2405971069

AF615891

Handbook of Laser Technology and Applications

Volume II: Laser Design and Laser Systems

Handbook of Laser Technology and Applications

Volume II: Laser Design and Laser Systems

Edited by

Colin E Webb

University of Oxford

and

Julian D C Jones

Heriot-Watt University

Institute of Physics Publishing
Bristol and Philadelphia

British Library Cataloguing-in-Publication Data

A catalogue record for this book is available from the British Library.

ISBN 0 7503 0960 1 (Vol. I)
0 7503 0963 6 (Vol. II)
0 7503 0966 0 (Vol. III)
0 7503 0607 6 (3 Vol. set)

Library of Congress Cataloging-in-Publication Data are available

Development Editor: David Morris
Production Editor: Simon Laurenson
Production Control: Sarah Plenty
Cover Design: Victoria Le Billon
Marketing: Nicola Newey and Verity Cooke

Published by Institute of Physics Publishing, wholly owned by The Institute of Physics, London

Institute of Physics Publishing, Dirac House, Temple Back, Bristol BS1 6BE, UK

US Office: Institute of Physics Publishing, The Public Ledger Building, Suite 929, 150 South Independence Mall West, Philadelphia, PA 19106, USA

Typeset in LaTeX 2_ε by Text 2 Text Limited, Torquay, Devon
Index by Indexing Specialists (UK) Ltd, Hove, East Sussex
Printed in the UK by MPG Books Ltd, Bodmin, Cornwall

Contents

VOLUME I: PRINCIPLES

PART A PRINCIPLES 1

A Principles 3
Richard Shoemaker
A1 Basic laser principles 5
Christopher C Davis
A2.1 Free-space laser resonators 81
Robert C Eckardt
A2.2 Waveguide laser resonators 115
Chris Hill
A3 Laser beam control 135
Jacky Byatt
A4 Nonlinear optics 161
Robert W Boyd
A5 Interference and polarization 185
Alan Rogers
A6 Optical waveguide theory 223
G Stewart
A7 Optical detection and noise 251
Gerald Buller and Jason Smith
A8 Introduction to numerical analysis for laser systems 281
George Lawrence

VOLUME II: LASER DESIGN AND LASER SYSTEMS

PART B LASER DESIGN, FABRICATION AND PROPERTIES 303

B1 Solid state lasers 305
R C Powell
B1.1 Transition metal ion lasers—Cr^{3+} 307
Georges Boulon
B1.2 Transition metal ion lasers other than Cr^{3+} 339
Stephen A Payne
B1.3 Rare earth ion lasers—Nd^{3+} 353
A I Zagumennyi, V A Mikhailov and I A Shcherbakov

B1.4 Lanthanide series lasers—near infrared 383
Norman P Barnes
B1.5 Rare-earth ions—miscellaneous: Ce^{3+}, U^{3+}, divalent, etc 411
Gregory J Quarles
B1.6 Lasers based on nonlinear effects 431
Fabienne Pellé
B1.7 Solid state Raman lasers 469
Tasoltan T Basiev and Richard C Powell
B1.8 Colour centre lasers 499
T T Basiev, P G Zverev and S B Mirov
B2 Laser diodes 523
Ian White
B2.1 Basic principles of laser diodes 525
N K Dutta
B2.2 Spectral control in laser diodes 561
Markus-Christian Amann
B2.3 High-speed laser diodes 585
Peter P Vasil'ev
B2.4 High-power laser diodes and laser diode arrays 605
Peter Unger
B2.5.1 Visible laser diodes: properties of III–V red-emitting laser diodes 619
Peter Blood
B2.5.2 Visible laser diodes: properties of blue laser diodes 641
Robert Martin
B2.6 Vertical-cavity surface-emitting lasers 659
B M A Rahman and K T V Grattan
B2.7 Long wavelength laser diodes 691
S Anders, G Strasser and E Gornik
B2.8 Semiconductor lasers and optical amplifiers for switching and signal processing 707
Hitoshi Kawaguchi
B3 Gas lasers 749
Julian Jones
B3.1 Carbon dioxide lasers 751
Denis R Hall
B3.2 Excimer, F_2, N_2 and H_2 lasers 791
W J Witteman
B3.3 Copper and gold vapour lasers 847
Colin Webb
B3.4.1 Chemical lasers: COIL 861
B D Barmashenko and S Rosenwaks
B3.4.2 Chemical lasers: HF/DF 881
Lee H Sentman
B3.5 Argon and krypton ion lasers 893
Malcolm H Dunn and Tony Gutierrez
B3.6 Helium–neon lasers 909
Alan D White and Lisa Tsufura

	B3.7	Helium–cadmium laser *William T Silfvast*	921
	B3.8	Optically pumped mid IR lasers: NH_3, C_2H_2 *Mary S Tobin*	929
	B3.9	Far-IR lasers: HCN, H_2O *Wilhelm Prettl*	951
	B4	Fibre and waveguide lasers *R C Powell*	961
	B4.1	Fibre lasers *David Hanna*	963
	B4.2	High power fiber lasers *Andreas Tünnermann and Holger Zellmer*	977
	B4.3	Cascaded Raman fibre lasers *Clifford Headley*	989
	B4.4	Soliton lasers *J R Taylor*	1007
	B4.5	Erbium and other doped fibre amplifiers *Kevin Cordina*	1025
	B4.6	High-power waveguide lasers *D P Shepherd*	1045
	B5	Other lasers *Colin Webb*	1063
	B5.1	Free electron lasers and synchrotron light sources *P G O'Shea and J B Murphy*	1065
	B5.2	X-ray lasers *Jorge J Rocca*	1087
	B5.3	Liquid lasers *David H Titterton*	1115
	B5.4	Solid-state dye lasers *David H Titterton*	1143
PART C	**LASER SYSTEM DESIGN**		**1163**
	C1	Optical components *Julian Jones*	1165
	C1.1	Optical components *Leo H J F Beckmann*	1167
	C1.2	Optical control elements *Alan Greenaway*	1183
	C1.3	Adaptive optics and phase conjugate reflectors *Michael J Damzen and Carl Paterson*	1193
	C1.4	Opto-mechanical parts *Frank Luecke*	1203
	C1.5.1	Power conditioning: supplies for driving semiconductor laser diodes *Ralph Savioli*	1211
	C1.5.2	Power conditioning: supplies for driving gas discharges (gas and solid state lasers) *I Smilanski*	1217

C1.5.3 Power conditioning: supplies for driving flash tubes and arclamps for solid state lasers 1237
Mark Greenwood and D W Miller

C2 Optical pulse generation 1247
Clive Ireland

C2.1 Quasi-cw and modulated beams 1249
K Washio

C2.2 Short pulses 1257
Andreas Ostendorf

C2.3 Ultrashort pulses 1273
Derryck T Reid

C3 Frequency conversion and filtering 1313
Terence A King

C3.1 Harmonic generation—materials and methods 1315
David J Binks

C3.2 Optical parametric devices 1347
M Ebrahimzadeh

C3.3 Laser stabilization for precision measurements 1393
G P Barwood and P Gill

C4 Beam delivery 1415
Julian Jones

C4.1 Basic principles 1417
D P Hand

C4.2 Free-space optics 1425
Leo H J F Beckmann

C4.3 Fibre optic beam delivery 1461
D P Hand

C4.4 Positioning and scanning systems 1475
Jürgen Koch

C5 Laser beam measurement 1499
Julian Jones

C5.1 Beam propagation 1501
B A Ward

C5.2 Detectors 1509
Kenny Weir

C5.3 Laser energy and power measurement 1523
Robert K Tyson

C5.4 Irradiance and phase distribution measurement 1527
B Schäfer

C6 Laser safety 1535
Colin Webb

C6.1 Laser safety 1537
J Michael Green and Karl Schulmeister

VOLUME III: APPLICATIONS

PART D APPLICATIONS: CASE STUDIES 1557

D1 Materials processing 1559
Clive Ireland
D1.1 Welding 1561
H Hügel and C Schinzel
D1.2 Cutting 1587
John Powell and Claes Magnusson
D1.3 Laser marking 1613
Terry J McKee
D1.4 Drilling 1633
S Williams
D1.5 Photolithography 1653
Shinji Okazaki
D1.6 Laser micromachining 1661
Malcolm Gower
D1.7 Rapid manufacturing 1693
Gary K Lewis
D1.8 Pulsed laser deposition of thin films 1705
Ian Boyd and D B Chrisey
D2 Optical measurement techniques 1721
Julian Jones
D2.1 Fundamental length metrology 1723
J Flügge, F Riehle and H Kunzmann
D2.2 Laser velocimetry 1749
C Tropea
D2.3 Laser vibrometers 1779
Neil A Halliwell
D2.4 Electronic speckle pattern interferometry (ESPI) 1805
Dave Towers and Clive Buckberry
D2.5 Optical fibre hydrophones 1839
Geoffrey A Cranch and Philip J Nash
D2.6 Optical fibre Bragg grating sensors for strain measurement 1881
David A Jackson and David J Webb
D2.7 High-speed imaging 1919
Adam Whybrew
D2.8 Particle sizing 1931
Nils Damaschke, Maurice Wedd, Adam Whybrew and Damien Blondel
D3 Medical 1951
Terence A King and Brian C Wilson
D3.1 Light–tissue interactions 1955
Steven Jacques and Michael Patterson
D3.2 Therapeutic applications: introduction 1995
Reginald Birngruber
D3.2.1 Therapeutic applications: ophthalmology 1999
Reginald Birngruber

D3.2.2 Therapeutic applications: refractive surgery 2009
Giovanni Cennamo and Raimondo Forte

D3.2.3 Therapeutic applications: photodynamic therapy 2019
Brian C Wilson and Stephen G Bown

D3.2.4 Therapeutic applications: thermal treatment of tumours 2037
Stephen G Bown

D3.2.5 Therapeutic applications: dermatology—selective photothermolysis 2045
Sean Lanigan

D3.2.6 Therapeutic applications: lasers in vascular surgery 2055
Mahesh Pai

D3.2.7 Therapeutic applications: hardtissue/dentistry 2065
Raimund Hibst

D3.2.8 Therapeutic applications: free-electron laser 2075
E Duco Jansen, Michael Copeland, Glenn S Edwards, William Gabella, Karen Joos, Mark A Mackanos, Jin H Shen and Stephen R Uhlhorn

D3.3 Medical diagnostics 2087
Brian C Wilson

D3.4 Laser applications in biology and biotechnology 2123
Sebastian Wachsmann-Hogiu, Alexander J Annala and Daniel L Farkas

D3.5 Biomedical laser safety 2155
Harry Moseley and Bill Davies

D4 Communications 2181
John Marsh

D4.1 The basic point-to-point communications system 2183
John Gowar

D4.2 High-capacity optical transmission systems 2231
Paul Urquhart

D4.3 Local area networks 2289
J Lehman and K L Johnson

D4.4 Fibre-to-the-chip: development of vertical cavity surface emitting laser arrays designed for integration with VLSI circuits 2321
A V Krishnamoorthy, L M F Chirovsky, K W Goosen, J Lopata and W S Hobson

D4.5 Optical satellite communications 2345
A Coello-Vera and M Maignan

D4.6 Smart pixel technologies and optical interconnects 2363
Marc P Y Desmulliez and Brian S Wherrett

D5 Optical information storage 2389
John Marsh

D5.1 Optical data storage 2391
Tom D Milster

D5.2 Lasers in printing 2421
Atsushi Kawamura, Seizo Suzuki and Yoshinori Hayashi

D6 Spectroscopy 2463
Colin Webb

D6.1 Laser cooling and trapping 2465
C S Adams and I G Hughes

D6.2	Ion trapping and laser applications to length and time metrology *P Gill and G P Barwood*	2485
D6.3	Time-resolved spectroscopy *Gavin D Reid and Klaas Wynne*	2507
D7	Earth and environmental sciences *Lance Thomas*	2529
D7.1	Satellite laser ranging *Roger Wood and Graham Appleby*	2531
D7.2	Lidar for atmospheric ozone remote sensing *Gérard Ancellet*	2563
D8	Lasers in astronomy *R C Powell*	2579
D8.1	Lasers in astronomy *Renaud Foy and Jean-Paul Pique*	2581
D9	Holography: holographic optical elements and computer-generated holography *Mohammad R Taghizadeh*	2625
D9.1	Holography: holographic optical elements—computer-generated holography—diffractive optics *Hans Peter Herzig*	2627
D10	High-intensity lasers for plasma studies *Colin Webb*	2643
D10.1	High-power lasers for plasma physics *M H R Hutchinson*	2645
D10.2	High-power lasers and the extreme conditions that they can produce *S J Rose*	2657
	Index	2665

PART B

LASER DESIGN, FABRICATION AND PROPERTIES

B1
Solid state lasers

R C Powell

Lasers can be categorized by whether their active medium is a solid, liquid or gas. Solid state lasers were the first type of laser to be invented (ruby lasers). They have the advantage of being rugged, compact systems that make them useful in non-laboratory environments. Technically both insulating host materials with optically active point defects and semiconductors are solid state lasers; however, the common terminology is to use the name 'solid state lasers' for the former and 'semiconductor lasers' for the latter. This chapter deals with solid state lasers while semiconductor lasers are treated in chapter B2.

There are a wide variety of laser host materials including oxide and fluoride crystals and glasses. Similarly, there are many types of optically active point defects including transition metal ions, rare earth ions and colour centres. It is the combination of specific optically active point defects with specific host materials that determine the fundamental properties of the laser. These combinations can operate as three-level, four-level and vibronic lasers spanning the near-ultraviolet, visible and near-infrared spectral regions. They can be pulsed or continuous wave (cw), discrete line or tunable, mode-locked or operate in Q-switched modes. The ruggedness and versatility of solid state lasers have lead to their use in many practical applications.

This section describes the fundamental science and technology of the most common solid state lasers and provides information on their systems design and laser operating parameters. One way to extend the spectral range of solid state lasers is through nonlinear optical effects. This section describes the use of several types of nonlinear optical effects in solid state lasers. The laser material for the systems described in this section is in the form of a rod or slab. Other configurations of solid state lasers such as fibres and waveguides are discussed in chapter B4. For further reading on the fundamental physics of solid state lasers the reader is referred to [1].

References

[1] Powell R C 1998 *Physics of Solid State Laser Materials* (New York: Springer)

B1.1 Transition metal ion lasers—Cr^{3+}

Georges Boulon

B1.1.0 Introduction

The most popular centres in inorganic solids are probably Cr^{3+} optical ones. As with many transition metal ions, Cr^{3+} is used as a colour agent in the glass industry and in jewellery where beautiful mineral stones are used such as ruby (Al_2O_3 sapphire), alexandrite ($BeAl_2O_4$ chrysoberyl), emerald [$Be_3Al_2(Si0_3)_6$], as well as several garnets containing aluminium or gallium. Colour characteristics can easily be connected to our light environment, with visible and strong broadband absorption. In solid-state laser sources, Cr^{3+} ions, which are dispersed in a host crystal as an impurity in the place of aluminium octahedral symmetry sites, can be pumped inside the usual absorption bands with an appropriate xenon flashlamp or with argon and krypton laser energy sources. The optical excitation energy is absorbed by raising electrons of $3d^3$ Cr^{3+} configuration from the ground level to one of the pump absorption bands. It is then transformed by fast radiation-less transitions into the lowest excited level. During this process, the energy lost by the electron of the excited Cr^{3+} ion is carried into the lattice by phonons. In the last step of the process, the electrons return to the ground state through spontaneous emission or fluorescence if pumping intensity is below the laser threshold. Laser emission, that is to say, stimulated radiation which produces the expected output beam, only occurs when the pump intensity is above the laser threshold, and then the necessary inversion of population begins to be reached between the ground level and the involved excited level. One of the advantages of Cr^{3+}-doped crystals is to show visible emission in the deep red spectral range where eyes are still sensitive. It is truly fascinating to detect the passage between spontaneous emission in all space and the directional stimulated emission in ruby at 694.3 nm (0.6943 μm). Theodore H Maiman [1] saw it for the first time in 1960 at the Hughes Research Laboratories in Malibu, California, by using a rod of synthetic ruby coated on the two opposite sides and pumped by a pulsed helical xenon flashlamp. Maiman's demonstration of the ruby laser, before the gas laser and the dye laser, opened the laser era.

The understanding of Cr^{3+}-doped crystals is thus connected with both the physical properties of Cr^{3+} activator ions, the host materials and relevant physical processes including:

(i) static electron–lattice interactions, to ascertain how the electronic energy levels of the free ions are perturbed by the electrostatic crystal field of the nearest neighbour host ions;

(ii) electron–photon interactions, to determine absorption and emission of light radiation between energy levels which can be classified as so-called allowed or forbidden transitions;

(iii) electron–phonon interactions, due to the thermal vibrations of the host ions which modulate the local crystal field at the Cr^{3+} site and then are strongly dependent on the temperature seen in shifts and broadening of spectral lines;

(iiii) ion–ion interactions, strongly dependent on the activator concentration leading to either non-radiative energy transfer accompanied by concentration quenching which is very well known in luminescent materials, or radiative energy transfer, so-called radiative trapping. An example of radiative trapping can be found in Cr^{3+} centres of ruby due to the presence of the resonant absorption-emission lines.

All these static and dynamic processes strongly affect the spectroscopic properties of Cr^{3+} ions due to the nature of the optically active electrons belonging to the $3d^3$ outer configuration. As a result, the absorption and emission spectra are characterized by the usual broadbands and sharp lines which are useful for laser operations, as we shall demonstrate.

This chapter has been divided into five parts. The first three summarize the general concepts of determining the energy levels for the $3d^3$ Cr^{3+} ion, first as a free ion and then as an impurity ion in the usual octahedral crystal field in which the stabilization energy is maximum. The last two explain the main physical properties, first of ruby as an example of a strong field Cr^{3+} laser crystal emitting $^2E \rightarrow {}^4A_2$ sharp line in the three-level scheme of Cr^{3+} laser crystals; second of alexandrite as an example of an intermediate field Cr^{3+} laser crystal emitting both $^2E \rightarrow {}^4A_2$ sharp line and $^4T_2 \rightarrow {}^4A_2$ broadband and, finally, of garnets and fluorides as examples of a weak crystal field Cr^{3+} laser crystal model emitting a $^4T_2 \rightarrow {}^4A_2$ broadband useful for tunable solid-state lasers in the four-level scheme.

B1.1.1 Free-ion energy levels

A detailed development of the theoretical description of electronic energy levels of atoms and ions can be found in several books [2–7]. The energy E and eigenfunction ψ of transition metal free ions can be obtained by solving Schrödinger's equation:

$$H_{\mathrm{fi}}\psi = E\psi \tag{B1.1.1}$$

$$H_{\mathrm{fi}} = H_1 + H_{\mathrm{e}} + H_{\mathrm{s-o}}. \tag{B1.1.2}$$

H_1 is the kinetic and potential energy of the optical active i electron in the spherically symmetric ion nucleus:

$$H_1 = \sum_i \frac{\hat{p}_i^2}{2m} - \frac{Ze^2}{4\pi\varepsilon_0 r_i}. \tag{B1.1.3}$$

H_{e} is the electrostatic Coulomb interaction between the electrons which configures the electrons into L and S terms:

$$H_{\mathrm{e}} = \sum_{i\rangle j} \frac{e^2}{4\pi\varepsilon_0 r_{ij}}. \tag{B1.1.4}$$

$H_{\mathrm{s-o}}$ is the spin–orbit interaction which is the relativistic effect that couples the magnetic moment of the spinning electrons and the magnetic field caused by the orbital motion of the electrons which forms J terms:

$$H_{\mathrm{s-o}} = \sum_i \xi(r_i)\hat{\boldsymbol{\ell}}_i \cdot \hat{\mathbf{s}}_i. \tag{B1.1.5}$$

The interelectron Coulomb interaction is simplified to a spherically-averaged potential term, by the central field approximation

$$H_{\mathrm{e}} = \langle \sum_{j\neq i} \frac{e^2}{4\pi\varepsilon_0 r_{ij}} \rangle = U_i(r_i). \tag{B1.1.6}$$

If the spin–orbit interaction is neglected and the central field approximation used, then the free ion Hamiltonian becomes

$$H_{\mathrm{fi}} = H_{\mathrm{o}} = \sum_i \frac{\hat{p}_i^2}{2m} - \frac{Z_e^2}{4\pi\varepsilon_0 r_i} + U_i(r_i) = V_i(r_i) \tag{B1.1.7}$$

Table B1.1.1. Quantum numbers associated with solutions to Schrödinger's equation for atoms and ions.

Quantum number	Allowed values	Purpose
Principal, n	1,2,3 ...	Indicates the energy and size of the orbital.
Orbital, l	$(n-1), (n-2), \ldots, 0$	Indicates the shape of the orbital and the electronic angular momentum.
Magnetic, l_z	$\pm l, \pm(l-1), \ldots 0$	Indicates the direction of an orbital and the electron's behaviour in a magnetic field.
Spin state, S_z	$\pm 1/2$	Indicates the axial angular momentum of the electron.

which is referred to as the central field Hamiltonian where $V_i(r_i)$ is the combined nuclear and interelectron potential energy term. The solutions to Schrödinger's equation are separable for each of the electrons, and the wavefunction, in polar coordinate form, can be written as:

$$u_i = R_{nili}(r_i) \cdot Y_{li}^{m_{zi}}(\theta_i, \Phi_i) \cdot S(s_{zi}) \tag{B1.1.8}$$

$$= |n\ell m_z\rangle_i |s_z\rangle_i = |n\ell m_z s_z\rangle_i \tag{B1.1.9}$$

where $R_{ni,li}(r_i)$ is the solution to the radial equation of the central field potential, $V_{(ri)}$ for the ith electron, $Y_{li}^{m_{zi}}(\theta_i, \Phi_i)$ are the spherical harmonics of the ith electron, $S(s_{zi})$ is the spin state of the ith electron. The solutions are parametrized by the four quantum numbers n, l, m_z and s_z as can be seen in table B1.1.1.

The Pauli exclusion principle states that no two electrons in an ion can occupy the same quantum state characterized by the four quantum numbers.

Since the electron is a spin $\frac{1}{2}$ particle, such as a fermion, the N electron wavefunction built from single particle states $u_1(1) \ldots u_n(N)$ must be antisymmetric and is given by the Slater determinant as linear combinations of products of the single-electron wavefunction. This wavefunction is antisymmetric because an interchange of the spatial and spin coordinates of two particles (say u_1 and u_2) is equivalent to interchanging two rows, so that the sign of the determinant changes. The energy of the state $\sum E_{n,e}$ depends on the set of n_i, ℓ_i values. This set, $n_i\ell_i$, is called the electron configuration. In each $n\ell$ orbital, there is a maximum of $2(2\ell + 1)$ electrons.

The electron configuration of a Cr atom with 24 electrons is: $1s^2 2s^2 2p^6 3s^2 3p^6 3d^5 4s$. Thus, the Cr^{3+} ion has given up three electrons from the outer shells leaving three unshielded electrons in the 3d orbitals. These are the electrons that play the dominant role in determining the optical properties of the ion, each electron having quantum numbers $n = 3$ and $\ell = 2$.

Finally, it can be observed that the energy of the states does not depend on the m_z and s_z values, and, as for the single electron functions, these eigenstates have a large energy degeneracy. Since H_0 commutes with ℓ_i the orbital angular momenta vectors, it will also commute with the total orbital angular momentum $L = \sum_i \ell_i$ and M_z. Likewise, H_0 commutes with s_i and hence the total spin angular momentum $S = \sum_i s_i$ and S_z, so that the eigenstates of H_0 can be classified by the set of quantum numbers $(n_i\ell_i)$ L S M_z S_z with corresponding ket$|(n_i\ell_i)LSM_zS_z\rangle$.

These L and S angular momenta vectors can then be coupled to give the total angular momentum for the ion, $J = L + S$. A second alternative to add up the angular momenta is first to sum the individual orbital and spin angular momenta for each electron to form the total $j_i = \ell_i + s_i$ and then to sum j_i over all electrons to give $J = \sum j_i$. Note that the size of the spin–orbit interaction determines which coupling scheme gives a better zero-order approximation to the actual spectrum. The first approach is more appropriate for transition

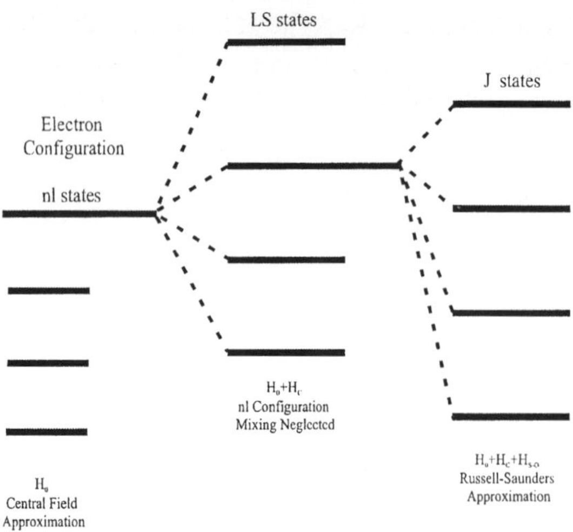

Figure B1.1.1. Splitting of free-ion states by LS coupling and spin–orbit coupling.

Table B1.1.2. Level energies of $3d^3$ (Cr^{3+}) spectral terms.

Spectral terms	Level energies
^{2}H	$3A - 6B + 3C$
^{2}G	$3A - 11B + 3C$
^{4}F	$3A - 15B$
^{2}F	$3A + 9B + 3C$
2D	$3A + 5B + 5C \pm (193B^2 + 8BC + 4C^2)^{1/2}$
4P	3A

metal ions such as Cr^{3+}. A diagram of the energy splittings due to the electronic and spin–orbit interactions is given in figure B.1.1.1.

The final degeneracy due to the $2J + 1$ different possible spatial orientations for the angular momentum can be partially or totally lifted only by an external perturbation such as an applied electric field (the Stark effect as in the octahedral symmetry crystallographic site of the hosts, as we shall see later) or magnetic field (the Zeeman effect). In the usual spectroscopic term designated by ^{2S+1}L, the superscript is called the multiplicity of the term. The spectroscopic notation S, P, D, F, ... is used for $L = 0, 1, 2, 3, \ldots$, respectively. When the spin–orbit interaction is taken into account, each term is split into states having specific values of J designated by $^{25+I}L_J$, called multiplets. In the Russell–Saunders approximation, or L–S coupling, L and S are still treated as good quantum numbers.

The energy value of the energy levels for each spectral term can theoretically be obtained by calculating the interaction of multi-electrons. Slater, Condon and Shortley used three parameters, F_0, F_2 and F_4 as radical integrators of static interaction, to denote the level position of the $3d^n$ electron configuration. These are known as the Slater parameters. Based on group theory, Racah used A, B and C to denote the parameters. These are known as the Racah parameters. The relationship between these two sets of parameters are $A = F_0 - 49F_4$, $B = F_2 - 5F_4$, $C = 35F_4$. A is the same for all the energy levels and only contributes as a shift. Table B1.1.2 gives level energies of $3d^3(Cr^{3+})$ spectral terms.

The eight terms have to be ordered with respect to their energies. Hund's rules require that the ground-state term has the greatest multiplicity possible and the greatest orbital angular momentum value consistent with this multiplicity. Therefore the ground state is the ^{4}F term. Finally, all energies measured relative to the ground state, yield:

$$\frac{E(^4\mathrm{F})}{C} = -15\frac{B}{C} \tag{B1.1.10}$$

$$\frac{E(^4\mathrm{P})}{C} = 0 \tag{B1.1.11}$$

$$\frac{E(^2\mathrm{H})}{C} = -6\frac{B}{C} + 3 \tag{B1.1.12}$$

$$\frac{E(^2\mathrm{P})}{C} = -6\frac{B}{C} + 3 \tag{B1.1.13}$$

$$\frac{E(^2\mathrm{G})}{C} = -11\frac{B}{C} + 3 \tag{B1.1.14}$$

$$\frac{E(^2\mathrm{F})}{C} = 9\frac{B}{C} + 3 \tag{B1.1.15}$$

$$\frac{E(^2\mathrm{D})}{C} = 5\frac{B}{C} + 5 - \sqrt{193\left(\frac{B}{C}\right)^{2} + 8\frac{B}{C} + 4} \tag{B1.1.16}$$

$$\frac{E(^2\mathrm{D}')}{C} = 5\frac{B}{C} + 5 + \sqrt{193\left(\frac{B}{C}\right)^{2} + 8\frac{B}{C} + 4} \tag{B1.1.17}$$

These expressions are plotted in units of E/C and B/C as shown in figure B1.1.2 [3]. By measuring the optical absorption and emission spectra of a specific type of 3d^3 ion such as Cr^{3+}, the energy-level splittings shown on the right-hand side of the figure have been found and identified with the designated terms. This fixes the value of B/C for Cr^{3+} at 0.222 as shown in the broken vertical line. Racah parameters for 3d^3 Cr^{3+} ion are $B = 918\ \mathrm{cm}^{-1}$, $C = 4133\ \mathrm{cm}^{-1}$ and $C/B = 4.50$.

The free-ion terms are further split by spin–orbit interaction. However, for 3d^3 ions in general and Cr^{3+} in particular, this is a smaller effect than the crystal-field splitting.

B1.1.2 Crystal-field splitting: ligand-field action

B1.1.2.1 Crystal field expression in the octahedral symmetry site

When a free-ion is put into a crystal host as an impurity, it is no longer in an environment of complete spherical symmetry. Instead it is surrounded by a set of nearest-neighbour ions, or ligands, in a geometric configuration. With the Cr^{3+} ion, the stabilization energy is maximum in the specific type of octahedral symmetry environment which can be described as a regular octahedron possessing a centre of inversion as shown in figure B1.1.3, and denoted by group theory as O_h point group symmetry.

All point groups are characterized by character tables giving the irreducible representations. The Mulliken notation is the one most commonly used: A and B denote one-dimensional representations, E denotes two-dimensional representations and T denotes three-dimensional representations. For groups with a centre of inversion, subscripts g and u are used to indicate representations that are symmetric (g for gerade) or antisymmetric (u for ungerade) with respect to the inversion operation respectively. For the O_h point group, the eight main representations are A_{1g}, A_{2g}, E_g, T_{1g}, T_{2g}, $E_{1/2g}$, $E_{3/2g}$ and G_g. In Bethe's notation, the first five representations are symbolised by Γ_1, Γ_2, Γ_3, Γ_4 and Γ_5.

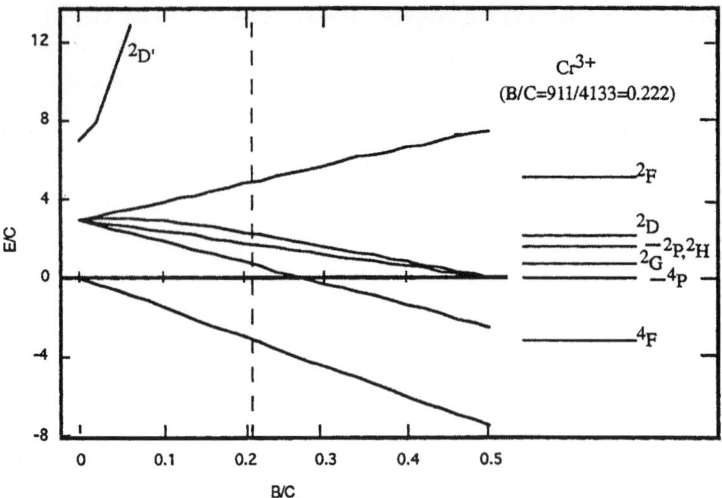

Figure B1.1.2. Free-ion terms of a d^3 ion and energy levels for Cr^{3+} [3].

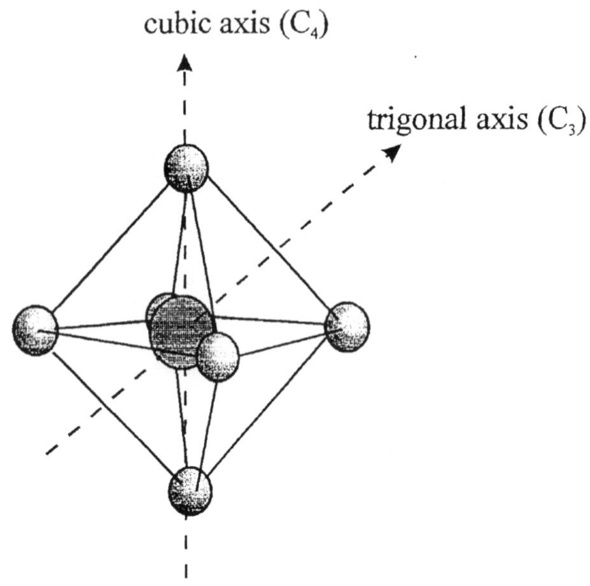

Figure B1.1.3. The Cr^{3+} ion in an octahedral ligand field. The cubic axis (C_4) and trigonal axis (C_3) are shown.

The Hamiltonian of the Cr^{3+} ion in an octahedral symmetry crystal field is given by the same expression [8] as well as the V_c crystal field energy term, assuming that all six neighbouring ions (oxygen or fluorine) can be approximated by point charges, each with the amount $-Z_p e$, fixed at their average positions $R_p(0) = (a_p, \theta_p, \Phi_p)$:

$$H = H_1 + H_e + H_{s-o} + V_c \tag{B1.1.18}$$

$$V_c = \sum_i^6 \sum_{p=1}^6 \frac{1}{4\pi\varepsilon_0} \frac{Ze}{|R_p(0) - r_i|}. \tag{B1.1.19}$$

The ith electron is at $\boldsymbol{r}_i = (r_i, \theta_i, \Phi_i)$. If we assume that $a_i < a_k$, we can expand V_c in terms of spherical harmonics according to the standard multi-pole expansion:

$$\frac{1}{r_{ij}} = \sum_k \sum_{p=1}^{+k} \frac{4\pi}{2k+1} \frac{r^k}{a^{k+1}} Y_k^m(\theta_2, \varphi_2). \tag{B1.1.20}$$

The potential function at the central ion due to the six ligands located at distance 'a' is then yielded by:

$$V_c = \sum_i \sum_{p=1}^6 \sum_{\ell=0}^{\infty} \sum_{m=-\ell}^{\ell} \frac{4\pi Ze}{2\ell+1} \cdot \frac{r_i^\ell}{a_p^{\ell+1}} \cdot Y_\ell^{m*}(\theta_i, \varphi_i) \cdot Y_\ell^m(\theta_j, \varphi_j). \tag{B1.1.21}$$

One obtains:

$$V_c = \frac{Ze}{4\pi\varepsilon_0}\left[\frac{6}{a_p} + \frac{7r_i^4}{2a_p^5}\left\{C_0^{(4)}(\theta_i\varphi_i) + \sqrt{\frac{5}{14}}(C_4^{(4)}(\theta_i, \varphi_i) + C_{-4}^{(4)}(\theta_i, \varphi_i))\right\}\right] = r_i^6\text{terms} + \cdots \tag{B1.1.22}$$

where

$$C_t^{(k)}(\theta, \varphi) = \left(\frac{4\pi}{2k+1}\right)^{1/2} Y_k^t(\theta, \varphi). \tag{B1.1.23}$$

This expression is of even parity as expected with the inversion symmetry of the local arrangement.

B1.1.2.2 Calculations of the Cr^{3+} energy levels

In the expression of the Hamiltonian H, the best way to calculate the energy level positions is to estimate the relative values of each term by applying the perturbation procedure. The perturbations are then treated from the highest to the lowest magnitude.

Usually, there are three types of treatment.

(1) *Weak crystal field* ($V_c < H_{s-o} < H_e$): the correct quantum numbers J and M_J are taken from the free-ion, and the crystal field acts as a perturbation that causes Stark splitting of the free-ion multiplets. This is the case for the 4f configuration of optically active rare earths shielded by the $5s^2 5p^6$ outer-shell electrons.

(2) *Intermediate crystal field* ($H_{s-o} < V_c < H_e$): this is the case of outer 3d electrons of optically active transition metal ions such as Cr^{3+}. Calculations of the energy levels can be made by neglecting both the crystal field Hamiltonian, V_c, and the spin–orbit Hamiltonian, H_{s-o}, and by considering only the LS terms of the Coulomb Hamiltonian, $H_1 + H_e$, representing the interaction between the electrons. The correct quantum numbers are L, S, M_L and M_S. The crystal field is a perturbation that splits the free-ion terms, and the H_{s-o} interaction is an additional perturbation that determines further splitting of the energy levels.

(3) *Strong crystal field* (H_{s-o}, $H_e < V_c$): this is the case with 4d or 5d optically active electrons. The intermediate crystal field procedure involved with Cr^{3+} ion is much more complicated than the other two. However, an illustrated model of d orbitals is very useful to interpret the spectroscopic data. As can be seen in figure B1.1.4, the angular forms of the five-fold degenerate d orbitals ($\ell = 2$ and $2\ell + 1 = 5$) of even parity in an octahedral crystal field can be divided into two types of levels labelled t_{2g} functions

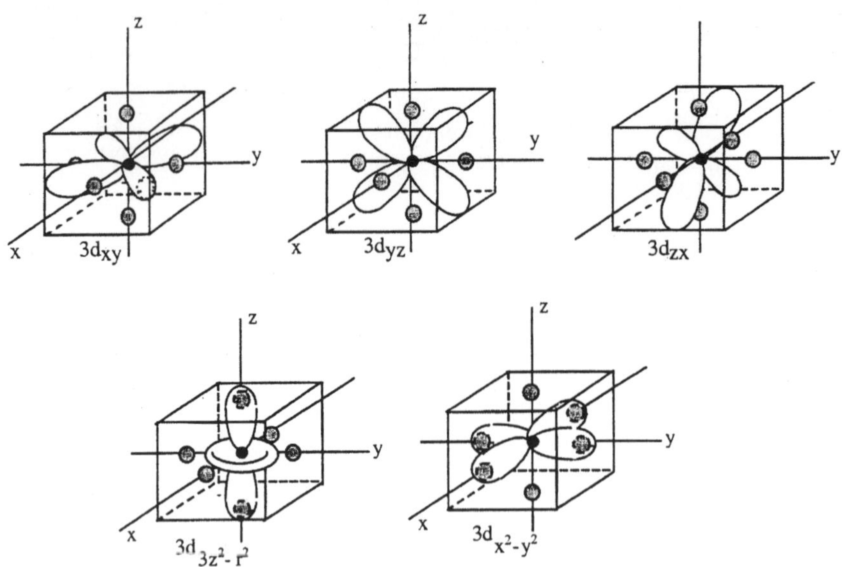

Figure B1.1.4. 3d orbital angular forms of even parity in an octahedral crystal field.

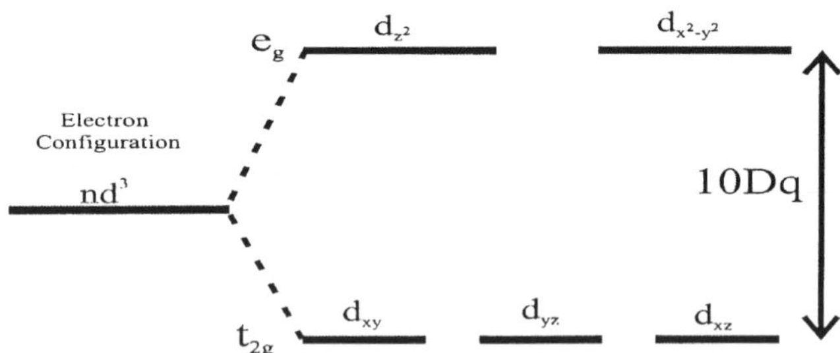

Figure B1.1.5. The two sets of d orbitals e_g (two-fold degenerate e level) and t_{2g} (three-fold degenerate t_2 level). $10D_q$ is the energy difference between the two sets.

for the three-fold degenerate t_2 level (d_{xy}, d_{yz}, d_{zx}) and e_g functions for the two-fold degenerate e level ($d^2_x - ^2_y$, d^2_z). It is easy to understand that the t_{2g} orbitals which are directed between the ligands are less strongly coupled to the lattice and have lower energy than the e_g orbitals, which are directed toward the ligands. The usual representation of t_{2g} and e_g is shown in figure B1.1.5 where the energy difference is designated by 10 D_q, D_q being the parameter characterizing the strength of the crystal field.

$\Delta = 10D_q$ has been estimated by using static electric theory

$$\Delta = \frac{5eq\bar{r}^4}{3R^5} \qquad \text{(B1.1.24)}$$

$$e/r$$

where e is the electric charge and r the average distance from the metal ion nucleus to the electron in its d orbital. q is the ligand charge and R is the distance of the metal ion to the centre of ligand.

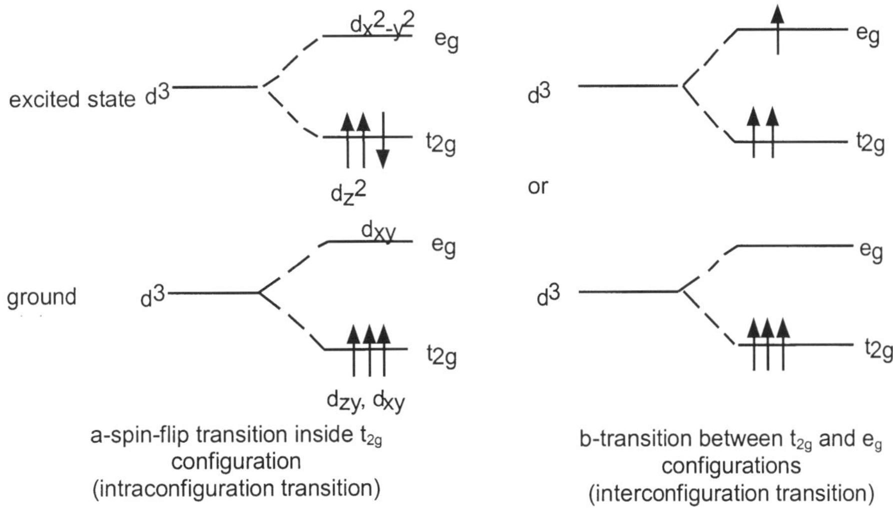

Figure B1.1.6. The two types of transitions of d^3 electrons in an octahedral crystal field.

Table B1.1.3. Irreducible representations of the O_h group.

Spectral terms (LS coupling)	Irreducible representations
^{2}H	$^2E + 2^2T_1 +^2 T_2$
^{2}G	$^2A_{ig} +^2 Eg + T_{1g} + T_{2g}$
^{4}F	$^4A_{2g} +^4 T_{1g} +^4 T_{2g}$
^{2}F	$^4A_{2g} +^4 T_{1g} +^4 T_{2g}$
^{2}D	$^2E_g +^2 T_{2g}$
^{2}D′	$^2E_g +^2 T_{2g}$
^{4}P	$^4T_{1g}$
^{2}P	$^4T_{1g}$

The Cr^{3+} ground state is t_{2g} with the maximum number of electrons (three for $3d^3$) and, according to Hund's rule, the maximum number of parallel spins (↑↑↑). We can get two kinds of transitions: intraconfigurational transition inside the t_{2g} configuration with a change of only one spin direction and interconfigurational transitions between t_{2g} and e_g with the same alignment of the three spin directions (figure B1.1.6).

This is a very important aspect of Cr^{3+} spectroscopy because it is needed to interpret absorption and emission spectra composed of broadbands and sharp lines as we shall see when we consider the coupling between the electrons of the Cr^{3+} ion and the lattice vibrations. In addition to these general considerations, we use the notation of the eight irreducible representations of the O_h octahedral symmetry group to label the eigenstates. The irreducible representations of the free-ion terms are reduced in terms of the irreducible representations of the O_h group as given in table B1.1.3.

The last step is to find the quantitative values of each term. Tanabe and Sugano [9, 10] calculated the energy states by considering the effect of the crystal field Hamiltonian V_c, before the Coulomb interaction between the outer electrons, H_e, that is to say in the strong field. This approach yields similar results to the

Table B1.1.4. Energies of the crystal field multiplets expressed in terms of the Racah parameters B and C and the the crystal field splitting $10D_q$.

t_2^3 configuration	t_2^2e configuration
$E(^4A_2) = 3A - 15B - 12D_q$	$E(^2A_1) = 3A - 11B + 3C - 2D_q$
$E(^2E) = 3A - 6B + 3C - 12D_q$	$E(^4T_1) = 3A - 3B - 2D_q$
$E(^2T_1) = 3A - 6B + 3C - 12D_q$	$E(^4T_2) = 3A - 15B - 2D_q$
$E(^2T_2) = 3A + 5C - 12D_q$	

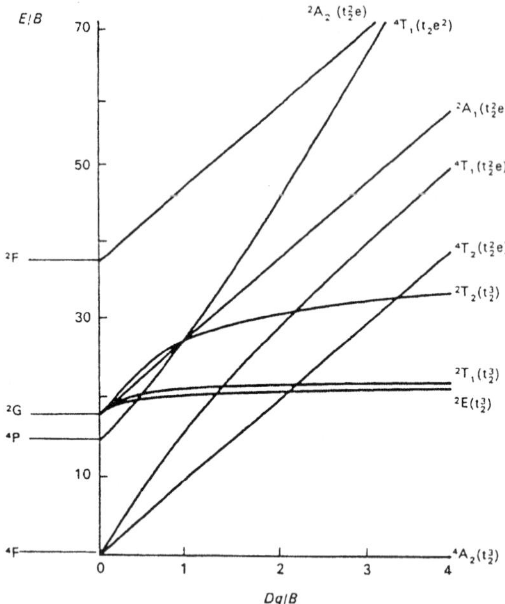

Figure B1.1.7. The Tanabe–Sugano diagram showing energy levels E/B of the Cr^{3+} ion in an octahedral crystal field as a function of the crystal field strength D_q/B. The theoretical free-ion levels are shown on the left by taking $D_q = 0$ [9, 10].

intermediate field approach since the effect of the interaction terms of the Hamiltonian, V_c and $H_1 + H_e$, are comparable.

The energies of the crystal field multiplets can be expressed in terms of the Racah parameters B and C and the crystal field splitting $10D_q$ previously introduced with e_g and t_{2g} orbitals. The results are shown in table B1.1.4 for the main LS term involved in the optical transitions.

The strength of the crystal field, $10D_q$, is characterized by the energy difference between the $^4A_2$ and $^4T_2$ states. Figure B1.1.7 is the Tanabe–Sugano diagram for Cr^{3+} which is very well known by spectroscopists. It can be seen that levels which are derived from the same electronic configuration t_{2g}^3, as the ground state $^4A_2$, ^2E, $^2T_1$, $^2T_2$ are almost independent of D_q for strong crystal fields and appear as horizontal lines. Figure B1.1.7 also shows that the lowest excited state is dependent on the strength of the crystal field: for strong field hosts it is the ^2E state giving $^4A_2 \Leftrightarrow {}^2E$ transition which is observed as a narrow spin-forbidden line ($\Delta S \neq 0$). For weak field hosts, the $^4T_2$ is the lowest giving the $^4A_2 \Leftrightarrow {}^4T_2$ transition which is observed

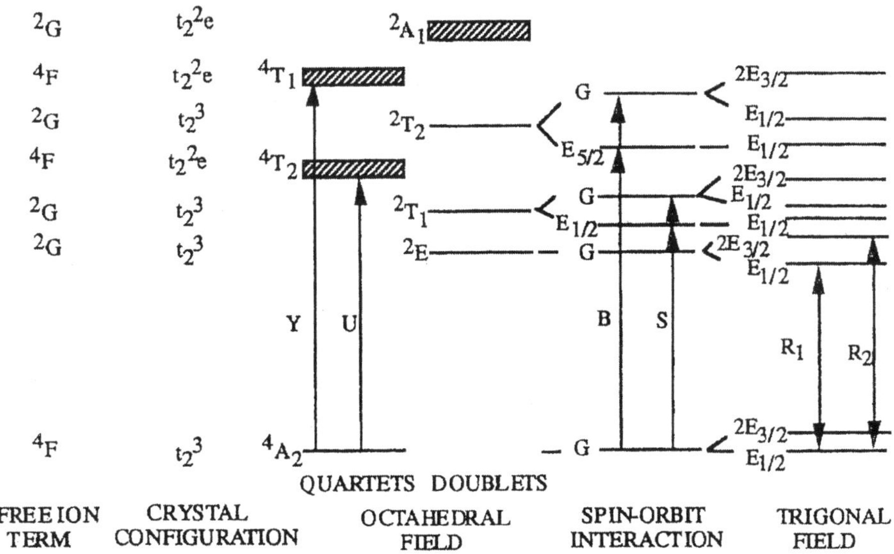

Figure B1.1.8. Cr^{3+} energy levels in the C_{3i} octahedral symmetry, spin–orbit interaction and trigonal field action are shown [7].

as spin-allowed broadband ($\Delta S = 0$). The separation in energy between the lowest vibrational levels of $^4T_2$ and ^2E determines the nature of the Cr^{3+} ion emission. $D_q/B = 2.3$ is the point separating weak and strong crystal fields.

In this theoretical part, we have assumed a perfect octahedral symmetry labelled as O_h point group. However, the oxygen octahedron is slightly disturbed along the threefold axis which is the c axis of the crystal (see figure B1.1.3) so that the site symmetry is lowered to C_{3v}. A more correct analysis of Cr^{3+} ion spectroscopy should be done with the C_{3v} character table and direct-product representations as well as with C_{3v} selection rules for forced electric dipole transitions. From these tables, the distinction between the two kinds of polarization, light polarized parallel and perpendicular to the c axis, can be made for each transition. Figure B1.1.8 summarizes all energy levels occurring within the C_{3v} symmetry site of Cr^{3+} ion in ruby. In addition to the octahedral Stark field within the two t_{2g} and e_g configurations, we have also mentioned the spin–orbit interaction and the trigonal field distortion which was described just above. The separation between the two sublevels is 0.38 cm^{-1} for the $^4A_2$ ground state and 29 cm^{-1} for the ^2E excited state in ruby. At room temperature, only ^2E splitting can be seen in the absorption spectrum with the use of a well resolved spectrophotometer.

B1.1.3 The dynamic effect of the ligand: electron–phonon interaction

B1.1.3.1 Configuration-coordinate diagram

The interaction of the optically active electrons of impurity ions with the lattice vibrations is most easily described using the single configuration coordinate diagram shown in figure B1.1.9 in which the configurational coordinate Q represents the collective position of the lattice relative to the optical ion. Although this is a great oversimplification of the true situation, it is still quite useful in explaining some aspects of the optical spectra of ions in solids. It is assumed that the lattice vibrates about the active ion in a totally symmetric 'breathing mode' such that the electronic states have additional energy due to the vibrations, which are represented by

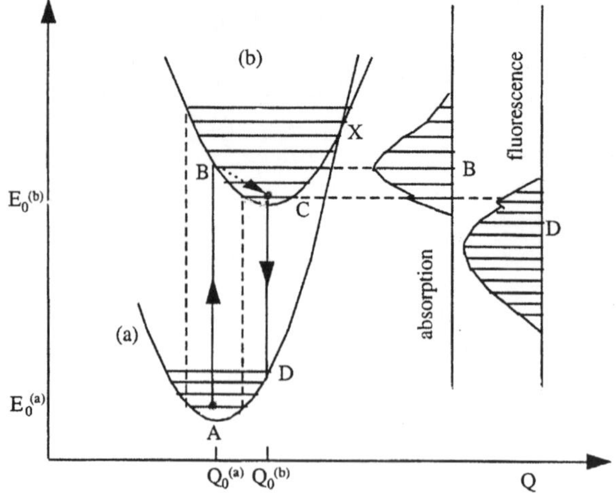

Figure B1.1.9. The usual single configuration coordinate diagram interprets absorption and luminescence properties of activator centres in solids. The 0-phonon line can be seen between $E_0^{(a)}$ and $E_0^{(b)}$ in both absorption and emission spectra.

a harmonic oscillator potential. In the configurational coordinate model, the harmonic oscillator energies are quantized so that:

$$E_{\text{vibr.}} = (n + 1/2)\hbar\omega \tag{B1.1.25}$$

where $\hbar\omega$ is the phonon energy and n is the phonon order.

The potential curves are described by parabolas in the harmonic approximation assuming the same frequency ω for the phonon energy of both ground state E_g and excited state E_e.

$$E_g = \tfrac{1}{2}k(Q - Q_o^g)^2 \tag{B1.1.26}$$

and

$$E_e = \tfrac{1}{2}k(Q - Q_o^e)^2 + E_o \tag{B1.1.27}$$

$$\omega = \left(\frac{k}{M}\right)^{1/2} \tag{B1.1.28}$$

where M is the mass of atoms involved in the vibration. The difference of configurational coordinates $Q_o^e - Q_o^g$ means the sensitivity difference of the two electronic levels to the host environment. This is expressed by the Huang–Rhys dimensionless parameter S:

$$\tfrac{1}{2}M\omega^2(Q_o^e - Q_o^g)^2 = S\hbar\omega \tag{B1.1.29}$$

The Huang–Rhys parameter can also be used to characterize the difference in the strength of vibronic coupling between the two states in terms of the Stokes shift between absorption and emission lines:

$$\text{Stokes shift} = 2S\hbar\omega \tag{B1.1.30}$$

B1.1.3.2 Optical band shapes

We find then that absorption and emission line shapes both become broadband by taking into account the vibrational state interaction. The quantum mechanical approach shows that the band shape is Pekarian evaluated as proportional to:

$$\frac{\exp(-S)S^m}{m!}$$

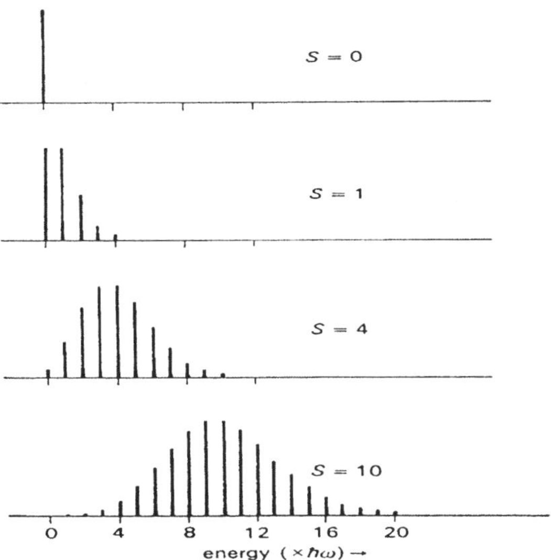

Figure B1.1.10. The relative intensities of the different electronic vibrational lines change with the strength of the coupling. S is the Huang–Rhys parameter.

which is plotted in figure B1.1.10 for various values of S. As S increases, the band shape becomes more Gaussian.

The previous parts of this chapter have shown that electronic transitions of Cr^{3+} ions involve two types of orbitals e_g and t_{2g} in an octahedral crystal field. The transitions between the same type of orbital (t_{2g}) which correspond to the spin-flip case, are not strongly coupled to vibrations so that S is small. The optical spectra are dominated by transitions that are purely electronic in nature, not involving phonons. This is the case for the so-called R_1 and R_2 lines associated with $^4A_2 \Leftrightarrow {}^2E$ transitions in absorption and emission occurring at the same wavelength (694.3 nm and 692.9 nm, respectively in ruby).

The transitions between the two types of orbitals e_g and t_{2g}, which involve a change in the spatial orbital, have strong electron–phonon coupling and hence large values of S. The absorption and emission spectra are composed of a 0-phonon line at the same energy in each and a vibronic broadband which is Stokes-shifted in the emission spectrum. An example of this is the $^4A_2 \Leftrightarrow {}^4T_2$ transition of the Cr^{3+} ion.

Finally, from this description, only three levels are needed to interpret absorption and emission spectra of the Cr^{3+} ion. Besides the $^4A_2$ ground state, the $\bar{E}$ component of ^2E and the $^4T_{2g}$ level which are the two lowest excited states, are systematically used to explain laser operations. Figure B1.1.11 shows a schematic representation of the three types of crystal field models usually met in Cr^{3+}-doped crystals. In ruby, which is a strong crystal field model shown in figure B1.1.9(a), $\bar{E}$ is the lowest excited state and the laser transition is interpreted as $\bar{E} \rightarrow {}^4A_2$ at 693.3 nm. Whereas in almost all other Cr^{3+}-doped crystals, $^4T_2$ is the lowest excited state giving rise to $^4T_2 \rightarrow {}^4A_2$ transition of a tunable solid-state laser between 700 and 1250 nm depending on the nature of the host. The parameter playing the main role is the distance R between the Cr^{3+} ion and the ligand occurring in the $10D_q$ formula as a result of the static electric theory, so that R increases progressively from a to b and c in figure B1.1.11. The crystal field dependence can be measured by the gap between $^4T_2$ and ^2E levels from either intensity or lifetime thermal dependence. For example, characteristics in Cr^{3+}-doped garnets are given in table B1.1.5 confirming the relationship between ε and the cell parameter a_o of the cubic structure in garnets.

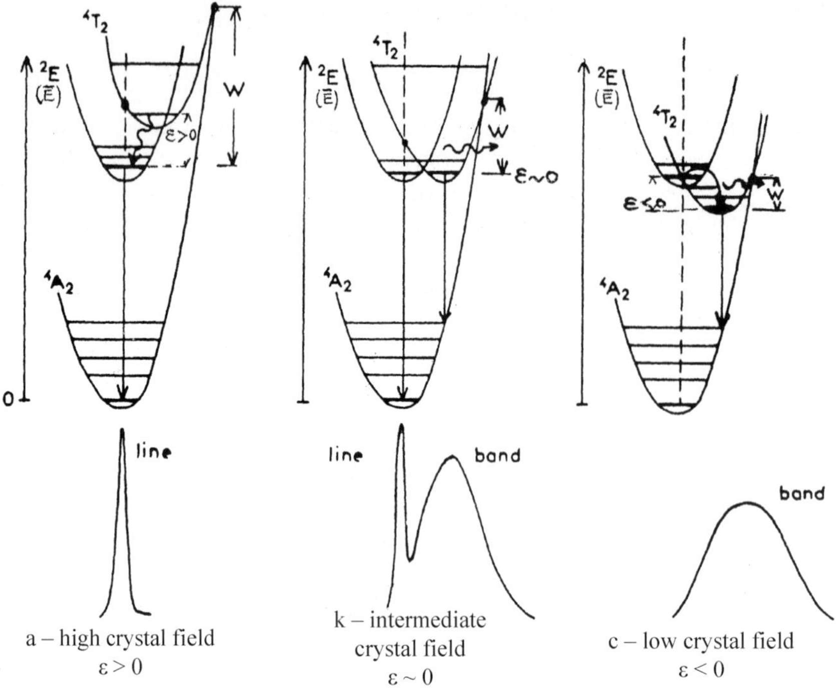

Figure B1.1.11. The three main types of crystal field strength ion. Cr^{3+}-doped crystals can be measured by the energy gap ε between the lowest vibrational levels of ^2E and $^4T_2$ electronic levels.

Table B1.1.5. Spectroscopic parameters and cell parameters in garnets.

Garnet host	ε (cm^{-1})	a_0 Å	R-line $^2E \rightarrow {}^4A_2$ (nm)	Wavelength of the maximum of $^4T_2 \rightarrow {}^4A_2$ (nm)	τ (μs)
$Y_3Al_5O_{12}$ [11, 12]	1000	12.005	680	720	9000
$Y_3Ga_5O_{12}$ [7]	600		693	720	240
$Gd_3Ga_5O_{12}$ [16, 17]	380	12.376		769	160
$Y_3(Sc,Ga)_2Ga_3O_{12}$ [15]	250		696.3–694.9	740	145
$(Gd,Ca)_3(Ga,Mg,Zr)_2Ga_3O_{12}$ [14]	100	12.50		760	106
$Gd_3(Sc,Ga)_2Ga_3O_{12}$ [15]	50		690	760	120
$Gd_3Sc_2Ga_3O_{12}$ [13]	50	12.567	696.4–695.0 694.2–693.2 (4K)	785	245–340 530–330 (4K)
$La_3Lu_2Ga_3O_{12}$ [16, 17]	−1000			830	68

B1.1.3.3 Radiative and non-radiative transitions. Energy transfer processes.

Radiative and non-radiative transitions

As the Cr^{3+} transition metal ion of $3d^3$ configuration is characterized by strong absorption associated with the ${}^4A_2 \rightarrow {}^4T_2$ and ${}^4A_2 \rightarrow {}^4T_1$ spin-allowed transitions of vibronic nature, there can be efficient pumping thanks to broad xenon flashlamp or laser visible sources. In the examples shown in figure B1.1.11, under ${}^4T_2$ pumping, populating the excited state of the laser transition requires efficient radiation-less relaxation down to the lowest state in each case. The dynamics of the non-radiative decays between ${}^4T_1$, ${}^4T_2$ and ${}^2E$ excited states have been analysed by applying the four-wave mixing technique in several hosts [16–18]. The time scale between vibrational levels is very fast, within tens of picoseconds. Therefore, radiative transition provides spontaneous emission and stimulated emission in lasers. The quantum efficiency η of the emitting level ($\eta = \frac{\tau\ \mathrm{exp}}{\tau\ \mathrm{rad}}$) is an important parameter in the search for efficient radiative emission. η depends on both the experimental fluorescence (τ exp) lifetime and the radiative lifetime (τ rad) which is very often difficult to measure. Calculations of both oscillator strength (f), absorption cross section σ abs and stimulated emission cross sections σ of the transition are detailed in many references [6, 7, 15]. However, such a description considers isolated ions in the host material, but experience reveals a strong concentration dependence of the optical properties due to ion–ion interaction, which is a function of the distance of two neighbour ions. The main consequence is the energy transfer process which takes place when distances are less than $\sim$2 nm. The interactions known are dipole–dipole, dipole–quadrupole and quadrupole–quadrupole. Two types of energy transfers have been observed: resonant and non-resonant or phonon-assisted energy transfers. The energy transfers may lead to energy diffusion through resonant radiative processes and to concentration quenching when a sensitizer ion transfers the excitation to an unexpected and unexcited activator ion. These processes have to be understood in doped-crystals. Especially in the high crystal field schema of Cr^{3+} ion in which ${}^2E \leftrightarrow {}^4A_2$ resonant transition occurs, lifetime measurements provide useful information.

Role of radiative and non-radiative energy transfers in the concentration quenching processes

Resonant radiative energy transfer is well known for acting as a radiation trap. Resonant transitions permit a long-range energy diffusion between identical ions by successive re-absorption/re-emission processes. This serial mechanism affects the experimental excited state decay because it leads to a lengthening of the fluorescence lifetime measured over the volume of the sample relative to the lifetime of a single isolated ion. Strong spectral overlap between the fluorescence and absorption spectra enhances fluorescence re-absorption. Examples of resonant laser transitions in this class are ${}^2E \leftrightarrow {}^4A_2$ of Cr^{3+} in crystals and glasses like in ruby at 694.3 nm, more generally any resonant 0-phonon lines of transition metal ions and rare earth ions such as ${}^5I_7 \leftrightarrow {}^5I_8$ in Ho^{3+} at 2.1 μm, ${}^3F_4 \leftrightarrow {}^3H_6$ in Tm^{3+} at 2.0 μm, ${}^4I_{13/2} \leftrightarrow {}^4I_{15/2}$ in Er^{3+} at 1.5 μm and ${}^2F_{5/2} \leftrightarrow {}^2F_{7/2}$ in Yb^{3+} at 980 nm. The size and the concentration of the sample are critical factors since the probability of a re-absorption/re-emission event depends on the distance that the emitted light has to travel in the sample. Depending on the nature of the sample, micro-crystal or bulky single crystal, large variations in lifetimes can be found [19, 20].

Figure B1.1.12 shows the observed experimental results lifetime *versus* the dopant concentration for the ruby. Identifying the origin of the observed lifetime variations allows the understanding of the excited state dynamics. The fluorescence lifetime increases from 4.2 ms at 10^{-3} weight% Cr_2O_3 up to 10.5 ms at 0.2 weight% Cr_2O_3 and decreases to 2 ms at 1 weight% Cr_2O_3 [19]. The curve can be divided into two regimes: in the lowest concentration range the experimental lifetime increases as the dopant concentration increases whereas for the higher concentration range, measured lifetime decreases when the doping rate increases.

Consequently, the two behaviours seen from curve (b) in figure B1.1.12 are, respectively:

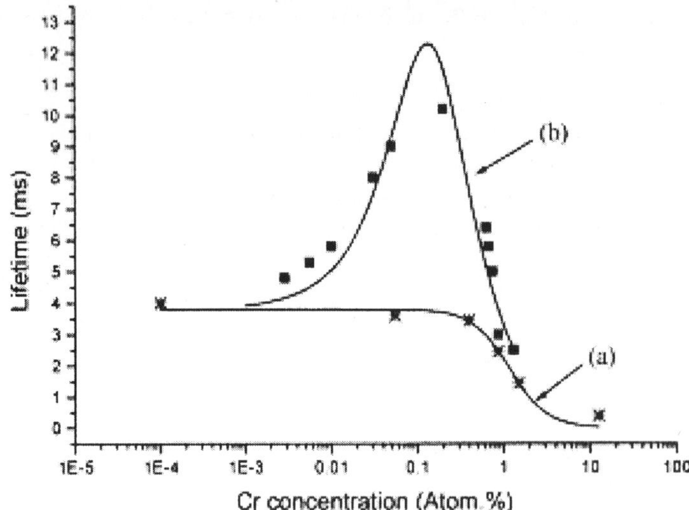

Figure B1.1.12. Self-quenching of ruby at 77 k. (*a*) Comparison of experimental points for powder from [19] with theory as given from [21] with $\tau_W = 3.8$ ms and $N_0 = 1.4\%$; (*b*) comparison of experimental points for a bulk crystal from [20] with theory given from [21] by with $\tau_W = 3.8$ ms and $N_0 = 0.2\%$.

- fluorescence re-absorption; the so-called self trapping process between resonant transitions which results from radiative energy transfer in the case of $^2E \rightarrow {}^4A_2$ and $^4A_2 \rightarrow {}^2E$ transitions. Experimentalists have to be careful to record purely radiative lifetime due to the strong influence of the concentration as observed in figure B1.1.12, since a wrong estimation of the lifetime can be the source of some error for the application to laser cavities;

- the usual quenching process as it can frequently be seen in luminescent materials; in a general way, fluorescence extinction is mainly due to energy migration through the excited level of the emitting ion to either the same neighbour ions (forming pairs and clusters) or to impurities such as transition metal ions for which oscillator strengths of the optical transition (3d configuration) are much higher than those of rare-earth ions (4f configuration) and even to unknown intrinsic defects.

We can say that the same behaviour is observed with a rare-earth ion, (Ho^{3+}, Tm^{3+}, Er^{3+} and Yb^{3+}) presenting resonant transitions between the first excited level and the fundamental state, by using concentration gradient fibres [22]. The originality of concentration gradient samples is that one fibre contains a collection of single crystals of different concentration spread along its axis. Such single-crystalline fibres present a dopant concentration varying continuously between two chosen compositions. Gradient concentration fibres are synthesized by the laser heated pedestal growth (LHPG) technique from a special ceramic feeding rod and the concentration distribution is analysed by EDX microprobe at regular intervals along the axis of the fibre. These procedures could be applied to Cr^{3+} doped crystals to evaluate the concentration quenching processes and the measurement of the intrinsic radiative lifetime.

It is worthwhile reminding readers that the linewidth of the transition at very low temperature should simply reflect the radiative lifetime of the state as is required by the uncertainty principle. Fluorescence line narrowing (FLN), optical hole burning and optical coherent transients have all been employed in the study of this problem in dilute systems such as ruby [23]. The high-resolution laser spectroscopy technique has determined a width of ~100 MHz which is therefore likely to contain residual inhomogeneous components.

B1.1.4 Emission sharp line of the Cr^{3+} ion in the ruby laser: a strong crystal field model and a three-level scheme

The ruby laser occupies a special place in laser science: a rod of synthetic ruby produced the first laser emission demonstrated by Maiman in 1960 [1] in the deep red at 694.3 nm and, even nowadays, it is one of the rare solid-state laser mediums emitting in the visible range. The ruby story will stay with us as one of the best examples of scientists being obstinate in their approach because, before 1960, it was claimed that the ruby laser would not work. Forty years later, however, such laser sources are still useful for certain applications such as pulsed holography, interferometry, non-destruction tests and plasma measurements. Ruby consists of a sapphire Al_2O_3 single crystal with corundum structure in which approximately 0.05 weight% of Al^{3+} cations have been replaced by Cr^{3+} active ions, i.e. 1.58×10^{19} ions cm^{-3}. This substitution is possible due to the same 3+ valence and an ionic radius (62 pm) similar to that of aluminium (53 pm). Depending on the initial concentration of Cr^{3+}, pink or red rods are available from 3–25 mm in diameter and up to 20 cm long. They can be grown either by the Verneuil method, used since the beginning of the century, or by the more recent Czochralski method which gives better optical quality samples. The ruby's success is associated with its excellent mechanical, chemical and thermal properties and, obviously, to the possibility of getting inversion of population between the $^4A_2$ ground state and the ^2E excited state of Cr^{3+} ions. Sapphire is, in fact, a hard, refractory, unalterable material. The optical transparency between UV fundamental absorption and usual infrared absorption of the sapphire host is large. The fundamental absorption occurs near 165 nm; i.e. a band gap of 6.2 eV so that excited-state absorption of the laser emission from the Cr^{3+} emitting levels to the conduction band is inefficient. The thermal conductivity is the largest of the laser crystals, 42 W m^{-1} K^{-1} at room temperature, much higher than YAG (11 W m^{-1} K^{-1}) which is considered as the reference. The main drawback of the ruby laser is the operation as a three-level scheme, the ground state also playing the role of the terminal level of the laser transition considerably limiting the laser output energy. Therefore, such a laser is characterized by a high threshold. Finally, only three levels are important for the usual interpretation of Cr^{3+} absorption and emission spectra. Besides the $^4A_2$ ground state, the $\bar{E}$ component of the ^2E and $^4T_{2g}$ levels, which are the two lowest excited states, will be used to understand laser operations. In ruby, $\bar{E}$ is the lowest excited state, and therefore the laser transition is interpreted as $^2E \rightarrow {}^4A_2$ at 694.3 nm. In almost all other Cr^{3+}-doped crystals, $^4T_2$ is the lowest excited state, giving rise to a $^4T_2 \rightarrow {}^4A_2$ tunable solid state laser between 700 and 1200 nm depending on the nature of the host.

The spectroscopic properties of ruby can be interpreted in figure B1.1.15 within Tanabe and Sugano's strong crystal field model which has been explained in the first part of this article. $^4T_2$ and $^4T_1$ visible broad absorption bands are pumped from the $^4A_2$ ground state with xenon flashlamps. The non-radiative processes between $^4T_2$, $^4T_1$ and ^2E are much higher in probability than the direct transition to the ground state. The transition $^2E \rightarrow {}^4A_2$, which appears as two sharp emission lines at 694.3 nm and 692.9 nm, belongs to the same t_{2g} configuration and represents a forbidden electronic dipole transition and even spin-forbidden transition. The ^2E excited state is then a metastable level with a lifetime of 3 ms.

Figures B1.1.13 and B1.1.14 show the absorption spectra of $^4A_2 \rightarrow {}^4T_1$, $^4A_2 \rightarrow {}^4T_2$ and $^4A_2 \rightarrow {}^2E$ transitions analysed as a function of the polarized pump beam, parallel or perpendicular to the optic axis c of the uniaxial sapphire crystal. $^4A_2 \rightarrow {}^4T_1$ and $^4A_2 \rightarrow {}^4T_2$ absorption broadbands which match nicely a large part of the visible spectral range except in the red are at the origin of the very well-known colour of the ruby crystal.

Such a long lifetime allows high-energy storage, much higher than with Nd^{3+}-YAG in which the $^4F_{3/2}$ excited state lifetime is only 0.25 ms. This is the main advantage of the ruby laser, producing high-energy Q-switched pulses. The ruby laser can operate in either an oscillator (figure B1.1.16) or oscillator-amplifier configuration in one of the three-pulse modes [26, 27]:

(i) the long-pulse mode called normal or conventional from 0.3 to a few milliseconds long depending on the length of the light pulse from the pump lamp. The generation of spikes during the pulse is known as

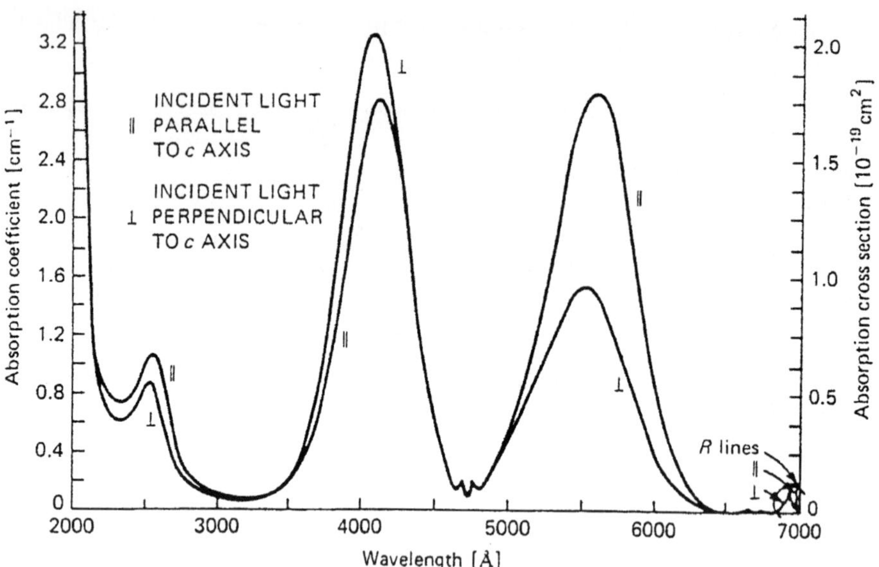

Figure B1.1.13. The absorption coefficient and absorption cross section of ruby for the incident beam polarized parallel and orthogonal to the c axis (Cr^{3+} concentration is $1.88.10^{19}$ at cm^{-3}) [24].

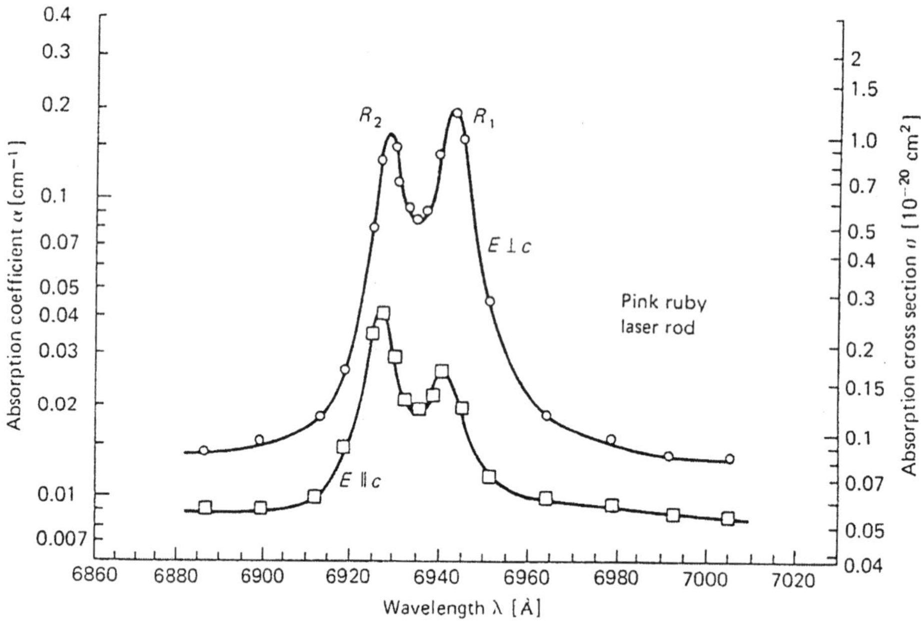

Figure B1.1.14. The absorption coefficient and absorption cross section of the R_1 and R_2 lines in ruby for an incident beam polarized parallel and orthogonal to the c axis (Cr^{3+} concentration is 1.58×10^{19} at cm^{-3}) [25].

a peculiar feature of the ruby laser since the first observation of stimulated emission in this system. One application of instantaneous pulses is for drilling holes in ruby samples, applied in watches for instance;

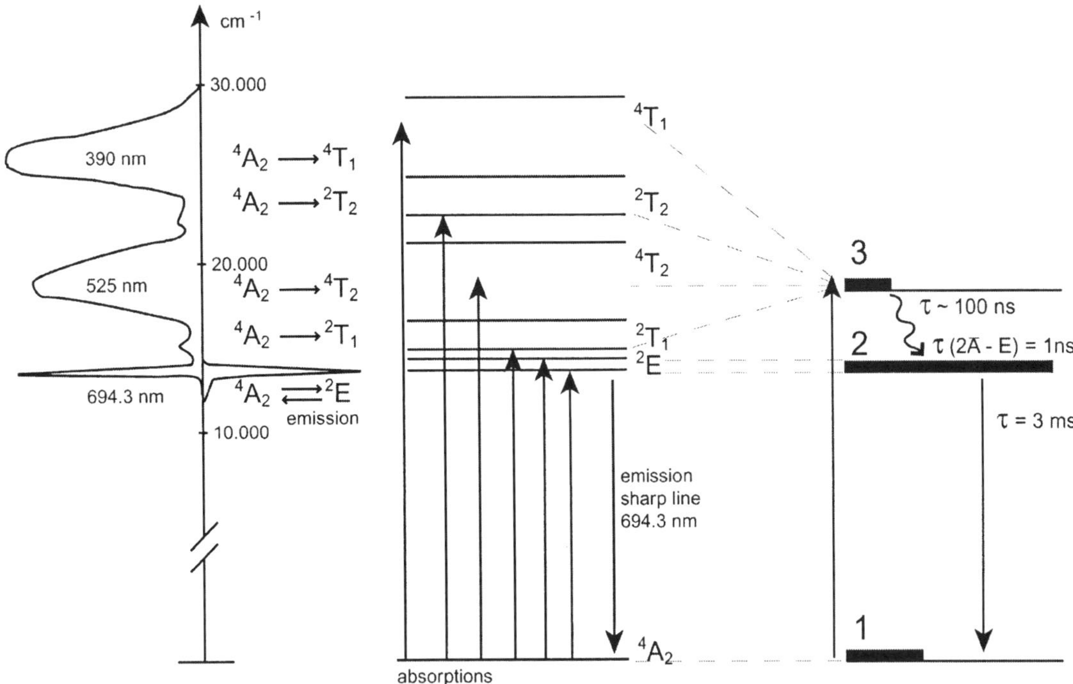

Figure B1.1.15. An interpretation of the absorption and emission spectra of ruby by the schematic three-level system.

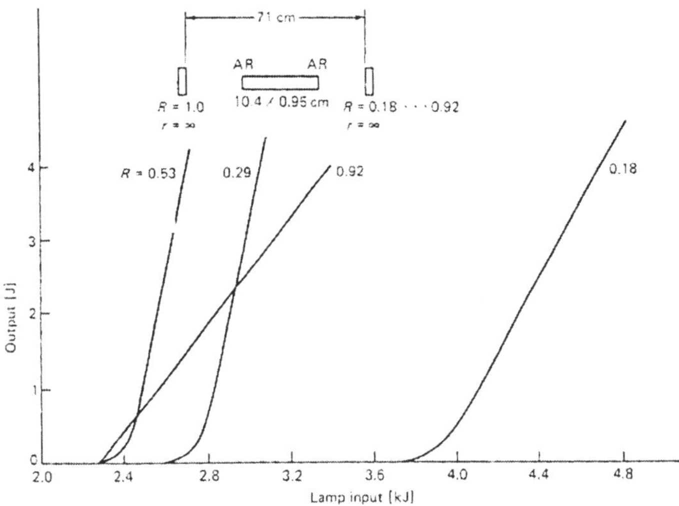

Figure B1.1.16. Oscillator output *versus* lamp input with output coupling as parameters [20, 21].

(ii) the Q-switched mode with a usual Pockels cell or even a passive saturable dye. Ten to 35 ns pulse durations can be obtained with peak power in the 100 MW range or 1 GW in the oscillator-amplifier system;

(iii) mode-locked, ruby lasers can produce a train of 20–30 pulses each as short as 3–4 ps. Each pulse has energy around 1 mJ, so peak power can reach hundreds of megawatts but are highly unstable.

Finally, we can say that, even within the most difficult three-level scheme operation, the ruby laser's initial success at room temperature was the result of favourable conjunctions. These are the two broad and intense absorption bands in the visible spectrum ($^4A_2 \rightarrow {}^4T_1$ at ~400 nm and $^4A_2 \rightarrow {}^4T_2$ at ~550 nm), a unique emission line in the red ($^2E \rightarrow {}^4A_2$ at 694.3 nm) with a relatively narrow linewidth (330 GHz = 0.53 nm), a very long excited-state lifetime (3 ms) and a high radiative quantum efficiency of 0.7. Within the visible part of the spectrum, characterizations were easier than in UV or IR with the use of sensitive photomultipliers. In addition, we recall that ruby is the simplest host for Cr^{3+} doping due to its natural substitution of Al^{3+} by Cr^{3+} in the desirable octahedral position. The growth of ruby by the pioneering experiments of Verneuil, who developed the so-called flame fusion technique one century ago, gave at one's early disposal, good optical quality samples and more recently, even better samples by using the Czochralski method. All of these features strongly contributed to the outcome of the first laser operation in 1960 by Maiman, which was followed by the impressive development of solid-state laser engineering and the discovery of new laser media.

B1.1.5 Emission broadband Cr^{3+} laser crystals: a weak crystal field model and a four-level scheme

B1.1.5.1 Spectroscopic approach

We have seen that the ruby laser emits only a sharp line associated with $^2E \rightarrow {}^4A_2$ emission which is resonant with the $^4A_2 \rightarrow {}^2E$ absorption line. ^2E and $^4A_2$ belong to orbitals within the t_{2g} configuration corresponding to lobes between ligand directions, as a spin-flip transition. As a result, configuration-coordinate diagrams therefore use parabolas centred on the same vertical energy axis, without any Stokes shift, and with a Huang–Rhys parameter $S \sim 0$ (figure B1.1.10). At room temperature, no trace of the $^4T_2 \rightarrow {}^4A_2$ emission broad band can be detected due to the large gap of 2300 cm^{-1} between the ^2E level and the $^4T_2$ upper level. Such a broad band is only seen at higher temperatures when ^2E high-vibrational levels occupy the same energy as those of the $^4T_2$ level. The only way to observe $^4T_2 \rightarrow {}^4A_2$ broadband variation is to control it with the crystal field strength between Cr^{3+} and neighbour ligands in different hosts, i.e. to adjust the distance R between Cr^{3+} to the centre of the ligand and therefore the parameter $\Delta = \frac{5eq\bar{r}^4}{3R^5}$ given previously. If R increases, Δ decreases and, therefore, for some Cr^{3+}-doped crystals, the $^4T_2$ parabola can be drawn at the same energy level as the ^2E level and even below the ^2E level. In this case, after relaxational processes between both the ^2E and $^4T_2$ levels, the observed emissions are associated with both the $^2E \rightarrow {}^4A_2$ sharp line as well as the transitions between the $^4T_2$ lowest vibrational levels and the $^4A_2$ vibrational levels. In addition, a Stokes shift is associated with the electron–phonon coupling in the $^4T_2$ level which is stronger than in ^2E, since $^4T_2$ belongs to orbitals transformed as e_g, corresponding to lobes along the ligand directions.

Besides ruby, other Cr^{3+}-doped crystals can show R laser lines in the strong crystal field model at room temperature, such as spinel $MgAl_2O_4$(14 660 cm^{-1}), $Y_3Al_5O_{12}$(YAG) (14 530 cm^{-1}) and alexandrite $BeAl_2O_4$(14 706 cm^{-1}), compared to 14 400 cm^{-1} in ruby. However, as we have just demonstrated, broadband fluorescence occurs from the $^4T_2$ level in some hosts where the crystal field is weaker. This can be used for vibronic tunable lasers. It has the advantage of a four-level scheme built up with vibrational levels of both $^4A_2$ and $^4T_2$ electronic states (figure B1.1.17) thanks to optical pumping within intense and broad bands in the visible range. Solid-state vibronic lasers, which work similarly to dye lasers, were considered to be much more simple, compact and robust. The development of this media is strongly dependent on the success of the single laser crystal growth techniques from melts, with severe requirements such as high optical quality of the crystal, homogeneous distribution of Cr^{3+} active ions, large sizes for rods and slabs, as well as low loss at laser wavelengths in the red and near infrared and absence of laser induced damage. Thus, since the seventies, research has been very keen to discover new Cr^{3+}-doped crystals characterized by a weak crystal field model. Table B1.1.6 gives the main crystals grown and shows slope efficiency measurements of the

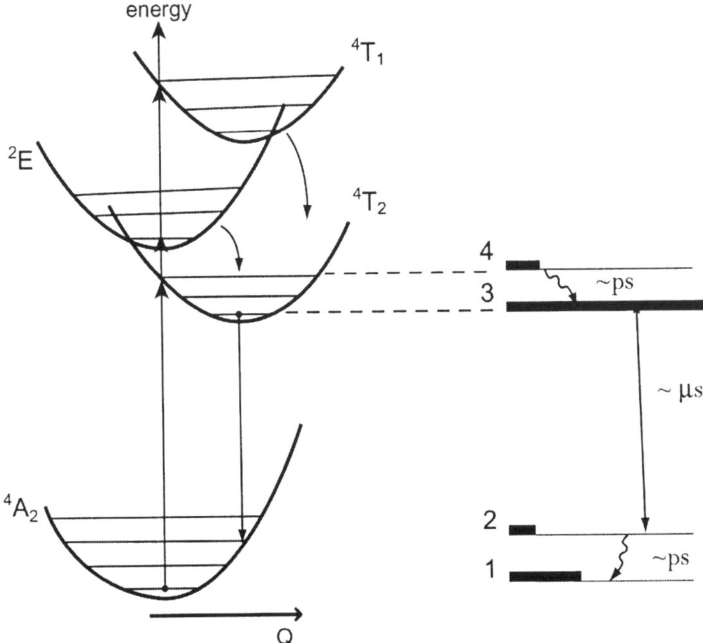

Figure B1.1.17. An interpretation of Cr^{3+} centres in the weak crystal field model by the schematic four-level system. Relaxation times have been shown.

$^4T_2 \rightarrow {}^4A_2$ peak lasing wavelength under the laser-pumped configuration [28]. It can be seen that emerald, LiCAF and LiSAF fluorides and alexandrite are at the top of the evaluation. Unfortunately, emerald is too difficult to grow for industrial development. We shall only illustrate this part with the two main crystal families under consideration today, alexandrite and LiCAF or LiSAF fluorides.

B1.1.5.2 Alexandrite laser crystal

The first vibronic lasers were reported in 1963 by Bell Labs researchers [29, 30] using the transition-metal ions Ni^{2+}, Co^{2+} and V^{2+} in MgF_2, MgO, ZnF_2 or $KMgF_3$ hosts which only work when cooled to cryogenic temperatures. This is why the first vibronic laser to reach the market, in 1979, working at room temperature, was the alexandrite crystal or Cr^{3+}-doped $BeAl_2O_4$ [31]. $BeAl_2O_4$, (also named chrysoberyl) has hexagonal close-packed structure isomorphous with olivine in the orthorhombic space group D^{16}_{2h}. The Al^{3+} ions are octahedral, coordinated by the oxygen ions and occupy two unequivalent crystallographic sites, one with mirror symmetry (C_s point group) occupied by 78% of Cr^{3+} ions and one with inversion symmetry (C_i point group) occupied by 22% of Cr^{3+} ions. The concentration of Cr^{3+} dopant is usually between 0.05 and 0.3 at% (0.1 at% represents 3.51×10^{19} ions cm^{-3}). Alexandrite is optically and mechanically similar to ruby. High thermal conductivity (23 W m^{-1} K^{-1}, two-thirds that of ruby and twice that of YAG) enables alexandrite rods to be pumped at high average powers without thermal fracture.

Absorption and emission spectra are shown in figure B1.1.18 for the three types of polarizations $E \parallel a$, $E \parallel b$ and $E \parallel c$ since alexandrite is biaxial (a, b and c are the usual crystallographic axes). Laser radiation is emitted from $E \parallel b$ as can be seen in figure B1.1.18. Both $^2E \rightarrow {}^4A_2$ sharp lines of the two crystallographic sites and $^4T_2 \rightarrow {}^4A_2$ broadbands are observed. By analogy with ruby, alexandrite can operate at room temperature as a three-level scheme with mirror symmetry Cr^{3+} ions: Rm lines at 680.4 nm with a lifetime

Table B1.1.6. The measured efficiencies of Cr^{3+} vibronic lasers utilizing the laser-pumped laser configuration.

Host material	Name	Peak lasing wavelength (nm)	Slope efficiency %
$Be_3Al_2(SiO_3)_6$	Emerald	768	64
$LiCaAlF_6$	LiCAF	780	54
$BeAl_2O_4$	Alexandrite	752	51
$LiSrAlF_6$	LiSAF	825	36
$ScBeAlO_4$	Scalexandrite	792	30
$ScBO_3$	Borate	843	29
$Gd_3Sc_2Ga_3O_{12}$	GSGG	785	28
$Na_3Ga_2Li_3F_{12}$	GFG	791	23
$Y_3Sc_2Al_3O_{12}$	YSAG	767	22
$Gd_3Sc_2Al_3O_{12}$	GSAG	784	19
$SrAlF_5$	Pentafluoride	932	15
$KZnF_3$	Perovskite	820	14
$ZnWO_4$	Tungstate	1035	13
$La_3Ga_5SiO_{14}$	LGS	968	10
$Gd_3Ga_5O_{12}$	GGG	769	10
$Ca_3Ga_{5.5}Nb_{0.5}O_{14}$	Niobate	1040	5
$Y_3Ga_5O_{12}$	YGG	(740)	5
$Y_3Sc_2Ga_3O_{12}$	YSGG	(750)	5
$La_3Lu_2Ga_3O_{12}$	LLGG	830	3

of 1.54 ms. The other crystallographic site which is occupied by 22% of Cr^{3+} ions is characterized by inversion symmetry and then optical transitions are strongly forbidden. As a result, the lifetime is 63 ms below 130 K, one of the longest known lifetimes. The stimulated emission cross section σ_{RM} of the Cr^{3+} ions in used mirror symmetry sites is ten times larger (3×10^{-19} cm^2) than in ruby. However, the main interest of alexandrite is associated with the $^4T_2 \rightarrow {}^4A_2$ broadband which allows laser tunability between 700–826 nm. The $^4T_2$ lifetime is 260 s and the stimulated emission cross section σ_m (band) is 1.0×10^{-20} cm^2. Alexandrite is pumped by either xenon flashlamp or mercury arc sources into the $^4T_1$ and $^4T_2$ absorption broadbands between 380–630 nm. The tunable output can reach 100 W average under pulsed pumping and 50 W under CW pumping [32].

An interesting spectroscopic feature has to be mentioned regarding the broadband emission of alexandrite crystal. Laser photons in the cavity can be absorbed by excited-state absorption of the ^2E long-lived metastable state [33]. Obviously excited-state absorption adds a supplementary loss to the photopopulation rates and additional heating of the crystal. However, in this compound the excited-state absorption spectrum exhibits a broad minimum area which exactly matches the $^4T_2 \rightarrow {}^4A_2$ broadband tunable laser emission. This effect represents 10% at the peak maximum and the only detected limitation is observed for the long-wavelength laser range.

Temperature dependences of the output laser energy in figures B1.1.19 and B1.1.20 show strong effects. As the temperature increases and the gain of alexandrite increases, the laser wavelength shifts to longer wavelength and the fluorescence lifetime decreases. This is connected with the Boltzmann distribution of the $^4A_2$, ^2E and $^4A_2$ states which produces an inversion population between the $^4T_2$ and $^4A_2$ states. The stimulated cross section increases since, in addition to the usual pumping *via* $^4T_1$ and $^4T_2$ vibrational levels, $^4T_2$ is also populated *via* the ^2E vibrational levels only separated by 800 cm^{-1} thereby increasing the population of the initial laser level. However, raising the temperature also populates the $^4A_2$ vibrational levels,

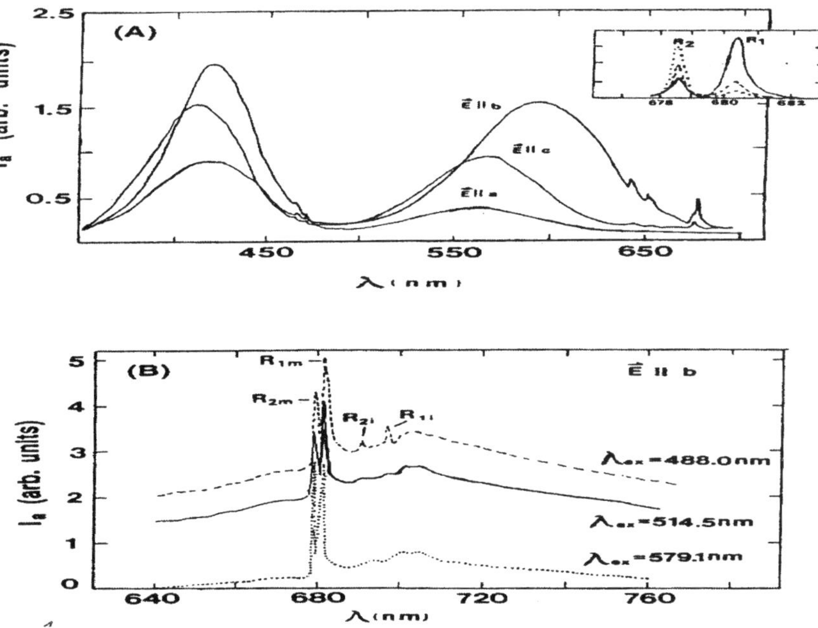

Figure B1.1.18. The absorption and emission spectra of alexandrite at room temperature for incident beam polarized parallel to the *a*, *b*, *c* axes, respectively [30].

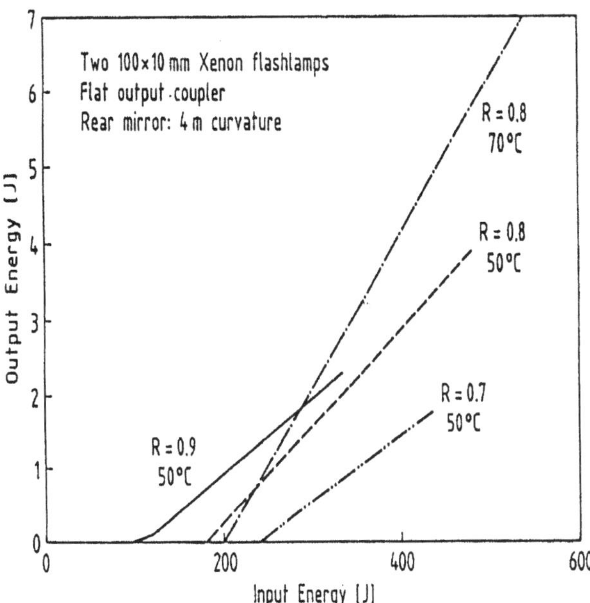

Figure B1.1.19. Output energy *versus* input energy for the alexandrite laser at 50 °C and 70 °C respectively [26].

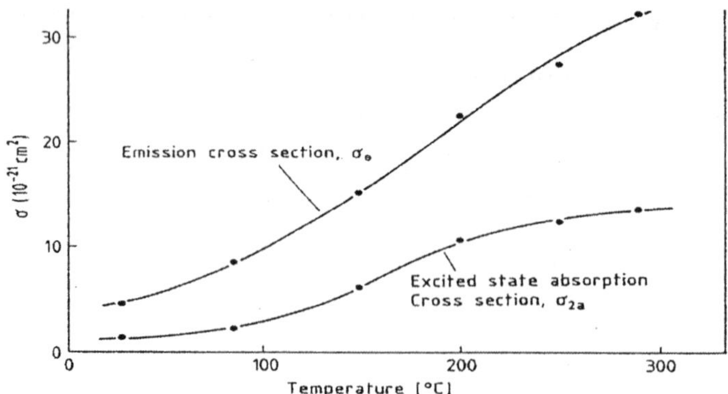

Figure B1.1.20. The temperature dependence of emission and excited-state absorption cross section at 790 nm [26].

and so conflicting effects occur. The result is that performance is only positively affected by temperature increases for wavelengths above 730 nm [26].

B1.1.5.3 Colquiriite solid-state laser

Cr^{3+}-doped fluoride crystals also show a vibronic emission which is used for tunable solid-state lasers. The optical transparency of fluorides is higher than oxides which is important to reduce any excited-state absorption of the laser emission from the Cr^{3+} emission level. Thus, non-radiative transitions *via* the conduction band can be neglected. An important advance has been made with the discovery of $LiCaAlF_6$ (LiCAF) and $LiSrAlF_6$ (LiSAF) of the colquiriite family by the Lawrence Livermore group [34, 35]. In particular, the Cr^{3+}-doped LiCAF laser appears to be as efficient as alexandrite when pumped by other lasers [36] (seen in tables B1.1.6 and B1.1.7), or by flashlamps [37]. The second of the colquiriite-structure crystals to be lased was Cr^{3+}-doped LiSAF [35]. Here, Sr^{2+} is used to replace the Ca^{2+} ion of the LiCAF host crystal. Absorption and emission cross sections are shown in figure B1.1.21 under polarized optical pumping.

Although the efficiency of Cr^{3+}-LiSAF was found to be slightly less than that of Cr^{3+}-LiCAF (see tables B1.1.6 and B1.1.7), it has, nevertheless, drawn more attention due to its higher emission cross section, broader tuning range and, in addition, the option of being doped to high Cr^{3+} concentrations giving the first Cr^{3+} stoichiometric laser crystal [38]. The tuning range of this laser crystal has been demonstrated to be from 780–1010 nm. The high emission cross section of 4.8×10^{-20} cm^2 and the emission lifetime of 67 s have resulted in successful Q-switching [39]. The absorption bands have also permitted laser action by pumping with visible AlGaInP diodes at 670 nm [40]; laser output performances are given in table B1.1.8, Cr^{3+}-Nd^{3+}:GSGG garnet has been listed for comparison with the neodymium solid-state laser.

Colquiriite crystals can be an attractive alternative to Ti^{3+}:Al_2O_3 sapphire since they can be pumped with a diode and exhibit greater energy storage and lower laser threshold. Obviously, diode laser pumping on the low-energy side of the $^4T_2$ broadband minimizes the so-called quantum defect between photon energies of excitation and emission transitions which is an important limitation of the efficiency. Since 1995, it is possible to find on the market CW diode-pumped, single frequency (resolution limited linewidth $<$20 MHz), Cr^{3+}-doped LiSAF micro-lasers and diode-pumped sub-100 femtoseconds (fs) passive mode-locked lasers. Due to the broadband absorption, which overlaps with the emission bands of xenon flashlamps, and longer luminescence lifetime in comparison with Ti^{3+}:Al_2O_3; Cr^{3+}:LiSAF and Cr^{3+}:LiCAF are suitable amplifier media for flashlamp pumping. A terawatt class system with a Cr^{3+}:LiSAF regenerative amplifier to produce 150 fs pulses with outputs of 15 mJ was designed as shown in the schematic diagram of figure B1.1.22 [41].

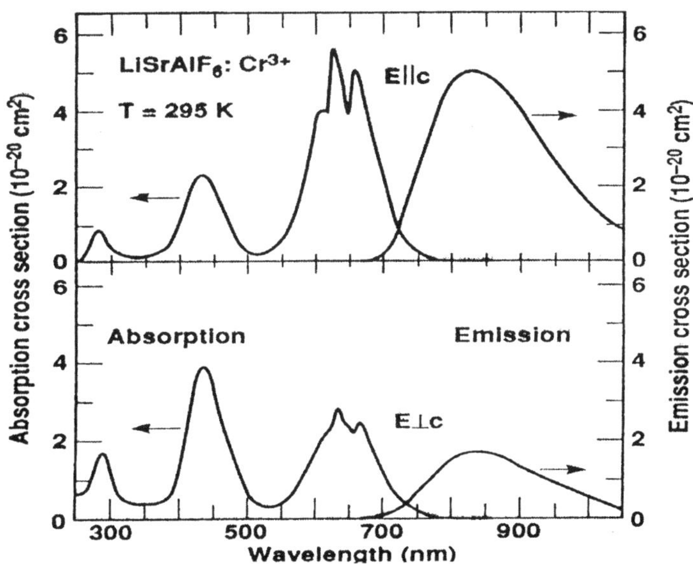

Figure B1.1.21. Absorption and emission spectra at room temperature of Cr^{3+}:LiSAF for two incident beam polarized parallel and orthogonal to the c axis [35].

Table B1.1.7. Laser efficiency optical and thermal properties of Cr^{3+}-doped colquiriite structure crystals.

Host	$LiCaAlF_6$	$LiSrGaF_6$	$LiSrAlF_6$
Name	LiCAF	LiSGAF	LiSAF
Intrinsic efficiency η_0(%)	67	52	53
Quantum intrinsic efficiency η_{0q}(%)	82	66	67
π emission cross section, σ_{em} (10^{-20} cm^{-2})	1.3	3.3	4.8
Emission peak nm	763	835	846
Emission lifetime τ (μsec)	170	88	67
Zero-phonon line (nm)	678	699	699
Expansion α_c coefficient	3.6	0	−10
(10^{-6}/°C) α_a	22	12	25
Melting point (°C)	810	716	766

Also a self-starting, self-mode-locked, femtosecond, diode-pumped Cr^{3+}:LiSAF laser has been demonstrated [42] with a Kerr-lens, mode-locked, diode-pumped crystal that starts with a simple mirror and produces 70 fs pulses at 850 nm with an average power of 50 mW, as shown in figure B1.1.23. The same group has developed

Table B1.1.8. Laser output performances of diode laser pumped Cr^{3+}:LiSAF and Cr^{3+}:LiCAF [15].

Laser	Output power (mW)	1 W-diode pumped slope efficiency (%)	Maximum slope efficiency (%)	Optical conversion efficiency (%)	Overall electrical efficiency (%)
Cr:LiCAF	158	28	50	19	1.7
Cr:LiSAF	176	25	45	21	1.9
Cr,Nd:GSGG	173	23	42	21	1.9

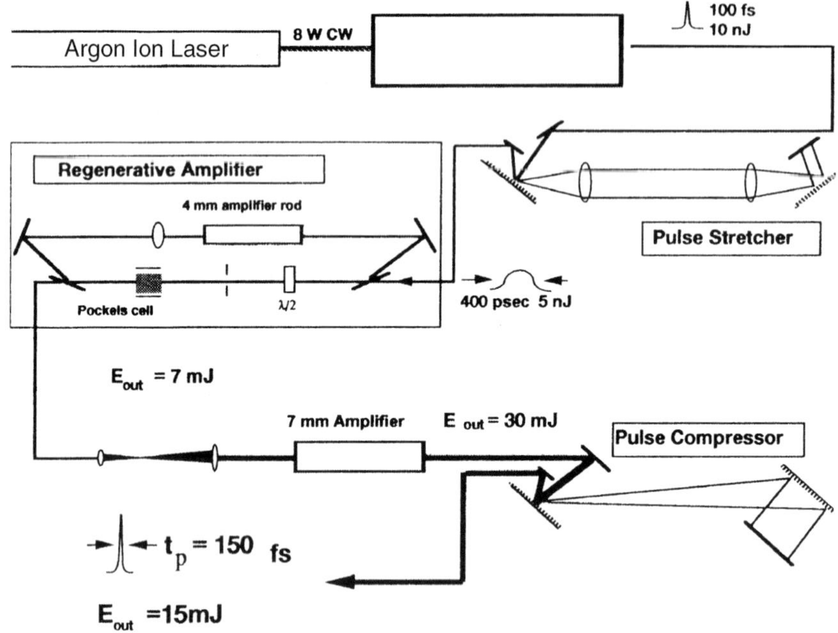

Figure B1.1.22. An example of the Cr^{3+}:LiSAF laser optical layout [41].

a Cr^{3+}:$LiSrGaF_6$ laser that produces, in Q-switched operation, four times as much energy (12 μJ at 10 kHz) as a Cr^{3+}:LiSAF laser under the same pumping conditions (four red diodes emitting 400 mW) [43].

We would like to mention that the development of the colquiriite family of laser crystals is the result of a proper scientific approach. Laser performances of Cr^{3+}-doped crystals tend to be best when Cr^{3+} ions are substituted for small cations such as Al^{3+} at octahedral sites and when the host is composed of small and more highly-charged cations constituting a stiff lattice. The comparison between $LiSrGaF_6$, $LiSrAlF_6$, $LiCaAlF_6$ and $LiCaGaF_6$ compounds shows that the nearest-neighbour (nn) cations (Sr^{2+} or Ca^{2+}) impact on the Cr^{3+} properties more significantly than the identity of the substitutional site (Al^{3+} or Ga^{3+}). Two qualitative explanations for this observation have been given: (i) lattice relaxation at the Cr^{3+} ion compensates for the change in the size of the substitutional site; (ii) the nn cation serves to displace the six fluorines surrounding Cr^{3+}. This latter change is not compensated for by lattice relaxation. The nn cations impose a t_{2u}-type distortion on the fluorines at the Cr^{3+} site. The t_{2u} twist angles θ_{tw} have been measured and may be related

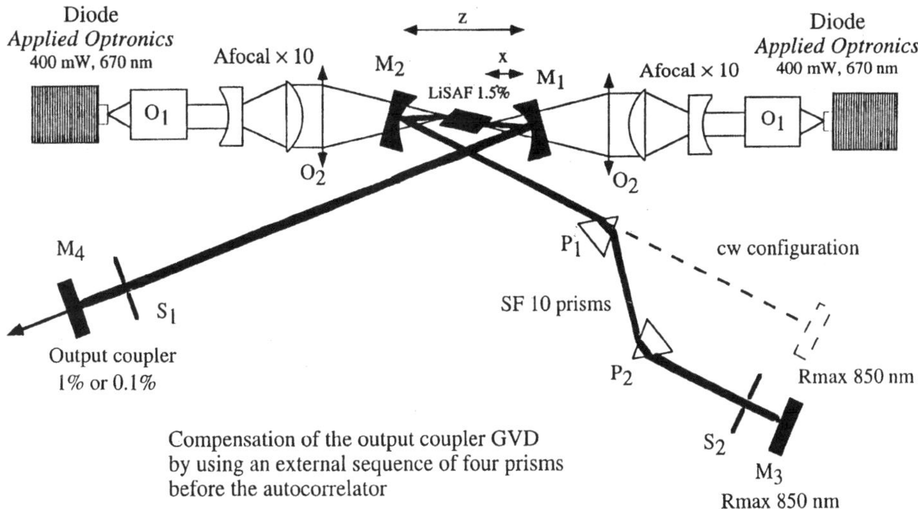

Figure B1.1.23. The x-fold cavity configuration used in a compact diode-pumped Kerr-lens mode locked Cr^{3+}:LiSAF laser [42] (M, mirrors; S, slits; P, prisms; O, objectives).

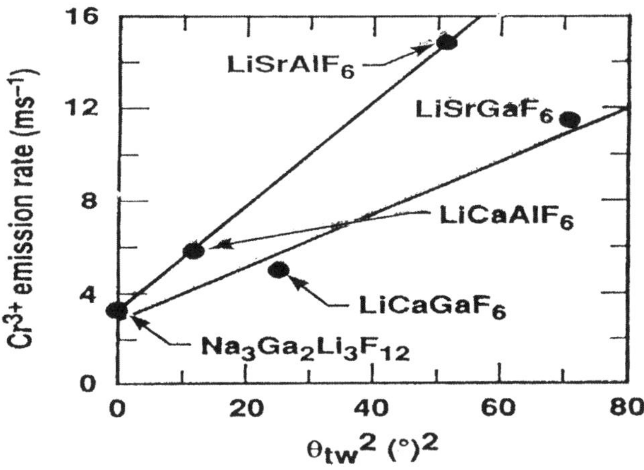

Figure B1.1.24. Cr^{3+} emission rate in m s^{-1} as a function of θ_{tw}^2 in $(°)^2$ for fluoride garnet and colquiriite crystals.

to the observed Cr^{3+} emission lifetime τ_{em} (figure B1.1.24).

The comparison with $Na_3Ga_2Li_3F_{12}$(GFG) garnet is interesting because the Cr^{3+} ion substitutes for the Ga^{3+} ion which is in a site of almost regular octahedral symmetry C_{3i}, with inversion-site symmetry. Consequently, the fluorescence lifetime is longer (310 μs) than in Cr^{3+}:LiSAF (67 μs) and the emission cross section is much lower (0.6×10^{-20} cm^2). In figure B1.1.24, two straight lines are observed for each family in accordance with θ_{tw}^2 so that the transition rate increases from 3.23 ms^{-1} to 14.93 ms^{-1} in the series GFG, LiCAF, LiSAF. Moreover, the temperature independence of the decay rate (14.93 ms^{-1} from $\tau = 67$ μs) of the $^4T_2$ state in Cr^{3+}:LiSAF implies that the transition probability is dominated by the static contribution of the trigonal distortion. Referring to the configurational coordinate diagrams in figure B.1.1.11, the temperature dependence of the $^4T_2$ state is dependent on an 'activation energy' W, which

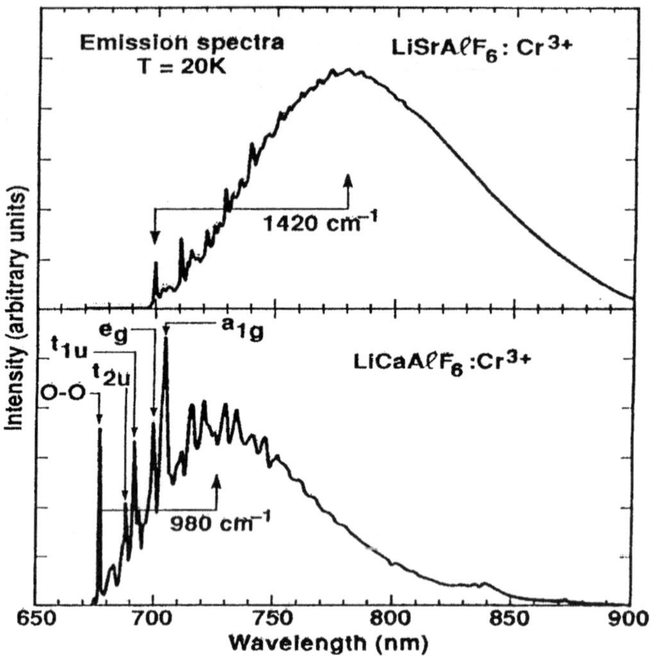

Figure B1.1.25. Emission spectra at 20 K of Cr^{3+}:LiSAF and Cr^{3+}:LiCAF. 0-phonon lines of the $^4T_2 \rightarrow {}^4A_2$ broadband are denoted by 0–0 (699 nm in LiSAF and 678 nm in LiCAF). The difference between 0-phonon lines and the emission peak is also noted (1420 cm^{-1} in LiSAF and 980 cm^{-1} in LiCAF) as an indication of the Stokes shift. The vibronic structure can be interpreted in the lower Stokes shift crystal by t_{2u}, t_{1u}, e_g and a_{1g}.

is determined by the crystal field strength. In this case, W is unusually high, quenching occuring only at higher than room temperatures, thus non-radiative decay processes are not efficient in the fluorescence quenching. Another feature of Cr^{3+}:colquiriite crystals is the larger Stokes shift of the Cr^{3+}:LiSAF emission compared to Cr^{3+}:LiCAF (figure B1.1.25), probably indicating that the Jahn–Teller effect is larger for LiCAF with, in addition, opposite shifts of the $^4T_2$ and $^4T_1$ excited states (the Jahn–Teller effect can be seen in crystals where strong electron–phonon coupling lifts electronic state degeneracy through a splitting of the vibronic energy levels). The result of such opposite distortion in $^4T_2$ and $^4T_1$ is that excited state absorption between $^4T_2$ and $^4T_1$ is unusually broadband but, fortunately, characterized by a low cross section. A theoretical model proves that up-conversion and thermal quenching of fluorescence strongly limit small-signal gain in a CW pumped Cr^{3+}:LiSAF laser [44]. The pumping configuration of a 3% doped Cr^{3+}:LiSAF pumped by two red laser diodes emitting 400 mW has been optimized. In Q-switched operation, this laser produces tunable nanosecond pulses with an energy up to 6.5 μJ. Finally, the emission broadband at 20 K in Cr^{3+}:LiCAF shows vibronic structure in figure B1.1.24 analysed in terms of the splitting of the $^4T_2$ level with t_{2u}, t_{1u}, t_{1u}, e_g and a_{1g} representations in the O_h point group. Recently, the shortest pulse ever produced of 15 fs from an all solid-state mode-locked Cr^{3+}:LiSAF laser has been demonstrated [45].

To summarize, Cr^{3+}:LiCAF and Cr^{3+}:LiSAF laser crystals can be efficiently pumped by flashlamps, diodes or lasers. Cr^{3+} has a unity distribution coefficient and crystal growth conditions are relatively easy at around 800 °C. The tunability is relatively large (~200 nm for Cr^{3+}:LiSAF) and the emission cross section is high (3.2×10^{-20} cm^2 for Cr^{3+}:LiSAF). Long energy storage lifetimes allow Q-switching and amplifier configurations.

Table B1.1.9. Typical operating characteristics of the main Cr^{3+}-doped laser crystals.

Crystals	Ruby Al_2O_3	Alexandrite $BeAl_2O_4$	Garnet $Gd_3Sc_2Al_3O_{12}$	Emerald $Be_3Al_2(SiO_3)_6$	Colquiriite LiCAF $LiCaAlF_6$	Colquiriite LiSAF $LiSrAlF_6$
Crystal field	Strong	Intermediate	Intermediate	Intermediate	Weak	Weak
$\varepsilon(^4T_2$-$^2E)(cm^{-1})$	2300	800	—	400	< 0	< 0
R emission line $^2E \rightarrow {}^4A_2$ (nm)	694.3	680 696	—	—	—	—
Emission broadband $^4T_2 \rightarrow {}^4A_2$ (nm)	—	752	785	768	780	825
Laser tunability (nm)	—	700–826	750–840	750–760	720–840	760–920
Slope efficency (%)		71	28	64	36	54
Quantum efficency	0.70	0.76	0.28	0.86	0.82	0.67
Lifetime of the excited level at RT	3 ms	240 μs and 63 ms	150 μs	65 μs	170 μs	67 μs
Emission cross section (cm^2)	2.5×10^{-19} 19	1.1×10^{-20}	0.70×10^{-20}	1.9×10^{-20}	1.3×10^{-20}	4.8×10^{-20}
Pump	Flashlamp	Flashlamp	Flashlamp	Flashlamp or Kr laser	Laser diode or Kr laser	Laser diode
Examples of performances	ms fixed Q pulses of 100 J	CW50 W	CW250 mW	CW320 mW	(see table B1.1.8)	15 fs pulses (see table B1.1.8)
Main drawback	three-level scheme	difficulties in growing the crystals	Sc high cost	Difficulties in growing the crystals	—	—

B1.1.6 Conclusion

In this chapter our main objective was to describe the concepts that allow us to understand the optical properties of the Cr^{3+} metal transition ion in inorganic oxide and fluoride crystals applied to laser sources. Typical operating characteristics of the main Cr^{3+} laser crystals are summarized in table B1.1.9. Laser science history retains the Cr^{3+} ion not only as the first demonstration of laser radiation in the famous experiment on the ruby, Cr^{3+}:Al_2O_3 laser crystal, by Maiman in June 1960 and then on alexandrite, Cr^{3+}:$BeAl_2O_4$ the first tunable vibronic solid-state laser, but also as the activator ion which triggered scientific discussions on the truest physical models which have to be adopted in laser source research. An important advantage of the Cr^{3+} ion is connected with the two sets of orbitals, e_g and t_{2g} belonging to the $3d^3$ configuration which offer the three types of crystal field strengths in the octahedral symmetry site of oxide and fluoride host crystals: strong crystal fields giving rise to the red sharp line $^2E \rightarrow {}^4A_2$; intermediate crystal fields for which the $^2E \rightarrow {}^4A_2$ and the $^4T_2 \rightarrow {}^4A_2$ broadband appear; and weak crystal fields in crystals in which only the $^4T_2 \rightarrow {}^4A_2$ broadband is observed and which has allowed tunable solid-state lasers to be developed. By playing with the crystal field strength, all possibilities are met with the Cr^{3+} ion under flashlamp, laser or diode pumping sources: CW or pulsed operations in the millisecond to femtosecond range.

Another important interest of research on the Cr^{3+} ion in host crystals has been to stimulate basic knowledge in spectroscopy on other chromium oxidation states and other metal transition ions in order to provide new laser radiation. Among them, we can mention the following:

- The Cr^{4+} ($3d^2$)-doped forsterite (Mg_2SiO_4) which was discovered while looking for a Cr^{3+} dopant and is now commercialized as a tunable solid-state laser emitting between 1167 and 1345 nm, and also Cr^{4+}-doped YAG as a tunable solid-state laser and a mainly saturable absorber crystal;
- Cr^{2+} ($3d^4$)-doped ZnSe which works between 2000 and 2800 nm;
- Cr^{5+} ($3d^1$) now known for its infrared fluorescence broadband around 1400 nm in the tetrahedral symmetry site of silicates and phosphates;
- Ti^{3+} ($3d^1$)-doped sapphire (Al_2O_3) now the most developed tunable solid-state laser emitting between 660 and 1060 nm.

These laser ions will be presented in the next chapters.

References

[1] Maiman H 1960 *Nature* **187** 493
[2] Condon E U and Shortley G H 1935 *The Theory of Atomic Spectra* (London: Cambridge University Press)
[3] Griffith 1961 *The Theory of Transition Metal Ions* (London: Cambridge University Press)
[4] Prather J L 1961 *Atomic Energy Levels in Crystals* (Washington, DC: Department of Commerce)
[5] Cotton F A 1963 *Chemical Applications of Group Theory* (New York: Wiley)
[6] Di Bartolo B 1968 *Optical Interactions in Solids* (New York: Wiley)
[7] Powell R 1998 *Physics of Solid-State Laser Materials* (New York: Springer)
[8] Imbush F 1987 *Spectroscopy of Solid-State Laser-Type Materials (Ettore Majorana International Science Series Physical Sciences 30)* ed B Di Bartolo (New York: Plenum) p 141
[9] Tanabe Y and Sugano S 1954 *J. Phys. Soc. Japan* **9** 753
[10] Sugano S, Tanabe Y and Kamimura H 1970 *Multiplets of Transition Metal Ions in Crystals* (New York: Academic)
[11] Nie W, Monteil A and Boulon G 1990 *Opt. Quantum Electron.* **22** S227–45
[12] Monteil A, Nie W, Madej C and Boulon G 1990 *Opt. Quantum Electron.* **22** S247–57
[13] Monteil A, Garapon C and Boulon G 1988 *J. Luminescence* **39** 167
[14] Boulon G, Garapon C and Monteil A 1987 *Optical Science and Engineering—Series 8—Advances in Laser Science—Conf. Proc.* vol 160 (Seattle, WA: American Institute of Physics) p 87
[15] Gan F 1995 *Laser Mater.* (Singapore: World Scientific)
[16] Hasmi F, Ver Steeg K, Durville F, Powell R C and Boulon G 1990 *Phys. Rev.* B **42** 3818
[17] Hasmi F, Powell R C and Boulon G 1992 *Opt. Mater.* **1** 281
[18] Boulon G 2001 Advances in energy transfer processes *Book of the International School of Atomic and Molecular Spectroscopy (Erice, June 1999)* ed B Di Bartolo and X Chen (Singapore: World Scientific)
[19] Imbusch G F 1967 *Phys. Rev.* **153** 326
[20] Tolstoi N A and Liu Shun'-Fu 1962 *Opt. Spectrosc. (URSS)* **13** 224
[21] Auzel F, Bonfigli F, Gagliari S and Baldacchini G 2001 *J. Luminescence* **94–95** 293–7
[22] Laversenne L, Goutaudier C, Guyot Y, Cohen-Adad M T and Boulon G 2002 *J. Alloys Compounds* **341** 214–19
[23] Yen W M and Selzer P M (ed) 1981 *Laser Spectroscopy of Solids (Topics in Applied Physics 49)* (Berlin: Springer-Verlag) pp 141–88
[24] Maiman T H, Hoskins R H, D'Haenens I T, Asawa C K and Eotuhov V 1961 *Phys. Rev.* **123** 1151
[25] Cronemeyer O C 1964 *J. Opt. Soc. Am.* **56** 1703
[26] Koechner W 1992 *Solid-State Laser Engineering* 3rd edn (New York: Springer)
[27] Hecht J 1986 *The Laser Guidebook* (New York: McGraw-Hill)
[28] Payne S A and Chase L L 1989 *J. Appl. Phys.* **66** 1051
[29] Johnson L F, Dietz R E and Guggenheim H J 1963 *Phys. Rev. Lett.* **11** 318–20
[30] Johnson L F, Dietz R E and Guggenheim H J 1966 *Phys. Rev. Lett.* **17** 13–15
[31] Boulon G 1989 *Disordered Solids—Structures and Processes (Ettore Majorana International Science Series 46)* ed B Di Bartolo (New York: Plenum) pp 317–41
[32] Walling J C, Jenssen H P, Morris R C, O'Dell E W and Peterson O G 1979 *Opt. Lett.* **4** 182
[33] Powell R C, Xi L, Gang X, Quarles G J and Walling J C 1985 *Phys. Rev.* B **32** 2788
[34] Payne S A, Chase L L, Neivkirk H W, Smith L K and Krupke W F 1988 *IEEE J. Quantum Electron.* **QE-24** 2243–52
[35] Payne S A, Chase L L, Smith L K, Kway W L and Neivkirk H N 1989 *J. Appl. Phys.* **66** 1051–6
[36] Payne S A, Chase L L, Atherton L J, Caird J A, Kway W L, Shinn M D, Hughes R S and Smith L K 1990 *SPIE* **1223** 84–93
[37] Payne S A, Chase L L, Smith L K and Chai B H T 1990 *Opt. Quantum Electron.* **22** S259–68
[38] Stalder M, Chai B H T and Bass M 1991 *Appl. Phys. Lett.* **58** 216–18

[39] Scheps R, Myers J F and Payne S A 1990 *IEEE Photon. Technol.* **2** 626–8
[40] Scheps R 1992 *Opt. Mater.* **1** 1
[41] Beaud P, Richardson M, Miesak E J and Chai B H T 1993 **18** 1550
[42] Falcoz F, Balembois F, Georges P and Brun A 1995 *Opt. Lett.* **20** 1874
[43] Falcoz F, Balembois F, Georges P and Brun A 1997 *Opt. Lett.* **22** 387
[44] Falcoz F, Balembois F, Kerboull F, Druon F, Georges P and Brun A 1997 *IEEE J. Quantum Electron.* **QE-33** 269
[45] Vemura S and Torizuka K 1999 *Advanced Solid-State Lasers (OSA TOPS 26)* (Washington, DC: OSA) pp 410–14

B1.2
Transition metal ion lasers other than Cr^{3+}

Stephen A Payne

B1.2.0 Introduction

In this chapter, the background and status of transition-metal-based solid-state lasers are discussed [1,2]. The transition metal ions which have been demonstrated to lase include Ti^{3+}, V^{2+}, Cr^{2+}, Cr^{3+}, Cr^{4+}, Mn^{5+}, Fe^{2+}, Co^{2+} and Ni^{2+}. All of these ions will be discussed with the exception of Cr^{3+}, which is reviewed in chapter B1.1. These transition metal ions have been demonstrated to laser oscillate in a cavity, by serving as the activators in various host media including oxide, fluoride, sulphide and selenide crystals. Laser oscillation has never been reported for any transition-metal-doped glass: all known laser systems are based on crystalline host media. They are incorporated onto sites of either fourfold or sixfold coordination, referred to as tetrahedral and octahedral sites, respectively. In terms of their practical use, Ti^{3+} in sapphire (Ti:Al_2O_3) is, by a wide margin, the most common transition metal laser. The next most common transition metal lasers include the Cr^{3+} lasers, alexandrite and Cr:LiSAF (see chapter B1.1 for further detail), forsterite (Cr^{4+} in Mg_2SiO_4) and Co:MgF_2. A number of other materials are still being explored. Media based on Cr^{2+} and Fe^{2+} have only recently been discovered, so their utility has not yet been evaluated in depth. Lasers based on V^{2+} and Ni^{2+} do not seem to have taken hold significantly, in spite of the many years that have elapsed since the initial demonstration of their lasing capability. Some ions have been reported to exhibit gain but an ability to oscillate has not yet been demonstrated, for example Rh^{2+} [3] and Cu^{+} [4], as discussed later.

B1.2.1 Survey of transition metal lasers

B1.2.1.1 Ti^{3+} lasers

The Ti:sapphire laser is, undoubtedly, the most ubiquitous solid state laser based on a transition metal ion dopant [5]. In a way, its emergence may be construed as the confluence of two essential discoveries: that of vibronic (broadband emitter) lasers and the ruby laser. By replacing the chromium ion in ruby with titanium (i.e. changing Cr^{3+}:Al_2O_3 to Ti^{3+}:Al_2O_3), broadband laser operation becomes possible while the favourable thermo-mechanical properties of the sapphire host can simultaneously be exploited. Ti:sapphire can be continuous wave (cw) pumped by powerful Ar ion lasers with 488 and 514 nm output lines, as well as with doubled Nd:YAG lasers. In fact, with the recent commercialization of high-power diode-pumped Nd:YAG and Nd:YVO_4 lasers, all-solid-state versions of these laser systems are now available. Ti^{3+} has also been shown to lase in chrysoberyl, or $BeAl_2O_4$, although inadequate information is available to assess the true potential of this material [6, 7].

Another critical discovery ensued, which was the discovery that Kerr lens mode-locking provided a simple, robust means of generating femtosecond pulses with Ti:sapphire [8–10]. The Kerr lens mode-locked oscillator inspired a fresh realization that transition-metal-doped solid state lasers could be unique as well as practical. Since then, Ti:sapphire has served as the workhorse of short-pulse lasers, often arranged with several stages of amplification to operate at average powers of more than 10 W [11].

The tuning range was found to be extraordinary, reaching beyond 0.66–1.18 μm, encompassing the 1.05 μm wavelength typical of Nd:glass lasers and permitting their use in hybrid systems (Ti:sapphire master oscillator with Nd:glass power amplifiers). The material properties of sapphire are outstanding, in light of its high thermal conductivity and high fracture toughness. Ti:sapphire lasers are most commonly pumped by argon ion and doubled Nd:YAG lasers, although other approaches such as flashlamps have been explored with some success. It is conceivable that Ti:sapphire may one day be pumped by nitride-based semiconductor lasers, if their brightness can be increased considerably beyond their current value.

B1.2.1.2 Transition metal lasers: V^{2+}, Co^{2+} and Ni^{2+} ions in octahedral sites

The V^{2+} ion [12, 13] has only been lased in fluoride hosts, including V:MgF_2, $KMgF_3$ and $CsCaF_3$, primarily because of the difficulty in stabilizing the divalent oxidation state of vanadium in oxide media. The efficiencies have generally been found to be quite low however, because of the presence of excited-state absorption (ESA) losses. Vanadium lasers operate in the range 1.1–1.3 μm.

Ni:MgF_2 has been demonstrated to lase [14–16] in several hosts including $KMgF_3$, MgF_2 and $CaY_2Mg_2Ge_3O_{12}$, although the sub-ambient temperatures together with the low efficiencies (due to low luminescent efficiencies and ESA losses) have limited their practical utility. Nickel lasers operate in the range 1.4–1.8 μm.

The most common Co^{2+} laser [17–19] is, undoubtedly, Co:MgF_2, which has been effectively pumped by a 1.31 μm Nd:YAG laser on a time scale shorter than the room temperature storage time of the medium ($\sim$37 μs). The tuning range for Co:MgF_2 is impressive, being 1.5–2.5 μm. The broad tuning range has enabled Nd:YAG-pumped Co:MgF_2 lasers to service a niche scientific market. Low-temperature operation of Co:$KZnF_3$ has also been reported [20, 21].

B1.2.1.3 Transition metal lasers: Cr^{4+}, Cr^{2+}, Fe^{2+} and Mn^{5+} ions in tetrahedral sites

It is interesting that, following much activity leading to the identification of over a dozen Cr^{3+} lasers, further exploratory work found that both Cr^{2+} and Cr^{4+} ions were also interesting and practical lasers. While Cr^{3+} is exclusively found in sixfold coordinated sites, the new divalent and tetravalent species appear only in fourfold sites. Cr^{2+} has been found to lase in the 2.1–2.9 μm range in ZnS, ZnSe, CdSe and $Cd_{0.85}Mn_{0.15}Se_2$ hosts [22–28]. Cr^{2+} lasers are actively being explored and developed and these have been pumped by Tm lasers, Co:MgF_2 and by 1.7 μm laser diodes. The first report of mode-locking has recently appeared in the scientific literature, as have output powers of the order of a watt. Fe^{2+} has recently been discovered to lase in ZnSe in the 4.0–4.5 μm region below about 200 K [29], although the first report of laser action dates back to 1983 [30], for Fe in the semiconductor InP and for 2 K (pumping above the bandgap).

Cr^{4+} lasers operate in the near infrared at 1.2–1.6 μm and have been reported in such hosts as Mg_2SiO_4, Ca_2GeO_4, Y_2SiO_5 and $Y_3Al_5O_{12}$ [31–39]. These lasers can be conveniently pumped with a 1.06 μm Nd:YAG laser. The original material representing this new class of gain media is known as forsterite (Cr:Mg_2SiO_4). Several watts of power are presently available from this class of materials and both Cr:YAG and forsterite have been mode-locked [40–43].

Mn^{5+} lasers operate as so-called three-level lasers, similar to the original ruby laser, where the laser transition at 1.18 μm terminates near the ground state of the ion. Mn^{5+} has been studied in the $Ba_3(VO_4)_3$ host crystal thus far, where a scientific observation of laser action has been reported [44].

Table B1.2.1. Laser properties of transitional metal lasers other than Cr^{3+}.

Laser ion	Laser crystal	Electronic Transition	Wavelength, μm (temp., if < RT)	References
Ti^{3+}	Al_2O_3	^2E–$^2T_2$	0.66–1.18	[5]
	$BeAl_2O_4$		0.80	[6, 7]
V^{2+}	$CsCaF_3$	$^4T_2$–$^4A_2$	1.24–1.33 (80 K)	[12]
	MgF_2		1.07–1.16 (80 K)	[13]
Cr^{4+}	Mg_2SiO_4	$^3T_2$–$^3A_2$	1.17–1.35	[31–33]
	Ca_2GeO_4		1.35–1.48	[34]
	$Y_3Al_5O_{12}$		1.33-1.57	[35, 36]
	$Y_3Sc_xAl_{5-x}O_{12}$		1.39–1.63	[37]
	Y_2SiO_5		1.25–1.35 (77–300 K)	[38, 39]
Cr^{2+}	ZnS	^5E–$^5T_2$	2.35	[23]
	ZnSe		2.1–2.8	[22, 23, 26] [28]
	$Cd_{0.85}Mn_{0.15}Se_2$		2.3–2.66	[24, 25]
	CdSe		2.3–2.9	[27]
Mn^{5+}	$Ba_3(VO_4)_3$	^1E–$^3A_2$	1.181	[44]
Fe^{2+}	n-InP	$^5T_2$–^5E	3.53 (2 K)	[30]
	ZnSe		3.98–4.54 (20-170 K)	[29]
Co^{2+}	MgF_2	$^3T_2$–$^3A_2$	1.5–2.5 (80–300 K)	[17–19]
	$KZnF_3$		1.63–2.08 (25–150 K)	[20, 21]
Ni^{2+}	$KMgF_3$	$^3T_2$–$^3A_2$	1.59 (80 K)	[15]
	MgF_2		1.63, 1.73–1.75 (20–200 K)	[14]
	$CaY_2Mg_2Ge_3O_{12}$		1.46 (80 K)	[16]

B1.2.1.4 Tabulation of transition metal lasers

A summary of the known laser materials based on transition metals other than Cr^{3+} appears in table B1.2.1. It should be noted that there is a diversity of host materials and ions and that there are certainly many more media yet to be discovered.

B1.2.2 Physics and engineering issues

B1.2.2.1 Electronic structure

Transition metals occupy three rows of the periodic table of the elements. Thus far, only the 'first-row' elements have demonstrated the ability to lase as an oscillator, although gain has been reported for Rh^{2+} [3] (second-row ion). The treatment of the electronic structure begins with the atomic species, which is described as $3d^n$ for the first row. Because the atomic interactions are the strongest, including the nuclear, repulsive and exchange interactions, the atomic states are the starting point of consideration [45].

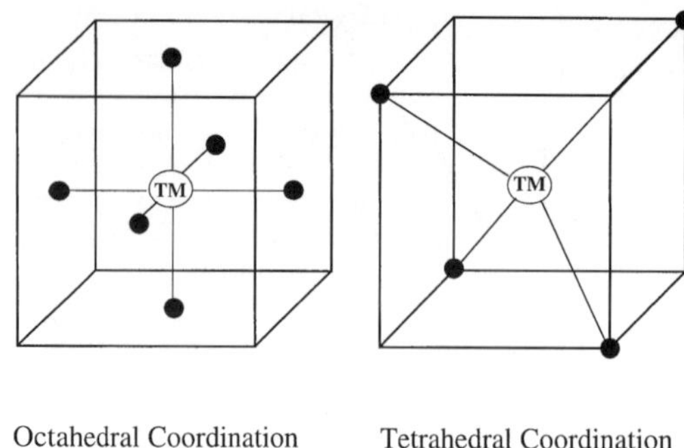

Octahedral Coordination Tetrahedral Coordination

Figure B1.2.1. Depiction of a transitional metal ion in octahedral and tetrahedral sites in a crystalline lattice.

For a $3d^1$ ion such as for Ti^{3+}, the *only* electronic state that arises from this configuration is the ^2D state. The upper left superscript is the multiplicity (or degeneracy) of the spin, given as $2S + 1$, while the 'D' stands for the $L = 2$ orbital angular momentum. Since there is only a single electron (with spin 1/2), then $2S + 1 = 2$, which is known as a *doublet*. If we now consider the Cu^{2+} ion which has a $3d^9$ configuration, we may regard this as being one electron short of the $3d^{10}$ fully symmetric closed shell. Therefore, the $3d^9$ structure is characterized as having one *hole* and also yields the ^2D atomic state. For more complicated transition metal ions, for example those with the $3d^2$ configuration, numerous states result including the ^1S, ^3P and ^1D states, where certain singlet and triplet states are excluded by the requirement of antisymmetrization for the overall electronic wavefunction. Again, d^8 ions give rise to these same states as a consequence of the electron–hole symmetry noted earlier.

Once the atomic state is specified, the crystal field in which the ion is situated must be accounted for. The crystalline environment of transition metal ions is however, simplified however, since they occur mainly in fourfold or sixfold coordinated sites, referred to as tetrahedral or octahedral sites, respectively, see figure B1.2.1. Furthermore, it is often the case that the sites do not deviate far from perfect sites (metal ion at the centre of the cube, with the anions at the centre of the six faces for an octahedron or at four of the corners of the cube for a tetrahedron).

The octahedral or tetrahedral sites further split the orbital component of the atomic state into *crystal field* states. For example, a D state is split into E and T_2 states, although the S and P states are not split further but become identified as A_1 and T_1 states, respectively. The rules governing crystal field splitting are encompassed within the discipline of Group Theory and the magnitude of the effect is encompassed in the well-known Tanabe–Sugano diagrams [45]. As an example of the result of this procedure, the d^8 configuration of Ni^{2+} in an octahedral coordination appears in figure B1.2.2. The abscissa of the Tanabe–Sugano diagram represents the magnitude of the crystal field, so at the furthest left-hand position the energy levels are simply the atomic states. As a final note, it should be mentioned that a sign reversal in crystal field splitting occurs in two situations: incorporating the ion into a tetrahedral as opposed to an octahedral site, and considering electrons instead of holes. In other words, the diagram in figure B1.2.2 also represents a d^2 ion in a tetrahedral environment.

B1.2.2.2 Absorption and emission

As an example of the type of absorption and emission spectra exhibited by all of ions of interest in this chapter, the data for Cr^{2+}:ZnSe are plotted in figure B1.2.3. One of the features of the spectra is the approximate 'mirror image' quality of the absorption and emission spectra, where they appear to be roughly symmetrically disposed about a line of reflection. In principle, the material can be pumped anywhere within the wavelength

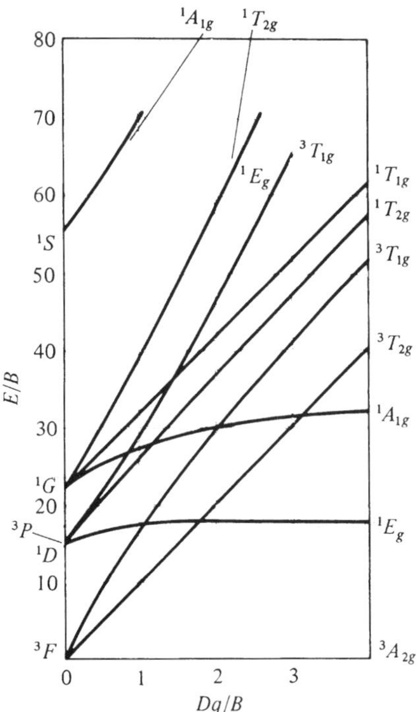

Figure B1.2.2. Tanabe–Sugano diagram for a d^8 ion (such as Ni^2) in anoctahedral coordination.

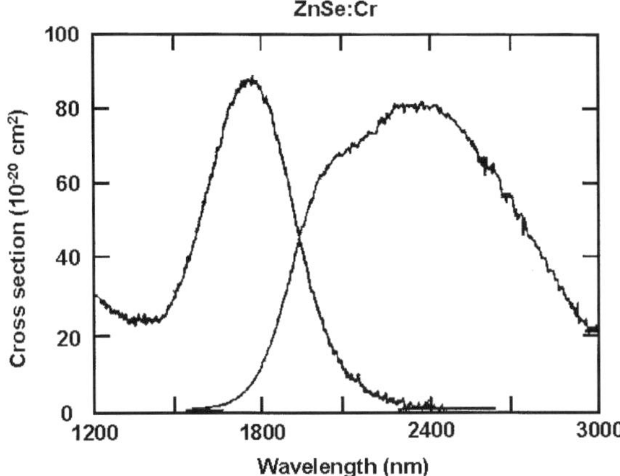

Figure B1.2.3. Absorption and emission spectra of Cr^{2+}:ZnSe, serving as an illustration of the vibronic nature of the emission band and revealing the appearance of a mirror image of the absorption and emission features.

range of the absorption band and the material may then be expected to emit photons according to the emission spectrum. Nearly all of the transition metal lasers operate in the vibronic mode by emitting vibrations during the emission process, with the exceptions of Mn^{5+} and high-field Cr^{3+} (as in for the ruby laser, covered in chapter B1.1). As a consequence, we will focus on broad-band vibronic emitters in this section.

A 'configuration coordinate' diagram description of absorption and emission is sketched in figure B1.2.4.

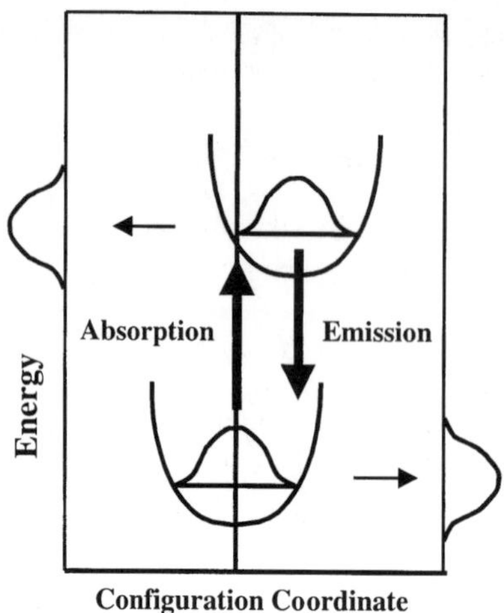

Figure B1.2.4. Idealized configuration coordinate diagram describing the absorption and emission processes for an ion embedded in a host medium.

The essence of this conceptual diagram is that the Gaussian distribution of interatomic distances in the initial state is unchanged in the course of the electronic transition to the other state, so the distribution is mapped via the vertical intersection with the other curve (up for absorption, down for emission) and then to the energy axis. The shape of the potential energy surfaces is regarded as approximately harmonic in nature, given as [46]

$$V = \tfrac{1}{2}k_{\mathrm{gnd}}Q^2 \qquad \text{(B1.2.1)}$$

for the ground state and as

$$V = \tfrac{1}{2}k_{\mathrm{ex}}(Q - \Delta Q)^2 \qquad \text{(B1.2.2)}$$

for the displaced excited state. Q is regarded as the embodiment of the generalized configuration coordinate representing the internuclear separation between the transition metal and the host anion and ΔQ as the displacement between the two electronic states. If, for the purposes of simplicity, we assume that the ground and excited states have the same shape (i.e. $k_{\mathrm{gnd}} = k_{\mathrm{ex}}$), then the 'Stokes shift' between the absorption and emission peaks is given as $2Sh\nu_{\mathrm{ph}}$, where S is known as the Huang–Rhys factor and $h\nu_{\mathrm{ph}}$ is the characteristic phonon energy of the host material. At low temperature, the spectral width of the absorption or emission feature is given as $\Delta = S^{1/2}h\nu_{\mathrm{ph}}$, so a greater ΔQ shift between the ground and excited potential energy surfaces leads to broader spectral features. To a higher degree of specificity, the lineshape can be described by a Pekarian function:

$$L_{\mathrm{ph}}(p) = S^p \exp(-S)/p! \qquad \text{(B1.2.3)}$$

where p is the number of phonons, which are converted to photon energy via the relationship $E = ph\nu_{\mathrm{ph}} + E_{\mathrm{ZPL}}$ by including the energy of the zero-phonon line (ZPL).

The emission bandshape is a critical aspect of the laser properties of the system since it determines the possible tuning range and because it has a major consequence on the magnitude of the emission cross section, which is given by [5]

$$\sigma_{\mathrm{em}} = \lambda^2/(8\pi n^2 \tau_{\mathrm{rad}} \Delta\nu_{\mathrm{em}}) \qquad \text{(B1.2.4)}$$

where λ is the wavelength, n is the refractive index, τ_{rad} is the radiative lifetime and $\Delta\nu_{\mathrm{em}}$ is the spectral width of the transition in s^{-1}. It is, therefore, easy to see that, with other factors held constant, the emission cross section must necessarily be reduced to capture the advantage of greater tuning range for the laser. In addition, the potential to generate short pulses also requires the availability of bandwidth, in that the shortest achievable pulsewidth is limited by the uncertainty principle

$$\tau_{\text{lase}} > 0.3/\Delta\nu_{\text{lase}} \tag{B1.2.5}$$

and the laser pulsewidth must naturally be less than that of the emission band ($\Delta\nu_{\text{lase}} \ll \Delta\nu_{\text{em}}$). For the case of an amplifier having gain G, the bandwidth is given by

$$\Delta\nu_{\text{lase}} = \Delta\nu_{\text{em}}/(\ln G)^{1/2}. \tag{B1.2.6}$$

For example, if a multi-pass regenerative amplifier is operating with $G = 10^6$, then the gain-narrowing factor is 3.7, indicating an attainable pulsewidth of 12 fs. Recent work with Ti:sapphire oscillators has yielded a pulsewidth of ~5 fs [10] (close to the limit imposed by the emission bandwidth of $\Delta\nu_{\text{em}} \sim 3$ fs).

B1.2.2.3 Excited-state absorption

Excited-state absorption (ESA) is one of the more subtle aspects of laser materials that needs to be understood to determine whether it is able to offer net gain at the laser wavelength. The ESA cross section directly subtracts from the emission cross section of equation (B1.2.4) to yield the net gain cross section:

$$\sigma_{\text{gain}} = \sigma_{\text{em}} - \sigma_{\text{ESA}} \tag{B1.2.7}$$

In other words, the light passing through the gain medium can either experience stimulated emission, pump-induced loss or both, depending on whether the ion undergoes a transition from the excited state to a lower lying level or to a higher level, respectively. For some ions such as Ti^{3+} there is no ESA, simply because there are no accessible excited states above the metastable pumped energy level of the ion. Other ways to avoid the negative impact of ESA for ions include situations where the ESA is either wavelength-shifted from the emission band such that it does not significantly overlap it or for cases where it possesses a significantly weaker cross section. We can consider several examples to compare these cases:

- no relevant higher lying excited states—Ti:sapphire (octahedral site);
- weak ESA that is spin-forbidden—Cr^{2+} in ZnSe (tetrahedral site); and
- strong ESA which reduces the laser efficiency—Ni^{2+} and V^{2+} ions (octahedral site) and Cr^{4+} (tetrahedral site) in certain crystals.

The energy levels for each of these cases are shown in figure B1.2.5. Ti^{3+} is the simplest case because the valence electronic configuration consists only of a single 3d electron. Then, we can imagine that the octahedral crystal field surrounding the ion splits the orbitals into two types: t_{2g} and e_g (t_{2g} orbitals 'point' between the oxygens while the e_g point at them and, therefore, have higher energy). So, since there is a single electron, the states become the $^2T_{2g}$ ground state and the ^2E_g excited state, and the laser transition is the transition between them as pictured in figure B1.2.5.

Cr^{2+} also does not suffer from extensive ESA losses [23], even though there are a great many excited states above the pumped metastable level. Cr^{2+} has four 3d electrons, so we can imagine that it is one 3d electron short of a half-filled shell—describable as a 3d *hole*. So with a 3d hole, the two lowest electronic states again become the T_{2g} and E_g states with a spin multiplicity of $2S + 1 = 5$. The laser transition occurs between the $^5T_{2g}$ and ^5E_g states. It is coincidentally the same orbital transition as for Ti^{3+} due to two inversions of its nature: because it is a hole rather than an electron; and because it is a tetrahedral rather than an octahedral arrangement of host anions. The important point relevant to ESA is that all of the other excited states (of which there are many), must necessarily be either spin triplets or singlets and are, therefore, spin-forbidden from the excited state. For this reason, Cr^{2+} ions in a tetrahedral coordination experience

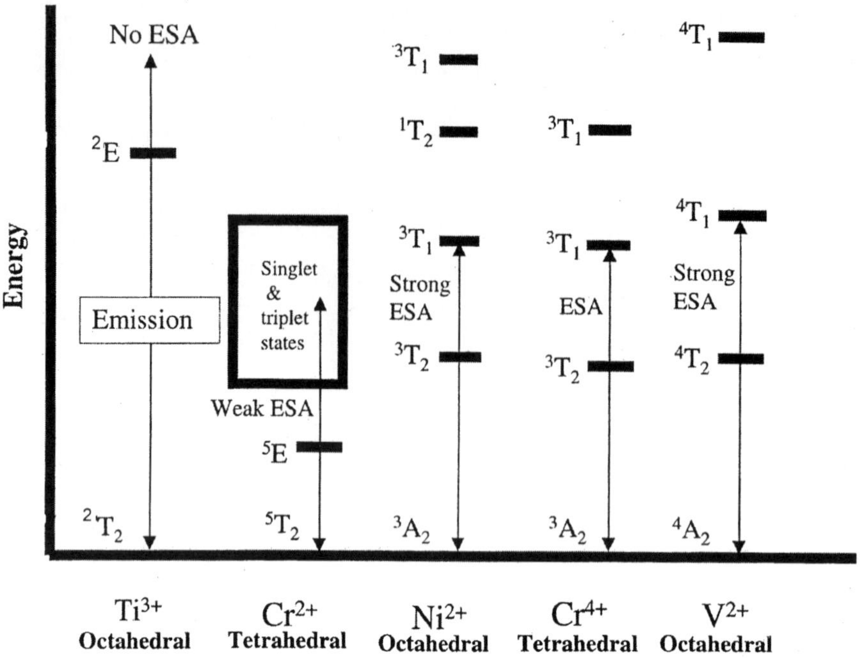

Figure B1.2.5. Energy levels of Ti^{3+} (octahedral); Cr^{2+} (tetrahedral); Ni^{2+} (octahedral); Cr^{4+} (tetrahedral); and V^{2+} (octahedral). The laser and ESA transitions are noted.

little ESA loss. In contrast, Mn^{2+} and Fe^{3+} (d^5) in octahedral sites do not lase because the ESA transitions are spin-allowed while the emission is forbidden—the opposite of the favourable condition that exists for Cr^{2+} in tetrahedral sites.

For the $3d^8$ Ni^{2+} laser ion situated in an octahedral environment [47, 48] or the $3d^2$ Cr^{4+} ion in a tetrahedral site, the emission transition ($^3T_2$–$^3A_2$) may be expected to have roughly the same transition strength as the ESA transition ($^3T_2$–$^3T_1$). Similar considerations apply to V^{2+} although the electronic states are spin quartets rather than triplets [49, 50]. In both cases, because the transitions are spin-allowed they derive their transition strength from vibrational interactions with odd-parity phonons that break the symmetry. (While there is sometimes a magnetic dipole contribution to the strength of d–d transitions, the vibrational component normally dominates.) It turns out that the σ_{em} and σ_{ESA} values are approximately comparable for these three systems, although, by serendipity, the emission cross section can sometimes be somewhat larger leading to efficient laser performance. For example, for forsterite (Cr^{4+}:Mg_2SiO_4), the emission cross section turns out to be significantly larger than that of the ESA for one of three polarizations. In contrast, Cr^{4+}:$LiAlO_2$ [51] does not exhibit any net gain since the crystal is isotropic and it appears that $\sigma_{ESA} > \sigma_{em}$. Another way in which the emission 'beats' the ESA is by reducing the temperature below ambient. This effect is apparent for Ni^{2+}:MgO where the emission features sharpen up and thereby rise above the ESA. The efficiency of V^{2+} lasers is also negatively impacted by ESA.

B1.2.2.4 Radiative rates and non-radiative losses

The emission rate experienced by a metal ion doped into a host medium is the sum of the radiative and non-radiative rates. The basic equation governing the radiative rate is known as Fermi's Golden Rule [2]:

$$W_{\mathrm{rad}} = (8\pi^3/h^2)|\langle I|H_{\mathrm{ph}}|F\rangle|^2 g(\nu) \tag{B1.2.8}$$

where the initial and final states, I and F, are linked via the photon field, H_{ph}, and $g(\nu)$ is the lineshape function in units of s. For transition metal ions, the relevant electronic states are all derived from the $3d^n$ electronic shell so the transitions are nominally symmetry-forbidden on the atomic basis. That is to say that the electronic states are both of even parity and, together in consideration of the odd parity of the photon operator, the overall integral is identically zero. However, for the case where the crystalline field is tetrahedral, the inversion symmetry of the ion is destroyed and the 3d–3d transition becomes allowed. This is the nature of the radiative transitions for Cr^{4+}, Cr^{2+}, Fe^{2+} and Mn^{5+} ions, all of which lase when they occupy the tetrahedral sites of certain host media. For Cr^{4+}, Cr^{2+} and Fe^{2+}, the radiative lifetimes are of the order of 1–100 μs for their spin-allowed laser transitions, while it is longer (about 1 ms) for Mn^{5+} because the laser transition is also spin-forbidden. In referring back to equation (B1.2.8), we see that the radiative rate can be seen as arising from the admixture of opposite parity states into the initial and final states, as a consequence of the crystal field.

For transition metals doped into an octahedral crystal field, the radiative rate can arise from the small residual acentric component of the crystal field and the dynamically induced transition strength resulting from the vibrations that destroy the centre of inversion. Magnetically-induced dipole strength can also play a role as in the $^4T_2$–$^4A_2$ emission transition of V^{2+}. Ti:sapphire is an example of the first case, since the sixfold coordinated site of Al_2O_3 has a significant asymmetric component. The dynamic component of the radiative rate is more important for materials where the transition metal site is most purely octahedral such as for $KMgF_3$.

Non-radiative decay is generally described as a thermally-activated process for transition metal ions, where the excited state is thermally promoted to a very high vibrational level of the lower ground state and then subsequently decays to the ground state. At this point, the excitation energy is deactivated into vibrational energy and then heat. The competition between radiative and non-radiative decay can be described by

$$k_{\mathrm{em}} = k_{\mathrm{rad}} + k_{\mathrm{nr}} \tag{B1.2.9}$$

where the non-radiative component is usually given by $k_{\mathrm{nr}} = A_{\mathrm{nr}} \exp(-E_{\mathrm{nr}}/kT)$. The main issue is the magnitude of the 'quenching temperature', T_{q}, where the radiative and non-radiative rates become comparable. As an example of this type of data, the emission lifetimes of Cr^{2+}:ZnSe and Fe^{2+}:ZnSe are plotted in figure B1.2.6. T_{q} is about 170 and 380 K for these two materials [23]. For comparison, it is 500 [52], 300 [12] and 370 K [53] for Ni^{2+}:MgO, V^{2+}:$CsCaF_3$ and Ti^{3+}:Al_2O_3, respectively.

The Cr^{4+} ion offers some insight into how it can sometimes be difficult to have both favourable emission efficiency and low ESA in a single material. The ion–host combination, Cr^{4+}:$LiAlO_2$, proved to be >30% radiative at room temperature [51] (while the standard forsterite laser material Cr^{4+}:Mg_2SiO_4 is only about 13% efficient) [31–33]. As a result, there was initially much excitement based on the hope that a more efficient Cr^{4+} laser was about to emerge. In spite of the enhanced emission, researchers were unable to lase Cr^{4+}:$LiAlO_2$. It was eventually recognized that ESA was more prominent in $LiAlO_2$ because it did not have the advantage of a strong polarization dependence which can favour low ESA in a particular polarization.

B1.2.2.5 Optical loss

Optical loss in laser crystals can be due to crystal imperfections and the presence of impurities that absorb at the laser wavelength. In addition, absorption can arise from the differing oxidation states of the doped

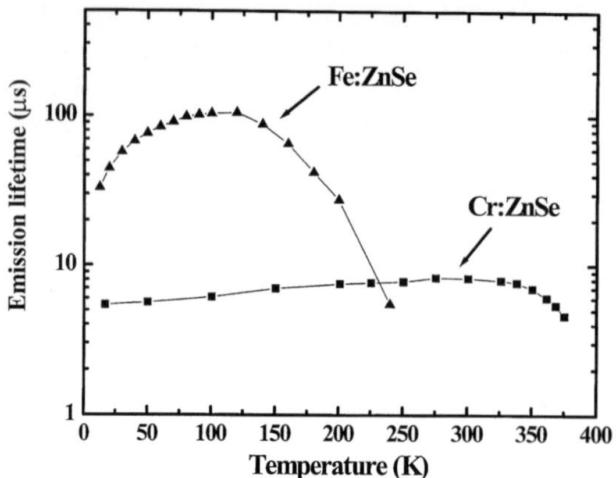

Figure B1.2.6. Emission lifetimes of Cr^{2+} and Fe^{2+} in ZnSe. The precipitous drop in the lifetime is due to the onset of non-radiative decay.

laser ion. Finally, poor optical quality of the gain element can, in effect, lead to diffraction of the light out of cavity mode so that it is registered as a loss (or a reduction in the beam quality of the oscillator or amplifier). Reducing the optical loss to $\sim$ 0.1%/cm often requires several years of crystal growth development and, therefore, must be warranted on the basis of the anticipated properties of the laser material. Such was the case, for example, for Ti:sapphire, where eventually Ti^{4+} was eliminated using carefully controlled growth conditions, which reduced the losses to a favourable level. Some host materials such as $Y_3Al_5O_{12}$, MgF_2 and ZnSe had been developed partly for other reasons and are available as high-quality optical materials.

B1.2.2.6 Laser gain and oscillation

Laser materials can naturally be arranged to operate in a wide variety of configurations, although we will consider the simple case of a cw longitudinally pumped material. The gain afforded by a laser material can be written as

$$G = T_{\mathrm{surf}} \exp[(\sigma_{\mathrm{gain}} N_{\mathrm{ex}} - \alpha_{\mathrm{loss}})d] \tag{B1.2.10}$$

where N_{ex} is the excited-state population, α_{loss} is the distributed loss in the material, d is the pathlength of the light in the material and T_{surf} is the surface transmission factor. The excited-state population is given as

$$N_{\mathrm{ex}} = \alpha_{\mathrm{abs}} I_{\mathrm{pump}} \tau_{\mathrm{em}} / h\upsilon_{\mathrm{pump}} \tag{B1.2.11}$$

where α_{abs} is the pump absorption coefficient, I_{pump} is the pump intensity, $h\upsilon_{\mathrm{pump}}$ is the energy of the pump photon and τ_{em} is the emission lifetime of the excited state. By employing equations (B1.2.10) and (B1.2.11) and allowing for the gain to just equal the loss in an oscillator, the threshold condition can be derived:

$$P_{\mathrm{th}} = \pi \omega^2 h\upsilon_{\mathrm{pump}} (T_{\mathrm{OC}} + T_{\mathrm{loss}}) / 4\sigma_{\mathrm{gain}} \tau_{\mathrm{em}} \tag{B1.2.12}$$

where P_{th} is the minimum threshold pump power for the oscillator, ω is the pump laser radius, T_{OC} is the output coupling, T_{loss} is the total double-pass transmission of the gain medium, which accounts for the passive losses, and the emission lifetime accounts for the negative impact of non-radiative decay since $\tau_{\mathrm{em}}^{-1} = k_{\mathrm{rad}} + k_{\mathrm{nr}}$. Thus, equation (B1.2.9) reveals that losses, ESA (recall $\sigma_{\mathrm{gain}} = \sigma_{\mathrm{em}} - \sigma_{\mathrm{ESA}}$) and non-radiative decay all lead to an increase in the laser threshold of the oscillator, by attenuating the light intensity, reducing the gain and reducing the inversion density, respectively.

Table B1.2.2. Thermal, mechanical and optical properties of host media.

	Thermo-mechanical		Optical	
Crystal	Thermal conductivity, κ ($\mathrm{W\,m\,K^{-1}}$)	Fracture toughness, R_T ($\mathrm{W\,m^{-1/2}}$)	Index change $\mathrm{d}n/\mathrm{d}T$ ($10^{-6}\,\mathrm{K^{-1}}$)	Nonlinear index n_2 (10^{-13} esu)
Al_2O_3	28	22	12	1.2
$Y_3Al_5O_{12}$	10	4.6	9	2.7
Y_2SiO_5	4.5	1.9	6, 7, 9 (anisotropic)	3.7
Mg_2SiO_4	8	—	—	—
MgF_2	21	7.6	0.3, 0.9	0.3
ZnSe	18	5.3	70	170

B1.2.2.7 Host properties

Solid state lasers operating in an 'average power' mode must appropriately manage the repercussions of the thermal gradients existing in the medium, which lead to thermal aberrations (sometimes mainly a spherical lens) and the potential to fracture the gain element. Assuming that the thermal gradients are fully developed in time (i.e. in equilibrium), then the allowed thermal flux that can safely be removed from the surface of a symmetrical thin slab without fracture is [54]

$$Q_{\mathrm{th}} = 6bR_{\mathrm{T}}/d \tag{B1.2.13}$$

where d is the thickness of the slab and b is the safety factor (usually taken as 20%). An overall measure of a material's ability to resist fracture is quantified with the parameter:

$$R'_{\mathrm{T}} = \kappa(1-\nu)\, K_{1\mathrm{c}}/\alpha E \tag{B1.2.14}$$

where κ is the thermal conductivity, ν is Poisson's ratio (0.2–0.3 for most materials), $K_{1\mathrm{c}}$ is related to the material's propensity to resist crack propagation, α is the expansion coefficient and E is Young's modulus. Finally, R_{T}, the fracture toughness, is given by

$$R_{\mathrm{T}} = R'_{\mathrm{T}}/a^{1/2} \tag{B1.2.15}$$

where a is the radius of the surface flaws, which is usually taken to be $a = 25\ \mu\mathrm{m}$. Some typical values of R'_{T} and κ are listed in table B1.2.2 [54, 55].

Another important impact of the thermal gradients within the optical material is the introduction of a thermal lens. This effect arises from the stresses, strains and temperature rise within the material. For the purposes of simplicity, we can consider the case of an unstressed optical element, where the pathlength due to temperature variation of ΔT is given by

$$\Delta p = [\alpha(n-1)\mathrm{d}n/\mathrm{d}T]L\Delta T. \tag{B1.2.16}$$

The magnitude of the change in refractive index with temperature $\mathrm{d}n/\mathrm{d}T$ is obviously an important parameter in determining the amount of thermal lensing exhibited by the material. From the data in table B1.2.2, it should be noted that MgF_2, for example, has a much lower $\mathrm{d}n/\mathrm{d}T$ than ZnSe. The thermal gradient that occurs in the material generally gives rise to a large spherical wavefront distortion, which can be partially mitigated with additional optics.

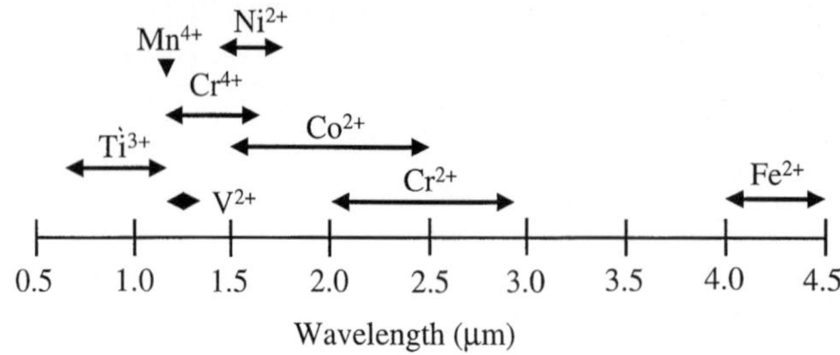

Figure B1.2.7. Summary of tuning ranges offered by known laser ions.

Lastly, the magnitudes of several nonlinear refractive indices are listed in table B1.2.2 [56]. The n_2 become important for pulsewidths of the order of nanoseconds or less, since it leads to an intensity-dependent refractive index and lensing effect. The nonlinear index, n_2, is defined by

$$n = n_0 + n_2 \langle E^2 \rangle \tag{B1.2.17}$$

where E is the electric field of the laser beam. The nonlinear indices generally follow the rule of fluorides $<$ oxides $<$ selenides. The nonlinear index is also is responsible for Kerr lens mode-locking by virtue of the slight difference in the oscillator stability that arises from the intensity-dependent focusing power, causing the mode-locked state to be preferred over cw output.

B1.2.3 Summary

In this chapter, the nature and scope of transition metal ion lasers (excluding Cr^{3+}) have been summarized, as they are embodied in today's solid state laser systems. Ti:sapphire is by far the most common transition-metal-based solid state laser, and it has led to many new opportunities for short-pulse lasers. Co^{2+}:MgF_2 lasers are commercially available and Cr^{4+} and Cr^{2+} lasers are growing in importance. Research in transition metal lasers is continuing, as the vibronic laser Fe^{2+}:ZnSe was first reported to lase in 1999 and many of the second- and third-row ions are yet to be explored in depth. The electronic structure, absorption and emission spectra, excited-state absorption, radiative properties, optical gain/loss and host properties have been described. The tuning ranges encompassed by the known transition metal ions are pictured in figure B1.2.7.

Work done under the auspices of the US Department of Energy by the University of California Lawrence Livermore National Laboratory under Contract W-7405-ENG-48.

References

[1] Caird J A and Payne S A 1991 *Handbook of Laser Science and Technology. Supplement 1: Lasers* ed M J Weber (Boca Raton, FL: Chemical Rubber Company) p 1
[2] Powell R C 1998 *Physics of Solid-State Laser Materials* (New York: Springer)
[3] Powell R C, Quarles G J, Martin J J, Hunt C A and Sibley W A 1985 *Opt. Lett.* **10** 212–14
[4] Barrie J D, Stafsudd O M and Nelson P 1987 *J. Luminesc.* **37** 303
[5] Moulton P F 1986 *J. Opt. Soc. Am.* B **3** 125–33
[6] Pestryakov E V, Trunov V I and Alimpiev A I 1987 *Sov. J. Quantum Electron.* **17** 585–6
[7] Segawa Y, Sugimoto A, Kim P H, Namba S, Yamagishi K, Anzai Y and Yamaguchi Y 1987 *Japan. J. Appl. Phys.* **36** L291–2
[8] Keller U 1994 *Appl. Phys.* B **58** 347–63
[9] French P M W 1996 *Contemp. Phys.* **37** 283–301
[10] Morgner U, Kaertner F X, Cho S H, Chen Y, Haus H A, Fujimoto J G, Ippen E P, Scheuer V, Angelow G and Tschudi T 1999 *Opt. Lett.* **24** 411–13
[11] Wang H, Backus S, Chang Z, Wagner R, Kim K, Wang X, Umstadter D, Lei T, Murnane M and Kapteyn 1999 *J. Opt. Soc. Am.* **16** 1790–4

[12] Brauch U and Durr U 1985 *Opt. Commun.* **55** 35–40
[13] Moulton P F 1982 *Appl. Phys.* **28** 233
[14] Moulton P F, Mooradian A and Reed T B 1978 *Opt. Lett.* **3** 164–6
[15] Johnson L F, Guggenheim H J and Bahnck D 1983 *Opt. Lett.* **8** 371–3
[16] Moulton P F 1982 *IEEE J. Quantum Electron.* **18** 1185–8
[17] Moulton P F 1985 *IEEE J. Quantum Electron.* **21** 1582–95
[18] Fox A M, Maciel A C and Ryan J F 1986 *Opt. Commun.* **59** 142–4
[19] Wedford D and Moulton P F 1988 *Opt. Lett.* **13** 975–7
[20] Kunzel W, Knierim W and Durr U 1981 *Opt. Commun.* **36** 383–6
[21] German K R, Durr U and Kunzel W 1986 *Opt. Lett.* **11** 12–14
[22] Page R H, Schaffers K I, DeLoach L D, Wilke G D, Patel F D, Tassano J B, Payne S A, Krupke W F, Chen K-T and Burger A 1997 *IEEE J. Quantum Electron.* **33** 609–19
[23] DeLoach L D, Page R H, Wilke G D, Payne S A and Krupke W F 1996 *IEEE J. Quantum Electron.* **32** 1–11
[24] Hoemmerich U, Wu X, Davis V R, Trivedi S B, Grasza K, Chen R J and Kutcher S 1997 *Opt. Lett.* **22** 1180–2
[25] Seo J T, Hoemmerich U, Trivedi S B, Chen R J and Kutcher S 1998 *Opt. Commun.* **153** 267–70
[26] Podlipensky A V, Shcherbitsky V G, Kuleshov N V, Mikhailov V P, Levchenko V I, Yakimovich V N, Postnova L I and Konstantinov V I 1999 *Opt. Commun.* **167** 129–32
[27] McKay J, Schepler K L and Catella G C 1999 *Opt. Lett.* **24** 1575–7
[28] Wagner G J, Carrig T J, Page R H, Schaffers K I, Ndap J-O, Ma X and Burger A 1999 *Opt. Lett.* **24** 19–21
[29] Adams J J, Bibeau C, Page R H, Krol D M, Furu L H and Payne S A 1999 *Opt. Lett.* **24** 1720–2
[30] Klein P B, Furneaux J E and Henry 1983 *Appl. Phys. Lett.* **42** 638–40
[31] Petricevic V, Gayen S K, Alfano R R, Yamagishi K, Anzai H and Yamaguchi Y 1988 *Appl. Phys. Lett.* **52** 1040–2
[32] Petricevic V, Gayen S K and Alfano R R 1989 *Opt. Lett.* **14** 612–14
[33] Carrig T J and Pollock C R 1991 *Opt. Lett.* **21** 1662–4
[34] Petricevic V, Bykov A B, Evans J M and Alfano R R 1996 *Opt. Lett.* **21** 1750–2
[35] Eilers H, Dennis W M, Yen M, Kueck S, Petermann K, Huber G and Jia W 1993 *IEEE J. Quantum Electron.* **29** 2508–12
[36] Sennaroglu A, Pollack C R and Nathel H 1995 *J. Opt. Soc. Am.* B **12** 930–7
[37] Kueck S, Petermann K, Pohlmann U, Schoenhoff U and Huber G, 1994 *Appl. Phys.* B **58** 153–6
[38] Deka C, Chai B H T, Shimony Y, Zhang X X, Munin E and Bass M 1992 *Appl. Phys. Lett.* **61** 2141–3
[39] Koetke J, Kueck S, Petermann K, Huber G, Cerullo G, Danailov M, Magni V, Qian L F and Svelto 1993 *Opt. Commun.* **101** 195–8
[40] Seas A, Petricevic V and Alfano R R, 1992 *Opt. Lett.* **17** 937–9
[41] Robertson A, Fuchs H, Ernst U, Wallenstein R, Scheuer V and Tschudi T 2000 *J. Opt. Soc. Am.* B **17** 668–71
[42] French P M W, Rizvi N H, Taylor J R and Shestakov 1993 *Opt. Lett.* **18** 39–41
[43] Sennaroglu A, Pollack C R and Nathel H 1994 *Opt. Lett.* **19** 390–2
[44] Merkle L, Pinto A, Verdun H R and McIntosh B 1992 *Appl. Phys. Lett.* **61** 2386–8
[45] Sugano S, Tanabe Y and Kamimura H 1970 *Multiplets of Transition Metal Ions in Crystals* (New York: Academic)
[46] DiBartolo B 1975 *Optical Properties of Ions in Solids* (New York: Plenum)
[47] Koetke J, Petermann and Huber G 1993 *J. Luminesc.* **60&61** 197–200
[48] Payne S A 1990 *Phys. Rev.* B **41** 6109–16
[49] Payne S A and Chase L L 1987 *J. Luminesc.* **38** 187–9
[50] Moncorgé R and Benyattou T 1988 *Phys. Rev.* B **37** 9177–85
[51] Kuck S, Hartung S, Petermann K and Huber G 1995 *Appl. Phys.* B **61** 33–6
[52] Iverson M V, Windscheif J C and Sibley W A 1980 *Appl. Phys. Lett.* **36** 183–4
[53] Albers P, Stark E and Huber *J. Opt. Soc. Am.* B **3** 134–9
[54] Albrecht G F and Payne S A 1994 *Electro-Optics Handbook* ed R Waynant and M Ediger (New York: McGraw-Hill)
[55] Krupke W F, Shinn M D, Marion J E, Caird J A and Stokowski S E 1986 *J. Opt. Soc. Am.* B **3** 102–14
[56] Adair R, Chase L L and Payne S A 1989 *Phys. Rev.* **39** 3337–50

B1.3
Rare earth ion lasers—Nd^{3+}

A I Zagumennyi, V A Mikhailov and I A Shcherbakov

B1.3.0 Spectroscopic and laser properties of Nd^{3+} ions

In this chapter the basic physical, spectroscopic and laser properties of Nd:$Y_3Al_5O_{12}$, Nd:ceramics, Nd:$YLiF_4$, Nd:YVO_4, Nd:$GdVO_4$, Nd:$YAlO_3$, Nd:Cr:GSGG, Nd:Cr:YSGG, Nd:LSB crystals and Nd:glass will be discussed.

B1.3.1 Nd:YAG laser crystal

The Nd:YAG laser is by far the most commonly used type of commercial solid state laser. Neodymium-doped yttrium aluminum garnet $Y_{3-x}Nd_xAl_5O_{12}$ (Nd:YAG) possesses a combination of properties uniquely favorable for laser operation. Undoped pure $Y_3Al_5O_{12}$ is a colourless, optically isotropic crystal which possesses a cubic structure characteristic of garnets. Its optical properties are isotropic: the refractive index does not depend on the direction of the light nor its polarization. The cubic structure of YAG favors a narrow fluorescence linewidth, which results in a high gain and a low threshold for laser operation. The YAG host is hard, of good optical quality and has a high thermal conductivity. The basic physical and chemical properties of YAG crystals are summarized in table B1.3.1.

The major laser activator ion for YAG is Nd^{3+}, which replaces isomorphic Y^{3+} ions. As the Nd^{3+} ion radius (0.98 Å) exceeds that of the Y^{3+} ion (0.90 Å), the entry of neodymium ions into the YAG lattice is limited, up to a concentration of a few percent. In Nd:YAG, trivalent neodymium substitutes for trivalent yttrium, so charge compensation is not necessary.

It is possible to grow a transparent Nd^{3+}:YAG crystal with an atomic concentration of Nd^{3+} of up to 1.5% with the substitution of Y^{3+} ions. The Nd doping level in YAG is sometimes expressed in different concentration units: a concentration of 1.0% Nd atoms in the lattice is equivalent to 0.727% Nd or 0.848 wt% Nd_2O_3, respectively. The concentration of Nd^{3+} sites in these cases is 1.386×10^{20} cm^{-3}. The growth is carried out by the Czochralski (CZ) method (pulling from a melt) using iridium crucibles. The high quality crystal must be grown with a very slow growth rate (0.5 mm hr^{-1}).

The difference in the ion radii of Nd^{3+} and Y^{3+} determines the distribution coefficient of neodymium ions in the melt–crystal system. The concentration of Nd ions in the growing crystal is much less than the concentration in the melt, because the distribution coefficient is only ~0.18. Because the Nd distribution coefficient is low, not all Nd:YAG crystal boules are uniformly doped. This problem arises as a result of the crystal-growth mechanism. In substituting the larger Nd^{3+} for a Y^{3+} in $Y_3Al_5O_{12}$, the Nd^{3+} is preferentially retained in the melt. The increase in concentration of Nd from the seed to the terminus of a 20 cm long boule is about 20–25%. For a laser rod 3–8 cm long, this end-to-end variation may be 0.05–0.10 wt% of Nd_2O_3.

In additional to CZ growth from iridium crucibles, crystals can also be grown by zone melting in molybdenum containers. However, the higest quality crystals are produced by CZ growth. In this method,

Table B1.3.1. Comparison of Nd:YAG, Nd:YLF and Nd:$YAlO_3$.

Crystal	YAG	$YAlO_3$	YLF
Crystal class	Cubic	Orthorhombic	Tetragonal
Space group	I_{a3d}	P_{bnm}	$I4_{1/a}$
Doping limit (at %)	0.3–1.6	1–3	1–2
Lattice parameters (Å)	12.01	$a = 5.176$ $b = 5.307$ $c = 7.355$	$a = 5.26$ $c = 10.94$
Melting point (°C)	1970	1870	825
Density (undoped)(g* cm^{-3})	4.56	5.35	3.99
Hardness (Mohs)	8.5	8.0	4–5
Thermal conductivity with 1 at% Nd (W m^{-1} K^{-1})	11.1	9.1	6.0
Thermal conductivity of undoped crystal at 300 K (W m^{-1} K^{-1})	13.0	11.0	6.3
Refractive index at laser wavelength	1.818	$n_\alpha = 1.929$ $n_\beta = 1.943$ $n_\gamma = 1.952$	1.470(e) 1.448(o)
Lasing wavelength (nm)	1064.2	1079.6 (*b*-axis) 1064 (*c*-axis)	1047.0 (π) 1053.0 (σ)
Linewidth (nm)	0.6	1.1	1.2
Radiative lifetime (μs)	255	190 (0.1 at%)	520
Lifetime for 1 at% Nd ions (μs)	220	170	460
Effective laser cross section (10^{-19} cm^2)	2.8	3.1 ‖ *a*-axis 3.9 ‖ *b*-axis 2.3 ‖ *c*-axis	1.87 (π) 1.25 (σ)
Diode pump peak wavelength (nm)	807.5	803 E ‖ *c*-axis 807 E ‖ *a*-axis	792
Peak abs. coefficient with 1 at% Nd (cm^{-1})	8.0	14	32.1 (π) 8.6 (σ)
FM75%—Absorption bandwidth at 1 at% Nd(nm)	2.5	2.8 ‖ *a*-axis	6 (π)
Thermal change of indices at 1064 nm (10^{-6} K^{-1})	9.0	9.7 (α) 14.5 (γ)	−4.3 (e) −2.0 (o)
Linear expansion coefficient (10^{-6} K^{-1})	7.3	9.5 ‖ *a*-axis 4.3 ‖ *b*-axis 10.8 ‖ *c*-axis	8.3 ‖ *c*-axis 13.3 ‖ *a*-axis

o, ordinary polarization.
e, extraordinary polarization.
π, $E \parallel C$-axis.
σ, $E \perp C$,-axis.

the melt is induction heated via the iridium crucible, which contains the starting materials. The growth process originates on the seed rod. Since the temperature of the seed crystal is lower than the temperature of the melt, the melt gradually crystallizes on the seed crystal during the slow lift and rotation rates (10–20 rev min^{-1}). The crystal orientation of the growing boule reproduces that of the seed crystal. For the crystal growth process, the melt temperature must be maintained above the melt temperature of YAG (1970 °C) with a very high degree of accuracy. At such high melt temperatures, one difficulty is the crucible material. For Nd:YAG, the best material is iridium, although its high cost and rarity are a serious problem in the manufacture

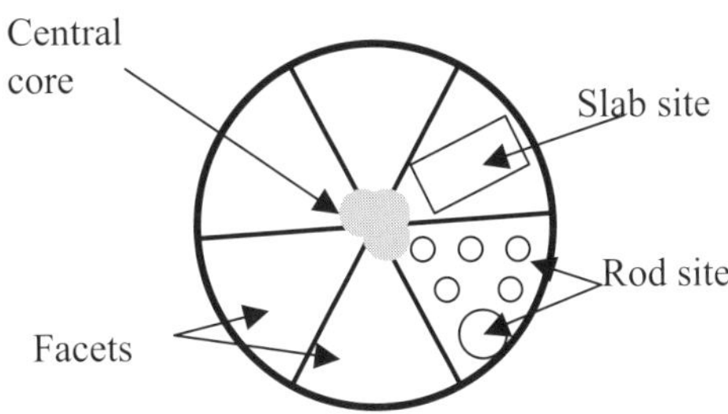

Figure B1.3.1. Nd:YAG boule cross section showing the facet regions and location of slab and rod sites.

of Nd:YAG crystals. CZ growth of Nd:YAG is a capital-intensive process. A single iridium crucible for growing large-sized crystals costs almost US $100 000. Even worse, a significant portion of the crucible is unrecoverable, as it is 'burned-up' during use, limiting its lifetime. The capital investment is increased by the additional need for protection systems to ensure a continuous supply of electrical power, cooling water and other utilities essential for maintaining a reasonable productivity level. Therefore, the Nd:YAG rod, particularly a slab-shaped crystal, is often one of the most expensive components in the laser system.

YAG crystals grow in an inert atmosphere of nitrogen or argon. At the optimum growth rate of 0.5 mm hr^{-1}, the manufacture of the average size Nd:YAG crystal boules (length about 150 mm with a diameter of about 100 mm) takes about 300 hr, i.e. about two weeks. CZ growth enables large Nd:YAG crystal to be produced. The biggest Nd:YAG crystal grown to date is 100 mm in diameter and 300 mm long.

During growth defects form in the central part of the crystal, giving rise to mechanical and optical distortions. Electron microprobe studies have revealed that, in the core region, the Nd concentration can be as much as twice as high as in the surrounding areas. The core originates from the presence of facets on the growth interface which have a different Nd^{3+} distribution coefficient from the surrounding growth surface. These compositional differences cause corresponding differences in the thermal expansion coefficients which, in turn, give rise to the stress patterns observed as the crystals cool down from the growth temperature. Annealing does not seem to eliminate the core and, thus far, no way has been found to prevent the formation of facets on the growth interface. Therefore, active elements are cut out from the peripheral parts of Nd:YAG crystals. As a rule, the active elements are made in the form of long thin cylinders with their axis directed along the growth of the crystal boule. By changing the orientation of the seed rod, it is possible to change the orientation of the crystallogrophic axes in an active element. Cylindrical elements produced from Nd:YAG have, as a rule, diameter of 3, 4, 5, 6.3, 8 or 10 mm and lengths 32, 40, 50, 65, 80, 100, 125, 150 or 200 mm.

In specifying Nd:YAG rods, the emphasis is on size, dimensional tolerance, doping level and passive optical tests of rod quality. Cylindrical rods with flat ends are typically finished to the following specifications: end flat to $\lambda/10$, ends parallel to ± 4 arc sec, perpendicularity to rod axis to ± 5 min, rod axis parallel to within $\pm 5°$ to the [111] direction. Dimensional tolerances typically are ± 0.5 mm in length and ± 0.025 mm in diameter. Most suppliers furnish laser crystals with a photograph showing the fringe pattern of the crystal as examined by a Twyman–Green interferometer. A double-pass Twyman-Green interferometer quickly reveals any strained areas, small defects or processing errors. The optical quality of such rods is normally quite good and comparable to the best quality of optical glass. For example, 6 mm by 100 mm rods cut from the outer sections of 20 mm by 150 mm boules typically may show only one to two fringes in a Twyman–Green interferometer.

Usually, lamp-pumped Nd:YAG lasers with an average output power up to 500 W use a single rod up to 10 mm in diameter and 150 mm long. Higher-power systems with outputs of 2–3 kW can have up to four or, less-commonly, six pump heads with a laser rod up to 200 mm long. More recently rectangular slab configuration crystals with up to 10 mm by 25 mm cross-sectional dimensions and lengths in excess of 200 mm have entered the market.

The spectroscopic properties of Nd:YAG are determined by the basic properties of an Nd^{3+} ion, as slightly modified by the YAG crystalline matrix. Neodymium is a rare earth metal. The optical and laser properties of three-valency neodymium, as well as other rare earth ions, are determined by the transitions inside the 4f-electronic shell (figure B1.3.2). Under the action of the electrostatic crystal field of the lattice, the energy levels split into Stark levels, which vary from crystal to crystal, and while the lines broaden as a result of interactions with the phonons of the crystal lattice. At room temperature, the main 1.06 μm line in Nd:YAG is homogeneously broadened by thermally-activated lattice vibrations.

The structure of the energy-level-free Nd^{3+} ion is determined by the electrons' Coulomb interactions with the nucleus and with each other, and also with the spin and orbital moments of the electrons. The lowest Nd^{3+} ion energy level belongs to a ^4I_J atomic state. This means that three electrons of the Nd^{3+} ion, being on the 4f shell, have a total spin moment S of 3/2 and orbital moment $L = 6$. As a result of the spin-orbital interaction, the 4f multiplet is split into four Stark levels with total moment J from $L - S = 6 - 3/2 = 9/2$ up to $L + S = 6 + 3/2 = 15/2$.

In figure B1.3.2, the splitting of the lowest ^4I levels of Nd^{3+} with various spin-orbits and Stark splits by the crystal field in a YAG crystal is shown. The energy levels of a free Nd^{3+} ion are also shown.

B1.3.1.1 Energy levels of Nd^{3+} ions in a YAG crystal.

A detailed diagram of the Nd^{3+} energy levels in YAG is given in [1]. All levels of the free ion in crystals split as a consequence of the Stark effect arising from the crystal field. Each term is split into $(2J + 1)/2$ components, in particular: $^4I_{9/2}$ into 5, $^4I_{11/2}$ into 6, $^4I_{13/2}$ into 7; $^4I_{15/2}$ into 8. Energy splitting between components lies in the ranges of about 10–100 cm^{-1}. In figure B1.3.2, the energy level diagram for the Nd^{3+} ion in a YAG crystal is shown at a temperature T of 300 K; however, the Stark splitting is only shown for levels which directly participate in laser generation. The number of levels from the Stark splitting is large, compared to those for the free ion. However, the weak splitting of the levels are grouped around the free ion energy. As a result of the low symmetry of the local crystal field of the Nd^{3+}ion in YAG, the degeneracy of levels in YAG is removed completely, except for Kramer's degeneracy of the J_z projection, which can be removed only by using an external magnetic field.

According to parity selection rules, optical transitions are forbidden between the energy levels in one electronic shell. In our case this means that either the probability of all optical transitions between levels is small or that the time for spontaneous optical transitions between levels is large (in comparison with the time constant of permitted optical transitions, being of the order of 10^{-8}–10^{-9} sec). For Nd^{3+} ions, all transitions between levels of the 4f shell are weak. Their intensity differs from zero as a consequence of mixing the 4f shell with a shell with opposite parity (for example, the 5d shell) under the action of odd harmonics arising from the crystal field. The intensity of transitions strongly varies from crystal to crystal and is defined by the specific properties of the matrix.

B1.3.1.2 The lifetime of Nd^{3+} ions in YAG

In the general case, the lifetime of the energy levels is determined, as noted earlier, by radiative and non-radiative relaxation between levels, which has been investigated by many research groups. We shall consider them more in detail.

According to the energy levels of Nd^{3+} ions in garnet crystals, optical transitions can occur from the metastable levels, including the $^4F_{3/2}$ level, downwards, to the $^4I_{9/2}$–$^4I_{15/2}$ multiplets. All these transitions are forbidden and their characteristic lifetimes are about 10^{-3}–10^{-4} sec. The energy distances between the pump levels above the $^4F_{3/2}$ level are rather small in comparison with the YAG phonon energy. Therefore,

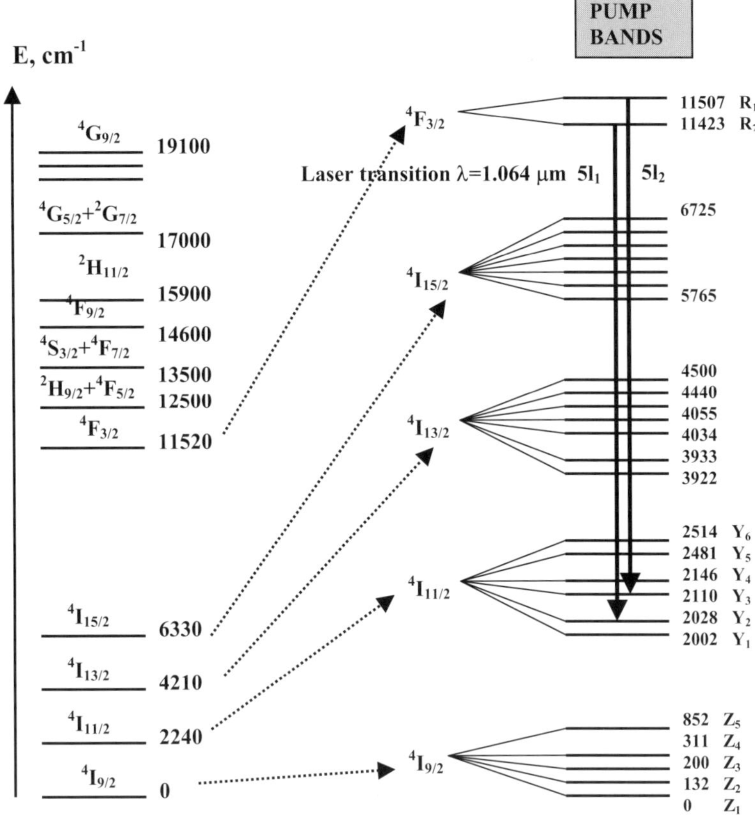

Figure B1.3.2. Energy level diagram of the free ion and the splitting of the Nd^{3+} ion inside the crystal field of YAG.

the probabilities for non-radiative transitions between them are great and the lifetimes accordingly small ($\ll 10^{-6}$ sec). Thus, the Nd^{3+} ions, upon being excited to pump levels, relax to the $^4F_{3/2}$ level quickly, in practice completely non-radiatively.

The $^4F_{3/2}$ level is metastable because the next lower level is separated from it by 4698 cm^{-1}. Therefore, non-radiative transitions downwards from the $^4F_{3/2}$ level have a small probability. They can be made only through multi-phonon relaxation requiring close to six phonons; the probability of such transitions is small and the lifetime is great, about $(1\text{–}3)\times 10^{-2}$ sec. As the radiative time for the $^4F_{3/2}$ level transition is also great (about 2.5×10^{-4} sec), this level is, therefore, metastable—a long lived level which is convenient for use as the upper level in a four-level scheme for lasers.

It is worthwhile noting that the ratio $\eta = (\tau_{em})^{-1}/[(\tau_{em})^{-1} + (\tau_{\sigma})^{-1}]$ determining the quantum yield of emission from the upper laser $^4F_{3/2}$ level in YAG is greater than 99.5% with a fluorescence lifetime of 230 μs.

For the lower $^4I_{9/2}$–$^4I_{15/2}$ multiplets, the energy distances within and between them are less or comparable to the maximum phonon energy (850 cm^{-1}). The transitions within and between these multiplets experience non-radiative relaxations with a short time constant of about 10^{-8} sec.

Thus, in view of the energy distances and lifetimes for Nd^{3+} ions in YAG, the laser scheme (see figure B1.3.2) will be close to the ideal four-level system. The pump level '4' includes all levels lying above the $^4F_{3/2}$ level. The metastable transition level '3' is served by the $^4F_{3/2}$ state, split into two Stark levels: R_1

Table B1.3.2. Wavelength, branching emission, β_{ij}, and cross sections, σ, for the strongest Nd^{3+} ion transitions in YAG at $T = 300$ K.

Transition	Bottom level (cm^{-1})	Wavelength (nm)	β_{ij}	σ, 10^{-20} cm^2
$^4F_{3/2} \rightarrow {}^4I_{13/2}$	3922	1318.4 (R_2)	0.021	1.5
	3933	1335.1 (R_1)	0.015	0.92
	4034	1338.1 (R_2)	0.023	1.5
	2028	1052.1 (R_2)	0.041	2.7
	2002	1051.5 (R_1)	0.091	4.7
$^4F_{3/2} \rightarrow {}^4I_{11/2}$	2110	1064.15 (R_2)	0.125	7.1
	2028	1064.4 (R_1)	0.041	1.9
	2146	1068.2 (R_2)	0.040	1.8
	2110	1073.7 (R_2)	0.062	2.6
	2146	1078.0 (R_1)	0.043	1.2
$^4F_{3/2} \rightarrow {}^4I_{9/2}$	852	938.5 (R_2)	0.041	0.81
	852	946.4 (R_1)	0.049	1.34

Note: The R_1, R_2 transitions arise according to 11 507 and 11 423 cm^{-1} Stark levels of the $^4F_{3/2}$ multiplet.

at 11 423 cm^{-1} and R_2 at 11 507 cm^{-1}. The '2' transitions can be served by any levels of the $^4I_{11/2}$–$^4I_{15/2}$ multiplets. And, at last, the lowest level is comprised from a set of Stark levels from the $^4I_{9/2}$ multiplet. Hence, laser generation on the four-level scheme can occur over the whole range of channels formed by the different Stark levels in the multiplets.

B1.3.1.3 Effective emission cross sections of laser transitions

The emission cross sections of the basic laser transitions of Nd:YAG are well investigated and are described in [1]. In table B1.3.2 the basic lines for Nd:YAG laser generation, the emission cross sections of transitions σ and factors of branching emission β are given. The ratio β shows the relative intensity of the emission for transitions between levels $i \rightarrow j$. The sum β_{ij} of all transitions between the $^4F_{3/2}$ multiplet to all multiplets is equal to one. Obviously, stronger transitions give a larger β_{ij} coefficient. As it can be seen from table B1.3.2, even among the strongest transitions there can be large distinctions in the emission cross sections observed, which are reflected in light amplification of the active laser rod and in the thresholds for laser generation. The branching ratio of emission from $^4F_{3/2}$ is as follows: $^4F_{3/2} \rightarrow {}^4I_{9/2} = 0.25$, $^4F_{3/2} \rightarrow {}^4I_{11/2} = 0.60$, $^4F_{3/2} \rightarrow {}^4I_{13/2} = 0.14$, and $^4F_{3/2} \rightarrow {}^4I_{15/2} < 0.01$. This means that almost all the ions transferred from the ground level to the pump bands end up at the upper laser level and 60% of the ions at the upper laser level cause fluorescence output to the $^4I_{11/2}$ manifold. Therefore, in commercial lasers in basic use only the strongest transitions with wavelengths of 1064.15, 1318.4 and 1338.1 nm are used.

B1.3.1.4 Absorption spectra and pump levels of Nd:YAG crystals

As previously noted and shown in figure B1.3.2, $^4F_{3/2}$ and higher levels are used to pump Nd:YAG crystals. These levels consist of Stark levels that are broadened by the influence of lattice fluctuations on the neodymium ions. It is obvious that the wider transitions absorb the pump light more strongly and, as a result, are more

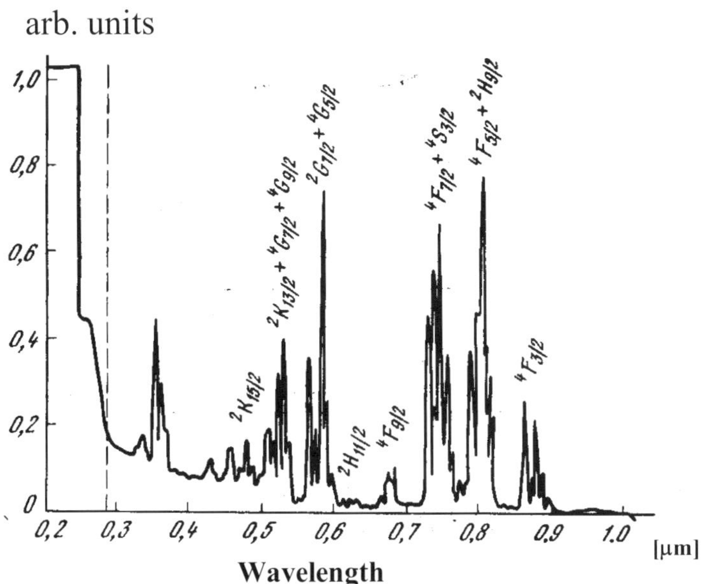

Figure B1.3.3. Absorption spectra of an Nd:YAG crystal.

effective for laser operation. The absorption coefficient of the pumping light is determined by the effective cross section of the transitions.

As an overwhelming number of the neodymium ions are located in the $^4I_{9/2}$ level (more precisely, the Z_1–Z_4 sublevels), light absorption practically only occurs from this level. The intensity of the absorption lines is dictated by the cross section of the transitions. In addition to the neodymium ions, the pumping light can be absorbed into the matrix of the YAG crystal. The YAG matrix is transparent in the range of 240–5500 nm. Therefore, in the range of visible and near-infrared radiation, where the strongest absorption lines are, the absorption by the matrix is not relevant. The matrix absorption spectrum may be caused by impurities or defects. In the light of these remarks, we shall consider the absorption spectrum of Nd:YAG crystals, shown in figure B1.3.3. As can be seen from the spectrum, the basic absorption is given by five main peaks, arising from transitions to the following pump levels:

(1) $^4F_{3/2}$: $\lambda \simeq 880$ nm;

(2) $^4F_{3/2} + {}^2H_{9/2}$: $\lambda \simeq 810$ nm;

(3) $^4F_{7/2} + {}^4S_{3/2}$: $\lambda \simeq 750$ nm;

(4) $^2G_{7/2} + {}^4G_{5/2}$: $\lambda \simeq 580$ nm;

(5) $^2K_{13/2} + {}^4G_{7/2} + {}^4G_{9/2}$: $\lambda \simeq 520$ nm.

The other absorption peaks have a small effect on the laser pumping process because of their small cross sections. The two most intense peaks are located in the spectral range of the basic pump source. The ultraviolet area is characterized by absorption peaks 360 and 260 nm while the wing of the matrix absorption begins at 400 nm and rises strongly to 240 nm (i.e. the absorption edge of the transparent crystal/matrix). As the pump light absorbed by the matrix does not contribute to useful laser radiation and only results in heating, the ultraviolet pumping radiation in real lasers is absorbed by a special light filter, to prevent it from impinging upon the Nd:YAG laser rod. The UV filter is necessary to avoid the generation of colour centres from residual impurities and defects and, in particular, from iron ions. Such colour centres absorb pump and laser radiation and sharply reduce the efficiency of the laser.

B1.3.1.5 Properties of the basic $^4F_{3/2} \rightarrow {}^4I_{11/2}$ *transition at 1064 mn*

The Nd:YAG emission line at 1064.15 nm offers the maximum emission cross section and efficiency. It is possible, however, to obtain oscillation at other wavelengths by inserting etalons or dispersive prisms in the resonator, by utilizing a specially designed resonant reflector as an output mirror or by employing highly selective dielectrically coated mirrors. These elements suppress laser oscillation at the undesirable wavelength and provide optimum conditions at the desired wavelength. With this technique over 20 transitions have been made to lase in Nd:YAG (table B1.3.2).

The 1.064 and 1.061 μm lines of the $^4F_{3/2} \rightarrow {}^4I_{11/2}$ provide the lowest threshold in Nd:YAG. At room temperature, the 1.064 μm line $R_2 \rightarrow Y_3$ is dominant, while at low temperatures the 1.061 μm line $R_1 \rightarrow Y_1$ has the lower threshold. If the flashlamp-pumped laser crystal is cooled, additional laser transitions are obtained, most notably the 0.946 μm line. Diode-pumped Nd:YAG lasing is very effective at the 0.946 μm line for a temperature of 300 K (table B1.3.8).

The 1.064 and 1.061 μm lines are caused by two transitions close in frequency: $5l_2$ between levels 11 507 cm^{-1}($^4F_{3/2}$) and 2110 cm^{-1} ($^4I_{11/2}$) and $4l_1$, between levels 11 423 cm^{-1} ($^4F_{3/2}$) and 2028 cm^{-1} ($^4I_{11/2}$) (figure B1.3.2, table B1.3.2). The energy and wavelengths of these transitions are: $5l_2$ 9397 cm^{-1} (1064.17 nm), $4l_1$ 9395 cm^{-1} (1064.4 nm). The linewidth for room temperatures is equal 6.5 cm^{-1}, which appreciably exceeds the distance between them (2 cm^{-1}). Therefore, the line contours are strongly blocked by each other, hence forming a total contour, which is only slightly distinguishable from a Lorentzian. The intensity of the $4l_1$ line is weaker than the $5l_2$ line ($\sigma_{4l_1} \simeq 1.9 \times 10^{-19}$ cm^2, $\sigma_{4l_2} = 7.1 \times 10^{-19}$ cm^2) and, consequently, the centre of the total line is closer to the centre of the stronger line corresponding to a wavelength of 1064.15 nm. The effective cross section slightly exceeds the cross section of the stronger line at 8.8×10^{-19} cm^2. At room temperature (300 K), the Maxwell–Boltzmann fraction in the upper Stark sub-level is 0.427 for Nd:YAG. The effective stimulated emission cross section is the spectroscopic cross section multiplied by the occupation of the upper laser level relative to the entire $^4F_{3/2}$ manifold population.

Heating the crystal displaces the line to a longer wavelength. For temperature intervals of $\pm 60\,°C$, which are of interest for practical applications, this displacement can be well described by a linear function with a slope of $d\lambda/dT = 5 \times 10^{-3}$ nm K^{-1} and the line centre is located at 1064.15 nm at room temperature (300 K).

B1.3.1.6 Concentration quenching of emission for high neodymium doping levels

For Nd lasers to function effectively, the activator-ions should absorb the maximum amount of pump radiation. As the absorption lines of rare earth ions in crystals are weak, increasing the activator concentration is of interest. However, at an increased activator concentration, we encounter the phenomenon of the concentration quenching of the emission, which appears as a short non-expotential fluorescence decay and a fall in the quantum efficiency. In figure B1.3.4(*a*) the dependence of the $^4F_{3/2}$ level lifetime on the Nd^{3+} concentration in the YAG crystal is shown. Up to 1 at%, the lifetime does not depend on concentration and is about 250 μs; at $C_{Nd} > 1$ at% there is a sharp reduction in lifetime. Similarly at $C_{Nd} = 1$ at% the maximum emission intensity is observed (figure B1.3.4(*b*)). This drop at very low concentrations is connected with a reduction in absorbed pump light and at large concentrations with the fall in the quantum efficiency due to non-radiative relaxation. Neodymium concentration (by at%) in YAG has been limited to 1.0–1.5%. Higher doping levels tend to shorten the fluorescence lifetime, broaden the linewidth and cause strain in the crystal, resulting in poor optical quality. There is an optimum concentration $C_{Nd} = 1$ at% corresponding to the minimum threshold for laser generation.

The nature of the concentration quenching of the emission, as previously specified, is connected to the interaction of Nd^{3+}ions with each other. Two closely located Nd^{3+} ions are coupled with each other by a dipole–dipole interaction, which strongly depends on distance ($W \simeq 1/R^{-6}$). A pair of Nd^{3+} ions is

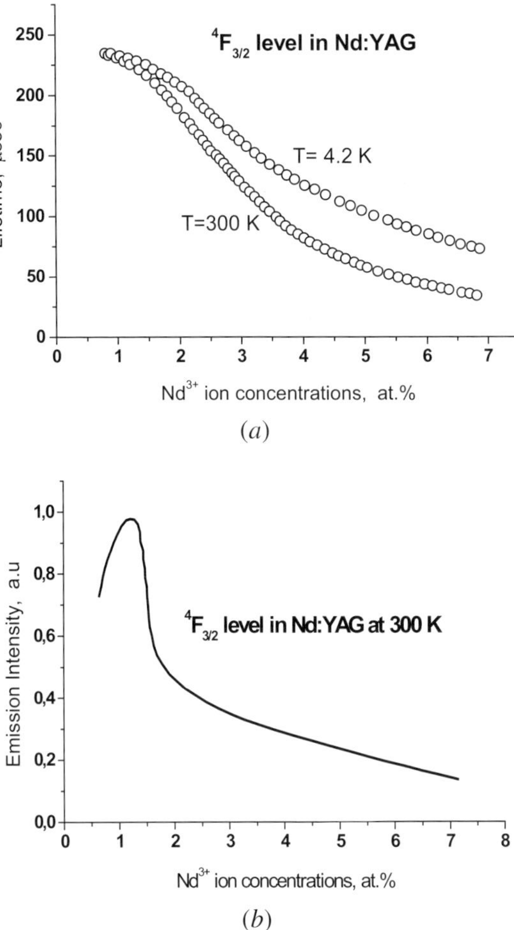

Figure B1.3.4. (*a*) Lifetime dependence of $^4F_{3/2}$ level on Nd^{3+} ions concentration in YAG crystals. (*b*) The concentration quenching of $^4F_{3/2}$ level emission.

capable of creating an additional channel for non-radiative relaxation of Nd^{3+} ions from the $^4F_{3/2}$ level. The performance of a Nd:YAG laser can be somewhat improved by the choosing of the optimum Nd concentration. As a general guideline, it can be stated that a high doping concentration (approximately 1.2 at%) is desirable for Q-switched operation because this will lead to high energy storage. For cw operation, a lower doping concentration (0.6–0.8 at%) is usually chosen to obtain good beam quality and the minimum lasing threshold.

The design, fabrication and properties of many lasers based on different Nd^{3+}-doped crystals are described in the papers presented at Advanced Solid State Lasers conferences [2–11]. The performance of a laser-diode-pumped solid state laser is superior to that of a flashlamp-pumped solid state laser including such characteristics as high efficiency, high beam quality, high output power stability. It is more difficult to develop a high-power laser with high output parameters. Diode-pumped solid state lasers operating at high cw power levels are attractive light sources for various applications ranging from material processing to sophisticated fundamental physics experiments.

A design for a diode-pumped cw Nd:YAG laser [12] with an output power up to 750 W is shown in

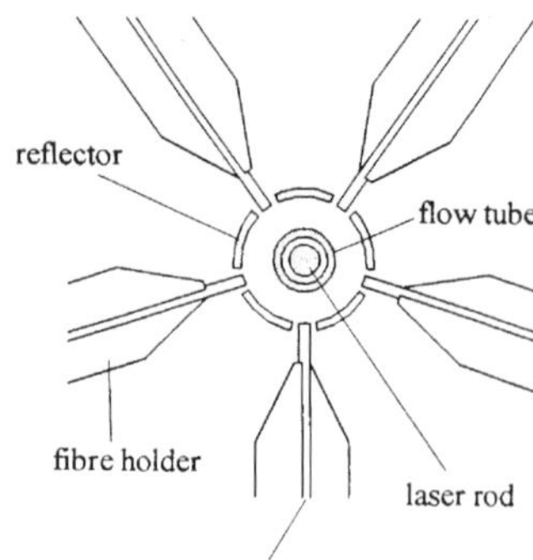

Figure B1.3.5. Diode-pumped Nd:YAG laser.

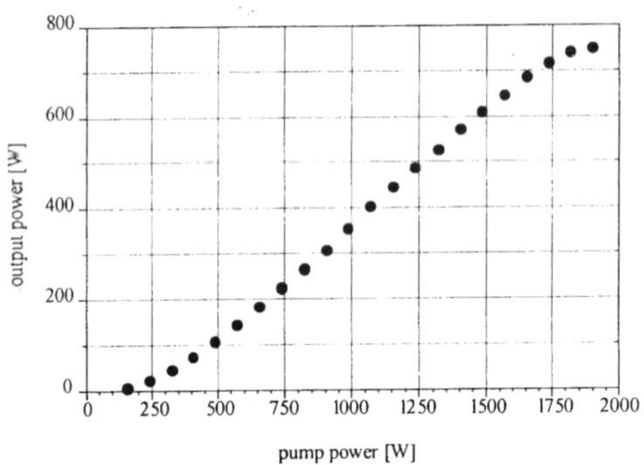

Figure B1.3.6. Output power at 1.06 μm.

Table B1.3.3. Parameters of a side-pumped Nd:YAG laser.

Design feature	Performance summary
Side pumped 0.5%Nd:YAG rod	Optical-to-optical efficiency = 40% for $P_{out} = 750$ W
6 mm diameter rod	Efficiency respect to electrical input = 10%
CW-fibre coupled diode.	Output power = 750 W in multi-mode
Power = 10 W, Core diam.= 800 μm,	Output power = 80 W in TEM_{00} mode
Numerical aperature = 0.20	
Maximum pump power = 2 kW	

figure B1.3.5. The output power of the laser at 1.06 μm is plotted in figure B1.3.6. The design features and performance are summarized in table B1.3.3.

A multi-kilowatt, high brightness, diode-pumped laser for precision laser machining [13] is presented schematically in figure B1.3.7 and table B1.3.4. The laser gain uses a 0.5 cm $\times$ 3.6 cm $\times$ 17 cm Nd:YAG slab. Figure B1.3.8 shows the output-input energy of a high-power module containing the zigzag Nd:YAG slab pumped by a stacked quasi-cw diode array. The zigzag geometry is already well known to minimize thermal lensing. The Brewster-cut slab allows nearly fulfilled minimizing regions of unextracted gain due to the slab geometry which means that no coating is required on the crystal end faces. The input faces of the slab are anti-reflection coated and cooled directly with water. An unstable resonator using a hyper Gaussian variable

Table B1.3.4. Multi-kilowatt, high brightness, diode-pumped Nd:YAG laser.

Design feature	Performance summary
Pump source = 30 diode laser arrays	Beam quality = 2–4 times diffraction limited
Each array = 1cm × 2cm package	Output = 1100 W with beam 2.4 × DL
Package = 16 of 1 cm quasi-cw bar	Output power = 3600 W with beam 3.5 × DL
Peak power per bar = 50 W and duty cycles = 20%	Optical efficiency = 25% for P_{out} = 3600 W
	Output power = 5000 W in multi-mode operation
Diode pulse duration from 100 μs to 1 ms	
Average power per arrays = 4800 W	Optical efficiency = 35% in multi-mode operation
Laser gain module = zigzag Nd:YAG slab	
Size = 0.5 cm × 3.6 cm × 17 cm	
Active gain volume = 3 cm × 0.5 cm × 11 cm	

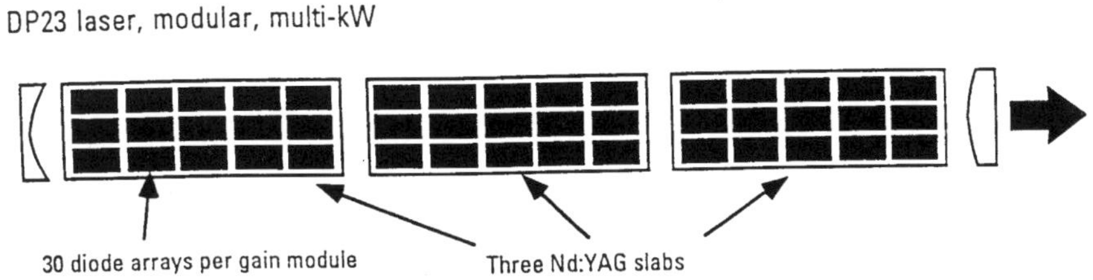

Figure B1.3.7. Laser diode pumping and resonator configuration of Nd:YAG laser.

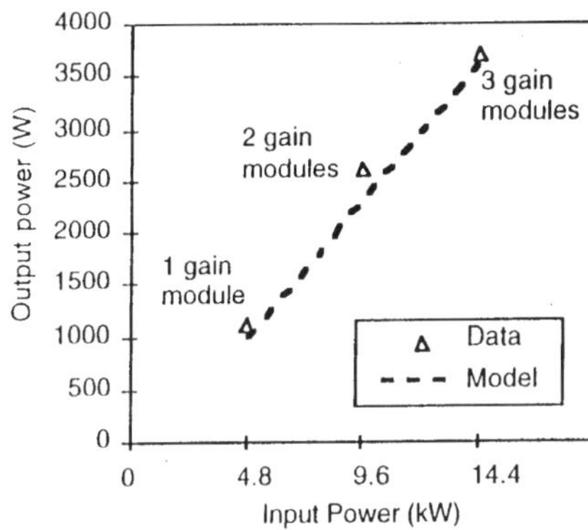

Figure B1.3.8. Laser output as function of diode power.

reflectance mirror enables the laser to run in a single transverse mode with the beam quality limited by a single pass of the gain modules. The resonator magnification was 1.5 and the cavity length was approximately 1 m. Figure B1.3.9 summarizes the beam quality of the laser as a function of output power.

A high brightness mode-locked Nd:YAG end-pumped zigzag slab laser has been developed with a mode-locked master oscillator and high-power amplifier using a novel architecture [14]. The laser emitted 150 W of linearly polarized power with an average M^2 beam quality of 1.25. A schematic layout of the laser system is shown in figure B1.3.10. The gain module uses a 5.6 mm × 1.7 mm × 67 mm (height × thickness ×

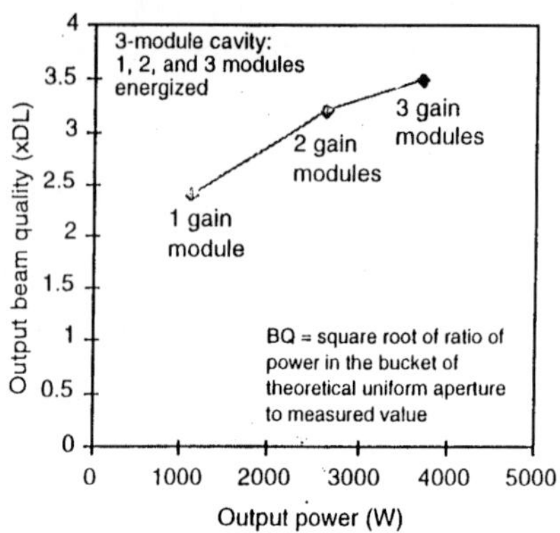

Figure B1.3.9. Laser beam quality as a function of output power.

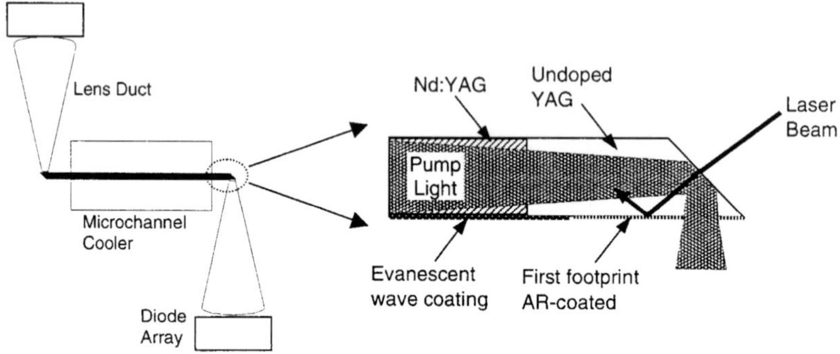

Figure B1.3.10. Schematic diagram of condition-cooled, end-pumped zigzag slab architecture.

length) composite zigzag slab. The central 49 nm length of the slab is 0.2% doped Nd:YAG with 9 mm long, diffusion-bonded, undoped YAG end caps on each side to reduce end effects by confining heat generation to the centre section. The slab ends are cut at 45° and anti-reflection-coated for 1064 nm. The total internal reflection (TIR) faces have 3 μm thick, SiO_2 evanescent wave coatings which allow the faces to be contacted and conduction cooled. Water-cooled copper microchannel coolers are thermally contacted to the slab's TIR faces using a low thermal resistance interface. The slab is pumped from both ends by 805 nm cw diode arrays. Microlenses collimate the fast axis of each diode bar and a lens duct with a 93% throughput concentrates the pump light to a 1.5×5 mm area. The diode light is injected through the first TIR footprint, where the evanescent wave coating also acts as an anti-reflection coating at 805 nm. The pump light undergoes a TIR reflection from the 45° input face of the slab and is guided down the length of the slab by the TIR for efficient and uniform end-pumping. Approximately 85% of the total pump light is absorbed in the slab. The total diode pump power is 1200 W and the Nd:YAG laser operated with an optical efficiency of 12.5%.

Efficient, cw TEM_{00} mode infrared beam sources are of interest in both industrial and scientific applications, e.g. optical communications, material processing, gravitational wave detection, and so on. Furthermore, cw-pumped TEM_{00} mode of 1064 nm beam sources are in high demand in the information technology field, where frequency-converted ultraviolet or visible high-repetition-rate Q-switched laser beams are used for drilling or trimming the printed circuit boards mounted in mobile phones and computers (see chapter D1.6). Various attractive schemes for TEM_{00} mode cw diode-pumped Nd:YAG lasers have been proposed [15–18]:

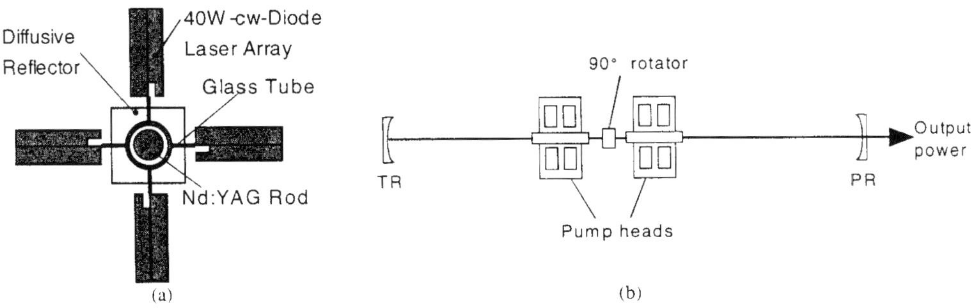

Figure B1.3.11. Schematic diagram of (*a*) the pump module and (*b*) the resonator configuration.

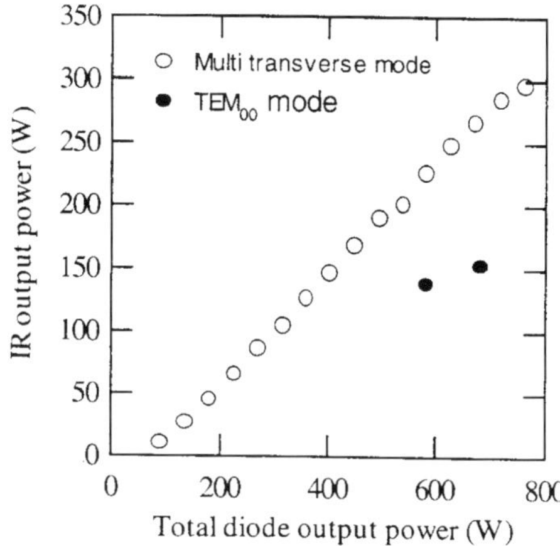

Figure B1.3.12. Multi-transverse and TEM_{00} mode laser performances.

their highest single-transverse-mode power was less than 100 W with an electrical efficiency of less than 10%.

CW 153 W output power in the TEM_{00} mode at 1064-nm generated by a laser-diode-pumped Nd:YAG rod laser an with an electrical efficiency of 10.9% and beam quality $M^2 = 1.18$ described in [19]. Figure B1.3.11 shows the laser design. A quartz 90° polarization rotator was placed between two uniformly pumped Nd:YAG rods for polarization-dependent bifocusing compensation [20]. The pump heat consisted of two modules. Each module contained four 1 cm long linear cw diode arrays (808 nm wavelength, 40 W average output power). One module was rotated 22.5° from the other around the optical axis to produce a uniform pump-light distribution within the rod cross section. Each Nd:YAG rod (4 mm in diameter, 70 mm in length, with 0.6% Nd doping) was surrounded by a flow tube and a diffusive reflector. In figure B1.3.12 high-order ($M^2 = 40$) laser performance in two mode-multi-transverse and TEM_{00}–at 1064 nm operation are compared. In both cases, the transmittance of the output coupler was 20–25%.

Based on the calculation reported in [18], the 950 mm long concave–concave resonator was designed to provide stable operation near the highest diode pump power of 700 W. A maximum TEM_{00} output power of 153 W was generated with 682 W total diode pump power and 1409 W electrical input power into the laser diodes, corresponding to electrical efficiency of 10.9% and an optical-to-optical conversion efficiency of 22.4%.

The use of diode-laser bars as efficient high-power pump sources for end-pumped Nd:YAG solid state lasers is now well established [21–24]. Slope efficiencies reaching 50–60% at 1064 nm have been reported for end-pumped Nd:YAG lasers by fibre-coupled diodes or a beam-shaped diode bar, the maximum attainable power was limited by the rod diameter and space for the diode. The side pumping has been favourable for scaling to a high output power, for it is convenient to use a large number of diode arrays and reduce the thermal loading problems, but the slope efficiency is reduced to 25–40%; the lower slope efficiency is usually caused by imperfect matching between the laser mode volume and the diode pumping volume.

B1.3.2 Polycrystalline Nd:YAG ceramic laser

Highly transparent nanocrystalline Nd:$Y_3Al_5O_{12}$ (Nd:YAG) is a new type of inorganic laser media for a new generation of solid state lasers. A polycrystalline ceramic laser material is an aggregate of crystalline grains, each randomly oriented with respect to neighbouring grains. Recently, polycrystalline Nd:YAG has received much attention since the quality of ceramics Nd:YAG has been improved greatly and highly efficient laser oscillation can be obtained [25–32]. Compared to single-crystal Nd:YAG , ceramic Nd:YAG laser material has several advantages:

(1) its ease of fabrication;
(2) it is less expensive;
(3) large sizes and high neodymium concentrations can be fabricated;
(4) it has a multi-layer and multi-functional ceramic structure and;
(5) it can be mass produced.

The first solid state laser based on a ceramic polycrystalline fluoride material (Dy:CaF) was reported in the middle of the 1960s. But only in the last decade have ceramic Nd:YAG laser materials received much attention, after highly transparent nanocrystalline $Y_3Al_5O_{12}$ (YAG) doped with Ln activators, in particular Nd ions, were developed (see, for example, [25–27]). The first Nd:$Y_3Al_5O_{12}$ (Nd:YAG) ceramic laser was reported in 1995 [28]. Quite recently [29–33], high-power and highly efficient lasers with laser-diode pumping were constructed from ceramics of Nd:YAG.

The room-temperature absorption spectra of ceramics and single-crystal Nd:YAG are shown in figure B1.3.13. For both single-crystal and ceramic Nd:YAG, the absorption peaks are centred at 808.6 nm.

The FWHMs (full width at half maximum) of the absorption coefficients are both 1.04 nm. Room-temperature absorption spectra for 0.6%, 2% and 4% ceramics Nd:YAG have also been measured. Not much difference has been observed except that the absorption coefficient increases with increase in Nd^{3+} concentration. Figure B1.3.14 shows the peak absorption coefficient of ceramic Nd:YAG around 808.6 nm *versus* neodymium concentration. From this figure, one can see that the relationship between the peak absorption coefficient and neodymium concentration is linear.

Figure 1.3.15 shows the room-temperature fluorescence spectrum for the $^4F_{3/2} \rightarrow {}^4I_{11/2}$ transition of a 1% Nd:YAG single crystal and ceramic, respectively. In order to have a clear comparison, the fluorescence spectra for the single crystal and ceramic were normalized and put together. From this figure, these two spectra are almost identical. The main emission peak is at 1064.18 nm. The FWHM is 0.78 nm.

Emission spectra for 0.6%, 2% and 4% Nd:YAG ceramics have also been measured [29]. A small wavelength redshift has been observed with increasing concentration because of a slight change in the crystal field. Figure B1.3.16(*a*) shows the main fluorescence peak spectra at 1064 nm for 0.6%, 1%, 2% and 4% Nd:YAG ceramics, respectively.

The four emission peaks are centred at 1064.15, 1064.18, 1064.24 and 1064.30 nm for 0.6%, 1%, 2% and 4% Nd:YAG ceramics, respectively. The redshift from 0.6% to 4% Nd:YAG ceramics is 0.12 nm. Because of the fluorescence quenching effect, the fluorescence emission linewidth at 1064 nm is also slightly broadened with concentration increases greater than 1%. Figure B1.3.16(*b*) shows the FWHM of the 1064 nm fluorescence peak. The FWHMs are 0.78, 0.78, 0.81 and 0.85 nm for 0.6%, 1%, 2% and 4% Nd:YAG ceramics, respectively. The linewidths for 0.6% and 1% Nd:YAG ceramics are identical. This means that the

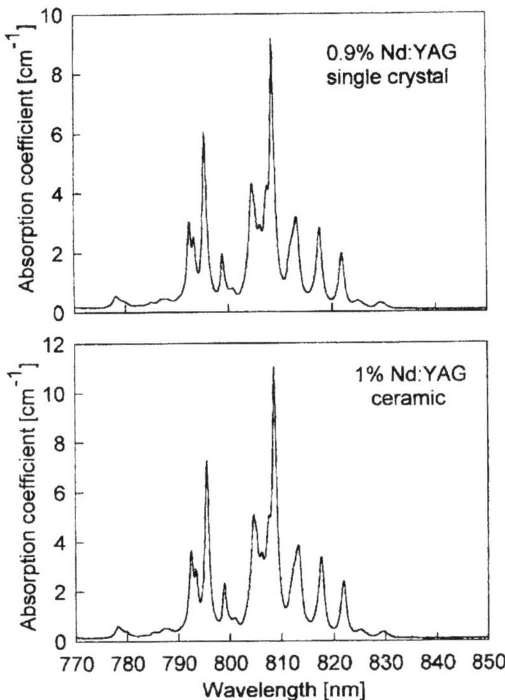

Figure B1.3.13. Absorption spectrum: single crystal 0.9% Nd:YAG and ceramic 1% Nd:YAG.

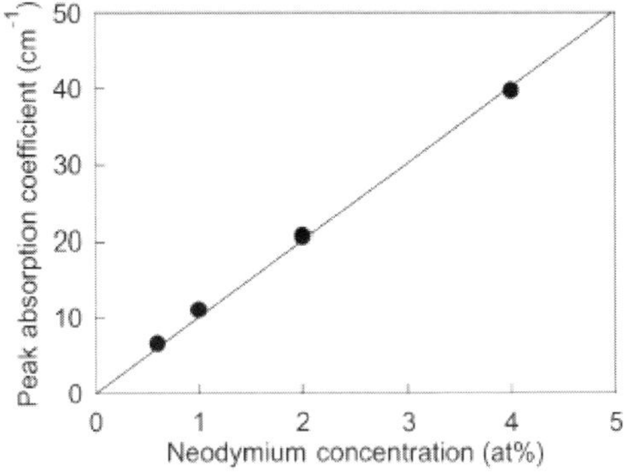

Figure B1.3.14. Peak absorption coefficient of ceramics Nd:YAG around 808 nm *versus* neodymium concentration.

fluorescence quenching effect is very weak for neodymium concentrations less than 1%, which is similar to that of a single crystal.

Figure B1.3.17 shows the fluorescence lifetime of single crystal and Nd:YAG ceramics *versus* neodymium concentration. The fluorescence lifetime for a single-crystal 0.6%Nd:YAG (procured from Litton-Airtron Inc) and a single-crystal 0.9% Nd:YAG are 256.3 and 248.6 μs, respectively.

Fluorescence lifetimes of 257.6, 237.6, 184.2 and 95.6 μs have been measured, respectively, for 0.6%, 1%, 2% and 4% ceramics Nd:YAG. These data agree well with the results in [1]. The fluorescence lifetime decreases dramatically when the neodymium concentration exceeds 1%. The fluorescence lifetimes for the

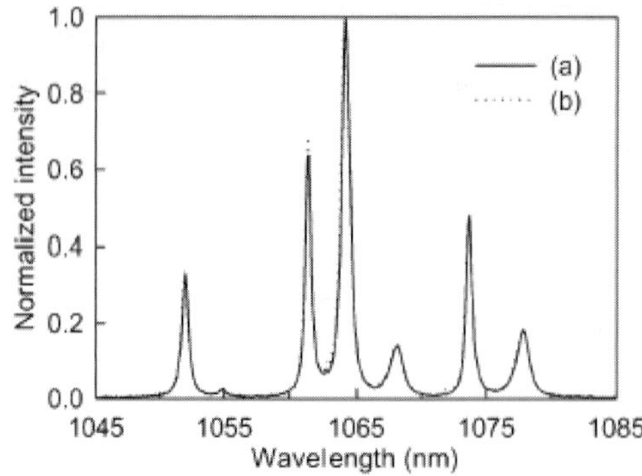

Figure B1.3.15. Fluorescence spectrum: (*a*) ceramic 1%Nd:YAG, (*b*) single crystal 0.9% Nd:YAG.

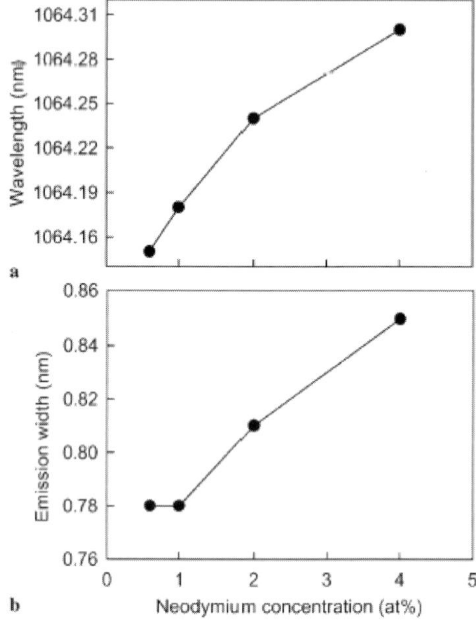

Figure B1.3.16. (*a*) Fluorescence redshift and (*b*) FWHM at 1064 nm *versus* neodymium concentration.

0.6% doped single crystal and the ceramics are almost identical (only 1.3 μs difference). The fluorescence lifetime difference between 0.9% Nd:YAG single crystal and 1% Nd:YAG ceramic is 11 μs. We could predict that for the same concentration for single-crystal and ceramic Nd:YAG, for example, a concentration of 0.9%, the lifetime difference should be less than 11 μs. From the fitted curve for the ceramic fluorescence lifetimes, the lifetime for the 0.9% Nd:YAG ceramic is 244.2 μs, which is only 4.4 μs different from that of the single-crystal 0.9% Nd:YAG [29].

Quite recently [34], a 1.46 kW Nd:YAG ceramic laser was demonstrated by the Toshiba Corporation, Japan. The optical-to-optical conversion efficiency is 42%. The dependence of output power on pump power is shown in figure B1.3.18. The data on the single-crystal Nd:YAG laser are also shown on the same graph to allow a comparison. An output power of 1.72 kW was obtained for the single-crystal laser, with an

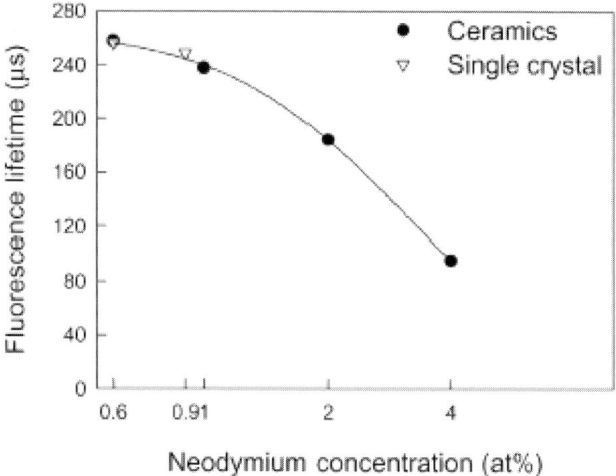

Figure B1.3.17. Fluorescence lifetime of ceramic and single-crystal Nd:YAG *versus* neodymium concentration. The full curve is the *fitted curve* for the ceramic fluorescence lifetime.

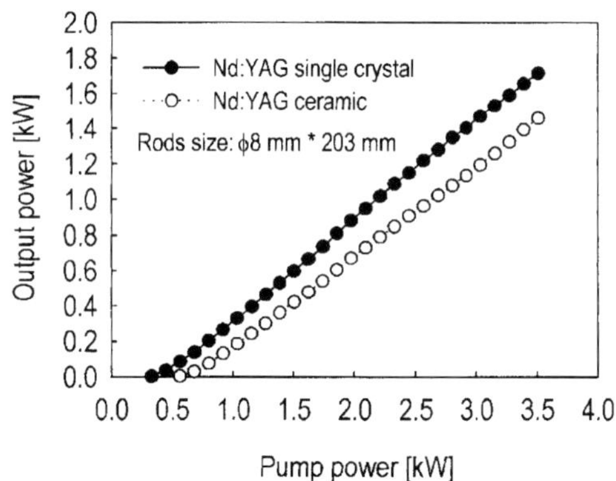

Figure B1.3.18. Laser outputs of ceramic and single crystal Nd:YAG lasers as a function of pump power.

optical-to-optical conversion efficiency of 49%. Although the laser efficiency of the ceramic Nd:YAG laser is still less than that of the single-crystal Nd:YAG laser, it shows that it could possibly be used in industry in the near future.

Nanocrystalline Nd:YAG ceramics, as the new generation of laser materials will before long become a very good alternative to the widely used Nd:YAG single crystals for different types of solid state lasers, including multi-kW industrial lasers.

B1.3.3 Nd:YLF laser crystal

Nd:YLF is a good candidate for certain specialized applications because the output is polarized and the crystal exhibits a lower thermal birefringence [35, 36]. Nd:YLF has a higher energy storage capability (due to its lower gain coefficient) compared to Nd:YAG and its output wavelength matches that of phosphate Nd:glass, therefore mode-locked and Q-switched Nd:YLF lasers [37] have become the standard oscillators for the large glass lasers employed in fusion research.

Some of the important physical properties of Nd:YLF are listed in table B1.3.1, together with its optical and laser parameters.

Nd:YLF is grown utilizing a modified CZ technique. The as-grown crystals are then processed into laser rods or slabs.

The favourable features of Nd:YLF are:

- a large product of the stimulated emission cross section and lifetime for a low cw threshold,
- high-power, low-beam-divergence, efficient single-mode operation,
- high average power Q-switched operation at a moderate repetition rate,
- linearly polarized resonators for Q-switching and frequency doubling,
- potential uniform mode for large diameter rods or slabs and
- the 1053 nm output matches the gain curves of Nd:glass and performs well as an oscillator and pre-amplifier for this type of laser.

B1.3.4 Nd:vanadate laser crystals

For small and efficient diode-laser-pumped solid state laser systems, it is desirable to decrease the requirement for temperature control of the diode-laser pump and to increase the tolerance in the wavelength selection. Therefore, it is necessary that the laser material features a high absorption coefficient and large absorption linewidth. Disordered crystals, for instance, offer broad absorption lines but have the disadvantage of low cross sections due to the inhomogeneous broadening of the transitions. Therefore, the advantage of low temperature sensitivity of the pumping process and the corresponding increased tolerance in pump-wavelength selection is correlated in such crystals with decreased inhomogeneous emission cross sections.

In vanadate crystals like Nd:YVO_4 [38–41] and Nd:$GdVO_4$ [42], both broad homogeneous absorption lines and homogeneous emission lines feature high peak cross sections (table B1.3.5). As we know, the accidental degeneracy of the $^4F_{3/2}$ level in $GdVO_4$ enhances the emission cross sections because the spectra condense into fewer emission lines [43].

B1.3.4.1 Preparation of the Nd:YVO_4 and Nd:$GdVO_4$ crystals

Laser quality crystals of vanadate doped with neodymium can be grown by the CZ technique. Yttrium vanadate and gadolinium vanadate belong to the group of oxide compounds crystallizing in a $ZrSiO_4$ structure with the tetragonal space group $I4_1/amd$. The fourfold symmetry axis is the crystallographic c-axis. Perpendicular to this axis are the two undistinguishable a- and b-axes. The rare earth ion, which is surrounded by eight oxygen atoms, has site symmetry 42 m.

In comparison with YVO_4, the replacement of Y^{3+} ions by the larger Gd^{3+} ions increases the distances between the dodecahedral lattice sites. This decreases the ion–ion interactions between neighbouring ions and also makes the segregation coefficient of Nd^{3+} ions closer to unity.

The YVO_4 and Nd:$GdVO_4$ crystals are grown by the CZ method along the c-axis or a-axis. As growth parameters it used a pulling rate of 1–2 mm hr^{-1}, a rotation rate of 10–30 rev min^{-1} and an atmosphere of nitrogen containing 2% oxygen by volume. The crystals were then annealed at 1200 °C for 10 hr in air. The melting point of Nd:$GdVO_4$ was measured with an optical pyrometer to be 1800 ± 20 °C.

Table B1.3.5. Comparison of Nd:YVO_4 and Nd:$GdVO_4$ [44–46].

General information		
Composition of 1 at% Nd doped crystal	$Y_{0.99}Nd_{0.01}VO_4$	$Gd_{0.99}Nd_{0.01}VO_4$
Concentration (for 1 at% Nd) ($\times 10^{20}$ Nd atom cm^{-3}	1.252	1.25
Nd doping limit (%)	3	3
Nd segregation coefficient	0.35	0.9
Crystal structure	Tetragonal	Tetragonal
Lattice parameters (Å)	$a = 7.12$ $a = 6.29$	$c = 7.21$ $c = 6.35$
Space group	$I4_1/amd$	$I4_1/amd$
Crystal type	Positive uniaxial	Positive uniaxial
Structure type	Zircon	Zircon
Optical properties		
Indices of refraction (n_e, n_o):	$n_e = 2.168$ at 1064 nm $n_o = 1.958$ at 1064 nm	$n_e = 2.192$ at 1063 nm $n_o = 1.972$ at 1063 nm
Birefringence ($n_e - n_o$):	+0.21	+0.24
Transparency range (μm)	0.4–5.0 μm	0.4–5.0 μm
Optical activity:	None	None
Absorption coefficient (cm^{-1}) (a-cut at 808 nm)	20 (0.7 at%) 31 (1.1 wt%) 54 (1.78 wt%) 72 (2.02 wt%) 110 (3.0 at%)	74 ($E \parallel c$-axis, 1.3 at% Nd) 10 cm^{-1} ($E \perp c$-axis, 1.3 at% Nd)
Thermal properties		
Melting point (°C)	1810	1800
Thermal conductivity (W $m^{-1}K^{-1}$ undoped crystal at 300 K	5.2 (∥ to c-axis) [47] 5.1 (⊥ to c-axis) [47] 9.6 (∥ to a-axis)	12.3 (∥ to c-axis) 9.9 (∥ to a-axis)
Thermal conductivity Nd doped crystal at 300 K	—	11.7 (along ⟨011⟩ 1.3 at% Nd) 9.63 (∥ to a-axis, 1.3 at% Nd)
Thermal expansion coefficient ($\times 10^{-6}$ K^{-1})	4.43(a-axis) 11.37 (c-axis)	1 (a-axis) 6.3 (c-axis)
Refractive index change ($\times 10^{-6}$ K^{-1})		
(dn_o/dT)	(8.5 ± 0.9)	4.7±
(dn_e/dT)	2.9	
Laser properties		
Emission cross sections in π-polarization ($\times 10^{-19}$ cm^2)	15.6 at 1064 nm 7.61 at 1342 nm	7.6 at 1064 nm 1.8 at 1342 nm
Lasing wavelength (nm)	1064.3	1062.9
Optical-to-optical efficiency (%)	>60	65
Polarized laser emission	Yes ($E \parallel c$), π-polarization	Yes ($E \parallel c$), π-polarization
Intrinsic loss at laser wavelength (cm^{-1})	0.04	0.01
Laser linewidth (nm)	1.3 (at 1.06 μm)	1.25 (at 1.06 μm) 2.15 (at 1.34 μm)
Excited-state absorption	None	None

Table B1.3.5. (Continued.)

Spectroscopic data		
Absorption bandwidth (FWHM) (nm)	748–760 803–813 877–881	748–760 803–813 877–881
FW75% (wavelength range where at least 75% of diode light is absorbed) (nm)	15.7 (1.1 at% Nd)	13.5 (1.0 at% Nd)
Branching ratios (experimental)	$\beta = 0.422$ at 0.9 μm $\beta = 0.464$ at 1.06 μm $\beta = 0.112$ at 1.34 μm	—
Metastable splitting at 300 K (cm^{-1})	14	~0
Fluorescence linewidth (cm^{-1})	7	6
Radiative lifetime (μs)	115	110
Fluorescence lifetime (μs)	100 (0.87 wt%Nd) 90 (1.1 wt%Nd) 50 (1.8 wt%Nd) 47 (2.2 wt%Nd)	90 (1.2 at% Nd) 40 (4 at% Nd) 20 (6.9 at% Nd)
Mechanical properties		
Hardness (Mohs)	5.5 ± 0.5	5
Knoop hardness (kg mm^{-2})	480	470
Density (g cm^{-3})	4.22	5.47

B1.3.4.2 Absorption and emission of Nd:GdVO$_4$ [42, 43]

The uniaxial crystal Nd:$GdVO_4$ shows strong polarization-dependent absorption transitions due to the anisotropic crystal field. At 808.4 nm, the peak absorption coefficient in π-polarization is 78 cm^{-1} for a 1.2% neodymium doping level, resulting in an effective absorption cross section of $5.2 \times 10^{-19} cm^2$. For σ-polarization, the absorption coefficient is only 17 cm^{-1}. The absorption spectra as a function of wavelength for both polarizations are shown in figure B1.3.19.

The half-width of the absorption at 808.4 nm is 1.6 nm. At the same neodymium concentration the absorption coefficient of Nd:$GdVO_4$ is seven times higher and the transition line is 80% broader in comparison with Nd:YAG.

Figure B1.3.20 shows an enlarged plot of the ${}^4F_{3/2} \rightarrow {}^4I_{11/2}$ in π-polarization in comparison with a corresponding spectrum of Nd:YAG. The total Stark splitting of the ${}^4F_{3/2}$ multiplet is about one-third of that in Nd:YAG. In addition, the number of transition lines is reduced in Nd:$GdVO_4$. Transition pairs, resulting from the ${}^4F_{3/2}$ splitting (indicated by arrows in the Nd:YAG spectrum), are not assignable in the spectrum of Nd:$GdVO_4$. This leads to the assumption that the upper laser level ${}^4F_{3/2}$ is completely degenerate, resulting in a reduction of transition lines by a factor of two, which could be verified by the measurements at 12 K. For completeness the detected energy levels derived from corresponding 77 K spectra of Nd:$GdVO_4$ are listed in table B1.3.6.

We do not know other laser crystals with ${}^4F_{3/2}$ degeneracy. Nd:YVO_4, which is a similar compound, has ${}^4F_{3/2}$ splitting of 18 cm^{-1}. Therefore, the effective emission cross section, in the case of Nd:$GdVO_4$, is identical to the atomic emission cross section. This is illustrated in figure B1.3.21, which shows the splitting

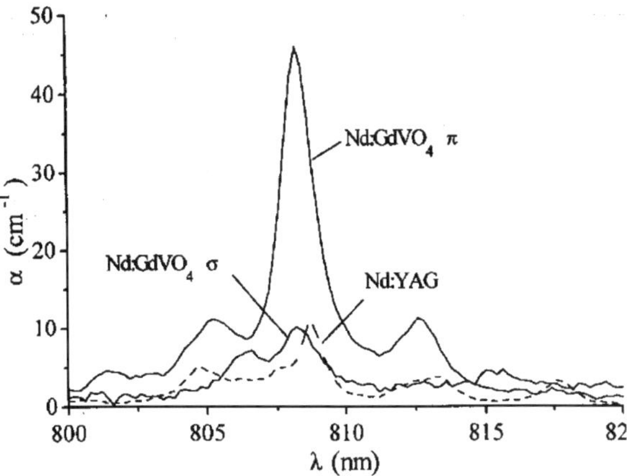

Figure B1.3.19. Absorption spectra for Nd:$GdVO_4$ in π- and σ-polarization. The absorption spectrum of Nd:YAG is shown for comparison.

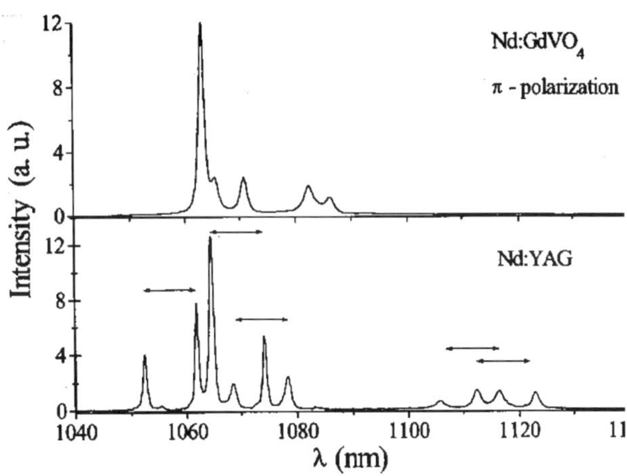

Figure B1.3.20. Fluorescence spectra of the $^4F_{3/2} \rightarrow {}^4I_{11/2}$ in π-polarization in comparison with Nd:YAG at 300 K.

Table B1.3.6. Splitting of the ^4I and ^4F terms of Nd^{3+} in $GdVO_4$ at 77 K.

Multiplets	Energy (cm^{-1})
$^4I_{9/2}$	0, 107, 173, 267, 409
$^4I_{11/2}$	1967, 1987, 2035, 2137, 2165
$^4I_{13/2}$	3915, 3932, 3970, 4022, 4075, 4146
$^4I_{15/2}$	5839, 5900, 5914, 5933, 6055, 6235, 6294
$^4I_{3/2}$	11 375 (degenerate)
$^4F_{5/2}$	12 372, 12 403, 12 413

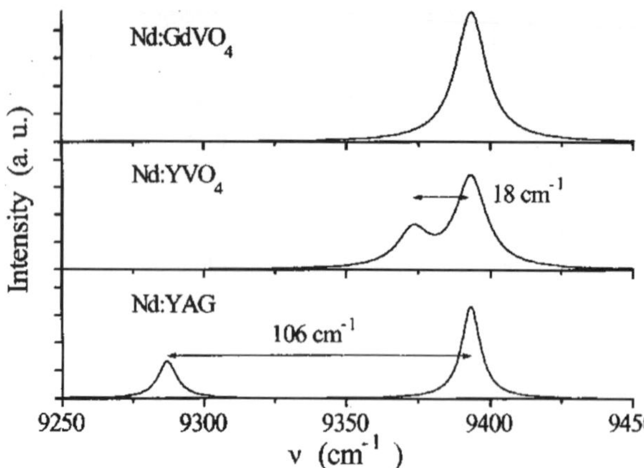

Figure B1.3.21. Merging of transition lines for the three crystals—Nd:YAG, Nd:YVO_4 and Nd:$GdVO_4$ at 300 K. A splitting of the $^4I_{3/2}$ level in Nd:$GdVO_4$ could not be detected.

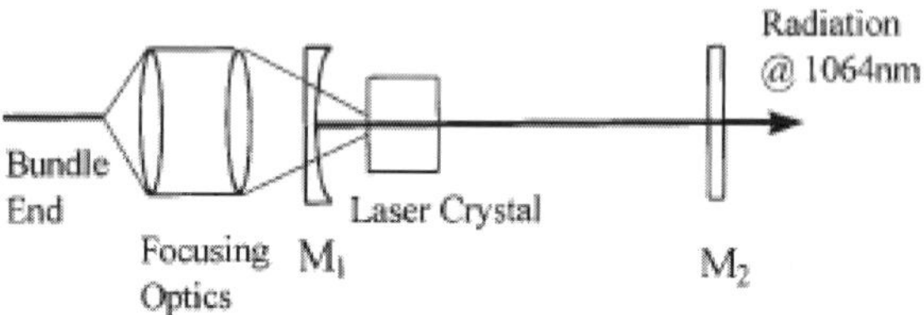

Figure B1.3.22. Schematic diagram of a diode-pumped Nd:$GdVO_4$ crystal lasing at 1063 nm. Mirrors: M_1 , M_2.

of the materials Nd:YAG, Nd:YVO_4 and Nd:$GdVO_4$, by the transition lines of the $^4F_{3/2}$ level, into a single level of the $^4F_{3/2}$ multiplet.

B1.3.4.3 Nd:YVO_4 and Nd:$GdVO_4$ lasers

Neodymium-doped gadolinium vanadate (Nd:$GdVO_4$) is an attractive and very efficient laser material for diode pumping and has been receiving considerable attention in recent years [42–46]. Compared with Nd:YAG crystal, Nd:$GdVO_4$ has a seven fold larger absorption cross section (5.2×10^{-19} cm^2, $E \parallel c$) and a three fold larger emission cross section at 1.06 μm (7.6×10^{-19} cm^2 , $E \parallel c$). More importantly, the Nd:$GdVO_4$ thermal conductivity (12.3 W m^{-1} K^{-1}, direction $\parallel$ to c-axis) is very good and comparable to that of YAG. Such unique spectroscopic and thermal properties make a Nd:$GdVO_4$ crystal a promising substitute for Nd:YAG diode-pumped compact solid-state lasers [48–50].

The laser output power [48] of a Nd:$GdVO_4$ at 1.06 mm was 14.3 W at the highest pump power of 26 W, the light–light conversion efficiency was 54.8%, the pumping threshold was 0.32 W and the slope efficiency was 62%. Figure B1.3.22 gives a schematic diagram of the laser and figure B1.3.23 shows its input–output power.

Passively Q-switched laser output of an Nd:$GdVO_4$ crystal at 1.06 μm has been demonstrated [49].

An actively acousto-optical Q-switched diode-pumping Nd:$GdVO_4$ laser (figure B1.3.24) is described in [48]. The average Q-switched (60 kHz) laser output power at 1.06 μm was 7.3 W at a pump power of

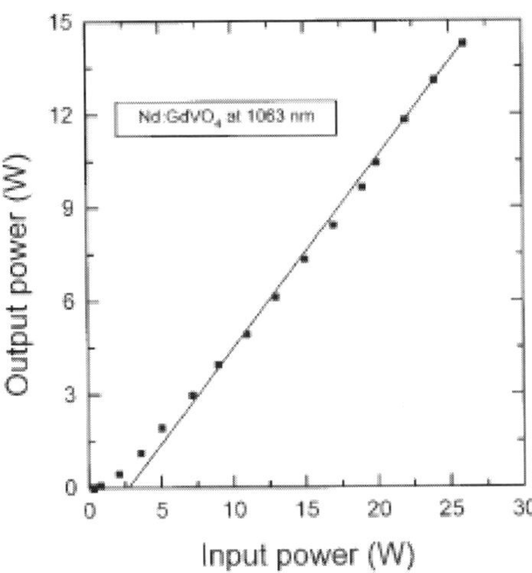

Figure B1.3.23. Output power of an Nd:$GdVO_4$ crystal at 1063 nm.

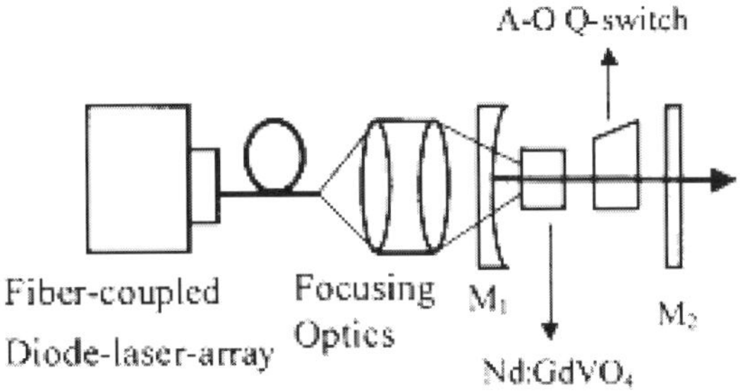

Figure B1.3.24. Schematic diagram of an end-pumped Q-switched 1.06 μm Nd:$GdVO_4$ laser: mirrors, M_1 ,M_2; A–O, acousto-optical.

19 W, the pumping threshold was 1.3 W, the optical-optical conversion efficiency was 38.1% and the average slope efficiency was 42.9%.

Some important parameters of an Nd:$GdVO_4$ crystal have been calculated or estimated [48]: $dn/dT = 4.7 \times 10^{-6}\ K^{-1}$; material constant, $3.5 \times 10^{-6}\ m^2\ s^{-1}$; FOM of the thermal stress resistance >5.73 W cm^{-1}, power per unit length at the stress limit 51.6 W cm^{-1}.

New generation diode pump sources allow end-pumped cw lasers to achieve gain levels previously reached only in quasi-cw diode-pumped systems. A very effective cw pumping oscillator layout is shown in figure B1.3.25 [50]. The short cavity lengths and high pump intensity ensured very short pulse Q-switched operation (table B1.3.7). The end-pumped Nd:YVO_4 (or Nd:YAG or Nd:YLF) lasers with single-pass small-signal gain $G = 10$, produce Q-switched output at the 16 W level in 6 ns pulses at 20 Hz (table B1.3.7).

One of the major advantages of Nd:YVO_4 (Nd:$GdVO_4$) is its ability to retain short pulse output even at very high peak rate frequency (prf). The Nd:YVO_4 demonstrated sub-20 ns pulses at 100 kHz prf. However, the usual laser scheme based on Nd:YVO_4 and Nd:$GdVO_4$ performs poorly at lower prf, while Nd:YAG and Nd:YLF perform well at 10 kHz, the Nd:YVO_4 drops in output power by 50% compared to operation at 100 kHz. The use of short cavity laser design on figure B1.3.25 allows highly efficient Q-switched operation with very low threshold at low and high prf to be obtained.

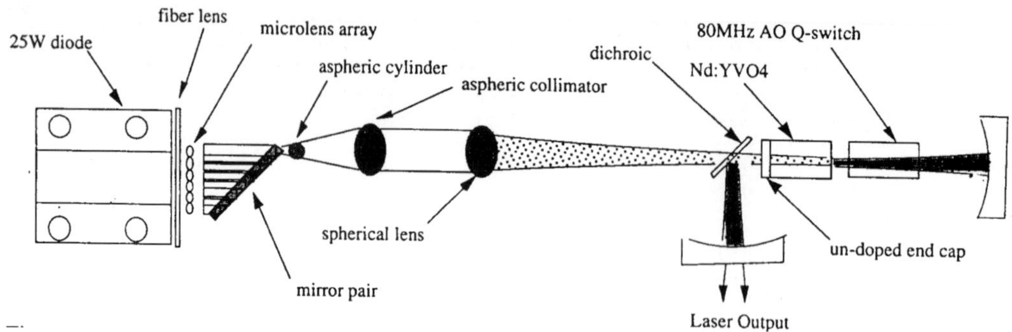

Figure B1.3.25. High-gain end-pumped Nd:YVO_4 laser.

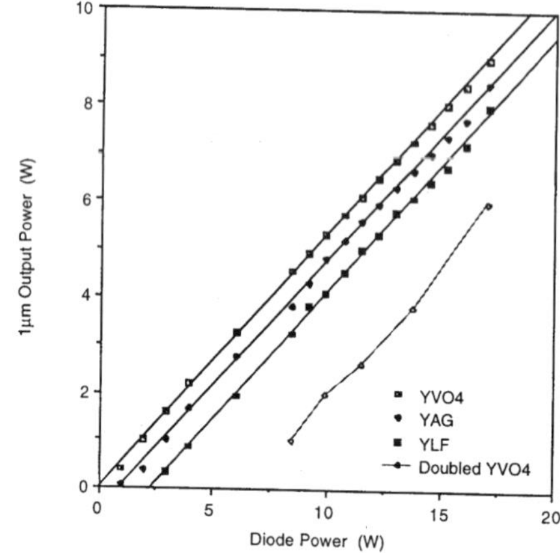

Figure B1.3.26. Performance characteristics of an actively Q-switched end-pumped laser for a short cavity design.

Table B1.3.7. TEM_{00} laser properties of Q-switched operation [50].

Crystal	Output power (W)	Pulselength (ns)		Efficiency (%)	
		10 kHz	20 kHz	Real	Optical
Nd:YVO_4	9	5	8	36	53
Nd:YAG	8.5	8	12	32	50
Nd:YLF	8	11	16	34	47

Now many companies in the world produce commercial low and middle-power lasers based on vanadate crystals. A diode-pumped laser with two Nd:YVO_4 rods and four diode bars produces an output power of about 35 W at 1064 nm, 20 W at 532 nm and 8 W at 355 nm—all with a beam quality $M^2 < 1.3$ (Spectra-Physics, Tucson, AZ, USA).

Diode-pumped Nd:$GdVO_4$ lasers at 1.06 μm have demonstrated an output power of ~25 W with a slope efficiency of 60% with beam quality $M^2 = 1.55$. In such lasers vanadate crystals about 3 mm × 3 mm × 4 mm in size and cut with the a-axis parallel to the optical axis of the laser system are used.

Table B1.3.8. Performance of cw diode-pumped quasi-three-level lasers.

Crystal	Lasing wavelength (nm)	Output power (mW)	Slope efficiency (%)	Reference
$Nd:GdVO_4$	912	2900	48	[46]
$Nd:YVO_4$	915	98	9	[46]
Ng:YAG	946	7400	32	[51]
$Nd:YAlO_3$	930	800	40	[52]

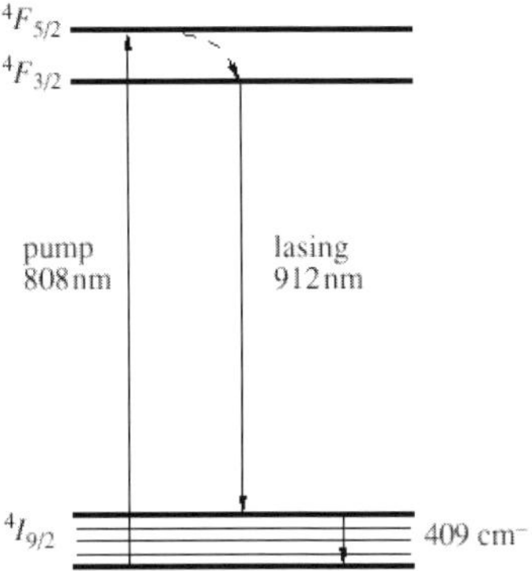

Figure B1.3.27. Diagram of the levels and energy transformation in $Nd:GdVO_4$ crystals during lasing at 912 nm upon the $^4F_{3/2} \rightarrow {}^4I_{9/2}$ transition.

A diode-pumped quasi-three-level Nd^{3+} laser operating on $^4F_{3/2} \rightarrow {}^4I_{9/2}$ has been realized for Nd:YAG, $Nd:YAlO_3$, $Nd:YVO_4$ and $Nd:GdVO_4$ crystals (table B1.3.8). The three-level scheme for Nd^{3+} lasing in a $GdVO_4$ host is shown in figure B1.3.27.

B1.3.5 Nd:Cr:GSGG and Nd:Cr:YSGG crystals

Soon after the invention of the Nd:YAG laser, attempts were made to increase the efficiency in transferring radiation from the pump source to the laser crystal by utilizing a second dopant called a 'sensitizer'. A particularly attractive sensitizer is Cr^{3+} [53, 54] because the broad absorption bands of chromium can efficiently absorb light throughout the whole visible region of the spectrum (figure B1.3.28). The concept of improving efficiency by co-doping a Nd:laser crystal with Cr^{3+} ions is based on transferring excitation [55,56], absorbed by the broad Cr^{3+} absorption bands, over to the Nd^{3+} ions. No improvement was achieved with the host crystal YAG because all the Cr^{3+} excitation was deposited in the ^2E level and the spin-forbidden nature of the $^2E \rightarrow {}^4A_2$ transition resulted in an inefficient transfer process.

However, measurements at the author's laboratory have shown that nearly 100% transfer efficiency could be achieved in the co-doped garnet crystal $Gd_3Sc_2Ga_3O_{12}$ (GSGG) and $Y_3Sc_2Ga_3O_{12}$ (YSGG). Unlike YAG, a large percentage of the Cr^{3+} excitation in GSGG and YSGG appears in the $^4T_2$ state, non-radiative transfer to Nd^{3+} ions can occur via the $^4T_2 \rightarrow {}^4A_2$ transition, which is spin-allowed and has a good spectral overlap with the Nd^{3+} levels (figure B1.3.29).

Experiments performed in [53–56] showed, with flashlamp-pumped operation, nearly a factor-of-three improvement in slope efficiency for the doubly doped garnet compared to a Nd:YAG crystal. For an

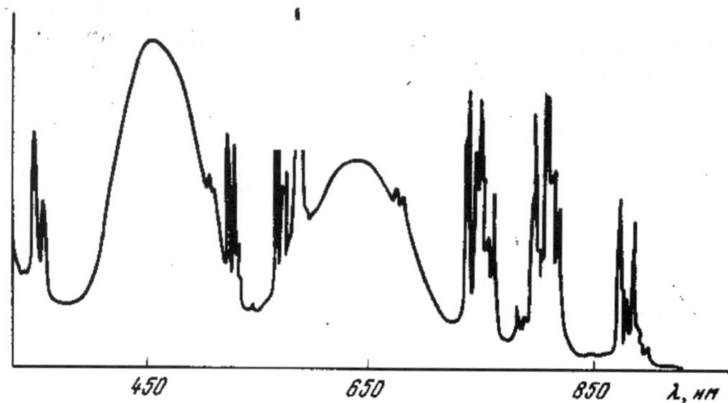

Figure B1.3.28. Absorption spectra of an Nd:Cr:GSGG crystal.

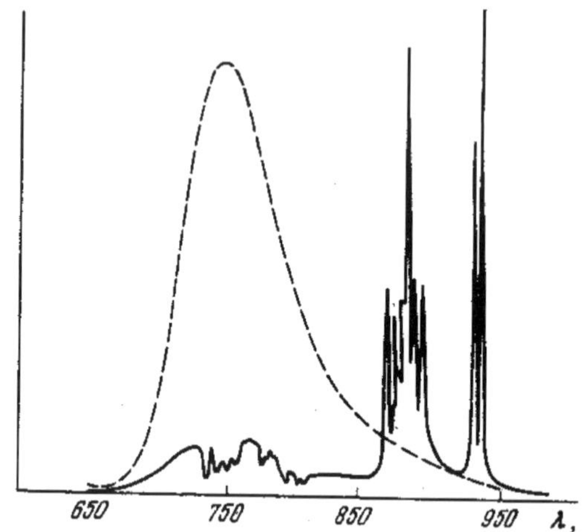

Figure B1.3.29. Emission spectra of Cr:GSGG crystals (dotted line) and emission spectra of Nd:Cr:GSGG crystal (full) at temperature 300 K.

Nd:Cr:YSGG laser rod 6.3 mm in diameter and 100 mm in length, we have achieved an output energy of 2 J electrical efficiency of 7.7% and a slope efficiency of 8.8%, which is the highest slope efficiency yet reported for flashlamp-pumped solid state lasers. It was found that the pumping efficiency improvement is not a strong function of the Cr concentration for the range $1–2 \times 10^{20}$ cm^{-3}. The higher pump efficiency of Nd:Cr:GSGG and Nd:Cr:YSGG does not automatically translate into better system performance because Nd:Cr:GSGG does exhibit much stronger thermal focusing and stress birefringence, compared to Nd:YAG (See tables B.1.3.1 and B.1.3.9). The absorption efficiency and heat deposition rate for the Cr:Nd:GSGG rod is almost three times those of Nd:YAG, a consequence of the broad red and blue absorption bands of the Cr^{3+} sensitizer. As a consequence, the thermal focusing power as a function of lamp input power has been reported to be several times larger in Cr:Nd:GSGG or Cr:Nd:YSGG than in Nd:YAG. Therefore, if beam brightness is the criterion, rather than output energy, some of the advantages of Cr:Nd:GSGG and Cr:Nd:YSGG are offset, particularly at medium average powers. Now Nd:Cr:GSGG and Nd:Cr:YSGG laser rods are widely used in compact military range-finders. Measurements have shown that under single-shot and low-repetition-rate operation, a factor of two higher beam brightness is achieved in a GSGG or YSGG systems compared to a

Table B1.3.9. Comparison of Nd:Cr:GSSG and Nd:Cr:YSSG and Nd:LSB crystals.

Crystal	Cr:Nd:GSSG	Cr:Nd:YSGG	Nd:LSB
Crystal class	Cubic	Cubic	Monoclinic
Space group	I_{a3d}	I_{a3d}	$C2/c$
Doping limit (at% or atom cm^{-3})	Cr:(1–2)10^{20} Nd:(1–3)10^{20}	Cr:(1–2)10^{20} Nd:(1–3)10^{20}	5–50 at%
Lattice parameters (Å)	12.57	12.42	$a = 9.778$ $c = 7.929$
Melting point (°C)	1825	1850	1600
Density (undoped)(g cm^{-3})	6.50	5.2	3.8
Hardness (Mohs)	~7	>7	7
Thermal conductivity at 300 K (W m^{-1} K^{-1})	6.0	7.9	2.8
Thermal conductivity undoped crystal at 300 K (W $m^{-1}K^{-1}$)	6.4	8.3	3.2
Refractive index at laser wavelength	1.943 (1060 nm) 1.955 (750 nm)	1.926 (1060 nm)	$n_x = 1.828$ $n_y = 1.827$ $n_z = 1.828$
Lasing wavelength (nm)	1061.2	1058.5	1800 1348 1062 905
Linewidth (nm)	1.8	1.6	4
Radiative lifetime (μs)	280	270	118 (10 at% Nd)
Effective laser cross section (10^{-19} cm^2)	1.5	1.5	1.3
Diode pump peak wavelength (nm)	808	808	808
Thermal change of indices at 1064 nm (10^{-16} K^{-1})	10.9	12.3	—
Linear expansion coefficient (10^{-6} K^{-1})	7.4	8.3	—

YAG system.

Active elements manufactured from Nd:Cr:GSGG and Nd:Cr:YSGG crystals are optimum materials for low- and medium-power flashlamp-pumped lasers with low repetition rates up to several cycles per second. The advantages of YSGG and GSGG crystals compared with YAG crystals are lost when large-size elements are used because of the worse thermal characteristics of Nd:Cr:GSGG and Nd:Cr:YSGG crystals [57, 58].

High-optical-quality Nd:Cr:GSGG and Nd:Cr:YSGG crystals up to 13 cm in diameter and 23 cm long are grown commercially in Russia and the USA by the Czochralski technique [57].

B1.3.6 Nd:LSB crystal

Neodymium-doped lanthanum scandium borate (Nd:$LaSc_3(BO_3)_4$ or Nd:LSB) is one of the most efficient laser crystals for diode pumping as well as for flashlamp pumping [58, 59]. It is a new laser crystal with improved spectral properties stipulated by the possibility of the introduction of high Nd^{3+} concentrations (up to 50 at%) into the crystalline host.

These properties make the Nd:LSB crystal indispensable in the design of low-power (up to 150 mW at 531 nm) highly efficient compact diode-pumped solid state laser systems because they allow the effect of temperature drift of the wavelength of the pumping laser diode to be reduced and the stability of the radiated laser frequency to be increased because of the wide absorption line of Nd ions at the 808 nm wavelength. The

Table B1.3.10. Optical, physical and spectroscopic properties of Nd-doped glasses.

	Glass type					
Properties	Silicate NS-0835 (Illinos)	Silicate Q-246 (Kiger)	Silicate LSG-91H (Hoya)	Phosphate LHG-5 (Hoya)	Phosphate LHG-6 (Hoya)	Silicate LG-630 (Schott)
Nd^{3+} concentration (10^{20} cm^{-3})	4.6	4.5	3.04	3.174	3.397	2.8
Peak wavelength (nm)	1060	1062	1062	1054	1054	1061
Cross section ($\times 10^{20}$)	2.8	2.9	2.7	4.1	4.2	2.7
Fluorescent lifetime (μs)	330	340	310	290	315	350
Linewidth FWHM (nm)	26	27.7	25.2	18.6	20.1	27.8
Thermal conductivity (W m K^{-1})	0.91	1.3	1.2	1.19	0.98	1.35
Density (gm cm^{-3})	2.63	2.55	2.81	2.68	2.83	2.54
Index of refraction	1.5196	1.568	1.5498	1.539	1.528	1.561
Nonlinear index n_2 [10^{-13}]	1.5	1.4	1.59	1.28	1.13	1.41
dn/dt (10^{-6} °C^{-1})	−2.2	2.9	1.8	8.6	−5.3	2.9
Optical thermal coefficient (10^{-6} °C^{-1})	8.5	8.0	7.9	4.6	0.6	8.0
Thermal expansion coeff. (20°–40 °C) (10^{-6} °C^{-1})	10.3	9.0	10.5	8.6	12.7	9.3
Softening point (°C)	661	518	505	455	485	468
Specific heat (J g^{-1} K^{-1})	0.9	0.93	0.91	0.71	0.75	0.92
Knoop hardness	490	600	590	497	321	490
Young's modulus (kg mm^{-2})	9910	8570	8890	6910	5109	6249

maximum absorption and radiation cross section are similar to those of Nd:YAG but the bands are five times wider. The absorption coefficient of 10 at% Nd^{3+}:LSB crystals is three times higher than that in Nd:YAG crystals. This allows Nd:LSB active elements to be used in end-pumped microchip configurations just about 1 mm long. The optical slope efficiency of the Nd:LSB laser at 1062 nm is about ~67%.

Investigation of Nd:LSB crystals shows a high efficiency for the use of these crystals for intracavity frequency doubling with potassium titanyl phosphate (KTP) crystals. The corresponding optical efficiency is greater than 25%, and the efficiency of conversion of the basic frequency into the second harmonic is 55% [60–62]. The comparative spectroscopic data of Nd:LSB (at room temperature) are given in table B1.3.9.

B1.3.7 Nd^{3+}-doped glass

There are a number of characteristics which distinguish Nd:glass from Nd-doped crystals [63–65]. A wide variety of Nd-doped laser glasses depending on the composition of the glass is known. Many compositions of oxide, fluoride and sulphide glasse have been developed. The commercial Nd:glasses are the oxide glasses, silicate and phosphate (i.e. SiO_2 and P_2O_5 based glasses). Table B1.3.10 listed some important spectroscopic, optical and thermal properties of typical silicate and phosphate glasses.

The physical, spectroscopic and laser properties of Nd:glass are isotropic. There are two important differences between Nd glasses and Nd crystals. First, the thermal conductivity of the glass is much lower than that of most crystal hosts. (The high thermal conductivity is a more important parameter for the Nd:YAG crystal: it is the reason this laser crystal has taken up to 70% of the laser market.) Second, the emission lines of Nd^{3+} in glasses are considerably broader than in crystals. A wider line increases the threshold for laser lasing and the amplification value. In contrast, this broadening has an advantage because the broader laser lines are used to obtain and amplify very short light pulses. In addition, a broader line permits larger amounts of energy to be stored in the amplifying Nd:glass medium for the same linear amplification coefficient.

A Nd:crystal host is a better material for cw or very high-repetition-rate operation, because the commercial crystals provide higher gain and greater thermal conductivity. Nd:glasses with their low thermal conductivity are more suitable for high-energy pulse operation because of their large size and broadened fluorescence line. Nd^{3+} doped glasses have been made in a variety of shapes and sizes, from a fibre a few micrometres in diameter, to rods 2 m long and 7.5 cm in diameter and discs up to 1 m in diameter and 10 cm thick.

The Lawrence Livermore National Laboratory (USA) has developed the world's largest Nd:glass laser system which comprises 192 beam lines that will deliver 1.8 MJ, 500 TW to the fusion target where internal fusion ignition and burn was demonstrated [65] (see also chapters D10.1 and D10.2). In the present design, the short pulse amplifiers comprise bundles of 4 cm thick Nd:glass slabs positioned at the Brewster angle, arranged in columns four slabs high, two columns wide and eleven deep in the 'regenerative' four-pass section, five deep in the single-pass 'booster' section. Each beam line has an aperture 40 cm × 40 cm.

References

[1] Kaminskii A A 1990 *Laser Crystal (Springer Series in Optical Science 14)* 2nd edn (Berlin: Springer)
[2] 1993 *OSA Proceedings of Topical Meeting (New Orleans)* vol 15 (Washington, DC: Optical Society of America)
[3] 1994 *OSA Proceeding on Advanced Solid State Lasers (ASSL'94)* vol 20 (Washington, DC: Optical Society of America)
[4] 1995 *OSA Proceedings on Advanced Solid State lasers (ASSL'95)* vol 24 (Washington, DC: Optical Society of America)
[5] 1996 *OSA Trends in Optics and Photonics, Advanced Solid State lasers (ASSL'96)* vol 1 (Washington, DC: Optical Society of America)
[6] 1997 *OSA Trends in Optics and Photonics, Advanced Solid State Lasers (ASSL'97)* (Washington, DC: Optical Society of America)
[7] 1998 *OSA Trends in Optics and Photonics, Advanced Solid State Lasers (ASSL'98)* vol 19 (Washington, DC: Optical Society of America)
[8] 1999 *OSA Trends in Optics and Photonics, Advanced Solid State Lasers (ASSL'99)* vol 26 (Washington, DC: Optical Society of America)
[9] 2000 *OSA Trends in Optics and Photonics, Advanced Solid State Lasers (ASSL)* vol 34 (Washington, DC: Optical Society of America)
[10] 2001 *OSA Trends in Optics and Photonics, Advanced Solid State Lasers (ASSL)* (Washington, DC: Optical Society of America)
[11] 2002 *OSA Trends in Optics and Photonics, Advanced Solid State Lasers (ASSL)* (Washington, DC: Optical Society of America)
[12] Schone W *et al* 1997 *Advanced Solid State Lasers (ASSL'97)* p 292
[13] Machan J, Moyer R, Hoffmaster D, Zamel J, Burchman D, Tinti R, Holleman G, Marabella L and Injeyan H 1988 *Advanced Solid State Lasers (ASSL'98)* pp 263–5
[14] Pealese S, Harkenrider J, Long W, Chui F, Hoffmaster D, Burt W and Injean H 2001 *Advanced Solid State Lasers (ASSL)* pp 300–22
[15] Tidwell S C, Seamans J F and Bowers M S 1993 *Opt. Lett.* **18** 116–18
[16] Golla D, Bode M, Knoke S, Schone S and Tunnerman A 1996 *Opt. Lett.* **21** 53–5
[17] Shine R J, Alfrey A J and Byer R L 1995 *Opt. Lett.* **20** 459–61
[18] Konno S, Fujikawa S and Yasui K 1997 *Appl. Phys. Lett.* **70** 2650–951
[19] Konno S, Fujikawa S and Yasui K 2001 *Advanced Solid State Lasers (ASSL)* pp 9–11
[20] Yasui K 1996 *Appl. Opt.* **35** 2566–96
[21] Shannon D and Wallacc R 1991 *Opt. Lett.* **16** 318–20
[22] Clarkson W A and Hanna D C 1996 *Opt. Lett.* **21** 869–71
[23] Kopf D, Keller U, Emanuel M A, Beach R J and Skidmore J A 1997 *Opt. Lett.* **22** 99–101
[24] Du. K, Liano Y and Loosen P 1997 *Opt. Commun.* **140** 53–6
[25] Sekita M, Haneda H, Yanagitani T and Shiransaki S 1990 *J. Appl. Phys.* **67** 453
[26] Sekita M, Haneda H, Shiransaki S and Yanagitani T 1991 *J. Appl. Phys.* **69** 3709
[27] Ikesue A, Kamata K and Yoshida K 1995 *J. Am. Ceram. Soc.* **78** 3545
[28] Ikesue A, Kinoshita T, Kamata K and Yoshida K 1995 *J. Am. Ceram. Soc.* **78** 1033
[29] Lu J, Prabhu M, Song J, Li C, Xu J, Ueda K, Kaminskii A A, Yagi H and Yanagitani T 2000 *Appl. Phys.* B **71** 469–73

[30] Lu J, Song J, Prabhu M, Xu J, Ueda K, Yagi H, Yanagitani T and Kudryashov A 2000 *Japan. J. Appl. Phys.* **39** L1048
[31] Lu J, Prabhu M, Xu J, Ueda K, Yagi H, Yanagitani T and Kaminskii A A 2000 *Appl. Phys. Lett.* **77** 3707
[32] Lu J, Prabhu M, Song J, Li C, Xu J, Ueda K, Yagi H, Yanagitani T and Kaminskii A A 2001 *Japan. J. Appl. Phys.* **40** L552
[33] Lu J, Murai T, Takaichi K, Uematsu T, Misiwa K, Prabhu M, Xu J, Ueda K, Yagi H, Yanagitani T, Kaminskii A A and Kudryashov A 2001 *Appl. Phys. Lett.* **78** 3586
[34] Lu J, Ueda K, Yagi H, Yanagitani T, Akiyama Y and Kaminskii A A 2002 *J. Alloys Compounds* **341** 220–5
[35] Harmer A L, Linz A and Gabbe D 1969 *J. Phys. Chem. Solids* **30** 1483–91
[36] Heinz P, Seilmeir A and Piskarskas A 1997 *Opt. Commun.* **136** 433
[37] Fluck R, Zhang G, Keller U, Weingarten K and Moser M 1996 *Advanced Solid State lasers (ASSL'96)* vol 1 (Washington, DC: Optical Society of America) pp 274–6
[38] O'Conner P R 1966 *Appl. Phys. Lett.* **9** 407–9
[39] Bagdasarov H S, Bogomolova G A and Popov A A 1968 *Dokl. Acad. Nauk. (DAN) SSSR* **180** 1347–50
[40] Fields R A, Birnbaum M, Fincher C L 1987 *Appl. Phys. Lett.* **51** 1885
[41] Chai B H, Loutts G, Lefaucheur J, Zhang X X, Hong P, Bass M, Shcherbakov I A and Zagumennyi AI 1994 *OSA Proc. Adv. Solid-State Lasers* **20** 41
[42] Zagumennyi A I, Ostroumov V G, Shcherbakov I A, Jensen T, Meyn J P and Huber G 1992 *Quantum Electron.* **22** 1071–2
[43] Jensen T, Meyn J P, Huber G, Ostroumov V G, Zagumennyi A I and Shcherbakov I A 1993 *OSA Proceedings of Topical Meeting (New Orleans)* vol 15 (Washington, DC: Optical Society of America) p 64
[44] Zagumennyi A I, Zavartsev Yu.D, Studenikin P A, Shcherbakov I A, Umyskov A F and Popov P A 1996 *SPIE* **2698** 182–92
[45] Zagumennyi A I, Mikhailov V A and Shcherbakov I A 1996 *Laser Phys.* **6** 582–8
[46] Zagumennyi A I, Mikhailov V A, Vlasov V I, Sirotkin A A, Podreshetnikov V I, Kalachev Yu L, Zavartsev Yu D, Kutovoi S A and Shcherbakov I A 2003 *Laser Phys.* **13** 1–8
[47] Klei P H and Croft W J 1967 *J. Appl. Phys.* **38** 1603
[48] Zhang H, Liu J, Wang J, Wang C, Zhu L, Shao Z, Meng X, Hu X and Jiang M 2002 *J. Opt. Soc. Am.* B **19** 18
[49] Song C, Shen D, Kim N and Ueda J 2000 *Appl. Phys.* B **70** 471–4
[50] Fuller M, Matthews D and Marshal L R 1998 *Advanced Solid State Lasers, Technical Digest* p 252
[51] Abraham M, Bar-Lev A, Epshtein H, Goldring A, Zimmerman Y, Lebiush E and Lavi R 2001 *Advanced Solid State Lasers, Technical Digest* p 357
[52] Kellner T, Heine F, Struve B, Ostroumov V, Peterman K and Huber G 1996 *OSA Tops Advanced Solid-State Lasers* p 356
[53] Zharikov E V, Zhitnyk V A, Zverev G V, Kalitin S R, Kuratev I I, Laptev V V, Osiko V V, Pashkov V A, Prokhorov A M, Shestakov A V and Shcherbakov I A 1982 *Sov. J. Quantum Electron.* **12** 1652–3
[54] Zharikov E V, Osiko V V, Prokhorov A M and Shcherbakov I A 1984 *Izv. Akad. Nauk SSSR Ser. Fiz.* **48** 1330
[55] Caird J A, Shinn M D, Kirchoff T A, Smith L K and Wilder R E 1986 *Appl. Opt.* **25** 4294–304
[56] Ostroumov V G, Privis Yu S, Smirnov V A and Shcherbakov I A 1988 *J. Opt. Soc. Am.* **3** 81
[57] Stokowski S E, Randles M H and Morris R C 1988 *IEEE J. Quantum Electron.* **24** 934–48
[58] Jackel S, Moshe I and Lallouuz P 1997 *Opt. Eng.* **36** 2031
[59] Kutovoi S A, Laptev V V and Matsnev S Yu 1991 *Quantum Electron.* **18** 149–50
[60] Meyn J P, Jensen T and Huber G 1994 *IEEE J. Quantum Electron.* **30** 913–17
[61] Laser Focus World, June 1997, p 70
[62] Ostroumov V G, Heine F and Kuck S *et al* 1997 *Appl. Phys.* B **64** 301–5
[63] Aus de Au J, Kopf D and Keller U 1997 *Opt. Lett.* **22** 307
[64] Stuart B C, Perry M D and Wharton K 1997 *Opt. Lett.* **22** 242
[65] Zapata L, Rotter M, Marshall C and Erlandson A 1998 *Tops Advanced Solid-State Lasers* p 206

B1.4
Lanthanide series lasers—near infrared

Norman P Barnes

B1.4.0 Introduction

Lanthanide series solid state lasers which operate in the near to mid infrared region of the spectrum are explored in this chapter. Lanthanide series lasers which produce wavelengths longer than ~1.4 μm are analyzed here, in particular, lasers having lanthanide series atoms of Dy, Ho, Er and Tm. Lasers producing wavelengths longer than 1.4 μm enjoy an advantage with regard to eye safety because these wavelengths are absorbed volumetrically in the vitreous humour of the eye. Therefore, the laser radiation is not focused onto the retina of the eye, which greatly increases the radiation level required for eye damage. These active atoms also often have other common properties including a dependence on energy transfer processes. Specifically the population of various manifolds or groups of closely spaced energy levels are populated through the transfer of quanta of energy from other manifolds. Energy transfer processes will also be covered in this chapter. Not covered here are lanthanide series lasers in glasses or fibre lasers (see chapter B4), including lasers utilized for communication purposes.

Lanthanide series solid state lasers have several advantages over most other types of lasers, advantages which are associated with their spectroscopic properties. A primary advantage of these lasers is their ability to store energy over relatively long time intervals. Virtually all solid state lasers are optically pumped, either with a lamp or laser diodes. With the capability to store energy, solid state lasers can behave like an optical integrator, storing the optical energy in the active atoms. Having stored the pump energy for time intervals up to many milliseconds, most of it can be released in a single pulse with characteristics tailored to the specific application. This optical integration ability is virtually unique with solid state lasers.

Solid state lasers also have the ability to concentrate the pump radiation into a nearly diffraction-limited beam, a narrow spectral bandwidth or both. While laser diode arrays are more efficient than solid state lasers, the ability of solid state lasers to produce high power or energy in a very narrow spectral bandwidth or diffraction-limited beam is often a reason to select them. Also, solid state lasers have the ability to generate a wide range of pulselengths and pulse repetition frequencies, ranging from a continuous beam to a pulse repetition frequency up to ~1.0 GHz while pulselengths can be as short as tens of femtoseconds.

Although Nd:YAG is useful for many applications (see chapter B1.3), particular wavelengths generated by other lanthanide series lasers may be sought because they interact with matter differently. A different interaction may be important, for example yielding increased eye safety. Some wavelengths are strongly absorbed by water, such as the 2.9 μm Er:YAG laser, and thus useful for medical applications. In contrast, lasers in the blue green region of the spectrum may be sought because they are transmitted by water. Other laser materials and active atoms are selected for particular properties including the ability to store energy, produce a polarized output or generate a minimum amount of heat internal to the laser.

In the following sections of this chapter, the physics and spectroscopy of lanthanide series solid state lasers are reviewed along with energy transfer processes and diffusion. Due to limited space, the review will be necessarily brief and other references are given for a more comprehensive review. Pertinent material parameters and their effect on laser design are presented in the next section (B1.4.1). Pumping mechanisms, including both lamps and laser diodes, as well as pumping geometries are reviewed in the following section. Basic laser physics and simplified laser equations are presented in section B1.4.3 along with methods for estimating the laser performance. Laser performance, in either continuous or pulsed modes, can usually be characterized by a threshold and slope efficiency. Possible laser transitions are presented next, along with a method for determining which particular transition will oscillate. Examples of Er, Tm, Ho and Dy lasers are presented in section B1.4.7 with thresholds and slope efficiencies reported in the literature. Finally, section B1.4.8 will summarize results.

B1.4.1 Quantum mechanics and spectroscopy

Lanthanide series lasers are characterized by the particular lanthanide series atom employed and, to a lesser degree, on the particular laser material in which the atom resides. Electron shells in the lanthanide series atoms are filled between 1s and 4d. Laser transitions for these elements occur between various energy levels of the 4f electrons. In addition, lanthanide series atoms have completely filled 5s and 5p shells. Usually electrons in the 6s shell as well as one electron from the 4f shell form the bonding electrons. Because the 5s and 5p shells are completely filled, they shield the 4f electrons from the electric field produced by the laser material. Thus, the electric field of the laser material exerts only a relatively weak effect on the 4f electrons. This insensitivity of the 4f energy levels on the particular laser material allows lanthanide series atoms to be arranged in an energy level chart for facile reference [1], often referred to as a Dieke chart. An abridged Dieke chart for the atoms of interest appears in figure B1.4.1.

Energy levels of lanthanide series lasers are determined primarily by the quantum mechanics of the individual lanthanide series atom. A spherically symmetric central attractive force is provided by the nucleus and the core electrons, that is the 1s–4d electrons. This attractive force is the dominant mechanism in lanthanide series atoms. Next, in order of importance, is the mutual repulsion of the electrons of the 4f electrons. Because of the mutual repulsion of the electrons, they couple to make the total angular momentum, L, and total spin, S, approximately good quantum numbers. Next, in order of importance, is the spin–orbit coupling. Spin–orbit coupling makes the total angular momentum, J, an approximately good quantum number. In many cases, the spin–orbit effect is nearly equal to the mutual repulsion term. Taken together, these interactions are not sufficient to remove all of the energy level degeneracy. Finally, the electric field of the laser material, or crystal field, splits the various free atom energy levels slightly, removing most of the remaining degeneracy. A set of energy levels associated with a particular L, S and J are referred to as a manifold. Manifold splittings are sufficiently small that they appear in a Dieke diagram only as a broadening of the line that represents the manifold. A figure showing the relative magnitude of the various interactions is given in figure B1.4.2.

Manifolds are designated in terms of the Russell–Saunders notation. Typically, the manifold has the notation $^{2S+1}L_J$ where S is the total spin quantum number, J is the total angular momentum quantum number while L designates the total angular momentum quantum number. With Russell–Saunders notation, the letters S, P, D, F, G, H, I,. . . represent the total angular momentum 0, 1, 2, 3, 4, 5 and 6. . .. If the total spin quantum number is integer, the maximum for the total number of levels in a manifold is $(2J + 1)$. If the total spin quantum number is half integer, in the crystalline field all of the levels are doubly degenerate, reducing the maximum number of levels to (2J +1)/2. Often this condition is referred to as Kramers' degeneracy.

Crystal field effects can be characterized by a set of crystal field parameters, B_{nm}. Crystal field parameters result from the electric field effects which neighbouring atoms have on the active atom. In crystalline laser materials, the crystal field parameters depend on the symmetry imposed by the neighbouring atoms. Crystals have an ionic character so that an ionic charge is associated with each neighbouring atom. Crystal field parameters can, in principle, be calculated given the position and charge of neighbouring atoms. Usually,

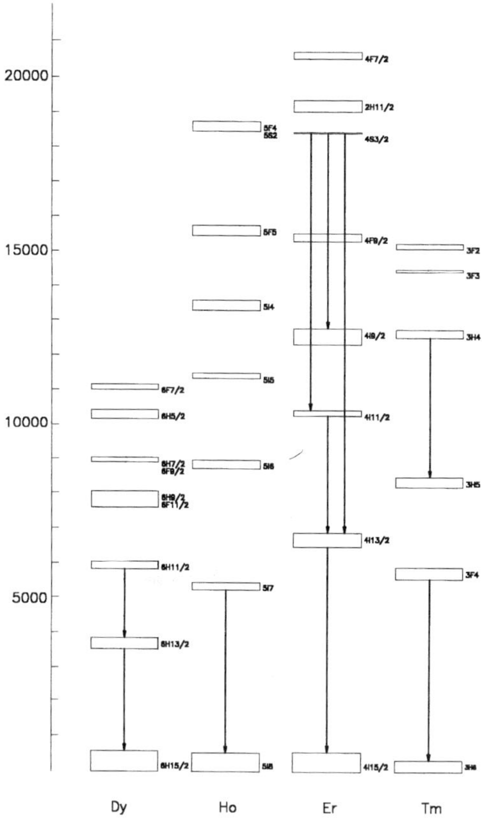

Figure B1.4.1. Abridged Dieke diagram for Dy, Ho, Er and Tm. Common laser transitions are indicated

however, the energy levels of a lanthanide series atom in a laser material are measured spectroscopically and the crystal field parameters are adjusted to reflect the experimental data.

Laser transitions are primarily electric dipole transitions. Transition probabilities between a set of levels can be forbidden and occur only because the specific levels are not pure levels but have some other levels included in the wavefunctions. While the contribution of other levels which are included can be relatively small, it is sufficient to allow the transition to occur. As a result of this mixing of levels, selection rules are rarely strictly observed. In general, however, the rule that the total spin of the level does not change during the transition does have some validity. Although magnetic dipole transitions are much weaker than electric dipole transitions under normal circumstances, in the case of lanthanide series atoms, magnetic dipole contributions to a transition can be noticeable.

Laser transitions in lanthanide series atoms are usually from a single level within a manifold to another single level in another manifold. For closely spaced energy levels, more than one pair of levels may contribute to a particular transition. An individual transition can be approximated by a Lorentzian lineshape, that is

$$g(\nu) = (\Delta\nu/2\pi)[(\nu - \nu_0)^2 + (\Delta\nu/2)^2]^{-1}$$

where $\Delta\nu$ is the linewidth and ν_0 is the centre frequency. However, many lines have some variation in the centre wavelength. Thus, a Voigt profile (see section A1.9) is often a better approximation. The linewidth

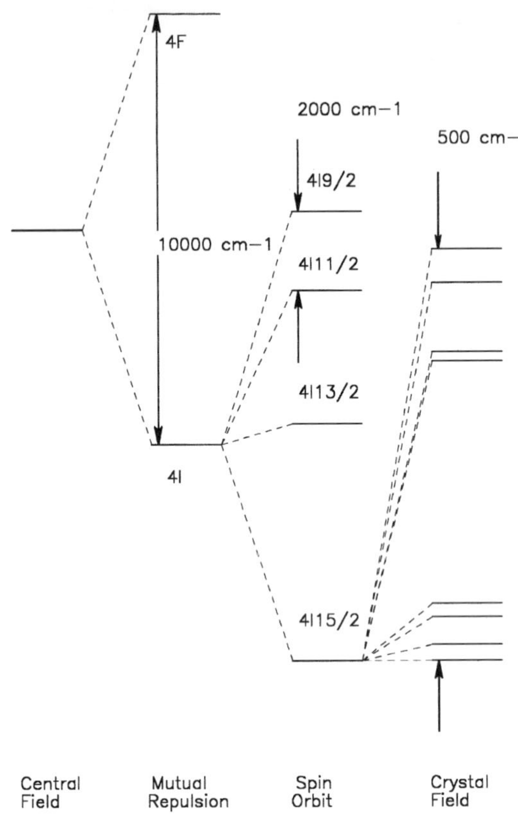

Figure B1.4.2. Magnitude of the various quantum mechanical interactions.

of the Lorentzian lineshape function is dependent on the temperature and is of the order of 10 cm^{-1} at room temperature. Linewidths become narrower as the temperature is reduced.

If the crystal does not have sufficient symmetry, the laser transitions are polarized. For example, in uniaxial crystals the laser transition is usually different if the electric field of the emitted radiation is polarized perpendicular to or parallel to the optical axis of the crystal. If a polarized laser is desired, the polarization dependence of the transition can be used to advantage.

B1.4.2 Energy transfer and diffusion

Lanthanide series lasers are often made efficient by energy transfer processes. If the energy difference between a pair of manifolds in one atom is nearly equal to the energy difference between a pair of manifolds of a neighbouring atom, the atom in the excited manifold can transition to a lower energy manifold while simultaneously raising the neighbouring atom from a low energy manifold to a higher energy manifold. A classic case of an energy transfer process involves the exchange between the first excited manifolds of Ho and Tm. A Tm atom which initially resides in the first excited manifold, $^3F_4$, transitions to the ground manifold, $^3H_6$, while a neighbouring Ho atom transitions from the ground manifold, $^5I_8$, to the first excited manifold, $^5I_7$, shown as process H in figure B1.4.3. No photons are emitted and reabsorbed, rather there is an energy

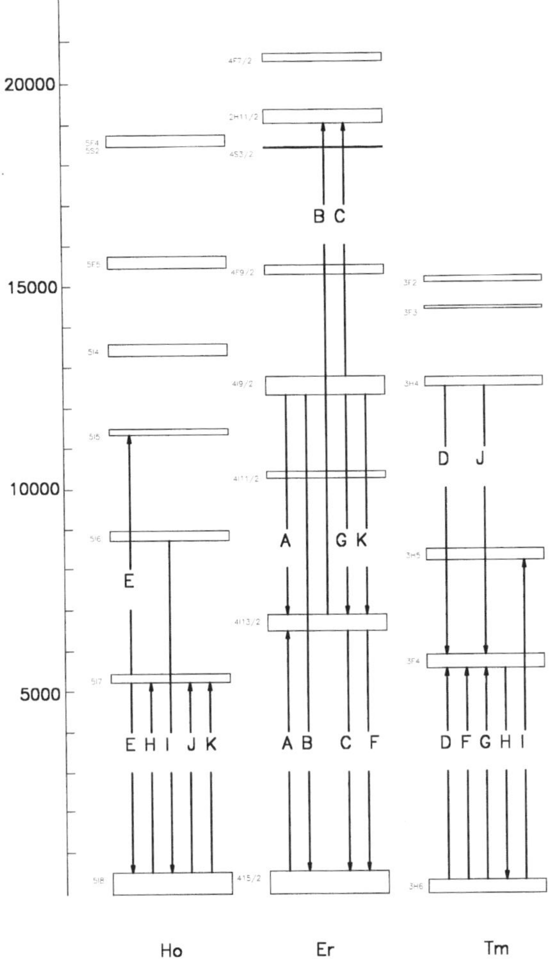

Figure B1.4.3. Abridged Dieke diagram for Ho, Er and Tm. Forward energy transfer interactions are indicated by a pair of lettered arrows. Reverse energy transfer interactions would have reversed arrows.

exchange between two atoms. For historical reasons, atoms donating the excitation are denoted as sensitizers while atoms receiving the excitation are denoted as active atoms.

Energy transfer usually occurs between neighbouring atoms by a dipole–dipole interaction. Dipole fields decrease as the inverse cube of the distance separating the dipoles. Consequently, the dipole–dipole interaction depends on the inverse sixth power of the distance. Dipole–quadrupole and quadrupole–quadrupole interactions are possible but they decrease as the inverse eighth and inverse tenth power of the distance. Because of the distance dependence and strength of the interactions, the dipole–dipole mechanism is usually responsible for energy transfer.

Historically, energy transfer was analyzed by Forster [2] and Dexter [3]. Energy transfer parameters were simply calculated by integrating over the product of the sensitizer emission spectra and the active atom absorption spectra. The strength of this interaction depends on the orientation of the atoms with respect to each other and the orientation of both dipoles with a vector connecting the atoms. In the classical approach, all possible orientations were averaged. In addition, all possible distances separating the interacting atoms were also averaged. Because of the inverse sixth power dependence of the interaction on the distance, the selection of the nearest possible distance for the dipoles becomes very critical. These approximations are valid for fluids, to which these models were initially applied. However, they are not strictly applicable to crystals where the distances and orientations are fixed by the crystal lattice. An energy transfer model has been developed which takes into account the restrictions set by the crystal lattice [4].

Energy transfer parameters have been modelled for crystalline laser materials by starting with a knowledge of the energy levels of the particular laser material. Utilizing a quantum mechanical model, level-to-level transition probabilities or lifetimes can be calculated. It is necessary to know the level-to-level lifetimes, rather than the manifold-to-manifold lifetimes, because energy transfer is greatly enhanced by a resonance between the interacting levels. Thus, pairs of energy levels which are in resonance or near resonance can contribute heavily when they have adequate transition probabilities. To calculate the energy transfer parameters, all possible transitions between all of the upper manifold and lower manifold levels must be known for both the sensitizer and active atom. Rather than integrating over the emission and absorption spectra, a summation over all possible pairs of transitions can be performed. In addition, a summation over all possible sites for the sensitizer and active atom can be performed. Possible sites and the allowed orientation of the dipole are both determined by the crystal lattice. When performing the summation over the sites, it was observed that for common laser materials, YAG and YLF, the nearest similar neighbours contributed about three-quarters of the total sum.

Active atoms or sensitizers which have a next nearest similar neighbour that is also an active atom or sensitizer usually have a substantially different lifetime from atoms which do not enjoy this condition. Because of the shielding of the 4f electrons from the crystal field and because the next similar neighbours are usually not very close, the occupancy of these sites has little effect on the energy levels and transition probabilities. However, the existence of a close cooperative neighbour can have a significant effect on the energy transfer process. An energy transfer process can shorten the fluorescent lifetime of neighbouring similar atoms. To describe the dynamics of the active atoms or sensitizers, the atoms can be divided into two or more groups depending on the distance to a cooperative atom.

Atoms can be divided into distinctive groups. An atom may have no nearest similar site occupied by a cooperative atom, called a singlet, or may have a single nearest similar site occupied by a cooperative atom, called a doublet. Or, the concentration of cooperative atoms may not be constant, as they prefer to form clusters. In fact, experimental data indicate that this occurs in some solid state lasers. In either case, the decay could be described with similar mathematics. In these cases, the decay of the manifolds may be described by a sum of exponentials, an exponential term for atoms having cooperative atoms close and another for atoms having cooperative atoms further away.

Many energy transfer processes are possible, as displayed in figure B.1.4.3. Energy transfer from Tm to Ho, described earlier, is an example. In addition, Ho in the $^5I_7$ manifold can also transition to the $^5I_8$ manifold while it raises a neighbouring Tm atom from the $^3H_6$ to the $^3F_4$ manifold. Obviously this is simply the reverse of the first process and is sometimes referred to as back transfer. In self-quenching, the sensitizer and the active atom are the same type of atom. As an example, a Tm atom in the $^3H_4$ manifold can transition to the $^3F_4$ manifold while raising a neighbouring Tm atom from the $^3H_6$ manifold to the $^3F_4$ manifold, process D. Finally, the opposite processes can occur, that is two Tm atoms both initially in the $^3F_4$ manifold can interact promoting one of the Tm atoms to the $^3H_4$ manifold while the other Tm atom transitions to the $^3H_6$ manifold. Because the final state of one of the initial Tm atoms is an energy manifold higher than the original excited manifold, this process is referred to as up-conversion.

Energy transfer can be described using a set of differential equations. For example, the differential equations for Tm and Ho forward and back energy transfer are given. Denoting the Tm $^3H_6$, Tm $^3F_4$, Ho $^5I_7$ and Ho $^5I_8$ manifolds as 1, 2, 7 and 8, respectively, the differential equations describing the population densities of the various manifolds, N_i, are:

$$\frac{dN_7}{dt} = -\frac{N_7}{\tau_7} - P_{71}N_7N_1 + P_{28}N_2N_8$$
$$\frac{dN_2}{dt} = -\frac{N_2}{\tau_2} + P_{71}N_7N_1 - P_{28}N_2N_8.$$

In this approximation, the energy transfer rate is proportional to the product of the number density of atoms in the manifold which donates the excitation and the number density of atoms in the manifold which receives the excitation. P_{ij}, is referred to as the energy transfer parameter. The subscripts, i and j, refer to the initial manifolds of the atom acting as the sensitizer and the atom acting as the active atom. If more than a single site for the cooperative atoms exists, similar differential equations can be written for each. A fraction of the population inversion can be associated with each site. The total manifold population is then described by the sum of the solutions.

Energy transfer and back energy transfer are related and their ratio is a function of temperature. In general, for every energy transfer process there is a reverse energy transfer process. Often the excited manifold of one atom has a higher average energy than the excited manifold of the other atom. Given these differences in energy, the interacting levels have different thermal occupation or Boltzmann factors. As a consequence, the difference in the thermal occupation factors favours the forward energy transfer over the back energy transfer. Again, the classical case is the energy transfer processes between the Tm $^3F_4$ and the Ho $^5I_7$ manifolds. If the partition function of a manifold is defined as

$$Z_i = \sum_j \exp(-E_{ij}/kT)$$

where the summation is over all energy levels in the manifold, k is Boltzmann's constant and T is the temperature. Using this, the ratio of the forward-to-backward energy transfer rates can be shown to be [5]

$$P_{71}/P_{28} = (Z_2Z_8/Z_1Z_7)\exp(-(E_{ZL2} - E_{ZL7})/kT)$$

where the subscripts 1, 2, 7 and 8 again refer to the Tm $^3H_6$, Tm $^3F_4$, Ho $^5I_7$ and Ho $^5I_8$ manifolds, respectively, and E_{ZL2} and E_{ZL7} are the energies of the lowest or zero level of manifolds 2 and 7, respectively.

Diffusion can affect the energy transfer of the various atoms by permitting the excitation to migrate among the same type of atoms. Diffusion is a special case of energy transfer where the sensitizer and active atom are the same type and the interacting manifolds are the same. As an example of diffusion, a Tm atom in the $^3H_4$ manifold can interact with a neighbouring Tm atom in the $^3H_6$ manifold. The first atom transitions to the $^3H_6$ manifold while promoting the second atom to the $^3H_4$ manifold. On average, no energy is lost in this process and the same number of Tm atoms remains in the $^3H_4$ manifold. Diffusion tends to be orders of magnitude faster compared with the more general energy transfer because all pairs of energy levels have at least one resonance. Diffusion can assist other transfer processes by allowing transfer to a pair of atoms where energy transfer can occur faster.

Several authors have proposed various models to describe the effects of diffusion on the decay of atoms affected by energy transfer processes. A static transfer model describes the decay of an initially excited manifold described by $\exp[-t/\tau - k_S(C_{DA}t)^{1/2}]$ where k_S is a constant and C_{DA} is the sensitizer to active atom transfer constant. However, this model neglects diffusion and back energy transfer. A diffusion-limited model describes decay by $\exp[-t/\tau - k_D(C_{DA}t)^{1/2}f(x)]$ where k_D is a constant and $f(x)$ is a ratio of polynomials in x while x is $Dt^{2/3}/C_{DA}^{1/3}$. This model assumes that diffusion is much slower than energy transfer, a condition usually contrary to the previous analysis, and neglects back energy transfer. A migration-assisted model describes decay by $\exp[-t/\tau - k_M(C_{DA}t)^{1/2} - Wt]$ where k_M and W are constants.

Table B1.4.1. Laser material properties.

Parameter	YAG	YLF(a)	YLF(c)	Units
Transparency	0.24–5.5	0.12–7.3		μm
Refractive index	1.819	1.450	1.471	
Sellmeier				
A	2.08745	1.38757	1.31021	
B	1.20810	0.70757	0.84903	
C	0.02119	0.00931	0.00876	
D	17.2049	0.01885	0.53607	
E	1404.45	50.997	134.957	
dn/dT	10.4	−0.7	−2.3	10^{-6} K^{-1}
Lattice constant	1201	516.8	10.736	pm
Density	4560	3980		kg m^{-3}
Lanthanide sites	1.38	1.39		10^{28} m^{-3}
Phonon maximum	860	560		cm^{-1}
Young's modulus	3.6	0.075		TPa^{-1}
Poisson's ratio	0.25	≈0.33		
Strength	0.20	0.033		GPa
Melting point	1940	819		°C
Specific heat	590	791		J/kg
Conductivity	13	5.8	7.2	W mK^{-1}
Thermal expansion	7.8	8.3	13.3	10^{-6} mK^{-1}
Thermal shock	790	180		W m^{-1}

This model assumes that migration is essentially a random walk fast enough to maintain the same relative distribution, but neglects back energy transfer.

Experimental decay curves can often be fitted almost equally well with several of these models. The sum of exponentials approach has the attraction of using simple rate equations to describe the processes and has proven to work rather well for practical laser applications [14].

B1.4.3 Properties of laser materials

Although the choice of the laser material makes only a small change in the position of the energy levels, it can make a large difference in other properties, such as cross sections and thermal properties. Properties of interest can be grouped into optical, mechanical and thermal properties. Each are discussed here along with some germane effects. A table of laser material properties is given in table B1.4.1 for an oxide, $Y_3Al_5O_{12}$ or YAG, and a fluoride, $YLiF_4$ or YLF (see also chapter B1.3).

Optical properties include: range of transparency, refractive index and variation in the refractive index with temperature. For efficient operation, the laser material must be transparent at both the pump and laser wavelengths. In most cases, pump wavelengths are shorter than laser wavelengths. The ultraviolet absorption features of laser materials do not have a sharp cut-off wavelength; thus, some laser material absorption may extend into the pump region. Because absorption by the laser material does not contribute to the laser process but does add to the heat deposited in the laser material, some short wavelengths may be filtered out, especially for flashlamp-pumped systems.

Refractive indices are usually best described by a Sellmeier equation. Specifically, the refractive index,

n, is related to the wavelength, λ, by

$$n^2 = A + B\lambda^2/(\lambda^2 - C) + D\lambda^2/(\lambda^2 - E)$$

where A, B, C, D and E are constants and λ is almost always in μm. Some materials are birefringent; that is the index of refraction is dependent on the polarization. In this case, a set of Sellmeier coefficients is associated with each different polarization. In addition, the variation of the refractive index with temperature is important because of thermal focusing effects which are discussed later. When a linearly polarized output is desired, a birefringent material is often selected. Birefringent materials usually produce a polarized laser output because the transition probability and, thus, the gain, is dependent on the polarization. Also, the natural birefringence, usually being much larger than the thermally induced birefringence, swamps any thermal variations which tend to scramble the polarization.

Lattice parameters set the size of the cell which forms the basic building block of the laser material lattice. In general, there are more atoms in a unit cell than can be found in one chemical formula. The number density of potential active atoms sites can be determined using the lattice parameters. If there are n chemical formulae of atoms in the unit cell and the chemical formula has x atoms for which the active atom could be substituted, the number density of potential sites, N_S, can be expressed as

$$N_S = nx/abc$$

where a, b and c are the lattice parameters along the axes of a rectangular lattice. For example, YAG has eight chemical formulae contained in a unit cell and there are three Y atoms in the chemical formula. Because YAG is a cubic lattice, a, b and c are equal and the number density of potential sites is $24/a^3$. The density is the mass per unit volume of the laser material and is also obviously dependent upon the lattice parameters.

The response of the laser material to an applied stress can be described by the mechanical parameters: Young's modulus and Poisson's ratio. Young's modulus, E, is the ratio of applied tensile stress, or force per unit area, to strain, or change in length per unit length. While the laser material elongates in response to the applied tensile stress, it simultaneously shrinks along the lateral dimensions. Poisson's ratio, ν, is the ratio describing the decrease in the lateral dimensions to the increase in the length. The strength of the laser material is the maximum stress that can be applied without inducing permanent changes. Crystalline laser materials usually fail in tension with the appearance of a crack originating on the surface. The strength of the laser material depends heavily on the surface finish, smoother surfaces producing a higher strength.

The lifetime of a laser manifold depends strongly on the maximum phonon energy. The lifetime of a manifold depends on the radiative decay and non-radiative quenching. In non-radiative quenching, an atom drops from a high manifold to the manifold directly below it with the emission of one or more phonons. If two laser manifolds are separated by an energy about five times the maximum phonon energy, non-radiative quenching of the upper lifetime is usually small. Manifolds which are not non-radiatively quenched often possess long lifetimes and are referred to as metastable manifolds. Fluoride laser materials tend to have lower maximum phonon energies than oxide laser materials.

Thermal parameters, including the coefficient of thermal expansion and thermal conductivity, may depend on the orientation with respect to the crystal axes. Because resonant frequencies depend on the optical length of the resonator, a change in the temperature can cause a change in the exact laser wavelength. The optical length of a laser material changes according to

$$\frac{\Delta(n\ell)}{n\ell} = \left[\frac{1}{n}\frac{\mathrm{d}n}{\mathrm{d}T} + \alpha\right]\Delta T$$

where n and α are the appropriate refractive index and coefficient of thermal expansion and ℓ is the length. In an open resonator, an expansion of the laser material implies a decrease in the air pathlength, which must be included in the resonator length calculations. Because laser materials are pumped volumetrically and cooled on the lateral surfaces, pumping causes thermal gradients. With circular symmetry, uniform pumping causes parabolic temperature profiles. When the laser material has different thermal conductivities in different lateral directions, even uniform pumping produces complex thermal profiles.

A temperature gradient, from the centre to the surface of the laser material, causes thermal lensing because of the variation in the refractive index with temperature and stress. (A temperature gradient causes a gradient in the index of refraction and thus in the optical pathlength, that is a lensing effect.) For isotropic laser rods, the thermally induced focal length, f, is approximated by [6, 7]

$$\frac{1}{f} = \frac{P_{\mathrm{H}}}{2\pi a^2 k_{\mathrm{c}}}\left[\frac{\mathrm{d}n}{\mathrm{d}T} + 2n^3\alpha C\right]$$

where P_H is the heat generation rate in the laser rod, k_c is the thermal conductivity, a is the radius of the laser rod, α is the coefficient of thermal expansion and C is a constant depending of the polarization and the photoelastic constants. In oxide laser materials such as YAG, both the first and second terms in the square brackets are positive. In this case, thermal lensing can be large and positive. In fluoride laser materials such as YLF, the first term is negative and the second term is positive. In the latter case, the terms tend to cancel and the thermal lensing can be small and possibly negative. Because of the decreased thermal lensing, fluoride laser materials may be selected for applications where thermal focusing is a concern. For diode-pumped lasers, the heat generation rate is roughly linearly proportional to the pump power. However, for flashlamp-pumped lasers, the heat generation rate may increase faster than the pump power.

Thermal gradients can cause depolarization in isotropic laser materials. In isotropic laser rods, depolarization is associated with the photoelastic term, the C term mentioned earlier. Linearly polarized radiation incident on a thermally stressed laser rod can be resolved into radial and tangential components on the laser rod ends. Because of small differences in the radial and tangential refractive indices, an initially linearly polarized beam becomes elliptical upon exiting the laser rod. Generally, a portion of the elliptically polarized beam will be rejected by a linear polarizer in the laser resonator, causing a loss. This depolarization loss has to be averaged over the laser beam cross section because the radial and tangential directions as well as the phase difference, $\Delta\phi$, vary over the cross section. For a material like YAG, the maximum phase difference is

$$\Delta\phi_0 = (2n_0^3\alpha P_{\mathrm{H}}/k_{\mathrm{c}}\lambda)[(p_{11} - p_{12} + 4p_{44})(\nu + 1)/48(1 - \nu)]$$

and the phase difference varies with radial position as $\Delta\phi_0(\rho/a_r)^2$. Losses have been calculated for both Gaussian and uniform beams [8] and plotted in figure B1.4.4. Losses for a Gaussian beam are smaller because the beam is concentrated in the centre of the laser rod where the phase difference is small. Conversely, losses for an uniform beam can be as high as 0.33.

B1.4.4 Pump and population inversion

Population inversions in solid state lasers are created almost exclusively using optical pumping. Optical pumping can be achieved by various methods including noble gas discharges, laser diodes and sometimes another laser. Optical pump radiation is absorbed by the laser material which uses the absorbed energy to promote active atoms to the upper laser level. Establishing a higher population density in the upper laser level rather than in the lower laser level, that is, a population inversion, may require several processes. When more active atoms reside in the upper laser level than in the lower laser level, laser action is possible. Solving the laser pumping problem is the first step in determining the laser output.

The laser process can be either continuous or pulsed. If it is pulsed, lasing can occur during the pumping process, referred to as normal mode, or lasing can occur in a single short pulse near the end of the pump pulse, referred to as Q-switched mode. In the latter case, a fast optical shutter, that is a Q-switch, is needed

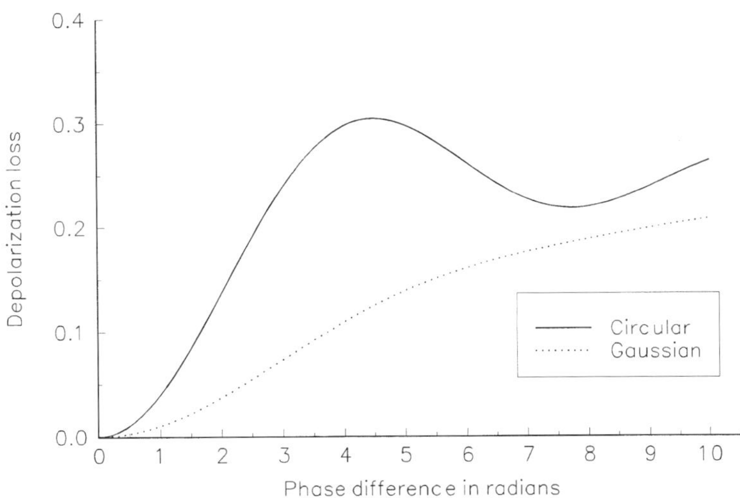

Figure B1.4.4. Depolarization loss *versus* phase difference.

to prevent lasing before the pumping is completed. Although the pumping and energy conversion processes are the same in all cases, different approaches are taken to calculate the laser output.

Noble gas discharges, similar to the flashlamps associated with photography, were the first optical pumps. In fact, they are still extensively used today with little modification from early configurations. While Kr and Ar flashlamps have been used, Xe is the most commonly used gas. Lamps are usually fabricated in a thin cylinder and laser rods are often configured in a similar manner to facilitate optical coupling between the lamp and laser rod. In a continuous discharge mode, the optical spectrum of Xe has a low-level blackbody spectrum with superimposed emission lines of Xe. In a pulsed discharge, the blackbody component becomes dominant but broad Xe emission features can still be recognized. Xe emission lines are grouped in the near infrared. Conversion of electrical energy to optical energy can be relatively efficient, approximately 0.3 to 0.5, depending on the lamp and the discharge parameters. But because the emission spectrum of the lamp and the absorption spectrum of the laser material usually do not match well, lamp pumping has a limited efficiency. Arc-lamp lifetimes are usually in the hundreds of hours and flashlamp lifetimes are in the range of 10^7–10^8 shots, depending on the discharge and the lamp design.

Laser diodes and laser diode arrays can improve both the efficiency and reliability of lasers and are becoming increasingly common (see chapter B2). Laser diodes can efficiently convert electrical energy into optical radiation, roughly 0.3 to 0.5. However, the spectral emission can be rather narrow, of the order of a few nm. By matching the laser diode emission spectrum with an absorption feature of the laser material, pump radiation can be efficiently absorbed. Initially, most laser diodes used GaAlAs technology and were confined to the 0.78–0.88 μm region. Presently, driven by communications applications, laser diodes can cover the spectral region between ~0.6 and 2.1 μm. Laser diode wavelengths and technologies are listed in table B1.4.2. High average power laser diodes and laser diode arrays are available primarily around 0.80, 0.98 and 0.67 μm. Laser diode lifetimes are often in the range of 10 000 h or more and pulsed diodes last for 10^8–10^9 shots, depending on the current pulses.

Laser diodes arrays for laser pumping applications are often fabricated from edge emitting bars with multiple emitting stripes. Bars are usually 10 mm in length and can be stacked to form arrays. Because the

Table B1.4.2. Wavelengths of commercial laser diodes.

Material system		Minimum	Maximum
Active region	Substrate		
InGaAlP	GaAs	0.61	0.69
GaAlAs	GaAs	0.78	0.88
InGaAs	GaAs	0.88	1.10
InGaAsP	GaAs	0.69	0.88
InGaAsP	InP	1.10	1.60
InGaAs	InP	1.60	2.10

emitting stripes of edge emitting laser diodes are only a few micrometres by tens of micrometres, the radiation pattern is not symmetric. The mode pattern perpendicular to the junction is often single mode but highly divergent, $\sim 35°$, because of the small size of the emitting stripe in this plane. The transverse mode pattern in the perpendicular plane is highly multi mode but not as divergent, $\sim 10°$. For pumping applications, glass fibre lenses are often mounted in front of each bar to collimate the highly divergent component of the beam effectively.

Laser pumping is often employed for lasers with short upper laser level lifetimes or if laser diodes with the proper wavelength are not available. Pumping a laser with another laser has other advantages associated with the relatively good beam quality of the pump laser. It is often possible to match the pumped volume with the TEM_{00} laser mode volume, thereby enhancing good beam quality. If the laser transition possesses a short upper laser lifetime, a pump laser can deliver the pump energy in a short Q-switched pulse. Because Q-switched pulses are typically shorter than most upper laser manifold lifetimes, energy storage losses can be obviated and a Q-switched-like pulse produced. This is referred to as gain switching.

Given a pump source, the optical pump radiation is used to create a population inversion. Pump radiation is coupled to the laser material, absorbed, transferred to the upper laser manifold and stored until the laser pulse evolves. Each process is assigned an efficiency and these efficiencies are often nearly independent of the pumping level. The efficiency of the laser system is roughly proportional to the product of these various efficiencies.

Lamps are coupled to the laser material utilizing specular or diffuse cavities. Specular cavities are usually in the form of a conic section, such as an elliptical cylinder. Lamps are usually near one focus of the ellipse and the laser rod is near the other focus. Radiation from the lamp is focused onto the laser material in this arrangement. The coupling efficiency, η_C, of specular cavities can be high, many above 0.9 [9, 10]. Conversely, diffuse cavities are constructed from extremely high reflecting material, such as $BaSO_4$. Pump radiation essentially fills the cavity and reflects until it is either absorbed by the laser rod, reabsorbed by the lamp or it finds another way to escape, such as propagating into a hole in the cavity wall. Diffuse cavities are often thought to produce a more nearly uniform absorbed pump distribution.

Laser diodes are coupled to the laser material in several ways including merely being in close proximity to the laser material. For laser materials with reasonable absorption, the efficiency of this method is difficult to surpass. For laser materials with low absorption, the high brightness of laser diodes can be used to good advantage. Pump radiation is admitted into a highly reflecting cavity through a small slit. On average, it makes many passes through the laser material before it can escape.

After being coupled to laser material, the pump radiation must be absorbed to be useful. The absorption efficiency relates energy stored in the upper laser manifold to incident pump energy. To calculate the absorption efficiency, the absorption coefficients must be known as a function of the pump wavelength. The absorption efficiency, η_A, can be approximated as [11]

$$\eta_A = \int_0^\infty \int_0^\pi \int_0^{2\pi} P(\lambda_P)(1-\exp(-\beta_{1A}Cl))\eta_Q(\lambda_P/\lambda_L)F(\theta,\phi)\sin(\theta)\,d\phi\,d\theta\,d\lambda \times \left(\int_0^\infty \int_0^\pi \int_0^{2\pi} P(\lambda_P)F(\theta,\phi)\sin(\theta)\,d\phi\,d\theta\,d\lambda\right)^{-1}$$

under the assumption of limited depletion of the ground manifold. $P(\lambda_P)$ is the pump power as a function of wavelength, β_{1A} is the absorption coefficient normalized to unity concentration, C is the concentration of absorbing atoms, η_Q is the quantum efficiency, l is the absorption length and $F(\theta,\phi)$ is the distribution of incident angles for the pump photons. A factor of (λ_P/λ_L) is needed because every absorbed photon creates one atom in the pump manifold. Obviously, longer pump wavelengths can be more efficient. This assumes that the ground manifold is not significantly depleted. If depletion occurs, the absorption coefficient will decrease as will the absorption efficiency.

The absorption coefficients, β_{1A}, for the lanthanide series atoms of interest appear in figures B1.4.5–B1.4.7. Although absorption spectra were taken for lightly doped material, the calculated absorption coefficient was divided by the concentration of active atoms to yield the normalized results which appear in the figures.

Sensitizers are sometimes used to increase the absorption efficiency of weakly absorbing active atoms. To be effective, the sensitizer must absorb the pump radiation and transfer the absorbed energy to the active atom with a high efficiency. For example, Tm absorbs pump radiation in the $^3H_4$ manifold from GaAlAs laser diodes and subsequently pumps the Ho $^5I_7$ manifold. As another example, Cr absorbs flashlamp radiation and subsequently pumps the Tm $^3F_4$ manifold. Because Cr is a transition metal atom, rather than a lanthanide series atom, the absorption features are broad, as shown in figure B1.4.8. By covering much of the visible region, Cr is a very good absorber of flashlamp radiation.

For complex systems, absorption transfer efficiency can be roughly approximated by assigning a quantum efficiency, which may be greater than unity, to each manifold [12]. In this approximation, the absorption efficiency integration over pump wavelengths is split into a sum of integrals, one over each pump manifold. Each pump manifold is assigned a quantum efficiency, η_Q. Because of the nature of the lanthanide series spectra, absorption features may be sufficiently separated making identification of the absorbing manifold straightforward. Manifolds which have beneficial energy transfer effects may have a quantum efficiency greater than unity.

In the Q-switched case, the atoms in the upper laser manifold must be stored until the laser operation commences. A storage efficiency, η_S, for a square pump pulse

$$\eta_S = (\tau_2/\tau_I)(1-\exp(-\tau_I/\tau_2))$$

where τ_I is the length of the pump pulse and τ_2 is the upper laser manifold lifetime. In essence, some active atoms excited to the upper laser manifold may decay away before the Q-switch opens. For a half sinusoid pump pulse, the storage efficiency is quite similar.

To approximate the energy transfer effects better, the pumping of every manifold and all subsequent energy transfer processes are described with a set of laser pumping equations. Pumping rates are calculated for every manifold by determining the pump power at the appropriate wavelengths and calculating the manifold absorption efficiency. Pumping rates for each manifold, R_i, are proportional to $\eta_{Ai}P_i$ where P_i is the pump power at the appropriate wavelengths. Lifetimes, τ_i, and energy transfer parameters, P_{ij}, must be known for each manifold.

To illustrate these processes, consider a Tm laser and, for the nonce, neglect the $^3H_5$ manifold, as shown in figure B1.4.9. If the $^3H_4$ manifold is pumped, the laser pumping equations describing the evolution of the

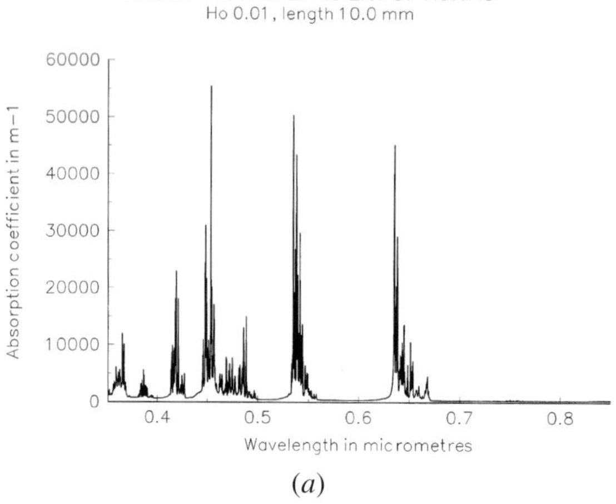

(*a*)

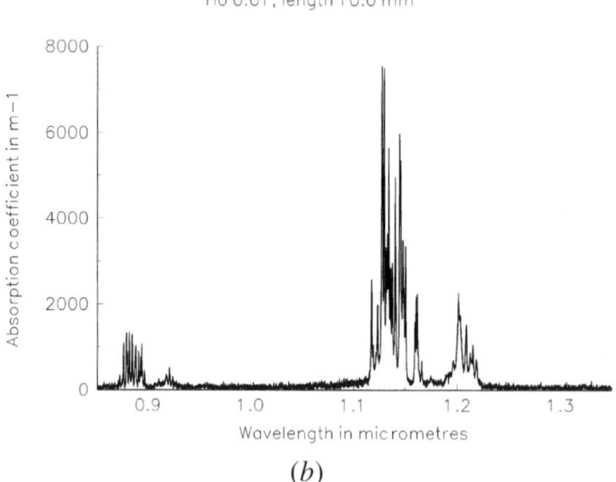

(*b*)

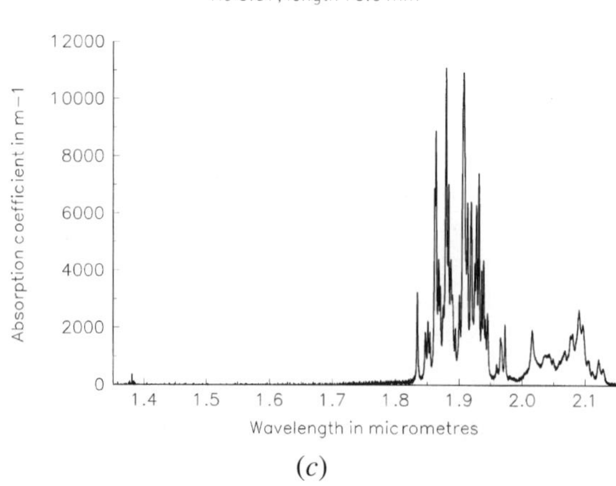

(*c*)

Figure B1.4.5. Normalized absorption coefficient for Ho:YAG.

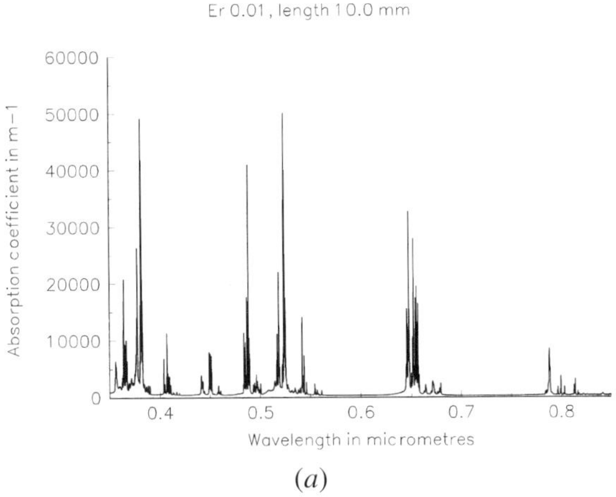

(a)

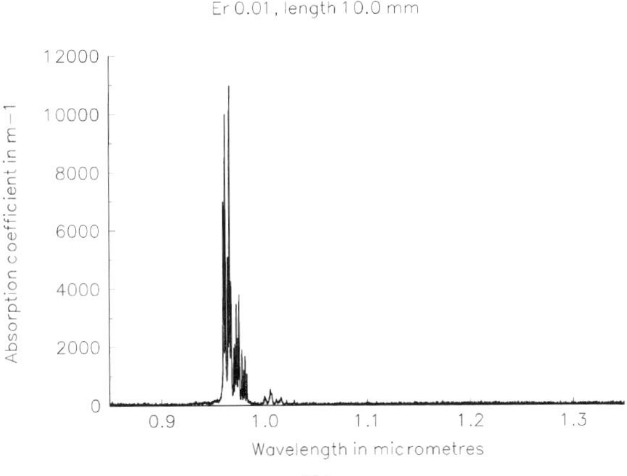

(b)

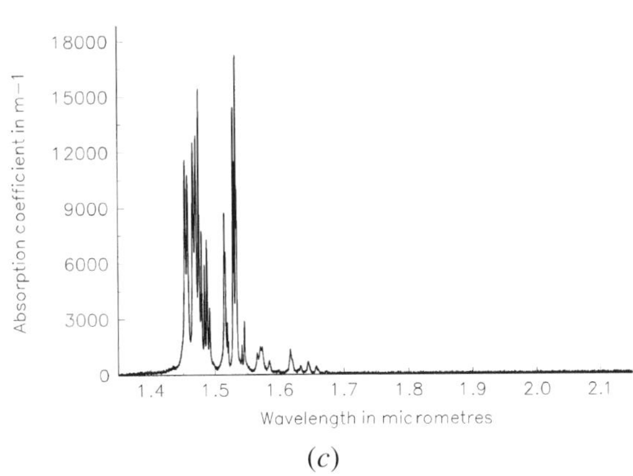

(c)

Figure B1.4.6. Normalized absorption coefficient for Er:YAG.

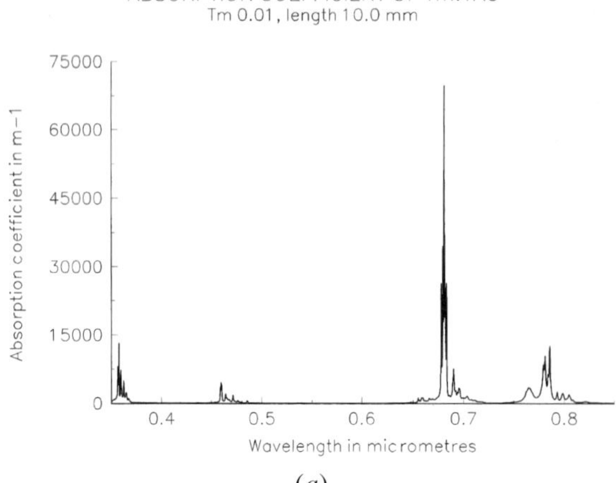

(*a*)

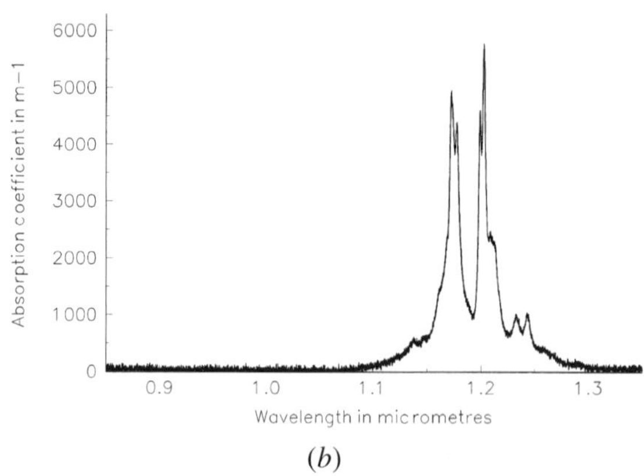

(*b*)

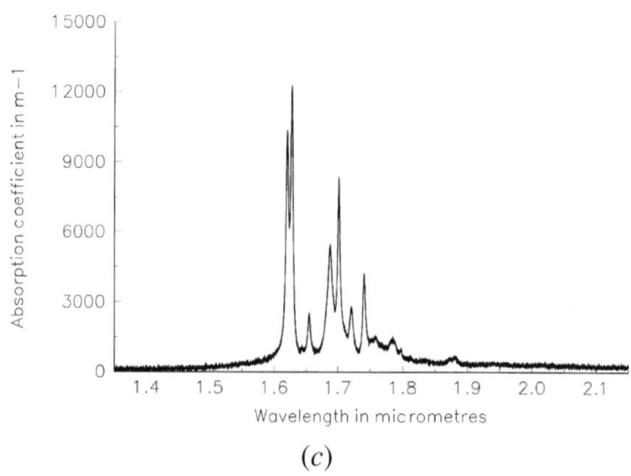

(*c*)

Figure B1.4.7. Normalized absorption coefficient for Tm:YAG.

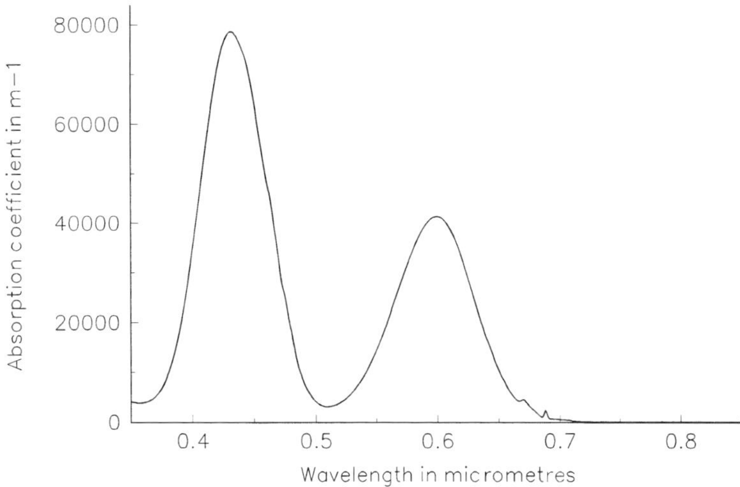

Figure B1.4.8. Normalized absorption coefficient for Cr:YAG.

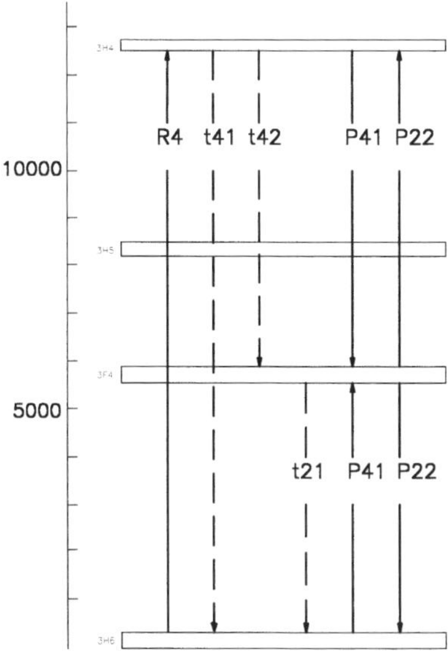

Figure B1.4.9. Simplified Dieke diagram for Tm showing pumping and dynamics.

relevant manifolds are:

$$\frac{dN_4}{dt} = -\frac{N_4}{\tau_4} - P_{41}N_4N_1 + P_{22}N_2^2 + R_4$$

$$\frac{dN_2}{dt} = -\frac{N_2}{\tau_2} + \frac{\beta_{42}N_4}{\tau_4} + 2P_{41}N_4N_1 - 2P_{22}N_2^2$$

$$N_1 + N_2 + N_4 = C_T N_S$$
$$1/\tau_4 = 1/\tau_{41} + 1/\tau_{42} = \beta_{41}/\tau_4 + \beta_{42}/\tau_4.$$

β_{ij} is the branching ratio, that is the fraction of the decaying population density of manifold N_i which ends up increasing the population density of manifold N_j. Energy transfer out of manifold 4, $^3H_4$, by self-quenching process P_{41}, depends on the number density in manifold 4, the manifold which donates the energy and the number density in manifold 1, $^3H_6$, the manifold which receives the energy. Conversely, the up-conversion, process P_{22}, depends on the number density in manifold 2 squared because atoms in this manifold both donate and accept the energy. A factor of 2 appears with the energy transfer processes in the second equation because 2 Tm atoms in the $^3F_4$ manifold are created at a time. Energy transfer processes, P_{41} and P_{22} are inverse process. Every energy transfer process has an inverse process although one of the energy transfer parameters, P_{ij}, may be significantly larger than its corresponding inverse energy transfer parameter. The third equation replaces the differential equation for the ground manifold by noting that the number of Tm atoms must remain constant, equal to $C_T N_S$ where C_T is the concentration of Tm atoms. Note that the storage efficiency is automatically included in the solution of the laser pumping equations.

B1.4.5 Basic laser equations

To calculate the laser output (see also chapters A1–A3 and A8), the differential equations describing the laser pumping and the laser dynamics are solved. Differential equations describe the manifold population densities rather than the level population densities. This approximation is usually valid because the levels within a manifold equilibrate in very short time intervals, on the order of a ps. Thus, even though energy is taken from only one level, the entire manifold contributes to the output. For different temporal operating modes, methods of solution will differ.

Strictly speaking, solutions for the different operating modes may not produce a linear relation between the laser output power or energy and the electrical power or energy. Nevertheless, a linear relationship is a rather good approximation. Therefore, laser performance can be approximated using a threshold power, P_{ETH}, or energy, E_{ETH}, and a slope efficiency, σ_S. The threshold power or energy is the electrical power or energy required for the laser to begin lasing. The slope efficiency is the rate at which the laser output power or energy increases as electrical power or energy increases above threshold. A plot of typical normal mode laser output energy *versus* electrical energy for a TEM_{00} Ho:Tm:Cr:YAG laser is shown in figure B1.4.10. A curve fit, also shown, yields the threshold and slope efficiency for various output mirror reflectivities.

Ho, Er and Tm can operate on transitions which have a small but non-negligible thermal population in the lower laser level and are referred to as quasi-four-level lasers. In three-level lasers, extraction of a photon decreases the population inversion by two, a decrease of one from the upper laser level and an increase of one into the lower laser level. In contrast, in four-level lasers, extraction of a photon decreases the population by one, a decrease of one from the upper laser level. The lower laser level has no thermal population and it relaxes quickly. Thus, there is virtually zero population in the lower laser level. For intermediate cases, the extraction of a photon decreases the population by a factor of γ. If γ is closer to one than two, the laser is termed a quasi-four-level laser, which is the case for the lasers considered here.

When laser pumping can be approximated by a pair of pumping rates, the differential equations (see section A8.3) governing the laser dynamics can be approximated as

$$\frac{\partial N_2}{\partial t} = -\frac{N_2}{\tau_2} - \frac{\sigma c}{n}(Z_2 N_2 - Z_1 N_1) N_P + R_2$$
$$\frac{\partial N_1}{\partial t} = -\frac{N_1}{\tau_1} + \frac{\beta_{21} N_2}{\tau_2} + \frac{\sigma c}{n}(Z_2 N_2 - Z_1 N_1) N_P + R_1$$
$$\frac{\partial N_P}{\partial t} + \frac{c}{n}\frac{\partial N_P}{\partial z} = \frac{\sigma c}{n}(Z_2 N_2 - Z_1 N_1) N_P$$

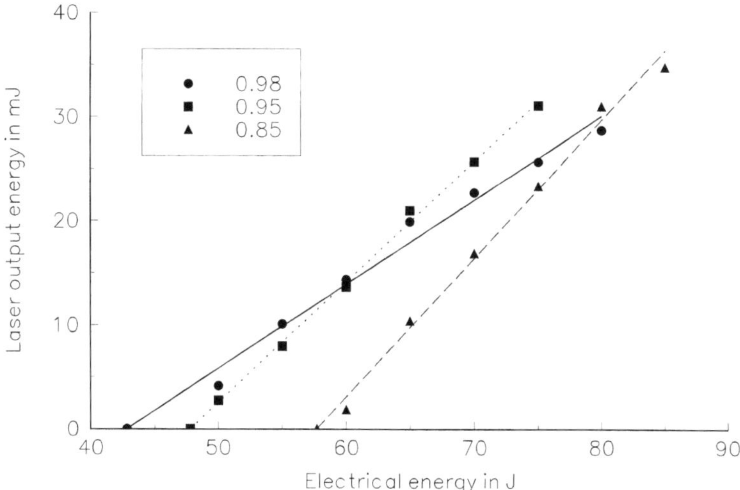

Figure B1.4.10. Typical laser output energy *versus* pump energy.

Here, N_2 and N_1 are the upper and lower laser manifold population densities, τ_2 and τ_1 are the lifetimes and R_2 and R_1 the pump rates of the upper and lower laser manifolds. Further, Z_2 and Z_1 are the thermal occupation or Boltzmann factors for the upper and lower laser levels, σ is the emission cross section, c is the speed of light, n is the index of refraction, and N_P is the photon density. The γ factor is

$$\gamma = 1 + Z_1/Z_2.$$

For quasi-four-level lasers, the lower laser level is in the ground manifold and the term $(-N_1/\tau_1)$ disappears. To a good approximation in many cases, the active atoms reside either in the upper or lower laser manifolds, that is

$$N_1 + N_2 \approx C N_S$$

where C is the concentration of the active atoms while N_S is the number density of potential active atom sites. Thus, CN_S is the number density of active atoms. Optical transparency is defined by the population density in the upper laser manifold necessary to overcome the thermal population density in the lower laser level. Without optical pumping, the laser material displays a loss, caused by the population density in the lower laser level. With optical pumping, the population density in the upper laser level increases, offsetting the loss. At optical transparency, the gain which results from the population density in the upper laser level, Z_2N_2, balances the loss caused by the population density in the lower laser level, Z_1N_1, so there is no net gain or loss. Using this relation, optical transparency occurs when the upper laser level population density, N_{2T}, is

$$N_{2T} = CN_S(\gamma - 1)/\gamma.$$

With quasi-four-level lasers, the differential equation for the population density of the lower laser level can be eliminated permitting closed-form approximations for the laser performance. When the active atoms are primarily in the upper or lower laser manifolds, the derivative of one quantity is simply the negative of the other. Integrating the laser dynamics equations along the length of the resonator and accounting for the photon losses,

$$\frac{\partial N_2}{\partial t} = -\frac{N_2}{\tau_2} - \frac{2\sigma_e \ell_c}{n}\frac{c}{2\ell_c}[\gamma N_2 - (\gamma - 1)CN_S]N_P + R_2$$
$$\frac{\partial N_P}{\partial t} = 2\sigma_e \ell \frac{c}{2\ell_c}(\gamma N_2 - (\gamma - 1)CN_S)N_P + \frac{c}{2\ell_c}N_P \ln(R_M R_L)$$

where σ_e is the effective emission cross section, that is $Z_2\sigma$, ℓ and ℓ_c are the lengths of the laser rod and the laser resonator, R_M is the reflectivity of the output mirror and R_L represents the losses. Integration along the length of the resonator is straightforward when the output mirror reflectivity is reasonably high, usually a good approximation for the lasers considered here.

Different methods are used to calculate the laser performance in different operating modes. For continuous operation, steady-state laser output is of interest. Thus, the temporal derivatives of the laser pumping and laser dynamics equations are set equal to zero and the resulting algebraic equations solved for the photon population density. For the other extreme, for Q-switching, the laser pumping equations are solved with no laser photons present. Population densities of the upper and lower laser manifolds at the instant of Q-switching serve as initial conditions in the solution of the laser dynamics equations for the photon population density. For normal mode operation, the laser pumping equations for all germane manifolds and the laser dynamics equations are solved simultaneously to obtain the laser output [4]. Normal-mode lasing occurs somewhat after the pump pulse commences. Initial temporal behaviour often appears as a series of damped short pulses while the laser output relaxes to quasi-continuous operation. These initial pulses are referred to as relaxation oscillations. When the laser operates in a single transverse mode, both the frequency and the exponential damping constant of the relaxation oscillations can be predicted [12, 13].

Solving the laser dynamics equations for the continuous case [12], the laser output power, P_{LO}, is related to the pump rate by

$$P_{LO} = \pi w^2 \ell \frac{hc}{\lambda_L}\frac{\ln(R_M)}{\ln(R_M R_L)}(R_2 - R_{2TH})$$
$$R_{2TH}\tau_2 = [(\gamma - 1)CN_S/\gamma - \ln(R_M R_L)/2\sigma_e \ell \gamma]$$

while the pump rate many times above threshold is approximated as

$$R_2 = \eta_E \eta_C \eta_A (P_E \lambda_L / hc\pi a_r^2 \ell)$$

where η_E is the efficiency with which electrical power is converted to optical pump power and λ_L is the laser wavelength . For operation close to threshold, the pump rate is approximated as being linearly related to the electrical power. To obtain some closed-form solutions, it is assumed that the laser material is a right circular cylinder with radius a_r and the laser output has a circular beam profile, that is a constant intensity to a radius of w. Gaussian beam profiles can also be handled but space does not allow its presentation. Using these equations, the threshold and slope efficiency are

$$P_{ETH} = \pi a_r^2 \ell (hc/\lambda_L \tau_2)[(\gamma - 1)CN_S/\gamma - \ln(R_M R_L)/2\sigma_e \ell \gamma]/\eta_E \eta_C \eta_A$$
$$\sigma_S = (\pi w^2/\pi a_r^2)[\ln(R_M)/\ln(R_M R_L)]\eta_E \eta_C \eta_A.$$

The threshold is a combination of overcoming the population in the lower laser level and the optical losses in the resonator. The slope efficiency is a product of the overlap factor, that is the ratio of the cross-sectional area of the mode to the cross-sectional area of the laser rod, multiplied by the ratio of the output to the total loss.

The threshold of quasi-four-level lasers is quite sensitive to the temperature, primarily through the factor $(\gamma - 1)$. Noting that this factor is Z_1/Z_2 and the nearly exponential nature of Z_1, the critical temperature dependence becomes apparent. Most initial experiments with quasi-four-level lasers were performed at reduced temperatures, down to cryogenic temperatures. But, with aggressive pumping, it was found that quasi-four-level lasers performed well at room temperatures even though some cooling still improves the performance.

One general closed-form solution for normal-mode operation is not practical because of the variety of particular systems under analysis. Typically lasing commences somewhat after the pump pulse begins. Initial lasing undergoes relaxation oscillations, characterized by a series of regularly spaced temporal pulses with exponentially decreasing amplitude. If the laser is confined to a single transverse mode, relaxation oscillations are often observed. If the laser is not confined to a single transverse mode, several sets of relaxation oscillations complicate the temporal picture. As the oscillations damp out, laser output may relax to a quasi-continuous power output. Both the period between and damping constants of the relaxation oscillations are calculated for quasi-four-level lasers [12].

For Q-switched operation, the laser pumping equations are solved with no energy extraction by laser photons. Solutions for population densities of the upper and lower laser manifolds at the instant the Q-switch opens are then used as the initial conditions for the laser dynamics equations. Because the Q-switched pulse evolution time interval and pulselength are short compared with energy transfer time intervals, energy transfer occurring during these time intervals can be ignored in most cases. Solving the laser dynamics equations for the population density of the upper laser manifold by eliminating the photon density [12],

$$\frac{N_{20}-N_{2F}}{N_{20}-N_{2T}}=\frac{\ln(R_M R_L)}{2\sigma_e \ell\gamma(N_{20}-N_{2T})}\ln\left(1-\frac{N_{20}-N_{2F}}{N_{20}-N_{2T}}\right)$$

where N_{20} and N_{2F} are the initial and final population densities of the upper laser manifold. Under the approximation that the upper laser manifold population density is directly proportional to the electrical pump energy and uniformly distributed

$$N_{20}=(\eta_E\eta_C\eta_A E_E)(\lambda_L/hc)/\pi a_r^2\ell.$$

Given the ratio of gain coefficient and length product to losses $2\sigma_e\ell\gamma(N_{20}-N_{2T})/\ln(R_M R_L)$, the normalized ratio of the difference between the initial and final population densities of the upper laser manifold, $(N_{20}-N_{2F})/(N_{20}-N_{2T})$, can be calculated using the previous transcendental equation above. The results of the solution are given in figure B1.4.11.

With the solution of the transcendental equation, the laser output energy, E_{LO}, and the pulselength, τ_P, can be determined using closed-form approximations. Laser output energy can be expressed as

$$E_{LO}=\frac{w^2}{a_r^2}\frac{\ln(R_M)}{\ln(R_M R_L)}(E_{SO}-E_{ST})\frac{N_{20}-N_{2F}}{N_{20}-N_{2T}}$$

where

$$E_{SO}=\pi a_r^2\ell hcN_{20}/\lambda_L$$
$$E_{ST}=\pi a_r^2\ell hcN_{2T}/\lambda_L$$

are the initial stored energy and the stored energy required to reach optical transparency.

The pulselength can be approximated as E_{LO}/P_{LOmax} where P_{LOmax} is the peak power in the laser pulse. The latter quantity is available through the examination of the laser dynamics equations [12]. With this,

$$\tau_P=-\frac{2\ell_c}{c\ln(R_M R_L)}\frac{N_{20}-N_{2F}}{N_{20}-N_{2T}}\left[1+\frac{\ln(R_M R_L)}{2\sigma_e\ell\gamma(N_2-N_{2T})}\left(1+\ln\left(\frac{-\ln(R_M R_L)}{2\sigma_e\ell\gamma N_2-N_{2T}}\right)\right)\right]^{-1}.$$

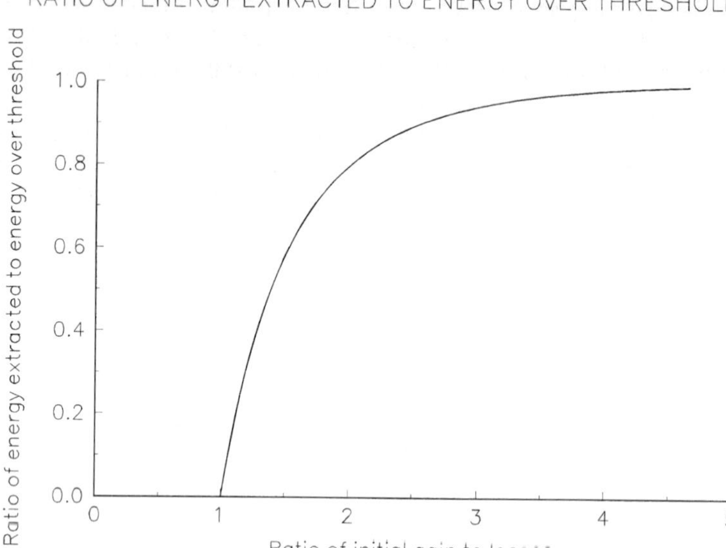

Figure B1.4.11. Differential population inversion *versus* ratio of loss to gain.

Thus, with a knowledge of some basic laser material parameters and the initial population density in the upper laser manifold, both the laser output energy and the pulselength can be calculated under the assumption of a uniform excitation density.

Quasi-four-level lasers can be effective laser amplifiers for Q-switched lasers. The laser pumping equations are solved as a function of time with no laser photons present. Upper and lower laser manifold population densities at the time of Q-switching are used as the initial conditions for the amplifier. The Frantz Nodvik equations, modified for quasi-four-level lasers, are then solved for the gain and energy extraction of the amplifier [13]. A diode-pumped Ho:Tm:YLF laser demonstrated a large signal gain greater than 12 following this procedure. In another case, the small signal gain of Ho:Tm lasers was measured *versus* pump energy and compared with the gain *versus* pump energy predicted by simultaneously solving the energy transfer equations for the four lowest Ho and four lowest Tm manifolds [14]. Good agreement was found for a variety of Ho and Tm concentrations and for both polarizations of this birefringent laser material.

B1.4.6 Tuning

While individual transitions of lanthanide series lasers are relatively narrow, the multiplicity of individual levels promotes virtually continuous tuning. For example, the upper laser manifold of Ho:YAG, the $^5I_7$ manifold, has 15 individual levels while the lower laser manifold of Ho:YAG, the $^5I_8$ manifold, has 17 individual levels. Of the 255 possible transitions, some may be forbidden or relatively weak, depending on the laser material. Nevertheless, transitions essentially cover the spectral range between 1.85 and 2.12 μm almost continuously, with the exception of two relatively small gaps near 2.0 μm. However, because of the thermal populations, only the longer wavelengths will operate.

The laser wavelength depends on the strength of allowed laser transitions, the temperature, and, for some modes, the level of pumping. For a lanthanide laser material like Nd:YAG operating on the $^4F_{3/2}$–$^4I_{11/2}$ transition (see section B1.3), there is essentially no population density in the lower laser level. Consequently, the wavelength of the laser tends to be independent of pumping. However, for quasi-four-level lasers, the transition with the highest gain depends on the strength of the transition and the population density of both the upper and lower laser levels. For resonator configurations with a high output mirror reflectivity, overcoming the thermal population in the lower laser level is often the largest threshold factor. In this case, typically the transition with the longest wavelength reaches threshold first. The longest wavelength usually indicates that the lower laser level is the highest level in the ground manifold and, therefore, has the lowest thermal population. However, for resonator configurations with a lower output mirror reflectivity, overcoming the optical losses becomes the largest threshold factor. In this case, the transition with a larger emission cross section may be favoured.

Which transition dominates depends on the level of pumping in the Q-switched mode while, in the normal mode, whichever transition reaches threshold first will dominate. In normal mode, at some instant during the pump pulse, one of the transitions will reach threshold. At this point, the population inversion will be limited to the threshold value of the operating transition. By limiting the population inversion, competing transitions will not achieve threshold. In the Q-switched mode, the Q-switch will prevent all transitions from reaching threshold during the pump pulse. When the Q-switch finally opens, pulses for all transitions will begin to evolve. The transition with the highest net gain will achieve a high power level first, extract much of the population inversion and limit the gain of competing transitions. Thus, the population inversion will be depleted before the other transitions can extract a significant amount of energy. A high net gain is a combination of a large emission cross section and population inversion and low losses. Because the population inversion is a function of both the wavelength and the level of pumping, the dominant wavelength may change with a change in the level of pumping.

Which transition will dominate in normal-mode operation also depends on the design of the laser resonator. The threshold occurs when the round trip gain is unity, that is

$$R_M R_L \exp[2\sigma_e(\gamma N_2 - (\gamma - 1)CN_S)\ell] = 1.$$

However, both σ_e and γ are dependent on the particular transition. Thresholds are competitive where [16]

$$[(\gamma_{\lambda 1} - 1)/\gamma_{\lambda 1} - (\gamma_{\lambda 2} - 1)/\gamma_{\lambda 2}]2CN_S\ell = [1/(\sigma_{e\lambda 1}\gamma_{\lambda 1}) - 1/(\sigma_{e\lambda 2}\gamma_{\lambda 2})]\ln(R_M R_L).$$

In a plot of the concentration and length product, $CN_S\ell$, *versus* the losses, $\ln(R_M R_L)$, lines defined by this equation separate the regions where the particular transitions would operate. For an example, Ho has transitions at 2.122, 2.097 and 2.091 μm. The effective emission cross section is the smallest for the longest wavelength transition and increases upon going toward the shorter wavelengths. For low loss cases, overcoming the thermal population density in the lower laser level is the dominant effect. Thus, in this case, the 2.122 μm transition is the strongest because it has the lowest population in the lower laser level. Conversely, for high loss cases, overcoming the losses is the dominant effect. In this case, the 2.091 μm transition is the strongest because it has the largest effective emission cross section. For intermediate losses, the 2.097 μm transition operates. In fact, measurements of the operating wavelengths for various Ho concentrations and output mirror reflectivities support this analysis [16].

Atmospheric conditions also affect the laser wavelength. In this region of the spectrum, many atmospheric absorption features exist. H_2O and CO_2 are the primary contributors to the atmospheric absorption in the near and mid infrared region of the spectrum. If strong atmospheric absorption features exist in the vicinity of the laser wavelength, continuous tuning of the laser may be hampered by the existence of these absorption features. When tuning is tried, the laser is apt to skip over the wavelengths where the absorption features are strong. In fact, this has been observed for Ho lasers [17].

B1.4.7 Laser performance

Laser operation utilizing several lanthanide series atoms was demonstrated soon after the discovery of the laser itself. A partial list of laser performance parameters appears in table B1.4.3. By cooling to liquid nitrogen temperatures, laser operation was demonstrated in Ho:YAG on the $^5I_7$–$^5I_8$ transition, in Er:YAG on the $^4I_{13/2}$–$^4I_{15/2}$ transition and in Tm:YAG on the $^3F_4$–$^3H_6$ transition using flashlamp pumping [18]. The importance of energy transfer processes and the possibility of using up-conversion to produce visible photons from infrared pump photons was identified [19].

Efficient operation on the Ho $^5I_7$–$^5I_8$ transition in multiply doped laser materials was demonstrated at cryogenic temperatures by several groups. With a Ho:Tm:Er:YAG laser rod with a low Ho concentration, a continuous laser demonstrated a threshold of ~200 W and a slope efficiency of 0.065 using an incandescent W–I lamp [20]. Both normal-mode and Q-switched operations were demonstrated in Ho:Tm:Er:YLF [21] and Ho:Tm:Er:YAG [22]. Normal-mode thresholds were 14 and 30 J while slope efficiencies were 0.029 and 0.012, respectively, in a good quality beam, often TEM_{00}. Because of Ho:Tm sharing, slope efficiencies for Q-switched operation were somewhat lower.

Despite the fact that Ho:YLF has a higher thermal population than Ho:YAG, better performance was achieved at room temperature with the former. Ho:Tm:Er:YLF and Ho:Tm:Er:YAG laser rods, with similar concentrations and dimensions, were evaluated in a flashlamp-pumped system. Normal-mode thresholds were 100 and 155 J while slope efficiencies were 0.0013 and 0.0004, respectively [23]. Substantially higher normal-mode efficiency was demonstrated by using Cr for a sensitizer in a Ho:Tm:Cr:YAG laser. Thresholds were reduced to ~45 J and slope efficiency was increased to 0.051 [24]. Because Cr substitutes for Al in YAG, Cr has not been tried in YLF.

Continuous laser operation near room temperature was obtained using laser diode pumping. With a laser diode at 0.785 μm to pump the Tm in an end-pumped configuration, a Ho:Tm:YAG laser had an optical threshold of ~52 mW and a slope efficiency of ~0.06 at −10 °C [25]. With a similar configuration, single wavelength operation was achieved [26]. Single wavelength operation is important for seeding purposes and heterodyne detection. By using laser diodes operating at 1.9 μm and end pumping, a Ho:YAG laser had a threshold of ~0.55 W and a slope efficiency of 0.35, optical to optical [27]. In this case, the laser was cooled to −53 °C and the optical power was the pump power incident on the laser material.

Normal-mode and Q-switched operation of diode-pumped Ho:Tm:YAG was demonstrated in a side-pumped arrangement at room temperature. The optical-to-optical threshold was ~2.4 J while the slope efficiency was ~0.026 and ~0.0023 for normal-mode and Q-switched operation, respectively. A Q-switched slope efficiency substantially closer to normal mode was obtained by permitting multiple Q-switched pulses [28]. Normal-mode and Q-switched operation of diode-pumped Ho:Tm:YLF in a side-pumped arrangement at 16 °C was achieved. Here, the optical-to-optical thresholds were 1.5 and 2.5 J while the slope efficiency was 0.124 and 0.060 for the normal-mode and Q-switched operations [29]. Using similar hardware, a Ho:Tm:YLF diode-pumped amplifier demonstrated 0.6 J/pulse [30].

Because up-conversion between Tm and Ho was recognized to be deleterious, a Tm:YAG laser pumping an Ho:YAG laser was tried. By placing both laser materials in the same resonator, a continuous threshold of 0.19 W and a slope efficiency of 0.42 was achieved where the input power is the absorbed optical power [31]. In a similar vein, a pulsed Ho:LuAG laser pumped by a Co:MgF_2 laser demonstrated a 4.3 mJ threshold and a slope efficiency of 0.82, absorbed optical to optical [32].

Although Er:YAG operated on the $^4I_{13/2}$–$^4I_{15/2}$ transition at liquid nitrogen temperatures, limited work has been performed on this transition. Operation on this transition at higher temperatures is somewhat difficult, partially because of excited-state absorption between the $^4I_{13/2}$ and $^4I_{9/2}$ manifolds. Using a low Er concentration and a Yb sensitizer, flashlamp pumped, room temperature operation was achieved with a threshold of ~75 J. Lasing which originates on the $^4I_{11/2}$ and $^4S_{3/2}$ manifolds was also reported.

Efficient operation on the $^4I_{11/2}$–$^4I_{13/2}$ transition in Er:YAG and Er:YSGG has been demonstrated. Although the lower laser level is not thermally populated, the lower laser manifold has a lifetime much longer than the upper laser manifold. Because the lower laser manifold does not decay quickly, lasing can be terminated as this manifold populates. Nevertheless, with similar Er:YAG and Er:$YAlO_3$ laser rods, normal-mode thresholds of 17.4 and 8.9 J were possible in a resonator having very low output coupling [33].

Utilizing Er:YSGG, a 30 J threshold and a 0.012 slope efficiency in a flashlamp-pumped system has been demonstrated [34]. Subsequently, work was done on optimizing the pump pulselength and a threshold of ~10 J and slope efficiency of ~0.01 was achieved [35]. Diode pumping of this transition has been achieved using a two-dimensional laser diode array at 0.973 μm. With 4.0 ms quasi-continuous pump pulses, an ~20 W threshold and 0.029 incident optical-to-optical efficiency were obtained [36].

Seeking room temperature operation on the $^4I_{13/2}$–$^4I_{15/2}$ transition, an Er:$YAlO_3$ laser with a very low Er concentration was aggressively pumped. Rather than observing this, the $^4S_{3/2}$–$^4I_{9/2}$ transition was observed at 1.663 μm instead. A 53 J threshold was obtained in a resonator with two highly reflecting mirrors [37]. On the same transition in Er:$YAlO_3$ but at 1.677 μm, a threshold of ~55 J and a slope efficiency of ~0.00015 was observed. In an Er:YLF laser, the $^4S_{3/2}$–$^4I_{9/2}$ transition at 1.73 μm achieved both normal-mode and Q-switched outputs. Thresholds of 14 and 16 J and slope efficiencies of 0.006 and 0.004, respectively, were achieved [39]. Also in an Er:YLF laser, the $^4S_{3/2}$–$^4I_{13/2}$ transition at ~0.85 μm was observed with a threshold of ~100 J and a slope efficiency of ~0.012 [40]. Operation of the $^4S_{3/2}$–$^4I_{11/2}$ transition at 1.23 μm was also observed in Er:YLF.

Er:YLF is suitable for up-conversion lasers as well. Using a pulsed 1.53 μm Er:glass laser as a pump, a three-step up-conversion process created a population inversion in the $^4S_{3/2}$ manifold. A three-step process excites the $^2H_{11/2}$ manifold which subsequently decays quickly to the $^4S_{3/2}$ manifold which serves as a metastable level. The most efficient operation was achieved at 1.73 μm where the total efficiency was 0.11, optical to optical, at 110 K [41]. Conversely, a 0.797 μm laser diode generated a population inversion on the $^4S_{3/2}$–$^4I_{15/2}$ transition at 48 K. At this temperature, a continuous 0.55 μm laser resulted in a threshold of ~0.4 W and a slope efficiency of 0.08 [42].

Efficient, flashlamp-pumped, normal-mode operation using Cr as a sensitizer has been demonstrated in Tm:Cr:YAG. Operating on the $^3F_4$–$^3H_6$ transition at 2.016 μm, a room temperature Tm:Cr:YAG laser achieved a threshold of 73 J and a slope efficiency of 0.045 [43]. Operation on the $^3H_4$–$^3H_5$ transition at 2.274 μm was demonstrated in Tm:Cr:$YAlO_3$ using flashlamp pumping. The threshold was ~70 J but the highly reflecting resonator provided little output to determine a slope efficiency [44].

Continuous operation of a diode-pumped Tm:YAG laser provided a high power output at 2.013 μm. Utilizing laser diodes in an end-pumped configuration, continuous operation had a threshold of 1.7 W and a slope efficiency of 0.33 at −15 °C [45]. Utilizing a high power laser diode array and a lens duct to concentrate the pump radiation into an internally reflecting laser rod, a threshold of ~50 W and a slope efficiency of 0.40 was achieved, where the pump power is considered to be the power delivered to the laser rod [46]. The power produced by this arrangement in a highly multi-mode beam was 115 W.

Using a Ti:Al_2O_3 laser in an end-pumped configuration, high slope efficiencies were achieved. Operating on the $^3F_4$–$^3H_6$ transition, thresholds of ~0.23 and ~0.1 W and slope efficiencies of 0.59 and 0.44 were achieved in Tm:YAG [47] and Tm:$YAlO_3$ [48], absorbed and incident optical to optical, respectively. The laser wavelengths were 2.105 and 1.94 μm for the two laser materials. High slope efficiencies verified that the quantum efficiency was greater than unity when pumping the $^3H_4$ manifold.

Continuous operation was also achieved in Tm:Y_2O_3 and Tm:Sc_2O_3 using laser diode pumping in an end-pumped scheme. Both lasers operated on the $^3F_4$–$^3H_6$ transition but the wavelengths were ~2.05 and ~2.15 μm in the two laser materials. Thresholds were 0.30 and 0.48 W and slope efficiencies 0.18 and 0.08, absorbed optical to optical, respectively. It was suggested that better quality laser material would benefit the operation [49].

Both normal-mode and Q-switched operation were demonstrated in Tm:LuAG using laser diode pumping. In a side-pumped configuration, the $^3F_4$–$^3H_6$ transition at 2.023 μm produced a threshold of 0.78 J and a slope efficiency of 0.144 and 0.050, optical to optical, for normal mode and Q-switched operation [50].

A Dy:BaY_2F_8 laser operating on the $^6H_{13/2}$–$^6H_{15/2}$ transition was achieved by cooling the laser rod to liquid nitrogen temperatures. The threshold for this device was 510 J and the wavelength was 3.022 μm [51]. By using an Er:YLF laser as a pump, a Dy:YLF laser operated at 4.340 μm on the $^6H_{11/2}$–$^6H_{13/2}$ transition

Table B1.4.3. Near to mid infrared laser performance.

Laser	Pump	Mode	Temperature	Threshold	Slope
$^5I_7$ to $^5I_8$					
Ho:Tm:Er:YAG	lamp	continuous	≈77 K	≈200 W	0.065
Ho:Tm:Er:YAG	lamp	NM, QS	≈77 K	30 J	0.012
Ho:Tm:Er:YLF	lamp	NM, QS	≈77 K	14 J	0.029
Ho:Tm:Er:YAG	lamp	NM	room	155 J	0.0004
Ho:Tm:Er:YLF	lamp	NM	room	100 J	0.0013
Ho:Tm:Cr:YAG	lamp	NM	room	≈45 J	0.051
Ho:Tm:YAG	diode	continuous	−10 °C	52 mW	≈0.06
Ho:Tm:YAG	diode	NM	room		0.026
Ho:Tm:YAG	diode	QS	room		0.0023
Ho:Tm:YLF	diode	NM	16 °C	1.5 J	0.124
Ho:Tm:YLF	diode	QS	16 °C	2.5 J	0.060
$^4I_{13/2}$–$^4I_{15/2}$					
Er:YAG	lamp	NM	≈77 K		
Er:Yb:YAG	lamp	NM	room	75 J	
$^4I_{11/2}$–$^4I_{13/2}$					
Er:YAG	lamp	NM	room	17.4 J	
Er:$YAlO_3$	lamp	NM	room	8.9 J	
Er:YSGG	lamp	NM	room	30 J	0.012
Er:YSGG	lamp	NM	room	≈10 J	≈0.01
$^4S_{3/2}$–$^4I_{9/2}$					
Er:$YAlO_3$	lamp	NM	room	55 J	0.00015
Er:YLF	lamp	NM	room	14 J	0.006
Er:YLF	lamp	NM	room	16 J	0.004
$^4S_{3/2}$–$^4I_{13/2}$					
Er:YLF	lamp	NM	room	≈100 J	≈0.012
$^3F_4$ to $^3H_6$					
Tm:Cr:YAG	lamp	NM	room	73 J	0.045
Tm:YAG	diode	continuous	−15 °C	17 W	0.33
Tm:YAG	diode	continuous	room	50 W	0.40
Tm:YAG	Ti:Al_2O_3	continuous	room	0.23 W	0.59
Tm:$YAlO_3$	Ti:Al_2O_3	continuous	room	0.1 W	0.44
Tm:Sc_2O_3	diode	continuous	room	0.30 W	0.18
Tm:Y_2O_3	diode	continuous	room	0.48 W	0.08
Tm:LuAG	diode	NM	room	0.78 J	0.144
Tm:LuAG	diode	QS	room	0.78 J	0.050
$^3H_{13/2}$–$^3H_{15/2}$					
Dy:BaY_2F_8	lamp	NM	≈77 K	510 J	
$^3H_{11/2}$–$^3H_{13/2}$					
Dy:YLF	Er:YLF	NM	room	0.015 J	0.07

[52]. The threshold was less than 0.015 J while the slope efficiency was greater than 0.07, optical to optical at room temperature.

B1.4.8 Summary

Solid state lasers can generate a variety of wavelengths in the near to mid infrared region of the spectrum with respectable efficiencies. Despite the fact that some of these lasers are quasi-four-level lasers, room temperature operation is practical. But to achieve room temperature operation, aggressive pump techniques are required.

These lasers are more complicated than their Nd counterparts primarily because of the importance of the energy transfer processes, which include: energy transfer to different atoms and back transfer, self-quenching, up conversion and diffusion. For some lasers, the energy transfer processes are vital to the operation and, in other cases, they are detrimental. Energy transfer processes are well described with a rate equation approach, similar to the laser dynamics rate equations. Solutions depend on the mode of operation: continuous, normal mode or Q-switched mode. Usually laser dynamics equations must include any nonnegligible population density in the lower laser level.

A number of examples of laser performance for Dy, Ho, Er and Tm have been presented. Performance is characterized here by the threshold and slope efficiency. While lamp pumping has been used to achieve the requisite high levels of pumping, laser diode pumping is becoming more common.

For further information on the physics of solid state lasers, the reader is encouraged to consult [1]. For further information of the engineering of solid state lasers, [54] may be consulted.

References

[1] Powell R C 1997 *Physics Of Solid State Laser Materials* (New York: Springer)

[2] Forster T 1947 Experimentalle Und Theoretische Untersuchung Des Zwischenmolerkularen Ubergangs Von Elektronenregungsenergie *Naturfurshung* **4A** 321–7

[3] Dexter D L 1953 A theory Of sensitizer luminescence in solids *J. Chem. Phys.* **21** 836–50

[4] Barnes N P, Filer E D, Morrison C A and Lee C J 1996 Ho:Tm lasers I: theoretical *IEEE J. Quantum Electron.* **QE-32** 92– 103

[5] Walsh B M, Barnes N P and DiBartolo B 1997 On the distribution of energy between the Tm $^3F_4$ And Ho $^5I_7$ manifolds in Tm sensitized Ho luminescence *J. Luminescence* **75** 89–98

[6] Foster J D and Osterink L M 1970 Thermal effects in a Nd:YAG laser *J. Appl. Phys.* **41** 3656–63

[7] Koechner W 1970 Thermal lensing in a Nd:YAG laser rod *Appl. Opt.* **9** 2548–53

[8] Williams-Byrd J A and Barnes N P 1990 Laser performance, thermal focusing and depolarization effects in Nd:Cr:GSGG and Nd:YAG *Proc. Solid State Lasers* **1223** 237–46

[9] Bowness C 1965 On the efficiency of single and multiple elliptical laser cavities *Appl. Opt.* **4** 103–7

[10] Schuldt S B and Aagard R L 1963 An analysis of radiation transfer by means of elliptical cylinder reflectors *Appl. Opt.* **2** 509–13

[11] Barnes N P, Storm M E, Cross P and Skolaut M W Jr 1990 Efficiency of Nd laser materials with laser diode pumping *IEEE J. Quantum Electron.* **QE-26** 558–69

[12] Barnes N P, Murray K E and Jani M G 1997 Flash lamp pumped Ho:Tm:Cr:YAG and Ho:Tm:Er:YLF lasers: modeling of a single, long pulse length comparison *Appl. Opt.* **36** 3363–74

[13] Singh U N, Yu J, Petros M, Barnes N P, Williams-Byrd J A, Lockard G E and Modlin E A 1998 Injection seeded, room temperature, diode pumped Ho:Tm:YLF laser with output energy of 600 mJ at 10 Hz *Proc. Advanced Solid State Laser Conf.* (Washington DC: Optical Society of America) pp 194–6

[14] Barnes N P, Rodriguez W J and Walsh B M 1996 Ho:Tm:YLF Laser amplifiers performance *J. Opt. Soc. Am.* B **13** 2872–82

[15] Jani M G, Barnes N P and Murray K E 1997 Flash lamp pumped Ho:Tm:Cr:YAG and Ho:Tm:Er:YLF lasers: experimental *Appl. Opt.* **36** 3357–62

[16] Lee C J, Han G and Barnes N P 1996 Ho:Tm Lasers II: experiments *IEEE J. Quantum Electron.* **QE-32** 104–11

[17] Barnes N P, Jani M G, Murray K E and Harrel S R 1994 Diode pumped Ho:Tm:YLF laser pumping an $AgGaSe_2$ parametric oscillator *J. Opt. Soc. Am.* B **11** 2422–6

[18] Johnson, L F, Geusic J E and van Uitert L G 1965 Coherent oscillation from Tm^{3+}, Ho^{3+}, Yb^{3+} and Er^{3+} in yttrium aluminum garnet *Appl. Phys. Lett.* **7** 127–9

[19] Johnson L F, Guggenheim H J, Rich T C and Ostermayer F W 1972 Infrared to visible conversion by rare earths in crystals *J. Appl. Phys.* **43** 1125–37

[20] Beck R and Gurs K 1975 Ho laser with 50 W output and 6.5% slope efficiency *J. Appl. Phys.* **46** 5224–5

[21] Barnes N P 1979 TEM_{00} mode Ho:YLF laser *Proc. SPIE* **190** 297–304

[22] Barnes N P 1981 Pulsed Ho:YAG oscillator and amplifier *IEEE J. Quantum Electron.* **QE-17** 1303–8

[23] Chicklis E P, Naiman C S, Folweiler R C and Doherty J C 1972 Stimulated emission in multiply doped Ho^{3+}:YLF and YAG-A comparison *IEEE J. Quantum Electron.* **QE-8** 225–30

[24] Quarles G J, Rosenbaum A, Marquardt C L and Esterowitz L 1989 High efficiency 2.09 μm flashlamp pumped laser *Appl. Phys. Lett.* **55** 1062–5

[25] Kintz G J, Esterowitz L and Allen R 1987 CW Diode-pumped Tm^{3+}:Ho^{3+}:YAG 2.1 μm room-temperature laser *Electron. Lett.* **22** 616–616

[26] Storm M E and Rohrbach W W 1989 Single longitudinal mode lasing of Ho:Tm:YAG at 2.09 μm *Appl. Opt.* **28** 4965–7

[27] Nabors C D, Ochoa J, Fan T Y *et al* 1995 Ho:YAG laser pumped by 1.9 μm diode laser *IEEE J. Quantum Electron.* **QE-31** 1063–5

[28] Bowman S R, Lynn J G, Searles S K *et al* 1993 High average power operation of a Q-switched diode pumped holmium laser *Opt. Lett.* **18** 1724–6

[29] Jani M G, Naranjo F L, Barnes N P *et al* 1995 Diode pumped, long pulse length Ho:Tm:YLF laser at 10 Hz *Opt. Lett.* **20** 872–4

[30] Singh U N, Williams-Byrd J A, Barnes N P *et al* 1997 Diode pumped 2.0 μm solid state lidar transmitter for wind measurements *Proc. SPIE* **3104** 173–8

[31] Stoneman R C and Esterowitz L 1992 Intracavity pumped 2.09 μm Ho:YAG laser *Opt. Lett.* **17** 736–8

[32] Hart D W, Jani M and Barnes N P 1996 Room temperature lasing of end pumped Ho:Tm:$Lu_3Al_5O_{12}$ *Opt. Lett.* **21** 728–30

[33] Breguet J, Umyskov A F, Semenkov S G *et al* 1992 Comparison of threshold energy of selectively excited $YAlO_3$:Er and YAG:Er lasers *IEEE J. Quantum Electron.* **QE-28** 2563–6

[34] Zharikov E V, Il'ichev N N, Kalitan S P *et al* 1986 Spectral, luminescence, and lasing properties of yttrium scandium gallium garnet activated with chromium and erbium *Sov. J. Quantum Electron.* **16** 635–9

[35] Moulton P F, Manni J G and Rines G A 1988 Spectroscopic and laser characteristics of Er,Cr:YSGG *IEEE J. Quantum Electron.* **QE-24** 960–73

[36] Waarts R, Nam N and Sanders S 1994 Two dimensional Er:YSGG microlaser array pumped with a monolithic two dimensional laser diode array *Opt. Lett.* **19** 1738–40

[37] Weber M J, Bass M and DeMars G A 1971 Laser action and spectroscopic properties of Er in $YAlO_3$ *J. Appl. Phys.* **42** 301–5

[38] Datwyler M, Luthy W and Weber H P 1987 New wavelength of the $YAlO_3$:Er laser *IEEE J. Quantum Electron.* **QE-23** 158–9

[39] Barnes N P and Allen R A 1986 Operation of an Er:YLF laser at 1.73 μm *IEEE J. Quantum Electron.* **QE-22** 337–43

[40] Chicklis E P, Naiman C S and Linz A 1972 Stimulated emission at 0.85 μm in Er:YLF, Dig. Tech. Papers VII International Quantum Electronics Conference, Montreal Canada

[41] Pollack S A, Chang D B and Birnbaum M 1989 Threefold upconversion laser at 0.85, 1.23 and 1.73 μm in Er:YLF pumped with a 1.53 μm Er:glass laser *Appl. Phys. Lett.* **54** 869–71

[42] Stephens R R and McFarlane R A 1993 Diode pumped upconversion laser with 100 mW output power *Opt. Lett.* **18** 54–7

[43] Quarles G J, Rosenbaum A, Marquardt C L and Esterowitz L 1990 Efficient room temperature operation of a flashlamp pumped Cr:Tm:YAG laser at 2.01 μm *Opt. Lett.* **15** 42–4

[44] Caird J A, DeShazer L G and Nella J 1975 Characteristics of room temperature 2.3 μm laser emission from Tm in YAG and $YAlO_3$ *IEEE J. Quantum Electron.* **QE-11** 874–80

[45] Kmetec J D, Kubo T S and Kane T J 1994 Laser performance of diode pumped thulium doped $Y_3Al_5O_{12}$, $(Y,Lu)_3Al_5O_{12}$, and $Lu_3Al_5O_{12}$ crystals *Opt. Lett.* **19** 186–8

[46] Honea E C, Beach R J, Sutton S B *et al* 1997 115 W Tm:YAG CW diode pumped solid state laser *Advanced Solid State Laser Conference (Orlando, FL)* (Washington DC: Optical Society of America)

[47] Stoneman R C and Esterowitz L 1990 Efficient, broadly tunable, laser pumped Tm:YAG and Tm:YSAG CW lasers *Opt. Lett.* **15** 486–8

[48] Stoneman R C and Esterowitz L 1989 Efficient 1.94 μm Tm:$YAlO_3$ laser *IEEE J. Selected Topics* **1** 78–81

[49] Fornasiero L, Berner N, Dicks B M *et al* 1999 Broadly tunable laser emission from Tm:Y_2O_3 and Tm:Sc_2O_3 *Advanced Solid State Laser Conference (Boston, MA)* (Washington DC: Optical Society of America)

[50] Barnes N P, Jani M G and Hutcheson R L 1995 Diode pumped, room temperature Tm:LuAG laser *Appl. Opt.* **34** 4290–4

[51] Johnson L F and Guggenheim H J 1973 Laser emission at 3 μm from Dy^{3+} in BaY_2F_8 *Appl. Phys. Lett.* **23** 96–8

[52] Barnes N P and Allen R E 1991 Room temperature Dy:YLF laser operation at 4.34 μm *IEEE J. Quantum Electron.* **QE-27** 277–82

[53] Powell R C 1998 *Physics of Solid State Laser Materials* (New York: Springer)

[54] Koechner W 1976 *Solid State Laser Engineering* (New York: Springer)

B1.5
Rare-earth ions—miscellaneous: Ce^{3+}, U^{3+}, divalent, etc

Gregory J Quarles

B1.5.0 Introduction

Major advances have been made during the last two decades in the development of rare-earth-activated solid state laser systems. Much of this effort has focused upon improving the efficiencies of near-infrared lasers utilizing trivalent neodymium or several of the other ions which directly exhibit laser action in the near infrared (see chapters B1.3 and B1.4). However, since the beginning of studies in the late 1950s of rare-earth-activated solid state materials for laser systems, ions other than just neodymium have played a key role in the evolution of this field. Much of the initial research and spectroscopy of lanthanide and actinide ions in solids was focused on investigations of hosts and activators for the phosphor industry. These investigations evolved to looking at ions other than those that emitted in the visible region and expanded into the ultraviolet and infrared regions of the electromagnetic spectrum.

This chapter will initially review the cerium ion, which exhibits the only directly-pumpable, tunable, ultraviolet 5d–4f stimulated emission in the rare-earth ion series. Also covered will be the trivalent uranium ion, which has brought about such historical trends as the second ion ever lased, the only actinide ion ever to exhibit stimulated emission and the first recorded diode-pumped solid state laser. Both of these ions have seen a revitalization in their interest to the solid state laser community in the 1990s, after virtually no research in the 1980s. New and efficient solid state lasers have been demonstrated for both cerium (in the 290 nm region of the spectrum) and uranium (at approximately 2.8 μm). The final section in this chapter will overview the history of the search for other rare-earth lasers utilizing divalent, rather than trivalent activator ions, including divalent dysprosium, europium, samarium and thulium.

The chapter is divided into three primary sections, with each section providing a historical overview of the spectroscopy, historical investigations, development and demonstrations of the stimulated emission for each of the ions in a variety of hosts. There has been an attempt in each section to provide an overview of the significance of the research, the applications of the laser in the industrial and scientific fields and the future directions for research with each ion. A beginning review will be presented of the energy levels, the absorption and emission spectra, the pertinent upper state lifetimes and the spectroscopic assets and deficiencies for each ion. Finally, a review of the laser performance demonstrated for each ion in the continuous wave (cw) and pulsed regimes will be summarized in each section.

There have been many excellent papers and texts written reviewing rare-earth-activated solid state laser research over the past 40 years. I have also referenced several books, manuscripts and chapters that are considered noteworthy with summaries covering the topics to be reviewed in this chapter in the Further Reading section. For any reader interested in a more in-depth review of these divalent and trivalent lanthanide and actinide ions, it is strongly suggested that these seminal and historical reviews be read. These reviews provide a comprehensive summary and bring together all aspects from basic materials science and crystallography, to stimulated emission spectroscopy and solid state laser design and engineering.

B1.5.1 Trivalent cerium lasers

Investigations involving the cerium ion have had several cyclical periods of intense interest during the past 40 years. The primary driver for this interest is the fact that the cerium ion exhibits direct ultraviolet (UV) absorption and emission and can provide tunable stimulated emission in the proper choice of host crystal. The interest in these ultraviolet sources has spanned a number of applications, including photolithography, medical, detection of biological and chemical agents and lidar. If the laser were to emit in the UV, and demonstrate tunability, then the applications become even more vast. Many groups have attempted for decades to develop a superior UV laser source but material limitations consistently seem to provide obstacles. These limitations can be as simple as the band edge of the host being too close to upper laser levels and limiting the UV transmission, to UV-induced colour centres and UV-activated defects that diminish or impede the stimulated emission.

Most of the previously studied tunable solid-state lasers have included transition metal ions such as trivalent titanium (Ti^{3+}) and chromium (Cr^{3+}) in a variety of oxide and fluoride hosts [1–3] (see chapters B1.1 and B1.2). These transition metal ions typically exhibit broadband emission in the visible to near-infrared (IR) regions of the spectrum due to vibronic transitions within the 3d shell. Most trivalent lanthanide rare-earth ions exhibit very sharp absorption and emission bands throughout the spectrum based upon 4f–4f transitions [4, 5]. However, with cerium in a trivalent state (Ce^{3+}), the excited state is a 5d level and the emission from this state can be broad and efficient. This is due to the electronic configuration of the trivalent cerium ion and the separation between the 5d and 4f energy levels [2–5]. The Ce^{3+} ion has been activated into a number of different fluoride and oxide hosts and the laser efficiencies are varied. The proximity of the band edge in some hosts has impacted the performance, as well as the fact that the allowed 5d $\rightarrow$ 4f transitions are very fast, typically with lifetimes of the order of nanoseconds. This severely limits the possibility of coupling flashlamp technologies as an excitation source. Recent research into hosts which have crystal fields that give rise to cerium absorptions which match well with commercially available gas and solid state laser sources has caused a tremendous increase in the past decade of cerium-activated solid state lasers.

The electronic configuration of Ce^{3+} consists of a palladium core with 46 electrons and with a $4f^1 5s^2 5p^6 5d^1 6s^2$ outer electron configuration. The inner 4f shell is unfilled, with a single electron, and the filled 5s and 5p outer shells shield this 4f shell. For the typical trivalent rare-earth ion, the Coulomb interactions and spin–orbit coupling are both larger than the magnitude of the crystal field, because of this shielding of the optically active 4f electrons [4, 6–8]. In a self-consistent field approximation, it is assumed that the electrons move in a centrally symmetric field, brought about by the other electrons and the nucleus. In this approximation, the spin–orbit coupling adheres to the rule of $L + S \geq J \geq |L - S|$. If $L \geq S$, then J can have the values of $2S + 1$, while for $L < S$, then J equals $2L + 1$. The designation for the energy levels is then derived from the form ${}^{2S+1}L_J$ [2].

Additional imposed criteria include that if the principal quantum number, n, is related to the orbital angular momentum quantum number, l, by $n = 2l + 1$, then there is no multiplet splitting of the energy levels. However, if $n > 2l + 1$ or $n < 2l + 1$, the multiplet splittings are inverted or normal, respectively. The standard selection rules state that electric dipole transitions are allowed within these rare-earth ions if the following angular momenta relations hold: $\Delta J = 0$, and $J + J' \geq 1$. For the Ce^{3+} ion, utilizing the Russell-Saunders coupling approximations [2], for the 4f shell, $n = 4, l = 3, s = \pm 1/2, m = 0, \pm 1, \pm 2, \pm 3$ and $j = l + s$, which yields $j = 7/2$ and 5/2. Utilizing these quantum numbers, it is possible to formulate the splitting of the 4f shell into its two components, the ${}^2F_{7/2}$ excited state and the ${}^2F_{5/2}$ ground state, as illustrated in figure B1.5.1.

The splitting of the ${}^2F_{7/2}$ and ${}^2F_{5/2}$ energy levels is influenced by the ligand field in which the ion resides. This changes from host to host and the splitting of these levels can be between 1500 and 3000 cm^{-1} for trivalent cerium [8]. The single electron for the trivalent cerium is located in the 5d shell and gives rise to the 5d level situated in figure B1.5.1 approximately 50 000 cm^{-1} above the ground state in fluoride

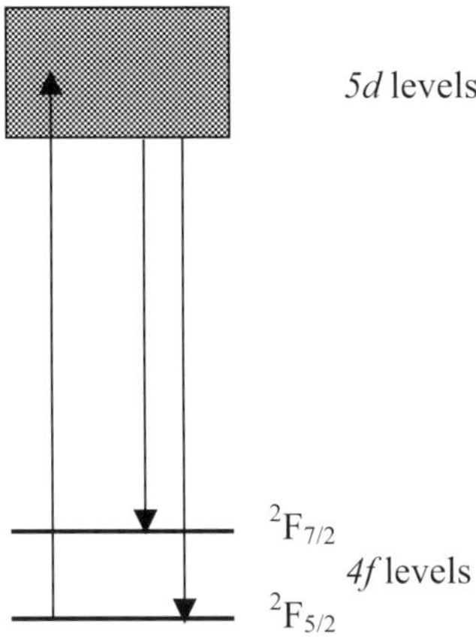

Figure B1.5.1. Partial Ce^{3+} energy level diagram (after Castillo [9]).

Table B1.5.1. Crystal properties of selected cerium-doped hosts, after Castillo [9].

Crystal	Space group	Local Site symmetry	Optical absorption (nm)	Refractive index	Reference
$Y_3Al_5O_{12}$	O_h^{10}	D_2	210–500	1.83	[11, 15]
LaF_3	D_{3d}^4	C_2	160–260	1.64	[25]
$LiYF_4$	C_{4h}^6	S_4	190–300	1.47	[26]
Rb_2NaYF_6	O_h^5	O_h	150–320	—	[39]
CaF_2	O_h^5	C_{4v}, O_h	180–320	1.47	[28]
BaY_2F_8	C_{2h}^3	C_{2h}	205–315	1.50	[38]
$LiCaAlF_6$	D_{3d}^2	D_3	250–280	1.39	[41, 53]
$LiSrAlF_6$	D_{3d}^2	D_3	250–280	1.41	[41, 53]

hosts and can be as low as 35 000 cm^{-1} in oxide hosts. The UV emission associated with the cerium ion originates from the 5d upper manifold and has electric-dipole-allowed 5d–4f transitions to the split ground-state energy levels. Fluorescence for nearly all cerium ions is composed of two peaks in the UV, with each of the peaks corresponding to a 5d $\rightarrow$ $^2F_{7/2}$ emission and a 5d $\rightarrow$ $^2F_{5/2}$ emission. Because of the nature of the 5d wavefunction, the resulting emission is generally a vibrationally broadened spectrum, as compared to those normally observed with the narrow emission of the electric-dipole-forbidden 4f–4f transitions in other rare-earth ions.

Over the past four decades, there have been numerous investigations of the spectroscopic and stimulated emission properties of the trivalent cerium ion in a variety of hosts. The trend that has been observed is that the fluorescence is generally very intense but the stimulated emission faces many potential loss mechanisms, including solarization, colour centre and defect formation and excited-state absorption processes. All of these can limit or even totally prevent the laser process. The oxide hosts have shown the greatest tendencies towards having multiple detrimental mechanisms inherent in their nature, although similar mechanisms have also been observed in the fluorides.

Table B1.5.2. Trivalent cerium crystal absorption, emission and flourescence lifetime properties in selected hosts after Hammons *et al* [19].

Host material	Peak absorption wavelength (nm)	Peak emission wavelength (nm)	Lifetime (ns)
Lu_2SiO_5	357	395	40
Y_2SiO_5	356	388	37
Gd_2SiO_5	340	402	60
$Y_3Al_5O_{12}$	457	527	65
$Lu_3Al_5O_{12}$	447	517	85
$LiSrAlF_6$	266	290	28
$SrAlF_5$	248	285	43
$LiYF_4$	287	310	40
$LiGdF_4$	289	310	41
KYF_4	287	218	38

The oxide host with the largest degree of characterization is most likely Ce^{3+}-activated $Y_3Al_5O_{12}$ (YAG) [7, 10–17]. Cerium doped in the YAG matrix was investigated heavily in the late 1960s for use as phosphors in cathode-ray tubes for the colour television market. In the YAG host, the trivalent cerium ion enters a distorted cubic site with a D_2 symmetry, causing the 5d levels to be very low lying, and it also provides a large crystal field to dramatically split the $^2F_{7/2}$ and $^2F_{5/2}$ energy levels. This low-lying 5d level causes the energy gap to be decreased between the upper and lower levels and shifts the emission into the visible (yellow) region of the spectrum in YAG. Table B1.5.1 lists the various crystal properties for Ce in YAG, as well as in other hosts. The absorption, emission and energy level schemes for Ce:YAG are illustrated in figure B1.5.2. The onset of the bandgap in YAG is visible with the tail of absorption beginning at 275 nm [17, 18], while the absorption peaks at nominally 350 and 450 nm correspond to the absorption between the ground state and the upper 5d manifolds. The band at 220 nm has been identified by Wong *et al* [14] as intrinsic lattice defects, as the same peaks can be observed in some undoped samples of YAG [16]. The broad emission spanning from approximately 500 to 700 nm arises from the direct transitions from the 5d upper level manifolds to the split $^2F_{7/2}$ and $^2F_{5/2}$ ground-state energy levels. The wavelengths for the absorption peaks, emission peaks and the fluorescent lifetimes for trivalent cerium in a variety of hosts are brought together for the reader in table B1.5.2.

The Ce ion in YAG has never been shown to exhibit stimulated emission, even with the significant quantity of research conducted in this field. The primary issue preventing the operation of Ce:YAG as a solid state laser is the excited-state absorption (ESA), which is observed through the overlap of the absorption and emission bands shown in figure B1.5.2. The ESA can be reduced through cooling the crystal but lasing is still not achieved [11–13, 15]. The radiation from the pump source, even at low temperature, excited the electrons into the phonon-coupled states and into the conduction band, and served as a primary loss mechanism for excitation energy [13, 15].

Several other oxide hosts have been activated with the cerium ion for investigation of the potential for stimulated emission. Studies showed that if the charge, ionic radius and the ionization potential of the activator ion (Ce^{3+}) and the host cation were similar, the chances for ESA would be minimized [13]. New oxide hosts studied included Ce^{3+}:$YAlO_3$ (YALO) [20], Ce^{3+}:Y_2O_3 [21] and Ce^{3+}:$LaPO_4$ [22]. The crystal field of the YALO shifted the upper-state manifolds to higher energy and gave broad emission from 325 to 400 nm but stabilization of the cerium ions was problematic. No laser action was ever observed in Ce:YALO.

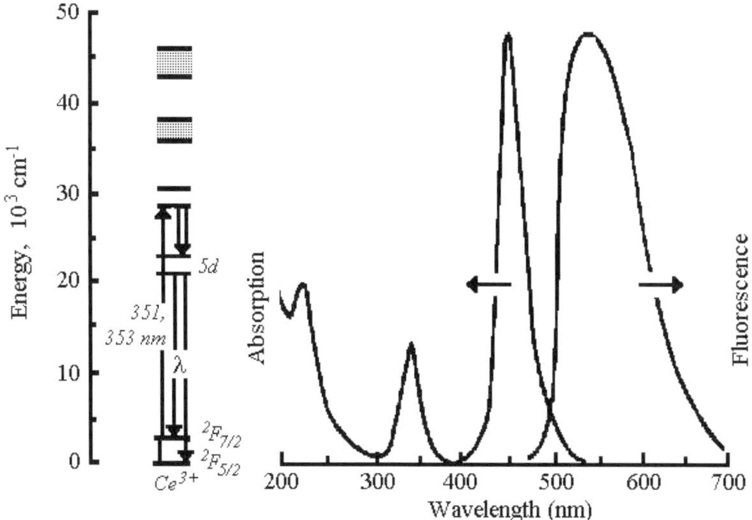

Figure B1.5.2. Ce:YAG absorption and emission spectra (after Jacobs *et al* [11]).

The cerium ion in Y_2O_3 and $LaPO_4$ proved to exhibit strong emission and tremendous phosphor potential; however, none of these oxide hosts could sustain stimulated emission either.

The failure of any of these oxide hosts to lase led researchers to concentrate most of their recent studies on the fluoride hosts. As compared to the oxides, the fluorides are more strongly insulating and characteristically have larger energy gaps between the various levels. They also tend to exhibit higher-energy conduction bands, resulting in better UV transmission. Optically, the fluoride host matrix is greatly influenced by the weak metal-to-fluorine bonds, the high electronegativity and small polarizability of the fluorine. These effects combine to give rise to typically lower refractive indices than the oxides, weaker crystal fields, wider transmission bands, lower melting temperatures and a general shift of the 4f levels to shorter wavelengths [9]. The earliest research with a fluoride host involved the LaF_3 fluoride matrix [23–25]. Researchers reported broadband fluorescence from 270–450 nm. The C_2 site symmetry causes the 5d state to split into five levels, and exhibit a broad emission with two peaks corresponding to the splitting of the ground state. The upper-state lifetime for this electric-dipole-allowed d–f transition was a rapid 20 ns. The first demonstration of laser performance was published by Ehrlich *et al* [25], in which they used a 40-mJ KrF excimer pump source at 249 nm to pump the Ce absorption bands at room temperature. Stimulated emission was measured at a peak wavelength of 285.5 nm, with a laser oscillation threshold of 3 mJ. Ehrlich *et al* also demonstrated tunability of the laser from 275–315 nm but the manuscript stated that the performance was inferior due to poor crystal quality. Minimal follow-up research has taken place with a LaF_3 host doped with trivalent cerium.

Another fluoride host investigated thoroughly was the $LiYF_4$ host (YLF). The initial report of laser operation in Ce:YLF was also reported at room temperature by Ehrlich and colleagues [26] using an identical pumping scheme with the KrF laser. The Ce:YLF laser operated at 325.5 nm and produced an estimated tuning range of 305–335 nm. In the YLF host, the S_4 site symmetry splits the 5d state into four levels, all of which can be seen in the ultraviolet absorption spectrum in figure B1.5.3.

The broken curve is the fluorescence spectrum, with two peaks indicating the transitions from the 5d manifold to the split ${}^2F_{7/2}$ and ${}^2F_{5/2}$ ground-state energy levels. This radiative lifetime is again very rapid, measured as 40 ns in the YLF host, and listed in table B1.5.2. It is interesting to note that, because of the allowed transitions, the stimulated emission cross-sections for trivalent cerium actually rival those found in

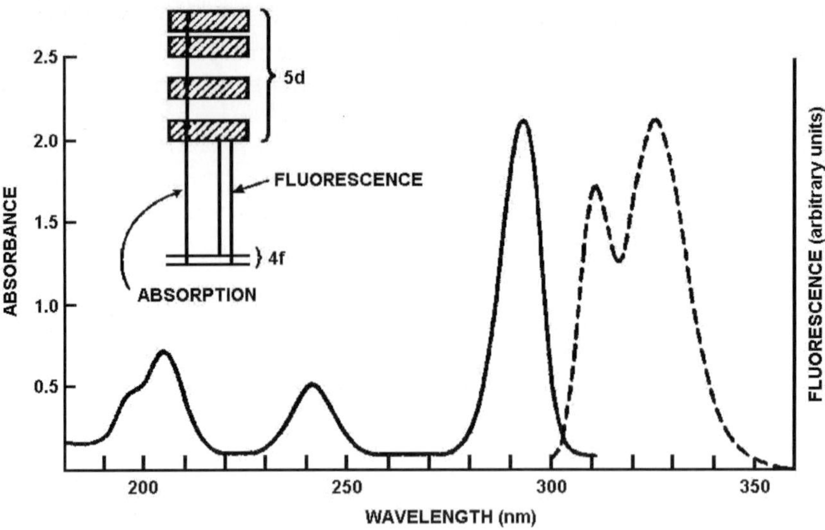

Figure B1.5.3. Absorption (full curve) and fluorescence (broken curve) spectra of Ce^{3+}:$LiYF_4$ (after Ehrlich *et al* [26]).

dye lasers, of the order of 7–9 $\times 10^{-18}$ cm^{-2}. Subsequent studies of the Ce:YLF laser material determined that the gain and pump rate were influenced by ESA of the pump light, which led to the formation of transient and stable colour centres [27]. The rate at which the defects and transients evolved impacted the gain saturation, slope efficiency and laser threshold. It is interesting to note that the ESA influenced the Ce:YAG system so strongly that there was no stimulated emission possible, whereas in the Ce:YLF system, tunable laser action was still feasible. Aside from this initial research with Ce:YLF, there has been very little investigation of the material during the last 20 years.

Spectroscopic and laser research was conducted during the same time period with Ce^{3+}:CaF_2 but similar ESA issues plague this host [28]. The fluorescence lifetime for this crystal is also around 40 ns. It was observed that the Ce^{3+}:CaF_2 crystal would actually turn from a clear water white to a reddish-brown colour during UV excitation, signifying the formation of the colour centres. This colouration could be reversed with heating of the crystal or by photo-bleaching at a wavelength absorbed by the colour centres [28]. Tables B1.5.1 and B1.5.2 list other fluoride and oxide hosts not discussed in detail here due to either a lack of stimulated emission or repeated incidences of ESA being problematic in the laser operation.

Two final hosts of note will be reviewed in the remainder of the section. One is an isomorph of YLF, namely $LiLuF_4$. This host doped with the cerium ion produces fluorescence peaks at 308 and 327 nm, and also exhibits a radiative lifetime of roughly 40 ns [29–37]. Compared to Ce^{3+}:$LiYF_4$, the Ce^{3+}:$LiLuF_4$, compound is more photo-chemically stable and has better opto-mechanical properties. This Ce^{3+}:$LiLuF_4$ gain material has been shown to be efficient as a 308 nm emitting solid state laser pumped by either a Ce:$LiSrAlF_6$ (LiSAF) pump source at 290 nm, a second-harmonic from a copper-vapour laser 289 nm, or as an amplifier for a KrF laser at 248 nm. The efficiencies for the LiSAF-pumped device exceed 55% [33], while output powers of greater than 1.25 W have been demonstrated by McGonigle *et al* [34–37]. Prism tuning of this source was demonstrated from 305.5 to 316 nm and a second band was tuned from 323 to 331 nm [35]. The only two noted drawbacks are that the cost of the LuF_3 for the growth of this material is generally very expensive and some colour centre formation, albeit slight, has been observed. More future research with Ce^{3+}:$LiLuF_4$ should be forthcoming, as this is a relatively new host and dopant combination.

The final Ce-activated hosts presented in this section are a series of isomorphs, based upon the colquiriite family of materials, including $LiSrAlF_6$ (LiSAF), $LiCaAlF_6$ (LiCAF) $LiSrGaF_6$ (LiSGaF) and the other isomorphs. LiCAF was first investigated and characterized at Lawrence Livermore National Laboratory in 1988 and was activated with trivalent chromium [40]. Although the colquiriite hosts have proven to be intriguing new laser materials when activated with chromium, success when doping with other transition metal ions or rare-earth ions has been limited, prior to the investigations with trivalent cerium. In a colquiriite host, the cerium takes on a charge of 3+ and carries an ionic radius of roughly 1.15Å. The trivalent aluminium ion in this type of structure has a much smaller ionic radius, approaching 0.67Å; thus, it is assumed that the Ce^{3+} might prefer to reside on the Sr^{2+} or Ca^{2+} sites, since their radii are more nearly equivalent at 1.27 and 1.14Å, respectively [41–45].

The initial investigations of UV laser properties of the Ce-doped LiSAF and LiCAF hosts were conducted by Dubinskii, *et al* [46–49] in 1993, followed up shortly thereafter by parallel research at the Naval Research Laboratory (NRL) [50, 51] and Lawrence Livermore National Laboratory (LLNL) [42, 52–54]. Utilizing well-oriented slabs of LiSAF and LiCAF, both the NRL and LLNL groups were able to demonstrate high efficiencies and broad tunability using two different resonator configurations. Efficiencies for Ce:LiSAF in a gain-switched, side-pumped configuration were approximately 17% when pumped at NRL with a fourth-harmonic Nd:YAG laser at 266 nm [50, 51]. In an end-pumped, nearly-confocal resonator configuration investigated at LLNL, efficiencies as high as 29% and 21% were measured with Ce:LiSAF and Ce:LiCAF, respectively [52, 53].

The research by Marshall, *et al* [42] was one of the initial works describing the energy levels and interactions within the Ce-colquiriites. The ground state in the colquiriites may be split into three nearly doubly-degenerate levels by the crystal field of the host. The crystal field also splits the 5d orbital into its two components, the t_{2g} and the e_g levels. Because of the sixfold coordinate octahedral field present at the Ca and Sr sites (potential sites for the Ce ion), the t_{2g} is assumed to be the lower of the split levels of the excited state. Figure B1.5.4 shows the absorption and emission cross-section spectra as a function of wavelength and polarization for these uniaxial materials. The Ce^{3+} absorption has a weak polarization dependence for both hosts. These peaks are also slightly shifted with each polarization, occurring at 266 nm for $\pi(E\|c)$ and at 272 nm for $\sigma(E\perp c)$ polarizations in LiSAF, and at 267 and 271 nm in LiCAF. These spectra are for 1 mol% CeF_3, and exhibit extremely high absorption and emission cross sections. The measured fluorescence lifetime of the 5d to 4f transition in LiSAF is nominally 28 ns at 290 nm [51–54]. Just as with many of the other Ce-activated solids, colour centres exist in this family of hosts, more strongly in LiSAF than in LiCAF [42, 54]. Co-doping and charge compensation have shown signs of minimizing these effects in the colquiriite isomorphs [9, 42, 44, 45, 54].

Figure B1.5.5 shows the strong polarization dependence observed in the output of both LiCAF and LiSAF, with the σ-polarization showing strongly decreased laser output [42]. The following diagram, (figure B1.5.6), shows the broad tunability of the LiSAF and LiCAF lasers from NRL data utilizing a prism in the cavity. Tunability has been investigated by a number of groups [42, 55–57] and the output follows the profile of the stimulated emissions cross sections plotted previously in figure B1.5.4. Finally, besides use as an oscillator material, Ce-doped LiCAF and LiSAF have been used by several groups as a high-peak -power amplifier for femtosecond pulses [58–62], which may open up a variety of UV-based applications, including high-field physics and materials processing.

Spectroscopic and laser investigations of the trivalent cerium ion have led to efficient pulsed lasers operating in the 280–330 nm region of the ultraviolet. To date, there have been no oxide-based hosts which have supported stimulated emission but all Ce-doped lasers have originated in the fluoride hosts. Lasers using these fluoride crystals are just starting to emerge from the laboratory and are being utilized in hardened systems for field experiments. These newly investigated hosts and dopants show promise for consideration for future commercial tunable solid state UV systems.

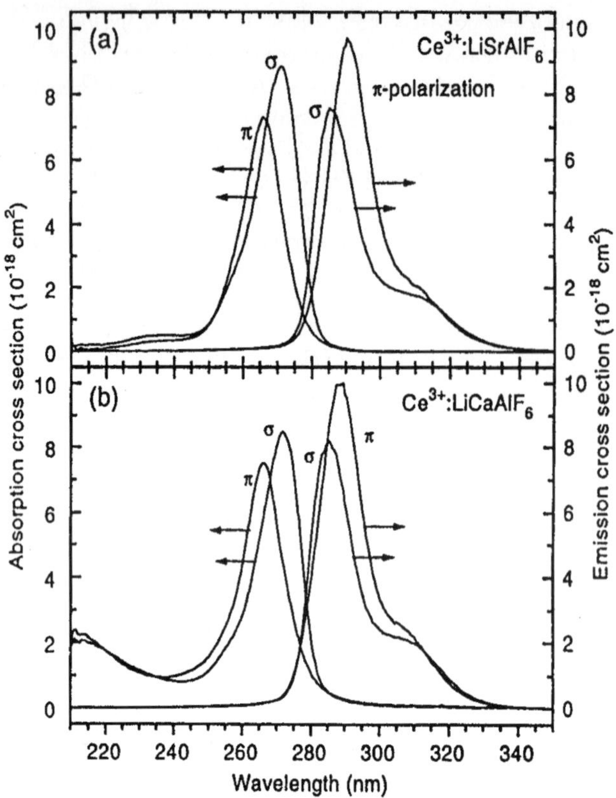

Figure B1.5.4. Ce-doped (*a*) LiSAF and (*b*) LiCAF absorption and emission cross section spectra, as a function of polarization (after Marshall *et al* [42]).

B1.5.2 Trivalent uranium lasers

Trivalent uranium (U^{3+}) activated into various laser hosts has played an interesting and important role in the history and evolution of solid state lasers. The spectroscopy of the uranium ion doped into CaF_2 was first investigated in the late 1950s by Galkin and Feofilov [63, 64] and simultaneously by Conway [65]. The intense early interest was due to the vast number of visible and near-infrared absorption bands and the strong fluorescence bands between 2.0 and 2.8 μm. Most of these early studies focused on the spectroscopy of CaF_2 [63–79], LiF^- [80], SrF_2 [63, 64, 67, 77], BaF_2 [63, 64, 77] and $LaCl_3$ [81, 82], while more recent spectroscopy has dealt with trivalent uranium in $LiYF_4$ [83–86]. These most recent studies of uranium in $LiYF_4$ involved researchers investigating materials and their spectroscopic properties to find crystals that would allow efficient diode- or laser-pumping of transitions in the near infrared, with a focus on lasing these transitions in the 2 to 3 μm wavelength region [83–87], competing with the Er^{3+} ion. Much of the recent focus on U^{3+}- activated hosts evolves from the search for materials for radar and medical applications operating in the infrared, with broad outputs which may give rise to tunability. It is interesting to note that after the advent of the ruby laser in 1960, a laser based upon U^{3+}:CaF_2 was the second laser ever demonstrated (also in 1960) [66]. One of the major notable differences was that ruby is a three-level laser, whereas the laser action of the actinide-based U^{3+} ion behaves as a four-level laser. It is also interesting to note that the uranium ion

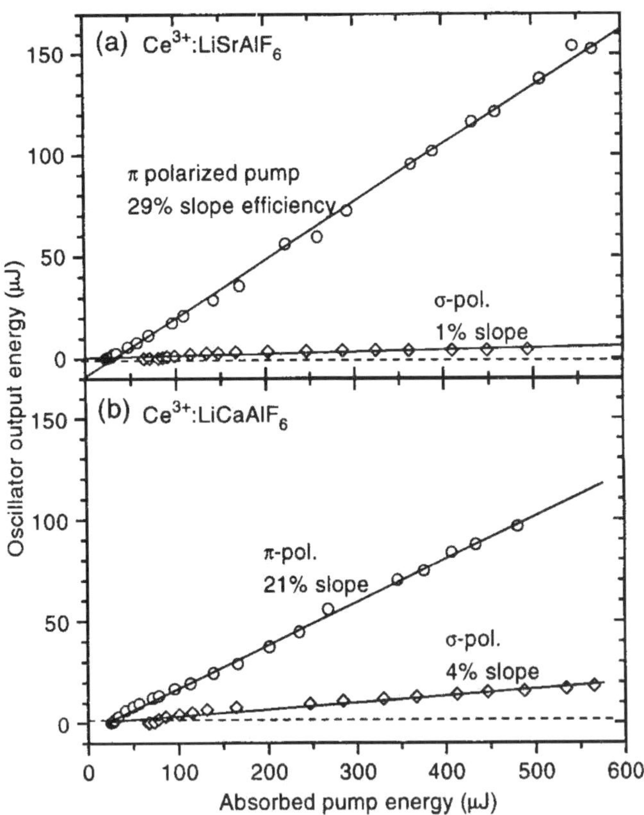

Figure B1.5.5. Laser output energies at ~292 nm as a function of absorbed 266 nm pump energy in Ce-doped (*a*) LiSAF and (*b*) LiCAF (after Marshall *et al* [42]).

is the only actinide ion ever to have stimulated emission reported, during these past 40 years of intense solid state laser research.

One issue that has always been of concern with spectroscopists and crystal growth scientists relative to the uranium ion is that of charge compensation. Since the desired valence state for the uranium ion for these infrared laser transitions is the trivalent state, and in CaF_2, SrF_2 and BaF_2 hosts the uranium ion resides on a divalent site, there is naturally a charge mismatch. This has often given rise to multiple valence states of uranium residing in the host lattice to provide a charge-neutral lattice. Uranium has been spectroscopically identified in the various hosts in the U^{2+}, U^{3+}, U^{4+} and U^{6+} valence states [67, 73, 74, 77–84].

The U^{3+} ion is a member of the actinide series of elements, having a partially filled 5f shell, making its $5f^3$ configuration of energy levels nearly analogous and actually iso-electronic to those found in its $4f^3$ counterpart trivalent neodymium (Nd^{3+}). Since there are three electrons in the unfilled 5f shell, by Hund's rules, the lowest term of an f^3 configuration is a ^4I_J term, with the $J = 9/2$ state being the lowest-lying state. The shielding of the 5f levels from the crystalline field is actually much weaker than that of the 4f shell but it has a greater spin–orbit interaction which gives rise to larger splittings between the ^4I_J levels than those found with Nd^{3+}. It has been determined quantitatively that the actinide ions exhibit crystal field strengths nearly twice as strong as those for the lanthanides, and that the cross sections for the absorptions

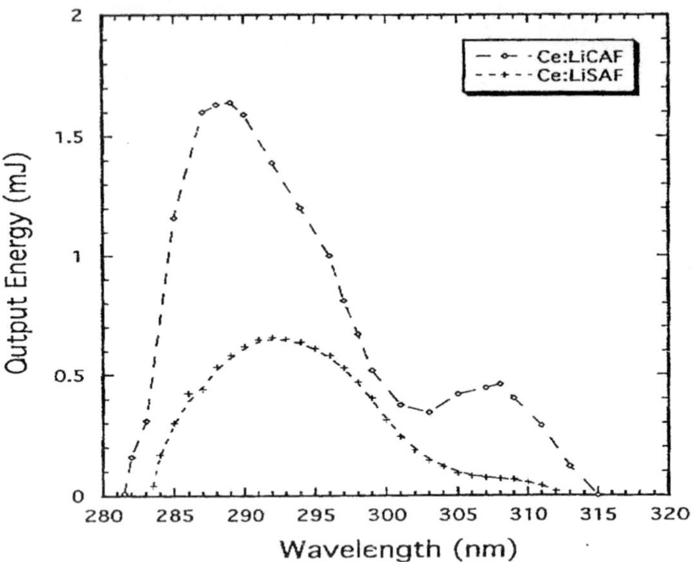

Figure B1.5.6. Laser tuning spectra and output energy of Ce-doped LiCAF and LiSAF, pumped at 266 nm at room temperature (after Pinto *et al* [55]).

of the 5f–5f transitions are sometimes two orders of magnitude larger than those of their lanthanide 4f–4f counterparts [87].

As with their lanthanide counterparts in the rare-earth series, the absorption and emission lines are sharp and well-defined. These sharp lines can be attributed to the transitions occurring between configurations of an unfilled inner f-electron shell that, as alluded to previously, is partially shielded. Transitions between these well-split ^4I_J levels involve a violation of the Laporte rule of $\Delta l = \pm 1$ for electric dipole transitions. This larger spacing between the ^4I_J levels also allows for a reduction in the non-radiative decay between the adjacent ^4I_J levels which results in the potential for laser transitions in the 2.5–2.9 μm range between the $^4I_{11/2}$ and the $^4I_{9/2}$ levels of U^{3+}. Figure B1.5.7 illustrates this energy level scheme and the wide splittings of the ^4I_J levels for the U^{3+} ion in CaF_2.

This diagram highlights the ^4I multiplets of the trivalent uranium ion in a typical fluoride host crystal field, identifying three of the major ^4I_J energy manifolds located throughout the visible and near-infrared portions of the spectrum. The upward arrows illustrate potential excitation paths in the 500 nm region (20 000 cm^{-1}) and in the 850–950 nm region (into the $^4I_{15/2}$ manifold at approximately 11 000 cm^{-1}). Representative absorption coefficient spectra as a function of wavelength for these manifolds in U^{3+}:$LiYF_4$ (YLF) are illustrated in figures B1.5.8 and B1.5.9. Because of the uniaxial nature of YLF, there are two distinct polarization-dependent spectra in each wavelength region, one corresponding to $E \perp c$ and the other for $E \parallel$ to the c-axis. These two figures, shown at room temperature, illustrate the breadth of the absorption spectrum which is due to the multiple transitions between the individual Stark levels.

As exhibited in figure B1.5.7, there is a broad band of energy levels in YLF corresponding to the $^4I_{11/2}$ manifold located at approximately 5000 cm^{-1}. The crystal-field splitting of this ground-state manifold is particularly large compared to that of the U^{3+} ion in CaF_2, measured as 1113 cm^{-1} *versus* 602 cm^{-1}, respectively. The optical absorption between the $^4I_{9/2}$ and the $^4I_{11/2}$ energy levels can give rise to absorptive losses on the high-energy side of the emission spectra which define the region of the laser transition from the $^4I_{11/2}$ upper laser level to the $^4I_{9/2}$ ground state. These absorption transitions between the $^4I_{9/2}$ and the

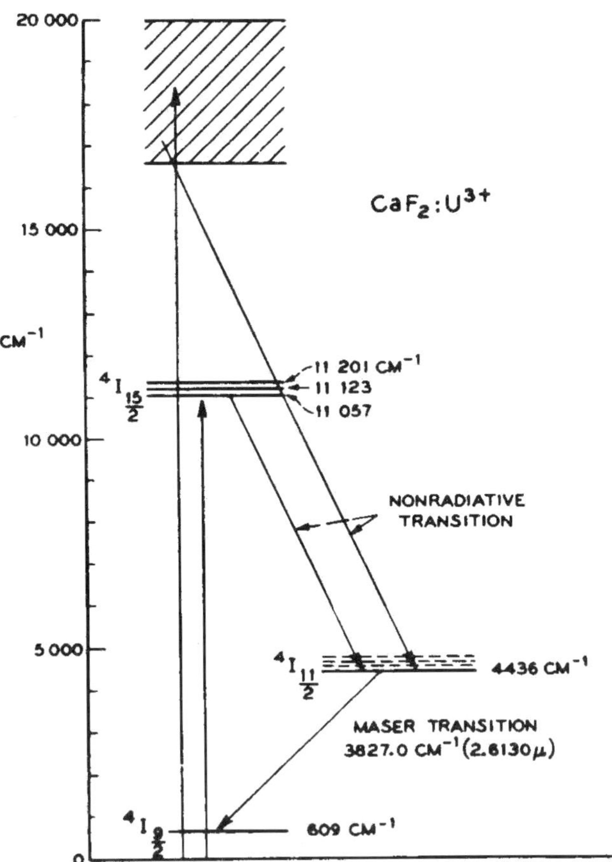

Figure B1.5.7. The absorption, fluorescence and laser transitions of the 2.613 μm transition in CaF_2:U^{3+} (after Boyd *et al* [88]).

$^4I_{11/2}$ levels are illustrated in figure B1.5.10, and are typical for trivalent uranium in a fluoride-based host. At room temperature, these absorption losses diminish past 2.6 μm, thus allowing any laser transitions at greater than 2.6 μm to be very efficient with minimal intrinsic or ground-state absorptive losses. Figure B1.5.10 illustrates these absorption transitions between the ground state and first excited state, while figure B1.5.11 is representative of the broad fluorescence spectrum originating from the $^4I_{11/2}$–$^4I_{9/2}$ transitions occurring at room temperature in U^{3+}:YLF.

As with any transition when trying to determine the probability for stimulated emission, the lifetime of the metastable upper laser level plays a key role. If pumping with a pulsed source, optimization is most efficient with an excitation source that is nearly matched to this lifetime. Since the $^4I_{11/2}$–$^4I_{9/2}$ emission arises as a Laporte forbidden transition, the lifetimes are reasonable and typically single exponential. For the CaF_2 family of hosts, the lifetimes measured at room temperature are of the order of 15 μs [88], whereas for the YLF host, these same transitions exhibit lifetimes an order of magnitude longer at 175–192 μs [83, 87]. The difference between these two reported $^4I_{11/2}$ lifetime values could be attributed to concentration effects. Compared to the lanthanides, the actinides (such as trivalent uranium) can exhibit concentration quenching of the upper-state lifetimes at much lower concentrations. This, as referenced earlier in this section, can be due

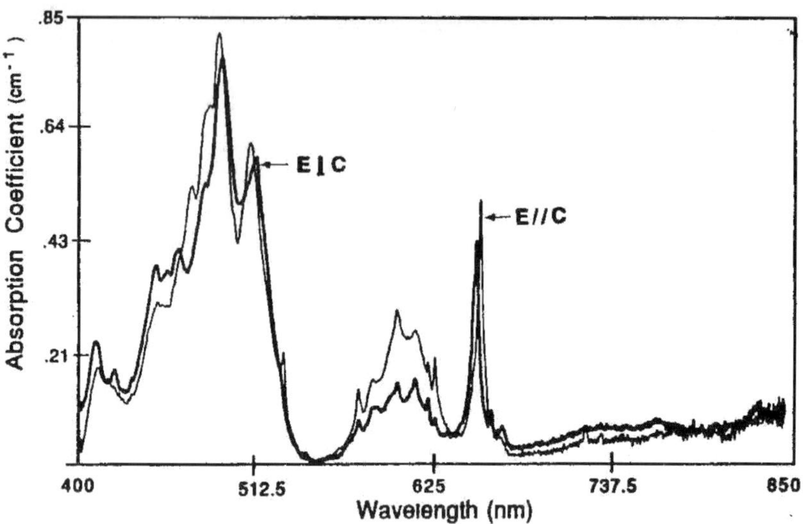

Figure B1.5.8. The polarized absorption spectra of U^{3+}:YLF in the visible region at 300 K (after Quarles *et al* [83]).

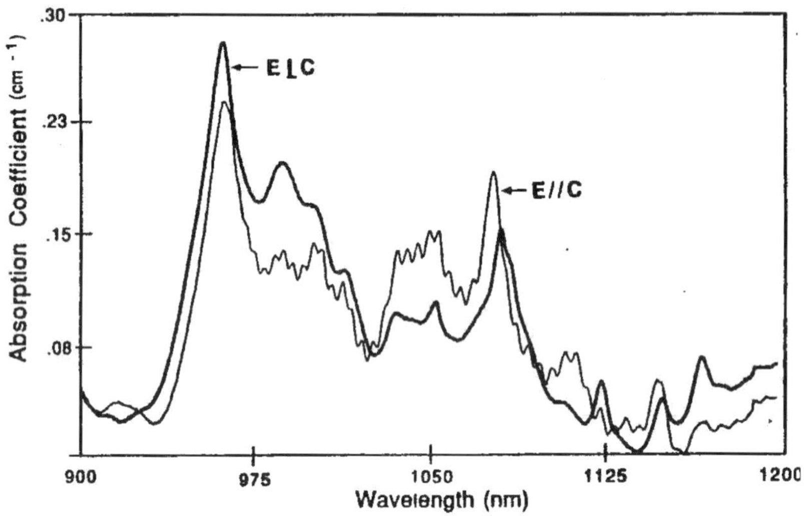

Figure B1.5.9. The polarized absorption spectra of U^{3+}:YLF in the near-infrared region at 300 K (after Quarles *et al* [83]).

to the fact that the oscillator strengths for the actinide ions are one to two orders of magnitude larger than those for the lanthanides and, thus, the probability for energy transfer from ion to ion can occur at a much lower concentration [39]. If there is ion-to-ion energy transfer, this could be exhibited by a measurable decrease in the upper-state lifetime and occur in concentrations as low as 0.1% in an actinide-activated material.

Laser action in the trivalent uranium ion was first observed by Sorokin and Stevenson in 1960 [66] in the CaF_2 host. A review of the literature shows that there were many simultaneous studies and various reports of different lasing conditions and different wavelengths. Several groups measured room temperature, pulse-pumped lasing in CaF_2 at 2.61 μm [66, 73, 90–92], at 2.57 μm [73], and 2.51 μm [73, 90, 91]. Others

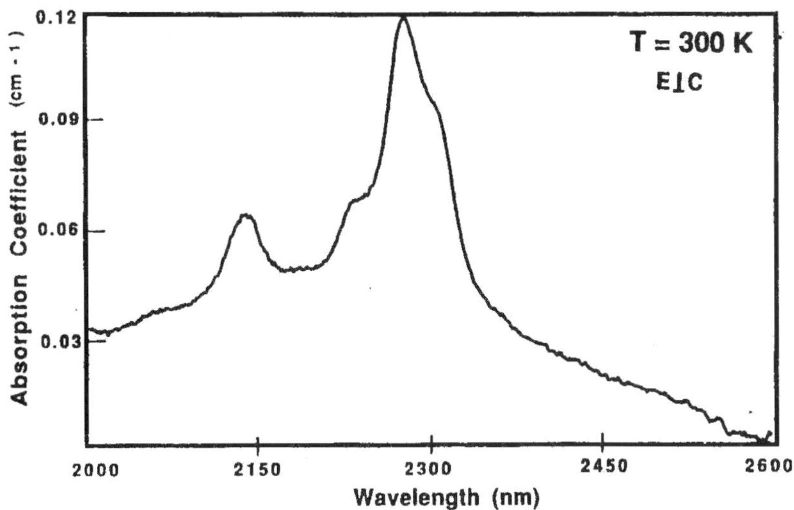

Figure B1.5.10. The $E \perp c$ polarized $^4I_{9/2}$–$^4I_{11/2}$ absorption spectra of U^{3+}:YLF at 300 K (after Quarles *et al* [83]).

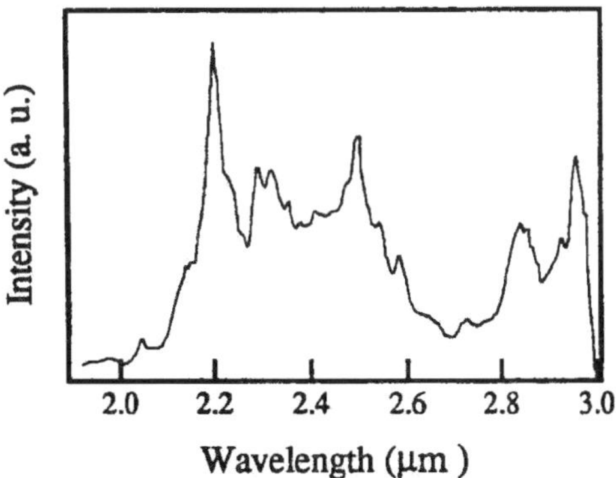

Figure B1.5.11. The $^4I_{11/2}$–$^4I_{9/2}$ fluorescence spectra of U^{3+}:YLF at 300 K (after Louis *et al* [87]).

have measured the pulse-pumped lasing at reduced temperatures (77 K) at wavelengths of 2.24 μm in CaF_2 [67–69], at 2.41 μm in SrF_2 [71] and at 2.56 μm in BaF_2 [70, 93]. In most of these cases, xenon pumping was utilized to provide a strong overlap between the xenon output and the high-quantum-efficiency infrared absorption bands. The various wavelengths of laser operation recorded can most likely be attributed to transitions oscillating on various components between the Stark-split manifolds. Likewise, since the crystals can be stressed during growth and during the pumping process, this may cause perturbations which may make some of the transitions less forbidden and, thus, give rise to different research groups measuring slightly different laser oscillation wavelengths. Finally, as exhibited in figures B1.5.10 and B1.5.11, any changes in the ground-state absorptive losses in the laser host crystal can impact the intracavity laser losses and, thus, the threshold operating conditions for the various transitions.

Continuous wave (cw) excitation also leads to lasing of the trivalent uranium ion in tests in the early 1960s. The first cw lasing was accomplished by Boyd *et al* [88] using a cw mercury lamp to excite the uranium ions at approximately 900 nm. The laser output was also measured at 2.61 μm for CaF_2 in a cw

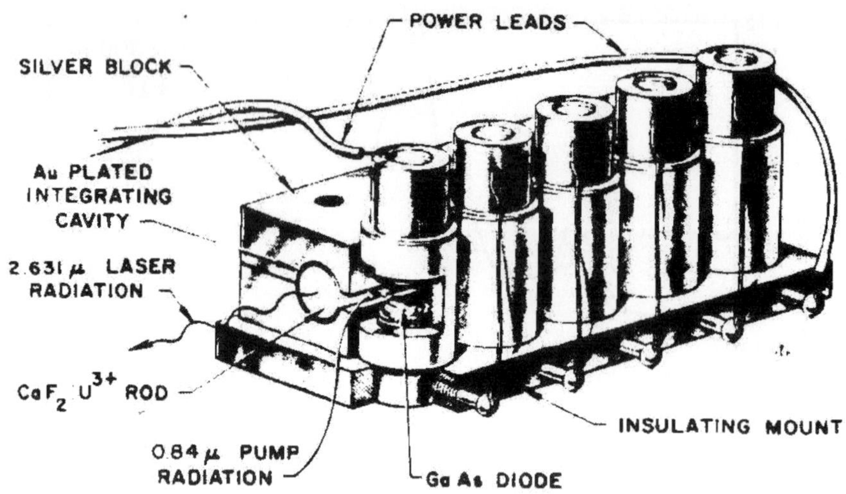

Figure B1.5.12. Artist's rendition of the first diode pumped solid state laser GaAs pumped U:CaF_2 (after Keyes *et al* [94]).

operating condition. One of the most significant breakthroughs recorded in solid-state laser history also occurred with cw excitation of the uranium ion. In 1963, Keyes and Quist became the first scientists to successfully excite a solid state laser with GaAs diodes [94]. An artist's rendition of this first diode-pumped solid state laser is illustrated in figure B1.5.12. They utilized cw diodes operating at 840 nm, and demonstrated that the $^4I_{9/2}$–$^4I_{15/2}$ transition in U^{3+}:CaF_2 could efficiently be excited to lase at 2.631 μm. Even though this lasing took place at 4.2 K, this was the first demonstration of the efficient use of semiconductor laser diodes as an excitation source for solid state lasers. More recent research has led to the demonstration of titanium:sapphire-pumped cw lasers based upon U^{3+}:YLF as the host material [87, 89]. These recent studies have shown that lasing could be achieved at 2.827 μm when pumped at room temperature between 810 and 890 nm. The efficiency of this most recent cw pumping approaches 0.3%. As can be seen in a comparison of figures B1.5.10 and B1.5.11, this lasing represents a four-level laser scheme when operating at the longer wavelengths of the $^4I_{11/2}$–$^4I_{9/2}$ transition.

Spectroscopic and laser investigations of the trivalent uranium ion have led to efficient pulsed and cw lasers operating in the 2.5–2.9 μm wavelength region. Significant historical and scientific studies have documented the stimulated emission of the U^{3+}-based 5f–5f transitions in various fluoride hosts. Of historical note is that U^{3+} in CaF_2 was the second solid state laser ever reported (the same year as the ruby laser); it is the first and only actinide ever reported to exhibit stimulated emission and it has set historical precedence in the now booming field of diode-pumped solid state lasers as the first solid ever lased using diode pumping. Even though there is not a proliferation of these lasers in the commercial marketplace, their place in the historical context of solid state lasers is noteworthy.

B1.5.3 Divalent rare earth ion lasers

In the previous sections, it has been shown how trivalent lanthanide rare-earth ions can exhibit stimulated emission on both the f -to- f transitions and the d-to- f transitions. There exists another group of rare-earth ions that carry a divalent charge which have also shown laser action in previous studies. Amongst these divalent ions are samarium (Sm^{2+}), dysprosium (Dy^{2+}) and thulium (Tm^{2+}). As with the trivalent ions, the electronic configuration consists of a palladium core with 46 electrons and with a $4f^n5s^25p^65d^16s^2$ outer

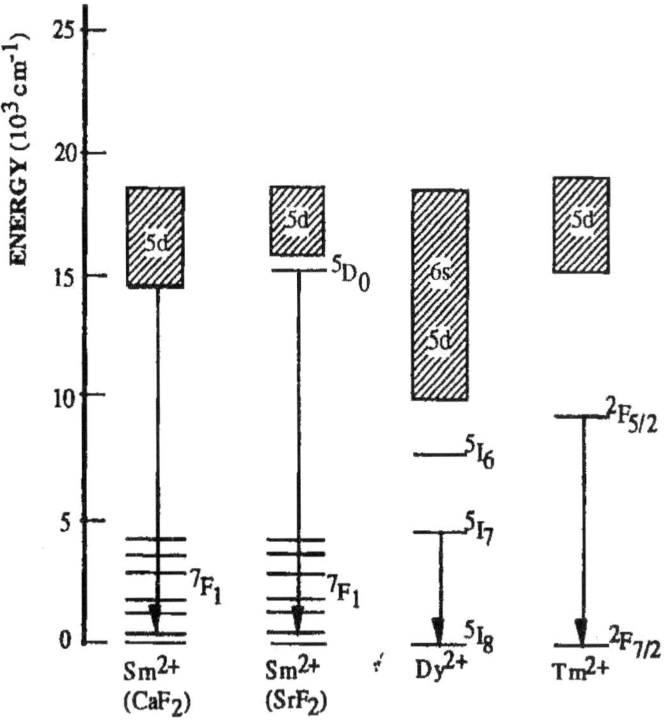

Figure B1.5.13. Partial energy level diagrams for the divalent rare earths: Sm^{2+}, Dy^{2+} and Tm^{3+} (after Powell [2]).

electron configuration, where for $4f^n$, n = 6 for Sm^{2+}, $n = 10$ for Dy^{2+} and $n = 13$ for Tm^{2+}. Because of this filling sequence for the $4f^n$ shell, this makes Sm^{2+} isoelectronic to Eu^{3+}, Dy^{2+} is isoelectronic to Ho^{3+}, and Tm^{2+} is isoelectronic to Yb^{3+}. Just as with the trivalent ions, this inner 4f shell is partially filled, and the 5s and 5p outer shells are completely filled, shielding this 4f shell. One important difference of the divalent lanthanide ions, compared with those of the trivalent lanthanides, is that the energy levels of the $4f^{n-1}5d$ configurations will typically occur at lower energies. Examples of the placement of these energy levels is shown in figure B1.5.13 for the divalent Sm (in two different hosts), Dy and Tm ions.

In some of the host materials investigated by various groups, the 4f–5d transitions occur in the visible region of the spectrum (such as in the Sm and Tm examples illustrated in figure B1.5.13), and can support visible to near-infrared stimulated emission. For ions such as divalent Sm, the 5d–4f laser transitions are spectrally broadened by the interaction of the 5d electron level and the crystalline host. These 5d–4f laser transitions are also electric-dipole-allowed transitions, similar to those described in the earlier section with the trivalent cerium ion and, thus, have much stronger oscillator strengths. As is clearly seen in figure B1.5.13, many divalent rare-earth ions, when activated in fluoride-based hosts, have 4f energy levels high enough in energy that radiationless decay processes can occur efficiently from the 5d to the 4f levels. The one exception to this case involves the Sm^{2+} ion, and the choice of host. The specific crystal host and, thus, the crystal field strength, determines whether the upper laser level belongs to the $^5D_0$ f-orbital (as shown in figure B1.5.13 for Sm^{2+} in SrF_2), or to the 5d level (as with CaF_2 in figure B1.5.13).

One divalent ion which has been studied extensively but has never exhibited stimulated emission is the divalent europium ion (Eu^{2+}), which is a close analogue of Sm^{2+}. Many researchers have anticipated being able to make Eu^{2+} lase, especially because of its very strong fluorescence, but all efforts have been hampered by extremely strong ESA [95,96]. Even though in most hosts the Eu^{2+} ion emits from an allowed-electric-dipole 5d–4f transition, the ESA has completely overwhelmed the possibility of reaching population inversion.

The first divalent rare-earth ion studied in great detail was the Sm^{2+} ion in CaF_2. The early pioneers of the trivalent uranium studies were also the pioneers of the divalent samarium ion, namely Feofilov [97, 98] and, separately, Sorokin and Stevenson [99]. Sorokin and Stevenson went on to be the first researchers to demonstrate stimulated emission with the Sm^{2+} ion in CaF_2, exciting with a pulsed Xe lamp and the crystal at 20 K [97–101]. The output wavelength was 708.5 nm, on the allowed 5d–4f transition. The measured upper-state lifetime for this transition was a very short 1.4 μs. Subsequent studies have shown that the Sm ion in CaF_2 can be lased on various lines between 708.3 and 745 nm when excited with a pulsed ruby laser excitation source, at temperatures ranging from 65 to 210 K [102]. Studies were also conducted on the SrF_2 host activated with divalent Sm. In 1962, Sorokin *et al* [103] followed up their Sm:CaF_2 laser studies of 1961, with evidence of stimulated emission of Sm in SrF_2. The divalent Sm ion in SrF_2 lased on the $^5D_0$–$^7F_1$ transition at 696.9 nm, at a temperature of 4.2 K. The upper state lifetime for this d–d transition was much longer than the transition in CaF_2, measured to be 10 ms. Following these initial publications, Sm:SrF_2, was also lased using a pulsed ruby excitation source [104, 105]. Subsequent research over the next three decades has determined that significant ESA, defect-induced losses and non-radiative decay processes severely limit the utility of the divalent Sm ion at other than cryogenic temperatures [106–110].

Another ion that was studied heavily during the 1960s was the divalent dysprosium ion [111–113]. Just as with its Sm^{2+} counterpart, the Dy^{2+} ion was initially excited with an Xe lamp in an attempt to induce stimulated emission. As is seen in figure B1.5.13, the Dy^{2+} ion lases on an f–f transition between the $^5I_7$ and $^5I_8$ manifolds, at nominally 2.358 μm. The Dy^{2+} ion was first lased in the CaF_2 host in 1962 under both cw [114, 115] and pulsed [116] conditions. The pulsed laser demonstration occurred with a Hg lamp as the excitation source and the crystal was held at 77 K [116], while for the cw experiments [114, 115], a Xe lamp was used and the crystal temperature was lower at 4 K. Other pumping sources included a tungsten (W) lamp (cw, between 4 and 77 K) [117, 118], and a Xe lamp at 20–77 K [119, 120]. One of the Xe lamp excited systems was with a SrF_2 host [120], which exhibited a slightly longer wavelength of 2.3659 μm. There were other pumping configurations, including a pulsed system [121], a cw system with giant pulses [122] and a giant pulsed system with a very high repetition rate [123], Finally, there were several unique and intriguing pumping schemes explored with the divalent Dy ion in CaF_2 and SrF_2 hosts, including ones utilizing sunlight focused in a cw scheme to achieve laser oscillation with the crystal temperature held at 77 K [124–126] and a scheme whereby the laser was excited in a quasi-cw manner using a pyrotechnic detonation [127]. All of these studies led to unique laser systems, but none of them exhibited stability or room temperature operation.

One final divalent rare-earth ion which has been moderately investigated is the thulium ion (Tm^{2+}). As mentioned previously, the Tm^{2+} ion is isoelectronic to the Yb^{3+} ion and operates as a laser between the $^2F_{5/2}$ and $^2F_{7/2}$ transitions at 1.116 μm. There has been no reported laser research involving the divalent thulium ion since 1980. The reports by Kiss and Duncan [117, 128, 129] were for pulsed pumping with a Xe lamp at 4.2 K, and for cw operation, pumped with a Hg lamp at crystal temperatures between 2.7 and 4.2 K. Neither of these operating schemes showed any promise for commercialization because of the low temperatures required to stabilize the ion in the fluoride host.

All of the divalent rare-earth ions investigated, and reported in this section, have exhibited deficiencies when considered for solid state lasers. Even though stimulated emission was possible with each ion in a variety of fluoride hosts, none of the combinations can overcome the limitations of ESA, optical instability, and defect-induced losses. It can be stated that the Tm^{2+}, Dy^{2+} and Sm^{2+} ions all indeed exhibit laser operation but none efficient or robust enough to warrant significant research or development over the past two decades.

B1.5.4 Summary

The purpose of this chapter was to highlight the historical and recent research involving several lesser-known, yet historically significant, trivalent and divalent rare-earth-activated solid state lasers. Laser action in the UV region of the spectrum has been observed in fluorides doped with the trivalent cerium ion. These lasers have been shown to be tunable and highly efficient, with some of the highest stimulated emission cross sections for any rare-earth-doped solids. The trivalent uranium ion holds several historically significant records in the evolution of solid state crystalline lasers. Trivalent uranium is the first and only actinide ion to exhibit stimulated emission, it is also the first ion ever to be lased by direct diode excitation and, finally, it is the second ion ever lased, immediately following the discovery of the ruby laser. This trivalent ion, like its cerium counterpart, is broadly tunable, except the uranium ion demonstrates stimulated emission in the 2.0–2.8 μm region of the infrared. Finally, a tremendous amount of research has gone into attempting to understand divalent rare-earth ions such as Tm, Dy and Sm. Each has exhibited stimulated emission and in some unusual configurations, such as solar-pumped or pyrotechnic pulsed, but because of instabilities, ESA and defects, none of the divalent systems has shown promise as a robust system for commercial applications.

References

[1] Koechner W 1988 *Solid-State Laser Engineering* (Berlin: Springer)

[2] Powell R C 1998 *Physics of Solid-State Laser Materials* (New York: Springer)

[3] Weber M J (ed) 1982 *CRC Handbook of Laser Science and Technology, Volume 1: Lasers and Masers* (Boca Raton, FL: Chemical Rubber Company)

[4] Deike G H 1968 *Spectra and Energy Levels of Rare-Earth Ions in Crystals* (New York: Wiley–Interscience)

[5] Wyborne B G 1965 *Spectroscopic Properties of Rare Earths* (New York: Wiley–Interscience)

[6] Hamilton D S 1985 *Tunable Solid State Lasers* ed P Hammerling, A B Budgor and A Pinto (New York: Springer) pp 80–90

[7] Weber M J 1973 *Solid State Commun.* **12** 741–4

[8] Dorenbos P 2000 *J. Lumin.* **91** 155–76

[9] Castillo V K 2002 *PhD Thesis* University of South Florida

[10] Blasse G and Bril A 1967 *J. Chem. Phys.* **47** 5139–46

[11] Jacobs R R, Krupke W F and Weber M J 1978 *Appl. Phys. Lett.* **33** 410–12

[12] Miniscalco W J, Pellegrino J M and Yen W M 1978 *J. Appl. Phys.* **49** 6109–11

[13] Owen J F, Dorain P B and Kobayasi T 1981 *J. Appl. Phys.* **52** 1216–23

[14] Wong S M, Rotman S R and Warde C 1984 *Appl. Phys. Lett.* **44** 1038–40

[15] Hamilton D S, Gayen S K, Pogatshnik G J, Ghen R D and Miniscalco W J 1989 *Phys. Rev.* B **39** 8807–15

[16] Rotman S R, Tuller H L and Warde C 1992 *J. Appl. Phys.* **71** 1209–14

[17] Doroshenko M E, Ivanov M A, Sigachev V B and Timoshechkin M I 1992 *Sov. J. Quantum Electron.* **22** 588–90

[18] Tomiki T, Akamine H, Gushiken M, Kinjoh Y, Miyazato M, Togokawa N, Hiraoka M, Hirata N, Ganaha Y and Futemma T 1991 *J. Phys. Soc. Japan* **60** 2437–45

[19] Hammons D A, Richardson M C, Chai B H T, Bass M, Quarles G J and Castillo V 1995 *OSA Annual Meeting Proc.* (Washington, DC: Optical Society of America) p 158

[20] Li G, Wu J, Chen Y, Chen J, Shi Z and Xiao C 1993 *Chinese J. Lasers* B **2** 91–6

[21] Nolas G S, Tsoukala V G, Gayen S K and Slack G A 1994 *Opt. Lett.* **19** 1574–6

[22] Benderskaya L P, Voloshinovskii A S, Novikova T P, Pidzyrailo N S and Pashchuk I P 1991 *Opt. Spectrosc. (USSR)* **70** 198–200

[23] Elias L R, Heaps W S and Yen W M 1973 *Phys. Rev.* B **8** 4989–95

[24] Yang K H and DeLuca J A 1977 *Appl. Phys. Lett.* **31** 594–6

[25] Ehrlich D J, Moulton P F and Osgood R M Jr 1980 *Opt. Lett.* **5** 339–41

[26] Ehrlich D J, Moulton P F and Osgood R M Jr 1979 *Opt. Lett.* **4** 184–6

[27] Lim K-S and Hamilton D S 1989 *J. Opt. Soc. Am.* B **6** 1401–6

[28] Pogatshnik G J and Hamilton D S 1987 *Phys. Rev.* B **36** 8251–7

[29] Dubinskii M A, Abdulsabirov R Yu, Korableva S L, Naumov A K and Semashko V V 1992 *18th Int. Quantum Electron. Conf., OSA Technical Digest* (Washington, DC: Optical Society of America) pp 548–50

[30] Sarukura N, Liu Z, Segawa Y, Edamatsu K, Suzuki Y, Itoh T, Semashko V V, Naumov A K, Korableva S L, Abdulsabirov R Y and Dubinskii M A 1994 *OSA Proc. Advanced Solid-State Lasers (TOPS 20)* (Washington, DC: Optical Society of America) pp 395–9

[31] Sarakura N, Liu Z, Segawa Y, Semashko V V, Naumov A K, Korableva S L, Abdulsabirov R Yu and Dubinskii M A 1995 *Opt. Lett.* **20** 599–601

[32] Sarakura N, Liu Z, Segawa Y, Edamatsu K, Suzuki Y, Itoh T, Semashko V V, Naumov A K, Korableva S L, Abdulsabirov R Yu and Dubinskii M A 1995 *Opt. Lett.* **20** 294–6
[33] Rambaldi P, Moncorgé R, Wolf J P, Pedrini C and Gesland J Y 1998 *Opt. Commun.* **146** 163–6
[34] McGonigle A J S, Coutts D W and Webb C E 1999 *Opt. Lett.* **24** 232–4
[35] McGonigle A J S, Coutts D W and Webb C E 1999 *OSA Proc. Advanced Solid-State Lasers (TOPS 26)* (Washington, DC: Optical Society of America) pp 123–9
[36] McGonigle A J S, Coutts D W and Webb C E 1999 *IEEE J. Selected Topics Quantum Electron.* **5** 1526–31
[37] McGonigle A J S, Coutts D W, Webb C E, Girard S and Moncorgé R 1999 *Electron. Lett.* **35** 1640–1
[38] Kaminskii A A, Kochubei S A, Naumochkin K N, Pestryakov E V, Trunov V I and Uvarova T V 1989 *Sov. J. Quantum Electron.* **19** 340–3
[39] Aull B F and Jenssen H P 1986 *Phys. Rev.* B **34** 6647–55
[40] Payne S A, Chase L L, Newkirk H W, Smith L K and Krupke W F 1988 *IEEE J. Quantum Electron.* **24** 2243–53
[41] Wannemacher R and Meltzer R S 1989 *J. Lumin.* **43** 251–60
[42] Marshall C D, Speth J A, Payne S A, Krupke W F, Quarles G J, Castillo V and Chai B H T 1994 *J. Opt. Soc. Am.* B **11** 2054–65
[43] Castillo V K and Quarles G J 1995 *UV and Visible Lasers and Laser Crystal Growth, San Jose (Proc. SPIE 2380)* (Bellingham, WA: SPIE) pp 43–50
[44] Castillo V K, Quarles G J and Chang R S F 2001 *J. Cryst. Growth* **225** 445–8
[45] Castillo V K and Quarles G J 1997 *J. Cryst. Growth* **174** 337–41
[46] Dubinskii M A, Semashko V V, Naumov A K, Abdulsabirov R Yu and Korableva S L 1993 *Laser Phys.* **3** 216–17
[47] Dubinskii M A, Semashko V V, Naumov A K, Abdulsabirov R Yu and Korableva S L 1993 *J. Mod. Opt.* **40** 1–5
[48] Dubinskii M A, Semashko V V, Naumov A K, Abdulsabirov R Yu and Korableva S L 1994 *Laser Phys.* **4** 480–4
[49] Dubinskii M A, Abdulsabirov R Y, Korableva S L, Naumov A K and Semashko V V 1993 *OSA Proc. Advanced Solid-State Lasers (TOPS 15)* (Washington, DC: Optical Society of America) pp 195–8
[50] Pinto J F, Esterowitz L, Rosenblatt G H and Quarles G J 1993 *LEOS '93 Conf. Technical Proceedings* Post-deadline paper PD1.4, pp 7–8
[51] Pinto J F, Rosenblatt G H, Esterowitz L, Castillo V and Quarles G J 1994 *Electron. Lett.* **30** 240–1
[52] Marshall C D, Payne S A, Speth J A, Tassano J B, Quarles G J and Castillo V 1994 *Presented at OE/LASE'94 Conf., Los Angeles* Session 2115, Paper 2115–02
[53] Marshall C D, Payne S A, Speth J A, Krupke W F, Quarles G J and Castillo V 1994 *OSA Proc. Advanced Solid-State Lasers (TOPS 20)* (Washington, DC: Optical Society of America) pp 389–94
[54] Bayramian A J, Marshall C D, Wu J H, Speth J A, Payne S A, Quarles G J and Castillo V K 1996 *J. Lumin.* **69** 85–94
[55] Pinto J F, Esterowitz L and Quarles G J 1995 *Electron. Lett.* **31** 2009–11
[56] Rambaldi P, Douard M and Wolf J P 1995 *Appl. Phys.* B **61** 117–20
[57] Pinto J F, Esterowitz L and Carrig T J 1998 *Appl. Opt.* **37** 1060–1
[58] Sarakura N, Liu Z, Ohtake H, Segawa Y, Dubinskii M A, Abdulsabirov R Yu, Korableva S L, Naumov A K and Semashko V V 1997 *Opt. Lett.* **22** 994–6
[59] Sarakura N, Dubinskii M A, Zhenlin L, Semashko V V, Naumov A K, Korableva S L, Abdulsabirov R Yu, Edamatsu K, Suzuki Y, Itoh T and Segawa Y 1995 *IEEE J. Selected Topics Quantum Electron.* **1** 792–804
[60] Sarakura N, Liu Z, Segawa Y, Semashko V V, Naumov A K, Korableva S L, Abdulsabirov R Yu and Dubinskii M A 1995 *Appl. Phys. Lett.* **67** 602–4
[61] Shimamura K, Baldochi S L, Mujilatu N, Nakano K, Liu Z, Sarukura N and Fukuda T 2000 *J. Cryst. Growth* **211** 302–7
[62] Liu Z, Kozeki T, Suzuki Y, Sarukura N, Shimamura K, Fukuda T, Hirano M and Hosono H 2001 *IEEE J. Selected Topics Quantum Electron.* **7** 542–50
[63] Galkin L N and Feofilov P P 1957 *Sov. Phys. Dokl.* **2** 255–7
[64] Galkin L N and Feofilov P P 1959 *Opt. Spectrosc. (USSR)* **1** 492–3
[65] Conway J G 1959 *J. Chem Phys.* **31** 1002–4
[66] Sorokin P P and Stevenson M J 1960 *Phys. Rev. Lett.* **5** 557–9
[67] Title R S, Sorokin P P, Stevenson M J, Pettit G D, Scardefield J E and Lankard J R 1962 *Phys. Rev.* **128** 62–6
[68] Yariv A 1962 *Phys. Rev.* **128** 1588–92
[69] Johnson L F, Boyd G and Nassau K 1962 *Proc. IRE* **50** 86
[70] Porto S P S and Yariv A 1962 *Proc. IRE* **50** 1542–3
[71] Porto S P S and Yariv A 1962 *Proc. IRE* **50** 1543–4
[72] Porto S P S and Yariv A 1962 *J. Appl. Phys.* **33** 1620–1
[73] Wittke J P, Kiss Z J, Duncan R C, Jr and McCormick J J 1963 *Proc. IEEE* **51** 56–62
[74] Meredith R E and Jenney J A 1963 *J. Chem Phys.* **39** 3127–30
[75] Bazhulin P A, Malyshev V I , Markin A S, Rakov A V and Bagdasarov Kh S 1964 *Opt. Spectrosc. (USSR)* **16** 291–2
[76] Porto S P S and Yariv A 1964 *Quant. Elect. III, Paris* vol 1 (New York: Columbia University Press) pp 717–23
[77] Hargreaves W A 1967 *Phys. Rev.* **156** 331–42
[78] Hargreaves W A 1970 *Phys. Rev.* B **2** 2273–84
[79] Hargreaves W A 1991 *Phys. Rev.* B **44** 5293–95

[80] Feofilov P P 1959 *Opt. Spectrosc. (USSR)* **1** 493–4

[81] Gruber J H, Morrey J R and Carter D G 1979 *J. Chem. Phys.* **71** 3982–7

[82] Crosswhite H M, Crosswhite H, Carnall W T and Paszek A P 1980 *J. Chem. Phys.* **72** 5103–17

[83] Quarles G J, Esterowitz L, Rosenblatt G H, Uhrin R and Belt R F 1992 *OSA Proc. Advanced Solid-State Lasers (TOPS 13)* (Washington, DC: Optical Society of America) pp 306–9

[84] Jenssen H P, Noginov M A and Cassanho A 1993 *OSA Proc. Advanced Solid-State Lasers (TOPS 15)* (Washington, DC: Optical Society of America) pp 463–7

[85] Simoni E, Louis M, Gesland J Y and Hubert S 1996 *J. Lumin.* **65** 153–61

[86] Hubert S, Simoni E, Louis M, Zhang W P and Gesland J Y 1994 *J. Lumin.* **65** 245–9

[87] Louis M, Simoni E, Hubert S, Auzel F, Meichenin D and Gesland J Y 1995 *OSA Proc. Advanced Solid-State Lasers (TOPS 24)* (Washington, DC: Optical Society of America) pp 141–5

[88] Boyd G D, Collins R J, Porto S P S, Yariv A and Hargreaves W A 1962 *Phys. Rev. Lett.* **8** 269–72

[89] Meichenin D, Auzel F, Hubert S, Simoni E, Louis M and Gesland J Y 1994 *Electron. Lett.* **30** 1257

[90] Sorokin P P and Stevenson M J 1961 *Advances in Quantum Electronics* ed J Singer (New York: Columbia University Press) pp 65–76

[91] Duncan R C, Jr, Kiss Z J and Wittke J P 1962 *J. Appl. Phys.* **33** 2568–9

[92] Miles P A 1961 *Advances in Quantum Electronics* ed J Singer (New York: Columbia University Press) p 76

[93] Bostick H A and O'Connor J R 1962 *Proc. IRE* **50** 219–20

[94] Keyes R J and Quist T M 1964 *Appl. Phys. Lett.* **4** 50–2

[95] Owen J F, Dorain P B and Kobaysi T 1981 *J. Appl. Phys.* **52** 1216–23

[96] Lawson J K and Payne S A 1992 *OSA Proc. Advanced Solid-State Lasers (TOPS 13)* (Washington, DC: Optical Society of America) pp 330–2

[97] Feofilov P P 1956 *Opt. Spektrosk.* **1** 992–9

[98] Feofilov P P and Kaplyanskii A A 1962 *Opt. Spektrosk.* **12** 493–500

[99] Sorokin P P and Stevenson M J 1961 *IBM J. Res. Dev.* **5** 56–8

[100] Kaiser W, Garrett C G B and Wood D L 1961 *Phys. Rev.* **123** 766–76

[101] Anan'yev Yu A, Grezin A K, Mak A A, Sedov B M and Yudina Ye N 1968 *Sov. J. Opt. Technol.* **35** 313–16

[102] Vagin Y U S, Marchenko V M and Prokhorov A M 1969 *Sov. Phys. JETP* **28** 904–9

[103] Sorokon P P, Stevenson M J, Lankard J R and Pettit G D 1962 *Phys. Rev.* **127** 503–8

[104] Konyukhov V K, Marchenko V M and Prokhorov A M 1966 *Opt. Spectrosc. (USSR)* **20** 299–300

[105] Konyukhov V K, Marchenko V M and Prokhorov A M 1966 *IEEE J. Quantum Electron.* **2** 541–2

[106] Forrester P A, Green G M and Sampson D F 1965 *Br. J. Appl. Phys.* **16** 1209–10

[107] Malkin B Z, Ivanenko Z I and Aizenberg I B 1970 *Sov. Phys. Solid State* **12** 1491–6

[108] Payne S A, Chase L L, Krupke W F and Boatner L A 1988 *J. Chem. Phys.* **88** 6751–6

[109] Chase L L, Payne S A and Wilke G D 1987 *J. Phys. C: Solid State Phys.* **20** 953–65

[110] Lawson J K, Lee H W H, Payne S A and Boatner L A 1991 *OSA Proc. Advanced Solid-State Lasers (TOPS 10)* (Washington, DC: Optical Society of America) pp 386–9

[111] Galaktionova N M, Egorova V F, Zubkova V S, Mak A A and Prilezhaev D S 1967 *Opt. Spektrosk.* **23** 949–53

[112] Voronko Yu K, Osiko V V, Prokhorov A M and Shcherbakov I A 1974 *Proc. P.N. Lebedev Phys. Inst.* vol 60, ed D V Skobeltsin (New York: Consultants Bureau) pp 1–30

[113] Zolotov E M 1974 *Proc. P.N. Lebedev Phys. Inst.* vol 60, ed D V Skobeltsin (New York: Consultants Bureau) pp 87–132

[114] Johnson L F 1962 *Proc. IRE* **50** 1691–2

[115] Yariv A 1962 *Proc. IRE* **50** 1699–1700

[116] Kiss Z J and Duncan R C 1962 *Proc. IRE* **50** 1531–2

[117] Kiss Z J 1964 *Quant. Elect. III, Paris* vol 1 (New York: Columbia University Press) pp 805–15

[118] Pressley R J and Wittke J P 1967 *IEEE J. Quantum Electron.* **QE-3** 116–29

[119] Hatch S E, Parson W E and Weagle R J 1964 *Appl. Phys. Lett.* **5** 153–4

[120] Zolotov E M, Osiko V V, Prokhorov A M and Shipulo G P J 1972 *Appl. Spectrosc. (USSR)* **8** 627–8

[121] Isaev S K, Kornienko L S, Lariontsev E G and Khritankov M S 1973 *Sov. J. Quantum Electron.* **2** 335–8

[122] Konyukhov V K, Kostin B B, Kulevskii L A, Murina T M and Prokhorov A M 1965 *Dokl. Akad. Nauk. SSSR* **165** 1056–8

[123] Kostin V V, Kulevsky L A, Murina T M, Prokhorov A M and Tikhonov A A 1966 *IEEE J. Quantum Electron.* **2** 611–12

[124] Kiss Z J, Lewis H R and Duncan R C 1963 *Appl. Phys. Lett.* **2** 93–6

[125] Kaminskii A A, Kornienko L S and Prokhorov A M 1965 *Dokl. Akad. Nauk. SSSR* **161** 1063–4

[126] Kozlov N A, Mak A A and Sedov B M 1966 *Opt. Mekh. Prom.* **N11** 25–9

[127] Bodretsova A I, Kaminskii A A, Levikov S I and Osiko V V 1967 *Sov. Phys. Dokl.* **12** 507–8

[128] Duncan R C and Kiss Z J 1963 *Appl. Phys. Lett.* **3** 23–4

[129] Kiss Z J and Duncan R C 1962 *Proc. IRE* **50** 1532–3

Further reading

Caird J A and Payne S A 1990 Crystalline paramagnetic ion lasers *CRC Handbook of Laser Science and Technology* ed M J Weber (Boca Raton, FL: Chemical Rubber Company)

Deike G H 1968 *Spectra and Energy Levels of Rare-Earth Ions in Crystals* (New York: Wiley–Interscience)

Di Bartolo B 1968 *Optical Interactions in Solids* (New York: Wiley)

Kaminskii A A and Osiko V V 1970 *Izv. Akad. Nauk. SSSR, Neorgan. Mat.* **6** 629–96

Kaminskii A A 1984 *Izves. Akad. Nauk. SSSR, Neorgan. Mat.* **6** 901–24

Kaminskii A A 1981 *Laser Crystals, Their Physics and Properties* (Berlin: Springer)

Koechner W 1988 *Solid-State Laser Engineering* (Berlin: Springer)

Mak A A, Anan'ev A and Ermakov 1968 *Sov. Phys.–Usp.* **92** 419–52

Moulton P F 1982 Paramagnetic ion lasers *CRC Handbook of Laser Science and Technology, Volume 1: Lasers and Masers* ed M J Weber (Boca Raton, FL: Chemical Rubber Company)

Penzkofer A 1988 *Prog. Quantum Electron.* **12** 291–427

Powell R C 1998 *Physics of Solid-State Laser Materials* (New York: Springer)

Reisfeld R and Jørgensen C K 1977 *Lasers and Excited States of Rare Earths* (Berlin: Springer)

Uitert L G 1966 *Luminesence of Inorganic Solids* (New York: Academic) pp 465–539

Walling J C 1987 Tunable paramagnetic-ion solid state lasers *Tunable Lasers* ed L F Mollenauer and J C White (Berlin: Springer)

Weber M J 2001 *CRC Handbook of Lasers* (Boca Raton, FL: Chemical Rubber Company)

Weber M J (ed) 1982 *CRC Handbook of Laser Science and Technology Volume 1: Lasers and Masers* (Boca Raton, FL: Chemical Rubber Company)

Wyborne B G 1965 *Spectroscopic Properties of Rare Earths* (New York: Wiley–Interscience)

Yariv A 1985 *Optical Electronics* (New York: CBS College Publishing)

B1.6
Lasers based on nonlinear effects

Fabienne Pellé

B1.6.1 Introduction

Recent advances in semiconductor laser diodes (see section B2) have given a new impulse in the research for visible solid state laser sources with the opportunity to realize compact devices. There is an important potential interest for compact visible sources emitting either in the blue, green or red in many applications. Low power compact continuous wave (cw) visible lasers are the key components in the development of undersea optical communications, high brightness displays, high-speed printers, high-density data storage, for medical applications and the environment (control of pollution), while high-power (cw or pulsed visible) devices are required for spectroscopy, materials treatment or surgery. Semiconductor laser diodes emitting in the near infrared operating in a stable way at room temperature require simple electric power supplies providing ideal excitation sources for miniaturized devices. Several optical processes such as parametric oscillation, stimulated Raman scattering or frequency mixing, or frequency conversion in a nonlinear crystal, allow laser emission in the infrared range to be converted into visible laser radiation. Frequency conversion or, more precisely, second harmonic generation in a nonlinear crystal will be the subject of this chapter. An infrared cw laser emission can be converted into a visible laser light either by intracavity frequency doubling or by self-doubling. These new approaches are promising in work to achieve compact all-solid-state lasers emitting in the blue-green spectral range. The basic principles of second harmonic generation will be given in the first part. Then intracavity frequency doubling will be compared to self-doubling in their principles and performances. For a deeper understanding, the reader can refer to several reference text books on nonlinear optics [1–5] laser materials [6] and engineering [7, 8].

B1.6.2 Physical background

B1.6.2.1 Nonlinear polarization (see chapter A4)

As an introduction, we will consider the propagation of electromagnetic radiation in a medium with nonlinear dielectric properties. In such a medium, when subjected to an electric field $\boldsymbol{E}$, the induced polarization $\boldsymbol{P}$ is not proportional to the applied field but should be expressed as an expansion in powers of the applied field $\boldsymbol{E}$:

$$\boldsymbol{P} = \varepsilon_0(\chi^{(1)}\boldsymbol{E} + \chi^{(2)}\boldsymbol{E}^2 + \chi^{(3)}\boldsymbol{E}^3 + \cdots) \qquad \text{(B1.6.1)}$$

with ε_0 the dielectric constant of the medium, $\chi^{(k)}$ the kth the susceptibility tensor of rank (k). In the following, we will restrict the nonlinearity to the second order since we are concerned only with second harmonic generation. Consider two electromagnetic waves incident on the nonlinear medium. They are taken as plane waves propagating along the z direction with wavevectors $\boldsymbol{k}_l$, amplitudes $\boldsymbol{E}_l$ and frequency ω_l ($l = 1, 2$).

$$\boldsymbol{E}_l^{\omega_l} = \tfrac{1}{2}\mathrm{Re}(\boldsymbol{E}_l \exp[\mathrm{i}(\omega_l t - \boldsymbol{k}_l \cdot \boldsymbol{r})]). \tag{B1.6.2}$$

The contribution of the quadratic term of equation (B1.6.1) is then given by:

$$\boldsymbol{P}^{(2)} = \varepsilon_0 \chi^{(2)}[\tfrac{1}{2}(\boldsymbol{E}_1^2 + \boldsymbol{E}_2^2) + \tfrac{1}{2}\boldsymbol{E}_1^2 \cos(2\omega_1 t) + \tfrac{1}{2}\boldsymbol{E}_2^2 \cos(2\omega_2 t) + \boldsymbol{E}_1 \boldsymbol{E}_2[\cos(\omega_1 + \omega_2)t + \cos(\omega_1 - \omega_2)t]]. \tag{B1.6.3}$$

The nonlinear response of the medium results in an exchange of energy between electromagnetic fields of different frequency. Then, the second-order susceptibility gives rise, in addition to a constant polarization, to oscillating components at the second harmonics, $2\omega_1$, $2\omega_2$, and components at the sum or difference frequencies ($\omega_1 \pm \omega_2$). The last components have important applications, i.e. parametric oscillation $\omega_3 = \omega_1 + \omega_2$ and the generation of tunable coherent sources $\omega_3 = \omega_1 - \omega_2$ developed in chapter A4.

B1.6.2.2 Second harmonic generation

In this section, we will focus our interest on second harmonic generation (SHG). The nonlinear polarization is usually written in a more compact way: the ith component of the induced polarization is determined by the tensor component $\chi_{ijk}^{(2)}$ with the components of the incident field following:

$$P_i^{\omega_3} = \varepsilon_0 \sum_{j,k=1}^{3} \chi_{i,j,k}^{(2)} E_j^{\omega_1} E_k^{\omega_2} \tag{B1.6.4}$$

where i, j, k run from 1 to 3 and stand for x, y, z, respectively, and $\omega_3 = \omega_1 + \omega_2$.

SHG is the limiting case of a three-wave interaction where the two fundamental harmonics have the same frequency ($\omega_1 = \omega_2$). An alternative notation $d_{ijk}^{(2)}$ for the nonlinear tensor is often used; $d_{ijk}^{(2)}$ is related to $\chi_{ijk}^{(2)}$ as follows.

$$d_{ijk}^{(2)} = \tfrac{1}{2}\chi_{ijk}^{(2)}. \tag{B1.6.5}$$

In the interaction process between the two fields, no physical significance can be attached to the exchange of E_j and E_k. Therefore, it is possible to replace the product of subscripts k and j by a single symbol, in order to reduce the number of subscripts in the formula. The Voigt notation is commonly adopted. In this convention, the first indices, x, y, z are replaced by 1, 2, 3 respectively, while the second and third indices are replaced by one index taking the following values:

$$xx = 1 \qquad yy = 2 \qquad zz = 3 \qquad yz = zy = 4 \qquad zx = xz = 5 \qquad xy = yx = 6. \tag{B1.6.6}$$

The contracted $d_{ij}^{(2)}$ tensor has 3×6 components instead of 27 for the $d_{ijk}^{(2)}$. The non-vanishing components of the induced polarization are determined by the form of the tensor and the components of the incident fields. The first condition for a crystal to exhibit a quadratic nonlinear susceptibility (i.e. a non-vanishing $d^{(2)}$), is the lack of an inversion centre in their structure. The form of this tensor (i.e. the non-vanishing components) is linked to the point group symmetry of the crystal. In addition to these symmetry-imposed restrictions on the number of non-vanishing d_{ij} elements, other restrictions arise, since the d_{ij} elements should be invariant under each symmetry operation. This introduces relationships between the components and further reduces the number of independent coefficients. As an example, one of the most popular nonlinear

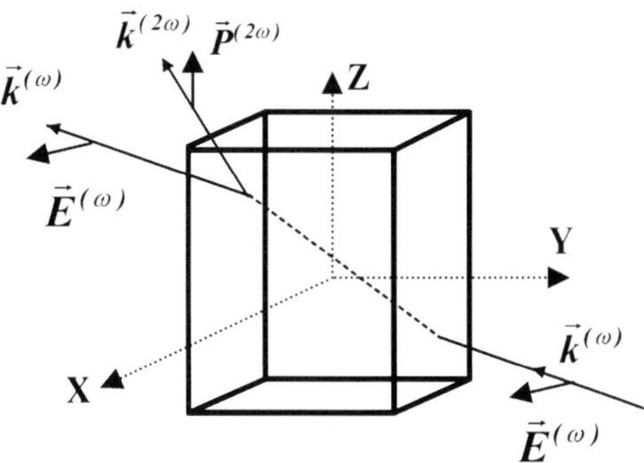

Figure B1.6.1. The effect of nonlinear susceptibility in an uniaxial crystal: an incident wave, propagating in a plane perpendicular to the optic axis, with electric field $\boldsymbol{E}^{(\omega)}(\mathrm{E}_x,\mathrm{E}_y,0)$ induces a polarization, at frequency 2ω, $\boldsymbol{P}^{2(\omega)}(0,0,\mathrm{P}_z^{(2\omega)})$ (i.e. parallel to the optic axis).

crystals, dihydrogen phosphate KH_2PO_4 (KDP) belongs to the $\bar{4}2m$ point group symmetry, then its nonlinear susceptibility is reduced to

$$d_{ij} = \begin{vmatrix} 0 & 0 & 0 & d_{14} & 0 & 0 \\ 0 & 0 & 0 & 0 & d_{25} & 0 \\ 0 & 0 & 0 & 0 & 0 & d_{36} \end{vmatrix}. \tag{B1.6.7}$$

In this case, only a few components of the tensor are non-vanishing and further symmetry considerations imply that $d_{14} = d_{25}$.

From this example and equation (B1.6.4), it is easy to deduce that the direction of the induced nonlinear polarization may be different from those of the applied electromagnetic field. Consider an incident wave propagating in a direction perpendicular to the optic axis $\boldsymbol{E}(E_x, E_y, 0)$. The induced nonlinear polarization will be $\boldsymbol{P}^{(2\omega)}(0, 0, P_z)$ with $P_z^{2\omega} = 2\varepsilon_0 d_{36} E_x^{(\omega)} E_y^{(\omega)}$ (see figure B1.6.1).

B1.6.2.3 Phase matching

The macroscopic SH wave will result from the summation of individual contributions generated by atoms at different positions (x, y, z) on the path of the fundamental beam in the nonlinear medium. Then, the SH wave will have an appreciable intensity if the vectors of the phase velocities of the fundamental wave and the polarization wave are properly matched, i.e. $v_{\mathrm{ph}}^{(\omega)} = v_{\mathrm{ph}}^{(2\omega)}$ with $v_{\mathrm{ph}} = \omega_i / k^{(\omega_i)} = c/n^{(\omega_i)}$. This phase-matching condition can be written as

$$\boldsymbol{\Delta k} = \boldsymbol{k}(\omega_1 + \omega_2) - \boldsymbol{k}(\omega_1) - \boldsymbol{k}(\omega_2) = 0 \tag{B1.6.8}$$

which may be interpreted as momentum conservation for the three photons participating in the mixing process (figure B1.6.2). In the limiting case of SHG, the phase-matching condition reduces to $\boldsymbol{k}(2\omega) = 2\boldsymbol{k}(\omega)$. The maximum exchange of energy between the fundamental and the second harmonic will be obtained for a collinear propagation of the waves, which implies that their phase velocity must be equal.

The phase-matching condition is not easily fulfilled in crystals because of phase velocity dispersion. Let us consider the general case $\Delta(\boldsymbol{k}) \neq 0$ which means $v_{\mathrm{ph}}(\omega) \neq v_{\mathrm{ph}}(2\omega)$. If the incident wave propagates in the z direction in a nonlinear medium, a phase difference between the fundamental wave and the wave generated at 2ω will increase and, after travelling a path l, it will be $\Delta k = |k(2\omega) - k(\omega)|l$. To simplify the analysis, we consider the general case of a low conversion efficiency (undepleted regime). In this case, the

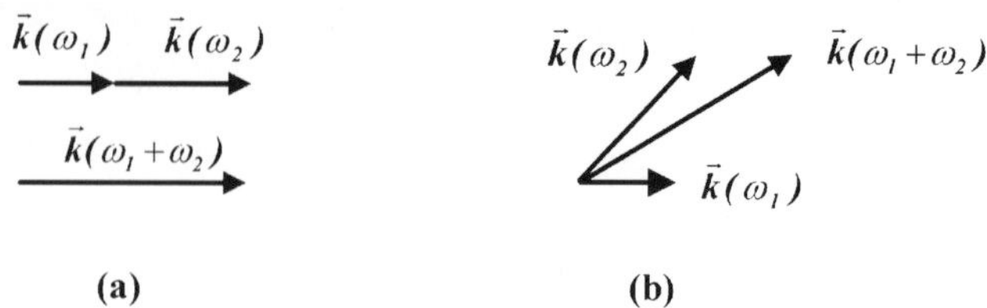

Figure B1.6.2. Phase-matching condition as momentum conservation for (*a*) collinear propagation and (*b*) non-collinear propagation of waves.

amplitude of the incident wave remains constant along the pathlength through the nonlinear crystal. The total amplitude of the SH wave for a crystal length L results from the summation of the microscopic contribution $\mathrm{d}E(2\omega, z)$ from $z = 0$ to $z = L$. The intensity of the SH wave at the output of the crystal is then

$$I(2\omega, L) = I(\omega)^2 \frac{2\omega^2 |d_{\mathrm{eff}}^2| L^2}{n^3 c^3 \varepsilon} \operatorname{sinc}^2(\Delta k L/2) \qquad \text{with } \operatorname{sinc}(x) = \sin(x)/x. \tag{B1.6.9}$$

A maximum in the intensity of the second harmonic is obtained for $\Delta kl/2 = \pi$. Half of the distance between two adjacent peaks of this interference pattern (also called Maker fringes) corresponds to the so-called *coherence length* l_{c}. Using $k^{\omega} = \omega n^{\omega}/c$, l_{c} is given by

$$l_{\mathrm{c}} = \frac{2\pi}{\Delta k} = \frac{\lambda}{4[n^{(2\omega)} - n^{(\omega)}]} \tag{B1.6.10}$$

where λ is the free-space wavelength of the fundamental wave.

The coherence length represents the maximum distance over which the fundamental and the SH waves are in phase and interfere constructively (i.e. the distance along which the second amplitude increases). For $l > l_{\mathrm{c}}$, destructive interference occurs which reduces the SH amplitude. From a practical point of view, the difference $n(2\omega) - n(\omega)$ should be as small as possible to provide a large coherence length compared to the crystal length. This can be achieved by taking advantage of the wave propagation in birefringent crystals and satisfying the condition in equation (B1.6.8).

Let $\boldsymbol{E}$ be the electric field incident on an anisotropic medium—from equation (B1.6.4) the induced polarization $\boldsymbol{P}$ is not necessarily parallel to $\boldsymbol{E}$. The electric displacement vector $\boldsymbol{D}$ is related to $\boldsymbol{E}$ through the dielectric tensor ε_{kl} as follows

$$D_k = \varepsilon_{kl} E_l \tag{B1.6.11}$$

where k, l refer to the Cartesian coordinates (x, y, z) fixed with respect to the crystal axes. An axis transformation is used to diagonalize the tensor ε_{kl}. In the new system, $D_i = \varepsilon_i \boldsymbol{E}_i$ with $i = x, y, z$, the principal dielectric axes. Defining the principal refractive indices n_i such as $n_i^2 = \varepsilon_i$, the equation of a general ellipsoid is then obtained:

$$\frac{x^2}{n_x^2} + \frac{y^2}{n_y^2} + \frac{z^2}{n_z^2} = 1. \tag{B1.6.12}$$

According to equation (B1.6.12), it is possible to determine the refractive index associated with a plane wave propagating along an arbitrary direction in a crystal.

Phase matching in uniaxial crystals

In *uniaxial* crystals, z is taken along the symmetry axis (also referred as the optic axis). The principal plane is defined by the z-axis and the wavevector $\boldsymbol{k}$ of the propagating wave. The *ordinary beam* ‘o’ propagates with

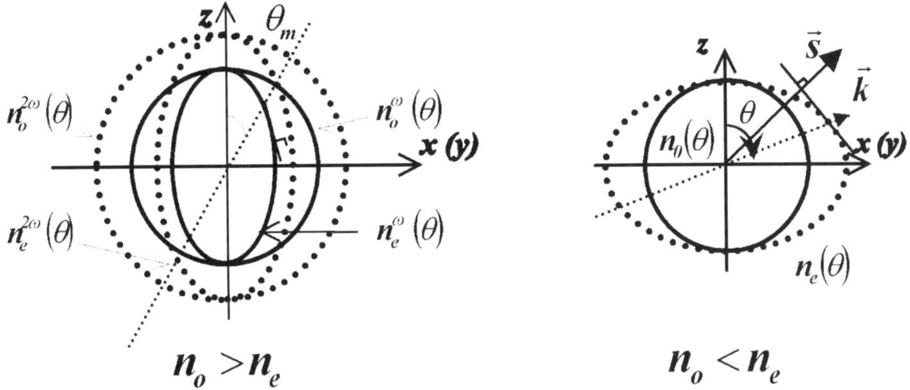

Figure B1.6.3. Uniaxial crystals: the dependence of the ordinary (n_o) and extraordinary (n_e) indices on the angle θ between the wavevector $\boldsymbol{k}$ and the optic axis. Phase matching is for SHG in a negative birefringent crystal ($n_o > n_e$). The illustrated energy and wave propagations are for a positive birefringent crystal ($n_o < n_e$).

a polarization perpendicular to the optic axis, the *extraordinary beam* 'e' is polarized in the principal plane. Thus, equation (B1.6.12) is simplified since $n_x = n_y = n_o$ and $n_z = n_e$: n_o is the ordinary index of refraction and n_e the extraordinary one. By definition, the optic axis is the direction in which the o- and e-rays propagate with the same velocity. For a *positive* uniaxial crystal $n_o < n_e$, while $n_o > n_e$ for *negative* uniaxial crystal (figure B1.6.3). Consider a wave propagating along an arbitrary direction, n_o remains constant whatever the beam polarization and its propagation direction, whereas n_e will depend on θ, the angle between the optic axis and the direction of propagation, following

$$\frac{1}{n_e(\theta)^2} = \frac{\cos^2(\theta)}{n_o^2} + \frac{\sin^2(\theta)}{n_e^2}. \qquad \text{(B1.6.13)}$$

Both n_o and n_e are functions of the frequency. There exists an angle which satisfies the phase-matching condition ($n_e^{2\omega} = n_o^{\omega}$). This *phase-matching angle* θ_m is determined by the intersection of the sphere corresponding to the index surface of the ordinary beam at ω with the index ellipsoid associated to the extraordinary beam at 2ω (see, for example, the case of the negative uniaxial crystal in figure B1.6.3). θ_m is then obtained by solving

$$\frac{1}{(n_o^{\omega})^2} = \frac{\cos^2\theta_m}{(n_o^{2\omega})^2} + \frac{\sin^2\theta_m}{(n_e^{\omega})^2}. \qquad \text{(B1.6.14)}$$

Different types of interaction processes in the harmonic generation can be encountered depending on the two possible orientations for the linear polarization of the fundamental beam. These are summarized in table B1.6.1. A complete calculation of the phase-matching angles in uniaxial crystals is developed in [5].

Phase matching in biaxial crystals

The propagation of light waves in *biaxial* crystals is much more complex. A complete description can be found in [5]. The dependence of the refractive indices with the light propagation direction is described by surfaces in a coordinate system (X, Y, Z) defined as *crystallophysical*, which differs from the crystallographic one (a, b, c). In the (X, Y, Z) system, (XY), (XZ) and (YZ) are defined as principal planes. The crystallophysical axes are obtained in such a way that the considered optic axis always belongs to the XZ principal plane. The

Table B1.6.1. Types of phase matching in uniaxial crystals.

ω_1	ω_2	SH (ω_3)	Crystal	SH type
o	o	e	negative	I
e	e	o	positive	I
o	e	e	negative	II
e	o	e	negative	II
e	e	o	positive	II
e	o	o	positive	II

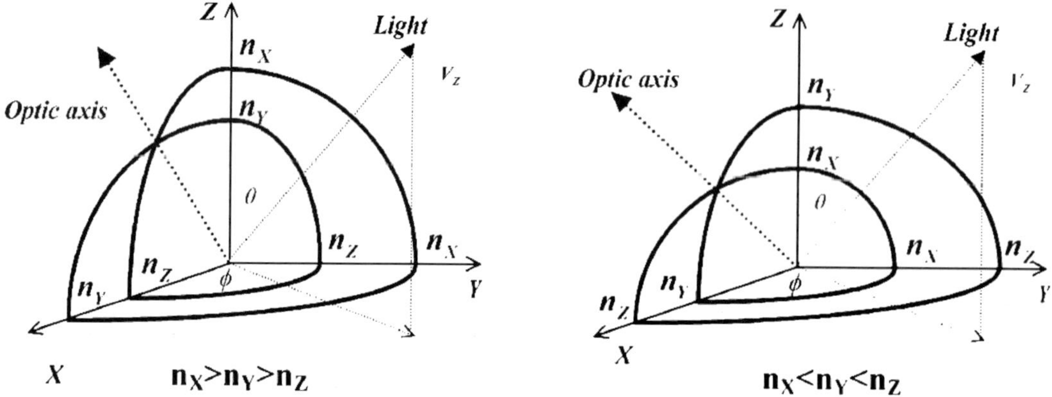

Figure B1.6.4. Indicatrices of refractive indices in biaxial crystals.

optic axis direction is given by the intersection of the ellipse and circle in XZ (figure B1.6.4). The angle V_Z between one of the optic axes and the Z-axis is given by

$$\sin V_Z = \frac{n_Z}{n_Y}\left(\frac{n_Y^2 - n_X^2}{n_Z^2 - n_X^2}\right)^{(1/2)} \quad \text{for } n_X < n_Y < n_Z \tag{B1.6.15}$$

$$\cos V_Z = \frac{n_X}{n_Y}\left(\frac{n_Y^2 - n_Z^2}{n_X^2 - n_Z^2}\right)^{(1/2)} \quad \text{for } n_X > n_Y > n_Z \tag{B1.6.16}$$

Different phase-matching types are possible depending on the direction of the light propagation in the principal planes (table B1.6.2). The corresponding equations for calculating phase-matching angles are detailed in [5]. A systematic review of phase matching in uniaxial and biaxial crystals is developed in [9].

The efficiency of SH conversion is not independent of the azimuthal angle ϕ measured between the x-axis and the projection in the xy-plane of the phase-matching direction. Thus, the effective nonlinear coefficient d_{eff} will be generally expressed as a combination of several d_{ij} components of the nonlinear tensor, θ_{m} and ϕ. As an example, in the case of BBO (a negative uniaxial crystal), phase matching is achieved for a type I interaction (see table B1.6.1) and d_{eff} is given by $d_{\text{ooe}} = d_{31} \sin\theta - d_{22} \cos\theta \sin 3\phi$.

Table B1.6.2. Types of phase matching in biaxial crystals.

	$n_X < n_Y < n_Z$		$n_X > n_Y > n_Z$	
Principal plane	Interaction	Phase matching	Interaction	Phase matching
XY	ooe	I	eeo	I
	eoe	II	oeo	II
	oee	II	eoo	II
YZ	eeo	I	ooe	I
	oeo	II	eoe	II
	eoo	II	oee	II
$ZX(\theta < V_Z)$	ooe	I	eeo	I
	eoe	II	oeo	II
	oee	II	eoo	II
$ZX(\theta > V_Z)$	eeo	I	ooe	I
	oeo	II	eoe	II
	eoo	II	oee	II

Limiting parameters of the doubling efficiency

For an extraordinary wave propagating in an anisotropic crystal, the directions of wave propagation ($\boldsymbol{k}$) and energy propagation ($\boldsymbol{s}$) are generally different. The energy propagation direction is defined by the normal to the tangent at the intersection of ($\boldsymbol{k}$) and the $n_o(\theta)$ curve (figure B1.6.3). Therefore, the angle ρ between $\boldsymbol{k}$ and $\boldsymbol{s}$, (called the *birefringence angle* or, more usually, the *walk-off angle*) is given by

$$\rho(\theta) = \pm\arctan[(n_o/n_e)^2 \tan\theta] \mp \theta \qquad \text{(B1.6.17)}$$

(the upper and lower signs refer to a negative and positive crystal respectively).

A laser beam with an arbitrary linear polarization entering a crystal will be split into two beams which propagate with orthogonal polarizations, i.e. an *ordinary* beam and an *extraordinary* beam. If L is the crystal length, at the output face of the crystal, the two beams will be separated by δ given by

$$\delta = L\tan\rho \qquad \text{(B1.6.18)}$$

with ρ (the *walk-off angle*) the angle between the two beams (figure B1.6.5).

The walk-off effect effectively decreases the interaction length and can, of course, be minimized if the beams are propagating in a direction along which $\boldsymbol{k}$ is parallel to $\boldsymbol{s}$.

However, if the phase matching is achieved for an angle θ_m other than 90° with respect to the optic axis of an uniaxial crystal, there will be double refraction: the incident beam will split into two beams with orthogonal polarizations. The beams will be completely separated after a distance $l_{d_f} = d_f\rho^{-1}$, with d_f the diameter of the fundamental beam. l_{d_f} is called the *aperture length*. This case corresponds to *critical phase matching*. Otherwise, the case $\theta_m = 90°$ corresponds to *non-critical phase matching* because the effect of limiting factors to the conversion efficiency is much smaller than in the previous case. Provided $n_o^{\omega} \sim n_e^{2\omega}$ and $\mathrm{d}(n_o^{\omega} - n_e^{2\omega})/\mathrm{d}T \neq 0$, non-critical phase matching can be achieved by tuning the temperature of the crystal (this is the case of $LiNbO_3$).

In real situations, other factors can affect perfect phase matching such as the divergence and spectral width of the fundamental beam or temperature variations in the nonlinear crystal. These factors are of fundamental importance for the applications in which we are interested in this chapter. Following the

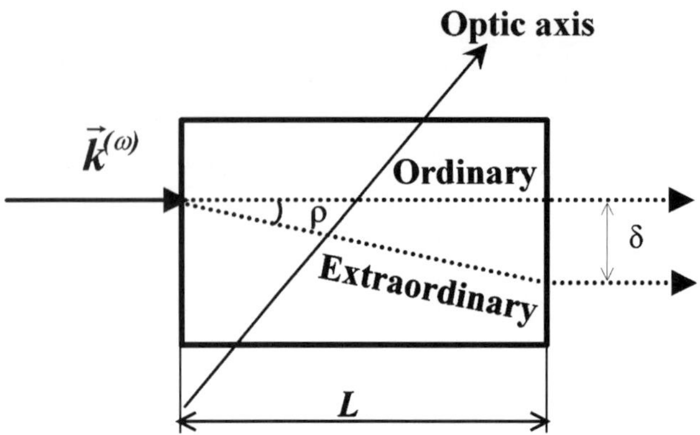

Figure B1.6.5. Walk-off effect in an uniaxial crystal.

analysis of Dmitriev *et al* [5], the dependence of Δk on these different effects can be expressed, in the linear approximation, as

$$\Delta k \simeq \Delta k(0) + \frac{\partial(\Delta k)}{\partial T}\Delta T + \frac{\partial(\Delta k)}{\partial(\delta\theta)}\Delta\theta + \frac{\partial(\Delta k)}{\partial\nu}\Delta\nu \qquad \text{(B1.6.19)}$$

where $\Delta k(0)$ is the mismatch to exact phase matching: the partial derivatives with respect to one argument are taken with the assumption that the other two are constant and $\Delta k(0) = 0$. From equation (B1.6.9), the conversion efficiency $(I_{2\omega}/I_{\omega})$ is reduced to half its value for $\mathrm{sinc}^2(\Delta k L/2)$=0.5 or $\Delta k = 0.886\pi L^{-1}$. Then, the angular $(\Delta\theta)$, thermal (ΔT) and spectral $(\Delta\nu)$ bandwidths can be written as

$$\Delta\theta = 1.772\pi L^{-1}\left[\frac{\partial(\Delta k)}{\partial(\delta\theta)}\right]^{-1}_{\theta=\theta_{\mathrm{pm}}} \qquad \text{(B1.6.20)}$$

$$\Delta T = 1.772\pi L^{-1}\left[\frac{\partial(\Delta k)}{\partial T}\right]^{-1}_{T=T_{\mathrm{pm}}} \qquad \text{(B1.6.21)}$$

$$\Delta\nu = 1.772\pi L^{-1}\left[\frac{\partial(\Delta k)}{\partial\nu}\right]^{-1}_{\nu=\nu_{\mathrm{pm}}} \qquad \text{(B1.6.22)}$$

In equation (B1.6.20), $[\partial(\Delta k)/\partial(\delta\theta)]_{\theta=\theta_{\mathrm{pm}}}$ defines the angular sensitivity (β_θ). Since $\Delta k = (4\pi/\lambda)(n^{2\omega} - n^{\omega})$, the different derivatives in equations (B1.6.20), (B1.6.21) and (B1.6.22) will depend on the dispersion of the refractive indices and on the type of phase matching. If, in the case of non-critical phase matching, the first derivative with respect to θ equals 0 then the second one becomes important, in this case, $\Delta\theta \propto [(\partial^2(\Delta k)/\partial(\delta\theta)^2)^{-1}]^{1/2}$. The equations for calculating $\Delta\theta$, ΔT and $\Delta\nu$ are given in [5] for all types of phase matching and interactions. The relevant parameters for nonlinear crystals of interest in this chapter are summarized in table B1.6.3. Furthermore, frequency doubling with focused Gaussian beams is discussed in [10, 11]. This is the case for *intracavity frequency doubling* and the principal results are given in section B1.6.3.1.

B1.6.3 Self-frequency doubled lasers *versus* intracavity frequency doublers

Optical harmonic generation has been widely developed in order to extend the spectrum of laser sources. In particular, in the field of solid state lasers, considerable efforts have been devoted to harmonic generation of Nd^{3+} lasers for visible and UV laser radiations. Large conversion efficiency requires high power densities not

Table B1.6.3. Frequency doublers: physical properties and tolerance parameters: angular $\Delta\theta$(mrad cm^{-1}), thermal ΔT(deg cm^{-1}), spectral $\Delta\lambda$ (nm cm^{-1}).

		$\lambda_{fund.}$ (μm)	θ_{PM} (deg)	d_{eff} (pm V^{-1})	n^{ω}	Transparency (μm)	$\rho^{(o)}$	Tolerance factors $\Delta\theta$	ΔT	$\Delta\lambda$	
Banana	ooe	1.06	74	13.2[a]		0.37–5	—	2.4	0.8	0.3	[5]
$(mm2)$[c]				14.6[b]							
KTP	eoe	1.06	24.3	3.18	1.74	0.35–4.5	0.26	25	25	0.56	[8]
$(mm2)$[d]											
BBO	ooe	1.06	22.8	1.94	1.65	0.198–2.6	3.19	0.5	55	0.66	[8]
(3m)[e]		0.946		2.08	1.658		3.5	0.42	23	0.43	[12]
$LiNbO_3$[f]	ooe	1.06	90	4.7	2.23	0.33–5.5	0	2.3	0.6	0.23	[8]
(3m)[e]								3.1	1.1	0.3	[5]
$LiIO_3$	ooe	1.06	30.2	4.1	1.85	0.3–6	4.26	0.7	—	—	[8]
(6)[e]		0.946	34.3	4.01	1.861		4.6	0.28	20	0.16	[12]
LBO	ooe	0.946	19.5	0.92	1.6077	0.16–2.6	0.7	2.29	8	0.63	[12]
$(mm2)$[e]											
$KNbO_3$	eeo	0.946	30	16.4		0.4–4.5	—	—	—	—	[13]
$(mm2)$[c]											

[a] along Y
[b] along X
[c] negative biaxial
[d] positive biaxial
[e] negative uniaxial
[f] Temperature tuning at 107 °C

available with cw sources. For this reason, Q-switched or mode-locked lasers delivering high peak powers were used as fundamental radiation, the harmonic generator being external to the cavity. In this way, visible and UV sources were available in the nanosecond and picosecond regimes. In 1968, the possibility of using the high intracavity power for efficient frequency conversion and a continuous 532 nm laser source (350 mW) was demonstrated with a nonlinear crystal ($Ba_2NaNb_5O_{15}$, also called 'Banana') inside the cavity of a Nd:YAG laser delivering 1 W at 1.064 μm [14]. Furthermore, with the recent appearance of semiconductor laser diodes (LD), LD-pumped cw lasers with intracavity SHG is a promising approach for the development of compact visible laser sources. More recently, another promising approach for compact all-solid-state lasers in the blue–green wavelength range was developed with the use of a gain medium which simultaneously acts as harmonic generator (self-frequency doubling). In a first step, we will focus our interest on intracavity frequency doubling, then on self-frequency doubling. Finally, we will compare the advantages and drawbacks of both systems.

B1.6.3.1 Intracavity frequency doubling

Nd^{3+}-crystals as gain medium

Up to now, intracavity frequency doubling has been developed with Nd^{3+}-doped crystals as the gain medium, most of them being LD pumped (see section B1.6.5). In addition to the well-known 1.064 μm laser emission (see chapter B1.3) due to the $^4F_{3/2} \longrightarrow {}^4I_{11/2}$ transition, the Nd^{3+} ion exhibits other laser transitions at

Table B1.6.4. Lasers with intracavity doubling frequency (A = three-mirror folded, B = standing-wave, C = ring, D = concentric, H = hemispherical (plane-concave), P = plane parallel, M = microchip).

Laser crystal	Length (mm)	Frequency doubler	Length (mm)	Cavity	λ_{SH} (nm)	Other data	Ref.
YAG		Banana $T = 83\,°C$	2.8	~H ($L = 450$ mm)	532		[14]
YAG	1	$LiIO_3$		H	473	$\omega_l = 40\ \mu m$	[17]
YAG[a]	3	$LiIO_3$	4	A ($L < 70$ mm)	473	$P_{th}^{(SH)} = 10$ mW	[12]
YAG[a]	3	BBO	4	A	473		[12]
YAG[a]	3	LBO	4	A	473		[12]
YAG (1.1 at.%)	10	KTP	15	C	532	$\omega_p = 290\ \mu m$ $\omega_l = 90\ \mu m$	[24]
YAG (0.6 at.%)	115	KTP	5	A ($L = 560$ mm)[b] EO Q-switch ($F = 20$ kHz, $\tau = 500$ ns)	532	Beam quality (Fund.) Function of P_{pump} $0.92 < \frac{M_x^2}{M_y^2} < 1.06$	[25]
YAG	8	KTP	5	A ($L = 330$ mm)	532		[26]
YAG	1	$KNbO_3$	3.7	H[c]	473	$\omega_l = 40\ \mu m$	[13]
YAG (1.1 at.%)	1	$KNbO_3$	1.5		473	$\omega_0 = 45\ \mu m$	[27]
YAG	3	$KNbO_3$	6.5	B	473	$\frac{\omega_p}{\omega_l} = 0.5$	[28]
YVO_4 (1 at.%)	0.5	LBO	2	M	671		[29]
YVO_4 (1.1 at.%)	1	KTP	5	D ($L = 21$ mm)	532	$M_x^2 = 1.2$ $M_y^2 = 1.3$	[30]
YVO_4 (1.1 at.%)	1	KTP	5	H ($L = 21$ mm)[c] AO Q-switch ($F = 100$ Hz)	532	$P_{th}^{(SH)} = 100$ mW	[31]
YVO_4 (3 at.%)	0.5	KTP	2	P			[32]
YVO_4 (1 at.%)	6	KTP	5	C ($L = 330$ mm)	532		[26]
YVO_4 (1 at.%)	1	KTP	5	H ($L = 100$ mm)[c]	532	Beam quality (?) $M^2(\parallel \mathbf{c}_{YVO_4}) = 1.1$ $M^2(\perp \mathbf{c}_{YVO_4}) = 1.2$	[33]
YVO_4 (1 at.%)	1	KTP	10	P ($L = 7$ mm) Q-switch[d]	532		[34]
S-FAP	3	KTP	3	H[c] ($L = 45$ mm)	529.5		[35]
LSB	1.2	KTP	1	P ($L = 3$ mm)	531		[36]
YLF (0.5 at.%)	15	LBO $T = 162.6°$ C		C	526.5	$TEM_{00} M^2 < 1.2$ $\omega_c = 60\mu m$	[37]
YAP	77	KTP	10	C ($L = 140$ cm)	540	$\frac{\omega_0^{inKTP}}{\omega_l^{inYAP}} = 0.061$	[38]

[a] Composite laser rod: 3 mm long Nd:YAG with two diffusion bonded 3 mm long undoped cap ends.
[b] Side-pumped.
[c] Front mirror = coated face of the crystal.
[d] Passive Q-switch with Cr^{4+}:YAG ($L = 1.5$ mm).

1.32 μm ($^4F_{3/2} \longrightarrow {}^4I_{13/2}$) and 946 nm ($^4F_{3/2} \longrightarrow {}^4I_{9/2}$) which can be used to achieve SHG for red and blue radiations, respectively. The fundamental wavelength is provided by Nd^{3+} emission in crystals commonly used for infrared laser sources: $Y_3Al_5O_{12}$ (YAG), YVO_4, $YAlO_3$ (YAP), $LiYF_4$ (YLF). Intracavity frequency doubling has also been achieved in $Sr_5(PO_4)_3F$ (S-FAP) and $LaSc_3(BO_3)_4$ (see table B1.6.4 and table B1.6.8).

Type of nonlinear crystals

The most important criterion for selecting the nonlinear crystal is the dependence of the effective nonlinear coefficient on the fundamental wavelength. As an example, KTP, widely used as a frequency-doubling crystal for the 1.064 μm laser transition, cannot be used for frequency doubling the 946 nm radiation since it exhibits a high nonlinearity at 1.064 μm but a low d_{eff} coefficient at 946 nm. In this case, $KNbO_3$, $LiIO_3$ or β-BaB_2O_4 (BBO) or LiB_3O_5 (LBO) are used as nonlinear crystals for harmonic generation in the blue [12, 15]. Other factors for the selection of a particular crystal have also to be considered such as spectral, temperature and, angular acceptance bandwidths and walk-off angle (table B1.6.3). For example, $KNbO_3$ exhibits a high nonlinearity and a possibility of non-critical phase matching but severe drawbacks limit its use : very small temperature and spectral acceptance bandwidths, photorefractivity leading to optical damage. Furthermore, the possibility of domain reversal has been reported particularly for non-critical phase matching where the crystal has to be heated up to 180 °C [12]. BBO and $LiIO_3$ exhibit a large temperature bandwidth, then temperature control is not necessary. $LiIO_3$ has the highest nonlinear coefficient; however, it has also the smallest spectral acceptance bandwidth. The small angular acceptance bandwidth and the large walk-off angle of these crystals limit their use. LBO is a promising candidate with respect to walk-off angle and spectral and angular acceptance bandwidths even if its nonlinear coefficient is less than that of BBO and $LiIO_3$.

Output power of the second harmonic

For intracavity frequency doubling, the nonlinear crystal is placed inside the cavity. To optimize the output power of the second harmonic, the output mirror of the cavity should reflect 100% of the fundamental radiation and be totally transparent at the SH wavelength. Therefore, in addition to the linear loss of the cavity due to absorption scattering, imperfect mirror coatings and so on, one must consider the nonlinear loss due to the SHG, since in this case, a non-negligible fraction of the power at ω is converted. The detailed theory of internal optical SHG is described at length in [16]. We will only briefly give the most important results. The pump and the laser beams are assumed to have a Gaussian profile and be collinear in the crystal. If the laser crystal length L is much shorter than the distance in which the beam section area doubles relative to its value at the waist, the waist radius of the pump beam is taken to be approximately constant within the crystal. The effect of the nonlinear crystal in the cavity is taken into account by introducing a power-dependent loss. Then, in steady-state conditions and for small-signal gain G_0, the round-trip saturated gain of the laser equals the loss of the cavity (see chapters A1–A2):

$$\frac{2G_0 l}{1+2SI} = \beta_{\mathrm{L}} + KI \tag{B1.6.23}$$

where the term on the right-hand side corresponds to the gain in the cavity, the term on the left to the losses (β_{L} being associated to the total round-trip linear loss due to the different components of the cavity, KI the round-trip nonlinear loss due to frequency conversion by the nonlinear crystal in the cavity). The factor 2 takes account of the saturation in the forward and backward directions in the gain medium of length l, S is the saturation term given by

$$S = \frac{1}{I_{\mathrm{sat}}} = \frac{\sigma_{\mathrm{em}}\tau}{\hbar\omega} \tag{B1.6.24}$$

where σ_{em} is the emission cross section, τ the emission lifetime and $\hbar\omega$ the laser frequency. The round-trip nonlinear loss of the cavity, which is the sum of the SHG power in the forward and backward directions, is given by [16]

$$P^{(2\omega)} = K P_c^{(\omega)} \frac{P_c^{(\omega)}}{\pi \omega_c^2} \tag{B1.6.25}$$

with K defined as the nonlinear coupling term. For simplicity, the direction of propagation is assumed to be along the principal axis, the effects of double refraction are then neglected. With this hypothesis, K is expressed as (MKS units):

$$K = \kappa \operatorname{sinc}^2\left(\frac{\Delta k L_n}{2}\right) \qquad \text{with } \kappa = 4\frac{\omega_c^2}{\omega_n^2}\left(\frac{\mu_0}{\varepsilon_0}\right)^{1/2} \frac{\omega_1^2 d_{eff}^2 L_n^2}{n_1^2 n_2 c^2} \tag{B1.6.26}$$

where ω_c and ω_n are the beam waist in the cavity and in the nonlinear crystal respectively, ω_1 is the radian frequency of the fundamental, $n_1 = n^\omega$ and $n_2 = n^{2\omega}$, d_{eff} is the relevant nonlinear coefficient and L_n is the length of the nonlinear crystal.

From equation (B1.6.25), the conversion efficiency ($P^{(2\omega)}/P_c^{(\omega)}$) depends on the nonlinear coupling term and the fundamental power density in the nonlinear crystal. The nonlinear coupling term includes the length of the nonlinear crystal and the phase mismatch. To optimize the cavity, the thermal lens effect (see chapters A2–A4) which appears at high power density should be taken into account especially with YAG and YVO_4 crystals.

When the conversion efficiency exceeds 20%, the depletion of the fundamental power should be taken into account. In the case of depleted input, equation (B1.6.25) should be replaced by [1]

$$\frac{P^{(2\omega)}}{P_c^{(\omega)}} = \tanh^2\left[\kappa^{1/2}\left(\frac{P_c^{(\omega)}}{\pi\omega_c^2}\right)^{1/2} \operatorname{sinc}(\Delta k l/2)\right]. \tag{B1.6.27}$$

Modelling the behaviour of a solid state laser with reabsorption as an additional loss: example of Nd^{3+} 946 nm laser emission (see section B1.3)

In practice, the output harmonic power can be lower than the expected value. This is explained by other processes such as reabsorption which introduce additional losses. Reabsorption is especially important for low gain lasers of interest for intracavity frequency doubling. This is the case for the 946 nm laser transition in which the effective emission cross section is lower by about one order of magnitude than that of 1.06 μm and corresponds to a quasi-four-level system. The lower laser level is the upper crystal field component of the $^4I_{9/2}$ ground state multiplet. Then, due to the small energy gap (for example, 857 cm^{-1} in YAG), there is a significant reabsorption loss owing to the population of this state at room temperature [17].

A room-temperature cw operation in a short gain medium with a low threshold can be achieved by using end pumping which allows a high inversion efficiency in a small volume that helps to overcome the reabsorption loss. The detailed modelling of the behaviour of a solid state laser with reabsorption loss is reported in [18] providing a comprehensive understanding of mode-overlap effects in longitudinally pumped lasers. In particular, the effects of reabsorption loss on the threshold and on the slope are analyzed in detail. The threshold as expressed in terms of incident pump power is given by [17, 18]

$$P_{th,inc} = \frac{\pi h \nu_p (\omega_c^2 + \omega_p^2)}{4\sigma f \tau (1 - \exp(-\alpha_p l))}(L_c + T + 2N_1^0 \sigma l) \tag{B1.6.28}$$

where σ is the gain cross section, N_1^0 the equilibrium population density of the lower laser level, ω_c the laser beam waist in the cavity, ω_p the pump beam radius, ν_p the pump frequency, τ the upper state lifetime, α_p

the absorption coefficient at the pump wavelength , l the laser crystal length, L_c the fixed cavity loss, T the output coupling and f the fraction of the total $^4F_{3/2}$ population that stays in the lower crystal field component of the $^4F_{3/2}$ multiplet.

From equation (B1.6.28), reabsorption is just like an additional loss leading to an increase in the threshold. It can also be deduced from equation (B1.6.28) that making the pump-beam waist ω_p and laser-beam waist ω_c as small as possible will reduce the threshold. Furthermore, it follows from equation (B1.6.28) that the optimum crystal length l_0 results from a compromise between large absorption of the pump and minimum reabsorption of the laser emission. The length of the laser crystal that minimizes the threshold can be calculated by differentiating equation (B1.6.28) providing that the other factors are known:

$$\alpha_p \exp(-\alpha_p l_0)\left(\frac{2\sigma N_1^0}{\alpha_p} + L + T - 2\sigma N_1^0 l_0\right) - 2\sigma N_1^0 = 0. \tag{B1.6.29}$$

As an example (see section B1.3), an optimum length l_0 equal to 1.2 mm is obtained for a 1.1% doped Nd:YAG ($N_1^0 = 1.1 \times 10^{18}$ ions cm^{-3}) and pumping by a dye laser (α_p=0.95 cm^{-1}), $\sigma = 4 \times 10^{-20}$ cm^2, assuming $L = 0.001$ and $T = 0.0085$ [17].

The key parameter for the internal slope efficiency (s_{int}) is $r_{waist} = \omega_p/\omega_c$ for a laser with reabsorption loss. There is an optimum value r^o_{waist} which maximizes the slope efficiency. r^o_{waist} depends on both the internal laser power and the reabsorption loss, i.e. it will increase as the reabsorption loss or the internal power increases. Anyway, for an appropriate value of r_{waist} and large internal laser powers, the reabsorption loss is saturated and the slope efficiency is comparable with that of a laser without reabsorption loss. Another parameter for the choice of the output coupler of the cavity is the external slope efficiency defined as [18]

$$\frac{dP_{out}}{dP_p} = \frac{T}{T+L}\frac{\nu_c}{\nu_p}\eta_a s_{int} \tag{B1.6.30}$$

where η_a is the efficiency with which incident pump photons are absorbed (corresponding to $[1-\exp(-\alpha_p l)]$ in equation (B1.6.28)), s_{int} is the efficiency with which absorbed pump photons are converted to laser photons, ν_c/ν_p is the quantum defect associated to this conversion, T is the output coupler transmittance, L the total optical round-trip loss of the cavity. Then, $T/(T+L)$ represents the fraction of laser photons that are emitted through the output coupler (see chapters A1–A2).

B1.6.3.2 Designs

Optimization of the diode end-pumped lasers (see chapter A2)

Several models have been proposed to optimize the design of diode end-pumped lasers. Simple design criteria were given for longitudinally pumped lasers operating in TEM_{00}, using a circular or elliptical Gaussian beam to simulate the laser diode beam distribution [19]. An analytical model has been developed for fibre-coupled laser diode end-pumped lasers, by including the effect of pump-beam quality into the mode overlap integrals [20, 21]. From this model, a simple procedure to optimize the design of the laser resonator and the optical coupling system has been derived. In another approach [22], the pump beam coming from the optical fibre is considered as Gaussian and its propagation is described by its M^2 factor [23]. This model allows a straightforward procedure to design the optimum laser resonator and the coupling optics. These models are briefly presented in section B1.6.5.

Optimization of second harmonic generation: conditions to achieve an efficient laser source

Another problem to be solved for intracavity doubling is the optimization of SHG. As we have previously noted, the walk-off effect reduces the efficiency of the frequency conversion. Taking this effect into account

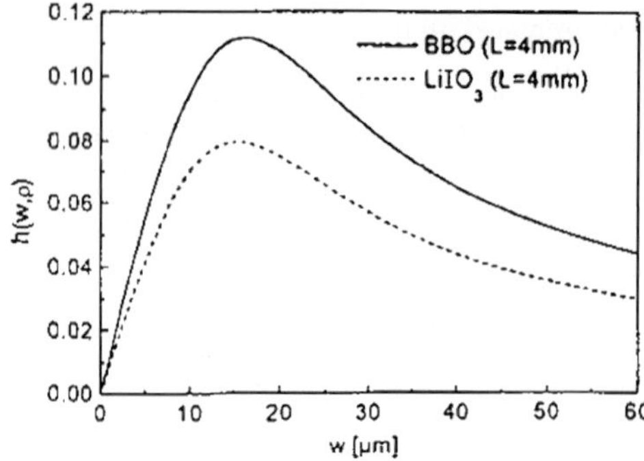

Figure B1.6.6. Dependence of the nonlinear coupling function h on the fundamental beam waist for $LiIO_3$ and BBO (length of both crystals $L = 4$ mm) (after [12]).

[11], and assuming a focused Gaussian beam, the nonlinear coupling term K is then given by

$$K = \frac{2\omega^2 d_{\text{eff}}^2 L_n h k_\omega}{\epsilon_0 c^3 \pi_\omega^3} \tag{B1.6.31}$$

where h is a nonlinear function depending on the walk-off angle ρ, the length of the nonlinear crystal (L_n), the fundamental laser beam waist (ω_c) and the fundamental wavevector k_ω inside this crystal. In the case of non-critical phase matching, $h = 1$. For non-critical phase matching, if the fundamental beam waist is such as $\lambda/\rho\pi n_\omega < \omega_c < \rho L_n/(24)^{1/2}$ then h is expressed as

$$h = \frac{\sqrt{\pi}}{2B\sqrt{\xi}} \arctan(\xi) \tag{B1.6.32}$$

with B the walk-off parameter and ξ the focusing parameter given by:

$$B = \rho/2\sqrt{\frac{\pi n L_n}{2\lambda}} \qquad \text{and} \qquad \xi = \frac{L_n \lambda}{2\pi\omega_0^2 n_\omega}. \tag{B1.6.33}$$

An example of the analytical solution of h for BBO and $LiIO_3$ crystals [12] is given in figure B1.6.6.

In conclusion, to achieve an efficient visible laser source by intracavity frequency doubling, the laser crystal should have a large emission cross section, a long emission lifetime, a low passive loss, an effective absorption bandwidth (important in the case of diode pumping) and an optimized length to reduce the reabsorption loss, if any. The nonlinear crystal should obviously have a large nonlinear coefficient but a small walk-off angle. Visible laser sources with intracavity frequency doubling have been realized by optimizing all the parameters important to achieve a large SH output power. Some examples are given in table B1.6.4 and table B1.6.8 (in section B1.6.6). As shown in table B1.6.8, the best optical conversion efficiency is obtained for the YAG-Banana system due to the large nonlinear coefficient of this frequency doubling crystal compared to the others (table B1.6.3). However, Banana has a poor optical quality and a low damage threshold which prevents its commercial use. The absorption cross section of the Nd^{3+} $^4I_{15/2} \rightarrow {}^4F_{3/2}$ transition is lower in YAG than in YVO_4 (see chapter B1.3). In addition, the absorption bandwidth is larger in YVO_4, thus this material is more appropriate to realize miniaturized devices such as diode-pumped microchips. Different designs and pumping devices are detailed in section B1.6.5.

B1.6.3.3 Self-frequency doubled lasers

There is another approach to realizing blue–green or red, compact, solid state laser sources. It is possible to combine the laser properties and the quadratic nonlinear characteristics into a single element. Self-frequency doubling is an attractive alternative to intracavity doubling. Though the first self-frequency-doubled laser was demonstrated in Tm^{3+}:$LiNbO_3$ [39], most of the systems have been developed with Nd^{3+}- or Yb^{3+}-doped crystals. Except for the evident interest of Nd^{3+} ions, the recent development of strained InGaAs diode lasers emitting around 980 nm has given renewed interest to Yb^{3+}-doped materials for compact solid state laser devices.

B1.6.3.4 Nature of crystals

Requirements in finding crystals

If self-doubling appears more simple than intracavity doubling, the problem lies in finding a crystal satisfying several requirements. These crystals should have a large nonlinearity, a broad transparency range and a high optical damage threshold. Large crystals of good optical and mechanical quality are required. The structure should offer suitable sites for rare earth ions (RE) and accept high RE concentrations. Furthermore, the doped laser crystal should exhibit good laser performance in the phase-matching orientation. As for intracavity doubling, other parameters such as walk-off, spectral, thermal and angular acceptances are of fundamental importance.

Up to now, only a few crystals have simultaneously exhibited good laser properties and quadratic nonlinear susceptibility. Self-frequency doubling has been demonstrated in MgO:$LiNbO_3$ [40–44] periodically poled (PP) [45], $LiTaO_3$ (PP) [46], $BaCaBO_3F$ [47], $YAl(BO_3)_4$ (YAB) [44, 48–50, 52–56], $YCa_4O(BO_3)_3$ (YCOB) [57–59], $GdCa_4O(BO_3)_3$ (GdCOB) [60–63], $Gd_2(MoO_4)_3$ [66], $LaBGeO_5$ [44, 67] and $Ce_{1-x}Gd_xSc_3(BO_3)_4$ (CSB) [68].

B1.6.3.5 Nd^{3+} and Yb^{3+} uses

A visible laser radiation is obtained by self-doubling the infrared laser emission provided by Nd^{3+} or Yb^{3+}. If, in both cases, the laser emission occurs in a quite similar spectral range, the optical properties of Nd^{3+} and Yb^{3+} are quite different. The energy levels of the $4f^3$ configuration of Nd^{3+} ions extend from 0 to, at least, 40 000 cm^{-1} while the level scheme of the $4f^{13}$ configuration of Yb^{3+} ions is very simple: only one excited state ($^2F_{5/2}$) $\simeq$10 000 cm^{-1} above the ground-state multiplet ($^2F_{7/2}$). So, the well-known Nd^{3+}1.06 μm laser emission is obtained in a four-level system while for Yb^{3+}, the infrared laser emission is obtained in a three- or quasi-four-level scheme depending on the splitting of the ground-state multiplet. This provides advantages and drawbacks which we will discuss in the following sections.

Nd^{3+} ion spectroscopy

First of all, in both cases, the three parameters (absorption and stimulated emission cross sections, σ_{abs} and σ_{em}, and the emission lifetime, τ) related to the potential performance of the laser material are evaluated as usual (see chapters B1.3 and B1.4). Moreover, the laser properties of the crystal should be investigated in the appropriate direction corresponding to the phase-matching conditions. σ_{abs} and σ_{em} are evaluated from polarized absorption and emission spectra and application of the Judd–Ofelt theory (in the case of Nd^{3+} ions) or using the Füchtbauer–Ladenburg equation (FL) or the method of reciprocity (for Yb^{3+}). So, absorption and emission spectra are recorded in σ ($\boldsymbol{E} \perp \boldsymbol{c}$, $\boldsymbol{H} \parallel \boldsymbol{c}$) and π ($\boldsymbol{E} \parallel \boldsymbol{c}$) polarizations for uniaxial crystals, $\boldsymbol{c}$ being the optic axis. For biaxial crystals, the spectra are recorded with absorption or emission light polarized along each of the three principal axes of the optical indicatrix, i.e. X, Y, Z.

Besides the assessment of laser performance, the study of the absorption spectra is an important step to determine the best pumping spectral range in order to reduce the energy difference between the pump and stimulated emission photons (i.e. a small quantum defect). A small quantum defect allows low thermal loading of the crystal to be achieved. Typically, the relevant absorption extend from 700 to 800 nm for the ($^4I_{9/2} \rightarrow {}^4F_{3/2}$) Nd^{3+} transition and from 900 to 1100 nm for the ($^2F_{7/2} \rightarrow {}^2F_{5/2}$) Yb^{3+} absorption. In both cases, pumping can be achieved using infrared laser diodes. However, due to additional 4f energy levels, Nd^{3+} ions suffer several drawbacks compared to Yb^{3+}. The pumping efficiency can be reduced by excited state absorption (ESA) of the pump from the $^4F_{3/2}$ manifold to upper levels. However, it is noteworthy that resonant absorption of the SH can occur from the ground state to upper levels thus lowering the self-doubling efficiency. Furthermore, the broadening of the spectral features due to the interaction between the 4f electrons and the vibrations of the host is more pronounced for Yb^{3+} due to a larger electron–phonon coupling than for other rare earths. Then, Yb^{3+} exhibits a large absorption bandwidth (for example, FWHM= 19 nm in $BaCaBO_3F$ [47]) compared to Nd^{3+} (FWHM = 5.5 nm in CSB [68]). Yb^{3+}-doped crystals appear to be more suitable for diode pumping in contrast to their Nd^{3+} counterparts; due to their wide absorption bandwidth they offer less sensitivity to the laser diode output wavelength affected by temperature changes. Yb^{3+} also exhibits a large emission bandwidth which opens the possibility of tunability and generation of ultra-short pulses.

In this section, we only briefly discuss the determination of the laser characteristics. The reader is invited to read chapters B1.3 and B1.4 for more detailed treatments.

The absolute radiative properties of Nd^{3+}-doped materials can be computed from simple and accurate absorbance measurements. The line strength $S_q^{\text{meas}}(J \rightarrow J')$ of an electric-dipole transition is related to its optical density through the following relation:

$$S_q^{\text{meas}}(J \rightarrow J') = \frac{3ch}{8\pi^3 e^2} \frac{(2J+1)}{N} \frac{9n}{(n^2+2)^2} \frac{1}{\bar{\lambda}} \int_{J \rightarrow J'} k_q(\lambda)\, \mathrm{d}\lambda \tag{B1.6.34}$$

where e is the electron charge, c the light velocity, N the Nd^{3+} concentration, $\bar{\lambda}$ the mean absorption wavelength of the transition, n the refractive index of the crystal at $\bar{\lambda}$, J (J') the total momentum quantum number of the initial (final) multiplet and $k_q(\lambda)$ the absorption coefficient at λ, q stands for the electric field polarization with respect to the optic axis or of the optical indicatrix X, Y or Z. According to Judd–Ofelt theory [69, 70], the line strength of an electric dipole transition between the initial manifold, $|(S, L)J\rangle$, and the final $|(S', L')J'\rangle$ state can be written in the form

$$S_q^{\text{meas}}(|(S, L)J\rangle \rightarrow |(S', L')J'\rangle) = \sum_{t=2,4,6} \Omega_t \, |\langle (S, L)J \parallel U^{(t)} \parallel (S', L')J' \rangle|^2 \tag{B1.6.35}$$

where $\langle \parallel U^{(t)} \parallel \rangle$ are the doubly reduced unit tensor operators calculated in the intermediate coupling approximation. The numerical values of the unit tensor operators have been calculated by Carnall *et al* [71]. The intensity parameters Ω_t (with $t = 2, 4, 6$) include the odd-symmetry crystal field terms, radial integrals and perturbation denominators. These phenomenological parameters, determined by least-square fitting between experimental and calculated line strengths, allow the different characteristics required to assess the potential laser performance of the doped crystal to be calculated.

- The probability for a transition from initial J' manifold to $\bar{J}$ final manifold is given by

$$A[(S', L')J'; (\bar{S}, \bar{L})\bar{J}] = \frac{64\pi^4 e^2}{3h(2J'+1)\bar{\lambda}^3} \frac{n(n^2+2)^2}{9} \sum_{t=2,4,6} \Omega_t |\langle (S', L')J' \parallel U^{(t)} \parallel (\bar{S}, \bar{L})\bar{J} \rangle|^2. \tag{B1.6.36}$$

- The fluorescence branching ratios for transitions originating from an initial $|(S', L')J'\rangle$ manifold are:

$$\beta[(S', L')J'; (\bar{S}, \bar{L})\bar{J}] = \frac{A[(S', L')J'; (\bar{S}, \bar{L})\bar{J}]}{\sum_{(\bar{S},\bar{L})\bar{J}} A[(S', L')J'; ((\bar{S}, \bar{L})\bar{J}]} \tag{B1.6.37}$$

  where the summation is over all the possible terminal states.
- The spontaneous radiative lifetime

$$\tau_{\text{rad}} = \left(\sum_{\bar{S},\bar{L},\bar{J}} A[(S', L')J'; (\bar{S}, \bar{L})\bar{J}] \right)^{-1}. \tag{B1.6.38}$$

- The radiative quantum efficiency of the $|(S', L')J'\rangle$ manifold is

$$\eta = \frac{\tau_{\text{f}}}{\tau_{\text{rad}}} \tag{B1.6.39}$$

  where τ_{f} is the measured fluorescence lifetime. η is usually less than 1 since τ_{f} is reduced by multi-phonon and ion–ion relaxation processes. Due to the small energy gap ($\Delta\epsilon \sim 4800\ \text{cm}^{-1}$) between the ${}^4F_{3/2}$ emitting level and the next lower ${}^4I_{15/2}$ multiplet, several phonons $p = \Delta\epsilon/\hbar\omega$ ($\hbar\omega$ being the effective phonon frequency) are required to bridge the gap, then the non radiative de-excitation by multi-phonon relaxation has a high probability. Thus, the lifetime is reduced, this obviously decreases the quantum efficiency. As examples, $\eta = 0.26$ for Nd:YAB as a result of a high multi-phonon relaxation probability involving three effective phonons of $1346\ \text{cm}^{-1}$ [73], $\eta = 0.15$ in GdCOB ($\hbar\omega = 1472\ \text{cm}^{-1}$). These are rather low compared to that in YAG (0.88-1) or in YAP (0.88)($\hbar\omega_{\text{YAG}} = 857\ \text{cm}^{-1}$, $\hbar\omega_{\text{YAP}} = 750\ \text{cm}^{-1}$). At high concentration, Nd–Nd interaction leads to concentration quenching through cross-relaxation, further reducing τ_{f}.
- Finally, the stimulated emission cross section can be calculated from the radiative rate and the lineshape of the fluorescence using the Füchtbauer–Ladenburg (FL) relation (see chapter A1):

$$\sigma_q^{(J\to J')}(\nu) = \frac{\lambda^2 g_q(\nu)}{8\pi n^2 \tau_{\text{rad}}} \tag{B1.6.40}$$

  where $g_q(\nu)$ is the normalized lineshape function obtained in polarization q. For anisotropic crystals, $g_q(\nu)$ is thus given by

$$g_q(\nu) = \frac{I_q(\nu)}{\frac{1}{3}\sum_q I_q} \tag{B1.6.41}$$

  where I_q is the integrated fluorescence intensity recorded with electric polarization along the optical indicatrix X, Y or Z.

However, it should be noted that Judd–Ofelt (JO) theory assumes that all the Stark levels of the initial manifold are equally populated. This assumption is usually not valid for temperatures near 300 K except in the case of low crystal field splitting [74]. In equation (B1.6.40), in addition to this approximation, it is assumed that all the transitions have the same strength regardless of the Stark components involved. However, the values determined this way are roughly not so far from the real ones.

Evaluation of Yb^{3+} ion emission parameters

The emission cross section of Yb^{3+} ions can be determined using either the method of reciprocity or the FL relation [75]. The reciprocity method is used if absolute polarized absorption cross sections, the energy and the degeneracy of the Stark components in the ground and excited manifolds are known, otherwise FL relation (equation (B1.6.40)) is employed. Since only one absorption transition is available within the $4f^{13}$ configuration, the radiative probability cannot be deduced from absorption and JO theory. However, due to the simple energy level scheme, the shortening of the lifetime is less probable in case of Yb^{3+} ions since up-conversion by energy transfer or a cross-relaxation process is unlikely to occur. Furthermore, due to the large energy gap between the excited manifold and the ground state ($\sim$10 000 cm^{-1}), a great number of phonons is required to bridge the gap even for materials which have a large phonon cut-off frequency, the multi-phonon process has then a low probability. Thus, in most materials, it is assumed that $\tau_{rad} = \tau_f$. Unlike Nd^{3+}, Yb^{3+} ions do not suffer from concentration quenching and the multi-phonon relaxation process that adds to the thermal loading of the crystal. In addition, the small energy difference between the pump and the emitted photons results in negligible heat generation, about four times less than that of Nd^{3+} for unit pump power. However, lifetime measurement requires some precautions due to the radiative trapping effect occurring in thick samples or highly concentrated ones due to a significant reabsorption of the emission. The trapping effect produces an apparent lengthening of the lifetime and then leads to an underestimation of the emission cross section. Lifetime measurements done on very thin samples or lightly doped crystals are more representative of the intrinsic lifetime.

Yb^{3+}-based lasers are expected to be quasi-four-level systems. Then, the major problem is a non-negligible ground-state absorption at the laser wavelength, as for the 946 nm Nd^{3+} laser transition we have previously discussed. One important parameter for a quasi-four-level system is β_{min}, the minimum fraction of ions that must be excited to balance exactly the gain with the ground state absorption at λ_{ext} (the extraction wavelength of the laser) and the Yb-doped crystal becomes transparent at λ_{ext}. β_{min} is related to the absorption and the emission cross sections at λ_{ext} following:

$$\beta_{min} = \frac{\sigma_{abs}(\lambda_{ext})}{\sigma_{abs}(\lambda_{ext}) + \sigma_{em}(\lambda_{ext})}. \tag{B1.6.42}$$

It is then possible to evaluate the minimum absorbed pump intensity I_{min} required to get a lossless material and to reach the threshold. I_{min} is related to β_{min} and the pump saturation intensity I_{sat} as given by equation (B1.6.24) in which σ_{em} is replaced by $\sigma_{abs}(\lambda_{ext})$:

$$I_{min} = \beta_{min} I_{sat} \tag{B1.6.43}$$

Low thresholds will be obtained for small β_{min} values corresponding to minimal resonant losses at λ_{ext}. Such a condition is achieved for large splitting of the $^2F_{7/2}$ ground state, otherwise cooling is necessary to reduce the ground state absorption at λ_{ext}, improving the laser performance. The splitting is linked to the Yb^{3+} site symmetry and the nature of the ligands as we will see in section B1.6.4. The relevant spectroscopic parameters of Nd^{3+} and Yb^{3+} in the different self-doubler crystals are summarized in tables B1.6.5 and B1.6.6.

Refractive indices measurements

Measurements of the refractive indices ($n_{o,e}$, for uniaxial, $n_{X,Y,Z}$ for biaxial crystals) and determination of the dispersion curves in the entire transparency range, especially in the range of the laser emission wavelengths selected for frequency doubling, are needed for further estimation of the phase-matching directions. The wavelength dependence of the refractive index is given by the Sellmeier equation:

$$n(\lambda)^2 = A + \frac{B}{\lambda^2 - C} - D\lambda^2 \tag{B1.6.44}$$

where A, B, C and D are obtained by least-square fit to experimental data. Using equations given by Dmitriev *et al* [5] for calculating the phase-matching angles for the group symmetry of the crystal and the Sellmeier fit

Table B1.6.5. Nd^{3+}-doped self-doubling crystals: spectroscopic parameters (peak absorption cross section, emission cross section, fluorescence lifetime and quantum efficiency).

	$LiNbO_3$:MgO [44]	YAB [44]	$LaBGeO_5$ [44]	GdCOB [61]	YAB [50][a]
Concentration	0.2 at.%	5.6 at.%	1.4 at.%	7 at.%	5.6 at.%
λ_{abs} (nm)	807.5 $^{(\sigma)}$ 812.19 $^{(\pi)}$	807.3 $^{(\sigma)}$ 807.4 $^{(\pi)}$	798.4 $^{(\sigma)}$ 798.4 $^{(\pi)}$	811	
$\sigma_{abs}(\times 10^{20} cm^2)$	2.3 $^{(\sigma)}$ 2.7 $^{(\pi)}$	2.58 $^{(\sigma)}$ 0.94 $^{(\pi)}$	2.36$^{(\sigma)}$ 2.2 $^{(\pi)}$	2.23 $^{(Z)}$ 1.57 $^{(Y)}$ 1.86 $^{(X)}$	
	0.08 (σ_{SH}) 0.058 (σ_{ESA})	0.31 (σ_{SH}) 0.018 (σ_{ESA})	0.438 (σ_{SH}) 0.003 (σ_{ESA})	0.23 (σ_{SH}) 0.008(σ_{ESA})	
λ_{em} (nm)	1094 $^{(\sigma)}$ 1085 $^{(\pi)}$	1062 $^{(\sigma)}$ 1060 $^{(\pi)}$	1049 $^{(\sigma)}$ 1049 $^{(\pi)}$	1060	1338.4 $^{(\sigma)}$ 1337.8 $^{(\pi)}$
$\sigma_{em}(\times 10^{20}$ $cm^2)$	7.8 $^{(\sigma)}$ 27 $^{(\pi)}$	9.7 $^{(\sigma)}$ 13.7 $^{(\pi)}$	7.6 $^{(\sigma)}$ 11.3 $^{(\pi)}$	4.2 $^{(Z)}$ 2.1 $^{(Y)}$ 1.9 $^{(X)}$	2.46 $^{(\sigma)}$ 2.56 $^{(\pi)}$
Fluorescence lifetime (μs)	95	53	280	98	
Quantum efficiency (η)	0.94	0.18	0.89	0.15	

[a] $\lambda_{em} = 1338$ nm.

of the refractive indices, one can find all the phase-matching configurations for SHG of fundamental radiation in the near- or mid-infrared range. As an example, from phase-matching curves calculated for GdCOB, a biaxial crystal, represented on figure B1.6.7, we can deduce that type I and II phase-matching are possible in all the principal planes for fundamental wavelengths ranging from 830 nm to 2.7 μm but frequency doubling at 1.064 μm is allowed only for type I in the XY (with $\theta_{PM} = 90°$ and $\phi_{PM} = 45.99°$) and the ZX (with $\theta_{PM} = 19.68°$ and $\phi_{PM} = 0°$) planes [60].

According to the linear and nonlinear properties of the self-frequency doubling crystal, a suitable orientation to get a good compromise between high cross sections and large nonlinear optic (NLO) coefficients is determined and the cavity design improved. Formulas (B1.6.28) and (B1.6.30) used to calculate the threshold and output power (given in section B1.6.3.1 and derived from [18]) also apply for self-frequency-doubled laser cavity. Furthermore, in this case also, if the SH is generated by a focused Gaussian fundamental input beam, then $P^{2\omega}$ is expressed by equation (B1.6.31). Theoretical models of a self-doubling laser have been developed taking into account end-pumping, radially varying laser gain, beam-waist data, walk-off effect and the reabsorption of the second harmonic waves by Nd^{3+} ions [76]. Technical details on cavity configuration are developed in section B1.6.5. In table B1.6.7, the characteristics of the more popular systems are summarized.

Table B1.6.6. Yb^{3+}-doped self-doubling crystals: spectroscopic parameters (peak absorption cross sections, emission cross section and fluorescence lifetime) and laser parameters defined in section B1.6.3.1 (I_{sat}, β_{min}, I_{min}).

	$LiNbO_3$:MgO [43] [a]	GdCOB [63]	BCBF [47]
Concentration	1 at.%	7 at.%	2–3 at.%
λ_{abs} (nm)	980	902	912
$\Delta\lambda_{abs}$ (nm)	5	15	19
$\sigma_{abs}(\times 10^{20}\ cm^2)$	1.13 $^{(\sigma)}$ 0.87 $^{(\pi)}$	0.41	1.1
λ_{em} (nm)	1063.9	1032 1052† 1082‡	1034
$\sigma_{em}(\times 10^{20}\ cm^2)$	0.55$^{(\sigma)}$ 0.24$^{(\pi)}$	0.55	1.53
Fluorescence lifetime (ms)	0.540	2.6	1.17
I_{sat} (KW cm^{-2})	33.2	25.5	17
β_{min}	0.106	0.06 0.03† 0.007‡	0.097
I_{min} (KW cm^{-2})	3.51	1.54 0.67† 0.19‡	1.064

[a] Since self-frequency doubling can only be achieved for the fundamental ordinary beam (i.e. σ polarization), only the laser parameters of interest in our purpose are given in this table.
† refers to 1052 emission line.
‡ refers to 1082 emission line.

B1.6.3.6 Comparison between both systems

From tables B1.6.8 and B1.6.9, it appears that intracavity frequency doubling (ICFD) allows us to achieve a more powerful visible laser source than self-doubling. However, a self-frequency-doubling (SFD) laser provides a more compact system with reduced losses due to reflection and scattering. Both suffer from the same drawback due to many oscillating longitudinal modes and large amplitude fluctuations which strongly modulate the SH output power. The instability of intracavity frequency doubled lasers arises from the coupling of longitudinal modes. In multi-longitudinal-mode lasers, the doubling process can occur either by direct doubling of individual longitudinal modes or by frequency mixing of two different modes. Then the coupling between modes can lead to chaotic intensity fluctuations which is often called the 'green problem'.

In addition to the restrictions due to transparency in the UV or poor acceptance angles, frequency-doubling crystals used for ICFD suffer from other limitations such as their maximum available size and hygroscopy. SFD crystals have a better chemical stability and can be grown as large crystals since most of them are obtained with congruent fusion (except YAB and CSB). Their limitations come from the stability

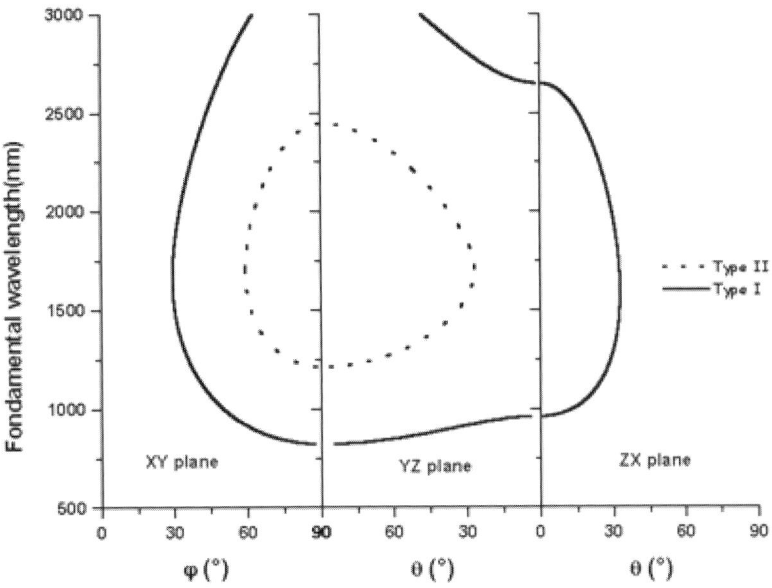

Figure B1.6.7. Phase-matching curves for propagation of the fundamental wavelength on the GdCOB XY, YZ and X principal planes: broken curve, type I phase matching; full curve, type II phase matching (after [60]).

Table B1.6.7. Self-doubling crystals: point group symmetry (S), phase matching (PM), effective nonlinear coefficient d_{eff}, phase-matching angle θ_{PM}, walk-off angle (ρ) and tolerance parameters: angular sensitivity β_θ, angular bandwidth $\Delta\theta$ and temperature sensitivity ΔT. (All data are given for a fundamental wave at 1.064 μm, data concerning single-domain $LiNbO_3$ are reported in table B1.6.3.)

	S	PM	d_{eff}	θ_{PM}	ρ(deg)	β_θ (cm^{-1} mrad)	$\Delta\theta$ (mrad cm)	ΔT (°C cm)	
YAB	R_{32}	Type I	1.27	$\theta = 33°, \phi = 0°$		8.69	8.7		[49, 64]
		Type II	0.73	$\theta = 46°, \phi = 90°$					
YAB[a]		Type I						50	[50]
GdCOB	C_m	Type I							
		ZX plane	1	$\theta = 19.7°, \phi = 0°$	0.87		2.6	> 50	[51]
		XY plane	0.5	$\theta = 90°, \phi = 46°$	0.87		1.7		
YCOB	C_m	Type I							
		ZX plane	1.1	$\theta = 33.95°, \phi = 0°$	0.61		1.3	65	[65]
		XY plane	0.5	$\theta = 90°, \phi = 35.3°$	0.86		1.3	65	[65]
$LaBGeO_5$	P_{321}	Type I	0.296[b]	$\theta = 54°$	1.15		1.47	10.08	[77]
CSB	R_{32}	Type I	2–3d(KDP)	$\theta = 33°$					[68]
		Type II	2–3d(KDP)	$\theta = 48°$					
$Gd_2(MoO_4)_3$	P_{ba2}	Type I	1.9	$\theta = 60°, \phi = 45°$					[72]
BCBF	P_{62m}	Type I	0.26	$\theta = 37°$		5.75			[47]
$LiNbO_3$:MgO		QPM	21.6						[45]

[a] at 1.048 μm.

of the structure with regard to the introduction of a high RE concentration. As an example, when the Yb^{3+} concentration exceeds 50% in YCOB, large crystals cannot be obtained due to incongruent melting.

B1.6.3.7 Periodically poled structures

Recent progress was reported with periodically poled (PP) structures with ferroelectric oxides such as $LaTiO_3$ [46] and $LiNbO_3$ [45] where ferroelectrics domains alternate. As an example, the use of quasi-phase matching (QPM) in PP structures offers the possibility of using π-polarized fundamental laser oscillation for SH generation. This represents two advantages compared to a single-domain Yb:$LiNbO_3$:MgO system, where the fundamental beam should have a σ-polarization to achieve SH generation in type I phase matching. First, the π polarized beam cross section is much higher than the σ-polarized one. In addition, QPM with $\boldsymbol{E} \parallel \boldsymbol{c}$ allows us to access the d_{33} component of the nonlinear tensor, increasing the effective nonlinear coefficient from 5.6 pm V^{-1} (type I) to 21.6 pm V^{-1}, and avoid the walk-off effect. Under non-optimal conditions, a conversion of 18% of infrared to visible output power was obtained, which is higher than in NYAB (14%) [52] or in Nd:YCOB (15%) [59]. If the problem linked to the accurate control of the domain period length remains to be solved, this should allow the SFD perfomances of lasers based on $LiNbO_3$:RE^{3+} to be enhanced.

In comparison with Nd^{3+}, Yb^{3+} has a simple energy level diagram which avoids detrimental complications such as ESA, up-conversion, concentration quenching effect and ground state resonant absorption of the SH (especially important in the case of NYAB). The long fluorescence lifetime of the $^2F_{5/2}$ emitting level and the low quantum defect offer long energy storage and a weak thermal loading of the system. Pumping with high power laser diode can result in local heating of the crystal. This local heating can induce a thermal lens effect but also a phase mismatching between the fundamental and SH beams which reduces the SH output power and affects the beam pattern [78]. Yb^{3+}-doped materials should be more suitable for high power laser applications than Nd^{3+}-doped one. Moreover, the drawback of Yb^{3+} compared to Nd^{3+} is a significant ground state absorption of the laser emission which can be avoided either by cooling or by choosing the host providing a large splitting of the ground state as we will discuss in the next section.

B1.6.4 Ion–host combination, nonlinear coefficients

B1.6.4.1 Ion–host combination: how to choose the best host with Nd^{3+} or Yb^{3+} dopants

For self-frequency doubling, the choice of the ion–host combination is a fundamental factor to achieve simultaneously good laser and NLO properties. The chemical problem of the substitution of Nd^{3+} or Yb^{3+} into a nonlinear crystal has to be taken into consideration. First of all, the structure of the host should allow RE ion doping at a high concentration without disorder of the substitution and distortion of the site. This is especially important for Nd^{3+}-doped materials where increasing the RE concentration results in the formation of clusters which favour energy transfer between RE ions leading to a lower quantum efficiency. The best condition to avoid the concentration quenching effect is to consider a crystal where Nd^{3+} or Yb^{3+} replaces another rare earth (La^{3+}, Y^{3+}, Sc^{3+} or Gd^{3+}) of similar ionic radius and chemical properties. This explains why the oxyborates have shown significant results compared to $LiNbO_3$ for which, in addition to photorefractive damage, RE^{3+} substitutes Li^+ resulting in multiple site symmetries for RE^{3+} due to charge compensation and RE^{3+} can only be introduced with a rather low concentration. Furthermore, $LiNbO_3$ suffers from optical damage due to the photorefractive effect. Doping $LiNbO_3$ with MgO (5%) allows this drawback to be overcome, reducing the photorefractive effect [79], but prevents the introduction of a high Re^{3+} concentration [44].

Other important factors to take into account when choosing a host material are its symmetry and the crystal field strength, which greatly influence the polarized spectroscopic features and the Stark splitting of the excited and ground state multiplets. The crystal field is a key parameter, especially for Yb^{3+} ions, in order to achieve a quasi-four-level system. The case of Yb: $BaCaBO_3F$ is a beautiful example to illustrate the role of charge compensation at the Yb^{3+} site [47]. In this case, Yb^{3+} occupies sevenfold coordinated Ca^{2+} sites. The coordination polyhedron consists of five O^{2-} from the coplanar borate groups and two F^-

located on the ***c***-axis above and below this plane. The O^{2-} replacing one of the F^- ions gives rise to a large crystal field splitting and a strong polarization dependence for the spectra which are both favourable for laser performance. A figure-of-merit has been developed which can be useful in predicting the laser performance of the crystal.

Finally, electron–phonon coupling has also to be considered since it prevails over the thermal shift and the broadening of the spectral lines, which is favourable for diode pumping, but also favours multi-phonon relaxation which competes with the radiative channel. The last effect would be detrimental to the emission efficiency especially in the case of Nd^{3+} due to the rather small energy gap between the emitting level and the next lower multiplet (~4800 cm^{-1}). The order of the multi-phonon process is mainly determined by the highest-energy vibrational mode; hence, crystals with a high phonon cut-off frequency should be avoided. As an example, in YAB structure, Nd^{3+} ions are shielded by the B–O–Al triangular prism and separated from each other. In this case, the Nd–Nd interaction is expected to be weak. However, the quantum efficiency is low resulting from the high non-radiative probability due to the high phonon cut-off frequency of borate compounds mediating the multi-phonon process.

B1.6.4.2 Nonlinear coefficients

Several approaches have been developed to understand the structure–property relationships in nonlinear materials. In oxides such as KTP, the optical nonlinearity is linked to O–Ti–O bonds running along extended chains, while in borate systems, the origin of the nonlinear properties is ascribed to boron–oxygen oxoanions which can be treated as molecular crystals. In the following, we will summarize these different approaches and illustrate them with two examples (see also chapter A4).

Case of KTP: analysis of the electronic hyperpolarizability tensor

A systematic analysis of the electronic hyperpolarizability of KTP has been performed since its nonlinear properties are well characterized and its structure is well suited for theoretical study using the techniques of quantum chemistry. The structure of KTP consists of distorted TiO_6 octahedra and PO_4 tetrahedra which are linked in a three-dimensional framework through shared corners. The orthorhombic structure that results contains chains of –Ti–O–Ti–O– atoms running parallel to the crystallographic ***a***-axis, one oxygen being shared between every two adjacent TiO_6 octahedra. Along these chains derived from distorted TiO_6 octahedron, long (~2.10 Å), intermediate (~1.96 Å) and short (~1.74 Å) Ti–O distances alternate. Early studies using bond-parameter models for the KTP nonlinear properties identified the distorted TiO_6 octahedron as a key contributor to the microscopic hyperpolarizability. The theoretical calculations of the hyperpolarizability tensor of KTP, developed in [80–82], were undertaken essentially to identify basic patterns in the nonlinear response and to estimate the influence of the Ti–O bond length, in a first step, of the Ti–O–Ti angle, in a second step, the magnitude and the direction of the changes in the nonlinear microscopic susceptibility when replacing K or Ti by other cations. The authors were more interested in drawing the main tendency following these parameter variations rather than calculating the exact value of the nonlinear coefficient.

At the microscopic scale, the components of the dipole moment, induced in a molecule by electric fields, are related to the microscopic polarizability tensors through a relation similar to equation (B1.6.1), where the macroscopic $\chi^{(n)}$ susceptibilities are replaced by microscopic parameters (keeping the usual summation convention for repeated indices):

$$p_i = \alpha_{ij} E_j(\omega) + \beta_{ijk} E_j(\omega_1) E_k(\omega_2) + \cdots . \qquad \text{(B1.6.45)}$$

Second-order nonlinear effects are governed by the hyperpolarizability tensor β_{ijk}. The standard perturbative expression of the individual components β_{ijk} of the hyperpolarizability tensor involves excited states, energies and dipole transition moments. Evaluation of β_{ijk} under the sum-over-state approach thus implies computing the matrix elements and supplying the energies of the molecular orbitals. Details of the sum-over-states formalism, including a description of its implementation using LCAO-MO can be found in [81].

First of all, to understand the origins and the structural dependence of the electronic hyperpolarizability tensor, the theoretical calculations of electronic hyperpolarizability β were performed using a perturbative sum-over-states formalism together with extended Hückel wavefunctions for TiO_2 and $(TiO_6)^{8-}$ entities. Indeed, these asymmetric species can be considered as the smallest sub-units of the crystal that are expected to be responsible for the nonlinear properties [81]. TiO_2 has a linear structure with long and short Ti–O distances along the z-axis; Δ, the asymmetry parameter, being defined as their difference. In TiO_6, the remaining four oxygens and the titanium belong to the xy-plane. In this frame, many of the orbitals identified in TiO_2 are found in $(TiO_6)^{8-}$ as well, appropriately modified to reflect the participation of the equatorial oxygen atoms in the xy-plane. To study the origin of the electronic hyperpolarizability, evaluation of β was done by considering separately the contribution from excitations in π and σ bonding and non-bonding orbital systems. State-by-state analysis reveals that NLO effects arise almost entirely from excitations within the π system. From all β_{ijk}, the β_{zzz} component has the largest magnitude in both TiO_2 and TiO_6, as an obvious consequence of an asymmetric O–Ti–O axis (β_{zzz} corresponds to both incident and second harmonic radiations polarized along the asymmetric O–Ti–O axis). To investigate the influence of bond lengths, extended Hückel/sum-over-states calculations of electronic hyperpolarizability were made as a function of Δ. In TiO_6, only the Ti–O axial distances are varying while equatorial ones are left constant. In both TiO_2 and $(TiO_6)^{8-}$, β_{zzz} is remarkably sensitive to bond length. The net hyperpolarizability results from the near cancellation of large contributions with opposite signs. Specific contributions to the hyperpolarizability classified into three major π families and three major σ families, typically pass through minima and maxima as Δ is increased. As we have already mentioned, the σ contribution to β is almost equal to zero. From the behaviour of individual π terms with Δ, one can understand that the SHG intensities observed for the KTP isomorphs vary over wide ranges, while the TiO_6 octahedra in the crystal structures appear to exhibit almost insignificant variations with composition. As examples, substitution of arsenic for phosphorus ($KTiOAsO_4$) seems to enhance SHG by up to approximately 15%, although replacing Ti by Sn leads to significant degradation of the performance. No net tendency is observed replacing K by other cations (Na, Rb, Cs, Tl, Ag or NH_4). Therefore, the last step of these theoretical investigations concerns the influence of the original electronic distribution on the hyperpolarizability as a result of the possible chemical substitutions [82]. A systematic assessment of the effect of changing the Ti–O–Ti angle and oxygen electronegativity on the electronic hyperpolarizability has been performed. So, a larger system made of three adjacent TiO_6 octahedra was considered. This system is the minimum necessary to explore the effect of the Ti–O–Ti angle. The analysis is focused on the local nonlinear response of the central TiO_6 unit placed in the context of an extended chain. This condition implies the determination of the molecular orbitals of the complete system. The microscopic hyperpolarizability was computed only for the central unit using the molecular orbitals truncated to include those atomic orbitals within the considered octahedron. Furthermore, the atoms within the central unit are not allowed to move, so any variation of the local β results from a purely electronic effect and cannot be due to geometrical effects. To compare this with experiments, isotropic averages of the squared hyperpolarizability tensor, $\langle\beta^2\rangle$, are calculated since they should be proportional to powder SHG intensity. An increase of 25% is obtained as the Ti–O–Ti angle in the chain goes from 125° to 145°. Most of the increase is taken by the β_{zzz} component. This nonlinear responsc enhancement follows from the perturbed local electronic structure at the different angles, as reflected by changes in the electronic third moment. This result is consistent with the correlation between SHG efficiency and Ti–O–Ti angle that has been observed experimentally for the KTP isomorphs. In $KTiOP_{(1-x)}As_xO_4$, the SHG powder signal increased by 17% for x going from 0 to 1, while the Ti–O–Ti angle increases by 5.5%. In the $K_{(1-x)}Na_xTiOPO_4$ system, the SHG signal is almost unchanged as x varies from 0 up to 0.5, but falls by an order of magnitude as x approaches 1. The Ti–O–Ti angle remains almost constant for $0 < x < 0.5$ and decreases for $x > 0.5$. All these results exhibit the same trends as predicted by theoretical calculations.

Modelling chemical substitutions

Modelling chemical substitutions can be achieved by changing the oxygen electronegativity. For the original system (i.e. KTP), contributions to the sum over states are distributed over the entire range of occupied orbitals. Increasing the electronegativity of the oxygens relative to the titanium, however, results in the formation of localized oxygen orbitals at lower energies. Most of the terms in the sum then originate from these perturbed levels and SHG in the model system is reduced as a result. When, by contrast, the oxygen electronegativity is decreased, the energies of the uppermost oxygen orbitals are raised leading to a significant near-resonant enhancement of the nonlinear response. However, it should be mentioned that, in real systems, there is a correlation between the Ti–O–Ti angle and the electronegativity of the substituent cations, although changes in the angle may have different origins depending on whether the site substitution is within the framework (phosphorus atom) or is outside the framework (potassium atom). This consideration means that the two effects—namely the variation of $\langle\beta^2\rangle$ with both the angle and the ionization potential of the chain oxygens—act cooperatively and, hence, cannot be examined separately in real systems. The changes observed in $\langle\beta^2\rangle$ should be larger than those predicted from either of the calculations taken in isolation. If the macroscopic coefficient cannot be derived from these calculations, it is possible to derive general trends that can help to synthesize new inorganic frequency-doubling crystals.

Case of borates: calculation of microscopic hyperpolarizability

In the case of borate systems, the optical nonlinear property arises from conjugate bonds within BO groups and the macroscopic χ^2 susceptibility can be calculated from the microscopic hyperpolarizability of the group and application of anion group theory [83]. SHG coefficients calculated, in this way, are in good agreement with experimental data. So, structural criteria were proposed to select, from the borate family, crystals with NLO properties. According to anionic group theory, the NLO susceptibility in a crystal results from the interaction between incident photons and electrons of anionic groups (for inorganic crystals) or molecules (for organics). The overall SHG coefficient is assumed to be the geometric superposition of all individual coefficients of the relevant groups, following

$$\chi_{i,j,k}(2\omega) = N \sum_{p=1}^{n} \sum_{i',j',k'} \alpha_{ii'}\alpha_{jj'}\alpha_{kk'}\beta^{(p)}_{i'j'k'}(2\omega) \qquad \text{(B1.6.46)}$$

where N is the number of cells per unit volume, n the number of groups in the unit cell, $\alpha_{ii'}$, $\alpha_{jj'}$ and $\alpha_{kk'}$ are the direction cosines of the Cartesian system axes of the group with the crystallographic axes, and $\beta^{(p)}_{i'j'k'}(2\omega)$ the microscopic hyperpolarizability of the pth group in the unit cell.

As previously, the microscopic hyperpolarizability is calculated using standard perturbative theory but, here, the wavefunctions are the localized wavefunctions of the group.

Due to the multiple discrete and condensed oxoanions, borate compounds can be synthesized with very different structures. D A Keszler *et al* [84] have recently proposed an overview of the structural chemistry of non-centrosymmetric borates selected for their NLO properties. This analysis of the structure–NLO properties relationship is based on anion group theory and provides important guidelines for the synthesis of new solid state inorganic crystals of interest in NLO applications. The widely used BBO and LBO have no derivative chemistry; hence, it is impossible to slightly alter their physical properties. However, new structural types that could be chemically derived would allow subtle control over both nonlinear and linear properties (i.e. susceptibility coefficients and the magnitude of birefringence) through atomic substitutions within a given structural type.

For example, YAB, BBO and LBO belong to the orthoborate type but their structure differs by the arrangement of the orthoborate anion: independent BO_3 groups in YAB, isolated B_3O_6 rings in BBO and condensed B_3O_7 rings in LBO. However, each of these anion groups belong to the D_{3h} symmetry. In [84], the orientational dependencies of the group hyperpolarizability coefficients are derived. In BBO, B_3O_6 planar rings are tilted from the local z-axis. Then, from the structural data and anion group theory, the NLO coefficient ($d_{22} = 2.3$ pm V^{-1}) is only 67% of the optimal value expected for planar groups parallel to the z-axis. Following the same analysis, from the BO_3 group alignment with respect to the z-axis in YAB, the NLO coefficient observed ($d_{11} = 1.5$ pm V^{-1}) represents 49% of the optimal value. The maximum nonlinearity of an orthoborate crystal is expected to be approximately 3 pm V^{-1}. Since the microscopic hyperpolarizability β_{zzz} is invariant to rotation about the B–O bond along the z-axis, a variety of orthoborate structural arrangements are expected to produce an optimal nonlinearity. In particular, it should be possible to synthesize an orthoborate with a nonlinearity that is comparable to that of BBO while maintaining a smaller birefringence, providing a smaller sensitivity to phase matching and an enhanced conversion efficiency for the production of UV light. In this goal, $BaCaBO_3F$ (BCBF) has been synthesized [84]. When doped with Yb^{3+} ions, this crystal behaves as self-doubler [47] as we mentioned in section B1.6.3.3. From the relative orientation of the BO_3 groups in the BCBF hexagonal structure, the number density of BO_3 groups and the hyperpolarizability β_{zzz} of that group derived from YAB, a nonlinear coefficient d_{11} equal to 0.4 pm V^{-1} is predicted. Furthermore, because of the coplanar arrangement of the BO_3 groups in that structure, a birefringence similar to that of BBO is expected. For type I phase matching, θ_m is expected between 20° and 30° which would produce $d_{eff} = 0.35$ pm V^{-1}. It is remarkable to note the good agreement between this predicted value and the experimental one (0.26 pm V^{-1}) measured on Yb:BCBF [47]. To improve the linear property, the lack of coplanarity of the BO_3 groups would allow the birefringence to be reduced. A smaller birefringence can be achieved if the coplanarity of the BO_3 groups is disrupted while maintaining a small tilted angle between each of the BO_3 group and the c-axis.

Other structures, such as pyroborates of the type $AMOB_2O_5$ (A≡K,Rb and Cs, M≡Nb and Ta), were demonstrated to be a unique means for examining the structural interrelationship of linear and NLO properties [84]. As the size of the alkali-metal decreases, the B_2O_5 groups adopt nonlinear geometries by twisting about the central B–O bond and the coplanarity of the group, observed in the Cs–Nb derivative, is severely disturbed. From consideration of the alignment of the B_2O_5 groups, the highest nonlinearity and birefringence are expected with the Cs derivative. Another interesting feature of these structures is the presence of chains of alternating short–long M–O interactions that extend approximately orthogonal to the B_2O_5 planes. These chains, of course, are reminiscent of the Ti–O distances in KTP that contribute, as we have already discussed, to its high nonlinearity. In the borates, however, the chains are not expected to contribute significantly to the nonlinearity as pairs of chains are related by an approximate centre of symmetry, i.e. the sequence long–short–long M–O distances in one chain alternates with a short–long–short one in a neighbouring chain. This analysis demonstrates the feasibility for controlling the linear and NLO properties and synthesizing new non-centrosymmetric borates for NLO applications.

B1.6.5 System configurations: resonator designs, pumping

B1.6.5.1 Modelling intracavity frequency doubled lasers

As we have already mentioned in section B1.6.3.1, a key parameter for efficient intracavity doubling is the beam waist in the nonlinear crystal, especially for critical phase matching (see equations (B1.6.32) and (B1.6.33)). The actual trend is to develop compact laser sources and then the pumping is achieved using laser diodes. Several models [19–22, 31, 85] were proposed and allow simple criteria to optimize coupling optics and the resonator geometry to be derived. These analyses demonstrate that the most important parameter, for laser efficiency and output power, is the spatial overlap between the pump beam and the laser mode in the active medium.

Diode end-pumped lasers

The first model [19], based on a space-dependent rate equation analysis for a four-level system, was investigated to study the influence of the optimum ratio between mode and pump spot size on input power. In this analysis, Gaussian and elliptic beam profiles of the pump are considered. This treatment is particularly well adapted for diode laser end-pumping since asymmetric pump distribution is taken into account. This model provides a convenient way to calculate the optimum spot size for the available input pump power and pumping configuration for a plano-concave resonator. Integrating equation (B1.6.30) gives

$$P_{\rm out} = \frac{T}{2\gamma}\frac{J_1^2}{J_2}\eta_{\rm p}[P_{\rm in} - P_{\rm th}] \qquad \text{(B1.6.47)}$$

with γ the total logarithmic loss per pass, J_1^2/J_2 a geometrical factor accounting for the overlapping efficiency, $\eta_{\rm p}$ the pumping efficiency, $P_{\rm in}$ and $P_{\rm th}$ the incident and threshold power, respectively.

The geometrical factor is expressed from overlap integrals between the normalized mode distribution $s_l(x,y,z)$ and the normalized pump distribution in the active medium $r_{\rm p}(x,y,z)$:

$$\frac{J_1^2}{J_2} = \frac{(\int\int\int s_l(x,y,z) r_{\rm p}(x,y,z)\,{\rm d}v)^2}{\int\int\int s_l^2(x,y,z) r_{\rm p}(x,y,z)\,{\rm d}v}. \qquad \text{(B1.6.48)}$$

The pumping efficiency $\eta_{\rm p}$ corresponds to the ratio between the power incident on the active medium to the power emitted by the source. It is related to the transfer efficiency η_t, $\eta_{\rm a}$ as defined in equation (B1.6.30) and the quantum defect by the following relation:

$$\eta_{\rm p} = \eta_t \eta_{\rm a}\frac{\nu_{\rm c}}{\nu_{\rm p}}. \qquad \text{(B1.6.49)}$$

In [19], two parameters, ζ and δ are introduced in order to obtain an explicit expression for the output power depending on the ellipticity of the pump beam:

- $\zeta = (\omega/\bar{\omega}_{\rm p})^2$ (ω is the mode spot size and $\omega_{\rm p}$ the minimum spot size allowed by the given configuration); and
- δ, the ratio between the minor and the major spot size ($0 \le \delta \le 1$), accounts for the ellipticity of the beam ($\delta = 1$ for a circular Gaussian beam).

Calculation of both J_1 and J_2 in a closed form allows $P_{\rm out}$ to be expressed as

$$P_{\rm out} = \frac{T}{2\gamma}\eta_{\rm p} P_{\rm tho} f_1(\zeta,\delta)[\xi - f_0(\zeta,\delta)] \qquad \text{(B1.6.50)}$$

where $P_{\rm tho} = \pi\gamma I_{\rm sat}\bar{\omega}_{\rm p}^2/2\eta_{\rm p}$ is the threshold limit obtained in the ideal case of vanishing mode area, its value is determined by the spectroscopic parameters of the crystal: $\xi = P_{\rm in}/P_{\rm tho}$ corresponds to the amount by which the threshold limit is exceeded. f_0 is the normalized threshold given by

$$f_0 = \frac{1}{\delta}[(\zeta\delta^2+1)(\zeta+1)]^{\frac{1}{2}}; \qquad \text{(B1.6.51)}$$

f_1 is the overlapping efficiency, expressed as

$$f_1 = \frac{\zeta\delta[(\zeta\delta^2+2)(\zeta+2)]^{\frac{1}{2}}}{[(\zeta\delta^2+1)(\zeta+1)]}. \qquad \text{(B1.6.52)}$$

From the plot of the normalized output power $P_{out}/[(T/2\gamma)\eta_p P_{tho}]$ as a function of ξ for different values of ellipticity δ, the optimum mode size can be easily calculated, for the available input pump power and pumping configuration which determines $\bar{\omega}_p$. The corresponding threshold power $P_{th}^{opt} = P_{th}(1+\zeta)$ and slope efficiency are also readily obtained. For a beam coming from the fibre with a circular Gaussian profile, equation (B1.6.47) becomes:

$$P_{out}(P_{in}, \zeta, \bar{\omega_p}) = \frac{T\eta_p}{2\gamma}\frac{\zeta(\zeta+2)}{(\zeta+1)^2}\left[P_{in} - \frac{\pi\gamma I_{sat}\bar{\omega}_p^2}{2\eta_p}(\zeta+1)\right]. \tag{B1.6.53}$$

Experimental results for an Nd:YAG laser pumped by a 500 mW diode array have shown a fairly good agreement with the predictions of this model. This analysis gives a straighforward procedure for designing the resonator but the effect of some parameters, such as the divergence and the focus position of the pumping light, are not considered. These parameters are important for fibre-coupled laser-diode end-pumped lasers. Therefore, an analytical model was proposed to design the laser resonator and to optimize the optical coupling system by taking into account the beam quality (see chapter B1.3).

Fibre coupled laser-diode end-pumped lasers:coupling and resonator

In this analysis [20, 21], the minimum value of $\bar{\omega}_p$ is limited, as previously, by the properties of the gain medium but also by the pump-beam quality and the focus position in the active medium; while, in the Laporta and Brussard treatment [19], the minimum spot size $\bar{\omega}_p$ was assumed constant and depended only on the resonator configuration. In the frame of this second analysis, $\bar{\omega}_p$ depends on the z-coordinate due to the divergence of the beam coming out from the fibre, then it is expressed, on the basis of the paraxial approximation, as:

$$\omega_p(z) = \omega_{p0} + \theta_p|z - z_0| \tag{B1.6.54}$$

where ω_{p0} is the radius at the waist, θ_p is the far-field angle and z_0 the focal plane in the active medium. The beam waist for a given beam is related to its associate far-field angle by the brightness theorem, therefore $\theta_p = C/n\omega_{p0}$ (C is a constant characterizing the beam quality). For a given value ω_{p0} of the pump beam waist in an active medium of length L and an absorption coefficient α at λ_p, the average spot size is simply expressed as:

$$\bar{\omega_p}(\beta, \omega_{p0}, \xi_0) = \omega_{p0} + \frac{\beta}{\omega_{p0}}(\xi_0 - 1 + 2\exp(-\xi_0)) \qquad \text{with } \xi_0 = \alpha z_0 \text{ and } \beta = C/n\alpha. \tag{B1.6.55}$$

The parameter β accounts for the pump-beam quality and spectroscopic properties of the active medium. With β and ω_{p0} fixed, the optimum focal position which minimizes $\bar{\omega_p}$ is simply derived by setting the first derivative of $\bar{\omega_p}$ with respect to ξ_0 equal to 0, then $z_{0,opt} = \ln(2)/\alpha$. This allows the minimum spot size $\omega_{p0,opt}$ to be determined using equation (B1.6.55) with $\xi_{0,opt}$ and the condition $\partial\bar{\omega_p}/\partial\omega_{p0} = 0$,

$$\omega_{p0,opt} = \sqrt{\frac{C\ln(2)}{n\alpha}} \qquad \text{then } \theta_{p,opt} = \sqrt{\frac{C\alpha}{n\ln(2)}}. \tag{B1.6.56}$$

Then, the minimized average spot size is simply derived: $\bar{\omega}_{p,min} = 2\sqrt{\beta\ln(2)}$. The output power in the condition of minimum average pump size is obtained by substituting $\zeta = (\omega_{p0}/\bar{\omega}_{p,min})^2$ (at fixed β) in equation (B1.6.47). These simple expressions are useful guidelines for designing the coupling system.

To design the resonator, all the optimized parameters are replaced in equation (B1.6.47) and the optimum mode size for a given β and input power P_{in} is obtained for $\partial P_{out}(P_{in}, \beta, \omega_{p0})/\partial\omega_{p,0})/\partial\omega_{p,0} = 0$:

$$\omega_{p,opt} = \sqrt{4\beta\ln(2)[g_+(\chi,\beta) - g_-(\chi,\beta) - 1]} \tag{B1.6.57}$$

where

$$g\pm(\chi,\beta) = \left[\frac{\sqrt{3}\sqrt{27(\chi/4\beta\ln(2))^2+1}}{9} \pm \frac{\chi}{4\beta\ln(2)}\right]^{1/3} \qquad \text{with } \chi = \frac{P_{\text{in}}}{(\pi\gamma I_{\text{sat}}/2\eta_{\text{p}})}. \tag{B1.6.58}$$

The value of χ is determined from the spectroscopic properties of the gain medium, the total resonator losses and the pumping efficiency. This analytical model provides a simple way to design the coupling optics and the resonator of fibre-coupled laser-diode end-pumped lasers.

These previous models considered the transversal pump beam's distribution as a Gaussian one with the longitudinal propagation described by its divergence. Another treatment has been proposed which makes use of the M^2 formalism (see section A2.1.4.3) to describe the longitudinally pump-beam propagation in the active medium [22, 85].

The output-to-input characteristics and mode-matching efficiency are defined by the overlap between the spatial distribution of the pump energy and the spatial distribution of the laser photons in the active medium as previously. Using the M^2-factor, the propagation of pump-beam spot size $\omega_{\text{p}}(z)$ is expressed as [23]:

$$\omega_{\text{p}}(z) = \omega_{\text{p0}}\left[1 + \frac{M^2\lambda_{\text{p}}l}{n\pi\omega_{\text{p0}}^2}\left(\frac{z}{l} - b\right)^2\right]^{1/2} \tag{B1.6.59}$$

where λ_{p} and ω_{p0} are the wavelength and beam waist of the pump respectively, the non-dimensional parameter $b = z/l$ is introduced to describe the focusing position in the active medium. From the mode-overlap integrals, the optimum focusing conditions that ensure a minimum threshold pump power and a maximum overlap efficiency, at a defined laser beam spot size, were determined by numerical simulations for different values of M^2, b and r^{waist} (the pump-beam waist ratio as defined in section B1.6.3.1). Analytical expressions could be derived for an optimum beam waist ratio, focusing positions, maximum overlap-efficiency.

- *Optimum beam waist ratio*

$$r_{\text{waist}}^{\text{opt}} = (A_1 - A_2\alpha_{\text{a}}l)(M^2 - 1)^{(A_1 - A_4\alpha_{\text{a}}l)}. \tag{B1.6.60}$$

- *Optimum focusing position*

$$b_{\text{opt}} = (B_1 - B_2\alpha_{\text{a}}l)(B_3 - B_4\alpha_{\text{a}}l)\log(M^2 - 1). \tag{B1.6.61}$$

- *Optimum overlap efficiency*

$$\eta_{\text{m}}^{\text{opt}} = (\eta_1 + \eta_2\alpha_{\text{a}}l)(\eta_3 + \eta_4\alpha_{\text{a}}l)M^2 \tag{B1.6.62}$$

where the different parameters A_i, B_i and η_i are functions of the laser beam volume $V_l = \pi\omega_{\text{c}}^2 l$. In the same way, an expression for the optimum threshold can be obtained $F_{\text{th}}^{\text{opt}} = P_1 + P_2(r_{\text{waist}}^{\text{opt}})^2$ where $P_i = f(V_l)$. Good agreements were obtained between predicted values and experimental measurements [22, 85]. This method provides a very simple procedure (schematized in figure B1.6.8) for the design of the coupling optics and the resonator.

B1.6.5.2 Resonator designs: examples of intracavity-frequency-doubled lasers (see chapter A2)

Intracavity doubling is achieved by introducing a nonlinear crystal into the laser resonator. Almost all intracavity-frequency-doubled (ICFD) crystals are longitudinally pumped. The input mirror has, as usual, a high transmission at the pump wavelength and high reflectance at the fundamental but also at the SH wavelength. Since high conversion efficiency requires a high power density, the output mirror should be 100% reflective at the pump and fundamental laser wavelengths and totally transmitting at the SH wavelength.

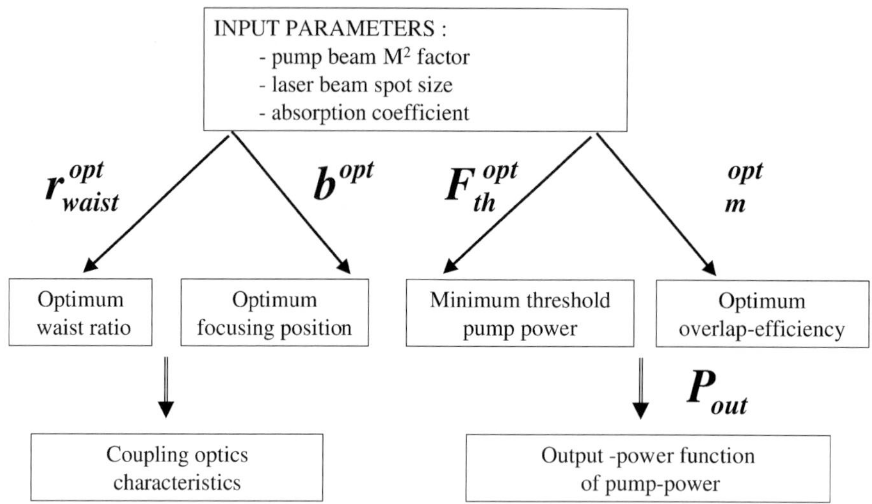

Figure B1.6.8. Schematic procedure for the design of the coupling optics and resonator for fibre-coupled end pumped lasers.

For a laser oscillation at 946 nm, the output mirror has a sufficient transmission at 1.06 μm to suppress the $^4F_{3/2} \rightarrow {}^4I_{11/2}$ laser transition. The pump side of the crystal is anti-reflection coated at the pump wavelength and high reflection coated at both the fundamental and the SH wavelengths to act a as an input mirror of the cavity. The opposite side is anti-reflection coated at the fundamental wavelength. The nonlinear crystal is anti-reflection coated at the fundamental to reduce the passive loss of the cavity. As we have already seen, from equation (B1.6.25) in section B1.6.3.1, the SH power is proportional to the fundamental power density and to the fundamental power. Then, the beam waist inside the nonlinear crystal should as small as possible in order to increase the fundamental power density. Concurrently, the intracavity beam must be large enough inside the laser rod to utilize the maximum volume that can contribute to the laser oscillation. This generally requires that the beam cross sectional area be at least one order of magnitude larger inside the laser rod than inside the nonlinear crystal.

Four major types of resonators are used to develop ICFD lasers. Ring and three-mirror folded cavities are more appropriate to generate high SH powers. Hemispherical cavity and microchip delivers lower SH power and is used to realize miniaturized visible laser sources. In the following, we will give the basic configuration of each of them. They operate in continuous wave mode, a pulsed regime can be achieved by either using the nonlinear crystal, which simultaneously acts as frequency doubler and electro-optic (EO) Q-switch (this is possible with KTP [31]), or by adding an acousto-optic (AO) Q-switch or a saturable absorber (Cr^{4+}:YAG [34], for example).

Ring cavity

High power visible sources were demonstrated in ICFD ring lasers (tables B1.6.4 and B1.6.8). A 'bow-tie' ring resonator provides a tight focus at the location of the frequency-doubler crystal and a good mode matching with the pump in the rod. With this configuration, the pump spot size in the rod can be adjusted in order to optimize the fundamental power while avoiding thermal effects (thermal lensing). Furthermore, the travelling-wave resonator eliminates the spatial hole burning due to the standing-wave distribution of the intensity in conventional oscillators. Compact pumping sources such as high-power, beam-shaped diode bars ([24,37]) can be used to give a well-confined pump spot, collimated over the length of the laser rod. A Faraday rotator (see section A5.3.9) is placed inside the cavity to enforce unidirectional single-frequency

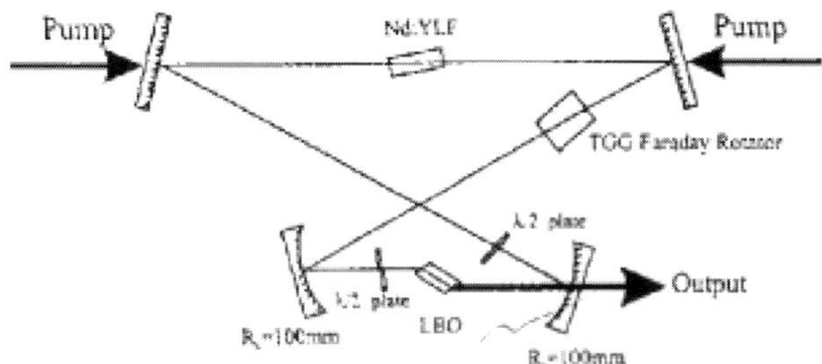

Figure B1.6.9. Schematic diagram of an Nd:YLF ring resonator, with aTGG Faraday rotator and LBO frequency-doubling crystal (after [37]).

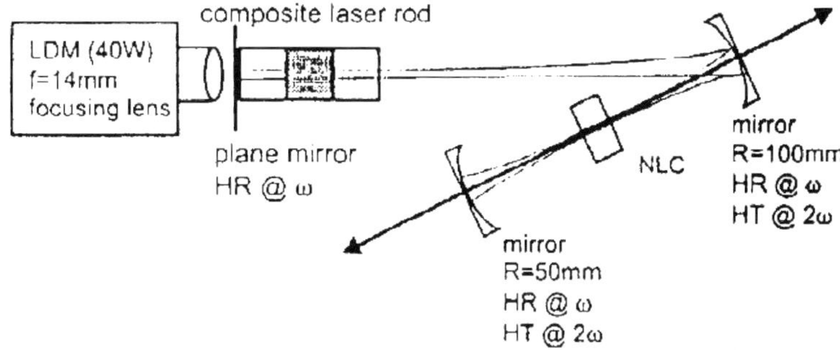

Figure B1.6.10. Schematic diagram of a three-mirror folded cavity (after [12]).

laser operation, thus suppressing the chaotic fluctuation behaviour known as the 'green problem'. Half-wave plates are added on each side of the frequency-doubler crystal to align the polarization in the proper direction for SHG before entering the crystal and one at the output face to rotate the laser polarization back to its initial orientation. A schematic diagram of an Nd:YLF ring laser resonator pumped from opposite ends by two beam-shaped diode bars is represented in figure B1.6.9.

A standing-wave laser cavity has been used instead of a ring cavity to simplify longitudinally pumping. A frequency-stabilized laser emission was obtained at 946 nm with harmonics at 473 nm and 237 nm [28].

Three-mirror folded cavity

A schematic diagram of a three-mirror folded cavity is represented in figure B1.6.10. The cavity is folded by a harmonic separator with a high reflectance and a high transmittance at the fundamental and SH wavelengths, respectively. The laser rod is in the first arm of the cavity, close to the front mirror and the nonlinear crystal in the second arm between the folded and the rear mirror. This configuration also allows the beam radius in the laser rod and in the nonlinear crystal to be adjusted independently. High SH powers are obtained by combining the forward and reverse SH beams into one single output beam.

Hemispherical cavity

A hemispherical cavity is appropriate to realize compact laser-diode-pumped visible sources. The input coupler can be a coated mirror or the coated front face of the laser crystal, the output coupler being a curved mirror with a high transmission at the SH wavelength. In this case, the beam radii cannot be adjusted

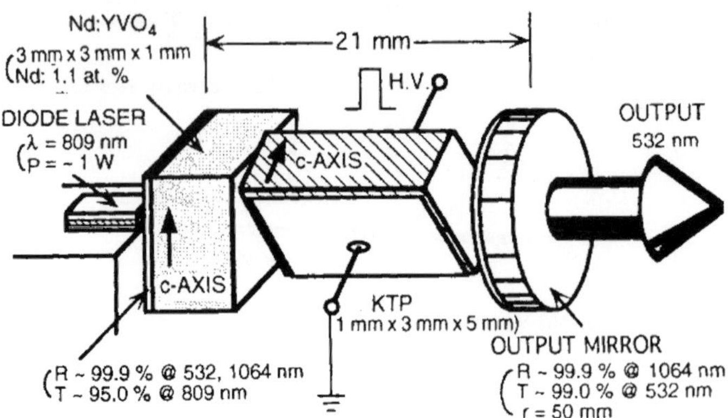

Figure B1.6.11. Schematic diagram of a diode-laser-pumped ND:YVO_4 laser. A single KTP crystal functioned as the intrafrequency doubler and the EO Q-switcher (HV high voltage) (after [31]).

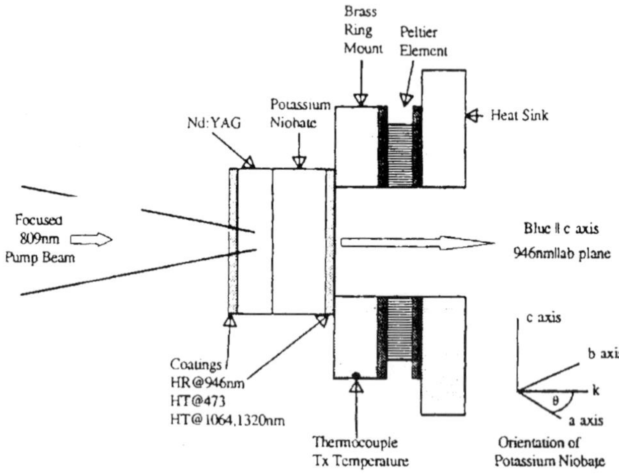

Figure B1.6.12. Composite-cavity microchip scheme (after [27]).

separately, the laser crystal is placed near the front mirror and the nonlinear crystal is placed close to the gain medium near the cavity waist. To achieve an efficient conversion, the pump beam is tightly focused ($\sim$40–60 μm). Due to the moderate pumping power, the thermal lens effect is not so critical as in previous designs [13, 31, 35]. In figure B1.6.11, a compact EO Q-switch ICFD laser is schematized.

Microchip

A composite-material microchip is the most compact device. The laser crystal is cut in a plane-parallel chip in close contact with the plane-parallel nonlinear crystal. The surfaces of the composite material are coated to have a high reflectance at λ_{fund}, a high transmittance at the SH wavelength for the ouptut face. An example of microchip laser is represented in figure B1.6.12.

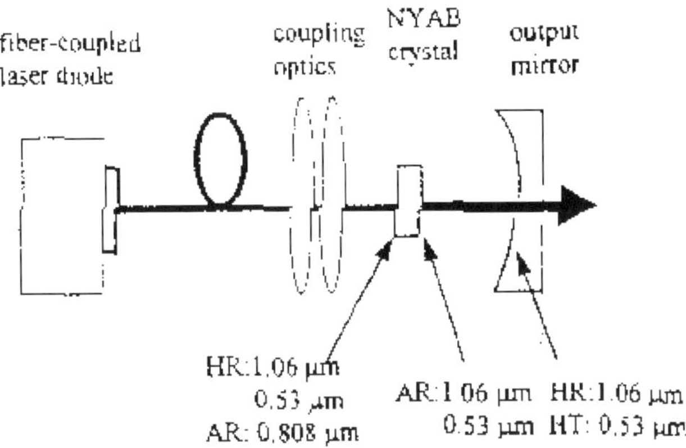

Figure B1.6.13. Fibre-coupled diode pumping experimental set-up (after [54]).

B1.6.5.3 Resonator designs: self-doubling lasers

In this case, the laser crystal is both the laser active medium and the nonlinear medium for frequency doubling, the waist size of the cavity mode as well as the focusing of the pump radiation into the crystal being the most critical parameters. Pump focusing and the configuration of the laser cavity has, thus, to be a compromise between the focusing required for optimum laser performance and the waist size of the TEM_{00} which provides optimum frequency doubling. The threshold and visible output power are given by equations (B1.6.28) and (B1.6.25), respectively, or equation (B1.6.25) using (B1.6.32) and (B1.6.33) for a focused Gaussian fundamental beam (see section B1.6.3.1). According to equation (B1.6.28), the minimization of P_{th} requires minimized radii for the pump (ω_p) and laser mode (ω_c). The pump mode radius is minimized by an appropriate choice of the focusing system, while the radius of the laser modes can be minimized by increasing the cavity length inside the stability region of the optical resonator. The hemispherical cavity is one of the designs which minimizes ω_c. The minimum threshold can be further decreased by the use of a concentric cavity to reduce the ω_c *and also anti-reflection coatings on the elements.* Modelling self-doubled lasers is simpler than for ICFD lasers and predictions of their performance can be achieved by numerical analysis based on the conventional rate equations with additional terms to account for the conversion of the fundamental beam (see [41, 52, 76, 86, 87], for example).

The first model was developed by Dmitriev *et al* [86, 87]. In this model, the spatial dependence of the pumping rate is not taken into account. Therefore, it is more appropriate for flashlamp side pumping. More recently, Brenier [76] has modified this theoretical investigation to describe cw laser diode end-pumped Nd^{3+} SFD lasers. Then, a forward and backward fundamental wave and a second harmonic wave are considered to propagate in the crystal and the nonlinear equations are solved numerically giving their spatial dependence. The laser operation is described by rate equations for a four-level system with additional terms to take counterpropagating fundamental and SH waves into account, together with the reabsorption of the SH wave by the laser emitting ions. This model was applied to NYAB and $LiMgO_3$:MgO:Nd^{3+} using the optical and spectroscopic data published in the literature. The influence of the cavity parameters (ω_p, ω_1, ω_2) and the length of the crystal on the green output power was investigated.

The model previously presented for fibre-coupled, laser-diode, end-pumped lasers [20, 21] also applies to determine the optimum pumping condition for SFD lasers taking into account the pump-beam quality. This analysis was performed for NYAB using spectroscopic data (τ, σ_p, α_p) and round-trip cavity losses [54]. The threshold was calculated as a function of the pump-beam waists for different fibre-coupled laser diodes. The coupling optics are chosen to match this optimum pumping condition. The experimental set-up is schematized

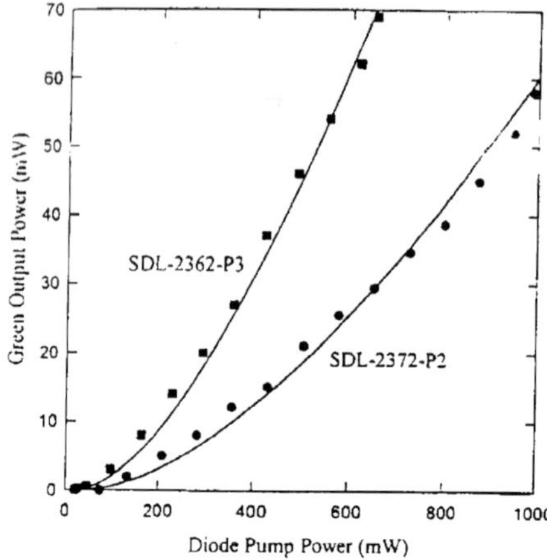

Figure B1.6.14. SH output power *versus* optical pump power of an NYAB laser: experimental (symbols) and theoretical results (full curves) (after [54]).

in figure B1.6.13. The theoretical predictions are in excellent agreement with the experimental data as shown on figure B1.6.14.

B1.6.6 Typical operating characteristics

Table B1.6.8. Lasers with intracavity doubling frequency (T_{out} is the output coupler transmission at λ_{SH} and ζ is the linear optical-optical conversion efficiency).

	Pump	λ_{pump} (nm)	T_{out} (%)	λ_{fund} (nm)	P_{SH}/P_{pump}	ζ (%)	Ref.
YAG:Banana				1064	0.350 W/1 W	35	[14]
YAG:$LiIO_3$	2×SDL-2420	808	50	946	50 μ W		[17]
YAG^a:$LiIO_3$	LDA (2 × 20 W)	804		946	0.520 W/2.4 W	21.7	[12]
YAG^a:BBO	LDA (2 × 20 W)	804		946	0.550 W/2.580 W	21.3	[12]
YAG^a:LBO	LDA (2 × 20 W)	804		946	0.372 W/2.4 W	15.5	[12]
YAG:KTP				1064	3 W/14 W	21.4	[24]
YAG:KTP	LDA (2 × 20 W)	808		1064	20 W^b/0.141 W	14.2	[25]
YAG:KTP	LDA (2 × 20 W)	808		1064	2 W/11.2 W	17.8	[26]
					4 W/27 W (unstable)		
YAG:$KNbO_3$	SDL-303V	809	85	946	3.1 mW/420 mW	0.74	[13]
YAG:$KNbO_3$	Ti-Sa(cw)	808.5		946	3.6 mW/870 mW	0.41	[27]
YAG:$KNbO_3$	SDL-2473 (2 W)		HT	946	100 mW/250 mW^{IR}		[28]
YVO_4:LBO	SDL-2362 (1.2 W)		60	1342	10 mW/750 mW^{abs}		[29]
YVO_4:KTP	810-CT100(1 W)	809	99	1064	230 mW/780 mW	29.5	[30]
YVO_4:KTP	ML 812-4D(1 W)	809	99	1064	105 mW/900 mW	11.5	[31]
YVO_4:KTP	SDL-2432(0.5 W)	809		1064	50 mW/380 mW	13	[32]
YVO_4:KTP	LDA (2 × 20 W)	808		1064	1 W/10 W	10	[26]
YVO_4:KTP	MDL-100(1 W)	808	> 95	1064	286.5 mW/881.4 mW	32.5	[33]
YVO_4:KTP	SDL-2372-P3	808	HT	1064	0.350 W/1.2 W	29.2	[34]

Passive Q-switch		808			240 W[b]/1.5 W		
S-FAPF-KTP	QW1000	805	HT	1059	40 mW/400 mW	10	[35]
LSB-KTP	SDL-2482(2 W)	808	M	1062.5	0.522 W/2.05 W	25	[36]
YLF LBO	LDA (2 × 20 W)	792	91	1053	6.2 W/28 W	22	[37]
YAP-KTP	High-pressure Kr lamp			1080	1 W$^{(SH)}$		[38]

[a] Composite laser rod: 3 mm long Nd:YAG with two diffusion-bonded 3 mm long undoped cap ends.
[b] Peak power.

Table B1.6.9. Self-frequency doubled lasers: performances (T^{SH}_{out} is the output coupler transmission at λ_{SH}, η is the slope efficiency for the infrared, P_{th} is the threshold power, ζ is the linear optical–optical conversion efficiency, H stands for hemispherical cavity (plane-concave), CC for a concentric resonator).

Nd-doped	Pump	λ_p (nm)	T^{SH}_{out} (%)	Cavity	λ_{fund} (nm)	η (%)	P_{th} (mW)	P_{SH}/P_{abs} (mW/mW)	ζ (%)	
YAB	Ti-Sa	807(o)	60	H	1338	16	98	0.6/750	0.1	[50]
YAB	SDL-2430-H1	800	> 85	H[c]	1060		40	6/420	1.4 20 W[d]	[94]
YAB	SDL-2362-P2	808	> HT	H	1063		40	67/650^{pump}	20 W[d]	[95]
YAB[a]	SDL-2362-P2	808	> 90	H	1063		~11	67 /650	10.3	[54]
YAB[a]	2×(SFH-487401) Ti-Sa	807	88	H	1062			225 /1610 450 /2200	14 20.5	[52]
YAB[a]	2×(SDL-2462-P1)	808	86	CC	1063			51 /1380	3.7	[49]
$Gd_2(MoO_4)_3$[a]	SDL-2460	807	85	SC	1060.6			0.48 /420^{pump}		[72]
YCOB[a]	POL-5100BW	812	> 96	H	1060	51		62 /900	7	[59]
$LiNbO_3$:MgO[a]	SDL-2482-P1	809	> 98	H	1093		220	8 /870	1	[42]
$LiNbO_3$:MgO[a]	SDL-5412-H1 Ti-Sa	813 809	> 90	H M[b]	1093	 2.4	5.2 7.7	0.27 /33 0.2 /100	0.8 0.2	[41]
Yb-doped										
YCOB 20 at.%	Ti-Sa	900	> 96	H	1090	35.8	370	1 /900		[58]
$LiNbO_3$:MgO 1 at.% (4 mm)	Ti-Sa	980	80	CC	1060	47	101	30 /560		[43]
$LiNbO_3$:MgO^{PP} 1 at.%	Ti-Sa	980	50	H	1060	7.5		10.5 /1400		[45]

[a] Face coated.
[b] Monolithic laser: the flat and spherical faces of the crystal are coated with dielectric mirrors.
[c] Acousto-optic Q-switch (repetition rate 16 kHz)
[d] Peak power (τ_{pulse} = 25 ns).

References

[1] Blombergen N 1965 *Nonlinear Optics* (New York: Benjamin)
[2] Yariv A 1975 *Quantum Electronics* 2nd edn (New York: Wiley)
[3] Shen Y R 1984 *The Principles of Nonlinear Optics* (New York: Wiley)
[4] Levenson M D and Kano S S 1988 *Introduction to Nonlinear Laser Spectroscopy (Quantum Electronics—Principles and Applications)* revised edn, ed P F Lia and P Kelley (New York: Academic)
[5] Dmitriev V G, Gurzadyan G G and Nikogosyan D N 1991 *Handbook of Nonlinear Optical Crystals (Springer Series in Optical Sciences)* ed A E Siegman (Berlin: Springer)
[6] Kaminskii A A 1990 *Laser Crystals: Their Physics and Properties (Springer Series in Optical Sciences)* 2nd edn, ed D L MacAdam (Berlin: Springer)
[7] Demtröder W 1996 *Laser Spectroscopy: Basic Concepts and Instrumentation* 2nd enlarged edn (New York: Springer)
[8] Koechner W 1996 *Solid State Engineering* 4th edn (New York: Springer)
[9] Hobden M V 1967 *J. Appl. Phys.* **38** 4365–72
[10] Kleinman D A, Ashkin A and Boyd G D 1966 *Phys. Rev.* **145** 338–79
[11] Boyd G D and Kleinman D A 1968 *J. Appl. Phys.* **39** 3597–639
[12] Kellner T, Heine F and Huber G 1997 *Appl. Phys.* B **65** 789-92
[13] Risk W P, Pon R and Lenth W 1989 *Appl. Phys. Lett.* **54** 1625–7
[14] Geusic J E, Levinstein H J, Singh S, Smith R G and van Uitert L G 1968 *IEEE J. Quantum Electron.* **QE-4** 352–3
[15] Fan T Y and Byer R L 1987 *Opt. Lett.* **12** 809–11
[16] Smith R G 1970 *IEEE J. Quantum. Electron.* **QE-6** 215–23
[17] Risk W P and Lenth W 1987 *Opt. Lett.* **12** 993–5
[18] Risk W P 1988 *J. Opt. Soc. Am.* B **5** 1412–23
[19] Laporta P and Brussard M *IEEE J. Quantum Electron.* **QE-27** 2319–26
[20] Chen Y F, Lio S T, Kao C F, Huang T M, Lin K H and Wang S C 1996 *IEEE J. Quantum Electron.* **QE-32** 2010–16
[21] Chen Y F, Kao C F and Wang S C 1997 *Opt. Commun.* **133** 517–24
[22] Pavel N, Kurimura S and Taira T 1999 *Advanced Solid State Lasers* vol 26, ed M M Fejer, H Injeyan and U Keller (Washington, DC: OSA TOPS) pp 253–9
[23] Sasnett M W 1992 Propagation in multimode laser beams—the M^2 factor *The Physics and Technology of Laser Resonator* ed D R Hall and P E Jackson (Bristol: IOP) pp 132–42
[24] Martin K I, Clarkson W A and Hanna D C 1996 *Opt. Lett.* **21** 875–7
[25] Konno S and Yasui K 1998 *Appl. Opt.* **37** 551–4
[26] Yelland C and Sibbett W 1996 *J. Mod. Opt.* **43** 893–901
[27] Matthews D G, Conroy R S, Sinclair B D and MacKinnon N 1996 *Opt. Lett.* **21** 198–200
[28] Hollemann G, Peik E and Walther H 1994 *Opt. Lett.* **19** 192–4
[29] Conroy R S, Kemp A J, Friel G J and Sinclair B D 1997 *Opt. Lett.* **22** 1781–3
[30] Pavel N, Taira T and Furuhata M 1998 *Opt. Laser Technol.* **30** 275–80
[31] Taira T and Kobayashi T 1995 *Appl. Opt.* **34** 4298–301
[32] MacKinnon N and Sinclair B D 1994 *Opt. Commun.* **105** 183–7
[33] Shen D, Liu A, Song J and Ueda K 1998 *Appl. Opt.* **37** 7785–8
[34] Chen Y F 1998 *IEEE Photon. Technol. Lett.* **10** 669–71
[35] Zhao S, Wang Q, Zhang X, Wang S, Zhao L, Sun L and Zhang S 1997 *Opt. Eng.* **36** 3107–10
[36] Meyn J P and Huber G 1994 *Opt. Lett.* **19** 1436–8
[37] Hardman P J, Clarkson W A and Hanna D C 1998 *Opt. Commun.* **156** 49–52
[38] Pan Q, Zhang T, Zhang Y, Li R, Peng K, Yu Z and Lu Q 1998 *Appl. Opt.* **37** 2394–6
[39] Johnson L F and Ballman A A 1969 *J. Appl. Phys.* **40** 297–302
[40] Fan T Y, Cordova-Plaza A, Digonnet M J F, Byer R L and Shaw H J 1986 *J. Opt. Soc. Am.* B **3** 140–7
[41] Ishibashi S, Itoh H, Kaino T, Yokohama I and Kubodera K 1996 *Opt. Commun.* **125** 177–85
[42] Zhang K, Xie C, Guo R, Wang J and Pen K 1996 *Appl. Opt.* **35** 3200–2
[43] Montoya E, Capmany J, Bausá L E, Kellner, Diening A and Huber G 1999 *App. Phys. Lett.* **74** 3113–15
[44] Jaque D, Capmany J, Sanz García, Brenier A, Boulon G and García Solé J 1999 *Opt. Mater.* **13** 147–57
[45] Capmany J, Montoya E, Bermúdez V, Callejo D, Diéguez and Bausá L 2000 *Appl. Phys. Lett.* **76** 1374–6
[46] Abedin K S, Tsuritani T, Sato M and Hiromasa I 1996 *Appl. Phys. Lett.* **70** 10–12
[47] Schaffers K I, DeLoach L D and Payne S A 1996 *IEEE J. Quantum. Electron.* **QE-32** 741–8
[48] Lu B S, Wang J, Pan H F, Jiang M H, Liu E Q and Hou X Y 1989 *J. Appl. Phys.* **66** 6052–4
[49] Hemmati H 1992 *IEEE J. Quantum Electron.* **QE-28** 1169–71
[50] Jaque D, Capmany J and García Solé J 1999 *Appl. Phys. Lett.* **74** 1788–90
[51] Aka G, Kahn-Harari A, Vivien D, Benitez J M, Salin F and Godart J 1996 *Eur. J. Solid State Inorg. Chem.* **33** 727–36
[52] Bartschke J, Knappe R, Boller K J and Wallenstein R 1997 *IEEE J. Quantum. Electron.* **QE-33** 2295–300
[53] Bartschke J, Boller K J, Wallenstein R, Klimov I V, Tsvetkov V B and Shcherbakov I A 1997 *J. Opt. Soc. Am.* B **14** 3452–6

[54] Chen Y F, Wang S C, Kao C F and Huang T M 1996 *Appl. Opt.* **37** 514–17

[55] Wang C L, Lin K H, Hwang T M, Chen Y F, Wang S C and Pan C L 1998 *Appl. Opt.* **37** 3282–5

[56] Jaque D, Capmany J, García Solé J, Luo Z D and Jiang A D 1998 *J. Opt. Soc. Am.* B **15** 1656–62

[57] Jang W K, Ye Q, Eichenholz J M, Richardson M C and Chai B H T 1998 *Opt. Commun.* **155** 332–4

[58] Hammons D A, Eichenholz J M, Ye Q, Chai B H T, Shah L, Peale R E, Richardson M and Qiu H 1998 *Opt. Commun.* **156** 327–30

[59] Eichenholz J M, Hammons D A, Shah L, Ye Q, Peale R E and Richardson M 1999 *Appl. Phys. Lett.* **74** 1954–6

[60] Aka G, Kahn-Harari A, Mougel F, Vivien D, Salin F, Coquelin P, Colin P, Pelenc D and Damelet J P 1997 *J. Opt. Soc. Am.* B **14** 2238–47

[61] Mougel F, Aka G, Kahn-Harari A, Hubert H, Benitez J M and Vivien D 1997 *Opt. Mater.* **8** 161–73

[62] Mougel F, Augé F, Aka G, Kahn-Harari A, Vivien D, Balembois F, Georges P and Brun A 1998 *Appl. Phys.* B **67** 533–5

[63] Mougel F, Dardenne K, Aka G, Kahn-Harari A and Vivien D 1999 *J. Opt. Soc. Am.* B **16** 164–72

[64] Fan Y X, Schlecht R, Qiu M W, Luo D, Jiang A D and Huang Y C 1992 *Proc. Adv. Solid State Lasers* **13** 371–5

[65] Iwai M, Kobayashi T, Furuya H, Mori Y and Sasaki T 1997 *Japan. J. Appl. Phys.* **36** 276–9

[66] Kaminskii A A, Ueda K, Bagaev S N, Pavlyuk A A, Song J, Nishioka H, Uehara N and Musha M 1996 *Quantum Electron.* **26** 379–80

[67] Capmany J, Jaque D, García Solé J and Kaminskii A A 1998 *Appl. Phys. Lett.* **72** 531–3

[68] Ostroumov V, Petermann K, Huber G, Ageev A A, Kutovoj S, Kuzmin O, Panyutin V, Pfeifer E and Hinz A 1997 *J. Luminesc.* **72/74** 826–8

[69] Judd B R 1962 *Phys. Rev.* **127** 750–61

[70] Ofelt G S 1962 *J. Chem. Phys.* **37** 511–20

[71] Carnall W T, Crosswhite H and Crosswhite H M 1977 Energy level structure and transition probabilities of the rare earth lanthanides in LaF^3 *Internal Report* (John Hopkins University, Baltimore, MD)

[72] Kaminskii A A, Bagaev S N, Ueda K, Pavlyuk A A and Musha M 1997 *Quantum Electron.* **27** 657–8

[73] Jaque D, Munõz J A, Cussó F and García Solé J 1998 *J. Phys.: Condens. Matter* **10** 7901–5

[74] Krupke W F 1974 *IEEE J. Quantum Electron.* **QE-10** 450–7

[75] DeLoach L D, Payne S A, Chase L L, Smith L K, Kway W L and Krupke W F 1993 *IEEE J. Quantum Electron.* **QE 29** 1179–91

[76] Brenier A 1997 *Opt. Commun.* **141** 221–8

[77] Capmany J and García Solé J 1997 *Appl. Phys. Lett.* **70** 2517–19

[78] Omatsu T, Kato Y, Shimosegawa M, Hasegawa A and Ogura I 1995 *Opt. Commun.* **118** 302–8

[79] Bryan D A, Gerson R and Tomaschke H E 1984 *Appl. Phys. Lett.* **44** 847–9

[80] Jarman R H, Munowitz M and Harrison J F 1991 *J. Crystal Growth* **109** 353–60

[81] Munowitz M, Jarman R H and Harrison J F 1992 *Chem. Mater.* **4** 1296–304

[82] Munowitz M, Jarman R H and Harrison J F 1993 *Chem. Mater.* **5** 1257–67

[83] Chen C, Wu Y C, Li R K 1985 *Chim. Phys. Lett.***2** 389–92

[84] Keszler D A, Akella A, Schaffers K I and Alekel T 1994 *New Materials for Advanced Solid State Lasers Symposium* ed B H T Chai, S A Payne, T Y Fan, A Cassanho and T H Allik (Pittsburgh P A: Mater. Res. Soc.) pp 85–94

[85] Pavel N and Taira T 1999 *Opt. Eng.* **38** 1806–13

[86] Dmitriev V G and Zenkin V A 1976 *Sov. J. Quantum Electron.* **6** 984–6

[87] Dmitriev V G, Zenkin V A, Kornienko N E, Ryzhkov A I and Strizhevskii 1978 *Sov. J. Quantum Electron.* **8** 1356–61

[88] Belabaev K G, Kaminskii A A and Sarkisov S E 1975 *Phys. Stat. Sol.* a **28** K17–20

[89] Risk W P, Baumert J C, Bjorklund G C, Schellenberg F M and Lenth W 1988 *Appl. Phys. Lett.* **52** 85–7

[90] Risk W P and Lenth W 1989 *Appl. Phys. Lett.* **54** 789–91

[91] Kretschmann H M, Huber G, Batchko R G, Meyn J P, Fejer M M and Byer R L 1999 *Laser Phys.* **9** 293–8

[92] Ye Q, Shah L, Eichenholz J M, Hammons D A, Richardson M, Chin A and Chai B H T 1999 OSA Tops vol 26 *Advanced Solid State Lasers* ed M Fejer, H Injeyan and U Keller pp 100–3

[93] Kaminskii A A, García-Solé J, Bagaev S N, Jaque D and Capmany J 1998 *Quantum Electron.* **28** 1031–3

[94] Li Z, Fan Q, hou F, Ma J and Xue Q 1994 *Opt. Eng.* **33** 1138–41

[95] Chen Y F, Wang S C, Kao C F and Huang T M 1996 *IEEE Photon. Techol. Lett.* **8** 1313–15

B1.7
Solid state Raman lasers

Tasoltan T Basiev and Richard C Powell

B1.7.1 Introduction

Almost 75 years ago, two research groups in Russia and India simultaneously and independently observed an important nonlinear optical process—Raman scattering. Since then the field of nonlinear optics has increased in importance and provided us with a significant amount of information about fundamental light–matter interactions and enabled the development of sophisticated new types of optical devices including optical parametric amplifiers and Raman lasers.

Raman scattering is a process involving the inelastic scattering of light by a material occurring simultaneously with the absorption or emission of thermal energy (phonons). Conservation of energy requires that the energy of the scattered photon differs from that of the incident photon by an amount equal to the energy of the phonon that is absorbed or emitted. The two Indian scientists who first observed Raman scattering, C V Raman and K S Krishnan, were performing light-scattering experiments on molecular liquids [1]. The two Russian scientists who simultaneously observed the same effect, G S Landsberg and L I Mandelshtam, were investigating light scattering in crystals [2]. Since then Raman light scattering has also been observed in many types of gases.

Initially the investigations of Raman scattering centred on trying to understand the fundamental physics associated with the process. This included the mechanism by which the incident light beam interacts with the atoms of the material and the electron–phonon interaction causing some of the light energy to be converted to thermal energy. The details of these interactions differ depending on the type of material being used as the Raman medium. Once the physics of the process was understood, Raman scattering became useful as a spectroscopic tool to study the vibrational, structural and chemical properties of molecules and solids. The Raman spectrum of different molecules is so distinct that Raman spectroscopy has become a common method used for chemical analysis. Forty years ago, Woodbury and Ng [3] reported frequency shifting of a ruby laser beam through Raman scattering. In this case, the intensity of the incident light beam was great enough to result in stimulated Raman scattering (SRS). Raman wavelength shifters made from gas cells with long pathlengths have become a normal accessory for lasers. Since stimulated scattering is a process similar to stimulated emission, it was a logical step to place a Raman gas cell in an optical resonator to obtain a Raman laser. Solid state Raman lasers are of greatest interest because of the ability to build them in a compact, rugged, low-maintenance format. Some of the properties of crystals used for Raman lasers and amplifiers are discussed here and then some potential Raman laser applications are described (see chapter B4.3 for fibre Raman lasers). Raman scattering is a third-order nonlinear optical process in a medium. It does not require phase matching. The resonance characteristics of the process allow it to be useful for investigating the microscopic physics of the atoms, ions or molecules of the medium and for the generation of new frequencies of light. Efficient Raman scattering requires high-power laser sources and materials that have very specific properties. Molecular gases such as H_2 are useful as Raman media because they have high-frequency vibrational modes and narrow spectral linewidths. These properties lead to large Raman frequency shifts and large Raman scattering cross sections as discussed later. However, gas media have a low particle concentration, $N \simeq 10^{20}$ cm^{-3} even for high pressures, and low thermal conductivity. This requires optical interaction lengths of up to a few metres to achieve a significant Raman signal or SRS gain and achieving a pump intensity that exceeds the SRS threshold ($>$1.0 GW cm^{-2}) along the entire length of a gas cell is difficult. The use of constant gas flow cells to overcome thermal problems is also cumbersome.

Liquid media have a higher density of active particles than gases but the broadening of the vibrational transitions due to fluctuations in bond lengths, bond angles and coupling constants as well as particle rotations and collisions is much larger. In general, the increased density outweighs the increased linewidth, resulting in higher Raman gain coefficients for liquid media compared to gas media. Low thermal conductivity is a major drawback of a liquid Raman medium since flowing cell techniques have problems associated with obtaining the homogeneous laminar flows that are required for laser transmission. Also, thermal lensing and self-focusing of the pump laser beam can be a major problem in liquid Raman cells.

Many of the problems associated with gas and liquid media are avoided by using crystalline solids as the Raman medium. The high density of solids ($N \simeq 10^{22}$ cm^{-3}) along with narrower spectral linewidths result in high Raman gains and this allows for a more precise resonance enhancement effect. A great variety of crystals have been investigated for Raman scattering and these provide Raman frequency shifts from 10 to 3000 cm^{-1} in the spectral range from the ultraviolet to the infrared.

Raman scattering in solids occurs because the light wave travelling through the medium is frequency modulated by optical phonons or molecular vibrations. This creates sidebands on the light wave shifted by an amount equal to the vibrational frequency at higher (anti-Stokes) and lower (Stokes) frequencies. The spectral width of the Raman sidebands depends on the lifetime of the vibrational excitation and their intensity depends on the effective scattering cross section describing the light–matter interaction. Conservation of the wavevector determines the direction of the Raman scattered beam of light. Monodirectional gain in a collinear pumping scheme results in SRS in the near forward or backward directions compared to the incident light beam.

B1.7.2 Theoretical background

The theory of SRS follows the usual treatment of third-order nonlinear optical processes [4] with the nonlinear polarization, in this case given by

$$P^{\mathrm{NL}} = N(\partial\alpha/\partial q)q_{\mathrm{av}}E \tag{B1.7.1}$$

where N is the density of scattering centres, q_{av} is the average value for the displacement operator of the normal vibrational mode, E is the electric field of the optical beam and α is the polarizability.

In the steady-state regime, when the pump duration t_{p} is much longer than the Raman mode dephasing time T_{R} ($t_{\mathrm{p}} \gg T_{\mathrm{R}}$), solving the coupled differential equations for q_{av} and the electric fields of the light beams propagating through the media leads to an expression for the gain coefficient g that causes the buildup of the Stokes beam I_{S}, due to SRS. After travelling a length L in the Raman medium, the Stokes beam intensity is given by

$$I_{\mathrm{S}}(L) = I_{\mathrm{S}}(0)\exp[gI_{\mathrm{p}}L] \tag{B1.7.2}$$

in the non-pump-depletion region, where I_{p} is the intensity of the incident laser pump beam and the gain coefficient is

$$g = \frac{\lambda_{\mathrm{p}}\lambda_{\mathrm{S}}^2 N}{\pi c\hbar n_{\mathrm{S}}^2 \Delta\nu_{\mathrm{R}}}\left(\frac{\mathrm{d}\sigma}{\mathrm{d}\Omega}\right). \tag{B1.7.3}$$

Here, λ_{p} and λ_{S} are the wavelengths of the laser pump beam and the Stokes Raman beam, respectively, and $\Delta\nu_{\mathrm{R}}$ is the full-width at half maximum of the Raman spectral line in units of s^{-1}. The gain coefficient is the most important parameter for solid state laser applications and its magnitude is usually given in units of cm GW^{-1}. These expressions show that the intensity of the Raman beam increases with the intensity of the pump laser and the interaction length and is greater for higher scattering cross sections. In addition, the gain is greatest for materials with a high density and a small Raman linewidth [5]. In the transient case, when the pump pulse duration is smaller than the dephasing time and the pump spectral width is broader than that of the Raman line ($t_p < 10T_R$; $\Delta\nu_p \gg \Delta\nu_R$), the analytical expression for the Stokes intensity for large amplification gain, in the non-pump-depletion region, can be written as [6]

$$I_S(L) = I_S(0)\exp\left(-\frac{t_p}{T_R}\right)\exp 2\left[I_p t_p L \frac{\lambda_p \lambda_S^2 N}{\hbar n_S^2}\left(\frac{d\sigma}{d\Omega}\right)\right]^{1/2}. \tag{B1.7.4}$$

Comparing expression (B1.7.4) with (B1.7.1) and (B1.7.2), one can see that the Raman gain has a slower (square root) dependence on the crystal length L and the total integral Raman scattering cross section, $d\sigma/d\Omega$, and does not depend on the Raman linewidth, $\Delta\nu_R$, at all. The linear pump intensity dependence for steady-state regimes changes to the pump pulse energy ($W = I_p t_p$) square root dependence for the transient regime described by equation (B1.7.4). The transient regime is very important when one tries to realize short-pulse picosecond and femtosecond Raman shifting and lasing.

The width of a Stokes Raman line is determined by the lifetime of the vibrational mode produced in the Raman interaction $\Delta\nu_R = (\pi C T_R)^{-1}$ and is strongly temperature dependent. This decay or dephasing of a phonon mode is generally associated with phonon–phonon interactions in which the initial phonon decays into two other phonon modes with conservation of energy and wavevector. This takes place through anharmonic coupling of the vibrational modes represented by the interaction Hamiltonian H_A. The expression for the width of a Stokes Raman line is [7]

$$\begin{aligned}\Delta\tilde{\nu}_R \sim \int d\kappa \sum_{s,s'} |\langle n_{s\kappa_1}+1, n_{s'\kappa_2}+1, n_0 = 0|H_A|n_{s\kappa_1}, n_{s'\kappa_2}, n_0 = 1\rangle|^2 \\ \times\, \delta(\omega_0 - \omega_{s\kappa_1} - \omega_{s'\kappa k_2})\delta(\kappa - \kappa_1 - \kappa_2)\end{aligned} \tag{B1.7.5}$$

where the phonon wavevectors and frequencies are given by κ and ω, respectively. The rate at which a phonon decay process occurs, thus, depends on the strength of the anharmonic coupling in the lattice and the phonon density of states. Thus, materials that have a small anharmonicity parameter will have a small Raman linewidth and, therefore, a high gain coefficient [5].

The second material parameter affecting the Raman gain is the scattering cross section, which depends on the strength of the photon–phonon coupling. This can be expressed in terms of the laser-induced modulation of the local polarizability through the relationship

$$\left(\frac{d\sigma}{d\Omega}\right) = \frac{\hbar\omega_S^4}{2m\omega_R c^4}\left(\frac{\partial\alpha}{\partial q}\right)^2. \tag{B1.7.6}$$

Here ω_R is the frequency of the vibrational mode. Thus, materials that have a high value of polarizability and vibrational modes that induce a large modulation of this polarizability will have a large Raman scattering cross section and, thus, a high Raman gain. Because of the resonant denominator appearing in the quantum mechanical expression for the dipole moment in the polarizability, materials with low-energy electronic transitions (such as molecular crystals with charge transfer bands) will have a high polarizability. In addition, the rate of the Raman scattering processes involves the matrix element of polarizability between initial and final states and this leads to symmetry selection rules with directional properties for each vibrational mode [7]. Totally symmetric 'breathing modes' of vibration produce the greatest degree of modulation of the polarizability. Also lattice structures such as diamond, zinc-blende and wurtzite are favourable for covalent bonding and allow for a greater vibrational modulation of the polarizability than ionic binding.

The two-photon nature of Raman transitions results in a very small scattering cross section for the process. In order to attain high efficiency for the stimulated Raman conversion starting from spontaneous noise of very low intensity, we must provide a very high increment $I_p g L \simeq 30$ in equation (B1.7.2). For a typical crystal length of about 3 cm, this leads to a very high gain value $I_p g \simeq 10\ \mathrm{cm}^{-1}$. This estimate means that, even for high gain materials with $g \simeq 10\ \mathrm{cm\ GW}^{-1}$, we must use a very high pump density at the level of 1 GW cm^{-2} without laser damage to the Raman crystal. For low gain materials, the pump density must be even higher. One of the best examples of a new synthetic solid-state Raman material is $Ba(NO_3)_2$ operating on the A_{1g} nitrate molecular ion breathing mode [8–14]. This provides a frequency shift of 1046 cm^{-1} with an extraordinarily small linewidth at room temperature of 0.4 cm^{-1}, leading to a Raman gain of as high as 47 cm GW^{-1} for green (0.532 μm) and 11 cm GW^{-1} for infrared (1.064 μm) pumping. $Ba(NO_3)_2$ crystals are now available in large sizes with high optical quality and a laser damage threshold higher than 1 GW cm^{-2}.

Measuring and analysis of the temperature dependencies of the spectral line broadening and dephasing time for the A_{1g} Raman active vibrational mode provide a better understanding of the unique narrow linewidth feature of $Ba(NO_3)_2$ crystals [13, 15, 16]. The forbiddance of fast three-phonon splitting processes in dephasing and line broadening (which is the main process for most other materials) together with very small inhomogeneous linewidth results in the extraordinarily small Raman linewidth, $\Delta\nu_R$, and strong temperature dependence of the linewidth. The latter is well explained by a four-phonon relaxation and dephasing process with a much smaller relaxation rate [13, 15, 16].

Another good synthetic crystal with practically the same Raman frequency shift ($\nu_R = 1069\ \mathrm{cm}^{-1}$) is $NaNO_3$. The broader linewidth of $NaN0_3$ ($\Delta\nu_R = 1.0\ \mathrm{cm}^{-1}$) results in a lower gain coefficient in the steady-state regime compared to $Ba(NO_3)_2$.

For natural crystals, the best Raman properties are exhibited by diamond and calcite ($CaCO_3$) crystals providing high-frequency shifts of 1333 and 1086 cm^{-1}, respectively. Unfortunately, size, optical quality and the price of natural crystals limit their applications. The quality of natural crystals causes the inhomogeneous widths of their Raman lines to vary greatly, from 1.5 to 2.7 cm^{-1} for diamond and from 1.2 to 1.6 cm^{-1} for calcite [17]. This leads to significant variations in gain.

$KGd(WO_4)_2$ synthetic crystals have a smaller Raman frequency shift and much broader Raman lines, 5–8 cm^{-1}. These have been useful for picosecond pulse applications.

Lithium iodate ($LiIO_3$) and niobate ($LiNbO_3$) synthetic crystals were among the most popular Raman materials at the beginning of the Raman laser era. However, due to the large linewidth, 5–20 cm^{-1}, and moderate laser damage threshold, 0.1–0.5 GW cm^{-2}, they did not find wide practical applications.

To estimate the potential of a specific crystal as a Raman laser material, it is important to compare the frequency and spectral width of the Raman line, as well as the peak and integrated cross sections for the two-photon Raman transition. For a great number of prospective Raman materials, a comparative study was undertaken [17–20] and some of the results are presented in table B1.7.1. The materials that were analyzed included carbonate, nitrate, phosphate, tungstate, niobate, iodate, bromide, borate and silicate crystals. The frequencies and linewidths of the most intense and narrow quasi-molecular Raman vibrations of $[CO_3]$, $[NO_3]$, $[WO_4]$, $[MoO_4]$, $[NbO_6]$, $[NbO_4]$, $[IO_3]$, $[PO_4]$, $[C1O_3]$, $[B_3O_6]$, $[BrO_3]$, $[SiO_2]$, $[SO_4]$ anion and $[NH_4]$ cation groups were compared. The relative intensities (integral and peak) of spontaneous Raman scattering (RS) lines were measured for more than 30 synthetic crystals in comparison with a reference diamond crystal (100%). The RS line intensities were studied for various orientations of the crystals with respect to the exciting and scattered beams (see table B1.7.1). As can be seen from the table, the highest gain for the steady-state regime (peak intensity) compared to diamond (100%) is found in $Ba(NO_3)_2$ (63%), $BaMoO_4$ (64%) and $BaWO_4$ (63%). For $SrMoO_4$ and other tungstates, iodates and niobates, the gain is between 50 and 20% and for calcite it is only about 10%. In the transient regime or for femto- and picosecond operation, where the integrated scattering cross section is the most important parameter, the record holder is expected to be $LiNbO_3$ (166%) and rutile TiO_2 (159%), the next is diamond (100%), then the molybdates,

Table B1.7.1. Parameters of spontaneous Raman scattering in crystals [17]

Material	Lattice space group	Molecular group	Raman frequency ν_R (cm^{-1})	Raman linewidth $\Delta\nu_R$ (cm^{-1})	Integral cross section (au)	Peak intensity (au)	Geometry of excitation scattering *K*	*E*
Diamond	O_h^7	—	1332.9	2.7	100	100	$\parallel C_3$	$\perp C_3$
Nitrates and calcites								
$Ba(NO_3)_2$	Th^6	$[NO_3]$	1048.6	0.4	21	63	$\parallel C_4$	$\parallel C_4$
$NaNO_3$	D_{3d}^6	$[NO_3]$	1069.2	1.0	23	44	$\parallel C_3$	$\perp C_3$
$CaCO_3$	D_{3d}^6	$[CO_3]$	1086.4	1.2	6.0	10.6	$\parallel C_3$	$\perp C_3$
Tungstates								
$CaWO_4$	C_{4h}^6	$[WO_4]$	910.7	6.95	52	18.6	$\perp C_4$	$\parallel C_4$
$SrWO_4$	C_{4h}^6	$[WO_4]$	921.5	2.74	50	41	$\perp C_4$	$\parallel C_4$
$BaWO_4$	C_{4h}^6	$[WO_4]$	926.5	1.63	47	64	$\perp C_4$	$\parallel C_4$
$KGd(WO_4)_2$	C_{2h}^6	$[WO_6]$	901	5.4	50	35	$\perp C_2$	$\perp C_2$
$KGd(WO_4)_2$	C_{2h}^6	$[WO_6]$	768	6.4	59	37	$\perp C_2$	$\parallel C_2$
Molybdates								
$CaMoO_4$	C_{4h}^6	$[MoO_4]$	879.3	5.0	64	34	$\perp C_4$	$\parallel C_4$
$SrMoO_4$	C_{4h}^6	$[MoO_4]$	887.7	2.8	55	51	$\perp C_4$	$\parallel C_4$
$BaMoO_4$	C_{4h}^6	$[MoO_4]$	892.4	2.1	52	64	$\perp C_4$	$\parallel C_4$
Iodate and niobates								
$LiIO_3$	C_6^6	$[IO_3]$	821.6	5.0	54	25	$\parallel C_2$	$\perp C_2$
$LiNbO_3$	C_{3v}^6	$[NbO_6]$	872	21.4	44	5	$\parallel C_3$	$\perp C_3$
$LiNbO_3$	C_{3v}^6	$[NbO_6]$	632	27	166	18	$\perp C_3$	$\parallel C_3$
$LiNbO_3$	C_{3v}^6	$[NbO_6]$	250	28	—	22	$\perp C_3$	$\parallel C_3$
$LaNbO_4$	C_{2h}^3	$[NbO_4]$	805	9	22	7.1	$\perp C_2$	$\parallel C_2$
Phosphates								
$Ca_5(PO_4)_3F$	C_{6h}^2	$[PO_4]$	964.7	2.8	3.4	3.8	$\perp C_6$	$\parallel C_6$
$Sr_5(PO_4)_3F$	C_{6h}^2	$[PO_4]$	950.3	2.8	3.4	3.8	$\perp C_6$	$\parallel C_6$
$Li_3PO_4^a$	—	$[PO_4]$	951	7.7	—	—	—	—
Other								
$Ba_3(B_3O_6)_2$	C_{3v}^6	$[B_3O_6]$	636	4.5	1	0.6	$\parallel C_3$	$\perp C_3$
TiO_2	D_{4h}^{14}		612	48	159	9.8	$\perp C$	$\parallel C$
ZnO	C_{6v}^4		438	6.0	7.14	2.7	$\parallel C$	$\perp C$
Y_2O_3			379	4.0	3.3	2.16		
SiO_2	D_3^6	$[SiO_4]$	464	7	2.2	1.2	$\perp C_3$	$\parallel C_3$
$Y_3Al_5O_{12}$			783	8	3	1.0		
Al_2O_3			419	2	0.05	0.07		
CdF_2			318	21	0.16	0.2		
CaF_2			323	8.4	0.07	0.21		

[a] Line with inhomogeneous splitting.

tungstates and iodates (70–54%). Nitrates $Ba(NO_3)_2$ and $NaNO_3$ show only 21–23%, and calcite $CaCO_3$ is even smaller at 6%.

An analysis of the data given in table B1.7.1 shows that the best crystals for low-threshold SRS frequency converters should have both a large integral RS cross section and a small RS transition broadening. Two promising classes for this purpose are tungstates and molybdates, which possess large integral RS cross sections (40–60%) and a great variety of line broadenings (1.6–15 cm^{-1}).

Since the line broadening (both due to the up-conversion and decay dephasing) of high-frequency SRS-active vibrational modes relates to their interaction with lattice phonons, the authors of [21–24] selected the crystals with the heaviest cations to reduce the frequencies of phonon modes. In [21–24], simple tungstates of alkaline-earth metals (Ca, Sr and Ba) with a scheelite structure were synthesized and investigated. As is evident from table B1.7.1, the $CaWO_4$ crystal frequently encountered in the scientific literature has a modest peak scattering cross section, equal to 18%, which is three times smaller than in the $Ba(NO_3)_2$ crystal and almost two times smaller than in $KGd(WO_4)_2$. This results from the significant broadening of its RS line, $\Delta\nu_R = 6.95\ cm^{-1}$, at room temperature. Replacement of the small-size and low-mass calcium ions for larger, heavier strontium and barium ions resulted in a much lower relaxation rate and the several times narrower spectral line (2.74 and 1.63 cm^{-1}) of $[WO_4]^{2-}$ Raman vibrations at room temperature.

As the cation mass increases in the sequence Ca, Sr and Ba, the maximum frequency ω_{lat} of lattice phonons decreases, from 274 to 194 cm^{-1}. In addition, the increase in the cation radius and lattice constant results in the higher frequency of the totally symmetric A_{1g} mode of the $[WO_4]^{2-}$ complex (see table B1.7.1). These two factors weaken the vibron–lattice interactions and considerably narrow the spectrum of Raman vibration, which should manifest itself in a growth of the peak scattering cross section and the steady-state SRS gain.

A detailed study of the RS fine structure of the A_{1g} mode in scheelites [23] shows that due to the largest lattice constants and the W–W distances for the $BaWO_4$ crystal (as well as for $Ba(NO_3)_2$), the Davydov splitting (DS) for these crystals is very small, $\Delta E_{DS} \approx 0.5–1\ cm^{-1}$ [23]. A very low acoustic phonon density at this frequency, $h\omega_{ph} = \Delta E_{DS} = 0.5–1\ cm^{-1}$, cannot provide fast dephasing and strong broadening of these Davydov-split components by means of direct single-phonon bridge processes. Conversely, in $CaWO_4$ and partly in $SrWO_4$ crystals with smaller lattice constants, the Davydov splitting was found to be much higher—50 and 28 cm^{-1}, respectively. This can explain the much faster dephasing and stronger broadening (6.9 and 2.74 cm^{-1}) in these crystals due to direct single-phonon bridge processes from one DS component to another with absorption or emission of acoustic phonons with 30–50 cm^{-1} frequencies, whose density (ρ) can be rather high ($\rho \approx \omega_{ph}^2$).

Similar regularities were observed in the series of alkaline-earth molybdates with a scheelite structure, $CaMoO_4$, $SrMoO_4$ and $BaMoO_4$. Table B1.7.1 demonstrates the relative values of integral and peak scattering coefficients. As is seen, even with the broader RS lines, the new crystals $BaWO_4$ and $BaMoO_4$ surpass the unique $Ba(NO_3)_2$ crystal in the peak cross section, while the $SrWO_4$ and $SrMoO_4$ crystals are only slightly below.

The data on the peak cross sections of spontaneous RS found by the comparative study were subsequently confirmed by direct measurements of SRS gain and thresholds [17, 21, 22, 24]. This proved the high potential of the newly developed crystals for applications in SRS amplifiers and lasers, which will be discussed later.

B1.7.3 Visible nanosecond SRS based on $Ba(NO_3)_2$ crystals

One of the first studies of visible SRS in synthetic nitrate crystals $Ba(NO_3)_2$, $Pb(NO_3)_2$ and $NaNO_3$ in comparison with natural $CaCO_3$ crystals was reported in [8]. SRS crystals 15 mm long, were studied in a single-pass scheme under 530 nm excitation. The pump laser, consisting of an Nd-glass master oscillator with passive Q-switching, a double-stage amplifier and a second harmonic generator, produced pulses of green radiation with energy 0.8 J and a duration of 15 ns. The pump energy density was increased from 1 J cm^{-2} to the point of laser damage of the SRS crystal, which is 6–15 J cm^{-2}. For the $Ba(NO_3)_2$ crystal, the minimum SRS threshold energies were about 1 J cm^{-2} for the first Stokes, and 3 J cm^{-2} for the second Stokes, and the highest total conversion efficiency was equal to 26%. The first Stokes maximum output energy was as high as 0.15 J at the pump energy density of about 8 J cm^{-2}, which was close to the surface damage threshold. For the $Pb(NO_3)_2$ crystal, the SRS thresholds were approximately double, while the maximum conversion efficiency was lower by about 20%, and the first Stokes output of about 70 mJ was reached at the pump density of 6–7 J cm^{-2}, starting the bulk damage of the sample. The $NaNO_3$ and $CaCO_3$ crystals showed even higher SRS thresholds of about 4–5 J cm^{-2} but also had a higher laser damage threshold of 14 J cm^{-2}. The output energies were measured to be 0.15 J with a conversion efficiency of 17% for $NaNO_3$ and 0.12 J with 14% efficiency for $CaCO_3$. Thus, by the first and second Stokes frequency shifts of 1045–1085 cm^{-1}, all these crystals can provide highly efficient yellow (560 nm) and orange (598 nm) visible radiation, but the best performance was shown by the $Ba(NO_3)_2$ crystal.

Raman lasers based on the $Ba(NO_3)_2$, $NaNO_3$ and $CaCO_3$ crystals under external pumping by nanosecond ($t_p = 10$ ns) 532 nm radiation were studied in [10]. In a flat resonator 26 cm long with the reflection coefficients of mirrors $R1 \approx 100\%$ and $R2 = 8\%$, the threshold pump densities were equal to 5–7 MW cm^{-2} for nitrate crystals and 20 MW cm^{-2} for calcites (the crystals were 5–8 cm long). These data correlate well with the directly measured SRS gain coefficients: $g_{Ba(NO_3)_2} = 47 \pm 5$ cm GW^{-1}, $g_{CaCO_3} = 13 \pm 3$ cm GW^{-1}, and $g_{NaNO_3} = 47 \pm 5$ cm GW^{-1}.

At a pump density exceeding the threshold by fivefold, the efficiency of the pump conversion to all the Stokes components reached a maximum of 50–65%. Since the variations in R2 from 4 to 20% did not change the conversion efficiency, the exit face of the Raman crystal can be used instead of the output mirror. The maximum conversion efficiency in the $Ba(NO_3)_2$ crystal was 40% for the first Stokes component and 25% for the second Stokes component at single-pass pumping. It was shown that an increase in the number of pump beam passes in the cavity resulted in a better selection and an increase in the conversion efficiency (up to 50% for the first Stokes component at two-pass pumping). The second Stokes component in a flat and unstable resonator can be efficiently suppressed by the on-coming pumping.

Visible SRS lasers with a much higher repetition rate were studied in [25]. The authors for the first time obtained SRS conversion under copper vapour laser pumping with a repetition rate of 16 kHz and a mean power of 1.9 W at a wavelength of 510.6 nm and 1.8 W at 578.2 nm. The beam divergence of the copper vapour laser was rather low, 2×10^{-4} rad, and the pulse duration was short, 14 ns. In a $Ba(NO_3)_2$ crystal 5 cm long, the SRS radiation at $\lambda_{Stokes} = 539.4$ nm had a power of 600 mW and a conversion efficiency of 38%, and the radiation at $\lambda_{Stokes} = 615.4$ nm was obtained with a power of 300 mW and a conversion efficiency of 18%. Using monochromatic pumping, the yellow band had a higher SRS threshold than the green band, which correlates with the spectral dependence of the gain coefficient. Using biharmonic pumping, the yellow band SRS threshold was reduced to that of the green band. The summed efficiency of the high order (second and third) Stokes components did not exceed 2–3%. The power of the backward SRS radiation was fourfold weaker than the power of the direct SRS. The authors observed thermally induced birefringence and offered methods to eliminate it. It was supposed that additional Stokes losses appear due to nonlinear absorption. The life of the SRS crystals before laser damage was found to depend on the quality and purity of the chemical reagents used for the synthesis of $Ba(NO_3)_2$ crystals. By summing the pump and SRS frequencies in the KDP and DKDP crystals, ten lines of UV radiation were obtained in the region 203–308 nm with a mean power of 1–80 mW.

Airborne or space-based LIDAR transmitters with two UV on-line and off-line outputs for ozone measurements can be developed on the basis of a $Ba(NO_3)_2$ Raman laser emitting the first (563 nm) and second (599 nm) Stokes frequencies under 532 nm pumping, with the following frequency doubling to 281 and 299 nm. This approach was proposed in [26]. The authors used a compact frequency-doubled Q-switched Nd:YAG laser with a pulse duration of 8 ns and a repetition rate of up to 30 Hz. The $Ba(NO_3)_2$ crystal 5 cm long was placed in a separate 15 cm Raman cavity, consisting of a flat high-reflecting end mirror and a 10 m radius concave output coupler. With a small pump beam spot diameter (1 mm), a pump energy of 10–30 mJ and a pump flux of 1–4 J cm^{-2}, the output energy for the first and second Stokes frequencies was up to 6–8 mJ with a conversion efficiency of 20–30%. There was only a small difference in the characteristics for the 1 and 10 Hz repetition rates. For a higher pump energy (20–70 mJ) and larger pump spot diameter (2.5 mm), at half the flux and with an optimized output coupler, first and second Stokes radiation with a power, respectively, of 25 and 32 mJ was obtained with a high conversion efficiency of 40–48%. The total conversion efficiency into the two Stokes frequencies was more than 65% at the repetition rate of 30 Hz.

Amplification of 563 nm laser radiation by a barium nitrate Raman oscillator/amplifier was measured in [27]. As the pump source, the authors used the second harmonic ($\lambda = 532$ nm) of an Nd:YAG laser with a pulse duration of 5–7 ns and a beam diameter of 1.8 mm. The dependence of the amplified first Stokes energy on the pump energy (5–25 μJ) was measured for the two seed energies (2 and 35 μJ) of the first Stokes beam. In both cases, the output energy was virtually the same and reached 3.5 mJ. Hence, the experiments demonstrated a high, from 100 to 1600, Raman gain of the single-pass $Ba(NO_3)_2$ Raman amplifier.

The experiments described here demonstrate the excellent performance of $Ba(NO_3)_2$ crystals as SRS materials for visible-frequency Raman shifting at nanosecond pumping of high peak power (the steady-state regime of SRS).

Transient SRS in $Ba(NO_3)_2$ crystals was first studied in [28] under pumping by 532 nm second harmonic radiation of an Nd:YAG laser. The pump pulse duration was about 22 ± 2 ps, which is less then the dephasing (transverse) relaxation time $T_2(= 28$ ps) of internal symmetrical vibrations of quasi-molecular $(NO_3)^-$ complexes of $Ba(NO_3)_2$ crystals. In the single-pass scheme, SRS threshold intensities of $Ba(NO_3)_2$ crystals 50 and 40 mm long were estimated to be 1.1 and 1.45 GW cm^{-2}, respectively, which is much higher than for the steady-state nanosecond oscillation regime. This corresponds to an approximately tenfold decrease in the picosecond Raman gain in comparison with steady nanosecond pumping. It should be noted that the overall conversion efficiency of a $Ba(NO_3)_2$ crystal at a pump density of 2 GW cm^{-2} was 25% for the first Stokes Raman shifting ($\lambda_{IST} = 563$ nm) and 5% for the second Stokes ($\lambda_{IIST} = 599$ nm).

These data, restrict application of $Ba(NO_3)_2$ crystals for picosecond SRS, especially in the infrared spectral region.

B1.7.4 Infrared SRS shifters and Raman lasers based on $Ba(NO_3)_2$

Solid state Raman lasers, discretely or continuously tunable, are of special importance for the IR spectral region, where no dye lasers with high quantum yield and good photo and thermal stability exist and only a few IR bands are covered by available tunable solid state lasers based on impurity-doped and radiatively-coloured crystals.

At the same time, SRS becomes a much more complicated problem when moving to the IR spectral region, because the SRS gain cross section decreases with an increase in the pump wavelength. For example, the $Ba(NO_3)_2$ crystal has $g = 47$ cm GW^{-1} for $\lambda = 532$ nm and $g = 11$ cm GW^{-1} for $\lambda = 1064$ nm [12, 13].

By the early 1980s, unique tunable lasers for the near-IR range were developed by employing LiF crystals with F_2^- and F_2^+ colour centres [29–31]. These lasers were tunable in the 0.82–1.28 μm IR region. To extend the tuning range further into the IR region, new SRS-active crystals, $Ba(NO_3)_2$ and $KGd(WO_4)_2$, were offered [32]. The investigations described in [32–34] have shown that $Ba(NO_3)_2$ crystals operate advantageously in the near-IR range up to 1.7 μm with efficiencies as high as 60% under nanosecond pumping, with power densities well below the laser damage threshold. Further extension into the IR is restricted by $Ba(NO_3)_2$ fundamental absorption beginning at approximately 1.8 μm. The $KGd(WO_4)_2$ crystals have a much wider transparency region, up to 5 μm, and also can convert the IR radiation to the Stokes region, but the four times smaller steady-state gain coefficient and low laser radiation resistance restrain their application in nanosecond lasers.

The development of SRS shifters based on $Ba(NO_3)_2$ crystals enabled the spectral range of tunable solid state lasers to be extended to the 1.3–1.7 μm IR region and, as a result, to develop a solid state laser spectrometer for a very wide near IR region (from 0.82 to 1.7 μm). This spectrometer also covered the visible (0.42–0.82 μm) and UV (0.25–0.45 μm) regions by using the second and higher harmonics of tunable IR radiation [31–34]. It was the first all-solid-state laser spectrometer operating in all the three spectral regions—UV, visible and IR.

Interest in SRS lasers was rekindled in the early 1990s, when solid state lasers based on impurity-doped crystals found so many applications that there was a need to prevent the human eye from being hurt by a laser beam. The laser wavelength determined to be safe for the human eye was standardized to be $\lambda = 1.54$ μm and the radiation of the majority of lasers was found to be in the prohibited wavelength area with $\lambda < 1.5$ μm (as an example, Nd:YAG laser crystals operate at 0.53, 1.06 and 1.32 μm, ruby at 0.69 μm, alexandrite at around 0.78 μm and Ti:sapphire at 0.7–1.1 μm). In this respect, the development of SRS frequency shifters for laser radiation available at the eye-safe wavelength of 1.54 μm is of great importance.

The small atmospheric absorption and the high transmittance of fibres at this wavelength also allow wide applications in such devices as LIDARs, free space communication and fibre links.

One of the simplest and most effective ways to obtain 1.54 μm radiation was proposed in 1993–94 in [35–42], where a $Ba(NO_3)_2$ Raman laser was pumped by the 1.064 and 1.320 μm radiation of a Q-switched neodymium laser. Under 1064 nm nanosecond pumping, a $Ba(NO_3)_2$ crystal 30–50 mm long in a compact cavity with dichroic mirrors, specifically designed for optimum lasing of the first, second and third Stokes waves, produced a single-mode radiation at new, including eye-safe, wavelengths (1.2–1.4 μm and 1.5–1.6 μm), with a light-to-light conversion efficiency of 20–50% [35–38].

Application of a single-pass $Ba(NO_3)_2$ SRS converter in combination with an amplifier on an LiF crystal with F_2^- colour centres is described in [43]. In this work, a SRS converter produced a weak (with an efficiency of 1%) nanosecond signal of the first Stokes component at 1197 nm, which was then amplified in an LiF: F_2^- crystal 12 cm long. The same Nd:YAG laser with an energy of 60 mJ and a pulse duration of 12 ns at 1064 nm, used as the pump source for the converter and amplifier, and a good spectral overlap of the generation and amplification regions provided a high total efficiency for the system, of up to 35% from the pump energy. No beam focusing to the SRS crystal was used and the second Stokes component was completely absent. In the intracavity pumping scheme, when both the SRS and LiF: F_2^- crystals were placed inside the Nd:YAG cavity, the conversion efficiency was improved by up to 50% [44]. When the Nd:YAG crystal was replaced with Nd:YAP, a conversion efficiency of 50% was recorded for frequency shifting in both the pump (1078 nm) and first Stokes (1216 nm) radiation. Using a KGW:Nd crystal in the same cavity, operating both as the active laser and SRS element, the authors [43] had the first Stokes radiation at 1180 nm with an efficiency of 50%. The oscillation train contained 1–15 pulses with an energy 45–10 mJ each depending on the initial transmittance of the LiF: F_2^- crystal, which worked as the passive Q-switch, converter to the first Stokes radiation and amplifier.

The 1.320 μm nanosecond intracavity pumping of a $Ba(NO_3)_2$ crystal, when the pump and Raman laser had one common cavity or two optically coupled cavities, enabled the conversion efficiency to the first Stokes frequency at 1.5 μm to be increased up to 90% [39–42]. In this case, the effect of nonlinear cavity dumping makes it possible to shorten the pump pulse a few times and to increase the peak power. An important additional advantage of SRS lasers is that their output beam can have a much better spatial quality and smaller beam divergence ('beam clean-up') and, hence, higher brightness than the pump beam [36–38, 45].

A diffraction-limited high-energy (0.25 J) eye-safe Raman laser with wavelengths of 1535 and 1560 nm was developed in 1995 [46, 47] when an intracavity $Ba(NO_3)_2$ crystal was pumped by a high-power nanosecond Nd:YAG laser operating at 1.318–1.338 μm. The two standing-wave resonators, a pump cavity 69 cm long and a Raman cavity 20 cm long were coupled by a dichroic beamsplitter. Using a 50 mm uncoated $Ba(NO_3)_2$ crystal, the authors [46, 47] had 12 mJ of Raman laser output with 48% conversion efficiency at a pump power of 25 mJ. The SRS pulse was about half that of the 10 ns pump pulse. Nonlinear Raman beam

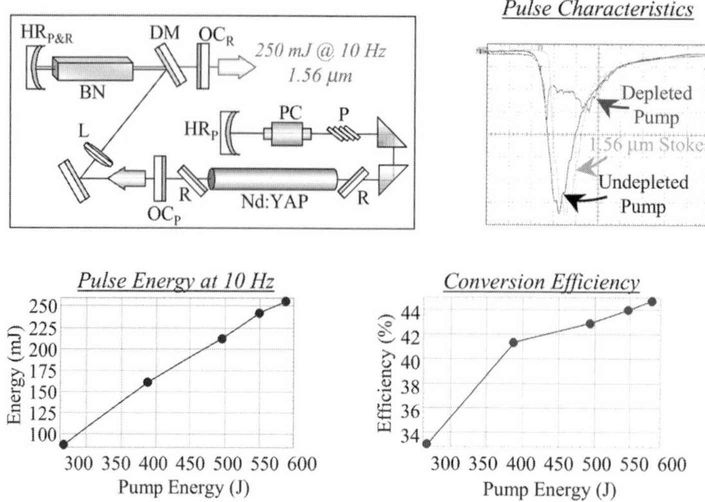

Figure B1.7.1. Optical scheme and typical characteristics of an eye-safe solid state Raman laser. (Data provided by J M Murray and W Austin, Lite Cycles, Inc.)

clean-up was observed in the near and far fields of the 1.5 μm beam profile. The second Stokes radiation at 1.820–1.860 μm was generated through a cascaded SRS process. In an enhanced (more powerful) system with a Nd:YAG rod ⌀7 mm × 110 mm and two flashlamp pump cavities, the output diffraction-limited eye-safe radiation had a power of 0.25 J at a repetition rate of 1 Hz. The repetition rate was limited by the thermal effects due to the low thermal conductivity of $Ba(NO_3)_2$ crystals.

The optical scheme and characteristics of a typical eye-safe Raman laser with increased pulse repetition rate are shown in figure B1.7.1.

An attempt to develop 20 Hz eye-safe cascade Raman lasers on $Ba(NO_3)_2$ crystals with 140 mJ, 1064 nm Nd:YAG laser pumping was done in [48]. The authors used an external pumping scheme with a 50 mm long $Ba(NO_3)_2$ crystal placed in a 20 cm, flat-mirror Raman laser cavity. In the double-pass pumping scheme with output coupling $T = 82\%$, they obtained a threshold energy of 50 mJ (1.3 J cm^{-2}) and an SRS output energy of 28 mJ for the first Stokes radiation at $\lambda = 1197$ nm. For the second Stokes 1369 nm laser cavity with an output coupler transmission $T = 62\%$, the output energy was 27 mJ. Using highly reflecting mirrors for the first and second Stokes wavelengths and the output coupler for the third Stokes with $T = 71\%$, the authors had 8.5 mJ of the third Stokes radiation at the eye-safe wavelength of 1.598 μm. The pulsewidth decreased from 30 ns of the pump pulse to 11 ns for the first and to 9.5 ns for the third Stokes. Substantial energy drops of 7, 46 and 76% for the first, second and third Stokes, respectively, were observed due to the thermal load increasing in the $Ba(NO_3)_2$ crystal at a repetition rate of 20 Hz and a pump energy of 140 mJ.

In order to develop a marine-based LIDAR laser transmitter (see chapter D7.2 on LIDAR), operating around 580 nm, for the very turbid water of a surf zone, the authors of [49] have studied and optimized a $Ba(NO_3)_2$ SRS laser with intracavity pumping by a Nd:YLF laser and with second harmonic generation by an LBO crystal. With an Nd:YAG rod ⌀7 mm × 76 mm and a T-shaped high-Q pump laser cavity coupled with a low-Q Raman cavity (400 and 85 mm in length, respectively), the authors of [49] realized nonlinear cavity-damping regimes. They use a 20 mm long MgF_2 AR-coated $Ba(NO_3)_2$ SRS crystal and a Raman cavity output coupler with $R = 25\%$. With 50 J of flashlamp pump energy and 10 Hz, the maximum output for a Raman wavelength of 1176 nm was 150 mJ, which was transferred into 90 mJ of yellow (588 nm) radiation by the LBO (type I) crystal 12 mm long. To stop pulse modulation and frequency cascading, the

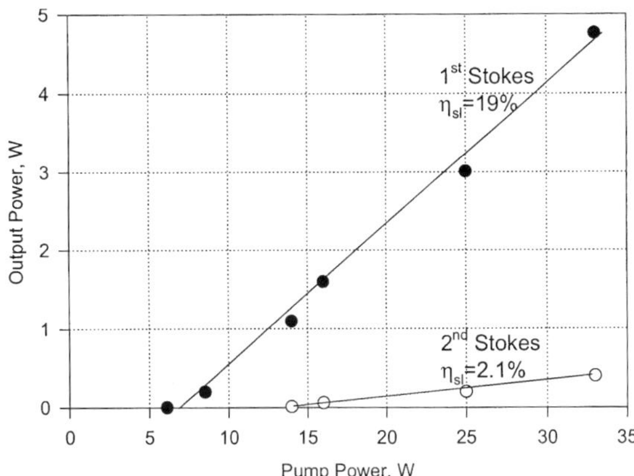

Figure B1.7.2. Pump-to-output average power dependences for the first and second Stokes radiation of $Ba(NO_3)_2$ Raman lasers under 1064 nm high-power Nd:YAG pumping.

authors included a prism in the Raman cavity. In this case, they had 1.5 ns pulsed radiation at 588 nm wavelength, which is well suited for coastal water LIDAR applications.

Application of a single-mode, single-frequency Nd:YAG laser with very high spectral and spatial brightness as a pump source allowed an average output power of 4.7 W to be reached by a $Ba(NO_3)_2$ Raman laser under 33 W pump power, which corresponds to a slope efficiency of about 19% (see figure B1.7.2) [33]. Simultaneously, the second Stokes radiation was observed with a power of 0.4 W at a slope efficiency of 2.1%. The pump laser beam produced five-pulse trains with a repetition rate of 30 Hz. The train energy was 1.1 J at a pulse energy of 0.22 J and peak power of 4.4 MW. The laser beam in the $Ba(NO_3)_2$ crystal was 2 mm in diameter. The Raman laser cavity 6.5 cm long was formed by flat mirrors: the first mirror had a high transmission for the pump wavelength and high reflection for the first Stokes radiation, while the output coupler had a reflectivity of 77% for 1.064 nm and 35% for the first Stokes radiation.

At the lower repetition rate of 10 Hz and a single-pulse train, the maximum conversion efficiency to the first Stokes radiation was 35% and the maximum output energy reached was 30 mJ. At the same time, the second Stokes energy was as high as 16 mJ with an efficiency of about 10%.

B1.7.5 IR-pumped $LiIO_3$ Raman laser

The $LiIO_3$ crystalline Raman laser operating in the IR region at a high repetition rate was first described in [50, 51]. The pumping Nd:YAG laser operated in a quasi-continuous-wave (cw) mode at 1080 nm with a repetition rate of 1–10 kHz and a pulse duration of 100 ns. In a shared two-mirror cavity with a $LiIO_3$ SRS crystal 25 mm long, cut at $\theta \approx 20.5°$, the first Stokes radiation (the Raman shift $\nu_R = 818\ \mathrm{cm}^{-1}$) had an average output power of 1.26 W and a pulse duration of 50 ns. In a separate experiment, a second Stokes radiation of 1310 nm was obtained with power 0.55 W and a pulse duration of 15 ns [50].

Using a three-mirror coupled cavity, the author of [51] had a strong pulse shortening, $t_{p(1St)} = 4$–8 ns with an average output power of 0.9 W, a peak power of 7 kW and a conversion efficiency of 77%. After optimization, the second Stokes pulse duration was even shorter, $t_{p(2St)} = 2$ ns with an average power of 0.2 W and a peak power of 100 kW.

A picosecond Raman laser on a $LiIO_3$ SRS crystal was studied in [52]. The authors used a Y-cut lithium iodate crystal 10 mm long and, as the source of synchronous pumping, a self-mode-locked neodymium phosphate glass laser operating at $\lambda = 1054$ nm. The pump pulses had a duration of 6–10 ps at an axial interval of 6.7 ns and the duration of the train envelope was about 200 ns.

A Raman laser cavity 1 m long was formed by an entrance edge mirror with $R_{1.05\mu} = 25\%$ and $R_{1.14c} = 98\%$, an output mirror with $R_{(1.05\mu,1.14\mu)} = 50\%$ and a two-lens intracavity telescope with a $LiIO_3$ crystal in between. The first Stokes component was observed at a 1.143 μm with a frequency shift of 760 cm^{-1}. In the vicinity of the SRS threshold (1 mJ), the output train consisted of 25 pulses and had a total duration of 170 ns. Varying the Raman cavity length, the authors found that the maximum shortening (by a factor of 7–8) of the SRS pulses in comparison with the pump pulses occurs when the Raman cavity length is equal to the length of the pump cavity. The minimum SRS pulse duration was 1 ps at an energy conversion efficiency of 30%. The maximum conversion efficiency reached 40%.

A series of studies [53–58] has been devoted to the practical development of Raman lasers based on the $LiIO_3$ crystal, operating in the yellow spectral region at a wavelength of around 580 nm, which is efficiently absorbed by haemoglobin and only slightly absorbed by the melanin of human skin.

With flashlamp quasi-cw (3 kHz) Nd:YAG intracavity pumping, the authors of [53] obtained a 770-cm^{-1} Raman frequency shift and a first Stokes radiation at the wavelength of 1179 nm with an average power of 1.8 W, a pulse duration of 40 ns and a conversion efficiency of 50%. After frequency doubling in an LBO crystal, the radiation had a wavelength of 580 nm and an average power of 550 mW. By using a transversely diode-pumped Nd:YAG laser, the authors [54–56] reached a power of 2.7 W for the first Stokes output radiation at 1155 nm with an electrical-to-optical efficiency of 10%. The second harmonic output power at 577 nm was 0.96–1.2 W at a repetition rate of 7–10 kHz, a pulse duration of 8 ns and an overall electrical-to-optical efficiency of 13%. The thermal lensing in the intracavity-pumped $LiIO_3$ Raman laser was measured in [55] under excitation by a cw lamp-pumped, acousto-optically Q-switched Nd:YAG laser operated at 1064 nm by 200 ns pulses with a typical power of 2.7 kW and a repetition rate of 3 kHz. The intracavity-pumped $LiIO_3$ crystal oscillated in the same cavity with an average power of 1 W at 1157 nm. The power of the negative thermal lens in the $LiIO_3$ crystal was observed to be linearly dependent on the first Stokes average power with a slope of -4.2 m^{-1} W^{-1} and almost independent of the output coupler reflectivity $R = 25$–50%. Longitudinal diode pumping of the Nd:YAG–$LiIO_3$ Raman laser and optimization of the thermal lensing resulted in an increase in the output average power of yellow radiation to 1.42 W at 20 ns pulse duration and an 8 kHz repetition rate with an electrical-to-optical efficiency of 7.5% [57].

Among the limitations and shortcomings of the $LiIO_3$ SRS crystal, we should mention its low thermal conductivity, moisture resistance and laser damage threshold.

B1.7.6 Picosecond Raman frequency shifters based on KGW ($KGd(WO_4)_2$) crystals

A specific feature of picosecond Raman converters is that they cannot successfully work in a compact cavity a few cm long, which is the order of a Raman crystal length. In addition, SRS has a non-stationary character when the pump pulse duration is less than or equal to the time of phase relaxation of the Raman oscillation ($t_p \leq 10T_R$). At the same time, the laser damage threshold of nonlinear crystals is much higher when the crystal is irradiated by picosecond laser pulses than by nanosecond ones, i.e. the picosecond operation mode allows pumping with a higher power density without crystal damage. As was shown earlier, with non-stationary pumping, the Raman gain coefficient sharply decreases in comparison with the steady-state regime, loses its dependence on the scattering peak cross section and pump power and keeps only the slow square root dependence on the integral Raman scattering cross section and picosecond pump pulse energy density.

The first results on SRS in Nd:$KY(WO_4)_2$ (Nd:KYW) and Nd:$KGd(WO_4)_2$ (Nd:KGW) crystals were obtained during 1984–85 [58] under picosecond pumping. At the same time, it was also the first study where the Raman material worked simultaneously both as an active and nonlinear material.

At about 20 J flashlamp pump energy, active elements 5 mm in diameter and 50 mm long oscillated at the fundamental Nd^{3+} wavelength of 1067 nm in a single-mode regime, producing a train of about 10 ultrashort pulses, each having a duration of 11–12 ps, at a spectral width of 1.8 cm^{-1} and an energy of 0.5 mJ. The efficiency of stimulated Raman self-conversion to the first Stokes component ($\lambda = 1182.7$ nm) for a single picosecond pulse from the beginning of the train was 6.14% of the total energy and reached 35% in the middle of the train. In the case when two Nd:KYW crystals were used (one as the lasing and self-converting Raman master oscillator and the second as the synchronously laser-pumped Raman amplifier (without flashlamp pumping), the authors had a maximum conversion efficiency $\eta_{1St} = 37\%$ to the first Stokes, $\eta_{2St} = 16\%$ to the second Stokes ($\lambda_{2St} = 1323.7$ nm) and $\eta_{aS} < 5\%$ to the anti-Stokes ($\lambda_{aS} = 947.9$ nm). When the Nd:KYW amplifier was additionally pumped by 30 J of flashlamp radiation, the total output energy increased to 11 mJ. For the two Nd:KGW crystals 50 mm long used in one laser cavity, the conversion efficiency of the whole train to the Raman scattering radiation was not so high, $\eta_{1St} = 3.84\%$ ($\lambda_{1St} = 1180.5$ nm), probably due to the quality of the crystals, and was improved with an SRS amplifier to $\eta_{1St} = 16.58\%$, $\eta_{2St} = 0.96\%$ and $\eta_{aS} = 0.03\%$.

The authors also briefly mentioned an oscillation observed in Nd:$K(MoO_4)_2$ and Nd:$NaLa(MoO_4)_2$ crystals. The pulse duration at the fundamental wavelength was 5–6 ps and, at the first Stokes component, 1.5–2.5 ps, with a conversion efficiency $\eta_{1St} = 40\%$.

Almost simultaneously with [58], intracavity SRS generation from Nd:$KGd(WO_4)_2$ crystals was reported in [59]. The authors studied the frequency doubling of the picosecond radiation of a Nd:KGW laser with passive mode-locking. New radiation wavelengths of 1180 and 1320 nm were found, corresponding, respectively, to the first and second Stokes components of the SRS conversion of the Nd^{3+} fundamental wavelength (1067 nm) in an Nd:KGW crystal with a frequency shift of 900 ± 10 cm^{-1}. This was verified by the SRS radiation from an Nd:KGW crystal that was placed in the picosecond Nd:YAG laser cavity and was not flashlamp pumped. Estimates of the SRS conversion efficiency showed that only 10% of the total energy belonged to the 1180 nm radiation and the other 90% falls to the 1067 nm radiation. The SRS pulse duration was 3.8 ± 0.4 ps, which was considerably shorter, by a factor of 2.5, than the pump pulse.

The SRS of undoped KGW crystals with different orientations in an external pumping scheme under 30 ps Nd:YAG laser pumping was studied later in [60]. In the 1.06–1.32 μm spectral region, the SRS radiation from 12 new spectral lines with different combinations of Stokes shifts of 84, 767 and 901.5 cm^{-1} was recorded. The normalized thresholds and gain coefficients for different polarization orientations were measured. The maximum gain coefficients were measured to be 6.4 cm GW^{-1} for the 901.5 cm^{-1} Stokes frequency shift and 4.3 cm GW^{-1} for 767 cm^{-1}. With increasing pump density, the SRS radiation of the higher Stokes components, up to the fourth component with a wavelength $\lambda_{4St} = 1730$ nm, was recorded.

The SRS pulse compression in Nd:KGW and Nd:KYW picosecond lasers was specially investigated in [61, 62]. In a cavity 1.3 m long formed by the back dichroic mirror with $R_{1064\ nm} = 99.94\%$ and $R_{1180\ nm} = 72.6\%$ and by a glass plate as the output coupler, the first Stokes radiation comprised 22–40% of the pump pulse energy at the middle of the pulse train, the first Stokes pulse duration was as short as $t_{1St} = 1.3$ ps, and the pump pulse duration was also shortened to 3.9 ps. Application of the SRS amplifier on an Nd:KGW crystal with a synchronous laser pumping in a passive regime (without flashlamp pumping of the Nd^{3+} ions) allowed the authors to shorten the Stokes pulse to a minimum value of 0.9 ps and increase the conversion efficiency to above 40%. With active amplifying, the Stokes pulse duration grew continuously from 1.1 to 6 ps as the energy of the Nd:KGW amplifier pumping E_p increased from 25 to 80 J. The duration of the Nd laser fundamental radiation pulse decreased from 10 to 4.4 ps.

A detailed review of the KGW crystal structure, lattice cell parameters, atomic positions and coordinates and interatomic distances can be found in [63, 64]. A spectroscopic study of spontaneous and stimulated Raman scattering in KGW crystals with different orientations of the pump beam direction and polarization with respect to the crystallographic axes was also presented. The Stokes components of the stimulated Raman scattering, from the first (1162.2 nm) to the fourth (1734.8 nm) ones, were observed under pumping by 1064 nm radiation from a Nd:YAG laser with a pulse duration of 30 ps. The threshold power was 4.8–24.3 GW cm^{-2}. The threshold pump density for the KGW first Stokes with shifts of 767 and 901.5 cm^{-1} was measured for different orientations and compared with those measured for $CaWO_4$ and $Ba(NO_3)_2$ crystals. The normalized SRS threshold for KGW was measured to be from 3.9 to 14.8 GW cm^{-2} depending on the crystal orientation, which was much less than 23–25 GW cm^{-2} for $CaWO_4$ crystals. The SRS threshold pump density measured for $Ba(NO_3)_2$ crystals of 6.2 GW cm^{-2} is much less than that for $CaWO_4$ but higher than the minimum value for KGW at the optimum orientation.

The non-stationary gain coefficient calculated from the measured thresholds turned out to be minimal for the $CaWO_4$ crystals ($g = 1$–1.1 cm GW^{-1}, four times higher than for the $Ba(NO_3)_2$ crystals ($g =$ 4 cm GW^{-1}), and the highest for KGW crystals for the optimal orientation *p[mm]p* ($g = 6.4$ cm GW^{-1} for the 901.5 cm^{-1} Stokes shift). For the KGW Stokes shift of 767 cm^{-1} and the orientation *q[pp]q*, this value was $g = 5.2$ cm^{-1}. These data on KGW and KYW Raman lasing properties demonstrate the very efficient operation of these crystals in the pico- and sub-picosecond regimes of oscillation, where high-density pumping (1–10 GW cm^{-2}) can be used without radiation damage to the crystal. These crystals demonstrate multi-wavelength frequency shifts of coherent radiation to the hard-to-reach IR region from 1 to 1.7 μm with a high conversion efficiency (about 40%) and very effective radiation pulse shortening (more than 10 times) to femtosecond durations.

Recently, the compression of pump laser pulses by backward SRS in KGW crystals has been studied [65]. The KGW crystals were pumped by 19 ps pulses of 532 nm radiation. At the optimized focusing conditions (a focusing angle of 75 mrad) and pumping exceeding the SRS threshold by two to three times, the authors obtained the 555 nm radiation of the first Stokes component with a total conversion efficiency of about 30% with a output pulse duration of 1.6–1.7 ps, which corresponds to the high, by a factor of 11–12, compression of 3–5 μJ pump pulses to 1 μJ output pulses.

B1.7.7 Nanosecond Raman lasers based on KGW crystals

The main problem associated with the use of KGW crystals as Raman shifters for nanosecond IR laser radiation is a rather high SRS threshold (due to the low Raman gain) in combination with a low laser damage threshold for nanosecond pumping. The narrow gap between the SRS and laser damage thresholds essentially restricts the SRS conversion efficiency and imposes strong requirements upon the pump radiation profile and the optical quality of the crystal and all the elements of the Raman cavity.

The first paper in which the authors succeeded in realizing nanosecond intracavity SRS generation with a Nd:KGW crystal was [66]. The pulse duration of the fundamental radiation was $t_p = 7.5$ ns and the duration of the first Stokes pulse at 1180 nm was measured to be threefold shorter, $t_{SRS} = 2.5$ ns. The intracavity SRS threshold was estimated to be 150–300 MW cm^{-2}. However, no data on the conversion efficiency and output energy were reported.

Nanosecond SRS in KGW crystal in a single-pass scheme with external pumping was also studied in [67, 68]. The SRS gain coefficients for both Stokes shifts of 901.5 cm^{-1} and 767.3 cm^{-1} were found to be 6 cm GW^{-1} when pumped by a Nd:YAG laser with a radiation wavelength of 1064 nm. For nanosecond ($t_p = 10$ ns) pumping by a Nd:KGW laser in a single-pass scheme, the total efficiency of SRS conversion to both (first and second) Stokes was 70% with an output energy of 10 mJ. These results did not depend on whether the Stokes shift frequency was 767.3 or 901.5 cm^{-1}.

Application of KGW crystals for SRS frequency shifting of nanosecond radiation with longer wavelengths, 1.1–1.4 μm, where the gain coefficient is much weaker, was first demonstrated in [32]. A KGW crystal 40 mm long was used for tunable nanosecond SRS frequency conversion of a solid state $LiF:F_2^-$ colour centre laser MALSAN-201 (see chapter B1.8). Using the single-pass Raman shifting scheme with a pump pulse duration of 8 ns, energy of about 20 mJ and a spectral width of 0.1 cm^{-1}, the authors observed the tunable radiation in new spectral regions, 1.23–1.37 μm (first Stokes) and 1.43–1.6 μm (second Stokes).

The $LiF:F_2^-$ colour centre laser radiation was tunable from 1.1 to 1.23 μm with a pump density of about 1 GW cm^{-2}. The maximum conversion efficiency reached 30% for the first Stokes radiation and 20% for the second. These results were among the first experiments on solid state Raman lasers operated in the eye-safe spectral region (1.54 μm). It should be noted that the laser damage threshold was only two to three times higher than the SRS threshold, which strongly limited the reported operation.

Ten years later, a compact flashlamp-pumped solid state laser for a range finder operating in the eye-safe spectral region at 1538 nm was developed in [68,69] with the application of SRS in a Nd:KGW active crystal self-pumped by 1350 nm (the fundamental wavelength) Nd^{3+} laser radiation. The authors used the Nd:KGW active nonlinear elements with dimensions of ⌀3 mm × 50 mm and ⌀4 mm × 65 mm. The cavity mirrors had the reflectivity $R1 = R2 = 99.99\%$ at the fundamental wavelength of 1351 nm and the exit mirror has a 35–40% reflectivity at the first Stokes wavelength of 1538 nm. Passive polymer and an electro-optical $LiNbO_3$ Q-switches were used.

In the case of a Nd:KGW crystal with small dimensions, ⌀3 mm × 50 mm, the threshold pump energy of the single-pulse mode changed from 2.5 to 5.5 J and the energy of the 40 ns SRS pulse increased from 1 to 6 mJ, as the passive Q-switch transmission changed from 45 to 33%. The KGW crystal with dimensions of ⌀4 mm × 65 mm allowed an increase in the output energy to 9 mJ at the pump energy of 7.5 J. With electro-optical Q-switching by a Brewster-cut $LiNbO_3$ crystal, the output energy of the 12 ns pulses reached 13.5–15 mJ at a pump energy of 10 J. The overall efficiency was quite high—0.15%.

Using diode laser pumping of Nd-doped laser crystals, one can build compact and efficient solid state lasers for IR and green spectral regions. A transversely diode-pumped $Nd:KGd(WO_4)_2$ Raman laser with self-frequency conversion was described in [70]. An AlGaAs/GaAs quasi-cw diode laser operating at 808 nm with a power of 300 W and a pulse duration of 300 μs was used for side pumping a $Nd:KGd(WO_4)_2$ crystal rod 4 mm in diameter and 22 mm in length. The plano-concave Raman laser cavity with an acousto-optical Q-switch inside had a high reflectivity for the fundamental wavelength (1067 nm) and an output coupler transmission of 5% for the first Stokes wavelength (1162 nm). As a result, the first Stokes pulsed radiation had an energy of 0.1 mJ for a pulse duration of 50 ns, while the pump pulse duration was 140 ns with a repetition rate of 47 Hz. The second harmonic radiation (581 nm) was generated with a 30% efficiency by a LiB_3O_5 crystal with non-critical phase matching at a temperature of 54 °C. The second harmonic pulses were threefold shortened to 16 ns.

Recently, intracavity self-frequency conversion in a microchip composed from Yb:KGW and Yb:KYW crystals was reported by two groups of authors [71,72]. They used 1.7 and 1.1 mm thick plates of Yb-doped laser crystals, pumped by a 980 nm laser diode and a Cr^{4+}:YAG plate as the saturable absorber. With an output coupler reflectivity of 7 and 5%, the authors obtained nanosecond pulses of the first Stokes radiation at a wavelength of 1139 nm with an average power of 7 and 2 mW and a repetition rate of 17 and 49 kHz, respectively.

This historical review of the best SRS materials for nano- and picosecond Raman lasers shows that, up until the mid 1990s, there was no clear understanding as to why one SRS material is better for nanosecond operation and another operates better in the picosecond regime. There was no complete comparative analysis of the major fundamental properties of SRS materials, which would allow one to select the most suitable materials among the known synthetic crystals or to develop SRS materials with the required properties.

B1.7.8 New $BaWO_4$ SRS crystals

The newly developed $BaWO_4$ crystals for SRS exhibit many advantages in comparison with the two previously discussed best synthetic crystals, $Ba(NO_3)_2$ and $KGd(WO_4)_2$ (the characteristics of these crystals are compared in table B1.7.1). As one can see, the $BaWO_4$ crystal has a much higher hardness, thermal conductivity and moisture resistance than $Ba(NO_3)_2$. The transparency range in the IR is also much wider for $BaWO_4$ than for $Ba(NO_3)_2$ crystals. The Raman scattering integral cross section of $BaWO_4$ is at least double, which is important for picosecond application, where $BaWO_4$ can provide a much higher gain than $Ba(NO_3)_2$. The extremely small Raman line broadening (0.4 cm^{-1}) in $Ba(NO_3)_2$ results in the record high gain in the nanosecond (steady-state) regime of SRS but at the same time restricts the steady-state regime to the picosecond time scale. As the result, the Raman gain of $Ba(NO_3)_2$ falls steeply down on shortening the picosecond pulse duration due to the transient character of SRS ($t_p < 10T_R$).

For nanosecond pulses, the $BaWO_4$ crystal exhibits only slightly smaller SRS peak cross sections and gain coefficients than those of $Ba(NO_3)_2$ crystals, but, due to the four times larger linewidth (1.6 cm^{-1}) and correspondingly four times shorter dephasing time T_R, its steady-state regime can be extended to the four times shorter laser pulses on the picosecond time scale.

Comparing the $BaWO_4$ crystal with $KGd(WO_4)_2$, one can see almost equally high hardness, thermal conductivity, moisture resistance and wide transparency range. The integral cross sections of these two crystals are also quite similar and high. A difference can be found only in the line broadening and the dephasing time. Due to the three times broader Raman line, the $KGd(WO_4)_2$ crystals have a much lower nanosecond (steady-state) SRS cross section and gain coefficient than $Ba(NO_3)_2$ and $BaWO_4$ crystals. Even for the picosecond region 10^{-11}–10^{-9} s, the new $BaWO_4$ crystal can show much better SRS operation than $KGd(WO_4)_2$. Only with few-picosecond or sub-picosecond pumping can KGW crystals overcome the $BaWO_4$ crystal in gain and efficiency, due to the deeper transient behaviour and faster gain decrease in $BaWO_4$ in comparison with KGW.

Similar properties are demonstrated by the $SrWO_4$ crystal, whose line broadening and dephasing time lie between those of $BaWO_4$ and KGW crystals and define the intermediate values of $SrWO_4$ SRS cross section and Raman gain, which are better than those for KGW and poorer than for $BaWO_4$ in the nano and picosecond steady-state regime.

B1.7.9 IR $BaWO_4$ and $SrWO_4$ nanosecond Raman lasers

IR SRS laser oscillation in the newly developed $BaWO_4$ crystal and two well-known nonlinear crystals was comparatively studied in [24], using a 9 cm long Fabry–Pérot cavity formed by two plane dielectric mirrors, an input mirror with $R_{1.064} < 5\%$ and $R_{1.1-1.25} > 98\%$ and output mirror with $R_{1.0-1.2} \approx 55\%$.

A pulsed single-mode Nd:YAG laser passively Q-switched by an $LiF:F_2^-$ crystal was used as the pump source. The pump laser operated by 12 ns pulses with a repetition rate of 10 Hz. The pump beam was focused on the centre of an SRS crystal by a lens with a focal length of 50 cm. The beam's focal diameter was measured by a CCD camera to be 90 μm at half maximum. The beam shape was close to Gaussian.

The real Raman laser efficiency reaches 26% for $Ba(NO_3)_2$ and 20.5% for $BaWO_4$. In both cases, the slope efficiency exceeds 75% and the laser thresholds are very close. At the same time, no efficiency saturation is observed for $BaWO_4$ with increasing pump power; hence, one can expect further improvement of the laser characteristics. The best results are obtained at the $BaWO_4$ orientation $E \parallel C_4$ with slightly poorer ones at $E \perp C_4$. An even longer KGW crystal demonstrates a much higher threshold and considerably lower slope efficiency (about 45%) in the same experimental scheme.

The Raman laser efficiency of $SrWO_4$ SRS crystal was studied in [80] under similar conditions. The laser scheme differed from the one considered earlier by having a shorter cavity (70 mm), lower reflectivity of the output coupler ($R_{1.05-1.32} = 30 \pm 5\%$), twice longer focal length of the focusing lens was twice as long ($f = 100$ cm) and the focal spot twice larger ($\varnothing = 200$ μm).

To prevent coupling between the SRS crystal (without any anti-reflection coatings), the Raman laser cavity and the pump laser cavity, they were misaligned by about 1° from each other.

Figure B1.7.3 shows the Raman laser output energy *versus* the pump energy for the new $SrWO_4$ and $BaWO_4$ crystals in comparison with KGW and $Ba(NO_3)_2$. One can see that the $SrWO_4$ crystal has a lower threshold and an efficiency which is approximately double ($\approx$ 12%) that of KGW. In spite of poorer threshold, gain and output characteristics of $SrWO_4$ in comparison with those of $BaWO_4$ and $Ba(NO_3)_2$ crystals, the

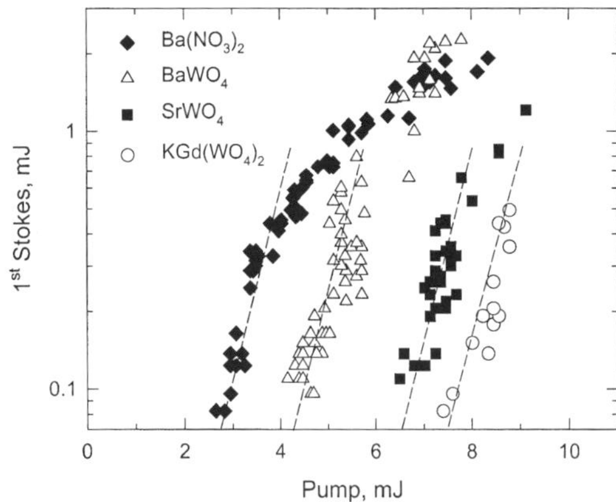

Figure B1.7.3. Dependences of the first Stokes output energy on the pump energy for $Ba(NO_3)_2$ 49 mm long (rhombs), $BaWO_4$ 43 mm long (triangles), $SrWO_4$ 47 mm long (squares) and $KGd(WO_4)_2$ 50 mm long (circles) crystals.

$SrWO_4$ crystal allows rather high levels of rare earth doping, which makes it promising for many applications as a multi-functional nonlinear (SRS) active material.

SRS self-conversion of laser radiation inside a Nd^{3+}:$SrWO_4$ laser crystal was first observed under flashlamp pumping in a Q-switched regime [73]. With high reflectivity cavity mirrors at the fundamental wavelength of 1.06 μm, the first Stokes oscillation at 1.18 μm wavelength was recorded with an output energy of 3 mJ. The first Stokes oscillation of this Raman laser combined two oscillation mechanisms: the broad-band stimulated emission of F_2^- colour centres and SRS inside the laser medium. This combination allowed SRS at lower intracavity energy density, i.e. it reduced the SRS threshold. Under alexandrite laser pumping such intracavity SRS conversion schemes provide the first Stokes radiation with a peak power of 440 kW at a pulse duration of about 3 ns [99].

B1.7.10 $BaWO_4$ yellow Raman shifters and lasers

The yellow SRS in a $BaWO_4$ crystal 33 mm long under green (532 nm) pumping was studied in [74] in comparison with the $KGd(WO_4)_2$ crystal. The pumping thresholds for different orientations of the beam propagation vector $\boldsymbol{K}$ and polarization vector $\boldsymbol{E}$ in respect to the C_4-axis of the $BaWO_4$ crystal were measured to be 240 (for $\boldsymbol{K} \perp C_4$ and $\boldsymbol{E} \parallel C_4$) and 200 MW cm^{-2} (for $\boldsymbol{K} \perp C_4$ and $\boldsymbol{E} \parallel C_4$). These values are much lower than 340 MW cm^{-2} for the KGW crystal (for $\boldsymbol{K} \perp C_2$ and $\boldsymbol{E} \perp C_2$ and $\boldsymbol{E} \parallel C_2$) and point to a much higher gain in the new $BaWO_4$ crystal for both orientations.

In the single-pass Raman shifting arrangements without laser cavities, the conversion efficiency for the KGW crystal was up to 15% in the first Stokes ($\lambda = 559$ nm) and up to 5% in the second Stokes ($\lambda = 589$ nm). The conversion efficiency of the $BaWO_4$ crystal was almost double (30%) for the first Stokes ($\lambda = 560$ nm) and 6% for the second Stokes ($\lambda = 590$ nm) under the same pumping conditions.

In the double-pass scheme [75] with one reflecting back mirror, which increased the length of the light–crystal interaction by a factor of two (to 66 mm), the SRS threshold was equal to 95 MW cm^{-2}, which is half as high as in the single-pass scheme and the conversion efficiency was increased up to 45%.

Measurements of the SRS temporal profile have shown that the SRS pulse is two to three (and more) times shorter than the pump pulse duration when the pump energy exceeded the threshold value by a factor of approximately three. With a 5.1 ns pump pulse, the first Stokes pulse duration was 2 ns and the second Stokes pulse had a duration of 1.5 ns. These measurements show that it is possible to have the Raman frequency shift virtually without loss of peak power.

Table B1.7.2. Comparative SRS and thermo-mechanical characteristics of $Ba(NO_3)_2$, $KGd(WO_4)_2$, $SrWO_4$ and $BaWO_4$ crystals.

Characteristic	Raman crystal			
	$Ba(NO_3)_2$	$KGd(WO_4)_2$	$SrWO_4$	$BaWO_4$
Raman frequency shift ν_R (cm^{-1})	1047	901; 767	922	924
Raman linewidth $\Delta\nu_R$ (cm^{-1})	0.4	5.4; 6.4	2.74	1.6
Dephasing time T_R (ps)	28	2.0	4	6.6
Gain coefficients:				
g(0.53 μm) steady-state, (ns)	47	11		36–40
g(0.53 μm) transient (20 ps)	4.7	11		14.4
g(1.06 μm) steady-state (ns)	11	4	5	8.5
g(1.06 μm) transient (30–50 ps)	1.1	3		3.8
g(1.3 μm) steady-state (ns)				5.8
Transparency range, (μm)	033–1.8	0.3–5	0.25–5	0.255–5
Moisture resistance	Low	High	High	High
Thermal conductivity at 25 °C, W K^{-1} m^{-1}	1.17	2.5–3.4	3	3.0
Thermal expansion coefficient α, 1 $°C^{-1}$	13×10^{-6}	$(1.6–8.5) \times 10^{-6}$		6×10^{-6}
Hardness	19.2 ($NaNO_3$ Knoop)	4–4.5 (Mohs)	4 (Mohs)	4(Mohs) 400 (Knoop)

A pulsed yellow $BaWO_4$ Raman laser operated at a high repetition rate was studied in [76] under Nd:YLF high-power green (527 nm) pumping. The first (544 nm) and second (583 nm) Stokes radiation were obtained with an output power of 130 and 22 mW at 70% and 50% output coupler reflectivity, respectively. Strong SRS pulse shortening, from 400 to 10 ns, was observed. The slope efficiency for the frequency conversion from the pumping to the first Stokes was measured to be as high as 15%.

B1.7.11 $BaWO_4$ picosecond Raman frequency shifters

A comparative study of picosecond SRS in this new $BaWO_4$ crystal and in $KGd(WO_4)_2$ and $KY(WO_4)_2$ crystals in a single-pass scheme was fulfilled in [74, 77–81] for 28 and 40 ps pulses with wavelengths of 532 and 1064 nm, respectively. With visible green light pumping, the picosecond Raman gain coefficient calculated for the $BaWO_4$ crystals from the pump threshold measurements by the formula $g_{exp} = 25/I_{pl}l$, was equal to 14.4 cm GW^{-1}, which is 30% higher than that of KGW (11.5 cm GW^{-1}) but 30% lower than that of KYW (18.7 cm GW^{-1}).

In comparison with the steady-state regimes, where the $BaWO_4$ crystals show approximately a three times higher gain than KGW, this ratio at picosecond pumping is moderate, though the $BaWO_4$ still has a higher gain than the KGW crystal. This is caused by the deeper transient behaviour of $BaWO_4$ due to its smaller linewidth and longer $T_R (= 6.6$ ps).

A much stronger difference can be found when comparing the $BaWO_4$ crystal with $Ba(NO_3)_2$. With nanosecond pumping (steady-state regime), $BaWO_4$ has only a 20–30% lower Raman gain (40 cm GW^{-1} against 52 cm GW^{-1}) [74, 75]) but in the picosecond mode its gain is about three times higher (see table B1.7.2).

For 1064 nm picosecond pumping, the $BaWO_4$ Raman gain reaches 3.8 cm GW^{-1}, which is slightly lower than that for KYW (4.7 cm GW^{-1}) but higher than for KGW (3 cm GW^{-1}). The maximum conversion efficiency for a 30 mm long $BaWO_4$ crystal under 532 nm picosecond pumping was 30% for the first Stokes frequency. This is much higher than the 18% efficiency of an even longer (40 mm) KGW crystal. For the second Stokes scattering, both crystals show a similar efficiency of about 15%.

Optimization of the pump-to-output conversion efficiency for the best pumping orientation gives a similar first Stokes (560 nm) efficiency ($\sim$50%) for $BaWO_4$ and KGW crystals and the $BaWO_4$ second Stokes (590 nm) efficiency up to 20% [74].

The study of picosecond SRS temporal behaviour reveals an approximately twofold shortening of the SRS pulse (18 ps) with respect to the pump pulse duration (35 ps). Hence, keeping in mind the 50% conversion efficiency of $BaWO_4$, we can conclude that the efficiency of peak power conversion from green to yellow radiation is 100%. A detailed study of $BaWO_4$ SRS under 1.06 μm, 50 ps pulsed pumping [80] allowed the authors to obtain the first Stokes radiation at 1180 nm with a conversion efficiency of 25% in the single-pass and 35% in the double-pass regimes. Placing the $BaWO_4$ crystal into a short (3.8 cm) cavity with an optimized output coupler provided a 55% pump-to-first-Stokes conversion efficiency with an output energy as high as 3 mJ.

Thus, we can conclude that the $BaWO_4$ is the first universal crystal, which shows an almost record high gain, both in the nanosecond and picosecond modes of operation.

B1.7.12 $PbWO_4$ SRS shifters and Raman lasers

Another interesting SRS crystal was recently introduced into Raman laser development. This is a well-known scintillating material, $PbWO_4$, with a scheelite structure similar to that of $BaWO_4$. Analogous to other tungstates, one can expect a high integral cross section for the $PbWO_4$ crystal, but the measured linewidth of the totally symmetrical A_{1g} 904 cm^{-1} Raman vibration shows a rather high value of 5.6 cm^{-1} for a polycrystalline sample [17] close to KGW. Detailed study of the A_{1g} Raman line broadening at different temperatures up to the melting point shows that, similar to $CaWO_4$ and $SrWO_4$, lead tungstate $PbWO_4$ has a rather high Davydov splitting ($\Delta E_{DS} = 32$ cm^{-1}) caused by a strong $[WO_4]^{2-}$–$[WO_4]^{2-}$ intermolecular interaction [23]. This value is one or two orders of magnitude higher than that in $BaWO_4$ and $Ba(NO_3)_2$ crystals (with a record high peak value of gain) $\Delta E_{DS} = 0$–1 cm^{-1}. The high density of acoustic phonon states leads to fast dephasing and large Raman line broadening $\Delta\nu_R(300\ K) = 4.7$ cm^{-1} (for a single crystal) due to the direct relaxation processes between the DS components with single-phonon absorption or emission.

The study of $PbWO_4$ crystals using picosecond SRS spectroscopy techniques and under nanosecond Raman laser pumping was first reported in [82–84]. The authors presented the measured data on the gain coefficient (8.4 cm GW^{-1} at 532 nm and 3.1 cm GW^{-1} at 1064 nm) and some other characteristics of the $PbWO_4$ single crystal. They observed 18 different spectral lines of SRS Stokes and anti-Stokes radiation in the 485.5–1117 nm spectral region under pumping by 1064 nm Nd:YAG laser pulses with a duration of 100 ps and by its second harmonic (532 nm) with a pulse duration of 80 ps.

Raman laser oscillation in $PbWO_4$ crystals under external pumping by 100 ps pulses of an Nd:YAG laser at 1064 nm was studied in [82, 83]. In the 200 mm long plane-concave laser cavity with a spherical ($r = 0.5$ m) output coupling mirror with a reflectivity of 78%, the first Stokes radiation energy of 1 mJ with a slope efficiency as high as 40% was obtained at a pump energy of 5 mJ. In these experiments, the pump beam was focused inside the 30 mm long $PbWO_4$ crystal by a lens with a focal length of 12 cm. Data are also presented on the steady-state Raman gain measured in visible light under pumping by 30 ps pulses of the frequency-doubled radiation of an Nd:YAG laser. The authors concluded that the gain coefficient of $PbWO_4$, measured to be 8.4 cm GW^{-1}, is slightly higher than the 8 cm GW^{-1} of KGW crystal, measured by the same technique.

A linear Raman cavity 110 mm long with one flat mirror serving as the pump input coupler and another concave mirror with a radius of curvature of 10 m as an output coupler with 46% transmission at the first Stokes wavelength were used. Two $PbWO_4$ crystalline samples, each 45 mm long, were placed inside the cavity. Under external pumping by nanosecond (5–7 ns) 532 nm radiation, collimated inside the crystal to the beam diameter of 1.1 mm, the output energy of the first Stokes beam (558.9 nm) was 1 mJ with a conversion efficiency of 13%. Laser damage on the crystal surface was observed with a pump fluence of 5.1 J cm^{-2}. The IR Raman lasing by one or two 45 mm long $PbWO_4$ crystals was studied in a ring Raman laser cavity 285 mm long with two highly reflecting mirrors and one output coupler with a reflectivity of 27% for the first Stokes wavelength [84]. The Raman crystals were pumped by a Nd:YAG laser (1064 nm) beam 2.1 mm in diameter with a pulse duration of 7 ns. At a pumping energy of 36 mJ (close to the Raman crystal laser damage threshold), the output energy of the first Stokes radiation was as high as 5.3 mJ for one $PbWO_4$ crystals. The highest conversion efficiency reached 20% for two SRS crystals in the cavity. The authors noted that the laser damage threshold for IR radiation was surprisingly low, about 2 J cm^{-2}, which is more than twofold lower than that for visible radiation.

The $PbWO_4$ crystal can also be easily activated by Nd^{3+} ions [82]. The first attempt to develop a diode-pumped passively Q-switched Nd^{3+}:$PbWO_4$ Raman laser with self-frequency conversion was described in [85]. An active and SRS Nd(0.5 at%):$PbWO_4$ crystal with the dimensions 3 mm × 3 mm × 15 mm was transversely pumped by a 100 W GaAlAs quasi-cw diode laser array ($\lambda = 808$ nm) with a pulse duration of 160 μs and a repetition rate of 20 Hz. In a highly-reflective cavity 24 mm long, passively Q-switched by a Cr:YAG saturable absorber with $T = 95\%$, under a pump energy of 4.5 mJ, the Raman output energy at 1170 nm was rather low, 2.5 μJ, with an efficiency of 0.6% and a threefold output pulse shortening to 8 ns.

Comparative data on the energy, pulse duration, power and efficiency of SRS radiation are given in table B1.7.3 along with the pump laser parameters, Raman shifter and laser optical schemes and SRS crystal types.

Table B1.7.3. Pump and output characteristics of solid-state SRS shifters and Raman laser.

SRS crystal, Raman shift	Optical scheme	Pump laser and wavelength (μm)	Pump pulse duration, Rep. rate	SRS efficiency	SRS energy, power, pulse duration	SRS wavelength (nm)	Refs
Green nanosecond pumping with low repetition rate							
$Ba(NO_3)_2$, 1047 cm^{-1}	SRS shifter	Nd glass, 0.532	15 ns single pulse	26%	150 mJ	560	8
$Pb(NO_3)_2$	”	”	”	20%	70 mJ	”	
$NaNO_3$, 1066 cm^{-1}	”	”	”	17%	150 mJ	”	
$CaCO_3$ 1084 cm^{-1}	”	”	”	14%	120 mJ	”	
$Ba(NO_3)_2$, 1047 cm^{-1}	Raman laser	Nd:YAG 0.532	10 ns	40–50%	60 mJ	560	10
				25%	20 mJ	598	
	SRS shifter	Nd:YAG 0.532	22 ps	25%		583	28
				5%		599	
	Raman laser, external pumping	Nd:YAG 0.532	8 ns (1–30 Hz)	40–48%	25 mJ	563, 281	26
					32 mJ, 5 ns	599, 299	
	Oscill. + amplifier	Nd:YAG 0.532	5–7 ns	14%	3.5 mJ	563	27
$BaWO_4$, 926 cm^{-1}	Single-pass SRS	Nd:YAG 0.532	5.1 ns	30%	2 ns	560	74, 75
				6%	1.5 ns	590	
	double-pass SRS	”	”	45%		560	75
	Raman laser, external pumping	Nd:GGG 0.531	29 ns	50%	6 mJ	559.5	87, 88
				30%	4 mJ	588.95	
$PbWO_4$, 904 cm^{-1}	Ring cavity Raman laser	Nd:YAG 0.532	5–7 ns	13%	1 mJ		91
KGW, 901.5 cm^{-1}	Backward single-pass SRS	Nd:YAG 0.532 μm	19 ps	30%	1 mJ, 1.6–1.7 ps	555	65
$BaWO_4$, 926 cm^{-1}	Single pass SRS	0.532 μm	28–35 ps	30–50%	18 ps	560	74
				15–20%		590	
$Ba(NO_3)_2$, 1047 cm^{-1}	Single pass SRS	Copper vapour laser 510.6; 578.2	15 ns (16 kHz)	38%	0.6 mW	539. 4	25
				18%	0.3 W	615.4	
					0.01–0.08 W	203–308	
$BaWO_4$, 925 cm^{-1}	Raman laser	Nd:YLF 2ω, 527	400 ns (1 kHz)		130 mW	544	76
					22 mW	583	
IR nanosecond low repetition rate pumping							
$Ba(NO_3)_2$, 1047 cm^{-1}	Single-pass shifter	LiF:F_2^- 1.08–1.28	15 ns, 20Hz	60%	0.1 J	1230–1370	32
				20%	0.03 J	1430-1600	
	Raman laser	Nd:YAG 1.064 μm	12 ns, 10 Hz	39%	6 mJ	1197	35–38
				27%	4.6 mJ	1363	
				21%	3.5 mJ	1598	
		Nd:YLF 1047	10 Hz		150 mJ (1.5 ns)	1176	49
					90 mJ	588 (2ω)	
	Raman laser, external pumping	Nd:YAG 1.064	30 ns, 20 Hz	20%	2.8 mJ, 11 ns	1197	48
				20%	2.7 mJ, 9.5 ns	1367	
				7%	8.5 mJ, 3.5 ns	1598	
			5 ns	30%	100 mJ, 3 ns	1197	86
					18.5 mJ	598.5	

			50 ns, 10–30 Hz	35%	32 mJ	1197	
				9%	16 mJ	1363	
			50 ns	19%	17.3 mJ, 2.6 W	1197	33
			5-pulse trains, 30 Hz	14.4%	32 mJ, 4.76 W	1197	
				8.5%	2.7 mJ, 0.4 W	1363	
	Raman laser, external pumping	Nd:YAG 1.318	64 ns	39%	5 mJ	1529	38
			64 ns	39%	5 mJ	1556	
			32 ns	52%	3 mJ	1556	
		Nd:YAP 1.341	60 ns	40%	5 mJ	1560	
	Raman laser, intracavity pumping	Nd:YAG 1.338	10–12 ns	90%	1.5 mJ	1556	39–42
		Nd:YAG 1.318–1.338	10 ns	48%	12 mJ	1529–1556	46, 47
						1820–1860	
		Nd:YAP 1.341	1–10 Hz	44%	250 mJ	1529–1556	
	Single-pass SRS	Nd:YAG+LiF:F_2^-	12 ns	35–50%	10–45 mJ, 20–45 ns	1197	43
		Nd:YAP+LiF:F_2^-				1216	
KGW, 901.5–767.3	Single-pass SRS	Nd:YAG 1.064	10 ns	70%	10 mJ	1St+2St	67
KGW, 901.5 cm^{-1}	Single-pass SRS	LiF:F_2^- 1.1–1.23	8 ns	30%	6 mJ	1230-1370	32
				20%	4 mJ	1430—1600	
	Intracavity Raman laser	Nd:YAG+LiF:F_2^-	12 ns	50%	10–45 mJ, 20–45 ns	1180	44
		Nd:YAP + LiF:F_2^-					
Nd:KGW, 901.5 cm^{-1}	Raman laser, intracavity self-conv.	Nd:KGW LD pumping	40 ns		1–6 mJ, 40 ns	1538	68–69
					13.5–15 mJ, 12 ns	1538	
	SRS self-conversion	Diode pumping 1.067	140 ns		0.1 mJ, 50 ns	1162	70
					0.03 mJ, 16 ns	581	
$BaWO_4$, 926 cm^{-1}	Raman laser, external pumping	Nd:GGG 1.062	20 ns	24%	15 mJ	1179.9	87, 88
		Nd:YAG 1.064	12 ns	30%	2.4 mJ	1180	24, 73
		Nd:YAG 1.318	50 ns, 3 Hz	12–20%	1.8 mJ (300 K)	1501	33
				19–35%	3 mJ (77 K)		
		Nd:GGG 1062	50 ns	20%	2 J	1178	98
			50 pul			1322	
			0.1 Hz				
$SrWO_4$, 921 cm^{-1}	Raman laser, external pumping	Nd:YAG 1.064	12 ns	15%	1.3 mJ		73
Nd:$SrWO_4$, 921 cm^{-1}	Raman laser, intracavity self-conv.	Nd:$SrWO_4$ 1.06			3 mJ	1180	73
$PbWO_4$, 904 cm^{-1}		7 ns	17–20%	5.3 mJ		84	
Nd:$PbWO_4$, 904 cm^{-1}	Raman laser intracavity self-conv.	Nd:$PbWO_4$ LD pumping		0.6%	2.5 μJ (8 ns)	1170	85
IR high repetition rate nanosecond pumping							
$LiIO_3$, 818 cm^{-1}	Raman laser, intracavity pumping	Nd:YAP 1.080	100 ns	77%	1.26 W (50 ns)	1180	50, 51
					0.55 W (15 ns)	1310	
					0.9 W (4–8 ns) 7 kW peak	1180	
					0.2 W (2 ns) 100 kW peak	1310	
$LiIO_3$, 770 cm^{-1}	"	Nd:YAG 1.064	100 ps	57%	1.8 W (40 ns)	1179	53
			100 kHz		0.55 W	580	53
			15 kHz	42%	2.7 W	1155	54–56
					0.96–1.2 W (8 ns)	577	
					1.42 W (20 ns)	577	57
$CaWO_4$, 911 cm^{-1}	"	Nd:YAG 1.064	10 kHz	8 ns	0.5 W	1178	89
			8 kHz	2 ns	0.062 W	1320	
$Ba(NO_3)_2$	Raman laser, intracavity pumping	Nd:YAG 1.064	3–7 kHz		0.15–1.4 W	1197	90
Yb:KGW, 901 cm^{-1}	Raman laser, intracavity self-conv.	Diode pumped Yb:KGW 1.03			7 mW (17 kHz)	1139	71, 72
					2 mW (49 kHz)		
IR picosecond low repetition rate pumping							
$BaWO_4$, 921 cm^{-1}	Raman laser, external pumping	Nd:YAG 1.064	20–100 ns	35–21%	6 W	1180	97
			150–300 Hz				
$LiIO_3$, 760 cm^{-1}	Raman laser	Phosphate glass 1.054	6–10 ns	40%	1 mJ (1 ps)	1143	52
Nd:KYW, 900 cm^{-1}	Raman laser, intracavity pumping	1.067	10–12 ps	<5%		947.9	58–64
	Self-conv. Raman laser	1.067	10–12 ps	37%	1.3 ps	1180.5	
				16%		1323.7	
	Raman amplifier		10–4 ps	40%	0.9 ps	1180	
			5–6 ps	40%	1.5–2.5 ps		
$BaWO_4$, 926 cm^{-1}	Single-pass SRS	1.06	50 ps	25%		1180	80
	Double-pass SRS	"	"	35%	3 mJ	"	
	3.8-cm laser cavity	"	"	55%		"	
$PbWO_4$, 904 cm^{-1}		Nd:YAG 1.064	100 ps	20%	1 mJ		82, 83

The basic optical schemes for SRS generation are single-pass SRS, double-pass SRS, Raman laser with external pumping, Raman laser with synchronous pumping and Raman laser with intracavity pumping and self-conversion.

One can see that the wavelength range of SRS shifters and Raman lasers covers a wide spectral region, 0.5–2 μm, limited only by the fundamental absorption of the SRS material and the wavelengths of the pump sources. The SRS conversion efficiency comprises tens of percents, reaching 90% for the intracavity scheme. For nanosecond pulses, SRS reaches the energy of hundreds of milliJoules and an average power of several watts. The SRS conversion of picosecond pulses is accompanied by pulse shortening by a factor of up to 10 and reaches subpicosecond durations (0.9 ps).

B1.7.13 Raman laser applications

As we discussed earlier, coupling an Nd-YAG pump with a barium nitrate Raman crystal results in Raman laser output at 1.56 μm. This is an optimum wavelength for atmospheric propagation and eye-safe operation. Therefore, this type of Raman laser is ideal as a transmitter for atmospheric LIDAR systems [46–48, 68, 69]. Since laser radars are used for many different purposes, it is important to be able to make use of the design flexibility of Raman lasers to provide a transmitter with optimized operational parameters. For example, detecting atmospheric pollutants through backscattering from aerosols can be done with direct detection LIDAR systems (see also section D7). If measuring distance is important, range-gating techniques can be used and the temporal pulsewidth of the laser becomes important. As mentioned previously, Raman lasers can provide output pulses of 1 ns or less in duration which results in a significant improvement in range resolution over normal Q-switched or cavity-damped lasers [45, 75]. It is difficult to obtain this pulsewidth from any other type of solid state laser. If it is important to identify the type of chemical species that is in the atmosphere, differential absorption (DIAL) techniques can be employed. In this case, the laser transmitter must operate at two different wavelengths, on and off a resonant absorption line of the chemical of interest. Raman lasers can easily be configured to meet this requirement through use of the pump and Raman wavelengths or through the use of first-Stokes and second-Stokes output. Depending on the position of the chemical absorption lines, this may require laser emission anywhere in the spectral range from the near ultraviolet to the near infrared. Using Raman lasers in conjunction with other frequency conversion techniques, all of these spectral regions can be reached. For applications requiring the measurement of the Doppler shift of the return signal due to wind or atmospheric turbulence, coherent LIDAR techniques are used. This requires a laser transmitter operating with high stability at a single frequency so heterodyne detection can be used. Injection seeding of Raman lasers has been demonstrated to show the ability of obtaining single frequency operation for use in this type of application [91].

Similar systems have been demonstrated for marine imaging LIDAR applications [49, 92]. For these systems the transmitter wavelength must be chosen to match the maximum transmission through the water. This changes from blue-green in open seawater to yellow or orange in the littoral zone due to the difference in amount of particulates and their effect on scattering light. In general, these systems are used to identify objects underwater such as types of fish or obstructions that might injure ships. High image quality and accurate range resolution requires a laser transmitter with a short temporal pulse width and good beam quality. This is ideal for a solid state Raman laser and excellent results have been obtained with marine imaging LIDAR systems employing Raman lasers [49, 92].

Semiconductor processing applications need efficient, compact, all-solid-state sources with a high repetition rate of 10 kHz, good pulse-to-pulse stability and diffraction-limited beam quality. One of the important wavelengths is close to but longer than the silicon band-to-band and excitonic absorption at 1.12–1.153 μm [89]. As a device satisfying these requirements, the authors offered an end-pumped intracavity Raman laser. They used a 2 mm $\times$ 10 mm Nd:YAG rod longitudinally pumped by a 5 W diode bar via a fibre bundle 600 mm in diameter. A cavity, high-reflecting at 1064 nm and low-reflecting at 1178 nm, with a $CaWO_4$ SRS crystal of dimensions 4 mm $\times$ 4 mm $\times$ 22 mm and acousto-optical Q-switching, was calculated and optimized. This resulted in about 500 mW for the first Stokes (1178 nm) single-mode radiation with a repetition rate of 10 kHz and a pulse duration of 8 ns. At a lower repetition rate, 8 kHz, the cascaded second Stokes radiation at 1320 nm was observed with the average power of 62 mW and a much shorter pulse duration of about 2 ns. External frequency doubling by a 12-mm-long LBO crystal provided about 250 mW of 589 nm (visible–yellow) and 27 mW of 660 nm (visible–red) radiation. The yellow 589 nm radiation is of interest for pumping the sodium resonance line (D2), which can be applied in adaptive optic systems. The red frequency can be applied for photodynamical therapy (see chapter D3.2.3).

An interesting application of SRS technology was proposed in [89]. The authors suggested using SRS radiation to form artificial guide stars in the 10 km deep mesoscopic layer situated 95 km from the earth's surface. In this case, a laser is projected through a telescope to the upper atmosphere where it is resonantly absorbed by a layer of sodium atoms. The sodium fluorescence acts like an artificial star and the telescope detects the atmospheric distortion of its image. Then, using adaptive optics techniques, one of the optical elements of the telescope system is altered in such a way as to null out the atmospheric distortion of the image. With the optical system in this configuration, images of astronomical objects that are free of atmospheric distortion can now be obtained. The requirements for the laser to create an effective sodium guide star are quite stringent. To create the maximum photon return, the wavelength, temporal pulsewidth, spectral pulsewidth, energy per pulse and pulse repetition rate must all be exactly correct to get the maximum amount of light absorbed by the sodium atoms and emitted as fluorescence without the effects of saturation, radiation pressure or interference from multiple pulses (see chapter D8.1). So far this has proven to be a major challenge for the laser community and no acceptable solution is currently working in a telescope. However, a solid state Raman laser appropriate for sodium guide star applications has been demonstrated in the laboratory [93, 94]. A second type of Raman guide star laser based on a fibre as the Raman shifting element is also being developed in the laboratory [95]. This has the advantage of power scaling for non-astronomy adaptive optics applications.

The sodium atomic transition corresponds to a wavelength of 589.2 nm and has a radiative lifetime of 16 ns. A laser source for pumping the sodium resonance should be adjusted to this wavelength with accuracy better than 1 GHz and must operate in the burst pulsed mode or pulsed-periodic mode. All solid state, high efficiency and reliable laser systems based on a diode-pumped crystalline neodymium laser, solid state Raman frequency shifter and nonlinear frequency doubler could be promising for this type of application.

Such a system could consist of an Nd:YAG pump laser and a $CaWO_4$ SRS frequency shifter with subsequent doubling. However, this frequency matches the Na resonance only when the Nd:YAG crystal is cooled to below room temperature. At the same time, the $CaWO_4$ SRS crystal is not the best candidate for frequency shifting due to its strong Raman line broadening, low gain and high SRS threshold.

A different scheme of sodium Raman laser, based on the new best SRS crystal, $BaWO_4$, which has a line several times narrower (1.6 cm^{-1}), higher gain and lower threshold, was proposed in [87, 88]. In comparison with the $Ba(NO_3)_2$ crystal, with its record high value of nanosecond Raman gain, the $BaWO_4$ has almost the same gain but operates better in the sub-nanosecond and picosecond modes. The $BaWO_4$ crystal also has a much higher moisture resistance and a thermal conductivity at least twice as good .

As the pump source, the authors of [87, 88] used a Nd:GGG solid state laser, because its operation wavelength, $\lambda = 1062.1$ nm, shifted by $\nu_R = 925\ cm^{-1}$ using the $BaWO_4$ crystal and then doubled, fits the Na resonance very well. The Nd:GGG laser crystal is well known as a good candidate for high-power solid state lasers. In comparison with Nd:YAG lasers, it demonstrates a little longer metastable level lifetime (280 *versus* 250 μs) and a little wider linewidth (7.2 *versus* 5 cm^{-1}), which results in a smaller emission cross section, $\sigma = 1.7 \times 10^{-19}\ cm^2$. The Nd:GGG crystal also has a 30% lower thermal conductivity and a 50% higher temperature derivative of the refractive index. At the same time, Nd:GGG crystals can be grown with a much higher speed and have a larger diameter and very high optical quality, free of central core defects. This is very important for the large disc, slab and zigzag geometry of high-power lasers.

The neodymium concentration in these crystals can be increased by a factor of three to five in comparison with Nd:YAG crystals. In addition, co-doping with Cr^{3+} and Ce^{3+} ions can provide much better sensitization and stability. As a result, the Nd:GGG lasers can have a doubled laser efficiency when pumped by flashlamps.

In order to prove this approach to the sodium resonance laser source, two laser schemes for an Na solid state yellow Raman laser were tested in [87, 88]. The first one was based on IR Raman shifting of the Nd:GGG laser fundamental wavelength $\lambda_p = 1062.1$ nm by the first Stokes $BaWO_4$ crystalline Raman laser to $\lambda_{1St} = 1177.9$ nm. This wavelength was then halved to $\lambda_{SHG} = 588.95$ nm by a second harmonic generator on a $LiNbO_3$, LBO or KTP nonlinear crystal. The results of the pump-to-output conversion in the $BaWO_4$ Raman laser are shown in figure B1.7.4(*a*), which demonstrates the more than 30% slope efficiency and the 15 mJ output energy. The second scheme was based on the Raman shifting of visible radiation. In

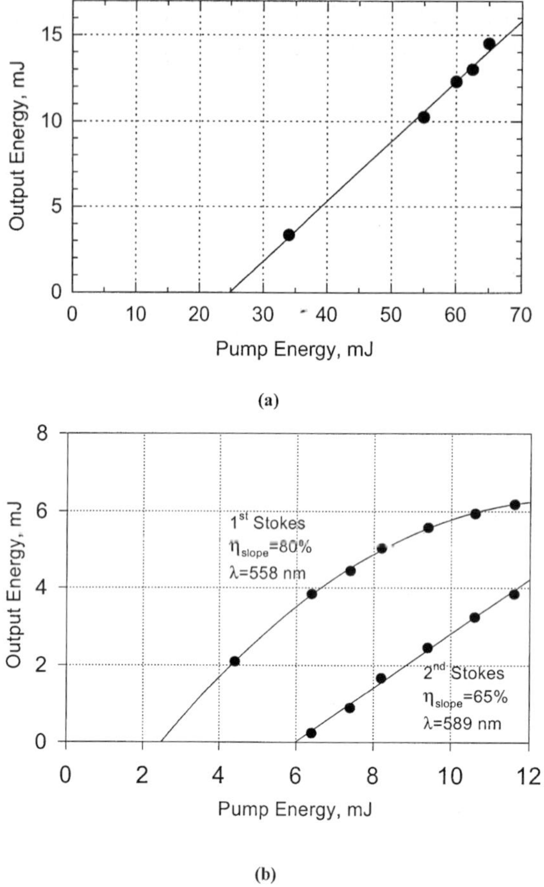

Figure B1.7.4. (*a*) The first Stokes (1179.9 nm) output energy of a $BaWO_4$ Raman laser *versus* 1062.1 nm pump energy of an Nd:GGG laser. (*b*) The first (559.5 nm) and second (588.95 nm) Stokes output energy of a Na-star Raman laser *versus* 531.05 nm pump energy of an Nd:GGG laser (second harmonic).

this scheme, the 1062.1 nm radiation of an Nd:GGG laser is first transformed by a nonlinear crystal, BBO or KTP, to the second harmonic, $\lambda_{SHG} = 531.05$ nm, which then serves as the pump radiation for the $BaWO_4$ visible Raman laser. In this case, the second Stokes output of the Raman laser oscillation was used to reach the Na resonance wavelength. As is seen from figure B1.7.4(*b*), this scheme gives more than 50 and 30% output-to-pump conversion efficiency of the first ($\lambda_{1St} = 559.5$ nm) and second Stokes ($\lambda_{2St} = 588.95$ nm), respectively. The 60% slope efficiency indicates that these characteristics can be improved. The better efficiency for the second scheme is caused by the higher Raman gain of the $BaWO_4$ crystal for green pumping ($g_{0.53} \approx 40$ cm GW^{-1}) than for IR pumping ($g_{1.06} \approx 8$ cm GW^{-1}). At the same time, as we discussed earlier, application of the intracavity IR pumping scheme of Raman laser allows one to increase the IR Raman conversion efficiency up to 100% (due to the use of the nonlinear intracavity dumping regime), to improve the spatial beam quality (beam clean-up), and to decrease the pulse duration with increasing peak power, which is important for the following second harmonic generation. Another advantage is that the $BaWO_4$ Raman laser and the Nd:GGG solid state pump laser permit some temperature wavelength tuning at around room temperature. This provides fine temperature adjustment of the oscillation wavelength in the vicinity of the sodium resonance without using a dispersive cavity. Recently, a long (up to 4 ms) pulse train of

50 giant pulses (50 ns each) of a high-energy (20 J in the train) Nd:GGG laser was used as a pump source for a $BaWO_4$ Raman Laser. As a result, a SRS train energy as high as 2 J was obtained, with a peak power up to 1 MW and conversion efficiency reaching 20% [98]. A high average power (up to 6 W) of the SRS radiation was realized under self-phase-conjugated Nd:YAG pumping at the conversion efficiency of 15-35% with a rather low thermal lensing (0.12 D) of the $BaWO_4$ SRS crystal [97]. The spectral range of output wavelengths and the variety of output configurations used in the applications described here demonstrate the potential for Raman lasers in military applications such as range finders, target designators and munitions guidance; in medical applications requiring specific wavelengths in the visible and near infrared spectral regions; and in material processing applications requiring high beam quality and precise wavelength selection [96].

Since Raman lasers can be based on a wide variety of combinations of pump lasers and active Raman crystals and they can be constructed in a variety of configurations using slabs or rods with normal Q-switching and mode-locking technology, it is possible to envision many more applications in the future.

B1.7.14 Conclusions

In summary, high gain solid state Raman materials provide infrared Raman shifting and lasing for tunable and fixed frequency operation in the 1.1–2.2 μm spectral range and allow the development of high efficiency intracavity solid state Raman lasers based on coupled resonator or shared resonator designs. These intracavity configurations take advantage of the third-order nonlinear nature of SRS to provide for special operating characteristics, such as spatial beam clean-up, leading to diffraction-limited output and pulsewidth control through nonlinear cavity damping. Important results have been reported on self-Raman shifting lasers in which the Raman crystal is also the host material for the rare earth ions that are active in lasing. Tunable Raman laser emission can be obtained by using glass fibres as the active medium, by scattering from polaritons in polar crystals such as $LiNbO_3$, and by parametric interactions among different types of stimulated emission processes.

In spite of strong Raman gain decrease in the IR spectral region ($g_{0.53\ \mu m}/g_{1.06\ \mu m} \approx 4$), the unique features of $Ba(NO_3)_2$ crystals allowed for the development of a single-pass SRS cell and then a Raman laser for extending the tuning range of $LiF:F_2^-$ solid state colour-centre lasers. With nanosecond pulse pumping in the range 1.10–1.25 μm, the steady-state SRS to the first and second Stokes components provided narrow line, broadly tunable laser oscillation in the 1.23–1.37 μm and 1.43–1.6 μm spectral regions with conversion efficiencies as high as 60 and 20%, respectively.

The special properties of solid state Raman lasers make them useful in many different types of applications. The appropriate intracavity design employing a high Fresnel number pump cavity with conversion to a single-mode Raman laser output can provide lasers with extremely high wall-plug efficiencies. The third-order nature of the nonlinearity allows for efficient wavelength conversion without the requirement of a high spatial quality pump beam and strict acceptance parameters for wavelength, angle and temperature, as is the case for second-order nonlinear effects used in harmonic generators and optical parametric oscillators. In fact, the stimulated Raman process in the cavity that acts to clean up the spatial quality of the laser beam is a mechanism similar to phase conjugation. High efficiency and excellent beam quality are important laser properties for any application. In addition, intracavity Raman laser configurations can be designed to operate with nonlinear cavity damping to control the temporal properties of pulsed output. This allows for laser emission with pulsewidths of 1 ns or less without the complexity of mode locking.

Acknowledgments

The research discussed in this article is based on work partially supported by ISTC and EOARD Partner project 2022p. The authors gratefully acknowledge the major contributions to the work described here by their many colleagues, especially P G Zverev, A A Sobol, L I Ivleva, M E Doroshenko, V V Osiko and

A M Prokhorov of the General Physics Institute and James T Murray and William L Austin of Lite Cycles, Inc.

References

[1] Raman C V and Krishnan K S 1928 *Nature* (London) **121** 501
[2] Landsberg G S and Mandel'shtam L I 1928 *Zh. Rus. Fiz. Khim. Obshch.* **60** 335
[3] Woodbury E J and Ng W K 1962 *IRE* **50** 2367
[4] Levenson M D 1982 *Introduction to Nonlinear Laser Spectroscopy* (New York: Academic)
[5] Murray J T, Powell R C and Peyghambarian N 1996 *J. Luminesc.* **66&67** 89
[6] Kaiser W and Maier M 1972 *Laser Handbook* vol 11, ed F T Arecchi and E O Shultz-Dubois (Amsterdam: North-Holland) p 1077
[7] Loudon R 1963 *Proc. R. Soc.* A **275** 218
[8] Eremenko A S, Karpukhin S N and Stepanov A I: 1980 Stimulated Raman scattering of the neodymium laser second harmonic in nitrate crystals *Kvantovaya Elektron.* **7** 196–7 (Engl. transl. 1980 *Sov. J. Quantum Electron.* **10** 113)
[9] Karpukhin S N and Yashin V E 1984 *Kvantovaya Elektron.* **10** 1992 (Engl. transl. 1984 *Sov. J. Quantum Electron.* **14** 1337)
[10] Karpukhin S N and Stepanov A I 1986 Stimulated emission from the cavity under SRS in $Ba(NO_3)_2$, $NaNO_3$, and $CaCO_3$ crystals *Kvantovaya Elektron.* **13** 1572–7 (Engl. transl. *Sov. J. Quantum Electron.*)
[11] Belevtseva L I, Voitsekhovskii V N, Nazarova N A, Romanova G I, Shvedova M V and Yakobson V E 1989 *Opt. Mekh. Promst.* **56** 38 (Engl. transl. *Sov. J. Opt. Technol.*)
[12] Voitsekhovskii V N, Karpukhin S M and Yakobson V E 1995 Single crystal barium nitrate and sodium nitrate as efficient materials for laser radiation frequency conversion based on stimulated Raman scattering *J. Opt. Technol.* **62** 770–6
[13] Zverev P G, Basiev T T, Osiko V V, Kulkov A M, Voitsekhovskii V N and Yakobson V E 1999 *Opt. Mater.* **11** 315–34
[14] Zverev P G, Basiev T T and Prokhorov A M 1999 *Opt. Mater.* **11** 335
[15] Zverev P G and Basiev T T 1995 *Kvantovaya Elektron.* **22** 1241 (Engl. transl. *Sov. J. Quantum Electron.*)
[16] Zverev P G, Jia W, Liu H and Basiev T T 1995 *Opt. Lett.* **20** 2378
[17] Basiev T T, Sobol A A, Zverev P G, Fedorov V V, Doroshenko M E, Ivleva L I, Polozkov N M, Osiko V V, Prokhorov A M and Hager G 1999 ed V J Corcoran and T A Goldman *Laser '98, Proc. Int. Conf. (Tucson AZ, 1998)* (McLean VA: SRS Press) p 712
Basiev T T, Sobol A A, Zverev P G, Fedorov V V, Doroshenko M E, Ivleva L I, Polozkov N M, Osiko V V and Prokhorov A M 1999 Solid state Raman materials characterization for high average power 1.3 μm laser frequency shift *Technical Report* LMTRC of GPI for EOARD, Moscow, January, p 85
[18] Basiev T T, Sobol A A, Zverev P G, Voron'ko Yu K, Osiko V V and Powell R C 1998 *Advanced Solid State Lasers* vol 19, ed W R Rosenberg and M M Fejer (Washington DC: OSA TOPS) p 546
[19] Basiev T T, Sobol A A, Zverev P G, Osiko V V and Powell R C 1999 *Appl. Opt.* **38** 594
[20] Basiev T T, Sobol A A, Zverev P G, Ivleva L I, Osiko V V and Powell R C 1999 *Opt. Mater.* **11** 307
[21] Basiev T T, Zverev P G, Sobol A A, Fedorov V V, Doroshenko M E, Skornyakov V V, Ivleva L I and Osiko V V 1999 *Novel Lasers and Devices—Basic Aspects Tech. Digest. Int. Conf. CLEO—Europe Focus Meeting '99 (Munich)* (Washington DC: Optical Society of America) p 160
[22] Basiev T T, Sobol A A, Zverev P G, Ivleva L I and Osiko V V 2000 Laser material for stimulated Raman scattering *Patent* RU 2178938
[23] Basiev T T, Sobol A A, Voron'ko Yu K and Zverev P G 2000 Spontaneous Raman spectroscopy of tungstate and molybdate crystals for Raman lasers *Opt. Mater.* **15** 205–16
[24] Zverev P G, Basiev T T, Sobol A A, Skornyakov V V, Ivleva L I, Polozkov N M and Osiko V V 2000 Stimulated Raman scattering in alkaline-earth tungstate crystal *Quantum Electron.* **30** 55–9
[25] S A Vitsinskii, Isakov V K, Karpukhin S M and Lovchil L L 1993 Stimulated Raman scattering of copper-vapour laser radiation in barium nitrate crystal *Kvantovaya Elektron.* **20** 1555–8 (Engl. transl. *Sov. J. Quantum Electron.*)
[26] He Chuan and Chyba T H 1997 Solid-state barium nitrate Raman laser in the visible region *Opt. Commun.* **135** 273–8
[27] McCray C L, Lee S W and Chyba T H 1998 Measurements of stimulated Raman amplification and Raman shifts in barium nitrate, *Advanced Solid State Lasers* vol 19, ed W B Bosenberg and M M Fejer (Washington DC: OSA TOPS) pp 550–4
[28] Zverev P G, Murray J T, Powell R C, Reeves R J and Basiev T T 1993 *Opt. Commun.* **97** 59–64
[29] Basiev T T, Voron'ko Yu K, Mirov S B, Osiko V V and Prokhorov A M 1982 *Izv. Akad. Nauk SSSR, Ser. Fiz.* **46** 1600–10
[30] Basiev T T 1985 Tunable color-center lasers and their application in selective spectroscopy of disordered media *Izv. Akad. Nauk SSSR, Ser. Fiz.* **49** 68–77 (Engl. transl. *Bull. Sov. Acad. Sci. USSR Phys. Ser.*)
[31] Basiev T T, Osiko V V and Mirov S B 1988 *IEEE J. Quantum Electron.* **24** 1052
[32] Basiev T T, Voitsekhovskii V N, Zverev P G, Karpushko F V, Lyubimov A V, Mirov S B, Morozov V P, Mochalov I V, Pavlyuk A A, Sinitsyn G V and Yakobson V E 1987 Conversion of tunable radiation from a laser utilizing an LiF crystal containing F_2^- color centers by stimulated Raman scattering in $Ba(NO_3)_2$ and $KGd(WO_4)_2$ crystals *Kvantovaya Elektron.* **14** 2452–3 (Engl. transl. 1987 *Sov. J. Quantum Electron.* **17** 1560–1)
[33] Basiev T T, Fedin A V, Gavrilov A, Osiko V V and Smetanin S N 2003 Stimulated Raman scattering of high average power Nd:YAG laser radiation *Laser Physics J.* accepted

[34] Basiev T T and Mirov S B 1994 *Room Temperature Tunable Color Center Lasers (Laser Science and Technology Books Series)* (New York: Gordon and Breach Science/Harwood Academic)
[35] Zverev P G and Basiev T T 1993 *Proc. All-Union Conference Optika Laserov (Leningrad)* **2** 363
[36] Zverev P G and Basiev T T 1994 *J. Physique* **4** C4-599
[37] Zverev P G, Basiev T T and Prokhorov A M 1994 *Tech. Digest CLEO/Europe'94* 94TH0614-8 (New Jersey: IEEE) p 154
[38] Zverev P G, Basiev T T, Ermakov I V and Prokhorov A M 1994 *Laser Methods of Surface Treatment and Modification (Proc. SPIE 2498)* (Bellingham: SPIE) p 164
[39] Basiev T T, Sigachev V B, Doroshenko M E, Zverev P G, Osiko V V and Prokhorov A M 1994 *Tech. Digest CLEOI Europe'94* 94TH0614-8 (New Jersey: IEEE) p 125
[40] Basiev T T, Sigachev V B, Doroshenko M E, Zverev P G, Osiko V V and Prokhorov A M 1994 *Laser Methods of Surface Treatment and Modification (Proc. SPIE 2498)* (Bellingham: SPIE) p 171
[41] Zverev P G and Basiev T T 1995 *Advanced Solid State Lasers* vol 24, ed B H T Chai and S A Pain (Washington DC: OSA) p 288
[42] Zverev P G, Basiev T T, Jia V and Liu H 1996 *Advanced Solid State Lasers* vol 1, ed S A Pain and C Pollok (Washington DC: OSA TOPS) p 554
[43] Khulugurov V M, Ivanov N A, Inshakov D V, Oleinikov E A, Voitsekhovskii V N and Yakobson V E 1992 Amplification in LiF:F_2^- crystal at SRS in $Ba(NO_3)_2$ crystal *Kvantovaya Electron.* **19** 162–3
[44] Khulugurov V M, Ivanov N A and Oleinikov E A 1994 Nanosecond lasers based on Raman converters with resonant excitation *Technical Digest CLEO'94* CtuK70 (Washington DC: Optical Society of America) p 126
[45] Murray J T, Austin W L, Powell R C and Peyghambarian N 1999 *Opt. Mater.* **11** 373
[46] Murray J T, Powell R C and Austin E L 1995 *Opt. Photon. News* **6** 32
[47] Murray J T, Powell R C, Peyghambarian N, Smith D, Austin W and Stolzenberger R A 1995 Generation of 1.5 μm radiation through intracavity solid-state Raman shifting in $Ba(NO_3)_2$ nonlinear crystals *Opt. Lett.* **20** 1017–19
[48] Kannari F and Takei N 1997 20-Hz Operation of an eye-safe cascade-Raman laser with a $Ba(NO_3)_2$ crystal *Proc. Int. Symp. On Lasers and Nonlinear Optical Materials (Singapore, 3–5 November)* pp 289–91
[49] Murray J T, Austin W L, Calmes L K, Powell R C and Quarles G J 1997 Nonlinear cavity-damped intracavity solid state Raman laser transmitters *OSA TOPs on Advanced Solid State Lasers* vol 20, ed St A Payne and C K Pollock (Washington DC: OSA) pp 72–6
[50] Ammann E O and Decker C D 1977 *J. Appl. Phys.* **48** 1973–5
[51] Ammann E O 1978 *Appl. Phys. Lett.* **32** 52–4
[52] Grigoryan G G and Sogomonyan S B 1989 A synchronously pumped picosecond SRS laser on a $LiIO_3$ crystal *Kvantovaya Electron.* **16** 2180–3 (Engl. transl. *Sov. J. Quantum Electron.*)
[53] Pask H M and Piper J A 1998 Practical 580 nm source based on frequency doubling of an intracavity-Raman-shifted Nd:YAG laser *Opt. Comm.* **148** 285–8
[54] Pask H M and Piper J A 1999 Efficient all-solid-state yellow laser source producing 1.2 W average power *Opt. Lett.* **24** 1490–2
[55] Revermann M, Pask H M, Blows J L and Omatsu T 2000 Thermal lensing measurements in an intracavity $LiIO_3$ Raman laser *Advanced Solid State Lasers* ed H Injeyan, U Keller and Ch Marshal (Washington DC: OSA TOPS) pp 506–9
[56] Pask H M and Piper J A 2000 Diode-pumped $LiIO_3$ intracavity Raman laser *IEEE J. Quantum Electron.* **36** 949–55
[57] Pask H M and Piper J A 2002 Design and operation of an efficient 1.4 W diode-pumped Raman laser at 578 nm *Advanced Solid-State Lasers Technical Digest Int. Conf. (Quebec City, 3–6 February)* (Washington DC: OSA) pp 1–3
[58] Andryunas K, Vishakas Yu, Kabelka V, Mochalov I V, Pavlyuk A A, Petrovskii G T and Syrus V 1985 *Pis. Zh. Ek. Teor. Fiz.* **42** 333 (Engl. transl. 1986 *JETP Lett.* **42** 410–12)
Andryunas K, Vishakas Yu, Kabelka V, Mochalov I V, Pavlyuk A A, Ionina N V and Syrus V Patent of USSR # 1227074
[59] Ivanyuk A M, Ter-Pogosyan M A, Shakhverdov P A, Belyaev V D, Ermolaev V L and Tikhonova H P 1985 Picosecond radiation pulses at intracavity stimulated Raman scattering inside the active element of neodymium laser *Opt. Spectrosk.* **59** 950–2 (Engl. transl. *Sov. Opt. Spectrosc.*)
[60] Mikhailov A V, Mochalov I V and Lyubimov A V 1987 Raman scattering in KGW crystals *Lasers and Optical Nonlinearity* ed E K Maldutis (Vilnyus: University) pp 265–9 (in Russian)
[61] Andryunas K, Barila A, Vishchakas Yu and Syrus V 1987 Investigation of pulse duration dynamics in a laser based on $KGd(WO_4)_2$:Nd^{3+} and $KY(WO_4)_2$:Nd^{3+} *Lasers and Optical Nonlinearity* ed E K Maldutis (Vilnyus: University) pp 43–50 (in Russian)
[62] Andryunas K, Barila A, Vishchakas Yu, Mochalov I V, Petrovskii G T and Syrus V 1988 Temporal characteristics of picosecond pulses of SRS self-conversion *Opt. Spektrosk.* **64** 397–401 (Engl. transl. *Sov. Opt. Spectrosc.*)
[63] Vishchakas Yu K, Mochalov I V, Mikhailov A V, Klevtsova R F and Lyubimov A V 1988 Crystal structure and Raman scattering in $KGd(WO_4)_2$ crystals *Lietuvos fizikos rinkinys* **28** 224–34
[64] Mochalov I V 1997 Laser and nonlinear properties of the potassium gadolinium tungstate laser crystal $KGd(WO_4)_2$:Nd^{3+} (KGW:Nd) *Opt. Eng.* **36** 1660–9
[65] Kurbasov S V and Losev L L 1999 Raman compression of picosecond microjoule laser pulses in $KGd(WO_4)_2$ crystal *Opt. Commun.* **168** 227–32
[66] Ivanyuk A M, Sandulenko V A, Ter-Pogosyan M A, Shakhverdov P A, Chervinskii V G, Lukin A V and Ermolaev V L 1987 Intracavity stimulated Raman scattering in a nanosecond neodymium laser based on potassium-gadolinium tungstate *Opt.*

Spektrosk. **62** 961–2 (Engl. transl. *Sov. Opt. Spectrosc.*)

[67] Berenberg V A, Karpukhin S N and Mochalov I V 1987 SRS of nanosecond pulses in $KGd(WO_4)_2$ crystals *Kvantovaya Electron.* **14** 1849–50 (Engl. transl. *Sov. Quantum Electron.*)

[68] Kaminskii A A, Ustimenko N S, Gulin A V, Bagaev S N and Pavlyuk A A 1998 Raman-parametric interactions in monoclinic $KGd(WO_4)_2$ and $KY(WO_4)_2{:}Nd^{3+}$ crystals: picosecond multicomponent Stokes and anti-Stokes generation and nanosecond self-SRS conversion in human eye-safe 1.5 μm range *Dokl. Akad. Nauk.* **359** 179–83 (Engl. transl. *Sov. Phys. Docl.*)

[69] Ustimenko N S and Gulin A V 2000 1.538 μm wavelength radiation of $KGd(WO_4)_2$:Nd3+ lasers based on stimulated Raman scattering with self-conversion *Pribory i Tekhnika Eksperimenta* **3** 99–101 (Engl. transl. *Sov. Instrum. Exp. Technol.*)

[70] Fideisen J, Eichler N J and Peuser P 2000 Self-stimulating, transversely diode pumped Nd:$KGd(WO_4)_2$ Raman laser *Opt. Commun.* **181** 129–33

[71] Lagatsky A A, Abdolvand A and Kuleshov N V 2000 Passive Q-switching and self-frequency Raman conversion in a diode-pumped Yb: $KGd(WO_4)_2$ laser *Opt. Lett.* **25** 616–18

[72] Grabtchikov A S, Kuzmin A N, Lisinetskii V A, Orlovich V A, Demidovich A A, Eichler N V, xBednarkiewics N V, Strek W and Titov A N 2002 Yb:KGW Microchip laser performance: fundamental frequency generation and Raman self-frequency conversion *Advanced Solid State Lasers, Technical Digest Int. Conf. (Quebec City, 3–6 February)* (Washington, DC: OSA) pp 1–3

[73] Basiev T T, Zverev P G, Sobol A A and Osiko V V 2002 Search and characterization of new crystals for Raman lasers *Advanced Solid State Lasers* MB10 (Washington DC: Optical Society of America) pp 1–3

[74] Cerný P, Jelínková H, Basiev T T and Zverev P G 2001 Properties of transient and steady-state stimulated Raman scattering in $KGd(WO_4)_2$ and $BaWO_4$ tungstate crystals *Growth, Fabrication, Devices and Applications of Laser and Nonlinear Devices (Proc. SPIE 4268)* (Los Angeles CA: SPIE) pp 101–8

[75] Cerný P, Jelínková H, Šulc J, Zverev P G and Basiev T T 2001 Evaluation of $BaWO_4$ Steady state Raman gain in single- and double-pass arrangement *Advanced Solid State Lasers (OSA TOPS, Seattle, WA)* (Washington DC: Optical Society of America) vol 50.

[76] Zverev P G, Basiev T T, Sobol A A, Ermakov I V and Gellerman W 2001 $BaWO_4$ crystal for quasi-cw yellow Raman laser *Advanced Solid-State Lasers (OSA Technical Digest)* (Washington DC: OSA) pp 124–5

[77] Cerný P, Zverev P G, Jelínková H and Basiev T T 2000 Efficient Raman shifting of picosecond pulses using $BaWO_4$ crystal *Opt. Commun.* **177** 397–404

[78] Cerný P, Zverev P G, Jelínková H, Basiev T T, Ivleva L I and Osiko V V 2000 Comparison of stimulated Raman scattering of picosecond pulses in tungstate crystals *Nonlinear Materials, Devices, and Applications (Proc. SPIE 3928)* (Los Angeles CA: SPIE) ed J W Pierce pp 124–31

[79] Cerný P, Jelínková H, Zverev P, Basiev T and Kubecek V 2000 Picosecond Raman gain of $CaCO_3$, $KGd(WO_4)_2$ and $BaWO_4$ crystals *Book of Abstract CLEO/Europe, Int. Quantum Electronics Conf. 2000 (Nice)* paper no CthE14 (New Jersey: IEEE) p 94

[80] Cerný P, Jelínková H, Miyagi M, Basiev T T and Zverev P G 2002 Efficient picosecond Raman lasers on $BaWO_4$ and $KGd(WO_4)_2$ tungstate crystals emitting in 1.15 to 1.18-μm spectral region *LASE 2002, part of Photonics West Symp. (San Jose)* paper no 4630-17 (Los Angeles CA: SPIE)

[81] Cherny P and Jelinkova H 2002 Picosecond stimulated Raman scattering in $BaWO_4$ crystal with close to quantum limit efficiency *Advanced Solid-State Lasers, Tech. Digest (Quebec City)* (Washington DC: Optical Society of America) pp 1–3

[82] Kaminskii A A, Eichler H J, Ueda K, Klassen N V, Redkin B S, Li L E, Findeisen J, Jaque D, Garcia-Sole J, Fernandez J and Balda R 1999 Properties of Nd^{3+}-doped and undoped tetragonal $PbWO_4$, $NaY(WO_4)_2$, $CaWO_4$, and undoped monoclinic $ZnWO_4$ and $CdWO_4$ as laser-active and stimulated Raman scattering-active crystals *Appl. Opt.* **38** 4533–46

[83] Findeisen J, Eichler H J and Kaminskii A A 1999 Efficient picosecond $PbWO_4$ and two-wavelength $KGd(WO_4)_2$ Raman lasers in the IR and visible *IEEE J. Quantum Electron.* **35** 173–8

[84] Kaminskii A A, McCray C L, Lee H R, Lee S W, Temple D A, Chyba T H, Marsh W D, Barnes J C, Annanenkov A N, Legun V D, Eichler H J, Gad G M A and Ueda K 2000 High efficiency nanosecond Raman lasers based on tetragonal $PbWO_4$ crystals *Opt. Commun.* **183** 277–87

[85] Chen W, Inagawa Y, Omatsu T, Tateda M, Takeuchi N and Usiki Y 2001 Self-stimulating, passively Q-switched, diode-pumped Nd3+:$PbWO_4$ Raman laser *Advanced Solid-State Lasers (Technical Digest Int. Conf., January, Seattle, WA)* (Washington, DC: OSA)

[86] Rawle C B, Ter-Mikirtychev V V, Kinnie I T and Sandle W J 2001 High-energy solid-state Raman laser based on barium nitrate crystal for near IR and visible spectral range *Advanced Solid-State Lasers (Technical Digest Int. Conf. 28–31 January, Seattle, WA)* (Washington, DC: OSA) 22/MB1-1–24/MB1-3

[87] Basiev T T, Doroshenko N E, Zverev P G and Prokhorov A M 2000 Solid-state laser of yellow spectral region *Russian Patent* Appl. No 2000110183/28(010974

[88] Zverev P G, Basiev T T, Doroshenko M E and Osiko V V 2000 Barium tungstate Raman laser—a new coherent source for sodium star experiments *OSA Trends Opt. Photon. Series* **34** 348-3-54

[89] Murray J T, Austin W L and Powell R C 1998 End-pumped intracavity solid-state Raman lasers *Advanced Solid-State Lasers (OSA Trends in Optics and Photonics 19)* ed W R Bosenberg and M M Feiger (Washington DC: OSA) pp 129–35

[90] Pask H M, Blows J L, Piper J A and Omatsu T 2001 Thermal lensing in a barium nitrate Raman laser *Advanced Solid-State Lasers*

(Technical Digest Int. Conf. 28–31 January, Seattle, WA) (Washington DC: OSA) 276/TuB15-1–278/TUB15-3
[91] Murray J T, Powell R C and Kock G 1996 *NASA Research Report*
[92] McLean J W and Murray J T 1998 *Laser Focus World (January)* (Los Angeles CA: SPIE) p 171
Murray J T, Austin W L, Calmes L K, Powell L K and Quarles G J 1997 *Advanced Solid State Lasers (OSA TOPS 10)* ed C R Pollock and W R Bosenberg (Washington DC: OSA) p 72
Murray J T, Austin W L, Calmes L K, Powell R C, McLean J W and Brian L 1997 *Proc. SPIE* (Orlando, FL)
[93] Murray J T, Roberts W T Jr, Austin W L, Powell R C and Angel J R 1998 *Proc. SPIE* (Los Angeles CA: SPIE)
[94] Murray J T, Powell R C, Austin W L, Roberts W T Jr, Angel J R, Shelton C T and Sandler D G 1997 *Proc. ESO Laser Technology and Laser Guide Star Workshop (Garching, June)* (LosAngeles CA: SPIE)
[95] Murray J T, Roberts W T Jr, Austin W L, Powell R C and Bonaccini D 1998 *Proc. SPIE* (Los Angeles, CA: SPIE)
[96] Murray J T, Austin W L and Powell R C 1998 *Advanced Solid-State Lasers (OSA TOPS 11)* ed W R Bosenberg (Washington DC: OSA) p 82
[97] Basiev T T, Danileiko Yu K, Doroshenko M E, Fedin A V, Gavrilov A, Osiko V V and Smetanin S N 2003 High-energy $BaWO_4$ Raman laser pumped by a self-phased-conjugated Nd:GGG laser *Optics of Lasers Int. Conf. June, St. Petersburg*
[98] Basiev T T, Fedin A V, Gavrilov A, Osiko V V and Smetanin S N 2003 High-average-power $BaWO_4$ Raman laser pumped by a self-phase-conjugated Nd laser *CLEO-Europe, June, Munich*
[99] Sulc J, Jelinkova H, Cerny P, Skornjakov V, Doroschenko M, Zverev P and Basiev T T 2003 Nd:$SrWO_4$ Raman Laser *Proceedings of Workshop 2003* Part A, CTU reports, Prague, volume 7, pp 174–4

Further reading

Loudon R 1964 *Adv. Phys.* **13** 423
Zubov V A, Sushchinskii M M and Shuvalov I K 1964 Stimulated Raman scattering *Usp. Fiz. Nauk* **83** 197 (Engl. transl. *Sov. Phys.–Usp.*) 7419
Erckhart G 1966 *IEEE J. Quantum Electron.* **QE-2** 1
Bloembergen N 1967 The stimulated Raman effect *Am. J. Phys.* **35** 989
Glass A 1967 7.4—Design consideration for Raman lasers *IEEE J. Quantum Electron.* **QE-3** 516–20
Lugovoi V N 1968 *Introduction to the Theory of Stimulated Raman Scattering* (Moscow: Nauka) in Russian
Kaiser W and Maier M 1972 *Laser Handbook* vol 11, ed F T Arecchi and E O Shultz-Dubois (Amsterdam: North-Holland) p 1077
Grasyuk A Z 1974 *Kvantovaya Elektron.* **3** 485 (Engl. transl. *Sov. J. Quantum Electron.*)
Murray J T, Powell R C and Peyghambarian N 1996 *J. Luminesc.* **66&67** 89
Mochalov I V 1997 Laser and nonlinear properties of the potassium gadolinium tungstate laser crystal $KGd(WO_4)_2$:Nd^{3+} (KGW:Nd) *Opt. Eng.* **36** 1660–9
Zverev P G, Basiev T T, Osiko V V, Kulkov A M, Voitsekhovskii V N and Yakobson V E 1999 *Opt. Mater.* **11** 315–34
Basiev T T 1999 Spectroscopy of new SRS-active crystals and solid-state SRS crystals *Sov. Phys.–Usp.* **42** 1051–62
Zverev P G, Basiev T T and Prokhorov A M 1999 *Opt. Mater.* **11** 335
Powell R C and Basiev T T 2000 Seventy years of Raman scattering *Advances in Laser Physics* ed V S Letokhov and P Meystre (Amsterdam: Harwood Academic) pp 55–66

B1.8
Colour centre lasers

T T Basiev, P G Zverev and S B Mirov

B1.8.1 Crystal hosts for active elements of colour centre lasers

Colour centres (CCs) exist in many types of crystalline solids, including both ionic and covalently bonded lattices. In table B1.8.1, the major physico-chemical, optical and mechanical characteristics of the crystal hosts intended for CC laser-active elements are considered and compared with each other and with other well-known laser media. The most promising materials for developing CCs are crystals of alkali-halide metals, alkali-earth fluorides, oxides (for instance, sapphire) and covalent crystals (at present the only available representative is diamond).

The most widespread media with CCs—alkali-halide crystals—are optically transparent in a wide range of spectra (0.11–35 μm) and are widely used as optical materials for instruments in UV, visible and IR optical technology, optics and spectroscopy. The structure of the alkali-halide crystals refers to the Fm3m spatial group with a general cubic symmetry of the crystal lattice. Thus, the alkali-halide crystals are inherently isotropic, i.e. their refraction index does not depend upon the polarization or direction of light propagation. A simple, fast and available Kyropoulos technique enables the commercial growth of high-quality crystals with large cross sections and volumes (for instance, ⌀400 mm × 200 mm) on a large scale. The crystal media so grown may be used for fabricating large active elements and Q-switches in powerful laser systems.

LiF and NaF crystals are the most outstanding among the alkali halides because of their low hygroscopicity. Their solubility in water is rather low and, hence, they are acceptable for practical applications in natural conditions. It is possible to use NaF crystals in a dry atmosphere. All the remaining alkali-halide crystals, without exception, cannot be used in damp air. Their facets become opaque without a special protective coating.

Of the alkali halides, LiF crystals have the smallest (close to $-1.2\times10^{-5}\,^\circ C^{-1}$) value for the temperature-derivative refractive index and a high thermal conductivity coefficient (14 W m^{-1} C^{-1} at 300 K, even higher than the value of 13 W m^{-1} C^{-1} for the YAG crystal). Thus, the critical thermal power $P_{cr}(P_{cr} = \lambda K|dn/dt|^{-1})$ of an LiF crystal [1, 2], for which the value of thermal aberration becomes comparable with the laser wavelength amounts to 1.25 W at $\lambda = 1.06\ \mu$m (for YAG:Nd^{3+} $P_{cr} = 1.3$ W). Therefore, LiF must be considered as a promising laser lattice for high average-power laser operations at room temperature. A comparatively small nonlinear refractive index, n_2, for LiF provides a high level of self-focusing critical power—0.46 MW, ($P_{cr}^{s} = \varepsilon_0 C\lambda^2/8\pi n_2$)—which exceeds the P_{cr}^{s} of YAG crystals (0.30 MW) and is slightly lower than that for ED-2 glass ($P_{cr}^{s} = 0.69$ MW). From this point of view as well as for its high (compared to the YAG crystal) laser-induced bulk damage threshold of 3.6 GW cm^{-2} (at 1.06 μm for 10 ns pulses), LiF crystals are also considered as prospective lattices for nonlinear saturable absorbers and Q-switchers in resonators in powerful laser systems.

As shown in table B1.8.1, alkali-halide crystals have low Knupp hardness values, 10–200 times lower than those for Al_2O_3 and YAG crystals which are the most widely used ones in laser technology. This makes it difficult to achieve and preserve highly polished surfaces during the process of preparation and application for all alkali-halide crystals except LiF and NaF.

The most durable of the alkali-halide crystals—LiF—has a tensile strength almost 20 times less than that for YAG, ten times less than for laser silica glasses and close to the strength of phosphate glasses. Moreover, alkali-halide crystals easily come to pieces along the [100] cleavage planes under impact. Therefore, they have to be carefully handled during exploitation and protected from appreciable thermal gradients and thermal and mechanical shocks.

Alkali-earth fluorides are of great interest as host crystals for active media with CCs. They are easily produced to have low hygroscopicity and high mechanical and thermal strength, in contrast to alkali-halide crystals.

Sapphire and diamond are also promising for application as CC active media. These crystals feature a unique combination of thermal and mechanical properties, which exceed (as seen in table B1.8.1) the corresponding values for YAG crystals. However, they are not so readily accessible for technology as alkali halides. Since LiF can be grown as large crystals and has remarkable physical and optical properties (the best material for room temperature operation) and because of the extensive research on this material we will primarily focus on it for the rest of this chapter.

B1.8.2 Types of CCs in ionic crystals

Table B1.8.2 shows the major types of CCs that can be used to achieve stimulated radiation. In general, the major crystals where laser active CCs can be developed are the pure and impurity-doped fluorides and chlorides of Li, Na, K and Rb alkali metals [2–6] as well as CaF_2 and SrF_2 [7]. Laser oscillation has also been reported using CCs in Al_2O_3 [8, 9], diamond [10], MgF_2 [11] and the compound fluorides, $KMgF_3$ [12] and $YLiF_4$ [13]. In the primal state, these crystals are optically transparent. Under irradiation with high-energy electrons, neutrons, γ-rays, x-rays or hard UV, or calcination in alkali metal vapour (additive colouration), anionic vacancies appear in the crystal lattice and serve to localize free electrons. The absorption bands of these (F) centres (vacancy + electron) give a typical colouring to the crystals.

The simplest of point defects is the anionic vacancy V, denoted in table B1.8.2 by an empty square (empty circle in figure B1.8.1) in place of an absent negative ion. The vacancy has an effective positive charge and, on capturing an electron, forms a defect whose electronic structure is similar to that of the hydrogen atom. Such an optical centre is referred to as an F centre and has absorption bands in the visible or UV spectral regions (for many reasons [6] F centres are not suitable for laser oscillation).

The other intrinsic electronic CCs on which oscillation has been obtained (see table B1.8.2 and figure B1.8.1) are complexes derived from F centres. For instance, the F_2^+ centre consists of two neighbouring anionic vacancies with one captured electron, the F_2 centre consists of two neighbouring vacancies with two captured electrons (an analogue of the hydrogen molecule) and the F_2^- centre is composed of a pair of vacancies with three captured electrons (see table B1.8.2 and figure B1.8.1).

The CCs denoted with a subscript A (besides their own defects) involve a neighbouring impurity ion replacing the positive ion in the nearest site of the cationic sublattice (for example, an Li impurity in Na or K sublattices or an Na impurity in K or Rb sublattices). The centres denoted by subscript B comprise an association of an anionic vacancy capturing one electron with two neighbouring impurity ions (see table B1.8.2).

The principles of colour centre laser (CCL) operation can be accounted for using the Frank–Condon configuration diagram, which describes the optical properties of an electron system interacting with molecular or lattice vibrations; in particular, the optical transition of CCs. Pump radiation is absorbed in the wide band of the electric-dipole electron–vibrational transition $1 \rightarrow 2$. In a time of the order of 10^{-12}–10^{-13} s, radiationless relaxation to the minimum of the potential curve of the excited electronic state, $2 \rightarrow 3$ occurs, accompanied by a mutual rearrangement of the neighbouring ions and by phonon emission. Then a radiative

Table B1.8.1. Main physico-chemical, mechanical and optical characteristics of the most promising crystal hosts with CCs [2]. The thermal shock parameter, R, characterizes the thermo-mechanical strength of the material. For instance, for a plate element with a thickness t, thermal power P_V, which may be absorbed by the medium with a parameter R_T without being damaged, $P_V = 12R_T/t^2$, where $R_T = S_T(1-\mu)K/\alpha E$.

Symbols: ρ, density (g cm^{-3}); d, lattice constant (Å); H, Knupp hardness (kg mm^{-2}); T_m, melting temperature (K); K, coefficient of thermal conductivity at 300 K (W m^{-} K^{-1}); α, coefficient of linear expansion at 300 K (K^{-1}); E, Young modulus (kg mm^{-2}); μ, Poisson's coefficient; S, compression (tension) or bending strength (kg mm^{-2}); R, thermal shock parameter (W m^{-1}); n, refractive index at $\lambda = 1\ \mu$m; dn/dT, temperature derivative refractive index at $\lambda = 1\ \mu$m (K^{-1}); I_{thr}, laser-induced bulk damage threshold at $\lambda = 1.06\ \mu$m for 10 ns pulse duration.

Crystal	ρ (g cm^{-3})	d (Å)	H (kg mm^{-2})	Solubility (g/100 g H_2O)	T_m (K)	K (W/mK)	$\alpha \times 10^6$ (K^{-1})	E (kg mm^{-2})	μ	S_T (kg mm^{-2})	R_T^* (W m^{-1})	n	dn/dT × 10^5 (K^{-1})	$I_{thr} \times 10^{12}$ (W m^{-2})	Transparency range at $\alpha = 1$ cm^{-1} level
LiF	2.64	4.03	99–102	0.12	1121	14.2	33.7	8820	0.28	1.2–4.0	43–143	1.387	−2.9	3600	0.11–6.6
NaF	2.79	4.62	60	4.2	1270	9.2	32.2	8780	0.19	—	—	1.321	−1.8	1400	0.16–11.2
NaCl	2.17	5.64	15.2–18.2	36.0	1074	6.4	39.6	4360	0.20	—	—	1.53	−3.7	200	0.17–18.0
KF	2.50	5.35	—	94.9	1130	—	—	—	—	—	—	—	—	—	0.2–15.0
KCl	1.99	6.29	7.2–9.3	37.4	1049	6.0	37.0	3810	0.14	—	—	1.48	−3.3	700	0.18–23.0
KBr	2.75	6.60	6–7	70.9	1007	4.8	38.4	3290	0.32	—	—	1.54	−3.2	500	0.21–28.0
KI	3.13	7.07	5	144.0	959	2.1	41.2	2550	0.14	—	—	1.64	−4.5	200	0.3–35.0
RbCl	2.76	6.58	—	94.2	717	—	32.8	—	—	—	—	—	—	—	—
MgF_2	3.18	$a = 4.64$	576	0.0076	1536	$21 \perp c$	$9.4 \perp c$	17 265	0.27	5.4	470	$n_0 = 1.373$	11.2	1000	0.1–7.0
		$c = 3.06$				$30 \parallel c$	$13.4 \parallel c$					$n_c = 1.385$	5.8		
CaF_2	3.18	5.46	120–163	0.0016	1676	9.7	18.8	14 000	—	—	—	1.429	−1.05	> 1000	0.13–9.4
SrF_2	4.24	5.79	144	< 0.1	1190	—	19.6		—	—	—	—	—		0.1–9.0
Al_2O_3	3.974	$a = 4.76$	2100	0	2313	35.0	$5.0 \perp c$	35 230	0.27	55.0	10 000	$n_0 = 1.765$	1.3	1900	0.18–5.1
		$c = 13.0$					$6.7 \parallel c$					$n_c = 1.757$	0.97		
Diamond	3.515	3.57	8820	0	3770	900.0	1.0	58 000	—	—	—	2.40	0.4	12 000	0.24–2.7
$Y_3Al_5O_{12}$	4.55	12.0	1350	0	2223	13.0	7.8	31 275	—	20.0	790	1.815	1.05	2000	0.21–5.3
ED-2 glass	2.539	—	—	—	582	1.4	8.0	9190	0.24	10.0	140	1.56	0.3		—

Table B1.8.2. Colour centre representations.

Type of centre	Schematic representation
F_2	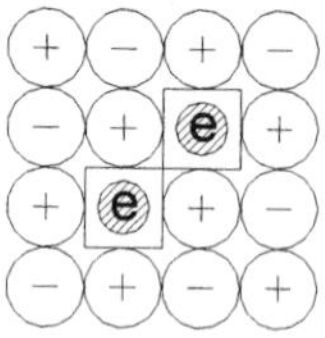
F_2^+	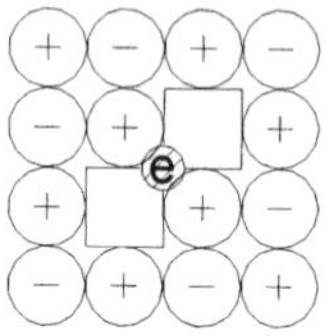
F_2^-	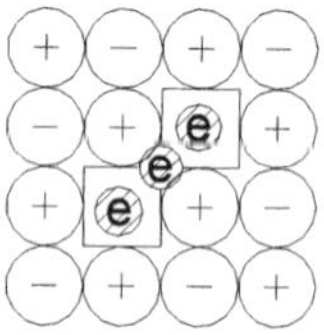
$(F_2^+)_A$	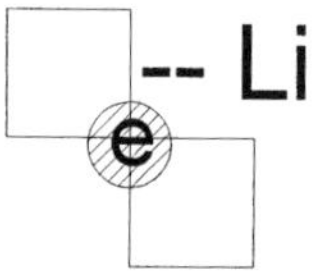 e -- Li
$(F_2)_A$	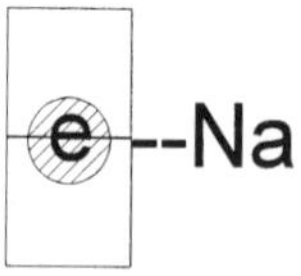 e --Na
$F_A(II)$	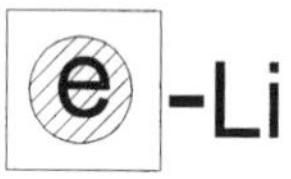 e -Li
$F_B(II)$	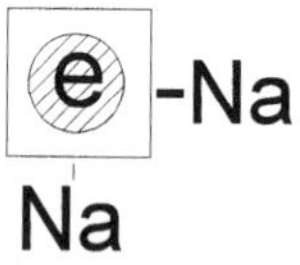 e -Na Na

electron–vibrational transition (3 → 4) occurs with a probability of $A = 10^7$–10^8 s^{-1}, followed by another rapid vibrational relaxation (4 → 1) to the potential curve minimum of the ground electronic state with a resetting of the spatial ion configuration. Disregarding the details, one may consider this scheme as the four-level laser scheme of oscillation.

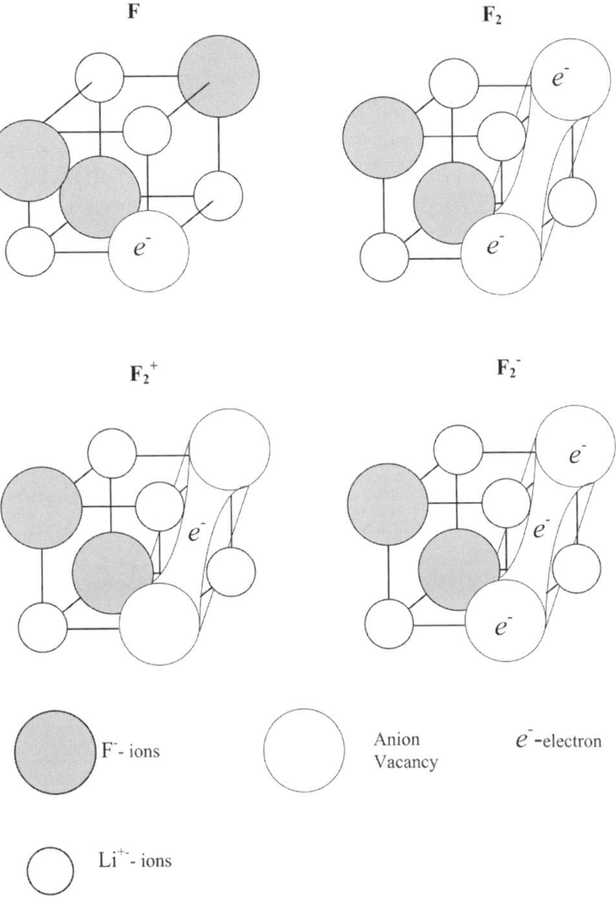

Figure B1.8.1. Structure of the simplest CCs in LiF crystals.

B1.8.3 Colour centre formation in ionic crystals

Colour centres are formed in originally transparent crystals through ionic displacement caused by high-energy irradiation or through the creation of a stoichiometric excess of an alkali or alkali-earth metal in the crystal at elevated temperature. The latter method is called additive colouration and provides CC with a longer lifetime.

The formation of aggregate CCs in alkali-halide crystals under ionizing radiation is a complicated process. It involves the emergence, separation and subsequent association of primary Frenkel defects into aggregate CCs as well as their recharging by electrons and band holes.

The necessary radiation damage can be produced by x-rays, γ-rays, neutrons, electrons or other charged particles. However, neutron beams are not readily available. The production of CCs using an electromagnetic field requires relatively long (days) exposure times but provides high homogeneity in the resultant CCs.

The readily available electron beam is a commonly used irradiation source for CC preparation. To penetrate through a crystal with a thickness of several millimetres, electrons with energies of 1–2 MeV are required and the crystal is usually exposed from both sides to provide sufficient homogeneity. The crystals are always irradiated at lower temperatures (77–177 K) with a current density of about 10 μA cm^{-2}. Irradiation time may vary from several minutes to a few hours depending on the crystal lattice and the type of centres to be produced [5].

The procedure for additive colouration consists of calcination of the crystal in metal vapour at a temperature between the crystal melting point and the temperature of colloid formation. For instance, for KCl the melting point is 768 °C and the colloid formation is initiated at a temperature of 400 °. As a result of this procedure, the alkali-metal atoms are captured and become ionized on the crystalline surface, attracting the halogen ions. The latter diffuse towards the surface, forming anion vacancies in the bulk of the crystal. The electrons appearing as a result of alkali-metal atom ionization also diffuse to be captured by the vacancies; hence forming F centres. The crystal is then quickly cooled down to room temperature in order to avoid colloid formation. Further accumulation of the F_A, F_B, F_2^+ centres is stimulated by additional light treatment of the crystal. The technology for additive colouration in alkali-halide crystals was developed a long time ago. A detailed description of the relevant techniques and apparatus can be found in [14].

The technology for electrolytic colouration was also available rather a long time ago but it was only in 1987 that the first laser crystals were produced by this technique [15]. The procedure is as follows. A crystal is placed in a crucible and heated to the temperature of anion mobility, at which point 'thermal' anion vacancies also appear. A dc high-voltage electric field is applied to the crystal to drive an ionic current: anions move towards the anode and the cathode emits electrons. The latter are captured by the anion vacancies to form F centres. The next part of the crystal treatment, as in the case of additive colouration, involves rapid cooling and light treatment. Additive and electrolytic colouration techniques are better than radiation techniques because they produce more temperature- and light-resistant laser crystals. The disadvantages of the previously-mentioned methods are the low colouring homogeneity, the poor optical quality of the crystals due to colloid formation and the inability to produce large-size active media.

B1.8.3.1 Colour centre formation in alkali-halide crystals under ionizing irradiation

The formation of aggregate CCs in alkali-halide crystals under ionizing radiation is a complicated process. It involves the emergence, separation and recombination of primary Frenkel defects, association into aggregate F_2, $(F_2)_A$, F_2^+, $(F_2^+)_A$, F_2^-, F_3, F_3^+, F_3^- and other CCs and recharging of the CCs by electrons and band holes. Defect formation may occur rapidly or slowly. Decomposition of self-localized excitons into primary radiational defects and the recharging of CCs are relatively fast processes (10^{-12}–10^{-7} s). Slow processes occur either because of spatial diffusion of the defects and their associates or diffusion of self-localized holes (resulting in CCs recharging).

All these processes determine the efficiency with which any type of CC is formed. They depend upon the irradiation temperature, and crystal storage, impurity composition of the initial material, ionizing radiation dose power and irradiation dose.

Separated electrons and holes, free and self-localized excitons are generated in alkali-halide crystals under ionizing irradiation. Self-localized excitons decompose with the emergence of pairs of F centres and interstitial halogen atoms (H). Rapid recharging of these pairs under a flux of electrons gives rise to the simultaneous formation of anion vacancies V_a^+ and interstitial halogen ions I_a [16]:

$$e_s^0 \Rightarrow F + H \qquad \text{(B1.8.1)}$$
$$e_s^0 \Rightarrow V_a^+ + I_a^- . \qquad \text{(B1.8.2)}$$

The mechanisms for aggregate centre formation through the migration of anion vacancies have been studied by Delbecq [17]. According to him, charged F_2^+ centres first appear and then capture electrons to produce neutral F_2 centres:

$$V_a^+ + F \Rightarrow F_2^+ \qquad \text{(B1.8.3)}$$
$$F_2^+ + e \Rightarrow F_2 . \qquad \text{(B1.8.4)}$$

Simultaneously with rapid electron capture, F_2^+ centres may take part in a slow temperature-dependent migration process. Colliding with the F, F_2^+ and F_2^- CCs, they form more complex CCs—F_3^+, F_4^+ and F_4, respectively:

$$F_2^+ + F \Rightarrow F_3^+ \tag{B1.8.5}$$

$$F_2^+ + F_2 \Rightarrow F_4^+ \tag{B1.8.6}$$

$$F_2^+ + F_2^- \Rightarrow F_4 \tag{B1.8.7}$$

whose further aggregation leads to the appearance of colloid particles in the crystal. The processes for F_2 CC formation by the scheme in (B1.8.3) and (B1.8.4) compete with those for ionization due to a rapid capture of free electrons:

$$F_2 + e \Rightarrow F_2^- \tag{B1.8.8}$$

or holes

$$F_2 + h \Rightarrow F_2^+ \tag{B1.8.9}$$

or due to the diffusion processes involving mobile anion vacancies

$$V_a^+ + F_2 \Rightarrow F_3^+ \tag{B1.8.10}$$

and self-localized holes

$$F_2 + V_k \Rightarrow F_2^+. \tag{B1.8.11}$$

These processes (B1.8.4, B1.8.8, B1.8.9, B1.8.11), leading to recharging in the group of F_2, F_2^+ and F_2^- CCs, are also inherent in the group of F_3, F_3^+ and F_3^- CCs.

The ratio of the contributions of different reactions depends on the irradiation procedure and impurity composition in the irradiated crystal. The optimization of irradiation treatment to create F_2^- and F_2^+ stabilized CCs in doped LiF crystals is discussed later.

B1.8.3.2 LiF:F_2^- active elements optimization

Here we use the LiF crystal with an F_2^- CC as an example to illustrate the way in which irradiation parameters optimization overcomes the technological difficulties faced when producing an optically dense active medium.

LiF:F_2^- crystals, with their unique combination of modulation, thermal and operational characteristics, are now widely used as active media in tunable lasers and as nonlinear elements with saturable absorption for neodymium laser Q-switches. However, the available techniques for the preparation of such laser crystals fail to produce the optically dense, high-contrast media required for a series of applications. Usually, the value of the active absorption at the working wavelength does not exceed 0.4–0.6 cm^{-1} at a contrast (the ratio of the active absorption coefficient to the loss coefficient at $\lambda = 1.06$ μm) of 10–20. Under the action of γ-quanta commonly used for producing LiF:F_2^- laser crystals, the F_2^+ CCs emerging by reaction (B1.8.3) may be involved in one of three processes: (a) rapid electron capture to produce F_2 (reaction (B1.8.4)) and F_2^- (reaction (B1.8.8)) CCs; (b) slow diffusion processes of aggregation of the type (B1.8.5)–(B1.8.7); and (c) dissociation on the F centre and anion vacancy $F_2^+ \Rightarrow F + V_a^+$.

The rates of the various processes and the final concentration of CCs depend, as mentioned previously, on the impurity composition of the initial material, irradiation dose, power and crystal temperature. By varying the ratio of these parameters, one can essentially influence the processes of F_2^- CC formation in LiF.

When the procedure of LiF crystal treatment under ionizing irradiation is optimized so that processes (B1.8.3) and (B1.8.4) are efficient and (B1.8.5)–(B1.8.7) negligible, the maximum active absorption of F_2^- CCs at the wavelength 1064 μm exceeds 1.5 cm^{-1} at a contrast as high as 20–40.

B1.8.3.3 LiF:F_2^+ active elements optimization

LiF:F_2^+ is a high-gain active medium with large absorption and emission cross sections and a wide near-IR tuning range [2, 6]. Unfortunately, the low thermal stability of 'pure' F_2^+ centres at room temperature (half decay time ~12 hr) made the LiF:F_2^+ laser medium very inconvenient for use, as it could be operated only at temperatures < 100 K [6].

To overcome the low thermostability of LiF:F_2^+ active media, a technique for the two-step photoionization of neutral F_2 centres in LiF was proposed [18]. However, in this case, the problem of low active element operational stability has only partially been solved. While the pumping beam is turned on, the F_2^+ laser exhibits stable operation but, after the pumping beam is switched off, the active zone of the crystal becomes exhausted due to thermal degradation of the F_2^+ centres after one day at $T = 300$ K. By thermoelectrically cooling an LiF:F_2^+ CC crystal to −30–0 °C the operation time can be extended to a few weeks or months.

In 1978, Khulugurov and Lobanov [19] discovered that the thermal stability of positively charged 'pure' F_2^+ centres can be increased by co-doping LiF crystals with suitable anion or cation impurities, with the formation of complex F_2^+-like centres. However, the improved thermostability of these centres at room temperature did not provide the appropriate photostability of the active media under powerful laser excitation at wavelengths ≤ 590 nm. For example, pumping the LiF crystals with F_2^+-stabilized CCs by powerful radiation of the second harmonic of a Nd:YAG laser (532 nm) results in a significant fading in the CCL output [20], although at low levels of pumping (several tens of mJ), the CCL will operate relatively well [21]. The fading under high power excitation is caused by the photo-chemical process in the pumping channel of the crystal, which involves the two-step photoionization of neutral F_2 centres ($F_2 + 2h\nu = F_2^+ + e$) and trapping of the released electron e by the positively charged F_2^{+**} centres ($F_2^{+**} + e = F_2^{**}$). The concentration of 'pure' F_2^+ centres which appeared due to photoionization of F_2 centres quickly reduces as F_2^+ centres migrate through the crystal with the creation of F_3^+ centres: $F_2^+ + F = F_3^+$ [22].

In [21,23], the development of a high-power room-temperature photo- and thermostable tunable LiF:F_2^{+**} laser was reported. The optimized technology of a thermostable F_2^{+**} CCs in LiF crystals and an appropriate choice for the pumping laser wavelength in [21, 23] allowed for the simultaneous solution of the problems of photo- and thermostability of laser materials based on LiF crystals with F_2^+-like CCs. The peculiarities of the technological preparation of LiF:F_2^{+**} crystals according to [21, 23, 24] are discussed later.

The crystals were grown by the Kyropulos method in an argon atmosphere from nominally pure raw materials and doped with LiOH, Li_2O and MgF_2. As-grown LiF crystals have an intense infrared absorption band at $\nu = 3730\,\mathrm{cm}^{-1}$ (absorption coefficient $\simeq 1.2\,\mathrm{cm}^{-1}$) due to OH^- ions, which substitute for the fluoride in the anion node of the crystalline lattice, and intense bands with frequency maxima at 3560 and 3610 cm^{-1}, due to $Mg^{2+}OH^-OH^-V_c^-$ complexes. The UV absorption spectra of the grown LiF crystals exhibit strong absorption bands in the region 200–270 nm that are due to the O^{2-}–V_a^+ as well as O^{2-}–Mg^{2+} dipoles.

In order to obtain a high concentration of F_2^{+**} centres and a small concentration of colloids and parasitic aggregate CCs, as-grown crystals are subjected to a multi-step γ-irradiation treatment with a Co^{60} source.

At the first stage, the crystals are subjected to an ionizing treatment (γ-irradiation, 2–5 × 10^7 Rad) at temperatures lower than temperature $T(V_a{}^+) = 240$ K of anion vacancy ($V_a{}^+$) mobility in LiF crystals.

The ionizing treatment creates a great number of free electrons and holes, which can form self-localized excitons after thermalization. These excitons localized near a specific anion node of the crystalline lattice may annihilate and the released energy can be used for shifting an anion from its node to the interstitial position, causing formation of an anion vacancy and an F centre according to the schemes (B1.8.1) and (B1.8.2).

Simultaneously with primary defect formation according to (B1.8.1) and (B1.8.2), two processes of $(OH)^-$ ions dissociation are taking place under the ionizing treatment [4]:

$$2(OH^-) \Rightarrow O_2^- + e + V_a^+ + 2H_i^0 \qquad (B1.8.12)$$

and

$$2(OH^-) + V_a^+ + e \Rightarrow O^{2-}V_a^+ + O^- + 2H_i^0 \qquad \text{(B1.8.13)}$$

where O_2^- is a molecular ion of oxygen, O^- is a single ionized atom of oxygen and H_i^0 is an interstitial atom of hydrogen.

It is very important that there is no formation of aggregate CCs during the irradiation of the crystal at temperatures below the temperature of mobility of anion vacancies. O^{2}–V_a^+ dipoles are accumulated in a high quantity, since they are photo- and thermostable under γ-irradiation. In addition to this, the ionizing treatment creates a high concentration of single ionized oxygen atoms and molecules due to a hydroxyl group dissociation. It is noteworthy to mention that one of the products of OH^- radiolysis—interstitial H_i^0 atoms—are efficient traps of electrons:

$$H_i^0 + e \Rightarrow H_i^- \qquad \text{(B1.8.14)}$$

which prevent the electrons from being captured by the F_2^+ centre at the subsequent stages of the crystal preparation, thereby increasing the efficiency with which stable F_2^{+**} centres accumulate. The role of the Mg impurity can be explained as follows. F-centre formation is likely to occur near impurity-cation-vacancy dipoles Mg^{2+}–V_c^-. In accordance with the following reaction:

$$Mg^{2+}V_c^- + F \Rightarrow Mg^+ + V_a^+V_c^- \qquad \text{(B1.8.15)}$$

an F centre can give its electron to a divalent metal impurity forming Mg^+ and a pair of vacancies $V_a^+V_c^-$. Anion–cation dipoles, $V_a^+V_c^-$, play an important role in the formation of F_2^+ perturbed CCs at the subsequent stages of the crystal preparation. However, an excessive concentration of the Mg dopant may play a negative role due to the decreasing concentration of the useful $O^{2-}V_a^+$ dipoles as they bond with oxygen and form $Mg^{2+}O^{2-}$ dipoles that do not take part in F_2^+-like CC formation.

At the second stage of the technological process, the crystals are heated up to $T_1 < T < T_2$, where T_1 (~240 K) is the threshold temperature of V_a^+ mobility and T_2 (~270 K) that of F_2^+, $V_a^+V_c^-$and $O^{2-}V_a^+$ mobility in an LiF crystal, and stored at this temperature for some time. This temperature interval is chosen within these specific limits so that anion vacancies are mobile in the crystal and F_2^+ centres as well as $V_a^+V_c^-$ and O^{-}–$V_a{}^+$ dipoles are immobile and cannot diffuse in the crystalline lattice. During this time, an important process in the formation of unperturbed F_2^+ CCs takes place due to the diffusion of mobile anion vacancies and their association with F centres:

$$V_a^+ + F \Rightarrow F_2^+. \qquad \text{(B1.8.16)}$$

Consequently, the concentration of F centres decreases due to their association with anion vacancies. This is a useful process which helps to increase the concentration of F_2^+ stabilized centres at the following stages and to decrease the concentrations of other aggregates and colloids. At the third stage, the crystals are re-irradiated at the temperature $T < T_1$ in order to increase the amount of anion vacancies (that have been exhausted at the second stage) by ionizing the neutral F centres. After that, the crystals are again subjected to the procedure described for stage 2 in order to increase the amount of pure F_2^+ centres due to process (B1.8.16).

At the fourth stage, the crystals are heated up to the room temperature and stored for some period of time, after which the crystals exhibit a stable concentration of the CCs of interest and are ready for use as a laser medium. The processes that take place at this stage are as follows.

(1) The $O^{2-}V_a^+$ complexes are mobile at room temperature and their migration leads to the development of F_2^{+**} centres stabilized with the O^{2-} ion ($F_2^+O^{2-}$):

$$O^{2-}V_a^+ + F \Rightarrow F_2^+O^{2-}. \qquad \text{(B1.8.17)}$$

(2) Another very important reaction occurs due to the intrinsic mobility of F_2^+ centres at room temperature. The migration of these centres may lead to a useful process of association of F_2^+ centres with a single

ionized oxygen atom O^- with the formation of $F_2^+O^{2-}$ centres:

$$F_2^+ + O^- + e \Rightarrow F_2^+O^{2-}. \tag{B1.8.18}$$

(3) Since at room temperature bi-vacancy dipoles $V_a^+V_c^-$ are also mobile ($V_a^+V_c^-$ start to be mobile at temperatures of about 273 K), they may associate with F centres and perturbed F_2^+ centres can occur:

$$V_a^+V_c^- + F \Rightarrow F_2^+V_c^-. \tag{B1.8.19}$$

Due to the described multi-stage LiF crystal treatment, the optimum conditions for useful processes (B1.8.17–B1.8.19) of F_2^+ stabilized CC formation are provided and the efficiency of the parasitic processes ($F_2^+ + e \Rightarrow F_2$, $F_2 + e \Rightarrow F_2^-$, $F_2^+ + F \Rightarrow F_3^+$, $F_2^+ + F_2 \Rightarrow F_4^+$, $F_2^+ + F_2^- \Rightarrow F_4$, $V_a^+ + F_2 \Rightarrow F_3^+$), using the techniques responsible for formation of other aggregate and colloid centres, is suppressed.

Using the techniques developed in [21, 23], LiF:F_2^{+**} crystals did not exhibit any photo-degradation when pumped by radiation in the 620–750 nm range at energy densities of up to ~5 J cm^{-2}. We estimate the lifetime of these CC crystals to exceed 10 years at room temperature. The real conversion efficiency of the LiF:F_2^{+**} laser exceeds 50% with a maximum output power of 100 mJ, limited only by the capabilities of the available pumping lasers.

Detailed descriptions of the laser oscillation in CCLs have been presented in several review papers [1, 2, 5, 6, 25]. Here, only the main features of specific operating regimes of CCLs will be described.

B1.8.4 CW and quasi-cw CCL operation

Stationary CC oscillation was realized for the first time at the beginning of the laser era, in 1965, by Fritz and Menke [26], who obtained CC lasing at a cryogenic temperature under flashlamp pumping. Now there are many cw lasers that can be used as pumping sources for tunable CCLs, but the most popular are the widespread Ar^+ and Kr^+ ion gas lasers oscillating in the visible range (see chapter B3.5) and solid state lasers on neodymium-doped yttrium aluminum garnet and perovskite, which effectively lase in the IR, at wavelengths of 1.06 and 1.3 μm (see chapter B1.3). These lasers in the cw regime can provide an output power of several tens of watts for pumping CCLs.

From the laser threshold expression (see, for example, equation (7) in [2]), one can see that a pump density of several kW cm^{-2} is needed to reach the threshold for CCL oscillation. This is easily achieved for pulsed pumping by a flashlamp emitting a few tens of joules during a millisecond pulse. In this case, the lasing volume of the CC crystal can be as large as 1 cm^{-3} or even more.

In order to get the same pump densities for cw laser pumping with a power of 0.1–10 W, the pump radiation has to be strongly focused and the CCL crystal placed at the focal point or cavity waist, 0.1–1 mm in diameter. The crystal length is limited by the confocal parameter of focused beam and does not usually exceed 1 cm (typically it is a few millimetres).

The most popular optical scheme now used for cw CCL operation is the Kogelnik scheme with a Brewster-cut active element and a V or Z-shaped folded cavity to compensate for spherical lens or mirror astigmatism [27].

The relationship between the CC crystal thickness t, focal length $f = r/2$ of the spherical mirrors with radius r and folded angle of the cavity is given by the expression

$$\sin\theta \tan\theta \frac{n^4}{(n^2-1)(n^2+1)^{1/2}} = \frac{t}{f}. \tag{B1.8.20}$$

For a typical thickness, $t = 2$ mm and the mirror curvature, $r = 50$ mm, the compensation angle 2θ is about 20 °C and depends slightly on the refractive index n, which is typically 1.35–1.5 for alkali halides (table B1.8.1).

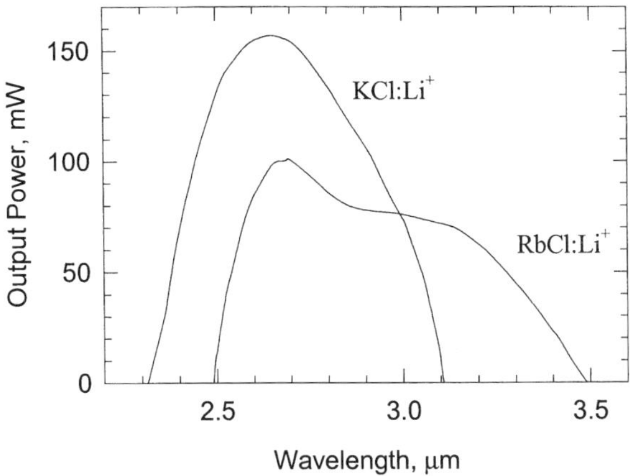

Figure B1.8.2. Tuning curves of KCL:Li$^+$ and RbCl:Li$^+$F$_A$(II) CCLs.

The beam waist diameter, w_0, and confocal length, b, can be found from the following formulae,

$$w_0 = f\sqrt{\lambda/2\pi l_2} \qquad t \le b = f^2/l_2 \qquad \left(\frac{b}{f_2} = \frac{f}{l_2}\right), \tag{B1.8.21}$$

As one can see, such a strong focusing condition leads to strong local heating of the CC crystal pumped zone due to Stokes losses and can result in a strong thermal lens. Fortunately, deep cooling to liquid nitrogen temperature greatly increases the thermal conductivity of CC active crystals and leads to a decrease in thermal lensing.

There are two most popular regimes for operating in the cw mode. The first is narrow-line (single-frequency) broadly tunable CCL oscillation used for visible and IR spectroscopy, analytical application and fundamental metrology. The second is the mode-locked pico- and femtosecond pulse operating regime used for time-resolved transient spectroscopy, high-peak-power nonlinear conversion and fast telecommunication systems.

Single-mode, single-frequency operation of a KCl:Li F_A(II) CCL under krypton ion laser pumping at 647 nm with a linewidth less than 260 KHz and tunability in the 2.5–2.9 μm spectral range using a pair of birefringent plates and an intracavity solid etalon was obtained in [28]. By active frequency stabilization, linewidths as low as 2 KHz were realized later [29]. To obtain single-mode operation and to improve linewidth and wavelength stability, ring cavity operation can also be applied [30, 31]. The typical output power of an F_A(II) CCL varies from tens to hundreds of milliwatts in the 2.3–3.5 μm spectral region (figure B1.8.2) [25].

With the same pump source, LiF:F_2^+, CCL operation with a very high slope efficiency of 60%, close to the physical limit (output-to-pump photon energy ratio) of 70%, was obtained. A wide tuning range, 0.84–1.04 μm, was covered with a cw output power of up to 1.1 W in the pure cw regime and under synchronous picosecond pumping [6]. In the last case, shortening by a factor of 25 from a 100 ps pump to a 4 ps CCL output was demonstrated. Optimal output coupling of the cw CCL varied from about 30% to 1%, depending on the gain. To work at the maximum of the tuning curve, where the gain is high, one usually uses low-output coupled laser cavities with a transmittance $T = 10$–30%, while; however, when the tuning curve widens, a high-output coupling of $T = 1$–3% should be chosen.

One of the main problems of the LiF(F_2^+) CCL is fast optical bleaching due to the orientational diffusion of F_2^+ CCs [1]. The maximum cw lasing output (2.7 W) was recorded in a KF(F_2^+) CCL under 1.064 μm Nd^{3+}:YAG pumping with a tuning range of 1.22–1.5 μm [25].

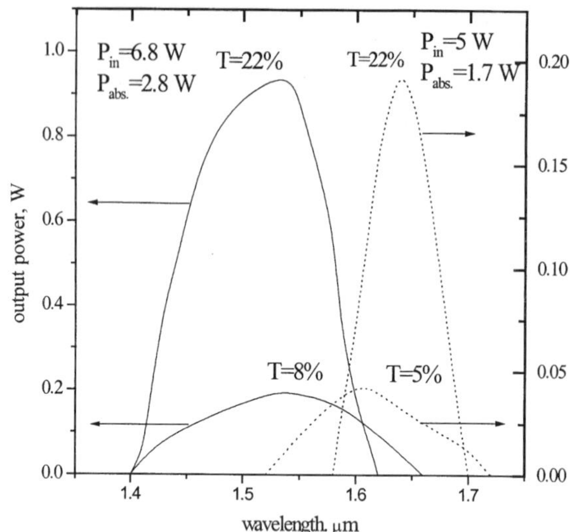

Figure B1.8.3. Tuning curves of an $F_A(Tl)$ colour centre laser in $KCl:Tl^+$ (full curves) and $KBr:Tl^+$ (dashes) pumped by Nd:YAG laser radiation (1.064 μm). The increase in output coupler transmission from 5% to 22% results in a considerable increase in oscillation power accompanied by a narrowing of the overall spectral range.

In the case of $(F_2^+)A$ CCs, the KCL:Li crystal showed a very good performance: $P_{out} = 400$ mW under 1.32 μm cw Nd laser pumping and $E_{out} = 2.3$ mJ under 1.41 μm pulsed pumping with a tuning range of 2–2.5 μm [25].

The operation of $(F_2^+)_H$ CCs in NaCl crystals doped by OH^- ions looks very promising. It was reported for the first time in [32] and then studied by many other groups [5,25]. An output power as high as 3.05 W at a slope efficiency of 33% was reported in [33] under 9 W pumping at 1.064 μm and 200 mW collinear auxiliary light of 514 nm. A very wide tuning range, 1.4–1.92 μm, important for telecommunications studies and tests with outputs of up to 1.3 W was obtained in [34] at 6 W 1.064 μm laser pumping and only 2 mW of 532 nm collinear auxiliary light. This laser can also easily operate in the mode-locked regime with picosecond pulses as short as 5 ps.

Stable cw laser operation without the use of auxiliary light was also obtained with the help of $F_A(Tl)$ CCs in $KCl:Tl^+$, $KBr:Tl^+$ and $KF:Tl^+$ crystals. The first of these crystals provides about 1.4 W output at the maximum of the tuning curve with tunability from 1.4 to 1.6 μm at the absorbed pump power of $\sim$3.5 W of a 1.064 μm Nd:YAG pump laser (figure B1.8.3) [5]. What distinguishes these CCs from the others is their partially forbidden optical transitions, resulting in a lower gain and the longest radiative lifetime equal to 1.6 μs.

B1.8.5 Pico- and femtosecond quasi-cw CCL operation ($T = 77$ K)

One of the most interesting and important regimes of CCL operation is the mode-locked regime, when many longitudinal modes of a broadband laser radiate with a ‘coordinated phase’. According to the Fourier transformation, this multi-frequency cw laser operation leads to short-pulse high-peak-power quasi-cw lasing with a pulse-to-pulse temporal interval equalling the laser cavity round-trip time and a pulse duration inversely proportional to the number of longitudinal modes operating in the mode-locking regime.

All types of the mode-locking techniques developed for dye lasers are applicable for developing mode-locked CCLs. CC mode-locked operation is quite similar to the dye laser mode-locking regime with only one distinction: the excited state lifetime of a CC τ_{CC} is about 15–1500 ns and is usually larger than the laser cavity round-trip time, $\tau_{cavity} = 6$–10 ns, while for the dyes it is smaller ($\tau_{dye} = 3$–5 ns).

One of the most popular and effective techniques for realizing of pico- and femtosecond lasing is synchronous pumping, when the pump laser mode-locking regime is transferred to a CCL. In this case, the cavity round-trip times of both pump and CCLs should be synchronized and the synchronized CCL gain modulation results in pulsed CCL lasing and pulse shortening to a duration much shorter than the pump pulse duration. The expression for the pulsewidth [35] shows that it should be proportional to the pump and CCL cavities mismatch parameter Δ and inversely proportional to the logarithmic gain loss product

$$\tau_p \simeq \frac{\Delta}{\ln(G \cdot R)}.$$

This means that the shortest pulse duration in the CCL can be realized at the highest gain, pump energy and power. When using 90 ps pulsed krypton laser synch-pumping, the LiF:F_2^+ CCL pulses had a duration of 2–3 ps in linear cavities and 0.7 ps in ring cavities [36].

Similar results were obtained in linear cavities of KF(F_2^+), NaCl(F_2^+) and KCL:FA(Tl^+) lasers under Nd:YAG synch-pumping and for KCl:Li(F_A) and RbCl:Li(F_A) CCL under mode-locked argon laser pumping (see references in [5]).

In the passive mode-locking regime with a dye saturable absorber, a pulse duration of 180 fs was achieved in a LiF:F_2^+ CCL ring cavity at a lasing bandwidth of about 60 cm^{-1} [37]. Using a multiple quantum-well saturable absorber, a NaCl(F_2^+) CCL demonstrated 260 fs pulse operation with 50 cm^{-1} bandwidth [38].

By developing a new additive pulse mode-locking technique for pulse shortening, a pulse duration of 127 fs instead of 23 ps from KCl:Tl CCL was realized [39]. In a similar regime, 50 fs pulses have been generated by the so-called 'soliton laser' on KCL:Tl at 1.5 μm [40].

B1.8.6 Room-temperature CCL operation

B1.8.6.1 CW and quasi-cw regimes

Room-temperature tunable CCLs have a special importance in comparison with dye and cryogenic CCLs due to their compactness, simplicity and ease of use both for laboratory and field conditions. Lithium fluoride with F_2^- CCs is one of the most reliable active media for room-temperature tunable operation.

During the last two decades only a few papers have been dedicated to the study of cw and quasi-cw laser operation on LiF:F_2^- crystals at $T = 300$ K [3, 41–44]. Most of them used neodymium solid state lasers as the pumping source. Due to the high absorption cross section of F_2^- CC in LiF crystals at Nd^{3+} wavelength (1.06 μm) which equals 2×10^{-17} cm^2 and which is much higher than the emission cross section ($10^{-19} \times 10^{-21}$ cm^2) of the Nd^{3+} ions wavelength, LiF:F_2^- CC crystals can be used not only as a gain medium but also as a crystalline saturable absorber for Nd laser Q-switching.

The first attempts to produce cw laser operation were made on a 4 mm long LiF:F_2^- crystal in a linear laser cavity. At a pump power of 6 W (cw YAG:Nd^{3+} laser), using a focusing lens with $f = 100$ mm, and a mechanical chopper, Gusev *et al* [3] attained an output power of 60 mW; but because of optical feedback between the two resonators of the LiF:F_2^- and YAG:Nd_3^+ lasers, a repetitive passive Q-switching mode with a peak power much higher than the average power level of 6 W was observed. As a result, only a quasi-cw high repetition rate nanosecond pulsed operation was achieved.

Using a pitchless YAG:Nd^{3+} laser operating in a free-running mode with smooth 150 μs pulses as a pump and a quarter-wave quartz plate with Glan polarizers to decouple the cavities of the YAG:Nd^{3+} and LiF:F_2^- lasers, the authors [1] produced pitchless quasi-cw lasing on a 5 cm long LiF:F_2^- crystal with a pulse duration of 150 μs and an efficiency of 0.5%.

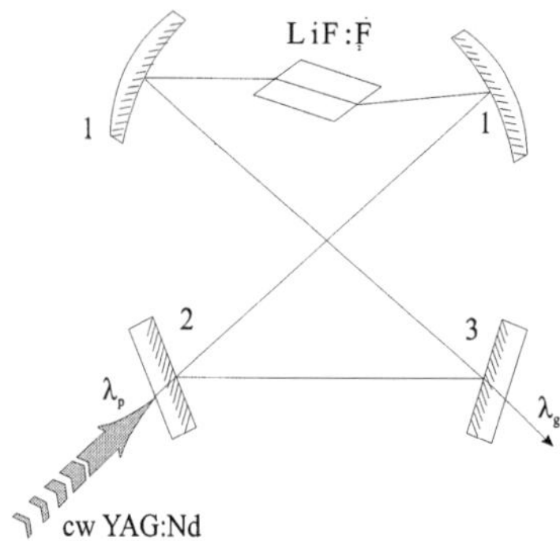

Figure B1.8.4. Typical cavity for cw or quasi-cw ring colour centre laser.

The application of a ring cavity for removing feedback enabled lasing on a 1 cm long LiF:F_2^- crystal with 60% absorption and a calculated diameter for the beam waist of 80 μm [3]. For a pump of 6 W and a tunable range of 1.14–1.19 μm, the output power was 80 mW. However, considerable thermo-optical distortions, which started manifesting themselves at an average pump power of more than 0.5 W, prevented the researchers from attaining a purely cw lasing mode without a mechanical chopper, because the lasing threshold was 2.5 W.

For the first time, pure cw LiF:F_2^- lasing has been attained using crystals with double the absorption coefficient; $k_{abs} = 1.3$ cm^{-1} at a pump wavelength of 1.064 μm and low losses ($K_L = 0.01$ cm^{-1}) in the lasing region of 1.1–1.2 μm [43] . A ring cavity with a 1 m base and astigmatism compensation incorporated two plane mirrors (2 and 3) with 99.7% and 99% reflectance in the lasing region and two 100% reflectance spherical mirrors (1) with a curvature radius of 10 cm (see figure B1.8.4). At a pump power higher than 0.5 W, lasing was disrupted. This can probably be attributed to a strong negative lens with a focal distance of less than 5 mm, which was induced in the sample during pumping. The induced lens was reduced by heating the sample surface to 20–40 °C, or by slowly rotating it at 6 turns/min. As a result, pure cw LiF:F_2^- lasing was realized. For a pump power of 3 W, the lasing output power was 60 mW. These results were further improved: the maximum output cw power of a LiF:F_2^- laser exceeded 100 mW at a pump power of 5 W. The LiF:F_2^- active element preserved all its properties after a 30 h performance.

The first results on frequency tuning and picosecond generation at room temperature of cw LiF:F_2^- laser were obtained under synchronous pumping with a continuous train of picosecond pulses of a cw-pumped Nd^{3+}:YAG laser. The use of a ring cavity like that in figure B1.8.4 and a Lyot filter as a dispersive element provided lasing over the 1.1–1.2 μm spectral region. LiF:F_2^- oscillation under Nd^{3+}:YAG laser synchronous pumping provided picosecond pulses with a duration of 10 ps and a 7 ns axial interval for a continuous train of pulses [44] (the pulsed mode-locking regime will be discussed later).

Because of a small Stokes shift between the energies of pump and lasing photons ($\nu_p - \nu_{osc} = 200$–300 cm^{-1} at room temperature), the investigation of the efficiency of LiF:F_2^- lasing under different pump wavelengths performed in [45] is very interesting. Pumping with a wavelength lying in the region of wide overlapping in the electronic-vibrational absorption and luminescence bands of the F_2^- CCs in addition to excitation or stimulated upward transitions gave rise to stimulated downward transitions and a corresponding deactivation of the excitation. As a result, the maximum concentration of the excited CCs in a crystal pumped by Nd^{3+}:YAG laser radiation ($\lambda_p = 1.064$ μm) could only reach 26% of the total concentration of these centres even in saturation conditions. Shortening the pump wavelength by replacing the Nd^{3+}:YAG pump laser with an Nd^{3+}:$YLiF_4$ laser ($\lambda_p = 1.047$ μm) increased the maximum concentration of the excited CCs up to 43% and, as calculations show, in the case of a Yb^{3+}:YAG laser ($\lambda_p = 1.029$ μm), this concentration

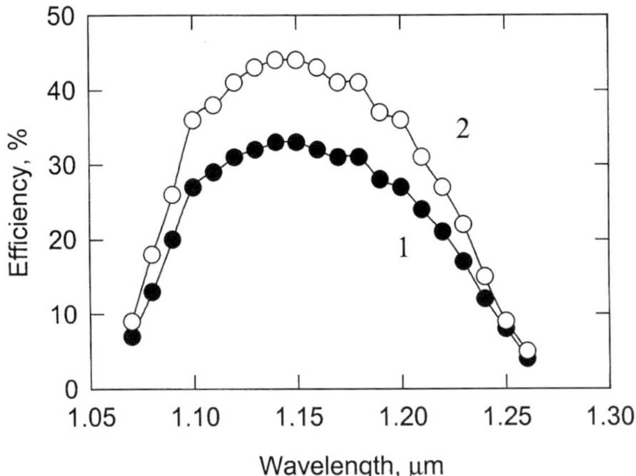

Figure B1.8.5. Tuning curves of quasi-cw oscillation of LiF:F_2^-CCL pumped by Nd^{3+}:YAG (1) and Nd^{3+}:$YLiF_4$(2) lasers.

rises to 62%. Therefore, the pump laser wavelength is an important parameter in attempts to increase CCL gain and efficiency.

In a non-selective cavity for the same peak pump power, the LiF:F_2^- laser was 2.5 times more efficient when pumped by the radiation of an Nd^{3+}:$YLiF_4$ laser [42]. Moreover, the efficiency did not reach its saturation value and could be further improved by increasing the peak pump power. The adoption of a shorter pump wavelength also resulted in a ten-fold reduction in the LiF:F_2^- lasing threshold.

Figure B1.8.5 shows the tuning curves of the LiF:F_2^- laser obtained in a selective cavity with a diffraction grating. The laser was pumped by an Nd^{3+}:$YLiF_4$ laser (1.047 μm) radiation at a pulse repetition rate of 1 kHz. The duration of the pump pulses was 450 ns and their peak power was 750 W. A tuning range of 1.07–1.27 μm was also obtained in a dispersive cavity with a high (45%) efficiency at the maximum of the tuning curve.

B1.8.6.2 High-peak-power room-temperature CCL operation in pico- and femtosecond regimes

Since the middle 1970s, there have been many studies on the picosecond operation of F_2^+, F_2^+ O^-, F_2^- and F_3^- CCs in LiF and NaF crystals at room temperature. Synchronous pumping by a train of picosecond pulses from ruby, Nd:YAG or Nd:glass lasers was used as the most effective scheme for CCL pulse shortening.

Fast CC gain switching (modulation) leads to an effective CCL pulse shortening depending on the degree to which the pump power exceeds threshold, the number of pulses and the length of the train. Here, we would mention only a few results on developing high peak power and high conversion efficiency LiF:F_2^- CCL with short (2–3 ps) picosecond and femtosecond pulse durations.

In the first case, a high optical density and high contrast LiF:F_2^- CC crystal was used in a semi-concentric cavity with a prism pair for dispersion compensation under pumping with an Nd-phosphate glass picosecond pulse train. The pump scheme was almost longitudinal with a transverse angle of 2–3° of beams. The Brewster-cut LiF:F_2^- crystals had lengths of 60 and 160 mm. Precise matching of the resonator's lengths gave rise to LiF:F_2^- femtosecond oscillation with a pulse duration of less then 500 fs, at a full train duration of 300–400 ns and a train energy of 250 μJ [46].

The durations of the subpicosecond pulses were measured directly on the screen of the streak camera (8) with a resolution of 0.6–0.7 ps (figure B1.8.6) as well as with an autocorrelator which provided an average

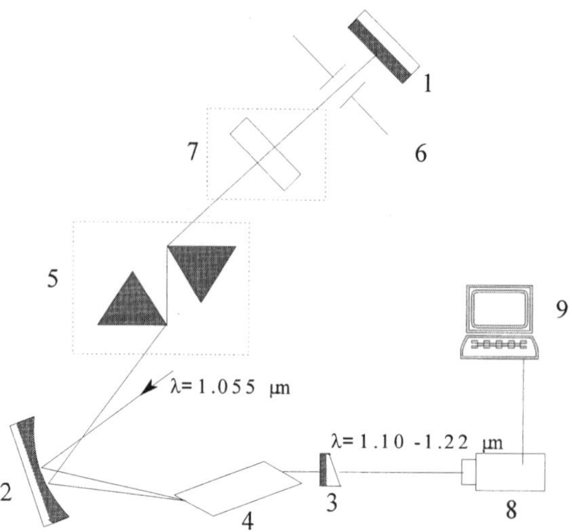

Figure B1.8.6. Experimental set-up of a picosecond LiF:F_2 CCL:(1,2,3), end, spherical and cavity mirrors; (4), active element (LiF:F_2^- crystal); (5) prism compensator of the group velocity dispersion; (6), aperture; (7), Lyot filter; (8), streak camera and (9), PC.

pulse duration in the LiF:F_2^- train of less than 0.5 ps. A conservative estimation of the peak power for 0.5 ps output pulses gives a minimum power of 10 MW, which greatly (by three to four orders) exceeds the standard peak power generated by ordinary quasi-cw femtosecond dye and CC lasers Stokes.

F_2^- CCs feature a long fluorescence decay time ($\tau = 55$ ns) that makes them very attractive for developing synchronously pumped mode-locked lasers. A highly efficient LiF:F_2^- lasing of high-power short picosecond pulses was realized under LiF:F_2^- synchronous pumping with a train of 50 pulses with 7 ps duration from a passively mode-locked neodymium phosphate glass laser ($\lambda = 1.055$ μm) [47]. The energy of the train was up to 6 mJ. The pump laser radiation was focused into an LiF:F_2^- CC crystal by mirror (2) (see figure B1.8.6). The Brewster-cut LiF:F_2^- CC crystal was 4 cm long. The initial absorption in this crystal at the pump wavelength was 0.7 cm^{-1}. The LiF:F_2^- laser cavity was formed by a plane mirror (1), a highly reflective spherical mirror (2) with a radius of curvature $r = 1500$ mm and a plane output mirror (3) whose reflection coefficient varied to ensure the optimal output energy. The group velocity dispersion was compensated for by introducing a pair of compensating prisms into one of the cavity arms.

The efficiency of LiF:F_2^- lasing *versus* the reflectivity of the output mirror is depicted in figure B1.8.7. The maximum efficiency of conversion of the pump radiation to CCL output and output energy reached 35% and 1.4 mJ, respectively. The temporal parameters of the output radiation were measured directly by an IMACON-501 streak camera, coupled to a linear photodiode array. The time resolution of the apparatus was 3 ps.

Figure B1.8.8 presents the duration of the output pulses of CC laser. The minimum pulse duration was close to the time resolution of the recording system and was about 2 ps. The half-width, $\Delta\lambda$, of the lasing spectrum was 8 nm. Tuning the output wavelength over 1.10–1.22 μm range was performed by a birefringent filter or by moving the slit aperture across the optical axis of the cavity. We estimate that, in this scheme, 35% efficiency can be obtained for 2 ps pulses with a peak power as high as 100 MW. As mentioned before, short wavelength (compared to 1.064 and 1.055 μm excitation) 1.047μm Nd:$YLiF_4$ pump laser radiation provides a much higher gain in LiF:F_2^- CC laser. In addition to this, Kerr lens mode-locking provided shorter picosecond pump pulses with a duration of about 4 ps each. This resulted in a cavity round-trip time of

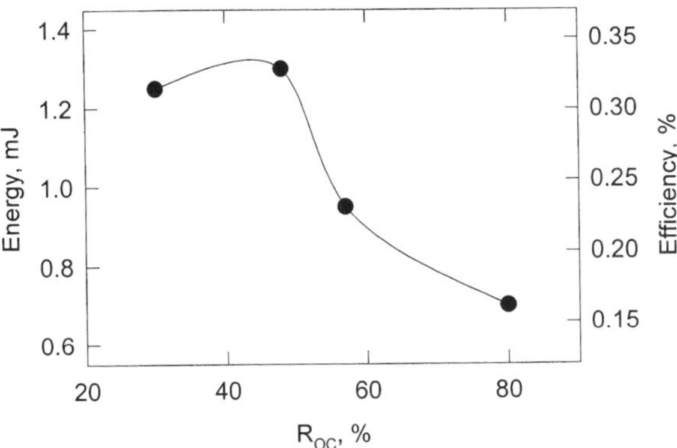

Figure B1.8.7. Energy and efficiency of a picosecond LiF:F_2^- CCL with respect to the reflectivity of the output coupler for 4 mJ energy in the pump pulse train.

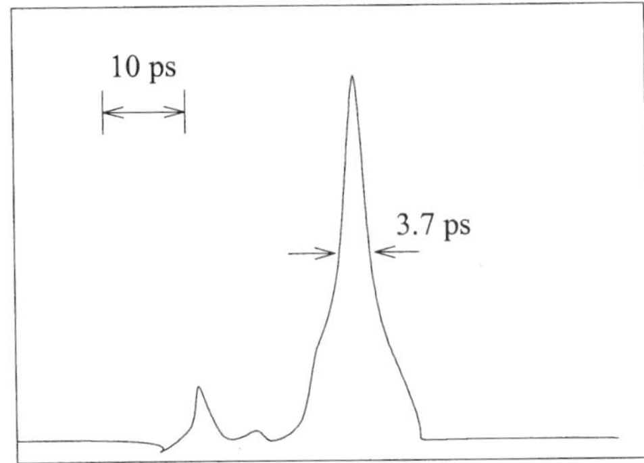

Figure B1.8.8. Temporal shape of the picosecond LiF:F_2^- CCL.

10 ns and a train duration of 7–10 μs, which was 20–30 times longer than that noted earlier [47]. Average pump energy in each pulse was about 3–4 μJ with a slow decrease at the end of the pulse train. The 40 mm long LiF:F_2^- Brewster-cut crystal with an absorption coefficient of 0.7 cm^{-1} was longitudinally pumped by a focused (500 mm focal distance) radiation through a dichroic mirror. The pulse duration was measured by means of non-collinear second harmonic generation with $LiIO_3$ nonlinear crystal.

The minimum pulse duration of LiF:F_2^- CCL was measured to 0.2–0.5 ps, assuming a Lorentzian pulseshape (figure B1.8.9) that constitutes a 15-fold output pulse compression with respect to the pump pulse duration [48]. These results were obtained at a wavelength of 1.14 μm near the maximum of the gain curve when a 75% reflectivity output coupler was used. Figure B1.8.10 presents a CCL pulse duration dependence on the degree to which the pump energy exceeds that of the threshold. The minimum pulse duration corresponded to the maximum pump energy, which was three times higher than the threshold value. The pulse duration was very critical to a mismatch of more than 10 μm. The total energy conversion efficiency for the shortest pulses was measured to be 6–8% with respect to the pump pulse energy. Peak power estimations show that, in this long pulse synchronous pumping scheme, one can obtain LiF:F_2^- femtosecond output pulses with a peak power exceeding 1 GW which is many orders of magnitude higher than that from an Al_2O_3:Ti femtosecond laser pumped under similar conditions.

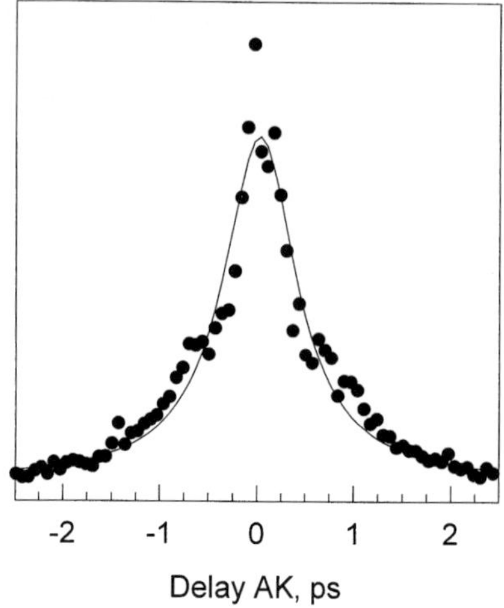

Figure B1.8.9. Auto correlation curve of a pulse duration of synchronously pumped LiF:F_2^- CCL with 75% output coupler.

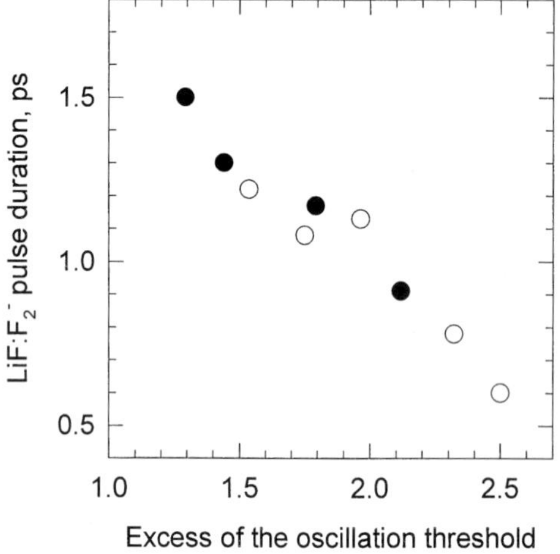

Figure B1.8.10. Dependence of the pulse duration of a synchronously pumped LiF:F_2^- CC laser with respect to pump threshold excess: filled circles, 50% reflectivity output coupler and hollow circles, 75%.

B1.8.6.3 High energy and power in colour centre lasers

Success in the growth of large size and good quality alkali-halide optical crystals and in the development of homogeneous, high-contrast colouration techniques gives rise to a unique opportunity to design large-scale high-power lasers and amplifiers based on CCLs. Using an LiF:F_2^- active crystal 50 mm × 50 mm × 50 mm in size and an initial transmittance of 8% for the 1.064 μm wavelength of an Nd-glass pump laser, an output energy as high as 20 J with a 23% pump-to-output conversion efficiency and a 100 ns pulse duration was obtained [49].

Developing large-scale homogeneously coloured LiF:F_2^- CC active crystals with a size of up to 40 mm × 80 mm × 200 mm with four large polished faces allows a huge (700 J in 120 ns pulse duration) pump energy from a multi-stage Nd:glass laser system. In a two-pass pumping scheme with a 5 J cm^{-2} pump energy density, transversely directed lasing of F_2^- CCs was realized in a non-selective 300 mm long Fabry–Pérot cavity with 50% output coupling. Under these conditions, the authors were able to reach the record output energy of 100 J and a peak power of 1 GW at 1.12–1.16 μm and 100 ns pulse duration.

For smaller (40 mm × 40 mm × 20 mm) LiF:F_2^- crystals in a transversal LiF and Nd:glass coupled cavity arrangement, the Nd laser pump-to-CCL output conversion efficiency was increased by up to 80% with a CCL output energy of 8 J at 1.15 μm [49].

B1.8.6.4 Narrow-line tunable and super-broadband CCL oscillation

The main advantage of a CCL is the possibility of obtaining coherent radiation tunable over a wide near-IR spectral range. The oscillating wavelength can be roughly selected by choosing an appropriate CC active medium. In a non-dispersive cavity, the oscillation occurs at wavelengths near the maximum of the fluorescence curve. The width of the oscillating spectra can be up to several hundreds of cm^{-1}. The radiation of CCLs in dispersive cavities is tunable when prisms, diffraction gratings, Fabry–Pérot etalons, Lyot filters and other wavelength-selective elements are utilized.

As an example, we consider a nanosecond LiF:F_2^+ CCL pumped by the second harmonic radiation of an Nd^{3+}:YAG laser. The use of a selective cavity with a single glass prism enables the radiation to be tunable over 0.84–1.1 μm with a spectral width of 20 cm^{-1}. Utilization of three dispersive prisms narrows the output spectrum to 5–7 cm^{-1}. A diffraction grating with 1200 lines mm^{-1} provides radiation with a 1–3 cm^{-1} spectral width tunable over the 0.84–1.1 μm spectral region. An additional intracavity Fabry–Pérot etalon provides a further tenfold reduction in the oscillating linewidth of up to 0.1–0.3 cm^{-1}. A narrow-line single-frequency laser oscillation regime is of special interest for numerous applications. The use of a grazing incidence scheme with a diffraction grating operating simultaneously as an intracavity telescope and dispersive element allows the linewidth of nanosecond high-peak-power lasers to be reduced to the level of 0.03–0.01 cm^{-1} and to realize a single-frequency oscillation [1].

Narrow line tunable LiF:F_2^- CCL system with nanosecond powerful amplifier with up to 50 mJ output pulse energy was developed in [56,57]. The master oscillator with synchronous scanning of autocollimation grating and dielectrically coated quarts Fabry–Pérot etalon (d=1 cm) provided stable (more than 20 min) single frequency operation (<300 MHz) smoothly and continuously tunable in 1 cm^{-1} frequency region. This system allowed the authors to investigate and optically pump HCl, HBr molecular gases in wide spectral range 1.1–1.25 μm. As an example, mid IR HCl gas laser operation at 3.88 μm at (3,2) P(5)–(2,1) P(6) optical transition was demonstrated under single frequency CC laser optical pumping [57].

Recently a new type of laser, specifically a super-broadband colour centre laser (SCCL) was proposed for LiF:F_2^- and LiF:F_2^+ CC active elements. It provides two types of laser radiation: simultaneous super-broadband lasing over the whole amplification band (about 1000 cm^{-1}) of the active medium (0.85–1.1 and 1.1–1.25 μm for LiF:F_2^+ and LiF:F_2^- lasers, respectively) and multi frequency or spectrally controlled laser oscillation (20 lines simultaneous oscillation) [50]. The basic idea of super-broadband lasing is the suppression of mode competition and the creation of a number of independent channels in the cavity, each lasing at different wavelengths within the amplification band of the active medium. As a result, super-broadband operation of an LiF CCL provided 'low-coherence' easily collimated high-intensity laser radiation with a spectral width of 1400 Å centred at 0.96 and 1.14 μm, correspondingly.

By simultaneous second harmonic generation in one nonlinear crystal, the oscillation in the visible (blue-green (0.43–0.53 μm) and green-red (0.55–0.62 μm)) spectral ranges was obtained in both regimes for LiF:F_2^+ and LiF:F_2^- CCLs and can be called rainbow or 'sun-colour' lasers. Super-broadband lasing can be of great interest for fast fluorescence and absorption spectroscopy, two-photon nonlinear spectroscopy, nanosecond colour photography, optical communication systems, information coding and other scientific and technical areas.

Free-running pulsed room-temperature tunable operation of a LiF:F_2^- laser pumped by InGaAs laser diode has been described in [51]. In the free-running mode, with a non-selective resonator, the LiF:F_2^- laser showed a 4% slope conversion efficiency and a measured 2.5% pumping-to-lasing efficiency at the maximum of the tuning curve for tunable narrow-band operation. The LiF:F_2^- oscillator was tunable in the 1122–1201 nm range.

Pulsed operation of a tunable LiF:F_2^{+*} CCL has been obtained at room temperature under pumping by a 680-nm fibre-coupled linear array of laser diodes [52]. Tunable operation in the 880–995 nm spectral region has been demonstrated with an efficiency of 2% in the maximum of the tuning curve when a birefrigent plate was utilized as a cavity dispersive element.

B1.8.7 CC energy and power amplifier

In many cases, it is necessary and convenient to amplify a weak laser radiation with special spectral, spatial or temporal properties to a high energy or peak power values. The four-level energy diagram describing the electronic–vibrational transitions of CCs together with a considerable Stokes shift and the quasi-homogeneous nature of their wide absorption and fluorescence bands, as well as a high quantum efficiency and a large amplification cross section, are very favourable factors for tackling this amplification task. CC crystal can be effectively used in master oscillator power amplifier (MOPA) schemes or for pico- and femtosecond pulse amplification in near-IR spectral regions.

The high emission cross section ($\sigma_{em} = 7 \times 10^{-17}$ cm^2) and long lifetime ($\tau = 100$ ns) of F_2^- CCs promises a high amplification gain and efficiency for an LiF:F_2^- amplifier. Compared to dye solutions, the concentration of an F_2^- CCs in LiF crystal is rather low. Due to this, a longitudinally pumped amplifier scheme is preferable for good spatial overlapping of the pump and probe radiation in the bulk CC crystal with a 40–80 mm length.

The experimental set-up included a pulsed Nd^{3+}:YAG pump laser with an output energy of up to 200 mJ, a 3 Hz pulse repetition rate and a 10 ns pulse duration. It was used to pump both a CC master oscillator and a CC power amplifier. A tunable LiF:F_2^- CCL provided a low divergence, tunable nanosecond probe beam with a spectral linewidth of less than 1 cm^{-1}.

The output energy and gain dependences of single-pass amplification in an LiF:F_2^- CC crystal 40 mm long with an initial absorption coefficient $K_0^{(1.064)} = 0.67$ cm^{-1} and 40 mJ pump pulse energy are shown in figure B1.8.11. The highest values of the amplification gain, $G = 12$, and amplification coefficient, $\alpha = 0.62$ cm^{-1}, were obtained for a low-energy probe pulse ($E_{input} = 0.06$ mJ). An increase in the probe pulse energy to 1 mJ resulted in gain saturation with a corresponding lowering of the amplification gain and gain coefficient to $G = 4$ and $\alpha = 0.35$ cm^{-1}, respectively, on increasing the amplified output energy to 4 mJ per pulse. This value corresponded to about 10% energy conversion efficiency for the LiF:F_2^- CC power amplifier [53].

Due to the limited concentration of F_2^- CCs in a LiF crystal, an increase in the interaction length by using longer crystals or multi-pass amplifier schemes is of great importance. A two-pass LiF:F_2^- CC amplifier scheme utilized the reflection from the back HR mirror located close to the LiF crystal. Figure B1.8.12 shows the results from the amplification in a LiF:F_2^- CC crystal 88 mm long with $K_0^{(1.064)} = 0.5$ cm^{-1} for a 40 mJ pump pulse. For a low input energy of 0.08 mJ, a high amplification gain of 56 ($\alpha = 0.26$ cm^{-1}) was realized with an output energy of 4.2 mJ and an efficiency of about 10%. For a higher energy input beam (2.2 mJ), the CCL amplifier output energy was increased to 12 mJ with a pump conversion efficiency of up to 30%, which is close to the best values for CCL conversion efficiency [53].

The utilization of a LiF:F_2^- double-pass amplifier with an 80 mJ Nd^{3+}:YLiF$_4$ laser with a more efficient, shorter pumping wavelength (1047 nm) provided the highest gain coefficient ($G = 300$) for a low-energy

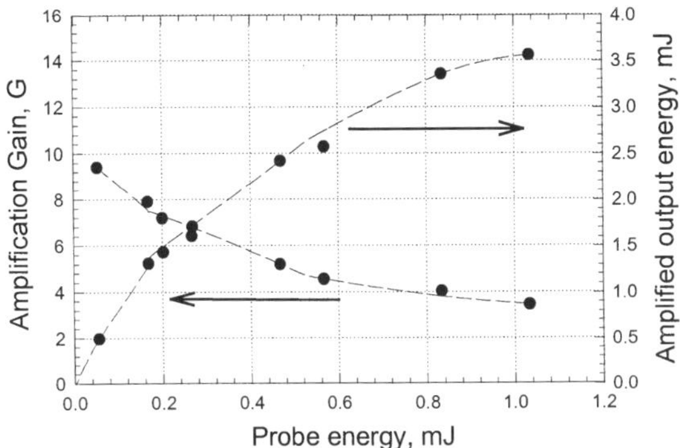

Figure B1.8.11. Energy dependences of the amplification gain and output energy of an LiF:F_2^- single-pass CCL amplifier with a 40 mJ pump pulse.

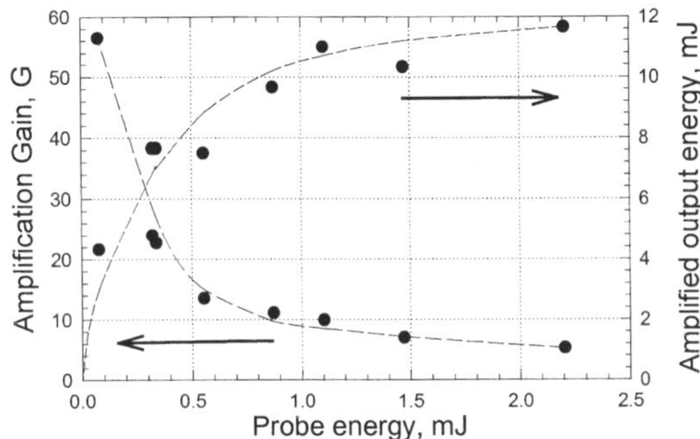

Figure B1.8.12. Energy dependences of the amplification gain and output energy of an LiF:F_2^- two-pass amplifier with a 40 mJ pump pulse.

probe pulse of 0.1 mJ. The maximum amplified output energy of 34 mJ with amplification gain $G = 90$ and an energy conversion efficiency of 45% was obtained for a high-energy probe pulse of 0.6 mJ.

The experimental results on LiF:F_2^- CC crystal amplification have shown that this crystal is a promising candidate for developing both high-gain and high-power optical amplifiers for low-divergence, short-pulse or narrow-line laser radiation in the 1.1–1.25 μm spectral region.

The feasibility of amplifying the broadband radiation of diode lasers and light-emitting diodes with the aid of LiF:F_2^+ CC crystals was demonstrated in [54]. A single-pass amplifier configuration was used. The active centres were excited by the second harmonic ($\lambda_p = 532\ nm$) of a multi-mode-pulsed Nd^{3+}:YAG laser emitting pulses of 10 ns duration. A GaAs laser diode generated 85 ns radiation pulses with a linear polarization at the wavelength 900 nm (close to the maximum of the fluorescence spectrum of F_2^+ CCs) with the spectral width being about 55 cm^{-1}.

In this experiment, single-pass gains (G) of 5 in the short crystal ($l = 0.8$ cm) and 20 in the long crystal ($l = 2$ cm) were reached with pump pulse energy of 1 mJ. No saturation of the gain relative to the energy of the amplified radiation from the laser diode was observed. The optical gain was constant when the pump energy was in the range 0.1–1 mJ and the energy of the amplified IR radiation was 0.4–33 nJ.

In the second case, the pump beam was split into two beams of equal intensities, which pumped the active element from opposite sides. An increase in the pump energy to 5 mJ per pulse and focusing of the pump radiation behind the exit end of the crystal enhanced the optical gain in the short 0.8 cm crystal up to eight per single pass. Hence, in this case, the gain coefficient, α, was at least 2.6 cm^{-1}.

The high stability and good laser performance of LiF CCLs with O^{2-} and OH^- stabilized F_2^+ CC has been demonstrated at room temperature [21, 23, 55]. Stabilized F_2^+ CCs have a broad absorption band with a spectral width of more than 3500 cm^{-1} with a maximum of about 600 nm. The room temperature fluorescence spectrum is centred at 890 nm with a bandwidth of more than 2300 cm^{-1}. The complexity of tunable cavity development for such broadly tunable CCLs arises from the necessity to use several sets of mirrors to obtain oscillation at the edges of the tuning curves. The use of different kinds of dichroic cavity mirrors in these CCLs reduced two-pass superfluorescence and provided an extremely wide tunable range (800–1200 nm) with a conversion efficiency of 12–25% in the maximum of the tuning curve.

B1.8.8 Conclusions

In this chapter, we have reviewed the progress in the physics of LiF CC formation and stabilization and demonstrated recent developments in dramatically improved LiF CC lasers providing stable, room temperature, broadly tunable, near infrared, optical output and operating efficiently in all known modes of operation. We believe that now, when the merits and shortcomings of tunable lasers based on impurity-doped crystals are clearly understood and the new merits of CC lasers are clearly demonstrated, the interest in CC lasers will be renewed.

LiF CC active elements have a unique combination of spectroscopic, oscillation, thermo-optic, and operational properties and the lasers based on them have many of the positive features of a solid state dye-like laser. These advantages include high gain coefficients, broad homogeneous gain profile, low threshold and highly efficient single-mode operation with extremely narrow spectral outputs, wide wavelength tunability, compactness, long operational lifetime, rigidity, ease of handling and insensitivity to the quality of the cavity's optical elements, and to the spatial angular and spectral characteristics of the pump source. The system also exhibits virtually no temporal delay between pump and output pulses. This temporal overlap between the pump and generated CC laser pulses for the entire range of tunability provides an easy extension to the UV, visible and middle-IR region through sum and difference frequency generation of the pump and CCL output. The described developments are expected to result in LiF CC lasers finding widespread acceptance in research laboratories throughout the world and generating an abundance of activity in different areas of science.

References

[1] Basiev T T, Mirov S B and Osiko V V 1988 Room-temperature colour centre lasers *IEEE J. Quantum Electron.* **24** 1052–69

[2] Basiev T T and Mirov S B 1994 *Room Temperature Tunable Colour Centre Lasers (Laser Science and Technology Book Series 16)* ed V S Letokhov *et al* (New York: Gordon and Breach Science/Harwood Acadademic) pp 1–160

[3] Gusev Yu L, Marennikov S I and Chebotaev V P 1980 *Bull. Acad. Sci. USSR, Phys. Ser.* **44** 15

[4] Parfianovich I A, Khulugurov V M, Lobanov B D and Maksimova N T 1979 Luminescence and stimulated emission of CCs in LiF *Bull. Acad. Sci. USSR, Phys. Ser.* **43** 20–7

[5] Gellermann W 1991 *J. Phys. Chem. Solids* **52** 249

[6] Mollenauer L F 1987 Colour centre lasers *Tunable Lasers* vol 59, ed L F Mollenauer and J C White (Berlin: Springer) pp 225–77

[7] Archandelskaya V A and Feofilov P P 1980 *Sov. J. Quantum Electron.* **7** 657

[8] Martynovich E F, Baryshnikov V I and Grigorov V A 1985 Visible lasing of Al_2O_3 colour centre crystals at room temperature *Opt. Commun.* **53** 257–8

[9] Voytovich A P, Grinkevich V A, Kalinov V S and Mikhnov S A 1988 Spectroscopic and oscillation characteristics of colour centre sapphire crystals in the range of 1.0 μm, *Kvant. Electron.* **15** 318–20 (Engl. transl. *Sov. J. Quantum Electron.*)

[10] Rand S C and DeShazer L G 1985 Visible colour-centre laser in diamond *Opt. Lett.* **10** 481–3

[11] Shkadarevich A P and Yarmolkevich A P 1985 New laser media on colour centre compound fluoride *Preprint* Inst. Phys. AN BSSR, Minsk, USSR, p 24

[12] Horsch G and Paus H J 1986 A new colour centre on the basis of lead-doped $KMgF_3$ *Opt. Commun.* **60** 69–73

[13] Basiev T T, Vakhidov F A and Mirov S B 1988 Radiational transformations in a new $LiYF_4$ laser crystal with CCs *Kratkie Soobsheniya po Fizike* **7** 3–5 (Engl. transl. *Sov. Phys. Lebedev Inst. Rep.* **7** 1–5)

[14] Mollenauer L F 1979 *Colour Centre Lasers in Quantum Electronics* part B, vol 15, ed C L Tang (New York: Academic) ch 6

[15] Rzepka R E, Bernard M, Lefrant S, Dubost H, Charneau R and Galaup Y P 1987 Laser performance of FA(II)Li centres prepared by electrolytic colouration in Li-doped KCl crystals *Opt. Commun.* **62** 174–8

[16] Lushchik Ch B, Vitol I K and Elango M A 1977 Decay on electronic excitation of radiational defects in ionic crystals *Uspehi Fizicheskih Nauk* **122** 223–52 (Engl. transl. *Sov. Achievements Phys. Sci.*)

[17] Delbecq C J 1963 A study of M centre formation in additively coloured KCL *Z. Phys.* **171** 560–81

[18] Basiev T T, Voron'ko Yu K, Mirov S B, Osiko V V and Prokhorov A M 1979 Kinetics of accumulation and oscillation of F_2^+ CCs in LiF crystals *Pis'ma v JETP* **30** 661–5 (Engl. transl. 1979 *Sov. JETP Lett.* **30** 626–9)

[19] Khulugurov V M and Lobanov B D 1978 Colour-centre lasing at 0.84–1.13 μm in a LiF–OH crystal at 300 K *Sov. Tech. Phys. Lett.* **4** 595–6

[20] Basiev T T, Konyushkin V A, Mirov S B and Ter-Mikirtychev V V 1992 Efficient tunable LiF:F_2^+ lasers utilizing F_2 and F_2^+ colour centres *Sov. J. Quant. Electron.* **22** 128

[21] Basiev T T, Ermakov I V, Fedorov V V, Konushkin V A and Zverev P G 1994 Laser oscillation of LiF:F_2^+—stabilized colour centre crystals at room temperature *Proc. Int. Conf. on Tunable Solid State Lasers (Minsk, Inst. Mol. and Atom. Phys. 64)*

[22] Nahum J 1967 Optical properties and mechanism of formation of some F-aggregate centres in LiF *Phys. Rev.* **158** 814

[23] Dergachev A Yu and Mirov S B 1998 Efficient room temperature LiF:F_2^{+**} colour centre tunable laser tunable over 820–1210 nm range *Opt. Commun.* **145** 107–12

[24] Mirov S B and Dergachev A Yu 1997 Powerful, room-temperature stable LiF:F_2^{+**} laser *Proc. SPIE* **2986** 162–73

[25] German K R 1989 *Handbook of Solid State Lasers* ed P K Cheo (New York: Marcel Dekker) ch 5

[26] Fritz B and Menke E 1965 *Solid State Commun.* **3** 61

[27] Kogelnik H W, Ippen E P, Dienes A and Shank C V 1972 *IEEE J. Quantum Electron.* **QE-8** 373

[28] Beigang R, Liftin G and Welling H 1977 *Opt. Commun.* **22** 269

[29] Breant Ch, Baer T, Nesbitt D and Hall J L 1983 *Laser Spectroscopy* vol VI, ed H P Weber and W Luthy (Berlin: Springer) p 138

[30] Johnston T F and Proffitt W 1980 *IEEE J. Quantum Electron.* **QE-16** 483

[31] Pollock C R and Jennings D A 1982 *Appl. Phys.* B **28** 308

[32] Rong F, Yang Y and Luty F 1989 *Cryst. Latt. Def. Amorph Mater.* **18** 1

[33] Beigang R and Wynne J J 1981 *Opt. Lett.* **6** 295

[34] Pinto J F, Stratton E and Pollock C R 1985 *Opt. Lett.* **10** 384

[35] Yasa Z A 1983 *Opt. Lett.* **8** 277

[36] Langford N, Smith K and Sibbett W 1987 *Opt. Commun.* **64** 274

[37] Langford N, Smith K and Sibbett W 1987 *Opt. Lett.* **12** 903

[38] Islam M N, Sundermann E R, Bar-Joseph I, Sauer N and Chang T Y 1989 *Appl. Phys. Lett.* **54** 1203

[39] Mark J, Liu L Y, Hall K L, Haus H A and Ippen E P 1989 *Opt. Lett.* **14** 48

[40] Mollenauer L F and Stolen R H 1984 *Opt. Lett.* **9** 13

[41] Gellermann W, Muller A, Wandt D, Wilk S and Luty F 1987 *J. Appl. Phys.* **61** 1297

[42] Basiev T T, Zverev P G, Papashvili A G, Konyushkin V A and Osiko V V 1997 *Quantum Electron.* **27** 759

[43] Basiev T T, Gusev Yu L, Kruzhalov S V, Mirov S B and Petrunkin V Ya 1988 *Sov. J. Quantum Electron.* **12**

[44] Basiev T T, Gusev Yu L, Kruzhalov S V, Mirov S B and Petrunkin V Ya 1988 Abst. Rep. XIII Inter. Conf. Coherent Nonlinear Opt., Minsk, USSR, Part 2, 256 (1988).

[45] Basiev T T, Zverev P G, Papashvili A G and Fedorov V V 1997 *Quantum Electron.* **27** 574

[46] Babushkin A V, Basiev T T, Vorob'ev N S, Mirov S B, Prokhorov A M, Serdyuchenko Yu N and Shchelev M Ya 1986 *Sov. J. Quantum Electron.* **16** 1492

[47] Basiev T T, Dergachev A Yu, Karasik A Ya, Fedorov V V and Shubochkin R L 1996 *Quantum Electron.* **26** 1042

[48] Zverev P G, Basiev T T and Laubereau A 1999 *Novel Lasers and Applications—Basic Aspects, OSA Technical Digest* (Washington DC: OSA) p 57

[49] Basiev T T, Dolzhenko S V, Ershov B V, Kravtsov S B, Mirov S B, Spiridonov V A Fedorov V B 1988 *Bull. Acad. Sci., USSR* **52** 164

[50] Basiev T T, Zverev P G, Fedorov V V and Mirov S B 1997 *Appl. Opt.* **36** 2515

[51] Ter-Mikirtychev V V 1998 Diode-pumped tunable room-temperature LiF:F_2^- colour-centre laser *Appl. Opt.* **37** 6442

[52] Ter-Mikirtychev V V 1998 Diode-pumped LiF:F_2^{+*} colour centre laser tunable in 880–995 nm region at room temperature *IEEE Photon. Technol. Lett.* **10** 1395

[53] Basiev T T, Doroshenko M E, Zverev P G, Skornyakov V V and Hager G 2001 Laser amplifier on LiF:F_2^- colour centre crystal for 1.1–1.25 μm spectral region *Advanced Solid State Lasers, OSA Technical Digest* (Washington DC: OSA) pp 54–5.

[54] Basiev T T, Mirov S B and Ter-Mikirtychev V V 1992 *Proc. SPIE Int. Soc. Opt. Eng.* **1839** 227

[55] Jenkins N W, Mirov S B and Fedorov V V 2000 Temperature-dependent spectroscopic analysis of F_2^{+**} and F_2^+ like CCs in LiF *J. Luminescence* **91** 147–53

[56] Basiev T T, Doroshenko M E, Kravtsov C B, Skornyakov V V, Zverev P G, Vasiliev S V, Alimpiev S S, Nikiforov S M and Hager G 2003 Oscillation of HC1 molecular gas laser at 4 μm spectral region under optical pumping with tunable color center laser

in the third oscillation obertone *Quantum Electronics* **33** 210

[57] Basiev T T , Papashvilli A G, Fedorov V V, Vasiliev S V and Gellermann W 2003 Single-longitudinal-mode pulsed LiF:F_2^- color center laser for high resolution spectroscopy *Laser Physics* (in press)

B2
Laser diodes

Ian White

The laser diode has proved to be an enduring device. Whilst at the most fundamental level its operation is very simple, using a two-level transition normally within a p–n junction, the device has proved to be complex to develop. As a result, despite being invented not long after the first laser, room-temperature continuous-wave operation took several years to develop and robust single-mode operation was only achieved using distributed feedback structures yet later.

Since then, a large number of variants of the basic laser diode structure have been developed. A wide range of materials has been used to enable lasing action across a range of wavelengths and the use of quantum wells and dots within active regions has generated many technological advances. Advanced laser structures have been developed for ensuring high-performance laser diode operation in terms of spectral and beam quality of the laser output. A key feature of the diode laser is its ability to allow high-speed direct modulation and, hence, a range of structures has been developed to allow both short-pulse and high-frequency modulation.

The result of this period of development is that laser diodes now find widely varying applications. They are the dominant optical sources in optical communications and storage and are increasingly being used in sensing and in high-power applications.

A full treatment of the laser diode must consider many aspects and, hence, in this section, following a treatment of the basic principles of the device, chapters concentrate on spectral control in laser diodes and their high-speed and high-power operation. Special attention is then given to laser devices able to operate in the red, in the blue and at long wavelengths where unipolar devices have been demonstrated. Finally, advanced laser diode devices, able to carry out signal-processing functions, are reviewed.

B2.1
Basic principles of laser diodes

N K Dutta

B2.1.1 Introduction

Since its invention in 1962 [1–4], the semiconductor injection laser has emerged as an important device in many optoelectronic systems such as optical recording, high-speed data transmission through fibres, optical sensors, high-speed printing and guided wave signal processing. Perhaps the most important impact of semiconductor lasers is in the area of lightwave transmission systems where the information is sent through encoded light beams propagating in glass fibres. These lightwave transmission systems, which are currently being installed throughout the world, offer a much higher transmission capacity at a lower cost than coaxial copper cable transmission systems.

The advantages of semiconductor lasers over other types of lasers—gas lasers, dye lasers and solid state lasers—lie in their considerably smaller size and lower cost and their unique ability to be modulated at gigahertz speeds simply by modulation of the injection current. These properties make the laser diode an ideal device as a source in several optoelectronic systems, especially optical fibre transmission systems.

B2.1.1.1 Basic laser diode

The concept of a semiconductor laser diode is a unique blend of semiconductor device physics and quantum electronics. The basic laser diode chip consists of two parallel cleaved facets, which form an optical cavity (figure B2.1.1). The other two edges are saw cut. The device has a pn junction near the light-emitting region (active region) for effective current injection. The typical cavity length is about 200–400 μm.

The light–current characteristics of a laser diode are shown in figure B2.1.2. As the injected current is increased beyond a certain value (threshold current), the light output from the facet increases dramatically. Associated with this is a narrowing of the spectral emission from ~600 Å in the below-threshold region to ~30–40 Å in the above-threshold region. The light below threshold is emitted spontaneously as the injected electrons and holes recombine in the active region. Above threshold, stimulated emission into a low divergence beam characteristic of laser action is observed. For digital lightwave systems, the light signal is encoded by modulating the injection current, which generally switches the laser between the 'on' state (stimulated emission regime) and the 'off' state (spontaneous emission regime).

A close relative of the injection laser is the light-emitting diode (LED) [5]. The LED is a spontaneous emission device: its emission characteristics are similar to those of lasers below threshold. LEDs generally emit lower power and have a broader spectral emission and lower modulation speed than lasers. The basic concepts, the physics of semiconductor lasers and semiconductor laser designs are described in this chapter. The semiconductor laser is essentially an oscillator where the feedback is provided by the cleaved mirror facets and the amplification (or gain) is provided by electron–hole radiative recombination in the active region. The condition for sustained oscillation (laser action) is that gain equals loss (see chapter A1).

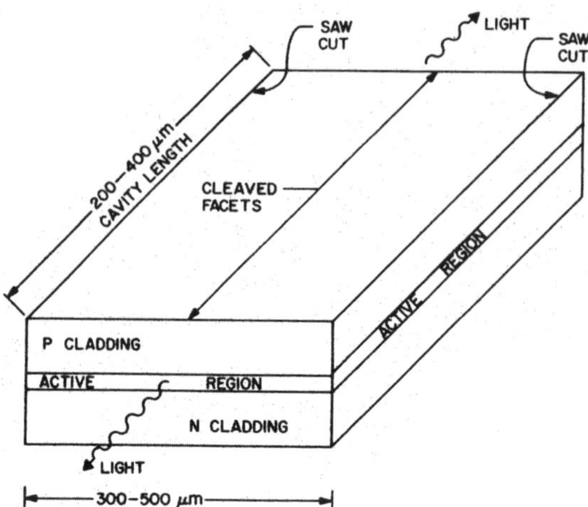

Figure B2.1.1. Schematic diagram of the basic semiconductor laser diode.

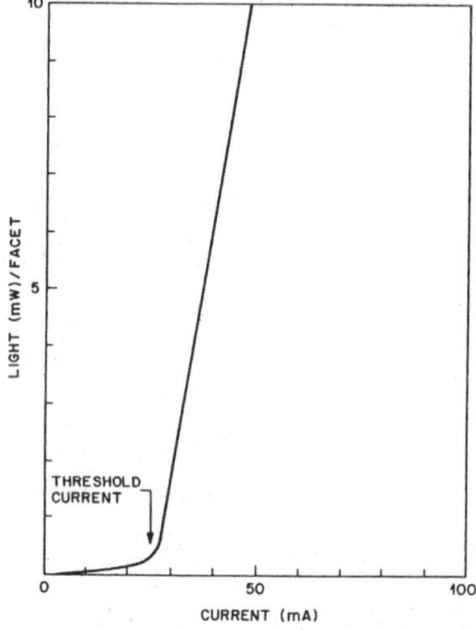

Figure B2.1.2. Light emitted from the facet *versus* the injected current for a semiconductor injection laser. The device is a proton-stripe GaAs laser.

B2.1.1.2 Materials

The choice of materials for semiconductor lasers is principally determined by the requirement that the probability of radiative recombination should be sufficiently high that there is enough gain to overcome the cavity losses. This is usually satisfied for 'direct-gap' semiconductors. The various semiconductor material systems along with their range of emission wavelengths are shown in figure B2.1.3. Many of these material systems are ternary (three-element) and quaternary (four-element) crystalline alloys that can be grown lattice-matched over a binary substrate.

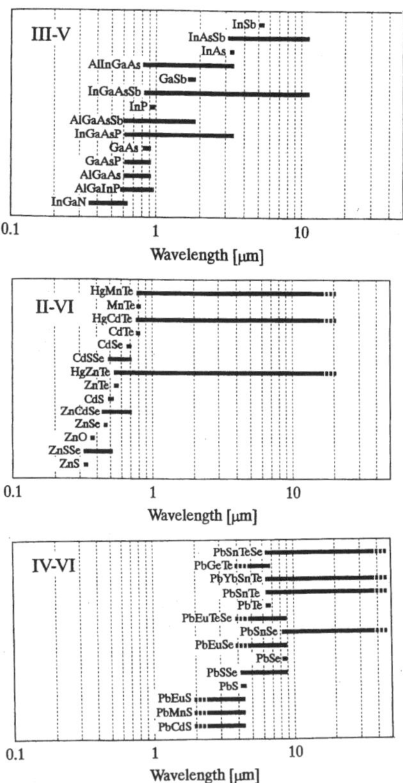

Figure B2.1.3. Semiconductor materials used in laser fabrication at different regions of the spectrum. The three figures refer to compound semiconductors formed using group III and group V elements (III–V), group II and group VI elements (II–VI) and group IV and group VI elements (IV–VI) of the periodic table (After Eliseev [6]). The quantum cascade lasers are fabricated using III–V semiconductors and they operate on intraband transitions in a quantum well.

B2.1.1.3 Epitaxy

The lattice-matched crystalline growth of one semiconductor over another is called epitaxy. The development of sophisticated epitaxial growth techniques has been of major significance in the development of high-quality reliable semiconductor lasers. The commonly used techniques are liquid-phase epitaxy (LPE) [7], vapour-phase epitaxy (VPE) [8] and molecular epitaxy (MBE) [9].

In LPE, the epitaxial layer is grown by cooling a saturated solution of the component of the layer to be grown while that solution is in contact with the substrate. In VPE, the epitaxial layer is grown by the reaction of gaseous elements or compounds at the surface of a heated substrate. The VPE technique has also been called chemical vapour deposition (CVD) depending on the constituents of the reactants. A variant of the technique is metal organic chemical vapour deposition (MOCVD) [10], which has been very successful for lasers in which metal alkyls are used as the compound source. In MBE, the epitaxial layer growth is achieved by the reaction of atomic or molecular beams of the constituent elements (of the layer to be grown) with a crystalline substrate held at high temperature in ultrahigh vacuum.

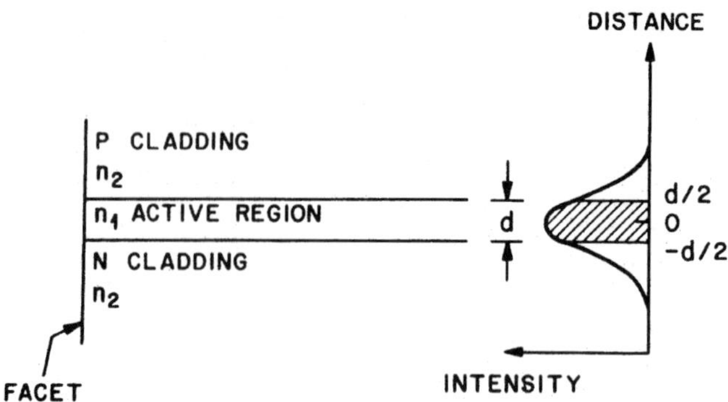

Figure B2.1.4. The dielectric waveguide of the semiconductor laser. n_2 is the refractive index of the cladding layers and n_1 that of the active region: $n_1 > n_2$. The cladding layers are of a higher bandgap material than the active region. The intensity distribution of the fundamental mode is shown. The cross-hatched region represents the fraction of the mode (Γ) within the active region.

B2.1.1.4 Lightwave system applications

As mentioned previously, one of the principal applications of semiconductor lasers is in the area of lightwave transmission systems. These systems have been made possible by two major technological advances: the development of low-loss silica (glass) fibres, which act as a transmission medium; and the development of high-performance reliable semiconductor lasers. Early lightwave systems used multi-mode fibres and an operating wavelength of about 0.85 μm. AlGaAs semiconductor lasers are used as sources for these systems. Current systems using single-mode fibres and an operating wavelength near 1.3 μm offer longer repeater spacing because of the lower silica fibre loss near 1.3 μm than at $\sim$0.85 μm. Even larger repeater spacing is allowed for an operating wavelength near 1.55 μm, where the fibre loss is minimum. Semiconductor lasers fabricated using the InGaAsP material system are, by and large, the only sources for commercial high-data-rate long-haul lightwave systems operating near 1.3 and 1.55 μm (see section D4). Because of its substantial impact on lightwave technology, a large portion of semiconductor laser research and development has been on A1GaAs ($\lambda \sim 0.85\ \mu$m) and InGaAsP ($\lambda \sim 1.3$ and 1.55 μm) lasers. For the same reason, we shall use AlGaAs and InGaAsP material systems as examples to illustrate the concepts in semiconductor lasers. Early work in the area of semiconductor lasers has been extensively reviewed in three books [11–13].

B2.1.2 Principles of operation

B2.1.2.1 The dielectric waveguide

Conceptually, stimulated emission in a semiconductor laser arises from electron–hole radiative recombination in the active region, and the light generated is confined and guided by a dielectric waveguide (figure B2.1.4). The active region has a slightly higher index than the p- and n-type cladding layers and the three layers form a dielectric waveguide. The energy distribution of the fundamental mode of the waveguide is also sketched in figure B2.1.4. A fraction of the optical mode is confined in the active region. Two types of fundamental transverse modes can propagate in the waveguide: the transverse electric (TE) and the transverse magnetic (TM) modes. The confinement factor (Γ), the fraction of mode in the active region, has been previously calculated [12, Part A, p 34]. Figure B2.1.5 shows the calculated Γ as a function of active layer thickness for the TE and TM modes for the $\lambda = 1.3\ \mu$m InGaAsP DH with p-InP and -InP cladding layers.

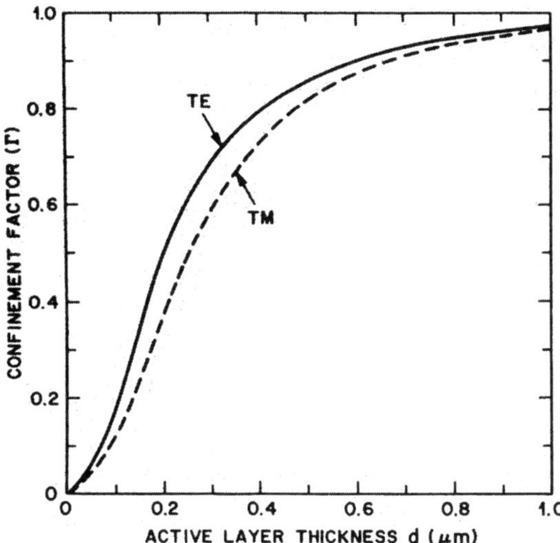

Figure B2.1.5. Confinement factor of the fundamental TE and TM modes for a waveguide with an InGaAsP ($\lambda = 1.3\ \mu$m) active layer and InP cladding layers as a function of the thickness of the active region. The refractive indexes of the active and cladding region are 3.51 and 3.22, respectively.

B2.1.2.2 Threshold condition

At threshold, the optical gain equals the total optical loss in the laser cavity. The condition for threshold (gain = loss) is

$$\Gamma g_{\mathrm{th}} = \alpha_{\mathrm{a}}\Gamma + (1 - \Gamma)\alpha_{\mathrm{c}} + \frac{1}{L}\ln\frac{1}{R} \tag{B2.1.1}$$

where g_{th} is the threshold gain in the active region; α_{a}, α_{c} are the absorption losses in the active and cladding regions, respectively: L is the cavity length; and R is the mirror facet reflectivity. Typically, $L \sim 300\ \mu$m and $R \sim 0.3$. For a 0.2 μm-thick active layer, Γ (TE) is ~0.47 (from figure B2.1.5) and using $\alpha_{\mathrm{a}} \simeq 30\ \mathrm{cm}^{-1}$, the calculated $g_{\mathrm{th}} \simeq 150\ \mathrm{cm}^{-1}$. This compares with a value of 10^{-2}–$10^{-3}\ \mathrm{cm}^{-1}$ for HeNe lasers [14].

B2.1.2.3 Condition for stimulated emission

Sufficient numbers of electrons and holes must be excited in the semiconductor for stimulated emission or net optical gain. The condition for net gain at photon energy E is given by [15]

$$E_{\mathrm{f_c}} + E_{\mathrm{f_v}} = E - E_{\mathrm{g}} \tag{B2.1.2}$$

where $E_{\mathrm{f_c}}$, $E_{\mathrm{f_v}}$ are the quasi-Fermi levels of electrons and holes, respectively, measured from the respective band edges (positive into the band) and E_{g} is the bandgap of the semiconductor. For zero net gain (transparency), this condition becomes $E_{\mathrm{f_c}} + E_{\mathrm{f_v}} = 0$. For undoped material at a temperature T, the quasi-Fermi energy $E_{\mathrm{f_c}}$ is related to the injected carrier (electron or hole) density n by

$$n = N_{\mathrm{c}}\frac{2}{\sqrt{\pi}}\int\frac{\mathrm{d}\varepsilon}{1 + \exp(\varepsilon - \varepsilon_{\mathrm{f_c}})} \tag{B2.1.3}$$

with $N_{\mathrm{c}} = 2(2\pi m_{\mathrm{c}}kT/h^2)^{3/2}$ and $\varepsilon_{\mathrm{f_c}} = E_{\mathrm{f_f}}/kT$ where k is the Boltzmann constant, h is Planck's constant, T is the temperature and m_{c} is the effective mass of the electrons in the conduction band. A similar equation

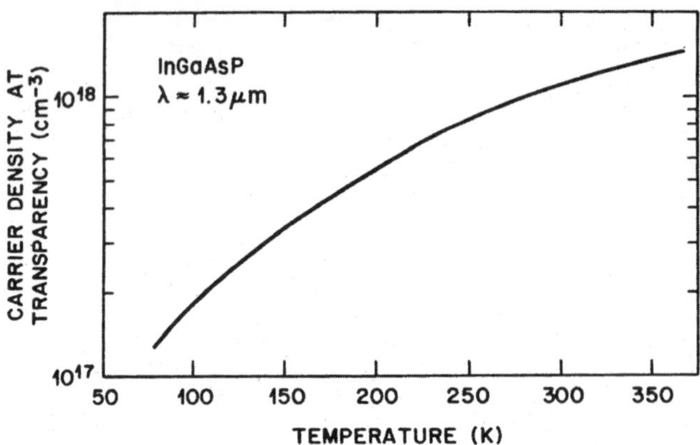

Figure B2.1.6. The calculated injected carrier density for transparency as a function of temperature for undoped $\lambda = 1.3\ \mu$m InGaAsP.

holds for holes. Figure B2.1.6 shows the variation of the injected carrier density for transparency n_t as a function of temperature for undoped $\lambda = 1.3\ \mu$m InGaAsP. The parameter values used in the calculation are $m_c = 0.061\ m_0$, $m_{hh} = 0.45\ m_0$, $m_{lh} = 0.08\ m_0$, where m_0, m_{hh} and m_{lh} are the free-electrons, heavy-hole, and the light-hole mass, respectively. Figure B2.1.6 shows that n_t is considerably smaller at low temperatures. The threshold carrier density is ~20–30% higher than n_t.

The high carrier densities needed for gain can be generated by optical excitation or current injection. The first injection laser operation was demonstrated in 1962 [1–3]. The injection laser utilized the unique characteristics of the pn junction to confine carriers near the depletion region under forward bias.

B2.1.2.4 The pn junction laser

Under high current injection through a pn junction, a region near the depletion layer can have a high density of electrons and holes. These electrons and holes can recombine radiatively if the interfaces are free of traps. The device will lase if the threshold condition (equation B2.1.2) is satisfied.

The energy-band diagram of a pn junction between two similar semiconductors (homojunction) at zero bias is shown in figure B2.1.7. The broken line represents the Fermi level. Under forward bias, both electrons and holes are present near the depletion region. This region can have net gain if the electron and hole densities are sufficiently high. However, the thickness of the gain region is very small (~100 Å), which makes the confinement factor (Γ) for an optical mode very small. Hence, from equation (B2.1.1), it follows that the threshold gain and, hence, the threshold current are high for a homojunction laser.

The threshold current of an injection laser was historically reduced using a DH for carrier confinement, which increased the size of the region of optical gain [16, 17]. The double heterostructure (DH) laser utilizes a pn heterojunction for carrier injection. A heterojunction is a junction in a single crystal between two dissimilar semiconductors. Thus, the fabrication of heterojunctions had to wait until the development of epitaxial growth techniques. The energy-band diagram of a DH laser at zero bias is shown in figure B2.1.8. The device consists of a narrow-gap semiconductor (p-type, n-type or undoped) sandwiched between higher-gap p- and n-type semiconductors. The narrow-gap semiconductor is the light-emitting region (active region) of the laser. The broken line represents the Fermi level. For the purpose of illustration, we have chosen the active region to be p-type in figure B2.1.8. The band diagram under forward bias is shown in figure B2.1.8(*b*). Electrons and holes are injected at the heterojunction and are confined in the active region. Thus, the region of optical gain is determined by the thickness of the active region in lasers. In addition, the refractive index

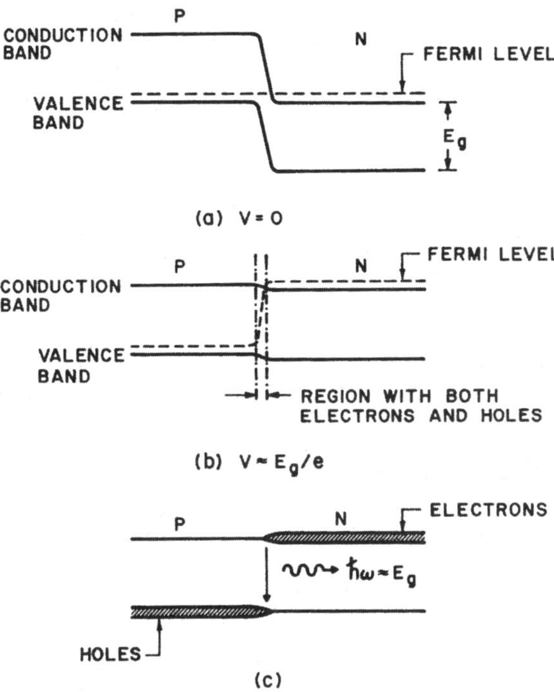

Figure B2.1.7. Energy-band diagram of a pn junction at (*a*) zero bias and (*b*) forward bias ($V \approx E_g/E$). (*c*) Schematic representation of electrons and hole densities under forward bias ($V \approx E_g/e$).

of the lower-gap active region is higher than that for the n- and p-type higher-gap confining (also known as cladding) layers. These layers form a waveguide for the lasing optical mode, as discussed previously.

The active region thickness of DH lasers is typically in the range 0.1–0.3 μm. Within the past two decades, with the development of MBE and MOCVD growth techniques, it has been possible to fabricate very thin epitaxial layers (<300 Å) bounded by higher-gap cladding layers. These double heterostructures are called quantum well double heterostructures because the kinetic energy for carrier motion along the thickness of the active region is quantized, similar to that for a one-dimensional potential well [18]. The modification of the electron–hole recombination characteristics (which is the basis for laser action) in quantum wells is described in section B2.1.5. For the purposes of this chapter, a laser device with an active layer thickness greater than 300Å is a DH laser, a laser with a smaller active layer thickness we shall call a quantum well laser.

B2.1.3 Radiative recombination

The basis of light emission in semiconductors is the recombination of an electron in the conduction band with a hole from the valence band; the excess energy is emitted as a photon (light quantum). The process is called radiative recombination, The energy *versus* wave–vector diagram of the electrons and holes in a cubic (zinc-blende type) semiconductor is shown in figure B2.1.9. For direct-gap semiconductors, the bottom of the conduction band and the top of the valence band are at the same point in momentum space or $\boldsymbol{k}$ space ($\boldsymbol{k} = 0$ in figure B2.1.9). This allows both energy and momentum conservation in the process of photon emission by electron–hole recombination. For indirect-gap semiconductors (e.g. silicon), the momentum

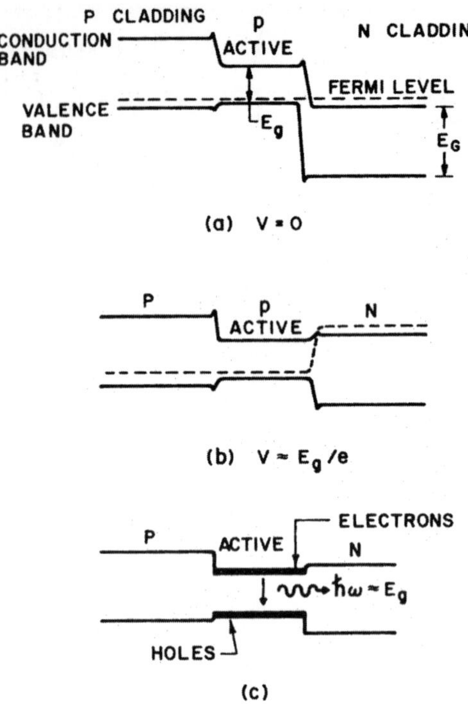

Figure B2.1.8. Energy-band diagram of a double heterostructure laser at (*a*) zero bias and (*b*) forward bias ($V \approx E_g/e$).

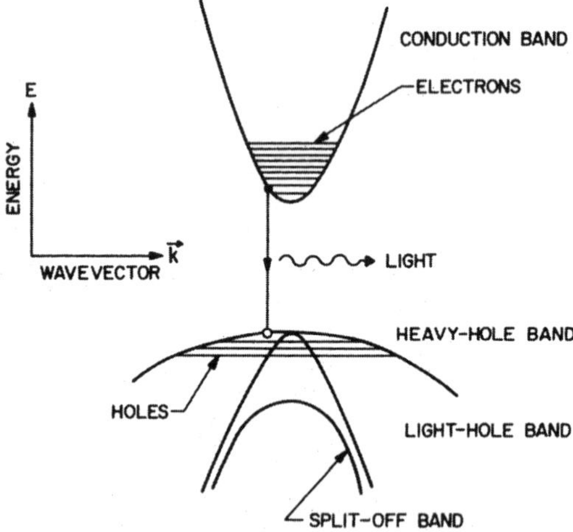

Figure B2.1.9. Energy *versus* wavevector of the four major energy bands for a zinc-blende type of direct-gap semiconductor.

conservation can be achieved with the assistance of a phonon (lattice vibration), which significantly decreases the probability of radiative recombination.

The valence band in many III–V semiconductors is represented by three major sub-bands. These are

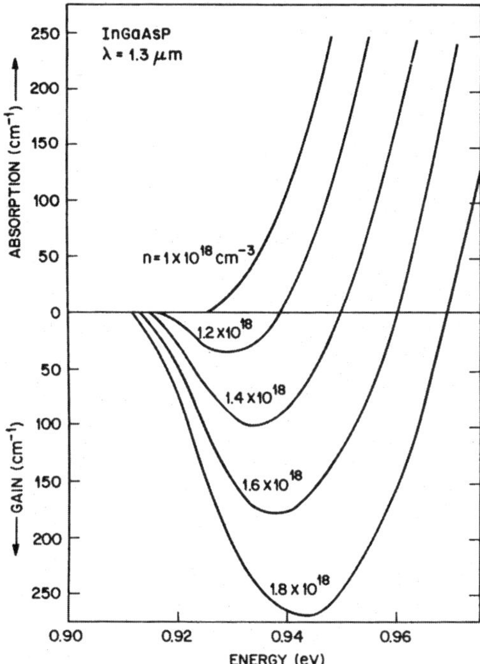

Figure B2.1.10. Calculated gain or absorption as a function of photon energy for $\lambda = 1.3$ μm ($E_g = 0.9$ eV) InGaAsP at various injected carrier densities (after Dutta [22]).

the heavy-hole band, the light-hole band and the spin–split-off band. The radiative transitions occurring near bandgap energies are due to the recombination of electrons with heavy holes and light holes. A heavy hole, as the name implies, has a larger effective mass than a light hole, which makes the density of states (and also the available number of heavy holes for a given Fermi level) larger than that for light holes.

B2.1.3.1 Gain calculation

The quantities associated with a radiative recombination are the absorption spectrum, emission spectrum, gain spectrum and total radiative emission rate. The optical absorption or gain for a transition between the valence band and the conduction band at an energy E is given by [12, part A, p 129]

$$\alpha(E) = \frac{e^2 h}{2\varepsilon_0 m_0^2 c n E} \int_{-\infty}^{+\infty} \rho_c(E')\rho_v(E'')|M(E', E'')|^2[f(E'' = E' - E) - f(E')]\,\mathrm{d}E' \qquad \text{(B2.1.4)}$$

where m_0 is the free-electron mass, e is the electron charge, ε_0 is the permittivity of free space, E is the photon energy, n is the refractive index at energy E and ρ_c and ρ_v are the densities of state per unit volume per unit energy in the conduction and valence band, respectively. $f(E')$ is the probability that a state of energy E' is occupied by an electron and M is the effective matrix element between the conduction-band state of energy E' and the valence-band state of energy . Hwang [19] has shown that the contribution of the band-tail impurity states can be significant for photon energies near the band edges. Several models for the density of states and matrix element for the transition between band-tail states exist (e.g. [11, part A, Chap 3]). The latest of such models that take into account the contributions of the parabolic bands and impurities is a

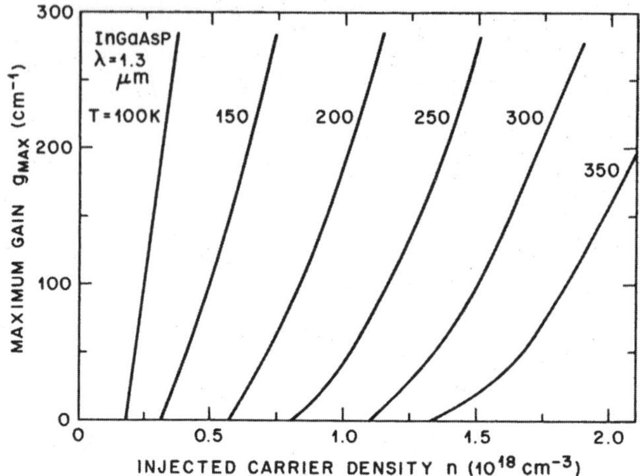

Figure B2.1.11. The maximum gain as a function of injected carrier density for undoped InGaAsP ($\lambda = 1.3\ \mu$m) at different temperatures (after Dutta and Nelson [27]).

Gaussian fit of the Kane form to the Halperin–Lax model of band tails. This was first proposed by Stern [20] and used to calculate the gain and recombination rate in GaAs.

The matrix element M may be expressed as a product of two terms: $M = M_{\mathrm{b}} M_{\mathrm{env}}$ The quantity M_{b} arises from the band-edge Bloch functions and M_{env} arises from the envelope wavefunctions. For III–V semiconductors using the Kane model [21],

$$|M_{\mathrm{b}}|^2 = \frac{m_{\mathrm{c}}^2 E_{\mathrm{g}}}{12 m_{\mathrm{c}}} \frac{E_{\mathrm{g}} + \Delta}{E_{\mathrm{g}} + \frac{2}{3}\Delta} \tag{B2.1.5}$$

where m_{c} is the conduction-band effective mass, E_{g} is the energy gap and Δ is the spin–orbit coupling. The envelope matrix element for the band-tail states have been calculated by Stern.

The calculated spectral dependence of absorption or gain at various injected carrier densities is shown in figure B2.1.10 for InGaAsP ($\lambda \simeq 1.3\ \mu$m) with acceptor and donor concentrations of 2×10^{17} cm^{-3}. The calculation was done using Gaussian Halperin–Lax band tails and Stern's matrix element. The material parameters used in the calculation are $m_{\mathrm{c}} = 0.061\ m_0$, $m_{\mathrm{hh}} = 0.45\ m_0$, $m_{lh} = 0.08\ m_0$, $\Lambda = 0.26$ eV, $E_{\mathrm{g}} = 0.96$ eV, and $\varepsilon = 11.5$. Figure B2.1.10 shows that the gain peak shifts to higher energies with increasing injection. Figure B2.1.11 shows the maximum gain as a function of injected carrier density at different temperatures. Note that considerably lower injected carrier density is needed at a lower temperature to achieve the same gain; this is the origin of the lower threshold current at low temperature. It is often convenient to use a linear relationship between the gain g and the injected carrier density n of the form

$$g = a(n - n_0) \tag{B2.1.6}$$

where a is the gain coefficient and n_0 is the injected carrier density at transparency.

B2.1.3.2 Spontaneous emission rate

At unity quantum efficiency, the total spontaneous radiative recombination rate R equals the excitation rate. The latter is usually expressed in terms of the nominal current density J_{n} [20, 22]:

$$J_{\mathrm{n}}[\mathrm{A\ cm^2\ \mu m^{-1}}] = eR \tag{B2.1.7}$$

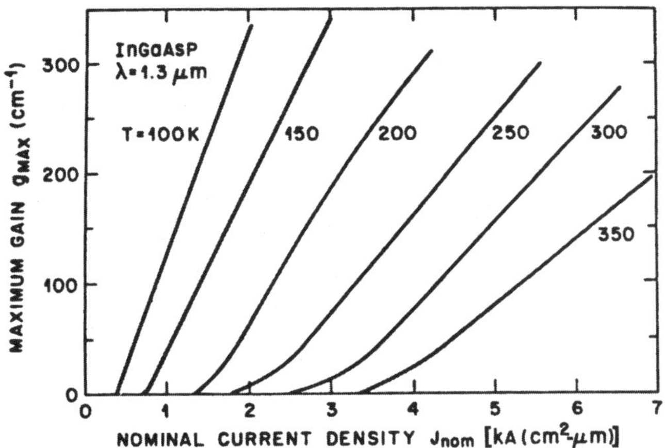

Figure B2.1.12. The maximum gain as a function of nominal current density for InGaAsP ($\lambda = 1.3\ \mu$m) at different temperatures (after Dutta and Nelson [27]).

with $R = \int r_{\mathrm{spon}}(E)\,\mathrm{d}E$ where e is the electron charge and the thickness of the active region is assumed to be 1 μm. $r_{\mathrm{spon}}(E)$ is the spontaneous emission rate at a photon energy E. It is given by [22]

$$r_{\mathrm{spon}}(E) = \frac{4\pi n e^2 E}{m^2 \varepsilon_0 h^2 c^3} \int_{-\infty}^{+\infty} \rho_{\mathrm{c}}(E')\rho_{\mathrm{v}}(E'')|M(E', E'')|^2 f(E')[1 - f(E'')]\,\mathrm{d}E' \qquad \text{(B2.1.8)}$$

The calculated maximum optical gain g as a function of the nominal current density at various temperatures is shown in figure B2.1.12 for InGaAsP ($\lambda \simeq 1.3\ \mu$m). The calculation is for an undoped lightly compensated material with 10^{17} cm^{-3} of acceptors and donors, respectively. Note that gain varies linearly with J_{n} above a certain gain.

The total spontaneous radiative recombination rate R can be approximated by

$$R = Bnp \qquad \text{(B2.1.9)}$$

where B is the radiative recombination coefficient and n, p are the electron and hole densities, respectively. For undoped semiconductors, equation (B2.1.9) becomes $R = Bn^2$. For GaAs, the measured $B = 1 \times 10^{-10}$ cm^3 s^{-1}. Calculation of the radiative recombination rate shows that B decreases with increasing carrier density [20]. This has been confirmed by Olshansky and coworkers [23] using carrier lifetime measurements.

B2.1.3.3 Threshold current calculation

In DH lasers, the injected carriers can recombine by both radiative and non-radiative recombination. The injected current density J is simply the sum of the radiative R and non-radiative R_{nr} recombination rates in the absence of carrier leakage:

$$J = e(R + R_{\mathrm{nr}})d = J_{\mathrm{r}} + J_{\mathrm{nr}} \qquad (J_{\mathrm{r}} = J_{\mathrm{n}}d) \qquad \text{(B2.1.10)}$$

where e is the electron charge, d is the active layer thickness and J_{r}, J_{nr} are the radiative and non-radiative components of the current density, respectively.

For a GaAs DH laser where the quantum efficiency is believed to be close to unity, i.e. most of the injected carriers recombine radiatively, $J \simeq J_r$. The relation between the optical gain and the nominal current density for a GaAs material system has been calculated by Stern [20]. It is given by

$$g = 0.045(J_{\text{nom}} - 4222)\ \text{cm}^{-1} \tag{B2.1.11}$$

where J_{nom} is expressed in amperes per square centimetre per micrometre. The threshold current of a 250-μm-long GaAs–A1GaAs DH laser with 2 μm $\times$ 0.2 μm active region can now be calculated using equations (B2.1.1) and (B2.1.11). The calculated $\Gamma \simeq 0.6$ for a 0.2 μm-thick GaAsAl$_{0.36}$Ga$_{0.64}$As double heterostructure. Assuming $R = 0.3$ and $\alpha_a = \alpha_c = 20\ \text{cm}^{-1}$, we get $g_{\text{th}} = 114\ \text{cm}^{-1}$, $J_{\text{th}} = 1.35\ \text{kA cm}^2$ and a threshold current of 6.7 mA. This compares well with the experimentally observed threshold currents in the range 5–10 mA for GaAs–A1GaAs DH lasers with good current confinement. Figure B2.1.12 shows that the current density needed to achieve a certain optical gain increases with temperature, that is the threshold current density of a laser is expected to increase with increasing temperature. It is experimentally observed that the threshold current density J_{th} of a laser varies with the temperature T as $J_{\text{th}} = J_0 \exp(T/T_0)$, where T_0 is a characteristic temperature determining the temperature sensitivity. For $g = 114\ \text{cm}^{-1}$, the calculated $T_0 \simeq 210$ K from figure B2.1.12 in the temperature range 300–350 K. This agrees well with the measured values for GaAs lasers.

For InGaAsP lasers, a significant amount of non-radiative recombination is believed to be present at carrier densities comparable to the lasing threshold near room temperature. Non-radiative recombination can increase the threshold current density as discussed in the next section.

B2.1.4 Non-radiative recombination

An electron-hole pair can recombine non-radiatively, meaning that the recombination can occur through any process that does not emit a photon. In many semiconductors—for example pure germanium or silicon—non-radiative recombination dominates radiative recombination. The measurable quantities associated with non-radiative recombination are internal quantum efficiency and carrier lifetime. The variation of these quantities with parameters such as temperature, pressure and carrier concentration are, by and large, the only way to identify a particular non-radiative recombination process.

The effect of non-radiative recombination on the performance of injection lasers is to increase the threshold current. If τ_{nr} is the carrier lifetime associated with the non-radiative process, the increase in threshold current density is given approximately by

$$J_{\text{nr}} = \frac{e n_{\text{th}} d}{\tau_{\text{nr}}} \tag{B2.1.12}$$

where n_{th} is the carrier density at threshold, d is the active layer thickness and e is the electron charge. The non-radiative recombination processes described in this section are the Auger effect, surface recombination and recombination at defects or traps.

B2.1.4.1 Auger effect

Since the pioneering work by Beattie and Landsberg [24], it is generally accepted that Auger recombination can be a major non-radiative recombination mechanism in narrow-gap semiconductors. Recent attention to the Auger effect has been in connection with the observed higher temperature of threshold current of long-wavelength InGaAsP lasers compared to short-wavelength AlGaAs lasers [25–29]. It is generally believed that the Auger effect plays a significant role in determining the observed high-temperature sensitivity of the threshold current in InGaAsP lasers emitting near 1.3 and 1.55 μm.

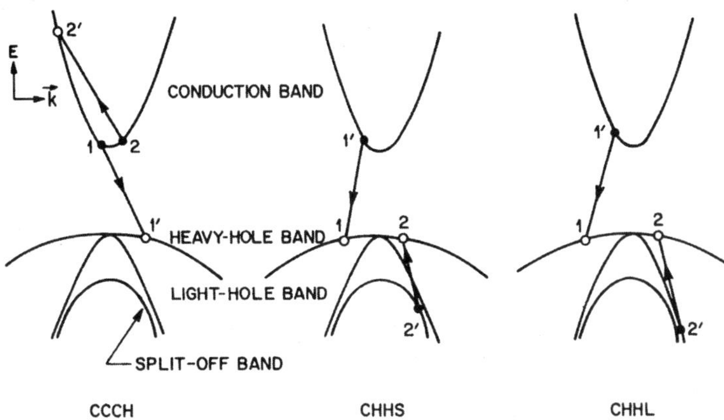

Figure B2.1.13. Band-to-band Auger recombination processes in a direct-gap semiconductor.

There are several types of Auger recombination processes. The three major ones are: the band-to-band process, the phonon-assisted Auger process, and the trap-assisted Auger process.

The band-to-band Auger processes in direct-gap semiconductors are shown in figure B2.1.13. The three processes are labelled CCCH, CHHS and CHHL, where C stands for the conduction band and H, L, S stand for heavy-hole, light-hole and split-off valence-band hole, respectively. The CCCH mechanism involves three electrons and a heavy hole and is dominant in n-type material. The process was first considered by Beattie and Landsberg [24]. The CHHS process involves one electron, two heavy holes and a split-off band hole. The CHHL process is similar to the CHHS process except that it involves a light hole. The CHHS and CHHL mechanisms are dominant in p-type material. Under the high injection conditions present in lasers, all these mechanisms must be considered.

Band-to-band Auger processes are characterized by a strong temperature and bandgap dependence, the Auger rate decreasing rapidly either for low temperature or for high bandgap materials. These dependencies arise from the energy and momentum conservation that the four free-particle states involved ($1, 2, 1', 2'$ in figure B2.1.13) must satisfy. This may be seen in the following way for the CCCH process. The momentum and energy conservation laws give rise to a threshold energy E_{T} for each of the processes. For the CCCH process, if we assume $E_1 \simeq E_2 = 0$, only holes with energies greater than $\sim(E_{\mathrm{T}} - E_{\mathrm{g}}) = \delta E_{\mathrm{g}}$ can participate (δ is a constant that depends on the effective masses). The number of such holes varies approximately as $\exp(-\delta E_{\mathrm{g}}/kT)$ for non-degenerate statistics and, thus, the Auger rate varies as $\exp(-\delta E_{\mathrm{g}}/kT)$ in the non-degenerate case. In the absence of momentum conservation, there is no threshold energy E_{T}. Thus, the strong temperature dependence does not appear if the individual particle states are not states of definite momentum, for example if they are trap states or if momentum conservation is satisfied through phonon assistance. Examples of a CCCH-type phonon-assisted process and a shallow-donor-trap-assisted process are shown in figure B2.1.14.

The Auger rate R_{A} in n-type semiconductors with a carrier concentration n_0 varies as [24]

$$R_{\mathrm{A}} = C n_0^2 \delta p \tag{B2.1.13}$$

where $\delta p \ll n_0$ is the injected minority carrier (hole) density and C is the Auger coefficient. Thus, the minority carrier lifetime is given by

$$\tau_{\mathrm{A}} = \frac{R_{\mathrm{A}}}{\delta p} = \frac{1}{C n_0^2}. \tag{B2.1.14}$$

The calculated τ_A for n-InGaAsP with a carrier concentration 10^{18} cm^{-3} is shown in figure B2.1.14. The phonon-assisted process dominates at high bandgap and the band-to-band processes dominate for low bandgap semiconductors. The active region of an InGaAsP laser is nominally undoped. Under high injection, the Auger rate R_A varies approximately:

$$R_A = Cn^3 \tag{B2.1.15}$$

where n is the injected carrier density. Calculations of the Auger coefficient using the Kane band model [21] yield a value of 10^{-28} cm^6 s^{-1} for $\lambda \simeq 1.3$ μm InGaAsP. Haug [29] has calculated a value of $\sim 2.5 \times 10^{-29}$ cm^6 s^{-1} for $\lambda \approx 1.3$ μm InGaAsP using a different model for the band structure. The experimental values of the Auger coefficient for this material are in the range 2–7 $\times 10^{-29}$ cm^6 s^{-1}. For GaAs, $C \simeq 10^{-31}$ cm^6 s^{-1}. From equations (B2.1.10) and (B2.1.15), the current lost to Auger recombination at threshold is given by

$$J_A = edCn^3. \tag{B2.1.16}$$

Thus, the effect of Auger recombination on the threshold current of GaAs lasers is small compared to that for InGaAsP lasers because the Auger coefficient in GaAs is smaller by two orders of magnitude. Since the threshold carrier density increases with increasing temperature, the carrier loss to the non-radiative Auger effect increases with increasing temperature, which results in a more rapid increase in threshold current with increasing temperature (low T_0) for long-wavelength InGaAsP lasers than for AlGaAs lasers. The measured variation in threshold current density as a function of temperature for InGaAsP lasers can be expressed by the relation $J_{th} = J_0 \exp(T/T_0)$ with $T_0 = 50$–70 K in the temperature range 300–350 K. Auger recombination plays a significant role in determining the smaller T_0 values of InGaAsP ($\lambda \simeq 1.3$ and 1.55 μm) lasers compared to the shorter wavelength ($\lambda \simeq 0.85$ μm) AlGaAs lasers.

B2.1.4.2 Surface recombination

In an injection laser, the cleaved facets are surfaces exposed to the ambient surroundings. In addition, in many index-guided laser structures, the edges of the active region can be in contact with curved surfaces, which may not be a perfect lattice. A surface, in general, is a strong perturbation of the lattice, creating many dangling bonds that can absorb impurities from the ambient. Hence, a high concentration of defects that can act as non-radiative recombination centres can occur. Such localized non-radiative centres, in addition to increasing the threshold current, can cause other performance problems (e.g. sustained oscillations) in lasers.

The recombination rate of carriers at the surface is expressed in terms of a surface recombination velocity S. If A is the surface area and n_{th} the threshold carrier density, then the increase in threshold current, ΔI_{th}, due to surface recombination, is given by

$$\Delta I_{th} = en_{th}SA \tag{B2.1.17}$$

where e is the electron charge. The surface recombination velocity for InP surface exposed to air is about two orders of magnitude smaller than that for GaAs.

B2.1.4.3 Recombination at defects

Defects in the active region of an injection laser can be formed in several ways. In many cases, they are grown in during the epitaxial growth process. They can also be generated multiply or propagated during a stress aging test [30]. Defects can propagate along a specific crystal axis in a strained lattice. The well-known dark line defect (DLD—dark region of linear aspect) is generally believed to be responsible for the high degradation rate (short life span) of early AlGaAs lasers.

Defects, in general, produce a continuum of states in a localized region. Electrons or holes that are within a diffusion length from the edge of the defect may recombine non-radiatively *via* the continuum of

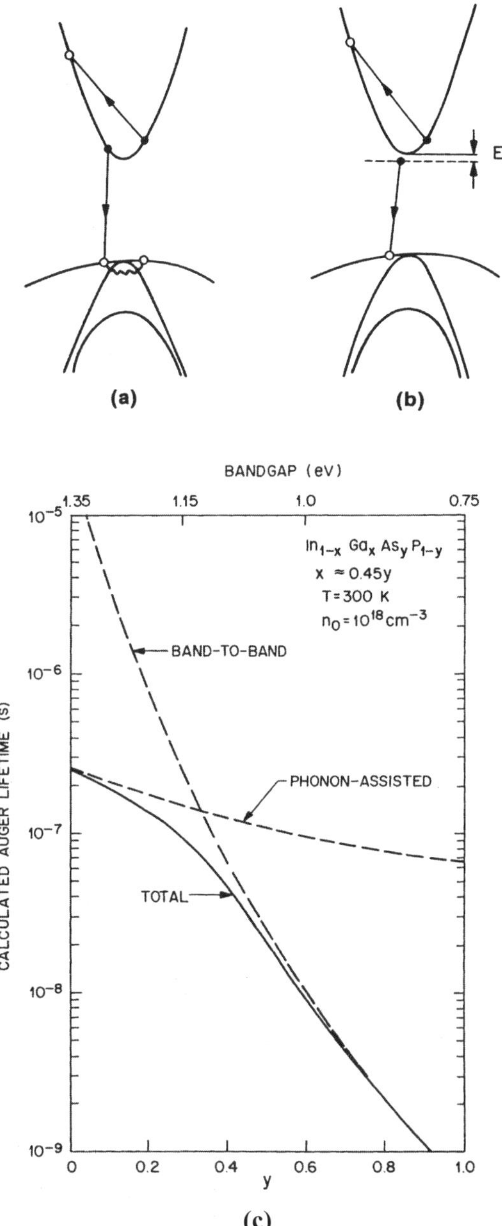

Figure B2.1.14. (*a*) Photon-assisted CCCH-type Auger recombination process. (*b*) Auger recombination process using a donor trap. (*c*) The calculated Auger lifetime for n-InGaAsP with a carrier concentration of 10^{18} cm^{-3} (after Dutta and Nelson [27]).

states. The rate of recombination at a defect or trap is usually written as

$$R = \delta \upsilon N_{\mathrm{t}} \tag{B2.1.18}$$

where δ is the capture cross section of the trap, N_t is the density of traps and υ is the velocity of the electrons or holes. A trap can preferentially capture electrons or holes. The study of recombination at defects in semiconductors is a vast subject—a detailed discussion is beyond the scope of this chapter.

B2.1.5 Laser designs and performance

A schematic diagram of a typical DH used for laser fabrication was shown in figure B2.1.1. It consists of n-InP, undoped $In_{1-x}Ga_xP_yAs_{1-y}$, p-InP and p-InGaAsP grown over a (100) oriented n-InP substrate. The undoped $In_{1-x}Ga_xP_yAs_{1-y}$ layer is the light-emitting layer (active layer). It is lattice matched to InP for $x \sim 0.45y$. The bandgap of the $In_{1-x}Ga_xP_yAs_{1-y}$ material (lattice matched to InP) which determines the laser wavelength is given by

$$E_g(eV) = 1.35 - 0.72y + 0.12y^2. \tag{B2.1.19}$$

B2.1.5.1 Laser designs

For lasers emitting near 1.3 μm, $y \simeq 0.6$. The DH material can be grown by the LPE, gas source molecular beam epitaxy (GSMBE) or MOCVD growth technique. The DH material can be processed to produce lasers in several ways. Perhaps the simplest is the broad area laser (figure B2.1.1) which involves putting contacts on the p- and n-sides and then cleaving. Such lasers do not have transverse-mode confinement or current confinement which leads to high threshold and nonlinearities in light *versus* current characteristics. Several laser designs have been developed to address these problems. Among them are the gain-guided laser, the weakly-index-guided laser and the buried heterostructure (strongly-index-guided) laser. Typical versions of these laser structures are shown in figure B2.1.15. The gain-guided structure uses a dielectric layer for current confinement. The current is injected in the opening in the dielectric (typically 6–12 μm wide), which produces gain in that region and, hence, the lasing mode is confined to that region. The weakly-index-guided structure has a ridge etched on the wafer and a dielectric layer surrounds the ridge. The current is injected in the region of the ridge and the optical mode overlaps the dielectric (which has a low index) in the ridge. This results in weak index guiding.

The buried heterostructure (BH) design shown in figure B2.1.15 has the active region surrounded (buried) by lower index layers. The fabrication process for a DCPBH (double channel planar buried heterostructure) laser involves growing a double heterostructure, etching a mesa using a dielectric mask and then regrowing the layer surrounding the active region using a second epitaxial growth step. The second growth can be a single Fe-doped InP (Fe:InP) semi-insulating layer or a combination of p-InP, n-InP and Fe:InP layers. Generally the MOCVD growth process is used for the growth of the regrown layer. Researchers have often given different names to the particular BH laser design that they have developed. For the structure in figure B2.1.16, the Fe-doped InP layer provides both optical confinement to the lasing mode and current confinement to the active region. BH lasers are generally used in communication system applications because a properly designed strongly-index guided BH design has superior mode stability, higher bandwidth and superior linearity in light *versus* current (*L versus I*)characteristics compared to the gain and weakly index guided designs. Early recognition of these important requirements for communication grade lasers led to intensive research on InGaAsP BH laser designs all over the world in the 1980s. It is worth mentioning that BH lasers are more complex and difficult to fabricate compared to the gain and weakly index guided lasers.

A scanning electron micrograph of a capped mesa buried heterostructure (CMBH) laser along with the laser structure is shown in figure B2.1.16. The current-blocking layers in this structure consist of i-InP (Fe-doped InP), n-InP , i-InP and n-InP layers. These sets of blocking layers have the lowest capacitance and are, therefore, needed for high-speed operation. An optimization of the thickness of these layers is needed for highest-speed performance. The laser fabrication involves the following steps. Mesas 1 μm wide are etched on the wafer and the current-blocking layers consisting of i-InP, n-InP i-InP and n-InP layers are grown on

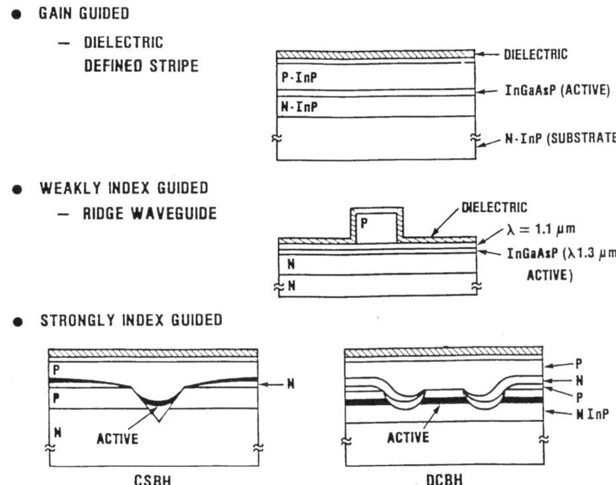

Figure B2.1.15. Schematic diagram of various laser designs using the InGaAsP/InP material system.

the wafer with an oxide layer on top of the mesa in place. The oxide layer is then removed and a third growth of a p-InP cladding layer and a p-InGaAs contact layer is carried out. The wafer is then processed using standard lithography, metallization and cleaving techniques to produce the lasers.

B2.1.5.2 Laser performance

The light *versus* current characteristics at different temperatures of an InGaAsP BH laser emitting at 1.3 μm is shown in figure B2.1.17. A typical threshold current for a BH laser at room temperature is in the 5–10 mA range. For gain and weakly index guided lasers, typical room temperature threshold currents are in the 25–50 and 50–100 mA range, respectively. The external differential quantum efficiency defined as the derivative of the L *versus* I characteristics above threshold is $\sim$0.25 mW mA^{-1} facet^{-1} for a cleaved uncoated laser emitting near 1.3 μm.

An important characteristic of the semiconductor laser is that its output can be modulated easily and simply by modulating the injection current. The relative magnitude of the modulated light output is plotted as a function of the modulation frequency of the current in figure B2.1.18 at different optical output powers. The laser is of the BH type (shown in figure B2.1.15), has a cavity length of 250 μm and the modulation current amplitude was 5 mA. Note that the 3 dB frequency to which the laser can be modulated increases with increasing output power and the modulation response is at a maximum at a certain frequency (ω_r). The resonance frequency, ω_r, is proportional to the square root of the optical power. The modulation response determines the data transmission rate capability of the laser, for example, for 10 Gb s^{-1} data transmission, the 3-dB bandwidth of the laser must exceed 10 GHz. However, other system level considerations, such as allowable error-rate penalty, often introduce much more stringent requirements on the exact modulation response of the laser.

A semiconductor laser with cleaved facets generally emits in a few longitudinal modes of the cavity. A typical spectrum of a laser with cleaved facets is shown in figure B2.1.19. The discrete emission wavelengths are separated by the longitudinal cavity mode spacing which is $\sim$10 A for a laser ($\lambda \simeq 1.3$ μm) with 250 μm

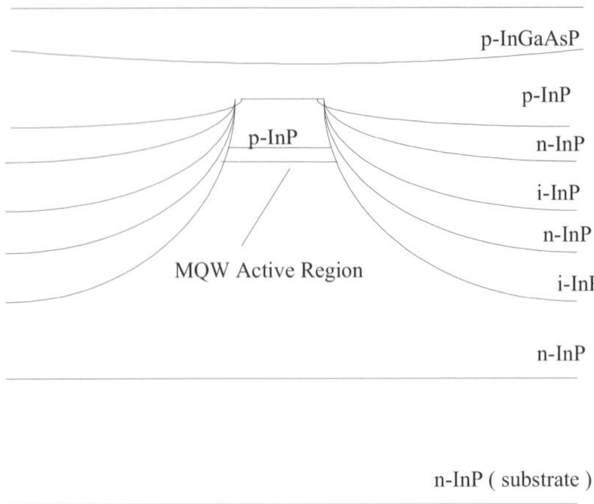

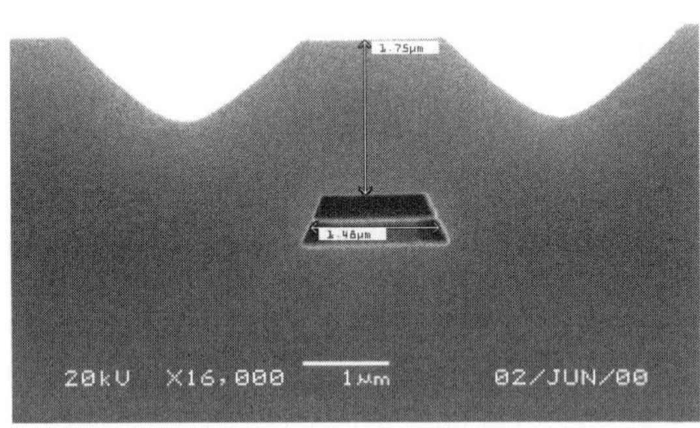

Figure B2.1.16. Schematic diagram and scanning electron photomicrograph of a buried heterostructure (BH) laser.

cavity length. Lasers can be made to emit in a single frequency using frequency-selective feedback, for example using a grating internal to the laser cavity as described later.

The active region waveguide dimensions of a semiconductor laser are typically comparable to or smaller than the wavelength of emission. This results in diffraction of the light as it emerges from the facet. The diffraction leads to a divergence of the emitted light. The magnitude of the divergence depends on the waveguide dimensions, index of refraction and wavelength of emission. The measured far-field patterns parallel and perpendicular to the junction plane at different output powers of a typical laser are shown in figure B2.1.20. The figure shows that the laser operates in the fundamental transverse mode in the entire operating power range from threshold to 60 mW. The full width at half maximum of the beam divergences parallel and normal to the junction plane are 40° and 30° respectively.

B2.1.6 Quantum well lasers

As shown in figure B2.1.1, a conventional DH laser consists of an active layer sandwiched between higher-gap cladding layers. The active layer thickness is typically 0.1–0.5 μm. Over the last decade, DH lasers have been fabricated with an active layer thickness of ~100 Å or less. The carrier (electron or hole) motion

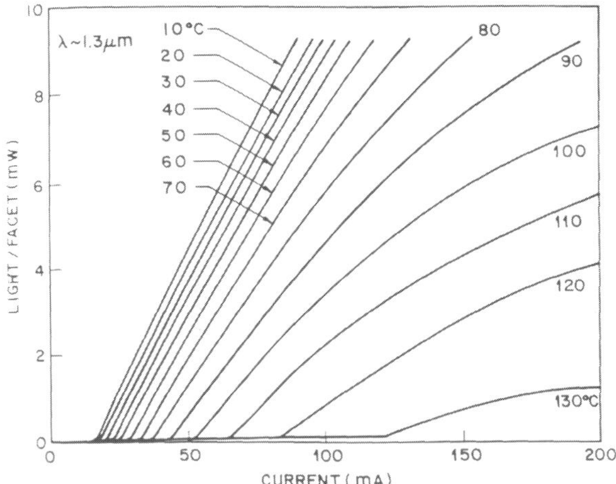

Figure B2.1.17. Light *versus* current characteristics at different temperatures of a typical buried heterostructure (BH) laser.

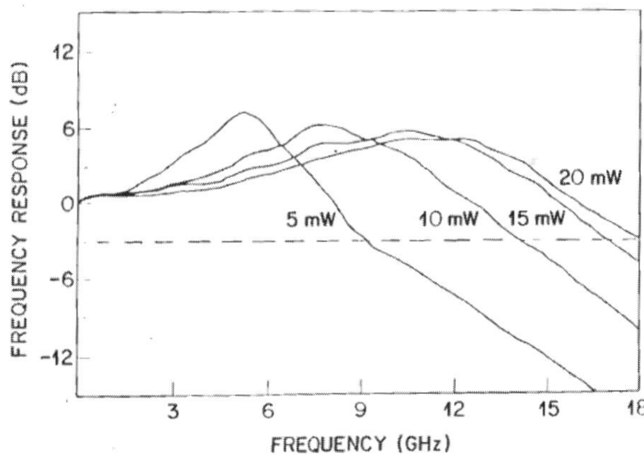

Figure B2.1.18. Frequency response, i.e. modulated, output power as a function of modulation frequency at different output powers.

normal to the active layer in these structures is restricted. This may be viewed as carrier confinement in a one-dimensional potential well and, hence, these lasers are called quantum well lasers [31, 32].

B2.1.6.1 Energy levels

When the thickness L of a narrow-gap semiconductor layer confined between two wide-gap semiconductors becomes comparable to the de Broglie wavelength ($\lambda \simeq h/p \simeq L_z$), quantum-mechanical effects are expected to occur. The energy levels of the carriers confined in the narrow-gap semiconductor can be determined by separating the Hamiltonian into a component normal to the layer (z component) and into the usual (unconfined) Bloch function components (x, y) in the plane of the layer. The resulting energy eigenvalues are:

$$E(n, k_x, k_y) = E_n + \frac{\hbar^2}{2m_n^*}(k_x^2 + k_y^2) \qquad \text{(B2.1.20)}$$

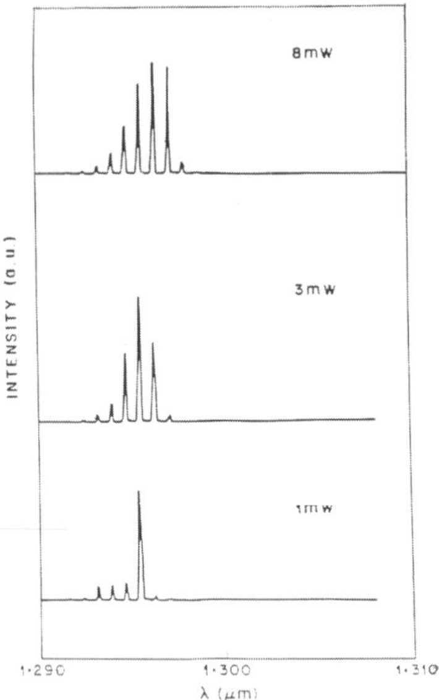

Figure B2.1.19. Typical Fabry–Pérot laser spectrum at different output powers.

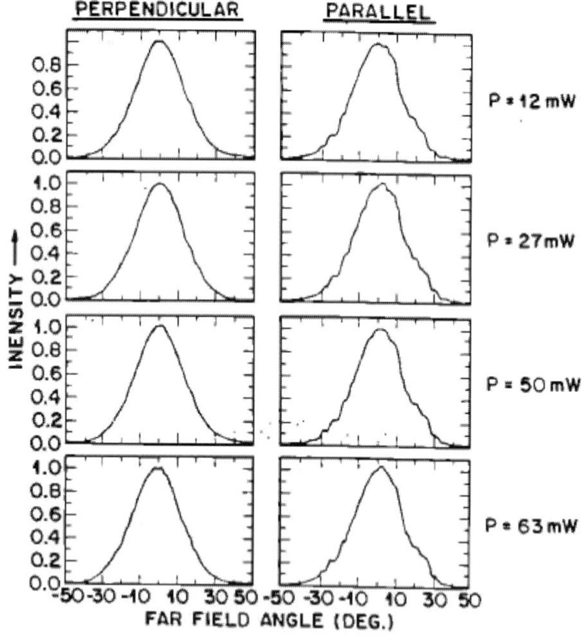

Figure B2.1.20. Typical beam divergence characteristics of a semiconductor laser parallel and perpendicular to the junction plane.

where E_n is the n_{th} confined-particle energy level for carrier motion normal to the well. m_n^* is the effective mass of the n_{th} level, $\hbar$ is Planck's constant and k_x, k_y are the usual Bloch wavevectors in the x and y

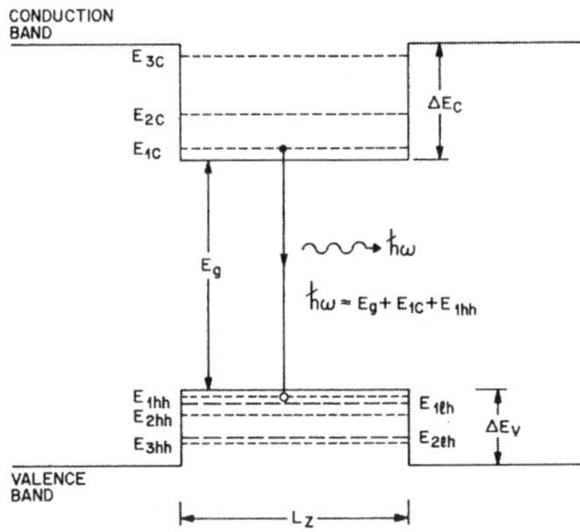

Figure B2.1.21. Energy levels in a quantum well structure (after Holonyak *et al* [32]).

directions. Figure B2.1.21 shows schematically the energy levels E_n of the electrons and holes confined in a quantum well. The confined particle energy levels E_n are denoted by E_{1c}, E_{2c}, E_{3c} for electrons; E_{1hh}, E_{2hh} for heavy holes; and E_{1lh}, E_{2lh}, for light holes. The calculation of these quantities is a standard problem in quantum mechanics for a given potential barrier (ΔE_c, ΔE_v). For an infinite potential well, the following simple result is obtained:

$$E_n = \frac{h^2 n^2}{8L_z^2 m_n^*}. \tag{B2.1.21}$$

Since the separation between the lowest conduction-band level and the highest valence-band level is given

$$E_q = E_g + E_{1c} + E_{1hh} \approx E_g + \frac{h^2}{8L_z^2}\left(\frac{1}{m_c} + \frac{1}{m_h}\right) \tag{B2.1.22}$$

it follows that in a quantum well structure, the energy of the emitted photons can be varied simply by varying the well width L_z. For the realistic case of a finite well, the Schrödinger equation for a finite well needs to be solved to get more accurate values for the sub-band energy levels in the (conduction and valence band) and, hence, the emission energy in a quantum well [36]. The infinite potential model overestimates the value of E_n in equation (B2.1.22). Figure B2.1.22 shows the experimental results [37] for an InGaAs quantum well bounded by InP cladding layers.

The discrete energy levels result in a modification of the density of states in a quantum well to a 'two-dimensional-like' density of states. This modification of the density of states results in several improvements in the laser characteristics such as lower threshold current, higher efficiency, higher modulation bandwidth and lower cw and dynamic spectral width. All of these improvements were first predicted theoretically and then demonstrated experimentally [23–34].

B2.1.6.2 Quantum well laser performance

The development of InGaAsP quantum well (QW) lasers were made possible by the development of MOCVD and GSMBE growth techniques. The transmission electron micrograph of a multiple QW laser structure is shown in figure B2.1.23. Shown are four InGaAs quantum wells grown over an n-InP substrate. The well thickness is 70 Å and they are separated by barrier layers of InGaAsP ($\lambda \simeq 1.1$ μm). Multi-quantum well (MQW) lasers with threshold current densities of 600 A cm^{-2} have been fabricated [33]. MQW BH lasers have also been fabricated. The composition of the InGaAsP material from the barrier layers to the cladding

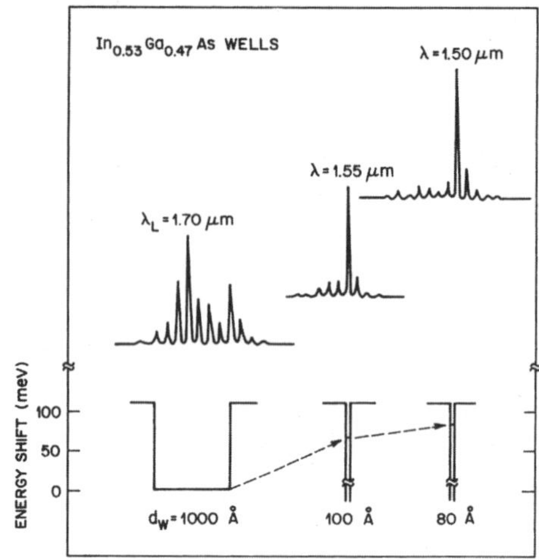

Figure B2.1.22. The laser spectrum for different well widths for InGaAs single quantum well structures.

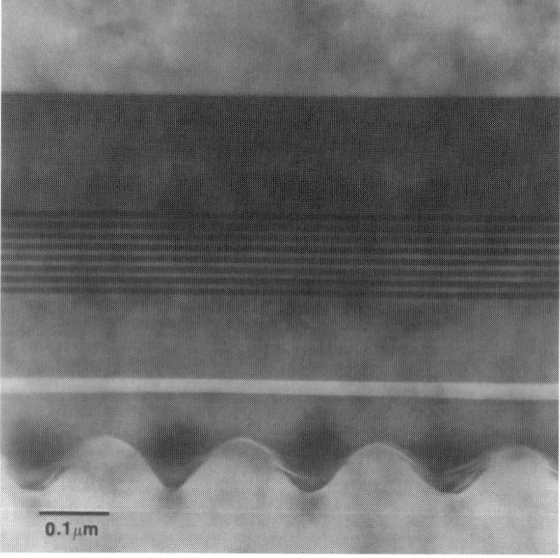

Figure B2.1.23. Transmission electron micrograph of a quantum well structure fabricated over a substrate with a grating.

layer (InP) is gradually varied in this structure over a thickness of $\sim 0.1\ \mu$m. This produces a graded variation in index (GRIN structure) which results in a higher optical confinement of the fundamental mode than that for an abrupt interface design. The larger mode confinement factor results in a lower threshold current. Many BH lasers utilize an MQW active region and Fe doped semi-insulating (SI) InP layers for current confinement and optical confinement. The light *versus* current characteristics of an MQW BH laser are generally similar to those of other BH lasers except for a lower threshold current. Also, the two-dimensional-like density of states of the QW lasers makes the transparency current density of these lasers significantly lower than that for regular DH lasers [30]. This allows the fabrication of very low threshold lasers using high-reflectivity coatings.

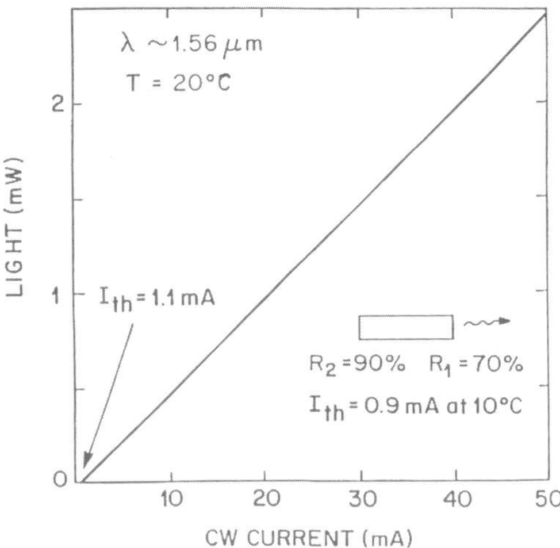

Figure B2.1.24. Light *versus* current characteristic of a low-threshold quantum well laser fabricated using high-reflectivity coatings.

The optical gain (g) of a laser at a current density J is given by

$$g = a(J - J_0) \tag{B2.1.23}$$

where a is the gain constant and J_0 is the transparency current density. Although a logarithmic dependence of gain on current density [35] is often used in order to account for gain saturation, a linear dependence is used here for simplicity. The logarithmic relationship between gain (g) and current density (J) for decoupled quantum wells is given by

$$g = N\Gamma B \ln(J/J_0) \tag{B2.1.23a}$$

where N is the number of quantum wells in the active region, Γ is the coupling factor, B is a constant and J_0 is the current needed for transparency.

The cavity loss α is given by

$$\alpha = \alpha_c + (1/L)\ln(1/R_1R_2) \tag{B2.1.24}$$

where α_c is the free carrier loss, L is the length of the optical cavity and R_1, R_2 are the reflectivities of the two facets. At threshold, gain equals loss; hence, it follows from (B2.1.23) and (B2.1.24) that the threshold current density (J_{th}) is given by

$$J_{th} = \alpha_c/a + (1/La)\ln(1/R_1R_2) + J_0. \tag{B2.1.25}$$

Thus for a laser with high-reflectivity facet coatings ($R_1, R_2 \simeq 1$) and with low loss ($\alpha_c \simeq 0$), $J_{th} \simeq J_0$. For a QW laser, $J_0 \simeq 50$ A cm^{-2} and for a DH laser, $J_0 \simeq 700$ A cm^{-2}; hence, it is possible to get a much lower threshold current using QWs as the active region.

The light *versus* current characteristic of a QW laser with high-reflectivity coatings on both facets is shown in figure B2.1.24 [36]. The threshold current at room temperature is ~1.1 mA.

The laser is 170 μm long and has 90% and 70% reflective coating at the facets. This laser has a compressively strained MQW active region. For lattice-matched MQW active regions, a threshold current of 2 mA has been reported [37]. Such low-threshold lasers are important for array applications. Recently QW lasers have been fabricated which have a higher modulation bandwidth than regular DH laser. Current and

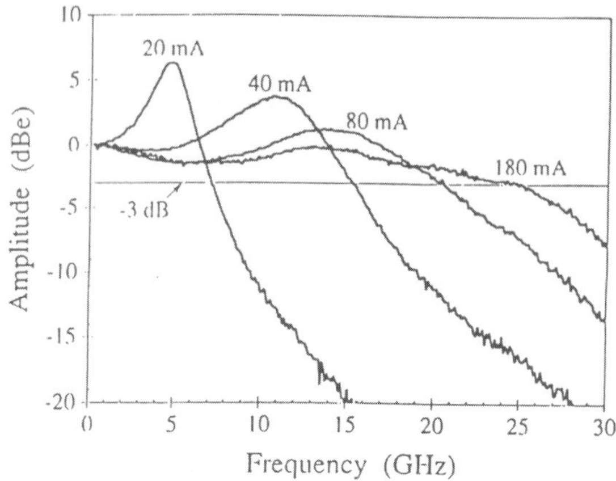

Figure B2.1.25. Modulation response of a quantum well buried heterostructure laser fabricated using Fe:InP current-blocking layers.

optical confinement in this laser is carried out using MOCVD-grown Fe-doped InP lasers. The laser structure is then further modified by using a small contact pad and etching channels around the active region mesa. These modifications are designed to reduce the capacitance of the laser structure. The modulation response of the laser is shown in figure B2.1.25. A 3-dB bandwidth of 25 GHz is obtained [38].

B2.1.6.3 *Strained quantum well lasers*

QW lasers have also been fabricated using an active layer whose lattice constant differs slightly from that of the substrate and cladding layers. Such lasers are known as strained QW lasers. Over the last few years, strained QW lasers have been extensively investigated all over the world [39–46]. They show many desirable properties such as (i) a very low threshold current density and (ii) a lower linewidth than regular MQW lasers, both under cw operation and under modulation. The origin of the improved device performance lies in the band-structure changes induced by the mismatch-induced strain [47, 48]. Strain splits the heavy- and light-hole valence bands at the Γ-point of the Brillouin zone where the bandgap is minimum in direct bandgap semiconductors.

Two material systems have been widely used for strained QW lasers: (i) InGaAs grown over InP by the MOCVD or CBE growth technique [39–43] and (ii) InGaAs grown over GaAs by the MOCVD or MBE growth technique [44–46]. The former material system is of importance for low-chirp semiconductor lasers for lightwave system applications, while the latter has been used to fabricate high-power lasers emitting near 0.98 μm, a wavelength of interest for pumping erbium-doped fibre amplifiers.

The alloy $In_{0.53}Ga_{0.47}As$ has the same lattice constant as InP. Semiconductor lasers with an $In_{0.53}Ga_{0.47}As$ active region have been grown on InP by the MOCVD growth technique. Excellent material quality is also obtained for $In_{1-x}Ga_xAs$ alloys grown over InP by MOCVD for non-lattice-matched compositions. In this case, the laser structure generally consists of one or many $In_{1-x}Ga_xAs$ QW layers with InGaAsP barrier layers whose composition is lattice matched to that of InP. For $x < 0.53$, the active layer in these lasers is under tensile stress, while for $x > 0.53$, the active layer is under compressive stress.

BH lasers have been fabricated using compressive- and tensile-strained MQW lasers. Lasers with compressive strain have a lower threshold current than that for lasers with tensile strain [49]. This can be explained by the light and heavy-hole bands splitting under stress [50, 51].

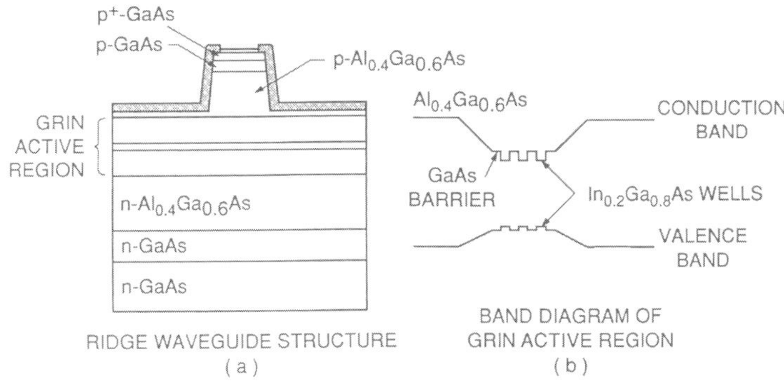

Figure B2.1.26. Schematic diagram of a ridge waveguide-type InGaAs/GaAs laser.

Strained QW lasers fabricated using $In_{1-x}Ga_xAs$ layers grown over a GaAs substrate have been extensively studied [44, 46, 52–57]. The lattice constant of InAs is 6.06 Å and that of GaAs is 5.654 Å. The $In_{1-x}Ga_xAs$ alloy has a lattice constant between these two values and, to a first approximation, can be assumed to vary linearly with x. Thus, an increase in the In mole fraction x increases the lattice mismatch relative to the GaAs substrate and, therefore, produces larger compressive strain on the active region.

A typical laser structure grown over the n-type GaAs substrate is shown in figure B2.1.26 [44] for this material system. It consists of a MQW active region with one to four $In_{1-x}Ga_xAs$ wells separated by GaAs barrier layers. The entire MQW structure is sandwiched between n- and p-type $Al_{0.3}Ga_{0.7}As$ cladding layers, and the p-cladding layer is followed by a p-type GaAs contact layer. Variations of this structure with different cladding layers or large optical cavity designs have been reported. Emission wavelength depends on the In composition, x. As x increases, the emission wavelength increases and, for x larger than a certain value (typically $\sim$0.25), the strain is too large to yield high-quality material. For $x \simeq 0.2$, the emission wavelength is near 0.98 μm, a wavelength region of interest for pumping fibre amplifiers [49]. Threshold current densities as low as 47 A cm^{-2} have been reported for $In_{0.2}Ga_{0.8}As$/GaAs strained MQW lasers [55]. High-power lasers have been fabricated using an $In_{0.2}Ga_{0.8}As$/GaAs MQW active region. Single-mode output powers of greater than 400 mW have been demonstrated using a ridge-waveguide-type laser structure.

B2.1.7 Distributed feedback lasers

Semiconductor lasers fabricated using the InGaAsP material system are widely used as sources in many lightwave transmission systems. One measure of the transmission capacity of a system is the data rate. Thus the drive towards higher capacity pushes systems to higher data rates where the chromatic dispersion of the fibre plays an important role in limiting the distance between regenerators. Sources emitting in a single wavelength help reduce the effects of chromatic dispersion and are, therefore, used in most systems operating at high data rates (>1.5 Gb s^{-1}).

The single-wavelength laser source used in most commercial transmission systems is the distributed feedback (DFB) laser where a diffraction grating etched on the substrate close to the active region provides frequency-selective feedback which makes the laser emit in a single wavelength. This section reports on the fabrication, performance characteristics and reliability of DFB lasers [58].

B2.1.7.1 Distributed feedback laser design and fabrication

A schematic diagram of a DFB laser structure is shown in figure B2.1.27. The fabrication of the device involves the following steps. First, a grating with a periodicity of 2400 Å is fabricated on a (100) oriented

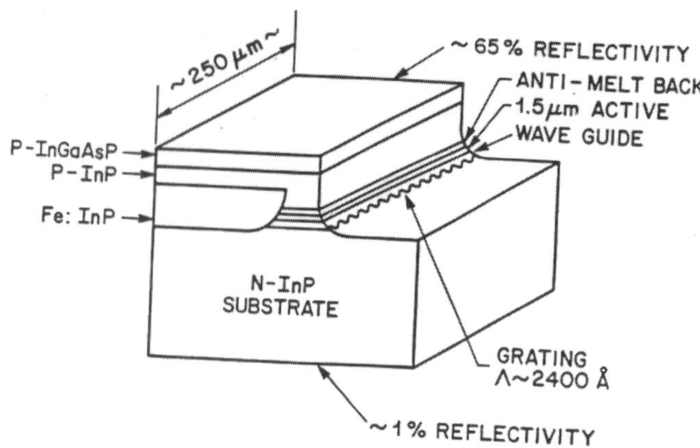

Figure B2.1.27. Schematic diagram of a distributed feedback (DFB) buried-heterostructure laser.

n-InP substrate using optical holography and wet chemical etching. Four layers are then grown over the substrate. These layers are: (i) an n-InGaAsP ($\lambda \simeq 1.3$ μm) waveguide layer, (ii) an undoped InGaAsP ($\lambda \simeq 1.55$ μm) active layer, (iii) a p-InP cladding layer and (iv) a p-InGaAsP ($\lambda \simeq 1.3$ μm) contact layer. Mesas are then etched on the wafer using a SiO_2 mask and wet chemical etching. Fe doped InP semi-insulating layers are grown around the mesas using the MOCVD growth technique. The semi-insulating layers help confine the current to the active region and also provide index guiding to the optical mode. The SiO_2 stripe is then removed and the p-InP cladding layer and a p-InGaAsP contact layer are grown on the wafer using VPE. The wafer is then processed to produce 250 μm long laser chips using standard metallization and cleaving procedures. The final laser chips have anti-reflection coating ($<1\%$) at one facet and high-reflectivity coating ($\sim 65\%$) at the back facet. The asymmetric facet coatings help remove the degeneracy between the two modes in the stop band.

B2.1.7.2 Laser performance

The cw light *versus* current characteristics of a laser are shown in figure B2.1.28. Also shown is the measured spectrum at different output powers. The laser emits near 1.3 μm. The threshold current of these lasers is in the 10–15 mA range. For high fibre coupling efficiency, it is important that the laser emits in the fundamental transverse mode.

Tunable single-wavelength semiconductor lasers are needed for many applications. Examples of application in lightwave transmission systems are: (i) wavelength-division multiplexing (WDM) where signals at many distinct wavelengths are simultaneously modulated and transmitted through a fibre; and (ii) coherent transmission systems where the wavelength of the transmitted signal must match that of the local oscillator. Several types of tunable laser structures have been reported in the literature [59–64]. The two principal schemes are (i) the multi-section DFB laser and (ii) the multi-section distributed Bragg reflector (DBR) laser. The multi-section DBR lasers generally exhibit higher tunability than that for the multi-section DFB lasers. The three sections of a multi-section device are: (i) the active region section that provides the gain, (ii) the grating section that provides the tunability and (iii) the phase-tuning section that is needed to access all wavelengths continuously. The current through each of these sections can be varied independently. The tuning mechanism can be understood by noting that the emission wavelength, λ, of a DBR laser is given by $\lambda = 2n\Lambda$ where Λ is the grating period and n is the effective refractive index of the optical mode in the grating section. The latter can be changed simply by varying the current in the grating section.

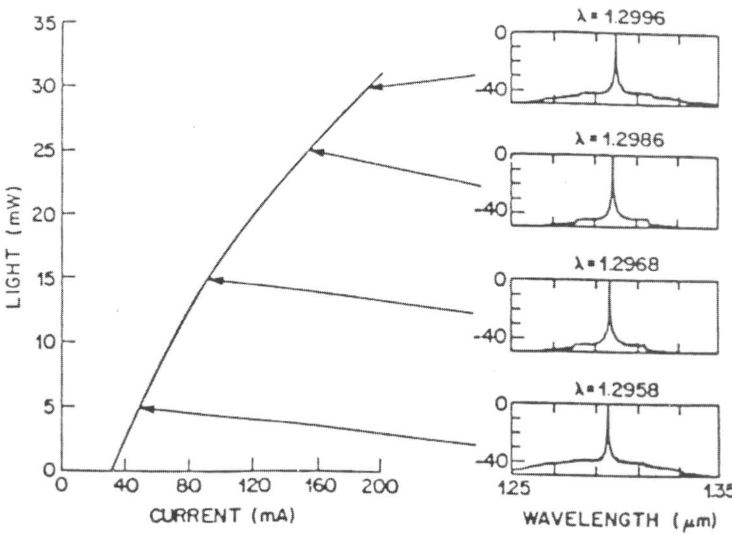

Figure B2.1.28. Light *versus* current characteristics and spectrum at different powers of a distributed feedback laser.

B2.1.7.3 Phase noise and intensity noise

The coupling of the spontaneous emission to the stimulated emission in a semiconductor laser leads to phase and intensity noise in the laser output. The phase noise manifests itself as a finite spectral width of the laser even for cw operation. Low phase noise is needed for lasers used in coherent transmission.

The intensity noise arises from the coupling of the electrons in the semiconductor and the photons in the electric field. The intensity noise is generally characterized by the relative intensity noise (RIN). The RIN value must be sufficiently small if the laser is used in analogue or subcarrier-based transmission systems. Typical RIN values for DFB lasers are in the range of -150 to -160 dB Hz^{-1}.

An important characteristic of lasers for applications requiring a high degree of coherence is the spectral width (linewidth) under cw operation. The cw linewidth depends on the rate of spontaneous emission in the laser cavity. For coherent transmission applications, the cw linewidth must be quite small. The minimum linewidth allowed depends on the modulation format used. For differential phase-shift keying (DPSK) transmission, the minimum linewidth is approximately given by $B/300$ where B is the bit-rate. The linewidth varies inversely with the output power at low powers (<10 mW) and shows saturation at high powers. Typical cw linewidths are in the 1–10 MHz range, the lowest reported value being ~200 kHz.

B2.1.8 Surface emitting lasers

Semiconductor lasers described in the previous sections have cleaved facets that form the optical cavity. The facets are perpendicular to the surface of the wafer and light is emitted parallel to the surface of the wafer. For many applications requiring a two-dimensional laser array or monolithic integration of lasers with electronic components (e.g. optical interconnects), it is desirable to have the laser output normal to the surface of the wafer. Such lasers are known as surface-emitting lasers (SEL). A class of surface-emitting lasers also have their optical cavity normal to the surface of the wafer [65–72]. These devices are known as vertical-cavity surface-emitting lasers (VCSEL) (see chapter B2.6) in order to distinguish them from other surface emitters.

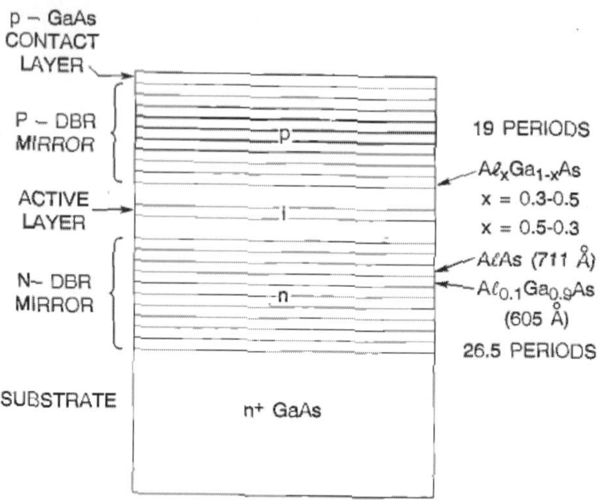

Figure B2.1.29. Schematic diagram of the layers of a AlGaAs/GaAs surface-emitting laser

B2.1.8.1 Surface emitting laser designs

A generic SEL structure utilizing multiple semiconductor layers to form a Bragg reflector is shown in figure 29. The active region is sandwiched between n- and p-type cladding layers which are themselves sandwiched between the two n- and p-type Bragg mirrors. This structure is shown using the AlGaAs/GaAs material system which has been very successful in the fabrication of SELs. The Bragg mirrors consist of alternating layers of low-index and high-index materials. The thickness of each layer is one quarter of the wavelength of light in the medium. Such periodic quarter-wave-thick layers can have very high reflectivity. For normal incidence, the reflectivity is given by [73].

$$R = \frac{(1 - n_4/n_1(n_2/n_3)^{2N})^2}{(1 + n_4/n_1(n_2/n_3)^{2N})^2} \tag{B2.1.26}$$

where n_2, n_3 are the refractive indices of the alternating layer pairs, n_4, n_1 are the refractive indices of the medium on the transmitted and incident sides of the DBR mirror and N is the number of pairs. As N increases, R increases. Also for a given N, R is larger if the ratio of n_2/n_3 is smaller. For a AlAs/$Al_{0.1}Ga_{0.9}As$ set of quarter-wave layers, typically 20 pairs are needed for a reflectivity of ~99.5%. Various types of AlGaAs/GaAs SELs have been reported [74–82].

For an SEL to have a threshold current density comparable to that of an edge-emitting laser, the threshold gains must be comparable for the two devices. The threshold gain of an edge-emitting laser is $\sim 100\ cm^{-1}$. For an SEL with an active-layer thickness of 0.1 μm, this value corresponds to a single-pass gain of ~1%. Thus, for the SEL device to operate with a threshold current density comparable to that of an edge emitter, the mirror reflectivities must be >99%. The reflectivity spectrum of an SEL structure is shown in figure B2.1.30. The reflectivity is >99% over a 10nm band. The drop in reflectivity in the middle of the band is due to the Fabry–Pérot mode.

The number of pairs needed to fabricate a high-reflectivity mirror depends on the refractive index of layers in the pair. For large-index differences, fewer pairs are needed. For example, in the case of CaF_2 and ZnS for which the index difference is 0.9, only six pairs are needed for a reflectivity of 99%. By contrast, for a InP/InGaAsP ($\lambda \simeq 1.3\ \mu$m) layer pair for which the index difference is 0.3, more than 40 pairs are needed to achieve a reflectivity of 99%.

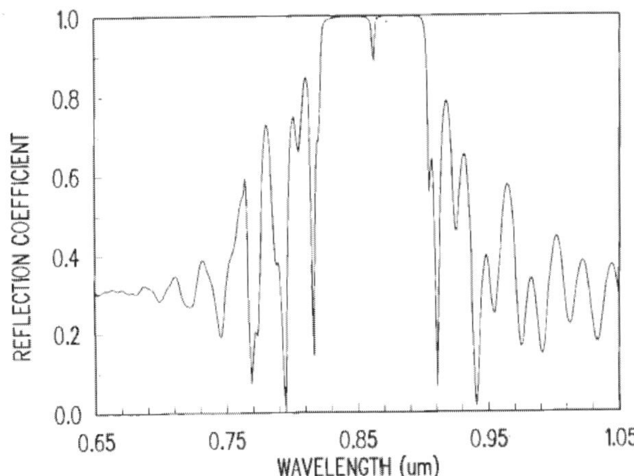

Figure B2.1.30. Reflection spectrum of a typical surface-emitting laser structure.

An important SEL structure for the AlGaAs/GaAs material system is the oxide aperture device [81] (figure B2.1.31). AlAs has the property that it oxidizes rapidly in the presence of oxygen to aluminium oxide, which forms an insulating layer. Thus, by introducing a thin AlAs layer in the device structure, it is possible to confine the current to a very small area. This allows the fabrication of very low threshold (<0.2 mA) and high bandwidth ($\sim$14 GHz) lasers.

Central to the fabrication of low-threshold SELs is the ability to fabricate high-reflectivity mirrors. In the late 1970s, Soda *et al* [83] reported on a SEL fabricated using the InP material system. The surfaces of the wafer form the Fabry–Pérot cavity of the laser. Fabrication of the device involves the growth of a DH on an n-InP substrate. A circular contact is made on the p-side using an SiO_2 mask. The substrate side is polished making sure that it is parallel to the epitaxial layer and ring electrodes (using an alloy of Au-Sn) are deposited on the n-side. The laser had a threshold current density of $\sim$11 kA cm^{-2} at 77 K and operated at output powers of several milliwatts.

B2.1.9 Visible and quantum cascade lasers

B2.1.9.1 Visible lasers

For visible and shorter wavelength operation, several material systems have been studied. Both III–V and II–VI-based material have been studied.

Most of the initial work on visible semiconductor lasers were based on the InAlGaP system. The typical laser design consists of an InGaP active region bounded by n- and p-type InAlGaP cladding layers and a p-GaAs contact layer. The entire structure is grown over a GaAs substrate. The typical emission wavelength is $\sim$630–670 nm [84]. Both gain and index guided lasers have been reported. The shortest wavelength that can be reached using this system is $\sim$600 nm at room temperature (see chapter B2.5.1).

Shorter wavelength operation using II–VI based materials such as Zns, ZnSe, ZnSSe and CdZnSe [85,86] has been reported. A typical laser design may consist of a CdZnSe active region bounded by n-ZnSe, n-ZnSSe and p-ZnSe, p-ZnSSe layers grown over an n-GaAs substrate. Laser operation at $\sim$490 nm has been observed. The reliability of II–VI lasers needs to be improved considerably.

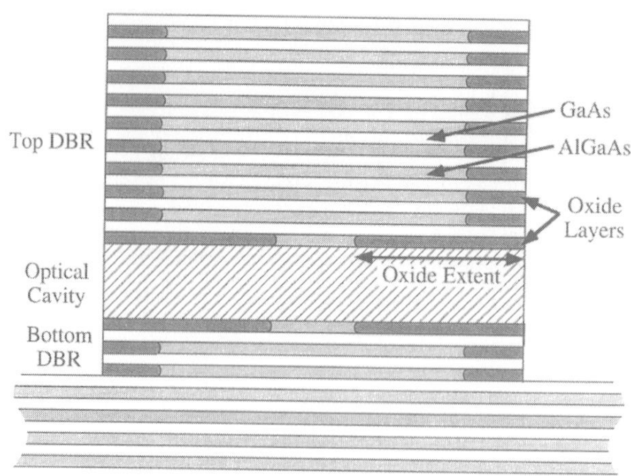

Figure B2.1.31. Schematic diagram of a oxide aperture AlGaAs/GaAs surface-emitting laser structure.

Nitride-based (GaN, AlGaN, InGaN, AlInGaN) laser technology has been developed extensively over the last seven years [87, 88]. Blue lasers and LEDS have been realized using a strained InGaN active region. The lasers are grown on an n-SiC substrate and have n- and p-AlGaN cladding layers (see chapter B2.5.2).

B2.1.9.2 Quantum cascade lasers

Quantum cascade lasers (QCLs), which emit in the infrared (see chapter B2.7), are based on intraband transitions in a quantum well. QCLs were first demonstrated by Faist and Capasso at Bell Laboratories in 1994 [89]. The operation of the QCL is illustrated with the aid of figure B2.1.32. The QCL relies upon unipolar injection of electrons into quantum states in a three-level system formed in coupled QW composed of AlInAs/InGaAs layers grown on an InP substrate by MBE. The emitted photon energy in the QCL is determined by the energy difference between the quasi-confined energy states in the quantum wells. These QW states can be tuned over a very broad range by 'band gap engineering' through appropriate selection of the thickness and bandgap of the constituent semiconductors that form the superlattice. The QCL typically contains multiple periods (typically 30) of the QW injection and active emission regions in order to obtain useful laser gains and low-threshold current densities. The operating wavelengths of the QCL has been demonstrated over the range of 3–17 μm [90–93]. The ability to specifically design the operating wavelength of the QCL for spectroscopic detection of gas molecules is the key reason why the QCL is the laser of choice for gas sensor applications.

The operation of unipolar electron injection in a QCL to obtain laser oscillations is quite different compared to the normal operation of pn diode semiconductor lasers. The photon energy in conventional semiconductor lasers results from direct radiative recombination of electrons and holes and is determined by the difference in the conduction- and valence-band energy levels which is simply the semiconductor bandgap. Therefore, the emission wavelength in pn junction lasers is fixed by the direct semiconductor bandgap.

MBE growth technology is generally used in synthesizing QCL structures to meet the precise layer thickness and interface abruptness that determines the QW electron states and, hence, the emission wavelength. Figure B2.1.32 shows a schematic diagram of a QCL active region grown by MBE. The material system is InGaAs/InAlAs grown over n-InP. These devices have been fabricated into wide-stripe lasers and have operated with peak powers of $\sim$200 mW at 200 K.

For improved performance, designs based on double-phonon resonance [92] and a bound to continuum resonance [93] have been used. The former has recently been used to demonstrate cw lasers [92]. The active

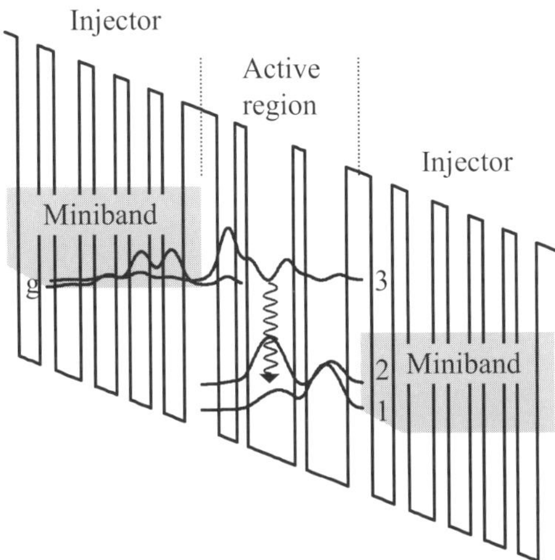

Figure B2.1.32. Conduction-band profile of one active region with the two adjacent injector regions under an applied electric field of 55 kV cm^{-1}. The moduli squared of the relevant wavefunctions are labelled 1, 2, 3, and g. The wavy arrow denotes the laser transition. The shaded region in the injectors represents a band-like manifold of states. This 'miniband' extracts electrons from levels 1 and 2 of a previous active region and injects them into state 3 of the following downstream active region. An energy range with a low density of states above the miniband prevents electrons from tunnelling out of level 3. Electrons traverse the structure from left to right. The QW material is InGaAs, the barrier material AlInAs; both are grown lattice matched to InP. The band offset is assumed as 520 meV. The length of one period of active region and injector is 43.4 nm; 32 periods were grown.

region consists of four InGaAs/InAlAs QWs designed so that the lasing transition takes place from level 4 to 3 in one QW and has three coupled lower energy levels separated from each other by one phonon energy (figure B2.1.33). The fast interband transition in the lowest subbands leads to a high population inversion at room temperature. The active also has a narrow QW barrier to increase the injection efficiency to the upper laser level (level 91).

The active region is fabricated by the MBE technique using InGaAs/InAlAs material over an n-InP substrate. The entire active region consists of 25–30 periods of the type shown in figure B2.1.33. Each period will have an n-type injector region and an undoped four QW light-emitting (active) region. The entire active region structure is bounded by an InGaAs layer followed by an InP layer. These layers form the waveguide. The high-speed performance and tunability of the QCL is described in [92].

B2.1.10 Laser reliability

The performance characteristics of injection lasers used in lightwave systems can degrade during their operation. The dominant mechanism responsible for the degradation is determined by any or all of the several fabrication processes including epitaxial growth, wafer quality, device processing and bonding [94–104] . In addition, the degradation rate of devices processed from a given wafer depends on the operating conditions, i.e. the operating temperature and the injection current. Although many of the degradation mechanisms are not fully understood, extensive empirical observations exist in the literature which have allowed the fabrication of InGaAsP laser diodes with extrapolated median lifetimes in excess of 25 years at an operating temperature of 20 °C [94].

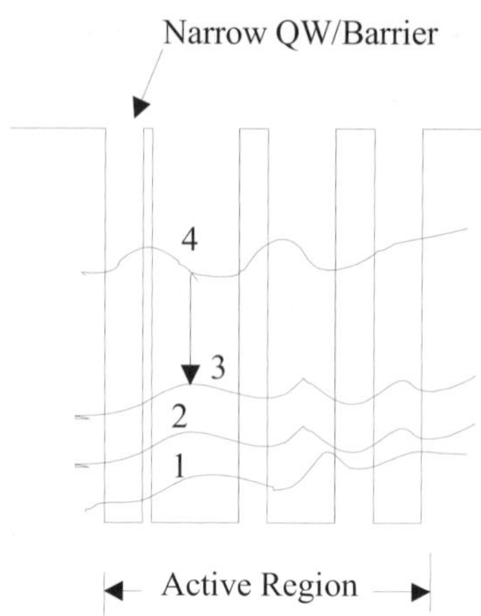

Figure B2.1.33. Schematic diagram of one period of a QCL active region that operated cw at room temperature. The laser transition is from level 4 to level 3. The moduli-squared wavefunctions in the four-QW active region is shown. The separation between levels 3 and 2 and that between 2 and 1 equals one phonon energy which helps depopulate the lower laser level (level 3).

The detailed studies of the degradation mechanisms of optical components used in lightwave systems have been motivated by the desire to have a reasonably accurate estimate of the operating lifetime before they are used in practical systems. Since for many applications, the components are expected to operate reliably over a period in excess of 10 years, an appropriate reliability assurance procedure becomes necessary, especially for applications such as an undersea lightwave transmission system where the replacement cost is very high. The reliability assurance is usually carried out by operating the devices under a high stress (e.g. high temperature) which enhances the degradation rate so that a measurable value can be obtained in an operating time of a few hundred hours. The degradation rate under normal operating conditions can then be obtained from the measured high-temperature degradation rate using the concept of an activation energy [94]. The light output *versus* current characteristics of a laser change after stress aging. There is generally a small increase in threshold current and a decrease in external differential quantum efficiency following the stress aging. Aging data for 1.3 μm nGaAsP lasers used in the first submarine fibre optic cable are shown in figure B2.1.34 [94].

Some lasers exhibit an initial rapid degradation after which the operating characteristics of the lasers are very stable. Given a population of lasers, it is possible to quickly identify the 'stable' lasers by a high stress test (also known as the purge test) [94, 105]. The stress test implies that operating the laser under a set of high stress conditions (e.g. high current, high temperature, high power) would cause the weak lasers to fail and stabilize the possible winners. Observations on the operating current after stress aging have been reported by Nash *et al* [94]. It is important to point out that the determination of the duration and the specific conditions for stress aging are critical to the success of this screening procedure. The expected operating lifetime of a semiconductor laser is generally determined by accelerated aging at high temperatures and using an activation energy. The lifetime (t) at a temperature T is experimentally found to vary as $\exp(-E/kT)$ where E is the activation energy and k is the Boltzmann constant [106]. The operating current of good lasers increases at a rate of less than 1%/kh of aging time at 60 °C operating temperature. Assuming a 50% change in operating current as the useful lifetime of the device and an activation energy of 0.7 eV, this aging rate corresponds to a light-emitting lifetime of greater than 100 years at 20 °C.

A parameter that determines the performance of the DFB laser is the side-mode suppression ratio

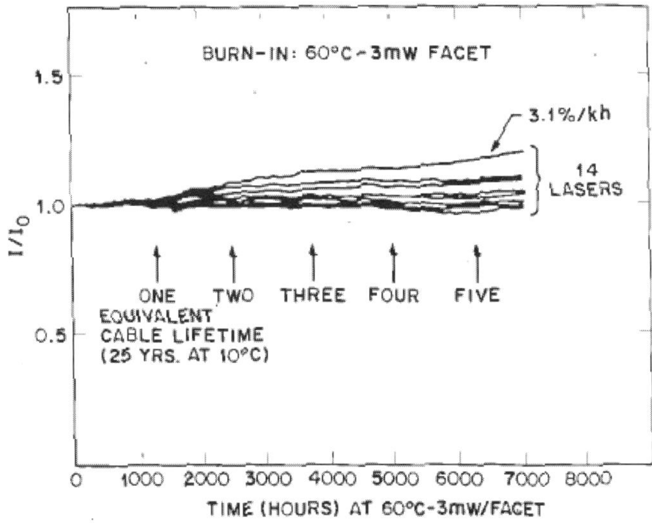

Figure B2.1.34. Aging data of lasers used in the first submarine cable application in 1988.

(SMSR), i.e. the ratio of the intensity of the dominant lasing mode to that of the next most intense mode [107]. The SMSR of good DFB lasers does not change significantly after aging, which confirms the spectral stability of the emission.

There have been a significant number of developments in the technology of optical integration of semiconductor lasers and other related devices on the same chip. These chips allow higher levels of functionality than that achieved using single devices. For example, laser and optical modulators have been integrated, serving as simple monolithic transmitters.

B2.1.11 Summary and future challenges

Tremendous advances in semiconductor lasers have occurred over the last decade. The advances in research and many technological innovations have led to the worldwide deployment of fibre optic communication systems that operate near 1.3 and 1.55 μm and compact storage discs that utilize lasers for read/write purposes. Although most of these systems are based on digital transmission, lasers have also been deployed for carrying high-quality analogue cable TV transmission systems. However, many challenges remain.

The need for higher capacity is pushing the deployment of WDM-based transmission which needs tunable or frequency settable lasers. The development of lasers with very stable and settable frequency will continue to be an important reasearch area. The integration of many such lasers on a single substrate would provide the ideal source for WDM systems.

Laser-to-fibre coupling is also an important area of research. Recent development in spot size converter integrated lasers are quite impressive but some more work, perhaps, needs to be done to make them easy to manufacture. This may require more process developments.

Although WDM technology is currently being considered for increasing transmission capacity, the need for sources with very high modulation capability still remains. Hence, research on new mechanisms for very high speed modulation is important.

The surface-emitting laser is very attractive for two-dimensional arrays and for single-wavelength operation. Several important advances in this technology have occurred over the last few years. An important challenge is the fabrication of a device with characteristics superior to that of an edge emitter.

Quantum well lasers have shown that lower dimensionality (two-dimensional-like density of states) leads to improved performance. Further performance improvement is expected and has been demonstrated to some extent using quantum wire and quantum dot structures. These lower dimensional systems require further investigation before they become commercial realities. Lasers based on inter-sub-band transitions in quantum well, quantum cascade lasers have been developed over the last decade. Further improvements, such as cw operation at high temperature, will continue to be important for wide commercial use.

Finally, much of the advances in laser development would not have been possible without the advances in materials and processing technology. The challenges of current laser research are intimately linked with the challenges in materials growth which include not only the investigation of new material systems but also improvements in existing technologies to make them more reproducible and predictable.

References

[1] Hall R N, Fenner G E, Kingley J D, Soltys T J and Grlson R O 1962 *Phys. Rev. Lett.* **9**

[2] Nathan M I, Dumke W P, Burns G, Dill F H Jr and Lasher G 1962 *Appl. Phys. Lett.* **1** 63

[3] Quist T M, Rediker R H, Keyes R J, Krag W E, Lax B, McWhorter A L and Ziegler H J 1962 *Appl. Phys. Lett.* **1** 91

[4] Holonyak N Jr and Bevacqua S F 1962 *Appl. Phys. Lett.* **1** 82

[5] Burrus C A and Dawson R W 1970 *Appl. Phys. Lett.* **17** 97

[6] Eliseev P G 1998 *Semiconductor Lasers* vol II, ed E Kapon (New York: Academic) ch 2

[7] Nelson H 1963 *RCA Rev.* **24** 603

[8] Finch W F and Mehal E W 1964 *J. Electrochem. Soc.* **111** 814

[9] Cho A Y 1971 *J. Vac. Sci. Technol.* **8** 531

[10] Dupuis R D 1981 *J. Crystal Growth* **55** 213

[11] Kressel H and Butler J K 1977 *Semiconductor Lasers and Heterojunction LEDs* (New York: Academic)

[12] Casey H C Jr and Panish M B 1978 *Heterostructure Lasers* (New York: Academic)

[13] Thompson G H B 1981 *Physics of Semiconductor Laser Devices* (New York: Wiley)

[14] Yariv A 1976 *Quantum Electronics* 2nd edn (New York: Wiley) p 306

[15] Bernard M G A and Duraffourg G 1961 *Phys. Status Solidi* **1** 699

[16] Alferov Zh I, Andreev V M, Portnoi E L and Trukan M K 1970 *Sov. Phys. Semicond.* **3** 1107

[17] Hayashi I, Panish M B, Foy P W and Sumski S 1970 *Appl. Phys. Lett.* **17** 109

[18] Dingle R, Wiegman W and Henry C H 1974 *Phys. Rev. Lett.* **33** 827

[19] Hwang C J 1970 *Phys. Rev.* B **2** 4117

[20] Stern F 1973 *J. Appl. Phys.* **47** 5382
Stern F 1973 *IEEE J. Quantum Electron.* **QE-9** 290

[21] Kane E O 1957 *J. Phys. Chem. Solids* **1** 249

[22] Dutta N K 1980 *J. Appl. Phys.* **51** 6095 (1980)
Dutta N K 1981 *J. Appl. Phys.* **52** 55

[23] Olshansky R, Su C B, Manning J and Powaznik W 1984 *IEEE J. Quantum Electron.* **QE-20** 838

[24] Beattie A R and Landsberg P T 1959 *Proc. R. Soc.* **249** 16
Beattie A R and Landsberg P T 1960 *Proc. R. Soc.* A **258** 486

[25] Horikoshi Y and Furukawa Y 1979 *Japan. J. Appl. Phys.* **18** 809

[26] Thomson G H B and Henshall G D 1980 *Electron. Lett.* **16** 42

[27] Dutta N K and Nelson R J 1981 *Appl. Phys. Lett.* **38** 407
Dutta N K and Nelson R J 1982 *J. Appl. Phys.* **53** 74 and references therein

[28] Sugimura A 1981 *IEEE J. Quantum Electron.* **QE-17** 627

[29] Haug A 1983 *Appl. Phys. Lett.* **42** 512

[30] DeLoach B C Jr, Hakki B W, Hartman R L and D'Asaro L A 1973 *Proc. IEEE* **61** 1042

[31] Dingle R 1975 *Festkorperprobleme* **XV** 21–47

[32] Holonyak N Jr, Kolbas R M, Dupuis R D and Dapkus P D 1980 *IEEE J. Quantum Electron.* **QE-16** 170

[33] Hess K, Vojak B A, Holonyak N Jr, Chin R and Dapkus P D 1980 *Solid State Electron.* **23** 585

[34] Dutta N K 1982 *Electron. Lett.* **18** 451
Dutta N K 1982 *J. Appl. Phys.* **53** 7211

[35] Arakawa Y and Yariv A 1985 *IEEE J. Quantum Electron.* **QE-21** 1666

[36] Agrawal G P and Dutta N K 1992 *Semiconductor Lasers* (Dordrecht: Kluwer Academic) ch 9

[37] Kazmierski C, Ougazzaden A, Blez M, Robien D, Landreau J, Sermage B, Bouley J C and Mirca A 1991 *IEEE J. Quantum Electron.* **27** 1794–7

[38] Morton P, Logan R A, Tanbun-Ek T, Sciortino P F Jr, Sergent A M, Montgomery R K and Lee B T 1993 *Electron. Lett.* **29** 1429–30
[39] Thijs P J A, Tiemeijer L F, Kuindersma P I, Binsma J J M and van Dongen T 1991 *IEEE J. Quantum Electron.* **27** 1426
[40] Thijs P J A, Tiemeijer L F, Binsma J J M and van Dongen T 1994 *IEEE J. Quantum Electron.* **QE-30** 477–99
[41] Temkin H, Tanbun-Ek T and Logan R A 1990 *Appl. Phys. Lett.* **56** 1210
[42] Tsang W T, Yang L, Wu M C, Chen Y K and Sergent A M 1990 *Electron. Lett.* 2033
[43] Laidig W D, Lin Y F and Caldwell P J 1985 *J. Appl. Phys.* **57** 33
[44] Fischer S E, Fekete D, Feak G B and Ballantyne J M 1987 *Appl. Phys. Lett.* **50** 714
[45] Yokouchi N, Yamanaka N, Iwai N, Nakahira Y and Kasukawa A 1996 *IEEE J. Quantum Electron.* **QE-32** 2148–55
[46] Beernik K J, York P K and Coleman J J 1989 *Appl. Phys. Lett.* **25** 2582
[47] Loehr J P and Singh J 1991 *IEEE J. Quantum Electron.* **27** 708
[48] Corzine S W, Yan R and Coldren L A 1993 Optical gain in III–V bulk and quantum well semiconductors *Quantum Well Lasers* ed P Zory (New York: Academic)
[49] Temkin H, Tanbun-Ek T, Logan R A, Coblentz D A and Sergent A M 1991 *IEEE Photon. Technol. Lett.* **3** 100
[50] Adams A R 1986 *Electron. Lett.* **22** 249
[51] Yablonovitch E and Kane E O 1986 *J. Lightwave Technol.* **LT-4** 50
[52] Wu M C, Chen Y K, Hong M, Mannaerts J P, Chin M A and Sergent A M 1991 *Appl. Phys. Lett.* **59** 1046
[53] Dutta N K, Lopata J, Berger J R, Sivco D L and Cho A Y 1991 *Electron. Lett.* **27** 680
[54] Kuo J M, Wu M C, Chen Y K and Chin M A 1991 *Appl. Phys. Lett.* **59** 2781
[55] Chand N, Becker E E, van der Ziel J P, Chu S N G and Dutta N K 1991 *Appl. Phys. Lett.* **58** 1704
[56] Choi H K and Wang C A 1990 *Appl. Phys. Lett.* **57** 321
[57] Dutta N K, Wynn J D, Lopata J, Sivco D L and Cho A Y 1990 *Electron. Lett.* **26** 1816
[58] Agrawal G P and Dutta N K 1992 *Semiconductor Lasers* (New York: Van Nostrand) ch 7 (1st edn 1985)
[59] Koch T L, Koren U, Gnall R P, Burrus C A and Miller B I 1988 *Electron. Lett.* **24** 1431
[60] Suematsu Y, Arai S and Kishino K 1983 *J. Lightwave Technol.* **LT-1** 161
[61] Koch T L and Koren U 1991 *IEEE J. Quantum Electron.* **QE-27** 641
[62] Dutta N K, Piccirilli A B, Cella T and Brown R L 1986 *Appl. Phys. Lett.* **48** 1501
[63] Liou K Y, Dutta N K and Burrus C A 1987 *Appl. Phys. Lett.* **50** 489
[64] Tanbun-Ek T, Logan R A, Chu S N G and Sergent A M 1990 *Appl. Phys. Lett.* **57** 2184
[65] Soda H, Iga K, Kitahara C and Suematsu Y 1979 *Japan. J. Appl. Phys.* **18** 2329
[66] Iga K, Koyama F and Kinoshita S 1988 *IEEE J. Quantum Electron.* **24** 1845
[67] Jewell J L, Harbison J P, Scherer A, Lee Y H and Florez L T 1991 *IEEE J. Quantum Electron.* **27** 1332
[68] Chang-Hasnain C J, Maeda M W, Stoffel N G, Harbison J P and Florez L T 1990 *Electron. Lett.* **26** 940
[69] Geels R S, Corzine S W and Coldren L A 1991 *IEEE J. Quantum Electron.* **27** 1359
[70] Geels R S and Coldren L A 1990 *Appl. Phys. Lett.* **5** 1605
[71] Tai K, Hasnain G, Wynn J D, Fischer R J, Wang Y H, Weir B, Gamelin J and Cho A Y 1990 *Electron. Lett.* **26** 1628
[72] Tai K, Yang L, Wang Y H, Wynn J D and Cho A Y 1990 *Appl. Phys. Lett.* **56** 2496
[73] Born M and Wolf E 1977 *Principles of Optics* (New York: Pergamon) section 1.6.5, p 69
[74] Jewell J L, Scherer A, McCall S L, Lee Y H, Walker S J, Harbison J P and Florez J P 1989 *Electron. Lett.* **25** 1123
[75] Lee Y H, Tell B, Brown-Goebeler K F, Jewell J L, Leibenguth R E, Asom M T, Livescu G, Luther L and Mattera V D 1990 *Electron. Lett.* **26** 1308
[76] Tai K, Fischer R J, Seabury C W, Olsson N A, Huo D T C, Ota Y and Cho A Y 1989 *Appl. Phys. Lett.* **55** 2473
[77] Ibaraki A, Kawashima K, Furusawa K, Ishikawa T, Yamayachi T and Niina T 1989 *Japan. J. Appl. Phys.* **28** L667
[78] Schubert E F, Tu L W, Kopf R W, Zydzik G J and Deppe D G 1990 *Appl. Phys. Lett.* **57** 117
[79] Geels R S, Corzine S W, Scott J W, Young D B and Coldren L A 1990 *IEEE Photon. Technol. Lett.* **2** 234
[80] Vakhshoori D, Wynn J D, Liebenguth R E 1994 *Appl. Phys. Lett.* **65** 144
[81] See, for example, Deppe D and Choquette K 2000 *Vertical Cavity Surface Emitting Lasers* ed J Cheng and N K Dutta (New York: Gordon and Breach) chs 1 and 2
[82] Dutta N K, Tu L W, Zydzik G J, Hasnain G, Wang Y H and Cho A Y 1991 *Electron. Lett.* **27** 208
[83] Soda H, Iga K, Kitahara C and Suematsu Y 1979 *Japan. J. Appl. Phys.* **18** 2329
[84] Valster A, van der Pol C J, Finke M N and Boermans M J B 1992 *Electron. Lett.* **28** 144
[85] Drenton R, Petruzzello J and Haberern K 1995 *Phillips J. Res.* **49** 225
[86] Nurmikko A V and Gunshor R L 1994 *IEEE J. Quantum Electron.* **QE-30** 619
[87] Nakamura S, Senoh M, Iwasa N and Nagahama S 1995 *Japan. J. Appl. Phys.* **34** L797
[88] Nakamura S and Fasol G 1997 *The Blue Laser Diode—GaN Based Light Emitters* (Berlin: Springer)
[89] Faist J *et al* 1994 *Science* **264** 553
[90] Capasso F *et al* 2002 *IEEE J. Quantum Electron.* **38** 511
[91] Gmachl C *et al* 2002 *IEEE J. Quantum Electron.* **38** 569
[92] Beck M *et al* 2002 *Science* **295** 301
[93] Faist J *et al* 2001 *Appl. Phys. Lett.* **78** 147

[94] Nash F R, Sundburg W J, Hartman R L, Pawlik J R, Ackerman D A, Dutta N K and Dixon R W 1985 *AT&T Tech. J.* **64** 809
[95] The reliability requirements of a submarine lightwave transmission system are discussed in a special issue of *AT&T Tech. J.* **64** 3
[96] DeLoach B C Jr, Hakki B W, Hartman R L and D'Asaro L A 1973 *Proc. IEEE* **61** 1042
[97] Johnston W D and Miller B I 1973 *Appl. Phys. Lett.* **23** 1972
[98] Petroff P M, Johnston W D Jr and Hartman R L 1974 *Appl. Phys. Lett.* **25** 226
[99] Matsui J, Ishida R and Nannichi Y 1975 *Japan. J. Appl. Phys.* **14** 1555
[100] Petroff P M and Lang D V 1977 *Appl. Phys. Lett.* **31** 60
[101] Ueda O, Umebu I, Yamakoshi S and Kotani T 1982 *J. Appl. Phys.* **53** 2991
[102] Ueda O, Yamakoshi S, Komiya S, Akita K and Yamaoka T 1980 *Appl. Phys. Lett.* **36** 300
[103] Yamakoshi S, Abe M, Wada M, Komiya S and Sakurai T 1981 *IEEE J. Quantum Electron.* **QE-17** 167
[104] Mizuishi K, Sawai M, Todoroki S, Tsuji S, Hirao M and Nakamura M 1983 *IEEE J. Quantum Electron.* **QE-19** 1294
[105] Gordon E I, Nash F R and Hartman R L 1983 *IEEE Electron. Device Lett.* **ELD-4** 465
[106] Hartman R L and Dixon R W 1975 *Appl. Phys. Lett.* **26** 239
[107] Joyce W B, Liou K Y, Nash F R, Bossard P R and Hartman R L 1985 *AT&T Tech. J.* **64** 717

B2.2
Spectral control in laser diodes

Markus-Christian Amann

B2.2.1 Introduction

The purpose of this chapter is to present an introduction into the various techniques for the control of the spectral properties of laser diodes and to investigate the most important device parameters that determine the spectral performance. The relevant characteristic hence is the laser spectrum and its variation with time. It should be noted that—strictly speaking—the laser spectrum should not be time dependent because it is defined as the Fourier transform of the time-dependent laser signal. Restricting the integration of the Fourier transform to finite time intervals, however, we may define time-dependent laser spectra.

It is well known that because of the large resonator length with respect to the optical wavelength, lasers in general but even laser diodes with their small dimensions tend to operate in several longitudinal modes. However, the transverse and lateral dimensions of laser diodes can be made small enough to provide the transverse and lateral single-mode operation. Let i denote the longitudinal mode number, the wavelength of the ith longitudinal mode is given as

$$\lambda_i = \frac{2n_{\mathrm{eff}}(\lambda_i)L}{i} \tag{B2.2.1}$$

where n_{eff} is the effective refractive index describing the transverse and lateral waveguide characteristics, L is the laser length and the longitudinal mode number i (typically 1000–10 000) equals the number of nodes of the intensity distribution along the laser axis.

The spacing $\Delta\lambda_m$ between adjacent longitudinal modes around mode i is

$$\Delta\lambda_m \simeq \frac{\lambda_i^2}{2n_{\mathrm{g,eff}}L} \tag{B2.2.2}$$

where $n_{\mathrm{g,eff}}$ is the effective group index that takes into account the wavelength dependence of n_{eff}

$$n_{\mathrm{g,eff}} = \left.\frac{\mathrm{d}\,(k_0 n_{\mathrm{eff}})}{\mathrm{d}k_0}\right|_{\lambda_i} = n_{\mathrm{eff}}(\lambda_i) - \lambda_i \left.\frac{\mathrm{d}n_{\mathrm{eff}}}{\mathrm{d}\lambda}\right|_{\lambda_i}. \tag{B2.2.3}$$

Since the effective refractive index usually decreases with increasing wavelength, the effective group index is commonly larger than the effective refractive index. In InGaAsP/InP laser diodes operated in the 1.3–1.55 μm wavelength regime, n_{eff} is typically of the order 3.3 and $n_{\mathrm{g,eff}} \simeq 4$. A schematic representation of a longitudinal mode spectrum of a Fabry–Pérot laser diode is shown in figure B2.2.1.

Accordingly, for a well-controlled laser spectrum it is most important to force the laser operation in one single longitudinal mode. The physical parameter defining the quality of the single-mode operation is the side-mode suppression ratio (*SSR*), which is the power ratio between the dominant and second strongest laser

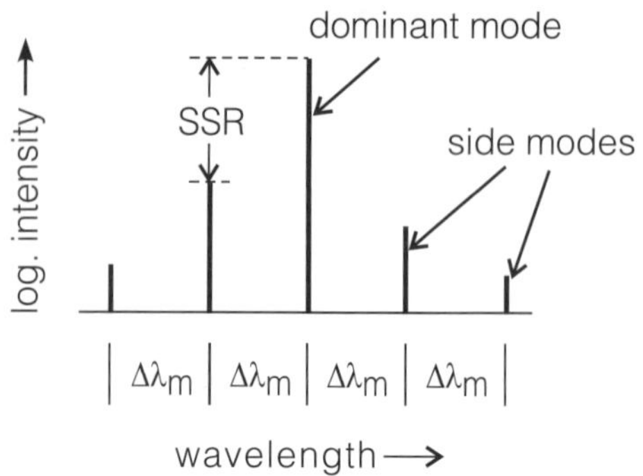

Figure B2.2.1. Longitudinal mode spectrum of a multi mode Fabry–Pérot laser diode.

mode. With P_i being the output power of the dominant mode i per mirror in a symmetric laser, the side-mode suppression ratio is

$$SSR = \frac{2P_i \Delta g}{h\nu v_g n_{sp} \alpha_{tot} \alpha_m} \qquad \text{(B2.2.4)}$$

where Δg is the mode gain difference between the dominant mode and the second strongest mode, h is Planck's constant, ν is the optical frequency ($h\nu$ is photon energy), n_{sp} is the spontaneous emisssion factor (typ. 2), α_{tot} is the total optical loss and α_m is the mirror loss. While single-mode operation may more easily be achieved in the vertical-cavity surface-emitting laser diodes (VCSELs, see chapter B2.6) because of their small cavity length, considerable effort has to be spent in the more usual edge-emitting laser diodes to achieve a sufficiently large suppression of side modes ($SSR > 30$ dB). The techniques to achieve single-mode operation in edge-emitting laser diodes are described in detail in the next section B2.2.2.

After accomplishing the transverse and longitudinal single-mode operation of a laser diode, the next relevant spectral characteristic is the spectral linewidth of the lasing mode, which usually is in the megahertz regime. The spectral linewidth represents the power spectrum of the laser output, which because of the carrier density dependence of the refractive index in semiconductor lasers may change during modulation and/or relaxation processes. In practice, this effect manifests itself as a parasitic frequency modulation (chirping of the laser frequency) during amplitude modulation. As with the laser linewidth, the amount of chirping strongly depends on the amplitude–phase coupling in laser diodes, which is usually described by the α-factor α_H [1].

Finally, the spectral purity of laser diodes is significantly affected by feedback of the laser output as caused, for instance, by reflections at coupling lenses or remote reflectors such as fibre couplers. Because of the marked sensitivity of the laser diodes to feedback and because of possible deterioriation of their static and dynamic performance in demanding applications, one usually applies optical isolators with isolation above 30 dB. The effect of optical feedback can be investigated by means of the simplified illustration in figure B2.2.2. Besides the remote reflectivity r_3, the amplitude external reflection coefficient r_{ext} includes the coupling efficiency η_c between laser diode and external components (e.g. a fibre), and the optical losses α_{ext} along the external cavity of length L_{ext}:

$$r_{ext} = \eta_c^2 r_3 e^{-\alpha_{ext} L_{ext}}. \qquad \text{(B2.2.5)}$$

The rate equation analysis of the external optical feedback shows that the different feedback regimes can conveniently be described by the feedback parameter

$$C = \frac{\tau_{ext}}{\tau_l}\frac{r_{ext}}{r_2}(1 - |r_2|^2)\sqrt{1+\alpha_H^2} \qquad \text{(B2.2.6)}$$

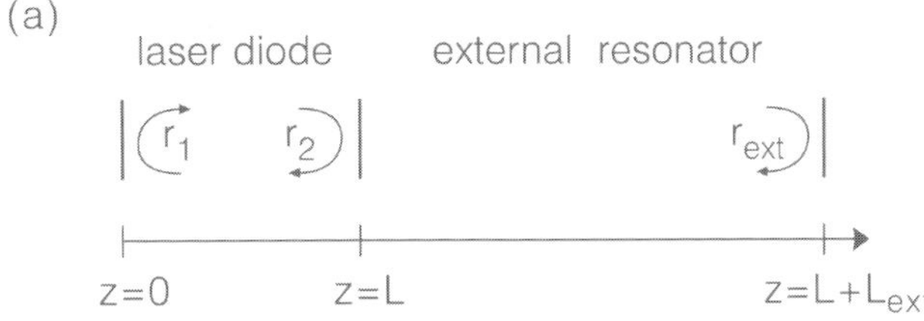

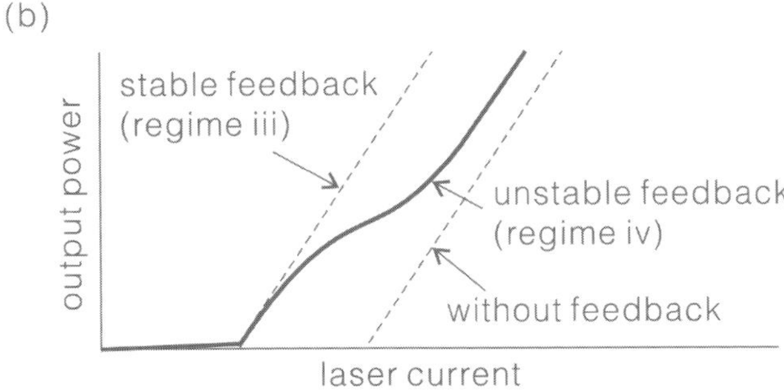

Figure B2.2.2. Schematic illustration of a laser diode with external optical feedback (*a*) and light–current characteristics for various degrees of external optical feedback (*b*).

where τ_l and τ_{ext} denote the round-trip times in the laser diode and external resonator. Depending on C and the power reflectivity given by the external reflection $R_{ext} = |r_{ext}|^2$, we may distinguish five regimes for a single-mode laser diode [2]:

(i) For $C < 1$, the laser operates in single mode and the external feedback may only affect the spectral linewidth.

(ii) For $1 < C \leq 2$ and $R_{ext} < 3 \times 10^{-4}$, mode jumps and hysteresis effects may occur.

(iii) For still larger feedback up to $R_{ext} = 10^{-4}$, the laser operates relatively stably with a *reduced* spectral linewidth.

(iv) For $R_{ext} > 10^{-4}$, strong instabilities occur (coherence collapse) with large linewidth broadening.

(v) For an external reflection that exceeds the reflection at the laser facet the instabilities cease and laser with external feedback operates like a laser with a combined length $L_{ext} + L$ with an internal perturbation by mirror 2 (r_2).

The light–current characteristics of laser diodes with different external optical feedback are shown in figure B2.2.2(*b*).

Once the single-mode operation has been established one may want to tune the laser wavelength over a certain wavelength range. This may either be required in wavelength division multiplex (WDH) applications, in measurements and sensing or in order to fix the laser wavelength precisely to a given value. Depending on the application, the wavelength tuning should be continuous, i.e. without mode jumps, or quasicontinuous, so that over the tuning range each wavelength may be addressed by at least one of the longitudinal modes. The established methods for electronic wavelength tuning of laser diodes using current or voltage variations at additional electrodes are compiled in section B 2.2.3 together with the corresponding device structures and performance.

Furthermore, laser diodes may also be suited for applications where simultaneous operation at several wavelengths is needed. The corresponding laser diode arrays are presented in section B 2.2.4.

B2.2.2 Single-mode laser diodes

Truly single-longitudinal-mode laser diodes with high performance were first realized in the InGaAsP/InP material system with anticipated applications in fibre optical communications. State-of-the-art single-mode laser diodes in the 1.55 μm wavelength range exhibit a side-mode suppression ratio (SSR) of at least 30 dB and spectral linewidth of the order of several MHz at a few ($\simeq$5) mW of optical output power [3].

The most successful technique to achieve single-mode operation is to introduce strongly wavelength dependent cavity loss or gain in order to favour one single longitudinal mode. In practice, this is usually done by integrating a Bragg grating into the laser cavity that replaces one or both mirrors by the distributed feedback along the periodic refractive index changes. The Bragg grating thereby acts as like a pile of pairs, each $\lambda/4$ thick, of alternating dielectric layers with different refractive index yielding a strong reflection at the Bragg wavelength

$$\lambda_B = 2\Lambda\overline{n}_{\text{eff}} \tag{B2.2.7}$$

where Λ is the grating period and $\overline{n}_{\text{eff}}$ is the average effective refractive index.

As shown in figure B2.2.3, the Bragg grating filters can be arranged either outside the active region at one or both ends of the laser cavity and act then as wavelength selective reflectors (distributed Bragg reflector (DBR)) [4], or can be collocated with the active region along the entire laser cavity yielding a distributed feedback (DFB) [5] structure. The longitudinal index variation is thereby introduced by periodic thickness variation of the grating layer with refractive index n_1, which covers a semiconductor region with different refractive index n_0. Since, in the type 1 DBR laser, the DBR acts as a wavelength selective reflector through which the outcoming light must pass, optimum reflectivity and output efficiency cannot be obtained at the same time. Consequently, the type 2 DBR laser is advantageous because output coupling and wavelength selective reflection occur at different ends of the laser cavity. In the DFB laser the distribution of the optical feedback along the entire laser length allows various designs with high output efficiency and wavelength selectivity. Furthermore, by the application of anti-reflection and/or high-reflection coatings onto the end facets the laser performance may be tailored to the anticipated application.

It should be noted that for an effective reflectivity of the grating it is *not* necessary that the index variations occur stepwise; even a soft periodic index transition, e.g. by a sine-like function, may provide strong and wavelength-selective feedback [5]. Since these index perturbations occur only within part of the transverse mode profile, their effect on the effective refractive index n_{eff} of the composite waveguide structure is weakened. The modulation of n_{eff} is determined by the product of the index difference n_1–n_0 and the confinement factor within the grating area.

In DBR and DFB lasers, the grating-induced refractive index perturbations lead to a coupling between the forward and backward propagating waves of the particular laser mode throughout the entire grating region. Therefore, the optical feedback is not localized at the cavity end facets but is distributed over the entire laser cavity (DFB) or only part of it (DBR). The analysis of these devices, therefore, requires the investigation of wave propagation and wave coupling in periodic structures. This is usually accomplished by means of coupled-mode theory.

B2.2.2.1 Distributed Bragg reflector (DBR) laser

First we will consider the DBR laser, which effectively represents a Fabry–Pérot laser, whose mirrors are replaced by DBRs so that the mirror reflectivities become wavelength dependent. Accordingly, the characteristics of the DBRs decisively determine the device performance, particularly the spectral purity.

A schematic model for a DBR is shown in figure B2.2.4 with a sine-like z-dependence of n_{eff}.

The DBR extends from $z = 0$ to $z = L_B$ and the z-dependent effective index reads as

$$n_{\text{eff}}(z) = \overline{n}_{\text{eff}} + \frac{\Delta n_{\text{eff}}}{2} \sin k_g z \tag{B2.2.8}$$

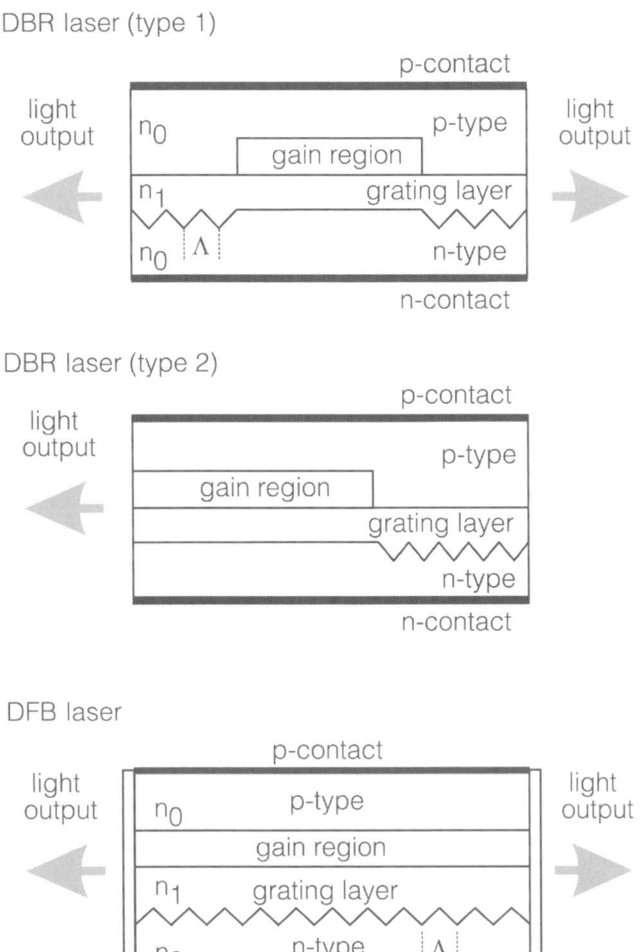

Figure B2.2.3. Schematic longitudinal sections of DBR and DFB laser diodes.

where $\overline{n}_{\mathrm{eff}}$ and Δn_{eff} denote the average effective refractive index and the effective refractive index difference (peak to peak). The grating wavevector k_{g} is related to the grating period Λ as

$$k_{\mathrm{g}} = \frac{2\pi}{\Lambda}. \tag{B2.2.9}$$

As indicated in figure B2.2.4, the periodic feedback leads to a decay in the incoming wave E_+ and produces a backward travelling (reflected) wave E_-, which starts with zero amplitude at the grating end $z = L_{\mathrm{B}}$. The reflection is not complete and a fraction of the forward travelling wave escapes from the Bragg grating at $z = L_{\mathrm{B}}$. The amplitude reflection at $z = 0$ is usually complex and is defined as

$$r = \frac{E_-(0)}{E_+(0)}. \tag{B2.2.10}$$

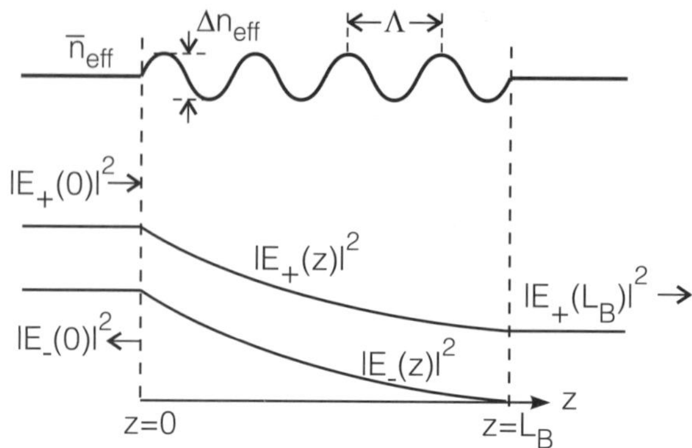

Figure B2.2.4. Forward and backward propagating waves in DBR.

Analysis by coupled-mode theory requires a coupling coefficient κ which describes the strength of the distributed feedback

$$\kappa = \frac{\pi\,\Delta n_{\mathrm{eff}}}{2\lambda}. \tag{B2.2.11}$$

The reflectivity of the Bragg grating is furthermore dependent on the grating length L_{B}, so that the product κL_{B} can be used to concisely describe the grating. The result of coupled-mode theory for the simple grating structure shown in figure B2.2.4 with no reflection at both ends of the grating yields for the amplitude reflectivity at wavelength $\lambda_{\mathrm{B}} + \Delta\lambda$ [4]:

$$r(\Delta\lambda) = \frac{-j\kappa\sinh(\gamma L)}{\gamma\cosh(\gamma L) + j\Delta\beta\sinh(\gamma L)} \tag{B2.2.12}$$

where

$$\Delta\beta = \frac{2\pi\,\Delta\lambda}{\lambda^2} \tag{B2.2.13}$$

$$\gamma^2 = \kappa^2 - \Delta\beta^2 \tag{B2.2.14}$$

and $\Delta\lambda$ is the deviation from the Bragg wavelength. At the Bragg wavelength, where $\Delta\lambda = 0$, the amplitude reflection is

$$r = -j\tanh(\kappa L_{\mathrm{B}}). \tag{B2.2.15}$$

Plots of the power reflectivity $|r(\Delta\lambda)|^2$ are shown in figure B2.2.5 with the κL_{B} product as parameter.

As can be seen, reasonable power reflection is achieved for κL_{B} larger than unity. At the same time, the filter bandwidth increases and the shape of the filter characteristic approaches that of a bandpass filter.

As mentioned earlier, the amplitude reflectivity at $z = 0$ is complex, which indicates a phase shift of the reflected wave E_- with respect to the incoming wave E_+ and can be interpreted by a longitudinal displacement of the reflecting plane from $z = 0$ to an effective penetration depth into the Bragg reflector $z = L_{\mathrm{eff}}$. Near the Bragg wavelength, therefore, (B2.2.12) can be written as equation

$$r(\Delta\lambda) \simeq \tanh(\kappa L_{\mathrm{B}})\mathrm{e}^{j4\pi\bar{n}_{\mathrm{eff}}L_{\mathrm{eff}}\Delta\lambda/\lambda_{\mathrm{B}}^2} \tag{B2.2.16}$$

with an effective penetration depth

$$L_{\mathrm{eff}} = \frac{\tanh(\kappa L_{\mathrm{B}})}{2\kappa}. \tag{B2.2.17}$$

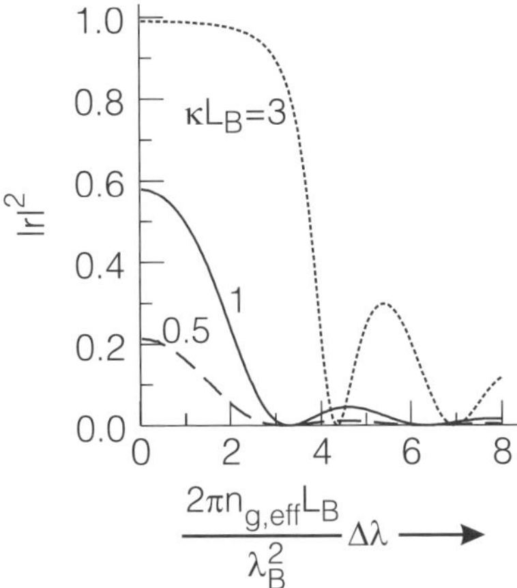

Figure B2.2.5. Power reflectivity $|r|^2$ of DBR as function of $\Delta\lambda$ with the κL_{B}-product as parameter.

The bandwidth $\Delta\lambda_{\mathrm{DBR}}$ of the DBR increases with κ as

$$\Delta\lambda_{\mathrm{DBR}} \simeq \frac{\kappa\lambda_{\mathrm{B}}^2}{\pi n_{\mathrm{g,eff}}}. \tag{B2.2.18}$$

A small bandwidth with a large reflectivity is obtained for long DBRs (large L_{B}) and small coupling coefficients κ. Even though the bandwidth can be made smaller than the longitudinal mode spacing $\Delta\lambda_{\mathrm{m}}$, DBR lasers have not been successfully commercialized as single-mode laser diodes. This is because the wavelengths of the longitudinal modes are still given by the phase condition (see equation (B2.2.1)), where the effective length of the DBR(s) has to be included in the total laser length L. Hence, the the wavelengths of the longitudinal laser modes are independent of the DBR parameters and any relative position of the longitudinal comb-mode spectrum may occur with regard to the Bragg wavelength, where the reflection is maximum. This is particularly crucial if the wavelength changes due to temperature and laser current changes are taken into account. Therefore, the unfortunate case in which two longitudinal modes are placed exactly symmetrically to the Bragg wavelength with a spacing of each $\Delta\lambda_{\mathrm{m}}/2$ so that both modes attain equal reflection and oscillate simultaneously may not be excluded. This results in a two mode laser with poor spectral behaviour and strong mode competition noise worse than those in multi-mode lasers with many longitudinal modes. For a truly single-mode laser diode with a stable single-mode emission, it is necessary that the longitudinal comb-mode spectrum is stably synchronized with the Bragg wavelength. As will be seen in the next section, this goal can conveniently be achieved with DFB lasers.

B2.2.2.2 Distributed feedback (DFB) laser

In contrast to the DBR laser, in a DFB laser (see figure B2.2.3) the gain and distributed feedback are not longitudinally separated but homogeneously extend over the entire laser axis. Technologically, therefore, DFB lasers are simpler to fabricate because no longitudinal integration of gain and DBR sections is required. Moreover, the longitudinal inhomogeneity of the DBR laser may yield unintentional reflections at the junctions of gain and DBR section that may deterioriate the spectral performance. However, the mathematical treatment of DFB lasers is more complicated because the amplitude and phase are not separable as in the case of the DBR laser.

Numerous types of DFB lasers have been proposed and these essentially differ in the grating structure and the treatment of the reflection at the end facets [6]. A selection of the most common DFB laser grating

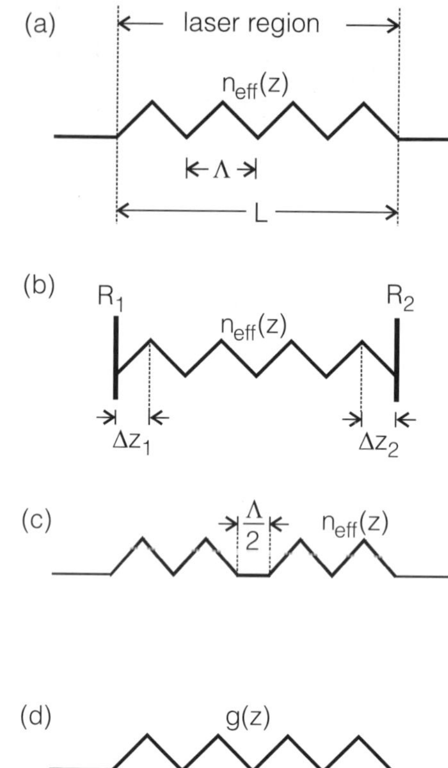

Figure B2.2.6. Various grating structures for single-mode DFB lasers: (*a*) 'pure' DFB laser, (*b*) DFB laser with two reflecting end facets with random phase offsets Δz_1 and Δz_2, (*c*) quarter-wave shifted DFB laser and (*d*) gain/loss-coupled DFB laser.

structures is compiled in figure B2.2.6. First of all the simplest 'pure' DFB laser is shown in figure B2.2.6(*a*). Practically, this structure can be realized by applying anti-reflection coatings onto both end facets. Depending on the facet-to-grating phase, however, even small residual reflections at the end facets may significantly affect the DFB laser spectral performance. Since a DFB laser structure without anti-reflection-coated end facets would be advantageous for technological reasons, we will also consider the DFB laser structure with two reflecting end facets as shown in figure B2.2.6(*b*). In this case, it is found to be technologically impossible to control the phase relationships between the two end facets and the grating, as indicated by the offsets Δz_1 and Δz_2. Accordingly, a random phase relationship has to be taken into account and the laser performance (*SSR*, yield of single-mode lasers) may be predicted only statistically [7, 8]. Optimum single-mode performance with a high device yield can be achieved with either a grating with a $\Lambda/2$-shift near the centre (see figure B2.2.6(*c*)) [9, 10] and with a gain/loss-grating (see figure B2.2.6(*d*)) [11] in which that is not the real part of the refractive index but its imaginary part, namely the mode gain or loss, that is longitudinally modulated.

Even for the relatively simple 'pure' DFB laser (figure B2.2.6(*a*)), coupled-mode solutions can only be obtained numerically [5]. These solutions give the threshold gain that is to be supplied by the active region for the various longitudinal modes. In the DFB laser, the mode with the smallest threshold gain will start lasing and the gain difference to the neighbouring modes determines the SSR. Due to the grating, a strong variation of the threshold gain can be observed around the Bragg wavelength. This is illustrated in figure B2.2.7, where

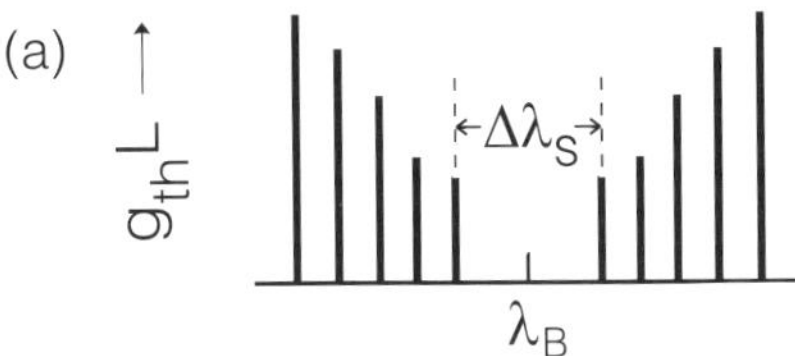

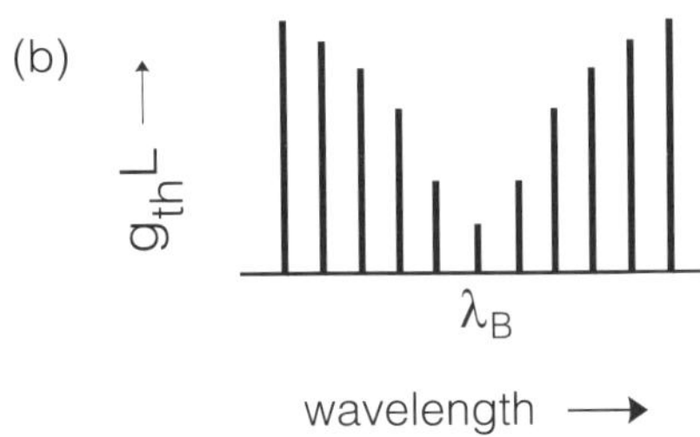

Figure B2.2.7. Threshold gain spectra for (*a*) a pure DFB laser and (*b*) quarter-wave-shifted and gain/loss coupled DFB laser.

the product of threshold gain and laser length is plotted *versus* the wavelength for the longitudinal modes nearest to the Bragg wavelength.

As can be seen, the threshold gain spectrum is symmetric with respect to λ_B and shows a minimum threshold gain for the modes nearest to λ_B as well as an effective gain discrimination. The spacing between the two modes nearest to the Bragg wavelength is called the stop band and has a width of

$$\Delta\lambda_s \simeq \Delta\lambda_m \sqrt{1 + \left(\frac{\kappa L}{\pi}\right)^2}. \qquad \text{(B2.2.19)}$$

In the limit of small κL products, the stop bandwidth approaches the longitudinal mode spacing of a Fabry–Pérot laser diode with equal length.

The threshold gain can be decreased for all modes by increasing the κL product. In the case of large κL, the threshold gain for the two modes nearest to the Bragg wavelength can be approximated by

$$g_{th} = \frac{2}{L}\left(\frac{\pi}{\kappa L}\right). \qquad \text{(B2.2.20)}$$

Theoretically, the symmetry of the threshold gain spectrum would result in exactly a two-mode operation of this type of DFB laser with bad spectral performance. Owing to spatial hole burning because of the longitudinal inhomogeneous intensity distribution, however, the degeneracy of the threshold gain breaks and one of the two modes becomes dominant to a certain degree. Nevertheless, this DFB laser type can hardly be considered as a reproducibly single-mode laser.

The threshold gain spectra strongly change if the end facets are reflecting (figure B2.2.6(*b*)). In this case the symmetry of the threshold gain spectrum is usually broken and one of the longitudinal modes is preferred. From the fabrication point of view this represents a considerable advantage, because no anti-reflection coating has been applied to the laser end facets, yielding an end facet reflectivity of the order 30% in InP-based devices. However, the random phase relationship(s) between mirror positions and grating require that, after laser fabrication, well-performing devices have to be selected using a spectral measurement.

Statistically, a high yield of single-mode lasers can be achieved if one facet is anti-reflection coated while the other one is high-reflectivity coated [7].

A deterministic approach to single-mode DFB lasers is the application of the quarter-wave-($\Lambda/2$)-shifted Bragg grating and no reflection at the end facets. In this structure the threshold gain spectra exhibit one mode with its minimum threshold gain located exactly at the Bragg wavelength [5, 12]. The field distribution within the laser peaks at the phase shift and exponentially decays towards the end facets. This means a large resonator quality factor and a narrow spectral linewidth; however, the optical output power may be small. Optimum laser performance therefore requires that the κL product is not too large (i.e. <2) in order to provide a good balance between spectral linewidth and output power and positioning of the phase shift near one of the end facets [13].

While the fabrication of the quarter-wave shift requires considerable effort (e.g. electron beam lithography), a similar performance can be achieved with the complex coupled DFB lasers in which a gain/loss Bragg grating or a mixed refractive index and gain/loss grating is incorporated [11, 14]. The technological realization usually employs a loss grating because the fabrication of a structured active region requires the partial removal of the active region with a subsequent overgrowth that may reduce the active region gain and induce degradation. The loss grating is usually made by placing a grating with an absorbing low-bandgap material (e.g. InGaAs with $E_g = 0.75$ eV in 1.55 μm lasers) near to the active region. As has already been indicated in the first DFB laser theory [5], the gain/loss Bragg grating shows an equivalent threshold gain spectrum to that of the quarter-wave-shifted DFB laser. More detailed theoretical work as well as experimental investigations furthermore showed that the axial field distribution is more strongly fixed yielding additional stability, particularly with respect to external optical feedback [15].

B2.2.2.3 Spectral linewidth

The spectral linewidth $\Delta\nu$ of a single-mode laser diodes is ultimately limited by the phase noise input due to spontaneous emission into the lasing mode [16, 17]. Here single-mode operation means a suppression of side modes (SSR) of at least 30 dB, otherwise mode competition effects may severely broaden the spectrum and dominate the spectral linewidth of the modes [18, 19]. Due to the gain clamping mechanism in lasers, the spontaneous emission saturates above threshold, while the photon number in the laser cavity as well as the optical output power increase further with current. Accordingly, the spectral linewidth decreases with optical output power as described by the Schawlow–Townes–Henry linewidth formula for laser diodes:

$$\Delta\nu = \frac{h\nu v_g^2 \alpha_{tot} \alpha_m n_{sp}}{8\pi P}(1 + \alpha_H^2). \tag{B2.2.21}$$

Here, $h\nu$ is the photon energy (e.g. 0.8 eV for 1.55 μm wavelength), v_g is the group velocity of light in the laser, α_{tot} is the total loss of the laser cavity, α_m is the mirror loss, n_{sp} ($\simeq 2$) is the spontaneous emission coefficient that takes into account the non-complete inversion, and P is the optical output power per facet by assuming a symmetric structure. The occurrence of α_H^2 in the linewidth formula is a speciality of laser diodes that, with α_H-values of 2–7, significantly affects the spectral linewidth of laser diodes. It originates from the fact that the gain characteristic of laser diodes is not symmetric with respect to frequency as is usually the case in other laser types, where the gain profiles are usually Lorentzian (homogeneous broadening) or Gaussian (inhomogeneous broadening).

The Kramers–Kronig relation shows that in the case of non-symmetric gain characteristics in laser diodes, gain changes during relaxation processes induce refractive index changes, that directly broaden the line as described by the α_H-factor. Since the α_H-factor is smaller for wavelengths smaller than the gain peak wavelength, negative detuning of the DFB laser wavelength by means of the grating pitch results in a significant improvement of the spectral linewidth [20]. A detuning of the order -30 nm is therefore usually accomplished in state-of-the-art DFB laser diodes in the 1.55 μm regime.

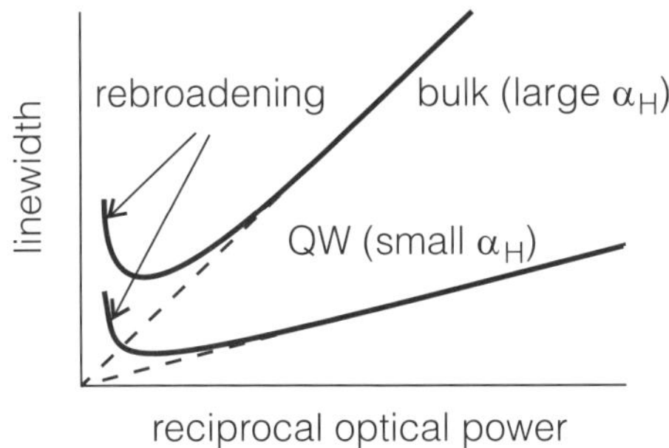

Figure B2.2.8. Schematic illustration of the spectral linewidth *versus* reciprocal optical power for bulk and QW single-mode lasers. The broken line is the theoretical relationship according to equation (B2.2.21).

Typical values of the spectral linewidth for laser diodes in the 1.3–1.55 μm wavelength range at a few mW of optical output power are of the order of several hundred kHz to a few MHz. The reciprocity between output power and linewidth usually ceases at very high output powers or high currents, respectively, as shown schematically in figure B2.2.8. As can be seen, a reduction of $\Delta\nu$ with increasing P changes into a steep increase of the linewidth, which is attributed to the appearance of side modes because of strong longitudinal hole burning by the dominant mode [21]. The smallest linewidth reported for monolithic semiconductor lasers is 3.6 kHz [22].

B2.2.3 Wavelength-tunable laser diodes

Electronically wavelength-tunable single-mode laser diodes are required for numerous applications in optical communications, measurement and sensing [23, 24]. In particular, advanced optical communication techniques such as optical heterodyning and wavelength division multiplexing rely on the availability of these sources [25]. A comparative review on widely tunable laser diodes for wavelength-division-multiplexed applications in the 1.55 μm region can be found in [26]. After the development of truly single-mode DFB and DBR lasers, therefore, the realization of electronically wavelength-tunable laser diodes represents an additional gain in device functionality.

The principal function of an idealized tunable single-mode laser is displayed in figure B2.2.9. The wavelength control requires at least one control current (I_t), while the output power should be adjusted by I_l. For practical reasons, a clear separation of the control functions for the output power and wavelength tuning is highly desirable, as indicated in the figure. However, since all of the available physical mechanisms for wavelength control also introduce varying (additional) optical losses, the output power usually changes during tuning. By a simultaneous adjustment of the power control current, the output power variations can, in principle, be eliminated. However, the output power variation via I_l changes the refractive index (e.g. because of temperature variations), which in turn affects the laser wavelength. As a consequence, the complete separate control of output power and emission wavelength is practically impossible and compromises must be found.

For most applications, continuous tuning with a high spectral purity is demanded [25, 27, 28]. In contrast to microwave oscillators, for instance, the continuous tuning of laser oscillators is rather difficult to achieve because of the large cavity dimensions compared to the light wavelength. This is true even for comparatively small laser diodes with cavity lengths of only several hundreds of micrometres.

Depending on the operating principle, three different tuning modes (shown in figure B2.2.10) may be distinguished. Besides continuous wavelength tuning, in which any wavelength within the covered wavelength band can be achieved, one may observe discontinuous tuning, in which discrete wavelength jumps occur, and

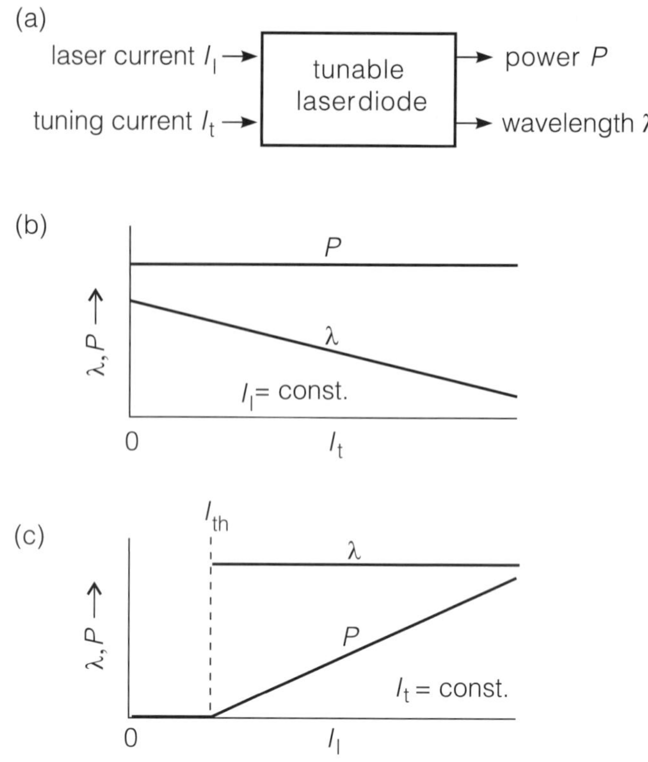

Figure B2.2.9. Schematic illustration of an idealized tunable laser diode (*a*). The wavelength and light power can be controlled independently by the two control currents I_t (*b*) and I_l (*c*), respectively.

quasicontinuous tuning, in which overlapping intervals of each continuously tunable regimes appear. The wavelength jumps in the discontinuous tuning mode stem from longitudinal mode jumps corresponding to the longitudinal mode spacing $\Delta\lambda_m$. The continuously tunable intervals in the quasicontinuous tuning mode each correspond to operation in a certain longitudinal mode, that can be continuously tuned over an interval exceeding $\Delta\lambda_m$. While in the discontinuous tuning mode not any wavelength within the tuning range can be addressed, the quasicontinuous mode allows access to all wavelengths by a proper choice of the corresponding control currents. Apparently, the tuning effort is largest in the discontinuous tuning mode requiring at least two current (or voltage) controls for the wavelength setting. Correspondingly, quasicontinuous tuning can be performed reasonably only by using a microprocessor control with look-ahead tables that are obtained by a detailed characterization of each device.

Electronic wavelength control requires the integration of additional controls into the laser cavity for the variation of the total wavelength-dependent cavity gain. This integration can either be performed longitudinally [29] or transversely [30] as shown in figure B2.2.11. As can be seen, in the longitudinal integration one replaces one (or both) of the laser mirrors, that usually are not wavelength selective, by wavelength-selective elements, the centre wavelength of which can be controlled electronically. In practice, these elements are DBR grating reflectors equipped with electronic control of the refractive index.

Likewise, the transverse integration scheme (figure B2.2.11(*b*)) comprises a DFB laser structure that provides single-mode operation and an additional electronic control mechanism for the effective refractive index of the transverse waveguide structure. In both approaches, the Bragg condition $\lambda_B = 2n_{\text{eff}}\Lambda$ yields the

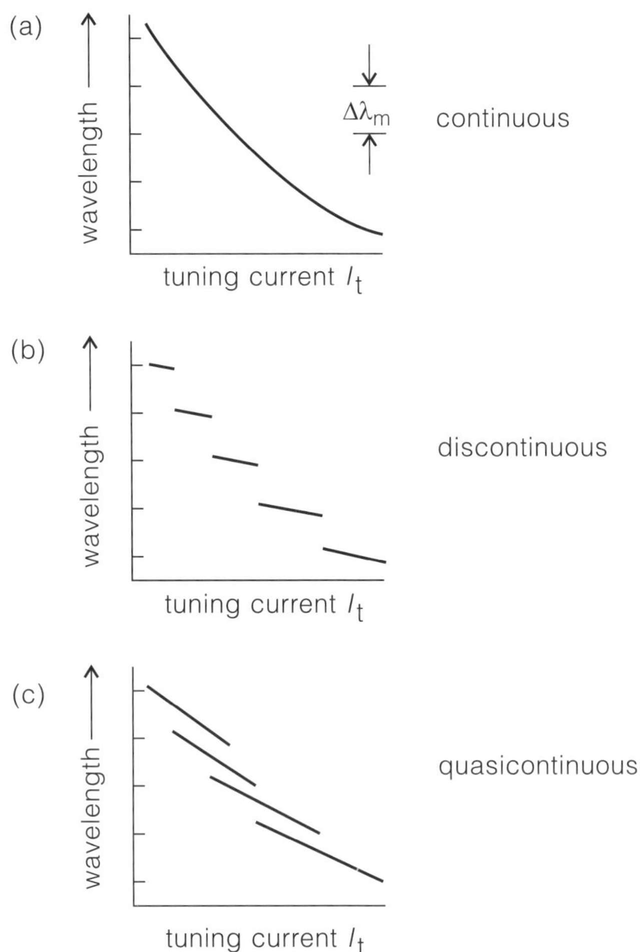

Figure B2.2.10. Tuning characteristics for the continuous (*a*), discontinuous (*b*) and quasicontinuous (*c*) tuning mode.

wavelength variation $\Delta\lambda$ due to the effective index change Δn_{eff}:

$$\Delta\lambda = \lambda_0 \frac{\Delta n_{\mathrm{eff}}}{n_{\mathrm{g,eff}}}. \tag{B2.2.22}$$

Whether the tuning is performed continuously or discontinuously depends on whether the optical cavity length $n_{\mathrm{eff}}L$ scales simultaneously with the centre wavelength of the wavelength-selective element. This is illustrated in figure B2.2.12(*a*), where discontinuous tuning via the shift of the filter curve of the Bragg reflector in a longitudinally integrated tunable laser is shown. Since the spectral positions of the longitudinal (Fabry–Pérot) modes remain unchanged, the otherwise continuous shift of the centre wavelength of the Bragg grating results in jumps from one longitudinal mode to the next. If, however, the longitudinal mode spectrum can be shifted synchronously (figure B2.2.12(*b*)) the continuous tuning mode can be achieved. For the transverse integration scheme of figure B2.2.10(*b*), an inherently continuous tuning behaviour is obtained because the shift of Bragg wavelength and optical cavity length, that determines the wavelengths of the longitudinal modes, occur synchronously.

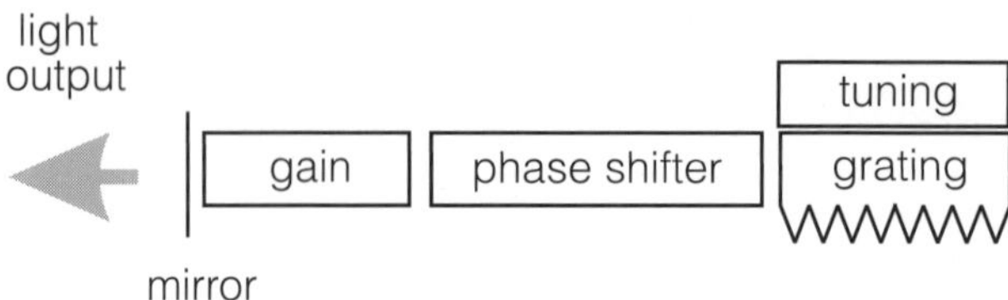

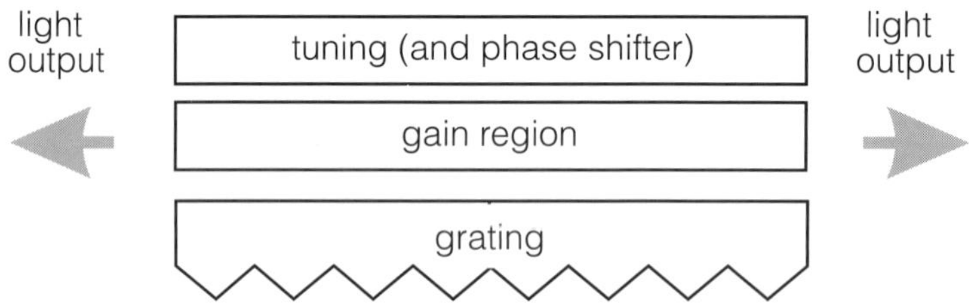

Figure B2.2.11. Longitudinally (*a*) and transversely (*b*) integrated wavelength-tunable laser diodes.

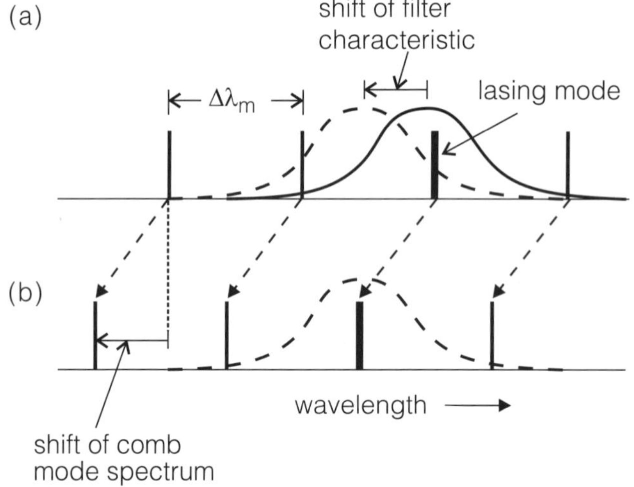

Figure B2.2.12. Discontinuous and continuous tuning. Tuning only the filter characteristic results in discontinuous tuning with mode jumps from one longitudinal mode to the next one (*a*). The simultaneous tuning of filter characteristic and comb mode spectrum results in continuous tuning (*b*).

The practically relevant physical mechanisms for electronic wavelength control are:

- the free-carrier plasma effect [31, 32],
- the quantum-confined Stark effect (QCSE) [33, 34] and
- the temperature dependence of the refractive index [35, 36].

For the plasma effect, the injection of an electron–hole plasma into a heterostructure reduces the refractive index leading to a red shift of the laser wavelength. The refractive index change Δn is proportional to the injected carrier density N:

$$\Delta n = \beta_{\mathrm{pl}} N \tag{B2.2.23}$$

where

$$\beta_{\mathrm{pl}} = -\frac{e^2\lambda^2}{8\pi^2 c^2 n\epsilon_0}\left(\frac{1}{m_{\mathrm{e}}} + \frac{1}{m_{\mathrm{h}}}\right) \tag{B2.2.24}$$

and m_{e} and m_{h} denote the effective masses of the injected electrons and holes, respectively, and n is the refractive index of the semiconductor. Note that β_{pl} is negative so that the refractive index and the wavelength both decrease by using the plasma effect for tuning. In addition to the real part, the imaginary part of the refractive index also changes, yielding an additional optical loss

$$\alpha_{\mathrm{pl}} = \frac{e^3\lambda^2}{4\pi^2 c^3 n\epsilon_0}\left(\frac{1}{m_{\mathrm{e}}^2\mu_{\mathrm{e}}} + \frac{1}{m_{\mathrm{h}}^2\mu_{\mathrm{h}}}\right) N \tag{B2.2.25}$$

where μ_{e} and μ_{h} are the mobilities of the electrons and holes, respectively. Δn can be as large as ~ -0.03 with an optical loss α_{pl} around 30 cm^{-1} at 1.5 μm wavelength in InGaAsP/InP heterostructures. While the plasma effect represents the strongest mechanism for electronic wavelength tuning, it suffers from the disadvantage that, due to recombination, a sustained current flow is needed that leads to significant heat generation. The latter not only degrades the laser performance (power) but also yields an inverse wavelength shift that markedly reduces the magnitude of the combined refractive index change.

The QCSE occurs if a reverse electric field is applied onto a pn-junction comprising a multiple quantum-well structure [33]. In this way the effective bandgap energy decreases and the electron and hole wavefunction overlap changes. Accordingly, the dipole matrix element changes as well resulting in a refractive index change. Either a red or blue shift may be obtained with the QCSE depending on the quantum-well design. The amount of index change is rather small; however, no sustained current flow is required and additional heat generation can be neglected. Also, the QCSE allows a fast change of the refractive index and is thus well suited for high-speed tuning or FM modulation in the GHz regime.

The refractive index variation by temperature change is essentially based on the temperature dependence of the bandgap energy of semiconductors. This shifts the absorption edge which, by means of the Kramers-Kronig dispersion relation, indirectly changes the refractive index. Thereby, the decrease of the bandgap energy with temperature yields an increase of the refractive index resulting in a red shift of the laser wavelength. In 1.55 μm wavelength InGaAsP/InP lasers the slope of the λ *versus* T characteristic is $\simeq$0.1 nm K^{-1} [35]. It should be noted that thermal tuning exhibits the slowest tuning speed (μs to ms regime) among the relevant wavelength control mechanisms. Admitting a maximum temperature change of 50 K, therefore, a tuning range of the order of 5 nm (0.3%) can be achieved by thermal tuning of InP-based laser diodes in the 1.55 μm wavelength region. However, heating laser diodes degrades many important laser parameters such as the output power, threshold current, spectral linewidth, modulation bandwidth and long-term reliability.

B2.2.3.1 Continuously tunable monolithic devices

The technologically simplest approach for a wavelength-tunable laser diode is the multi-section (typically two or three sections) DFB laser [37–39] shown in figure B2.2.13. Referring to figure B2.2.11, the three-section DFB laser is a longitudinally integrated device. With respect to reproducibility and single-mode yield, the symmetrical three-section DFB laser with a quarter-wave-shifted DFB grating and anti-reflection-coated end facets performs particularly well with largest continuous tuning ranges around 5 nm [40]. The anti-reflection coating is important because of the unknown and uncontrollable phase relation between the end facet and grating, which results in random device performance and reduced device yield.

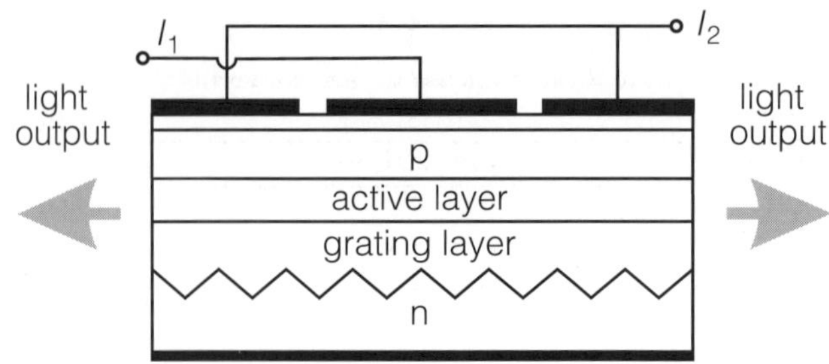

Figure B2.2.13. Schematic longitudinal view of a three-section DFB laser.

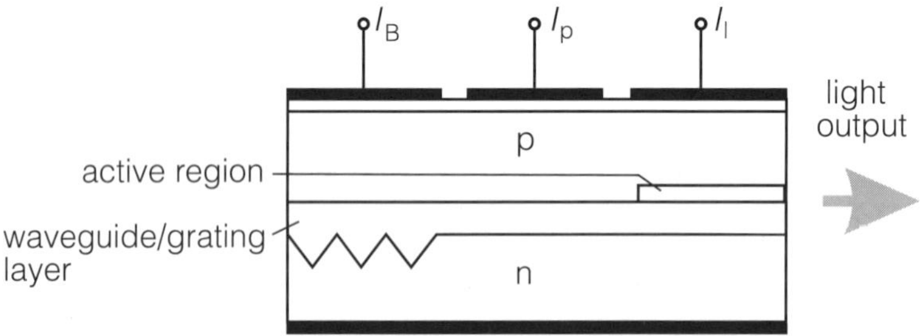

Figure B2.2.14. Schematic longitudinal view of three-section DBR laser.

Wavelength tuning is induced by an inhomogeneous biasing of the centre and side regions. Considering the gain-clamping mechanism in laser oscillators that keeps the total mode gain constant above threshold, it appears surprising that inhomogeneous biasing may lead to a wavelength shift. In fact, at equal total current no tuning occurs in an inhomogeneously biased Fabry–Pérot laser. In the case of DFB lasers, however, the longitudinally varying complex-field amplitudes yield longitudinally varying amplitude phase coupling. As a consequence, constant total gain may be maintained by a change of the round-trip phase that performs the wavelength shift. In summary, wavelength tuning in all-active multi-section DFB lasers is a rather complex mechanism, also comprising spatial hole-burning effects.

A distinct disadvantage of this laser type is that the controls of output power and wavelength are not separated, i.e. both currents I_1 and I_2 simultaneously affect output power and oscillation wavelength. The result is an increased complexity of the device control, particularly because of possible longitudinal-mode jumps for certain I_1–I_2 combinations. Despite their relative simple device technology, therefore, the multisection DFB lasers have become less important as continuously tunable laser diodes than other approaches.

A more convenient handling with an effective separation between power and wavelength control offers the three-section DBR laser [29, 41, 42] shown in figure B2.2.14, which is also a longitudinally integrated device. In this laser, the current into the active region (I_l) mainly determines the output power, while wavelength tuning is essentially accomplished with currents I_B and I_p. Referring to figure B2.2.14, the Bragg region current I_B determines the Bragg wavelength, i.e. the wavelength where the total cavity gain exhibits its maximum. The phase current I_p controls the effective refractive index in the phase section and thus the optical cavity length ($n_{eff}L$ product). The latter is relevant for the comb spectrum of the longitudinal modes and small changes of $n_{eff}L$ essentially affect the positions of the modes, rather than the spacings. Accordingly, the positions of the comb modes can be controlled by the phase current, so that continuous

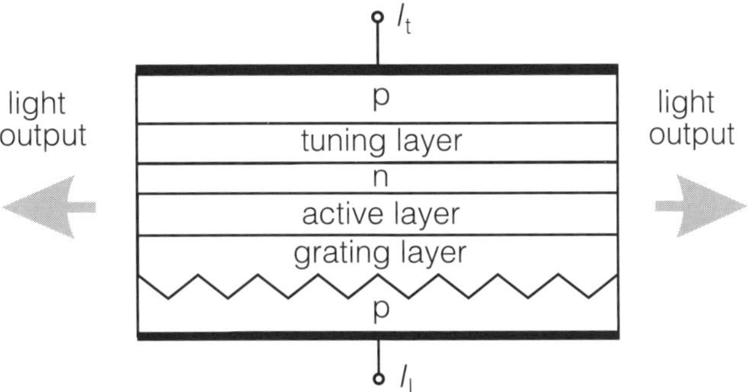

Figure B2.2.15. Schematic longitudinal section of transversely integrated tunable twin-guide (TTG) laser diode.

tuning can be achieved in the way shown in figure B2.2.12 by a proper mutual adjustment with the Bragg region. Otherwise, by a solitary change of either I_B or I_p the tuning behaviour is essentially discontinuous.

The continuous tuning range of the three-section DBR laser is usually limited by the tunability of the phase section, because the tuning of the phase section has to match the optical cavity changes including the active section that is not tuned. With well-designed devices, a continuous wavelength tuning over some 5 nm [42, 43] and a discontinuous tuning up to 8 nm [44] can be achieved in InP-based lasers at 1.55 μm wavelength. Again, however, the wavelength setting requires considerable effort with respect to device characterization and control of the three currents and temperature. In particular, it is important that mode jumps are avoided, which usually requires device control via microprocessors and look-ahead tables.

The most convenient handling performance can be achieved with the transversely integrated structures, such as the tunable twin-guide (TTG) laser [45–47], shown in figure B2.2.15, or the striped heater DFB lasers [48]. This is because in these devices the refractive index change is applied homogeneously along the entire laser axis so that the optical cavity length changes are automatically synchronized with the Bragg wavelength changes. Consequently, the tuning behaviour of these devices is inherently continuous. As can be seen for the case of the TTG laser, carriers injected into the tuning layer lead to a longitudinally homogeneous reduction of the refractive index resulting in a blue shift. Also, only two control currents I_l and I_t are required for the output power and wavelength control. As indicated by the subscripts, current I_l controls the output power and current I_t essentially controls the wavelength. Both controls, however, are not strictly orthogonal to each other, so that the output power is affected also by I_t because of the additional loss by the injected carriers and the wavelength is influenced also by I_l because of temperature changes.

Since the tuning layer extends over the entire laser length, and not only part of it as in the phase control section in the three-section DBR laser, the tuning effect is stronger in the transversely integrated devices. Accordingly, the largest continuous tuning ranges reported so far ($\simeq$13 nm) have been achieved with TTG laser diodes [49] while the thermal tuning of DFB lasers yields up to some 4 nm of continuous tuning at 1.55 μm wavelength [48].

Besides the laser wavelength, the electronic wavelength tuning influences other laser characteristics. First of all, the optical power usually decreases because of the additional loss (plasma effect) or the larger threshold current (thermal tuning). The reduction of the optical power, in turn, leads to an increased spectral linewidth according to the Schawlow–Townes–Henry formula (equation (B2.2.21)). In case of the most often used plasma effect the linewidth generally shows distinct additional broadening, denoted $\Delta\nu_t$, because of shot-noise in the tuning region(s) [50]. This contribution to the linewidth amounts to

$$\Delta\nu_t = 4\pi e \frac{c^2}{\lambda^4} \left(\frac{d\lambda}{dI_t}\right)^2 I_t \tag{B2.2.26}$$

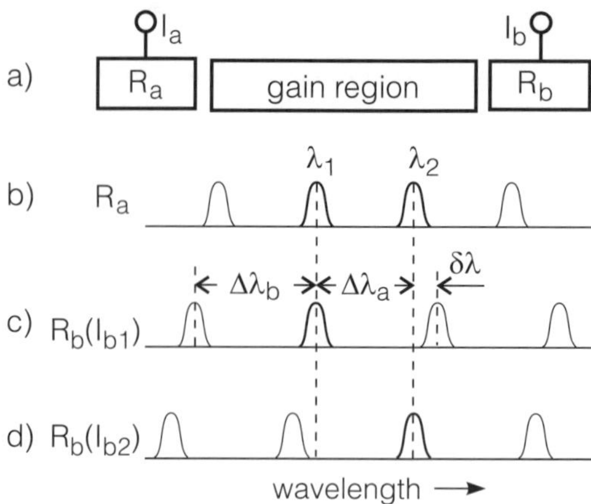

Figure B2.2.16. Schematic illustration of widely tunable laser diodes exploiting the vernier effect (*a*). Both mirrors exhibit comb reflection spectra (*b,c*) with different pitches $\Delta\lambda_a$ and $\Delta\lambda_b$. Lasing can occur at wavelength λ_1, where both mirrors yield reflection. By a small shift $\delta\lambda$ of the reflection spectrum of mirror B, the laser wavelength shifts by a much larger amount, $\Delta\lambda_a$, to λ_2, where now both mirrors reflect (*d*).

in TTG laser diodes and

$$\Delta\nu_t = 4\pi e \frac{c^2}{\lambda^4}\left\{\left(\frac{\partial\lambda}{\partial I_p}\right)^2 I_p + \kappa L_B\left(\frac{\partial\lambda}{\partial I_B}\right)^2 I_B\right\} \tag{B2.2.27}$$

in three-section DBR laser diodes. Apparently, this broadening is proportional to the tuning efficiencies squared so that the wider tunable devices show the larger broadening. In TTG laser diodes with tuning ranges of the order of 7 nm at 1.55 μm wavelength, $\Delta\nu_t$ can be up to several ten MHz, while three-section DBR lasers with their smaller continuous tuning ranges, $\Delta\nu_t$ can generally be kept below about 10 MHz [50].

B2.2.3.2 Wavelength-tunable vertical-cavity surface-emitting lasers (VCSELs)

Continuous wavelength tuning can also be accomplished with vertical-cavity surface-emitting lasers (VCSELs). Longitudinal-single-mode operation and the efficient suppression of other longitudinal modes because of the short cavity length eases the mode control considerably. VCSELs may be tuned either externally by using a piezoelectric transducer actuator [51] or with an integrated micromechanically moveable top mirror [52]. Tuning ranges around 40 nm have been reported for the wavelength ranges 950 nm [53] and 1.3 μm [51].

B2.2.3.3 Discontinuously tunable monolithic devices

The continuous electronic wavelength tuning is ultimately limited by the maximal achievable effective refractive index change (see equation (B2.2.22)). With maximum refractive index changes of the order of 1% and maximum confinement factors of the tuning regions around 50%, therefore, the continuous tuning range is limited to about 0.5–1% of the wavelength. Significantly larger tuning ranges can be achieved if discontinuous or quasicontinuous tuning can be accepted. In this case the tuning range is enhanced by mode jumps from one longitudinal mode to the next one. Again, electronic tuning is induced by refractive index changes exploiting the physical mechanisms discussed in the preceding section. Thereby, largest discontinuous and/or quasicontinuous tuning range around 22 nm was achieved by combining the free-carrier plasma effect and thermal heating [54].

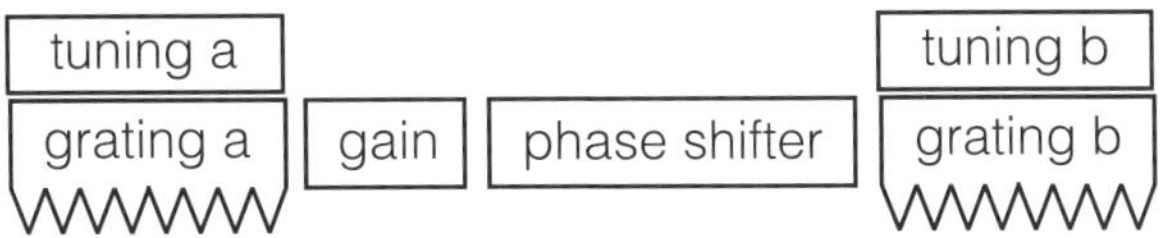

Figure B2.2.17. Schematic longitudinal section of longitudinally integrated tunable laser diode exploiting the vernier effect.

In order to obtain a large tuning effect one exploits the vernier effect, by which small relative refractive index changes can be used to yield large relative wavelength changes. The application of the vernier effect in tunable laser diodes is illustrated in figure B2.2.16 for the case of the longitudinal integration technique. The essential device components are the two mirrors with comb reflection spectra exhibiting different pitch $\Delta\lambda_a$ and $\Delta\lambda_b$, respectively. Typically, $\Delta\lambda_a$ and $\Delta\lambda_b$ differ only by 3–10%. The comb reflection spectra may be achieved, for instance, by spatially modulated Bragg gratings (sampled gratings, super structure gratings, chirped gratings). As shown in figures B2.2.16(*b*) and (*c*), lasing can occur only at wavelength λ_1, where both mirrors exhibit non-vanishing reflection. By a small shift $\delta\lambda$ of the comb reflection spectrum of mirror B, using the free-carrier plasma effect, the 'resonance' at λ_1 breaks and lasing can now take place at wavelength λ_2. Consequently, the wavelength tuning by $\Delta\lambda_a$ from λ_1 to λ_2 is much larger than $\delta\lambda$ because of the vernier effect provided by the two comb reflection spectra with slightly different pitch.

The schematic structure of the corresponding laser in the longitudinal integration technique [55–57] is shown in figure B2.2.17 and includes a phase-shift section [58] to enable the quasi-continuous tuning mode. This type of device may have up to four control currents including the current into the gain section for power control, two tuning currents into the mirrors for adjusting the comb reflection spectra, and a phase control current into the phase shift section to adjust the Fabry–Pérot modes in order to enable quasicontinuous tuning.

Practically, the amplification factor due to the vernier effect,

$$T = \frac{\Delta\lambda_a}{\Delta\lambda_a - \Delta\lambda_b} \tag{B2.2.28}$$

can be of the order of 10–100, so that the maximum discontinuous tuning range is now finally limited by the gain bandwidth of the laser gain region. The latter is of the order of 10% of the laser wavelength for the usual III/V laser diodes and discontinuous and quasicontinuous tuning ranges around 100 nm were achieved in InGaAsP/InP laser diodes at 1.55 μm wavelength [59, 60].

The vernier effect can also be applied to transversely integrated structures such that the small differences in optical pathlength of the interfering modes of a twin waveguide are tuned [61, 62]. In this way, the wavelength for constructive interference, i.e. the laser wavelength, can be tuned widely by small changes of the optical path difference

$$(n_{\mathrm{eff},1} - n_{\mathrm{eff},2})L \tag{B2.2.29}$$

where $n_{\mathrm{eff},1}$ and $n_{\mathrm{eff},2}$ are the effective refractive indices of the participating transverse modes. The resulting wavelength change is

$$\Delta\lambda = \lambda\frac{\Delta(n_{\mathrm{eff},1} - n_{\mathrm{eff},2})}{n_{\mathrm{g,eff},1} - n_{\mathrm{g,eff},2}} \tag{B2.2.30}$$

where $\Delta(n_{\mathrm{eff},1} - n_{\mathrm{eff},2})$ denotes the change of the difference of the effective refractive indices of the two modes. The comparison with equation (B2.2.22) shows that the amplification factor of the transversely integrated structures due to the vernier effect is

$$T = \frac{n_{\mathrm{g,eff},1}}{n_{\mathrm{g,eff},1} - n_{\mathrm{g,eff},2}}. \tag{B2.2.31}$$

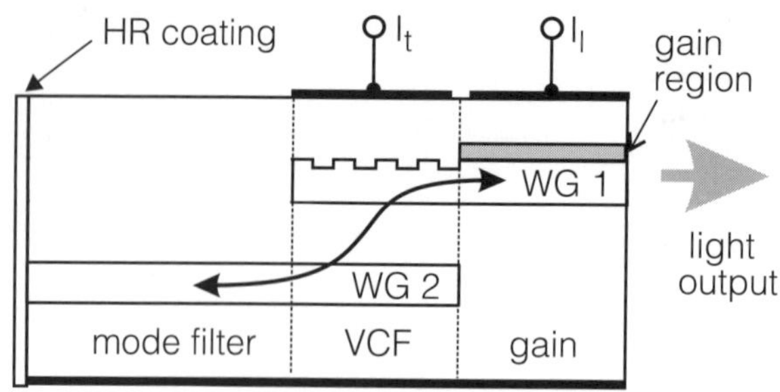

Figure B2.2.18. Schematic longitudinal section of a discontinuously tunable laser diode in the transverse integration technique (VCF laser: vertical cavity filter).

While $n_{g,eff,1}$ is of the order of 3–4, $n_{g,eff,1} - n_{g,eff,2}$ is typically less than 0.1, T is larger than 30 so that even small changes $\Delta(n_{eff,1} - n_{eff,2})$ may cause large changes of the laser wavelength. The schematic longitudinal section of a representative device is shown in figure B2.2.18. The tuning in this device is done by the vertical coupler filter (VCF), in which the superposition of the two guided transverse mode gives a power transfer from waveguide 1 (WG 1) to waveguide 2 (WG 2) only at the coupling wavelength. The latter can be tuned by current I_t which, by the plasma effect, dominantly changes the effective refractive index of the mode in the upper waveguide and thus the difference $n_{eff,1} - n_{eff,2}$ occurring in equation (B2.2.30). The tuning ranges achievable with the transversely integrated devices is again of the order of 50–100 nm for 1.55 μm lasers [61]. Particular performance with tuning ranges exceeding 100 nm at 1.55 μm wavelength [63] and spectral linewidth below 4 MHz [64] can be obtained by combining the longitudinal integration scheme with the transverse one.

Only few data have been reported on the spectral properties of discontinuously tunable laser diodes. This is mainly because the linewidth strongly changes during tuning, showing singularities near the mode jump regimes. In linewidth-sensitive applications, therefore, care has to be taken to avoid operation of discontinuously as well as quasicontinuously tunable devices near to longitudinal mode jumps.

B2.2.3.4 External cavity tunable laser diodes

The use of an external cavity for a tunable laser diode [65, 66] may yield significant advantages; the efforts in adjustment and control, however, may be considerable. For a proper operation of a laser diode in an external cavity it is necessary that the reflection of at least one of the end facets be sufficiently well suppressed in order to avoid multiple cavity effects or, still worse, to establish an optical feedback configuration with reduced or unpredictable stability. The main advantages of external cavity diode lasers with respect to tuning are the smaller spectral linewidth due to the smaller spontaneous emission input into the laser mode, and the possibility to tune the wavelength precisely by means of mechanical movements. However, mechanical and thermal stability of the external cavity may be a problem in practice.

The principal configuration of a simple external cavity tunable laser diode is shown in figure B2.2.19 In this device the laser wavelength is determined by the inclination of the grating reflector, because only at wavelength

$$\lambda = 2\Lambda \sin\phi \qquad \text{(B2.2.32)}$$

does reflection back into the diode occur. Changing the angle ϕ allows discontinuous wavelength tuning with mode jumps equal to the (small) longitudinal mode spacing of the external cavity. With this technique large tuning ranges can be achieved: the maximum tuning range reported so far is around 240 nm in the 1.3–1.55 μm wavelength regime [67]. Continuous tuning is also possible by choosing the axis for the grating

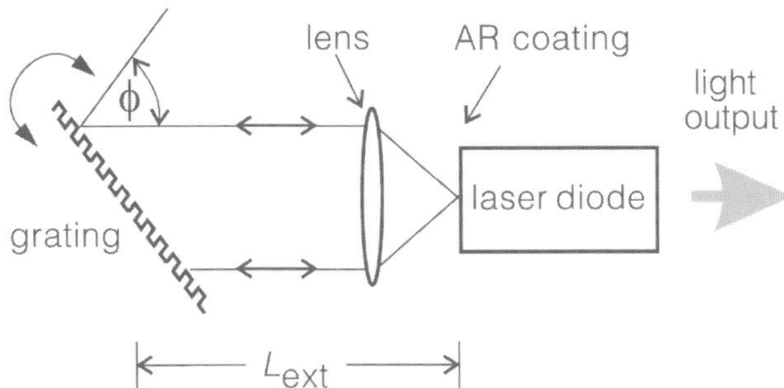

Figure B2.2.19. Principal setup of external cavity tunable laser diode with a wavelength selective grating reflector.

rotation appropriately [68]. In this way the length of the external cavity is changed simultaneously with the laser wavelength which is the prerequisite for continuous tuning.

B2.2.4 Multiwavelength laser diodes

In many applications, such as WDM, an array of laser diodes is required in which each laser operates at a well-defined wavelength, usually in equidistant wavelength channels. For proper device operation it is necessary for each laser to be tunable over a certain wavelength range. If the tuning range of each element exceeds the wavelength difference between the elements these devices may cover a huge wavelength range without a gap, making these devices particularly attractive as spare light sources. Simple multiwavelength laser arrays exhibit separate outputs for each laser. Most often, however, it is necessary to combine the outputs of all elements into one output, e.g. a single optical fibre. In the latter case the multiwavelength laser array represents a considerably integrated photonic device comprising a combiner. Multiwavelength laser diodes can essentially be of the following four types:

- simple arrays [69, 70],
- laser arrays integrated with a combiner [71],
- multi-stripe grating cavities [72] or
- phased arrays [73].

Multiwavelength laser diodes are also useful in applications where multiple wavelengths in a fixed channel spacing must rapidly be accessed. Devices with 40 wavelengths spaced by 100 GHz in the 1.55 μm wavelength range based on a chirped waveguide grating router have successfully been demonstrated [74].

References

[1] Osiński M and Buus J 1987 Linewidth broadening factor in semiconductor lasers—an overview *IEEE J. Quantum Electron.* **23** 9–29

[2] Petermann K 1995 External optical feedback phenomena in semiconductor lasers *IEEE J. Selected Topics Quantum Electron.* **1** 480–9

[3] Kojima K and Kyuma K 1990 Multi-quantum well distributed feedback and distributed Bragg reflector lasers *Semicond. Sci. Technol.* **5** 481–93

[4] Suematsu Y, Arai S and Kishino K 1983 Dynamic single-mode semiconductor lasers with a distributed reflector *IEEE J. Lightwave Technol.* **1** 161–76

[5] Kogelnik H and Shank C V 1972 Coupled-wave theory of distributed feedback lasers *J. Appl. Phys.* **43** 2327–35

[6] Carroll J, Whiteaway J and Plumb D 1998 *Distributed Feedback Semiconductor Lasers* (Stevenage: IEE)

[7] Buus J 1986 Dynamic single-mode operation of DFB lasers with phase shifted gratings and reflecting mirrors *IEE Proc.* **133** 163–4

[8] Matsuoka T, Yoshikuni Y and Nagai H 1985 Verification of the light phase effect at the facet on DFB laser properties *IEEE J. Quantum Electron.* **21** 1880–6

[9] Utaka K, Akiba S, Sakai K and Matsushima Y 1986 λ/4-shifted InGaAsP/InP DFB lasers *IEEE J. Quantum Electron.* **22** 1042–51

[10] Sasaki T, Takano S, Henmi N, Yamada Y, Kitamura M, Hasumi H and Mito I 1988 1.5 μm $\lambda/4$ shifted multiple quantum well distributed feedback laser diodes *Electron. Lett.* **24** 1408–9

[11] Luo Y, Nakano Y, Tada K, Inoue T, Hosomatsu H and Iwaoka H 1990 Purely gain-coupled distributed feedback semiconductor lasers *Appl. Phys. Lett.* **56** 1620–2

[12] Utaka K, Akiba S, Sakai K and Matsushima Y 1984 Analysis of quarter-wave-shifted DFB laser *Electron. Lett.* **20** 326–7

[13] Usami M and Akiba S 1989 Suppression of longitudinal spatial hole-burning effect in λ/4-shifted DFB lasers by nonuniform current injection *IEEE J. Quantum Electron.* **25** 1245–53

[14] Borchert B, David K, Stegmüller B, Gessner R, Beschorner M, Sacher D and Franz G 1991 1.55 μm gain-coupled quantum-well distributed feedback lasers with high single-mode yield and narrow linewidth *IEEE Photon. Technol. Lett.* **3** 955–7

[15] Hui R, Kavehrad M and Makino T 1994 External feedback sensitivity of partly gain-coupled DFB semiconductor lasers *IEEE Photon. Technol. Lett.* **6** 897–9

[16] Henry C H 1982 Theory of the linewidth of semiconductor lasers *IEEE J. Quantum Electron.* **18** 259–64

[17] Henry C H 1986 Phase noise in semiconductor lasers *IEEE J. Lightwave Technol.* **4** 298–311

[18] Pan X, Olesen H and Tromborg B 1990 Spectral linewidth of DFB lasers including the effects of spatial hole burning and nonuniform current injection *IEEE Photon. Technol. Lett.* **2** 312–15

[19] Pan X, Tromborg B and Olesen H 1991 Linewidth rebroadening in DFB lasers due to weak side modes *IEEE Photon. Technol. Lett.* **3** 112–14

[20] Ogita S, Kotaki Y, Kihara K, Matsuda M, Ishikawa H and Imai H 1988 Dependence of spectral linewidth on cavity length and coupling coefficient in DFB laser *Electron. Lett.* **24** 613–14

[21] Kitamura M, Yamazaki H, Sasaki T, Kida N, Hasumi H and Mito I 1990 250 kHz spectral linewidth operation of 1.5 μm multiple quantum well DFB-LDs *IEEE Photon. Technol. Lett.* **2** 310

[22] Okai M, Suzuki M, Taniwatari T and Chinone N 1994 Corrugation-pitch-modulated distributed feedback lasers with ultranarrow spectral linewidth *Japan. J. Appl. Phys.* **33** 2563–70

[23] Park C A and Williams P J 1989 Single-mode laser sources for FSK systems *IEE Proc.* **136** 18–21

[24] Strzelecki E M, Cohen D A and Coldren L A 1988 Investigation of tunable single frequency diode lasers for sensor applications *IEEE J. Lightwave Technol.* **6** 1610–18

[25] Koch T L and Koren U 1990 Semiconductor lasers for coherent optical fiber communications *IEEE J. Lightwave Technol.* **8** 274–93

[26] Delorme F 1998 Widely tunable 1.55-μm lasers for wavelength-division-multiplexed optical fiber communications *IEEE J. Quantum Electron.* **34** 1706–16

[27] Kotaki Y and Ishikawa H 1991 Wavelength tunable DFB and DBR lasers for coherent optical fibre communications *IEE Proc.* **138** 171–7

[28] Dieckmann A and Amann M C 1995 FMCW-LIDAR with tunable twin-guide laser diode *Trends in Optical Fibre Metrology and Standards* (Dordrecht: Kluwer) pp 791–802

[29] Coldren L A and Corzine S W 1987 Continuously-tunable single-frequency semiconductor lasers *IEEE J. Quantum Electron.* **23** 903–8

[30] Amann M C and Thulke W 1990 Continuously tunable laser diodes: longitudinal versus transverse tuning scheme *IEEE J. Selected Areas Commun.* **8** 1169–77

[31] Bennet B, Soref R and DelAlamo J 1990 Carrier-induced change in refractive index of InP, GaAs, and InGaAsP *IEEE J. Quantum Electron.* **26** 113–22

[32] Weber J P 1994 Optimization of the carrier-induced effective index change in InGaAsP waveguides—application to tunable Bragg filters *IEEE J. Quantum Electron.* **30** 1801–16

[33] Miller D A B, Chemla D S, Damen T C, Gossard A C, Wiegmann W, Wood T H and Burrus C A 1984 Band edge electroabsorption in quantum well structures: the quantum confined Stark effect *Phys. Rev. Lett.* **53** 2173–6

[34] Susa N and Nakahara T 1992 Enhancement of change in the refractive index in an asymmetric quantum well *Appl. Phys. Lett.* **60** 2457–9

[35] Chinen K, Gen-Ei K, Suhara H, Tanaka A, Matsuyama T, Konno K and Muto Y 1987 Low-threshold 1.55 μm InGaAsP/InP buried heterostructure distributed feedback lasers *Appl. Phys. Lett.* **51** 273–5

[36] Woodward S L, Koren U, Miller B I, Young M G, Newkirk M A and Burrus C A 1992 A DBR laser tunable by resistive heating *IEEE Photon. Technol. Lett.* **4** 1330–2

[37] Horita M, Tsurusawa M, Utaka K and Matsushima Y 1993 Wavelength-tunable InGaAsP-InP multiple-λ/4-shifted distributed feedback laser *IEEE J. Quantum Electron.* **29** 1810–16

[38] Tohyama M, Onomura M, Funemizu M and Suzuki N 1993 Wavelength tuning mechanism in three-electrode DFB lasers *IEEE Photon. Technol. Lett.* **5** 616–18

[39] Ishii H, Kondo Y, Kano F and Yoshikuni Y 1998 A tunable distributed amplification DFB laser diode (TDA-DFB-LD) *IEEE Photon. Technol. Lett.* **10** 30–2

[40] Tohyama M, Funemizu M, Onomura M, Takakuwa C and Suzuki N 1995 Mechanism of wavelength tuning and frequency modulation in three-electrode DFB lasers *IEEE J. Selected Topics Quantum Electron.* **1** 416–26

[41] Caponio N P, Goano M, Maio I, Meliga M, Bava G P, Destefanis G and Montrosset I 1990 Analysis and design criteria of three-section DBR tunable lasers *IEEE J. Selected Areas Commun.* **8** 1203–13

[42] Murata S, Mito I and Kobayashi K 1987 Over 720 GHz (5.8 nm) frequency tuning by a 1.5 μm DBR laser with phase and Bragg wavelength control regions *Electron. Lett.* **23** 403–5

[43] Murata S, Mito I and Kobayashi K 1988 Tuning ranges for 1.5 μm wavelength tunable DBR lasers *Electron. Lett.* **24** 577–9

[44] Koch T L, Koren U, Gnall R P, Burrus C A and Miller B I 1988 Continuously tunable 1.5 μm multiple-quantum-well GaInAs/GaInAsP distributed-Bragg-reflector lasers *Electron. Lett.* **24** 1431–3

[45] Amann M C, Illek S, Schanen C and Thulke W 1989 Tunable twin-guide laser: A novel laser diode with improved tuning performance *Appl. Phys. Lett.* **54** 2532–3

[46] Amann M C 1995 Broad-band wavelength-tunable twin-guide lasers *Optoelectronics—Devices and Technologies* **10** 27–38

[47] Yamamoto E, Hamada M, Suda K, Nogiwa S and Oki T 1992 Wavelength tuning characteristics of tunable twin-guide lasers with improved current-injection structures *Appl. Phys. Lett.* **60** 805–6

[48] Sakano S, Tsuchiya T, Suzuki M, Kitajima S and Chinone N 1992 Tunable DFB laser with a striped thin-film heater *IEEE Photon. Technol. Lett.* **4** 321–3

[49] Wolf T, Illek S, Rieger J, Borchert B and Amann M C 1994 Extended continuous tuning range (over 10 nm) of tunable twin-guide lasers *Conf. on Lasers and Electro-Optics (CLEO '94) (Anaheim, CA)* p CWB 1

[50] Amann M C and Schimpe R 1990 Excess linewidth broadening in wavelength-tunable laser diodes *Electron. Lett.* **26** 279–80

[51] Hsu K, Miller C M, Babic D, Houng D and Taylor A 1998 Continuously tunable photopumped 1.3 μm fiber Fabry–Pérot surface-emitting lasers *IEEE Photon. Technol. Lett.* **10** 1199–201

[52] Tyebati P, Wang P, Vakhshoori D, Lu C C, Azimi M and Sacks R N 1998 Half-symmetric cavity tunable micromechanical VCSEL with single spatial mode *IEEE Photon. Technol. Lett.* **10** 1679–81

[53] Wang P, Tayebati P, Vakhshoori D, Lu C C, Azimi M and Sacks R N 1999 Half-symmetric cavity micromechanically tunable vertical cavity surface emitting lasers with single spatial mode operating near 950 nm *Appl. Phys. Lett.* **75** 897–8

[54] Öberg M, Nilsson S, Klinga T and Ojala P 1991 A three-electrode distributed Bragg reflector laser with 22 nm wavelength tuning range *IEEE Photon. Technol. Lett.* **PTL-3** 299–301

[55] Tohmori Y, Yoshikuni Y, Tamamura T, Ishii H, Kondo Y and Yamamoto M 1993 Broad-range wavelength tuning in DBR lasers with super structure grating (SSG) *IEEE Photon. Technol. Lett.* **5** 126–9

[56] Mason B, Fish G A, DenBaars S P and Coldren L A 1998 Ridge waveguide sampled grating DBR lasers with 22 nm quasi-continuous tuning range *IEEE Photon. Technol. Lett.* **10** 1211–13

[57] Jayaraman V, Chuang Z M and Coldren L A 1993 Theory, design, and performance of extended tuning range semiconductor lasers with sampled gratings *IEEE J. Quantum Electron.* **29** 1824–34

[58] Ishii H, Tanobe H, Kano F, Tohmori Y, Kondo Y and Yoshikuni Y 1996 Broad-range wavelength coverage (62.4 nm) with superstructure-grating DBR laser *Electron. Lett.* **32** 454–5

[59] Tohmori Y, Yoshikuni Y, Tamamura T, Yamamoto M, Kondo Y and Ishii H 1993 Over 100 nm wavelength tuning in superstructure grating (SSG) DBR lasers *Electron. Lett.* **29** 352–4

[60] Jayaraman V, Heimbuch M E, Coldren L A and DenBaars S P 1994 Widely tunable continuous-wave InGaAsP/InP sampled grating lasers *Electron. Lett.* **30** 1492–4

[61] Kim I, Alferness R C, Koren U, Buhl L L, Miller B I, Young M G, Chien M D, Koch T L, Presby H M, Raybon G and Burrus C A 1994 Broadly tunable vertical-coupler filtered tensile-strained InGaAs/InGaAsP multiple quantum well laser *Appl. Phys. Lett.* **64** 2764–6

[62] Amann M C and Illek S 1993 Tunable laser diodes utilizing transverse tuning scheme *IEEE J. Lightwave Technol.* **11** 1168–82

[63] Rigole P J, Nilsson S, Bäckbom L, Klinga T, Wallin J, Stålnacke B, Berglind E and Stoltz B 1995 114 nm wavelength tuning range of a vertical grating assisted codirectional coupler laser with a super structure grating distributed Bragg reflector *IEEE Photon. Technol. Lett.* **7** 697–9

[64] Saavedra A A, Rigole P J, Goobar E, Schatz R and Nilsson S 1998 Relative intensity noise and linewidth measurements of a widely tunable GCSR laser *IEEE Photon. Technol. Lett.* **10** 481–3

[65] Lee B L and Lin C F 1998 Wide-range tunable semiconductor lasers using asymmetric dual quantum wells *IEEE Photon. Technol. Lett.* **10** 322–4

[66] Mehuys D, Mittelstein M, Yariv A, Sarfaty R and Ungar J E 1989 Optimized Fabry–Pérot (AlGa)As quantum-well lasers tunable over 105 nm *Electron. Lett.* **25** 143–5

[67] Bagley M, Wyatt R, Elton D J, Wickes H J, Spurdens P C, Seltzer C P, Cooper D M and Devlin W J 1990 242 nm continuous tuning from a GRIN-SC-MQW-BH InGaAsP laser in an extended cavity *Electron. Lett.* **26** 267–9

[68] Favre F and Le Guen D 1991 82 nm of continuous tunability for an external cavity semiconductor laser *Electron. Lett.* **27** 183–4

[69] Okuda H, Hirayama Y, Furuyama H, Kinoshita J I and Nakamura M 1987 Five-wavelength integrated DFB laser arrays with quarter-wave-shifted structures *IEEE J. Quantum Electron.* **23** 843–8

[70] Nakao M, Sato K, Nishida T and Tamamura T 1990 Distributed feedback laser arrays fabricated by synchrotron orbital radiation lithography *IEEE J. Selected Areas Commun.* **8** 1178–82

[71] Young M G, Koren U, Miller B I, Chien M, Koch T L, Tennant D M, Feder K, Dreyer K and Raybon G 1995 Six wavelength laser array with integrated amplifier and modulator *Electron. Lett.* **31** 1835–6

[72] Asghari M, Zhu B, White I H, Seltzer C P, Nice C, Henning I D, Burness A L and Thompson G H B 1994 Demonstration of an integrated multichannel grating cavity laser for WDM applications *Electron. Lett.* **30** 1674–5

[73] Zirngibl M, Joyner C H, Doerr C R, Stulz L W and Presby H M 1996 An 18-channel multifrequency laser *IEEE Photon. Technol.*

Lett. **8** 870–2
[74] Doerr C R, Joyner C H and Stulz L W, 1999 40-wavelength rapidly digitally tunable laser *IEEE Photon. Technol. Lett.* **11** 1348–50

Further reading

Amann M C and Buus J 1998 *Tunable Laser Diodes* (Norwood, MA: Artech House)

A survey over fundamentals, device structures and performance, and applications of monolithic tunable laser diodes.

Buus J 1991 *Single Frequency Semiconductor Lasers* (Bellingham, WA: SPIE)

A comprehensive introduction into theory and device structures of single mode laser diodes.

Carroll J, Whiteaway J and Plumb D 1998 *Distributed Feedback Semiconductor Lasers* (Stevenage: IEE)

Covering all aspects of DFB lasers including theory, modelling, system aspects. Provides also a useful set of simulation programs (MATLAB).

Coldren L A and Corzine S W 1995 *Diode Lasers and Photonic Integrated Circuits* (Chichester: Wiley)

An extensive textbook proceeding from the fundamentals of diode lasers to advanced photonic devices.

Ikegami T, Sudo S and Sakai Y 1995 *Frequency Stabilization of Semiconductor Laser Diodes* (Norwood, MA: Artech House)

A rather complete compilation of the various techniques for frequency stabilization of laser diodes and their practical performance.

Morthier G and Vankwikelberge P 1997 *Handbook of Distributed Feedback Laser Diodes* (Norwood, MA: Artech House)

A detailed monography on distributed feedback laser diodes with particular emphasis on the theoretical decription and design.

Ohtsu M 1996 *Frequency Control of Semiconductor Lasers* (Chichester: Wiley)

A useful book on theory and performance of frequency control techniques for laser diodes.

B2.3
High-speed laser diodes

Peter P Vasil'ev

B2.3.1 Introduction

Semiconductor lasers are now considered to be the optimum optical sources for a wide range of optical fibre communication links and ultrafast optical data processing systems. Diode lasers capable of being modulated at very high bit rates are key requirements for high-speed optical communication links. The major feature of the laser is its ability to generate modulated optical signals as a result of the direct injection of electrical currents. Hence the dynamic response and modulation behaviour of lasers are of great importance for many applications and have been extensively studied by many researchers.

High-speed modulation can be realized either internally or externally. Directly modulated laser diode systems are simpler to implement compared with external modulation. They can be fabricated in conjunction with the drive electronics of the transmitter circuitry on the same wafer to make up optoelectronic integrated circuits. A number of external modulators, including electroabsorption and electrorefractive modulators, can be used for ultrafast external modulation and data encoding which can often provide less frequency chirp than that under direct modulation.

In principle, two aspects to designing high-speed diode lasers exist. The first one is the physical aspect which involves the optimization of the laser active area, taking into account an enhancement of the differential gain, decreasing the threshold current, minimizing the carrier transport effects in quantum well (QW) lasers, etc. The other is the device aspect which involves the fabrication of a structure and a laser cavity, minimization of the parasitic capacitance and series resistance, etc. Both these aspects are thought to be of equal importance for the realization of practical high-speed diode lasers and are discussed here.

Very high bit rate communication systems with a capacity of 100 Gbit s^{-1} and higher require sources of ultrashort optical pulses. Laser pulses shorter than 10 ps with minimum possible bandwidths (bandwidth-limited pulses) are necessary to ensure such speeds of data transmission and processing. Both gain/Q-switching and mode-locking techniques have been widely used to generate picosecond and femtosecond optical pulses from diode lasers. A number of methods for external optical pulse compression, including the soliton effect, have been often utilized to produce even shorter pulses than lasers can generate themselves.

B2.3.2 Dynamics of laser diodes

B2.3.2.1 Review of rate equations for study of dynamics

The dynamics of diode lasers, including laser modulation and ultrashort pulse generation, have been conventionally modelled using a set of a few coupled first-order differential equations [1–4]. The simplest rate equations consist of just two equations, one for the carrier density n in the active area and the other for the photon density S in the laser cavity

$$\frac{\mathrm{d}n}{\mathrm{d}t} = \frac{j(t)}{ed} - g_0(n - n_\mathrm{t})S - \frac{n}{\tau_\mathrm{s}} \tag{B2.3.1}$$

$$\frac{\mathrm{d}S}{\mathrm{d}t} = \Gamma g_0(n - n_\mathrm{t})S - \frac{S}{\tau_\mathrm{ph}} + \frac{\beta\Gamma n}{\tau_\mathrm{s}} \tag{B2.3.2}$$

where n_t is the transparency density, g_0 is the differential gain coefficient, β is the spontaneous coupling factor, e is the electron charge, d is the active layer thickness, $j(t)$ is the current density, Γ is the optical confinement factor and τ_s and τ_ph are the carrier and photon lifetimes. Here, the optical gain is assumed to be proportional to the carrier concentration. These simple equations are written on the assumption that the laser operates in a single longitudinal and transverse mode, that is, they do not take into account the distribution of S between any multiple cavity modes. τ_ph is given as [1]

$$\tau_\mathrm{ph} = 1/[v_\mathrm{g}(\alpha_\mathrm{m} + \alpha_\mathrm{i})] \tag{B2.3.3}$$

where v_g is the group velocity of light and α_m and α_i are the mirror and internal loss, respectively.

In order to take the frequency dependence of the optical gain into account and to simulate multi-mode laser dynamics, the rate equations can be modified to the form [2, 4],

$$\frac{\mathrm{d}n}{\mathrm{d}t} = \frac{j(t)}{ed} = \Sigma g_m S_m - \frac{n}{\tau_\mathrm{s}} \tag{B2.3.4}$$

$$\frac{\mathrm{d}S_m}{\mathrm{d}t} = \Gamma g_m S_m - \frac{S_m}{\tau_\mathrm{ph}} + \frac{\beta\Gamma n D_m}{\tau_\mathrm{s}} \tag{B2.3.5}$$

where $g_m = g_0(D_m n - n_\mathrm{t})$, S_m is the photon density in the mth laser mode and D_m is the line shape factor which is defined, say, for Lorentzian shape as

$$D_m = \frac{\delta\lambda/(\pi\Delta\lambda_\mathrm{g})}{1 + [(\lambda_m - \lambda_0)/\Delta\lambda_\mathrm{g}]^2} \tag{B2.3.6}$$

where $\delta\lambda$ is the longitudinal mode spacing, $\Delta\lambda_\mathrm{g}$ is the gain bandwidth, λ_0 and λ_m are the wavelengths of the central and mth laser cavity mode. The total photon density is thus given by $S(t) = \Sigma S_m$. The evaluation of these multi-mode rate equations often gives more accurate results than those obtained using the single-mode equations (B2.3.1)–(B2.3.2).

The linear dependence of the optical gain on the carrier density which is assumed in equations (B2.3.1)–(B2.3.2) and (B2.3.4)–(B2.3.5) is an approximation. In fact, an additional nonlinear dependence of the optical gain upon the photon density is observed. In particular, the gain nonlinearity manifests itself in long-wavelength InP/InGaAsP lasers and in QW lasers. A number of physical processes, the most important of which are spectral hole burning, spatial hole burning and carrier heating, are generally thought to be responsible for gain nonlinearity in diode lasers [2, 3]. An analytical expression of gain, accounting for the effects of photon density, is generally written as

$$g(n, S) = \frac{g_0(n - n_\mathrm{t})}{(1 + \varepsilon S)^m} \tag{B2.3.7}$$

where ε is the nonlinear gain compression coefficient [2] and m is a fitting constant which is often thought to take a value of either unity or a half. Nonlinear gain effects play a significant role in laser diode dynamics. They impact on the modulation bandwidth of lasers and other modulation properties and determine the minimum achievable pulsewidths and maximum peak power levels of ultrashort pulses from diode lasers (see later).

The carrier density variations across the laser active region and the associated effects of lateral carrier diffusion on the dynamic properties have been neglected in all the rate equations presented so far. This is thought to be a good approximation in high-speed diode lasers because they have typically active regions which are sufficiently narrow in the lateral dimension. In contrast, carrier transport effects in the direction of current injection (the transverse direction) are very important in modern QW high-speed lasers and should be included [3, 5]. QW lasers are different from bulk ones where the entire undoped region is the active area and the carrier transport effects are not significant. For instance, electron and hole transport in QW lasers with a separate confinement heterostructure (SCH) consists basically of two parts. The first is the transport of carriers across the SCH from the doped cladding layers to the QW region, whereas the second is carrier capture by the QW. Thermionic emission or the thermally activated carrier escape from the QW and tunnelling between the QWs are other transport mechanisms which strongly affect high-speed laser dynamics.

The rate equation model derived for high-speed multiple QW (MQW) lasers, taking into account all carrier transport effects, is based upon the equations describing the dynamics of the carrier densities in the QW N_{w} and in the SCH layer N_{b}, and the photon density dynamics in the laser cavity. They can be written as follows [5]

$$\frac{\mathrm{d}N_{\mathrm{b}}}{\mathrm{d}t} = \frac{I(t)}{eV_{\mathrm{SCH}}} - \frac{N_{\mathrm{b}}}{\tau_{\mathrm{n}_1}} + \frac{N_{\mathrm{w}}}{\tau_{\mathrm{e}}}\frac{V_{\mathrm{w}}}{V_{\mathrm{SCH}}} \tag{B2.3.8}$$

$$\frac{\mathrm{d}N_{\mathrm{w}}}{\mathrm{d}t} = -\frac{G(N_{\mathrm{w}})S}{1+\varepsilon S} - \frac{N_{\mathrm{w}}}{\tau_{\mathrm{e}}} - \frac{N_{\mathrm{w}}}{\tau_{\mathrm{n}_2}} + \frac{N_{\mathrm{b}}}{\tau_{\mathrm{tr}}}\frac{V_{\mathrm{SCH}}}{V_{\mathrm{w}}} \tag{B2.3.9}$$

$$\frac{\mathrm{d}S}{\mathrm{d}t} = \frac{\Gamma G(N_{\mathrm{w}})S}{1+\varepsilon S} - \frac{S}{\tau_{\mathrm{ph}}} + \frac{\beta\Gamma N_{\mathrm{w}}}{\tau_{n_2}} \tag{B2.3.10}$$

where V_{SCH} is the volume of the SCH, V_{w} is the volume of the QW, τ_{n_1} and τ_{n_2} are the carrier recombination lifetime in the SCH and active regions, respectively, τ_{tr} is the carrier transport time across the SCH region which also includes the local capture time at the QW, τ_{e} is the carrier thermionic emission time from the QW into the SCH layer and $G(N_{\mathrm{w}})$ is the carrier-density-dependent gain. Note that the nonlinear gain has been taken into consideration here using equation (B2.3.7) with $m = 1$.

Some dynamic regimes of high-speed diode lasers, including Q-switching and mode locking of multiple contact lasers in particular, require travelling-wave rate equations rather than 'space-averaged' ones. Indeed, those lasers have a non-uniform distribution of driving current along the laser cavity axis or unpumped passive sections which result in inherently non-uniform distributions of both carrier and photon densities along the cavity axis. To account for these space distributions, one needs to include spatial coordinates into the equations. This is particularly important when the generation of ultrashort light pulses with a spatial length smaller than or about the cavity length is described. Travelling-wave rate equations, which become partial differential equations, are often written as follows [2]

$$\frac{\partial S^+}{\partial t} + v_{\mathrm{g}}\frac{\partial S^+}{\partial z} = \frac{\Gamma g_0(n-n_{\mathrm{t}})S^+}{1+\varepsilon(S^+ + S^-)} - v_{\mathrm{g}}\alpha_{\mathrm{i}}S^+ + \frac{\beta n}{\tau_{\mathrm{s}}} \tag{B2.3.11}$$

$$\frac{\partial S^-}{\partial t} - v_{\mathrm{g}}\frac{\partial S^-}{\partial z} = \frac{\Gamma g_0(n-n_{\mathrm{t}})S^-}{1+\varepsilon(S^+ + S^-)} - v_{\mathrm{g}}\alpha_{\mathrm{i}}S^- + \frac{\beta n}{\tau_{\mathrm{s}}} \tag{B2.3.12}$$

$$\frac{\partial n}{\partial t} = \frac{j(z,t)}{ed} - g_0(n-n_{\mathrm{t}})(S^+ + S^-) - \frac{n}{\tau_{\mathrm{s}}} \tag{B2.3.13}$$

where $S^+(z,t)$ and $S^-(z,t)$ are forward and backward travelling waves (photon densities), respectively.

Boundary conditions are now necessary to solve the partial differential equations (B2.3.11)–(B2.3.13). The boundary conditions should include reflections of light at the laser facets and look like

$$S^+(0,t) = R_1 S^-(0,t) \qquad S^-(L,t) = R_2 S^+(L,t)$$

where R_1 and R_2 are the power reflectivities at the chip facets, L is the cavity length. Since all the rate equations are nonlinear ones, it is very difficult to solve them analytically and numerical computations are generally required. However, equation linearization and small-signal analysis can be used to study laser dynamics in some cases. For instance, the modulation properties of high-speed lasers can be readily established analytically using the latter approach.

B2.3.2.2 Modulation bandwidths and damping

The overall dynamic response of a laser includes the response of the package parasitics, the parasitics associated with the laser chip and the intrinsic laser, i.e. the active region and the cavity. The parasitics cause a high-frequency roll-off in the small-signal response of the laser which slows down the fast transients of drive current waveforms. The small-signal response of the intrinsic laser is obtained by linearizing equations (B2.3.1)–(B2.3.2) using

$$I(t) = I_0 + i\mathrm{e}^{\mathrm{i}\omega t} \qquad N(t) = N_0 + n\mathrm{e}^{\mathrm{i}\omega t} \qquad S(t) = S_0 + s\mathrm{e}^{\mathrm{i}\omega t}$$

with

$$i \ll I_0 \qquad n \ll N_0 \qquad s \ll S_0$$

where $I(t)$ is the small-signal current modulation and N_0 and S_0 are the steady-state (average) carrier and photon densities, respectively. The small-signal analysis of the rate equations (B2.3.6)–(B2.3.8) gives the intrinsic frequency response $R_{\mathrm{int}}(\omega) = |M(\mathrm{j}\omega)/M(0)|$ as

$$R_{\mathrm{int}}(\omega) = \frac{\omega_0^4}{(\omega^2 - \omega_0^2)^2 + \omega^2\gamma^2} \qquad \text{(B2.3.14)}$$

where $\gamma = g_0 S_0$ is the damping rate and the resonant frequency ω_0 is given by

$$\omega_0^2 = \frac{g_0 S_0}{\tau_{\mathrm{ph}}}. \qquad \text{(B2.3.15)}$$

Figure B2.3.1 illustrates the general form of $R_{\mathrm{int}}(\omega)$ for two values of S_0. A resonance in modulation response at ω_0 is known as the relaxation oscillation and its strength depends on the spontaneous emission factor, carrier diffusion, carrier transport effects, etc. Typical values of $\omega_0/2\pi$ are in a range of 1–10 GHz. Above the resonance peak, the magnitude of the intrinsic response approaches asymptotically a slope of -40 dB/decade [6]. A parameter $f_{3\ \mathrm{dB}} = \omega_{3\ \mathrm{dB}}/2\pi$ is widely referred to as the maximum modulation bandwidth of the laser. The 3 dB bandwidth is defined by setting $R(\omega) = 1/2$.

There is a universal relationship between the resonant frequency and the damping rate. It has been demonstrated both experimentally and theoretically that γ varies linearly with $f_0^2 = (\omega_0/2\pi)^2$. The proportionality constant is called the K-factor [9]. The K-factor was originally defined given by $\gamma = Kf_0^2$. However, a dc offset is always observed and, hence, a better definition is: $\gamma = Kf_0^2 + \gamma_0$ [5, 10]. From this definition and using the small-signal solution of the rate equations, the K-factor is given by $K = 4\pi^2\tau_{\mathrm{ph}}$. The K-factor is often taken as a figure of merit for high-speed lasers since the maximum possible intrinsic modulation bandwidth f_{max} (3-dB bandwidth) is determined solely by this factor [9]:

$$f_{\mathrm{max}} = \sqrt{2}(2\pi/K) \approx 8.88/K. \qquad \text{(B2.3.16)}$$

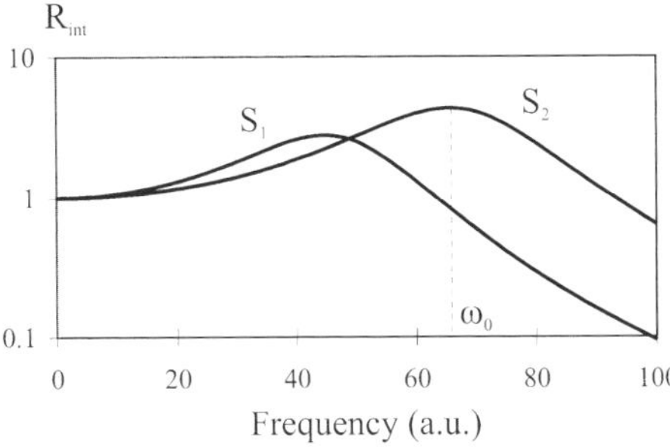

Figure B2.3.1. Small-signal AM response of a diode laser for two values of the photon density S_1 and S_2, $S_2 > S_1$.

In fact, the K-factor depends not only on the photon lifetime but it is also significantly affected by nonlinear gain effects and carrier transport effects (in QW lasers). From the small-signal analysis of the equations (B2.3.8)–(B2.3.10) and for negligible τ_{n_1} and $\tau_n = \tau_{n_2}$ one obtains, for the resonant frequency and damping rate, the following analytical approximations [5]

$$\omega_0^2 = \frac{(g_0/\chi)S_0}{\tau_{\mathrm{ph}}(1+\varepsilon S_0)}\left\{1+\frac{\varepsilon}{g_0\tau_{\mathrm{n}}}\right\} \quad \text{(B2.3.17)}$$

$$\gamma = \frac{(g_0/\chi)S_0}{1+\varepsilon S_0} + \frac{\varepsilon S_0}{\tau_{\mathrm{ph}}(1+\varepsilon S_0)} + \frac{1}{\chi\tau_{\mathrm{n}}} \quad \text{(B2.3.18)}$$

where the carrier transport factor χ is given by $\chi = 1 + \tau_{\mathrm{tr}}/\tau_{\mathrm{e}}$. The K-factor in this case is given by

$$K = 4\pi^2\left(\tau_{\mathrm{ph}} + \frac{\varepsilon}{g_0\chi}\right). \quad \text{(B2.3.19)}$$

There is a number of laser parameters to optimize if one wants to design a laser with the maximum possible modulation bandwidth. More specifically, the differential gain g_0, the photon density S_0 and the carrier escape time τ_{e} should be maximized, whereas the carrier transport time τ_{tr}, the photon lifetime τ_{ph}, the gain compression factor ε and device parasitics should be as small as possible. The use of quantum wells, wires and quantum dots instead of bulk materials for the active layers of high-speed diode lasers has been predicted to increase the differential gain strongly. It has been demonstrated [2, 5, 10] that, in contrast to bulk lasers, g_0 in QW lasers is a structure-dependent parameter, i.e., it is determined not only by the material parameters of the active layer but is also influenced by the dimensions of the QW and barrier layers, optical confinement factor and carrier injection and confinement processes. Wavelength detuning is considered to be a promising method of differential gain enhancement in laser structures where the lasing wavelength can be fixed independently of the peak wavelength of the material gain, for instance in distributed feedback (DFB) lasers. The differential gain is larger on the 'blue' wavelength side of the gain distribution. Thus, negative detuning of the lasing wavelength from the material gain peak is often exploited in both bulk and QW DFB lasers.

Strained-layer QW structures have recently received a lot of attention since an in-plane comprehensive strain can reduce the valence-band effective masses, which leads to further enhancement of differential gain. P-doping and modulation doping of the active layer have been predicted and demonstrated to enhance the differential gain and decrease the K-factor of high-speed QW lasers. P-doped lasers demonstrate very efficient high-speed direct modulation. A maximum possible intrinsic bandwidth of 63 GHz (a K-factor of 0.14 ns and the value of $\varepsilon = 4.6 \times 10^{-17}$ cm^3) has been reported in strained InGaAs/GaAs MQW lasers [11]. In 1.5 μm InGaAsP/InP MQW strained p-doped lasers, an $f_{\max}$ as high as 40 GHz has been demonstrated [12]. The actual 3 dB modulation bandwidth of 25 GHz has been limited by the parasitics of the devices.

The photon density S_0 is related to the driving current I_0 (by the total output optical power) as follows

$$S_0 = \left(\frac{\Gamma}{V}\right) \frac{\eta(I_0 - I_{\mathrm{th}})}{e v_{\mathrm{g}}(\alpha_m + \alpha_{\mathrm{i}})} \tag{B2.3.20}$$

where V is the volume of the active region, η is the injection efficiency and I_{th} is the threshold current. Although high output power levels of a high-speed diode laser are highly desirable for large modulation bandwidths, it is the optical mode volume V/Γ which determines the photon density within the laser resonator. As a result, a large modulation bandwidth laser structure requires tight optical confinement within the laser cavity. Facet passivation and application of special coatings on device facets can substantially increase the power level of the optical damage of the facets and allows for an enhancement of values of S_0.

The effects of carrier transport on the modulation bandwidth are evident from equations (B2.3.17)–(B2.3.19). The first effect consists in a parasitic-like roll-off in the modulation response which would significantly limit the 3 dB modulation bandwidth for large τ_{tr}. This limitation which originates from the finite transport time across the SCH is indistinguishable from the parasitics [5]. The second effect is the reduction of the effective differential gain by a factor of χ. This leads to the reduction of the resonance frequency for the same power levels. To optimize the high-speed performance of QW lasers, the transport time across the SCH has to be minimized. A narrower SCH should result in shorter τ_{tr}. However, decreasing the SCH width below an optimum point will also lead to a drop in the optical confinement factor and an increase of the internal loss. This, in turn, results in a higher carrier density at threshold and, consequently, a lower differential gain which originates from rapid gain saturation in QW lasers.

It has been demonstrated that VCSELs have a very large modulation bandwidth of over 70 GHz. The physical reason for such enormous increase in the bandwidth is that VCSELs have very short cavity lengths which result in a substantial reduction of the photon and electron lifetimes.

B2.3.2.3 Frequency chirp and spectral control

Carrier density variations which occur under high-speed modulation result in a corresponding change in the refractive index in the laser cavity. This change, in turn, causes a variation in the lasing wavelength (or the oscillation frequency) which is referred to as wavelength chirping. Enhanced linewidth and frequency chirp under dc modulation can also originate from the finite difference in the carrier density between the 'on' and 'off' states of the laser. The chirping response of a laser can be determined by means of the rate equations with an additional differential equation describing the evolution of the optical phase $\phi(t)$ which is given by [9]

$$\frac{\mathrm{d}\phi}{\mathrm{d}t} = \frac{1}{2}\left\{\frac{\Gamma g_0(N - N_{\mathrm{t}})S}{1 + \varepsilon S} - \frac{1}{\tau_{\mathrm{ph}}}\right\} \tag{B2.3.21}$$

where α is the linewidth enhancement factor. The linewidth enhancement factor, describing the coupling between the carrier-induced variation in the real $\mu(n)$ and imaginary $g(n)$ parts of the optical susceptibility, can be defined as [13, 14]

$$\alpha = -2k\frac{\mathrm{d}\mu/\mathrm{d}n}{\mathrm{d}g/\mathrm{d}n} \tag{B2.3.22}$$

where $k = 2\pi/\lambda$ is the free-space wavenumber. Since $\mathrm{d}\mu/\mathrm{d}n$ is negative, α is positive.

The frequency chirp is defined as $\Delta\nu(t) = (1/2\pi)\mathrm{d}\phi/\mathrm{d}t$. Using equation (B2.3.21), the chirp can be related directly to the output laser power $P(t)$ as

$$\Delta\nu(t) = \frac{\alpha}{4\pi}\left\{\frac{\mathrm{d}\ln P(t)}{\mathrm{d}t} + \frac{2\Gamma\varepsilon}{V\eta\hbar\omega}P(t)\right\} \tag{B2.3.23}$$

where η is the laser differential quantum efficiency. The magnitude of the chirp represented by the first term in equation (B2.3.23) may be large during high-speed turn-on and turn-off transients, while being relatively small at the on and off levels. By contrast, the second term results in frequency offset between the on and off levels owing to a difference in the steady-state values of the carrier density. Under digital pulse modulation when zeros of the pulse sequence are close to threshold, shoulders can be apparent on both the long wavelength and short wavelength side of the optical spectrum. The former is caused by the relaxation oscillation and light present in the trailing edges and nominal zeros of the pulse stream, while the latter results from the relaxation oscillation at turn on.

From equation (B2.3.23) it is clear that the α-factor is a key parameter in obtaining low-chirp emission from high-speed directly modulated lasers. Hence, lasers with lower values of α should exhibit lower wavelength chirp under high-speed modulation. Typical values of α in most lasers lie in the range 3–5 [14]. The development of laser structures with $\alpha < 2$ has attracted considerable attention in recent years. According to equation (B2.3.22), α is inversely proportional to the differential gain of the laser medium. Since the differential gain is enhanced in QW lasers, the α parameter has been predicted and observed to be reduced in this type of laser. Consequently, QW lasers are the most promising for low-chirp operation under high-speed modulation. However, α is a structure-dependent parameter [14] and a parameter α_{eff} should be used in equation (B2.3.23) for the chirp evaluation. In the case of QWs, laser carrier transport, particularly carrier density variation in the SCH region, has a significant impact on α_{eff} and frequency chirping leads to enhanced values of it. An expression for α_{eff} can be written as

$$\alpha_{\text{eff}} = \frac{4\pi}{g_0\lambda}\left[\frac{\partial n}{\partial N_{\text{w}_0}} + \left(\frac{1}{\Gamma} - 1\right)\left(\frac{V_{\text{w}}}{V_{\text{SCH}}}\right)(\chi - 1)\left(\frac{\tau_{\text{b}}}{\tau_{\text{b}} + \tau_{\text{tr}}}\right)\frac{\partial n}{\partial N_{\text{b}_0}}\right] \tag{B2.3.24}$$

where N_{w_0} is the steady-state carrier density in the quantum well, N_{b_0} is the steady-state carrier density in the SCH region, and τ_{b} is the barrier transport time.

Strained-layer QW lasers exhibit lower values of α_{eff} when compared with their unstrained counterparts. For example, the α_{eff} parameter of an unstrained MQW DFB laser was estimated to be 3.1 at the 1.55 μm wavelength, while a similar laser structure with compressive strain of 1.9% had a linewidth enhancement factor as low as 1.1 [15]. P-type modulation doping of QWs has also been demonstrated to reduce the linewidth enhancement factor. A reduction of α_{eff} by a factor of two due to modulation doping in both 1.3 and 1.55 μm InGaAsP/InP strained MQW DFB lasers has been observed [16]. Reduced wavelength chirp is thus expected in these lasers as a result of a reduction in the α_{eff} parameter. Since α_{eff} depends on wavelength, the frequency chirp of a laser can be controlled by detuning the lasing wavelength with respect to the gain profile peak.

B2.3.2.4 Noise

Noise in diode lasers originates from both carrier density fluctuations and photon density fluctuations. The relative intensity noise (RIN) spectrum is commonly used to analyse the noise properties of high-speed diode lasers and to determine the modulation bandwidth [5, 10]. The RIN is defined as the ratio of the intensity fluctuations to the averaged intensity. The RIN spectrum of a laser can be derived using the same formalism of the rate equations as before. The RIN spectrum can be written as

$$\text{RIN}(f) = \frac{4}{\pi}(\delta f)_{\text{ST}}\frac{f^2 + (\gamma^*/2\pi)^2}{(f_0^2 - f^2)^2 + (\gamma/2\pi)^2 f^2} \tag{B2.3.25}$$

where $(\delta f)_{\text{ST}}$ is the intrinsic Schawlow–Townes linewidth which derives from spontaneous emission and γ^* is given by

$$\gamma^* = \frac{(g_0/\chi)S_0}{1 + \varepsilon S_0} + \frac{1}{\chi\tau_n}.$$

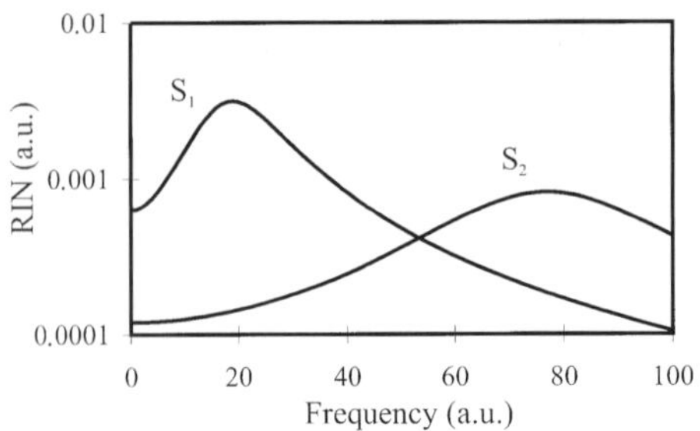

Figure B2.3.2. Typical RIN spectra for a diode laser.

The γ term is very close to γ^* at sufficiently low powers and small ε. It is commonly believed that the RIN spectrum measurement is a parasitic-free means of determining the actual modulation performance of high-speed diode lasers. However, significant transport effects can affect strongly results of RIN spectrum measurements.

Typical RIN spectra for two values of S_0 are shown in figure B2.3.2. The RIN spectra have been plotted on a relative scale and do not give the actual noise level of the laser. The resonance frequency f_0 and damping γ can be defined from the RIN spectrum with a high degree of accuracy.

B2.3.3 High-speed laser diode design

The device aspects of the requirements and techniques employed to design high-speed diode lasers comprise a variety of means, including decreasing the photon lifetime, decreasing the device size to eliminate microwave propagation effects, reduction of parasitic capacitance and series resistance, using semi-insulating substrates, etc. Device structures with low parasitics are considered to be of great importance since, in practice, the dominant limit to the modulation response of high-speed diode lasers is the device parasitics.

B2.3.3.1 Device optimization

To disclose the effects of parasitics on device optimization, one should treat the laser as an electrical element and establish an equivalent circuit for the intrinsic laser in conjunction with the bond wire, the mount elements, the transmission line used for impedance matching, etc. A simplified equivalent circuit of a diode laser is shown in figure B2.3.3(*a*) [2]. It includes the bond wire inductance L_p, the series resistance of the layers above and below the active layer R_s, the capacitance between the top and the bottom contact in the area outside the lasing region C_p and the intrinsic laser diode.

A more practical equivalent circuit should include additional R, L, C elements which describe the parasitic elements of each particular device accurately. Figure B2.3.3(*b*) shows an example of an equivalent circuit of a double-heterostructure laser with an oxide isolation layer and blocking layers. The circuit includes a contact resistance R_c, an effective resistance in series with the active layer R_a and the blocking layer R_b, a parallel effective capacitance comprising the oxide isolation layer in series with either blocking layer C_i, a capacitance of the blocking layer C_b and the diffusion capacitance of the device C_j.

The input resistance of the intrinsic laser is approximately given by $2kT/eI_0$ and is typically less than 1 Ω [3] at currents I_0 above threshold due to clamping of the Fermi levels. Typical values of R_s for high-speed diode lasers can be in the range of 2–4 Ω [4]. The use of p-type substrates instead of the typical n-type has been proposed for the fabrication of lower resistance high-speed lasers. The actual device resistance is constrained by the maximum doping in the cladding for acceptable levels of loss due to free-carrier absorption and the minimum cladding layer thickness required for good transverse optical confinement.

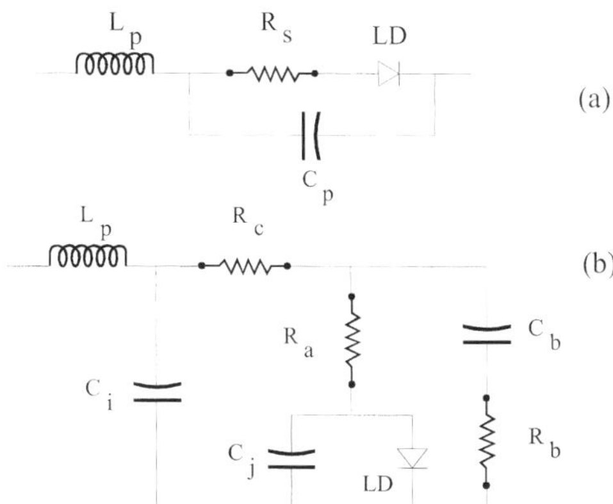

Figure B2.3.3. Equivalent circuits of a laser diode: (*a*) simplified circuit; (*b*) small-signal circuit of a double-heterostructure laser.

The RC constant characterizing the parasitic roll-off in the overall response is given by the product of R_s and the effective shunt capacitance C_i. In some lasers C_i was measured to be as large as 40 pF while in low-parasitic devices it can be as low as 1-3 pF. To minimize the parasitic capacitance, a number of structures has been developed [3,4,6,7]. They include polyimide buried ridge waveguide geometry structures, undercut 'mushroom' mesa-type structures, devices with isolation channels in a buried heterostructure configuration and Fe-doped semi-insulating InP-based structures.

If the bond wire inductance L_p is less than 0.2 nH, it has little effect on the laser response up to 20 GHz. L_p depends on the length of the wire, its diameter and its proximity to the ground plane. Low inductance can be achieved by using a short (<1 mm) wire or with a tape or mesh instead of the usual thin wire. An inductance L_p in excess of 1 nH results in significant roll-off in the laser response at frequencies above 6 GHz. The R_sC_i product of the chip dominates in the case of small L_p and these parasitics must be minimized if high-speed operation is to be achieved.

B2.3.3.2 Microwave propagation effects

As the modulation speed increases above 20–30 GHz, the laser size becomes very critical. Conventional designs that treat a laser diode as a lumped electrical element as presented in figure B2.3.3 are valid only when the laser cavity length is much smaller than the modulation wavelength. This is not the case for modulation frequencies of 25–30 GHz and higher and cavity lengths of about 300 μm. In addition, at very high frequencies, the time delay from the feed point of the modulation current to the far end of the laser can be a significant fraction of the modulation period. The high-frequency current injection and the optical modulation response are, thus, modified by the distributed microwave nature of the laser. It is thought that microwave propagation effects put an ultimate limit on the modulation bandwidth.

The distributed circuit laser model shown in figure B2.3.4 was proposed as a first-order estimation of the propagation effects [17]. The total current flowing into the device is a sum of all currents flowing through the distributed resistors R_s (injection current) and through the distributed capacitors C_s (displacement current). The active region laser impedance is Z_d. A sizeable frequency-dependent current loss and change of the phase velocity caused by the microwave propagation effects which must be taken into consideration in laser design at very high frequencies exist. As the frequency increases the loss increases as well. This, in turn, leads to a roll-off in the current injection into the active region. The estimated 3 dB roll-off frequency happens at approximately 25 GHz for a 300 μm long laser when the device is fed at the centre of the laser stripe.

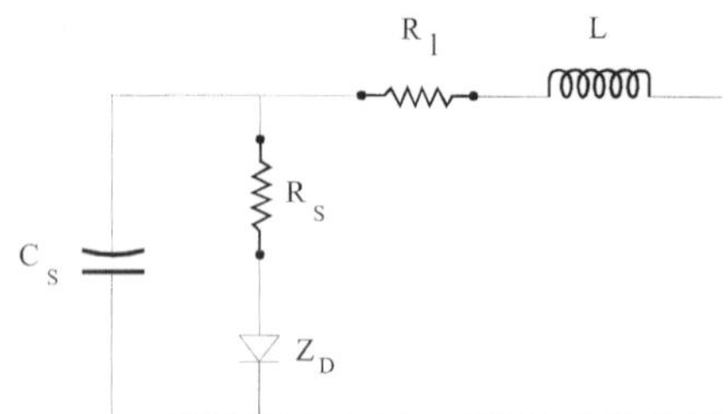

Figure B2.3.4. Distributed equivalent circuit of a diode laser.

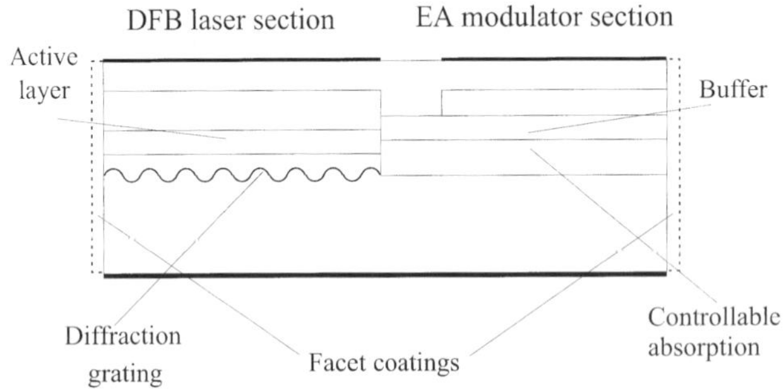

Figure B2.3.5. Electroabsorption modulator integrated with a DFB laser.

This is considered to be the optimal feeding condition since waves can propagate in either direction from the centre of the cavity.

B2.3.4 External modulation

An alternative approach to high-speed modulation of semiconductor laser emission is to use an external modulator instead of internal modulation. An external modulator can be either monolithic (integrated) or a separate device. External modulation of laser emission has significant advantages over direct modulation because frequency chirping is low and modulation at very high frequencies (>40 GHz) is possible. Electroabsorption (EA) and electrorefraction (ER) modulators are the most commonly utilized types of external modulators for high-speed modulation and data encoding [18, 19]. Among various designs of external modulators for practical applications, waveguide (in-plane) and surface-normal (transverse) devices are the most widely used configurations. Waveguide devices tend to provide low drive voltages since the electric field is applied perpendicular to the light propagation direction. At the same time, the coupling of the light into the modulator waveguide adds some complexity to practical implementation. However, surface-normal devices exhibit much higher drive voltages due to their shorter interaction lengths [19].

Highly efficient modern EA and ER modulators exploit either the Franz–Keldysh effect in bulk semiconductor materials or the quantum-confined Stark effect in MQW heterostructures. The latter refers to the red shift of the exciton absorption peak of the QW when an electric field is applied in the perpendicular direction. Extremely large absorption changes of 5×10^3–10^4 cm^{-1} are readily achievable in MQW waveguide modulators at reasonably low electric fields and small drive powers [20]. By virtue of their small dimensions, the speed of devices may be very high because the RC time constant is small.

In an EA waveguide modulator, the transmitted optical power coupled into the waveguide is attenuated by a factor of $\exp\{-\alpha(V)L\}$ which is dependent on the absorption coefficient α at the applied voltage V and the length of the device L. The efficiency of the modulator depends on how close the optical wavelength is to the EA peak. However, there will always be substantial insertion loss in well-designed EA waveguide modulators. A typical EA modulator structure is schematically shown in figure B2.3.5. It consists of an EA modulator section with a semi-insulating buried heterostructure and a DFB laser section. Both the active layer and absorption layer consist of either bulk or SCH MQW structures with slightly different material compositions and thicknesses. An additional buffer layer is often introduced between the upper cladding layer and the modulator waveguide layer in order to reduce the hole pile-up effect at high optical power levels. The modulating electric field is created by applying a voltage to the pin structure of the modulator section, the MQWs being enclosed in the i-region. Lengths of the EA modulator section and the DFB laser section are typically 100–250 and 200–450 μm [5], respectively. Anti-reflection coatings on the facets are used for optimizing the output power. The electrical isolation resistance between the two sections usually exceeds 1 MΩ [6]. Typical driving voltages for a 10 dB extinction ratio can be as low as 1.0–3.0 V_{p-p}. In principle, RC-limited bandwidths of over 30–40 GHz with a very low drive voltage ranging from 1–2 V can be achieved by optimizing both material and waveguide design simultaneously. It has been predicted [20] that an EA modulator with a large contrast ratio of 10–20 dB, a very high modulation bandwidth of 60–100 GHz, a small drive voltage (<3 V) and a small insertion loss of less than 4–5 dB can be obtained.

B2.3.5 Short pulse generation

The generation of ultrashort optical pulses is another important field of applications of semiconductor lasers. A wide gain bandwidth of a semiconductor medium potentially allows optical pulses as short as 50–100 fs to be generated [2]. A number of methods for the generation of picosecond and femtosecond pulses in diode lasers has been developed. The main techniques include gain-switching, Q-switching and different types of mode locking.

B2.3.5.1 Gain and Q-switching

The idea of gain switching was initially conceived as a result of early observations of the relaxation oscillations generated by a diode laser simply being turned on from below threshold by electrical pulses with a fast leading edge. Gain switching therefore consists of exciting the first pulse of a relaxation oscillation and then terminating the electrical pulse before the onset of the next optical relaxation pulse. Typical pulsewidths of electrical pulses are in the range of 50–1000 ps. Alternatively, modulation of a laser biased below threshold with a large sinusoidal signal at sub-GHz or GHz frequencies can result in gain-switched optical pulses.

Dc bias I_0 and pulsed I_1 currents are assumed to be applied to the laser under gain switching. I_0 can be set to be below the laser threshold level. The lasing does not thus occur until the electrical pulse I_1 causes the carrier density in the laser to increase to above the device threshold density. Above this level, the carrier density reaches its maximum value n_i. Then lasing starts and represses the increase in carrier density and, consequently, $n(t)$ is pulled down to its minimum value n_f. Using the rate equations (B2.3.1)–(B2.3.2), one can approximately describe a gain-switched pulse as a combination of two exponential curves with time constant τ_r on the rising edge of $S(t)$ and τ_f on the trailing edge. According to equation (B2.3.2), for a sufficiently small value of S, gain compression is negligible and $S(t)$ increases exponentially as a result of stimulated emission:

$$\frac{1}{S}\frac{dS}{dt} = \frac{1}{\tau_r} = g_0\Gamma(n_i - n_{th}). \qquad \text{(B2.3.26)}$$

The trailing edge of the optical pulse also exhibits an exponential decay constant and, normally, τ_f approximately equals (equations (B2.3.2), (B2.3.3)) τ_r or $\tau_f = \tau_r + \tau_{ph}$. The full width half maximum (FWHM) of the pulse can be roughly estimated by the sum $\tau_f + \tau_r$. As a result, the pulse duration is limited by the rate of energy transfer between electron and photon populations. From equation (B2.3.26) it is clear that there is a corresponding improvement in the temporal response of gain switched pulses if g_0 increases.

In practice, the generated optical pulsewidths are of the order of several cavity round-trip times. It has been observed that the pulsewidth decreases and the peak power increases with increasing rf power until a point at which a second relaxation oscillation pulse appears. For most standard devices, pulsewidths as low as 7–12 ps can be achieved at repetition rates of the order of, 1 GHz for GaAs/AlGaAs lasers, and 25–30 ps at the 1.5 μm wavelength. It is thought that gain-switched optical pulses generated in InGaAsP lasers are longer than pulses in the 0.85 μm wavelength region due to enhanced nonlinear gain compression in quaternary lasers.

The peak powers of gain-switched pulses typically lie in the range 0.1–1 W at the approximately 850 nm wavelength and from 10 to 100 mW for 1.3 or 1.5 μm InGaAsP lasers. Again, the peak powers for the InGaAsP systems are lower because of enhanced nonlinear gain compression. More recently, gain switching of VCSELs has been demonstrated, the shortest pulse durations being 15 ps.

Pulse-to-pulse timing jitter is an important parameter of gain-switched pulses as, for many applications, the timing jitter may affect dramatically the performance of a system where the laser is used. The timing jitter of gain-switched pulses originates from random fluctuations of the photon density in the laser cavity caused by the spontaneous emission. Typical values of timing jitter lie in the range 1–10 ps. The jitter decreases as the ratio I_1/I_{th} increases and it can be less than 1 ps in multi-mode Fabry–Pérot lasers, where large levels of spontaneous emission act to reduce the turn-on jitter effect.

The optical bandwidth of picosecond pulses is another key parameter that plays a crucial role in a number of applications of gain-switched lasers. For instance, an excessive emission bandwidth results in a significant dispersion penalty in optical communications systems and can be the limiting factor in their performance. In Fabry–Pérot gain-switched lasers, a larger number of longitudinal modes are generated under gain switching than in steady-state operation as a consequence of the dynamic overshoot of electron concentration. This can occur even if the laser emission is single mode at steady state. If a single-frequency laser is used for gain switching, multi-mode operation can be avoided. However, the bandwidth under gain switching exceeds the bandwidth under cw operating conditions. Due to wavelength chirp, gain-switched pulses have a time-bandwidth product $\Delta\tau\Delta\nu$ which is often far from the theoretical limit for bandwidth-limited pulses (see chapter C2.3).

Picosecond optical pulses may also be generated in a semiconductor laser by a Q-switching technique (see chapter C2.2). Here, current is applied to the gain section of a multi-section laser in which at least one section acts to switch the Q-factor of the cavity. While the laser population inversion is building up, the Q-switch is held in the 'off' or high-loss state, ensuring a low Q. As the carrier density in the laser reaches its peak, the cavity is suddenly switched to the high-Q condition and the stored energy is rapidly emitted in an output optical pulse. If the switching is caused by external influence, the technique is called active Q-switching. However, pulse generation can also be achieved by passive Q-switching by using an optically saturable absorber as the Q-switching element. Unlike other types of lasers it is possible simply to divide the diode into several sections and apply different currents to them. The parts of the laser with high current injection act as the gain medium, while the unpumped sections or sections with a low pumping rate act as absorbers. Active Q-switching can be realized by modulation of the loss in the cavity. The modulation of loss in diode lasers is realized by shifting the band edge in the modulator section using the Franz–Keldysh effect or the quantum-confined Stark effect in MQW lasers caused by an applied electric field.

It can be shown that the Q-switched pulse parameters are drastically dependent on the photon lifetime τ_{ph} and on the initial inversion level n_{i}. One can deduce this from consideration of the basic rate equations (B2.3.1)–(B2.3.2). When the ultrashort pulse is emitted, stimulated emission is the dominant mechanism and the contribution of all other processes, including spontaneous emission, carrier pumping and recombination, are negligible. Thus, using equations (B2.3.1)–(B2.3.2), one can derive the following relation between carrier and photon densities in the laser cavity [2]:

$$S(t) = n_{\rm i} - n(t) - \frac{1}{g_0\tau_{\rm ph}} \ln \frac{n_{\rm i} - n_{\rm t}}{n(t) - n_{\rm t}}. \tag{B2.3.27}$$

By introducing the inversion ratio r and the energy extraction efficiency η as [1,8]

$$r = g_0\tau_{\rm ph}(n_{\rm i} - n_{\rm t}) \qquad \text{and} \qquad r = \frac{1}{\eta} \ln \frac{1}{1-\eta}$$

the peak power of the Q-switched pulse is given by

$$S_0 = \frac{r - 1 - \ln r}{g_0\tau_{\rm ph}} \tag{B2.3.28}$$

and the pulsewidth of the pulse is

$$\tau_{\rm Q} = r\eta\tau_{\rm ph}/(r - 1 - \ln r). \tag{B2.3.29}$$

To obtain the shortest pulses, the initial inversion should be maximized whereas the photon lifetime should be minimized. For instance, for $r = 1.1$, the pulsewidth $\tau_{\rm Q}$ is about $50\tau_{\rm ph}$ and $\tau_{\rm Q} = 5\tau_{\rm ph}$ for $r = 2.0$ [23].

Passive Q-switching is known as one of the simplest techniques for generating picosecond optical pulses in lasers. A difference between active and passive Q-switching is that, in contrast to active Q-switching, no external electrical or optical modulation is required for self-Q-switching to produce ultrashort optical pulses. Additionally, in the case of active Q-switching, the pulse repetition frequency is set by an external rf drive signal, whereas for passive Q-switching the frequency is governed by the laser parameters and pumping conditions. As a result of the nonlinear transmission characteristics of the absorber, the absorption of the leading edge of the pulse is typically larger than that of the trailing edge. This results in the generation of asymmetric pulses.

To generate self-Q-switched pulses, the saturable absorber in the laser cavity must satisfy a number of conditions. The most important parameters which affect the dynamic behaviour of the laser are the ratio of the differential loss, $\mathrm{d}a/\mathrm{d}n$, of the absorber to the differential gain, $\mathrm{d}g/\mathrm{d}n$, of the amplifier section and the ratio of the carrier lifetime in the gain $\tau_{\rm g}$ and the absorber $\tau_{\rm a}$ sections. It can be shown [2] that for the generation of high-quality Q-switched pulses the condition

$$\frac{\mathrm{d}a/\mathrm{d}n}{\mathrm{d}g/\mathrm{d}n} > 1 \tag{B2.3.30}$$

must be satisfied. Owing to the dependence of gain and absorption on carrier concentration in lasers, the values of $\mathrm{d}g/\mathrm{d}n$ and $\mathrm{d}a/\mathrm{d}n$ can be readily controlled by using different currents applied to different sections of the laser. As a result, the output performance is also dependent on the driving currents. The parameters of Q-switched pulses are dependent on the photon lifetime and the absorber recovery rate. Although the photon lifetime can be set by cavity design, the recovery of the absorber is due to the spontaneous recombination of excited electron–hole pairs. A number of techniques for decreasing $\tau_{\rm a}$ have been developed. For example, proton or ion bombardment has been used to introduce a small region of saturable absorption having a very short carrier lifetime. Alternatively, a reverse bias can be applied on the absorber section of a multiple contact Q-switched laser to reduce $\tau_{\rm a}$ [21,22]. Optical loss in the cavity can be changed by a reverse dc bias applied to the absorber section. The location of the absorber section at the middle between the gain regions and application of the reverse bias on the absorber results in the improvement of the pulse parameters generated by three section lasers as compared with results for two section devices. Using the technique, optical pulses as short as 5 ps have been generated, with peak powers as large as around 10 W. The pulse repetition rate can readily be varied from 0.8 up to 18.5 GHz by a variation of dc bias and the driving currents.

Passive Q-switching exhibits poor timing jitter. This is due to the influence of random spontaneous photons during the build-up of an ultrashort pulse in the laser cavity. The turn-on of a Q-switched pulse relies upon amplification of random spontaneous photons. In addition, the repetition rate of Q-switched pulses is very sensitive to variation in the applied bias currents; this also causes enhanced timing jitter. To reduce the pulse-to-pulse jitter, rf modulation can be applied to the absorber section of the device for locking of self-Q-switched pulses [23]. This approach is significantly more effective than that using direct modulation of the optical gain. The required rf power is modest (10–20 dBm) and pulse train locking over a wide range of frequencies has been achieved.

Many longitudinal modes of a Q-switched laser cavity may be excited as a consequence of the variation of carrier density during the emission of optical pulses. Self-Q-switched lasers normally have a spectral bandwidth as wide as 3–5 nm and thus generate picosecond pulses with time–bandwidth products that exceed, by several orders of magnitude, those for bandwidth-limited pulses. To reduce the spectral bandwidth of Q-switched pulses, single-frequency lasers such as distributed Bragg reflector (DBR) and DFB devices should be used. Using such devices, bandwidth-limited Q-switched operation can be achieved. For instance, optical bandwidths as narrow as 0.18 nm can be achieved by Q-switching of quarter-wavelength-shifted MQW triple section DFB lasers [24].

High-peak-power ultrashort optical pulses with good spatial mode quality are increasingly required for many scientific and industrial applications, including space communications and optical sensing. The simplest method of increasing the output power of Q-switched pulses is to increase the active region area of the laser. This can be done by increasing the laser volume in all space dimensions: thickness, width and length. There are a number of approaches for achieving this: (a) to fabricate a laser structure with a thicker active layer; (b) to use a broad area structure with a straight waveguide; (c) to use bow-tie devices; (d) to use structures with flared or tapered waveguides; or (e) to fabricate an array of several individual devices. Alternatively, post-amplification in a semiconductor optical amplifier may be used.

There exists another type of self-Q-switching in multiple-contact DFB lasers. It is called dispersive self-Q-switching because the Q-factor of the laser cavity is varied by dispersive effects in this case. Self-Q-switching can occur when the lasing wavelength coincides with a negative slope of the reflectivity of a DFB section. Through the linewidth enhancement factor α, an amplitude modulation causes a wavelength and feedback modulation. This type of Q-switching results in very high repetition rates of optical pulses in the range 10–100 GHz [25].

B2.3.5.2 Mode-locking of laser diodes

It is well known [1,2] that laser emission can be considered as a set of resonator modes ω_m, $m = 1, 2, 3 \ldots$, separated by $\delta\omega = \pi c/(\mu_g L)$, where μ_g is the group refractive index. The number of modes which oscillate is limited by the spectral bandwidth $\Delta\omega_g$ over which the laser gain exceeds the cavity loss. The output therefore consists of the sum of oscillating modes and its electric field is given as (without taking into account the spatial distribution)

$$E(t) = \sum_m A_m \exp\{\mathrm{i}[(\omega_0 + m\delta\omega)t + \phi_m]\} \tag{B2.3.31}$$

where A_m and ϕ_m represent the amplitude and phase of the mth mode. In general, the relative phases between the modes are randomly fluctuating. If nothing fixes the phases ϕ_m, the laser output varies randomly in time, Alternatively, if the modes are forced to maintain a fixed phase and amplitude relationship then the output of the laser has a periodic function of time, i.e.

$$E(t) = A_0 \frac{\sin[(k+1)\delta\omega t_1/2]}{\sin[\delta\omega t_1/2]} \mathrm{e}^{\mathrm{i}\omega_0 t} \tag{B2.3.32}$$

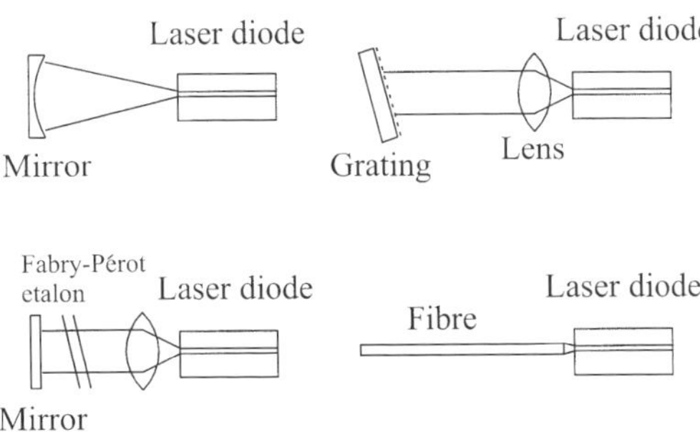

Figure B2.3.6. External cavity configurations of mode-locked diode lasers.

where k is the number of locked modes and $t_1 = t + \delta\phi/\delta\omega$. The operating conditions given by equation (B2.3.32) result in the generation of a train of regularly spaced optical pulses with a width of $\Delta\tau$ which is proportional to $1/\Delta\omega_g$. The pulse train has a temporal periodicity of $T = 2L\mu/c$ and the ratio of the period T to the pulsewidth is equal to the number of locked modes k. This dynamic regime is called mode locking (ML). It is also possible to produce ML with several pulses inside the cavity spaced by a multiple of $c/2L\mu$. The basic advantage of ML is that much shorter pulses are generated than those produced using gain/Q-switching. ML can be realized using a number of techniques including active, passive and hybrid methods.

Active ML is achieved by modulation of the loss or gain of a laser at a frequency which is equal to the mode spacing $\delta\upsilon$ of the laser cavity. As a result, the phases of the modes are locked by the external modulation. Because of the very small dimensions of the lasers (the length of a cavity is typically 250-400 μm), $\delta\upsilon$ is very large (more than 100 GHz). Consequently, it is quite difficult to modulate the gain at such a frequency. A straightforward solution to this problem is, therefore, to increase the cavity length. If the cavity length is increased by placing the laser in an external resonator or using an extended monolithic cavity up to several millimetres long, the mode spacing is decreased and active ML can easily be realized.

The simplest method for realizing active ML is to place a laser in an external cavity, the most commonly used configurations of external resonators being shown schematically in figure B2.3.6. ML in semiconductor lasers can also be achieved using phase-conjugate optical feedback. Phase-conjugate feedback, in contrast to ordinary feedback, directly couples pairs of longitudinal modes of a multi-mode laser. For certain strengths of phase-conjugate reflectivities, the phases and beat frequencies of various longitudinal modes become locked and the laser produces ultrashort pulses even though the laser is pumped continuously.

Unless special precautions are taken, the time–bandwidth product of ML pulses exceeds considerably the theoretical values for bandwidth-limited pulses. A number of methods for the restriction of the optical spectrum to a single mode of the laser have been used in external cavity lasers. Placing a Fabry–Pérot etalon in the cavity or using a diffraction grating as the external reflector, as shown in figure B2.3.6, restricts the lasing spectrum considerably and leads to the generation of bandwidth-limited pulses. In this case, values of $\Delta\tau\Delta\nu$ lie in the range of 0.3–0.6. As mentioned earlier, Bragg reflectors can be integrated in the laser cavity, both in monolithic extended-cavity lasers and single-mode fibre composite cavity lasers for the same purpose.

In a similar manner to gain or Q-switched pulses, ML pulses exhibit wavelength chirping. The chirp leads to excessive values of the time–bandwidth product for ML pulses. Although the temporal variation of the instantaneous wavelength of the ML pulses is complex, the chirp is approximately linear near the peak

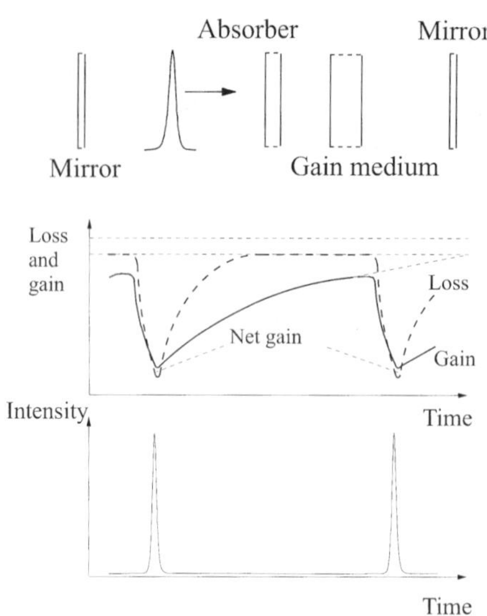

Figure B2.3.7. Passive mode locking with a saturable absorber: cavity configuration, gain and loss dynamics, and optical output.

of the pulses. The frequency chirp observed in actively ML lasers is of the opposite sign to that exhibited by passively ML lasers. This has been accounted for in terms of the different magnitudes and signs of the gain/refractive index coupling in the gain and absorber sections of the lasers. The frequency chirp, however, has been widely used for further pulse compression (see later).

Ultrashort pulse generation by a passively ML laser is schematically illustrated in figure B2.3.7. The cavity configuration and the time-dependent gain and absorption loss in the laser are shown. The laser is driven by a dc current and no external modulation is used. Before the arrival of the optical pulse, the loss and gain are approaching their steady-state values (broken curves in figure B2.3.7), the loss being larger than the gain. Thus, the leading edge of the pulse experiences loss. In a passively ML laser the loss must saturate more quickly than the gain. When the gain becomes saturated below the unsaturated loss (due to mirror transmission, internal losses, etc), the trailing edge of the pulse again experiences a loss. If the recovery time of the absorber is faster than that of the amplifier, then the loss remains greater than the gain everywhere except near the peak of the pulse. Spontaneous emission is not able to build up between pulses. The unsaturated gain must also be greater than the unsaturated loss so that the laser oscillation can build up when the laser is turned on. On each round trip in the cavity, the combined action of the gain and loss saturation is to shorten the pulse by amplifying the peak of the pulse and attenuating the leading and trailing edges. In the steady state, this shortening is balanced by broadening mechanisms (detuning, dispersion and self-phase modulation). It is necessary to have substantial changes in both gain and loss during the passage of a pulse. Thus it is desirable for the pulse repetition rate to be of the order of the gain or loss recovery time. Passive ML is a self-starting process and output pulses are formed from spontaneous fluctuations. The fundamental difference between passive ML and passive Q-switching is that in contrast to Q-switching where the pulse repetition rate is set by amplitudes of applied currents, the repetition rate of passive ML pulses is determined by the round-trip time of the laser cavity and is not dependent on the pumping conditions.

Colliding pulse mode (CPM) locking is a very effective technique for generating ultrashort pulses in the femtosecond range [1, 2]. This method is based upon the interaction of two counter-propagating optical pulses inside the saturable absorber of a laser with a linear or ring resonator. Coherent interaction of the pulses builds up a transient standing-wave optical field and a periodic saturation modulation in the absorber. This, in principle, should result in shorter pulses when compared with passive ML pulses obtained from the same laser. The counter-propagating pulses form transient standing waves in the absorber. The carrier generation rate is high at the peak of the standing wave while it is low at the nulls of the standing wave. The carrier density absorption grating facilitates the saturation of the absorber. The rigorous theory of the pulse collision effects in CPM lasers predicts the pulsewidth to be four times shorter at best than that for passive ML with the same absorber and gain medium and without pulse collisions [2, 26]. It should be noted that the pulse collision effectively shortens ML pulses in thin saturable absorbers only, i.e. when the length of the absorber is less than the spatial dimension of the pulses. CPM can be obtained with a laser cavity configuration where a saturable absorber is placed near a highly reflecting mirror. In this case the transient absorber grating is formed by the interference of the incident and reflected pulses (self-colliding pulse configuration). Pulses as short as 500–600 fs can be produced in external cavity CPM diode lasers.

Monolithic cavity ML laser structures are very compact and do not exhibit the mechanical instabilities associated with optical elements in an external cavity [27]. Overall cavity lengths of monolithic ML lasers range typically from 0.25 to 7.0 mm, the corresponding repetition rates of the ML pulses varying from 5.5 to 350 GHz. Passive ML of cw monolithic lasers results in generating optical pulses with typical FWHM pulsewidths of 0.6–2 ps ($\Delta\tau\,\Delta\nu$ of around 1) and peak powers in the range of 10–30 mW at repetition rates above 100 GHz [28].

Hybrid ML of a monolithic laser can be achieved with the addition of a separate gain modulation section. Because of parasitics of the device it is difficult to couple the modulation signal to a laser at very high frequency. As a result, the cavity length of hybridly ML lasers is longer than that of passive ML and CPM lasers. Since saturable absorption is the dominant mechanism for obtaining very short pulses, the requirements on the electrical modulation waveform are reduced in hybridly ML lasers. Here it is not as important to have an extremely short electrical modulation waveform as in actively ML lasers where the pulse-shortening mechanism is conditioned by the curvature of the electrical waveform. Harmonic passive monolithic ML can produce transform-limited ultrashort pulses at repetition rates from 500 GHz to over 1.5 THz with an output power in excess of 15 mW [29].

Peak power of ultrashort pulses is another key parameter which strongly affects potential applications of these pulses in various fields. The main means of enhancing peak power currently consists of increasing the emitting area of diode lasers. This can be done in a number of ways, including fabrication of structures with flared (tapered) waveguides, bow-tie devices or using laser arrays. A flared waveguide expands the optical mode from a narrow region which gives a single lateral optical mode to a wider multi-mode region for higher pulse saturation energy. A promising approach for increasing the output power of optical pulses is the use of tapered-strip travelling-wave semiconductor amplifiers in conjunction with ML or gain/Q-switched lasers. Tapered single-pass amplifiers allow for increasing energies of ultrashort pulses because the saturation energy can be made relatively large at the flared output end of the device. A master oscillator power amplifier (MOPA) configuration has been considered one of the most promising for high-power generation in diode lasers [30].

B2.3.5.3 Pulse compression of laser diode pulses

It is well known that ultrashort laser pulses can be compressed by a number of methods exploiting self-phase modulation (SPM) and group velocity dispersion (GVD) effects. The fundamental idea of optical pulse compression was borrowed from radar physics, where chirped pulses at microwave frequencies were compressed with the use of dispersive delay lines. The principle of pulse compression consists of cancelling the initial chirp of the pulse by passing it through a medium which exhibits a dispersion-induced chirp of the opposite direction. This results in pulse narrowing. Positively chirped pulses, i.e. pulses with the instantaneous frequency increasing towards the trailing edge, require anomalous (negative) GVD in order to

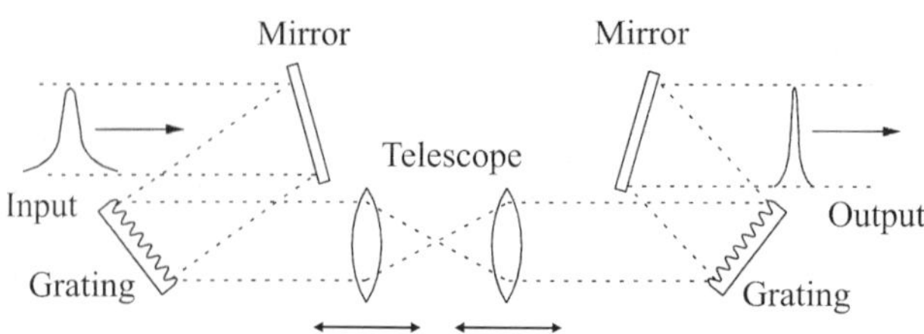

Figure B2.3.8. Schematic diagram of a grating compressor for the compression of both positively and negatively chirped ultrashort optical pulses.

slow down the red-shifted leading edge. By contrast, normal (positive) GVD is necessary for compression of negatively chirped pulses.

There are three main types of pulse compressors that are commonly used for compression of ultrashort pulses from diode lasers, namely fibre, grating and soliton-effect compressors. Owing to intrinsic negative chirping of picosecond pulses generated by gain/Q-switched and actively mode-locked diode laser, it is possible to use a medium with normal dispersion for pulse compression. A piece of a fibre of appropriate length is the simplest pulse compressor. The minimum (transform-limited) value of the pulsewidth after compression [7] is determined by the initial chirped bandwidth $\Delta\lambda$ [8] as

$$\tau_{\min} = \frac{\lambda_0^2}{\Delta\lambda\pi c}. \tag{B2.3.33}$$

For a chirped bandwidth of initial pulses of about 1 nm, $\tau_{\min}$ [10] as short as 2–3 ps can be obtained using 300–700 m long fibres with a typical dispersion parameter D of 20–30 ps nm^{-1} km^{-1}. It is possible to get the compression factor in the range of 5–10 with a single-stage fibre compressor. To get transform-limited compressed pulses, single-mode initial pulses are required.

Grating compressors are adjustable and can provide larger compression factors. A conventional grating compressor consists of a pair of diffraction gratings. In grating compressors the GVD arises from the angular spectral dispersion. A grating-pair compressor that can provide both normal and anomalous dispersion is schematically shown in figure B2.3.8 [31]. It consists of a pair of diffraction gratings and two lenses in a telescope arrangement. By using the telescope between the two gratings the sign of the GVD could be either positive or negative. The sign inversion is achieved if the gratings are placed between the focal planes of the lenses. The compressor dispersion in the range of -100 to $+100$ ps nm^{-1} can be obtained by using 50 cm focal-length lenses. A compression factor of 15.8 was demonstrated [32] using such a compressor. Pulses of 5.2 ps from a passively mode-locked external cavity GaAs MQW two-section laser with the chirped bandwidth of 1.1 THz were compressed to 320 fs pulses having a time–bandwidth product of 0.39. Compressed pulses were obtained with the single-pass compressor in the region of negative GVD, corresponding to positively chirped input pulses.

Soliton-effect compression of diode laser pulses has been demonstrated both at 1.3 μm [11] and 1.5 μm [12] wavelengths. Ultrashort pulses obtained from a diode laser are compressed by a fibre or grating compressor at first. Then they are normally amplified by a travelling-wave semiconductor amplifier or by an Er-doped fibre amplifier. Finally, the pulses are launched into a fibre with anomalous dispersion for soliton-effect compression. The fibre length is normally chosen to be a few (1–3) km. Extremely large compression factors using a soliton-effect compressor have been reported [33]. Initial pulses were generated by a gain-switched InGaAs/InP MQW laser. The pulses at 1.55 μm had a FWHM of 5 ps. The compressor had four fibre stages with different dispersions. By careful management of the fibre dispersion, the minimum pulsewidth of compressed pulses was as short as 20 fs.

B2.3.6 Future trends in high-speed diode lasers

The ultrafast picosecond and femtosecond diode systems seem, at present, to be in an advanced state. Extensive experimental and theoretical work is in progress in order to take full advantage of all the possibilities offered by ultrafast diode lasers. At present extremely high repetition rates of several hundred GHz of femtosecond pulses can be generated by CPM monolithic diode lasers. However, the weaknesses of ultrashort pulse generation by diode lasers are small pulse energies and longer pulsewidths when compared with CPM dye and mode-locked solid state lasers. Thus, the goal is to increase peak pulse power and to decrease pulsewidth without using non-semiconductor (solid state, dye, etc) components.

A promising way for decreasing pulsewidth on the femtosecond time scale in diode lasers is to utilize coupled-cavity ML [34]. Fourier synthesis of optical harmonics is another good candidate for producing ultrashort optical pulse trains. Realization of super-radiance or cooperative spontaneous emission in semiconductor laser structures is a novel promising way for the generation of very high-power femtosecond pulses from diode lasers [35, 36].

Another potentially important technique for the generation of ultrashort pulses with an ultrahigh repetition rate is Fourier synthesis of the laser emission. Indeed, optical pulses can be Fourier-synthesized using independent cw semiconductor lasers. Because there is no fundamental limitation on a wider frequency separation or on the laser material, Fourier synthesis is particularly suitable for the generation of ultrafast optical pulse trains. For instance, it is possible to produce pulses having a FWHM as short as 170 fs at a repetition rate of around 1.8 THz using three cw diode lasers and a nonlinear phase-locking scheme [37].

The maximum modulation bandwidth achieved so far (30–40 GHz) can be increased by putting forward and realizing novel ideas on how to beat the present limits set by the device parasitics and device size. New fabrication techniques and device structures with cavity lengths less than 50–100 μm are required to solve the problem. Development of potentially chirpless high-speed laser structures is also very important for applications in long-haul high-bit-rate optical fibre communication systems. Different types of optical TDM and WDM systems utilizing all the advantages of high-speed diode lasers can provide very high bit rates ranging from 40 to 100 Gb s^{-1}.

References

[1] Siegman A E 1986 *Lasers* (Mill Valley CA: University Science Books)

[2] Vasil'ev P 1995 *Ultrafast Diode Lasers: Fundamentals and Applications* (Norwood MA: Artech House)

[3] Agrawal G P and Dutta N K 1993 *Semiconductor Lasers* (New York: Van Nostrand Reinhold)

[4] Petermann K 1988 *Laser Diode Modulation and Noise* (Dordrecht: Kluwer Academic)

[5] Nagarajan R, Ishikawa M, Fucushima T, Geels R S and Bowers J E 1992 High speed quantum-well lasers with carrier transport effects *IEEE J. Quantum Electron.* **28** 1990–2007

[6] Tucker R S 1985 High speed modulation in semiconductor lasers *J. Lightwave Technol.* **3** 1180–92

[7] Bower, J E, Hemenway B R, Gnauck A H and Wilt D P 1986 High-speed InGaAsP constricted mesa lasers *IEEE J. Quantum Electron.* **22** 833–44

[8] Lau K Y 1989 Short pulse and high-frequency signal generation in semiconductor lasers *J. Lightwave Technol.* **7** 400–19

[9] Olshansky R, Hill P, Lanzisera V and Powazinik W 1987 Frequency Response of 1.3 μm InGaAsP high speed semiconductor lasers *IEEE J. Quantum Electron.* **23** 1410–18

[10] Tatham M C, Lealman I F, Seltzer C P, Westbrook L D and Cooper D M 1992 Resonance frequency, damping, and differential gain in 1.5 μm multiple quantum-well lasers *IEEE J. Quantum Electron.* **28** 408–14

[11] Ralston J D, Weisser S, Esquivias I, Larkins E C, Rosenzweig J, Tasker P J and Fleissner J 1993 Control of differential gain, nonlinear gain, and damping factor for high-speed application of GaAs-based MQW lasers *IEEE J. Quantum Electron.* **29** 1648–59

[12] Morton P A, Logan R A, Tanbun-Ek T, Sciortino P F Jr, Sergent A M, Montgomery P K and Lee B T 1992 25 GHz bandwidth 1.55 μm InGaAsP p-doped strained multiple quantum well lasers *Electron. Lett.* **28** 2156–7

[13] Henry C H 1982 Theory of the linewidth of semiconductor lasers *IEEE J. Quantum Electron.* **18** 259–64

[14] Osinski M and Buus J 1987 Linewidth broadening factor in semiconductor lasers—an overview *IEEE J. Quantum Electron.* **23** 9–29

[15] Yasaka H, Takahata K and Naganuma M 1992 Measurement of gain saturation coefficient in strained-layer multiple quantum-well distributed feedback lasers *IEEE J. Quantum Electron.* **28** 1294–304

[16] Kano F, Yamanaka T, Yamamoto N, Yoshikuni Y, Mawatari H, Tohmori Y, Yamamoto M and Yokoyama K 1993 Reduction of linewidth enhancement factor in InGaAsP/InP modulation-doped strained multiple-quantum-well lasers *IEEE J. Quantum Electron.* **29** 1553–9

[17] Tauber D A, Spickermann R, Nagarajan R, Reynolds T, Holms A L and Bowers J E 1994 Inherent bandwidth limits in semiconductor lasers due to distributed microwave effects *Appl. Phys. Lett.* **64** 1610–12

[18] Agrawal G P 1997 *Fiber-Optic Communication Systems* (New York: Wiley)

[19] Wakita K 1997 *Semiconductor Optical Modulators* (Dordrecht: Kluwer Academic)

[20] Chin M K and Chang W S C 1993 Theoretical design optimization of multiple-quantum-well electroabsorption waveguide modulators *IEEE J. Quantum Electron.* **29** 2476–88

[21] Vasil'ev P P 1988 Picosecond injection laser: a new technique for ultrafast Q-switching *IEEE J. Quantum Electron.* **24** 2386–91

[22] Vasil'ev P P 1993 High-power high-frequency picosecond pulse generation by passively Q-switched 1.55 μm diode lasers *IEEE J. Quantum Electron.* **29** 1687–92

[23] Vasil'ev P P, White I H, Burns D and Sibbett W 1993 High-power, low-jitter encoded picosecond pulse generation using an rf-locked self-Q-switched multicontact GaAs/GaAlAs diode laser *Electron. Lett.* **29** 1593–4

[24] Vasil'ev P P, White I H and Fice M J 1993 Narrow line high power picosecond pulse generation in a multicontact distributed feedback laser using modified Q-switching *Electron. Lett.* **29** 561–3

[25] Sartorius B, Mohrle M, Reichenbacher S, Preier H, Wunsche H-J and Bandelow U 1997 Dispersive self-Q-switching in self-pulsating DFB lasers *IEEE J. Quantum Electron.* **33** 211–18

[26] Vasil'ev P P, Morozov V N, Popov Yu M and Sergeev A B 1986 Subpicosecond pulse generation by a tandem-type AlGaAs DH laser with colliding pulse mode locking *IEEE J. Quantum Electron.* **22** 149–52

[27] Vasil'ev P P and Sergeev A B 1989 Generation of bandwidth-limited 2 ps pulses with 100 GHz repetition rate from multisegmented injection laser *Electron. Lett.* **25** 1049–50

[28] Chen Y K and Wu M C 1992 Monolithic colliding-pulse mode-locked semiconductor lasers *IEEE J. Quantum Electron.* **28** 2176–85

[29] Arahira S, Matsui Y and Ogawa Y 1996 Mode locking at very high repetition rates more than terahertz in passively mode-locked distributed-Bragg-reflector laser diodes *IEEE J. Quantum Electron.* **32** 1782–90

[30] Mar A, Helkey R, Bowers J, Mehuys D and Welch D 1994 Mode-locked operation of a master oscillator power amplifier *IEEE Photon. Technol. Lett.* **6** 1067–9

[31] Martinez O E 1987 3000 times grating compressor with positive group velocity dispersion: application to fiber compensation in 1.3–1.6 μm region *IEEE J. Quantum Electron.* **23** 59–64

[32] Schrans T, Salvatore R A, Sanders S and Yariv A 1992 Subpicosecond (320 fs) pulses from CW passively mode-locked external cavity two-section multiquantum well lasers *Electron. Lett.* **28** 1480–2

[33] Igarashi K, Kishi M and Tsuchiya M 2001 Higher order soliton compression of optical pulses from 5 ps to 20 fs by a 1.5 m long single-stage step-like dispersion profiled fibre *Jap. J. Appl. Phys.* **40** 6426–9

[34] Dianov E M and Okhotnikov O G 1991 *IEEE Photon. Technol. Lett.* **3** 499–501

[35] Vasil'ev P P 1997 Superfluorescence in semiconductor lasers *Quantum Electron.* **27** 860–5

[36] Vasil'ev P P 1999 A role of high gain of the medium in the superradiance generation and observation of coherent effects in semiconductor lasers *Quantum Electron.* **29** 4–8

[37] Hyodo M, Abedin K S and Onodera N 2001 Fourier synthesis of 1.8 THz optical-pulse trains by pulse locking of three independent semiconductor lasers *Opt. Letts.* **26** 340–2

B2.4
High-power laser diodes and laser diode arrays

Peter Unger

B2.4.1 Introduction

Laser diodes providing high optical output power show numerous advantages in comparison to conventional high-power laser systems. Superior features are high electrical-to-optical power conversion efficiency, small size, reliable operation and low price. The main uses of high-power laser diodes include optical pumping for solid state lasers and fibre amplifiers, optical data storage, free-space data communication, material treatment, medical applications and infrared illumination. Depending on the specific application, the laser diodes are optimized with regard to high optical output power, spectral and spatial beam quality or emission wavelength. In this chapter, the basic criteria for the design of high-power laser diodes and laser diode arrays are discussed together with a presentation of state-of-the-art device implementations.

B2.4.2 Mechanisms limiting output power and beam quality of laser diodes

There are three important mechanisms which limit the output power of semiconductor laser diodes. The first effect is the heating of the device during operation leading to an increase in laser threshold current and a decrease in the differential efficiency. The second problem, usually called filamentation, is the distortion of the beam profile at high output power levels. The third phenomenon is catastrophic optical mirror damage (COMD) which is an irreversible destruction of the laser mirror facet.

B2.4.2.1 Thermal issues

The electrical-to-optical conversion efficiency of laser diodes is typically around 50% approaching record values close to 70% [4, 20]. Thus, a significant amount of heat is generated in the device. Depending on the thermal resistance R_{therm}, which is defined by

$$R_{\text{therm}} = \frac{\Delta T}{P_{\text{el}} - P} \tag{B2.4.1}$$

where P_{el} is the electrical input power and P is the optical output power, the active region of the laser diode experiences a temperature rise ΔT. The thermal resistance, also called the thermal impedance, can be influenced be the geometry and the thermal conductivity of the different materials and layers in the device. Another factor that governs thermal resistance is the mounting and packaging technology (see section B2.4.6). The elevated temperature leads to a degradation of the output power P *versus* operating current I characteristic of the device. This characteristic is usually characterized by the threshold current I_{th} where laser operation

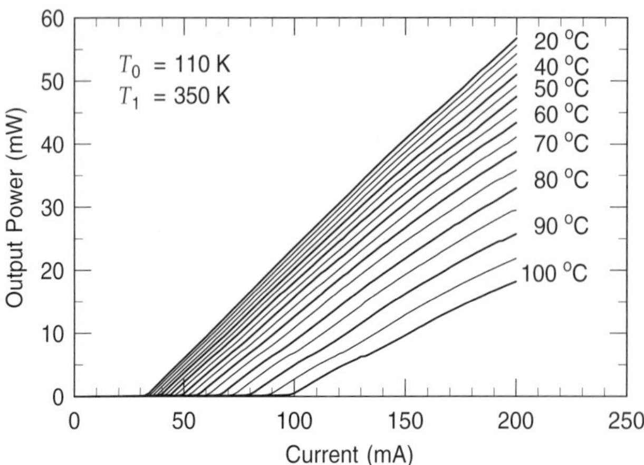

Figure B2.4.1. Continuous-wave (cw) output characteristics of a ridge-waveguide laser diode at different temperatures. At higher temperatures, the threshold current increases and the differential efficiency decreases. This behaviour is phenomenologically described by the characteristic temperatures T_0 and T_1.

starts and the differential quantum efficiency η_d. The differential quantum efficiency can be determined from the slope $(\mathrm{d}P)/(\mathrm{d}I)$ of the linear part of the characteristic above threshold using the equation

$$\eta_d = \frac{e}{\hbar\omega}\frac{\mathrm{d}P}{\mathrm{d}I} = \frac{(\mathrm{d}P)/(\hbar\omega)}{(\mathrm{d}I)/e} \tag{B2.4.2}$$

where $e = 1.602\,189 \times 10^{-19}$ C is the elementary charge and $\hbar\omega$ is the photon energy of the emitted light. Thus, the differential efficiency η_d can be interpreted as the differential increase in emitted photons per unit time $(\mathrm{d}P)/(\hbar\omega)$ divided by the differential increase in injected electrons per unit time $(\mathrm{d}I)/e$ above laser threshold [33].

Increasing the temperature T of the device leads to an increase in the non-radiative recombination processes and a reduced carrier confinement in the active region, which results in a higher threshold current and a lower differential efficiency. Usually, the temperature behaviour of the output characteristic of a laser is phenomenologically described by two characteristic temperatures T_0 and T_1 for the threshold current I_{th} the differential efficiency η_d, respectively, according to the equations

$$I_{th} \propto \exp\left(\frac{T}{T_0}\right) \qquad \text{and} \qquad \eta_d \propto \exp\left(-\frac{T}{T_1}\right). \tag{B2.4.3}$$

As an example, the output power characteristics of a ridge-waveguide laser are shown in figure B2.4.1 for different temperatures. If the active region gets extremely hot, thermal rollover is observed. In this case, a further increase in threshold current results in a decrease in output power, as shown in the left-hand side diagram of figure B2.4.3.

B2.4.2.2 Filamentation

At high optical output levels, the beam profile of semiconductor laser diodes deteriorates. The reason for this phenomenon, which is usually called filamentation, is the interaction between the optical wave and the electronic carriers in the active region of the device. A high optical power density leads to spatial hole burning,

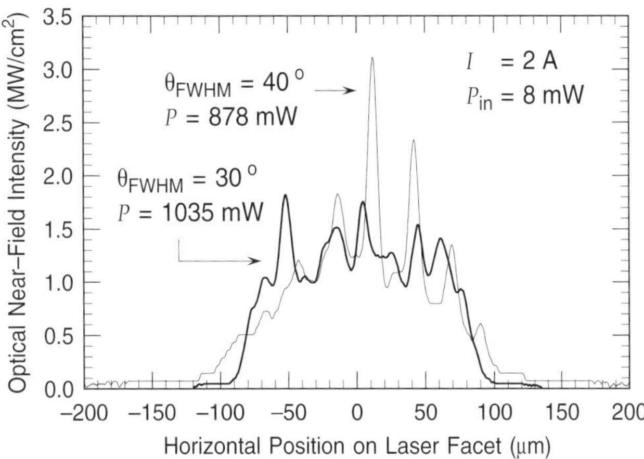

Figure B2.4.2. Optical near field intensity at the laser facet of two broad-area laser amplifiers having different optical power densities inside the active region. The device with low optical density has a smaller full-width at half maximum angle Θ_{FWHM} of the vertical far-field pattern and shows less filamentation although the overall optical output power P is larger. The laser amplifier is driven by an electrical current of $I = 2$ A and an optical input power of $P_{\mathrm{in}} = 8$ mW.

which locally increases the refractive index, resulting in self-focusing of the optical wave. Figure B2.4.2 shows an example of these filamentations for two broad-area laser amplifiers having different power densities inside the active region.

A detailed discussion of the filamentation effect is provided by the articles of Lang *et al* [22, 23], Hess *et al* [19], Dai *et al* [8] and Mullane and McInerney [29]. According to these publications, the complex refractive index $n(N)$ in the active region as a function of the electronic carrier density N is the sum of the carrier independent refractive index n_0 and a complex part originating from the modal optical gain $g_{\mathrm{m}}(N)$.

$$n(N) = n_0 - \frac{1}{2k_0} g_{\mathrm{m}}(N)\alpha_{\mathrm{H}} + \frac{\mathrm{i}}{2k_0}[g_{\mathrm{m}}(N) - \alpha_{\mathrm{i}}]. \tag{B2.4.4}$$

In this equation, $k_0 = 2\pi/\lambda$ is the vacuum wavenumber, α_{i} is the intrinsic optical modal loss inside the cavity and α_{H} is the Henry factor, which considers the influence of the optical gain on the refractive index via the Kramers–Kronig transformations [17, 18]. To reduce filamentation effects in high-power lasers, it is therefore necessary to enlarge the optical mode volume and to reduce the modal gain [5, 26].

B2.4.2.3 Facet damage

Another very important effect limiting the output power is catastrophic optical mirror damage (COMD). An example of this phenomenon is shown in figure B2.4.3. When a device is properly cooled, the output power limitation is caused by an irreversible and spontaneous destruction of the facet. A scanning electron micrograph of a facet after COMD is shown in figure B2.4.4. The dynamics of this type of degradation is described in detail in Moser *et al* [27], Brugger and Epperlein [6], Moser and Latta [28] and Fukuda [13]. The COMD can be reduced by lowering the optical power density on the laser facet using a suitable design of the vertical optical waveguide. With the help of sophisticated facet coatings, the COMD effect can be drastically reduced or even completely eliminated.

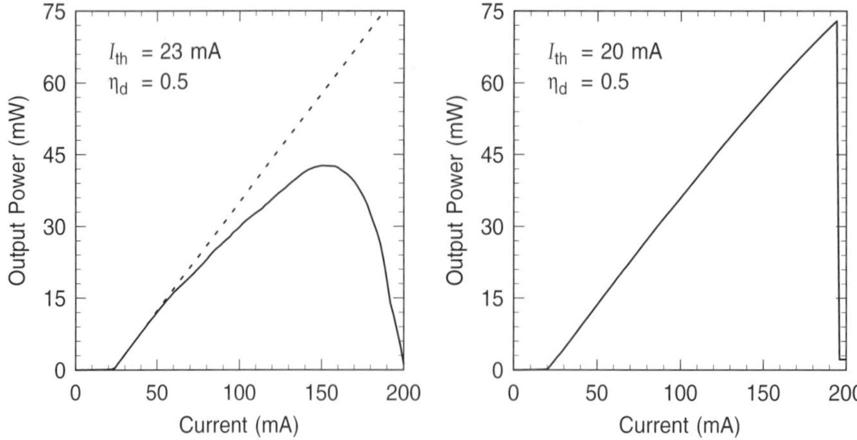

Figure B2.4.3. Optical output power *versus* operating current for two ridge-waveguide lasers. The output power of the device on the left-hand side is limited by thermal rollover, whereas the device on the right-hand side is destroyed by catastrophic optical mirror damage (COMD). Unlike COMD, thermal rollover is a reversible process, since it is normally not associated with a destruction or degradation phenomenon.

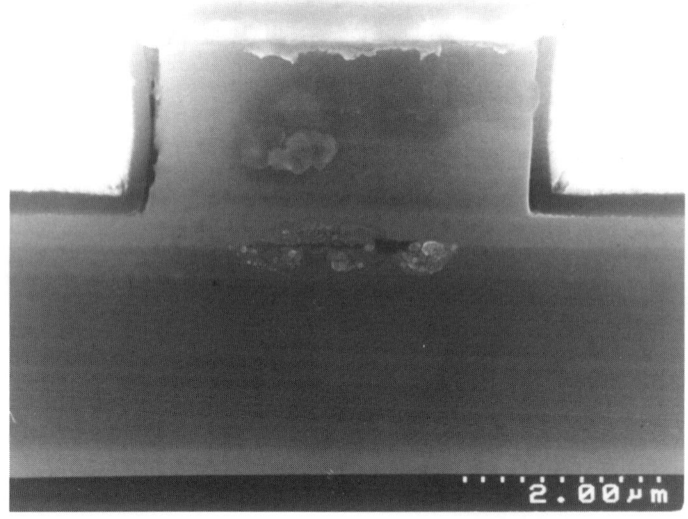

Figure B2.4.4. Scanning electron micrograph of the facet region of a ridge-waveguide laser where the facet has been destroyed by catastrophic optical mirror damage. The pattern of the single-mode optical near-field distribution can be clearly identified.

B2.4.3 Vertical waveguide design for high-power operation

B2.4.3.1 Material systems

The vertical multilayer structures of laser diodes are fabricated using epitaxial growth techniques. The following list gives an overview of material systems commonly used for high-power operation.

- $Al_xGa_{1-x}As$ grown on GaAs is the classical material for high-power laser diodes. Since the radii of gallium and aluminium ions are nearly equal, $Al_xGa_{1-x}As$ can be grown lattice-matched for any composition x, and laser emission in the wavelength range 700–870 nm can be realized.
- Using an InGaAs quantum well, the emission range of the AlGaAs/GaAs system can be extended to longer wavelengths. Since an indium ion has a larger radius than a gallium ion, the lattice parameter of GaInAs is higher than the lattice parameter of GaAs. Therefore, only thin GaInAs quantum-well layers containing compressive mechanical strain can be grown keeping lattice-matching. The emission wavelength can be adjusted in the range 800–1100 nm by varying the thickness and indium content of the strained quantum well.
- The same wavelength range can be covered using the strained GaInAs quantum well in combination with GaInAsP separate-confinement layers and $Ga_{0.51}In_{0.49}P$ cladding layers. Like GaInAs-AlGaAs/GaAs, this system is also lattice-matched to GaAs but is completely aluminium free [4, 25].
- The visible-red short-wavelength range 600–700 nm can be accessed using $(Al_xGa_{1-x})_{0.5}In_{0.5}P$. Again, this material is lattice-matched to GaAs for any aluminium concentration x since the ion radii of aluminium and gallium are approximately equal.

As mentioned previously, it is highly recommended to use a vertical epitaxial waveguide structure providing a low optical power density and a reduced optical modal gain. There are two major approaches for the vertical waveguide design to fulfil these requirements, a graded-index separate-confinement heterostructure (GRINSCH) having a narrow waveguide or a structure with a broad-width vertical waveguide. Both waveguide designs, which are discussed in the next two sections in more detail, provide a number of advantages compared to conventional vertical waveguide structures which are used for low-power devices like laser diodes for fibre communication.

- The beam quality at high output power is improved due to reduced filamentation effects caused by spatial hole burning and self-focusing (see figure B2.4.2).
- The device reliability is improved since regions with increased optical density and increased temperature (hot spots) leading to degradation effects are avoided.
- The localized power density at the laser facet is reduced resulting in less spontaneous device failures caused by COMD.
- A waveguide with a low optical power density in the active region has a rather broad optical mode profile. Since the optical far field is the Fourier transform of the optical mode profile, the vertical far-field angle Θ_{FWHM} is smaller (see figure B2.4.5). As a consequence, lens and fibre coupling is easier due to the lower numerical aperture in the fast optical axis.

B2.4.3.2 Graded-index separate-confinement heterostructure

An example of a GRINSCH structure having a narrow vertical waveguide is shown in figure B2.4.5. The active region of the device is a single compressively strained quantum well with a typical thickness of 5–10 nm, allowing laser emission in the range of 900–1000 nm. The quantum well is surrounded by $Al_xGa_{1-x}As$ graded-index (GRIN) regions. This core-waveguide GRINSCH region is sandwiched between p- and n-doped cladding layers. Since the core region of the vertical waveguide is rather narrow, the resulting optical near-field intensity distribution is broad. The near-field distribution shown in figure B2.4.5 has been calculated from the refractive-index profile using the one-dimensional Helmholtz equation. As a consequence, the full-width at half-maximum angle Θ_{FWHM} of the vertical far-field pattern is reduced compared to conventional vertical waveguide structures.

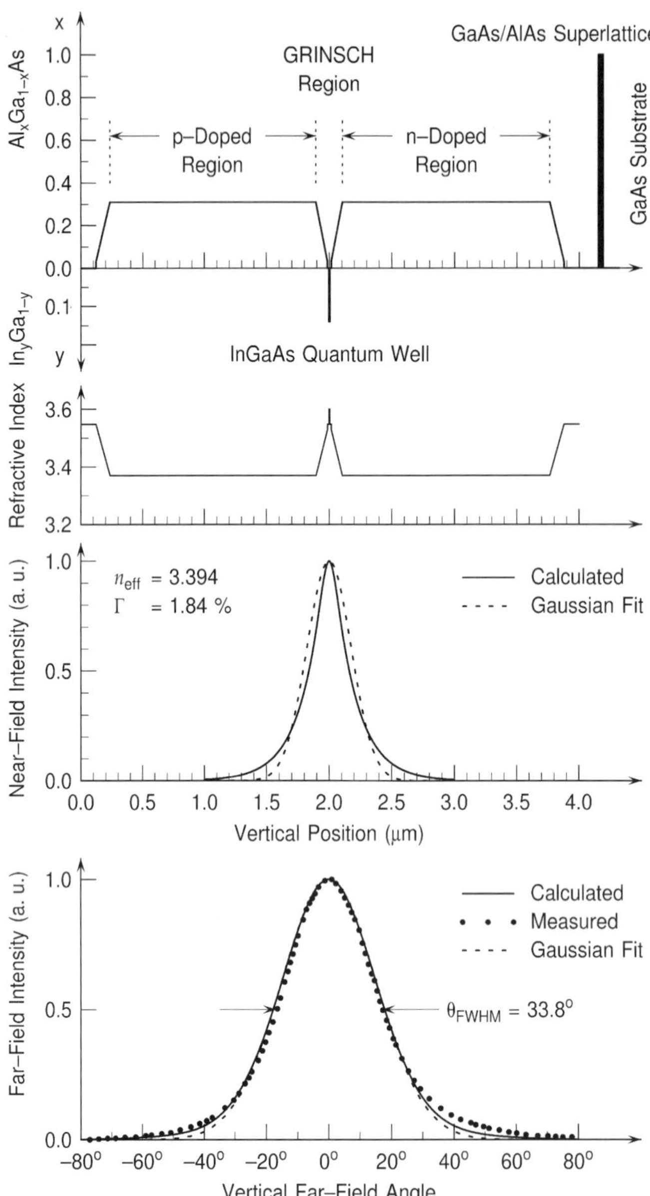

Figure B2.4.5. Vertical GRINSCH laser structure for high-power operation. The material composition is shown in the top diagram. The centre diagrams show the vertical refractive-index profile and the calculated optical near-field intensity together with a Gaussian fit. Listed also are the effective refractive index n_{eff} of the vertical waveguide and the optical confinement factor Γ which is the overlap of the near-field intensity with the quantum well. In the diagram at the bottom, the calculated and measured far-field patterns and a Gaussian fit are plotted. Θ_{FWHM} is the full-width at half-maximum angle of the far-field distribution.

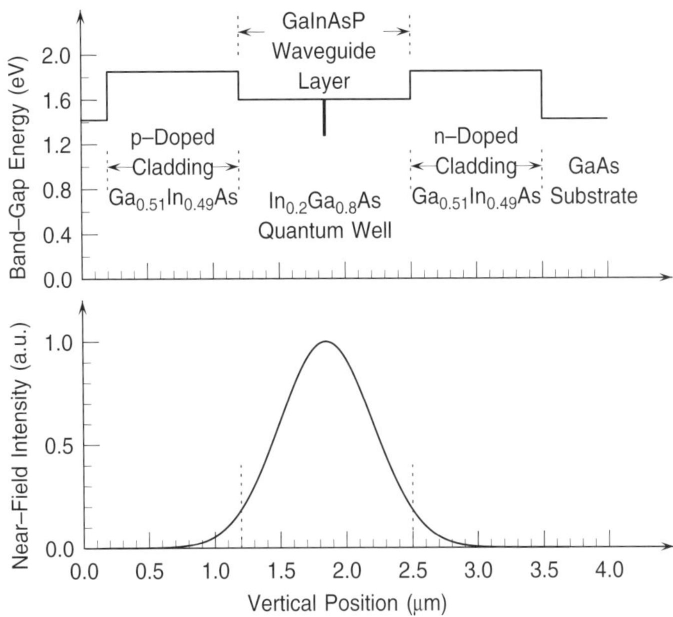

Figure B2.4.6. Aluminum-free broad vertical waveguide structure for 970 nm emitting high-power laser diodes. The waveguide consists of a 1.3-μm-wide undoped GaInAsP waveguide having a band-gap energy of 1.6 eV. In the centre of the broad-waveguide layer, one or two compressively-strained InGaAs quantum wells are located. The cladding layers consist of $Ga_{0.51}In_{0.49}P$ having a band-gap energy of 1.85 eV. In the diagram at the bottom, the fundamental optical vertical mode intensity of the waveguide is plotted. The full-width at half-maximum vertical far-field angle of this structure was reported to be $\Theta_{FWHM} = 36°$ (after Al-Muhanna *et al* [1]).

B2.4.3.3 Broad-width vertical optical waveguide

An alternative approach to the epitaxial structure of high-power laser uses a broad-width vertical optical waveguide to achieve a broad near-field intensity distribution. These structures, which are also called large-optical-cavity (LOC) structures are mainly used in the InGaAs/GaInAsP/GaInP material system where a grading of the refractive index in the optical waveguide is hard to achieve. In such a broad-width waveguide, the first- and second-order modes can also propagate. Since the first-order mode has a node at the quantum well, its confinement factor Γ is close to zero and thus the modal optical gain is negligible. The second-order mode is drastically reduced because its near-field distribution has a significant intensity in the GaInAsP cladding layers and therefore it suffers from radiation losses due to the width of these layers and of absorption losses at the p-side metal contact. A detailed discussion of the design of this type of structure can be found in Botez [5].

B2.4.4 Lateral design of high-power laser diodes

A lateral fundamental-mode output beam can be achieved by choosing a narrow-stripe index- or gain-guided lateral waveguide having a typical width in the range of 4–8 μm. For this kind of device, a maximum single-mode output power of a few 100 mW has been reported limited by a deterioration of the beam profile or by COMD. Applications for lateral-waveguide single-mode lasers are mainly optical data storage and pumping

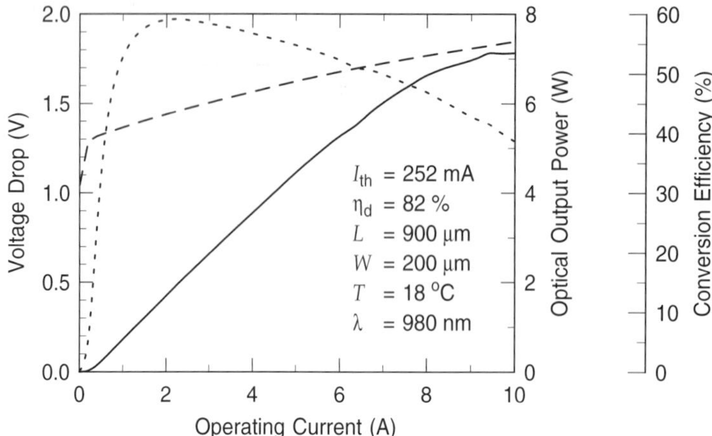

Figure B2.4.7. CW characteristics of a broad-area InGaAs/AlGaAs semiconductor laser diode having a vertical epitaxial layer structure as shown in figure B2.4.5. Plotted are the optical output power (full curve), the voltage drop (dashes) and the electrical-to-optical power conversion efficiency (dots). Listed are the threshold current I_{th}, the differential quantum efficiency η_d (determined according to equation (B2.4.2) from the slope of the output power characteristic), the cavity length L, the lateral width of the active region W, the temperature of the heat sink T and the emission wavelength λ (after Deichsel *et al* [10]).

of fibre amplifiers. To achieve higher output power levels, the lateral aperture width of the laser output facet has to be enlarged. In the following sections, different types of wide-aperture devices are discussed.

B2.4.4.1 Broad-area laser diodes

A rather simple approach to achieve high output powers uses broad-area laser diodes having a broad stripe design which allows the propagation of a very large number of different lateral modes. These devices are widely employed for applications where single-mode beam profiles are not required, such as infrared illumination, material treatment and solid state laser pumping. Electrical-to-optical conversion efficiencies of more than 50% can be achieved, making broad-area lasers the most efficient technical light sources. Record output powers of around 10 W for lasers with cavity lengths of 2 mm and lateral stripe widths of 100 μm have been reported for the InGaAs/AlGaAs material system [16, 31] as well as for aluminium-free InGaAs/GaInAsP/GaInP lasers [1]. An example for the characteristics of a broad-area InGaAs/AlGaAs laser diode is shown in figure B2.4.7. For pumping solid-state lasers, several broad-area lasers are usually combined to a monolithically integrated bar having a typical width of 1 cm. When properly cooled, such bars yield continuous output power levels in the range of 50–200 W.

B2.4.4.2 Tapered laser devices

With tapered laser devices, the beam quality of wide-aperture laser diodes at high output powers can be significantly improved in comparison to broad-area devices. In figure B2.4.8, three examples for tapered lateral designs are presented.

A tapered travelling-wave laser amplifier has a narrow-aperture input facet and a wide-aperture output facet. Both mirror facets are covered with antireflection coatings to suppress any optical resonances. The basic function of the device is the amplification of a low-power beam profile from an external master-oscillator

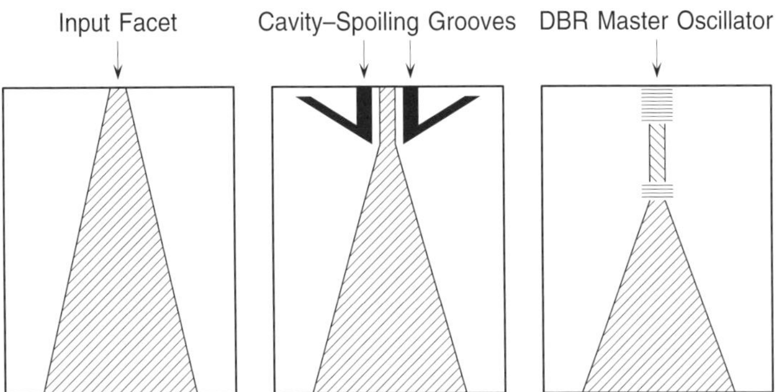

Figure B2.4.8. Lateral design of the active area for different types of tapered laser diodes with wide-aperture output facets. At the left-hand side, a tapered travelling-wave amplifier is shown (after Goldberg *et al* [15]). The drawing in the centre illustrates a tapered laser oscillator having cavity-spoiling grooves at the narrow end of the taper (after Choi *et al* [7]). The device sketched at the right-hand side is a monolithically integrated master-oscillator power amplifier (MOPA) (after O'Brian *et al* [30]).

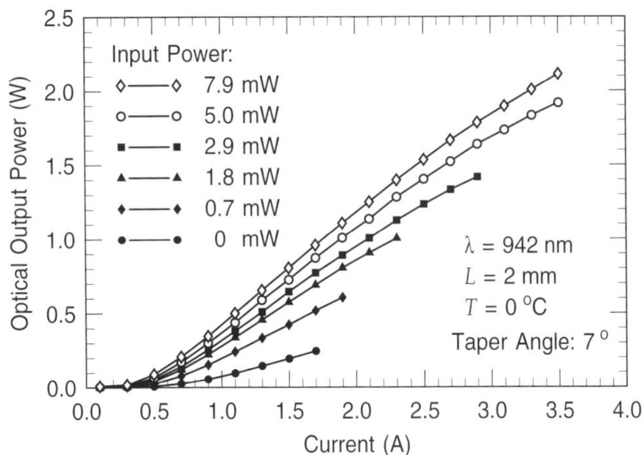

Figure B2.4.9. CW characteristics of a tapered semiconductor laser power amplifier with a taper angle of 7° for different input powers from the master-oscillator laser. Listed are the emission wavelength λ, the propagation length of the travelling-wave amplifier L and the heat-sink temperature T.

laser while maintaining its beam properties. The master oscillator can be any type of laser having an excellent beam profile, usually another low-power single-mode laser diode. During amplification inside the tapered travelling-wave amplifier area, the beam propagates freely in the lateral direction. Therefore, the taper angle must be properly adjusted to the diffraction angle of the propagating beam. According to Gehrig and Hess [14], the mode pattern of the output beam is very sensitive to the coupling accuracy of the input beam, the quality of the antireflection coatings and the taper geometry. Figure B2.4.9 shows a typical example for the output power characteristics of a tapered amplifier.

Tapered laser oscillators with wide output apertures consist of a tapered gain region and a lateral waveguide section at the narrow-aperture mirror facet. The output facet is typically antireflection coated to

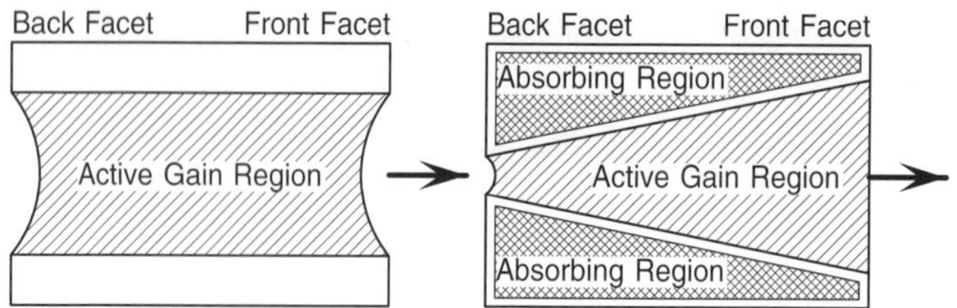

Figure B2.4.10. Two possible examples for lateral designs of unstable-resonator semiconductor lasers with dry-etched curved mirrors. On the left-hand side, a broad-area laser with curved mirrors is sketched, the right-hand side shows a design having a tapered active area and lateral absorber regions to suppress parasitic laser oscillations. Front and back facets of the devices are equipped with low- and high-reflectivity coatings, respectively.

a reflectivity of less than 1%, therefore only a small part of the output light is reflected back towards the waveguide section which works like a spatial mode filter. Cavity-spoiling grooves around the waveguide, which are tilted against the output facet suppress any further Fabry–Pérot oscillations. A detailed discussion on the design of tapered lasers can be found in Williams *et al* [37].

The third example for a tapered laser design shown at the right-hand side of figure B2.4.8 is a monolithically integrated master-oscillator power amplifier (MOPA). Again there is a tapered gain region with an antireflection coating at the wide-aperture output facet. At the narrow end of the taper, a master laser with a single-mode waveguide is monolithically integrated into the chip. The resonator mirrors of the master laser are realized using distributed Bragg reflectors (DBRs).

B2.4.4.3 Unstable resonator lasers

In recent years, considerable interest has been generated in semiconductor device research to realize unstable-resonator designs. The most promising approaches use curved mirror facets [3, 10, 34, 35]. To produce these devices, a reliable dry-etching process for vertical, flat and smooth mirror facets is required [9, 36]. In figure B2.4.10, two examples for unstable resonator designs using curved mirrors are shown. The curved mirrors create a high-loss resonator through the introduction of lateral divergences in the oppositely propagating wavefronts reflected by the curved mirror surfaces. The design of the mirror curvature has to be optimized so that the lowest-loss mode will produce a single-lobed output beam and all higher-order modes are strongly suppressed because of their substantially higher losses.

B2.4.4.4 α-DFB lasers

Recently, a new class of broad-area lasers called grating-confined broad-area lasers has been demonstrated in which angled gratings are used to support broad-area laser ring modes. The most promising candidate of this class is the angled-grating distributed-feedback (α-DFB) laser [24] which provides near-Gaussian near and far fields, a high side-mode-suppression ratio and a strong suppression of the amplified spontaneous emission. The device which is illustrated in figure B2.4.11 uses an angled DBR grating to generate ring modes. The active region consists of a broad-area gain stripe with a lateral distributed-feedback (DFB) grating embedded between the DBR gratings. The angled grating supplies feedback and spatial filtering, enforcing single-spatial-mode and single-longitudinal-mode oscillation.

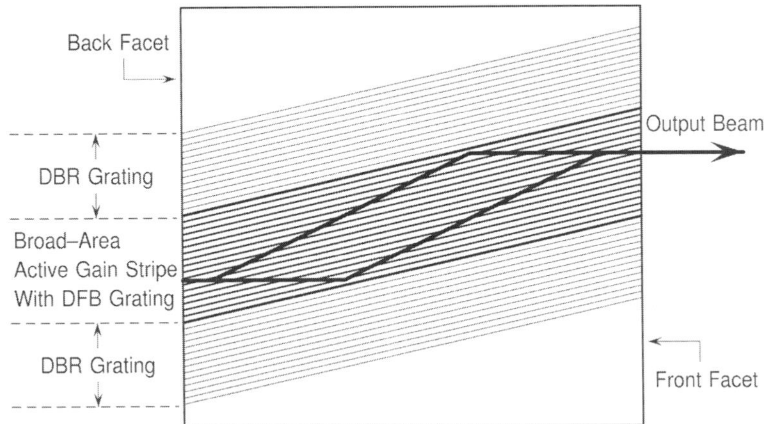

Figure B2.4.11. Schematic illustration of an α-DFB laser (after Lang *et al* [24]). The broad-area active gain stripe is misorientated against the cleaved facets. The gain stripe is covered by a lateral DFB grating and is embedded between DBR gratings orientated parallel to the stripe. The front and back facets of the device are equipped with a low- and high-reflectivity coating, respectively.

B2.4.5 High-power laser diode arrays

Monolithic integration of laser diodes can be easily performed by arranging the individual devices onto the chip in the form of an one-dimensional array. If the distance between the individual emitters is large enough, no optical coupling is observed and the emission of each laser is independent of the emission of its neighbours. Typical array devices of this type are the 1-cm-wide bars which are used for the pumping of solid state lasers. Common emission wavelengths are 808 and 940 nm for the pumping of Nd:YAG and Yb:YAG, respectively. Commercially available 1-cm-wide laser bars with a cavity length of 900 μm are specified for continuous optical output powers of 50 W at operating currents of approximately 70 A. Although such devices can be driven to output powers of more than 100 W, the lifetime is drastically reduced under such conditions. When operated under specified conditions, lifetimes up to 10 000 h have been reported. Another example of broad-area laser arrays are individually addressable arrays of single-mode lasers used for laser printers.

If the single emitters are arranged in close proximity, the optical field generated by each element of the array is coupled to other elements. Typical arrays of this type consist of single-mode index- or gain-guided waveguides which are closely spaced. These devices are called phase-locked laser arrays, because there is a fixed phase correlation between the individual emitters. Optical coupling can occur either via evanescent waves or leaky modes (radiation modes). The nature of the coupling strongly depends on the geometry of the waveguides and the effective refractive index step from the waveguide regions to the regions that mediate the optical coupling between the waveguides. For evanescent-type array modes, the fields are peaked in the high-effective-index regions, whereas for leaky-type array modes, the fields are peaked in the low-index array regions. Evanescent-wave arrays have propagation constants between the low and high refractive-index level, while leaky modes have values for the propagation constant below the low effective refractive index. The modes are said to be ‘in-phase’ when there is no phase shift for the fields of each waveguide and are called ‘out-of-phase’ when the fields of adjacent waveguides show a phase shift of π.

Generally, evanescent-wave arrays tend to operate in the out-of-phase mode resulting in a double-lobed far field. The reason for this behaviour is the better field overlap of the out-of-phase mode with the gain regions because this mode has zero intensity in the regions between the waveguides. Another disadvantage of evanescent-wave arrays is the limitation in the effective refractive-index step between the waveguides and the inter-element regions since a high index step will allow the propagation of higher-order modes in the individual waveguide elements. This low index step makes the device sensitive to spatial hole burning resulting in deteriorated beam properties at high output-power levels.

In leaky-mode arrays, the optical gain is provided in the low-index regions. Unlike evanescent-wave arrays, there is no limitation in the refractive index step resulting in stable operation at high output power. When properly designed, leaky-mode arrays operate in phase and exhibit a single-mode beam profile at output power levels in the range of 1–2 W. A detailed treatment of phase-locked arrays can be found in the book edited by Botez and Scifres (1994).

B2.4.6 Mounting of high-power laser diodes and diode arrays

A proper cooling of laser diodes is essential to obtain high output power and reliable operation. To provide a low thermal resistance, the devices are usually mounted junction-side down, which is a rather sophisticated process because of the requirements in positional accuracy, the risk of injury to the laser facet and reliability problems attributed to the solder material. For small chips having dimensions below 2 mm, eutectic Au/Sn solder can be used which provides a rather durable connection. In many cases, the chip is first soldered onto a heat spreader consisting of a low-thermal-resistivity material like diamond, boron nitride, silicon carbide or copper before being soldered into the device package. For larger chips like laser bars having a typical width of 1 cm, either a heat sink having the same thermal expansion coefficient as the chip or a flexible and soft solder material like indium has to be implemented to avoid cracks in the chip caused by the thermal expansion during operation [12]. Laser bars are mostly mounted onto water-cooled heat sinks having microchannels of some 100 μm dimension directly below the heat-sink surface to provide an efficient and homogeneous heat transfer [2].

B2.4.7 Reliability of high-power laser diodes

Depending on the application, different specifications for the device reliability of high-power lasers are expected by the customer. For consumer electronics, typical lifetimes are some thousands of hours. For pumping solid state lasers, devices with lifetimes in the range of 5000–10 000 h are commercially available. For telecommunications applications, guaranteed lifetimes of 300 000 h are required.

For single-mode ridge-waveguide or narrow-stripe lasers, which have output powers of some 100 mW, the lifetime-limiting effect is a COMD at the mirror facet. Since the internal temperature is low compared to broad-area laser diodes and laser bars, degradation of the bulk material is normally not observed. To achieve the device reliability required in the telecommunications market, e.g. for pumping of fibre lasers, sophisticated processes are used to passivate and coat the laser facet [13].

For broad-area lasers and laser bars, the active fraction of the device area is much larger in comparison to ridge-waveguide or narrow-stripe lasers. In this case, the device reliability is governed by the excess heat produced in the device and a gradual degradation is typically observed. Important factors influencing the reliability and degradation behaviour are the defect density of the epitaxial material, the diffusion stability of the doping profiles at elevated temperatures and the mounting technology.

B2.4.8 Conclusion and future trends

Semiconductor diode lasers providing high optical output powers are attractive devices for applications where a small size, a high electrical-to-optical power conversion efficiency, reliable operation and a low price is desired. Broad-area laser devices and laser diode bars exhibit optical output powers up to 100 W and are widely used for pumping solid state lasers, material treatment, medical applications and infrared illumination. Due to filamentation effects, the beam quality of these devices is rather poor. State-of-the-art single-mode devices like ridge-waveguide and narrow-stripe lasers are limited to output power levels of several 100 mW.

A considerable part of the ongoing reseach activities on high-power lasers is attributed to approaches which allow high optical output powers and diffraction-limited beam profiles at the same time. Devices of this kind have tapered lateral designs, implement unstable resonator concepts, stabilize the mode by lateral angled gratings or use coupled arrays of single-mode waveguides.

In future, the wavelength range of 600–1100 nm for commercially available high-power laser diodes will be extended towards longer wavelengths, e.g. using InGaAsN quantum wells [11]. An example of a new and promising device concept is the optically pumped semiconductor disc laser [21, 32]. This type of laser, which is also called vertical external cavity surface-emitting laser (VECSEL), has excellent beam properties, is scalable in output power and allows intracavity frequency doubling.

References

[1] Al-Muhanna A, Mawst L J, Botez D, Garbuzov D Z, Martinelli R U and Connolly J C 1998 High-power (>10 W) continuous-wave operation from 100-μm-aperture 0.97-μm-emitting Al-free diode lasers *Appl. Phys. Lett.* **73** 1182–4

[2] Beach R, Benett W J, Freitas B L, Mundinger D, Comaskey B J, Solarz R W and Emanuel M A 1992 Modular microchannel cooled heatsinks for high average power laser diode arrays *IEEE J. Quantum Electron.* **28** 966–76

[3] Biellak S A, Fanning C G, Sun Y, Wong S S and Siegman A E 1997 Reactive-ion-etched diffraction-limited unstable resonator semiconductor lasers *IEEE J. Quantum Electron.* **33** 219–30

[4] Botez D, Mawst L J, Bhattacharya A, Lopez J, Li J, Kuech T F, Iakovlev V P, Suruceanu G I, Caliman A and Syrbu A V 1996 66% wallplug efficiency from Al-free 0.98 μm-emitting diode lasers *Electron. Lett.* **32** 2012–13

[5] Botez D 1999 Design considerations and analytical approximations for high continuous-wave power, broad-waveguide diode lasers *Appl. Phys. Lett.* **74** 3102–4

[6] Brugger H and Epperlein P W 1990 Mapping of local temperatures on mirrors of GaAs/AlGaAs laser diodes *Appl. Phys. Lett.* **56** 1049–51

[7] Choi H K, Walpole J N, Turner G W, Eglash S J, Missaggia L J and Connors M K 1993 GaInAsSb–AlGaAsSb tapered lasers emitting at 2 μm *IEEE Photon. Technol. Lett.* **5** 1117–19

[8] Dai Z, Michalzik R, Unger P and Ebeling K J 1997 Numerical simulation of broad-area high-power semiconductor laser amplifiers *IEEE J. Quantum Electron.* **33** 2240–54

[9] Deichsel E, Eberhard F, Jäger R and Unger P 2001 High-power laser diodes with dry-etched mirror facets and integrated monitor photodiodes *IEEE J. Selected Topics Quantum Electron.* **7** 106–10

[10] Deichsel E, Jäger R and Unger P 2002 High-brightness unstable-resonator lasers fabricated with improved dry-etching technology for ultra-smooth laser facets *Jpn. J. Appl. Phys. Part I* **41** 4279–82

[11] Egorov A Y, Bernklau D, Livshits D, Ustinov V, Alferov Z I and Riechert H 1999 High power cw operation of InGaAsN lasers at 1.3 μm *Electron. Lett.* **35** 1643–4

[12] Endriz J G, Vakili M, Browder G S, DeVito M, Haden J M, Harnagel G L, Plano W E, Sakamoto M, Welch D F, Willing S, Worland D P and Yao H C 1992 High power diode laser arrays *IEEE J. Quantum Electron.* **28** 952–65

[13] Fukuda M 1996 Reliability of high power pump lasers for erbium-doped fibre amplifiers *Int. J. High Speed Electron. Syst.* **7** 55–84

[14] Gehrig E and Hess O 2001 Spatio–temporal dynamics of light amplification and amplified spontaneous emission in high-power tapered semiconductor laser amplifiers *IEEE J. Quantum Electron.* **37** 1345–55

[15] Goldberg L, Mehuys D, Surette M R and Hall D C 1993 High-power, near-diffraction-limited large-area traveling-wave semiconductor amplifiers *IEEE J. Quantum Electron.* **29** 2028–43

[16] He X, Srinivasan S, Wilson S, Mitchell C and Patel R 1998 10.9 W continuous wave optical power from 100 μm aperture InGaAs/AlGaAs (915 nm) laser diodes *Electron. Lett.* **34** 2126–7

[17] Henry C H, Logan R A and Merritt F R 1980 Measurement of gain and absorption spectra in AlGaAs buried heterostructure lasers *J. Appl. Phys.* **51** 3042–50

[18] Henry C H, Logan R A and Bertness K A 1981 Spectral dependence of the change in refractive index due to carrier injection in GaAs lasers *J. Appl. Phys.* **52** 4457–61

[19] Hess O, Koch S W and Moloney J V 1995 Filamentation and beam propagation in broad-area semiconductor lasers *IEEE J. Quantum Electron.* **31** 35–43

[20] Jäger R, Heerlein J, Deichsel E and Unger P 1999 63% wallplug efficiency MBE grown InGaAs/AlGaAs broad-area laser diodes and arrays with carbon p-type doping using CBr_4 *J. Crystal Growth* **201/202** 882–5

[21] Kuznetsov M, Hakimi F, Spargue R and Mooradian A 1999 Design and characteristics of high-power (>0.5-W CW) diode-pumped vertical-external-cavity surface-emitting semiconductor lasers with circular TEM_{00} beams *IEEE J. Selected Topics Quantum Electron.* **5** 561–73

[22] Lang R J, Hardy A, Parke R, Mehuys D, O'Brian S, Major J and Welch D 1993 Numerical analysis of flared semiconductor laser amplifiers *IEEE J. Quantum Electron.* **29** 2044–51

[23] Lang R J, Mehuys D, Welch D F and Goldberg L 1994 Spontaneous filamentation in broad-area diode laser amplifiers *IEEE J. Quantum Electron.* **30** 685–94

[24] Lang R J, Dzurko K, Hardy A A, Demars S, Schoenfelder A and Welch D F 1998 Theory of grating-confined broad-area lasers *IEEE J. Quantum Electron.* **34** 2196–210

[25] Mawst L J, Bhattacharya A, Lopez J, Botez D, Garbuzov D Z, DeMarco L, Connolly J C, Jansen M, Fang F and Nabiev R F 1996 8 W continuous wave front-facet power from broad-waveguide Al-free 980 nm diode lasers *Appl. Phys. Lett.* **69** 1532–4

[26] Mikulla M, Chasan P, Schmitt A, Morgott S, Wetzel A, Walther M, Kiefer R, Pletschen W, Braunstein J and Weimann G 1998 High-brightness tapered semiconductor laser oscillators and amplifiers with low-modal gain epilayer-structures *IEEE Photon. Technol. Lett.* **10** 654–6

[27] Moser A, Latta E E and Webb D J 1989 Thermodynamics approach to catastrophic optical mirror damage of AlGaAs single quantum well lasers *Appl. Phys. Lett.* **55** 1152–4

[28] Moser A and Latta E E 1992 Arrhenius parameters for the rate process leading to catastrophic damage of AlGaAs–GaAs laser facets *J. Appl. Phys.* **71** 4848–53

[29] Mullane M P and McInerney J G 1999 Minimization of the linewidth enhancement factor in compressively strained semiconductor lasers *IEEE Photon. Technol. Lett.* **11** 776–8

[30] O'Brian S, Lang R, Rarke R, Major J, Welch D F and Mehuys D 1997 2.2-W continuous-wave diffraction-limited monolithically integrated master oscillator power amplifier at 854 nm *IEEE Photon. Technol. Lett.* **9** 440–2

[31] O'Brian S, Zhao H, Schoenfelder A and Lang R J 1997 9.3 W CW (In)AlGaAs 100 μm wide lasers at 970 nm *Electron. Lett.* **33** 1869–71

[32] Schiehlen E, Golling M and Unger P 2002 Diode-pumped semiconductor disk laser with intracavity frequency doubling using lithium triborate (LBO) *IEEE Photon. Technol. Lett.* **14** 777–9

[33] Smowton P M and Blood P 1997 The differential efficiency of quantum-well lasers *IEEE J. Selected Topics Quantum Electron.* **3** 491–8

[34] Unger P, Boegli V, Buchmann P and Germann R 1993 Fabrication of curved mirrors for visible semiconductor lasers using electron-beam lithography and chemically assisted ion-beam etching *J. Vacuum Sci. Technol.* B **11** 2514–18

[35] Unger P, Boegli V, Buchmann P and Germann R 1994 High-resolution electron-beam lithography for fabricating visible semiconductor lasers with curved mirrors and integrated holograms *Microelectron. Eng.* **23** 461–4

[36] Vettiger P, Benedict M K, Bona G L, Buchmann P, Cahoon E C, Dätwyler K, Dietrich H P, Moser A, Seitz H K, Voegeli O, Webb G J and Wolf P 1991 Full-wafer technology—a new approach to large-scale laser fabrication and integration *IEEE J. Quantum Electron.* **27** 1319–31

[37] Williams K A, Penty R V, White I H, Robbins D J, Wilson F J, Lewandowski J J and Nayar B K 1999 Design of high-brightness tapered laser arrays *IEEE J. Selected Topics Quantum Electron.* **5** 822–31

Further reading

Botez D and Scifres D R (ed) 1994 *Diode Laser Arrays* (Cambridge: Cambridge University Press)

A multi-author book providing a comprehensive overview of the fundamental principles and applications of semiconductor diode laser arrays.

Carlson M M 1994 *Monolithic Diode-Laser Arrays* (New York: Springer)

This book presents and analyzes the results of work performed over the last 20 years in the development of high-brightness diode-laser arrays.

Casey H C and Panish M B 1978 *Heterostructure Laser—Part A: Fundamental Principles, Part B: Materials and Operating Principles* (Orlando, FL: Academic)

A comprehensive treatment of the general aspects of modern semiconductor lasers including theory, fabrication technology and characterization of these devices.

Coldren L A and Corzine S W 1995 *Diode Lasers and Photonic Integrated Circuits* (New York: Wiley)

A modern textbook on semiconductor lasers primarily written for students.

Diehl R (ed) 2000 *High-Power Diode Lasers—Fundamentals, Technology, Applications (Topics in Applied Physics 78)* (Berlin: Springer)

A multi-author book covering various aspects of semiconductor diode lasers including theory, microscopic dynamic effects, epitaxial growth, beam properties and packaging. Also applications of these devices like frequency conversion and pumping of solid state lasers are discussed.

B2.5.1
Visible laser diodes: properties of III–V red-emitting laser diodes

Peter Blood

B2.5.1.1 Introduction

B2.5.1.1.1 Background

In a semiconductor, light is produced by electrons making *radiative* transitions across the *direct* energy gap (E_g), with the energy lost by each electron being given up by the creation of a photon. The wavelength of the emitted light is determined by the magnitude of the direct energy gap and for short-wavelength emission we require a wide-gap semiconductor: 650 nm corresponds to a photon energy of 1.91 eV. However, as illustrated in figure B2.5.1.1, other energy gaps exist which are termed *indirect* because the electron has to change its momentum as well as its energy in making a transition. The 'momentum' associated with a photon is small and cannot compensate for the momentum difference between initial and final states of a transition across the indirect gap. The radiative transition rate across the indirect gap is therefore very small because another 'particle' such as a phonon must also be involved in the process to conserve momentum, consequently the dominant indirect recombination processes are non-radiative, producing heat rather than light. If the conduction band minimum of the indirect gap is lower than that of the direct gap then most of the electrons will reside in the indirect minimum and will not be available to make direct radiative transitions. Thus, the direct energy gap can only be exploited for light emission where it lies below the indirect gap. The indirect gap has a major influence on the design of red-emitting devices.

In the elemental semiconductors Si and Ge the indirect gap has the lowest energy so these are not effective light-emitting materials. The most common direct gap, light-emitting semiconductors are formed from elements in columns III and V, or II and VI of the periodic table. Although compounds such as GaAs and ZnS have a defined composition and energy gap ($E_g = 1.41$ eV and 3.6 eV, respectively) it is possible to obtain a range of energy gaps by alloying compounds together. For example, by alloying AlAs with GaAs, effectively replacing the group III element Ga on some sites with Al, it is possible to vary the direct gap from 1.42 eV to 3.03 eV by changing the composition, however, the indirect gap at about 2 eV restricts the useful range for light emission. Semiconductor alloys are the key component in engineering laser diode structures, providing the basis for tailoring the emission energy and enabling multiple-layer heterostructures to be built in which the band gap changes with distance to confine the carriers and provide a waveguide for the emitted light. The double heterostructure (see [7]) is the essential element in realizing continuous operation of diode lasers at room temperature and its inventors, Kroemer and Alferov, were recipients of the Nobel Prize for Physics in 2000.

In principle, it is possible to fabricate devices emitting at 700 nm using the $Al_xGa_{1-x}As$ alloy system but, at the Al content required ($x \cong 0.28$), the radiative efficiency of the material is poor and the threshold

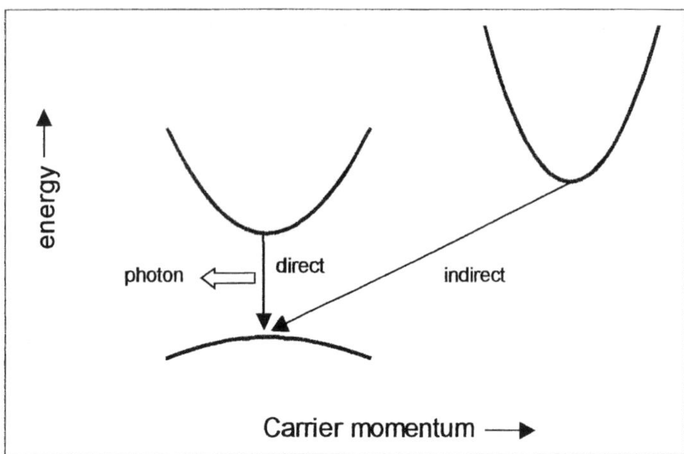

Figure B2.5.1.1. Diagram of energy *versus* momentum for electrons in a crystalline semiconductor illustrating the direct and indirect energy gaps. An indirect transition requires the electron to change both energy and momentum. Because photons have negligible momentum they are produced most effectively by direct transitions.

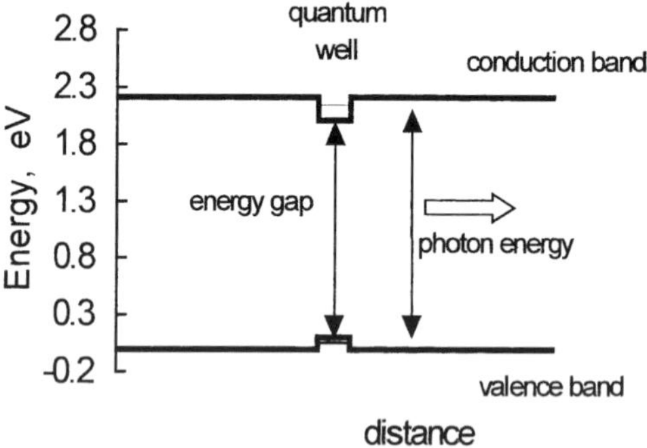

Figure B2.5.1.2. Illustration of the energy bands of a quantum well. As the well width is reduced the confined states in the well move toward the band edges of the barrier material and the energy of the emitted photon is increased above the band gap of the well material. The maximum photon energy is set by the band gap of the barrier material. (This diagram is drawn to the energy scale of the GaInP well in figure B2.5.1.5.)

current for laser action is prohibitively high. In a given material system the energy of recombination can be increased by confining electrons in a thin quantum well as illustrated in figure B2.5.1.2 (the thinner the well the greater the electron energy). A quantum well offered a more promising means of reducing the emission wavelength from GaAs. Lasers operating at about 710 nm have been made in this way [36] but the threshold currents were large and strongly temperature-sensitive due to the leakage of carriers from the quantum well to the adjacent barrier material. This is because the confined states are near the top of the quantum well [4] in very short wavelength devices and the reduction in wavelength is ultimately limited by the band gap of the barrier material (figure B2.5.1.2). To fabricate low threshold red-emitting lasers it is necessary to use a material system which provides a wider gap barrier to enable the photon energy to be increased while still providing strong localization of carriers within the quantum well. These requirements have been met using the AlGaInP system, which is more correctly termed a ternary alloy of the III–V compounds AlP, GaP and InP. Further general details of quantum well lasers are given in section B2.1.6.

Before describing this material system in greater detail we briefly consider general trends in the properties of semiconductor lasers with band gap to put the particular features of red-emitting lasers into a wider context.

B2.5.1.1.2 Fundamental trends with wavelength

Properties of different III–V materials follow systematic trends with band gap, for example, as the gap increases from one material to another the effective mass of the electrons increases and the dielectric constant decreases. Consequently there are trends in laser characteristics which apply to devices from the near infra-red (e.g. 1.55 μm communications lasers) through GaAs-based lasers (e.g. 780 nm CD lasers) to red (650 nm) and blue (430 nm) emitting devices. Due to the increase in effective mass, the carrier density (n_{th}) at laser threshold increases with increasing gap. There are two intrinsic processes which contribute to the threshold current: radiative recombination across the direct gap producing incoherent spontaneous emission and Auger recombination, in which the recombining electron excites another electron which then gives up the excess energy as heat (see [1]). Auger recombination is undesirable because it reduces the fraction of the current which is used to produce light. The Auger coefficient is large in narrow-gap materials and Auger recombination is responsible for up to 80% of the threshold current in 1.55 μm lasers, but it decreases with increasing gap and the Auger current is negligible in the wide gap materials used for red- and blue-emitting devices. Although the Auger recombination rate varies as $(n_{th})^3$ the coefficient has the dominant effect. However, the spontaneous recombination current varies as $(n_{th})^2$ and its coefficient is only weakly dependent on band gap; consequently, the threshold current increases with increasing gap because n_{th} increases. There is, therefore, an increase in intrinsic threshold current going from 870 nm to 650 nm lasers. These trends are illustrated in figure B2.5.1.3. The important message is that we should not expect red- and blue-emitting lasers to have threshold currents as low as 870 nm lasers, for equivalent structures.

Most of the theories used to design laser diodes treat the electrons as independent, non-interacting particles. However, the Coulomb interactions between electrons become stronger as the band gap is increased due to the higher mass and lower dielectric constant. In addition to an increase in optical gain for a given carrier density, these interactions increase the rate of recombination per electron contributing a further increase in threshold current density. At a fundamental level the 'free electron' laser theories based on non-interacting electrons are poor approximations in red-emitting lasers and totally inappropriate for blue-emitting lasers. We should therefore expect to adopt more complex theories for short wavelength devices.

B2.5.1.1.3 The AlGaInP material system

Red-emitting lasers are made using the $(Al_xGa_{1-x})_yIn_{1-y}P$ material system which provides variation in the direct band gap from 1.9 eV to about 2.5 eV. To avoid formation of unwanted crystal defects it is necessary that the layer structure of the laser is grown with the same lattice parameter as the substrate material. The lattice parameter is determined chiefly by the ratio of [Al+Ga] to [In], the parameter y. The band gap can be changed by substituting [Al] for [Ga] with very little change in lattice parameter, or by changing y, which also causes the lattice parameter to change. (For a survey of AlGaInP band-gap parameters, see [29].) This material can be grown lattice-matched to a GaAs substrate with $y = 0.51$. If the device structure cannot be grown lattice-matched to the substrate the extended defects produced by the mismatch have a deleterious effect on both the operating current and the lifetime of the devices. The availability of a suitable substrate material is the major factor which determines the viability of a material system for laser diode manufacture. Blue-emitting AlGaInN laser diodes have been plagued with problems of short operating life because there is no lattice-matched substrate available.

Figure B2.5.1.4 shows the variation of the direct and indirect band gaps as functions of the Al/Ga ratio (the composition x) for AlGaInP material lattice-matched to GaAs. The variation of the direct gap is given by

$$E_g(x) = 1.91 + 0.61x \qquad [\text{eV}]. \tag{B2.5.1.1}$$

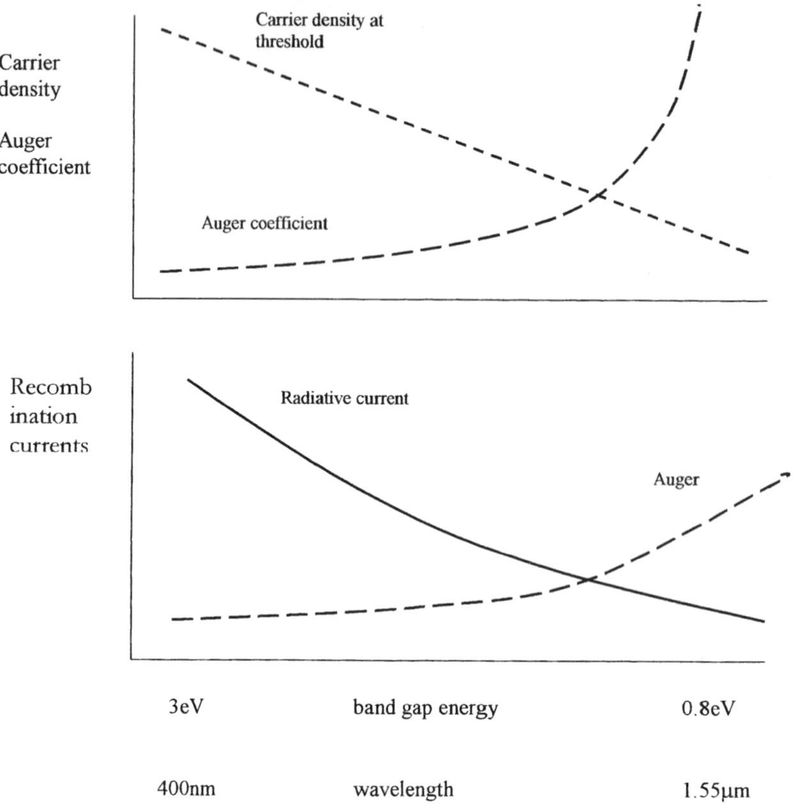

Figure B2.5.1.3. The upper plot illustrates the variation of threshold carrier density and Auger coefficient with band gap and wavelength. The lower plot shows the resulting variation in intrinsic recombination currents due to radiative and non-radiative Auger processes.

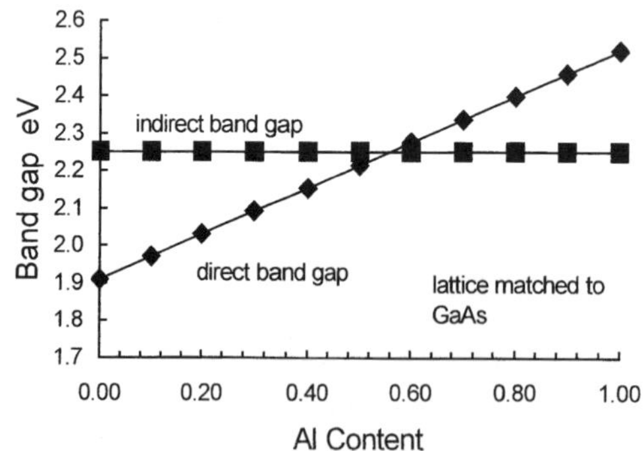

Figure B2.5.1.4. Direct and indirect energy gaps of $(Al_xGa_{1-x})_yIn_{1-y}P$ as a function of Al composition (x) for $y = 0.51$ corresponding to the lattice matched to the GaAs substrate.

The maximum useful direct gap is limited to about 2.25 eV ($x < 0.55$) because the minimum gap becomes indirect at higher compositions. In practice this material has been used to make laser diodes of reasonable

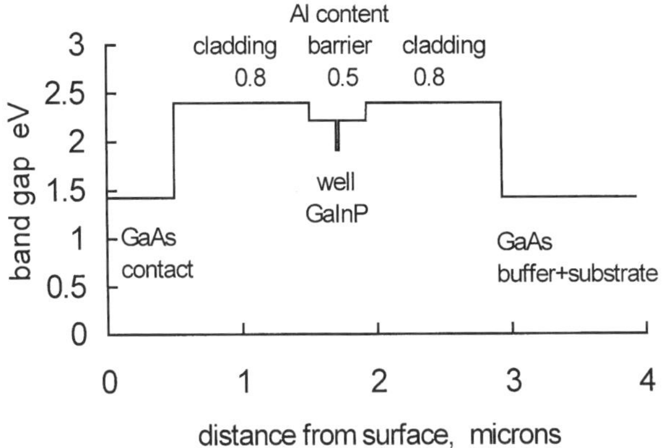

Figure B2.5.1.5. Schematic diagram of the energy gap through a typical AlGaInP laser structure with a single GaInP quantum well. There are GaAs contact layers at the surface and adjacent to the substrate. The Al content of the cladding and barrier layers is indicated.

performance down to about 625 nm (photon energy of about 2 eV).

It is assumed that the different atomic species of a semiconductor alloy are distributed randomly over the appropriate lattice sites. However, in AlGaInP there is a tendency for the distribution of In on the group III sites to be ordered such that a periodic In distribution is produced along the ⟨111⟩ direction (see for example, [16, 20]). In GaInP there is a tendency to form alternating (111) planes of InP and GaP. This has the effect of reducing the band gap below that given by equation (B2.5.1.1). The degree of ordering depends upon the growth conditions and, unless these are well controlled, there is poor control over the emission wavelength. In developing red-emitting lasers, considerable effort was devoted to suppressing this ordering process by the choice of the growth temperature or by growth on substrates with orientation other than ⟨100⟩ [8]. By suppressing ordering the maximum band gap is obtained and the emission wavelength is reproducible.

The layer structure of a typical lattice-matched AlGaInP laser ($y = 0.51$) is shown in figure B2.5.1.5 as a plot of the variation of band gap of the layers with distance from the surface. In this example there is a single GaInP quantum well (5–8 nm thick) within a barrier layer with $x = 0.5$ and of overall width 200 nm, surrounded by outer cladding layers each about 1 μm thick and having $x = 0.8$. Since the cladding layers have a lower refractive index than the inner barrier material they provide a waveguide which serves to localize the laser mode in the region of the well which provides the optical gain. The structure has GaAs contact and buffer layers. The substrate is n-type and the structure is doped n-type through the lower cladding layer. The barrier and well region is undoped and the upper cladding and contact is p-type to form the junction necessary for carrier injection into the well to achieve population inversion and laser action. The emission wavelength can be tailored by choice of the width of the well.

While it is necessary that the overall device structure is lattice-matched to the substrate, it is possible to accommodate a limited amount of strain elastically in the thin GaInP quantum well. This enables the composition (y) of the $Ga_yIn_{1-y}P$ well to be varied in addition to the width so that a wider range of emission wavelengths can be engineered. The composition dependence of the band gap of 'bulk' (i.e. unstrained) $Ga_yIn_{1-y}P$ is given by:

$$E_g(y) = 1.35 + 0.73y + 0.70y^2 \qquad [\text{eV}]. \qquad \text{(B2.5.1.2)}$$

The strain also affects the band gap and, in an epitaxial layer, the gap is increased above values given by this equation by compressive strain ($y < 0.51$) and decreased below these values for tensile strain (> 0.51).

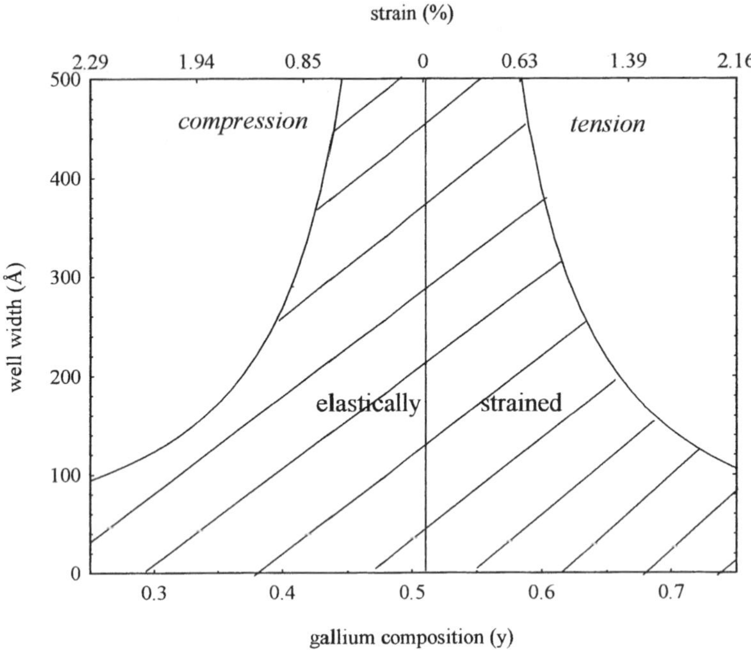

Figure B2.5.1.6. For the combinations of Ga composition (y) and well width within the shaded area of this diagram the strain of $Ga_yIn_{1-y}P$ layer can be accommodated elastically. Outside this region dislocations are formed [29].

The shaded area in figure B2.5.1.6 defines the combinations of well composition (y) and width (L_z) which satisfy the criterion for elastic deformation according to the Mathews–Blakeslee condition [17]. Outside the shaded region the strain cannot be accommodated elastically and extended defects form. As well as allowing greater freedom in the choice of well composition, the presence of strain can also reduce the threshold current of the device through strain-induced modification of the electronic structure of the well material [21].

AlGaInP is used to make diode lasers over the red region of the spectrum, from 620 nm to 690 nm, and we concentrate on devices with the generic structure shown in figure B2.5.1.5. Their design exploits the wider parameter space made available by strained quantum wells, with attendant benefits in performance. The proximity of the indirect and direct gaps in the cladding layer places a serious restriction on the realization of short wavelength lasers because the indirect gap is responsible for the leakage of electrons from the active region. The high electrical and thermal resistance of the wide-gap cladding layer also pose serious thermal problems for high power devices (see chapter B2.4). The device structures are similar to other device families and are not discussed in the limited space available here.

B2.5.1.2 III–V red-emitting laser diodes

B2.5.1.2.1 Design of the quantum well structure

With a lattice-matched $Ga_{0.51}In_{0.49}P$ quantum well emission can be obtained at wavelengths from the practical lower limit of about 625 nm in a thin well to 650 nm in a wide well, set by the band gap of the well material itself (1.91 eV). However, by using compressively-strained GaInP it is possible to reduce the band gap of the well to extend this to longer wavelengths. For example, the Ga composition can be reduced to about 0.25 giving a minimum direct gap of about 1.63 eV (760 nm) in the well. However, figure B2.5.1.6 shows that at

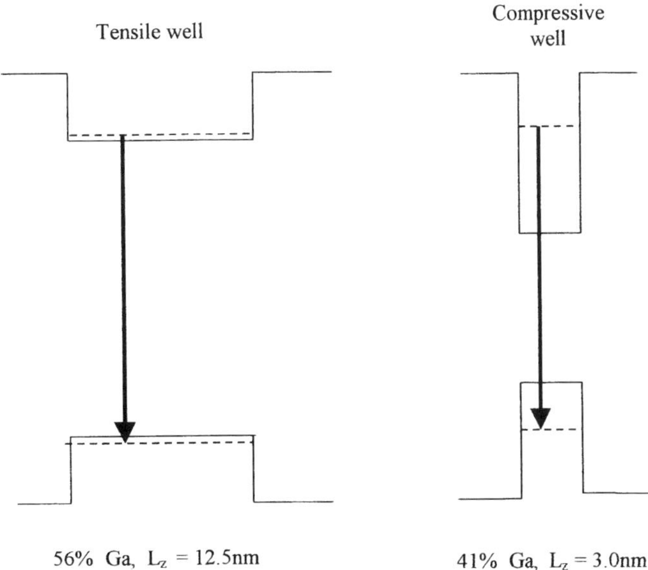

Figure B2.5.1.7. Illustration of compressive and tensile strained quantum well structures, both of which provide a transition energy of 633 nm [27]. The wide tensile-strained well will have further subbands at higher electron and hole energies.

this composition the well must be thinner than 10 nm and the resulting quantum confinement increases the transition energy, resulting in a wavelength of about 725 nm. In principle it is possible to use thin strained layers of GaInP to obtain wavelengths as long as 750 nm, where the compressive strain is about 2%. In practice, devices operating at room temperature can only be made using strains of up to about 1%. For higher compressive strain the GaInP well material becomes inhomogeneous, forming clusters rather than a continuous epitaxial layer. These clusters may be on a dimensional scale similar to the wavelength of light and scatter light out of the laser mode leading to a high internal optical loss which ultimately prevents the laser working [19]. In using tensile-strained GaInP layers to increase the photon energy and reduce the wavelength, quantum confinement arising from the need to keep the layer thin to accommodate the strain serves to contribute a further increase to the transition energy. In this case the lower wavelength limit is set by the loss of carriers to the indirect energy gap and this is discussed in detail in a subsection of section B2.5.1.2.

As well as using strain to access a wider range of wavelengths, it is also possible to use different combinations of well width and well composition to achieve the same transition energy. Figure B2.5.1.7 shows two possible quantum well structures for emission at 633 nm. The compressively-strained well has $y = 0.41$ and a width of 3 nm while the tensile strained well has $y = 0.56$ giving a much larger band gap so the well is wider at 12.5 nm to reduce the effects of quantum confinement. The question then arises as to which structure gives the 'best' device performance. Valster and coworkers [33] have investigated the effect of strain at the fixed wavelength of 633 nm (figure B2.5.1.8) and these data shows that about 0.5% compressive and tensile strain reduces the threshold of these devices, with tensile strain giving the greatest effect. Smowton and coworkers [25] have investigated the choice of well width/strain combinations for optimum 670 nm lasers. From a practical standpoint, it is difficult to achieve wavelength reproducibility in thin wells because of the sensitivity of the transition wavelength to monolayer changes in well width (one monolayer changes the transition energy of a 3 nm wide well by about 5 nm) therefore wider tensile strained wells may be preferred.

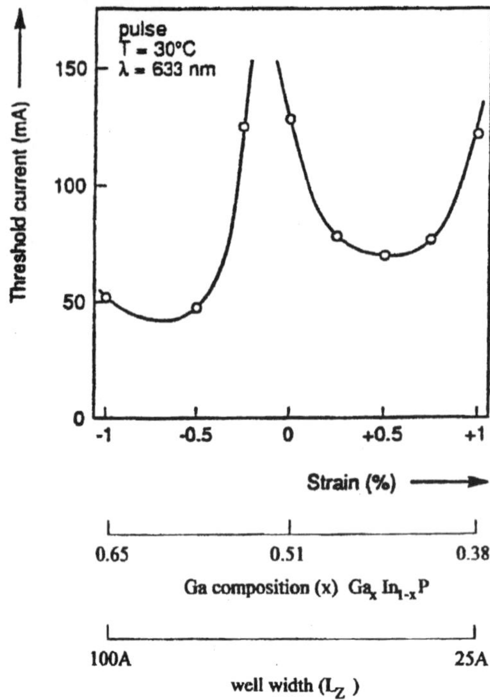

Figure B2.5.1.8. Experimental variation of the room temperature threshold current of 633 nm lasers with different combinations of well widths and compositions/strain [33].

B2.5.1.2.2 Intrinsic gain and recombination characteristics

The modal optical gain (G) is defined as the fractional increase in energy in a specific optical mode per unit path length (see [3, 11]). Because the mode is only partially coupled to the gain-generating region of the well it is usual to define a local gain (g) for the well (the well is much thinner than one optical wavelength) which is related to the modal gain by the optical confinement factor (Γ). The gain necessary for laser action is produced by stimulated transitions of electrons from the conduction band to the valence band across the direct gap. A detailed general account of the theory of optical gain, treating the electrons as non-interacting particles, is given in textbooks such as [11]. Gain is produced over a range of photon energies and, for a given quantum well, its magnitude is determined by the separation of the quasi Fermi levels for electrons and holes, which is equal to the internal voltage of the device. In an edge-emitting device, laser action occurs when the optical gain just matches the optical losses from the cavity (due to scattering of light out of the guided mode and due to light emitted from the laser for external use). Laser operation first occurs at the wavelength on the gain spectrum where the gain achieves this value. The threshold is thus determined by the peak value on the gain spectrum. Through the gain process the optical loss determines the position of the quasi Fermi levels at threshold and these are important in controlling the carrier leakage from the active region.

Current must be supplied to the well to maintain the threshold carrier density and internal voltage against recombination processes across the gap. In the absence of extrinsic non-radiative processes and carrier leakage the current at and below threshold is due to spontaneous emission (figure B2.5.1.9); it is usual to suppose that the photon density at threshold is sufficiently small that the stimulated recombination rate is negligible *at threshold*. Consequently, the intrinsic threshold current density is given by the integral of the

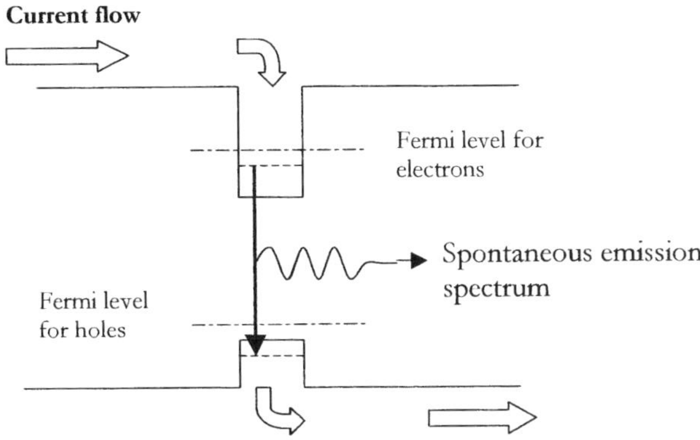

Figure B2.5.1.9. Band diagram of a quantum well showing the recombination current path at a given Fermi level separation.

spontaneous emission spectrum:

$$J_{\rm spon}^{\rm th} = -e \int R_{\rm spon}(h\nu)\, {\rm d}h\nu \tag{B2.5.1.3}$$

where $R_{\rm spon}(h\nu)$ is the spontaneous recombination rate per unit area at photon energy $h\nu$, at threshold.

It is necessary to take account of the broadening of the individual transitions due to dephasing by carrier collisions and the effects of monolayer variations in thickness along the quantum well. The former process is often incorporated as spectral broadening by a lineshape function (often a Lorentzian) defined by a fixed dephasing lifetime or, more realistically, by a lifetime which is calculated for the specific carrier density and transition energy. The variation in well width within a device can have a significant effect when the wells are thin and it is an important factor in determining the optimum well width [25]. The effect of broadening processes is to widen both the gain and emission spectra, reducing their peak values, consequently it is necessary to drive the device to a higher internal voltage and carrier density to generate the required peak gain, increasing the threshold current compared to that calculated in the absence of broadening.

This is illustrated in figure B2.5.1.10 which shows calculations of peak local gain as functions of recombination current for two different 670 nm GaInP wells with compositions and well widths of $Ga_{0.41}$ 2.5 nm and $Ga_{0.25}$ 6.8 nm. Calculations are shown with and without well width variations; the variations occur over ±2 monolayers with a standard deviation of 1 monolayer in both cases. It is clear that the current needed to achieve a given value of gain is increased by these fluctuations by about 25% for the thin well, whereas the effect is negligible for the thicker well.

Within the framework of a non-interacting free electron picture the intrinsic characteristics of AlGaInP quantum wells are not fundamentally different from those of other materials. However, these calculations do not provide a fully accurate description of the gain process. Using a simple lifetime to describe the collision dephasing process it is not possible to obtain agreement between theoretical and experimental gain spectra. However, as shown in figure B2.5.1.11, when the collisions are treated using the quantum kinetic equations the spectra are in good agreement with theory [10].

A general characteristic of the variation of peak gain with current for quantum wells is that it saturates at high current for recombination between the lowest pair of subbands. [18] suggested that the relation between optical gain g and current density J for a single pair of subbands could be written in the form

$$g = g_{\rm t} \ln\left(\frac{J}{J_{\rm t}} + 1\right) \tag{B2.5.1.4}$$

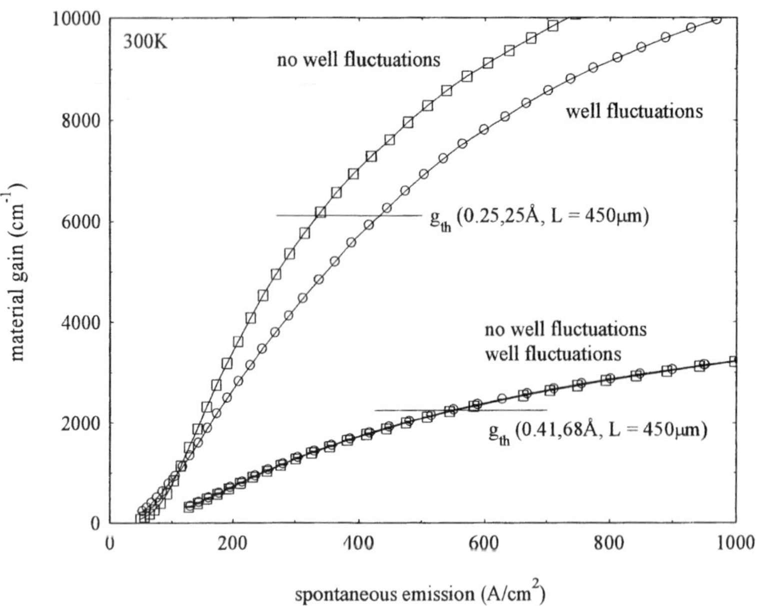

Figure B2.5.1.10. Calculated curves for peak local (or material) gain as a function of the integrated spontaneous recombination current for two different GaInP wells both of which emit at 670 nm. The wells are 2.5 nm ($y = 0.25$) and 6.8 nm wide ($y = 0.41$) and the calculations have been done with and without well width fluctuations. The horizontal lines indicate the threshold local gain, $g_{\rm th}$, required for a 450 μm long laser at each well-width. The effect of well width fluctuations is negligible for the wider well [29].

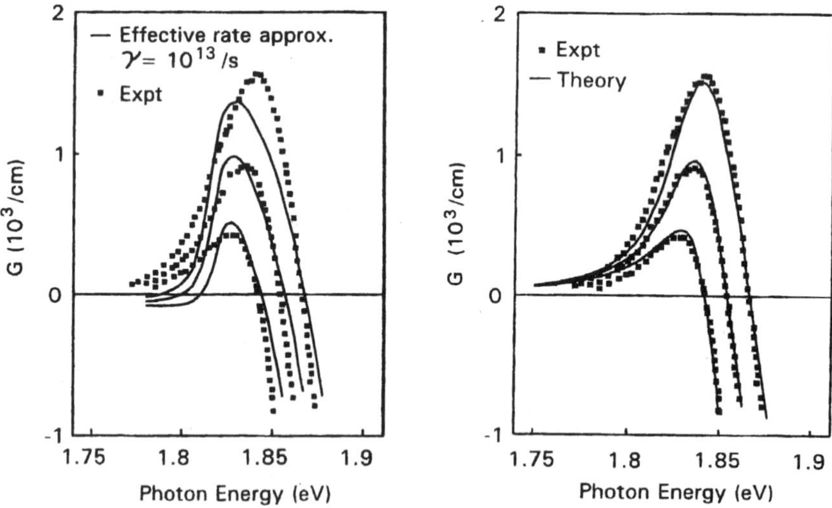

Figure B2.5.1.11. Experimental data (points) and theoretical gain spectra (lines) for a 6.8 nm-wide 670 nm $Ga_{0.41}In_{0.59}P$ quantum well. The calculation for the left-hand figure used a fixed carrier scattering time of 10^{-13} s. For the right-hand figure the dephasing was calculated from first principles using a quantum kinetic theory [10].

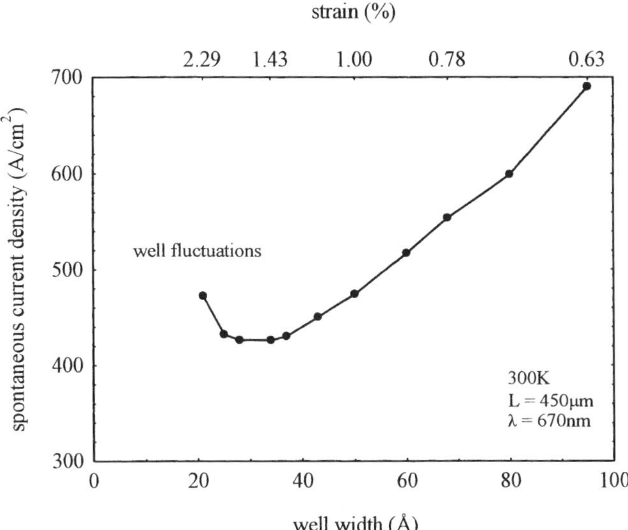

Figure B2.5.1.12. Calculations of the threshold current density of 450 μm long 670 nm GaInP lasers as a function of well width with the composition/strain adjusted to keep the same transition energy. The calculation uses an energy- and density-dependent dephasing time and includes well-width variations. These results have been derived from gain-current curves such as those in figure B2.5.1.10. The up-turn in threshold current for the very thin well is due to the broadening of the gain spectrum by well-width variations [29].

where g_t and J_t are parameters which describe the $g(J)$ relation of specific structures and are defined as the point where a line through the origin is a tangent to the $g(J)$ curve. If these parameters are known from theoretical calculations then these relations enable the gain-current relations to be reconstructed. They also enable analytic expressions to be derived for threshold current as a function of cavity length for optimization of device designs with respect to cavity length and mirror reflectivity. Because the gain-current curve is nonlinear there is a minimum in the threshold *current* as a function of cavity length [2, 18 and figures 26 and 27 of 37]. Typical values for the gain parameters for GaInP quantum wells are given in table B2.5.1.1.

A strained structure offers considerable choice of well width and composition/strain for emission at a specific wavelength and one use of gain-current curves is to provide guidance in the choice of an appropriate structure. As an example, figure B2.5.1.12 shows calculations of the intrinsic threshold current density of 450 μm long, 670 nm lasers as a function of well width (lower axis), with the composition/strain being adjusted as indicated on the upper axis to maintain the same transition energy. The curve has a minimum at a well width of about 3.5 nm indicating that there is a specific structure which will provide a minimum threshold current. The details of optimization of 670 nm devices have been discussed by [25]. The upturn of this curve for narrow wells is due to the spectral broadening by well-width fluctuations. This 'U-shaped' variation dependence of threshold current density at fixed wavelength is characteristic of strained quantum well active regions [5].

B2.5.1.2.3 Threshold current and power

The gain-current curves discussed in the previous section enable the *intrinsic* threshold current to be calculated for a specific device. However, in real lasers there are further *extrinsic* current paths which also contribute to the total threshold current. In red lasers the chief extrinsic processes are non-radiative recombination via

Table B2.5.1.1. Theoretical and experimental data for gain-current parameters (equation (B2.5.1.4)) of GaInP quantum wells. The experimental data provides a comparison between tensile and compressive strained structures at 635 nm.

Well width nm	Composition Ga, y	Wavelength nm	Comments	g_{th} cm^{-1} (material gain)	J_t A cm^{-2} (spontaneous current)
2.5	0.25	670	No well fluctuations (figure B2.5.1.10)	5100	275
2.5	0.25	670	With well fluctuations (figure B2.5.1.10)	4800	335
4.0	0.41	650	No well fluctuations [note 1]	4000	485
6.8	0.41	670	From figure B2.5.1.10. [note 2]	1600	371
6.8	0.41	670	[note 3]	1850	470
Experimental data at 160 K [note 4]				Modal gain cm^{-1}	(Total current)
12.5	0.56	637	Tensile strain	62 ± 4	344 ± 12
3.0	0.41	637	Compressive strain	69 ± 9	426 ± 34
12.0	0.42	675	Compressive strain	60 ± 7	399 ± 30

1. Data from [29]. The effect of well width fluctuations is small at this well width.
2. As figure B2.5.1.10 shows the effect of well width fluctuations is negligible at this well width.
3. This is the result of a full 'many-body' calculation as described by [30]. As figure B2.5.1.10 shows, the effect of well width fluctuations is negligible at this well width and fluctuations have not been included in the calculation.
4. Experimental data from [27]. These results are for modal gain (local gain $\times$ optical confinement factor) and the current is that due to all paths through the structure. The measurements are at 160 K where carrier leakage is thought to be negligible but there may be a non-radiative contribution. The intrinsic current parameter J_t increases approximately linearly with temperature.

defects in the quantum well, which provides a parallel current path (J_{nr}) to the spontaneous recombination in figure B2.5.1.9, and thermally activated carrier leakage (J_l) through the cladding layer (see figure B2.5.1.5); this process is discussed in more detail below. Thus, the total current at threshold is

$$J_{th} = J_{spon}^{th} + J_{nr} + J_l. \tag{B2.5.1.5}$$

The extrinsic currents depend upon the material quality and the details of the specific structure and the non-radiative current in particular is difficult to predetermine. Sometimes these currents are lumped together into an unknown radiative quantum efficiency (η) defined as the ratio of the radiative current to the total current so that

$$J_{th} = \frac{J_{spon}^{th}}{\eta}. \tag{B2.5.1.6}$$

We issue a word of warning that the term 'quantum efficiency' is used in a variety of inconsistent ways in the literature, for example it is often erroneously interchanged with the differential quantum efficiency (or slope efficiency). The interested reader should refer to the paper by [28] for a discussion of these issues in red-emitting lasers.

Experimental results for threshold current, plotted as a function of temperature, are shown in figure B2.5.1.13 for devices operating at different wavelengths. The wells have different composition/strain and are of similar width [31]. At low temperatures the threshold current increases linearly with temperature, characteristic of the intrinsic radiative recombination current in the quantum well. At higher temperatures

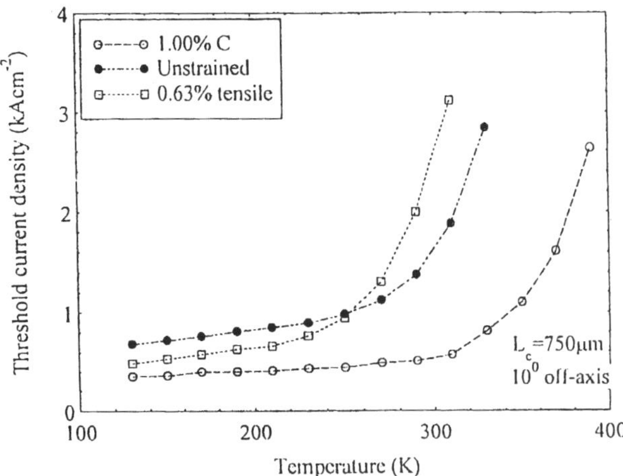

Figure B2.5.1.13. Experimental measurements of the threshold current as a function of temperature for AlGaInP edge-emitting lasers with different strained active regions of similar well width. The emission wavelengths range from 664 nm for the 1% compressive structure to 623 nm for the 0.63% tensile strain structure. At low temperatures both compressive and tensile strain reduce the threshold current. However, the tensile strained structure has a strong temperature dependence of threshold current due to carrier leakage from the shallow well (see [31]).

the threshold increases in an exponential-like manner due to thermally-activated loss of electrons to the indirect conduction band minimum of the p-type cladding layer where they are transported away from the active region to the contact [6, 26]. At low temperatures, where carrier leakage is negligible, compressive and tensile strain reduce the threshold current relative to the unstrained device due to the beneficial effects of strain on the intrinsic radiative current. However, at high temperature the tensile strain device has the highest threshold current because in this short wavelength device the well is shallow, the confined states are close to the top of the well and carriers are easily excited thermally to the barrier. From observations made only at room temperature it could be concluded, erroneously, that tensile strain is deleterious: these data show the problem arises due to carrier loss from a confined state near the top of the well. Carrier leakage can be eliminated in 670 nm devices for operating temperatures up to about 80 °C, but it makes a substantial contribution to the threshold current of 630 nm devices at room temperature and is the chief factor which prevents manufacture of shorter wavelength devices.

While thermally-activated leakage is a major contribution to the threshold current of short-wavelength GaInP lasers, there is evidence that non-radiative recombination in the wells themselves, probably via defects in the adjacent barrier material, also contributes to the current and is the dominant non-radiative carrier loss in longer wavelength devices. A comparison between the quantum kinetic theory and experimental data for peak gain as a function of current density suggests the quantum efficiency at threshold is only about 55%. [30]. This indicates that the threshold current density can be reduced by attention to the quality of the material, probably the Al-containing barrier material. Nevertheless, thermally-activated carrier leakage remains the major issue in the development of AlGaInP lasers at short wavelength and we now examine this mechanism in detail.

Many applications of red-emitting lasers demand high output power and this is an area where there are continuing steady improvements in performance. As will be clear from our discussion, achieving high output power at short wavelength poses a major challenge and power performance is very much related to the operating wavelength. At the time of writing, single stripe devices have been reported with output powers in

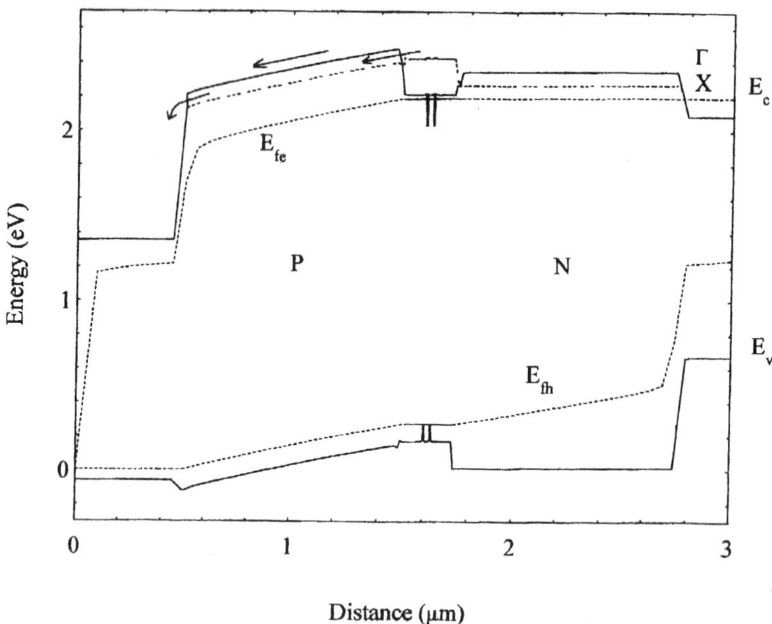

Figure B2.5.1.14. The calculated profile of the energy bands through a complete GaInP laser diode heterostructure for a 250 μm-long, edge-emitting, double well device at threshold. The full curves are the conduction (E_c) and valence band (E_v) edges corresponding to the direct gap and the broken curve is the indirect conduction band minimum (labelled X). The quasi Fermi levels are shown as dotted curves (E_{fe} and E_{fh}) The cladding material is $(Al_{0,70}Ga_{0.30})_{0.52}In_{0.48}P$ and the waveguide core and barrier material is $(Al_{0.30}Ga_{0.70})_{0.52}In_{0.48}P$. In the cladding layer the X-minimum is lowest and electrons are lost by conduction through this band [12].

excess of 100mW (e.g. 140 mW at 660 nm in a single lateral mode for DVD read and write) and arrays of devices have been reported with output of several watts (e.g. 3 W at 643 nm). At all power levels the output spectra and the field patterns depend upon the details of the device structure. Simple lasers usually show closely spaced multiple longitudinal modes within the envelope of the gain spectrum while ridge waveguide devices provide control of the lateral modes.

B2.5.1.2.4 Carrier leakage

The strong temperature dependence of threshold current in GaInP lasers is due to the leakage of electrons through the p-cladding layer [6,26,35]. Figure B2.5.1.14 shows self-consistent calculations of the energies of the conduction and valence bands as functions of distance through the complete device structure at threshold for a double-well 670 nm device. The band edge of the n-type cladding layer is flat, whereas there is a significant electric field across the p-cladding layer due to its high electrical resistivity caused by a low hole mobility and difficulties in achieving high concentrations of electrically active acceptors in the cladding material. This field is generated primarily by the majority carrier (hole) current which flows through the cladding layer to the active region. In the p-cladding layer the indirect X conduction band edge lies below the direct minimum (see figure B2.5.1.4) and represents the minimum energy gap. Electrons in the active region are distributed thermally above the confined state in the well and those with energies above the X-minimum are able to leak through the p-cladding material. This occurs by both diffusion (due to their concentration gradient) and drift (due to the electric field) and the electrons eventually recombine with holes

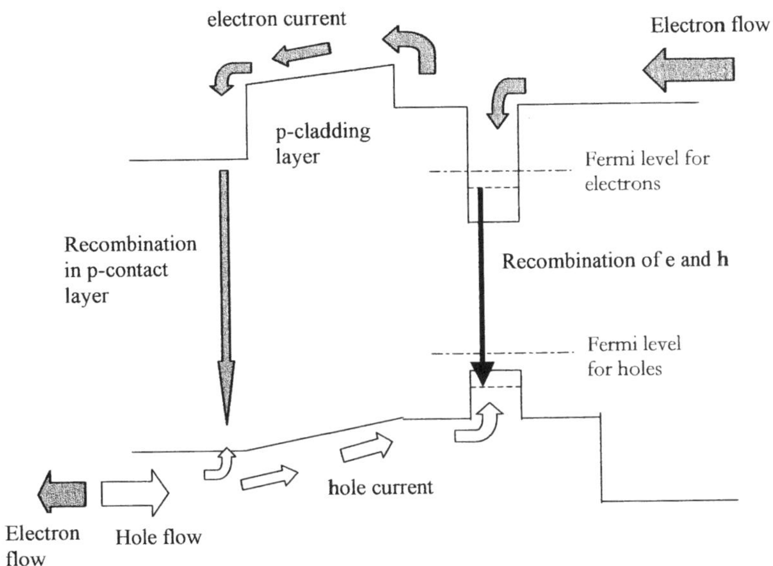

Figure B2.5.1.15. Schematic diagram showing current flow through a laser structure. Electrons leak as minority carriers through the p-cladding layer to recombine with holes in the p-contact layer. The hole current to the well generates an electric field across the resistive p-cladding layer which assists the flow of electrons away from the active region.

in the p-contact layer to complete the circuit through the structure, as illustrated in figure B2.5.1.15. As the temperature increases, the number of electrons able to escape increases exponentially. Furthermore, the intrinsic current through the well needed to maintain a fixed gain increases linearly with temperature and this also increases the leakage by increasing the field in the p-cladding, adding to the increase in total current with temperature.

Electron leakage is substantial in this material because of the low energy of the indirect conduction band edge (which determines the number of electrons able to escape), and because the field in the cladding layer is large (which increases the rate at which they can escape). The field arises for three reasons: it is difficult to achieve high densities of active acceptors, the acceptor ionization energy is larger than in narrower gap materials so thermal ionization may not be complete and the hole mobility is low due to the higher hole mass.

There are three possible strategies to decrease this thermally activated current.

(1) To reduce the resistivity of the p-cladding material by increasing the p-doping density in order to reduce the field. Improvements have been made in growth techniques and hole densities in excess of 10^{18} cm^{-3} can now be achieved.

(2) To reduce the field by keeping the majority carrier current through the cladding layer as small as possible by minimizing the gain requirement by use of a long cavity and/or by coating the mirrors. The local gain can also be minimized by designing the waveguide to maximize coupling between the well and the optical field.

(3) To increase the height of the potential barrier seen by the electrons. The leakage process is controlled by the energy difference between the quasi-Fermi energy of electrons in the well and the X-conduction band edge and this can be increased in three ways.

- Increasing the p-doping density reduces the spacing between the valence band edge and the hole quasi Fermi level which raises the energy of the X-conduction band edge of the p-cladding material relative to the electron quasi-Fermi level in the well.

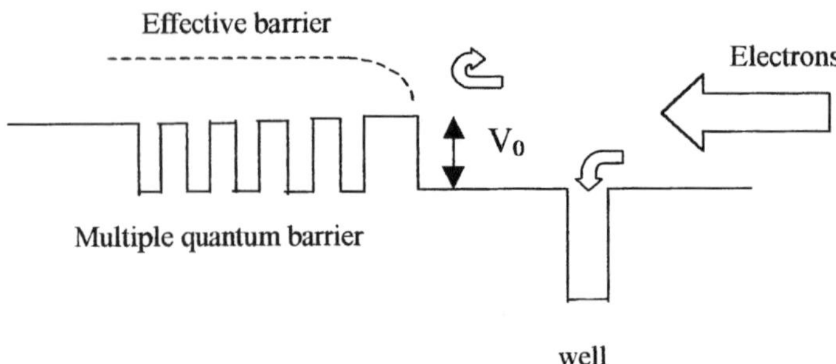

Figure B2.5.1.16. Illustration of the operation of a multiple quantum barrier to reduce carrier leakage. The effective height of the barrier is greater than the electrostatic barrier height V_0 due to quantum mechanical reflection of electrons by the multilayer structure.

- If the gain requirement is reduced the quasi-Fermi level for electrons at the well is lowered relative to the barrier. This can be done by increasing the number of wells, cavity length and mirror reflectivity, and by maximizing coupling to the optical field.
- The band gap of the cladding layer should be as wide as possible, though there is little opportunity to control this because the indirect gap is insensitive to alloy composition (figure B2.5.1.4). There have been attempts to increase the effective height of the barrier by making a quantum mechanical reflector as described later.

It must be remembered that these actions have other consequences for the performance of the laser. For example, designing the waveguide to maximize coupling between the optical field and the well usually increases the divergence of the external beam, and increasing the number of wells may increase the threshold current while reducing its temperature sensitivity. Because of these trade-offs between different characteristics it is necessary to optimize the design of red-emitting lasers for the particular requirements of a given application.

B2.5.1.2.5 Multiple quantum barriers (MQB)

While the strategies outlined go some way towards reducing the thermally activated leakage current, the benefits they provide are modest because they do not overcome the fundamental limitation of the inadequate barrier height determined by the band gaps of the constituent materials. It has been proposed [13] that the effective barrier height as seen by the electrons can be increased above the height of the electrostatic barrier by placing a multi-layer quantum mechanical reflector in the vicinity of the interface with the p-cladding layer (figure B2.5.1.16). This comprises alternating layers of materials of different band gap which reflect electrons quantum mechanically in a manner analogous to Bragg reflection of light by a mirror made up of alternating layers with different dielectric constant. It has been shown that such a mirror could provide an effective barrier up to twice the height of the electrostatic barrier [32]. If such effects could be achieved the benefits for short wavelength lasers would be considerable, however there are practical difficulties. The epitaxial layers required are as thin as 4–6 monolayers (about 2 nm) and the barrier must have accurate periodicity over about 10–15 periods. The structure must not introduce additional electrical resistance so the multilayers must be doped without inducing interdiffusion of the layers. The reflection occurs by quantum mechanical interference and it is not certain that the electronic wavefunctions maintain sufficient coherence at room temperature for the effect to be significant. There have been a number of attempts to employ such structures in GaInP lasers (e.g. [22]) and, although reductions in threshold current have been achieved, it is difficult to attribute these unambiguously to an increase in the effective barrier height. However, the

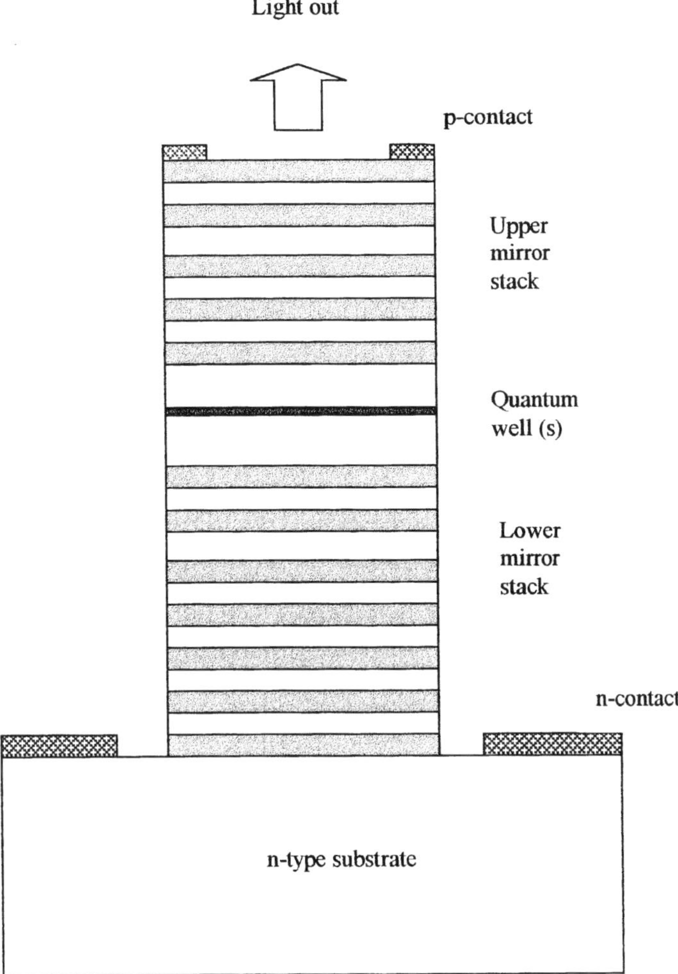

Figure B2.5.1.17. Schematic diagram of a vertical cavity surface emitting laser (VCSEL).

MQB is the only means devised as yet which could enable the leakage to be substantially reduced to make low-threshold lasers operating at wavelength down to 600 nm.

B2.5.1.2.6 Vertical cavity surface emitting lasers (VCSELs)

Conventional diode lasers are in the form of a chip typically 300 μm square and with an active stripe region about 5 μm wide. To realize a laser device it is necessary to cleave the wafer to produce reflecting end facets, consequently, the light is emitted from the end of the chip. These are termed edge-emitters. This geometry has several disadvantages: special mounting arrangements are needed because the light is emitted in a direction perpendicular to the plane on which the device is bonded, it is not easy to produce arrays of devices (for example, for high-power sources) and it is necessary to go through the labour intensive process of cleaving the wafer into chips before the device performance can be assessed. Edge-emitters also have an eliptical far-field pattern.

Iga *et al* [14] proposed the vertical cavity geometry, where the light is emitted in a direction normal to the plane of the wafer, which has since evolved to use multilayer epitaxial mirrors to take the form illustrated in figure B2.5.1.17. This has many advantages by allowing on-wafer testing before dicing into discrete chips, easier mounting due to the planar structure, fabrication of arrays and a circular far field. The device is usually made by growing a lower epitaxial multilayer dielectric Bragg reflector of alternating layers of semiconductor with different dielectric constant, then the laser cavity with a quantum well active region followed by an upper Bragg mirror. The cavity is very short, usually a few optical wavelengths, so the mirrors must have a high reflectivity (typically 0.996 or higher) so that the optical loss can be matched by the gain of the active region. These are termed vertical cavity surface emitting lasers (VCSELs). An extensive account of the current status of VCSELs is given in [34] and in section B2.1.8 and chapter B2.6. While there is a general drive to utilize the generic advantages of VCSELs in red-emitters, the ability to make arrays is particularly attractive for high-speed lasers printers.

The major issue which has had to be addressed in making AlGaInP red-emitting VCSELs is fabrication of the mirrors [23]. The variation of refractive index with composition is weaker in AlGaInP compared to AlGaAs, consequently more layer pairs are needed to achieve similar values of reflectivity, and the band of high reflectivity is narrower. This puts greater demands on the control of the growth process and introduces more electrical and thermal resistance into the structure. For this reason some of the best red-emitting VCSELs have combined a AlGaInP well and barrier system with AlGaAs multi-layer mirrors [24]. As with edge-emitters, thermally-activated carrier loss is an important issue and it is desirable to achieve a low electrical resistance in the mirror stack [24] however, because of the strong coupling to the optical field this introduces a higher free-carrier optical loss, increasing the gain requirement. In early devices lateral current confinement was provided by ion implantation, but more recently confinement by lateral oxidation of high-Al-content layers has been shown to be a better approach [9, 15].

B2.5.1.2.7 Summary

The design and performance of red-emitting AlGaInP laser diodes are distinguished by the following features.

- The availability of a ternary alloy system which can be lattice matched to GaAs substrates enables device designs to be optimized through the independent choice of well width and well composition, with the latter determining the degree of elastic strain. There are many width/composition combinations which give emission at the same wavelength and a detailed knowledge of the physics of the gain medium is needed to exploit these choices to the full.
- The indirect band gap at about 2.25 eV imposes a fundamental limit on the range of wavelengths which can be engineered in this system, even though the direct gap can be increased to about 2.5 eV.
- Carrier leakage through the indirect conduction band minimum is exacerbated by the electric field caused by the relatively high electrical resistivity of the p-type cladding layer due to the low hole mobility and hole density. This introduces a strong temperature dependence to the threshold current and sets the lowest wavelength achievable in practice to about 635 nm. Although multiquantum barriers offer a means to overcome this problem this technology has not yet been clearly demonstrated.
- This alloy system permits only small changes in refractive index which means that many periods are required for effective AlGaInP Bragg reflectors for VCSELs.
- AlGaInP has a lower thermal conductivity than AlGaAs, causing a higher temperature rise in the active regions with implications for the design of high power devices.

Future developments are likely to be concerned with further efforts to reduce the operating wavelength and increase output power. Fundamental to progress in this area is the reduction of carrier leakage, thereby gaining improvements in threshold current and efficiency. Further advances will also be made in vertical cavity lasers. Performance improvements will result in wider applications in medicine and displays.

Acknowledgments

The author wishes to thank his colleagues in the Optoelectronics Group in Cardiff for their contributions to research on GaInP lasers. The contributions of collaborators Weng Chow (Sandia Labs), David Bour (Agilent, Palo Alto) and Steve Bland (IQE, Cardiff) are also acknowledged. Funding for the Cardiff research activities has been provided by the Engineering and Physical Sciences Research Council.

References

[1] Agrawal G P and Dutta N K 1986 *Long Wavelength Semiconductor Lasers* (New York: Van Nostrand Reinhold)

[2] Blood P 1991 Heterostructures in lasers *Physics and Technology of Heterojunction Devices* ed D V Morgan and R H Williams (London: Peter Peregrinus)

[3] Blood P 2000 On the dimensionality of optical absorption, gain and recombination in quantum-confined structures *J. Quantum Electron.* **36** 354–62

[4] Blood P, Fletcher E D, Woodbridge K, Heasman K C and Adams A R 1989 Influence of the barriers on the temperature dependence of threshold current in GaAs/AlGaAs quantum well lasers *IEEE J. Quantum Electron.* **25** 1459

[5] Blood P and Smowton P M 1995 Strain dependence of threshold current in fixed-wavelength GaInP laser diodes *IEEE J. Selected Topics Quantum Electron.* **1** 707–11

[6] Bour D P, Treat D W, Thornton R L, Geels R S and Welch D F 1993 Drift leakage current in AlGaInP quantum well lasers *IEEE J. Quantum Electron.* **29** 1337–43

[7] Casey H C Jr and Panish M B G 1978 *Heterostructure Lasers* (San Diego, CA: Academic)

[8] Chen G S, Jaw D H and Stringfellow G B 1990 Effects of substrate misorientation on ordering $InGaAs_{0.5}P_{0.5}$ grown by organometallic vapor phase epitaxy *Appl. Phys. Lett.* **57** 2475–7

[9] Choquette K D, Schneider R P, Crawford M H, Gleib K M and Figiel J J 1995 Continuous room temperature operation of 640–660 nm selectively oxidized AlGaInP vertical cavity lasers *Electron. Lett.* **31** 1145–6

[10] Chow W W, Smowton P M, Blood P, Girndt A, Jahnke F and Koch S W 1997 Comparison of experimental and theoretical GaInP quantum well gain spectra *Appl. Phys. Lett.* **71** 157–9

[11] Coldren L A and Corzine S W 1995 *Diode Lasers and Photonic Integrated Circuits* (New York: Wiley)

[12] Foulger D L, Smowton P M, Blood P and Mawby P A 1997 Self-consistent simulation of AlGaInP visible lasers *IEE Proc. Optoelectron.* **144** 23–9

[13] Iga, K, Uenohara H and Koyama F 1986 Electron reflectance of multiquantum barrier (MQB) *Electron. Lett.* **22** 1008–9

[14] Iga K, Koyama F and Kinoshita S 1988 Surface emitting semiconductor lasers *IEEE J. Quantum Electron.* **24** 1845–55

[15] Jager R, Grabherr M, Jung C, Michalzik R, Reiner C, Weigl B and Ebeling K J 1997 57% wall-plug efficiency oxide confined 850 nm wavelength VCSELs *Electron. Lett.* **33** 330–1

[16] Kondow M, Kakibayashi H, Tanaka T and Minagawa S 1989 Ordered structure in $Ga_{0.7}In_{0.3}P$ alloy *Phys. Rev. Lett.* **56** 884–6

[17] Matthews J W and Blakeslee A E 1974 *J. Crystal Growth* **27** 118

[18] McIlroy, P W A, Kurobe A and Uematsu Y 1985 Analysis and application of theoretical gain curves to the design of multiple quantum well lasers *IEEE J. Quantum Electron.* **21** 1958–63

[19] Mogensen P C, Hall S A, Smowton P M, Bangert U, Blood P and Dawson P 1998 The effect of high compressive strain on the operation of AlGaInP quantum well lasers *J. Quantum Electron.* **34** 1652–9

[20] Okuda H, Anayama C, Tanahashi T and Nakajima K 1989 X-ray investigation of the ordered structure of AlGaInP quaternary alloys *Appl. Phys. Lett.* **56** 731–3

[21] O'Reilly E P 1989 Valence band engineering in strained-layer structures *Semicond. Sci. Technol.* **4** 121–37

[22] Raisch P, Winterhoff R, Wagner W, Kessler M, Schweitzer H, Reidl T, Wirth R, Hangleiter A and Scholz F 1999 Investigations on the performance of multiquantum barriers in short wavelength (630 nm) AlGaInP laser diodes *Appl. Phys. Lett.* **74** 2158–60

[23] Schneider R P Jr and Lott J A Jr 1993 InAlP/InAlGaP distributed Bragg reflectors for visible vertical cavity surface emitting lasers *Appl. Phys. Lett.* **62** 2748–50

[24] Schneider R P Jr, Hagerott Crawford M, Choquette K D, Lear K L, Kilcoyne S P and Figiel J J 1995 Improved AlGaInP-based red (670–690 nm) surface emitting lasers with novel C-doped short cavity epitaxial design *Appl. Phys. Lett.* **67** 329–31

[25] Smowton P M, Summers H D, Rees P and Blood P 1994 Threshold current of 670 nm AlGaInP strained quantum well lasers *IEEE Photon. Technol. Lett.* **6** 910–12

[26] Smowton P M and Blood P 1995, GaInP–(Al_yGa_{1-y})InP 670 nm quantum well lasers for high temperature operation *IEEE J. Quantum Electron.* **31** 2159–64

[27] Smowton P M, Blood P, Mogensen P C and Bour D P 1995, Role of sublinear gain-current relationship in compressive and tensile strained 639 nm GaInP lasers *Int. J. Optoelectron.* **10** 383–91

[28] Smowton P M and Blood P 1997 The differential efficiency of quantum well lasers *IEEE J. Special Topics Quantum Electron.* **3** 491–8

[29] Smowton P M and Blood P 1997 Visible-emitting AlGaInP laser diodes *Strained Layer Quantum Wells and their Applications* ed M O Manasreh (New York: Gordon and Breach) pp 431-87

[30] Smowton P M, Blood P and Chow W W 2000 Comparison of experimental and theoretical gain-current relations in GaInP quantum well lasers *Appl. Phys. Lett.* **76** 1522–4

[31] Summers H D and Blood P 1994 Strain effects in AlGaInP lasers operating at fixed threshold gain *Electron. Lett.* **30** 236–8

[32] Takagi T, Koyama F and Iga K 1991 Design and photoluminescence study on a multiquantum barrier *IEEE J. Quantum Electron.* **27** 1511–19

[33] Valster A, van der Poel C J, Finke M N and Boermans M J B 1992 Effect of strain on the threshold current of GaInP/AlGaInP quantum well lasers emitting at 633 nm *13th Int. IEEE Semiconductor Laser Conf. (Takamatsu, Japan) (Conference Digest)*

[34] Wilmsen C W, Temkin H K and Coldren L A 1999 *Vertical-Cavity Surface Emitting Lasers* (Cambridge: Cambridge University Press)

[35] Wood S A, Molloy C H, Smowton P M, Blood P and Button C C 2000 Minority carrier effects in GaInP laser diodes *J. Quantum Electron.* **36** 742–50

[36] Woodbridge K, Blood P, Fletcher E D and Hulyer P J 1984 Short wavelength (Visible) GaAs quantum well lasers grown by molecular beam epitaxy *Appl. Phys. Lett.* **45** 16–18

[37] Zory P S, Reisinger A R, Mawst L J, Costrini G, Zmudzinski M A, Emanuel M A, Givens M E and Coleman J J 1986 Anomalous length dependence of threshold for thin quantum well AlGaAs diode lasers *Electron. Lett.* **22** 475–6 (Use of the word anomalous in the title of this paper is misleading. The effect reported was certainly ususual at the time, but as understanding of quantum wells advanced it became clear that this is exactly what should happen in such structures as discussed by [2])

Further reading

Bastard G 1988 Wave mechanics applied to semiconductor structures *Les Editions de Physique* (New York: Halstead)

An advanced quantum mechanical treatment of quantum well structures at undergraduate/graduate level.

Blood P 1991 Heterostructures in lasers *Physics and Technology of Heterojunction Devices* ed D V Morgan and R H Williams (Peter Peregrinus)

An introductory review of quantum well heterostructure lasers at undergraduate level.

Bour D P 1993 AlGaInP quantum well lasers *Quantum Well Lasers* ed P S Zory (San Diego, CA: Academic)

A book chapter providing an introductory research review on GaInP lasers.

Casey H C Jr and Panish M B 1978 *Heterostructure Lasers* (San Diego, CA: Academic)

A useful introduction to the physics of diode lasers with further chapters on heterostructures, basic p-n junction theory relevant to laser diodes, properties of semiconductor alloy systems (band gaps, lattice parameters), and crystal growth methods. Though it predates quantum well lasers, this is a good introduction at the advanced undergraduate level.

Chow W W and Koch S W 1999 *Semiconductor Laser Fundamentals* (Berlin: Springer)

A serious text on the theory of semiconductor lasers at a post graduate level. Deals with electronic structure of bulk materials and quantum wells and many body theories of gain.

Chuang S L 1995 *Physics of Optoelectronic Devices* (New York: Wiley)

A comprehensive text at advanced graduate and postgraduate level. In this book the derivations of equations are given in detail: no recourse to ‘it can be shown that’ in this text. Covers the basics of semiconductors and quantum mechanics and waveguiding, as well as full accounts of the physics of lasers.

Coldren L A and Corzine S W 1995 *Diode Lasers and Photonic Integrated Circuits* (New York: Wiley)

A comprehensive account of the physics of quantum well laser structures at fourth-year undergraduate or first-year postgraduate level, including the basic physics of quantum well structures.

Fox M 2001 *Optical Properties of Solids* (Oxford: University Press)

A useful introduction to the basic physics at undergraduate level.

Smowton P M and Blood P 1997 Visible-emitting AlGaInP laser diodes *Strained Layer Quantum Wells and their Applications* ed M O Manasreh (New York: Gordon and Breach) pp 431–87

A research review of GaInP red-emitting laser diodes at a postgraduate level.

Weisbuch C and Vinter B 1991 *Quantum Semiconductor Structures* (San Diego, CA: Academic)

A research review of the fundamental physics and applications of semiconductor quantum well structures.

Wolfe C M, Holonyak N Jr and Stillman G 1989 *Physical Properties of Semiconductors* (Englewood Cliffs, NJ: Prentice-Hall)

A valuable account of the physics of semiconductors including crystal structure, electronic band structure, carrier distributions, minority carrier behaviour, optical properties and heterostructures at final year undergraduate level.

B2.5.2
Visible laser diodes: properties of blue laser diodes

Robert Martin

B2.5.2.1 Introduction

That the demonstrations of red and infrared laser diodes (LDs) in the early 1960s were not quickly followed by the development of similar shorter wavelength devices was not due to a lack of desirability and commercial attractiveness but rather due to the difficult nature of the wide bandgap semiconductors. Despite early work on II–VI compounds, GaN and ZnO, the technology and quality of epitaxial structures composed from these materials lagged far behind that of narrower gap materials such as GaAs. High brightness blue light emitting diodes and lasers were not developed until the 1990s and followed major advances in the fabrication of group III-nitride (AlGaInN) structures. This chapter presents an outline of these developments and then proceeds to a description of the current status and performance of GaN-based blue LDs. The extension of the lasing wavelength in GaN-based lasers to both shorter (UV) and longer (blue/green) wavelengths is described along with the competing technologies: namely ZnO-based structures and frequency doubled systems for the blue and II–VI semiconductor structures for the green/yellow spectral region.

Note that the performance, range and applications of short wavelength LDs (and LEDs) are still improving rapidly with time and this chapter represents the situation in spring 2002. For a deeper understanding of the development and properties of blue-emitting LEDs and lasers, the reader is directed to several books [1, 2] and conference proceedings [3, 4].

B2.5.2.2 The development of blue LEDs and laser diodes

B2.5.2.2.1 InGaN-based LEDs

The commercialization of the first short wavelength LDs (violet, 400–410 nm emission) by the Nichia Corporation in February 1999 was the culmination of a dramatic series of research successes. The work within Nichia, under the leadership of Shuji Nakamura, spanned almost a decade and has been well documented (e.g. [1]). It included demonstrations of ultra-bright UV, blue, green and amber InGaN/GaN LEDs and rapid progress in improving the operating characteristics and longevity of violet LDs.

These achievements built on notable research successes by Akasaki and others in overcoming the main problems that had thwarted the development of GaN-based devices following the demonstration of metal–insulator–semiconductor LEDs in the 1970s [5]. The two main problems were the poor crystalline quality of GaN and the lack of a viable route to p-type material. High-quality GaN crystalline films were first grown in the late 1980s by metal organic vapour phase epitaxy (MOVPE) using low-temperature grown buffer layers of AlN [6] or GaN [1]. These layers helped to isolate a region of high-quality GaN from the nitride–substrate interface, where lattice and thermal mismatches generate severe dislocations and defects. Nakamura made a further significant improvement in GaN epitaxy through the development of a novel

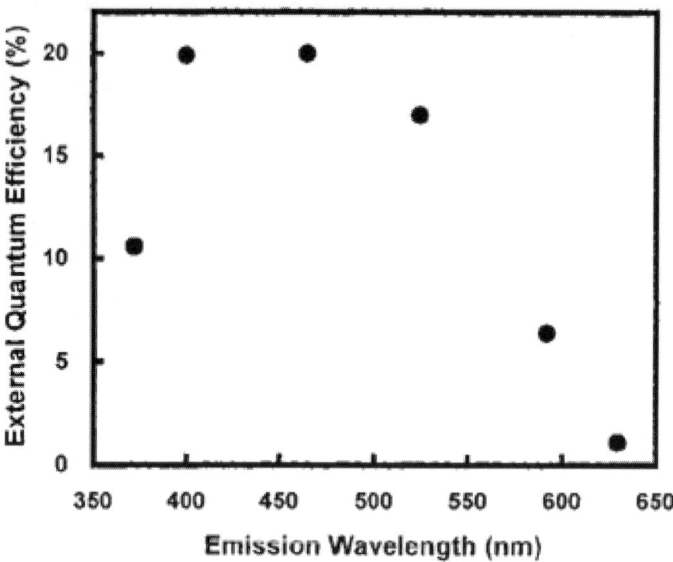

Figure B2.5.2.1. The dependence of the external quantum efficiency of InGaN LEDs on the emission wavelength. From [9], reproduced with permission from *Physica Status Solidi*.

two-flow MOVPE reactor [1]. The quality of MOVPE-grown III-nitrides has continued to improve at a rapid rate. As discussed later, additional dramatic reductions in dislocation densities were essential for laser development and were achieved using lateral epitaxial overgrowth techniques. The doping problem was also solved by Akasaki's group who produced p-type GaN by irradiating Mg:GaN with low-energy electrons and fabricated the first electroluminescent GaN pn diode GaN [7]. Nakamura subsequently showed that thermal annealing in a nitrogen ambient was a better method of p-doping, with advantages including speed, reliability and homogeneous conversion of the full depth of a layer [1]. By early 1991, Nakamura was fabricating homojunction GaN LEDs emitting 430 nm blue light with output powers and external quantum efficiencies approximately ten times greater than commercial SiC blue LEDs at that time [8]. The light emission originated from an impurity transition related to the magnesium dopant within the p-GaN and rapid progress was made on changing to the use of band-edge emissions in InGaN. The Nichia Corporation were the first to succeed in the commercial production of high-brightness blue LEDs, starting in November 1993. Toyoda Gosei Co., Ltd also pioneered similar devices, starting commercial production in October 1995.

The performance of GaN-based LEDs improved rapidly with the incorporation and improvement of InGaN quantum wells in the active region. The main features of GaN-based LEDs on sapphire substrates are as follows: following the deposition of GaN nucleation layers, several micrometres of n-type Si-doped GaN are grown. Next comes the active region incorporating InGaN quantum wells (typically 1–4 nm) with either GaN or low-InN fraction InGaN barriers and, finally, a p-type GaN layer. The Mg-dopants must be activated and are contacted with Ni/Au. For LEDs on sapphire substrates, the underlying n-type material is exposed by etching through the top layers and contacted using Ti/Al.

Nichia has LEDs incorporating multiple InGaN quantum wells with room-temperature external quantum efficiencies of approximately 20% in the blue (11 mW with a forward current of 20 mA, 50 A cm^{-2}) and green (8 mW, current as before) [9]. Figure B2.5.2.1 plots the external quantum efficiency against LED emission wavelength, showing the fall-off either side of the blue/green region. Measurements as a function of temperature for a 400 nm emitter indicate a considerable (more than double) improvement in efficiency on cooling to ~100 K [9]. These LEDs contain remarkably high densities of dislocations threading through the active region (~10^{10} cm^{-2}) [10]. Comparisons with LEDs fabricated on laterally overgrown GaN with considerably reduced dislocation densities (<10^7 cm^{-2}) show virtually no improvement for blue LEDs [11] and only a slight improvement (~25%) for UV LEDs [9, 12]. Such results are indicative of the origin of the InGaN luminescence, which results from radiative recombination within localizing InN-rich nanometre-scale regions [13–15]. This spontaneous extreme localization isolates the carrier recombination from the non-radiative recombination centres at the dislocations and leads to a surprising independence of luminescence efficiency from dislocation density. The low InN mole fraction in UV InGaN LEDs does allow for some improvement due to the limited number and depth of the localized states.

Several other groups soon followed Nichia and Toyoda Gosei in the fabrication of high-brightness blue/green LEDs, including a number of variations on the general scheme outlined earlier. Use of conducting substrates, such as 6H-SiC, offer a number of advantages: principally, the use of a vertical geometry with a back contact. Different approaches to MOCVD growth reactors have been successfully employed and InGaN LEDs have also been fabricated using molecular beam epitaxy, although not of commercial quality.

Currently (spring 2002) the best InGaN LEDs emit 30 lu W^{-1} but the performance continues to improve at a rate indicating that over 100 lumen W^{-1} is achievable (this is to be compared with 15 lumen W^{-1} for conventional incandescents and 50 lumen W^{-1} for conventional fluorescents). One of the leaders in the development of high-power LEDs, Lumileds Lighting, has advanced flip-chip Luxeon devices showing record external quantum efficiencies of ~30% and optical powers in excess of 1 W (at 425 nm) in a 2×2 configuration of LEDs [16, 17]. CREE Inc. has high-brightness 'Xbright' LEDs on SiC emitting at 470 nm, 505 nm (11 mW) and 525 nm (9 mW) [18]. White GaN-based LEDs can also be fabricated using a number of routes. Devices emitting ~20 lumen W^{-1} are available using a blue LED coated with a yellow phosphor, which partially transmits the blue light. The impressive performance of III-nitride LEDs is exemplified by the demonstration of headlights using white Lumileds Luxeon LEDs on a concept car unveiled at the 2002 Geneva Motor Show.

B2.5.2.2.2 The first GaN-based laser diodes

The bright InGaN/GaN LEDs produced by Nakamura's MOCVD process and with p-type material resulting from thermal activation of Mg donors were the forerunners of the violet LDs. However, additional problems would have to be overcome to achieve reliable lasers. The fabrication of mirrors for a diode laser's optical cavity is generally achieved by cleaving, which is not practical for the c-face sapphire substrates used for InGaN LEDs. Early InGaN lasers contained relatively rough mirror facets created by reactive ion etching [19]. Shortly afterwards, Nakamura also demonstrated InGaN lasers on a-face (11$\bar{2}$0) sapphire substrates, where mirror facets could be formed by cleaving, although no reduction in threshold current was achieved [20].

The implementation of two additional features was significant for the extension of the laser lifetime and a reduction in the operating voltage. First was the inclusion of modulation-doped strained layer superlattice (MDSLS) cladding layers above and below the active region of the lasers. These GaN/AlGaN MDSLSs reduced cracking within the laser structure, improved optical confinement and lowered the operating voltage. The second was the implementation of lateral overgrowth on patterned silica substrates [21, 22]. Stripes of silica, typically 4 μm wide and 12 μm apart, are deposited on a GaN surface prior to growth of a thick GaN layer. Above the mask, the GaN grows laterally and soon coalesces to form a flat surface with, most importantly, approximately 10^4 fewer dislocations cm^{-2} in the regions above the silica stripes. In simple terms, the silica mask has almost eradicated the memory of the GaN/substrate interface for these regions. The structure of one of these layer diodes is represented schematically in figure B2.5.2.2. Further improvement of the lasers was achieved by removing the sapphire substrate from beneath several micrometres of overgrown GaN to create 'GaN substrates' which are easily cleaved and have the extra benefit of higher thermal conductivity [23, 24].

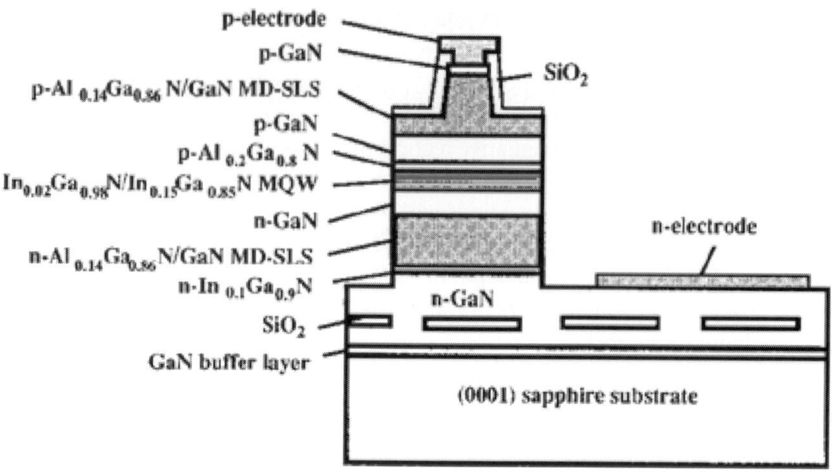

Figure B2.5.2.2. Schematic diagram of the layer structure of one of the early ELOG GaN-based laser diodes from the Nichia Corporation. From [22], reproduced with permission from the American Institute of Physics.

The first reports of violet LD operation at room temperature under pulsed conditions [19] and then with continuous wave (cw) operation [25] came from Nakamura in 1996. Using the steps previously described, the lasing lifetime was improved dramatically over a short period of time; with 30 hr in December 1996, 300 hr in May 1997 and accelerated lifetime tests indicating achievement of the 'magical' 10 000 hr lifetime in December 1997 [22]. The dramatic progress made by Nakamura in the late 1990s clearly shows the importance of minimizing dislocations for increasing the lasing lifetime. The degradation of InGaN LDs is also linked to the generation of heat within the p-cladding and contact layers and there is a report suggesting a contribution of catastrophic optical damage to the failure mechanism [26].

B2.5.2.2.3 The structure and performance of state-of-the-art GaN-based LDs

Following the previously described development, the Nichia Corporation proceeded to commercialize violet (~405 nm)-emitting GaN LDs with ~5 mW output in February 1999. Subsequently higher power (30 mW) versions (see section B2.5.2.2.4) and extended wavelength (450 nm) LDs (see section B2.5.2.2.5) were added to their product range.

Following Nichia's lead, violet GaN-based lasers have been demonstrated by a growing list of other companies and universities, with early players including Toyoda Gosei, University College of Santa Barbara (UCSB), CREE Inc., Samsung and Xerox. These devices share a number of similarities—they all employ MOCVD crystal growth, with some approach to a reduction in threshold density (lateral growth, low temperature interlayers, etc), they are grown p-side up due the to relative difficulty in doping and require some approach to lateral current spreading within the p-layer. Differences between them include the use of a variety of substrates, with the most notable variations being SiC (e.g. CREE Inc), bulk GaN crystals (UNIPRESS [27]) or conductive 'free-standing GaN' (e.g. NEC [28]) which all allow use of a back-side contact. Note that Nakamura has reported lasing on a variety of substrates (including a-face sapphire [20], $MgAl_2O_4$ [29] and 'free-standing GaN' [23]). There are a number of variations in the lateral overgrowth technique for which a range of names and acronyms exist—ELOG, lateral epitaxial overgrowth (LEO), pendeoepitaxy^{TM} [30], facet-induced epitaxial lateral overgrowth (FIELO) [28], lateral overgrowth from trenches (LOFT), etc.

The structure and performance of the Nichia devices will now be summarized. Operating specifications and illustrative figures for Nichia's devices are available on the company's website [31]. All of their devices

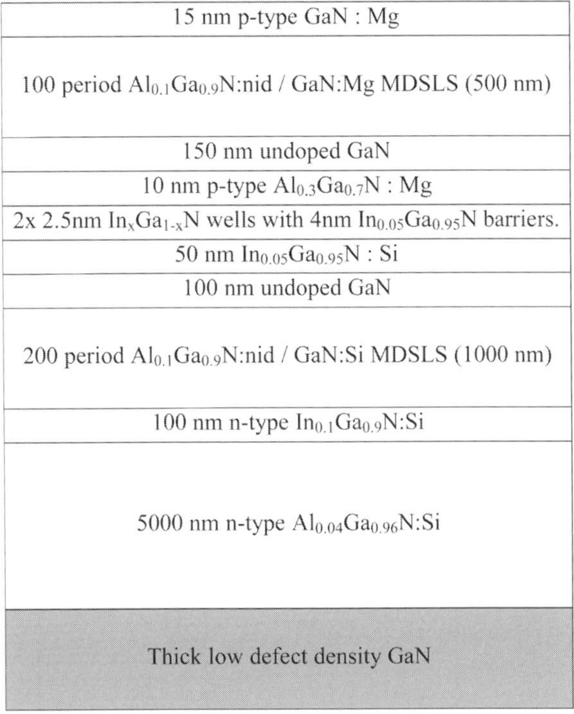

Figure B2.5.2.3. Schematic diagram of the layer structure of recent GaN-based LDs from the Nichia Corporation.

are based on a similar epitaxial layer structure, which can be generalized as follows. Starting with the reduced-defect-density GaN layer, a thick n-type GaN (or low-Al AlGaN) layer is followed by an n-type MDSLS, then an undoped GaN (or low-InN InGaN) layer containing InGaN quantum wells, then a p-type MDSLS and, finally, a thin p-type GaN layer. A more detailed layer structure reported for recent lasers [9] is represented schematically in figure B2.5.2.3. These layers are deposited on a low-defect density GaN layer produced by a lateral overgrowth of GaN. Nichia has detailed the fabrication of LDs by ELOG on sapphire and also 'ELOG on GaN' [32]. The ELOG on sapphire utilizes 100 nm thick SiO_2 masks on $\sim$2.5 μm of GaN. The masks are patterned into 8 μm stripe windows with 20 μm periodicity. Approximately 15 μm of GaN is then grown, giving a flat surface with very low defect density above the SiO_2 stripes and a so-called ELOG substrate. A thick (200 μm) layer of GaN is deposited on this, using hydride vapour phase epitaxy (HVPE) with a high growth rate of 40 μm hr^{-1}. The sapphire substrate, ELOG GaN and some HVPE GaN were then removed by polishing to obtain a free-standing GaN substrate. LDs are fabricated using the low-defect-density regions above the silica stripes. For the 'ELOG on GaN' substrates, the ELOG process is repeated to give a further reduction in the defect density. The dark spot density is reduced to 1×10^6 cm^{-2} for ELOG on sapphire and 7×10^5 cm^{-2} for 'ELOG on GaN'. The n-InGaN serves to counteract the build-up of strain and prevents crack formation within the structure. The MDSLSs act as cladding layers to confine carriers and also serve to maintain crystal quality and reduce operating voltage.

The LDs have a ridge geometry (typically 2.5 μm $\times$ 650 μm). In Nichia's case, a ZrO_2 coating is used to define the ridge by refractive index confinement of the laser beam [9]. For LDs fabricated on 'ELOG on GaN' substrates, the facets are formed by cleaving, whilst for ELOG on sapphire, the facets are formed by reactive

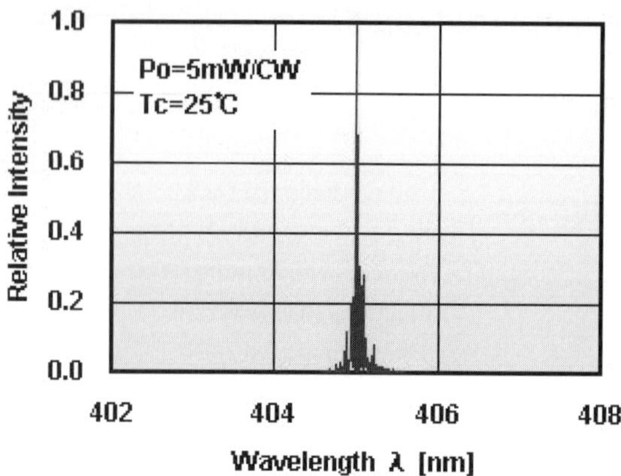

Figure B2.5.2.4. Lasing spectrum for the Nichia Corporation's 5 mW diode. Copyright Nichia Corporation [31].

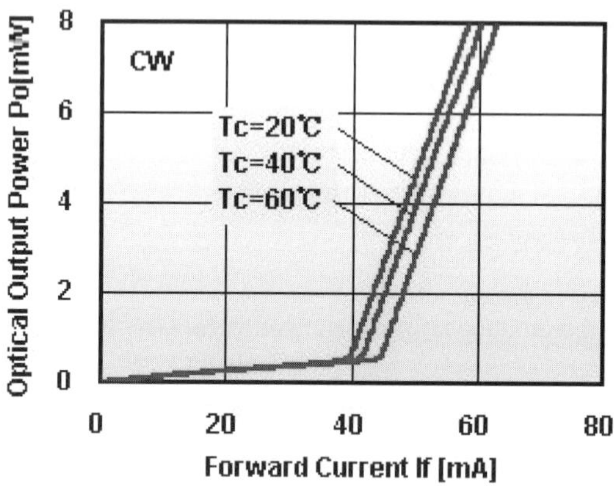

Figure B2.5.2.5. Output power against forward current for the Nichia Corporation's 5 mW diode. Copyright Nichia Corporation [31].

ion etching. Coating the facets, for example with TiO_2/SiO_2 multilayers, enhances device performance. Ni/Au is used for the p-contact and Ti/Al for the n-contact.

The laser emission wavelength is tuned mainly by control of the InN fraction within the wells, with a lesser effect from the width of the wells (see [33]). An optimization process for violet LDs resulted in a preferred well width of 2.5 nm, with wider wells resulting in increasing forward leakage currents and rising threshold current. With the choice of 2.5 nm well and 4 nm barrier layers, a second optimization showed a minimum threshold current for two wells (also a peak in emission wavelength) [32].

For the 405 nm, 5 mW LD, the quoted performance parameters are: operating voltage (4.5 ± 1.0) V, operating current (50 ± 20) mA, slope efficiency 0.7 W A^{-1}. The beam divergence (FWHM) is 10° and 28° for directions perpendicular and parallel to the growth direction. The LD spectrum is reproduced in figure B2.5.2.4 and figure B2.5.2.5 shows the output optical power plotted against forward current. The operating temperature is quoted as -10 to 60°C and figure B2.5.2.6 plots the dependence of the peak wavelength on case temperature.

CREE Inc have been developing blue/violet GaN lasers using SiC substrates and in February 2002 announced that their 405 nm, 3 mW blue LDs exhibited a projected lifetime exceeding 10 000 hr at room temperature.

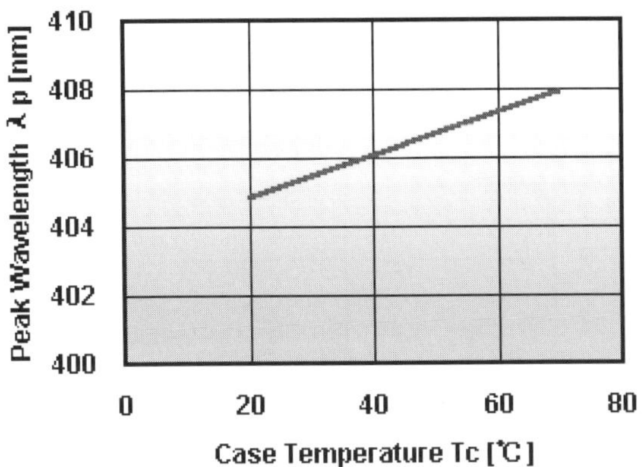

Figure B2.5.2.6. The dependence of laser peak wavelength on case temperature for the Nichia Corporation's 5 mW diode. Copyright Nichia Corporation [31].

All the demonstrations of laser action in the III-nitrides feature efforts to reduce the number of dislocations resulting from the lack of an ideal substrate material. As described earlier high-performance LDs have been fabricated using free-standing GaN substrates produced by the removal of an alien substrate (usually sapphire) following growth of a thick low-defect-density GaN layer. The availability of high-quality large crystals of bulk GaN promises dislocation densities that are orders of magnitudes lower than in any other GaN and, consequently, a possible significant improvement in laser performance. However, the growth of bulk crystals of GaN is an immense challenge, due to the extremely high melting point (~2500 °C) and equilibrium pressure at melting (~45 000 atm!). The team at the High Pressure Research Centre, Polish Academy of Sciences (UNIPRESS) are alone in producing such crystals and have recently demonstrated a blue LD fabricated by MOCVD on a substrate cut from a GaN crystal [27]. The laser is a separate confinement heterostructure device containing five InGaN quantum wells. Dislocation densities in the n-type-conducting substrates are as low as 10–100 cm^{-2} and fully strained MOCVD structures can be grown with no evidence for the formation of additional dislocations. Pulsed operation of the first laser at −40 °C showed a threshold at 1.7 A. The preparation of the bulk GaN substrates is clearly a highly technical challenge but the rewards for overcoming it are similarly great. This is evidenced by reports of large improvements in performance of lasers grown at the Nichia Corporation in 1999 when Nakamura was provided with a small number of Polish bulk substrates (×10 increase in lifetime, ×2 increase in power over corresponding devices on sapphire) [27]. A number of alternative routes to 'bulk-like' GaN are actively being researched and show promise for the future.

B2.5.2.2.4 Advancing GaN-based LDs to higher power

The first commercial violet LDs (see section D5) had an output power of 5 mW but for some major applications, e.g. DVD writers, higher power and longer lifetimes are necessary. (In February 2002, nine companies jointly announced the establishment of the basic specifications for a next-generation optical disc dubbed the 'Blu-ray Disk', designed for the recording, rewriting and play back of 27 Gbits of data on a single-sided single-layer CD/DVD size disc using a 405 nm blue-violet laser.) For optical disc writers, laser powers of 30 mW or above are required and further increases in power are necessary for other applications, including laser-based display technologies, printing and certain medical applications. Additional desirable characteristics for high-power LDs include low operating voltage, the elimination of 'kinks' in the L–I curves, improved far-field patterns by reduction of the aspect ratio of the emitting region, low noise, long life, etc.

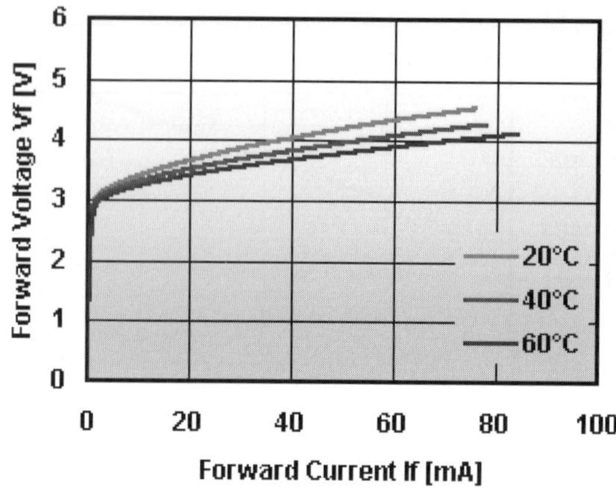

Figure B2.5.2.7. Forward voltage against forward current for the Nichia Corporation's 30 mW diode. Copyright Nichia Corporation [31].

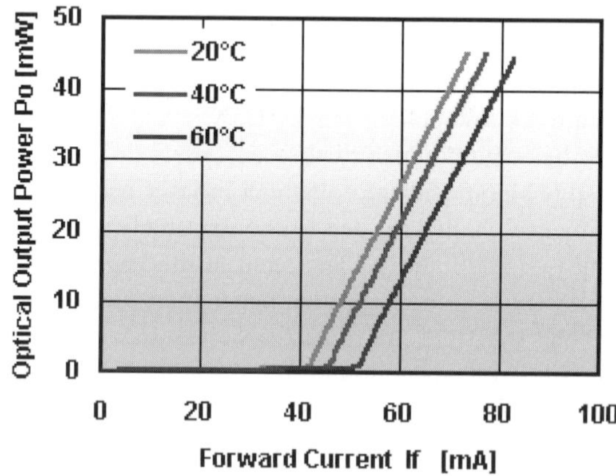

Figure B2.5.2.8. Output power against forward current for the Nichia Corporation's 30 mW diode. Copyright Nichia Corporation [31].

The Nichia Corporation has improved the reliability of their high-power LDs by use of free-standing GaNs and have commercially available 30 mW lasers. These higher power ~408 nm devices have the same operating voltage (4.5 V) as their commercial 5 mW devices and the typical operating current is only slightly higher, at 70 mA. Plots of the forward voltage and output power against forward current are shown in figures B2.5.2.7 and B2.5.2.8. The slope efficiency has increased to 1.2 W A^{-1} and the lateral spread (FWHM) decreased to 8° and 23° for directions perpendicular and parallel to the epitaxial growth direction (figure B2.5.2.9). Results for 55 mW lasers at 60 °C show lifetimes (defined as a 50% increase in current) exceeding 4000 hr for devices with both cleaved and etched facets, although productivity is improved by the latter.

NEC have developed high-power (>30 mW) devices using a ridge by a selective re-growth (RiS-) LD structure [34, 35] on a low-dislocation laterally grown GaN (FIELO) [28], as shown schematically in figure B2.5.2.10. The RiS-LD initially suffered from internal losses caused by light absorption in the region of

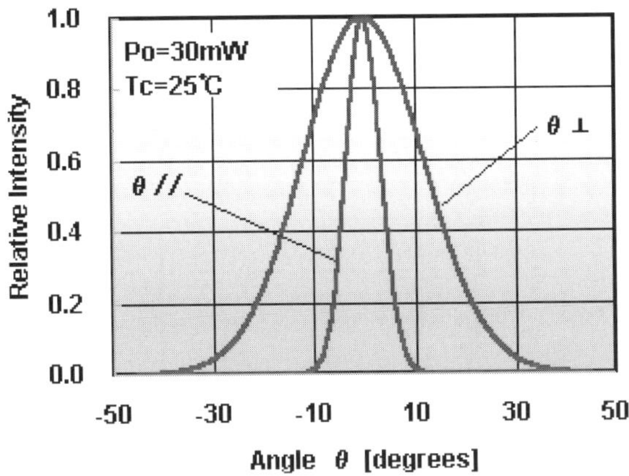

Figure B2.5.2.9. Beam spread for the Nichia Corporation's 30 mW diode. Copyright Nichia Corporation [31].

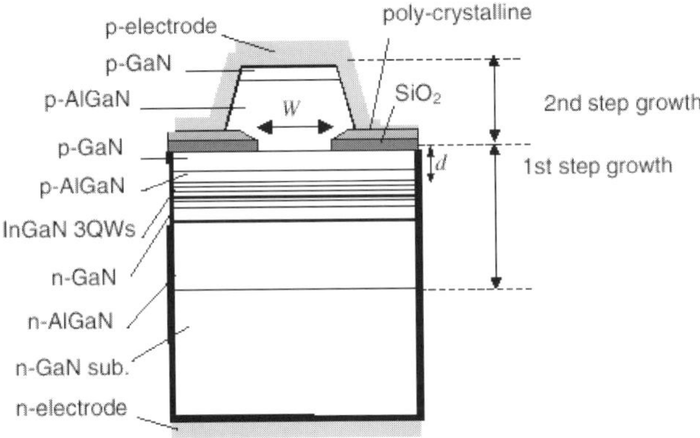

Figure B2.5.2.10. Schematic diagram of the layer structure of NEC's RiS laser diode fabricated on a FIELO n-GaN substrate. From [35], reproduced with permission from *Physica Status Solidi*.

the re-growth boundary for the upper ridge but performance has been improved by reducing the internal loss to as low as 26 cm^{-1}. The main source of loss was identified as Si contamination at the re-growth boundary, the effect of which was minimized by shifting the optical field deeper into the structure using a thick low-Al content n-AlGaN cladding layer. The result was superior laser performance with a threshold current as low as 10 mA ($J_{th} = 1.6$ kA cm^{-2}) with highly reflective-coated facets, as shown in figure B2.5.2.11. The large differential gain and suppressed non-radiative recombination in these structures may be associated with the lack of a dry etching step in the fabrication process. An impressively low aspect ratio, 2.0, is achieved in the 30 mW lasers also benefitting from the extension of the optical mode vertically in the lower cladding layer [35].

Extremely impressive progress in high-power GaN-based lasers has recently been reported by Sony [36–38]. They have developed blue LDs emitting in excess of 100 mW and demonstrated a 4.2 W LD array consisting of 44 individual lasers (11 laser chips each with four monolithically integrated laser stripes). Deviation between the individual emitters is the key limitation here but the power performance is impressive with a wall-plug efficiency approaching 10%. The 1 W cw lifetime of these laser arrays exceeded 1000 hr

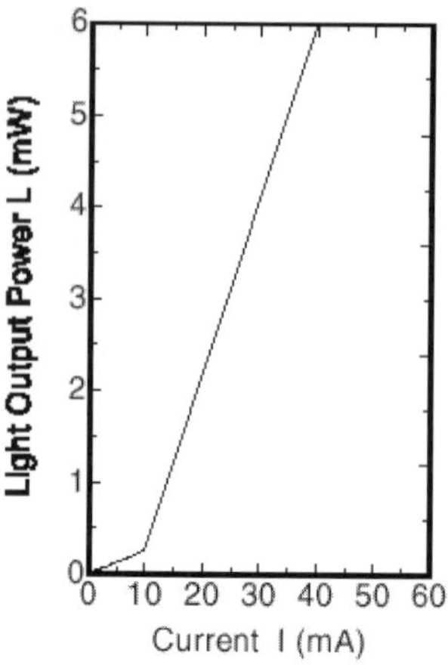

Figure B2.5.2.11. Light output power plotted against forward current for NEC's RiS laser diode with highly reflective facet coatings. From [35], reproduced with permission from *Physica Status Solidi*.

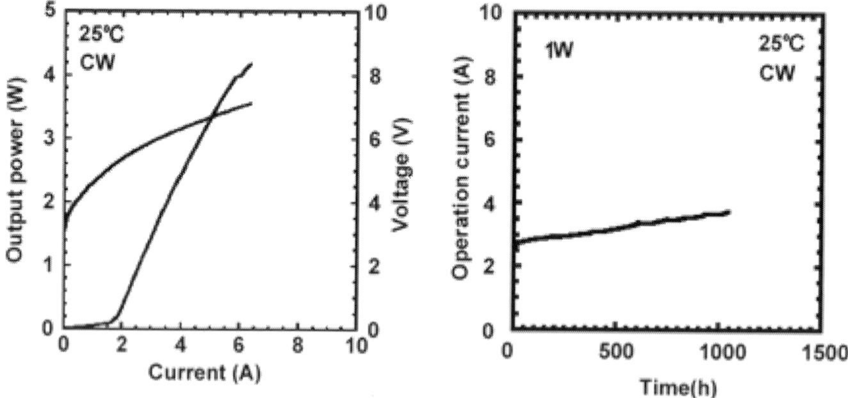

Figure B2.5.2.12. Output power and forward voltage as a function of forward current for Sony's 44 stripe GaN LD and lifetime data for an array running at 1 W. From [38], reproduced with permission from *Physica Status Solidi*.

at the time of writing. Figure B2.5.2.12 reproduces the L–I and V–I characteristics and lifetime data as presented at the ISBLLED2002 symposium in March 2002 [38]. It is noteworthy that the power consumption of an individual 100 mW blue laser ($>$ 20% wall-plug efficiency) is less than half that of an equivalent red LD. The Sony lasers use a licensed lateral overgrowth technique, based on pendeoepitaxy [30], to reduce the dislocation density. 'Kink-free' L–I curves have been demonstrated for powers in excess of 100 mW with lateral beam spreads of 8° and lifetimes (defined by a 20% increase in operating current) in excess of 15 000 hr. The kinks are associated with an increase in higher transverse modes, whose suppression is discussed in section B2.5.2.2.6 below.

Improving the laser output aspect ratio by increasing the angular spread parallel to the layers ($\theta_{\|}$) will lead to the kinks discussed earlier and, thus, any improvement in the far-field pattern can preferably be achieved by reducing the angular spread along the growth direction ($\theta_{\perp}$). Raising the thickness of the guiding layer is associated with an increased threshold current (due to less confinement) and so the reduction of $\theta_{\perp}$

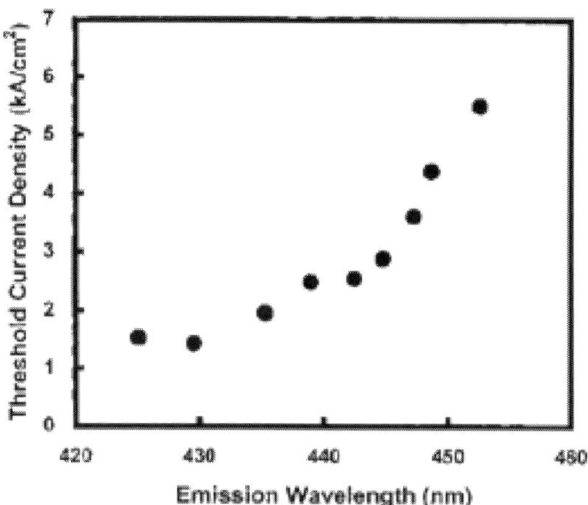

Figure B2.5.2.13. Emission wavelength dependence of the threshold current density of blue GaN-based laser diodes. From [9], reproduced with permission from *Physica Status Solidi*.

has been achieved by setting the active layer back from the p-AlGaN using an undoped spacer layer. An aspect ratio of 2.3 has been reported with no degradation in threshold current. The spacer layer is also found to have a beneficial impact on the internal losses (down to 10 cm^{-1}) due to the reduced absorption within the Mg-doped layers [38].

B2.5.2.2.5 Extending the wavelength range of GaN-based blue LDs

Figure B2.5.2.1 illustrated the falling luminescence efficiency for LEDs emitting outside of the violet–blue–green spectral range. For LDs, this characteristic of InGaN emitters imposes more severe limitations on the achievable emission wavelengths. Early GaN-based LDs were limited to the 390–420 nm spectral range, which includes the 405 nm selected for next-generation optical discs. However, longer and shorter wavelengths are very desirable and considerable progress is being made to this end.

'True blue' and longer wavelengths

Several years ago Nichia reported results on GaN-based lasers with emission wavelengths ranging from 425–454 nm, achieved by changes in the InGaN well-growth temperature [9]. The variation in threshold current is reproduced in figure B2.5.2.13 and highlights the problems to be overcome in raising the lasing wavelength. At this time, cw lasing was not achieved above 455 nm, due to the reduced quality of the layers with a higher indium nitride fraction and the damaging effects of band-tail states.

Spectra for LDs emitting at 402, 422 and 447 nm are illustrated in figure B2.5.2.14 as a function of excitation current density [32]. The 402 nm laser shows a red-shifting peak (~5 nm) due to heat generation. However, the 422 and 447 nm samples both show a blue shift (~10 and 15 nm), attributed to screening of the in-built piezoelectric fields for current densities below 0.25 kA cm^{-2} and thereafter dominated by the band-tail filling. Reference [32] also shows microscope images revealing deterioration of the quality of the ELOG material as the wavelength increases from 440 to 460 nm, which may be related to dissociation of the InGaN during growth of the p-type cladding layers.

Despite the difficulties, progress has been steady and the Nichia Corporation has been offering a commercial LD emitting at a wavelength of 440 nm for some time. The specifications are not dissimilar to

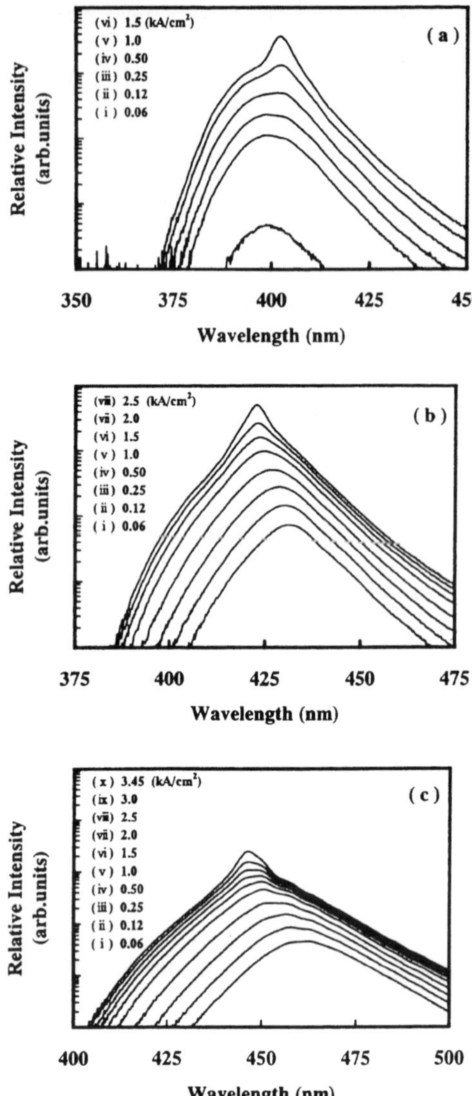

Figure B2.5.2.14. Emission spectra of lasers diodes with lasing wavelengths of (*a*) 402 nm, (*b*) 422 nm and (*c*) 447 nm at different forward current densities. From [32], reproduced with permission from the *Japanese Journal of Applied Physics*.

those of the commercial 405 nm 5 mW LD, although the typical operating voltage and current are increased to 5 V and 55 mA, respectively [31].

By mid-2001, Nichia had published details of InGaN LDs with emission wavelengths in excess of 460 nm [39]. The room temperature threshold current density of the device emitting at 460 nm was 3.3 kA cm^{-2} and the estimated lifetime approximately 3000 hr. The improvement in material quality using the 'ELOG on GaN' substrates described earlier has proved significant for enhancing the performance of the longer wavelength GaN-based LDs [32]. Recently the Nichia Corporation announced that they had achieved

GaN-based room-temperature cw lasing at 488 nm and further extensions of wavelength are sure to follow, albeit with declining performance characteristics [40].

UV laser diodes

The Nichia Corporation has also demonstrated laser emission at shorter wavelengths using either GaN [41] or AlInGaN [42] active regions. Lasing at 367 nm (pulsed) or 369 nm (cw) was demonstrated using a 10 nm well of binary GaN. For $Al_{0.03}In_xGa_{(0.97-x)}N$ wells, room-temperature cw lasing was achieved below 370 nm. In both cases, threshold current densities as low as 3.5 kA cm^{-2} and lifetimes up to 2000 hr were reported. For the AlInGaN wells, the threshold current rises with the increasing AlN fraction, required for shorter wavelengths, due to deteriorating crystal quality. Improvements in the performance of these and similar UV diode lasers will doubtless follow in the near future.

B2.5.2.2.6 Mode structure and control

Early GaN lasers with ridge widths above 3 μm gave outputs with multiple transverse modes for moderate powers. These are highly undesirable for optical storage and high-resolution spectroscopic applications but by narrowing the ridge widths, Nakamura achieved lasers with only the fundamental transverse mode for variable operating current [43].

When increasing the power of GaN-based lasers kinks are sometimes seen in the L–I traces, which are generally associated with a change in the transverse mode structure. 'Kink-free' operation has been demonstrated for powers in excess of 100 mW by suppression of higher transverse modes. For example, the conventional ridge structure has been replaced with one including layers of silicon or 'spin-on-glass' that preferentially absorb the first mode [37, 38].

Similar suppression of higher modes can be achieved by fabrication of distributed feedback (DFB) gratings, as described in [44] and [45]. Such gratings are defined by periodic variations in the refractive index, such as by etching trenches, laterally coupled to the optically active region. This also enables the emission wavelength to be tuned across the gain spectrum. For example, the use of DFB gratings with periods between 160 and 170 nm results in a 10 nm tuning range for a ~400 nm GaN-based LD, with single-mode behaviour [44].

B2.5.2.3 Stimulated emission mechanisms in GaN and related materials

As previously mentioned, the remarkable luminescence efficiency of the dislocation-ridden InGaN LEDs appears to be related to the spontaneous formation of strongly localizing In-rich regions [13–15, 33]. These regions result from segregation of InN and GaN and have a quantum-dot-like nature, which concentrates excitonic recombination away from the non-radiative centres. Strong strain variations and intense piezoelectric fields are associated with these composition variations and have major effects on the characteristics of the luminescence. However, many or all of the localized states will be filled in the stimulated emission (SE) regime controlling the operation of the LDs and different mechanisms will become dominant.

B2.5.2.3.1 Stimulated emission in GaN

Stimulated emission from GaN was first reported as early as 1971 by Dingle *et al* [46], using single-crystal GaN needles at 2 K. More recent studies have demonstrated SE in GaN at temperatures up to 700 K [47]. The large excitonic binding energy in GaN means that exciton effects persist to high temperatures as can be observed, for example, in measurements of optical absorption at elevated temperatures [48]. Stimulated emission, however, involves much higher excitation densities than in these demonstrations and the contribution of excitons will depend on the degree of screening of the Coulomb interaction by the carriers. Bidnyk *et*

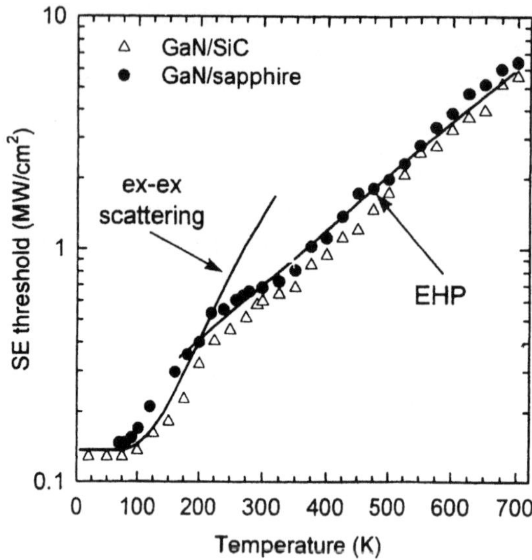

Figure B2.5.2.15. Threshold of stimulated emission as a function of temperature for GaN thin films. From [47], reproduced with permission from the American Institute of Physics.

al [49] reported a temperature dependence (20–700 K) of the SE threshold in GaN epilayers grown on sapphire and SiC substrates. Figure B2.5.2.15 shows the measured SE threshold increasing rapidly up to ~200 K and, in this range, inelastic exciton–exciton scattering appears to be the dominant gain mechanism. Exciton effects can, thus, still operate above the carrier densities required for SE at low temperatures. At higher temperatures, however, the dominant gain mechanism is electron–hole plasma (EHP) recombination, characterized by a strong red-shift of the stimulated emission peak relative to that of spontaneous emission due to bandgap renormalization.

Note that in ZnO-based devices (see section B2.5.2.4.1 below) exciton–exciton scattering persists up to higher temperatures due to the stronger exciton binding energy and ZnO-stimulated emission due to bi-excitonic processes has also been demonstrated [50, 51].

B2.5.2.3.2 Stimulated emission in InGaN

In comparison to the situation in GaN a discussion of the origin of SE in the InGaN layers present in almost all of the LDs discussed previously in this chapter is complicated by InN–GaN segregation, disorder, composition fluctuations, piezoelectric fields, strain effects, etc. Furthermore, measurements of InGaN SE have sometimes revealed two distinct SE peaks with quite different characteristics that compete for gain [52, 53]. Further work is required to explain the mechanism of SE in InGaN.

B2.5.2.4 Non-GaN blue lasers

B2.5.2.4.1 ZnO

ZnO is a wide bandgap (E_g = 3.37 eV at 300 K) semiconductor with similarities to GaN (e.g. wurtzite structure) but also a number of potential advantages. These include the extremely large excitonic binding energy (at ~60 meV, considerably larger even than for GaN), there is the possibility that lattice-matched heterostructures using $Mg_yZn_xCd_{1-x-y}O$ compounds and the availability of more closely matched substrates will be available. For example, homoepitaxy is possible using ZnO substrates [54] and $ScAlMgO_4$ substrates are almost lattice-matched to ZnO [55]. ZnO is naturally n-type but, as in the development of GaN devices, progress towards ZnO-based light emitters has been held up by the difficulty of doping p-type materials. Recently, however, Eagle Picher Technologies have demonstrated p-ZnO layers doped with Mg giving carrier densities of $\sim 10^{18}$ cm^{-3} (although at very low mobility) [54]. Given the advantages listed here, further progress with p-type ZnO could be expected to result in the emergence of high-performance LEDs and, possibly, lasers. Nevertheless it remains to be seen whether the spontaneous segregation and large polarization fields within the III-nitrides are the key to success in these areas.

B2.5.2.4.2 Frequency doubling

An alternative approach to semiconductor-based blue lasers involves frequency doubling of high-power infrared diodes. Both Matsushita and Coherent have second harmonic generation (SHG) blue lasers emitting 20–30 mW. Matsushita has demonstrated a 31% power-conversion efficiency using SHG in a periodically poled $MgO{:}LiNbO_3$ waveguide by frequency doubling an AlGaAs laser diode [56] and have a 410 nm emitter suitable for writing to optical discs. Coherent's devices employ an 810 nm diode to optically pump a vertical external cavity surface emitting GaAs-based laser and use different gain regions for devices emitting at 460 and 488 nm.

The size of the resulting device is an issue but Matsushita has reduced the package for their SHG laser to an impressive 0.3 cm^3, which may be suitable for optical disc-writers [57].

B2.5.2.4.3 II–VI materials

ZnSe-based blue lasers have been demonstrated with room temperature lasing at 463 nm (pulsed mode, binary ZnSe quantum well [58]) and 490 nm (cw mode, CdZnSe quantum well [59]). A major obstacle to the development of II–VI lasers remains the limited lifetime, largely resulting from the formation of dislocation networks in the active region, and the significantly higher lifetimes demonstrated for GaN-based lasers mean that ZnSe-based structures are very unlikely to compete in the blue region. However, as discussed in section B2.5.2.5 the II–VIs retain pole position for diode lasing at wavelengths above 500 nm, where InGaN proves more problematic.

B2.5.2.5 Extension of lasing emission beyond the blue/UV; green and yellow laser diodes

As described in section B2.5.2.2.5, the longest lasing wavelength demonstrated for GaN-based LDs is currently 488 nm. Researchers on III-nitride lasers will continue to work on increasing this value but experience thus far indicates that progress will be hard won. Thus far, diode lasers operating in the 510–520 nm green spectral region have only been demonstrated in ZnSe-based devices, including CdZnSe quantum wells. For these devices optimization of the point defect density within the active layer enabled researchers at Sony to report a ~400 hr cw lifetime at 20 °C in 1998 [60], although that result still retains the record for longevity.

Conventionally, ZnSe-based LDs are grown on conductive (n-type) GaAs substrates allowing back contacting [60, 61]. The active region is composed of CdZnSe or CdZnSSe quantum well(s), typically 3–4 nm wide. A separate confinement structure is employed using MgZnSSe cladding layers and ZnSSe waveguiding layers. The p-type contact can be formed by a ZnTe/ZnSe multi-quantum-well structure although there are other schemes with potential advantages. Indeed, achieving a reliable low-voltage ohmic p-contact remains one of the challenges for ZnSe-based lasers and appears to limit the lifetime [62]. The degradation of II–VI lasers is mostly due to a gradual process involving the development of dark-spot defects which align and propagate as dark-line defects, although failure by catastrophic optical damage has been observed.

As mentioned earlier ZnSe-based lasers are unable to compete with the III-nitrides for blue emission but are currently superior in the green/yellow area. Careful optimization of CdZnSSe quantum wells has led to the demonstration of yellow laser emission (560 nm [63]). The increased Cd content of the wells required for such emission tends to lead to high strains when using GaAs substrates, which would be expected to impair lasing severely. The addition of S to give the quaternary wells has allowed high structural and optical quality to persist even with Cd contents of up to >40% [63]. In some respects, the characteristics (e.g. higher resistance to catastrophic optical damage with 560 nm, powers exceeding 1 W in pulsed mode) of the yellow lasers exceeds that of the shorter wavelength ZnSe-based structures. This reflects the improved carrier confinement at the longer wavelengths since the bandgap of the MgZnSSe cladding material is similar in both cases.

The development of ZnSe substrates offers an alternative homoepitaxial route to ZnSe-based diode lasers with a number of potential advantages, including reduced defect densities and absorption losses. Very low thresholds (current densities as low as 176 A cm^{-2} [64]) have been demonstrated for II–VI gain guided LDs on ZnSe substrates and attributed to the reduction in the internal losses occurring in a non-transparent substrate such as GaAs. However, Wenisch *et al* [65] describe a more recent comparison of the characteristics of devices grown on GaAs and ZnSe substrates and report some similarities in the propagation of defects, although they have a relatively high defect density in the homoepitaxial lasers.

B2.5.2.6 Conclusion and outlook

The rapid development of III-nitride semiconductor devices in the last decade has produced high-performance LDs covering the UV to blue/green spectral region with more than 30 mW available in commercial 400–420 nm emitting devices. The rate of progress remains high, both in increasing power and extending the wavelength range. Schemes to minimize dislocations have been shown to be important in all of the GaN-based LDs. The most attractive of these would be the use of true bulk GaN substrates, for which the demonstration of lasers has been described. The difficulty of producing such GaN substrates remains extremely hard, however. ZnO-based devices may yet compete with the III-nitrides for blue LDs, with the availability of matched substrates a considerable advantage, but the development is at a very early stage. Laser diodes based on the II–VIs will almost certainly not be able to compete in the blue spectral region but they still hold the ground for use in green/yellow LDs. It remains to be seen whether the success of the III-nitrides in the UV and blue can be extended into these longer wavelength regions.

References

[1] Nakamura S, Pearton S and Fasol G 2000 *The Blue Laser Diode* 2nd edn (Berlin: Springer)
[2] Gil B (ed) 2002 *Low Dimensional Nitride Semiconductors* (Oxford: Oxford University Press)
[3] Proceedings of the International Conferences on Nitride Semiconductors; e.g. ICNS3 Montpellier; 1999 *Phys. Status Solidi* a **176**; 1999 *Phys. Status Solidi* b **216**; 1999 and ICNS4 Denver; 2001 *Phys. Status Solidi* a **188**; 2001 *Phys. Status Solidi* b **228**
[4] Proceedings of the International Symposia on Blue LEDs and lasers; e.g. Berlin 2000 *Phys. Status Solidi* a **180**; and Cordoba 2002 *Phys. Status Solidi* a **192**
[5] Pankove J I, Miller E A, and Berkeyheiser J E 1971 *RCA Rev.* **32** 383
[6] Amano H, Sawaki N, Akasaki I and Toyoda Y 1986 *Appl. Phys. Lett.* **48** 353
[7] Akasaki I, Amano H, Kito M and Hiramatsu K 1989 *Japan. J. Appl. Phys.* **28** L2112
[8] Nakamura S, Mukai T and Senoh M 1991 *Japan. J. Appl. Phys.* **30** L1998
[9] Nagahama S *et al* 2001 *Phys. Status Solidi* a **188** 1
[10] Lester S D, Ponce F A, Craford M G and Steigerwald D A 1995 *Appl. Phys. Lett.* **66** 1249
[11] Mukai T, Takekawa K and Nakamura S 1998 *Japan. J. Appl. Phys.* **37** L839
[12] Mukai T and Nakamura S 1999 *Japan. J. Appl. Phys.* **38** 5735
[13] Chichibu S, Azuhata T, Sota T and Nakamura S 1996 *Appl. Phys. Lett.* **69** 4188
[14] Narukawa Y, Kawakami Y, Funato M, Fujita Sz, Fujita Sg and Nakamura S 1997 *Appl. Phys. Lett.* **70** 981
[15] O'Donnell K P, Martin R W and Middleton P G 1999 *Phys. Rev. Lett.* **82** 237

[16] Kim A Y, Götz W, Steigerwald D A, Wierer J J, Gardner N F, Sun J, Stockman S A, Martin P S, Krames M R, Kern R S and Steranka F M 2001 *Phys. Status Solidi* a **188** 15
[17] Krames M R, Bhat J, Collins D, Gardner N F, Gotz W, Lowery C H, Ludoweise M, Martin P S, Mueller G, Mueller-Mach R, Rudaz S, Steigerwald D A, Stockman S A and Wierer J J 2002 *Phys. Status Solidi* a **192** 237
[18] http://www.cree.com/products/led/
[19] Nakamura S *et al* 1996 *Japan. J. Appl. Phys.* **35** L74
[20] Nakamura S *et al* 1996 *Japan. J. Appl. Phys.* **35** L217
[21] Zheleva T S *et al* 1997 *Appl. Phys. Lett.* **71** 2472
Sakai A *et al* 1997 *Appl. Phys. Lett.* **71** 2259
[22] Nakamura S *et al* 1998 *Appl. Phys. Lett.* **72** 211
[23] Nakamura S, Senoh M, Nagahama S, Iwasa N, Yamada T, Matsushita T, Kiyoku H, Sugimoto Y, Kozaki T, Umemoto H, Sano M and Chocho K 1998 *Appl. Phys. Lett.* **72** 2014
[24] Nakamura S *et al* 1998 *Japan. J. Appl. Phys.* **37** L309
[25] Nakamura S *et al* 1997 *Appl. Phys. Lett.* **70** 868
[26] Cohen D A, Margalith T, Abare A C, Mack M P, Coldren L A, DenBaars S P and Clarke D R 1998 *Appl. Phys. Lett.* **72** 3267
[27] Grzegory I *et al* 2001 *Acta Phys. Polon.* A **100** 57, 229
[28] Usui A *et al* 1997 *Japan. J. Appl. Phys.* **36** L899
[29] Nakamura S *et al* 1996 *Appl. Phys. Lett.* **68** 2105
[30] Zheleva T S, Smith S A, Thomson D B, Gherke T, Linthicum K J, Rajagopal P, Carlson E, Ashmawi W M and Davis R F 1999 *MRS Internet J. Nitride Semicond. Res.* **4S1** G3.38
[31] http://www.nichia.co.jp/
[32] Nagahama S, Yanamoto T, Sano M and Mukai T 2001 *Japan. J. Appl. Phys.* **40** 3075
[33] Martin R W, Edwards P R, Pecharroman-Gallego R, Liu C, Deatcher C J, Watson I M and O'Donnell K P 2002 *J. Phys. D: Appl. Phys.* **35** 604
[34] Kuramoto M *et al* 2001 *Japan. J. Appl. Phys.* **40** L925
[35] Kuramoto M, Sasaoka C, Futagawa N, Nido M and Yamaguchi A A 2002 *Phys. Status Solidi* a **192** 329
[36] Miyajima T *et al* 2001 *J. Phys.: Condens. Matter* **13** 7099
[37] Kijima S *et al* 2001 *Phys. Status Solidi* a **188** 55
[38] Takeya M, Tojyo T, Asano T, Ikeda S, Mizuno T, Matsumoto O, Goto S, Yabuki Y, Uchida S and Ikeda M 2002 *Phys. Status Solidi* a **192** 269
[39] Nagahama S *et al* 2001 *Appl. Phys. Lett.* **79** 1948
[40] Mukai T, Nagahama S, Yanamoto T and Sano M 2002 *Phys. Status Solidi* a **192** 261
[41] Nagahama S *et al* 2001 *Japan. J. Appl. Phys.* **40** L785
[42] Nagahama S *et al* 2001 *Japan. J. Appl. Phys.* **40** L788
[43] Nakamura S *et al* 1998 *Japan. J. Appl. Phys.* **37** L1020
[44] Hofman R *et al* 1996 *Appl. Phys. Lett.* **69** 2068
[45] Hofstetter D, Thornton R L, Kneissl M, Bour D P and Dunnrowicz C 1998 *Appl. Phys. Lett.* **73** 1928
[46] Dingle R, Shaklee K L, Leheny R F and Zetterstrom R B 1971 *Appl. Phys. Lett.* **19** 5
[47] Bidnyk S, Little B D, Schmidt T J, Krasinski J and Song J J 1998 *Proc. SPIE* **3419** 35
[48] Fischer A J *et al* 1997 *Appl. Phys. Lett.* **71** 1981
[49] Bidnyk S, Schmidt T J, Little B D and Song J J 1999 *Appl. Phys. Lett.* **74** 1
[50] Sun H D *et al* 2000 *Appl. Phys. Lett.* **77** 4250
[51] Sun H D *et al* 2000 *Appl. Phys. Lett.* **78** 3385
[52] Deguchi T, Azuhata T, Sota T, Chichibu S, Arita M, Nakanishi H and Nakamura S 1998 *Semicond. Sci. Technol.* **13** 97
[53] Schmidt T J *et al* 1999 *Appl. Phys. Lett.* **73** 3689
[54] http://www.epcorp.com/press/64.asp
[55] Ohtomo A *et al* 2000 *Appl. Phys. Lett.* **77** 2204
[56] Sugita T, Mizuuchi K, Kitaoka Y and Yamamoto K 1999 *Opt. Lett.* **24** 1590
[57] Kitaoka Y, Yokoyama T, Mizuuchi K and Yamamoto K 2000 *Japan. J. Appl. Phys.* **39** 3416
[58] Grillo D C *et al* 1994 *Electron. Lett.* **30** 2131
[59] Nakayama N *et al* 1993 *Electron. Lett.* **29** 2194
[60] Kato E, Noguchi H, Nagai M, Okuyama H, Kijima S and Ishibashi A 1998 *Electron. Lett.* **34** 282
[61] Klude M *et al* 2000 *Phys. Status Solidi* a **180** 21
[62] Kijima S *et al* 1998 *Appl. Phys. Lett.* **73** 235
[63] Klude M *et al* 2002 *Phys. Status Solidi* b **229** 935
[64] Katayama K *et al* 1998 *Appl. Phys. Lett.* **73** 102
[65] Wenisch H *et al* 1999 *Japan. J. Appl. Phys.* **38** 2590

B2.6
Vertical-cavity surface-emitting lasers

B M A Rahman and K T V Grattan

B2.6.1 Introduction

Since the development of semiconductor lasers in 1962 and the subsequent demonstration of low-loss silica fibre, the last four decades have seen rapid progress in photonics, with applications in optical fibre communications, medical instrumentation and such consumer products as high-volume CD and DVD players for the entertainment sector, laser printers for PCs and the barcode scanners used in supermarkets. The semiconductor laser diode has emerged, particularly over the last decade, as a key device that complements the transistor to form the two major technological features of today's information technology revolution.

The semiconductor laser is the backbone of modern typical optoelectronic systems and has played an important enabling role in the development of the modern era of optical communications since the early 1970s. Given the wider commercial applications of photonics, the importance of semiconductor lasers continues to increase rapidly. The semiconductor laser diode has gone beyond being just a conventional edge emitter where photons simply emerge from one edge of the semiconductor wafer. In the vertical-cavity surface-emitting laser (VCSEL), photons oscillate vertically between mirrors grown into the structure, and then emerge on a path at a right angle to the wafer surface. These differences seem simple but their consequences for higher manufacturing efficiencies and the wider applications of these lasers are tremendous.

Following the explosive growth in research and development in recent years, an expanding range of markets have been identified, closely associated with the wide wavelength coverage offered by these lasers, in which VCSELs show key advantages over conventional edge-emitting lasers. With today's precise epitaxial technology, VCSELs can be made simpler than conventional edge-emitting lasers for higher volume production and some of their performance features are superior to those of edge-emitting lasers. Low-cost broad-band optical interconnects based on VCSELs have begun to emerge from the laboratory and into the marketplace. Serial optical data links using VCSELs are already commercially available, with a performance that easily extends into the multi-gigabit-per-second regime and parallel optical data links are expected to follow shortly.

B2.6.2 Advantages

VCSELs are relatively new members of the semiconductor lasers family. However, they have generated considerable interest in recent years due to their inherent single longitudinal mode of operation as a result of their short cavity length. VCSELs possess several other unique features providing major advantages that distinguish them from conventional edge-emitting lasers. A number of their important features are discussed in the following.

VCSELs have a fundamental advantage in their beam geometry as they naturally emit 'clean' round beams. The beam of light that emerges from a typical edge-emitting laser through a narrow non-symmetrical aperture is usually both highly elliptical and divergent. If this beam is directly butt-coupled to a single-mode fibre (SMF) which has a larger round mode profile, nearly 80–90% of the launched power is lost. This laser beam profile is rather difficult to correct and often complex and costly microlenses are used to improve the coupling to an SMF. However, a circularly symmetric VCSEL beam has no astigmatism, unlike that of an edge-emitting laser. The surface-emitting aperture can be shaped to give the beam the ideal circular cross section and its diameter can be made large enough to minimize the divergence of the light rays highly compatible for coupling light into optical fibres with circular cross sections.

Most critically, VCSELs that emit light from their upper surface are suitable for integration into densely packed two-dimensional (2D) laser arrays and can be fabricated side-by-side on a wafer in vast numbers. However, in edge-emitting lasers, the light is emitted from the edge of the chip and this limits the integration of the lasers onto chip since they must be placed at the edges, which restricts them to one-dimensional (1D) arrays only. Surface-emitting lasers can form a 2D array whose power and direction can be carefully controlled, enabling new applications in optical computing and optical interconnections. Devices can be integrated monolithically and packing them so densely enhances the manufacturing efficiency and millions of these lasers can fit onto a 10 cm diameter wafer making them suitable for many applications, such as high-resolution miniature displays and 2D scanners.

In VCSEL production, the initial probe test can be performed automatically on the wafer before it is separated into chips for packaging, which reduces costs, as packaging is one of the principal costs in both edge and surface-emitting laser diode manufacture. For cleaved-edge emitting lasers, such an initial probe test is impossible before separation into chips and testing is done after packaging and the low yield of such devices significantly increases the overall cost of the few successful devices that are produced. By contrast, each VCSEL can also be conventionally screened at the wafer level, thanks to its built-in mirrors, before separation. Since wafer-level testing identifies defective devices before packaging, only those meeting the required performance standards are packaged and production costs are significantly reduced. Because packaging represents most of the cost of diode lasers, a reduction in the packaging cost can give VCSELs a real advantage in mass production of everyday consumer products such as compact-disc players, barcode scanners, laser pointers and laser printers.

Furthermore, since a VCSEL can have an ultrashort cavity length of the order of microns, it can lase in a single longitudinal mode. Edge-emitting lasers with relatively long cavity lengths (hundreds of laser wavelengths) have many cavity modes that overlap their gain bandwidth, although distributed feedback (DFB)-type edge emitters may have a smaller number of viable longitudinal modes. In contrast to this, dynamic single longitudinal mode operation is expected in VCSELs because the cavity mode spacing ($\Delta\lambda$ = 10–20 nm) is larger than the gain bandwidth of the active material and, thus, the device can be inherently single longitudinal mode. VCSELs can be produced by a fully monolithic process, such as the standard fabrication process used in silicon integrated-circuit (IC) fabrication. Although silicon IC fabrication uses a different processing technique, the concept is similar in that many devices may be fabricated in parallel. Even more important is that because of the way they are made, they can be monolithically integrated on a chip with transistors and other devices. There is no need to connect each of them individually to a circuit, as in edge emitters, and they can even be optically linked to overhead elements.

Today VCSELs are regarded as a key enabling technology for low-cost broadband optical interconnects, and they are finding a wide range of applications as attractive light sources for optical interconnections in highly parallel architectures, optical recording, image processing, optical pattern recognition, computing and free space optical communications. Optical interconnection is an important technology for both current and future computing systems and large switching systems. In recent years, VCSEL-based optical transmitters have been introduced into multi-gigabit sec^{-1} local area networks (LANs). Parallel links will penetrate more and more into areas such as intercabinet connections and even down to intraboard data communication links, which nowadays are mostly dominated by electrical interconnect solutions. As the clock speed of today's processor is increasing rapidly, the interconnections for a large cluster of workstations are becoming more difficult to achieve with conventional coaxial cable. Fibre optic transmission is an important alternative for such systems and is currently being implemented in very large systems. Using parallel fibre optic links, high-speed data transmission can be achieved with low electromagnetic-radiation-induced crosstalk, low skew and low noise. Future applications are likely to include, for example, not only LAN backbones but also links to remote peripherals, interconnections for workstation clusters, cabinet-to-cabinet links within supercomputers, backplanes for massive parallel processors, telecommunication switching and links from storage systems to computer hosts.

B2.6.3 Historical overview

Vertical-cavity lasers have existed both in concept and in prototype since their demonstration by Iga and co-workers at the Tokyo Institute of Technology since the late 1970s. The first surface-emitting semiconductor laser with a vertical cavity was reported in a paper in 1979 by Soda *et al* [1]. In it the authors describe an InGaAsP/InP laser with a cavity normal to the crystal growth surface using one mirror formed directly on the epitaxial crystal surface, and a second formed on the opposite side of the InP substrate. In that report, the cavity design was novel compared to established edge-emitting lasers and the design incorporated a 1.8 μm GaInAsP active region with a simple gold–zinc alloy mirror providing around 80% reflectivity. Such a low mirror reflectivity required a rather high (900 mA) threshold current (with an equivalent threshold current density of 44 kA cm^{-2}) under a pulsed condition at 77 K, emitting on a 1.2 μm wavelength.

The initial device performance was so poor in terms of the high threshold current density and low output power that it attracted very little attention. The Japanese group continued to work on the concept and further research work by Iga and his co-workers showed steady improvements in performance, eventually leading to the first room-temperature pulsed GaAlAs/GaAs VCSEL in 1984 [2]. They later produced the first VCSEL arrays [3], introduced an Au/SiO_2 mirror [4], a dielectric multi-layer reflector [5] and produced devices with reasonably low thresholds [6]. It was not until 1988, after continuous-wave (cw) low-threshold-current VCSELs were demonstrated at room temperature [7], [8], that practical applications were envisaged. Industrial adoption of this technology produced the first low-threshold room-temperature cw VCSEL [9] in 1989 and subsequently room temperature cw operation of 1.3 μm VCSELs was achieved [10] but under circumstances where the operating temperature range and output power were limited.

In parallel with the Japanese research at the Tokyo Institute of Technology, considerable research work was being carried out on both the concept and technology in many other places. The previous generation of VCSELs had very thick active layers, of the order of microns, which implied extremely high current densities. To make a low-threshold-current laser, the volume of the active region must be made very small: however, the corresponding mirror reflectivity must also be very high. From the early 1980s, several research groups, particularly at Bell Laboratories and at the Optical Sciences Center of the University of Arizona, had been working on the development of high-finesse cavity resonators for nonlinear optical applications. Around that time, other important innovations brought growing attention to the potential of the VCSEL. The use of epitaxially grown dielectric Bragg reflectors with high-finesse cavity mirrors allowed the creation of lasers which were thinner, as a small number of quantum wells (even as low as one) could be used in the active region, just as in edge-emitting lasers.

Probably the most important development following the early Japanese work was that of high-reflectivity semiconductor distributed Bragg reflectors (DBRs) [11], made possible with the advancing crystal growth techniques of molecular beam epitaxy (MBE) and metal organic chemical vapour deposition (MOCVD). Semiconductor DBRs were incorporated into optically pumped VCSELs by Gourley and co-workers at Sandia National Laboratories [12, 13]. Similar AlAs/GaAs DBRs grown by MBE played a critical role in the AT&T Bell Laboratories–Bellcore joint demonstration of the all-epitaxial, current injection VCSEL in 1989 [14, 15]. The achievement of submicron lateral dimensions [16] allowed device densities to be increased to more than 10^7 cm^{-2}. The addition of dopants and electrical contacts to this kind of structure produced more than one million low-threshold VCSELs on a single GaAs chip [14, 17–19].

One of the major innovations in VCSEL technology was the discovery of a stable oxide in the AlGaAs material system by Dallesasse *et al* [20] by selective wet oxidation of $Al_xGa_{1-x}As$ layers with a high AlAs content. This has enabled the fabrication of many high-performance devices [21–23]. Selective oxidation of $Al_xGa_{1-x}As$ is used to convert the thin layers to AlO_x. Owing to the low index of refraction of AlO_x ($n = 1.55$) and its high breakdown field ($E_{br} > 10^7$ V cm^{-1}), insulating AlO_x can be used to confine the current to a small active area while also providing index confinement within the VCSEL cavity. This technique allows very small microcavity devices with a very low (sub-microampere) threshold current to be realized by eliminating the high carrier loss that occurs due to defects created by proton implantation, while providing current confinement as well as some optical confinement for the lasing modes. This technique has resulted in the creation of devices with a threshold current in the tens of microampere regime. Since their introduction, oxide apertures have revolutionized the VCSEL field in a way unparallelled by any other single advance and they are being rapidly adopted by industry as they generally outperform implanted VCSELs.

Two recent review articles by Iga *et al* [24] and Towe *et al* [25] give excellent overviews of the historical development of VCSELs. There are also a number of books with excellent coverage of VCSELs and several books entirely on VCSELs [147–150].

B2.6.4 Basic Structure

There are basically two ways to achieve a surface-emitting lasers (SEL). The first approach is to form a normal in-plane laser and deflect the light upwards with a mirror or a grating. The radiation emitted by the horizontal cavity structure [26,27] with a higher-order grating shows extremely small divergence. An essential drawback of this structure however, is its low conversion efficiency. Another drawback with grating-coupled devices is that the beam angle is sensitive to any change of wavelength. Another proposed horizontal cavity structure possessing a 45° deflecting mirror [28–30] had difficulties in making a good quality flat reflector at an appropriate angle. Two turned-up cavity structures [31] suffer additional cavity losses and, thus, their threshold currents are difficult to reduce. Moreover, because of their relatively long resonators, none of these structures can, in principle, be used for constructing densely packed 2D arrays.

In an alternative approach pioneered by Iga and his colleagues, the cavity can be placed vertically and the mirrors made parallel to the wafer surface (the so-called vertical-cavity SEL or VCSEL). As their name indicates, VCSELs emit light perpendicular, rather than parallel, to the wafer surface. Thus, a fundamental difference between VCSELs and conventional edge-emitting lasers is the elimination of facet mirrors fabricated by either cleaving or dry etching. This simple change in the cavity orientation produces radical differences in the beam characteristics and the device scalability, optoelectronic design, fabrication and array configurability.

The basic structure could be considered as two Bragg reflectors separated by spacer layers, as shown in figure B2.6.1. In the centre of the cavity, an active region with one or more quantum wells is sandwiched between two highly reflective mirrors so that the laser oscillation and, thus, the output occurs normal to the wafer. Since the active regions are grown by an epitaxial technique, high-quality pn junctions are easily produced in the vertical direction, while they are rather more difficult to obtain horizontally, as for edge emitting lasers. The entire structure of the VCSEL—mirrors and all—is built layer by layer onto a substrate with atomic precision. By means of MBE and metal–organic vapour phase epitaxy, hundreds of layers of semiconductor materials can be grown on top of each other where, in essence, thin layers are used to confine electrons and holes, while thicker layers serve as mirrors.

In VCSELs, one period of the standing-wave field is of the order of 100–200 nm, so the incorporation of a thicker active layer would be partly wasted by having portions of it in regions of very low optical intensity. Thus, a better design would be to have an active-region thickness of less than one-quarter wavelength in

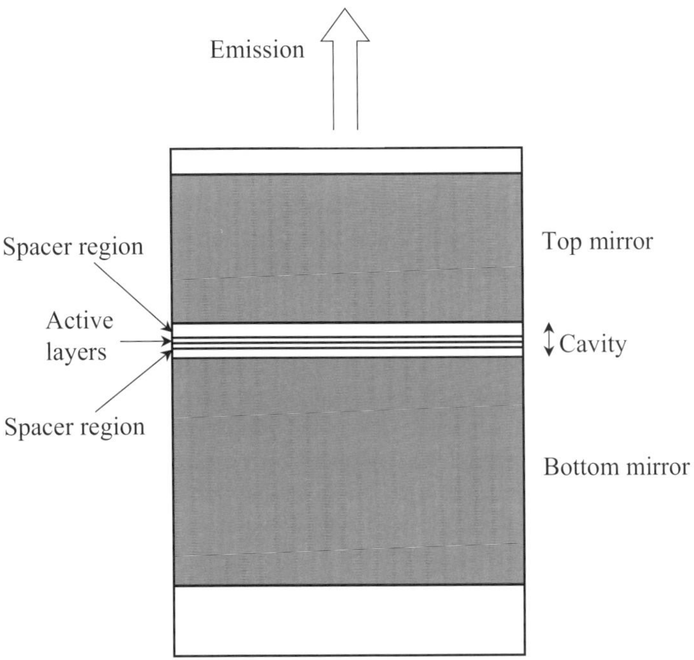

Figure B2.6.1. Schematic cross-sectional view of the VCSEL structure.

the material, with the peak of the electric field standing-wave pattern aligned with the gain region. Another key issue is to realize an effective current injection scheme and keep the carriers confined in the active region, by using an effective current-confining structure. For the purpose of effectively confining current in such an active region, some types of current-confining structures have been introduced, i.e. a round low mesa, round high mesa/polyimide buried and circular buried heterostructure [32]. By introducing a circular buried heterostructure (CBH), the threshold current was dramatically reduced and low-threshold-current room-temperature pulsed operation was obtained in a GaAlAs/GaAs system [33, 34]. On the top surface and the underside, metal is deposited to form electronic contacts. A circular aperture is left in one contact as an exit for the laser light. When a voltage is applied to the contacts, electrons and holes are injected through the opposing mirrors and collected in the active region of the laser.

It is recognized that a smaller active region is needed to reduce the threshold current. However, for a smaller volume, the optical gain per pass of such an active region could be of the order of 1%; hence, a very high mirror reflectivity, often greater than 99%, is also required for the device to reach threshold. The distributed Bragg reflector (DBR) mirrors in a VCSEL are grown into the laser structure itself. Each is composed of paired layers of two different materials, such as GaAs and AlAs: one material has a high index of refraction and the other a low index. Because the light waves from successive layers interfere constructively, together the pairs of layers can reflect more than 99% of the photons at the laser emission wavelength. The mirrors are known as quarter-wave stacks because each layer's thickness is equal to the quarter-wavelength of the optical wave inside the material. The index of refraction is the ratio of the speed of light in vacuum to the speed of light in the material. Therefore, the wavelength in the medium is shorter than in vacuum by a ratio equal to the refractive index. For example, for a lasing wavelength of 980 nm, as the refractive indices of GaAs and AlAs are 3.5 and 2.9 respectively, the respective layers need to be 70 and 80 nm thick to create an effective mirror.

There are several ways to achieve this high reflectivity including the use of metal mirrors [1], DBR mirrors [7, 9] and semiconductor DBR mirrors [14, 35–37]. These various techniques can be combined to obtain high reflectivity with a small number of dielectric or semiconductor layers by adding a metal layer on top of the DBR [14, 35–37].

Each quarter-wave stack, in fact, serves two functions: that of a highly reflecting mirror for the photons and as a current path for the electrons and the holes. Choosing materials for these two purposes is a balancing act. Material combinations with the largest possible difference in refractive indices make the best mirrors, since they give large reflections at each interface and so require fewer interfaces. For physical reasons, the indices of refraction and the bandgap energy are inversely correlated, so material pairs with large changes in index also show large changes in the bandgap energy. To achieve a specific output power, the external differential quantum efficiency must also be considered. As the mirror reflectivity is reduced, the output coupling from the cavity increases, resulting in a higher external quantum efficiency. However, reducing the reflectivity also increases the total loss in the cavity, tending to increase the threshold current. It may be noted that for the intra-cavity contact design, discussed later, its carriers bypass the DBR mirrors.

Semiconductor multi-layer reflectors have been successfully used in GaAs or GaInAs systems. However, long-wavelength VCSELs, unlike their short-wavelength counterparts, lack a good natural mirror. In order to improve the performance features of VCSELs, the development of high-reflectivity mirrors is one of the crucial issues for long-wavelength VCSELs. For long-wavelength lasers in lattice-matched GaInAsP/InP, a multi-layer reflector needs a large number of pairs because of its smaller refractive index difference. It may be difficult to form a vertical cavity using a GaInAsP/InP reflector on both sides because the total cavity thickness may exceed 10 μm. The InP-based Bragg mirrors often cannot reach a high enough reflectivity and the dielectric stack mirrors suffer from very poor thermal conductance and exhibit a rapid rise in lasing threshold with increasing temperature. Furthermore, the excessive losses due to inter-valence band absorption and non-radiative Auger recombination makes it hard for the device to reach the lasing threshold. To overcome these difficulties, researchers around the world are looking for new mirror technologies and more efficient laser gain media.

A hybrid mirror technology has been considered for long-wavelength VCSELs. Semiconductor/dielectric reflectors such as the Si/SiO_2 dielectric mirror, can be formed by chemical beam epitaxy (CBE), which provides a very high reflectivity even with a few pairs. However, the low thermal conductivity of SiO_2 may be a problem for cw operation. MgO/Si with its larger thermal conductivity would be a good choice for a binding side mirror and shows an improvement in the maximum cw operating temperature [38]. An alternative approach could be the epitaxial bonding of a quaternary/GaAs/AlAs mirror. An epitaxially bonded mirror made of GaAs/AlAs was introduced by Dudley *et al* [39] into SELs operating at 1.3 μm providing a 9 mA room-temperature pulsed threshold. Babic *et al* [40] have used bonding technology to fuse InP/InGaAsP to a GaAs/AlAs Bragg mirror to form 1.3 and 1.55 μm VCSELs. Because a GaAs/AlAs Bragg mirror can easily reach 99% reflectivity and have a superior thermal conductivity, the approach of using wafer-bonded hybrid mirrors solves the mirror problems for 1.3/1.55 μm VCSELs. Wafer-bonding techniques have also been considered by Liau *et al* [41] with one ZnSe/MgF dielectric mirror and another GaAs/AlAs mirror for InP/AlGaInAs cavity layers and by Qian *et al* [42] for a 1.3 μm VCSEL with two GaAs/AlAs Bragg mirrors. The high thermal conductivity of GaAs/AlGaAs DBRs has made wafer fusion a potentially attractive means of fabricating long-wavelength VCSELs that meet commercial performance requirements.

To make long-wavelength VCSELs with a performance comparable to that of the best edge-emitting lasers, a more effective gain medium other than bulk InGaAsP and regular quantum wells can be used. Because of the intrinsic material properties, more quantum wells are required for long-wavelength VCSELs, compared to their short-wavelength counterparts. Strained quantum wells have been well recognized as the most efficient gain medium for lasers because of the high optical gain and low transparency current. However, because of the critical thickness limitation, it is not possible to grow any desired number of strained quantum

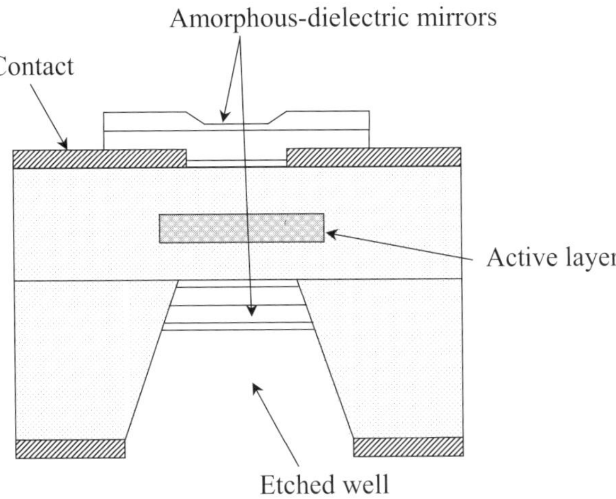

Figure B2.6.2. Etched-well VCSEL.

wells without compromising the material quality and device reliability. Miller *et al* at Cornell [43] used strain-compensated multiple quantum wells to overcome this problem for semiconductor lasers. Strain in a quantum well is cancelled by the opposite strain in its barrier. The alternating compressive and tensile strain between wells and barriers effectively offsets strain in any of the individual layers, thus allowing a large number of defect-free strained wells to be created.

B2.6.5 Device Geometries

There are several possible geometries for a VCSEL. Those which will be discussed below have demonstrated the necessary optical and/or electrical confinement. The choice may often depend on the desired wavelength, the material system used, the transparencies of the various layers, the fabrication of suitable reflectors and the conductivity of the reflectors. The preferred geometry may also depend on the device application.

B2.6.5.1 Etched-well devices

The first structure, as shown in figure B2.6.2, depicts an etched-mesa structure, where a 2D round or square post-like mesa is formed by stop etching, just above the active layer. Here, the current is confined in the lateral dimensions of the mesa but the carrier is free to diffuse in the active region. The etched-mesa structure also provides a lateral index of refraction step over the upper etched portion of the cavity. However, because of the larger refractive index, these devices will support many higher-order transverse modes unless submicron mesa dimensions are used. However, since the etched surface is not perfect, the higher-order modes will suffer higher scattering loss than the fundamental mode. A problem with the simple etched-mesa structure, as well as with most other VCSELs that employ conduction through the mirrors, is the series voltage that develops across the mirrors due to the potential barriers at the numerous hetero-interfaces. This is the simplest method of device fabrication, although non-radiative recombination at the outer wall may deteriorate performance. An alternative could be to avoid conduction through the mirrors, as described later.

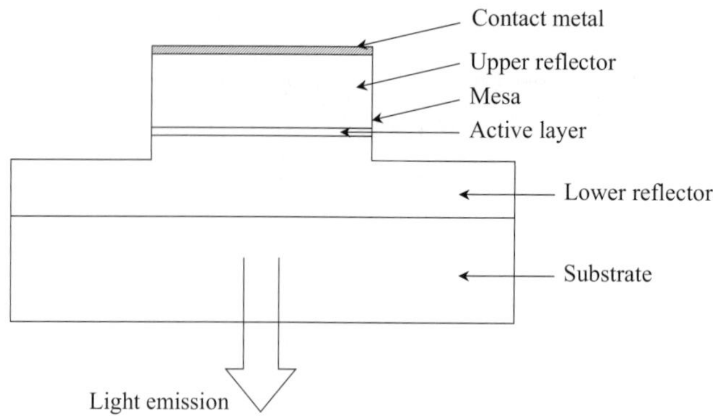

Figure B2.6.3. Bottom-emitting VCSEL.

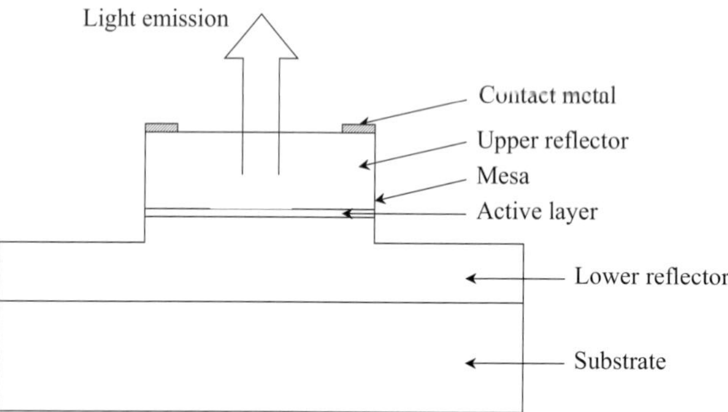

Figure B2.6.4. Top-emitting VCSEL.

B2.6.5.2 Top-and bottom-emitting mesas

VCSELs may also be either top- or bottom-emitting types. If the substrate is transparent, a bottom-emitting geometry is possible and the upper mirror reflectivity may be augmented by that of the contact metal. In the top-emitting device, the current must either be injected through a windowed contact such as an annulus or a grid pattern or through a transparent contact such as a very thin metallic film or indium tin oxide [44]. The top-emitting geometry may be desirable for arrays of devices to be mounted on conventional headers, such as those used for VLSI chips. The bottom-emitting geometry is useful for solder bump mounting of arrays onto driving circuitry or to provide intimate thermal contact to a heat sink (e.g. patterned diamond) without substrate thinning. The bottom-emitting devices have two minor drawbacks: first, a windowed lower electrode is required for light emission and, second, the reflections from the uncoated substrate–air interface can cause modal instability in the laser. Figures B2.6.3 and B2.6.4 show schematic diagrams of bottom- and top- emitting VCSELs.

B2.6.5.3 Proton/ion Implanted Devices

Implantation of ions into the top DBR mirror can be used to render the material around the laser cavity non-conductive and, thus, to concentrate the injected current into the active medium. Various ion species have been employed (H^+, O^+, N^+, F^+), although proton implants are the most common. These high-energy

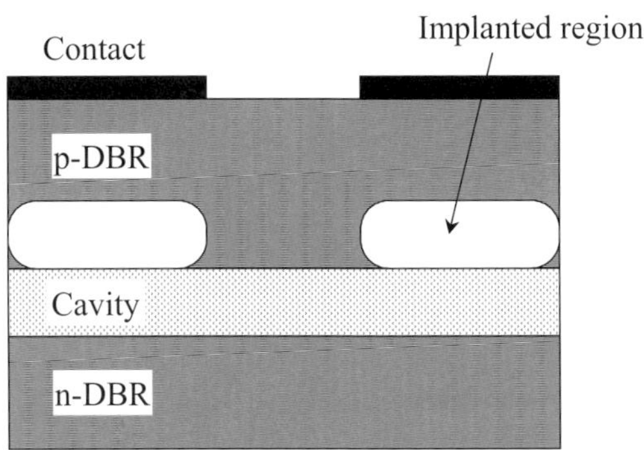

Figure B2.6.5. Ion-implanted VCSEL.

species can be used to produce selectively a buried current-blocking layer to funnel current through a small area of the active layer (figure B2.6.5). An insulating layer created by proton (H^+) irradiation limits the current spreading toward the surrounding area. Lateral carrier confinement is achieved by means of proton implantation with varying ion energies. The implantation energy and dosage are carefully chosen to minimize the damage in the active layer, while achieving efficient current confinement. Proton-implanted small-area VCSELs can operate in single transverse mode over the whole bias area. This structure is very simple to manufacture and early commercialized devices were made by this method. Therefore, it has been suggested by some that it is one of the most important VCSEL structures. However, so far Honeywell is the only major producer to have continued to pursue this type of device as the lack of in-built index contrast (for DBR mirrors) limits the dynamic performance and their applications have been mainly restricted to very slow data rate applications.

As with both top- and bottom- emitting mesas, ion implantation is quite appropriate for devices with epitaxially grown upper and lower reflectors and is applicable to both top- and bottom- emitting devices. Neighbouring devices may be isolated by mesa etching or by further implantation. Top-emitting devices may be produced in which the patterned upper electrode does not obscure the light output at all. The surrounding material acts as passivation and improves the thermal and ohmic heating properties of the devices. It is generally desirable to passivate optoelectronic devices. Mesa etched devices of both geometries may be suitably passivated with a number of materials e.g. polyimide, oxides, nitrides or by semiconductor overgrowth. Large contact pads or power traces may be applied over arrays of small passivated devices.

B2.6.5.4 Oxide confined

VCSELs have been the topic of intense research for several years and, although different schemes can be used to introduce a dielectric aperture into the planar microcavity, the recent VCSEL demonstrations using the selective oxidation technique have been the most impressive. Huffaker and co-workers [21] have reported the first native-oxide-confined VCSELs and demonstrated low room-temperature cw threshold currents of 220 μA in an 8 μm square device. An oxide aperture acts as a current window, as shown in figure B2.6.6, enabling the active region to be pumped more effectively. Apart from this electrical property, oxide apertures can also have beneficial optical effects, counteracting diffraction and improving the modal stability of oxide-confined VCSELs. For well-designed oxide-confined VCSELs, the optical loss can be considered to arise from two mechanisms: transmission loss through the DBRs due to normal propagation of the field to the cavity and diffraction loss in the DBR due to lateral propagation of the field within the cavity. Recent characterization data suggest that the oxide aperture effectively controls the diffraction losses for lateral device sizes >4 μm.

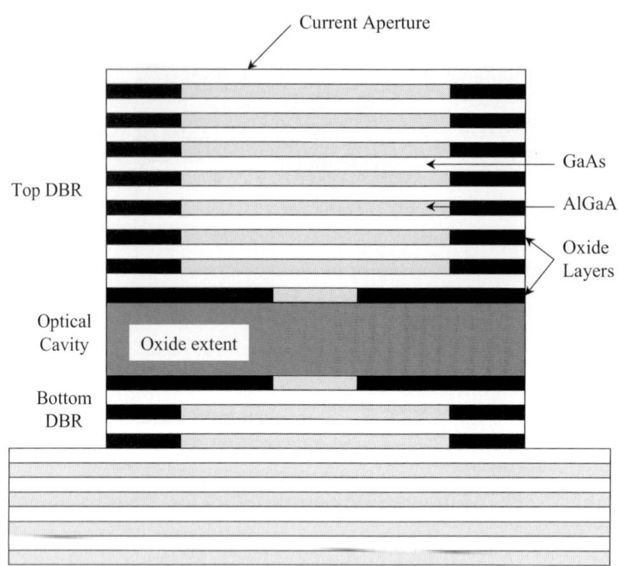

Figure B2.6.6. Oxide-confined VCSEL.

Oxide-confined VCSELs, with extremely low threshold currents e.g. below 100 μA have been demonstrated by several groups [45–50]. An extremely low threshold current of 8.7 μA for a single QW VCSEL [51] and a threshold current density (90 A cm^{-2} per quantum well) [52] have been reported. Choquette and co-workers have also reported a record low threshold voltage, at 1.33 V, only 50 mV above the photon energy [53] and also very high fabrication yield devices [54] grown by MOCVD. In terms of achieving high differential and power conversion efficiencies, impressive results have been reported by workers both at Sandia Laboratories and the University of Southern California [48, 55, 56]. Lear and co-workers demonstrated a record power conversion efficiency of more than 50% in a similar all-epitaxial structure [55, 57] and Yang *et al* [56] have shown 55% slope efficiency. Modulation bandwidths in excess of 15 GHz have also been reported [58].

These milestones in VCSEL development, together with the inherent possibility of wafer-scale testing, easy packaging and very good output beam quality make VCSELs the ideal light source for short-haul optical communications and for optical interconnects. In the few years since its first demonstration, the selectively oxidized VCSEL has become a major research topic in the area of coherent light emitters. Yet despite its vast popularity and impressive performance record, the physics of oxide apertures is still not completely understood and the stability and reproducibility issues of the oxides as a manufacturing process have not yet been fully determined.

B2.6.5.5 Intra-cavity contacted devices

Epitaxially grown reflectors can present a large resistance to the flow of current and limit the laser output by causing the device temperature to rise. The emphasis in the design of VCSELs of this size is in minimizing electrical series resistance through p-type semiconductor DBRs and optimizing the transmission through the output coupling efficiency. Choquette *et al* [53] have pointed out that the electrical series resistance in the oxide-confined VCSEL benefits from constricting the current flow only just above the QW active region, so that an extremely low threshold voltage is achieved. By employing a two-step mesa etch (figure B2.6.7), the current may be injected so as to avoid the upper reflector [59], which may be an oxide multi-layer or

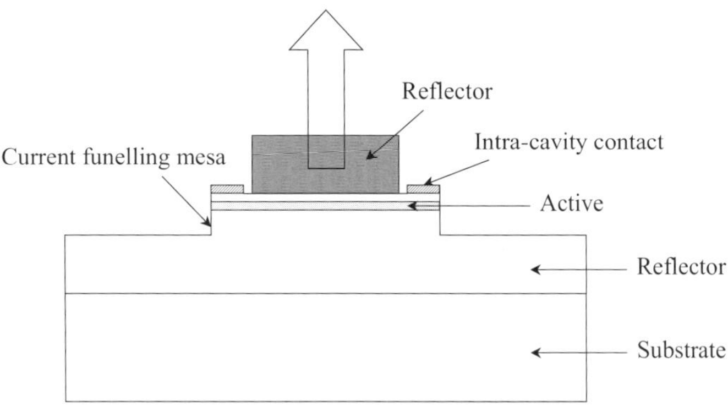

Figure B2.6.7. Intra-cavity VCSEL.

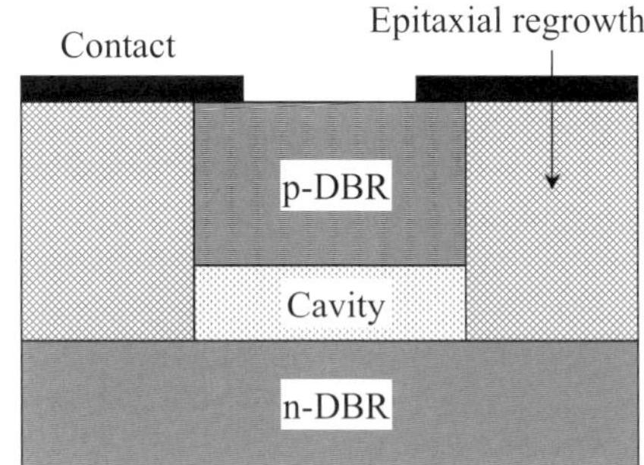

Figure B2.6.8. Buried-heterostructure VCSEL.

an oxide–semiconductor hybrid. Jewell and his colleagues have demonstrated a low-voltage vertical cavity method that allows the carrier to bypass the mirrors altogether, which is similar to one pioneered by Iga and his colleagues at the Tokyo Institute of Technology for indium-phosphide-based VCSELs. They injected current into the active region from the sides of the device, underneath mirrors that can even be made from insulating materials and that can be applied like coatings instead of being grown epitaxially. Although such an approach offers flexibility, it does not yet offer the same performance as devices based on epitaxially layered conducting mirrors. Poor current spreading from this type of contact may give rise to higher threshold currents and favour higher-order transverse optical modes.

B2.6.5.6 Buried heterostructure (BH) Devices

To achieve index-guiding in a planar VCSEL topology requires some manner of lateral variation of the refractive index around the laser. The refractive index difference can be small in the surrounding region, resulting in the formation of an index-guiding structure. The mesa, including the active region, is buried with a wide-gap semiconductor to limit the current. This is an ideal structure in terms of both current and optical confinement.

One method for fabricating such devices is to use an etch/regrowth process to deposit a semiconductor with a different composition (and, thus, refractive index) around the laser cavity, as shown in figure B2.6.8. This can be accomplished by depositing a robust etch mark, etching a pillar and then regrowing new material around the etched cavity. Regrowth VCSELs have the advantages of engineering index-guiding, transverse mode control, current confinement, passivation of the active region side walls and restored heat sinking. In addition, the epitaxially regrown semiconductor material could allow monolithic integration of other optoelectronic devices or microelectronic devices. The disadvantages are associated with the challenges, complexity and expense involved in the epitaxial regrowth and, in particular, in making 3D devices.

B2.6.6 Material systems

There are a wide range of III–V alloys that may be grown on GaAs or InP substrates. One of the basic requirements for alloy semiconductors is that they must be reasonably closely lattice matched to the readily available binary alloy substrates, GaAs or InP.

Various material systems may be considered, their choice primarily depending on the operating wavelength of interest, where III–V materials are ideal for complex VCSEL devices. The crystal lattice of AlGaAs matches the spacing of the lattice of GaAs substrates, so it can be grown epitaxially on GaAs wafers without fear of crystal defects and dislocations. AlGaAs lasers are attractive for optical discs, optical sensing and optical parallel processing. The ternary AlGaAs system matches the lattice closely and has a large bandgap difference at the two end-point binaries AlAs ($EgX = 2.16$ eV at 1.0 μm) and GaAs ($EgX = 1.42$ eV at 1.0 μm). $Al_xGa_{1-x}As$ has a direct bandgap for $x < 0.42$ which varies from 1.42 to 1.9 eV. To date, AlGaAs has been the most technologically relevant material system for VCSELs, enabling operation in the range 780 to 980 nm but it is commonly used in the range of 800–880 nm. Emission at these wavelengths is of interest for pumping erbium fibre lasers and other solid state lasers. There are also applications for short-haul communications. GaAs substrates are not transparent to the output light so a top-emission geometry is required. There have been some reports, however, of lasers operating at wavelengths as short as 670 nm [151]. This limitation is due to the high reactivity of the aluminium atom. The sources used in epitaxy (metallic Al in MSE or trimythaylaluminium (TMA) in MOVPE or metal organic molecular beam epitaxy (MOMBE)) are easily contaminated with OH groups or atomic oxygen. The material grown is often contaminated by O or C, which seriously degrades the non-radiative carrier lifetime in optoelectronic devices containing more than a few percent of aluminium [60]. However, by employing ultra-pure source materials for all the precursors, MOVPE growth of high-purity AlGaAs could also be possible.

It is advisable for a long-haul optical fibre communication system to use a GaInAsP/InP VCSEL single-mode laser emitting at the wavelengths of 1.3 and 1.55 μm which correspond, respectively, to zero dispersion and minimum absorption in mono-mode silica fibres. In fact, the first ever demonstrated VCSEL in 1979 [1] was created by using the GaInAsP/InP system. $Ga_xIn_{1-x}As_yP_{1-y}$ with $x = 0.47y$ is lattice matched to InP substrates. By varying the composition, both tensile and compressive strained quantum wells may be produced. Only the latter, however, provides any advantage for VCSELs. However, the GaInAsP/InP system has some substantial difficulties for making SELs for the following reasons: (a) Auger recombination and inter-valence band absorption (IVBA) are noticeable; (b) the index difference between GaInAsP and InP is relatively small; and (c) the valence band offset is large.

For many years it was believed that InGaAsP on InP was the only material system that met lattice-matching criteria for long-wavelength lasers. While InGaAsP met the needs of edge-emitting lasers, it is almost impossible to meet the requirements for growth of DBR quarter-wave VCSEL mirrors. This is because the refractive index contrast of InP/InGaAsP is insufficient and the thermal and electrical conductivities are too low to realize the required combination of high reflectivity and low thermal and electrical resistances. An alternative to InGaAsP is AlInGaAs lattice-matched to InP and active regions containing AlInGaAs may offer improved high-temperature performance compared to InGaAsP. The possibility of ternary (InGaAs) substrates [61] will also improve the choice of materials available for long-wavelength VCSELs but the inevitable higher cost of such substrates may only be justified by meeting the potential market.

Visible SELs are extremely important for display applications. In particular, red, green and blue surface emitters may provide much wider technical applications: in that respect GaInAlP/AlGaAs SELs have been developed and room-temperature operation exhibiting sub-milliampere threshold and a few mW output have been reported [62]. Schneider *et al* have used strained $Ga_{0.54}In_{0.46}P$ quantum wells and a $(Al_{0.5}Ga_{0.5})_{0.5}In_{0.5}P$ barrier to obtain 780 nm emission and they have also reported 650 nm emission [63]. Recently Knigge *et al* [64] reported a cw room-temperature emission of 10 mW power at 670 nm for an oxide-confined VCSEL.

Generally, the light-emitting device in the short-wavelength region is more difficult to operate than in longer-wavelength regions, since the photon energy is large and achieving p-type doping is technically harder to perform. If aluminium is included in the system, the degradation due to Al oxidation is appreciable. The II–VI compound ZnSe has a close lattice match to GaAs and the addition of a small amount of sulphur means a perfect match can be obtained. The ZnSe system is still the only material to provide cw operation of green–blue semiconductor lasers. However, these materials have been shown to have a very poor lifetime. In the quaternary MgZnSSe system a range of direct bandgaps, 2.9–4.3 eV is possible, all lattice-matched to GaAs it but is more limited in its range of bandgap. Laser diodes produced in these systems have been able to produce emission in the range 470 (blue) to 525 (green) nm. The cw lifetime at room temperature of laser diodes is rather limited; the failure mechanism appears to be due to the propagation of dark line defects.

It has been shown recently that the huge band bowing in the GaAsN system, arising from the addition of a small percentage of N, leads to a strong suppression of the GaAs bandgap. This allows the incorporation of AlGaAs-based DBRs and for them to be integrated with the 1300 nm active regions in a single epitaxial run, which has a distinct advantage over the wafer-fusion processes. GaAs-based dilute nitride VCSELs have shown excellent light quality at 1300 nm [152], which is, perhaps, the most important wavelength for VCSEL devices. The wide bandgap III–V materials based on GaN and related materials is currently arousing much interest and may cover a wide spectral range from green to UV. GaN are typically grown on GaAs or sapphire but often a strain relief layer is needed, as GaN does not form a close lattice match to these common substrates. Nevertheless, the reported reliability of GaN- based LEDs grown on GaAs substrates appears to indicate a good material potential for lasers as well. The InGaAlN system covers both the full visible spectrum and also extends considerably into the UV region. LEDs have been produced to operate in the range 400–500 nm [65]. Sandia National Laboratory has reported that they are developing InGaN/GaN multiple quantum well VCSELs emitting up to 3 mW at 383 nm when optically pumped with a 355 nm Nd:YAG laser source. The VCSEL reported to operate at room temperature in a quasi-cw mode was grown using MOVPE on a sapphire substrate with prospective applications in chemical and biochemical sensing. Another interesting system is GaInNAs, lattice-matched to GaAs and this is expected to be a new material [66], for long-wavelength VCSELs.

The energy-band diagram and lattice constants are shown in figure B2.6.9 and the most commonly used material systems for VCSEL is shown in figure B2.6.10.

B2.6.7 Mirror design

To reduce the threshold current of a VCSEL, it is necessary to have mirrors with very high reflectivities at each of the two ends. Generally, DBRs provide the necessary reflectivity for the high-finesse cavity of a low-threshold VCSEL. The reflectance spectrum of a simple metal mirror shows a reasonably high reflectivity, around 97–98%, over a very broad band. However, DBR mirrors can exhibit even higher reflectivity, which can reach a value of more than 99.5%, but over a narrow band. For the design of VCSEL devices, it is necessary to obtain the reflection spectra, the Bragg wavelength and the maximum reflectivity at this Bragg wavelength and the corresponding bandwidth.

The reflection coefficient of transverse electromagnetic (TEM) waves impinging at normal incidence onto planar reflector stacks can be calculated using a simple analytical approach. In a VCSEL, the lasing mode propagates in the vertical direction so the other angle of incidence may not need to be considered.

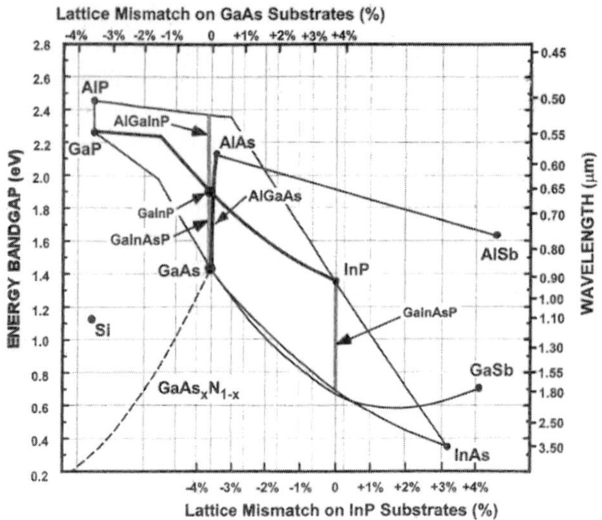

Figure B2.6.9. Dependence of bandgap energy and emission wavelength on lattice mismatch relative to GaAs and InP substrates [67].

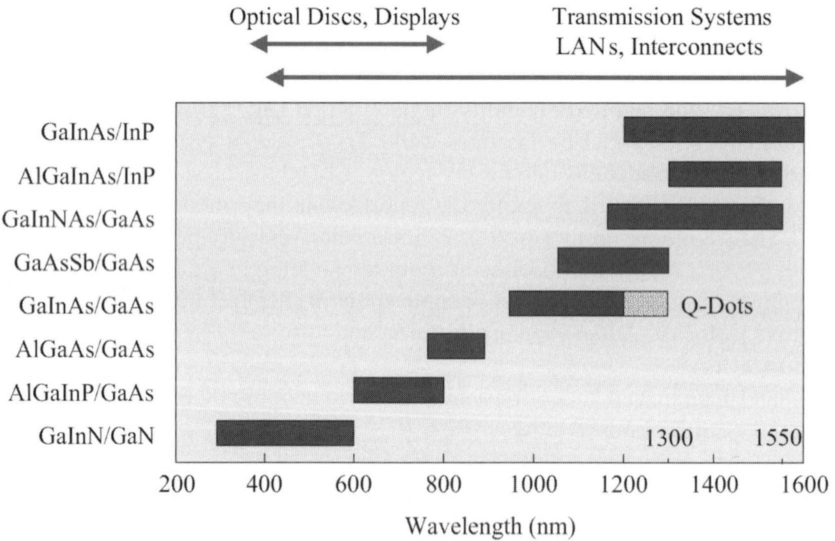

Figure B2.6.10. Materials used for VCSELs [24].

Devices are of finite dimension so the transverse modes are never truly TEM. However, in virtually all of the electrically pumped devices, the lateral dimension is many times the wavelength in the material, so the transverse component of the wavevector will be very small for the fundamental mode and, in an approximation, the TEM approach may be satisfactory; however, for mesa diameters below 5 μm, this approach ceases to be appropriate. The reflection coefficient of a plane wave incident normally on a dielectric interface, as shown in figure B2.6.11, could be calculated after matching the continuity of the transverse electric and magnetic

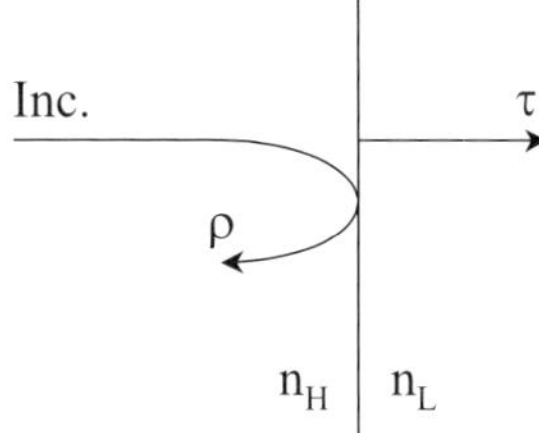

Figure B2.6.11. Reflection at a single interface.

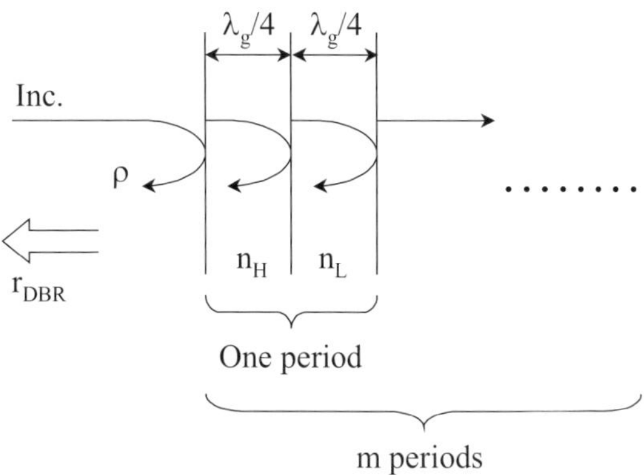

Figure B2.6.12. Multi-layer Bragg reflector.

field components, where

$$r = (n_H - n_L)/(n_H + n_L). \tag{B2.6.1}$$

In order to calculate the field reflectivity, r_{DBR}, of a multi-layer dielectric stack at an arbitrary wavelength, as shown in figure B2.6.12, it is necessary to resort to numerical techniques [68]. However, for the case where the interfaces are abrupt and each layer is exactly a quarter wavelength thick (only possible at a fixed wavelength, known as the Bragg wavelength), it is possible to formulate the results analytically.

$$R_{DBR} = |r_{DBR}|^2 = ((1 - (n_H/n_L)^{2m}/(1 + (n_H/n_L)^{2m})^2 \tag{B2.6.2}$$

where R_{DBR} is the reflectance (intensity) for m number of periods or $2m$ layers. The surprisingly simple formula for r_{DBR} comes about because the reflection from each interface adds exactly in phase with that from every other interface. When the total number of layers in each DBR mirror is odd, the overall reflection coefficients can also be obtained by using similar equations.

However, these simple equations give reflection coefficients only at the Bragg wavelength. The wavelength-dependent reflection spectrum of a multi-layer dielectric stack may be found in several ways. One is using the coupled mode theory (CMT) discussed by Kogelnik and Shanks [69] which considers the coupling per unit length between the forward and reverse travelling waves in a perturbed waveguide by introducing a coupling coefficient $K(\lambda)$. A Fourier analysis of the refractive index variation with depth can find the coupling coefficient, K. For a periodic guide with length L, the maximum reflectivity at the Bragg wavelength may be given by

$$R_{DBR} = \tanh^2(KL) \tag{B2.6.3}$$

Besides the maximum reflectance, the variation of the reflectance over the wavelength range can also be obtained by using the CMT approach. The CMT is generally valid for weakly coupled Bragg gratings and has been proved to be very effective for the designs of Bragg gratings in silica fibres and distributed Bragg gratings for semiconductor lasers. However, in the layered mirror structure of a VCSEL, the index perturbation is significantly large and CMT may not be adequate for many design applications. An alternative is the transfer matrix method TMM [70], which has been extensively used in the design of strong Bragg gratings [71]. This approach is particularly versatile being a discrete method and is, therefore, applicable to numerical solutions for arbitrary finite structures.

As mentioned, the guided modes in index- and gain-guided VCSELs are never exactly TEM but have often been considered so. A number of authors have observed the reduced mirror reflectivity as seen by the guided modes due to diffraction effects. In calculating the reflectivity spectra and the bandwidth, the material dispersion also needs to be included. If the period of the stack is chirped with depth then the bandwidth can effectively broaden, the side lobe can also be reduced, but usually to the detriment of the peak reflectivity. It has also been observed that a DBR with a smaller transverse shape shows a distinctive blue shift in the resonating wavelength. More rigorous numerical approaches to characterize DBR mirrors, including the use of the matrix method, will be described later, in section B2.6.9.

B2.6.8 Thermal Problems

Thermal problems have played a significant role since the very beginnings of semiconductor laser history and they continue to play an even more important role in the operation of VCSEL devices and arrays. This further complication is due to the fact that VCSEL devices usually have a higher threshold current density than the normal edge-emitting lasers because of their short cavity length and, hence, this generates a greater problem with laser heating. Most of the VCSEL designs that have been reported consist of an active region sandwiched between two DBR mirrors. The device is biased through two electrodes attached to the outside of these two DBR mirrors where the current is injected into the active region through the DBR mirrors and resistive heating in the mirrors thus increases the temperature of the device. The DBRs comprise the bulk of the current VCSEL designs and must be able to dissipate heat from the resistive and the lasing process effectively.

The thermal power densities generated inside the active region are extremely high. This leads to a substantial increase in temperature and a corresponding increase in the threshold current density. It has been observed that diodes of these types have operated in cw mode at up to an ambient temperature of 120 °C [72]. There have also been reports of operation as high as 210 °C [153], indicating a considerable temperature rise over room temperature. As a result of the increase in device temperature, the parameters of the radiation beam, such as the radiation power, the near-field and far-field patterns, the spectral characteristics, the current–voltage characteristics and the output-current characteristics are strongly affected. Heating is clearly manifested in the sharp rollover which eventually extinguishes the output characteristics and it is certainly the most performance-limiting factor at present. In the case of gain-guided structures, which avoid a strong index-guided pillar, the phenomena of thermal lensing and spatial hole burning can have important effects on the transverse mode profile due to the lack of a strong index discontinuity. It has been shown that for a gain-guided laser, both the near- and far-field intensities would be altered significantly with a quasi-uniform temperature rise as small as 10 °C.

The alternating high and low refractive indices needed in the Bragg reflectors also introduce corresponding high and low potential energies in the conduction bands. When the adjacent materials have very different bandgap energies, the carriers (electrons and holes) have to surmount these bands as they move from one mirror layer to the next. The impediment to carrier flow is evidenced in high resistances, with increased voltage drop and lowered power conversion efficiency and the generation of unwanted heat. Using graded layers at the interface between high and low index layers (graded interface) can reduce the electrical resistance [35]. The resistance of the doped DBRs has been greatly reduced from several tens of kilo-ohms to a few tens of ohms by the use of stepped or graded interfaces [35, 73, 74].

Thus, while resistive heating can be reduced by introducing parabolic-gap grading, the tailoring of the bandgap and doping profiles can decrease the quantum efficiency due to increased free-carrier absorption and a much increased thermal resistance, due to the presence of ternary alloys. Alternatively, the charge carriers can bypass the mirror altogether by injecting current into the active region from the sides of the device [75]. Furthermore, resistive joule heating sources such as non-radiative recombination and reabsorption of spontaneous emission add, combining to produce a destructive effect on the performance of the laser device.

Heat generated in the DBR region must be dissipated effectively to reduce the temperature rise in the VCSEL. In the design of Bragg mirrors, the thermal properties of the DBR layers are also very important parameters. Dielectric mirrors typically exhibit low thermal conductivity. Ternary and quaternary alloys also have substantially lower thermal conductivities than binary mirrors, due to alloy scattering.

Significant heat generation and the associated large temperature rise limit the number of devices that can be integrated onto a chip. Extremely high thermal power densities inside the cavity [76] present an obstacle to the development of a high-density solitary laser and a densely packed 2D array. Great efforts have been made to reduce heating so that the operating temperature could be increased. Thermal dissipation for cw operation is very important, especially in the long wavelength region. Efficient heat dissipation, along with an ultralow threshold, is, therefore, critical to applications requiring massive integration. By improving the heat sinking using a diamond submount and a highly reflective mirror, Dudley *et al* [39] have achieved room temperature cw operation for a wafer-fused VCSEL.

B2.6.9 Modelling VCSELs

In a relatively large device, the threshold current is proportional to the volume of the electrically pumped active region. Since the longitudinal dimension of this volume is fixed by the thickness of the active quantum wells, attention has focused on transverse current confinement. Both etched-post and oxide-apertured structures have been introduced in an effort to restrict the current path and the threshold currents have declined accordingly. It is important to realize, however, that these lateral patterning techniques will also affect the transverse optical confinement. As the transverse dimensions shrink to the order of the lasing wavelength, the spatial profile of each cavity mode changes from a plane wave to a three-dimensionally confined 'waveguide-like' mode which also alters the modal reflectivity.

In order to understand the input–output characteristics of VCSELs and optimize such devices, analytical and numerical modelling of these structures including such features as the self-consistent transport of charge carriers and heat and light generation are essential. Some of the simple models developed earlier have helped research workers to understand the fundamentals of many device parameters. However, as the device designs become more complex, the design and analysis of VCSELs presently depends heavily on the use of numerical simulation, due to the high cost of empirical optimization and the continual increase in the complexity of efficient devices.

To support further VCSEL development, modelling efforts are needed to understand and optimize several key parameters, such as single transverse mode and single polarization operation, low operating voltage, higher output power and a higher modulation bandwidth. An important way to tackle the design and characterization problem is by modelling, and this provides an effective tool for better designs and thus reduced development time and cost. Modelling VCSELs can be based on three sub-models: optical, electronic and thermal.

B2.6.9.1 Optical modelling

In the complete process of design, analysis and optimization of VCSELs, the optical analysis aspect is the most difficult task, when compared to the electronic modelling of carrier and gain profiles or thermal modelling. Because VCSELs are generally designed as open resonators where only a part of the cavity confines a mode (the other part is left unguided), it is very difficult to determine the electromagnetic eigenmodes of such structures. Although the axial length of the VCSEL is much shorter than that of an edge-emitting laser, due to structural complexity, a full 3D optical model is often necessary. At the heart of such simulations is the computation of the shape of the optical field, which in a 3D structure is vectorial in nature, together with its complex refractive index profile. The calculation of the field profile, the resonating wavelength and the modal gain or loss have proven particularly difficult when compared to their counterparts for edge-emitting lasers, because of the distributed nature of the Bragg mirrors and the high Q of the resulting cavity.

Modes in 3D optical waveguides are not truly TE or TM types but hybrid in nature and more particularly so in semiconductor waveguides with a high index contrast. Finding rigorous solutions to Maxwell's equations beyond the scalar approximation is a non-trivial problem. The steady-state time harmonics satisfy the 3D vector Helmholtz equation. This equation can be written either in terms of the electric, $\boldsymbol{E}$, or magnetic, $\boldsymbol{H}$, field. However, for use with various numerical approaches, the $\boldsymbol{H}$-field-based formulations [77] are more suitable for optical guided-wave devices as all three components of the magnetic field are naturally continuous across the dielectric interfaces. The vectorial Helmholtz equation in terms of the magnetic field, $\boldsymbol{H}$, in three dimensions can be written as follows:

$$\nabla^2 \boldsymbol{H} + k^2 \boldsymbol{H} = 0 \tag{B2.6.4}$$

where k is the wavenumber, ($k = \omega/c$), where for the time-harmonic wave a temporal dependence of $\exp(j\omega t)$ has been assumed. To find the electromagnetic field profile and the resonating wavelength for each cavity mode, the vector wave equation must be solved along with the appropriate boundary conditions. However, solving this equation for a truly 3D structure with arbitrary shape and index distribution with material gain and loss is not easy. Strictly speaking, a practical VCSEL geometry is not separable in any coordinate system, so the optical field cannot be found in a closed form by using any of the available analytical approaches and, hence, a numerical approach is essential. The effort to develop a robust VCSEL optical model has been ongoing for several years with, as yet, no universally embraced result.

As a fully 3D optical model is always computationally expensive, irrespective of the numerical approach used, there have been various approaches to simplifying the problem by assuming various approximations. The first is the use of the effective index method [78] that seeks to reduce the dimensionality of the problem and this has proven invaluable in increasing the understanding of waveguiding in VCSELs. It is accurate for certain geometries; however, it fails badly for other geometries, most of which are of current interest, including the oxide-confined devices that have recently set a record for both low threshold and high efficiency.

If the structure possesses cylindrical symmetry, such as the circular mesa, another simplification could be to exploit this symmetry ($\partial/\partial\phi = 0$), and then the wave equation can be written in terms of r and z only. The Helmholz equation in cylindrical (r, z) coordinates, as given here, can be solved for the optical mode:

$$\nabla^2 \boldsymbol{E}(r, z) + n(r, z)^2 k_0^2 \boldsymbol{E}(r, z) = 0 \tag{B2.6.5}$$

where $\boldsymbol{E}(r, z)$ is the space-dependent electric field, $n(r, z)$ the space-dependent refractive index and k_0 the wavevector in vacuum. In VCSEL modelling, so far this is the most widely used approximation [80, 81]. Besides the cylindrical symmetry, some authors have also considered a given higher-order cylindrical symmetry by considering the $\exp(jm\phi)$ dependence [79, 82]. For many VCSELs, the air post may not be designed deliberately to have the exact cylindrical symmetry to achieve the polarization selection. Even when a cylindrical post is used, most of the remaining structures as well as the post do not have cylindrical-symmetry. Again for a VCSEL array with many elements, cylindrical symmetry may not be available. In fact, in most cases, this cylindrical-symmetry approximation is not valid. Even for a structure with cylindrical symmetry, the modes may not obey the symmetry condition due to non-symmetrical current injection or the material nonlinearity. Hence, the imposition of this symmetry could restrict the build-up of the optical wave with all possible spatial distributions, such as for multi-moded structures, for high-power operation or during the transient phase. However, when such symmetry is available, even as an approximation, it is worthwhile exploiting it unless the stability of the modes is being studied.

Although a full vectorial solution of the optical modes is more desirable, one approach to simplifying the problem could be to solve the scalar Helmholtz equation for the cavity modes. Numerical methods based on the scalar wave equations have been presented in the literature [79, 83], but in this case the polarization issues were left untreated. For structures with dimensions of the order of the wavelength and where strong index contrasts are present (e.g. oxide-apertured devices), a vectorial approach seems to be more appropriate, especially for studying the polarization properties of higher-order transverse modes.

The common approach so far for rigorous modelling has been the use of beam propagation methods (BPMs) to find the longitudinal and lateral mode profiles. These BPM approaches could be based on the Fourier technique [84], as originally developed; however, at present the finite-difference-based BPM is more widely used as it is more flexible. The BPM approach can also be based on the finite element method (FEM) [85], which has been developed more recently, as it is numerically more efficient than the finite difference approach because the transverse cross section can be more accurately and efficiently represented. As with many other numerical approaches, the BPM approach could be either scalar or vectorial. Previously reported BPM formulations were mostly scalar [80]; however, semi-vectorial [86] or full-vectorial [85] BPM approaches which could be used to obtain the optical modes have also been reported. Recently, Rao *et al* [87] reported a blue-shift phenomenon in small oxidized VCSELs using a bi-directional BPM.

In general, the computation of the stationary (cw) characteristics of a laser using the BPM involves considering the iterative, back-and-forth propagation fields inside the cavity until the steady state is reached. The iteration procedure is started with an arbitrary field profile. It is solved by using the finite differences approach in the transverse plane and a Crank–Nicolson scheme for the longitudinal (z) coordinate. At each mirror, the forward-and-backward propagation waves are connected by the boundary conditions. The field is propagated until the desired convergence of the field profile variation is achieved after several round-trip propagations in the cavity. The fundamental and higher-order modes can be separated during the calculations because the fundamental mode has an even symmetry and the first higher-order mode has an odd symmetry. There are various types of boundary conditions that need to be implemented around the computational window. The natural boundary conditions are often of Dirichelet or Neumann type depending on the variational formulation being used. However, various artificial boundary conditions such as the absorbing boundary condition or the transparent boundary condition [88] can be introduced at the edges of the computational window in order to minimize reflections at the boundaries. At the moment, the most advanced boundary condition that has been used with the BPM is the perfectly matched layer approach [89], which can absorb unwanted optical power at the computational window most effectively.

The effects of the alternating dielectric layers may be approximated as an effective mirror instead of simulating the wave propagation in the DBR regions. Shimizu *et al* [90] proposed a BPM model for VCSELs and studied the transverse modes. However, they ignored the effects of the Bragg reflector structures and used two effective plane mirrors instead. This approximation may be adequate for simple geometries (such as implanted gain-guided devices) but it is questionable for complicated geometries where the distributed nature of the end mirrors plays an important role, since a fundamental assumption of many such models is reflection from a single plane. The more elaborate models do not suffer from any of these problems but have the common disadvantage of requiring extensive computing resources for their implementation. This might be a serious limitation of the BPM, especially when more sophisticated gain models including many-body effects [91] or other complications, such as electrical and thermal currents, are considered [92, 93].

The transfer matrix method (TMM) is another approach, which is particularly suitable for characterizing layered structures. Piprek *et al* [94] used the TMM to solve Maxwell's equations by matching the transverse field components at the dielectric interfaces. The tangential electric and magnetic field components must be continuous at each of these interfaces and recursive calculations for all the layers give the overall transfer matrix of the device. The propagation of forward-and-backward travelling waves through the layers can be performed by matrix multiplication. Burak and Binder [95] detailed the characteristics of the vector modes of the empty cavity—cold cavity modes that are assumed to have no gain or absorption—all refractive indices are real. They assumed cylindrical symmetry and considered an analytical approximation in the transverse direction for the hybrid modes and the use of the TMM in the longitudinal direction to find the cold cavity modes. They have also noted a change of the resonant wavelength with cavity radius and of a blue-shift of about 30 nm when the radius, R, is reduced from 3 to 1 μm. However, Yu *et al* [96] have reported an approach in which they have combined the TMM with the BPM.

However, in the design, analysis and optimization of optical cavity modes, besides using a fully 3D model to study the polarization dependence, a vectorial approach is mandatory. Some work has already been done in developing vectorial optical models for VCSEL structures. Noble *et al* [81] reported an approach based on the weighted index method but this method was unable to handle diffraction effects, which become predominant when dealing with small devices. Similar methods have been outlined by Kuszelewicz [97], who calculated the mode spectrum of the cold cavity, based on an expansion into hybrid modes of optical fibres. However, no optical gain or absorption is taken into account in these models, so the threshold gain cannot be evaluated directly. Demeulenaere *et al* [98] considered a vectorial junction analysis approach, which can handle diffraction and calculate the threshold but they enclosed the optical cavity inside an artificial cylindrical metal cavity to discretize the radiation modes. Conradi *et al* [82] used the method of lines in cylindrical coordinates to model VCSELs. Valle *et al* [99] obtained the modes of the circularly symmetric guide and Burak *et al* [100] subsequently studied the polarization properties of the passive cavity mode in detail. The vector FEM, a numerical technique based on the variational approximations to Maxwell's equations which yields a quasi-exact result, has been used by Noble *et al* [101].

B2.6.9.2 Electronic modelling

The electronic model may be considered to be the most important and represents the most basic model needed for the design and analysis of VCSELs. Many semi-analytical or numerical models have been reported, including some comprehensive studies at the energy band or circuit level [102].

The equation which describes the electric potential and ultimately links the carrier transport inside the structure is the Laplace equation, given here for a 3D structure:

$$\nabla.(\sigma\nabla V) = 0 \tag{B2.6.6}$$

where σ is the electrical conductivity. For a structure with cylindrical symmetry, the 3D del operator, ∇, can be written in terms of the transverse (r) and axial (z) directions. Often the electrical conductivity in layered materials, as encountered in the VCSEL structures, is anisotropic and then the electrical conductivity can be represented by a tensor:

$$\boldsymbol{\sigma} = \begin{bmatrix} \sigma_t & 0 \\ 0 & \sigma_z \end{bmatrix} \tag{B2.6.7}$$

It has been reported that the σ_t and σ_z, the conductivities in the radial and longitudinal directions, respectively, can differ substantially ($\sigma_r/\sigma_z = 100$) because of a reduced conductivity in the longitudinal direction caused by the band discontinuities in the DBR mirror. In the electronic model, the relation between the potential and the current density can be written as follows.

$$J = -\sigma\nabla V \tag{B2.6.8}$$

The axial component of the current density, J_z, can be written as

$$J_z = -\sigma \mathrm{d}V/\mathrm{d}z. \tag{B2.6.9}$$

The electron and hole densities inside the quantum well are computed by a rate equation, which considers the diffusion and recombination effect inside the active material. The strongly nonlinear rate equation for the electron and holes of the QW couples the carrier density to the emitted optical field of the laser. The carrier transport in the undoped active region (the quantum well) is governed by radial diffusion and is described by the ambipolar diffusion equation for the carrier density N, as

$$D\nabla_t^2 N + J_{\mathrm{inj}}(r)/de - N/\tau - BN^2 - CN^3 - g(r, N, f_m)\Sigma|E_m(r)|^2 = 0 \tag{B2.6.10}$$

where D is the diffusion coefficient of the carrier, J_{inj} is the current density, N is the carrier density, τ_s is carrier lifetime, which is a function of the carrier density and to meet the first-order approximation is taken to be constant, B is the quadratic recombination coefficient, C is the Auger recombination, E_m is the modal field of the mth mode, g is the gain, d the thickness of the active layer, e the electron charge and the summation is carried out for all the modes. However, in order to determine the temporal evolution of the optical power, it is necessary to introduce the time-dependent carrier and field equations [103]. The coupling between the injected current and density of the electrons and holes N within the active region is achieved by the previous rate equation. With the computed current density $\boldsymbol{J}$ within the active region, the carrier distributions in the QW can be calculated.

The general oscillation condition has the requirement that the field amplitudes reproduce themselves after a round trip. The optical loss for the resonant mode must balance the gain to reach the threshold; hence, the amplitude condition at the threshold of oscillation must satisfy the following round-trip relation:

$$R_1 R_2 \exp[2(g_{\mathrm{th}} - \alpha_{\mathrm{i}})] = 1 \tag{B2.6.11}$$

where R_1 and R_2 are the reflection coefficients of the mirrors, g_{th} is the gain at threshold and α_{i} is the total loss in the cavity. However, for a VCSEL the gain is not distributed over the entire effective cavity length but lies in the active region of thickness, d. The general condition is the modified equation for a VCSEL, i.e.

$$\Gamma g_{\mathrm{th}} d = \Gamma \alpha_{\mathrm{a}} d + \alpha_{\mathrm{c}}(L - d) - 2\ln(R_1 R_2) \tag{B2.6.12}$$

$$g_{\mathrm{th}} = \alpha_{\mathrm{ac}} + \alpha_{\mathrm{cs}}(L/d - 1) - (1/2d)\ln(R_1 R_2) + \alpha_{\mathrm{d}} \tag{B2.6.13}$$

where d is the active layer thickness, L is the effective length of the cavity, Γ is the confinement factor, α_{ac} and α_{cs} are the absorption loss in the active and cladding layers, respectively, and α_{d} is the diffraction loss.

The gain profile $g(r)$ in the active region may be approximated to a linear function of carrier concentration:

$$g(r) = a(N(r) - N_{\mathrm{tr}}) \tag{B2.6.14}$$

where a is the gain coefficient and N_{tr} is the carrier concentration for transparency. However, in reality it is not usually a simple analytic function of the carrier density but a better approximation could be obtained by using a logarithmic function, in the case of quantum wells (QWs):

$$g(r) = \xi g_0 \log(N(r)/N_{\mathrm{tr}}) \tag{B2.6.15}$$

where g_0 is the gain coefficient and ξ the longitudinal filling factor calculated from the standing-wave pattern [104]. More generally, the effect of the standing-wave pattern may be included by the use of a properly weighted complex dielectric constant (i.e. refractive index and the gain or loss) for the different regions.

The term J_{inj} is the injected current density distribution, which often varies along the transverse direction and can be obtained by solving the Poisson equations describing the current flow inside a laser. However, to simplify the problem, it may quite often be assumed that the active layer is excited by a uniform current density in the transverse direction and then the electrical effects, such as the current crowding at the edge of the aperture, are neglected. In the simplest form, the current density can be taken to be

$$J(r) = J \qquad r < R \qquad \text{(B2.6.16)}$$
$$J(r) = 0 \qquad r > R$$

where R is the radius of the active region. In several other models [105], it has also been assumed that the current density decays exponentially outside the active region. Dutta [106] and Nakwaski [107] have presented more elaborate modelling for current spreading in VCSELs.

B2.6.9.3 Thermal modelling

The thermal problem is recognized as one of the most important aspects for effective VCSEL design for devices and arrays. Thermal lensing provides additional transverse confinement, which affects the mode profiles. This is an important effect where the thermally induced modification of the mode profiles changes the overlap with the surface relief and must, therefore, be taken into account to obtain reliable results. Computer simulation is very important here, not only as a means to understand but also to solve this problem in the design of VCSEL devices and arrays, in particular in a cost-effective way.

The injected current flux leads to an increase in the temperature inside the laser, which has a significant influence on the optical power. In the steady state, the temperature within the laser satisfies the Poisson equation

$$(1/\kappa)\nabla(\kappa(\nabla T) = -(1/\kappa)Q \qquad \text{(B2.6.17)}$$

where κ is the thermal conductivity of the material and T is the temperature. The heat source is associated with the electrical and optical loss. For a structure with cylindrical symmetry, this equation can be written in the (r, z) plane:

$$\nabla(\kappa\nabla T(r, z)) = -Q_{\mathrm{t}}(r, z) \qquad \text{(B2.6.18)}$$

where Q_{t} is the net local heat generation rate. However, in the transient analysis, a time-dependent thermal diffusion equation needs to be considered. In contrast to edge-emitting diode lasers, where the non-radiative recombination of the charge carriers in the active region is the dominant heat source, VCSELs have a much more complicated distribution of heat sources. Two important heat sources are non-radiative recombination and reabsorption of spontaneous emission in the active region and Joule heating. The main source of Joule heating lies in both the confinement layers and in the p-contact region, and in the cladding layers and in contact heating.

$$Q_{\mathrm{t}} = Q_{\mathrm{J}} + Q_{\mathrm{nr}}. \qquad \text{(B2.6.19)}$$

The Joule heat dissipation in such anisotropic materials can be expressed as [108]

$$Q_{\mathrm{J}} = \sigma_z(\partial V/\partial z)^2 + \sigma_r(\partial V/\partial r)^2 \qquad \text{(B2.6.20)}$$
$$Q_{\mathrm{nr}} = eV_{\mathrm{active}}N/\tau \qquad \text{(B2.6.21)}$$

where V_{active} is the voltage drop across the active region.

The thermal conductivities of the semiconductor materials depend not only on the compositions but also strongly depend on the operating temperature. As an example, the thermal conductivities of the $\mathrm{Al}_x\mathrm{Ga}_{1-x}\mathrm{As}$ and the GaAs layers are obtained from the following relationship [109]:

$$\kappa(x) = 0.44/(1 + 12.70x - 13.22x^2)\ \mathrm{W\ cm^{-1}\ K^{-1}}. \qquad \text{(B2.6.22)}$$

The temperature dependence of κ, determined for GaAs [110], is assumed to be valid for any composition of ternary $Al_xGa_{1-x}As$ [111, 112] and is given by

$$\kappa(x, T) = \kappa(x, 300K)(300/T)^{5/4}. \tag{B2.6.23}$$

For a layered structure, such as that encountered in VCSELs, the thermal conductivity, κ, often depends on the heat flux direction and it can be represented by a tensor. As an example, for a AlAs/GaAs lattice, a vertical value of $k_v = 0.3$ W cm^{-1} K^{-1} has been measured [113], which is less than half the value of $k_l = 0.69$ W cm^{-1} K^{-1}.

By solving Poisson's equation, the temperature profile inside the VCSEL can be obtained. However, the spatial temperature profile (x, y, z or r, z) depends on the heat sources and thermal conductivity profiles but the thermal conductivity profile also depends on the temperature profile. This makes the thermal problem highly nonlinear in nature and only an iterative procedure [114] can be used to solve such a nonlinear problem.

Technically, solving the heat-transfer problem is equivalent to solving the Poisson equation with appropriate boundary conditions. At present, many researchers either ignore the thermal effects or use very approximate analytical techniques. To work in a proper thermal modelling regime, Joyce and Dixon [115] used the separation of variables and Fourier series expansion for thermal modelling. A similar method [116] was used for thermal modelling edge-emitting lasers and it was subsequently adapted for SELs [107] with the later additional incorporation of spatial transformation to study VCSEL structures [117] with cylindrical symmetry. Sarzala and Nakwaski [118] have also developed an FEM-based model for thermal analysis and, subsequently, Rahman *et al* [114] used a versatile FEM-based approach to obtain the temperature profile inside a VCSEL. As an alternative, general purpose commercial packages, such as ANSYS, based on the FEM, can be used for thermal modelling. Piprek *et al* [94] used this ANSYS package for both thermal and electronic simulations, and, in particular, the 3D finite-element code, which allows anisotropic material parameters to be considered for thermal simulations was employed. Often a simple semi-analytical model, in which it is assumed the VCSEL is a disc-shaped heat source with semi-infinite claddings, can provide a reasonably accurate thermal picture of the device. However, the flexibility of the FEM to represent the heat source profile accurately simulated can provide convincing results. The FEM-based approach has the potential to simulate all three physical models, i.e. the electronic, optical and thermal models, iteratively, with the important advantage of using the same structural mesh discretization for consistent modelling of the device.

B2.6.9.4 Consistent modelling

The thermal and electrical conductivity and refractive index, i.e. the parameter needed for the optical model, are dependent on the temperature profile. Thus, the thermal model relates back to the electrical and optical models due to the dependence on the electrical conductivity and the permittivity distribution. The heating effects on the optical behaviour are connected to the temperature dependence of the refractive index and, in particular, the refractive index of a semiconductor strongly depends on its operating temperature. The temperature variation of the refractive index can be given as

$$n(T) = n(0) + B\Delta T \tag{B2.6.24}$$

where B is the thermal coefficient and as, an example, the thermal dependence of the GaAs is known to be about 4×10^{-4} K^{-1} [119].

The refractive index also depends on the carrier distribution. The carrier dependence of the refractive index may be given as [120]

$$dn/dN = -q^2/(2n_0\varepsilon_0\omega^2 m_n) \tag{B2.6.25}$$

where m_n is the effective mass of the electron. Hence, the built-in refractive index profile is modified by both the carrier profile and the temperature rise in the device. The spatial distribution of the electrical conductivity also depends on the local temperature and the temperature dependence can be approximated as [82]

$$\sigma = \sigma_0 (T_0/T)^{1.25}. \tag{B2.6.26}$$

The heat source inside a VCSEL depends on the photon intensity, the injected carrier density and also the carrier density in the active region. However, the process is more complex, because of the coupling between the space-dependent optical field and the carrier density (through stimulated emission) which, in turn, changes the gain and, hence, the propagation coefficients and the self-consistent field. The temperature dependence of the gain arises due to the temperature dependencies of the gap energy and the quasi-Fermi levels.

Temperature not only changes the refractive indices: by considering just the thermal expansion of the semiconductor materials, it can be shown that a 100 °C temperature rise generates an increase in layer thickness of around 0.06%. The emission wavelength, λ_0, rises with increasing temperature, due to the enlarged optical length of the cavity. The gain peak also changes rapidly with increasing temperature in a way that is much faster than the change in emission wavelength.

The VCSEL threshold may be calculated self-consistently. First, a field profile (Gaussian or eigenmode) may be injected into the cavity and propagated for several round-trips until its shape is stabilized, that is until the cavity mode is reached. Then the power variation after one round trip in the cavity is calculated and the injection current is modified to compensate for this variation. For a VCSEL operating above threshold, the space-dependent carrier–photon interaction (stimulated emission) should be included in the carrier rate equation through the term involving the optical modal field intensities. The propagation equation could be solved for a given laser current above threshold, where the carrier density and the optical field profile should be calculated self-consistently. The output power may be calculated from the profile of the outgoing field at the facets of the hard mirrors, after taking into account the transmission of the DBR mirrors.

Michalzik and Ebeling [121] considered a simpler model to study the optical thermal and electronic characteristics for gain-guided proton-implanted VCSELs, self-consistently. Vukusic *et al* [122] used the effective index method and the finite-difference approach after considering cylindrical symmetry. Conradi *et al* [82] considered the method of lines for all three sub-models for a structure with cylindrical symmetry. Previously, Zhang and Peterman [80] and Hadley *et al* [79] had reported consistent modelling by combining all three individual models by using the more powerful BPM. Typically, 2000 round trips may be needed for good convergence. Zhang and Peterman [80] used the BPM to calculate the threshold properties, including the Bragg reflector in the structure they modelled. However, they considered only the fundamental mode and did not model the above-threshold behaviour. Subsequently, Hadley [83] and Bissessur *et al* [123] considered the use of the more versatile BPM based on the finite-difference method, but only examined the scalar approach. Bissessur *et al* [123] presented a steady-state self-consistent, quasi-3D model for the analysis of VCSELs. The model was based on a forward BPM scheme and included light absorption, diffraction and scattering as well as carrier diffusion; however, the thermal effects were not taken into account. Recently, Li *et al* [124] presented a comprehensive self-consistent model for oxide-aperture VCSELs using the BPM.

Comprehensive numerical modelling of VCSELs requires a full vectorial optical model and appropriate sub-models to include the thermal and electronic physical properties. A full vectorial optical model would be mandatory for polarization studies and also single transverse mode operation for those VCSEL structures with narrow optical apertures, large index contrast or operating near to the modal cut-off region. However, a full 3D optical treatment of a VCSEL, including the effect of the distributed mirrors in the structure, may prove too time- and computer-resource-consuming, if it is necessary to account for all the different phenomena involved. Depending on the specific needs, appropriate simplifications may be made to improve computational efficiency.

B2.6.10 Conclusions and the Future Trends

With several decades of development, the VCSEL has clearly emerged from the laboratory and is being used in a wider range of applications, ranging from short- to long-distance optical communications, consumer products such as CD and DVD players and laser printers, as an optical pump for other sources as well as for optical sensing applications.

VCSELs are satisfying a growing number of existing commercial markets and those which may emerge in the future. The number and size of these markets will depend on the improvements in VCSEL design and the development of reliable manufacturing processes. If sufficient advances can be made, then edge-emitting diode lasers could be confined to niche markets, while VCSELs take the lead in major markets. Recently, several companies have reported supplying sampling devices to partners operating at 1500 and 1600 nm with power levels between 1 and 2 mW. For some of these based on InP, using MBE, the process yield has been higher than that for DFB lasers and these may be the dominant sources for dense and coarse WDM applications.

The incredible growth of the internet and data transmission is pushing the bandwidth requirement of long-haul, metro and local area networks at an unprecedented pace. One of the key requirements for the expansion of optical networks and ultimately 'fibre-to-the-home' is low-cost high-performance lasers that are easily coupled to fibres. Various vendors have been reporting that the time is nearly ready to mass produce commercially viable 1.55 μm devices, without major problems, for the 'last-mile' to bring 'fibre-to-the-home'.

It has been shown [154] that carrier transport and hot carriers have a significant influence on the relaxation-oscillation frequency and the modulation bandwidth of VCSELs. Relaxation oscillation frequency of VCSELs could be as high as 71 GHz [125]. However, the maximum modulation bandwidth has so far been found to be limited to below 20 GHz [58, 126]. VCSELs offer viable directly modulated sources for DWDM applications [155] and Chang-Hasnain [156] has recently reviewed the status of tunable VCSELs. As a result, it is necessary to have a thorough understanding of the dynamic behaviour of VCSELs for which a time-domain travelling-wave model would be useful. Ultra-low threshold operation can be expected, with a reduced active volume until the limit is reached, this being expected to be due to the non-radiative recombination and optical and electrical confinements effects.

To achieve what is required, a number of developments have already occurred. The selectively oxidized VCSEL, under development since 1994, has rapid and significant performance advantages; however, issues pertaining to control, uniformity, reliability and reproducibility still need to be fully understood. Long-distance communication over silica fibre dictates operation in the 1.3 and 1.55 μm windows. The development of InP-based, longer-wavelength VCSELs, suitable for long-haul fibre-based optical communications, has met with slower progress because of the inherent difficulties involved. VCSEL performances at these wavelengths have been limited by some substantial difficulties: higher diffraction loss at the longer wavelengths, Auger recombination, inter-valence band absorption and the large valence band offset in GaInAsP/InP. The relatively small index difference in GaInAsP/InP has made it difficult to design appropriately thick (>10 μm) semiconductor mirrors to achieve the required reflectivity. Advanced designs in which one or both epitaxially grown semiconductor mirrors are replaced with conventional dielectric mirrors could open the way to long-wavelength VCSEL diodes. Wafer bonding has been successfully demonstrated for long-wavelength operation; however, this technique also adds to the electrical resistance and the combination of the yield and the triple epitaxy required for the bonding is generally regarded as too costly.

One of the main problems is the substantial elimination of the heat associated with the light generation process and the poor thermal conductivity of quaternaries. Since the inception of VCSELs, thermal problems have plagued their development and, to date, they represent a major hurdle that must be overcome if massive integration into 2D arrays is ever to be realized.

Understanding the transverse mode and polarization properties is important, particularly for fibre-optic applications. In contrast to the commonly claimed single-mode operation, a VCSEL often emits multiple transverse modes if the laser cavity is a multi-mode waveguide. For numerous applications in optical interconnects, optical recording and optical communications, single transverse mode operation is essential. VCSEL transverse mode dynamics are complex and can affect the level of mode partitioning noise in network applications. Several control strategies for the fundamental transverse mode have been proposed, including the use of mode-suppressing oxide layers [127], anti-guide structure [128], a reduced optical aperture [105], zinc diffusion and impurity-induced disordering [129], spatial mode filtering with contact metal [127] and localized modification of the mirror reflectivity by using focused ion-beam etching [130].

The VCSEL generally shows linear polarization but the direction of the beam cannot be tightly controlled: occasionally it switches over due to spatial hole burning or temperature variations. Polarization selectivity appears to be the key weakness—due to rotational symmetry, two orthogonal polarized modes are degenerate and VCSELs show no intrinsic polarization selectivity. At present, the polarization control of VCSELs presents one of the greatest challenges since its polarization state must be stable against the injection current, external feedback and the operating temperature variations. Random polarization fluctuations also cause excess noise, resulting in the degradation of the signal-to-noise ratio and this limits the bit-error-rate performance of optical transmission systems. Thus, polarization control is one of the most important subjects to be resolved, especially for polarization-sensitive applications such as magneto-optic discs and coherent detection systems.

Several approaches, including the use of anisotropy cavity geometry [132], amorphous inscription of subwavelengths [133], metal/dielectric gratings on the top mirror [134], trench etching [135], elliptic surface etching [136], cavity tilting [137], anisotropic gain medium of a fractional-layer superlattice [138] and anisotropic stress from a hole in the substrate [139] have been considered to stabilize the polarization states. Usually, most of the VCSELs which are grown on a (100) substrate show an unstable polarization state. Fabricating optically active quantum well devices grown on (n11) substrates [140–146], which creates large polarization selectivity in the active region on a misoriented substrate, may be more promising than other approaches. Intrinsic anisotropy in optical gain can create large polarization selectivity and the VCSEL fabrication process is also similar to that for fabrication on a (100) substrate.

Within the last decade, VCSELs have become successful industrial products. Many convincing demonstrations of effective devices have been reported, due in most part to trial and error or intuition on the part of their designers. However, as VCSEL designs have reached a reasonable maturity, more and more advanced issues need to be tackled for high-performance high-speed VCSELs, including beam quality, polarization stability, polarization switching, enhanced modulation bandwidth and wavelength tunability but also at a reasonable (inevitably) lower cost. In this respect a comprehensive theoretical model to understand polarization selection in realistic VCSELs would be very valuable commercially. Ultimately, commercial VCSEL development will be driven by the requirements for their applications and their manufacturing costs: low cost and high volume should be an inherent VCSEL attribute of benefit for a wide range of future applications.

References

[1] Soda H, Iga K, Kitahara C and Suematsu Y 1979 GaInAsP/InP surface emitting injection lasers *Japan. J. Appl. Phys.* **18** 2329–30

[2] Iga K, Ishikawa S, Ohkouchi S and Nishimura T 1984 Room temperature pulsed oscillation of GaAlAs/GaAs surface emitting laser *Appl. Phys. Lett.* **45** 348–50

[3] Uchiyama S and Iga K 1985 Two-dimensional array of GaInAsP/InP surface-emitting lasers *Electron. Lett.* **21** 162–4

[4] Uchiyama S and Iga K 1986 Low threshold GaInAsP/InP surface emitting lasers *Trans. IECE Japan* **E69** 587–8

[5] Kinoshita S, Kobayashi K, Sakaguchi T, Odagawa T and Iga K 1986 Room temperature pulse operation of GaAs surface emitting laser using TiO_2/SiO_2 dielectric multilayer reflectors *Trans. IECE Japan* **J69-C** 412–20

[6] Iga K, Kinoshita S and Koyama F 1987 Microcavity GaAlAs/GaAs surface emitting laser with $I_{th} = 6$ mA *Electron. Lett.* **25** 134–6

[7] Koyama F, Kinoshita S and Iga K 1988 Room-temperature cw operation of GaAs vertical cavity surface emitting lasers *Trans. IEICE Japan* **E71** 1089–90

[8] Koyama F, Kinoshita S and Iga K 1989 Room-temperature continuous wave lasing characteristics of GaAs vertical cavity surface emitting laser *Appl. Phys. Lett.* **55** 221–3

[9] Ibariki A, Kawashima K, Furusawa K, Ishikawa T, Yamagushi T and Niina T 1989 Buried heterostructure GaAs/AlGaAs distributed Bragg reflector surface emitting laser with very low threshold (5.2 mA) under room temperature cw condition *Japan. J. Appl. Phys.* **28** L667–8

[10] Baba T, Yogo Y, Suzuki K, Koyama F and Iga K 1993 First room temperature cw operation of GaInAsP/InP surface emitting lasers *IEICE Trans. Electron.* **EL76-C** 1423–4

[11] Gourley P L and Drummond T L 1986 Single-crystal, epitaxial multilayers of AlAs, GaAs, and $Al_xGa_{1-x}As$ for use as optical interferometric elements *Appl. Phys. Lett.* **49** 489–91

[12] Gourley P L and Drummond T J 1987 Visible room temperature surface-emitting lasers using an epitaxial Fabry–Pérot resonator with AlGaAs/AlAs quarter wave high-reflectors and AlGaAs/GaAs quantum wells *Appl. Phys. Lett.* **50** 1225–127

[13] Gourley P L, Drummond t J, Zipperian T E, Reno J L and Plut T A 1988 Threshold characteristics of epitaxial Al(Ga,As) surface-emitting lasers with integrated quarter-wave high reflectors *J. Appl. Phys.* **64** 6578–80

[14] Jewell J L, Scherer A, McCall S L, Lee Y H, Walker S J, Harbison J P and Florez L T 1989 Low threshold electrically pumped vertical cavity sutface emitting micro-lasers *Electron. Lett.* **25** 1123–4

[15] Jewell J L, Harbison J P, Scherer A, Lee Y H and Florez L T 1991 Vertical cavity surface emitting lasers: Design, growth, fabrication, characterization *IEEE J. Quantum Electron.* **27** 1332–45

[16] Jewell J L, McCall S L, Scherer A, Huch H H, Whitaker N A, Gossard A C and English J H 1989 Transverse modes, waveguide dispersion and 30 ps recovery in submicron GaAS/AlAs microresonator *Appl. Phys. Lett.* **55** 22–4

[17] Lee Y H, Jewell J L, Scherer A, Bell B, Brown-Goebeler K, Harbison J and Florez L T 1989 Effect of etch depth and ion implantation on surface emitting microlasers *Electron. Lett.* **25** 225–6

[18] Lee Y H, Jewell J L, Scherer A, McCall S L, Harbison J P and Florez L T 1989 Room-temperature continuous wave vertical cavity single quantum well microlaser diodes *Electron. Lett.* **25** 1377–8

[19] Scherer A, Jewell J L, Harbison J P, Florez L T, Lee Y H and Sandroff C J 1989 Fabrication of low threshold cw electrically pumped surface emitting microlasers *Appl. Phys. Lett.* **55** 2724–6

[20] Dallesasse J M, Holonyak Jr N, Sugg A R, Richard T A and El-Zein N 1990 Hydrolyzation oxidation of $Al_xGa_{1-x}As$-AlAs-GaAs quantum well heterostructure and superlattices *Appl. Phys. Lett.* **57** 2844–6

[21] Huffaker D L, Deppe D G, Kumar K and Rogers T J 1994 Native-oxide defined ring contact for low threshold vertical-cavity lasers *Appl. Phys. Lett.* **65** 97–9

[22] Hu S Y, Hegblom E R and Coldren L A 1998 Multiple-wavelength top-emitting vertical-cavity photonic integrated emitter arrays for direct coupled wavelength-division multiplexing applications *Electron. Lett.* **34** 189–90

[23] Sjolund O, Louderback D A, Hegblom E R, Ko J and Coldren L A 1998 Monolithic integration of microlensed resonant photodetectors and vertical cavity lasers *Electron. Lett.* **34** 1742–3

[24] Iga K 2000 Surface emitting laser- its birth and generation of new optoelectronic field *IEEE J. Selected Topics Quantum Electron.* **6** 1201–15

[25] Towe E 2000 A historical perspective of the development of the vertical-cavity surface-emitting laser *IEEE Selected Topics Quantum Electron.* **6** 1558–63

[26] Reinhert R K and Logan R A 1975 GaAs–AlGaAs double heterostructure lasers with taper-coupled passive waveguides *Appl. Phys. Lett.* **26** 516

[27] Defreez R K, Bossert D J, Yu N, Hunt J M, Ximen H, Elliot R A, Carlson N W, Lurie M, Evans G A, Hammer J M, Bour D P, Palfrey S L, Amantea R, Winful H G and Wang S S 1989 Picosecond optical properties of a grating surface emitting two-dimensional coherent laser array *IEEE Photon. Technol. Lett.* **1** 209–11

[28] SpringThorpe A J 1977 A noble double-heterostructure p-n junction laser *Appl. Phys. Lett.* **31** 524–5

[29] Saito H and Noguchi Y 1989 A reflection-type surface-emitting 1.3 μm InGaAsP InP laser array with microcoated reflector *Japan. J. Appl. Phys.* **28** L1239–41

[30] Liau Z L and Walpole J N 1985 Surface emitting GaInAsP/InP laser with low threshold current and high efficiency *Appl. Phys. Lett.* **46** 115–17

[31] Wu M C, Ogura M, Hsin W, Whinnery L R and Wang S 1987 Surface emitting laser diode with bent double heterostructure *Proc. CLEO'87* (Piscataway, NJ: IEEE) paper WG4

[32] Uchiyama S, Ohmae Y, Shimizu S and Iga K 1986 GaInAsP/InP surface emitting lasers with current confining structure *J. Lightwave Technol.* **4** 846–51

[33] Kinoshita S, Odagawa T, Sakaguchi T and Iga K 1986 Low threshold circular buried heterostructure (CBH) GaAlAs/GaAs surface emitting lasers *Proc. IEEE Int. Semiconductor Laser Conf.* (Piscataway, NJ: IEEE) Paper H-2, pp 106–7

[34] Iga K, Kinoshita S and Koyama F 1986 Micro-cavity GaAlAs/GaAs surface emitting lasers with I_{th} = 6 mA, *Proc. 10th Semiconductor Laser Conf.* (Piscataway, NJ: IEEE) paper PD-4, p 12

[35] Geels R S, Corzine S W, Scott J W, Young D B and Coldren L A 1990 Low threshold planarized vertical cavity surface emitting lasers *IEEE Photon. Technol. Lett.* **2** 234–6

[36] Hasnain G, Tai K, Wynn J D, Wang Y H, Fisher R J and Hong M 1990 Continuous wave top surface emitting quantum well lasers using hybrid metal semiconductor reflectors *Electron. Lett.* **26** 1590–2

[37] Geels R S and Coldren L A 1990 submilliamp threshold vertical-cavity laser diodes *Appl. Phys. Lett.* **57** 1605–7

[38] Tanobe H, Osikiri M, Araki M, Koyama F and Iga K 1992 A preliminary study on MgO/Si multilayer reflectors for improving thermal conductance in surface emitting lasers *Dig. LEOS'92* (Piscataway, NJ: IEEE) DLTA 12.2

[39] Dudley J J, Babic D I, Mirin R, Yang L, Miller B I, Ram R J, Reynolds T, Hu E L and Bowers J E 1994 Low threshold, wafer-fused long wavelength vertical cavity lasers *Appl. Phys. Lett.* **64** 1463–5

[40] Babic D I, Piprek J, Streubei K, Mirin R P, Margalit N M, Mars D E, Bowers J E and Hu E L 1997 Design and analysis of double-fused 1.55 μm vertical-cavity lasers *IEEE J. Quantum Electron.* **33** 1369–83

[41] Liau Z L and Mull D E 1990 Wafer fusion—a novel technique for optoelectronic device fabrication and monolithic integration *Appl. Phys. Lett.* **56** 737–9

[42] Qian Y, Zhu Z H, Lo Y H, Hou H Q, Wang M C and Lin W 1997 1.3 μm vertical-cavity surface emitting lasers with double-bonded GaAs/AlAs Bragg mirrors *IEEE Photon. Technol. Lett.* **9** 8–10

[43] Miller B I, Koren U, Young M G and Chien M D 1991 Strain-compensated strain-layer super lattices for 1.5 μm wavelength lasers *Appl. Phys. Lett.* **58** 1952–4

[44] Matin M A, Jezierski A F, Bashar A F, Lacklison D E, Benson T M, Cheung T S, Roberts J S, Sale T E, Orton J W, Foxon C T and Rezazadeh A A 1994 Optically transparent indium-tin-oxide (ITO) ohmic contacts in the fabrication of vertical cavity surface emitting lasers *Electron. Lett.* **30** 318–20

[45] MacDougal M H, Dapkus D P, Pudikov V, Zhao H and Yang G M 1995 Ultralow threshold current vertical cavity surface emitting lasers with AlAs oxide-GaAs distributed Bragg reflectors *IEEE Photon. Technol. Lett.* **7** 229–31

[46] Hayashi Y, Mukaihara T, Hatori N, Ohnoki N, Matsutani A, Koyama F and Iga K 1995 Lasing characteristic of low threshold oxide confinement InGaAs-GaAlAs vertical-cavity surface-emitting lasers *IEEE Photon. Technol. Lett.* **7** 1234–6

[47] Bond A E and Dapukus P D 1998 Monolithically integrated surface and substrate emitting vertical cavity laser with spatially selected distributed Bragg reflectors *Appl. Phys. Lett.* **73** 19–21

[48] Yang G M, MacDougal M H and Dapkus P D 1995 Ultralow threshold current vertical-cavity surface-emitting lasers obtained with selective oxidation *Electron. Lett.* **31** 886–8

[49] Huffaker D and Deppe D G 1997 Improved performance of oxide-confined vertical-cavity surface-emitting lasers using a tunnel injection active region *Appl. Phys. Lett.* **71** 1449–51

[50] Deppe D G, Huffaker D L, Shin J and Deng Q 1995 Very-low-threshold index-confined planar micocavity lasers *IEEE Photon. Technol. Lett.* **7** 965–7

[51] Yang G M, MacDougal M H and Dapkus P D 1995 Ultralow threshold VCSEL's fabricated by selective oxidation from all epitaxial structure *CLEO'95 (Baltimore, MD)* (Piscataway, NJ: IEEE) postdeadline paper CPD4-1

[52] Choquette K D, Lear K L, Schneider Jr R P, Geib K M 1995 Cavity characteristics of selectively oxidized vertical-cavity lasers *Appl. Phys. Lett.* **66** 3413–15

[53] Choquette K D, Schneider R P, Lear K L and Geib K M 1994 Low threshold voltage vertical cavity lasers fabricated by selective oxidation *Electron. Lett.* **30** 2043–4

[54] Choquette K D, Lear K L, Schneider R P, Geib K M, Figiel J J and Hull R 1995 Fabrication and performance of selectively oxidized vertical cavity lasers *IEEE Photon. Technol. Lett.* **7** 1237–9

[55] Lear K L, Choquette K D, Schneider R P, Kilcoyne S P and Geib K M 1995 Selectively oxidized vertical cavity surface emitting lasers with 50% power conversion efficiency *Electron. Lett.* **31** 208–9

[56] Yang G M, MacDougal M H and Dapkus P D 1995 Ultralow threshold current vertical-cavity surface-emitting lasers obtained with selective oxidation *Electron. Lett.* **31** 886–8

[57] Jager R, Grabherr M, Jung C, Michalzik R, Reinner G, Weigl B and Ebeling K J 1997 57% wallplug efficiency oxide-confined 850 nm wavelength GaAs VCSEL's *Electron. Lett.* **33** 330–1

[58] Lear K L, Mar K D, Choquette K D, Kilocoyne S P, Schneider R P Jr and Geib K M 1996 High-frequency modulation of oxide-confined vertical cavity surface emitting lasers *Electron. Lett.* **32** 457–8

[59] Rogers T J, Lei C, Deppe D G and Streetman B G 1992 Low threshold voltage continuous wave vertical surface-emitting lasers *Appl. Phys. Lett.* **62** 2027–9

[60] Roberts J S, David J P R, Sale T E and Tihanyi P I 1993 High purity AlGaAs from methyl-based precursors using in situ gettering of alkoxides *J. Cryst. Growth* **143** 135–40

[61] Shoji H, Uchida T, Kusunoki T, Matsuda M, Kurakake H, Yamazaki S and Ishikawa H 1994 Fabrication of $In_{0.25}Ga_{0.75}As$/InGaAsP strained SQW laser on $In_{0.05}Ga_{0.95}As$ ternary substrate *IEEE Photon. Technol. Lett.* **6** 1170–2

[62] Schneider R P Jr, Choquette K D, Lott J A, Lear K L, Figiel J J and Malloy K J 1994 Efficient room-temperature continuous-wave AlGaInP/AlGaAs visible (670 nm) vertical-cavity surface-emitting laser diodes *IEEE Photon. Technol. Lett.* **6** 313–16

[63] Lott J A and Schneider R P Jr 1993 Electrically injected visible (639–661 nm) vertical-cavity surface emitting lasers *Electron. Lett.* **29** 830–2

[64] Knigge A, Zorn M, Weyers M and Trankle G 2002 High-performance vertical-cavity surface-emitting laser with emission wavelength between 650 and 670 nm *Electron. Lett.* **38** 882–3

[65] Nakamura S, Mukai T and Soneh M 1994 Candella-class high brightness InGaN/AlGaN double-heterostructure blue-emitting diodes *Appl. Phys. Lett.* **64** 1687–9

[66] Harris J S 2002 GaInNAs long-wavelength lasers: progress and challenges *Semicond. Sci. Technol.* **17** 880–91

[67] Schneider R P Jr and Houng Y H 1999 Eptaxy of vertical-cavity lasers *Vertical Cavity Surface-emitting Lasers, Design,*

Fabrication, Characterization and Applications ed C Wilmsen, H Temkin and L A Coldren (Cambridge: Cambridge University Press)

[68] Born M and Wolff E 1993 *Principles of Optics* 6th edn (New York: Pergamon)

[69] Kogelnik H and Shank C V 1972 Coupled wave theory of distributed Bragg reflector lasers *J. Appl. Phys.* **43** 2327–35

[70] McLeod 1986 *Thin Film Optical-Filters* (Bristol: Adam Hilger)

[71] Makino T 1992 Effective index matrix analysis of distributed feedback semiconductor lasers *IEEE J. Quantum Electron.* **28** 434–40

[72] Catchmark J M, Morgan R A, Kojima K, Leibenguth R E, Guth G D, Focht M W, Asom M T, Luther L C and Przybylek G P 1993 High temperature cw operation of vertical-cavity top surface-emitting lasers *Conf. Lasers and Electroptics, CLEO'93 (Baltimore, MD)* (Piscataway, NJ: IEEE) pp 138–9

[73] Jewell J L, Lee Y H, Scherer A, McCall S L, Olsson N A, Harbison J P and Florez L T 1990 Surface-emitting microlasers for photonic switching and interchip connections *Opt. Eng.* **29** 210–14

[74] Tai K, Yang L, Wang Y H, Wyan J D and Cho A Y 1990 Drastic reduction of series resistance in doped semiconductor distributed Bragg reflector for surface emitting lasers *Appl. Phys. Lett.* **56** 2496–8

[75] Scott J W, Thibeault B J, Young D B, Coldren L A and Peters F H 1994 High efficiency submillimeter vertical cavity lasers with intra-cavity contacts *IEEE Photon. Technol. Lett.* **6** 678–80

[76] Nakwaski W 1990 Correspondence: Consideration on geometry design of surface-emitting laser diodes *IEE J. Optoelectron.* **137** 129–31

[77] Rahman B M A and Davies J B 1984 Finite element solution of integrated optical waveguides *J. Lightwave Technol.* **2** 682–8

[78] Hadley G R 1995 Effective index model for vertical-cavity surface-emitting lasers *Opt. Lett.* **20** 1483–5

[79] Hadley G R, Lear K L, Warren M E, Choquette K D, Scott J W and Corzine S W 1996 Comprehensive numerical modelling of vertical-cavity surface-emitting lasers *IEEE J. Quantum Electron.* **32** 607–16

[80] Zhang J -P and Peterman K 1994 Beam propagation model for vertical cavity surface emitting lasers: Threshold properties *IEEE J. Quantum Electron.* **30** 1529–36

[81] Noble M J, Loehr J P and Lott J A 1998 Analysis of microcavity VCSEL lasing modes using a full-vector weighted index method *IEEE J. Quantum Electron.* **34** 1890–903

[82] Conradi O, Helfert S and Pregla R 2001 Comprehensive modelling of vertical-cavity laser-diodes by the method of lines *IEEE J. Quantum Electron.* **37** 928–35

[83] Hadley G R 1998 Low-truncation-error finite difference equations for photonic simulation II: Vertical-cavity surface-emitting lasers *J. Lightwave Technol.* **16** 142–51

[84] Feit M D and Fleck J A 1980 Computation of mode properties in optical fiber waveguide by a beam propagation method *Appl. Opt.* **19** 1154–64

[85] Obayya S S A, Rahman B M A and El-Mikathi H 2000 New full-vectorial numerically efficient propagation algorithm based on the finite element method *J. Lightwave Technol.* **18** 409–15

[86] Whuang W P, Xu C L and Choudhuri S K 1992 A finite-difference vector beam propagation method for three dimensional waveguide structures *IEEE Photon. Technol. Lett.* **4** 148–151

[87] Rao H L, Steel M J, Scarmozzino R and Osgood R M 2001 VCSEL design using the bi-directional beam propagation method *IEEE J. Quantum Electron.* **37** 1435–40

[88] Hadley G R 1991 Transparent boundary condition for the beam propagation method *Opt. Lett.* **16** 621–6

[89] Berenger J P 1994 A perfectly matched layer for the absorption of electromagnetic waves *J. Comput. Phys.* **114** 185–200

[90] Shimizu M, Koyama F and Iga K 1991 Transverse mode analysis for surface emitting laser using beam propagation method *IEICE Trans.* **E-74** 3334–41

[91] Chow W W 1995 Nonequlibrium and many-body Coulomb effects in the relaxation oscillation of a semiconductor laser *Opt. Lett.* **20** 2318–20

[92] Sarzala R P, Nakwaski W and Osinski M 1995 Effects of carrier diffusion on thermal properties of proton-implanted top-surface emitting lasers *Physics and Simulation of Optoelectronic Devices (Proc. SPIE 2393)* vol III, ed M Osinski and W W Chow (Bellingham, WA: SPIE) pp 583–604

[93] Osinski M and Nakwaski W 1994 Thermal effects in vertical cavity surface emitting lasers *Int. J. High Speed Electron. Systems* **5** 667–730

[94] Piprek J, Wenzel H and Sztefka G 1994 Modelling thermal effects on light vs current characteristics of gain-guided surface-emitting lasers *IEEE Photon. Technol. Lett.* **6** 139–42

[95] Burak D and Binder R 1997 Cold-cavity vectorial eigenmodes of VCSEL's *IEEE J. Quantum Electron.* **33** 1205–15

[96] Yu S F 1998 An improved time-domain travelling-wave model for vertical-cavity surface-emitting lasers *IEEE J. Quantum Electron. Lett.* **34** 1938–48

[97] Kuszelewicz R and Aubert G 1997 Modal matrix theory for light propagation in laterally restricted stratified media *J. Opt. Soc. Am.* A **14** 3262–72

[98] Demeulenaere B, Bienstman P, Dhoet B and Baets R 1999 Detailed study of AlAs-oxidized apertures in VCSEL cavities for optimized modal performance *IEEE J. Quantum Electron.* **35** 358–67

[99] Valle A, Pesquera L and Shore K A 1997 Polarization behaviour of birefringent multitransverse mode vertical-cavity surface-emitting lasers *IEEE Photon. Technol. Lett.* **9** 557–9

[100] Burak D, Moloney J V and Binder R 2000 Macroscopic versus microscopic description of polarization properties of optically anisotropic vertical-cavity surface-emitting lasers *IEEE J. Quantum Electron.* **36** 956–70

[101] Noble M J, Loehr J P and Lott J A 1998 Quasi-exact optical analysis of oxide-apertured microcavity VCSELs using vector finite elements *IEEE J. Quantum Electron.* **34** 2327–39

[102] Mena P V, Morikuni J J , Kang S M, Harton A V and Wyatt K W 1999 A comprehensive circuit-level model of vertical cavity surface-emitting lasers *J. Lightwave Technol.* **17** 2612–32

[103] Valle A 1998 Selection and modulation of high-order transverse modes in vertical-cavity surface-emitting lasers *IEEE J. Quantum Electron.* **34** 1924–32

[104] Corzine S W, Geels R G, Scott J W, Yan R H and Coldren L A 1989 Design of Fabry–Perot surface emitting lasers with a periodic gain structure *IEEE J. Quantum Electron.* **25** 1513–21

[105] Zhao Y G and McInerney J G 1996 Transverse-mode control of vertical-cavity surface-emitting lasers *IEEE J. Quantum Electron.* **32** 1950–8

[106] Dutta N K 1990 Analysis of current spreading, carrier diffusion, and transverse mode guiding in surface emitting lasers *J. Appl. Phys.* **68** 1961–3

[107] Nakwaski W and Osinski M 1991 Thermal properties of etched-wall surface-emitting semiconductor lasers *IEEE J. Quantum Electron.* **27** 1391–401

[108] Chen G 1995 A comparative study on the thermal characteristics of vertical-cavity surface-emitting lasers *J. Appl. Phys.* **77** 4251–8

[109] Adachi S 1983 Lattice thermal resistivity of II–V compound alloys *J. Appl. Phys.* **54** 1844–8

[110] Amith A, Kudman I and Steigmeier E F 1965 Electron and photon scattering in GaAs at high temperature *Phys. Rev.* **138** A1270–6

[111] Garel-Jones P and Dyment J C 1975 Calculation of the continuous-wave lasing range and light output power for double heterostructure lasers *IEEE J. Quantum Electron.* **11** 408–13

[112] Blakemore J S 1982 Semiconductor and other major properties of galium arsenide *J. Appl. Phys.* **53** R123–81

[113] Yao T 1987 Thermal properties of AlAs/GaAs superlattice *Appl. Phys. Lett.* **51** 1798–800

[114] Rahman B M A, Lepkowski S P and Grattan K T V 1995 Thermal modelling of vertical-cavity surface emitting lasers using the finite element method *IEEE J. Proc. Optoelectron.* **142** 82–6

[115] Joyce W B and Dixon R W 1993 Thermal resistance of heterostructure lasers *J. Appl. Phys.* **46** 855–62

[116] Nakawaski N 1987 Thermal properites of buried hetero-structure laser-diodes *IEE J. Optoelectron.* **134** 87

[117] Nakwaski W and Osinski M 1993 Thermal analysis of GaAs-AlGaAs etched-well surface-emitting double-hetero structure lasers with dielectric mirrors *IEEE J. Quantum Electron.* **29** 1981–95

[118] Sarzala P R and Nakwaski W 1990 An application of usability of the finite element method for the thermal analysis of stripe-geometry diode lasers *J. Thermal Anal.* **36** 1171–89

[119] Casey H C and Panish M B 1978 *Heterostructure Lasers* part A (New York: Academic) p 31

[120] Cook D D and Nash F R 1975 Gain-induced gain and stigmatic output beam of GaAs lasers *J. Appl. Phys.* **46** 1660–72

[121] Michalzik R and Ebelling K J 1993 Modelling and design of proton-implanted untralow threshold vertical cavity laser diodes *IEEE J. Quantum Electron.* **29** 1963–74

[122] Vukusic J A, Martinsson H, Gustavsson J S and Larsson A 2001 Numerical optimization of the single fundamental model output from a surface modified vertical cavity surface emitting lasers *IEEE J. Quantum Electron.* **37** 108–17

[123] Bissessur H K, Koyama F and Iga K 1997 Modelling of oxide-confined vertical-cavity surface-emitting lasers *IEEE J. Selected Topics Quantum Electron.* **3** 344–52

[124] Li X, Sale T E and Knowles G 2002 Self-consistent modelling of oxide aperture visible vertical-cavity surface-emitting laser devices *J. Mod. Opt.* **49** 905–12

[125] Tauber D, Wang G, Geels R S, Bowers J E and Coldren L A 1993 Large and small signal dynamics of vertical cavity surface emitting lasers *Appl. Phys. Lett.* **62** 325–7

[126] Shtengel G, Temkin H, Brusenbach P, Uchida T, Parsons C, Quinn W E and Swirhun S E 1993 High speed vertical cavity surface emitting laser *IEEE Photon. Technol. Lett.* **5** 1359–62

[127] Nishiyama N, Arai M, Shinada S, Suzuki K, Koyama F and Iga K 2000 Multi-oxide layer structure for single-mode operation in vertical-cavity surface-emitting lasers *IEEE Photon. Technol. Lett.* **12** 606–8

[128] Wu Y A, Chang-Hasnain C J and Nabiev R 1994 Transverse mode selection with a passive antiguide region in vertical cavity surface emitting lasers *IEEE Photon. Technol. Lett.* **6** 924–6

[129] Floyd P D, Peters M G, Coldren L A and Merz J L 1995 Suppression of higher-order transverse modes in vertical-cavity lasers by impurity-induced disordering *IEEE Photon. Technol. Lett.* **7** 1388–90

[130] Morgan R A, Hibbs-Brenner H K, Lehman J A, Kalweit E L, Walterson R A, Marta T M and Akinwande T 1995 Hybrid dielectric/AlGaAs mirror spatially filtered vertical cavity top-surface emitting laser *Appl. Phys. Lett.* **66** 1157–9

[131] Dowd P, Raddatz L, Sumaila Y, Asgharai M, White I H, Penty R V, Heard P J, Allen G C, Schneider R P, Tan M R T and Wang S Y 1997 Mode control in vertical-cavity surface-emitting lasers by post-processing using focused ion-beam etching *IEEE Photon. Technol. Lett.* **9** 1193–5

[132] Choquette K D and Leibenguth R E 1994 Control of vertical-cavity laser polarization with anisotropic transverse cavity geometry *IEEE Photon. Technol. Lett.* **6** 40–2

[133] Schablitsky S J, Zhuang L, Shi R C and Chou S Y 1996 Controlling polarization of vertical-cavity surface-emitting lasers using amorphous silicon subwavelength transmission gratings *Appl. Phys. Lett.* **69** 7–9

[134] Mukaihara T M, Ohnoki N, Hayashi Y, Hatori N, Koyama F and Iga K 1995 Polarization control of vertical-cavity surface emitting lasers using a birefringent mela/dielectric polarizers loaded on top distributed Bragg reflectors *IEEE Selected Topics Quantum Electron.* **1** 667–73

[135] Dowd P, Heard P J, Nicholson J A, Raddatz L, White I H, Penty R V, Allen G C, Corzine S W and Tan M R 1997 Complete polarization control of GaAs gain-guided top-surface emitting vertical cavity lasers *Electron. Lett.* **33** 1315–17

[136] Unold H J, Reid M C, Michalzik R and Ebeling K J 2002 Polarization control in VCSELs by elliptic surface etching *Electron. Lett.* **38** 77–8

[137] Chu H Y, Yoo B -S, Park M S and Park H -H 1997 Polarization characteristics of index-guided surface-emitting lasers with tilted pillar structure *IEEE Photon. Technol. Lett.* **9** 1066–8

[138] Chavez-Pirson A, Ando H, Saito H and Kanbe H 1993 Polarization properties of a vertical cavity surface emitting lasers using a fractional layer superlattice gain medium *Appl. Phys. Lett.* **62** 3082–4

[139] Mukaihara T, Koyama F and Iga K 1993 Engineered polarization control of GaAs/AlGaAs surface-emitting lasers by anisotropic stress from elliptical etched substrate hole *IEEE Photon. Technol. Lett.* **5** 133–5.

[140] Choquette K D, Schneider J R P and Lott J A 1994 Lasing characteristics of visible AlGaInP/AlGaAs vertical-cavity lasers *Opt. Lett.* **19** 969–71

[141] Kaneko Y, Nakagawa S, Takeuchi T, Mars D E, Yamada N and Mikoshiba N 1995 InGaAs/GaAs vertical-cavity surface-emitting laser on (311)B substrate *Electron. Lett.* **31** 805–6

[142] Numai T, Kurihara K, Kuhn K, Kosaka H, Ogura I, Kajikawa M, Saito H and Kasahara K 1995 Control of light-output polarization for surface-emitting laser type device by strained active layer grown on misoriented substrates *IEEE J. Quantum Electron.* **31** 636–42

[143] Takahashi M, Vaccaro P, Fujita K, Watanabe T, Mukaihara T, Koyama F and Iga K 1996 An InGaAs-GaAs vertical-cavity surface emitting laser grown on GaAs (311)A substrate having low threshold and stable polarization *IEEE Photon. Technol. Lett.* **8** 737–9

[144] Taneo K, Ohiso Y, Amano C, Wakatsuki A and Kurokawa T 1997 Growth of vertical-cavity surface-emitting laser structures on GaAs (311)B substrates by metalorganic chemical vapor deposition *Appl. Phys. Lett.* **70** 3395–7

[145] Mizutani A, Hatori N, Ohnoki N, Nishiyama N, Ohtake N, Koyama F and Iga K 1998 InGaAs–GaAs vertical-cavity surface-emitting laser on GaAs (311)B substrate using carbon auto-doping *Japan. J. Appl. Phys.* **37** 1408–12

[146] Nishiyama N, Arai M, Shinada S, Azuchi M, Miyamoto T, Koyama F and Iga K 2001 Highly strained GaInAs-GaAs quantum-well vertical-cavity surface-emitting laser on GaAs (311)B substrate for stable polarization operation *IEEE Selected Topics Quantum Electron.* **7** 242–7

[147] Lee T P (ed) 1995 *Current Trends in Vertical Cavity Surface Emitting Lasers* (London: World Scientific)

[148] Sale T E 1995 *Vertical Cavity Surface Emitting Lasers* (Baldock, Herts: Research Studies Press)

[149] Wilmsen C, Temkin H and Coldren L A 1999 *Vertical-Cavity Surface-Emitting Lasers: Design, Fabrication, Characterization and Applications* (Cambridge: Cambridge University Press)

[150] Cheng J and Dutta N K (ed) 2000 *Vertical-Cavity Surface-Emitting Lasers: Technology and Applications* (Amsterdam: Gordon and Breach Science)

[151] Knigge A, Zorn M, Weyers M and Trankle G 2002 High-performance vertical-cavity surface-emitting laser with emission wavelength between 650 and 670 nm *Electron. Lett.* **38** 882–3

[152] Jouhti T, Peng C S, Pavelescu E M, Konttinen J, Gomes L A, Okhotnikov O G and Pessa M 2002 Strain-compensated GaInNAs structures for 1.3 μm lasers *IEEE J. Selected Topics Quantum Electron.* **8** 787–94

[153] Otsubo K, Nishijima Y and Ishikawa H 1998 Long-wavelength semiconductor lasers on InGaAs ternary substrates with excellent temperature characteristics *Fujitsu Scientific Tech. J.* **34** 212–22

[154] Yu S F 1996 Dynamic behaviour of vertical-cavity surface-emitting lasers *IEEE J. Quantum Electron.* **32** 1168–79

[155] Dutta N K, Tayahi M and Choquette K D 1997 Transmission experiments using oxide confined vertical cavity surface emitting lasers *Electron. Lett.* **33** 1147–8

[156] Chang-Hasnain C J 2000 Tunable VCSEL *IEEE J. Selected Topics Quantum Electron.* **6** 978–87

B2.7
Long wavelength laser diodes

S Anders, G Strasser and E Gornik

B2.7.1 Introduction

Optoelectronics relies on emitters and detectors of electromagnetic radiation. The most widely used mechanism to generate light from a semiconductor is the recombination of electrons and holes across the bandgap of a semiconducting material. Soon after the invention of the laser, this photon-emitting process was used to demonstrate the first semiconductor lasers based on GaAs [1–3]. The bandgap of GaAs (1.4–1.5 eV) sets the emission wavelength in the near infrared (~800 nm). Considerable effort has been undertaken to develop semiconductor lasers that operate at longer and at shorter wavelengths. For short wavelengths, cheap and efficient diode lasers based on GaAs and InP are available. A breakthrough has occurred by solving the technological problems of wide-gap semiconductors. Green, blue and ultraviolet current injection lasers based on II–VI compounds (ZnSe [4]) as well as on III–V compounds (GaN [5]) have been developed. For wavelengths in the mid-infrared (MIR, also called 'long wavelengths'), narrow gap semiconductor lasers were demonstrated in the sixties [6]. The most prominent example is lead-salt compounds with an emission range from below 3 μm to about 30 μm. A gap exists in the far-infrared, while wavelengths beyond 1 mm can be readily generated by high-frequency oscillators. Figure B2.7.1 presents an overview of the emission ranges of various types of lasers. While the near-infrared range is well covered by commercially available lasers, quantum cascade lasers (QCLs) are very promising sources for the mid- and far-infrared range.

In recent years, steady progress has been made to fill gaps in the emission wavelengths that are not conveniently accessible. The advance in epitaxial-layer growth and bandstructure engineering led to improvements of MIR sources that have been available for decades. In particular, these are lead-salt lasers, antimonide-based and mercury-cadmium-telluride-based lasers. Most notably, molecular beam epitaxy allowed the realization of quantum cascade lasers [7]. QCLs are unipolar lasers based on transitions between quantum levels in ultrathin layers.

The emission range of the MIR (wavelength range: 3–30 μm) is interesting for several industries. Compact, cheap and reliable coherent sources are useful for a wide spectrum of applications ranging from IR astronomy to the automobile and telecommunication industries and to chemical sensing applications.

Many small molecules have their fundamental absorption bands in the MIR. Therefore, this frequency range is sometimes called the fingerprint region of molecules. Applications in gas sensing have been shown as early as 1998 [8]. A recent overview of near- and mid-infrared lasers as optical sensors for gas analysis is presented in the work of Werle *et al* [9]. The designable wavelength of quantum cascade lasers makes it possible to install a detection system for literally any organic molecule. Environmental control, the food industry, pollution and contamination control applications will most probably make use of these coherent sources once the product is established. A number of dedicated QCL 'mini-spectrometers' installed at strategic points in any process line (like, e.g., semiconductor industries) would ensure that any process failures are rapidly detected, thereby reducing redundant processing and minimizing product loss.

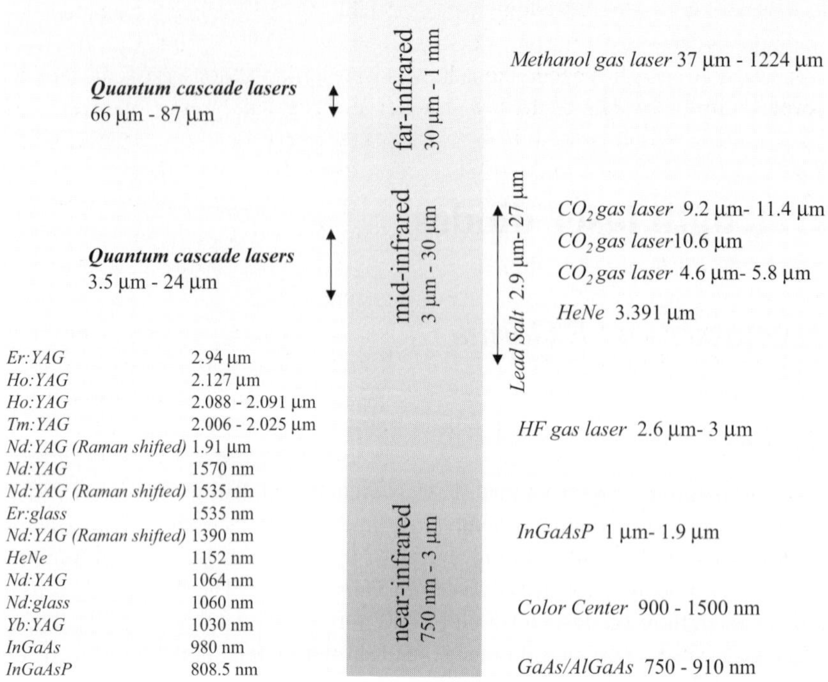

Figure B2.7.1. Overview of the emission ranges of various types of lasers. The wavelength axis is not to scale. (Source: *The Photonic Spectrum Reference Wall Chart* Photonics Spectra 2000.)

Other potential applications are in IR astronomy and remote sensing collision avoidance radars. The atmospheric windows in the MIR spectral range—from 3–5 and 8–12 μm—make coherent MIR sources favourable to telecommunications [11, 12], because these wavelengths are not strongly scattered by fog, smog and dust particles suspended in the atmosphere. This would allow a replacement of microwave RF links by MIR systems that are smaller, directional and can be easily focused. Further, high-speed applications of actively and passively mode-locked QCLs are being developed [13]. Emission in the mid- to far-infrared is well suited for law enforcement (concealed weapons detection) and applications in medicine and biology [14].

This chapter will describe selected MIR emitters, namely lead-salt lasers, antimonide-based and mercury-cadmium-telluride-based lasers, and quantum cascade lasers. Most of the space is dedicated to quantum cascade lasers. They are relatively new and considered most exciting by researchers in the field because an extended quantum mechanical structure is grown in a controlled fashion, and because of their potential for applications. For more extensive reviews about the established lasers, the reader may refer to the literature suggested as further reading.

B2.7.2 Lead salt lasers

Lead-salt lasers cover the emission range from below 3 μm to almost 30 μm [15–17]. Continuous emission is observed at temperatures that can be achieved with thermoelectric coolers, while pulsed operation is possible above room temperature [18]. For more than two decades, lead-salt lasers have been commercially available. Their high sensitivity and selectivity makes them well suited for MIR gas analysis. Once a single-mode emitter with an emission near the target absorption energy has been selected and tuned, it can be operated for years with reproducible performance.

In principle, emitters with long wavelengths are possible by decreasing the bandgap of the material. However, as the bandgap is decreased, the non-radiative Auger recombination drastically increases. Especially at the high electron and hole densities needed to achieve lasing action at non-cryogenic temperatures, Auger recombination severely limits the gain of the active material. Lead-salt lasers circumvent this hindrance to some extent. The amount of Auger recombination depends, among other things, on the bandstructure. The direct bandgap minimum of lead-salt compounds is not located at the Γ point like in III–V band shapes, but rather there are four minima located at the L-points. The bandstructure influences energy and momentum conservation of an Auger process. Thus, it becomes clear why Auger processes are less likely in some bandstructures. The inhibited Auger recombination is the reason why conventional III–V diode lasers cannot compete with lead-salt lasers for wavelengths longer than about 3.5 μm.

Examples of lead-salt components are PbS (one of the few naturally occuring semiconductors), PbTe, PbSe, PbSnTe, PbSnSe, PbCdSe and PbSSe [19]. The infrared dielectric constant of these materials is quite high and, as usual, increases with decreasing bandgap. Thus, one can combine the materials to obtain a structure that confines the carriers and guides the light. A disadvantage for lasing operation is that the lead-salt components have a low thermal conductance. Lead-salt lasers were made early on by chemical or vapour deposition with a curing step. Further, epitaxially grown (liquid phase, hot wall and molecular beam epitaxy is used) and single crystal lead-salt lasers have been fabricated. Vertical-cavity surface-emitting lasers based on lead-salt components have also been demonstrated [20, 21].

B2.7.3 Antimonide-based lasers

Lasers fabricated from antimony-containing III–V compounds are developed for emission wavelengths of 2–5 μm. Before 1990, antimonide-based lasers were mostly grown by liquid phase epitaxy. Room-temperature continuous-wave operation was reported at 2.3 μm [22], however, at longer wavelengths lasing operation was restricted to cryogenic temperatures. Liquid phase epitaxy cannot be used to grow sophisticated structures such as strained quantum well lasers that can potentially reduce Auger recombination. Therefore, the use of molecular beam and organometallic vapour phase epitaxy improved the performance of antimonide-based lasers.

So far, GaInAsSb/AlGaAsSb quantum well lasers have shown the best performance between 1.9 and 3 μm. Room temperature pulsed operation has been observed up to 2.9 μm [23], while continuous-wave operation has been obtained at 2 μm [24, 25]. Using InAsSb as the active layer allows emission beyond 3 μm, however, well below room temperature because of Auger recombination.

In addition to these type I lasers, type II lasers have been demostrated. In type II lasers, conduction and valence band minima are physically separated. In principle, the wavelength can now be extended beyond the values given by the bandgap of each alloy. Also, Auger recombination can be reduced by proper design. Type II lasers with GaInSb/InAs superlattices or quantum wells have emission wavelengths between 2.7 and 5.2 μm. Optically pumped type II lasers function at higher temperatures (350 K [26]) than electrically pumped type II lasers (255 K [27]). Both maximum operating temperatures are for an emission wavelength of 3.2 μm. Operation at 27 μm is proposed [28].

B2.7.4 Mercury-cadmium-telluride-based lasers

Mercury-cadmium-telluride compounds have been studied for 40 years [29,30]. The $Cd_xHg_{1-x}Te$ alloy is the only semiconductor that can be grown with any bandgap of 0–1.48 eV. Zero bandgap is reached for $x = 0.15$. The lattice parameter varies only by 3×10^{-3} for this composition range, and the energy band structure is of the direct type. Even though these are attractive properties for MIR emitters, so far this compound has been mostly used for the fabrication of infrared detectors and detector arrays.

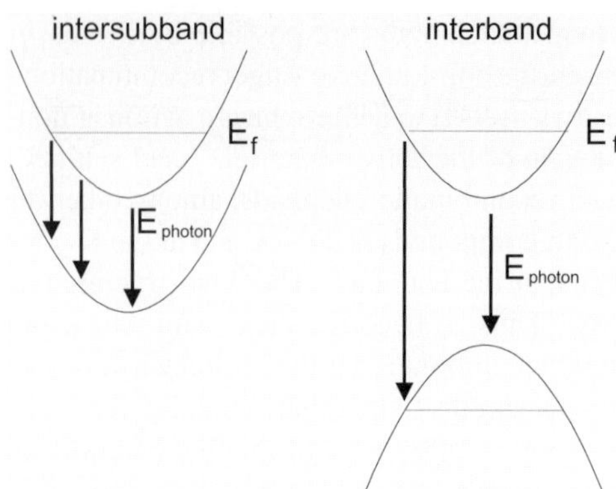

Figure B2.7.2. Schematic diagrams of interband and intersubband optical transitions.

The best optically pumped lasers were grown by MBE. They were designed to emit at 3.2 μm by using a composition of $x = 0.38$ and a quantum well thickness of 20 nm. The cladding layers had a composition of $x = 0.5$ and a thickness of 1 μm. The devices worked up to 154 K and had an external quantum efficiency of 7.5% at 80 K [31]. For electrical pumping, the active material has to be inserted into a p–i–n junction with doping levels in excess of 10^{17} cm^{-3} for the doped regions. Non-radiative recombination (mostly Auger recombination) appears to limit the performance of the electrically pumped lasers produced so far to operating temperatures below 100 K [32–34]. Improvements are expected by incorporating quantum wells into the recombination region because carrier confinement strongly modifies the Auger recombination [35, 36].

B2.7.5 Quantum cascade lasers

In contrast to bandgap lasers, which are based on electron-hole recombination, unipolar lasers are based on *intersubband* transitions. The lasing transition occurs between sublevels of the same polarity (see figure B2.7.2). Electronic levels are used for InP and GaAs based QCLs, because the conduction band offset in these materials is larger than the valence band offset. Because the electrons that are involved in the optical transition are still available after the transition, the operation of several light emitting stages in series is possible. One electron passing through the device can thus emit several photons ('electron recycling'). This boosts the optical output to several hundred mW, depending on the temperature, the type of operation (pulsed or continuous mode) and the material. *Interband* cascade lasers are also possible; here, we focus on intersubband cascade lasers. The unipolar nature of these lasers means that, as long as non-parabolicity effects are excluded, all initial and final states have the same curvature. The joint density of states is sharp; the gain linewidth does not directly depend on the temperature. The gain is only limited by the amount of current that can be driven through the structure.

The light emitting cascades consist of a multiple-quantum well heterostructure. Figure B2.7.3 shows the conduction band with an applied field for two QCL designs. The wavefunctions are obtained by solving Schrödinger's equation. To illustrate the transition probabilities, the squared wavefunctions are plotted at the energy of the electronic states. In a superlattice design (figure B2.7.3(*a*)), the active region consists of several wells and barriers such that the lower lasing level is emptied by tunnelling of the electrons into a miniband. For three-well designs, the thicknesses of the wells and barriers in the active region is chosen such that the lasing transition occurs between the wavefunctions labelled '3' and '2' in figure B2.7.3(*b*). The lower lasing level '2' is emptied efficiently because the energy difference between level '2' and '1' corresponds to one longitudinal optical phonon energy, 30–40 meV depending on the material. For both three-well and superlattice designs, the active regions are separated by injector regions. These serve as charge reservoirs and

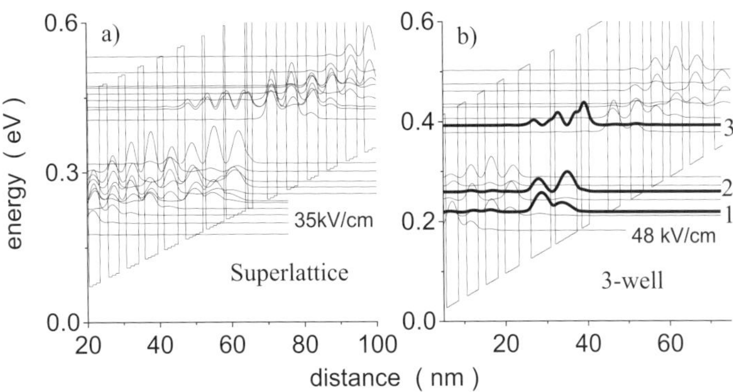

Figure B2.7.3. Band diagram and squared wavefunctions of a superlattice and a three-well quantum cascade structure.

contain doped layers. Injectorless QCLs have been fabricated by using an appropriately designed chirped superlattice active region [38].

The main difficulty in the realization of an intersubband laser is the achievement of population inversion, which is not an intrinsic property like in interband lasers, but a result of properly designed scattering times. The dominant scattering mechanism between subbands separated by more than one optical phonon energy is the emission of optical phonons [39, 40]. This limits the non-radiative lifetimes of the excited states to the order of a picosecond. Lifetime measurements [41] on InGaAs/InAlAs show that, in general, the lifetime of an excited state has a minimum ($\sim$0.25 ps) for a subband spacing equal to the energy of the optical phonons. It increases monotonically up to a few ps for subband spacings of about 300 meV. In the far-infrared (or terahertz) regime, where the subband spacing is well below an optical phonon energy, a different mechanism to establish population inversion had to be found.

For a three-well intersubband emitter, the active region consists of a three-level scheme, designed to ensure population inversion between level 3 and level 2. In a simple model, the inversion condition is reduced to $\tau_{32} > \tau_2$. Here, $1/\tau_{32}$ denotes the non-radiative scattering rate from level 3 to level 2, and $1/\tau_2$ denotes the total scattering rate out of level 2. The transition $3 \rightarrow 1$ is, in coupled quantum wells, not negligible. This leads to an inversion condition of $\tau_3 > \tau_2$ (total lifetimes). A characterization for the $3 \rightarrow 2$ transition is given by the strength of the wavefunction overlap of these two states. For a strong overlap, the transition is called vertical; in a regime with a weak overlap, the term diagonal transition is used.

Generating infrared radiation based on the electronic transitions between minibands in a superlattice has some advantages compared to coupled quantum well systems. The main advantage is the high oscillator strength of the radiative transitions between two minibands. The oscillator strength has a maximum at the mini-Brillouin zone boundary ($k_z = \pi/d$). The injected electrons are located in the upper miniband, while in the lower miniband in properly designed and carefully doped structures the states at the zone boundary are empty. This is called 'natural inversion'.

In weakly coupled superlattices, transport proceeds by sequential resonant tunnelling, and is usually accompanied by the formation of electric-field domains. Applying external electric fields breaks superlattice structures into high- and low-field domains, where the energy levels lock in resonances [42]. In a laser, it is necessary to have a homogenous and stable electric field distribution. Therefore, doping is inserted into the QCL to compensate the total negative charge in the active region by positive ions (impurities). This prevents charge buildup even for strong injection. Also, the operating point of a QCL structure has to be stable to avoid field domain formation. The active region is normally left undoped, since impurity atoms in the active region reduce the lifetimes of the excited states. Dopants in the active region broaden and red-shift the lasing

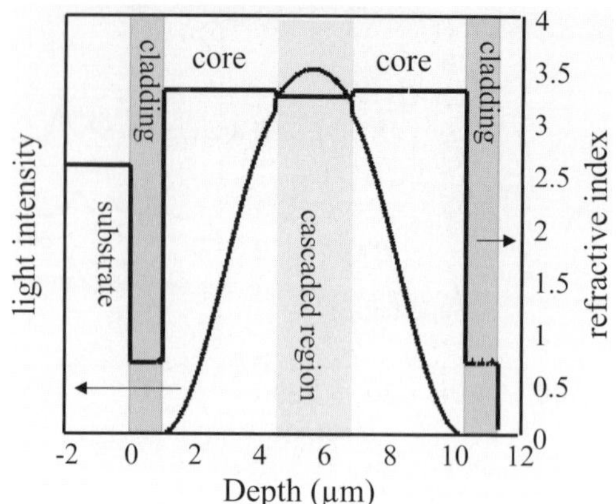

Figure B2.7.4. Refractive index and light intensity of a double plasmon enhanced waveguide.

transitions [43]. The doping is usually inserted in the centre of the injectors, where scattering increases the relaxation probability and, thus, enhances the injection efficiency.

In addition to a radiative transition combined with a population inversion in a quantum cascade structure, a cavity is necessary to achieve lasing. In the simplest approach, the cleaved side facets of a laser bar act as mirrors. Optical confinement within these mirrors is, like in bandgap lasers, achieved by dielectric waveguides. The main requirement for an optical waveguide, apart from guiding the light, is to have low optical losses. In QCLs, the waveguide losses are predominantly controlled by free carrier absorption and can be described by a Drude model [44,45]. In contrast to near-infrared lasers, mid-infrared QCLs have to deal with high internal losses and rather thick cladding layers. The total thickness of the waveguide structure is limited because the growth time should not be excessive, and because the defect density should be kept minimal. Also, the thermal transport across the layers needs to be as efficient as possible.

A plasmon enhanced waveguide may be used for mid-infrared emitting lasers. The first plasmon enhanced waveguide for a QCL was demonstrated in the InGaAs/InAlAs system in 1995 [46] and in the GaAs/AlGaAs system in 1999 [47]. In this waveguide design, the refractive index variations are based on changes in the doping concentration n or on interfaces between different materials. The plasma frequency ω_{p} of the highly doped n^{++} layers is shifted close to the laser frequency. Especially for the GaAs system, plasmon enhanced GaAs waveguides are preferred to AlGaAs cladding layers for a variety of reasons. First, the refractive index of GaAs is higher than the effective refractive index of the active material. Hence, in contrast to the InGaAs/InAlAs/InP system, the substrate does not support the waveguiding. From the technological point of view, the growth of thick GaAs layers is much more reliable than the growth of thick AlGaAs layers with high Al contents, where residual strain limits the total thickness. Further, ternary alloys like AlGaAs have a poor thermal conductivity compared to binary materials like GaAs. Figure B2.7.4 shows the refractive indices of a double plasmon enhanced waveguide together with the optical intensity. For the plasmon enhanced waveguides, the overlap between the n^{++}-layers and the optical mode needs to be minimized to avoid very high waveguide losses. This is done with 3–4 μm thick low doped (some 10^{16} cm^{-3}) GaAs layers.

A QCL consists of several tens of cascades, each containing an active region and an injector, embedded in the waveguide. The total thickness of MBE-grown material adds up to several micrometres. Figure B2.7.5 shows a TEM image of a cross section of several layers. Figure B2.7.6, a $\omega/2\theta$ scan, shows an x-ray scan along the (004) plane of GaAs.

Compared to bandgap lasers, the major advantage of QCLs is that their emission energy is not fixed

Figure B2.7.5. TEM image of a cascaded structure.

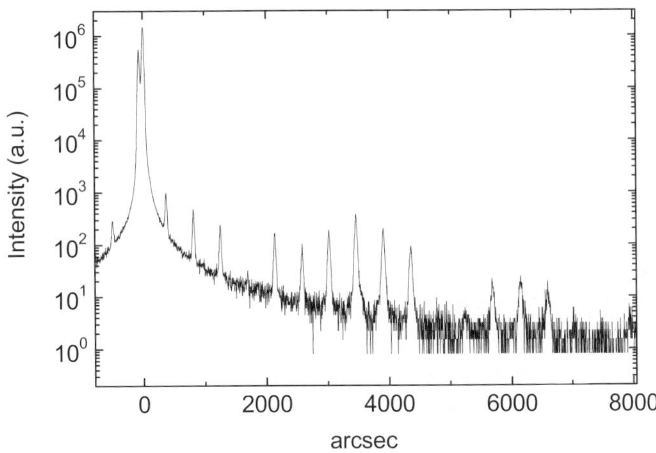

Figure B2.7.6. X-ray scan of a cascaded structure.

by the bandgap of the utilized materials, but can be tailored by bandstructure engineering. This allows a wide emission range in the mid-infrared, currently from 3.5 μm [48] to 24 μm [49] and from 66 μm [50] to 87 μm [51].

Lasing in a cascaded structure was first proposed in 1971 [52]. In this work, electrons tunnel from a quantum well ground state to an excited state of the neighbouring well by photon-assisted tunnelling. After the photon emission, the electrons non-radiatively relax into the ground state and are injected into the next stage. The population inversion necessary for lasing action is obtained here because of a long non-radiative transition time ($\geq$10 ps) in the 'diagonal' transition compared to the short intrawell transition time. The first intersubband emission data were observed in an electron gas at a Si/SiO_2 interface [53]. The first intersubband luminescence spectrum in a GaAs/AlGaAs superlattice pumped by resonant tunnelling was demonstrated in 1988 [54]. Because of the stringent requirements on the layer thicknesses and interface quality, the growth of QCLs was only possible after the advent of molecular beam epitaxy (MBE) [55]. In 1994, the first QCL

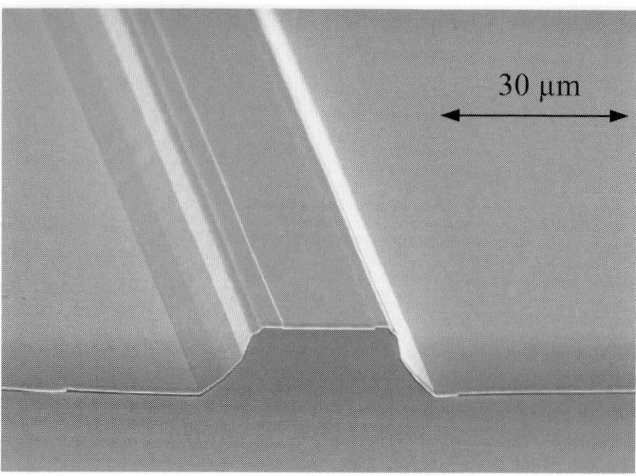

Figure B2.7.7. SEM image of a Fabry–Pérot laser cavity.

was built [7]. It was based on the InGaAs/InAlAs material system. Four years later, the AlGaAs/GaAs system was the second system that showed lasing [56]. This material system potentially has a number of key advantages, including a more mature materials technology, amenability to dry etching for improved device fabrication, good lattice matching over the entire range of alloy compositions, and the possibility to realize extremely long wavelength emission. Disadvantages arise due to the lower conduction band offset in this material system [57], which places a lower limit on the lasing wavelength ($\sim$7 μm) and decreases the confinement of the upper lasing state. However, mostly because the InGaAs/InAlAs material system is better explored, continuous-wave operation at room temperature has been demonstrated only for this system so far [58]. Products based on the InP/InGaAs/AlInAs technology are already commercially available [59]. While there is some effort to develop SiGe-based QCL structures [60], to date InP and GaAs remain the only two material systems that show quantum cascade lasing.

The processing of electrically pumped laser cavities from the MBE-grown heterostructure is done by photolithography. Typically, the laser cavity consists of a ridge that is about 1 mm long and several tens of micrometres wide (see figure B2.7.7 for an SEM image). The lateral confinement can be done by either wet-chemical or, if a straight profile is desired, by reactive-ion etching. Due to the unipolar nature of QCLs, surface recombination is, in contrast to bandgap lasers, not an issue. To complete the most simple device, metallization is deposited on top of the ridge as well as on the back of the device. Alternatively, extended top contacts for improved current homogeneity can be fabricated. SiN or SiO may be used as isolation. The metal contacts can be processed as either Schottky contacts or, if less heat dissipation is desired, as alloyed ohmic contacts. Here, care must be taken not to alloy too deep into the waveguide, since a high charge concentration close to the light field is detrimental to the light output power.

The end facets are frequently left as-cleaved. For higher reflectivity, a coating of, for example, ZnSe and Au may be applied to one of the mirrors. Another approach is to process a Bragg mirror with a focused ion beam. This can decrease the threshold current by up to 40% [61]. Microcylinder-shaped cavities do not need mirrors because the light propagates mostly by total internal reflection (see figure B2.7.8 for an example of a ring-shaped laser) [62].

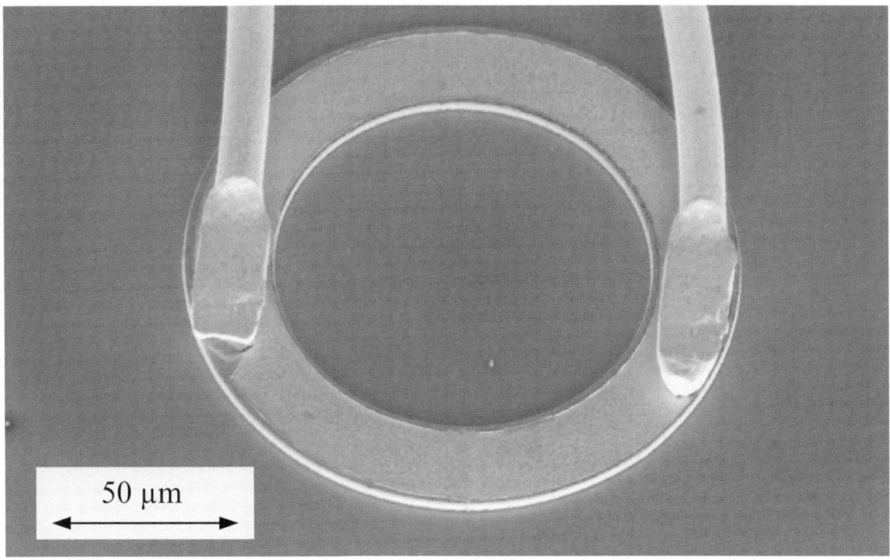

Figure B2.7.8. SEM image of a ring-shaped laser cavity.

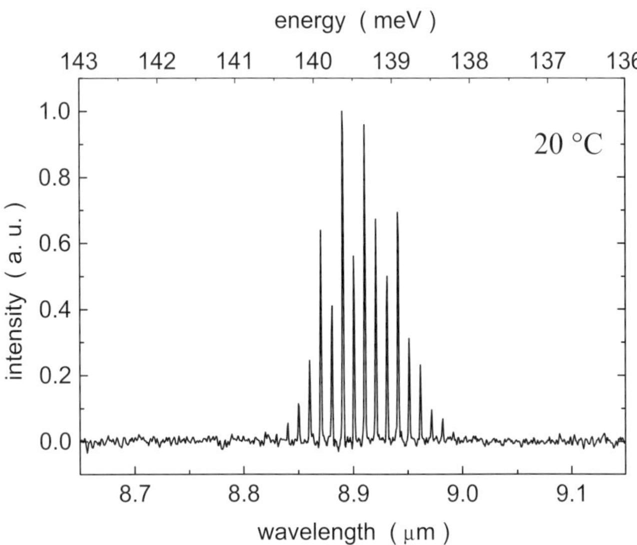

Figure B2.7.9. Emission spectrum of a QCL with Fabry–Pérot cavity.

B2.7.5.1 Distributed feedback QCLs

For several applications it is desirable to have a single-mode emitter. While conventional ridge cavities emit a Fabry–Pérot spectrum (see figure B2.7.9), the fabrication of a grating on top of the laser ridge fixes the emission wavelength to a desired value. The first 'distributed feedback (DFB) laser' was introduced in 1997. Figure B2.7.10 shows a SEM image of a grating that was defined by photolithography. For gratings with a period of less than about 1 μm, holographic lithography may be used. Figure B2.7.11 demonstrates the

Figure B2.7.10. SEM image of a DFB grating. The grating extends across the trench that defines the laser cavity. The SiN isolation (300 nm thick) is visible.

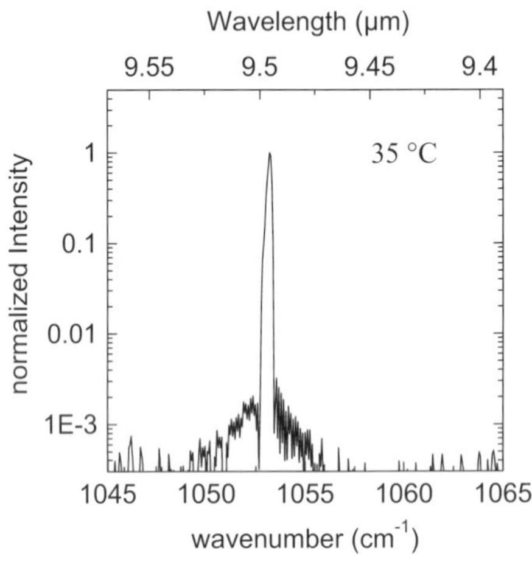

Figure B2.7.11. Single mode emission spectrum of a DFB laser.

good sideband suppression of a DFB laser. Figure B2.7.12 shows that the emission energy can be tuned by temperature, thus allowing it to sweep across, for example, a chemical absorption spectrum.

B2.7.5.2 Terahertz (far-infrared) QCLs

Recently, the emission range of QCLs has been extended into the terahertz range [64]. This is a particularly important feat since it is the first versatile approach to bridge the 'terahertz gap'. Already in 1984 lightly doped germanium has been shown to show lasing action under crossed electric and magnetic fields [65]. However, the properties of this semiconductor laser were fixed by the bulk material. The great advantage of a QCL terahertz emitter is that the design is flexible. Terahertz sources are beneficial for applications in astronomy, telecommunications, medical imaging and concealed weapon detection. The terahertz gap is approached from the high-energy side by semiconductor lasers that can operate at frequencies no lower

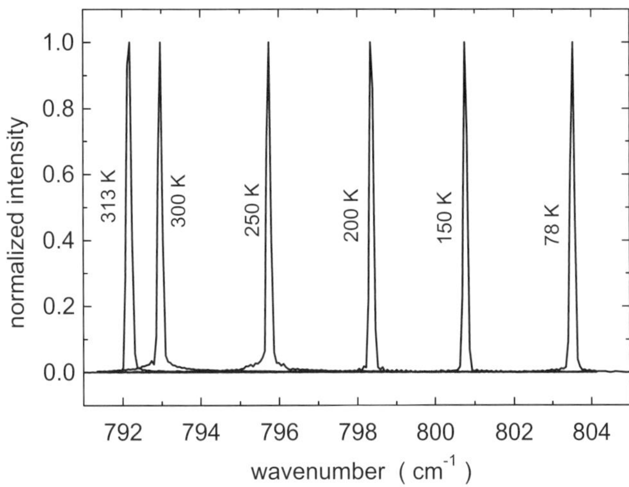

Figure B2.7.12. Temperature tuning behaviour of a DFB laser.

than 30 THz (about 10 μm). On the low-energy side, transistors deliver frequencies not higher than about 300 GHz. Between the two technologies, the conversion of electrical power into electromagnetic radiation poses several difficulties. Free-carrier absorption losses increase quadratically with the wavelength, and the optical mode scales with the wavelength so that a sufficiently large (or clever) waveguide must be designed. Further, the lifetime of the upper lasing level is short because of much electron–electron scattering between the closely spaced energy levels. For the first terahertz QCL, Köhler *et al* developed a hybrid waveguide between a 'metallic microwave strip-line' and a 'surface plasmon waveguide'. A chirped superlattice heterostructure has been used to achieve fast depopulation of the lower lasing level as well as efficient depletion of the upper level. Figure B2.7.13 shows a lasing spectrum of the first functioning terahertz QCL together with spontaneous emission spectra at currents below threshold. The spectra were taken at 8 K. Improvements of terahertz lasers with power output and maximum operating temperature moving towards commercial terahertz technology are already underway [50, 66].

B2.7.5.3 Continuous wave operation

Considerable progress has been made as the first QCL operating continuous wave at room temperature was introduced [58]. This QCL emits at 9.1 μm. The output power is 17 mW at 20 °C. Because no cooling is needed and the linewidth can be narrower than the 0.1 cm^{-1} that limits pulsed lasers, such a laser is a milestone for applications in high-resolution spectroscopy and free-space optical communication.

Compared to bandgap lasers, the thresholds of even the best QCLs are rather high because of the large fraction of non-radiative recombination. Threshold current densities of several kA cm^{-2} dissipate several tens of kW cm^{-2}. This is enough heat load to substantially raise the laser temperature above the ambient temperature. If the coupling to the heat sink is poor, the laser temperature of a continuous-wave-driven QCL can easily be 100 K above the heatsink temperature. Pulsed laser operation circumvents this problem by allowing the laser to cool in between pulses. Figure B2.7.14 shows the phase shift of a detection laser beam that illuminates the laser ridge. The phase shift is proportional to the temperature change. Thus, figure B2.7.14 illustrates how the temperature in the laser ridge increases during a 100 ns long pulse. After the pulse, the temperature decreases exponentially. The decrease can be fitted by assuming heat conductivities of 0.25 W K^{-1} cm^{-1} parallel to the layers and 0.015 W K^{-1} cm^{-1} perpendicular to the layers [67].

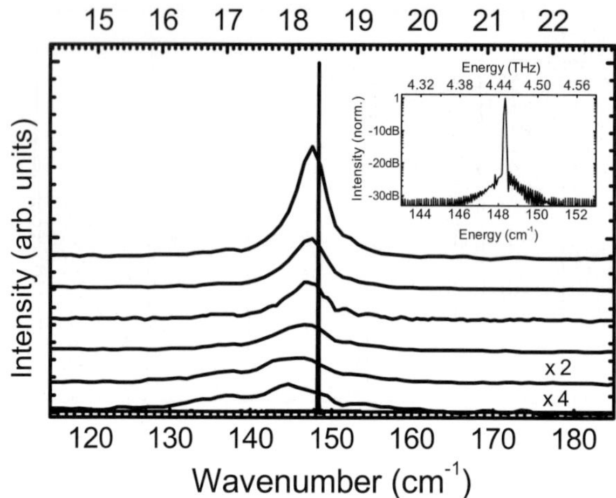

Figure B2.7.13. Terahertz emission from a 1.24 mm long and 180 μm wide device recorded at 8 K. Shown are spectra below and above lasing threshold. Currents are 300, 450, 600, 700, 750, 850 and 1240 mA. Inset: single-mode emission with a side-mode suppression ratio of 20 dB. (Reproduced by permission of Nature Publishing Group from Köhler R, Tredicucci A, Beltram F, Beere H E, Linfield E H, Davies A G, Ritchie D A, Iotti R C and Rossi F 2002 *Nature* **417** 156.)

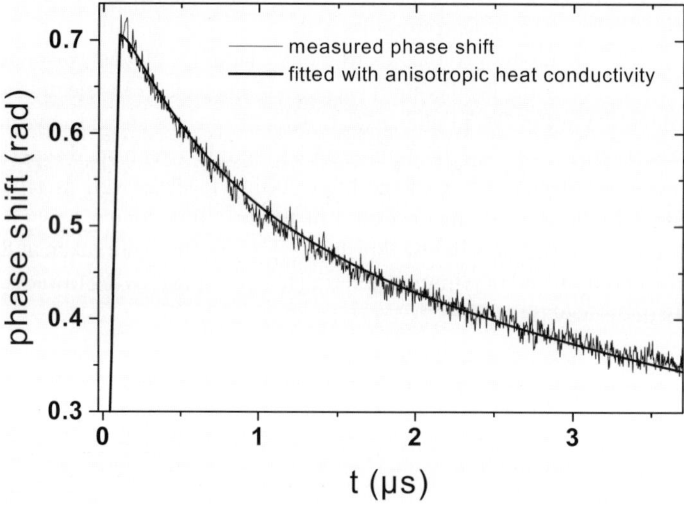

Figure B2.7.14. Phase shift of a detection laser beam that illuminates the QCL ridge from the back. The phase shift is proportional to the temperature increase of the QCL while it is driven with a 100 ns long pulse.

For room-temperature continuous-wave operation, sufficient thermal coupling is thus essential. Further, the characteristic temperature T_0, which describes the temperature dependence of the threshold $J_{th} = J_0 \exp(T/T_0)$, should be large. In the approach of Beck *et al* [58], the active region consisted of four quantum wells (see figure B2.7.15). The lower three energy states are separated roughly by one longitudinal optical phonon energy each. A narrow quantum well and barrier separate the injector region from the active region.

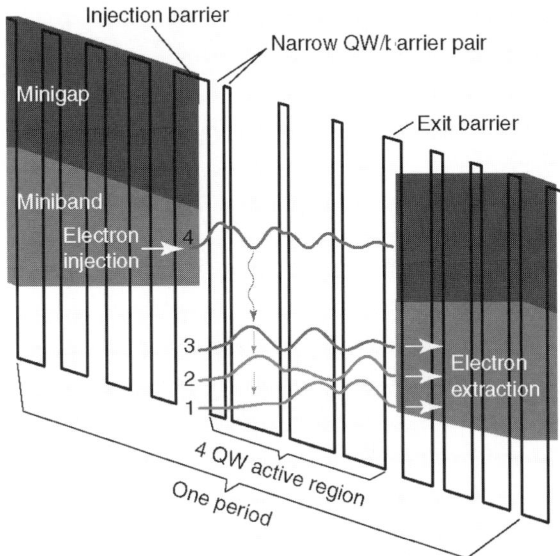

Figure B2.7.15. Four-well design of the structure that emitted continuous wave at room temperature. (Reprinted with permission from Beck M, Hofstetter D, Aellen T, Faist J, Oesterle U, Ilegems M, Gini E and Melchior H 2001 *Science* **295** 310. Copyright 2002 American Association for the Advancement of Science.)

This increases the magnitude of the upper state wavefunction locally and enhances the injector efficiency. The devices were processed in a narrow-stripe, planarized and buried heterostructure geometry, all of which serve to enhance the heat flow. Continuous wave operation was observed up to 39 °C, which is quite a jump from the previously observed record of −30 °C where junction-down-mounted devices were used. The threshold current density at room temperature was 4.3 kA cm^{-2} at a voltage bias of 7.6 V. Figure B2.7.16 shows high-resolution spectra at various currents and temperatures.

B2.7.6 Summary

A remarkable number of new concepts and contributions demonstrate the scientific and industrial relevance of new or improved mid- and far-infrared sources. Thinking in a time scale of one to five years, steady process in the performance of the already existing mid- and far-infrared sources will lead to an increased wavelength range as well as to improved temperature and power output performance. The continuing efforts put into the understanding and the technology of crystal growth techniques like MBE and MOCVD will increase the number of possible material combinations.

Improvements of material knowledge and technology will allow further development of all the discussed laser types. For lead-salt lasers, there is no theoretical argument that makes continuous-wave room-temperature operation impossible. These lasers, widely commercially available, will probably remain, for a while, the workhorse of mid-infrared gas analysis with high-resolution spectroscopy. For mercury-cadmium-telluride-based lasers, the use of sophisticated heterostructures recently led to better performance. Better material quality of antimonide-based lasers resulted in larger output power and higher operating temperatures.

In the 5–12 μm range, InP-based quantum cascade laser structures currently show the best performance in terms of high output power, aging, single mode emission and temperature performance. Nevertheless, there is room for improvement. First, the design of the active region and of the waveguide can be further improved, including new solutions for low loss optical confinement. To extract the heat from the active region

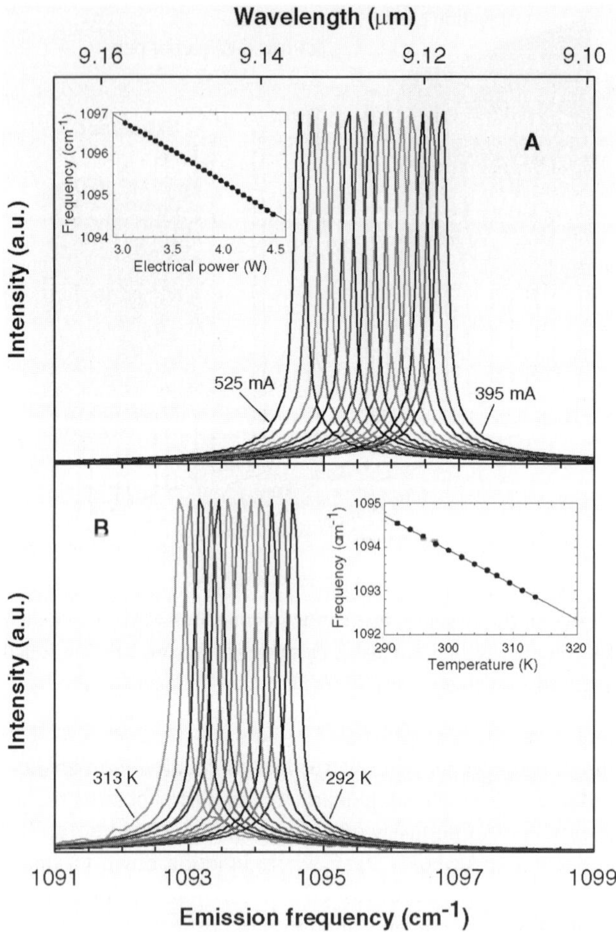

Figure B2.7.16. (A) High-resolution continuous-wave spectra for various injection currents. Inset: emission frequency as a function of the electrical input power. (B) Spectra of a continuous-wave QCL at various temperatures around room temperature. Inset: temperature tuning behaviour. (Reprinted with permission from Beck M, Hofstetter D, Aellen T, Faist J, Oesterle U, Ilegems M, Gini E and Melchior H 2001 *Science* **295** 310. Copyright 2002 American Association for the Advancement of Science.)

as effectively as possible, regrowth of InP, most likely by a MOCVD technique, will become state of the art. Junction-down mounting and high reflectivity mirrors (done by, for example, coatings, Bragg reflectors or photonic bandgap structures) will further improve the high-temperature performance.

Similar improvements are feasible for GaAs-based quantum cascade lasers. Continuous wave operation at room temperature has yet to be demonstrated. Comparing GaAs- to InP-based QC lasers, the lower band offset of the GaAs-based lasers may well be compensated by the higher degree of freedom in the design of the active region, as well as by the amenability to RIE techniques. Limiting the losses at long wavelengths allowed the realization of a GaAs/AlGaAs terahertz QCL. Further improvements in the far-infrared emission range with respect to output power and maximum operating temperature are certainly just around the corner.

References

[1] Hall R N, Genner G E, Kingsley J D, Soltys T J and Carlson R O 1962 *Phys. Rev. Lett.* **9** 366
[2] Nathan M I, Dumke W P, Burns G, Dill F J Jr and Lasher G J 1962 *Appl. Phys. Lett.* **1** 62
[3] Quist T M, Rediger R H, Keyes R J, Krag W E, Lax B, McWorther A L and Zeigler H J 1962 *Appl. Phys. Lett.* **1** 91
[4] Gaines J M, Drenten R R, Habern K W, Maeshall T, Mensz P and Petruzello J 1993 *Appl. Phys. Lett.* **62** 2462
[5] Nakamura S, Mukai T and Senoh M 1993 *Appl. Phys. Lett.* **64** 1687
[6] Horikoshi Y 1985 *Semiconductors and Semimetals* vol 22C, ed W T Tsang (Orlando, FL: Academic)
[7] Faist J, Capasso F, Sivco D L, Sirtori C, Hutchinson A L and Cho A Y 1994 *Science* **264** 553
[8] Namjou K *et al* 1998 *Opt. Lett.* **23** 219
[9] Werle P, Slemr F, Maurer K, Kormann R, Muecke R and Jaenker B 2002 *Opt. Lasers Eng.* **37** 101
[10] Faist J and Capasso F 1997 *Quantum Cascade Laser (McGraw-Hill Yearbook of Science and Technology)* (New York: McGraw-Hill)
[11] Martini R, Gmachl C, Falciglia J, Curti F G, Bethe C G, Capasso F, Whittaker E A, Paiella R, Tredicucci A, Sivco D L and Cho A Y 2001 *Electron. Lett.* **37** 111
[12] Blaser S, Hofstetter D, Beck M and Faist J 2001 *Electron. Lett.* **37** 778
[13] Paiella R, Capasso F, Gmachl C, Sivco D L, Baillargeon J N, Hutchinson A L, Cho A Y and Liu H C 2000 *Science* **290** 1739
[14] Waynant R W, Ilev I K and Gannot I 2001 *Phil. Trans. R. Soc.* A **359** 635
[15] Tacke M 2001 *Phil. Trans. R. Soc.* A **359** 547
[16] Bauer G, Kriechbaum M, Shi Z and Tacke M J 1995 *Nonlinear Opt. Phys. Mater.* **4** 283
[17] Katzir A, Rosman R, Shani V, Bachem K H, Boettner H and Preier H M 1989 Tunable lead-salt lasers *Handbook of Solid State Lasers* ed P K Cheo (New York: Dekker) pp 227–347
[18] Schiessl U P and Rohr J 1999 *Infrared Phys. Technol.* **40** 328
[19] Holloway H and Walpole J N 1979 *Prog. Cryst. Growth Charac.* **2** 49
[20] Heiss W, Schwarzl T, Springholz G, Biermann K and Reimann K 2001 *Appl. Phys. Lett.* **78** 862
[21] Zhao F, Wu H, Jayasinghe L and Shi Z 2002 *Appl. Phys. Lett.* **80** 1129
[22] Bochkarev A E, Dolginov L M, Drakin A E, Eliseev P G and Sverdlov B N 1988 *Sov. J. Quantum Electron.* **18** 1362
[23] Lee H, York P K, Menna R J, Martinelli R U, Garbuzov D Z, Naryan S Y and Connolly J C 1995 *Appl. Phys. Lett.* **66** 1942
[24] Garbuzov D Z, Martinelli R U, Lee H, York P K, Menna R J, Connolly J C and Naryan S Y 1996 *Appl. Phys. Lett.* **69** 2006
[25] Garbuzov D Z, Martinelli R U, Lee H, Menna R J, York P K, DiMarco L A, Matarese M G, Naryan S Y and Connolly J C 1997 *Appl. Phys. Lett.* **70** 2931
[26] Malin J I, Felix C L, Meyer F R, Hoffmann C A, Pinto J F, Lin C H, Chang P C, Murry S J and Pei S S 1996 *Electron. Lett.* **32** 1593
[27] Miles R H and Hasenberg T C 1997 *Antimonide-Related Strained-Layer Heterostructures* ed O Manasreh (New York: Gordon and Breach)
[28] Vurgaftman I and Meyer J R 1999 *Appl. Phys. Lett.* **75** 899
[29] Willardson R K and Beer A C (ed) 1981 *Mercury Cadmium Telluride, Semiconductors and Semimetals* vol 18 (New York: Academic)
[30] Capper P (ed) 1994 *Properties of Narrow Gap Cadmium-Based Compounds (EMIS Data Review Series 10)* (London: INSPEC)
[31] Le H Q, Arias J M, Zandian M, Zucca R and Singh J 1994 *Appl. Phys. Lett.* **65** 810
[32] Zucca R, Zandian M, Arias J M and Gill R V 1992 *J. Vac. Sci. Technol.* B **10** 1587
[33] Million A, Colin T, Ferret P, Zanatta J P, Bouchut P, Destefanis G L and Bablet J 1993 *J. Cryst. Growth* **127** 291
[34] Arias J M, Zandian M, Zucca R and Singh J 1993 *Semicond. Sci. Technol.* **8** S255
[35] Jiang Y, Teich M C and Wang W I 1991 *J. Appl. Phys.* **69** 6869
[36] Vurgaftman I and Meyer J R 1998 *Opt. Express* **2** 137
[37] Yang R, Yang B H, Zhang D, Lin C-H, Murry S J, Wu H and Pei S S 1997 *Appl. Phys. Lett.* **71** 2409
[38] Wanke M C, Capasso F, Gmachl C, Tredicucci A, Sivco D L, Hutchinson A L, Chu S N G and Cho A Y 2001 *Appl. Phys. Lett.* **78** 3950
[39] Price P J 1981 *Ann. Phys.* **133** 217
[40] Ferreira R and Bastard G 1989 *Phys. Rev.* B **40** 1074
[41] Faist J, Capasso F, Sirtori C, Sivco D L, Hutchinson A L, Chu S-N G and Cho, A Y 1993 *Appl. Phys. Lett.* **63** 1354
[42] Grahn H T, Schneider H and von Klitzing K 1990 *Phys. Rev.* B **41** 2890
[43] Faist J, Capasso F, Sirtori C, Sivco D L, Hutchinson A L and Cho A Y 1994 *Appl. Phys. Lett.* **65** 94
[44] Drude P 1900 *Ann. Physik* **1** 566
Drude P 1900 *Ann. Physik* **3** 369
[45] Jensen B 1985 *Handbook of Optical Constants* ed E D Palik (Orlando, FL: Academic)
[46] Sirtori C, Faist J, Capasso F, Sivco D L, Hutchinson A L and Cho A Y 1995 *Appl. Phys. Lett.* **66** 3242
[47] Sirtori C, Kruck P, Barbieri S, Page H, Nagle J, Beck M, Faist J and Oesterle U 1999 *Appl. Phys. Lett.* **75** 3911
[48] Faist J, Capasso F, Sivco D L, Hutchinson A L, Chu S-N G and Cho A Y 1998 *Appl. Phys. Lett.* **72** 680
[49] Colombelli R, Capasso F, Gmachl C, Hutchinson A L, Sivco D L, Tredicucci A, Wanke M C, Sergent A M and Cho A Y 2001

Appl. Phys. Lett. **78** 2120
[50] Rochat M, Ajili L, Willenberg H, Faist J, Beere H, Davies G, Linfield E and Ritchie D 2002 *Appl. Phys. Lett.* **81** 1381
[51] Williams B S, Callebaut H, Kumar S, Hu Q and Reno J L 2003 *Appl. Phys. Lett.* **82** 1015
[52] Kazarinov R and Suris 1971 *Fiz. Tekh. Poluprov.* **5** 797
[53] Gornik E and Tsui D C 1976 *Phys. Rev. Lett.* **37** 1425
[54] Helm M, England P, Colas E, DeRosa F and Allen S J Jr 1988 *Phys. Rev. Lett.* **63** 74
[55] Cho A Y (ed) 1994 *Molecular Beam Epitaxy* (Woodbury, NY: American Institute of Physics)
[56] Sirtori C, Kruck P, Barbieri S, Collot P, Nagle J, Beck M, Faist J and Oesterle U 1998 *Appl. Phys. Lett.* **73** 3486
[57] Adachi S 1985 *J. Appl. Phys.* **58** R1
Adachi S 1989 *J. Appl. Phys.* **66** 6030
[58] Beck M, Hofstetter D, Allen T, Faist J, Oesterle U, Ilegems M, Gini E and Melchior H 2002 *Science* **295** 301
[59] http://www.alpeslaser.ch
[60] Dehlinger G, Diehl L, Gennser U, Sigg H, Faist J, Ensslin K, Gruetzmacher D and Mueller E 20002 *Science* **290** 2277
[61] Hvozdara L, Lugstein A, Finger N, Gianordoli S, Schrenk W, Unterrainer K, Bertagnolli E, Strasser G and Gornik E 2000 *Appl. Phys. Lett.* **77** 1241
[62] Anders S, Schrenk W, Gornik E and Strasser G 2002 *Appl. Phys. Lett.* **80** 4094
[63] Faist J, Gmachl C, Capasso F, Sirtori C, Sivco D L, Baillargeon J N and Cho A Y 1997 *Appl. Phys. Lett.* **70** 2670
[64] Köhler R, Tredicucci A, Beltram F, Beere H E, Linfield E H, Davies A G, Ritchie D A, Iotti R C and Rossi F 2002 *Nature* **417** 156
[65] Andronov A A *et al* 1984 *JETP Lett.* **40** 804
[66] Menon V M, Goodhue W D, Karakashian A S, Naweed A, Plant J, Ram-Mohan L R, Gatesman A, Badami V and Waldman J 2002 *Appl. Phys. Lett.* **80** 2454
[67] Pflügl C, Litzenberger M, Schrenk W, Pogany D, Gornik E and Strasser G 2003 *Appl. Phys. Lett.* **82** 1664
[68] Hofstetter D, Beck M, Aellen T, Faist J, Oesterle U, Ilegems M, Gini E and Melchior H 2001 *Appl. Phys.* **78** 1964

Further reading

Capasso F, Paiella R, Martini R, Colombelli R, Gmachl C, Myers T L, Taubman M S, Williams R M, Bethea C G, Unterrainer K, Hwang H Y, Sivco D L, Cho A Y, Sergent A M, Liu H C and Whittaker E A 2002 Quantum cascade lasers: Ultrahigh-speed operation, optical wireless communication, narrow linewidth, and far-infrared emission *IEEE J. Quantum Electron.* **38** 511

Gmachl C, Capasso F, Sivco D L and Cho A Y 2001 Recent progress in quantum cascade lasers and applications *Rep. Prog. Phys.* **64** 1533

Helm M (ed) *Long Wavelength Infrared Emitters Based on Quantum Wells and Superlattices* (New York: Gordon and Breach)

Liu E H C and Capasso F 2000 *Intersubband Transitions in Quantum Wells (Physics and Device Applications II)* (San Diego, CA: Academic)

Tacke M 2001 Lead-salt lasers *Phil. Trans. R. Soc.* A **359** 547

B2.8
Semiconductor lasers and optical amplifiers for switching and signal processing

Hitoshi Kawaguchi

B2.8.1 Introduction

This section focuses on the fundamental characteristics of semiconductor optical amplifiers (SOAs) and applications of diode lasers and SOAs for switching and signal processing.

Switching is one of the most important functions in telecommunications and computing. Today, nearly all switching systems in operational networks are realized with electronics. The bit rate in systems for video, image and data transmission keeps increasing. It is difficult for electronics alone to satisfy the demand of increasingly higher bit rates. The use of optical fibres as a transmission medium has become common in local networks and long-haul terrestrial or transoceanic systems. An optoelectronic conversion is necessary in these systems, as the switching function is realized with electronics while the transmission is optical. The introduction of optics in switching systems would increase the capacity and flexibility of these systems.

Optical transmission through optical fibres is now a mature technology, while photonic switching is in an early stage of development. Important progress in optoelectronic devices such as laser diodes and SOAs has been made in the development of optical fibre transmission systems, making possible the use of these devices and technologies for switching and signal processing applications. Laser diodes and SOAs are very attractive for optical signal processing in wavelength division multiplexing (WDM) and optical time division multiplexing (OTDM) networks due to their fast and efficient interaction between optical input signals and gain medium.

In this section, we review their switching functions and possible applications in photonic switching systems and signal processing. In section B2.8.2, after briefly reviewing the basic characteristics of SOAs, much of this section deals with ultrafast dynamics in SOAs and their use as switches. Section B2.8.3 describes wavelength conversion. Optical bistability is reviewed in section B2.8.4. Future trends are described from the viewpoints of both device level and system applications in section B2.8.5.

B2.8.2 Semiconductor optical amplifiers

B2.8.2.1 Basic characteristics of semiconductor optical amplifiers [1]

We first investigate the essential characteristics of SOAs. Mukai *et al* [2] identified four important parameters which can be used to describe the performance of SOAs in an optical fibre communication system. They include: the signal gain, the frequency bandwidth, the saturation output power and the noise figure.

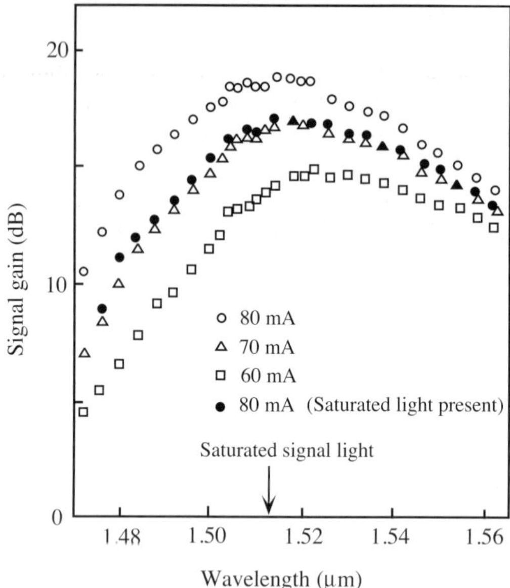

Figure B2.8.1. Signal gain spectra measured for a weak probe beam with and without a strong injected saturating beam. Circle ● denotes the saturated signal gain for a TWA with a bias current of 80 mA. Symbols ○, △ and □ are unsaturated signal gains measured at bias currents of 80, 70 and 60 mA, respectively, with no saturating beam [4].

B2.8.2.1.1 Signal gain

The most important property of an optical amplifier is its ability to amplify the power of light. The gain of the optical amplifier is of primary interest, as it determines many other essential factors, like the signal-to-noise (S/N) ratio, when they are incorporated into systems. The signal gain, G, of an optical amplifier is given, in decibels, by

$$G = 10\log[P_{\mathrm{out}}/P_{\mathrm{in}}] \tag{B2.8.1}$$

where P_{out} is the light power measured at the output of the optical amplifier and P_{in} is that measured at the input end.

We can further refine the definition of signal gain G by considering the light path. If equation (B2.8.1) describes the input and output light power due to a single light path from input to output of the optical amplifier, the resulting gain is known as the signal pass gain G_{s}. If positive feedback is provided (i.e. by reflections from end-facets in SOAs), the signal gain G becomes

$$G = \frac{G_{\mathrm{s}}}{1 + F_{\mathrm{B}} G_{\mathrm{s}}} \tag{B2.8.2}$$

where F_{B} is the proportion of output signal which is fed back to the input. When $G = G_{\mathrm{s}}$, it corresponds to amplifiers without feedback known as travelling wave amplifiers (TWAs). If the reflectivities are finite so that some sort of optical feedback is provided to the optical amplifier, then F_{B} is finite and the resulting amplifier structure becomes a Fabry–Pérot amplifier.

B2.8.2.1.2 Frequency bandwidth [3]

The gain in an optical amplifier is not the same for all frequencies of the input signal. This is more commonly known as the optical gain spectrum of the amplifier. Figure B2.8.1 gives an example [4]. The points to note here are that there is a practical overlap between the open triangles and solid circles, and that no hole burning (see section A1.15.1) due to the intense cw saturated light is observed. This means that the carriers will interact similarly with light of any wavelength. This signifies that the SOA is effectively a homogeneous gain medium. Whilst the curves shown in figure B2.8.1 are for cw light, an experiment using short light pulses indicates that the SOA can be treated as a homogeneous gain medium for optical pulses of the picosecond order also.

Figure B2.8.1 shows several characteristics in addition to those already mentioned. The maximum signal gain at an injection current of 80 mA is 19 dB, and the 3 dB gain bandwidth is approximately 50 nm. This broad gain bandwidth is a major advantage of SOAs. Whilst the gain spectra given in figure B2.8.1 are for a bulk-type amplifier, it is known that multiple quantum well (MQW) amplifiers can give an even broader gain spectrum due to their peculiar band structure [5].

Another feature seen in the figure is that the gain peak wavelength shifts towards the shorter wavelength as the injection current increases. This is caused by the effects of 'band filling'. Because of this band filling effect, attention must be paid to wavelength selection when an SOA is actually manufactured. When a normal diode laser is antireflection-coated to be an amplifier, the gain peak in the amplifier will be at a shorter wavelength than the original laser wavelength since it is used at a higher injection state than the original laser oscillator. It is therefore necessary to start with a laser oscillating at a wavelength some 50 nm, for example, longer than the intended amplification wavelength. Another effect of band filling is its influence on signal gain saturation characteristics.

Two major reasons for the finite bandwidth of SOAs can be identified. The first reason is that the material gain itself has a finite bandwidth as mentioned. The second reason is because of the waveguiding action of the amplifier (or dispersive effects). Optical waveguides possess cut-off frequencies and, hence, a finite bandwidth. Many proposed models of SOAs have neglected the dispersive effect in deriving the amplifier gain spectra $F(f)$, because experimental work showed that material gain effects dominate in SOAs. However, it can be shown that, for many SOA structures, the full material gain bandwidth cannot be utilized because of the presence of resonant behaviour in the device, which significantly reduces the actual bandwidth of the amplifier.

B2.8.2.1.3 Saturation output power

The signal gain of an optical amplifier is limited also by a finite range of input and, hence, output power. Experimentally, it is observed that, once the input power is increased to a certain level P_s, the gain G starts to drop. The output power at the -3 dB points is known as the saturation output power P_{sat}, and the corresponding saturation output intensity I_{sat} can be used to describe the gain saturation effect quantitatively. When the amplifier gain G is measured against the output light intensity I_{out} a similar phenomenon of saturation can be observed, which can be described in an optical amplifier with the gain given by

$$G_s = \exp(g_m L) = \exp[g_0 L/(1 + I_{out}/I_{sat})] \qquad \text{(B2.8.3)}$$

where g_m, measured in cm^{-1} is the material gain coefficient, L is the length of the amplifier and g_0 is the unsaturated value of g_m.

This behaviour of optical amplifiers is known as gain saturation. A qualitative explanation for this phenomenon can be obtained from the two-level system model. The pumping source creates a fixed amount of population inversion at a particular rate and, on the other hand, the amplification process is continuously draining the inverted population by creating stimulated emissions. As we increase the input power, a point arrives where the rate of draining due to amplification is greater than the rate of pumping, such that the population inversion level can no longer be maintained at a constant value and starts to fall. Thus, the gain of the system starts to fall. In other words, gain saturation arises simply because of the conservation of energy. Because of this saturation phenomenon, when an optical amplifier is used to amplify several channels of light signals as in a WDM system, if the total optical power input to the amplifier (not the optical power of an individual channel) from all the channels exceeds P_{sat}, the amplifier will be saturated. The saturation behaviour is complicated by the fact that, in real optical amplifiers, the mechanisms involved are far more complex than in the two-level system described, and often the nonlinear effects of the material gain will also play a role.

B2.8.2.1.4 Noise figure

Optical amplifiers are not free from noise. The noise process in optical amplifiers is due to spontaneous emissions. Because these spontaneous emissions are random events, the phases of the emitted photons are also random. If the spontaneous emission photons emit, they will interact with the signal photons through a change of carrier population and cause both amplitude and phase fluctuations. In addition, these spontaneous emissions will be amplified as they travel across the optical amplifier towards the output. Hence, at the output of the amplifier the measured power consists of both the amplified signal power GP_{in} and the amplified spontaneous emissions (ASE) power P_N. This means that

$$P_{out} = GP_{in} + P_N \tag{B2.8.4}$$

A figure of merit can be attributed to an optical amplifier to describe its noise performance. In an SOA, its noise performance is measured by a noise figure F, which describes the degradation of the signal-to-noise S/N ratio due to the addition of amplifier noise. Mathematically, it is given as

$$F = \frac{(S/N)_{in}}{(S/N)_{out}} \tag{B2.8.5}$$

where $(S/N)_{in}$ is the signal-to-noise ratio at the amplifier input and $(S/N)_{out}$ is that at the amplifier output. It can be shown either by quantum mechanics or semiclassical arguments that, because spontaneous emission is unavoidable in any optical amplifier, the minimum noise figure of an ordinary optical amplifier is 3 dB [6] (unless the optical amplification process is due to parametric amplification, for which a noise figure of 0 dB can be achieved).

Noise in an optical amplifier is the most important parameter. It will not only limit the signal-to-noise ratio in systems incorporating optical amplifiers, but it will also impose other limitations on various applications of optical amplifiers in optical fibre communications. For example, consider several optical amplifiers cascaded in tandem along a transmission span as linear repeaters to compensate fibre losses. The ASE noise power P_N contributes a part of the output power P_{out} of a particular amplifier in the chain, and becomes the input to the next amplifier. Therefore, P_N can be further amplified by subsequent amplifiers. Because gain saturation depends on the total amount of power input to the amplifier, the ASE noise from the output of the earlier stages in the optical amplifier chain can be so large that it will saturate the following ones. If the reflectivities on both input and output ends of the amplifier are low, backwardly-emitted ASE from the amplifiers of the later stages can also input to amplifiers of the earlier stages, enhancing saturation due to ASE. To minimize this effect, optical isolators can be installed along the fibre link to cut off backward-emitted ASE, but this will prevent the system being used for bidirectional transmission.

In addition to the degradation of performance in terms of power, the phase contamination of the signal due to spontaneous emissions is also manifest as additional amplitude and frequency noise, especially due to stray reflections from optical interfaces. As the input signal to optical amplifiers already has a finite amount of phase noise due to the finite spectral spread of the laser source [7, 8], further enhancement of noise from the amplifier is possible. This will further degrade the performance of optical communication systems using phase modulation and coherent detection.

B2.8.2.1.5 Polarization dependency of gain

Small signal gain (unsaturated signal gain) in an ideal TWA can be written as follows if we take into account the light confinement factor Γ:

$$P_{\text{out}} = P_{\text{in}} \exp[(\Gamma g_0 - \alpha)L]. \tag{B2.8.6}$$

Here P_{out} is the output signal light power, P_{in} the input signal light power, g_0 the gain coefficient of the medium for the sufficiently low light intensity and α the optical loss.

According to equation (B2.8.6) it can be seen that the value of the signal gain depends on Γ. In a conventional semiconductor laser, Γ is different for TE and TM polarized lights. Therefore, an amplifier fabricated from a conventional semiconductor laser will give different signals for differently polarized lights. When an optical amplifier is applied to a practical system, signal gain with no polarization dependence is desired. Efforts have been therefore made to eliminate polarization dependence of Γ. It should be mentioned here that, in an MQW amplifier, the gain coefficient g_0 itself is polarization-sensitive.

Strategies for low polarization dependence can broadly be classified into two groups:

(i) hybrid configurations of plural optical amplifiers or of an optical amplifier and bulk optical components; and

(ii) seeking to achieve low polarization dependence of the amplification gain through the design of the active layer structure in the SOA.

Reduction of polarization dependence may be achieved by arranging two SOAs in series or in parallel [9]. From the viewpoint of practical application, however, it is desirable to achieve low polarization sensitivity in the amplifier itself by optimizing the device structure. In the semiconductor laser an active layer thickness of around 0.1 μm is used to reduce the oscillation threshold. When fabricating an SOA, however, the need for low polarization sensitivity means that the active layer should be given a square cross section. Using a relatively square active layer, values below 1 dB have been reported [10, 11]. Because there are limits on the reduction in active layer width that can currently be achieved, there remains some difference in the confinement factor between the two modes.

Therefore new research has studied the introduction of tensile strain into the barrier layer [12]. By compensating for the difference in Γ_{TE} and Γ_{TM} in the confinement factor to increase the TM-mode gain coefficient, it is possible to effect an overall reduction in the mode dependence of the amplification gain. The amplification characteristics in this SOA are shown in figure B2.8.2. These results are for a specific wavelength, but they show perfect agreement between the TE-mode and TM-mode gains. SOAs using a tensile-type superlattice in the barrier would appear to be an extremely effective approach towards achieving polarization-independent amplification whilst maintaining some of the characteristics of the MQW structure.

B2.8.2.2 Gain clamped SOA [13]

Crosstalk is one of the most severe limitations for the introduction of SOAs in systems. Gain nonlinearity arises from the very short carrier lifetime (100–500 ps) that allows the gain to change in a lapse of time shorter than the bit width of the signal. In a conventional amplifier, the amplified signal output power has to be maintained well below (6–10 dB) the maximum available output power of the SOA, so as to prevent signal distortions such as extinction ratio degradation. This considerably limits the practical dynamic range of power at the input of the SOA.

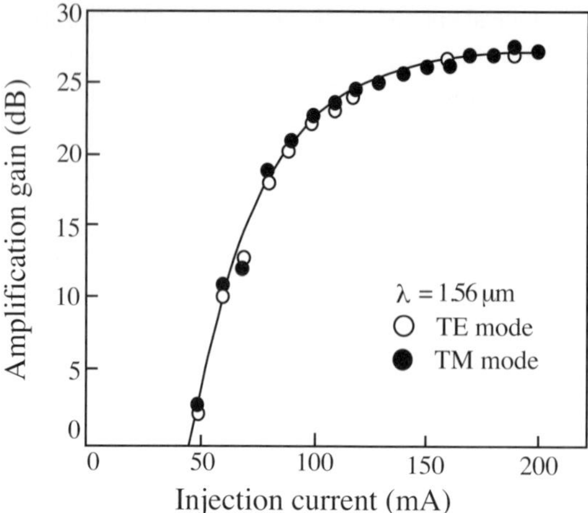

Figure B2.8.2. Amplification characteristics in an SOA with a tensile strained-layer superlattice [12].

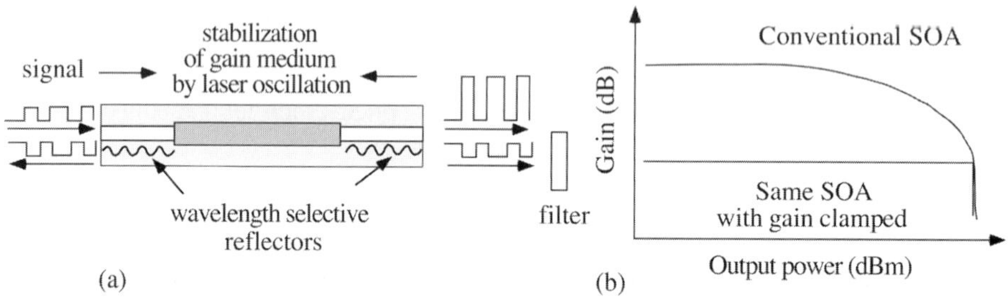

Figure B2.8.3. (*a*) Structure of gain clamped SOA for large dynamic range of input power [13]. (*b*) Typical gain *versus* signal output power characteristics for conventional SOA and the same SOA with gain clamped.

The gain clamping concept in SOA was investigated [14] and has been recognized [15] as an attractive approach to reduce the nonlinearities induced by gain saturation that plagues SOA. The principle is based on the locking of the carrier population by means of a laser oscillation detuned from the amplification bandwidth (see figure B2.8.3(*a*)). The gain then becomes insensitive to the fluctuations of the amplified signal power as long as this one does not exceed the oscillating power. As shown in figure B2.8.3(*b*), the gain curve, with respect to the output power, of such an amplifier is quite flat and saturates abruptly without the gentle saturation region as for a conventional SOA. Consequently, all the available power of the amplifier can be extracted without introducing distortion of the amplified signal.

One other important feature of the gain-clamped concept is that it offers the flexibility to control separately the gain and the available output power. The gain is fixed by the feedback level whereas the available output power is given by the driving current. In contrast, for a conventional SOA, the gain and the output power are not separately controlled. Therefore, if the output power is increased when increasing the driving current, the dynamic range is unchanged because the gain increases in the same time. With the gain-clamped concept, it becomes possible to increase considerably the dynamic range of input power by maintaining the gain at a moderate value. This has been well demonstrated using an external reflector.

Optical switching in a WDM-based system is one of the most attractive applications for gain-clamped SOA [16]. First, coherent crosstalk imposes a high on/off ratio and second, a large number of channels imposes a large dynamic range of input power. Thanks to their high absorption in the off-state, SOAs are able to provide on/off ratios much higher than 40 dB. The problem of the dynamic range of input power

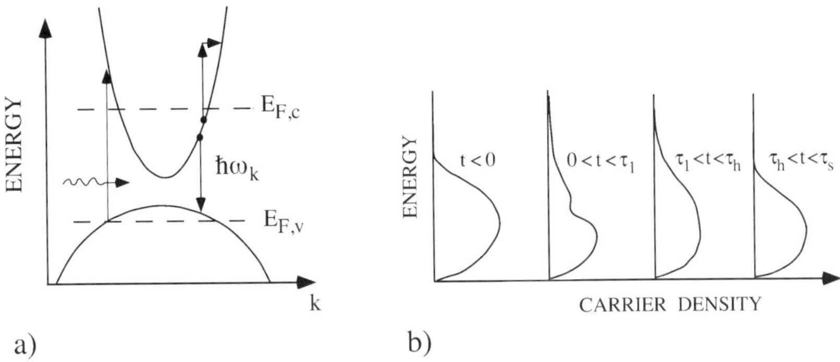

Figure B2.8.4. (*a*) Schematic representation of the interactions between the semiconductor and the electromagnetic field [18]. (*b*) Carrier density evolution in the semiconductor after the interaction with a short optical pulse.

is particularly stringent with high bit rate transmissions. Increasing bit rate requires higher signal-to-noise ratios and thus imposes a higher signal power at the input of the SOA so as to avoid the penalty induced by the noise of the amplifier.

B2.8.2.3 Ultrafast dynamics in SOAs

B2.8.2.3.1 Spectral hole burning and carrier heating [17]

Contributors to the nonlinear gain (and absorption) of active semiconductor materials are fundamental to the material itself. One such nonlinearity is spectral hole burning (SHB). In an inhomogeneous gain medium, light at a specific wavelength causes stimulated transitions only between specific energy levels, not across the entire gain or absorption spectrum (see section A1.15.1). In a semiconductor material, this implies that stimulated transitions tend to distort—i.e. burn a hole in—the Fermi distribution. The width of the spectral hole depends on k-vector selectivity and the homogeneous dephasing rate, thought to be extremely fast (<10 fs) at high carrier densities. Figure B2.8.4 shows a pictorial representation of this behaviour at the time of $0 < t < \tau_l$ [18]. Here, τ_l is the lifetime of SHB. In the gain region, an optical beam with centre wavelength $\lambda(\lambda = (2\pi c/\omega_k))$ will stimulate emission, but will tend to deplete the carrier density over only a limited range. Thus, this effect can be neglected for cw input light as mentioned in section B2.8.2.1. In the absorption regime, too, an optical beam generates carriers over only a limited energy range, bleaching a hole in the absorption. At the transparency point, where there are no net stimulated transitions, there is no SHB. These localized non-uniformities or holes in the carrier distributions persist, while carrier–carrier scattering redistributes the total carrier density into a new Fermi distribution that contains greater or fewer total carriers depending on whether carriers were added or removed by the stimulated transitions.

Another nonlinearity that is fundamental to the semiconductor gain medium results from non-equilibrium carrier heating. Even when the carriers have achieved a Fermi distribution, the temperature of this distribution may differ from that of the lattice. (see figure B2.8.4(*b*), $\tau_l < t < \tau_h$) When electron and hole distributions are heated relative to the lattice, the gain of the active semiconductor medium is reduced. Heating the carrier distribution compresses the gain and shifts the gain peak to longer wavelengths (lower energy). Calculations suggest that significant changes in the gain coefficient are possible for moderate changes in the electron and hole distribution temperatures. For example, heating the distributions to 40 K above room temperature reduces the diode gain by a factor of 2.

The carrier distribution can be heated by several mechanisms. Free carrier absorption (FCA) can create highly energetic carriers in both the conduction and valence bands. Through carrier–carrier scattering, these

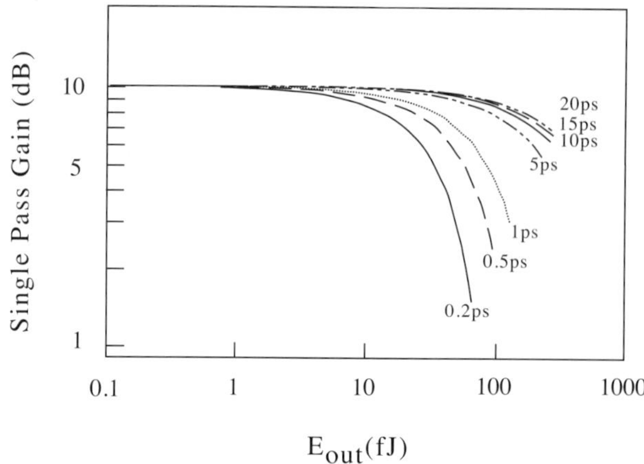

Figure B2.8.5. Predicted gain for an SOA as a function of output energy for different input pulse widths [17].

'hot' carriers share their energy with the rest of the distribution. As a result, the carrier distribution heats up. This distribution cools back to the lattice temperature on a slower timescale by coupling to the lattice vibrations (phonon emission). Notice that heating the carrier distribution reduces the gain across the entire range of energies indicated. This effect occurs because the electron quasi-Fermi level lies at higher energy, further in the absorption regime. The carrier distributions may also be heated by higher-order effects, such as two-photon absorption (TPA), but this effect is much smaller than other heating effects in active semiconductor materials biased in the gain regime.

Besides FCA, the carrier distribution can also be heated by energy changes resulting from stimulated transitions [17]. Carriers occupying energy levels below the average energy are 'cold', whereas those occupying levels above the average energy are 'hot'. Photons with energies between the bandgap energy (E_g) and the transparency energy (E_{trans}), incident on this material system, stimulate emission, removing cold electrons and holes from the distribution. Removing cold carriers effectively heats the distribution. Photons with energies between E_{trans} and the average energy $\langle E \rangle$ stimulate absorption, creating cold carriers. Photons with energies greater than $\langle E \rangle$ stimulate absorption and create hot carriers. This process heats the distributions.

The relative importance of stimulated transitions and FCA depends on the way the device is biased and on the photon energies of the optical beams that are perturbing the device.

The saturated output energy for ultrashort pulses is less than for longer pulses because of the nonlinear gain already mentioned. Partial gain recovery following such pulses is very fast ($\sim$1 ps). Thus, a rapid sequence of ultrashort pulses can be used to extract the same net energy as that of a long pulse. Figure B2.8.5 shows the calculated saturation curves for a bulk amplifier and for pulsewidths ranging from 200 fs to 20 ps [17]. Note that for pulses longer than approximately 10 ps, the spectral hole burning and carrier heating nonlinearities have negligible influence on the gain. For these longer pulses, the saturation behaviour of the diode becomes pulsewidth independent. Clean, pedestal-free pulses, even those as long as 5 ps, have reduced output saturation energies owing to the carrier heating nonlinearity.

B2.8.2.3.2 Propagation characteristics of short optical pulses [19]

For optical pulses in the tens of picosecond range, a theoretical model which assumes that the gain saturation of the SOA is caused only by depletion of carriers due to stimulated emission was used to explain spectral broadening and temporal distortion. When the optical pulse width is shorter than several picoseconds, gain

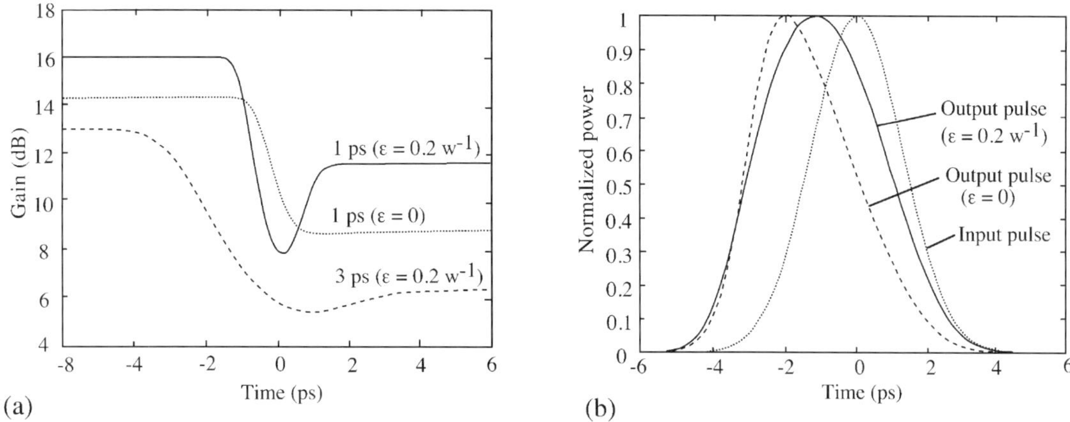

Figure B2.8.6. (*a*) Dynamic gain response of the SOA to different pulse widths (input pulse peak power, 1 W). (*b*) Normalized pulse shapes after amplification for different gain compression factors (input pulse width, 3 ps; peak power, 1 W) [19]

compression, mainly governed by intraband processes of CH and SHB, has been found to play an important role.

In this subsection, we discuss the influence of gain compression on the optical pulse properties in the SOA subject to optical pulses with a pulse duration of a few picoseconds. To outline the effects of gain compression on chirp and spectral properties of amplified picosecond pulses, the following simple phenomenological equation is used in order to model the relevant phenomena

$$g_{\mathrm{m}}(z,t) = \frac{g(N)}{1+\varepsilon|A(z,t)|^2} \tag{B2.8.7}$$

here, g_{m} is the gain, ε is the phenomenological gain compression factor and $A(z,t)$ is the slowly varying envelope of the optical pulse.

Figure B2.8.6 shows how the gain compression influences the amplified pulse characteristics in the temporal domain. The dynamic gain responses are shown in figure B2.8.6(*a*) for 1 ps and 3 ps. For comparison, the result for the case when $\varepsilon = 0$ with 1 ps pulse duration is also given. The salient feature of the gain dynamics is the rapid gain depletion near $t = 0$ for the 1 ps pulse. Increasing the pulse width to 3 ps significantly reduces the rapid depletion, whilst in the case when $\varepsilon = 0$ no such rapid process appears even if a 1 ps pulse is utilized. The rapid depletion is due to the effects of carrier heating and spectral hole burning. A relaxation time of almost 1 ps for carrier heating results in the fast partial recovery of the gain to a lower gain level value taking about 1 ps. The magnitude of the gain outside the transient region is shown to be dependent on the pulse energy and peak power; large pulse energy and peak power lead to low gain.

The normalized output pulse shapes for different gain compression factors are given in figure B2.8.6(*b*). The pulse width broadens significantly and the pulse shape becomes less asymmetric compared with the case when $\varepsilon = 0$. Such behaviours result because the gain compression reduces the pulse peak power (see equation (B2.8.7)) and decreases the gain experienced by the leading edge of the pulse (shown in figure B2.8.6(*a*)). Equation (B2.8.7) also shows that the large pulse peak power can enhance the effect of gain compression. Therefore, the pulse width broadens while increasing the pulse energy.

The influence of gain compression on chirp and spectral properties of the amplified optical pulses are shown in figure B2.8.7. The frequency chirp is negative across the pulse for 3 ps pulse duration and, for a 1 ps pulse, positive chirp grows considerably in the trailing edge of the pulse. By contrast, no positive

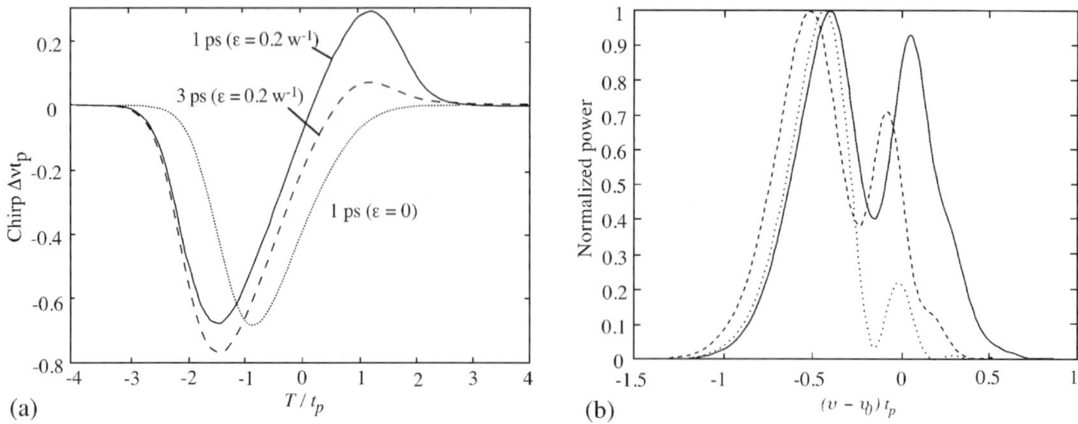

Figure B2.8.7. (*a*) The chirp profiles of the amplified pulses for different pulse widths (input pulse peak power, 1.0 W) where $\Delta\nu = \nu - \nu_0$ is the frequency difference and t_p and t_0 are given by the relation $t_0 \approx 1.665\ t_\mathrm{p}$ for a Gaussian pulse [19]. (*b*) The pulse spectra corresponding to (*a*).

frequency chirp occurs if the gain compression is not included. The reduction in carrier density associated with the saturation of the gain increases the refractive index, which does not recover significantly during the passage of the pulses because of the much longer recovery time of the carrier density than the pulse width. This gives rise to the instantaneous frequency shifting towards the end. In the same way, the positive chirp appearing for the 1 ps pulse can be understood by considering figure B2.8.6(*a*). Owing to gain compression the fast recovery of the gain located in the trailing edge of the pulse reduces the refractive index. Thus, the instantaneous frequency in the trailing edge shifts towards the blue.

The corresponding spectra given in figure B2.8.7(*b*) show multipeak structures. For 3 ps pulses or in the case when $\varepsilon = 0$, there is a dominant spectral peak shifting to the low frequencies side, which is apparent also in figure B2.8.7(*a*). The multipeak structures result from the interference phenomenon. On the other hand, for the 1 ps pulse an interesting feature of the spectrum shown in figure B2.8.7(*b*) is two dominant peaks; one is far red-shifted, and the other is blue-shifted. The frequency difference between the two dominant peaks exceeds 500 GHz. The spectral features can again be deduced from figure B2.8.7(*a*).

B2.8.2.4 Use as optical switches

B2.8.2.4.1 Optical gate array

An SOA is one of the most promising options for gate elements because it provides a fibre-to-fibre lossless operation and a high extinction ratio over a wide wavelength range [20]. The switching time is limited by the carrier lifetime, but is sufficiently fast for several photonic asynchronous transfer mode (ATM) switching systems.

Research is now focused on the application of optical gate arrays due to progress in process technique, such as fine fabrication, and in good reproducibility for multichannel arrays. An SOA optical gate array needs to operate at a low carrier density to have a low gain in a fibre-to-fibre lossless operation. It also needs low-energy consumption. Thus, an SOA gate array has a different structural design from an SOA optical amplifier.

To develop an SOA gate array, several inherent problems need to be overcome as follows [21].

(1) Connecting optical fibres to an SOA gate array. It is difficult to obtain good coupling through a

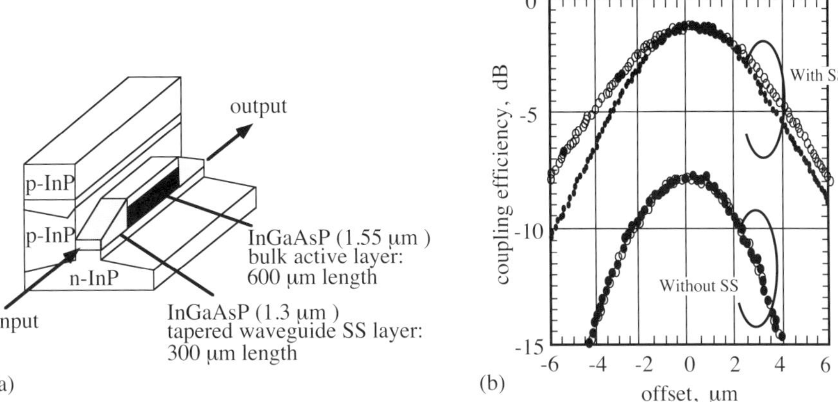

Figure B2.8.8. The SS-SOA gate. (*a*) Device structure [20]. (*b*) Horizontal (○) and vertical (●) alignment tolerances of SS-SOA gate [22].

multichannel gate array because of spot-size mismatching. Hybrid integration with a silica-based planar-lightwave-circuit (PLC) platform is a promising approach to solve this problem. Spot-size converters (SSs) at both ends of the SOA reduce the mismatching. Figure B2.8.8(*a*) shows the structure of an SS-SOA gate [20]. The SSs must have both good coupling efficiency and a large alignment tolerance between the PLC waveguide and the SOA gate array. Figure B2.8.8(*b*) shows the coupling efficiency and its offset tolerances [22]. The typical coupling loss was 2.6 dB. A 1 dB increase in coupling loss corresponds to alignment tolerances of ± 2.1 μm (horizontal) and ± 1.9 μm (vertical). The alignment tolerance of the optical axis direction was 12.0 μm.

(2) Reducing polarization sensitivity. To achieve low polarization sensitivity over a wide wavelength region, an almost square bulk active structure can be used. However, adding an SS structure to an SOA creates additional polarization sensitivity.

(3) Reducing the operating current. A large confined active region enables fibre-to-fibre lossless operation with a low driving current and a high extinction ratio.

(4) Increasing uniformity. Many optical gates are needed to construct a photonic network system, and it is important that these gates have a uniform width (and thus uniform performance) to enable multichannel operation.

A 4-ch SS-SOA gate array with high performance has been fabricated by solving these problems [21].

B2.8.2.4.2 Optical switches with integrated amplifiers [23]

In this subsection, we review the optical switches with amplifiers. To compare device characteristics, such as loss compensation, crosstalk and spontaneous emission we categorize the switches into the four types listed in figure B2.8.9. The first (a) is a static switch and the other three categories of switches are monolithic dynamical routing switches. Here we use the terms gate TWA and booster TWA to contrast the dynamical gate effect of a TWA to improve crosstalk with the continuous gain of a TWA only to compensate for losses. The terms passive and active denote static and dynamic routing functions of the switches, respectively. Switches for large-scale switching systems should be lowloss, low crosstalk and low spontaneous emission noise characteristics. To reduce the noise from the TWAs, the coupling efficiency for the TWAs should be high and the number of TWA stages in a device should be reduced.

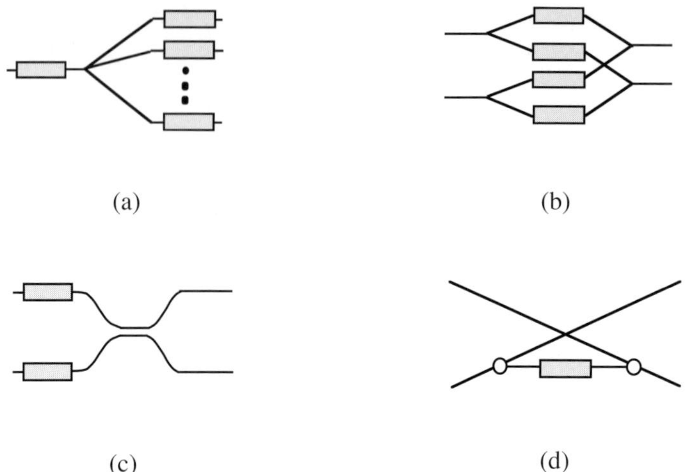

Figure B2.8.9. Optical switches with integrated SOAs [23]. (*a*) Passive switch with booster TWA; (*b*) passive switch with gate TWA; (*c*) active switch with booster TWA; and (*d*) active switch with gate TWA.

The passive switches with booster TWAs (a) are static switches, which distribute an optical signal from one port to N ports and/or combine optical signals from N ports to one port, and in which the booster TWAs compensate for losses in the device. The static switch is a key device for WDM systems and information broadcasting systems. A 1×16 switch with 17 TWAs integrated into both ends of a 1×16 passive splitter was demonstrated [24].

The passive switches with gate TWAs (b) enable us to obtain an extremely low crosstalk using the gate function of TWA. The switching of this type of switch is divided into the following three stages: splitting light from an incoming port, selective gating in a TWA and light combining for an outgoing port. The passive switches consume no power and are easy to fabricate, since these are branching waveguides. However, the passive switch has intrinsic loss: at least 3 dB per switch. The loss of the passive switch is undesirable for the spontaneous emission noise from the TWA and loss compensation by the gain of the TWA, since the loss decreases the coupling efficiency for the TWA and increases the gain load for the TWA. As one example, Ikeda *et al* have expanded the 2×2 switch to a rearrangeable non-blocking 4×4 switch array consisting of six 2×2 switches in the Benes network configuration [25]. To switch an optical path, seven stages of TWAs were turned on, and fibre-to-fibre lossless switching for all sixteen switching paths and an extinction ratio greater than 30 dB were atttained.

The third category of switches is an active switch with booster TWA (c), and can minimize the coupling efficiency for the TWA when the booster TWA is integrated into the input end of the device. This is advantageous for the spontaneous emission noise, but the booster TWA does not improve the crosstalk. The crosstalk is limited by the performance of the active switches. The active switches with booster TWAs were demonstrated by using a carrier-injection type 2×2 directional coupler switch [26].

The last active switches with gate TWAs (d) also enables us to obtain extremely low crosstalk. To contrast these passive switches with gate TWAs, the splitting and combining of the passive Y-branches are done by the two active switches where the active switches and the gate TWA are synchronously turned on or off. To introduce the gate function of TWA into switches, the TWA has to integrate between two switching components. Since the active switches can reduce the intrinsic loss of the passive switches, the coupling efficiency for the TWA is better than that for the passive switches with gate TWAs and the dimension of the switch array without booster TWAs is larger. Inoue *et al* proposed a combination of two total internal reflection switches and a gate TWA called a carrier-injection-type optical single-slip structure (S^3) switch with travelling-wave amplifier (COSTA) and they expanded the switch array to 4 $\times$ 4 [27] in the crossbar configuration.

B2.8.2.5 Designs for high power operation [28, 29]

Many applications require SOAs with large values of the output saturation power. Guides to increasing saturation intensity are as follows. First, increasing the carrier density to shorten τ_s is an effective step. This can be done simply by increasing the injection current. However, if the injection current is increased, then the signal-pass gain also increases, leading to a situation where resonance effects due to residual reflectivity appear. Therefore, a reduction in reflectivity ultimately increases saturation output. Meanwhile, in another method, the carrier density is increased whilst the single-pass gain remains the same. As equation (B2.8.6) shows, the single-pass gain is determined by $\Gamma g_0 L$. If the carrier density is increased, then g_0 increases, but if the confinement factor Γ or the amplifier length L is reduced by the same degree, then the single-pass gain will be unchanged, thus suppressing resonance effects. In other words, by lowering Γ or L the maximum injectable carrier density can be increased, leading to an increase in saturation output. Doping with impurities to reduce τ_s and boost the saturation intensity is an alternative strategy.

As shown in figure B2.8.1, when carrier number decreases the gain peak wavelength shifts towards the longer wavelengths. This means that at long wavelengths, even if the carrier number decreases, there is little reduction in the gain value and saturation does not readily occur. This is particularly notable in an MQW-type amplifier. In an MQW device, there exist conditions, for a long wavelength signal, where signal gain scarcely changes with variation in the carrier density because of its peculiar band structure. High saturation output can therefore be achieved through the introduction of an MQW structure.

Many applications of semiconductor lasers require high power levels in excess of 1 W. Such power levels are difficult to obtain from a single semiconductor laser directly mainly because of a relatively small active-region volume. We have looked at the issue of saturation intensity, but in an actual amplifier the main concern is the optical power, which is given by the product of the light intensity and the light mode cross section. Accordingly, increasing the mode cross section is also of benefit. An approach based on this concept is to unfold the waveguide width towards the output facet (see figure B2.8.10(*a*)). The mode field area is large at the output and a high saturation output power is obtained.

Master-oscillator power-amplifier MOPAs have demonstrated high output power and still have a good beam quality [29]. Their power can be as large as 5 W in pulsed or cw operation. The output of a low-power semiconductor laser, acting as a master oscillator, is amplified in a TW-type SOA acting as a power amplifier (see figure B2.8.10(*b*)). If the master oscillator and the SOA are integrated monolithically on the same chip, the resulting device can provide power levels in the range 1–10 W while maintaining single-mode operation and a good beam quality.

Most of the studies conclude that filament formation is the most severe obstacle for achieving high-power semiconductor lasers and amplifiers [28]. The physics of the filamentation has been attributed to several causes. First, fluctuation of beam intensity leads to non-uniform gain depletion and so spatial variation of the carrier concentration, which then further causes the optical gain to vary spatially. Second, the variation of carrier concentration also causes the refractive index to change spatially, so the optical field is further distorted due to the self-focusing effect. Third, experiments have shown that the temperature distribution across the waveguide in tapered and broad-area semiconductor lasers is not uniform. While most works study the reasons for the filamentation, Lai and Lin discuss the possibility of reducing the filamentation by changing the intrinsic property of the semiconductor material [28].

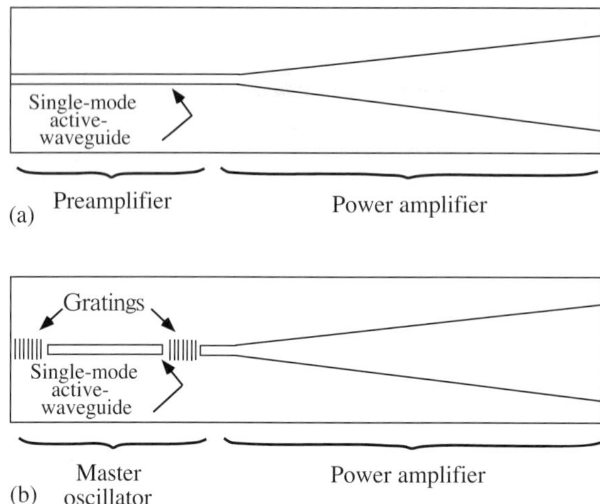

Figure B2.8.10. Two variations of tapered structures: (*a*) a single-mode preamp section coupled to a tapered gain region; and (*b*) the MOPA structure discussed in the text. Minor changes in both devices can be made to produce simple laser structures [29].

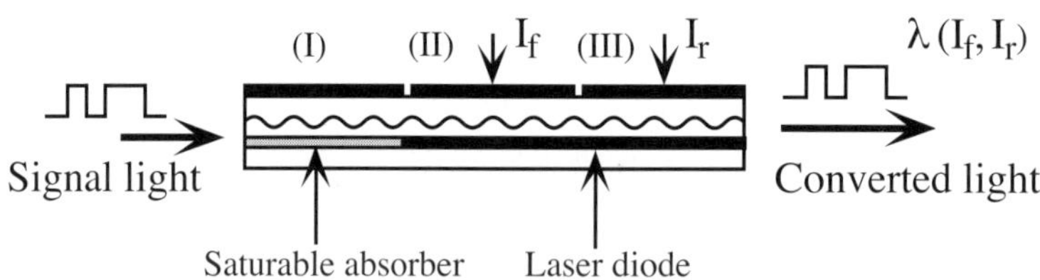

Figure B2.8.11. Wavelength converter constructed by monolithic integration of a tunable wavelength LD and an optically triggered gate.

B2.8.3 Wavelength conversion [30, 31]

Wavelength converters transform information on one wavelength to another [32]. Conventional wavelength conversion techniques used second-harmonic generation or frequency shifting and were found to be difficult to apply to optically functional systems. Wavelength conversion using an LD, however, has wide tunability and fast operation for multiwavelength photonic systems. The converters fall into four groups: optoelectronic converters, laser converters, coherent converters (four-wave mixing) and converters based on optically controlled optical gates.

Optoelectronic conversion is a straightforward solution that is already used in WDM point to point links. The optoelectronic (O/E/O) converter is conceptually simple since it consists of a detector followed by amplification or regeneration and transmitter stages. The approach has the big advantage that it can be implemented with existing technology. Most optoelectronic converters reported till now operate at 2.5 Gbit s^{-1}. For wavelength conversion in high-speed networks operating at 10 Gbit s^{-1} and in the future possibly at 100 Gbit s^{-1}, the power consumption of the optoelectronic converter will be high and bandwidth limitations of electronic circuitry may be encountered. Therefore, all-optical converters are highly interesting.

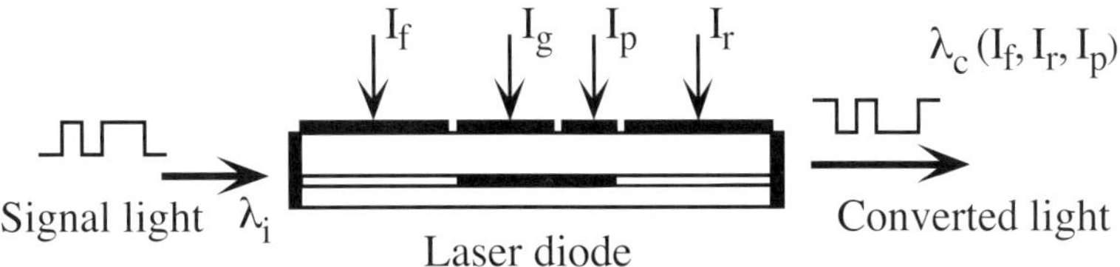

Figure B2.8.12. Wavelength converter based on gain saturation in LD.

B2.8.3.1 Laser converters

B2.8.3.1.1 Saturable absorber gates [32]

A tunable wavelength conversion device can be constructed by monolithic integration of a tunable wavelength LD and an optically triggered gate. Basic device structure is shown in figure B2.8.11. The function was first demonstrated using a tandem-type bistable DFB LD with three divided sections [33]. The device has two main functions: (1) the device emits coherent light only when light is injected and (2) the output wavelength can be continuously tuned by changing the driving current. Therefore, this device is used for intensity modulation (IM)/IM conversion. The function of the first is explained as follows. The driving current injected into the electrode located on one end is set at zero or at a low level. The region acts as a saturable absorber whose absorption coefficient decreases with increasing input optical power. When optical input is injected into the saturable absorber, the absorption coefficient decreases, light intensity in the laser resonator increases, and positive feedback is established. Therefore, light output power increases abruptly at the laser threshold. On the other hand, when the light input intensity decreases, the output decreases abruptly through the reverse process. The current-light output curve shows a bistable or a differential gain characteristic, depending on the level of saturable absorption. Therefore, the device emits coherent light only when light is injected, provided that the bias current is set just below the laser turn-off threshold. The wavelength tunability can be realized by changing the driving-current distribution in a manner similar to that of an ordinary multielectrode DFB LD. Independent adjustment of the driving current to electrodes (II) and (III) results in a non-uniform distribution of carriers in the active layer. This changes the refractive index of the different sections of the active layer and increases the difference between the effective grating pitches of each region. The larger difference increases the threshold current of the laser. As a result, the output wavelength becomes shorter through the carrier-density-dependent refractive index as the current non-uniformity becomes more pronounced.

B2.8.3.1.2 Gain quenching of LDs

All-optical wavelength conversion can also be performed by optical control of single frequency lasers, as shown in figure B2.8.12. The input signal (λ_i) to be converted is launched into the laser where it causes gain saturation that controls the oscillation of the laser. The result can be either intensity modulation (IM) or continuous-phase frequency-shift keying (CPFSK) output formats depending on the operation of the laser. The lasing wavelength (λ_c) is either fixed or electronically tunable depending on the system requirements.

With steep input-output transfer functions (see figure B2.8.12) the IM output format mode can achieve fine signal waveforms that even allow for cascading of a few converter stages. The IM output mode is, however, associated with chirp that will limit transmission on non-dispersion-shifted optical fibres. The problems related to chirp could be overcome by operating the laser well above threshold in the CPFSK output mode. In this case a frequency discriminator/filter is, however, needed at the output to obtain an IM signal format.

Note that the laser converter basically consists of a single component, so it is conceptually simple. Its

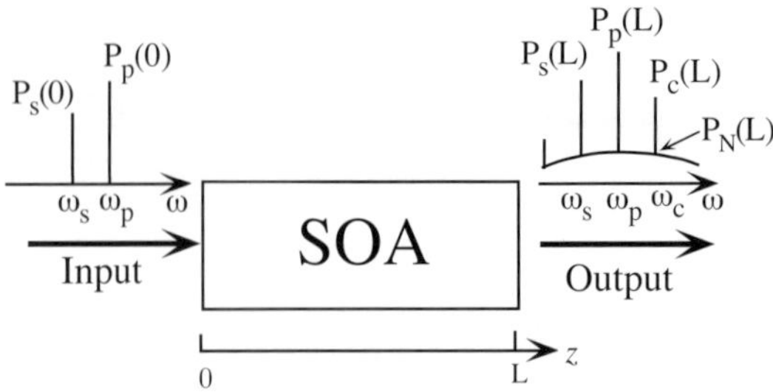

Figure B2.8.13. FWM in SOAs. Here z denotes the spatial coordinate and L is the amplifier length.

optical input power requirements are 0–10 dBm, but unless a special waveguide design is implemented for the gain section of the laser, it is polarization-dependent. Moreover, the maximum bit rate, determined by the laser's resonance frequency, is limited to around 10 Gbit s^{-1}.

B2.8.3.2 Four-wave mixing [34]

B2.8.3.2.1 Principle of four-wave mixing in SOAs

Figure B2.8.13 shows a typical four-wave mixing (FWM) arrangement. Two optical waves, a pump and a signal wave, are injected into the SOA with the same state of polarization. The angular frequencies are $\omega_{\rm p}$ and $\omega_{\rm s}$ and the optical input powers are $P_{\rm p}(0)$ and $P_{\rm s}(0)$ for pump and signal wave, respectively. The beating of both waves influences the active medium, thereby generating index and gain modulation. The interaction of the injected waves with these modulations leads to new frequency components, the FWM signals that can be observed in the SOA output spectrum. Of main interest is the converted FWM signal $P_{\rm c}(L)$, the optical angular frequency of which is $\omega_{\rm c} = \omega_p + \Delta\omega$, with $\Delta\omega = \omega_{\rm p} - \omega_{\rm s}$ denoting the detuning of pump and signal wave angular frequencies. The nonlinear gain dynamic in SOAs, responsible for the process of FWM, is based on inter- and intraband effects. Whereas the former refers to transitions between the conduction and the valence band, the latter are the carrier distribution within one band. The interband effects change the carrier density due to the carrier depletion caused by the stimulated emission, which is referred to as carrier density pulsation (CDP). The characteristic time of CDP is the effective carrier lifetime, i.e. the lifetime due to all recombination processes, including stimulated emission. The intraband effects are associated with basically two phenomena. First, the stimulated emission burns a hole in the carrier distribution causing a deviation from the Fermi distribution. This is referred to as spectral hole burning (SHB). The time constant associated with this process is the time needed to restore the Fermi distribution (carrier–carrier scattering time), and is typically several tens of femtoseconds. Additionally, free carriers at low-energy levels are removed by stimulated emission or transferred to higher levels due to free carrier absorption. This phenomenon leads to a heating of the electron and hole distributions and is called carrier heating (CH). The characteristic time needed to cool down to the lattice temperature, due to carrier–phonon scattering, is of the order of several hundred femtoseconds. It has been found that four-wave mixing is mainly mediated by CDP, CH and SHB, for pump-signal frequency detuning $\Delta\nu$ up to a few THz. At higher detunings the two photon absorption and the Kerr effect can have a significant role.

B2.8.3.2.2 Efficiency and signal-to-background ratio

Optical signal processing using the nonlinearities of a SOA offers several advantages over the use of passive nonlinear devices, such as nonlinear crystals or optical fibres. One advantage is the high FWM conversion efficiency of the SOA, which enables short interaction lengths of the injected waves. Phase matching is therefore not a crucial problem for copropagating signals in SOAs. Moreover, only low optical input powers (in the range of 1 mW) are required for the pump and signal waves. The conversion efficiency

$$\eta_{\mathrm{FWM}} = \frac{P_{\mathrm{c}}(L)}{P_{\mathrm{s}}(0)} \tag{B2.8.8}$$

is defined as the ratio of the FWM signal output power to the signal input power. η_{FWM} therefore describes the efficiency by which an optical signal is converted from the angular frequency ω_{s} to the angular frequency ω_{c}. However, one drawback of using an active component rather than a passive one is the amplified spontaneous emission (ASE) noise, which is added by the SOA. In order to quantify the noise performance of an SOA, which is of crucial interest for practical applications, it is convenient to introduce the signal-to-background ratio SBR in addition to η_{FWM}. The signal-to-background ratio

$$\mathrm{SBR} = \frac{P_{\mathrm{c}}(L)}{P_{\mathrm{N}}(L)} \tag{B2.8.9}$$

is the ratio of the FWM signal power to the noise power $P_{\mathrm{N}}(L)$ measured at the optical frequency ω_{c} within a certain optical bandwidth at the SOA output. Since the ASE noise is the main contribution to the noise in an SOA, $P_{\mathrm{N}}(L) \approx P_{\mathrm{ASE}}(L)$ is often used to determine the SBR. Note that the SBR is not equivalent to the signal-to-noise ratio SNR, which is important for data transmission. However, supposing that the ASE generated noise is dominant, a relation exists between these two quantities.

FWM, however, is polarization dependent; it requires that signal and pump have the same polarization. This feature limits the use of FWM in SOAs as a frequency converter because of the random polarization of incoming signals. Several methods have been proposed to obtain polarization-independent frequency conversion [35–40]. The use of two pumps [35–38], both at higher frequencies than the signal, provides good polarization independence, but, at least for the method proposed in [35–37] does not allow the accommodation of broadband signals. A different two-pump configuration shows better polarization independence, with potential high-speed transmission, but allows only single-interval frequency-conversion [39]. Lacey and coworkers [40], proposed a polarization-diversity scheme, i.e. the use of two amplifiers in parallel, with only one pump.

All methods cited suffer from the strong dependence of the FWM efficiency on the conversion interval. Greco *et al* have shown that if we inject into a SOA two pumps (P1 and P2) with orthogonal polarization and a signal, S, with polarization parallel, say, to P1, the mixing process will produce two sidebands on P2 (C2 and S2). They have shown that C2 and S2 have intensity independent of the conversion interval within the SOA gain bandwidth.

They experimentally demonstrate that the two-orthogonal pump configuration used in a polarization-diversity scheme, like that proposed in [40], leads to a polarization-independent frequency conversion with constant efficiency over a large conversion interval. Figure B2.8.14 shows the ideal scheme [41]. The signal, S, randomly polarized, and P1 are mixed and sent to one input port of a polarization beam-splitter (PBS1). The polarization of P1 is set linear and at 45° with respect to the principal axes of PBS1. P2 is injected in the other input port of PBS1, again with linear polarization at 45° with respect to the principal axes of PBS1. P1 and P2 have perpendicular polarization at each output port of PBS1, reciprocal with respect to the other port. PBS1 then routes the signal to the two SOAs, which independently convert the two orthogonal polarizations. At the output of the two amplifiers, the sum of the intensity of the C2 (as for S2, not shown here) fields generated in both amplifiers is constant. In the ideal scheme, a second PBS (PBS2) suppresses P1, S and the noise associated with their polarization, including that originated by EDFA upstream of this device, if any.

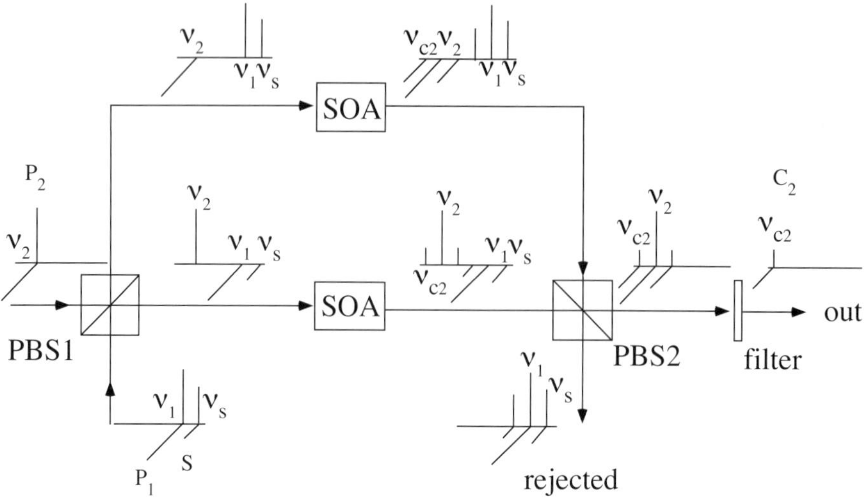

Figure B2.8.14. Scheme of the two-pump polarization-diversity configuration. For a detailed explanation, refer to the text [41].

B2.8.3.2.3 Four-wave mixing in LDs

Wavelength conversion by single devices is desirable for its practicality. If we use the lasing beam of a DFB laser for the pump beams and generate a conjugate beam by non-degenerate four-wave mixing (NDFWM) within the DFB laser itself, the use of external pump lasers will be unnecessary. As the lasing mode of DFB lasers is stable, it is not necessary to use a master laser for injection locking as is the case of Fabry–Pérot lasers. NDFWM in a lasing $\lambda/4$-phase-shifted DFB laser with a long cavity and antireflection (AR) coating on both facets was demonstrated [42]. (see figure B2.8.15). Lasing beams working as pump beams are stabilized in the middle of the stopband in this laser. On the other hand, the laser works as a TWA for signal beams or conjugate beams outside the stopband. A high conversion efficiency and a wide conversion range were demonstrated.

B2.8.3.3 Converters based on optically controlled gates

As illustrated in figure B2.8.16, the principle is to let the input power at λ_i control the gating of cw light at λ_c. Thereby the data are converted from λ_i to λ_c. The cw light originates from a light source with either a fixed or a tunable output wavelength depending on the application of the converter. Clearly, the transfer function of the gate should be as steep as possible and, depending on its positive or negative slope, the converted signal will be in-phase or inverted relative to the input.

B2.8.3.3.1 Cross gain modulation gates

A simple optical gate is realized by an SOA. Its gain saturation due to an optical input signal is controlling the gain and thereby the state of the gate. The resulting converter, also called a cross gain modulated (XGM) converter, is extremely simple to assemble. It is polarization-insensitive because of polarization-independent SOA gain and is very power efficient. It has, however, a number of shortcomings. The signal is inverted relative to the input signal (negative slope) and the extinction ratio for the converted signal may degrade going from shorter to longer wavelength. Moreover, the converted signal has a relatively large frequency chirp.

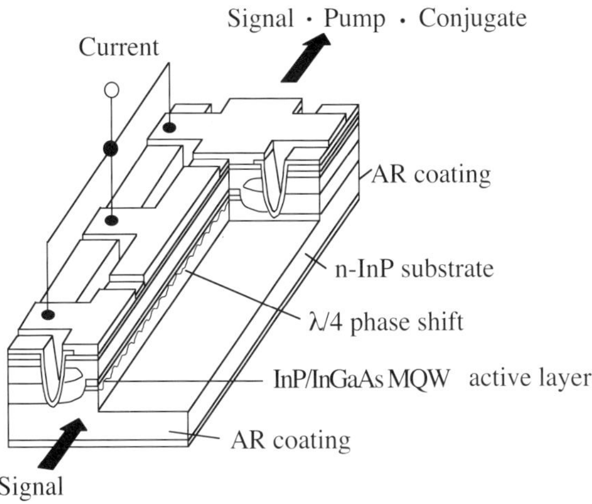

Figure B2.8.15. Laser structure used for non-degenerate FWM [42].

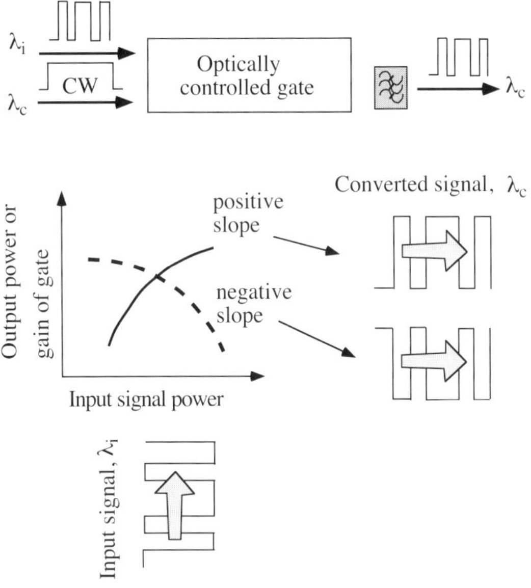

Figure B2.8.16. Schematic illustration of a wavelength converter based on optically controlled gate and the dependency of the converted signal on the transfer characteristic of the gate [30].

Despite the non-ideal properties, the converter has been used with fine results in a number of switch-block experiments. The shortcomings may also be eliminated if an interferometric converter follows the XGM converter.

The conversion speed is determined by the carrier dynamics that are governed by the relatively slow interband carrier recombinations. The detailed analyses show that gain saturation plays a significant role for obtaining a high bit rate. Simple guidelines for achieving high bit rate conversion are to operate the SOAs

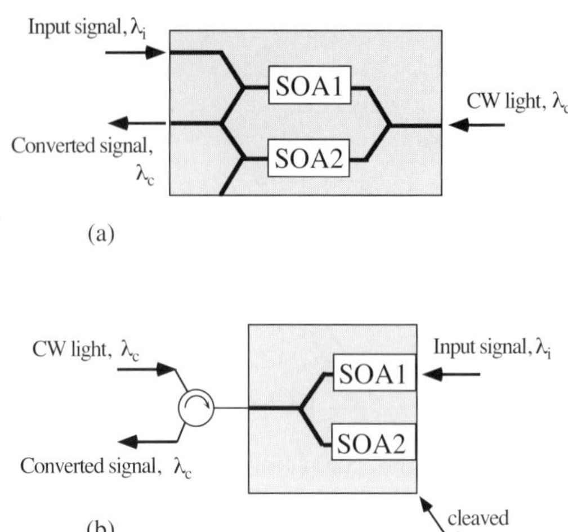

Figure B2.8.17. Wavelength conversion principle using MZI (*a*) and MI structures (*b*) with SOAs [30].

with: (1) large current injection; and (2) high optical power levels. Moreover, the SOA waveguide should have: (3) large optical confinement factors; and (4) large differential gain.

Since the allowed injection current is limited, the conversion speed can also be increased using longer cavity lengths. Recently 100 Gbit s^{-1} wavelength conversion has been achieved by XGM conversion in a 2 mm long SOA followed by a grating for FM to AM conversion of the chirped output to extend the conversion bandwidth [43]. Obviously, the SOA converters add spontaneous emission noise to the converted signals. Still, with noise figures of 8–10 dB (including 2–4 dB input coupling loss), it is possible to cascade a few SOA-XGM converters as shown in, e.g. [44].

Briefly, it should also be noted that very promising results have been achieved for cross absorption modulation (XAM) in electro-absorption modulators [45]. It has been shown that the technique works up to at least 40 Gbit s^{-1}, that extinction ratio enhancement may be achieved and that 30 nm operation bandwidths are possible.

B2.8.3.3.2 Cross phase modulation gates

Better wavelength converters are obtained by the integration of two or more SOAs into interferometric waveguide configurations (figure B2.8.17). In these converters, the optical input signal controls the phase difference between the interferometer arms through the relation between the carrier density and the refractive index in the SOAs (cross phase modulation, XPM); thereby a cw light is modulated [46]. For stable operation the XPM converters must be integrated. The Mach–Zehnder interferometer (MZI) structures have been developed into structures with a separate waveguide for coupling the input signal into only one of the SOAs thereby creating the needed phase difference between the two interferometer arms in a very efficient way.

Michelson interferometric (MI) converters have a simpler structure since they offer direct access for the input signal to the SOAs [47], see figure B2.8.17(*b*). The high optical power levels due to the direct coupling will enable a high-modulation bandwidth. The MI converter has a reflective facet making it a folded version of the MZI converter. Therefore, it works very similarly. It should be noted, however, that the MZI allows the cw light and the input signal to be launched counter-directionally, whereas, for the MI, a part of the original signal will always pass through the converter and be transmitted along with the converted signal. This prohibits the input and output wavelengths from being identical due to interference crosstalk.

The interferometric converters have the advantage of very steep transfer functions enabling extinction ratio regeneration of the converted signal. The most attractive point of this device is that the converted signal is non-inverted. Only small input signals are needed to introduce a π-phase difference between the interferometer arms, so very efficient conversion is obtained almost independent of wavelength. Because of the small modulation associated with the π-phase shift, the frequency chirp of the output signal will also be small compared to, e.g. the XGM converter. Moreover, for conversion on the positive slope the resulting chirp compensates against the influence of fibre dispersion.

Besides signal waveform and spectral reshaping, the interferometric converters have excellent noise properties with optical signal-to-noise ratios for the converted signals as high as 30 dB (measured in 1 nm spectral bandwidth). Moreover, the noise is redistributed due to the transfer function. As a result, the noise accumulates less rapidly than for a chain of optical amplifiers. This allows for cascading of several converters.

The interferometric converters will typically have <1 dB penalty for bit rates of 10 Gbit s^{-1}. Like the XGM converters, the interferometric XPM converters can also achieve 40 Gbit s^{-1}. The MI converter used in the 40 Gbit s^{-1} experiment is realized using multiquantum-well based layer structures with a 10 well tensile strained InGaAs/InAsP waveguide core [48].

Progress toward development of practical interferometric converters has been achieved with the realization of packaged all-active MZI structures [49]. These packaged converters have been tested in field trials. A practical obstacle for the interferometric converters is their small input power dynamic range of typically 2–3 dB. This can, however, be significantly improved by simple compensation schemes [50]. In 10 Gbit s^{-1} experiments it was shown that scheme A with simple power monitoring and electronic feed forward gives an 8 dB dynamic range, while the use of an EDFA preamplifier operated in saturation, as shown in scheme B, can give a 40 dB dynamic range. It may be a disadvantage that the EDFA has a response time in the millisecond range. Use of a SOA preamplifier with electronic gain control as shown in scheme C is much faster and give a 28 dB dynamic range.

Another interferometric arrangement which has been widely evaluated is the nonlinear Sagnac interferometer, or nonlinear optical loop mirror (NOLM) [51]. In the first of such devices, an asymmetric input coupler and the nonlinear refraction of optical fibre provided the basis for the switching function. A SOA inserted within the loop mirror can provide nonlinear refraction in a more compact device: the SLALOM (semiconductor laser amplifier in a loop mirror) [52]. The SLALOM performs several nonlinear functions, including demultiplexing from 9 Gbit s^{-1} to 3 Gbit s^{-1}.

A further modification of the SLALOM device, in which the SOA is offset only slightly from the midpoint of the loop, is the terahertz optical asymmetric demultiplexer (TOAD) device [53]. The TOAD may include coupling ports for a control pulse, and is shown generically in figure B2.8.18. A pulse train incident on port 1 is split into clockwise (cw) and counter-clockwise (ccw) pulse trains at coupler A. After traversing the loop, the pulse trains interfere at coupler A. A polarization controller in the loop (not shown in figure B2.8.18) can be adjusted so that the pulse trains are reflected out port 1, or transmitted out port 2 (or some combination). As the pulse intensities increase so that they are sufficient to compress the gain of the SOA in the loop, the gain and phase delay experienced by the second pulse to reach the SOA, the ccw in figure B2.8.18, are different from those of the first pulse, so the interference of the pulse trains at coupler A and, hence, the transmission of the output port 2, will become intensity-dependent. This is the basis of the SLALOM device. In the operation of this device, the control pulse saturates the amplifier. The control pulse is timed to arrive in the SOA after the cw pulse has passed the amplifier but before the ccw pulse has arrived. Thus, the ccw pulse will experience lower gain and a raised refractive index in the SOA. When the pulses interfere at coupler A, switching will occur. The switching time window is determined by twice the delay between the centre of the loop and the position of the SOA, corresponding to time shift between the response of the SOA probed by the two counter-propagating pulses, as shown at the bottom of figure B2.8.18. With the TOAD device, a switching window of 4 ps has been demonstrated.

With an arrangement similar to that in figure B2.8.18, an OTDM data stream applied at port 1 has been

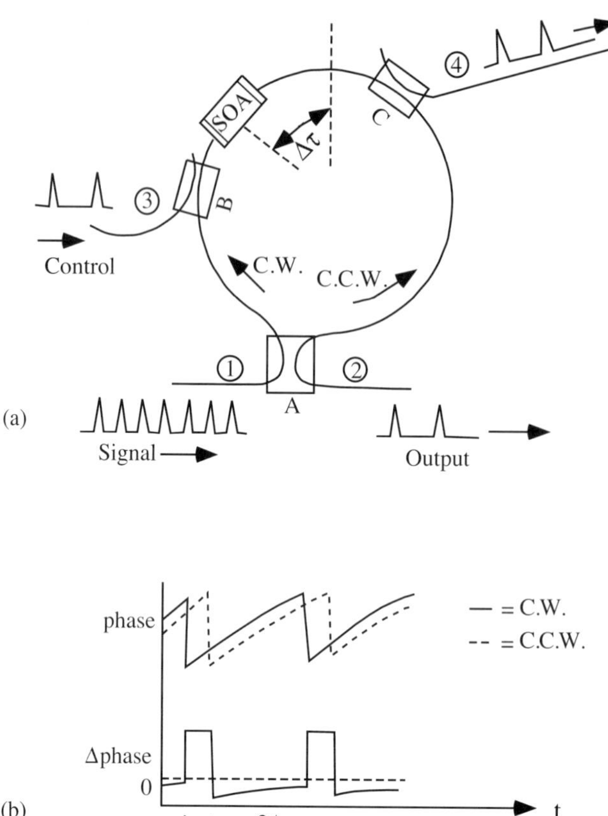

Figure B2.8.18. (*a*) SLALOM/TOAD demultiplexer [31]. Clockwise (cw) and counterclockwise (ccw) pulses interfere at coupler A after travelling in opposite directions around the loop mirror. The control pulse can switch the output of the signal pulse from ① to ②. (*b*) Temporal variation of the phase experienced by the two signal pulse streams and a representation of the switching window.

demultiplexed by a clock pulse stream that has been applied at port 3. Demultiplexing has been demonstrated from 160 Gb s^{-1} to 10 Gb s^{-1} [54]. Because the SOA is the nonlinear element, the control pulse (clock stream here) must have pulse energies such that after amplification in the gain-compressed SOA the pulse energies are comparable to the saturation energies of a few pJ. Thus, input control energies of a few tenths of pJ can be used. In the preceding results, the clock period is shorter than the gain recovery time of the SOA. The demultiplexer can function because the clock signal is periodic and the phase change has a sawtooth shape, as shown in the bottom of figure B2.8.18. The switching window is the displacement between the sawtooth curves.

B2.8.4 Optical bistability in laser diodes

B2.8.4.1 Optical bistability [55]

Optical bistability, as the term implies, refers to the situation in which two stable optical output states are associated with a single optical input state (figure B2.8.19(*a*,*c*)). Two general requirements must be satisfied for optical bistability to occur. The first is that there must exist an appropriate system parameter, such as

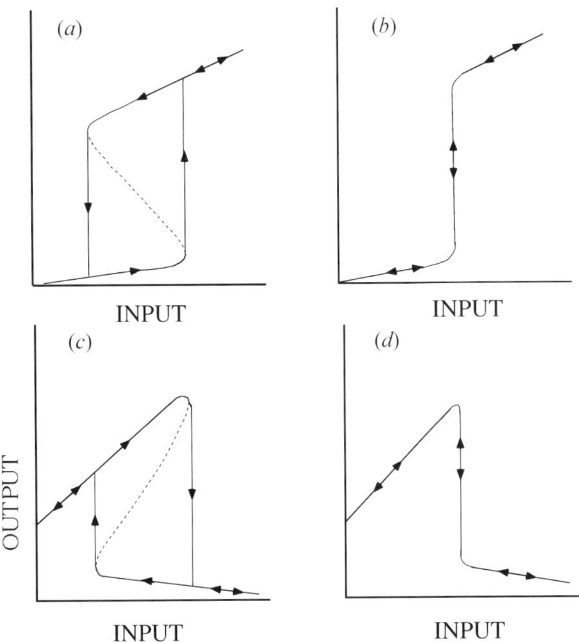

Figure B2.8.19. Illustrations of the various types of optical bistability.

the absorption or gain coefficient or refractive index, which depends on optical input intensity. The second is the existence of a feedback mechanism. By setting the parameters to proper values, we can also obtain differential-gain characteristics (figure B2.8.19(*b,d*)). Various types of bistability appear in LDs, which depend on their configurations and operation conditions. Hysteresis occurs both in a counterclockwise sense (figure B2.8.19(*a*)) and in a clockwise sense (figure B2.8.19(*c*)). These types of hysteresis are called S-shape bistability.

Pitchfork bifurcation is usually defined as a form represented by the following differential equation, which depends on a single parameter μ:

$$\frac{\mathrm{d}x}{\mathrm{d}t} = \mu x - x^3. \qquad \text{(B2.8.10)}$$

The bifurcation diagram for this equation is depicted in figure B2.8.20, in which the branches of equilibria are shown in (μ, x) space. Here, the only bifurcation point is $(\mu, x) = (0, 0)$. It is easy to check that the unique fixed point $x = 0$ existing for $\mu \leq 0$ is stable, that it becomes unstable for $\mu > 0$, and that the new bifurcating fixed points at $x = \pm\sqrt{\mu}$ are stable. As will be described later, many types of bistability that have shapes similar to pitchfork bifurcation appear in two-mode LDs (figure B2.8.20(*b*)). These are called pitchfork bifurcation bistability.

B2.8.4.2 Absorptive bistability

LDs that include saturable absorbers in their cavity show bistability in the optical output *versus* current curve and in the optical output *versus* optical input curve. A saturable absorber is defined as a material whose absorption decreases with the increase in incident radiation intensity. In the OFF state of a BLD, there will be no laser action. In the ON state, the device operates as a laser. The population in the absorber is inverted by optical pumping from the gain region so that it is essentially transparent to the laser radiation. The electron

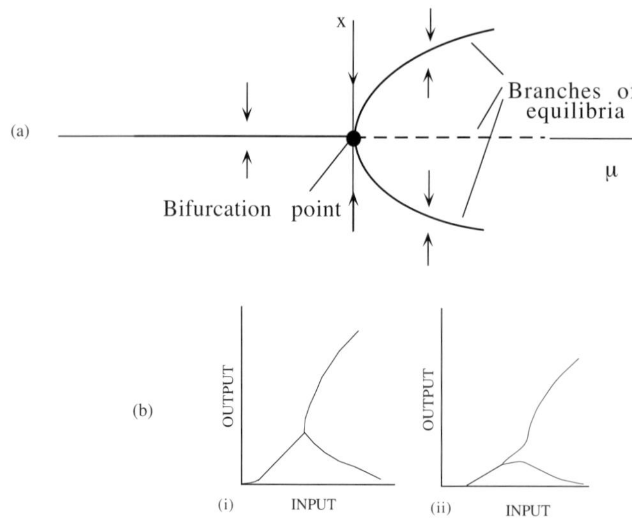

Figure B2.8.20. (*a*) Pitchfork bifurcation; (*b*) some forms of pitchfork bifurcation bistability [55].

quasi-Fermi level in the gain region decreases as it goes from OFF state to ON state, while, in the absorber, it increases. The two states of the device are stable, as will be shown later.

The simplest model for absorptive bistability is the mean-field and single-mode one in the two-section LD with current biasing. The output characteristics can be analysed by the following rate equations

$$\frac{\mathrm{d}n_{\mathrm{e1}}}{\mathrm{d}t} = P_1 - Bn_{\mathrm{e1}}^2 - g_1(n_{\mathrm{e1}})n_{\mathrm{p}}v_{\mathrm{g}} - \frac{n_{\mathrm{e1}}}{\tau_{\mathrm{nr1}}} \tag{B2.8.11}$$

$$\frac{\mathrm{d}n_{\mathrm{e2}}}{\mathrm{d}t} = P_2 - Bn_{\mathrm{e2}}^2 - g_2(n_{\mathrm{e2}})n_{\mathrm{p}}v_{\mathrm{g}} - \frac{n_{\mathrm{e2}}}{\tau_{\mathrm{nr2}}} \tag{B2.8.12}$$

and

$$\frac{\mathrm{d}n_{\mathrm{p}}}{\mathrm{d}t} = n_{\mathrm{p}}\upsilon_{\mathrm{g}}[\gamma_1 g_1(n_{\mathrm{e1}}) + \gamma_2 g_2(n_{\mathrm{e2}})] - \frac{n_{\mathrm{p}}}{\tau_{\mathrm{p}}} \tag{B2.8.13}$$

where $\gamma_1 + \gamma_2 = 1$. Subscripts 1 and 2 refer to parameters in regions I and II, respectively, while n_{e} and n_{p} are the numbers of injected carriers and photons. The pump rate per unit volume is $P = I/qV$, where I is the total input current and V is the volume. B is the recombination coefficient, υ_{g} is the group velocity, β_{sp} is the spontaneous emission coefficient, γ_{i} is a ratio of the length of region i to the whole cavity length. We can obtain characteristics related to bistable operation, such as ON and OFF thresholds, hysteresis width and switching dynamics between ON and OFF states using the above rate equations.

As one example of absorptive BLDs, the InGaAsP DFB BLD structure operating at 1.55 μm is shown in figure B2.8.21(*a*) [56]. The p-type electrode was divided into three parts, and the divided regions can be excited independently through the electrodes. If the current of region I_1 is set at zero or a low value, region I_2 acts as a saturable absorber. It is then possible to obtain bistable characteristics in both the light output current and optical output input curves.

A typical measured light output *versus* current curve is shown in figure B2.8.21(*b*)(i). A noticeable hysteresis loop is seen in the curve. The current range over which bistability exists decreases with an increase in the current into the saturable absorber (not shown in the figure).

When the bias current is set at a value just below the turn-off threshold, the optical output input curve exhibits bistability resulting from an injection of optical power into the saturable absorber, as shown in fig-

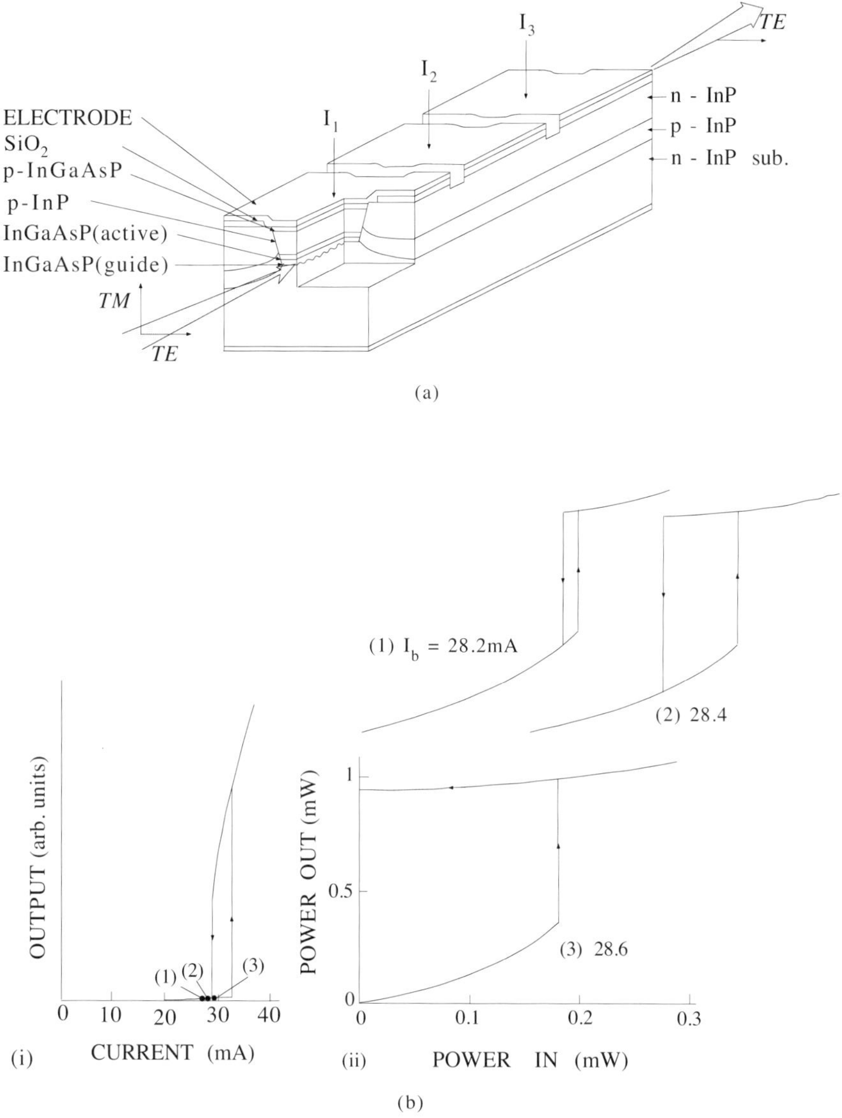

Figure B2.8.21. (*a*) Structure of absorptive bistable DFB LD [56]. When I_1 is set at a low value; region I_2 acts as a saturable absorber. (*b*) Output characteristics of absorptive bistable DFB LD: (i) light output *versus* current curve; (ii) optical output *versus* input curve. Bias current values are indicated by the closed circles (1)–(3) in (i).

ure B2.8.21(*b*)(ii). The bias current values are indicated by the closed circles (1) to (3) in figure B2.8.21(*a*)(i). An increase in the bias current markedly increases the bistability range of the input optical power. When the bias current is within the bistability range, the BLD emits coherent light when input optical power is reset to zero. This means that the BLD acts as an optical memory device.

Generally, in absorptive BLDs, the output wavelength differs from the input wavelength. This characteristic suggests the use of an absorptive BLD as a wavelength converter (shown in section B2.8.3).

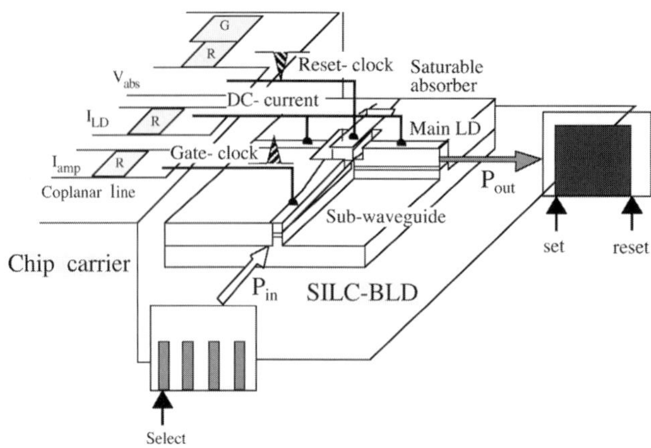

Figure B2.8.22. Schematic diagram of SILC-BLD and chip-mount [58].

The side-injection light-controlled bistable laser diode (SILC-BLD) which consists of a main waveguide laser for output and an orthogonally crossed subwaveguide gate for input was demonstrated [57]. The switching power dependence on input light wavelength is very small yielding a 28 nm wide 3 dB sensitivity. However, the highest operation speed was limited to 1 Gbit s^{-1}. For decreasing parasitic capacitance of the laser, a SILC-BLD buried with semi-insulating InP was developed as shown in figure B2.8.22 [58]. The input signal and output laser cavity are completely isolated, so there is no back reflection or signal leakage. 2.5 Gbit s^{-1} demultiplexed output signals are each selected once every 4 bits from 10 Gbit s^{-1} NRZ optical input signals using this SILC-BLD as described section B2.8.4.5.2.

B2.8.4.3 Dispersive bistability

A resonant-type LD amplifier, consisting of a normal LD biased below the laser oscillation threshold, can act as a nonlinear cavity and shows bistability in the optical input-output characteristics. This is because the active layer refractive index changes due to gain saturation by light injection. The analysis is carried out by using the relationships derived by Adams [59] to describe laser amplifier behaviour. The characteristics of dispensive BLDs strongly depend on their bias current and the detuning from the resonant frequency. Figure B2.8.23(*a*) shows plots of the normalized transmitted output intensity as a function of the normalized input intensity for a bistable amplifier [60]. The input power for switching increases as the detuning from the resonance frequency is increased. In the reflected light, a rich variety of behaviour may be observed [61]. Figure B2.8.23(*b*) shows the three main forms of hysteresis loop exhibited by a Fabry–Pérot amplifier in reflection. For values of the gain close to the threshold, the loop is the same as that for transmission, and the hysteresis occurs in a counterclockwise sense (figure B2.8.23(*b*) (i)). In contrast, for lower values of the gain, the loop is described in a clockwise sense and is rather similar to that for reflection from a passive Fabry–Pérot etalon (figure B2.8.23(*b*) (ii)). For intermediate values of gain, the loop is traversed in the clockwise direction for increasing input, and in the counterclockwise direction for decreasing input (figure B2.8.23(*b*)(iii)).

B2.8.4.4 Two-mode bistability through gain saturation

B2.8.4.4.1 General considerations

If a laser is oscillating in two different modes, the cross-effect between the two modes arises through gain saturation [62]. To the lowest order in the nonlinearity, the saturated gain $\bar{g}_i(I_i)$ of a particular mode of (i), an

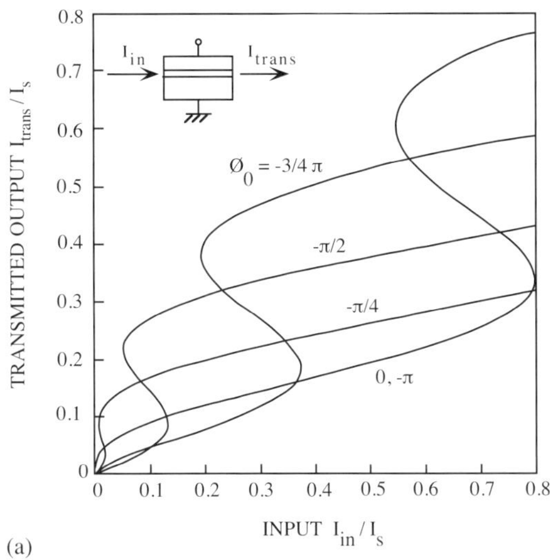

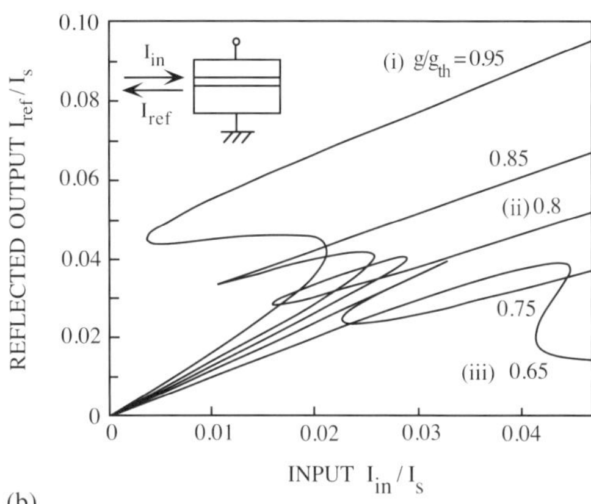

Figure B2.8.23. (*a*) Transmission output *versus* input curve with initial detuning ϕ_0 as the parameter for a Fabry–Pérot amplifier: $b = 3$; $g_0 = 0.95 g_{th}$ [60]. (*b*) Reflected output *versus* input curve with g/g_{th} as the parameter: $\phi_0 = -0.3\pi$ [61].

idealized laser, is related to the unsaturated gain g_{i0} through the mode intensity I_i and saturation parameter $1/\varepsilon_{ii}$ as follows;

$$\bar{g}_i(I_i) = g_{i0}(1 - \varepsilon_{ii} I_i). \tag{B2.8.14}$$

In the case of two modes, the rate of change of one-mode intensity also depends on the intensity of the other:

$$\frac{dI_1}{dt} = g_1 I_1(1 - \varepsilon_{11} I_1 - \varepsilon_{12} I_2) \tag{B2.8.15}$$

$$\frac{dI_2}{dt} = g_2 I_2(1 - \varepsilon_{21} I_1 - \varepsilon_{22} I_2) \tag{B2.8.16}$$

where ε_{11} and ε_{22} are self-saturation and ε_{12} and ε_{21} are cross-saturation coefficients.

Two possible characteristic situations, weak coupling and strong coupling, as shown in figure B2.8.24 can occur. If $\varepsilon_{11}\varepsilon_{22} > \varepsilon_{12}\varepsilon_{21}$ (weak coupling), in general, there can be stable one- or two-mode operation as shown in figure B2.8.24(*a*)–(*c*). Bistability will occur if $\varepsilon_{21}\varepsilon_{12} > \varepsilon_{11}\varepsilon_{22}$ (strong coupling). In this strong-coupling case, only one of the modes oscillates, the other mode being suppressed. In the resulting steady state, therefore, the laser does not oscillate simultaneously in two modes, but it oscillates in only one of the modes. At the stable point P_1 in figure B2.8.24(*d*), it oscillates in mode 1, while at P_2 it oscillates in mode 2. Which of the bistable states is reached is dependent on the initial conditions, as can be seen from the curves in figure B2.8.24(*d*).

B2.8.4.4.2 Nonlinear optical gains [63]

We calculate self- and cross-saturation coefficients by solving the equation of motion for the density matrix perturbationally, where the optical field is described by Maxwell equations, while the electronic structure is calculated quantum mechanically by diagonalizing the Luttinger's Hamiltonian.

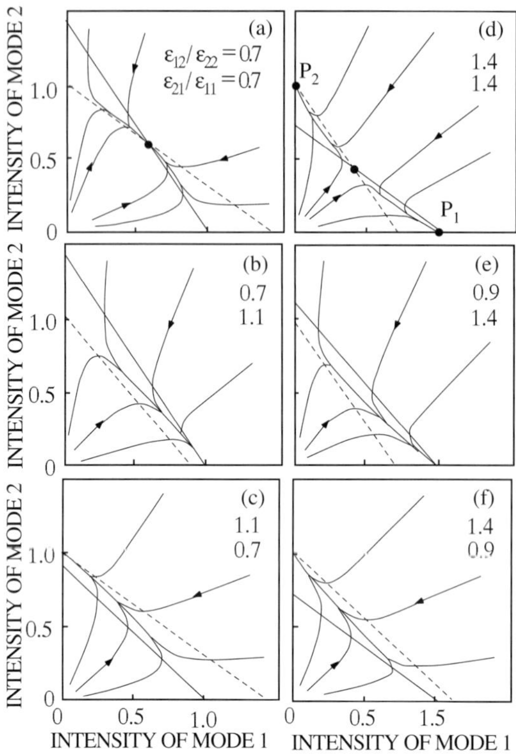

Figure B2.8.24. Examples of trajectories of the intensities of the two modes, I_1 and I_2 (abscissa and ordinates in units of $1/\varepsilon_{11}$ and $1/\varepsilon_{22}$, respectively), for various initial values in the phase plane: (*a*)–(*c*) $\varepsilon_{11}\varepsilon_{22} > \varepsilon_{21}\varepsilon_{12}$; (*d*)–(*f*) $\varepsilon_{21}\varepsilon_{12} > \varepsilon_{11}\varepsilon_{22}$; g_1 is assumed equal to g_2 for all cases; solid straight line: $\mathrm{d}I_2/\mathrm{d}I_1 = \infty$; dashed straight line: $\mathrm{d}I_2/\mathrm{d}I_1 = 0$ [62].

When the polarization of lasing field is in the TE direction, the rate equation for the photon density is given by

$$\frac{\mathrm{d}I_\mathrm{e}}{\mathrm{d}t} = \upsilon_\mathrm{e} g_\mathrm{e} \qquad \text{(B2.8.17)}$$

where υ_e is the light velocity, g_e is the linear gain coefficient, ε_se is the self-saturation coefficient, and ε_xee and ε_xem are the cross-saturation coefficients for parallel and orthogonal polarizations, respectively. A similar expression is obtained when the laser is operating TM mode with frequency ω_m in the presence of two injected fields.

The gain and saturation coefficients are defined in terms of imaginary parts of linear and nonlinear susceptibility.

The density matrix formulation is used in general for nonlinear optical systems in order to obtain nonlinear susceptibility tensors. We solve the (8 × 8) Luttinger–Kohn Hamiltonian to obtain energy dispersions and dipole transition matrix elements at each k states. We solve the quantum mechanical Liouville equation by perturbation.

We have numerically calculated linear gains and saturation coefficients for semiconductor lasers composed of bulk $\mathrm{In}_{1-x}\mathrm{Ga}_x\mathrm{As}_y\mathrm{P}_{1-y}$ lattice matched to InP, which emits in the 1.3 μm region.

We show in figure B2.8.25 the photon energy dependence of the self-saturation coefficients ε_se for TE

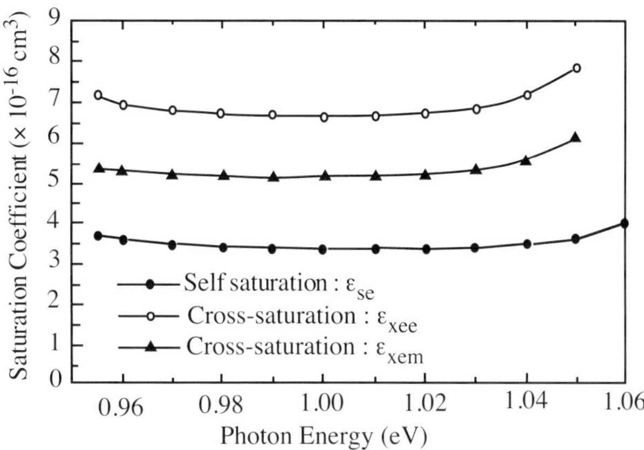

Figure B2.8.25. Self-saturation coefficients $\varepsilon_{\mathrm{se}}$ and the peak magnitudes of cross-saturation coefficients $\varepsilon_{\mathrm{xee}}$ and $\varepsilon_{\mathrm{xem}}$ at 300 K with an injection carrier density of 4×10^{18} cm^{-3} [63].

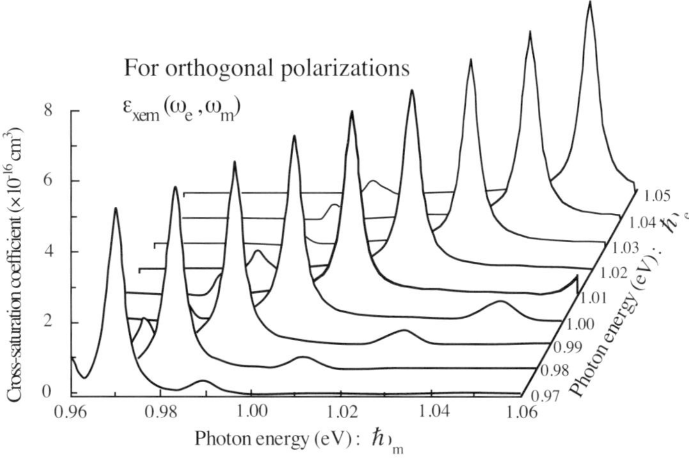

Figure B2.8.26. Cross-saturation coefficients for orthogonal polarizations, $\varepsilon_{\mathrm{xem}}(\omega_{\mathrm{e}}, \omega_{\mathrm{m}})$ at 300 K with the carrier density 4.0×10^{18} cm^{-3} [63].

mode. The self-saturation coefficients barely depend on photon energies. Self saturations in TM mode, $\varepsilon_{\mathrm{sm}}$ are very similar to $\varepsilon_{\mathrm{se}}$.

There are four types of cross-saturation coefficients; $\varepsilon_{\mathrm{xee}}$ is the saturation coefficient for two optical fields with parallel polarizations, that is, it represents the gain saturation of laser light I_{e} (with a frequency ω_{e} in TE mode) caused by the injected light I_{e}' (with a frequency ω_{e}' in TE mode); $\varepsilon_{\mathrm{xem}}$ is the saturation coefficient for two optical fields with orthogonal polarizations, representing the gain saturation of the laser light I_{e} caused by the injected light I_{m} in TM mode. (The other two parameters, $\varepsilon_{\mathrm{xmm}}$ and $\varepsilon_{\mathrm{xme}}$, have a similar meaning.) Thus, cross-saturation coefficients are the function of two optical frequencies, ω_{e} and ω_{e}', or ω_{e} and ω_{m}.

We plot the photon energy dependence of $\varepsilon_{\mathrm{xem}}$ in figure B2.8.26. (We have the similar plot for $\varepsilon_{\mathrm{xee}}$ with its magnitude slightly larger as discussed later.) The cross-saturation coefficients are significant only around

$\omega_e = \omega_{e'(m)}$ when both of the photon energies are resonant with e-hh transitions or e-lh transitions, and they vanish otherwise.

We show in figure B2.8.25 the magnitudes of ε_{xee} and ε_{xem} at peak ($\omega_e = \omega_{e'(m)}$). They show similar trends to the self-saturation coefficients; the magnitudes are almost independent of photon energies.

The relative magnitude of ε_{se}, ε_{sm} and ε_{xem}, ε_{xme} is very important since it determines the polarization bistability properties of semiconductor lasers. As described in section B2.8.4.4.1, the polarization bistability occurs when $\varepsilon_{xem}\varepsilon_{xme} > \varepsilon_{se}\varepsilon_{sm}$. We can state several characteristics of those parameters from the symmetry properties of the crystal. When the crystal is isotropic, $\varepsilon_{se} = \varepsilon_{sm}$. In the present case of InGaAsP with cubic symmetry, the degeneracy is lifted, resulting in the slight difference of ε_{se} and ε_{sm}. For cross-saturation coefficients with orthogonal polarizations, it can be shown generally from the symmetry properties that $\varepsilon_{xem} = \varepsilon_{xme}$ in isotropic and in cubic symmetry.

There are other important relationships between self-saturation coefficients and cross-saturation coefficients for the parallel polarizations, which hold generally independent of the details of the material structure,

$$\varepsilon_{xee}(\omega_e = \omega_0, \omega'_e = \omega_0) = 2 \times \varepsilon_{se}(\omega_0),$$
$$\varepsilon_{xmm}(\omega_m = \omega_0, \omega'_m = \omega_0) = 2 \times \varepsilon_{sm}(\omega_0).$$

The present numerical calculations confirm this relation within the numerical precisions as can be seen in figure B2.8.25.

The present numerical calculations for this laser material InGaAsP show,

$$\varepsilon_{se} < \varepsilon_{xem} < \varepsilon_{xee} \qquad \text{and} \qquad \varepsilon_{se} < \varepsilon_{xem} < \varepsilon_{xee}\varepsilon_{sm} < \varepsilon_{xme} < \varepsilon_{xmm},$$

at $\omega_e = \omega_{e'(m)}$, thus satisfying the condition $\varepsilon_{xem}\varepsilon_{xme} > \varepsilon_{se}\varepsilon_{sm}$ for polarization bistability.

The numerical calculations already mentioned take into account the effect of spectral hole burning. However, it is believed that carrier heating also has considerable contribution to the gain saturation in laser diodes. We should point out that the gain saturation caused by spectral hole burning is a coherent process while the saturation by carrier heating is an incoherent process. Thus in the former process, the system retains the 'memory' of how the spectral hole was created, resulting in the frequency and polarization-direction-dependent saturation coefficients. However, in the latter process, carriers are relaxed in energy-momentum space and are in quasithermal equilibrium within the electron system. In other words, carriers already lost the 'memory' of how they were created. Thus, the spectra of saturation coefficients by carrier heating should be smooth and flat functions of photon energies, and independent of the direction of optical polarizations. Then the contribution from carrier heating should be to add some constant background component to the saturation coefficient caused by the carrier heating, ε_{ch}, from the experimentally observed carrier heating in semiconductor lasers, which yielded 3×10^{-17} cm^3. ε_{ch} is of the same order as our calculations for saturation coefficients. The relative magnitudes of self- and cross-saturation coefficients in the present study are unaffected even when the constant components of carrier heating are combined.

B2.8.4.4.3 Rate equation analysis

When a cross section of an optical waveguide is rectangular, there are two possible polarization modes. Usually a laser diode oscillates with TE mode because of its strong optical confinement into the active layer and high reflectivity at the facets. However, if the structure of a laser diode is optimized, the device becomes polarization-insensitive and we can see a bistability between the TE mode oscillation and TM mode oscillation.

We can analyse the static and the transient behaviour of the polarization bistability in LDs using the following rate equations including the gain saturation

$$\frac{dn}{dt} = P - \frac{n}{\tau_s} - v_g g(n)(1 - \varepsilon_{EE}S_E - \varepsilon_{EM}S_M)S_E \tag{B2.8.18}$$

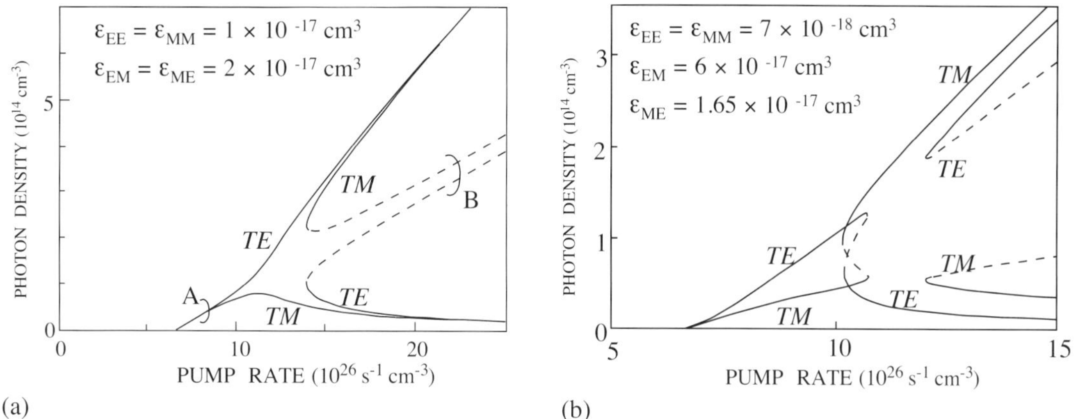

Figure B2.8.27. (*a*) Pitchfork bifurcation polarization bistability arising from the inherent intraband relaxation processes [55]. (*b*) Ordinary S-shaped polarization bistability arising from enhancement of the cross-saturation effect.

$$\frac{\mathrm{d}S_{\mathrm{E}}}{\mathrm{d}t} = \upsilon_{\mathrm{g}}\Gamma_{\mathrm{E}}g(n)(1 - \varepsilon_{\mathrm{EE}}S_{\mathrm{E}} - \varepsilon_{\mathrm{EM}}S_{\mathrm{M}})S_{\mathrm{E}} \tag{B2.8.19}$$

$$\frac{\mathrm{d}S_{\mathrm{M}}}{\mathrm{d}t} = \upsilon_{\mathrm{g}}\Gamma_{\mathrm{M}}g(n)(1 - \varepsilon_{\mathrm{ME}}S_{\mathrm{E}} - \varepsilon_{\mathrm{MM}}S_{\mathrm{M}})S_{\mathrm{M}}. \tag{B2.8.20}$$

Equation (B2.8.18) is for the electron density. Equations (B2.8.19) and (B2.8.20) are the photon densities of TE and TM polarization. In these equations, S_{E} and S_{M} are the photon densities for the TE and TM modes, respectively. The carrier density in the active layer is n. The material gain is given by $g(n) = an - b$, P is the pump rate, τ_{s} is the carrier lifetime, υ_{g} is the group velocity. The optical confinement factors are Γ_{E} and Γ_{M} where the subscripts E and M refer to the TE and TM modes, respectively. The coefficients for self-saturation are $\varepsilon_{\mathrm{EE}}$ and $\varepsilon_{\mathrm{MM}}$, and $\varepsilon_{\mathrm{EM}}$ and $\varepsilon_{\mathrm{ME}}$ are those for cross-saturation from the nonlinear gain, τ_{pE} and τ_{pM} are the photon lifetimes in the laser cavity, β is the spontaneous emission factor and B is the recombination coefficient. We assumed the following relations from the consideration mentioned in the previous subsection: $\varepsilon_{\mathrm{ME}} = \varepsilon_{\mathrm{EM}} = 2\varepsilon_{\mathrm{EE}} = 2\varepsilon_{\mathrm{MM}}$. The enhancement of cross-saturation effect can be expected with the use of additional transverse mode effects. When a coherent optical signal $S_{\mathrm{iE(M)}}$ (E(M) refers to the TE(TM) mode) at a wavelength within the injection locking band is injected into the BLD, an additional term $\sqrt{S_{\mathrm{E(M)}}\eta S_{\mathrm{iE(M)}}}/\tau_{\mathrm{pE(M)}}$ is included in (B2.8.19) or (B2.8.20) depending on its polarization. Here, η is the coupling coefficient. If we take into account a detuning of incident optical beam from the cavity resonant frequency of an LD, rate equations should be modified [64].

We can obtain two kinds of polarization bistability, conventional S-shaped bistability and pitchfork bifurcation bistability depending on the values of self- and cross-gain saturation coefficients. Only the intraband relaxation processes need to be considered to establish the pitchfork bifurcation bistability, as shown in figure B2.8.27(*a*). In this case, as the pump rate is increased from zero, the output intensities of TE and TM polarization bifurcate at a critical point, and form the branch A. Branch B is obtained by polarization switching caused by the injection of the TM trigger input. In order to attain S-shape bistability, an enhanced cross-saturation effect has to be included (figure B2.8.27(*b*)). Here, the strong gain-saturation for TE mode caused by a TM mode power is assumed. S-shape bistability has been found, for example, in low temperature LDs and ridge waveguide LDs, due to the stress-induced nonlinear gain effect. Again, we can see the branches caused by the pitchfork polarization bistability in the region of high pump rate.

An all-optical flip-flop operation is realized with the help of trigger optical pulses. The bistable LD

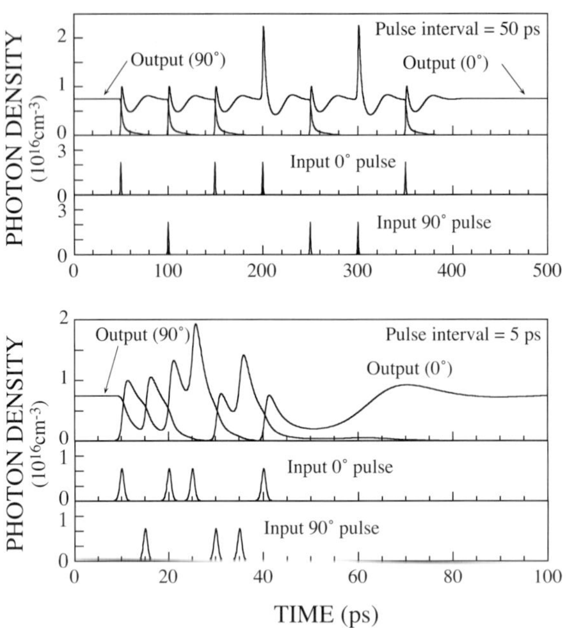

Figure B2.8.28. Ultrafast all-optical flip-flop operation in pitchfork bifurcation polarization bistability [65].

is biased in the regime of the bifurcated solution, and is injected with trigger pulses of either TE or TM polarization. Figure B2.8.28 shows calculated time variations of two orthogonal polarization outputs of a vertical-cavity surface-emitting laser (VCSEL) [65] (see chapter B2.6). The ultrafast flip-flop operation is realized at a rate of 150 Gbit s^{-1}. The main reason for such fast switching is that the flip-flop operation can be obtained by using only ON switching for both modes.

B2.8.4.4.4 Two-mode bistability in stripe lasers

A pitchfork bifurcation polarization bistability in an LD with a two-armed polarization-selective external cavity was examined [66]. All-optical flip-flop operation has been successfully demonstrated.

TE/TM mode switching in GaAsP/AlGaAs tensilely strained quantum-well (QW) LDs with two electrodes was demonstrated [67]. Ridge-stripe LDs with tandem electrodes were fabricated. The LD oscillated in the TM mode when current was injected into both electrodes, and oscillated in the TE mode when current was injected only into the longer electrode. The mode was switched by controlling the injection current of the two electrodes. The modulation efficiency of the TM mode output selected by a polarizer was very high (15 W A^{-1}) in this mode switching.

B2.8.4.4.5 Polarization bistability in vertical-cavity surface-emitting lasers

In this subsection, some of the experimental results displaying pitchfork bifurcation polarization bistability and the all-optical flip-flop operation in VCSELs are reviewed [65].

The structure of the device used in the experiment is shown in figure B2.8.29. The VCSEL consists of $In_{0.2}Ga_{0.8}As$ active layers and AlAs-GaAs DBR mirrors. The top mesa is 6 μm square. The small size of the device ensured its stable single transverse mode operation. The VCSEL used in this experiment shows a polarization switching between 0° and 90° at certain operation conditions, such as temperature and current.

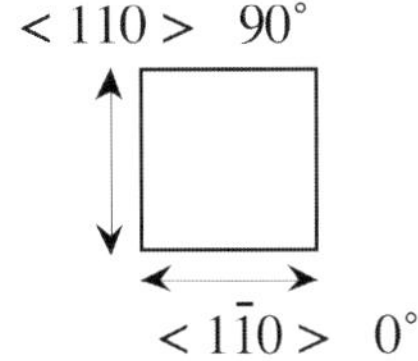

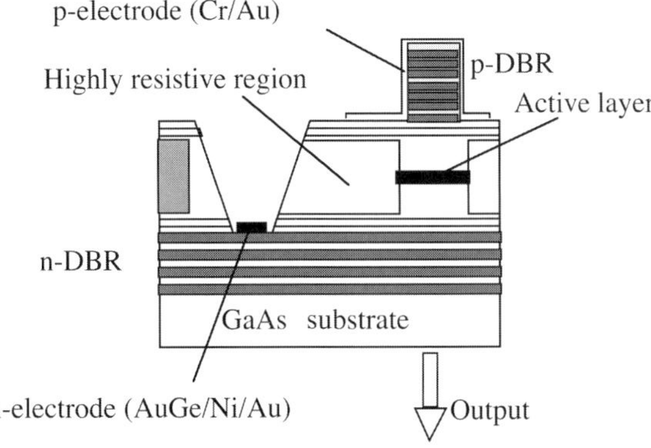

Figure B2.8.29. Schematic structure of the VCSEL [64].

Here we call these polarization modes '0° mode' and '90° mode', respectively. One device of a 10 × 8 independently addressable VCSEL array was used in the experiment.

The polarization resolved L–I curve measured at 18.5 °C is shown in figure B2.8.30(a). No polarization switching was observed. The free-running output light was linearly polarized at 100°. Pitchfork bifurcation polarization bistability was observed, when light input having a wavelength close to the oscillation wavelength of the VCSEL was injected as shown in figure B2.8.30(b). When the injection current was increased from zero, the VCSEL oscillated with the 100° polarization. The injection current was set at 3.52 mA. A 10° polarized trigger light pulse was then injected into the VCSEL. The polarization of the VCSEL changed to the same polarization (10°) as that of the incident light. No change in the oscillating wavelength of the VCSEL was observed. Even when the incident light was blocked, the polarization of the VCSEL remained unchanged (10°). When the injection current was increased up to 3.9 mA, the dominant polarization remained at 10°. On the contrary, however, when the injection current was decreased, the polarization was switched back to the initial polarization state at 3.1 mA.

Switching between the two polarization states was accomplished by injecting the trigger optical inputs (figure B2.8.31(a)). The VCSEL oscillated with 100° polarization at the initial condition. When a 10° polarization trigger pulse was injected, the VCSEL changed its polarization from 100° to 10°. At the 10° polarization state, when a 100° polarization trigger pulse was injected, the polarization of the VCSEL changed from 10° to 100°. If an optical trigger input having the same polarization as the VCSEL was injected, the polarization state of the VCSEL did not change.

All-optical flip-flop operation was obtained at 5.3 GHz, which is the highest repetition rate of flip-flop operation using LDs. To measure the switching between the orthogonal polarizations with much faster speed, modelocked pulses from a Ti:sapphire laser were used as optical injection pulses. Figure B2.8.31(b) shows the ultrafast polarization bistable switching. The polarization switching time can be estimated to be about 7 ps. Even for the ultrafast bistable switching obtained by a picosecond trigger input, we can see stable

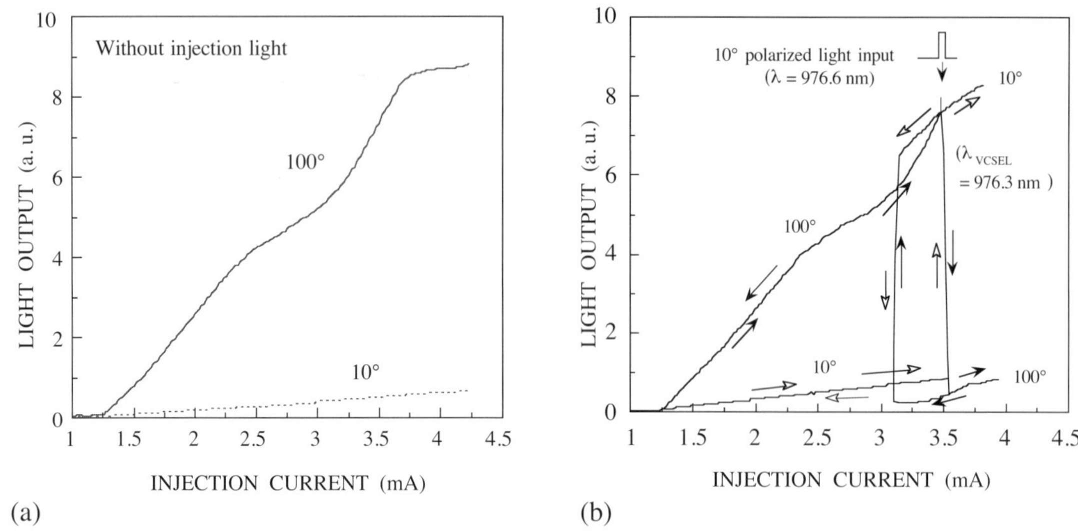

Figure B2.8.30. (*a*) Polarization resolved L–I curves measured at 18.5° [64]. (*b*) Polarization resolved L–I curves for when 10° polarization light pulse is injected into the VCSEL at $I = 3.52$ mA.

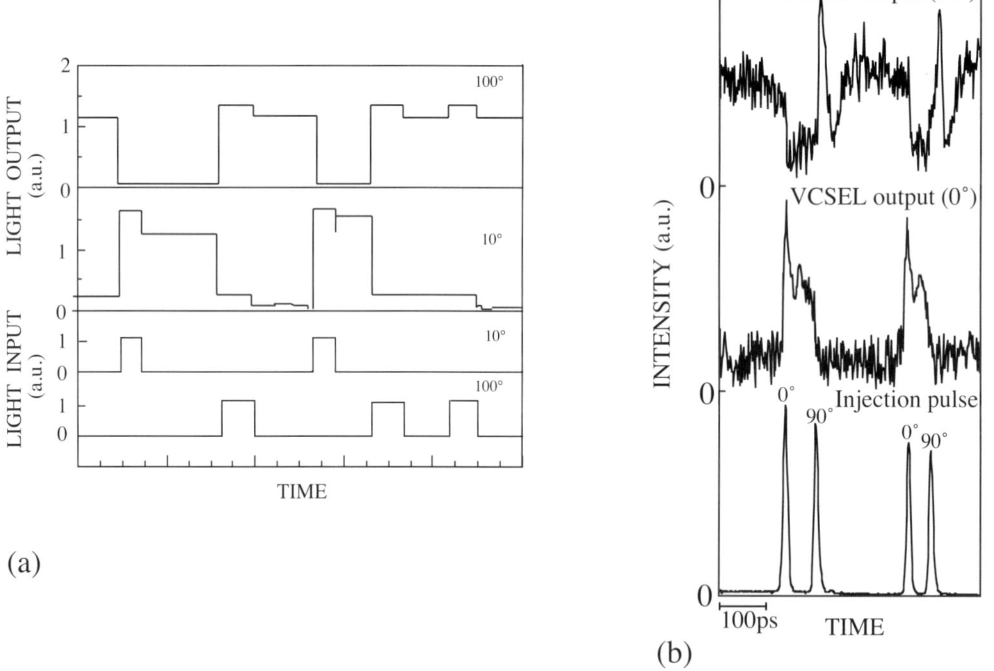

Figure B2.8.31. (*a*) Time variability for all-optical flip-flop operation [64]. (*b*) Time variability for ultrafast polarization bistable switching.

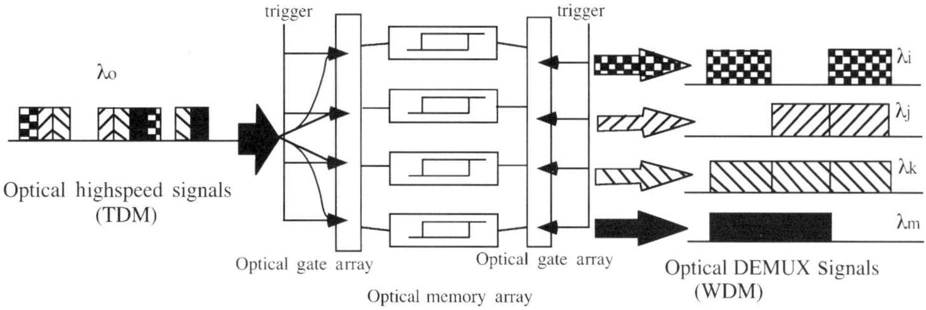

Figure B2.8.32. Optical demultiplexing scheme with an SILC-BLD in a TDM/WDM switching network [71].

polarization bistable switching with a holding time of 12 ns that is determined by the input pulse repetition rate.

B2.8.4.5 Applications

B2.8.4.5.1 Digital signal regeneration

Nonaka *et al* have developed a SILC-BLD that works as wavelength converter, as well as digital signal regeneration device, due to its threshold amplification characteristics [68]. The SILC-BLD is used to digitally regenerate optical signals with; (1) low extinction, (2) intensity fluctuation and (3) waveform deformation. The SILC-BLD consists of a main waveguide laser for output and an orthogonally crossed subwaveguide for input. The subwaveguide amplifies input light and conducts it to the saturable absorption region in the main laser cavity. The optical input-output characteristic shows thresholding. The thresholding characteristics act as a discriminator, suppressing the noise level and amplifying the on-set signal to a constant level, so the output signal level is stable and signal extinction is improved to 17 dB for 622 Mb s^{-1} and 1 Gb s^{-1} bit rates. 2.5 Gb s^{-1} error-free digital regeneration has been demonstrated with the semi-insulating InP buried BLD [69].

B2.8.4.5.2 Optical demultiplexing

Nonaka *et al* [70, 71] have demonstrated direct optical demultiplexing of NRZ signals with simultaneous wavelength conversion by using a semi-insulating InP buried SILC-BLD module (figure B2.8.32). High-bit-rate multiplexed signals from backbone networks are distributed to WDM local area networks in the switching nodes, where signals are converted to low bit rate and to several different wavelengths corresponding to the end-terminals. In this situation, the high-speed optical gating and signal bit width and wavelength conversion function are required simultaneously. To demultiplex NRZ optical signals simultaneously with wavelength conversion, the photonic device must have integrated functions for gate operation, memory and wavelength conversion. This could be applied to the photonic switching nodes of TDM/WDM hybrid transmission networks.

In the SILC-BLD, input light signal from a fibre are selected by the sub-waveguide that operates as an optical gate and are memorized in the main BLD. The output light signals with longer pulsewidths and the converted wavelength are read from the output fibre. Clock pulses for the voltage control part of the saturable absorption region are produced by the trigger pulses obtained from the pin-photodiode in the module. In their former experiment, the 250 Mb s^{-1} demultiplexed output signals are selected once every four bits from

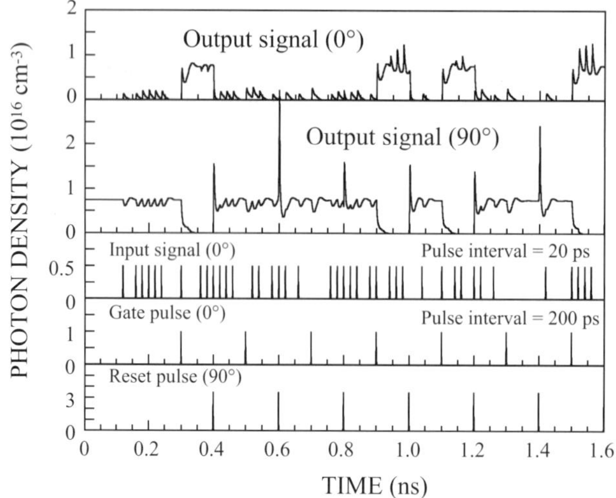

Figure B2.8.33. Computed all-optical demultiplexing from 50 Gbit s^{-1} to 5 Gbit s^{-1} signal [72].

1 Gb s^{-1} NRZ optical input signals [70]. Each signal is regenerated and its bit width and wavelength were converted on one chip.

Recently, input and output switching speeds of this device has been improved with the InP buried heterostructure and chip carriers having 50 Ω matched coplanar lines [71]. The 2.5 Gb s^{-1} demultiplexed output signals were selected and bit width was converted in the same chip from 10 Gb s^{-1} NRZ optical input-signals. The newly designed SILC-BLD consists of a main waveguide laser for output and an orthogonally crossed sub-waveguide gate for input as shown in Figure B2.8.22.

Using the memory function and thresholding function of the polarization bistable VCSEL, all-optical demultiplexing with a signal-bit width conversion function can be obtained from high bit rate multiplexed signals [72]. Only when the input signal and the gate pulse are injected simultaneously into the polarization bistable VCSEL with 0° polarization, the polarization of the VCSEL is set to 0° polarization due to its AND gate function. Then laser oscillation in 0° polarization is continued until the reset pulse with 90° polarization is injected. Thus, optical demultiplexing with a signal-bit width conversion function is achieved in the 0° polarization output signal. Figure B2.8.33 shows one example of computed time variations of all-optical demultiplexing. The 5 Gbit s^{-1} demultiplexed output signals are selected once every 10 bits from the 50 Gbit s^{-1} optical input signal. All-optical demultiplexing has been successfully demonstrated using the ultrafast polarization bistable VCSEL [72].

B2.8.5 Future trends

Optical nonlinearities in LDs and SOAs and their applications to optical functional devices for switching and signal processing are studying extensively. Here, three topics, i.e. bandgap-engineering for enhancement of optical nonlinearities, devices utilizing light polarization states and monolithic integration with other devices, are selected, which may become important research fields in the device level. WDM applications, mid-span spectral inversion and high-speed signal processing are selected as important system applications in the future.

Bandgap engineering consists of the tailoring of an association of materials in order to custom design the structure for some desired properties unattainable in homostructures. Third-order optical nonlinear susceptibilities $\chi^{(3)}$ in compressively strained and non-strained InGaAs–InGaAsP quantum wells (QWs) under the population inversion condition have been discussed [73]. The small effective mass of compressively strained QWs increases the contribution of the carrier density pulsation effect and the carrier heating effect of $\chi^{(3)}$. The hole burning effect is also increased due to the decrease of the carrier–carrier scattering rate. The calculation including these effects shows an enhancement of factor 3 due to 0.8% compressive strain. $\chi^{(3)}$ in 0.8% compressively strained QWs is three times larger than that in non-strained QWs with the same linear gain.

Quantum dot (QD) SOAs are extensively studied for the applications of demultiplexing, regenerating and wavelength conversion from the perspective of its large and ultrafast optical nonlinearity [74]. The enhancement in the optical nonlinearity is expected to originate from the slower relaxation time of SHB in QDs in comparison with QWs, because relaxation occurs not via intraband carrier–carrier scattering as in QWs, but via carrier capture by QDs. The ultrafast response is expected due to the absence of slow gain recovery component as produced by carrier density pulsation in QWs.

Spectral broadening and cross relaxation of an aggregate of highly carrier-injected quantum dots have been studied experimentally by evaluating the gain saturation and optical-wave mixing characteristics of SOAs that include self-assembled 0.98 μm InGaAs dots [75]. Experimental results suggest that cross relaxation is so fast that the dot aggregate has a homogeneously broadened gain spectrum. Self-saturation and cross-saturation characteristics, and the detuning characteristics of highly non-degenerate FWM were also investigated. The latter revealed that the characteristic times of two cross-relaxation processes are about 1 ps and 100 fs.

The optical gain saturation characteristics and the ultrafast carrier dynamics of quantum dots in the gain region were measured [74]. The quantum dot layers are closely stacked Stranski–Krastanov InAs dots on GaAs substrate with few monolayers-thick intermediate layers in the growth direction (columnar dot). $\chi^{(3)}$ value was measured to be 5.5×10^{-16} m^2 V^{-2} with linear gain g_0 of 285 cm^{-1}, and the gain recovery time was measured to be $\sim$1.8 ps. Nonlinearity expressed by $\chi^{(3)}/g_0$ was 1.9×10^{-5} m^3 V^{-2} for a bias current of 50 mA, which is comparable to that of quantum wells. This result is contrary to the expectation that the $\chi^{(3)}$ value should become large in quantum dots. Further investigation is required.

The sensitivity of SOA-based devices to the state of polarization of input waves is of prime interest in most transmission and network applications. A great deal of effort has been concentrated in making SOAs with polarization-insensitive gain. However, if we can use light polarization states as information, optical devices will have a variety of new functions and much flexibility.

As described in section B2.8.4.4, the switching speed of bistable laser diode has been greatly improved by using the pitchfork bifurcation polarization bistability in VCSELs. The fastest switching speed and the highest repetition rate against year are shown in figure B2.8.34 [76]. It may operate from the few hundreds of GHz range to even the THz range.

The difference in the group velocity for TE and TM modes in SOAs allows the implementation of an all-optical switch based on a non-interferometric principle that only uses a SOA and does not require integrated optics. Two kinds of birefringence effects have been observed in a SOA [77, 78]. The first is attributed to different refractive indices for two principal axes. The second effect is observed when a strong beam (control beam) is introduced in the amplifier. Due to the quasi-Gaussian intensity profile of the control beam in the amplifier, the carriers on the plane perpendicular to the propagation direction are not depleted uniformly. This produces a wavefront whose polarization state depends on the position of the wavefront in the transverse plane. This phenomenology affects in a similar way a linearly polarized signal beam simultaneously introduced in the SOA with the strong control beam, both with polarizations coinciding with the TE axis. The change in the signal polarization at the amplifier output is increased as the control beam power is increased. Therefore, by turning on and turning off the control beam it is possible to obtain a signal at the amplifier output with two different polarization states. An extinction ratio of about 30 dB obtained with a 51 dBm off-state has been confirmed.

Another important research field is monolithic integration of photonic devices. Monolithic integration leads to a drastic reduction in volume of complex optical devices. Compact WDM channel selectors which

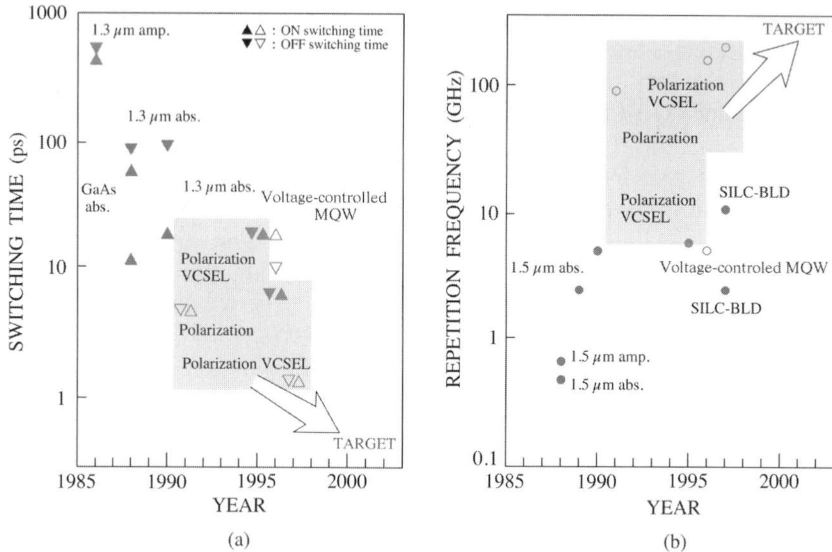

Figure B2.8.34. Progress in performances of BLDs [76]. Switching on (▼) and off (▲) time and repetition frequency (●) of BLDs. ▼, ▲ and ● are experimental values and △, ▽ and ○ are theoretical ones.

are monolithically fabricated on InP substrates has been reported and will be described later in detail [79]. A deep-ridge structure with a small bending radius is used for passive waveguides to reduce the device size. In addition, a buried waveguide structure, whose reliability has been confirmed in conventional laser production, is used in the SOA regions. The hybrid waveguide structure enables us to use the optimum design for each region, therefore, compact high-performance WDM channel selectors can be made.

One of the most exciting developments in the field of optical communication (see section D4) is the explosive growth of wavelength division multiplexing (WDM) systems [80]. Today's commercial systems are developed for use in point-to-point transmission; the next step will be the use of WDM in advanced network applications. WDM systems require devices with a greater complexity than systems operating at a single wavelength, such as tunable or multiwavelength transmitters and receivers, WDM add-drop multiplexers and WDM-crossconnects.

With the technology of integration of SOAs with passive components a variety of functions can be realized. Figure B2.8.35 shows a WDM channel selector [79] consisting of a demultiplexer and a multiplexer with a SOA gate array in between, that can pass or block each of the eight wavelengths applied to the input demultiplexer. The device operates with zero-loss between fibres for the transmitted channels and an extinction of 40 dB for unwanted channels.

For ultrahigh-speed optical transmission in standard single mode fibre (SSMF), the transmission length will be limited by chromatic dispersion. This limitation can be overcome by installing dispersion-shifted fibres (DSF). However, to upgrade the capacity of the large existing net of SSMF, compression techniques are required, and one technique, which has attracted much attention, relies on spectral inversion of the transmitted signal. It is based on four-wave mixing (FWM) in a nonlinear medium, situated in the middle of the transmission span, mid-span spectral inversion. Using this scheme, e.g. 5×40 Gbit s^{-1} over 105 km SSMF fibre has been reported [81].

Recently, polarization-independent FWM in an SOA used for mid-span spectral inversion was implemented and introduced only 0.9 dB penalty compared to polarization-dependent mid-span spectral inversion [82]. The polarization dependence in receiver sensitivity is 1 dB. An SOA monolithically integrated with

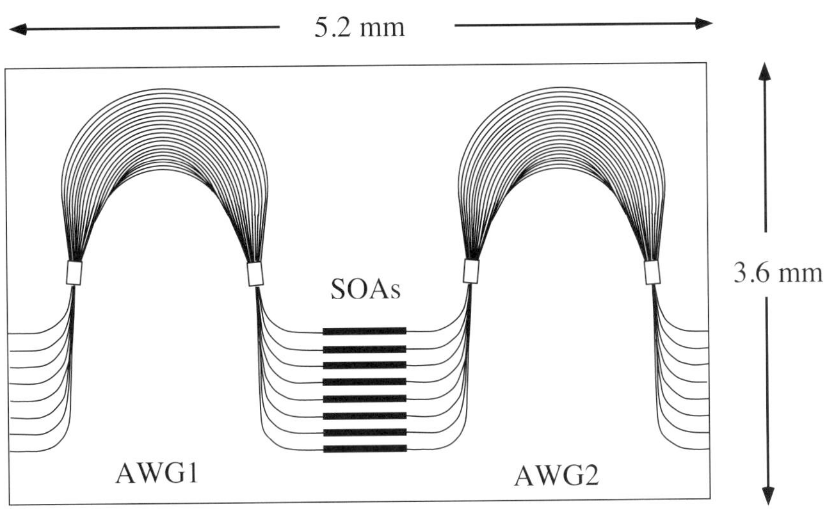

Figure B2.8.35. An eight-channel WDM channel selector fabricated on an InP substrate [79].

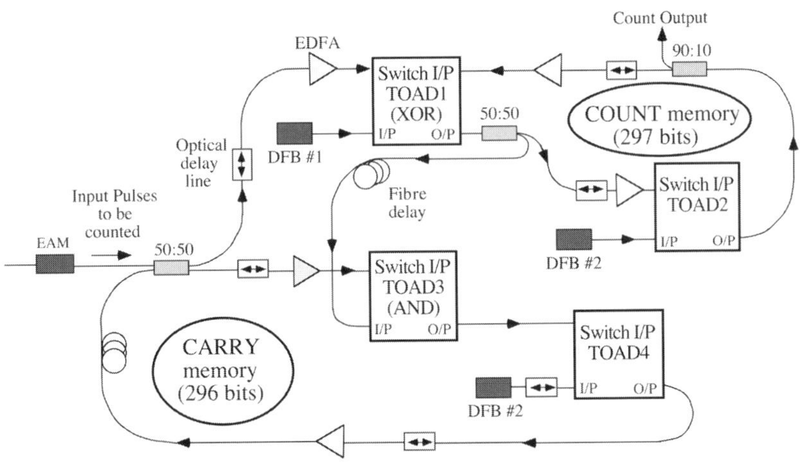

Figure B2.8.36. Schematic diagram of the experimental all-optical counter [84]. EDFA = erbium-doped fibre amplifier, I/P = input, O/P = output, DFB = gain switched distributed feedback laser diode, EAM = electroabsorption modulator.

a distributed feedback pump laser is used for the first time to provide dispersion compensation at 40 Gbit s^{-1} using mid-span spectral inversion over more than 100 km of standard fibre [83].

The deployment of all-optical digital processing in future high-speed optical network architectures is dependent upon being able to add functionality directly at the optical layer. In photonic networks based on binary modulated packets of optical pulses, all-optical processing directly at the bit level can be used to regenerate the packets and perform functions such as parity checking. Poustie *et al* describe an all-optical binary counter [84]. Within optical packet networks, a binary counter can be used at the packet level to verify network performance within a node. Ultimately, an optical pulse counteroperating at the bit level could enable functions such as header extraction and payload processing to be realized. In the experiment

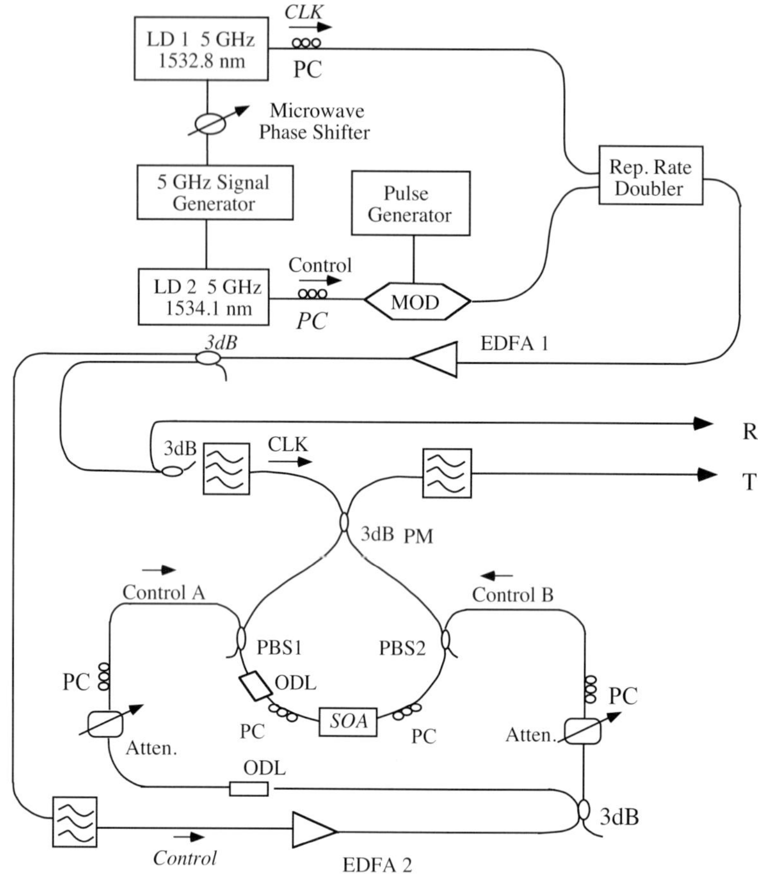

Figure B2.8.37. Experimental set-up of Boolean XOR logic [85].

two coupled optical regenerative memories are used to form the optical counter, as shown in figure B2.8.36.

Boolean XOR logic was demonstrated on a pseudo-data pattern using an SOA three-terminal fibre Sagnac gate by the experimental configuration shown in figure B2.8.37 [85]. Bit pattern switching with low pattern dependence is achieved at 10 Gb s^{-1}.

References

[1] Ghafouri-Shiraz H 1996 *Fundamentals of Laser Diode Amplifiers* (New York: Wiley)

[2] Mukai T, Yamamoto Y and Kimura T 1985 Optical amplification by semiconductor lasers *Semiconductors and Semimetals* vol 22, part E, ed W T Tsang (New York: Academic Press)

[3] Shimada S and Ishino H (ed) 1994 *Optical Amplifiers and their Applications* (New York: Wiley) chs 3 and 4

[4] Mukai T, Inoue K and Saitoh T 1987 *Appl. Phys. Lett.* **51** 381–3

[5] Magari K, Kondo S, Yasaka H, Noguchi Y, Kataoka T and Mikami O 1990 *IEEE Photon. Technol. Lett.* **2** 792–3

[6] Jopson R M and Darcie T E 1991 Semiconductor laser amplifiers in high-bit-rate and wavelength-division-multiplexed optical communication systems *Coherence, Amplification, and Quantum Effects in Semiconductor Lasers* ed Y Yamamoto (New York: Wiley) pp 323–66

[7] Spano P, Piazzola S and Tamburrini M 1983 *IEEE J. Quantum Electron.* **QE-19** 1195–9

[8] Henry C H 1986 *IEEE J. Lightwave Technol.* **LT-4** 298–311

[9] Grosskopf G, Ludwig R, Waats R G and Weber H G 1987 *Electron. Lett.* **23** 1387–8

[10] Olsson N A, Kazarinov R F, Nordland W A, Henry C H, Oberg M G, White H G, Garbinski P A and Savage A 1989 *Electron. Lett.* **25** 1048–9
[11] Mersali B, Gelly G, Accard A, Lafragette J-L, Doussiere P, Lambert M and Fernier B 1990 *Electron. Lett.* **26** 124–5
[12] Magari K, Okamoto M, Suzuki Y, Sato K, Noguchi Y Mikami O 1994 *IEEE J. Quantum Electron.* **30** 695–702
[13] Doussière P 1996 *TOPS* **V** 170–88
[14] Simon J C, Doussière P, Lamouler P, Valiente I and Riou F 1994 *Electron. Lett.* **30** 49–50
[15] Tiemeijer L F, Thijs P J A, Dongen T v, Binsma J J M, Jansen E J and van Helleputte H R J R 1995 *IEEE Photon. Technol. Lett.* **7** 284–6
[16] Soulage G 1996 *CLEO* CMA2
[17] Hall K L, Thoen E R, Ippen E P 1999 Nonlinear optics in semiconductors *Nonlinear Optics in Semiconductors* vol II, ed E Garmire and A Kost (New York: Wiley) pp 83–160
[18] Spano P 1995 Nonlinear optical effects in active semiconductor devices *Nonlinear Optical Materials and Devices for Applications in Information Technology* ed A Miller *et al* (Dordrecht: Kluwer Academic) pp 183–205
[19] Tang J M, Spencer P S and Shore K A 1998 *J. Mod. Opt.* **45** 1211–18
[20] Ito T, Yoshimoto N, Mitomi O, Magari K, Ogawa I, Ebisawa F, Yamada Y and Hasumi Y 1998 *IEICE Trans. Electron.* **E81-C** 1237–44
[21] Magari K 1999 *CLEO Pacific Rim'99* pp 1153–4
[22] Yoshimoto N, Magari K, Ito T, Kawaguchi Y, Mitomi O and Tohmori Y 1999 *IEE Proc. Optoelectron.* **146** 71–6
[23] Kirihara T and Inoue G 1996 InP-based switch arrays using semiconductor optical amplifiers *Current Trends in Optical Amplifiers and Their Applications* ed T P Lee (Singapore: World Scientific) pp 85–151
[24] Koren U, Young M G, Miller B I, Newkrik M A, Chien M, Zirngibl M, Dragone C, Glance B, Koch T L, Tell B, Brown-Goebeler K and Raybon G 1992 *Appl. Phys. Lett.* **61** 1613–15
[25] Ikeda M, Oku S, Shibata Y, Suzuki T and Okayasu M 1992 *Proc. Topical Meeting on Photonic Switching '92 (Minsk, Belarus)* paper 2B1
[26] Glastre G, Rondi D, Enard A, Lallier E, Blondeau R and Papuchon M 1993 *Electron. Lett.* **29** 124–6
[27] Kirihara T, Ogawa M, Inoue H, Kodera H and Ishida K 1994 *IEEE Photon. Technol. Lett.* **6** 218–21
[28] Lai J-W and Lin C-F 1998 *IEEE J. Quantum Electron.* **34** 1247–56
[29] Walpole J N 1996 *Opt. Quantum Electron.* **28** 623–45
[30] Stubkjaer K E, Kloch A, Hansen P B, Poulsen H N, Wolfson D, Jepsen K S, Clausen A T, Limal E and Buxens A 1999 *IEICE Trans. Electron.* **E82-C** 338–48
[31] For a review of wavelength conversion in SOAs see, e.g. Wiesenfeld J M 1996 Current trends in optical amplifiers and their applications *Gain Dynamics and Associated Nonlinearities in Semiconductor Optical Amplifiers* ed T P Lee (Singapore: World Scientific) pp 179–222
[32] Kawaguchi H 1994 *Bistabilities and Nonlinearities in Laser Diodes* (Norwood, MA: Artech House)
[33] Kawaguchi H, Magari K, Yasaka H, Fukuda M and Oe K 1988 *IEEE J. Quantum Electron.* **QE-24** 2153–9
[34] For reviews of four-wave mixing in SOAs see, e.g. Vahala K J, Zhou J, Geraghty D, Lee R, Newkirk M and Miller B 1996 Four wave mixing in semiconductor travelling-wave amplifiers for wavelength conversion in all-optical networks *Current Trends in Optical Amplifiers and Their Applications* (Singapore: World Scientific) pp 153–77
Guekos G (ed) 1999 *Photonic Devices for Telecommunications* (Berlin: Springer)
[35] Jopson R M and Tench R E 1993 *Electron. Lett.* **29** 2216–17
[36] Hunziker G, Paiella P, Gerharty D F, Vahala K J and Koren U 1996 *IEEE Photon. Technol. Lett.* **8** 1633–5
[37] Lin L Y, Wiesenfeld J M, Perino J S and Gnauck A H 1998 *IEEE Photon. Technol. Lett.* **10** 955–7
[38] Schnabel R, Hilbk U, Hermes Th, Meißner P, Helmolt Cv, Magari K, Raub F, Pieper W, Westphal F J, Ludwig R, Küller L and Weber H G 1994 *IEEE Photon. Technol. Lett.* **6** 56–8
[39] Mecozzi A, Contestabile G, Graziani L, Martelli F, D'Ottavi A, Spano P, Dall'Ara R and Eckner J 1998 *Appl. Phys. Lett.* **72** 2651–3
[40] Lacey J P R, Madden S J and Summerfield M A 1997 *IEEE Photon. Technol. Lett.* **9** 1355–7
[41] Greco C M, Martell F, D'Ottavi A, Mecozzi A, Spano P and Dall'Ara R 1999 *IEEE Photon. Technol. Lett.* **11** 656–8
[42] Kuwatsuka H, Shoji H, Matsuda M and Ishikawa H 1997 *IEEE J. Quantum Electron.* **33** 2002–10
[43] Ellis A D, Kelly A E, Nesset D, Pitcher D, Moodie D G and Kashyap R 1998 *Electron. Lett.* **34** 1958–9
[44] Wiesenfeld J M and Glance B 1992 *IEEE Photon. Technol. Lett.* **4** 1168–71
[45] Edagawa N, Suzuki M and Yamamoto S 1998 *IEICE Trans. Electron.* **E81-C** 1251–7
[46] Durhuus T, Joergensen C G, Mikkelsen B, Pedersen R J S and Stubkjaer K E 1994 *IEEE Photon. Technol. Lett.* **6** 53–5
[47] Mikkelsen B, Durhuus T, Joergensen C, Pedersen R J S, Danielsen S L, Stubkjaer K E, Gustavsson M, van Berlo W and Janson M 1994 *Proc. ECOC '94 (Firenze, Italy)* vol 4, pp 67–70
[48] Schilling M, Idler W, Laube G, Daub K, Dütting K, Lach E, Baums D, Wünstel K 1996 *Tech. Digest of OFC'96 (San Jose, CA)* paper WG2
[49] Jourdan A, Berthelon L, Bonno P, Bruyér F, Chbat M, Coeurjolly C, Emery J Y, Gavignet P, Grard E, Janz C, Noury A, Soulage G and Zami T 1997 *Proc. ECOC'97* 3 We1D
[50] Hansen P B, Danielsen S L, Joergensen C G and Stubkjaer K E 1997 *Tech. Digest of Photonics in Switching '97* pp 132–5
[51] Doran N J and Wood D 1988 *Opt. Lett.* **13** 56–8

[52] Eiselt M, Pieper W and Weber H G 1995 *J. Lightwave Technol.* **13** 2099–112
[53] Sokoloff J P, Prucnal P R, Glesk I and Kane M 1993 *IEEE Photon. Tech. Lett.* **5** 787–90
[54] Suzuki K, Iwatsuki K, Nishi S and Saruwatari M 1994 *Electron. Lett.* **30** 660–1
[55] Kawaguchi H 1994 *Bistabilities and Nonlinearities in Laser Diodes* (Norwood, MA: Artech House)
[56] Kawaguchi H, Magari K, Yasaka H, Fukuda M and Oe K 1988 *IEEE J. Quantum Electron.* **QE-24** 2153–9
[57] Nonaka K, Tsuda H, Uenohara H, Iwamura H and Kurokawa T 1993 *IEEE Photon. Technol. Lett.* **5** 139–41
[58] Nonaka K, Kobayashi F, Kishi K, Tadokoro T, Itoh Y, Amano C and Kurokawa T 1998 *IEEE Photon. Technol. Lett.* **10** 1484–6
[59] Adams M J 1985 *IEE Proc. Pt. J* **132** 343–8
[60] Kawaguchi H 1987 *Opt. Quantum Electron.* **19** S1–36
[61] Adams M J 1987 *Opt. Quantum Electron.* **19** S37–45
[62] Tang C L, Schremer A and Fujita T 1987 *Appl. Phys. Lett.* **51** 1392–4
[63] Takahashi Y, Neogi A and Kawaguchi H 1998 *IEEE J. Quantum Electron.* **34** 1660–72
[64] Kawaguchi H 1997 *IEEE J. Selected Topics Quantum Electron.* **3** 1254–70
[65] Kawaguchi H 1998 *5th Int. Workshop on Femtosecond Technology* p 15
[66] Kawaguchi H, Irie T and Murakami M 1995 *IEEE J. Quantum Electron.* **31** 447–55
[67] Tanaka H, Shimada J and Suzuki Y 1994 *Appl. Phys. Lett.* **64** 158–60
[68] Nonaka K, Noguchi Y, Tsuda H and Kurokawa T 1995 *IEEE Photon. Technol. Lett.* **7** 29–31
[69] Tadokoro T, Kobayashi F, Kishi K, Nonaka K, Amano C, Itoh Y and Kurokawa T 1997 *Appl. Phys. Lett.* **70** 2946–8
[70] Nonaka K and Kurokawa T 1995 *Electron. Lett.* **31** 1865–6
[71] Nonaka K, Kobayashi F, Tadokoro T, Kishi K, Amano C, Itoh Y and Kurokawa T 1997 *Proc. 8th Eur. Conf. Integrated Optics* paper JFA3, pp 470–3
[72] Kawaguchi H, Yamayoshi Y, Tamura K 1999 *ECOC'99* pp II-268–9
[73] Kuwatsuka H, Simoyama T and Ishikawa H 1999 *IEEE. J. Quantum Electron.* **35** 1817–25
[74] Akiyama T, Shimayama T, Kuwatsuka H, Nakata Y, Mukai K, Sugawara M, Wada O and Ishikawa H 1999 *ECOC'99* pp II-76–7
[75] Nambu Y, Tomita A, Saito H and Nishi K 1999 *Japan. J. Appl. Phys.* **38** 5087–95
[76] Kawaguchi H 1999 *ICTON '99* pp 67–70
[77] Soto H, Erasme D and Guekos G 1999 *IEEE Photon. Technol. Lett.* **11** 970–2
[78] Soto H, Erasme D and Guekos G 1999 *CLEO Pacific Rim'99* p P2.29
[79] Ishii H *et al* 1998 *Proc. 24th Eur. Conf. on Opt. Comm. ECOC'98* pp 329–30
[80] Smit M K 1999 *ECOC'99* p I-98
[81] Watanabe S *et al ECOC'97* vol 5, pp 1–4
[82] Clausen A T, Buxens A, Poulcen H N, Oxenløre L and Jeppesen P 1999 *ECOC'99* pp I-146–7
[83] Stephens M F C, Nesset D, Williams K A, Penty R V, White I H and Fice M J 1999 *ECOC'99* pp I-148–9
[84] Poustie A J, Blow K J, Kelly A E and Manning R J 1999 *ECOC'99* pp I-246–7
[85] Houbavlis T, Hatziefremidis A, Avramopoulos H, Occhi L, Guekos G, Hansmann S and Burlchard H 1999 *ECOC'99* pp I-252–3

Further reading

For other reviews of SOAs see, e.g. Agrawal G P (ed) 1995 *Semiconductor Lasers Past, Present, and Future* (New York: AIP)

Saitoh T and Mukai T 1991 Travelling-wave semiconductor laser amplifiers *Coherence, Amplification, and Quantum Effects in Semiconductor Lasers* ed Y Yamamoto (New York: Wiley) pp 257–322

For another review of optical bistability in LDs/SOAs see, e.g. Duan G-H 1995 Semiconductor lasers and amplifiers for optical switching *Semiconductor Lasers Past, Present, and Future* ed G P Agrawal (New York: AIP) pp 321–57

B3
Gas lasers

Julian Jones

For many readers, perhaps their first sight of laser radiation was from a helium–neon lasers, the subject of Alan White and Lisa Tsufura's chapter, B3.6. The helium–neon laser was first demonstrated in 1961; it was the prototype of all gas lasers, and the first laser of any kind to be operated cw, first in the near-IR at 1.15 μm and then—by White (co-author of this chapter) and Rigden—at the familiar red wavelength of 633 nm. Production of the helium–neon laser peaked in the 1980s at more than 750 000 units per annum, for applications from holography to printing and barcode scanning. Since then it has been displaced in some areas by semiconductor diode lasers (see section B2), but still has sales exceeding 200 000 each year, more than for any type other than the diode laser.

The helium–neon laser is one of a large class of devices in which the pumping mechanism is by means of an electrical discharge in a gas mixture. It was soon joined by the argon ion and krypton ion lasers, the subject of chapter B3.5 (Malcolm Dunn and Tony Gutierrez). Ion lasers have fundamental wavelengths spanning the UV to near-IR, from 275 nm to 1.09 μm, and with frequency-doubling can be used efficiently in the deep-UV, from 229 to 264 nm. Their range of applications is exceptional: holography; printing; imaging; spectroscopy; and non-destructive testing. A major application has been in the pumping of other lasers, such as the Ti:sapphire (see chapter B1.8), although there has often been displaced by solid-state pump sources. The frequency-doubled Nd:YAG laser (chapter B1.3) has become a convenient substitute for applications e.g. in instrumentation, requiring a visible laser source. Nevertheless, ion lasers are attractive in the UV, violet and red, having excellent coherence properties, and where cw solid-state lasers are not available.

There has always been a strong drive towards the development of shorter-wavelength lasers. One reason was to enable better spatial resolution in photolithographic applications, notable in the manufacture of microelectronics (chapter D1.5). UV gas lasers are the subject of chapter B3.2 by W J Wittemann. The excimer lasers (notably the rare-gas halide lasers XeCl, KrF, ArF, etc) and the similar F_2 lasers became the sources of choice for high-resolution photolithography, at progressively shorter wavelengths as feature sizes reduced. The chapter covers also the related N_2 and H_2 lasers.

The highest-average-power gas laser is the carbon dioxide laser, operating at a wavelength of 10.6 μm, described in Denis Hall's chapter, B3.1. The CO_2 laser has many applications, but is perhaps best know for its use in laser material processing (D1), in which its high power, high efficiency and reliability are decisive advantages. Only the Nd:YAG laser (chapter B1.3) is used in a comparable volume of laser processing applications, where an advantage that it has over the CO_2 laser is its shorter wavelength, 1.06 μm, compatible with delivery by fused silica optical fibres (chapter C4.3). The chapter also covers the related CO and N_2O lasers.

The development of far-IR lasers opened up a new range of spectroscopic applications, previously hampered by the inadequate performance of the only sources previously available, which were thermal. HCN and H_2O lasers are the subject of chapter B3.9, by Wilhelm Prettl. The 337 μm line of the HCN laser is the strongest of any discharge-pumped far-IR laser, with powers typically of a few mW. The H_2O laser offers over 100 emission lines, from 7 μm to 220 μm.

The concentration of molecular-vibrational spectra in the mid-IR led to an obvious need for sources in that region suitable for spectroscopy, satisfied by the optically-pumped NH_3 and C_2H_2 lasers described by Mary Tobin in chapter B3.8, with wavelengths in the 8 μm to 20 μm range. The ammonia laser is usually pumped by CO_2 lasers, and frequency-doubled CO_2 lasers are suitable for pumping C_2H_2. The ammonia laser dominates in applications, one of which is for differential absorption LIDAR, for atmospheric monitoring (chapter D7.2).

A sub-class of the discharge-pumped gas lasers are those in which the active medium is a metal. The helium–cadmium laser, the subject of William Silfvast's chapter (B3.7) is one of the most widely-used metal–vapour lasers, with emission at 325, 354 and 442 nm and with principal applications in stereolithography and printing. The stereolithographic technique involves forming three-dimensional solid objects by scanning a laser beam to produce photopolymerization of a liquid monomer. Photosensitivity is more easily achieved at the short wavelength of the helium–cadmium laser.

Copper and gold vapour lasers are the subject of Colin Webb's chapter, B3.3. They are examples of cyclic, or self-terminating lasers, efficiently generating short pulses. The copper vapour laser is the more common, with fundamental wavelengths in the green (511 nm) and yellow (578 nm), typically generating pulses of a few mJ, 10 ns duration and—most importantly—at very high repetition rates of perhaps 10 to 30 kHz. Copper vapour lasers first found use in isotope separation. Since then, their visible wavelength and short, high repetition rate pulses have made them very useful in high speed imaging (chapter D2.7) and in material processing (section D1). The gold vapour laser is in many respects similar, with wavelengths in the red (628 nm) and UV (312 nm).

Quite different from all other lasers in this book are those pumped by chemical reaction, rather than by electrical or optical means, described in section B3.4. The chemical iodine oxygen, COIL, laser (chapter B3.4.1, Boris Barmashenko and Samit Rosenwaks) is the only known example of a high-power chemically-driven electronic transition laser. The reaction of gaseous chlorine with basic hydrogen peroxide solution produces enormous (multi MW) power at 1315 nm and 30% chemical efficiency. Its development was driven by military applications, but it is also suitable for some material processing applications, such as in the heavy cutting needed in some nuclear decommissioning work. In chapter B3.4.2, Lee Sentman describes HF and DF chemical lasers, with wavelengths of 2.8 μm and 3.8 μm respectively, or in the range 1.3 to 1.4 μm for the HF overtone laser. Small-scale lasers with power outputs in the range of tens to hundreds of W are available commercially.

B3.1
Carbon dioxide lasers

Denis R Hall

B3.1.1 Introduction

The study of the carbon dioxide laser and its spectacular portfolio of applications has been the topic of near continuous research and development for most of the past 40 years. Although the physical principles underlying the basic operation of the CO_2 laser itself had become known and well understood within about a decade of its invention in 1964 [1], the ensuing three decades has seen huge activity world-wide in the development of diverse technological implementations of the basic laser physics. Many different types of excitation technology have been developed, as well as a wide variety of optical resonator designs required to accommodate the vastly different gain medium architectures and scales which have been invented to exploit the basic excitation physics. This wide diversity of laser technologies developed for CO_2 lasers is capable of generating a *range* of laser beam characteristics that is probably unmatched by any other laser. These have been applied in many important applications to produce considerable practical impact over 40 years in key areas of science, industry, surgery and in the military sector. Moreover, despite the evident reduction in device research on CO_2 lasers over recent years, the scale and diversity of these applications, particularly in industrial manufacturing, continues to expand year-on-year despite the growth in competitive technologies such as solid state and power diode array lasers. As a result, despite such 'competition', the CO_2 laser remains a basic work-horse of the laser industry, particularly where powerful and flexible beam sources are required for industrial applications (see section D1, Materials processing).

Not only can the CO_2 laser be constructed in many different configurations, it also has the virtue of being a highly efficient laser capable of $\sim$20% electrical-to-optical conversion with near-diffraction-limited beam quality and able to be tuned to deliver more than $\sim$200 discrete transitions in the 8–18 μm wavelength range. It is at its most powerful and useful in the 9–11 μm spectral region in two bands centred at 10.4 μm and 9.4 μm, where many important engineering materials absorb well and where there is a good transmission window in the atmosphere. Many different designs of CO_2 laser are commercially available with average powers in the range from a few watts to tens of kilowatts, and with MW peak power levels and with pulse duration in the nanosecond to microsecond range. Moreover, much more extreme performance levels in terms of peak power (TW in master oscillator power amplifier (MOPA) systems), pulse energy ($\sim$100 kJ) and average power levels (100s of kW) have been produced in the laboratory. Many different designs of CO_2 have been developed by companies large and small to match the attractive versatility in CO_2 laser beam characteristics, and although there has been considerable consolidation in the industry in recent years, this is a still a market where the scale and diversity of applications, particularly in the industrial sector, has sustained a highly competitive and expanding commercial environment. Over the years, there have also been very important CO_2 laser applications in certain surgical procedures, where compact, sealed laser sources have played an important role (see sections D3.1.5.4.3, D3.2.4.1, D3.2.5.1.1, D3.2.5.5, D3.2.7.2 and D3.4.6.1.4). In the military and aerospace sectors, CO_2 lasers have been the focus of very substantial research and development efforts as the active source for applications such as direct detection or coherent laser radar (see sections D7.2.2.5, D7.2.3.4) including laser ranging, designation, Doppler-based velocity measurement systems and directed energy. In addition, the atmospheric window, beam properties and eye safety features of CO_2 lasers make them suitable for many applications in atmospheric remote sensing. However, the most important market and key technology driver, particularly in recent years, has been the industrial application sector. This has continued to grow at about 10% per annum throughout the last decade or so notwithstanding the well-documented travails of other parts of the photonics industry. There are many factors that have contributed to the growth of the CO_2 laser as a viable and productive modern industrial tool. Aside from the beam diversity and the strength of the basic laser device technologies, research on key laser–material interaction processes has been strong, often combining a sound science base with effective connections to industrial needs (see section D.1, Materials processing). In addition, certain key industries and notably machine tool and innovative automotive manufacturers embraced laser-based techniques at an early stage and integrated them into their production lines. This has stimulated significant growth in recognition of the commercial advantages of laser tools in manufacturing generally with respect to product design for manufacture, process optimization and just-in-time methodologies.

As a consequence of this high level of activity, there is considerable literature covering CO_2 laser physics device technology and applications, and many authoritative reviews (for example [2–8]) have been published over the years on most aspects of the topic. In an effort to minimize unnecessary repetition, this chapter will concentrate on CO_2 device physics and technology covering only the past 20 years or so, while providing a brief coverage of the essential basic physics and a historical review of the key milestones that occurred in the first 20 years of technology development. The limitations of space inevitably mean that not all interesting and relevant topics can be adequately covered, or even mentioned.

In section B3.1.2 which covers the basic physics of the excitation of CO_2 laser, the aim is to provide a brief summary of the 'rules of the game' that one must appreciate in order to facilitate an understanding of the extraordinary developments in device technology that have occurred over the past 30 years. Section B3.1.3 provides a rather brief historical review of the main developments over the first two decades of this technology development, covering the main events up until about 1980, while section B3.1.3 covers the modern era of CO_2 lasers, which, as we will see, is based largely on the device technology opportunities made possible by the exploitation of high power radiofrequency discharges.

B3.1.2 Basic physics of excitation

B3.1.2.1 Introduction

Aside from the primary literature, many comprehensive treatments of the physics of CO_2 laser excitation have been published both in textbooks and in detailed review articles, e.g. [2–8]. The brief outline presented here is intended to provide sufficient background information on the fundamental excitation physics of the CO_2 laser to provide a basic 'underlay' for the device physics and technology covered in remainder of the chapter.

The CO_2 laser operates by virtue of energy exchanges between pairs of the low-lying rotational–vibrational energy states of the molecule. Molecules are excited to the upper energy state via collisions with energetic electrons in a gas discharge, leading subsequently to a transition that is induced or 'stimulated' by the radiation field to the lower energy state, accompanied by the generation of an IR photon with energy corresponding to the difference in energy between the two states. The rate at which such transitions can occur, and the corresponding laser output power is dependent on *both* the rate at which molecules can be excited to the upper state *and* the rate at which the lower level can be de-populated. The excitation rate of the upper level is dependent on the efficacy of the gas discharge in providing electrons with appropriate kinetic energy to preferentially excite ground-state CO_2 molecules to the upper level. As will be seen later, the efficiency of

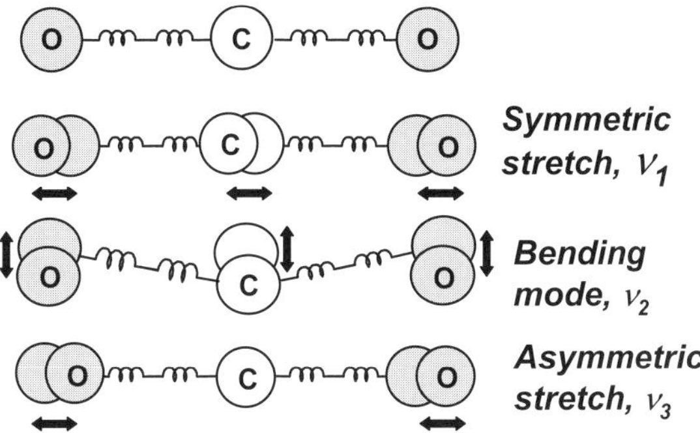

Figure B3.1.1. CO_2 molecular structure and three fundamental modes of vibration.

this process is hugely increased if nitrogen is added to the CO_2 laser gas mix. However, the second process essential for efficient laser action, namely the de-excitation of the lower level may be seriously impeded if the discharge gas temperature is allowed to rise too high, because of the thermally-induced population of this level from the ground state. Thus, for gas temperatures above about 550 K, the rate of depopulation of the lower energy level becomes seriously impeded resulting in reduced laser output power levels. Although the basic electrical-to-optical energy conversion efficiency of the CO_2 laser is high (for a laser!), with a basic quantum defect of 40% and demonstrated devices efficiencies of >20%, the difference between laser input power and laser output power appears as gas heating. It is, therefore, clear that the magnitude of the power that can be transferred to the laser discharge capable of high efficiency conversion to optical output is dependent on the rate at which waste heat can be removed from the system.

These two factors, namely the excitation of the upper level usually by some form of gas discharge and the de-excitation of the lower level mediated via effective gas cooling, are at the very core of much of the technological innovation that has characterized different types of CO_2 laser developed over the past four decades. For excitation, various forms of gas discharge have been employed, including longitudinal dc and ac discharges, transverse rf discharges, pulsed dc discharge, electron beam discharges and a variety of hybrid excitation schemes. *Three* main options characterize the gas cooling issue. At a basic level, assuming no net heat removal from the system, the maximum input energy that can be achieved for *pulsed* systems is limited by the heat capacity of the laser gas itself to values of approximately $300\ \mathrm{J \cdot l^{-1}}$ of laser gas at atmospheric pressure, with pulsed laser output capacity at the $\sim 40\ \mathrm{J \cdot l^{-1}}$ atmosphere level. The *second* and by far the simplest gas-cooling method for cw or low-repetition-rate operation is passive and relies on the thermal conductivity of the gas and thermal transfer to the (cooled) vessel containing the discharge. However, much higher output power levels may be achieved by an active heat removal method where the hot gas itself is removed from the laser cavity by some form of 'fast gas flow' technique.

B3.1.2.2 CO_2 molecular structure and laser excitation issues

Carbon dioxide is a linear polyatomic molecule which probably because of its relative simplicity and importance in basic sciences has been studied extensively for more than 100 years [9]. Its principal axis of symmetry lies along the line of the nuclei and it has a plane of symmetry perpendicular to this axis, as illustrated in figure B3.1.1. The molecule has three vibrational degrees of freedom with vibrational modes corresponding to symmetric stretch (v_1), where the oxygen atoms vibrate in a straight line in opposition to each other, the bending mode (v_2), which is doubly degenerate because the oxygen atoms can oscillate in two planes perpendicular to the principal molecular axis, and the asymmetric stretch (v_3), where the oxygen atoms are always moving in the same direction along the axis. The vibrational levels are characterized in

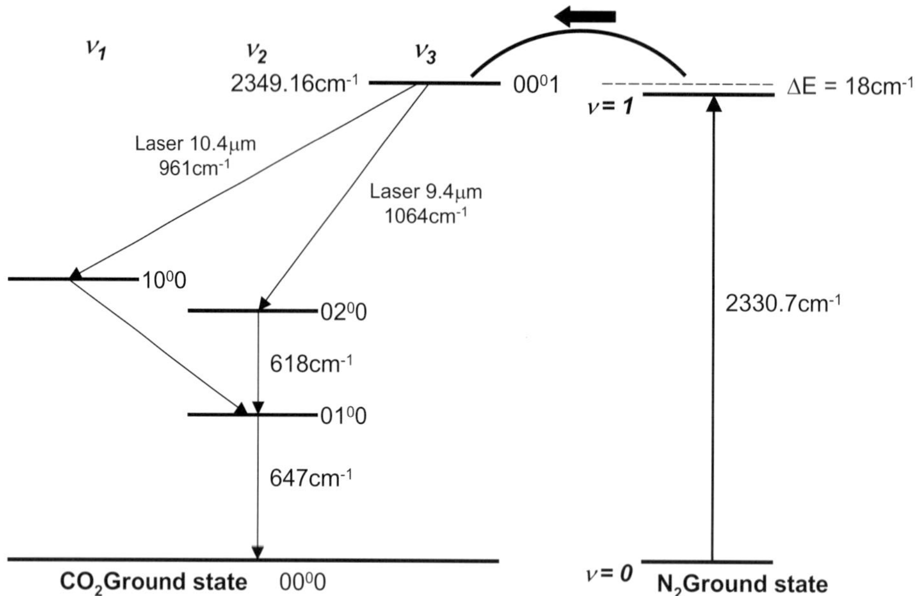

Figure B3.1.2. CO_2 laser energy level diagram showing the lowest vibrational levels of the electronic ground state of the CO_2 and N_2 molecules.

terms of the quantum numbers, n_1, n_2, n_3, which refer to the number of quanta of energy in each of the vibrational modes. In addition, since the bending mode, υ_2, is twofold degenerate, the quantum number, n_2, has a superscript, l, which specifies the angular momentum quantum number. The relevant section of the energy level term diagram for CO_2 is shown in figure B3.1.2, along with the lowest two energy levels for nitrogen. This simplified CO_2 diagram omits much of the rich spectroscopic structure, for example that which gives rise to 'hot bands' and 'sequence bands', where laser action is possible with some extra effort [10], because here we concentrate only on the molecular energy states which contribute to the most important laser emission bands centred at 9.4 μm and 10.4 μm. The upper laser level is the 00^01 excited state corresponding to a single quantum in the υ_3 asymmetric stretch mode with an energy, υ_3, of 2349 cm^{-1}. As indicated in figure B3.1.2, there are two possible lower laser vibrational energy levels corresponding to the two emission bands referred to earlier. The first is the 00^01 energy state with a single quantum in the symmetric longitudinal mode υ_1 with a potential energy of 1388 cm^{-1}, where the energy difference give rise to laser lines centres at 10.4 μm, while the second lower energy level is the 02^00 state, corresponding to two quanta in the bending mode and energy of 1285 cm^{-1}, with emission lines centred at 9.4 μm.

So far we have only considered the vibrational character of the molecular motion, but it is clear that molecules are also free to rotate. The rotational energy of the molecule is characterized by a series of quantized rotational sub-levels designated by the rotational angular momentum quantum number J, which may be either zero or a positive integer. The quantum mechanical selection rules for transitions between rotational–vibrational levels require that in emission $\Delta J = +1$ (describing the so-called P-branch) or $\Delta J = -1$ (giving the R-branch), as illustrated in figure B3.1.3. Individual rot–vib transitions are then designated by the letter P or R for each branch followed by the J value of the *lower* level of the transition. For example, the strongest transition line under typical laser operating conditions is the line P(20) (at a wavelength of 10.59 μm) denoting a P-branch transition from the $J = 19$ upper level (of 00^01) and terminating on the $J = 20$ level of the 10^00 state – where for the P branch $\Delta J = +1$.

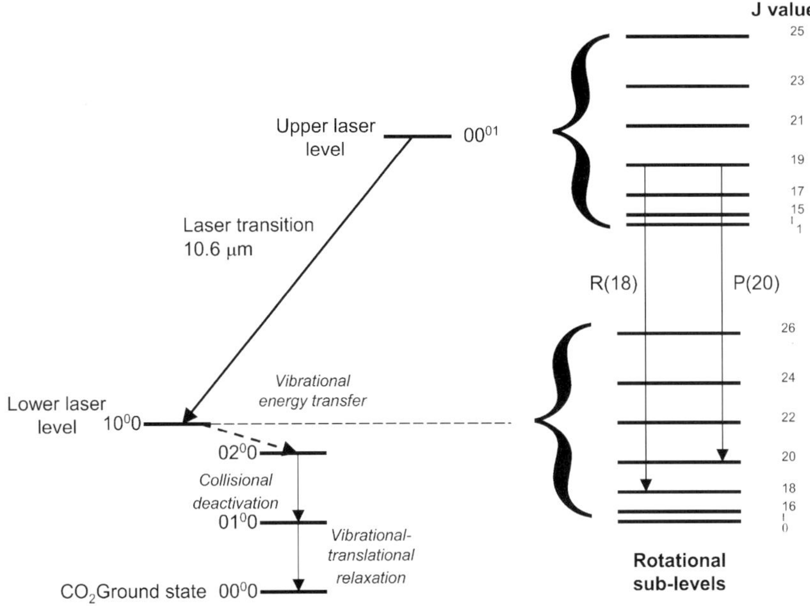

Figure B3.1.3. Some detail of the CO_2 laser (00^01–10^00) transition rotational spectrum.

B3.1.2.2.1 Upper laser level excitation

Here we consider the most important class of CO_2 lasers where the basic laser inversion is produced by energetic electrons produced in suitable electric *discharges*. The process of molecular excitation to the upper laser state, 00^01 may be achieved by three main 'routes', either direct excitation of the CO_2 molecules or via intermediate states in nitrogen or carbon monoxide.

B3.1.2.2.2 Electron excitation of N_2 and resonant energy transfer to CO_2

Nitrogen is a diatomic molecule and therefore only has a single vibrational mode of which the two lowest states are shown in figure B3.1.2. It can be seen that the potential energy of the first excited state ($v = 1$) of N_2 is 2330.7 cm^{-1} above the ground state and is in near-resonance ($\Delta E = 18$ cm^{-1}) with the 00^01 state of CO_2. The energy difference between the two states is small compared to the value of kT of about 350cm^{-1} for a typical discharge temperature of ~500 K. It follows that the probability of resonance energy transfer to the 00^01 state of CO_2 is high in the reaction

$$N_2(v = 1) + CO_2(00^00) \leftrightarrow N_2(v = 0) + CO_2(00^01) - \Delta E(18\ \mathrm{cm}^{-1}). \qquad \text{(B3.1.1)}$$

Moreover, since radiative decay to the ground state is a forbidden process (metastable state) for a symmetric diatomic nitrogen molecule whereas the $CO_2(00^00)$ molecule can decay radiatively, the reaction proceeds predominantly in the *forward* direction, providing an efficient route for excitation of the upper laser level. However, the CO_2 lower energy level (10^00) is some 900 cm^{-1} below the highly populated N_2 ($v = 1$) level, which is too large an energy decrement for efficient collisional energy transfer to occur. Consequently, very little vibrational energy transfer occurs between these two levels, to the considerable benefit of CO_2 laser pumping efficiency. So far we have seen that the coincidence of the energy states of CO_2 and nitrogen provide a potentially very efficient upper level pumping route. But mother nature's contribution to the creation of a

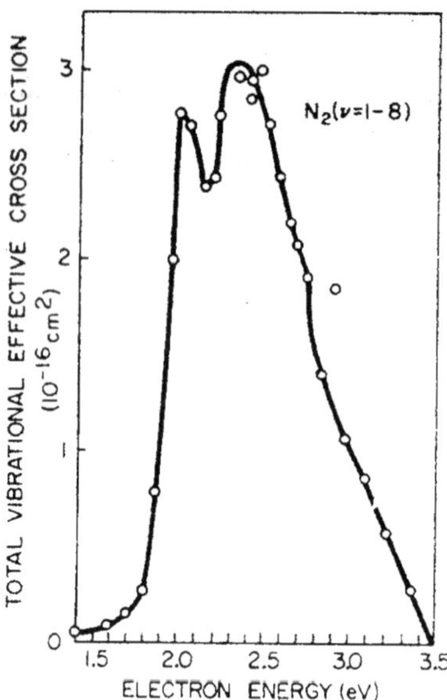

Figure B3.1.4. Total effective cross section for vibrational excitation by electron impact of N_2 for the $\nu = 1$–8 vibrational levels. After [1].

very efficient high power CO_2 laser does not end here. In addition, as first determined by Schultz *et al* [11], the excitation cross sections of individual higher-lying vibrational states of nitrogen are comparable to each other and rather high for the first to the fourth level falling to 10% at the seventh and eighth levels. The net excitation effect can thus be reasonably approximated by looking at the sum to the eighth state and, as shown on figure B3.1.4, the cross section for electron excitation to these levels is large ($>2 \times 10^{-16}$ cm^2) for electron energies in the approximate range 2.0–2.7 eV. Moreover, the upper-lying vibrational levels N_2 decay rapidly to the $\upsilon = 1$ state, so completing an efficient set of excitation channels that funnel into the first vibrational level in N_2, as required for CO_2 laser pumping by this route.

B3.1.2.2.3 Direct electron collisional excitation of CO_2

The measured cross sections for direct electron collisional excitation of vibrational levels in CO_2 show a series of resonances associated with the three lowest asymmetric stretch, (υ_3) levels which occur for electron energies of <1 eV [12]. However, although the value peak of the cross section for excitation of the 00^01 level is 3×10^{-16} cm^2 and occurs at an electron energy of 0.9 eV [13], it also has a high energy tail which extends to > 1.5 eV with $\sigma \sim 10^{-16}$ cm^2. Consequently, the 00^01 upper laser level can be directly excited from the ground state via electron impact, according to the inelastic collision process described by

$$CO_2(00^00) + e^-_{\text{fast}} \rightarrow CO_2(00^01) + e^-_{\text{slow}}. \qquad \text{(B3.1.2)}$$

In fact the electronic excitation of both N_2 and CO_2 are actually quite complex processes, which proceed via a series of intermediate steps involving negatively charged ions, and which have a significant effect on the energy threshold for the processes.

B3.1.2.2.4 Electron excitation of CO and resonant energy transfer to CO_2

The coincidence of the near-resonance of the $v = 1$ vibrational state in nitrogen with the 00^01 state in CO_2 (and many of the corresponding excitation processes) are mirrored by a similar near-resonance involving carbon monoxide, since the energy of the $v = 1$ vibrational state in CO is within $\sim$206 cm^{-1} of the CO_2 upper laser level. This has some impact on the excitation kinetics since, in practice, in typical gas discharges a dissociation equilibrium is established very rapidly after switch-on in which a significant fraction of the CO_2 molecules dissociate into carbon monoxide and oxygen, so that CO molecules are available to participate in resonance energy transfer reactions. The physics of the excitation via CO is very similar to the processes mentioned earlier for nitrogen, including the involvement of a significant contribution to the excitation of $CO_2(00^01)$ via higher-lying states in CO. Excitation cross sections for low lying-levels in CO were also measured by Schultz [11] and Boness [13] and values observed to be $>4 \times 10^{-16}$ cm^2 for electron energies in the range 1.5–2.3 eV. A major difference, however, is that the resonance transfer probability from CO is lower than from N_2, due to the much larger value of ΔE for CO.

Considering the overall situation for electron excitation of the upper laser level via each of the three channels indicated, it is clear that for efficient laser pumping, electron collision processes are required where the electron energies lie in the approximate range 1.5–2.5 eV. Fortunately for the creation of CO_2 laser excitation technologies, there is considerable choice in that the properties of several different types of self-sustained discharges are such that the energy range of electrons can be optimized to be in the appropriate range

A very practical consequence of the effect of CO_2 and N_2 on CO_2 laser kinetics has been observed in research aimed at reducing the time constant for upper state excitation in order to maximize the repetition rate achievable with direct pulsing of the laser discharge to achieve repetitively pulsed laser operation. The time constant for resonant energy transfer to the upper-state population from nitrogen imposes a repetition rate limit of a few tens of kHz in typical gas mixtures due to the time taken to 'exhaust' the energy 'trapped' in the vibrational states of nitrogen and CO. Efforts to 'beat' this limit by using nitrogen-less gas mixes and aiming to rely only on direct excitation of the upper level (at some cost in terms of maximum power) are partly thwarted by the creation of CO in the discharge and 'locking-up' of upper-state population in CO near-resonant transfer.

It is important to also consider possible de-excitation routes from the upper laser level other than the desired one via either of the two possible lower laser levels as shown in figure B3.1.2, since this would represent a net loss of inversion. One obvious possibility is radiative decay from the 00^01 level to the 00^00 ground-state. However, for gas densities used in typical CO_2 lasers, the concentration of ground-state molecules is sufficient to ensure that radiation trapping effectively negates this loss possible mechanism. Radiative lifetimes for molecular vibrational transitions are, in fact, very long, varying from milliseconds to seconds at typical gas pressures in the tens of torr range. Perversely the lifetime of the upper level laser in CO_2 is actually *shorter* than that of the lower laser level, so that under conditions where radiative decay dominates, laser action would be impossible. However, the important mechanisms in CO_2 laser kinetics are all collisional. Collisional lifetime effects dominate since, with values in the range 10^{-7}–10^{-8} s, they are many orders of magnitude shorter than those for spontaneous emission. Thus spontaneous emission effects are usually ignored in setting up the kinetic rate equations and in kinetic process modelling.

B3.1.2.2.5 Lower laser level de-excitation processes

Radiative decay transitions from the lower laser levels $10^00 \rightarrow 01^00$ and $02^00 \rightarrow 01^00$ are both allowed but, as mentioned previously, the corresponding spontaneous emission decay lifetimes are very long and the level population densities are dominated by collision effects. So far we have considered the two lower laser levels, 10^00 and 02^00, as though they are completely independent of each other. In fact, the energy difference between the two levels is very small compared to kT and, furthermore, there is a coupling between the two levels (Fermi resonance) which results from the fact that the motion of the oxygen atoms in the bending mode vibration produces a degree of linear displacement of the oxygen molecules, which corresponds to symmetric stretch. The populations of the two lower laser levels are, thereby, effectively close-coupled together and to the 01^00 state via near-resonant collisions with ground state molecules as follows:

$$CO_2(10^00) + CO_2(00^00) \rightarrow CO_2(01^10) + CO_2(01^10) + \Delta E \quad \text{(B3.1.3)}$$

$$CO_2(12^00) + CO_2(00^00) \rightarrow CO_2(01^10) + CO_2(01^10) + \Delta E'. \quad \text{(B3.1.4)}$$

These two relaxation processes have a very high degree of probability since both energy decrements ΔE and $\Delta E'$ are much smaller than kT so, as a result, the three levels reach a state of thermal equilibrium in a very short time. The necessary process of de-population of the lower laser levels thus reduces to processes which effect the decay of the 01^00 state to the ground state, if we are to avoid the build-up of the population and, hence, an effective 'bottleneck' in this state. However, this is the least energetic transition of any in the term diagram, and, therefore, decay from the 01^00 state can only occur by means of the transfer of vibrational energy to the translational energy of the partners participating in the collision, the so-called vibrational-translational (VT) relaxation process. Since the probability of VT energy transfer in a collision is increased for light-atom collision partners, the use of helium in the laser gas mix, as noted later, is highly advantageous for efficient de-population of the lower laser level. The use of He in the gas mix has additional significant benefits for CO_2 laser action, as we will see later.

B3.1.2.3 Optimization of CO_2 laser operating parameters

A key component of laser design is the optimization of operating conditions to maximize laser power and conversion efficiency. With the CO_2 laser, there are two particular parameter choices where optimization may be achieved, namely in the use of gas additives to the basic gas mixture and the selection of the operating conditions (particularly the E/N ratio) of the gas discharge.

B3.1.2.3.1 Selection of gas additives

The use of selective gas additives in CO_2 laser gas mixtures is motivated by a combination of factors that unite to produce operational benefits and enhanced laser performance. These include the desire to increase the excitation rate of the upper level, to mediate more effective de-population of the lower level, to reduce gas temperature and, thereby, produce a positive effect on the gas kinetics, to modify the electron energy distribution to produce increased population inversion and to reduce molecular decomposition and to stabilize the gas chemistry for stable long-life operation. We have already seen how the addition of *nitrogen* provides an alternative and highly effective pumping route for upper laser level excitation via electron impact to the $v = 1$ and higher-lying vibrational N_2 states which channel their energy to the 00^01 state via collisional resonant energy transfer. Also, it has already been noted that the presence of *helium* in the gas mix can significantly increase laser gain and output power, as first observed by Moeller and Rigden [14]. Helium plays an important role in the relaxation of the lower laser levels via its role in mediating efficient VT collisional energy transfer processes and also in reducing electron temperature. Most importantly, helium has a crucial part to play in gas temperature control. Because its thermal conductivity is more than an order of magnitude greater than the values for CO_2 or N_2 its presence in the mix significantly enhances the rate of heat transfer to the discharge vessel walls, thus minimizing increases in temperature and, thereby, reducing the probability of a thermally induced population of the 01^00 intermediate energy state. Helium may also be used to good effect as a useful gas component in the optimization of the discharge E/N ratio for maximum fractional input power transfer to the resonant vibrational states of N_2 and CO_2, as described later. The presence of helium also cools the rotational levels as pointed out by Patel [15] and so enables these closely spaced levels to come to equilibrium with the gas temperature, thereby helping to determine the J line with the highest gain. This fast thermalization of the rotational manifold has the important effect of feeding energy into the lasing transition from all other neighbouring rotational transitions, which has the very important consequence that most of the energy in the upper vibrational state is extracted in a single line.

Xenon gas is frequently used as a CO_2 laser gas mix additive usually in small concentrations and for economic reasons, only in sealed-off devices. The addition of a few percent Xe has been observed to produce increases in laser power of ~30% in both dc discharge [16] devices and rf lasers [17, 18]. The presence of Xe in the gas mix is also observed to have beneficial effects on the operational lifetime of sealed-off lasers. To explain these effects, we first note that xenon has no energy levels that are resonant with vibrational levels in CO_2, CO or N_2 to produce any direct selective effects on population inversion and as, a heavy atom it does not produce positive effects on gas cooling via its thermal conductivity properties or on thermalization of the rotational manifold. It is now widely believed that the observed benefits of the use of Xe in CO_2 lasers is a consequence of its effects on the discharge plasma properties, in particular its impact on driving the electron energy distribution to one that is more favourable for electron impact excitation of vibrational levels. For example, experiments [19] indicate that the presence of Xe produces fewer energetic electrons with energy above 2.5 eV and an increase in those with energy below this value, which explains the reduced dissociation observed, the increased pumping rates and higher gain. With rf discharges, the presence of Xe is observed to produce higher values of gain and lower measured electron temperatures and [18] and reduced discharge sheath thickness.

B3.1.2.3.2 Optimization of discharge operating conditions

In the previous discussions of optimization, we have referred rather loosely to the idea of average electron temperature. However, this is an oversimplification and the distribution of electron energies in a molecular gas discharge are neither Maxwellian not Druyvesteyn [20] and cannot be characterized by a single electron temperature. Here, we consider the main opportunities that exist to optimize laser performance by appropriate selection of the operating parameters of the gas discharges. In particular, we consider the optimization of the parameter E/N, which is the ratio of the electric field ($V \cdot cm^{-1}$) to the total neutral particle number density (cm^{-3}) and which is the most useful and frequently quoted optimization parameter. A number of authors have calculated the distribution electron energies for a range of relevant laser gas mixtures. However, perhaps the most useful results are those of Nighan and Bennett [21], who determined the fractional rate of energy transfer to the various excited states of CO_2 and nitrogen relevant to CO_2 laser operation as a function of E/N, as shown in figure B3.1.5. They used an energy balance procedure that equates the rate of energy input to electrons from the electric field of the discharge to the rate of energy transfer from the electrons to the various molecular states that are excited by electron impact collisions. The calculations shown are for a 8:1:1 mixture of He, CO_2 and N_2 and demonstrate in a striking manner the sensitivity of the energy transfer processes to the value of E/N. In particular, it can be seen that for an E/N value of 10^{-16} $V \cdot cm^2$, nearly 65% of the available electron energy goes directly into pumping the CO_2 00^01 upper laser level and less than 8% into the 01^00 lower level. Virtually all energy available goes into vibrational states and nearly none into electronic states or ionization, so that discharges with these conditions are potentially extremely efficient for converting electricity into light. The calculated fractional power transfer values of ~50% coupled with the 41% CO laser quantum efficiency are in very good agreement with the experimentally observed laser efficiency values of >20% [22]. However, the selection of discharge conditions to provide the value of E/N that gives optimum laser excitation conditions is secondary to the primary requirement to produce a stable (self-sustaining) discharge and, in practice, one is usually faced with a complex iterative design and optimization process, since the conditions for ideal stable discharges with the desired physical scale are not always compatible with the requirements to optimize E/N.

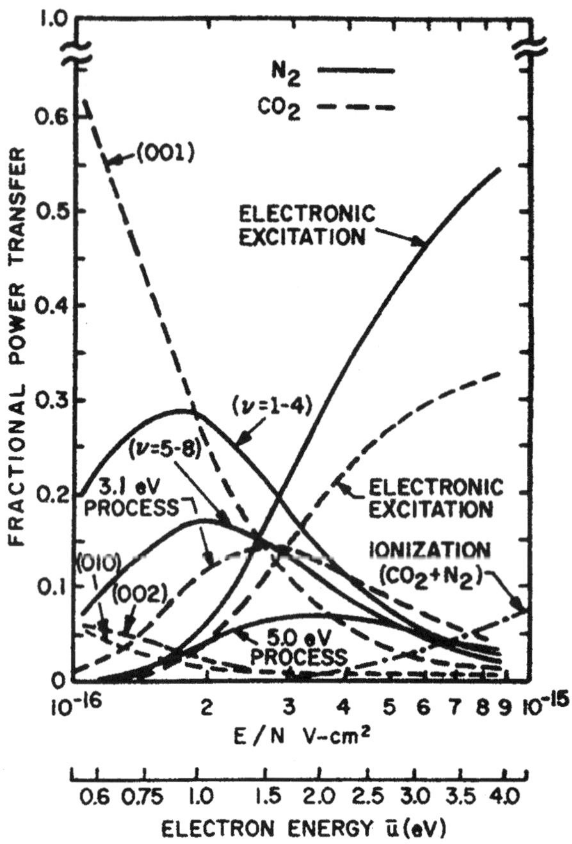

Figure B3.1.5. Fractional power transferred from electrons to vibrational and electronic excited states of CO_2 and N_2 as a function of E/N for a CO_2–N_2–He, 1:1:8 gas mixture. After [21].

B3.1.3 Events in the life of the carbon dioxide laser

Although most of the key excitation physics underpinning the operation of the CO_2 laser were quite well understood after the first 10 years or so of its life, the ensuing 30 years or so has seen the growth of a very extensive literature devoted to laser device technologies and applications. As mentioned previously, the technology developments have been driven by a diverse application sector, dominated by industrial processing and military applications but also including applications in surgery, science and remote sensing. However, it is clear that most of the leading edge research particularly prior to 1980 was driven by military and aerospace requirements. To meet these needs, CO_2 lasers have been developed with the capability of producing a very wide range of laser beam characteristics, and a correspondingly diverse range of disparate device *technologies*. In fact, of all lasers, the CO_2 laser probably has the widest diversity of laser output beam formats. Thus, different CO_2 laser technologies may operate in cw mode from mW to the MW range, in pulsed mode with pulse-widths from millisecond to sub-nanosecond, with repetition rates to hundreds of kHz with pulse energies to the MJ range, and with >200 discrete lines in the 8–18 μm wavelength regions, that may be frequency stable to a few parts in 10^{-12}. Moreover, many of these laser technologies have been developed for commercial production for a very wide range of applications and proven to be very successful in the market.

Although, as will be seen later, technologies for CO_2 lasers may appear to be extremely dissimilar, they all have in common the key constituent elements of a gain medium excitation technique, a gas-cooling method, an optical resonator design, a gas mixture control sub-system and a mechanical support system. Each of these may be realized in very different ways but all are necessary for effective laser operation. Virtually all gain excitation techniques for CO_2 lasers involve some form of gas discharge, though excitation via rapid gas expansion through a nozzle is the basis of the gas dynamic laser, which had a relatively transient 'life' in the 1970s as a feasible but technically demanding route to very high average power levels [121]. Thus, discharges involving dc, ac or rf power generators, with cw or pulsed outputs have all been developed as practical and effective pumping techniques, though only rf techniques have seen significant development in the 'modern era' since about 1980. In section B3.1.2 the importance of controlling gas temperature rise has been emphasized, so effective gas cooling is a necessity. Techniques have included 'passive' cooling based on diffusion of the laser gases to water or air-cooled surfaces; selection of appropriate gas mixtures with effective transport properties is key here and high helium content is usually important. To achieve higher power levels, gas transport cooling techniques have been developed which employ high-speed gas flow, which may be either transverse or co-axial with the laser optic axis.

The range of optical resonators and intracavity optical components selected for use with CO_2 lasers is also extensive and it is usually the physical scale of the gain medium *transverse* aperture which is the key factor determining the design. Conventional stable free-space Gaussian resonators in linear or folded designs are an excellent match to many excitation schemes involving modest diameter (usually less than a few cm) cylindrical tubes to contain the gas discharge. For very small apertures, free-space resonators cannot be used and a distinct class of 'waveguide resonators' that incorporate a waveguide section are appropriate (see chapter A2.2, Waveguide resonators) while for large aperture devices such as TEA lasers or electron beam excited lasers [122], unstable resonators (see section A2.1.7) are usually required. Hybrid designs, e.g. waveguide-unstable or stable–unstable are often found to be suitable for lasers with large-aspect-ratio gain media, where the transverse and lateral dimensions are very different. Most CO_2 lasers operate in the 9–11 μm infrared wavelength range where there are few high transmission materials for use as substrates for optical components. Of these zinc selenide has the lowest absorption loss and is the most frequently used material for output couplers, lenses etc, although germanium and KCl have been used in the past. Optical coating technology is well developed for both high damage threshold, low-loss high reflectors where reflectance values of >0.99 may be achieved and partially transmitting mirrors with low losses and effective anti-reflection coatings. For high reflectors, glass, semiconductor (silicon on GaAs) or metals (copper, beryllium–copper etc) are in common use depending on the power levels. For high average-power lasers, water cooling must be used to prevent surface distortion or damage in resonator mirrors and beam-forming optics [123]. Large aperture mirrors for unstable resonators have been developed for high-power lasers, and there has been considerable progress in the development of high damage-threshold coatings. Reflective diffraction gratings with acceptably low losses are also available with the capacity to resolve adjacent rotational–vibrational lines and facilitate the design of single-line tunable lasers. Overall, the industry supplying optical components for use in CO_2 lasers has became very mature with a strong element of consolidation in recent years, as befits a well-established commercial marketplace for these lasers. Control of the gas chemistry and, in particular, the level of dissociation of the CO_2 gas is also a crucial requirement. In sealed lasers, such control also demands effective vacuum engineering and several companies have developed proprietary processing regimes capable of producing equilibrium mixtures with adequate CO_2 concentration for particular discharge regimes to produce stable laser power levels sustained over lifetimes exceeding 10 000 hr. The use of 'slow- flow' of the laser gas to remove dissociation products and to ensure constant gas composition is technically less demanding to achieve but requires the use of a consumable gas supply and vacuum pump and is usually unacceptable in modern commercial lasers. In some slow-flow systems and many 'fast-flow' convectively cooled laser designs, catalysts are employed to stabilize CO_2 dissociation and ensure constant gas composition. For all lasers, a suitable mechanical system is required to support

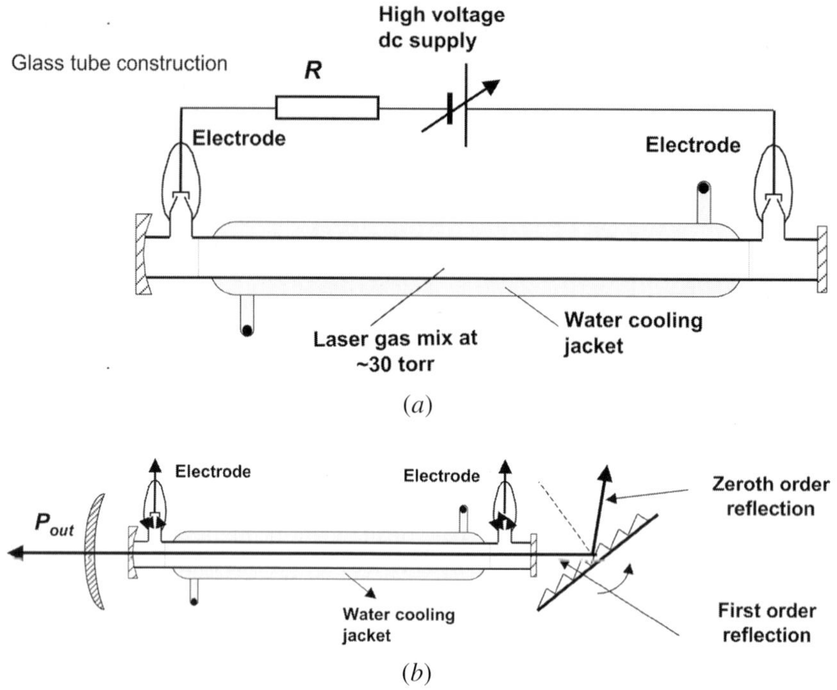

Figure B3.1.6. (*a*) Longitudinal dc discharge excited CO_2 laser tube and (*b*) line tunable selectable CO_2 laser using diffraction grating tuning laser.

and package the key operational elements of the device in a manner appropriate for the overall design. In particular, a detailed opto-mechanical design is an important (but sometimes neglected) requirement to ensure stable beam properties over the range of thermal and mechanical conditions that it may be anticipated that the laser will encounter.

B3.1.3.1 DC discharge excited lasers

The earliest form of CO_2 laser in common use employed longitudinal dc discharge excitation. A typical laser of this type involves the use of a cylindrical glass tube with a co-axial water jacket with anode and cathode structures often in side arms, as illustrated in figure B3.1.6(*a*). The dc discharge behaves as a negative impedance element (see also section B3.6.2), so the high voltage dc power supply is required to provide discharge circuit stabilization, most simply via a series resistor as shown in the figure. The gas mixture may be static in a sealed-off configuration or operated in slow-flow mode with a mixture of CO_2, N_2 and He (with Xe included in sealed mode mixtures) at a total pressure of a few tens of torr. As a consequence of the law of 'similar discharges' (pd = Constant) [23, 24], the optimum gas pressure is inversely proportional to the tube diameter, which is typically in the 10–20 mm range with gas pressures of a few tens of torr (760 torr = 1 bar = 1.013×10^5Pa). The gas is cooled here by radial diffusion to the glass tube walls, with modest levels of heat transfer through the glass to the circulating water coolant. In general, the laser power output increases with current density until gas heating effects exceed the capacity of the cooling system and a thermal bottleneck in the lower laser levels occurs (corresponding to an axial gas temperature of ~600 K) whereupon it falls. Thus, the optimum value of current density is inversely proportional to tube diameter, though as dictated by discharge similarity principles, the output power of the laser depends only on tube length. In practice, laser powers of 50–100 W m^{-1} can be achieved, depending on the laser constructional details and degree of control of the gas chemistry. The development of sealed-off cw CO_2 lasers was a topic of major practical interest for many years, driven by application needs in military and space-borne systems. It was shown that careful attention is necessary in the choice of material and design for the cold-cathode structure coupled with the choice of gas additives [25].

Commercial versions of slow-flow lasers of this type began to be available in the late 1960s at power levels of tens of watts and were used for some industrial applications (e.g. ceramic scribing), as well as scientific and government research. The wavelength selectivity feature of the CO_2 laser was soon exploited by the use of a planar diffraction grating in the Littrow configuration as one of the laser mirrors, as illustrated in figure B3.1.6(*b*). Lasers with this basic tuning design are still in use for spectral applications that require unambiguous single-line operation at wavelengths away from the band centres. Frequency tuning within a given rotational–vibrational emission line may also be achieved via resonator length tuning, for example using a piezoelectric length transducer fitted to one of the resonator mirrors. However, for these lasers operating pressures are low at a few tens of torr, so that the tuning range is limited to $\sim 50 - 100$ MHz.

Much higher power levels were soon achieved in experimental systems by geometric folding of the resonator. In one design developed at MIT Lincoln Laboratory, an average power of 1 kW was obtained from a device 230 ft in length, employed in such a planar folding system [27], while a year later in 1969 Horrigan reported a 750 ft-long folded laser which produced 8.8 kW [83]. Driven by demands from applications in industrial materials processing, commercial folded lasers at power levels to a few kW became available from the major laser suppliers during the early 1970s. Notable among these was a highly innovative and compact design from Ferranti Ltd in the UK, that employed a 3D folding system utilizing a Herriott cell which produced near-diffraction-limited beam quality at power levels of 400 W—and later to 2 kW [26]. A great deal of development work on longitudinal discharge lasers was carried by numerous companies in Europe and North America, especially during the 1970s, to meet the rising demands for reliable lasers for the industrial processing market. During this period, considerable practical advances were made in discharge technology, folded resonator design, gas-handling systems and power supplies, so that this generation of lasers, though rather bulky, became reliable and important commercial workhorses for several generations of laser-based production systems.

B3.1.3.2 Convectively cooled lasers

Although the CO_2 laser has an intrinsic capacity for high electrical-to-optical energy-conversion efficiencies, as mentioned previously the output power is nevertheless limited by excess temperature rise of the gain medium. For diffusion-cooled dc excited lasers, the average power extraction limit of <100 W m^{-1} limits laser power to a few kW usually in rather bulky slow-flow lasers that may be inconvenient for integration into processing systems. An alternative approach to achieving multi kW laser outputs is to significantly improve the heat removal ability of the laser beyond the limited capacity of passive diffusion-cooling in centimetre-scale dielectric tubes. This may be achieved by flowing the gas mixture out of the gain region at high speed ($\sim$100 m s^{-1}), so that the heat is removed from the active laser discharge gain region by simply removing the hot gas itself and replacing it with equivalent cool gas in a process termed *convective cooling*. The hot gas mixture removed by the flow system can then be cooled by a suitable heat exchanger unit within an external flow system and the cooled gas subsequently returned to the discharge gain region. This approach provides much more effective cooling than the passive diffusion technique described earlier and specific power output values of several kW m^{-1} can be achieved in practice. The maximum input power P^{i}_{max} to the discharge under convectively cooled conditions is given by

$$P^{\mathrm{i}}_{\max} = (1-\eta)\dot{M}c_{\mathrm{p}}\Delta T \qquad \text{(B3.1.5)}$$

where η is the laser conversion efficiency, $\dot{M}$ is the mass flow rate, c_P is the specific heat and ΔT is the rise in the stagnation temperature (about 300 K). The laser efficiency is ~15–20% and the mass flow rate depends on the operating gas pressure and the volume rate of the gas blowers.

As illustrated in figure B3.1.7, two types of convective cooling (fast-flow) systems have been developed for use in multi-kilowatt lasers: *axial-flow* systems involving gas flow along the optic axis of the laser and *transverse-flow* lasers, where the gas flow is transverse to the optical axis [28]. Longitudinal flow designs are much more commonly employed due to the simpler constructional arrangements that can be used, involving single or multiple discharge sections in cylindrical glass tubes a few centimetres in diameter, with in-line cylindrical electrodes. In addition, for axial-flow cooling it is usually less complex to design lasers with high beam quality since any variations in gain medium particle density and refractive index are approximately radial. Since dc discharges in large bore tubes operate at gas pressures of only a few tens of torr, high mass flow rates demand large volume rates and, therefore, (large) high capacity blowers are required for the production of high laser powers. However, such blowers are not available with very high volume rates. In fast *axial*-flow devices, Rootes blowers must be used in order to overcome the large pressure drop across the discharge tube but unfortunately such blowers have limited volume flow capacity. Nevertheless, in practice, there are relatively few serious problems in extending the power levels of axial-flow lasers to ~5–10 kW range by multiple 'U' or 'V' folds and many companies have developed such lasers for the industrial market, notably for high-speed flat-bed cutting (see section D1).

By contrast for cross-flow lasers, gain medium density fluctuations are one-dimensional and occur in a direction transverse to the laser axis as shown in figure B3.1.7(*b*): optical resonators that provide effective compensation for these effects are, therefore, rather more difficult to design. However, blowers with simpler designs (such as single-stage compressors or radial fans) are available which are capable of high volume rates (and, therefore, higher laser powers) because of the much smaller pressure drops involved.

To achieve the stable long-term operation that is needed for most applications of fast-flow lasers, it is necessary to take account of molecular dissociation. To achieve this, it is usual to design flow-through catalysis units into the external gas-flow loop in order to control the level of dissociation and the concentration levels of the key gases. Over the years, considerable effort has gone into producing highly effective proprietary designs for flow loops for convectively cooled CO_2 lasers, although some degree of 'gas make-up' is often required. For obvious geometric reasons relating to the transit time of the laser gas through the gain region (see figure B3.1.7(*a*) and (*b*)), the transverse-flow technique provides, *at least in principle* a larger value of heat removal capacity than for axial flow, so that transverse-flow systems potentially offer advantages for very high average power lasers. For such lasers it is advantageous to operate at higher values of gas pressures (~100 torr), that is five to ten times larger than the value typically used for axial lasers. In order to minimize the required inter-electrode separation, as well as other practical reasons, it is usual for cross-flow lasers to employ discharges that are maintained in a direction transverse to both the flow direction *and* the optic axis, as shown in figure B3.1.7(*b*). Transverse-flow lasers offering power levels in excess of 10 kW have been developed for material-processing applications, such as thick section welding that require very high power levels and where the somewhat reduced beam quality is less important in practice.

B3.1.3.3 TEA CO_2 lasers

Although it is possible to operate longitudinal dc discharges of the type discussed earlier in a *pulsed* mode and thereby to produce pulsed laser radiation, the attainable peak powers and pulse energies are severely restricted by limitations in scaling the discharge in volume, gas pressure and power. However, beginning in about 1969 a series of discoveries occurred within a few years of each other which completely changed the situation with regard to CO_2 laser scaling in peak power and pulse energy because of significant advances in scaling lasers to larger discharge volumes at high (atmospheric) pressures. It was realized that the key to success here is use a transverse pulsed dc discharge architecture with essentially parallel-plate electrodes.

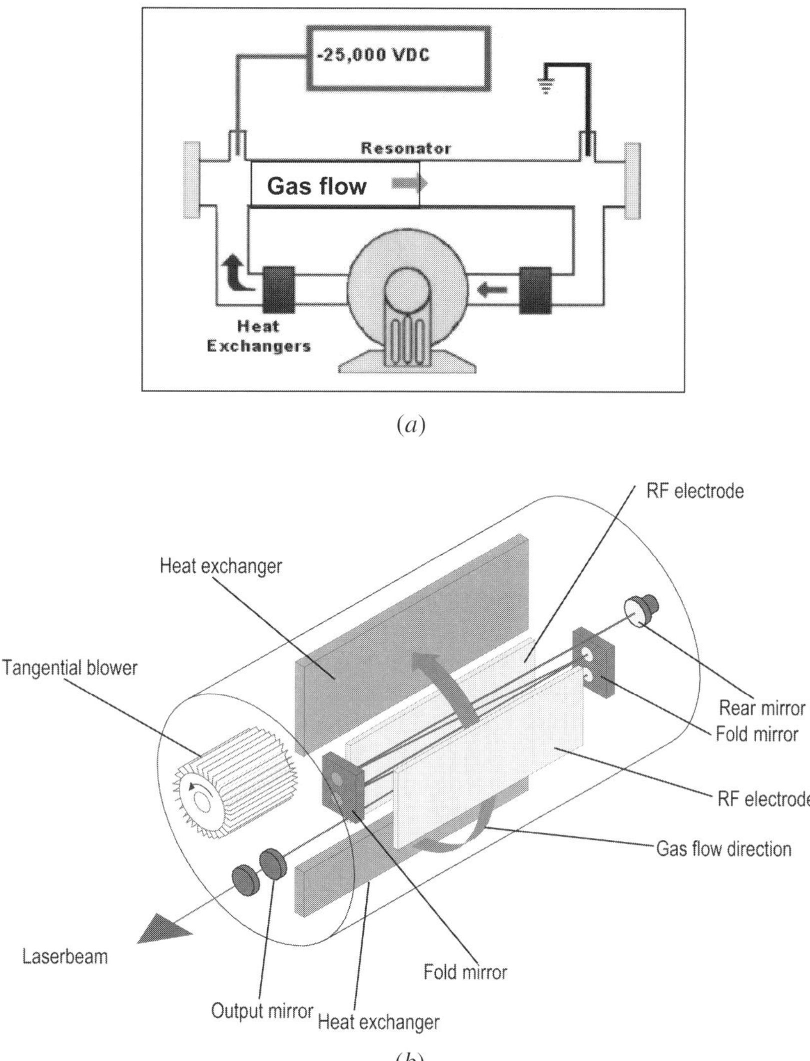

Figure B3.1.7. (*a*) Convectively cooled CO_2 laser employing fast axial gas flow and (*b*) convectively cooled rf excited CO_2 laser employing fast transverse gas flow.

However, for several years the missing technical advance was the invention of some technique to 'volume-stabilize' the discharge. First Beaulieu in Canada [29], then Dumanchin and Rocca Serra in France [30] and Lamberton and Pearson in the UK [31] succeeded with novel methods for controlling a transverse avalanche discharge in a high (atmospheric) pressure mixture of CO_2, N_2 and CO_2, and preventing the onset of instabilities and the triggering of arcs within the discharge. Thus was born the TEA laser, the acronym for transverse electrically-excited, atmospheric pressure laser, a major landmark in CO_2 technological history. These inventions triggered a period of intense world-wide research activity in the race to 'control' larger and larger discharge volumes and to develop practical technologies for repetitively pulsed lasers [32, 33]. This was seen as the route to scaling TEA lasers for very high peak powers and pulse energies, and to produce

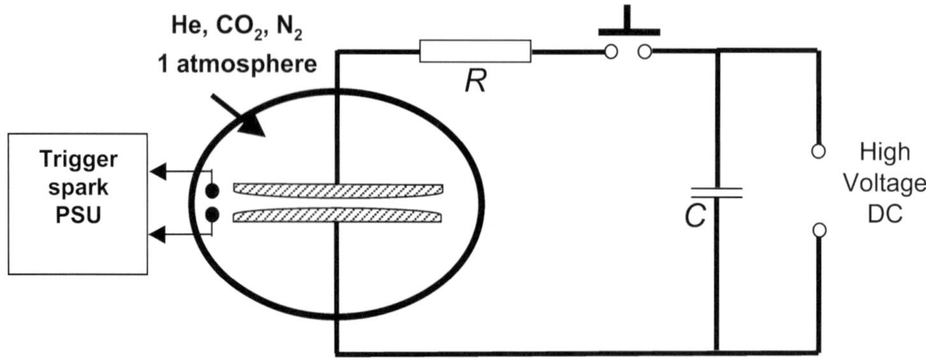

Figure B3.1.8. Schematic diagram of a TEA (Transverse–electric discharge–atmospheric pressure) CO_2 laser.

effective and practical technologies for both low and high repetition rate laser devices.

A schematic diagram of a typical TEA laser of the modern era is shown in figure B3.1.8. A pair of shaped metal electrodes is arranged in parallel and separated by ~1–5 cm with lateral dimension up to about 10 cm defining a (usually) rectangular cross-sectional discharge with length up to ~1 m. The electrode shape is determined by the requirement to control the discharge width and to ensure uniformity in the electric field and, thereby, in the discharge. In this design, pre-ionization of the main discharge volume is provided by the UV radiation emanating from two rows of running arc discharges between pairs of subsidiary pin electrodes positioned at intervals of 5–10 cm along the length and on each side of the main discharge. These arc discharges are supplied with electrical energy stored in the ceramic capacitors that support the pin electrodes, as illustrated in the schematic diagram. The deep UV produced by the sparks produces the necessary ionization in the gas volume by both photoionization of constituents of the laser gas mixture and also by photoionization of the electrode surface. Once the pre-ionization has become established within the whole laser volume, a fast high voltage switch (spark gap or hydrogen thyratron) is used to rapidly switch the energy from the main (~10–50 nF) capacitor, that can be repetitively charged to ~25–40 kV via appropriate pulse-forming circuitry. Other forms of pre-ionization that have been used include trigger wires, x-ray pre-ionization, and electron beams, as described in [32, 33].

Much of the early interest in TEA lasers was for military applications and as a source of high-peak power, short-pulse radiation for research on laser-generated plasmas. However, interest in the use of TEA laser systems for industrial applications and, in particular, for product marking began to increase in the late 1970s. The discharge electrode geometry is suitable for the generation of large relatively uniform multi-mode rectangular beams, which are used to illuminate a thin mask, usually of copper in which the mark/code/message to be marked has been etched. The image is then projected onto the product to be marked, for example as it passes on a production line. Typical TEA lasers for this application emit pulse energies of ~5 J in 1 μs and with pulse repetition rates up to ~50 Hz.

B3.1.3.4 Waveguide CO_2 lasers

Waveguide lasers can be distinguished from those employing 'conventional' resonators by the fact that over at least some part of the intra-cavity path, the laser radiation is guided and does not obey the laws of free-space Gaussian propagation. In the early 1970s, the demands of one particular NASA application led to a surge of interest in lasers with gain cells with cross-sectional dimensions which are too small for Gaussian beam propagation, and where optical guiding can be exploited. The optical propagation laws [34] and resonator physics [35] for such lasers differ considerably from that of free-space devices and have been

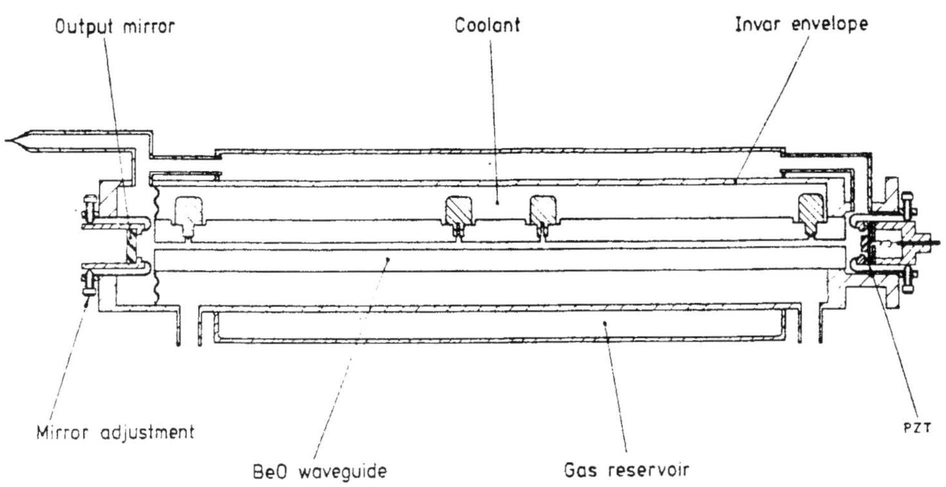

Figure B3.1.9. Schematic diagram of a 6 W dc excited waveguide CO_2 laser. After [43].

the subject of considerable subsequent research—see [8, 36, 37] and chapter A2.2. The NASA requirement was to produce continuous frequency tunability of ±500 MHz in heterodyne receiver local oscillator laser for use in wideband telecommunication links between satellites in earth orbit [38]. In typical CO_2 laser gas mixtures, collision effects produce line broadening at ~5 MHz torr^{-1}, so gas pressures of >150 torr are required to achieve the required continuous tuning range. Recalling the similarity rules gas discharges, this implies the use of tube diameters of <2 mm, which is not possible for conventional resonators of the required length. Earlier work with helium neon lasers [39] had demonstrated the concept of lasers based on hollow dielectric waveguides and it was not long before the CO_2 waveguide lasers became a reality [23, 40, 41]. The potential of waveguide lasers to provide compact, stable sources of infrared laser light for applications in laser radar and other military/aerospace applications was rapidly recognized and considerable development took place in the mid and late 1970s. A typical sealed–off dc waveguide laser of the time, capable of 5 W cw is shown in figure B3.1.9. As with conventional diffusion-cooled free-space lasers, tubes for dc discharge excited waveguide lasers are usually cylindrical but may be fabricated from alumina (or, in extreme cases, beryllia [42]) instead of Pyrex glass, for improved thermal conductivity, reduced waveguiding losses and to meet demands for more rugged construction. The use of small-bore higher-pressure discharges requires the use of longitudinal high dc voltages, and the need for series ballast resistance means that twin 'central' electrodes are needed to permit electrically grounded mirror mounts as shown in the schematic diagram. Considerable research was devoted to investigating laser scaling laws in the spirit of discharge similarity principles [23, 43], and to developing device technology for practical (field) applications. Technology limitations such as the maximum available length of (very straight) dielectric waveguide, excessively high dc voltages and/or the complexity of multiple high voltage dc discharges in restricted volumes put quite practical restrictions on cavity folding strategies and limited power levels to a few tens of watts. This matched several important military requirements but was generally too low for many key industrial applications of the time. Nevertheless, much of the research conducted on dc waveguide laser device physics laid the groundwork for future significant developments in the later life of the CO_2 laser. In particular, it prepared the way for the subsequent utilization of rf discharge excitation, which effectively solved most of the technical problems of dc discharge devices but, even more significantly, it opened the door to *planar waveguide* architectures and the huge device technology potential of *area-scaling* to produce ultra-compact high-power high-brightness lasers, which is a main topic of the next section (B3.1.4).

B3.1.4 Modern era: CO_2 lasers excited by rf discharges

B3.1.4.1 Introduction

The onset of the 'modern era' in the progress of CO_2 laser technologies to their present state can be traced to an intensive burst of research activity that began to gain momentum around 1980, and which lasted for about a dozen years, underpinning a new generation of ultra-compact high-performance commercial lasers. The key factor that sparked this interest was the growing realization that transverse rf discharges offer significant advantages over dc discharges as an effective technique for the excitation of CO_2 in both diffusion-cooled and fast-flow formats. Aside from offering a series of important but incremental technology improvements over dc lasers, the major impact of rf technology has been in enabling *new laser device architectures*. Thus, the ability to control rf discharges over very large areas of *all-metal electrodes* at unprecedented power densities in relatively simple and inexpensive structures provides the basis for high average power lasers in ultra-compact sealed packages, using the planar waveguide or annular configuration. This feature, which has spawned a whole new generation of ultra-compact high-power sealed-off lasers, is not feasible with dc discharge excitation due to the onset of arcs and discharge filamentation. In addition, rf discharges can be modulated or switched at much higher frequencies than for dc discharges, so facilitating considerable benefits in terms of more flexible temporal formats, which is crucial for many important material-processing applications. Also, at a practical technological level, there are several important device engineering advantages of rf *versus* dc discharge excitation. These include the avoidance of high voltages and cathode processes for a reduced impact of gas chemistry effects on the sealed lifetime, an all-metal construction for more effective cooling, the avoidance of ohmic losses in series ballast resistors and simpler, more rugged and lower-cost construction opportunities. Convectively cooled lasers share many of these advantages.

Research on high-frequency discharges can be traced back to the end of the 19th century [44] and, indeed, rf discharge excitation was involved in a number of landmark experiments in the early days of gas laser physics. For example the first gas (HeNe) laser and the first CO_2 laser were excited by rf discharges [45,46] and the first waveguide laser also employed rf fields [39]. However, despite several apparently promising results [47,48] obtained with rf and combinations of rf and dc fields, high-frequency excitation was largely ignored in favour of dc discharge pumping until the end of the 1970s. The resurgence of interest owes most to the pioneering work of Laakmann and co-workers at Hughes Aircraft Company [49–52] and of Chenausky *et al* at United Technologies Research Center [53,54] in the USA and the Hull University, UK, group [55,56] following the first published report by Lachambre and colleagues at DREV in Quebec, Canada [57]. At about the same time, several Russian researchers including Raizer [58], Myshenkov [59] and Yatsenko [60] were active in fundamental studies of the structure and properties of transverse high-frequency discharges and their work was to have a profound impact on the development of practical laser excitation technologies over the following decade.

After a few years, it became apparent that two distinct strands could be identified in the international research efforts in the area of rf excitation of CO_2 lasers. Thus, while most of the initial work was focused on rf discharges in hollow waveguide gain structures, aiming at higher performance sealed lasers cooled by *diffusion*, a different orientation soon became apparent where researchers, mostly in Germany, began developing rf excitation technologies for fast-flow, *convectively* cooled lasers. This research led to the development of rf excited lasers involving axial gas flow in glass tubes of diameter $\sim$2–10 cm, with transverse rf discharges maintained between shaped external metal electrodes. In this configuration, the dielectric nature of the glass tube walls provides additional capacitative ballast, which *combines* with the internal stabilizing influence of the rf discharge sheaths to permit stable discharge operation at high power densities. These values are substantially higher than for equivalent dc discharge lasers (even when 'stabilized' by turbulent flow) where thermally induced arcs are the limiting factor, though they are significantly lower than for the all-metal 'thin' discharges of diffusion-cooled planar waveguide lasers mentioned earlier. However, the use of glass tubes and external electrodes is conveniently compatible with fast-flow cooling techniques, and offers relatively simple laser engineering opportunities for folded cavity designs typical of those that had previously been developed for dc lasers. These designs were soon undergoing commercial development and were the forerunners of several successful families of commercial rf lasers.

This section of the chapter endeavours to provide a summary of the key device physics/technology features of each of these two types of transverse rf discharge laser. As a necessary introduction to this, we first introduce the underlying physics and operational characteristics of transverse rf discharges, since it is their distinct properties that are the essential enabling factor in the most significant developments in CO_2 laser engineering technologies over the past 20 years.

B3.1.4.2 Rf gas discharges

B3.1.4.2.1 Basic rf discharge structures

AC discharges differ from dc discharges because the polarity of the applied electric field reverses every half cycle, though at low frequencies the properties of ac discharges do not differ appreciably from those which are uni-directional. However, when the oscillation frequency increases sufficiently so that the electron ambit is smaller than the discharge dimensions, the situation is transformed and the discharge structure and physics for high-frequency discharges is very different from the dc case. Moreover, as discussed later, rf discharge excitation offers considerable fundamental and practical advantages for high-performance CO_2 lasers. The use of the terms 'high' or 'radio' frequency are somewhat imprecise but, in the present context, they refer to frequencies in the approximate range 1–200 MHz, though for practical reasons most research and all commercial rf lasers employ power generators with frequencies below $\sim$100 MHz. Both 13.56 MHz and 27.12 MHz are preferred industrial frequencies for larger electrode separation, non-waveguide lasers where frequencies at the low end of this range are required. At a very basic level, the physical phenomena involved in dc and ac discharges are not dissimilar. The main differences derive from the temporal interaction of the frequency of the applied field, the oscillation frequencies, collision frequencies and drift velocities of the charged and neutral particles, on the one hand, and the scale and geometry of the electrodes and/or containing vessel, on the other. Electrons gain energy from the applied field and transfer it to atoms and molecules via collisions and collisions may result in electronic/molecular excitation or ionization of atoms or molecules or electron loss from the discharge depending on conditions. For example, electron multiplication may be produced by ionization processes in the bulk of the gas or following secondary emission following ion bombardment of the electrodes/walls or by a combination of both processes. For laser excitation, we look for electron collisions to produce excitation of atoms (molecules), while recognizing that as well as electron multiplication in the field, electron loss will occur by recombination with positive ions, attachment to neutrals to form negative ions or simply by collision with the walls.

We consider a simple parallel-plate metal electrode structure typical of rf lasers as illustrated in figure B3.1.10(*a*). A transverse rf electric field is applied between the plates in a vacuum enclosure containing gas at intermediate pressure ($\sim$10–300 torr). Assuming that only electron motion need be considered due to the high relative mass of atoms/molecules, the initiation of the discharge will result in potentially large electron loss to the electrodes due to the frequency-dependent and finite oscillatory electron motion in the applied field. This leads very rapidly (microseconds) to the creation of electron depleted regions of high positive ion concentration or 'space charge layers' near each electrode. The resulting combination of space charge fields and applied rf field determines the spatial structure, dynamics and frequency dependence of the discharge for a given inter-electrode spacing. As described later, the space charge regions play a crucial stabilizing role in the role in rf discharge operation. Two stable forms of the rf discharge were identified by Levitskii [61] for this transverse-field situation. The first of these was designated the alpha (α) discharge, reflecting the fact that, in this case, the ionization is maintained largely by electron multiplication in the main volume, where the Townsend α- coefficient applies. The α- discharge is usually, though not always, characterized by low current

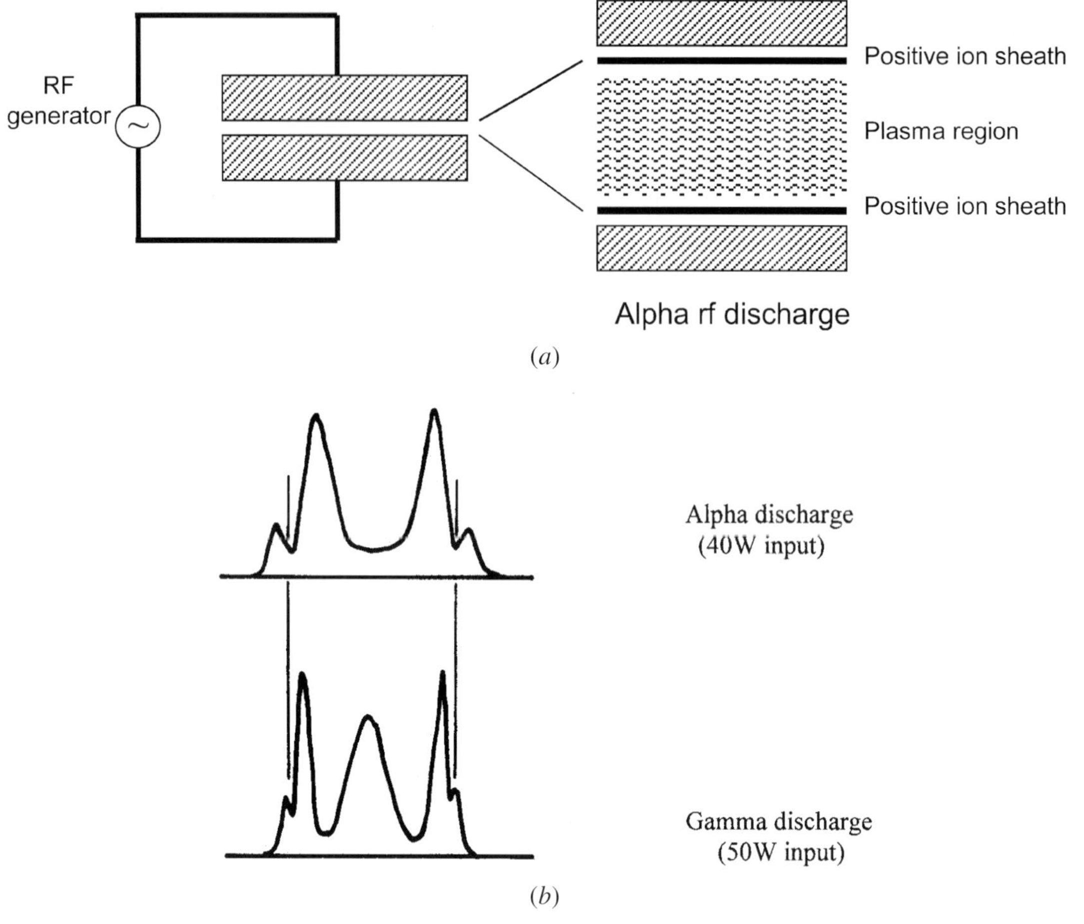

Figure B3.1.10. (*a*) Schematic diagram of transverse rf electrodes and α-discharge structure and (*b*) TV line scans of the visible emission from an rf discharge in the α and γ modes. After [93].

densities, it has a positive net impedance characteristic and, in practice, is the only one of the two discharge types that is relevant to the pumping of molecular lasers such as CO_2. The second type, designated the gamma (γ) discharge, is associated with higher current densities and involves secondary electron emission from the electrode surfaces, hence, the γ-designation consistent with the Townsend γ-coefficient describing electrode secondary emission. Gamma discharges have near-electrode regions which are rf analogues of the near-cathode regions (cathode sheath, negative glow and Faraday dark space) of the dc glow discharge (see section B3.6.2). Although the γ-discharge may have a diffuse plasma column that may be quite uniform, the high power dissipation in the sheaths together with the low electron temperature in the plasma glow renders this type of discharge useless for molecular laser excitation [59, 62]. Thus, in practice, the transition from an α-discharge to a γ-discharge may be considered as an 'instability' process, to be avoided at all costs.

Observation of the visible light emission from alpha rf discharges reveals significant transverse structure, with intense visible light emission ('striations') evident in the high-field 'sheath' regions close to but not immediately adjacent to each electrode [63], as illustrated in figure B3.1.10(*a*). The striations are caused by electronic excitation by energetic electrons accelerated by the high fields in the boundary layers, with the sheath regions effectively isolating the more diffusely emitting discharge plasma column from the electrodes. The α-discharge is characterized by a stable and positive impedance characteristic and its plasma column can provide efficient laser excitation, while suffering low dissipative power losses in the sheath regions. The scale of the sheath regions depends on the operating conditions and, ideally, they occupy only a small fraction of the inter-electrode spacing since they do not contribute directly to laser excitation. Given the strong parametric dependencies of rf discharge properties, an issue of major practical importance is the optimization of rf discharge conditions for efficient laser operation. This applies both for the case of narrow-discharge 'waveguide-type' diffusion-cooled devices and also for larger electrode separation laser designs that are cooled by fast gas flow. However, first, we consider the more fundamental question of the instability processes to which alpha rf discharges are subject particularly under conditions of high power loading. The aim is to design devices such that the probability of instabilities occurring is minimized. From the earliest days of rf laser investigations, it had been observed that although in certain situations, the magnitude of the power density that can be maintained in the plasma column of a spatially uniform rf discharge may be large, nevertheless it is limited. Above certain values, depending on the conditions, it is observed that the alpha rf discharge becomes unstable, producing significant observable changes in its structure and properties. In particular, for laser pumping such instabilities result in the destruction of the inversion and laser action ceases. Therefore, the challenge was to understand the key instability processes so that inappropriate operating conditions could be avoided. Perversely in this context, the basic question that confronted researchers for many years was to determine the origin of the high degree of *stability* that characterized rf discharges maintained between *metallic* electrodes, rather than specifically to understand how the *in*stability occurs. In the early 1980s Myshenkov and Yatsenko [59] had shown that the instability threshold depends critically on macro-parameters such as generator frequency, electrode separation and gas pressure and composition. Moreover Kolesnychenko *et al* [64] and Wester *et al* [65] among others noted that the insertion of dielectric sheets (or glass tube walls) between the metal electrodes and the low-pressure gas produced an increased instability threshold. They correctly attributed this to the additional ballasting (current stabilizing) effects of the series capacitance of the dielectric sheet. However, the problem remained to explain the relatively high level of stability of α-discharges with no added series dielectric/capacitative ballast and to determine the instability threshold for situation where parallel all-metal electrodes are used. This is particularly relevant for laser design since the avoidance of dielectric sheets in an all-metal structure is highly desirable to enable efficient cooling and the attainment of high-power density excitation.

With a detailed analysis and a careful set of experiments on alpha rf discharges, Vitruk *et al* [66] were able to determine for the first time the electrical properties of the sheath and plasma glow regions *separately*. Using the results of Smirnov [67], the thickness and, thereby, the capacitance of the sheath was determined using vector analysis of the equivalent circuit of the discharge, including the electrodes and associated structure. This allowed the determination for the first time of the current–voltage characteristics of the sheath and plasma regions separately, as shown in figure B3.1.11(a). Analysis of these data leads to important conclusions concerning the structure and stability of the α-discharge. For example, it is clear from this figure that the positive slope of the current–voltage characteristics of the α-type discharge is a consequence of the properties of the *sheath* rather than of the plasma glow region, and moreover that the high current densities and relatively low plasma voltages compared to the ion sheath voltages indicate that the α-discharge behaves almost as a capacitative load. Taking account of the thickness of the two sheaths, one can now compute the value of E/N averaged across the plasma region of the discharge and this is plotted in figure B3.1.11(b). The values of the reduced field in the plasma are seen to be pressure dependent and to be comparable with the near-optimum values observed by Abrams and Bridges [23] for dc discharge waveguide lasers. This is not unexpected since for typical rf discharge laser conditions the electron collision frequency is much higher than the excitation frequency [69].

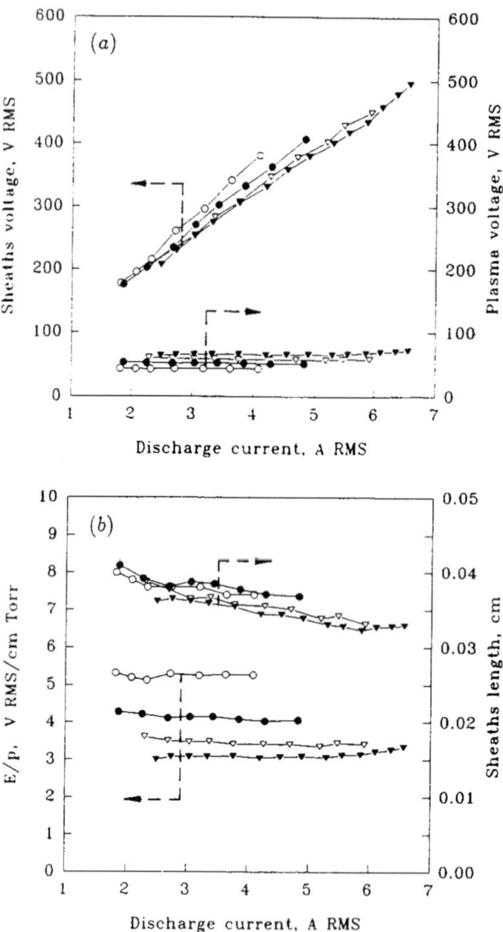

Figure B3.1.11. (*a*) Voltage–current characteristics of the sheaths and plasma region. (*b*) Plasma-reduced electric field and sheath length *versus* discharge current for different pressures. From [66].

B3.1.4.2.2 RF discharge instability processes

The susceptibility of rf discharges to instability processes is shared by many other types of gas discharges and such instabilities can be generally categorized as either thermal or ionic in origin. For most designs of diffusion-cooled laser that employ all-metallic electrode systems, the alpha-to-gamma transition which is an example of an ionic instability, is the most important type to consider. However, for fast-flow rf lasers which have a larger electrode separation, the thermal instability process is usually dominant since series dielectric ballast (e.g. the glass tube structure) is usually employed and heat removal is much less efficient than for narrow-gap diffusion cooled lasers [70]. Let us consider the sequence of events in the initiation of an α-type discharge and the transition instability to the γ-type. For an all-metal parallel-electrode geometry with a separation of a few mm, the application of rf power via a suitable impedance-matching network triggers the initiation of an α-type discharge, the area of which may be less than the full electrode area. Increase in the rf power supplied causes the discharge to expand to cover the entire electrode surface area, after which further increases cause the discharge current density to increase and the visible discharge glow to intensify, reflecting the positive impedance characteristic of the discharge under these conditions. As the rf power is further increased, a threshold is reached where a sudden transition to the γ-type discharge occurs, accompanied by a reduction in discharge area and an increase in current density. The visible transverse profile of the α- and the γ-type discharges are shown in figure B3.1.10(*b*). The γ-discharge has a bright central region with spectral emission from both ionized and dissociated molecules being present in the typical CO_2 laser gas mixture and the electrodes are covered with a bright glow, similar in appearance to that of a dc-voltage excited cold cathode surface. The electrical changes occurring in the α/γ transition in a small test cell are illustrated in figure B3.1.12 [71]. Following ignition, the positive impedance characteristic of the α-type discharge with inter-electrode voltage increasing with rf power is observed, and large areas of discharge can be uniformly excited so long as the electrode voltage is uniform across the electrode area. At a well-defined value of the power density, which is dependent on the generator frequency, gas mixture and pressure and the electrode material and gap size, a discontinuous switch to the γ-type discharge occurs. The transition is usually accompanied by a slight decrease in the power coupled to the discharge due to the changed impedance characteristic of the gamma discharge, and the I–V curve is observed to be slightly negative. Figure B3.1.12 also shows the 'bistable effect' where a considerable reduction in applied rf power must be made in order to return the discharge to the α-mode. The conditions in the central region of the γ-discharge are akin to those in the dc discharge negative glow, with high ion densities being sustained by hot electrons being 'injected' from the boundary layers, high gas temperatures and significant molecular dissociation. As mentioned, these conditions are highly inimical to CO_2 laser excitation. Consequently, a good understanding of this instability process is required to facilitate laser designs, that for good cooling and low cost reasons, employ all-metallic electrodes, so that the transition and latching into the γ-discharge can be avoided. Figure B3.1.11(*a*) shows that the net positive slope of the I–V characteristic is a consequence of the behaviour of the *sheath* regions while the slope of the plasma column is never observed to be positive [66]. We can conclude from this that the capacitative sheaths perform the crucial role in stabilizing the plasma in an α-type rf discharge, and behave as an *internal* ballast element in series with the plasma 'load' which has a non-positive I–V characteristic. Thus, the onset of the α/γ transition instability is very unlikely to occur when the plasma voltage is small compared to the sheath's voltage. Moreover, experiments [66] indicate that the threshold for the α/γ transition instability is the point when the sheath voltage exceeds a critical threshold value linked to the α/γ transition, and this value increases with increased frequency as explained by the model proposed in [66].

B3.1.4.2.3 Similarity and parameter scaling in rf discharge lasers

The experimental optimization of discharge parameters usually aimed at maximizing the laser power extraction involves a complex interplay between many intrinsic and extrinsic factors. The study of parametric dependencies or scaling for dc discharge lasers has been investigated within the framework of the so-called *similarity laws* for such discharges [72, 73]. Gas heating in the discharge plasma is seen to play an important role in the definition of a series of invariant parameters of the system. Comparing the situation for rf discharges, the principal additional parameter where optimization is required is the generator frequency. Vitruk *et al* [68] have extended the similarity principles for dc discharges to include frequency effects and have included the key structural features of the alpha rf discharge. Bringing together earlier data on the optimization of the generator frequency with inter-electrode spacing indicates that these parameters are linked by the simple relationship that the (frequency $\times$ spacing) product $fd{\sim}280$ mm MHz. The results of their theoretical analysis of similarity in alpha rf discharges are compared with experimental data on the current–voltage–power characteristics of these discharges over a wide range of generator frequencies, f, gas pressures, p, and inter-electrode distances, d for a typical CO_2 laser gas mix. The analysis results in a set of invariant parameters and the formulation of a set of scaling and similarity laws for rf discharges. These relations extend the laws developed for dc discharges to include the generator frequency and the impact of

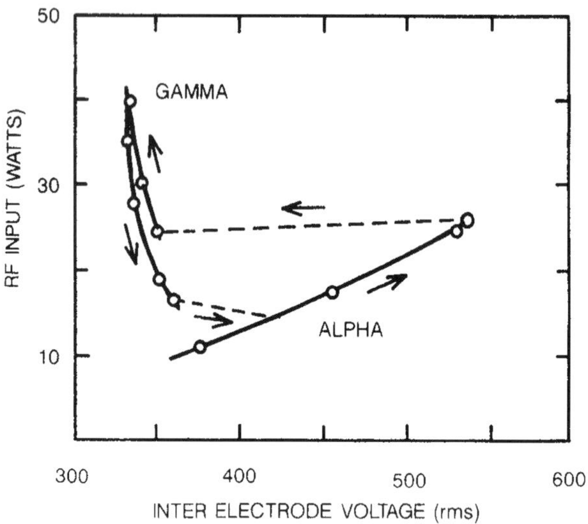

Figure B3.1.12. Electrical characteristics of a 6 mm gap rf discharge, showing the alpha and gamma regions and the bistable switching effect: He, N_2, CO_2, 3:1:2 + 5% Xe at 15.5 MHz. After [93].

varying f on key parameters relevant to the stability of the alpha rf discharge. These rules may be applied to the design of discharge structures and impedance-matching networks. Moreover, the scaling laws for plasma and sheath voltages are particularly useful in defining the frequency and power limits of discharge scalability, as illustrated by the operational 'map' [68] shown in figure B3.1.13. Thus, for large-gap discharges at low excitation frequencies, the sheath voltage may become large enough such that there is significant risk that the α/γ transition instability may occur. The frequency at which the α-discharge sheath voltage becomes comparable with the α/γ transition sheath voltage defines the low frequency limit of stability for unballasted metal electrodes. For the typical CO_2 gas mixture used, the transition voltage is about 400 Vrms [66] and, hence, the frequency scalability limit is predicted to be ~40–50 MHz, as shown in figure B3.1.13. These predictions are in excellent agreement with experimental observations. A second reason to avoid low excitation frequencies arises from a consideration of the rf power losses in the sheaths. The ratio of power dissipation in the sheaths (which produces unwanted gas heating) to desirable plasma power dissipation is proportional to $1/f^2$. The relatively high observed values of lasing efficiency at high frequencies with high power loading may be due to very low power losses in the sheaths at such frequencies. There is a high frequency limit on excitation that occurs here at about 300 MHz when the sheath and plasma voltage curves cross and the net discharge impedance is then only barely positive. This leads to increasing current density non-uniformities along the electrodes even for cases where the applied inter-electrode voltage uniformity is constant. Finally, for high-power pulsed rf discharges with metal electrodes that are required for generating high peak power laser pulses from planar waveguide devices [102] (see later), figure B3.1.13 indicates that high frequencies (>100 MHz) are preferable for stable discharges, though a compromise at, say, 80 MHz provides very good performance. This is because the value of V_s can be maintained below the α/γ transition voltage even for electrical power density inputs of ~1 kW cm^{-3}. This is key information for the development of enhanced pulsed (so-called 'ultra-super-pulse' (USP)) lasers [102].

As previously mentioned, the modern era of CO_2 laser research has spawned two parallel technology development tracks that have been driven, to a large extent, by the requirements of different laser market

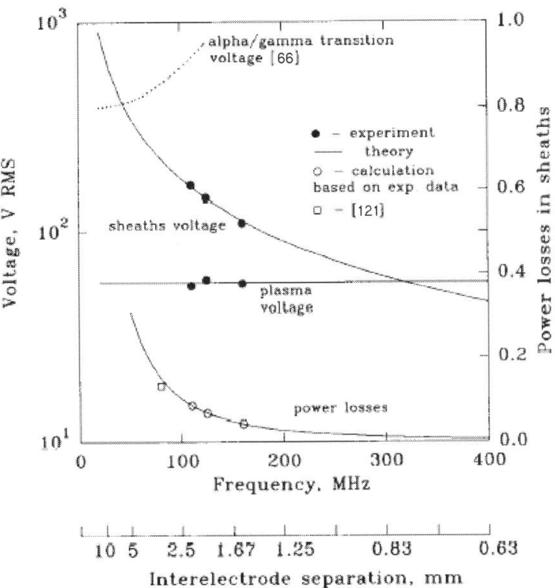

Figure B3.1.13. Frequency dependencies of some of the alpha 'similar discharge' parameters for He:N_2:CO_2 = 3:1:1 + 5% Xe gas mixture.

sectors. These two topic areas, namely (i) diffusion-cooled rf waveguide lasers and (ii) convection-cooled free-space lasers, are the main subjects for the remainder of this chapter.

B3.1.4.3 Diffusion-cooled rf-excited waveguide lasers

B3.1.4.3.1 Linear rf waveguide lasers

In many ways a *linear* rf waveguide laser represents a direct progression from the dc discharge waveguide lasers developed in the early 1970s, employing similar optical cavity designs (typically dual Case I resonators, see [35–37] and chapter A2.2) but with rf excitation technology providing a number of important performance advantages. A typical rf waveguide laser discharge cross section shown in figure B3.1.14(*a*) is designed around a metal/ceramic sandwich structure where the inter-electrode spacing is typically ~2 mm and waveguide length is less than 0.5 m. The lower electrode is grounded and the entire electrode structure and vacuum envelope is maintained near room temperature by water-cooling or forced air. A pair of planar dielectric coated mirrors is used in a simple waveguide case I resonator with output coupling usually <10%. A typical gas mixture is 3:1:1 helium, nitrogen and CO_2 with 5% xenon added at a total pressure of ~100 torr (the optimum value depends on electrode separation). Power is delivered from an rf power generator often via a 50 Ω cable, though lasers with integrated rf power supplies have been developed commercially. The laser can be considered as an electrical load comprising a complex impedance consisting of the mechanical structure combined with the rf discharge. Power transfer efficiency to the discharge is a key feature for laser operation and this is made more complicated by the fact that the rf discharge behaves as a *power-dependent* impedance element. Moreover, at the high frequencies (>50 MHz) where rf waveguide laser performance is optimized, unacceptable longitudinal variations of the inter-electrode voltage occur due to transmission-line effects caused by the rf electrical properties of the laser structure of parallel metal conductors (electrodes) separated by high dielectric ceramic spacers. To solve this problem, compensation techniques producing distributed impedance matching have been developed [8, 52, 54, 74], and such designs are used to minimize linear variations in the inter-electrode voltage and the consequent triggering of the α/γ transition, that would otherwise limit the discharge power density that can be achieved. The schematic diagram in figure B3.1.14(*b*) shows one approach that involves distributed resonant matching with an array of inductors that is designed to compensate for the effects of the structure capacitance by achieving parallel resonance at, or near, the generator frequency. Beginning with the first reported rf waveguide laser of Lachambe *et al* in 1978 [57], laser performance in terms of power extraction and conversion and efficiency increases significantly over the next few years. This is evident from the increase achieved over a 5-year period in a frequently-quoted figure of merit for linear lasers, the specific power, P_L, measured in W cm^{-1} of the discharge length. Reported values of P_L showed an increase from the earliest value of 0.04 W cm^{-1} achieved by Lachambre to the value of 0.83 W cm^{-1} reported by He and Hall who in 1983 obtained a cw power output of 30.6 W and a maximum efficiency of 13% from a 37 cm rf discharge operated at 125 MHz in a 3:1:1 mixture of He, CO_2 and N_2 with the addition of 5% xenon at a total pressure of 110 torr [56]. These power increases were due in the main to increased recognition of the design requirement to optimize the generator frequency [75], to ensure negligible longitudinal variation in the transverse rf voltage and to maximize heat transfer to the all-metal electrodes. The use of trace quantities of Xe in CO_2 laser mixtures is well known to have several beneficial effects including increased laser power and increased sealed-off lifetime [16, 76]. A 24% increase in laser power was observed with 5% Xe [77] and this observation is linked to gain increases [78] by spatially resolved measurements of electron temperature, T_e, using an electrostatic double-probe technique. These measurements show T_e falling to 1.5 eV for 5% Xe concentration, a value where vibrational pumping of N_2 and CO (formed by CO dissociation) has a peak cross section with corresponding maximum gain, and under near-optimized condition, Vidaud *et al* [79] achieved efficiency values 20%. Diffusion-cooled rf lasers became available commercially as sealed-off devices in the early 1980s at power levels of ~5–50 W from several companies including Laakmann Electro-optics Inc. in the USA and Laser Applications Ltd in the UK, who developed a folded rf laser design producing >50 W for applications in surgical and industrial systems. These new laser technologies produced a rapid increase in diverse applications of laser-based ‘table-top’ systems in medicine, science and industry. In addition, there was a parallel series of rf laser device developments that were aimed largely at military, aerospace and remote-sensing applications, for example at United Technologies Research Center and Ferranti Ltd in the UK. With these rapid developments under way, it became apparent by the mid-1980s that there were very large potential markets for sealed CO_2 lasers at power levels of tens to hundreds of watts, particularly for surgical and industrial manufacturing applications.

B3.1.4.3.2 Power scaling—folded waveguide lasers

Because of its volumetric power efficiency and technical convenience, rf waveguide laser technology was seen as a prime candidate for further development but the key question was how best to achieve the necessary power scaling. Many laboratories around the world became interested in the challenge of how to scale CO_2 lasers to the kilowatt power range while maintaining beam brightness, sealed-off operation, compactness and ease of usage, given that the intrinsic efficiency and power extraction capability of CO_2 lasers are limited by discharge-dependent excitation kinetics, gas chemistry and gas heating. Thus, the specific laser power, P_L, is limited to an optimum value of <100 W m^{-1} for linear sealed or slow-flow rf lasers, so the only realistic line of attack for power scaling is to somehow increase the gain volume. Clearly, the simplest conceptual approach is to employ longer gain lengths. However, there are problems with fabricating very long waveguides with sections in the range ~1–3 mm, especially if a machined ceramic is used to define all or part of the waveguide, since the imperative of low waveguiding losses imposes quite stringent fabrication conditions of waveguide surface straightness and finish. In practice, this limits single-section waveguides to maximum lengths of ~0.5 m. A linear series of multiple sections is possible in principle but mutual alignment problems and excessive overall device length makes this an unattractive option. An alternative is to deploy

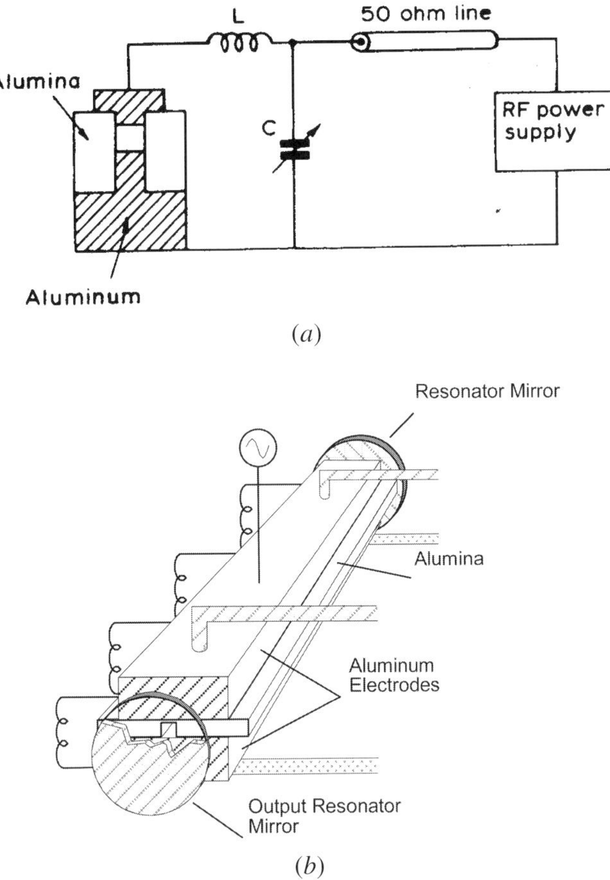

Figure B3.1.14. (*a*) Cross section and electrical schematic diagram of an rf excited waveguide CO_2 laser. (*b*) Schematic diagram of an rf excited waveguide CO_2 laser showing inductors in parallel resonance configuration.

an optical folding system using multiple parallel waveguides with a series of 'U' folds or else 'V' folds with multiple waveguides in a folded zigzag arrangement. When repeated, these folding arrangements can be used to concatenate several metres of an active gain medium into a compact structure, with the advantage that a single pair of planar metal electrodes and a single rf power generator can be employed. However, the design of low-loss optical folds is, of course, complicated by the problem of the beam passing repeatedly from one waveguide section to another via a folding mirror positioned a non-negligible distance away, so that mode-coupling losses and beam effects associated with multiple waveguide/free-space transitions becomes a major issue. Jackson *et al* [80] compared several types of waveguide folding designs and produced a compact folded waveguide laser of 60 W in a compact design. Newman and Hart [81] developed the folding concept considerably further and this work became the basis for a successful commercial range of rf excited folded waveguide lasers developed by DeMaria Electrooptic Systems Inc. and operated either cw or with an internal Q-switching element [82].

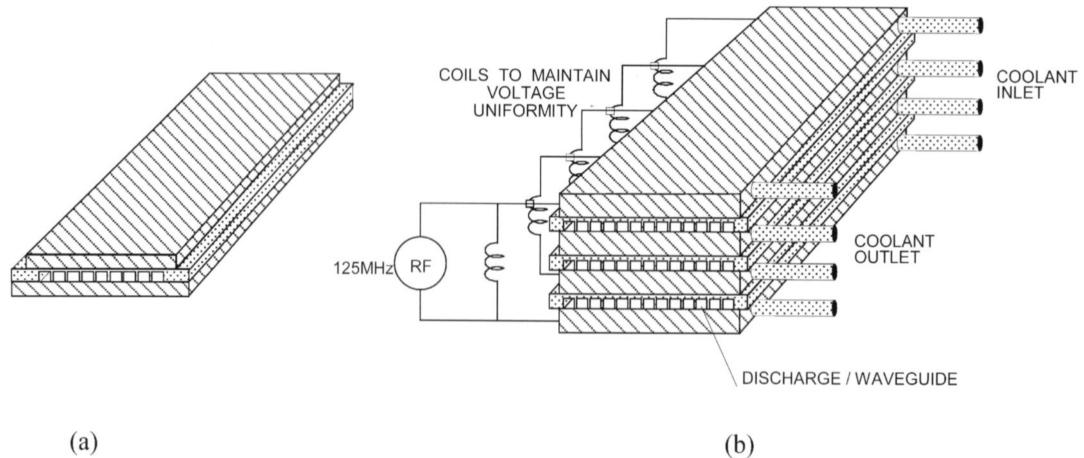

Figure B3.1.15. Schematic diagram of rf excited CO_2 waveguide array laser gain media: (*a*) one-dimensional array and (*b*) two-dimensional array. After [91].

B3.1.4.3.3 Power scaling—waveguide laser arrays

However, it was apparent that linear resonator waveguide lasers with multiple folds were likely to be too complex to permit effective power scaling to the desirable levels of 100 W–1 kW. Nevertheless, the potential power-scaling capacity of multiple parallel waveguides in a monolithic *array* structure had been noted earlier and proposed as a route for the realization of high-power dc waveguide lasers [84]. The additional potential benefit of successfully phase-locking the elements of such arrays for the production of a coherent output and the possibility of high-speed beam scanning via differential phase modulation, by analogy with radar antenna technology was also recognized. However, the practical limitations of high-voltage dc technology prevented the development of effective high-power laser arrays. By contrast, the benefits of transverse rf excitation techniques employing a single rf generator to excite multiple waveguides in a planar structure, cooled efficiently by diffusion to cooled metal/ceramic surfaces (as illustrated in figure B3.1.15(*a*)) appeared as a most attractive option, at least in terms of 'raw' power. The parallel quest for the necessary high beam quality drove considerable research on phase-locking techniques for rf waveguide laser arrays. Youmans [85] showed the principle of phase-locking between a pair of adjacent waveguides separated by a plate of optically transparent zinc selenide and Newman *et al* [86] soon demonstrated higher power phase-locked operation, using a hollow-bore ridge waveguide structure. Subsequently other techniques were devised and demonstrated by the Heriot Watt University group for coherent coupling (phase-locking) of linear waveguide laser arrays, including intra-cavity four-wave mixing [87], distributed coupling [88] and the use of the waveguide-confined Talbot effect [89]. However, the realization of a *practical* phase-locking technique capable of extending array power towards the kilowatt region was seriously hindered by the difficulties in ensuring adequate phase uniformity across the array in the face of variations in element-to-element rf power deposition and cooling across the array. Thus, heterodyne measurements indicated [90] that these effects produce frequency shifts of 1 MHz W^{-1}, thereby imposing severe challenges to the locking range of any candidate coherent coupling technique. Nevertheless, large planar and two-dimensional array lasers (see figure B3.1.15(*b*)), deliberately designed to be incoherent with respect to each other to avoid interference effects, have been demonstrated with power levels >2 kW with the excellent individual element beam quality and the capacity for array beam reformatting to produce useful beam shapes at high average power levels for industrial applications [91, 92].

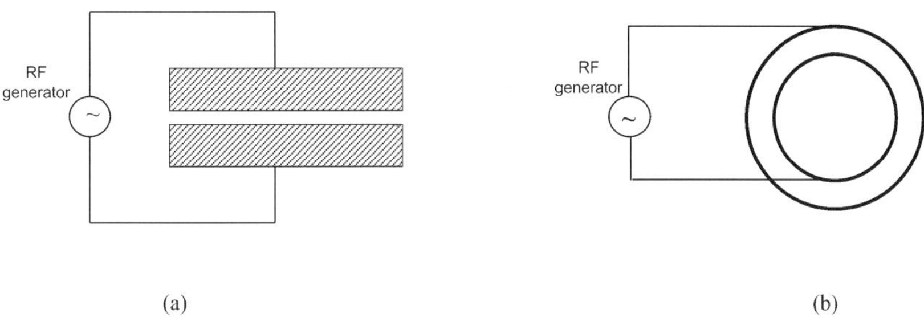

Figure B3.1.16. Schematic diagram of a large area diffusion-cooled laser using (*a*) planar or (*b*) annular electrode geometry.

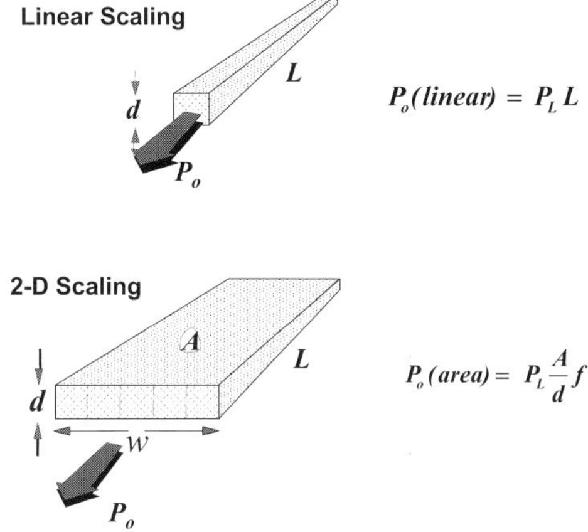

Figure B3.1.17. Schematic diagram illustrating two-dimensional (area) power scaling for planar lasers.

B3.1.4.3.4 Power scaling—area scaling in planar waveguide lasers

The planar waveguide laser concept can fully exploit the enormous (power scaling) benefit that rf discharges can be *area* rather than length scaled. Large-area rf discharges can be operated in a stable manner at high power densities and efficiently cooled by diffusive heat flow to concentric or plane parallel metallic electrodes, as illustrated in figure B3.1.16. If we consider a uniform rf discharge maintained between metallic electrodes of width, w, length, L, and spacing, d, then the total power that can be extracted from the volume is given by [94]:

$$P = f P_L (w/d) L = f P_L (A/d) \tag{B3.1.6}$$

where P_L is the specific power (~ 80 W m^{-1}) and A is the electrode area, as illustrated in figure B3.1.17. The numerical factor f is a number <1 which takes account of the difference between two-dimensional heat flow in a linear laser and one-dimensional flow in planar devices. Geometric arguments suggest that $f \sim 2/\pi$ and this is confirmed by experiment [94]. We can thus arrive at a very simple *area*-scaling law and define a figure of merit, P_A, for area-scaled lasers:

$$P_L = P/A = f P_L/d. \tag{B3.1.7}$$

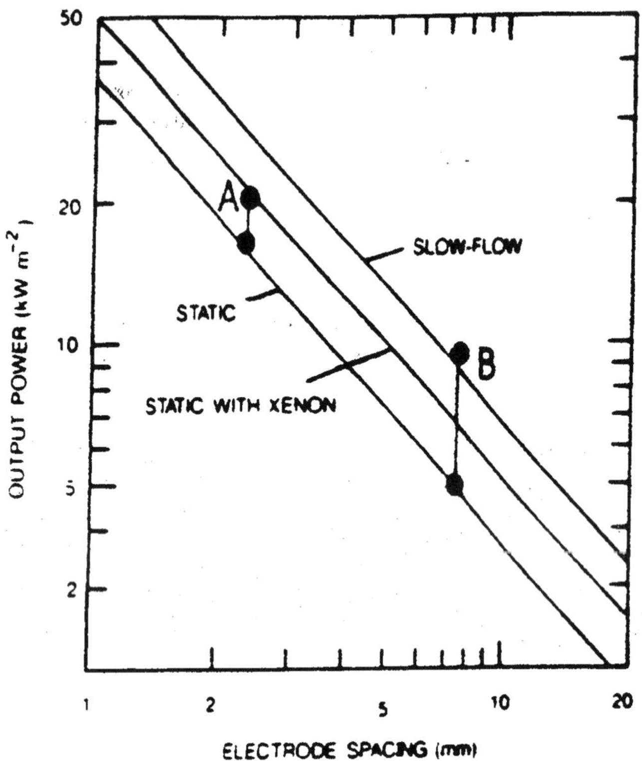

Figure B3.1.18. Scaling of laser power per unit area of large-area discharges with an electrode gap. Experimental points A are from a 2.25 mm device with a static gas fill. Points B are for a 7.5 mm device with slow flow and a static gas. After [93].

The huge potential of the area-scaling concept, and the importance of the $(1/d)$ factor was demonstrated by a simple plot [93, 94] based on equation (B3.1.7) and reproduced in figure B3.1.18. This shows a plot of laser output power per unit area of electrode, P_A, measured in kW m^{-2} that can be expected from a large-area discharge with one-dimensional heat flow, plotted against electrode separation for three slightly different operating modes. The upper line relates to a slow-flow mode where helium-rich mixtures can be used to give maximum thermal conductivity and provide the highest output for all relevant electrode spacings. For static gas /sealed mode operation, the higher molecular gas content in the mixture of helium, nitrogen and CO_2 necessary to stabilize dissociation for a long gas lifetime reduces the gas conductivity and laser power output falls by about 50% compared to slow flow. The middle line indicates the expected power levels obtained when ~5% xenon is added to the mix. These projections were extremely exciting, predicting laser power generation scaling at $1/d$ and reaching levels in excess of 20 kW m^{-2} for inter-electrode spacings of ~2 mm, characteristic of optical waveguiding for 10 μm radiation. These projections were based on the assumption that the discharge is optimized in terms of the rf drive frequency and the achievement of efficient power transfer and uniform power deposition throughout the structure. The experimental data points shown are from measurements of multi-mode laser output from two different types of planar structure that confirm the power-scaling behaviour of large-area discharges. The first set was obtained with a discharge gap of 2.25 mm forming a one-dimensional waveguide with an electrode area of 66 cm^2, while the second was measured using an electrode pair with area 320 cm^2, and a gap of 7.5 mm operating as a free space resonator device [105].

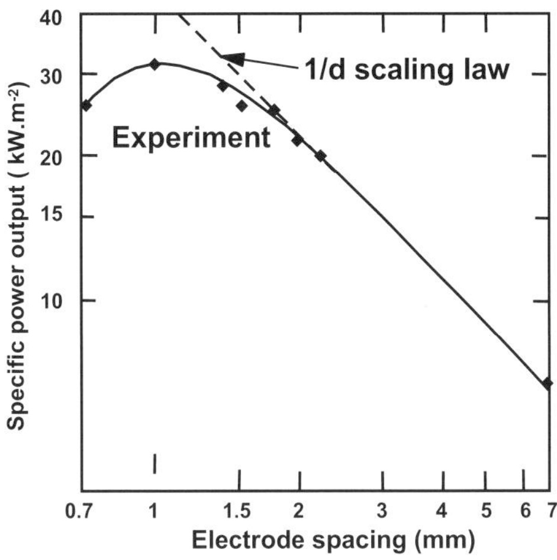

Figure B3.1.19. Dependence of specific laser output power (power per unit area of electrode (kW m^{-2})) on electrode spacing, demonstrating the $1/d$ scaling law. After [95].

The power scaling data for the more interesting small gap (waveguide) data are presented in greater detail in [94], along with experimental evidence for the key feature which is the transition from two-dimensional to one-dimensional discharge cooling. The data in figure B3.1.18 confirm that these power-scaling laws based on thermal conduction limits are, indeed, realistic and demonstrated quite dramatically the power-scaling opportunities available with this 2-D laser concept. The attainment [94] of 20 kW m^{-2} from a large-area discharge under conditions that are suitable for long-term sealed-off operation was a particularly exciting result, which heralded the subsequent development of a new generation of ultra-compact lasers with average powers in the multi-kilowatt range. Later research [95] further investigated the validity of the $(1/d)$ scaling law down to electrode gap sizes of 0.5 mm. This research showed that for values of d below about 1.5 mm, the power per unit value was seen to depart from the $1/d$ dependency and to show a peak value of 30 kW m^{-2} for a gap of $\sim$1.2 mm as shown in figure B3.1.19. The departure from $1/d$ can be expected because of the increased waveguiding losses of the machined aluminium alloy electrodes which scale as $\sim d^{-3}$ and increasing mirror–guide coupling losses, as well as an increased impact of ion sheath losses in the discharge for very small gaps [68].

The extraordinary potential benefits of the 'thin' laser idea for producing ultra-compact laser designs could perhaps be best realized by considering the input power density map, first presented at the 1988 GCL Conference, and shown in figure B3.1.20 [93, 96]. This map shows—as the diagonal band—the input power density which the laser gain medium can 'tolerate' while still maintaining high laser conversion efficiency (corresponding to an axial gas temperature <600 K), as a function of the gain medium thickness. Also shown for comparison are the 'allowed' input power density values that are representative of both (passive) diffusion-cooled non-waveguide lasers and (active) convectively cooled devices of the day. What emerges from considering figure B3.1.20 is that the planar waveguide laser technology is capable of operating, with high electrical-to-optical power conversion efficiency, even at power density levels which are more than 100 times greater than diffusion-cooled, conventional geometry lasers and 10–50 times higher than even the much more complex and bulky fast-flow convectively cooled lasers. Much more recently, it has been shown [102] that stable discharges at power density values to 1 kW cm^{-3} can be used to produce high peak powers in pulsed operation in the so-called USP mode (see later).

In the quest for high-*brightness* lasers, a challenge of equal importance to the development of these

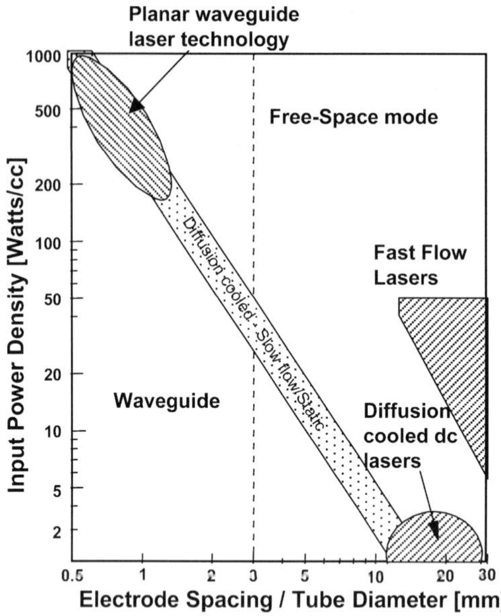

Figure B3.1.20. Dependence of discharge electrical power density for efficient laser power extraction on electrode spacing/tube diameter output power, indicating the ×10 advantage of rf planar waveguide lasers. After [93].

new diffusion-cooled gain medium structures was the design of effective optical resonators capable of combining diffraction-limited beam quality with high power extraction to produce high-brightness beams in ultra-compact devices. The benefits of the very high values of P_A that are associated with the use of narrow electrode gaps offer particular rewards for diffraction-limited resonators that can efficiently incorporate waveguide gain structures. The most suitable resonator so far developed for such planar waveguide gain structures was proposed by Jackson *et al* [97]. This is a *hybrid* design based on the use of classical waveguide resonators [35] for mode control in the narrow (transverse) dimension, and a confocal unstable resonator (see section A2.1.7) for beam control in the wide (lateral) direction. The resonator they proposed is illustrated in figure B3.1.21(*a*). It consists of a pair of spherical mirrors of radius R_1 and R_2 spaced by a distance, L, and arranged in a confocal setting with $R_1 - R_2 = 2L$. This design provides a magnification factor $M = R_1/R_2$ and a geometric output coupling fraction, $\delta = 1 - 1/M$. The concave (back) mirror R_1 completely 'covers' the discharge width, a_0, while the 'front' mirror R_2 has a diameter and position selected to project a distance, d_0, into the active discharge region such that the magnified edge-coupled output beam matches the discharge width, a_0 such that $Md_0 = a_0$. This resonator design permits the use of quite large radius of curvature mirrors (5 and −4.22 m) for the confocal unstable resonator to produce lateral beam control. Simultaneously, such reflectors behave as 'pseudo-planar' mirrors in a near-case I waveguide resonator design to produce low-loss waveguide–mirror coupling in the transverse direction. In the experiment, a stable α-type discharge at 125 MHz was maintained in a static 3:1:1 He, N_2 CO_2, gas mixture with 5% Xe between water-cooled electrodes (380 mm × 45 mm) separated by 22.25 mm. A maximum laser power of 240 W was obtained at a conversion efficiency of 12% and a value of specific power extraction, P_A of 14 kW m^{-2}. This reduced value of P_A is a consequence mostly of the reduced volume overlap in the unoptimized resonator. The *negative* branch hybrid waveguide-confocal unstable resonator shown in figure B3.1.21(*b*) has the advantage of a higher optical stability against the effects of mechanical misalignments, though the fill factor is usually lower. The x–y hybrid nature of the resonator design means that beam divergence in the X and Y directions are usually

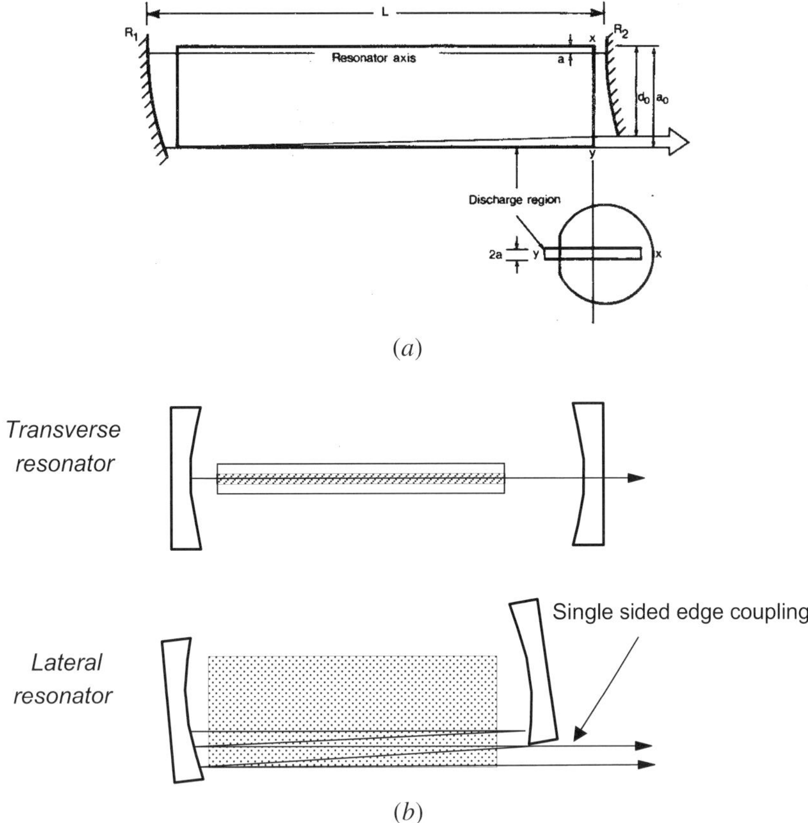

Figure B3.1.21. (*a*) Hybrid waveguide–positive-branch confocal unstable resonator. After [97]. (*b*) Hybrid resonator for a planar waveguide CO_2 laser waveguide in the transverse direction and negative-branch confocal unstable in the lateral direction.

not identical. Moreover, the laser near-field beam pattern shows small ripples deriving from diffraction at the edge of the coupling mirror. However, subsequent experiments showed that a simple extra-cavity optical transformation with spherical lenses or mirrors and the judicious positioning of a one-dimensional spatial filter produces a beam which is circular and has identical divergence in the X and Y directions. Optical losses as low as 5% are achievable using fixed passive optics and the net result is a near-perfect beam with values of $M^2 < 1.2$.

As mentioned already, the use of a positive-branch type of unstable resonator [97, 98] permits the use of large-radius mirrors, which are most convenient for rapidly planned experiments but which provided only a modest degree of waveguide mode discrimination. A further refinement of this resonator design involving the use of the negative-branch confocal unstable resonator has several useful advantages. First, as pointed out for example by Seigmann [99], the negative-branch resonator produces considerable enhanced optical beam pointing stability against resonator mirror misalignment due to thermal/vibration effects or other mechanical factors. In addition, it is possible to select radii of curvature that which satisfying the conditions for the confocal negative-branch design, simultaneously match the radius of curvature of the fundamental waveguide mode at a specified distance from the end of the discharge waveguide. Following this demonstration of planar waveguide sealed lasers at the 250 W level, area-scaling was applied to take the laser power to the

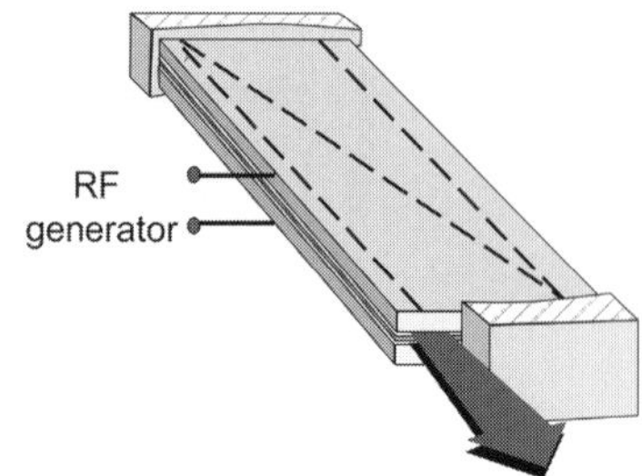

Figure B3.1.22. Photograph and schematic diagram of a sealed 1.1 kW CO_2 planar waveguide CO_2 laser (Heriot-Watt University). After [101].

1 kW as reported by Nowack *et al* [100] from the DLR lab in Stuttgart, and Colley *et al* [101] at Heriot-Watt University with the demonstration of 1.1 kW in the device shown in figure B3.1.22. At the same time, parallel development of commercial versions of these lasers occurred at Rofin Sinar GmbH in Hamburg, who aimed largely at the flat-bed cutting market with lasers at >1 kW, and Coherent Inc. in Hull, UK, and California, where focus was on compact sealed technology at power levels to 500 W. Considerable development has occurred in the ensuing years, with compact hard-sealed long-life planar waveguide lasers with all solid state power supplies being available at power levels up to >600 W for 'table-top' industrial and medical systems, while static gas, refillable planar waveguide lasers with power levels up to 6 kW being supplied into the material-processing industrial markets by Rofin Sinar GmbH. These lasers also feature excellent (near-diffraction-limited) beam quality and attractive modulation characteristics.

One final planar waveguide laser development is worthy of note. Although offering many advantages as power sources for industrial processing, two particular limitations of 'conventional' planar waveguide laser 'operating space' are evident for some applications. First, conventional planar waveguide lasers offer only a very modest value (of ~3) in the ratio of attainable peak-to-average power; second, pulsewidths are restricted to values larger than ~100 μs at full operating efficiency. However, some applications are best served with shorter pulses with higher peak power levels and this problem was addressed by Villarreal *et al* , who demonstrated the so-called 'ultra-super-pulse (USP)' mode of operation with peak power of 3.8 kW and pulses to 10 μs at kHz repetition rates, while maintaining high efficiency and a slight *increase* in average output power from 100–120 W in compact sealed devices [102]. Such lasers offer clear performance advantages for applications including microvia drilling for advanced PCB fabrication [119] (see also section D1.6.3.1) and rapid melting of metal surfaces for increased corrosion resistance [120].

B3.1.4.3.5 Annular discharge lasers

Annular gain medium lasers with radial rf discharges between coaxial electrodes offer the key advantage of large cross-sectional areas (large gain volumes) without the need for excessive inter-electrode distances that would otherwise necessitate low gas pressures. Thus coaxial lasers provide the opportunity to employ large electrode areas in compact (rolled-up planar) structures for efficient diffusion cooling, as well as utilizing cylindrical geometries that are convenient for manufacture. An early and prescient example of rf CO_2 laser design (one of the first rf CO_2 laser on record) was that of Crocker and Wills at SERL, UK who used an annular discharge to produce what appeared to be a very promising result in 1969 with a device at 200 W [47]. However, the design of high beam quality resonators with good azimuthal mode discrimination for the annular geometry, low misalignment sensitivity and efficient volumetric power extraction represented a significant challenge. Several approaches were investigated including toric resonators [103] and multipass resonators based on the Herriott cell optical folding system [104] to produce a single-mode Gaussian beam, though the discharge fractional utilization and corresponding value of P_A was quite low. The Herriott multi-fold system has also been applied to the *planar* electrode design, offering good beams at reasonable values of P_A [26, 105]. Subsequently a compact annular *waveguide* coaxial discharge laser was developed at the 500 W level [106] but such designs are seriously hampered by the extreme mechanical precision and tight tolerances demanded of annular waveguide structure and resonator mirrors, so that the full benefit of $1/d$ scaling that waveguiding requires cannot easily be achieved and, to date, it appears not to have been possible to develop practical resonators for annular *waveguide* devices that are capable of the high beam quality necessary for modern lasers.

In the more tractable non-waveguide stable resonator regime, promising designs of transverse rf excited annular laser construction at the kilowatt level were demonstrated at ILT, Aachen, in Germany [107]. This led to the design and demonstration of a novel hybrid resonator that is stable in the radial direction and unstable in the azimuthal direction that uses a specially manufactured helical mirror. With this resonator, near-diffraction-limited beam output was obtained at power levels of 2 kW and 10% efficiency from a coaxial gain length of 1.8 m [108].

B3.1.4.4 Convectively cooled rf-excited free-space rf lasers

As mentioned earlier, the exploitation of transverse rf discharges for innovation in diffusion-cooled CO_2 laser technologies in the 1980s was parallelled by important new developments in convectively cooled (fast-flow) lasers. The main driver for research in combining fast-flow gas cooling techniques with transverse rf discharges was the relatively buoyant market for multi-kilowatt lasers required for an increasing range of industrial material-processing applications. Although dominated numerically by metal-cutting applications, the development of higher power lasers (>5 kW) has resulted in a steady increase in the use of industrial CO_2 lasers for welding and, to a lesser extent, for the surface treatment of metals (see section D1, Materials processing). As workpiece handling and system control hardware for laser-cutting systems have improved and competition intensified, the industrial market has demanded ever higher processing speeds and better finished product quality. This has driven industrial laser manufacturers to seek ever higher average laser powers to match required production rates, combined with better optical quality beams in ever more reliable and low-cost laser systems. For instance, whereas in the past lasers for cutting typically operated at the 1.0–1.5 kW range, modern requirements, for example, for fine cutting applications are for lasers at 5 kW and above with near-diffraction-limited beam quality.

By 1980, fast-flow dc /ac lasers had been under development for about ten years, with axial-flow lasers meeting many industrial laser needs at power levels of up to ~5 kW. However, as previously mentioned, rf discharges are stable at higher values of power density and gas pressure than dc discharges. In addition, they can be modulated at higher frequencies and operated with electrodes external to the gas discharge vessel, thereby avoiding electrode sputtering reducing gas consumption. dc excited lasers with *transverse* flow are often capable of higher average power levels than axial devices but are usually viewed as being technically more complex, often with inferior beam quality and lower levels of reliability. Today, after 20 years or so of further development, the situation is not dissimilar. Partly because rf technology is highly compatible with longitudinal flow structures, axial fast-flow lasers now dominate the industrial laser scene and power outputs from folded cylindrical designs have kept pace with industrial market demand for power levels up to ~20 kW.

B3.1.4.4.1 Fast axial flow (FAF) transverse rf excited lasers

Most fast-axial-flow rf discharge excited lasers are based on the use of metal/dielectric electrodes, where the metal electrode is separated from the discharge by a sheet of dielectric material as illustrated in figures B3.1.23(*a*) and (*b*) which shows the version of this concept that is commonly used in rf-excited FAF lasers [70]. The rf electric field is applied to the metal electrodes and there is a voltage drop that is pro-

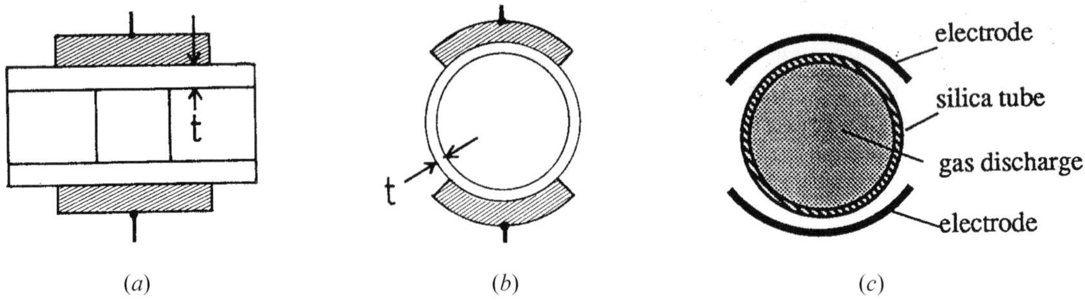

Figure B3.1.23. Schematic cross section of electrode system for fast axial flow lasers, showing reactive ballast of dielectric layers: (*a*) planar structure, (*b*) glass tube structure and (*c*) illustration of optimized metal electrode shape.

portional to the current density across the capacitative dielectric layer. As previously mentioned, the use of dielectric sheets in series with the discharge acts to supplement the stabilizing influence of the internal rf discharge sheaths [65] and to increase the power density threshold for the onset of α-to-γ instabilities. This added stabilizing effect of the dielectric layers is particularly important given that tube diameters of several centimetres are called for (values of ~2–8 cm are common) to achieve effective gas flow and such inter-electrode distances are associated with *low* frequency operation for optimized operation, where α-to-γ transition instabilities are more probable (see figure B3.1.13 and [68]). The use of series dielectric ballast permits the use of desirable industrial frequencies at 13.56 MHz or 27.12 MHz, and these are the two most commonly used generator frequencies for fast-flow lasers. The impact of the selected frequency, f, (and material) on the effectiveness of the dielectric ballast can also be seen from the voltage drop across the glass wall ballast in figure B3.1.23(*b*), which is given by $V_b = 2tJ/\varepsilon\varepsilon_0\omega$ where t is the dielectric thickness and J is the current density [93]. For a 1 mm wall thickness in a typical FAF laser, the voltage drop across the glass walls is ~2000 V at 13.6 MHz, compared with only ~30 V at 126 MHz. Thus, for the low-frequency case, the ballast voltage is significantly larger than the discharge voltage which helps to inhibit the onset of gamma discharges.

However, in order to maintain the high level of transverse homogeneity of the rf discharge needed to produce high beam quality, more complex electrodes designs than the simple matched-diameter cylindrical shape shown in figure B3.1.23(*b*) are necessary. Viewed from the tube end, the discharge with electrodes that are shaped and positioned such as in figure B3.1.23(*b*) shows evidence of the expected electrode sheath regions (of thickness ~a few mm) close to to each electrode. There is also clear evidence of localization of the transverse current density in the highest field regions corresponding to the shortest inter-electrode distances. Electrode optimization using electric field calculations and experiments [109] have produced shapes such as that shown schematically in figure B3.1.23(*c*) which have been shown to produce very uniform visible luminosity across the tube diameter at right angles to the field direction [110]. For the (usual) case of multiple discharges that are optically in series, residual non-uniformities can be further reduced by rotating the electrode pairs by an appropriate angle about the tube axis. In practice, proper attention to these issues enables excellent phase homogeneity in the gain medium and permits circularly symmetric high-quality laser beams to be obtained. However, there is a remaining problem that the spatially averaged gain profile may have a minimum value on axis which makes the achievement of TEM_{00} mode operation rather difficult. Such lasers often are designed to operate in the TEM_{01} 'doughnut' mode.

One has also to be concerned with non-uniformities in the longitudinal, axial flow direction since as the gas proceeds through the discharge region (usually some tens of cm in length) its temperature rises, resulting in a density (and discharge impedance) gradient in the direction of the flow. Thus, assuming a constant inter-electrode electric field, the transverse current density will vary in the axial direction if the

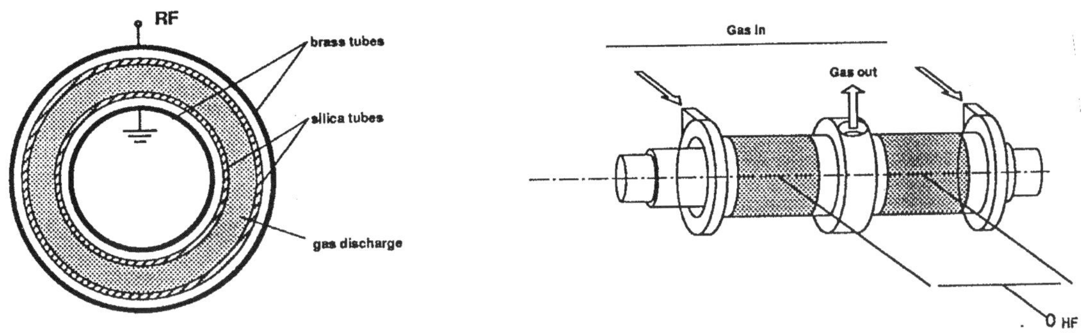

Figure B3.1.24. (*a*) Coaxial electrode structure and annular rf discharge. (*b*) Schematic diagram of a two-section coaxial fast flow laser. After [114].

electrode spacing is constant along the tube. This effect may be countered, for example, by tilting the electrodes so that the separation increases appropriately in the direction of the flow [111]. With the design refinements mentioned, rf excitation of FAF lasers can operate very effectively. For rf lasers this is true even without the highly turbulent flow and consequent high values of pressure drop that are characteristic of dc lasers, so that turbo-blowers can be substituted for Rootes pumps allowing the design of very compact lasers with integrated blowers [112]. Nevertheless, it is necessary to fully understand power deposition uniformity issues, including the effects of generator frequency, gas conductivity and gas flow conditions since, as with all discharges, hot spots can occur and lead to thermal instabilities, discharge filamentation and destructive arcs [65]. The demand for higher power levels has stimulated a need to understand power scaling in terms of gas pressure, tube length and diameter and electrical issues where the limit is usually the onset of discharge instabilities [113]. Considerable effort has been devoted to theoretical and experimental research on discharges and gas flow systems for the optimization of spatial uniformity of gain in FAF rf lasers, for example using the CARS techniques [110]. The resulting technology combining rf excitation and FAF gas cooling can provide an excellent 'building block' for scaling to very high powers in a modular fashion. As with diffusion-cooled lasers, such scaling is multi-stage involving multiple (as many as 16) discharge sections and optically folded resonators.

Since heat is removed by convection, a second type of power scaling is available which seeks to extract higher powers per unit length of discharge by using larger tube diameters ($\sim$65 mm) and higher gas pressures (100 mbar). This approach exploits the flow conditioning and discharge homogenization techniques developed by modelling and experiment [114], and leads to designs that are capable of average power levels to $>$20 kW, [115, 116]. In one experiment at ILT Aachen, a large-diameter high-pressure laser was constructed with eight-discharge sections that incorporated four radial compressors for gas flow with a volume rate of 3300 m^3 hr^{-1}. With a power input of 160 kW from a pair of 100 kW generators operating at 27.12 MHz, a laser power of $>$25 kW was obtained, with the expectation that power could be scaled further towards 40 kW [114].

The need for process quality and competition in the industrial laser sector means that near-perfect beam quality is an indispensable requirement in modern industrial lasers. The key to achieving this for FAF rf lasers is the achievement of a high degree of optical homogeneity in the gain medium by attention to optimized electrode design and flow conditioning. Failure to do this results in strong negative lensing and large aberration effects and laser beam quality that is inadequate for modern lasers. However, with an optimized electrode configuration, only small negative lens effects remain due to transverse heat conduction gradients and, for normal operation, their influence on beam quality is quite low [117]. Optical transient effects may also occur.

For example, transient statistical phase distortions associated with turbulent flow in the medium may cause optical effects which may have to be taken into account. Nevertheless, a high degree of uniformity of the gain medium refractive index can be achieved, excellent laser beam quality at multi-kilowatt power has been demonstrated using either stable or unstable cavities and folded designs depending on the overall laser length and aspect ratio. This type of FAF laser technology is available from several companies including Trumpf GmbH and Rofin Sinar GmbH.

As has been shown for diffusion-cooled lasers, an alternative to discharge-length scaling as a route to very high power lasers is area scaling and the annular rf discharge with coaxial electrodes offers a very promising route to very high average power lasers since it offers a large effective cross section for gas flow coupled with a relatively small gap, high gas pressure and, therefore, high mass flow, as illustrated in figure B3.1.24 [114]. This system uses two silica tubes as dielectric layers, with a grounded inner metal electrode and an outer metal cylinder to which the rf is applied, to provide a discharge with inner and outer diameters of 95 and 117 mm, respectively, and a total discharge length (two sections) of 0.4 m. A Rootes blower with a 1700 m^2 hr^{-1} maintain the fast gas flow, and a hybrid stable azimuthal unstable resonator was employed [109]. In initial experiments, more than 2 kW of laser power was obtained at $\sim$10% efficiency with a high optical quality beam which, after extra-cavity beam transforming optics, was measured as $M^2 < 1.2$.

B3.1.4.4.2 Transverse flow rf lasers

In the renaissance period of rf laser development since 1980, there has been considerable work on transverse-flow rf excited lasers. The main thrust of much of this work has concentrated on the integration of the requirements of transverse rf discharges with the technology for handling high-speed gas flow in the transverse direction, gas recirculation and cooling. Although transverse systems are superficially attractive in providing very short transit distances through the active discharge medium, as mentioned earlier the main (intrinsic) deficiency relates to the transverse density (refractive index) gradients which have deleterious effects on the optical beam quality and pointing stability. Although these effects may be compensated to some degree for cw lasers, for repetitively pulsed or modulated lasers which are required for some applications, the transient moving density gradients produce non-stationary phase-front distortions and steering which it is not possible to correct. Nevertheless, the use of rf technology permits significantly increased discharge power density ≤ 30 W cm^{-3} for pseudo-cw excitation within and up to 100 W cm^{-3} for 'super-pulsed' operation [118]. Such lasers have been demonstrated at power levels of >10 kW but the limitations in terms of achievable beam quality limit the use of these lasers to some of those uses (e.g. some welding and heat treatment applications) where imperfect beam quality can be sometimes tolerated. The relative decline in interest in transverse-flow laser technologies has been particularly accentuated by the recent development of highly compact and lower cost FAF technologies at power levels to >25 kW with superior beam quality. The latter have become highly competitive in the commercial market for high-power CO_2 lasers.

References

[1] Patel C K N 1964 *Phys. Rev. Lett.* **13** 617–19

[2] Witteman W J 1986 *The CO_2 Laser (Springer Verlag Series in Optical Sciences)* (Berlin: Springer)

[3] Cheo P K 1971 CO_2 lasers *Advance in Lasers* vol 3, ed A K Levine and A J De Maria (New York: Dekker) ch 2

[4] Sobelev N N and Sokovikov V V 1967 *Sov. Phys.–Usp.* **10** 153

[5] Willett C S 1974 *An Introduction to Gas Lasers: Population Inversion Mechanisms* (Oxford: Pergamon)

[6] Tyte D C 1970 Carbon dioxide lasers *Advances in Quantum Electronics* vol 1, ed D W Goodwin (London: Academic)

[7] De Maria A J 1976 Review of high power CO_2 lasers *Principles of Laser Plasmas* ed G Bekefi (New York: Wiley) ch 8

[8] Hall D R and Hill C A 1987 Radiofrequency discharge excited CO_2 lasers *Handbook of Molecular Lasers* ed P Cheo (New York: Dekker) ch 3

[9] Herzberg G 1964 *Molecular Spectra and Molecular Structure* vol 2 (Princeton, NJ: Van Nostrand)

[10] Reid J and Siemson K 1976 *Appl. Phys. Lett.* **29** 250

[11] Schultz G J 1959 *Phys. Rev.* **116** 1141

[12] Hake R D and Phelps A V 1967 *Phys. Rev.* **158** 70
[13] Boness M J W and Schultz G J 1968 *Phys. Rev. Lett.* **21** 1031
[14] Moeller G and Rigden G D 1965 *Appl. Phys. Lett.* **7** 274
[15] Patel C K N 1964 *Phys. Rev.* A **136** 1187
[16] Clark P O and Wada J Y 1968 *IEEE J. Quantum Electron.* **QE-4** 263
[17] He D and Hall D R 1984 *J. Appl. Phys.* **56** 856
[18] Vidaud P and Hall D R 1985 *J. Appl. Phys.* **57** 1757
[19] Novgorodov M Z, Sviridov A G and Sobolev N N 1971 *IEEE J. Quantum Electron.* **QE-7** 508
[20] Willett C S 1974 *An Introduction to Gas Lasers: Population Inversion Mechanisms* (Oxford: Pergamon) p 280
[21] Nighan W L and Bennett J H 1969 *Appl. Phys. Lett.* **14** 240
[22] Vidaud P, He D and Hall D 1985 *Opt. Commun.* **56** 185
[23] Abrams R L and Bridges W B 1973 *IEEE J. Quantum Electron.* **QE-9** 940
[24] Vitruk P, Baker H J and Hall D R 1994 *IEEE J. Quantum Electron.* **QE-30** 1623
[25] Hochuli U and Sciacca T P 1974 *IEEE J. Quantum Electron.* **QE-1** 239
[26] Baker H J 1989 Multifold laser resonators *The Physics and Technology of Laser Resonators* ed D R Hall and P E Jackson (Bristol: Adam Hilger) ch 5
[27] Miles P A and Lotus J W 1968 *IEEE J. Quantum Electron.* **QE-4** 811
[28] Tiffany W B, Targ T R and Foster J D 1969 *Appl. Phys. Lett.* **15** 91
[29] Beaulieu A J 1970 *Appl. Phys. Lett.* **16** 504
[30] Dumanshin R and Rocca-Serra J 1970 *C. R. Acad. Sci.* **269** 916
[31] Lamberton H M and Pearson P R 1971 *Electron. Lett.* **7** 141
[32] Brown R T 1987 CO_2 TEA Lasers in *Handbook of Molecular Lasers* ed P Cheo (New York: Dekker) ch 2
[33] Figueira J 1987 High energy short-pulse CO_2 lasers *Handbook of Molecular Lasers* ed P Cheo (New York: Dekker) ch 4
[34] Marcatili E A J and Schmeltzer R A 1964 *Bell. Syst. Tech. J.* **43** 1783
[35] Degnan J J and Hall D R 1973 *IEEE J. Quantum Electron.* **QE-8** 901
[36] Degnan J J 1976 *Appl. Phys. Lett.* **11** 1
[37] Hill C A 1989 Theory of waveguide laser resonators *The Physics and Technology of Laser Resonators* ed D R Hall and P E Jackson (Bristol: Adam Hilger) ch 3
[38] Hall D R, Peruso C, Johnson E, Schiffner G, McElroy J H and McAvoy N 1972 *Opt. Eng.* **11** 77
[39] Smith P W 1971 *Appl. Phys. Lett.* **19** 132
[40] Bridges T J, Burkhardt E G and Smith P W 1972 *Appl. Phys. Lett.* **20** 403
[41] Abrams R L 1974 *Appl. Phys. Lett.* **25** 304
[42] Burkhardt E G, Bridges T J and Smith P W 1972 *Opt. Commun.* **6** 193
[43] Hall D R, Jenkins R M and Gorton E K 1978 *J. Phys. D: Appl. Phys.* **11** 859
[44] Thomson J J 1891 *Phil. Mag.* **32** 321
[45] Javan A, Bennett W R Jr and Herriott D R 1961 *Phys. Rev.* **6** 106
[46] Patel C K N 1964 *Phys. Rev.* A **136** 1187
[47] Crocker A and Wills M S 1969 *Electron. Lett.* **5** 33
[48] Brown C O and Davis J W 1972 *Appl. Phys. Lett.* **21** 480
[49] Laakmann K D 1978 *Proc. Lasers '78 Conf. (Orlando, FL)*
[50] Laakmann K D 1979 *Proc. Lasers '79 Conf. (Orlando, FL)* pp 741–3
[51] Sutter L V 1980 *Proc. SPIE 227 (Washington, DC)*
[52] Griffiths G A 1980 *Proc. SPIE 227 (Washington, DC)* pp 6–11
[53] Chenausky P P 1980 Paper TUKK4 *Conf. on Lasers and Electro-optic Systems (San Diego, CA)*
[54] Chenausky P P, Drinkwater E H and Laughman L 1982 US Patent 4,363,126
[55] Allcock G and Hall D R 1981 *Opt. Commun.* **37** 49
[56] He D and Hall D R 1983 *Appl. Phys. Lett.* **43** 726
[57] Lachambre J L, MacFarlane J, Otis G and Lavigne P 1978 *Appl. Phys. Lett.* **32** 652
[58] Raizer Y P 1979 *Sov. J. Plasma Phys.* **5** 232
[59] Myshenkov V I and Yatsenko N A 1981 *Sov. J. Quantum Electron.* **11** 1297
[60] Yatsenko N A 1982 *Sov. Phys. Tech. Phys.* **27** 741
[61] Levitskii S M 1958 *Sov. Phys. Tech. Phys.* **2** 887
[62] Vidaud P, Durrani S M and Hall D R 1988 *J. Phys. D: Appl. Phys.* **21** 57
[63] He D, Baker C J and Hall D R 1984 *J. Appl. Phys.* **55** 4120
[64] Kolesnychenko F Yu, Matyukhin V D, Murv'ev V F and Smaznov S I 1979 *Dokl. Akad. Nauk. SSSR* **246** 1091
[65] Wester R, Siewert S and Wagner R 1991 *J. Phys. D: Appl. Phys.* **24** 1796
[66] Vitruk P, Baker H J and Hall D R 1992 *J. Phys. D: Appl. Phys.* **25** 1767
[67] Smirnov A S 1984 *Sov. Phys.–Tech. Phys.* **29** 34
[68] Vitruk P, Baker H J and Hall D R 1994 *IEEE J. Quantum Electron.* **QE-30** 1623
[69] Baker C J, Hall D R and Davies A R 1984 *J. Phys. D: Appl. Phys.* **17** 1597

[70] Wester R, Herziger G and Schulke H 1987 *SPIE* **81** 14
[71] Durrani S M A 1988 Optimization of rf discharges for CO_2 laser excitation *PhD Thesis* (Edinburgh: Heriot Watt University)
[72] Konyukhov V K 1971 *Sov. Phys.—Tech. Phys.* **15** 1283
[73] Muehe C E 1974 *J. Appl. Phys.* **45** 82
[74] He D and Hall D R 1983 *J. Appl. Phys.* **54** 4367
[75] He D and Hall D R 1984 *IEEE J. Quantum Electron.* **QE-20** 509
[76] Siemson K J 1980 *Appl. Opt.* **19** 818
[77] He D and Hall D R 1984 *J. Appl. Phys.* **56** 856
[78] Vidaud P and Hall D R 1985 *J. Appl. Phys.* **57** 1757
[79] Vidaud P and Hall D R 1985 *Opt. Commun.* **56** 185
[80] Jackson P E, Hall D R and Hill C A 1989 *Appl. Opt.* **28** 935–41
[81] Newman L A and Hart R A 1987 *Laser Focus* **18** 80
[82] Spacht D P and DeMaria A J 2001 *Laser Focus World* **37** 397
[83] Horrigan F 1969 *Microwaves* **8** 68
[84] Corcoran V J and Crabbe I A 1974 *Appl. Opt.* **13** 1755
[85] Youmans D G 1984 *Appl. Phys. Lett.* **44** 365
[86] Newman L A, Hart R A, Kennedy J T, Cantor A J, DeMaria A J and Bridges W B 1986 *Appl. Phys. Lett.* **48** 1701
[87] Abramski K, Colley A D, Baker H J and Hall D R 1992 *Opt. Commun.* **90** 61
[88] Abramski K, Colley A D, Baker H J and Hall D R 1992 *Appl. Phys. Lett.* **60** 530
[89] Baker H J, Hornby A and Hall D R 1993 *Appl. Phys. Lett.* **63** 2591
[90] Abramski K M, Colley A D, Baker H J and Hall D R 1991 *IEEE J. Quantum Electron.* **QE-27** 1939
[91] Abramski K M, Baker H J, Colley A D and Hall D R 1996 *IEEE J. Quantum Electron.* **32** 304
[92] Villarreal F, Baker H J, Abram R, Jones D and Hall D R 1999 *IEEE J. Quantum Electron.* **QE-35** 267
[93] Hall D R and Baker H J 1988 *Proc. Gas Flow and Chemical Laser Symposium (Vienna) (Proc SPIE 1031)* p 60
[94] Abramski K M, Colley A D, Baker H J and Hall D R 1989 *Appl. Phys. Lett.* **54** 1833
[95] Shackleton C, Baker H J and Hall D R 1992 *Opt. Commun.* **89** 423
[96] Hall D R and Baker H J 1989 *Laser Focus World* **25** 77
[97] Jackson P E, Baker H J and Hall D R 1989 *Appl. Phys. Lett.* **54** 1950
[98] Tulip J 1988 US Patent 4,719,639
[99] Siegman A E 1986 *Lasers* (Oxford: Oxford University Press) ch 22
[100] Nowack R, Opower H, Schaefer U and Wessel K 1990 *Proc. SPIE* **1276** 18
[101] Colley A D, Baker H J and Hall D R 1992 *Appl. Phys. Lett.* **61** 136
[102] Villarreal F, Murray P, Baker H J and Hall D R 2001 *Appl. Phys. Lett.* **78** 2276
[103] Ferguson T R 1984 *Appl. Opt.* **23** 2122
[104] Xin J G and Hall D R 1987 *Appl. Phys. Lett.* **51** 469–71
[105] Zhang X, Baker H J and Hall D R 1993 *J. Phys. D: Appl. Phys.* **26** 359
[106] Bethel J W, Baker H J and Hall D R 1998 *Opt. Commun.* **125** 352
[107] Erlichmann D, Habich U and Plum H-D 1993 *J. Phys. D: Appl. Phys.* **26** 183
[108] Erlichmann D, Habich U, Plum H-D, Loosen P and Herziger G 1994 *IEEE J. Quantum Electron.* **30** 1441
[109] Schwede H 1989 *Proc. 9th Int. Laser Congress (Munich)* (Berlin: Springer)
[110] Loosen P 1992 *Proc. Gas and Chemical Laser Conf. (Crete) (SPIE 1810)* p 26
[111] Hage H, Martinen H and Northemann T 1987 *Proc. High Power Lasers Conf. (The Hague) (Proc. SPIE 801)* p 58
[112] Beck R 1987 *Appl. Phys.* B **42** 233
[113] Loosen P and Wester R 1995 *J. Phys. D: Appl. Phys.* **28** 849
[114] Habich U, Du K, Erlichmann D, Jarosch U, Niehoff J, Plum H-D, Meyer R, Wolf N, Loosen P, Beck T, Hertzler C and Wollermann-Windgasse R 1994 *Proc. Gas and Chemical Laser Conf. (Friedrichshafen)*
[115] Habich U, Jarosch U, Maly H, Meyer R, Loosen P, Wolf N, Beck T, Hertzler C and Wollermann-Windgasse R 1996 *Proc. Gas and Chemical Laser Conf. (Edinburgh) (SPIE)*
[116] Habich U, Loosen P, Hertzler C and Wollermann-Windgasse R 1996 *Proc. SPIE* **2702** 374
[117] Loosen P 1992 *Proc. Gas and Chemical Laser Conf. (Madrid) (SPIE 1810)* p 26
[118] Schock W, Hall T, Wildermuth E, Wessel K, Gehringer E, Schnee P and Hadinger F 1988 *Proc. Gas and Chemical Laser Conf. (Vienna) (SPIE 1031)* p 76
[119] Baker H J, Villarreal F J and Hall D R 2002 *The 3rd Int. Symp. on Laser Precision Microfabrication LPM2002 (Osaka)*
[120] Tsagkarakis M, Villarreal F J, Baker H J and Hall D R 2003 *J. Laser Appl.* in press
[121] Anderson J D 1976 *Gas Dynamic Lasers: An Introduction* (New York: Academic)
[122] Fenstermacher C A, Nutter M J, Leyland K and Boyer K 1972 *Appl. Phys. Lett.* **20** 56
[123] Dyson D J 1989 Resonators for high power CO_2 lasers *The Physics and Technology of Laser Resonators* ed D R Hall and P E Jackson (Bristol: Adam Hilger) ch 13

B3.2
Excimer, F_2, N_2 and H_2 lasers

W J Witteman

B3.2.1 Introduction

In the field of laser development, researchers have always been challenged to move to the shorter wavelength part of the spectrum. In the period shortly after the invention of the laser, research activities were mainly concentrated on systems operating in the visible and near-infrared part of the spectrum. This choice of spectrum is understandable if we look at the required instrumental conditions: in particular, the optical components and technology for the required input powers were already available. The appropriate excitation method and input power could be realized at that time with classical instruments available in any well-equipped laboratory. Futhermore, the excitation power for reaching the laser threshold decreases rapidly with longer wavelength. Some near-infrared lasers like the CO_2 and Nd:YAG lasers could, therefore, be developed energetically as mature high-power instruments for large-scale applications in roughly one decade. Being so successful and driven by potential new applications, similar systems in the UV, VUV and even XUV were desired.

However, world-wide research efforts to unlock UV and VUV laser systems revealed the limitations of the available suitable materials, the complexity of excitation technologies and, last but not least, the threshold conditions. Up to now, the availability of these lasers has been rather limited. This is partly related to the required low absorption of the laser medium and optical components in this part of the spectrum. Nearly all solids and liquids strongly absorb VUV radiation. Fortunately, the situation for some selected gases is much more favourable.

Another limiting factor is the required high input power of those lasers. This can be understood by the following analysis. The cross section for stimulated emission σ_{st} is described by

$$\sigma_{st} = \frac{c^2}{8\pi\nu^2} A g(\nu) \qquad \text{(B3.2.1)}$$

where c is the speed of light, ν the transition frequency, A the decay rate of the spontaneous emission and $g(\nu)$ the lineshape function of the transition. To obtain laser action the small signal gain α_0 should exceed the minimum required threshold gain α_{th} which is equal to the losses in the resonator (mainly mirror and medium absorption). Thus we have the condition

$$\alpha_0 \geq \alpha_{th} = \sigma_{st} \bigtriangleup N_{th} \qquad \text{(B3.2.2)}$$

where $\bigtriangleup N_{st}$ is the threshold inversion. For each photon the input power has to produce at least one excited state with energy exceeding that of the transition. Let us assume, for the moment very conservatively, an excitation efficiency of 100%. The minimum input power P_{th} to produce the threshold inversion density is then

$$P_{th} > h\nu A \bigtriangleup N_{th}. \qquad \text{(B3.2.3)}$$

Substituting equations (B3.2.1) and (B3.2.3) into (B3.2.2) we have the condition

$$P_{\mathrm{th}} > \frac{8\pi h\nu^3\alpha_{\mathrm{th}}}{c^2 g(\nu)} \tag{B3.2.4}$$

Usually the lineshape $g(\nu)$ is expressed in terms of the linewidth $\Delta\lambda$ of the transition with the relation $g(\nu) = \lambda^2/c\,\Delta\,\lambda$ so that equation (B3.2.4) becomes

$$P_{\mathrm{th}} > \frac{8\pi hc^2\alpha_{\mathrm{th}}\,\Delta\,\lambda}{\lambda^5}. \tag{B3.2.5}$$

It can be seen that the minimum required input power density increases very rapidly with decreasing wavelength and it does not depend on the spontaneous decay rate of the transition. In most VUV lasers the values for cavity losses and $\Delta\lambda/\lambda$ are higher—sometimes by several orders of magnitude—than for visible or infrared lasers. But even assuming comparable values for these quantities the power density increases with the fourth power of the wavelength ratio of the compared systems. For example, at 100 nm a threshold gain of 1% per cm and a linewidth of $\Delta\lambda/\lambda = 10^{-2}$, we need at least 36 kW cm^{-3}. Usually the excitation efficiency is not 100% but a small fraction of it so that the minimum input power must be considered at least an order of magnitude higher. Thus, it is not the fast decay rates of the upper laser transitions but, in most cases, the broad line profiles and short wavelengths that are the principal reasons for very high input powers. These high powers are not realistic for continuous operation and can only be reached in pulsed systems.

For VUV lasers with very short lifetimes in the upper laser level, of the order of 1 ns or less, an additional complication arises if the system is self-terminating, i.e. the lifetime of the lower laser level is larger than that of the upper level. For such systems time-independent spatial homogeneity of the excitation pulse, even at the required high density, is catastrophic. Gain is only present for a period equal to the lifetime of the upper state. Stimulated emission is, therefore, only feasible if a high-power excitation pulse moves with the speed of the generated stimulated emission. The initially produced spontaneous emission will then be amplified by the generated gain and this stimulated emission thus follows the propagating excitation pulse. The pulse duration is given by the lifetime of the upper state. The output power is amplified spontaneous emission (ASE). The molecular hydrogen laser is a typical example.

Successful pulsed operation of gas systems is based either on high-energy beams that penetrate the laser gas or a self-sustained glow discharge. In special cases a combination of both techniques turns out to be advantageous.

As mentioned already, the transparency for UV and VUV is often another serious handicap even for reaching threshold. Although there have been successful attempts at purifying suitable materials, this has so far only contributed to the improvement of such optical components as mirrors. For gases the situation is more promising. The absorption can often be circumvented or reduced to an acceptable level when the composition, density and purity are well chosen. This, together with the high potential of discharge technology, are the main reasons why short-wavelength gas lasers have been successfully developed.

Of all known short wavelengh systems the excimer lasers can be considered, with respect to output power and efficiency, as the champions. They represent an interesting and important class of molecular lasers involving electronic transitions between an excited bound state and an unbound or weakly bound ground state. They use rare gas molecules like Xe_2 (172 nm), Ar_2 (126 nm) and Kr_2 (146 nm) or, more successfully, rare gas halogen molecules, for example ArF. This is schematically shown in figure B3.2.1 with the potential energy curves for the upper and repulsive lower state of a diatomic molecule. The upper potential curve has a minimum so that the excited state exists, i.e. a vibrating molecule in an electronic state. Decaying to the lower state the molecule dissociates immediately because of the repulsive potential. For the vibrating molecule there is a broad band of transitions with different energies. The attractive feature for laser action is the absence of the lower level so that absorption from the ground state is absent. The system does not saturate by the laser process itself. Such systems are, in principle, good candidates for high output powers.

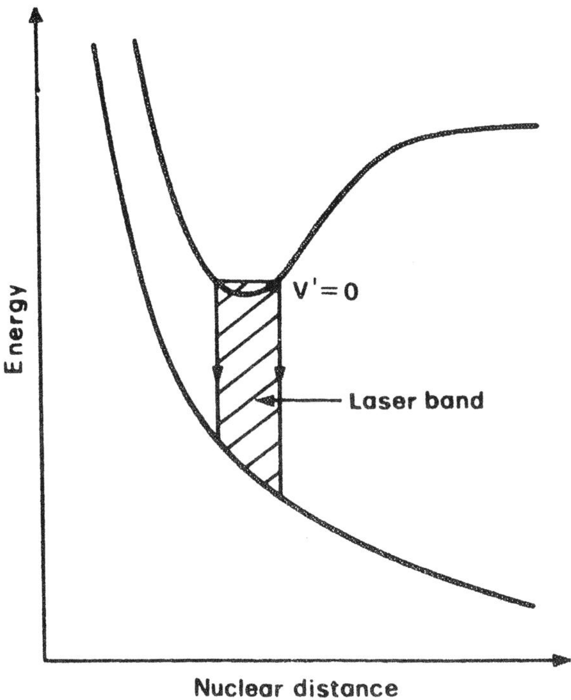

Figure B3.2.1. Energy states of an excimer molecule.

Electronic transitions with an unbound lower state have been known for decades in the field of gaseous electronics and spectroscopy of high-pressure rare gases. Gas discharges in multi-atmospheric rare gases producing excited molecules like He_2, Ne_2 and Ar_2 are often used as high-intensity UV light sources. Since these molecules exist only in an excited electronic state, they are called excimers, a contraction from excited dimer. Later on, especially when other gases were studied as candidates for UV lasers, it was found that rare gas and halogen mixtures show similar behaviour with an unbound ground state. For that reason they were also called excimers although, strictly speaking, we are not dealing with a dimer because they are composed from different atoms.

B3.2.2 Historical perspectives

The broadband emission of excimer molecules was already known around 1930 when the UV emission of He_2 was described [1]. Much later, other rare-gas molecules like Ar_2, Kr_2 and Xe_2 as well as metal vapours [3] like Hg_2 and Cd_2 were reported [2]. Rare gas halogen compounds were probably observed for the first time around 1974 when the quenching of metastable states of rare gases was studied [4, 5]. The idea to select excimers as an attractive medium for laser action because of the absence of a ground level came shortly after the invention of the laser [6]. However, at that time the instrumental developments, i.e. the technology for high-pressure discharges to fulfil the laser conditions, were inadequate to realize excimer lasers. It took more than ten years before the first excimer laser was reported [7]. It used stimulated emission from Xe_2 molecules obtained by electron-beam excitation of cryogenically cooled xenon.

The first experimental studies with excimer lasers go back to 1972 when a high-pressure xenon gas was excited by an electron beam [8]. Laser action in rare gas halogen excimers was reported for the first time in 1975 for XeBr and XeF [9, 10]. Shortly after the first realizations of rare gas halogen excimer lasers all possible combinations of rare gases and halogens were thoroughly investigated as laser candidates. Practically all combinations were successful, although some of them were superior with respect to output power and efficiency. This is especially true for XeCl, KrF, ArF and XeF.

It became clear at that time that, in general, the laser physics of excimers was understood to a great extent but not the technology for realizing good performing systems. Further development was, therefore, seriously hampered by the absence of the technical knowledge to realize stable discharges in large volumes of multi-atmospheric rare gas halogen mixtures. During the subsequent decade there was a worldwide effort to find new technologies to improve discharge performance. This was also stimulated by the awareness of several potential applications of high-power UV lasers in industry, especially in the field of material processing and lithography. The found solutions were promising and it even became a challenge to build very large systems with average output powers in the range of kilowatts. In order to pay for the costly developments, international collaboration was then strongly stimulated by national governments. This approach has, for instance, been very successful for the kilowatt XeCl laser. In a sense, this laser development has also had a strong technological push on the progress in high-density gas discharges containing species with negative ion formation.

B3.2.3 Excitation technique

Pulsed excitation technology for excimer lasers has developed along two different paths. The first, electron-beam (e-beam) excitation, which can be successfully applied to all systems, is based on direct excitation with relativistic electrons. It offers the advantage of uniformly depositing large amounts of energy into any high-density gas mixture. The attractive feature is the possibility of separately studying the input energy transfer and the laser kinetic process for various gas compositions. Studies with this excitation technique have, therefore, contributed tremendously to a better understanding of excimer formation and its lasing potential. Moreover, this technique has been developed successfully to reach the highest output powers and efficiencies for excimer systems. The disadvantages, however, are its complexity, high costs and limited repetition rate.

The other more practical excitation technology is based on the self-sustained discharge of a laser gas. To limit the applied voltage for high-density gases, transverse discharge geometries were chosen. Large electrodes are parallel to the laser axis. The discharge electrons gain their kinetic energy from the applied electric field and lose it to the gas by excitation, ionization and elastic collisions. The discharge's behaviour depends strongly on the gas density and its composition. In most cases it is strongly inclined to unstable behaviour so that pulse durations are inherently limited. Such a discharge can be kept stable for a longer period with the help of a simultaneous excitation with a rather weak e-beam, say a combination of e-beam and gas discharge. Nevertheless, for practical reasons, one likes to avoid the complexity of e-beams. Therefore, many more attractive alternatives have been successfully worked out. Those techniques are very specific and refined. They are relatively cheap compared with e-beams and, in principle, suitable for high repetition rates so that high average output powers become available.

B3.2.3.1 Electron-beam excitation

High-energy electrons to excite the laser gas are generated outside the laser in a vacuum chamber or diode compartment. The electrons are emitted from the cathode in this compartment, usually by field emission, and are accelerated by a high voltage. The high-energy electrons are then directed towards the anode and penetrate through a thin foil that separates the high-pressure gas from the vacuum chamber. The e-beam, with energies usually in the range of several 100 kV, excites and ionizes the laser gas. This direct transfer is then followed by a chain of kinetic reactions involving energy exchange between the ions and excited states with other gas constituents eventually leading to the excited excimer molecules. For efficient excitation the current density must be at least 10 A cm^{-2} over a length of at least 20 cm. The realization of such devices was energetically carried out thanks to the pioneering work in the AWRE Aldermaston Laboratory [11] in the 1960s. The required high voltage of several hundreds of kilovolts is usually achieved by means of a fast switching Marx generator that charges a coaxial oil or water capacitor as a pulse-forming line. The principle of the Marx generator is based on the parallel charging of its capacitors with a relatively low-voltage power supply followed by decharging in series through spark gaps into the coaxial capacitor. The transfer to the pulse-forming coaxial capacitor is done because the high-voltage pulse of the switched Marx, especially in the case of large capacitors, is too slow to obtain a high current density. When the voltage of the coaxial capacitor has risen to a sufficiently high value, a low-inductance spark gap closes spontaneously transmitting the charge to the load, i.e. the laser chamber. In the case of impedance matching with the diode compartment, a short, clean, square pulse of high current density is obtained. The current pulselength is determined by the length of this coaxial capacitor and ranges, in practical systems, between 20 and 200 ns.

The high current density of the cathode is mostly obtained by field emission on a cold cathode. (Conventional thermionic electron emitters cannot be used for currents above 1 A cm^{-2} because of their limited yield.) The construction is very special to ensure high current fields. It contains tantalum knife-edges, needles, carbon wedges or carbon felt. This is based on the strong enhancement of the electric field on a microscopic scale. The anode can be a grid supporting a thin metal foil often made of titanium, a metal with a high ratio of tensile strength and mass density. This foil separates the diode compartment from the laser vessel. For very high gas densities, say above 5 bar, some constructions have an additional foil with a support structure that contains holes or slots with a transparency of 70 to 80%.

To achieve an optimum incoupling of the input energy the diode impedance must be matched as well as possible with the pulse-forming line. The maximum current density in the diode is space-charge limited according to the Child–Langmuir law [12]:

$$J = 2.37 \times 10^2 \frac{V^{3/2}}{d^2} \tag{B3.2.6}$$

where J is the current density [A cm^{-2}], V the diode voltage [MV] and d the anode-diode distance [cm]. The maximum obtainable pulselength depends on d, because at the onset of the vacuum current a plasma is formed that moves across the gap with a velocity of a few cm per microsecond, decreasing the actual impedance [13]. Furthermore, for high current densities, pinching effects in the diode caused by the self-magnetic field or by the magnetic field of the return current can limit the freedom in designing diodes adapted to a more or less homogeneous excitation current in the laser vessel.

B3.2.3.1.1 Electron-beam geometries

The most simple and commonly used diode configuration has a one-sided transversal geometry. A cross-sectional view is shown in figure B3.2.2(*a*). The electrons emitted from a cathode wedge in the vacuum diode (VD) penetrate through the supporting foil into the laser compartment (LC). The broken circle in LC indicates schematically the cross section of the laser beam in the cavity. The main drawback of this construction is its poor energy utilization by the excitation inhomogeneity of the laser beam. A more uniform deposition can be obtained with a guiding magnetic field [14] or with a symmetric construction where two identical beams on both sides are applied as shown in figure B3.2.2(*b*). The best way to obtain beam uniformity is a coaxial symmetry [15], see figure B3.2.2(*c*). The anode is a self-supporting cylindrical foil tube with a high electron transmittance. The electrons are focused to the laser axis so that a uniform high excitation density can be performed. Using titanium foil 50 μm thick, the laser gas can be pressurized to at least 14 bar in cylinders of 1 cm diameter [16]. Scaling up this geometry to larger diameters and higher current densities, say 4 cm and 200 A cm^{-2}, the self-magnetic field of the current may cause current contractions near the cavity ends.

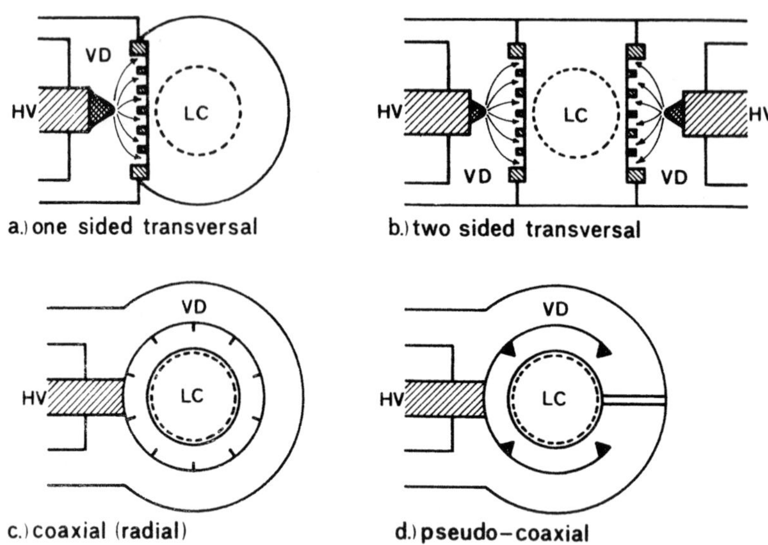

Figure B3.2.2. Different e-beam diode configurations.

This problem can be circumvented by an additional current return, parallel to the laser tube, as indicated in figure B3.2.2(*d*).

A relatively simple cylindrical construction for a laser tube 20 cm in length with a diameter of 1 cm has been successfully used to study various excimer systems. A cross section of this laser is shown in figure B3.2.3 [17]. The coaxial vacuum diode is driven directly by a low-inductance ten-stage Marx generator. The spark gaps are UV-triggered, except for the first gap, which is triggered externally. Each stage is formed by eight pairs of $BaTiO_3$ capacitors of 1.8 nF each. The total capacitance per stage is 7.2 nF. It is seen that the generator is enclosed in a cylindrical vessel 1 m in height and with a diameter of 25 cm, filled with 4 bar SF_6 to prevent flash-over. The compact arrangement provides a low inductance and a fast risetime. The Marx generator is capable of providing 100 J at a maximum voltage of 600 kV. The spark gaps are flushed with dry air. The voltage is monitored by means of a differentiating voltage sensor [18] and by means of a self-integrating Rogowski coil [19]. The cathode consists of a 20 cm long cylinder of diameter 5 cm with rounded ends to prevent flash-over to the vacuum chamber walls. Inside this cylinder two graphite felt strips of 20×0.7 cm^2 are mounted parallel to the axis. In this way, a two-sided transversal excitation in a radial symmetric electric field is obtained. The anode or laser tube is made of 25 μm thick titanium foil with a single electric weld along the axis. Titanium end-pieces are laser welded to this tube. The tube can stand gas pressures of up to 10 bar. This cathode configuration provides a good impedance matching to the Marx generator. The corresponding voltage and current waveforms are illustrated in figure B3.2.4. The current pulsewidth is about 50 ns. Since most halogen components in the gas mixture are very aggressive the total reflector is an MgF_2-coated aluminium mirror. As outcoupler, an uncoated plane-parallel suprasil or MgF_2 window is used.

B3.2.3.2 Self-sustained discharge technology

The chain of kinetic reactions leading to the formation of excimer molecules in gas discharges is based on three-body collisions. For those collisions to be effective, it is requisite to consider multi-atmospheric gas densities. This fundamental condition has a strong impact on the discharge technology for the relevant

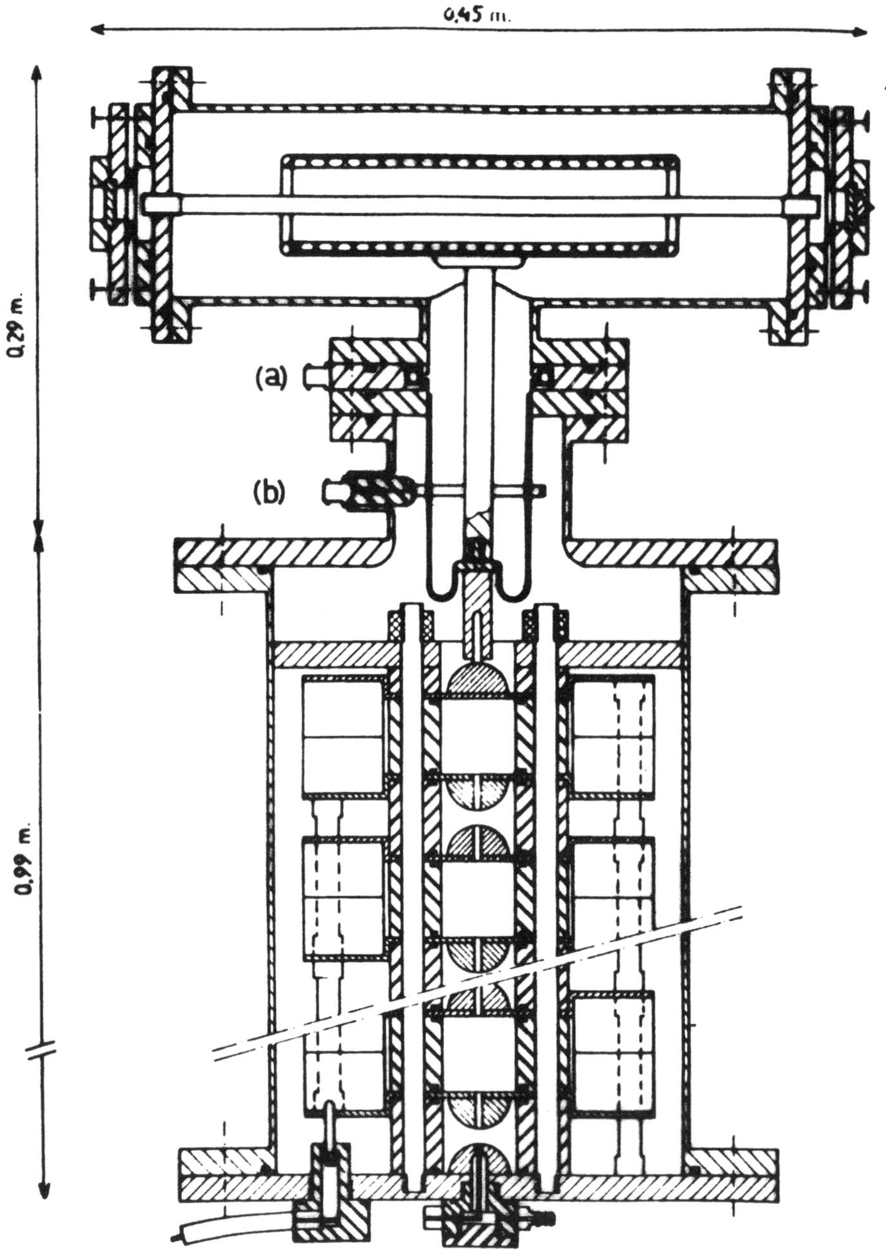

Figure B3.2.3. Cross section of an e-beam pumped coaxial laser with Marx generator. (*a*) Rogowski coil, (*b*) voltage sensor.

gas mixtures. Most multi-atmospheric self-sustained discharge systems are, by nature, unstable and the presence of the halogen constituent with its negative ion formation intensifies the instability process. As pointed out earlier the laser conditions can only be fulfilled with pulsed systems. Therefore, the technological development of the discharge is focused on slowing down or temporarily suppressing instability formation in

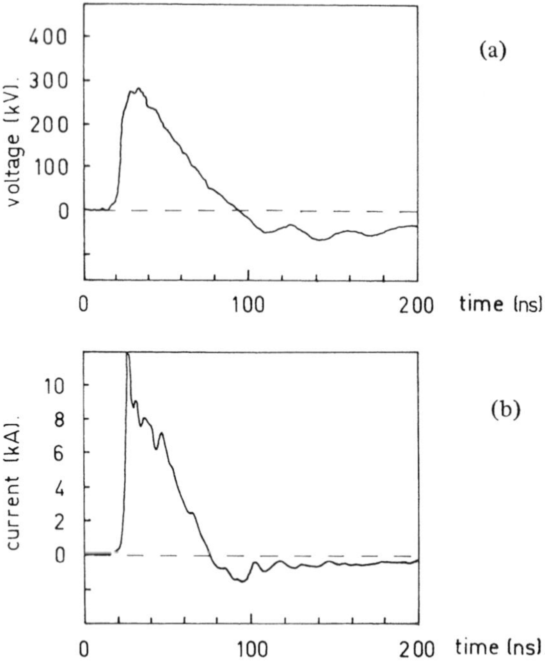

Figure B3.2.4. Voltage and current pulse of the vacuum diode as a function of time.

an initially stable pulsed discharge. For this purpose several innovations have been pioneered. They all have in common that the quality of the discharge of any excimer laser depends, among other things, strongly on the initial electron density and its distribution by the pre-ionization source. Furthermore, all these developments are based on a fast transverse discharge geometry (perpendicular to the laser axis). The average electron energy in this kind of discharge is only a few electronvolts so that the production of excited states and ions and the subsequent formation of excimer molecules is less efficient than in e-beam systems. However, this discharge technique can operate at a high repetition rate. The systems are then also capable of delivering high average powers.

B3.2.3.2.1 Pre-ionization

Operation of high-pressure gas lasers in a self-sustained discharge regime demands a proper pre-ionization technique to obtain a uniform glow discharge. This can be performed by direct e-beam interaction, by UV or by x-ray irradiation of the laser gas. Both UV and e-beam have their own shortcomings for high-pressure gases. Those for UV pre-ionization arise mainly due to high absorption of the UV radiation in the laser gas. The main limitation of e-beam pre-ionization is its window. Since it must be transparent for the electrons a thin metal foil is used. The strength requirements for this thin foil separating the vacuum diode from the high-pressure laser chamber are not compatible with high transmission. For x-ray pre-ionization, the situation is much more favourable. Medium-energy x-rays (50–100 kV) already have a high penetration depth and relatively thick windows can be constructed. Therefore, the x-ray technique has been successfully applied to excimer lasers.

In general, medium-energy x-rays with high peak powers and short risetimes are required. This is especially the case for x-ray switched single discharges [20, 21] where the applied discharge voltage is

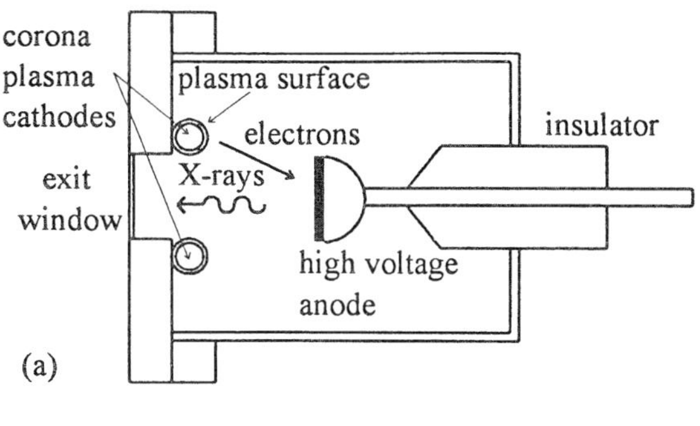

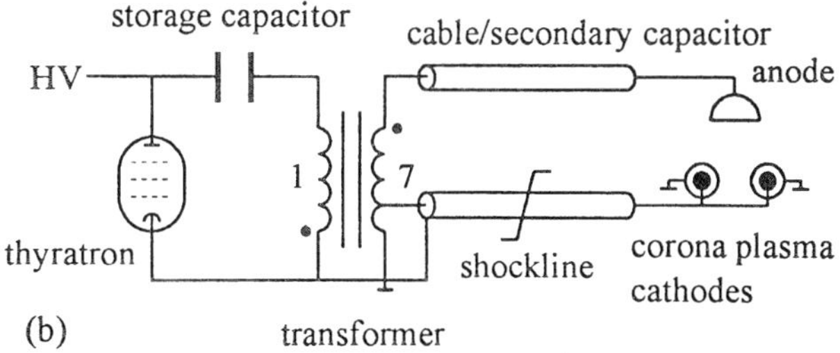

Figure B3.2.5. Cross-sectional view of the x-ray generator and its electrical circuit.

relatively high compared to the quasi-steady-state discharge voltage. After switching a fast electron avalanche, a uniform glow discharge free from streamers is obtained, provided the x-ray peak power is sufficiently high and uniform with a short risetime. A premature electron avalanche due to a long risetime of the x-ray pulse causes a non-uniform unstable discharge.

B3.2.3.2.2 X-ray source with corona plasma

A very successful development for obtaining a short intense x-ray pulse with a fast risetime is based on a corona plasma cathode that supplies the x-ray-producing electrons [22]. A cross-sectional view and its electrical circuit are shown in figure B3.2.5. The anode consists of a tantalum foil and the cathode contains two corona plasma tubes each 80 cm in length. These tubes are made of quartz tubes of 1 cm diameter with the inside metal bars acting as trigger electrodes and outside thin tungsten wires wound around them. When a fast-rising positive voltage pulse reaches these trigger electrodes, a negative charge accompanied by a corona plasma along these quartz tubes is formed. The plasma is produced by the interaction of the accelerated electrons along field lines on the quartz surface with (impurity) molecules adsorbed to the quartz surface. The plasma electrons are then extracted from the surface and accelerated to the anode by the anode voltage. The distance between the cathode and the anode is 3.5 cm. The anode is connected to a pulse transformer by means of a 80 cm long, 25 Ω coaxial cable. The primary storage capacitor of 17 nF is charged to 20 kV. The trigger voltage for the corona on the quartz tube surface taken from a tab of the pulse transformer passes a 180 cm long shock line to compress and to delay the pulse [23]. Figure B3.2.6 shows the waveforms of the input and output pulses. It is seen that the risetime is reduced from 100 ns to less than 20 ns and delayed by about 100 ns. During this delay the acceleration voltage pulse reaches the anode. Thus, as soon as the trigger pulse is applied, the plasma electrons are accelerated and bombard the tantalum foil producing x-rays. Figure B3.2.7 shows the waveforms of the anode voltage, current and x-ray pulses. These x-ray pulses with fast risetimes are successfully applied to pre-pulse–main pulse excited excimers.

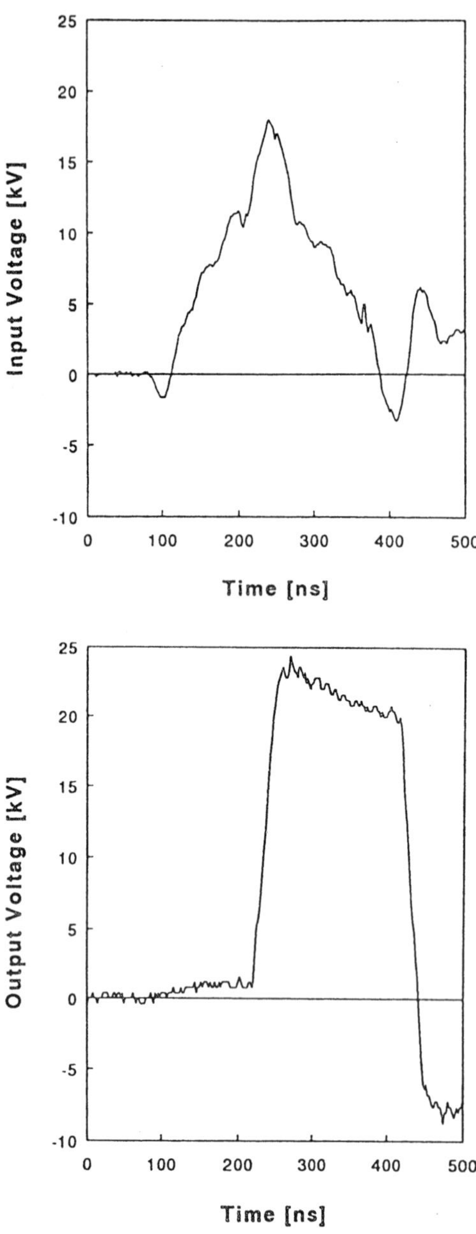

Figure B3.2.6. Input and output voltages of the ferrite shock line.

B3.2.3.2.3 Single discharge pulse

After the initial low electron density obtained by the x-ray pulse, the discharge implies the sudden application of a strong electric field that causes electron multiplication in the laser gas. In order to avoid the formation of filamentary arcs, i.e. an unstable irregular discharge, it is necessary to start with an initially homogeneous electron density of at least of the order of 10^8 electrons cm^{-3} and a fast risetime, of the order 10 ns, of the

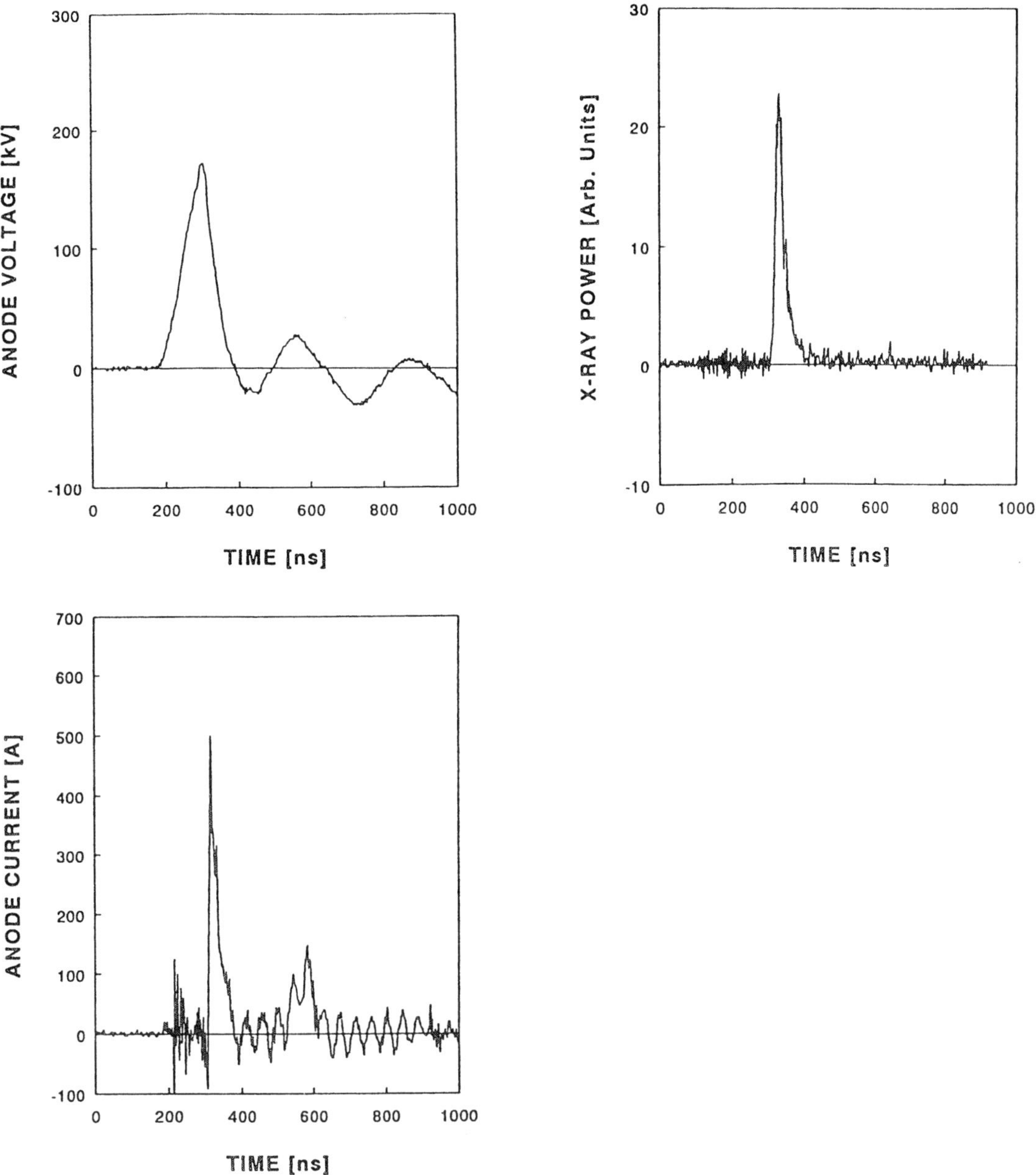

Figure B3.2.7. Waveforms of the anode voltage, current and x-ray power of the x-ray generator.

applied electric field. The fast risetime leads to the formation of a dense hot-spot structure on the cathode which enhances the homogeneity of the discharge [24]. If the discharge is formed by an applied electric field, the plasma resistivity decreases rapidly with increasing plasma density. In this way the discharge will reach the so-called steady-state conditions for which the voltage U_B across the discharge gap varies, from now on, relatively little with increasing current. The impedance across the discharge gap will then become lower

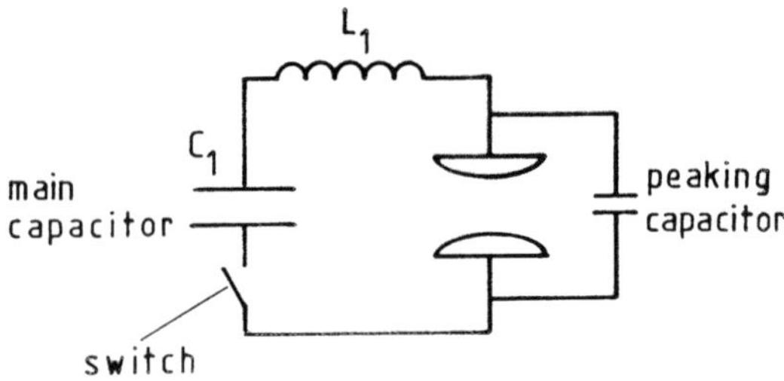

Figure B3.2.8. Scheme for a single discharge.

than the driving circuit's output impedance. When this occurs, the discharge will go into an oscillatory mode which immediately destroys the stability necessary for the laser kinetic process. The circuit is schematically shown in figure B3.2.8 where the electrodes are shown in a cross-sectional view perpendicular to the laser axis. The energy storage capacitor is connected to the electrodes in series with a switch [25–28]. To increase the initial electric field of the discharge a small peaking capacitor is used so that at the onset twice the charging voltage is applied to the electrodes. The first phase before the current reversal of an initially stable discharge is most effective for the laser process. In the appendix it is shown that for efficient energy transfer from the capacitor into the discharge the charging voltage should be twice the steady-state discharge voltage. This condition is not consistent with the required high-voltage breakdown of the discharge. Hence, for such a system, the voltage and current show circuit ringing and only during the short period of the first oscillation is laser power generated.

B3.2.3.2.4 Pre-pulse–main pulse

Since the impedance before breakdown of the discharge is high compared to its steady-state value and the voltage required for rapid breakdown is not consistent with efficient impedance matching of the electric circuit, using separate electric circuits for the avalanche breakdown (pre-pulse) and the quasi-steady-state discharge (main pulse) respectively has been proposed [29–31]. The energy of the pre-pulse is kept low, in most cases only a few joules at a voltage, depending on the specific excimer, of roughly six times the steady-state value U_B. This low pre-pulse energy facilitates the realization of the desired fast current rise for obtaining a dense hot-spot structure on the cathode, as discussed earlier, resulting in a high overvoltage and a rapid breakdown of the gas. The sustainer capacitor for the main pulse will then be charged to twice the steady-state voltage U_B for optimum impedance matching (appendix). With such a circuit there are, in principle, two modes of operation: the switch mode and the diode mode as will be described in the next paragraph. The diode mode is not attractive because the polarities of the two pulses are equal and, consequently, there will be a time delay between the two pulses due to the inductance of the main pulse circuit. Since the mechanism for forming streamers or filamentary arcs is always present and time-dependent any delay should be avoided. The situation is much more favourable for the switch mode.

An attractive way to separate the two circuits is by the use of a saturable magnetic isolator. The principle is shown in figure B3.2.9. For the essential requirement of fast discharging into the laser gas, during their discharge the inductances of the circuits are as low as possible. Therefore, the sustainer capacitor C_1 charged at $2U_B$, is connected to the laser head by a metal plate surrounded by a race track made of low-loss ferrite bricks. The pre-pulse circuit consists of a small peaking capacitor C_2 in series with a capacitor C_3 of equal size and a spark gap. The capacitor C_3 can be positively (switch mode) or negatively (diode mode) charged with respect to the main capacitor. For the switch mode, one side of C_3 is connected to C_1 with the voltage $2U_B$ whereas the other side is connected with a voltage source of about $6U_B$. Directly after firing the x-ray

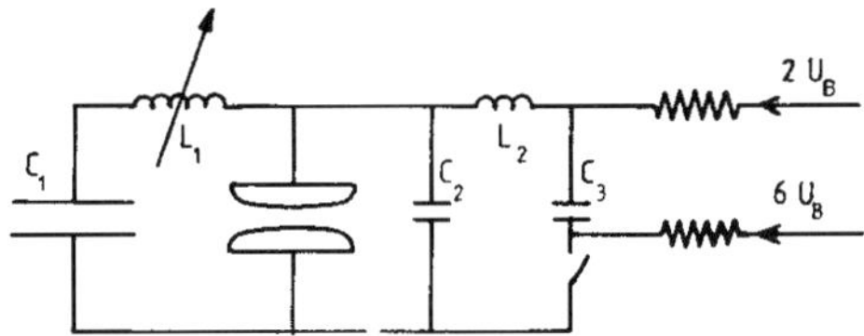

Figure B3.2.9. Scheme for a pre-pulse–main pulse discharge in switch mode.

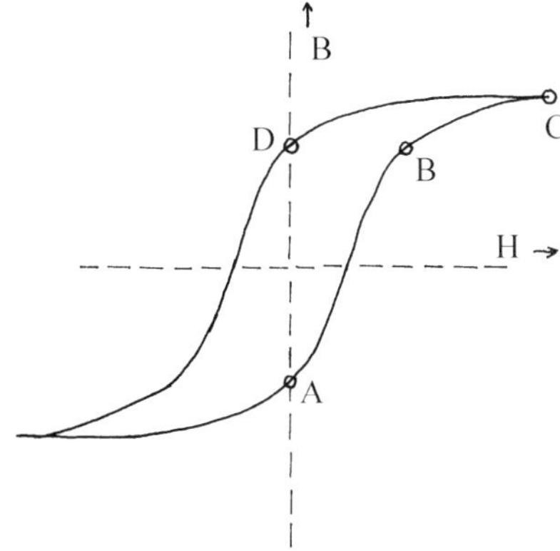

Figure B3.2.10. *BH* curve of the saturable ferrite inductor showing the inductive states of the pulsed current operation.

pulse the pre-pulse circuit is switched so that the voltage across C_3 becomes $-4U_B$ which will be directly transferred to the peaking capacitor C_2. Then this voltage is also across the electrodes causing, during several tens of nanoseconds, the breakdown of the discharge. In the meantime, this pre-pulse also saturates the magnetic switch so that after the avalanche process of the pre-pulse, the main pulse will automatically maintain the discharge by delivering, in principle, all its energy.

The inductive state of the saturable ferrite inductor of the main switch can be described with the *BH* curve as shown in figure B3.2.10. After charging the main capacitor C_1, the ferrites are, for increasing *H*-values, in a high inductive state, indicated by point A. During the firing of the pre-pulse the ferrites remain in this high inductive state while passing the section from A to B. When the gas breaks down at point B the ferrites are almost saturated and the main current with still increasing *H* values starts immediately at point C in a low inductive state. At the end of the pulse, the current is zero and the state of the ferrite reaches point D. Charging C_1 again the left-hand part of the loop with negative *H*-values is followed, finally reaching point A again for zero current.

It should be noted that when the main pulse starts there is a current reversal in the discharge and that the saturating voltage across the switch is $6U_B$. It may happen that in order to obtain a faster avalanche and a stabler discharge, depending on the specific excimer, the charging voltages, especially of the pre-pulse circuit, are somewhat different.

Since there is a current reversal through the laser, the anode and cathode function of the electrodes are interchanged. It is well known in discharge physics that the electron emission of the cathode is accompanied by material sputtering. With x-ray pre-ionization, the thin plate that transmits the x-rays is, therefore, chosen as the anode for the main pulse. However, during the pre-pulse, this electrode is the cathode so that material is sputtered. Therefore this mode of operation limits the lifetime of this thin electrode. This mode of operation with the present construction is therefore not suitable for high repetition operation.

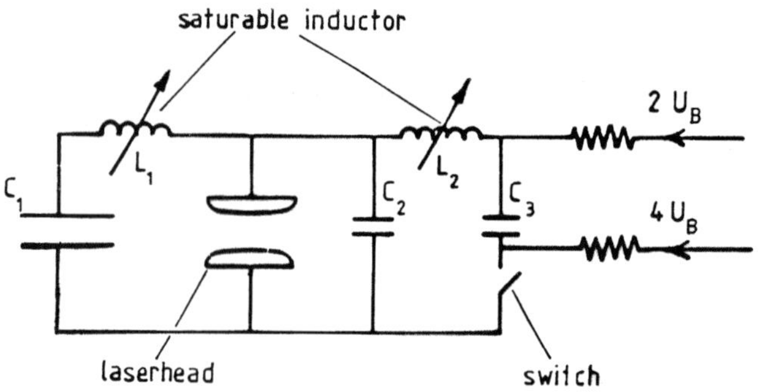

Figure B3.2.11. Resonant overshoot circuit scheme.

B3.2.3.2.5 Resonant overshoot mode

The resonant overshoot mode [32] is, at first glance, very similar to switch mode operation. The schematic diagram is shown in figure B3.2.11. It is essential for obtaining fast current rise to replace L_2 by a saturable inductor [33]. Furthermore, the difference in operation is set by the specific values of the chosen components. We have the additional requirement that the resonance time τ_1 of the current flow from C_1 to C_2 just after the saturation of L_1 is shorter than the resonance time τ_2 of a reversed current flow from C_3 to C_2.

The principle of operation is as follows. The capacitors C_1 and C_2 are also charged to $2U_B$ and the connected voltage to C_3 is, in this case, only $4U_B$ as shown in the figure. After pre-ionization the pre-pulse is again fired within a few microseconds after charging C_1, C_2 and C_3. After switching, the voltage drop over C_3 is now $-2U_B$ so that the voltage over C_2 will subsequently change from $2U_B$ to $-2U_B$ which is too low to start the avalanche. Thus the laser gas does not break down at this phase. In the meantime the magnetic switch L_1 saturates by its voltage drop of $4U_B$ and allows a current flow from the main capacitor C_1 to the small peaking capacitor C_2 which results in an overshoot by reaching, at most, $6U_B$. The increasing voltage on the peaking capacitor in this charge transfer will now reach such a level that a fast avalanche breakdown in the discharge will occur to be directly followed by the main discharge.

The inductive state of the saturable ferrite inductor L_1 can also be described with the BH curve shown in figure B3.2.10. Following the BH curve we start after the charging of C_1 at point A. After switching C_3 and charging C_2 at $-U_B$ we reach point B. Next the ferrites will further saturate as C_2 is now charged by the main capacitor C_1. After breakdown, a large current flows through the inductor and, for increasing current, point C will be reached. At the end of the discharge when the current again becomes zero point D of the upper part of the curve is reached. By charging C_1 again the current flows in the opposite direction, bringing the ferrites back in the high inductive state indicated by point A.

It is important that, just after switching the pre-pulse, the resonance charging of C_2 from $2U_B$ to $-2U_B$ occurs very rapidly because the size of the magnetic main switch L_1 is proportional to the switching time. The inductor L_2 is, during this resonance charging of C_2, in a saturated state so that for a current reversal between C_2 and C_3 the inductor L_2 is unsaturated and the condition that τ_1 is shorter than τ_2 is easily fulfilled.

We note that, because of a voltage drop of only $4U_B$ over the main magnetic switch, less ferrite bricks are needed for the race-track construction and that the current to be switched by the pre-pulse between C_3 and C_2 is only two-thirds of that described earlier for the switch mode. We also note that, in this case, there is no current reversal in the discharge. This means that the thin x-ray transmitting anode will, in principle, not be damaged. This all makes this mode of operation attractive for high repetition operation.

B3.2.3.2.6 Photo-triggered mode

For the photo-triggered mode, an x-ray beam starts the discharge and, hence, this mode is often called the x-ray triggered mode. In this mode of operation the laser gas discharge itself is used as the switch in a single discharge circuit. Thus, the laser discharge is triggered by the pre-ionization pulse instead of a pre-pulse as in the previous case. This breakdown, now initiated by the pre-ionization pulse after the high voltage pulse has been applied over the electrodes, has a strong impact on the discharge physics. The closing time of the laser discharge as switch, which is equal to the formative time of the discharge, depends on the voltage applied to the electrodes and on the electron production rate of the pre-ionizer. Usually the charging voltage of the circuit capacitor is considerably higher than twice the steady-state voltage, in order to obtain a rapid avalanche in the electron multiplication process from the pre-ionization density of, say, 10^8 cm^{-3} to a steady-state value of about 10^{14} cm^{-3}. The high charging voltage leads to a current oscillation of which only the first phase before the current reversal will be stable. Consequently, the conversion efficiency will be low. However, each photo electron will immediately gain sufficient ionization energy so that the multiplication remains, in principle, spatially homogeneous, while in the previous case some pre-ionization electrons drifted away from the cathode due to the presence of the main voltage across the electrodes thereby making the discharge less homogeneous.

The absence of a pre-pulse, of course, also has several advantages. The system becomes simpler. The thyratron switch, power supply and trigger circuit of the pre-pulse as well as the magnetic switches of the main circuit and pre-pulse compressor are not needed. This mode of operation has been successfully developed for XeCl lasers [20, 21].

B3.2.3.2.7 Instrumental design

The most promising discharge technique for the rare gas halide excimer lasers is the pre-pulse–main pulse technology with resonant overshoot, although it is more complicated than the single-discharge technique. Most attractive are the better conversion efficiency, the longer output pulses, less system poisoning by electrode sputtering and wear out of the thin x-ray transmitting electrode and the suitability for high repetition rate operation.

A successful pre-pulse–main pulse design for a self-sustained excimer laser requires spatial homogeneities, pulse compressions and fast risetimes for all input quantities of the discharge. A cross-sectional view of such a system and a scheme for its electrical circuit are shown [34] in figures B3.2.12 and B3.2.13, respectively. A pulsed collimated x-ray beam (described in section B3.2.3.2.2) enters the discharge chamber through a 2 cm × 60 cm aluminium window 1 mm thick. The electrode gap is 2.5 cm. The pulse compression network of the pre-pulse also serves to keep the current of the thyratron below its maximum allowable value. The saturable inductor of the pre-pulse circuit consists of six rings of CMD 5005 high-frequency ferrite from Ceramic Magnetics with a total cross section of 26.4 cm^2. The pulse-forming network (PFN) capacitor of the main pulse circuit consists of eight double-plate transmission lines connected in parallel to the laser head. Each line contains 20 TDK capacitors of 2.7 nF each resulting in a total capacity of 432 nF. The saturable inductor of the main switch is constructed in a rectangular shape in order to reach minimum inductance after saturation. It has a 2 mm slit in the middle to be used for the conducting plate that connects the laser chamber with the PFN capacitor. The inductor is made of CMD 5005 ferrite blocks with a total cross section of 12 cm^2. The peaking capacitor is 2.8 nF. In general, the pre-pulse will be chosen as high as possible without a discharge breakdown during its negative voltage swing. The risetime of the resonance voltage just before breakdown is given by $\pi(C_p L_{p,sat})^{1/2} = 24$ ns, where $L_{p,sat}$ is the residual inductance of the saturated ferrite blocks.

Applying this system, for example, to a XeCl laser system by using a mixture of 1 mbar HCl, 20 mbar Xe and 4 bar Ne and charging voltages of 10 and 20 kV for the PFN and pre-pulse capacitor respectively;

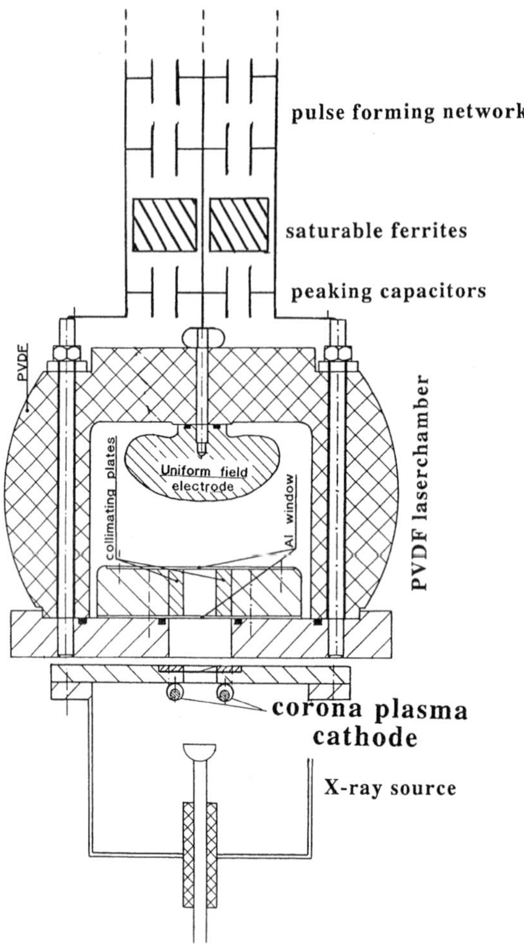

Figure B3.2.12. Cross-sectional view of a pulsed excimer laser system

typical waveforms are shown [34] in figure B3.2.14. Thanks to fast risetimes and impedance matching, the discharge stability for this XeCl system of 4 bar is 500 ns, whereas the output pulse follows the current pulse closely.

The stability of the inherently unstable excimer systems depends on both gas composition and instrumentation parameters. For usual gas compositions, the stability deteriorates very rapidly even after appropriate starting conditions with increasing risetimes of the various circuits and with improper matching. Stated differently, the stability period of an excimer discharge can be extended by shorter risetimes in the circuits. It should be noted that the discussed instrumental principles apply to practically all self-sustained discharge excimers, although the values for circuit components and electrical parameters may differ widely.

B3.2.4 Rare gas halogen excimers

The rare gas halogen excimer has received most attention in the field of excimer lasers. They are highly developed and are currently the most powerful lasers in the UV. Several systems are well engineered and there is a great deal of understanding about the discharge physics and the kinetic chain for the formation of

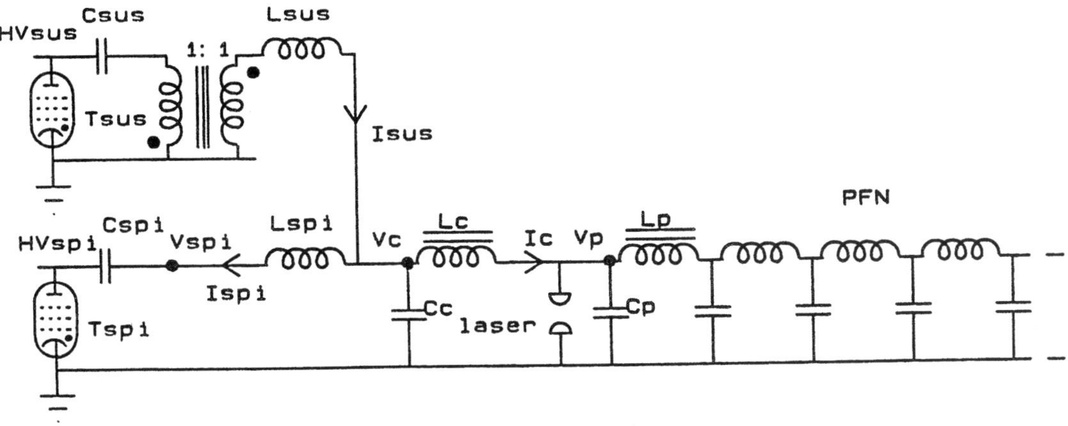

Figure B3.2.13. Scheme of the pre-pulse–main-pulse circuit with thyratron switches. The subscript 'sus' refers to the sustainer and the subscript 'spi' to the pre-pulse or spiker pulse.

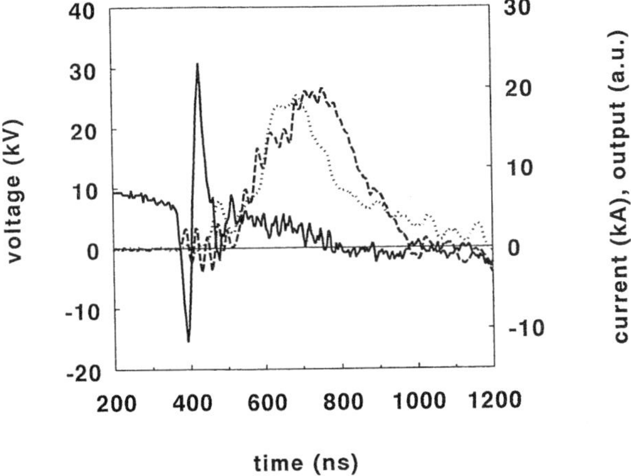

Figure B3.2.14. Waveforms of the discharge voltage (full), current (dashes) and optical output (dots) for the resonant overshoot mode of an XeCl laser.

the lasing molecules. In particular the properties of XeCl (308 nm), KrF (248 nm), ArF (193 nm) and XeF (351 and 353 nm) have been studied extensively.

The spectroscopy of the rare gas halide molecules exhibits a general structure which is shown in figure B3.2.15 [35]. The ground state is covalently bonded and correlates to ground state ^{1}S rare gas and ^{2}P halogen atoms at infinite internuclear separation. The ground state manifold of the halogen atom with the orbital angular momentum of one consists of two states. The $^2\Sigma$ state has the lowest energy since in this configuration the single occupied halogen orbital is directed toward the rare gas atom. Since it forms the ground state it is referred to as the X state [36]. This state is generally nearly flat or, at most, weakly bound as in the case of XeCl (255 cm^{-1}). The exception is XeF which has a binding of approximately 1065 cm^{-1} [37]. The $^2\Pi$ state of the ground-state manifold is always repulsive. Since it forms the first state above the ground state it is referred to as the A state. The upper state is a charge transfer state between the positive rare gas

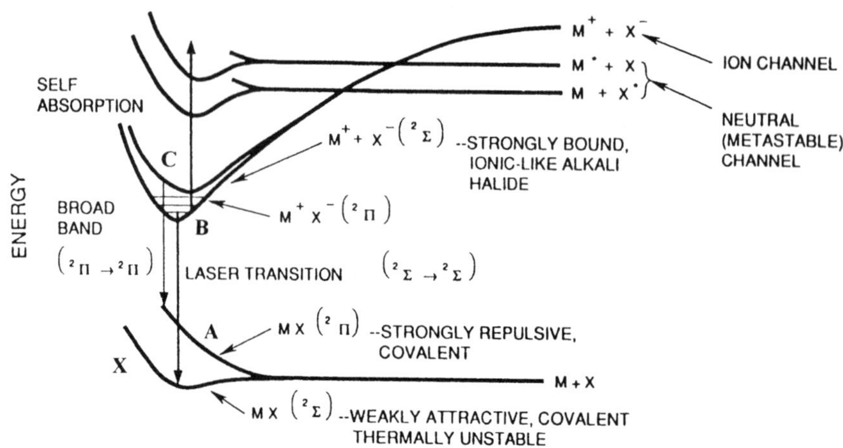

Figure B3.2.15. Schematic potential energy diagram for a diatomic rare-gas halide molecule showing the dominant B→X transition and the less efficient C→A transition.

atom and the negative halogen atom with the ^{2}P and ^{1}S state, respectively, at infinite separation distance. Beginning at an energy equal to the ionization potential of the rare gas less the electron affinity of the halogen, the potential of the state of this system formed by the two particles follows a Coulomb curve which crosses, in the adiabatic sense, the covalent curves correlating to excited states of the rare-gas atom and the halogen atom. The crossings generally occur at large internuclear separation where the overlap between orbitals centred on the rare gas and the halogen is small, so that the interactions will be small at the crossing points. At close internuclear separation, the potential energy curve splits into $^2\Sigma$ and $^2\Pi$ states. The structure of the upper level manifold is similar to that of the ground-state manifold, again with the $^2\Sigma$ state lying lowest as indicated. By convention this is referred to as the B state. In the interesting cases, the Coulomb curve crosses all the excited covalent curves, and the ionic state forms the lowest excited state. The emission spectrum of the rare gas halides consists of several bands. The strongest band is the $B(^2\Sigma) \rightarrow X(^2\Sigma)$ transition and gives rise to a laser transition. Other transitions have, in general, a much higher input energy threshold value if they will lase at all.

B3.2.4.1 Kinetics

The kinetics of the rare-gas halide lasers are rather complex. The laser gas mixtures are at multi-atmospheric densities and consist of a high percentage of a lighter rare gas, called a buffer gas (He, Ne or Ar), a small amount, less than 10%, of a lighter donor rare gas (Ar, Kr or Xe) and about 1% of a halogen or halogen-bearing gas. The electrical excitation of the gas mixture results mainly in the formation of excited atoms and atomic ions of the dominant lighter rare gas. The formation of the desired excimer molecule occurs at the end of the kinetic reaction chain. This reaction chain consists of charge transfer processes and recombination processes such as collisions between charge carriers of opposite sign and the dimerization of excited atoms with neutral atoms. The internal energy must decrease as a result of each reaction, otherwise the reaction is not possible. For instance, in the case of atomic ions forming the excimer molecule, the energy decrease is equal to the sum of the ionization energy of the positive ion and the electron affinity of the negative ion minus the excitation energy of the excimer. Once the oppositely charged ions or excited particles begin to interact, the ability of the system to dispose of this excess energy determines the probability that recombination occurs or reaction products are obtained. The requirement that linear and angular momentum be conserved in the reaction makes it almost impossible to dispose of the recombination energy by an increase in the kinetic energy of the recombining ions. However, the energy release can be achieved through an increase in the kinetic energy of a neutral particle that is also involved in this interaction process. Then we are dealing with three-body collisions. The three-body collisions take place in several reactions and are essential in the formation of the lasing excimer molecules. This means that for obtaining sufficient fast recombination or energy transfer, the density of neutral particles (buffer gas) must be at least at atmospheric density.

The predominantly excited and ionized buffer gas atoms may then lead rapidly to the excimer molecule. Although the excited buffer atoms are initially in different states, at high gas density these states are rapidly quenched to lower states and by far most of the excited atoms will then be found in the lowest metastable state. These atoms will dimerize by three-body collisions:

$$X^* + 2X \rightarrow X_2^* + X \tag{B3.2.7}$$

where X is the buffer atom. Then the energy of the excited states are transferred between the buffer gas and the heavier donor rare gas (Y). Reactions of the type $X^* + Y \rightarrow X + Y^*$ are relatively slow unless there is a near resonance between the initial and final states. The rate depends very sensitively on the closeness of the resonance. Reactions of the type

$$X_2^* + Y \rightarrow 2X + Y^* \tag{B3.2.8}$$

are much more rapid because the excess energy can be disposed of in the dissociation products.

Also, the ions formed by the electrical excitation rapidly dimerize by three body collisions of the type

$$X^+ + 2X \rightarrow X_2^+ + X \tag{B3.2.9}$$

where X^+ is directly produced by the high-energy electrons of e-beam pumping or, in the case of discharge pumping, in a two-step process. The first step is the formation of the metastable state X^* followed in the next step by the ionization of X^*. This happens because the average electron energy in the discharge is low (a few eV), only relatively few electrons are able to ionize directly and much more can reach the metastable level. The metastable state density can be relatively large in a gas discharge. Furthermore, the cross section for ionization of a metastable state is orders of magnitude larger than the cross section for ground-state ionization.

Direct charge transfer between the buffer gas ion and the neutral-donor rare-gas atom is generally very slow like the reaction of the excited buffer atom with the donor gas. However, charge transfer from the dimer ion to the donor atom is generally very fast in a reaction like

$$X_2^+ + Y \rightarrow 2X + Y^+. \tag{B3.2.10}$$

The excited-donor rare-gas atom reacts directly with the halogen (Z_2) or halogen-bearing molecule (ZM) as

$$Y^* + Z_2(\text{or ZM}) \rightarrow YZ^* + Z(\text{or M}) \tag{B3.2.11}$$

where YZ^* is the excited rare gas halide molecule. These kinds of reaction have large cross sections and are often called *harpooning* in analogy with reactions between alkali atoms and halogen molecules. If other reaction products do not occur, reaction (B3.2.11) is said to have a unit branching ratio. This formation process is often called the metastable or neutral channel.

The forming of an excited rare-gas halide may also follow the ion channel. The negative halide ion is produced in a fast exothermic dissociative attachment reaction such as

$$Z_2 + e^- \rightarrow Z^- + Z \tag{B3.2.12}$$

or

$$ZM + e^- \rightarrow Z^- + M. \tag{B3.2.13}$$

In spite of the strong Coulomb attraction between the negative halogen atom and the positive donor rare gas atom, a recombination in the absence of collisions with a third body is not possible because linear and angular momentum are not conserved. The recombination is, therefore, mainly achieved by a three-body process such as

$$Y^+ + Z^- + X \rightarrow YZ^* + X \tag{B3.2.14}$$

which is very fast at high buffer gas densities.

So far we have discussed, in general, the main kinetics in the formation process. A full description, including all species produced in the discharge, may deal, depending on the specific excimer, with more than 60 different reactions and, as numerical simulation has shown, most of them have a negligible effect on the production rate of the lasing excimer molecule. For instance, the reaction of the excited and ionized buffer gas atoms with the halogen molecules and halogen ions, similar to reactions (B3.2.11) and (B3.2.14), are much less likely with an appropriate choice for the partial pressures of the system. There are, however, fast quenching processes of the excited excimer. Direct quenching takes place by all gas components and they are most effective by the halogens. This may result in the dissociation of the excited excimer or in the formation of triatomic species of the type

$$YZ^* + Y + X \rightarrow Y_2Z^* + X. \tag{B3.2.15}$$

This process is found to be very effective.

Although several of the quenching processes are very rapid, they do not, in general, give serious problems for well-engineered laser systems. In such systems, the excited excimer molecule is also subjected to the radiation field in the laser cavity, so there is a finite stimulated emission lifetime which, if the optical field is strong enough, can be made shorter than the quenching lifetime. The quenching losses are then negligible. However, the intracavity absorption may become a dominant loss mechanism. For optimizing the output, there will be a trade-off between these processes

The significance of these kinetics depends strongly on the initial excitation of the gas. When we are dealing with e-beam pumping, the energy of the incident electrons is so high that they mainly ionize the particles and transfer high energy to the secondary electrons which, in turn, further ionize the gas by slowing down. They also create excited atoms, although these atoms will be ionized as well. This means that in e-beam systems, the excimer production rate is mainly given by the ionization channel, i.e. reactions (B3.2.9), (B3.2.10) and (B3.2.14). The metastable route is less significant here. In contrast, the discharge-pumped systems produce mainly along the metastable channel described by reactions (B3.2.7), (B3.2.8) and (B3.2.11). This is because the average electron energy in the discharge is much smaller than the first excited state and certainly not enough to reach direct ionization. Only a small fraction of the discharge electrons are capable of exciting the gas and a relative small amount of ions is then produced in a two- or multi-step process. This also counts for the much lower efficiency that is obtained with discharge systems compared with e-beam systems.

B3.2.5 KrF Laser (248 nm)

The first reported KrF laser was based on e-beam pumping [10,38–40]. Around this time, discharge pumping for this system was demonstrated [41–43]. The high electron energy density and plasma stability that is imposed by e-beam excitation resulted in the successful development of this laser. The usual gas mixture of an e-beam system is 2–6 mbar F_2 or NF_3, 100–300 mbar Kr and 2–4 bar Ar. In the formation kinetics, both metastable atoms and ions are produced by the e-beam electrons interacting mainly with the buffer gas. When Ar is the buffer gas, on average, approximately 26 eV is deposited in the gas for each produced electron–ion pair. Some of it is transferred to the emitted secondary electron which, in turn, excites atoms and finally, through successive atomic collisions, rapidly cools to about a few eV. Of all e-beam pumped excimer lasers, the highest intrinsic efficiency (about 10% [44–47]) and an output power density of more than 60 J l^{-1} [48] have been reported for KrF. The reason for the high efficiency comes from the relatively low ionization energy of the buffer gas, a few potential energy crossings of the Coulomb curve of the Kr^+ and F^- ions covalent curves correlating to excited states of Kr and F atoms and the relatively low intracavity radiation absorption.

The KrF emission spectrum consists of several bands. The strongest band with the B→X transition is observed as the lasing transition. The linewidth is about 2 nm and the X state is weakly repulsive. The upper state has a large number of overlapping vibrational and rotational states at the usual high gas density. This homogeneous line broadening also has a favourable effect on the efficiency. The lifetime of the upper state is found to be 9 ns. [49]. The cross section becomes 2.4×10^{-16} cm^2 as follows from equation (B3.2.1).

The Kr ions for forming KrF* are either produced directly by the interaction of the e-beam with Kr or by the reaction with Ar_2^+ according to reaction (B3.2.10). The secondary electrons from the ionization process undergo easily attachment with F_2 to form F^- ions according to reaction (B3.2.12). The ions recombine in three-body collisions:

$$Kr^+ + F^- + Ar \rightarrow KrF^* + Ar \qquad (B3.2.16)$$

This process is rapid because the ions attract each other with the long-range Coulomb interaction. Although for e-beam pumping there is also, in principle, a metastable channel, it is not important because only a fraction (15–20%) of the electron energy is transferred to the metastable atoms. Moreover, the sequence reactions in this neutral channel are much slower than those with ions so that an even smaller part of the electron energy goes through the metastable channel. The KrF* molecules also undergo quenching reactions with the laser gas and electrons [50]. These quenching losses in the laser are in competition with the stimulated emission process and are determined by the optical quality of the cavity. However, intracavity absorption by discharge species such as Ar_2^+, Kr_2^+, F^- and F_2 increase with the cavity field. For an optimized system, for the outcoupling mirror of the laser there is a trade-off between these competing processes.

Analysing the kinetic process, it is found that a limited number of reactions plays a critical role in calculating the laser parameters. The quenching of KrF* determines critically the optimized gas densities, especially the Ar density. F_2 is less critical and the additional F_2 concentration that compensates for the burn-up loss during the excitation has practically no effect on the output power when the excitation pulse length is increased. At optimized conditions, of the KrF* produced about 50% is lost by quenching, about 20% is lost by absorption and about 30% is correlated to the outcoupled beam [51]. Experiments have confirmed that the output power follows the input power at least up till 500 ns [52]. In the figures B3.2.16 and B3.2.17, the output powers are plotted as a function of, respectively, the Ar and Kr density with an average ionization rate of an Ar atom, P[s^{-1}], as parameter [53]. It is seen that the optimized densities of Ar and Kr increase with the pumping power as indicated by P. For instance at $P = 4 \times 10^3$ which corresponds to an e-beam current density of 70 A cm^{-2} the observed optimized densities for Ar and Kr are respectively, 4 bar and 160 mbar and the output power density 100 MW l^{-1}.

Higher output power densities for a given input current of the e-beam is obtained when the buffer gas is increased with a mixture of Ar and Ne. This effect can be attributed to the fact that the quenching by Ne is less effective so that the optimization occurs at higher density and that the energy deposited in the Ne buffer is efficiently transferred to the Kr species. The metastable Ne atoms are formed in so-called Penning reactions with Ar^+ and Kr^+ [54]. The effect of Ne as a buffer gas becomes even more pronounced at higher input powers. At an e-beam current of 400 A cm^{-2} the optimized Ar density is 5 bar in the Ar–Kr–F_2 mixture. Replacing Ar by Ne the laser output scales linearly with the Ne density over the range 2–14 bar [55]. For the high density regime, the addition of Ar has no beneficial effect. This range was limited by the strength of the anode foil of the e-beam. At 14 bar the output with the Ne–Kr–F_2 mixture exceeds the optimized Ar–Kr–F_2 mixture by a factor of 4.5. The dimerization of Ne^+ and Ne* proceeds at a speed three to twenty times slower than the corresponding processes in a mixture with Ar as the buffer gas [56]. So it is expected that the dimer

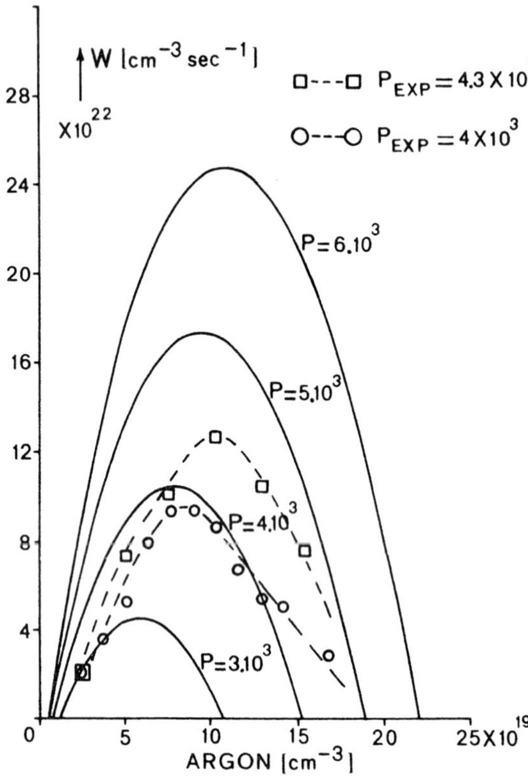

Figure B3.2.16. The measured and calculated extracted photon density rate as a function of the Ar density for Kr and F_2 densities of 3×10^{18} and 6×10^{16} cm^{-3}, respectively. The parameter P is the average number of ionizations of an Ar atom per second.

densities in the Ne mixture are relatively low which is favourable since the absorption coefficient of Ne_2^+ at 248 nm is at least an order of magnitude higher than for Ar_2^+.

The impressive performance of the e-beam-pumped KrF with its highest output density and efficiency at relatively short wavelengths was a strong encouragement for developing huge systems with energies above 10 kJ to study laser fusion by means of inertial confinement [57].

For discharge KrF lasers, the kinetic processes and system technology are very similar to those in ArF systems to be discussed later. For KrF the same gas mixture is used except that Ar is replaced by Kr. The buffer gas is again He or Ne. Usually He is used as the buffer gas [28, 58–60]. However, it is expected that when the risetime of the pre-pulse discharge is small, say 10 or less nanoseconds, the use of Ne will lead to longer pulses and higher output energies, as will be discussed for ArF. A discharge-pumped KrF laser needs less pumping power and yields more output than the ArF laser under identical circumstances.

B3.2.6 XeCl Laser (308 nm)

The first reported XeCl laser was based on e-beam pumping [38]. The high electron energy is, as mentioned earlier, advantageous for obtaining an efficiency as high as 8% [61]. The more interesting development in this field is based on gas discharge excitation for reasons explained in section B3.2.3. However, the efficiency of discharge pumping is lower than that of e-beam pumping. The highest reported value is 5% [32].

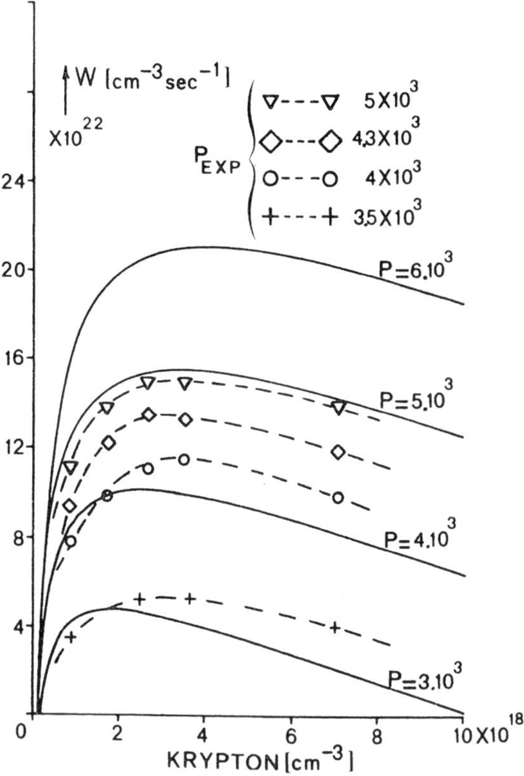

Figure B3.2.17. The measured and calculated extracted photon density rate as a function of the Kr density for Ar and F_2 densities of 6.5×10^{19} and 6×10^{16} cm^{-3}, respectively. The parameter P is the average number of ionizations of an Ar atom per second.

For the laser process the main upper levels are the B and C states which have a strong bond. The lower A state is strongly repulsive whereas the X state is loosely bound. The energy needed to break the X state is so small that the thermal vibrational energy of the molecule is sufficient to break it. The strongest laser lines at 307.9 and 308.2 nm involve the B$\rightarrow$X transition with a bandwidth of $\Delta\lambda = 0.5$ nm [36]. The B state has a rotational–vibrational structure. In the XeCl formation process, the inversion is spread over these levels. Between these levels the vibrational energy can be transferred by collisions. There is, however, competition between this relaxation process and the stimulated emission process. The depleted strongest laser line will be filled with less populated levels if the collision process is faster than the stimulated emission process of these levels. So it may happen that one strong transition is fed by all excited vibrational levels. If, however, the radiation field is weak (lower pumping power) and the relaxation process sufficient slow (lower gas density), several lines oscillate simultaneously. Also, the loosely bound X state has some vibrational levels so that transitions can, in principle, take place between different upper levels and different lower levels. The strongest transition is then determined by the population inversion as well as by the corresponding Franck–Condon factor. It was found that, for a non-selective cavity, the $0 \rightarrow 1$ and the $0 \rightarrow 2$ vibrational transitions are the strongest [62, 63]. The C$\rightarrow$A transition which emits at $\lambda = 345$ nm has, due to its strong repulsive lower state, a large linewidth of 20 nm so that the cross section for stimulated emission is much smaller than for the B→X transition. Since the upper states are also strongly coupled normally only the B→X transition will oscillate.

A typical gas mixture of the discharge laser contains 1 mbar HCl, 10 mbar Xe and Ne as the buffer gas with a total pressure of 5 bar. In the past He has also been used as a buffer gas but less successfully. The halogen bearing molecules BCl_3, CCl_4, CH_2Cl_2, CCl_2F_2, $CFCl_3$ and $CHCl_3$ were also tried, but the best results are obtained with HCl. The partial pressures of HCl and Xe can be very controllably kept constant by means of a cold trap gas purifier. Since the optimized HCl density depends on the discharge current and its pulselength, it is convenient to regulate the partial pressure by the cold trap temperature. Furthermore, the purifier controls the vapour pressures and the surplus is frozen in the cold trap. In particular, this buffer takes care that the HCl consumed during operation will be supplied. This happens because the discharge dissociates the HCl and the chlorine ions attach themselves to the electrodes and other metal parts of the laser head. The H ions eventually form H_2 molecules. Thus there is a loss of HCl. Another advantage of a cold trap gas purifier is that it freezes in contaminants such as the H_2O and CO_2 released from the walls. A typical temperature for the purifier is 122 K.

Detailed kinetic modelling of the laser process requires a thorough understanding of the reactions and their cross sections. The main kinetics for the inversion production are:

$$\mathrm{Xe^+ + Cl^- + Ne \rightarrow XeCl^* + Ne} \quad \text{(B3.2.17)}$$

$$\mathrm{Xe_2^+ + Cl^- \rightarrow XeCl^* + Xe} \quad \text{(B3.2.18)}$$

$$\mathrm{Xe^* + HCl_{\upsilon=1,2} \rightarrow XeCl^* + H} \quad \text{(B3.2.19)}$$

where Xe^+, Xe_2^+ and Xe^+ are formed in reactions with dimer ions and metastable dimers of the buffer gas as pointed out in section B3.2.4.1

The ionization of Xe occurs *via* excited states because the average electron energy of only a few electron volts is too small. The discharge instability is strongly enhanced by the dissociative attachment of electrons to HCl by the reaction

$$\mathrm{HCl + e^- \rightarrow Cl^- + H.} \quad \text{(B3.2.20)}$$

The higher the HCl concentration is, the higher the attachment rate of the electrons and the faster the growth of instability will be. The formation reaction for HCl from H and Cl is very slow so that the HCl must be replenished. During the discharge many other reactions take place [64].

B3.2.6.1 The discharge

Shortly after a homogeneous pre-ionization, with an initial electron density of at least 10^7 cm^{-3}, is obtained, a quasi-stable discharge can be maintained for a period up to 1 μs depending on the discharge parameters. During this period, instabilities are building up and these will eventually short-circuit the discharge. Thus there is a transit time during which the discharge remains uniform and useful for the laser process.

The discharge physics is as follows. The high electron density of about 10^{14} cm^{-3} for reaching the quasi-steady state is obtained by a fast high voltage pre-pulse. The initial electrons are accelerated in an avalanche process. The shorter the risetime is, the longer the acceleration voltage will remain much higher than the quasi-steady-state voltage and the faster the avalanche will be. The short risetime also has a favourable effect on the formation of a dense hot-spot structure on the cathode which enhances the discharge homogeneity [24]. The electrons move toward the positive electrode and leave the slow ions behind. In this way each initial electron is creating a cone of positive charges behind it. If there is sufficient pre-ionization the created cones will overlap and will form a homogeneous plasma moving to the anode. If, however, the pre-ionization is insufficient, an individual cone will reach the anode much faster than the overlapping cones forming a narrow channel (called a streamer) with a high plasma density between the anode and cathode. The streamer

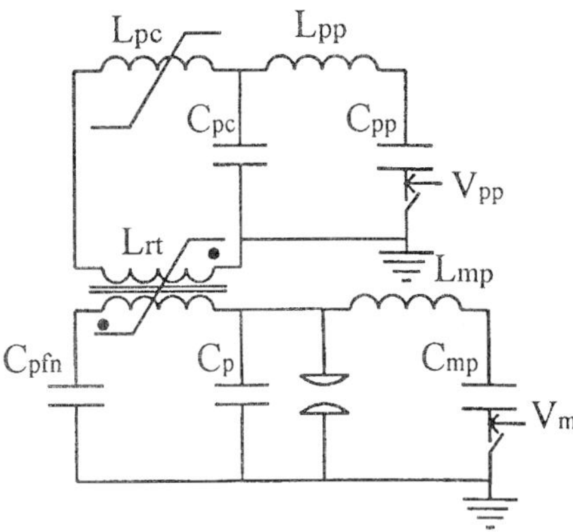

Figure B3.2.18. Scheme for the electrical circuit of an XeCl laser with a race-track transformer.

short-circuits the discharge. In the case of sufficient pre-ionization and a fast avalanche process, the ions shield the acceleration field for the electrons by forming a positively charged ion layer which increases its own acceleration field so that the ions acquire sufficient energy to bombard the cathode and thereby release fresh electrons (secondary emission) to maintain the discharge. It may also happen that some emission from the cathode is due to field emission by microscopic protrusions on the surface. Anyhow a quasi-steady state might be produced. However, the cathode will not emit uniformly and, after a short time, so-called hot spots are formed on the cathode surface, which are small areas where the electron emission is very dense and which develop to filamentary arcs destroying the homogeneity of the discharge. This process is strongly enhanced by the electron attachment to the HCl molecules. The quasi-steady time period, more or less the time needed for the filamentary arcs to reach the anode, must be used for the laser process. The question then is how much pumping energy can be supplied. Since the discharge voltage in the quasi-steady state is more or less independent of the current, the inversion production is then proportional to the current. Therefore, the highest output power and pulse energy will be obtained for the highest current that can be reached in the stable period. This is possible either by increasing the charging voltage of the circuit or by decreasing the inductance of the circuit. The former has been successfully performed in the photo-triggered mode (x-ray triggering). High output energies of 1.2 J l^{-1} [20], 0.5 J l^{-1} [65] and 3.5 J l^{-1} [66] at efficiencies of 1.7%, 0.9% and 1.6%, respectively, have been obtained. Increasing the voltage to double the steady-state value leads to current ringing and, because only the first half current wave is stable, this is not the most efficient energy conversion. Higher efficiencies with no current ringing and better performance are obtained with pre-pulse–main-pulse excitation [29, 30]. With the so-called switch mode, an output energy of 3.3 J l^{-1} at 4% efficiency [31] has been reported. For the resonant overshoot mode, the best reported results are 3.6 J l^{-1} at 5% efficiency [32]. The disadvantage of the pre-pulse–main-pulse technique is the high voltage of the main pulse across the electrodes which may initiate a slow avalanche of the pre-ionization electrons and therewith hampering an effective avalanche process by the pre-pulse. A fast risetime for the pre-pulse and a low inductance for the main pulse to obtain high current density are, therefore, the best conditions for high output energy.

B3.2.6.2 Parametric studies

We consider the resonant overshoot mode. The effects of various parameters have been studied for a system 60 cm in length and with an electrode gap of 2.5 cm [67]. The discharge width at optimum performance is 2 cm. The laser construction is shown in figure B3.2.12. The gas mixture used is 1 mbar HCl, 10 mbar Xe and Ne as buffer gas with a total pressure of 5 bar. The scheme of the electrical circuit is shown in

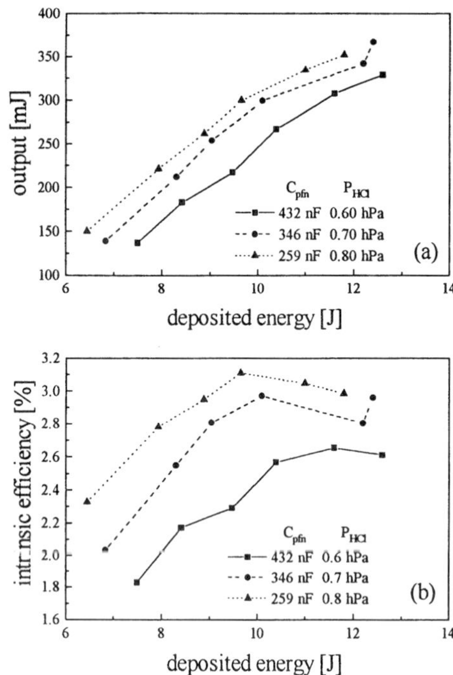

Figure B3.2.19. Optical output and intrinsic efficiency as a function of the deposited energy for different values of the pulse-forming network (PFN) capacitor.

figure B3.2.18. It consists of a pre-pulse and a main-pulse circuit. The saturable inductor of the main-pulse circuit is a race-track transformer to conduct the pre-pulse to the laser head. The pre-pulse circuit contains the capacitors $C_{pp} = 5.1$ nF and $C_{pc} = 3.4$ nF. For the main circuit: $C_p = 3.6$ nF and $C_{pfn} = 389$ nF. After charging C_{pp} its energy is switched to C_{pc} while it saturates L_{pc} in order to compress the pre-pulse passing through the transformer. The cross sections of both the ferrites of the magnetic pulse compressor and the main switch must be adjusted to hold off the required amount of volt-seconds. The relatively low current of the pre-pulse compressor is switched with eight ferrite rings having a total cross section of 35.2 cm^2. The low inductance race-track transformer, built from ferrite bars, has a length equal to the discharge and a cross section of 24 cm^2. It is observed that these cross sections are not critical for optimum performance.

The energy of the storage capacitor C_{pp} is transferred in about 130 ns to C_{pc}. Then, after saturating the ferrites, the energy is rapidly transferred to C_p in about 25 ns. Because of energy losses by transfer and saturation of the ferrites, the subsequent capacitors become smaller each time. The energy needed for the pre-pulse can be estimated by calculating the minimum ionization energy that is required to produce an ion density of 10^{13}–10^{14} cm^{-3}. Having an ionization potential of 22 eV for Ne and a discharge volume of 300 cm^3 we find 0.01 to 0.1 J. Since the efficiency for ionization is low the required energy will be in the range of 1 J. This has indeed been observed. Using peaking capacitors in the range 2.8–4 nF and charged in the range 20–30 kV, satisfactory results are obtained.

As already mentioned, the main capacitor is charged to twice the quasi-steady-state voltage in order to avoid ringing of the discharge current. If the voltage is higher only the first half-cycle of the current pulse is effective for the laser process and the intrinsic efficiency, defined as the output energy divided by the deposited energy of both pre-pulse and main pulse, will be less (see appendix). The dependence of output energy and intrinsic efficiency on the deposited energy are plotted in figure B3.2.19. The plotted results are

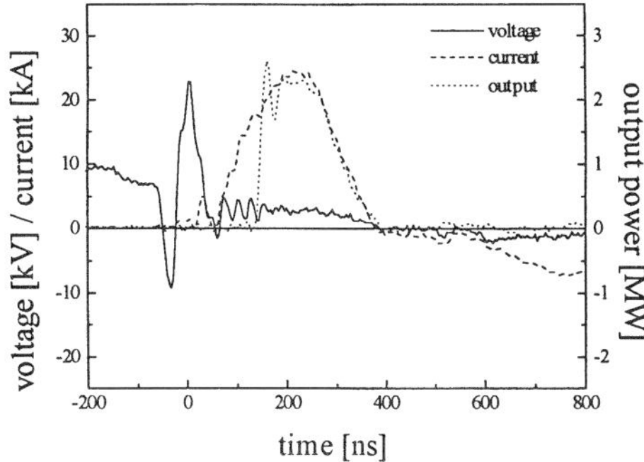

Figure B3.2.20. Voltage, current and output power of the XeCl laser with a PFN of 346 nF containing 17 J stored energy (10 kV).

optimized with respect to the HCl density. The smaller the value of C_{pfn} is the smaller the current pulse and the higher the HCl density become. This can be understood from the fact that for a shorter current pulse a faster growth for the instabilities caused by HCl, i.e. a higher HCl density, is allowed. A higher HCl density gives higher output power. Thus, for optimized conditions, the HCl density is adjusted in such a way that the discharge remains stable up till the end of the current pulse. The optical pulse will then closely follow the current pulse. Typical waveforms of the discharge current, voltage and output power for optimized conditions are shown in figure B3.2.20. It is seen that the output power does not terminate before the current pulse. If we increase the current pulse by increasing the inductance of the main circuit the optical pulse will follow this extension, provided the HCl density is sufficiently reduced. Laser pulses up to 1 μs can be obtained in this way. However, the total output energy will always be reduced by extending the pulse. As a general observation, it is found that a shorter current pulse gives a higher HCl density and a higher output energy. The maximum energy for short pulses is above 1 J l^{-1}.

B3.2.6.3 High average power XeCl laser

The development of high-repetition-rate XeCl lasers in the kilowatt range can be considered as a logical extension of successful single-pulse experiments. In principle, it can be performed with a fast-flowing circulating laser gas with a homogeneous velocity profile and thin boundary layers near the electrodes of the laser discharge. The flow speed is then determined by the condition of gas renewal after each pulse. However, although for repetition rates up to 1 kHz the pulse discharge time is much shorter than the time interval in between pulses, the flow pattern can be disturbed by discharge shock waves which, in turn, may have a detrimental effect on the homogeneity of the discharge and output beam. Furthermore, the optical elements and circuit components of the discharge have to be compatible with high repetition rate and long lifetime. For instance, fast-switching circuits with low heat production are required. The heat consumption of the saturable magnetic switches and its cooling is critical since the operational temperature range for high quality ferrites is about 50 °C above room temperature This may set a limit to the repetition rate. Also the reliability of a short-pulse, high-repetition rate x-ray pre-ionizer requires new developments in its high voltage performance and materials with long lifetimes.

The efforts to realize a high average power system were driven by various potential applications. In the middle of the 1980s, large national and international projects were started to reach the kilowatt power level with discharge systems. Depending on the application, the interest was in high pulse energy of the order of 10 J at low frequency or about 1 J at a much higher frequency. In Europe several companies in close collaboration with university groups or institutes participated in the Eureka programs. In Japan it

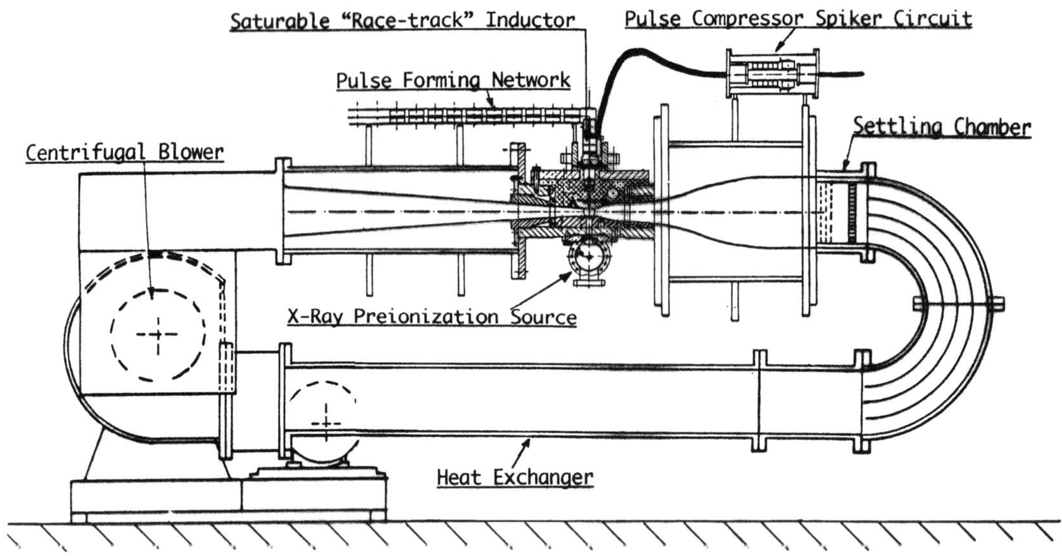

Figure B3.2.21. Schematic drawing of a 1 kW average power XeCl laser.

was coordinated by the Advanced Material Processing and Technology Research Association (AMMTRA). The Kurchatov Institute of Atomic Energy in Moscow and Rand Afrikaans University in Johannesburg also became involved.

The development carried out by the Nederlands Centrum voor Laser Research (NCLR) in collaboration with the University of Twente involves a repetition rate of 1 kHz, a pulse energy of 1 J and a pulse duration of 250 ns. [68]. The gas mixture consists of 2 mbar HCl, 20 mbar Xe and 4 bar Ne. The distance between the electrodes is 2.5 cm, the discharge width and length are, respectively, 2 and 80 cm. On both sides of the discharge region there are spaces to prevent mirror damage. For a general performance, the stable resonator has a 5 m radius of curvature end mirror with maximum reflectivity and a flat outcoupling mirror with 70% reflectivity. An output energy of 1 J is reached. The system also operates with an unstable resonator.

The laser system includes a gas circulation unit with a heat exchanger to remove the discharge and flow friction heat of the circulating gas. A cross-sectional view of the system is shown in figure B3.2.21. In order to minimize flow inhomogeneity and turbulence in the discharge chamber a settling chamber containing honeycombs and screens is placed in front of the contraction section. Passing the contraction section the flow speed is increased to reach 60 m s^{-1} in the discharge section. With a discharge width of 2 cm, the volume flow between two pulses is then three times the discharge volume. This clearing ratio of three is conservative and, with the chosen impedance-matched discharge technique, it may be reduced. After the discharge region the flow passes the diffuser to reduce the speed and therefore the friction losses. The contraction and diffusor walls connect smoothly to the uniform-field [69] high-voltage electrode and to the flat thin ground electrode that also acts as the entrance window for the x-ray pre-ionization. Current returns are placed both upstream and downstream to minimize the self-inductance of the laser chamber. The insulator of the high-voltage feed-through and the parts of the contractor and diffusor near the high voltage electrode were made of alumina. A gas purification unit with a cold trap and a hot reactor are incorporated into the gas-handling system to remove contaminations mainly produced by interaction with long-living discharge species and the walls. A constant HCl density is maintained by regulating the temperature of the cold trap. In this way the HCl losses by wall adsorption are replenished.

The x-ray source used (discussed in section B3.2.3.2.2) is based on a corona plasma cathode. An additional advantage of the chosen x-ray illumination through the thin anode is the separation of the source and the laser system. The operation condition requires a homogeneous short x-ray pulse with a very short risetime. The short pulse is not only beneficial for the discharge build-up, it also reduces the heat consumption of the plasma cathode. The lifetime of the plasma cathode at high repetition rate is limited to the order of magnitude of 10^8 pulses, mainly caused by material wear-out and surface sputtering. An alternative may be a thermionic x-ray source with high-yield (scandate) dispenser cathodes [70], which are well known for their long lifetime.

The discharge uses the resonant overshoot mode with the pre-pulse–main-pulse circuitry schematically shown in figure B3.2.13. As pointed out before, this technique requires the lowest amount of energy that can be switched as fast as possible. Furthermore, the main pulse is impedance matched to the laser discharge so that current oscillations do not occur. Sputtering from the x-ray transmitting electrode is also avoided. Thyratrons are the best choice for high voltage switching at high repetition rates, provided the average current and its change in time are limited. This is no problem for charging the pulse-forming network (PFN) of the main discharge up to 10 kV, about twice the steady-state discharge voltage. The PFN consists of eight double-plate transmission lines connected in parallel to the laser chamber. Using 160 ceramic capacitors of 2.7 nF each, resulting in a total capacity of 432 nF, the charged energy is 20 J. In the resonant overshoot mode, the voltage of the pre-pulse capacitor after the swing becomes $6U_B$. With a capacity of 4.5 nF the pre-pulse energy is 2 J.

For a pulse duration of 100 ns the average current is about 1.5 kA and dI/dt of the order of 10^{10} A s^{-1} which are within the specifications of the thyratron. This pre-pulse is further compressed to 25 ns with a saturable inductor. A good choice for a saturable high-frequency material is ferrite CMD 5005 from the company Ceramic Magnetics. The flux swing is about 0.5 T, the magnetic field intensity to reach saturation, H_s, is 2 kA m^{-1} and the field intensity to achieve μ–$5\mu_0$ is about 5 kA m^{-1} whereas the dissipation losses are reported as only 100 J m^{-3}. This inductor contains six rings with a total flux area of 21.6 cm^2.

The saturable inductor that connects the PFN with the laser chamber is, due to the large current, made from eight blocks with the dimensions 2.5 cm $\times$ 2.4 cm $\times$ 40 cm of the same ferrite material. They are arranged in a rectangular configuration with a 2 mm central slit for the high voltage conductor. Optimum performance of the laser was achieved with a cross section of 12 cm^2. Both saturable inductors are reset after each pulse by charging the PFN for the next one. Because of the considerable amount of dissipated saturation energy at 1 kHz, both inductors must be effectively cooled with a flowing oil circuit. Since the ferrite material itself is a poor heat conductor, copper-supporting structures with forced convection flows reaching the inner volume were made.

It should be noted that the application of a very short pre-pulse, limited by the smallest inductance of the laser head, is not only a requirement for pulse stability, it also reduces the amount of saturable ferrite material. For high repetition rates, it is crucial to minimize the amount of ferrite material because of costs, saturation energy losses and effective cooling.

B3.2.7 KrCl laser (222 nm)

The first lasing of this system was again observed with e-beam pumping [71, 72]. The performance of the e-beam-pumped KrCl system may be compared to that of KrF. The gas mixture also contains 100–300 mbar Kr, 2-4 bar Ar and 2–6 mbar Cl_2 instead of F_2. Ar as the buffer gas again gives the highest output energy. However, the observed output energy and efficiency of KrCl are both considerably lower than in the case of KrF under equal e-beam-pumping conditions. This is due to the different loss processes such as relatively high medium absorption at the laser wavelength and quenching of KrCl by the gas components.

The discharge pumping is straightforward. It has been studied for both single-discharge short pulses [73] and double-discharge long pulses with pre-pulse–main-pulse discharges [74]. The behaviour of the KrCl laser operating in the long-pulse discharge mode may also be compared to the XeCl laser. Both the kinetics of the laser process and the system technology are very similar. Self-sustained discharge pumping and the technology described for XeCl are appropriate. In fact, if for a XeCl pre-pulse–main-pulse system the Xe is replaced by Kr and the mirror reflectivities are adjusted for a 222 nm wavelength with an outcoupling percentage about half the value used for XeCl only the pulse energies have to be increased to reach the optimum performance. A typical gas mixture contains 3–4 bar Ne, 100–150 mbar Kr and 1–2 mbar HCl. Because of the relatively mild attachment rate of HCl compared with other halogen donor gases, the discharge stability allows also for relatively long pulses up to 500 ns without substantial loss of total energy. The gain, output energy density and the intrinsic efficiency are all 30–40% lower than that of the XeCl laser.

B3.2.8 ArF Laser (193 nm)

The first ArF was realized with e-beam pumping [40]. Shortly afterwards the first paper on discharge pumping appears [41]. The intrinsic efficiency, defined as the optical output energy divided by the electrical energy delivered to the laser medium, of the e-beam-pumped system is reported to be as high as 8% [75], whereas discharge pumping is less than 4% [76]. This system has also been examined very thoroughly in the past.

The development of the e-beam system was very similar to that of the KrF laser and its prospects for high energy systems to be used for inertial confinement studies were also considered [77–79]. A mixture of only Ar and F_2 where Ar is both donor and buffer gas does not work [16]. The best results are obtained with Ar and F_2 as donors and Ne as the buffer gas. Like other excimer lasers, it is found that a high e-beam current density gives higher optimized gas densities. At a high current density of 375 A cm^{-2} in a small coaxially pumped e-beam system, the optimum output does not saturate even at 14 bar, similar to KrF. Using a mixture of 28 mbar F_2, 280 mbar Ar and 14 bar Ne an output energy of 8.1 J l^{-1} was obtained for a pulse of 40 ns [16].

In spite of the successful e-beam studies, much more attention is, like several other excimer systems, given to discharge pumping because of its relative simplicity, higher development potential and more promising applications. A laser gas of a discharge system with densities in the range 2–5 bar contains about 1% F_2 or fluorine-bearing gas and about 5% Ar in a mixture with He or Ne as the buffer gas. The laser has been operated with NF_3 as the fluorine donor. As in the KrF system, the main kinetics for the formation of the ArF^* molecule are:

$$Ar^+ + F^- + X \rightarrow ArF^* + X \tag{B3.2.21}$$

where Ar^+ is formed in the reaction of Ar with the dimer ion of the buffer gas as given by equation (B3.2.10), X is the buffer gas atom and along the metastable channel

$$Ar^* + F_2 \rightarrow ArF^* + F. \tag{B3.2.22}$$

The F^- ions are formed in the dissociative attachment of electrons to F_2 i.e.

$$F_2 + e^- \rightarrow F^- + F. \tag{B3.2.23}$$

For e-beam pumping the ionization channel followed by reaction (B3.2.21) is dominant; whereas for discharge pumping with an average electron energy of only a few electron volts, the metastable channel and the formation by reaction (B3.2.22) is most productive. In modelling the kinetic process many reactions have been considered for e-beam [80] as well as for discharge systems [75, 81, 82]. Parallel to the formation channel there are also several reactions that channel energy away from the desired path or which quench the ArF^* molecule itself. The quenching is in competition with the stimulated emission and can be minimized by a cavity design that imposes a stimulated emission lifetime that is shorter than that of the quenching. Besides the losses by quenching, laser photons are absorbed by F^-, Ar^+ and rare-gas dimers in the discharge plasma [75, 83]. Furthermore, photons are also absorbed by impurities like O_2. However, this can be avoided

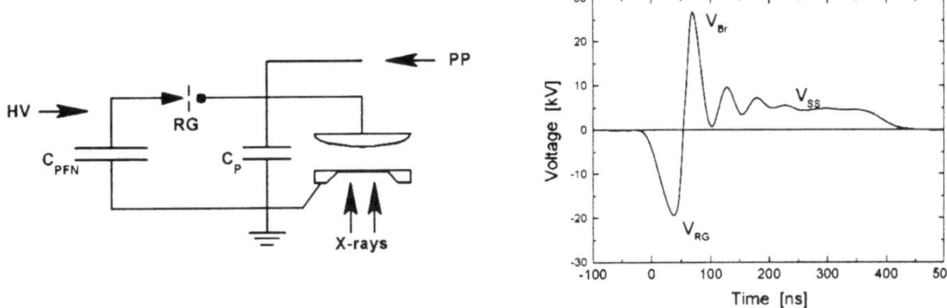

Figure B3.2.22. Scheme for a pre-pulse–main-pulse excitation circuit for a field-free medium during the pre-ionization: pp; pre-pulse; RG; rail gap switch, V_{RG}; breakdown voltage of rail gap; V_{br}: breakdown voltage of laser; V_{ss}; steady-state voltage of discharge.

by using a cold trap gas purifier and a thorough passivation of the laser vessel. Satisfactory purification is obtained by continuously flowing the gas mixture through a liquid nitrogen gas purifier with a cold trap of 90 K to freeze out the impurities [84].

The ArF* molecule exists in two very closely lying B and C upper states. The mixing between these states is so strong that they act as a single state with a lifetime of 4.2 ns [86]. The lower states, A and X, are also strongly mixed. Therefore, the radiative transitions between the upper and lower levels are a structureless continuum [86].

B3.2.8.1 The discharge

A major problem for maintaining a quasistable discharge is the dissociative electron attachment to fluorine. The attachment rate is of the order of 10^{-9} cm^3 s^{-1} so that, for an F_2 concentration of 10^{17} cm^{-3}, the free electron lifetime is only 10 ns. This means that the discharge pulse or pre-pulse must have a overlap with the pre-ionization pulse by starting the multiplication process. Furthermore, as pointed out for the XeCl excimer, the continuity of the discharge current is governed by both the field emission of microscopic protrusions on the surface of the cathode and dominantly by secondary emission of the cathode. Near the cathode a positively charged ion layer is formed which is accompanied by a high electric field near the cathode so that ions from this layer will bombard the cathode with sufficient energy to release electrons for the current continuity. In the used gas mixture with electronegative fluorine, the discharge instabilities develop in a few nanoseconds from minor pre-ionization inhomogeneities and cathode hot spots. This instability process depends strongly on the initial electron density near the cathode and the risetime of the acceleration voltage. A fast risetime for the current leads to the formation of a dense hot spot structure which enhances the homogeneity and favours stability. The absence of electrons or a low density near the cathode is very favourable for instabilities. This happens, for instance, in the resonant overshoot mode where the main voltage is across the electrodes before the pre-ionization pulse starts. Some of the pre-ionization electrons will then drift away from the cathode before the pre-pulse of the discharge is applied and, since the attachment rate in F_2 is much faster than in HCl, the electron depletion near the cathode will become too strong. Thus, for mixtures containing F_2 the stability period becomes shorter, if present at all, than in mixtures containing HCl. There will be at least a loss of potential laser energy. This has, indeed, been observed when the resonant overshoot mode has been applied to mixtures containing F_2 in ArF and KrF lasers. The cathode depletion problem can be circumvented with the discharge scheme shown in figure B3.2.22 [84]. It is very similar to the resonant overshoot mode except for the absence of the main voltage across the electrodes before pre-ionization.

Even if good pre-ionization conditions exist, the fast attachment rate still requires a very fast electron

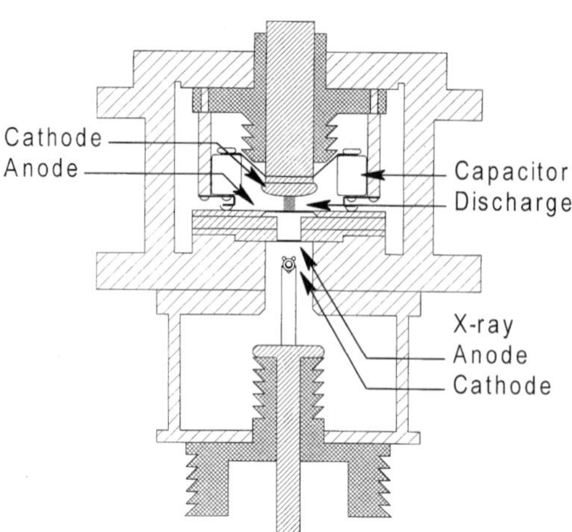

Figure B3.2.23. Cross-sectional view of a laser head with peaking capacitors and current return rods inside the laser head to reduce self-inductance.

multiplication process or avalanche process. This can be performed by minimizing the current risetime of the pre-pulse. The self-inductance of the pre-pulse circuit should, therefore, be as short as possible. A successful construction is shown in figure B3.2.23 where the peaking capacitance is placed within the laser vessel [84]. The decrease in current risetime results in a denser hot spot structure on the cathode surface and therewith to a more homogeneous discharge, longer pulses and more output.

B3.2.8.2 Laser performance

In the past several authors have discussed the ideal buffer gas option for optimizing the ArF laser. Some of them found the best results with He, while others had better results with Ne or a mixture of these gases. However, the general impression has, for a long time, been that it does not make much difference which choice is made. It has been suggested that, with Ne, the formation kinetics of the lasing molecule is much more efficient, whereas the optical extraction efficiency, limited by intracavity absorption, is favoured by a He gas medium with the result that both options are effectively more or less equal [85, 86].

In a He buffer gas mixture, the electron multiplication process and breakdown are faster than in a Ne mixture, because of the lower inductance by less scattering losses and therewith higher current. However, the more effective electron losses by three-body volume recombination in Ne buffered mixtures such as

$$\mathrm{ArNe}^+ + \mathrm{e}^- \rightarrow \mathrm{Ar} + \mathrm{Ne} \qquad \text{(B3.2.24)}$$

offer a current control mechanism which helps to stabilize the discharge. In the case of He, the production rate of ArHe^+ is much smaller because of the higher ionization energy and the lower E/p value which is related to the average electron energy. Since instability growth is very fast in mixtures containing electronegative fluorine, it follows that the choice of buffer gas depends on the current risetime of the pre-pulse discharge. Using a system with minimized inductance, as shown in figure B3.2.23, the difference between the two buffer gases is substantial. Experiments have been performed with the following parameters [84]. The discharge length is 60 cm, the electrode gap is 1.2 cm and the discharge width, determined by the pre-ionization area, is about 0.7 cm so that the active medium is about 50 cm^3. The peaking circuit is formed by 10 ceramic capacitors with a total value of $C_p = 460$ pF. To reduce the self-inductance of the main circuit the laser vessel is fitted with additional current returns. These peaking capacitors and current returns are distributed evenly

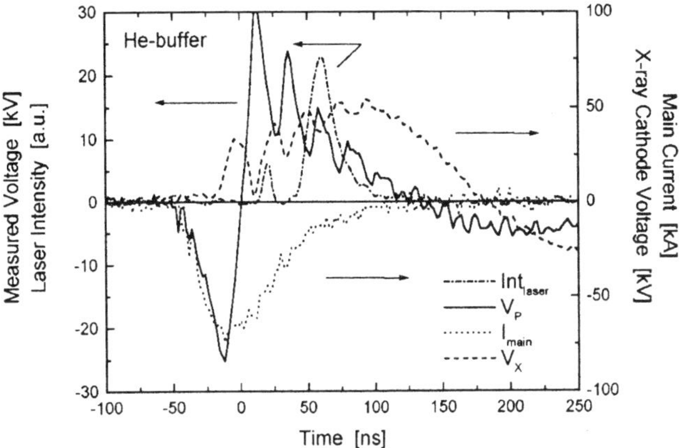

Figure B3.2.24. Waveforms of the ArF laser using He as the buffer gas.

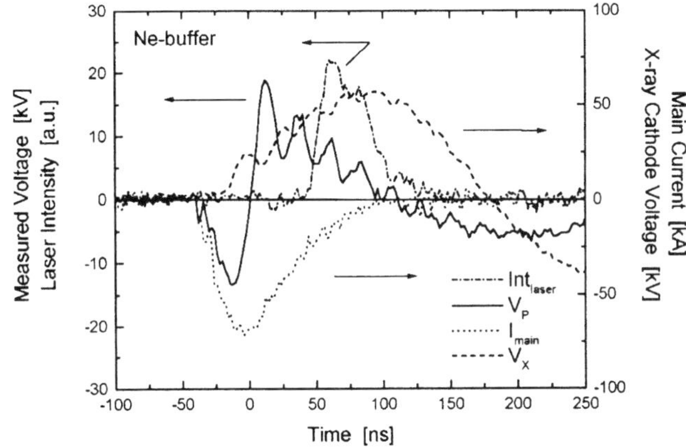

Figure B3.2.25. Waveforms of the ArF laser using Ne as the buffer gas.

along the cathode length. In this way the inductance of the laser vessel is about 8 nF. The main circuit has a capacitance of $C_{pfn} = 162$ nF and is charged to 32 kV. The switch gap together with C_p is very fast charged by means of a mini-Marx generator with an effective output capacitance of 0.6 nF delivering a 120 kV pulse with a risetime of 2 ns. For a gas mixture that consists of 3 mbar F_2, 250 mbar Ar and 4.75 bar He the optimum pulse length of this system is 32 ns (FWHM), as shown in figure B3.2.24, and the output energy is 1.9 J l^{-1}. In a mixture of 3 mbar F_2, 250 mbar Ar, 60 mbar He and 4.7 bar Ne the pulselength becomes 48 ns (FWHM), as shown in figure B3.2.25, and the output is 3.6 J l^{-1}. These output energies are in the range 2–5 J l^{-1} as is also obtained with single-discharge systems producing pulselengths of no more than 12 ns. [87–90].

B3.2.8.3 Long pulses

Since the fast growth of the instabilities are mainly induced by the electronegative fluoride gas, it is expected that pulse elongation can be obtained by reducing the fluoride content, i.e. with lean mixtures. This has indeed been observed. However, it has also been observed that the strongly reduced self-inductance of the laser

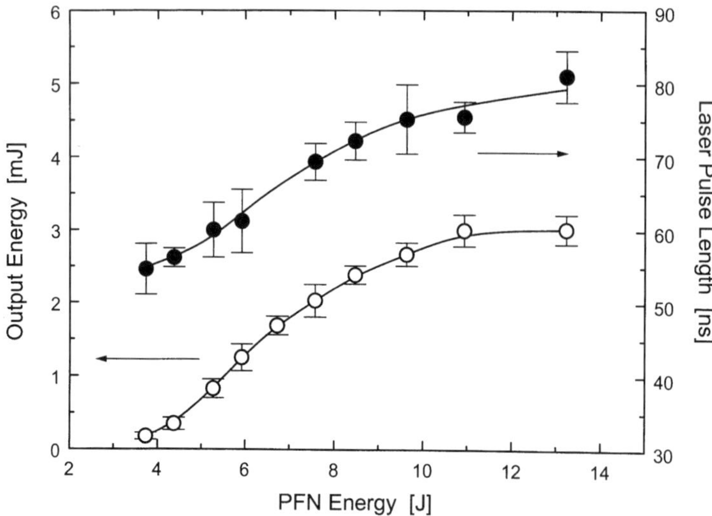

Figure B3.2.26. Output energy and pulse duration of an ArF laser as a function of the energy stored in the PFN.

vessel, as shown in figure B3.2.23, is also essential [84]. The usual systems with circuit components outside the laser vessel performing much longer current risetimes are not appropriate for minimizing the instabilities. Unfortunately, the fluorine reduction also affects the laser power and pulse energy and an extension of the quasi-stable discharge period is found at the expense of laser power and total energy. The reduced gain requires a higher output coupler reflector that has to increase to at least 90% instead of the usual 60–70% for the previously described high-energy systems.

An advantage of operational systems with long pulses and lean mixtures is that, due to the reduced electron attachment rate, the more convenient resonant overshoot mode with a saturable magnetic ferrite switch can be successfully applied. The experiments have been performed with the following parameters [84]. The gas mixture consists of 1 mbar F_2, 20 mbar He, 50 mbar Ar and 5 bar Ne. The main circuit capacitance $C_{pfn} = 427$ nF. The peaking capacitance $C_p(= 460$ nF) placed inside the laser vessel is again charged very fast by the mini-Marx with an output capacitance of 0.6 nF and delivering a 120 kV pulse with a risetime of 2 ns. If the charging voltage of C_{pfn} is varied between 4.1 and 7.9 kV with a corresponding energy variation between 3.7 and 13.2 J, both the pulselength and output energy increase as shown in figure B3.2.26. Having a discharge volume of about 50 cm^3, the output reaches 60 mJ l^{-1}.

The pre-ionization timing is a very critical parameter in this laser system. The first condition for discharge stability is a homogeneous pre-ionization with an electron density of at least 10^8 cm^{-3}. If this pre-ionization is too early, too many electrons are lost by attachment and fast instabilities will develop when the discharge starts. If the pre-ionization is too late, there is too little time left for electron multiplication. In figure B3.2.27, the output energy and corresponding pulse duration(FWHM) are shown as a function of the pre-ionization time delay, defined as the time between the onset of the X-ray pre-ionization pulse and the discharge breakdown. The pulselength is, as expected, dependent on the fluorine content. This is plotted in figure B3.2.28 where the other gas components are as previously described.

B3.2.9 XeF laser (351, 353 and 486 ± 35 nm)

Like most excimer systems the first XeF(B→X) excimer laser was realized with e-beam pumping [39]. Although the system is complex, this was an attractive way to get some understanding of the potentials of

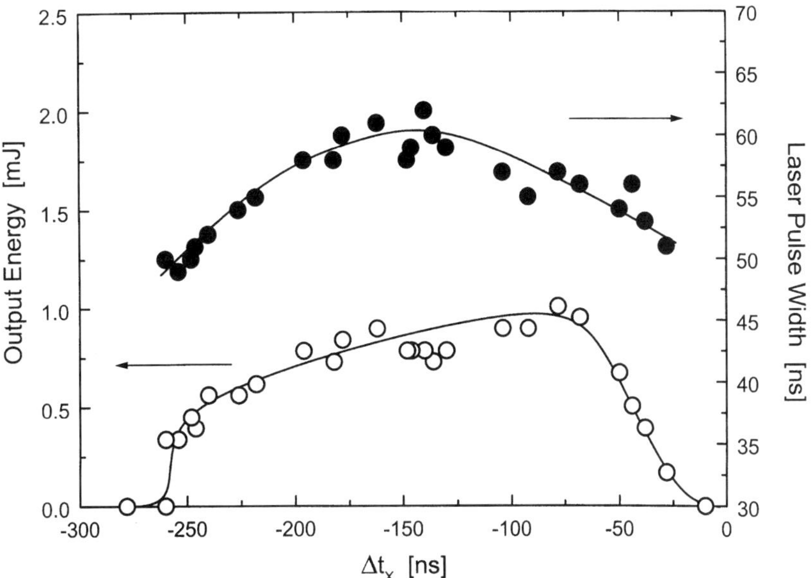

Figure B3.2.27. Pulsewidth (FWHM) and output energy of an ArF laser as a function of the pre-ionization delay time.

this excimer [91, 92]. The instabilities which are inherent to the simpler discharge systems can be avoided so that the choice of gas composition can be totally based on kinetic considerations. Shortly afterwards papers on discharge-pumped XeF(B→X) lasers appeared [93, 94]. Although the efficiency and output energy density are much less than with e-beam pumping, the continuing research is mainly carried out on discharge systems because of their relative simplicity, higher development potential and more promising applications. The gas mixture consists of a few mbar F_2 or NF_3, a few mbar Xe and several bars of He or Ne as buffer gas. The discharge stability sets stringent limitations on gas composition and densities, especially for the halogen donor gas. Detailed kinetic modelling of the laser process requires a thorough understanding of the reactions and their cross sections. The most important reactions are similar to those of other excimers. For the ionic channel we have

$$\mathrm{Xe^+ + F^- + X \rightarrow XeF^* + X} \tag{B3.2.25}$$

where X is the buffer gas atom. As described in section B3.2.4.1, Xe^+ is formed in a charge transfer reaction with dimer ions of the buffer gas. The metastable channel has the formation reaction

$$\mathrm{Xe^* + F_2 \rightarrow XeF^* + F} \tag{B3.2.26}$$

or

$$\mathrm{Xe^* + NF_3 \rightarrow XeF^* + fragments} \tag{B3.2.27}$$

where Xe^* is mainly formed in a reaction with the metastable dimer of the buffer gas as indicated in section B3.2.4.1. The F^- are formed by the dissociative attachment of the electrons to NF_3 or F_2.

For e-beam pumping the ionic channel followed by reaction (B3.2.25) is dominant, whereas for discharge pumping the metastable channel with the reaction (B3.2.26) or (B3.2.27) is most productive.

The attachment rate has a great impact on the stability of the discharge. This rate is different for NF_3 and F_2 [95] which means that the discharge behaviour depends on the choice of halogen donor gas. It is found that for the XeF system NF_3 is the better choice. The main reason is the rather large absorption cross

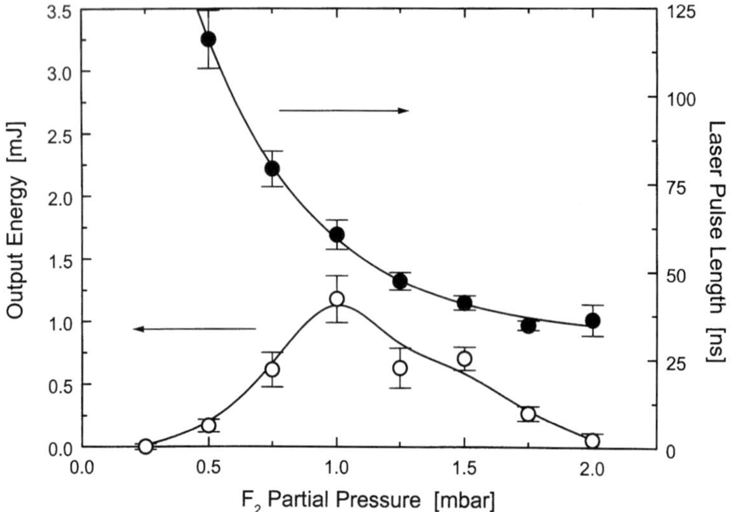

Figure B3.2.28. Pulsewidth (FWHM) and output energy of an ArF laser as a function of the F_2 density.

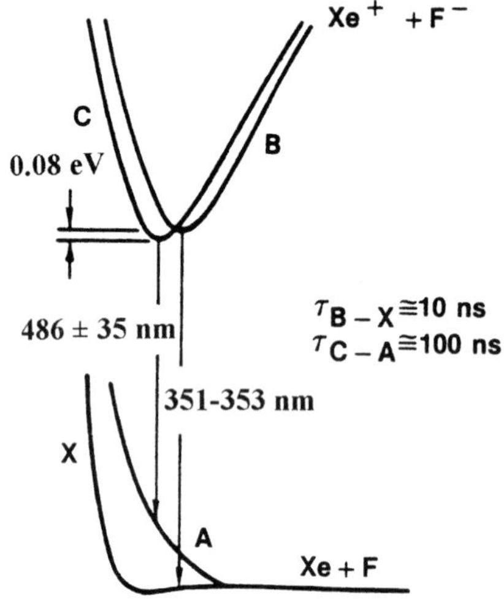

Figure B3.2.29. Schematic potential energy diagram of XeF showing both the B→X and the C→A transitions.

section of the F_2 molecules for the laser radiation. Furthermore, with F_2 the discharge becomes unstable more quickly compared to NF_3-based gas mixtures, although the total output energy remains more or less equal [96].

Once the XeF^* is formed, it can easily be quenched by all gas constituents and discharge electrons. The quenching process will be in strong competition with the stimulated emission so that a high radiation field density in the cavity is required. However, there is also optical absorption, mainly photoionization of the dimer states Xe_2^+, Xe_2^* and Ne_2^* as well as photo-dissociation of F^- and F_2. For a optimized laser system, there will be a trade-off between quenching and absorption losses.

B3.2.9.1 The spectroscopy

The bottom of the C state of the excimer potential curve as a function of the internuclear separation distance is approximately 0.08 eV lower than that of the B state as indicated in figure B3.2.29. The A state is repulsive while the X state is weakly bound. The binding energy of the X state of 0.15 eV is rather strong compared to other excimers. The bottom of this potential curve is situated at shorter internuclear separation than the bottom of the potential curve of the B state. Because of the weakly bound lower state the emission in the B→X band has a complex structure [97], starting on the lower vibrational levels of the B state and terminating on high vibrational levels of the X state. The strongest lines are at 351.1 and at 353.1 nm which are due to the 1–4 and 0–3 vibrational bands respectively. The linewidths of these bound–bound B→X transitions are of the order of 0.1 nm. The bound-free C→A transition is much broader with a bandwidth of about 70 nm centred at 486 nm. Tunability, as will be discussed later, reaches from 454 to 525 nm. The radiative lifetime of the B state is about 10 ns, while the C state has a radiative lifetime of 100 ns. The effective lifetime of the X state is about 1 ns for the usual gas composition and densities. The cross sections for stimulated emission for the strongest transitions of the XeF excimer laser are, for the usual gas conditions, estimated as 4.5×10^{-16} cm^2 for the B→X band and 7.4×10^{-18} cm^2 for the C→A band, respectively [98]. Because of the much smaller cross section for stimulated emission of the C→A band, which is due to its large linewidth, the laser action will always occur in the B→X band unless special measures are taken.

B3.2.9.2 The discharge

A straightforward approach to a XeF(B→X) laser is the use of a single self-sustained discharge; with a conventional single-discharge circuit with a spark gap switch and UV pre-ionization laser pulses of 20 ns with a specific output of 1.2 J l^{-1} have been reported [99]. A further improvement is obtained with x-ray pre-ionization and a single-discharge circuit that includes a low-inductance multi-channel spark gap for fast switching. By increasing the charging voltage of the circuit capacitor, the discharge current can be enhanced to obtain higher output energy. In this way laser pulses of 85 ns and a specific output energy of 4.7 J l^{-1} have been obtained for a gas density of 6 bar [100]. In such a discharge the driving voltage of the circuit is much higher than twice the steady-state voltage so that the discharge will go into an oscillatory mode, of which only the first phase before the current reversal will be stable. The efficiency, in this case 0.5%, is therefore inherently low. Furthermore, this technique is not suitable for high repetition rates.

A much better performance is obtained with the overshoot mode of the pre-pulse–main-pulse technique [101]. The system has the following parameters. The gas mixture used contains 1 mbar NF_3, 5 mbar Xe and 4 bar Ne. The cavity discharge length is 60 cm, the electrode gap is 1 cm and the discharge width 1.25 cm. The PFN capacitor of the main circuit of 108 nF has a charging voltage of 17.4 kV. The capacitance of the pre-pulse circuit of 5.6 nF is charged to 20 kV. The output energy and pulsewidth have broad maxima with respect to the delay time between the pre-ionization and the pre-pulse [101], as shown in figure B3.2.30. For good performance, the delay of the pre-pulse after the onset of the pre-ionization should be more than the attachment lifetime of the electrons in the gas mixture and less than the pre-ionization pulsewidth. The maximum output energy is 180 mJ or 2.4 J l^{-1} with a pulse length of 70 ns. It should be noted that the pulselength, similar to that of the ArF laser, can be extended at the expense of the output energy if the halogen content is reduced.

The risetime of the pre-pulse taking care of the avalanche electron multiplication must be, as pointed out in section B3.2.3.2.7, as short as possible. In the previously mentioned results, this risetime is about 50 ns. By placing a magnetic pulse compression circuit (MPC) just before the charging of the pre-pulse this risetime can be further reduced to 20 ns. The output enhancement is remarkable. This is seen in figure B3.2.31 where the effect of the pre-pulse risetime is shown by plotting the output energy as a function of the charging voltage of the main capacitor (PFN). The highest output energy for the mentioned gas mixture is 275 mJ or 3.7 J l^{-1}.

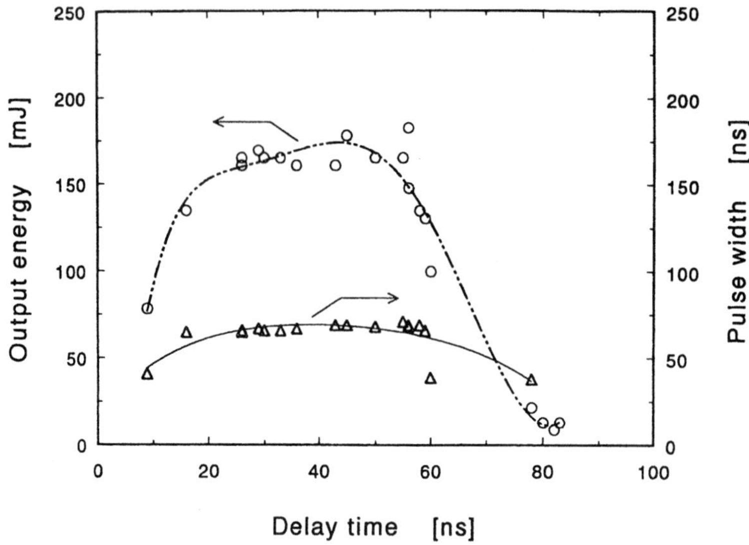

Figure B3.2.30. Pulsewidth (FWHM) and output energy of a XeF laser as a function of the pre-ionization delay time.

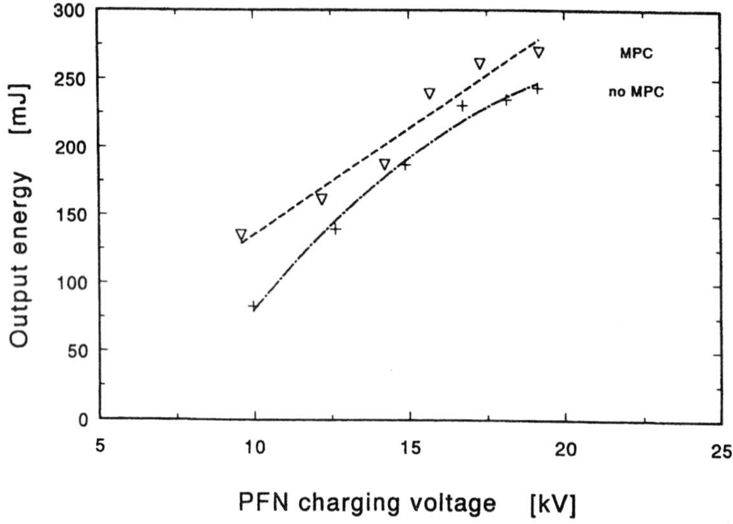

Figure B3.2.31. Output energy as a function of the charging voltage with and without magnetic pulse compression (MPC) of the pre-pulse.

B3.2.9.3 Broadband blue–green C→A transition

Most excimer lasers operate on the relatively narrow-band transitions between the B and the nearly flat or at most weakly bound X state. Transitions between the C and the strongly repulsive A state are relatively broad because, even with a small deviation from the bottom of the potential well of the C-state energy curve, a considerable wavelength shift will occur and, therefore, this transition exhibits a much smaller cross section for stimulated emission than the B→X transition. Moreover, most excimers have a C state sufficiently higher than the B state so that, at room temperature for the gas, it is essentially empty. (During the discharge and laser pulse the gas temperature does not increase substantially. The discharge energy is initially mainly transferred to electronic excitation and ionization processes. Some of this energy will then be transferred to thermal motion after the laser pulse.) This means that the B→X transition is the only one observed for most excimers.

The XeF system is, in this respect, exceptional because the C state lies 0.08 eV below the B state as indicated in figure B3.2.29. At high buffer gas density, the B and C states are coupled by collisional mixing. The buffer gas will thermalize the state densities in times shorter than the radiative lifetime of the B state. The C state may therefore have a considerably larger population density which is favourable for satisfying the laser threshold condition. Nevertheless, the small signal gain equal to the product of the upper state density and its cross section of stimulated emission is smaller than for the B→X transition so that, in order to get laser oscillations on the C→A transition, special measures must be taken. This can be done with mirrors that are highly reflective for the region around 486 nm and less reflective in the region of 351 nm. To make the system tunable over the entire blue–green region, highly reflective dispersive optical elements like a Littrow prism or a diffraction grating can be used as end reflectors. The output coupler can then be a highly reflective mirror with a few per cent transmission.

Laser oscillation on the C→A transition has been reported by a number of investigators with e-beam excitation [102, 103]. Typical gas mixture are 1–3 mbar NF_3, 3–10 mbar Xe and 2–8 bar Ar. The addition of 0.1–1 bar Kr has a favourable effect on the laser performance by suppressing the UV of the B→X transition. Small amounts of F_2 are also beneficial. Using a mixture of 6 bar Ar, 0.5 bar Kr, 8 mbar NF_3, 10 mbar Xe and 2 mbar F_2 excited by 1 MeV electrons of 250 A cm^{-2} and a pulse duration of 10 ns, a pulse output energy of more than 2 J l^{-1} at an intrinsic efficiency of 1.5% has been obtained [104]. Scalable and continuously tunable e-beam-pumped systems covering the blue–green region of the spectrum has been described by several authors [105–109].

Nearly simultaneously discharge-pumped systems operating on the C→A transition have been obtained [110, 111]. Because of the small cross section for stimulated emission, intense electric pumping is needed. The first report [110] described an output density of 0.56 mJ l^{-1} which was achieved in a short pulse UV pre-ionized single discharge with an input power of 11 MW cm^{-3} and a gas mixture of 3 bar He, 4 mbar Xe and 2.7 mbar NF_3. Using a mixture of 2.2 bar He, 7 mbar Kr, 3.5 mbar NF_3 and an input energy of 2.7 MW cm^{-3} an output energy of 9 mJ l^{-1} was reached [112]. These intense discharges with high electron densities contribute to the mixing of the B and C states which has a detrimental effect on the C-state density because the average kinetic electron energy is much higher than that of the laser gas. Strong pumping may also increase the production of absorbers of the C→A radiation. Furthermore, the applied short single-pulse UV pre-ionized discharge leads to considerable difficulties in optical energy extraction because the available build-up time for stable cavity oscillation is short for this excitation technique. For e-beam devices, one has the advantage of generating higher gain to prolong its period and to minimize the absorbers by choosing the gas composition and its density independently from the discharge parameters so that laser oscillations on the C→A transition have been successfully generated even for short e-beam pulses. An approach to solve this problem for short-pulse discharges is to apply external injection of a suitable dye laser pulse into a discharge laser to provide an enhanced optical flux in the cavity that is much higher than the C→A spontaneous emission flux. In this way, the entire blue–green line can be tuned by tuning the injection source. This has been successfully demonstrated by injection into both a stable [113] and an unstable [114] XeF discharge system.

A very promising principle [115] for obtaining XeF(C→A) lasers with long pulses and high output energy is optical pumping by photodissociation of XeF_2 vapour. The XeF_2 vapour can be conveniently produced from solid crystalline XeF_2 kept in a separate container closely attached to the laser tube. In this photolysis process, the components XeF(B) and F are formed. Efficient energy transfer from the B to the C state occurs by atomic and molecular collisions. In the first experiments it was demonstrated by using e-beam excited fluorescence of Xe_2^* at 172 nm as the photodissociation source. Nearly simultaneously, broadband optical pumping of XeF_2 with the exploding wire technique was successfully applied [116]. Further development has led to multi-joule laser pulses of several microseconds [117,118]. For obtaining repetitively operating devices, surface discharges are used [119,120]. By means of ferrite surface plasmas [121], selective radiation production can be obtained in the spectrum region below 220 nm that coincides with the dissociation absorption continuum of XeF_2 which lies between 140 and 200 nm. With high-current surface discharges on ferrite rods placed near the wall of the laser tube effective illumination is obtained. Each rod is initially conditioned by burning a conductive path into its surface by discharging a current pulse through a thin aluminium wire taped to the surface of each rod. The thus formed narrow track on each surface then becomes a line light source from the narrow high-current surface discharge when a capacitor is fired in a fast circuit. Using two parallel rods of 76 cm length energized by a 6 μF capacitor at 50 kV and a gas filling of 450 torr Ar, 50 torr N_2 and 12 torr XeF_2 in an active volume of 1 l an output energy of 8 J at 351 nm has been obtained [122]. A compact device with an output pulse energy of 225 mJ based on this ferrite surface plasma has also been described [123].

B3.2.10 F_2 laser (157 nm)

The halogen molecular laser F_2^* operating at 157 nm has, so far, been the only high-power laser operating in the VUV. As pointed out in the introduction, solid-state-based VUV laser systems are no alternative as the thermal loading of the laser crystals and the absorption of the frequency conversion crystals prohibit operation at comparable power levels. The F_2^* laser, like other halogen lasers, operates on transitions between molecular electronic states. At the usual gas densities, the lifetime of the lower state is smaller than that of the upper state. Strictly speaking, this laser does not belong to the class of excimer lasers as they do not radiate to an unbound ground state. However, the formation kinetics and laser parameters resemble that of the rare-gas halogen excimer lasers. The F_2 laser has, like the XeF laser, a weakly bound ground state and the depopulation rate is, even at high output power densities, fast enough to prevent saturation of the radiation production. The gas composition is also very similar. The halogen lasers also have a multi-atmosphere rare-gas density with a very small fraction of the lasing halogen gas.

A great deal of the F_2^* laser research was carried out with e-beam pumping [124–130]. The great advantage of e-beam pumping is, in principle, the detailed study of the kinetics for various input powers and gas densities without loss of homogeneity and input stability as is the case for discharge pumping. The observed very high gain is one of the most striking phenomena. In the absence of resonator mirrors, strong amplified spontaneous emission (ASE) will be observed and optimized lasers may have output couplers with a few per cent reflection. For e-beam-pumped systems a gain of more than 50% per cm has been measured for an input power density of 6 MW cm^{-3} in a gas of 8 bar He containing only 0.1% F_2. The gain even increases to more than 60% per cm if, for the same gas density, 60% of the He is replaced by Ne and the input power density with the same input current density changes to 11 MW cm^{-3} [131]. It is observed that the optical pulse duration is only limited by the e-beam duration, even at high input power. The lifetime of the stable ground state is, for the usual gas conditions, only a few nanoseconds. Any self-termination by the accumulation of the lower level population has not been observed for pulses up to 160 ns. Both He and Ne have been used as buffer gases. Since the deposition of the e-beam energy in Ne is higher than in He the gain and output power increase slightly if, for the same e-beam current, the Ne concentration is between 10 and 60%. With an e-beam pulse of 160 ns delivering an input energy of 400 J l^{-1} an output energy of 1.8 J in a volume of 157 cm^3, corresponding to 11.4 J l^{-1}, was realized with an intrinsic efficiency of 2.8% for an Ne–F_2 mixture of 8 bar containing only 0.15% F_2 [131]. E-beam pumping yields, as for all excimer lasers, much higher power densities and efficiencies than discharge pumping. Nevertheless, the much simpler discharge pumping has attracted much more research efforts.

For discharge systems, the best performance is obtained with He as the buffer gas because of discharge stability. The development of this type of laser challenged previous successful advances in discharge technology for rare-gas halogen excimer systems. Furthermore, it may also have been stimulated by its application potential. The high photon energy combined with high beam power discloses a new area of application which could not be reached by the rare-gas halogen excimers.

The first step of the reaction channel for reaching the excited electronic state of the upper laser level is the formation of the rare-gas halogen excimer molecule by the Coulomb attraction between the F^- ions and the He^+ or He_2^+ ions. The F^- ions are produced by the dissociative attachment of discharge electrons to F_2 molecules according to reaction (B3.2.23). The helium ions are produced by the interaction of discharge electrons with the buffer gas as described in section B3.2.4.1. The initially highly vibrational excited excimer molecule relaxes towards its lower vibrational levels in collisions with the buffer gas. During this relaxation several curve crossings with repulsive He+F* states occur, giving the excimer molecule a high probability to pre-dissociate into ground-state helium and an excited fluorine atom [132] according to the reactions

$$He^+ + F^- + He \rightarrow HeF^* + He \rightarrow F^* + 2He \qquad (B3.2.28)$$

and

$$He_2^+ + F^- \rightarrow HeF^* + He \rightarrow F^* + 2He. \qquad (B3.2.29)$$

When Ne is used as the buffer gas, the reactions are similar.

The formation of the $F_2(D', \upsilon)$ molecule in the upper laser level occurs during a three-body collision of F^* with a fluorine molecule and a buffer atom like:

$$F^* + F_2 + He \rightarrow F_2(D', \upsilon) + F + He. \qquad (B3.2.30)$$

Excited F^* atoms will also be ionized in the discharge and subsequently recombine with F^- in three-body collisions to form the excited fluorine molecule according to

$$F^+ + F^- + He \rightarrow F_2(D', \upsilon) + He. \qquad (B3.2.31)$$

Highly vibrational excited $F_2(D', \upsilon)$ may relax to lower or lowest vibrational state in collisions with buffer atoms before the stimulated emission takes place. If the stimulated emission process is not fast enough, because of a low optical quality of the resonator, the $F_2(D', \upsilon)$ molecules are quenched in collisions with the gas and the discharge electrons, of which F_2 is the dominant quencher. The amount of F_2 is, therefore, critical. In principle, it should be enough for the production of the upper laser state during the excitation pulse. For lower concentrations, the output will also be relatively lower. (The regeneration process of fluorine molecules is too slow during the excitation pulse.) Moreover, the electron density will be higher due to the reduced loss of electrons by attachment, so that electron quenching becomes higher. However, too high a fluorine concentration results in too high a quenching rate by fluorine. With a longer excitation pulse, the trade-off with respect to the fluorine content becomes more critical.

The kinetics of the laser process are rather efficient. Some energy is lost by the radiative decay of HeF^* forming F^*. Minor losses in the discharge take place by radiative decays of various excited states of atoms and dimers. Although the radiative lifetime of F^* to the ground state is only a few nanoseconds, its radiation is trapped by ground-state fluorine atoms which are produced in various reactions. The effective lifetime in the usual laser gas is estimated to be 100–200 ns [133]. The radiative decay of F^* runs parallel to the formation of the upper laser state $F_2(D', \upsilon)$. For the usual gas mixtures the decay losses are relatively low.

There are also some minor absorption losses. The main losses are the photoionization of He^* and F^* and the electron photo-detachment of F^-. However, the absorption cross sections are small. There may be some additional absorption from impurities like O_2, HF and CF_4 that are present in the fluorine gas. Furthermore, the resonator mirrors are usually made of coated or uncoated MgF_2 which also gives some absorption losses for the laser wavelength.

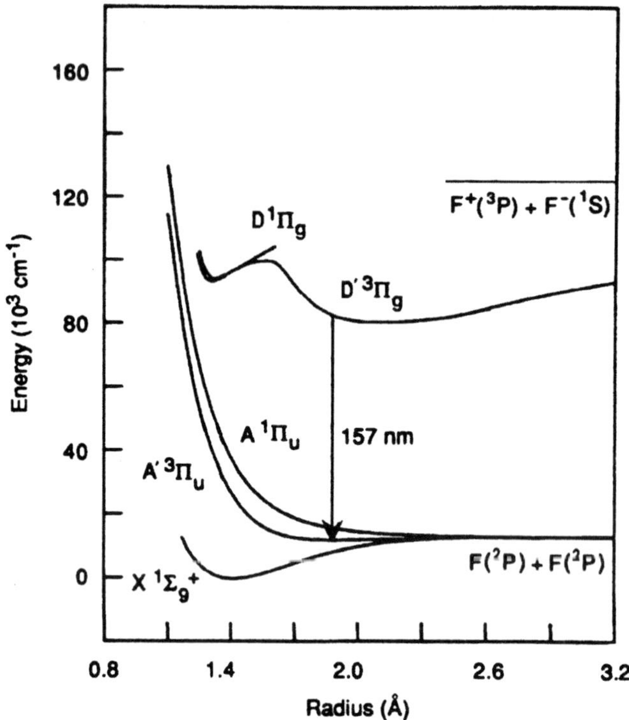

Figure B3.2.32. Schematic potential energy diagram for electronic levels in F_2 showing the laser transition.

B3.2.10.1 The spectroscopy

The laser transition takes place between the $D'^3\Pi_g$ to the $A'^3\Pi_u$ state in molecular fluorine, as shown [134] in figure B3.2.32. The spontaneous emission spectrum shows a complex structure containing many rotational–vibrational bands. At the usual high pressure of the laser gas, the vibrational and rotational relaxations are fast enough to assume a homogeneously broadened transition. The observed spontaneous emission is predominantly from its lowest vibrational level. The linewidth is approximately 0.3 nm. Laser oscillations are observed on two lines: a strong one at 157.63 nm which is always accompanied by a weaker line at 157.52 nm [62, 135]. The energy gap between the two lines is attributed to a rotational splitting in the lower laser level. An additional weak line at 156.71 nm is observed in e-beam-pumped systems [136]. When helium is used as the buffer gas, the laser pulse is also accompanied by a red laser pulse which originates from transitions of the higher energy levels in the atomic F* manifold [124]. This is not observed when neon is used as the buffer gas and may be due to the fact that the predissociation of NeF* mainly populates the lower energy levels of this manifold which prevents laser inversion [125].

The lower laser level of the F_2^* system is a covalently bound electronic state. It has practically the same small binding energy of about 0.15 eV as the lower laser level of the XeF* laser. Such a small binding energy leads under the usual high gas density to an effective lifetime of the lower laser level of about 1 ns. This small lifetime prevents a substantial population build-up of the lower laser level that might lead, at high power densities, to an effective reduction in the inversion density. The laser process, similar to that in excimer lasers, is not hampered by the ground state. From a radiation lifetime of 3.7 ns for the upper state [134] and a linewidth of 0.3 nm the stimulated emission cross section according to equation (B3.2.1) becomes $\sigma_{st} = 6.8 \times 10^{-16}$ cm^2. This relatively large cross section, compared with those found for most rare gas halogen excimer lasers, is due to its relatively narrow linewidth and explains the relatively high gain of the F_2^* laser.

B3.2.10.2 Laser performance

Discharge-pumped F_2^* lasers are commonly excited by a fast single-discharge circuit that allows a high pumping power density in a short pulse [127, 137]. The output pulse durations are typically in the range 10–30 ns (FWHM). The limit on the pulse duration is imposed by the very fast growth of discharge instabilities which are inherent for the high-pressure gas mixtures used. Filaments formed in the discharge absorb and scatter the laser field and the laser oscillation may terminate before the end of the discharge pulse. Since the time is very limited, researchers have been looking to optimize the input power density. Performed systems with high pumping power exhibit very high gain, of the order of 10% per cm, and output beams that are dominated by amplified spontaneous emission. The beam quality is rather poor with strong divergence up to a few mrad [138].

For good laser performance, it is crucial to use a short-pulse high-intensity x-ray pre-ionization source. This can be successfully done with a corona plasma cathode from which the electrons are emitted [22]. They are accelerated towards the anode with voltages ranging from 40 to 110 kV and are stopped in a 25 μm thick tantalum foil to produce the x-rays [131]. The x-ray pulse has a sharp peak with a width of 25 ns [FWHM]. The pre-ionization electron density in the laser gas is proportional to the x-ray dose entering the gas. A uniformity within 8% over the full entrance window can be obtained. It is essential that the inductance of the discharge circuit of the laser is minimized in order to get a fast current risetime and a high current density. This can be reached by placing a series of capacitors along the cathode as close as possible to the discharge. Following these principles, the following experiment can be described for a laser gas containing 0.1% F_2 in 5 bar He [131]. Using a discharge length of 42 cm, an electrode spacing of 1.5 cm and a discharge width of 2.4 cm, the inductance is 1.7 nH for a capacitance of 14 nF and 1.1 nH for 30 nF. The discharge width depends on the electrode spacing. Reducing the spacing to 0.8 cm the width becomes 1.5 cm. Setting the latter spacing and installing a capacitance of 14 nF charged at 20 kV, the observed waveforms are shown in figure B3.2.33. While the capacitors are charged, the x-rays are injected. After typically 30–100 ns, the gas breaks down and the discharge starts. The applied discharge voltage drops while the current rises very fast. Both the pump power pulse and the laser pulse are as short as 7.5 ns [FWHM]. It was found that the onset of the x-ray pulse must occur during the voltage rise on the electrodes. If the pre-ionization is earlier, the electrons are already lost by attachment and if it occurs near the end or after the full charging of the capacitors, the avalanche electron multiplication is not homogeneous.

The output energy density depends on the electrode spacing. If, for the same capacitance of 14 nF and inductance of 1.7 nH, the spacing is increased from 0.8 to 1.5 cm the discharge width increases from 1.5 to 2.4 cm, whereas the maximum output as a function of the charging voltage increases from 10 mJ at 20 kV to 26 mJ at 35 kV. The maximum specific output energy density is 0.3 J l^{-1} at an intrinsic efficiency of 0.3%. By increasing the electrode spacing to 1.5 cm, nearly a factor of two, the breakdown voltage is expected also to increase by a factor of two so that the stored energy in the capacitor is about four times more. The simultaneously observed discharge width changes by a factor of 1.6. Then the input power density, within experimental error, was more or less constant. The output density seems to saturate only with the current density.

B3.2.10.3 Long pulses

The poor beam quality of the described short-pulse single-discharge systems can be improved when the number of cavity round-trips can be increased. This means that the gain duration of the medium should be increased. The question is then to look for discharge parameters and technology that extend the stability

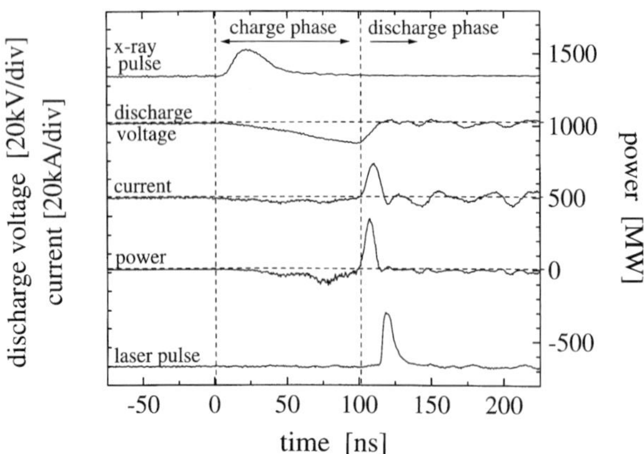

Figure B3.2.33. Typical waveforms of the F_2 laser showing the x-ray pulse, discharge voltage, laser current and output pulse.

duration of the discharge. One possibility is a reduction in the fluorine content which is accompanied by a reduction in the output. This reduction of output can be limited by the application of the resonant overshoot mode in a pre-pulse–main-pulse configuration which has already been shown to improve the performance of rare-gas halogen excimer lasers. Applied to the F_2^* laser system, it also resulted in relatively long output pulses. The system then may run without current reversal and opens the way to the development of high repetition systems. To obtain fast risetimes, the pre-pulse capacitors are again placed as close as possible to the electrodes. This technique with magnetic switching has been successfully demonstrated with the following parameters [131]. Using the same laser head construction as described in the previous section except that the main capacitor of 153 nF is now built as a pulse-forming network (PFN) consisting of nine rows of five low inductance ceramic capacitors, 1.7 nF each. The ten pre-pulse capacitors with a total capacitance is 3.6 nF are equally distributed along either side of the cathode. Its circuit inductance is about 3 nH. The series inductance of the saturated magnetic switch together with the main capacitance is about 12 nH. Typical waveforms for a discharge in a gas mixture of 2 bar He and 3 mbar F_2 are shown in figure B3.2.34. The charging voltages of the main and pre-pulse capacitors are 12 and 20 kV, respectively. At the peak of the charging voltage a positive pre-pulse is applied to the peaking capacitors. This creates a large voltage difference across the magnetic switch driving it from a high to a low inductance state. Next, the peaking capacitors are resonantly charged by the main capacitor. The voltage reverses and rises rapidly until the laser gas breaks down. The low inductance of the laser allows a fast rise in the current and, thus, a high power deposition from the pre-pulse capacitors during the start of the discharge. The fast risetime of the current leads to the formation of a dense hot-spot structure on the cathode which enhances the homogeneity of the discharge [24] and favours the generation of long laser pulses. After this breakdown, the current from the main capacitor starts flowing through the low-impedance discharge driving the magnetic switch deep into saturation so that the stored energy in the main capacitor is very rapidly dissipated in the discharge. The oscillations on the voltage waveform are due to current oscillations in the circuit that is formed by the discharge, the laser head and the peaking capacitor.

The timing of the start of the pre-ionization is most effective between 5 and 85 ns after the start of the pre-pulse. For the described system, an output of a few mJ and a pulse duration of 70 ns have been demonstrated [131].

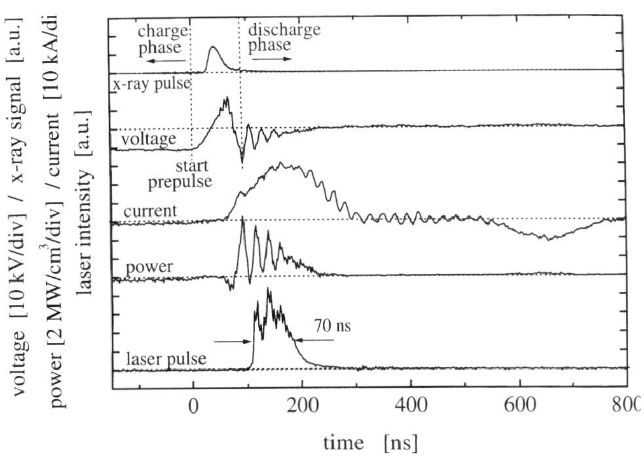

Figure B3.2.34. Typical waveforms of the long pulse F_2 laser showing the x-ray pulse, discharge voltage, laser current and output pulse.

B3.2.11 Other rare-gas halide lasers

Apart from the previously discussed rare-gas halide lasers some more systems have shown stimulated emission. These are XeBr (281.8 nm) [9], KrBr (206.5 nm) [140], ArCl (169 and 175 nm) [141] and NeF (108 nm) [139]. These systems have been demonstrated with self-sustained discharges, except for NeF which was pumped by e-beam. The energies and efficiencies are low. This may explain the low interest in further development.

Next to diatomic excimers, poly-atomic excimers have also been investigated. The broadband emission spectra, mostly in the visible part of the spectrum, observed in e-beam excitation of rare-gas halide mixtures are assigned to triatomic excimers [142]. The large linewidths of those spectra with $\Delta\lambda/\lambda$ of the order of 10% provoked the development of high power lasers, tunable over these relatively broad spectral ranges. However, the experiments showed that it was difficult to reach threshold because of the low gain and high absorption of the emitted radiation. Using a high e-beam current density of 800 A cm^{-2} and a short pulse of 10 ns lasing have been found for Xe_2Cl at 520 nm with $\Delta\lambda = 30$ nm [FWHM] [143] and for Kr_2F in the range 380–480 nm [144]. The cross sections for stimulated emission are very small due to the large linewidths of the transitions. The studies have indicated that these complex molecules have a high laser threshold, high working pressure, low gain and low efficiency.

B3.2.12 N_2 laser (337–406 nm and 741–1230 nm)

Nitrogen is also an attractive medium for obtaining high output laser energy in the UV region of the spectrum. This laser was introduced as a simple operating self-sustained discharge system shortly after the invention of the first lasers [145, 148]. More than ten years later, when suitable e-beam technology became available, additional studies were carried out to obtain improved performances and a better understanding of the system [146, 147]. Its high gain obtained with a simple *LC* circuit in the easily available N_2 gas made the system very popular as a home-made system to be used for various applications where intense UV radiation was required, for instance dye laser pumping and spectroscopy. The lasing scheme has some similarities with the excimers although the formation kinetics of the lasing molecule is different.

Laser action occurs on electronic vibrational transitions of the N_2 molecule. The electronic energy states with vibrational levels of the N_2 molecule are shown in figure B3.2.35. The main UV transition takes place

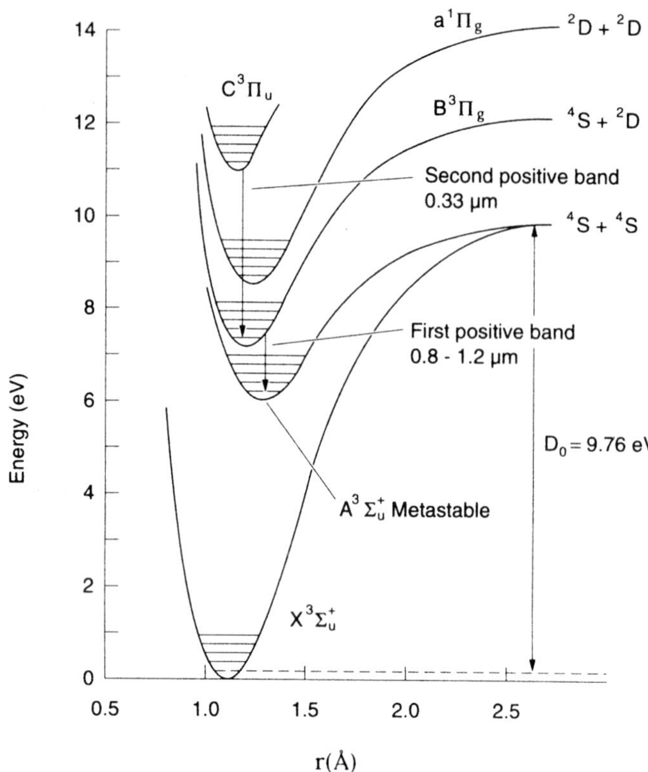

Figure B3.2.35. Potential energy diagram for electronic-vibrational levels in N_2 showing the first and second positive bands; D_0 is the dissociation energy.

in the so-called second positive band, i.e. in the transition from the $C^3\Pi_u$ to the $B^3\Pi_g$ state at 337.1 nm. The upper state is formed by electronic excitation from the ground state of N_2. Since the lifetime of the lower state, about 10 μs, is much longer than the 40 ns of the upper state, the laser is self-terminating and, therefore, it operates only in the pulsed mode. As a consequence, during the laser process there will be a strong population build-up of the $B^3\Pi_g$ state so that cascade laser transitions in the first positive band between the $B^3\Pi_g$ and $A^3\Sigma_u^+$ states in the near infrared (741–1230 nm) may also occur [148, 149], which are also self-terminating because the lifetime of the $A^3\Sigma_u^+$ state is about 1 ms. Since the lifetime of the $C^3\Pi_u$ state is 40 ns, it follows that the excitation pulse is only effective within this period and that the radiation build-up time in the cavity must be much shorter than this lifetime. This means that the discharge pulse must be as short as possible to produce the high gain for a fast radiation build-up.

Initially, the N_2 molecules are at room temperature and practically all in the ground vibrational state of the ground electronic $X^3\Sigma_u^+$ state. The probability of exciting a vibrational state of a higher electronic state from this initial ground state and the subsequent radiative transition probability between states can be calculated quantum mechanically. It depends on the overlap of the wavefunctions of the respective states and is described by the Franck–Condon (FC) factor. The FC principle states that during an electronic transition in a molecule induced by the influence of an external perturbation as in electronic impact, absorption or emission of radiation the internuclear separation of the molecule and the velocity of the relative nuclear motion alter to a negligible extent. The transition takes place so quickly that the nuclei do not have time to move any appreciable distance. This is due to the fact that the impulse transfer to the relatively heavy nuclear mass of a molecule is very small. The transitions are then indicated by vertical lines in the molecular potential energy diagram. According to this FC principle the electronic excitation of the N_2 molecules to the ground vibrational state of the $C^3\Pi_u$ manifold is more likely than to that of the $B^3\Pi_g$ manifold. This follows from the energy state diagram in figure B3.2.35 where the potential minimum of the $B^3\Pi_g$ state relative to that of the ground state is shifted to a larger internuclear separation distance than in the case of the $C^3\Pi_u$ state.

B3.2.12.1 The discharge

The discharge system has a typical gas mixture of 1 bar He and 50–100 mbar N_2. The He contributes to the initial stability of the principally unstable discharge and not so much to collisional pumping of the laser. The discharge requires high electric fields of the order of 10 kV cm^{-1} or E/p values of about 10 V cm^{-1} mbar^{-1}. This is obtained by transverse electrodes similar to excimer laser constructions. The energy needed to excite the upper laser level is about 11.7 eV so that discharge electrons with higher energy are desired. The initial quasi-stability of the discharge can be obtained with e-beam, x-ray or UV pre-ionization, similar to that in excimer lasers. The simplest constructions are based on UV pre-ionization by means of a row of spark plugs parallel to the laser axis, surface flashes or corona discharges. Once the more or less uniform pre-ionization is obtained, the discharge follows immediately. Since the discharge instabilities will build up rapidly and the lifetime of the upper level is only 40 ns it is advantageous to have the current pulse as short as possible. The inductance of the circuit must then be as low as possible to obtain the required fast current risetime. This can be done by mounting parallel capacitors as close as possible to the discharge electrodes.

The initial voltage that is applied across the discharge gap just after switching is several times higher than the quasi-steady-state voltage that follows after breakdown. A faster switch gives a higher initial voltage and a higher electron energy of the initial discharge. Due to fast switching, a high electric field in the laser gas is maintained during the delay time to break down the discharge. This delay time is typically in the range 100 ns. Just after the break down, the E/p value drops very rapidly, say in about 5 ns, to its steady-state value. During this short period, the electrons have sufficient energy for laser pumping. The duration of the fast fall of the electric field depends on the gas components and their densities. It determines the laser pulse duration. The following quasi-steady-state discharge is not suitable for laser pumping because of its low average electron energy. Small amounts of electronegative gases like SF_6, NF_3 and F_2 added to the laser gas increase the delay time before breakdown so that the maximum voltage just before breakdown is increased. The field strength during the transient time for pumping will increase and higher output energies are obtained [150, 151].

The gain of the N_2 discharge lasers is, in general, very high so that the short output pulse may be considered as amplified spontaneous emission. Output powers above 1 MW are feasible. The optical beam quality is poor as a consequence of the short pulse durations which do not allow sufficient cavity passes to reach coherence. Often a single mirror is used at one end to provide a unidirectional output beam. High repetition rate performance requires adequate electric circuitry. Spark gaps have limited repetition rates and lifetimes. Thyratrons are more suitable, although their current is limited. This can be circumvented by applying magnetic pulse compression before the current passes the discharge.

Higher output energies, longer pulses and higher intrinsic efficiencies [146, 152] have been obtained with e-beam pumping where the discharge parameters are independent from the gas composition. Most successful is the addition of Ar as a buffer gas in multi-atmospheric systems. The accumulated excitation energy of Ar is transferred to the upper laser level of N_2 during molecular collisions in multi-atmospheric systems. It was also observed that the addition of He and Ne to the Ar–N_2 mixture has a further enhancing effect on the output energy, efficiency and pulse duration. Next to 337.1 nm, several additional wavelengths, e.g. 357.7, 380.5 and 405.9 nm, have been observed [152, 153].

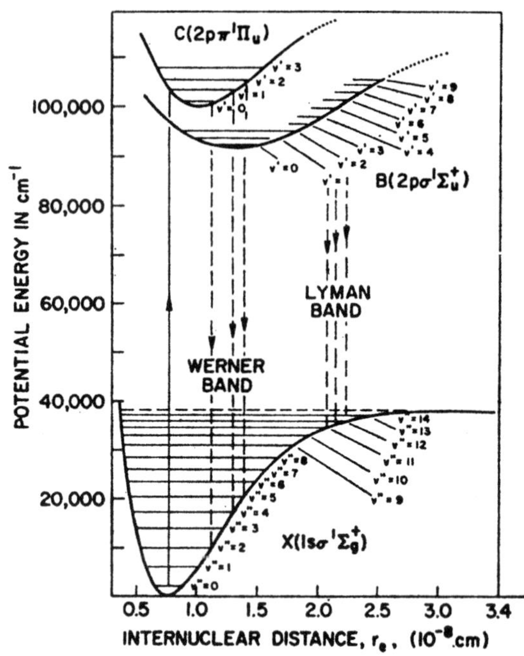

Figure B3.2.36. Potential energy diagram for electronic-vibrational levels in H_2 showing the Lyman and Werner bands and the electron impact excitation path [164].

B3.2.13 H_2 laser (140–165 nm and 100–120 nm)

The molecular hydrogen laser has much in common with the nitrogen laser. The system is attractive for producing laser power in the vacuum ultraviolet region of the spectrum. The system was also suggested shortly after the invention of the laser [154]. Since the gain needed for laser action decreases drastically with wavelength, as we discussed in the section B3.2.1, extremely high pumping power densities are required. Fortunately, this condition can still be fulfilled for laser action in the Lyman band (140–165 nm) as well as for the Werner band (100–120 nm). Both high current electron beams [155, 156] and very intense transverse discharges [157–159] have been applied successfully. The lasing scheme is very similar to that of the N_2 laser and has also similarities to the excimer lasers. Laser actions occur between the B and X states in the Lyman band and between the C and X state in the Werner band. The X state is the ground state and the B and C states are, respectively, the first and second excited electronic state of molecular hydrogen. Each electronic state can be further described by its vibrational and rotational energy distributions. The system is also self-terminating because the lifetime of the lower laser level, with its long rotational and vibrational relaxation times, is much longer than that of the excited B or C state. The selective kinetic paths in the laser processes are mainly determined by the FC principle, similar to what we have discussed for the N_2 laser. Figure B3.2.36 shows the potential energy diagram of molecular hydrogen with electronic and vibrational energy levels, excitation paths and radiative decay paths in the Lyman and Werner bands. The transitions according to the FC principle are indicated by vertical lines. The upper vibrational levels being initially empty at room temperature have a very low probability of being excited because of the FC principle.

Table B3.2.1. Wavelengths and vibrational transitions of observed laser lines in the Lyman band [157].

Transition	Wavelength (nm)		
B(υ_1)–X(υ_2)	P(1)	P(3)	R(1)
2–9	157.18	—	
3–10	159.15	159.61	
4–11	160.47	160.86	
5–12		161.33	
6–13	160.74		
7–13		158.05	157.71
8–14	156.73		

Table B3.2.2. Wavelengths and vibrational transitions of the strongest observed laser lines in the Werner band [156]

Transition	Wavelength (nm)	
C(υ_1)–X(υ_2)	Q(1)	P(3)
1–4	116.136	116.617
2–5	117.586	118.050
1–5	120.668	
2–6	121.900	122.358
3–7	123.004	
4–8	123.956	

B3.2.13.1 Lyman band (140–165 nm)

The light hydrogen molecules are at room temperature in their ground vibrational state and have only a few rotational levels occupied, mainly the levels $J_2 = 0, 1, 2$ and 3 of which $J_2 = 1$ has the highest density. The excited B state will then, according to the FC principle, have the vibrational levels $\upsilon_1 = 2$–11 and preferentially $\upsilon_1 = 6$ and 8. The electronic excitation of these molecules results in even-numbered rotational levels in the B state ($J_1 = 0, 2$) with $J_1 = 2$ preferred. Since the upper vibrational levels of the X state are empty, the laser process is mainly observed between B($\upsilon_1 = 2$–11; $J_1 = 0, 2$) and X($\upsilon_2 = 9$–14; $J_2 = 1, 3$) states with intensities proportional to the FC factors. See table B3.2.1.

B3.2.13.2 Werner band (100–120 nm)

In this laser process, the electronic excitation from the ground state occurs according to the FC principle to the low-lying vibrational levels ($\upsilon_1 = 0$–4) of the C state. Laser oscillation can then be obtained between these low-lying vibrational states and the high vibrational levels of the X state. The intensity distribution depends on the FC factors. The Q(1) line with $\Delta J = -1$ of the C($\upsilon_1 = 1$) to X($\upsilon_2 = 4$) transition is the strongest. Observed lines are shown in table B3.2.2.

The isotopes of hydrogen can also be successfully used. The observed stimulated emission spectrum of H_2, HD and D_2 shows tens of lines in both Lyman and Werner bands [160].

B3.2.13.3 Excitation technique

As pointed out in section B3.2.1, the main difficulty encountered in obtaining laser action in the VUV spectral region is the required high pumping power density. Because of the very short radiative lifetime of the upper states of about 0.8 ns and the relatively long lifetime of the lower state, stimulated emission of this self-terminating system may occur on the order of 1 ns. The shortness of the upper state lifetime means that both optical feedback by cavity mirrors and excitation pulses longer than this lifetime are of no use. It even means that time-independent spatial homogeneity of the excitation pulse is catastrophic which imposes an additional constraint on the applied excitation technology. Stimulated emission is only feasible when the medium is excited by an excitation pulse that moves with the speed of stimulated emission. The first reports on observed stimulated emission in H_2 describe an intense e-beam pulse propagating parallel to the stimulated emission system [155, 156]. A short e-beam pulse of a few nanoseconds penetrates longitudinally, parallel to the axis of a cylinder containing H_2. The e-beam generator consists of a Marx circuit for obtaining high voltages followed by a Blumlein circuit for pulse compression, voltage doubling and pulsefront sharpening. A 3 ns e-beam pulse of 10 kA cm^{-2} with an acceleration voltage of 600 kV excites longitudinally this cylinder containing 20–100 torr H_2. In this way stimulated emission in both Lyman and Werner bands has been observed. The short high-power e-beam pulse produces a large inversion density before the spontaneous emission can populate the upper vibrational levels of the ground electronic state. The gain is then high enough for stimulated emission in the direction of the propagating e-beam. The system then operates in the amplified spontaneous emission mode, just as the molecular nitrogen laser does. The laser pulses, estimated to be about 1 ns, have power densities in the range 50–500 W cm^{-2}.

An attractive alternative technique for pumping the H_2 system is the short-risetime travelling-wave discharge. This type of discharge can be performed with a Blumlein transmission line connected to the discharge channel [161–163]. The required high input power density is further increased by a narrow channel for the laser discharge. The Blumlein construction consists, in principle, of two parallel plates, one is a continuous conducting plate and the other is interrupted by the laser discharge channel. The two plates are charged like a capacitor. If on one end of the two parallel plates a fast switch is used to short-circuit the plates, a short square voltage pulse of twice the charging voltage is applied to the channel. The switching time determines the voltage risetime; the length of the plates with the dielectric the pulse duration; the dielectric with the distance between the plates the current per unit length. The trick in obtaining the required travelling excitation-current pulse wave is to position the switch in a corner of the parallel plates and to shape the conducting plates in such a way by the angle relative to the discharge channel that a travelling-wave current pulse reaches the discharge channel. With the correct angle, the excitation pulse seen by the discharge moves with a velocity equal to that of amplified spontaneous emission. Pulse durations of 1 ns are recorded at the end toward which the excitation pulse proceeds [157]. Using discharge channels with thicknesses of 50–1000 μm and H_2 densities 30–60 torr, pulse energies up to 20 μJ with a duration of 0.5 ns have been obtained at a high repetition rate [158].

Appendix

During the stable period of a pulsed discharge, there is an equilibrium between the production and loss of free electrons i.e. a glow discharge is formed. This discharge is characterized by a more or less constant steady-state voltage between the electrodes, independent of the discharge current. The steady-state voltage depends strongly on the gas composition and density. The main pulse maintaining the steady state of the discharge can then be described by a simple LC circuit as shown in figure B3.2.37. Starting with the main capacitor C_{pfn} charged at V_{pfn} and the constant voltage V_{ss} across the electrodes, the circuit current is described by the differential equation

$$LC_{\text{pfn}}\frac{\mathrm{d}^2 i}{\mathrm{d}t^2}+i=0$$

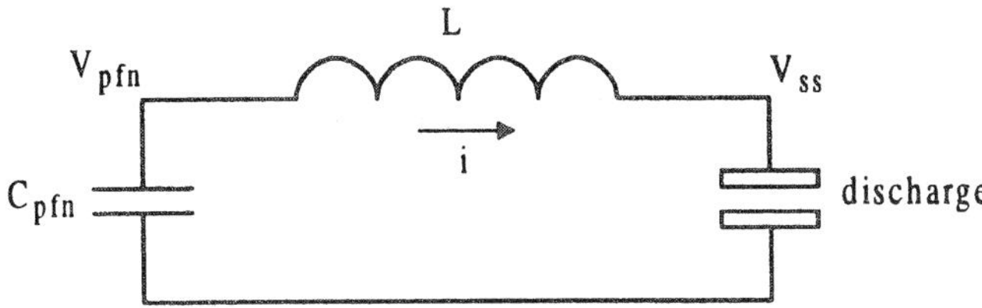

Figure B3.2.37. Single LC circuit of the main discharge.

with the initial conditions

$$\frac{\mathrm{d}i}{\mathrm{d}t} = \frac{V_{\mathrm{pfn}} - V_{\mathrm{ss}}}{L}$$

and $i = 0$ for $t = 0$. The solution of this equation is

$$i(t) = (V_{\mathrm{pfn}} - V_{\mathrm{ss}}) \left(\frac{C_{\mathrm{pfn}}}{L}\right)^{1/2} \sin\left(\frac{t}{\sqrt{LC_{\mathrm{pfn}}}}\right).$$

If we integrate the current over the first half-cycle, $\Delta t = \pi(LC_{\mathrm{pfn}})^{1/2}$, the charge transfer from the capacitor becomes $2C_{\mathrm{pfn}}(V_{\mathrm{pfn}} - V_{\mathrm{ss}})$. It is seen that for $V_{\mathrm{pfn}} = 2V_{\mathrm{ss}}$ the capacitor is empty and no current reversal can occur. This is the so-called matched discharge situation. The energy delivered to the discharge during the first half-cycle becomes, by integrating $i(t)V_{\mathrm{ss}}$, equal to $2V_{\mathrm{ss}}C_{\mathrm{pfn}}(V_{\mathrm{pfn}} - V_{\mathrm{ss}})$.

The efficiency η defined as the energy delivered to the discharge during the first half-cycle divided by the energy stored on the capacitor is then

$$\eta = 4\frac{V_{\mathrm{ss}}(V_{\mathrm{pfn}} - V_{\mathrm{ss}})}{V_{\mathrm{pfn}}^2}.$$

The efficiency reaches its maximum, $\eta = 1$, for $V_{\mathrm{pfn}} = 2V_{\mathrm{ss}}$.

References

[1] Hopfield, J J 1930 Absorption and emission spectra in the region 60–110 nm *Phys. Rev.* **35** 1133–4

[2] Tanaka, Y 1955 Continuous spectra of rare gases in the vacuum ultraviolet region *J. Opt. Soc. Am.* **45** 710–13

[3] Funkelnburg W 1938 *Kontinuierliche Spektren* (Berlin: Springer)

[4] Golde M F and Thrush B A 1974 Vacuum UV emission from reactions of metastable inert gas atoms: Chemiluminescence of ArO and ArCl *Chem. Phys. Lett.* **29** 486–90

[5] Velazco J E and Setser D W 1975 Bound free emission spectra of diatomic xenon halides *J. Chem. Phys.* **62** 1990–1

[6] Houtermans F G 1960 *Helv. Phys. Acta* **33** 933

[7] Basov N G, Danilychev V A, Popov Y M and Khodkevich D D 1970 Laser operating in the vacuum region of the spectrum by excitation of liquid xenon with an electron beam *JETP Lett.* **12** 329–31

[8] Koehler H A, Ferderber L J, Redhead D L and Ebert P J 1972 Stimulated VUV emission in high pressure xenon excited by high current relativistic electron beams *Appl. Phys. Lett.* **21** 198–200

[9] Searles S K and Hart G A 1975 Stimulated emission at 281.8 nm from XeBr *Appl. Phys. Lett.* **27** 243–5

[10] Brau C A and Ewing J J 1975 354 nm laser action on XeF *Appl. Phys. Lett.* **27** 435–7
Brau C A and Ewing J J 1975 Emission spectra of XeBr, XeCl, XeF and KrF *J. Chem. Phys.* **63** 4640–7

[11] Fleischmann H H 1975 High current electron beams *Phys. Today* **28** 35–43

[12] Langmuir I 1913 Effect of space charge and residual gases on thermionic currents in high vacuum *Phys. Rev.* **II-6** 450–86

[13] Bradley L P 1976 *High Power Gas Lasers 1975 (Conference Series 29)* (Bristol: IOPP)

[14] Hsia J C, Jacob, J H Mangano, J A and Rokni M 1977 Avco Everett Research Lab., Semiannual Report, DARPA 3125, NTIS Intermediar

[15] Bradley D J, Hull, D R, Hutchinson, M H R and McGeoch M W 1975 Coaxially pumped, narrow band, high power VUV xenon laser *Opt. Commun.* **14** 1–3

[16] Peters P J M, Fierkens I H T and Witteman W J 1987 Effect of Ne and Ar on the performance of a high pressure ArF laser pumped by a small coaxial electron beam *Appl. Phys. Lett.* **51** 883–5

[17] Kleikamp B M H H and Witteman W J 1984 High energy extraction of electron beam pumped KrF lasers at multi atmospheres *Opt. Commun.* **49** 345–8

[18] Ekdahl C A 1980 Voltage and current sensors of high density Z-pinch experiments *Rev. Sci. Instrum.* **51** 1645–8

[19] Pellinen D G and Spence P W 1971 A nanosecond risetime megampere current monitor *Rev. Sci. Instrum.* **42** 1699–701

[20] Lacour, B and Vannier C 1987 Phototriggering of a 1 J excimer laser using either UV or x-rays *J. Appl. Phys.* **62** 754–8

[21] Bollanti S, Letardi T and Zheng C 1991 Effect of pre-ionization on uniformity of photo-triggered XeCl laser discharges: modelling and comparison with experimental results *IEEE Trans. Plasma Sci.* **19** 361–8

[22] van Goor F A 1993 Fast risetime x-ray pre-ionization source using a corona plasma cathode *J. Phys. D: Appl. Phys.* **26** 404–9

[23] Seddon N and Thornton E 1988 A high voltage short risetime pulse generator based on a ferrite pulse sharpener *Rev. Sci. Instrum.* **59** 2497–8

[24] Makarov M 1995 Effect of electrode processes on the spatial uniformity of the XeCl laser discharge *J. Phys. D: Appl. Phys.* **28** 1083–93

[25] Levatter J I AND Bradford R S Jr 1978 Water dielectric Blumlein driven fast electric discharge KrF laser *Appl. Phys. Lett.* **33** 742–4

[26] Wang C P 1978 High repetition rate transverse flow XeF laser *Appl. Phys. Lett.* **32** 360–2

[27] Taylor R S, Sarjeant W J, Alcock A J and Leopold K E 1978 Glow discharge characteristics of a 0.8 J multi-atmosphere rare gas halide laser *Opt. Commun.* **25** 231–4

[28] Sze R C and Loree T R 1978 Experimental studies of a KrF and ArF discharge laser *IEEE J. Quantum Electron.* **14** 944–50

[29] Long W H, Plummer M J and Stappaerts E A 1983 Efficient discharge pumping of an XeCl laser using a high voltage pre-pulse *Appl. Phys. Lett.* **43** 735–7

[30] Taylor R S and Leopold K E 1985 Magnetically induced pulser laserexcitation *Appl. Phys. Lett.* **46** 335–7

[31] Fisher C H, Kushner M J, DeHart T E, MCDaniel J P, Petr R A and Ewing J J 1986 High efficiency XeCl laser with spiker and magnetic isolation *Appl. Phys. Lett.* **48** 1574–6

[32] Gerritsen J W, Keet A L, Ernst G J and Witteman W J 1990 High efficiency operation of a gas discharge XeCl laser using a magnetically induced resonant voltage overshoot circuit *J. Appl. Phys.* **67** 3517–19

[33] Witteman W J, Ekelmans G B, Trentelman M and van Goor F A 1990 Discharge technology for excimer lasers of high average power *Eighth Int. Symp. on Gas Flow and Chemical Lasers (SPIE Volume 1397)* ed J M Orza and C Domingo (Bellingham, WA: SPIE), pp 37–45

[34] Timmermans J C M, van Goor F A and Witteman W J 1993 A new mode to excite a gas discharge XeCl laser *Appl. Phys.* B **57** 441–5

[35] Brau Ch A 1984 Rare gas halide excimers *Excimer Lasers (Topics in Applied Physics 30)* ed Ch K Rhodes (Berlin: Springer) p 88

[36] Tellinghuisen J, Hoffman J M, Tisone G C and Hays A K 1976 Spectroscopic studies of diatomic noble gas halides: analysis of spontaneous and stimulated emission from XeCl *J. Chem. Phys.* **64** 2484–90

[37] Tellinghuisen J, Tisone G C, Hoffman J M and Hays A K 1976 Analysis of spontaneous and laser emission from XeF *J. Chem. Phys.* **64** 4796–7

[38] Ewing J J and Brau C A 1975 Laser action on the ${}^2\Sigma^+_{1/2} \rightarrow {}^2\Sigma^+_{1/2}$ bands of KrF and XeCl *Appl. Phys. Lett.* **27** 350–2

[39] Ault E R, Bradford R S and Bhaumik M L 1975 High power xenon fluoride laser *Appl. Phys. Lett.* **27** 413–15

[40] Hoffman J M, Hays A K and Tisone G C 1976 High power noble gas halide lasers *Appl. Phys. Lett.* **28** 538–9

[41] Burnham R and Djeu N 1976 Ultraviolet pre-ionized discharged pumped lasers in XeF, KrF and ArF *Appl. Phys. Lett.* **29** 707–9

[42] Mangano J A and Jacob J H 1975 Electron beam controlled discharge pumping of the KrF laser *Appl. Phys. Lett.* **27** 495–8

[43] Sutton D G, Suchard S H, Gibb O L and Wang C P 1976 Fast discharge initiated KrF laser *Appl. Phys. Lett.* **28** 522–3

[44] Jacob J H, Hsia J C, Mangano J A and Rokni M 1979 Pulse shape and lasing energy extraction from e-beam pumped KrF *J. Appl. Phys.* **50** 5130–4

[45] Rice J K, Tisone G C and Patterson P L 1980 Oscillator performance and energy extraction from a KrF laser pumped by a high intensity relativistic electron beam *IEEE J. Quantum Electron.* **16** 1315–26

[46] Edwards C B, O'Neil F and Shaw M J 1980 60 ns e-beam excitation of rare gas halide lasers *Appl. Phys. Lett.* **36** 617–20

[47] Swingle J C, Schlitt L G, Rapopart W R, Goldhar J and Ewing J J 1981 Efficient narrow band electron beam pumped KrF laser for pulse compression studies *J. Appl. Phys.* **52** 91–6

[48] Peters P J M, Bastiaens H M J, Witteman W J and Gerber T 1987 A study of the electron quenching of excimers in a KrF laser *Appl. Phys. B* **43** 253–61

[49] Burnham R and Searles S K 1977 Radiative lifetime of KrF *J. Chem. Phys.* **67** 5967–8

[50] Mangano J A, Jacob J H, Rokni M and Hawryluk A 1977 Three body quenching of KrF by Ar and broadband emission at 415 nm *Appl. Phys. Lett.* **31** 26–7

[51] Witteman W J and Kleikamp B M H H 1984 On the electron-beam pumped KrF laser *J. Appl. Phys.* **55** 1299–307

[52] Peters P J M, Bastiaens H M J and Witteman W J 1984 Compact coaxially excited high energy density KrF laser *J. Appl. Phys.* **55** 1410–12

[53] Witteman W J and Oomen G L 1980 On the performance of an e-beam pumped KrF laser *Opt. Commun.* **32** 467–72

[54] Oomen G L and Witteman W J 1978 Electron beam pumped KrF laser with a neon-argon diluent *Appl. Phys. Lett.* **33** 878–80

[55] Gerber T, Peters P J M, Bastiaens H M J and Witteman W J 1984 Enhancement of the specific output energy of an electron-beam pumped KrF laser using Ne as the main buffer gas *Appl. Phys. Lett.* **45** 356–7

[56] Ewing J J 1979 *Laser Handbook* vol 3, ed M K Stitch (Amsterdam: North-Holland) p 163f

[57] Inertial Confinement Fusion at Los Alamos 1989 *Progress in Inertial Confinement Fusion Since 1985* **1** 262

[58] Greene A E and Brau C A 1978 Theoretical studies of UV pre-ionized transverse discharge KrF and ArF lasers *IEEE J. Quantum Electron.* **14** 951–7

[59] Loree T R, Butterfield K B and Barker D L 1978 Spectral tuning of ArF and KrF discharge lasers *Appl. Phys. Lett.* **32** 171–3

[60] Zhupikov A A and Razhev A M 1998 Excimer KrF laser with He buffer gas, 0.8 J energy and 2% efficiency *Sov. J. Quantum Electron.* **28** 667–9

[61] Moody S E, Levin L A, Center R E, Ewing J J and Klosterman E L 1981 Measurements of laser performance and efficiency of e-beam pumped xenon chloride *IEEE J. Quantum Electron.* **17** 1856–61

[62] McKee T J 1985 Spectral narrowing techniques for excimer oscillators *Can. J. Phys.* **63** 214–19

[63] Efimovskii S V, Kurbasov S V, Novichkov A V and Palchikov K K 1996 Influence of the parameters of a Fabry–Pérot etalon on the output characteristics of a narrow band long optical pulse XeCl laser *Sov. J. Quantum Electron.* **26** 599–603

[64] Sorkina R, van Goor F A and Witteman W J 1992 Simulation studies of the pre-pulse main pulse XeCl discharge lasers with magnetic switching *Appl. Phys.* B **55** 478–84

[65] Bollanti S, Di Lazzaro P, Giordano G, Hermsen T, Letardi T and Zheng C E 1990 Performance of a ten liter electron avalanche discharge XeCl laser device *Appl. Phys.* B **50** 415–23

[66] Sentis M L, Delaporte P, Forestier B M and Fontaine B L 1991 Design and characteristics of high pulse repetition rate and high average power excimer laser systems *IEEE J. Quantum Electron.* **27** 2332–9

[67] Timmermans J C M 1995 Double discharge XeCl laser *PhD Dissertation* (University of Twente, Enschede, The Netherlands)

[68] Witteman W J, van Goor F A, Ekelmans G B, Trentelman M and Ernst G J 1990 Design studies of a high pulse repetition rate excimer Eurolaser *High Power Gas Lasers (SPIE 1225)* ed P V Avizonis, C Freed, J J Kim and F K Tittel (Bellingham, WA: SPIE), pp 132–41

[69] Ernst G J 1984 Uniform field electrodes with minimum width *Opt. Commun.* **49** 275–77

[70] van Goor F A and Witteman W J 1994 High average power XeCl laser with x ray pre-ionization and spiker sustainer excitation *High-Power Gas and Solid State Lasers (SPIE 2206)* (Bellingham, WA: SPIE) pp 30–40

[71] Murray J R and Powell H T 1976 KCl laser oscillation at 222 nm *Appl. Phys. Lett.* **29** 252

[72] Eden J G and Searles S K 1976 Observation of stimulated emission of KrCl *Appl. Phys. Lett.* **29** 350–2

[73] Panchenko A N and Tarasenko V F 1995 Maximum performance of discharge pumped exciplex laser at $\lambda = 222$ nm *IEEE J. Quantum Electron.* **31** 1231–6

[74] Hueber J M, Fontaine B L, Bernard N, Forestier B M and Sentis M L 1992 Long pulse KrCl excimer laser at 222 nm *Appl. Phys. Lett.* **61** 2269–71

[75] Boichenko A M, Derzhiev V I, Zhidkov A G and Yakovlenko S I 1992 Kinetic model of an ArF laser *Sov. J. Quantum Electron.* **22** 444–8

[76] Lo D, Shchedrin A I and Ryabtsev A V 1996 The upper energy limit of a self sustained discharge pumped ArF laser *J. Phys. D: Appl. Phys.* **29** 43–9

[77] Suda A, Obara M and Noguchi A 1986 Performance characteristics of the ArF excimer laser using a low pressure argon-rich mixture *J. Appl. Phys.* **60** 3791–3

[78] Mandle A 1986 ArF short pulse extraction studies *J. Appl. Phys.* **59** 1435–45

[79] Klementov A D, Morozov N V and Sergeev P B 1986 Electron beam excited ArF laser *Sov. J. Quantum Electron.* **16** 1139–42

[80] Maeda M, Nishitarumizu T and Miyazoe Y 1979 Formation and quenching of excimers in low pressure rare gas halogen mixtures by e-beam excitation *Japan. J. Appl. Phys.* **18** 439–45

[81] Akashi H, Sakai Y and Tagashira H 1994 Modelling of a self sustained discharge excited ArF excimer laser *J. Phys. D: Appl. Phys.* **27** 1097–106

[82] Akashi H, Sakai Y and Tagashira H 1995 Modelling of a self sustained discharge excited ArF excimer laser: the influence of photo-ionization and photodetachment by laser light on the discharge development *J. Phys. D: Appl. Phys.* **28** 445–51

[83] Nagai S, Sakai M, Furuhashi H, Kono A, Goto T and Uchida Y 1998 Effects of F^- ions and F_2 molecules on the oscillation process of a discharge pumped ArF excimer laser *IEEE J. Quantum Electron.* **34** 40–6

[84] Feenstra L 1999 On the long pulse operation of a discharge pumped ArF excimer laser *PhD Dissertation* (University of Twente, Enschede, The Netherlands)

[85] Nagai S, Furuhashi H, Uchida Y, Yamada J, Kono A and Goto T 1995 Formation dynamics of excited atoms in an ArF laser using He and Ne buffer gases *J. Appl. Phys.* **77** 2906–11

[86] Ohwa M and Obara M 1988 Theoretical evaluation of the buffer gas effects for a self sustained discharge ArF laser *J. Appl. Phys.* **63** 1306–12

[87] Saito T, Ito S and Tada A 1996 Long lifetime operation of an ArF excimer laser *Appl. Phys.* B **63** 229–35

[88] Andrew, Dyer, P E and Roebuck P J 1984 Improved energy output from discharge pumped ArF and KrCl lasers *Opt. Commun.* **49** 189–94

[89] Miyazaki K, Hasama T, Yamada K, Fukatsu T, Eura T and Sato T 1986 Efficiency of a capacitor transfer type discharge excimer laser with automatic pre-ionization *J. Appl. Phys.* **60** 2721–8

[90] Zhupikov A A and Razhev A M 1997 Excimer ArF laser with an output energy of 0.5 J and He buffer gas *Sov. J. Quantum Electron.* **27** 665–9

[91] Mandle A 1992 XeF(B-X) long pulse length laser studies *J. Appl. Phys.* **71** 1630–7

[92] Nishida N, Takashima T, Tittel F K, Kannari F and Obara M 1990 Theoretical evaluation of a short pulse electron beam excited XeF(B-X) laser using a low pressure, room temperature Ar/Xe/F_2 mixture *J. Appl. Phys.* **67** 3932–40

[93] Burnham R, Harris N W and Djeu N 1976 Xenon fluoride laser excitation by transverse electric discharge *Appl. Phys. Lett.* **28** 86–7

[94] Sarjeant W J, Alcock A J and Leopold K E 1977 A scalable multi-atmosphere high power XeF laser *Appl. Phys. Lett.* **30** 635–7

[95] Nighan W L 1978 Influence of electron F_2 collisions in rare gas halide laser discharges *Appl. Phys. Lett.* **32** 297–300

[96] Mei Q C, Peters P J M, Trentelman M and Witteman W J 1995 Optimisation of the pulse duration of a discharge pumped XeF(B $\rightarrow$ X) excimer laser *Appl. Phys.* B **60** 553–6

[97] Tellinghuisen J *Appl. Atomic Collision Phys.* vol 3, ed E C McDaniel and W L Nighan (London: Academic)

[98] Sze R C, Sakai T, Vannini M and Sentis M L 1991 Time resolved studies of the C-A transition in avalanche discharges: fluorescence and gain-loss studies *IEEE J. Quantum Electron.* **27** 77–89

[99] Kumagai H and Obara M 1989 Output energy enhancement of discharge pumped XeF(B to X) lasers with the two component halogen donor mixtures *IEEE J. Quantum Electron.* **25** 1874–8

[100] Peters P J M, Trentelman M, Mei Q C and Witteman W J 1994 Gas discharge XeF(B $\rightarrow$ X) laser with high specific output power *Appl. Phys.* B **59** 533–5

[101] Trentelman M, Peters P J M, Mei Q C and Witteman W J 1995 Gas discharge XeF(B $\rightarrow$ X) laser excited by a pre-pulse main pulse circuit with magnetic switching *J. Opt. Soc. Am.* B **12** 2494–501

[102] Ernst W E and Tittel F K 1979 A new electron beam pumped XeF laser at 486 nm *Appl. Phys. Lett.* **35** 36–7

[103] Cambell J D, Fisher C H and Center R E 1980 Observations of gain and laser oscillation in the blue–green during direct pumping of XeF by microsecond electron beam pulse *Appl. Phys. Lett.* **37** 348–50

[104] Nighan W L, Tittel F K, Wilson W L, Nishida N, Zhu Y and Sauerbrey R 1984 Synthesis of rare gas halide mixtures resulting in efficient XeF(C $\rightarrow$ A) laser oscillation *Appl. Phys. Lett.* **45** 45–7

[105] Liegel J, Tittel F K and Wilson W L Jr 1981 Continuous broadband tuning of an electron beam pumped XeF(C $\rightarrow$ A) laser *Appl. Phys. Lett.* **39** 369–71

[106] Bischel W K, Ekstrom D J, Walker H C Jr and Tilton R A 1981 Photolytically pumped XeF(C-A) laser studies *J. Appl. Phys.* **52** 4429–34

[107] Mandle A and Litzenberger L N 1988 Efficient long pulse XeF(A $\rightarrow$ C) laser at moderate electron beam pump rate *Appl. Phys. Lett.* **53** 1690–2

[108] Hofmann Th, Yamaguchi S, Dane C B, Wilson W L Jr, Sauerbrey R, Tittel F K, Rubino R A and Nighan W L 1991 Wavelength agile operation of an injection controlled XeF(C $\rightarrow$ A) laser system *Appl. Phys. Lett.* **58** 565–7

[109] Peters P J M, Bastiaens H M J, Witteman W J, Sauerbrey R, Dane C B and Tittel F K 1990 Efficient XeF(C $\rightarrow$ A) laser excited by a coaxial electron beam at intermediate pumping rates *IEEE J. Quantum Electron.* **26** 1569–73

[110] Fisher C H, Center R E, Mullaney G L and McDaniel J P 1979 A 490 nm XeF electric discharge laser *Appl. Phys. Lett.* **35** 26–8

[111] Burnham R 1979 A discharge pumped laser on the C $\rightarrow$ A transition of XeF *Appl. Phys. Lett.* **35** 48–50

[112] Cheng Yong-kang, Yang Shao-peng, Wang Min-xiao and Ma Zu-guang 1993 Free running XeF(C-A) lasing at 2.2 atm in a commercial discharge excimer laser *Appl. Phys. Lett.* **62** 4–6

[113] Fisher C H, Center R E, Mullaney G J and McDaniel J P 1979 Multipass amplification and tuning of the blue-green XeF(C $\rightarrow$ A) laser *Appl. Phys. Lett.* **35** 901–3

[114] Voges H and Marowsky G 1988 Injection control of a discharge excited XeF(C $\rightarrow$ A) laser *IEEE J. Quantum Electron.* **24** 827–32

[115] Bischel W K, Nakana H H, Eckstrom D J, Mill R M, Huestis D L and Lorents D C 1979 A new blue–green excimer laser in XeF *Appl. Phys. Lett.* **34** 565–7

[116] Basov N G, Zuev V S, Kanaev A V, Mikheev L D and Stavrovskii D B 1979 Laser action due to the bound free C(3/2)-A(3/2) transition in the XeF molecule formed by photodissociation of XeF_2 *Sov. J. Quantum Electron.* **9** 629–32

[117] Eckstrom D J and Walker H C 1982 Multi-joule performance of the photolytically pumped XeF(C $\rightarrow$ A) laser *IEEE J Quantum Electron.* **18** 176–81

[118] Zuev V S, Mikheev L D and Stavrovskii D B 1984 Efficiency of an optically pumped XeF laser *Sov. J. Quantum Electron.* **14** 1174–8

[119] Zuev V S, Kashnikov G N, Kozlov N P, Mamaev S B, Orlov V K, Protasov Yu S and Sorokin V A 1986 Characteristics of an XeF(C-A) laser emitting visible light as a result of optical pumping by surface discharge radiation *Sov. J. Quantum Electron.* **16** 1665–7

[120] Zuev V S, Kashnikov G N, Kirilenko V V, Mamaev S B, Sorokin V A and Sukhorukov V F 1989 Photodissociation XeF laser emitting visible and ultraviolet radiation when pumped with radiation from a sectioned surface discharge *Sov. J. Quantum Electron.* **19** 748

[121] Watanaba K, Kashiwabara S and Fujimoto R 1987 Development of a formed ferrite flash plasma light source for gas laser applications *Appl. Phys. Lett.* **50** 629–31

[122] Gross R W F, Schneider L E and Amimoto S T 1988 XeF laser pumped by high-power sliding discharges *Appl. Phys. Lett.* **53** 2365–7

[123] Mitko S V, van Goor F A, Witteman W J and Ochkin V N 1996 Optical pumping and ferrite flash discharges *Proc. NATO Advanced Research Workshop ARW 950443 on Gas Lasers—Recent Developments and Future Prospects (Moscow, Russia, July 2–5, 1995) (Kluwer Academic Publishers Series 3: High Technology 10)* (Dordrecht: Kluwer Academic) p 185

[124] Kovacs M A and Ultee C J 1970 Visible laser action in fluorine *Appl. Phys. Lett.* **17** 39–40

[125] Huestis D L, Hill R M, Nakano H H and Lorents D C 1978 Quenching of Ne, F and F_2 in Ne/Xe/NF_3 and Ne/Xe/F_2 mixtures *J. Chem. Phys.* **69** 5133–9

[126] Bastiaens H M J, van Dam B M C, Peters P J M and Witteman W J 1993 Small signal gain measurements in an electron beam pumped F_2 laser *Appl. Phys. Lett.* **63** 438–40

[127] Bastiaens H M J, Peters P J M and Witteman W J 1997 *Eleventh Int. Symp. on Gas Flow and Chemical Lasers (SPIE 3092)* ed H J Baker (Bellingham, WA: SPIE), pp 374–7

[128] Kakehata M, Uematsu T, Kannari F and Obara M 1991 Efficient characterization of vacuum ultraviolet molecular fluorine laser (157 nm) excited by an intense electric discharge *IEEE J. Quantum Electron.* **27** 2456–64

[129] Yamada K, Miyazaki K, Hasama T and Sato T 1989 High power discharge pumped F_2 molecular laser *Appl. Phys. Lett.* **54** 597–9

[130] Lankhorst F T J L, Bastiaens H M J, Peters P J M and Witteman W J 1993 High specific laser output energy at 157 nm from an electron beam pumped He–Ne–F_2 gas mixture *Appl. Phys. Lett.* **63** 2869–71

[131] Bastiaens H M J 2000 On the long pulse operation of the molecular fluorine laser *PhD Dissertation* (University of Twente, Enschede, The Netherlands)

[132] Rokni M and Jacob J H 1982 *Appl. Atomic Collision Phys. Gas Lasers* vol 3, ed H S W Massey, E W McDaniel and B Bederson (New York: Academic) p 274

[133] Asinovskii E I, Vasilyak L M and Markovets V V 1983 Wave breakdown of gas filled gaps *High Temperature* **21** 293–305

[134] Diegelmann M, Hohla K, Rebentrost and Kompa K L 1982 Diatomic interhalogen laser molecules: Fluorescence spectroscopy and reaction kinetics *J. Chem. Phys.* **76** 1233–47

[135] Ishchenko V N, Kochubei S A and Razhev A M 1986 High power efficient vacuum ultraviolet F_2 laser excited by an electric discharge *Sov. J. Quantum Electron.* **16** 707–9

[136] Woodworth J R and Rice J K 1978 An efficient, high power F_2 laser near 157 nm *J. Chem. Phys.* **69** 2500–4

[137] Kakehata M, Hashimoto E, Kannari F and Obara M 1990 High specific output energy operation of a vacuum ultraviolet molecular fluorine laser excited at 66 MW cm^3 by an electric discharge *Appl. Phys. Lett.* **56** 2599–601

[138] Kakehata M, Yang C H, Ueno Y and Kannari F 1993 Output characteristic of a discharge pumped F_2 laser (157 nm) with an injection seeded unstable resonator *J. Appl. Phys.* **74** 2241–6

[139] Rice J K, Hays A K and Woodworth J K 1977 VUV emissions from mixtures of F_2 and the nobel gases. A molecular F_2 laser at 157.5 nm *Appl. Phys. Lett.* **31** 31–3

[140] Waynant R W 1978 Inert gas halide VUV lasers *Sov. J. Quantum Electron.* **8** 1002–4

[141] Waynant R W 1977 A discharge pumped ArCl superfluorescence laser at 175 nm *Appl. Phys. Lett.* **30** 234–5

[142] Tittel F K, Marowsky G, Wilson W L and Smayling M C 1981 Electron beam pumped broadband diatomic and triatomic excimer lasers *IEEE J. Quantum Electron.* **17** 2268–81

[143] Tittel F K, Wilson W L, Stickel R E, Marowsky G and Ernst W E 1980 A triatomic Xe_2Cl excimer laser in the visible *Appl. Phys. Lett.* **36** 405–7

[144] Tittel F K, Marowsky G, Smayling M C and Wilson W L 1980 Blue laser action by rare gas halide trimer Kr_2F *Appl. Phys. Lett.* **37** 862–4

[145] Heard H G 1963 Ultraviolet gas laser at room temperature *Nature* **220** 667

[146] Searles S K and Hart G A 1974 Laser emission at 357.7 and 380.5 nm in electron beam pumped Ar–N_2 mixtures *Appl. Phys. Lett.* **25** 624–6

[147] Bychkov Yu I, Losev V F, Mesyats G A and Tarasenko V F 1977 Stimulated emission of three lines from an N_2+Ar laser *Sov. J. Quantum Electron.* **7** 789–90

[148] Mathias L E S and Parker J T 1963 Stimulated emission in the band spectrum of nitrogen *Appl. Phys. Lett.* **3** 16–18

[149] Tarasenko V F and Bychkov Yu I 1973 Nitrogen laser with a transverse discharge *Instrum. Exp. Techniques* **16** 221–2

[150] Junichi I, Kiichiro K and Yoshitake K 1975 Intense laser emission on 337.7 nm using N_2-SF_6 mixtures in a TE N_2 laser *Appl. Phys. Lett.* **27** 503–4

[151] Losev V F and Tarasenko V F 1976 Effect of SF_6 on characteristics of the N_2 laser *Sov. Phys. Technical Phys.* **21** 1295–7

[152] Chou M S and Zavarzkas G 1981 Long pulse N_2 UV lasers at 357.7, 380.5 and 405.9 nm in N_2–Ar–Ne–He mixture *IEEE J. Quantum Electron.* **17** 77–81

[153] Derzhiev V I, Losev V F and Skakun V S 1986 Effect of neon and helium impurities on the lasing energy of an Ar–N_2 laser *Opt. Spectrosc.* **60** 499–500

[154] Bazhulin P A, Knyazev I N and Petrash G G 1965 *Sov. Phys.–JETP* **21** 649

[155] Hodgson R T 1970 *Phys. Rev. Lett.* **25** 494

[156] Hodgson R T and Dreyfus R W 1972 Vacuum-UV laser action observed in H_2 Werner bands *Phys. Rev. Lett.* **28** 536–9

[157] Waynant R W, Shipman J D, Elton R C and Ali A W 1970 Vacuum ultraviolet laser emission from molecular hydrogen *Appl. Phys. Lett.* **17** 383–4

[158] Goldsmith J E M and Knyazev I N 1977 A simple compact high repetition rate hydrogen VUV laser for scientific applications *J. Appl. Phys.* **48** 4912–21

[159] Waynant R W 1972 Observations of gain by stimulated emission in the Werner band of molecular hydrogen *Phys. Rev. Lett.* **28** 533–5

[160] Dreyfus R W and Hodgson R T 1974 Molecular hydrogen laser: 109.8–161.3 nm *Phys. Rev.* A **9** 2635–48

[161] Shipman J D 1967 Travelling wave excitation of high power gas lasers *Appl. Phys. Lett.* **10** 3–4

[162] Schwab A J and Hollinger F W 1976 Compact high power N_2 laser: circuit theory and design *IEEE J. Quantum Electron.* **12** 183–8

[163] Wang C P 1976 Simple fast discharge device for high power pulsed lasers *Rev. Sci. Instrum.* **47** 92–5

[164] Waynant R W 1971 A travelling wave vacuum ultraviolet laser *Proc. Technical Program, Electro-Optical Systems Design Conference (Anaheim)* pp 1–5

B3.3
Copper and gold vapour lasers

Colin Webb

B3.3.1 Atomic physics and excitation mechanisms of copper vapour lasers

Historically, the first of the 'self-terminating' or 'cyclic' class of laser systems was the 723 nm atomic lead laser [1] demonstrated by Fowles and Silfvast in 1965. In 1966 Walter *et al* [2] reported oscillation on the 511 and 578 nm transitions of the copper atom Cu, a system belonging to the same generic type as the Pb laser, and the one which has become the best known of the entire class. The operating principles can best be understood by reference to figure B3.3.1, which shows the low-lying energy levels of atomic copper.

The $^2S_{1/2}$ ground state of Cu arises from the configuration $3d^{10}4s$. Promotion of the outer electron to successively higher orbitals $3d^{10}4p$, $3d^{10}5p$, etc, gives rise to a series of ^2P levels connected to the ground state by strong resonance transitions (whose first members are 328 and 325 nm) analogous to the principal series in alkali atoms.

However, in contrast to the alkalis, in Cu there exists an alternative configuration ($3d^94s^2$) that happens to be almost as low in energy as the ground state—indeed it misses being the ground state by only about 1.5 eV. This configuration gives rise to $^2D_{3/2}$ and $^2D_{5/2}$ levels which are fully metastable because radiative transitions from them to the $^2S_{1/2}$ ground state are rigorously forbidden by the parity rule for electric dipole radiation. Emission from the ^2P levels of the $3d^{10}4p$ configuration to the ^2D levels of $3d^94s^2$ would be forbidden by the electric dipole selection rule that allows only one electron at a time to change orbital. However, a slight mixing of other configurations in Cu is sufficient to allow this rule to be violated to some extent, because the levels of the real atom are not rigorously described by the quantum numbers implied by the pure LS description.

Thus, while free Cu atoms in the $3d^{10}4p$ ^2P levels can relax to the $^2S_{1/2}$ ground state with untrapped lifetimes of the order of 7 ns, they do have a finite probability (with $A_{21} \sim 2 \times 10^6$ s^{-1} [3]) for decay to the ^2D levels in these 'forbidden' transitions. Further, at the vapour densities appropriate to laser action in the copper vapour laser (CVL), decay from the $3d^{10}4p$ ^2P upper laser levels direct to the ground state does not represent a significant loss since the 328 and 325 nm transitions are subject to strong radiation trapping. Thus, the only effective radiative decay route is *via* the laser transitions to the ^2D levels.

Now, if a fast rising current pulse is applied to a discharge containing copper vapour, Cu atoms are more likely to be promoted to the levels of the $3d^{10}4p$ configuration than they are to the $3d^94p^2$. This is because the 4s outer electron of the ground state Cu atom presents a bigger target than one of the $3d^{10}$ electrons, at least for electrons of more than a few eV energy. The result is that, provided T_e (the temperature of the electron energy distribution) exceeds about 3.5 eV the $3d^{10}4p$ levels are populated faster than the $3d^94p^2$ ^2D levels and a large population inversion can be created transiently.

In Cu vapour the wanted excitation of the $3d^{10}4p$ ^2P upper laser levels provides by far the largest inelastic collision cross section for plasma electrons, just as the wanted process of excitation of the 3p ^2P resonance levels from the 3s $^2S_{1/2}$ ground state of the Na atom dominates all other electron energy-loss processes in

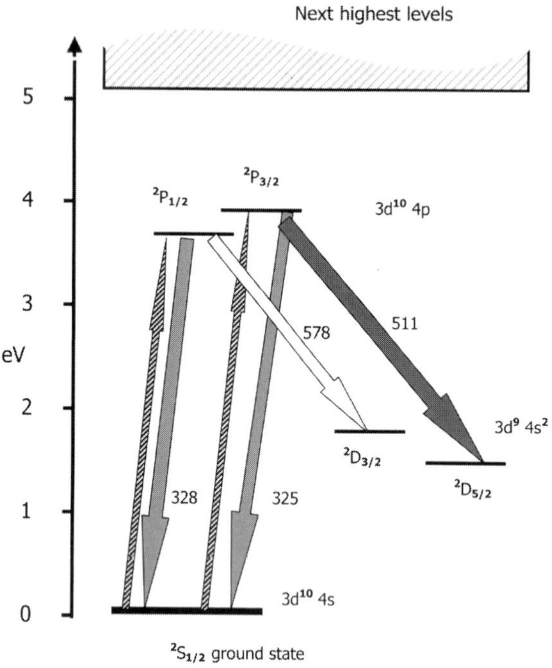

Figure B3.3.1. First few levels of atomic copper.

the sodium vapour lamp. For the same reason that sodium vapour makes an efficient lamp, the CVL is an efficient laser—more than ten times as efficient as the argon ion laser, with over 1% true wall-plug efficiency available from present commercial systems.

Electric dipole optical transitions from the ^{2}D lower laser levels back to the ^{2}S ground state are rigorously forbidden since both the configurations $3d^94s^2$ and $3d^{10}4s$ are of even parity. Because of this the lower levels are truly metastable, and the Cu vapour laser fails to satisfy the minimum condition for CW lasing:

$$A_{21}^{-1} > (g_2/g_1)\tau_1 \tag{B3.3.1}$$

in which A_{21} is the Einstein coefficient for spontaneous emission on the laser transition and g_2 and g_1 are the statistical weights of upper and lower levels respectively and τ_1 is the lower level lifetime. The gain on the 511 and 578 nm transitions is sufficiently strong (e.g. ~50 000 per pass in the case of 511 nm) that the laser can produce substantial output power even without an optical cavity. With a cavity, oscillation builds up in a very small number of cavity round trips. By the time this has occurred, stimulated emission is dumping population into the ^{2}D metastable lower levels which have no means of disposing of it on the timescale of 100 ns. This, together with the decrease in electron temperature as the current pulse develops, precipitates a rapid switch in the behaviour of the active medium—from high gain to high loss—and the laser action self-terminates well before the end of the current pulse. It is then necessary to wait until the lower level population has been reduced to a much smaller value before the next excitation pulse can be applied and the cycle of events repeated.

It is important to point out that the processes acting to de-excite atoms in the ^{2}D lower laser levels by collisions with buffer gas atoms or with the tube wall are not very effective in emptying the lower level of population. If there were no other mechanism, the rate at which laser pulses could be repeated would be very slow. Experiments by Lewis [4] showed unambiguously that the population of the ^{2}D levels in Cu during the 100 μs interval between excitation pulses become clamped after about 10 μs to that of the ^{2}S ground state. The clamping mechanism is the kinetically balanced two-way exchange of population brought about by inelastic collisions of electrons with ground state ^{2}S atoms which excites them to ^{2}D and the thermodynamic inverse process of de-excitation from ^{2}D to ^{2}S in superelastic collisions:

$$\mathrm{Cu}(^2\mathrm{S}_{1/2}) + \mathrm{e(fast)} \Leftrightarrow \mathrm{Cu}(^2\mathrm{D}_{3/2}, {}^2\mathrm{D}_{5/2}) + \mathrm{e(slow)}. \qquad \text{(B3.3.2)}$$

Although the cross section for these optically forbidden processes tends to be much smaller than those for the allowed processes, which excite the upper laser levels from the ground state, the electron density in the interpulse period remains high enough to ensure that the forward and reverse rates in equation (B3.3.2) are the dominant processes affecting the ^{2}D population in elemental CVLs without gas additives.

The pulse repetition frequency limitation in practical CVLs is due not so much to the population remaining in the metastable lower laser levels from one excitation current pulse to the next—the superelastic de-excitation rate is usually sufficient to ensure this. Rather it is the remanent electron density which causes the discharge to remain conducting after the discharge pulse and prevents the electron temperature rising to the required high values at the onset of the next pulse of applied voltage.

The characteristic features of the term diagram of a 'self terminating' or 'cyclic' laser are thus: an upper laser level connected by strong transitions to the ground state but by less allowed transitions to the lower levels which are themselves optically forbidden to decay the ground state. These features are exemplified by Pb, Au, Cu and several other atomic systems.

B3.3.2 Development of the copper vapour laser

The first CVL of Walter in 1966 employed an alumina discharge tube of 10 mm internal diameter and 800 mm length. The tube contained metallic copper beads and helium at a few mbar pressure. To achieve its working temperature of 1500 °C (sufficient to give enough ground-state copper density for trapping on the resonance lines to be effective) it employed an external oven.

The next major advance in CVL development was reported by workers in Russia. In 1972, Isaev *et al* [5], using a discharge tube surrounded by a thick layer of thermal insulator, showed that the heat generated by the pulsed discharge was sufficient to maintain the elemental copper at the CVL working temperature and obtained 15 W at 1% true efficiency.

Development of elemental copper lasers continued in the late 1970s and 1980s in the USA, prompted largely by the requirements of the Atomic Vapor Laser Isotope Separation (AVLIS) program at the Lawrence Livermore National Laboratory (LLNL).

The scaling-up of CVL power was based on the demonstration in 1978 by Smilanski *et al* [6] in Israel that the CVL could be operated in tubes of relatively wide bore (up to 40 mm) and with neon as buffer gas at a pressure of 180 mbar. The surprising feature of Smilanski's experiments was the excellent uniformity with which the Ne+Cu discharge filled the cross section of the discharge tube. Discharges in rare gases above a few mbar pressure in wide bore tubes tend to be self-constricted columns following tortuous and irreproducible paths between electrodes. Later work by Gabay and Smilanski [7] has provided the explanation by showing that ionization remaining from one laser pulse to the next is able to provide enough seeding to make the discharge current flow in a uniform fashion across the tube cross section.

With the proof that the CVL power could be scaled by increasing the size of the discharge device, rapid progress was made at LLNL so that by 1991 that laboratory was able to report that CVL amplifiers of 80 mm tube diameter with extractable powers of 650 watts per unit were entering production and were qualified to install in a full scale isotope enrichment facility at LLNL [8].

By 1980, the development of elemental copper vapour lasers was well under way in the Clarendon Laboratory at the University of Oxford and in 1982 a commercial 25 W CVL was introduced by Oxford Lasers Ltd based on the Clarendon Laboratory design.

Table B3.3.1. Properties of a typical commercial CVL.

Average power on 511 578 nm combined 45 W	Green/yellow power ratio 2:1
Pulse energy (max) 3 mJ	Pulse width (FWHM) 25 ns (nominal)
Peak power 200 kW	Pulse repetition frequency 10–30 kHz
Timing jitter $<$ 2 ns RMS	Beam diameter 23 mm
Beam divergence 4 mradians (plane-plane cavity) 75 μradians (unstable resonator fitted)	
Power consumption for normal running 4 kW	Cooling water requirement 5 litre min^{-1}
Gas supply neon/HCl premix	Gas consumption 0.5 litre atm hr^{-1} (running)

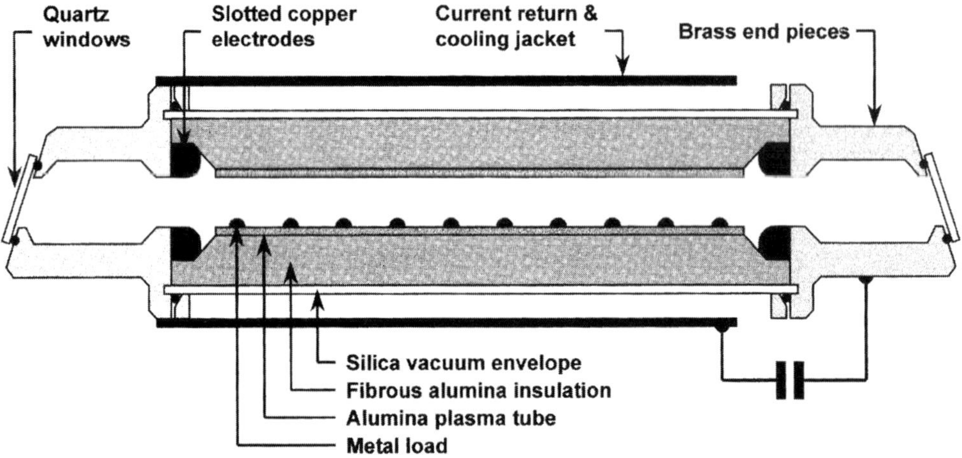

Figure B3.3.2. CVL discharge tube, schematic illustration.

Typical performance figures for a modern commercial 45 W CVL (Oxford Lasers model LM100-X) are shown in table B3.3.1.

B3.3.3 Elemental CVLs—practical considerations

Most commercial CVLs for scientific and industrial (non-nuclear) applications operate with output powers in the range 10–100 W. Figure B3.3.2 shows a schematic diagram of the construction of a typical elemental CVL, i.e. one employing thermal evaporation of metallic copper as the source of the vapour. The discharge tube (typically an alumina ceramic tube of 20–40 mm internal diameter and 1 m length) is surrounded by a layer of ceramic fibre material for thermal insulation. The entire discharge tube and thermal insulator assembly is contained within a gas envelope of pyrex or silica closed at each end by silica windows set tilted a few degrees from the normal to the laser tube axis. One of the two cylindrical electrodes at either end of the discharge tube is electrically connected to a metal tube enclosing the discharge tube which acts both as a heat sink (to absorb the heat radiated by the discharge tube) and as a coaxial current return path. Small pellets of copper are placed at intervals down the length of the discharge tube that also contains a buffer gas, usually neon, at a pressure of 20–200 mbar. Both the efficiency and beam quality are improved if a small amount of hydrogen (typically ~0.5–3%) is added to the buffer gas (see section B3.3.5).

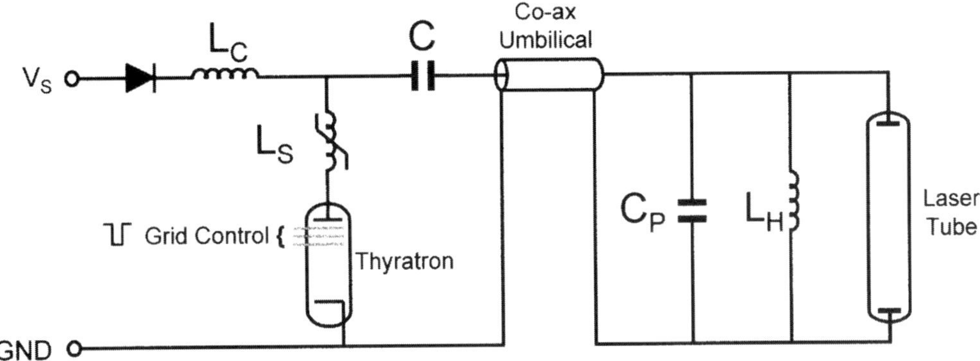

Figure B3.3.3. CVL discharge circuit.

During the period from initial switch-on, the heat generated by the repetitively pulsed discharge (carried by the buffer gas) raises the temperature in the discharge tube until it reaches its stable working temperature of about 1500 °C. This occurs when the heat lost by radiation through the end windows and conduction radially outwards is sufficient to balance the power input to the discharge. At this temperature the vapour pressure of copper is 0.3 mbar, and the discharge in the heated zone is dominated by the copper vapour component with the buffer gas playing only a minor role.

During hundreds of hours of operation, the copper fill slowly migrates from the hot region to condense in the cold regions at each end of the tube, and laser power will thus eventually decrease. For this reason, if the laser tube is permanently sealed off (as in the case of 3–10 W devices manufactured for many years by the Russian firm, Istok) it is necessary to replace the complete tube assembly at roughly 1000 hr intervals of operating time. An alternative approach is to provide demountable window sections on the tube so that one or both can be removed at appropriate servicing intervals to remove the copper condensed in the catcher regions at each end of the tube, and to recharge the central region of the tube with a new load of metallic copper. In the CVL, unlike the Ar^+ laser, there is no problem in exposing the electrodes to air since the electrodes are not coated with chemically reactive materials. The CVL tube can simply be connected permanently to a source of low-pressure neon and supplied with a trickle flow of the gas to purge continuously any impurities. The copper refill procedure also provides the opportunity to clean the laser tube windows if necessary, although the buffer gas in the end regions very effectively protects the windows from becoming coated with copper.

A basic version of the circuit which powers the laser is shown in figure B3.3.3 (see also chapter C1.5.2). It comprises a high voltage supply capable of resonantly charging the storage capacitor C *via* the inductor L_H to a voltage between 10 and 20 kV. When a low-voltage trigger pulse is applied to the control grid of the thyratron T, the tube conducts and allows the charge on the storage capacitor C to be transferred to the peaking capacitor C_P mounted close to the laser tube assembly and connected directly across the two electrodes. To avoid damage to the thyratron, the initial switching current through the thyratron is limited by the saturable inductor L_S. As the voltage on the peaking capacitor rises, the gas in the main discharge tube undergoes breakdown and the peaking capacitor rapidly discharges through the low-inductance circuit comprising the gas discharge and the external coaxial current return path.

The thyratron is a gas-filled hot-cathode tube containing hydrogen. Once conducting, it can only be restored to the non-conducting state by removing the voltage across it or even applying a slightly negative voltage for a short period while the hydrogen ions recombine. With a suitable choice of thyratron, operation at several tens of kilohertz is possible. The upper limit on frequency of operation is set by kinetic processes within the laser medium itself. Power supplies incorporating all solid-state switching elements and magnetic pulse compression techniques offer an alternative to the use of gas-filled tube technology (see section C1.5.2.2), but the efficiency, power handling capability, and low-timing jitter associated with thyratron-switched circuits make them still highly competitive.

B3.3.4 Copper halide lasers

In the years following the first demonstration of the CVL, the warm-up time needed to bring a self-heated CVL into operation (usually about 1 hr, depending on the size of the laser) and the fact that the tube operated at high temperature were seen as problems which would inhibit the widespread adoption of the technology. In fact, the high tube temperature, although it does restrict the choice of construction materials, has not proved to be a barrier to the attainment of long-term reliability. Modern elemental CVLs operate for many tens of thousands of hours without experiencing tube failure. However, the idea of making a CVL which could operate at lower temperature (even room temperature) was an important motivation for testing many molecular compounds of copper which could be used as donors of copper atoms.

As early as 1966, Walter *et al* [9] had suggested the use of volatile compounds as a source of atomic copper. Although in 1973–4 Liu *et al* [10] reported observation of 'super-radiance' (i.e. amplified spontaneous emission) in discharges containing copper halide compounds, the first successful operation of a laser containing a He–CuCl mixture at temperatures in the range of 500–600 °C was reported by Chen *et al* [11] in 1974. Their device, in common with many in the years immediately following, used a system of double pulse excitation in which the first pulse of the pair served to dissociate the molecular copper donor compound in the volume of the tube. Laser action was achieved on the second pulse of each pair. This strategy avoided the problems of discharge stability to which longitudinally excited discharges in gases containing electronegative components are prone. The tendency of such discharges to knot themselves into self-constricted and tortuous filaments associated with the build-up of excess free halogen compounds in the tube were ameliorated by operating the devices either with low repetition rate pulse trains of double pulses in short bursts of high repetition rate pulses.

Steady progress in the development of copper halide lasers (particularly the CuBr system which they pioneered) was made in the years from 1974 onwards by the group headed by Sabotinov [12] at the Institute of Solid State Physics in Sofia, Bulgaria. The characteristic feature of many of the laser tube designs they developed which could be run in a truly pulse repetitive mode was, as indicated in figure B3.3.4(*a*), a series of discharge confining silica discs spaced at intervals along the wide bore silica gas envelope. Narrow apertures in discs served to confine the discharge along the axis of the tube, while the wide pockets between neighbouring discs allowed the copper bromide density to be regulated independently from discharge heating.

By providing traps containing copper filings in the electrode regions, as in figure B3.3.4(*b*), or by other means, the build-up of excess bromine can be avoided, and these lasers can be run in a truly pulse repetitive fashion at pulse repetition frequencies typically in the range 10–20 kHz. Although powers of over 120 W have been reported from CuBr lasers [13], most commercial CVL such as those marketed by the Australian company, Norseld (whose designs are based on the Bulgarian work) have typically been for devices in the power range up to 10 W on 511 and 578 nm combined. Tube lifetimes are of the order of 1000 hr for these sealed-off tubes.

B3.3.5 Effect of additive hydrogen

As early as 1980, Bokhan *et al* [14] reported evidence of the beneficial effect of adding a small proportion (typically 0.3–0.6%) of hydrogen to the neon buffer gas in an elemental CVL. Other workers at first found the effect difficult to reproduce. This was largely because the amount of hydrogen needed to obtain a positive effect is small enough that the optimum concentration was already provided by the dissociation of water vapour from the atmosphere permeating the flexible plastic hoses connecting the laser tube to the buffer gas supply. Another source of water vapour is the alumina used both for the thermal insulation and for

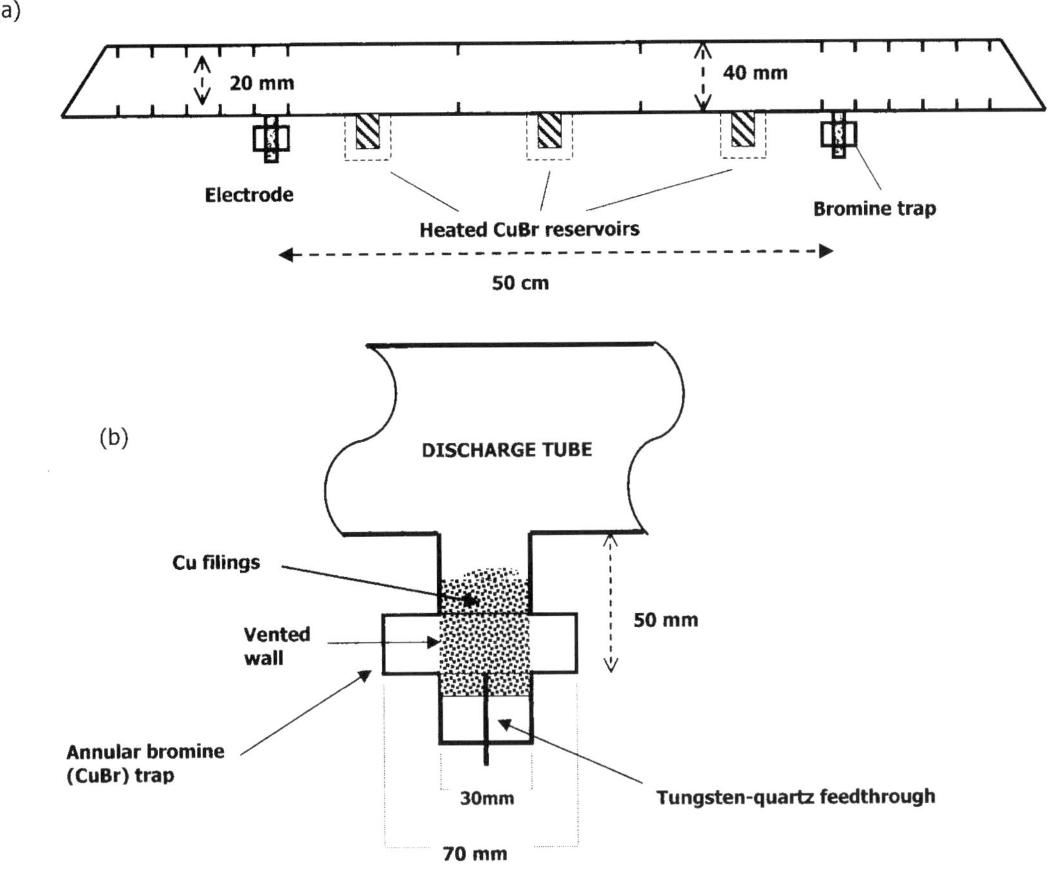

Figure B3.3.4. Copper-bromide laser tube—after Sabotinov [12].

the discharge tube itself. One form of alumina is $Al_2O_3(5H_2O)$, and the amount of this compound in the construction materials of the laser depends on the manufacturing process. At elevated temperatures the water leaves the alumina and enters the discharge where it is dissociated. The oxygen forms oxides and leaves the laser gas mixture, while the hydrogen concentration continuously builds up in the tube unless a steady purge of gas is maintained.

However the beneficial effect of adding hydrogen is certainly real, and results in a increase in output power by some 50–70% in discharge gas mixtures containing 0.5% H_2 in neon when compared to the power obtained with a buffer gas of pure neon. A similar power enhancement with addition of hydrogen is also observed in copper halide lasers [15].

The beneficial effect of adding hydrogen is threefold. Not only does it allow the laser to be run at a higher pulse repetition frequency, but the energy per pulse is increased. Even more importantly, as shown by Withford *et al* [16], by optimizing both the total neon buffer gas pressure and the hydrogen concentration, the radial dependence of gain early in the gain period can be manipulated to advantage. Without hydrogen, the gain and emission rises first in the outer regions of the discharge tube near the tube walls. This annular pattern of radiation seeds the growth of the beam by stimulated emission and causes the final output beam to have a highly undesirable doughnut shape in the far field. (For this reason some early CVL designs [17] employed off-axis unstable resonators to take advantage of the rise in gain near the discharge wall in the early part of the pulse to build up a low-divergence beam which could extract power from the axial regions later in the pulse.) However, with optimized neon and hydrogen concentrations, the gain rises first on the tube axis, and a much more desirable axially peaked output beam results. CVLs of modern design employ on-axis unstable resonators, with beneficial effects on the symmetry and divergence properties of the output beam.

The mechanisms by which added hydrogen affects the discharge kinetics is discussed in detail in the encyclopaedic treatise on metal vapour lasers by Little [18]. The H_2 molecules, with their myriad of closely spaced rotational and vibrational levels, act to provide a rapid means of cooling the electrons in the period immediately following the discharge current pulse. This rapid decrease of electron temperature not only reduces the population in the lower lasers levels more rapidly than would otherwise be the case, but even more importantly, promotes rapid recombination of the electrons and ions because of the sensitive dependence (typically $T_e^{-9/2}$) of recombination rate upon electron temperature.

B3.3.6 Copper HyBrID lasers

A radical innovation in CVL design was introduced by Maitland, Little, Jones and co-workers at St Andrews University in the early 1990s [19]. Although these (Ne–HBr–CuBr) lasers employ a copper halide to achieve adequate densities of copper atoms, they differ from the copper bromide (Ne–H_2–CuBr) lasers described in section 4 by generating the CuBr *in situ* in the discharge tube by the reaction of HBr gas with annular strips of metallic copper placed at intervals along the bore of the discharge tube. The copper component is transported from the cooler regions near the tube wall into the axial region by the same kind of mechanism operating in tungsten–halogen filament lamps. The halogen scavenges copper by forming CuBr at the lower temperature of the wall region, and then migrates to the axis of the tube where it is dissociated into its atomic components.

These so-called HyBrID (Ne–HBr–CuBr) take their name from the slightly awkward acronym **hy**drogen **br**omide **i**n **d**ischarge. They have proven capable of high efficiency (3%), high repetition rate (20 kHz in large bore devices) and average powers of up to 280 W [20], at the same time operating at moderate temperatures in the range 600–700 °C. The lower operating temperature range brings advantages not only in shortening the warm-up period before full laser power is obtained, but also gives the laser designer greater freedom in the choice of materials for laser tube construction. In particular, the entire laser tube assembly can be made from fused silica by conventional glassblowing techniques.

To date, CVLs of the HyBrID design do not appear to have been made commercially. Perhaps one reason for this is the susceptibility of the lasers to form 'dendrites'—tree-like accumulations of metallic copper—which hang from the walls in the cooler regions at the ends of the tube and may ultimately block the optical path of the beam inside the cavity.

B3.3.7 CVL cavities and beam quality

When oscillating in the usual plane-plane resonator cavity the seed radiation determining the form of the output beam from a CVL typically diverges with an angle determined by the geometric value representing the aspect ratio of the discharge tube. Matters can be improved significantly by providing a confocal unstable resonator such as that shown schematically in figure B3.3.5 (see section A2.1.7).

The large concave high-reflectance mirror at one end of the cavity returns incident spontaneous emission back through the amplifying medium towards the beam splitter—a high-reflectance flat plate at 45° to the cavity axis with a small hole at its centre. The radiation returning towards the large mirror *via* this hole after reflection at the small concave mirror (of short focal length) therefore appears to diverge from a point close to where the two mirrors have their common focal point. After making a further double transit of the discharge tube most of this beam exits from the cavity in the form of an annular beam having been reflected from the outer region of the beam splitter plate. The remaining central part of the beam continues through

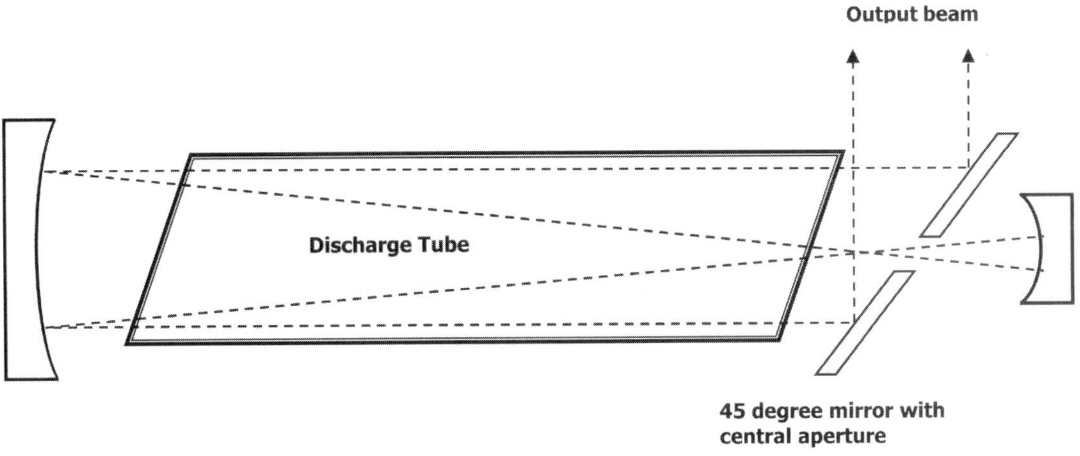

Figure B3.3.5. Confocal unstable resonator.

the hole and is reflected again at the smaller mirror. However, now the returning beam appears to diverge from a source point even closer to the common focal point of the mirrors. Thus, the beam which exits from the cavity after a further round trip (reflection at the large mirror and the beam splitter) is of much lower divergence. This iterative process continues as long as the gain pulse lasts, with the apparent source point approaching arbitrarily close to the common focus and the divergence correspondingly decreasing until (ideally) diffraction sets a limit. In a cavity of magnification 100, approximately 0.01% of the output beam is fed back from the small mirror and the flux divergence decreases by a factor of about 100 on each round-trip.

For the 45 W laser described in section B3.3.2, the output beam divergence with a plane-plane cavity is some 4 milliradians, but use of an unstable resonator reduces this below 75 microradians. In conventional CVLs with hydrogen addition, the time taken to build up the low divergence signal corresponds to two round trips. However, the maximum gain occurs early in the pulse. The result is that only the last part of the laser pulse, containing about 30–60% of the total energy, is of the low divergence form.

B3.3.8 MOPA CVLs

The problem of extracting power over the full duration of the gain pulse in a stand-alone CVL oscillator is associated with the finite time taken to build up a radiation distribution inside the unstable resonator cavity corresponding to the steady-state low divergence. One method of solving this problem is to inject low-divergence laser light into an amplifier unit just as its excitation pulse rises. The optical layout of such a master oscillator–power amplifier (MOPA) scheme used by Coutts *et al* [21] is shown in figure B3.3.6.

The narrow bore (12 mm) master oscillator laser tube is enclosed in its own unstable resonator. Its unpolarized output is transmitted *via* a rotatable polarizing beam splitter and an adjustable iris to be focused by a 500 mm achromatic lens through the 1 mm central hole in the 45° coupling mirror into the double-pass amplifier. After traversing the gain tube once, a mirror of 2 m radius of curvature produces a collimated beam of 25 mm diameter that is returned for a second pass through the gain medium, this time extracting power efficiently. The double-pass output exits by reflection at the 45° coupling mirror and is finally transmitted through a second, fixed, polarizing beam-splitter cube. By rotating the first polarizing beam splitter, full control over the power output of the MOPA CVL is obtained without variation of any other beam parameter. The divergence of the final beam for an output power of 19 W was measured to be 27 μradians (FWHM), only 1.3 times the diffraction-limited value.

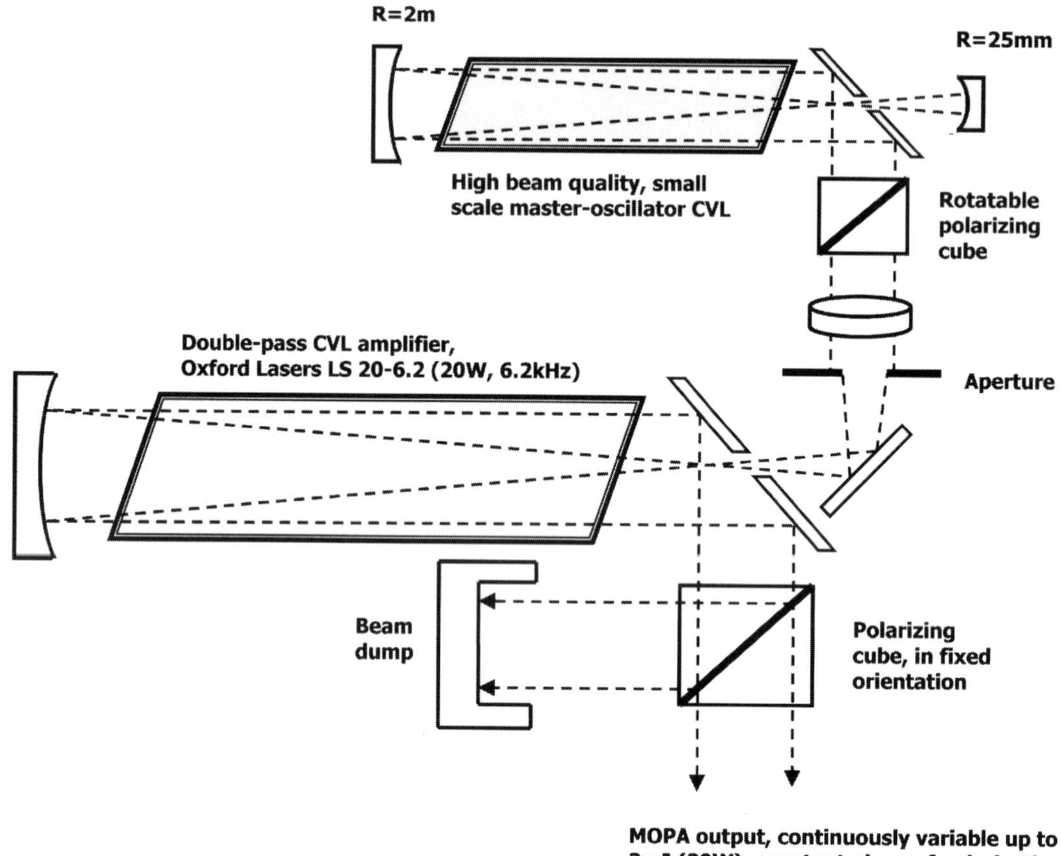

Figure B3.3.6. Master oscillator–power amplifier (MOPA) CVL configuration.

B3.3.9 Kinetically enhanced CVLs

The recent development of kinetically enhanced CVLs (KE-CVLs) by Piper and co-workers [22] at the Centre for Lasers and Applications at Macquarie University, Australia has resulted in a doubling of the power output capability of elemental CVL of a given size. The enhancement is brought about by adding a mixture of H_2 and HCl to the neon buffer gas of a high-temperature elemental CVL. The philosophy that led to this advance was the idea of decoupling the effect of the halide additives on the enhanced relaxation of electron and metastable densities from its role as the provider of ground-state copper which it has in the low-temperature HyBrID concept. The behaviour of KE-CVLs suggest that besides the increased cooling rate the electrons experience in the interpulse period due to the presence of H_2, the process of dissociative attachment of electrons with the hydrogen halide plays a dominant role in reducing the prepulse electron density [23]. The increased decay rate of the electron density combined with the efficient reformation of HCl in the plasma region has allowed very dramatic improvements in repetition rate capability as well as specific output power to be demonstrated. Operation of a KE-CVL with 9 W of output power at a pulse repetition rate of 100 kHz has been reported by Marshall *et al* [24]. Another benefit is that mutual neutralization of (positive) copper ions with negative hydrogen and halide ions benefit the replenishment of ground state copper atoms near the axis of the tube, which otherwise would be depleted due the migration of copper ions to the wall by ambipolar diffusion. The

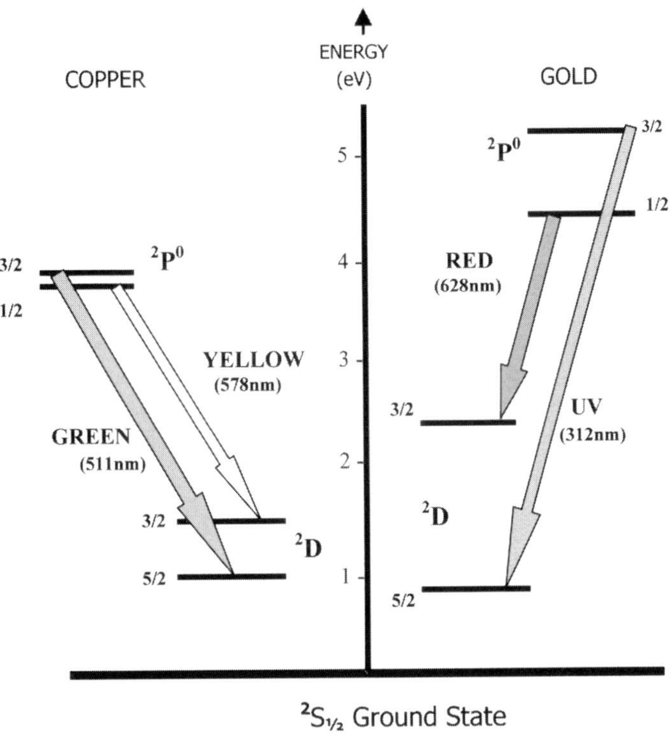

Figure B3.3.7. Low-lying energy levels of atomic copper and gold.

result is that, early in the discharge pulse, the gain rises to high values on axis. This is most important for the beam quality that the laser is able to achieve, since the effects of axial depletion of copper ground-state density combined with the skin effect tend to produce a gain profile which rises first near the walls with the deleterious effects noted in section B3.3.5.

B3.3.10 The gold vapour laser

The gold vapour laser is another member of the class of self-terminating neutral metal vapour lasers. It is closely analogous to the copper system, indeed gold is the only element in the whole periodic table which has exactly the same arrangement of low-lying energy levels as those of the copper atom. As shown in figure B3.3.7 where the term diagrams of copper and gold are drawn side by side, the ground state of Au is $^2S_{1/2}$ and arises from the configuration $5d^{10}6s$. The ^2P upper laser levels lie about 5 eV above the ground state, and the $J = 3/2$ and $J = 1/2$ levels are separated by a much greater interval than in the case of Cu, as are the ^2D lower levels, owing to the much greater spin-orbit splitting in the heavier element. Thus, while the $^2D_{3/2}$ lower level of the red 628 nm transition (analogous to the yellow 578 nm transition of Cu) at 2.8 eV is higher in energy than the corresponding state of Cu (1.5 eV), the opposite is true of the $^2P_{5/2}$ level which, in Au, lies only 1 eV above the ground state.

This has important consequences for the power capabilities of the ultraviolet 312 nm laser transition of Au. Because gold is less volatile than copper, the typical running temperature of a GVL is 1650 °C as against 1500 °C for the CVL. Even at these temperatures, thermal population of the $^2D_{5/2}$ is a problem to such an extent that relaxation of its population between pulses is not very effective and the 312 nm does not show oscillation at all in wide bore tubes at high repetition rates.

These same considerations also explain the lack of strong dependence of pulse energy with repetition rate displayed by the 628 nm transition of Au. The thermalization of the lower laser-level population to the Boltzmann fraction $\exp(-E_1/kT_e)$ of the ground-state population at T_e (the electron temperature) in the interpulse period and the fact that T_e can decay no further than T_{gas} is sufficient to explain the observed behaviour .

The $^2P_{1/2}$ upper level of the 628 nm transition in Au is significantly higher in energy than its counterpart in Cu. This may go some way to explaining why the GVL at 628 nm is not as efficient as the CVL operating on the analogous transition at 578 nm. In a laser of similar dimensions, the 628 nm transition of Au produces only 50% of the power available on the 578 nm transition of Cu. To date, the highest power recorded for the GVL is the 16 W of 628 nm output achieved by Lewis [4] at the Clarendon Laboratory in 1985. However, since to obtain a vapour pressure of Au comparable to that of Cu in the CVL the temperature of the discharge tube must be raised by a further 100 °C, it is possible that simply problems associated with preventing discharge tube sag have limited GVL performance to date. Little development of GVLs has been reported in the 1990s, however, in view of the improvements made to CVL technology and performance over the past decade it is interesting to speculate whether, by applying similar concepts to the GVL, significant improvements in performance might be realized in this laser system.

References

[1] Fowles G R and Silfvast W T 1965 High-gain laser transition in lead vapor *Appl. Phys. Lett.* **6** 236–7

[2] Walter W T, Piltch M, Solimene N and Gould G 1966 Pulsed laser action in atomic copper vapor *Bull. Am. Phys. Soc.* **11** 113

[3] Bielski A 1975 A critical survey of atomic transition probabilities for CuI *J. Quantum Spectrosc. Radiat. Transfer* **15** 463–72

[4] Lewis R R 1985 *DPhil Thesis* University of Oxford

[5] Isaev A A, Karayan M A and Petrash G G 1972 Effective pulsed copper-vapour laser with high average generation power *JETP Lett.* **16** 27–9

[6] Smilanski I, Kerman A, Levin L A and Erez G 1978 Scaling of the discharge heated copper vapour laser *Opt. Commun.* **25** 79–82

[7] Gabay S and Smilanski I 1980 Effect of preionization on a copper vapor laser *IEEE J. Quantum Electron.* **QE-16** 598–601

[8] Warner B E 1991 Status of copper vapor laser technology at Lawrence Livermore National Laboratory *Paper CFH4 in CLEO 1991 Technical Digest* (Washington DC: Optical Society of America) pp 516–18

[9] Walter W T, Piltch M, Solimene N and Gould G 1966 Efficient pulsed gas discharge lasers *IEEE J. Quantum Electron.* **QE-2** 474–9

[10] Liu C S, Sukov E W and Weaver L A 1973 Copper superradiant emission from pulsed discharges in copper iodide vapor *Appl. Phys. Lett.* **23** 92–3

[11] Chen C J, Nerheim N M and Russell G 1973 Double-discharge copper vapor laser with copper chloride as lasant *Appl. Phys. Lett.* **23** 514–15

[12] Sabotinov N V 1996 Copper bromide lasers *Pulsed Metal Vapour Lasers (NATO ASI Series 1)* vol 5, ed C E Little and N V Sabotinov (Dordrecht: Kluwer) pp 113–24

[13] Astadjov D N, Dimotrov K D, Jones D R, Kirkov V K, Little C E, Sabotinov N V and Vuchkov N K 1997 Copper bromide laser of 120 W average output power *IEEE J. Quantum. Electron.* **33** 705–9

[14] Bokhan P A, Silant'ev V I and Solomonov V I 1980 Mechanism for limiting the repetition frequency of pulses from a copper vapour laser *Sov. J. Quantum Electron.* **10** 724–7

[15] Sabotinov N V, Vuchkov N K and Astadjov D N 1993 Effect of hydrogen in the CuBr and CuCl-vapour lasers *Opt. Commun.* **95** 55–6

[16] Withford M J, Brown D J W, Coutts D W and Piper J A 1995 Increased efficiency of high-quality beam extraction from a copper vapor laser with H_2–Ne admixtures *IEEE J. Quantum Electron.* **31** 898–902

[17] Webb C E 1991 High power dye lasers pumped by copper vapour lasers *High Power Dye Lasers (Springer Series in Optical Sciences 65)* ed F J Duarte (Berlin: Springer) ch 5

[18] Little C E 1999 *Metal Vapour Lasers: Physics, Engineering and Applications* (Chichester: Wiley)

[19] Livingstone E S, Jones D R, Maitland A and Little C E 1992 Characteristics of a copper bromide laser with flowing Ne–HBr buffer gas *Opt. Quantum Electron.* **24** 73–83

[20] Le Guyadec E, Coutance P, Bertrand G and Peltier C 1999 A 280 W average power Cu–Ne–HBr laser amplifier *IEEE J. Quantum Electron.* **35** 1616–21

[21] Coutts D W, Wadsworth W J and Webb C E 1998 High average power blue generation from a copper vapour laser pumped titanium sapphire laser *J. Mod. Opt.* **45** 1185–97

[22] Withford M J, Brown D J W, Carman R J and Piper J A 1998 Enhanced performance of elemental copper-vapor lasers by use of H_2–HCl–Ne buffer gas mixtures *Opt. Lett.* **23** 706–8

[23] Webb C E and Hogan G P 1996 Copper laser kinetics—a comparative study *Pulsed Metal Vapour Lasers—Physics and Emerging Applications in Industry, Medicine and Science (NATO ASI Series 1)* vol 5, ed C E Little and N V Sabotinov (Dordrecht: Kluwer Academic) pp 29–42

[24] Marshall G D and Coutts D W 2000 Repetition rate scaling up to 100 kHz of a small scale (50 W) kinetically enhanced copper vapor laser *IEEE J. Selected Topics Quantum Electron.* **6** 623–8

Further reading

Little C E 1999 *Metal Vapour Lasers: Physics, Engineering and Applications* (Chichester: Wiley)

This is a truly encyclopaedic work on the subject of metal vapour lasers. It covers the history, science, practical construction details, performance, and applications of all types of metal vapour lasers, together with an exhaustive list of literature references on the subject.

Little C E and Sabotinov N V (ed) 1996 *Pulsed Metal Vapour Lasers—Physics and Emerging Applications in Industry, Medicine and Science (NATO ASI Series 1)* vol 5 (Dordrecht: Kluwer Academic)

This book is the proceedings of the NATO Advanced Research Workshop on Pulsed Metal Vapour Lasers held at the University of St Andrews in August 1995. The meeting brought together many of the pioneers of the subject and leading research specialists from around the world to discuss the latest advances in the science and applications of metal vapour lasers.

B3.4.1
Chemical lasers: COIL

B D Barmashenko and S Rosenwaks

B3.4.1.1 Principle of operation

The chemical oxygen iodine laser (COIL) [1] is the only known example of a high-power chemically-driven electronic transition laser. Unlike other chemical lasers (e.g. HF/DF lasers described in section B3.4.2), which operate on rotational–vibrational transitions of molecules, the COIL operates on an atomic electronic transition. At present the COIL is the shortest wavelength chemical laser system [2, 3]. Figure B3.4.1.1 shows the electronic states of the species participating in the pumping process of the COIL. The laser transition, at 1315 nm, takes place between the spin-orbit levels of the ground state configuration of the iodine atom, $I(5p^5\,{}^2P_{1/2}) \rightarrow I(5p^5\,{}^2P_{3/2})$. Basic chemical reactions taking place in the COIL are presented in table B3.4.1.1 [4, 5]. The iodine atoms are pumped by a near-resonant energy transfer from oxygen molecules in the excited singlet-delta state, $O_2(a\,{}^1\Delta_g)$ (reaction 1, table B3.4.1.1). $O_2({}^1\Delta)$ is a metastable molecule with a very long radiative lifetime (76 min) [6] and a pronounced stability against collisional deactivation processes in many environments. Hence it can be used as the energy carrier for the COIL. Mixing of chemically produced $O_2({}^1\Delta)$ with I_2 molecules results in dissociation of molecular iodine to atomic iodine which is subsequently excited *via* reaction 1 (table B3.4.1.1).

$O_2({}^1\Delta)$ is produced in a chemical generator by the reaction of gaseous chlorine with basic hydrogen peroxide (BHP) solution. BHP is prepared by mixing an aqueous solution of H_2O_2 with aqueous alkali metal hydroxide MOH (M≡Na or K). The mixing results in formation of HO_2^- ions in reaction 2 (table B3.4.1.1). Due to the large value of the equilibrium constant of reaction 2 the BHP is composed of HO_2^-, H_2O_2 and H_2O with very little OH^-. The BHP solutions used in the COIL are 6–8 molar in HO_2^-, 1–3 molar in H_2O_2 and about 50% wt H_2O [2, 7]. Since reaction 2 is extremely exothermic ($\Delta H = -50$ KJ mole^{-1}), the BHP has to be cooled down during preparation and is kept at -10 to $-30\,°C$. The chlorine molecules are transferred from the gas–phase to the gas/liquid interface and then enter the liquid. $O_2({}^1\Delta)$ is then produced in the liquid (mainly in a layer that extends 100 nm from the interface) *via* reactions 3a–c (table B3.4.1.1) with the overall reaction 3 between the chlorine and HO_2^- [8, 9]. The $O_2({}^1\Delta)$ product diffuses back to the interface and emerges from the liquid, some fraction of $O_2({}^1\Delta)$ being quenched in the liquid by reaction 4. Key parameters of the $O_2({}^1\Delta)$ chemical production are the $O_2({}^1\Delta)$ yield Y and chlorine utilization U defined as

$$Y = [O_2({}^1\Delta)]/[O_2]_{total} \qquad \text{(B3.4.1.1)}$$

and

$$U = [O_2]_{total}/[Cl_2] \qquad \text{(B3.4.1.2)}$$

respectively, where and $[O_2]_{total} = [O_2({}^1\Delta)] + [O_2({}^1\Sigma)] + [O_2({}^3\Sigma)]$ and $[Cl_2]$ is the chlorine density prior to its reaction with the BHP. Although the $O_2({}^1\Delta)$ yield in reaction 3 is near one [8, 10], the yield obtained at the exit of the singlet oxygen generator (SOG) is usually smaller than unity, ranging from 0.5 to 0.7 [11, 12]

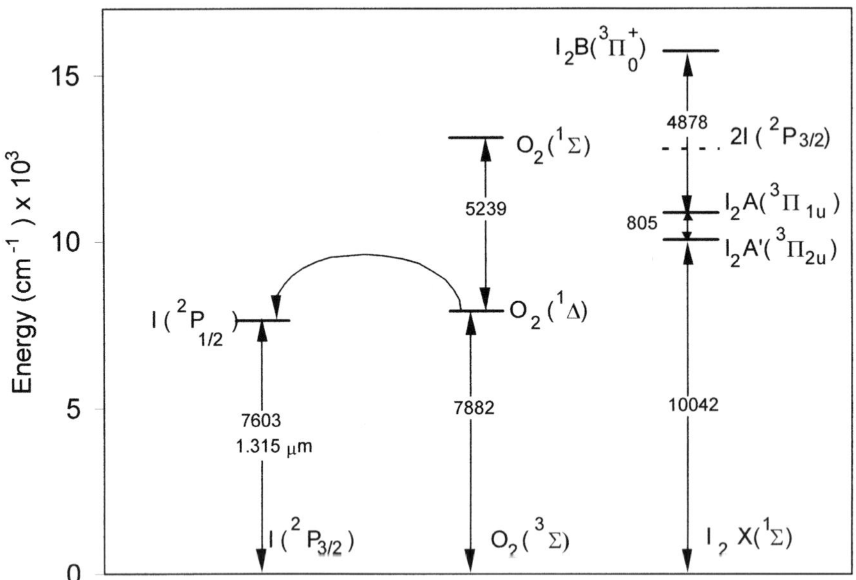

Figure B3.4.1.1. Low-lying electronic states of O_2, I_2 and I. Energy gaps are given in cm^{-1}.

for the-high pressure SOGs used in the modern COIL devices (see section B3.4.1.2). $O_2(^1\Delta)$ losses are due to both liquid-phase and gas-phase (reaction 5) quenching. For high pressures in the SOG, the liquid-phase losses are enhanced by HO_2^- depletion near the gas–liquid interface, which results in the reaction zone shifting into the bulk of the liquid and an increase in the $O_2(^1\Delta)$ transport time to the interface. To estimate the values of Y and U, taking into account all the aforementioned processes, one can use simple analytical expressions [13–15]. Unlike Y, the utilization U is close to unity for almost all types of the SOG.

The mechanism by which $O_2(^1\Delta)$ dissociates I_2 is one of the most complex and least understood aspects of COIL chemistry. Since the loss of excited $O_2(^1\Delta)$ molecules in the iodine mixing–dissociation zone results in a decrease of laser power, this process is very important for the COIL operation. Three mechanisms have been suggested for iodine dissociation. The first [16] involves I_2 dissociation by energy transfer from $O_2(^1\Sigma)$(reaction 8), which has sufficient energy to dissociate I_2 (see figure B3.4.1.1). $O_2(^1\Sigma)$ is initially formed by the energy-pooling reaction 5 and then (once an appreciable density of I is formed) strongly supplemented by the pooling reaction 6, where I* ($\equiv I(^2P_{1/2})$ is formed by the pumping reaction 1. However, dissociation in COIL devices, where significant water concentrations severely reduce the steady-state density of $O_2(^1\Sigma)$ by the quenching reaction 7, is too fast to be attributed solely to this mechanism. Hence, reaction 8 is important only at the initiation stage of dissociation when a small fraction of I_2 is dissociated to atoms.

The second and most celebrated mechanism of iodine dissociation was suggested by Heidner *et al* [4]. They found that iodine dissociation is an autocatalytic chain reaction accelerated by I*. A schematic representation of the dissociation mechanism is shown in figure B3.4.1.2. The dissociation is a two-step process. In the first step, I_2 is excited to an intermediate state $I_2^\dagger$ by reaction 9 and the more rapid reaction 11. Then excited $I_2^\dagger$ molecules are dissociated by $O_2(^1\Delta)$ in reaction 10. The dissociation rate is controlled by the rates of the excitation reactions 9 and 11. The slow reaction 9 initiates the process but after the build up of I atoms (excited to I* in reaction 1), the much faster reaction 11 dominates. Consequently, the rate of iodine dissociation is dependent on the number density of I_2 in the flow. Although the iodine intermediate $I_2^\dagger$ was not observed directly, there is evidence that it is vibrationally excited $I_2(X, v > 20)$ and not one of the

Table B3.4.1.1. Basic chemical reactions in the COIL [4, 5]. $I^* \equiv I(^2P_{1/2})$, $I \equiv I(^2P_{3/2}) \cdot I_2^\dagger$ is the intermediate state of I_2.

No	Reaction	Rate constant ($cm^3\ s^{-1}$) or equilibrium constant (K_e)
	Energy transfer pumping reaction	
1	$O_2(^1\Delta) + I \leftrightarrow O_2(^3\Sigma) + I^*$	$k_{1f} = 7.8 \times 10^{-11}, k_{1b} = k_{1f}/K_{e1}$, $K_{el} = 0.75 \exp(402/T)$
	Liquid phase $O_2(^1\Delta)$ production	
2	$OH^- + H_2O_2 \leftrightarrow HO_2^- + H_2O$	$K_{e2} = 4 \times 10^4$
3a	$HO_2^- + Cl_2 \rightarrow HOOCl + Cl^-$	$k_{3a} = 4.5 \times 10^{-14}$
3b	$HO_2^- + HOOCl \rightarrow ClO_2^- + H_2O_2$	$k_{3b} = \infty$
3c	$ClO_2^- \rightarrow Cl^- + O_2(^1\Delta)$	$k_{3c} = 10^7\ s^{-1}$
3	$Cl_2 + 2H_2O^- \rightarrow O_2(^1\Delta) + 2Cl^- + H_2O_2$	
4	$O_2(^1\Delta) \rightarrow O_2(^3\Sigma)$	$k_4 = 5 \times 10^5\ s^{-1}$
	I_2 dissociation by $O_2(^1\Sigma)$	
5	$O_2(^1\Delta) + O_2(^1\Delta) \rightarrow O_2(^1\Sigma) + O_2(^3\Sigma)$	$k_5 = 2.7 \times 10^{-17}$
6	$O_2(^1\Delta) + I^* \rightarrow O_2(^1\Sigma) + I$	$k_6 = 1.1 \times 10^{-13}$
7	$O_2(^1\Sigma) + H_2O \rightarrow O_2(^1\Delta) + H_20$	$k_7 = 6.7 \times 10^{-12}$
8	$I_2(X) + O_2(^1\Sigma) \rightarrow I + I + O_2(^3\Sigma)$	$k_8 = 4 \times 10^{-12}$
	Auto catalytic mechanism of I_2 dissociation	
9	$O_2(^1\Delta) + I_2 \rightarrow O_2(^3\Sigma) + I_2^\dagger$	$k_9 = 7 \times 10^{-15}$
10	$O_2(^1\Delta) + I_2^\dagger \rightarrow O_2(^3\Sigma) + 2I$	$k_{10} = 3 \times 10^{-10}$
11	$I^* + I_2 \rightarrow I + I_2^\dagger$	$k_{11} = 3.5 \times 10^{-11}$
12	$O_2(^1\Delta) + I_2^\dagger \rightarrow O_2(^3\Sigma) + 2I$	$k_{12} = 3 \times 10^{-10}$
13	$I^* + H_2O \rightarrow I + H_2O$	$k_{13} = 2 \times 10^{-12}$
14	$I_2^\dagger + M \rightarrow I_2 + M(M = H_2O, O_2, He)$	$k_{14H_2O} = 2 \times 10^{-12}, k_{14O_2} = 5 \times 10^{-11}$, $k_{14He} = 4 \times 10^{-12}$

low-lying triplet I_2 electronic states shown in figure B3.4.1.1 [4, 17]. However, very recently a new model of iodine dissociation has been suggested [18] where both $I_2(A')$ and vibrationally excited $I_2(X)$ are significant dissociation intermediates. This model requires at least three $O_2(^1\Delta)$ molecules to dissociate I_2.

The condition for the existence of population inversion between the levels involved in the $I(^2P_{1/2}) \rightarrow I(^2P_{3/2})$ transition is

$$\frac{[I(^2P_{1/2})]}{[I(^2P_{3/2})]} > 0.5. \tag{B3.4.1.3}$$

Taking into account that fast equilibrium between I^* and $O_2(^1\Delta)$ is established in reaction 1, one can determine the yield Y_{th} necessary to achieve threshold population inversion in the O_2–I system:

$$Y_{th} \equiv \frac{1}{2K_{el} + 1} \tag{B3.4.1.4}$$

where K_{el} is the equilibrium constant. At room temperature ($T = 300$ K) $Y_{th} = 0.15$.

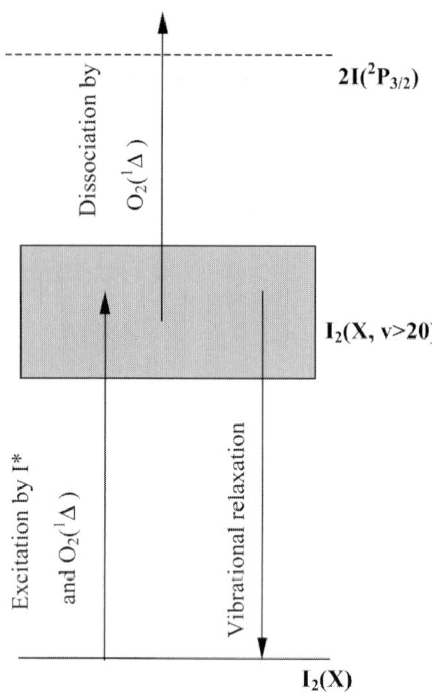

Figure B3.4.1.2. Schematic diagram of iodine dissociation mechanism in the COIL.

To find the small signal gain it is necessary to take into account that the upper laser level $^2P_{1/2}$ is split into two hyperfine sublevels with total angular momentums of the I atom $F = 2$ and 3 and the lower level $^2P_{3/2}$ consists of four hyperfine sublevels with $F = 1, 2, 3$ and 4. Possible transitions between these sublevels and their wavelengths in vacuum are shown in table B3.4.1.2. The largest value of the Einstein coefficient $A = 5\ \mathrm{s}^{-1}$ corresponds to the $F = 3 \rightarrow F = 4$ which is the lasing transition. The gain coefficient for this transition is given by

$$g = \tfrac{7}{12}\sigma_{34}([I(^2P_{1/2})] - 0.5[I(^2P_{3/2})]) \qquad \text{(B3.4.1.5)}$$

where $\sigma_{34} = 1.29 \times 10^{-17}(300/T)^{1/2}\ \mathrm{cm}^2$ is the stimulated emission cross section for Doppler broadening [19]. Taking into account the aforementioned assumption of the equilibrium between I* and $O_2(^1\Delta)$, the following equation for g is derived [19]:

$$g = \frac{7}{12}\sigma_{34}\frac{I_{\mathrm{total}}}{2}\frac{(2K_{\mathrm{el}} + 1)Y - 1}{(K_{\mathrm{el}} - 1)Y + 1} \qquad \text{(B3.4.1.6)}$$

where $[I]_{\mathrm{total}} = [I(^2P_{1/2})] + [I(^2P_{3/2})]$. Hence the gain increases with increasing iodine density and decreasing temperature and weakly depends on the yield Y. At room temperatures and $Y = 0.5$–0.8, $g\ (\mathrm{cm}^{-1}) = (5\text{–}7) \times 10^{-18}[I]_{\mathrm{total}}$, where $[I]_{\mathrm{total}}$ is in cm^{-3}. Hence, to get gain around 10^{-3}–$10^{-2}\ \mathrm{cm}^{-1}$, the iodine densities in the COIL active medium should be around 10^{14}–$10^{15}\ \mathrm{cm}^{-3}$.

The lasing power P can be estimated with the aid of an energy book-keeping methodology such as the COIL heuristic equation [20]:

$$P = 91(\mathrm{kJ\ mole}^{-1})nCl_2U(Y_{\mathrm{plen}} - Y_{\mathrm{diss}} - Y_{\mathrm{th}})\eta_{\mathrm{mix}}\eta_{\mathrm{ext}} \qquad \text{(B3.4.1.7)}$$

where 91(kJ mole^{-1}) is the energy of I*, nCl_2 is the chlorine molar flow rate, Y_{plen} is the $O_2(^1\Delta)$ yield just

Table B3.4.1.2. Wavelengths of transitions between the hyperfine sublevels of $^2P_{1/2}$ and $^2P_{3/2}$ states of I [21].

Transition	Wavelength in vacuum (nm)
$3 \to 4$	1315.2463
$3 \to 3$	1315.2220
$3 \to 2$	1315.2106
$2 \to 3$	1315.3360
$2 \to 2$	1315.3246
$2 \to 1$	1315.3204

upstream of the iodine injection,

$$Y_{diss} = NF\frac{nI_2}{nCl_2\eta_{mix}} \tag{B3.4.1.8}$$

is the $O_2(^1\Delta)$ loss during iodine dissociation, η_{mix} is the mixing efficiency defined as the fraction of $O_2(^1\Delta)$ mixed with iodine, η_{ext} is the optical extraction efficiency of the resonator, F is the iodine dissociation fraction and N is the number of $O_2(^1\Delta)$ molecules lost in the region of iodine dissociation per I_2 molecule. From energy considerations, N cannot be less than two. However, analytical and numerical computations [20, 22] and experimental measurements [23, 24] show that N can be appreciably larger, typically in the range 4–6. The expressions for η_{ext} for different types of optical resonators are presented later in section B3.4.1.3. A very important parameter of the COIL is the chemical efficiency η_{chem} defined as the number of emitted photons per number of chlorine molecules passed through the generator of $O_2(^1\Delta)$ and, hence, given by

$$\eta_{chem} \equiv \frac{P}{91(\text{kJ mole}^{-1})nCl_2} = U(Y_{plen} - Y_{diss} - Y_{th})\eta_{mix}\eta_{ext}. \tag{B3.4.1.9}$$

High values of the chemical efficiency can be achieved for efficient SOG with large values of U and Y_{plen}, good mixing and optical extraction corresponding to large η_{mix} and η_{ext}, small dissociation losses Y_{diss} and low temperature in the resonator corresponding to small Y_{th}.

There are two types of COILs: subsonic and supersonic. The first COIL [1] used subsonic flow of the reagents in the resonator. Due to small flow velocities, low densities of iodine are needed to provide for the total dissociation of I_2 and to extract energy $O_2(^1\Delta)$ while maintaining a reasonable laser mode volume [25]. Small values of the iodine density and, hence, of the small signal gain forces the laser beam region to become unbearably long. For example, the first 100–500 W class subsonic devices [26, 27] had a gain length of 0.5 to 2 m and an output power of 4.6 kW was eventually obtained from a 4 m gain-length device [2]. In the supersonic COILs, the flow is brought to supersonic velocity *via* expansion in a converging–diverging nozzle. The supersonic COILs, which have been intensively studied during the last decade, have several advantages over the subsonic devices. Most importantly, the low temperature achieved by supersonic expansion (predicted to be 150 K for typical devices) displaces the equilibrium of reaction 1 (table B3.4.1.1) in favour of I^* production. The equilibrium constant increases from $K_{e1} = 2.88$ at room temperature to 10.9 at 150 K. Using equation (B3.4.1.4) one obtains that the threshold yield Y_{th} falls to 0.044; hence, both the gain and power, given by equations (B3.4.1.6) and (B3.4.1.7), respectively, rise. Also, supersonic flow stretches the gain region, which facilitates power extraction from the optical cavity. In addition, the low temperature reduces the Doppler linewidth of the lasing transition, thereby increasing the gain at the line-centre. These advantages were demonstrated in a supersonic COIL with a 25 cm length nozzle, which yielded powers comparable to the 4 m subsonic device mentioned earlier [2].

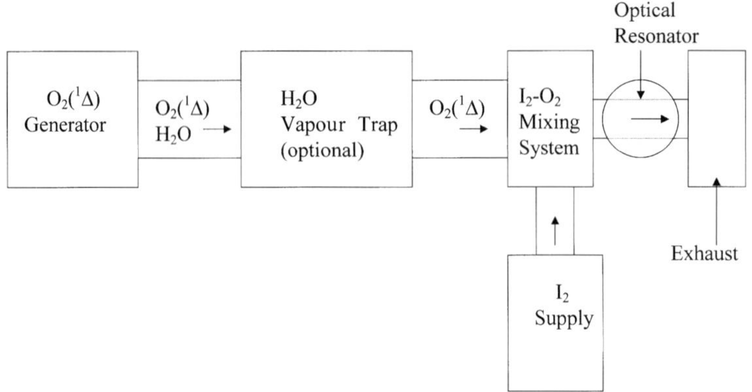

Figure B3.4.1.3. Block diagram of a typical COIL.

B3.4.1.2 Construction and components

A block diagram of a typical COIL is shown in figure B3.4.1.3. The major components of the COIL are the SOG, water vapour trap (WVT) for H_2O removal (in most modern COILs this trap is not used), iodine supply system, iodine–oxygen mixing system, optical resonator and reactant exhaust system.

Figure B3.4.1.4 shows four main types of SOGs: bubbler (or sparger), wetted wall with rotating discs, jet and aerosol generators. In a bubbler or sparger generator (figure B3.4.1.4(*a*)) $O_2(^1\Delta)$ is produced by bubbling (sparging) of the Cl_2 gas through a batch of pre-cooled BHP. Cl_2 is injected into the BHP through small orifices in a plate located in the lower part of the generator. Chlorine bubbles formed in this way float up and react with the BHP. The chlorine is absorbed by the solution and reacts in the liquid phase generating $O_2(^1\Delta)$, which in its turn is transferred from the liquid to the interior of the bubble. This generator has a very simple configuration and a very high specific surface area for the gas–liquid interface, S (the area per active unit volume in the generator), which can reach 300 cm^{-1} and ensures high utilization of chlorine (close to 100%). However, the yield is sufficiently high (above $\sim$0.3) only when the partial pressure of oxygen in the generator is relatively low $\sim$1–2 torr.

In a wetted wall SOG, the gaseous Cl_2 reacts with a thin film of the BHP solution on the surface of packing placed inside the SOG. This generator evolved from falling film generators [28] to the current rotating-disc generator shown in figure B3.4.1.4(*b*) [29]. In this generator, closely spaced discs rotate through a pool of BHP producing a set of parallel-plate surfaces coated with a thin BHP film. The Cl_2 gas is forced through the discs, producing $O_2(^1\Delta)$. This generator operates better at higher pressures, the chlorine utilization and $O_2(^1\Delta)$ yield being about 0.9 and 0.5–0.55 [12], respectively, at a partial oxygen pressure of 11.5 torr. The specific surface S for disc generators reaches 10 cm^{-1}.

The third type of the SOG shown in figure B3.4.1.4(*c*) is a jet generator [30]. The BHP is pushed through holes of 0.6–0.8 mm diameter in a perforated plate into the generator, producing jets at a velocity of 6 m s^{-1} in the reaction zone. Cl_2 is delivered into the reaction zone at the lower part of the generator and flows at a velocity of 10 m s^{-1} between the BHP jets, against their flow direction, to the generator exit. Most of the (80–90%) reacts with the BHP to produce singlet oxygen at a yield of 0.6–0.7 [11] and a partial oxygen pressure of 20–30 torr. An increase in the velocity of the jets to 20 m s^{-1} and of the chlorine to 30 m s^{-1} enables the oxygen pressure to be increased to 100 torr and the yield is then about 0.6 [31].

In aerosol SOGs, the gaseous Cl_2 reacts with the BHP delivered in the form of droplets. The best design is the uniform droplets SOG [32] shown in figure B3.4.1.4(*d*). Just as in the jet SOG, BHP is sprayed through a large number of holes in an injector plate located at the top of the generator. The plate is vibrated

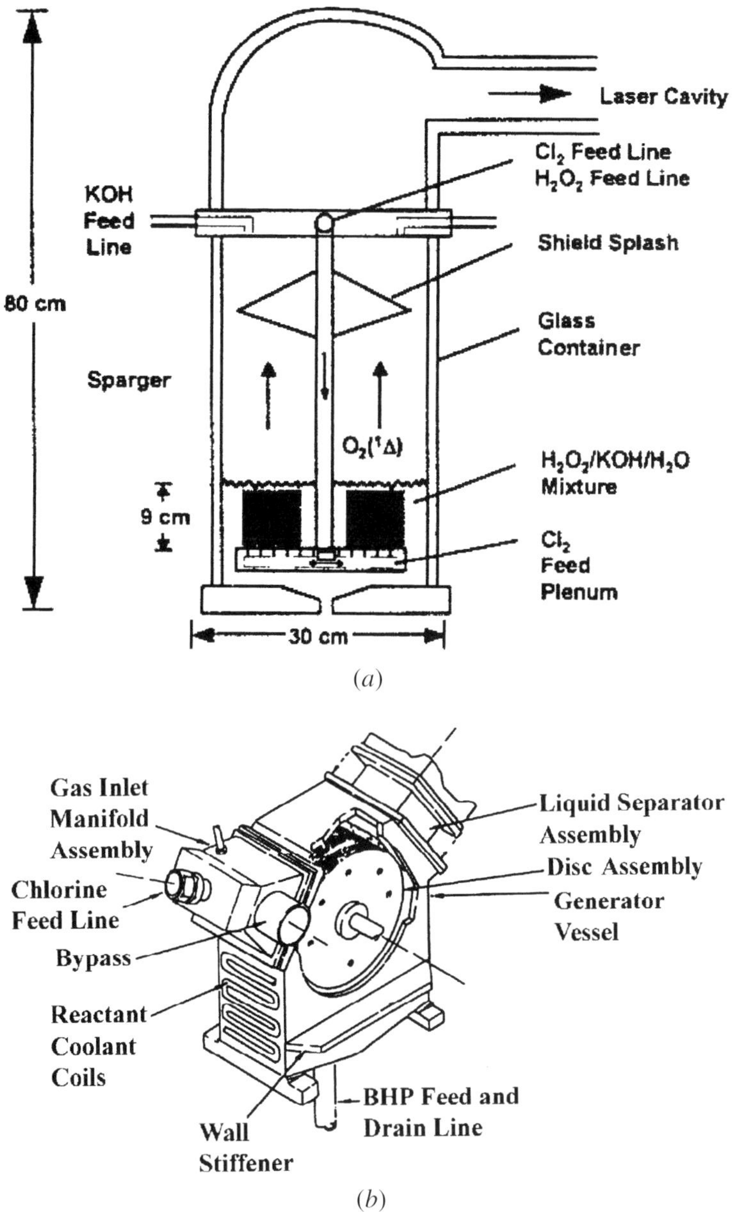

Figure B3.4.1.4. Different types of singlet oxygen generators: (*a*) bubble (reprinted with permission of SPIE from [5]); (*b*) wetted wall with rotating discs (reprinted with permission of SPIE from [5]); (*c*) jet [30]; (*d*) aerosol with uniform droplets (originally published in [32]. Copyright© 1994/2000 by the Am. Inst. Aeronaut. and Astronaut, Inc. Reprinted with permission.)

by a piezoelectric actuator at a frequency of 4 kHz to generate a field of uniformly sized 0.5 mm diameter droplets at a velocity 12 m s^{-1}. Mutual orientation of the chlorine and droplet flows of 90°, as suggested and analysed by Barmashenko *et al* [13], is employed in this generator. Application of a uniform droplet field

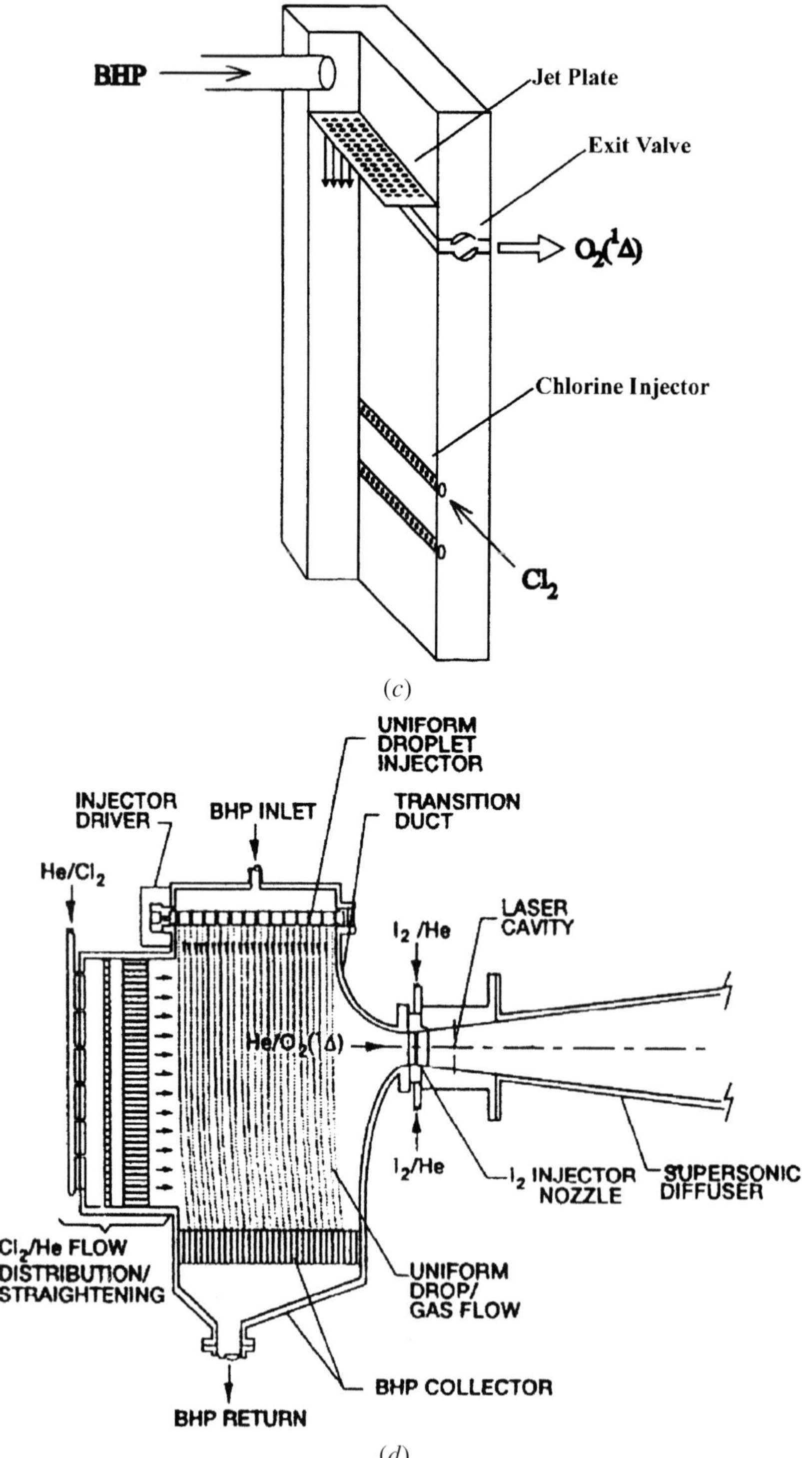

(*c*)

(*d*)

Figure B3.4.1.4. (Continued.)

and 90° orientation in this SOG enables the separation of the droplets from the gas flow to be precluded. At the nominal conditions of 10 torr partial oxygen pressure the chlorine utilization and $O_2(^1\Delta)$ yield are 0.8 and 0.7, respectively. The uniform droplet SOG is used in the multi-megawatt COIL for the US Air Force Airborne Laser [33].

The flow emerging from the SOG contains water vapour that strongly quenches both I^* and $I_2^\dagger$ by reactions 13 and 14 (table B3.4.1.1) and, hence, decreases laser power. To minimize this effect, two approaches have been used: (1) WVT to freeze the water downstream of the SOG; and (2) lower BHP temperature to minimize water vapour formation [34]. The first approach works well for the bubble SOG used in subsonic COILs, where pressure is low and losses of $O_2(^1\Delta)$ due to the pooling reaction 5 (table B3.4.1.1) are small. Different types of WVTs are used: cold-finger-style WVTs [5], finned WVT [5] and rotational/scraping finned cold traps [35] to remove the ice forming on the trap walls. The WVT temperature is usually about 200 K. For the disc, jet and aerosol SOGs used mainly in supersonic COILs, the pressure is significantly higher and, therefore, the WVT cannot be used due to large $O_2(^1\Delta)$ losses. In this case the second approach is employed and the BHP is kept at −15–30 °C. It is worth noting that the harmful effect of the water on the COIL operation is determined by the ratio $[H_2O]/[O_2(^1\Delta)]$ rather than by the absolute H_2O vapour density [9]. In the case of high $O_2(^1\Delta)$ partial pressure this ratio just after the generator is not larger than that after the WVT in the case of low $O_2(^1\Delta)$ pressure. That means that COILs using high-pressure SOG can work without WVT.

The I_2 supply system uses the following two methods of iodine vapour production: (1) sublimation of solid I_2 into a carrier gas (Ar, He or N_2) by heat lamps or circulating oil heater [5, 36, 37] and (2) iodine boilers where I_2 is swept out of molten iodine by heated carrier gas [38, 39]. The tubes delivering the iodine vapours to the iodine–oxygen mixing system are heated to avoid I_2 condensation.

Subsonic and supersonic COILs use different kinds of iodine–oxygen mixing systems. For subsonic COILs the primary stream emerging from the WVT, and typically carrying the oxygen and a diluent (buffer) gas, flows in a rectangular duct with constant cross section. The secondary stream carrying the iodine and additional diluent is injected into the primary stream through either small (~0.5 mm diameter) holes, uniformly distributed across the primary flow (with a distance of several millimetres between adjacent holes) or a narrow slit in metal (brass or stainless steel) tubes. Due to low pressure and flow velocity, almost total mixing is achieved several centimetres downstream of the injection location.

In supersonic COILs, as previously mentioned (section B3.4.1.1), the primary stream is expanded in a supersonic nozzle with ~2:1 area ratio and ~10:1 pressure ratio [2]. Both slit mono-nozzles [39] (see figure B3.4.1.5) and grid nozzles, i.e. arrays of several supersonic nozzles [2, 38] (see figure B3.4.1.6), are used. The secondary flow is injected into the primary flow at some location in the nozzle, i.e. in the subsonic, transonic or supersonic part of the flow (see figure B3.4.1.5). Usually one or two rows of small injection holes, uniformly distributed across the flow, are drilled in each side of the nozzle wall. Most of the supersonic COILs operating with He primary diluent use a subsonic mixing scheme [2, 39]. In subsonic mixing, He is preferred due to the high velocity of flows diluted with He. For conditions where a nitrogen buffer gas or no primary buffer gas is used, this scheme is not optimal and transonic [37] or quasi-transonic (with injection holes located several millimetres upstream of the critical cross section [40, 41]) mixing schemes are employed. A supersonic mixing scheme uses either iodine injection in the diverging section of a slit nozzle [7, 42], or supersonic mixing of parallel jets of O_2 and of I_2/N_2 with Mach number $Ma \simeq 1.5$ [43] created by a two-dimensional conical nozzle array, referred to as a Cassady grid nozzle [44].

Very recently a new concept of ejector COIL has been suggested which also uses supersonic mixing [45, 46]. In order to achieve good pressure recovery and effective exhaust of the laser gas into the atmosphere, it is necessary to increase the stagnation pressure at the supersonic section of the flow, which usually is less than 100 torr. To this end special grid nozzles were designed where a transonic ($Ma = 1$) oxygen flow is mixed with hypersonic ($Ma = 4$–5) I_2/N_2 flow, the N_2 flow rate being five to ten times larger than that of the oxygen. If the total momentum of the N_2 stream is much larger than the momentum of oxygen stream, the latter is pulled into the laser cavity by the nitrogen flow and the stagnation pressure of completely mixed

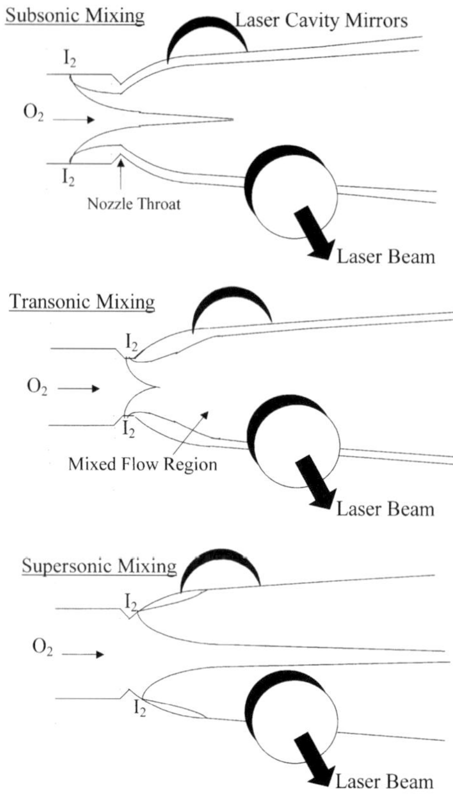

Figure B3.4.1.5. Subsonic, transonic and supersonic injection schemes (in slit nozzles).

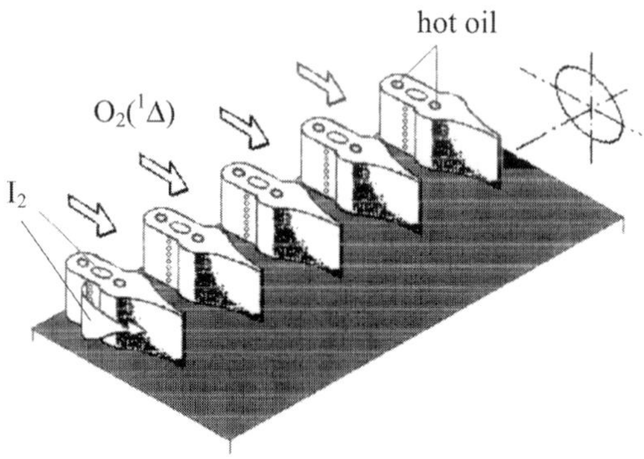

Figure B3.4.1.6. Grid nozzle with subsonic injection. Either hot oil or electrical heating is used to heat the nozzle blades to preclude I_2 clogging.

stream is determined mainly by the $N_2 + I_2$ flow. Figure B3.4.1.7 shows one such mixing nozzle with separate supplies of N_2 and I_2 [45]. The ejector COIL operates at stagnation pressure in the cavity of 200 torr and a Mach number of about three and, hence, is able to ensure a good pressure recovery using small pumping rates.

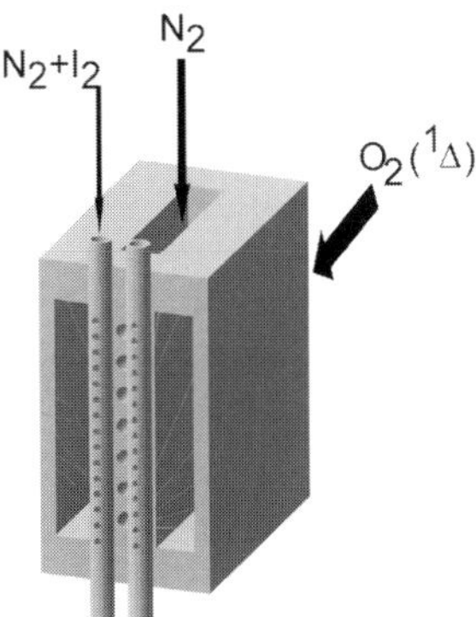

Figure B3.4.1.7. Fragment of the nozzle bank of the ejector COIL. (Originally published in [45]. Copyright© 1994/2000 by Am. Inst. Aeronaut. and Astronaut. Inc. Reprinted with permission.)

The optical resonator where the energy of $O_2(^1\Delta)$ is converted into the laser radiation is located several centimetres downstream of the iodine injectors. The optical axis of the resonator is usually transverse to the flow direction. The design and extraction efficiency of the resonators are discussed later in section B3.4.1.3.

The laser effluent is then exhausted into a diffuser-pumping system where the pressure is recovered to the atmospheric pressure [47]. The pumping system usually consists of one or two Roots stages on the vacuum side followed by a rotary pump [38]. Sometimes a cryogenic trap is installed upstream of the pumps to remove iodine and residual chlorine from the exhaust gas in order to prevent corrosion in the pumps. For high stagnation pressures in supersonic COILs ejectors can be used for pressure recovery [47].

Considerable heat release downstream of the supersonic nozzle (and thus in the optical cavity) driven by the water deactivation of the excited iodine atoms (reaction 13, table B3.4.1.1) has a direct impact on the hardware design of the supersonic part of the flow channel and on the performance of the gas recovery system. To preclude a temperature rise in the gas the floor and the ceiling of the supersonic flow duct usually diverge at an angle of 4–8 ° [20, 48]. Calculations of the shape of the flow duct for the supersonic diffuser should also take into account the aforementioned heat release [49].

B3.4.1.3 Cavity design and extraction efficiency

The measured values of the small signal gain are typically smaller than 1–1.5 m^{-1} and 0.15 m^{-1} in supersonic [7, 39, 50] and subsonic [51, 52] COILs, respectively. Hence, stable resonators (see chapter A2.1) have been commonly used in these lasers. The mirror separation depends on the gain length L of the COIL and is usually in the range of 0.5–3 m, the mirror radius of curvature being 2–10 m. The outcoupling mirror transmissions for supersonic COILs with large gain length are about 10-20%, whereas for subsonic COILs a typical transmission coefficient does not exceed 5%. The mirrors are isolated from the gas flow by ducts purged with He or N_2 to protect the mirror coatings from BHP contamination as well as to prevent iodine from entering the ducts. To extract most of the energy stored by $O_2(^1\Delta)$, the aperture of the mirrors in the flow direction should be 3–10 cm, hence, 5 or 10 cm diameter mirrors are typically used in COILs. Since each iodine atom is repumped and cycled many times throughout the flow field during the lasing process, the saturation characteristics of the COIL are different than in other lasers. To find the extraction efficiency η_{ext} appearing in equation (B3.4.1.7) for the laser power, simple models for gain saturation and power extraction in the COIL should be used [22, 53, 54]. In particular, if the mirror's aperture is much larger than the $O_2(^1\Delta)$

energy extraction distance along the flow, the efficiency η_{ext} for Fabry–Pérot resonator is given by

$$\eta_{ext} = \frac{t}{t+a} \frac{\left(1 - \frac{g_{th}}{g}\right)}{\left[1 - \frac{g_{th}}{g}\frac{Y - 1/(2K_{el}+1)}{Y + 1/(K_{el}+1)}\right]} \tag{B3.4.1.10}$$

where t and a are the transmission and loss of the mirrors, respectively, and $g_{th} \equiv (t+a)/2L$ is the threshold gain of the resonator. The values of η_{ext} of stable and unstable resonators are shown to be higher than those of a Fabry–Pérot resonator [22].

The COIL using stable resonator with mirror aperture of several centimetres has a large Fresnel number and, hence, oscillates on high-order transverse modes. Therefore, it has a large mode volume and is well suited for the extraction of high power. However, this advantage is associated with a large beam divergence and, thus, with a beam quality that is far from what is required for the general applications of such a laser. This problem can be solved by using an unstable resonator (see section A2.1.7) which, on the one hand, has a large mode volume and, on the other hand, is capable of discriminating the high-order transverse modes [55]. A conventional confocal unstable resonator consists of either concave and convex (positive branch) or two concave (negative branch) spherical mirrors (see chapter A2.1). The most important parameter of this kind of resonator is the magnification M, which is equal to the ratio between the mirror apertures. M describes the outcoupled fraction of the mode volume by simple geometry, being also a measure of the amount of power (intensity) that can be concentrated in the central peak of the far-field mode pattern. In addition to the central peak, this pattern has also side diffraction maxima. Low power in the side maxima is obtained only for large M. However, for small- and medium-sized COILs, only a small fraction of the intracavity power can be coupled out without a serious loss in the total lasing power. For small values of M (close to unity), the radiation field forms a narrow ring with only a small fraction of power in the central peak and the beam divergence is large. This means that a conventional unstable resonator is unable to provide for high lasing power and good quality of the laser beam. In experiments carried out in China [56], an output power of 7.1 kW was achieved at a chlorine flow rate of 450 mmole s^{-1} for a 1 m gain-length COIL a using confocal, conventional unstable resonator with $M = 1.39$. However, the experimental beam divergence angle was almost six times larger than the beam diffraction angle, defined as the divergence of a filled-in beam with uniform amplitude and phase; and the aforementioned ring intensity profile pattern was observed in the near field. One way for improving this situation was suggested by Anan'ev *et al* [57] and investigated for the COIL by Latham *et al* [58]. Their resonator utilizes an oblique roof-top reflector to turn the phase by $90°$ with each path (figure B3.4.1.8). This resonator (named UR-90) was used to obtain a uniform intensity profile pattern in the near field. The same output power as for a conventional unstable resonator was obtained with a beam divergence angle only twice the beam diffraction angle. Another unstable double-pass resonator of the hybrid, confocal type was designed for outcoupling the radiation of a 10 kW class COIL with a gain length of 20 cm developed at the Deutsche Zentrum fur Luft und Raumfahrt (DLR) in Germany [55]. This positive branch unstable resonator uses two cylindrical mirrors with $M = 1.17$. While being unstable in the flow direction, it is of the common Fabry–Pérot type in the other direction. Predicted mode field distribution shows excellent beam quality in the far field with over 95% of the outcoupled power in the central peak with 0.3×0.5 mrad divergence.

B3.4.1.4 Operating characteristics, performance and modes of lasing

Operating characteristics and performance of different cw COIL devices are shown in table B3.4.1.3. The first lasers, starting with the 4 mW COIL (No 1, table B3.4.1.3) demonstrated in 1977 [1], used subsonic velocity gas in the laser cavity. After an initial scaling from the milliwatt level to 100 W by Benard *et al* [26], the subsonic COIL was further scaled to 4.6 kW (No 2, table B3.4.1.3 [5]). The chemical efficiencies of

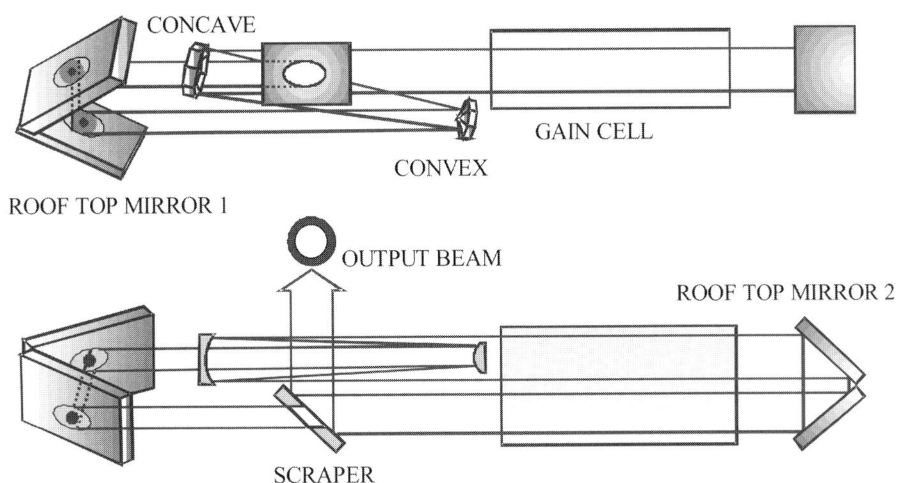

Figure B3.4.1.8. Unstable resonator with 90 ° phase rotation [56].

Table B3.4.1.3. Operating characteristics and performances of different cw COILs.

No	Name of the COIL and comment	Type of the SOG	Gain length (cm)	Cl_2 flow rate, (mmole s^{-1})	Power, (W)	Chemical efficiency (%)	Reference
			Subsonic COILs				
1	The first COIL	Bubble	70	4	0.004	10^{-5}	[1]
2	COIL-IV	Bubble	400	600	4600	8.4	[5]
3	Extremely efficient COIL	Bubble	35	5	192	42	[59]
4	Industrial COIL	Bubble	100	69	1015	16	[61]
5	Compact subsonic COIL	Bubble	10	0.4	5	13.7	[62]
6	High pressure subsonic COIL	Jet	7.5	19.7	448	25	[63]
			Supersonic COILs				
7	VertiCOIL with primary He	disc	5	70.8	1730	27	[39]
8	RADICL	disc	25	500	12 790	28.1	[69, 70]
9	RADICL	Jet	25	509	13 690	29.6	[70]
10	RotoCOIL	disc	54.1	1800	39 000	24	[2]
11	COIL for Airborne Laser	Uniform droplets	—	—	$>10^6$	—	[33]
12	DLR (Germany) 10 kW class COIL	disc	20	500	9700	21.3	[66]
13	Dalian (China) 10 kW class COIL	Jet	100	450	8800	21.4	[71]
14	VertiCOIL with primary N_2	disc	5	23.8	490.6	23	[40]
15	Compact COIL with primary N_2	Jet	5	75	1408	20.7	[68]
16	Ejector COIL	Jet	5	39.2	700	19.7	[45]
17	COIL without primary buffer gas	Jet	5	17.4	517	32.7	[42]

these COILs were smaller than 10%. Much higher values of the efficiencies were achieved at the end of the 1980s in Japan. In particular, a COIL with extremely high efficiency of 42% (No 3, table B3.4.1.3), was demonstrated by Yoshida *et al* [59]. The highest reported chemical efficiency of both subsonic and supersonic COILs is 42%. The same authors developed the first subsonic industrial COIL (No 4, table B3.4.1.3) with a power of 1 kW and a highest efficiency of 23% at the same power level. Long-term stable operation was demonstrated with the output power kept within 3% at 1 kW for over 1 hr [60].

As has already been mentioned in section B3.4.1.1, most subsonic devices have a very large gain length and, therefore, are inconvenient for operation and upscaling. The first compact, 10 cm gain length subsonic COIL [61] was developed in Israel (no 5, table B3.4.1.3). In spite of relatively high efficiency, low (~0.5 torr)

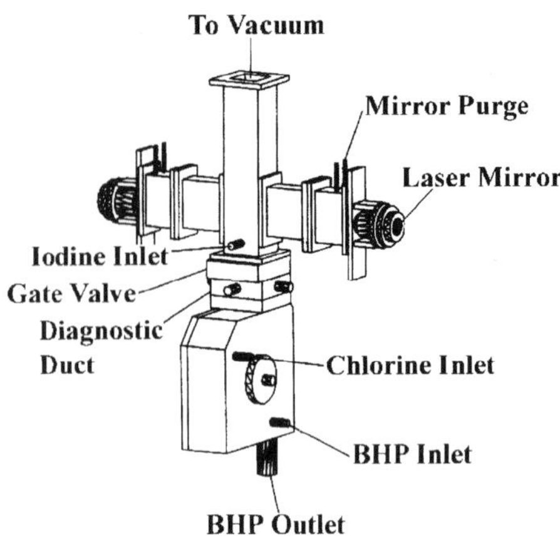

Figure B3.4.1.9. Three-dimensional view of VertiCOIL generator, duct and cavity hardware. (Reprinted with permission from [39]. Copyright© 1999 IEEE.)

pressure in the cavity resulted in a very small output power. Very recently 5–7.5-cm gain-length compact subsonic COILs with an output power of several hundreds watts (No 6, table B3.4.1.3) have been developed in Russia and Japan [63, 64]. Unlike the first COIL devices, these COILs are energized by jet SOGs and have chemical efficiencies larger than 20%. To increase the efficiency of these compact COILs, pre-cooling of the primary buffer gas, previously demonstrated by Blayvas *et al* [65] for supersonic COIL, was employed.

From the end of the 1980s, the emphasis of COIL research switched to the supersonic devices first demonstrated in 1982 in the Phillips Laboratory in Albuquerque, NM [2]. For the VertiCOIL (shown in figure B3.4.1.9), RADICL and RotoCOIL devices (Nos 7, 8, 9 and 10, respectively, table B3.4.1.3) power larger than 1, 10 and 30 kW, respectively, were achieved. Both VertiCOIL and RADICL used slit supersonic nozzles, whereas a grid nozzle array was employed in the 54-cm gain-length RotoCOIL. A chemical efficiency of 29.6%, obtained for the RADICL device energized by the jet SOG, is the highest reported chemical efficiency for supersonic COILs (No 9, table B3.4.1.3). These COILs served as the prototypes for the multi-megawatt supersonic COIL (No 11, table B3.4.1.3) employed in the US Air Force Airborne Laser [18]. Two 10 kW class supersonic COIL devices operating with a grid nozzle array were developed in China (Dalian) and at the DLR in Germany (Nos 12 and 13, respectively, table B3.4.1.3). The aforementioned supersonic COILs used He as the diluent gas and, hence, a subsonic mixing scheme (see section B3.4.1.2). In order to obtain high gas velocity and a low partial stagnation pressure for $O_2(^1\Delta)$, the He molar flow rates were three to five times higher than that of the chlorine.

For industrial applications, the use of large amounts of He is expensive. Therefore, supersonic COILs with a nitrogen diluent (which is more readily available and cheaper than He) or without a primary buffer gas were developed at the end of the 1990s. A supersonic COIL with N_2 buffer gas using subsonic mixing scheme was first demonstrated by Blayvas *et al* [65], the chemical efficiency being only about 13%. Much higher chemical efficiencies (over 20%) were achieved by moving the mixing location downstream to the nozzle throat, i.e. by using quasi-transonic mixing scheme [40, 41]. The highest (for compact COILs) reported chemical efficiency with a room-temperature nitrogen diluent—23%—was obtained by Carroll *et al* [40] using a VertiCOIL with modified iodine injector blocks (No 14, table B3.4.1.3). A higher efficiency, with room-temperature N_2, 24.6%, was obtained very recently by Endo *et al* [67]; however, in these experiments a much larger gain length, 22.5 cm, was employed. The highest power, with room-temperature nitrogen, 1.4 kW, was achieved by Zagidullin *et al* [68] using a 5-cm gain-length supersonic COIL (No 15, table B3.4.1.3). A highly efficient ejector COIL (No 16, table B3.4.1.3) described earlier in section B3.4.1.2 has been demonstrated by Nikolaev *et al* [45]. Efficient supersonic COILs without any primary diluent applying either transonic or supersonic mixing schemes have been demonstrated in Ben-Gurion University, Israel [37,42,48]. The maximum chemical efficiencies were 18% and 21% (No 17, table B3.4.1.3) for transonic [48] and supersonic [42] mixing schemes, respectively.

The range of applications of the COIL can be extended considerably by pulsed or pulse-periodic operation. A cw COIL can be transformed into a pulse-periodic laser by using either gain-switching (which is similar to Q-switching) or mode-locking methods.

The design of a pulsed COIL is practically identical to that of a cw laser. This means that all the characteristic features of the operation of a COIL associated with dissociation of molecular iodine, relaxation, etc are still retained. The advantage of this method is that there is no need to alter the laser design in going over from pulsed to cw operation, i.e. both may be achieved in the same device. The principle of gain switching, first suggested by Schmiedberger *et al* [7], is based on magnetic suppression or modulation of the gain by means of applying strong (hundreds of Gauss) magnetic field directly to the COIL cavity. Increasing the magnetic field in the range 1–500 G causes splitting of the hyperfine levels of the I atom due to the Zeeman effect and a significant monotonic decrease in the gain coefficient on the strongest $F = 3 \rightarrow F = 4$ transition. Modulation of the gain of a kilowatt-class COIL was demonstrated in the Phillips Laboratory in Albuquerque, NM [73]. Initially, a static magnetic field of 400 G was applied to the cavity of a supersonic COIL with an output power of 3 kW. An alternating magnetic field, directed opposite to the static field, compensated the latter at certain moments and the lasing pulse appeared at these moments. The peak power in the pulse, 39 kW, was about 13 times larger than the average power corresponding to cw operation, the pulse duration and repetition rate being of 10–20 μs and 500 Hz, respectively.

The mode-locking experiments were carried out at the Phillips Laboratory using the RADICL device [74]. To achieve mode locking, an acousto-optic modulator was inserted into the resonator and lasing was demonstrated on the $TEM_{(00)}$ mode with a small intracavity aperture. A dc magnetic field was used to increase the number of axial modes. The field broadened the gain by weakly splitting the Zeeman components of the $I(^2P_{1/2}, F = 3) \rightarrow I(^2P_{3/2}, F = 4)$ hyperfine transition. A peak power of 2.5 kW was reached with a pulse width of 2.1 ns at a repetition rate of 43.68 MHz.

Pulsed operation in COILs can also be obtained in a flowing gas mixture containing $O_2(^1\Delta)$ and an iodide (CH_3I, CF_3I or C_3F_7I) exposed to a pulse (e.g. a light flash) which generates iodine atoms throughout the whole volume [3]. Optimization of the laser performance made it possible to generate 4.4 J pulses of 15 μs duration, which corresponds to a pulse power of 300 kW [3]. This is the highest value reported so far. These results were obtained at a chlorine flow rate of 17 mmole s^{-1}, when under the best circumstances the cw output power could be 680 W (assuming that the laser chemical efficiency is 40%). Thus, the power of a pulsed COIL (with flash formation of iodine) was over 400 times higher than the cw power. Different methods of the pulsed operation of the COIL have been comprehensively analysed by Yuryshev [3].

The high power of both the cw and pulsed COIL devices makes it possible to generate second harmonics (see section C3) in these devices. The frequency doubling of 1315 nm radiation is very important since chemical lasers in the visible wavelength region have not yet been constructed. Different research groups have carried out both intra- and extra-cavity frequency-doubling experiments. The maximum second-harmonic power was obtained in a series of extracavity frequency-doubling tests performed using the output from a RotoCOIL at the Phillips Laboratory [75]. The diffraction-limited radiation of 6.8 kW power was focused into an $LiIO_3$ crystal. Conversion efficiencies of 8% were achieved resulting in visible (657 nm) cw outputs of nearly 700 W. It was found that the optical strength of the nonlinear crystals was insufficient: it fractured after 1 s exposure to a focused beam. Intracavity frequency-doubling experiments was carried out using both $LiIO_3$ [76] and LBO crystals [77], the maximum conversion efficiency being about 8% [76].

B3.4.1.5 Conclusions

Since the invention of the COIL in 1977, this laser has made impressive progress and is now one of the most powerful and efficient lasers. It can operate at multi-megawatt power levels with 30% chemical efficiency at a wavelength of 1315 nm, favourable for fibre and atmospheric transmission. Employment of unstable resonators offers excellent beam quality. Development of COILs with a small consumption of a nitrogen diluent makes their operation inexpensive. These characteristics make COIL an ideal laser for both military and industrial applications, in particular for laser cutting, decommissioning of nuclear power facilities and the removal of space debris [24]. COILs have higher investment and operation costs than the classical CO_2 and Nd:YAG lasers if they are compared on the basis of equal power [78]. However, the advantages of the COIL, which has a much better beam quality than the Nd:YAG laser and a much shorter wavelength and, hence, a much higher material processing speed than the CO_2 laser, may more than compensate for these higher costs.

In spite of the progress in laser technology, some fundamental processes in the COIL active medium are not still properly understood. In particular, one of the more critical issues, the kinetics of iodine dissociation, is still obscure. Other important issues are understanding the hydrodynamics in the supersonic COIL and finding new mixing schemes resulting in minimal losses of singlet oxygen and ensuring a good mixing efficiency. An additional problem is how to increase the chemical efficiency of COILs, using a nitrogen diluent. It would also be very interesting to find alternative methods for generating singlet oxygen and atomic iodine and, in particular, to develop discharge SOGs [79].

References

[1] McDermott W E, Pchelkin N R, Benard D J and Bousek R R 1978 An electronic transition chemical laser *Appl. Phys. Lett.* **32** 469–70

[2] Truesdell K A, Helms C A and Hager G D 1995 A history of COIL development in the USA *SPIE* **2502** 217–37

[3] Yuryshev N N 1996 Chemically pumped oxygen–iodine laser *Quantum Electron.* **23** 583–600

[4] Heidner R F III, Gardner C E, Segal G I and El-Sayed T M 1983 Chain reaction mechanism for I_2 dissociation in the $O_2(^1\Delta)$-I atom laser *J. Phys. Chem.* **87** 2348–60

[5] Truesdell K A and Lamberson S E 1993 Phillips laboratory COIL technology overview *SPIE* **1810** 476–92

[6] Newman S M, Lane I C, Orr-Ewing A J, Newnham D A and Ballard J 1999 Integrated absorption intensity and einstein coefficients for the O_2 (0, 0) transition: a comparison of cavity ringdown and high resolution fourier transform spectroscopy with a long-path absorption cell *J. Chem. Phys.* **110** 10 749–57

[7] Furman D, Bruins E, Rybalkin V, Barmashenko B D and Rosenwaks S 2001 Parametric study of a small-signal gain in a slit nozzle, supersonic chemical oxygen–iodine laser operating without primary buffer gas *IEEE J. Quantum Electron.* **37** 174–82

[8] Held A M, Halko D J and Hurst J K 1978 Mechanisms of chlorine oxidation of hydrogen peroxide *J. Am. Chem. Soc.* **100** 5732–40

[9] Storch D M, Dymek C J and Davis L P 1983 MNDO study of the mechanism of $O_2(^1\Delta)$ formation by reaction of Cl_2 with basic H_2O_2 *J. Am. Chem. Soc.* **105** 1765–9

[10] Avizonis P V and Truesdell K A 1995 Chemical oxygen–iodine laser (COIL) *SPIE* **2502** 180–203

[11] Furman D, Barmashenko B D and Rosenwaks S 1999 Diode-laser based absorption spectroscopy diagnostics of a jet-type generator for chemical oxygen-iodine lasers *IEEE J. Quantum Electron.* **35** 540–7

[12] Kendrick K R, Helms C A and Quillen B G 1999 Determination of singlet-oxygen generator efficiency on a 10-kW class supersonic chemical oxygen iodine laser *IEEE J. Quantum Electron.* **35** 1759–64

[13] Barmashenko B D, Kochelap V A and Mel'nikov L Yu 1985 Singlet oxygen generator of the atomizer type (for O_2–I chemical laser) *Sov. J. Quantum Electron. (USA)* **15** 1346–52

[14] Barmashenko B D and Rosenwaks S 1993 Theoretical modelling of chemical generators producing $O_2(^1\Delta)$ at high pressure for chemically pumped iodine laser *J. Appl. Phys.* **73** 1598–611

[15] Copeland D A, McDermott W E, Quan V and Bauer A H 1993 Exact and approximate solutions of the utilization and yield equations for $O_2(^1\Delta)$ generators *Am. Inst. Aeronaut. Astronaut.* **93** 3220

[16] Arnold S J, Finlayson N and Ogryzlo E A 1966 Some novel energy-pooling process involving $O_2(^1\Delta_g)$ *J. Chem. Phys.* **44** 2529–30

[17] Van Benthem M H and Davis S J 1986 Detection of vibrationally excited I_2 in the iodine dissociation region of chemical oxygen–iodine lasers *J. Phys. Chem.* **90** 902–5

[18] Kommisarov A V, Goncharov V and Heaven M C 2001 Chemical oxygen iodine laser (COIL) kinetics and mechanisms *SPIE* **4184** 7–12

[19] Avizonis P V, Hasen G and Truesdell K A 1990 The chemically pumped oxygen–iodine laser *SPIE* **1225** 448–74

[20] Hon J, Plummer D N, Crowell P G, Erkkila J, Hager G D, Helms C A and Truesdell K A 1996 Heuristic method for evaluating COIL performance *Am. Inst. Aeronaut. Astronaut.* **43** 1595–603

[21] Engelman R Jr, Keller R A and Palmer B A 1980 Hyperfine structure and isotope shift of the 1.3 μm transition of ^{129}I *Appl. Opt.* **19** 2767–70

[22] Barmashenko B D and Rosenwaks S 1996 Analysis of the optical extraction efficiency in gas-flow lasers with different types of resonator *Appl. Opt.* **35** 7091–101

[23] Churassy S, Bacis R, Bouvier A J, Pierre dit Mery C, Erba B, Bachar J and Rosenwaks S 1987 The chemical oxygen–iodine laser: comparison of a theoretical model with experimental results *J. Appl. Phys.* **62** 31–5

[24] Heaven, M C 2001 Chemical dynamics in chemical laser media *Chemical Dynamics in Extreme Environments (Advanced Series in Physical Chemistry)* ed R A Dressler (Singapore: World Scientific) ch 4, pp 138–205

[25] Scott J E, Shaw J L R, Truesdell K A, Hager G D and Helms C A 1994 Design considerations for the chemical oxygen–iodine supersonic mixing nozzle *Am. Inst. Aeronaut. Astronaut.* **94** 2436

[26] Benard D J, McDermott W E, Pchelkin N R and Bousek R R 1979 Efficient operation of a 100-W transverse-flow oxygen–iodine chemical laser *Appl. Phys. Lett.* **34** 40–1

[27] Wiswall C E, Bragg S L, Reddy K V, Lilenfeld H V and Kelley J D 1985 Moderate-power cw chemical oxygen–iodine laser capable of long duration *J. Appl. Phys.* **58** 115–18

[28] Richardson R J, Wiswall C E, Carr P G A, Hovis F E and Lilenfeld H V 1981 An efficient singlet oxygen generator for chemically pumped iodine laser *J. Appl. Phys.* **52** 4962–9

[29] Harpool G M, English W D, Berg G D and Miller D J 1992 Rotating disc oxygen generator *Am. Inst. Aeronaut. Astronaut.* **92** 3006

[30] Zagidullin M V, Kurov A Y, Kupriyanov N L, Nikolaev V D, Svistun M I and Erasov N Y 1991 Highly efficient jet generator *Sov. J. Quantum Electron.* **21** 747–53

[31] Azyazov V N, Zagidullin M V, Nikolaev, V D, Svistun M I and Khvatov N A 1994 Jet $O_2(^1\Delta)$ generator with oxygen pressures up to 13.3 kPa *Quantum Electron.* **24** 120–3

[32] Thayer W J III and Fisher C H 1994 Comparison of predicted and measured output from a transverse flow uniform droplet singlet oxygen generator *Am. Inst. Aeronaut. Astronaut.* **94** 2454

[33] Lamberson S E 2001 The airborne laser *SPIE* **4184** 1–6

[34] Takeshita K, Kikuchi T and Uchiyama T 1988 Laser operation of chemical oxygen–iodine laser without water vapor trap *J. Appl. Phys.* **63** 1785–6

[35] Fuji H, Yoshida S, Iizuka M and Atsuta T 1989 Long-term stability in the operation of a chemical oxygen–iodine laser for industrial use *J. Appl. Phys.* **66** 1033–7

[36] Elior A, Barmashenko B D, Lebiush E and Rosenwaks S 1995 Experiment and modelling of a small-scale, supersonic chemical oxygen–iodine laser *Appl. Phys.* B **61** 37–47

[37] Furman D, Barmashenko B D and Rosenwaks S 1997 An efficient supersonic chemical oxygen–iodine laser operating without buffer gas and with simple nozzle geometry *Appl. Phys. Lett.* **70** 2341–3

[38] Handke J, Werner A, Bohn W J and Schall W O 1995 Multikilowatt supersonic chemical oxygen–iodine laser *SPIE* **2502** 266–71

[39] Rittenhouse T L, Phipps S P and Helms C A 1999 Performance of a high efficiency 5-cm gain length supersonic chemical oxygen–iodine laser *IEEE J. Quantum Electron.* **35** 857–66

[40] Carroll D L, King D M, Fockler L, Stromberg D, Solomon W C, Sentman L H and Fisher C H 2000 High-performance chemical oxygen–iodine laser using nitrogen diluent for commercial applications *IEEE J. Quantum Electron.* **36** 40–51

[41] Zagidullin M V, Nikolaev V D, Svistun M V, Hvatov N A and Ufimtsev N I 1997 Highly efficient supersonic chemical oxygen–iodine laser with a chlorine flow rate of 10 mmole/s *Quantum Electron.* **27** 195–9

[42] Rosenwaks S, Rybalkin V, Katz A and Barmashenko B D 2003 Comparative studies of different schemes of iodine injection in a high efficiency supersonic COIL *Am. Inst. Aeronaut. Astronaut.* **2003** 302

[43] Azyazov V N, Zagidullin M V, Nikolaev V D and Safonov V S 1997 Chemical oxygen–iodine laser with mixing of supersonic jets *Quantum Electron.* **27** 491–4

[44] Cassady P E, Newton J F and Rose P H 1976 A new mixing gasdynamic laser *Am. Inst. Aeronaut. Astronaut.* **76** 343

[45] Nikolaev V D, Zagidullin M V, Hager G D and Madden T J 2000 An efficient supersonic COIL with more than 200 torr of the total pressure in the active medium *Am. Inst. Aeronaut. Astronaut.* **2000** 427

[46] Yang T T, Hsia Y C, Moon L F and Dickerson R A 2000 Advanced mixing nozzle concepts for COIL *SPIE* **3931** 116–30

[47] Walter R F and O'Leary R A 1993 Pressure recovery in supersonic gas lasers *SPIE* **1810** 328–33

[48] Furman D, Barmashenko B D and Rosenwaks S 1998 Parametric study of an efficient supersonic chemical oxygen–iodine laser/jet generator system operating without buffer gas *IEEE J. Quantum Electron.* **34** 1068–74

[49] Malkov V M, Boreysho A S, Savin A V, Kiselev I A and Orlov A E 2001 About choise of working parameters of the pressure recovery systems for high power gas flow chemical lasers *SPIE* **4184** 419–22

[50] Gruenwald K M, Handke J and Duschek F 2001 Small signal gain and temperature profiles in supersonic COIL *SPIE* **4184** 75–8

[51] Schmiedberger J, Kodymova J, Kovar J, Spalek O and Trenda P 1991 Magnetic modulation of gain in a chemical oxygen–iodine laser *IEEE J. Quantum Electron.* **27** 1262–4

[52] Watanabe K, Kashiwabara S, Sawai K, Toshima S and Fujimoto R 1983 Small-signal gain and saturation parameter of a transverse-flow cw oxygen–iodine laser *IEEE J. Quantum Electron.* **19** 1699–2609

[53] Rybalkin V, Katz A, Barmashenko B D and Rosenwaks S 2003 A 33% efficient chemical oxygen–iodine laser with supersonic

mixing of iodine and oxygen *Appl. Phys. Lett.* **82** 3838–40

[54] Hager G D, Helms C A, Truesdell K A, Plummer D, Erkkila J and Crowell P 1996 A simplified analytic model for gain saturation and power extraction in the flowing chemical oxygen–iodine laser *IEEE J. Quantum Electron.* **32** 1525–36

[55] Schall W O, Hall T and Handke J 2001 Unstable resonator for COIL *SPIE* **4184** 461–4

[56] Yang Bailing 1998 Latest advances in COIL in Dalian *SPIE* **3574** 281–9

[57] Anan'ev Yu A, Kuprenyuk V I and Sherstobitov V E 1979 Properties of unstable resonators with field rotation *Sov. J. Quantum Electron.* **9** 1105–10

[58] Latham W P, Paxton A H and Dente G C 1990 Laser with 90 degree rotation *SPIE* **1224** 265–82

[59] Yoshida S, Endo M, Sawano T, Amano S, Fuji H and Fujioka T 1988 Chemical oxygen iodine laser of extremely high efficiency *J. Appl. Phys.* **65** 870–2

[60] Fuji H and Atsuta T 1992 Current status of industrial COIL development *SPIE* **1980** 148–52

[61] Shimizu K, Sawano T, Tokuda T, Yoshida S and Tanaka I 1991 High-power stable chemical oxygen–iodine laser *J. Appl. Phys.* **69** 79–83

[62] Bachar J and Rosenwaks S 1982 An efficient, small scale chemical oxygen-iodine laser *Appl. Phys. Lett.* **41** 16–18

[63] Wani F, Endo M and Fujioka T 1999 High-pressure, high-efficiency operation of a chemical oxygen–iodine laser *Appl. Phys. Lett.* **75** 3081–3

[64] Zagidullin M V, Nikolaev V D, Svistun M I and Khvatov N A 1998 Comparative characteristics of subsonic and supersonic oxygen–iodine lasers *Quantum Electron.* **28** 400–2

[65] Blayvas I, Barmashenko B D, Furman D, Rosenwaks S and Zagidullin M V 1996 Power optimization of small-scale chemical oxygen–iodine laser *IEEE J. Quantum Electron.* **32** 2051–7

[66] Handke J, Grunewald K M, Schall W O and Entress-Fursteneck L 1998 Power extraction investigations for a 10 kW-class supersonic COIL *SPIE* **3574** 309–14

[67] Endo M, Kazuyoku T, Sugimoto D, Nanri K, Uchiyama T and Fujioka T 2001 Development of a prototype COIL for decomissioning and dismantlement *SPIE* **4184** 23–6

[68] Zagidullin M V, Nikolaev V D, Svistun M I and Khvatov N A 2000 Supersonic oxygen–iodine 1.4-kW laser with a 5 cm gain length and a nitrogen-diluted active medium *Quantum Electron.* **30** 161–6

[69] Helms C A 2001 Review of the US Air Force research laboratory's 10-kW RADICL Llaser *SPIE* **4184** 13–18

[70] Yang T T, Copeland D A, Bauer A H, Quan V, McDermott W E, Cover R A and Smith D M 1997 Chemical oxygen–iodine laser performance modelling *Am. Inst. Aeronaut. Astronaut.* **97** 2384

[71] Sang Fengting, Yang Bailing and Zhaung Qi 2001 Chemical oxygen–iodine laser research in China *SPIE* **4184** 27–31

[72] Schmiedberger J, Kodymova J, Spalek O and Kovar J 1991 Experimental study of gain and output coupling characteristics of a cw chemical oxygen–iodine laser *IEEE J. Quantum Electron.* **27** 1265–70

[73] Hager G D, Kopf D, Plummer D, Salsich T and Crowell P 1993 Demonstration of a repetitively pulsed magnetically gain-switched chemical oxygen–iodine laser chemical *Phys. Lett.* **204** 420–9

[74] Phipps S P, Helms C A, Copland R J, Rudolph W, Truesdell K A and Hager G D 1996 Mode locking of a cw supersonic chemical oxygen–iodine laser *IEEE J. Quantum Electron.* **32** 2045–50

[75] Hager G D, Hanes S A and Dreger M A 1992 Continuous wave frequency doubling of a high-energy 1315 nm laser *IEEE J. Quantum Electron.* **28** 2573–6

[76] Frolov M P, Ishkov D V, Kryukov P G, Pazyuk V S and Yurishev N N 1992 A frequency-doubled pulsed chemical oxygen–iodine laser *Appl. Phys.* B **54** 490–1

[77] Baba T, Tezuka T, Ito D, Uchiyama T and Fuji H 1995 Intracavity 2nd-harmonic generation of chemical oxygen–iodine laser-emission using a LBO crystal *Appl. Phys.* B **60** 369–73

[78] Von Bulow H and Schall W O 1995 Oxygen–iodine lasers for industrial applications *SPIE* **2502** 258–65

[79] Schmiedberger J and Fujii H 2001 Radio-frequency plasma jet generator of singlet delta oxygen with high yield *Appl. Phys. Lett.* **78** 2649–51

Further reading

Avizonis P V and Truesdell K A 1995 Chemical oxygen–iodine laser (COIL) *SPIE* **2502** 180–203

Basic principles of the COIL operation are presented; the main equations for the gain and power are derived.

Basov N G 1990 *Chemical Lasers* 1st edn (Berlin: Springer)

Main principles of the chemical lasers operation are described, theoretical description of different processes in the COIL is given.

Heaven M C 2001 Chemical dynamics in chemical laser media *Chemical Dynamics in Extreme Environments (Advanced Series in Physical Chemistry)* ed R A Dressler (Singapore: World Scientific) ch 4, pp 138–205

Energy transfer and reaction kinetics in the COIL active medium are described in detail. Special attention is paid to the iodine dissociation kinetics and hyperfine relaxation in the COIL.

Truesdell K A, Helms C A and Hager G D 1995 A history of COIL development in the USA *SPIE* **2502** 217–37

Historical overview of the COIL development in the USA. Key technical developments are reviewed beginning in 1960 and culminating in 1977 with the first COIL lasing demonstration. Particular emphasis is placed on how the singlet oxygen generator and iodine–oxygen mixing nozzle technologies evolved.

Yuryshev N N 1996 Chemically pumped oxygen–iodine laser *Quantum Electron.* **23** 583–600.

A review is given of the principles of operation of the COIL. Various types of chemical generators of singlet oxygen are considered. Experimental investigations of cw and pulsed COILs are described.

B3.4.2
Chemical lasers: HF/DF

Lee H Sentman

The HF/DF chemical lasers are based on the cold and hot pumping reactions:

$$
\begin{array}{lll}
F + H_2 \rightarrow HF(v, J) + H & v = 1, 2, 3 & 34\ \text{Kcal mol}^{-1}\ \text{‘cold reaction’} \\
H + F_2 \rightarrow HF(v, J) + F & v = 1\text{–}8 & 102\ \text{Kcal mol}^{-1}\ \text{‘hot reaction’} \\
F + D_2 \rightarrow DF(v, J) + H & v = 1, 2, 3, 4 & 34\ \text{Kcal mol}^{-1}\ \text{‘cold reaction’} \\
D + F_2 \rightarrow DF(v, J) + F & v = 2\text{–}9 & 102\ \text{Kcal mol}^{-1}\ \text{‘hot reaction’}
\end{array}
$$

which produce the excited species $HF(v, J)$ or $DF(v, J)$ in a population inversion from which energy is extracted as a laser beam at 2.8 μ for HF or 3.8 μ for DF. These lasers have been extensively developed since their inception in 1967. Both the continuous wave (cw) and pulsed versions of these lasers were pursued. Since the cw device is the laser of choice for high-power applications, this article will concentrate on this type of HF/DF laser.

A schematic diagram of a typical HF/DF laser is shown in figure B3.4.2.1. The HF/DF laser consists of an F-atom generator, a nozzle array to accelerate the F-atom stream and reduce its pressure and temperature, a device to inject H_2 or D_2 and promote their mixing with the F atom stream, an optical resonator to extract the laser beam and a pressure recovery system because these lasers operate at subatmospheric pressures. The power produced by these lasers scales with the flow rates of the reactants: to double the power, the flow rates are doubled.

B3.4.2.1 F-atom generators

At the present time, small-scale devices that operate at tens to hundreds of watts are commercially available from Helios Inc. (1822 Sunset Place, Longmont, CO 80501, USA). These lasers use a low pressure arc to partially dissociate SF_6 to produce F atoms. If a high-pressure, high-temperature arc was used, these lasers would produce several kilowatts at the same flow rates. The hundreds of kilowatts to several megawatts lasers react to NF_3 with a small amount of D_2 or H_2 to raise the temperature in the combustor high enough to thermally dissociate the remaining NF_3. This results in F atoms, N_2 and some DF or HF flowing through the primary nozzles at the exit of which H_2 or D_2 is mixed with the primary stream to form the lasing species, HF or DF.

B3.4.2.2 Mixing nozzles

Once the F atoms have been produced, the power of the laser is determined by the effectiveness with which the F atoms and the H_2 or D_2 are mixed. The physical-chemical characteristics of the HF/DF systems determine the design of the mixing nozzles. The pumping reactions that produce the population inversions are very fast, of the order of microseconds. If the chemicals were premixed, the width of the laser beam would be of the

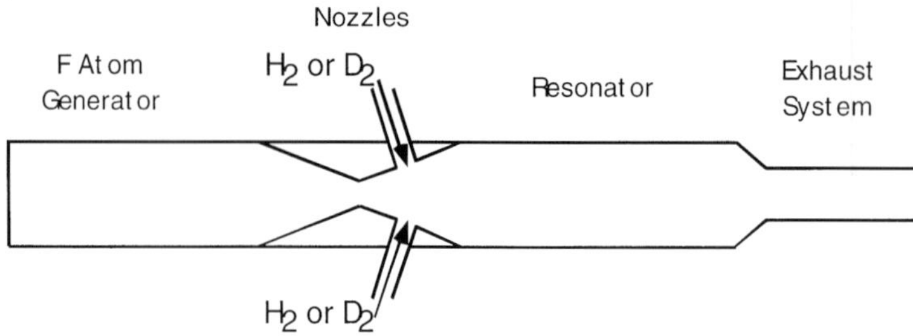

Figure B3.4.2.1. A schematic diagram of a typical HF/DF chemical laser.

order of a millimetre which results in very large intensities which windows and mirrors could not survive. The exothermicity of the pumping reactions raises the temperature which increases the rates of deactivation of the excited HF/DF. The problems of the heat release and the width of the lasing zone are solved by introducing He and expanding the F atom/He stream through a nozzle. For combustor and high-pressure, high-temperature arc-produced F-atoms, the nozzles must be supersonic nozzles. Typically these will be Mach 5 nozzles which produce 2–10 torr exit pressures and gain zones that are several centimetres long, for example [1]. For low-pressure, high-voltage arc-produced F-atom flows, either supersonic or subsonic nozzles may be used. These nozzles typically produce exit pressures of 2–10 torr and gain zones that are 3–12 mm long. In the supersonic case, they are typically Mach 2 nozzles [2].

The original HF/DF lasers employed alternating primary (F atom) and secondary (H_2 or D_2) slit nozzles at the exits of which the two streams begin mixing [3]. Because the pressures are so low, the mixing is two dimensional and diffusive with the mixing layer growing as $x^{1/2}$, where x is the distance downstream measured from the nozzle exit plane. Flow field photography showed that the mixing was slow and that the primary and secondary streams were not fully mixed before the fluid exited the resonator. One of the most successful schemes developed to increase the rate of mixing injected interleaved jets of He near the exits of the primary and secondary nozzles. These jets, called trip jets, caused a rapid increase in the rate of mixing and about a factor of two increase in power. With laser induced fluorescence, Driscoll [4–6] showed that the trip jets introduced fluid element stretching which dramatically increased the surface area of contact between the primary and secondary streams which increased the rate at which they mixed. Driscoll showed that the same effect could be obtained by putting alternating, interleaved solid ramps at the exits of adjacent nozzles. With this understanding of the fluid dynamic mechanism responsible for the increased mixing due to side-wall injection in these low-pressure nozzles, the logical development of the trip jet concept led to eliminating the secondary nozzle and injecting the H_2 directly into the primary flow near the exit of the nozzle. To shield the H_2 from the primary flow until it exited the nozzle, a He jet was placed immediately upstream of the H_2 jets. This is illustrated by the TRW HYLTE nozzle [7]; see figure B3.4.2.2.

Another efficient mixing scheme is called the double axisymmetric nozzle. This is a conical hole primary nozzle aound the exit of which the H_2 or D_2 is injected from a circular slit. This is illustrated by the Bell Aerospace Textron BCL-13 nozzle, figure B3.4.2.3 [8]. These nozzles are excellent performers.

The choice of nozzle depends upon the mission of the laser. If space and volume constraints do not drive the design, rectangular nozzle banks are usually used. The flow system is linear, with the fluid flowing from the combustor through the nozzle bank to the laser cavity and out through the exhaust system. When space and volume constraints are critical, cylindrical gain generators are used. An example is the TRW Alpha laser which is a prototype for a space based laser system [9]. In this case, the nozzle bank is wrapped around the surface of a cylinder, see figure B3.4.2.4. The centre of the cylinder is the combustor which produces the F atoms. The fluid flows radially out through the nozzles and the resonator to the exhaust ducts

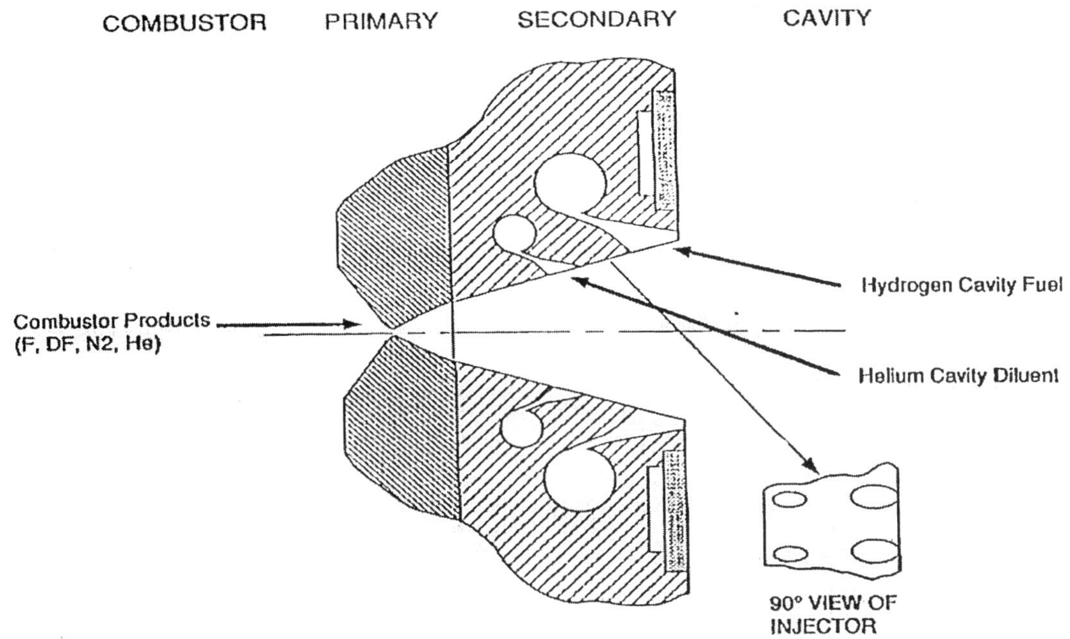

Figure B3.4.2.2. A schematic diagram of the TRW HYLTE nozzle. This figure was originally published in [7], copyright© 1991 by the AIAA, reprinted with permission.

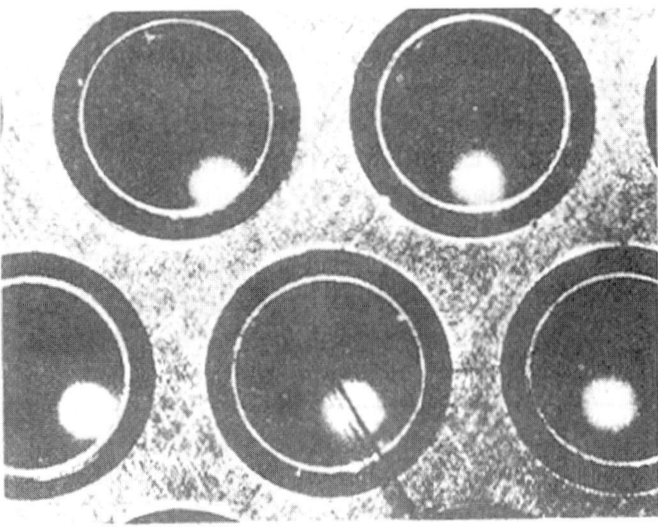

Figure B3.4.2.3. A photograph of the face of the Bell Aerospace Textron double axisymmetric BCL-13 nozzle. The primary conical nozzle is surrounded by the conical secondary nozzle. The diameter of the exit of the primary nozzle is of the order of 0.12 inches.

and out to space. The choice of nozzle geometry for the cylindrical gain generator is heavily influenced by manufacturing considerations.

HF/DF laser nozzles typically have throat widths that are 0.10 inch or less. The secondary He and H_2

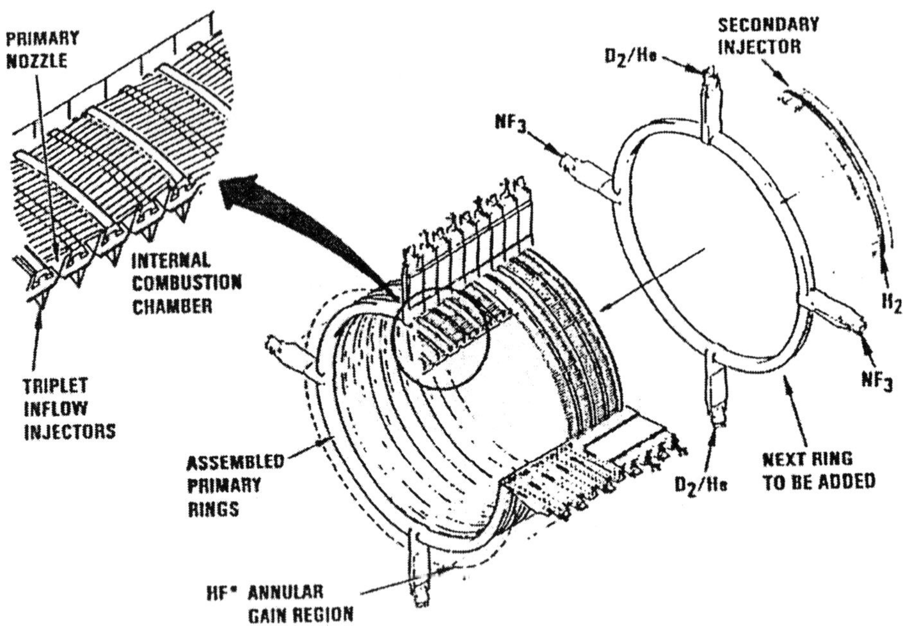

Figure B3.4.2.4. TRW hypersonic wedge nozzles on the surface of the ALPHA laser cylindrical gain generator. This figure was originally published in [9], copyright© by the AIAA, reprinted with permission.

injection holes are of the order of 0.020 inch diameter spaced several millimetres apart and are staggered on opposite nozzle walls. Since these are low pressure nozzles, the boundary layers are quite thick, typically up to 50% of the nozzle exit area. The side wall injection of the H_2 and He partially blocks the primary nozzle, further reducing its effective area ratio. These effects must be taken into account when designing the nozzle so that the desired exit pressure will be obtained.

With side-wall injection of the secondary He and H_2, the manifolds for these fluids usually determine the thickness of a nozzle blade. The size of the base area between adjacent primary nozzles may become large enough to provide a recirculation region for ground-state HF/DF, see figure B3.4.2.5. In this case, the base region must be purged with He to prevent the build up of ground-state HF/DF. If this is not done, up to 50% of the laser power may be lost. The He base purge flow rates are determined experimentally to maximize the power.

B3.4.2.3 Optical resonators

Each of the three types of resonators, Fabry–Pérot (two plane, parallel mirrors), stable and unstable (see chapter A2.1), have been used to extract power from the HF/DF chemical laser. Since it generally produces the least power, the Fabry–Pérot resonator is only used for special studies. Normally, the stable or unstable resonator is used to extract power. Since the stable resonator extracts power through a partially transmitting mirror, the stable resonator is usually used when the power is less than tens of kilowatts. At higher power levels, partially transmissive mirrors cannot handle the radiative fluxes that occur. For powers of tens of kilowatts and larger, power is extracted with an unstable resonator (see section A2.1.7) which employs all reflective optical elements. The outcoupled beam from an unstable resonator generally has a hole in it so that some of the radiation can be fed back into the resonator to keep the lasing process going. The problem is to design the unstable resonator to produce an outcoupled beam with a uniform phase so it can be focused to a

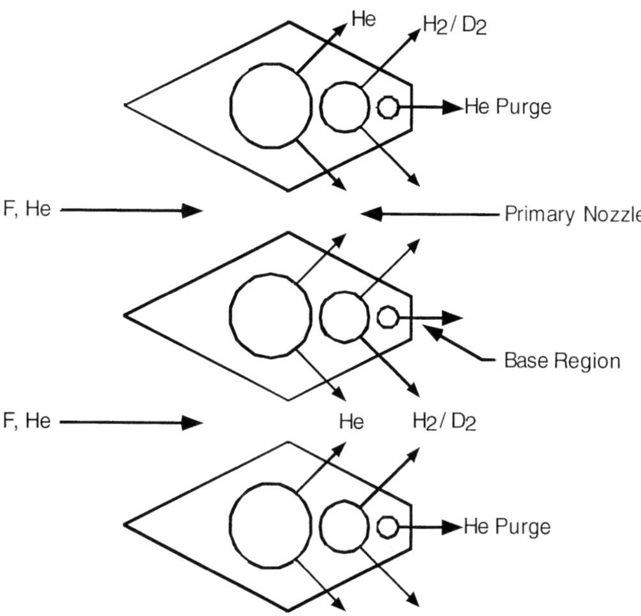

Figure B3.4.2.5. A schematic diagram of a nozzle bank showing the side wall injection of H_2 and He and the base region He purge.

spot in the far field. Since diffraction effects play a major role in the performance of an unstable resonator, the resonator must be designed with a wave optics code. As a minimum, these models are two dimensional, and in most cases, a three-dimensional wave optics model is used.

The rotational non-equilibrium kinetic-fluid dynamic model must be coupled with the wave optics model of the unstable resonator or the geometric optics model of the stable resonator. Since the flow and optical axes are perpendicular to each other, the computer models (see chapter A8) are coupled in an iterative fashion [10]. An initial guess at the intensity distribution on each line is used to run the fluid dynamic-kinetic model to obtain the gain distribution on each line. Then the optics model propagates the initial guess of the intensity distribution on each line one round trip through the resonator. Each time the optical wave passes through the gain medium, the intensity distribution is modified by the gain distribution from the preceding fluid dynamics-kinetics calculation. After the round trip through the resonator, the new intensity distribution is used in the fluid dynamic-kinetic model to recalculate the gain. When the changes in I and α are less than some small number ε, the procedure has converged. The resulting power, spectra and intensity distributions agree very well with data. With appropriate fluid dynamic, kinetic and optical models, it is possible, in theory at least, to predict laser performance as various aspects of the laser design are changed.

When choosing the outcoupling fraction and the location for the optical axis for an unstable resonator, knowledge of the gain distribution on each of the lasing lines is important. If the optical axis is placed so that the saturated gain distribution of some of the lasing lines does not fill the resonator, a time-dependent oscillation may occur on those lines [11]. The period of the oscillation increases as the magnification of the resonator decreases and the amplitude increases as the fraction of the resonator filled by the line decreases [12]. The oscillations do not occur if the gain medium is strongly coupled to the optical fields diffractively, the Fresnel number $N_F = (D^2/4\lambda L) < 3$, or geometrically, the number of round trip passes required for a wave to exit the resonator after leaving the Fresnel core $n_p = [\ln\{MD_s/2(\lambda L)^{1/2}\}]/\ln M > 3.8$, where D is the diameter of the large mirror, D_s is the diameter of the small mirror, λ is the wavelength, M is the resonator

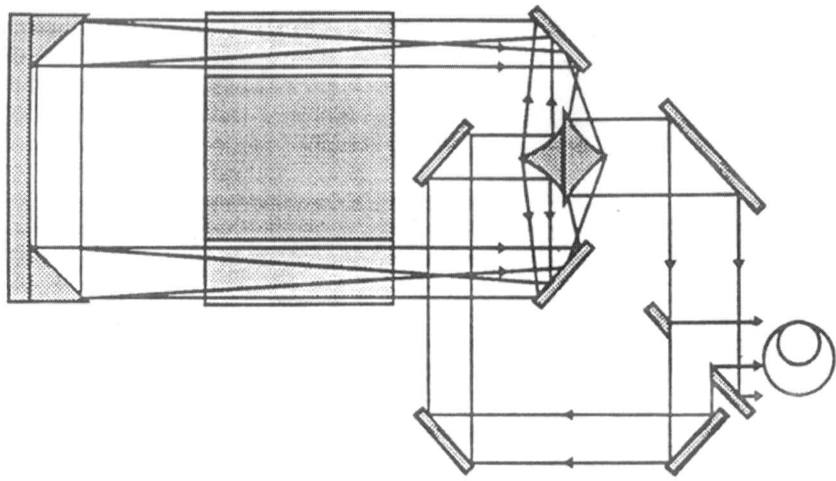

Figure B3.4.2.6. A schematic view of the HEXDARR resonator used to extract power from the cylindrical gain generator of the ALPHA laser. The dark shaded region is the cylindrical combustor and the light shaded region is the gain zone of the cylindrical laser. This figure was originally published in [14], copyright© 1993 by SPIE, reprinted with permission.

magnification and L is the distance between the mirrors.

Prior to the development of the cylindrical gain generator, HF/DF lasers were rectangular and the optical resonators were developed to extract power from rectangular gain regions. The cylindrical gain region produced by the cylindrical gain generator required the development of an entirely new resonator to extract the power. This resulted in a class of resonators denoted the high extraction efficiency decentred feedback annular ring resonator (HEXDARR) [13, 14]. A schematic view of the HEXDARR resonator is shown in figure B3.4.2.6. Testing of the TRW Alpha laser has demonstrated good performance for the HEXDARR resonator [15].

For all laser geometries, the mirrors that form the resonator must be protected from the lasant flow. This is accomplished with He purges (jets of He) that are positioned to prevent the laser fluids from entering the mirror cavity. Improperly placed or missing purges will result in ground-state HF/DF building up in the mirror boxes which reduces the laser power and degrades the mirror coatings. The mirror purge flow rates are experimentally optimized to maximize the power.

B3.4.2.4 Exhaust systems

The exhaust system depends upon the application for the laser. For power levels up to tens of kilowatts, blowers backed by mechanical pumps are used. The pump exhaust is scrubbed before discharge to the atmosphere. For hundreds of kilowatts power levels, the laser effluent is pumped with a steam ejector jet pump system [16–18] which also scrubs the exhaust.

B3.4.2.5 Laser performance

HF/DF laser performance is characterized by the power of the laser beam and by the spectral content of the laser beam or the power spectral distribution. The power of the laser is determined by the flow rates of the reactants and the effectiveness of the mixing nozzles. The power scales linearly with the flow rates of the reactants. For fixed flow rates, an efficient mixing nozzle may double the power obtained with an inefficient mixing nozzle.

The HF laser operates on several (v, J) transitions (wavelengths) simultaneously, with the spectra generally peaked around $J = 6, 7, 8$ in both $v = 2 \rightarrow 1$ and $v = 1 \rightarrow 0$ vibrational bands. Since rotational relaxation is the fastest collisional deactivation mechanism, the original models of laser performance, (e.g. [19–21]), assumed the HF was in rotational equilibrium. These models were capable of predicting the correct power in each vibrational band. However, they allowed only one (v, J) transition in each vibrational band to lase at a time. They predicted a sequential shift as lasing progressed from low-J to high-J lines. In the early 1970s, Polanyi and Woodall, and Polanyi and Sloan performed a comprehensive set of experiments on the hydrogen halides, including HF, that measured the fraction of product molecules that were formed in each (v, J) state for both the cold and hot reactions [22, 23]. The data showed that the product molecules were produced in a decidedly non-equilibrium distribution over both rotational and vibrational states. A kinetic model that treats each (v, J) state of the HF molecule as a separate species where the nascent distribution was that measured by Polanyi and coworkers was set up. This approach was implemented by Hough and Kerber in 1975 for the pulsed laser and by Sentman in 1975 for the cw laser [24, 25]. For the cw case, 21 J states in $v = 0, 1, 2$, that is 63 states for the lasing molecule, are followed. The results of these models were the prediction of simultaneous lasing on many (v, J) lines in each vibrational band, in agreement with experiment. When the rotational relaxation rate constants in these models are increased by a factor of 10^6 above the measured values, the rotational non-equilibrium models reproduce the results of the rotational equilibrium models.

The computer models still predicted significant power in the higher vibrational bands populated by the hot reaction. Particularly in the HF cw case, under certain conditions some power is observed in the $3 \rightarrow 2$ band, but generally no power occurs in the $4 \rightarrow 3$, $5 \rightarrow 4$ or $6 \rightarrow 5$ bands. Bartoszek, Manos and Polanyi [26] performed a set of chemiluminescence depletion with mass spectrometry experiments that allowed them to measure the relative rates for the deactivation reactions:

$$\mathrm{HF}(v, J) + \mathrm{D} \rightarrow \mathrm{F} + \mathrm{HD} \qquad \text{and} \qquad \rightarrow \mathrm{H} + \mathrm{DF} \qquad \text{for } v = 3, 4, 5, 6.$$

These measured rates showed that these reactions deactivated the HF (v = 3–6) as fast or faster than the pumping reactions produced it. When these reactions were incorporated into the kinetic models, the predictions of which vibrational bands lased were in agreement with data.

Before the collisional decomposition reactions were incorporated into the computer models of the HF laser, these models predicted large increases in performance if the laser could be operated to emphasize the hot reaction by increasing F_2 in the flow (e.g. [27]). Unfortunately, the experimental data did not come close to the theoretical predictions. When the collisional decomposition reactions are included in the computer models, the predicted performance is in reasonable agreement with the experimental data [28].

The power spectral distribution of the HF laser is primarily determined by the rotational non-equilibrium produced by the pumping reactions and the fact that the lasing process is faster than rotational relaxation which results in many (v, J) transitions lasing simultaneously in each vibrational band.

The essential rotational non-equilibrium kinetics required to model the HF laser are:

pumping reactions

$$\begin{aligned}
\mathrm{F} + \mathrm{H_2} &\rightarrow \mathrm{HF}(1, J) + \mathrm{H} \\
\mathrm{F} + \mathrm{H_2} &\rightarrow \mathrm{HF}(2, J') + \mathrm{H} \\
\mathrm{F} + \mathrm{H_2}(0) &\Leftrightarrow \mathrm{HF}(3) + \mathrm{H} \\
\mathrm{F} + \mathrm{H_2}(1) &\rightarrow \mathrm{HF}(3) + \mathrm{H} \\
\mathrm{F_2} + \mathrm{H} &\rightarrow \mathrm{HF}(3) + \mathrm{F}
\end{aligned}$$

The H_2 in the first two pumping reactions is $H_2(0) + H_2(1)$.
collisional deactivation reactions

$$\begin{aligned} \mathrm{HF}(1, J) + M &\Leftrightarrow \mathrm{HF}(0, J''') + M \\ \mathrm{HF}(2, J') + M &\Leftrightarrow \mathrm{HF}(1, J) + M \\ \mathrm{HF}(3) + M &\Leftrightarrow \mathrm{HF}(2, J') + M \\ \mathrm{H_2}(1) + \mathrm{H} &\Leftrightarrow \mathrm{H_2}(0) + \mathrm{H} \end{aligned}$$

where M = HF, F, H, H_2 and DF
multiquantum deactivation reactions

$$\begin{aligned} \mathrm{HF}(2, J') + M &\rightarrow \mathrm{HF}(0, J''') + M \\ \mathrm{HF}(3) + M &\rightarrow \mathrm{HF}(1, J) + M \\ \mathrm{HF}(3) + M &\rightarrow \mathrm{HF}(0, J''') + M \end{aligned}$$

where M = HF, F, H and DF
collisional decomposition reaction

$$\mathrm{H} + \mathrm{HF}(3) \rightarrow \mathrm{H_2} + \mathrm{F}$$

VV transfer reactions

$$\begin{aligned} \mathrm{H_2}(1) + \mathrm{HF}(2, J') &\Leftrightarrow \mathrm{H_2}(0) + \mathrm{HF}(3) \\ \mathrm{H_2}(1) + \mathrm{HF}(1, J) &\Leftrightarrow \mathrm{H_2}(0) + \mathrm{HF}(2, J') \\ \mathrm{H_2}(1) + \mathrm{HF}(0, J''') &\Leftrightarrow \mathrm{H_2}(0) + \mathrm{HF}(1, J) \\ \mathrm{HF}(1, J) + \mathrm{HF}(2, J') &\Leftrightarrow \mathrm{HF}(0, J''') + \mathrm{HF}(3) \\ \mathrm{HF}(2, J') + \mathrm{HF}(2, J') &\Leftrightarrow \mathrm{HF}(1, J) + \mathrm{HF}(3) \\ \mathrm{HF}(1, J) + \mathrm{HF}(1, J) &\Leftrightarrow \mathrm{HF}(0, J''') + \mathrm{HF}(2, J') \end{aligned}$$

rotational relaxation reactions

$$\begin{aligned} \mathrm{HF}(2, J') + M &\Leftrightarrow \mathrm{HF}(2, J) + M \\ \mathrm{HF}(1, J) + M &\Leftrightarrow \mathrm{HF}(1, J'') + M \\ \mathrm{HF}(0, J''') + M &\Leftrightarrow \mathrm{HF}(0, J') + M \end{aligned}$$

where M = HF, F, H, H_2, F_2, DF, He, Q

The species denoted by Q is included to take account of any other combustion or dissociation products that may be present in the mixture and which would contribute to the rotational relaxation but not the collisional deactivation of the lasing species.

B3.4.2.6 HF overtone laser

The cw HF overtone laser operates on the first overtone $\Delta v = 2$ ($v = 2 \rightarrow v = 0$) transitions between 1.3–1.4 μ [29,30]. An extensive series of experiments [7,29,31], demonstrated that 60%–90% of the fundamental power is obtainable on the overtone, the overtone is optimized by the same flow rates as the fundamental and overtone efficiency is independent of mode volume and whether the mixing is slow or fast. Since the overtone is a low gain system, $\alpha_{\mathrm{OT}} \approx (1/80)\alpha_{\mathrm{FUND}}$, to suppress fundamental lasing, mirror coatings must be less than 1% reflective over the fundamental wavelengths and highly reflective (>99%) over the overtone wavelengths.

Optimization of the output power [32] requires careful selection of the reflectivity of the resonator mirrors in terms of their absorption/scattering losses. The overtone mirror design problem led to the development of uncooled silicon optics. Measurements of the fundamental gain while lasing on the overtone [$P_{20}(7)$, $P_{20}(8)$, $P_{20}(9)$, $P_{20}(10)$] [33] showed that lasing on the overtone suppressed the gains of the low J lines $P_1(4–6)$ and $P_2(4–6)$ by 41–96% and suppressed the gains of the high J lines $P_1(7–9)$ and $P_2(7–9)$ by 3–44% for a well-saturated overtone laser. The high J lines are suppressed because their upper or lower levels are directly involved in overtone lasing. The upper levels of the $P_2(7–9)$ lines are depopulated and the lower levels of the $P_1(7–9)$ lines are populated by overtone lasing, which decreases their gains. The low J P_2 lines are suppressed [34] because overtone lasing depopulates the high $J v = 2$ states which blocks the rotational relaxation that populates the low $J v = 2$ states which decreases the gains of the $P_2(4–6)$ lines. The low J P_1 lines are suppressed because overtone lasing populates the high $J v = 0$ states which increases the rotational relaxation that populates the $v = 0$ low J states which decreases the gains of the $P_1(4–6)$ lines.

B3.4.2.7 Line selected performance of the HF laser

To operate the HF laser in a multiple line selected mode, a grating is incorporated into the resonator in the off-Littrow orientation [35]. In an unstable resonator, the feedback mirrors are placed at the locations of the positive first-order diffraction peaks of the desired lines. In a stable resonator, one line can be selected in the Littrow orientation and the remaining lines selected by placing mirrors at the locations of the positive first-order diffraction peaks of the desired lines. The possible line combinations are dictated by the grating equation. The grating characteristics, ruling and blaze angle, are selected to permit only first-order diffraction and to ensure that the grating incident angle does not exceed the limit for high grating efficiency [36]. For first-order diffraction peaks of the HF wavelengths of interest, the maximum grating incident angle must be less than 57°. For incident angles greater than this, the grating begins to behave as a grazing optic. Experiments by Chodzko, Gordon *et al* and Sentman *et al* [37–39] have shown that 55%–80% of the multiline power can be obtained by operating the HF laser in the line selected mode on two to four lines.

B3.4.2.8 Concluding remarks

The fundamental processes responsible for the performance, power, spectra and beam quality of cw HF/DF chemical lasers are reasonably well understood. An HF/DF laser can be designed with confidence to meet the requirements of an application. The rotational non-equilibrium distribution produced by the pumping reactions is primarily responsible for the power spectral distribution of these lasers. The basic kinetic processes are fairly well understood. The rate constants for the major kinetic processes are known although some of the energy transfer/redistribution rate constants are less well known. Mixing is the mechanism primarily responsible for the power of the HF/DF laser and the basic fluid dynamic mechanism that controls the mixing is fluid element stretching. Resonator designs that efficiently extract the energy from the gain medium in a beam that can be focused to a spot in the far field have been developed for rectangular and cylindrical gain media. Efficient HF overtone lasing has been demonstrated. The processes responsible for the suppression of the fundamental gain while lasing on the overtone have been identified. The grating characteristics required for efficient line selected operation of the HF laser have been identified. Line selected operation of the HF laser has been demonstrated for both stable and unstable resonators.

References

[1] Spencer D J, Mirels H and Durran D A 1972 Performance of cw HF chemical laser with N_2 or He diluent *J. Appl. Phys.* **43** 1151–7

[2] Sentman L H, Theodoropoulos P T, Nguyen T, Carroll D L and Waldo R E 1989 An economical supersonic CW HF laser testbed *AIAA Preprint 89-1898, AIAA 20th Fluid Dynamics, Plasma Dynamics and Lasers Conf. (June 12–14)*

[3] Durran D A and Spencer D J 1970 Axisymmetric mixing nozzle for supersonic diffusion laser *Technical Report* TOR-0059 (6756-02)-1, The Aerospace Corporation, Los Angeles, CA
[4] Driscoll R J 1984 Effect of reactant-surface stretching on chemical laser performance *AIAA J.* **22** 65–74
[5] Driscoll R J 1986 Mixing enhancement in chemical lasers, part I: Experiments *AIAA J.* **24** 1120–6
[6] Driscoll R J 1987 Mixing enhancement in chemical lasers, part II: Theory *AIAA J.* **25** 965–71
[7] Duncan W, Patterson S, Graves B and Holloman M 1991 Recent progress in hydrogen fluoride overtone chemical lasers *AIAA Preprint 91-1480, AIAA 22nd Fluid Dynamics, Plasma Dynamics and Laser Conf. (June 24–26, Honolulu, Hawaii)*
[8] Raymonda J W, Subbiah M, Schimke J T, Zelazny S W and Sentman L H 1979 Advanced HF and DF chemical laser performance modelling vol I The CNCDE/Blaze rotational equilibrium code *Technical Report* DRCPM-HEL-CR-79-7, Bell Aerospace Textron, Buffalo, NY
[9] Horkovich J A and Pomphrey P J 1997 Recent advances in the alpha high power chemical laser program *AIAA Preprint 97-2409, AIAA 28th Plasmadynamics and Lasers Conf. (June 23–25, Atlanta, GA)*
[10] Sentman L H 1978 Chemical laser power spectral performance: a coupled fluid dynamic, kinetic and physical optics model *Appl. Opt.* **17** 2244–9
[11] Sentman L H, Nayfeh M H, Townsend S W, King K, Tsioulos G and Bichanich J 1985 Time-dependent oscillations in a cw chemical laser unstable resonator *Appl. Opt.* **24** 3598–609
[12] Sentman L H, Gilmore J and Carroll D 1989 Mechanism for time-dependent oscillations in a cw HF chemical laser unstable resonator *AIAA Preprint 89-1897, AIAA 20th Fluid Dynamics, Plasma Dynamics and Laser Conf. (June 12–14, Buffalo, NY)*
[13] Wade R C 1998 Chemical lasers with annular gain media *Optical Resonators-Science and Engineering* ed R Kossowsky *et al* (Dordrecht: Kluwer Academic) pp 211–23
[14] Wade R C 1993 Annular resonators for high-power chemical lasers *Proc. of Laser Resonators and Coherent Optics: Modelling, Technology and Applications* vol 1868, ed A Bhowmik (Bellingham, WA: SPIE)
[15] Horkovich J A, Nefzger C L, Platt B C, Hyver G A, Pomphrey P J, Jacoby J L and Loomis D N 1998 Space-based laser programs high power testing progress *AIAA Preprint 98-2661, AIAA 29th Plasmadynamics and Lasers Conf. (June 15–18, Albuquerque, NM)*
[16] Emanuel G 1976 Optimum performance for a single-stage gaseous ejector *AIAA J.* **14** 1292–6
[17] Emanuel G 1982 Comparison of one-dimensional solutions with Fabri theory for ejectors *Acta Mech.* **44** 187–200
[18] Taylor D and Toline F R 1968 Summary of exhaust gas ejector-diffuser research, AEDC-TR-68-84, Arnold Engineering Development Center, Tullahoma, TN
[19] Emanuel G 1971 Analytical model for a continuous chemical laser *J. Quant. Spectrosc. Radiat. Trans.* **11** 1481–520
[20] Emanuel G, Adams W D and Turner E B 1971 RESALE-1: A chemical laser computer program, TR-0172(2776)-1, The Aerospace Corporation, July
[21] Mirels H, Hofland R and King W S 1973 Simplified model of CW diffusion-type chemical laser *AIAA J.* **11** 156–64
[22] Polanyi J C and Sloan J J 1972 Energy distribution among reaction products, VII. $H + F_2$ *J. Chem. Phys.* **57** 4988–98
[23] Polanyi J C and Woodall K B 1972 Energy distribution among reaction products, VI. $F + H_2, D_2$ *J. Chem. Phys.* **57** 1574–86
[24] Hough J J T and Kerber R L 1975 Effect of cavity transients and rotational relaxation on the performance of pulsed HF chemical lasers: a theoretical investigation *Appl. Opt.* **14** 2960–70
[25] Sentman L H 1975 Rotational nonequilibrium in cw chemical lasers *J. Chem. Phys.* **62** 3523–37
[26] Bartoszek F E, Manos D M and Polanyi J C 1978 Effect of changing reagent energy. x. vibrational threshold energies for alternative reaction paths $HF(v) + D \rightarrow F + HD$ and $\rightarrow H + DF$ *J. Chem. Phys.* **69** 933–5
[27] Cummings J C, Broadwell J E, Shackleford W L, Witte A B, Trost J E and Emanuel G 1977 H_2–F_2 chain reaction rate investigation *J. Quant. Radiat. Spectrosc. Trans.* **17** 97–116
[28] Sentman L H and Detweiler G L 2002 The possibility of hot reaction enhancement of CW HF laser performance, AAE TR 02-02, UILU ENG 02-0502, Aeronautical and Astronautical Engineering Dept., University of Illinois, Urbana, IL, March
[29] Jeffers W Q 1989 Short wavelength chemical lasers *AIAA J.* **27** 64–6
[30] Jeffers W Q 1988 United States patent 4,760,582 July 26
[31] Carroll D L, Sentman L H, Theodoropoulos P T, Waldo R E and Gordon S J 1993 Experimental study of continuous wave hydrogen-fluoride chemical laser overtone performance *AIAA J.* **31** 693–700
[32] Carroll D L and Sentman L H 1993 Maximizing output power of a low-gain laser system *Appl. Opt.* **32** 3930–41
[33] Theodoropoulos P T, Sentman L H, Carroll D L, Waldo R E, Gordon S J and Otto J W 1996 Continuous wave hydrogen fluoride overtone lasing saturation effects on fundamental gain suppression *AIAA J.* **34** 1216–23
[34] Theodoropoulos P T and Sentman L H 1996 Fundamental gain suppression mechanisms in a continuous wave hydrogen fluoride overtone laser *AIAA J.* **34** 1589–94
[35] Chodzko R A 1974 Multiple-selected-line unstable resonator *Appl. Opt.* **13** 2321–5
[36] Petit R 1980 *Electromagnetic Theory of Gratings* (Berlin: Springer)
[37] Chodzko R A 1993 Multiple line selection in cw HF/DF chemical lasers, Report Number ATR-92(2732)-1, The Aerospace Corp., Los Angeles, CA
[38] Gordon S J, Sentman L H, Jenkins D S and Eyre A J 1997 A study of line selection in cw HF lasers, AAE TR 97-02, UIUL ENG 97-0502, Aeronautical and Astronautical Engineering Dept, University of Illinois, Urbana, IL, March
[39] Sentman L H, Cassibry J T, Wootton B P and Eyre A J 1999 Influence of Grating Design on cw HF laser line selected performance,

AAE TR 99-08, UILU ENG 99-0508, Aeronautical and Astronautical Engineering Dept., University of Illinois, Urbana, IL

Further reading

Bott J F and Gross R W F (ed) 1976 *Handbook of Chemical Lasers* (New York: Wiley)

Excellent review of all aspects of chemical lasers as of the date of publication. This is probably the best reference for an introduction to the field of chemical lasers.

Ultee C J 1979 chemical and gas dynamic lasers *Laser Handbook* vol 3, ed M L Stitch (Amsterdam: North-Holland)

Meinzer R A 1987 HF/DF chemical lasers *Handbook of Molecular Lasers* ed P K Cheo (New York: Marcel Dekker) pp 393–493

Solomon W C 1991 Chemical lasers *Encyclopedia of Lasers and Optical Technology* ed A Meyers (New York: Academic) pp 1–8

Shellan J B, Smith W and Wade R C 1990 Novel resonators for high power HF overtone lasers *Proc. Optical Resonators (SPIE 1224)* ed D A Holmes (Bellingham, WA: SPIE) pp 302–11

Powers R B 1963 Steam-Jet Air Ejectors: Specification, Evaluation and Operation, ASME Paper 63-WA-143

This is a comprehensive review of steam ejectors with an extensive reference list.

Lacour B 2003 Recent advances in pulsed HF/DF lasers in *Proc. XIV Int. Symp. Gas Flow Chemical Lasers and High Power Lasers* ed K Abramski, E Plinski and W Wolinski SPIE vol 5120 (Bellingham, WA: SPIE) *in press*

B3.5
Argon and krypton ion lasers

Malcolm H Dunn and Tony Gutierrez

B3.5.1 Introduction

Since its discovery some 37 years ago in 1964, the argon ion laser (and the closely related krypton ion laser) has continued to be a valued 'powerhouse' of coherent radiation in the ultraviolet (UV), visible and near infrared (NIR). Continuous-wave (cw) powers of several tens of watts in the visible (450–530 nm) and of several watts in the ultraviolet (229–363 nm) are available from commercial devices providing both high beam quality and spectral refinement. In addition devices have been demonstrated that generate hundreds of watts of cw power in the blue–green spectral region.

Despite being gas lasers and ones of poor electrical to optical conversion efficiency, the ion lasers have continued to hold their own on the technological scene despite the rapid development in recent years of diode-laser-pumped solid state lasers. This is particularly the case with regard to their high-power capabilities in the UV and deep-UV (DUV) and their coverage of spectral ranges (e.g. red, violet) not accessible at present to solid state cw sources.

This review covers the early history of ion lasers, particularly the argon ion laser (section B3.5.2), basic processes involved in their operation (section B3.5.3) and device development to the present day including contemporary capabilities and applications (section B3.5.4).

A word on notation may be appropriate here. The conventional spectroscopic notation for neutral Ar is Ar I. The singly ionized species (which discharge physicists denote as Ar^+) in spectroscopic notation is denoted by Ar II and the doubly ionized species (Ar^{++}) in spectroscopic notation is denoted by Ar III, etc.

B3.5.2 Early history

In the spring of 1964, Bridges [1] and Convert *et al* [2] reported pulsed laser oscillation on ten transitions of singly ionized argon in the range 450–530 nm; all but one of these transitions being from the levels of the (^{3}P) 4p to those of the (^{3}P) 4s configuration. This was soon followed by reports from Bennett *et al* [3] and Gordon *et al* [4] of quasi-continuous and true continuous laser oscillation on these transitions.

The active medium of the argon ion laser is the positive column region of a low-pressure (<1 mbar) DC discharge in argon. In the first generation of argon lasers, the positive column was constricted by a quartz capillary tube (bore 1–4 mm) to produce the high current densities (100 A cm^{-2}) necessary to create population inversions. The capillary was usually between 25 cm and 1 m long.

Since high currents have to be passed into the gas, hot oxide-coated cathodes were employed to keep discharge voltages and cathode sputtering to a minimum. Voltage gradients in the positive column are around 10 V cm^{-1} and so the power dissipation in the capillary, even at an operating current as low as 1 A, is around 10 W cm^{-1}. A power dissipation of 25 W cm^{-1} is sufficient to heat quartz discharge tubes (with these dimensions) to incandescence within a few seconds, so that, for prolonged running at high currents, water cooling of the constricted region of the discharge is essential.

For a 2-mm bore tube the optimum argon filling pressure is about 0.5 torr (1 torr = 1.32 mbar) and varies only slightly between the different transitions. The E/p ratio for the positive column is therefore around 20 V cm^{-1} torr^{-1}.

Considerable progress was made in improving the power outputs available from a fixed volume of active medium, both through the development of improved discharge tubes and through the use of longitudinal magnetic fields.

Bridges and Halstead [5] reported total output powers of up to 8.5 W from a 46 cm-long 4-mm bore quartz discharge tube at a current of 45 A (350 A cm^{-2}) in the presence of a longitudinal magnetic field of 0.1 T. In 1965 Labuda *et al* [6] established that laser power output increases monotonically with increasing discharge current and that the power output is limited through the current limitation set by the destruction of the quartz capillary under positive ion bombardment (maximum power dissipation in quartz being approximately 140 W cm^{-2}).

By using a ceramic discharge tube (3.5 mm bore, 50 cm long) Paananen [7] was able to overcome the current limitation set by the quartz capillary, and reported a total output of 14 W at a current of 75 A (800 A cm^{-2}) in the presence of a magnetic field. Using a similar discharge to one given earlier, Paananen [7] also reported cw oscillation in Ar III at 351.1 nm. An output power of 13 mW was obtained at a discharge current of 72 A.

Labuda *et al* [6], Rigden [8], Hernqvist and Fendley [9] and McMahan [10] investigated a range of materials in attempts to replace quartz as the confinement material for the discharge. These included molybdenum, aluminium, tantalum and pyrolytic graphite. Since the wall of the discharge tube is electrically conducting when such materials are used, it must be constructed in the form of short segments, each segment being insulated by a suitable spacer material from its neighbours.

By going to larger bore quartz discharge tubes (10 mm) Boersch *et al* [11] in July 1967 succeeded in producing continuous output powers of the order of 100 W at currents of the order of 300 A (400 A cm^{-2}) in the absence of a magnetic field.

In order to reach the current densities that produce saturation of power output in the large-bore tubes, Boersch *et al* [12] devised an argon ion laser where the capillary was constructed in anodized aluminium segments (bore 12 mm). Power output was found to depend on the product of current density with tube radius (*JR*) and to reach a maximum value at around $JR = 170$ A cm^{-1} (280 A cm^{-2}). Maximum power output on all transitions of Ar II was 105 W m^{-1} at filling pressures around 1 torr. The saturation of power output for transitions of Ar II was found to correspond to the threshold for Ar III oscillation, and for JR values around 250 A cm^{-1} (420 A cm^{-2}) two Ar III transitions (351.1, 363.8 nm) produced a total coherent output power of 3.0 W.

If large total output powers (100 W) are required then large-bore tubes with large discharge currents are used but if more moderate powers are sufficient (tens of watts) the narrow-bore tube with axial magnetic field is more convenient.

Fourteen cw laser transitions of Ar II are listed in table B3.5.1 together with level designations.

B3.5.3 Basic processes

B3.5.3.1 Energy levels

The configurations of Ar II of relevance to the argon ion laser are of the form $3s^2 3p^4 nl$. The parent configuration (Ar III), $3s^2 3p^4$, is essentially governed by LS coupling so that the designation of the parent levels is $^3P_{2,1,0}$, $^1D_2$ or $^1S_0$. When the outer electron of Ar II is added to the Ar III parent its associated

Table B3.5.1. Wavelengths of cw laser transitions in Ar II (all levels have a ^{3}P core except where otherwise indicated).

Wavelength in air (nm)		Designation
437.1	(^{1}D)	4p $^2D^0_{3/2}$–3d $^2D_{3/2}$
448.2		4p $^2D^0_{5/2}$–3d $^2D_{5/2}$
454.5		4p $^2P^0_{3/2}$–4s $^2P_{3/2}$
457.9		4p $^2S^0_{1/2}$–4s $^2P_{1/2}$
472.7		4p $^2D^0_{3/2}$–4s $^2P_{3/2}$
476.5		4p $^2P^0_{3/2}$–4s $^2P_{1/2}$
488.0		4p $^2D^0_{5/2}$–4s $^2P_{3/2}$
488.9		4p $^2P^0_{1/2}$–4s $^2P_{1/2}$
496.5		4p $^2D^0_{3/2}$–4s $^2P_{1/2}$
501.7	(^{1}D)	4p $^2F^0_{5/2}$–3d $^2D_{3/2}$
514.2	(^{1}D)	4p $^2F^0_{7/2}$–3d $^2D_{5/2}$
514.5		4p $^4D^0_{5/2}$–4s $^2P_{3/2}$
528.7		4p $^4D^0_{3/2}$–4s $^2P_{1/2}$
1090		4p $^2P^0_{3/2}$–3d $^2D_{5/2}$

coupling energies are larger than the spin–orbit energy of the $3p^4$ core for the configurations of principal interest here (3d, 4s, 4p, 4d, 5s) and, hence, are describable by LS coupling.

In the argon ion laser the configurations of particular interest are all associated with the ^{3}P parent term. The following are of particular interest (figure B3.5.1): the $3p^4(^3P)4p$ configuration since it contains the majority of the upper laser levels of Ar II; the $3p^4(^3P)4s$ configuration since it contains the short-lived lower laser levels (doublet levels) as well as metastable levels of interest in relation to the pumping cycle (quartet levels); the $3p^4(^3P)3d$ configuration since it also contains metastable levels of relevance to the pumping cycle; the $3p^4(^3P)$ configurations since this is a source of cascade pumping to the 4p levels; and finally the $3p^5$ ground configuration with levels $^2P_{1/2,3/2}$. The excited configurations contain both doublet and quartet terms.

B3.5.3.2 *Transition probabilities and level lifetimes*

Shortly after the discovery of the Ar II laser transitions in 1964, Statz *et al* [13] calculated transition probabilities for transitions between levels of the $3p^4(^3P)4p$ and $3p^4(^3P)4s$ configurations and the $3p^4(^3P)4s$ and $3p^5$ configurations of Ar II. This was an extension of earlier work by Garstang [14]. In 1967, Rudko and Tang [15] extended the calculations of Statz *et al* [13] to cover all transitions from the 4p levels (transitions to 3d levels now being included), as well as transitions from the 5s and 4d configurations to the 4p configuration (with the aim of assessing the part played by the 4d and 5s levels in cascade pumping of the 4p levels). In 1968 Koster *et al* [16] published corrections to the original transition probabilities of Statz *et al* [13] for transitions from the $3p^4(^3P)4s$ to the $3p^5$ configuration; finding that the transition probabilities for transitions from the 4s lower laser levels were a factor of five larger than originally thought. These calculations have been confirmed by Kitaeva *et al* [17].

Experimental investigations on relative intensities of transitions in Ar II were first carried out by Olsen [18] and, subsequently, Rudko and Tang [15] carried out detailed spectroscopic measurements of absolute spontaneous intensities. Bennett *et al* [19] have determined experimentally total radiative lifetimes for the 4p states.

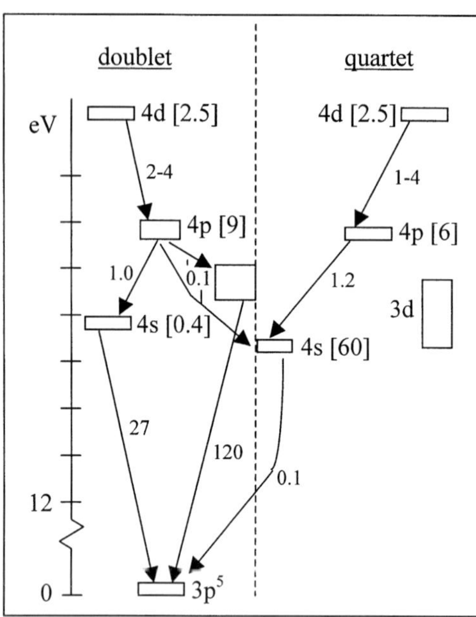

Figure B3.5.1. Electronic configurations in Ar II. All have the ^3P as parent term. The vertical extension of each block indicates the energy range associated with the levels of that configuration (units shown are electronvolts). The numbers at the arrows designate the radiative transition probabilities (spin conserved) in units of 10^8 s^{-1}. Level lifetimes (in ns) are given in square brackets.

Average values of the lifetimes for the different configurations associated with Ar II are summarized in figure B3.5.1 (for a more detailed consideration the reader is referred to the original references cited earlier), where both transition probabilities between the configurations and level lifetimes within the configurations are indicated.

The following general statements regarding transition probabilities (see chapter A1) in the argon ion can be summarized from a consideration of figure B3.5.1:

(i) For transitions between the same pair of configurations, transition probabilities are roughly an order of magnitude smaller for transitions involving spin changes compared with transitions not involving spin changes. (The former only occur at all because of deviations from pure LS coupling.)

(ii) In the $3p^4(^3P)4s$ configuration, the levels are $^2P_{1/2,3/2}$ and $^4P_{1/2,3/2,5/2}$. Since the ion ground states are doublet states, and since LS coupling is valid approximately, the decays of the quartet 4s states to ground states are slow compared to the decays of the doublet 4s states to the ground states. As a consequence there are two orders of magnitude difference in radiative lifetime between quartet and doublet 4s states; the latter, but not the former, being short-lived enough to be lower laser levels for transitions from the 4p configuration.

(iii) Transition probabilities for 4d–4p transitions are generally about half an order of magnitude greater than for 4p–4s transitions. (The influence of spin change considered in (i) applies to both these sets of transitions.)

It will be recognized from a comparison of table B3.5.1 and figure B3.5.1 that all of the laser transitions terminate on the short-lived doublet 4s states where lifetimes (0.4 ns) are considerably shorter than those (6–9 ns) associated with the upper (4p) states of these transitions, whereas none terminate on the much

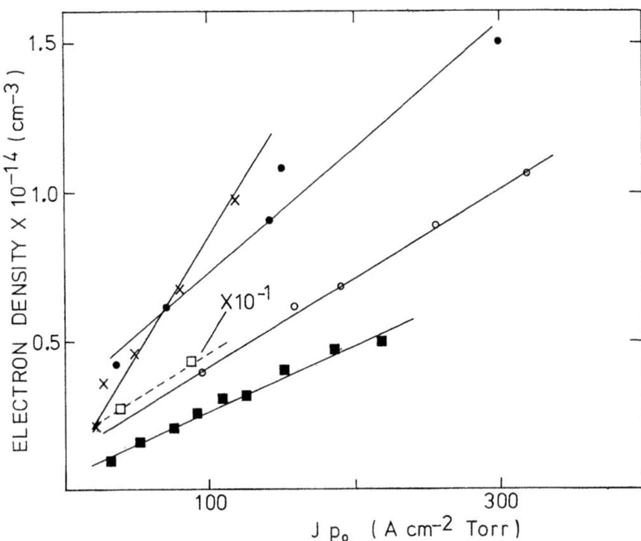

Figure B3.5.2. Electron density (cm^{-3}) as a function of Jp_0 (A cm^{-2} torr): ○, Labuda *et al* [21] (2 mm bore); ●, Pleasance and George [23] (2 mm bore); ■, Kitaeva *et al* [20] (1.6 mm bore); □, Sze and Bennett [24] (2 mm bore); ×, Kitaeva *et al* [22] (1 cm bore). (Reproduced from Dunn M H and Ross J N 1976 The argon ion laser *Prog. Quantum Electron.* **4** 233–69, with permission.)

longer lived quartet 4s states whose lifetimes (30–90 ns) far exceed those of the 4p states. This is because population builds up on the latter of the 4s states, hence destroying population inversions but not on the former 4s states. The laser transitions at 514.5 and 528.7 nm are particularly interesting in that they involve spin changes. Despite the lower transition probabilities associated with these transitions, as a result, both show laser oscillation—indeed the transitions at 514.5 nm delivers the highest power output of all the Ar II laser transitions and is probably the one most often selected for usage in applications of the laser.

B3.5.3.3 Plasma parameters of the discharge

The basic plasma parameters of the argon discharge operating as the active medium of a laser are particularly important in evaluating excitation processes. It is appropriate to divide measurements into two groups: those carried out on small-bore laser discharges (diameter less than 6 mm), which were the first type to be investigated; and those carried out on larger bore systems (diameter greater than 6 mm). The former type of discharge has been extensively investigated by three groups: that under the direction of V F Kitaeva at the Lebedev Physics Institute, Moscow; that under the direction of W R Bennett Jr at Yale University, New Haven; and that at Bell Telephone Laboratories, Murray Hill, involving A N Chester, E I Gordon, E F Labuda, R C Miller and C E Webb. Investigations on the large-bore systems have mainly been by G Herziger and collaborators at the University of Bern. In this brief review we restrict our attention to electron density and electron temperature measurements alone since these are of particular significance in the context of excitation mechanisms.

Electron densities in the argon discharge have been determined by (i) measuring the Stark broadening of the H_α and H_β lines (Kitaeva *et al* [20], Labuda *et al* [21]); (ii) probe measurements (Labuda *et al* [21], Kitaeva *et al* [22]); (iii) microwave cavity techniques (Pleasance and George [23]); (iv) Voigt analysis of the Ar I line profiles (Sze and Bennett [24]). In figure B3.5.2 electron density has been plotted as a function

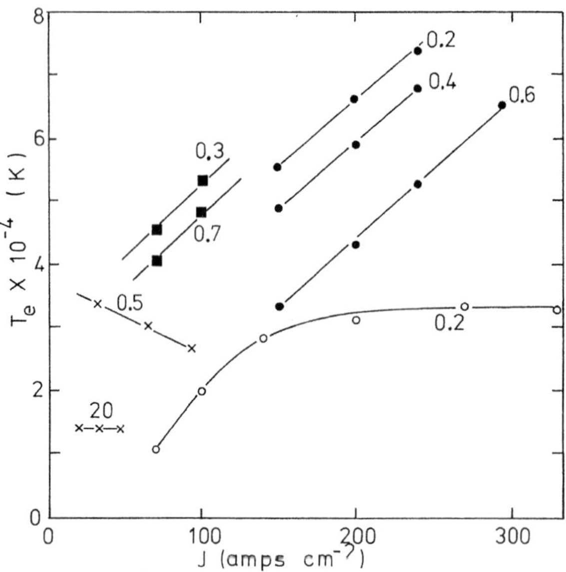

Figure B3.5.3. Electron temperature (K) as a function of discharge current (A cm^{-2}) with pressure as parameter (torr): ×, Labuda *et al* [21] (2 mm bore, probe); O, Sze and Bennett [24] (2 mm bore, line-profile); ●, Kitaeva *et al* [20] (1.6 mm bore, line-profile); ■, Kitaeva *et al* [22] (10 mm bore, probe). (Reproduced from Dunn M H and Ross J N 1976 The argon ion laser *Prog. Quantum Electron.* **4** 233–69, with permission.)

of Jp_0, where J in the operating current density and p_0 is the argon filling pressure in torr, to allow the measurements of different investigations to be compared. A linear dependence on Jp_0 is to be noted. This observation is of particular importance in the subsequent discussion on excitation mechanisms since it means that an experimentally observed dependence of a level population on discharge current can immediately be interpreted as a similar dependence on electron density. There are, however, discrepancies in absolute magnitudes of electron density, most notably in the case of Sze and Bennett [24], where results are about a factor of ten greater than those of the other investigators.

Radial dependence of electron density has also been investigated using the probe technique developed by Kitaeva *et al* [22].

Electron temperatures have been determined by (i) probe measurements (Labuda *et al* [21], Kitaeva *et al* [22]), (ii) linewidth measurements on Ar II transitions (Sze and Bennett [24]). Results are summarized in figure B3.5.3, where electron temperature is plotted as a function of discharge current with pressure as a parameter.

In order to deduce electron temperatures by linewidth or line-profile techniques, the broadening of Ar II transitions when viewed transverse to the discharge tube axis is measured. The additional broadening so observed is a consequence of a drift velocity superimposed on the ion random velocity by the strong radial electric field in the plasma. This radial electric field may be related to electron temperature—roughly speaking the radial potential drop between tube axis and the beginning of the wall sheath is of the order of kT_e/e, where e is the electronic charge but the exact details depend on the discharge model assumed.

As may be seen from figure B3.5.3, there is reasonable agreement between the probe and linewidth measurements of Kitaeva *et al.* Sze and Bennett repeated the linewidth measurements using a Fabry–Pérot system of much higher finesse, so that detailed line profiles were obtained for the Ar II transitions. These measurements, also shown in figure B3.5.3, indicate that although electron temperature increases initially with increasing current, it soon reaches a constant value for current densities above about 200 A cm^{-2}, at variance with the observations of Kitaeva *et al* discussed previously. However, the findings of Sze and Bennett are more consistent with the probe measurements of Labuda *et al*, both with regard to the magnitude of electron temperature and its independence of current density.

The behaviour of electron temperature with discharge parameters, in particular current density, is therefore a region of some uncertainty. This is particularly unfortunate as such knowledge is required to identify unequivocally the excitation mechanisms associated with the argon laser, as will become apparent shortly.

B3.5.3.4 Level populations

The dependence on discharge parameters of the population of 4p, 4s and 3d levels of Ar II and of neutral argon have been experimentally investigated by Gordon *et al* [21, 25], Webb [26], Rudko and Tang [15], Vladimirova *et al* [27] and Ross [28]; radiating levels by emission spectroscopy and metastable levels by the Ladenburgh–Reiche line-absorption technique. The general conclusions that can be drawn about level populations from these observations are as follows.

(i) The populations of the radiating levels of neutral argon are only weakly dependent on current density, showing a tendency to decrease slowly with increasing current density.

(ii) Metastable level populations of neutral argon are only weakly dependent on current density, again showing a tendency to decrease with increasing current density.

(iii) Populations of radiating levels of Ar II are a quadratic function of discharge current density.

(iv) Metastable level populations of Ar II (4s ^{4}P, 3d ^{4}D, 3d ^{4}F) are a linear function of discharge current density.

By a consideration of simple rate equations describing the steady-state populations of the excited levels, several important conclusions regarding excitation and relaxation mechanisms for these levels may be drawn. These may be summarized as follows.

(i) Since the populations of both radiating and metastable levels of neutral argon effectively saturate with discharge current, electron collisional processes (including both superelastic collisional de-excitation and collisions which lead to ionization [26]) can be identified as the dominant loss mechanism for both. In the case of radiating levels it predominates over losses due to radiative decay and, in the case of metastable levels, it predominates over losses due to diffusion to the walls. These conclusions are based on the assumption that the electron temperature and, hence, also electronic excitation and de-excitation rates are independent of discharge current. (We have previously discussed the uncertainties in the dependence of the electron temperature with the current density in the discharge.)

(ii) The linear dependence on discharge current of the populations of both the ion ground state and the ion metastable levels is consistent with the metastable levels being created by electron impact excitation from the ion ground level and being destroyed by electron collisions. In other words, the populations of the ion metastable levels are saturated relative to the population of the ion ground level, in just the same way as the populations of the neutral metastable levels are saturated relative to the population of the neutral ground level.

(iii) The quadratic dependence of the populations of the radiating levels of the ion on discharge current is consistent with these levels being populated by electron impact excitation from the ion ground level, as before, but being destroyed by radiative decay. Since, like those of the ion ground level, the population of ion metastable levels is linearly dependent on current, the radiating levels could equally be created by electron impact excitation from metastable levels and still show a quadratic current dependence.

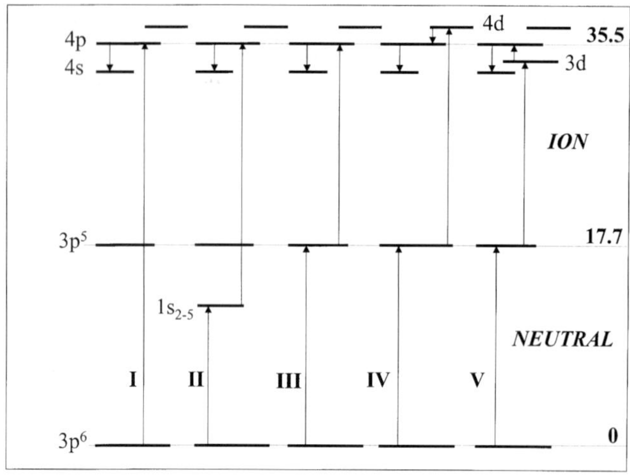

Figure B3.5.4. Excitation pathways for the 4p states of Ar II.

(iv) Webb [27] has investigated the radial variation of Ar I and Ar II excited state populations using a novel 'end-on' collimation technique. Measurements were made on tubes of 2 and 3 mm bore. His findings confirmed many of the previous observations while providing additional insights.

B3.5.3.5 Excitation mechanisms

Numerous excitation pathways have been proposed to explain the 4p–4s population inversions in the argon ion laser. It is convenient to consider five distinct mechanisms, although the active medium excitation probably involves more than one of these mechanisms. The proposed excitation pathways may be summarized as follows (figure B3.5.4):

(i) direct electron excitation from the $3p^6$ neutral argon ground state to the 4p upper laser level (mechanism I);

(ii) electron excitation to an intermediate Ar I metastable state ($1s_{2-5}$) followed by electron impact excitation from this state to the 4p state (mechanism II);

(iii) electron impact excitation to the Ar II ground state ($3p^5$) followed by electron impact excitation from this ion ground state to the 4p state (mechanism III);

(iv) electron impact excitation to the Ar II ground state followed by electron impact excitation to a 4d or higher state, followed by radiative cascade to the 4p state (mechanism IV); and

(v) electron impact excitation to the Ar II ground state, followed by electron impact excitation to an intermediate ion metastable state (4s, 3d) followed by electron impact excitation to the 4p state (mechanism V).

Obviously, combinations of these excitation processes are possible (for example, IV and V).

As a consequence of the large electron energy required to reach threshold in the direct excitation process (35 eV) compared to the mean electron energy in the cw laser discharge (see figure B3.5.3), the direct excitation process can be shown to make only a negligible contribution. (It is, however, much more significant in the case of the pulsed argon ion laser.)

Webb [26] has analysed excitation and destruction rates for the $1s_{2-5}$ metastable levels of Ar I, pertinent to pathway II. Under discharge conditions appropriate to the operation of the argon laser (electron temperature of the order of 1.5 eV) the predominant loss mechanism for these metastable levels is through ionization to the ground state of Ar II by electron collisions. Since the metastables are excited by electron collisions in the first place, the metastable populations of Ar I saturate with current, and indeed this can be shown to occur even at very low currents. It can further be shown, using arguments relating to the dependences of relevant cross sections on electron temperature, that this two-step electron excitation of the Ar II ground state dominates by at least an order of magnitude over the single-step process.

Although excitation of the ion ground state is, therefore, predominantly a two-step process, its population is still expected to be a linear function of discharge current since the intermediate populations ($1s_{2-5}$ metastables) are saturated relative to the neutral ground state. The approximately linear dependence of electron density (and hence ion density) on discharge current that is observed experimentally is consistent with this excitation mechanism for the ion ground state (see figure B3.5.2).

Excitation processes from the ion ground state can be either directly to the 4p upper laser levels or through intermediate levels. The populations of the upper laser levels (4p levels of Ar II) show a quadratic dependence on discharge current. However, it is not possible from this dependence alone to conclude, therefore, that the 4p levels are excited solely by direct electron excitation from the Ar II ground state, as opposed to *via* intermediate excitation to Ar II metastable states. This is because the metastable level populations of Ar II (quartet 4s, metastable 3d) are themselves saturated with respect to the ion ground state and hence show the same dependency on current as this ground state itself.

For both direct and indirect excitation processes, a quadratic dependence of 4p populations on discharge current is, hence, to be expected. However, Dunn [29, 30] and Ross [31] have applied the technique of perturbation spectroscopy to a study of the excitation processes in argon, in particular to assess the roles of the quartet 4s and metastable 3d states. By measuring the changes in the populations of other 4p levels when the laser oscillates on a particular 4p–4s transition, Dunn has shown that an important excitation pathway for the 4p states is through the quartet 4s states. Ross has shown that under typical laser conditions 57% of the loss rate from the 3d $^4F_{7/2}$ and 75% of the loss rate from the 3d $^4F_{5/2}$ level is by electron excitation to the 4p levels. Circulation of populations between 4s, 3d and 4p states of Ar II, therefore, plays a significant role in creating the population inversions of the 4p–4s transitions.

Excitation by radiative cascade from above has been investigated experimentally by Rudko and Tang [32]. They have shown by comparing relative spontaneous emission intensities that, for example, 30% of the 4p $^2D_{5/2}$ state population appears to be due to cascade from above.

Process II can be ruled out by the observed quadratic dependence of the 4p populations on discharge current—direct excitation from neutral metastable states ($1s_{2-5}$) implies a linear dependence on current, since these metastable populations are themselves saturated with respect to current.

In conclusion, all of the processes III–V are generally considered to play significant roles in determining the populations of the upper laser states of Ar II (4p states of argon II) and, hence, in determining the population inversions associated with the laser transitions.

B3.5.4 Device development to the present day

The ion laser tube represents the most costly component to the end-user. Therefore, improvements in laser tube lifetime have been a major focus of device development in the last 25 years. Coherent's Innova® Series V™ metal/ceramic plasma tube incorporates many improvements in tube design, as well as tube processing techniques that have evolved since 1981 when Coherent Inc. introduced metal/ceramic tube technology [33]. An example of a modern large-frame laser of this type is shown in figure B3.5.5.

In the 1960s, the first commercially available plasma tubes were fabricated from either BeO or quartz/graphite. These plasma tubes had significant limitations in terms of performance and lifetime. However, hard sealing by metal/ceramic technology [34] eliminated a source of contamination and thus led to monumental increases in performance and lifetime.

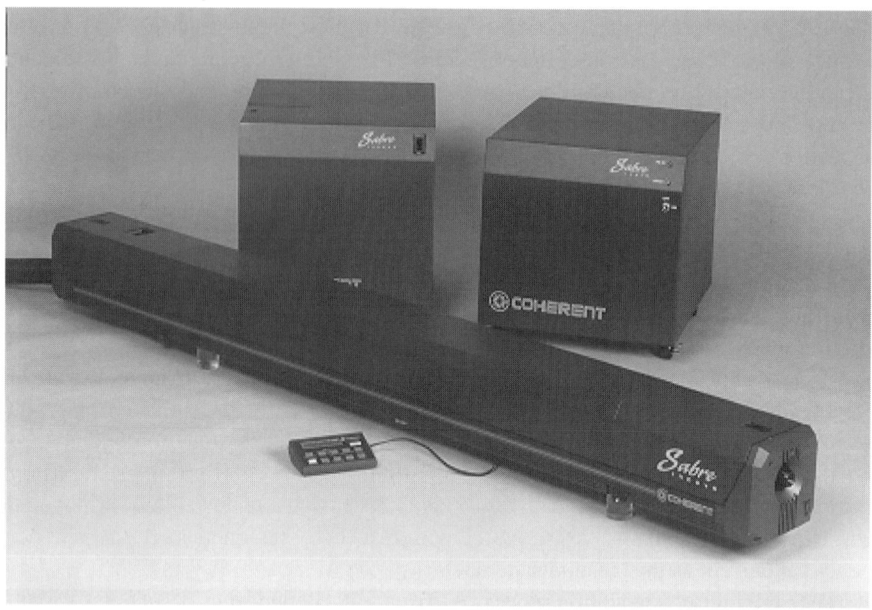

Figure B3.5.5. A large frame (25 W) Innova® Series V™ argon ion laser.

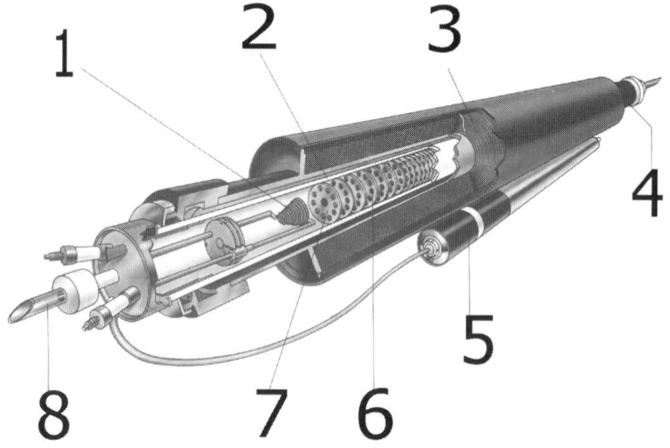

Figure B3.5.6. Innova ion laser tube: 1, thermionic cathode; 2, gas envelope (high conductivity ceramic); 3, solenoid magnet; 4, conduction-cooled anode; 5, gas reservoir for automatic pressure control; 6, internal gas return apertures; 7, tungsten discs supported in conduction-cooled copper cups; 8, hard-seal crystalline quartz Brewster windows.

In 1975, crystalline quartz windows replaced fused silica as Brewster window material. Crystalline quartz is more resistant than fused silica to changes in optical thickness due to densification [35] as a result of prolonged exposure to VUV radiation from the plasma discharge. In addition, crystalline quartz is less prone to colour-centre formation. Both of these factors led to increases in available UV output power [36].

In 1981, Innova (metal/ceramic) plasma tube technology was introduced. In plasma tubes of this design, the discharge path is defined by a series of apertures in thin tungsten discs aligned to an accuracy better than 100 μm. Tungsten is used because of its resistance to sputtering and stable crystalline structure at temperatures up to 1400 °C. Supporting the tungsten discs (which define the bore) are oxygen-free copper cups [37], to which the tungsten discs are brazed as shown in the cutaway view of figure B3.5.6. The high thermal conductivity of copper makes it ideal for transferring heat to the ceramic tube envelope. It is this feature, removal of the heat flux generated by the discharge by conduction to the water-cooled alumina tube gas envelope, which differentiates the Innova design from the pioneering segmented-metal-disc plasma confinement structures developed by Labuda and Gordon in the 1960s which were cooled by radiative transfer via a water-cooled silica tube gas envelope [6, 37].

The next major breakthrough in plasma tube development, however, occurred in 1985, when coated Brewster windows were introduced. The coated crystalline quartz windows further reduced colour-centre formation and led to increases in UV tube lifetimes of more than 1000 hr [38]. Also in 1985, gas pressure modifications were made inside the tube, increasing cathode lifetime and decreasing bore erosion. Increased pressure at the cathode allowed material sputtered off the cathode to redeposit onto it, thus increasing the usable lifetime of the cathode itself [39]. In 1989, sealed mirror tubes with no anode window were introduced. Elimination of the anode window increased lifetime by an additional 500 to 1000 hr, depending on the type of plasma tube. UV plasma lifetimes also increased to more than 2000 hr. Shortly thereafter, in 1991, new higher strength ceramic materials were introduced. This improved the plasma tube's ability to withstand thermal stresses, thereby reducing a potential source of catastrophic failure. In 1994, a new braze alloy with better thermal uniformity and joint strength was introduced, further reducing sources of catastrophic failure due to thermal stresses within the plasma tube.

In 1995, Coherent, Inc. introduced the Series V™ plasma tubes with internal re-engineered gas pathways, which substantially improved cathode performance over a wide range of operating currents. With this improvement in cathode lifetime, cathode degradation was no longer considered a primary failure mechanism. UV plasma lifetimes also increased to more than 3500 hr [33]. The working lifetime of tubes designed for output in the visible region increased to more than 5000 hr [40].

Lifetime data taken from actual in situ field installations provide a useful metric for measuring tube lifetimes in conjunction with life-tested lasers. Life-tested lasers are usually operated 24 hr a day at maximum currents while measurements are made of the output power, mode quality and other electrical and optical parameters. Although this type of life test data is useful, actual field data are more pertinent. This is true because the number of life-tested lasers that can be tested is much less than the actual number of lasers in the field. Also, lasers in the field are subject to actual usage patterns that are variable, depending on the installation. This variability in operating conditions, such as on-and-off cycling, operating powers and operating currents and extreme ambient conditions provide useful information about the reliability and lifetimes of plasma tubes.

A summary (shown in table B3.5.2) of tube lifetimes from 1990 to the present, based on the size, quasi-air-cooled cartridge, small frame, or large frame, tube type, dual Brewster window (DBW) or tunable sealed mirror (TSM) and wavelength, provides a useful guideline for laser selection.

Along with improvements in laser tube technology, productivity enhancements, such as PowerTrack™, an active power stabilization, mode optimization and noise-reduction system that operates by feedback control of the output coupler or the high-reflector mirror, have extended both ion laser performance and lifetime. PowerTrack also shortens the warm-up time for ion laser systems since users can operate the laser immediately upon turn-on. PowerTrack also maintains the optimal alignment of the optical beam through the resonator. This extends tube lifetime by minimizing bore erosion of the tube. Similarly, v-Track™, an active cavity, single-frequency stabilization system, enhances applications that require mode-hop-free operation. v-Track operates by a feedback circuit that monitors a temperature-controlled etalon, allowing the etalon to track a single longitudinal mode of the laser resonator and provide mode-hope-free operation.

Device development has also been focused on extending the fundamental wavelength regimes available

Table B3.5.2. Argon-ion-laser tube lifetimes.

Small or large frame/ output wavelengths	Tube type	Mean lifetime (hr)
Small/visible lines	Cartridge TSM	7100
Small/visible lines	DBW	6200
Small/UV lines	DBW	3500
Small/UV and visible	TSM	6200
Large/UV and visible	DBW	2500
Large/UV and visible	TSM	3300

Table B3.5.3. Output power and wavelengths from commercially available argon ion lasers (courtesy Coherent Inc.).

	Output power			
Wavelength (nm)	25 W main frame (DBW25) Multi-line visible 25.0 W (W)	Frequency-doubled 25 W main frame (Sabre MotoFRED) Multi-line ultraviolet 3.0 W (nm)[a]	(W)	Small-frame Laser 610 (612) Multi-line Visible 1.00 W (mW)
528.7	1.8	(264.3)	0.10	
514.5	10.0	(257.2)	1.00	350
501.7	1.8			
496.5	3.0	(248.2)	0.30	
488.0	8.0	(244.0)	0.50	200
476.5	3.0			
472.7	1.3			
465.8	0.8			
475.9	1.5	(229.0)	0.04	(80)[b]
454.5	0.8			
333.6–363.8	7.0			(Multi-line UV 100 mW)[b]
350.7				(40)[b]
351.1	1.8		1.0	
363.8	1.7		1.0	
351.1–385.8	4.4			
300.3–335.8	3.0			
275.4–305.5	1.6			
334.5	0.5			
302.4	0.38			
275.4	0.18			

[a] Frequency-doubled output wavelengths in brackets.
[b] Available with appropriate UV mirror cavity.

from ion lasers. This is done by doubling the frequency of the visible lines to the corresponding DUV wavelengths by intracavity critically phase-matched BBO crystals [41] (see chapters A4 and C3.1).

Table B3.5.4. Output power and wavelengths from commercially available krypton ion lasers (courtesy Coherent Inc.).

	Output power	
Wavelength (nm)	Main-frame krypton (Model 302) Multi-line red 1.0 W (W)	Mixed-gas krypton/argon (Model 70—spectrum) Multi-line 'whitelight' 2.5 W (W)
793.2–799.3	0.03	—
752.5–799.3	0.25	—
752.5	0.10	0.03
676.4	0.15	—
647.1	0.80	0.25
568.2	0.15	0.15
530.9	0.20	0.13
520.8	0.07	0.13
514.5 (Ar)	—	0.25
488.0 (Ar)	—	0.25
482.5	0.03	—
476.5 (Ar)	—	0.10
476.2	0.05	—
457.9 (Ar)	—	0.03
413.1	0.30	—
	Multi-line violet 0.60 W	
406.7	0.20	—
356.4	0.12	—
350.7	0.25	—
	Multi-line UV 0.5 W	Multi-line UV 0.05 W

B3.5.5 Present systems and capabilities

Presently, air-cooled ion lasers with powers up to 500 mW are available commercially. Water-cooled units with multi-line powers ranging from 1 to 25 W are also on the market. Tables B3.5.3 and B3.5.4 show output power as a function of wavelength for a representative set of commercial water-cooled ion laser systems. These output power/wavelength tables include argon, krypton and mixed-gas (argon/krypton) ion laser systems. Table B3.5.5 lists a number of plasma tube parameters for typical small-frame and large-frame ion lasers [33].

Ion lasers are commercially available in a number of configurations. Examples include systems designed to provide multi-line operation, single-line operation and single-frequency operation (where a suitable etalon assembly is installed). Similarly, a number of active stabilization systems, diagnostics and computer-interfacing capabilities are commercially available. The stabilization systems can maintain constant power, constant frequency and constant transverse mode. Ion laser diagnostics currently available include power measurement, TEM_{00} transverse-mode detectors and wavelength indicators. The digital computer interfacing

Table B3.5.5. Typical plasma tube parameters of high-power large-frame and small-frame argon-ion lasers. The typical powers are for TEM_{00} or a combination of TEM_{01} and TEM_{01} modes [33].

Laser parameter	Large frame	Small frame
Maximum tube current	65 A	60 A
Nominal tube voltage	535 V	235 V
Active tube length	1.11 m	0.41 m
Bore diameter	3.25 mm	3.15 mm
Current density	785 A cm^{-2}	779 A cm^{-2}
Heat load per cm	300 W cm^{-1}	320 W cm^{-1}
Typical output power (457.9–514.5 nm)	30 W	10 W
Typical output power (333.3–363.8 nm)	10 W	2 W
Efficiency (457.9–5145 nm)	0.08%	0.07%

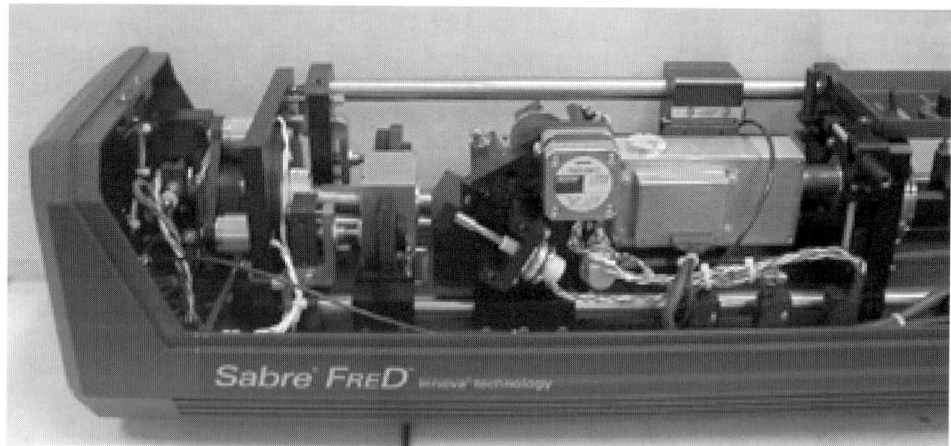

Figure B3.5.7. Motorized mount for doubling crystal in a frequency-doubled argon ion laser cavity.

available today includes RS-232/RS-422 and/or IEEE-488. Digital remote-control modules that allow remote operation of the laser systems are also available.

Ion lasers are currently available with fundamental wavelengths that range from 275.4 to 1090 nm. Lasers generating DUV wavelengths, obtained by intracavity frequency doubling the fundamental wavelengths, are commercially available in the 229.0–264.3 nm range.

Figure B3.5.7 shows the intracavity frequency-doubling crystal section of a modern laser of this type (Sabre MotoFRED). The average tube lifetimes of today's ion lasers operating in the visible regime is greater than 6000 hr. In contrast, an ion laser operating directly at fundamental wavelengths in the DUV region will have an average tube lifetime greater than 3500 hr. Because frequency-doubled UV systems are designed so that the tube is exposed only to the visible fundamental laser wavelengths, average tube lifetimes are well over 6000 hr.

B3.5.6 Ion laser applications

Key applications currently for ion laser systems are holography (see chapters D9.1 and D9.2), printing (see chapter D5.2), digital imaging, non-destructive testing, spectroscopy (see chapter D6.2), optical pumping, confocal microscopy, flow cytometry, compact disc and DVD mastering (see chapter D5.1), basic R&D, photomask direct imaging, printed circuit board direct imaging, writing fibre Bragg gratings (see chapters D4.3 and D2.6) laser art and precision optics inspection. These applications require the use of argon, krypton and argon/krypton mixed-gas lasers, as well as frequency-doubled argon ion systems.

The technology of ion lasers is a mature one. Ion lasers are used in both scientific and commercial applications. At present the scientific use of ion lasers accounts for approximately 20% of all new ion laser deployments. Solid state lasers have replaced ion lasers in many traditional applications, such as laser-pumping cw and pulsed titanium:sapphire lasers (see chapter B1.2), and pumping dyes (see chapter B5.3). However, solid state pumping has yet to provide a viable solution for pumping dyes in the UV. This is an application where the ion laser remains the laser of choice. Similarly, ion lasers are used in scientific and commercial applications, such as laser spectroscopy in the UV, DUV and red wavelengths, and violet wavelengths, all where solid state cw solutions are not readily available.

References

[1] Bridges W B 1964 *Appl. Phys. Lett.* **4** 128

[2] Convert G, Armand M and Martinot-Lagarde P 1964 *C. R. Acad. Sci.* **258** 4467

[3] Bennett W R, Knutson J W, Mercer G N and Detch J L 1964 *Appl. Phys. Lett.* **4** 180

[4] Gordon E I, Labuda E F and Bridges W B 1964 *Appl. Phys. Lett.* **4** 178

[5] Bridges W B and Halstead A S 1966 *IEEE J. Quantum Electron.* **QE-2** 84

[6] Labuda E F, Gordon E I and Miller R C 1965 *IEEE J. Quantum Electron.* **QE-1** 273

[7] Paananen R 1966 *Appl. Phys. Lett.* **9** 34

[8] Rigden J D 1965 *IEEE J. Quantum Electron.* **QE-1** 221

[9] Hernqvist K G and Fendley J R Jr 1967 *IEEE J. Quantum Electron.* **QE-3** 66

[10] McMahan W H 1968 *Appl. Phys. Lett.* **12** 383

[11] Boersch H, Herziger G, Seelig W and Volland I 1967 *Phys. Lett.* A **24** 695

[12] Boersch H, Boscher J, Hoder D and Schafer G 1970 *Phys. Lett.* A **31** 188

[13] Statz H, Horrigan F A, Koozekanani S H, Tang C L and Koster G F 1965 *J. Appl. Phys.* **36** 2278

[14] Garstang R H 1954 *Mon. Not. R. Astron. Soc.* **114** 118

[15] Rudko R I and Tang C L 1967 *J. Appl. Phys.* **38** 4731

[16] Koster G F, Statz H, Horrigan F A and Tang C L 1968 *J. Appl. Phys.* **39** 4045

[17] Kitaeva V F, Osipov Yu I, Rubin P L and Sobolev N N 1968 On oscillation mechanisms in the CW ion argon laser, private communication

[18] Olsen H N 1963 *J. Quantum Spectroc. Radiat. Trans.* **3** 59

[19] Bennett W R Jr, Kindlmann P J, Mercer G N and Sunderland J 1964 *Appl. Phys. Lett.* **5** 158

[20] Kitaeva V F, Osipov Yu I and Sobolev N N 1967 *Sov. Phys. Dokl.* **12** 55

[21] Labuda E F, Webb C E, Miller R C and Gordon E I 1968 A study of capillary discharges in noble gases at high current densities. Bell Telephone Laboratories, private communication

[22] Kitaeva V F, Osipov Yu I, Sobolev N N, Shelekhov A L and Agheev V P 1974 *IEEE J. Quantum Electron.* **QE-10** 803

[23] Pleasance L D and George E V 1971 *Appl. Phys. Lett.* **18** 557

[24] Sze R C and Bennett W R Jr 1972 *Phys. Rev.* A **5** 837

[25] Miller R C, Labuda E F and Webb C E 1967 Excited level populations in high current density argon discharges *Bell Syst. Tech. J.* **281**

[26] Webb C E 1968 *J. Appl. Phys.* **39** 5441

[27] Vladimirova N M, Konkov I D, Rovinskii R E and Cheburkin N V 1971 *Opt. Spectrosc.* **31** 91

[28] Ross J N 1974 *J. Phys. D: Appl. Phys.* **7** 1426

[29] Dunn M H 1968 *IEEE J. Quantum Electron.* **QE-4** 43

[30] Dunn M H 1974 *PhD Thesis* University of St Andrews, Fife, Scotland

[31] Ross J N 1975 *J. Phys. B: At. Mol. Phys.* **8** 529

[32] Rudko R I and Tang C L 1996 *Appl. Phys. Lett.* **9** 41

[33] Arrigoni M, Spinelli L and Barker G 1991 Experimental methods for designing, testing, and producing high-power ion lasers *SPIE 1620 Laser Testing and Reliability* (Bellingham, WA: SPIE)

[34] US Patent No 4,376,328.
[35] Ginouves P T 1996 Longer lifetimes revive water-cooled ion lasers *Laser Focus World* (Tulsa, OK: Penwell) pp 69–74
[36] US Patent No 3,993,965
[37] US Patent No 3,531,734
[38] US Patent No 5,101,415
[39] US Patent No 4,378,600
[40] US Patent No 4,736,379
[41] Asher S A, Bormett R W, Chen X G, Lemmon D H, Cho N, Peterson P, Arrigoni M, Spinelli L and Cannon J 1993 UV resonance Raman spectroscopy using a new cw laser source: convenience and experimental simplicity *Appl. Spectrosc.* **47** pp 628–33

Further reading

At introductory level:

Davis C C 1996 *Lasers and Electro-Optics—Fundamentals and Engineering* (Cambridge: Cambridge University Press) section 9.7
Silfvast W T 1996 *Laser Fundamentals* (Cambridge: Cambridge University Press) section 13.2

At research level:

Dunn M H and Ross J N 1976 The argon ion laser *Prog. Quantum Electron.* **4** 233–69
Bridges W B 1982 Ionized gas lasers *Handbook of Laser Science and Technology* vol II, ed M J Weber (Boca Raton, FL: Chemical Rubber Company)
Bridges W B 2000 Ion lasers—the early years *IEEE J. Selected Topics Quantum Electron.* **6** 885–98

B3.6
Helium–neon lasers

Alan D White and Lisa Tsufura

B3.6.1 Introduction

The helium–neon gas laser was the first gas laser to be developed and the first truly continuous wave (cw) laser of any kind. It is a textbook example of light amplification by stimulated emission of radiation from free atoms, in which population inversion occurs by selective excitation of the upper level through resonant energy transfer.

The initial demonstration in 1961 by Javan, Bennett and Herriott of Bell Telephone Laboratories [1] of laser oscillation on the infrared transition of Ne at 1.15 μm was followed shortly afterwards by the demonstration of laser oscillation on the 632.8 nm red transition by White and Rigden [2] also at Bell Labs. From that time to the present day, the helium neon (HeNe) laser has been a mainstay of scientific, industrial, medical and commercial laser applications. At its peak in the late 1980s, HeNe laser manufacturers were shipping more than 750 000 units a year for applications as diverse as holography (see section D9) and barcode scanning, biological cell counting (see chapter D3.4), metrology (see chapters C3.3, D2.1, D6.2), construction alignment (see chapter C4.4), printing and reprographics (see chapter D5.2), medical and industrial pointing, and light shows. The overall numbers have dropped considerably, but even today, nearly ten years after the 'death' of the HeNe was predicted as a result of advances in semiconductor diode laser technology (see chapter B2.1) approximately 200 000 are sold annually—more than all other lasers combined (excluding, of course, diode lasers).

The oscillating fields in the HeNe laser match very closely the classical description of light modes in a resonant optical cavity (see chapter A1, section A2). Its output beam when operating in its lowest transverse mode (see section A2), approximates very closely that of a coherent Gaussian beam with an M^2 value (see section A2), typically smaller than 1.02. This is largely a consequence of the optical homogeneity of gas-phase light-amplifying media at low-pressure and low-power dissipation. In precision scanning applications such as recording, imaging, defect measurement, reprographics and in biotechnology applications including microscopy and cytometry it is critical for the beam to have a known, repeatable energy distribution and an M^2 value close to unity in order to properly calculate, create or measure artifacts with extremely high accuracy and resolution. A particular advantage over many types of diode lasers that HeNe lasers possess is the polarization extinction ratio. This is of value in applications like ellipsometry where precise measurements of relative phase changes are made. HeNes can obtain extinction ratios on the order of 500:1, 1000:1 or higher without additional cost of polarization elements.

Plasma tube life is typically greater than 10 000 hr, and many systems in the field have been operating 50 000 hr or more without a tube replacement. The lasers are small in size (relative to other gas-laser sources), convection cooled and can be operated on normal household power. Lower-power lasers can be operated easily from batteries, making them ideal sources for field equipment including theodolites, laser levels and other construction alignment equipment. HeNe lasers are low in cost; in single-unit quantities, purchase

Table B3.6.1. Output power characteristics of commercial HeNe lasers.

Laser wavelength (nm)	Power range (mW)
1523 near infrared	0.25–1.0
633 red	0.25–40
612 orange	0.25–7.0
594 yellow	0.25–4.0
543 green	0.10–3.0

prices range from a few hundred to a few thousand dollars. For original equipment manufacturer quantities, the prices are much lower. Finally, in addition to the familiar red (633 nm) output, HeNes with orange (612 nm), yellow (594 nm), green (543 nm) and infrared (1523 nm) output became commercially available in the early 1980s finding fluorescence applications in microscopy, cytometry, genomics and other applications.

HeNe lasers are low-power devices, with output ranging from 0.1 mW or less to about 40 mW for the largest practical tube. This limitation makes them unsuitable for most materials-processing applications or for pumping other laser sources (e.g. dye lasers, see chapter B5.3).

About a dozen visible transitions and perhaps 15 times as many infrared transitions of the neon atom have been reported as showing either laser oscillation or optical gain [3]. However, the most common wavelengths for which HeNe lasers are commercially available are listed in table B3.6.1.

In this review we shall deal primarily with the 633 nm red HeNe laser. With some obvious changes, the discussion applies to all the visible and infrared transitions which originate from the same upper level (3s in Paschen notation) of the neon atom and therefore compete with it. Similar considerations apply to a group of infrared transitions which originate from the 2s group of levels about 0.9 eV below the 3s levels of Ne (see section B3.6.3).

B3.6.2 The glow discharge in rare gases

The active medium of the HeNe laser is a glow discharge in a mixture of the 'rare' or 'noble' gases helium and neon. For that reason it is worthwhile at this point to review briefly the main features of the discharge physics relevant to the excitation and inversion mechanisms of the laser [3–5].

The term glow discharge refers to a specific manifestation of electrical conduction in low-pressure gas, in which sufficient charge carriers and excited states are created in the gas by electron impact to establish a low current self-sustaining discharge. Figure B3.6.1 illustrates such a discharge in a long narrow tube provided with a planar cold cathode and anode, and filled with one of the rare gases at a pressure of a few torr (1 torr = 1.32 mbar).

The region adjacent to the cathode, termed the cathode fall or sheath, is strongly accelerating for electrons leaving the cathode. Within this region, and extending into the light and dark regions just beyond, fast electrons make ionizing collisions with gas atoms, creating large numbers of ion–electron pairs. Ions drifting back to the cathode under the influence of the field produce secondary electrons in an Auger process. Each secondary electron must generate enough ions (usually 3–5) in traversing the cathode sheath to produce its Auger replacement at the cathode, and the cathode fall voltage (typically 100–300 V) adjusts until this criterion is met. Beyond the cathode region and extending to the anode, is the uniform, strongly luminous region known as the positive column. Though characterized by a much smaller electric field than the cathode region, a similar criterion applies: enough ions, electrons and excited atoms must be generated by electron impact within the positive column to supply the loss of energetic particles through diffusion and recombination at the wall, and to radiation and gas heating.

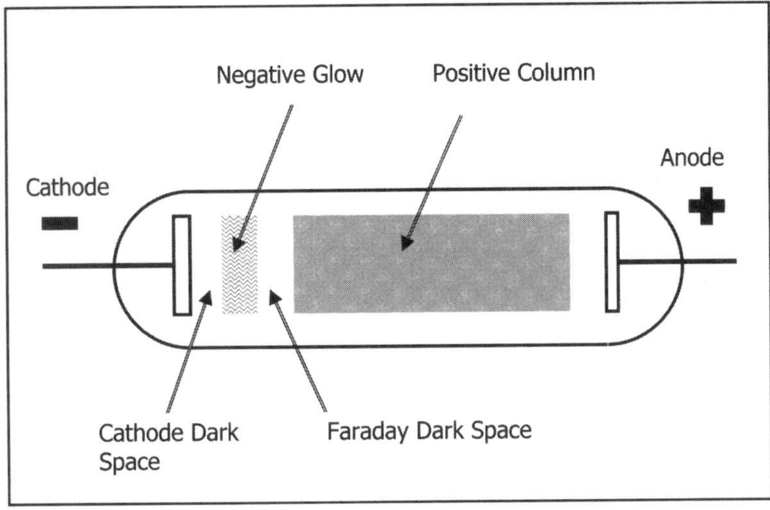

Figure B3.6.1. The glow discharge.

Electrons gain energy continuously from the electric field in the positive column since they make only elastic collisions with gas atoms until they acquire sufficient energy to excite or ionize. The electron population tends to thermalize as a group (electrons colliding with other electrons share their energy) and, to an approximation, may be characterized by a Maxwellian energy distribution at some high temperature, T_e. Ions, on the other hand, share their energy on collision with other ions and neutral atoms, the net effect of which is to raise the gas temperature slightly. Except at the tube wall, where a negative sheath is created by fast electrons, charge equality is closely maintained. One way of looking at the positive column, which relates it to the external discharge parameters, is to note that an energy balance exists throughout its length, such that the power input per unit length (product of tube current and voltage gradient) equals the power dissipated per unit length [4].

Electrons at the end of the column are collected at the anode. Depending on whether the electron diffusion current to the anode is larger or smaller than the tube current, the (usually small) anode sheath becomes negative or positive with respect to the adjacent positive column. The dynamic resistance of a glow discharge is negative, which, if discovered empirically, may lead to an entirely different regime, termed the destructive arc discharge.

The source of light in neon signs, and the seat of HeNe laser action, is the positive column. In order to examine some of the properties of a specific positive column in detail, we choose as an example the typical 1–2 mW HeNe laser shown in figure B3.6.2.

Here the cold cathode is an aluminium cylinder, close-fitting within the tube envelope, which surrounds the 1 mm precision bore capillary tube defining the positive column. The mirrors are hard-sealed to metal end caps, one of which serves as an anode. The tube voltage and current are 1450 V and 6 mA, respectively.

The cathode sheath voltage drop is 200 V. The length of the positive column is typically 15 cm, the positive column voltage gradient is 90 V cm^{-1} and the current density in the capillary is 0.7 A cm^{-2}. The gas fill is a 5:1 mixture of helium 3 (see section B3.6.3 below) and neon to a total pressure of 4 torr. With these operating parameters, the electron and the positive ion densities in the positive column plasma are in the range 10^{11} cm^{-3}, representing fractional ionization of less than one part in 10^6. With some significant exceptions, discussed later, all excited states, including ions, are created by single or multiple step inelastic collisions with hot electrons, characterized by a temperature, $T_e \sim 80\,000$ K (10 eV average electron energy).

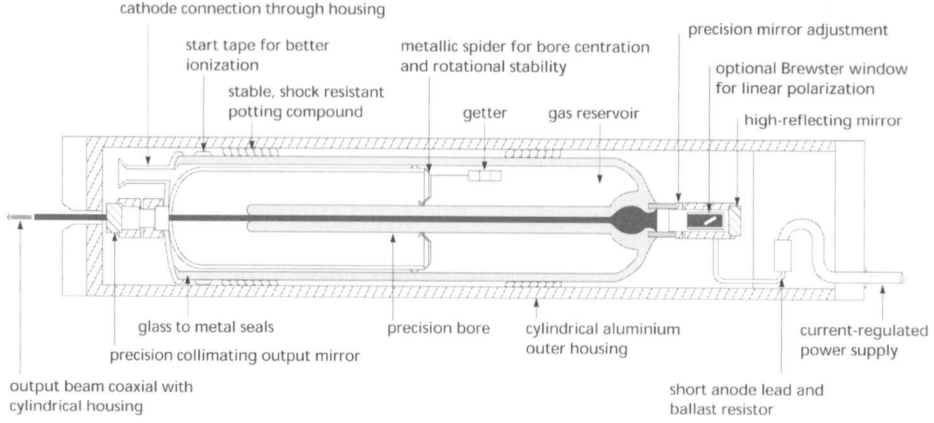

Figure B3.6.2. Construction of a modern commercial HeNe laser (courtesy Melles Griot Inc.).

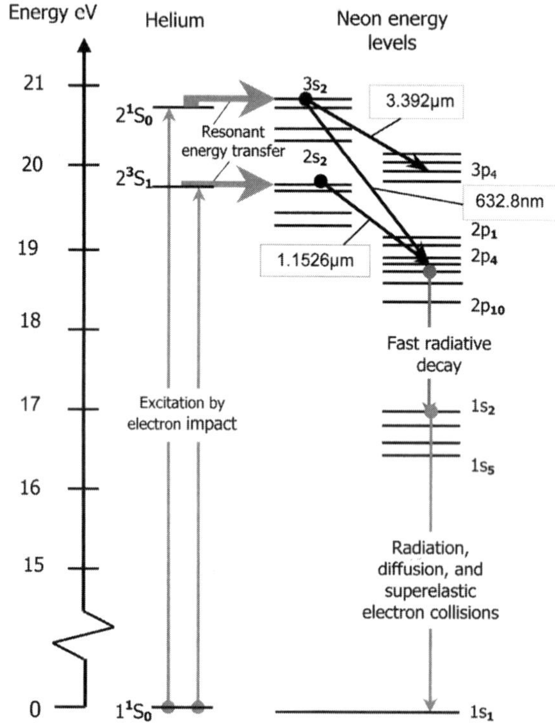

Figure B3.6.3. Energy level diagrams of He and Ne.

B3.6.3 Population inversion, optical gain and oscillation

In the discharge, both helium and neon excited state populations are comparable to electron and ion populations. However, the helium metastable populations are likely to be an order of magnitude higher because they are unable to relax to the ground state by photon emission. Decay from the 1s2s $^3S_1$ and 1s2s $^1S_0$ first excited levels of He to the $1s^2\,{}^1S_0$ ground state by electric dipole transitions is rigorously forbidden by the parity selection rule.

In order to realize population inversion and optical gain, some departure from electron thermal excitation equilibrium must occur. In the HeNe discharge this departure from equilibrium is a result of fortuitously close energy coincidence between the triplet 2^3S and singlet 2^1S metastable levels of helium and the $2s_2$ and $3s_2$ levels of neon, respectively [6].

The two upper laser levels in Ne are $3s_2$ and $2s_2$ and have quite short radiative lifetimes (10–20 ns) since both can decay via fast transitions in the vacuum ultraviolet back to the ground state. However, at any neon pressure above 0.05 torr the probability of resonance radiation photons in these transitions being reabsorbed by ground state atoms before they can escape discharge is so high that the decay paths in the transitions from $3s_2$ and $2s_2$ to the ground state may be considered as fully blocked. Thus, the *effective* lifetimes of $3s_2$ and $2s_2$ are determined entirely by their slower decay in visible transitions ($\tau_{eff} \sim 96$ ns for $2s_2$, $\tau_{eff} \sim 110$ ns for $3s_2$). The two lower levels $3p_4$ and $2p_4$ can decay by reasonably fast UV transitions to levels in the ls_2–ls_5 block with lifetimes of order 20 ns. Thus, the He–Ne system has a $\tau_{upper} : \tau_{lower}$ ratio very much in its favour on all three laser transitions.

The other desirable factor, selective excitation of the *upper* laser levels is ensured by the presence of the helium. By two completely unrelated accidents of nature, the energy of the He 2^3S lies within 0.049 eV of $2s_2$ Ne (a coincidence of 1 part in 500) and that of 2^1S He coincides with $3s_2$ Ne to within 0.039 eV (better than 1 part in 400). When a discharge is run in pure He at low pressure (1 torr, say) electron collisions with ground-state He will excite some atoms directly (and by cascade from higher levels) to both metastable levels from which they can only escape by collision with the wall or by making an ionizing collision with an electron.

A reservoir of excited He atoms in these two long-lived levels builds up to steady values of the order of 10^{11}–10^{12} atoms cm^{-3}. If now a small quantity of Ne is added to the discharge (not enough to change its character as a discharge governed by the properties of He) selective transfer of energy to the Ne $2s_2$ and $3s_2$ levels can occur in the reactions

$$\mathrm{He}(2\,{}^3S_1) + \mathrm{Ne}(1s_1) \rightarrow \mathrm{He}({}^1S_0) + \mathrm{Ne}(2s_2) + 0.04\ \mathrm{eV} \tag{B3.6.1}$$

and

$$\mathrm{He}(2\,{}^1S_0) + \mathrm{Ne}(1s_1) \rightarrow \mathrm{He}({}^1S_0) + \mathrm{Ne}(3s_2) - 0.05\ \mathrm{eV} \tag{B3.6.2}$$

Note that because atom–atom interactions occur via short-range forces, transfer of excitation will be appreciable only for those states of Ne close enough to exact resonance that thermal energy ($kT \sim 0.03$ eV at 350 K) can make up the difference. Because of the accidental energy resonances already mentioned, the transfer of excitation is selective in exciting predominantly the $3s_2$ and $2s_2$ levels of neon and creating inversions of population with respect to the $3p_4$ and $2p_4$ lower levels.

The reactions (B3.6.1) and (B3.6.2) can also go backwards, but since neon atoms in $3s_2$ and $2s_2$ either radiate spontaneously or are stimulated out of these levels fairly promptly, the backward reaction rate is not serious.

Excitation through energy transfer from the helium 2^1S metastable is the dominant process populating the $3s_2$ level. There is experimental evidence that it is 70–90 times the excitation rate to the same level by direct electron impact. The three other levels in the 3s group (despite similarly close energy coincidence) are excited to a much smaller extent by energy transfer, a circumstance which is attributed to the fact that those interactions do not conserve spin. The transfer process reaches a broad maximum with the helium to neon ratio in the range 7:1. By employing the lighter helium isotope, ^{3}He, a 10% to 20% increase in excitation transfer is realized. Several factors contribute to this increase, but the most important appear to be increased frequency of energy transfer collisions and increased electron temperature, T_e.

With such strong pumping to the upper $3s_2$ level, population inversion and optical gain is obtained for all nine permitted visible transitions to the 2p group, as well as all the permitted transitions to the 3p group. Small signal gain for the $3s_2$–$2p_4$ red line in a typical 1 mm bore capillary at optimum conditions is 0.4–0.6% per cm. For all other visible lines, the gain is smaller. The highest gain line is the $3s_2$–$3p_4$ transition at 3.39 μm, where a gain of 50 dB m^{-1} has been measured.

Excitation energy transfer from the triplet 2^3S helium metastable level to the $2s_2$ neon level, though not as dramatic as the 2^1S to $3s_2$ transfer, is 10–20 times the electron excitation rate. However, this transfer is distributed more uniformly over the 2s group, so that inversion also occurs between other 2s levels and various 2p levels. The strongest oscillation occurs on the $2s_2$–$2p_4$ transition at 1.1529 μm.

Given a suitable optical cavity and gain medium, optical transitions with sufficient gain to overcome cavity loss will oscillate. When several transitions having a common upper level show gain, the line with the largest excess gain will dominate and, through oscillation, deplete the upper level to the point where competing lines are no longer inverted. This situation exists with the $3s_2$–2p group of lines, where the 632.8 nm line dominates. By introducing a wavelength selective device, such as a prism, into the cavity to suppress all wavelengths but one, other visible lines in the $3s_2$–2p group can be made to oscillate. As mentioned before, the $3s_2$–$3p_4$ line at 3.39 μm has an extremely high gain approaching 50 dB m^{-1}. This gain is high enough to achieve oscillation with uncoated mirrors. To make the stronger point, it is almost impossible to suppress oscillation at 3.39 μm without the use of special mirrors having very low reflectivity at this wavelength.

Oscillation will build up in the cavity until the net gain (gain minus loss and transmission) approaches zero at the frequency of oscillation. The steady-state output power depends in a complex manner on cavity properties, atomic lifetimes and the rate at which the upper level is pumped. Unlike homogeneously broadened lasers, which 'burn' the gain profile uniformly, the HeNe laser is Doppler-broadened and each axial mode burns two 'holes' in the gain profile (corresponding to atoms moving in opposite directions whose frequency is Doppler-shifted to match the frequency of the cavity resonance–see chapters A1, C3.3, D2.1, D6.2). In the case of a single axial mode at atomic line centre, the two holes coincide, resulting in a power dip (Lamb dip). The hole width depends on the intensity of the field within the cavity; higher fields extract power from a wider spectrum of atoms. For this reason, reducing the mirror loss so as to increase the field intensity is crucial for maximizing the output power of single axial mode lasers.

There are several similarity relations [5] pertinent to the HeNe laser discharges which have some theoretical justification and which have, to a considerable degree, received empirical substantiation by many years of engineering practice. For optimum power output, changes in capillary diameter requires an inverse change in filling pressure such that the product $p \cdot D$ of pressure (p in torr) and diameter (D in mm) remains about 3.6 torr mm. Optimum discharge current is nearly proportional to tube diameter throughout the range 1.5–8 mm when $p \cdot D$ is held constant. Optical gain is found to be inversely proportional to tube diameter and proportional to discharge length.

B3.6.4 HeNe laser tube construction

Until the late 1960s, the capillary bore and glass bottle were constructed from 'soft' glasses. Unfortunately, these were permeable to helium, limiting life to a few hundred operating hours. Modern HeNe lasers are made from 'hard' borosilicate glass, which is impermeable to helium.

Many cathode designs have been tried, but virtually all HeNe lasers now use the interior surface of a cylindrical aluminium cold cathode that has been oxidized by running an oxygen discharge during the initial processing of the plasma tube. The overall life of the plasma tube is determined by the quality of the cathode processing and the configuration of the plasma tube. To maximize life, the discharge should utilize the maximum possible cathode area. A tightly confined discharge, or a discharge that reaches a sharp edge of the cathode, will significantly degrade operating life. To keep the discharge away from edges, the cathodes are usually quite long. Likewise, the diameter of the cylinder should be as large as practicable to allow the discharge to expand as it leaves the bore. Since the only task of the anode is to collect electrons, the drop in potential near the anode is much smaller that that near the cathode, and the impact of electrons is insufficient to cause sputtering. Thus, unlike the cathode whose concave geometry acts to confine sputtered material, the configuration of the anode is usually not critical.

As was true with most gas lasers, the resonator (mirror) structure that enables lasing action was mounted external to the plasma tube on early HeNe lasers, with light entering and exiting the plasma tube through Brewster windows. However, with the development of damage-resistant laser coatings, it quickly became apparent that the mirrors could be mounted directly to the ends of the plasma tube as part of the gas-confinement structure. This had two major advantages: the cost of the external resonator support structure was eliminated, and the problem of mirror and Brewster-window contamination was eliminated. If linearly polarized output was required, it could be obtained by inserting a piece of glass, in the beam path, at Brewster's angle, inside the tube itself. Today, most HeNe lasers incorporate internal mirrors. External-mirror lasers with Brewster windows are available, however, that provide access to up to 100 W of recirculating laser power within the laser cavity. Such lasers are used extensively in high-speed small-particle counting applications because of their predictable energy profiles and excellent reliability.

HeNe lasers are operated by a current-controlled dc supply. The operating characteristics vary with the size and configuration of the plasma tube, but typical configurations would have an output current ranging from 4–8 mA and an output voltage of 1000–3000 V. The tube is ignited by a voltage multiplier (Crockcroft–Walton) circuit which quickly rises to 8 kV or more to break down the discharge, then immediately drops to operating tube voltage.

The illustration of figure B3.6.2 shows a cross section of a commercial HeNe laser head. The entire plasma tube assembly is mounted in an aluminium cylinder using a shock-resistant potting compound. Inside the tube, a metallic support (spider) is attached to the cathode to support the narrow bore discharge capillary tube, preventing droop and reducing the effects of shock and vibration. A contaminant-absorbing flashless getter is installed in the tube to maximize shelf and operating life.

The glow discharge in a HeNe laser operates at negative impedance, meaning that, as the tube current increases, the voltage across the tube decreases. Since the power supplies require positive impedance across their output for proper current regulation, a ballast resistor is added to the package to compensate for the effects of the laser tube.

B3.6.5 The evolution of expectations

One of the reasons for the continued success of the HeNe laser has been its ability to evolve over time to meet the ever-increasing demands of sophisticated customers. In the early years of the laser (the first commercial HeNe lasers became available in 1963), obtaining red light was sufficient. The main concern was lifetime, because early HeNes lasted only a few hundred hours at most as the helium leaked out through the glass envelopes. Once this problem was solved (circa 1965), customers began to focus on the output parameters—mode, divergence, power fluctuations and optical noise. This meant that manufacturers had to increase attention to bore straightness and diameter; precisely define and control mirror curvature, mirror transmission properties and mirror spacing; as well as improve structural designs to provide uniform heat loading (to minimize power drift) and improve the control electronics in the power supplies (to decrease noise).

In the mid-1970s, the intensity and location of satellite beams, the extraneous beams caused by reflections between the front and back surfaces of the output coupler, became a major issue for many customers. Although the total power in all of the satellite beams was rarely more than a few per cent of the power in the main beam, the increased sensitivity of detection electronics, coupled with cosmetic and quality perceptions of customers, demanded that these extraneous effects should be eliminated. This required laser manufacturers to work with optics manufacturers to improve dramatically the quality of the antireflection coatings used on output couplers and put stringent requirements on the centration and wedge angle of the optics themselves.

In the early 1980s, HeNe manufacturers had to meet their greatest challenge—the requirements of the barcode scanning industry. Although flatbed scanner installations in supermarkets were becoming commonplace, the major growth opportunity was in handheld barcode scanners for general-purpose merchandising, an extremely cost-sensitive application. Because traditional HeNes were too large and too expensive, the manufacturers literally went back to the drawing board to develop new designs amenable to semiautomated processing techniques. They were successful. The volume of lasers sold annually went from approximately 150 000 per year to well over 750 000, and average unit prices dropped from approximately $500 to $150. In high volume, some HeNe plasma tubes were selling for less than $50.

By 1990 the barcode tide was receding as HeNe lasers were replaced with red diode lasers operating at 650 nm. However, the requirements for shorter-wavelength HeNes were growing strong, particularly for the 543 nm (green) wavelength, which was especially suited to a wide variety of biotechnology applications. Because of the low gain of these lasers, processing requirements were much more stringent than for the red HeNes, and severe demands were made on optics vendors to develop extremely low-loss, scatter-free, damage-resistant coatings at these wavelengths. Once again, the industry came through and, today, 30 per cent of unit HeNe sales are for non-red lasers.

B3.6.6 HeNe laser manufacturing processes and technology

The basic techniques used to manufacture HeNe lasers have not changed significantly over the last 40 years, although processes and testing procedures have been refined extensively in order to meet the ever-more-stringent customer demands. The bulk of the manufacturing process is ‘hands-on’. Only tube processing and some testing procedures are semi-automated.

The manufacturing procedure begins with the cleaning and degreasing of all the plasma tube components. This step is particularly important because even the smallest particulate, lodged in the laser bore, can cause severe power or beam-quality degradation, particularly in the lower-gain lasers.

Once cleaned, the plasma tube (sans mirrors) is assembled. Historically, some manufacturers have used frits (mixtures of powdered glasses and ceramic that melt to form a vacuum-tight seal) to attach the metallic components to the glass envelopes, but these have proved to be susceptible to vibration. Today, most HeNe laser plasma tubes are assembled with traditional glass-to-metal seals.

In the next step, mirrors are applied to the metallic mirror mounts on each end of the tube with a frit. The seal is made by placing the components in an oven with a highly controlled multistage heating cycle designed for the specific frit. At this point, the tube is ready to be placed on the processing station.

Processing of the plasma tube occurs in three stages. In the first stage, the tube is baked, under vacuum, to remove any impurities that could cause outgassing. In the second stage, the tube is filled with oxygen, and a discharge is struck. This develops a stable aluminium oxide layer on the cathode which will act as an electron emitter in the finished tube. It also oxidizes any residual contaminants in the tube. Finally, the tube is evacuated and filled with the helium and neon lasing mixture, and a discharge is ignited.

Depending on the tube configuration and the process requirements, these steps may be repeated several times. Overall processing time can vary from three to eight hours or more. Once processing is completed, the getter is activated and the tube is pinched off from the processing station. At this point, the mirrors are aligned so that the plasma tube can produce laser light. A technician deforms the metallic end pieces with a tool to move the mirrors into the proper position. Because the laser cavity is designed with relatively short-radius mirrors, this can be accomplished easily. A skilled technician can fully align a plasma tube in one or two minutes.

Figure B3.6.4. Testing HeNe lasers.

Finally, the aligned plasma tube is 'burned in' for 24–48 hr to allow the tubes to stabilize and to isolate any premature failures. At the end of this period, the tubes are tested (see figure B3.6.4) to determine their operating parameters and then sorted for output power.

B3.6.7 'Single-frequency' HeNe lasers

Some of the earliest uses of HeNe lasers were in holography (see section D9), laser ranging and metrology (see sections D2.1) applications requiring very narrow linewidths ($\Delta\nu$) and long coherence lengths ($c/\Delta\nu$). Consequently, various schemes were devised to narrow the frequency spread of these lasers. The red HeNe laser discharge has a gain (amplification) curve that is approximately 1.4 GHz wide (full width at half maximum) centred at 632.8 nm. If the laser is operating in the fundamental Gaussian mode, the laser resonator (mirror structure) can only support output at specific, equally spaced, frequencies (longitudinal modes) which are determined uniquely by the mirror spacing (see chapter A1, section A2). The actual output of the laser is a comb of frequencies that coincide with the intersection of the resonator modes and the laser gain curve. The greater the separation of the resonator mirrors, the more frequencies will be present in the laser output. Conversely, the shorter the distance between the mirrors, the fewer frequencies will be present. It is possible to make the laser short enough that only a single frequency will be present, with an instantaneous linewidth of a few kilohertz.

In a gas laser, the laser gain is fixed by the parameters of the discharge, but the frequency of a longitudinal mode is determined by the effective mirror spacing, which can vary significantly as a function of alignment, ambient temperature, vibration, barometric pressure and many other environmental factors. Since it takes only 0.6 μm of change in spacing to sweep a single longitudinal mode completely through the HeNe gain curve, it is imperative to control the laser cavity length precisely in a stabilized single-frequency laser. The precision of control will determine the effective linewidth.

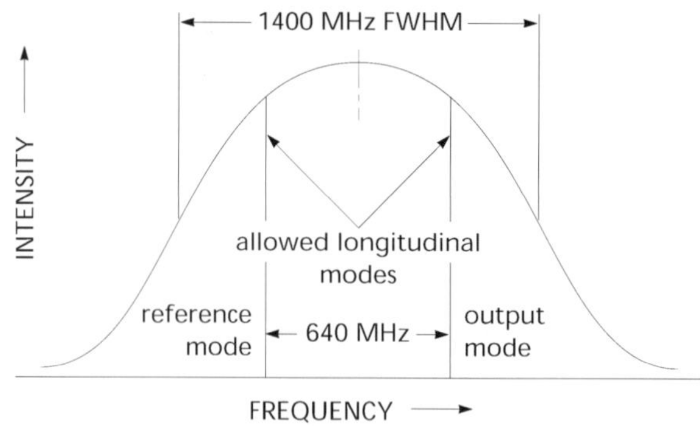

Figure B3.6.5. HeNe gain curve showing two orthogonally polarized modes.

Several techniques have been developed to determine the frequency of a laser and to provide feedback information to adjust the cavity length. Early methods with true single-frequency lasers focused on stabilizing the cavity frequency around the Lamb dip found in the gain curve which occurs when single-isotope gases are used (see chapters A1, C3.3, D2.1, D6.2). However, the very short lasers required for true single-mode operation limited output power to a small fraction of a milliwatt.

An approach embodied in commercially available stabilized single-frequency lasers takes advantage of the fact that adjacent longitudinal modes are cross polarized. The overall laser length is increased to allow two adjacent modes to lase, as shown in figure B3.6.5. As the laser cavity length changes, the modes move through the gain curve, changing in both frequency and amplitude. As one mode increases in power, the other decreases and *vice versa*. To stabilize the system, the two modes are separated into two beams by polarization components, and their amplitudes are compared electronically. The cavity length is then adjusted to maintain the proper relationship between the modes. Only one beam is allowed to exit the system. Even though as much as half the power is not utilized, the increase in overall gain obtained from lengthening the laser provides significantly more output in the remaining beam than could be obtained from a laser that only oscillates in a single mode.

A significant advantage of the two-beam system is that, by adjusting the ratio of power in one beam with respect to the other, the output frequency can be swept 300 MHz or more, while maintaining a 1 MHz linewidth. Since the ratio of power in the two beams is completely independent of the output power itself, excellent frequency stability can be maintained over long periods of time, regardless of power fluctuations or long-term power degradation. Since cavity fluctuations are a major contributor to power fluctuations in a HeNe laser, this stabilization technique has the added advantage of reducing overall system noise. Typical frequency and noise characteristics of a stabilized HeNe laser are shown in figure B3.6.6.

The cavity length of a stabilized HeNe laser can be controlled by placing the discharge tube in an oven and relying on the expansion properties of the glass plasma tube to change the cavity length, or by providing a dynamically adjustable mirror on one end of the tube. The advantage of the oven approach is that it uses a standard, off-the-shelf plasma tube, but its overall response is slower, due to the thermal mass of the system.

The adjustable-mirror approach is somewhat more complex and it requires a special plasma tube, but it can reach the frequency lock point in about 10 min compared to approximately 30 min for the temperature-controlled systems. In addition to the aforementioned applications, single frequency HeNe lasers are used for wavelength calibration and for a wide variety of interferometric applications (see also chapters C3.3, D2.1, D6.2).

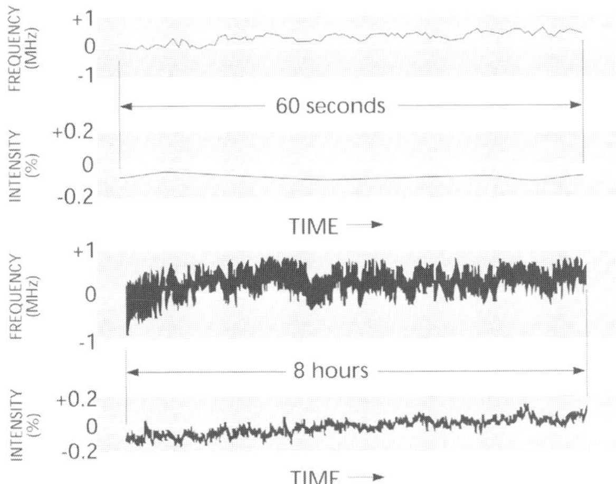

Figure B3.6.6. Short- and long-term frequency and intensity stability of a commercial frequency-stabilized (courtesy Melles Griot Inc.).

B3.6.8 HeNe lasers: typical performance characteristics

As mentioned before, the chief operational characteristics of a HeNe laser are superb beam quality, long life and low power consumption. Depending on the laser, output power ranges from 0.25 mW to more than 40 mW. Other characteristics of these lasers bear mentioning and include:

- Polarization. The basic HeNe laser is randomly polarized, but by inserting a Brewster's angle window into the beam path between the resonator mirrors, the laser can be linearly polarized with a 500:1 polarization ratio. If the laser is to be used with any polarizing elements in the optical path, it is recommended that a linearly polarized laser is used. Randomly polarized lasers can cause unexpected shifts and amplitude fluctuations when used with beamsplitters and other polarizing elements.
- Mode sweeping. As already mentioned, when the length of the laser changes as a result of environmental conditions or changes in temperature, the longitudinal modes sweep across the gain curve causing fluctuations in output power. A very short laser with a single longitudinal mode can fluctuate from maximum power to zero power and back to maximum as the temperature changes—the longer the laser, the smaller the fluctuations. A seven-inch-long laser typically will have mode-sweeping fluctuations of 10%. A fourteen-inch-long laser will normally have a mode-sweeping fluctuation of less than 2%. These fluctuations are particularly noticeable while the laser is warming up. Once equilibrium is reached, the frequency of the fluctuations decreases dramatically. By stabilizing the temperature of the laser, fluctuations can be eliminated completely.
- No automatic power control. HeNe lasers operate as normal glow discharges with negative impedance. Unfortunately, with these lasers, there is not a direct, linear relationship between output power and discharge current. Increasing discharge current may increase output power, decrease output power or have no effect. This makes it very difficult to provide an economically attractive method to adjust or control output power.

B3.6.9 HeNe lasers: the future

For virtually all consumer applications (e.g. DVD and CD players) and many commercial applications (e.g. barcode scanning) the red semiconductor diode laser (see chapter B2.1) with its small size, low cost and low power consumption, is the better choice. On the other hand, for applications requiring exceptional mode quality, narrow linewidth or output at shorter wavelengths (orange, yellow, green), the HeNe laser is the only low-cost choice, and there exist no other technologies that will supplant it in the foreseeable future.

In addition to the demand for lasers for new systems, there is a huge replacement market. Today, there are several million HeNe lasers still in use around the world. In many cases, when these units fail, it is more cost effective to replace the laser plasma tube for a few hundred dollars than to replace the instrument for many thousands of dollars.

References

[1] Javan A, Bennett W R Jr and Herriott D R 1961 *Phys. Rev. Lett.* **6** 106
[2] White A D and Rigden J D 1962 *Proc. IRE* **50** 1697
[3] Willett C S 1974 *Introduction to Gas Lasers: Population Inversion Mechanisms* (Oxford: Pergamon) ch 4
[4] Webb C E 1975 The fundamental discharge physics of atomic gas lasers *High Power Gas Lasers 1975 (IOP Conf. Series 29)* ed R Pike (Bristol: IOPP) ch 1
[5] Francis G 1956 The glow discharge at low pressure *Handbuch der Physik Vol XXII Gas Discharges* ed S Flugge (Berlin: Springer)
[6] Moore C E 1949 *Atomic Energy Levels (NBS Circular 467)* vol 1 (Washington, DC: US Government Printing Office)

Further reading

For introductory level texts on gas lasers see, e.g.

Davis C C 1996 *Lasers and Electro-Optics—Fundamentals and Engineering* (Cambridge: Cambridge University Press) Section 9.7 ISBN 0-521-30831-3

Silfvast W T 1996 *Laser Fundamentals* (Cambridge: Cambridge University Press) Section 13.2 ISBN 0-521-55424-1

For a more advanced text, with an extensive and detailed coverage of HeNe lasers see:

Willett C S 1974 *Introduction to Gas Lasers: Population Inversion Mechanisms* (Oxford: Pergamon) ch 4 ISBN 0-08-017803-0

B3.7
Helium–cadmium laser

William T Silfvast

B3.7.1 General description

The cw He–Cd laser is perhaps the best known and most widely used metal vapour laser. It produces visible laser output in the blue at 441.6 nm [1] and ultraviolet output at 354.0 nm and 325.0 nm [2]. The blue transition produces the most power output [3] while the 354 nm transition is the weakest. A number of other lasers in the green and red portions of the spectrum have been observed in various types of He–Cd discharges. However, these lasers are not available in the commercial He–Cd laser that operates in a positive column type of dc discharge, similar to that of the He-Ne laser. Hence these other visible transitions will not be dealt with further in this discussion.

B3.7.2 Energy levels and laser transitions

Laser action occurs on transitions within the singly ionized (Cd^+) species. To understand the nature of these transitions we must first consider the neutral Cd atom. The ground state has an outer electronic configuration of $4d^{10}5s^2\ {}^1S_0$. The removal of a 5s electron leaves the atom in the ground state of the ion ($4d^{10}5s\ {}^2S_{1/2}$) as indicated in figure B3.7.1. The electronic structure of the upper laser levels is relatively unique in that it involves the removal of an inner shell d electron from the Cd atom leaving the atom ionized in the $4d^9 5s^2$ ${}^2D_{5/2}$ and ${}^2D_{3/2}$ levels as also seen in figure B3.7.1. The laser transition involves decay from the upper laser level to the $4d^{10}5p\ {}^2P^o_{3/2}$ and ${}^2P^o_{1/2}$ excited ion levels which then rapidly decay to the ion ground state as shown in figure B3.7.1. The relevant transition probabilities are shown for the laser transitions as well as for the transitions on which the lower levels decay. It can easily be seen how the population inversion is established due to the much more rapid decay rates of the lower laser levels compared to those of the upper laser levels.

B3.7.3 Pumping or excitation techniques

There are three excitation mechanisms that have been proposed for the excitation of the Cd laser as indicated in figure B3.7.2. The first one involves the transfer of energy *via* Penning ionization from He metastable levels to the Cd upper laser levels [4]. This is similar to the process for the excitation of the He–Ne laser except that, since the Cd upper laser level is an ion level, the exact energy coincidence is not necessary. The electron emitted in the ionization process takes up the energy difference between the He metastable energy and the Cd upper laser level energy (approximately 12 eV). The second process is electron excitation from the Cd ion ground state to the upper laser level [5]. The third process, which can be shown to represent only a few per cent of the total excitation for the conditions of the He–Cd laser discharge, involves photoionization of the Cd atoms to populate the upper laser level *via* emission from the 50–60 nm transitions of the excited

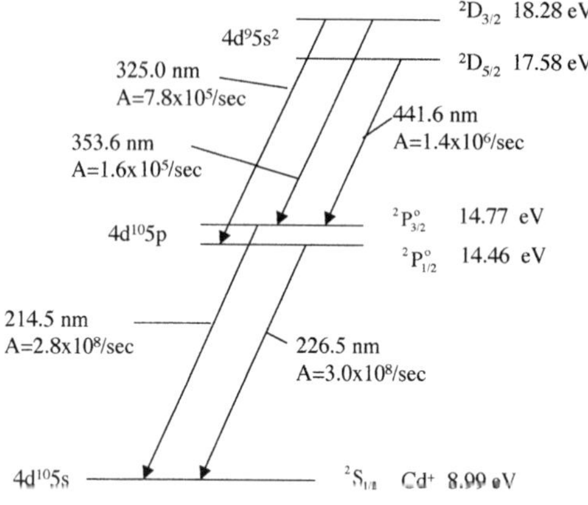

Figure B3.7.1. Energy level diagram for the He–Cd laser.

He atoms radiating within the discharge [6]. A recent paper with additional experimental results supports the dominance of Penning ionization [7].

B3.7.4 Uniform vapour distribution—cataphoresis

Normally a uniform vapour is obtained within a chamber by placing the vapour source (either liquid or solid) within the chamber and heating the entire chamber uniformly. In the case of Cd, the chamber would have to be heated to a temperature of approximately 260 °C. However, in the He–Cd laser, this would be very impractical. The only region that is necessary to be heated is the region of the discharge capillary bore where the excitation and amplification occur. It is much more desirable to have the laser windows, mirrors, electrodes, etc at or near room temperature. This is achieved by using a process known as cataphoresis to control the Cd vapour distribution [8]. In this process the Cd metal is heated and vapourized at the anode (positive potential) end of the discharge and is transported towards the cathode end (negative potential) of the discharge by the electric field acting upon the Cd ions that are produced by the discharge current. The Cd atoms then condense in the cathode region. The laser species is essentially ‘flowing’ down the capillary, albeit at a very slow rate of approximately one to two grams per 5000 hr. The discharge current heats the gain region sufficiently so that Cd does not condense in that region while it is being transported down the capillary.

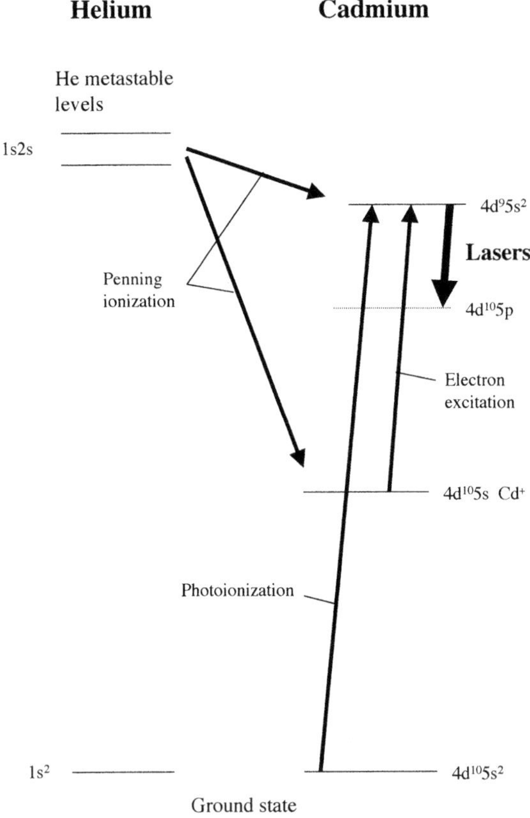

Figure B3.7.2. Excitation process for the He–Cd laser.

The cataphoresis process is sensitive to the fractional degree of ionization of the desired species (in this case Cd) to be transported through the discharge region [8]. A fractional ionization of the order of 1% is required to produce the uniform distribution throughout the entire length of a long discharge capillary such as that used in the He–Cd laser. In the case of Cd, this unusually high fractional ionization is produced primarily due to the process of Penning ionization; previously mentioned as one of the excitation processes. The highest Penning ionization cross section to produce a specific Cd^+ level is to the ion ground state $4d^{10}5s$ $^2S_{1/2}$ [9]. The second highest Penning ionization cross section is to the upper laser level species ($4d^95s^2$) in the $^2D_{5/2}$ and $^2D_{3/2}$ levels. Penning ionization cross sections to all other Cd ion levels are significantly lower since the process would require the removal of one electron and the change of another electron, whereas the two states mentioned require only the removal of one electron with no change of another electron.

Hence, whether Penning ionization from He metastable atoms or electron excitation from the Cd ion ground state is the dominant excitation mechanism to the Cd upper laser levels, the Penning ionization process plays a major role in producing the high degree of ionization of the Cd atom. This allows either of the excitation mechanisms to occur by providing the necessary uniform vapour distribution along the gain region of the capillary bore. It also provides a high-density source of Cd ions in the ion ground state from which electron collisional excitation could occur.

B3.7.5 Role of helium

The importance of the helium metastable atoms involved in the Penning ionization process was described both by producing excitation directly to the Cd upper laser levels, as well as providing the high percentage (1%) of ionization of Cd necessary to obtain a uniform vapour distribution *via* cataphoresis. The He metastable density population is of the order of 10^{12} cm^{-3} within the discharge bore region whereas the population of most excited levels in both He and Cd are of the order of 10^{9}–10^{10} cm^{-3}. Hence, the He metastables are the dominant excited species within the discharge gain region and consequently the rate of energy transfer from them *via* Penning ionization can be quite large. In addition, the presence of He serves to maintain a relatively high electron temperature in the discharge, favourable for all of the mentioned excitation processes.

B3.7.6 Laser gain—isotope distribution

The laser is typically operated with a naturally occurring mixture of Cd isotopes which includes Cd 106, 108, 110, 111, 112, 113, 114 and 116. This natural isotope mixture favours Cd 112 and Cd 114, each contributing approximately 1/4 of the Cd atoms within the discharge. For most gas laser transitions, Doppler broadening dominates over natural broadening, as is the case for the He–Cd laser. However, when using the Cd natural isotope mixture, isotope broadening becomes the dominant broadening process. Each of the even isotopes has a 1.1 GHz emission linewidth (FWHM) on the laser transitions and the series of even isotopes are separated by approximately 1.5 GHz to produce an emission spectrum as indicated in figure B3.7.3. Also included in that spectrum are the distributed spectra of the odd isotopes Cd 111 and Cd 113. The odd isotopes have a nuclear spin and thus each of them divides into three spectral components with a distribution spread out enough that those isotopes do not have a major impact upon the emission spectrum at any specific wavelength as indicated in figure B3.7.3. With the natural isotope mixture, the laser operates over a bandwidth of approximately 3 GHz, the width of Cd 112 and Cd 114 combined (as shown in figure B3.7.3), since the other isotopes have a gain too low to contribute significantly to laser output. However, the laser can also be operated with a single even isotope of Cd (at a significant additional expense) to produce approximately a factor of two more power output and a much narrower emission bandwidth than that of the natural mixture. The higher power is obtained since the gain is higher and the energy is concentrated in fewer longitudinal modes.

B3.7.7 Gain, saturation intensity, and output power

The laser gain coefficient for the 441.6 nm laser transition is of the order of 0.3 m^{-1} for a laser that uses naturally occurring Cd as the Cd vapour source, as indicated in table B3.7.1. Hence, for a discharge tube having a gain length of 30 cm, the single pass gain within the discharge would be approximately 11%/pass. The saturation intensity of this laser is 0.4 W cm^{-2}. Thus, assuming a laser system that has losses of the order of 0.2%/pass, and assuming that a high reflecting mirror is located at the rear of the laser tube, the optimum mirror transmission for the output mirror would be approximately 2.2%. Inserting these values into a well-known expression for laser power output suggests an optimum output power of 24 mW, which is approximately what is experimentally achieved.

B3.7.8 Long-life operation—He cleanup

In the He–Cd laser, helium is gradually lost within the laser system both by leaking through the glass structure of the laser tube and by being buried underneath the Cd as the Cd condenses in the form of a metallic coating at the cathode end of the laser tube [10]. Thus, a He reservoir (ballast) is provided as part of the laser system. At the end of the ballast is a separate reservoir (bottle) that is initially filled with helium at approximately atmospheric pressure and is separated from the ballast region by a thin membrane that allows He gas to leak slowly through from the high-pressure He bottle into the low pressure ballast region. The amount of He

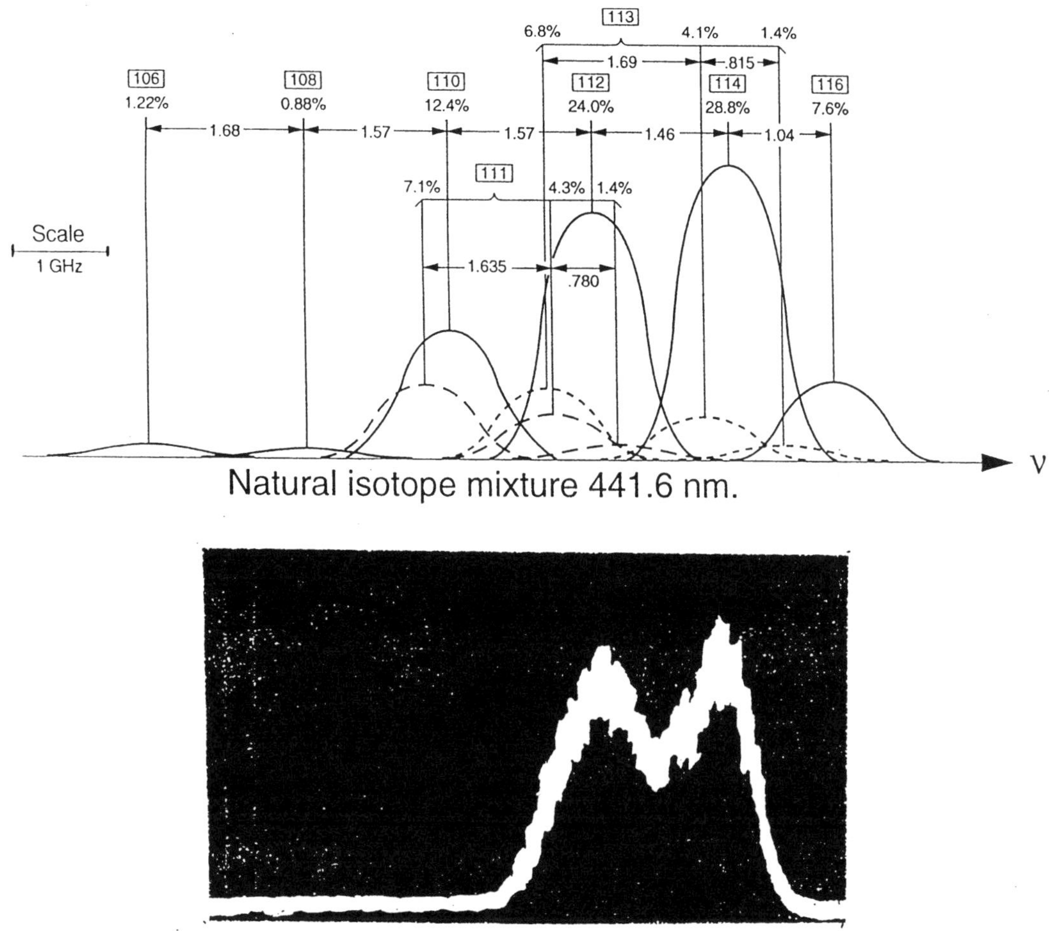

Figure B3.7.3. Natural isotope distribution and typical laser emission spectrum for 441.6 nm Cd laser transition.

leakage through the membrane is temperature sensitive and is controlled with a pressure sensor and an active feedback control system. This system is able to maintain a constant He pressure within the laser tube over the lifetime of the laser.

B3.7.9 Structural arrangement

There are typically two versions of this laser including a long discharge tube with a bore length (gain region) of the order of 60–75 cm, and a shorter version with a bore length of the order of 25–35 cm. The longer version comprises either one long bore or two shorter bores within the same laser cavity. The lasers are installed within air-cooled housings (some use cooling fans) with lateral dimensions of the order of 15 cm and an overall length that is approximately 40 cm longer than the bore length. A small power supply is separate from the tube assembly. The larger lasers produce up to 200 mW multimode at 441.6 nm and 100 mW multimode at 325 nm. A 10 mW single TEM_{00} mode blue laser is also produced from a 25–30 cm bore length. A diagram of the small version of a commercial He–Cd laser is shown in figure B3.7.4. The heated Cd reservoir is located at the anode end of the laser discharge bore region. A metal vapour trap is located

Table B3.7.1. Typical helium–cadmium laser parameters.

Laser wavelengths	441.6 nm, 353.6 nm, 325.0 nm
Laser transition probability	$1.4 \times 10^6\ s^{-1}$, $1.6 \times 10^5\ s^{-1}$, $7.8 \times 10^5\ s^{-1}$
Upper laser level lifetime	7.1×10^{-7} s (1.1×10^{-6} s)
Stimulated emission cross section	$8 \times 10^{-18}\ m^2$
Emission linewidth and laser gain bandwidth for each isotope	$1.5 \times 10^9\ s^{-1}$, $1.4 \times 10^9\ s^{-1}$, $1.5 \times 10^9\ s^{-1}$
Inversion density	$4 \times 10^{16}\ m^{-3}$
Small signal gain coefficient	$0.3\ m^{-1}$
Laser gain medium length (L)	0.25–1.5 m
Single pass gain	0.07–0.45
Gas pressure	5–10 Torr He
Gas mixture	100:1, He:Cd
Operating temperature	Tube bore 350 °C, Cd temp. 260 °C
Pumping method	Electrical discharge
Electron temperature	5–7 eV
Gas temperature	300 °C
Output power	10–200 mW
Mode	TEM_{00} or multimode

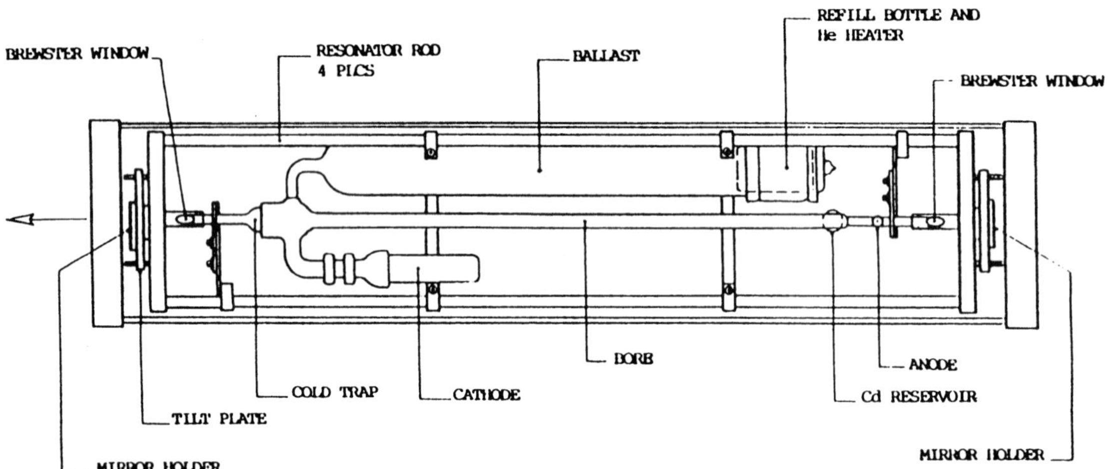

Figure B3.7.4. Structural arrangement of the He–Cd laser (courtesy of LiCONix, Inc.).

beyond the cathode to prevent metal vapour from condensing on the laser window or mirror. The ballast is a region within the tube that provides a large reservoir of He gas to help maintain the proper He pressure. Beyond the reservoir is a bottle containing He at high pressure. The He in that bottle is diffused into the reservoir region as needed to maintain the proper He pressure. The rigid cavity structure is used to maintain the mirror alignment with respect to the fixed location of the laser bore.

B3.7.10 Discharge tube design

The lasers operate within a 1–2 mm bore glass discharge tube. The tube contains He gas at a pressure of several torr and a partial pressure of Cd that is approximately 1% of that of He. A dc discharge is operated within the tube at currents in the range of 60–100 mA. The lasers typically operate for a lifetime of the order of 4000–5000 hr with a loading of one or two grammes of Cd metal in the anode reservoir. The general properties of the laser are summarized in table B3.7.1.

B3.7.11 Applications

Applications of these lasers include stereolithography, printing, making CD masters, microchip inspection, flow cytometry, lithography and fluorescence analysis. Stereolithography is a process that uses the ultraviolet laser to make computer-generated models in a plastic material. The blue wavelength is used for printing on photosensitive materials in which the short wavelength blue photons react more readily with many photosensitive materials. In flow cytometry, the He–Cd laser is most favourable in some situations where the short wavelength is desired from either the standpoint of scattering or absorption of the light by the material. Also, for fluorescence analysis, the short wavelengths of the He–Cd laser are often more useful than for the other visible lasers. The lasers are also used extensively for inspection of electronic circuit boards.

References

[1] Silfvast W T, Fowles G R and Hopkins B D 1966 Laser action in singly ionized Ge, Sn, Pb, In, Cd, and Zn *Appl. Phys. Lett.* **8** 318–19

[2] Silfvast W T 1969 New cw metal–vapor laser transitions in Cd, Sn, and Zn *Appl. Phys. Lett.* **15** 23–5
Goldsborough J P 1969 A continuous laser oscillation at 3250 A in the Cd ion *IEEE J. Quantum Electron.* **QE-5** 133

[3] Silfvast W T 1968 Efficient cw laser oscillation at 4416 A in cadmium *Appl. Phys. Lett.* **13** 169–71

[4] Silfvast W T 1971 Penning ionization in a He–Cd dc discharge *Phys. Rev. Lett.* **27** 1489–92

[5] Goto T, Shimizu Y, Hattori S and Sakurai T 1983 Quantitative interpretation of radial profiles of Cd(II) lines in the positive column He–Cd laser discharge *J. Phys. D: Appl. Phys.* **16** 261–8

[6] Silfvast W T, Macklin J J and Wood O R II 1983 High gain inner-shell photoionization laser in Cd vapor pumped by soft-x-ray radiation from a laser-produced plasma *Opt. Lett.* **8** 551–3

[7] Ivanov I G, Latush E L and Sem M F 1966 *Metal Vapor Ion Lasers—Kinetic Processes and Gas Discharges* ed C Little (Chichester: Wiley)

[8] Sosnowski T P 1969 Cataphoresis in the helium–cadmium laser discharge tube *J. Appl. Phys.* **40** 5138–44

[9] Cermak V 1971 Penning ionization electron spectroscopy III, ionization of cadmium *Coll. Czechoslovakian Chem. Commun.* **36** 948–50

[10] Sosnowski T P and Klein M B 1971 Helium cleanup in the helium–cadmium laser discharge *IEEE J. Quantum Electron.* **QE-7** 425–6

B3.8
Optically pumped mid IR lasers: NH_3, C_2H_2

Mary S Tobin

B3.8.1 Introduction

The development of mid-infrared (MIR) optically pumped gas lasers (MIROPLs) has been motivated by the concentration of molecular vibrational–rotational spectra in the MIR, with the consequential need for sources for spectroscopy and gas identification. Another source need is for selective excitation of vibrational modes to produce photochemical reactions, e.g. for isotope separation. The objective here is to provide an overview of the MIROPLs NH_3 and C_2H_2 and their isotopic variations, primarily in the 8–20 μm range. We differentiate MIR from far-infrared (FIR) lasers in that the MIROPL transitions occur between different vibrational bands whereas FIROPL transitions generally occur on pure rotational transitions within a given vibrational mode at wavelengths mainly from 50 μm to 1 mm [1]. Much of the MIROPL development occurred in the 1970s and 1980s, and several reviews are available [2–6].

Electrical discharge pumping has produced MIR pulsed laser action in NH_3 [7, 8]. Electrical excitation of H_2 with He or CO, followed by vibrational energy transfer, has produced laser action near 8 μm in C_2H_2 [9–11]. Optical pumping has advantages in that it does not dissociate the molecules, which in many cases allows cw (as opposed to pulsed only) and sealed-off operation. Further, it is energy-level selective and more efficient than electrical pumping. As in the FIR, the CO_2 laser is the major pump choice. A fundamental requirement of optical pumping is that the molecule must have an absorption transition in the spectral region of the pump laser. NH_3 is the most efficient, powerful and important of all the MIROPLs because of a combination of its fundamental properties and the near-coincidence of absorption lines with the CO_2 laser. There has been one report of an HF laser pumping an NH_3 laser [12]. The use of an isotopic species, e.g. $^{15}NH_3$, can sometimes provide a better match with the pump and produce many new laser lines. A complete list of close matches of CO_2 laser lines with ammonia lines is given by Frank *et al* [13]. Whereas the CO_2 laser cannot directly pump C_2H_2, it can pump C_2D_2, with a performance that rivals NH_3 for some lines. In this chapter we discuss these lasers including their simplest spectroscopy, the principles of operation, laser construction and current topics of interest.

B3.8.2 Basic spectroscopy of NH_3 and C_2H_2

The absorption and emission spectra of molecules in the MIR are related to vibrational and rotational motions of the electronic ground state. For a fundamental mode of vibration at frequency ν, the vibrational energy is $E_v^{\mathrm{vib}} = h\nu(v + 1/2)$ + anharmonic terms, where v is the vibrational quantum number. Acetylene is a linear molecule belonging to point group $D_{\infty h}$, which means it has a plane of symmetry perpendicular to the internuclear axis. Figure B3.8.1 indicates the five normal vibrational modes and fundamental frequencies for C_2H_2 and C_2D_2; ν_1, ν_2, and ν_3 are non-degenerate stretching modes and ν_4 and ν_5 are degenerate bending modes. Vibrational states of C_2H_2 can be denoted by the quantum numbers $(v_1, v_2, v_3, v_4^{l4}, v_5^{l5})$. Bending

Mode	Frequency (cm^{-1})	Description	Symmetry	Normal Mode H C C H
ν_1	3374 (2700)	Sym. C-H stretch	Σ_g^+	
ν_2	1974 (1762)	C-C stretch	Σ_g^+	
ν_3	3287 (2427)	Asym. C-H stretch	Σ_u^+	
ν_4	612 (505)	Sym. bend	Π_g	
ν_5	729 (539)	Asym. bend	Π_u	

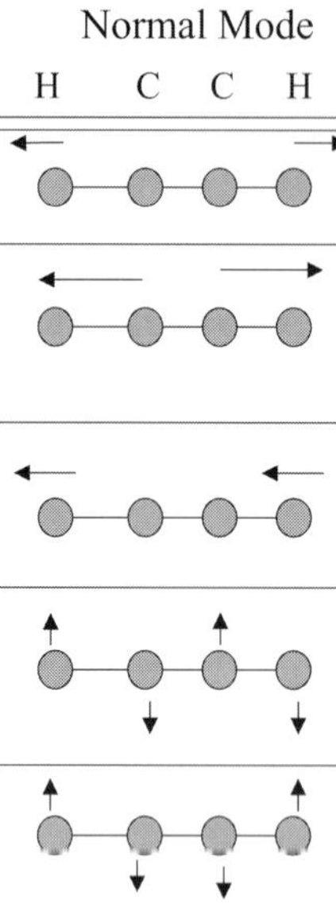

Figure B3.8.1. The fundamental modes and frequencies of C_2H_2 (frequencies of C_2D_2 in parentheses).

modes have a vibrational angular momentum about the internuclear axis characterized by quantum number l_i, which are combined to give a total vibrational angular quantum number l. The notation Σ, Π, Δ refers to $l = 0$, 1, 2, respectively. The superscripts (+ and −) indicate the parity of the wavefunctions when the vibrational normal coordinates are reflected in a plane containing the molecular axis. The subscripts, g and u, indicate that the vibrational wave function is +1 or −1, i.e. even (gerade) or odd (ungerade) upon inversion of the normal coordinates with respect to the molecule-fixed axis. The only infrared-active fundamental modes are ν_3 and ν_5; neither of these modes for C_2H_2 (or C_2D_2) overlaps the CO_2 laser emission.

Each vibrational level consists of many rotational levels. In the rigid-rotor harmonic oscillator approximation, the energy of a molecule can be considered as a sum of vibrational and rotational energy. For a linear molecule $E_v^{\mathrm{rot}}(J) = B_v[J(J+1) - l^2] - [D_v J^2(J+1)^2]$ where l is included for degenerate levels, B_v and D_v are vibrational–rotational parameters, and J is the angular momentum quantum number. Selection rules for transitions depend on whether the oscillating dipole moment is parallel ($\Delta l = 0$, $\Delta v = \pm 1$, $\Delta J = \pm 1$) or perpendicular ($\Delta l = \pm 1$, $\Delta v = \pm 1$, $\Delta J = 0, \pm 1$) to the molecular axis.

For a symmetric top molecule such as NH_3, the rotational energy for a non-degenerate vibrational mode including centrifugal and Coriolis effects in the non-rigid molecule is

$$E_v^{\mathrm{rot}}(J) = B_v[J(J+1) - l^2] - [D_v J^2(J+1)^2] \qquad \text{(B3.8.1)}$$

where B_v, C_v and the Ds are vibrational-rotational parameters, K is the projection of the rotational angular momentum number J on the molecular symmetry axis ($-J \leqslant K \leqslant J$). Radiative transitions follow dipole-like selection rules, $\Delta J = \pm 1,0$; $\Delta K = 0$. The bands corresponding to $\Delta J = -1$, 0 and +1 are referred to as the P, Q and R branches, respectively, where the sign of the change in J is taken from lower level to upper level, for either absorption or emission. Therefore, the R, Q and P branches extend from higher to lower energy, respectively. Degenerate vibrations are more complicated to treat and further details can be found in [14, 15]. As in the case of FIROPLs, the gain for MIROPLs is preferentially parallel or perpendicular to the pump polarization depending on the type of pump and MIR transition [16].

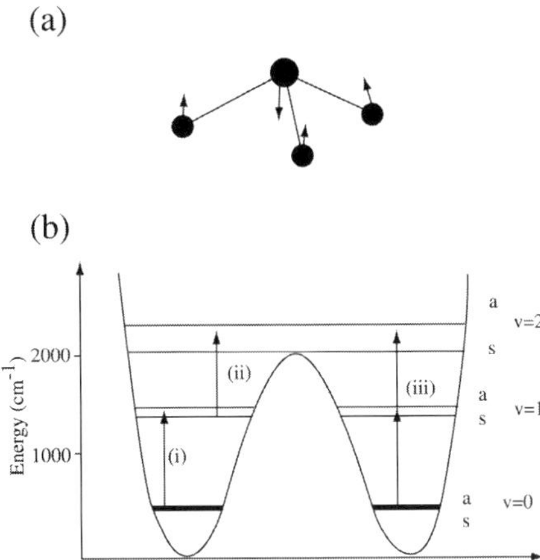

Figure B3.8.2. Schematic diagrams of (*a*) the ν_2 mode of NH_3 and (*b*) the potential function of the ν_2 mode indicating a double minimum and showing the possible pumping processes in NH_3: {i}, fundamental $0 \rightarrow 1$ {ii}, hot band $1 \rightarrow 2$ {iii}, two-photon (or two-step) pumping $0 \rightarrow 2$.

NH_3 is a pyramidal molecule with threefold rotational symmetry (point group C_{3v}). It has four fundamental vibrational modes: $\nu_1 = 3337$ cm^{-1} and $\nu_2 = 950$ cm^{-1} are totally symmetric (species A_1), and $\nu_3 = 3414$ cm^{-1} and $\nu_4 = 1628$ cm^{-1} are degenerate (species E) [14]. The only fundamental mode that the CO_2 laser can pump is ν_2 (figure B3.8.2). The lines are doubled by the inversion of the molecule, which corresponds to the two states that the N atom can take on opposite sides of the plane formed by the H atoms, and 950 cm^{-1} is the average of the two doublets due to passage through the potential barrier. The inversion removes the degeneracy in K, and separates the levels into symmetric (s) for the lower level and antisymmetric (a) for the upper level. Vibrational transitions can only occur between levels of opposite parity i.e. $a \leftrightarrow s$. Another concept important in understanding the NH_3 laser is the concept of ortho- and para-NH_3. Rotational levels with $K = 3n$, so called ortho-NH_3, species A, have twice the statistical weight of the levels with $K = 3n \pm 1$, para-NH_3, species E; ortho-NH_3 cannot convert to para-NH_3 or *vice versa*, by collisions.

The strength of a vibrational transition is proportional to the square of the vibrational transition dipole moment. The ν_2 transition dipole moments for NH_3 are about $0.24D$ for the fundamental and $0.28D$ and $0.42D$ for the overtone $2\nu_2$s and the $2\nu_2$a band, respectively [17]. These are relatively large and contribute to the success of the NH_3 laser.

B3.8.3 Principle of operation

Figure B3.8.3(*a*) indicates the basic principle of operation of the MIROPL. The pump laser radiation is absorbed on a vibrational–rotational transition $1 \rightarrow 2$, followed by laser action on vibrational–rotational transition $2 \rightarrow 3$ at a longer wavelength. Several types of transitions can occur. For the strongest NH_3 lines, 1 and 3 represent rotational levels of the ground vibrational state and 2 is of the fundamental ν_2 vibrational band. Pumping to an overtone band followed by lasing on a hot band can also occur, e.g. two-step pumping to the $2\nu_2$ band of NH_3. Pumping is also possible to a combination band (as for C_2D_2) with lasing on a hot band. The merits of various pumping schemes, based on the relative level populations and transition strengths, are compared in Harrison and Gupta [4]. The FIROPL shares the same excitation mechanism but lasing occurs on a pure rotational transition. There are fewer MIROPL molecules and laser lines compared to FIROPLs. One reason is that the MIR transition is often back to the same (ground) vibrational state and

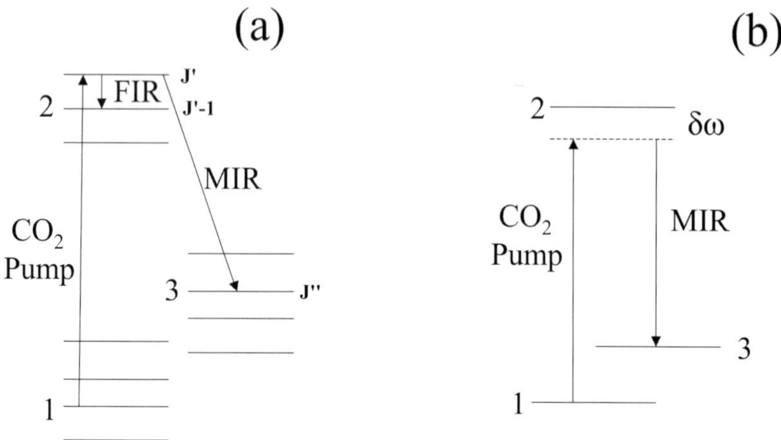

Figure B3.8.3. Energy-level schematic diagram to show (*a*) the basic principle of operation for MIROPLs (and FIROPLs) and (*b*) the Raman process, where the pump and emission frequencies are off-resonance from the molecular transitions.

the relatively small population difference between the initial and final states can result in significant MIR self-absorption [18]. This absorption can be reduced by pumping far off line centre (Raman lines) as was done for the first cw NH_3 laser [19] but this requires more pump power.

The dynamics of NH_3 lasers show the presence of two gain mechanisms, inversion and Raman [19–22]. For inversion gain, the pump is absorbed by a specific ammonia transition and the excitation may be transferred to other states via collisions: buffer gases such as N_2 and Ar can aid this process. The frequency of the pump does not determine the frequency of MIR emission. In contrast, Raman gain is a coherent process simultaneously involving two radiation fields and three molecular levels (figure B3.8.3(*b*)). A pump photon at frequency $\omega_{21} - \delta\omega$ is converted to an emission photon with a corresponding offset at frequency $\omega_{23} - \delta\omega$, and the molecule is transferred directly from level 1 to level 3 where $\omega = 2\pi\nu$. In general, the gain can be expressed as a sum of population inversion and Raman gain:

$$g(\omega) = \sigma_{01}[S_I(n_2 - n_1) + S_R(n_1 - n_3)] \qquad \text{(B3.8.2)}$$

where σ_{12} is the absorption cross section at frequency ω_{21}, S_I and S_R are lineshape factors for the inversion and Raman gain, respectively, and n_1, n_2 and n_3 represent populations of levels 1, 2 and 3 respectively. Raman gain only appears on laser transitions having one level in common with the pumped transition. Gain models are discussed further in section B3.8.6.

B3.8.4 Experimental aspects

The major pump source for MIR- and FIROPLs is the CO_2 laser (see chapter B3.1) because of its large number of lines (approximately 80—see figure B3.8.4) in the 9 and 10 μm molecular 'finger print' region. Isotopic variations and sequence band CO_2 lasers offer additional pump frequencies. The cw CO_2 laser is relatively inexpensive, easy to operate, has tunability of about 70 MHz, produces up to 50 W per metre length and is about 20% efficient. The laser tube is water cooled and a flowing gas mixture of about 10–30 torr of a mixture of CO_2, N_2 and He is used. The laser can be sealed off, often with the addition of xenon. Waveguide cw CO_2 lasers (see section B3.1.3.4) exhibit higher operating pressure and greater tunability and have produced new FIROPL lines [23]. A cw CO_2 waveguide laser at 52 MHz above the 10R(24) line was

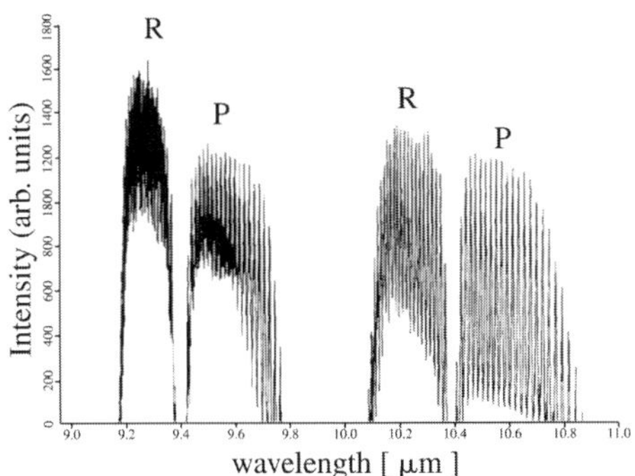

Figure B3.8.4. CO_2 laser tuning profile showing the R- and P-branches of the 00^01–02^00 and 00^01–10^00 bands in the 9 and 10 μm regions, respectively [112]. (Reproduced by permission of Elsevier.)

used to pump an NH_3 waveguide MIROPL on the aP(4,0) transition but severe feedback problems, in the absence of an optical isolator, discouraged this approach [25].

Pulsed MIROPLs mainly use CO_2 pumps with transverse excitation at atmospheric pressure (TEA). The short transverse discharge paths of TEA lasers allow excitation of large volumes of gas at high pressures at reasonable voltages (20–50 kV). Pulse energies in the fraction to multi-joule range are obtained depending on gas and excitation conditions. Tunability ranges from about 1 GHz for a one-atmosphere to the entire R and P branches for a multi-atmosphere CO_2 TE laser [26]. Frequency-selective elements (e.g. gratings, etalons, apertures) or injection locking can be used to achieve single longitudinal and transverse mode operation. The TEA laser output typically consists of a 100–150 ns pulse followed by a tail, which contains most of the energy, of several microseconds duration. The initial peak can saturate the MIR absorption so that the pulse-tail transmits with little contribution. Biswas *et al* [27] improved MIROPL efficiency 30% by redistribution of the energy within the pump pulse through choice of gas mixtures.

The MIR gas cell can be an oversized waveguide with dielectric or metallic walls with a cooling jacket. The cell is evacuated by a standard mechanical pump with a liquid nitrogen (LN_2) cold trap, then filled and sealed off. A superfluorescent laser consists of the pump laser and a single- or multi-pass gas cell [26, 28]. Otherwise, the pump radiation must be admitted into the MIR cavity while maintaining good MIR resonator efficiency. Line selection, prevention of feedback into the pump laser and separation of the residual pump from the MIR output are all important. Optical alignment can be complicated by the different material optical properties in the MIR *versus* visible, e.g. germanium and silicon are opaque in the visible. Some MIR materials are toxic and some are hygroscopic. ZnSe optical windows and elements are frequently used, although they may be more expensive. These various requirements present challenges, as evidenced by the many papers that have addressed MIROPL designs over the years and continue to do so [29, 30].

The pump beam can be injected into the MIR cavity off-axis through the Brewster angle window [31, 32], or side-window of the gain tube [33] (as shown in figure B3.8.5) and reflected down the MIR resonator by an oversized waveguide [34]. Such a design is relatively simple and inexpensive to make, the resonators can be independently optimized, there is little pump feedback and the pump beam is separated from the MIR output. However, for off-axis pumping, much of the gas can remain unexcited and contribute significant MIR loss, especially for lines terminating in the ground state, e.g. the NH_3 $\nu_2 \rightarrow 0$ lines. Hole coupling through the MIR mirror is inexpensive but far from optimal with regard to power output and mode quality. Improvements are achieved by on-axis pumping and matching the pump with the MIR cavity mode. Dichroic optical elements, which can selectively transmit the pump but strongly reflect the generated MIR radiation, are frequently used as mirrors for collinear pumping (figures B3.8.6 and B3.8.7).

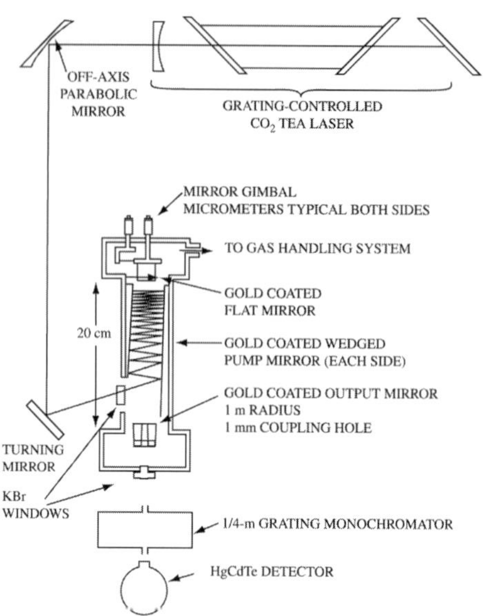

Figure B3.8.5. Experimental set-up for off-axis pumping. Note the MIR elements of KBr windows and HgCdTe fast detector [33]. (Reproduced by permission of Elsevier.)

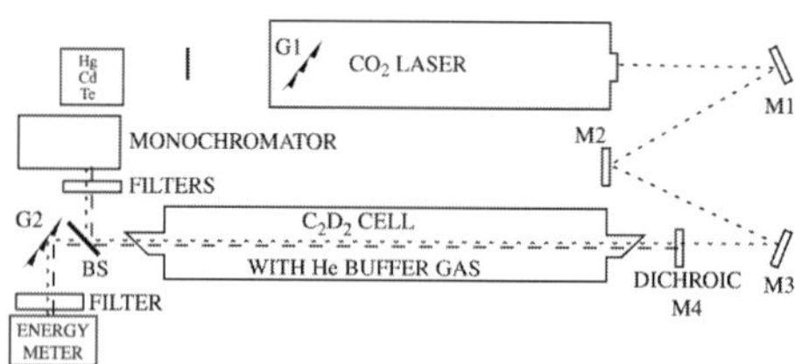

Figure B3.8.6. Experimental set-up for on-axis pumping. Note use of dichroic mirror for input of pump into MIR cavity and use of the Littrow grating with output in zeroth order for MIR resonator [35]. (Reproduced by permission of the American Institute of Physics.)

Gratings are used in both pump and MIR lasers for wavelength selection. For light of wavelength λ, the grating equation is $m\lambda = d(\sin I + \sin R)$, where m is the diffraction order, d is the the grating constant and I and R are the incident and reflection angles, respectively Gratings are blazed and d can be selected for optimum performance at a given wavelength. Figures B3.8.5–B3.8.7 demonstrate the Littrow configuration, whereby the grating replaces a CO_2 pump laser end mirror, in which case $I = R, m\lambda_L = 2d \sin I$ and wavelength λ_L is reflected directly back on itself. Turning the grating does line selection. Some light is always reflected in zeroth order, and this can be used for the MIR output (figure B3.8.6). Note this provides for inefficient pumping as only one pass of the CO_2 results and has the inconvenience that the output direction varies at twice the grating tuning angle. Radiation-damage-resistant original gratings can be used to reflect the pump in first order for on-axis pumping while simultaneously providing for MIR line selection (figure B3.8.8). A variety of other configurations have been reported [36, 37] and reviewed by Harrison and Gupta [4].

Acousto-optic modulators (AOMs) have application in frequency-shifting pump lines into coincidence with absorption lines [38]. In figure B3.8.7 the full line demonstrates double pass through a 90 MHz AOM to achieve a 180 MHz downshift. The disadvantage of AOMs is that they are lossy ($\sim$40% loss per pass);

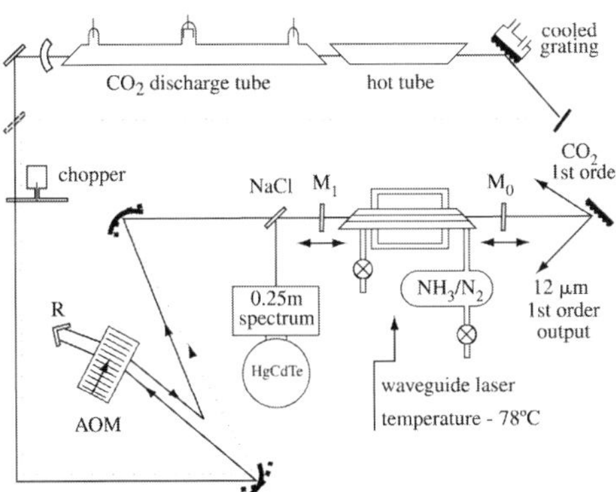

Figure B3.8.7. Experimental set-up for on-axis pumping of a cooled MIR waveguide laser. The hot cell in the CO_2 cavity enables sequence-band operation. The full line shows the pump path when an acousto-optic modulator (AOM) in the double-pass configuration is used. The dotted line is without AOM. M_1 and M_0 are dichroic mirrors. The grating is used to separate the 12 μm MIR output from the residual pump radiation [35]. (Reproduced by permission of the Optical Society of America.)

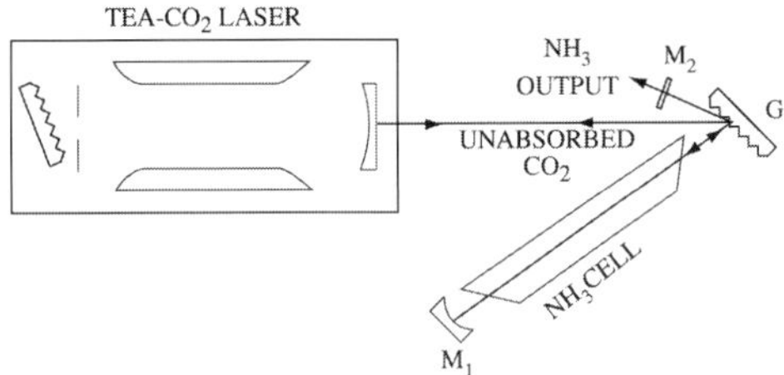

Figure B3.8.8. Schematic diagram of a TEA CO_2 laser pumped NH_3 laser. Grating G diffracts the pump energy in first order into NH_3 cell and selects MIR frequency [27]. (Reproduced by permission of Elsevier.)

however, a benefit is pump-feedback isolation. Polarization isolators, ring cavity configurations and single-pass configurations also provide isolation. Recently, an efficient cavity configuration was proposed [30] which solves many problems including feedback, although it requires accurate focusing of the pump beam through a hole in the input mirror to the MIR cavity.

B3.8.5 NH_3 laser

The first MIROPL was NH_3 pumped by the 9R(16) CO_2 TEA laser, producing 5 kW at 12.81 μm for a 1.5 MW pump [34]. Many improvements have since been made through resonator design, gas cooling and buffer gas use. There are now about 200 MIR laser transitions known for $^{14}NH_3$ and $^{15}NH_3$. Pulse energies have reached the 1 J range with peak powers in the MW range and energy conversion efficiencies greater than 20% [39]. Repetitively pulsed operation at 100 Hz can give 20 W average power [40]. CW operation over 10 W at 28% conversion efficiency has been reported [41]. Table B3.8.1 classifies the NH_3 laser in terms of the vibrational bands. Table B3.8.2 lists the pump wavelength, pump transitions and the pump offset for some of the more important NH_3 lines. Table B3.8.3 lists some specific results. The best-known NH_3 lines are associated with the CO_2 9R(30) and 9R(16) lines pumping the ν_2-fundamental with lasing at 12.06 and 12.82 μm, respectively, on the same band (figure B3.8.9). In both cases, pumping occurs on an NH_3 R-branch line with lasing on a P-branch line, which is ideal for a molecule with a small moment of inertia [4].

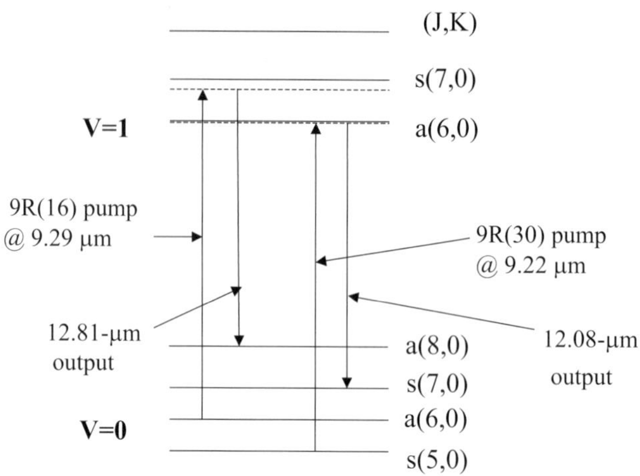

Figure B3.8.9. Energy-level schematic diagram showing two major NH_3 laser transitions. The $CO_2$9R(16) line pumps the aR(6,0) transition; lasing occurs on the aP(8,0) transition at 12.08 μm. Also the $CO_2$9R(30) line pumps the sR(5,0) transition with emission on the sP(7,0) transition at 12.08 μm.

Table B3.8.1. Summary of NH_3 and $^{15}NH_3$ lasing bands and wavelengths: SPP, single-photon pumping; TPP, two-photon (includes two-step) pumping.

Molecule	Pump information	No lines	Lasing band	Wavelength range (μm)	References
$^{14}NH_3$	SPP	116	$\nu_2 \rightarrow 0$	9.3–13.8	(a)
	TPP 10P(24)&10P(18)	2	$\nu_4 \rightarrow 0$	6.27, 6.69	(b)
	TPP (several combinations)	38	$2\nu_2 \rightarrow \nu_2$	11.5–21.3	(c)
$^{15}NH_3$	SPP–various lines	22	$\nu_2 \rightarrow 0$	10.8–13.5	(d)
	True TPP/single pump line 9R(46)	1	$2\nu_2 \rightarrow \nu_2$	13.9	(e)

(a) [42, table 3.2.3.4a], [43, table 3.2.3.4a], [16, 33–35, 38, 44–51].
(b) [42, table 3.2.3.4a], [54].
(c) [43, table 3.3.2.3c], [17, 52–57].
(d) [43, table 3.3.2.3d], [58].
(e) [43, table 3.3.2.3d], [59].

Buffer gases and temperature are important in MIROPLs. Individually or in some combination, He, H_2, N_2 and Ar can increase the output and the lasing spectrum for the CO_2 TEA 9R(30) pumped line [33, 49] by several mechanisms. First, pump absorption increases substantially, e.g. threefold [28], by pressure broadening of the NH_3 line. With 350 torr He, the sR(5,K), $K = 0.5$ multiplet lines merge into a single absorption band that can be pumped by the 9R(30) CO_2 TEA laser [48]. Figure B3.8.10 indicates the transition strengths of this multiplet and some relevant linewidths. Buffer gases also provide rotational relaxation, which allows tuning over much of the ν_2 band. With NH_3/Ar, White *et al* [39] reported more than 70 vibrational

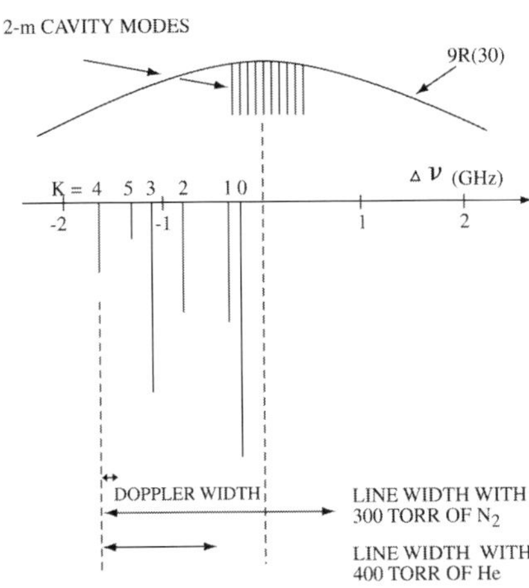

Figure B3.8.10. Pressure-broadened linewidths for comparison with CO_2 laser gain bandwidth. The position of the sR(5,K) multiplet is shown [60]. (Reproduced by permission of Springer.)

Table B3.8.2. CO_2 pump line/NH_3 pumped transition/pump offset for well-known NH_3 lines. Unless otherwise designated, the normal isotopic species are implied. The numbers in the square brackets are the relevant references.

CO_2 pump/λ(μm)	NH_3 pumped transition(s)	Pump offset ($\nu_{NH_3}-\nu_{pump}$) (MHz)	NH_3 emission (μm)
9R(40)/9.107	$^{15}NH_3$ sR(6,6)	−72	sP(8,6) at 12.476159130 [61, 62]
9R(30)/9.22	sR(5,0)	−184 [63]	sP(7,0) at 12.08 μm
	sR(5,1)	−336 [63]	>70 lines 10.08–14.14 [39]
9R(16)/9.29	aR(6,0)	1355 [64]	aP(8,0) at 12.812 [34]
9P(24)/9.59	2sR(4,3)	−475 [64]	13 lines 15.9–21.3 [17]
9P(34)/9.67	aR(4,2)	−5164 [65]	aP(6,2) at 12.26 [66]
10R(42)/10.11	$^{15}NH_3$ aR(2,0)	52.65 [63]	aP(4,0) at 11.76 [66]
			12 lines total 10.79–12.74 [35, 44]
$^{13}CO_2$ 9P(8)/8.89	aR(3,3)	76.8 [63]	aP(4,0)/7 lines total [35]

band transitions between 10.08 and 14.14 μm for the single 9R(30) CO_2 TEA pump line. The optimum NH_3 concentration and pressure are transition dependent. Cooling the NH_3 tube with dry ice or LN_2 vapour increases power for some lines as indicated in table B3.8.3. The response to temperature can be explained through the Boltzmann distribution, which gives the populations of the rotational levels as a function of $E(J, K)$, the statistical weights, as well as the temperature. The buffer gas N_2 has ground-state rotational energy levels that are nearly resonant to those of NH_3, providing for possible collisional deactivation of the lower laser level [37], and results of Grasiuk *et al* [2]. show a correlation between those lines with close coincidences and best performance. The results from modelling the effects of buffer gases are discussed in section B3.8.6.

In contrast to the 9R(30)-CO_2 pumped transition, buffer gases generally have little effect on the 9R(16) pumped 12.82 μm line. The pump laser is 1.355 GHz from the NH_3 aR(6,0) line so that pressure-broadening effects on absorption are insignificant. The limited tuning achieved with He is attributed to that fact the

Table B3.8.3. Some of the strongest results for cw and pulsed NH_3.

Pump	MIR lambda (μm)	MIR output	Parameters	Reference
TEA 9R(30) $\sim$5 J pulse^{-1}	Over 70 lines 10.08–14.14	1 J pulse^{-1} on strongest lines	For $J \leq 7$, 0.2% NH_3/Ar at 170 torr/T = 200 K $J > 7$, 0.2% NH_3/Ar at 270 torr/T = 300 K	White *et al* [39]
TEA 9R(16) $\sim$120 mJ 80–160 ns	12.8	18 mJ pulse^{-1} from their figure 3	For 250 cm cell	Gupta *et al* [67]
CW9R(30) 37 W	12.08	10.2 W	Raman. Linear dry-ice cooled waveguide 1.2 m/2.5 mm bore	Rolland *et al* [41]
CW 9R(30) (AOM/-180 MHz) 18 W	41 total lines strongest are: 11.52–12.31	3.5 W Max	3.0 torr-NH_3/N2 1.0 m/2.5 mm bore waveguide at 200 K	Kroeker and Reid [47]
CW 9R(30) (AOM/-360 MHz) $\sim$6 W	para-NH_3 24 total lines strongest are: 11.98–12.25	90 mW Max	1 or 2% NH_3/Ar 1.0 m/1.5 mm bore waveguide at 200 K	Kroeker and Reid [47]
CW 10R(42) 8.8 W	12 lines 11–13	1.9 W max power	NH_3/Ar dry-ice cooled 1.0 m/2.5 mm waveguide	Chakrabarti and Reid [44]

12.82 μm line is Raman for a broad range of pressures and pump intensities [49]. However, with the high repetition rate pulsed laser, thermal effects become important. Baranov *et al* [40] reported $\sim$100% increase in output with He (or H_2) for the 9R(16) pumped line but no effect with N_2. These results are explainable by the high thermal conductivity of He, which is effective in removing heat. For comparison, output for 9R(30) pumped line increased 700%.

The initial cw NH_3 lasers were Raman type [19]. The offsets for the sR(5,0) and sR(5,1) NH_3 lines are 184 and 336 MHz, respectively, from the CO_2 9R(30) pump line (table B3.8.2). The cw CO_2 laser can be frequency shifted with an AOM to obtain population inversion on either transition. Thereafter, collisions with buffer gas (either N_2 or Ar) thermalize the NH_3 rotational population, and gain can occur throughout the band for line tunable emission. If the $K = 0$ transition is pumped, gain occurs only on the ortho ($K = 3n$) transitions [38]. If the $K = 1$ transition is pumped, gain occurs only on the para ($K = 3n \pm 1$) transitions. Using appropriate AOMs to pump the sR(5,0) and sR(5,1) transitions and dry-ice cooling, Kroeker and Reid [47] reported line tunable cw emission on a total of 65 laser lines (41 in ortho-NH_3 and 24 in para-NH_3) in the 10.3–13.8 μm region.

CW operation, with thresholds below 1 W, is obtained without modulators with $^{15}NH_3$. The aR(2,0) line is offset 52.65 MHz from the CO_2 9R(42) line [35], which allows for direct pumping within the linewidth.

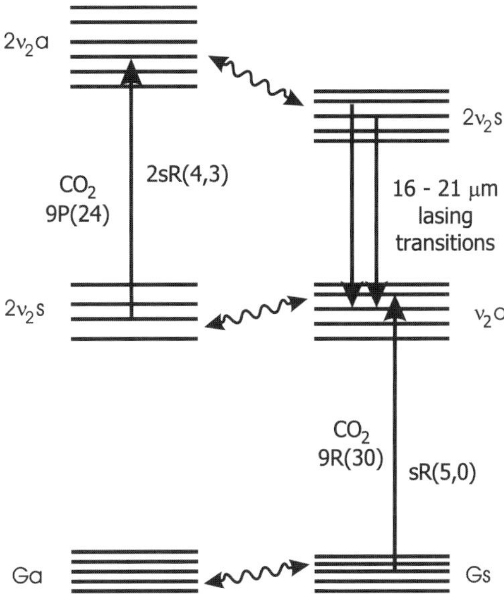

Figure B3.8.11. Partial energy level schematic diagram of NH_3 indicating two-step pumping that leads to efficient emission in the 16–21 μm region [17]. (Reproduced by permission of the Optical Society of America.)

Twelve line-tunable transitions in the 10.8–12.8 μm region were obtained using buffered $^{15}NH_3$/Ar or $^{15}NH_3/N_2$ [44]. Using a cooled hybrid waveguide and Fox-Smith mode selector, Wazen and Bourdet [68] showed 550 MHz tunability for the cw-pumped $^{15}NH_3$ aP(4,0) emission at 11.76 μm and 430 MHz for the collisionally pumped aP(4,2) emission at 11.76 μm. Both CO_2 sequence band and isotopic $^{13}CO_2$ lasers can also cw pump NH_3 and $^{15}NH_3$ [35]. Using a 3.4 W maximum power cw $^{15}NH_3$ MIROPL, Tachikawa and Evenson [69] sequentially pumped $^{15}NH_3$ to produce new FIROPLs.

Simultaneous pumping with two pulsed TEA lasers, either as two-step or two-photon pumping, has extended the NH_3 laser operation to the 16–21 μm region ($2\nu_2 \rightarrow \nu_2$ band) as indicated in table B3.8.1. Two-photon pumping with a single pump line has also been used [59]. Pumping with the combination of the 9R(30) and the 9P(24) TEA CO_2 lines (figure B3.8.11) for buffered NH_3 has led to line tunable emission on 13 lines from 15.9 to 21.3 μm with 100 mJ maximum pulse energy [17].

Still more NH_3 lines can be pumped with a multi-atmosphere CO_2 TE laser, which can tune continuously $\sim$10 cm^{-1} on each branch, compared to about 0.1 cm^{-1} per line for the usual TEA laser [46]. With a 20-atmosphere pump laser and a multi-pass Raman cell, Tillert *et al* [26] obtained 52 emissions in the 11–50 μ region including 37 MIR lines (25 new) in normal NH_3 and 15 MIR lines (14 new) in NH_3. These are not included in the number of lines in table B3.8.1. The cavity was not optimal for wavelengths below 20 μm, leaving the possibility of more lines.

Optically pumped NH_3 produces amplification over a large bandwidth. White *et al* [70] measured small signal gain near 7% and pump conversion efficiencies of 46% under saturated gain conditions for the CO_2 TE 9R(30) pumped NH_3 transition aQ(3,3) at 10.7 μm. Good performance is expected for many other transitions as well. CW amplifiers have also been investigated. Wazen *et al* [71] investigated the cw 12 μm Raman amplifer pumped by the CO_2 9R(30) line, and obtained amplifier power conversion efficiencies of 30%. Scaling relations indicate the practicality of a 150 W system.

The self-absorption of NH_3 at many lasing frequencies may be disadvantageous to the gain but can be used to good advantage for stabilization. The NH_3 laser output beam can be re-absorbed in a low-

Table B3.8.4. Comparison of buffer gases on aR(1,1) for equal absorption linewidth.

	NH_3	Buffer gases N_2	Ar	He
$\Delta\nu/p$ (MHz torr^{-1})	19.8 ± 0.6[a]	4.34 ± 0.28[a]	1.91 ± 0.20[a]	1.15 ± 0.20[a]
k_{VT} (ms torr^{-1})	$(1.3 \pm 0.2) \times 10^{3}$[b]	12 ± 1.3[b]	5.9 ± 0.7[b]	9.3 ± 1.0[b]
Partial pressure (torr)	1.0[a]	75[a]	170[a]	283[a]
Peak gain (% cm^{-1})		0.93[a]	0.90[a]	0.50[a]

[a] For 10R(14)-pump intensity of 1.0 MW cm^{-2} [21].
[b] Hovis and Moore [83].

pressure NH_3 cell to produce strong saturation dips, which have been used to stabilize the laser frequency and accurately measure the line-centre frequencies [63]. The dispersion associated with this absorption (Kramers–Kronig) can lead to a sub-kilohertz spectral linewidth of the NH_3 laser under certain purely passive operating conditions [72, 73].

Other interesting aspects of the NH_3 laser are self-pulsing instabilities [74] leading to chaos in the 12 μm emission, as predicted for a homogeneously broadened system by Haken [75]. Chaotic and periodic pulsation behaviour for the NH_3 12.8 μm aP(8,0) transition have been observed [76] and, under normal low-power conditions, for the sP(7,0) transition [77].

B3.8.6 Gain models

Rate-equation models [21, 78–80], which treat level populations, can describe the dynamics of inversion gain. Such computer models successfully describe laser action in NH_3 in cw and pulsed operation particularly where buffer gases are used and particularly where population occurs through collisional transfer. It is assumed that (1) rotational populations are always thermalized except for the two levels of the pumped transition and (2) only one species ortho-NH_3 ($K = 3n$) or para-NH_3 ($K = 3n \pm 1$) is pumped, with no conversion between. Consider three levels, n_0, n_1 and n_2 are the ground, pumped state and final state populations, respectively, and $n = n_0 + n_1 + n_2$. Input to the equations [79] includes the pumping rate from rotational level r_0 in n_0 to r_1 in n_1: $W_p\Delta r = (I_p/h\nu_p)\sigma[r_0 - r_1(g_0/g_1)]$ where I_p is the pump intensity at frequency, σ is the absorption cross section and g_0 and g_1 are the degeneracies of r_0 and r_1.

Other input includes the V–T relaxation rate $(\tau_{VT})^{-1}$, which is given by $(k_{NH_3}p_{NH_3} + k_B p_B)$ where k is the rate and p is pressure of NH_3 or gas B. The rotational relaxation rates τ_0 and τ_1 are related to the pressure-broadened absorption linewidths. For steady-state pumping, the time derivatives to the rate equations are set to zero to obtain a numerical solution for the gain spectrum. Figure B3.8.12 compares theory and experiment for some cw lines for the aR(6,0)-pumped line [79]; the gain spectrum represents a thermalized distribution for the buffered NH_3/N_2 mixture. Morrison *et al* [21] compared the effects of the three buffer gases N_2, Ar and He by a rate-equation analysis on the aR(1,1) system. Their results are included in table B3.8.4. They conclude that the major effect of N_2 is the large pressure broadening relative to its V–T rate contribution, rather than the resonance deactivation as discussed in section B3.8.5.

For off-resonant pumping involving two-photon Raman processes, a quantum mechanical theory based on the density formalism is required. The original model developed for FIROPLs [24] is readily extended to MIROPLs [22, 66, 81, 111]. This theory treats a coupled three-level system with mixed broadened transitions interacting with two travelling-wave fields of arbitrary strength. A numerical solution can be obtained for the small signal and saturated gain. Silva *et al* [82] have extended the model to include the effect of a standing-wave MIR as would occur in a Fabry–Pérot cavity.

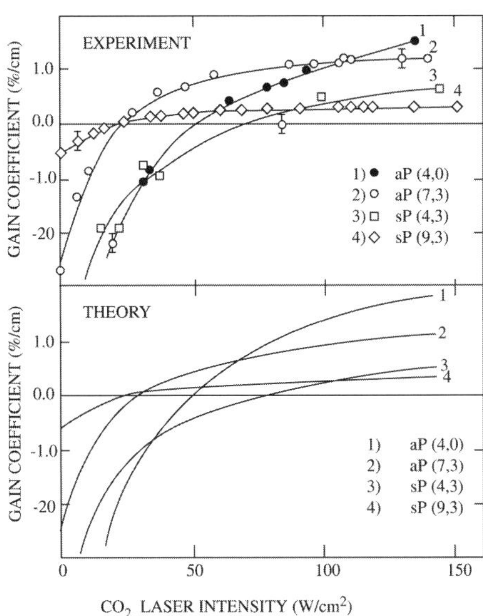

Figure B3.8.12. Comparison of gain coefficients: experiment (top) and rate-equation model (lower): The experiments and calculations are for a 1%NH_3/N_2 mixture at 5 torr. The NH_3 cw-pumped transition is aR(6,0) [79]. (Reproduced by permission of the Optical Society of America.)

Morrison *et al* [66] performed calculations and experiments for the 12.26 μm NH_3 aP(6,2) transition under experimental conditions that avoid confounding effects such as population transfer, power broadening and ac Stark shifts. Their results show typical Raman gain characteristics: at low pressures, the gain is higher for the co-propagating case than for the counter-propagating case. Further, the Raman linewidth is substantially narrower for the co-propagating case. For high pump intensities, the lasing-transition gain shape is strongly influenced by the coherent pump field through the dynamic Stark or Autler–Townes effect [84]. The pump levels are split by an amount equal to the Rabi frequency ($\omega_R = |\mu_{20}|E_2/\hbar$); this lead to a splitting of the MIR gain in the Doppler broadened regime for the case of co-propagation of MIR and pump beams. Although we tend to associate Raman lasers with large pump-frequency offsets, this is not necessarily the case. Wazen and Bourdet [85] investigated the Doppler broadened 11.76 μm $^{15}NH_3$ line optically pumped in a ring laser and found that, although the pumping occurs inside the Doppler width, Raman processes appear to be responsible for the emission. Figures B3.8.13 and B3.8.14 show experimental and calculated tuning curves for the co- and counter-propagating waves, respectively, for three different laser gas pressures for the cw 11.76 μm $^{15}NH_3$ line. For the (+) wave, the two-photon effect dominates, while for the (−) wave, population inversion dominates. Conditions can be chosen so that the laser operates uni-directionally in either direction. Co-propagating gain spikes were observed by Siemsen *et al* [86] in a simple straight gain cell at low NH_3 pressure.

The very complicated dynamics is demonstrated by the NH_3 aR(6,0) pumped transition, which has a 1.355 GHz offset (table B3.8.3) from the 9R(16) CO_2 pump but produces efficient lasing at 12.8 μm. Under conditions of single-mode 1 MW cm^{-2} pump intensity [22], the associated Rabi frequency is comparable to the offset of the pump from the absorbing transition. A variety of processes including population transfer occur. An interesting aspect of pumping this transition is that there are actually two different observed Raman transitions: the aP(8,0) shares the upper level with the pump and the aP(6,0) shares the lower level with the pump. However, the aP(6,0) is pumped indirectly because it first requires a large population transfer to the

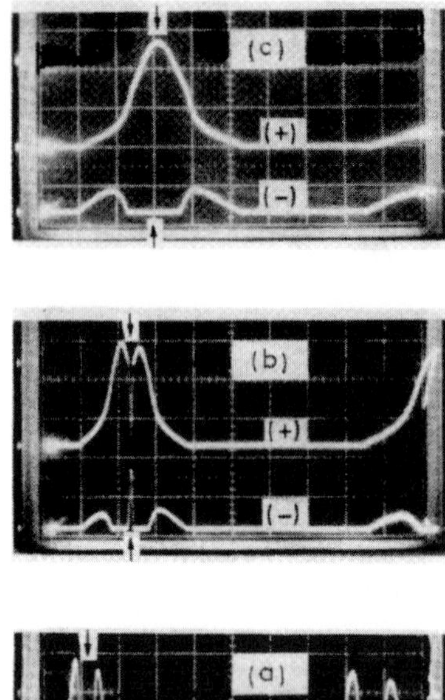

Figure B3.8.13. Experimentally observed tuning curves for $^{15}NH_3$ at 11.76 μm for co-propagation (+) and counter-propagation (−) of MIR and pump beams for total pressure of (*a*) 0.75 (*b*) 1.3 and (*c*) 3 torr, and for $I_p = 150$ W cm^{-2} [87] (©1991 IEEE). (Reproduced by permission of the IEEE.)

upper vibrational level. Raman gains of greater than 10% cm^{-1} and inversion gain greater than 2% cm^{-1} are observed and compared to a combined density-matrix and rate-equation model [22].

B3.8.7 Optically pumped laser action in C_2H_2 and its isotopes

The CO_2 laser cannot directly optically pump C_2H_2. However, the frequency-doubled 9P(24) TEA CO_2 laser can pump the CO (0 → 1) P(14) transition. Energy is then vibrationally transferred by collisions to the ν_2 level of C_2H_2. Lasing occurs on four lines [1243.08, 1243.75, 1244.12 and 1244.52 ± 0.04 cm^{-1} (vacuum)] of the $\nu_2 \to \nu_5$ band near 8 μm [88]. The addition of hydrogen improves performance through depopulation of the lower laser level. Recently, C_2H_2 was optically pumped with a tunable optically pumped oscillator (OPO) [89]. The 3 μm OPO pumps C_2H_2 from the ground state to $(\nu_2 + \nu_4 + \nu_5)$; laser action occurs on transitions $(\nu_2 + \nu_4 + \nu_5) \to (\nu_2 + \nu_5)$ and $(\nu_2 + \nu_4 + \nu_5) \to (\nu_4 + \nu_5)$ at 13.6 and 15.7 μm, respectively. Super-radiant emission in both directions was obtained with total unidirectional output energy up to 20 μJ for 1 mJ pump energy for 100 mtorr C_2H_2.

The CO_2 9.6 μm laser can optically pump the C_2D_2 $(\nu_4 + \nu_5)$ band (figure B3.8.15) to produce MIR lasing for 29 total lines of the $(\nu_4 + \nu_5) \to \nu_4$ transition (table B3.8.5); the strongest lines are at 20 and 17.8 μm. Nine additional lines associated with other high-order bands have been reported. The C_2D_2 laser can be categorized according to type of pump laser: (1) TEA, (2) repetitively-pulsed TEA and (3) multi-atmosphere TE. Only pulsed, not cw, operation has been reported. A total of seven lines from a tunable TEA laser produced 15 C_2D_2 emissions ranging from 17.4 to 20.5 μm [90]. The 9R(12) CO_2 pump

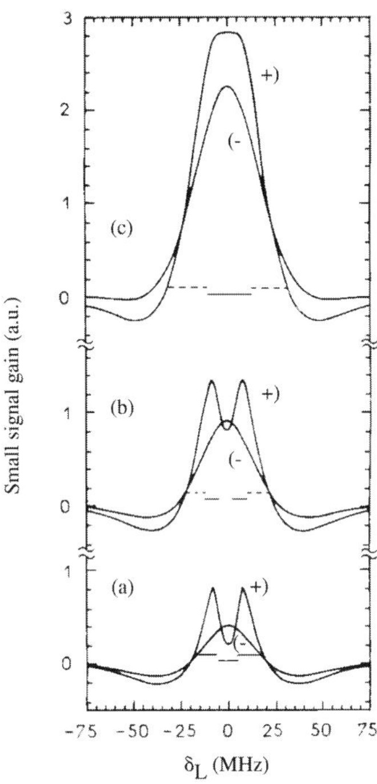

Figure B3.8.14. Calculated small-signal gain for $^{15}NH_3$ at 11.76 μm for co-propagation (+) and counter-propagation (−) of MIR and pump beams for total pressure of (*a*) 0.75 (*b*) 1.3 and (*c*) 3 torr, and for $I_p = 150$ W cm^{-2} [87] (©1991 IEEE). (Reproduced by permission of the IEEE.)

produces the highest C_2D_2 pulse energies, which are 380 mJ for the combined 20 and 17.8 μm lines with energy efficiencies up to 15% for a cooled cell at 160 K [91]. At 300 K, 130 mJ was obtained. Fischer and Wittig [92] obtained rotationally relaxed, grating-tuned oscillation on seven C_2D_2 P-branch lines, in the 17.79–20 μm region, was obtained with a C_2D_2/He mixture at 50 torr for the 9R(12) TEA CO_2 pumped line. To achieve tunability, superfluorescence and Q-branch oscillation had to be quenched by less than ideal conditions, such as short tube lengths and high pressure. When pumped with the 9R(12) CO_2 TEA laser repetitively pulsed at 100 Hz, ~1 MW peak power and ~5 W average power were obtained for the C_2D_2 17.8 μm line for a liquid-nitrogen-vapour cooled cell [93]. The dependence on pressure and repetition rate is shown in figure B3.8.16. Lasing on the 20 μm line produced about one-fifth as much power. The addition of buffer gases (He, N_2 and CF_4) all led to a monotonic sharp drop in the output energy of the C_2D_2 laser [93]. Smith *et al* [94] investigated the relaxation mechanisms of C_2D_2 but the modelling of this MIROPL has not been reported.

Similarly as for NH_3, [95] used a continuously tunable multi-atmosphere CO_2 TEA laser to pump C_2D_2 and produced 16 additional lines, including a very strong line at 20 μm. The isotope $^{13}C_2D_2$ has produced a total of four lines in the 17–21 μm region for four different CO_2 pump lines [91]. Finally, 15 lines in the 17.7–20.4 μm region were obtained with C_2HD using eight pump lines from a 9 μm region CO_2 laser [96].

B3.8.8 Applications and future prospects

The future prospects for the NH_3 and C_2H_2 MIROPLs and their isotopes will depend on the needs associated with applications and what alternative sources exist. Gas lasers have the attributes that they can be constructed

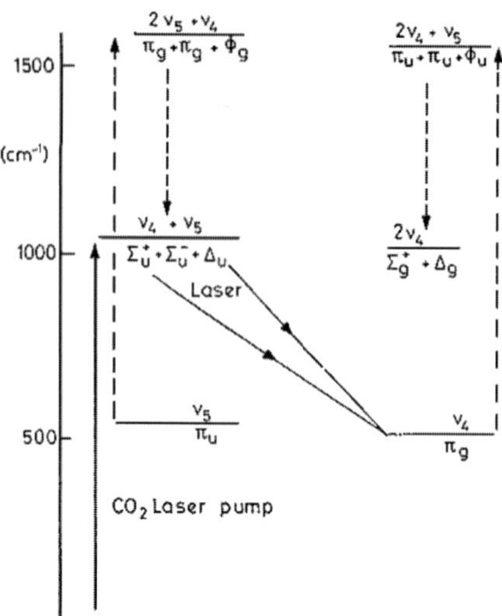

Figure B3.8.15. Partial energy level diagram of C_2D_2 indicating bands leading to MIROPL emission [90]. (Reproduced by permission of Elsevier.)

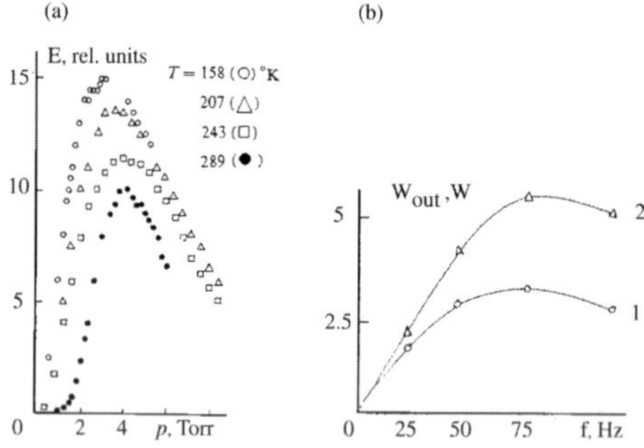

Figure B3.8.16. C_2D_2 laser results: (*a*) pulse output energy *versus* pressure at four temperatures [93]; (*b*) dependence of average MIR power on pump repetition frequency [97]. Curve 1 corresponds to a CO_2 short pump pulse of ~150 ns FWHM. Curve 2 corresponds to a CO_2 pulse consisting of a 150 ns spike and ~3 μs tail that contains more than 50% of the energy. (Figures reproduced by permission of the American Institute of Physics).

in a laboratory with glass and vacuum capabilities. However, gas lasers suffer from their size and are replaced by solid state lasers if one exists that meets the power and spectral requirements. Recently the mid-infrared quantum cascade (QC) semiconductor laser (see chapter B2.7) has come to the forefront and offers great promise as an efficient source for applications not requiring large powers, e.g. chemical sensing [98]. Such QC lasers based on GaInAs/AlInAs structures emit at wavelengths in the 3.4–20 μm region with peak output powers in the fraction of a watt range.

The NH_3 laser covers a spectral region not covered by the CO_2 laser or any other intense source. The very high spectral purity of an NH_3 laser has been used in spectroscopic investigation of the fine structure of a single trapped barium ion [99, 100]. Other applications of the NH_3 MIROPL include gas absorption measurements [101], laser spectrometers for analysis of gaseous mixtures [102], remote gas analyser [103], lidar [105, 106], (sequential) optical pumping of the NH_3 FIR laser, i.e. using the 11–13 μm NH_3 laser to

Table B3.8.5. Summary of optical pumping results in C_2H_2 and its isotopes.

Molecule	No lines	Lasing band	Wavelength range (μm)	Reference
C_2H_2	4	$\nu_2 \rightarrow \nu_5^1$	About 8.04	Optical transfer pumping [88]
C_2H_2	2	$(\nu_2 + \nu_4 + \nu_5) \rightarrow (\nu_2 + \nu_5)$	13.6	OPO pump 3 μm [89]
		$\rightarrow (\nu_4 + \nu_5)$	15.7	
C_2D_2	21	$(\nu_4 + \nu_5)\Sigma_u^+ \rightarrow \nu_4$	17.5–20.3	[42, table 3.2.3.4b], [90, 95]
C_2D_2	2	$(\nu_4 + \nu_5)\Delta_u^+ \rightarrow \nu_4$	17.61, 19.67	[42, table 3.2.3.2b], [90]
C_2D_2	9	Other higher bands	17.5–20.5	[42, table 3.2.3.2b], [90]
C_2D_2	6	$(\nu_4 + \nu_5)\Sigma_u^+ \rightarrow \nu_4$	17.9–19.9	[43, table 3.2.3.2b], [92]
$^{13}C_2D_2$	4		17.2–20.9	[91]
C_2HD	15 total	$(2\nu_4)\Sigma^+ \rightarrow \nu_4$ $2\nu_4(\Delta_c) \rightarrow \nu_4$ and hot bands.	17.7–20.4	[43, table 3.3.2.3b], [96]

pump a FIR laser in the range of 60–400 μm [69], optical pumping an InSb spin-flip Raman laser [32], and multi-photon dissociation of molecules [107]. Mixing an ammonia laser with a CO_2 laser and microwave radiation has generated tunable FIR radiation for high-resolution spectroscopy [108]. High-repetition-rate pulsed gas lasers have their applications in chemistry and isotope separation [109], e.g. separation of tritium from deuterium [110].

Multi-frequency lasing makes the NH_3 laser attractive for differential-absorption lidar systems. The NH_3 laser has the unique effect that when it is pumped by a TEA CO_2 laser operating on the 9R(30) line, it can generate up to 30 lines *simultaneously* and co-linearly in the 12–14 μm region [102] with efficiencies of 20% [105]. Numerical simulations have been carried out to understand the simultaneous multi-frequency operation aspect of NH_3 [111], but this aspect has not been fully investigated [102].

Ammonia has clearly become the leading MIROPL; in comparison, there has been relatively little application reported for the C_2H_2 or C_2D_2 MIROPL; nonetheless these sources offer alternative wavelengths should a need develop. Deuterated acetylene, in particular, has strong lines for applications in the 16 μm region; NH_3 also emits in this region but most of the lines require two pump sources.

References

[1] Chang T Y 1977 Optical pumping in gases topics *Nonlinear Infrared Generation (Applied Physics 16)* ed Y R Shen (New York: Springer) pp 215–72

[2] Grasiuk A Z, Letokhov V S and Lobko V V 1980 Molecular infrared lasers using resonant laser pumping *Prog. Quantum. Electron.* **6** 245–93

[3] Gupta P K and Mehendale S C 1987 Mid-infrared optically pumped molecular lasers *Hyperfine Interact.* **37** 243–74

[4] Harrison R G and Gupta P K 1983 Optically pumped mid-infrared molecular gas lasers *Infrared and Millimeter Waves, Vol 7, Coherent Sources And Applications* part II, ed K J Button (New York: Academic)

[5] Harrison R G and Gupta P K 1984 Mid infrared optically pumped molecular gas lasers *NATO Advanced Study Institute on Physics of New Laser Sources (NATO ASI Series 132)* (San Miniato) (New York: Plenum Press) pp 201–16

[6] Jones C R 1978 Optically pumped mid-IR lasers 1978 *Laser Focus* **14** 68–74

[7] Akitt D P and Wittig C F 1969 Laser emission in ammonia *J. Appl. Phys.* **40** 902–3

[8] Mathias L E S, Crocker A and Wills M S 1965 Laser oscillations at wavelengths between 21 And 32 μm from a pulsed discharge through ammonia *Phys. Lett.* **14** 33–4

[9] Nelson L Y, Fisher C H, Hoverson S J, Byron S R, O'Niell F and Whitney W T 1977 Electron-beam -controlled excitation of a CO–C_2H_2 energy transfer laser *Appl. Phys. Lett.* **30** 192–5

[10] Shelton C F and Byrne F T 1970 Laser emission near 8 μm from a H_2_C_2H_2-He mixture *Appl. Phys. Lett.* **17** 436–7

[11] Stregack J A, Wexler B L and Hart G A 1976 CW CO-CS_2, CO-C_2H_2, CO-N_2O Energy-transfer lasers *Appl. Phys. Lett.* **28** 137–9

[12] Jones C R, Buchwald M I, Gundersen M and Bushnell A H 1978 Ammonia laser optically pumped with an HF laser *Opt. Commun.* **24** 27–30

[13] Frank E M, Weiss C O, Siemsen K, Grinda M and Willenberg G D 1982 Predictions of far-infrared laser lines from $^{14}NH_3$ and $^{15}NH_3$ *Opt. Lett.* **7** 96–8

[14] Herzberg G 1945 *Molecular Spectra and Molecular Structure, II. Infrared and Raman Spectra of Polyatomic Molecules* (New York: Van Nostrand)

[15] Hollas J M 1998 *High Resolution Spectroscopy* 2nd edn (New York: Wiley)

[16] Chang T Y and McGee J D 1976 Off-resonant laser action in NH_3 and C_2H_4 without population inversion *Appl. Phys. Lett.* **29** 725–7

[17] White J D and Reid J 1993 Efficient NH_3 laser operation in the 16 μm to 21 μm region *Appl. Opt.* **32** 2053–7

[18] Rolland C, Reid J, Garside B K, Jessop P E and Morrison H D 1983 Tunable-diode-laser measurements of gain in optically pumped NH_3 *Opt. Lett.* **8** 36–8

[19] Rolland C, Garside B K and Reid J 1982 CW optically pumped 12 μm NH_3 laser *Appl. Phys. Lett.* **40** 655–7

[20] Julien F, Wazen P, Lourtioz J-M and DeTemple T A 1983 The cw optically pumped 12.08 μm Raman laser—theory and experiments for high power conversion efficiency *IEEE J. Quantum Electron.* **QE-19** 1654–62

[21] Morrison H D, Garside B K and Reid J 1984 Dynamics of the optically pumped midinfrared NH_3 laser at high pump power—part I: inversion gain *IEEE J. Quantum Electron.* **QE-20** 1051–9

[22] Morrison H D, Garside B K and Reid J 1984 Dynamics of the optically pumped midinfrared NH_3 laser at high pump power—part II: Raman gain and ac Stark shifts *IEEE J. Quantum Electron.* **QE-20** 1060–4

[23] Tobin M S, Sattler J P and Daley T W 1982 New SMMW laser transitions optically pumped by a tunable CO_2 waveguide laser *IEEE J. Quantum Electron.* **QE-18** 79–86

[24] Heppner J, Weiss C O, Hubner U and Schinn G 1980 Gain in cw laser pumped FIR laser gases *IEEE J. Quantum Electron.* **QE-13** 392–401

[25] Siemsen K J 1986 Private communication

[26] Tillert T, Zimnas P and Renk K F 1996 Partly tunable mid-infrared NH_3 laser *Int. J. Infrared Millimeter Waves* **17** 1011–21

[27] Biswas D J, Sarkar S K and Nayak A K 1992 Improved performance of an NH_3 laser by temporal shaping of the CO_2 laser pulse *Opt. Commun.* **89** 249–52

[28] Midorikawa K, Matsuda I, Obara M and Fujioka T 1980 Efficient energy extraction by superradiant emission in NH_3 *Opt. Commun.* **32** 447–50

[29] Makowe J, Boyarkin O V and Rizzo T R 1998 Highly efficient optically pumped NH_3 laser with near diffraction limited output *Rev. Sci. Instrum.* **69** 4041–3

[30] Nilaya J P and Biswas D J 2001 Versatile cavity for optically pumped molecular lasers *Rev. Sci. Instrum.* **72** 1343–5

[31] Harrison R G and Al-Watban F A 1977 High power laser emission at 12.812 μm from an optically pumped ammonia laser *Opt. Commun.* **20** 225–8

[32] Miyazaki K, Kasada H and Ohtsuka M 1994 Tunable laser for sensing atmospheric transmission in the infrared region of 11–16 μm *Int J. Infrared Millimeter Waves* **15** 1669–74

[33] Fry S M 1976 Optically pumped multi-line NH_3 laser *Opt. Commun.* **19** 320–4

[34] Chang T Y and McGee J D 1976 Laser action at 12.812 μm in optically pumped NH_3 *Appl. Phys. Lett.* **28** 526–8

[35] Siemsen K J, Reid J and Danagher D J 1986 Improved cw lasers in the 11–13 μm wavelength region produced by optically pumping NH_3 *Appl. Opt.* **25** 86–91

[36] Shaw E D and Patel C K N 1978 Improved pumping geometry for high power NH_3 lasers *Opt. Commun.* **27** 419–22

[37] Vasil'ev B I, Grasyuk A Z, Dyad'kin A P, Sukhanov A N and Yastrebkov A B 1980 High-power efficient optically pumped NH_3 laser, tunable over the range 770–890 cm^{-1} *Sov. J. Quantum Electron.* **10** 64–8

[38] Rolland C, Reid J and Garside B K 1984 Line-tunable oscillation of a cw NH_3 laser from 10.7 to 13.3 μm *Appl. Phys. Lett.* **44** 380–2

[39] White J D, Chakrabarti A and Reid J 1990b High-power, high-efficiency optically pumped NH_3 lasers *Appl. Phys.* B **51** 371–3

[40] Baranov V Y, Kazakov S A, Pis'menny V. D, Starodubtsev A I, Velikhov E P, Gorokhov Y A and Letokhov V S 1978 Multiwatt optically pumped ammonia laser operation in the 12–13 μm *Appl. Phys.* **17** 317–20

[41] Rolland C, Reid J and Garside B K 1984 10 W cw optically pumped 12 μm laser *Appl. Phys. Lett.* **44** 725–7

[42] Weber M J (ed) 1985 *CRC Handbook of Laser Science and Technology, Volume 1: Lasers and Masers* (Boca Raton, FL: Chemical Rubber Company)

[43] Weber M J (ed) 1991 *CRC Handbook of Laser Science and Technology, Supplement 1: Lasers* (Boca Raton, FL: Chemical Rubber Company)

[44] Chakrabarti A and Reid J 1987 Simple cw NH_3 laser operating in the 11–13 μm region *Rev. Sci. Instrum.* **58** 1413–16

[45] Danielewicz E J, Malk E G and Coleman P D 1976 High-power vibration–rotation emission from 14 NH_3 optically pumped off resonance *Appl. Phys. Lett.* **29** 557–9

[46] Deka B K, Dyer P E and Winfield R J 1980a optically pumped NH_3 laser using a continuously tunable CO_2 laser *Opt. Commun.* **33** 206–8

[47] Kroeker D F and Reid J 1986 Line-tunable cw ortho- and para-NH_3 lasers operating at wavelengths of 11 to 14 μm *Appl. Opt.* **25** 2929–33

[48] Tashiro H, Suzuki K, Toyoda, K and Namba S 1980 Wide range line-tunable oscillation of an optically pumped NH_3 laser *Appl. Phys.* **21** 237–40
[49] Tiee J J and Wittig C 1978 Optically pumped molecular lasers in the 11–17 μm region *J. Appl. Phys.* **49** 61–4
[50] Yamabayashi N, Yoshida T, Miyazaki K and Fujisawa K 1979 Infrared multi-line NH_3 laser and its application for pumping an InSb laser *Opt. Commun.* **30** 245–8
[51] Znotins T A, Reid J, Garside B K and Ballik E A 1980 12 μm NH_3 laser pumped by a sequence CO_2 laser *Opt. Lett.* **5** 528–30
[52] Bobrovskii A N, Vedenov A A, Kozhevnikov A V and Sobolenko D N 1979 NH_3 laser pumped by two CO_2 lasers *JETP Lett.* **29** 536–9
[53] Eggleston J, Dallarosa J, Bischel W K, Bokor J and Rhodes C K 1979 Generation of 16 μm radiation in $^{14}NH_3$ by two-quantum excitation of the $2\nu_2^-$ (7,5) state *J. Appl. Phys.* **50** 3867–70
[54] Jacobs R R, Prosnitz D, Bischell W K and Rhodes C K 1976 Laser generation from 6 to 35 μm following two-photon excitation of ammonia *Appl. Phys. Lett.* **29** 710–2
[55] Lee W, Kim D, Mark E and Leap J 1979 Hot-band lasing in NH_3 *IEEE J. Quantum Electron.* **QE-15** 838–9
[56] Morrison H D, Reid J and Garside B K 1984c 16–21 μm line-tunable NH_3 laser produced by two step optical pumping *Appl. Phys. Lett.* **45** 321–3
[57] Pinson P, Delage A, Girard G and Michon M 1981 Characteristics of two-step and two-photon-excited emissions in $^{14}NH_3$ *J. Appl. Phys.* **52** 2634–7
[58] Akhrarov M, Vasil'ev B I, Grasyuk A Z and Soskov V I 1984 Middle infrared laser utilizing isotopically substituted $^{15}NH_3$ ammonia molecules *Sov. J. Quantum Electron.* **14** 572–3
[59] Akhrarov M, Vasil'ev B I, Grasyuk A Z, Soskov V I and Yastrebkov A B 1986 $^{15}NH_3$ laser with two-photon optical pumping *Sov. J. Quantum Electron.* **16** 1016–19
[60] Pinson P, Girard G and Michon M 1982 Optically pumped $^{14}NH_3$ laser and its Stark tuning at 10.78 μm *Appl. Phys.* B **28** 55–62
[61] Siemsen K J, Madej A A and Whitford B G 1995a Frequency stabilized ammonia laser system for probing the 24 THz clock transition of a trapped barium ion *IEEE J. Quantum Electron.* **31** 1764–73
[62] Siemsen K J, Madej A A and Whitford B G 1995b Absolute frequency measurement of the sP(8,6) transition of $^{15}NH_3$ *J. Mol. Spectrosc.* **174** 613–14
[63] Siemsen K J and Reid J 1985 Heterodyne frequency measurements of $^{14}NH_3$ and $^{15}NH_3$ ν_2-band transitions *Opt. Lett.* **10** 594–7
[64] Sattler J P, Miller L S and Worchesky T L 1981 Diode-laser heterodyne measurements on (NH-3)-N-14 *J. Mol. Spectrosc.* **88** 347–51
[65] Sattler J P and Worchesky T L 1981 Additional diode-laser heterodyne measurements on ammonia *J. Mol. Spectrosc.* **90** 297–301
[66] Morrison H D, Garside B K and Reid J 1982 Raman gain in 12 μm NH_3 lasers *Opt. Lett.* **7** 520–2
[67] Gupta P K, Kar A K, Taghizadeh M R and Harrison R G 1981a 12.8 μm NH_3 laser emission with 40–60% power conversion and up to 28% energy conversion efficiency *Appl. Phys. Lett.* **39** 32–4
[68] Wazen P and Bourdet G L 1989 CW MIR optically pumped ammonia laser with large frequency tunability *Opt. Commun.* **71** 81–4
[69] Tachikawa M and Evenson K M 1996 Sequential optical pumping of a far-infrared ammonia laser *Opt. Lett.* **21** 1247–9
[70] White J D, Bruce D M, Beckwith P H and Reid J 1990a Efficient optically pumped NH_3 amplifiers in the 10–12 μm region *Appl. Phys.* B **50** 345–54
[71] Wazen P, Lourtioz J-M, Julien F and DeTemple T 1987 High-power optically pumped cw 12 μm NH_3 amplifier systems *IEEE J. Quantum Electron.* **23** 623–32
[72] Siemsen K J, Williams E and Reid J 1987 Alternative approach to stable lasers: the optically pumped laser *Opt. Lett.* **12** 879–81
[73] Siemsen K J, Madej A A, Hanes G R and Reid J 1992 Improved frequency stability of the ammonia laser *Appl. Phys.* B **54** 126–31
[74] Harrison R G and Al-Saidi I A 1985 Experimental-evidence of self-pulsing and chaos in an optically pumped 12 μm NH_3 laser *Opt. Commun.* **54** 107–11
[75] Haken H 1975 Analogy between higher instabilities in fluids and lasers *Phys. Lett.* A **53** 77–8
[76] Biswas D J and Harrison R G 1985 Observation of optical turbulence in a single mode homogeneously broadened optically pumped molecular laser *Opt. Commun.* **54** 112–16
[77] Siemsen K J, Reid J and Weiss C O 1987 Self pulsing and chaos of a vibrational NH_3 laser *Opt. Commun.* **63** 415–16
[78] Gupta P K and Harrison R G 1981b Rate equation model for mid IR OPML having common pump and upper lasing level; application to 12.8 μm emission from NH_3 *IEEE J. Quantum Electron.* **QE-17** 2238–44
[79] Sinclair R L, Reid J, Morrison H D Garside B K, and Rolland C 1985 Dynamics of the line-tunable 12 μm continuous-wave NH_3 laser as measured with a tunable-diode laser *J. Opt. Soc. Am.* B **2** 800–5
[80] Morrison H D, Garside B K and Reid J 1985 Gain dynamics in pulsed 12 μm NH_3 lasers *J. Opt. Soc. Am.* **2** 535–40
[81] Luo X and Qui R 1997 Optimized operation of optically pumped NH_3 laser emission at 12.08 μm and 12.81 μm *Int. J. Infrared Millimeter Waves* **18** 641–52
[82] Silva F, Vilaseca R, and Corbalán R 1992 Theroretical gain spectrum of coherently pumped mid-infrared Fabry–Pérot lasers *Opt. Commun.* **94** 589–98
[83] Hovis F E and Moore C B 1980 Temperature dependence of vibrational energy transfer in NH_3 and $H_2^{18}O$ *J. Chem. Phys.* **72** 2397–402
[84] Delsalt C and Keller J-C 1978 The optical Autler–Townes effect in Doppler broadened three-level systems *J. Physique* **39** 350–60
[85] Wazen P and Bourdet G L 1989 Experimental investigation of the MIR optically pumped ammonia bidirectional ring laser *Appl.*

Phys. B **49** 377–81

[86] Siemsen K J, Madej A A and Reid J 1991 Narrow gain spikes and frequency pulling in the midinfrared ammonia laser *IEEE J. Quantum Electron.* **27** 1199–206

[87] Wazen P and Bourdet G L 1991 Gain anisotropy and simultaneous bidirectional emission of a Doppler-broadened MIR optically pumped ammonia ring laser *IEEE J. Quantum Electron.* **27** 152–7

[88] Kildal H and Deutsch T F1975 Optically pumped infrared V–V transfer lasers *Appl. Phys. Lett.* **27** 500–2

[89] Tapalian H C, Michaels C A and Flynn G W 1997 Midinfrared molecular gas lasers optically pumped by a continuously tunable infrared optical parametric oscillator *Appl. Phys. Lett.* **70** 2215–17

[90] Rutt H N and Green J M 1978 Optically pumped laser action in dideuteroacetylene *Opt. Commun.* **26** 422–6

[91] Rutt H N 1980 Development of the C_2D_2 laser: high energy, line-tuned, and 13C isotopic operation *Topical Meeting on Infrared Lasers: A Digest of Technical Papers Presented at the Topical Meeting on Infrared Lasers (Los Angeles, CA)* (Washington, DC: OSA)

[92] Fischer T A and Wittig C 1982 Rotationally relaxed, grating tuned laser oscillations in optically pumped C_2D_2 *Appl. Phys. Lett.* **41** 107–9

[93] Baranov V Y, Dyad'kin A P, Pigul'skii S V and Starodubtsev A I 1989 Laser-pumped C_2D_2 laser *Sov. J. Quantum Electron.* **19** 715–18

[94] Smith N J G, Davis C C and Smith W M 1985 Relaxation of C_2D_2 (ν_4, ν_4) by vibration-rotation, translation energy transfer *J. Chem. Soc. Faraday Trans.* **2** 417–32

[95] Deka B K, Dyer P E and Winfield R J 1980b New 17–21 μm laser lines in C_2D_2 using a continuously tunable CO_2 laser pump *Opt. Lett.* **5** 194–5

[96] Rutt H N 1984 Optically-pumped laser action in monodeutero-acetylene (HCCD) *Infrared Phys.* **24** 535–41

[97] Baranov V Y, Dyad'kin A P, Kazakov S A, Pigul'skii S V and Starodubtsev A I 1987 Optically pumped pulse-periodic C_2D_2 laser *Sov. J. Quantum Electron.* **17** 771–2

[98] Capasso F, Gmachl C, Tredicucci A, Sivco D L, Baillargeon J N, Cho A Y and Liu H C 2000 New frontiers in quantum cascade lasers and applications *IEEE J. Selected Topics Quantum Electron.* **6** 931–46

[99] Madej A A, Siemsen K J, Sankay J D, Clark, R F and Vanier J 1993 High-resolution spectroscopy and frequency measurement of the midinfrared $5d^2D_{3/2}$–$5d^2D_{5/2}$ transition of a single laser-cooled barium ion *IEEE Trans. Instrum. Meas.* **42** 234–41 a review *Infrared Phys. Technol.* **36** 465–73

[100] Whitford B G, Siemsen K J, Madej A A, and Sankey J D 1994 Absolute-frequency measurement of the narrow-linewidth 24 THz D-D transition of a single laser-cooled barium ion *Opt. Lett.* **19** 356–8

[101] Benzerhouni K, Meyer F and Lourtioz J M 1986 Absorption-measurements using cw MIR NH_3 laser *Infrared Phys.* **26** 377–80

[102] Baranov V Y, Khakhlev A A, Maluta D D, Mezhevov V S, Petrushevich Y V and Poliakov G A 2000 Analysis of gaseous mixtures with multifrequency NH_3 laser spectrometer *Infrared Phys. Technol.* **41** 97–113

[103] Banakh V A, Ponomarev Y N, Smalikho I N, Firsov K M, Maluta D D and Poliakov G A 2000 Simulation of operation of multiwave remote gas-analyser based on NH_3-laser *Infrared Phys. Technol.* **41** 115–31

[104] Anan'ev V Y, Vasil'ev B I, Lobanov A N, Lytkin A P, Whang C C and Sung K J 2000 two-frequency lidar based on an ammonium laser *Quantum Electron.* **30** 535–9

[105] Vasil'ev B I and Whan C C 2000 NH_3 Laser as a radiation source for a two-frequency lidar *Quantum Electron.* **30** 1105–6

[106] Makowe J, Boyarkin O V and Rizzo T R 2000 Collision-free infrared multiphoton dissociation of silane *J. Phys. Chem.* A **104** 11 505–11

[107] Odashima H, Tachikawa M, Zink L R and Evenson K M 1997 Extension of tunable far-infrared spectroscopy to 7.9 THz *Opt. Lett.* **22** 822–4

[108] Baranov V Y 1983 High repetition rate pulsed gas lasers and their applications in chemistry and isotope separation *IEEE J. Quantum Electron.* **QE-19** 1577–87

[109] Nayak A K, Sarkar S K, Biswas D J, RamaRao K V S and Mittal J P 1991 Separation of tritium from deuterium by pulsed NH_3 laser-selective multiple photon dissociation of $CTCl_3$ *Appl. Phys.* B **53** 246–9

[110] Petrushevich Y V 1998 Theoretical computational investigation of an NH_3 laser subjected to intensive pulsed pumping *Quantum Electron.* **28** 3–8

[111] Rolland C, Reid J, Garside B K, Morrison H D and Jessop P E Investigation of cw optically pumped 12-μm NH_3 lasers using a Tunable Diode Laser *Appl. Opt.* **23** 87–93

[112] Urban W 1995 Physics and spectroscopic applications of carbon monoxide lasers, a review *Infrared Phys. Technol.* **36** 465–73

Further reading

Barrow GM 1975 *Molecular Spectroscopy* (New York: McGraw-Hill)

This book is particularly recommended for the beginning spectroscopist.

Brown R T 1987 *CO_2 TEA lasers Handbook of Molecular Lasers* ed P K Cheo (New York: Marcel Dekker) ch 2, pp 93–164

This gives information concerning TEA lasers.

Cheo P K 1987 *Emission spectra of molecular lasers Handbook of Molecular Lasers* ed P K Cheo (New York: Marcel Dekker) ch 1, pp 1–92

This chapter describes the spectroscopy of the CO_2 molecule and lists the wavelengths for the various 12C16O2 laser transitions and the isotopic variations.

Gordiets B F, Osipov A I, Stupochenko E V, and Shelepin L A 1973 Vibrational relaxation in gases and molecular lasers *Sov. Phys.–Usp.* **15** 759–85

This paper discusses the vibrational kinetics and mechanisms of lasers that are based on vibrational-rotational transitions and is the foundation for most of the rate-equation models for the NH_3 MIROPL.

Marcatili E A J and Schmeltzer R A 1964 Hollow metallic and dielectric waveguides for long distance optical transmission and lasers *Bell Syst. Tech. J.* **43** 1783–809

This paper gives the theory of oversized waveguides that are frequently used in MIROPLs.

Tobin M S 1985 A review of optically pumped NMMW lasers *Proc. IEEE* **73** 61–85

This paper reviews FIROPLs but addresses optical pumping issues that also relate to MIROPLs.

B3.9
Far-IR lasers: HCN, H_2O

Wilhelm Prettl

B3.9.1 Introduction

This chapter deals with electric discharge pumped far-infrared lasers. The discovery of the H_2O and the HCN laser by Gebbie *et al* [8, 13, 14] opened up a new age of spectroscopy in the far-infrared, a spectral range which had been notoriously hampered by the low power of thermal radiation sources. Many laboratories took up research on glow discharge excited molecular lasers and suddenly a substantial number of fairly strong laser lines in the far-infrared were available. Many measurements of fundamental importance could be carried out in various fields. In particular in solid state physics a new technique, magneto-spectroscopy, evolved which made use of fixed frequency laser lines, tuning the energy levels of the object of investigation by an external magnetic field. For instance, after its feasibility was first demonstrated [5], most effective masses and band parameters of semiconductors were determined by this method [27, 28].

Besides HCN, H_2O and the isotopic derivatives DCN and D_2O, a number of other polyatomic molecules have been found to be suitable for laser action in an electric gas discharge. Among them are NH_3, OCS and SO_2. However, the lifetime of these molecules in a high voltage electric discharge is very short. Electrons with average kinetic energy in the range of $\sim$10 eV break chemical bonds and decompose molecules. Therefore most polyatomic molecule lasers can only be operated in a pulsed mode with a duty cycle of the order of 1 cycle s^{-1}.

Among the many laser lines observed, there are a few strong laser lines in the cw mode lying in atmospheric windows. Today only these lines are of practical importance. These lines are currently used as an analytical tool in plasma diagnostics [20], to heat electrons in semiconductors [15] and in Stark spectroscopy on molecules [19]. Finally, it is interesting to note that one of the HCN laser lines belongs to the about 100 cosmic laser transitions observed so far [33].

In this article, the basic physics, techniques and construction of HCN and H_2O lasers will be summarized.

B3.9.2 Laser lines

HCN and H_2O lasers may be electrically pumped in a longitudinal configuration of electrodes in pulsed and cw mode of operation as well as transversely excited [42]. Laser action of HCN by HF pumping and for both lasers by flash lamp excitation has also been observed. However, today only longitudinally pumped cw lasers are in practical use.

Under longitudinal excitation, the gain bandwidth of both lasers is for practically all lines smaller than the frequency separation of the subsequent fundamental longitudinal modes or longitudinal and higher transverse modes. Hence, the output intensity as a function of the resonator length is modulated with sharp peaks at resonance between the molecular transition frequency and the resonances of the resonator. Therefore a good estimate of the wavelength of an emission can easily be performed by monitoring the laser intensity as a

Table B3.9.1. Strong cw laser lines of the HCN and the H_2O laser. The highest reported power is given. (a) [18], (b) [2], (c) [37], (d) [12], (e) [11].

	Wavelength approximate (μm)	Wavelength high resolution (μm)	Frequency (GHz)	Power (mW)
HCN	337	336.5578 (a)	890.7607 (a)	250 (b)
H_2O	28	27.971 (c)	10,718.073 (e)	150 (c)
	78	78.442 (c)	3,821.755 (4)	12 (c)
	119	118.59 (c)	2,527.9528 (d)	52 (c)

function of the resonator length, a technique which has been termed resonator interferometry [40]. In many cases more than one laser line or more than one mode can oscillate under the same operational conditions. Tuning the resonator length here is also a useful tool with which to obtain single-line and single-mode emission.

The laser wavelengths were determined with high accuracy by long path-difference Michelson interferometry [9]. Very high precision measurements of laser line frequencies were carried out by heterodyning the laser emission with the high harmonics of coherent microwave sources like the klystron [17].

The gain in energy per pulse due to transverse excitation is not much larger than that at longitudinal excitation; however, discharge instabilities occurring at high longitudinal currents are avoided. In addition, transverse excitation increases the gain bandwidth, which makes tuning and stabilization of the resonator length unnecessary.

With HCN as a laser medium essentially two groups, each of four or five laser lines, may be obtained. One group is around the strongest HCN laser line at $\lambda = 337$ μm (see table B3.9.1) and the other is in the range of 130 μm. With DCN a similar set of lines may be obtained around 190 μm. The longest wavelengths with an electric discharge pumped molecular laser were observed by Steffen *et al* with ICN at 538 μm [38] and at 774 μm [39]. These lines had not been assigned until then.

The 337 μm line of HCN is the strongest far-infrared laser emission of an electrically pumped molecular laser. It may yield cw power around 250 mW [2].

With the water vapour laser in pulsed mode more than 100 laser lines and over a dozen lines in cw mode have been found in the wavelength range 7–220 μm. In addition to these lines attributed to the isotopic species $H_2{}^{16}O$, a comparable number of laser lines have been obtained with $H_2{}^{18}O$ and $D_2{}^{16}O$. The molecular transitions of most but not of all of these lines have been identified. A list of the wavelengths is given in [3,24].

The wavelengths of the strong lines in cw mode of the H_2O laser are 28, 78 and 119 μm, as summarized in table B3.9.1. Only these lines are of practical importance. An exhaustive list of experimentally observed lines of glow discharge pumped far-infrared lasers is given by [22].

B3.9.3 HCN laser

B3.9.3.1 Laser mechanism

The history of the identification of the energy levels involved in the far-infrared transitions of the HCN and the H_2O laser and even the recognition of the laser active molecules illustrates the complexity of molecular spectra and molecular kinetics. The striking feature of these triatomic molecular lasers is that there is no regularity in the emission-line spectra as in the case of the CO_2 or CO laser. It took three years of controversial discussions in the literature until Lide and Maki [23] succeeded in assigning the strong 337 μm line and the neighbouring line at 311 μm of the HCN laser correctly.

The essential physical background of laser transitions comprises the resonance perturbations in the vibrational–rotational system due to the Coriolis force acting in the frame of the rotating molecule and anharmonic resonances like the Fermi resonance [26]. The electric discharge creates a population inversion between low-lying predominantly stretching modes and bending modes. Laser action occurs by rotational-vibrational inter-mode transitions. In the harmonic approximation, inter-mode transitions are forbidden; therefore, anharmonicity of the molecular vibrations is important for laser action[1]. The transition probabilities between different vibrational modes are usually too small to give sufficient gain for laser oscillation. However, if two molecular states belonging to the same symmetry species come accidentally close in energy their wavefunctions may mix. The mixing of vibrational states of identical symmetry occurs due to anharmonicity. A coupling of vibrational states of differing symmetry may be accomplished by the Coriolis interaction of nearly degenerate rotational–vibrational states if the overall symmetry species are the same. In particular, the angular momentum quantum number J and the parity of the interacting rotational states must be identical. In some sense the inter-mode vibrational transition borrows oscillator strength from rotational transitions. If laser oscillation occurs between two different modes, rotational states in the individual modes are populated and depopulated. Hence, laser action may also occur by cascades of purely rotational transitions. Therefore, HCN as well as H_2O lasers have a tendency towards simultaneous multi-line emission.

In the case of the linear molecule HCN laser, the mixing of different vibrational modes due to the Coriolis force is possible between almost degenerate states of the same angular momentum J and parity. The mixing of these states allows optical inter-mode transitions. In the electric discharge of a HCN laser the molecule is formed with a high amplitude for the stretching mode $(\nu_1\nu_2\nu_3) = (100)$ which is mostly made up of CN vibrations. This is due to the fact that the (100) state is practically metastable [16]. Vibrational transitions from (100) to the (000) ground state are almost three orders of magnitude slower than the (010)–(000) and (001)–(000) transitions [21]. In the gas discharge HCN laser, molecules are distributed over all vibrational states which are quickly depopulated by spontaneous emission cascading down the vibrational levels to the ground state. The metastable (100) state and all combinations based on it become overpopulated compared to other vibrational modes. Such combinations are obtained by adding one or two quanta of the bending mode ν_2 yielding a combination of the vibrational states (110) and (120) which are very close in energy to the pure bending modes (040) and (050), respectively. The bending and stretching combination states give the upper laser levels whereas the purely bending states represent the lower laser levels. This is due to faster relaxation of the bending modes compared to stretching–bending equilibrations [25].

The pumping mechanism is not very effective as the majority of the molecules relax back to the ground state without passing through the laser transitions. This correlates to the low small-signal gain of the order of 0.1 dB m^{-1} and the correspondingly moderate output power of the order of magnitude of 10 to 100 mW.

In the case of HCN transitions around the strong 337 μm line, two rotational levels of the combined stretching and bending mode (11^10) and the bending mode (04^00) having the same rotational quantum number and parity, $J^+(10)$, lie sufficiently close together that the Coriolis interaction relaxes the selection rule of forbidden inter-mode transitions. In addition, the inter-vibrational transitions pump purely rotational transition cascades at 284, 310, 335 and 373 μm. High-precision measurements of the frequencies of these laser lines support this level assignment. The sum of the frequencies of the 311 and 335 μm lines differ by less than one part in 10^6 from the sum of the frequencies of the 310 and 337 μm lines [18]. A schematic energy level diagram and the corresponding laser transitions are displayed in figure B3.9.1.

The same mechanism can be assumed for other groups of laser lines. In the HCN laser, the coupling of $J = 26$ rotational states of the 12^00 and the 05^10 vibration ladders leads to an emission of laser lines in the

[1] In fact, all electric discharge pumped molecular infrared gas lasers are based on inter-mode transitions made possible by anharmonicity.

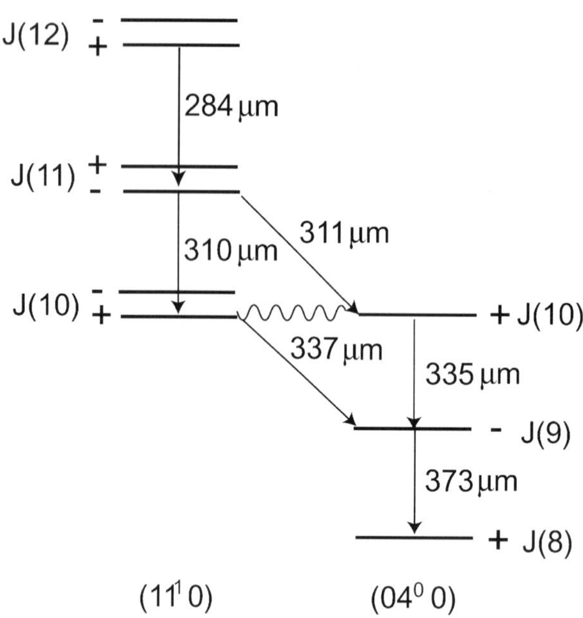

Figure B3.9.1. Rotation–vibration energy level diagram around the Coriolis resonance (curly line) of (11^10) and (04^00) vibrational states and laser transitions (arrows) leading to the strong 337 μm line and the lines in the vicinity of 337 μm. The energy levels are about 2900 cm^{-1} above the ground state. J and $\pm$ denote the angular momentum and the overall parity of the states. The plotted l-type doubling of the rotational levels of (11^10) is largely exaggerated.

region of ~130 μm and in the DCN laser, a similar perturbation occurs between $J = 21$ states of the (22^00) and (09^10) vibrations generating laser emission in the vicinity of 190 μm.

B3.9.3.2 Laser fuels

HCN molecules are not stable in the glow discharge. They are decomposed by the impact of high-energy electrons and form again in excited vibrational–rotational states. Thus the HCN laser is a chemical laser. In fact it is not necessary to use poisonous HCN as a laser medium but almost any compound or mixture of molecules containing C, N and H, with a sufficient excess of H, may produce HCN in the gas discharge and generate laser action. Several chemical reactions have been discussed, making population inversion conceivable [29].

Gas mixtures are selected from the point of view of high output power and small polymer deposition on the resonator wall (see later). A standard fuel is the gas mixture $N_2 + CH_4$ in volume ratios ranging from 1:1 to 1:3. Other laser media are CH_3N, C_2H_5CN and $CH_4 + NH_3$. Remarkably, $C_2H_2 + N_2$ is laser inactive but may be activated by adding H_2 [34].

B3.9.3.3 Wall effects

Physical and chemical processes at the inner walls of the discharge tube are of central importance for the operation of an HCN laser. A characteristic feature of the HCN laser is the deposit of a polymer layer on the discharge tube and on the mirrors of the resonator. The deposited film, coloured between yellow and dark brown, causes a substantial reduction in output power and, if the layer gets too thick, laser emission ceases.

The deposit can be removed by running the discharge with H_2O for several hours. Adding oxygen speeds up the process of converting the deposit into gaseous products.

Chemical analysis and infrared spectroscopy showed that the main constituent of the electric discharge polymerized material is $(CN)_x$ [35]. Laser emission of the 337 μm HCN line may be obtained in a discharge tube with the polymer deposit running the discharge with pure H_2O, H_2 or NH_3. Laser action occurs because HCN molecules are formed in sufficient concentration by the discharge from the wall deposit.

The generation of laser active molecules at the walls leads to a surprising dependence of the gain on the laser tube as a function of the distance from the tube axis. In typical low-pressure gas lasers the gain drops rapidly as it approaches the discharge tube walls. Atoms or molecules excited in the volume diffuse to the wall and lose their energy there. In contrast, the gain in the HCN laser is almost constant over the whole cross section of the discharge tube [41]. This unusual behaviour was explained by taking the volume and wall processes in the plasma into account simultaneously [35].

B3.9.4 H_2O laser

The physical mechanisms and operating conditions of the water vapour laser are very similar to those of the HCN laser. Both Coriolis and Fermi resonances yield the transition probability between different vibrational modes. The coupling occurs between rotational levels of the (100) and (001) vibrational states of the ν_1 and ν_3 modes, respectively, and the first overtone (020) of the symmetric bending mode of frequency ν_2. The strong interaction is caused by large anharmonicity and the fact that the frequencies of the symmetric (ν_1) and antisymmetric (ν_3) OH stretching modes are almost exactly twice the frequency of the bending fundamental vibration (ν_1). A symmetry analysis [16] shows that Fermi resonances cause perturbations of rotational levels in (100) and (020) whereas the Coriolis force is responsible for the interaction between (020) and (001).

As an asymmetric top molecule the rotational energy level structure of H_2O is considerably more complex than that of the linear molecule HCN. A very detailed discussion of the water vapour laser and an assignment of a large number of observed laser lines has been given by [3]. In figure B3.9.2, part of the energy level scheme of H_2O is shown which accounts, among other lines, for the stronger laser lines given in table B3.9.1. A concise discussion of the inversion mechanism was given by [31].

The H_2O laser is very convenient to operate as there are no problems of wall depositions in the laser tube and the lifetime of the molecule is quite long. In contrast to the HCN laser, H_2O is directly excited in the glow discharge without being dissociated. This was concluded from the observation that, in pulsed mode of operation, H_2O lases during the exciting current pulse and decays rapidly when the excitation terminates. HCN and other molecular species lase with characteristic temporal delay in the afterglow of the current pulse [7].

B3.9.5 Resonator

The glow discharge and the resonator mirrors of electrically pumped far-infrared lasers are usually confined in glass or Pyrex tubes. Typical dimensions of laser resonators are several centimetres in radius a and a few metres in length b. These units result in Fresnel numbers $N = a^2/\lambda b \approx 1$ where λ is the laser wavelength. This is in contrast to lasers in the visible or near infrared which have Fresnel numbers much larger than one. Thus, taking into account the low gain, diffraction losses cannot limit the laser threshold. This fact was first recognized by Schwaller *et al* [36] who pointed out the discrepancy between diffraction-selected resonator modes and actually observed modes. Steffen and Kneubühl [40] showed that, in fact, the resonator works like an oversized leaky waveguide. Instead of diffraction the losses are due to reflection at the dielectric resonator walls. The attenuation in a low-index leaky guide decreases with increasing radius like $\propto(\lambda/a)^3$ [45]. The loss of fundamental longitudinal modes is vanishingly small in oversized resonators due to the almost grazing

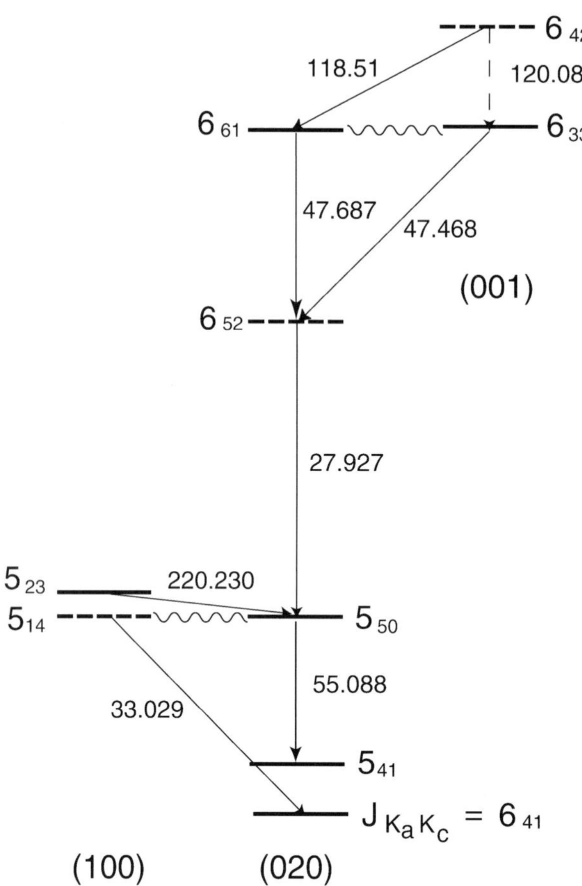

Figure B3.9.2. Part of the rotation–vibration energy level diagram and laser transitions of $H_2{}^{16}O$, among them the lines listed in table B3.9.1. The laser levels shown are between 2500 and 4500 cm^{-1} above the ground state. The transitions are indicated by their wavelength in μm. Energy levels are shown for the case of strong transitions of statistical weight 3 (parallel orientation to the hydrogen proton spins). The Fermi resonance ((100) ⇔ (020)) and the Coriolis resonance ((020) ⇔ (001)) are indicated by curly lines. J is the total angular momentum, K_a and K_c are the K values in the limit of prolate and oblate tops, respectively, where K is the angular momentum along the symmetry axis of a symmetric top. Levels of symmetry classes $++$ and $+-$ are plotted as full lines, classes $-+$ and $--$ are shown as dashes. The $\pm$ are parities of two two-fold rotations in physical space around two independent principal axes of the moment of inertia. For more details see [3].

incidence of radiation on the wall. The curvature of the mirrors is still important as it affects the mode volume [1].

Partially reflecting loss-free solid state mirrors, such as those used as output reflectors in optical lasers, do not exist in the far-infrared due to phonon absorption at room temperature. Therefore there is no standard scheme to couple radiation from the resonator. Several coupling systems have been developed and applied. The most used among them include hole coupling, coupling with a thin dielectric film, coupling with a Michelson interferometer and coupling by a metallic mesh.

Coupling through a hole in the centre of one of the resonator mirrors is technically simple and thus mostly applied. The disadvantage is the diffraction widening of the beam and the need for an individual hole diameter for each laser line in order to optimize output. The latter requirement is in fact nowadays no real restriction as most discharge-pumped lasers are used with one line only.

Coupling is performed by mounting a thin (5–20 μm) Mylar or polyethylene film with a tilt angle of 45° to the resonator axis. Absorption losses in the thin films can be ignored. The reflection coefficient is independent of wavelength in a wide wavelength range but it is polarization dependent. Thus the output beam is polarized. Two output ports may be used yielding two low-divergence beams of the same cross section as the resonator.

A Michelson interferometer (see section A5.2.5) with a dielectric film, usually Mylar, as beamsplitter is very reliable making the operation of a laser on different lines easier. By varying the length of one interferometer arm, the output can be tuned and optimized for different laser lines.

Thin metallic meshes represent low-loss partial reflectors in the far-infrared [43] and have been frequently used [4, 32]. The advantage is that they retain the laser beam and the linear structure of the resonator. The pitch has to be adapted to the particular laser line. Using two parallel metallic meshes with adjustable spacing gives the arrangement of a Fabry–Pérot interferometer. As an output coupler it allows the reflectivity and selection of single laser lines to be tuned [40].

Films of Mylar or polyethylene as well as sheets of TPX or crystalline quartz are used as window materials.

B3.9.6 Glow discharge and additive gases

The optimum radiation power of a longitudinally pumped cw laser is obtained with a pressure in the range of 0.2–0.8 mbar. The output power of the laser increases with rising current until it assumes a maximum and drops on further increase in current. The optimum current is in the range 1–3 A, yielding a voltage drop of about 2–4 kV. The peak output is found with a wall temperature between 100 and 150 °C. The wall temperature has been controlled by thermostatically temperature-stabilized oil flowing in a jacket around the laser tube. In the case of the HCN laser, a higher temperature is advantageous even if the power is not optimal because it reduces the wall deposit rate. The glow discharge of a cw laser is in the diffusion-dominated regime where the positive column fills almost the whole plasma tube (see section B3.6.2). In the positive column, striations are formed which become unstable at higher currents causing large fluctuations of the laser output power. This process limits the current and radiation power of stable laser operation. Stronger striations affect the lower frequency lines of HCN more than those of H_2O at higher frequencies as the refractive index fluctuations of the plasma decrease with the distance of the laser frequency to the plasma frequency.

A common method for improving stability is to introduce additive gases. Among the different chemical species investigated in the past [30] He, H_2 and H_2O are of importance. Due to its small mass and its high mobility, He reduces the temperature and the radial temperature gradiant. This supports the maintenance of inversion and improves the homogeneity and stability of the discharge. Addition of He shifts the current of maximum output power to higher values or, in turn, lowers the required discharge voltage for a given current. H_2 and H_2O have a similar effect on the discharge. In addition, H_2 may compensate for a shortage of hydrogen in the laser fuel and it helps to relax the energy of the lower laser level bending mode by hydrogen bonding effects.

B3.9.7 Design and operational characteristics

In figure B3.9.3 a standard design of a cw HCN laser is shown. The characteristic data are given in table B3.9.2.

The laser oscillates on the 337 μm line only. The gain of the laser is too small to obtain other HCN lines. The length of the laser tube is made of glass (Schott Duran 50) and is thermally stabilized by four invar rods. Sufficient thermal stability is achieved after running the laser for about 1 hr. The temperature at the wall

Table B3.9.2.

Resonator:	
length	2.4 m
curvature of movable mirror	4 m
curvature of coupling mirror	∞
hole coupling, bore	15 mm
output window	TPX sheet, 3 mm thick
thermal stabilization	4 invar rods, 25 mm diameter
Discharge:	
laser gas	CH_4 and N_2, 1:1
optional additive gases	H_2O, He
pressure	0.4. . . 0.6 mbar
flow rate	500 $cm^3\ min^{-1}$
anode	stainless steel ring, 70 mm wide
cathode	hollow copper structure 50 mm diameter, water cooled
voltage	1.0. . . 2.5 kV
current	0.5. . . 1.5 A
load resistor	600 Ω
Power:	
gain	0.5 dB m^{-1}
output power	5 mW

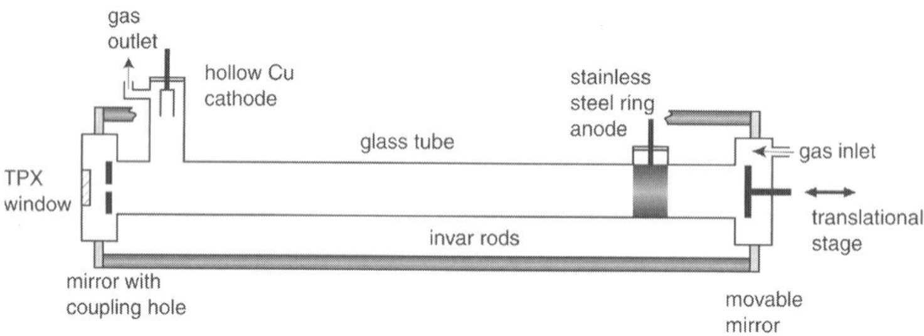

Figure B3.9.3. Design of a cw HCN laser.

of the tube is then 150 °C. The stability of the laser also depends critically on the electrode configuration. As an anode, a stainless steel ring 7 cm wide attached concentrically onto the inner wall of the discharge tube and, as a cathode, a hollow copper cylinder in a side arm of the tube, have proved to provide a stable discharge. The cathode is cooled by water flow. Another important parameter with respect to stability is the gas pressure. The discharge runs smoothly below about 0.6 mbar forming stable mushroom-like striations. Above this pressure the striations start irregular motions and mixing until any regular pattern in the discharge disappears. In this state the output power strongly fluctuates; however, the average power is two to three times higher than in the smoothly running mode. Adding small amounts of water vapour to the discharge increases stability, adding He does not have much effect.

Making full use of the waveguide effect of rather thin discharge tubes, similar but more compact cw laser designs have been reported by Belland and Veron [1], Véron *et al* [46] and Bruneau *et al* [4]. In these devices the temperature of the tube wall could be actively controlled by the flow of oil in an outer jacket around the tube. This gives another parameter at hand to optimize laser operation.

B3.9.8 Summary

Glow discharge lasers were coherent sources of substantial intensity which bridged the gap between the microwave range and near infrared. The revolution in the field of coherent far-infrared sources was concluded by the discovery of the CO_2 laser pumped molecular laser by Chang and Bridges [6] (see section B3.8) and approached maturity with the advent of the free-electron laser [10] (see section B5.1). Optically pumped molecular lasers deliver thousands of laser lines in the far infrared, they may be operated cw or pulsed, they produce short pulses or extremely high intensities when transversely excited atmospheric pressure (TEA)-CO_2 lasers (see section B3.1.2) are applied as the pump source. To all these the free-electron laser adds tunability in a wide wavelength range. Thus, these laser concepts superseded the electrically pumped far-infrared laser in most cases. However, there are still areas of research where it is reasonable to apply H_2O and HCN lasers. Both lasers are low cost, reliable and of simple construction. In cw mode, they are easy to handle due to direct electrical excitation and deliver on a few lines with a rather high power.

References

[1] Belland P and Veron D 1973 A compact cw HCN laser with high stability and power output *Opt. Commun.* **9** 146–8

[2] Belland P, Veron D and Whitbourn L B 1976 Scaling laws for cw 337-μm HCN waveguide lasers *Appl. Opt.* **15** 3047–53

[3] Benedict W S and Pollack M A 1969 The water-vapor laser *IEEE J. Quantum Electron.* **QE-5** 108–24

[4] Bruneau J L, Belland P and Veron D 1978 A cw DCN waveguide laser of high volumetric efficiency *Opt. Commun.* **24** 259–64

[5] Button K J, Gebbie H A and Lax B 1966 Cyclotron resonance in semiconductors with far infrared laser *IEEE J. Quantum Electron.* **QE-2** 202–7

[6] Chang T Y and Bridges T J 1970 Laser action at 452, 496, and 541 μm in optically pumped CH_3F *Opt. Commun.* **1** 423–6

[7] Coleman P D 1973 Far-infrared molecular lasers *IEEE J. Quantum Electron.* **QE-9** 130–8

[8] Crocker A, Gebbie H A, Kimmit M F and Mathias L E S 1964 Stimulated emission in the far-infrared *Nature (London)* **201** 250–1

[9] Daneu V, Hocker L O, Javan A, Ramachandra Rao D, Szöke A and Zernike F 1969 Accurate laser wavelength measurment in the infrared and far infrared using a Michelson interferometer *Phys. Lett.* A **29** 319–20

[10] Elias L R, Ramian G, Hu J and Amir A 1986 Observation of single-mode operation in a free-electron-laser *Phys. Rev. Lett.* **57** 424–7

[11] Evenson K M, Wells J S, Matarrese L M and Elwell L B 1979 Absolute frequency measurement of the 28- and 78-μm cw water vapor laser lines *Appl. Phys. Lett.* **16** 159–62

[12] Frenkel L, Sullivan T, Polack M A and Bridges T J 1967 Absolute frequency measurement of the 118.6-μm water-vapor laser transition *Appl. Phys. Lett.* **11** 344–5

[13] Gebbie H A, Stone N W B and Findlay F D 1964a Interferometric observations on far infra-red stimulated emission sources *Nature (London)* **202** 169–70

[14] Gebbie H A, Stone N W B and Findlay F D 1964b A stimulated emission source at 0.34 millimetre wave-length *Nature (London)* **202** 685

[15] Golubev V G and Prettl W 1991 FIR photoconductivity at filamentary current flow in n-GaAs in a magnetic field *Solid State Commun.* **79** 1035–8

[16] Herzberg G 1945 *Molecular Spectra and Molecular Structure II: Infrared and Raman Spectra of Polyatomic Molecules* (Princeton, NJ: Van Nostrand Reinhold) p 279

[17] Hocker L O, Javan A and Ramachandra Rao D 1967 Absolute frequency measurement and spectroscopy of gas laser transitions in the far infrared *Appl. Phys. Lett.* **10** 147–9

[18] Hocker L O and Javan A 1967 Absolute frequency measurements on the new cw HCN submillimeter laser lines *Phys. Lett.* A **25** 489–90

[19] Jackson M, Sudhakaran G R and Gansen E 1996 Lees R M 1999 Far-infrared laser Stark spectroscopy of $^{13}CD_3OD$ *J. Mol. Spectrosc.* **176** 439–41

[20] Jie Y X, Gao X, Cheng Y F, Yang K and Tong X D 2000 Multi-channel FIR HCN laser interferometer on HT-7 Tokamak *Int. J. IR MM-Waves* **21** 1375–80

[21] Kim K and King W T 1979 Intergrated intensities in hydrogen cyanide *J. Chem. Phys.* **71** 1967–72

[22] Kneubühl F K and Sturzenegger C 1980 Electrically excited submillimeter-wave lasers *Infrared and Millimeter Waves* vol 3, ed K J Button (New York: Academic) pp 220–60

[23] Lide D R and Maki A G 1967 On the explanation of the so-called CN laser *Appl. Phys. Lett.* **11** 62–4

[24] Mathias L E S and Crocker A 1964 Stimulated emission in the far-infrared from water vapour and deuterium oxide discharges *Phys. Lett.* **13** 25–36

[25] McGarvey J A Jr, Friedmann N E and Cool T A 1977 Vibrational energy transfer in HF-HCN, DF-HCN, and H_2-HCN mixtures *J. Chem. Phys.* **66** 3189–96

[26] Mills I M 1972 Vibration–rotation structure in asymmetric- and symmetric-top molecules *Molecular Spectroscopy: Modern Research* ed K N Rao and C W Mathews (New York: Academic) pp 115–40

[27] von Ortenberg M 1980 Submillimeter magentospectroscopy of charge carriers in semiconductors by use of the strip-line technique *Infrared and Millimeter Waves* vol 3, ed K J Button (New York: Academic) pp 275–345

[28] Otsuka E 1980 Cyclotron resonance and related studies of semiconductors in off-thermal equilibrium *Infrared and millimeter waves* vol 3, ed K J Button (New York: Academic) pp 347–416

[29] Pichamuthu J P 1974 The excitation mechanism of the HCN laser *J. Phys. D: Appl. Phys.* **7** 1096–100

[30] Pichamuthu J P 1983 Submillimeter lasers with electrical, chemical, and inciherent optical excitation *Infrared and Millimeter Waves* vol 7, ed K J Button (New York: Academic) pp 166–241

[31] Pollack M A 1969 Far-infrared laser gain resulting from rotational perturbations *IEEE J. Quantum Electron.* **QE-5** 558–62

[32] Prettl W and Genzel L 1966 Notes on the submillimeter laser emission from cyanic compounds *Phys. Lett.* **23** 443–4

[33] Schilke A, Mehringer D M and Menten K M 2000 A submillimeter HCN laser in IRC +10216 *Astrophys. J.* **528** L37–40

[34] Schötzau H J and Kneubühl F K 1975 Mass spectroscopy of the HCN-laser plasma *IEEE J. Quantum Electron.* **QE-11** 817–22

[35] Schötzau H J and Veprek S 1975 Wall processes and the radial gain of the cw HCN laser *Appl. Phys.* **7** 271–7

[36] Schwaller P, Steffen H, Moser J F and Kneubühl F K 1967 Interferometry of resonator modes in submillimeter wave lasers *Appl. Opt.* **6** 827–9

[37] Sentz A 1983 Operating conditions of a cw water vapour laser at 28, 47, 78, 79, and 119 μm *Infrared Phys.* **23** 9–14

[38] Steffen H, Steffen J, Moser J F and Kneubühl F K 1966 Stimulated emission up to 0.535 mm wavelength from cyanic compounds *Phys. Lett.* **20** 20–1

[39] Steffen H, Steffen J, Moser J F and Kneubühl F K 1966 Comments on a new laser emission at 0.744 mm wavelength from ICN *Phys. Lett.* **21** 425–6

[40] Steffen H and Kneubühl F K 1968 Resonator interferometry of pulsed submillimeter-wave lasers *IEEE J. Quantum Electron.* **QE-4** 922–1008

[41] Strafsudd O M and Yeh Y C 1969 The CW characteristics of several gas mixtures at 337 μm *IEEE J. Quantum Electron.* **QE-5** 377

[42] Sturzenegger Ch, Vetsch H and Kneubühl F 1979 Transversely excited double-discharge HCN laser *Infrared Phys.* **19** 277–96

[43] Ulrich R 1967 Far-infrared properties of metallic mesh and its complementary structure *Infrared Phys.* **7** 37–55

[44] Ulrich R, Bridges T J and Pollack M A 1970 Variable metal mesh coupler for far infrared lasers *Appl. Opt.* **9** 2511–16

[45] Ulrich R and Prettl W 1973 Planar leaky light guides and couplers *Appl. Phys.* **1** 55–68

[46] Véron D, Belland P and Beccaria M J 1978 Continuous 250 mW gas discharge DCN laser at 195 μm *Infrared Phys.* **18** 465–8

Further reading

Kneubühl F K and Sturzenegger C 1980 Electrically excited submillimeter-wave lasers *Infrared and Millimeter Waves* vol 3, ed K J Button (New York: Academic) pp 220–60

Kneubühl pioneered submillimetre glow discharge lasers. This review was written after understanding of the physics background was settled and after the technology had approached a certain technological maturity. Coverage of all excitation mechanisms (cw, pulsed longitudinal, transversely excited).

Pichamuthu J P 1983 Submillimeter lasers with electrical, chemical, and inciherent optical excitation *Infrared and Millimeter Waves* vol 7, ed K J Button (New York: Academic) pp 166–241

A review with emphasis on the kinetic and transport processes in the laser plasma. A very thorough discussion of chemical processes, excitation and relaxation in the plasma and rate equation models.

Chantry G W 1984 *Long-Wave Optics* (London: Academic)

A concise textbook presentation of the inversion mechanism and energy level scheme of the HCN and H_2O laser (Vol 1: Basics, pp 150–3) and technical performance (Vol 2: Applications, pp 570–3) in the context of other infrared lasers.

B4
Fibre and waveguide lasers

R C Powell

One of the major technological advances in laser systems was the development of fibre optics [1]. This allowed laser light to be accurately and efficiently delivered to distant targets over ranges from several millimetres to many kilometres. The combination of laser sources and fibre optic delivery systems has lead to important applications such as modern telecommunication systems and non-invasive laser surgery. Sections D2 and D4 describe some important application of fibre optics laser systems. The next step in this technology development is to combine the fibre delivery system and the laser into one device. Since fibres are made out of glass, they can act as the host material for rare earth laser ions to make fibre lasers. All of the fundamental science and technology of rare-earth-doped glass laser materials discussed in section B1 are relevant for fibre lasers. The major difference is that the length of a single pass of light in a fibre laser is much longer than in a rod or slab laser. This requires significantly different resonator design techniques.

The basic spectroscopic properties discussed in section B1 and the waveguide theory discussed in chapter A6 form the basis for fibre lasers. Because of the length of the gain media, nonlinear optical effects play an important role in fibre lasers. The basic principles of nonlinear optics are described in chapter A4. Nonlinear optical processes can be the limiting factor in determining the power output of a fibre laser. At the same time, nonlinear processes such as stimulated Raman scattering can form the basis of the fibre laser itself.

The development of the technology to produce fibres with special shapes and properties has lead to the ability to make fibre lasers with very low loss over very long distances. By controlling the core size and shape, and using Bragg gratings for distributed feedback, the power, mode structure, dispersion and wavelength of the laser can be designed. Rare-earth-doped fibre amplifiers have become standard devices in optical telecommunications systems. All of these topics are discussed in this section.

References

[1] Yeh C 1990 *Handbook of Fiber Optics* (San Diego, CA: Academic)

B4.1
Fibre lasers

David Hanna

B4.1.1 Introduction

Amongst fibre lasers there is a wide variety of fibre geometries, resonator designs and pumping arrangements. Some of these variants will be described later in this chapter and in companion chapters. First, we describe the canonical fibre laser design, since this illustrates the salient characteristics of fibre lasers in general.

This basic structure (see figure B4.1.1) is essentially the same as that of a conventional optical fibre as used in telecom applications, i.e. with a central glass core surrounded by a glass cladding, whose refractive index, n_{cl}, is less than that of the core, n_{co} thus ensuring guidance of light within the core. Unlike in conventional fibres, the core contains laser-active dopant ions which, when suitably pumped, provide amplification at wavelengths corresponding to a transition of the dopant ion. So far, laser action has been confined to the rare earths (RE) (see chapters B1.3 and B1.7). Pumping is achieved optically, e.g. by launching the pump light directly into the exposed core at the end of the fibre (see figure B4.1.4). Thus, the core not only guides the amplified laser light but also the pump light. The final ingredient, needed to complete the laser, is provision of a resonator for feedback, thus converting the amplifier to an oscillator (see chapters A1 and A2). In its simplest form, relying on the very high gains that can be achieved in active fibres, this feedback can be provided by Fresnel reflections from the fibre ends. Often a more sophisticated feedback arrangement is used, with the laser light emerging from a fibre end and then passing through a region of free space before being fed back, (see figure B4.1.1). This then allows a range of different components, such as filters and modulators, to be inserted into the resonator for control of the spectral and temporal characteristics of the laser output. We now briefly survey some of the main features of fibre lasers, before elaborating on these and giving them a quantitative basis in section B4.1.3.

B4.1.2 Features of fibre lasers

High gain and low threshold. The basis for these two connected features, is the small cross-sectional area of the core, typically in the range 10^{-11}–10^{-10} m^2, corresponding to core diameters in the range 4–10 μm. This area is a factor of 2–3 orders of magnitude smaller than the typical pumped area involved in an end-pumped bulk laser. Hence this same factor increases the gain for a given pump power, or conversely decreases the threshold for a given resonator loss. Submilliwatt threshold pump powers and gains of 10 dB mW^{-1} are not uncommon for active-fibre devices.

These high gains led to a number of important possibilities, of which high-gain amplifiers for telecom have been the most prominent [1–3], while more generally it provides very convenient master oscillator–power amplifier (MOPA) schemes for power-scaling of a wide variety of laser sources, including diode lasers and fibre lasers [4].

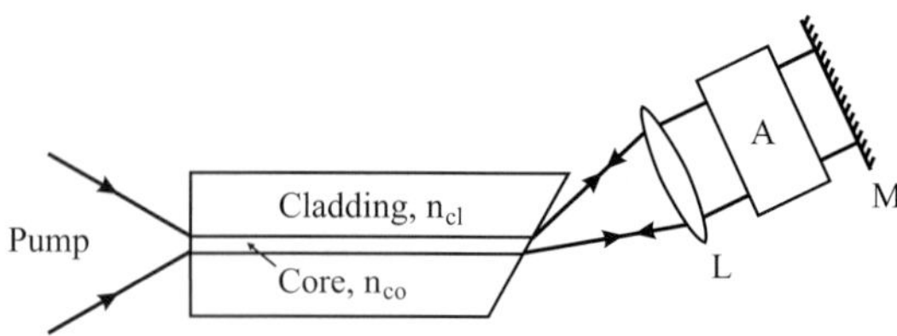

Figure B4.1.1. An end-pumped fibre laser. The resonator shown has feedback provided on the left by Fresnel reflection from the end-face of the fibre, and on the right by a separate mirror, M. Light emerging from the right-hand end of the fibre is collimated by a lens, L. Intracavity components can be inserted at A. Unwanted feedback from the right-hand end-face of the fibre can be reduced by cleaving or polishing the end of the fibre at an angle, as shown.

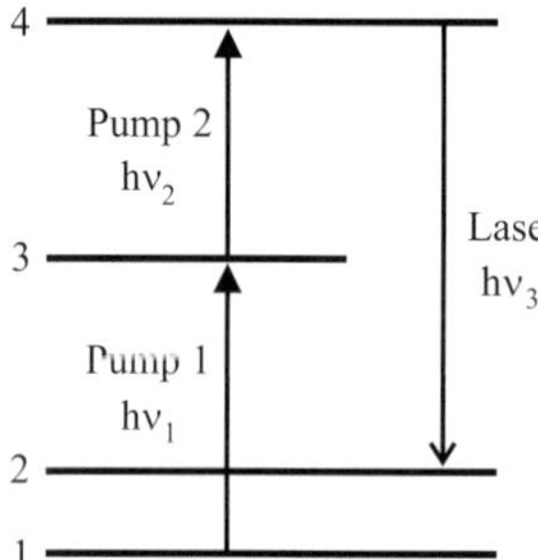

Figure B4.1.2. An up-conversion laser (see chapter B1.6). Sequential absorption of two pump photons of energy $h\nu_1$, $h\nu_2$, (these may be identical for a suitable energy-level scheme) takes ions to the excited level 4, and subsequent laser emission to level 2 produces laser-photons that are more energetic than the pump-photons.

Related to the fact that modest pump powers can lead to high pump intensities in the core, is the result of very strong saturation (or 'bleaching') of the absorbing ions, i.e. most of the ions can be pumped out of the ground state into an excited state. This population of excited ions can then easily be subjected to a second or even a third pumping-stage that takes them to yet higher excited states. Laser action can then take place from these highly excited states with the emission of photons that are more energetic than the pump-photons (see figure B4.1.2). This is referred to as an up-conversion laser, and while up-conversion lasers were first demonstrated in a bulk-laser geometry, a dramatic improvement in performance and practicality resulted when a fibre laser geometry was adopted, leading to cw visible lasers directly pumped by near-infrared diode lasers [5].

High efficiency. The excellent modal overlap between pump and laser modes over the entire length of the laser medium, contributes to the typically very high efficiency. Also the high gain and the low background losses of the host (for silica, a few dB km^{-1}), allow one to ensure that the output coupling is the dominant loss, i.e. essentially no laser photons are wasted. The ability to use a three-level laser effectively, by virtue of the high pump intensity in the core, has given prominence to the ytterbium (Yb^{3+}) ion. This ion offers no significant channels for parasitic loss of excitation from the upper laser level and essentially 100% quantum efficiency (~80% power efficiency) can be achieved [6].

Broad gain-linewidth. In a fibre, the host medium being glass (silica is commonly used), the dopant ions occupy a wider range of environments than in the ordered lattice of a crystal host. This results in a significant broadening of the transition linewidths, including that of the amplifying transition. As an example,

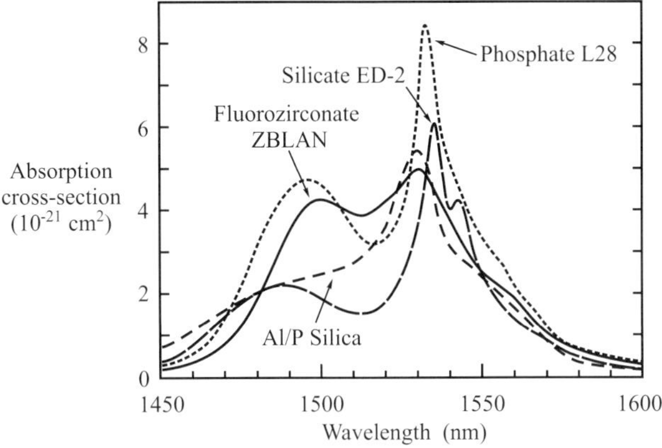

Figure B4.1.3. The spontaneous-emission spectrum of erbium at ~1500 nm in various glass hosts [7].

figure B4.1.3 shows a typical spontaneous-emission spectrum of an erbium-doped silica fibre in the 1.5 μm spectral region, indicating a bandwidth of ~4 THz or 25 nm. This is more than an order of magnitude broader than the same transition would exhibit in a typical crystalline host. Other REs show similarly broad, and in the case of the important ions Yb^{3+} and Tm^{3+} even broader, linewidths.

The broad gain-profile implied by figure B4.1.3. has an important application in optical telecom. The fact that the gain for a given pump power varies inversely with the emission bandwidth, is easily accommodated by the gain-enhancement conferred by the waveguide geometry. More generally, the broad linewidths offer significant scope for tuning of fibre lasers and for ultrashort-pulse mode-locked operation. The Er gain-bandwidth is sufficient to support mode-locked pulses of ~100 fs duration, as has been widely demonstrated [8] and Yb has produced even shorter pulses [9]. In the context of mode-locked operation, one should also note two other features of the fibre that play an important role. These are the dispersion, which can be significant as a result of a long length of medium, and self-phase modulation (SPM) [10], which can be strong as a result of the combination of high peak-intensity and long path length.

Robust modal characteristics. If the core diameter is chosen small enough, so as to support propagation of just the lowest-order-mode, the fibre laser will ensure single-spatial-mode output despite the fibre being subjected to major disturbances, such as bending of the fibre and thermal variations either of the environment or due to pump-induced heat-load. This contrasts with the much more critical behaviour of typical bulk lasers

Convenient heat-dissipation. As well as having the benign modal response to thermal disturbance, the long, thin geometry of the fibre makes heat dissipation very simple [11], particularly in comparison to bulk-lasers. It is interesting to note that the fibre has essentially the same limit of tolerable heat generation per unit length, determined by its liability to stress-fracture, as a bulk-laser-rod, since this is independent of diameter. To make a fibre long enough to circumvent this fracture limit is, however, trivially easy. On the other hand it should be noted that efforts to scale up the power of fibre lasers by scaling the core diameter will ultimately be confronted by the same thermal constraints as found in bulk lasers.

Availability of robust and versatile fibre components. The practicality of fibre lasers and their further prospects benefit enormously from the availability of a range of components and techniques flowing from developments in telecom. A striking example is that of the fibre-Bragg grating [12] (see chapter B4.3). This can be optically written directly into the core of the fibre (usually silica) and can serve as a mirror or filter. Not only does it have superior characteristics of stability and robustness compared to bulk counterparts, but it has much greater functionality, e.g. allowing it to fulfill sophisticated pulse-shaping functions that enhance the fibre laser's role as a source of ultra-short pulses. The grating can be made to extend over the full length of the gain medium, thus resulting in a distributed feedback (DFB) laser [13, 14], a very convenient source with an output that is robustly single-frequency (single-longitudinal-mode). Another example of a simple and stable component is the fibre-coupler, obviating the need for bulk beam-splitters. Figure B4.1.4 shows a typical example of how fibre lasers can have an all-fibre construction by incorporating various fibre components.

Power limitation. The small core size, necessitated by the requirements for supporting a single spatial-mode, inevitably imposes some limits on the power-handling capability of fibre lasers (see chapter B4.2). Despite this ultimate constraint, considerable power-scaling has been achieved by increasing the core-diameter while maintaining lowest-order-mode operation by reducing the index-difference, $n_{co} - n_{cl}$ (see section B4.2). Also it is found that even when a fibre is, in principle, capable of supporting higher-order-modes, lowest-order-mode operation can be maintained by suitable design [16]. Such techniques have elevated fibre lasers from their previously perceived status as low-power devices (a few tens of milliwatts) to the region of tens, and even hundreds, of watts. Multimode operation, which may be acceptable in some applications, offers yet further scope for power-scaling. A very important element in this scaling lies in the choice of geometry, both of the fibre and of the pump source, to enable high pump power to enter the core.

B4.1.3 Basic principles and formulae

For simplicity we assume that the fibre has a so-called step-index profile (uniform refractive index, value n_{co} in the core, n_{cl} in the cladding). The numerical aperture, NA, of the fibre defines the maximum incidence angle θ_m ('acceptance-angle') at the end-face of the fibre core, for subsequent guidance to take place in the core, with θ_m given by, $\sin\theta_m = NA$ (see chapter A6). The NA is given by [17]

$$NA = (n_{co}^2 - n_{cl}^2)^{1/2}. \tag{B4.1.1}$$

It is customary to define a quantity referred to as the V value, given by

$$V = 2\pi a(NA)/\lambda \tag{B4.1.2}$$

where a is the core-radius and λ is the (vacuum) wavelength of the guided wave. For the fibre to be monomode at wavelength λ, (i.e. for the LP_{11} and higher-order-modes to be cut off, allowing only the LP_{01} mode to propagate), V must be less than 2.4 (see section A6.4.1). Thus, substituting in equation (B4.1.2), a wavelength of 1 μm and a typical NA of 0.15, one finds that for $V = 2.4$, a has the value 2.5 μm, implying a maximum diameter of 5 μm for monomode propagation at this wavelength. One can see from equation (B4.1.2) that monomode operation can be maintained at larger diameters provided the NA is appropriately reduced to keep V below 2.4. There are however, practical limits to this procedure since, e.g., the reduced index difference, $n_{co} - n_{cl}$, makes the fibre more liable to bending losses.

Conversely, to increase gain for a given pump power, or equivalently to reduce the threshold power, one needs to minimize the mode areas for pump- and laser-modes. This can be achieved, up to a point, by reducing the core size. However, it is important to note that, while it is a reasonable approximation to identify the LP_{01} mode area with core area for V values around 2.4, this is not in general true and, for small V values, ($V < 2$) the mode size increases as V is further reduced. Hence, reduction in core size must be accompanied by a corresponding increase in NA (to keep V from decreasing below $\sim$2) if reduced mode volume is to result. Thus, assuming $V \sim 2.4$, so that mode-area $\cong$ core-area, then equation (B4.1.2.) implies that gain is proportional to $(NA)^2$.

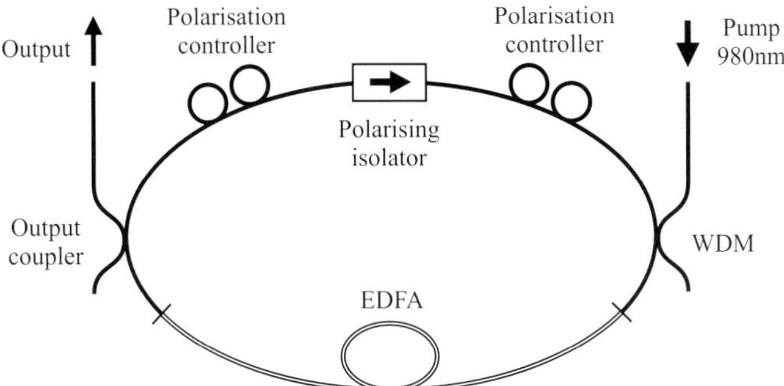

Figure B4.1.4. A stretched-pulse mode-locked Er-doped fibre laser (after [18]), using an all-fibre construction, in a unidirectional ring configuration. Pump light enters at the wavelength division multiplexer (WDM) and pumps the erbium-doped fibre amplifier (EDFA). Output is taken via the fibre output coupler. Tightly wound fibre loops are used to control the polarization state. The constituent fibre lengths and dispersion values, positive and negative, are chosen to achieve the desired stretching and compression of the pulses.

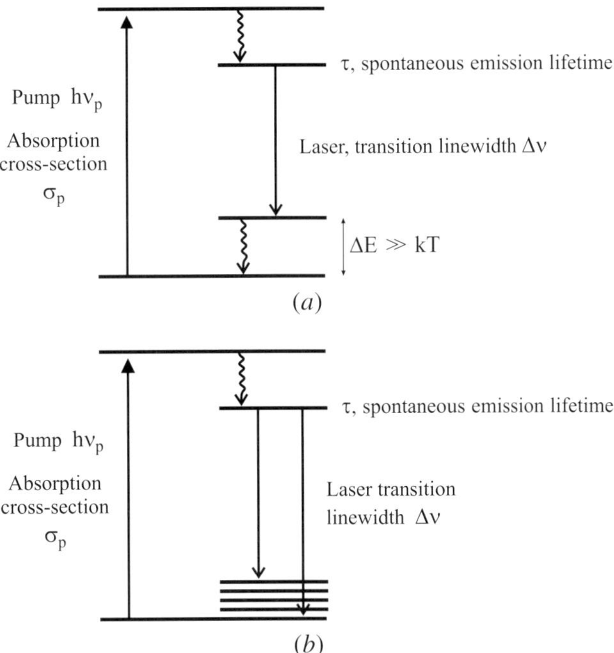

Figure B4.1.5. (*a*) A four-level transition, in which the lower laser level is several kT above the ground level. (*b*) A three-level transition, in which the lower laser level is the ground level. Wiggly arrows indicate non-radiative decays. The multilevel nature of the ground manifold is explicitly indicated in (*b*) illustrating the fact that longer wavelength emission has some 'four-level' character as it involves higher-lying and therefore less populated sublevels.

If one now considers a four-level laser, i.e. with an empty lower-laser level, (see figure B4.1.5(*a*)), assuming $V = 2.4$ and equating core area to mode area, then the gain, expressed in dB mW^{-1} of absorbed pump power, is given by

$$\text{Gain} = (NA)^2\phi/(10^4\Delta\nu h\nu_p). \tag{B4.1.3}$$

Here, $\Delta\nu$ is the linewidth (FWHM) of the amplifying transition, ν_p is the pump frequency and ϕ is the product of ϕ_p, the pump-quantum efficiency, i.e. the fraction of absorbed pump photons that produce an ion in the upper-laser level, and ϕ_r, the radiative quantum efficiency, i.e. the fraction of spontaneous decays of ions in the upper-laser level that occur as radiative emission via the amplifying transition. Insertion of typical numbers for the case of Nd^{3+} ions (see chapter B1.3) (0.8 μm pump, $\lambda_p \sim 1$ μm, $\phi_p = 1$, $\phi_r = 0.5$ (radiative decay also occurs via the 0.94 μm and 1.3 μm channels), $\Delta\nu = 3$ THz) and assuming an NA of 0.15, gives a gain of 1.5 dB mW^{-1}. A somewhat higher figure has been achieved in Er-doped fibre at $\sim$1.5 μm, where $\phi \sim 1$, and by deliberately using a larger NA, a gain of 11 dB mW^{-1} has been achieved [19]. The expression of equation (B4.1.3) cannot be applied directly to the case of the Er^{3+} ion, however, as this exhibits a three-level scheme, i.e. the lower-laser level is also the ground level (see figure B4.1.5(*b*)). An approximate analytical treatment of the three-level case leads to equation (B4.1.3) being modified by insertion of the factor $(I/I_{sat} - 1)/(I/I_{sat} + 1)$ into the right-hand side, where I_{sat} is the saturation intensity of the pump transition, $I_{sat} = h\nu_p/\sigma_p\tau$. Here σ_p is the pump-absorption cross section and τ is the lifetime of the upper-laser-level. When $I \gg I_{sat}$, a condition that is easily achieved in a fibre since it may typically need only $\sim$milliwatt power levels, then the three-level case approaches that of the four-level, as expected, since this corresponds to most of the ground-state population having been promoted to the upper-laser-level. In general, however, the treatment of the three-level case is more complicated analytically than that of the four-level, since the ratio I/I_{sat} decreases progressively along the fibre as the pump intensity is decreased by absorption (for $I < I_{sat}$, this factor indicates that the medium has negative gain, i.e. is absorbing). Furthermore, a number of laser transitions involve a lower-laser-level that is somewhat above the ground level, although still containing a significant thermally-excited population. This is often referred to as a quasi-three-level transition. Further complications arise from the fact that the nature of a transition may in fact be continuously graded from effectively three-level at the short-wavelength end of the emission spectrum, to four-level at the long-wavelength end, since the longer emission wavelengths correspond to emission into higher-lying sublevels of the ground manifold (see figure B4.1.5(*b*)).

To complete this brief survey of basic principles an indication is given of the role played by amplified spontaneous emission (ASE) in limiting the maximum achievable gain and thus also limiting the maximum energy that can be stored (and subsequently extracted, e.g. in the form of a Q-switched pulse). Photons created by spontaneous emission at one end of the fibre and then guided over the fibre's length will be amplified, thereby cause some reduction of the population-inversion and hence of the gain. This ASE process therefore acts as a limiter, reducing the achievable gain even in the absence of any feedback, to a level that, in practice, is typically 40 dB over a single pass. ASE for example, provides the ultimate limit on tuning range achievable from a fibre laser. Thus, when tuning to the extreme wings of a transition, where the gain is low, while one can, in principle, compensate the low gain by pumping harder, this can raise the gain at line-centre to a level that induces strong ASE. This therefore limits the tuning range that can be covered. ASE also limits the energy stored in the population inversion. The relation between gain and stored energy is

$$\text{Gain[dB]/(Stored energy)} = 4.34/E_{sat} \qquad \text{(B4.1.4)}$$

where $E_{sat} = A_{core}F_{sat}$, with A_{core} being the area of the core and F_{sat} the saturation fluence, given by $h\nu_l/\sigma_l$, where ν_l and σ_l are the laser frequency and emission cross section respectively. The maximum stored energy, limited by the onset of ASE, is found by inserting a gain of $\sim$40 dB in equation (B4.1.4). The effect of a small core-diameter is immediately apparent as the stored-energy, for a given gain, is proportional to A_{core}. Typical values for Nd^{3+} for a 5 μm core-diameter are $\sim$3 dB μJ^{-1}, thus the ASE-limited energy is $\sim$10 μJ. For Er^{3+}, with a much smaller σ_l, corresponding to the much longer τ_l than for Nd^{3+} ($\sim$30$\times$), there is a correspondingly greater energy-storage capability. Further increases in the energy storage are achievable by scaling up the core area.

B4.1.4 Dopants and fabrication

Fabrication. So far only rare earth dopants have produced laser oscillation in a glass host. A number of techniques have been used to introduce the dopant [3, 20, 21]. In one of the standard methods for fabricating silica-fibres, modified chemical vapour deposition (MCVD), the constituents that make up the core are introduced as gases, flowed through a hollow tube of silica and deposited as a solid on the inner surface of the tube as a result of heat-induced decomposition. By using a volatile compound containing the RE, this MCVD process was extended to allow incorporation of REs into the core material. Finally, strong heating of the tube makes its hollow core collapse so that the deposited material now forms the solid core. Introducing the dopant in this way gives accurate control of its concentration profile, since one can build up the core material with a sequence of depositions of different composition.

The volatile RE compounds are costly however, so an alternative technique is more widely used, the so-called solution-doping technique [22]. Here the core material, again produced by MCVD, is deposited as an open structure with many voids (a soot). The tube is then immersed in an aqueous solution containing RE salts. Heating in various stages then evaporates the water, leaving deposited RE in the voids; the frit is compacted and the core is collapsed.

The rod with its solid, doped core (called the preform) can now be heated in a furnace and drawn down into a fibre. This can be done in two stages, a partial drawing-down, to form a 'cane', and then this cane is inserted in a close-fitting tube ('sleeve') of silica and the combination drawn down together. This gives independent control over the core diameter and the outer fibre diameter. The latter is often chosen to have a standard dimension (125 μm), for convenience of handling and also to facilitate common procedures such as fusion splicing, used to join the ends of different fibres.

A wide variety of other fibre-fabrication techniques offer an extensive set of variations on the basic fibre-geometry [3], which translates into versatility in their role as laser- and amplifier-media. These include the so-called double-clad fibre where an inner cladding is surrounded by an outer cladding of lower index-value [23]. This allows highly multimode pump light, e.g. from a high-power diode-laser to be coupled into and guided within the inner cladding and thus eventually absorbed into a monomode core, resulting in a monomode laser output. This is an elegant scheme for achieving an output with brightness greatly enhanced over that of the pump, and is also a basis for high-power operation of fibre lasers where a major challenge is that of coupling high-power, multimode light into the fibre core. Variants on the inner cladding shape include rectangular cross section, D-shaped cross section etc, achieved by shaping the preform and then relying on the preservation of this form as the fibre is drawn. Other techniques include drilling a hole in a preform and inserting a doped rod to serve as the core when drawn down. Mention should also be made of so-called holey fibres, made, for example, by stacking hollow tubes around a solid rod (of doped material in the case of a laser), then drawing this down to a fibre, while preserving the air-holes of the original tubes [24, 25]. In this way, one can achieve a large NA for the core. since its surrounding is largely air [26]. The reduced average-index provided by the holey region gives an extra control over the refractive index and can be used to fashion a guiding-interface, e.g. at the outer boundary of the inner-cladding in a double-clad-fibre.

Dopants. Laser action in a fibre geometry has been reported for eight of the RE^{3+} ions. For several of the ions there are multiple laser transitions. In general, for a host-glass whose most energetic lattice vibrations are of low energy (low-phonon-energy glasses) there is a reduced probability of an excited ion decaying non-radiatively, by means of multiphonon emission. This probability of multiphonon decay depends strongly on the number of phonons (less probable for more phonons) that need to be emitted to bridge the energy-gap between the excited state and the nearest lower-lying state. Typically, for a silica host (with its relatively high value of maximum phonon energy, $\sim$1150 cm^{-1}) and for laser wavelengths longer than $\sim$1.5 μm, the corresponding energy gap is narrow enough for multiphonon decay to produce significant shortening of the upper-laser-level lifetime. For wavelengths beyond 2 μm, a low-phonon-energy glass is needed,

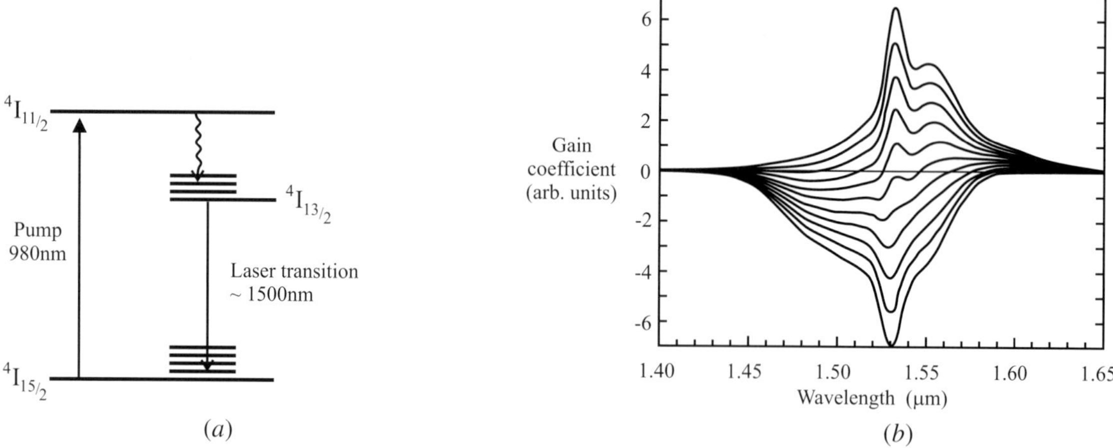

Figure B4.1.6. (*a*) An energy-level diagram of Er^{3+}, showing the lower energy levels relevant to pumping at 980 nm and lasing at ~1500 nm. The Stark sublevels of the upper and lower laser levels are indicated schematically. (*b*) Absorption loss/gain *versus* wavelength for the laser transition, with pump-intensity as a parameter, ranging from zero on the lowest curve (corresponding to unpumped absorption) to infinity on the top curve (corresponding to a fully-inverted medium). (© John Wiley & Sons, Inc. 1994. This material is used by permission of John Wiley & Sons, Inc.)

both to reduce the multiphonon decay rate and to extend the infrared-absortion edge to longer wavelengths, since silica shows significant absorption beyond 2 μm. The heavy-metal-fluoride glass, ZBLAN, is the best-developed material for this purpose [27] with a maximum phonon energy about half that of silica. This glass also plays an important role for up-conversion lasers, since it slows the multiphonon decay rates from many of the highly-excited states, reached by sequential multiple-photon absorption. Consequently, an abundance of up-conversion lasers have been demonstrated in ZBLAN fibre, none however in silica. Nevertheless, those laser transitions that operate in silica have particular importance, since they can fully exploit the benefits of silica. These benefits include low loss, high mechanical and optical strength, highly developed technologies such as Bragg-grating technology, and the variety of fabrication technologies referred to before. As illustrative examples of fibre laser transitions in silica, we now give details for the (~1.5 μm) Er^{3+} and (~1 μm) Yb^{3+} transitions, these representing the most developed systems, and some information is also given on the potentially important (~1.9 μm) Tm^{3+} and (0.94, 1.05 and 1.35 μm) Nd^{3+} transitions.

Erbium. Figure B4.1.6(*a*) shows the lower energy levels of Er^{3+}, relevant to its operation as a laser, pumped at 980 nm and lasing around 1530 nm. Its many, higher levels can also play a significant role, in a variety of parasitic processes, such as excited-state absorption (ESA) and ion–ion interactions such as cooperative energy transfer (CET), [2]. The lifetime of the upper laser level, $^4I_{13/2}$, is typically ~10 ms, essentially determined by radiative decay, without any significant non-radiative contribution. Figure B4.1.6(*b*) shows the typical characteristics of the gain/absorption behaviour, as a function of wavelength, with pump-intensity as a parameter, increasing from zero (unpumped medium, negative gain corresponds to absorption) for the bottom curve, to infinite intensity (fully-inverted medium) for the top curve. Clearly visible in these curves is the fact that gain appears at the longer wavelengths for low intensities (~4-level behaviour at long wavelengths) and spreads to shorter wavelengths for higher pump-intensities (~3-level behaviour at short wavelengths). This behaviour relates to the fact that longer wavelengths correspond to emission to higher (and therefore less populated) Stark levels of the ground-manifold, $^4I_{15/2}$. The Stark levels of the $^4I_{15/2}$, $^4I_{13/2}$ manifolds (8, 7 in number, respectively) are shown schematically in figure B4.1.6. Pumping can also

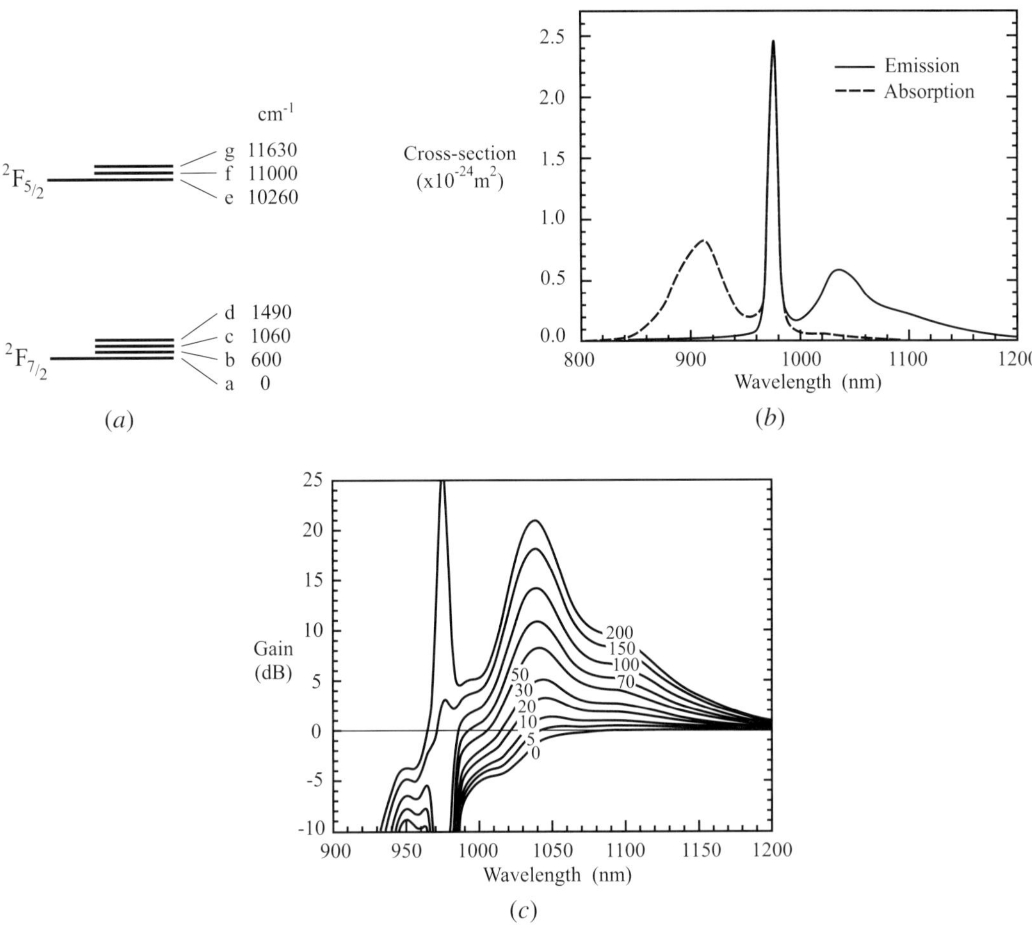

Figure B4.1.7. (*a*) The energy levels of the Yb^{3+} ion. The Stark sublevels (3, 4, respectively) of the upper and lower manifolds are indicated schematically. (*b*) Absorption and emission spectra for Yb^{3+} in fused silica (after [6]). (*c*) Calculated gain/absorption loss *versus* wavelength for a 1 m fibre, core-diameter 3.75 μm, $[Yb^{3+}] = 550$ ppm with launched pump power (mW) as parameter (after [6]).

be carried out directly into the upper Stark levels of $^4I_{13/2}$, (so-called 'in-band pumping'), at a wavelength of $\sim$1470 nm.

Ytterbium. Figure B4.1.7(*a*) shows the relevant energy levels of the Yb^{3+} ion, with the Stark levels (4, 3 in number, respectively) of the ground ($^2F_{7/2}$) and upper ($^2F_{5/2}$) manifolds. There are no higher energy levels before the UV region, thus the Yb^{3+} ion is free from ESA. While CET can still occur, its effect is only to transfer excitation of the $^2F_{5/2}$ level from one ion to another. The energy gap between upper and lower manifolds is also sufficient to ensure that the observed fluorescence lifetime of $\sim$1 ms is essentially radiative, with negligible non-radiative contribution. This absence of channels for decay of excitation from the upper-laser-level makes the Yb^{3+} system attractive for efficient high-power fibre lasers [6]. The well-spaced Stark levels give broad absorption- and emission-bands (see figure B4.1.7(*b*)). The simplicity of its energy-level scheme has also given Yb^{3+} an important role acting as a 'sensitizer' or 'donor', whereby it transfers its

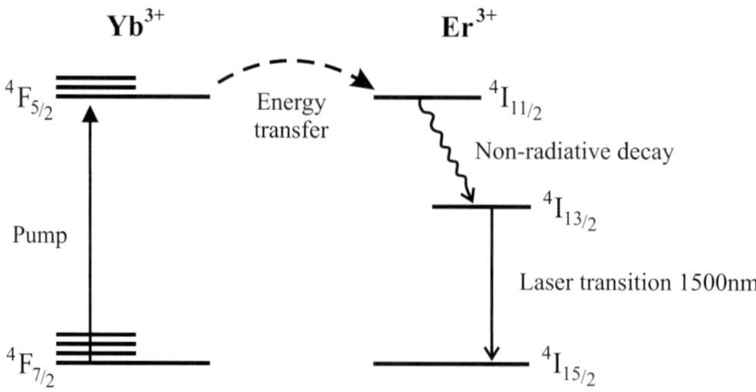

Figure B4.1.8. The energy transfer from Yb^{3+} to Er^{3+}. Reverse transfer can also take place, however if the non-radiative decay from $^4I_{11/2}$ to $^4I_{13/2}$ is sufficiently rapid, the population of $^4I_{11/2}$ remains small, and this reverse transfer is correspondingly reduced.

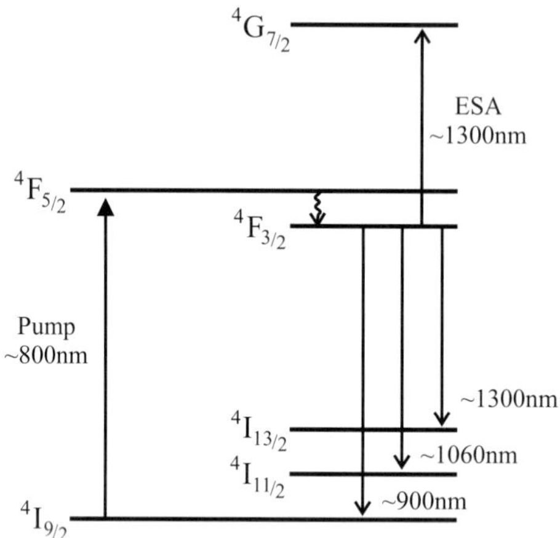

Figure B4.1.9. The relevant energy levels of Nd^{3+}. The emission at ~900 nm corresponds to a three-level transition, whereas emissions at ~1060 nm and ~1300 nm correspond to four-level transitions. Excited state absorption (ESA) from $^4F_{3/2}$ to $^4G_{7/2}$ at ~1300 nm reduces the range around 1300 nm over which gain can occur.

excitation to the excited state of a neighbouring ion [28]. The principle is illustrated in figure B4.1.8, for the Yb–Er co-doped system. An Yb^{3+} ion, having been pumped to the $^2F_{5/2}$ level, can pass this excitation to a neighbouring Yb^{3+} ion (initially in the ground-state) and so on until the excitation reaches an Yb^{3+} ion whose neighbour is an Er^{3+} ion in its ground state. The excitation then passes to the $^4I_{11/2}$ level of the Er^{3+} ion, and thus, by non-radiative decay, to its upper laser level. This can be a rather effective way of pumping Er^{3+} ions, with the population of Yb^{3+} ions offering a larger effective absorption cross section than via direct pumping of the Er^{3+} ions. This sensitizer role of ytterbium is also used to good effect in the Pr^{3+} visible up-conversion laser, using ZBLAN fibre with Yb–Pr co-doping.

Neodymium. The relevant energy levels of the Nd^{3+} ion, see figure B4.1.9, are perhaps the most familiar of any RE, with the transition at ~1060 nm being widely exploited on account of its 4-level nature. This transition has also been widely used in fibres [23], often serving as an initial test-bed before moving on to a

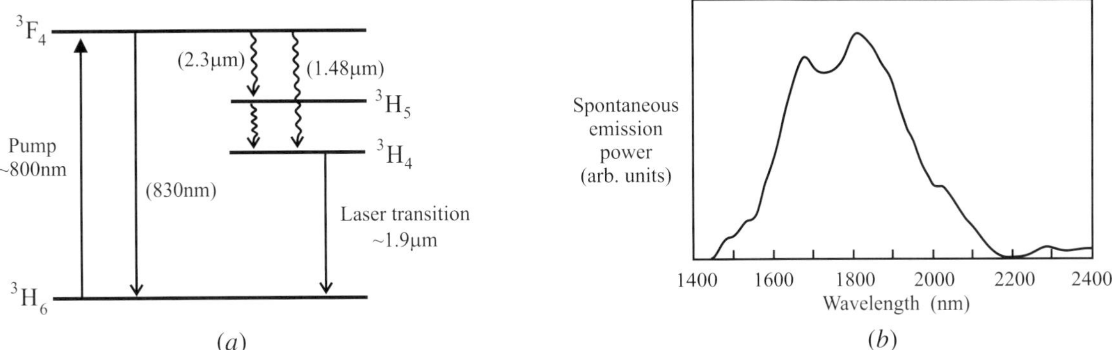

Figure B4.1.10. (*a*) Lower energy levels of Tm^{3+} relevant to its infrared laser transitions. In a silica host the non-radiative (multiphonon) decay from $^3F_4$ is rapid (wiggly arrows) and laser action is from $^3H_4$ to $^3H_6$ at $\sim$1.9 μm. In a ZBLAN host, besides up-conversion laser action in the blue (relevant upper levels not shown) the longer lifetime of the $^3F_4$ level allows laser action at $\sim$830 nm, 1.48 μm and 2.3 μm as shown. (*b*) The broad emission spectrum (after [29]) has permitted laser oscillation to be tuned over the range $\sim$1.75–2.1 μm.

more challenging transition. Two other laser transitions in Nd^{3+} (1.3 μm and 0.9 μm) have also attracted attention, although so far not resulting in significant applications. The 1.3 μm transition has been targeted as a possible amplifier for the second telecom window. However, it suffers from a small gain in comparison with 1060 nm (whose ASE limits the achievable gain), and from ESA at 1.3 μm, which biases the available gain towards longer wavelengths that are of less relevance in a telecom context.

Thulium. Figure B4.1.10(*a*) shows the lower energy levels of Tm^{3+}, relevant to its operation as a laser at $\sim$1.9 μm, on the $^3H_4$–$^3H_6$ transition (some authors refer to $^3H_4$ as $^3F_4$, reflecting the fact that total orbital angular momentum does not constitute a 'good' quantum number [23]). The lifetime of the $^3H_4$ level, typically $\sim$200 μs, is significantly shorter than its radiative lifetime ($\sim$2 ms), a consequence of the proximity between upper and lower laser levels which allows a high rate of multiphonon decay. While a reduced radiative quantum efficiency raises the threshold, it does not affect the slope efficiency. Thus, highly efficient performance has been obtained from the thulium-doped silica fibre lasers [29], with broad tunability, over the range 1.75 μm to 2.1 μm (see figure B4.1.10(*b*) for the emission spectrum), and high-power operation ($\sim$14 W) [30]. In a ZBLAN host, the many other energy levels of the Tm^{3+} ion offer wide scope for laser operation, with efficient operation demonstrated on a number of transitions, including visible up-conversion lasers [31].

B4.1.5 Laser performance

Fibre-lasers have been operated in many regimes, e.g. cw, Q-switched, mode-locked, single-frequency, frequency-tuned, etc on a wide variety of transitions, in various hosts and in a variety of geometrical configurations. Some examples have been alluded to in the preceding sections, however, for an extensive coverage of these topics, the reader is referred to [23] and, for other specific aspects of fibre lasers, to companion chapters in this handbook. Here, the discussion is restricted to a brief mention of three aspects of fibre laser operation, which emphasize the fibre laser's versatility: these are the fibre-DFB-laser, mode-locked fibre lasers and fibre-MOPA configurations.

MOPA configurations are now widely used with bulk lasers as a route to power scaling, while preserving desirable attributes in the output, such as specific temporal and spatial characteristics, which are most conveniently achieved in a low-power oscillator or perhaps by filtering and modulating the output of such an oscillator. One can then exploit sophisticated filters and modulators that would be unable to tolerate the high power of the amplified output. The MOPA configuration is particularly well suited to fibre lasers, since amplification in a fibre is very convenient, and very high gain is readily available, using if necessary, a chain of pre-amplifiers as well as a power amplifier. Besides using a fibre laser as the master oscillator, there are also attractive possibilities for other oscillator sources, such as semiconductor diode lasers. As a representative example of a MOPA scheme, one can cite the case of the mode-locked Yb-doped fibre laser, followed by a high gain Yb fibre amplifier [32], a scheme which is scaleable to the multiwatt average power level.

Mode-locked fibre lasers have been demonstrated in a bewildering array of variations on the mode-locking mechanism and on the geometrical configuration. Out of this plethora there have emerged robust and reliable designs [18, 33], which, as indicated before, when incorporated into a MOPA make the mode-locked fibre laser, a potential rival to the Ti–Sapphire laser, the standard workhorse for the femtosecond arena. With its prospects for lower costs than the Ti–Sapphire laser and for extensive power scalability, the mode-locked fibre laser seems set to be both a catalyst and a future workhorse for the burgeoning applications in femtosecond science and technology.

The distributed-feedback (DFB) fibre laser [13, 14] provides an illustration of the effectiveness that can result from the combination of fibre-Bragg-grating technology with an active fibre [12]. The achievement of single-frequency operation, a prerequisite for applications in many fields, including telecom, appeared initially to be doubly problematic for fibre lasers, on account of their broad gain linewidth and their typically rather longer resonator length than for miniature bulk lasers, thus leading to very many longitudinal modes within the gain bandwidth. Various routes to single-frequency operation have been demonstrated despite these hurdles [34], however, the DFB laser has the particular merit of being robustly constrained to single-frequency operation by virtue of the entire length of its gain medium being covered by a single Bragg-grating. This technology has depended on the successful development of a silica composition that is both photosensitive, thus allowing optical writing of the grating structure, and also suitable as a host for the laser-active ion. The ease with which the output power from the DFB-laser can be boosted by a MOPA arrangement is a further attraction, and there exists, as a result, the possibility of extensive spectral ranges in the near-infrared within which high-power single-frequency operation at an accurately-defined frequency can be accessed. By interposing a modulator between the DFB laser and the fibre amplifier one can create a high-power amplified output with any desired pulse shape and sequence.

B4.1.6 Future prospects

Developments in the area of fibres in general and in fibre lasers in particular, are both very rapid, not only in an incremental fashion but also in terms of radical innovations. In incremental terms a steady progress in power scaling is now under way and in the areas of cw lasers and femtosecond lasers it seems likely that fibre lasers will begin to displace some of the existing bulk laser systems. A recent, radical development has been the appearance of microstructured ('holey') fibres [24, 25] offering not only the attractive feature of smaller core-sizes; hence access to nonlinear effects at much reduced power levels, but also a new approach to engineering large changes in the local, average, value of the refractive index, a valuable ingredient for versatility in the design of guiding structures. Silica has maintained a dominant place in fibre laser developments, due in large part to its superior mechanical and optical (transparency) characteristics, thus making it eminently suitable for the subsequent major deployment in optical telecom systems. There, its present position seems unassailable. In the general area of fibre lasers, however, when free from the imperatives of the telecom area and, in particular, free from the requirement of very long fibre lengths, there are strong motivations for finding alternative hosts with significantly different characteristics from those of silica, for example with reduced phonon energy, higher nonlinearity, different photosensitivity etc. The long history of ZBLAN, mooted originally as an alternative to silica for low-loss transmission, must serve as a reminder that the search for alternative materials is arduous, expensive, and of uncertain outcome. Possibly a hybrid material, such as a ceramic will emerge, with the desirable mechanical and optical/spectroscopic properties contributed by different constituent parts. Whatever the route to their emergence, it is clear that new materials could have a strong effect on future developments in fibre lasers. At the same time, the capacity of silica to offer radical new capability should not be discounted. It has happened before, with the development of fibre-Bragg gratings in silica, and more recently with holey fibres.

References

[1] Mears R J, Reekie L, Jauncey I M and Payne D N 1987 Low noise, erbium-doped fibre amplifier operating at 1.54 μm *Electron. Lett.* **23** 1026–8

[2] Desurvire E 1994 *Erbium-Doped Fibre Amplifiers* (New York: Wiley)

[3] Sudo S 1997 *Optical Fibre Amplifiers: Materials, Devices and Applications* (Boston MA: Artech)

[4] Höfer S, Liem P, Zellmer H and Tünnermann A 2001 Single-frequency master-oscillator fibre power amplifier system emitting 20 W of power *Opt. Lett.* **26** 1326–8

[5] Baney D M, Rankin G and Kok-Wai Chang 1996 Blue Pr^{3+}-doped ZBLAN fibre up-conversion laser *Opt. Lett.* **21** 1372–4

[6] Pask H M, Carman R J, Hanna D C, Tropper A C, Mackechnie C J, Barber P R and Dawes J M 1995 Ytterbium-doped silica fiber lasers: versatile sources for the 1–1.2 μm region *IEE J. Selected Topics Quantum Electron.* **1** 2–13

[7] Miniscalco W J 2001 Optical and electronic properties of rare earth ions in glasses *Rare-Earth Doped Fibre Lasers and Amplifiers* 2nd edn, ed M J F Digonnet (New York: Marcel Dekker) pp 17–112

[8] Tamura K, Nelson L E, Haus H A and Ippen E P 1994 Soliton versus nonsoliton operation of fibre ring lasers *Appl. Phys. Lett.* **64** 149–51

[9] Cauterts V, Richardson D J, Paschotta R and Hanna D C 1997 Stretched pulse Yb^{3+}: silica fibre laser *Opt. Lett.* **22** 316–18

[10] Agrawal GP 1995 *Nonlinear Fibre Optics* 2nd edn (San Diego, CA: Academic)

[11] Brown D C and Hoffman H J 2001 Thermal, stress and thermo optic effects in high average power double-clad silica fibre lasers *Quantum Electron.* **37** 207–17

[12] Kashyap R 1999 *Fibre Bragg Gratings* (San Diego, CA: Academic)

[13] Asseh H, Storoy H, Kringlebotn J T, Margulis W, Sahlgren B and Sandgren S 1995 10 cm long Yb^{+} DFB fibre laser with permanent phase shifted grating *Electron. Lett.* **31** 969–70

[14] Loh W H and Laming R I 1995 1.55 μm phase-shifted distributed feedback fibre laser *Electron. Lett.* **31** 1440–2

[15] Richardson D J, Britton P E and Taverner D 1997 Diode pumped, high energy, single transverse mode Q-switch fiber laser *Electron. Lett.* **33** 1955–6

[16] Fermann M E 1998 Single-mode excitation of multimode fibres with ultrashort pulses *Opt. Lett.* **23** 52–4

[17] Snyder A W and Love J D 1983 *Optical Waveguide Theory* (London: Chapman and Hall)

[18] Tamura K, Ippen E P, Haus H A and Nelson L E 1993 77-fs pulse generation from a stretched-pulse mode-locked all-fibre ring laser *Opt. Lett.* **18** 1080–2

[19] Shimizu M, Yamada M, Horiguchi M, Takeshita T and Okayasu M 1990 Erbium-doped fiber amplifiers with an extremely high gain coefficient of 11.0 dB mW^{-1} *Electron. Lett.* **26** 1641–2

[20] Simpson J R 2001 Rare-earth doped fibre fabrication: techniques and physical properties *Rare-Earth Doped Fibre Lasers and Amplifiers* 2nd edn, ed M J F Digonnet (New York: Marcel Dekker) pp 1–15

[21] Craig-Ryan S P and Ainslie B J 1991 Glass structure and fabrication techniques *Optical Fibre Lasers and Amplifiers* ed P W France (Glasgow: Blackie) pp 50–78

[22] Townsend J E, Poole S B and Payne D N 1987 Solution doping technique for fabrication of rare-earth doped optical fibres *Electron. Lett.* **23** 329–31

[23] Digonnet M J F 2001 Continuous wave silica fibre lasers *Rare-Earth Doped Fibre Lasers and Amplifiers* 2nd edn, ed M J F Digonnet (New York: Marcel Dekker) pp 113–70

[24] Knight J C, Birks T A, Russell P St J and Atkin D M 1996 All-silica single-mode optical fibre with photonic crystal cladding *Opt. Lett.* **21** 1547

[25] Bennett P J, Monro T M and Richardson D J 1999 Toward practical holey fibre technology: fabrication, splicing, modelling and characterization *Opt. Lett.* **24** 1203

[26] Ranka J K, Windeler R S and Stentz A J 2000 Optical properties of high-delta air–silica microstructure optical fibers *Opt. Lett.* **25** 796

[27] France P W *et al* 1990 *Fluoride Glass Optical Fibres* (Glasgow: Blackie)

[28] Fermann M E, Hanna D C, Shepherd D P, Suni P J and Townsend J E 1988 Efficient operation of an Yb-sensitized Er fibre laser at 1.56 μm *Electron. Lett.* **24** 1135–6

[29] Hanna D C, Percival R M, Smart R G and Tropper A C 1990 Efficient and tunable operation of a Tm-doped fibre laser *Opt. Commun.* **75** 283

[30] Hayward R A, Clarkson W A, Turner P W, Nilsson J, Grudinin A B and Hanna D C 2000 Efficient cladding-pumped Tm-doped silica fibre laser with high power singlemode output at 2 μm *Electron. Lett.* **36** 711–12

[31] Paschotta R, Barber P R, Tropper A C and Hanna D C 1997 Characterization and modelling of thulium: ZBLAN blue up-conversion fiber lasers *J. Opt. Soc. Am.* B **14** 1213–18

[32] Fermann M E 2000 Self-similar propagation and amplification of parabolic pulses in optical fibres *Phys. Rev. Lett.* **84** 6010–13

[33] Fermann M E and Höfer M 2001 *Mode-Locked Fibre Lasers* Rare-Earth Doped Fibre Lasers and Amplifiers 2nd edn, ed M J F Digonnet (New York: Marcel Dekker) pp 395–447

[34] Langford N 2001 Narrow-linewidth fibre lasers *Rare-Earth Doped Fibre Lasers and Amplifiers* 2nd edn, ed M J F Digonnet (New York: Marcel Dekker) pp 243–312

Further reading

Agrawal G P 1995 *Nonlinear Fibre Optics* 2nd edn (San Diego CA: Academic)

Covers a range of nonlinear optical processes that occur in fibres. Includes a chapter on fibre lasers, with discussion of the principles of mode locking.

Desurvire E 1994 *Erbium-doped Fibre Amplifiers* (New York: Wiley)

Although focused on erbium-doped fibre, and on amplifiers rather than lasers, it gives a very thorough and extensive treatment of general principles relevant to any RE-doped fibres

Digonnet M J F (ed) 2001 *Rare-Earth Doped Fibre Lasers* 2nd edn (New York: Marcel Dekker)

The most comprehensive treatment of fibre lasers, comprising a multi-author collection of chapters each written by acknowledged experts, covering fabrication, spectroscopy, fibre components, silica and fluoride fibre lasers in various modes of operation and amplifiers.

France P W *et al* 1990 *Fluoride Glass Optical Fibres* (Glasgow: Blackie)

A useful compendium of spectroscopic data, and of details on fabrication of fluoride glass fibres. Discussion of device performance is now rather dated.

France P W (ed) 1991 *Optical Fibre Lasers and Amplifiers* (Glasgow: Blackie)

A multi-author work, with emphasis on silica fibres. A reasonable coverage of devices, although now dated. A useful compendium of spectroscopic and other data.

Kashyap R 1999 *Fibre Bragg Gratings* (San Diego CA: Academic)

A comprehensive treatment of fibre gratings, from basic principles to applications, with a useful chapter on fibre lasers incorporating gratings, including the DFB fibre laser.

Snyder A W and Love J D 1983 *Optical Waveguide Theory* (London: Chapman and Hall)

A weighty tome with a very thorough coverage of the theory of light propagation in fibres. A key source of essential formulae for an in-depth study of fibres.

Svelto O 1998 *Principles of Lasers* 4th edn (New York: Plenum)

Covers the basic principles of lasers. While not explicitly dealing with fibre lasers it gives good coverage of material needed for an understanding of fibre lasers, such as the principles of optically-pumped solid-state lasers and mode-locking of lasers.

Sudo S 1997 *Optical Fibre Amplifiers: Materials, Devices and Applications* (Boston MA: Artech)

While aimed predominantly at amplifiers and their applications in telecom, the book gives a comprehensive introduction to material relevant to fibre lasers, such RE spectroscopy and fibre fabrication.

B4.2
High power fibre lasers

Andreas Tünnermann and Holger Zellmer

Fibre lasers are based on silica fibres consisting of a rare-earth-doped core with a refractice index n_1, generally surrounded by a pure silica-glass cladding of refractive index $n_2 < n_1$, so that waveguiding occurs due to total internal reflection on the interface between core and cladding (see chapter A6 and B4.1). The fibre core serves both as active medium and as waveguide for the pump light and the generated laser light. Due to the long interaction length of pump and laser light the fibre design offers high gain and efficiency [24].

In contrast to conventional solid state lasers, in these lasers the beam quality is determined solely by the refractive index profile, i.e. the geometry and the numerical aperture of the laser core. The intensity distribution of the low-order LP_{lm} modes in a cylindrical optical waveguide (see section A6.4.1) is depicted in figure B4.2.1. The azimuthal mode number l and the radial mode number m correspond to the number of minima in the intensity distribution in the azimuthal and radial direction, respectively. The fundamental transversal mode LP_{01} possesses an intensity maximum on the fibre axis [8, 25].

The fibre laser oscillates without significant adverse effects from thermo-optical effects such as thermally induced lens formation or stress-induced birefringence even at high power levels [3, 5, 19]. Furthermore, for fibre lasers the thermal load caused by the pumping process is spread over a long region as well as, due to the greater surface to volume ratio, heat is more efficiently removed and, thus, the observed temperature increase in the laser core is small compared to bulk solid state lasers operating at the same power level [28]. Therefore, the reduction of the quantum efficiency of the active medium with increasing temperature during laser operation plays a subordinate role in fibre laser geometries.

Commonly, to excite the fibre laser optically, the pump radiation is coupled through the fibre facet into the laser core. However, in the case of longitudinal pumping of a fibre laser with mono-mode output the pump radiation has to be coupled to a waveguide with dimensions of a few micrometres. Therefore, to excite mono-modal fibre lasers pump radiation sources of high brightness are required.

B4.2.1 Cladding-pumping technique

The cladding-pumping technique overcomes the difficulty of efficiently coupling light from powerful low brightness pump sources—such as high-power diode laser arrays—into a laser waveguide that supports the operation only in a single transverse mode. In this technique the rare-earth-doped single-mode core is embedded in a multi-mode pump core. In order to realize the pump core which also acts as a cladding for the active core, the surrounding coating must have a lower refractive index. Normally a fluorine-doped silica glass or a highly transparent polymer with a low refractive index such as silicone or Teflon is used. The diameter of the pump core is typically a few hundred micrometres and its numerical aperture $\simeq$0.3–0.7, which is compatible to fibre-coupled high-power diode laser arrays. A schematic diagram of a double-clad fibre laser with a circular pump core and centred active core is shown in figure B4.2.2. The pump light which is launched into the multi-mode pump core is coupled into the active monomode core over the entire fibre length, where it is partly absorbed by the active ion, resulting in a population inversion. Using this

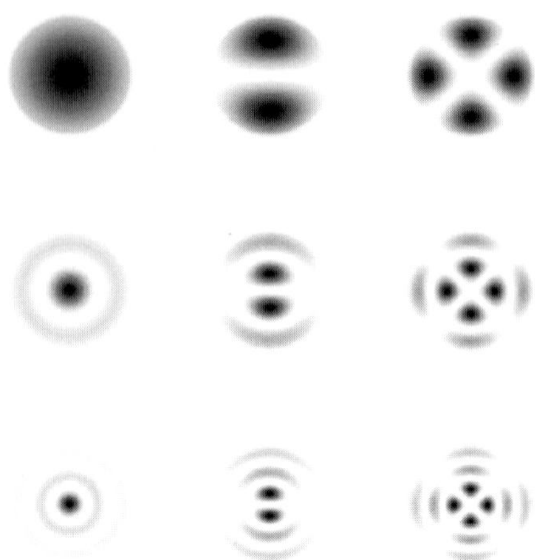

Figure B4.2.1. Intensity distribution of low-order fibre modes calculated with the WKB approximation; dark, high intensity; left column, LP_{01}, LP_{02}, LP_{03}; centre column, LP_{11}, LP_{12}, LP_{13}; right column, LP_{21}, LP_{22}, LP_{23}.

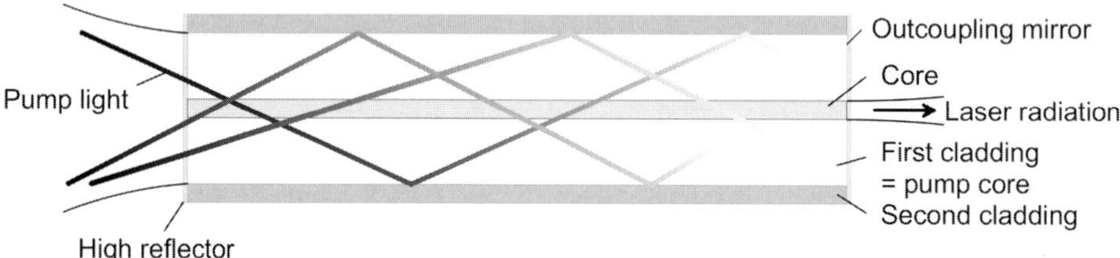

Figure B4.2.2. Double-clad fibre principle. The pump light is launched into an inner cladding which surrounds the active core and acts as a waveguide, too. The pump light propagates in the inner cladding and is absorbed in the active core over the entire fibre length.

technique, multi-modal pumping radiation can be converted into laser radiation with excellent beam quality. This technique was invented by E Snitzer [20]. Applying this concept, the operation of a fibre laser with an output power of several watts was possible for the first time [17, 26].

However, within a double-clad structure as shown in figure B4.2.2, the effective pump-light absorption coefficient is significantly reduced compared to conventional single-clad fibres, resulting in a worse optical efficiency. In general, for fibres with 100 μm pump-core diameter and 5 μm active-core diameter, only about 30% of the pump radiation can be absorbed. This reduced pump-light absorption results from the imperfect overlap between the intensity of the transversal modes in the pump core with the active core area. It can easily be seen from figure B4.2.1 that fibre modes with azimuthal mode number $l > 0$—which are by far the majority of fibre modes—possess no intensity on the fibre axis. Only the fibre modes LP_{01}, LP_{02}, $LP_{03}, \ldots$ can be absorbed completely. Therefore, within this cylindrical double-clad geometry the reduced pump-light absorption coefficient can not be compensated by increasing the rare-earth-doping concentration and/or the length of the fibre. To counteract this, fibre designs have to be applied that either convert high-order azimuthal modes into $l = 0$ modes or do not support stable modes, i.e. fibres with broken symmetries in which the pump light propagates in a chaotic way. In order to study the propagation and absorption of the pump light in double-clad fibres in detail, ray-tracing methods have been applied by several groups. These

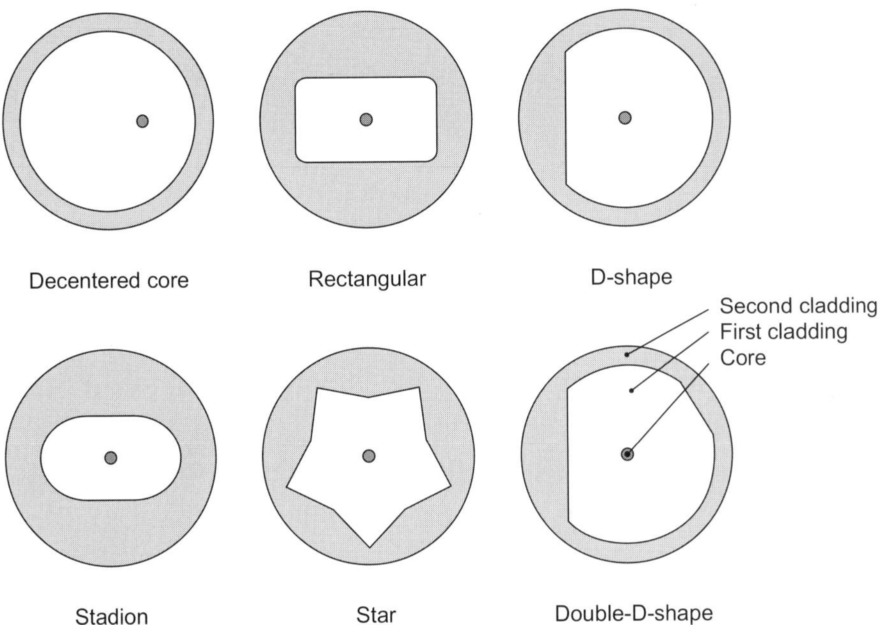

Figure B4.2.3. Possible cross sections of double-clad fibres for efficient pump absorption.

ray-tracing methods are valid as long as the dimensions of the optical waveguide are very large compared to the wavelength of the radiation. By applying ray-tracing methods, a distinct improvement in the pump-light absorption has been made by the development of novel pump-core geometries which utilize a disturbance of the rotational symmetry of the fibre along the fibre axis to hamper the propagation of helical rays. Different pump-core cross sections, which enforce a chaotic spreading of the pump radiation and ensure that the pump radiation coupled into the pump core crosses the active core during the propagation are shown in figure B4.2.3. Within these geometries almost complete pump-light absorption occurs even for short fibre length [16, 23]. A calculation for the normalized pump absorption in double-clad fibres with different geometries is depicted in figure B4.2.4. The D-shaped and double-D-shaped fibres have, in addition, the advantage of their quasi-cylindrical shape, which means that they can be cleaved and handled in almost the same way as conventional fibres and they can be polished into standard fibre connectors.

Although these novel fibre designs show an excellent laser performance, they have the disadvantage of costly production. As an alternative to the costly manufacturing of fibres with non-cylindrical symmetry the technique of mode mixing by periodic bending of the fibres can be applied. The curvature in the course of the fibre causes pump radiation from the outer regions of the pumping core to be transferred to the centre of the pump core. Coiling the fibre in a kidney shape is especially favourable for pump-radiation absorption in centred double-clad fibres (figure B4.2.5) [27]. The successive fibre parts with different radii and directions of curvature give rise to excellent pump-radiation absorption.

B4.2.2 High-power laser operation

Applying rare-earth-doped double-clad fibres, continuous-wave (cw) powers of up to 100 W have been achieved in ytterbium-doped fibres [3] with optical efficiencies as high as 80% [9]. Typically, the fibres made from fused silica have an inner cladding diameter of 200–400 μm and numerical aperture of 0.3–0.4, which allows an efficient coupling of diode laser bars with a maximum beam parameter product of up to

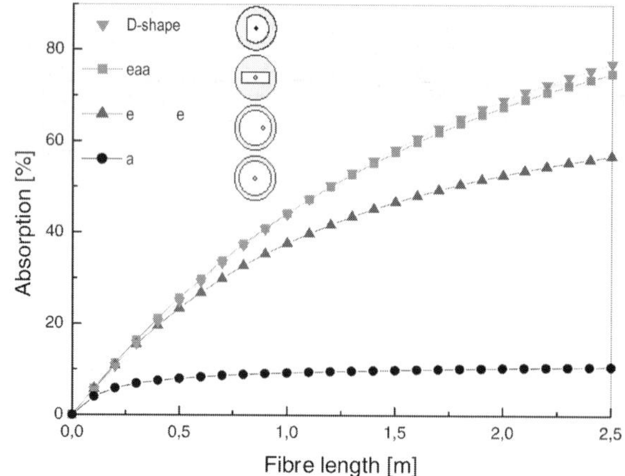

Figure B4.2.4. Ray-tracing calculation of the pump absorption in fibres with different shapes, but the same area of inner cladding.

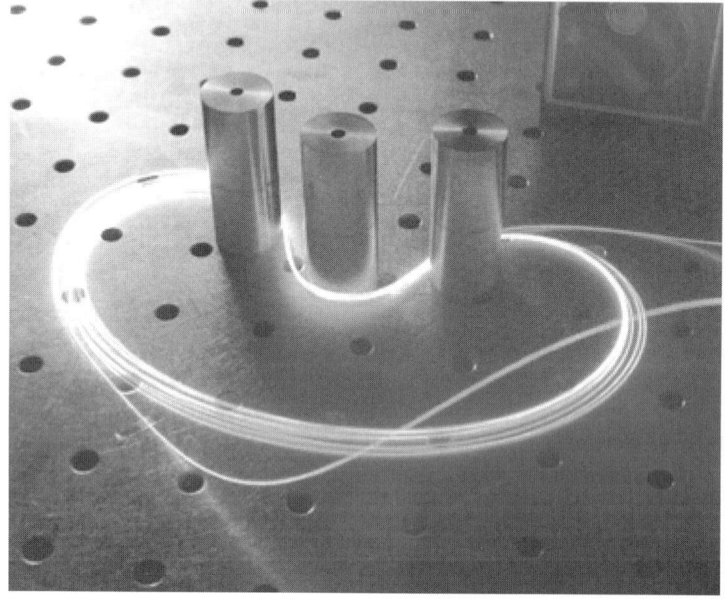

Figure B4.2.5. Kidney-shaped fibre coil which enables mode mixing and thus efficient pump-light absorption in circular double-clad fibres.

80 mm mrad. The active core of the fibres is typically slightly multi-mode with 10–12 μm diameter and a numerical aperture in the order of 0.15. The active fibre core is doped with several 1000 ppm (mol) of rare earth ions. A codoping consisting of germanium, aluminium and phosphor ensures the desired refractive index profile and a homogeneous distribution of the rare earth ions which reduces the background losses. The typical attenuation of the fibres resulting from scattering and residual absorption is in the range of 5 dB km^{-1} at 1150 nm. A photograph of the end facet of such a fibre is shown in figure B4.2.6.

The experimental set-up of a high-power Nd-doped fibre laser is shown in figure B4.2.7. The core of

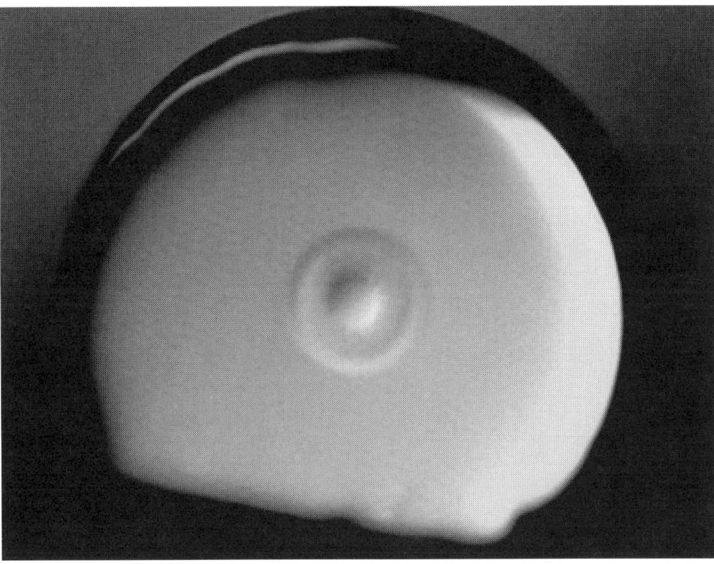

Figure B4.2.6. End face of a high-power laser fibre with extended mode-field diameter (large-mode-area fibre).

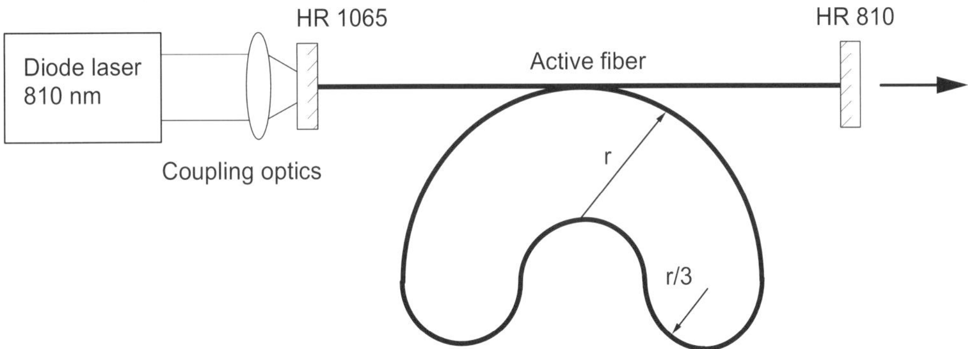

Figure B4.2.7. Setup of a neodymium-doped fibre laser with kidney-shaped fibre geometry.

this fibre is doped with 1300 ppm (mol) Nd^{3+} ions and has a diameter of 10 μm at a numerical aperture of approximately 0.17. The pump-core diameter is 400 μm, its numerical aperture approximately 0.38. The pump radiation of a diode laser operating at 808 nm is coupled into the pump core of the fibre through the resonator mirror. This mirror is highly transmittive for the pump wavelength and highly reflective for the laser wavelength. In order to reflect non-absorbed pump radiation back into the fibre, the outcoupling mirror is highly reflective for the pump light. For optimum laser output, the transmission of the mirror at the laser wavelength is about 90%. For a kidney-shaped coiling geometry, a cw output power of 32.5 W was achieved at a pump power level of 72 W [27]. The optical efficiency as a function of the pumping power is about 46%. A D-shaped fibre with the same fibre parameters pumped at the same power level yields 30 W with a slope efficiency of 43%. The beam quality is given by $M^2 = 1.3$, corresponding to mono-modal laser emission. Higher-order modes, which are guided in the fibre core due to its geometry, are effectively suppressed in the laser operation due to higher losses compared to the fundamental mode. The spectral properties of fibre lasers differ from those of conventional solid state lasers. As a result of spectral broadening of the laser transition

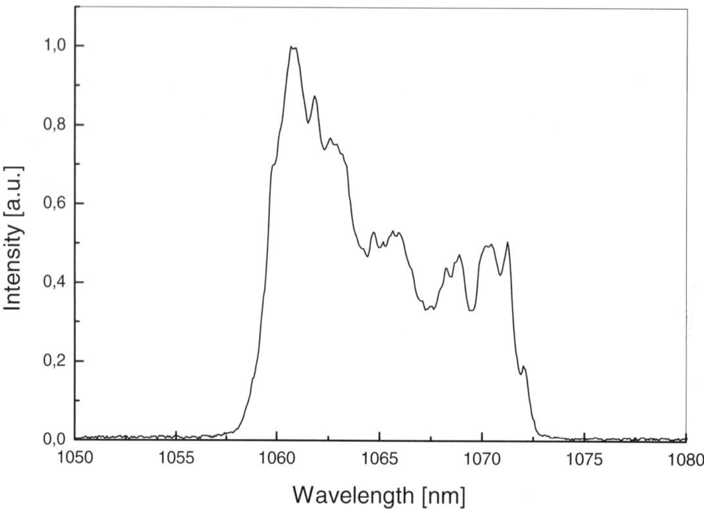

Figure B4.2.8. Laser spectrum of a high power neodymium fibre laser.

of the neodymium ions in silica host material, the emission linewidth is typically 15 nm at the output power level (figure B4.2.8).

Even at output powers of 100 W in cw operation, the linear input–output characteristics of fibre lasers indicates that, although the power densities are in the range of 10 MW cm^{-2}, no saturation effects from nonlinearities or thermally induced effects occur. However, due to the brightness of high-power diode laser arrays, the power scaling capability in an end-pumping geometry is limited [3].

An approach to overcome this problem is the use of a double-doped fibre. By codoping an ytterbium-doped fibre with neodymium, both the pump-bands of the ytterbium between 915 and 980 nm and that of the neodymium around 808 nm can be used to excite the upper laser states of the active ytterbium ions, either directly or by means of an energy transfer from neodymium to ytterbium. With superposition of two high-power diode lasers (808 and 940 nm), a fibre laser was operated at a power level of cw 150 W [14].

For further power scaling, novel side-pump geometries for double clad fibre lasers have been developed. One promising approach is the v-groove technique [9], which allows one to superimpose the output power of several diode lasers. In this technique the output of broad-stripe diode lasers is transversely reflected into the pump core by using V-shaped gold-coated grooves in the pump core of the fibre (figure B4.2.9). Alternative techniques for diode laser fibre side-pumping are based on fibre-coupled diode lasers attached to the pump core by fused splices or transverse coupling of pump light applying reflective diffraction gratings. The transverse pump light coupling techniques allow one to launch pump power levels up to the fundamental limits of the fibre laser operation.

The principal limits to the power scalability of fibre lasers in cw and pulse operation at single transverse mode are defined by the damage threshold of the fibre facets, as well as nonlinear effects like stimulated Raman and Brillouin scattering caused by the interaction of the high optical power density and fibre length [1]. Due to the linewidth of fibre lasers, the threshold for stimulated Brillouin scattering is at least several kilowatts. The Raman threshold for fibre core diameters in the range of 10 μm and a fibre length of 50 m can be estimated to be about 150 W in cw operation. In pulsed operation the pulse energy in single-mode fibres is limited to values in the range of 100 μJ for 1 ns pulses. To avoid catastrophic optical damage of the fibre end facets, a short piece of coreless fibre can be spliced to the fibre end. The length of this cylinder is limited by the diameter of the inner cladding divided by the numerical aperture to avoid reflection at the barrel face.

In order to scale the extractable output powers, the fibre core diameter can be increased. However, an

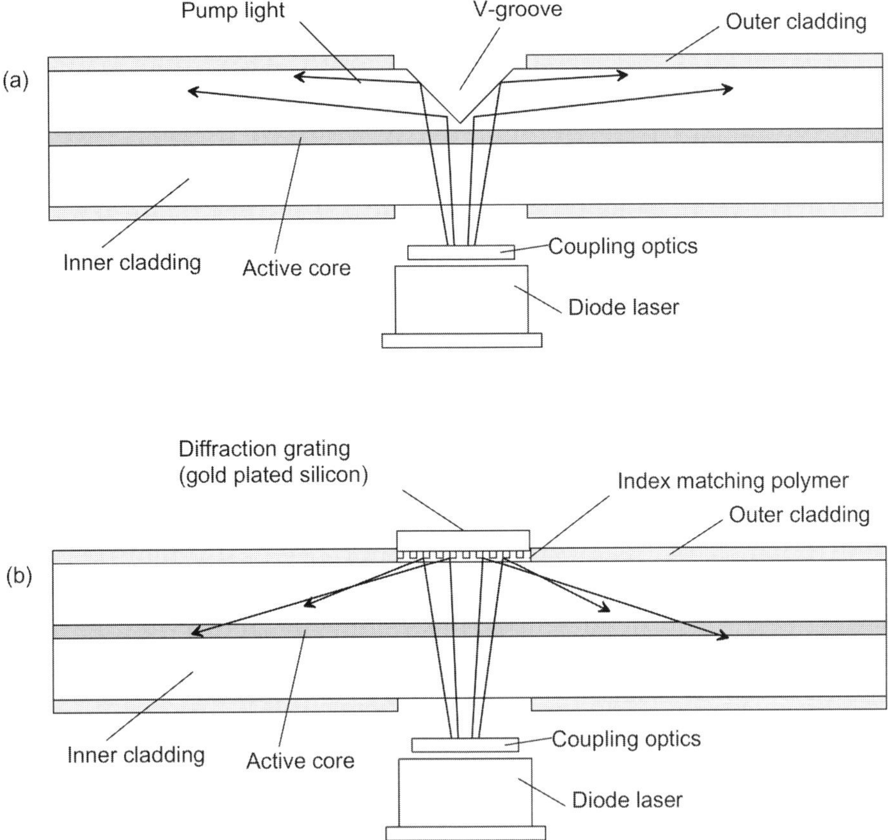

Figure B4.2.9. Side-pumping of a fibre laser applying v-grooves (*a*) and diffraction gratings (*b*).

increased fibre diameter permits multi-mode operation, resulting in reduced beam quality. Even for fibres which theoretically can guide several transverse modes, mono-modal fibre laser operation can be enforced by a narrow coiling of these multi-mode fibres or a partial tapering [10]. Alternatively, large-mode-area (LMA) fibres can be applied, which have been developed in the recent years [18]. Within these fibres the single transverse-mode diameter is significantly increased compared to conventional fibres by accurate control of the numerical aperture, which is limited to about 0.05. The refractive index profile of a Yb-doped fibre with a core diameter of 30 μm and a numerical aperture of $\sim$0.06 is shown in figure B4.2.10. For this fibre an M^2 value of 1.3 has been measured in cw operation at an emission wavelength of 1.06 μm [13]. With a core diameter of 30 μm, the scaling limit approaches the kW-output power level in cw operation. Applying LMA fibres, output pulse energies of Q-switched fibre lasers in the millijoule range have been achieved [2], which impressively demonstrates the capability of fibre lasers and amplifiers for extracting high pulse energies.

Advanced fibre designs consider a multi-mode fibre core that is doped with active ions only in its centre. The diameter of this active area is designed to have a good overlap with the fundamental mode but almost none with higher-order modes (figure B4.2.11). To enhance the mode selective effect, the outer region of the waveguide can be doped with an absorber that causes high losses to higher-order modes but almost no losses for the fundamental mode. In gain-loss-managed fibres, fundamental mode operation in fibres with more than 50 μm core diameter can be achieved [15].

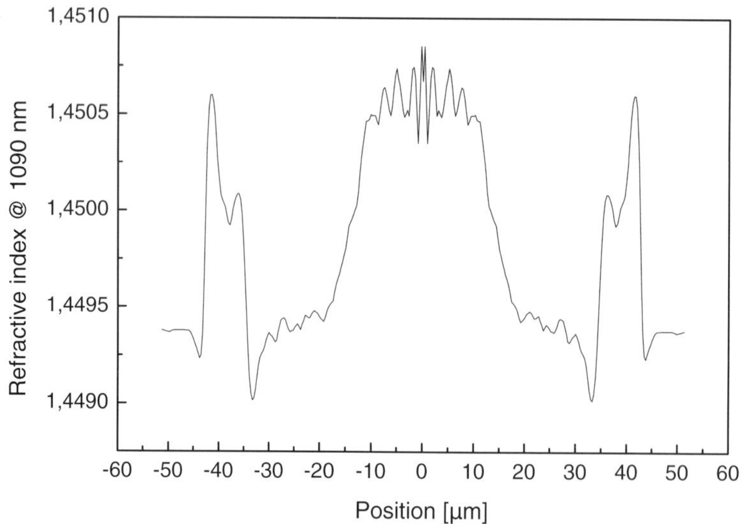

Figure B4.2.10. Refractive index profile of a large-mode-area fibre. The refractive index in the core is only slightly higher than in the cladding, which leads to an increased mode-field diameter. To minimize the bending losses in this low numerical aperture fibre, a guiding ring surrounding the active core is applied.

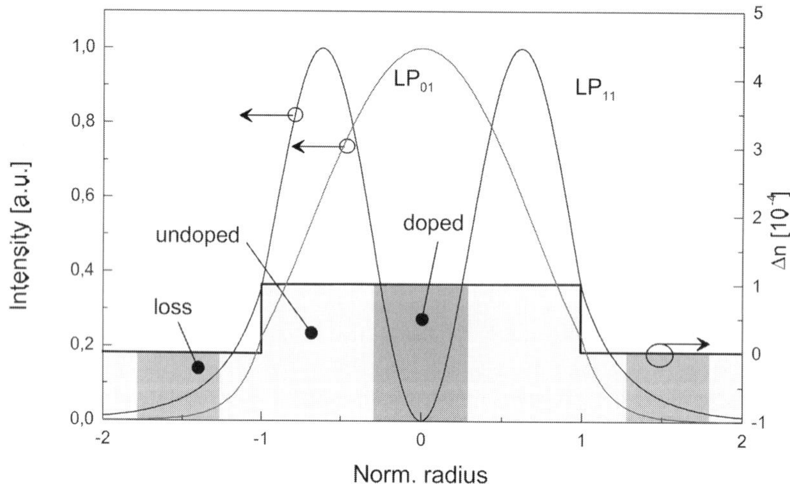

Figure B4.2.11. Gain-loss management in a fibre amplifier. In a low-order multi-mode fibre only the central region is doped with active ions. The doped area has a good overlap with the fundamental mode but almost none with higher-order modes. The mode selective amplification can be enhanced by an absorbing layer near the boundary between core and cladding. This layer induces high loss to higher-order modes. However, the fundamental mode is almost unattenuated.

Broad spectral gain bandwidth, high optical gain, pulse energy and pumping efficiency, therefore, predestine fibre amplifiers for the generation of high-energy ultrashort pulses using the chirped pulse amplification (CPA) technique [22]. This technique is based on a dispersive delay line, which is used to stretch an ultrashort pulse temporally to a pulsewidth of several 100 ps. After amplification, the pulse is then compressed in a

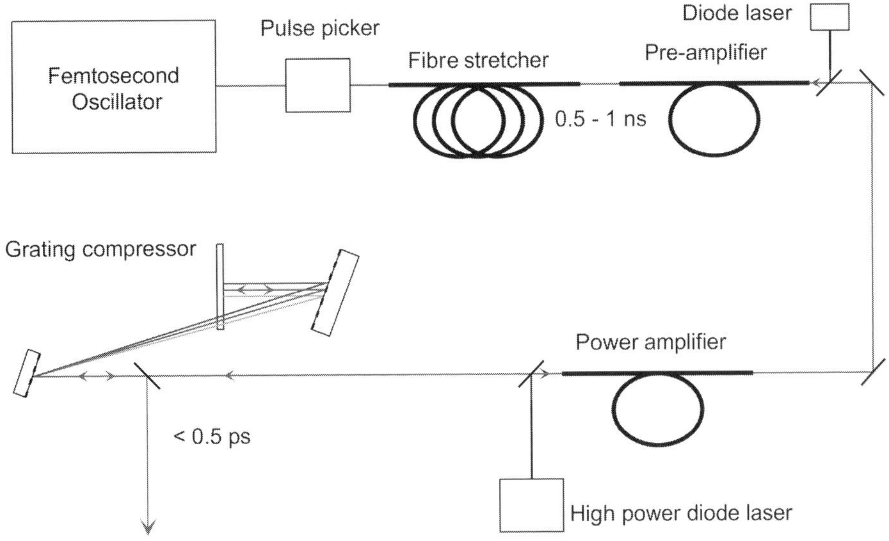

Figure B4.2.12. Set-up of a fibre chirped pulse amplification system.

second dispersive delay line. Nonlinear effects and optical damage can be avoided in the fibre amplifier stage [4, 21].

Applying Yb-doped LMA-fibre amplifiers sub-picosecond pulses with energies in the millijoule range have been demonstrated [6, 7, 11]. The principle of the set-up of a fibre CPA system is shown in figure B4.2.12. The system consists of a passively mode-locked, diode-pumped solid state laser, a (fibre-)stretcher, Yb-doped fibre power amplifiers and a diffraction-grating compressor. The total gain in energy within these systems is of the order of 10^6, therefore, in fibre CPA systems, the commonly used regenerative amplifier can be avoided, resulting in a robust and reliable arrangement of ultrashort pulse high-energy laser systems which opens up novel applications for ultrafast optics apart from basic science.

Due to cladding pumping, fibre lasers and amplifiers are no longer restricted to low-power operation. A femtosecond fibre CPA system with 125 W average output power has already been demonstrated [12]. The fundamental limits of the power scaling are given by nonlinear effects in the fibre, the energy stored in the fibre and the damage threshold of the fibre endfaces. In figure B4.2.13 the basic limits are shown for an ytterbium-doped fibre with 30 μm active core operating with 1 ns pulses. The damage threshold of the fibre ends can be significantly increased by application of coreless fibre endcaps. The expanded-mode-field diameter reduces the power density at the fibre ends by a factor of 10. With an expanded-mode-field diameter of 100 μm a pulse energy of at least 6.5 mJ can be obtained for 1 ns pulses. For short fibres the pulse energy is limited by the energy stored in the active material. A 30 μm fibre core with a doping concentration of 500 ppm (mol) Yb_2O_3 can store approximately 1–1.5 mJ per metre fibre length if the amplified spontaneous emission (ASE) can be suppressed. For fibres with a length of more than 3 m the pulse energy is limited by the threshold of stimulated Raman scattering (SRS). Other nonlinear effects like stimulated Brillouin scattering (SBS) or self-phase modulation (SPM) are negligible for 1 ns pulses. However, they have to be considered for other operating conditions. For further power scaling, e.g. in fibre CPA systems, longer pulse durations are required. With the expansion of pulse duration the maximum intensity and hence nonlinear effects that limit the pulse energy can be reduced. Latest results show that 10 mJ from a fibre laser are feasible.

In summary, average output powers in the kW range in single-transverse-mode operation with optical efficiencies as high as 80% with respect to a low brightness diode laser pump source are feasible. Novel

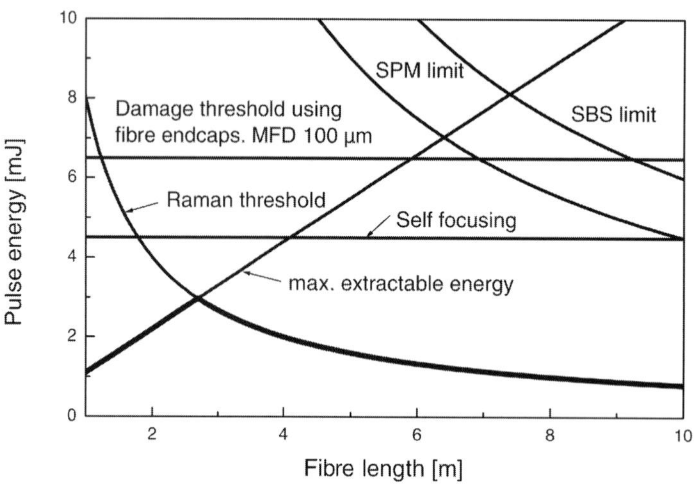

Figure B4.2.13. Power scaling limits calculated for 1 ns pulses in a fibre with 30 μm core diameter. For short fibres the pulse energy is limited by the energy stored by the active ions. For long fibres, the output energy is limited by the threshold of stimulated Raman scattering (SRS).

fibre designs open up the operation at high output energies. This, as well as the inherent reliability and the existence of numerous low-cost all-fibre elements, permits the construction of novel lasers and amplifiers in fibre form and promises more applications of laser technology in basic science and engineering.

References

[1] Agrawal G P 1995 *Nonlinear Fiber Optics* 2nd edn (San Diego, CA: Academic)

[2] Alvarez-Chavez J A, Offerhaus H L, Nilsson J, Turner P W, Clarkson W A and Richardson D J 2000 High-energy, high-power ytterbium doped Q-switched fiber laser *Opt. Lett.* **25** 37

[3] Dominic V, MacCormack S, Waarts R, Sanders S, Bicknese S, Dohle R, Wolak E, Yeh P S and Zucker E 1999 110 W fibre laser *Electron. Lett.* **35** 1158

[4] Fermann M E, Galvanauskas A and Harter D 1994 All-fibre source of 100 nJ subpicosecond pulses *Appl. Phys. Lett.* **64** 1315

[5] France P W 1991 *Optical Fiber Lasers and Amplifiers* 1st edn (Glasgow: Blackie)

[6] Galvanauskas A and Fermann M E 2000 13 W average power ultrafast fiber laser *Conf. on Lasers and Electro-Optics CLEO (San Francisco, USA)* postdeadline paper CPD 3

[7] Galvanauskas A, Sartania Z, Bischoff M 2001 Millijoule femtosecond fiber CPA system *Advanced Solid-State Lasers* OSA TOPS vol 50 ed C Mashall (Washington DC: OSA) p 679–81

[8] Geckeler S 1990 *Lichtwellenleiter für die Optische Nachrichtenübertragung* 3rd edn (Berlin: Springer)

[9] Goldberg L, Koplow J P and Kliner D A V 1999 Highly efficient 4 W Yb-doped fiber amplifier pumped by a broad-stripe laser diode *Opt. Lett.* **24** 673

[10] Koplow J P, Kliner D A V and Goldberg L 2000 Single-mode operation of a coiled multi-mode fiber amplifier *Opt. Lett.* **25** 442

[11] Liem A, Nickel D, Limpert J, Zellmer H, Griebner U, Unger S, Tünnermann A and Korn G 2000 High average power ultrafast fiber chirped pulse amplification system *Appl. Phys.* B **71** 889

[12] Liem A, Limpert J, Höfer S, Nolte S, Zellmer H, Tünnermann A, Reichel V, Unger S, Jetschke S and Müller H-R 2002 High average power femtosecond fiber CPA system *Advanced Solid-State Lasers* OSA TOPS vol 68 (Washington DC: OSA) pp 128–32

[13] Limpert J, Liem A, Höfer S, Gabler T, Zellmer H, Tünnermann A, Unger S, Jetschke S and Müller H-R 2001 High average power ultrafast Yb-doped fiber amplifier *Advanced Solid-State Lasers* OSA TOPS vol 50 ed C Mashall (Washington DC: OSA) pp 348–54

[14] Limpert J, Zellmer H, Tünnermann A, Pertsch T and Lederer F 2002 Suppression of higher order modes in a multimode fiber amplifier using efficient gain-loss-management (GLM) *Advanced Solid-State Lasers* OSA TOPS vol 68 (Washington DC: OSA) pp 112–4

[15] Limpert J, Liem A, Höfer S, Zellmer H, Tünnermann A, Unger S, Jetschke S and Müller H-R 2002 150 W Nd^{3+}/Yb^{3+} codoped fiber laser at 1.1 μm *Advanced Solid-State Lasers* OSA TOPS vol 73 (Washington DC: OSA) pp 590–1

[16] Liu A P and Ueda K 1996 The absorption characteristics of circular, offset, and rectangular double-clad fibers *Opt. Commun.* **132** 511–18
[17] Po H, Cao J D, Laliberte B M, Minns R A, Robinson R F, Rockney B H, Tricca R R and Zhang Y H 1993 High power neodymium-doped single transverse mode fibre laser *Electron. Lett.* **29** 1500
[18] Renaud C C, Offerhaus H L, Alvarez-Chavez J A, Nilsson J, Clarkson W A, Turner P W, Richardson D J and Grudinin A B 2001 Characteristics of Q-switched cladding-pumped ytterbium-doped fibre laser with different high-energy fiber designs *IEEE J. Quant. Electron.* **37** 199
[19] Richardson D, Minelly J and Hanna D 1997 Fiber laser systems shine brightly *Laser Focus World* **33** 87
[20] Snitzer E, Po H, Hakimi F, Tumminelli R, McCollum B C 1988 Double clad, offset core Nd fiber laser, Conference on Optical Fiber Sensors, New Orleans, January 27–29, 1988, Postdeadline Paper PD5
[21] Stock M and Mourou G 1994 Chirped pulse amplification in an erbium-doped fiber oscillator/erbium-doped fibre amplifier system *Opt. Commun.* **106** 249
[22] Strickland D and Mourou G 1985 Compression of amplified chirped pulses *Opt. Commun.* **56** 219
[23] Tünnermann A and Zellmer H 2000 New concepts for diode-pumped solid-state lasers *High-Power Diode Lasers, Topics Appl. Phys.* vol 78, ed R Diehl (Berlin: Springer) p 369
[24] Urquhart P 1988 Review of rare earth doped fibre lasers and amplifiers *IEE Proc. J.* **6** 135
[25] Yeh C 1990 *Handbook of Fiber Optics* (San Diego, CA: Academic)
[26] Zellmer H, Willamowski U, Tünnermann A, Welling H, Unger S, Reichel V, Müller H-R, Kirchhof J and Albers P 1995 High-power cw neodymium-doped fiber laser operating at 9.2 W with high beam quality *Opt. Lett.* **20** 578
[27] Zellmer H, Tünnermann A, Welling H, Reichel V 1997 Double-clad fibre laser with 30 w output power *Optical Amplifiers and their Applications (OSA Trends in Optics and Photonics Series TOPS 16)* ed A Willner, M Zervas and S Sasaki (Cambridge: Cambridge University Press) p 137
[28] Zenteno L 1993 High power double-clad fibre lasers *J. Lightwave Technol.* **11** 1435

Further reading

Agrawal G P 1995 *Nonlinear Fibre Optics* 2nd edn (San Diego, CA: Academic)

Thorough coverage of linear and nonlinear characteristics of optical fibres

Koechner W 1999 *Solid-State Laser Engineering* 5th edn (Berlin: Springer)

Thorough coverage of characteristics, design, construction, and performance of solid state lasers

Diehl R (ed) 2000 *High-Power Diode Lasers* (Berlin: Springer)

B4.3
Cascaded Raman fibre lasers

Clifford Headley

B4.3.1 Introduction

A cascaded Raman fibre laser (RFL) [1–12] uses the nonlinear process of stimulated Raman scattering (SRS) [13] (see chapter B1.7) in optical fibres to shift the wavelength of light from an input pump laser out to another desired wavelength. Devices at almost any wavelength can be made by proper choice of a pump wavelength and by cascading the pump through several Raman shifts. Fibre Bragg gratings (FBGs) are the reflectors most commonly used to form the multiple Fabry–Pérot resonator cavities. These devices were developed in the telecommunications field to fill a need for lasers that could emit in the fundamental mode of an optical fibre at higher powers than that which could be obtained from diode lasers.

There are three elements needed to form a laser cavity: gain, feedback and a pump. In this section the gain medium and feedback mechanism are described. The cladding-pumped fibre lasers typically used to pump RFLs are described in chapter B4.2. In the following section, an overview of SRS is given concluding with a schematic diagram of an exemplary device. Section B4.3.3 then discusses the design of the optimum fibre for an RFL. Section B4.3.4 describes FBGs and how they are fabricated. The effect of various design parameters on RFL performance is detailed in section B4.3.5. Finally, multiple-wavelength RFLs are presented in section B4.3.6.

B4.3.2 Overview: stimulated Raman scattering and a Raman fibre laser

Raman scattering is a process by which light incident on a medium is converted to a lower frequency [13]. This is shown schematically in figure B4.3.1. A pump photon, $\nu_{\rm p}$ excites a molecule up to a virtual level (non-resonant state). The molecule quickly decays to a lower energy level emitting a signal photon, $\nu_{\rm s}$, in the process. The difference in energy between the pump and signal photons is dissipated by the molecular vibrations of the host material. These vibrational levels determine the frequency shift and shape of the Raman gain curve. Due to the amorphous nature of silica, the Raman gain curve is fairly broad in optical fibres. The frequency (or wavelength) difference between the pump and the signal photon is called the Stokes shift.

For high enough pump powers, the scattered light can grow rapidly with most of the pump energy converted into scattered light. This process is called SRS and it is the gain mechanism for RFLs. An important point is that, because the pump photon is excited to a virtual level, Raman gain can occur for a pump source at any wavelength. This is unlike rare-earth-doped fibre lasers whose lasing wavelengths are defined by their resonant levels. Therefore, with the proper choice of a pump wavelength, an RFL can emit light at any wavelength.

An RFL consists of a spool of enhanced Raman gain single-mode fibre with a grating on either side. In the example shown in figure B4.3.2 using Ge-doped optical fibre, light is shifted from 1117 to 1480 nm in the Raman fibre. Four nested pairs of high-reflector (HR) FBGs, with reflectivities of approximately 100%,

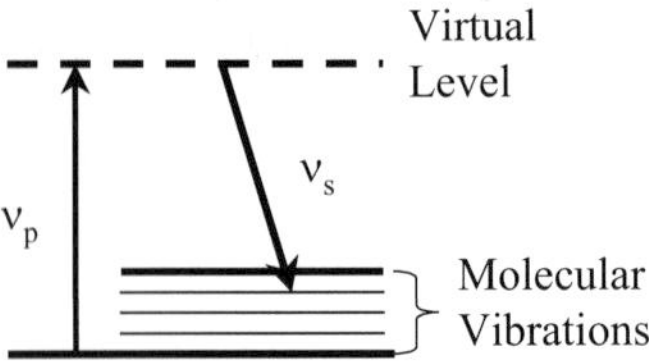

Figure B4.3.1. Schematic diagram of the quantum mechanical process taking place during Raman scattering.

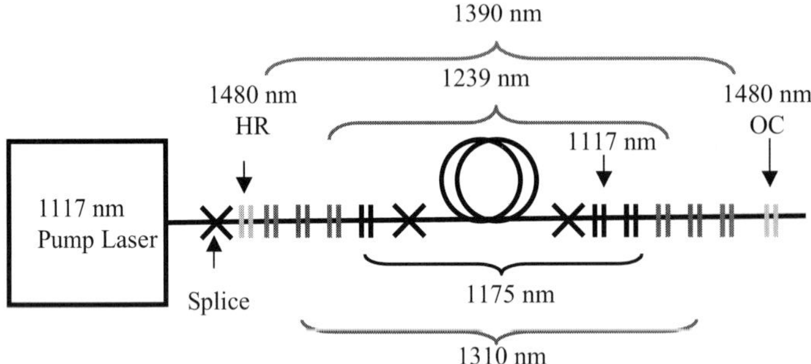

Figure B4.3.2. Schematic diagram of a typical FBG-RFL. In this example a 1117–1480 nm device is shown. The 1480 nm is five Stokes shifts away from the 1117 nm input pump.

at the intermediate Stokes orders prevent light at these wavelengths from leaving the cavity. In addition, the input grating set contains a 1480 nm HR and the output grating set a HR for the 1117 nm pump and an output coupler (OC, reflectivity $<100\%$) at 1480 nm. In the following section, it is shown that the number of pairs of gratings used will depend on the dopants used in the Raman gain fibre.

B4.3.3 The gain medium: optical fibre design

The trade-offs necessary to design a fibre optimized for an RFL can be understood by looking at a single-pass SRS system. Coupled equations describing the continuous wave (cw) evolution of the pump (P_p) and signal (P_s) power along the fibre can be written as [13]

$$\frac{dP_s}{dz} = \frac{\omega_p}{\omega_s}\frac{g_R}{A_{eff}} P_p P_s - \alpha_s P_s \tag{B4.3.1}$$

and

$$\frac{dP_p}{dz} = -\frac{g_R}{A_{eff}} P_p P_s - \alpha_p P_p \tag{B4.3.2}$$

The subscripts p and s designate the pump and Stokes light, g_R is the Raman gain coefficient, α is the fibre loss, ω is the angular frequency and A_{eff} is the effective area of the mode field of the light and, for simplicity, is assumed to be equal for the pump and Stokes light. The first term in equations (B4.3.1) and (B4.3.2) represents the loss (gain) due to SRS, the second term the intrinsic fibre loss in the pump (signal) light. These equations are readily solved for a fibre of length L if it is assumed that the depletion of the pump by the signal is negligible, to yield

$$P_s(L) = P_s(0)\exp(g_R P_{in} L_{eff}/A_{eff} - \alpha_s L) \tag{B4.3.3}$$

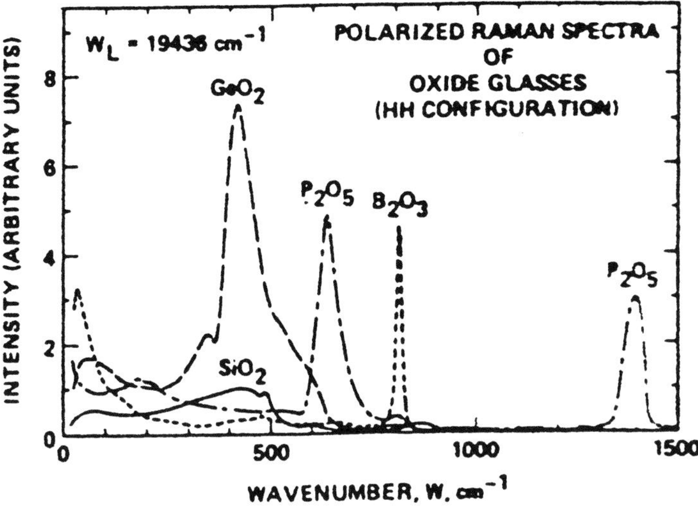

Figure B4.3.3. Relative Raman spectra of the primary glass formers SiO_2, GeO_2, P_2O_5 and B_2O_3 showing their relative strengths (after [14]).

Table B4.3.1. List of relative peak Raman cross sections of the primary glass formers, and the frequency shift at which the peak occurs.

Dopant	Relative intensity	Frequency shift (THz)
SiO_2	1	13.2
GeO_2	7.4	12.6
B_2O_3	4.6	24.2
P_2O_5	4.9	19.2
	3.0	41.7

where P_{in} is the input pump power and L_{eff}, the effective interaction length, is given by

$$L_{eff} = \frac{1}{\alpha_p}[1 - \exp(-\alpha_p L)]. \tag{B4.3.4}$$

It is clear from equations (B4.3.3) and (B4.3.4) that, in order to optimize a fibre for Raman gain, (i) g_R should be increased, (ii) A_{eff} should be reduced and (iii) $\alpha_{p/s}$ must be minimized. In addition, the fibre should either be photosensitive to allow low-loss gratings to be written into it or have a low splice loss to allow gratings to be spliced on. These parameters are all interrelated in the fibre design.

The largest influence on the Raman gain coefficient is the co-dopant used and the quantity of the dopant [14–19]. Figure B4.3.3 (from [14]) shows the relative strength of the Raman gain spectrum for different glass formers SiO_2, GeO_2, B_2O_3, and P_2O_5. The results are summarized in table B4.3.1, with corrections to the data shown in figure B4.3.3 to account for Fresnel reflections. The results show that in fused vitreous pure forms, GeO_2 has the largest Raman gain coefficient.

RFLs built to date have used either Ge- or P-doped fibres. Figure B4.3.4(*a*) is a diagram of the relative gain spectrum of P- and Ge-doped fibre. There are slight differences in the gain spectra when the dopants are incorporated with Si compared to that shown in figure B4.3.3. Germanium has been the traditional dopant

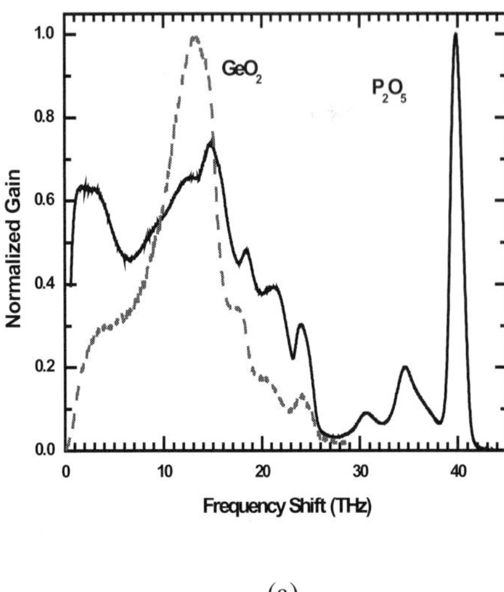

(a)

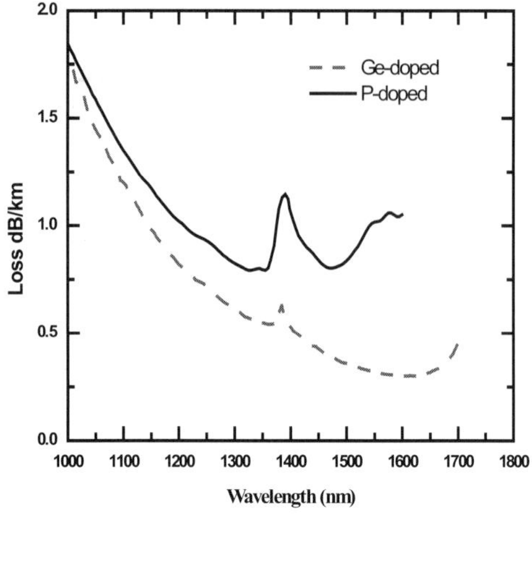

(b)

Figure B4.3.4. (*a*) Gain and (*b*) loss spectra of a P-doped fibre (full) and Ge-doped fibre (dashes). (P-doped fibre data courtesy of The Fiber Optics Research Center at the General Physics Institute, Russian Academy of Sciences, Ge-doped data courtesy of OFS Fitel Specialty Photonics Devices.)

used in optical fibres, and the techniques for incorporating Ge into fibres are well understood. However, P-doped fibre with a larger Stokes shift (39.9 THz) reduces the number of Raman shifts needed (e.g. two shifts from 1060 to 1480 nm) as compared to Ge-doped fibre (six shifts from 1060 to 1480 nm). In addition, [10] also points out that the gain coefficient of P_2O_5 fibre appears to increase by a factor of 2.6 compared to that given in table B4.3.1 when incorporated with Si.

The next goal of the fibre design is to reduce $A_{\rm eff}$. This can be accomplished by reducing the core area. While the reduction in the core radius, r, does not have a one-to-one correlation with the mode diameter ρ, it is an effective means of controlling ρ. There is a lower limit on r as eventually ρ will increase and the bend loss becomes excessive. This loss can be reduced by increasing the index difference, (Δn), between the core and the cladding, which also confines the mode more tightly, reducing $A_{\rm eff}$. However, there is another design constraint and that is the cut-off wavelength, $\lambda_{\rm c}$. It is intuitive that the most efficient Raman pumping will take place if the pump and signal wavelengths spatially overlap. Therefore, the light at all wavelengths in the fibre should guide in a single mode. The pump wavelength for an RFL then determines the design $\lambda_{\rm c}$. An expression for $\lambda_{\rm c}$ for step-index fibres is given as [20] (see section A6.4)

$$\lambda_{\rm c} = \frac{2\pi r}{V_{\rm c}}[(n_1 + n_2)\Delta n]^{1/2} \qquad \text{(B4.3.5)}$$

where n_1 and n_2 are the refractive indices of the core and cladding, respectively, and $V_{\rm c} = 2.405$ is the normalized frequency at the cut-off wavelength. It is now clear that in order to reduce $A_{\rm eff}$, r should be reduced and Δn increased with the changes constrained by equation (B4.3.5). However, there are additional problems.

The use of high-index difference fibres is both beneficial and detrimental. The dopants such as Ge and P added to increase the core index also increases the Raman gain coefficient as outlined earlier. However, an undesirable effect is that the fibre loss is also increased [16–19]. This is the third fibre design parameter. For low values of GeO_2 in the core, it is straightforward to predict the fibre loss based on knowing the

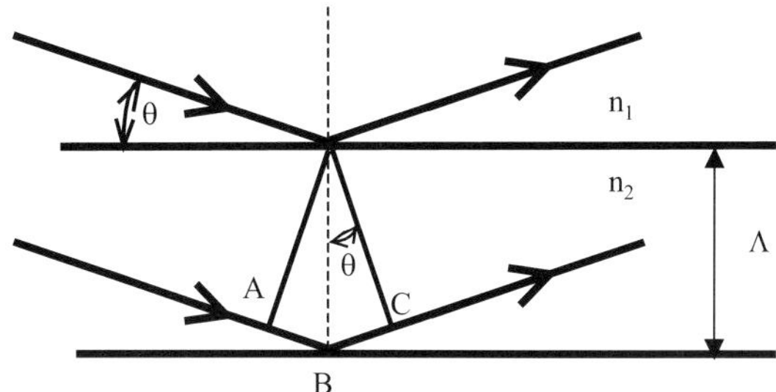

Figure B4.3.5. A schematic illustration of Bragg scattering from two dielectric layers.

concentration of GeO_2. It is well documented that there is an 'excess' loss in highly doped GeO_2 fibres. This loss is called excess because it goes beyond the loss predicted by merely extrapolating out loss calculations at lower concentrations. Three reasons are put forth for this excess loss. The first explanation is that it results from increased Rayleigh scattering. The origin of Rayleigh scattering is fluctuations in the refractive index due to concentration fluctuations frozen into the core. It is speculated that the inhomogeneous incorporation of a large amount of GeO_2 leads to increased Rayleigh scattering. A second cause that has been offered is the stress at the core–cladding interface due to the differences in the viscosity of the core and cladding as more GeO_2 is added to the core. This leads to waveguide imperfections. The final explanation is an increase in the UV absorption peak of the fibre due to the increased number of germanium oxygen-deficient centers that absorb UV light. It should be noted that the loss in these fibres depends more on drawing conditions than standard single-mode fibres.

A final design consideration for a Raman fibre is the ability to get low-loss intra-cavity splices in the RFL. The easiest route towards this is for the Raman fibre to allow the low-loss gratings to be written. The Raman fibre is then spliced to itself, which typically is a low splice loss. Alternatively, the Raman fibre should have a low splice loss to a photosensitive fibre; however, the lower A_{eff} results in a significant mode-field mismatch between the Raman gain fibre and other fibres. The different melting points and diffusion rates between high-index difference fibres and standard fibres also presents splicing challenges [21].

B4.3.4 The feedback mechanism: fibre Bragg gratings

With an understanding of the gain medium, the feedback mechanism used in RFLs is now described. At the interface between two dielectric materials scattering occurs. The fraction of light reflected depends on the angle of incidence and the refractive index of the materials. For a multilayered dielectric medium such as that shown in figure B4.3.5, where a small index difference has been assumed, scattering will occur at each interface. In order for there to be a significant amount of reflection from the multiple layers the reflections from each layer should add in phase. The condition under which this is satisfied is when the pathlength shown as AB+BC is an integer number, m, of wavelengths (λ/n) and is given as

$$2\Lambda \sin\theta = m\frac{\lambda_{\mathrm{B}}}{n_2} \tag{B4.3.6}$$

where Λ is the layer thickness, θ is defined in figure B4.3.5, and λ_{B} is called the Bragg wavelength and is the wavelength at which the phase-matching condition occurs. When this condition is satisfied, this is known as Bragg reflection. For multiple layers with a period Λ, a significant amount of light is reflected backward. Due to the wave-guiding nature of optical fibres, $\theta = 90°$ and the Bragg condition becomes $\Lambda = m\lambda_{\mathrm{B}}/2n_2$.

A multilayered structure is reproduced in optical fibres by inducing periodic changes in the refractive index of the core of the fibre. The coupling between the forward (transmitted) and backward (reflected) travelling waves in FBGs is represented by coupled mode equations. From these equations a reflection coefficient, R, can be extracted and is given by [22–26]

$$R = \frac{\kappa^2 \sinh^2(|S|L)}{\delta\beta^2 \sinh^2(|S|L) + |S|^2 \cosh^2(|S|L)} \qquad \kappa^2 > \delta\beta^2 \tag{B4.3.7a}$$

and

$$R = -\frac{\kappa^2 \sin^2(|S|L)}{\delta\beta^2 - \kappa^2 \cos^2(|S|L)} \qquad \kappa^2 < \delta\beta^2. \tag{B4.3.7b}$$

In these equations $\delta\beta = 2\pi n_{\mathrm{eff}}/\lambda - m\pi/\Lambda$ (from equation (B4.3.6)) measures the mismatch at the free-space wavelength λ from the Bragg condition (n_{eff} is the mode effective index) and Λ is the period of the index change. κ is the coupling coefficient between the transmitted and reflected fields. $S = [\delta\beta^2 - \kappa^2]^{1/2}$ is the dispersion relationship for the backward propagating wave determined from the boundary conditions of the coupled mode equations. Finally, L is the length of the grating. For a uniform sinusoidal modulation of the refractive index throughout the core, κ can be expressed as

$$\kappa = \frac{\pi\,\delta n}{\lambda_{\mathrm{B}}}\eta \tag{B4.3.8}$$

where δn is the periodic index difference and η is the overlap of the mode of the electric field with the core. Note that equation (B4.3.7a) describes the case when the coupling of the grating at λ is larger than the detuning of the grating from Bragg resonance whereas equation (B4.3.7b) is the reverse case.

Equations (B4.3.7) shows how the grating properties can be manipulated in order to obtain the reflectivities and spectral widths needed to make an RFL. The peak reflectivity is obtained from equation (B4.3.7) when the Bragg condition is satisfied and $\delta\beta = 0$ is given by

$$R_{\max} = \tanh^2(\kappa L). \tag{B4.3.9}$$

Hence, for $\kappa L = 3$ or larger, the reflectivity approaches 100%. The magnitude of κ and, hence, $R_{\max}$ is primarily determined by the index difference induced between consecutive sections of the grating.

Another important property of a grating is its spectral bandwidth $\Delta\lambda$. There are several measures of bandwidth. A simple one to derive is the difference between the first minimum occurring on either side of the reflection peak. This occurs when $\sinh^2(|S|L)$ equals zero and an expression for it is given by

$$\Delta\lambda = \frac{\lambda_{\mathrm{B}}^2}{\pi n_{\mathrm{eff}} L}[(\kappa L)^2 + \pi^2]^{1/2} \tag{B4.3.10}$$

where it has been assumed that $\lambda \approx \lambda_{\mathrm{B}}$.

B4.3.4.1 Grating writing

There are several different techniques for writing gratings in optical fibres. The two techniques that are primarily used are the dual-beam holographic method and the phase-mask approach [23–29] (see also chapter D2.6).

A diagram for the dual-beam holographic method is shown in figure B4.3.6. The bare fibre is exposed to two beams originating from one ultraviolet source. The interference pattern produced by the two beams leads to a periodic modulation of the refractive index of the glass core. Cylindrical lenses are used to elongate the beam, along the fibre length. The grating period is given by the expression

$$\Lambda = \frac{n_{\mathrm{eff}}\lambda_{\mathrm{laser}}}{n_{\mathrm{laser}}\sin\theta} \tag{B4.3.11}$$

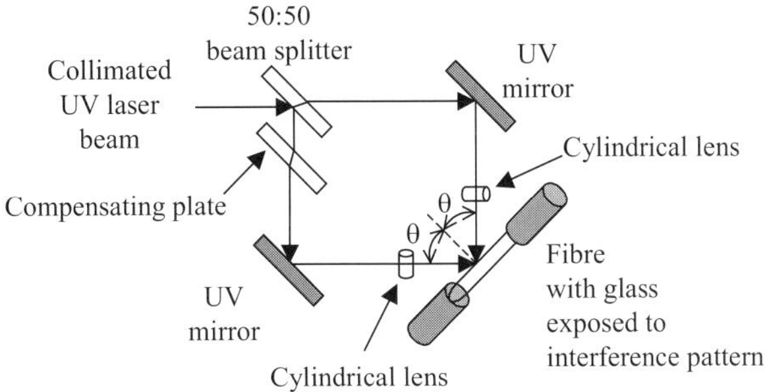

Figure B4.3.6. Schematic diagram of a dual-beam holography method for writing Bragg gratings.

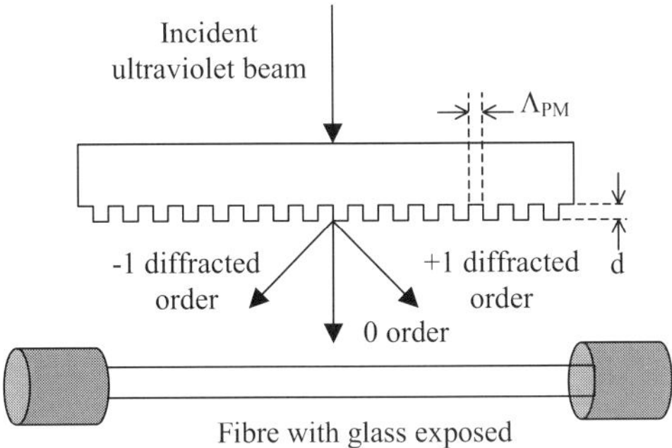

Figure B4.3.7. Schematic diagram of a Bragg grating writing using a phase mask.

where λ_{laser} and n_{laser} are the laser wavelength and index in the fibre respectively, and θ is the angle between the two beams as shown in figure B4.3.6. By adjusting θ and, hence, Λ, the wavelength at which the peak reflectivity occurs can be varied. The dual-beam holography method was one of the first approaches to remove the limitations imposed by the ultraviolet source on the obtainable grating wavelengths. This technique, however, has several disadvantages. The source needs a high degree of spatial and temporal coherence. This is because mechanical vibrations and air turbulence can affect the quality of the interference pattern generated in the fibre affecting grating quality. For low coherence sources, it is typical to insert a glass plate in the arm of the interferometer path that undergoes a reflection at the beamsplitter in order to compensate for the path difference. This makes this method ideally suited for gratings that can be written with high-energy pulsed sources with a very short exposure time.

The second approach, shown in figure B4.3.7, which overcomes some of the disadvantages of the dual-beam holographic approach, is to use a phase mask. The phase mask is obtained using photolithography techniques commonly employed for fabrication of integrated electronic circuit devices. The majority of the light incident on the phase mask (typically normally) is diffracted into the $m = 0$ and ± 1 orders. The interference between the ± 1 orders sets up a standing-wave pattern which then induces periodic changes in the index of a fibre placed in the field. In this approach the grating period Λ is truly independent of the ultraviolet source and is given by

$$\Lambda = \Lambda_{PM}/2 \tag{B4.3.12}$$

where Λ_{PM} is the period of the phase mask. For efficient diffraction into the first order, the phase mask should be written to have a groove–space ratio of 1:1. The depth of the grooves, d, affects the amount of light diffracted into the first order and is chosen so as to minimize the light in the zero order, with the expression for d with minimum transmission given by

$$d = \frac{\lambda_{\rm laser}}{2(n_{\rm laser} - 1)}. \tag{B4.3.13}$$

Hence, the zero-order minimization can only be done for a given laser wavelength. The advantages of this technique are reduced demand on the temporal and spatial coherence of the ultraviolet source, the gratings are very reproducible, gratings with a variation in period (chirped) can be made and very long gratings can be fabricated. Some disadvantages to this approach are that the quality of the grating is almost completely mask dependent and, beyond a limited wavelength range (~5 nm), a new phase mask is needed for each wavelength.

B4.3.4.2 Photosensitivity

The ability to write gratings in a fibre is due to the changes in the absorption spectrum of the fibre when exposed to ultraviolet (UV) light [30]. These changes are analogous to those light induces on a photographic plate. During the fabrication of optical fibres under ideal circumstances all the bonds formed would be SiO_2 (silica dioxide) bonds. When dopants such as germanium (Ge) are added they would then replace the Si atoms. However, Si–Si, Si–Ge and Ge–Ge bonds may form. These bonds are defects in the fibre and, when exposed to UV light around 244 nm, they break freeing an electron. It is believed that the increase in the number of this type of defect changes the absorption spectrum of the light resulting in a change in the refractive index of the fibre through the Kramers–Kronig relationship. A second explanation is that there is an increase in the density of the glass through the creation of these defects. The removal of the electron changes the shape of the molecule. This densification results in index changes. While the exact mechanism by which this occurs is not fully agreed upon, it is likely that both mechanisms play a role, with the determination of which dominates dependent on the fibre type, writing power and wavelength. While Ge is the dopant most generally used, other dopants such as P, Ce, Pb, Sn–Ge, Sb and Ge–B have been shown to be photosensitive at UV wavelengths ranging from 350 to 155 nm. An important aspect of UV-induced index changes is that they require thermal annealing to ensure long-term stability.

The quality of a grating can be improved by increasing the photosensitivity of the fibre. This is done by a high-pressure exposure of the fibre to hydrogen (H_2) [25] or deuterium (D_2) [26]. The increase in photosensitivity has two effects. It increases the amount of index change that can be induced and it reduces the exposure time needed to achieve the index change. It appears that these elements can produce Si–OH or Si–OD groups as well as oxygen-deficient Ge sites. These are the sites that causes index changes when exposed to UV light mentioned in the previous paragraph. The photosensitivity of fibres treated in this way can be two orders of magnitude higher than that of untreated fibre. Any unreacted H_2 or D_2 diffuses out of the fibre.

The discussion of FBGs in this section has provided insight into the process but it is not indicative of the rich possibilities with gratings (see chapter D2.6). Nevertheless, it is clear how both high reflectors and output couplers can be written in order to provide feedback to RFLs.

B4.3.5 Raman fibre lasers

With the understanding of the components that comprise an FBG–RFL obtained from the previous sections, a complete device is now described. The behaviour of an RFL depends on such parameters as fibre length, OC reflectivity and splice losses. In order to understand the trade-offs in making these choices, numerical simulations are used. In what follows the mathematical basis for the numerical model is given and the results of the numerical simulations are used to elucidate the effect of some design parameters on a RFL. Finally, experimental results are presented.

The evolution of the pump and Stokes power inside an RFL can be described by a set of nonlinear ordinary differential equations [33–37]. These are:

$$\frac{\mathrm{d}P_{\mathrm{p}}^{\pm}}{\mathrm{d}z} = \mp \alpha_{\mathrm{p}} P_{\mathrm{p}}^{\pm} \mp \frac{\nu_{\mathrm{p}}}{\nu_1} \frac{g_{\mathrm{R}}^{1}}{A_{\mathrm{eff}}^{1}} (P_1^{+} + P_{\mathrm{s}1}^{-}) P_{\mathrm{p}}^{\pm} \quad \text{(B4.3.14a)}$$

$$\frac{\mathrm{d}P_{i}^{\pm}}{\mathrm{d}z} = \mp \alpha_{i} P_{i}^{\pm} \mp \frac{\nu_{i}}{\nu_{i+1}} \frac{g_{\mathrm{R}}^{i}}{A_{\mathrm{eff}}^{i}} (P_{i+1}^{+} + P_{i+1}^{-}) P_{i}^{\pm} \pm \frac{g_{\mathrm{R}}^{i-1}}{A_{\mathrm{eff}}^{i}} (P_{i-1}^{+} + P_{i-1}^{-}) P_{i}^{\pm} \quad \text{(B4.3.14b)}$$

$$\frac{\mathrm{d}P_{n}^{\pm}}{\mathrm{d}z} = \mp \alpha_{n} P_{n}^{\pm} \pm \frac{g_{\mathrm{R}}^{n}}{A_{\mathrm{eff}}^{n}} (P_{n-1}^{+} + P_{n-1}^{-}) P_{n}^{\pm} \quad \text{(B4.3.14c)}$$

where the superscript $\pm$ designates the power, P, in the forward and backward travelling waves respectively. The forward direction is from the pump to the OC as shown in figure B4.3.2. The index p designates the pump wave, i the intermediate Stokes orders and n the lasing wavelength. The first term on the right-hand side of each equation represents the effect of intrinsic fibre loss on the power in that wave, the next term describes the depletion of that wave by the forward and backward travelling wave of the next highest Stokes order through the Raman effect. The final term in equations (B4.3.14b) and (B4.3.1c) represent the gain of ith wave through Raman pumping by the $(i-1)$th wave. The boundary conditions for these equations are given by

$$\begin{aligned} P_{\mathrm{p}}^{+}(0) &= P_{\mathrm{in}} & P_{\mathrm{p}}^{-}(L) &= R_{\mathrm{p}} P_{p}^{+}(L) \\ P_{i}^{+}(L) &= R_{i} P_{p}^{-}(L) & P_{i}^{-}(L) &= R_{i} = P_{p}^{+}(L) \\ P_{n}^{+}(L) &= R_{n} P_{n}^{-}(L) & P_{n}^{-}(L) &= R_{\mathrm{oc}} P_{n}^{+}(L) \end{aligned} \quad \text{(B4.3.15)}$$

where R is the reflectivity of the FBG, L is the length of the RFL cavity, and R_{oc} is the reflectivity of the OC. For notational simplicity $P_{\mathrm{out}} = (1 - R_{\mathrm{oc}}) P_{n}^{+}(L)$ is used.

Numerical solutions to equations (B4.3.14) provide insight into the wavelength conversion of light in a RFL. They neglect, however, several effects worth mentioning that are included in the model used to generate the results described here. Spontaneous emission has been omitted and the interaction between non-sequential Stokes lines (e.g. the first Stokes line interacting with the third) together with the possibility of generating the next Stokes line beyond the nth wave have been neglected. No results published to date predict the shape of the output spectrum of an RFL. Nonetheless, these models provide an excellent qualitative description of the effect of various parameters on an RFL.

In the simulations presented here, a 1117–1480 nm RFL such as that shown in figure B4.3.2 was modelled. Experimentally measured values of the Raman gain coefficient and the fibre loss for a Ge-doped fibre such as that shown in figure B4.3.4 were used. Unless otherwise stated the splice loss between fibres was assumed to be zero. The three measures of the RFL performance are the slope efficiency, η_{s}, pump threshold power, P_{th}, and the overall (total) efficiency, η_{T}. These quantities are defined from the linear fit of a graph of the launched pump power P_{in} *versus* output power P_{out} as

$$\begin{aligned} P_{\mathrm{out}} &= \eta_{\mathrm{s}}(P_{\mathrm{in}} - P_{\mathrm{th}}) \\ \eta_{\mathrm{T}} &= \frac{P_{\mathrm{out}}}{P_{\mathrm{in}}}. \end{aligned} \quad \text{(B4.3.16)}$$

Figure B4.3.8(a) is a plot of η_{s} for a 1480 nm RFL as a function of L for different R_{oc}. As L increases, η_{s} decreases almost linearly. This decrease is due to the increased loss as a function of L. The effect of changing R_{oc} can also be extracted from figure B4.3.8(a). By moving down a vertical line at a constant value of length, it is seen that η_{s} decreases fairly linearly as R_{oc} increases. The slightly closer spacing of the lines

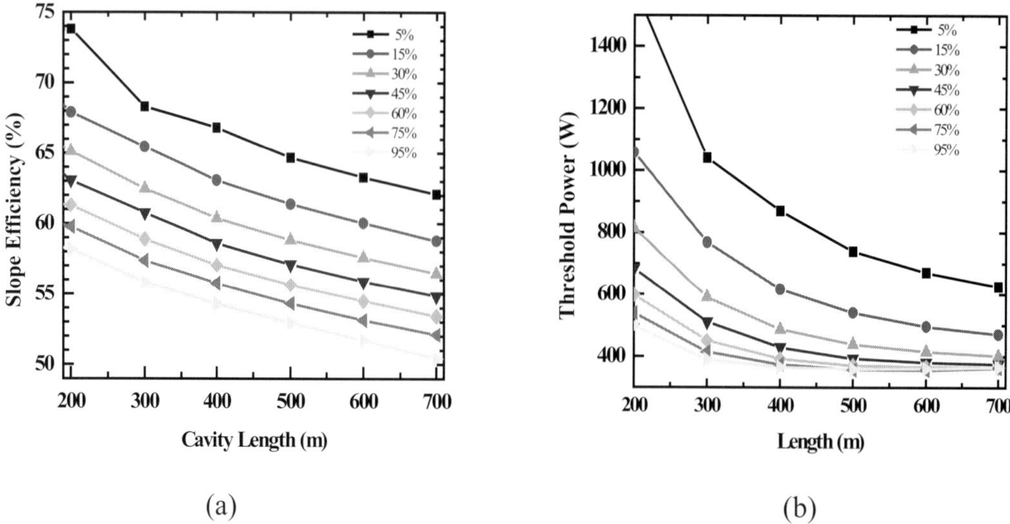

Figure B4.3.8. Simulation results showing the effect of fibre length on (*a*) slope efficiency and (*b*) threshold power for different output coupler reflectivities.

Table B4.3.2. Values of L and R_{oc} optimized to produce the maximum η_T for a given P_p.

P_{in} (W)	L (m)	R_{oc} (%)	P_{out} (W)	η_T (%)
1	500	70	0.4	40
2	400	45	0.9	46
3	400	25	1.5	51
4	400	15	2.1	53
5	300	15	2.7	55

at higher reflectivities suggests a slight curvature to the fall off. The decrease in η_s is expected since a higher R_{oc} means less power is extracted from the cavity.

The trade-offs in designing a RFL become apparent in comparing figure B4.3.8(*a*) with (*b*). All of the changes that had a negative effect on η_s now have a positive effect on P_{th}. Increasing the length of the fibre reduces P_{th}. This is simply because the length-integrated Raman gain increases with increasing fibre length lowering the threshold. Eventually the benefit on P_{th} of increasing fibre length saturates as the length integrated gain exceeds fibre loss. Simulations show P_{th} can eventually increase as the increased intrinsic fibre loss exceeds any increased benefit from the length-integrated Raman gain. As with η_s, the effect of R_{oc} on P_{th} can also be obtained from figure B4.3.8(*b*). By moving down a vertical line at a constant value of length it is seen that P_{th} decreases nonlinearly as R_{oc} increases, eventually saturating. For longer L a lower value of R_{oc} is needed to saturate P_{th}. This behaviour is expected since a larger R_{oc} increases the intercavity power and it is this power that determines the threshold. Since lasing occurs when gain equals loss, the lower R_{oc} needed for saturation at longer lengths is due to the increased integrated Raman gain at these lengths. Saturation occurs since, beyond the point at which gain equals loss, little value is obtained by retaining more power in the cavity (by increasing R_{oc}).

The behaviour described in the previous paragraphs points to the need to optimize L and R_{oc}. The

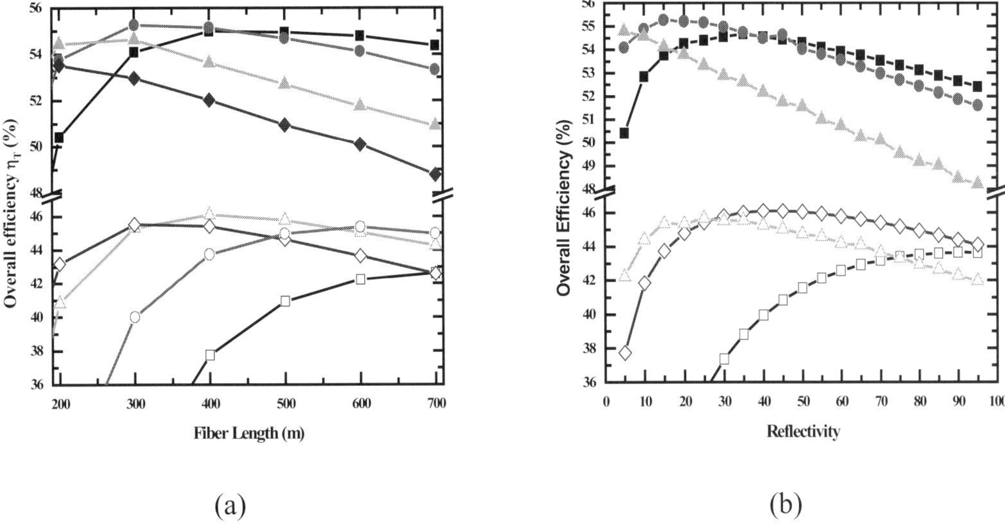

Figure B4.3.9. Overall efficiency as a function of (*a*) fibre length and (*b*) output coupler reflectivity for an input pump power of 5 W (filled symbols) and 2 W (open symbols). Plot (*a*) is for $R_{oc} = 5\%$ (squares), 15% (circles), 45% (triangles) and 70% (diamonds). Plot (*b*) is for $L = 200$ m (squares), 300 m (diamonds), 400 m (circles) and 600 m (triangles).

overall efficiency is a measure of this trade-off. The results of simulations to maximize η_T by varying L and R_{oc}, for different values of P_{in} are summarized in table B4.3.2 which shows that the optimized parameters depend on the desired operating point. For low P_{out}, it is more important to design the cavity to minimize P_{th} since it represents a large percentage of P_{in}. Increasing L and R_{oc} does this. For higher powers, η_s is more important since the device is operating far above threshold; hence, smaller values of L and R_{oc} are required. It is also seen that the maximum η_T increases with P_p. This is again a reflection of the decrease in the percentage of P_{in} light used to reach P_{th}.

Further information can be obtained by examining the behaviour of η_T as a function of L and R_{oc} for a fixed P_p. Plots for $P_p = 5$ W and 2 W are shown in figures B4.3.9(*a*) and (*b*) and these indicate the sensitivity of the optimum design points listed in table B4.3.2. Qualitatively, it is seen η_T is fairly insensitive to L and R_{oc} as long as the RFL is designed around the optimum parameters. It is noteworthy, however, that a laser designed to operate at 2 W(5 W) will not necessarily perform well at 5 W(2 W). Therefore, a laser required to operate over a wide power range will have to be a compromise between the two designs for optimum performance.

In section B4.3.2 the design trade-off between increasing fibre gain and loss was described. The practical effect of this trade-off is now examined. Table B4.3.3 shows the optimized L and R_{oc} for three different fibre types at 5 W and 2 W. For the prior simulations, a fibre specifically designed to enhance Raman gain (RF) was used. Note that the high slope dispersion compensating (HSDK) fibre has a higher gain and loss coefficient whereas the reverse is true for the Truewave® RS (TWRS) fibre. A figure of merit (FOM) is defined as [18, 38]

$$\mathrm{FOM} = \frac{g_R}{A_{\mathrm{eff}}\alpha_{1550\ nm}} \qquad \text{(B4.3.17)}$$

It is seen that the best performance is obtained for the highest FOM, which is the RF fibre. Other trends emerge. A higher gain means a shorter fibre for the maximum η_T. The difference is especially noticeable at lower power levels. The optimum reflectivity increases for the HSDK (TWRS) because of the higher loss

Table B4.3.3. Values of L and R_{oc} optimized to produce the maximum η_T for a given P_p and for fibres with various Raman gain and loss coefficients.

Fibre type	g_R/A_{eff} (km^{-1} W^{-1})	$\alpha_{1550\ nm}$ (dB km^{-1})	FOM (W^{-1} dB^{-1})	P_{in} (W)	L (m)	R_{oc} (%)	P_{out} (W)	η_T (%)
HSDK	3.3	0.64	5.1	5	200	25	2.6	51
RF	2.4	0.30	8.0	5	300	15	2.7	55
Truewave® RS fibre	0.7	0.21	3.4	5	600	55	2.4	48
HSDK	3.3	0.64	5.1	2	250	60	0.8	40
RF	2.4	0.30	8.0	2	400	45	0.9	46
Truewave® RS fibre	0.7	0.21	3.4	2	800	95	0.7	35

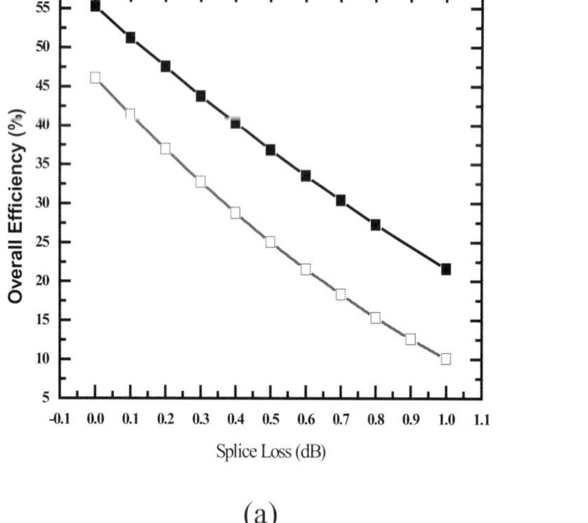

(a)

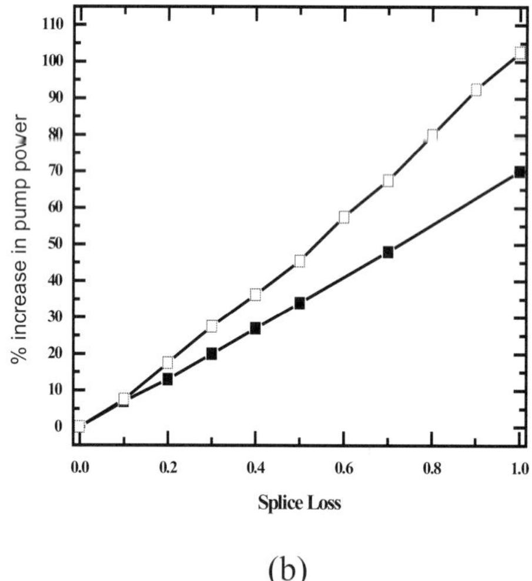

(b)

Figure B4.3.10. (a) Change in η_T as a function of splice loss for a 5 W (filled squares) and a 2 W (open squares) optimized RFL as described in table B4.3.1. (b) The percentage increase in pump power required to maintain the same output power as with 0 dB splice loss for a given splice loss, under the same conditions as those described in (a).

(lower gain); hence, more power must be retained in the cavity in order to reach threshold. Finally, the best η_T is more sensitive to fibre design at lower powers since P_{th} is a larger percentage of the total power.

As a final design parameter the effect of splice loss is examined in figures B4.3.10(a). For these simulations, the loss indicated is divided evenly between the splices indicated in figure B4.3.2. As indicated earlier, a splice loss of 0 dB was used in the previous simulations. Just adding 0.05 dB splice loss (considered an excellent splice loss) to each end of the cavity produces a predicted decrease in η_T of 4%(5%) for an RFL optimized for 5 W(2 W). The drop in overall efficiency is worse at lower powers since the increased cavity loss raises P_{th}, which represents a larger percentage of the pump power. Since in most applications a constant output power is what is required, figure B4.3.10(b) shows the percentage pump power increase required to maintain the same output power for a 0.1 dB splice loss compared to a 0 dB splice loss. In order for a 5 W (2 W) device to maintain the same operating power a 7% (7.5%) increase in pump power is needed. In applications such as telecommunications, this represents a significant amount of power. A summary of

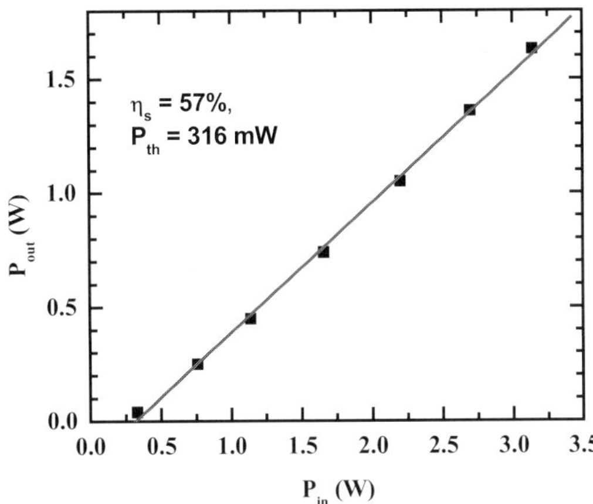

Figure B4.3.11. Plot of slope efficiency and threshold of a 1117–1480 nm Ge-doped RFL. (Courtesy of OFS Fitel Specialty Photonics Devices.)

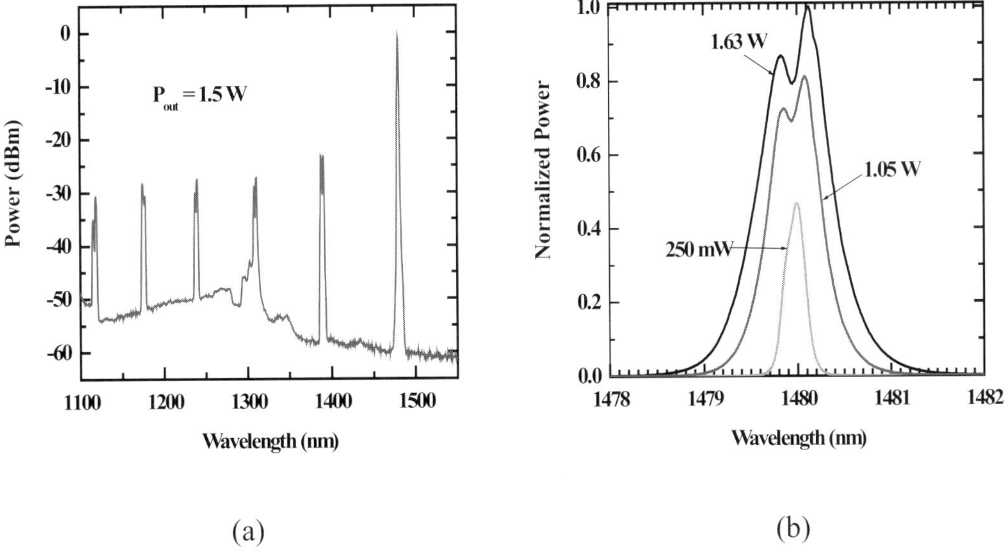

Figure B4.3.12. The output spectrum of an 1117–1480 nm (*a*) RFL showing the pump and intermediate Stokes orders and (*b*) a close-up of the spectrum of the 1480 nm line.

the design issues is, therefore, to choose the desired output power of the RFL, optimize the design for L and R_{oc}, then work hard to reduce splice losses.

To conclude this section experimental measurements of an RFL will be discussed. Figure B4.3.11 is a plot of the input *versus* output power for a Ge-doped RFL. This device is pumped with a 1117 nm Yb-doped cladding-pumped fibre laser. The cavity uses 500 m of Raman enhanced fibre and a 30% peak reflector. The slope efficiency for this device is 54%, with $P_{th} = 656$ mW. For a P-doped RFL, the best η_s in the literature is 48%, though it should be pointed out that this type of laser is in an earlier stage of development.

The output spectrum is shown for the same device in figures B4.3.12(*a*) and (*b*). The intermediate

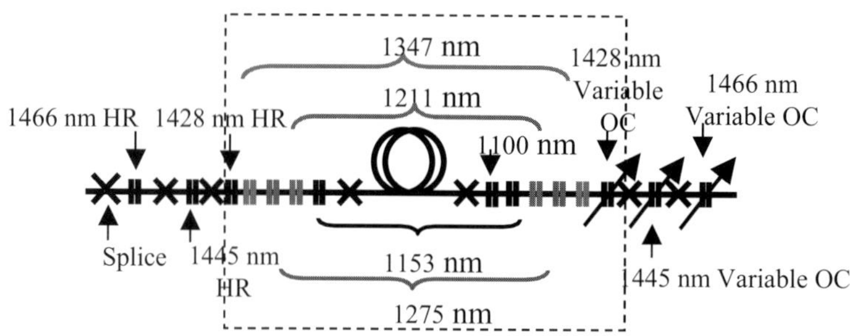

Figure B4.3.13. Schematic diagram of a 3λRFL lasing pumped at 1100 nm and lasing at 1427 nm, 1445 and 1466 nm.

Raman orders can be seen in the spectral plot. A term often used to characterize an RFL is the suppression ratio. This describes the difference in dB between the peak power of the desired output wavelength and the intermediate Stokes order with the highest output power. It should also be noted that the output spectrum of an RFL significantly broadens as the power is increased. To date, the reason for this has not been explained in the literature.

In optimizing an RFL, consideration should be given to the pump source. The best design should include the overall efficiency of the pump and RFL. The usual pump source used is a Yb-doped cladding-pumped fibre laser (CPFL), and the efficiency of this device will depend on the output wavelength selected. As an example, in considering whether to use P-doped or Ge-doped fibre the starting wavelength for a 1480 nm device is 1060 and 1117 nm, respectively. Because the CPFL operates more efficiently at 1117 nm this should be considered in designing a device [39].

B4.3.6 Future trends in Raman fibre laser designs

One emerging trend in the literature is the use of multi-wavelength RFLs ($n\lambda$RFL where n is the number of wavelengths emitted). These devices takes advantage of the inhomogeneously broadened Raman gain spectrum to produce multiple wavelengths from one RFL cavity [40–43]. A schematic diagram of such a device is shown in figure B4.3.13. In order to make the cavity emit at multiple wavelengths, for example a 3λRFL, high-reflector gratings at two additional wavelengths must be added to the input grating set and output couplers at the corresponding wavelengths on the output grating set. The resultant output spectrum for such a device emitting at 1428, 1445 and 1466 nm is shown as a full curve in figure B4.3.14.

In order to make a practical multi-wavelength RFL, it is essential that the distribution of power among the output wavelengths be variable. This is because the tolerance imposed on the distribution of power among the wavelengths by an end user is tighter than the possible manufacturing tolerance. Moreover, this distribution will vary with changing power. These reasons point to the need to have a multi-wavelength RFL where the wavelength power ratio can be controlled. Two approaches for accomplishing this have been described in the literature. The first involves strain tuning the output coupler grating of one of the wavelengths; hence, deliberately misaligning the laser cavity reducing the power at that wavelength. In a second approach, OC gratings with a variable reflectivity are used [41, 42, 44–46]. In this approach there is a minimum change in wavelength as the power is adjusted nor any net loss in the total power in the cavity. Figure B4.3.14 shows the output spectra of a 3λRFL with different OC reflectivity settings. To date, a source emitting up to six wavelengths has been demonstrated.

The ability of the source to be reconfigured was also considered [44]. A 3λRFL was interfaced with a computer such that the voltages controlling the output coupler reflectivities could be varied, and data on the resultant pump power distributions or gain in a transmission system recorded. The results showed that for

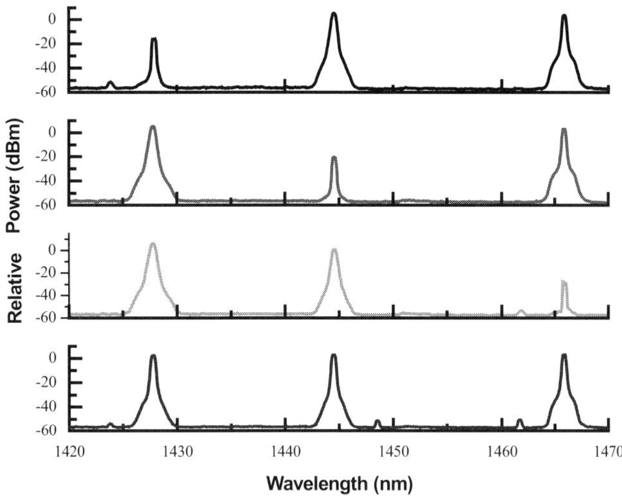

Figure B4.3.14. Output spectra of a configurable 3λRFL for different settings of the OC reflectivities.

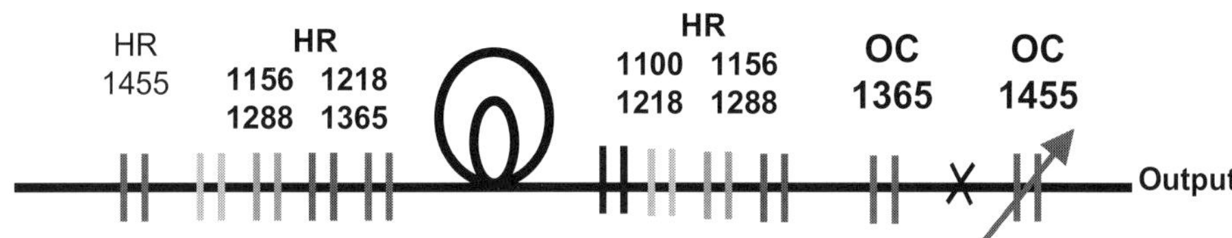

Figure B4.3.15. Schematic diagram of a dual-order RFL.

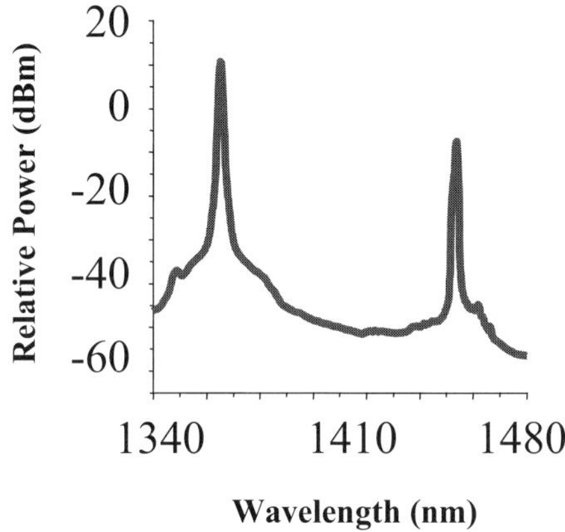

Figure B4.3.16. Output spectrum of a dual-order RFL.

a given pump power level, the total output power was conserved and varying the reflectivity of the output coupler resulted in the redistribution of the power among the different wavelengths. It also showed that locating a desired operating point can be easily implemented through a search routine since the operating point is a stable minimum.

Finally, the temporal stability of a multi-wavelength RFL needs to be addressed. A concern was that the power could fluctuate between the different wavelengths or that the coupled cavities would increase the noise of the laser. One way to characterize the noise performance of a laser is to measure its relative intensity noise (RIN). This is defined as the ratio of the mean square optical intensity fluctuation to the square of the average optical power in a 1 Hz bandwidth. A comparison of the RIN spectra of a 3λRFL compared to that of a 1λRFL shows that the performance of the 3λRFL was equal to or better than that of the 1λRFL.

A second trend is the use of multiple-order Raman fibre lasers [47, 48]. In these devices the wavelength separation between the emitted wavelengths is approximately one Stokes shift: the shorter wavelength also provides gain to the longer wavelength. A schematic diagram of this along with a plot of the spectrum is shown in figures B4.3.15 and B4.3.16. In this example, the OC of the longer wavelength is variable to allow the power ratio between the two emitted wavelengths to be controlled. Again the driver for this technology is telecommunications where the multiple orders provide an advantage during the Raman amplification of transmission spans [48].

References

[1] Stolen R H, Ippen E P and Tynes A R 1972 Raman oscillation in glass optical waveguide *Appl. Phys. Lett.* **20** 62–4

[2] Hill K O, Fujii Y and Johnson D C 1976 Low-threshold cw Raman laser *Appl. Phys. Lett.* **30** 647–9

[3] Lin C, Cohen L G, Stolen R H, Tasker G W and French W G 1977 Near-infrared sources in the 1–1.3 μm region by efficient stimulated Raman emission in glass fibers *Opt. Commun.* **20** 426–8

[4] Innis D *et al* 1997 Ultrahigh-power single-mode fiber lasers from 1.065 to 1.472 μm using Yb-doped cladding-pumped and cascaded Raman lasers *Conf. Lasers Electrooptics* Paper CPD-31 (Washington, DC: OSA)

[5] Grubb S *et al* 1994 1.3 μm cascaded Raman amplifier in germanosilicate fibers *Proc. Optical Amplifiers and their Applications* Paper PD3-1 (Washington, DC: OSA) p 187

[6] Grubb S G *et al* 1995 High-power 1.48 μm cascaded Raman laser in germanosilicate fibers *Proc. Optical Amplifiers and their Applications* Paper SA4 (Washington, DC: OSA) pp 197–9

[7] Karpov V I, Dianov E M, Kurkov A S, Paramonov V M, Protopopov V N, Bachynski M P and Clements W R L 1999 LD-pumped 1.48 μm laser based on Yb-doped double-clad fiber and phosphorosilicate-fiber Raman converter *Proc. Optical Fiber Communication Conf.* Paper WM3 (Washington, DC: OSA) pp 202–4

[8] Dianov E M *et al* 1999 CW highly efficient 1.24 μm Raman laser based on low-loss phosphosilicate fiber *Proc. Optical Fiber Communication Conf.* Paper PD25 (Washington, DC: OSA)

[9] Karpov V I *et al* 1999 Laser-diode-pumped phosphosilicate-fiber Raman laser with an output power of 1 W at 1.48 μm *Opt. Lett.* **24** 13 887–9

[10] Dianov E M, Bufetov I A, Bubnov M M, Grekov M V, Vasiliev S A and Medvedkov O I 2000 Three cascaded 1407-nm Raman laser based on phosphorus-doped silica fiber *Opt. Lett.* **25** 402–4

[11] Kurkov A S, Paramonov V M Medvedkov O I, Vasiliev S A and Dianov E M 2000 Raman fiber laser at 1.45 μm: comparison of different schemes *Optical Amplifiers and their Applications* Paper OMB5 (Washington, DC: OSA) pp 16–18

[12] Dianov E M and Prokhorov A M 2000 Medium-power cw Raman fiber lasers *IEEE J. Selected Topics Quantum Electron.* **6** 1022–8

[13] Agrawal G P 1995 *Nonlinear Fiber Optics* (San Diego, CA: Academic) ch 8

[14] Galeener F L, Mikkelsen J C Jr, Geils R H and Mosby W J 1978 The relative Raman cross sections of vitreous SiO_2, GeO_2, B_2O_3, and P_2O_5 *Appl. Phys. Lett.* **32** 34–6

[15] Shibata N, Horigudhi M and Edahiro T 1981 Raman spectra of binary high-silica glasses and fibers containing GeO_2, P_2O_5 and B_2O_3 *J. Non-Crystalline Solids* **45** 115–26

[16] Mashinsky V H, Dianov E M, Neustruev V B, Lavrishchev S V, Guryanov A N, Khopin V F, Vechkanov N N and Sazhin O D 1994 UV absorption and excess optical loss in preforms and fibers with high germanium content *Fiber Optic Materials and Components (Proc. SPIE 2290)* ed H H Yuce, D K Paul and R A Greenwell, (Bellingham, WA: SPIE) pp 105–12

[17] Dianov E M, Mashinsky V M, Neustruev V B, Sazhin O D, Guryanov A N, Khopin V F, Vechkanov N N and Lavrishchev S V 1997 Origin of excess loss in single-mode optical fibers with high GeO_2-doped silica core *Opt. Fiber Technol.* **3** 77–86

[18] L Grüner-Nielsen 1998 High index fibers *Thesis for Industrial PhD* EF 546/PhD No 94-0146-ATV Danish Academy of Technical Sciences

[19] Lines M E, Reed W A, DiGiovanni D J and Hamlins J R 1999 Explanation of anormalous loss in high delta singlemode fibres *Electron. Lett.* **35** 1009–10

[20] Marcuse D 1982 *Light Transmission Optics* (New York: Van Nostrand Reinhold) ch 8

[21] Kashima N 1995 *Passive Optical Components for Optical Fiber Transmission* (Boston, MA: Artech)

[22] Kogelnik H and Shank C W 1972 Coupled wave theory of distributed feedback lasers *J. Appl. Phys.* **43** 2327–35

[23] Bennion I, Williams J A R, Zhang L, Sugden K and Doran N J 1996 UV-written in-fibre Bragg gratings *Opt. Quantum Electron.* **28** 93–135

[24] Agrawal G P 1995 *Nonlinear Fiber Optics* (San Diego, CA: Academic) ch 10

[25] Kashyap R 1999 *Fiber Bragg Gratings* (San Diego, CA: Academic)

[26] Othonos and Kalli K *Fiber Bragg Gratings: Fundamentals and Applications in Telecommunications and Sensing* (Norwood, MA: Artech)

[27] Meltz G, Morey W W and Glenn W H 1989 Formation of Bragg gratings in optical fibres by transverse holographic method *Opt. Lett.* **14** 823

[28] Hill K O, Malo B, Bilodeau F, Johnson D C and Albert J 1993 Bragg gratings fabricated in monomode photosensitive optical fiber by UV exposure through a phase mask *Appl. Phys. Lett.* **62** 1035–7

[29] Anderson D Z, Mizrahi V, Erdogan T and White A E 1993 Production of in-fiber gratings using a diffractive optical element *Electron. Lett.* **29** 566–8

[30] Hill K O, Fujii Y, Johnson D C and Kawasaki B S 1978 Photosensitivity in optical fiber waveguides: Applications to reflection filter fabrication *Appl. Phys. Lett.* **32** 647–9

[31] Lemaire P J, Atkins R M, Mizrahi V and Reed W A 1993 High pressure H_2 loading as a technique for achieving ultrahigh UV photosensitivity and thermal sensitivity in Ge doped optical fibres *Electron. Lett.* **29**

[32] Mizrahi V, Lemaire P J, Erdogan T, Reed W A, DiGiovanni D J and Atkins R M 1993 Ultraviolet laser fabrication of ultrastrong optical fiber gratings and of germania-doped channel waveguides *Appl. Phys. Lett.* **63** 1727–9

[33] Reed W A, Coughran W C and Grubb S G 1995 Numerical modelling of cascaded cw Raman fiber amplifiers and lasers *Optical Fiber Communication Conf.* Paper WD1 (Washington, DC: OSA) pp 107–9

[34] Rini M, Cristiani I and Degiorgio V 2000 Numerical modelling and optimization of cascaded cw Raman fiber lasers *IEEE J. Quantum Electron.* **QE-36** 1117–22

[35] Bertoni A and Reali G C 1998 1.24 μm cascaded Raman laser for 1.31 μm Raman fiber amplifiers *Appl. Phys.* B **67** 5–10

[36] Vareille G, Audouin O and Desurvire E 1998 Numerical optimization of power conversion efficiency in 1480 nm multi-Stokes Raman fibre lasers *Electron. Lett.* **34** 675–6

[37] Jackson S D and Muir P H 2001 Theory and numerical simulation of nth-order cascaded Raman fiber lasers *J. Opt. Soc. Am.* B **18** 1297–306

[38] Qian Y, Povlsen J H, Knudsen S N and Grüner-Nielsen L 2000 On Rayleigh backscattering and nonlinear effects evaluations and Raman amplification characterizations of single-mode fibers *Proc. Optical Amplifiers and their Applications Conf.* Paper OMD18 (Washington, DC: OSA)

[39] Kurkov A S, Paramonov V M, Medvedkov O I, Vasiliev S A and Dianov E M 2000 Raman fiber laser at 1.45 μm: comparison of different schemes *Proc. Optical Amplifiers and their Applications Conf.* Paper OMB5 (Washington, DC: OSA)

[40] Do Il Chang, Dong Sung Lim, Min Yong Jeon, Hak Kyu Lee, Kyong Hon Kim and Taesang Park 2000 Dual-wavelength cascaded Raman fibre laser *Electron. Lett.* **36** 1356–8

[41] Papernyi S B, Karpov V I and Clements W R L 2001 Efficient dual-wavelength Raman fiber laser *Proc. Optical Fiber Communication Conf.* Paper WDD15 (Washington, DC: OSA)

[42] Mermelstein M D, Headley C, Bouteiller J-C, Steinvurzel P, Horn C, Feder K and Eggelton B J 2001 *Proc. Optical Fiber Communication Conf.* Paper PD3-1 (Washington, DC: OSA)

[43] Do Il Chang, Dong Sung Lim, Min Yong Jeon, Kyong Hon Kim and Taesang Park 2001 *Electron. Lett.* **37** 740–1

[44] Mermelstein M D, Headley C, Bouteiller J-C, Steinvurzel P, Horn C, Feder K and Eggleton B J 2001 Configurable three-wavelength Raman fiber laser for Raman amplification and dynamic gain flattening *IEEE Photon. Technol. Lett.* **13** 1286–8

[45] Mermelstein M D, Horn C, Huang Z, Steinvurzel P, Feder K, Luvalle M, Bouteiller J-C, Headley C and Eggleton B J 2002 Configurability of a three-wavelength Raman fiber laser for gain ripple minimization and power portioning *Proc. Optical Fiber Communication Conf.* Paper TuJ2-1 (Washington, DC: OSA)

[46] Mermelstein M D, Horn C, Bouteiller J-C, Steinvurzel P, Feder K, Headley C and Eggleton B J 2002 Six wavelength Raman fiber laser for C + L-band Raman amplification *Proc. Conf. on Lasers and Electro-Optics* Paper CThJ1 (Washington, DC: OSA)

[47] Prabhu M, Kim N S and Ueda K 2000 Simultaneous double-color continuous wave Raman fiber laser at 1239 nm and 1484 nm using phosphosilicate fiber *Opt. Rev.* **7** 277–80

[48] Bouteiller J-C, Brar K, Radic S, Bromage J, Wang Z and Headley C 2002 Dual-order Raman pump providing improved noise figure and large gain bandwidth *Proc. Optical Fiber Communication Conf.* Postdeadline Paper FB3 (Washington, DC: OSA)

B4.4
Soliton lasers

J R Taylor

B4.4.1 Introduction

Following the introduction of techniques for the mode-locking (see, for example chapter C2) of solid-state lasers in the mid 1960s, pulses of the order of a few picoseconds were readily available. The nonlinear problems caused by the relatively high intracavity flux of such pulses were soon recognized [1] and techniques were developed [2] to counteract the frequency chirping associated with this nonlinearity, primarily self-phase modulation (see chapter A4). Despite this, throughout the first two decades of technological short-pulse laser development, little effort was directed towards the production of transform-limited pulse operation. In 1973, Hasegawa and Tappert [3], first proposed the generation of solitons in single mode optical fibre, through a subtle mechanism that balanced the intensity-dependent spectral broadening and anomalous dispersion, that was a property of silica-based fibres in the spectral region above about 1.27 μm. It was not until 1980, however, when a suitable laser source was developed, that Mollenauer [4] reported the first demonstration of optical solitons. By the mid 1980s, with pulse durations from bulk lasers approaching the 100 fs regime, techniques for controlling both nonlinearity and intracavity dispersion were necessarily introduced (see chapter C2.2) and the first 'soliton laser' was reported [5]. With the introduction of the Kerr lens mode-locking technique [6] for reliable ultrashort pulse generation, methods of dispersion control have become vital for the production of pulses of a few tens of femtoseconds and below. The term soliton laser is now very widely applied to the generic dispersion-compensated ultrashort pulse laser although the pulses generated are generally not solitons in the strict mathematical definition, even in fibre-based systems.

In this chapter, only fibre-based systems will be considered and the emphasis will be placed on 'all fibre' formats rather than systems incorporating bulk elements and fibre gain media. In the twenty years since Mollenauer's soliton laser the devices have become compact, efficient, exhibiting pulsewidth and wavelength versatility. Ultrashort pulse fibre lasers are readily commercially available and innovative advances are still being made. Through the incorporation of holey fibre for dispersion compensation many of the techniques that will be described later for soliton pulse generation in the near-infrared can be applied to wavelengths below 1.27 μm, offering even more versatility to the palette of the soliton fibre laser.

B4.4.2 Solitons in optical fibre

The interaction of nonlinearity and dispersion, within certain bounds, gives rise to optical soliton formation. The (Kerr) nonlinearity results from the third order nonlinear susceptibility (chapter A4) which gives rise to an intensity dependent refractive index. This is generally written as

$$n = n_0 + n_2 I \tag{B4.4.1}$$

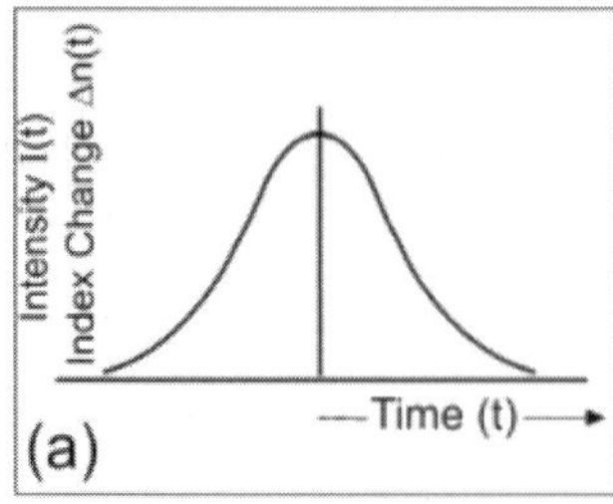

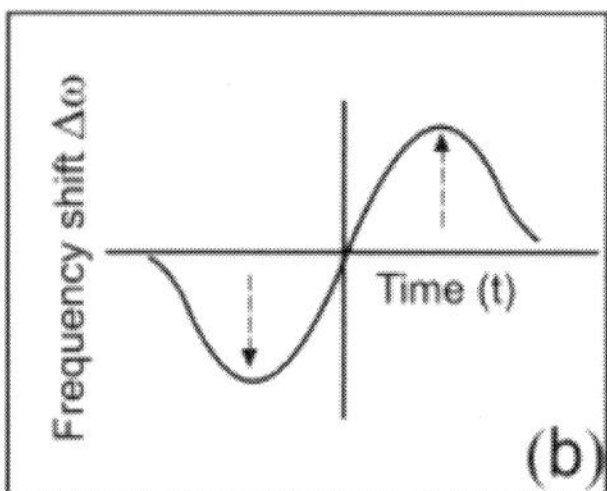

Figure B4.4.1. (*a*) The intensity profile and accompanying intensity-dependent index change of a pulse and (*b*) the associated frequency shift due to self-phase modulation.

where n_0 is the refractive index at zero intensity I and n_2 is the nonlinear index coefficient. For a standard telecom fibre, n_2 typically has a value of 2.2×10^{-20} m^2 W^{-1}, however, depending on the dopants used for modifying the refractive index of the respective silica-based fibre types, the measured nonlinear index coefficient can have values in the range 2.2–3.2 $\times 10^{-20}$ m^2 W^{-1} [7]. In silica, the nonlinear contribution to the refractive index has a response time of the order of a few femtoseconds, which is effectively instantaneous over the time scale of the pulses generated, hence, the induced index change will directly map the pulse intensity profile. On propagating over a length L of fibre, an accumulated intensity-dependent phase shift across an input pulse arises, given by

$$\Delta\phi(t) = dn \cdot k \cdot L = n_2 I(t) k L \tag{B4.4.2}$$

where k is the wavenumber $2\pi/\lambda$. This is illustrated in figure B4.4.1(*a*) which shows the instantaneous refractive index change across an input pulse with an intensity profile $I(t)$. Such a time-dependent phase shift gives rise to a time-dependent frequency shift (or chirp) given by

$$\Delta\omega(t) = -\frac{\mathrm{d}\Delta\phi(t)}{\mathrm{d}t} = -n_2 k L \frac{\mathrm{d}I(t)}{\mathrm{d}t}. \tag{B4.4.3}$$

As can be seem in figure B4.4.1(*b*) this gives rise to a frequency downshift on the leading edge of the pulse and a frequency up-shift on the trailing edge. In the absence of dispersion, the intensity-dependent refractive index would give rise only to spectral broadening through this self-phase modulation. The nonlinear length L_{NL}, the length scale over which nonlinearity becomes significant is

$$L_{\mathrm{NL}} = \frac{1}{\gamma P_0} \tag{B4.4.4}$$

where P_0 is the peak power and γ is $n_2\omega_0/cA_{\mathrm{eff}}$, where ω_0 is the carrier frequency and A_{eff} is the effective area of the core.

Group velocity dispersion (GVD), in the absence of nonlinearity, always gives rise to pulse temporal broadening irrespective of it being normally or anomalously dispersive. A dispersive lengthscale can be defined over which dispersion is significant, given by

$$L_{\mathrm{D}} = \frac{T_0^2}{\beta_2} \tag{B4.4.5}$$

where T_0 is the input pulse width and β_2 is the group velocity dispersion coefficient. Where the fibre length is longer or comparable to both the dispersion length and the nonlinear length, the simultaneous interaction of nonlinearity and dispersion can give rise to soliton formation if the GVD is anomalous. In conventional, silica-based fibres anomalous dispersion can only be achieved for wavelength above 1.27 μm. In the anomalously dispersive region, up-shifted frequencies travel faster than downshifted frequencies, consequently, empirically one can see that from figure 4.4.1(*b*) the rear of the chirped pulse (which is up-shifted) on propagation would move relatively towards the front, counteracting dispersive broadening. This simultaneous interaction is the basis of soliton formation. The treatment and solution of the nonlinear Schrödinger equation that describes soliton propagation is beyond the scope of this short review and the reader is referred to [8–10]. With the balance of the dispersion and nonlinear lengths a sech-like profile solution of the NLS occurs at a fundamental soliton power given in practical units by

$$P_1 = 0.776\frac{\lambda^3|D|A_{\text{eff}}}{\pi^2 c n_2 \tau^2} \tag{B4.4.6}$$

where D is the dispersion parameter $(2\pi c\beta_2/\lambda^2)$ and τ is the full width half maximum of the pulse. Assuming a value of 2.5×10^{-20} m^2 W^{-1} for n_2, the fundamental soliton power can be simplified to:

$$P_1 = 0.01\frac{D[\text{ps (nm km)}^{-1}]\lambda^3[\mu\text{m}]a_{\text{eff}}[\mu\text{m}^2]}{\tau^2[\text{ps}]}. \tag{B4.4.7}$$

Similarly, the soliton period, $(\pi/2)L_{\text{D}}$, is given by

$$z_0(m) = \frac{953.4\tau^2[\text{ps}]}{\lambda^2[\mu\text{m}]D[\text{ps (nm km)}^{-1}]}. \tag{B4.4.8}$$

Therefore, to establish a 5 ps soliton pulse at 1.55 μm in a fibre with an effective core area of 80 μm^2 and a dispersion of -10 ps(nm km)$^{-1}$, a peak power of approximately 1.2 W is required. At a pulse repetition rate of 1 GHz, this requires an in-fibre average power of 6 mW. If, however, the requirement is for 100 fs pulses, the peak power increases to approximately 3 kW and the average power to 300 mW. It can be seen that for defined fibre parameters, the average power-pulse duration product for solitons is a constant. In these examples, the associated soliton period, the fibre length required to establish soliton operation is 992 m and 0.4 m, respectively. From (B4.4.7) it is clear that by reducing the dispersion, the power requirement can be reduced. This, on the other hand, necessitates increased fibre lengths. Similarly, long pulse operation significantly reduces the power budget while requiring significantly increased fibre lengths. Consequently, the fibre parameters and available saturation power of amplifiers to be deployed will strongly affect output pulse performance of any soliton laser. Other effects, such as soliton interaction [11], additionally place limits on repetition periods to approximately ten pulse widths. Similarly, under certain conditions, as will be described later, solitons become unstable with periodic amplification, as experienced in a laser or in a transmission system, causing radiation to be shed from the soliton and giving rise to associated temporal and spectral instability.

B4.4.3 General fibre soliton laser

Numerous configurations have been adopted in the realization of fibre soliton lasers, some of which will be described in later sections, however, despite the obviously different adaptations, the basic conceptual building blocks for pulse generation, maintenance and shaping remain reasonably similar in each manifestation. Figure B4.4.2 is a representation of a generic fibre soliton laser containing the essential elements. The greatest body of work has been associated with Er-doped or Yb-Er co-doped silica-based amplifiers that operate broadly in the 1.53–1.60 μm range (see chapter B4.5). The reason for this being the ready availability of fibre pigtailed components primarily aimed at the telecommunications market. Where alternative wavelength ranges have

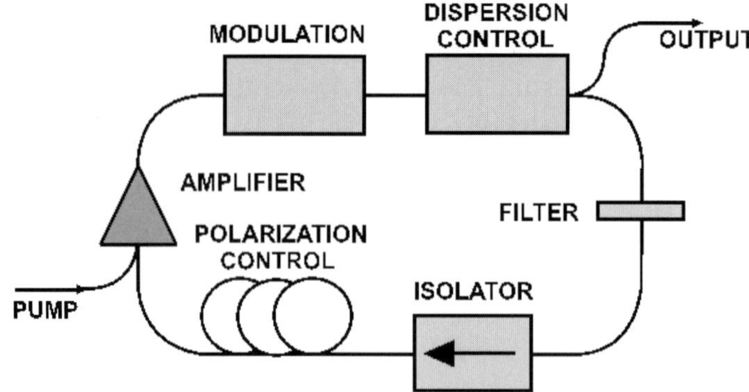

Figure B4.4.2. The components and schematic of a generic fibre soliton laser.

been investigated, the technological support of components has greatly affected all-fibre manifestations. For compact configurations, single stripe 980 nm or 1480 nm diode pumping of single clad Er-amplifiers has been the preferred technique, although this has set limitations on efficiency and power levels available. For increased efficiency and power scaling double clad geometries are preferable [12].

Other rare earth doped fibre amplifiers have allowed alternative gain bands to be accessed, for example in Pr around 1.3 μm [13], although this was with a low phonon, fluoride-based glass for efficiency at this wavelength. Femtosecond generation in a 786 nm diode-pumped Tm-silica fibre operating at 1.9 μm has been achieved [14]. Pulses as short as 53 fs have been obtained from a Nd-fibre laser operating on the 920 nm transition, although the gain element was the only fibre-based component in the cavity and anomalous dispersion was achieved through the use of bulk optical prism-pairs [15]. Recently, femtosecond operation has been reported for an all-fibre based Yb-doped, double clad amplifier, around 1065 nm, where the soliton shaping was achieved using anomalously dispersive holey fibre [16]. Typically, as a result of the relative low absorption coefficients, gain lengths in the rare earth doped based oscillators are in the range of a few metres up to many tens of metres, depending on the pump wavelength and whether double clad fibres are deployed (which generally require longer absorption lengths). In addition, the upper state lifetimes tend to be relatively long, (many hundreds of microseconds to milliseconds), which allows large inversion to be achieved and, although this permits good noise performance of the amplifiers, the high energy storage and the tendency to Q-switch can be catastrophic for waveguide modulators that may be placed intracavity to initiate pulse formation and mode-locking.

The optical fibre itself can be used as the gain medium, relying on stimulated Raman scattering (see section A4.11.1) as the amplifying process. In early, manifestations of fibre soliton lasers [17, 18] Raman was used as the gain mechanism. An advantage of the Raman gain process in silica is that broad gain bandwidths are achievable, about 30 nm (3 dB width) for a single wavelength pump but with multiple wavelength pumping this can be significantly increased (see chapter B4.3). The other major advantage is that the operational centre wavelength of the gain bandwidth is simply determined by the pump wavelength, with a Stokes shift of approximately 440 cm^{-1}, which corresponds to a shift of about 90 nm for a pump around 1.4 μm. Consequently, Raman gain can be achieved at any wavelength where an adequate pump power can be provided.

Mode-locking can only be achieved through the imposition of a modulation precisely at the fundamental or some harmonic of the frequency of the mode separation of the cavity. Both active and passive techniques have been used with fibre soliton lasers. Mollenauer's original soliton laser was based upon the concept of coupled cavity or additive pulse mode-locking, which, in fact, does not require soliton operation in the external cavity that contains the fibre to achieve femtosecond pulse operation, nor does it even need the fibre in certain circumstances. The general concept of the coupled cavity [19] is illustrated schematically

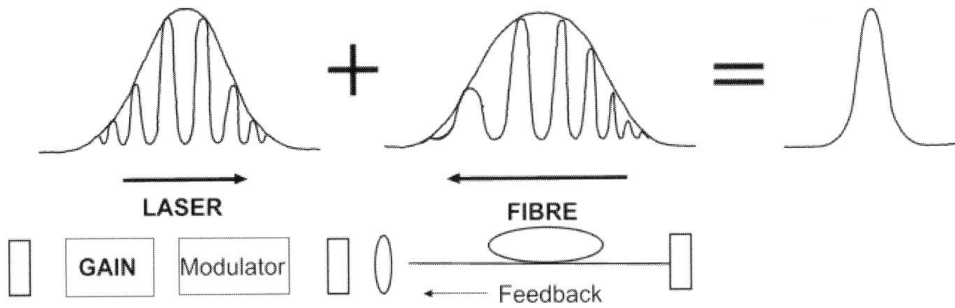

Figure B4.4.3. A schematic illustration of the concept of pulse shortening in additive pulse or coupled cavity mode-locking.

in figure B4.4.3. A pulse from a mode-locked laser experiences a phase perturbation, such as self phase modulation, in an external interferometrically-matched cavity which, in the example shown and in the original realizations, contained a single mode fibre. The returning phase modulated pulse adds interferometrically with the circulating pulse in the laser cavity, giving rise to pulse shortening. The pulse shortening process will continue until balanced by some additional mechanism such as dispersion or the bandwidth limitation of a filter. This concept has been extensively deployed both in linear and loop fibre lasers exploiting active and passive modulation techniques [20].

A laser system provides an environment that is considerably different from that of the ideal for stable soliton generation. The effects of nonlinearity and dispersion can be manipulated to be separate and significant only in certain sections of the laser system, as for example in early schemes where anomalous dispersion was provided by prism pairs. Dispersion management or control is essential in all-fibre formats where the mixing of fibres with both negative and positive dispersion is necessary to optimize performance and to provide an overall anomalous dispersion, which, although necessary for soliton generation, is not a prerequisite for short pulse performance. Consequently, on propagation around the laser, the generated solitons will experience perturbations through interaction with the differing regions of dispersion. Similarly, where rare earth amplifiers are used to provide lumped gain, a considerable perturbation to the soliton amplitude is experienced and also at the point of pulse extraction where a rapid decrease in the recirculating pulse amplitude is also experienced. Theoretically, it has been shown, using the guiding centre [21] or an average soliton concept [22], that stable soliton pulse propagation in the presence of a periodic gain (or loss or dispersion map) is possible when the length scale of the generated soliton is long compared to the length scaling of the periodic amplification or any periodic perturbation within the laser. This is schematically illustrated in figure B4.4.4. which maps the periodic amplitude fluctuations of a soliton over a cavity length Z_A. The configuration consists of a passive section, exhibiting a small loss (this section may be used for dispersion control) before entering a lumped, high gain amplifier, where rapid exponential gain is experienced. After exiting the amplifier, a considerable percentage of the pulse power is coupled from the cavity and the recirculated pulse again experiences the periodic perturbation cycle. A perturbed soliton will react to the perturbation by adjusting its associated pulse width and by shedding excess radiation as a dispersive wave. If, as in the example of figure B4.4.4, the cavity length Z_A is short compared to the associated soliton period, then the soliton does not have sufficient length to react to change in, for example, amplitude, before experiencing further periodic perturbations. Under these conditions, the description of the soliton can be undertaken in terms of the average energy or, for example, the average dispersion (should it exhibit variations). It is possible in this regime for the average soliton energy to be substantially less than the extracted pulse energy.

In the presence of no temporally limiting processes, the pulse durations in the fibre soliton laser will reduce to the point where the associated soliton period (length) is comparable to the cavity round-trip length and consequently the perturbed soliton sheds radiation. An enhanced mixing takes place between the soliton

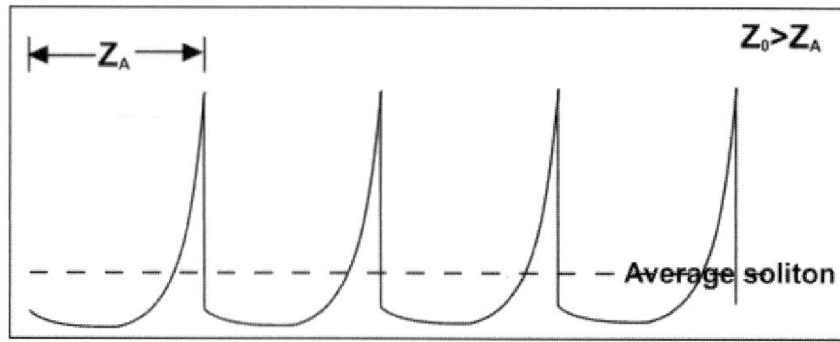

Figure B4.4.4. Intracavity periodic amplitude perturbations and the 'average' soliton condition.

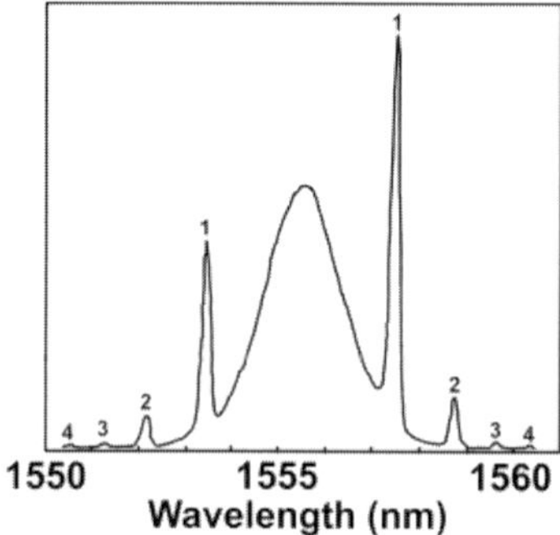

Figure B4.4.5. A representative spectrum, on an arbitrary linear intensity scale, of the output from a fibre soliton laser exhibiting characteristic sidebands, orders indicated.

and this dispersive wave, which, through a phase matching with the spatial harmonics of the amplifier length, gives rise to instability and characteristic spectral resonances, as shown in figure B4.4.5. The wavelength separation $\Delta\lambda_n$ of the resonance of order n, from the central λ wavelength is given by [22]

$$\Delta\lambda_n = \pm\frac{\lambda^2}{2\pi c t_0}\sqrt{\frac{8nz_0}{z_A} - 1} \qquad \text{(B4.4.9)}$$

where z_A is the periodic amplification length, and t_0 is the soliton pulse width, related to τ the pulse full width at half maximum by $\tau = 1.76t_0$.

The production of the dispersive continuum with the associated spectral sideband signature presents a limitation to stable short pulse generation in all fibre soliton lasers [23]. For a given average dispersion, dispersive wave generation limits the minimum pulse width and, hence, the peak power, but this will eventually always be the penalty for operation with solitons.

By including a spectral filter intracavity, in addition to selective wavelength tuning and reduction of problems with ASE, the spectral bands associated with the dispersive wave can be eliminated or reduced [24] and optimization procedures have been undertaken, concluding that to obtain maximum sideband reduction with minimal pulse broadening a filter of approximately twice the bandwidth of the unfiltered spectrum should be used [25].

Also included in the cavity of many soliton fibre lasers is some degree of polarization control. This can be used to compensate for unwanted birefringence caused by bends or stress. As will be seen later, controlled, intensity-dependent polarization rotation can be used to initiate soliton pulse formation. There have been few reports on all-polarization preserving cavities that would improve environmental stability, [26], however the sigma configuration [27] introduced a hybrid solution. This uses a loop of polarization preserving fibre

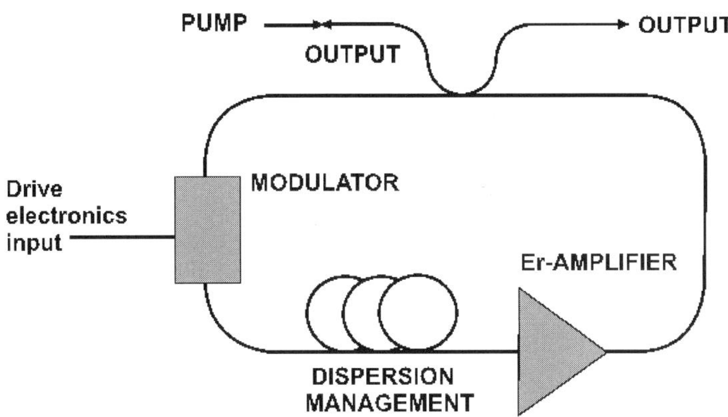

Figure B4.4.6. An actively mode-locked Er-doped soliton laser, after [31].

(see section A5.3.9) with an integrated waveguide amplitude modulator and coupled to a linear cavity via a polarizing beamsplitter. The linear section contained a 10 m Yb:Er gain fibre and 14.9 m of dispersion compensating fibre, which were non-polarization preserving, however, it was terminated by a Faraday rotator mirror assembly such that the effects of birefringence were totally compensated in each round-trip of the linear section. This system produced 1.3 ps soliton-like pulses at 10 GHz with a pulse drop out ratio of less than 10^{-12}.

Finally, output coupling in the vast majority of fibre loop soliton lasers is provided by fused couplers, and these devices are also used to define sub-cavities within the lasers if required for processes such as stabilization [28]. In linear cavities, Bragg grating reflectors (see section B4.3.4.1 and chapter D2.6) are now most commonly deployed [29].

B4.4.4 Actively mode-locked soliton lasers

The two principal methods of active mode-locking fibre lasers are amplitude and phase modulation (chapter C2). Generally, acousto-optic modulators in bulk-fibre hybrid formats have been used or preferably pigtailed lithium niobate integrated optic modulators for all fibre configurations. The greater modulation depth and the generally considerably greater bandwidth of integrated optical modulators also favours shorter pulse generation. With a simple acousto-optic modulator and without soliton shaping, the generated pulse widths are strongly limited by the modulation frequencies and pulse durations obtained are very well described by theory [30] as

$$\tau_{\mathrm{p}} = \frac{(2\ln 2)^{1/2} g_0^{1/4}}{\pi \delta^{1/2}} \left(\frac{1}{f_{\mathrm{m}} \Delta f_{\mathrm{A}}} \right)^{1/2} \tag{B4.4.10}$$

where g_0 is the single pass gain, δ is the modulation parameter, Δf_{A} is the amplification bandwidth and f_{m} is the modulator drive frequency. For a gain bandwidth of 50 nm at 1.55 μm and a modulator driven at 100 MHz providing a 35% modulation in a laser with a 12 dB round-trip gain, the predicted pulse width is around 50 ps, which is fairly typical for mode-locking with amplitude modulation.

A schematic diagram of the first totally fibre integrated actively mode-locked Er doped soliton laser is shown in figure B4.4.6. This system incorporated a 70 m long Ed doped amplifier, pumped by a fused coupler at 980 nm (from a Ti:sapphire laser) and was modulated at 90 MHz using a $LiNbO_3$ waveguide modulator driven by a step recovery diode delivering a comb of 100 ps pulses. The laser also contained 2 km of standard telecommunications fibre that exhibited an anomalous dispersion of 15 ps $(\mathrm{nm\ km})^{-1}$ around the lasing wavelength of 1.53 μm. This ensured an overall anomalously dispersive cavity. Without this dispersion management fibre, pulses of approximately 50 ps were obtained, while with it, soliton shaping dominated and pulses of 4 ps were generated. Average powers of around 1 mW were achieved that were consistent with soliton generation.

Using a phase modulator, the time varying frequency introduced can be compensated by fibre dispersion and together with soliton shaping was shown to be capable of sub-picosecond pulse generation in bulk element plus fibre configurations [32] and all-fibre formats were also realized [33].

Improved technology associated with low loss waveguide modulators readily allows tunable picosecond pulse generation at 40 GHz repetition rates and in polarization preserving format, while Er-based schemes still currently dominate the literature. In addition, actively mode-locked schemes that incorporate prism pairs for amomalous dispersion control tend to support the generation of shorter soliton pulses [34] since the problems associated with dispersive wave generation in relative long amplification periods, associated with all-fibre formats, can be negated.

Increased repetition rates have been achieved using the rational active mode-locking technique. In this an optical pulse train at a repetition rate of $(np \pm 1)\Delta f$ is produced when the modulation frequency is set at $(p \pm 1/n)\Delta f$, where Δf is the longitudinal mode spacing and p and n are integers. 80–200 GHz pulse trains have been reported by Yoshida [35], by rationally detuning a 40 GHz modulator in an all-polarization preserving fibre cavity containing 90 m of p–m dispersion-shifted fibre. Pulse durations of 1 ps and average powers of 6 mW were reported at 80 GHz. A problem with rational mode-locking, however is that a slow modulation on the output pulse train is often obtained.

Actively mode-locked fibre lasers are notoriously susceptible to environmental perturbation that change the effective length of the fibre cavity (either through thermal expansion or, for example change in birefringence), such that a fixed frequency modulation drive shifts off resonance, causing pulse drop out and instability. Even with polarization preserving fibre, temperature changes cause drift; consequently, length stabilizing schemes are required. By detecting the phase difference of the modulator drive frequency together with the output mode-locked pulse stream and using the difference correction signal to drive a piezo stretcher to adjust the fibre cavity length, Shan [36] demonstrated long-term stability and used the actively mode-locked laser for error free (BER $<10^{-9}$) operation at received power levels of -26 dBm.

B4.4.5 Passively mode-locked soliton lasers

The simplification of laser configurations through the removal of external drive circuitry and intracavity modulators makes passive mode-locking attractive if not from an engineering perspective at least from economic considerations. Passive mode-locking has been extensively studied in bulk laser schemes for several decades. It relies on the intensity-dependent transmission properties of a saturable absorber and the pulse formation and evolution of short pulses is critically dependent on the response time of the absorber used. Additionally other pulse shortening mechanisms can place additional restrictions and complications or the mode-locking process (chapter C2). In fibre based lasers, two principal mechanisms are used, based either on the Kerr effect which can be used, for example, in intensity dependent polarization rotation to mimic an ultrafast saturable absorber or the more traditional approach of a bulk material, with a saturable transition and a response time of the order of the pulse durations required. In the past decade semiconductor materials have been the choice in this latter category, rather than the more traditional organic dye or doped glass. The use of saturable absorbers also ensures that the mode-locking process is self starting, which in other passively mode-locked systems can be problematic.

B4.4.5.1 Passive mode-locking—saturable absorbers

The first report of an InGaAs/GaAs on GaAs superlattice saturable absorber for mode-locking an Er fibre ring laser [37] used an absorber with 120 periods on a polished 30 μm thick GaAs substrate and a transmission geometry was employed. Pulses of 1.2 ps were achieved at 50 MHz repetition rate at an average power of

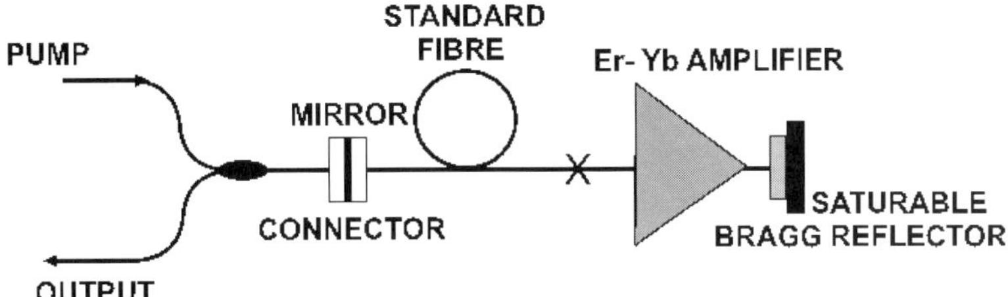

Figure B4.4.7. A schematic diagram of a semiconductor saturable absorber passively mode-locked Er/Yb fibre laser after [38].

0.6 mW, however, the authors attributed the generated picosecond pulses simply to the similar picosecond absorption recovery time, with no soliton shaping occuring. Femtosecond pulses were achieved by DeSouza employing a simple linear cavity with a 1.2 m polarization maintaining Er gain fibre using a 2 μm thick epitaxial layer of InGaAsP on InP that was directly mounted on a bulk cavity mirror [39]. The 320 fs pulses were also solely attributed to the fast recovery component of non equilibrium carrier redistribution. Figure B4.4.7 is a schematic diagram of a simple integrated, passively mode-locked, cavity design of Collings [38, 40]. The gain fibre which was typically around 15 cm was butt-coupled to a saturable Bragg reflector with InGaAs/InP quantum wells on an AlAs/GaAs reflector. Additional fibre ensured operation in the anomalous dispersion regime and the shortest pulses were observed when soliton shaping was effective. Pulses of 410 fs were achieved and the associated spectra exhibited spectral sidebands, however, by increasing the net dispersion (decreasing the soliton period) the pulse durations remained effectively the same but with no evidence of sidebands. Average output powers were 250 μW at 50 MHz repetition rate. Self-organized harmonic mode-locking was observed at 2.6 GHz repetition rate in cavities with shorter overall fibre length with 270 fs pulses and 1.6 mW average power.

B4.4.5.2 Passive mode-locking—figure of eight laser

The introduction of the nonlinear optical loop mirror (NOLM) presented a mechanism for fibre integrated all optical switching [41]. Based on the operation of a Sagnac interferometer (see section A5.2.5), asymmetry is introduced to the device through the use of a split ratio fused coupler such that counter-propagating beams in the fibre experience differing phase perturbations which, on recombining at the coupler, give rise to a power dependent transmission characteristic, thus mimicking a saturable absorber with the effectively instantaneous (femtosecond) response time of $\chi^{(3)}$. Because of the unequal splitting ratio at the coupler, complete switching is not possible and required fibre lengths tend to be long, otherwise associated power levels would also be too large. This was solved by the introduction of the concept of the NALM or nonlinear amplifying loop mirror [42], where a lumped amplifier is placed asymmetrically within the Sagnac fibre loop that is constructed using a 3 dB coupler. Switching in this case arises simply because nonlinearity and dispersion do not commute and counter-propagating pulses in the loop experience differing phase shifts. The advantage of the NALM is that practically full switching can be achieved and because solitons have uniform phase throughout their temporal profile, complete pulse switching with solitons is possible. As an amplifier is within the loop, input switching powers are consequently low making a very practical, fibre-based saturable absorber.

The figure of eight configuration, using a NALM, see figure B4.4.8 [43, 44] provided a simple all-fibre passively mode-locked laser for the production of femtosecond pulses. The additive pulse mode-locking mechanism can be seen from figure B4.4.8 where pulses selected from the noise profile of the intracavity flux are switched and transmitted by the NALM, recirculated in the passive loop and fed back again into the

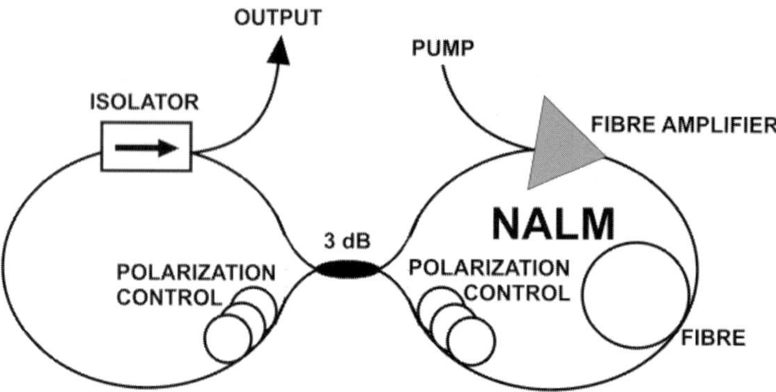

Figure B4.4.8. The components of a figure of eight fibre soliton laser.

NALM, leading to pulse evolution, amplification and narrowing. Any reflected radiation at the NALM is blocked by the isolator. The figure of eight laser was extensively investigated as a source of femtosecond pulses, primarily using Er or Yb:Er amplifiers in the 1.55 μm range, although operation at 1.3 μm using Pr doped fibre has been reported [13].

The output parameters are strongly dependent on fibre parameters (determining the soliton power), the saturation power of the amplifier and the output coupling. Typically, average output powers are in the several milliwatt range, depending on soliton power and number of pulses per round trip. Stable sub-100 fs pulse generation is difficult to achieve, although 98 fs has been reported [45] and short pulse generation is always accompanied by characteristic spectral sideband generation. Through optimization of the cavity birefringence through the polarization controllers, long, square pulse generation is also possible, with durations in the ns regime [46].

There are many problems associated with the passively mode-locked figure of eight laser that affect its experimental applicability. Not the least of these is that generally the mode-locking is not self-starting and a mechanical perturbation, usually through optimization of the intracavity polarization controllers is essential to start pulse formation. As the spontaneous passive mode-locking is essentially noise driven, the mode-locked outputs tend to consist of bunches of pulses, randomly distributed, but repetitive at the cavity round-trip time, which (depending on fibre lengths deployed) is in the range of 10–100 ns. Attempts have been made to stabilize the repetition rate and this has most simply been achieved by the insertion of a sub-cavity in the passive loop of the figure of eight configuration [35]. A self-organizing, repetition rate stabilization has also been observed in ring soliton fibre lasers where the pulse repetition rate varied approximately over the range 200 MHz to 1 GHz, at precise harmonics, as the pump power (and hence the output power) increased in a laser cavity with a fundamental resonant frequency of 3.2 MHz [47]. This behaviour is most likely attributable to the long-range interaction of solitons due to electrostriction. Hybrid techniques can also be applied by insertion of modulators to precisely determine single pulse repetition rates and this hybrid cavity has the advantage of being self-starting.

As the average soliton dynamics plays a dominant role, the control of the net dispersion is vital in determining achievable pulse durations [48]. The operational power level is also critically dependent on the fundamental soliton power. The generated solitons are limited by the switching characteristics of the NALM with spectral bandwidth limited durations. Once solitons have been established in the cavity, any further increase in the pump power will simply lead to an increase in the number of solitons circulating, and these will be randomly spaced in time [49].

The problems of reliable self-starting and essential stabilization of pulse repetition rates has limited the application of the figure of eight soliton fibre laser, despite it being an economic source of femtosecond pulses and its simple construction. Recently, wavelength diversity has been achieved using a Yb-doped silica-based

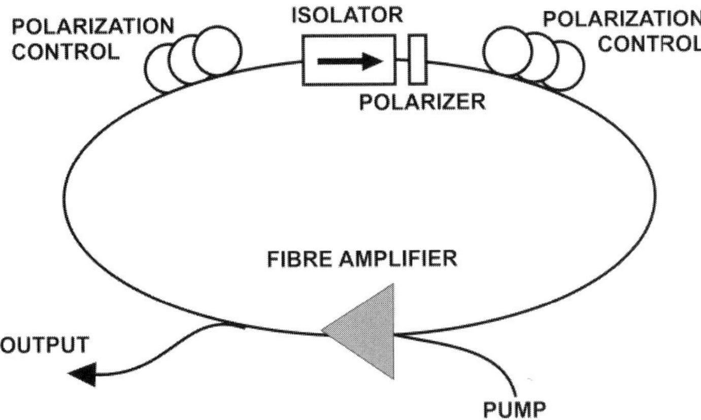

Figure B4.4.9. A schematic diagram of a soliton fibre ring laser passively mode-locked using intensity-dependent polarization rotation.

fibre amplifier scheme in a figure of eight geometry [16] where average soliton dynamics was obtained by deploying holey fibre in the ring to provide sufficient anomalous dispersion [50] at the operational wavelength of 1065 nm, allowing femtosecond pulse generation in an all-fibre format.

B4.4.5.3 Passive mode-locking—polarization rotation laser

The Kerr effect, manifesting itself through intensity dependent polarization rotation can also be used to simulate the effect of an ultrafast saturable absorber and has been extensively employed to mode-lock fibre lasers, generating femtosecond soliton pulses, again primarily in Er- and Yb:Er-based systems. A schematic diagram of such a laser is shown in figure B4.4.9, from which a qualitative understanding of the mode-locking mechanism can be understood. A polarizer placed within the cavity preferentially discriminates a linear state of polarization. This polarizer could be within the isolator that is used to define unidirectional propagation within the fibre loop. The state of polarization is changed to elliptical by the following polarization controller and, on amplification and propagation, the state of polarization evolves as a result of intensity dependent polarization rotation by self-phase and cross-phase modulation on the orthogonal polarization components. Across the intensity profile of a pulse, differing phase shifts will be experienced. At the second polarization controller, the polarization state is adjusted such that the following polarizer allows transmission of a selected section of the pulse. For mode-locking, preferentially, the wings of the pulse are rejected while the peak intensity is transmitted. Initial realizations containing several hundred metres to a few kilometres of fibre allowed nanosecond [51] and picosecond [52] pulse generation. Through optimizing fibre length and dispersion, femtosecond pulses [24] were obtained. In a fibre loop of 4.8 m total length, self-starting, single pulse operation per round-trip was achieved for a 980 nm pump power of 53 mW with average powers up to approximately 300 μW and pulse durations of 452 fs [53].

Generally the polarization rotation, mode-locked loop soliton laser displays all the negative characteristics of the figure of eight laser, such as irreproducibility in self-starting, random pulse bunching, energy quantization, spectral sideband instability and problems with environmentally indiuced instabilities, particularly in long fibres. Many of these can be easily solved. However, soliton-based systems because of the system pre-determined pulse energy will tend to show spectral instability and multiple pulsing. Consequently the considerable advances in high power fibre lasers are not reflected by soliton-based systems. This has been overcome through dispersion management in the loop laser employing a technique termed stretched

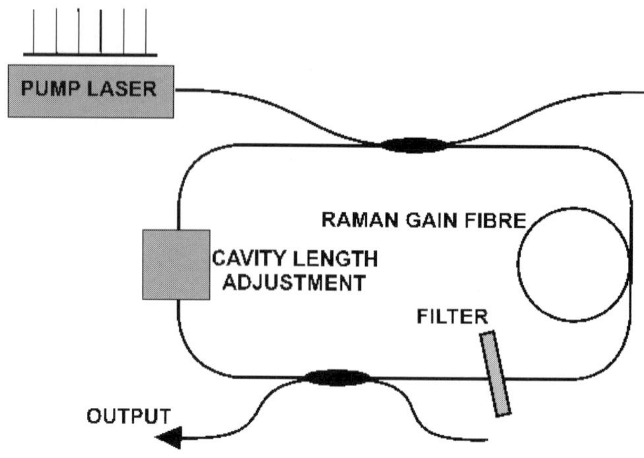

Figure B4.4.10. The cavity configuration of a fibre Raman soliton laser.

pulse mode-locking [54]. The experimental configuration is similar to that of figure B4.4.9, with the cavity configured from an erbium fibre amplifier that is normally dispersive while the remaining fibre is anomalously dispersive. The output being taken directly after passage through the normally dispersive amplifier. With a net positive dispersion, soliton formation is not possible and the output pulses are strongly chirped. This, however, allows significantly increased pulse energy extraction. Through optimization of the dispersion via the length of the output coupling fibre, the generated chirped pulses compressed to 77 fs and pulse energies of 90 pJ were measured from this self-starting mode-locked laser operating around 1560 nm. The overall cavity is compact with only 1.13 m of Er-doped amplifier and a further 3.37 m of standard (anomalously dispersive) fibre. The overall dispersion was normal and estimated to be $\sim$0.040 ps^2. Recently, stretched-pulse techniques have been applied to a Er-doped system incorporating a NOLM and 100 fs pulse with energies of 1 nJ have been demonstrated [55].

B4.4.6 Soliton Raman fibre lasers

The broad gain profile of stimulated Raman scattering in silica based optical fibre (chapter B4.3) provides a medium for femtosecond pulse generation over a spectral range that is simply limited by the transmission window of silica and available pump wavelengths. The first femtosecond soliton fibre lasers were based upon Raman scattering in standard single mode fibres [18]. Mode-locking was achieved through synchronous pumping using mode-locked, bulk solid-state lasers. A schematic diagram of a synchronously pumped, fibre Raman soliton laser is shown in figure B4.4.10. Typical pump pulses are in the range of 100 ps with peak powers of about 50–100 W in order to reach threshold in the deployed fibre Raman gain lengths of around a few hundred metres, which are limited by dispersive walk-off between pump and signal. The broad gain profile means that dispersive tuning, is possible. Alternatively, tuning can be achieved with an intracavity wavelength filter such that micron sensitive matching is required between the pump and slave cavities. Soliton shaping occurs for wavelengths above approximately 1.3 μm. From these lasers the pulse durations are in the range 150–250 fs and average output powers, limited by output coupling, soliton power levels and repetition rates are of the order of 10s of mW. By selecting the fibre dispersion such that the dispersion zero lies between the first and second Stokes orders, soliton shaping does not take place for the first Stokes order, hence, on cascading, the pulses in the first order remain long and act as the pump for the second order, where femtosecond pulses can evolve as a result of nonlinear shaping [56].

The efficiency of Raman conversion can be high (up to 70%), particularly in a region of low dispersion and single pass femtosecond pulse generation is possible. Where long pump pulses are used, particularly in a region of low anomalous dispersion, modulational instability initially occurs, leading to short pulse formation. Subsequent Raman gain gives rise to a short pulse continuum which evolves to longer wavelengths with increasing pump and interaction lengths. Through spectral selection in the continuum, pulses of 100–200 fs can be generated, however, the nature of the multisoliton continuum and presence of a dispersive wave component means that autocorrelation traces exhibit significant pedestals. It is not necessary for the pump radiation to be in the anomalous dispersion regime. Femtosecond pulse generation is possible with the pump in the normal dispersion region when the generated Stokes signal is in the anomalously dispersive regime. Pump power dependent, broad continua ~100 nm, supportive of pulses of ~200–300 fs can be generated, but at higher threshold levels than when modulational instability is present.

A contribution to soliton continuum generation is possible when launched pulses, on entering a negatively dispersive fibre, have peak power levels (P_N) many times that of the fundamental soliton (P_1). High-order solitons can give compression ratios of approximately 4.1 N where N is the soliton order ($P_N = N^2 P_1$). Short pulse propagation is unstable as a result of other nonlinearities such that the high-order solitons compress, break up and give rise to a multisoliton continuum. The generated continuum can extend over 100 nm and again, spectral selection within the continuum can select solitons as short as 100 fs [57] in simple, single pass geometries.

With the introduction of the Er-doped fibre amplifier, Raman based sources were not so attractive since pump sources were inefficient and laser geometries were primarily bulk element and fibre composites. In the past years, with the introduction of efficient high power diode pumps and efficient cascaded fibre Raman lasers (chapter B4.3), the efficiency and wavelength diversity offered by Raman-based soliton fibre lasers may offer a unique opportunity for simple compact femtosecond sources. The application of holey fibre intra-cavity for dispersion management has meant that femtosecond Raman soliton pulses are now possible in the 1.1 μm region (pumped by a compact mode-locked Yb fibre laser) and the same technique can be clearly applied to other wavelength regions, with the potential of fibre-based, femtosecond, visible lasers.

B4.4.7 Soliton self frequency shift sources

The spectrum of a 200 fs pulse at 1.55 μm has a bandwidth of approximately 50 cm^{-1}. As has been previously described, the gain profile of Raman scattering in silica-based fibre is at a peak for shifts of approximately 440 cm^{-1}, however, for shifts of around 50 cm^{-1} the gain is about 10% that of the peak value (chapter B4.3). Therefore, on propagation over moderate distances, the short wavelength components of, for example a 200 fs fundamental soliton pulse, can pump and provide considerable gain for the longer wavelength components. On transmission, this gives rise to a continuous spectral red shift of solitons [58]. The process was termed the soliton self-frequency shift [59] and this frequency shift with propagation distance was shown to follow a roughly linear dependence on the inverse fourth power of the soliton pulse duration [60]. Therefore, the effect can be used to provide a wavelength tunable source of femtosecond soliton pulses [61]. Usually, the process is self-terminating. In conventional fibre in the anomalous dispersion regime the dispersion increases with increasing wavelength, consequently, the required soliton power (equation (B4.4.7)) increases. As normally no additional amplification process is present (which would tend to suppress the soliton self-frequency shift anyway), the frequency-shifted solitons eventually have insufficient energy and as a consequence temporally broaden. This broadening reduces the effect of the self-frequency shift and eventually the pulses will decay through dispersion.

It is possible, through choice of fibre optimization, for example by employing a dispersively flattened fibre, to obtain significant tunability. Nishizawa [62] reported that on launching 200 fs 1.56 μm soliton pulses, at 48 MHz repetition rate into 200 m of dispersion shifted polarization maintaining fibre, distinct tuning of the soliton was observed from 1.56 μm to 1.75 μm on increasing the average launch power up to

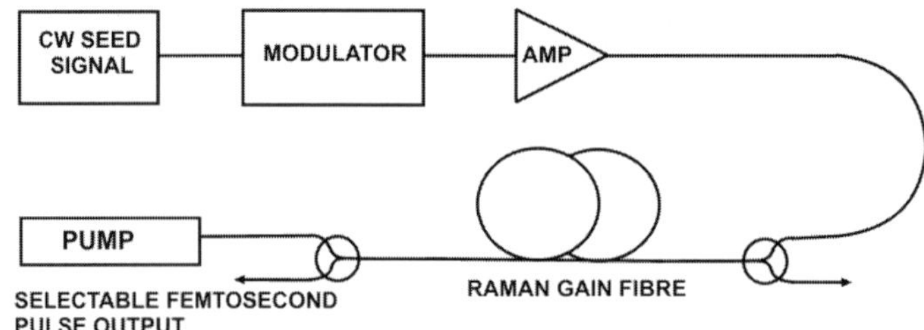

Figure B4.4.11. The experimental arrangement of a wavelength tunable, repetition rate and pulse width selectable, fibre soliton source based upon adiabatic, single pass Raman amplification.

a maximum 25 mW (~2.6 kW peak power). This technique has recently been extended to the near-infrared, to a region where conventional fibre is normally dispersive. Through launching 110 fs pulses at 806 nm into a 1.7 m length of holey fibre with a zero dispersion at 765 nm soliton generation, tunable from 850–1050 nm was achieved for launched peak powers up to 870 W [63]. By employing a doped holey fibre amplifier, the technique has been extended to cover the range 1.06–1.33 μm [64].

B4.4.8 Single pass, adiabatic compression soliton sources

On propagation, as a result of the distributed loss of a fibre, albeit relatively small, the resultant accumulated loss of energy in a soliton will cause the soliton to temporally broaden and eventually decay. It was suggested that distributed Raman gain could exactly counteract the distributed loss and lead to propagation without distortion, effectively over unlimited distances [65]. The Raman gain profile through the fibre is not uniform, exhibiting an exponential growth and, by using the distributed geometry, the amplification process can be made adiabatic, such that no soliton energy is radiated as dispersive waves and the soliton property of (amplitude × pulse-width) product remains constant. In addition to simply allowing the distributed Raman gain to compensate the loss, a net gain can be given to the soliton, which, if the soliton area is conserved by adiabatic amplification, will give rise to pulse compression. For this to occur, the gain experienced over a soliton period must be kept low, of the order of 10%. The Raman gain process, using a counter-propagating pump geometry is ideal for control of the net gain experienced over the associated soliton period of an adiabatically amplified soliton. For example, for a 20.4 km length of standard telecom fibre, pumped at 1.48 μm with an in-fibre power of 1.7 W and an input 10 ps soliton pulse, the net gain experienced at the input side is 1.7% over the first 2.48 km (the soliton period of the input 1.55 μm, 10 ps pulses). With adiabatic amplification, the pulse compresses and as the signal nears the output end of the fibre, the net gain per length exponentially increases, with a gain of approximately 3.4% over the final 24.8 m (the soliton period for an output 1 ps pulse). Figure B4.4.11 shows a typical experimental arrangement. The advantage of this adiabatic compression scheme over other single pass techniques, for example using dispersion decreasing fibre, is that the input wavelength and pulse length are not so strictly constrained by the fibre parameters [66]. In the example the seed signal is a cw fibre integrated laser source that can be wavelength tunable. This source is directly modulated to provide pulsed inputs to the Raman gain fibre. The modulator can be an electro-absorption modulator, in which case the input pulse profiles are effectively sech or soliton-like [67], however, it can be a simple fibre-integrated waveguide Mach–Zehnder amplitude modulator providing sine-like inputs [68]. These are quite adequate for soliton compression, as a pulse of any reasonable shape will form solitons, with the caveat that there will be a small percentage of dispersive wave present, and this can be removed by a NOLM or NALM. In certain circumstances, a pre-amplifier may be necessary to boost the input signal pulses to soliton power levels. A counter-propagating pump geometry in the Raman amplifier, a 20 km length of STF, reduces problems of noise transfer from the pump to the signal. For significant Raman

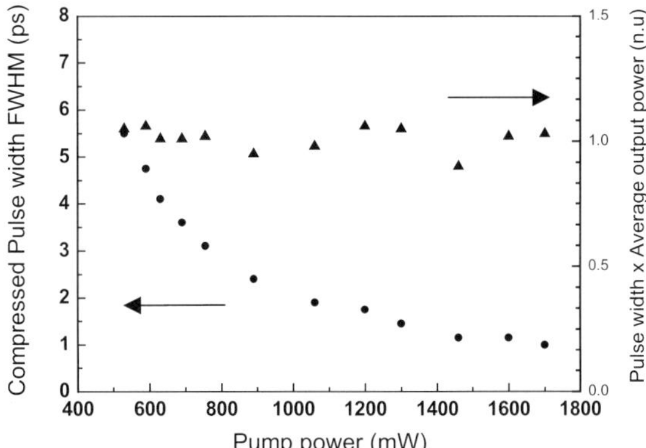

Figure B4.4.12. The variation of single pass Raman-amplified, adiabatically-compressed soliton output with pump power for a 10 ps input pulse. The associated product of (output average power × pulse width), in normalized units, is also indicated.

gain and compression, the required average pump powers are in the region of 2 W. This can be supplied by a cascaded fibre Raman laser (chapter 4.3). For peak gain around 1.55 μm a pump at 1.45 μm was used. Figure B4.4.12 shows the variation of the adiabatically compressed output 10 GHz repetition rate soliton pulses with pump power. For 10 ps input pulses, minimum pulses of 1 ps were achieved for a 1.7 W pump power. The average output power was 650 mW corresponding to the expected soliton power per pulse. The soliton nature of the output was confirmed by the fact that the product of the average power and the pulse width remained constant despite the amplification (see equation (B4.4.6)). Complete tunability throughout the Raman amplification band is possible and average powers and pulse widths are ultimately limited by the saturated power of the amplifier. The soliton power and dynamics of the amplification can be changed by varying the fibre type used as the distributed Raman amplifier. Femtosecond output pulses have been achieved using a dispersion-shifted fibre [68,69]. However, with the generation of femtosecond pulses, other competing nonlinearities, such as self-frequency shifting, affect the quality of the output pulses, through, for example the generation of dispersive waves that contribute to the background in pulse autocorrelations, and the pulse duration times average power factor not remaining constant. On the other hand, this simple, single pass experimental configuration does allow wavelength tunable, repetition rate tunable, pulse length selectable operation over extensive wavelength regions.

References

[1] Shimizu F 1967 Frequency broadening in liquids by a short light pulse *Phys. Rev. Lett.* **19** 1097–9

[2] Treacy E B 1968 Compression of picosecond light pulses *Phys. Lett.* A **28** 34–5

[3] Hasegawa A and Tappert F 1973 Transmission of stationary nonlinear optical pulses in dispersive dielectric fibers I—anomalous dispersion *App. Phys. Lett.* **23** 142–4

[4] Mollenauer L F, Stolen R H and Gordon J P 1980 Experimental observation of picosecond pulse narrowing and solitons in optical fibres *Phys. Rev. Lett.* **45** 1095–8

[5] Mollenauer L F and Stolen R H 1984 The soliton laser *Opt. Lett.* **9** 13–5

[6] Spence D E, Kean P N and Sibbett W 1991 60 fs pulse generation from self-mode-locked Ti:sapphire laser *Opt. Lett.* **16** 42–4

[7] Boskovic A, Chernikov S V, Taylor J R, Gruner-Nielsen, L and Levring O A 1996 Direct continuous-wave measurement of n_2 in various types of telecommunication fiber at 1.55 μm, *Opt. Lett.* **21** 1966–8

[8] Agrawal G P 2001 *Nonlinear Fiber Optics* 3rd edn (London: Academic)

[9] Hasegawa A 1989 *Optical Solitons in Fibers (Springer Tracts in Modern Physics 116)* 1st edn (Berlin: Springer)

[10] Hasegawa A and Kodama Y 1995 *Solitons in Optical Communications (Oxford Series in Optical and Imaging Science 7)* 1st edn (Oxford: Clarendon)

[11] Mitschke F M and Mollenauer L F 1987 Experimental observation of interaction forces between solitons in optical fibers *Opt. Lett.* **12** 355–7

[12] Fermann M E, Harter D, Minelly J,D and Vienne G G 1996 Cladding-pumped passively mode-locked fiber laser generating femtosecond and picosecond pulses *Opt. Lett.* **21** 967–9

[13] Guy M J, Noske D U, Boskovic A and Taylor J R 1994 Femtosecond soliton generation in a praseodymium fluoride fibre laser *Opt. Lett.* **19** 828–30

[14] Sharp R C, Spock D E, Pan N and Elliot J 1996 190-fs passively mode-locked thulium fiber laser with a low threshold *Opt. Lett.* **21** 881–3

[15] Hofer R, Hofer M, Reider G A, Cernusca M and Ober M H 1997 Mode-locking of a Nd-fibre laser at 920 nm *Opt. Commun.* **140** 242–4

[16] Avdokhin A V, Popov S V and Taylor J R 2003 Totally fiber integrated, figure-of-eight, femtosecond source at 1065 nm *Opt. Express* **11** 265–9

[17] Gouveia-Neto A S, Gomes A S L and Taylor J R 1987 Femtosecond soliton Raman ring laser *Electron. Lett.* **23** 537–8

[18] Kafka J D and Baer T 1987 Fiber Raman soliton laser pumped by a Nd:YAG laser *Opt. Lett.* **12** 181–3

[19] Blow K J and Wood D 1988 Mode-locked lasers with nonlinear external cavities *J. Opt. Soc. Am.* B **5** 629–32

[20] Haus, H A, Ippen E P and Tamura K 1994 Additive-pulse mode-locking in fiber lasers *IEEE J. Quantum Electron.* **30** 200–8

[21] Hasegawa A and Kodama Y 1990 Guiding center soliton in optical fibers *Opt. Lett.* **15** 1443–5

[22] Kelly S M J, Smith K, Blow K J and Doran N J 1991 Average soliton dynamics of a high-gain erbium fiber laser *Opt. Lett.* **16** 1337–9

[23] Noske D U, Pandit N and Taylor J R 1992 Source of spectral and temporal instability in soliton fiber lasers *Opt. Lett.* **17** 1515–17

[24] Noske D U, Pandit N and Taylor J R 1992 Subpicosecond soliton pulse formation from self-mode-locked erbium fibre laser using intensity dependent polarization rotation *Electron. Lett.* **28** 2185–6

[25] Tamura K, Ippen E P and Haus H A 1994 Optimization of filtering in soliton fiber lasers *Opt. Lett.* **6** 1433–5

[26] Takara H, Kawanishi, M Saruwatari and Nogushi K 1992 Generation of highly stable 20 GHz transform-limited optical pulses from actively mode-locked Er doped fibre lasers with an all-polarization maintaining ring cavity *Electron. Lett.* **28** 2095–6

[27] Carruthers T F and Duling III I N 1996 10 GHz, 1.3 ps erbium fiber laser employing soliton pulse shortening *Opt. Lett.* **21** 1927–9

[28] Dennis M L and Duling III I N 1992 High repetition rate figure eight laser with extracavity feedback *Electron. Lett.* **28** 1894–6

[29] Davey R P, Fleming R P E, Smith K, Kashyap R and Armitage J R 1991a Mode-locked erbium fibre laser with wavelength selection by means of fibre Bragg grating reflector *Electron. Lett.* **27** 2087–8

[30] Kuizenga D J and Siegman A E 1970 FM and AM mode-locking of the homogeneous laser - part 1 theory *IEEE J. Quantum Electron.* **QE 6** 694–708

[31] Kafka J D, Baer T and Hall D W 1989 Mode-locked erbium-doped fiber laser with soliton pulse shaping *Opt. Lett.* **14** 1269–71

[32] Davey R P, Langford N.and Ferguson A I 1991b Subpicosecond pulse generation from erbium doped fibre laser *Electron. Lett.* **27** 726–8

[33] Smith K, Greer E J, Wyatt R, Wheatley P and Doran N J 1991 Totally integrated erbium fibre soliton laser pumped by laser diode *Electron. Lett.* **27** 244–6

[34] Fermann M E, Andrejo M J, Silberberg Y and Weiner A M 1993 Generation of pulses shorter than 200 fs from a passively lode locked Er-fiber laser *Opt. Lett.* **18** 48–50

[35] Yoshida E, Kimura Y and Nakazawa M 1992 Laser diode-pumped femtosecond erbium-doped fiber laser with a sub-ring cavity for repetition rate control *Appl. Phys. Lett.* **60** 932–4

[36] Shan X, Cleland D and Ellis a 1992 Stabilizing Er fibre soliton laser with pulse phase locking *Electron. Lett.* **28** 182–4

[37] Zirngibl M, Stulz L W, Stone J, Hugi J, DiGiovanni D and Hansen P B 1991 1.2 ps pulses from passively mode-locked laser diode pumped Er-doped fibre ring laser *Electron. Lett.* **27** 1734–5

[38] Collings B C, Bergman K and Knox W H 1998 Stable multigigahertz pulse train formation in a short-cavity passively harmonic mode-locked erbium/ytterbium fiber laser *Opt. Lett.* **23** 123–5

[39] DeSouza E A, Soccolich C E, Pleibel W, Stolen R H, Simpson J R and DiGiovanni D J 1993 Saturable absorber mode-locked polarization maintaining erbium fibre laser *Electron. Lett.* **29** 447–9

[40] Collings B C, Bergann K, Cundiff S T, Tsuda S, Kutz J N, Cunningham J E, Jan W Y, Koch M and Knox W H 1997 Short cavity erbium/ytterbium fiber lasers mode-locked with a saturable Bragg reflector *IEEE J Selected Topics Quantum Electron.* **3** 1065–75

[41] Doran N J and Wood D 1988 Nonlinear-optical loop mirror *Opt. Lett.* **13** 56–8

[42] Fermann M E, Haberl F, Hofer M and Hochreiter H 1990 Nonlinear amplifying loop mirror *Opt. Lett.* **15** 752–4

[43] Duling I N III 1991 Subpicosecond all-fibre erbium laser *Electron. Lett.* **27** 544–5

[44] Duling I N III 1991 all-fiber soliton laser mode-locked with a nonlinear mirror *Opt Lett.* **16** 539–41

[45] Nakazawa M, Yoshida E and Kimura Y 1993 Generation of 98 fs optical pulses directly from an erbium-doped fibre ring laser at 1.57 μm *Electron. Lett.* **29** 63–4

[46] Richardson D J, Laming R I, Payne D N, Matsas V and Phillips M W 1991 Selfstarting passively mode-locked erbium fibre ring laser based on the amplifying Sagnac switch *Electron. Lett.* **27** 542–4

[47] Grudinin A B, Richardson D J and Payne D N 1993 Passive harmonic mode-locking of a fibre soliton ring laser *Electron. Lett.* **29** 1860–1

[48] Dennis M L and Duling III I N 1993 Role of dispersion in limiting pulse width in fiber lasers *App. Phys. Lett.* **62** 2911–13

[49] Grudinin A B, Richardson D J and Payne D N 1992 Energy quantization in figure eight fibre laser *Electron. Lett.* **28** 67–8

[50] Ferrando A, Silvestre E, Andrés, Miret J J and Andrés M V 2001 Designing the properties of dispersion-flattened photonic crystal fibers *Opt. Express* **9** 687–97

[51] Matsas V J, Newson T P and Zervas M N 1992 Self-starting passively mode-locked fibre ring laser exploiting nonlinear polarization switching *Opt. Commun.* **92** 61–6

[52] Matsas V J, Newson T P, Richardson D J and Payne D N 1992 Selfstarting passively mode-locked fibre ring soliton laser exploiting nonlinear polarization rotation *Electron. Lett.* **28** 1391–3

[53] Tamura K, Haus H A and Ippen E P 1992 Self-starting additive pulse mode-locked erbium fibre ring laser *Electron. Lett.* **28** 2226–7

[54] Tamura K, Ippen E P, Haus H A and Nelson L E 1993 77 fs pulse generation from a stretched-pulse mode-locked all-fiber ring laser *Opt. Lett.* **18** 1080–2

[55] Ilday F Ö, Wise F W and Sosnowski 2002 High-energy femtosecond stretched-pulse fiber laser with a nonlinear optical loop mirror *Opt. Lett.* **27** 1531–3

[56] Gouveia-Neto A S, Gomes A S L, Taylor J R, Ainslie B J and Craig S P 1987 Cascade Raman soliton fiber ring laser *Opt. Lett.* **12** 927–9

[57] Gouveia-Neto A S, Gomes A S L and Taylor J R 1987 High-efficiency single pass soliton like compression of Raman radiation in an optical fiber around 1.4 μm *Opt. Lett.* **12** 1035–7

[58] Dianov E M, Karasik A Ya, Mamyshev P V, Prokhorov A M, Serkin V N, Stel'makh M F and Fomichev A A 1985 Stimulated-Raman conversion of multisoliton pulses in quartz optical fibers *Pis'ma Zh. Eksp. Teor. Fiz.* **41** 242–4

[59] Mitschke F M and Mollenauer L F 1986 Discovery of the soliton self frequency shift *Opt. Lett.* **11** 659–61

[60] Gordon J P 1986 Theory of the soliton self frequency shift *Opt. Lett.* **11** 662–4

[61] Gouveia-Neto A S, Sombra A S B and Taylor J R 1988 A Nd:YAG laser pumped soliton self-frequency shifter *Opt. Commun.* **68** 139–42

[62] Nishizawa N, Okamura R and Goto T 2000 Widely wavelength tunable ultrashort soliton pulse and anti-Stokes pulse generation for wavelengths of 1.32–1.75 μm *Japan. J. Appl. Phys.* **39** L409–11

[63] Washburn B R, Ralph S E, Lacourt P A, Dudley J M, Rhodes W T, Windeler R S and Coen S 2001 Tunable near infrared femtosecond soliton generation in photonic crystal fibres *Electron. Lett.* **37** 1510–3

[64] Price J H V, Furusawa K, Monro T M, Lefort L and Richardson D J 2002 Tunable, femtosecond pulse source operating in the range 1.06–1.33 μm based on an Yb^{3+}-doped holey fiber amplifier *J. Opt. Soc. Am.* B **19** 1286–94

[65] Hasegawa A 1983 Amplification and reshaping of optical solitons in a glass fiber—IV: Use of the stimulated Raman process *Opt. Lett.* **8** 650–2

[66] Chernikov S V, Kashyap R, Guy M J, Moodie D G and Taylor J R 1996 Ultrahigh-bit-rate optical sources and applications *Phil. Trans. R. Soc.* A **354** 719–31

[67] Reeves-Hall P C, Lewis S A E, Chernikov S V and Taylor J R 2000 Picosecond soliton pulse-duration-selectable source based on adiabatic compression in a Raman amplifier *Electron. Lett.* **36** 622–4

[68] DeMatos C J S, Chestnut D A and Taylor J R 2002 Wavelength- and duration-tunable soliton source based on a 20 GHz Mach–Zehnder modulator and adiabatic Raman compression *Appl. Phys. Lett.* **81** 2932–4

[69] Reeves-Hall P C and Taylor J R 2001 Wavelength and duration tunable sub-picosecond source using adiabatic Raman compression *Electron. Lett.* **37** 417–8

Further reading

Duling I N III (ed) 1995 *Compact Sources of Ultrashort Pulses* 1st edn (Cambridge University Press)

This is an excellent review of short pulse sources, providing theoretical and experimental detail and a comprehensive review, extensively referenced, with several chapters on on fibre based systems. Highly recommended.

Agrawal G P 2001 *Applications of Nonlinear Fiber Optics* (London: Academic)

The chapter on fibre lasers contains good coverage of short pulse lasers with an emphasis on passive mode-locking techniques and the role of soliton shaping.

Taylor J R (ed) 1992 *Optical Solitons Theory and Experiment* 1st edn (Cambridge: Cambridge University Press)

A multi-author review of solitons in optical fibre, less emphasis is placed on short pulse lasers but a good introduction to the field with practical applications.

Fermann M E 1994 Ultrashort-pulse sources based on single-mode rare-earth-doped fibers *Appl. Phys.* B **58** 197–204

An early review of the principal mode-locking mechanisms, with extensive referencing.

Fermann M E, Galvanauskas A, Sucha G and Harter D 1997 Fiber-lasers for ultrafast optics *Appl. Phys.* B **65** 259–75

Review of passive and hybrid mode-locking, chirp pulse amplification of fibre lasers and nonlinear conversion.

Nelson L E, Jones D J, Tamura K, Haus H A and Ippen E P 1997 Ultrashort-pulse fiber ring lasers *Appl. Phys.* B **65** 277–94

Reviews the passive mode-locking technique of polarization additive pulse mode-locking with soliton and stretched pulse lasers, this is essential reading. See also Haus [20].

B4.5
Erbium and other doped fibre amplifiers

Kevin Cordina

B4.5.1 Introduction

The erbium-doped fibre amplifier (EDFA) has become a vital part of optical communications systems, enabling the economic transmission of multiple wavelengths through a single fibre over very long distances.

The EDFA is a member of the doped fibre amplifier (DFA) family, which extends to cover any amplifier which achieves gain *via* excitation and decay of electrons in ions within the material through which the light is passing. Typically these ions are rare earth elements which have been included as dopants within the material during its production. Different ions and materials provide gain at different wavelengths, figure B4.5.1 shows the dopants and amplification bands that have been demonstrated for use in telecoms applications.

The EDFA has become the most commonly utilized DFA due to the fact that its gain spectrum coincides with the third transmission window (*circa* 1550 nm) of silica optical fibre. Whilst this chapter will focus directly on EDFAs all of the principles discussed are applicable to other dopants and types of DFA. Where there are important differences these are cited within the text.

Fibre amplifiers in telecommunications are desirable for a number of reasons:

- no requirement for optical to electrical conversion;
- amplification is achieved within the fibre;
- high efficiencies can be attained;
- low noise figures can be achieved;
- broad gain window enabling wavelength division multiplexed (WDM) transmission;
- low cost;
- compact.

Optical to electrical conversion within a system is undesirable as it adds significant cost and complexity to the system. Therefore, the ability to amplify signals without this conversion is very important in communication systems.

The fact that amplification is achieved within the fibre is a key performance benefit—first no coupling is required from the input fibre, into the amplifier and then in reverse at the output, minimizing losses, costs and construction difficulties associated with the transition from fibre to other amplification media. Since the light is principally confined within the core of the fibre, very high cross-section overlaps can be achieved between the amplifying medium and the signal to be amplified, hence leading to high efficiencies.

It has been demonstrated that amplifiers with a noise figure tending towards the ‘quantum limit’ [1] of 3 dB can be achieved in the laboratory [2]. This high level of performance is achieved without paying a premium in other aspects of the device. The fact that a single amplifier can be used to amplify many channels in a wavelength division multiplexed (WDM) system is a key commercial consideration, and it contrasts with the fact that an independent electrical repeater would be needed for every channel.

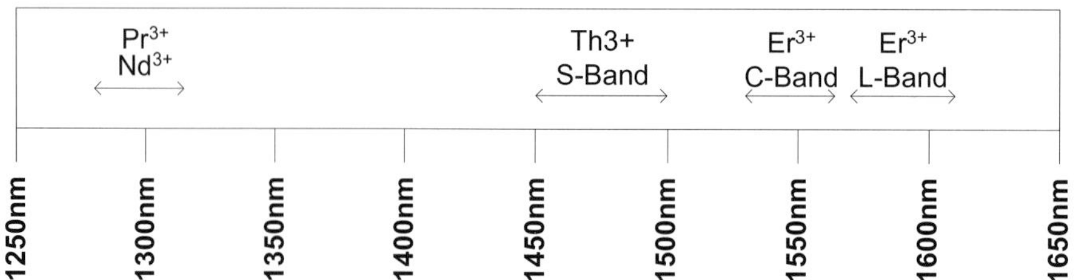

Figure B4.5.1. Rare earth dopants and their gain bands for telecom applications.

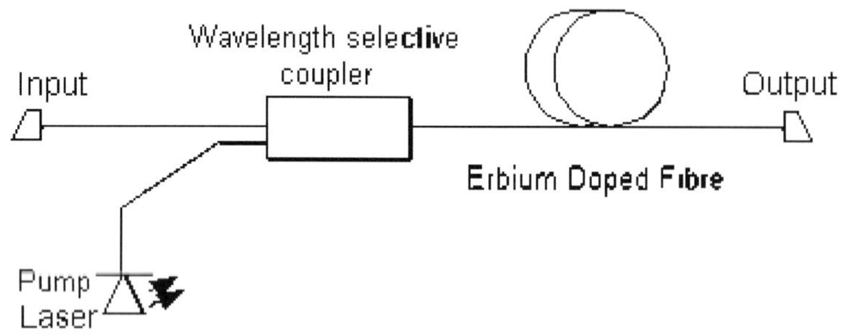

Figure B4.5.2. Topology of a simple EDFA.

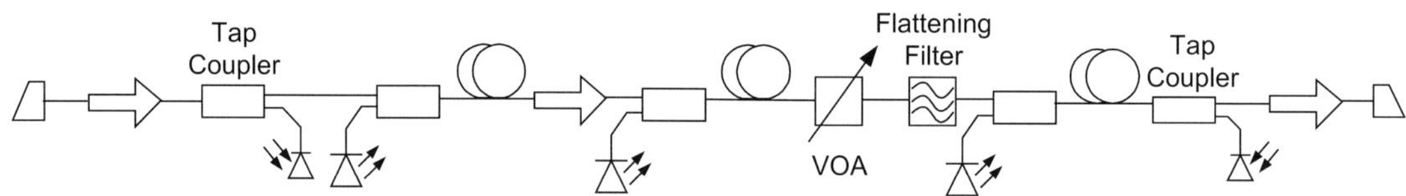

Figure B4.5.3. Topology of a typical commercial EDFA.

B4.5.2 Amplifier topology

Figure B4.5.2 shows the topology of a very basic EDFA.

A simple amplifier actually only requires three components—doped fibre, a pump multiplexer and a source of pump light. The pump multiplexer is generally a wavelength selective device which couples the signal and pump lights together into a single fibre, with only a small loss to both signal and pump (throughout we assume operation with single mode fibres only). Within the doped fibre the pump and signal lights interact with the dopant ions to provide gain to the signal. At the output of the amplifier there will be a small amount of residual pump light (typically $<0.1\%$ of input pump light), the amplified signal and a broadband noise radiation referred to as amplified spontaneous emission (ASE). As will be discussed later there are merits to utilizing a more complicated structure in order to improve the performance of the amplifier. The structure of a typical commercial telecoms EDFA is shown in figure B4.5.3. Multiple pump sources are used to achieve superior noise performance, and also due to the fact that there is a limit to the optical power that can be obtained from a single laser diode.

Optical isolators are utilized in the design for a number of reasons. The isolator at the input to the amplifier prevents backwards travelling noise from leaving the amplifier, which can degrade system performance and the isolator at the output prevents backwards travelling light entering the amplifier and

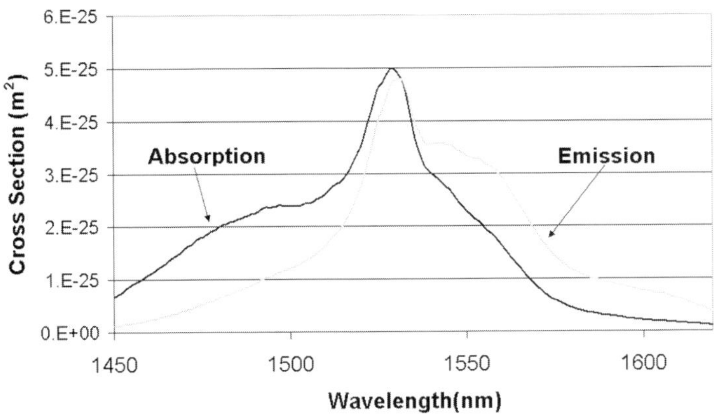

Figure B4.5.4. Typical absorption and emission cross sections.

leading to a degradation of its performance. The isolators within the amplifier break up the length of doped fibre and prevent accumulation of backwards travelling noise which may degrade performance. Detailed consideration to these requirements will be given later. The inherent gain spectrum of erbium-doped fibre is not flat and, therefore, for operation in WDM systems filters are required to flatten the gain spectrum.

B4.5.3 Gain spectra

The exact gain spectrum of an EDFA is dependent ultimately on the net cross section of the amplifier, which is defined by the inversion level and the absorption and emission cross sections of the doped fibre being used. If we know the net cross section we can calculate the gain spectrum that will be obtained from a given amplifier. Whilst the cross sections can be measured for a given fibre type, calculating the inversion level is very difficult—figure B4.5.4 shows typical absorption and emission cross sections. The doped fibre cannot be considered to have zero length, and therefore the flow of signal, pump and ASE power in both forward and backward directions must be considered. All of these powers are also a function of wavelength, which must be taken into account. The problem cannot be solved analytically and must be tackled numerically. Section B4.5.12 gives consideration to the modelling approaches commonly used.

Whilst it is very hard to determine the inversion level, we can use the cross section and inversion level as a starting point to discuss gain shapes that can be achieved with an EDFA. Figure B4.5.5 shows the net cross section *versus* wavelength for a range of inversion levels for a typical erbium-doped silica glass. The net cross section of a fibre is given by equation (B4.5.1)

$$\sigma_{\text{net}} = \frac{\sigma_{\text{e}} N_2}{N} - \frac{\sigma_{\text{a}} N_1}{N} \qquad \text{(B4.5.1)}$$

where σ_{net} is the net cross section; σ_{e} is the emission cross section; σ_{a} is the absorption cross section; N_2 is the population of the upper energy level; N_1 is the population of the lower energy level; and N is the total ion population.

The upper and lower lines (labelled 1 and 0) are the emission and absorption cross sections, respectively. The net cross section can be considered to be equivalent to gain per unit length, with a net cross section of 0 being a gain of unity, negative values being losses and positive values being gains. The net cross section therefore offers a route to predicting an amplifier's performance as a function of inversion level.

Figure B4.5.5 suggests that gain may be achieved from 1450 nm to beyond 1650 nm, however, in practically realizable amplifiers a sufficiently high inversion level cannot be obtained. The band from

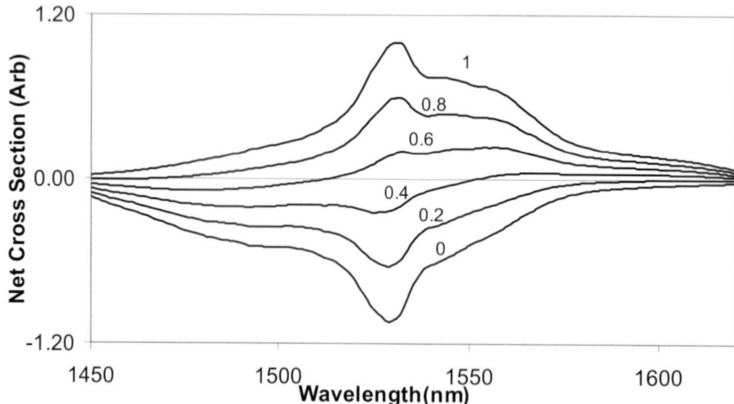

Figure B4.5.5. Net cross sections for varying inversion levels.

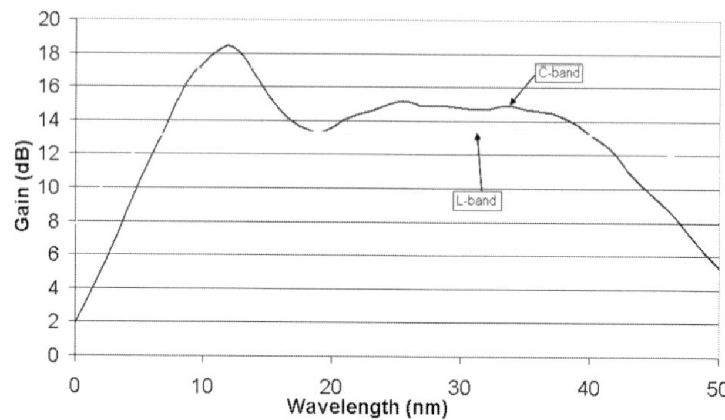

Figure B4.5.6. C-band and L-band gain spectra.

1528 nm to 1560 nm is known as the C-band, and amplifiers can be constructed in this region with an inversion level of around 0.8. Whilst an amplifier with this inversion does provide gain to longer and shorter wavelengths the discrepancy from the peak gain is such that the gain is not useful.

Another area of interest is the L-band extending from 1565 nm to 1610 nm, which can be accessed using inversion levels of around 0.4. The net cross section in this case is significantly smaller than for the C-band, hence the gain per unit length is much lower in an L-band amplifier and, to achieve useful gains, much longer lengths of fibre are used (up to 100 m). The maximum wavelength at which gain can be attained is generally limited by excited state absorption (ESA) of the signal. In conventional silica fibre this limits the edge of the L-band to approximately 1610 nm, however, special glass compositions have pushed the band out to beyond 1620 nm [3].

The natural gain spectrum obtained in the L-band is substantially flatter than the natural shape in the C-band, for comparison, figure B4.5.6 shows typical unflattened C- and L-band gain shapes. The flatter shape is desirable as it makes gain flattening easier, however unfortunately L-band amplifiers have yet to rival C-band amplifiers in terms of performance and cost.

The long lengths of fibre required for L-band amplifiers (up to 100 m) are undesirable as they add cost and reduce amplifier performance (due to background loss and other fibre impairments). In an effort to improve the performance of L-band amplifiers, highly doped fibre has been utilized [4]. These have

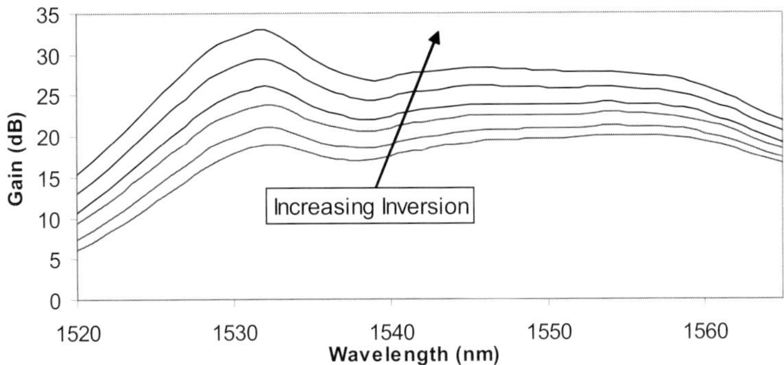

Figure B4.5.7. Gain spectrum demonstrating gain tilt.

successfully reduced fibre length by a factor of two, and, consequently, performance improvements have been realized. The level of doping that can be achieved in a fibre is limited by cooperative energy transfer (CET) effects that become significant when doping densities are high (due to the closer proximity of rare earth ions and the formation of clusters within the glass). The CET effects compete with the gain process and hence limit amplifier performance.

B4.5.3.1 Dynamic gain tilt

Dynamic gain tilt (DGT) is a method of describing changes in the gain spectrum of a given amplifier as the inversion level changes. Figure B4.5.7 shows a set of gain curves obtained by varying the inversion level.

As can be seen each of the curves is comparable, having been 'tilted' around a point to the long wavelength edge of the gain spectrum. This predictable dependence of gain spectrum on inversion level (and hence on signal and pump powers) can be used to model the behaviour of an amplifier under changing conditions. A set of values for DGT *versus* wavelength are calculated from gain spectra measurements and can then be used to predict gain spectra under different conditions. The DGT parameter is defined in equation (B4.5.2):

$$\mathrm{DGT}(\lambda) = \frac{\Delta g(\lambda_1)}{\Delta g(\lambda_2)} \tag{B4.5.2}$$

if $\lambda_1 < \lambda_2$ then DGT > 1.

B4.5.3.2 Variable gain amplifiers

If we wish to change the gain of an amplifier for a given set of input conditions, the only parameter we have control over is the inversion level (*via* the pump power). However, from figure B4.5.7 (and the fact that DGT $\neq 1$) we know that changing the inversion level also changes the gain spectrum, which implies that it is not possible to produce a variable gain amplifier that maintains a flat gain spectrum as its gain is changed.

In practice we obtain a variable gain amplifier by operating the amplifier at a fixed inversion level and including a variable optical attenuator (VOA) in the design. There are a number of locations within the amplifier in which the VOA can be placed, and here we consider the tradeoffs involved.

To describe the operation of a variable gain amplifier, we introduce the terms 'internal gain' and 'external gain'. The internal gain is the gain spectrum provided by the erbium-doped fibre (EDF), whilst the external gain is the gain from the input to the output, including all other optical components. Since we have said that the inversion level must be fixed, the internal gain spectrum must also be fixed. If a VOA is placed at the output of the amplifier, its loss can be adjusted to directly vary the external gain of the amplifier without affecting the operation of the doped fibre. The drawback of this approach is that it is inefficient, as the attenuation is performed where the powers are highest. The most efficient position for the VOA would be at the input of the amplifier where powers are at a minimum. The input, however, is the worst position from a noise figure (NF) perspective—any loss before gain adds directly to the NF.

A compromise is therefore reached whereby the VOA is 'buried' along the length of EDF. This allows a significant amount of gain to be achieved before the VOA, but reduces signal powers at the VOA and hence improves efficiency. The key consideration in selecting the position of the VOA is that the gain seen by the signals before the VOA is much larger than the maximum loss of the VOA to avoid any impact on the NF.

B4.5.3.3 Gain flattening

Amplifiers in a WDM system require a flat gain spectrum, which is not provided by EDF and therefore gain flattening filters (GFFs) must be included to flatten the gain spectrum. These filters are usually buried along the length of the doped fibre, according to the same compromises as the VOA discussed previously.

Many technologies are employed for the flattening filters, however dielectric components are the most common. The erbium gain shape is a complex one and, hence, the filter technology chosen must be capable of generating arbitrary shapes. Also, since the filters are often placed within a gain medium, it is important that the technology employed is not reflective (e.g. Bragg gratings), as the reflected light would result in a degradation of the amplifier performance.

Dielectric filters are cheap when purchased in large quantities, however, the design and prototyping phase is expensive. Other technologies such as blazed gratings and tapered fibre filters have been developed to provide a cheaper alternative to dielectric filters, however, the performance of dielectric filters means that they have yet to be replaced in commercial amplifiers.

Although it is possible to produce an amplifier with a very flat gain spectrum under one set of operating conditions, the gain spectrum is affected by a number of factors. The cross sections of erbium fibre are affected by temperature and can change over the lifetime of the fibre. Other components within the amplifier may also change their spectral shape, further affecting the gain shape. The most significant change is that of gain spectrum with temperature, and this has been tackled in a number of ways, for example controlling the temperature of the erbium fibre, or manufacturing filters whose shape changes in the same manner as the erbium fibre.

As the performance of optical communication systems has advanced the requirements for flat gain have become tighter, and this has led to the development of dynamic gain flattening technology. This technology utilizes filters whose shapes can be changed on a continual basis, which, when combined with spectral monitoring, can be used to provide a continuously flat amplifier.

Two approaches have been employed—continual filter shapes and per channel power control. The former utilizes a filter whose shape can be changed to flatten the gain spectrum. This can be accomplished utilizing a set of sinusoidal filters to generate a Fourier series [5]. By changing the characteristics of the filters the resulting filter shape can be defined. Figure B4.5.8 shows a typical C-band gain spectrum being successively flattened with up to five sinusoidal filters.

The per channel power control system controls the power of each channel by de-multiplexing the channels, passing them through a VOA and then re-multiplexing them. This is less desirable as the demultiplexing-multiplexing process degrades system performance, however the devices are cheaper and easier to control.

B4.5.4 Noise in EDFAs

Gain is achieved in an EDFA *via* a two- or three-level laser system (see section B4.1.4), in which electrons are excited to an upper energy level by a pump source and are then available for stimulated emission when

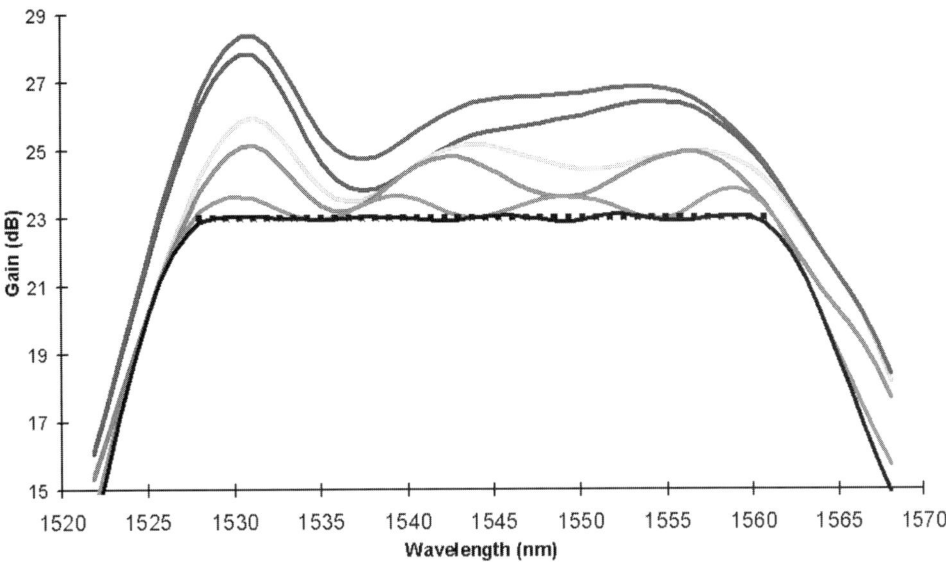

Figure B4.5.8. C-band gain spectrum successively flattened by up to five sinusoidal Fourier filters.

they interact with the light to be amplified. Electrons in the upper state are also available for spontaneous emission, and it is this process that leads to the creation of noise in an EDFA.

The photons emitted spontaneously from an ion will be travelling in a random direction and a small proportion of them will be captured by the numerical aperture (NA) of the fibre, and hence are guided along the fibre. These photons experience gain as they propagate along the fibre giving rise to ASE. As well as providing the amplifier with a finite NF, ASE also degrades the amplifier performance, as discussed later.

It can be shown that the NF of an EDFA is given by equation (B4.5.3)

$$\mathrm{NF}(\lambda) = \frac{\rho_{\mathrm{ASE}}(\lambda)}{h\nu G(\lambda)} + \frac{1}{G(\lambda)} \tag{B4.5.3}$$

where ρ is the ASE power spectral density; G is the linear gain; h is Planck's constant; and ν is the optical frequency at λ. Furthermore, in the case of high gain amplifiers we can show that the NF is approximated by:

$$\mathrm{NF} \approx 2N_{\mathrm{sp}} \tag{B4.5.4}$$

where N_{sp} is the spontaneous emission factor. For the case of three-level pumping, as occurs at 980 nm, the emission cross section is zero and, hence, the minimum value of N_{sp} is unity. This gives a minimum achievable NF for a 980 nm pumped EDFA of 3 dB [6]. This is borne out in simulation and experiment, where amplifiers tending to a 3 dB NF have been demonstrated [2]. The figure of 3 dB is often referred to as the quantum limited noise figure.

In the case of two-level pumping (around 1480 nm) the minimum value of N_{sp} is increased and is ultimately determined by the pump and signal cross sections. For signals in the C-band typical values of $N_{\mathrm{sp}}^{\mathrm{min}}$ are 1.3–1.5, which correspond to minimum NF of 4.1–4.7 dB.

In an ideal case the ASE is small at all points in comparison to the signals and does not affect the inversion level and, hence, gain spectrum, of the amplifier. Unfortunately, this is not the case and steps must be taken to manage the ASE within an amplifier to avoid degradation of performance. The first point to note is that since spontaneous emission occurs in all directions from an ion, the captured spontaneous

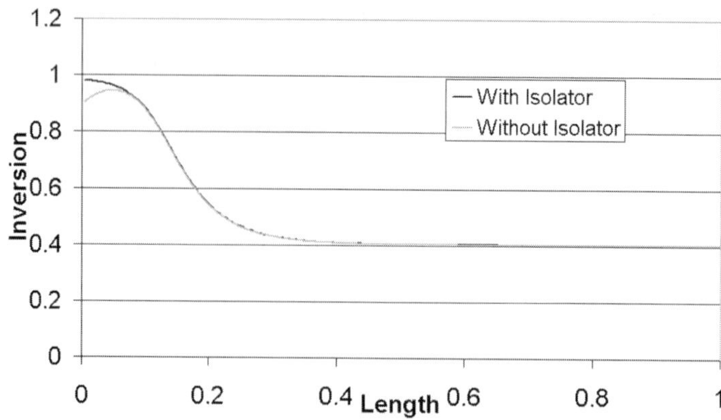

Figure B4.5.9. Inversion level with and without a buried optical isolator.

emission, and hence propagating ASE, will be travelling both forwards and backwards through the amplifier. Assuming that a sensible design of amplifier has been chosen, the power of the forwards propagating ASE will be substantially smaller than the signal at all points in the amplifier and hence will have little effect on the performance of the amplifier.

Counterpropagating ASE, on the other hand, behaves very differently to the signals—in fact the power profile of counter propagating ASE is broadly the opposite of the signals—it is smallest at the output end of the amplifier, and grows towards the input of the amplifier. This can mean that the power of the ASE at the input end of the amplifier is significant in comparison to the signal and, hence, may decrease the inversion of the amplifier. This will lead to an increase in the NF of the amplifier as the gain per unit length at the input of the amplifier will be lower. The conversion efficiency of the amplifier will also be reduced as power will be preferentially transferred to ASE, rather than the signals.

The backwards travelling ASE must therefore be managed along the length of the fibre to ensure that it never becomes significant with respect to the signal. This is achieved by the insertion of optical isolators along the length of doped fibre which serve to block the backwards propagating light, whilst allowing the forwards propagating signals (and ASE) to pass though. Figure B4.5.9 shows inversion level *versus* length for amplifiers with and without an isolator placed at the peak of the backwards propagating ASE. All other parameters for the amplifiers remained consistent.

A consequence of inserting isolators in the doped fibre is that they themselves may decrease performance. The loss of the isolator must be compensated for by a higher gain, and the loss may also degrade the NF if the gain before the isolator is not much larger than the loss. Isolators designed to work at signal wavelengths will not pass 980 nm pump light and, hence, a system to bypass the pump around the isolator must be introduced. This adds further loss (to both the signals and pump), and increases complexity and cost. To obtain the benefits of an isolator it is often, therefore, placed at a point where the vast majority of the pump light has been absorbed and so a pump bypass is not required. Another pump is then inserted after the isolator. There is often a trade-off between position for optimum noise performance and that required to minimize complexity and pump loss.

B4.5.5 Amplifier pumping

Erbium in a silica host glass has a number of absorption bands which can be used to pump an EDFA to provide gain in the 1550 nm optical transmission window. Figure B4.5.10 shows the absorption cross section with

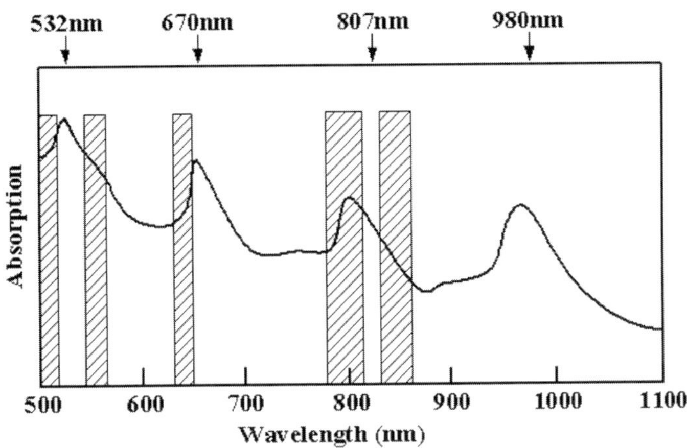

Figure B4.5.10. Erbium absorption spectra, regions of significant ESA (hatched boxes) and possible pumping wavelengths.

potential pumping wavelengths highlighted, along with regions of ESA. Of the possibilities, two wavelengths have emerged as the favoured ones—980 nm and 1480 nm (not shown on diagram).

The energy levels ($^4I_{11/2}$) of the 980 nm absorption band reside in a different multiplet to those corresponding to the 1550 nm amplification band ($^4I_{13/2}$) and we therefore obtain a true three-level laser system (see section B4.1.4). The non-radiative spontaneous decay time from the $^4I_{11/2}$ to the $^4I_{13/2}$ level is of the order of 10 μs giving effectively no spontaneous emission to the ground state from the $^4I_{11/2}$ level. This allows the creation of very high inversion levels using a 980 nm pump. In contrast, the energy levels of the 1480 nm band ($^4I_{13/2}$) reside in the same multiplet as the amplification levels and the amplification system can be considered a two-level system. Due to the thermal equilibrium within each multiplet the population of the pump level will not necessarily be zero and is dependent upon the distance of the given level from the bottom of the multiplet. There will, hence, be spontaneous and stimulated emission at the pump wavelength. The weaker absorption cross section at 1480 nm, and the emission that occurs means that it is not possible to achieve arbitrarily high inversion levels. As has been discussed previously this limits the minimum NF that can be achieved with 1480 nm pumping.

B4.5.5.1 Influence on amplifier performance

A plot of inversion level *versus* length can be used to understand the relative merits and issues of the two pumping wavelengths. Figure B4.5.11 shows this plot for a simple amplifier with a single copropagating pump. As can be seen the inversion levels peak near the input end of the amplifier and then tail off towards the output end, where there is little pump power left. This shape of inversion curve is desirable for good noise performance—the signals entering the amplifier immediately see a high gain, which is key to achieving a low noise figure. The decrease in inversion at the very front of the amplifier is due to counterpropagating noise and, as shown in figure B4.5.9 by can be dealt with the use of a buried isolator. Whilst the 980 nm and 1480 nm curves follow the same pattern there are quantitative differences between them, these are discussed in the following sections.

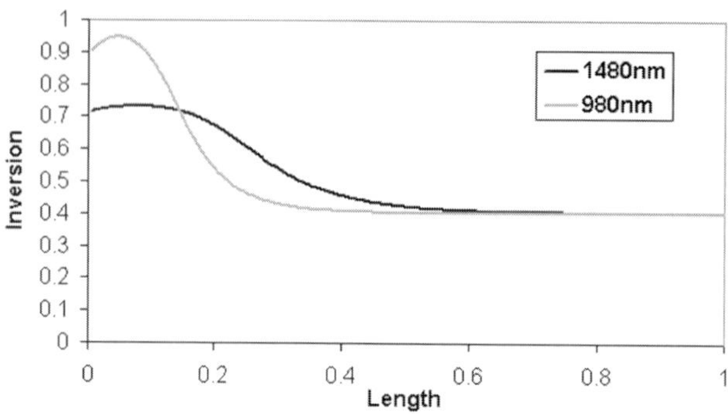

Figure B4.5.11. Inversion *versus* length for copropagating 980 nm and 1480 nm pumps.

B4.5.5.2 980 nm pumping

Due to the high absorption cross section and zero emission cross section a 980 nm pump can achieve a very high level of inversion in the fibre, which will provide a high level of gain per unit length for the optical signals. This is desirable at the input of the amplifier as it means the signals will experience gain immediately upon entering the amplifier and, hence, the noise performance is optimized.

Since the 980 nm absorption band is very strong, the pump light is absorbed very quickly and, hence, the inversion level of the amplifier decreases rapidly after the initial region where almost all of the pump is absorbed. Since the gain spectrum is dependent on the average inversion of the fibre, gain shapes which require low average inversion (e.g. L-band amplifiers) are hard to achieve with good performance using a 980 nm pump, as they require a long length of fibre with a very low inversion level, to counteract the high inversion at the input end. This fibre will absorb signal photons, which are then either spontaneously emitted as noise, or are lost non-radiatively, thus reducing efficiency.

The efficiency of a pumping scheme can be quantified using either power conversion efficiency (PCE), or quantum conversion efficiency (QCE). For practical amplifier design the PCE is the more useful figure as we are interested in the amount of pump power that will be required to produce an amplifier with a given signal output power.

Pumping at 980 nm is less efficient than at 1480 nm and an experimental figure of 50% PCE has been reported [7], in contrast to 75.6% for a 1480 nm pumped amplifier. Interestingly, the vast majority of this difference is due to the differences in photon energy as the QCE figures are similar (78.9% *versus* 79.4%).

An effect cited as pump mediated gain inhomogeneity has been reported for pumping in the 980 nm band. It has been shown [8] that significant changes in the gain spectrum occur with only small changes in the pump wavelength, when an amplifier is pumped near to 980 nm. Figure B4.5.12 shows results for the short and long wavelength edges of a C-band amplifier. Changes of over 2 dB are seen at the short wavelength end of the spectrum when the pump wavelength is changed over a range of 4 nm.

The gain spectrum of commercial EDFAs is specified very tightly and, therefore, due to pump mediated gain inhomogeneity, it is vital that if 980 nm pumps are utilized, their wavelength is specified tightly. This will add cost to the devices and is therefore a factor in the choice of pump wavelength for commercial amplifiers.

The main advantage of 980 nm pumping is the superior noise performance, which has been discussed in section B4.5.4.

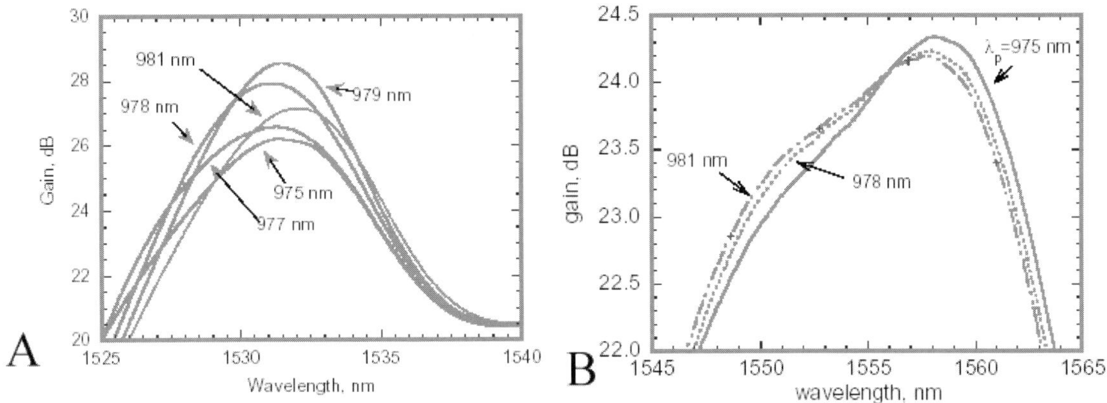

Figure B4.5.12. Pump mediated gain inhomogeneity. (*A*) Short wavelength edge of the C-band; (*B*) long wavelength edge of the C-band. (From [3]).

B4.5.5.3 1480 nm pumping

As can be seen from figure B4.5.11 the 1480 nm pump does not achieve such a high peak inversion level as the 980 nm pump, however due to the lower absorption the pump penetrates further into the amplifier giving a more even inversion level along the amplifier.

The lower inversion level at the input to the amplifier causes the noise performance of the 1480 nm pumped amplifier to be poorer than that of the 980 nm pumped amplifier, however, there are a number of other advantages to 1480 nm pumping. As mentioned in the previous section, the efficiency of 1480 nm pumping is substantially higher than 980 nm and so less pump power is required to achieve an amplifier with a given output power. 1480 nm pumps are therefore utilized where efficiency is a more important parameter than noise performance.

The more even inversion profile of a 1480 nm pumped amplifier is desirable for L-band amplifiers which require a low average inversion. It is also possible to pump away from the peak of the absorption band to achieve still lower absorption which may be useful in some situations, such as amplifiers requiring gain out to ultralong wavelengths (beyond *ca* 1610 nm).

Amplifiers with mixed pump wavelengths are commonly used—a 980 nm pump is used at the input to the amplifier to provide good noise performance, with 1480 nm pumps being employed towards the output to provide efficiency.

B4.5.5.4 Pump distribution schemes

Pump distribution schemes can be broadly categorized into three families—copropagating, counterpropagating and bidirectional pumping. In the small signal regime where the signal powers are too small to influence the inversion level, the co- and counterpropagating pump schemes are equivalent, however, the bidirectional pumping scheme can deliver higher gains. Since all practical amplifiers are operated in a saturated regime we will only consider that situation in detail here.

In the saturated regime the signal powers have a significant impact on the inversion level of the amplifier at any given point along the fibre and, hence, the relative pump, ASE and signal powers at each point through the amplifier play a role in determining the gain available to the signal. The two key performance factors that are optimized by the choice of pumping scheme are gain (and hence output power) and NF. Unfortunately, these two factors often demand opposite actions to optimize them and hence any design is ultimately a compromise.

A copropagating pump is invariably the preferred choice when optimizing an amplifier for noise performance, since the inversion and, hence, gain per unit length, is highest at the input of the amplifier. In the counterpropagating case the inversion is lowest at the input of the amplifier, and the lower gain per unit length is the equivalent of having some loss before the signal undergoes amplification, resulting in an increased NF. The difference in NF for co- and counterpropagating pumps is lessened for 1480 nm pumping as the pump penetrates further towards the input to the amplifier, increasing the inversion there and reducing the NF.

As a signal propagates through an amplifier its power increases such that, in a well designed amplifier, it is at a maximum level at the end of the doped fibre. With a copropagating pump, the pump power follows the inverse pattern to the signal, whereas with a counterpropagating pump, the pump power follows the same pattern as the signal powers. The counterpropagating pump is therefore more efficient.

B4.5.6 Doped fibre amplifiers for other telecoms windows

Doped fibre amplifiers have also been developed in other areas of the optical spectrum of interest to optical communications. This section summarizes two of the areas which have received attention and generated positive results.

B4.5.6.1 1450–1500 nm

Thulium-doped fibre amplifiers (TDFAs) have been widely investigated as a source of amplification in the 1450–1500 nm range, which is an area of interest to telecoms, being known as the S-band.

The performance of TDFAs has yet to reach that of EDFAs, however, practical amplifiers have been constructed. Different glass hosts are utilized for thulium-doped fibre, as the high phonon energy of silica makes thulium-doped silica impractical. This has led to a number of challenges in the production of useable amplifiers since the glass used as the host must be mechanically reliable and capable of joining efficiently to silica based fibres.

A further difference is that the energy levels of interest are part of a quasi-four-level laser system, as opposed to the two- or three-level system of erbium amplifiers. Figure B4.5.13 shows the energy levels of a thulium ion along with typical lifetimes when a fluoride-based glass is used as the host.

As can be seen the lifetime of the lower energy level is substantially longer than the upper level, which is undesirable in a lasing or amplification material and has implications for the pumping scheme employed. A number of pumping schemes have been demonstrated, typically utilizing an upconversion pumping process [9–11].

B4.5.6.2 1300 nm

Two rare earth dopants have been proposed in this wavelength range—neodynium and praeseodynium. Both of these dopants require non-silica hosts and work has concentrated on fluoride glasses. Whilst practical amplifiers have been developed, a number of mechanical issues with fluoride fibres remain, and the lack of commercial interest in the 1300 nm region has prevented their further development.

B4.5.7 Inhomogeneous effects

Inhomogeneous effects are those that result from restricting the set of ions available for interaction with signal photons to a subset of the total population. This occurs in both the polarization and wavelength domains. The principle inhomogeneous effects are summarized in this section.

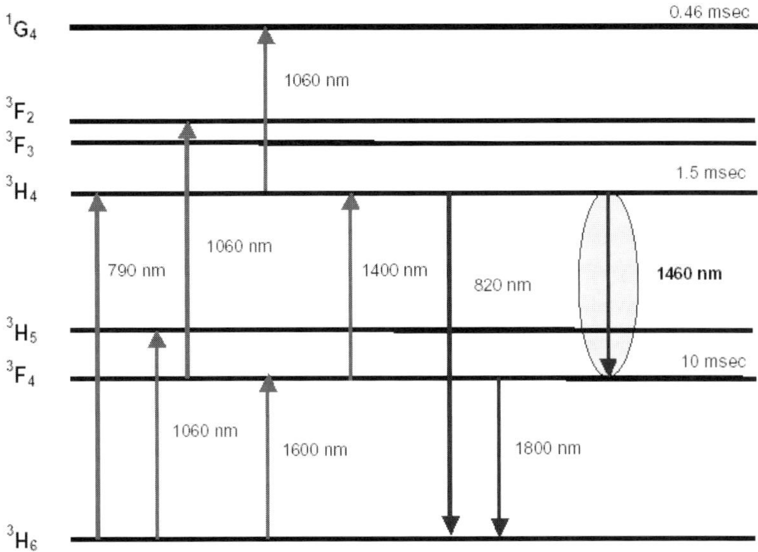

Figure B4.5.13. Energy levels in thulium, with the transition of interest to telecoms applications highlighted.

B4.5.7.1 Polarization hole burning

The absorption and emission cross sections of an ion are a function of the polarization of light in relation to the symmetry axis of the sites around the ions. Polarized light will therefore preferentially interact with a subset of the total ion population, causing preferential saturation in that subset.

ASE propagating in the orthogonal polarization state to the signal will interact with the orthogonal subset of ions, which have been depleted less by the signal and will therefore see a higher gain than would normally be expected. This results in a poorer NF than would be predicted if PHB is not taken into account.

PHB is a small effect of the order of 0.1 dB in a single amplifier [12], however, the underestimate of ASE will accumulate between amplifiers and hence may be significant in systems with many amplifiers.

B4.5.7.2 Polarization dependant gain

The relative polarizations of pump and signal photons in an amplifier can cause a change in the gain experienced by the signal. The gain is maximum when the polarizations are aligned and minimum when they are orthogonal. This is a very small effect and reports in the literature have shown dependencies of <0.1 dB [13]. The effect is generally masked by the polarization dependent loss (PDL) of other components in the amplifier and is very hard to measure experimentally.

B4.5.7.3 Spectral hole burning

Energy level broadening within the erbium ion in glass hosts has an inhomogeneous component which can affect the gain spectrum of the amplifier under certain loading conditions. The result of inhomogeneous broadening (see section A1.8) is that only a subset of the total ion population is available to provide gain at a certain wavelength, and hence the inversion level is a function of wavelength. If a high power wavelength is present in an amplifier, the inversion level local to that wavelength may be reduced and, hence, the channel will see a reduced gain. This reduction in inversion will also affect wavelengths close to the high power

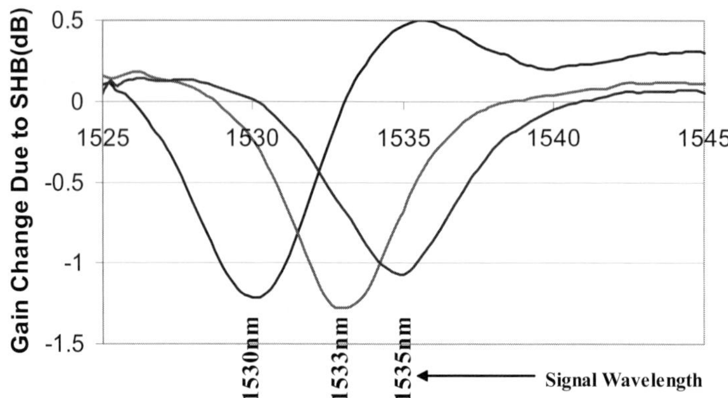

Figure B4.5.14. Gain change due to spectral hole burning (SHB).

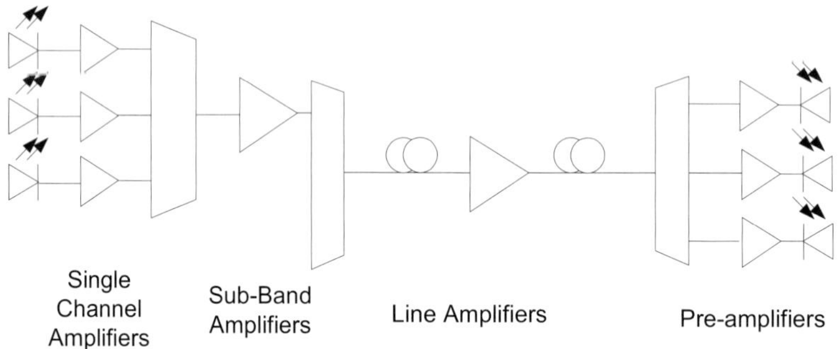

Figure B4.5.15. Amplifiers in an optical communication system.

wavelength and the effect will appear as depressions in the gain spectrum—this effect is known as spectral hole burning (SHB) [14] (see section A1.10.2).

Figure B4.5.14 shows the changes in the gain spectrum where a spectral hole has been created. The characteristics of the holes are a function of wavelength—the holes become deeper and narrower to the short wavelength edge of the C-band (1530 nm) and wider and shallower to longer wavelengths. Non-localized SHB has also been reported, whereby a spectral hole is burnt at a wavelength different to the saturating signal wavelength [15].

B4.5.8 The EDFA in a communication system

When deploying amplifiers in a system detailed consideration must be given to the specification of the devices to ensure that the system performs as desired.

As has been shown the performance of EDFAs can be optimized around a number of parameters (gain, noise figure, efficiency, bandwidth), each of which assumes a different level of importance depending upon the amplifier's purpose in the network. For example, in the case of an optical pre-amplifier the requirement is for a low noise device, whereas the optical bandwidth is not important as only a single channel passes through the amplifier. Figure B4.5.15 shows the typical locations of amplifiers in a network. Each of the locations has a set of distinctive requirements that govern the specification of the amplifier.

B4.5.8.1 Line amplifiers

These amplifiers are responsible for compensating for the loss of the fibre within the communication system, thus ensuring that sufficient signal gets to the receiver to enable data transmission. The amplifier's performance must be such that it can amplify all of the channels to a relatively high power level for transmission through the next fibre span. The bandwidth of the amplifier is therefore of key importance, as is the gain flatness. Many line amplifiers are cascaded along the length of the system and therefore any penalties associated with each one are multiplied through the system.

Typical span losses require line amplifiers to have a gain of approximately 20 dB, and current channel counts may require output powers over +17 dBm. In order to avoid excessive heat dissipation the efficiency of the amplifier is very important.

The line amplifier is generally the most challenging amplifier to design as every facet of its performance is important to the system.

B4.5.8.2 Pre-amplifiers

It is widely known that an optically pre-amplified receiver can give superior performance to a receiver with an electrical post amplifier and, therefore, the optical pre-amplifier plays a vital role in the system.

The main requirement of this amplifier is that it is a low noise device. Only one channel will be propagated through the amplifier and, hence, the bandwidth and output power requirements are not significant. This also removes the need for gain flattening filters. Since one amplifier is needed for every receiver (sometimes pre-amplifiers are shared but it is not desirable) the cost of the amplifier is also important.

B4.5.8.3 Narrow band amplifiers

Narrow band amplifiers are used to amplify signal levels at switch nodes and multiplexing sites before they are combined onto the transmission fibre. Typically, these amplifiers will be used to amplify 4–8 channels.

Since many of these amplifiers may be used in a network the cost and physical size of them is important. The output power and bandwidth requirements are limited by the fact that they only have to amplify a small number of channels. By careful design of these amplifiers it may be possible to remove the need for gain flattening filters and this has an impact on the cost of the devices.

B4.5.9 Amplifier control

In order to operate reliably and with good performance the amplifier must be controlled correctly throughout its lifetime. In an ideal situation an amplifier will be set up at the time of installation and left running with the defined pump powers until the system is switched off. In reality this is not possible as many parameters are likely to change during the lifetime of the system. Consideration must be given to changes in the number of channels operated on the system, in span losses and in variation in the operating environment of the amplifier.

Some form of autonomous control is required to enable amplifiers to operate over a long period of time. Two main forms of control system are used in optical amplifiers—total output power (TOP) control and gain control (GC).

B4.5.9.1 Total output power control

In TOP control the amplifier monitors its output power and adjusts the pump powers to achieve the desired output power. As parameters change the amplifier will continually update the pump powers to maintain the required output power. This method of control works well under steady-state conditions as the performance of a system is generally defined using the power launched into each span.

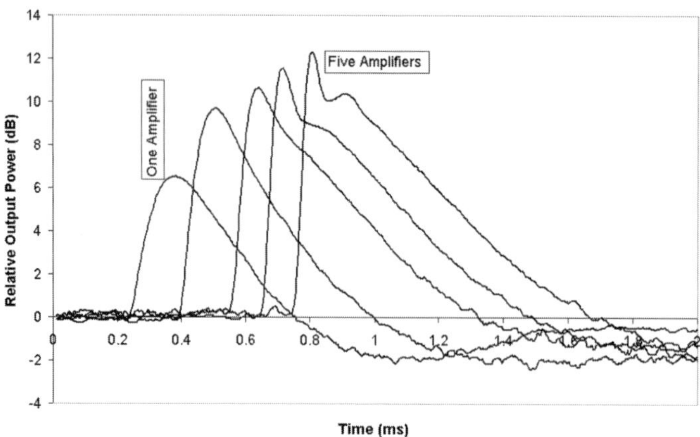

Figure B4.5.16. Transient propagating through six gain-controlled optical amplifiers.

Problems arise however when the number of channels passing through an amplifier changes. Since we have fixed the total output power, the per channel power (which is what is important to system performance) is dependent upon the number of channels passing through the amplifier. Thus, if channels are added, the per channel power is reduced and the system may stop operating, the amplifier set points must therefore be modified each time the channel count in a system is changed.

TOP control is achieved by using a broadband photodiode at the output of the amplifier to monitor the total power, and then the pump powers are controlled by an analogue or digital controller. At the output of the amplifier both signals and ASE are present and, in order that the signal powers are set to the desired level, an adjustment is often provided in the control system to account for the ASE power.

B4.5.9.2 Gain control

In this mode, the amplifier's operating point is controlled based on the optical gain provided by the amplifier. This is generally achieved by monitoring the total input and output powers and maintaining a constant ratio between them. This mode is preferential to TOP as it ensures that the channel powers are independent of the number of channels present.

B4.5.9.3 Transient control

It is possible that the channel count or channel powers will change very quickly due to failures within the optical network. If this occurs it is undesirable for the changes to impact other channels unconnected with the original failure.

If amplifiers are being operated in TOP mode and some channels rapidly disappear, the other channels will rapidly increase in power. In addition to failure of the system, damage to components can also occur.

Gain control solves transient control issues—the per channel powers are not dependent on the number of channels and, therefore, the remaining channels are not affected by the failure. Failures can occur on very short timescales, down to 1 μs for certain types of fibre cut and so control systems must be capable of tracking these changes. If the control loop does not respond fast enough, changes in the power of surviving channels will occur.

A key factor in transient control is that behaviour can accumulate as the transient passes through successive amplifiers, and can therefore be very large by the time the signals get to the receiver. Figure B4.5.16

shows an example of a transient passing through up to five gain controlled amplifiers. As can be seen the gain control is not fast enough to control the amplifier and hence some deviation is seen.

With the move towards optical switching in networks the control of transients will become more important as rapid changes in channel count will occur as the network is reconfigured.

B4.5.10 Alternatives to EDFAs

This section gives brief consideration to alternative types of amplifiers.

B4.5.10.1 Erbium-doped waveguide amplifiers

Erbium-doped waveguide amplifiers (EDWAs) utilize the same amplification mechanism as EDFAs, however the dopant is placed in a planar waveguide grown on a substrate as opposed to within an optical fibre. The advantages of this are seen as the ability to integrate many amplifiers onto a single substrate, and the possibility of constructing all parts of the amplifier in one growth process, creating significant production cost savings.

EDWAs have yet to match the performance of EDFAs, however progress is being made continuously. The main limitations are low output powers due to poor efficiency obtained in waveguide glasses, and also poor NFs due to coupling losses. EDWAs are unlikely to form a competitor to line amplifiers, however may find a market in narrow band amplifiers where many low power devices are needed in one location.

B4.5.10.2 Semiconductor optical amplifiers

Semiconductor optical amplifiers (SOAs) utilize the same amplification mechanism and structure as semiconductor laser diodes, however without the mirrors that cause lasing.

High output powers are hard to achieve and, again, coupling losses limit the NF. There is also a significant issue due to channel–channel cross talk which can occur due to the very short carrier lifetimes. SOAs have found application as pre-amps and as single channel booster amplifiers for transmitter lasers, where only modest output powers are required.

B4.5.10.3 Raman fibre amplifiers

The Raman effect can be utilized to obtain gain for signal wavelengths within standard transmission fibre (see chapter B4.3). Raman amplifiers simply consist of a high power pump of a shorter wavelength than the signals. This pump is inserted into the optical fibre *via* a wavelength selective coupler and allowed to propagate with the signals (often counterpropagating Raman pumps are used as they have performance benefits over copropagating pumps), the signal wavelengths then experience gain *via* energy transfer from the pump wavelength(s).

Two types of Raman amplifier have been utilized—discrete and distributed. Discrete amplifiers are operated in a very similar way to EDFAs, as stand alone amplification modules between spans. In contrast, a distributed Raman amplifier utilizes the fibre spans as the gain media and, thus, the span losses are reduced. Distributed Raman amplifiers are advantageous in extending the distances over which systems can operate as the reduced span loss improves signal to noise ratios.

Discrete Raman amplifiers have not seen widespread deployment due to poorer performance and efficiency than EDFAs. An advantage of Raman amplifiers is that their gain bandwidth is not restricted by the availability of suitable rare earth dopants, therefore gain can be produced at any wavelength. This may be important as the demand for capacity pushes signal wavelengths outside of the wavelengths accessible with DFAs.

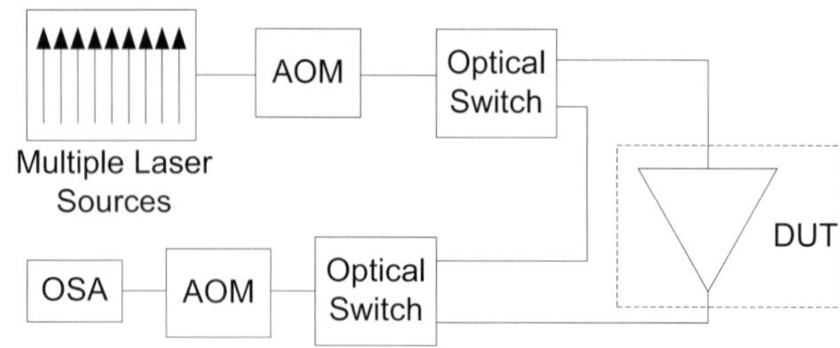

Figure B4.5.17. The EDFA measurement system.

B4.5.11 Amplifier measurement

Here we will consider the measurement of the gain and noise figure spectra of EDFAs.

The measurement of the gain spectrum would appear to be a relatively simple task, however, a complication arises due to the presence of ASE in the output spectrum. A direct measurement of the output spectrum will include the ASE and hence we are not measuring the true signal powers. We must therefore measure the ASE and subtract this from the output spectrum to obtain the true signal powers.

ASE measurement can be performed utilizing a method known as time domain extinction (TDE). EDFAs have a relatively long time constant (*ca* 10 ms) and, hence, the input signals can be modulated at a high frequency (>50 kHz) without affecting the inversion level and, hence, operating point, of the amplifier. The input signals are modulated, and then a gating method is used to govern when the output spectrum is sampled. Figure B4.5.17 shows such a measurement system. The optical switches allow the device under test (DUT) to be bypassed such that the system can be calibrated.

Two measurements are taken—one when the signals are present and one when the signals are switched off, measuring the full output spectra and the ASE respectively. We now have a measure of the ASE at the signal wavelengths and can hence calculate the actual signal output powers, gain and noise figure. Alternative methods utilizing interpolation have been used however the TDE method has been shown to give the most accurate results [16].

B4.5.12 EDFA modelling

An accurate method of modelling EDFAs is essential to their successful development. A brief introduction to two methods is given here, however no actual mathematical details are provided can be found in the referenced literature.

B4.5.12.1 Gain tilt modelling

The DGT parameter discussed previously can be used to predict the gain shape at different operating points as signal powers, pump powers and fibre lengths are changed. The method, however, does not provide any information on the noise properties of the resulting amplifier and only considers homogeneous effects.

Despite these limitations the method is very useful for understanding how a given amplifier will perform under changing conditions, and is often used for predicting gain shape changes as conditions change. The model is a very simple analytical expression and, hence, generates results rapidly.

B4.5.12.2 Rate equation modelling

In order to accurately model an EDFA, including noise and inhomogeneous effects, one must develop a numerical model based on the rate equations applicable to the amplification process. The solution process is complex and must take into account the wavelength and directions of power at each wavelength.

Data on the fibre being modelled can be captured either as absorption and emission cross sections, or as part of a 'Giles' dataset which contains the fibre data in a more readily measurable form. (See Desuvires' book, referenced in the further reading section, for an in-depth study of EDFA modelling.)

References

[1] Simon J C 1982 Semiconductor laser amplifiers for single-mode optical fiber communications *J. Opt. Commun* **4** 51

[2] Cordina K, Jolley N and Mun J 1999 Ultra low noise long wavelength EDFA with 3.6 dB external noise figure *OFC*

[3] Ellison A J G 2001 Extending the L-band to 1620 nm using MCS fiber *OFC* TuA2

[4] Ishikawa S *et al* 1998 High gain per unit length Silica based Erbium doped fiber for 1580nm band amplification *OAA TuC4*

[5] Parry S, King J P, Roberts K B, Jolley N E, Keys R and Mun J 1999 Dynamic gain equalization of EDFAs with Fourier filters *OAA*

[6] Oleshanky R 1988 Noise figure for erbium doped optical fibre amplifiers *Electron. Lett.* **24** 1363

[7] Massicott J F, Wyatt R, Ainslie B J and Craig-Ryan S P 1990 Efficient, high power, high gain, Er^{3+} doped silica fibre amplifiers *Electron. Lett.* **26** 1038

[8] Bennet K, Davis, F, Jakobson P A, Jolley N, Keys R, Newhouse M A Sheih S-J and Yadlowsky M J 1997 980 nm band pump wavelength tuning of the gain spectrum of EDFAs *OAA*

[9] Roy F, Leplingard F, Lorcy L, Le Sauze A, Baniel P and Bayart D 2001 48% power conversion efficiency in a single-pump gain-shifted thulium-doped fiber amplifier *OFC PD2*

[10] Roy F, Le Sauze A, Baniel P and Bayart D 2001 0.8 mm and 1.4 mm pumping for gain-shifted TDFA with power conversion efficiency exceeding 50% *OAA PD4*

[11] Aozasa, Masuda H, Ono H, Sakamoto Y, Kanamori T, Ohishi Y and Shimizu M 2001 1480-1510 nm band Tm doped fiber amplifier (TDFA) with a high power conversion efficiency of 42% *OFC PD1*

[12] Taylor M G 1993 Observation of new polarization dependence effect in long haul optically amplified system *IEEE Phot. Tech. Lett.* **5** 1244

[13] Mazurczyk V J and Zyskind J L 1994 Polarisation dependent gain in EDFAs *IEEE Phot. Tech. Lett.* **6** 616

[14] Aizawa Y, Sakai T, Wada A and Tamauchi R 1999 Effect of spectral hole burning on multi channel EDF gain profile *OFC WG1-1*

[15] Rudkevich E, Baney M, Simple J, Derickson D and Wang G 1999 Nonresonant spectral hole burning in erbium doped fiber amplifiers *IEEE Photon. Tech. Lett.* **11** 542–4

[16] Jolley N and Cordina K 1999 A modification of the time domain extinction measurement technique for erbium doped fibre amplifiers with improved accuracy *OAA*

Further reading

Desurvire E 1994 *Erbium Doped Fibre Amplifiers—Principles And Applications* 1st edn (New York: Wiley)

A thorough and complete reference on EDFAs. Mainly theoretical work, covering all areas of EDFAs.

Becker P C, Olsson N A and Simpson J R 1999 *Erbium Doped Fiber Amplifiers—Fundamentals and Technology* 1st edn (New York: Academic)

Covers all aspects of EDFAs, particularly considering the practical implications and impact on real-world applications of each aspect.

B4.6
High-power waveguide lasers

D P Shepherd

B4.6.1 Introduction

Planar waveguides can be fabricated in a wide range of crystal and glass materials by many methods. These include modification of the material refractive index, typically at or near the surface, through such techniques as ion-exchange [1], ion-implantation [2], ion-diffusion [3] and optical writing [4]. Alternatively, a waveguide may be formed by adhesion [5], deposition [6] or growth [7] of one material on another of different refractive index. These techniques generally lead to waveguides in the form of a planar thin film or, if some patterning technique is used, a planar array of channel waveguides (see figure B4.6.1).

The optical confinement afforded by these structures, even if only in one plane in the case of the thin film, leads to smaller pump and laser mode sizes than would be the case in an equivalent bulk laser. Thus, high optical gains can be achieved and, if the additional propagation losses due to the waveguide are not too high, low lasing thresholds are possible. For most planar devices the lengths involved are on the centimetre scale and the propagation losses are just a few tenths of a decibel. The optical confinement also tends to give a good spatial overlap of the population inversion and the laser mode, leading to high efficiency for low-loss waveguides.

The very large range of fabrication techniques means that many different laser hosts can be used in the planar waveguide format. These include high-gain materials such as rare-earth-doped YAG [7] (see chapter B1.3); broadly tunable materials such as Ti:sapphire [8] (see chapter B1.2); nonlinear and electro-optic materials such as $LiNbO_3$ [9]; low-phonon-energy materials such as LaF_3 [10]; and many different glasses, some of which may be difficult to produce in the form of an optical fibre.

Recently, further advantages of the planar geometry have been applied to the production of high-power laser devices. Firstly, there is a natural compatibility with high-power diode pump lasers, as these also have a planar format, simplifying the coupling optics required and allowing very compact laser devices [11]. Secondly, the planar waveguide can be seen as an extreme case of the slab geometry, commonly used in high-power bulk lasers [12] and consequently has similar excellent thermal power handling capabilities. The aim of this section is to describe how these geometric advantages can be combined with those of the optical guidance to make compact and robust, high-average-power, diode-pumped laser devices.

B4.6.2 Coupling of high-power diode pump sources to planar waveguides

The spatial properties of a typical high-power diode pump source are asymmetric, non-diffraction-limited and of high divergence. This makes coupling to a solid-state laser medium a major issue and many different techniques have been used in bulk and fibre lasers. These have included preconditioning optics such as fibre-coupling [13], lens ducts [14] and beam-shaping [15], as well as beam-confining techniques such as cladding-pumping [16], tapers [17], total internal reflection [18] and reflective pumping chambers [19]. Many

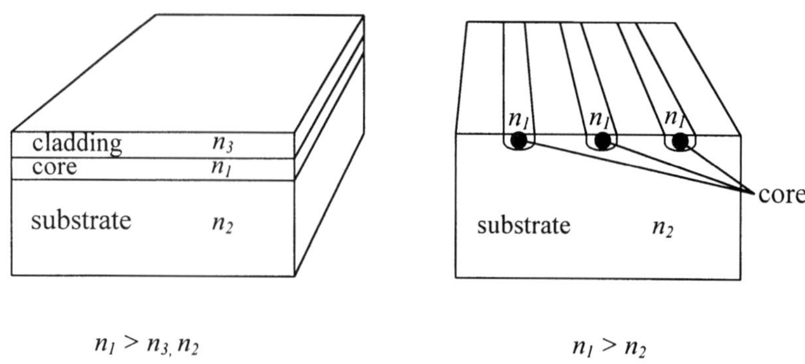

Figure B4.6.1. Typical planar and channel waveguide structures.

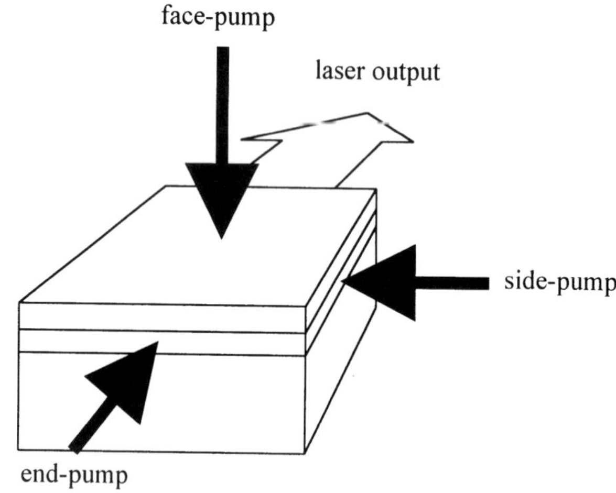

Figure B4.6.2. Pumping configurations for a planar waveguide.

of these techniques may also be used in the case of a waveguide, but it is also possible to use straightforward cylindrical focusing optics or direct proximity coupling.

The geometry of the planar waveguide suggests three possible pumping arrangements (see figure B4.6.2), each suited to different types of diode pump laser and waveguide structure.

In-plane pumping. Both side- and end-pumping of waveguides fall into the category of in-plane pumping techniques and so have some common requirements. If a waveguide laser is to be in-plane pumped by a beam with a divergence in the guiding axis M^2 times greater than the diffraction limit (see section A2.1.4.3), then the depth, D, and the numerical aperture, NA, of the waveguide must be chosen such that

$$NA \geq \frac{\lambda M^2}{D}. \qquad \text{(B4.6.1)}$$

This condition is shown graphically in figure B4.6.3, assuming a pump wavelength of $\lambda = 1\ \mu$m. For single diode units (broad stripe or diode bar) it is possible to arrange to have the near-diffraction-limited fast axis coincide with the guiding axis of thc waveguide so that relatively low NA structures can be used. The larger

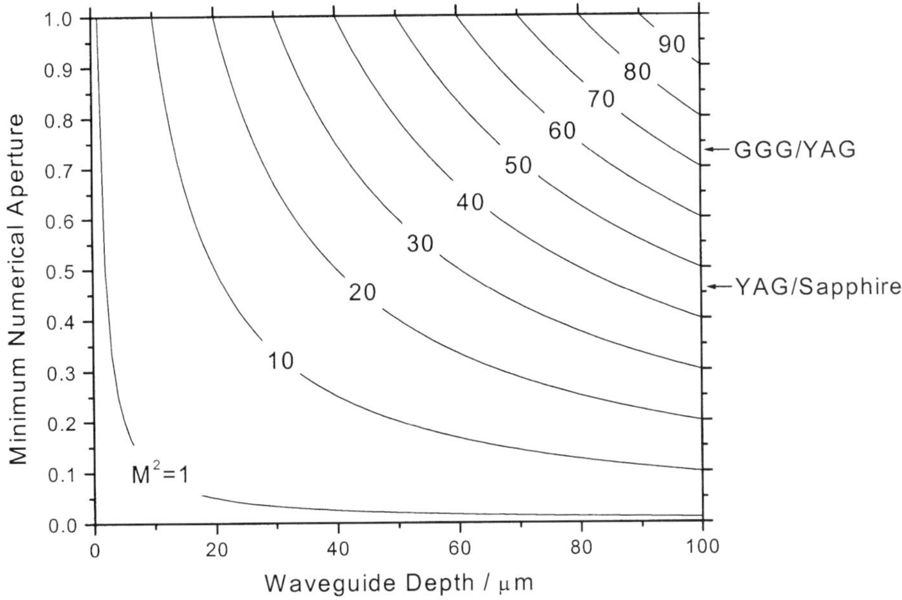

Figure B4.6.3. Graph of minimum waveguide NA, for a given depth D, required to confine beams of M^2 up to 90. Example waveguide NAs are also given.

NA waveguides can be used to confine higher M^2 beams, such as that from a beam-shaped or fibre-coupled diode bar, the slow axis of a broad-stripe diode, or the fast axis of a diode stack (a vertical array of diode bars). The high NA waveguides can also allow direct proximity coupling of a single diode unit to the waveguide with no focusing optics, allowing a very compact system. However, the slow axis of a diode bar or diode stack can be >1000 times diffraction-limited and so is normally aligned to be in a non-guided axis.

To make high NA guides of >10 μm depth it is usually necessary to combine two different materials through a deposition, growth or adhesion technique. The example waveguides given in figure B4.6.3 can both be fabricated by the adhesion technique of direct bonding [5], as this relies on the intermolecular forces between two highly polished surfaces and is thus very broadly applicable. In contrast, epitaxial growth techniques are more limited in the materials that they can combine, as a near lattice match is required between the two crystals. Nevertheless, some examples of high NA epitaxial waveguides have been successfully demonstrated, such as GGG on YAG structures fabricated by pulsed-laser-assisted deposition [20].

End-pumping. End-pumping allows focusing in the plane of the waveguide, as well as in the guided axis, and so tends to give the most intense pumping and the best spatial overlap of the inversion profile and the laser mode. Thus, low thresholds and high efficiency can be achieved. The relatively small pumping mode sizes also make it easier to obtain diffraction-limited output. However, the number of pumping units is limited and so this scheme is not as easily scalable in power as side or face pumping. Broad-stripe diodes are well suited to end-pumping, having powers of several watts, near-diffraction-limited fast-axis divergence and ~10–40 times diffraction-limited divergence in the slow axis. In general, only one of the pumping and/or lasing axes is guided and reported configurations have included guiding of only the diode slow-axis in relatively large waveguides [21], guidance of only the diode and laser fast-axes in smaller waveguides [22] (see figure B4.6.4), and one-axis guiding of both the pump and the laser using a beam-shaped diode source [23]. In all cases,

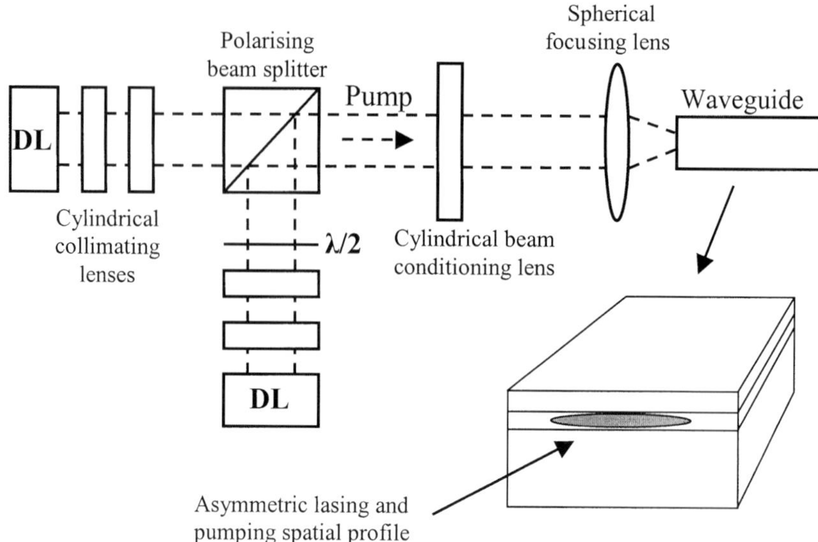

Figure B4.6.4. Typical configuration for end-pumping waveguide lasers with two polarization coupled broad-stripe diode lasers (DL) with guidance of the pump and laser modes in one axis (reproduced from [22] with permission from the Optical Society of America).

relatively simple cylindrical focusing can be used to couple the diode light into the waveguide and pumping power of $\sim$ 10 W is possible.

Side-pumping. Side-pumping is more suited to high-power pumping, as it is compatible with the use of high-power diode bars and can be scaled by adding further pumping units, vertically or horizontally, and on both sides. It also allows independent optimization of the lasing, pumping and heat removal, as these three processes occur along different axes.

Side-pumping of high NA guides can be achieved in a very compact fashion by simply proximity coupling the diode bars to either side of the waveguide (see figure B4.6.5). The delivery efficiency, η_{del}, of such a coupling scheme is given by [24],

$$\eta_{\mathrm{del}} = \frac{\displaystyle\int_{-\min(\vartheta_{\mathrm{inc}},\sin^{-1}(NA))}^{+\min(\vartheta_{\mathrm{inc}},\sin^{-1}(NA))} \exp\left[-\ln(16)\left(\frac{\theta}{\theta_{\mathrm{FWHM}}}\right)^2\right] \mathrm{d}\theta}{\displaystyle\int_{-90}^{+90} \exp\left[-\ln(16)\left(\frac{\theta}{\theta_{\mathrm{FWHM}}}\right)^2\right] \mathrm{d}\theta} \tag{B4.6.2}$$

where θ is the angle of propagation (in degrees) of the divergent diode beam with respect to the axis of the planar waveguide and θ_{FWHM} is the full-width half-maximum divergence of the diode fast-axis. If the half-angle subtended by the waveguide core to the diode emission aperture, θ_{inc}, is smaller than the angular acceptance of the waveguide, $\sin^{-1}(NA)$, then equation (B4.6.2) describes the overlap of the diverging pump beam with the physical aperture of the core. Any light incident on the waveguide outside this aperture is lost. If the angular acceptance of the waveguide is less than θ_{inc} then equation (B4.6.2) describes the fraction of the diverging pump beam that is contained by the numerical aperture of the waveguide. Figure B4.6.6 shows how the delivery efficiency varies with the core depth, D, and waveguide NA for typical experimental values

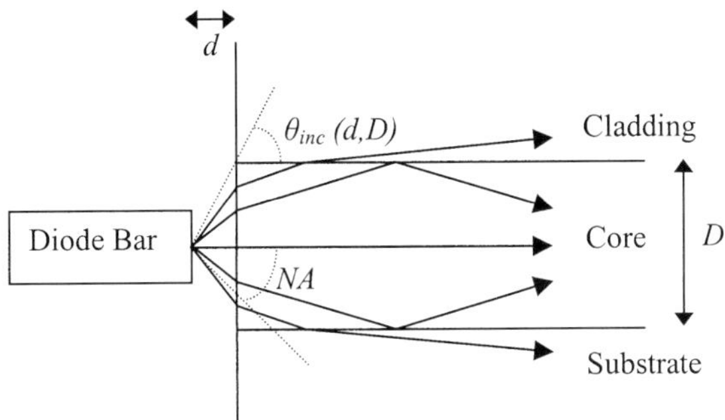

Figure B4.6.5. Side-pumping with proximity-coupled diode bars.

of coupling distance $d = 20\ \mu$m and diode divergence $\theta_{FWHM} = 35°$. It can be seen that high efficiencies (>80%) are obtained for waveguides with $NA \geq 0.4$ and $D \geq 15\ \mu$m. As diode bars with powers of ~60 W are now available, total pumping power of ~100 W cm^{-1} length of waveguide is quite feasible.

If focusing optics are used it can be seen from figure B4.6.3 that a high NA waveguide, such as a YAG/sapphire structure, of a few tens of microns thickness could easily be pumped by a ~20 times diffraction-limited diode beam. This is compatible with the fast-axis of a microlensed vertical diode stack of ~3 bars, i.e. pumping powers of ~300 W cm^{-1} length of waveguide.

Face-pumping. Face-pumping avoids the restrictions placed on the waveguide NA imposed by in-plane pumping and is an area-scalable scheme compatible with pumping by large diode stacks. As the pumping direction is perpendicular to the plane of the waveguide, a relatively large waveguide core ($D > 100\ \mu$m) and a reflective pumping chamber are required to achieve high-absorption efficiency. Current demonstrations have used a pumping chamber with narrow slots through which the divergent light from the individual bars of a diode stack could pass with high efficiency and then be effectively trapped within the chamber, multipassing through the thin bulk or waveguide slab to achieve efficient and uniform pump absorption [19]. Later developments included the integration of water-cooling in the same package [25], as shown in figure B4.6.7. In this way pumping with over 400 W has been demonstrated with a 10 bar diode stack and a 6 cm long, 1.1 cm wide waveguide consisting of a 200 μm thick Nd:YAG core and 400 μm thick YAG claddings [26].

B4.6.3 Thermal management properties

The thermal loading associated with high-power operation of solid-state lasers can lead to optical distortions such as thermal lensing and birefringence, and eventually to thermal stress fracture. High-power pumping can be achieved with face or side-pumping, and both can give a relatively uniform thermal loading of the slab-like doped-core. For an ideal, uniformly pumped slab (of infinite extent), a one-dimensional heat-flow is obtained by cooling both large faces, leading to well-defined stress-induced birefringence axes, and light polarized either in the plane, or at 90° to it, will not suffer from depolarization loss [27]. In a bulk slab, the focusing effects of the strong, birefringent, cylindrical lens associated with the stress and temperature profiles can be eliminated by taking a zigzag optical path, using total internal reflection off of the slab faces [28]. In the case of a waveguide, the thermally induced focusing is overcome by the guidance properties of the structure, even for the thicker waveguides used in the case of face-pumping [29, 30]. For the slab waveguide shown in figure B4.6.8, the temperature at the centre of the waveguide, T, and the maximum thermal power

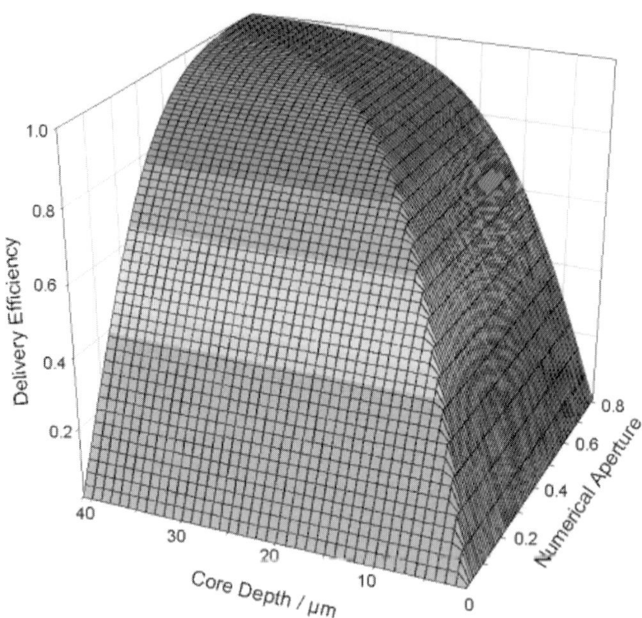

Figure B4.6.6. Theoretical delivery efficiency for a proximity-coupled diode bar with $\theta_{\mathrm{FWHM}} = 35°$ and $d = 20\ \mu$m.

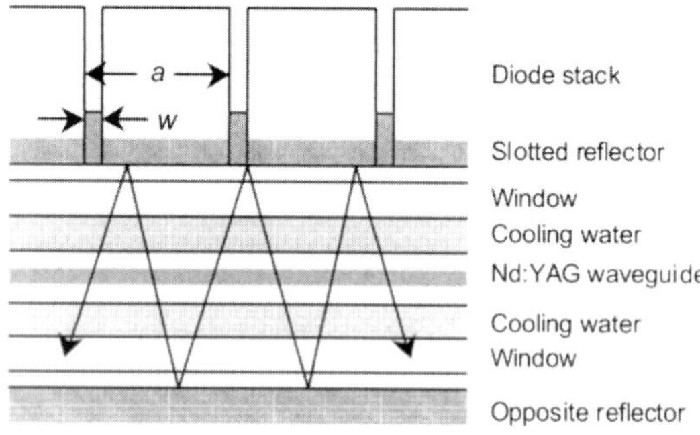

Figure B4.6.7. Schematic diagram of face pumping and cooling (reproduced from [26] with permission from the Optical Society of America).

per unit cooling area before stress fracture occurs, $P_{a\,\max}$, are given by [31],

$$T = \frac{P_a}{2}\left(\frac{1}{\lambda_t} + \frac{(2t - D)}{4k}\right) + T_c \tag{B4.6.3}$$

$$P_{a\,\max} = \frac{8R_s}{\left(t - \frac{D^2}{3t}\right)} \tag{B4.6.4}$$

where λ_t is the surface heat transfer coefficient, k is the thermal conductivity (for simplicity assumed to be the same in all regions of the waveguide), T_c is the coolant temperature and R_s is the thermal shock parameter [27]. These equations reduce to those for a uniformly pumped slab when $D = t$ [28]. Thus, a 100 μm thick YAG waveguide ($R_s = 790$ W m^{-1}) with a 50 μm thick doped region can handle a thermal

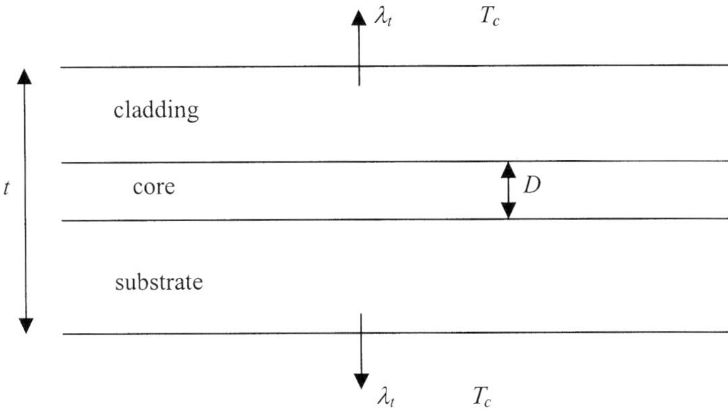

Figure B4.6.8. Symmetric slab waveguide used for thermal model (reproduced from [31] with permission from IOP publishing).

load per unit cooling area of up to 6.9k W cm^{-2}. However, it should be noted that the same waveguide has a maximum temperature rise of 85 °C for a thermal load per unit area of just 100 W cm^{-2} (using $\lambda_t = 6000$ W m^{-2} K^{-1} and $k = 13$ W m^{-1} K^{-1}). Assuming that the main contribution to the thermal load is due to the energy difference between the pump and laser photons, then a thermal load per unit area of 100 W cm^{-2} corresponds to over 1 kW cm^{-2} of actual pump power for the Yb:YAG system. Thus, very compact kW power level devices would appear to be a possibility. A theoretical comparison of power scaling in face- and side-pumped slabs has been made for the case of quasi-three-level bulk lasers [32], which include the important high-power material Yb:YAG, and it was found that the greater freedom of choice over the thickness and doping level in the case of side-pumping is advantageous.

B4.6.4 Spatial-mode selection

The need to confine non-diffraction-limited diode pump sources in the case of in-plane pumping, and the need to use relatively thick cores in the case of face-pumping, both lead to the use of multimode waveguides. In the non-guided plane both face and side-pumping (and the need for effective cooling) tend to lead to relatively large pumping mode sizes of 0.5–1 cm, which again tends to lead to multimode laser output. Various methods of spatial-mode control have thus been applied in both axes.

Double-clad waveguides. A standard solution to the conflicting requirements of needing a multimode waveguide for confinement of the pump, while simultaneously desiring a single-spatial-mode laser output, is to embed a single-mode lasing core within a much larger multimode pump guide [16]. Widely used in high-power fibre lasers, this cladding-pumping technique also increases the length of active medium required to efficiently absorb the pump. A similar double-clad structure can be used for mode selection in planar waveguides, but the design must be such that the increase in the absorption length is relatively small, as typical device dimensions are on the centimetre scale. By doping only the central portion of a multimode waveguide core (see figure B4.6.9) mode selection can be achieved by the gain profile rather than by the index profile. For a uniformly pumped medium, the gain exponent for a particular mode with a spatial profile $\phi_p(x, y, z)$, relative to that for a lasing mode, $\phi(x, y, z)$, is given by [33]

$$G_{rel} = \iiint \frac{d(x, y, z)\phi_p(x, y, z)}{(1 + S\phi(x, y, z))\left[\iiint \dfrac{d(x, y, z)\phi(x, y, z)}{1 + S\phi(x, y, z)\,dV}\right]}\,dV \qquad \text{(B4.6.5)}$$

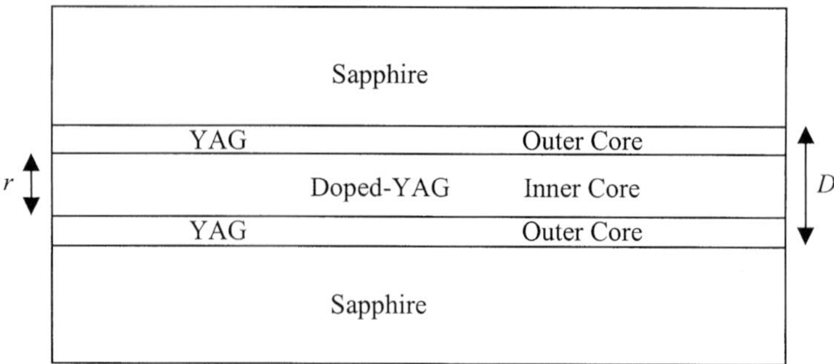

Figure B4.6.9. Double-clad planar waveguide for spatial-mode selection.

where $d(x, y, z)$ is the spatial profile of the doping and S gives a measure of the degree of saturation due to the intracavity laser power. For a centrally doped waveguide the fundamental mode reaches threshold first as it has the best spatial overlap with the doped region. Figure B4.6.10 then shows how G_{rel} varies for the ten lowest order modes of a step-index waveguide for conditions of high saturation. It can be seen that doping up to 60% of the core of a multimode step-index waveguide gives preferential gain for the fundamental mode and that, when lasing on the fundamental mode, the higher-order modes cannot reach threshold, even under conditions of high saturation [33]. In this way the absorption length is increased by less than a factor of two and fundamental-spatial-mode output is robustly selected. Coupling between the lasing fundamental mode and higher-order modes is very low due to the short and straight nature of the planar waveguide. One limitation to this technique is that any refractive index change caused by the rare-earth doping of the inner-core must be kept below the level at which this region becomes an independent waveguide. Thus for deeper waveguides compensatory doping of the outer-core to reduce the index difference may be necessary.

Self-imaging waveguides. Another technique of use for mode selection in the guided axis is that of self-imaging [34, 35]. This technique relies on the fact that an optical beam of a given spatial profile will periodically re-image within a multimode step-index waveguide with a period,

$$L = \frac{4n_1 D^2}{\lambda}. \tag{B4.6.6}$$

Given laser device lengths up to 10 cm and noting that axially symmetric profiles re-image for the first time at $L/4$ (see figure B4.6.11), implies effective use for core depths of up to a few hundred microns. This is an order of magnitude larger in scale than double-clad structures have currently reached and the design also maintains the normal pump absorption length. Self-imaging can be applied in practice to amplifiers [25, 36] and laser oscillators [37] by launching an undersized, diffraction-limited Gaussian beam at one end of the waveguide, which will couple to higher-order modes and efficiently sweep out the uniformly pumped core, before re-imaging to a diffraction-limited output at the end of the guide. Some degradation of the re-imaging process may occur in high-power systems due to the effects of gain saturation and thermal lensing, as well as simple misalignment of the input beam, but with careful design these effects can be minimized [38].

Tapered waveguides. One technique that can provide diffraction-limited output in both axes is to use a tapered waveguide. The basic principle (see figure B4.6.12) is that a multimode broad-stripe guide, compatible with diode pumping, is connected to a single-mode channel *via* a taper (in the lateral axis) designed to give minimal mode coupling between the fundamental mode and higher-order modes [39]. Thus, low-loss laser

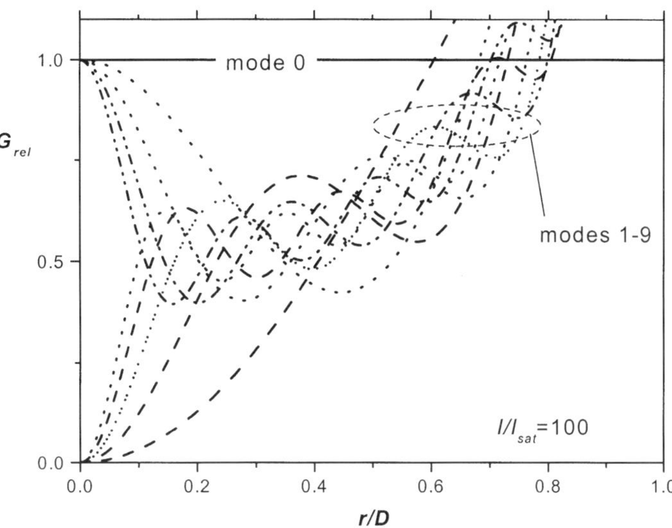

Figure B4.6.10. Relative gain for the ten lowest order modes against doping fraction, for a heavily saturated gain region (reproduced from [33] with permission from the Optical Society of America).

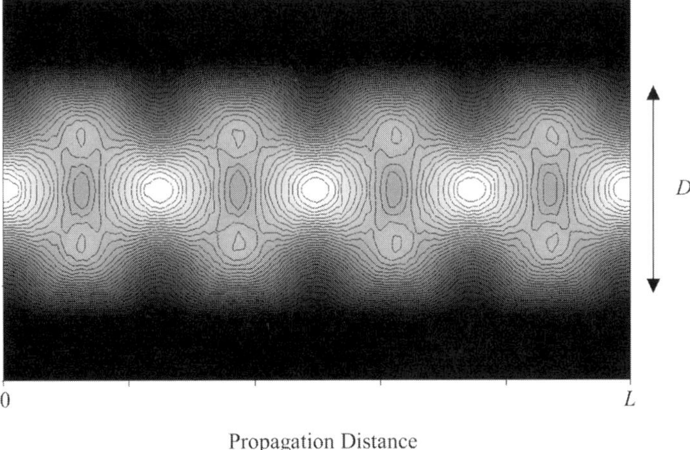

Figure B4.6.11. Theoretical plot for the field contours of a self-imaging, multimode, step-index waveguide.

operation is obtained in the fundamental mode, while efficiently pumping with multimode diode sources. A design condition, set to minimize mode coupling in a parabolic-shaped horn, has been given as [40]

$$L = \frac{n_1 D_{\text{max}}^2}{2\lambda} \tag{B4.6.7}$$

where D_{max} is the maximum width that the taper can open out to, in a length L. This has a similar form to the self-imaging condition of equation (B4.6.6) and imposes much the same sort of restriction, i.e. if the device is a few centimetres long then the taper can only open to a few hundred microns. Thus, this sort of waveguide is most suited to end-pumping with broad-stripe diodes, with the diode slow axis arranged

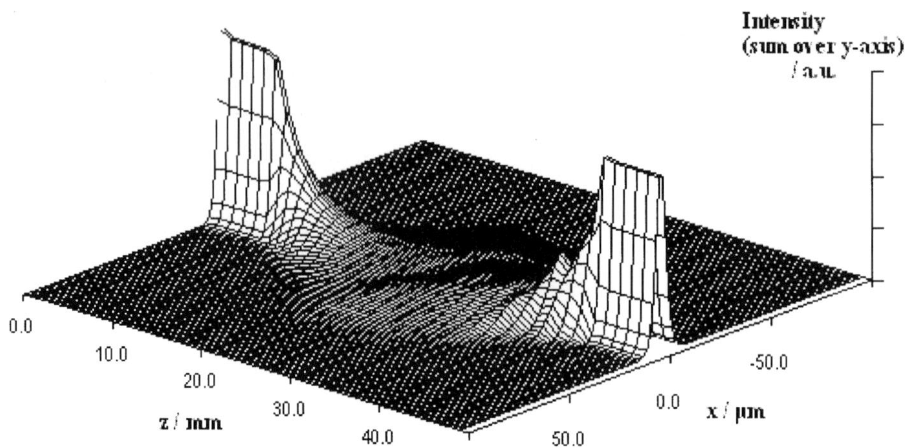

Figure B4.6.12. Theoretical plot of the mode propagation in a double-pass of a tapered planar waveguide laser (reproduced from [31] with permission from IOP publishing).

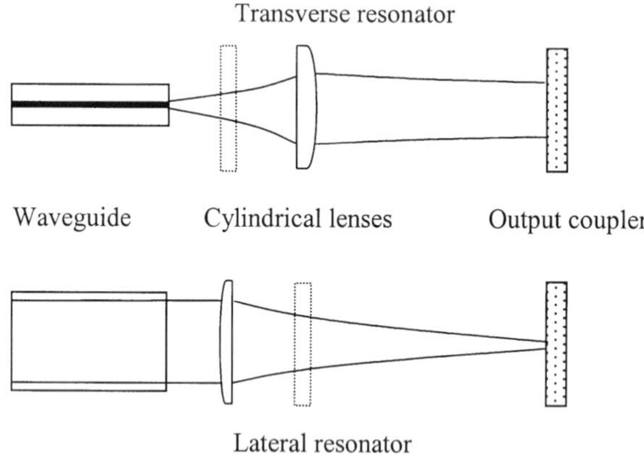

Figure B4.6.13. Schematic diagram of an extended stable cavity waveguide laser.

to coincide with the broad width of the taper and the diffraction-limited axis coinciding with the depth. It should be noted that end-pumping of monolithic thin-film waveguides with broad-stripe diode sources can give multiwatt diffraction-limited output without any need for structuring in the non-guided plane, as the focused pump can be kept small enough to give a good match to the fundamental mode size [23]. Thus, the use of a tapered structure is perhaps most useful for the production of an integrated laser source fabricated on the same substrate as, and connecting to, further waveguide circuits.

External resonator waveguides. An external resonator can be useful in controlling the mode of the waveguide laser in both axes, but is especially useful for the non-guided axis in face- or side-pumped waveguide lasers, where the large pump mode size leads to multimode operation in monolithic devices. Figure B4.6.13 shows a stable extended cavity applied to a side-pumped, double-clad, Nd:YAG waveguide laser. In the fast axis of the waveguide the fundamental mode selected by the double-clad structure is simply collimated onto an external plane mirror and diffraction-limited output ($M^2 < 1.2$) is easily obtained. In order

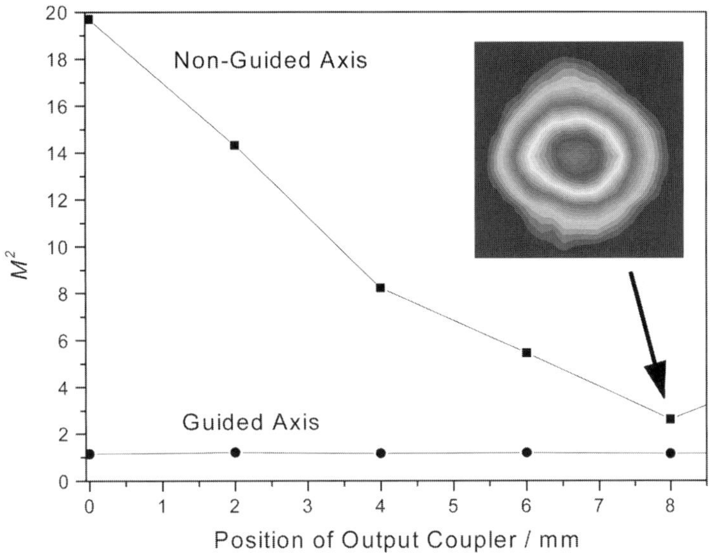

Figure B4.6.14. Plot of M^2 against position of output coupler for the extended stable cavity and CCD image of the collimated output spatial profile.

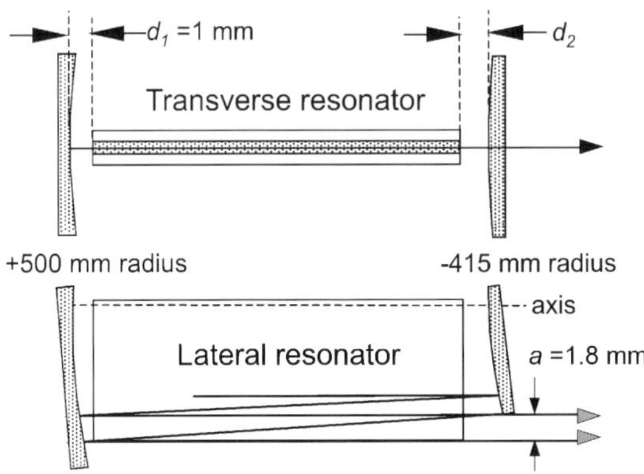

Figure B4.6.15. Hybrid waveguide-unstable resonator (reproduced from [29] with permission from Elservier).

to achieve a large fundamental mode size that efficiently sweeps out the gain in the 5 mm wide, non-guided axis of the waveguide, the cavity has to operate relatively close to instability. Nevertheless, as shown in figure B4.6.14 the M^2 of the output can be reduced to <3 in this axis. Stable extended resonators have also been applied to side-pumped, multimode, channel waveguide lasers [41], where the free-space propagating resonator mode is closely matched to the fundamental mode size of the large channel, again leading to low M^2 output.

Alternatively, a hybrid waveguide-unstable resonator design can be used [29] as shown in figure B4.6.15. In this scheme, low-order guided (transverse) mode operation is selected, as the low-order modes have the lowest recoupling loss from the waveguide to the mirrors and back. In the non-guided (lateral) axis the cavity corresponds to an off-axis, positive branch, confocal unstable resonator (see section A2.1.7). For this

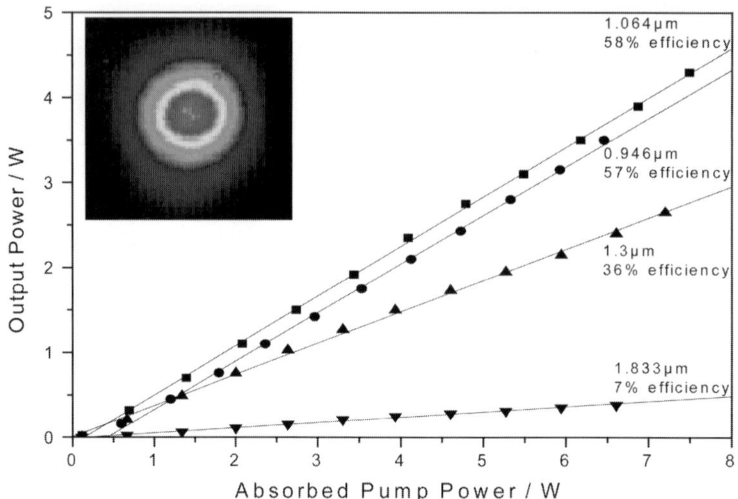

Figure B4.6.16. Output power characteristics of an end-pumped double-clad Nd:YAG waveguide laser, and CCD image of the collimated, near-diffraction-limited, output profile.

configuration the mirror radii are related by

$$R_1 - R_2 = 2\left(\frac{L}{n_1} + d_1 + d_2\right) \tag{B4.6.8}$$

and the effective output coupling is given by $(1 - R_1/R_2)$. This type of resonator has been used at output powers >100 W, sweeping out a gain width of ~ 1 cm, and delivering M^2 values of 4.4 in the unguided axis and 8.6 in the guided axis [26].

B4.6.5 Waveguide laser performance

End-pumping. Fully diffraction-limited, multiwatt performance is most easily achieved *via* the end-pumped configuration. Using a 10 W beam-shaped broad-stripe diode source at 807 nm ($M^2 = 17$ in both axes), a monolithic, Nd:YAG double-clad waveguide laser has delivered the performance shown in figure B4.6.16 [23]. The high efficiency ($\sim$60%) for the 1.064 μm and 946 nm transitions and the low M^2 values (1.0 by 1.3) are characteristic of end-pumped waveguides and the fact that the same double-clad structure can provide fundamental-mode selection across a wide range of wavelengths is an interesting feature. It can also be seen that the high gain provided by the waveguide structure allows laser operation of rather weak transitions, such as 1.833 μm in Nd:YAG. In this particular case the output coupling was not optimized and it was calculated that over 1 W of power could be available at this wavelength.

Passively Q-switched operation has also been demonstrated in the end-pumped configuration in both Nd and Yb-doped YAG, monolithic, double-clad waveguide lasers [22]. Using a similar direct-bonded structure to that shown in figure B4.6.9, but with a small part of the doped inner core being replaced by a Cr^{4+}-doped section to act as a saturable absorber, the results shown in figure B4.6.17 were obtained for the Yb:YAG guide. Although the double-clad design is strictly based upon cw operation the output was still found to be near-diffraction-limited with M^2 values of 1.3 (guided-axis) by 1.5. In this way peak powers of up to 18 kW were available (30 μJ pulses with 1.6 ns duration) in a linearly polarized beam.

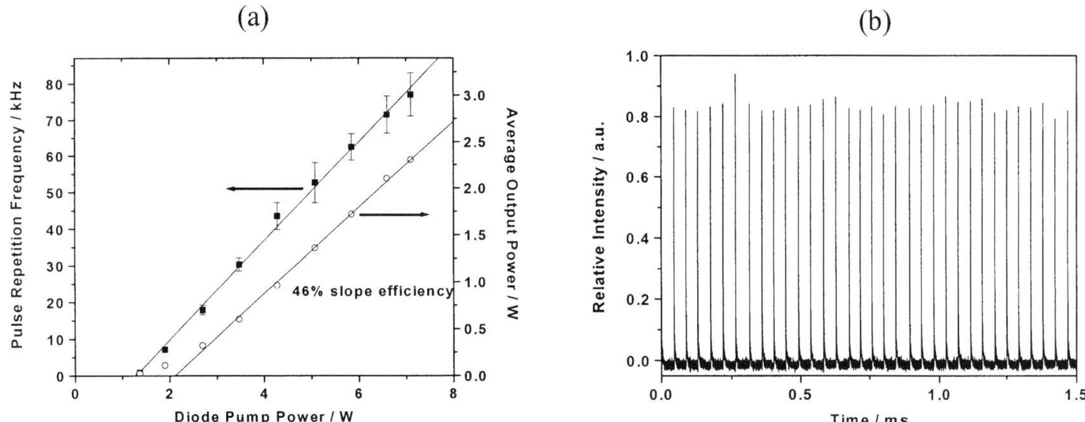

Figure B4.6.17. Output performance characteristics of an end-pumped passively Q-switched Yb:YAG waveguide laser (reproduced from [22]with permission from the Optical Society of America): (*a*) repetition rate and output power against diode pump power, (*b*) typical output pulse train.

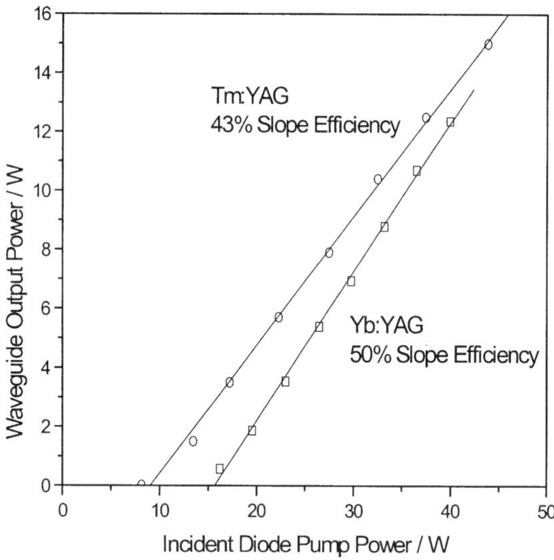

Figure B4.6.18. Output power against incident diode pump power for proximity-coupled, side-pumped double-clad waveguide lasers (reproduced from [31] with permission from IOP Publishing).

Side pumping. To date, side pumping has been restricted to pumping with diode bars rather than diode stacks. Thus, pump powers of up to 40 W, pumping both sides of 1 cm long waveguides with 20 W diode bars, have been achieved. Initial demonstrations of side pumping were carried out using Nd:YAG guides fabricated by liquid-phase epitaxy [42], but since then most work has used double-clad waveguides fabricated by direct bonding. Side pumping has been demonstrated in Nd, Yb and Tm-doped YAG double-clad waveguides in various configurations. The proximity-coupled results for Tm at 2.02 μm [43] and Yb at 1.03 μm [44] are shown in figure B4.6.18, with both demonstrating high efficiencies with respect to incident diode power. Using a monolithic cavity, the output is diffraction-limited in the guided axis but >100 times diffraction

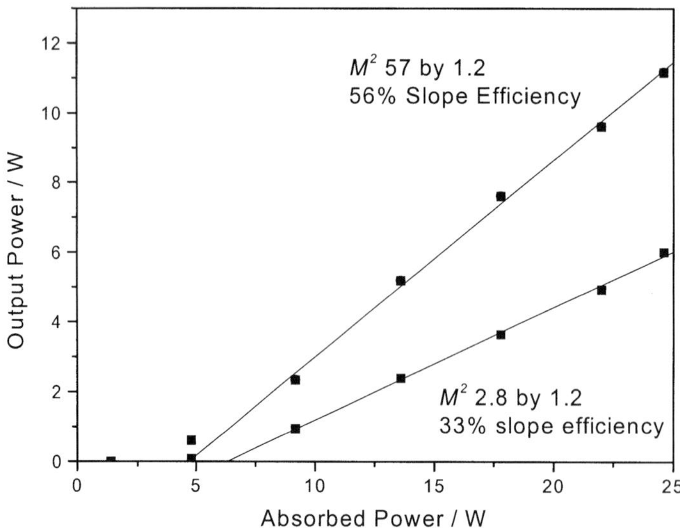

Figure B4.6.19. Output power against absorbed diode pump power for an extended stable cavity, side-pumped, double-clad waveguide laser, with the cavity length set for multimode and single-mode operation.

limited in the non-guided axis. Side-pumped passively Q-switched operation has also been demonstrated using a similar guide to that discussed in the previous section but based on Nd rather than Yb-doped YAG. Once again using proximity coupling of two diode bars and a monolithic laser cavity, average powers up to 8 W were obtained with peak pulse powers of nearly 30 kW [44].

An extended stable cavity of the design shown in B4.6.13, based on a side-pumped Nd:YAG double-clad waveguide operating at 1.064 μm, gave the performance shown in figure B4.6.19. These results show how implementing near-diffraction-limited performance can reduce the slope efficiency due to the fact that the now near-Gaussian laser mode has a worse spatial overlap with the uniformly pumped slab than the highly multimode case. In order to avoid this effect a cavity could be designed that takes a zigzag path through the unguided axis to efficiently sweep out the gain, as has been demonstrated in bulk systems [45]. Similar slope efficiencies and low output M^2 have been obtained with side-pumped, Yb:YAG, multimode channel [41] and thin-film waveguides [46] using extended stable cavities.

Face pumping. The highest output powers obtained to date from a planar waveguide have been from the face-pumped configuration shown in figure B4.6.7. Using plane resonator mirrors placed near to the end faces of the Nd:YAG waveguide (6 cm long, 1.1 cm wide and 200 μm-thick core) a maximum of 150 W of multimode output was obtained for 430 W of pump power [26]. Employing a hybrid waveguide-unstable resonator cavity, as shown in figure B4.6.15, the M^2 values were reduced to 8.6 (guided) by 4.4 for output powers of up to 110 W. A comparison of the multimode and unstable resonator output power performance is shown in figure B4.6.20 indicating that ~88% of the multimode output power can be converted into the high-brightness beam. In more recent work, using a hybrid-unstable resonator of the negative branch confocal type, M^2 values as low as 1.8 by 1.1 have been obtained at powers of 110 W [30].

B4.6.6 Future prospects

Rapid development in the power scaling of planar slab waveguides is currently in progress. Output powers in the kilowatt regime appear to be obtainable in the near future from either face-pumped or side-pumped

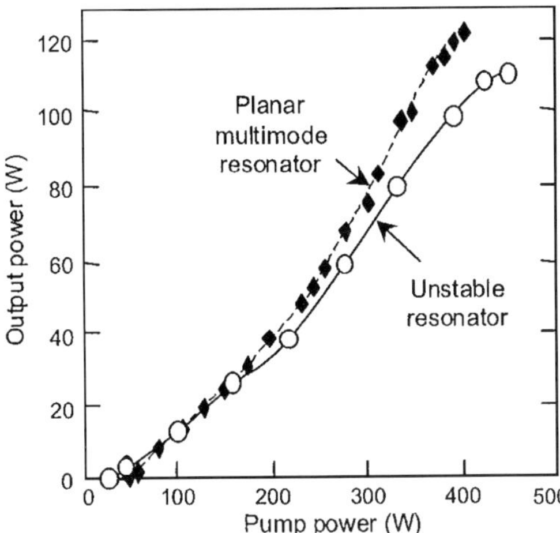

Figure B4.6.20. Comparison of the output power for a multimode plane mirror and unstable resonator configurations for a face-pumped Nd:YAG waveguide laser (reproduced from [26] with permission from the Optical Society of America).

configurations due to the excellent power handling capabilities of ultra-thin, uniformly pumped slabs. Despite the requirement of using multimoded waveguides, several technologies exist that can deliver high-brightness output beams with good optical efficiencies. These include waveguiding techniques such as tapers, double-cladding and self-imaging, as well as the use of stable and unstable external resonators. The high gains generated in the thin planar films can cause problems with amplified spontaneous emission and parasitic lasing and so these structures would appear to be more suited to cw or high-repetition-rate pulsed outputs than for high-energy storage. The natural compatibility of the planar waveguide geometry with high-power diode bars and stacks allows rather compact devices and relatively simple coupling optics. Thus, waveguide lasers operating in the kilowatt power regime should find numerous uses in scientific, industrial and military applications, while waveguides of a few watts to a few hundred watts may also find applications delivering efficient operation of unusual low-gain transitions, such as 1.833 μm operation of Nd:YAG, and integrated laser sources for telecommunications.

References

[1] Ramaswamy R V and Srivastava R 1988 Ion exchanged glass waveguides: a review *IEEE J. Lightwave Technol.* **6** 984–1002

[2] Chandler P J, Zhang L and Townsend P D 1991 Ion-implanted waveguides in laser host materials *Nucl. Instrum. Methods* B **59** 1223–7

[3] Schmidt R V and Kaminow I P 1974 Metal-diffused optical waveguides *Appl. Phys. Lett.* **25** 458–60

[4] Svalgaard M, Poulsen C V, Bjarklev A and Poulsen O 1994 Direct UV writing of buried single mode channel waveguides in Ge-doped silica films *Electron. Lett.* **30** 1401–3

[5] Brown C T A, Bonner C L, Warburton T J, Shepherd D P, Tropper A C and Hanna D C 1997 Thermally bonded planar waveguide lasers *Appl. Phys. Lett.* **71** 1139–41

[6] Kawachi M 1990 Silica wave-guides on silicon and their application to integrated-optic components *Opt. Quantum Electron.* **22** 391–416

[7] Chartier I, Ferrand B, Pelenc D, Field S J, Hanna D C, Large A C, Shepherd D P and Tropper A C 1992 Growth and low-threshold laser oscillation of an epitaxially grown Nd:YAG waveguide *Opt. Lett.* **17** 810–12

[8] Anderson A A, Eason R W, Hickey L M B, Jelinek M, Grivas C, Gill D S and Vainos N A 1997 Ti:sapphire planar waveguide laser grown by pulsed laser deposition *Opt. Lett.* **22** 1556–8

[9] Becker C *et al* 2000 Advanced Ti:Er:LiNbO$_3$ waveguide lasers *IEEE J. Quantum Electron.* **6** 101–13

[10] Bhutta T, Chardon A M, Shepherd D P, Daran E, Serrano C and Muñoz-Yagüe A 2001 Low phonon energy, Nd:LaF$_3$ channel waveguide lasers fabricated by molecular beam epitaxy *IEEE J. Quantum Electron.* **37** 1329–37

[11] Bonner C L, Bhutta T, Shepherd D P and Tropper A C 2000 Double-clad structures and proximity coupling for diode-bar-pumped planar waveguide lasers *IEEE J. Quantum Electron.* **36** 236–42

[12] Hodgson N, Ter-Mikirtychev V V, Hoffman H J and Jordon W 2002 Diode-pumped, 220 W ultra-thin slab Nd:YAG laser with near-diffraction-limited beam quality *Advanced Solid-State Lasers (OSA Trends in Optics and Photonics Series 68)* ed M E Fermann and L R Marshall (Washington, DC: Optical Society of America) pp 552–7

[13] Graf Th and Balmer J E 1993 High-power Nd:YLF laser end pumped by a diode-laser bar *Opt. Lett.* **18** 1317–19

[14] Beach R J 1996 Theory and optimization of lens ducts *Appl. Opt.* **35** 2005–15

[15] Clarkson W A and Hanna D C 1996 Two-mirror beam-shaping technique for high-power diode bars *Opt. Lett.* **21** 375–7

[16] Snitzer E, Po H, Hakimi F, Tumminelli R and McCollum B C 1988 Double-clad offset core Nd fibre laser *Dig. Conf. On Optical Fibre Sensors* paper PD5 (New York: IEEE)

[17] Minelly J D, Zenteno L A, Dejneka M J, Miller W J, Kuksenkov D V, Davis M K, Crigler S G and Bardo M E 2000 High power diode-pumped single-transverse-mode Yb fiber laser operating at 978 nm *Dig. Optical Fiber Communication Conf.* paper PD2 (Washington, DC: Optical Society of America)

[18] Bibeau C, Beach R J, Mitchell S C, Emanuel M A, Skidmore J, Ebbers C A, Sutton S B and Jancaitis K S 1998 High-average-power 1-μm performance and frequency conversion of a diode-end-pumped Yb:YAG laser *IEEE J. Quantum Electron.* **34** 2010–19

[19] Faulstich A, Baker H J and Hall D R 1996 Face-pumping of thin, solid-state slab lasers with laser diodes *Opt. Lett.* **21** 594–6

[20] Bonner C L, Anderson A A, Eason R W, Shepherd D P, Gill D S, Grivas C and Vainos N 1997 Performance of a low-loss pulsed-laser-deposited Nd:Gd_3 Ga_5O_{12} waveguide laser at 1.06 and 0.94 μm *Opt. Lett.* **22** 988–90

[21] Griebner U, Grunwald R, Schönnagel H, Huschke J and Erbert G 2000 Laser with guided pump and free-propagating resonator mode using diffusion-bonded rectangular channel waveguides *Appl. Phys. Lett.* **77** 3505–7

[22] Mackenzie J I and Shepherd D P 2002 End-pumped, passively Q-switched Yb:YAG double-clad waveguide laser *Opt. Lett.* **27** 2161–3

[23] Mackenzie J I, Li C and Shepherd D P 2003 Multi-watt, high-efficiency, diffraction-limited Nd:YAG planar waveguide laser *IEEE J. Quantum Electron.* **39** 493–500

[24] Mackenzie J I, Li C, Shepherd D P, Beach R J and Mitchell S C 2002 Modelling of high-power continuous-wave Tm:YAG side-pumped double-clad waveguide lasers *IEEE J. Quantum Electron.* **38** 222–30

[25] Baker H J, Chesworth A, Pelaez Millas D and Hall D R 1998 Power scaling of thin Nd:glass and Nd:YAG slab lasers, face-pumped by laser diodes *Advanced Solid-State Lasers (Trends in Optics and Photonics Series 19)* ed W R Besenburg and M M Fejer (Washington, DC: Optical Society of America) pp 407–10

[26] Lee J R, Baker H J, Friel G J, Hilton G J and Hall D R 2002 High-average-power Nd:YAG planar waveguide laser that is face pumped by 10 laser diode bars *Opt. Lett.* **27** 524–6

[27] Koechner W 1996 *Solid-State Laser Engineering* 4th edn (Berlin: Springer)

[28] Eggleston J M, Kane T J, Kuhn K, Unternahrer J and Byer R L 1984 The slab geometry laser, part I: theory *IEEE J. Quantum Electron.* **20** 289–301

[29] Baker H J, Chesworth A A, Pelaez Millas D and Hall D R 2001 A planar waveguide Nd:YAG laser with a hybrid waveguide-unstable resonator *Opt. Commun.* **191** 125–31

[30] Lee J R, Baker H J and Hall D R 2002 Thermal guiding, birefringence and brightness scaling of high average power, planar waveguide devices *Advanced Solid-State Lasers (OSA Trends in Optics and Photonics Series 68)* ed M E Fermann and L R Marshall (Washington, DC: Optical Society of America) pp 529–34

[31] Shepherd D P, Hettrick S J, Li C, Mackenzie J I, Beach R J, Mitchell S C and Meissner H E 2002 High-power planar dielectric waveguide lasers *J. Phys. D: Appl. Phys.* **34** 2420–32

[32] Rutherford T S, Tulloch W M, Gustafson E K and Byer R L 2000 Edge-pumped quasi-three-level slab lasers: design and power scaling *IEEE J. Quantum Electron.* **36** 205–19

[33] Bhutta T, Mackenzie J I, Shepherd D P and Beach R J, 2002 Spatial dopant profiles for transverse-mode selection in multi-mode waveguides *J. Opt. Soc. Am.* B **19** 1539–43

[34] Bryndhal O 1973 Image formation using self-imaging techniques *J. Opt. Soc. Am.* **63** 416–18

[35] Ulrich R and Ankele G 1975 Self-imaging in homogeneous planar optical waveguides *Appl. Phys. Lett.* **27** 337–9

[36] Lee J R, Friel G J, Baker H J, Hilton G J and Hall D R 2001 A Nd:YAG planar waveguide laser operating at 121W output with face-pumping by diode bars, and its use as a power amplifier diodes *Advanced Solid-State Lasers (OSA Trends in Optics and Photonics Series 50)* ed C Marshall (Washington, DC: Optical Society of America) pp 36–40

[37] Pelouch W S, Smith D D, Koroshetz J E, McKinnie I T, Unternährer and Henderson S 2002 Self imaging in waveguide lasers and amplifier diodes *Advanced Solid-State Lasers (OSA Trends in Optics and Photonics Series 68)* ed M E Fermann and L R Marshall (Washington, DC: Optical Society of America) pp 6–9

[38] Baker H J, Lee J R and Hall D R 2002 Self-imaging and high-beam-quality operation in multi-mode planar waveguide optical amplifiers *Opt. Express* **10** 297–302

[39] Hettrick S J, Mackenzie J I, Harris R D, Wilkinson J S, Shepherd D P and Tropper A C 2000 Ion-exchanged tapered waveguide laser in neodymium-doped BK7 glass *Opt. Lett.* **25** 1433–5

[40] Milton A F and Burns W K 1977 Mode coupling in optical waveguide horns *IEEE J. Quantum Electron.* **13** 828–35

[41] Griebner U and Schönnagel H 1999 Laser operation with nearly diffraction-limited output from a Yb:YAG multi-mode channel waveguide *Opt. Lett.* **24** 750–2

[42] Hanna D C, Large A C, Shepherd D P, Tropper A C, Chartier I, Ferrand B and Pelenc D 1992 A side-pumped Nd:YAG epitaxial waveguide laser *Opt. Commun.* **91** 229–35

[43] Mackenzie J I, Mitchell S C, Beach R J, Meissner H E and Shepherd D P 2001 15W Diode-side-pumped Tm:YAG waveguide Laser at 2 μm *Electron. Lett.* **37** 898–9

[44] Beach R J, Mitchell S C, Meissner H E, Meissner O R, Krupke W F, McMahon J M, Bennett W J and Shepherd D P 2001 Continuous wave and passively Q-switched cladding-pumped planar waveguide lasers *Opt. Lett.* **26** 881–3

[45] Snell K J, Lee D, Wall K F and Moulton P F 2000 Diode-pumped, high-power CW and modelocked Nd:YLF lasers *Advanced Solid-State Lasers (OSA Trends in Optics and Photonics Series 34)* ed H Injeyan, U Keller and C Marshall (Washington, DC: Optical Society of America) pp 55–9

[46] Griebner U, Grunwald R and Schönnagel H 1999 Thermally bonded Yb:YAG planar waveguide laser *Opt. Commun.* **164** 185–90

Further reading

Lee D L 1986 *Electromagnetic Principles of Integrated Optics* (New York: Wiley)

Develops the concepts and principles of guided wave optics starting from Maxwell's equations and includes a useful introduction to many waveguide fabrication techniques and devices.

Koechner W 1996 *Solid-State Laser Engineering* 4th edn (Berlin: Springer)

Thorough coverage of characteristics, design and performance of a wide variety of solid-state lasers, including a good coverage of thermo-optic effects and heat removal.

Siegman A E 1986 *Lasers* (Sausalito: University Science Books)

A comprehensive textbook giving a detailed treatment of laser physics.

B5
Other lasers

Colin Webb

To explore the spectral regions where most atoms and molecules have their strongest absorption transitions—e.g. those due to electrons changing orbitals—has required the development of sources which are tunable in the UV, XUV and x-ray regions. Likewise to explore the regions where transitions between different vibrational and/or rotational levels of molecules cause absorption requires sources tunable in the infrared and far-infrared regions.

One type of source which has the potential to provide radiation tunable over this entire range of wavelengths is the synchrotron and its closely related laser counterpart, the free electron laser (FEL). Although high-power synchrotron radiation sources have been available at various national facilities for the past half century, it is only in the last 15 years or so that FELs have begun to make their mark. Because both synchrotrons and FELs depend on fairly large and expensive electron accelerator technology, they tend to be built as shared national or regional facilities although there is some prospect that more affordable versions may become available in the future.

The ability to generate x-rays in the form of short pulses of narrow-band laser output on specific transitions of highly ionized atoms, pumped by radiation from conventional high-power near-visible laser sources has been realized at several national laboratories. With the advent of commercial ultrashort pulsed laser and amplifier systems, the prospect of tabletop versions of these systems, within the financial reach of many university laboratories, is now becoming a reality.

For the past 30 years or more, the work-horse tunable laser, available in both cw and pulsed forms, has been the liquid dye laser. The dye laser does not convert electrical power directly into laser radiation but relies on radiation from a fixed wavelength laser or a flashtube for excitation. Though reviled by many for the messiness involved in making up and replacing the dye solution, the fact that the thermally perturbed laser medium is bodily transported out of the lasing region to be replaced by optically uniform, cool medium means that the power-handling capability of the dye laser can be scaled up to several kW of average power. For the past 20 years or so, dye lasers of different types have been available to cover the spectral region from near UV through the visible to the near infrared, and the ability of a single dye–pump laser combination to tune across a spectral region 40–100 nm wide enabled them to gain an early hold as the laser of choice for many physical chemistry laboratories. However, this dominant position is now being challenged by various forms of tunable solid state laser (such as Ti:Sapphire) whose fundamental and harmonically generated tunable ranges cover most (but not all) of the same spectral region.

The idea of encapsulating the dye medium in a solid host (plastic or porous glass) which is clean to handle and does not require the circulating pump and plumbing associated with liquid dye systems is attractive from several points of view. True, the inability of the solid state hosts to dissipate heat does tend to limit both the repetition rate capability and useful lifetime of such devices compared to their liquid dye counterparts. However, there are particular applications in which these features are not overwhelming considerations, and for which the solid state dye laser provides a simple low-cost answer.

B5.1
Free electron lasers and synchrotron light sources

P G O'Shea and J B Murphy

B5.1.1 Introduction

For many years it has been a goal of optical researchers to produce coherent radiation at wavelengths where existing sources are weak. Free-electron lasers (FELs) are rapidly filling voids in the electromagnetic spectrum left by more conventional sources. FELs are attractive because of their potential to produce both high-peak and high-average power and because of their wavelength flexibility. When compared to conventional lasers, FELs are neither simple nor inexpensive. FELs can, however, produce intense coherent light at wavelengths where other sources are either weak or non-existent. This makes them the ideal choice when one moves away from the visible region of the spectrum to the far IR, or to the vacuum UV, x-ray and beyond. At present FELs operate over a broad spectrum from millimetre waves to the vacuum ultraviolet and at average power levels from watts to several kilowatts. No single device spans this entire range of wavelength or power. In spite of having the same underlying physical principles, the technology behind mm-wave FELs is quite different from that of UV devices.

The development of synchrotron radiation (SR) light sources has, until recently, been independent of the development of FELs. Therefore, for the most part, FELs and SR sources are treated independently in this chapter. Heretofore, FELs and synchrotron sources have not been in direct competition because they have occupied different regions of the electromagnetic spectrum. X-ray FELs currently proposed will offer direct competition to synchrotron sources (see sections B5.1.5 and B5.1.6). Such x-ray FELs are now often referred to as *fourth-generation synchrotron light sources*. Therefore, it is likely that the distinction between FELs and SR sources will become increasingly blurred, particularly at x-ray wavelengths.

B5.1.1.1 History and basic principles of FELs

FELs are a marriage of accelerator and laser technology. The essential technical elements are an electron accelerator, a magnetic undulator (or wiggler)[1] and photon optics. FEL operation is based on a fundamental radiative mechanism—that of an oscillating electron in a sinusoidal magnetic field of an undulator magnet—which we call spontaneous emission (see figure B5.1.1). This radiation is enhanced or amplified by the modulation of the electron beam density. In low-gain systems, an optical cavity provides the feedback and the optical mode selection necessary for proper performance; in high-gain FELs an optical cavity is often unnecessary. Typical undulators in use today are 1–5 m long and have less than 100 magnetic periods, each several centimetres long. Future undulators for x-ray FELs may be as long as 100 m.

The history of FELs can be traced back to microwave tubes, which were the first sources of coherent radiation from free electrons in vacuum. Such early devices relied on slow-wave structures with closely

[1] The terms *wiggler* and *undulator* are often used interchangeably in the FEL community; however, they have quite specific meaning in the synchrotron radiation community, see figure B5.1.7.

coupled boundary conditions and were limited to long wavelengths. The possibility of producing coherent radiation from electrons bending in a magnetic field without the necessity of boundary structures was first postulated by Schwinger in relation to synchrotron radiation [1]. In the 1950s, Motz [2] performed a series of experiments with a high-energy electron beam passing through an undulator-like device. These experiments produced incoherent emission in the visible. Between 1957 and 1964, Phillips demonstrated an FEL-like microwave tube called the ubitron (for 'undulating beam interaction') in which a bunched electron beam in an undulator produced 150 kW of coherent radiation at a wavelength of 5 mm [3]. In 1971, Madey [4] proposed an optical wavelength device, subsequently referred to as an FEL, which included a conventional optical resonator. Madey's proposal led directly to a series of seminal experiments that demonstrated FEL amplification [5] and oscillation [6] at near-IR wavelengths.

Is the FEL a laser? This question was originally addressed by Motz [7]. There is no doubt that the FEL mechanism is quite different from that of a conventional laser. FEL light output is, however, narrow band, transversely coherent and has well-defined spatial and temporal mode structures. In the optical region of the spectrum, FEL light looks very similar to conventional laser light. Madey originally coined the term free-electron laser to describe a device operating in the optical region with an optical resonator [8]. Today, the use of the term has been extended to include almost any device that produces coherent radiation from free electrons at any wavelength and which does not use closely coupled resonant structures.

B5.1.2 Physics of FELs

In an FEL, electron energy transitions occur in a continuum. Consequently, the discrete level transitions that exist in conventional lasers are not a factor. The decoupling of the FEL mechanism from the constraints of atomic and molecular media allows for a light source that is continuously tunable and potentially capable of both high peak and high average power. Note that in a conventional laser the thermal energy associated with heating of the lasing medium is carried away at acoustic velocities ($\approx 10^3$ m s^{-1}), whereas in an FEL the heat is carried away at the speed of the electrons ($>10^8$ m s^{-1}). Therefore, if sufficient electron beam power is available, FELs should be capable of extremely high average power operation.

The key step in an FEL is the generation of *spontaneous emission*. In an FEL the *lasing medium* is the electron bunch (10^8–10^{10} electrons typically). Consider a relativistic electron travelling in the $\hat{z}$ direction encountering a region with a transverse sinusoidal magnetic field $B = B_0 \sin(k_u z)\hat{x}$ as shown in figure B5.1.1. The undulator period ($\lambda_u = 2\pi/k_u$) is typically a few cm and B_0 is usually in the range 0.1–0.5 T. In this chapter, we consider only planar undulators such as that shown in figure B5.1.1. Planar undulators made with permanent magnets or electromagnets are by far the most common variety. The electrons wiggle in the $\hat{y}$ direction as they move through the field region. In the electron rest-frame, the undulator field is transformed into an intense virtual photon field with orthogonal electric and magnetic fields.

These virtual photons become real photons by Compton scattering off the relativistic electrons and are Doppler shifted to short wavelength when viewed in the laboratory frame. In the laboratory frame the real photons have wavelength

$$\lambda_r \approx \frac{\lambda_u}{2h\gamma^2}(1 + a_u^2) \tag{B5.1.1}$$

where $h = 1, 2, 3, \ldots$ is the harmonic number, $a_u = (eB_0/\sqrt{2}mck_u)$ is known as the rms normalized vector potential of the undulator (a_u is usually of order unity) and γ is the electron energy divided by the rest energy of the electron. For a 50 MeV electron $\gamma \approx 100$ and the light wavelength is a factor of 10^4 smaller than the undulator period. Most FELs operate at harmonic number $h = 1$; however, a number of FELs have operated successfully at the third ($h = 3$) and fifth ($h = 5$) harmonic. A second-harmonic FEL has recently been demonstrated but because the spontaneous emission and gain functions are zero on the electron beam axis, lasing was quite difficult to achieve [9].

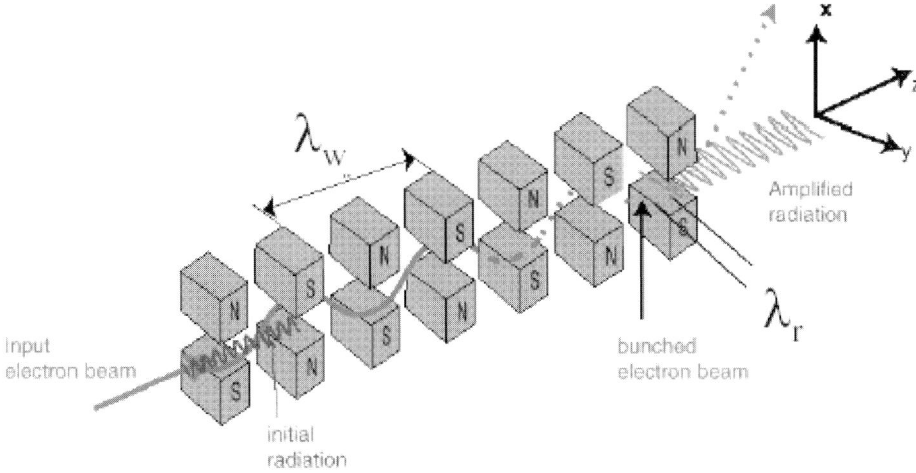

Figure B5.1.1. The free-electron laser radiative process: As the electron beam moves in a magnetic undulator it radiates light (full line with long undulations). Initially, the electron beam is unstructured on a scale comparable to the wavelength of emission. The electron emitters are randomly phased with respect to the optical wave and so incoherent or spontaneous emission results. As the electrons progress along the wiggler, their interaction with the optical field in the presence of the undulator field results in microbunching of the electrons (broken line). At saturation, the electrons become tightly bunched in clumps spaced one optical wavelength (λ) apart. Once the electrons are bunched, the emission becomes coherent and laser-like. After passage through the undulator, the electrons are separated from the light by a bending magnet (not shown).

B5.1.2.1 Spontaneous emission

The spontaneous emission is generally not very intense. A 50 MeV electron beam with a peak current of 100 A in a pulse several picoseconds long will produce a peak spontaneous emission power of the order of 1 W in a typical undulator. The intensity of the spontaneous radiation *on axis* near the fundamental ($h = 1$) resonant wavelength is given by [10]

$$\frac{\mathrm{d}^2 I}{\mathrm{d}\omega\,\mathrm{d}\Omega} = \frac{N_{\mathrm{u}}^2 e^2 \gamma^2}{2\pi\varepsilon_0 c} \frac{a_{\mathrm{u}}^2}{\gamma(1+a_{\mathrm{u}}^2)} \left(\frac{\sin(\pi N_{\mathrm{u}} \Delta\lambda/\lambda_{\mathrm{r}})}{\pi N_{\mathrm{u}} \Delta\lambda/\lambda_{\mathrm{r}}} \right)^2 F^2[1,\xi]. \tag{B5.1.2}$$

where $F[1,\xi] = J_0(\xi) - J_1(\xi)$, J_x is a Bessel function of order x and $\xi = a_{\mathrm{u}}^2/2(1+a_{\mathrm{u}}^2)$.

The bandwidth of the emission ($\Delta\lambda/\lambda$) is approximately $1/2N_{\mathrm{u}}$. The effective angular aperture is $\theta = \sqrt{\lambda_{\mathrm{r}}/\lambda_{\mathrm{u}} N_{\mathrm{u}}}$ and the source radius is $\omega = \sqrt{\lambda_{\mathrm{r}}\lambda_{\mathrm{u}} N_{\mathrm{u}}}/4\pi$. The number of photons per electron, per pass through the undulator within the angular aperture and bandwidth is given by

$$N_{\mathrm{p}} = \frac{\pi}{2}\alpha \frac{a_{\mathrm{u}}^2}{(1+a_{\mathrm{u}}^2)} F^2[1,\xi] \tag{B5.1.3}$$

where α is the fine structure constant. For the case of $a_{\mathrm{u}} = 0.5$, $N_{\mathrm{p}} = 2 \times 10^{-3}$ photons are emitted per electron per pass.

For a bunch of electrons, whose bunch length is much longer than λ, the total intensity is proportional to the number of electrons N_{e} in a bunch; typically, $N_{\mathrm{e}} \approx 10^9$. If the electrons were grouped into bunches approximately λ apart in the longitudinal direction, the number of photons emitted by the bunch would be proportional to N_{e}^2. Such a process results in *coherent emission* of photons.

B5.1.2.2 FEL lasing

The essence of the FEL mechanism is the conversion of spontaneous emission into narrow-band coherent emission by means of a feedback mechanism involving the interaction of the spontaneous photons and the electrons in the presence of the undulator magnetic field. The electrons either gain or lose energy depending on their phase with respect to the optical field. This results in a longitudinal clumping of the electrons. In the typical case where the electron pulselength (l) is much greater than the optical wavelength, the electron pulse will become modulated into l/λ_r micropulses spaced approximately λ_r apart. This modulation results in a narrowing of the spectrum and an increase in the output power, i.e. lasing. In low-gain systems, the optical feedback is provided by a resonator cavity, just as in conventional laser oscillators. In high-gain FELs, self-amplified spontaneous emission (SASE) occurs without the necessity of an optical cavity.

In an FEL the gain in the optical field depends on the peak current and the average optical power output depends on the average current. The quality of the electron beam is critical to FEL performance. For optimum gain the electron beam must overlap well with the optical mode. Poor electron beam quality will result in poor overlap and reduced FEL performance.

In order to go from spontaneous emission to saturation a total gain of the order of 10^8 is often required. Typically, low-gain oscillators can take up to several tens of passes to saturate, while high-gain amplifiers can achieve saturation in a single pass. In all cases, there is an incentive to have a large electron beam brightness, i.e. high peak current and small emittance (see section B5.1.3), so as to optimize gain.

The efficiency of FELs depends on the details of the undulator used. For a low-gain FEL with a simple undulator with constant period and constant magnetic field amplitude and N_u magnetic periods, η, the efficiency of conversion of electron energy into light, is approximately $\eta \approx 1/(4N_u)$, a number comparable to the fractional width of the spontaneous emission spectrum. Typically, for simple FEL systems, $\eta \approx$ 1–2%. If the undulator period or magnetic field amplitude is tapered to compensate for the reduction in electron-beam energy as power is extracted from the electrons, then much higher efficiencies can, in principle, be obtained. The present record for FEL efficiency at optical wavelengths is 5% at a wavelength of 10 μm [11]. In principle, efficiencies up to 20% appear feasible. The effective system efficiency can be enhanced by energy recovery of the electron beam in storage rings or in recirculating linac systems.

B5.1.2.3 FEL gain and saturation

We present here a set of formulae that will allow the reader to estimate FEL performance in two particular cases: (a) low-gain oscillators and (b) high-gain SASE or amplifiers. In both cases we deal only with the Compton regime where plasma oscillations of the beam are not important. This regime is representative of almost all short-wavelength FELs. When plasma oscillations are important, the term Raman regime is used. The Raman regime is important for long-wavelength FELs driven by low-energy electron beams.

Figure B5.1.2 shows idealized FEL spontaneous emission and gain functions. The independent variable is $\Omega = N_u\pi(\omega - h\omega_r)/\omega_r$, where $\omega = 2\pi c/\lambda$. The spontaneous emission peaks at the FEL resonant wavelength λ_r and the width of the spectrum to the first zero is $2/(hN_u)$. The gain function is given by the derivative of the spontaneous emission function. The peak optical gain occurs at a wavelength detuned on the long wavelength side of resonance by

$$\frac{\Delta\lambda}{\lambda} = \frac{1}{hN_u}. \tag{B5.1.4}$$

The negative side of the gain function corresponds to a situation where electrons gain energy from the optical field. An FEL operating on the negative side of the gain function is called an inverse FEL (IFEL) and IFELs have been proposed as electron accelerators.

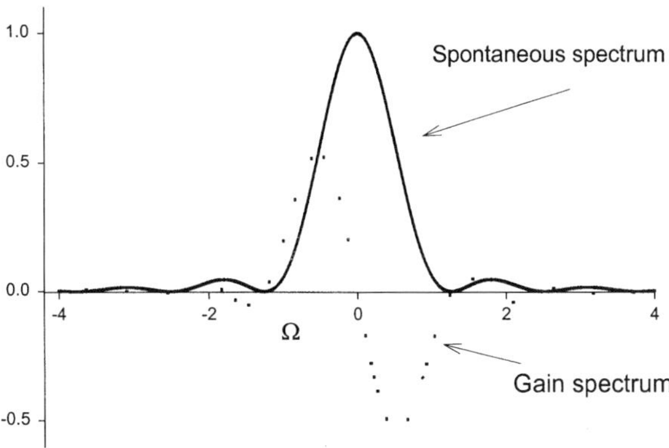

Figure B5.1.2. Distribution of the spontaneous emission and gain functions *versus* $\Omega = N_u\pi(\omega - h\omega_r)/\omega_r$, where $\omega = 2\pi c/\lambda$. Note that the peak of the gain function occurs at a longer wavelength than the FEL resonant wavelength.

B5.1.2.3.1 Low-gain oscillators

In a low-gain oscillator, optical power (P) builds up through many passes of the electron beam through the undulator. The gain length (L_G) for the growth of the optical power is such that $L_G \gg N_u\lambda_u$ (the undulator length). The small signal power gain per pass is denoted by δ, with $\delta \ll 1$. In the small signal regime where $P \ll P_{sat}$, where P_{sat} is the saturation power, we have

$$P = P_0(1 + \delta - \Gamma)^n, \tag{B5.1.5}$$

where P_0 is the spontaneous emission power and Γ represents the fractional loss of power per pass resulting from out coupling and internal cavity losses.

The peak value of $\delta = \delta_0$ is given by

$$\delta_0 = 1.3 \times 10^{-3}\frac{h^2 a_u^2 f}{\gamma(1 + a_u^2)} I_e N_u^2 F^2[h, \xi] \tag{B5.1.6}$$

where I_e is the peak electron current, f is a unitless quantity called the *filling factor* that varies between 0 and 1 and depends on the degree of overlap between the optical mode and electron beam (with $f = 1$ corresponding to perfect overlap); and

$$\begin{aligned} F[h, \xi] &= J_0(\xi) - J_1(\xi) \qquad \text{for } h = 1 \\ &= J_{\frac{(h-1)}{2}}(h\xi) - J_{\frac{(h+1)}{2}}(h\xi) \qquad \text{for } h = 3, 5, 7, \ldots \end{aligned} \tag{B5.1.7}$$

The spectral width of the gain function is shown in figure B5.1.2.

The expression of δ_0 is valid in the case of a perfect electron beam with zero emittance and no energy spread. In practice, however, imperfections in the electron beam, such as finite emittance and energy spread, reduce the gain so that the small-signal gain is given approximately by [12]

$$\delta_s \approx \frac{g_0}{\left(1 + \frac{4N_u a_u \varepsilon_n}{\gamma^2\lambda_r}\right)(1 + 27h^2 N_u^2 \sigma_\gamma^2)} \tag{B5.1.8}$$

where ε_n is the rms normalized emittance and σ_γ is the rms fractional energy spread of the electron beam. Lasing on the higher harmonics is difficult because of the increased sensitivity of the terms in the denominator of the expression for δ_s to ε_n and σ_γ as the harmonic number h increases.

In the low-gain regime, the optical mode structure is largely governed by the details of the optical cavity design. Many FELs use linear near-concentric cavities (see section A2.1.3.1), with the Rayleigh range (z_R) (see section A2.1.2.2) being chosen to match the electron and optical beam sizes at the centre of the undulator. To maximize the output power, the optimum outcoupling is usually chosen to be approximately one-third of σ_s. The maximum saturated power output is given by $P_{sat} \approx P_{ebeam}/N_u$.

B5.1.2.3.2 High gain FELs

The behaviour of high-gain FELs is quite different from that of low-gain devices. In contrast to the low-gain regime, the gain may be sufficiently high to saturate the FEL in a single pass through the undulator [13]. The high-gain regime is of particular interest in regions of the spectrum where high reflectivity mirrors do not exist, such as in the x-ray region. The optical power grows exponentially over most of the undulator distance z, except near the beginning and close to saturation. For an ideal beam, in the exponential gain regime at the fundamental wavelength,

$$P(z) = P_0 \exp(z/L_G) \tag{B5.1.9}$$

where $P_0 \approx \rho^2 mc^3(\gamma/\lambda_r)$ and $L_G = 0.046\lambda_u/\rho$ and the parameter

$$\rho = 5.7 \times 10^{-3} \left\{ \left(\frac{a_u F_1 \lambda_u}{\gamma} \right)^2 \frac{I_e}{\gamma R_e^2} \right\}^{\frac{1}{3}} \tag{B5.1.10}$$

is referred to as the *Pierce (or FEL) parameter*, with R_e the electron beam rms radius. The saturated power is given by $P_{sat} \approx \rho P_{ebeam}$. Typically, saturation is achieved after a length approximately equal to 20 L_G. For the FEL to perform as described by the simple 1D equations above the following conditions must be met:

(i) $L_G/z_R \ll 1$ (little diffraction over a gain length)
(ii) $4\pi\varepsilon_n/\lambda_r\gamma \ll 1$ (small electron beam emittance)
(iii) $\sigma_\gamma/\rho \ll 1$ (small energy spread).

When these conditions are not met, the gain is reduced and a more complicated set of equations is required [14] to calculate the performance. For short wavelengths, e.g. UV and x-ray, the requirement on the electron-beam emittance is the most difficult to meet (see later).

B5.1.3 Technology for FELs

The type of accelerator used to drive an FEL depends very much on the lasing wavelength. Because the wavelength scales inversely as the square of the electron energy, short-wavelength devices tend to be driven by large and sophisticated accelerators, whereas long-wavelength FELs are driven by smaller and simpler accelerators. It is interesting to note that this size scaling is opposite to that of conventional microwave and rf generators, i.e. in an FEL the wavelength of emission scales inversely with the square of the electron energy; therefore, short-wavelength FELs are generally bigger than longer-wavelength devices. Note that the gain of an FEL scales approximately inversely with electron energy.

The type of FEL configuration, e.g. oscillator or amplifier, also scales with wavelength. Table B5.1.1 gives a summary of accelerator technology in relation to FELs.

At the longest wavelengths, the FEL gain is usually so high that the devices can operate from spontaneous emission to saturation in the amplifier mode (self-amplified spontaneous emission or SASE). At intermediate wavelengths (near-IR, visible and near-UV), the gain is lower, therefore, oscillators with optical cavities dominate. At present, storage-ring-driven FEL oscillators are leading the way to short UV wavelengths. At the shorter wavelengths (<100 nm), mirrors become impractical and so one has to consider linear-accelerator-driven SASE amplifiers with very long undulators designed to achieve saturation in a single pass. Linear

Table B5.1.1. FEL accelerator technology is wavelength dependent.

Wavelength range	Accelerator type	FEL configuration
Far-IR(mm)	Induction linac, van de Graaff, Pulse-line accelerator	SASE amplifier, oscillator
Near-IR	rf linac, microtron	Oscillator
Visible	rf linac, storage ring	Oscillator
UV > 100 nm	Storage ring, rf linac	Oscillator
XUV	rf linac, storage ring	Amplifier, harmonic generation
1 keV–MeV	rf linac, storage ring	SASE (keV), Compton backscatter (keV–MeV)

accelerators are better suited as drivers for very-short-wavelength FELs rather than storage rings. This is because in a high-energy linear accelerator, higher electron beam peak current and lower beam emittance can be achieved than in a ring of equivalent energy.

At hard x- and γ-ray wavelengths even SASE systems are likely to become impractical. Such photons can be produced by Compton backscattering of FEL-generated photons against the electron beam [15, 16].

A measure of electron beam quality is the rms normalized emittance

$$\varepsilon_{\mathrm{n}} \approx \gamma R_{\mathrm{e}} \theta_{\mathrm{e}} \tag{B5.1.11}$$

where R_e is the rms electron beam radius and θ_e is the rms electron beam divergence angle. High electron beam brightness ($B \approx I_e/\varepsilon_n^2$) corresponds to high peak current and low emittance. One of the goals of accelerator development for FELs is to have I_e as high as possible while keeping ε_n as low as possible. In order to work well, the electron and optical beams must overlap effectively. Note that the product of the optical beam radius w and the emission angle θ_0 is given by $w\theta_0 = \lambda_r/4\pi$. Effective overlap of the beams requires $R_e\theta_e < w\theta_0$, or in terms of the emittance,

$$\frac{\varepsilon_{\mathrm{n}}}{\gamma} < \frac{\lambda_{\mathrm{r}}}{4\pi}. \tag{B5.1.12}$$

The current state of the art corresponds to $I_e \approx 100$ A with $\varepsilon_n \approx 2\ \mu$m. The requirements of x-ray FELs are such that this low emittance value will have to be maintained at currents of several kA. The generation and transport of an electron beam from its source to the undulator involves a considerable amount of manipulation of the electrons. Maintaining the quality of the electron beam (i.e. keeping the emittance low and current high) is the subject of extensive research in the FEL community.

At optical wavelengths, considerable effort has been put into developing compact cost-efficient FELs, and also toward improving the quality of electron beams, particularly in regard to electron sources for rf linacs [17]. The technology of choice, which both enhances the performance and reduces the size of the FEL, is the rf electron gun, operated either in the thermionic mode or in the laser switched mode. Typically FELs require peak currents (I_e) in the range of 100 A for low-gain oscillators, to several kA for high-gain amplifiers. In the case of low average power devices, increasing the electron beam brightness has the effect of making the accelerator more compact by allowing short-wavelength lasing at lower electron energy than would otherwise be possible, e.g. 370 nm using a 45 MeV electron beam [18].

B5.1.4 FEL applications

Because of their cost and complexity, FELs are most competitive at wavelengths where other sources are weak. In order for FELs to be of continued interest to users, they must deliver a performance that far surpasses

Table B5.1.2. A selection of FELs intended as user facilities. The performance numbers are subject to change. The asterisk indicates facilities that are proposed or under construction, numbers in parentheses indicate design goals.

Country	Institution	Device name	Wavelength (μm)	Micropulse length (ps)	Peak power (MW)	Average power (W)
China	Institute for High Energy Physics	—	10–16	4	20	2
France	Université de Paris-Sud	SuperACO	0.3–0.69	20	12	0.8
		CLIO	3–50	1.5–6	10	9
Germany	DESY	TESLA-FEL*	(0.0001–0.001)	(0.2)	(37 000)	(210)
The Netherlands	FOM-Institute for Plasma Physics	FELIX-2	5–30	0.5–10	2	1
		FELIX-1	25–110	1–10	2	0.5
Japan	iFEL	FELI 1	5–22	2	5	2
		FELI 2	1–6	2	5	0.5
		FELI 3	0.23–1.2	2	5	0.5
		FELI 4	20–80	2	5	1
		FELI 5	50–100	2	2	1
USA	Duke University	OK-4	0.19–0.4	10	1000	1
		Mark III	3–10	3	2	3
	Jefferson Laboratory	IR Demo	3–6.6	1–2	10	2000
		IR-Upgrade*	(1–15)			(10 000)
		UV-FEL*	(0.3–1)			(3000)
	University of Maryland*	MIRFEL	(10–150)	(4)	(4)	(0.5)
	Stanford University	FIREFLY	19–65	1–5	0.3	0.4
		SCA	3–10	0.7	10	1.2
	Univ. California Santa Barbara		150–2000	10^6	0.004	0.08
	Vanderbilt University	MKIII	2–10	2	10	10
	SLAC*	LCLS-2*	(0.00015–0.0015)	(0.3)	(20 000)	(1.5)
	BNL	DUVFEL	0.266	2	50	N/A

conventional sources in brightness, power and wavelength range. At present, most applications-orientated FELs operate in the near- to mid-IR and near-UV regions of the spectrum.

In the following description of some FEL applications, we highlight applications that have great practical utility but which still require work to reach maturity. Any application requiring average optical power in excess of a few tens of watts should be considered speculative at this time. Table B5.1.2 gives a list of FELs that are operated as user facilities. Further details of these and other operating and proposed FELs can be found elsewhere [19, 20].

Medicine is a very rich field for FEL applications [21]. Medical applications generally require rather modest average power levels of the order of several watts (see section D3.1.5).

In the area of laser surgery, there is considerable interest in finding wavelengths at which tissue can be cut cleanly with little collateral damage. Ablation of tissue with high peak power laser pulses in the IR and UV is of particular interest. There is some preference for the use of IR wavelengths because of the potential reduced risk of photochemical effects (with possible persistent collateral damage or genetic modifications) compared to UV wavelengths.

Studies at the Vanderbilt University FEL have shown that wavelengths near 6.45 μm, corresponding to the amide II absorption band of proteins, are particularly well suited for soft tissue surgery [22]. Remarkable results have been achieved in ocular and neural tissue (see sections D3.1.5 and D3.2.2), where extremely clear cuts were made with little collateral damage. In some cases, collateral damage extended only to a depth of 10 μm into the tissue from the cut boundary. Human studies are now underway.

The FEL is particularly suited for bulk materials processing because of its tunability for optimum wavelength, potential for scaling to high average power and low cost per unit of light energy. Because of the large investment of capital equipment and operating personnel, FELs are not cost effective for industrial processing at low average power levels. As the average power level increases, however, the cost effectiveness of the FEL as a photon source improves dramatically. Studies have shown that at average power levels in the 100–200 kW range, the cost of UV photons delivered should be less than \$0.0002/kJ [23].

A general category of material processing involves surface modification (see section D1.6.1). An example is the alteration of the surface characteristics of various fibres to make fabric less shiny, more hydroscopic, softer or more accepting of dyes [24]. Such an application would require approximately 100 kW of 193 nm light per min per 1000 m^2. of fabric. Other examples include improving the adhesion properties of material to reduce the requirement for glues and enhancing the antimicrobial properties of nylon by converting the amide groups to amines.

In the gas phase, infrared photochemistry by multiple-photon dissociation has been the subject of promising FEL research [25]. The IR excitation process that leads to molecular dissociation occurs by the successive absorption of a few tens of IR photons near a strong IR resonance. The particular features of the FEL make it a very appropriate device as a driver for bond-selective photochemistry. The FEL offers easy tunability to match molecular absorption resonances and the high peak power necessary for multi-photon processes (see section B3.1 and B3.8).

In general, the wavelength range covered by existing FELs is not far removed from that of conventional lasers. In many cases, FEL applications have been extensions of work already begun with other lasers. New opportunities exist at very short wavelengths (i.e. very high photon energies), where the only existing sources are incoherent.

The development of tunable, hard x-ray FELs such as the linac coherent light source (LCLS) at the Stanford Synchrotron Radiation Laboratory and the TESLA FEL facility at the Deutsches Elektronen Synchrotron (DESY) in Germany are commonly being referred to as *fourth-generation light sources* [26]. This contrasts with the previous three generations of incoherent synchrotron light sources. The major advantage of FELs is that they produce a coherent beam with a peak brightness many orders of magnitude higher than that possible from synchrotrons (see figure B5.1.12). Consequently, applications of light sources in the x-ray region will undergo an explosion similar to that experienced in the visible region brought on by the invention of the laser (see section B5.2).

The proposed x-ray FELS will have the spatial and temporal resolution to observe features on the angstrom length scales of molecular bonds and the sub-picosecond time scales of molecular vibrations. This would allow chemists to directly observe reaction processes such as the photo-dissociation of isolated gas-phase molecules, photochemically induced bond breakage and recombination and structural transformations in photosynthetic processes. Studies of nanoscale dynamics involve the overlap of different time and length scales. Techniques that will be possible include x-ray photon correlation spectroscopy; x-ray transient grating spectroscopy; and the study of the dynamics of entangled polymers, glassy dynamics and collective mode dynamics in liquids and gases.

Very high-energy ($\gg$10 keV) photons can be generated by Compton backscattering of FEL photons

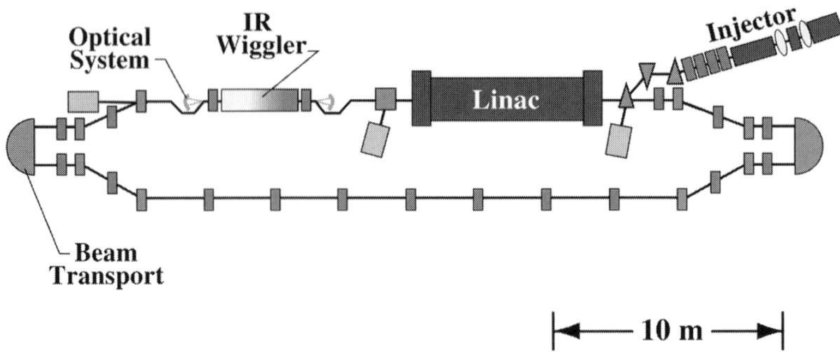

Figure B5.1.3. Thomas Jefferson National Accelerator Facility 50 MeV energy recovery linac and infrared FEL light source. (Courtesy Lia Merminga, TJNAF.)

against high-energy electrons. The head-on collision results in a pencil-like beam of hard photons whose energy is closely correlated with scattering angle. The highest energy photons are on the scattering axis, with energy falling monotonically as the scattering angle increases. Such a source is quite different in character from a conventional bremsstrahlung source in which photon energy is only loosely correlated with angle. Though the photon beam is incoherent, the correlation of photon energy with angle allows the collimation of the photon beam into a narrow beam with a small energy spread. Experiments have shown that a quasi-monochromatic beam of multi-MeV γ-rays could be produced by scattering near-UV FEL photons against 500 MeV electrons circulating in a storage ring [16]. Such beams will have many applications [27] in nuclear physics studies [28] and radiographic imaging [29].

B5.1.5 Future directions for FELs

Recent developments in both high average power and short wavelength have fueled a tremendous increase in interest in FELs in recent years.

The demonstration of 2 kW average optical power in the near-IR by the Jefferson laboratory group using a superconducting energy recovery linac (figure B5.1.3) has opened the door for high average power applications in materials processing (see section D1.6). Continued development of high-power FEL technology in the UV should produce considerable interest from industry.

Ultra-short-wavelength FELs will be an important component in the next generation of light sources that will produce coherent, short-pulse radiation at soft and hard x-ray wavelengths. The technical challenges are particularly great in this region where high-reflectivity optics are not available. Recent developments in linear accelerator and storage ring technology and synchrotron radiation research, however, have made the new frontiers both alluring and reachable [26,30]. An example of a proposed device is the linac coherent light source [26]. The LCLS is designed to operate from 0.15 to 2.5 nm and will require a 15 GeV electron beam with a peak current of $I_e > 3000$ A and a normalized emittance of 1–2 μm feeding a 100-m long undulator. The LCLS will produce an x-ray beam with a peak brightness of 10^{33} photons (s mm^2 mrad^2 0.1% bandwidth)$^{-1}$ in sub-picosecond pulses (see figure B5.1.12). Such a device would have a peak spectral brightness several orders of magnitude greater than existing synchrotron sources. Another x-ray FEL designed to operate at longer wavelengths than the LCLS is under construction at DESY in Germany [26]. Figure B5.1.12 shows a comparison of the performance of these proposed FELs and existing synchrotron light sources.

Another spectral region that is ripe for FEL development is the far-IR at wavelengths of several tens to several hundred micrometres (see sections B3.8 and B3.9). In this region lie many molecular resonances of biological interest that have not yet been investigated. Unlike the x-ray FELs, far-IR devices can be very

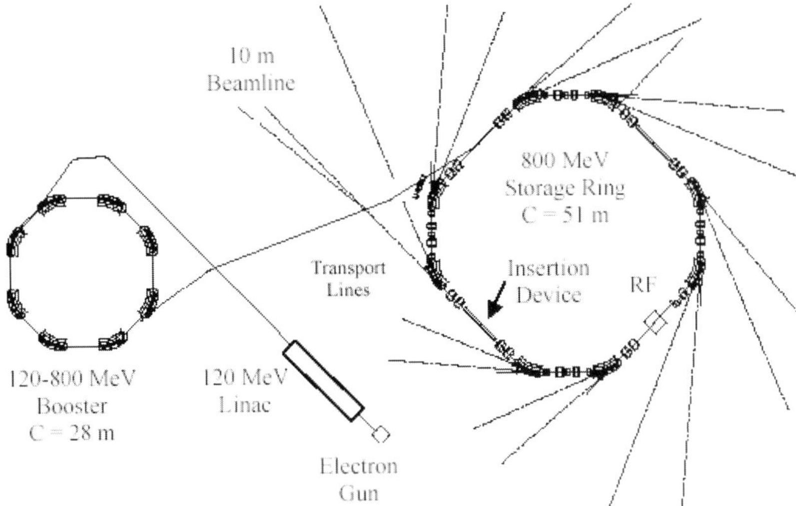

Figure B5.1.4. Schematic layout of a generic synchrotron light source.

compact and can be driven by electrons that are a few MeV in energy. It is anticipated that far-IR devices will fit on an optical bench, and will be suitable for individual investigator use. For further details on FELs the reader is referred to [31–35].

B5.1.6 Synchrotron light sources

B5.1.6.1 Introduction

A synchrotron light source typically consists of an electron storage ring which produces both synchrotron radiation (SR) in the bending magnets of the ring and wiggler or undulator radiation in the insertion devices installed in the straight sections of the ring. A detailed discussion of the properties of synchrotron, wiggler and undulator radiation will be given in a later section. As an example of a synchrotron light source the schematic layout of part of the National Synchrotron Light Source at Brookhaven National Laboratory is shown in figure B5.1.4.

The key elements of a synchrotron light source are the electron storage ring and the electron injection system:

The *electron storage ring* stores 10^{12} electrons for 5 hr and has the following components:

- dipole magnets which bend electrons in a circular orbit and produce SR;
- quadrupole magnets which provide focusing of the electron beam;
- rf cavity which replenishes the electron energy lost to SR;
- a vacuum system which reduces ring pressure to $\sim 10^{-7}$ Pa;
- straight sections which are typically ~5 m of empty space to accommodate insertion devices;
- beamlines providing an optical transport line (~5–30 m) to take the SR from the ring to the experimental end stations.

The *electron injection system* has the following components:

- electron gun which is the electron source;
- linac which pre-accelerates the electrons to 50–100 MeV; and
- a booster ring which accelerates the electrons to the energy of the storage ring (500–8000 MeV).

Table B5.1.3. Approximate number of synchrotron light source rings worldwide as a function of ring energy.

Region	0–1.5 GeV	1.5–3 GeV	6–8 GeV	Approximate total
Asia	13	5	1	19
Europe	3	7	1	11
India	1	1		2
Middle East	1			1
North America	5	5	2	12
Russia	2	3	1	6
South America		1		1

B5.1.6.2 Brief history of synchrotron light sources

In the literature one often reads that light sources are classified in terms of 'generations' as follows:

- *First generation*: SR is obtained parasitically from bending magnets in a high-energy physics ring (*circa* 1970s);
- *Second generation*: dedicated storage rings producing SR from bending magnets with a only a few insertion devices of moderate brightness (*circa* 1980s);
- *Third generation*: dedicated storage rings optimized for many insertion devices yielding very high brightness (*circa* 1990s); and
- *Fourth generation*: an open research topic and possibly not a storage ring at all but an FEL on a linear accelerator (*circa* 2000s)!

At present most of the first-generation light sources have been replaced by dedicated storage rings. There are more than 50 second- and third-generation rings in operation around the world with new rings being brought on line each year (see table B5.1.3).

The rings in table B5.1.3 are grouped into three ranges according to the energy of the electron beams. Using the formulae in the next section for synchrotron and undulator radiation, it can be shown that the rings of a particular energy produce photon beams as follows:

- *0–1.5 GeV rings*: infrared, ultraviolet photon beams (less than 2 keV);
- *1.5–3 GeV rings*: soft x-ray photon beams (1–10 keV); and
- *6–8 GeV rings*: hard x-ray photon beams (10 keV and beyond).

B5.1.6.3 Operational characteristics of synchrotron light sources

Typical stored currents in an electron storage ring are 750 mA for low energy rings and 200 mA for high-energy rings; this corresponds to of the order of 10^{12} electrons circulating in the ring. It should be noted that the electrons are not distributed uniformly around the ring; they are formed into circulating bunches with free space in between. The length of the electron bunch depends on the arrangement of the magnets, the energy of the ring and the voltage and frequency of the rf system. Bunch lengths typically range from 1 to 5 cm (rms). The number of electron bunches in a particular ring can range from one to several hundreds depending on the design of the ring and desires of the experimenters.

The operation of a generic ring proceeds as follows. All the hardware (magnets, rf, etc) is powered to the appropriate setpoints. An electron beam is emitted from a thermionic cathode and accelerated in the linear accelerator to low energy (~50–100 MeV). This beam is injected into a booster ring that cycles at a rate of

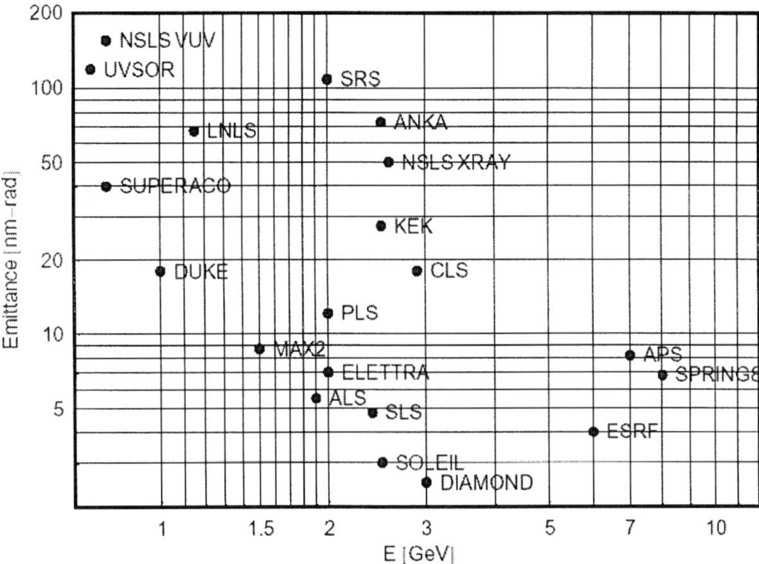

Figure B5.1.5. Horizontal electron beam emittance *versus* electron energy for several existing and planned synchrotron light sources worldwide.

1–30 Hz and the energy of the electron is raised to the level of that in the storage ring (~500–8000 MeV). The electron beam is ejected from the booster ring into a transport line and injected into the storage ring. The circulating electron beam generates radiation when it is bent in the dipole magnets of the ring or in one of the insertion devices in a straight section of the ring. The electron beam is stored for roughly 5–12 h. The electrons are lost from the ring due to Coulomb collisions among electrons in a bunch and also due to collisions with residual gas molecules in the ring. To prolong the storage time the vacuum in the storage ring should be very good ($\sim 10^{-7}$ Pa). This operation cycle is simply repeated throughout the life of the ring.

One of the key figures of merit of a synchrotron light source is the emittance of the electron beam. The Holy Grail is to produce a light beam without any measurable effects on the electron beam. Since a zero emittance electron beam is not possible, the desire is to reduce the emittance, such that $\varepsilon_{\rm n}/\gamma \sim \lambda/4\pi$, the so-called 'diffraction limited emittance'. Figure B5.1.5 displays the horizontal emittance of the electron beam *versus* energy of the electron beam for many existing and planned synchrotron light sources.

The main factors determining the emittance of a light source are the energy of the ring and the layout of the quadrupole and dipole magnets that make up the ring, the so-called 'lattice'. The lattice is usually constructed from basic cells as displayed in figure B5.1.6; the number of cells in the full ring is termed the superperiodicity, $N_{\rm s}$. The emittance, $\varepsilon_{\rm n}$, of a ring depends on the energy of the ring and the number of superperiods: $\varepsilon_{\rm n} \propto E^2/N_{\rm s}^3$ [36].

Thus, for a ring of particular energy, one varies the number of cells to obtain the desired emittance. Of course the circumference of the ring increases as the superperiodicity is increased. A ring with an energy of 1 GeV will have a circumference of roughly 75 m, a 3 GeV ring ranges from 170–350 m, while a 6–8 GeV ring grows to 1–2 km.

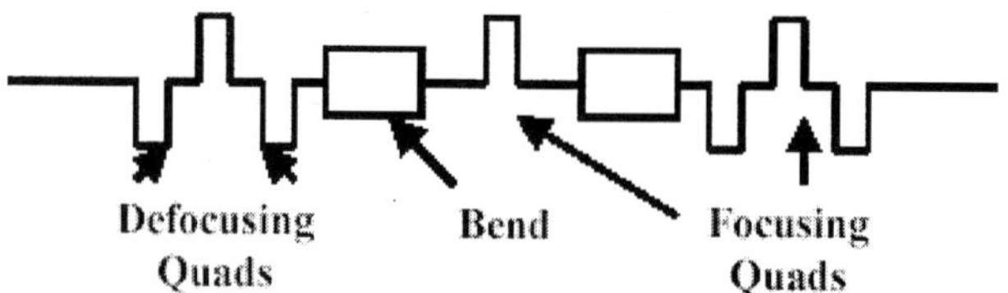

Figure B5.1.6. Basic building block of a light source lattice: the Chasman-Green double-bend achromat cell.

B5.1.6.4 Electromagnetic radiation from charged particles

When an electron circulates in an electron storage ring designed as a synchrotron light source, there are three types of radiation that are of interest to the experimentalist who makes use of this light:

- synchrotron or bending magnetic radiation,
- wiggler magnet radiation, or
- undulator magnet radiation.

In figure B5.1.7 we give a schematic diagram of the three radiation processes showing the electric field as a function of time and the intensity as a function of frequency for radiation from a single electron. In general, the radiation from the electrons is not coherent so the total intensity is simply given by the sum of the intensity for each electron. In the succeeding sections, we shall discuss the properties of each type of radiation and provide some formulas in 'practical units' to facilitate rapid calculations.

B5.1.6.4.1 Synchrotron or bending magnetic radiation

When an electron moves in a transverse magnetic field, the accelerated electron emits electromagnetic radiation known as synchrotron radiation (SR) as shown schematically in figure B5.1.8. SR has a very broad spectrum for low photon energies and falls off exponentially above a critical photon energy, $E_c = \hbar\omega_c = 3\hbar c\gamma^3/2\rho$, where $\gamma = E/m_e c^2 = E[\text{MeV}]/0.511$ [37–40]. ρ is the bending radius of the electron in the dipole magnetic field and is related to the electron energy and the dipole magnetic field, $\rho[\text{m}] = E[\text{GeV}]/0.3B[\text{T}]$. SR was first experimentally observed at the General Electric Corporation in 1947.

In practical units, the critical photon energy or wavelength can be written in terms of the electron energy and the dipole magnetic field as

$$E_c[\text{keV}] = 0.665B[\text{T}]E^2[\text{GeV}] \tag{B5.1.13}$$

$$\lambda_c[\text{Å}] = \frac{18.64}{B[\text{T}]E^2[\text{GeV}]} \tag{B5.1.14}$$

Half of the total synchrotron radiation power is radiated above the critical photon energy and half below. Due to the relativistic motion of the electron, the synchrotron radiation is highly collimated in the forward direction with a characteristic vertical opening angle [37],

$$\psi(\omega) \approx \begin{cases} \dfrac{1}{\gamma}\left(\dfrac{\omega_c}{\omega}\right)^{1/3} & \omega \ll \omega_c \\[2ex] \dfrac{1}{\gamma} & \omega \approx \omega_c \\[2ex] \dfrac{1}{\gamma}\left(\dfrac{\omega_c}{\omega}\right)^{1/2} & \omega \gg \omega_c. \end{cases} \tag{B5.1.15}$$

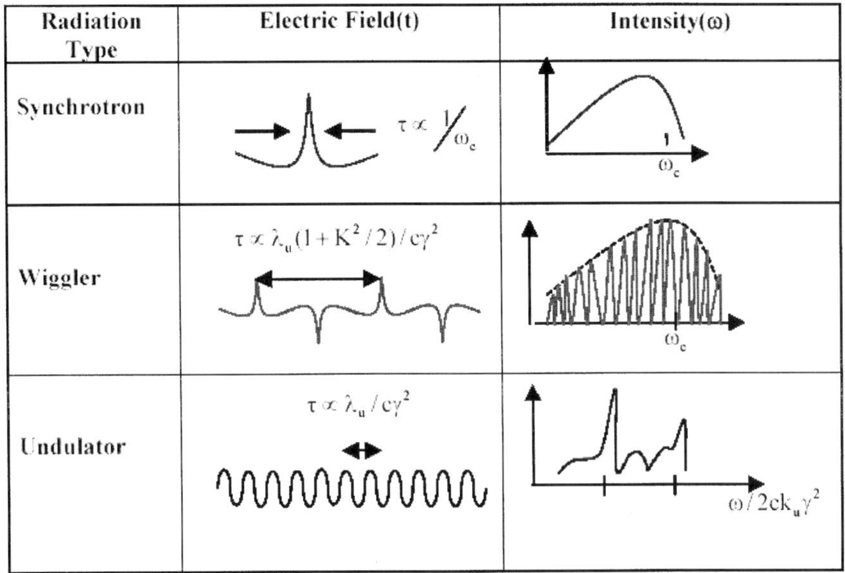

Figure B5.1.7. Schematic diagram of the three radiation processes of interest in a synchrotron light source. The second column displays the electric field as a function of time and the third column displays the intensity as a function of frequency.

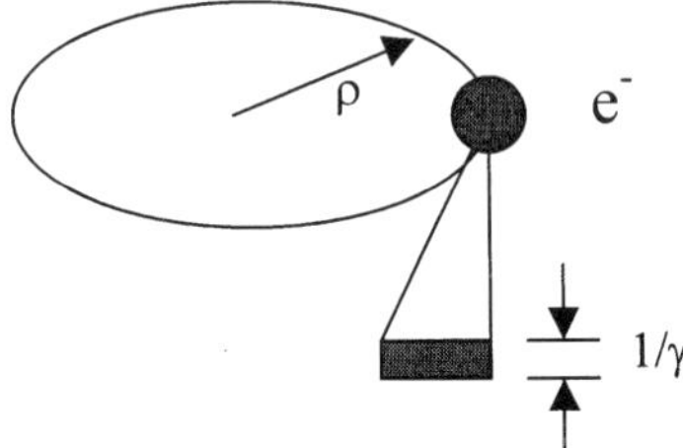

Figure B5.1.8. Schematic diagram for the emission of synchrotron radiation by a relativistic electron on a curved trajectory.

The spectral and angular distribution of the number of photons (flux) is given by [37–40]

$$\frac{\mathrm{d}N}{\mathrm{d}\Omega} = \frac{3\alpha\gamma^6}{4\pi^2}\left(\frac{\omega}{\omega_\mathrm{c}}\right)^2\left(\frac{1}{\gamma^2}+\psi^2\right)^2\left[K_{2/3}^2(\chi) + \frac{\psi^2}{1/\gamma^2+\psi^2}K_{1/3}^2(\chi)\right]\frac{I}{e}\frac{\Delta\omega}{\omega} \qquad \text{(B5.1.16)}$$

with units of photons $\mathrm{s}^{-1}\ \mathrm{sr}^{-1}$, $\alpha = 1/137.04$, I in amperes, $e = 1.6 \times 10^{-19}$ C, $K_{2/3}$ is a Bessel function of fractional order and where $\chi \equiv \omega(1+\gamma^2\psi^2)^{3/2}/2\omega_\mathrm{c}$. The first term in the square brackets corresponds to radiation with the electric field vector polarized in the plane of the orbit and the second term to radiation polarized perpendicular to the plane.

In the forward direction ($\psi = 0$), the previous equation can be written in practical units as

$$\frac{\mathrm{d}N}{\mathrm{d}\Omega} = 1.325 \times 10^{16} E^2[\mathrm{GeV}]I[\mathrm{A}]H_2(y, 0)\frac{\Delta\omega}{\omega} \qquad \text{(B5.1.17)}$$

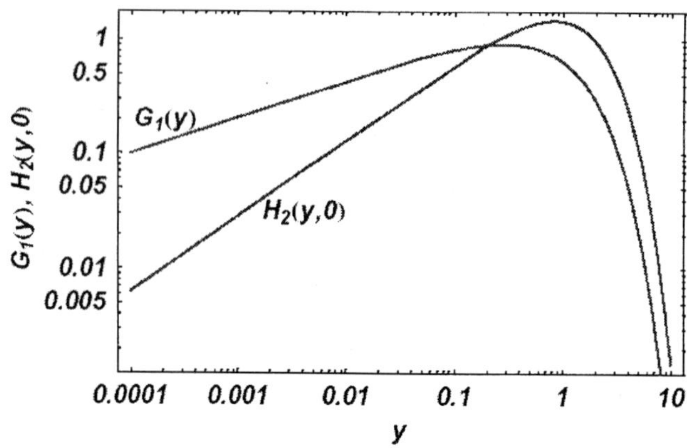

Figure B5.1.9. Plot of $G_1(y)$ and $H_2(y, 0)$ *versus* y.

with units photons s^{-1} $\mathrm{mrad}\theta^{-1}$ $\mathrm{mrad}\psi^{-1}$, $y \equiv \omega/\omega_c = \lambda_c/\lambda$ and $H_2(y, 0) = y^2 K^2_{2/3}(y, 2)$ is plotted in figure B5.1.9.

If the angular distribution is integrated over all vertical angles, one obtains the flux per milliradian of horizontal arc (θ),

$$\frac{\mathrm{d}N}{\mathrm{d}\theta} = \frac{\sqrt{3}\alpha\gamma}{2\pi \times 10^3} \frac{\Delta\omega}{\omega} \frac{I}{\mathrm{e}} \left(\frac{\omega}{\omega_c}\right) \int_{\omega/\omega_c}^{\infty} K_{5/3}(x)\,\mathrm{d}x \tag{B5.1.18}$$

with units of photons s^{-1} $\mathrm{mrad}\theta^{-1}$, I in amperes and $e = 1.6 \times 10^{-19}$ C.

In practical units, this can be written as

$$\frac{\mathrm{d}N}{\mathrm{d}\theta} = 2.457 \times 10^{16} E[\mathrm{GeV}] I[\mathrm{A}] G_1(y) \frac{\Delta\omega}{\omega} \tag{B5.1.19}$$

with units of photons s^{-1} $\mathrm{mrad}\theta^{-1}$ and where

$$G_\mathrm{n}(y) = y^\mathrm{n} \int_\mathrm{y}^{\infty} K_{5/3}(x)\,\mathrm{d}x \tag{B5.1.20}$$

Figure B5.1.9 is a plot of the functions $H_2(y, 0)$ and $G_1(y)$. Figure B5.1.10 is a plot of $\mathrm{d}N/\mathrm{d}\theta$ as a function of photon energy for radiation from a representative range of storage rings.

The spectral power per horizontal angle but integrated over all vertical angles is given in practical units by

$$\frac{\mathrm{d}P}{\mathrm{d}\theta}\left[\frac{\mathrm{mW}}{\mathrm{mrad}\,\theta}\right] = 8.73 \times 10^3 \frac{E^4[\mathrm{GeV}]}{\rho[\mathrm{m}]} I[\mathrm{A}] G_2(y) \frac{\Delta\omega}{\omega} \tag{B5.1.21}$$

Finally, the total power radiated by a circulating current I is given in SI units by

$$P_\mathrm{T} = \frac{4\pi r_e m c^2}{3e} \frac{\gamma^4}{\rho} I \tag{B5.1.22}$$

or in practical units by

$$P_\mathrm{T}[\mathrm{kW}] = \left[\frac{88.5 E^4[\mathrm{GeV}]}{\rho[\mathrm{m}]}\right] I[\mathrm{A}]. \tag{B5.1.23}$$

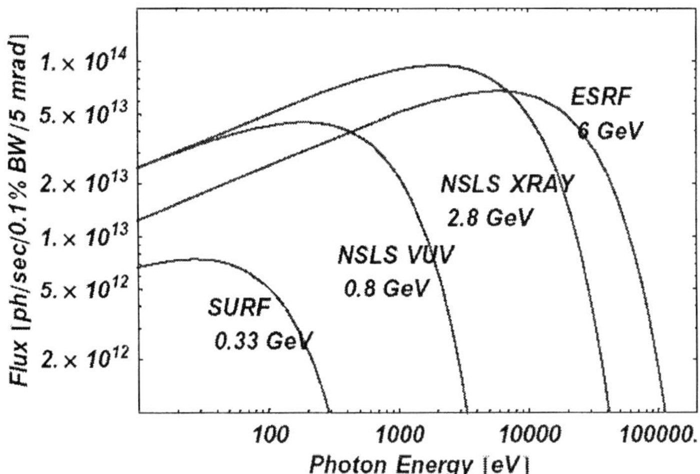

Figure B5.1.10. SR flux integrated over all vertical angles and for 5 mrad of horizontal angle *versus* photon energy for several synchrotron light sources of different energies.

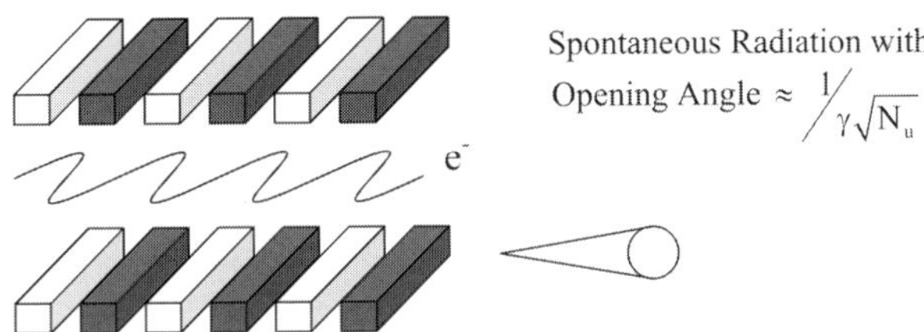

Figure B5.1.11. Schematic diagram of an electron undulating as it traverses an insertion device and emitting spontaneous radiation.

B5.1.6.4.2 Insertion devices

To enhance the performance of synchrotron light sources beyond what is available from the dipole magnet radiation so-called insertion devices are used in the straight sections of the electron storage ring. As mentioned in section B5.1.2, an insertion device is characterized by a periodic magnetic field $B_x[z] = B_0 \sin[2\pi z/\lambda_u]$ and it is constructed from alternating polarity permanent magnets as shown in figure B5.1.11. Although free electron lasers do exist in a few storage rings, the radiation from insertion devices in synchrotron light sources is simply spontaneous emission.

The figure of merit of an insertion device is the 'K parameter' given by [40],

$$K = \frac{eB}{mck_u} = 0.934B[\mathrm{T}]\lambda_u[\mathrm{cm}]. \tag{B5.1.24}$$

Note that the K parameter is simply related to the rms normalized vector potential introduced in section B5.1.2 as $a_u \equiv K/\sqrt{2}$. In the literature on synchrotron light sources, 'K' is the mostly widely used notation; however, we shall use 'a_u' to make the notation consistent throughout this chapter.

For $a_u \leq \sqrt{2}$, the insertion device is termed an undulator and for $a_u > \sqrt{2}$, the device is termed a wiggler. Note that $a_u = 2$ is not a threshold: there is a continuous transition from an undulator to a wiggler.

Schematically, the difference between an undulator and wiggler is shown in figure B5.1.7 where the electric field as a function of time and the intensity as a function of frequency are given.

Wiggler radiation: For a wiggler the electric field as a function of time is simply a periodic train of pulses similar to those of ordinary synchrotron radiation. In the frequency domain the intensity is then given by a large number of lines with an overall envelope given by ordinary SR but increased by twice the number of periods in the insertion device. This is the key advantage of a wiggler. By wiggling the particle back and forth many times one can greatly enhance the amount of flux compared to that available from, say, 5 mrad of arc in a dipole magnet. Another use of a wiggler is to shift the critical photon energy to higher values by using superconducting coils instead of permanent magnets to produce fields of 5–10 T.

The total power radiated in a wiggler is given by

$$P_{\mathrm{T}}[\mathrm{W}] = \frac{8\pi^2 r_{\mathrm{e}} m_{\mathrm{e}} c^2 a_{\mathrm{u}}^2 \gamma^2 N_{\mathrm{u}}}{3\lambda_{\mathrm{u}}} \frac{I}{e} \tag{B5.1.25}$$

$$= \frac{14.5 E^2[\mathrm{GeV}] I[\mathrm{A}] N_{\mathrm{u}} a_{\mathrm{u}}^2}{\lambda_{\mathrm{u}}[\mathrm{cm}]} \tag{B5.1.26}$$

where N_{u} is the number of periods in the wiggler.

Undulator radiation: For $a_{\mathrm{u}} \leq \sqrt{2}$, the electric field as a function of time is roughly sinusoidal giving rise to an intensity with a single peak in the frequency domain at

$$\omega = \frac{2ck_{\mathrm{u}}\gamma^2}{(1 + a_{\mathrm{u}}^2)} \tag{B5.1.27}$$

with an intensity proportional to N_{u}^2 and a bandwidth given by $\Delta\omega/\omega \approx 1/N_{\mathrm{u}}$ as given in section B5.1.2.1. As a_{u} is increased, the number of lines in the frequency domain increases but for $a_{\mathrm{u}} \leq \sqrt{2}$ the spectrum is still characterized by a few distinct lines. The narrow line structure arises from the interference between light emitted in successive periods. The very high intensity in a narrow spectral bandwidth is the signature feature of an undulator source. The formula given for the total power radiated in a wiggler also applies for an undulator. Thus, one additional feature of an undulator is that the total power incident on the beamline optical elements is greatly reduced as compared to a wiggler since $P \propto a_{\mathrm{u}}^2$.

Radiation in the forward direction occurs at odd harmonics of the frequency ω of equation (B5.7.27),

$$\omega_{\mathrm{h}} = \frac{2hck_{\mathrm{u}}\gamma^2}{(1 + a_{\mathrm{u}}^2)} \tag{B5.1.28}$$

with a linewidth of

$$\frac{\Delta\omega_{\mathrm{h}}}{\omega_{\mathrm{h}}} = \frac{1}{hN_{\mathrm{u}}}. \tag{B5.1.29}$$

In contrast to most FELs, the higher harmonics of undulator and wiggler radiation are widely used in synchrotron light sources. For completeness we generalize the expression for the spectral angular intensity in the forward direction given in section B5.1.2.1 to include higher *odd* harmonics and an electron beam with a current I. The intensity in units of photons $\mathrm{s}^{-1}\ \mathrm{sr}^{-1}$ is then given by

$$\left.\frac{\mathrm{d}N_{\mathrm{h}}}{\mathrm{d}\Omega}\right|_{\theta=0} = \alpha N_{\mathrm{u}}^2 \gamma^2 \frac{I}{e} \frac{\Delta\omega}{\omega} \frac{2h^2 a_{\mathrm{u}}^2}{(1 + a_{\mathrm{u}}^2)^2} F^2[h, \xi] \frac{\sin^2(x_{\mathrm{h}}/2)}{(x_{\mathrm{h}}/2)^2} \tag{B5.1.30}$$

with I in amperes, $h = 1, 3, 5, \ldots, x_{\mathrm{h}} = 2\pi h N_{\mathrm{u}}((\omega - \omega_{\mathrm{h}})/\omega_{\mathrm{h}})$ and $F[h, \xi]$ was given in section B5.1.2.3.1.

Again generalizing the results in section B5.1.2.3, the most useful part of the radiation is confined to a central cone about the forward direction within an angular divergence given by [40],

$$\sigma_{\rm r}' \approx \frac{1}{\gamma}\left[\frac{(1+a_{\rm u}^2)}{2hN_{\rm u}}\right]^{1/2} = \left[\frac{\lambda_{\rm h}}{L}\right]^{1/2} \tag{B5.1.31}$$

Using this definition of the radiation opening angle, the spectral and angular distribution in the forward direction at $\omega = \omega_{\rm h}$ can be rewritten as [40],

$$\left.\frac{{\rm d}N_{\rm h}}{{\rm d}\Omega}\right|_{\omega=\omega_{\rm h}} = \frac{N_{\rm h}}{2\pi\sigma_{\rm r}'^2} \tag{B5.1.32}$$

where

$$N_{\rm h} = \pi\alpha N_{\rm u}\frac{I}{e}\frac{\Delta\omega}{\omega}\frac{2ha_{\rm u}^2}{(1+a_{\rm u}^2)}F^2[h,\xi] \tag{B5.1.33}$$

is the intensity of photons in the central cone. It is interesting to note that this quantity does not depend explicitly on the energy of the electron!

A critical figure of merit for an undulator source is the on-axis peak brightness of the undulator radiation at $\omega = \omega_{\rm h}$ defined as [40]

$$B_{\rm h} = \frac{N_{\rm h}}{4\pi^2\Sigma_x\Sigma_x'\Sigma_y\Sigma_y'} \tag{B5.1.34}$$

where $\Sigma_{x,y} = \sqrt{\varepsilon_{x,y}\beta_{x,y} + \lambda L/(4\pi)^2}$ is the effective rms beam size of the combined electron beam, $\sigma_{x,y} = \sqrt{\varepsilon_{x,y}\beta_{x,y}}$ and photon beam $\sigma_{x,y}^{\rm ph} = \sqrt{\lambda L/(4\pi)^2}$ and, similarly, for the rms angular divergence $\Sigma_{x,y}' = \sqrt{\varepsilon_{x,y}/\beta_{x,y} + \sigma_{\rm r}'^2}$.

Figure B5.1.12 displays the average and peak brightness of several synchrotron light sources and FELs.

B5.1.7 Future directions for synchrotron light sources

With more than 50 rings in existence and more under construction and in the planning stages, storage-ring-based synchrotron light sources will continue to serve the largest user base for decades to come. For high average brightness, from the infrared portion of the spectrum to the hard x-ray regime, with pulse-lengths in the range of 50–500 ps, electron storage rings are the source of choice at present. If higher peak brightness and/or sub picosecond pulses are required then the source of choice is an FEL.

It must be noted that the underlying physics of electron storage rings means it will be difficult, if not impossible, to continue increasing the brightness significantly as the rings are simply becoming too large [41, 42]. In fact, a recent conference proceedings contained a paper entitled ‘Towards the Ultimate Storage Ring-Based Light Source’ [43]. New technologies will be needed to continue the trend of ever increasing brightness.

A candidate for a future light source that is beginning to appear on the drawing boards of machine designers is the so-called energy recovery linac (ERL) shown schematically in figure B5.1.3 [44–49]. This device uses a superconducting linear accelerator and a pair of half arcs to return the accelerated beam back through the linac with a 180-degree phase shift to recover the energy in a bunch before the spent beam is dumped at an energy of the order of 10 MeV. At present, the ERL at the Thomas Jefferson National Accelerator Facility is the only machine in operation. It has an energy of 50 MeV, a circulating current of 5 mA, and it also contains a high average power infrared FEL [50]. Machines of higher energy (0.6–6 GeV) and average currents similar to that in ring (100 mA) are being explored for feasibility and should be possible in the next decade.

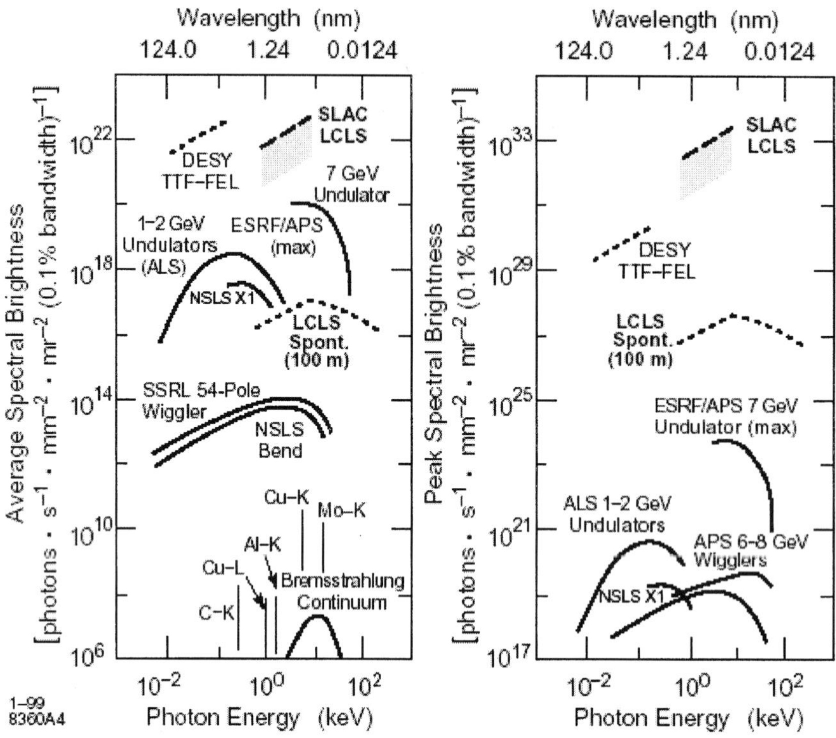

Figure B5.1.12. Average and peak brightness *versus* photon energy for several synchrotron light sources and FELs (courtesy of J Galayda/SLAC).

Potential advantages of an ERL-based light source are as follows [46]:

- diffraction limited electrons beams for 10 Å and below,
- average brightness exceeding that of existing storage rings,
- sub-picosecond electron bunches,
- round electron beams,
- virtual 'top off' operation and
- SC linac can serve as an FEL driver.

References

[1] Schwinger J 1949 *Phys. Rev.* **75** 1912–25
[2] Motz H, Thon W and Whitehurst R 1953 *J. Appl. Phys.* **24** 826–33
[3] Phillips R M 1988 *Nucl. Instrum. Methods* A **272** 1–9
[4] Madey J M J 1971 *J. Appl. Phys.* **42** 1906–13
[5] Elias L R, Fairbank W M, Madey J M J, Schwettmen H A and Smith T I 1976 *Phys. Rev. Lett.* **36** 717–20
[6] Deacon D A G, Elias L R, Madey J M J, Ramian G J, Schwettmen H A and Smith T I 1977 *Phys. Rev. Lett.* **38** 892–436
[7] Motz H 1979 *Phys. Lett.* A **71** 41–3
[8] Madey J M J, Fairbank W M, Schwettman H A 1973 *IEEE Trans. Nucl. Sci.* **NS-20** 980–3
[9] Neil G R *et al* 2001 *Phys. Rev. Lett.* **87** 084801
[10] Brau C A 1990 *Free-Electron Lasers* (San Diego, CA: Academic)
[11] Feldman D W *et al* 1989 *Nucl. Instrum. Methods* A **285** 11–16

[12] Dattoli G, Letardi T, Madey J M J and Renieri A 1984 *IEEE J. Quantum Electron.* **QE-20** 637–46
[13] Bonifacio R, Pellegrini C and Narducci L M 1984 *Opt. Commun.* **50** 359–78
[14] Kim K J and Xie M 1993 *Nucl. Instrum. Methods* A **331** 359–64
[15] Carroll F E 1990 *Invest. Radiol.* **25** 465–72
[16] Litvinenko V N *et al* 1997 *Phys. Rev. Lett.* **7** 4569–72
[17] O'Shea P G and Bennett H E (ed) 1997 Free-electron laser challenges *Proc. SPIE* **2988**
[18] O'Shea P G *et al* 1993 *Phys. Rev. Lett.* **71** 3661–4
[19] Freund H P and Granatstein V L 1999 *Nucl. Instrum. Methods* A **429** 37–40
[20] Colson W B 1999 *Nucl. Instrum. Methods* A **429** 33–6
[21] Carroll F and Brau C 1995 Medical applications of the free-electron laser *Tunable Laser Applications* ed F J Duarte (New York: Marcel Decker)
[22] Edwards G *et al* 1994 *Nature* **371** 416–19
[23] Neil G R 1996 *Proc. 1995 Particle Accel Conf.,* (Piscataway, NJ: IEEE) pp 137–9
[24] Kelley M J *Free-Electron Laser Challenges (Proc. SPIE 2988)* ed P G O'Shea and H E Bennett (Bellingham, WA: SPIE) pp 240–4
[25] Petrov A K *et al* 1998 *Nucl. Instrum. Methods* B **144** 203–10
[26] Nuhn H-D and Rossbach J 2000 *Synchrotron Radiation News* **13** 18–32
[27] O'Shea P G 1996 *Nucl. Instrum. Methods* A **375** 530–4
[28] Carman T S *et al* 1996 *Nucl. Instrum. Methods* A **378** 1–20
[29] Schreiber E C and O'Shea P G 1997 *Proc. SPIE* **3154** 94–102
[30] Poole M 2000 *Synchrotron Radiation News* **13** 4–17
[31] Freund H P and Neil G R 1999 *Proc. IEEE* **87** 782–803
[32] Brau C A 1990 *Free-Electron Lasers* (San Diego, CA: Academic)
[33] Freund H P and Antonsen T M 1996 *Principles of Free-Electron Lasers* 2nd edn (London: Chapman and Hall)
[34] O'Shea P G and Freund H P 2001 *Science* **292** 1853
[35] Colson W B *et al* 2002 *Phys. Today* **55** 35
[36] Sands M 1970 *SLAC Report* 121
[37] Jackson J D 1998 *Classical Electrodynamics* 3rd edn (New York: Wiley)
[38] Green G K 1976 *BNL Internal Report* 50522
[39] Hofmann A 1986 *SSRL ACD Note* 38
[40] Kim K J 1989 *AIP Conf. Proc.* **184** (New York: American Institute of Physics) p 565
[41] Laclare J L 1998 *Proc. EPAC 1998* (Bristol: Institute of Physics) p 78
[42] Wrulich A F 1999 *Proc. IEEE PAC 1999* (Mulhouse: European Physical Society) p 192
[43] Ropert A *et al* 2000 *Proc. EPAC 2000* (Piscataway, NJ: IEEE) p 83
[44] Kulipanov G N, Skrinsky A N and Vinokurov N A 1998 *J. Synchrotron Rad.* **5** 176
[45] Bazarov I *et al* 2001 *Proc. IEEE PAC 2001* (Piscataway, NJ: IEEE) p 230
[46] Ben-Zvi I *et al* 2001 *Proc. IEEE PAC 2001* (Piscataway, NJ: IEEE) p 350
[47] Corlett J *et al* 2001 *Proc. IEEE PAC 2001* (Piscataway, NJ: IEEE) p 2635
[48] Clarke D T *et al* 2001 *Proc. IEEE PAC 2001* (Piscataway, NJ: IEEE) **253**
[49] Levi B G 2002 *Phys. Today* **55** 23
[50] Neil G R *et al* 2000 *Phys. Rev. Lett.* **84** 662–5

B5.2
X-ray lasers

Jorge J Rocca

B5.2.1 Introduction

Specific proposals for excitation schemes for x-ray lasers date back to 1967, when the possibility of achieving amplification by photoionization pumping was first suggested by Duguay and Rentzepis [1]. This was followed by proposals for x-ray lasers based on the two population inversion mechanisms on which the present day's soft x-ray lasers are based: electron impact excitation schemes [2–5] and collisional recombination [6–9].

Several experiments realized during the 1970s and early 1980s yielded the observation of population inversions and gain [10–16]. Nevertheless, the experimental demonstration of large amplification at soft x-ray wavelengths was not realized until 1984, when Matthews *et al* [17] and Suckewer *et al* [18] observed amplification from the generation of population inversions in plasmas by collisional electron excitation and collisional electron–ion recombination, respectively. Subsequent experiments greatly expanded the number of laser transitions, achieved soft x-ray laser operation in the saturated regime for the first time [23, 31–36] and realized proof-of-principle demonstrations in several applications [37–41]. However, the large size, cost and low repetition rate of these lasers were barriers to their widespread utilization. Much progress has been achieved in recent years in the development of compact soft x-ray lasers. This is the result of advances in pump sources that include the development of compact multi-terawatt ultrashort pulse optical laser systems based on chirped pulse amplification [42–46], and fast capillary discharges capable of generating highly ionized plasma columns with high axial uniformity and length-to-diameter aspect ratios of 1000 to 1 [47–57]. Also contributing to the progress towards practical tabletop x-ray lasers is the implementation of excitation mechanisms that take full advantage of the high intensity and ultrashort pulsewidth of new optical laser systems. The field is rapidly approaching the stage at which soft x-ray lasers sufficiently compact to fit onto a normal optical table (frequently described as 'tabletop' lasers) will be routinely utilized in science and technology.

Several articles have reviewed the progress in x-ray laser development [59–64]. The book by Elton discusses the developments up to 1990 [65]. A detailed account of the progress realized since that time[1] can be found in the proceedings of two bi-annual conferences that focus on soft x-ray lasers [66, 74]. The advances in tabletop lasers have also been recently reviewed [75].

B5.2.2 Amplification and pump power requirements

In contrast to most optical lasers, the duration of the gain in soft x-ray lasers is usually shorter than the time required for the effective use of an optical cavity. This is the result of either the rapid self-terminating

[1] The field of x-ray laser research is a rapidly advancing one. The present chapter includes only material published before the date of submission of the first draft (September 2000).

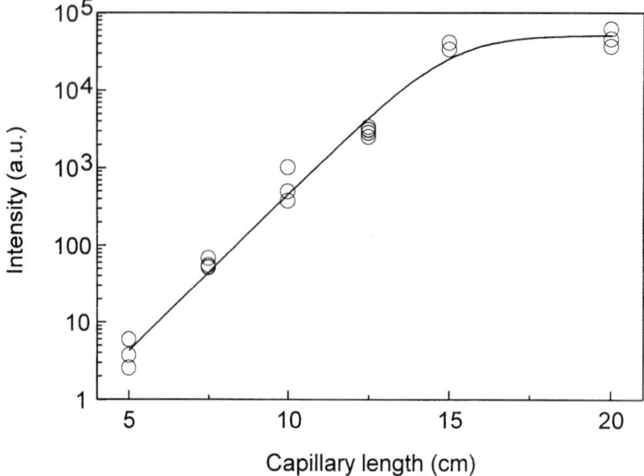

Figure B5.2.1. Integrated intensity of the 46.9 nm line of Ne-like Ar as a function of plasma column length in a capillary discharge amplifier. The exponential increase of the intensity ceases at $l \simeq 15$ cm, where the saturation intensity is reached (from J J Rocca *et al* [50].

nature of the population inversion in transient schemes (see section B3.3.1, equation (B3.3.1)) [76–86] or the difficulty in maintaining the stringent plasma conditions necessary for amplification in quasi-cw schemes for a sufficient period of time. Consequently, soft x-ray lasers normally operate with single- or double-pass amplification of the spontaneous emission through the gain medium. In such amplifiers, I, the spectrally integrated intensity of the laser line, increases as a function of l, the plasma length, as

$$I = \frac{E}{g}(e^{gl} - 1)^{3/2}(gle^{gl})^{-1/2} \tag{B5.2.1}$$

where g is the small-signal gain coefficient (assumed to be constant along the plasma column) and E is a constant proportional to the emissivity [87]. The product gl of the small-signal gain coefficient and the length of the amplifier (known as 'the gain–length product') has been widely used in the literature as a figure of merit to describe the level of amplification achieved.

A gain–length product greater than five, commonly measured by monitoring the growth of the laser line intensity as a function of plasma length, is usually a clear indication that lasing is taking place. However, there are, in the literature, quite a large number of x-ray amplification experiments that measured products less than 4. In these cases the data should be treated with significant caution, as such, phenomena such as for example plasma column end effects, could be responsible for the apparent enhancement of the line intensity. Ideally, the nearly exponential amplification continues until the intensity approaches the saturation intensity, as illustrated in figure B5.2.1 for a 46.9 nm Ne-like Ar capillary-discharge-pumped laser [48]. At this intensity, the stimulated emission is sufficiently strong to extract the majority of the energy stored in the population inversion, depleting the population inversion and the gain. In collisionally excited lasers, gain saturation typically occurs [32–35, 83] with $gl \simeq 14$–20. Consequently, to achieve saturated soft x-ray amplification in a single- or double-pass amplifier, the gain coefficient must be, depending on the length of the plasma, one to three orders of magnitude larger than those typically encountered in visible gas lasers which use optical cavities [88]. Moreover, the physics of the generation of amplification by stimulated emission determines a dramatic upward scaling of the power density deposition required to obtain substantial gain at ultrashort wavelengths. Consider a transition at wavelength λ from an upper level of degeneracy g_2 and population N_2 to a lower level of population N_1 and degeneracy g_1. The small-signal gain coefficient is the product of σ, the stimulated emission cross section and the population inversion density, N^*:

$$N^* = \left(N_2 - \frac{g_2}{g_1} N_1\right) \tag{B5.2.2}$$

and

$$g = \sigma N^* \cong \frac{A_{21}\lambda^2}{8\pi \Delta\nu} \left(N_2 - \frac{g_2}{g_1} N_1\right) \tag{B5.2.3}$$

in which A_{21} is the Einstein coefficient for spontaneous emission and $\Delta\nu$ is the linewidth of the transition.

Now, A_{21} scales along an isoelectronic sequence as $A_{21} \propto \lambda^{-2}$. Considering, as an example, the case of a naturally broadened unbranched transition with linewidth $\Delta\nu \propto A_{21}$ we obtain:

$$g \propto N\lambda^2. \tag{B5.2.4}$$

In turn, P_{min}, the minimum pump power density required to maintain a certain upper laser level population density N_2, scales as

$$P_{\text{min}} = N_2 A_{21} \frac{hc}{\lambda}. \tag{B5.2.5}$$

It results from the two previous equations that the power density required to obtain a certain gain coefficient in a naturally broadened transition scales as

$$P_{\text{min}} \propto g\lambda^{-5}. \tag{B5.2.6}$$

Consideration of the case of Doppler broadening, the line-broadening mechanism most frequently dominant under soft x-ray laser operating conditions, leads to a different but also dramatic scaling of the power requirement. In this case,

$$P_{\text{min}} \propto g\lambda^{-4} \sqrt{\frac{k_B T_i}{M_i}} \tag{B5.2.7}$$

in which T_i is the kinetic temperature of the emitting species which has atomic mass M_i.

Nevertheless, using the case of natural broadening for purposes of illustration, the operation of a saturated, mirrorless laser at 50 nm would require an estimated pump power density 10^7 times larger than that necessary for a 500 nm blue–green laser, while one operated at 5 nm would require an estimated pump power density 10^{12} times larger than that for the blue–green. In making these estimates, it is assumed that the use of the cavity for the 500 nm laser reduces the gain requirement for lasing by a factor of 100. In addition, the pump power required for lasing at soft x-ray wavelengths is often further increased by refraction losses that reduce the effective gain coefficient.

B5.2.3 Soft x-ray laser characteristics

B5.2.3.1 Wavelength

To date, saturated soft x-ray lasers have been developed at wavelengths ranging from 5.8 [89] to 46.9 nm [48, 55]. Significant amplification has been observed at wavelengths as short as 3.56 nm [31].

B5.2.3.2 *Output pulse energy*

The output pulse energy generated by soft x-ray lasers exceeds, by several orders of magnitude, that of other sources of coherent short wavelength radiation. Saturated collisional lasers pumped by large optical lasers have achieved energies of several mJ (e.g. 8 mJ at 15.5 nm [35]) and an 18.2 nm recombination laser has been reported to emit 3 mJ pulses [18]. Recently a highly saturated 46.9 nm tabletop discharge pumped laser has produced millijoule-level pulses at a repetition rate of 4 Hz [56]. Transient collisional lasers pumped by relatively compact optical lasers have produced picosecond laser pulses with energies of the order of 10 μJ [84].

B5.2.3.3 *Pulse duration*

The pulsewidth of presently available soft x-ray lasers ranges from a fraction of a nanosecond to several nanoseconds for lasers operating in a quasi-steady-state regime, to several picoseconds for transient collisional excitation lasers [79–86] and recombination lasers in transitions to the ground state [76–78]. Future photoionization x-ray lasers will generate femtosecond pulses.

B5.2.3.4 *Peak power*

Quasi-cw collisional electron excitation lasers have produced peak output powers of several tens of MW [90] and recombination lasers have reached peak powers of the order of 0.1 MW [18]. Very compact capillary discharge tabletop collisional lasers have reached peak powers up to 0.6 MW [56], and transient collisional lasers have generated powers of up to $\sim$1 MW [84].

B5.2.3.5 *Average power*

Most soft x-ray lasers demonstrated to date have produced low average power due to the low repetition rate used in the majority of the experiments that have produced significant output pulse energies (typically less than one shot every few minutes). An exception is a 4 Hz tabletop capillary discharge laser operating 46.9 nm, that produced an average power of 3.5 mW [56]. The average power of soft x-ray lasers can be expected to increase dramatically during the next decade as a result of the development of efficient high-repetition-rate tabletop pump sources.

B5.2.3.6 *Beam divergence*

In the presently available soft x-ray lasers, the beam divergence is dominantly determined by refraction and typically ranges from 1 to 10 mrad. Refraction bends the amplified x-rays out of the gain volume as a result of the transverse variation of the index of refraction due to the large electron density gradients.

B5.2.3.7 *Temporal coherence*

Since presently available soft x-ray lasers do not employ optical resonators, their temporal coherence depends on the atomic linewidth of the laser transition. In most of these single- or double-pass amplifiers, the dominant line-broadening mechanism is the Doppler effect. Taking into account the fact that amplification narrows the output by a factor of $(gl)^{-1/2}$ below that of spontaneous emission from a thin source, the value of the linewidth $\Delta\nu_{\mathrm{FWHM}}$ can be estimated as

$$\Delta\nu_{\mathrm{FWHM}} = \nu_0 \sqrt{\frac{8k_{\mathrm{B}}T_i \ln 2}{M_i c^2}} \frac{1}{\sqrt{gl}}. \tag{B5.2.8}$$

For values of gl above saturation ($gl \simeq 14$–20 in collisional lasers), some line re-broadening may occur. This linewidth determines l_{coh}, the longitudinal coherence length, and τ_{coh} the coherence time, via

$$l_{coh} = c\tau_{coh} = \frac{c}{\Delta\nu_{FWHM}}\sqrt{\frac{2\ln 2}{\pi}} = \frac{0.66c}{\Delta\nu_{FWHM}}. \tag{B5.2.9}$$

A few measurements of the linewidth of saturated collisional excitation lasers have been reported [91,92] and correspond to $l_{coh} \simeq 0.1$–0.2 mm .

B5.2.3.8 Spatial Coherence

Several authors [93–95] have theoretically analysed the spatial coherence of soft x-ray amplifiers. Single-pass soft x-ray amplifiers can sustain a large number of spatial modes and, therefore, have poor spatial coherence [96, 97]. However, significant improvements can be obtained by increasing the length of the plasma column [98] or by operating in a double-pass amplification [99]. The spatial coherence of a capillary discharge laser was measured to increase monotonically with plasma column length to reach nearly fully spatial coherence for plasma columns 36 cm in length [100]. This is the result of a decreasing number of spatial modes along the plasma column due to gain guiding and refraction anti-guiding.

B5.2.3.9 Peak spectral brightness

Soft x-ray lasers are among the brightest soft x-ray sources available. Both laser-excited [84] and discharge-pumped collisional excitation lasers have reached brightness values that exceed 1×10^{25} photons mm^{-2} mrad^{-2} s^{-1} (0.01%BW)$^{-1}$ [100]. Figure B5.2.2 compares the peak spectral brightness of some of the brightest soft x-ray lasers with other sources of coherent short-wavelength radiation.

B5.2.3.10 Average spectral brightness

Due to their low repetition rate, most of presently available saturated soft x-ray lasers produce low average brightness. An exception is a capillary discharge 46.9 nm laser [100] that produced an estimated spatially coherent average output power of nearly 1 mW within a relative spectral bandwidth $(\Delta\lambda/\lambda) < 1 \times 10^{-4}$, a spectral purity that exceeds that of some undulators at third generation synchrotrons. The average spectral brightness of laser-pumped soft x-ray lasers can be expected to increase significantly in the near future as compact high repetition rate optical laser pumps become available and more efficient target configurations are used to couple the laser pump energy to the plasma.

B5.2.4 Other methods for the generation of coherent x-ray radiation

The direct amplification of radiation in plasmas is not the only means by which coherent soft x-ray radiation can be generated. Alternative methods include high-order harmonic generation from the ouput of high-power optical lasers [103–107] (see chapter C3.1), synchrotron sources [108–111], free electron lasers (FELs) and up-conversion of optical laser pulses by scattering with energetic electron beams [112–115]. Synchrotron sources are large multi-user facilities based on the radiation emitted by high-energy electron beams. They have the advantages of broad tunability and high average power (see section B5.1.6). However, they fall short of the high peak brightness needed in applications such as the study of nonlinear phenomena at ultrashort wavelengths [116, 117] and the diagnostics of dense plasmas [40, 41, 101, 102]. Another alternative, also based on accelerator technology, is the use of self-amplified spontaneous emission in an FEL [111, 112] (see section B5.1.3). In a soft x-ray FEL, an electron beam would radiate at much higher powers and with better coherence than it does due to spontaneous synchrotron radiation. Several single-pass ultrashort-wavelength

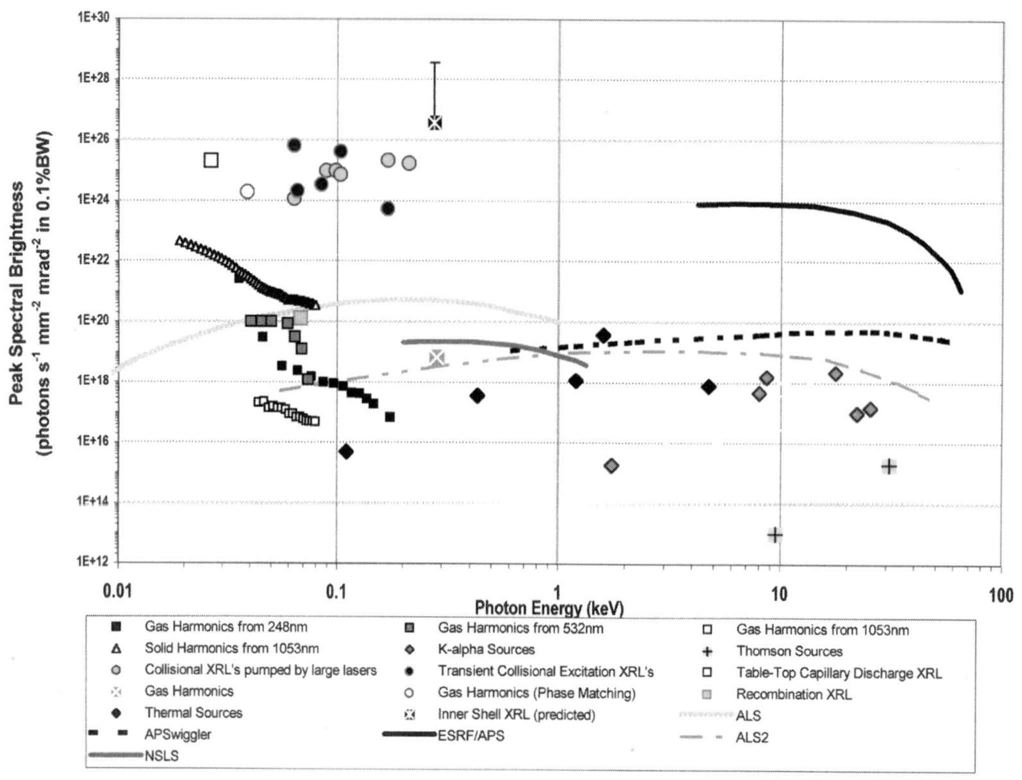

Figure B5.2.2. Peak spectral brightness of x-ray lasers compared to other x-ray sources. (The data are courtesy of R Smith and M H Key [211].)

FELs have been proposed or are under construction [111, 113]. Nevertheless, these lasers will also constitute large facilities that will not be tabletop ones.

Alternatively, soft x-ray coherent radiation can also be generated using nonlinear optical techniques for frequency up-conversion of optical laser radiation. Harmonic up-conversion of intense ultrashort pulse tabletop optical laser systems has been reported to generate radiation at wavelengths as short as 2.7 nm [103]. Presently, fairly optimized conditions in non-phase-matched configurations typically yield a conversion efficiency of about 10^{-6} in the range of 10–40 eV (of the order of 10^9 photons per pulse) and 10^{-8} in the 40–150 eV range (10^6–10^7 photons per pulse) [104]. The highest energy reported for a high-order harmonic pulse, 60 nJ at about 50 eV, was obtained using a powerful glass laser [105]. Recently, the demonstration of phase-matched harmonic conversion of visible light into soft x-rays with a generation efficiency of 10^{-5}–10^{-6} in the 40–70 eV spectral region was reported [106, 107]. Soft x-ray pulses with an energy >0.2 nJ pulse^{-1} per harmonic order were produced at a repetition frequency of 1 kHz utilizing 20 fs optical pulses [106].

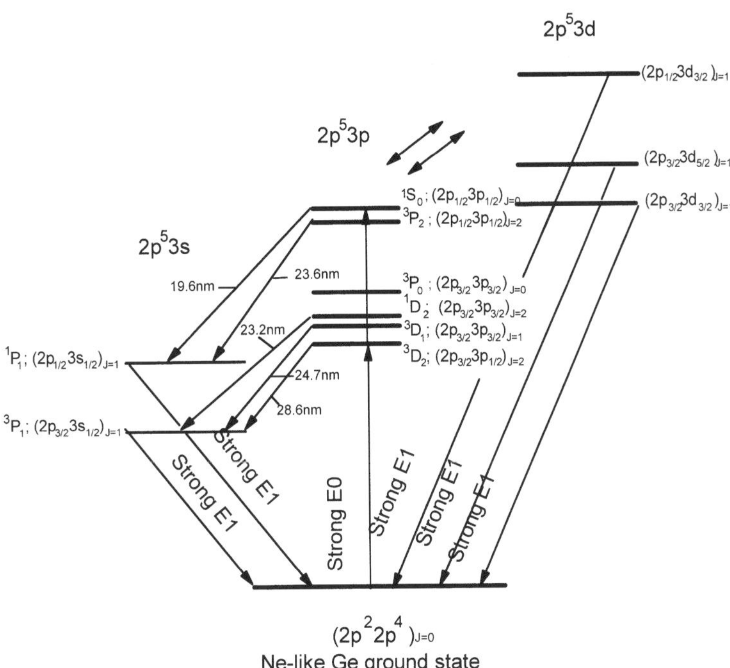

Figure B5.2.3. Simplified energy level diagram of Ne-like Ge showing the laser transitions of interest and the dominant excitation and de-excitation processes responsible for the creation of population inversions between the 3p and 3s levels. The laser levels are labelled in both LS and jj notations (C Keane, [62]).

B5.2.5 X-ray laser schemes

B5.2.5.1 Collisional electron excitation lasers

B5.2.5.1.1 Quasi-steady-state lasers

Collisional electron impact excitation of Ne-like and Ni-like ions has resulted in some of the most robust soft x-ray lasers available. In the traditional implementation of these lasers, the generation of a population inversion occurs in a quasi-cw regime by strong collisional monopole electron excitation of the upper level aided by the very favorable ratio between the radiative lifetime of the upper and lower levels. The upper levels are metastable with respect to radiative decay to the ground state and the lower levels are depopulated by strong dipole-allowed transitions. The first successful demonstration of a collisional soft x-ray laser was realized at Lawrence Livermore National Laboratory and involved the $2p^5 3p \rightarrow 2p^5 3s$ transitions in Ne-like Se and Ne-like Y [17]. Lasing has been extended to nearly all of the Ne-like ions with an atomic number between Si [26] and Ag [27] at wavelengths ranging from 87 to 9.93 nm.

Figure B5.2.3 shows a simplified energy level diagram for a typical Ne-like system (GeXXIII) illustrating the laser transitions and the dominant processes involved in the generation of amplification. The 3p upper levels are dominantly populated by electron monopole collisional excitation (E_0) from the Ne-like ion ground state and by cascades from higher energy levels. The population inversions are maintained by the very rapid radiative decay of the 3s lower levels to the ground state of the ion through strong dipole-allowed transitions (E_1). Therefore, operation of these lasers in a quasi-cw regime requires the plasma to be optically thin for the transitions originating from the lower level.

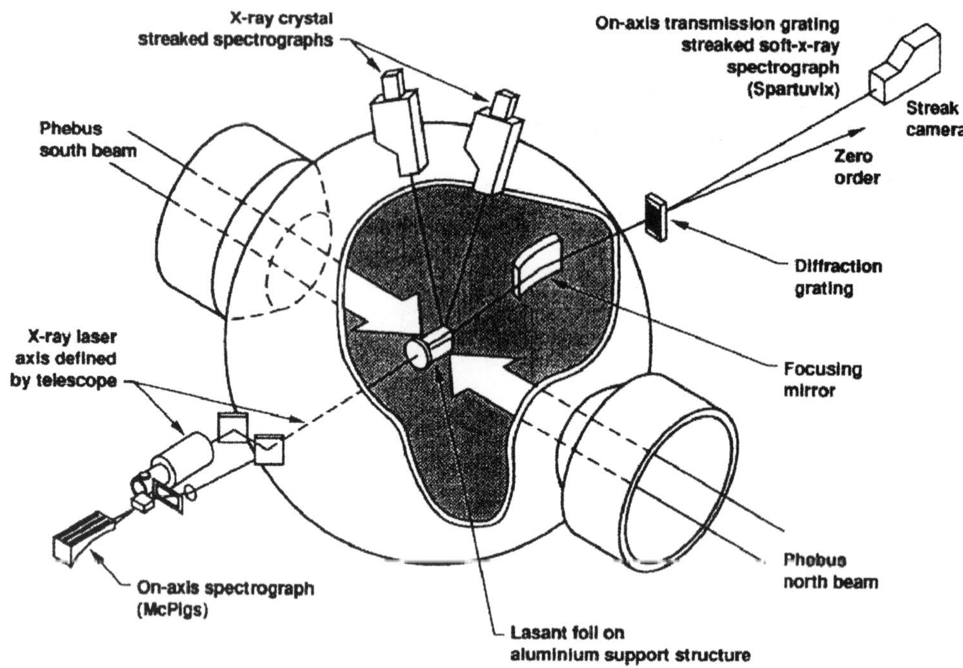

Figure B5.2.4. Schematic representation of a typical experimental set-up used to produce and diagnose a collisionally excited soft x-ray laser of the first generation. Two laser beams from an optical laser are line-focused onto a target to create the plasma column that constitutes the amplification medium. (From C J Keane *et al* [30].)

Lasers in the $3d^94d \rightarrow 3d^94p$ transitions of Ni-like ions are direct analogues to lasers in $2p^53p \rightarrow 2p^53s$ transitions in closed-shell Ne-like ions but have the advantage of producing amplification at shorter wavelengths for a given state of ionization. This higher quantum efficiency significantly reduces the pumping energy required to achieve lasing of a selected wavelength. Ni-like soft x-ray lasers were first demonstrated in 1987 in an Eu-laser-created plasma, producing amplification with a $gl \simeq 4$ at 7.1 nm [28]. Subsequently, the scheme was isoelectronically extrapolated to other ions with laser wavelengths as short as 3.56 nm in Ni-like Au [29]. Recently, gain-saturated operation has been obtained at wavelengths as short as 5.8 nm [32]. Hagelstein first proposed the use of low-Z Ni-like ions to develop tabletop collisional lasers at wavelengths near 20 nm and computed a significant gain for overheated plasma conditions [30]. Gain has also been observed in Co-like ions [29] and the use of the Nd-like sequence has also been proposed [30].

The collisionally excited soft x-ray lasers developed before 1990 consisted of line focus plasmas generated by optical laser pulses with energies ranging from several hundred joules to several kilojoules and, therefore, involved the use of very large facilities [17, 19–29]. One such laser is schematically illustrated in figure B5.2.4. In recent years, much progress has been made in improving laser efficiency. Strong refraction of the amplified beam caused by the large electron density gradients in the amplifier was early recognized as a major obstacle to the generation of efficient soft x-ray lasers with good beam quality. Methods implemented to mitigate refraction have included the use of foil targets [60, 124–127], opposite-gradient and curved targets and, in particular, the use of one or multiple pre-pulses [128–147]. In the pre-pulse technique, the first pulse is used to create a plasma with an optimized density for amplification and with reduced density gradients for improved beam propagation. The subsequent pulses which are absorbed more efficiently in the gain region to heat the plasma to lasing conditions lead to the observation of large amplification in a large number of elements with dramatic increase in the $(J = 0) \rightarrow (J = 1)$ line intensity and reduced excitation energy.

Another important step in the reduction of the pump energy has been the use of shorter excitation pulses. Daido *et al* [118, 119] reported lasing in several lanthanide ions in the spectral range 5.8–14 nm using a three-pre-pulse sequence and a main excitation pulse of 100 ps and ~250 J of energy. A subsequent series of experiments conducted at the Rutherford Laboratory obtained saturation in several transitions [120–123] with wavelengths as short as 5.86 nm (Ni-like Dy [58]) using sequences of 75 ps pulses with about 100 J of energy saturation. Balmer *et al* have used a relatively compact Nd:glass laser to demonstrate saturated lasing in Ne-like Fe (25.5 nm), Ni-like Ag (14.0 nm) and Ni-like Pd (14.7 nm) with driver energies below 30 J in a 100 ps pulse [148, 150]. In terms of improving the soft x-ray output beam characteristics, double-pass amplification experiments have demonstrated increased spatial coherence and reduced beam divergence.

B5.2.5.1.2 Transient collisional lasers

The collisional electron excitation scheme described earlier is intrinsically a quasi-steady-state scheme in which lasing can occur for as long as the plasma conditions necessary for the generation of a population inversion can be maintained. It was first recognized by Afanasiev and Shlyaptsev [151] that one to two orders of magnitude larger gain coefficients could be produced for a short period of time (typically subpicosecond to tens of picoseconds) by heating the plasma at a rate faster than the relaxation rate of the excited states. The larger gain coefficients are mainly the consequence of the larger rate of excitation of the upper level from the ion ground state by electron collisions, that results in a large population inversion before collisions have time to redistribute the populations. Another phenomenon that contributes to an increased gain is the increased rate of electron excitation in an overheated plasma. Transient gains in excess of 100 cm^{-1} have been predicted theoretically [151, 152]. In the transient regime, there is no need to limit the transverse dimension of the plasma in order to ensure optical transparency of the laser lower level radiation. A main advantage of the transient excitation scheme for the realization of tabletop x-ray lasers is the greatly reduced pump energy required for excitation.

The recent availability of multi-terawatt ultrashort pulsed optical laser systems with output energies of several joules (see chapter D10.1) has provided the opportunity to demonstrate soft x-ray lasing by transient electron collisional excitation [80–86]. The implementation is based on a two-step excitation sequence. First, a long pulse (typically of nanosecond duration) is used to produce a plasma containing the desired active ions, which are usually closed-shell Ne-like or Ni-like. The temperature of this pre-plasma must be sufficiently high to produce a ground-state population of these ions but it does not need to reach the values necessary to populate the upper level. The plasma is allowed to expand hydrodynamically to reach the desired degree of ionization, optimum electron density and minimum possible electron density gradient. Second, the plasma is rapidly heated with a picosecond or subpicosecond laser pulse to increase the electron temperature rapidly to values that exceed the excitation energy of the upper level. The ionization balance is not significantly altered and a large transient population inversion is generated by electron excitation.

The first demonstration of amplification by transient inversion was realized by Nickles *et al* in the 32.6 nm line of Ne-like Ti [80]. The experiment used a hybrid chirped-pulse amplification (CPA) Ti:sapphire/Nd:glass pump delivering synchronized long and short pulses of ~1.2 ns and 0.7 ps duration and energy of 3 and 2 J respectively. The soft x-ray pulse duration was measured to be less than 20 ps. The results were analysed to correspond to an average gain of 19 cm^{-1} and to $gl \simeq 9.5$ [80]. Weaker lasing was also observed in a line near 30 nm, identified as the $3d(J = 1) \rightarrow 3p(J = 1)$ transition of Ne-like Ti. These results were improved and extended to other Ne-like ions such as Fe and Ge in subsequent experiments conducted at several laboratories [81–86].

The transient collisional scheme has also been extended to shorter wavelengths utilizing the Ni-like sequence [81–86]. Dunn *et al* reported a gain of up to 35 cm^{-1} and $gl \simeq 12.5$ in the $4d^1S_0 \rightarrow 4p^1P_1$ line of

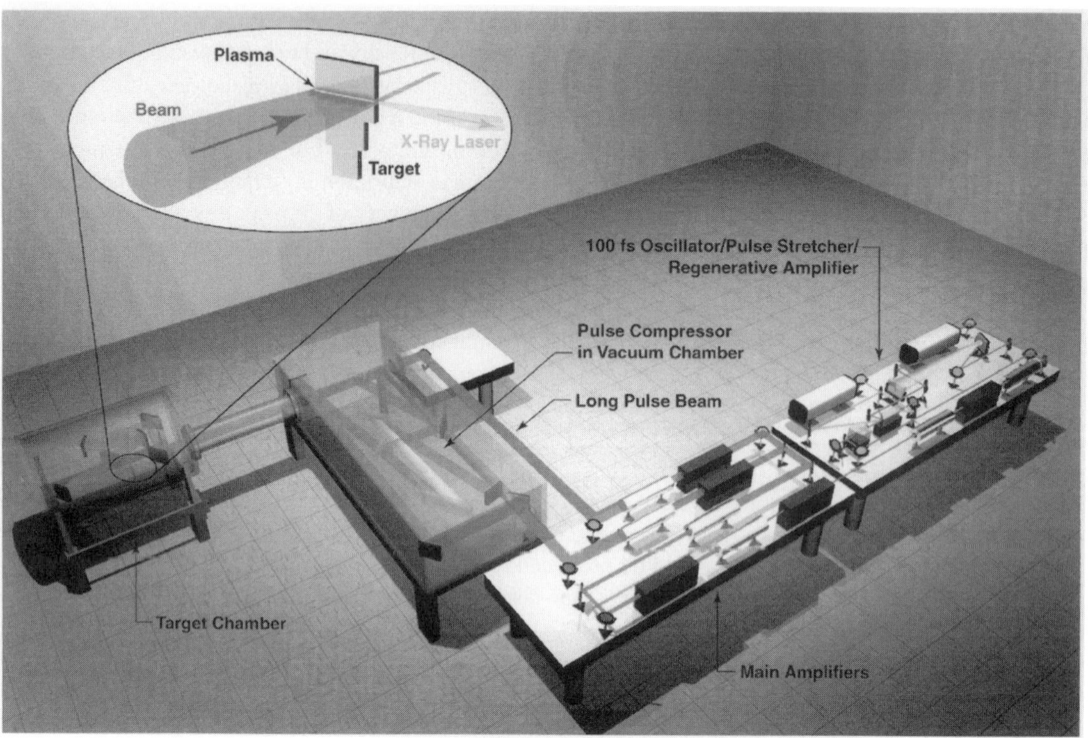

Figure B5.2.5. Schematic illustration of the CPM pump laser system and target chamber used to produce transient inversion collisional soft x-ray lasers at Lawrence Livermore National Laboratory. The pump laser occupies two standard optical tables of dimensions 1.2 m × 3.6 m with total area less than 10 m^2. The system can be fired once every 4 min. (Courtesy of J Dunn, Lawrence Livermore National Laboratory.)

Ni-like Pd at 14.7 nm and large amplification in several other Ni-like ions [81] using the relatively compact laser system illustrated in figure B5.2.5. The pump laser occupies two 1.2 m × 3.6 m optical tables with a total area of less than 10 m^2. This 15 TW hybrid laser system consists of a Ti:sapphire oscillator tuned to 1050 nm, and four-stage Nd-phosphate glass amplifiers. It generates two synchronized laser pulses of 7.5 J in 500 fs and 15 J in 600 ps at a repetition rate of one shot every 4 min. Typically 50 shots on target are achieved in a day. The full circles in figure B5.2.6(*b*) show the measured increase of the Ni-like Pd 14.7 nm line intensity for target lengths between 2 and 9 mm. The gain coefficient for lengths between 1 and 2 mm is about 35 cm^{-1} but continuously decreases to reach a value of 3.9 cm^{-1} for target lengths above 7 mm. This smooth decrease in gain with target length has been observed in all non-travelling-wave transient collisional excitation experiments and resembles gain saturation. However, it is mainly caused by the short duration of the gain and by refraction. Gain saturation with the transient excitation scheme was first demonstrated on the Ne-like scheme for the 32.6 nm line of Ne-like Ti and the 19.6 nm line of Ne-like Ge at the Rutherford Laboratory [83]. However, these experiments utilized a total reported excitation energy of 32 J and 60 J, respectively, which is not available in smaller facilities. To achieve gain saturation with smaller excitation energy, travelling-wave excitation schemes have been implemented at several laboratories.

B5.2.5.1.3 Travelling-wave excitation

In these transient systems, the shortness of τ_g, the gain duration, limits the amplification length to $l < c\tau_g$ (where c is the speed of light in the plasma), unless travelling-wave excitation is used to maintain the excitation

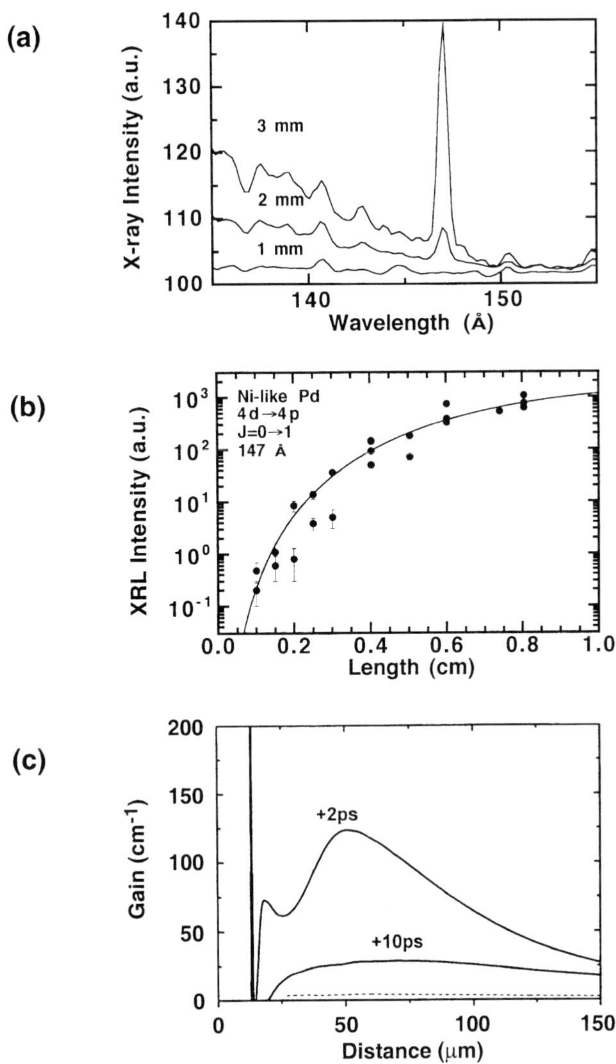

Figure B5.2.6. Results of transient inversion soft x-ray amplification experiment in the 4d–4p$J = 0$–1 line of Ni-like Pd at 14.7 nm. (*a*) Spectra showing the increase of the intensity of the laser line as a function of target length. The intensity increases by more than three orders of magnitude when the target length is increased from 1 to 3 mm. (*b*) Measured intensity of the laser line as a function of target length. The continuous line is a guide through the experimental points. The data show the decrease of the transient gain as the x-ray laser propagates along the plasma column. (*c*) Computed transient gain profiles as a function of time measured from the arrival of a 1 ps excitation pulse. (From J Dunn *et al* [81].)

in phase with the amplified x-ray pulse. The travelling-wave system was used to demonstrate gain saturation in Ni-like Pd at 14.7 nm with 1.8 J of long pulse energy and 5.2 J of short pulse. Figure B5.2.7 illustrates the advantage of travelling-wave excitation and corresponds to a 20–100-fold increase in x-ray laser intensity compared to the configuration without travelling-wave excitation [85]. The soft x-ray laser output energy with travelling-wave excitation was estimated at $\sim$10 μJ. Strong amplification was also obtained in Ni-like ions ranging from Mo (18.9 nm) to Sn (11.9 nm) [83]. A different travelling-wave method was also used at

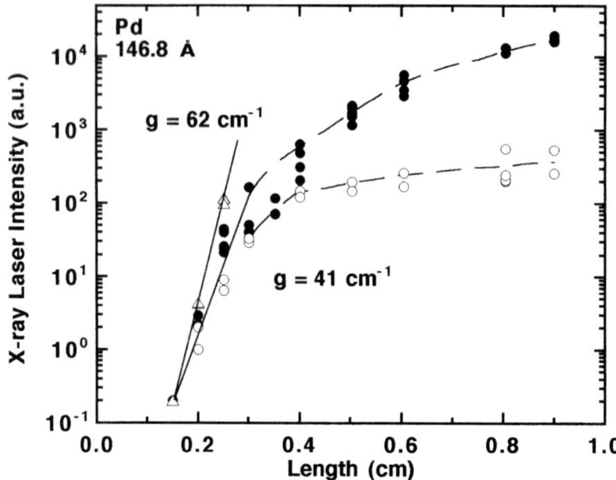

Figure B5.2.7. Comparison of the x-ray laser intensity *versus* plasma column length for the Ni-like Pd 4d–4p laser line at 14.7 nm with travelling wave (closed circles) and without travelling wave (open circles). The excitation energy consisted of a 1 .8 J long pulse and a 5.2 J short pulse. (From J Dunn *et al* (2000) [84].)

Table B5.2.1. Characteristics transient inversion collisional laser pumped by a compact multi-terawatt laser, as compared with quasi-steady-state collisional laser pumped by NOVA (after J Dunn *et al* [67].)

	COMET	NOVA
Size (sq. ft)	100	> 40 000
Cost ($)	1 M	> 100 M
Pump energy (J)	< 10	$5–10 \times 10^3$ (2 beams, 1ω)
XRL gain (cm^{-1})	30–65	1–8
XRL output (J)	$> 10 \times 10^{-6}$ (at 146.8 Å)	5×10^{-3}(at 155 Å)
Shot rate (per day)	50–100	4–6
XRL wavelength (Å)	119–330	35–330
Pulse duration (ps)	5–10	45–200
Brightness (ph mm^{-2}mrad^{-2}s^{-1}(0.01% BW)$^{-1}$	10^{24}	10^{24}
Cost/shot ($)	50–100	10 000–20 000

the Rutherford Laboratory to obtain gain saturation in several Ne-like ions and in Ni-like Sm at 7.3 nm [85]. Experiments conducted at CEA-Limeil yielded saturation of the 13.9 nm laser line of Ni-like Ag and also observed strong amplification at 16.05 nm in a 4f–4d transition in the same ion [86]. Table B5.2.1 compares the characteristics of soft x-ray lasers driven by the transient collisional scheme at a facility with a relatively compact multi-terawatt laser with those of the quasi-steady-state collisional lasers pumped by the Nova laser facility [84]. Future improvements in pumping efficiency resulting from optimized target configurations and travelling-wave excitation can be expected to lead to saturated transient inversion soft x-ray lasers occupying a single optical table and to tabletop lasers that operate at shorter wavelengths.

Another method for pumping collisional lasers in which travelling-wave excitation is intrinsic was demonstrated by Lemoff *et al* in Pd-like Xe utilizing optical field ionization. In this scheme, an intense circularly-polarized femtosecond laser pulse is used to simultaneously create closed-shell ions and hot

Table B5.2.2. Characteristics of capillary-discharge-pumped tabletop 46.9 nm laser in Ne-like Ar.

Laser parameters		Ref.
Pulse energy	0.88 mJ at 4 Hz	57
Average pulse power	3.5 mW	57
Peak pulse power	0.6 MW	57
Divergence	$\approx$ 4.6 mrad	56, 57
Pulsewidth	1.2–1.5 ns	56, 57
Peak spectral brightness	2×10^{25} photons s^{-1} mm^{-2} $mrad^{-2}$ (0.01% bandwidth)$^{-1}$	100

pumping electrons by tunnelling ionization [79]. An amplification of $gl \simeq 11$–12 in the 41.8 nm line of Pd-like Xe was measured by axially pumping a Xe gas cell up to 8.4 mm in length with 70 mJ pulses from a Ti:sapphire laser operating at 10 Hz.

B5.2.5.1.4 Collisionally excited capillary discharge lasers

Direct excitation of plasma columns with an electrical discharge has the advantage of generating soft x-ray lasers that are very efficient and compact. Fast discharge excitation of capillary plasmas has produced the highest soft x-ray laser average power to date [55, 56]. In these lasers the electromagnetic forces of a fast rising current pulse flowing through a capillary channel rapidly compresses the plasma to a size of about 300 μm diameter. The fast current risetime minimizes the amount of material that is ablated from the capillary walls before the magnetic field detaches it from the walls [47–50]. The first observation of large soft x-ray amplification in a discharge-created plasma was realized in 1994 by Rocca *et al* in the 46.9 nm line of Ne-like Ar [48, 49]. In that initial experiment, illustrated in figure B5.2.8, the fast capillary discharge set-up (generating current pulses of 60 ns half-cycle duration and ~40 kA peak current) was used to excite Ar plasma columns in 4 mm diameter capillary channels up to 12 cm in length. As indicated in figure B5.2.9, the capillary was placed in the axis of a 3 nF liquid- dielectric capacitor that was pulse-charged by a Marx generator. The capillary loads were excited by discharging the capacitor through a spark-gap switch pressurized with SF_6. A small gain was also observed in the $(J = 2) \rightarrow (J = 1)$ line of Ne-like Ar at 69.8 nm [49, 58]. The optimum conditions for lasing occur several nanoseconds before stagnation, when the first compression shock wave reaches the axis. Lasing by collisional electron excitation of Ne-like Ar ions takes place at a time when the electron density is rapidly increasing and reaches (0.3–1 $\times$ 10^{19} cm^{-3}), when the electron temperature is 60–80 eV [51]. The pulsewidth is approximately 1 ns [51, 54]. Subsequent experiments employing longer plasma columns under better optimized discharge conditions yielded an effective gain–length product of 27 and resulted in the first observation of gain saturation in a tabletop soft x-ray amplifier [51]. Saturation of the laser intensity was observed at products of about 14 [51].

A very compact, high repetition rate, saturated 46.9 nm laser of a size comparable to that of many widely utilized visible and ultraviolet gas lasers has been developed [55, 56]. Figure B5.2.10 illustrates the size of this capillary discharge soft x-ray laser in comparison to a 1 J Nd:YAG laser. The soft x-ray laser occupies a surface space of about 1 m by 0.4 m on an optical table. This laser was operated with capillaries 18.2 cm in length at a repetition rate of 7 Hz to produce an average output pulse energy of 135 μJ, corresponding to an average laser power of 1 mW [55]. Increasing the plasma column length to 35.4 cm resulted in the generation of 0.88 mJ laser pulses at a repetition rate of 4 Hz [56], corresponding to an average power of 3.5 mW as shown in figure B5.2.11. The spatially coherent average output power per unit bandwidth emitted by this compact laser at 26.5 eV is comparable to that generated by a beam-line at a third generation synchrotron

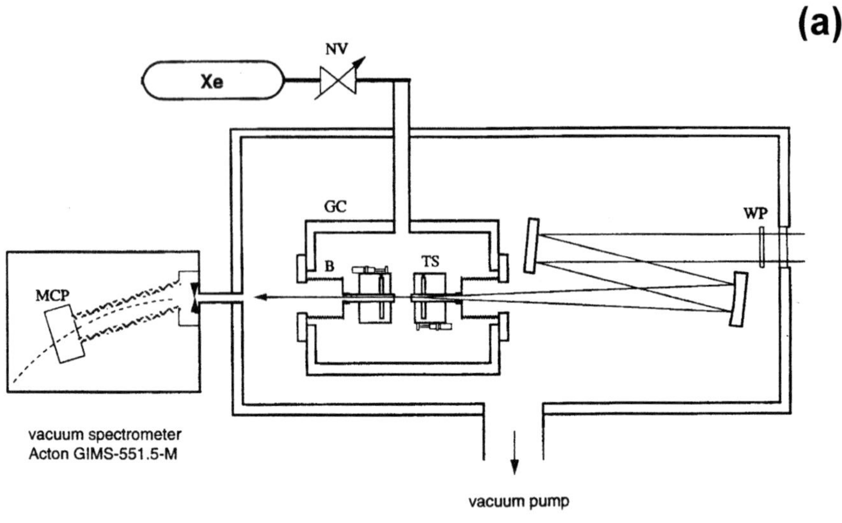

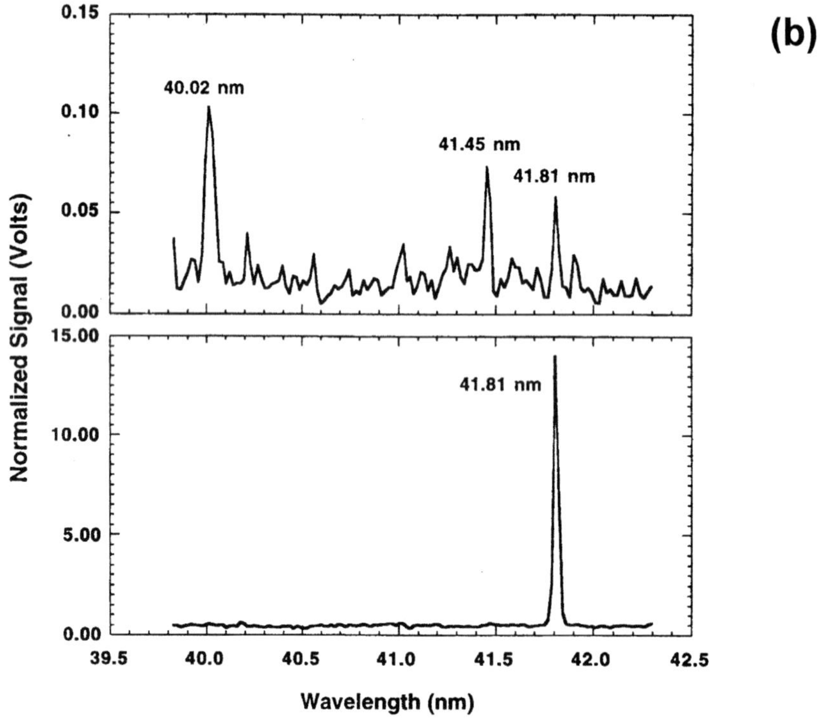

Figure B5.2.8. (*a*) Schematic illustration of the gas target set-up used to obtain lasing in Pd-like Xe at 41.8 nm in a plasma created by optical-field-induced ionization. Excitation was provided by circularly-polarized laser pulses with an energy of 70 mJ and a duration of 40 fs from a CPA Ti:sapphire laser operating at 10 Hz. (*b*) Spectra illustrating the dramatic increase in the intensity of the 41.8 nm laser line observed when the pressure is increased from 3 torr (top spectrum) to 12 torr (lower spectrum). (From B E Lemoff *et al* [79].)

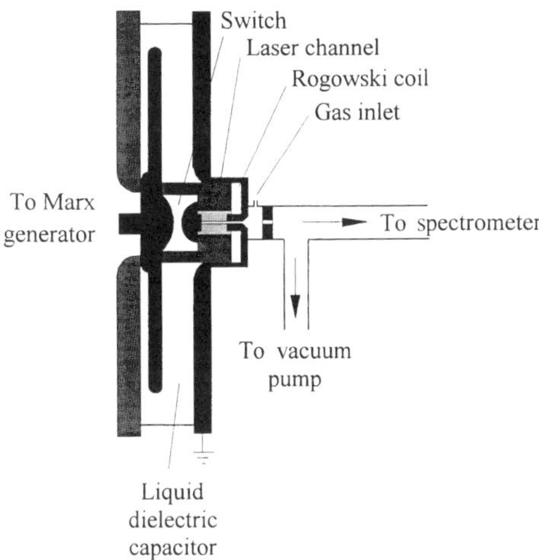

Figure B5.2.9. Schematic diagram of the capillary discharge set-up used to obtain saturated amplification of the 46.9 nm line of Ne-like Ar. (From J J Rocca *et al* [48, 49, 51].)

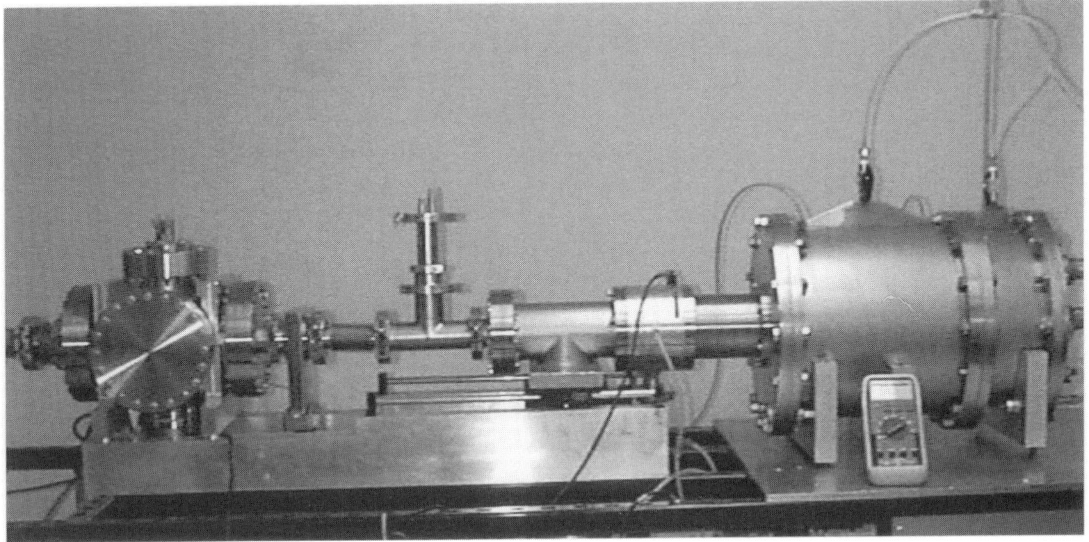

Figure B5.2.10. Photograph a of high average power tabletop 46.9 nm capillary discharge (right) and applications chamber (left). The multimeter provides a reference for the size of the laser. (From Colorado State University.)

facility, while its peak coherent power per unit bandwidth exceeds that of the synchrotron by several orders of magnitude [55, 56]. Table B5.2.2 summarizes the characteristics of the 46.9 nm Ne-like capillary discharge laser. These capillary-discharge-pumped tabletop lasers were successfully used in several applications, including high-resolution soft x-ray laser interferometry and shadowgraphy of plasmas [101, 102, 153, 154], the measurements of XUV optical constants of materials [155] and the demonstration of laser ablation with a focused soft x-ray beam [156].

The Ne-like Ar results were extended to Ne-like S (60.8 nm) [53] and Ne-like Cl (52.9 nm). To obtain

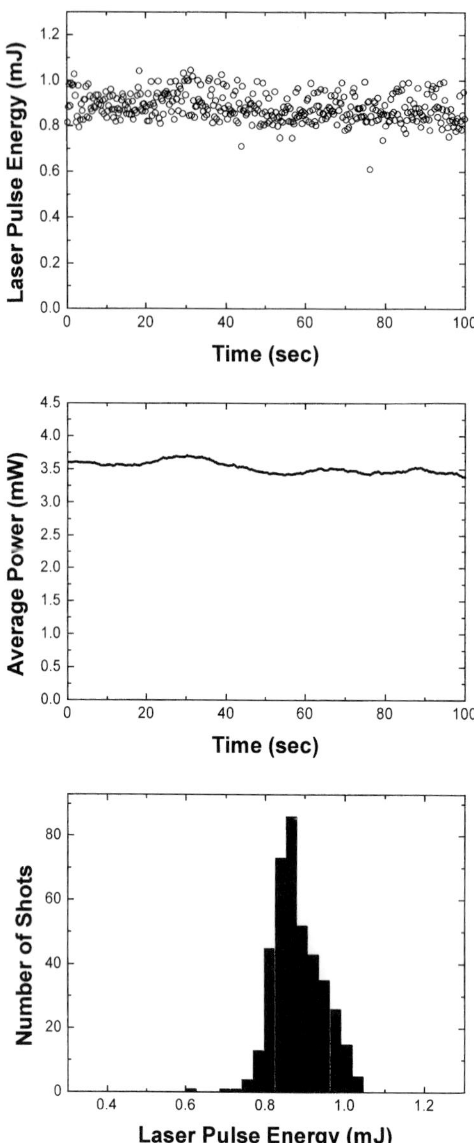

Figure B5.2.11. Measured output pulse energy and average output power of a tabletop capillary discharge 46.9 nm laser operating at a repetition frequency of 4 Hz. (*a*) Shot-to-shot laser output pulse energy. (*b*) Average output power computed as a walking average of 60 contiguous laser pulses. (*c*) Distribution of the output pulse energy. The average pulse energy is 0.88 mJ, corresponding to an average power of 3.5 mW. (From C Macchietto *et al* [57].)

amplification in Ne-like S, the discharge set-up illustrated in figures B5.2.8 and B5.2.9 was modified to allow for the injection of the sulphur vapour into the capillary channel through a hole in the ground electrode. The sulphur vapour was produced, ablating the wall of an auxiliary capillary channel, drilled in a sulphur rod, with a slow current pulse. Model computations indicate that, in this case, transient population effects increase the gain by 20–40% [53]. The scaling of plasma conditions and pump requirements for lasing at wavelengths as

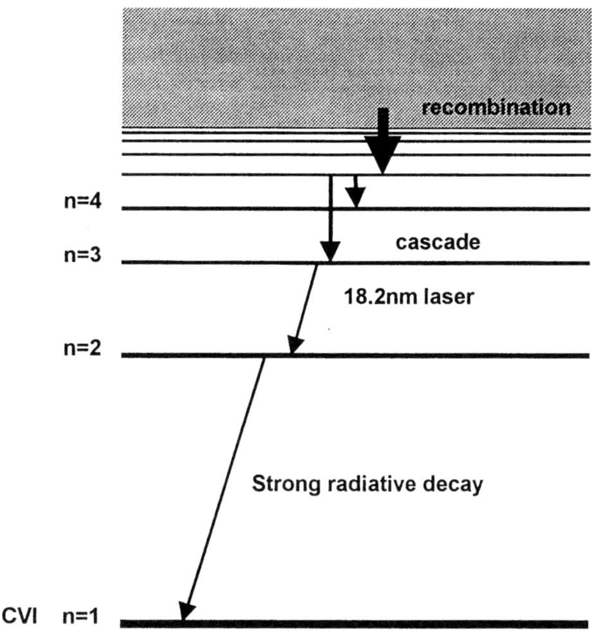

Figure B5.2.12. Simplified Grotrian diagram of H-like C showing the processes responsible for the generation of population inversion between the $n = 3$ and $n = 2$ levels in the 18.2 nm CVI recombination laser.

short as 10 nm have been estimated [49]. Recently, current pulses of up to 200 kA peak amplitude and ~10 ns risetime were used to explore the generation of capillary plasma columns with the conditions necessary for amplification at shorter wavelengths [157].

B5.2.5.2 Collisional recombination lasers

B5.2.5.2.1 Mechanisms

In this population inversion scheme, first suggested by Gudzenko and Shelepin [158], the upper level is populated following the recombination of ions of atomic species A with a charge $Z + 1$ with an electron, through a three-body interaction described as collisional or three-body recombination:

$$(\mathrm{A}^{Z+1}) + 2\mathrm{e} \rightarrow (\mathrm{A}^{Z})^{*} + \mathrm{e}. \tag{B5.2.10}$$

This process preferentially populates highly excited bound levels $(\mathrm{A}^Z)^*$ of the ion of charge Z, favouring the generation of population inversion. The collisional recombination rate is proportional to the square of the electron density. The recombination rate is also extremely sensitive to T_e, the electron temperature, following a dependence on $T_\mathrm{e}^{-4.5}$. Therefore, the generation of large population inversions by recombination requires a dense and relatively cold plasma.

Hydrogen-like ions have a very favourable energy-level structure for the generation of population inversions by collisional recombination. In principle, several transitions can be inverted but, initially, much of the attention focused in the very favourable $3 \rightarrow 2$ transition. Figure B5.2.12 shows the atomic processes involved in the generation of population inversion in the $3 \rightarrow 2$ transition of H-like C at 18.2 nm. When the plasma cools, collisional recombination populates highly excited states and electron collisions rapidly transfer the population to levels of lower energy. Since collisional electron de-excitation is inversely proportional to the square root of the energy difference between the levels, this electronic cascade reaches a level at which electron de-excitation is no longer dominant over radiative decay. At this level, the bottleneck that is produced in the cascade creates a population inversion with respect to a lower level that is de-excited by very rapid radiative decay to the ion ground state.

The amplification of soft x-rays by plasma recombination would seem to require conflicting plasma conditions: a very highly ionized plasma and a very cold electron temperature. In practice, the problem has been traditionally solved utilizing a two-step process. First, a hot and dense plasma with a large population of A^{Z+1} ions is generated by a heating pulse. Second, the plasma is rapidly cooled by an adiabatic expansion [10, 11, 16, 159–166], by electron heat conduction to a nearby wall or colder neighbouring plasma [167] or by radiation from high-Z ions introduced as impurities into the plasma [168, 169]. All three cooling mechanisms, or combinations of them, have been utilized experimentally to generate gain at soft x-ray wavelengths by collisional electron–ion recombination.

Important efforts have been devoted to amplification in the 18.2 nm $3 \rightarrow 2$ line of H-like C. Initial experiments observed population inversions in plasmas generated by ablating solid carbon targets or carbon fibres with high-power laser pulses and cooled by adiabatic expansion [10, 11, 14–16]. Large amplification was first observed by Suckewer *et al* in an experiment in which a 300 J pulse from a CO_2 laser with about 75 ns pulsewidth was used to generate a nearly totally ionized carbon plasma column by bombarding a carbon solid target immersed in a strong solenoidal magnetic field [18]. The magnetic confinement allowed the maintenance of a high electron density while the plasma was cooled by radiation and electron heat conduction to adjacent cooling blades. Laser pulses of about 3 mJ were generated at an efficiency of 10^{-5}. Amplification has also been demonstrated in numerous experiments in which line focus plasmas were cooled by free adiabatic expansion [41, 160–165]. Gain has been reported in H-like Na ions [163], in the 4f$\rightarrow$3d and 5f$\rightarrow$3d lines of Li-like ions [164, 165] and in Be-like and Na-like ions [170].

Several experiments have been conducted during the last ten years toward the realization of a small-scale recombination laser [171–178]. In 1989, Kim *et al* reported a gain of about 8 cm^{-1} in the 18.2 nm line of CVI using a 25 J pump pulse of 3 ns duration, line-focused into a 5 mm-long solid carbon target [171]. Later experiments obtained similar gain coefficients target with only 4 J of laser excitation energy using a multi-fin carbon target [173]. However, difficulties were found in increasing the amplification over $gl \simeq 5$. One of the latest attempts to increase the product in this line utilized a polyethylene microcapillary as target [174]. A Nd:YAG laser (2.5 J, 1.5 ns pulsewidth) was focused at the entrance of the microcapillary with the motivation of guiding the pump beam, plasma confinement and plasma cooling by heat conduction to the walls. A product of about 5 was inferred from the intensity ratio of lines in the axial and perpendicular direction. A similar experiment recently conducted in B_2O_3 resulted in a gain of $gl \simeq 5$ in the 3$\rightarrow$2 transition of H-like B at 26.2 nm [178]. A pre-plasma was created by a 20 ns, 0.2 J KrF laser and the excitation was produced by a 0.4 J, 8 ns Nd:YAG laser operating at 1 Hz. Hara *et al* have utilized multiple-pulse excitation from a pulse train with the objective of developing a compact soft x-ray recombination laser based on a train of 100 ps pulses produced by a tabletop Nd:YAG laser [175–177]. Experiments conducted utilizing 1.5–2 J of laser pulse energy to excite an 11 mm long Al slab target were reported to produce a gain coefficient of 3.2 cm^{-1} in the 15.5 nm line of Li-like Al [177] but the amplification was small.

An advantage of the recombination scheme with respect to the collisional excitation scheme involving $\Delta n = 0$ transitions is its more rapid scaling to shorter wavelengths with nuclear charge Z. However, recombination lasers have suffered the problem of not scaling adequately with plasma column length, a problem possibly related to the very high sensitivity of the gain in recombination schemes to the variation of the plasma parameters. The largest product achieved to date is eight [18].

B5.2.5.2.2 *Gain in discharge-pumped recombination systems*

The possibility of obtaining amplification by collisional recombination in capillary discharge plasmas was proposed by Rocca *et al* in 1988 [179]. Evidence of gain has been reported in several discharge-pumped recombination laser experiments [180–186]. However, in all cases only small exponentiation ($gl<4$) has been observed to date, with the line intensity remaining of the same order as that of surrounding non-lasing lines emitted by the plasma.

B5.2.5.2.3 *Recombination lasing in transitions to the ground state*

Jones and Ali [187] demonstrated theoretically that large transient gains could be generated in the $2 \rightarrow 1$ transitions of H-like ions following recombination of a totally ionized plasma of arbitrarily low temperature. This kind of laser in a transition to the ground state is very attractive to rapidly scale recombination lasers to very short wavelengths. However, for lasing to occur in a transition to the ground state, recombination has to be very rapid to allow for the generation of large upper level populations before the ion ground-state level is significantly populated. The recombination time must then be shorter than the radiative lifetime of the upper level which is, for example, 26 and 1.6 ps for the $n = 2$ level of H-like Li and H-like C, respectively. These time scales are shorter than the time required to cool a hot plasma before the ground state is significantly populated. Burnett and Corkum [188] recognized that a cold plasma of highly stripped ions in which recombination can occur in a very short time scale can be created by optical-field-induced ionization (OFI) using linearly polarized light and proposed the realization of ultrashort wavelength lasers utilizing this approach. Several theoretical studies analysed different aspects of the implementation of soft x-ray recombination lasers based on OFI [189, 190].

The first report of amplification following plasma recombination of an OFI plasma corresponds to an experiment 13.5 nm in the $2 \rightarrow 1$ transition of H-like Li realized by Nagata *et al* in 1993 utilizing the set-up illustrated in figure B5.2.13(*a*). [76, 77]. A cold plasma of totally ionized Li ions was generated in two steps. First, a plasma of singly-ionized Li ions was produced by line-focusing a 20 ns KrF laser pulse with an energy of 200 mJ onto a rotating lithium target. Second, after a selected time delay, the plasma was optical-field ionized with 50 mJ pulses from a linearly polarized subpicosecond (0.5 ps) KrF laser pulse focused to an intensity of 1×10^{17} cm^{-2}. Figure B5.2.13(*b*) shows the measured variation of the 13.5 nm laser line intensity with plasma length, from which a gain coefficient of 20 cm^{-1} was deduced. The 13.5 nm radiation pulse was measured to have a FWHM duration of less than 20 ps, with the peak intensity occurring about 20 ns after the pump pulse. However, the maximum resulting product only reached four because the plasma length was limited to 2 mm by ionization-induced refraction. This limitation is the result of the confocal geometry that causes the beam to refract due to the convex electron density profile resulting from the maximum ionization on axis. Approaches to overcome this problem have been suggested and include the generation of very flat transverse intensity profiles [190] and the generation of plasma waveguides with a minimum density on axis [78, 191–196].

Korobkin *et al* [78] used the set-up illustrated in figure B5.2.14(*a*) to increase the amplification length to 5 mm by pre-forming a plasma waveguide in an LiF microcapillary tube. A plasma with a minimum electron density on axis was generated by ablation of the capillary walls with a 100 mJ, 5 ns pulse of 1.06 μm wavelength radiation from a Nd:YAG laser loosely focused on the entrance of the microcapillary. Following a time delay of several hundred nanoseconds, the plasma was further ionized with a laser pulse of 50 mJ energy and 250 fs duration from a tightly focused KrF laser. The observed increase of the intensity of the 13.5 nm line of H-like Li as a function of length is shown in figure B5.2.14(*b*), and corresponds to $gl = 5.5$ [76]. These results were obtained at a repetition rate of 2 Hz. Gain by recombination in an OFI plasma has also been reported in the 2p2s ^{3}P$\rightarrow$2p^2 ^{3}P transition of OIII ($\lambda = 37.4$ nm) and in the 2p^{2}3s ^{2}P $\rightarrow$2p^3 ^{2}D line of OII ($\lambda = 61.7$ nm) [197]. The data suggest gain coefficients as high as 25 cm^{-1} but again the overall amplification was small due to the small length of the gas targets.

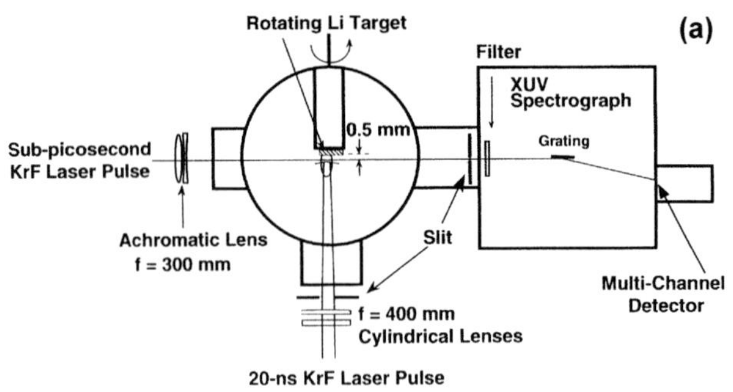

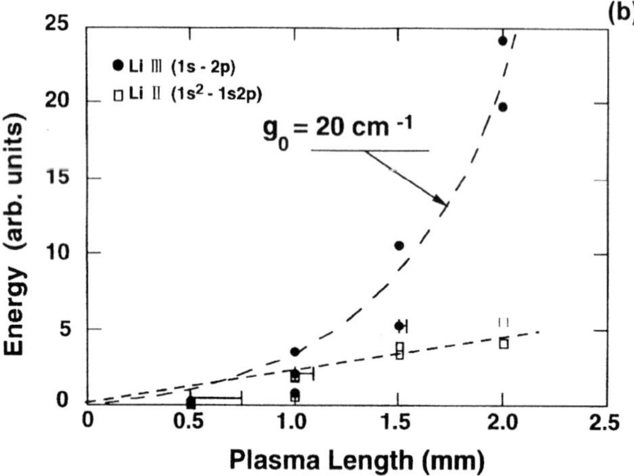

Figure B5.2.13. (*a*) Schematic illustration of the experimental set-up used at RIKEN in the observation of amplification in the 13.5 nm line of H-like Li in a plasma created by optical field induced ionization. (*b*) Results of amplification experiment in the $2 \rightarrow 1$ line of H-like Li in a plasma created by optical-field-induced ionization. Intensities of the 13.5 nm laser line of H-like Li and 19.9 nm line of He-like Li as a function of plasma column length. The broken curve is a calculated increase corresponding to a small-signal gain coeffient of 20 cm^{-1}. (From Y Nagata *et al* [77].)

The small excitation energies used in OFI soft x-ray laser experiments are compatible with the requirement for the generation of high-repetition-rate tabletop soft x-ray lasers. However, an increase in the product from the reported values of $\sim$6.5 to the larger values is still required to achieve practical output energies. Predictions of the output pulse energies that can be expected from OFI lasers range from 1 μJ [190] to 10–20 μJ [198]. The OFI recombination scheme can, in principle, also be scaled to shorter wavelengths by using ions with higher charge and more intense optical fields, provided that parametric electron heating effects (mainly stimulated Raman backscattering) can be overcome.

B5.2.5.3 Photoionization lasers

The generation of amplification following the selective x-ray photoionization of inner shell electrons was originally proposed by Duguay and Rentzepis in 1967 [1]. The generation of population inversions by this mechanism exploits the fact that at photon energies just above the threshold for inner-shell photoionization, the cross section is an order of magnitude larger for inner-shell electrons as compared to outer-shell electrons.

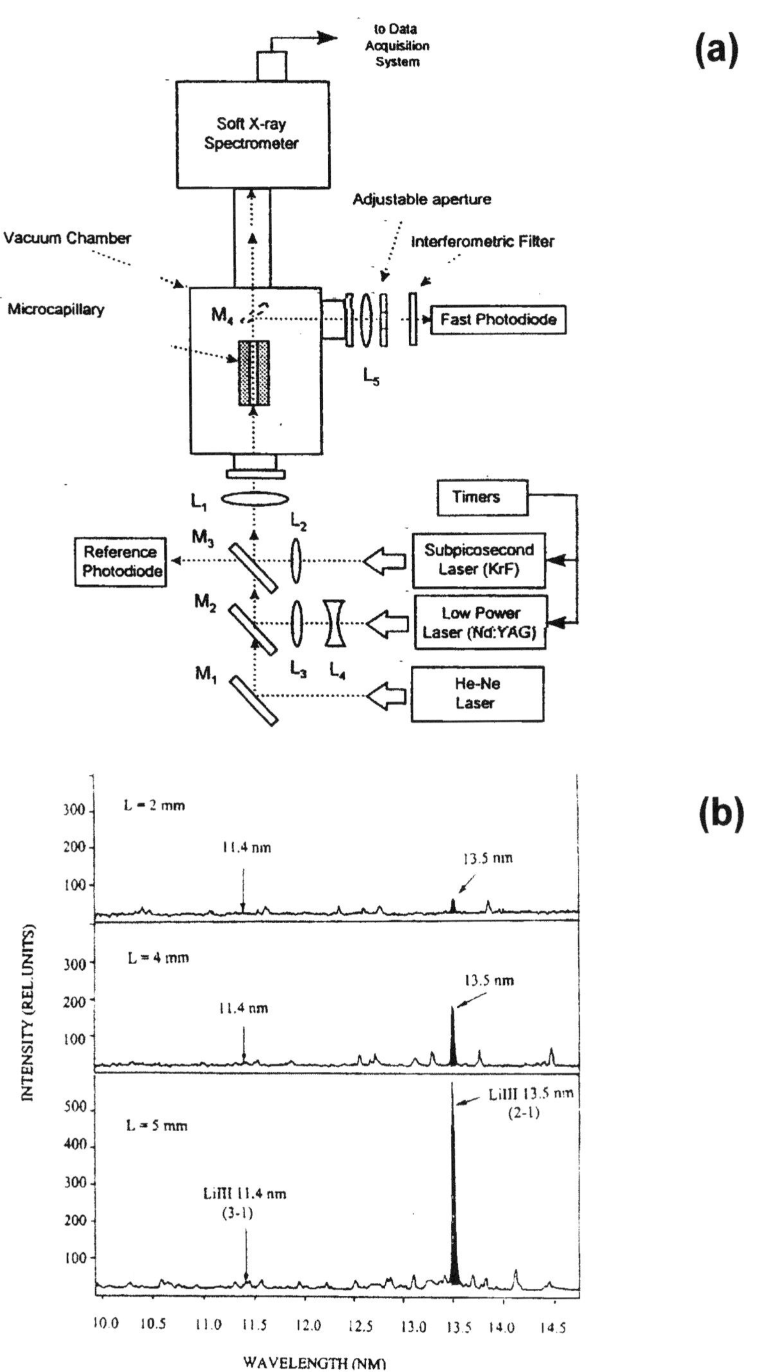

Figure B5.2.14. (*a*) Schematic illustration of the experimental set-up used at Princeton in the creation of OFI plasmas in LIF microcapillaries for the observation of amplification in the 13.5 nm laser line of H-like Li. (*b*) Axial spectra of an OFI plasma created in a LiF microcapillary using the set-up shown in (*a*). The $n = 2$–1 transition of H-like Li at 13.5 nm is observed to increase as a function of plasma length with a gain coefficient of 11 cm^{-1}. (From D Korobkin *et al* [78].)

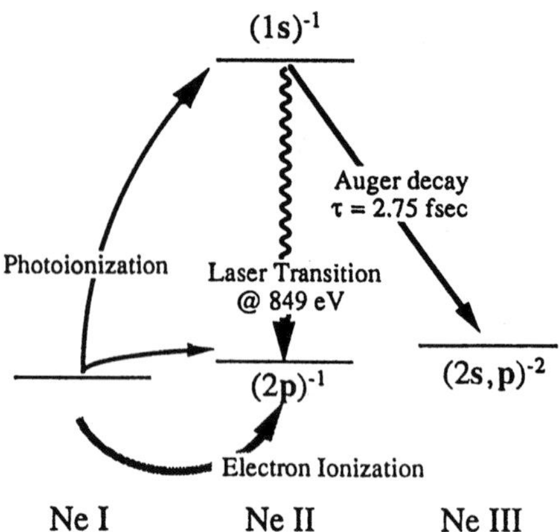

Figure B5.2.15. Simplified energy level diagram of a proposed inner-shell photoionization laser in the Kα transition of Ne at 1.5 nm. X-rays photoionize the 1s electrons of the Ne atoms creating a population inversion and gain in the $(1s)^{-1} \rightarrow (2p)^{-1}$ transition of singly ionized Ne. Rapid Auger decay depletes the $(1s^{-1})$ upper level. Electron impact ionization populates the $(2p)^{-1}$ lower level limiting the duration of the gain. (From H C Kapteyn [200].)

This scheme has the potential advantage of leading to compact lasers with wavelengths shorter than 1.5 nm [199, 200]. In principle, it can allow for operation at a low plasma temperature (<1 eV) with consequently less Doppler broadening and large gain coefficients. The incoherent x-ray photons that pump the laser media can be produced by a nearby plasma created by heating a high-Z target material, such as gold, with an intense ultrashort laser pulse. Photons at energy below the inner-shell binding energy can be removed from the pump with an appropriate filter to avoid pumping of the lower level.

The first proposal for inner-shell photoionization lasers at x-ray wavelengths [2] preceded by decades the development of sufficiently powerful ultrashort pulse laser drivers. McGuire proposed the production of population inversions by selective Auger decay [201]. In 1986 Kapteyn *et al* demonstrated lasing at 108.9 nm in doubly-ionized Xe by Auger decay following the photoionization of neutral Xe [202]. Output energies of several μJ in this transition were subsequently obtained at a 2 Hz repetition rate by Sher *et al* utilizing travelling-wave excitation [203]. Amplification in the equivalent transition in Kr at 90.7 nm was also observed [204]. However, no demonstration of a photoionization laser at x-ray wavelengths has yet been realized. Nevertheless, the recent development of lasers capable of producing peak powers of up to 100 TW with ~20 fs pulse duration increases the likelihood that this type of x-ray lasers will soon be realized.

Kapteyn has analysed the possibility of obtaining lasing by preferentially photoionizing the K-shell electrons of low-Z elements [199], focusing in particular on the Kα line of Ne ($\lambda = 1.5$ nm). A main advantage of the Kα scheme is its scalability to very short wavelengths. The energy-level diagram of the Kα Ne laser is illustrated in figure B5.2.15. The filtered x-rays from the pump primarily photoionize the inner-shell electrons, producing population inversion and gain in the allowed radiative transition between the $(1s)^{-1}\ {}^2S$ and $(2p)^{-1}\ {}^2P$ levels of the singly charged ion. Rapid Auger decay with a rate of $(2.7\ \text{fs})^{-1}$ is the dominant mechanism of depopulation of the $(1s)^{-1}$ upper level. However, the self-terminating lifetime is not the dominant process that limits the duration of the gain. Instead, the magnitude and duration of the gain is limited by electron collisional ionization of neutral neon atoms. The energetic electrons created in the lasing material by photoionization and Auger decay create predominantly ground-state Ne ions that are the lower level, destroying the gain. Therefore, this is intrinsically an ultrashort pulse x-ray laser. It has been estimated that lasing in the Kα transition of Ne at 1.5 nm could be produced by travelling-wave excitation using a pump laser generating ~10 J pulses of approximately 50 fs duration [199]. Eder *et al* [190, 200, 205] calculated a scaling law relating the necessary laser pump energy E_L (J) to the wavelength λ (nm) that can be achieved. Assuming 20 fs x-ray pump pulses, they obtained $E_L = (4.5/\lambda)^2$ [206]. Calculations for lasing in C at 4.5 nm suggest that driving laser energy of the order of 1 J should be sufficient to produce a gain coefficient of 10 cm^{-1} and a large product [205].

The possibility of using inner-shell electrons to obtain lasing in 2p→2s Ne-like ions and 3d→3p Ni-like ion transitions has also been suggested [207, 208]. Barty *et al* and Kim *et al* have recently discussed systems in which dominant Coster–Kronig decay of the laser lower level can overcome electron collisional filling of this state [209]. These systems are, therefore, more immune to secondary electron ionization. Computations show that, under certain conditions, inversions can be obtained with pumping by energetic electrons as well as x-rays. Conditions for inversion by Coster–Kronig lower level decay were analysed in several transitions in elements with nuclear charge up to $Z = 90$, and detailed computations of gain were published for the L_2M_1 transition in Ti ($Z = 22$) at 3 nm.

References

[1] Duguay M A and Rentzepis P M 1967 *Appl. Phys. Lett.* **10** 350
[2] Molchanov A G 1972 *Sov. Phys.–Usp.* **15** 124
[3] Elton R C 1975 *Appl. Opt.* **14** 97
[4] Zherikin A N, Koshelev K N and Letokhov V S 1976 *Sov. J. Quantum Electron.* **6** 82
[5] Vinogradov A V, Sobel'man I I and Yukov E A 1977 *Sov. J. Quantum. Electron.* **7** 32
Vainshtein L A, Vinogradov A V, Safranova V I and Skolev I Vu 1978 *Sov. J. Quantum Electron.* **8** 239
Vinogradov A V and Shlyaptsev V N 1980 *Sov. J. Quantum Electron.* **10** 753
Vinogradov A V and Shlyaptsev V N 1980 *Sov. J. Quantum Electron.* **13** 303
[6] Pert G J and Ramsden S A 1974 *Opt. Commun.* **11** 270
[7] Bohn W L 1974 *Appl. Phys. Lett.* **25** 15
[8] Konokov E Ya and Koshelev K N 1975 *Sov. J. Quantum Electron.* **4** 1340
[9] Jones W W and Ali A W 1975 *Appl. Phys. Lett.* **26** 450
[10] Irons F and Peacock N J 1974 *J. Phys. B: At. Mol. Phys.* **7** 1109
[11] Dewhurst R J, Jacoby D, Pert G J and Ramsden S A 1976 *Phys. Rev. Lett.* **37** 1265
[12] Kononov Y A, Koshelev K N, Levykin Yu A, Sidel'nikov Yu V and Churilov S S 1976 *Sov. J. Quantum Electron.* **6** 308
[13] Dixon R H and Elton R C 1977 *Phys. Rev. Lett.* **38** 1072
[14] Key M H, Lewis C L S and Lamb M J 1979 *Opt. Commun.* **28** 331
[15] Jaeglé P, Jamelot G, Carillon A and Wehenkel C 1978 *Japan. J. Appl. Phys.* **17** 483
[16] Jacoby D, Pert G J, Ramsden S A, Shorrock L D and Tallents G J 1981 *Optics Commun.* **37** 193
Jacoby D, Pert G J, Shorrock L D and Tallents G J 1982 *J. Phys. B: At. Mol. Phys.* **15** 3557
[17] Matthews D L 1985 *et al Phys. Rev. Lett.* **54** 110; Rosen M D, Hagelstein P L, Matthews D L, Campbell E M, Hazi A U, Whitten B L, MacGowan B, Turner R E and Lee R W 1985 *Phys. Rev. Lett.* **54** 106
[18] Suckewer S, Skinner C H, Milchberg H, Keane C and Voorhees D 1985 *Phys. Rev. Lett.* **55** 1753 and Suckewer S, Skinner C H, Kim D, Valeo E, Voorhees D and Wouters A 1986 *Phys. Rev. Lett.* **57** 1004
[19] MacGowan B J *et al* 1987 *J. Appl. Phys.* **61** 5245
[20] MacGowan B J, Maxon S, Hagelstein P L, Keane C J, London R A, Matthews D L, Rosen M D, Scofield J H and Whelan D A 1987 *Phys. Rev. Lett.* **59** 2157
[21] Lee T N, McLean E A and Elton R C 1987 *Phys. Rev. Lett.* **59** 1185
[22] MacGowan B J, Maxon S, Keane C J, London R A, Matthews D L and Whelan D A 1988 *J. Opt. Soc. Am.* B **5** 1858
[23] Carillon A *et al* 1992 *Phys. Rev. Lett.* **68** 2917
[24] O'Neill D M, Lewis C L S, Neely D, Uhomopbhi J, Key M H, MacPhee A, Tallents G J, Ramsden S A, Rogovski A and McLean E A 1990 *Opt. Commun.* **75** 406
[25] MacGowan B J *et al* 1992 *Phys. Fluids* B **4** 2326
[26] Li Y, Lu P, Pretzler G and Fill E E 1997 *Opt. Commun.* **133** 196
[27] Fields D J *et al* 1992 *Phys. Rev.* A **46** 1606
[28] MacGowan B J, Maxon S, DaSilva L B, Fields D J, Keane C J, Matthews D L, Osterheld A L, Scofield J H, Shimkaveg G and Stone G F 1990 *Phys. Rev. Lett.* **65** 420
[29] MacGowan B J *et al* 1990 *Phys. Rev. Lett.* **65** 2374
[30] Hagelstein P L 1988 *Proc. OSA Meeting on Short Wavelength Coherent Radiation: Generation and Applications* ed R W Falcone and J Korz (Washington DC: OSA)
Shlyaptsev V N 1987 P N Lebedev Physical Institute, *PhD Thesis* Preprint FIAN #40
[31] MacGowan B J *et al* 1992 *Phys. Fluids* B **4** 2326

[32] Smith R, Tallents G J, Zhang J, McCabe S, Pert C J and Wolfrum E 1999 *Phys. Rev.* A **59** R47

[33] Koch J A *et al* 1992 *Phys. Rev. Lett.* **68** 3291

[34] Rus B *et al* 1994 *X-Ray Lasers* ed D C Eder and D L Matthews (New York: AIP)

[35] DaSilva L B, MacGowan B J, Rowka S M, Koch J A, London R A, Matthews D L and Underwood J 1993 *Opt. Lett.* **18** 1174

[36] Wang S, Zhou G, Zhang G, Sheng J and Chunyu S 1994 *X-Ray Lasers* ed D Eder and D Matthews (New York: AIP) p 293

[37] Skinner C H, Dicicco D S, Kim D, Rosser R J, Suckewer S, Gupta A P and Hirshberg J G 1990 *J. Microsc.* **159** 51

[38] DaSilva L B *et al* 1992 *Science* **258** 269

[39] Trebes J E *et al* 1987 *Science* **238** 517

[40] Da Silva L B *et al* 1995 *Phys. Rev. Lett.* **74** 3991
Wan A S, Barbee T W Jr, Cauble R, Celliers P, Da Silva L B, Moreno J C, Rambo P W, Stone G F, Trebes J E and Weber F 1997 *Phys. Rev.* E **55** 6293

[41] Cauble R, Da Silva L B, Barbee T W Jr, Celliers P, Moreno J C and Wan A S 1995 *Phys. Rev. Lett.* **74** 3816

[42] Backus S, Durfee C G III, Murnane M and Kapteyn H C 1998 *Rev. Sci. Instrum.* **1207** 69

[43] Barty C P J *et al* 1996 *X-Ray Lasers* ed S Svanberg and C G Wahlstrom (Bristol: Institute of Physics Publishing) p 282

[44] Umstadter D P, Barty C, Perry M and Mourou G A 1998 *Opt. Photon. News* **9** 41

[45] Kato Y 1996 *X-Ray Lasers* ed S Svanberg and C G Wahlstrom (Bristol: Institute of Physics Publishing) p 274

[46] Yamakawa Y, Aoyama M, Matsuoka S, Kase T, Akahane V and Takuma H 1999 *Opt. Lett.* **23** 146

[47] Rocca J J, Cortázar O D, Szapiro B, Floyd K and Tomasel F G 1993 *Phys. Rev.* E **47** 1299

[48] Rocca J J, Shlyaptsev V, Tomasel F G, Cortázar O D, Hartshorn D and Chilla J L A 1994 *Phys. Rev. Lett.* **73** 2192–5

[49] Rocca J J, Tomasel F G, Marconi M G, Shlyaptsev V N, Chilla J L A, Szapiro B T and Guidice G 1995 *Phys. Plasmas* **2** 2547

[50] Rocca J J, Marconi M C, Chilla J L A, Clark D P, Tomasel F G and Shlyaptsev V N 1995 *IEEE J. Selected Topics Quantum Electron.* **945** 1

[51] Rocca J J, Clark D P, Chilla J L A and Shlyaptsev V N 1996 *Phys. Rev. Lett.* **77** 1476–9

[52] Tomasel F G, Rocca J J and Shlyaptsev V N 1996 *IEEE Trans. Plasma Sci.* **49** 24

[53] Tomasel F G, Rocca J J, Shlyaptsev V N and Macchietto C D 1997 *Phys. Rev.* A **55** 1437

[54] Benware B R, Moreno C H, Burd D J and Rocca J J 1997 *Opt. Lett.* **22** 796

[55] Benware B R, Macchietto C D, Moreno C H and Rocca J J 1998 *Phys. Rev. Lett.* **81** 5804

[56] Macchietto C D, Benware B R and Rocca J J 1999 *Opt. Lett.* **24** 1115

[57] Moreno C H, Marconi M C, Shlyaptsev V N, Benware B R, Macchietto C D, Chilla J L A, Rocca J J and Osterheld A 1998 *Phys. Rev.* A **58** 1509

[58] Frati M, Seminario M and Rocca J J 2000 *Opt. Lett.* **25** 1022

[59] Waynant R W and Elton R C 1976 *Proc. IEEE* **64** 1059

[60] Hagelstein P L 1983 *Plasma Phys.* **25** 1345
Hagelstein P L 1984 *Atomic Physics* vol 9, ed R S Van Dyck Jr and E N Fortson (Singapore: World Science) p 382

[61] Keane C J 1991 *SPIE* **1551** 2

[62] Kapteyn H C, Da Silva L B and Falcone R W 1992 *Proc. IEEE* **342** 80

[63] Matthews D L 1995 *Nucl. Instrum. Methods Phys. Res.* B **98** 91

[64] Skinner C H 1991 *Phys. Fluids* B **3** 2420

[65] Elton R C 1990 *X-Ray Lasers* (Boston, MA: Academic)

[66] Rocca J J and Da Silva L B (ed) 1999 *Soft X-Ray Lasers and Applications (Proc. SPIE 3776)* vol III (Bellingham, WA: SPIE)

[67] Kato Y, Takuma H and Daido H (ed) 1999 *X-Ray Lasers 1998* (Bristol: Institute of Physics Publishing)

[68] Rocca J J and Da Silva L B (ed) 1997 *Soft X-Ray Lasers and Applications (Proc. SPIE 3156)* vol II (Bellingham, WA: SPIE)

[69] Svanberg S and Wahlstrom C G (ed) 1996 *X-Ray Lasers 1996* (Bristol: Institute of Physics Publishing)

[70] Rocca J J and Hagelstein P L (ed) 1995 *Soft X-Ray Lasers and Applications (Proc. SPIE 2520)* (Bellingham, WA: SPIE)

[71] Eder D C and Matthews D L (ed) 1994 *X-Ray Lasers 1994* (New York: AIP)

[72] Suckewer S (ed) 1993 *Ultrashort Wavelength Lasers (Proc. SPIE 2012)* vol II (Bellingham, WA: SPIE)

[73] Fill E E (ed) 1992 *X-Ray Lasers 1992* (Bristol: Institute of Physics Publishing)

[74] Suckewer S (ed) 1991 *Ultrashort Wavelength Lasers (Proc. SPIE 1551)* (Bellingham, WA: SPIE)

[75] Rocca J J 1999 *Rev. Sci. Instrum.* **70** 3799

[76] Nagata Y, Midorikawa K, Kubodera S, Obara M, Tashiro H and Tokoda K 1993 *Phys. Rev. Lett.* **71** 3774

[77] Nagata Y, Midorikawa K, Kubodera S, Obara M, Tashiro H and Tokada K 1993 *Phys. Rev. Lett.* **71** 3774

[78] Korobkin D, Nam C H, Suckewer S and Golstov A 1996 *Phys. Rev. Lett.* **77** 5206

[79] Lemoff B E, Yin G Y, Gordon C L III, Barty C P J and Harris S E 1995 *Phys. Rev. Lett.* **74** 1576 and 1996 *J. Opt. Soc. Am.* B **13** 180

[80] Nickles P V, Shlyaptsev V N, Kalashnikov M, Schnurer M, Will I and Sandner W 1997 *Phys. Rev. Lett.* **78** 2748

[81] Dunn J, Osterheld A L, Shepherd R, White W E, Shlyaptsev V N and Stewards R E 1998 *Phys. Rev. Lett.* **80** 2825

[82] Dunn J, Osterheld A L, Shepherd R, White W E, Shlyaptsev V N, Bullock A and Stewart R E 1997 *Soft X-Ray Lasers and Applications (Proc. SPIE 3156)* vol II, ed J J Rocca and L B Da Silva (Bellingham, WA: SPIE) p 114

[83] Kalachnikov M P *et al* 1998 *Phys. Rev.* A **57** 4778

[84] Dunn J, Li Y, Osterheld A L, Nilsen J, Hunter J R and Shlyaptsev V N 2000 *Phys. Rev. Lett.* **84** 4834

Dunn J, Li Y, Osterheld A L, Nilsen J, Moon S J, Fournier K B, Hunter J R, Faenov A Ya, Pikuz T A and Shlyaptsev V N 1999 *Soft X-Ray Lasers and Applications (Proc. SPIE 3776)* vol III, ed J J Rocca and L B Da Silva (Bellingham, WA: SPIE) p 2

[85] Lewis C L S *et al* 1999 *Soft X-Ray Lasers and Applications (Proc. SPIE 3776)* vol III, ed J J Rocca and L B Da Silva (Bellingham, WA: SPIE) p 292

[86] Miquel J L *et al* 1999 *Soft X-Ray Lasers and Applications (Proc. SPIE 3776)* vol III, ed J J Rocca and L B Da Silva (Bellingham, WA: SPIE) p 24

[87] Linford G J, Peresini E R, Soor W R and Spaeth M L 1974 *Appl. Opt.* **13** 379

[88] For typical gains in visible gas lasers, see for example, Silfvast W T 1996 *Laser Fundamentals* (Cambridge: Cambridge University Press)

[89] Smith R, Tallents G J, Zhang J, Eker G, McCabe S, Pert G J and Wolfrum E 1999 *Phys. Rev.* A **59** R47

[90] Jamelot J *et al* 1999 *X-Ray Lasers 1998* ed Y Kato, H Takuma and H Daido (Reading: IOPP) p 25

[91] Koch J A, MacGowan B J, DaSilva L B, Matthews D L, Underwood J H, Batson P J and Mrowka S 1992 *Phys. Rev. Lett.* **68** 3291
Koch J A, MacGowan B J, DaSilva L B, Matthews D L, Underwood J H, Batson P J and Mrowka S 1994 *Phys. Rev.* A **50** 1877

[92] Kato Y *et al* 1991 *Ultrashort Wavelength Lasers (Proc. SPIE 1551)* ed S Suckewer (Bellingham, WA: SPIE) p 56

[93] London R A, Strauss M and Rosen M D 1990 *Phys. Rev. Lett.* **65** 563

[94] Feit M D and Fleck J J A 1991 *Opt. Lett.* **16** 76
Feit M D and Fleck J J A 1991 *J. Opt. Soc. Am.* B **7** 2048

[95] Hazak G and Bar A 1989 *Shalom Phys. Rev.* A **40** 7055

[96] Trebes J *et al* 1992 *Phys. Rev. Lett.* **68** 588

[97] Burge R E, Slark G E, Browne M T, Yuang X C, Charalambous P, Cheng X H and Lewis C L S 1997 *J. Opt. Soc. Am.* B **14** 2742
Burge R E, Slark G E, Browne M T, Yuang X C, Charalambous P, Cheng X H and Lewis C L S 1998 *J. Opt. Soc. Am.* B **15** 1920
Burge R E, Slark G E, Browne M T, Yuang X C, Charalambous P, Cheng X H and Lewis C L S 1998 *J. Opt. Soc. Am.* B **15** 2515

[98] Marconi M C, Chilla J L A, Moreno C H, Benware B R and Rocca J J 1997 *Phys. Rev. Lett.* **79** 2799

[99] Jamelot G *et al* 1995 *Soft X-Ray Lasers and Applications (Proc. SPIE 2520)* ed J J Rocca and P L Hagelstein (Bellingham, WA: SPIE) p 2

[100] Liu Y, Seminario M, Tomasel F G, Rocca J J, Chang C and Attwood D T 2001 *Phys. Rev.* A **63** 033802

[101] Rocca J J, Moreno C H, Marconi M C and Kanizay K 1999 *Opt. Lett.* **24** 240
Moreno C H, Marconi M C, Kanizay K, Rocca J J, Unspenskii Yu A, Vinogradov A V and Pershin Yu A 1999 *Phys. Rev.* E **60** 911

[102] Filevich J, Marconi M, Kanizay K, Chilla J L A and Rocca J 2000 *Opt. Lett.* **25** 356

[103] Chang Z, Rundquist A, Wang H, Murnane M M and Kapteyn H C 1997 *Phys. Rev. Lett.* **79** 2967

[104] L'Huillier A 1996 *X-Ray Lasers 1996* ed S Svanberg and C G Wahlstrom (Bristol: Institute of Physics Publishing) p 444

[105] Ditmire T, Crane J K, Nyugen H, Da Silva L B and Perry M D 1995 *Phys. Rev.* A **51** R902

[106] Rundquist A, Durfee C G III, Chang Z, Herne C, Backus S, Murnane M and Kapteyn H C 1998 *Science* **280** 1412

[107] Tamaki T, Nagata Y, Obara M and Midorokawa K 1999 *Phys. Rev.* A **59** 4041

[108] Attwood D T, Halbach K and Kim N J 1985 *Science* **228** 1264

[109] Attwood D T, Sommargren G, Beguiristain R, Nguyen K, Boker J, Ceglio N, Jackson K, Koike M and Underwood J 1993 *Appl. Opt.* **32** 7022

[110] Coisson R 1995 *Appl. Opt.* **34** 904

[111] Attwood D T *et al* 1999 *IEEE J. Quantum Electron.* **35** 709

[112] Rossback J P 1996 *X-Ray Lasers 1996* ed S Svanberg and C G Wahlstrom (Bristol: Institute of Physics Publishing) p 436

[113] Feldhaus J, Krzywinski J, Saldin E L, Schneidmiller E and Yurkov M V 1999 *X-Ray Lasers 1998* ed Y Kato, H Takuma and H Daido (Bristol: Institute of Physics Publishing) p 553

[114] Schoemlein R W *et al* 1996 *Science* **274** 236

[115] Ditmire T, Cowan T E, LeSage G, Allen M, Hays G, Wharton K B, Perry M D, Powell H T and Rosenzweig J B 1999 *IEEE LEOS Annual Meeting Conf. Proc. (San Francisco, CA)* (Piscataway, NJ: IEEE) p 47

[116] Shkolnikov P L and Kaplan A E 1996 *X-Ray Lasers 1996* ed S Svanberg and C G Wahlstrom (Bristol: Institute of Physics Publishing) p 512

[117] Muendel M H and Hagelstein P L 1991 *Phys. Rev.* A **44** 7573

[118] Daido H *et al* 1995 *Phys. Rev. Lett.* **75** 1074

[119] Daido H *et al* 1996 *Opt. Lett.* **21** 958

[120] Zhang J *et al* 1997 *Phys. Rev. Lett.* **78** 3856

[121] Zhang J *et al* 1997 *Science* **276** 1097

[122] Zhang J *et al* 1997 *Soft X-Ray Lasers and Applications (Proc. SPIE 3156)* vol II, ed J J Rocca and L B Da Silva (Bellingham, WA: SPIE) p 53

[123] Zhang J *et al* 1996 *Phys. Rev.* A **54** R4653

[124] Lunney J G 1986 *Appl. Phys. Lett.* **46** 891

[125] Lewis C L S 1992 *et al Opt. Commun.* **91** 71

[126] Kodama R, Neely D, Kato Y, Daido H, Murai K, Yuan G, MacPhee A and Lewis C L S 1994 *Phys. Rev. Lett.* **73** 3215

[127] Wanshiji *et al* 1992 *J. Opt. Soc. Am.* B **9** 360

[128] Boehly I T *et al* 1990 *Phys. Rev.* A **42** 6962
[129] Nilsen J, MacGowan B J, Da Silva L B and Moreno J C 1993 *Phys. Rev.* A **48** 4682
[130] Basu S, Hagelstein P L, Gooberlet J G, Muendel M H and Kaushik S 1993 *Appl. Phys.* B. **57** 203
[131] Nilsen J, Moreno J C, DaSilva L B and Barbee T W Jr 1997 *Phys. Rev.* A **55** 827
[132] Moreno J C, Nilsen J, Li Y L and Fill E E 1996 *Opt. Lett.* **21** 866
[133] Nilsen J and Moreno J C 1995 *Phys. Rev. Lett.* **74** 3376
[134] Jamelot G *et al* 1995 *Soft X-Ray Lasers and Applications (Proc. SPIE 2520)* ed J J Rocca and P L Hagelstein (Bellingham, WA: SPIE) pp 2–12
[135] Cairns G F *et al* 1996 *Opt. Commun.* **124** 777–89
[136] Li Y, Pretzler G and Fill E E 1995 *Phys. Rev.* A **52** R3433, and Fill E E, Li Y, Schlögl D, Steingruber J and Nilsen J 1995 *Soft X-Ray Lasers and Applications (Proc. SPIE 2520)* ed J J Rocca and P L Hagelstein (Bellingham, WA: SPIE) p 134
[137] Daido H, Kodama R, Murai K, Yuan G, Takagi M, Kato Y, Choi I W and Nam C H 1995 *Opt. Lett.* **20** 61
[138] Rus B, Carillon A, Dhez P, Jaeglé P, Jamelot G, Klisnick A, Nantel M and Zeitoun P 1997 *Phys. Rev.* A **55** 3858
[139] Daido H, Ninomiya S, Imani T, Kodama B, Takagi M, Kato Y, Murai K, Zhang J, You Y, and Gu Y 1996 *Opt. Lett.* **21** 958
[140] Fill E E, Li Y, Schlögl D, Steingruber J and Nilsen J 1995 *Opt. Lett.* **20** 374
[141] Tallents G J 1997 *Soft X-Ray Lasers and Applications (Proc. SPIE 3156)* vol II, ed J J Rocca and L B Da Silva (Bellingham, WA: SPIE) p 30
[142] Moreno J C, Nilsen J and DaSilva L B 1994 *Opt. Commun.* **110** 585
[143] Nilsen J, Li Y, Lu P, Moreno J C and Fill E E 1996 *Opt. Commun.* **124** 287
[144] Zhang J 1996 *et al Phys. Rev.* A **54** R4653
[145] Li Y L, Pretzler G, Lu P X, Fill E E and Nilsen J 1997 *Phys. Plasmas* **4** 479
[146] Yuang G, Murai K, Daido H, Kodama R and Kato Y 1995 *Phys. Rev.* A **52** 4861
[147] Zhang J *et al* 1996 *Phys. Rev.* A **53** 3640
[148] Prag A R, Loewenthal F and Balmer J E 1996 *Phys. Rev.* A **54** 4585
[149] Tomassini R, Lowenthal F and Balmer J E 1999 *Phys. Rev.* A **59** 1577
[150] Lowenthal F, Tommasini R and Balmer J E 1998 *Opt. Commun.* **154** 325
[151] Afanasiev Yu A and Shlyaptsev V N 1989 *Sov. J. Quantum Electron.* **19** 1606
[152] Shlyaptsev V N, Nickles P V, Schlegel T, Kalashnikov M R and Osterheld A L 1991 *SPIE* **1551** 111
Shlyaptsev V N, Rocca J J, Kalashnikov M P, Nickles P V, Sandner W, Osterheld A L, Dunn J and Eder D C 1997 *Soft X-Ray Lasers and Applications (Proc. SPIE 3156)* vol II, ed J J Rocca and L B Da Silva (Bellingham, WA: SPIE) p 93
Decker C D and London R A 1997 *Soft X-Ray Lasers and Applications (Proc. SPIE 3156)* vol II, ed J J Rocca and L B Da Silva (Bellingham, WA: SPIE) p 94
Nilsen J 1997 *Soft X-Ray Lasers and Applications (Proc. SPIE 3156)* vol II, ed J J Rocca and L B Da Silva (Bellingham, WA: SPIE) p 86
[153] Moreno C H, Marconi M C, Shlyaptsev V N and Rocca J J 1999 *IEEE Trans. Plasma Sci.* **27** 6
[154] Rocca J J, Moreno C H, Marconi M C and Kanizay K 1999 *Opt. Lett.* **24** 420
Filevich J, Kanizay K, Marconi M C, Chilla J L A and Rocca J J 2000 *Opt. Lett.* **25** 356
[155] Artioukov I A, Benware B R, Rocca J J, Forsythe M, Uspenskii Yu A and Vinogradov A V 2000 *IEEE. J. Selected Topics Quantum Electron.* **30** 328
[156] Benware B R, Ozols A, Rocca J J, Artioukov I A, Kondratenko V V and Vinogradov A V 1999 *Opt. Lett.* **24** 1714
[157] Gonzalez J J, Frati M, Rocca J J and Shlyaptsev V N 1999 ed Y Kato, H Takuma and H Daido 1999 *X-Ray Lasers 1998* (Bristol: Institute of Physics Publishing) p 163
Gonzalez J J, Frati M, Rocca J J and Shlyaptsev V N 1999 *SPIE* **3776** 159
[158] Gudzenko G A and Shelepin L A 1963 *Zh. Ensp. Fiz.* **45** 1445 (Engl. transl. 1964 *Sov. Phys.–JETP* **18** 998)
[159] Pert G J 1976 *J. Phys. B: At. Mol. Phys.* **9** 3301
Dave A K and Pert G J 1985 *J. Phys. B: At. Mol. Phys.* **18** 1027
[160] Seely J F, Brown C W, Feldman U, Richardson M, Yaakobi B and Behring W E 1985 *Opt. Commun.* **54** 289
[161] Grande M *et al* 1990 *Opt. Commun.* **74** 309
[162] Boswell B, Shavarts D, Boehly T and Yaakobi B 1990 *Phys. Fluids* B **2** 436
[163] Azumma H *et al* 1990 *Opt. Lett.* **15** 1011
[164] Jamelot G, Klisnick A, Carillon A, Guennov H, Sureau A and Jaeglé P 1998 *J. Phys. B: At. Mol. Opt. Phys.* **18** 4647
[165] Jaeglé P, Jamelot G, Carrillon A, Klisnik A, Sureau A and Guennov H 1987 *J. Opt. Soc. Am.* B **4** 563
Carillon A *et al* 1990 *J. Phys. B: At. Mol. Opt. Phys.* **23** 147
[166] Chenais-Popovics C *et al* 1987 *Phys. Rev. Lett.* **59** 2161
[167] Milchberg H, Skinner C H, Suckewer S and Voorhees D 1985 *Appl. Phys. Lett.* **47** 1151
[168] Suckewer S and Fishman H 1980 *J. Appl. Phys.* **51** 1922
[169] Keane C J and Suckewer S 1991 *J. Opt. Soc. Am.* B **8** 201
[170] Steingruber J, Chen S S and Fill E E 1992 *X-Ray Lasers 1992* ed E E Fill (Bristol: Institute of Physics Publishing) p 115
[171] Kim D, Skinner C H, Umesh G and Suckewer S 1989 *Opt. Lett.* **14** 665
[172] Kim D, Skinner C H and Suckewer S 1991 *Proc. SPIE* **1551** 295

[173] Polonsky L, Park C O, Krushelnick K and Suckewer S 1976 *Proc. IEEE* **64** 75
[174] Morozov A, Polonsky L and Suckewer S 1995 *Soft X-Ray Lasers and Applications (Proc. SPIE 2520)* ed J J Rocca and P L Hagelstein (Bellingham, WA: SPIE) p 180
[175] Hara T, Ando K, Aoyagi Y, Kusakabe N and Yashiro H 1989 *Japan. J. Appl. Phys.* **28** L1010
[176] Yamaguchi N, Hara T, Fujikawa C and Hisada Y 1997 *Japan. J. Appl. Phys. Lett.* **36** L1297
[177] Hirose H, Hara T, Ando K, Negishi F and Aoyagi Y 1993 *Japan. J. Appl. Phys. Lett.* **32** L1538
Hara T, Ando K, Negishi-Tsuboi F, Aoyagi Y, Hirose H and Yashiro Y 1992 *Proc. SPIE.* **1627** 377
[178] Korobkin D, Goltsov A, Morozov A and Suckewer S 1998 *Phys. Rev. Lett.* **81** 1607
[179] Rocca J J, Beethe D C and Marconi M C 1988 *Opt. Lett.* **13** 565
[180] Steden C and Kunze H J 1990 *Phys. Lett.* A **151** 534
[181] Rocca J J, Marconi M C, Szapiro B T and Meyer J 1976 *Proc. IEEE* **64** 275
[182] Shin H J, Kim D E and Lee T N 1994 *Phys. Rev.* E **50** 1376
[183] Wagner T, Eberl E, Frank K, Hartmann W, Hoffmann D H H and Thotz R 1996 *Phys. Rev. Lett.* **76** 3124
[184] Glender S and Kunze H J 1994 *Phys. Rev.* E **49** 1586
[185] Böss T, Neff W, Boboc T, Weigand F, Bischoff R and Langhoff H 1998 *J. Phys. D: Appl. Phys.* **31** 2472
[186] Dussart R, Rosefeld W, Richard N, Hong D, Cachoncinlle C, Fleurier C and Pouvesle J M 1999 *X-Ray Lasers 1998* ed Y Kato, H Takuma and H Daido (Bristol: Institute of Physics Publishing) p 171
[187] Jones W W and Ali A W 1975 *Appl. Phys. Lett.* **26** 450
[188] Burnett N H and Corkum P B 1989 *J. Opt. Soc. Am.* B **6** 1195
[189] Amendt P E and Eder D C 1991 *Phys. Rev. Lett.* **66** 2589
Eder D C, Amendt P and Wilks S C 1992 *Phys. Rev.* A **6761** 45
[190] Eder D C *et al* 1994 *Phys. Plasmas* **1744** 1
[191] Durfee C G III, Lynch J and Milchberg H M 1995 *Phys. Rev.* E **51** 2368
[192] Milchberg H M, Durfee C G III and Lynch J 1995 *J. Opt. Soc. Am.* B **12** 731
[193] Durfee C G III, Lynch J and Milchberg H M 1994 *Opt. Lett.* **19** 1937
[194] Ehrlich Y, Cohen C, Zigler A, Krall J, Sprangle P and Erasay E 1996 *Phys. Rev. Lett.* **77** 4186
[195] Rocca J J, Tomasel F G, Marconi M C, Chilla J L A, Moreno C H, Benware B R, Shlyaptsev V N, Gonzales J J and Macchietto C D 1997 *Soft X-Ray Lasers and Applications (Proc. SPIE 3156)* vol II, ed J J Rocca J J and L B Da Silva (Bellingham, WA: SPIE) p 164
[196] Goltson A *et al* 1999 *IEEE Sel. Top. Quant. Elect.* **5** 1453
[197] Chichkov B N, Egbert A, Eichmann H, Momma C, Nolte S and Wellegehausen B 1995 *Phys. Rev.* A **52** 1629
[198] Goltsov A, Korobkin D, Morozov A and Suckewer S 1998 *Proc. 1998 Int. Cong. on Plasma Physics and 25th EPS Conf. on Controlled Fusion and Plasma Physics (Prague)*
[199] Kapteyn H C 1992 *Appl. Opt.* **31** 4931
[200] Strobel G L, Eder D L, London R A, Rosen M D, Falcone R W and Gordon S P 1993 Short pulse-high-intensity lasers and applications II *SPIE* **1860** 157
[201] McGuire E J 1975 *Phys. Rev. Lett.* **35** 844
[202] Kapteyn H C, Lee R W and Falcone R W 1986 *Opt. Phys. Rev. Lett.* **57** 2939
[203] Sher M H, Benerofe S J, Young J F and Harris S E 1991 *J. Opt. Soc. Am.* B **8** 114
[204] Kapteyn H C and Falcone R W 1988 *Phys. Rev.* A **37** 2033
[205] Moon S J, Eder D C and Strobel G L 1994 *X-Ray Lasers 1994* ed D C Eder and D L Matthews (New York: AIP) p 262
[206] Matthews D L 1996 *X-Ray Lasers 1996* ed S Svanberg and C G Wahlstrom (Bristol: Institute of Physics Publishing) p 32
[207] Lewis C L S *et al* 1999 *X-Ray Lasers 1998* ed Y Kato, H Takuma and H Daido (Bristol: Institute of Physics Publishing) p 1
[208] Moribayashi K, Sasaki A and Tajima T 1998 *Phys. Rev.* A **58** 2007
[209] Kim D, Toth C and Barty C P J 1999 *Phys. Rev.* A **59** R4129

Further reading

Attwood D T 1999 *Soft X-ray and Extreme Ultraviolet Radiation* (New York: Cambridge University Press)
X-ray lasers 2000 *J. Physique* IV

B5.3
Liquid lasers

David H Titterton

B5.3.1 Introduction

The first liquid laser was demonstrated in 1966 [1], using a solution of organic dye molecules[1] as its gain medium: it was pumped by a ruby laser. There have been research studies, using rare earth ions, such as neodymium, in various solvents, but the emission performance of these lasers has been disappointing [2]. Consequently, the remainder of this chapter will focus on liquid-phase dye laser technology.

Operation of a liquid dye laser relies on the optical excitation of organic dye molecules dissolved in a suitable solvent. The optical excitation can be coherent from another laser or incoherent, using the energy from a flashlamp or arc lamp. An alternative approach involves the use of light from a surface discharge [3]. More than 50 laser dyes are available for use in the gain medium, enabling energy to be emitted from the ultraviolet to the near infrared, simply by using a different dye—this is wavelength diversity. Additionally, many dyes can be tuned over many tens of nanometres. A dye molecule requires a number of characteristics, such as strong fluorescence and low loss, to be suitable for laser operation. It is often possible to have a choice of solvents, as many laser dyes can be dissolved in a number of liquids. In fact, the solvent can have a very strong influence on the emission, governing the output power, as well as influencing the rate of degradation of the dye solution [4].

The optical set-up of liquid dye lasers can be extremely simple [5], which leads to their flexibility and general versatility. However, this does have a cost in mechanical complexity if sustained output is required, along with some of their other attributes, such as narrow linewidth, high repetition rate and high power. In this case, a circulation pump will be required to circulate the dye solution from a reservoir and additional elements in the cavity, such as an etalon, birefringent filter or set of prisms for reducing the spectral width, as well as a dispersive element for tuning, in addition to the resonant cavity.

Dye lasers are attractive for many applications, owing to their simplicity, versatility, very broad spectral emission capability, including simultaneous multicolour emission, and energy scalability. A big advantage of this technology, compared with solid-state devices, is the ease of removal of heat from the gain medium. However, care is required in the design of the circulation system to achieve a compact arrangement that allows an easy and clean procedure for changing the dye solution, otherwise it can become a messy configuration.

These features have made this source the laser of choice for a wide range of applications, and not only as a research tool in the laboratory. Industrial applications include inspection and laser isotope separation [6–8] to enrich uranium, with less emphasis on use as a manufacturing tool [9]. In medicine, the scope of their application appears to be growing rapidly: embracing skin treatment, including tattoo removal, diagnostic measurements, lithotripsy and activation of photosensitive drugs for photodynamic therapy [10, 11] (see

[1] Dye molecules are characterized by having a backbone of carbon atoms connected by alternating single and double bonds, with an overlapping set of π-orbitals running the length of the carbon atom backbone, in which the electrons are effectively de-localized.

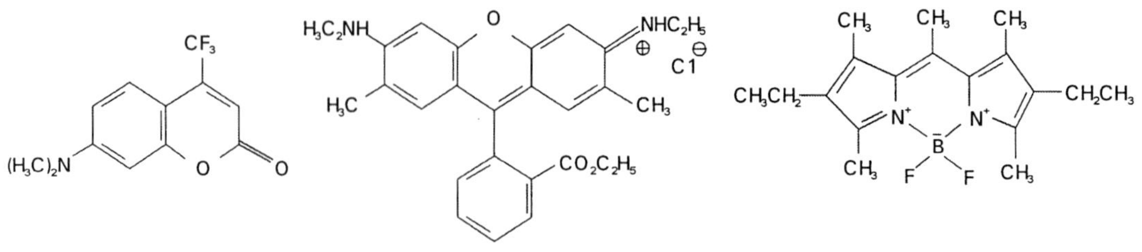

Coumarin 485 Rhodamine 6G Pyrromethene 567

Figure B5.3.1. Structures of dye molecules.

chapter D3.2.3). Spectroscopy is an application where dye lasers have been employed extensively in the past, but optical parametric oscillators (see chapter C3.2) now offer an alternative source. Other applications include communications, particularly underwater [12], bathymetry [13] and creation of guide stars [14, 15] for reference in an adaptive optical system (see chapter D8.1).

B5.3.2 General characteristics

B5.3.2.1 Gain medium

It is quite easy and cheap to make a gain medium with high optical quality by forming a laser dye solution. The laser head will have a cell, containing this solution, for use as the active medium. The great flexibility of this laser is provided by the range of organic compounds suitable for use in the gain medium. Many techniques have been demonstrated [16–21] to give simultaneous multicolour operation by direct laser action. Alternatively, a number of cells can be 'ganged together' [22], one for each colour. Furthermore, if energy from a flashlamp is used to create the population inversion in the organic molecules, then the laser head may also have an additional cell for fluorescence conversion to enhance efficiency [23].

Generally, laser dyes are quite large and complex molecules, containing a number of ring structures [24], which lead to complex absorption and emission spectra [25]. Many dyes are suitable for use in a laser: they can be categorized into classes or families by virtue of their structures that are chemically similar. Common examples are the coumarins, xanthenes and pyrromethenes. Xanthenes, for example, have a three-ring structure, the central ring containing oxygen. The differences between the absorption and emission spectra, associated available tuning ranges and physical behaviour of members within these broad classes, result from the substitution of specific chemical groups on the edges of the basic ring structures. Examples of three classes of dye molecules are shown in figure B5.3.1.

Optical energy from an external source is absorbed by the organic molecules raising them to an excited state. To a first approximation the wavelength absorbed by the molecules is proportional to the length of the 'backbone' of carbon atoms that form the chain structure [26]. Owing to the complex nature of the structures of these organic compounds, the energy level structure is very complex. However, a simplified generic diagram can be drawn to illustrate the principal interactions. Such a diagram is shown in figure B5.3.2.

It is essentially a four-level system where the upper and lower energy levels have an extended span, which leads to the broad absorption and tunability. The electronic states form the basic structure and transitions between these states lead to emission in, or close to, the visible part of the spectrum. The energy difference between the states, which leads to the visible band emission range from ~3.1 eV to 1.65 eV. The diagram shows that these electronic levels have other substates superimposed on them, resulting from the vibrational and rotational motion within the molecule. Energy differences caused by rotational motion are small and, when the other line-broadening mechanisms (see sections A1.7 and A1.8) are included [27], a continuum

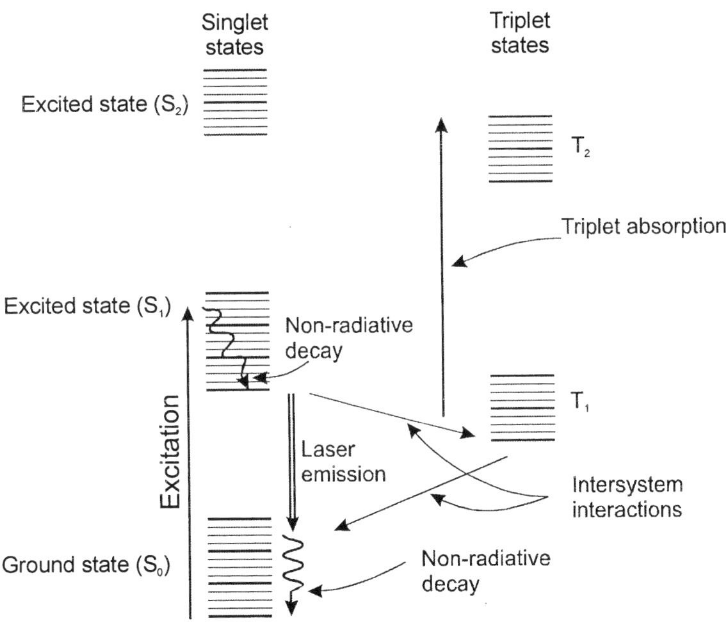

Figure B5.3.2. Simplified energy level structure diagram of a generic laser dye.

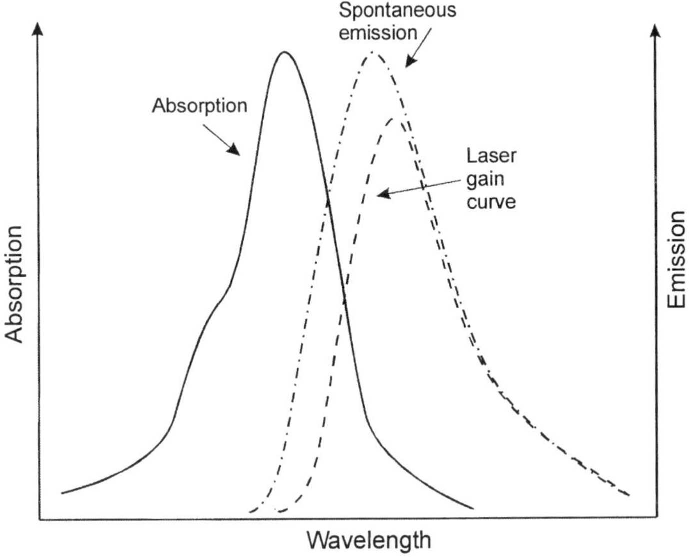

Figure B5.3.3. Absorption and emission spectra of a generic laser dye.

of states emerges, both for absorption and emission. The diagram shows a continuum of energy levels that enables a laser dye to absorb energy, as well as emit light over a broad range of wavelengths, particularly in comparison with other types of laser. Emission bands of 10 to 20 nm or more are possible, as shown schematically in figure B5.3.3.

Figure B5.3.2 also shows the processes leading to laser action, with the population inversion being created in the bottom group of 'excited states' (see chapter A1). A stimulated transition to a lower energy state provides the coherent light in the singlet manifold of states. Figure B.5.3.3 shows that a typical molecule can emit spontaneously through fluorescence over a broad band of wavelengths, the highest probability being at the centre of the curve. In the case of a free-running resonant cavity, the excess of gain is greatest where the difference between absorption and emission is largest and determines the position of the peak of the laser gain curve. Use of a tuning element (see section A3.4) controls the loss over a narrow spread of wavelengths and consequently determines the band where amplification can occur.

The lifetime of these states is very short, of the order of a few nanoseconds, so there is very little energy storage, hence Q-switching (see section A3.6) is not a viable option for generating high-intensity short-duration pulses. Consequently, the temporal-emission profile is closely related to that of the excitation source, particularly for short-duration excitation pulses. However, the broad bandwidth of the lasing emission does lend itself to the generation of ultrashort pulses via the mode-locking technique (see chapters C2.2 and C2.3).

In addition to the singlet states, there is a parallel set of energy bands, known as triplets, which have a slightly lower energy than the corresponding singlets. Triplets arise because the bonding of electrons in organic molecules occurs in pairs. In a singlet state the electron pair occupies the same energy level with their spins aligned in exactly opposite directions, as happens in the ground state. Triplet states are characterized by a pair of electrons having parallel spins. Since the electric field associated with a light wave does not act upon the magnetic dipole represented by a spin, the change from singlet to triplet cannot be induced easily by light (or put another way, the transition is forbidden by electric dipole selection rules). The build-up of population in any triplet state is detrimental since, owing to their long lifetimes, triplet states act as traps for molecules taking them out of the laser population circulation loop. Furthermore, the detrimental processes are compounded by the fact that the absorption process $T_1 \rightarrow T_2$ overlaps the emission $S_1 \rightarrow S_0$ and so can reduce the efficiency of both pumping and lasing.

Because the build-up of population in the lowest triplet state takes a finite time (see section B5.3.3.1), triplet-state losses tend not to cause too many problems during 'short-pulse' operations in the range up to a few microseconds. In the case of 'long-pulse' operations, a triplet-quenching agent, such as cyclo-octatetraene, will need to be added to the dye solution to enhance the performance of the gain medium. 'Triplet quenchers' [28] reduce the lifetime of the triplet states through collisional de-excitation.

The very large gain bandwidth of laser dye solutions makes them very attractive candidates for ultrashort pulse generation (see chapter C2.3). The large tuning range of laser-pumped rhodamine 6G implied that pulses of less than 10 fs duration are possible. Furthermore, the large homogeneously broadened bandwidth of these dyes suggests that linear amplification is feasible.

B5.3.2.2 Pumping options

Many different pump sources have been used to excite the gain medium, ranging from flashlamps to lasers, and recently surface discharges [3, 29, 30]. Tables B5.3.1 and B5.3.2 show some of the common options.

The incoherent sources tend to offer a cheaper system with greater flexibility to exploit the full range of emissions from various dyes with a single pump-source. Laser-pumped configurations can offer a simpler dye laser arrangement, as the pump source is directed into the dye cell, so the dye laser is essentially a wavelength conversion module. Longer wavelength laser-pumping also offers the prospect of greater longevity of the laser-dye solution, as there is no UV energy to cause photodegradation.

B5.3.2.3 Dye solutions

Many solvents are available for creating a laser-dye solution: alcohols and ethylene glycol are commonly used, although some are soluble in water, but a surfactant may be needed to prevent dimerization. Dimerization

Table B5.3.1. Incoherent pump-source characteristics.

Incoherent	λ (μm)	τ (μs)	PRF (kHz)	
Flashlamp–linear	UV to IR	∼5 to 10^4	To ∼1	Broadband, high flexibility, limited PRF, high energy, high average power, lifetime up to 10^6 pulses.
Flashlamp–coaxial	UV to IR	∼0.5 to 5	To ∼0.001	Broadband, very limited PRF, high energy, lifetime up to 10^4 pulses, longer duration pulses possible.
Surface discharge	UV to IR	∼0.15	∼1	Discrete lines, high efficiency, high PRF, high energy, high average power, long life, non-thermal.
Explosive	—	Long	—	Broadband, single shot!, high energy.

Table B5.3.2. Coherent pump-source characteristics.

Coherent	λ (nm)	τ (ns)	PRF (kHz)	
Copper vapour	511 and 578	∼10 to 60	3 ∼ 20	511 and 578 nm, high PRF, high average power, longer wavelengths.
Nd:YAG (2ν)	532	∼5	To cw	Variable PRF to cw, short-pulse operation.
Nd:YAG (3ν)	355	∼5	To cw	Variable PRF to cw, short-pulse operation.
Ion (Ar, Kr, . . .)	UV to green and red	∼0.18 (mode locked)	To cw	Various lines, high PRF to cw, short wavelength emission, low average power, low efficiency.
Nitrogen	337	∼2 to 10	0.01 to ∼0.3	Low average power, low efficiency, short-pulse operation.
Excimer	Various in UV (157^+)	∼10 to 200	∼0.5 to 3	Various lines in UV (e.g. 308 nm), high PRF, high average power, short-pulse operation, large pump source.
Diode	Red to IR	Long	To cw	Long wavelength pump, high PRF to cw.

occurs when there is a strong affinity between two or more molecules, leading to large structures with suppressed fluorescence. The solvent used to make the solution can have a significant influence on the optical properties of the laser gain medium and the longevity or usefulness of the dye solution. For example, solvents containing the deuterium isotope of hydrogen tend to have a lower transmission loss than those based on ‘common hydrogen’.

As noted earlier, the pump light can cause the dye molecules to decompose, but for broadband pumping the use of additives in the dye can ameliorate the effects of unwanted shorter wavelength energy. Furthermore, the use of a dye solution between the pump source and the gain medium can absorb the harmful shorter wavelength photons and convert their wavelengths, via fluorescence conversion [23], to match the absorption band of the active dye. Caffeine can also be added to this solution to block the UV.

Photodegradation-causing mechanisms are complex. Studies [4, 31] have shown that formation of

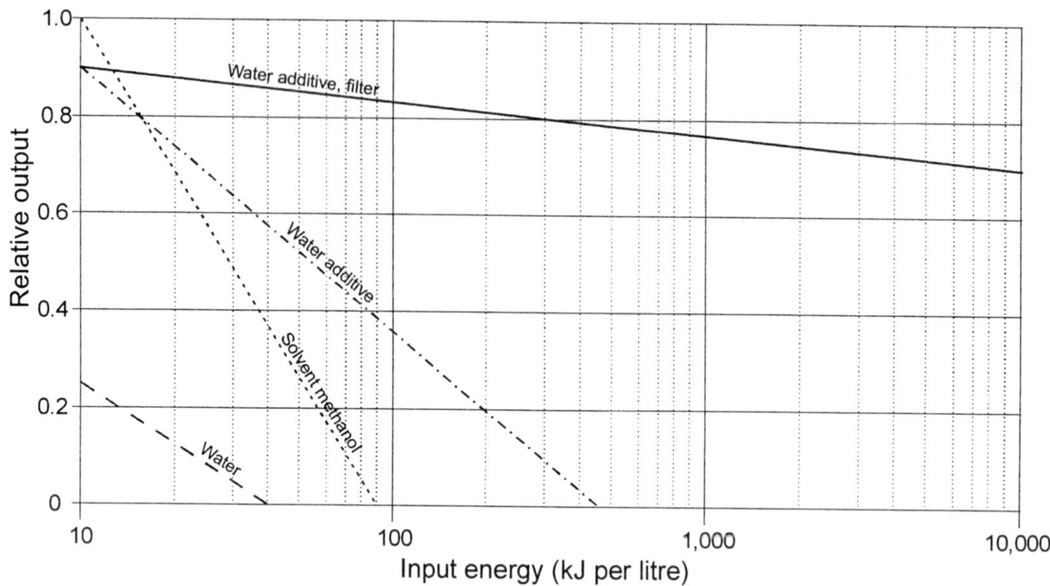

Figure B5.3.4. Dye solution longevity enhancement.

singlet oxygen in the solution is particularly detrimental. Elimination of oxygen enhances the longevity of the solution, but an alternative triplet-quenching agent will be needed. Figure B5.3.4 indicates how a solution's longevity varies with different operational conditions [4].

Many flashlamp-pumped dye lasers operate efficiently with dye solutions, having a concentration of the order of 10^{-4} to 10^{-5} molar. Laser-pumped systems often use a higher concentration, around 10^{-3} molar. The aim is to optimize the use of the pump photons in the gain volume, so that all the pump photons are absorbed by the dye solution throughout the dye cell. There is the flexibility to trade-off concentration, pumping efficiency and beam quality.

Some organic dyes are considered to be toxic, so data sheets must be consulted to establish whether or not special handling precautions are necessary. It is normal to wear a 'dust mask' and prepare the dye solution in a fume cupboard (or hood). Great care is recommended during handling and its use in the laser system. When the dye solution is no longer required, it must be disposed of as a toxic waste by a method approved by the local 'Health and Safety' service. Further guidance is given in section B5.3.2.14 on the safe operation of flashlamp-pumped dye lasers.

B5.3.2.4 Spectral control

Spectral control of the emission is possible because of the broad gain bandwidth and the homogeneous broadening of the transition, which enables the power (or energy) to be extracted at a single wavelength. A number of approaches can be applied, which rely on increasing the cavity losses at the unwanted wavelengths, resulting in only one or a few wavelengths oscillating. There is a subtle difference between narrowing an existing line and forcing the laser to oscillate on a single line. In the latter case, gain is extracted over a narrow spectral region, rather than discarding the unwanted wavelengths, hence offering superior efficiency.

Schematic representations of spectral control schemes are shown in figure B5.3.5. In the case of the filter technique, this could use a range of devices: examples include a birefringent (Lyot) filter (see section A3.4), a Fabry–Pérot etalon (see section A5.2.3), a dielectric filter, coloured glass or a dielectric mirror with a

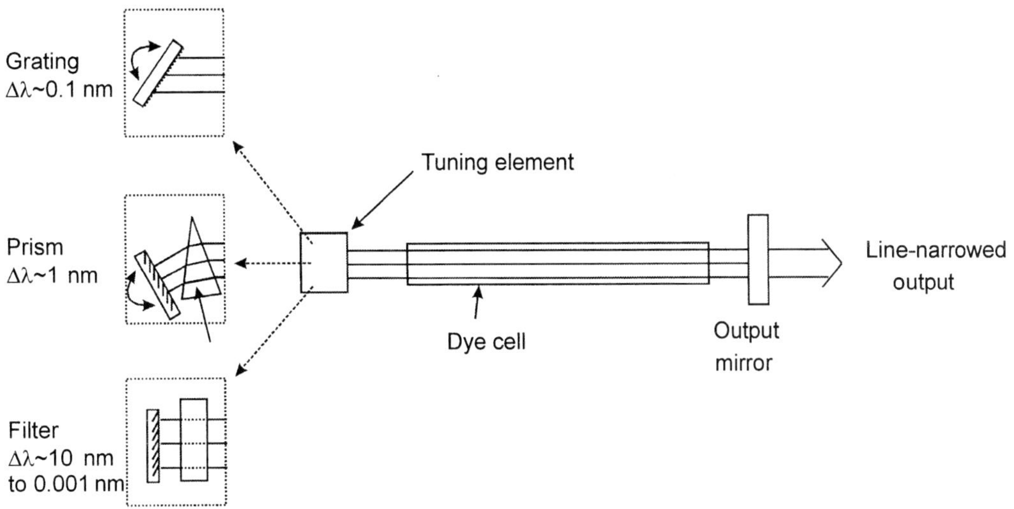

Figure B5.3.5. Tuning schemes.

narrow-band coating. It is vital that low loss devices, such as 'birefringent filters' (also known as 'Lyot' filters) and etalons, are used in continuous wave sources, owing to their lower gain. In contrast, high gain (pulsed) systems can use many different types of more lossy tuning elements, including gratings with amplified spontaneous emission suppression.

Full exploitation of the spectral properties of this technology requires the use of wavelength-selective elements in the resonant cavity, such as prisms, a diffraction grating, a birefringent filter, or Fabry–Pérot etalons. Prisms are attractive as they are relatively cheap and durable, but a single prism gives limited line narrowing. Use of an etalon, which is relatively expensive, with the appropriate finesse (equivalent to quality) gives enhanced line narrowing. A diffraction grating can provide excellent spectral control, including tunability, but great care is needed during high-power operation.

An effective and low-cost configuration, which is suitable for a flashlamp-pumped system, is to use a series of prisms for selection of a very narrow range of wavelengths to pass along the optical axis of the cavity; the remaining wavelengths are dispersed. An etalon can then be used for further line narrowing, usually with a finesse of about five. Additional line narrowing may be achieved by the use of more etalons; in this case the finesse may be up to 10, depending on the beam quality and laser aperture. This approach is inefficient, but can be very effective, particularly if the parasitic reflections from the dye-cell wall are controlled, for example, by inserting a helical spring inside the cell [32].

Tuning and line narrowing can be accomplished with a diffraction grating by using it as the 'high reflector' (Littrow grating configuration) [33, 34], as shown in the 'grating arrangement' of figure B5.3.5, with equal angles of incidence and diffraction on the same side of the normal. Adjusting the orientation of this grating gives the desired resonant condition. Many commercial systems use this technique and a series of prisms to expand the beam prior to incidence on a diffraction grating. The original demonstration [35] of this approach used a Cassegrain telescope to provide the beam expansion needed to take advantage of the available resolving power, but commercial systems now favour a prismatic system [33]. An alternative technique involves a fixed prismatic arrangement to expand and a near-grazing-incidence mirror to disperse the beam, movement of the feedback mirror providing tunable emission [36], shown schematically in figure B5.3.6.

Similar results can be achieved with a birefringent element or a tuning wedge. In the former case, a birefringent plate is mounted at Brewster's angle inside the laser cavity and oriented to act as a waveplate.

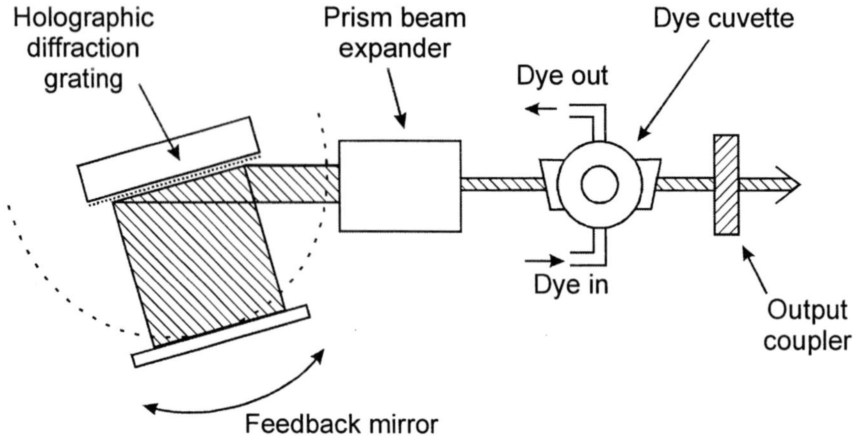

Figure B5.3.6. Tuning configuration (courtesy of Lumonics Inc).

There is no reflection at the plate surfaces for suitably polarized light. However, at most wavelengths a fraction of the light passing through the filter is transformed into the orthogonal polarization and then partially reflected out of the cavity. At the resonant wavelength, the retardation of the plate is an integer number of waves, the polarization is unchanged and the plate is loss-free. The resonant wavelength can be tuned by rotating the plate in its own plane. The tuning wedge uses interference as the discriminator by creating a Fabry–Pérot cavity. Translation of the wedge alters the separation of the plates of the interferometer.

Use of line-narrowing elements[2] has enabled linewidths as small as 5 GHz (sub-0.01 nm) to be achieved with a flashlamp-pumped dye laser. A laser-pumped system is capable of achieving 100 MHz, but with the application of special techniques this linewidth can be reduced to better than 1 MHz from a single-mode pulsed source. In the case of an injection-locked (see section A2.9) cw source, values of 1 kHz or better have been demonstrated; it is projected that 1 Hz could be achieved [33].

Some applications need a broad spectral output and this may be achieved by the use of dispersive elements to make different parts of the dye cell oscillate at different wavelengths within a dye's gain curve [37]. This has been used to extend the linewidth to beyond 30 nm, using the scheme shown in figure B5.3.7.

B5.3.2.5 Control of ASE

One of the common problems of dye lasers is the control of amplified spontaneous emission (ASE), which produces an incoherent 'pedestal' of broadband energy around the laser line. This is shown in figure B5.3.8. Use of multiple-prism beam-expanding configurations, in a closed-cavity arrangement with a Littrow grating [38], led to a significant reduction in ASE. A further attraction of prismatic cavities is the stability of alignment and also the prospect of a compact laser configuration.

B5.3.2.6 Temporal control

As noted earlier, the upper state lifetime of dye molecules is a few nanoseconds, so unlike the rare earth ion lasers (see chapter B1.3), there is very little scope for modifying the temporal characteristics by Q-switching (see chapters A3 and C2.2). If very short duration pump pulses cannot be achieved, as may be the case with flashlamp excitation, or a suitable pump pulse is not available, then the cavity-dumping technique [39, 40]

[2] The spread in frequency ($\Delta\nu$) and wavelength ($\Delta\lambda$), at a given wavelength (λ) is related by: $\Delta\nu = -c\Delta\lambda/\lambda^2$.

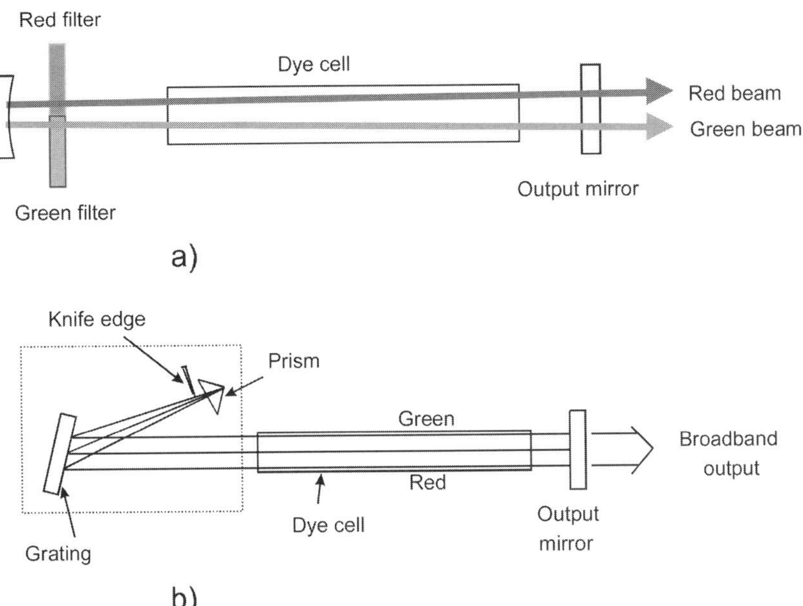

Figure B5.3.7. Broadband laser scheme. (*a*) Basic broadband laser scheme; (*b*) broadband laser cavity design.

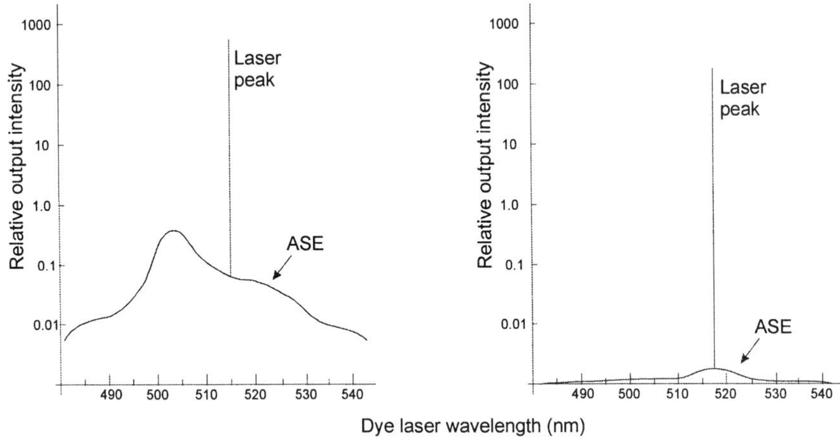

Figure B5.3.8. ASE reduction (from [38]).

approach can be applied. In this case, the energy is stored in the optical cavity[3] and released by altering the polarization of the travelling wave. Creation of a high-Q cavity during the millisecond pump-pulse operation, where many thousands of cavity round-trips occur, demands mirrors with very high reflectivity (better than 99.999%). Hence, unwanted optical activity must be eliminated to avoid leakage of energy during this storage phase. A cavity-dumping scheme for the high reflecting element is shown in figure B5.3.9: the remainder

[3] The pulse length in seconds (τ), assuming two passes of the cavity of length (l metres), is given by: $(2/3) \times l \times 10^{-8}$, hence for τ expressed in nanoseconds, this reduces to $(20l)/3$.

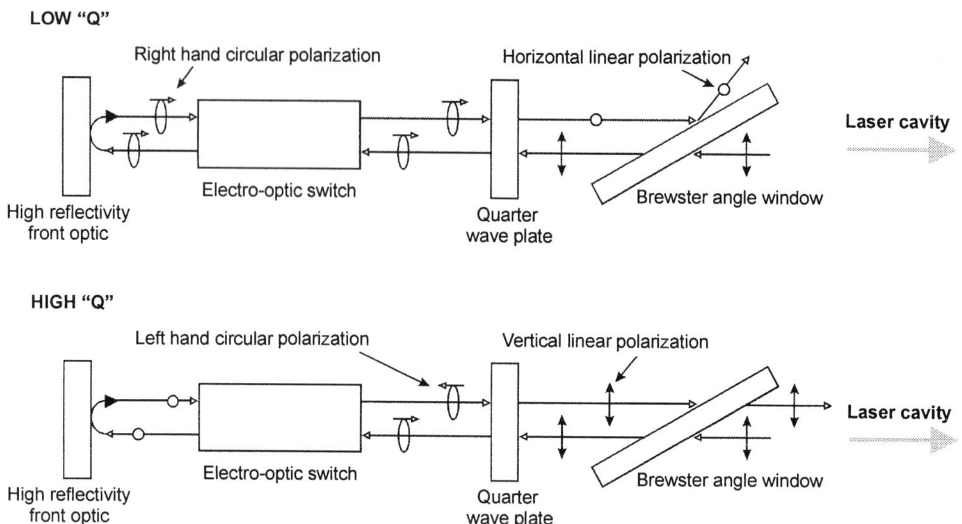

Figure B5.3.9. Cavity-dumping scheme (high reflector arrangement).

of the laser cavity is to the right, but is not shown. Pulses with a duration as short as 8 ns [40] have been demonstrated from a flashlamp-pumped source.

Ultra-short pulses have been generated from a self-mode-locking (see section C2.3.1.2) flashlamp-pumped dye laser with a saturable absorber [41], using an organic dye solution. Continuous wave dye lasers have been passively mode-locked [42]. The colliding-pulse concept has been applied to a ring dye laser [43] by including a saturable absorber in the ring cavity to give pulses of a duration of 90 fs. This concept may also be realized by placing the saturable absorber close to the high reflector. Further reduction in pulse length has been achieved by the application of dispersion compensation to produce a pulse length of the order of 50 fs [44]. Further enhancement and compensation with extra-cavity compression has led to the generation of a pulse length of 6 fs [45].

B5.3.2.7 Flashlamp-pumped devices

The first flashlamp-pumped devices were demonstrated in 1967 [46, 47] and have seen steady development to enable mature systems to be manufactured.

B5.3.2.8 Laser head

Flashlamps are essentially broadband emitters that are isotropic radiators, which can produce pulses of energy of a duration from about a microsecond to tens of milliseconds, depending on the impedance of the pulse-forming network [48]. In the case of a multi-lamp system, the flashlamps may be connected in series or parallel. A parallel arrangement tends to give short-duration pulses and longer lamp-life [49]. Schematic circuit diagrams of pulse-forming networks (PFN) are shown in figure B5.3.10.

It is important to have a rapid rise in the radiated power from the flashlamp in order to avoid unwanted triplet state formation. The following argument illustrates the problem for rhodamine 6G. Once the laser starts to oscillate, the value of the upper level population N_1 is clamped at N_{1CR},the critical value it has at the threshold of laser oscillation, since stimulated emission depletes the entire S_1 level at a rate faster than the spontaneous emission rate. During laser oscillation, the rate at which the triplet state is fed depends on

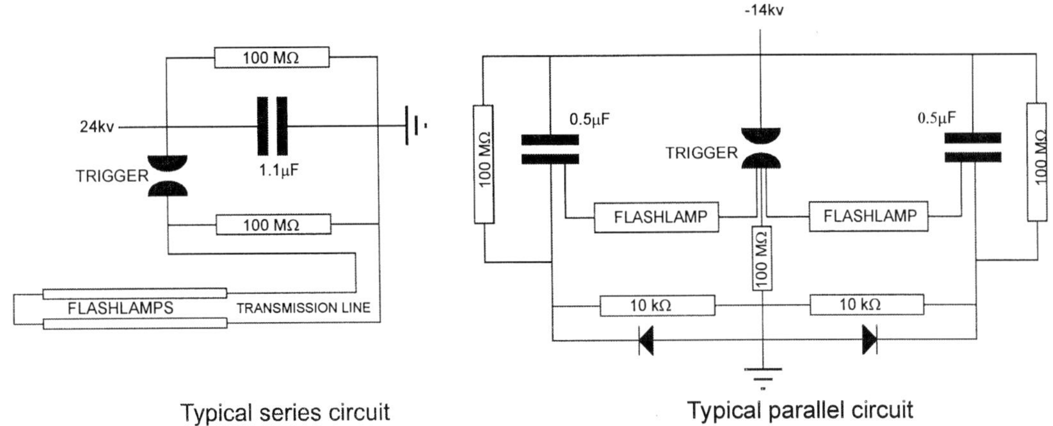

Figure B5.3.10. Flashlamp drive circuits.

k_{ST}, the rate constant for singlet to triplet conversion and the value of N_{1CR}:

$$\frac{dN_T}{dt} = N_{1CR}k_{ST} - \frac{N_T}{\tau_T} \cong N_{1CR}k_{ST} \tag{B5.3.1}$$

because the timescale for triplet decay (τ_T) is effectively infinite compared with other timescales. Within a single pulse, the triplet population at time t is given by:

$$N_T(t)\sigma_T(\lambda_L) = N_{1CR}k_{ST}t \tag{B5.3.2}$$

Now, lasing at wavelength λ_L will extinguish at t_f, such that for triplet absorption and emission cross sections σ_T and σ_e respectively,

$$N_T(t_f)\sigma_T(\lambda_L) = N_{1CR}\sigma_e(\lambda_L). \tag{B5.3.3}$$

For a typical dye molecule, such as rhodamine 6G, the triplet absorption cross section σ_T is related to σ_e as $\sigma_T \sim 0.1\sigma_e$ for wavelengths in the spectral band of interest, the laser pulse will end when

$$N_T(t_f) \sim 10N_{1CR} \tag{B5.3.4}$$

so

$$t_f < 10/k_{ST} \sim (3.3\ \mu\text{s for rhodamine 6G}) \tag{B5.3.5}$$

Consequently, the pulse-forming network should be designed to provide a rapid rise time of the order of a microsecond, which was a significant challenge in the early days of lasers and encouraged the development of alternative switches and new flashlamp designs. Failure to achieve full radiated power in more than a few microseconds, as described, results in energy not being used for pumping, but merely heating the dye solution and degrading the optical quality. Triplet quenching agents [28] are required to avoid these effects if long-pulse operation is required and are essential if efficient operation is required over a sustained period.

A flashlamp's envelope can be doped to modify its emitted spectrum, in order to match the spectral emission to the absorption spectrum of the dye. This can provide a modest enhancement to the longevity of the laser-dye solution.

A linear flashlamp is essentially a cylindrical envelope with electrodes at either end: a discharge is created between them in the gas. The other common form of flashlamp is the co-axial device, which has an annular-section active region, with the gain medium inserted along its hollow core. In the latter case, there is little need for any structure to direct the optical energy into the dye cell, as many of the photons are travelling towards the dye cell, whilst the remainder would need to cross the plasma in the discharge tube in order to pump the dye solution. Little will be gained from attempting to couple this energy from the discharge into the cell.

The situation with linear flashlamps is somewhat different: there is a smaller barrier created by the plasma. Two distinct pump-chamber designs have been developed and are in common use. They are distinguished by the techniques used to couple the excitation energy from the discharge into the dye cell, namely, specular or diffuse reflecting surfaces. These alternative approaches have an impact on the design of the pumping chamber.

Use of specular reflections from the outer walls of the pumping chamber usually dictates the need for a curved reflector 'behind' each flashlamp. In the case of a single flashlamp, the pumping chamber can be elliptical. A flashlamp is located at one focus and the dye cell at the other. The situation is normally more complex for a multiple lamp configuration [50]. It is common for a pumping chamber to be faceted to make a compact design and to avoid directing the reflected energy back into the plasma. However, in this case the 'rearward directed' energy may make multiple bounces before entering the dye cell, which may modify the duration of the pump pulse slightly, depending on the configuration. This type of pumping chamber tends to be quite expensive to produce and could be relatively heavy.

It is possible to make the entire pumping chamber from a diffuse-reflecting material, such as the resinous substance known as spectralon. In this case, the flashlamps can be placed very close to the dye cell in what is known as a close-coupled configuration, enabling a compact and lightweight design to be achieved. Alternatively, a metal structure can be fabricated and the inner surface sprayed with a diffuse-reflecting material with a high albedo, such as barium sulphate. Attempts have also been made to include dyes in this layer to provide some fluorescence conversion of the diffusely reflected energy, but this appeared to add little to the pumping efficiency. The shape of the inner surface of the pumping chamber is less critical than that for a specular reflector, so a curved surface has tended to be used with a cylindrical dye cell and a plane reflector with a rectangular-section dye cell. The close-coupled approach has been used extensively in dye lasers developed in the UK [5].

An alternative arrangement, known as the inverted configuration, has been investigated recently. A linear flashlamp is at the core of the pumping chamber and the dye cell has an annular cross section, which surrounds the flashlamp. The need for a reflector to enhance the pumping efficiency is less critical in this configuration, as the concentration of the dye solution may be manipulated to ensure optimum absorption of the energy radiating outwards from the flashlamp. Alternatively, a reflecting sleeve can be placed over the outer surface of the dye cell. This design is shown schematically in figure B5.3.11, along with the other approaches previously described.

B5.3.2.9 Dye cell configuration

Design of the dye cell is an important aspect of achieving optimum performance: it is made of transparent material, such as quartz or sapphire. To minimize self-absorption losses in the dye solution, resulting from the overlap of the absorption and emission spectra (shown in figure B5.3.3), its length is matched to that of the flashlamps. Normally, less than 5% of the gain-length is unpumped. Fresnel losses are minimized by using broadband anti-reflection coatings, matched to the anticipated band of emitted wavelengths, on the ends of the cell (windows): it is also possible to use 'Brewster windows' (see chapter C1.1). The dye is shielded from acoustic shocks by surrounding each flashlamp with a water-cooling jacket, which also helps to provide a uniform thermal environment, needed for good beam quality.

The size of the dye cell is an important parameter in determining the output power of the laser, along with the emission efficiency of the chosen dye and the optical energy coupled into the cell from the flashlamp(s).

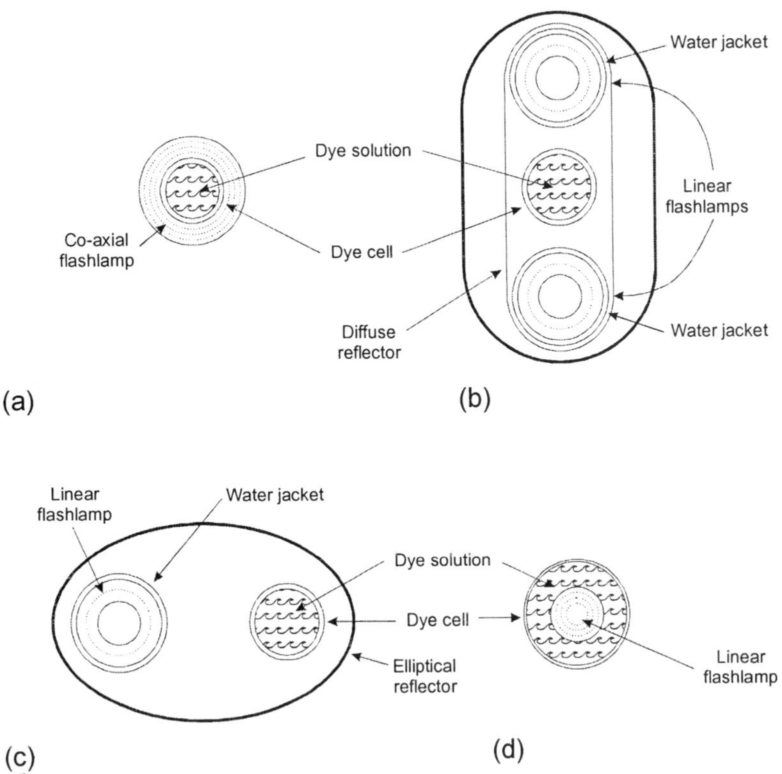

Figure B5.3.11. Cross-sectional schematic diagrams of laser head designs. (*a*) Co-axial flashlamp laser; (*b*) close-coupled linear flashlamp laser; (*c*) specular linear flashlamp; and (*d*) inverted laser.

An empirical rule-of-thumb is that a good dye, such as a xanthene or a pyrromethene, will allow about 0.25 J cc^{-1} to be extracted. Jensen [51] describes analytical techniques for accurate assessment of the size required for a given output performance.

If the laser is required to operate for more than a few 'shots', then it will be necessary to circulate the dye solution through the dye cell to maintain consistent laser efficiency and beam quality. As the temperature of the solution rises, efficiency degrades rapidly, owing to the thermal population of the lower laser level. Additionally, thermal gradients across the dye cell degrade the beam quality. Another empirical rule-of-thumb indicates the need for a 'flushing ratio (flush factor) of 2:1' of fluid in the cell between laser pulses to give an optimum trade-off between emission characteristics (particularly beam quality and dye solution longevity) and the volume of the circulated solution. The magnitude of the pumped volume therefore determines the size of the circulation pump required to meet the flushing ratio. Two common forms of flow geometry are used to pass the dye solution through the dye cell, namely, longitudinal and transverse flow.

Longitudinal flow allows a reasonably simple coupling of the circulating dye solution into the dye cell, and creation of a uniform flow throughout the cell. However, it can take a relatively long time for a given volume to be 'changed', particularly for a long dye cell. Hence, this geometry tends to be used when the repetition rate requirement is low, typically 15 Hz or less, in order to achieve the optimum flushing-ratio easily. Cylindrical-section cells are common. A filter is included in the circulation system to suppress the formation of bubbles, which would degrade the optical quality, generating increased optical scatter. Typically,

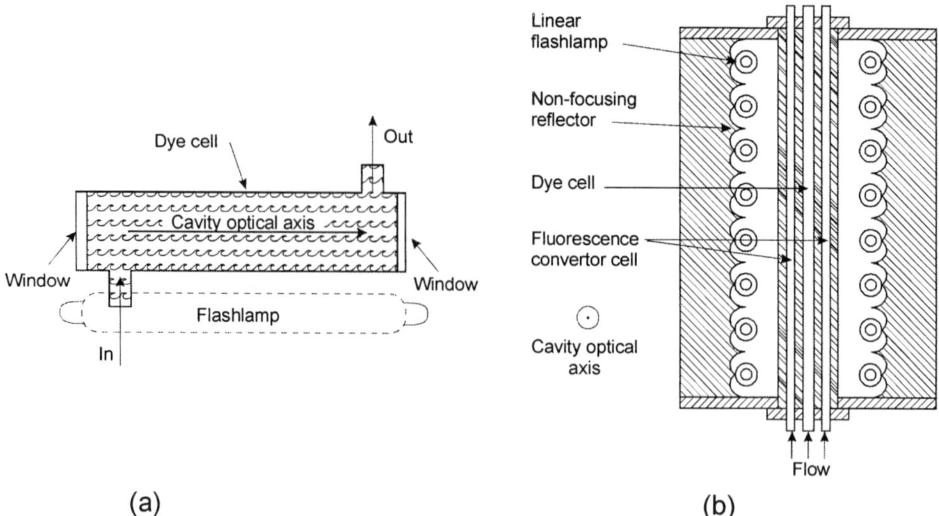

Figure B5.3.12. Dye cell flow schemes. (*a*) Longitudinal flow close-coupled cavity, and (*b*) transverse flow with faceted specular cavity (from [59]).

a 0.2 μm pore filter is used, as this provides a good compromise between flow resistance and required pump capacity.

In contrast, transverse flow across a cell is commonly used with higher repetition rates, as it allows large volumes of dye solution to be rapidly flushed through the cell. The one drawback of a transverse-flow arrangement is the design of the manifolds to ensure that the fluid maintains a uniform flow across the dye cell. These cells usually have a rectangular cross section. The formation of eddy currents or the creation of any non-uniform flow degrades the optical quality of the gain medium, thus having a direct impact on laser performance. Similarly, the formation of stagnation zones rapidly diminishes the performance of the laser head. It is normal in a transverse flow scheme to pump the solution vertically up through the cell to prevent the formation of air locks and similar voids in the system. A bubble-suppression filter is usually incorporated into the circulation system, as already noted, for longitudinal flow. Laminar-flow boundary layers along the dye cell walls must be minimized: fully-developed, small-scale pipe flow is required to produce an even flow distribution [52].

Repetition rates of the order of a kilohertz are possible with a transverse scheme, the limit being determined by the recovery time of the flashlamp and PFN. This restriction could well be eliminated by the use of a surface discharge pump source. These flow schemes are illustrated in figure B5.3.12.

B5.3.2.10 Multicolour operation

There have been many demonstrations [16–22] of multicolour output. In the case of a longitudinal-flow scheme, this can readily be achieved by having a series of annular cells co-axial within the central cell. Managing the separate circulation systems and the respective inlet and outlet ports becomes a significant design issue with more than three cells, although more are feasible. The approach is to arrange the dyes so that the fluorescence from one dye will pump the dye in its inner cell, so the longest wavelength emission is at the centre. A schematic diagram of a three-colour source is shown in figure B5.3.13.

A stack of rectangular-section dye cells can achieve a similar result for a transverse-flow laser head, as shown in figure B5.3.12(*b*). A photograph of a 'multicolour' laser head is shown in figure B5.3.14.

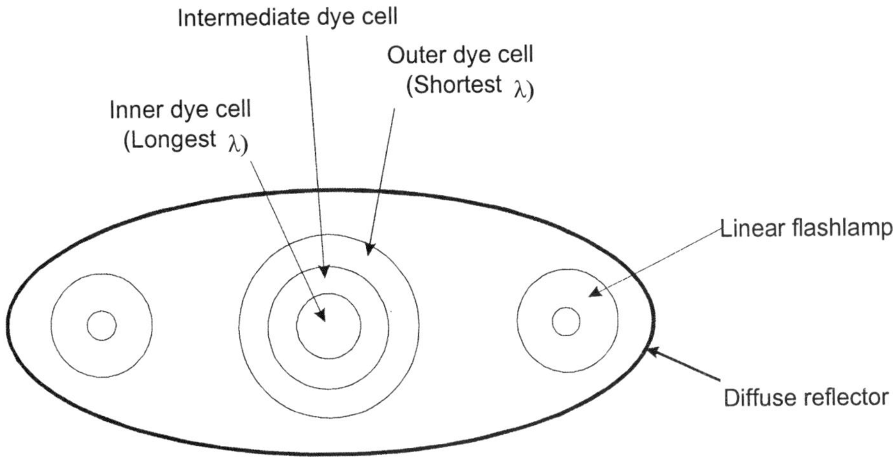

Figure B5.3.13. Multicolour longitudinal flow laser head schematic diagram.

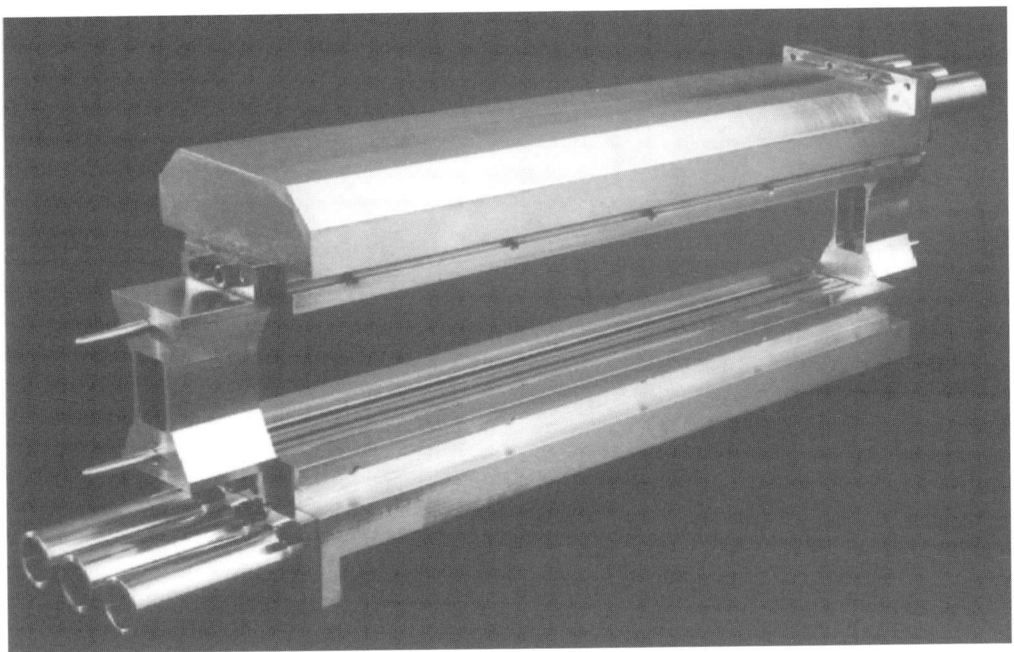

Figure B5.3.14. A transverse-flow multicolour dye laser head.

Many applications find it easier to exploit the wavelength diversity of a laser head by simply using a different dye in the solvent. Provided suitable broadband anti-reflection coatings have been applied to the optical elements in the resonator, then the same laser head can be used to generate light from the UV to the near IR. Carbon filters have been used to remove one organic laser dye from solution [4], leaving a few parts per million in solution after a few minutes' normal circulation. At this level of contamination the solvent appears 'clear'. A dye concentrate can then be injected to create a new solution with the appropriate

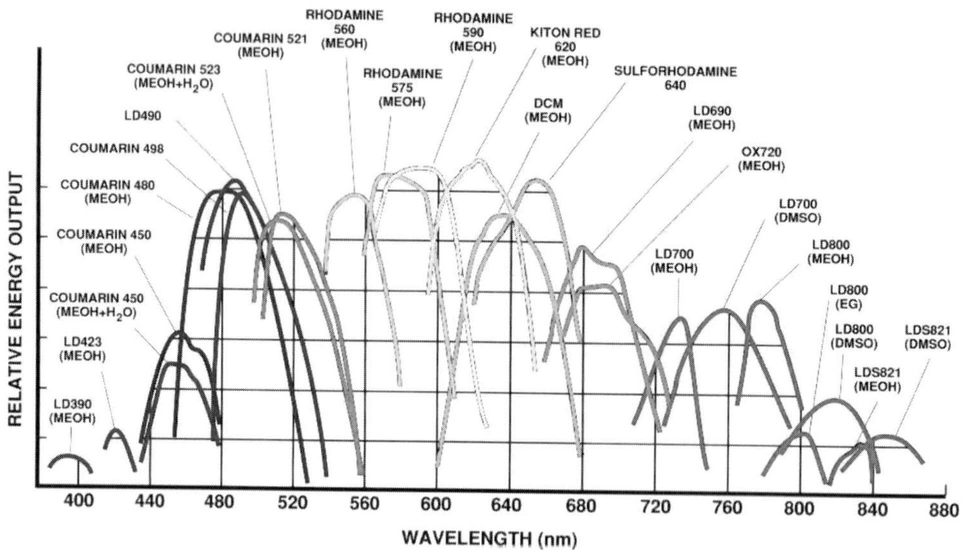

Figure B5.3.15. Dye spectral emission characteristics (courtesy of Exciton Ltd).

concentration of laser dye to provide the desired new spectral properties. The range of laser dyes offered by a single dye manufacturer is shown in figure B5.3.15.

B5.3.2.11 Laser-resonator design

The choice of resonator configuration (see section A2) has a significant impact on the key performance parameters: beam divergence and extracted energy. Consequently, different arrangements have been investigated [53, 54] in order to enhance some aspect of performance, especially beam quality in the case of a flashlamp-pumped dye laser. One approach is to have a long laser so that the Fresnel number[4] is small, thus only the lowest order modes can resonate. However, this is often not practical, as many dye cells have a diameter of about 10 mm; moreover it usually leads to inefficient light generation.

Positive branch unstable resonators (see section A2) with a magnification of about two have been exploited to optimize laser power extraction and simultaneously achieve high beam quality. In this case, the magnification reduces the (modified) Fresnel number[5], allowing optimum power extraction from the dye cell. Recently, graded-reflectivity mirrors [55] have been available to form resonators, permitting better beam quality to be maintained over a range of performance parameters.

The transmission of the output coupler (mirror) needs to be matched to the emission performance of the laser dye in the gain medium. With a high-gain lasing medium, it is beneficial to use a low-reflectivity output coupler. The converse is also true, but in this case if the energy extraction is too large, then the oscillator may fail to operate! Figure B5.3.16 shows the improvements that can be achieved by optimizing the output coupler reflectivity of a flashlamp-pumped dye laser.

Phase conjugation techniques (see chapter C1.3) offer another approach for enhancing beam quality. A self-starting resonator scheme with a primary and secondary cavity has been used to enhance beam

[4] Fresnel number $(F) = a^2/(\lambda \cdot L)$, where a is the radius of the limiting aperture, λ is the wavelength of operation and L is the length of the resonator.

[5] Modified Fresnel number $= \{(M - 1)/2\} \cdot F$, where M is the magnification.

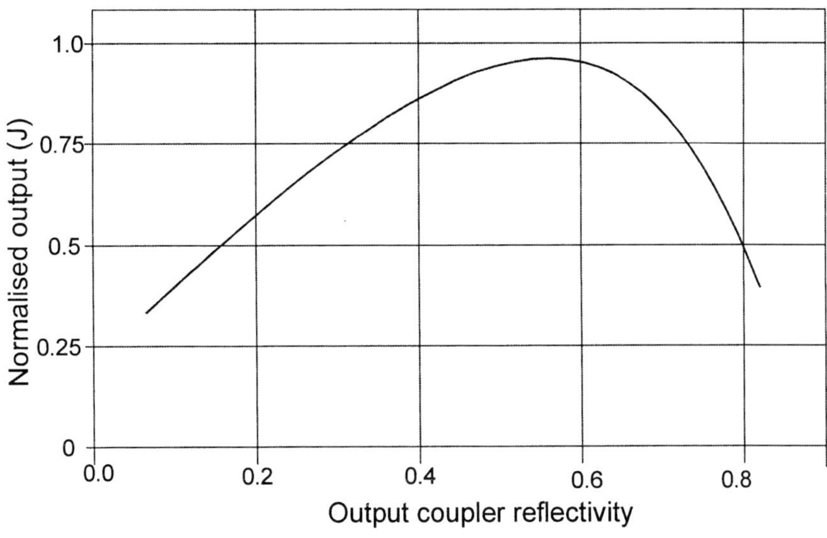

Figure B5.3.16. Output energy optimization.

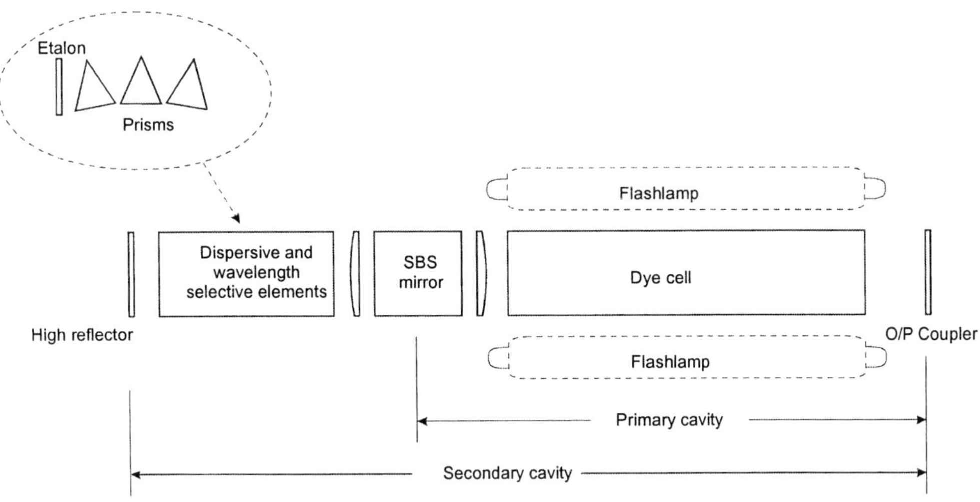

Figure B5.3.17. Schematic of self-starting phase conjugate resonator.

brightness [5]. This scheme, shown schematically in figure B5.3.17, improved the brightness by a factor of 100, compared with a plane-plane cavity.

B5.3.2.12 Other beam-quality enhancement techniques

A number of factors, other than the size of the gain volume, contribute to the degradation of the optical quality of the gain medium, causing excessive divergence of the emerging beam. The principal factors contributing to the reduction in beam quality are thermal gradients across the dye cell and non-uniform flow, including boundary-layer effects. The former effect can be reduced by a suitable choice of solvents with

Figure B5.3.18. A 35 J laser in operation.

a high thermal conductivity and matched thermal characteristics, similar to those of the cell. Table B5.3.3 shows the thermo-optical parameters [56] of a few common solvents and their influence on beam brightness.

Table B5.3.3. Thermo-optical parameters of typical dye solvents.

Solvent	(Wm^{-1} K^{-1}) Thermal cond'y	(K^{-1}) (dn/dT) $\times$ 10^{-3}	Rel. divergence	Rel. brightness
Water	0.561	0.08	1	1
Methanol	0.223	0.36	1.75	0.45
50% mixture	0.392	0.44	1.5	0.4

A number of different approaches, using standard 'fluid-dynamical-flow principles', have been suggested for reducing the boundary-layer effects of the fluid and parasitic reflections as it flows through a dye cell. A helical insert was proposed for use in gas lasers [32] to reduce these effects, consequently improving beam quality, and may be applied here. It is also anticipated that short duration pulses will reduce induced thermal effects, thereby enhancing beam quality. Finally, zigzag-pumping configurations have been demonstrated [57, 58], which average out thermally-induced effects in a dye cell.

B5.3.2.13 Flashlamp-pumped laser performance

These lasers are usually operated as simple oscillators, because high average-power performance is readily achievable. Average output power in excess of several kilowatts has been demonstrated [50, 59]. In some cases, this was from a low repetition rate (of approximately 10 Hz) string of high-energy pulses. Repetition rates of the order of a kilohertz have been demonstrated [6, 60] for this power class. Devices to produce one kilojoule per shot have been designed, but it has not been possible to confirm achieved performance. In principle this should be possible as flashlamp-pumped dye lasers are average-power devices. Hence, a 100 J/10 Hz device is equivalent to one emitting 1 kJ at 1 Hz. Figure B5.3.18 shows a high-energy source in operation. This performance is available from the UV to the near IR. However, it is dependent on the efficiency of the dyes, with the most efficient dyes tending to emit in the visible.

Miniature devices have been demonstrated using the same technology with average power of the order of a watt [60]. Scaling rules have been devised, enabling maximum use to be made of this technology. In

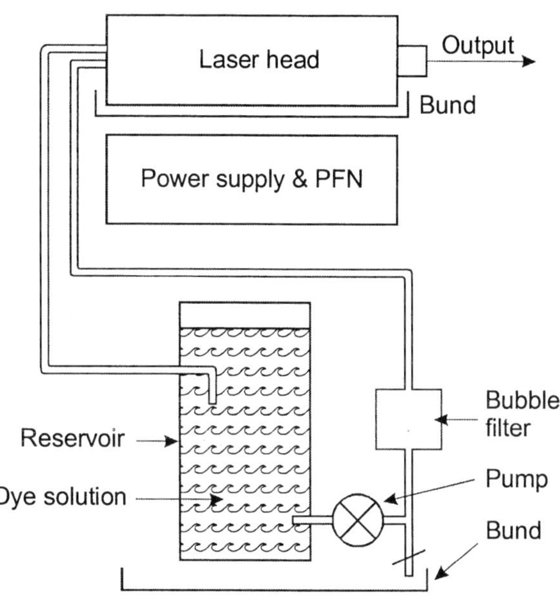

Figure B5.3.19. Scheme for safe operation.

general, it is easier to 'scale-up' a device than to reverse the process. One of the reasons for the scalability of this technology is the ease of removal of waste heat from the gain medium via a cooling circulation system.

Many techniques can be applied to modify the spectral and temporal characteristics. A formula[6] [61] relating the trade-off between the shortest possible duration of a pulse ($\Delta\tau$) and its linewidth ($\Delta\nu$), has been derived. It is not possible to combine a narrow linewidth in a very short pulse.

Linewidths of the order of 1 GHz have been demonstrated from flashlamp-pumped dye technology, whilst maintaining efficient operation. Pulse lengths are usually in the range of 1–5000 μs, although pulses as short as 1 ps have been reported [62], using passive mode-locking at the end of the pulse.

The efficiency of these lasers is in the region of 1–2% (wall plug) [59, 63] for the best dyes in medium- to high-power systems.

The *bête noire* of this technology has been the beam quality and the short useful life of the dye solution, particularly during high-power operation, which means the dye reservoir needs frequent replenishment. Use of unstable resonators with graded-reflectivity mirrors has enabled high-brightness beams to be produced. The beam-size is dominated by the diameter of the cell: it is typically of the order of 10 mm or more. The use of carefully designed cocktails of solvents, blocking of the unwanted hard-UV, preventing singlet oxygen generation and having optimum pump emission matched to a range of dyes, has enabled longevity to be enhanced by many orders of magnitude [4], as shown in figure B5.3.4.

Virtues of this technology are its simplicity, reliability and inherent low cost. The performance of this laser technology, particularly efficiency, is likely to be revolutionized by the application of surface discharge pump sources [3, 30].

[6] The relationship between pulse length ($\Delta\tau$) and linewidth ($\Delta\nu$) is given by: $\Delta\tau \cdot \Delta\nu \geq 0.441$.

B5.3.2.14 Safe operation

Some dyes may need to be dissolved in a volatile flammable solvent. Additionally, large voltages with rapid rising currents are used to pulse the flashlamp, thus there is a danger of an explosion under some combinations of circumstances. For example, if the solvent is very volatile with low viscosity, enabling it to penetrate joints in the circulation system, this risk can be mitigated by using 'metal plumbing', with compression fittings, and physical separation of the hazards. This is possible, because many solvents are heavier than air. However, use of efficient water-based solvents eliminates this hazard [4]. A schematic arrangement for safe operation with a hazardous solvent is shown in figure B5.3.19.

Safe handling, use and disposal of laser dye solutions, as well as the constituent chemicals, is discussed in section B5.3.2.3.

B5.3.3 Dye-lasers pumped by pulsed lasers

Radiation emitted by a visible- or near-UV pulse laser is an ideal technique for exciting dye lasers and was used in the first liquid laser demonstration [1,64]. Many different types of lasers, as indicated in table B5.3.2, are capable of delivering a very high intensity pulse of radiation, the peak power of which can reach 100 kW or more in a pulse lasting a few nanoseconds. The use of very short pumping pulses avoids the inefficiencies associated with triplet formation. Additionally, this method avoids distortion of the gain medium from acoustic effects.

In this case the dye laser becomes a wavelength-conversion module, enabling the user to select the emission wavelength. Clearly, there is an optimum pump wavelength for each type of laser-dye molecule, as the pump-radiation's wavelength must be within the region of high absorption, not too far away from the fluorescence region, although its exact wavelength is not critical. In some circumstances, with a pump laser that has a multiple-wavelength emission, it may be necessary to select one wavelength to achieve efficient use of the pump light. For example, with a copper-vapour pump laser, the green line could be used to excite one dye and the yellow line used for another.

The laser-head for this type of excitation is usually very much simpler than that used for incoherent (flashlamp) pumping, although many of the techniques described relate directly to achieving the desired spectral output.

B5.3.3.1 Dye cells

Laser pumping offers high optical conversion efficiency, owing to the large absorption and emission cross sections of many laser-dye molecules, which leads to a small interaction volume. Hence, the dye cells can be very small: the lower limit is often set by the damage threshold of dye-cell windows, particularly with oscillator-amplifier configurations.

The shape of a dye cell [2,62] can have a significant effect on the spectral characteristics of the beam generated in the gain volume. For example, a broadband output can be generated from a rectangular cell that is excited transversely. In contrast, narrow linewidth emission from a similar configuration would require a reduction in parasitic reflections. This can be achieved by tilting the emission windows at an angle to an axis in the plane of propagation normal to the resonator axis.

Alternative approaches include those with a parallelogram-shaped section and rectangular cells skewed at an angle on the plane formed by the optical axis of the resonator, and the axis normal to the plane of laser-light propagation. Prismatic-shaped cells have been used for uniform mode extraction, as have conical cells. These cells offer more uniform pumping of the dye solution by taking advantage of total internal reflections within the cell. Figure B5.3.20 illustrates some of the cell shapes often used.

As described before for the flashlamp-pumped dye laser, the effects of parasitic reflection may be reduced further by the use of anti-reflection coatings (see section C1.1.3.1.).

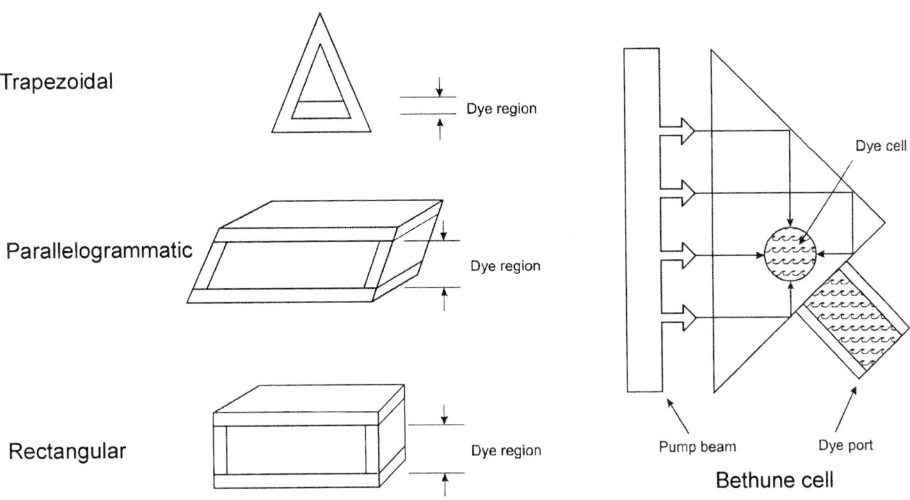

Figure B5.3.20. Dye cell shapes.

The pulse repetition rate (PRF) of the desired output and required duration of the emission determine whether it will be necessary to circulate the dye solution through the cell. A low-PRF emission, sustained over a lengthy period, will require the solution to flow through the cell with a low velocity. In this configuration there is a choice of longitudinal or transverse flow. A typical dye cell cross-section for low average-power emission is of the order of 1 cm^2.

A high-PRF laser output will demand rapid circulation of the solution through the cell, according to the 2:1 flushing ratio 'rule-of-thumb' described in section B5.3.2.9. For very high PRFs, it may be necessary to reduce the cross-sectional area to achieve the desired flow velocity with a pump of reasonable proportions and to avoid the phenomenon of cavitation. For example, cells with cross-sectional dimensions of 15×0.5 mm have been used for multikilohertz emissions. It is quite common to use longitudinal pumping with this operational regime, as the dye solution flowing through a cell moves faster in the centre. Judicious shaping of components in the cell can minimize the boundary layer effect and produce a flat, turbulent velocity profile necessary for good optical homogeneity in the cell, leading to good beam quality.

B5.3.3.2 Pump angle

The pump beam may be directed into the dye cell at a range of angles, depending on pump power and the laser configuration [65]. Use of high peak-power pulses of energy enables 'transverse pumping' to be used, i.e. the pump beam illuminates the dye cell orthogonal to the resonator's optical axis. This technique simplifies alignment requirements, but there is some sacrifice of efficiency. Provided that the pump powers are high, then it will not be necessary to use any optical elements to focus the beam into the dye cell.

Transverse-pumping is usually employed in an oscillator-amplifier configuration, which is commonly used to reduce the effects of ASE. In this case, the beam from a single-pump laser can be divided and directed transversely into both the oscillator and one or more amplifiers [62]. The use of a delay line may be required to achieve the optimum time-of-arrival of the pump pulses throughout the system, particularly at the amplifier to compensate for the build-up time in the oscillator.

Directing the pump beam along the optical axis of the cavity, or close to it, is known as longitudinal pumping. This approach tends to be used to improve the efficiency of low-gain systems and it also helps to improve beam quality by providing a good match between the pump and gain modes. An additional advantage of longitudinal pumping is that optimum use can be made of the faster (Newtonian) flow along the centre of the dye cell, so enhancing the 'flushing ratio' for a given size of circulation pump. A practical configuration of this pumping technique has the pump beam directed at a small angle to the 'resonator axis', just avoiding one of the cavity mirrors, hence dispensing with the complication of using dichroic mirrors in the laser resonator.

B5.3.3.3 Pumping configuration

Laser-pumping allows many configurations to be implemented, as the gain of the dye medium in a pulse-pumped system can be very high, owing to the peak intensity of the pump beam. Hence, the use of dispersive optic elements within the cavity, which have appreciable losses, can be tolerated without significantly reducing performance. Oscillator-amplifier configurations have been demonstrated: in the case of the laser isotope separation (LIS) facility in America [8], chains of amplifiers were used, pumped by banks of copper-vapour lasers.

The LIS system demonstrated the use of master oscillator power amplifier (MOPA) chains with three or four amplifiers in series: the first two being high gain pre-amplifiers to give sufficient power to saturate the following power amplifiers. This system showed that several copper-vapour pump-lasers (CVL) could be combined, with staggered timing to elongate the oscillator pulse prior to amplification. This system also demonstrated the multiplexing of CVLs to increase the PRF of the output. Additionally, it was necessary to control the path lengths through the various banks of amplifiers to maintain temporal coincidence between the pump and dye pulses at each amplifier [8].

Other arrangements have seen the use of optical fibres [66–68] to deliver the pump energy to the dye cell. This approach is particularly useful for operation in a hazardous environment, or to ease the alignment requirements of the pump beam. Furthermore, an anamorphic telescope can be used to transform a circular beam emerging from a fibre into an elongated footprint, leading to optimum overlap of the pump beam with the resonated beam.

B5.3.3.4 Linear cavities

This configuration offers the simplest arrangement, but can be used to generate a tunable emission. Figure B5.3.21 shows the arrangement for generating a tuned output from a dye laser pumped with a nitrogen laser.

B5.3.3.5 Efficiency

One important consideration with the laser-pumped configuration is to match the emission of the pump to the absorption bands of the laser dyes. With dyes such as the pyrromethenes and a well-matched pump-laser, it is possible to achieve conversion efficiencies well in excess of 50%, but this does not take into account the efficiency of the pump source. In some cases, this can be as high as 40% with a diode-pumped frequency-doubled neodymium laser, whereas other lasers are comparatively inefficient, of the order of 1 or 2 %.

B5.3.3.6 Mode locking

Generation of ultrashort pulses from a dye laser, operated in a broadband mode, is possible with some form of mode-locking (see chapter C2.3). Synchronous pumping of a dye laser is usually undertaken with a dye-jet configuration, enabling a string of very short pulses to be generated. This is considered further in section B5.3.5.1.

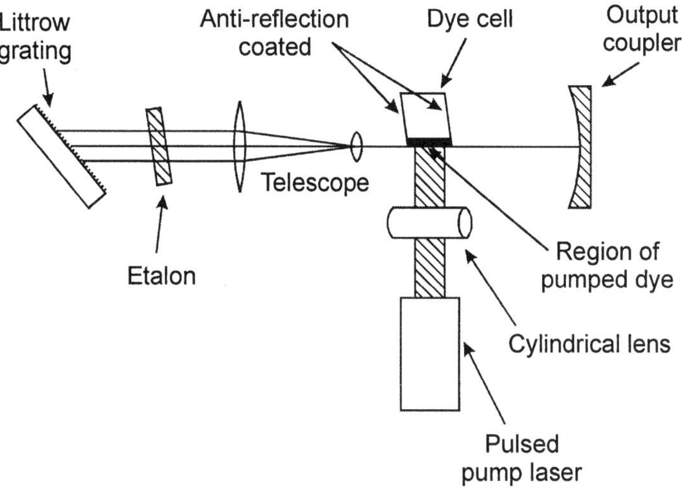

Figure B5.3.21. Pulse-pumped dye laser.

B5.3.3.7 Performance of laser-pumped devices

In theory, a very broad spectral emission is possible by merely changing the dye in the solution and, where necessary, the pump source or using a tunable source. However, the spectral coverage capability of a given configuration may well be limited by the resonator's optics. Usually the low-loss coatings on mirrors cover a broader spectral range than the gain bandwidth of any single dye, but the advantage of the AR coatings may be lost when a substantially different colour is generated. However, broadband coatings are available. Similarly, the use of wavelength-selecting elements for limiting the linewidth of the output can restrict the efficient emission to the spectral range of a single dye, or to dyes whose wavelength is close to the primary dye wavelength. Generally, the simpler the configuration, the greater the possible range of wavelengths that can be generated using a range of dyes, without the need to change optical elements.

Laser-pumped devices can generate a tunable output over a spectral range of many nanometres from a given dye. Increasing the reflectivity of the output mirror increases the tunable bandwidth. Use of additional elements in the resonator enables a beam of high spectral purity to be emitted of better than 1 MHz with single-mode emission, but more commonly of the order of 100 MHz, ($\Delta\lambda \sim 10^{-4}$ nm) [25, 33, 62]. Alternatively, use of mode-locking techniques has enabled very short pulses to be generated, of the order of 6 fs.

Pulsed lasers can give a few tens of watts of average power [65] close to the peak of the gain curve for efficient dyes, such as the xanthenes or the pyrromethenes. However, this value decreases as the wavelength is 'tuned' away from the peak of the gain curve. A master oscillator power amplifier (MOPA) system developed for laser isotope separation in the USA has demonstrated many kilowatts of average power with high spectral purity, from what is probably the largest dye laser in the world [8].

The temporal profile tends to follow, or be marginally shorter than, the pump pulse. This type of laser will give a very broad spectrum of pulse lengths from the very short to continuous, depending on the pump laser. Similar remarks apply to the pulse repetition rate. A range of pump lasers is available as discussed in section B5.3.2.2 to provide the desired output. For example, copper vapour lasers have pulse lengths of about 20 ns and a PRF of 3–20 kHz, whereas nitrogen lasers have a pulse length of a few nanoseconds and a typical PRF of up to 300 Hz: the characteristics of frequency-doubled solid-state lasers is similar to the latter, although PRFs of many tens of kilohertz are possible.

The beam quality of this class of dye laser is generally somewhat better than that from a flashlamp-pumped device, owing to less waste heat being generated in the laser head. In this case, the pump source and the design of the resonant cavity have a strong influence on the performance. However, in general, beam quality is often a secondary consideration. The divergence of the emitted beam is usually substantially less than 10 mrad, often better than 1 mrad. The beam size is usually less than 10 mm, typically in the range 1–3 mm.

Linearly polarized energy is generated if a tuning grating is included in the cavity. This also occurs if any component is oriented at Brewster's angle. Additional components can be included, either intracavity or extracavity, to enhance the polarized output.

The coherence length can be defined in various ways, but can be estimated from:

$$\text{Coherence length} = \text{Speed of light/frequency bandwidth.}$$

B5.3.4 Dye lasers pumped by cw lasers

The first cw dye laser was demonstrated in 1970 [69] using a cell with a flowing dye solution and excited by an ion laser (see chapter B3.5). This was followed in 1972 by the introduction of the dye-jet laser [70], also pumped by an ion laser. Recent developments have seen these lasers pumped with frequency-doubled cw neodymium sources (Nd:YAG and Nd:YLF—see chapter B1.3). Dye-jet lasers have also been excited using diodes (see chapter B2.4): a number of studies [71–73] report the use of red or near infrared diodes for producing compact cw dye lasers. In this case, the laser emission is generally in the near infrared.

Population of triplet states dominates the design of continuous wave dye lasers. There is also the problem of the pump laser sources, as noted in table B5.3.2, which are of modest power typically of the order of 10 W. Consequently, as the gain of a cw dye laser is very low, of the order of a few percent per pass, so the cavity must be designed to keep optical losses to an absolute minimum. Gratings, therefore cannot be used, but a combination of a birefringent filter and one or more tilted etalons is a feasible spectral control configuration.

The gain region of these lasers is in the form of a parallel-sided, thin stream of dye solution forced as a laminar-flow jet through a slit-like nozzle by a high-pressure pump. Longitudinal-pumping is often used in dye-jet lasers. In this case, optical elements are needed to focus the pump energy to a spot, typically a few micrometres in diameter, to broach the laser threshold. In order to avoid causing optical damage to windows, the dye usually flows from a nozzle to a receiver in an open configuration.

The use of a high-quality pump beam enables a common waist to be shared by both the pump beam and the dye-laser cavity. The diameter of such a waist is of the order of 15 μm, so with a dye-flow velocity of 10 ms^{-1} it means that a molecule only spends about 1.5 μs in the pump beam, hence the triplet-state population cannot build up to impair the efficiency of the laser action. If necessary, additional quenching agents can be used to augment the effects of atmospheric oxygen entrained in the dye jet.

B5.3.5 Cavity designs

Several cavity designs have been used for generating tunable cw operation. The diagrams in figure B5.3.22 show a simple two-way linear cavity and a ring cavity. Both contain frequency-selective elements (FSE), such as birefringent filters, to give coarse control of the oscillation and tilted etalons to restrict the spread of oscillation even further.

Line narrowing of the oscillation in a linear cavity is limited by the characteristics of the standing wave in a two-way cavity. The standing wave is spatially pinned with respect to the cavity mirrors, and the upper-level population is depleted more strongly at the antinodes of the electric-field pattern than at the nodes. This leads to the possibility of unwanted cavity modes developing, by scavenging the population inversion not accessed by the 'spatial hole burning' (see section A1.15.1) of the dominant mode.

The ring cavity, introduced in the late 1970s [74], avoids this drawback by creating travelling waves, so there are no nodes with unused gain that could support additional (unwanted) modes. In a simple

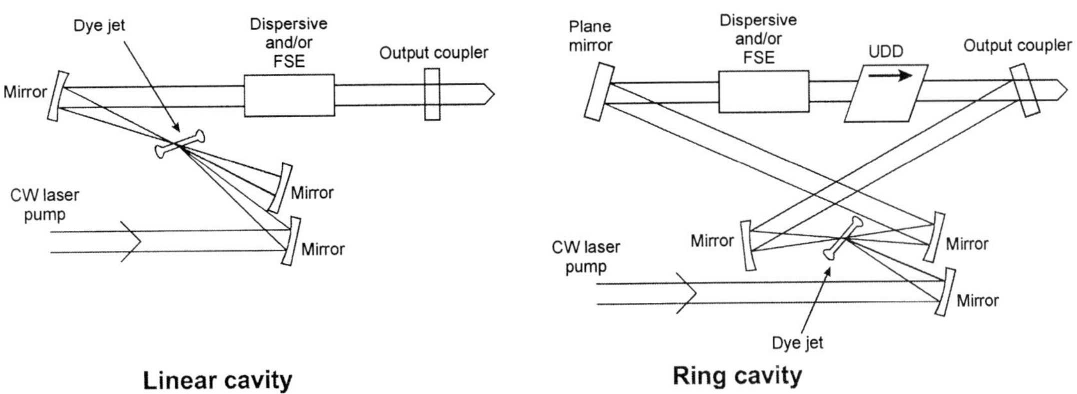

Figure B5.3.22. cw dye-laser arrangements.

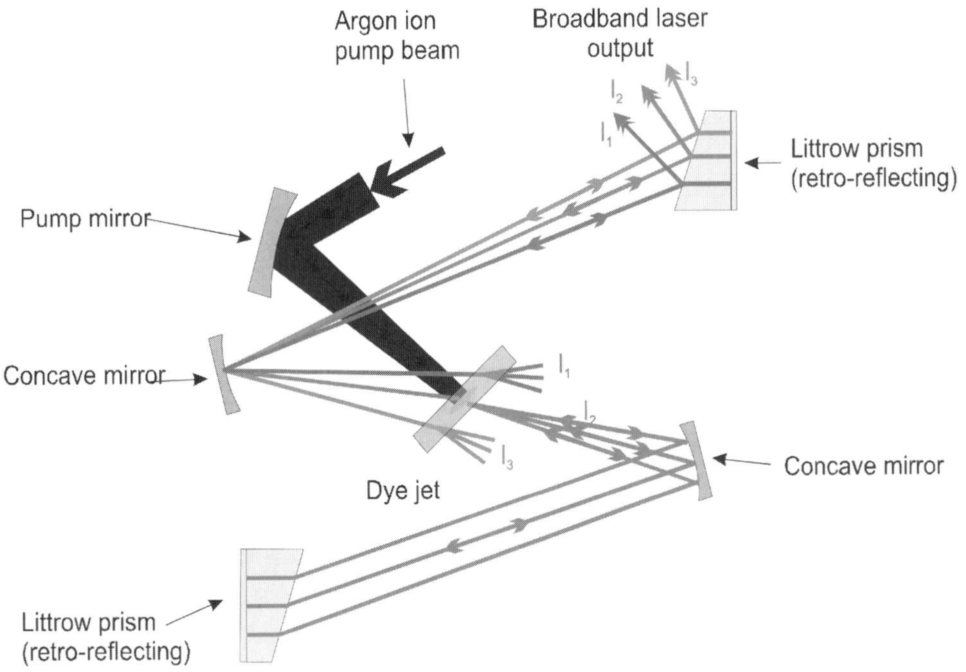

Figure B5.3.23. cw broadband dye-laser configuration.

configuration, laser energy can be generated in either of the two directions around the ring. The first to resonate will oscillate at the expense of the other, thus suppressing the second resonance. However, the direction of preference is random, which is unsatisfactory, as brief fluctuations in the cavity loss can cause the direction of propagation to be reversed. This presents problems for extracting the cavity flux, so an additional element is placed in the cavity that has a direction-dependent loss, giving a preferred propagation direction. A common technique is to use a Faraday rotator and standard polarizer, acting as a unique direction device (UDD). Consequently, only one travelling wave is established as the dominant mode and avoids spatial hole burning, leading to single mode emission.

B5.3.5.1 Mode locking

Synchronous pumping of a dye-jet laser enables a string of ultrashort pulses, in the femtosecond regime, to be generated, provided sophisticated techniques are applied to the laser configuration [43–45, 75] (see chapter C2.3). An absorber is used for providing the non-linearity to phase-lock the cavity modes. [43]. The general rule is the shorter the pulses, the broader the emission bandwidth needs to be.

The basic arrangement is to focus a continuous wave laser on to a dye laser jet in a linear cavity to provide the gain. The beam inside the cavity has another focal spot on an absorber dye, such as 3,3-diethyloxadicarbocyannine iodide (DODCI). Saturation of a dye in the absorbing element is used as the nonlinearity to phase-lock the cavity modes. Use of the colliding pulse technique [75] has enabled pulses of the order of 300 fs to be generated. Use of ring cavity configurations enabled this duration to be reduced to about 65 fs. More sophisticated techniques and compensation enabled this to be reduced by an order of magnitude [45, 76].

B5.3.5.2 Broadband emission

Broadband emission is feasible from a dye-jet laser, and a scheme is shown in figure B5.3.23. A mode-narrowing control aperture is required at the concave mirror, i.e. where the various beams intersect. The angles between the colours have been exaggerated for the purposes of illustrating the technique.

B5.3.5.3 Performance

The performance of cw dye lasers is comparable with laser-pumped sources in terms of the beam quality and spectral characteristics, with frequency stability or linewidths of the order of 1 kHz being demonstrated by injection seeding. The spectral emission range is comparable with other dye lasers, stretching 365–1000 nm. This range can be extended further into the UV and near IR by the use of nonlinear techniques.

This class of laser is capable of emitting beams with power in the 'watt class'. For special applications, the power has been extended to give tens of watts of tunable light. Recent developments with diode-pumping and the use of diode-pumped solid-state lasers have seen the efficiency of the complete cw dye laser system increase substantially, compared with the classic ion-laser pump source. Diode-pumping has been used for near infrared emission [71].

The development of ring laser cavities has led to the generation of beams with high spectral purity. Mode locking has enabled very short duration pulses to be emitted in the femtosecond regime.

References

[1] Sorokin P P and Lankard J R 1966 Stimulated emission observed from an organic dye, chloro-aluminum phthalocyanine *IBM J. Res. Develop.* **10** 162–3

[2] Samelson H 1982 *Handbook of Laser Science Technology* vol 1, ed M L Weber (Boca Raton, FL: Chemical Rubber Company) pp 397–422

[3] Fulker D J *et al* 1999 The evolution of pulsed discharges over PTFE in the presence of various cover gases *Proc. 12th IEEE Int. Pulsed Power Conf. (Monterey)* (Piscataway, NJ: IEEE) pp 1071–4

[4] Winstanley P A *et al* 1998 Enhancements achieved in the performance of liquid and solid-state dye laser technology *Proc. SPIE Solid State Lasers Conference (San Jose, CA)* vol 3265 (Bellingham, AW: SPIE) pp 42–50

[5] Titterton D H *et al* 1994 Studies to enhance the beam quality from flashlamp-pumped dye lasers *Proc. SPIE Solid State Lasers Conference (Los Angeles)* vol 2115 (Bellingham, WA: SPIE) pp 219–30

[6] Morton R G and Draggoo V G 1981 Reliable high average power high pulse energy dye laser *J. Quantum Electron.* **17** 222

[7] Ackerman D A 1990 Dye-laser isotope separation *Dye Laser Principles with Applications* ed F J Duarte and L W Hillman (Boston, MA: Academic) ch 9

[8] Bass I L *et al* 1992 The high-average-power dye laser at Lawrence Livermore National Laboratory *Appl. Opt.* **31** 6993–7006

[9] Klick D 1990 Industrial applications of dye lasers *Dye Laser Principles with Applications* ed F J Duarte and L W Hillman (Boston, MA: Academic)

[10] Wheeler M D 1999 Photonics in medicine-cancer in the crosshairs *Photon. Spectra* **33** 86

[11] Itzkan I and Izatt J A 1994 Medical uses of lasers *Encyclopaedia Appl. Phys.* **10** 33–59

[12] Pacheco D P *et al* 1990 A high-average-power blue-green laser for underwater communications *Proc. Int. Conf. on Lasers '89* (McLean, VA: STS) pp 376–82

[13] Estep L 1993 A review of airborne lidar hydrographic (ALH) systems *Hydrographic J.* **67** 25–43

[14] Hogan G P *et al* 1997 Designing a scaleable laser guide star system: the Oxford approach *Proc. ESO Workshop on Laser Technology for Laser Guide Star Adaptive Optics Astronomy Garching bei München, Germany* ed N Hubin, no 55, pp 92–8

[15] Friedman H W *et al* 1997 Design of a Laser Guide Star System for the Keck II Telescope *Proc. ESO Workshop on Laser Technology for Laser Guide Star Adaptive Optics Astronomy Garching bei München, Germany* ed N Hubin, no 55, pp 138–44

[16] Alden M *et al* 1984 Application of a two colour dye laser in CARS experiment *Appl. Opt.* **23** 2053–5

[17] Burlamacchi P *et al* 1979 Characterization of a multicolour dye laser *Opt. Commun.* **34** 185–8

[18] Hilborn R C and Brayman H C 1974 Simultaneous two-wavelength output from multiple-dye pulsed tunable dye lasers *J. Appl. Phys.* **45** 4912

[19] Nair L G 1979 A double wavelength nitrogen-laser-pumped dye laser *Appl. Phys.* **20** 97–9

[20] Saito Y *et al* 1985 Simultaneously tunable three-wavelength dye laser *Appl. Opt.* **24** 2477–8

[21] Williams S W *et al* 1983 Simultaneous two-wavelength tunable dye laser with no mode competition and with a wavelength separation of more than 200 nm *Opt. Commun.* **45** 112–4

[22] Malikechi N and Allen L 1987 A two wavelength dye laser with broadband tunability *J. Phys. E: Sci. Instrum.* **20** 558–9

[23] Pacheco D P *et al* 1989 The use of special conversion to increase the efficiency of flashlamp-pumped dye lasers *Proc. Int. Conf. on Lasers '88* (McLean, VA: STS) pp 410–19

[24] Griffiths J 1976 *Colour and Constitution of Organic Molecules* (London: Academic)

[25] Hecht J 1992 Dye lasers *The Laser Guidebook* 2nd edn (New York: McGraw-Hill) ch 17

[26] Schäfer F P 1973 Principles of dye laser operation *Dye Lasers* (Berlin: Springer) ch 1

[27] Koechner W 1992 *Solid-State Laser Engineering* (Berlin: Springer)

[28] Marling J B 1970 Chemical quenching of the triplet state in flashlamp-excited liquid organic lasers *Appl. Phys. Lett.* **17** 527–30

[29] Tuema F A *et al* 1999 The spectral properties of UV radiation from pulsed surface discharges *Proc. 12th IEEE Int. Pulsed Power Conf. (Monterey)* (Piscataway, NJ: IEEE) pp 1106–9

[30] Fulker D J 2002 The development and application of surface discharges for the pumping of dye lasers *PhD thesis* University of Strathclyde

[31] Miller A W *et al* 1995 Investigation into the photo-degradation products affecting the dye life in flashlamp pumped dye laser *Proc. SPIE Conf. on UV and Visible Lasers and Laser Crystal Growth (San Jose)* vol 2380 (Bellingham, WA: SPIE) pp 364–273

[32] Bradley D J *et al* 1974 Megawatt vuv xenon laser employing coaxial electron-beam excitation *Opt. Commun.* **11** 335–8

[33] Duarte F J 1990 Narrow-linewidth pulsed dye laser oscillators *Dye Laser Principles with Applications* ed F J Duarte and L W Hillman (Boston, MA: Academic) ch 4

[34] Hansch T W 1972 Repetitively pulsed tunable dye laser for high resolution spectroscopy *Appl. Opt.* **11** 895–8

[35] Beiting E J and Smith K A 1979 An on-axis reflective beam expander for pulsed dye laser cavities *Opt. Commun.* **28** 355–8

[36] Titterton D H and Stokoe C 1995 Progress with a flashlamp-pumped titanium sapphire laser *Proc. SPIE Conf. on UV and Visible Lasers and Laser Crystal Growth (San Jose)* vol 2380 (Bellingham, WA: SPIE) pp 115–26

[37] Danilov M B and Kristov I P 1986 Novel method of ultra-broadband laser generation *Opt. Commun.* **73** 235

[38] Duarte F J and Piper J A 1980 A double-prism beam expander for pulsed lasers *Opt. Commun.* **35** 100–4

[39] Morton R G *et al* 1978 Efficient cavity dumped dye laser *Appl. Opt.* **17** 3268–75

[40] Leslie S G *et al* 1984 Efficient cavity dumped HgBr laser *Appl. Opt.* **23** 36–9

[41] Schmidt W and Schäfer F P 1968 Self-mode-locking of dye-lasers with saturable absorbers *Phys. Lett.* A **26** 558–9

[42] Ippen E P *et al* 1972 Passive mode locking of the cw dye laser *Appl. Phys. Lett.* **21** 348–50

[43] Fork R L *et al* 1981 Generation of optical pulses shorter than 0.1 psec by colliding pulse mode locking *Appl. Phys. Lett.* **38** 671–2

[44] Dietel W *et al* 1983 Intracavity pulse compression with glass: a new method of generating pulses shorter than 60 fs *Opt. Lett.* **8** 4–6

[45] Fork R L *et al* 1987 Compression of optical pulses to six femtoseconds by cubic phase compression *Opt. Lett.* **12** 483–5

[46] Sorokin P P and Lankard J R 1967 Flashlamp excitation of organic dye lasers: a short communication *IBM J. Res. Dev.* **11** 148

[47] Schmidt W and Schäfer F P 1967 Blitzlampengepumpte Farbstofflaser *Z. Naturf.* a **22** 1563–6

[48] Markiewicz J P and Emmett J L 1966 Design of (a) flashlamp driving circuit *J. Quantum Electron.* **2** 707–11

[49] Miller A W *et al* 1998 Factors affecting the life of dye laser flashlamps *Proc. SPIE Solid State Lasers Conference (San Jose)* vol 3265 (Bellingham, WA: SPIE) pp 21–8

[50] Klimek D E and Aldag H R 1994 10 Hz kilowatt-class dye laser system *Proc. SPIE Conf. on Intense Laser Beams and Applications (Los Angeles)* vol 1871 (Bellingham, WA: SPIE) p 11

[51] Jensen C C 1991 Pulsed dye laser gain analysis and amplifier design *High-Power Dye Lasers* ed F J Duarte (Berlin: Springer)

[52] Schlichting H 1960 *Boundary Layer Theory* (New York: McGraw-Hill)

[53] Hall D R and Jackson P E 1989 *The Physics and Technology of Laser Resonators* (Bristol: Hilger)

[54] Siegman A W 1986 *Lasers* (Oxford: Oxford University Press)

[55] Morin M 1998 National Optics Institute (Canada) GRM resonator design: a tutorial *Proc. SPIE Conf. on Laser Resonators (San*

Jose) vol 3267 (Bellingham, WA: SPIE) p 6
[56] Kaye G W C and Laby T H 1986 *Tables of Physical Constants* (Singapore: Longman)
[57] Klimek D E and Mandl A 1995 Single mode, diffraction-limited operation of a zig-zag dye laser *Proc. SPIE Conf. on UV and Visible Lasers and Laser Crystal Growth (San Jose)* vol 2380 (Bellingham, WA: SPIE) pp 256–63
[58] Dearth J J *et al* 1988 A flashlamp-pumped zig-zag slab dye laser *Proc. Int. Conf. on Lasers '87* (McLean, VA: STS) pp 320–9
[59] Aldag H R 1996 Efficient high average power dye laser *Proc. SPIE Conf. on Gas Lasers (laser Optics '95) (St Petersburg)* vol 2771 (Bellingham, WA: SPIE)
[60] Sierra R 1988 Flashlamp-excited dye lasers achieve new performance levels *Laser Focus/Electro-Opt.* 76–92
[61] Svelto O 1982 *Principles of Lasers* 2nd edn (New York: Plenum)
[62] Duarte F J 1990 Technology of pulsed dye lasers *Dye Laser Principles with Applications* ed F J Duarte and L W Hillman (Boston, MA: Academic) ch 6
[63] Everett P N *et al* 1986 Efficient 7-J flashlamp-pumped dye laser at 500 nm wavelength *Appl. Opt.* **25** 2142–7
[64] Schäfer F P *et al* 1966 Organic dye solution laser *Appl. Phys. Lett.* **9** 306–9
[65] Hecht J 1992 Versatility keeps dye lasers alive *Laser Focus World* 59–74
[66] Anderson D J *et al* 1996 Fiber-optic-bundle delivery system for high peak power laser particle image velocimetry illumination *Rev. Sci. Instrum.* **67** 2675–9
[67] Anderson D J *et al* 1995 An optical-fiber delivery system for pulsed-laser particle image velocimetry illumination *Meas. Sci. Technol.* **6** 804–14
[68] Booth H 1998 Guide star lasers *High Power Fibre Optic Laser Beam Delivery* University of Oxford, DPhil Thesis, ch 4
[69] Peterson O B 1970 cw operation of an organic dye solution laser *Appl. Phys. Lett.* **17** 245–7
[70] Runge P K and Rosenburg R 1972 Unconfined flowing-dye films for cw dye lasers *IEEE J. Quantum Electron.* **8** 910–1
[71] Scheps R 1995 Near-IR dye laser for diode-pumped operation *IEEE J. Quantum Electron.* **31** 126–34
[72] Scheps R 1993 Low threshold dye laser pumped by visible diodes *IEEE Photonics Tech. Lett.* **5** 1156–8
[73] Scheps R 1995 Laser diode-pumped dye laser *Proc. SPIE Conference on UV and Visible Lasers and Laser Crystal Growth (San Jose)* vol 2380 (Bellingham, WA: SPIE) pp 274–84
[74] Schröder H W *et al* 1977 A high-power single-mode cw dye ring laser *Appl. Phys.* **14** 377–80
[75] Ruddock I S and Bradley D J 1976 Bandwidth-limited sub-picosecond pulse generation in mode-locked cw dye lasers *Appl. Phys. Lett.* **29** 296–7
[76] Diels J-C 1990 Femtosecond dye lasers *Dye Laser Principles with Applications* ed F J Duarte and L W Hillman (Boston, MA: Academic) ch 3

Further reading

Schäfer F P 1973 *Dye Laser* (Berlin: Springer)
Griffiths J 1976 *Colour and Constitution of Organic Molecules* (London: Academic)
Hecht J 1992 *The Laser Guidebook* 2nd edn (New York: McGraw-Hill)
Duarte F J and Hillman L W (ed) 1990 *Dye Laser Principles with Applications* (Academic)
Duarte F J (ed) 1991 *High-Power Dye Lasers* (Berlin: Springer)
Maeda M 1984 *Laser Dyes* (New York: Academic)
Förster T 1951 *Fluoreszenz Organizcher Verbindungen* (Göttingen: Vanderhöck and Ruprecht)
Pavlopoulos T G 2000 Laser dyes *Encyclopaedia of Materials Science and Engineering, Section on Colors, Paints and Dyes* (New York: Wiley) ch 7

B5.4
Solid-state dye lasers

David H Titterton

B5.4.1 Introduction

The broad spectral emission from a single laser cavity containing a dye-based gain medium makes these sources attractive for many applications, as described in chapter B5.3 on liquid lasers. A single laser configuration can be used to produce light from the ultraviolet to the near-infrared by merely changing the laser dye in the gain medium. In addition, it is possible to tune the wavelength of the light from a given laser dye over a narrow spectral range, often over a few tens of nanometres.

Traditionally, dye lasers have used a liquid gain medium but dye-doped materials, such as polymers and glasses, are now becoming established [1–24] with flashlamp-pumped configurations [15–24] as well as laser-pumped devices. The development of new laser dyes [25–30] over the last 15 years, offering greater photostability and potential for enhanced gain, has provided the catalyst for the development of dye-doped solid-state lasers. During this period there has also been a significant improvement in the optical quality as well as the range and stability of host materials. However, other limitations in the physical characteristics of the gain materials, such as the thermal degradation of the dye, along with thermal expansion of the gain element and its impact on the cavity stability, will limit the linewidth performance that can be achieved from this class of laser. Hence, solid-state dye lasers cannot be expected to supersede the liquid laser in all of its established applications, particularly those requiring high spectral purity, such as spectroscopy. For many other applications such as medical or military ones, which generally require energetic pulses of coloured light, the solid-state dye laser is most attractive owing to its greater simplicity.

Laser action from a dye gain medium was first demonstrated in 1966 [31,32]. In these studies, solutions of the phthalocyanine class of organic laser dyes formed the gain medium, which was pumped by photons from a ruby laser. Soon afterwards flashlamp-pumped dye lasers were demonstrated [33,34] and the same year saw the first laser action from a dye-doped polymer [35]. The following year, 1968, saw the demonstration of a flashlamp-pumped dye-doped polymer [36]. The conversion efficiency of these early devices was poor but was improved with the use of transfer dyes to improve conversion efficiency in dye-doped poly-(methyl methacrylate) [37].

During the latter part of the 1980s, there was an upsurge in interest in solid-state dye lasers [1, 2, 15], particularly for medical applications [10]. The combination of better dyes being incorporated into host matrices with lower optical losses resulted in the dye-doped polymer hosts being very effective sources. Various hosts have been developed, including polymers and glass-like materials. The laser-pumped devices have routinely exhibited conversion efficiencies of the order of 70%, or more, for some dyes [2, 11–13, 17], and recent results with flashlamp-pumped rods have seen emissions with energy in excess of 1 joule per pulse [22, 24]. Currently, there appears to be little prospect of sustained continuous-wave (cw) emission, although quasi-cw devices have been demonstrated [38].

Research has also been directed towards increasing the so-called 'service life' of the dye-doped solid-state gain medium [7, 12, 14], which relates the number of useful pulses of light that can be obtained from a gain element. Various criteria can be used to express the 'useful performance': a typical cut-off point is when the current output performance (such as energy per pulse) has fallen to 70% of its initial value.

A major *bête noire* of a liquid-phase laser is the need to circulate the laser-dye solution in order to remove heat, molecules in their triplet state and any dye molecules that have degraded from the gain volume. A dye-doped gain medium experiences similar performance-limiting effects, resulting from the trapping of pump light but their impact can be mitigated through careful design of the gain element and choice of laser dye. Several schemes have been designed to simulate dye flow by moving the dye-doped matrix in the resonant cavity, either by rotation, translation or a combination of both, which are somewhat simpler than a dye-circulation system. This approach has enabled enhanced 'service life' to be achieved, resulting in many tens of thousands of laser pulses to be emitted from a single dye-doped gain element without significant loss in performance. In some cases, well in excess of 500 000 shots have been reported from a small disc, of the order of 20 mm diameter. Recent developments indicate that this performance may be achievable without the need to move the gain element [14].

B5.4.2 Solid-state dye-doped host preparation

A few companies produce dye-doped materials for sale. However, good quality materials can be made relatively easily in a laboratory equipped for polymer or glass technology experiments. No particularly sophisticated equipment is required and the techniques that have been developed by various laboratories are quite straightforward; hence, research groups and other source developers can produce their own samples. Many different types of host [39] have been demonstrated and techniques for the production of various hosts are outlined in the following sections, the references providing details on experimental procedures.

B5.4.2.1 General characteristics

The use of conventional solid glass materials doped with laser dyes is quite attractive, as they have low optical scatter, low distortion, are hard and stable and can be polished to be smooth and flat. However, it is usually impossible to use such a host, owing to the need to heat the constituent materials to temperatures in excess of 1000 °C during production, thereby destroying the laser dye. Consequently, alternative host materials and techniques for fabrication at a lower temperature have been sought.

There are two major approaches: synthesizing polymer or 'glassy materials' at low temperature or producing a glass with a porous structure. The former technique allows a laser dye to be added to the constituent chemicals and is known as pre-doping. A porous structure allows a dye solution to be introduced into the structure to impregnate it by diffusion and is termed post-doping of the solid matrix.

Pre-doped hosts usually produce solid materials with good optical quality and uniform dye concentration. These dye-doped solids can have properties approaching those of glass but tend to be somewhat softer, which can have an adverse impact on performance, particularly if high-energy performance is required. In contrast, post-doped materials are usually more 'glass-like', with many of the excellent bulk properties of glass but the material can be susceptible to losses owing to excessive optical scatter from the pores in the host matrix. Clearly, there is a trade-off between ensuring that the average pore size is large enough to absorb the dye solution, to achieve a good concentration, and still have very good optical quality with low losses, particularly optical scatter.

B5.4.2.2 Pre-doped hosts

In this case, a solution of the chosen laser dye is mixed with one of the constituent chemicals used to form the host matrix, which is normally a thermoplastic polymer with a linear molecular structure. Consequently, successful application of this technique is dependent on limiting the temperatures used in the polymerization, condensation or curing processes invoked, to produce the solid material, to less than a few hundred degrees Celsius. This process has been used successfully by many teams to produce dye-doped sol-gel [40], poly-(methyl methacrylate) (PMMA) [6], ormosil [41–43] and poly-acrylamide gels [44], with dye concentrations of the order of 2×10^{-4} molar.

One of the snags with the pre-doping technique is that the laser dye may not be soluble in the basic constituent chemicals, such as the monomer—use of a co-polymer solves this problem [45]. An example is the popular dye rhodamine 6G, which is insufficiently soluble in methyl methacrylate. In this case, it is normal to use a hydroxylic co-monomer, e.g. 2-hydroxypropyl acrylate/methyl methacrylate, in order to produce rhodamine 6G-doped PMMA.

The curing process can take from a few tens of minutes for poly-acrylamide gels to many weeks for ormosils: PMMA takes five to 10 days. Careful thermal control is necessary in order to prevent the build-up of stresses during the formation of the host matrix as the volume can shrink by about 80% in some cases, for example with a sol-gel. Various strategies have been developed to alleviate the stress formation, including the use of drying control chemical additives; however, some additives can be difficult to remove.

B5.4.2.2.1 Sol-gel-derived silicates

This matrix is produced [40] by a low temperature (less than 100 °C) hydrolysis of tetramethoxysilane with water in a solvent, such as methyl alcohol, with subsequent poly-condensation. Following hydrolysis of the tetramethoxysilane, the poly-condensation reaction occurs to produce a sol of colloidal silica particles in suspension. This reaction continues to inter-link the sol particles to produce a rigid gel matrix. Further strengthening of the matrix results from additional reaction time and drying to remove the solvent and excess water. Such matrices are commonly known as xerogels.

The xerogel structure usually consists of a network of pores, their sizes being of the order of 1–10 nm. The pore surface is populated by silanol groups, Si–OH and is very polar. These silanol groups are suitable sites for hydrogen bonding with dye molecules inside the pore system.

B5.4.2.2.2 Poly-(methyl methacrylate) (PMMA)

The fundamental unmodified host is produced from the polymerization of the monomer methyl methacrylate [6], in which a suitable dye is dissolved. This reaction is generally started by a radical initiator, such as benzoyl peroxide, or azo bis-isobutyronitrile. Again, this is a low-temperature reaction with the mixture heated to about 50 °C; it is also quite quick, producing samples within a few days. This host is often regarded as the standard for pre-doped solid-state dye media.

The use of co-polymer mixtures [3, 8, 17, 45], such as methyl methacrylate/hydroxy propylacrylate and methyl methacrylate/hydroxy ethyl methacrylate, to produce host matrices has been successful. This modified PMMA material (m-PMMA) has some attractions as, for example, it allows polar dyes to be used such as rhodamine 6G.

A common modification is to use a low molecular weight additive [46–48], which plasticizes the polymer and increases the laser-induced damage threshold. The modest damage threshold of the unmodified host is a serious disadvantage for high average-power operation.

B5.4.2.2.3 Organically modified silicates

This host material is commonly known as ormosil, which has some similarities in its fabrication with sol-gels but produces an organic–inorganic composite substance on the molecular scale [49, 50]. The silicate structure is prepared by a low-temperature sol-gel technique and it is possible to produce materials where the PMMA is covalently bonded to the silicate network, producing a non-porous structure.

During the reaction tetramethoxysilane (TMOS) undergoes acid catalyzed hydrolysis and condensation reactions to form a silicate structure with loss of methyl alcohol. A second precursor, trimethoxysilylpropylmethacrylate (TMSPM), undergoes partial hydrolysis and condensation to allow it to link into the developing silicate structure but its Si–C group is not hydrolyzed, so it retains a methacrylate tail. This methacrylate tail allows the structure to bond to the third precursor, methyl methacrylate, which itself undergoes polymerization reactions, to form PMMA. This matrix is doped with a dye by adding a solution of laser dye in methyl alcohol to the reactants, for example after the initial rapid hydrolysis under acid catalysis.

The fabrication process can be somewhat lengthy, taking up to several weeks to produce a large fully-cured matrix with negligible stress. The use of carefully controlled cooling methods (typically about 1 °C per day) and curing techniques has led to the production of samples with very good optical quality, comparable with that of PMMA. Generally, this host is somewhat harder than PMMA.

B5.4.2.2.4 Optical epoxy

Optical epoxy polymer samples are fabricated [51,52] from two constituent parts: usually an epoxy oligomer and a polyamine hardener. These two components are mixed together according to the manufacturer's instructions and cured to give a thermoset epoxy polymer matrix. A suitable laser-dye solution is added to the two components prior to the curing process, which is conducted in an appropriate mould. These cured polymers tend to have a yellow colouration, because of the absorption of the matrix in the blue–violet region of the spectrum.

A Ukrainian group developed a matrix [53] formed by the radical photo-polymerization of oligomers, such as oligourethane acrylate. In this case, they used benzophenone as the initiator and it was introduced into the initial oligomer composition.

Epoxy materials are cheap and easy to manufacture and have high thermal resistance and damage threshold. Other advantages that are often cited, compared with PMMA, are a higher elastic modulus, a lower temperature coefficient of refractive index and better resistance to photochemical degradation. Moreover, during the curing process this type of host matrix only shrinks by a small percentage of the initial volume; hence, the dye concentration can be carefully controlled.

B5.4.2.2.5 Polyacrylamide gel

This laser gain medium, with high optical quality, can be produced [44] in about 20 minutes and is made up from three constituents: gel buffer, an acrylamide gel and an ammonium persulphate solution. The dye solution is added to the gel buffer prior to adding to the other constituent solutions, mixed quickly and allowed to set at room temperature in a quartz cuvette.

This dye-doped host degrades over time owing to the loss of water and, eventually, it becomes opaque. Loss of water from the matrix can be retarded by sealing the cuvette with a silicon rubber compound.

B5.4.2.2.6 Fluorophosphate glasses

A family of glasses in the lead-tin fluorophosphate system, which has viscosity characteristics that enable melting to occur at a temperature which does not cause thermal decomposition of many organic dyes has been reported [54]. The dye-doped samples are prepared by melting batch materials at about 400 °C in a vitreous carbon crucible and allowed to cool to 300 °C. The composition is such that the viscosity at this temperature allows dyes to be thoroughly mixed with the host. The samples are then allowed to cool in air followed by annealing at around 50 °C. In this case, the dye is incorporated into the matrix and is unaffected by the glass transition process.

Glasses within this class appear to be an ideal host for organic laser dyes. The hosts are transparent in the visible and near-infrared bands, with the onset of the ultraviolet absorption edge at around 350 nm.

B5.4.2.2.7 Other media

A range of other materials has been investigated, such as polystyrene, lavsan, di- and tri-acetate cellulose, polyvinylethial and polycarbonate. In general, these hosts have tended to be less successful for dye-laser applications than the matrix materials just described in detail, or the glass-like materials considered later.

B5.4.2.3 Post-doped hosts

The use of this technique enables genuine glass-like materials to be produced, using elevated temperatures to strengthen the matrix. A porous structure can either be generated through a chemical reaction or by selective etching of a solid glass matrix. The resulting porous material can then be doped with a suitable laser-dye solution; however, it is usually difficult to calculate the exact dye concentration in the host matrix. Thus, it can be difficult to optimize the dye concentration for a given pump source. One common method is to estimate the amount of dye solution taken up by the host matrix from the difference in dye solution at the beginning and end of the doping or impregnation process. Hence, the doping concentration can be estimated. The other major drawback is the problem of achieving an optical-grade polish on the surfaces after the doping process, owing to the porous structure.

B5.4.2.3.1 Densified sol-gel

An undoped xerogel matrix, produced by the hydrolysis process described in section B5.4.2.2.1 may be modified by heating to about 800 °C [6]. This produces a stronger and denser matrix with smaller pores, which has thicker pore walls. During this process, the xerogel shrinks to about 50% of its original volume. Heating to higher temperatures results in a further reduction in pore size until, at about 1200 °C, a non-porous structure is produced. This host is usually doped [6] by standing the partially densified matrix in a solution of laser dye for a couple of weeks at an ambient temperature of around 45–50 °C. Subsequently, the material is dried, using vacuum techniques.

B5.4.2.3.2 Microporous glass

The porous glass host can be created through selective etching of a particular glass matrix, which has at least two skeletal structures. It is necessary for each to have a different reaction to a given solvent, so that one structure can be removed to leave the other intact, thus creating a porous matrix.

Vycor is a sodium borosilicate glass which, when heated above its glass transition temperature, forms a liquid–liquid phase separation. The sodium borate glass phase can be leached out at room temperature in acid to leave an interconnected sponge-like structure of more or less pure silica glass. The matrix can be impregnated with laser dye: this has been demonstrated in wet and dry forms [55–57]. In the former case, the dye is in solution in a suitable solvent, such as an alcohol or an ester, and the dye-doped micro-porous matrix is sealed between quartz plates, for example. In the latter case, the solvent is removed by warming. Generally, the ‘wet fill’ has superior optical performance, owing to reduced optical scatter.

The overall porosity (pore volume density) and the average pore size can be controlled by varying the ratio of the two phases in the original matrix to give a structure optimized to receive a dye solution. Pore sizes in the range 1–10 nm are usually achieved and a pore volume density of 0.15–0.2 $cm^3\ g^{-1}$. The presence of a continuous glass structure, together with the relatively small pores, allows efficient heat dissipation throughout the entire structure and its subsequent extraction during laser action.

B5.4.2.3.3 Polycomposite glass

In this case, a sol-gel matrix has its pores impregnated with a refractive-index-matched material [6, 58] to reduce the optical scatter from the pores and, hence, improve the optical quality of the material. It is possible to use PMMA to 'index match' with silica glass, using an *in situ* polymerization of methyl methacrylate [6], producing a sol-gel–PMMA composite known as polycomposite (or polycom) glass. In this case, a solution of the monomer methyl methacrylate containing the laser dye and an initiator is permitted to diffuse into the pore network, then the polymerization is initiated at an ambient temperature of 40 °C for a couple of days.

One snag with this approach is the progressive micro-fracturing of the host owing to the mismatch in the thermal properties of the two components within the composite structure.

B5.4.2.3.4 Polymer-filled micro-porous glass

This has some similarities with polycomposite glass; in this case, a micro-porous glass [14, 59] forms the fundamental matrix of the host. As previously noted, the host-fabrication process can be controlled to give a porous structure optimized to receive the monomer containing the dye dopant and initiator. Moreover, the porous structure can filter the monomer during the impregnation process, which leads to purification and enhances the damage threshold of the polymer content.

B5.4.2.4 Additives

Studies have shown [46] that it has been possible to enhance the damage threshold of the host matrix and increase the resilience of the laser dye by clearing the initial monomer compositions of foreign inclusions and by doping PMMA with low-molecular-weight additives. It is recommended that the methyl methacrylate and the modifying additives are purified, by use of standard methods, to not less than 99.8 wt% of the main product.

Dye photostability [46] can be increased by doping a polymer with modifying additives that suppress the generation and branching of radical chains through cross-relaxation between the vibrational levels of the polymer macromolecules and those in the introduced additives. The function of these additives is to restrict the formation of radicals, which have a far lower ionization potential than the macromolecules. Elimination of these radicals and their products is vital for good 'service life' as they interact readily with the dye molecules in their first excited singlet (S_1) states (see section B5.3.2.1) and subsequently leads to the destruction of the dye molecules. These radical interactions have been postulated as the principal destruction mechanism.

B5.4.3 Gain media

B5.4.3.1 General considerations

One of the attractions of this class of laser source is that effective gain media, with good optical quality and low cost, can be made easily to match the requirement or application. With the exception of the poly-acrylamide gels, the dye-doped material requires careful cutting and polishing before it is suitable for use in a laser cavity. As noted in table B5.4.1, the shape of the gain medium is determined by the pump source configuration.

An important requirement of any gain medium is that its optical properties, such as refractive index, should be stable with varying environmental conditions, particularly temperature, humidity or water intake. Stress in the host matrix can not only lead to fracture but also to the creation of a lens within the material, which can be equally disruptive. As the material relaxes, this can lead to a variation in the focal length of the lens [24], which can prevent the effect being compensated or corrected.

Poor thermal conductivity, exhibited by some hosts, contributes to the bleaching of the dye and also distorts the gain medium, thus this class of gain medium will not be suitable for applications requiring high

Table B5.4.1. Characteristics of solid-state dye laser gain media.

Gain shape	Typical dimensions	Excitation	Host
Disc	Thickness: few to 20 mm Diameter: <10–100s of mm	Pulsed laser	Most pre-doped hosts except poly-acrylamide gels. All post doped.
Cuboids	Side: few to 10s of mm	Pulsed laser	All.
Rods	Diameter: 4–15 mm Length: 20–250 mm	Short rods—pulsed laser Long rods—flashlamp	As discs. Mainly pre-doped to date.
Slabs	Up to 150 mm × 50 mm × 3 mm	Pulsed laser, flashlamp and surface discharge [60]	Most, except poly-acrylamide gels.

Table B5.4.2. Thermo-optical properties of some hosts [from 14].

Host	$10^5 \times dn/dT$ ($^\circ C^{-1}$)	K ($W\ m^{-1}\ {}^\circ C^{-1}$)
PMMA	−0.85	0.21
Epoxy-filled MPG	−0.75	≈0.80
Silicate glass	0.30	1.3
Phosphate glass	−0.05	0.84

stability in the output beam. Hence, materials with poor thermal characteristics need careful heat control and management to avoid quickly changing optical properties. Changes to the optical properties can result in rapid fluctuations in laser performance, which may lead to a premature termination of laser action. Some hosts do have better thermo-optical characteristics than others, as shown in table B5.4.2. A combination of hosts with good thermal properties and suitable geometrical shape can be used to mitigate some of the effects of waste heat deposited in the matrix. Rotation or translation of the gain medium can also help to reduce the effects of thermal gradients and the bleaching effects of the pump energy [48]. Rotation of the gain medium can help to increase the 'service life' of dye-doped media. Recent reports [7, 13, 48] suggest that more than 60 000 pulses can be emitted from a small rotating disc, of the order of 25 mm (diameter), with the possibility of up to an order of magnitude more at low pump fluences.

B5.4.3.2 Fabrication of laser media

The fundamental approach is to mould the dye-doped material to the appropriate size, using a container of the approximate shape allowing for shrinkage and cutting to the required shape and size. The container can be made of any suitable inert substance, usually polythene or glass. Care is required during the curing or condensation processes to avoid introducing stress into the host and achieving a uniform density. It is essential to ensure that the polymer is fully cured before the machining of the pre-form is undertaken, which is dependent on the host. A range of techniques can be used to cut the sample to the approximate size and finish by diamond turning and/or polishing. A photograph of various dye-doped gain media is shown in figure B5.4.1.

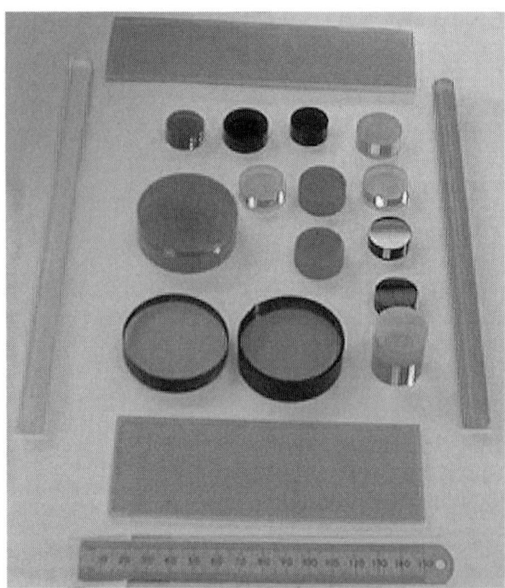

Figure B5.4.1. Dye-doped gain media.

B5.4.3.3 Optical coatings

Application of anti-reflection (AR) coatings (see chapter C1.1) to the dye-doped gain medium is important if high performance is to be achieved from the gain medium. The durability of 'coatings' is a problem with many materials, especially if the material to be 'coated' is soft or deforms easily. Sol-gel coatings have been spun on to a polymer [12], which enhanced the performance substantially. Motheye coatings [61, 62] developed by the Defence Evaluation Research Agency (Malvern, UK) have been applied to PMMA to control parasitic reflections. This form of coating is attractive as it is effective over a wide range of incidence angles. A magnified photograph of this modification to the surface texture is shown in figure B5.4.2 in which the pitch of the pattern is 400 nm.

B5.4.3.4 Damage threshold

The damage threshold of the plastic materials in an unmodified form is usually of the order of 2 J cm^{-2} but this can be increased by a factor of 5 or so by the use of additives [10], as noted earlier. The porous glasses tend to have an inherent laser-induced damage threshold that is in the region of 26 J cm^{-2}, with the value for ormosils of the order of 20 J cm^{-2}. Polycom glass is quoted [13, 63] as having a damage threshold of about 26 J cm^{-2}. Polyacrylamide gels potentially have a high damage threshold as they have some mobility within the matrix, so can be 'self-healing' [44].

Bulk damage effects can be mitigated by careful purification of the constituent chemicals used to create the host, particularly the exclusion of dust and other foreign particles, including moisture and bubbles.

B5.4.3.5 Photostability

The 'service life' of a dye-doped element is dependent on the photostability of the laser dye, which depends on the number of photons absorbed by each laser dye molecule [10], and the damage threshold of the host: both aspects are a function of the incident pump fluence. Some molecules are more resilient than others, for example pyrromethene 597 is much better than pyrromethene 650, which, in turn, is much more resilient than rhodamine 11B [10]. Photo-destruction results from excited dye molecules in a triplet state (see figure B5.3.2)

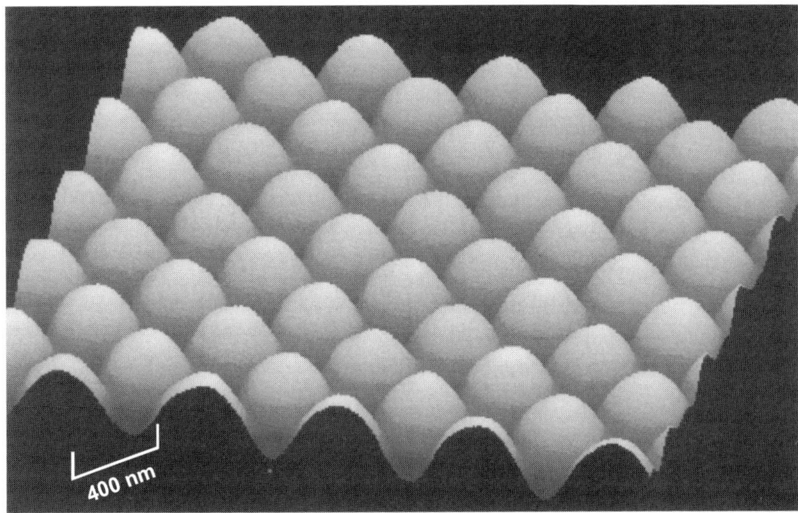

Figure B5.4.2. Motheye AR coating.

being excited from the lowest triplet energy state (T_1) into higher states (T_n). Having absorbed two light quanta, these molecules undergo chemical reactions, which may involve rearrangement of their chemical bonds, other reactions with dissolved impurities or residual solvent or oxygen from the atmosphere. The laser efficiency can be reduced significantly because the 'photo-products' often absorb in the same part of the spectrum as the laser emission.

Good photostability may be achieved if the laser–host combination has high laser efficiency. Recent studies have reported [10] that dye-doped modified PMMA was superior to a modified acrylic co-polymer, which was much better than 'pure' PMMA. In the case of an inefficient combination, a significant proportion of the pump light will be trapped in the matrix, which leads to local heating and stresses resulting in destruction of laser dye molecules in this zone. Thus the trapped energy assists the photo-degradation processes.

The photostability of a laser dye in a sol-gel matrix can be enhanced by increasing the organic character of the surrounding silica cage [50]. Additives to other hosts have also demonstrated enhanced service life. For example, use of 1,4-diazobicyclo[2,2,2] octane (DABCO) more than doubled the lifetime of pyrromethene 567 in PMMA and resulted in more than half a million shots being emitted from a small disc for a pump fluence well below the damage threshold [64].

B5.4.3.6 Health and safety

The procedures described in chapter B5.3 (sections B5.3.2.3 and B5.3.2.14) on liquid dye lasers are also recommended for the preparation, use and disposal of these materials. In general, a laser dye is considered to present a negligible hazard once it is encapsulated in a host.

B5.4.4 Laser dyes

B5.4.4.1 Absorption/fluorescence characteristics

It is natural to assume that the spectral characteristics of the laser dyes in a solid matrix are dependent not only on the intermolecular interactions between individual dye molecules but also on intermolecular interactions between dye molecules and macromolecules. The intermolecular interactions may crucially affect the lasing characteristics of the dye molecules. These characteristics are determined by the degree of overlap between a laser-dye's absorption and fluorescence bands, as well as the luminescence quantum yield. The quantum yield may depend on the tendency of the dyes in weakly polar media to form dimers or other higher-order aggregates, which are usually weakly fluorescent.

A number of studies [65, 66] have been conducted comparing the spectroscopic and photophysical properties of dyes in the liquid and solid states. The absorption and fluorescence spectra of dyes in the solid state are found to be similar to their analogous liquid counterparts (see figure B5.3.3). The spectral properties of the dye pyrromethene 567 have been found [65] to be governed by polarity and specific hydrogen bonding ability of the solvent. There is a bathochromic (red) shift in less polar solvents.

It has been reported [66] that the fluorescence lifetime of dyes in solid hosts is higher than in a liquid solution. In common with viscous solvents, the fluorescence emission from dyes in solid-state hosts is anisotropic. This is because the rotational diffusion of the dye molecules embedded in the solid is slow in comparison with the lifetime of the singlet S_1 state.

A study reported [15] similar fluorescence quantum yields for a number of dyes in solution and in a PMMA host, with the pyrromethene dyes having a superior performance in the solid host, compared with liquid hosts.

B5.4.4.2 Laser action

The physical processes, which lead to laser action in dye molecules, are described in chapter B5.3 on liquid dye lasers (see section B5.3.1). Similarly, the techniques for spectral and temporal control which may be applied to dye lasers are described in sections B5.3.2.4 and B5.3.2.6.

B5.4.5 Laser-pumped solid-state dye lasers

A number of pump sources are available for this type of laser and their characteristics are reviewed in table B5.3.2 of chapter B5.3 on liquid dye lasers. Pump lasers providing Q-switched pulses (see section C2.2.2) with durations of order 10 ns permit pump fluxes of the order of 100 MW cm^{-2} to be achieved from an energy density of about 1 J cm^{-2}. This is well within the operational limits of most host materials but is well in excess of that needed for fluorescence saturation ($\sim$1 MW cm^{-2}) of most dyes, so enabling the efficient extraction of laser energy from the gain medium. For these short pump pulses, the triplet state population is minimal and thermal-lensing effects should be small, particularly at low repetition rates.

A relatively simple configuration can be used for achieving laser action from a dye-doped disc, using an off-axis pumping arrangement. This approach avoids the need to use a dichroic mirror, which will transmit the pump energy but reflect the laser emission. A small penalty is paid in conversion efficiency, owing to the poor overlap of the pump energy and the resonant beam. In this arrangement the reflectivity of the output coupler is usually in the range 30–50%. End pumping is common but transverse pumping [7] is also feasible.

The modern dye-doped materials offer high optical gain, which can lead to parasitic laser action establishing both transverse and longitudinal modes: these effects are illustrated in figure B5.4.3. These parasitic effects result in loss of energy and poor tunability. Longitudinal parasitic laser action can be suppressed by using a 'motheye' anti-reflection coating [61, 62] on the faces of the disc. Transverse parasitic laser action can damage the cylindrical surface of the gain medium. This action can be suppressed by roughening and 'blackening' the curved surface of the disc.

B5.4.5.1 Performance

Table B5.4.3 shows the performance of three pyrromethene dyes in different hosts, when longitudinally-pumped with a frequency doubled Q-switched Nd:YAG source. 'Uncoated' discs were used with typical

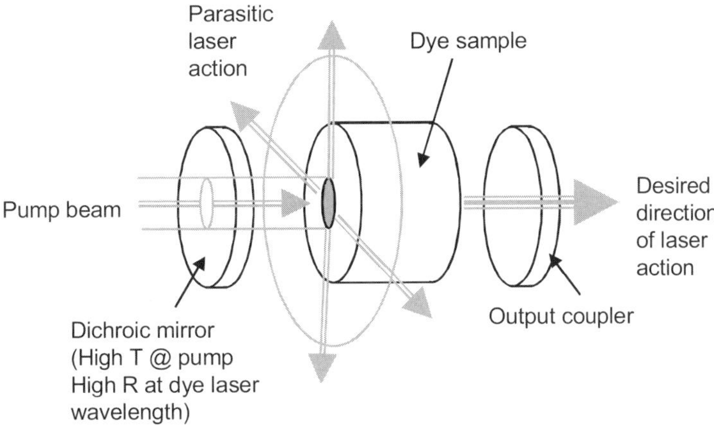

Figure B5.4.3. Parasitic effects in a dye-doped disc.

Table B5.4.3. Solid-state dye laser performance data.

	Laser performance data	
Laser material	Peak λ (nm)	Slope efficiency (%)
Pyr 567 in PMMA	554	85 (64)
Pyr 567 in ormosil	552	80 (67)
Pyr 567 in MeOH	557	>80
Pyr 580 in PMMA	563	86 (69)
Pyr 580 in MeOH	556	>80
Pyr 597 in PMMA	581	86 (58)
Pyr 597 in ormosil	578	80 (62)
Pyr 597 in MeOH	585	~80

sizes of about 25 mm diameter and 6–8 mm thick: the dye concentration was of the order of 2×10^{-4} molar. The figures in brackets are for off-axis pumping of 'uncoated' discs with 5 ns pulses. Data are also included for dyes in a methanol solution for comparison.

The threshold energy for laser action is typically in the range of 100 μJ of pump energy for the range of dyes in the hosts quoted in table B5.4.3. Owing to the small threshold energy there is little difference between the slope efficiency and the overall efficiency with a high energy per pulse output.

In the short-pulse regime, pyrromethene dyes show better conversion efficiency than xanthenes or other dyes. A recent review [13] indicated that pyrromethene dyes showed a similar performance in solid-state and liquid hosts. Furthermore, those laser dyes, which show good efficiency, also tend to show good photostability, as this avoids detrimental thermally-induced stress, as described earlier. An additional observation was that there had been little improvement in the performance of solid hosts doped with xanthenes over the last 15 years, despite the improvement in the host materials. This feature indicates that the limitation is dominated by the dye rather than the host: excited state absorption is probably the limiting loss mechanism with this class of dye. Optical epoxy hosts doped with pyrromethene dyes were found to have poorer laser efficiency and lifetime in comparison with analogous modified PMMA samples when tested [11].

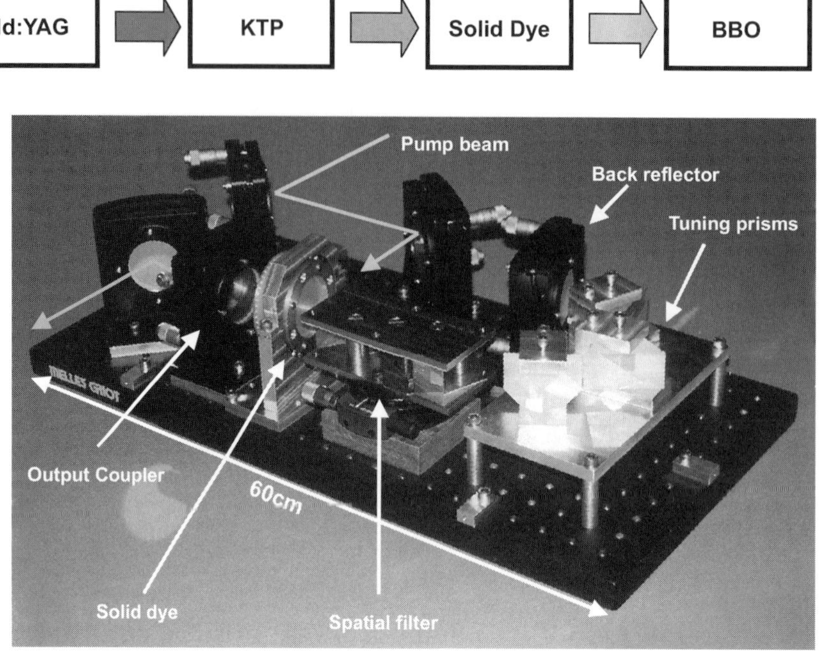

Figure B5.4.4. Laser-pumped dye laser.

Laser pumping of dye-doped materials with short pulses (nanoseconds) can provide output pulses with energy of 500 mJ pulse^{-1} when using high-performance dyes. The spectral range for this high performance is typically 550–650 nm. The performance when emitting pulses of microsecond duration tends to be inferior, owing to the increased triplet-state losses and the low peak power of the excitation source. The spectral emission range can extend from the UV to the near IR. An efficient route to the UV is frequency-doubling (see section C3.1.2) of yellow light [67]. A photograph of this form of arrangement is shown in figure B5.4.4.

The pulselength of the emitted energy is closely related to that of the pump, in the short-pulse regime up to about a microsecond. For longer pulselengths, triplet state population becomes the dominant factor, which can lead to premature cessation of laser action.

The spectral emission bandwidth from a dye-doped host can extend over many tens of nanometres, often significantly more than is observed with the identical dye in a solution [6, 13]. Figure B5.4.5 shows the emission bandwidth of two dyes in a PMMA host.

Recent studies [14] report a 'service life' of more than 60 000 pulses from a pyrromethene-doped microporous glass when pumped with a power density of 25 MW cm^{-2}. This value reduced to about 45 000 shots when the fluence increased to 50 MW cm^{-2}. The identical host doped with rhodamine 11B dye gave a service life of about 110 000 pulses. However, this dye has a somewhat lower efficiency than the pyrromethene dyes. For all of these studies, the 'service life' was defined as the point where the output energy fell to 70% of its initial value. The tests were undertaken by irradiating a single site in the gain medium and were limited by the photo-bleaching of the dye in the host and not by damage to the host. 'Service lifetimes' of many tens of thousands of shots have been recorded from dyes in ormosil and PMMA hosts [17, 20].

The pulselength of the excitation influences the 'service life' as a high fluence is required to achieve laser action in the long-pulse regime, which, of course, is energy dependent. Consequently, the 'service life'

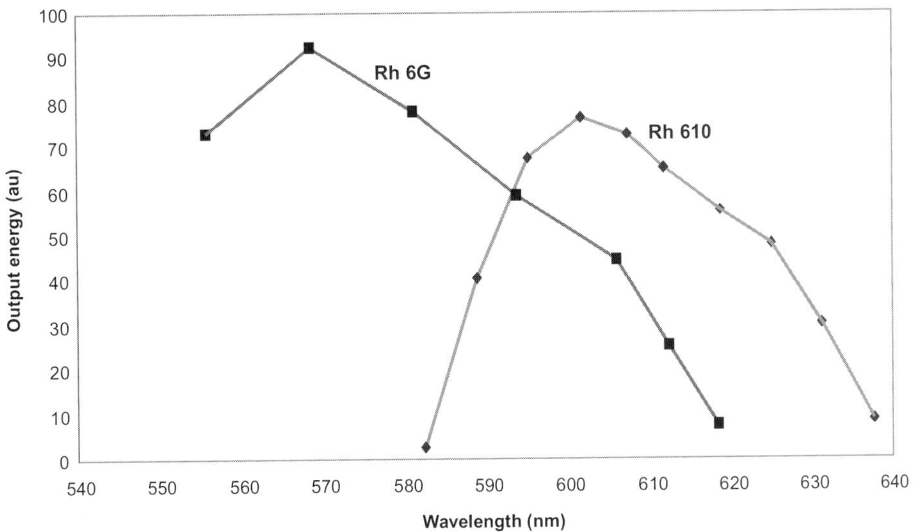

Figure B5.4.5. Laser dye emission bandwidth in PMMA.

for long-pulse operation is often in the realm of a few thousand pulses from a given site, although this value can be increased by moving the sample.

Short-pulse operations [68] with pulse lengths of 30 ps have been demonstrated with dye-doped gain media pumped with the second harmonic from a mode-locked neodymium laser (Nd:YAG). Single pulse and pulse trains were obtained from materials doped with pyrromethene 580 and rhodamine 11B.

Studies [6] have shown that the peak of the laser output from a given dye experiences a hypsochromic (blue) shift as the host is changed from an organic polymer, such as PMMA, to a partially organic matrix (ormosil) and finally to an inorganic glass host. These wavelength shifts reflect changes in the dye–host interactions, with increased interaction in inorganic hosts with hydrogen bonding sites, leading to dye absorption occurring. Some laser dyes, such as the perylenes, are relatively non-polar and in their ground state and more polar in an excited state, experience a redshift in sol-gel glass, as their interaction with the inorganic host is very weak [6]. Wavelength shifts can also occur during the degradation of a dye-doped host during operation, owing to the creation of photo-degradation products, which have an absorption profile coinciding with the laser emission band. This acts in a similar way to the spectral control filter described for liquid lasers (see section B5.3.2.4), by creating an additional loss at the emission wavelength and forcing the emission to a different wavelength with a better excess gain. Of course, this does not occur if the degradation products are transparent. Figure B5.4.6 shows comparative output spectra of a pyrromethene dye in various hosts, reflecting the modified dye–host interactions.

Dye-doped solid-state materials can be used in place of liquid phase dye cells in a range of laser-pumped systems, with the exception of the very high power devices, where bulk damage would be a serious issue and where very high spectral purity is required. For many applications this gain medium offers a more convenient alternative to the liquid-phase device, particularly for those with low power requirements.

B5.4.6 Flashlamp-pumped solid-state dye lasers

Progress with the performance of flashlamp-pumped dye-doped rods has been significant in the last few years [17, 19, 21–24]. One of the major problems with flashlamp-pumping has been producing rods which are

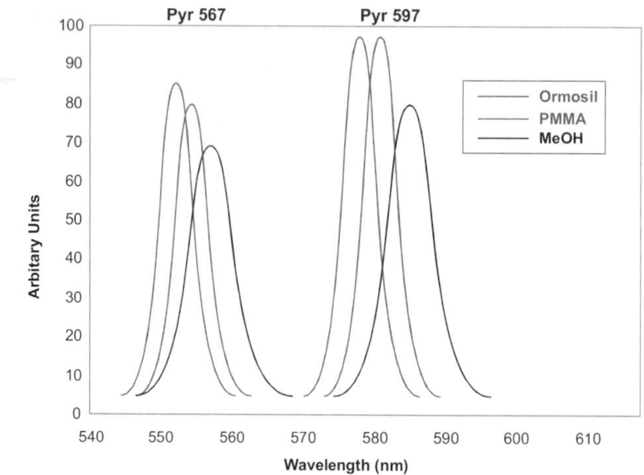

Figure B5.4.6. Lasing spectra of a pyrromethene dye in various hosts.

large enough to match the extended pumped volume, whilst exhibiting good optical quality and low losses. It has been estimated [16] that bulk losses should not exceed 0.5% cm^{-1} for satisfactory laser performance. The major breakthrough has been a refinement of the production method that has led to the reduction in optical scatter in the rods and control of the refractive index profile which, in the past, has led to strong lensing within the rod.

A relatively simple arrangement with a coaxial flashlamp can be used to create laser action. For example, a dye-doped rod may be inserted into the annulus of the lamp, with suitable mirrors arranged to form the resonant cavity. A schematic arrangement is shown in figure B5.4.7. For the best performance, it is necessary to match the length of the dye-doped rod to that of the flashlamp, as is common with liquid dye lasers.

The use of a flashlamp source presents a number of issues that need careful consideration, such as the triplet-excitation problem and the creation of a dynamical thermal lens in the rod. The triplet problem can be mitigated by use of relatively short flashlamp pulses or energy from a surface discharge and additives that collide with the excited dye molecules in the host. Consequently, the additives will only solve this problem if they can be arranged to be in very close proximity to the dye. The thermal lens can be controlled through careful design to avoid a thermal gradient. An additional problem is generating a sufficiently large flux in the cavity to ensure fluorescence saturation without exceeding the damage threshold of the host matrix. As previously noted, fluxes of the order of 1 MW cm^{-2} are required for efficient energy extraction from the cavity. However, when using long-pulse excitation, this would require pulses with a fluence of the order of 10 J cm^{-2}. Unfortunately, this fluence is well beyond the damage threshold of most polymeric and some glass-like hosts; hence, lower efficiency operation is necessary to achieve reliable and repeatable performance.

B5.4.6.1 Performance

In addition to thermally-induced focusing characteristics, a dynamic lens has been noted in polymer rods [24]: figure B5.4.8 shows how the stress-induced lens of a pyrromethene-doped PMMA rod changed over a 40-day observation period.

The number of high-energy pulses that can be generated from a rod is still limited and the output decreases with the number of pulses, as shown in figure B5.4.9. In the case of the low repetition rate the interval between pulses was between 1 and 8 minutes. As can be seen for the 1 Hz case, the decline in performance was more rapid, owing to the heat not being able to be dissipated and, hence, allowing the

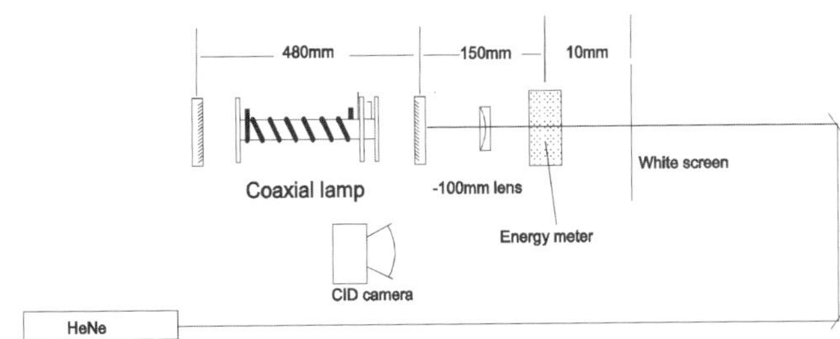

Figure B5.4.7. A schematic diagram of a flashlamp-pumped polymer rod test.

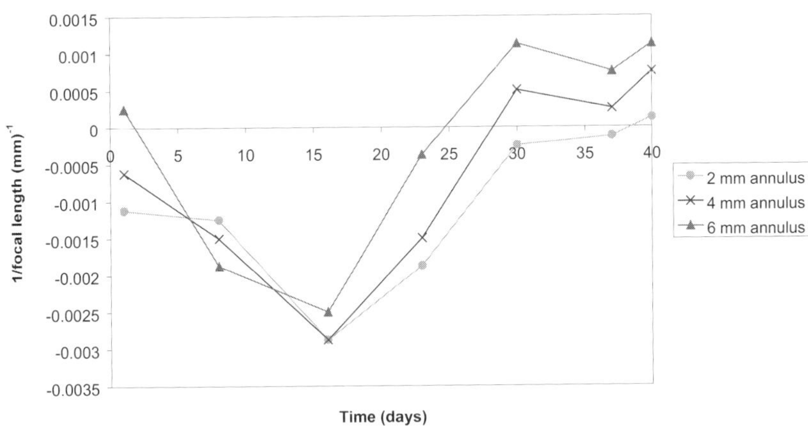

Figure B5.4.8. Variation of focal length of a polymer rod.

creation of a thermally-induced lens, in addition to triplet excitation. If the gain medium is not excited for a few minutes, the gain medium recovers and the original performance is re-established.

The temporal characteristics recorded from a similar dye-doped PMMA rod closely follow the flashlamp pulse, as shown in figure B5.4.10. The output trace is quite symmetrical and illustrates that there is no obvious indication of a premature termination of lasing, such that may result from dynamic thermal lensing in the rod or excessive triplet-state absorption.

Figure B5.4.11 shows the performance of three polymeric rods (PMMA), each doped with a different pyrromethene dye. The laser threshold was of the order of 100 J for each of them, with a slope efficiency of up to ~0.5% and an overall conversion efficiency of ~0.3%. More recent results [22, 24] have reported

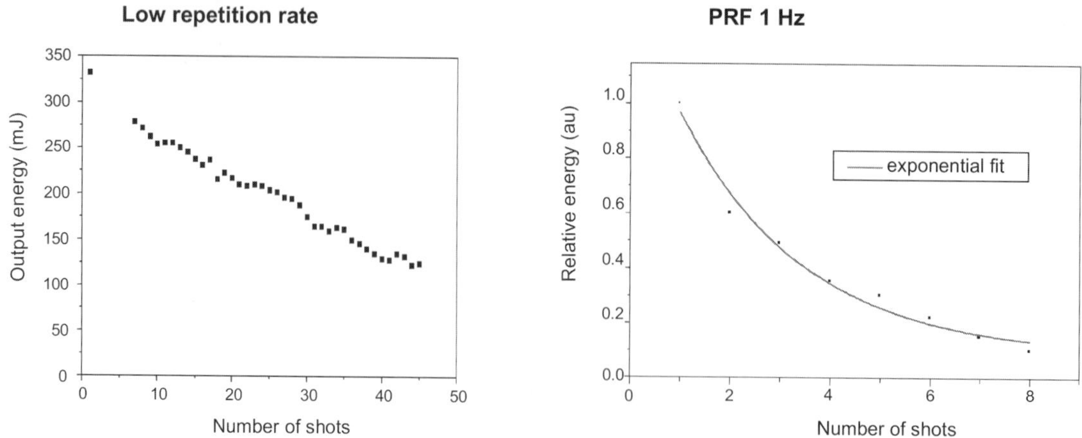

Figure B5.4.9. Degradation of output with shot number.

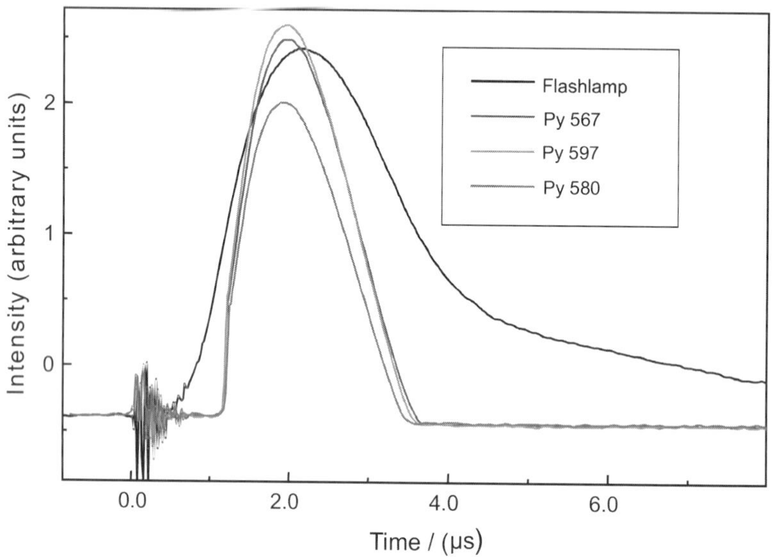

Figure B5.4.10. Temporal output of flashlamp and laser.

an output energy of 1.1 J per pulse with a conversion efficiency close to 0.5%: in this case the dye dopant was pyrromethene 567. The same study reports that pyrromethene 567 exhibits better laser efficiency than pyrromethene 580, which, in turn, is better than pyrromethene 597. All of the pyrromethene-doped rods out-perform those doped with rhodamine 6G. The divergence from these dye-doped laser rods is of the order of 10 mrad.

Long-pulse operation presents issues with the gain–loss balance and the need for a high fluence to generate a peak intensity above threshold. This can result in more damage to plastic hosts than other media, as the main damage mechanisms are energy related rather than power dependent. In addition, management of the triplet-state build-up is a considerable challenge for long-pulse operation. This process acts not only as a

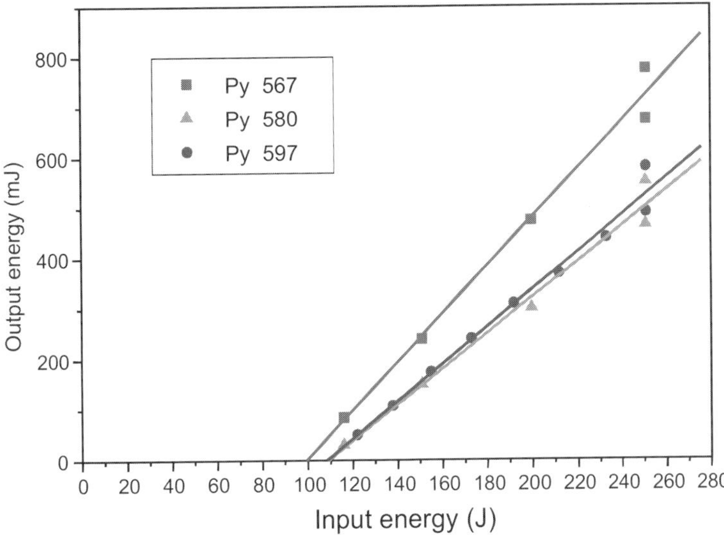

Figure B5.4.11. Comparative laser performance of dye-doped PMMA rods.

loss mechanism but can also provide a route to singlet oxygen generation, leading to rapid photo-degradation of the laser dye.

Spectral emission is readily achieved in the visible band. The most energetic pulses are in the green to orange part of the spectrum. Pulselengths are of the order of a microsecond and the spectral width is about 4 nm.

B5.4.7 Choice of host

As previously noted, excited-state absorption is an important loss mechanism in laser dyes but it should be independent of the host matrix, as it is only dependent on the energy levels in the dye molecule. Hence, there should be little difference between using a given laser dye in a polymer or glass-like host, except for small changes in the peak lasing wavelength. Other considerations are likely to determine the choice of host, such as the laser-induced damage threshold of the host, thermal environment or the repetition rate of the pulses. Thermal degradation is independent of the generation of excited states, although an increase in ambient temperature may increase the photodegradation rates as many processes are diffusion controlled; hence, it is important to match the thermo-optical parameters of the host to the operational environment.

Owing to the very low efficiency of flashlamp-pumped systems, it is critical that the gain medium has low loss and negligible optical distortion. To date, these characteristics have been most easily achieved with polymeric hosts, although it should be possible to use other hosts, such as the epoxy-filled micro-porous glasses. Other considerations tend to dominate the choice of host for short-pulse laser-pumped devices as a result of the ease with which fluorescence saturation may be achieved leading to high laser emission efficiency. Some of those considerations will include potential 'service life', desired operational configurations and laser damage threshold of the matrix.

Polymeric hosts have a number of advantages over the porous materials, such as better chemical compatibility with the organic dye molecules [50] and their suitability for inexpensive fabrication. However, dye-doped silica gels may offer superior photostability and better laser performance.

The geometry of the gain medium is usually dictated by mechanical mounting or service/replacement requirements for laser-pumped sources. In the case of high-performance laser-pumped systems, the shape of the gain medium becomes important for controlling the thermal balance of the system. Rods are popular for flashlamp pumping and slabs may be used for either form of excitation. Moreover, slabs may also be ideal for surface-discharge-pumped devices [60].

References

[1] Salin F *et al* 1989 Efficient tunable solid-state laser near 630 nm using sulforhodamine 540-doped silica gel *Opt. Lett.* **14** 785–7

[2] Hermes R E *et al* 1993 High-efficiency pyrromethene doped solid-state dye lasers *App. Phys. Lett.* **63** 877–9

[3] Hermes R E 1994 Lasing performance of pyrromethene-BF2 laser dyes in a solid polymer host *Conference on Visible and UV Lasers Los Angeles (Proc. SPIE 2115)* (Bellingham, WA: SPIE) pp 178–83

[4] Sastre R and Costela A 1995 Polymeric solid-state dye lasers *Adv. Mater.* **7** 1998–2002

[5] Canva M *et al* 1995 Perylene- and pyrromethene-doped xerogel for a pulsed laser *Appl. Opt.* **34** 428–31

[6] Rahn M D and T A King 1995 Comparison of laser performance of dye molecules in sol-gel, polycom, ormosil and poly (methyl methacrylate) host media *Appl. Opt.* **34** 8260–71

[7] Rodriguez M *et al* 1995 A simple rotating system to avoid early degradation of solid-state dye lasers *Meas. Sci. Technol.* **6** 971–8

[8] Costela A *et al* 1996 Polymeric solid state dye lasers: new rhodamine 6G copolymers and terpolymers *Conf. on Solid State Lasers V San Jose (Proc. SPIE 2698)* (Bellingham, WA: SPIE) pp 17–22

[9] Faloss M *et al* 1996 Pyrromethene-doped xerogels for solid state dye laser systems *Conf. on Solid State Lasers V San Jose (Proc. SPIE 2698)* (Bellingham, WA: SPIE) pp 2–7

[10] Anderson R S *et al* 1998 Photo-stability of dye doped modified polymers at extremely high intensities: Medlite TM laser systems *Proc. Solid State Lasers San Jose VII (Proc. SPIE 3265)* (Bellingham, WA: SPIE) pp 13–20

[11] Pacheco D P and H R Aldag 1998 Comparison of the laser performance of solid-state dye media under long- and short-pulse excitation *Proc. Solid State Lasers VII San Jose (Proc. SPIE 3265)* (Bellingham, WA: SPIE) pp 2–12

[12] Nechitailo V S *et al* 1999 Polymer dye lasers with exceptional performance *Proc. Solid State Lasers VIII San Jose (Proc. SPIE 3613)* (Bellingham, WA: SPIE) pp 105–11

[13] Rahn M D and T A King 1999 Comparison of solid-state dye laser performance in various host media *Proc. Solid State Lasers VIII San Jose (Proc. SPIE 3613)* (Bellingham, WA: SPIE) pp 94–105

[14] Aldag H R *et al* 2000 Efficient solid-state dye lasers based on polymer-filled micro-porous glass *Proc. Solid State Lasers XI San Jose (Proc. SPIE 3929)* (Bellingham, WA: SPIE) pp 133–44

[15] Pacheco D P *et al* 1988 A solid-state flashlamp-pumped dye laser employing polymer hosts *Proc. Int. Conf. on Lasers '87* ed F J Duarte (McLean, VA: STS) pp 330–7

[16] Minnigh S W *et al* 1996 Performance of a flashlamp-pumped solid-state dye laser *Proc. Solid State Lasers V San Jose (Proc. SPIE 2698)* (Bellingham, WA: SPIE) pp 55–64

[17] Titterton D H 1997 Performance of plastic dye lasers in the UK *Conf. of Solid State Lasers VI San Jose (Proc. SPIE 2986)* (Bellingham, WA: SPIE) pp 132–40

[18] Calderon O G *et al* 1997 Laser emission of a flash-lamp pumped Rhodamine 6 G solid copolymer solution *Appl. Phys. Lett.* **70** 25–7

[19] Finlayson A J *et al* 1998 Flashlamp pumped polymer rod dye laser *Proc. Int. Conf. on Lasers '97* (McLean, VA: STS) pp 252–5

[20] Finlayson A J *et al* 1998 Flashlamp pumped polymer dye laser containing Rhodamine 6G *Appl. Phys. Lett.* **72** 2153–5

[21] Winstanley P A *et al* 1998 Enhancements achieved in the performance of liquid and solid-state dye laser technology *Conf. on Solid State Lasers VII San Jose (Proc. SPIE 3265)* (Bellingham, WA: SPIE) pp 42–50

[22] Finlayson A J *et al* 1999 Flashlamp pumped solid-state dye laser incorporating pyrromethene 597 *Appl. Phys. Lett.* **75** 457–9

[23] Mandl A and Klimek D E 1999 Flashlamp pumped thin slab solid state plastic dye zig-zag laser *Proc. Solid State Lasers VIII San Jose (Proc. SPIE 3613)* (Bellingham, WA: SPIE) pp 119–24

[24] Finlayson A J and Venner M R W 2000 Flashlamp pumped polymer dye laser studies *Conf. on Solid State Lasers IX San Jose (Proc. SPIE 3929)* (Bellingham, WA: SPIE) pp 154–62

[25] Pavlopoulos T G *et al* 1988 Laser action from a tetramethylpyrromethene-BF2-complex *Appl. Opt.* **27** 4998–9

[26] Pavlopoulos T G *et al* Efficient laser action from 1,3,5,7,8-penta-methylpyrromethene-BF2-complex and its 2,6-disulphonate derivative *Opt. Commun.* **70** 425–7

[27] Pavlopoulos T G *et al* Laser action from 2,6,8-trisubstituted 1,3,5,7-tetramethylpyrromethene-BF2-complex: part 1 *Appl. Opt.* **29** 3885–6

[28] Boyer J H *et al* Laser action from 2,6,8-trisubstituted 1,3,5,7-tetramethylpyrromethene-BF2-complex: part 2 *Appl. Opt.* **30** 3788–9

[29] Pavlopoulos T G *et al* 1998 Laser action from a 2,6,8-trisubstituted 1,3,5,7-tetramethylpyrromethene-BF2-complex: part 3 *Appl. Opt.* **37** 7797–800

[30] Pavlopoulos T G 1999 Photostability of some pyrromethene laser dyes *Conf. on Solid State Lasers VIII San Jose (Proc. SPIE 3613)* (Bellingham, WA: SPIE) pp 112–18

[31] Sorokin P P and Lankard J R 1966 Stimulated emission observed from an organic dye, chloro-aluminum phthalocyanaine, *IBM J. Res. Dev.* **10** 162–3

[32] Schäfer F P *et al* 1966 Organic dye solution laser *Appl. Phys. Lett.* **9** 306–9

[33] Sorokin P P and Lankard J R 1967 Flashlamp excitation of organic dye lasers: a short communication *IBM J. Res. Dev.* **11** 148

[34] Schmidt W and Schäfer F P 1967 Blitzlampengepumpte Farbstofflaser *Z. Naturforsch.* a **22** 1563–6

[35] Soffer B H and McFarland B B 1967 Continuously tunable, narrowband, organic dye lasers *Appl. Phys. Lett.* **10** 266–7

[36] Peterson O G and Snavely B B 1968 Stimulated emission from flashlamp-excited organic dyes in polymethyl methacrylate *Appl. Phys. Lett.* **12** 238–40

[37] Drake J M *et al* 1972 The use of light converters to increase the power of flashlamp-pumped dye lasers *IEEE J. Quantum Electron.* **QE-8** 92–4

[38] Bank P W *et al* 1995 High average power quasi-cw tunable polymer laser *Conf. on Visible Lasers and Laser crystal Growth San Jose (Proc. SPIE 2380)* (Bellingham, WA: SPIE) pp 292–7

[39] Duarte F J 1995 Opportunity beckons for solid-state dye lasers *Laser Focus World* **31** 187–9

[40] Hench L L and West J K 1990 The sol gel process *Chem. Res.* **90** 32–72

[41] Schmidt H 1984 *Better Ceramics through Chemistry* ed C J Brinker, D E Clarck and D R Ulrich (New York: North-Holland) p 327

[42] Schmidt H 1989 Organic modification of glass structure—new glasses or new polymers *J. Non. Cryst. Solids* **112** 419–23

[43] Altman J C *et al* 1991 Solid-state laser using a rhodamine-doped silica gel compound *IEEE Photon. Technol. Lett.* **3** 189–90

[44] Kessler W *et al* Novel solid state dye laser host *Conf. on Visible and UV lasers Los Angeles (Proc. SPIE 2115)* (Bellingham, WA: SPIE) pp 190–5

[45] Maslyukov A *et al* 1995 Solid-state dye laser with modified poly(methyl methacrylate)-doped active elements *Appl. Opt.* **34** 1516–18

[46] Dyumaev A A *et al* 1992 Dyes in modified polymers: problems of photostability and conversion efficiency at high intensities *J. Opt. Soc. Am.* B **9** 143–51

[47] Dyumaev K M *et al* 1981 Transparent polymers as a new class of optical materials for lasers, laser induced damage in optical materials *NBS Spec. Publ.* **638** 31–40

[48] Nechitailo V S 1996 Recent advances in polymer dye lasers *Proc. Solid State Lasers V San Jose (Proc. SPIE 2698)* (Bellingham, WA: SPIE) pp 8–16

[49] Knobbe E T *et al* 1990 Laser behaviour and photostability characteristics of organic dye doped silicate gel materials *Appl. Opt.* **29** 2728–33

[50] Seddon A B 1997 Potential of organic-inorganic hybrid materials, derived by sol-gel, for photonic applications. *Sol-Gel and Polymer Photonic Devices (Critical Reviews of Optical Science and Technology 68)* ed M P Andrews and S I Najofi (Bellingham, WA: SPIE) pp 143–71

[51] Jones G and Pacheco D P 1998 Fluorescence and lasing properties of pyrromethene dyes in epoxy solid host media *Conf. on Solid State Lasers VII San Jose (Proc. SPIE 3265)* (Bellingham, WA: SPIE) pp 51–8

[52] Cazeca M J *et al* 1997 Epoxy matrix for solid-state dye laser applications *Appl. Opt.* **38** 4965–8

[53] Bezrodny V I *et al* 1982 Active and passive laser elements on the basis of polymers with organic dyes *Kvantovaya Elektron.* **9** 2455 (in Russian)

[54] Tompkin W R 1987 Non-linear optical properties of lead tin fluorophosphate glass containing acridene dyes *J. Opt. Soc. Am.* **4** 1030–4

[55] Al'tshuler G B *et al* 1983 output characteristics of a laser utilizing rhodamine 6G in micro-porous glass *Quantum Electron.* **10** 784–8

[56] Efendiev T S *et al* 1984 Tunable distributed-feedback laser-based on dye-doped micro-porous quartz glass *Appl. Phys.* B **33** 167–9

[57] Al'tshuler G B *et al* 1985 Field-induced transport of laser dyes in porous glass matrices *Sov. Phys. Tech. Phys.* **30** 941–2

[58] Rahn M D and King T A 1994 Lasers based on dye-doped sol-gel composite glasses *Proc. Sol-Gel Optics III San Diego (Proc. SPIE 2288)* (Bellingham, WA: SPIE) pp 382–91

[59] Meshkovskii I K *et al* 1995 Solid state dye laser *Conf. on Visible Lasers and Laser Crystal Growth San Jose (Proc. SPIE 2380)* (Bellingham, WA: SPIE) pp 292–305

[60] Fulker D J 2002 The development and application of surface discharges for pumping dye lasers *PhD thesis* University of Strathclyde

[61] Bernhard C G 1967 Structural and functional adaptation in a visual system *Endeavour* **26** 79–84

[62] Wilson S J and Hutley M C 1982 The optical properties of moth eye antireflection surfaces *Opt. Acta* **29** 993–1009

[63] Rahn M D and King T A 1998 High-performance solid-state dye laser based on perelene-orange doped polycom glass *J. Mod. Opt.* **45** 1259–67

[64] King T A *et al* 2000 Dye triplet state and singlet-oxygen quenching effects in solid-state dye lasers *Conf. on Solid State Lasers IX San Jose (Proc. SPIE 3929)* (Bellingham, WA: SPIE) pp 145–53

[65] Arbeloa F L *et al* 1998 Photophysical and lasing properties of pyrromethene 567 dye in liquid solution. Environment effects *Chem. Phys.* **236** 331–41

[66] Yariv E and Reisfeld R 1999 Laser properties of pyrromethene dyes in sol-gel glasses *Opt. Mater.* **13** 49–54

[67] Chandra S *et al* 1997 Tunable ultraviolet laser source based on solid-state dye laser technology and $CsLiB_6O_{10}$ harmonic generation *Opt Lett.* **22** 209–211

[68] Garnov S V 1998 Ultra short pulse generation in solid state dye lasers *Proc. Solid State Lasers VII San Jose (Proc. SPIE 3265)* (Bellingham, WA: SPIE) pp 306–8

PART C

LASER SYSTEM DESIGN

C1
Optical components

Julian Jones

Chapters C1.1–C1.4 are concerned with the optical and optomechanical components used to form and transport laser beams, building on the basic concepts of laser optics expressed in part A (notably chapters A1–A3) and as a foundation to the description of laser beam delivery systems to be found in section C4. Meanwhile, section C1.5 is devoted to the electronic components that go together to make up the circuits used to power practical lasers: C1.5.1 for diode laser power supplies; C1.5.2 for driving the electric discharges which are the basis of gas lasers; and C1.5.3 for the flash and arclamps used to pump solid-state lasers.

The topic of Leo Beckmann's chapter C1.1 is the passive components of lenses and mirrors that are used to condition laser beams. His starting point is to contrast the properties of coherent laser beams with the familiar formalism of incoherent optics and the formation of images, developing the theory of Gaussian beams (first encountered in chapter A3). Armed with the theoretical foundations, he goes on to consider the characterization of laser beams, introducing the M^2 parameter and more sophisticated means for describing the wavefront (these topics are covered in greater detail in section C5). The most simple components are those with plane surfaces, causing reflections, and whose properties can be modified by thin films, either singly or multiply. The most familiar optical component is surely the lens; nevertheless, the characteristics required for laser beam transmission are different from those in imaging systems, for reasons of high spatial coherence and power density. Methods for lens design and multiple lens systems are described, and extended to mirror systems. The section closes by considering thermal effects, optical specifications and manufacturing techniques.

In chapter C1.2, Alan Greenaway considers the optical components used for dynamic control of lasers, in scanning and positioning the beam, changing its size and shape, and modulating it in time and space. Relevant components for scanning are mechanical, acousto-, electro- and magneto-optic, and the use of diffractive optical effects. For controlling beam shape, spatial light modulators and various adaptive optical components are appropriate. Adaptive optics are the subject of Michael Damzen and Carl Paterson's chapter C1.3. Here are described the techniques first developed for astronomical imaging through the turbulent atmosphere for the correction of wavefront distortion. More recently, such techniques have been considered for intra- and extra-cavity use in lasers. They describe the relevant sensors (such as the Shack–Hartmann wavefront sensor), actuators (adaptive mirrors, faceplates, electrostatic membranes and biomorphs) and control systems. A special case of adaptive optics, in which there is no closed-loop control, is the phase-conjugate mirror, also capable of restoring distorted wavefronts to their original shape, and the use of, for example, four-wave mixing, stimulated Brillouin scattering, photorefractivity and self-intersecting loops are all discussed.

Optical components are of little use unless they can be held in the right place (statically or dynamically) and kept there; such is the function of the optomechanical components that form the subject of Frank Leucke's chapter C1.4, which considers the formalism for optomechanical design, practical approaches and testing, with special attention to precision positioning.

All laser systems require power supplies but the needs are very different for the various laser types that were the subject of part B. For the diode lasers of section B2, Ralph Savioli in chapter C1.5.1 describes the design criteria for some simple driver circuits. For the gas lasers of section B3, Israel Smilanski provides chapter C1.5.2, concerned with the components and circuits required to drive gas discharges, with descriptions of the various energy storage techniques, switch technologies and circuit designs that have all-round practical application. For the solid-state lasers of section B1, Mark Greenwood in chapter C1.5.3 describes the power supplies for the arc and flashlamps that are often used as the primary pumping mechanism.

C1.1
Optical components

Leo H J F Beckmann

C1.1.1 Introduction

This chapter will deal with the design, manufacture and handling of the components of optical systems used for manipulating laser radiation. Attention will focus on those aspects which are specific to laser-based optics. These occur mainly in two quite unrelated fields: (1) the propagation of laser beams due to spatial coherence; and (2) the thermal aspects of laser optics caused by the (often) high power of laser beams. Commonalities and analogies between classical and laser-based optics, which are plentiful, will be addressed if needed for completeness or clarity.

As a general rule, optical components for laser applications have to meet high quality standards and must be handled with care and the utmost cleanliness. At high power densities, any organic contaminants on optical surfaces will readily carbonize, become unremovable and cause absorption and local heating resulting in thermal gradients and stress. Mechanical damage, such as scratches, may cause unwanted local power concentrations, thermal stress and rupture, aside from stray radiation which may reach sensitive parts of the equipment. Substandard or damaged anti-reflection coatings will make a part of the incident radiation pursue unintended optical paths with unpredictable results. Some but not all of these risks can be minimized by careful design of the overall system, including suitable cooling provisions to ensure thermal stability.

C1.1.2 Optical design aspects of laser optics

C1.1.2.1 What and where are the object and image?

In a classical radiation source (an incandescent lamp etc), each spot on the source emits radiation independently of all others and is the origin of a spherical wave, which propagates into a hemisphere. The centre of this diverging hemispherical wave locates the source as an object and its radius of curvature is proportional to the distance from the object. By means of suitable optics, the divergent wavefront can be made convergent in the direction of propagation to form a region of smallest extent called an image. As the object and image are located at the centres of the respective wavefronts, we may, and customarily do, consider just the object and image distances and their relations to the 'wavefront-bending-capability' of the optical system, also known as its optical power. This well-known relation can be written [1]:

$$1/s + 1/s' = 1/f \tag{C1.1.1}$$

where s is the object distance, s' the image distance and $1/f$ the optical power of the optical system (we refer to its reciprocal, the quantity f, as the 'focal length').

The radiation from a laser is spatially coherent and propagates accordingly (see chapter A3). There is a point of minimum beam radius W_0, called the beam waist but it does not have the characteristics of an object in the sense of classical optics: it does not form the centre point of a spherical wave. Instead, the wavefront of laser radiation is flat at the waist (like that from a classical object at infinity) and its radius of curvature varies nonlinearly with the axial distance z from that waist according to

$$R(z) = z[1 + (z_R/z)^2] \tag{C1.1.2}$$

while the beam radius expands according to

$$W(z) = w_0\sqrt{[1 + (z/z_R)^2]}. \tag{C1.1.3}$$

As both relations are nonlinear in z, the simple relation (C1.1.1) is clearly not applicable to laser beam waists. The difference is most dramatic if z is smaller than z_R and vanishes only for z much larger than z_R. z_R, the so called Rayleigh range, with

$$z_R = \pi w_0^2/\lambda \tag{C1.1.4}$$

marks the distinction between the near field ($z < z_R$) and the far field ($z > z_R$) and is the axial position, where $R(z)$ attains a minimum of $2z = 2z_R$.

Yet, intuitively, one would like to consider laser beam waists as 'objects' and their optical conjugates as 'images'. This can be formalized by writing equation (C1.1.1) in expanded form:

$$1/(s+q) + 1/(s'+q') = 1/f \tag{C1.1.5}$$

where the beam parameters q and q' at the object and image waists, respectively, are purely imaginary [2]:

$$q = j\pi w_0^2/\lambda \qquad q' = j\pi w_0'^2/\lambda. \tag{C1.1.6}$$

As shown in section C4.2.3, from this we can derive an expression for the axial position of the image waist (s') as a function of the object waist position (s) and optical power ($1/f$) [1]:

$$1/[s + (z_R)^2/(s-f)] + 1/s' = 1/f. \tag{C1.1.7}$$

If we rewrite these equations in normalized form [1]:

$$(s'/f) - 1 = \frac{(s/f) - 1}{[(s/f) - 1]^2 + (z_R/f)^2} \tag{C1.1.8}$$

and compare this with the equally rewritten classical formula:

$$(s'/f) - 1 = \frac{1}{(s/f) - 1} \tag{C1.1.9}$$

we see that the appearance of the term $(z_R/f)^2$ removes the pole at $s = f$ in the classical formula, the case for which the object is at the (forward) focal point and for which the image moves to infinity. Thus, the position of the image waist of a laser is always (!) finite, its maximum distance is

$$s'(\max) = f[1 + (f/z_R)/2] \tag{C1.1.10}$$

for a source beam-waist distance of $s = f + z_R$. There is also no minimum distance between the source and image beam waists, unlike in classical optics. Finally, if the source distance becomes zero—which means a flat wavefront entering the optics and is the equivalent of a classical object at infinity—the image-waist distance is

$$s'(s \to 0) = f/[1 + (f/z_R)^2] \tag{C1.1.11}$$

that is the image waist is always closer to the imaging optics than the classical image. It is useful to note, however, that in most cases the focal length is much smaller than the Rayleigh range, making the resulting deviation negligibly small for most practical considerations.

All these formulas for laser beams reduce to the simpler relations for classical optics when z_R approaches zero, which emphasizes that classical optics is merely a special case of the more general (=coherent) laser optics, where the Rayleigh range shrinks to zero.

C1.1.2.2 Size of the image waist

From equations (C1.1.3) and (C1.1.4) it is seen that the divergence of a laser beam is related to its waist radius w_0. Conversely, the size of an image waist w_0' is related to the convergence of the beam in image space

$$w_0' = \lambda/(\pi \sin U'). \qquad \text{(C1.1.12)}$$

U' is the half-angle of the radiation cone that emerges from the optical system and is given by the beam radius $w(z)$ at the source-waist distance $z = s$ from the optical system (to be taken from (C1.1.3)) and the image-waist distance s' (to be taken from (C1.1.7)). w_0' is a minimum value. It assumes that virtually all radiant power from the source laser beam contributes to the (diffraction) image. This requires that the clear optical diameter of the imaging optics be substantially larger than $w(z)$, typically by a factor of 1.5, to avoid 'aperturing' (beam truncation). An even larger factor would theoretically be advantageous but is rarely effective in practice since some aperturing has always already taken place within the laser itself.

C1.1.2.3 Real laser beams

While the previous formulas assume an ideal (or 'diffraction-limited') laser beam and ideal images, they can be extended to real beams, by multiplying the wavelength by a 'times diffraction-limited factor', M^2, introduced by Siegman [3] (see also section C4). The quantity $M^2\lambda$ may be seen as an 'effective' wavelength. Real beams have a stronger divergence and a shorter Rayleigh range than an ideal laser beam of equal waist radius.

C1.1.2.4 Multiple optical elements and the use of ray tracing

The previous calculation of image-to-object relations applies to a single (thin) optical element and the distances from it. In practice, however, optical systems typically consist of a succession of optical components as needed to perform a given task. These relations can be expanded to cover this situation but, for all practical purposes, it is preferable to take an approach that is closer to the ray-tracing procedures which have been well developed for classical optical design. When dealing with laser beams, it is obviously necessary to take into account the specific rules of laser-beam propagation, equations (C1.1.2) and (C1.1.3). The laws of refraction (and reflection) apply unchanged. Details are discussed in chapter C4.2.3.

C1.1.2.5 Evaluation of aberrations and diffraction patterns

Standard ray-tracing algorithms include the computation of the optical path difference (OPD), the difference between the optical pathlengths along the ray and a reference (chief-)ray. By taking a large number of ray samples, typically arranged as a grid raster over the entrance pupil, one obtains a model of the actual wavefront in image space. The aberrations caused by the passage through the individual surfaces will be mapped into that emerging wavefront and will show up as deformations with respect to its ideal (i.e. spherical) shape. There are several ways in which such a computed wavefront can be evaluated. One is to express its shape by means of Zernike polynomials (see chapter C1.3), another is to compute the pattern of the power density, which will be created by diffraction in some reference plane. The latter involves addition of the (amplitude) contributions from samples, properly distributed and weighted for the type of beam, and the amplitude distribution associated with its wavefront. Diffraction calculations give the most relevant information for practical purposes, in particular for systems which approach the theoretical diffraction limit. Care has to be taken to ensure that the samples are taken over the full entrance aperture of the optical system and, thus, represent the true extent of the wavefront, which is admitted by the system's clear optical diameter. In practice, as already mentioned, this must well exceed the beam diameter to avoid undue beam aperturing. In the diffraction image, any beam aperturing will show up as a broadening of the central peak and a shift of power from that central peak to outer rings.

The calculation of the true diffraction pattern for a real beam requires full knowledge of the amplitude distribution within the source beam, which is usually known only for low-order modes. The characterization of a real beam by its M^2 value does not include that information and its diffraction calculation will yield the power distribution of a Gaussian beam, broadened in accordance with M^2 and including the effects of aberrations as applicable.

A treatment of laser beams in full accordance with this is not necessarily implemented in commercial optical design and analysis software and the respective instruction manuals must be consulted for any details about the actually used methods of computation of Gaussian beams and eventual simplifying assumptions or shortcuts.

C1.1.3 Surface phenomena and thin layer coatings

C1.1.3.1 Reflection at the surface of a dielectric

When radiation travelling in air hits the surface of a dielectric material, it undergoes several important changes:

(1) A part of it enters the dielectric, where it may be transmitted or absorbed.

(2) Its speed is slowed down by a factor equal to the refractive index n of the material and, as a result, it is refracted.

(3) The remainder of the incident radiation is reflected.

(4) The distribution between entering and reflected radiation depends on the angle of incidence and on the polarization (see chapter A5) and is governed by Fresnel's equations, which can be written as follows.

$$\frac{R_s}{E_s} = -\frac{\sin(I - I')}{\sin(I + I')} \qquad \frac{R_p}{E_p} = \frac{\tan(I - I')}{\tan(I + I')}$$
$$\frac{E'_s}{E_s} = -\frac{2\sin I' \cos I}{\sin(I + I')} \qquad \frac{E'_p}{E_p} = \frac{2\sin I' \cos I}{\sin(I + I')\cos(I - I')} \tag{C1.1.13}$$

where E, R and E' are the amplitudes of the electric vectors of the incident, reflected and refracted radiation, respectively. The subscripts p and s denote the two planes of polarization, I and I' are the angles of incidence and of refraction. As is immediately seen from these equations, light of different polarization behaves quite differently, but the differences vanish at zero angle of incidence ($I = 0$, which also means $I' = 0$), and when I approaches 90 degrees (grazing incidence, $\cos(I) = 0$), where all light is reflected. For p-polarized light, no light is reflected at the angle, where $\tan(I + I')$ passes through infinity, and for which $\tan(I) = n$, and since $\tan(I + I')$ changes sign at that angle, the reflected radiation also undergoes a phase shift of 180 degrees beyond that angle.

The angle, for which $\tan(I) = n$, and where the refracted and the reflected rays form a right angle, is known as Brewster's angle. Since, as mentioned already, no p-polarized light is reflected under that angle, a (plano-parallel) plate of dielectric material placed under that angle will fully transmit p-polarized light and is then called a 'Brewster window'. This has important applications in laser technology: if included in a laser cavity, (1) it offers a loss-free means to contain a laser gas and (2) it causes a higher cavity gain for p-polarized radiation and, thereby, incites the laser to oscillate exclusively in that plane of polarization, usually the preferred mode of operation.

For low-index glass, with a refractive index near 1.5, Brewster's angle is 56.3 degrees. Since the tangent function is relatively steep in that range, Brewster's angle runs up to just 61 degrees for a refractive index of 1.8. To avoid losses (from stray radiation etc) at these angles, the surface quality requirements for Brewster windows are much higher than those for near-vertical incidence.

C1.1.3.2 Vertical incidence and effect of a single thin layer

In many situations, including practically all applications of lenses, radiation impinges (almost) vertically on the surface of an optical component. To obtain equations for vertical incidence ($I = 0$), we evaluate the sines of $(I - I')$ and $(I + I')$, set the cosine terms to 1, use Snell's law to replace $\sin(I)$ by $(N/n)\sin(I')$, multiply everything by n and divide by $\sin I'$ giving

$$R_s/E_s = R_p/E_p = (N - n)/(N + n) \qquad \text{(C1.1.14)}$$

where N is the refractive index of the component and n that of air and, as the reflected power R is given by the square of the electric vector, we find that

$$R = [(N - n)/(N + n)]^2 \qquad \text{(C1.1.15)}$$

Thus, at each interface between air ($n = 1$) and a component of refractive index N, a sizable percentage of the incident radiation is lost by reflection. The loss is already 4% per surface for low-index glasses ($N = 1.5$) and reaches 17% per surface for a material like ZnSe ($N = 2.4$).

In general optical imaging, (dielectric) reflection losses are not only a nuisance because they reduce image brightness but—mostly more important—because the reflected radiation will, to some extent, find its way to the image plane and act like stray light, which reduces image contrast. In high-power laser systems, radiation lost through reflection is particularly troublesome in that the reflected power will typically hit the interior walls of the lens housing, which is thereby heated up with unpredictable results (misalignment, stress, thermal damage etc).

Fresnel's equations in the simplified form for vertical incidence (C1.1.15) also show the way in which to apply interference in order to reduce reflection. If the component (with refractive index N_s) is covered by a thin layer of a material with refractive index N_t, there will be two reflections:

(1) at the interface between the optical component and the thin layer with

$$R_1 = [(N_s - N_t)/(N_s + N_t)]^2$$

(2) at the interface between the thin layer and air with

$$R_2 = [(N_t - 1)/(N_t + 1)]^2.$$

The condition for the two reflections to become equal is easily found to be:

$$N_t = \sqrt{N_s} \qquad \text{(C1.1.16)}$$

If the (optical) thickness of the thin layer is $\lambda/4$, the radiation reflected at the interface between the thin layer and the component has travelled twice that distance, meets that reflected at the air–layer interface with a phase difference of $\lambda/2$ wavelength and is, thus, cancelled by interference.

Thus a single thin layer can, in principle, completely suppress reflection losses but the necessary condition (C1.1.16) is often difficult or even impossible to meet, as there is a lack of low-index materials: the lowest available are MgF_2 ($N = 1.38$) and kryolite ($N = 1.36$). Single layers are adequate for anti-reflection coating of materials with refractive indices above, say, $N = 1.8$, such as yttrium aluminium garnet (YAG), the most common host material for Nd lasers ($N = 1.82$) or ZnSe ($N = 2.4$) the most common material for optics for CO_2 laser radiation.

C1.1.3.3 Multilayer coatings and their applications

Very substantial design freedom for reducing reflection losses is gained by expanding the thin film coatings on an optical component to two or more layers. In a typical situation, the component (N_s) is coated with a first layer of a high-index material (N_1) and a second layer of low-index material (N_2), chosen such that

$$N_s(N_2)^2 = (N_1)^2 \tag{C1.1.17}$$

where both layers have an optical thickness of $\lambda/4$. This condition requires a high-index material of about 1.70 and a low-index material of 1.38 for a component of $N = 1.5$, both available.

In the cases discussed so far, thin-layer coatings were used to reduce reflection losses at only one wavelength—that of the laser radiation. The principle of manipulating reflection at the interface with a dielectric by means of thin-film coatings is, however, much more flexible and extremely powerful. It allows for the provision of low or high reflection, in a small or wide band of wavelengths and, if designed for finite angles of incidence, to create surfaces with highly selective polarization behaviour. In essence, thin-film coatings exploit the wave nature of light and, as Fresnel's equations already suggest, there is an almost infinite variability of complexity, given the dependencies of the splitting of power at a dielectric interface on refractive index, angle and polarization. The design of such coatings, which go much beyond the examples given here, requires advanced computer methods. Suitable software is offered by several vendors. The subject, which developed somewhat late (by Smakula in 1936 [4]), is treated in a large body of literature and a number of textbooks. The relevant chapter in the *Handbook of Optics* [5] includes 532 literature citations.

The technology of thin-film coatings is difficult to master and has become the foundation of highly specialized enterprises whose viability critically depends on a substantial body of experience. Technical challenges arise in practically every single issue, as the adherence of thin films critically depends on the method of deposition, which must be performed in high vacuum, and the cleanliness of the substrate. Layer thicknesses have to be adhered to—and thus monitored—with high accuracy while the materials for the successive layers have to be chosen to not only meet the required refractive indices but also to avoid undue thermal stress caused by differences in the expansion coefficients.

In general, one has to expect that coatings of any type will be much more sensitive to damage by high laser energy densities (of short laser pulses) and high power densities (of cw laser radiation) than the respective bulk materials. It is, therefore, mandatory to avoid focusing (high peak power) laser radiation onto optical surfaces and this includes secondary foci ('ghost images') created by unwanted reflections from refractive optical surfaces.

C1.1.4 Elementary lens forms

C1.1.4.1 The singlet

The singlet is a stand-alone lens made from a material which is transparent for the wavelengths of its intended use and of (mostly) homogeneous refractive index. It may have spherical or non-spherical surfaces, the former preferred for several reasons, including ease of use and alignment and cost of manufacturing to high precision. If r_1 and r_2 are the radii of curvature of the two lens surfaces, t the centre thickness and N the refractive index, the lens power P, which is, by definition, the reciprocal of the focal length f, is

$$P = 1/f = (N-1)\left[1/r_1 - 1/r_2 + t\frac{(N-1)}{N(r_1 r_2)}\right] \tag{C1.1.18}$$

The sign convention is such that light travels from left to right and a radius is considered positive for a surface with the centre of curvature to the right. A lens of positive power makes an incoming light beam more convergent. If the lens is thin, i.e. $t \ll f$, as is usually the case, the thickness term becomes small and may often be neglected, giving

$$P = 1/f = (N-1)[c_1 - c_2] \qquad \text{(C1.1.19)}$$

where we have also replaced the surface radii r_1, r_2 by the respective curvatures $c_1 = 1/r_1$, $c_2 = 1/r_2$.

For a lens of given material (N), the sum of the curvatures is thus fixed by the required lens power (or focal length) and the only design parameter left free is the ratio of the curvatures, that is the shape of the lens. This is typically taken so as to optimize the imaging quality, i.e. to minimize the 'aberrations' of the lens. If a (near) parallel beam of radiation (e.g. from a classical object at infinity or from a laser beam near the beam waist) is to be focused, the best lens shape depends on N and is approximately symmetrical (biconvex) for low-index materials ($N < 1.5$). It becomes plano-convex for $N = 1.8$ and is meniscus-shaped for higher values of N. The stronger curved surface is always in the direction of the incoming radiation. With increasing refractive index, the absolute values of the curvatures become smaller for a given lens power. This has the highly desirable result that the primary aberrations, spherical aberration and coma are also reduced. This means that the same level of performance is maintained at higher numerical aperture (*NA*) (which is the sine of the angle under which the beam radius is seen from the focal point) or lower 'f-number' (focal length divided by beam diameter). A factor of about five is gained by switching from a refractive index of 1.4 to 2.5 for a lens of equal focal length, which explains why optical designers prefer to work with high-index materials. As ZnSe, one of the very few materials which are highly transparent for CO_2 laser radiation, has a refractive index of 2.403, singlets made from ZnSe perform satisfactorily in many CO_2 laser applications. Details, examples and limits are discussed in chapter C4.2.5.

Singlets of the plano-convex and biconvex type can be purchased as catalogue items from a number of vendors. Cost and—often more important—much time can be saved if a given requirement can be (nearly) met with a catalogue lens. The steps between available focal lengths are, however, rather large and materials are mostly limited to BK7 and quartz but occasionally include some high-index glasses for lenses with high apertures.

C1.1.4.2 The dialyte and the achromat

Any combination of two lenses offers much extended design freedom over the singlet as there are four additional parameters: two curvatures, one refractive index and one lens separation. The classical use of this design freedom is to 'achromatize', i.e. to compensate the wavelength dependence of the refractive index (i.e. dispersion) of one lens by choosing a material with different dispersion for the second lens and arranging the curvatures for simultaneous correction of other aberrations. The subject is discussed extensively in all textbooks on optical design (see, for example, [6, 7]). In a very general way, 'splitting' a single lens into two elements will reduce the aberrations of that single lens and, thus, can, for instance, greatly improve the monochromatic performance and this is often the primary goal when handling (monochromatic) laser radiation. It should be noted that the rules for the best shape of a singlet almost never apply to combinations of lenses. Optimization by computer program, guided by the previous experience of the designer, is applied in practice.

For example, we consider the improvement gained by a two-lens arrangement (called a 'dialyte') over a single biconvex lens made from BK7 glass as discussed earlier (figure C1.1.1). If the singlet [a], which has acceptable performance at $NA = 0.062$ (or $f/8$) is replaced by two identical biconvex lenses (each having half the optical power) [b], the same performance is obtained at (a marginally better) $NA = 0.071$ ($f/7.0$). If replaced by two plano-convex lenses, arranged such that the curved surfaces are facing each other [c], the performance is just a bit better and allows use at $NA = 0.073$ ($f/6.8$). If the two lenses have the same orientation, both with their curved sides facing the entering radiation [d], the same performance is obtained at $NA = 0.087$ ($f/5.7$). If the second lens is given the optimum shape, which is a meniscus in case [e], the system is further improved to allow use at $NA = 0.104$ ($f/4.8$). This turns out to be the best that can be

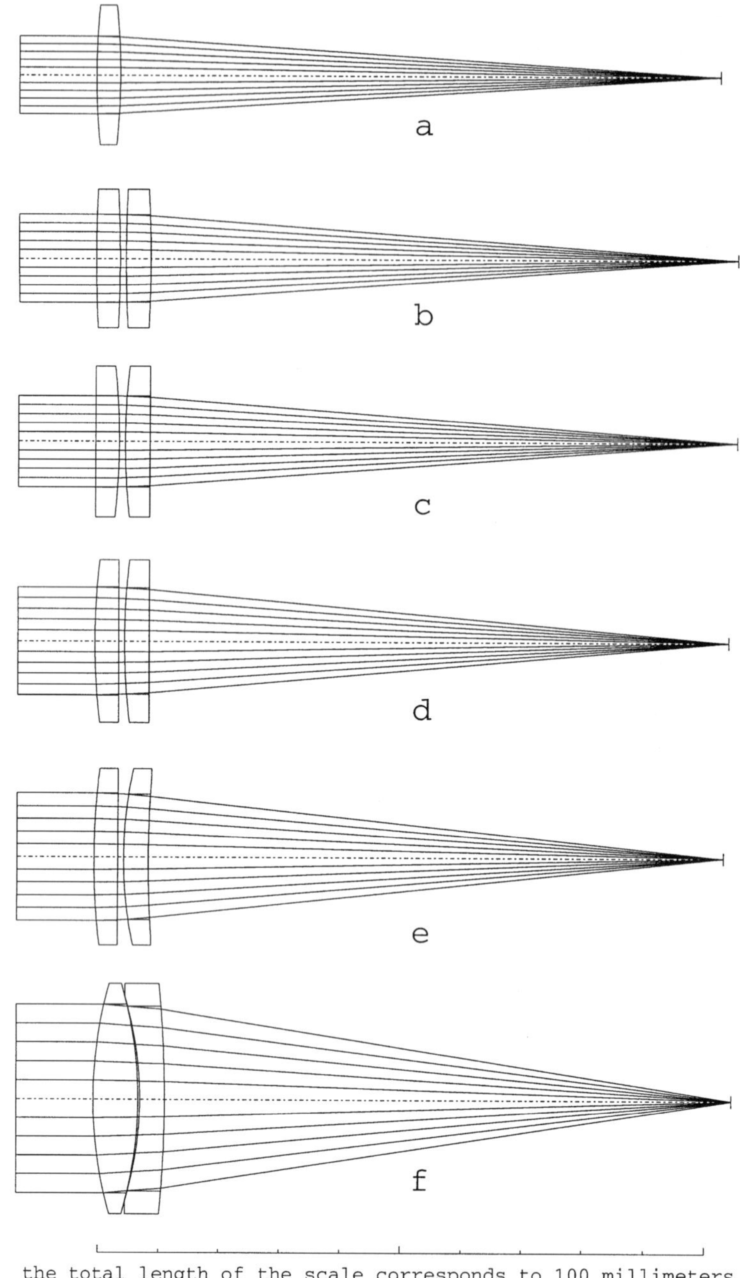

Figure C1.1.1. Comparison of six lens types of equal (diffraction-limited) performance. (*a*) singlet at $NA = 0.062$ ($f/8$); (*b*) two equal symmetrically bi-convex lenses of half-power at $NA = 0.071$ ($f/7$); (*c*) two equal plano-convex lenses facing each other at $NA = 0.073$ $f/6.8$); (*d*) the same lenses as in (*c*) both facing the incoming beam at $NA = 0.087$ ($f/5.7$); (*e*) one plano-convex lens plus one positive lens of same power bent to optimum (= meniscus) shape at $NA = 0.104$ ($f/4.8$); (*f*) one positive, symmetrically bi-convex lens plus one negative lens of optimum shape in edge contact with the first lens at $NA = 0.15$ ($f/3.3$).

achieved with two positive (!) lens elements (with that refractive index). If, however, the design is forced to have a positive (biconvex) first, and a negative meniscus second, with a small air gap between the lenses [f], it becomes substantially better and can be used at $NA = 0.15$ ($f/3.3$). This type of dialyte is particularly useful for such laser applications, where the wavelength dependence of the optics is no issue.

Proper shaping (also known as 'bending') of lens elements is by far the most powerful method to achieve best performance (i.e. low monochromatic aberrations). Correcting the wavelength dependence of lens properties however, can only be achieved by selecting suitable materials in conjunction with proper distribution of the optical power over these materials, whereas lens shape has very little effect.

A special form of the two-lens system, not a dialyte but an 'achromat', consists of a low-dispersion positive and a high-dispersion negative lens element, cemented together at the common interface. With properly selected materials (one popular combination for visible light uses the Schott glasses BK7 and SF2), the resulting system is corrected for spherical aberration and coma and has the same focal length at two wavelengths in the visible (e.g. C and F light) band. Cementing lenses has the advantage of creating a single mechanical unit where the mutual alignment between the elements has already been done by the manufacturer. It also removes two air–glass interfaces (and the need for anti-reflection coating of these surfaces). The use of cemented lenses must, however, be strongly discouraged for use with laser radiation of any appreciable power. The thermal properties of the constituents of an achromat tend to differ enough to create thermal stress upon exposure to high radiant powers and this will readily break up the cemented assembly.

The well-known glass combinations, which allow simultaneous correction of colour and monochromatic aberrations for visible light, are not transferable to other spectral regions. When moving towards the near infrared, the differences in the dispersion between the optical glasses become progressively smaller [8]. Colour correction, if at all possible, then demands different glass combinations. A typical example is discussed in section C4.2.6.3.

Building a dialyte from catalogue lenses is only marginally successful. As the previous example showed, replacing a single lens by a pair of identical lenses (of, accordingly, double focal length) will give some, but not much, improvement in performance and the best solution is not accessible with catalogue lenses, as it requires a very special negative lens element. If lowest aberrations are not mandatory, however, combinations of cleverly chosen catalogue lenses can help to achieve an otherwise not available focal length, since lens powers are, in a first approximation, additive.

C1.1.4.3 Lens systems with more than two elements

Much more complex designs than those already discussed are needed if a lens system has to provide imaging over an appreciable angular field onto a flat image plane, a requirement which is typical for all photographic applications but not normally for laser optical systems. In applications with direct focusing of a laser beam, we usually need to cover only small angular fields of less than, say, 5 mrad—just enough to make them insensitive to alignment errors. But there are exceptions: one is scanner optics (which often require f–θ lenses), another is the use of mask imaging for micromachining with excimer lasers. The latter subject is treated in chapter C4.2.7, the former in C4.4. More complex designs than those already discussed are also needed to achieve very high numerical apertures and/or a distance between the last optical surface and the image, which is substantially larger than the focal length. The latter may be desirable to protect a lens from spatters and fumes from a laser machining process (see section C4.2.5.2.2).

The art and science of designing optical systems for just about any application is highly developed. It is treated in a number of textbooks [6–9]. Excellent computer software for optical design and analysis is offered by a number of vendors and covers a broad price range. While such software is of great help to the educated designer (it alleviates the tediousness of manual calculations, provides excellent visualization of system layouts and (aberration) performance and always includes powerful optimization routines), it is no

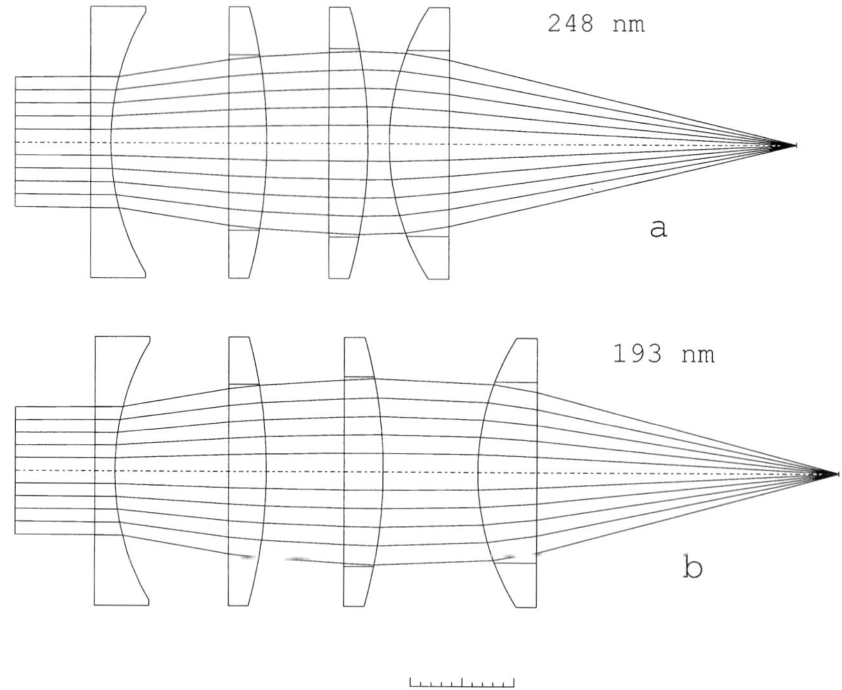

Figure C1.1.2. Example of a (UV) lens system assembled from (quartz) catalogue lenses, with lens distances adapted for optimum—diffraction limited—performance at wavelengths of 248 and 193 nm, respectively. For further details see text.

replacement for insight into the optical problem to be solved and the most adequate approach to its solution, which basically come from experience.

Designing multi-lens systems with the exclusive use of catalogue lenses is an interesting design challenge and can be quite successful. One will often end up with more lens elements than are necessary for a truly optimized system but this should not be a major concern. Figure C1.1.2 shows, as an example, a lens for 30:1 imaging of a small object in excimer laser light of [a] 248 nm (a KrF-laser) and [b] 193 nm (an ArF laser). Both versions use the same four (quartz) lenses, which can be found in most catalogues: one plano-concave, three plano-convex, of which two are identical. The respective focal lengths (for visible light) are: −50, 100 and 50 mm. Only the lens separations are different (and have been used for optimizing performance). Both versions have excellent imaging qualities up to, respectively, $NA = 0.22$ ($f/2.17$) at 248 nm and $NA = 0.25$ ($f/1.94$) at 193 nm wavelength, albeit for small fields.

C1.1.5 Use of (curved) mirrors

The optical power of a lens can also be provided by a curved mirror but with important differences, some advantageous, some not; mirrors function independently of wavelength (except for properties caused by surface coatings) and a single concave spherical mirror has much smaller aberrations than a comparable lens. In fact, a lens would have to be made of very high-index material ($N = 4.4$) to match the spherical aberration of a single concave spherical mirror of equal aperture! The problem with mirror optical systems lies in the overlap between the incoming and outgoing beams which, particularly in laser optics, requires the mirrors to be placed in some 'off-axis' arrangement, with the individual mirror elements tilted with respect to the optical axis to avoid obscuration. This is in sharp contrast to axially symmetrical mirror systems such as those used in astronomical instruments where a central obscuration is acceptable. Tilted non-flat optical surfaces are inevitably inflicted with off-axis aberrations, notably coma and astigmatism, which, in turn, must be corrected by additional (and again tilted) optical surfaces, resulting in more complex set-ups with at least three curved mirrors [10]. Thus, mirror optical systems lack the ease of alignment along a common optical axis, which makes lens-based systems attractive. This and the lower number of degrees of freedom have made mirror systems much less popular with optical designers. Extended mirror designs with non-spherical surface shapes do not really offer much relief: the improvement is typically limited to an extremely small angular field and at the expense of a large coma beyond. This is discussed for the off-axis paraboloid in section C4.2.5.3.2. Not surprisingly, mirror systems are discussed much less extensively in textbooks on optical design than lenses. A systematic study of systems with three and four spherical mirrors, respectively, was published recently [11, 12]. For use at very high laser power levels, mirrors are the only choice, as they offer excellent thermal stability due to the high thermal conductivity of the more common substrate materials (copper and silicon) and the ease with which efficient cooling can be implemented. These aspects are discussed in conjunction with high-power CO_2 laser applications in section C4.2.5.3.

C1.1.6 Non-focusing optical laser-beam handling and relaying

Large distances often have to be bridged between the location of a high-power stationary laser installation and that of the workpiece on which a process has to be performed. This situation is typical for (and limited to) CO_2 lasers. Since the laser beam waist is usually located near the output window of the laser, the beam will diverge while travelling the distance to the workpiece which, for instance, in the case of large gantry-type workstations is large and variable (see chapter C4.4). The beam radius at the location of the focusing optics, as well as its wavefront curvature, will vary accordingly (equations (C1.1.2) and (C1.1.3)). This is unwanted for two reasons: (1) it affects the axial position of the focus (equation (C1.1.7)) and (2) as the beam radius controls the numerical aperture of the image beam, the size of the focal radius (equation (C1.1.12)) will not be constant over the working envelope. A numerical example is given in table C1.1.1. Both drawbacks are avoided by relaying the primary beam so that an image waist is formed near the median of the range of distances to the workpiece and, thus, the range of waist distances with respect to the focusing optics. It is good practice to combine this operation with creating an image waist of optimum radius and Rayleigh range. This will keep the distances to the focusing optics to well within the near field and minimize focal distance variations as well as variations of the effective numerical aperture and resulting focal spot size. For the example of table C1.1.1, this would mean the use of a telescope which creates a waist of 12 mm radius at a distance of, say, 10 m. As a result, the focus remains constant within less than 1% over the same range of distances (±2 m).

Table C1.1.1. Effect of lens-to-waist distance on the focus of a lens with 100 mm focal length. The laser-beam characteristics are: $M^2 = 2$; $W_0 = 6$ mm; wavelength, 10.6 μm. The diffraction image includes the effects of the varying beam truncation at the lens diameter of 36 mm.

Lens-to-waist distance [m]	Beam radius on lens [mm]	Diffraction image diameter [mm]
8.0	10.8	0.132
10.0	12.8	0.119
12.0	14.8	0.112

Optical systems for performing this relaying function are, in essence, telescopes but with a capability of adjusting the exiting beam for (weak) convergence. In this application, the telescope must enlarge the beam enough to allow formation of a waist of suitable radius at the desired (forward) distance (cf equation (C1.1.10)). It is often referred to as a 'beam expander'. Telescopes can, of course, be built with lenses or mirrors, and can be of the 'Keplerian' or 'Galilean' type, i.e. with or without an intermediate focus. Unless needed for some special function (such as mode filtering), an intermediate focus is best avoided. Examples of telescopes based on (off-axis) mirrors and for use with (high power) CO_2-laser radiation are given in C4.2.5.1. Lens-based beam expanders are more commonly used in conjunction with shorter wavelength lasers. Beam adaptation—expansion or compression—in only one dimension is often needed for diode laser applications. This is discussed in section C4.2.6.4.3.

Planar mirrors play an important role in beam positioning for whatever purpose (relaying, alignment, scanning a workpiece etc). The high speeds with which some laser-based processes (scribing, marking etc, see section D1) can be performed, call for equally high speed of beam deflection by mirrors (see chapter C4.4). Low inertia and high mirror rigidity are then primary requirements, to be met by selection of (1) a suitable material with a high modulus of elasticity and low specific gravity and (2) a mirror geometry that minimizes inertia. For best results—highest scan speeds, fast settling to the end position with minimum overshoot, accurate positioning—matching of the scan mirror to the drive motor is essential. Several vendors offer integrated (and optimized) mirror-drive assemblies for a range of mirror sizes.

By their very nature, weakly curved mirrors form an essential component of laser resonators (see section A2). In this application, the local reflectivity of a mirror is occasionally adapted such that some desirable mode of oscillation is promoted within the resonator. If used in a folded cavity, (dielectric) mirrors will favour one direction of polarization, sufficient to induce a nearly completely (linearly) polarized output beam.

Mirror functions in systems for visible and near-infrared light are often implemented by means of reflecting prisms, usually exploiting total internal reflection for enhanced performance over surface mirrors. Reflective (and non-dispersing) prisms are often designed for multiple reflections. They, thus, allow not only for the direction of a beam to be changed but also for it to be deviated, displaced, reverted and/or inverted. The more common types of reflective prisms are treated in all textbooks on optics. A more complete yet concise overview can be found in the relevant chapter of the *Handbook of Optics* [13].

C1.1.7 Thermal effects in optical materials

The optical properties of materials are temperature dependent. The refractive index typically varies of the order of 10^{-4}–10^{-5} per °C. As long as the temperature distribution within an optical component is homogeneous, this variation is of concern only in situations where an optical instrument has to meet very high performance standards (e.g. an aerial camera) or where very wide temperature ranges are envisaged (as in certain military equipment). Specific cases have been discussed by Rogers [14, 15]. Quite a different problem arises in optical systems using high-power lasers. If a lens is exposed to high-power radiation and absorbs some of the incident radiation, the lens centre is heated but the heat will typically flow to the lens rim where it is eventually taken up by the lens mounting. This heat flow will, thus, cause a thermal gradient and, consequently, a radial gradient of refractive index in the lens material even under stationary conditions. Its optical effects are threefold:

(1) a change of the lens shape due to stronger thermal expansion in the lens centre area,

(2) an angle of refraction at the air–lens interfaces depending on ray height and

(3) a propagation of radiation within the lens along an inwardly curved path, leading to a shortening of the focal length.

A quantitative computational model based on these effects reveals that, in any realistic situation, the third phenomenon accounts for the largest contribution. If the actual temperature rise in the lens centre is known, the resulting shortening of the focal length can be predicted reliably [16]. The effect is of substantial importance—and this is largely underestimated—-for the ZnSe lenses used extensively for focusing CO_2 laser radiation. As a rule of thumb, the shortening of the focal length of a ZnSe lens will be about one Rayleigh range if, in thermal equilibrium, the lens centre runs 25 K warmer than the lens rim. Interestingly, the corresponding effect on aberrations is small; in fact, a thermal gradient reduces the spherical aberration of a single lens [17, 18].

C1.1.8 Specifying optics for laser applications

C1.1.8.1 Tolerances

The performance of optics in practice depends on many factors. While the parameters as designed will set an upper limit to what may be achieved under ideal circumstances, unavoidable manufacturing tolerances and imperfections cause real systems to perform below that limit. Thus, in the first place, the design of an optical system for a given application should always well exceed the acceptable minimum performance parameters by a reasonable margin to account for tolerances. Second, designs which, due to their basic principles, are less sensitive to tolerances should be preferred if possible. This typically involves a search for several, sometimes radically different, alternative possible solutions. Third, and usually quite complex, the tolerance budget has to be allocated to the individual components of a system to minimize the cost of manufacturing while staying within a given performance degradation. It is quite common for the effect of the tolerances of individual optical surfaces on the overall performance of the system to vary considerably, which makes it important to locate critical parts of a design and to cleverly balance the tolerances between all components. Modern commercial optical design software offers help for this task.

C1.1.8.2 Surface imperfections: shape deviations

Imperfections in an optical component can be divided into two parts: shape deviations and surface (and bulk material) quality imperfections. Shape deviations, also called form deviations, of an optical surface can be of different types. If the surface is perfectly spherical but its curvature deviates from the design value, the deviation is a 'sagitta error'. Deviations from the spherical shape, which are, in essence, higher-order shape deviations, are generally referred to as 'irregularities'. In practice, allowable surface shape tolerances are computed and specified in terms of sagitta errors and the corresponding tolerances for irregularity are typically specified at a substantially lower value.

C1.1.8.3 Surface imperfections: surface quality

Optical surfaces are supposed to be 'optically smooth', which means that their microscopic profile varies gradually with steps well below the wavelength for which the optics is destined. This is achieved by a polishing process. Polishing must both remove the irregularities left by prior grinding of the surface and precisely shape the surface to conform to the specified curvature within the specified form tolerances. If the polishing ends prematurely, surface damage caused by the grinding process may still be present; alternatively, the polished surface may be damaged by contaminated polishing agents or by inadequate handling. The resulting faults are termed 'surface imperfections' if they are localized small individual pits, scratches etc and referred to as 'long scratches' if they exceed 2 mm in length, because of their visibility. All surface imperfections which are of the order of, or larger than, the wavelength cause losses by stray radiation and in classical optical systems the acceptability of such losses is the basis for setting a limit for the accumulated sizes of the surface imperfections. For laser applications, the situation is generally much more severe. Surface imperfections of wavelength size or beyond may cause local power concentration and thereby lower damage thresholds. It is, therefore, customary to require the highest standards for surface quality imperfections for optical components to be used with laser radiation, particularly if pulsed laser operation with accordingly high peak powers is envisaged.

C1.1.8.4 Communication of specifications—ISO 10110

The communication of design data to a manufacturer is invariably by technical drawings which, to avoid misinterpretation, should be prepared according to a common standard. With the increasing globalization of the production of optical parts and systems during the last decades, a need was felt to agree upon an international standard. Issued by ISO, the International Organization for Standardization (Geneva, Switzerland), ISO 10110: (*Optics and Optical Instruments—Preparation of Drawings for Optical Elements and Systems*), has, since 1996, replaced the various national standards which were in effect before. ISO 10110 is now the preferred way of drawing optical elements and systems. Conformity to the standard must be explicitly embodied in the respective drawings.

ISO 10110 consists of 12 parts. An originally envisaged part 13 on laser damage thresholds has been cancelled but is still (and maybe confusingly) referred to in several other parts. Each part is a separate document and subject to revision. ISO 10110 is intended and has been written to standardize both the specification and measurement of the parameters of optical components.

Part 1 (General) explicitly regulates the drawing of optical components and systems, their dimensions and tolerances (linear dimensions always in millimetres, angular dimensions in degrees, minutes and seconds of arc). Parts 2, 3 and 4 are short documents describing how to specify three different types of bulk material imperfections: stress birefringence (part 2), bubbles and inclusions (part 3) and inhomogeneity and striae (part 4). Part 5 addresses the definition, specification, measurement and analysis of surface form tolerances, with particular attention to interferometric measurements and their interpretation. In part 6 the complex problem of specifying and indicating centring tolerances of single lenses and lens combinations are treated. Part 7 addresses surface imperfection tolerances and includes a choice of techniques for measurement or visual inspection. In part 8, surface textures as created by grinding or polishing are treated, whereas part 9 deals with (protective) surface treatment and (functional) coatings. Part 10 indicates how to optionally specify the data of an optical lens element in a table rather than on the drawing itself. Part 11 shows which tolerances apply, in case explicit indications are not given in a drawing. These tolerances then depend on the size of the component. Part 12 deals exclusively with aspheric surfaces, their mathematical description and presentation in drawings.

C1.1.9 Manufacture of optical components

The manufacture of lenses from brittle materials—which include optical glasses and most crystals—involves two basic steps: grinding and polishing. Grinding, which may be preceded by some coarse machining of the part to approximate shape, is mostly done with a slush of loose grit of a hard material such as emery or carborundum. The actual mechanism is that the localized pressure exerted by a hard particle on the glass surface causes compressive stress in the glass surface area, which, upon release, lets a small part of the glass surface separate from the bulk. A ground surface, therefore, retains compressive stress in a surface layer of thickness roughly proportional to the size of the grinding particles and decaying exponentially. Grinding is typically done in steps using successively smaller particles as the intended final shape is approximated and damage caused by the coarser particles is alleviated.

Polishing is the final process for generating an optical surface of the required smoothness and conformity to the specified shape. A measurement to determine surface conformity to the required shape by interferometric methods can only be done on a polished surface (see chapter A5). Polishing must also remove the stressed part of the surface layer and as much more of this layer as is affected by damage (digs, scratches) from previous grinding. Polishing, in contrast to grinding, is a slow process, basically involving chemical interaction between the glass and water, enhanced by a polishing agent. The degree of polishing of a surface is characterized by one of four grades, which specify the number of microdefects—irregularities of less than 1 μm size—per 10 mm of sampling length (ISO 10110-8). What is acceptable for optics for laser applications depends largely on the laser wavelength and on the peak power density, which will be encountered by the surface. The highest grade (P4 with less than three microdefects per 10 mm of sampling length) is typically required for optical components for use with pulsed lasers in the near-infrared (Nd:YAG), visible (ruby) and ultraviolet (excimer lasers) spectral region. Much less stringent requirements apply to optics for the longer wavelengths of (cw) CO_2 lasers.

The quality of polish affects that of a coating, both in terms of adherence and blemishes and, therefore, strongly affects the onset of damage induced by laser irradiation. Coated optical surfaces are much more prone to damage by high-peak-power (pulsed) laser radiation than bulk materials but, if produced by the best manufacturing techniques, coatings have approached the bulk thresholds to within one order of magnitude. Laser-induced damage to optical materials is a subject of extensive current research and results are periodically reported in conferences, many of which have been organized or, at least, supported by the US National Bureau of Standards. Proceedings have been published by 1999 *SPIE* **3902**, 1998 **3578**, 1997 **3244**, 1994 **2428**, 1992 **1848**. Sixty-two 'classical' papers on the subject are found in Vol. MS24 of the SPIE Milestone Series [19].

The manufacture of mirrors is similar to that of lenses in that brittle substrate materials (glasses, zerodur, silicon) are used. Metal mirrors for high laser power applications based on copper bodies with integrated cooling channels are mostly made by single point diamond turning on precision machines, a process which readily produces aspheric shapes.

C1.1.10 Summary and conclusions

When used with laser radiation, optical components perform similar tasks as in classical optics and, thus, benefit from the achievements in optical design and manufacturing, which have accumulated over centuries. The coherent nature of laser radiation has to be considered when planning and evaluating laser beam imaging. The possibly high (peak and or average) powers of laser radiation requires utmost care in the manufacturing and handling of optical components for laser applications and attention to achieving thermal stability in laser optics.

References

[1] Self S A 1983 Focusing of spherical Gaussian beams *Appl. Opt.* **22** 658–61

[2] Kogelnik H and Li T 1966 Laser beams and resonators *Appl. Opt.* **5** 1550–67

[3] Siegman A E 1990 New Developments in laser resonators *Proc. SPIE* **1224** 2–14

[4] Smakula 1936 German patent 685 767

[5] Dobrowolski J A 1995 Optical properties of films and coatings *Handbook of Optics* 2nd edn, ed M Bass (New York: McGraw-Hill)

[6] Kingslake R 1978 *Lens Design Fundamentals* (New York: Academic)

[7] Smith W J 1990 *Modern Optical Engineering* 2nd edn (New York: McGraw-Hill)

[8] Shannon R R 1997 *The Art and Science of Optical Design* (New York: Cambridge University Press)

[9] Smith W J 1989 Optical design *The Infrared Handbook* 3rd edn, ed W L Wolfe and G I Zissis (Washington, DC: Office of Naval Research) ch 8

[10] Beckmann L H J F and Ehrlichmann D 1994 Three-mirror off-axis systems for laser applications *Proc. Int. Optical Design Conf. (Rochester)* (Washington DC: OSA) pp 340–8

[11] Howard J M and Stone B D 2000 Imaging with three spherical mirrors *Appl. Opt.* **39** 3216–31

[12] Howard J M and Stone B D 2000 Imaging with four spherical mirrors *Appl. Opt.* **39** 3232–42

[13] Wolfe W L 1995 Nondispersive prisms *Handbook of Optics II 2nd edn* ed M Bass (New York: Mcgraw Hill)

[14] Rogers P J 1996 The selection of IR optical materials for low-mass applications *Proc. SPIE* **2774** 301–10

[15] Rogers P J 1993 Optics in hostile environments *Proc. SPIE* **1780** 36–48

[16] Tangelder R J, Beckmann L H J F and Meijer J 1992 Influence of temperature gradients on the performance of ZnSe-lenses *Proc. SPIE* **1780** 294–302

[17] Beckmann L H J F 1994 Modelling of, and design for, thermal radial gradients in lenses for use with high power laser radiation *Proc. Int. Optical Design Conf. (Rochester)* (Washington, DC: OSA) pp 11–15

[18] Beckmann L H J F and Meijer J 1998 Modelling of and design for thermal radial gradients in lenses for use with high power laser radiation *Welding in the World* **41** 97–104

[19] Wood R M (ed) 1990 *Laser Damage in Optical Materials (SPIE Milestone Series MS24)* (Bellingham, WA: SPIE)

C1.2
Optical control elements

Alan Greenaway

C1.2.1 Introduction

The extracavity control of laser beams includes requirements for scanning and positioning the beam, changing the size and shape of the beam, modulation of the beam in both time and space (also in amplitude and phase) and, in this increasingly safety-conscious age, ensuring the safe disposal of stray beams. Each of these requirements will be considered briefly in this section.

In many cases the most suitable technology will depend on the power and wavelength of the laser beam that is to be controlled. Other considerations are cost and the environment in which the control function is to be exercised (for example, when used in a laboratory by experienced researchers a degree of control and access to the beam may be required that would be wholly unsuited to a commercial environment in which the operators may be much less aware of the potential risks).

Note that the intracavity use of adaptive optics is treated in chapter C1.3. As used in this section the term 'adaptive optics' will refer exclusively to linear optical systems and discussion will refer to extracavity uses of such systems. A treatment of beam positioning and scanning systems is presented in chapter C4.4. The treatment of beam positioning and scanning technologies here will, therefore, include discussion of such applications as flat-bed scanners only briefly.

C1.2.2 Amplitude modulation

The most basic level of beam control is achieved through use of a shutter. Most laser systems (but especially solid state systems) operate most effectively when producing output at their design power output. The resonator design can, under these circumstances, take full account of optical defects and distortions due to the operating temperature of the laser. It follows that the use of a shutter is preferable to variation of the output power level (including power-down) in most situations. Various forms of shutter are available from a multitude of commercial sources and these are widely used in safety interlock systems on lasers.

Amplitude modulators are also widely used with laser systems at all power levels. In the case of low-power lasers, the amplitude modulation is most often a simple means to ensure that the power level delivered to a detector can easily and quickly be varied. In this case a simple neutral density filter wheel or variable attenuator is generally satisfactory (but if these are of the reflective type care may be required with reflected beam). With low-power beams the use of absorptive filters or a polarizer–analyser system provide effective beam attenuators. In all cases it is advisable to locate filters between a laser source and any beam-conditioning optics (such as an optical fibre or a spatial filter) so that optical imperfections in these elements do not lead to any degradation of beam quality.

C1.2.2.1 Electro-optic modulators

Electro-optic modulators using the Kerr effect or the Pockels effect are available, although most modern electro-optic modulators are based on Pockels cells. These modulators exploit the voltage-dependent birefringence induced in some crystalline materials when they are subjected to an external electric field. The Pockels effect has a linear relationship and the electro-optic Kerr effect a quadratic relationship between the indicatrix of the material and the applied electric field. Birefringence is a difference in refractive index dependent on the alignment of the electric field vector with crystal lattice directions.

A linearly polarized light beam can be resolved into two orthogonal polarizations of equal strength (see section A5.3.9). Suppose the beam enters a birefringent crystal along the crystalline optic axis (the axis of crystal symmetry denoted as the z-axis). If the linear polarization direction is at 45° to the other principal dielectric axes (x- and y-axes) in the crystal, one of the orthogonal polarization components aligns with the fast axis and the other with the slow axis within the birefringent material. As a result of the different refractive indices associated with the fast and slow axes, a phase change develops between the two polarization components as they propagate through the crystal and this, in turn, causes a rotation of the plane of polarization of the radiation.

In an electro-optic device the electro-optic element is positioned between a polarizer and analyser with a mutual orientation of 90°. In the absence of birefringent effects, no radiation will be passed by the analyser. However, if a voltage is applied that makes the crystal birefingent and the fast and slow axes of the induced birefringence are oriented at 45° to the input polarization plane, the plane of polarization is rotated as the beam passes through the crystal. If the plane of polarization is rotated by 90° the polarization will align with the analyser and the radiation will pass through the system. If the voltage is removed and the plane of polarization is unchanged by passage through the crystal, the polarization will be orthogonal to the analyser and the radiation cannot pass through. If the applied voltage results in a rotation of the polarization plane between 0° and 90°, the fraction of the energy passed will follow a cosine-squared law.

Suitable materials for electro-optic modulators include lithium niobate, lithium tantalate (for visible and near-IR regions) and, for use in the thermal IR, cadmium telluride. Electro-optic modulators may use various orientations of the crystalline axes. Dependent on the orientation of the electric field relative to the direction of propagation of the light, these are referred to as longitudinal (electric field and propagation direction are parallel and aligned with the crystalline optic axis) or transverse (electric field and propagation direction are orthogonal).

With longitudinal modulators the voltage must be applied across the entire propagation path within the crystal and, to maintain the required field strength, high voltages (in the kilovolt range) must be used. For KDP (potassium dihydrogen phosphate) approximately 8.7 kV is required at a wavelength of 633 nm to retard one polarization component by half a wave compared to the other (thus to obtain an electrically programmable half-wave plate and rotate the plane of polarization by 90°). It is generally difficult to switch high voltages at high frequency and thus longitudinal devices are used for lower-frequency applications. An additional feature is that the radiation must enter and leave the crystal through transparent (or wire grid) electrodes required to apply the electric field. As the applied voltage increases from zero to that required to achieve a half-wave delay, the polarization changes from the linear input to elliptical with the major axis aligned with the original polarization direction, through circular polarization, to elliptical polarization with the long axis aligned normal to the original direction of polarization and, finally, to linear polarization at 90° to the original direction. The interaction of these polarization states with the analyzer is what provides the desired amplitude modulation. Longitudinal modulators are used when large acceptance angles and wide fields of view are required.

Transverse modulators, by comparison, use an orientation that requires the electric field to be applied along the z-axis which is now orthogonal to the direction of propagation. This construction makes it possible to achieve a long interaction length at a given field strength. Lower applied voltages are, therefore, required

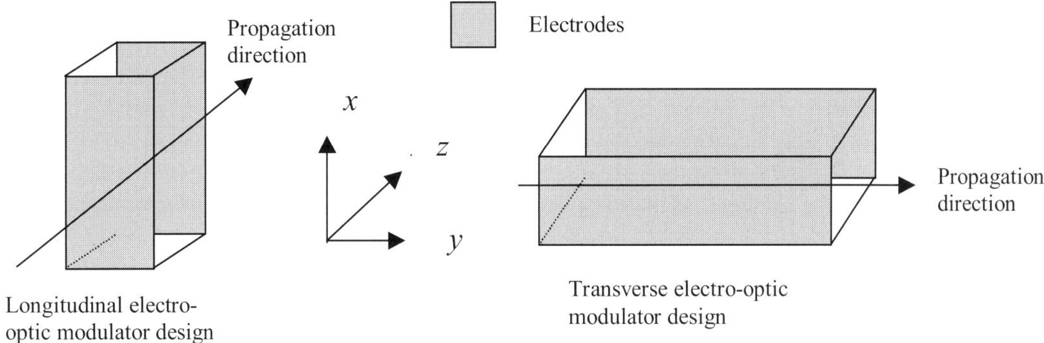

Figure C1.2.1. Schematic diagram showing axis directions and electrode location for longitudinal and transverse electro-optic modulators.

across the short dimension of long thin crystals, because the potential is applied across the material and orthogonal to the direction of propagation. Additionally, the input and output beams are no longer required to pass through the electrodes. Because of the long, thin structure only small acceptance angles are suitable. This design is ideally suited to modulators in waveguide configurations for integrated optics applications and has the added benefit of using lower electric voltages.

The voltage required to induce a given phase change between the polarization components (and thus rotation of the plane of polarization) increases with the wavelength of the radiation to be controlled.

Material exhibiting intrinsic birefringence can be used in electro-optic devices by cutting the active crystal into two segments and rotating the segments so that the fast and slow axes of the first segment align with the slow and fast axes in the second segment. Such an arrangement cancels the natural birefringence of the material and increases the range of materials that can be exploited to include birefringent materials such as ADP.

Pockels cells with extinction ratios (ratio between maximum and minimum transmission) of 10^3 are available and bandwidths of $>10^9$ Hz are available. This extinction ratio is not always adequate for use as a shutter but two Pockels cells may be used in series if required. Electro-optic devices offer a higher switching speed than magneto-optic and acousto-optic devices but are generally of smaller aperture and have the disadvantage of requiring relatively high voltage drivers. If the polarizer and analyser in an electro-optic device are absorbing, it is clear that the energy deposited in these elements will limit the power that can be handled.

C1.2.2.2 Acousto-optic modulators

Acousto-optic modulators rely on the variation of refractive index with position that is produced by pressure (acoustic) waves in various media. Acousto-optic materials require good optical quality, good optical transmission and good acoustic properties, but suitable materials exist across the UV, visible and near IR (e.g. fused silica/quartz) and in the IR (germanium). A typical acousto-optic modulator will consist of a material with suitable elasto-optic properties and a carefully-designed and coupled piezoelectric transducer bonded to one side of the material and an acoustic sink well coupled to the opposite side (to prevent acoustic reflection and the generation of standing waves).

The (acoustic) pressure waves give rise to a pattern of periodic variations in refractive index that is moving slowly compared to the electromagnetic wave. The light, therefore, sees a near-stationary Bragg grating structure and can be scattered by this grating (note, however, that Doppler effects lead to a slight frequency change, desirable e.g. in generating heterodyne carrier frequencies in some forms of interferometry).

For a given transducer frequency and a material with a given speed of sound, the acoustic wavelength (and thus the period of the pressure-induced refractive index changes) will be defined by the usual relationship between frequency, speed and wavelength. The angle at which an optical wave is scattered from the acoustically-generated grating is dependent on the ratio of the wavelength of the light and the wavelength of the Bragg grating. The fraction of the light energy scattered by the Bragg grating will be dependent on the amplitude of the Bragg grating and that, in turn, is determined by the acoustic power coupled into the material from the transducer.

By keeping constant the frequency of excitation but varying the acoustic power, the energy in the scattered beam can be varied from zero (no acoustic signal) to some maximum level. Acousto-optic modulators, therefore, have a high extinction ratio. Acousto-optic materials are compared through a 'figure of merit' that relates the optical diffraction efficiency to the acoustic power coupled into the material. The 'figure of merit' is not the only important parameter and the speed of operation (modulation bandwidth) also needs to be considered.

The pressure-induced Bragg grating propagates through the acousto-optic cell at the speed of sound in the given material and it follows that the reaction time of the device cannot significantly exceed the time taken for the acoustic wave to transit the optical beam diameter. The speed of operation of opto-acoustic devices can, therefore, be improved through the use of smaller light-beam diameters. However, by the principle of Fourier optics the beam spread induced by the Bragg grating is dependent on the number of grating periods across the beam. This effect, and the power-density level that the material can support, may both limit the extent to which the bandwidth may be improved by focusing the beam.

Acousto-optic devices have the merit of requiring only low electrical power and delivering good extinction ratios (from 90% of input to zero) and being physically compact. They are intrinsically slower than electro-optic devices and have the disadvantage that the beam used has been deviated from its original direction (the undeviated beam is dumped).

C1.2.2.3 High-power beams

Amplitude modulators for higher-power systems may be based on the use of diffractive or polarizing effects to divide the beam, after which one beam is used as the required modulated output and the other beams are directed to beam dumps. For high-power applications, these approaches have the merit that the absorption of the beam energy occurs on the beam stop and not within the diffractive or polarizing element.

At their simplest these elements can consist of a circular disc having phase gratings of various modulation depths in various segments around the disc. The grating diffracts the input beam into various diffraction orders 0, ± 1, ± 2, ... An aperture then passes one diffraction order and blocks the remaining orders. In a simple device of this form, it is generally the zero diffraction order (undeviated beam) that will be used downstream. These are low-cost elements that can be designed to specific user requirements and suitable materials for fabrication of the phase gratings can be found for wavelength ranges from the UV to the IR. In general, such elements would not be used where accurately calibrated attenuation ratios are required.

Variants on the same principle exploit polarization or acousto-optic effects to divide the input beam in amplitude and to dump all but one output beam safely whilst providing the required modulation on the remaining beam. In general, a larger dynamic range (i.e. a greater range in the extinction ratio) can be achieved through use of a deviated beam as the output beam (since it is generally possible to switch the output energy to zero in the deviated beam). For all of these devices maximum continuous power levels of a few hundred W cm^{-2} appear to be widely available.

If a sample of a high-power beam is required for laser diagnostic purposes, the use of a low-amplitude diffractive structure, or the Fresnel reflection from one or more accurately figured transparent wedges, may be used.

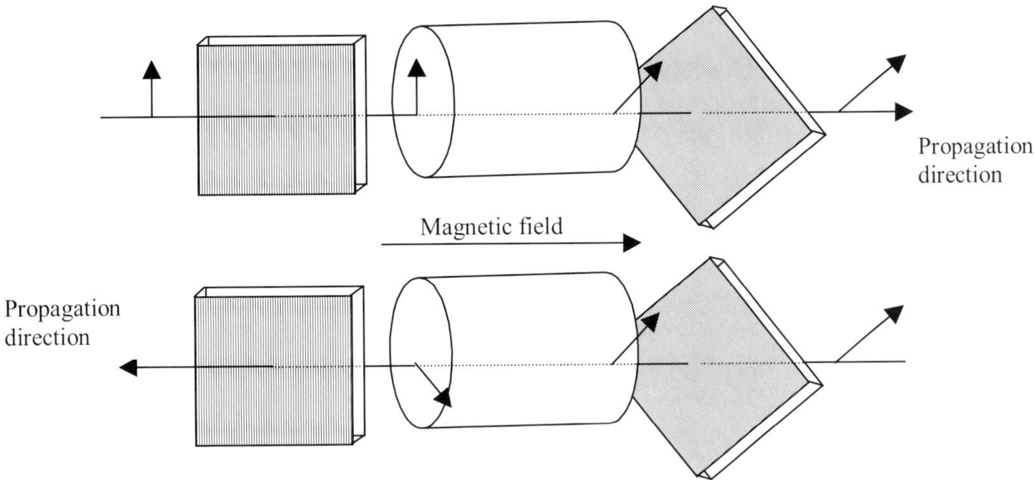

Figure C1.2.2. A vertically polarized beam propagating left to right is rotated clockwise by the Faraday rotator and passes unhindered through the analyzer. A reflected beam of the same polarization passes through the Faraday rotator and is rotated anticlockwise, resulting in an orientation normal to the polarizer. The backscattered beam is thus blocked.

C1.2.2.4 Magneto-optic isolators

The distinctive fact that the rotation of the polarization plane of an optical beam induced by the magneto-optic effect does not reverse direction if the direction of propagation is reversed gives rise to the unique application using optical isolators (see section A5.3.9). A reflected beam of initially weak amplitude re-entering a laser cavity can be amplified within the cavity and lead to serious effects in terms of the resonant properties of the laser cavity. Isolators offer a mechanism that can be used to prevent such unwanted feedback.

Many materials become optically active if they are immersed in an axial magnetic field. The rotation due to this Faraday effect depends on the Verdet constant, the length of the material and the modulus of the applied magnetic field. The vital feature here is that, if the rotation of the polarization is clockwise when the light propagates in the same direction as the magnetic field vector, it is anticlockwise if the direction of propagation is reversed with respect to the magnetic field. Thus, a beam reflected back through the system suffers twice the rotation. The basic configuration of a Faraday isolator is shown in figure C1.2.2.

The maximum achievable isolation from a Faraday isolator is likely to be limited by inhomogeneities in the crystal. It is, however, possible to square the attenuation of the backscattered beam by placing two isolators in series and arranging the magnetic field for each isolator to be reversed with respect to the other. Such an arrangement leaves the output polarization of the laser system plus isolators unchanged in addition to providing a high level of isolation.

C1.2.3 Scanning and positioning the beam

In all scanning operations it should be recalled that the focus of the beam scanned by any device may vary significantly in quality across the scan when the output plane is flat. Discussion of these issues is more appropriate to chapter C4.4.

C1.2.3.1 *Mechanical beam-directing systems*

At the simplest level, a system for re-directing a laser beam may be required in a research laboratory to provide a safe method for sharing the output from a laser between several experiments or simply to provide a convenient and easily re-arranged coherent source. Flat mirrors for angular deflection can be obtained from a variety of commercial sources. These can range from simple mirrors with defined mechanical positions (e.g. Thorlabs' 'indexing optical mount' provides 15 positions to allow a laser to be shared at very low cost) to controllable scanning mirrors of various degrees of sophistication and cost. In each case, care is likely to be required to intercept and to contain all stray beams generated, not only in the positions required but also during transitions from one beam position to the next.

For raster scans it is normal to use two single-axis deflectors arranged so that the scan directions are orthogonal. This arrangement provides a high degree of flexibility in terms of the scan direction sequence. For repeated raster scans, rotating systems using rotating polygonal mirrors are useful and are widely used in the IR [2].

For applications that require an ability to actively re-position the beam under computer control and over relatively short periods of time various systems are available. Fast steering mirrors of the type used in military applications can provide steering of high-power laser beams at high frequencies and with a 'go to' capability. Such devices do not come cheaply—a high-performance model from Ball Aerospace may cost in excess of \$50 000 (cheaper commercial devices are available through Newport Corp)—and are, therefore, not well suited to many applications where cost is an important factor. Such devices may offer kHz bandwidths and better than μrad resolution. Other, more modestly priced, devices can offer accurate 'go to' capability but generally with lower angular range, speed or power-handling ability.

The range of devices available includes galvanometer mirrors, voice-coil scanners, acousto-optic deflectors, electro-optic deflectors and various forms of adaptive optics mirrors. Mirrors with various forms of actuation are produced, the choice of actuation being determined by speed of operation, positional accuracy and repeatability required, cost, power-handling capability and angular deflection required. Moving magnet scanners are widely available and offer low-inertia, reliable and low-cost scanning systems capable of optical deflections of tens of degrees with good linearity and stability. Those available from Laser 2000, for example, have small-angle response times of a few tenths of a millisecond and can include position sensors to make them suitable for feedback control. Moving coil systems are likely to be more accurate but to have slower response. Repeatability ranging from a few 10s of μrad to a few μrad are offered over these scan ranges and, in both cases, optical scans over wide angles (up to 80 degrees) are offered in the commercial literature. Gimballed mirrors driven by voice-coil or piezoelectric actuation with more modest range and price are available from a wide variety of sources.

Wavefront modulator systems used in adaptive optics (for example, deformable membrane mirrors (see chapter D8.1)) can be used for scanning the beam over small deflections in addition to their use for changing the beam shape (this latter use will be discussed later). However, such adaptive optics wavefront modulators are generally designed for the shape deformation that they can produce. Since the amplitude available for wavefront deformation is very limited, it makes little sense in adaptive optics applications to compromise the wavefront shape change that can be achieved by using any of the limited modulator stroke for control of beam direction. For this reason it is general practice in adaptive optics to include within the system a 'tip-tilt' mirror of exactly the type that can be used for general laser-beam steering applications.

C1.2.3.2 *Acousto-optic and electro-optic scanners*

When considering scanners, one is generally interested in the format of the raster scan that can be produced, i.e. how many resolvable spots can be produced. This is dependent both on the total scan angle that can be achieved and on the beam divergence (spot size). A second important consideration is the time taken to 'go

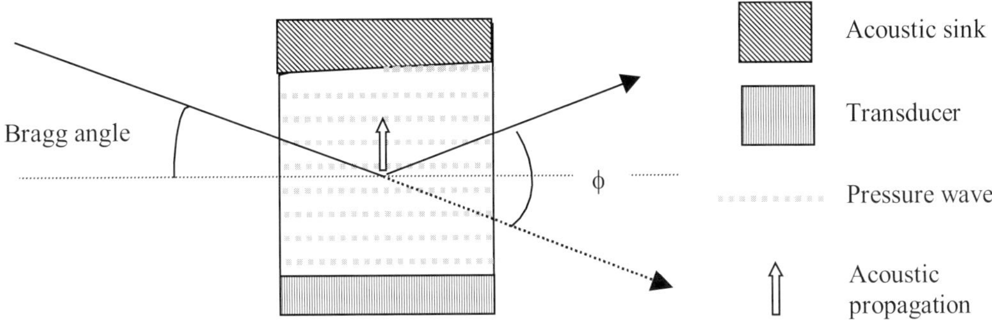

Figure C1.2.3. Illustrating the principles of acousto-optic deflection through angle ϕ.

to' a specific location within the scan area and the repeatability with which that spot can be re-acquired. For some applications, however, (e.g. image display), the time required to complete the sequential scan may be of more interest.

A range of electro-optic and acousto-optic beam deflectors is in use. These devices are essentially the same as those discussed in the previous section and used as modulators and Q-switches (see section C2). Electro-optic devices exploit the change in refractive index that occurs in electrically-induced birefringence. The acousto-optic effect exploits the change in refractive index that occurs as a result of the passage of sound waves (i.e. pressure waves) through a material.

Acousto-optic diffraction occurs from the pressure-induced grating and is maximum when the diffracted beam emerges at exactly twice the Bragg angle relative to the undeviated position. This angle depends on the acoustic frequency and, hence, can be controlled electronically through the drive signal fed to the transducer. The Bragg angle is given by $\sin\theta_B = \frac{\lambda}{2n\Lambda}$, where θ_B is the Bragg angle, λ the optical wavelength, Λ the acoustic wavelength and n the optical refractive index (see, e.g. [6]). The principle is shown schematically in figure C1.2.3.

Acousto-optic deflectors offer simple and fast control of the beam deflection for low electrical power input. The devices are much faster than mechanical scanners, simple in structure and offer a larger scan format than the electro-optic deflectors (but are slower in operation than electro-optical scanners). In some applications the small Doppler shift of the deflected beam may be a disadvantage.

Electro-optic deflectors use prisms of electro-optic material. Simple scanners may be based on the use of a single prism where the electric field is applied across the triangular ends of the prism to change the bulk refractive index and thus vary the beam deflection. However, a more common arrangement is to cement together two prisms whose optic axes (z-axis) are reversed in direction (figure C1.2.4). By applying an electric field along the z-axis the increase in refractive index in the upper prism is balanced by an equal and opposite refractive index decrease in the lower prism. Thus the upper ray is retarded with respect to the lower ray and the optical beam is deflected. Unfortunately the deflection is rather small for modest voltages and electro-optic deflectors are not as widely used as their acousto-optic counterparts.

The principal advantage of acousto- and electro-optic scanning systems is their speed, but this is delivered over small apertures and small scan angles (typically only a few degrees). It can be harder to control the properties of acousto-optic and electro-optic materials than to control mirror surface properties (especially emission) and for these reasons mechanical scanning systems tend to be more widely used in IR systems, although this is less true of laser-beam scanning applications.

In general, galvanometer systems are slowest but offer the highest format and largest overall scan angles. Acousto-optic scanners offer significantly higher access time but with smaller scan angles. Electro-optic scanners offer the fastest performance but the smallest scan formats.

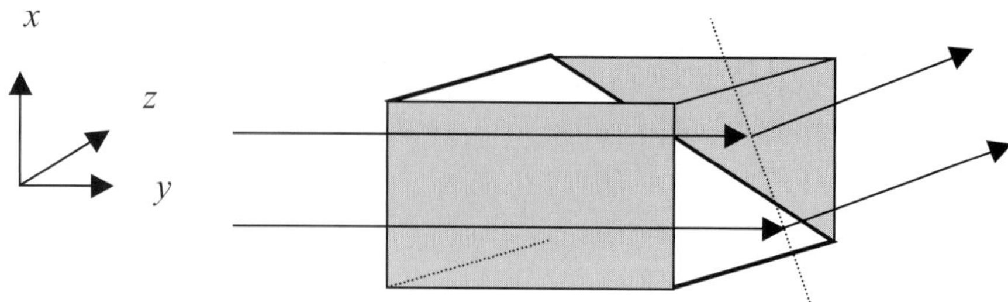

Figure C1.2.4. Schematic diagram of a double prism beam deflector. The upper and lower prisms have the orientation of their optic axis reversed. The shaded sides represent the electrodes across which the input voltage is applied.

C1.2.3.3 Diffractive beam steering

Programmable diffractive optical elements can be used with laser (and other monochromatic) beams for beam steering and spatial multiplexing. This is particularly useful since small-amplitude devices can exploit the intrinsic modulo 2π nature of the diffraction process to achieve beam division and relatively wide-angle beam-steering performance. Low-cost, widely-available components such as liquid crystal spatial light modulators with large spatial format can be exploited here, as can waveguide-type devices using electro-optic effects. Micro-machined mirrors, now widely used as display drivers, will offer an extremely flexible device for such applications when better control of the flatness of the individual mirror elements has been achieved—at present the curvature of the individual segments is an impediment to such application of these devices.

C1.2.3.4 Positioning the beam

When several sources (multiple lasers or both lasers and broadband sources) are to be used with a single optical system, the use of an optical waveguide (fibre) provides a method by means of which this may be achieved. The fibre end provides a well-defined source that may be left in position and energized sequentially with a wide variety of sources or energized with several sources simultaneously through the use of a cascade of fibre-optic couplers.

In general, the coherent interaction between the waveguide modes will mean that monomode fibres will be used here. Thus, the laser system may be used with a lens (usually a microscope objective) to couple the laser beam into the fibre core and the laser and coupling optics may be safely boxed. Such a procedure has the advantage that all stray beams are confined within the box and that the only laser hazard then comes from the fibre output. It has the additional advantage that if the fibre used is mono-mode at the laser wavelength, the beam is also conditioned by the fibre. All modes except the fundamental fibre mode leak into the fibre cladding and ultimately out of the fibre, thus, at the expense of the optical efficiency, a well-defined Gaussian mode can be obtained after mode stripping by even relatively short propagation lengths within the fibre. As a result, the output from a monomode fibre is a well-defined Gaussian mode irrespective of the quality of the laser beam itself. This represents a simple method to achieve a high-quality laser beam that may be coupled with precision into a laboratory experiment. The use of a mono-mode fibre is more effective, often has a lower cost and is easier to use than a conventional spatial filter based on a pinhole. Fibre-optic delivery systems also provide a mechanism for beam attenuation through the simple mechanism of deliberately misaligning the coupling optics at the fibre feed. A discussion of fibre-optic delivery systems is given in chapter C4.3.

Within limits imposed by transport efficiency and modes supported, the use of fibres and fibre-couplers offers a convenient method for combining several wavelengths of laser output in a single source that may be conveniently and easily positioned in both location and angle. The different frequencies in the output beam are then easily selected by using shutters at the points where the lasers are coupled to their individual fibres. Pre-packaged couplers for injecting laser beams into both mono-mode and multi-mode fibres are commercially available and many laser diodes are available with fibre pig-tails.

For situations where pure modal quality of the beam is less critical and some over-moded behaviour can be tolerated, the use of a pinhole filter offers an alternative to the use of fibres for spatial filtering and beam transport. The principal disadvantages associated with use of a pinhole are the bulk optics required to couple the laser light through the pinhole filter but the advantage can be more efficient use of the available laser output.

Applications using higher-power laser beams (such as material-processing applications discussed in section D1) are generally less critical in terms of the beam quality, although the beam M^2 (i.e. the beam quality; see chapter C4.1) can have a significant influence on the quality and efficiency of laser percussion drilling, for example (see chapter D1.4). Here the use of fibres with large core diameter can provide a useful mechanism for transport of the beam. Fibres with smaller cores (e.g. mono-mode fibres) may suffer physical damage due to the high input brightness over the small core area, as explained in chapter C4.3).

C1.2.4 Controlling the size and shape of the beam

To expand (or to reduce) the beam diameter, zoom lens expanders suitable for both manual and computer control can be obtained (see chapter C4.2). These systems can be capable of handling substantial power (500 MW cm^{-2} is quoted in the commercial literature) for operation over the 0.45–1.1 μm wavelength range. Such systems are of good optical quality (distortion less than $\lambda/4$) and offer either variable expansion or, with somewhat better optical performance, fixed expansion ratios. Designs using a combination of a negative lens and a positive lens may be preferred in high-power applications because the intense focused spot in a beam expander using two positive lenses can cause breakdown.

The beam shape (as opposed to diameter) can be controlled using linear adaptive optics, spatial light modulators of various descriptions [6] and diffractive optical elements (see section D9). Linear adaptive optics technology [4] for controlling light has been developed primarily for military and astronomical applications (see section C1.3.1 and chapter D8.1). It was declassified by the USAF in May 1991. Linear adaptive optics is also used in laser applications, where it may be confused with nonlinear optical effects, also described as 'adaptive optics' in the laser literature (e.g. [3]). Linear adaptive optics technologies are used for real-time measurement and correction of dynamic and stochastic aberrations due, for example, to turbulence in the earth's atmosphere or thermal blooming in the atmospheric propagation of high-power laser beams. Thermal blooming, caused by local heating of the atmosphere by a high-power beam, leads to defocusing and, in the presence of wind, causes the beam to swing upwind towards the higher density air. Used intracavity, an adaptive mirror can be used to modify the resonator properties to alter the output beam profile (usually to produce a super-Gaussian beam [1], i.e. one whose brightness profile matches a 'top-hat' more closely than the usual Gaussian beam modes). Used extracavity, adaptive optical wavefront modulators can produce variable focus control and/or small-angle beam re-direction, functions of considerable use in laser-beam materials processing. By changing the relative phase across a laser beam, adaptive mirrors (or other wavefront modulation devices) can be used to produce particular beam patterns in the diffracted beam.

C1.2.5 Safe disposal of unwanted beams

Stray beams are a safety hazard and 'beam dumps' are available to dispose of them. Such systems may range from a simple curtain screen used in a laboratory to confine low-power stray beams to a work area used only by personnel aware of the hazards, to the use of specially designed conical systems designed to ensure the minimum back leakage of any 'dumped' beam. The latter are particularly useful where beamsplitters are used to produce many beams, not all of which may be required during a particular experiment. For a discussion of laser safety, see chapter C6.

References

[1] Cherezova T Y, Chesnokov S S, Kaptsov L N and Kudryashov A V 1998 Super-Gaussian laser intensity output formation by means of adaptive optics *Opt. Commun.* **155** 99–106

[2] Montagu J and DeWeerd H 1996 *Optomechanical Scanning Applications, Techniques, and Devices (The Infrared and Electro-Optical Systems Handbook 3)* 2nd edn (Bellingham, PA: SPIE Optical Engineering Press)

[3] Pepper D M 1986 Applications of optical phase conjugation *Sci. Am.* **254** 56–65

[4] Tyson R K 2000 *Introduction to Adaptive Optics* (Bellingham, PA: SPIE)

[5] Yariv A and Yeh P 1984 *Optical Waves in Crystals* (New York: Wiley)

[6] Yu F T S and Khoo I C 1990 *Principles of Optical Engineering* (New York: Wiley)

Further reading

Davis C C 1996 *Lasers and Electro-Optics* (Cambridge: Cambridge University Press)

Written to be suitable for 'graduate students in electrical engineering and physics. It should also be useful to mechanical engineers or chemists who use lasers and electro-optic devices in their research'. Thorough, including much on crystal symmetries, tables of properties and many diagrams. Chapters 17 *Optical Fibres and Waveguides*, 18 *Optics of Anisotropic Media*, 19 *The Electro-Optic and Acousto-Optic Effects and Modulation of Light Beams* are particularly relevant here.

Smith S D 1995 *Optoelectronic Devices* (London: Prentice-Hall)

Written for 'graduate students in physics and engineering, as well as those working in research and in industry'. A balanced treatment with detailed comparisons of operational parameters for some specific (if now out of date) commercial devices. Chapter 7 *Fast optical modulators* is particularly relevant to this material.

Yariv A 1991 *Optical Electronics* 4th edn (Fort Worth, TX: Harcourt Brace Jovanovich College)

A standard text aimed at those 'interested in the generation and manipulation of optical radiation'. Contains much of interest on lasers. Chapter 9, *Electro-Optic Modulation of Laser Beams*, is particularly relevant to the material here. A thorough treatment with examples.

Yariv A and Yeh P 1984 *Optical Waves in Crystals* (New York: Wiley)

Written 'to present a clear physical picture of the propagation of laser radiation in various optical media and to teach the reader how to analyze and design electro-optical devices'. Chapters 4 *Electromagnetic Propagation in Anisotropic Media*, 7 *Electro-Optics*, 8 *Electro-Optic Devices*, 9 *Acousto-Optics* and 10 *Acousto-Optic Devices*, are particularly relevant to the material presented here. Well explained and thorough with examples and tables of material properties.

The LEOT Laser Tutorial (http://www.dewtronics.com/tutorials/lasers/leot/) is a good source of background material on many topics discussed here. Course 4 Module 7 *Electro-Optic and Acousto-Optic Devices* is particularly relevant.

There is a wealth of introductory and general background material available on-line in the web sites of suppliers of these forms of instrumentation. Anyone interested should examine the sites of all major suppliers (a few were noted in the text here) in order to get up-to-date information on the newer commercially available devices.

C1.3
Adaptive optics and phase conjugate reflectors

Michael J Damzen and Carl Paterson

Adaptive optics techniques, originally developed in the context of ground-based astronomy to compensate dynamically for optical aberrations arising from turbulence in the Earth's atmosphere, have found application in a number of aspects of laser system design. Intracavity adaptive optics allows fine-tuning of the laser cavity, for example, to compensate for thermal lensing effects and drift or to optimize efficiency automatically as operating parameters change. Extracavity adaptive optics systems are used widely in beam-shaping, beam-steering and temporal pulse shaping applications. The use of lasers in astronomy is discussed in chapter D8.

At the core of an adaptive optical system is an active element such as a mechanically deformable mirror having a number of externally controllable actuators that are used to alter the shape of its optical surface. An error signal derived from elsewhere in the system is used to control this adaptive mirror in a closed feedback loop. The error signal, which may in fact consist of a number of different parallel measurements in spatially complex systems, can be obtained from a wavefront sensor or by monitoring some other property of the laser output such as power.

Phase-conjugate reflectors differ significantly from conventional adaptive mirrors in that they do not have externally controllable actuators. Therefore, requiring no external error signal, their practical use in adaptive laser systems is markedly different from conventional adaptive mirrors and will be treated separately.

C1.3.1 Adaptive mirrors

A key component to an adaptive optical system is the adaptive mirror, for which a number of different technologies are available. The suitability of a given type of mirror for the application is determined primarily by the spatial properties of the deformations that it can produce and by its temporal response.

To understand the basic principles involved, consider first one of the simplest types of adaptive mirror, the continuous faceplate mirror. This consists of a single plate with piston actuators (such as piezoelectric stacks) attached at regularly spaced points across the back surface. Using the actuators, the front surface of the mirror can be deformed in a controllable fashion (figure C1.3.1(*a*)). Each actuator produces its own characteristic deformation to the mirror, referred to as an influence function, the amount of the deformation depending on the value of the control signal given to that actuator. Treating the mirror as a linear system, which, given the small deformations involved, is usually a very good assumption, for a mirror with N_{A} actuators, the total deformation of the mirror surface $\Phi_{\mathrm{M}}(x, y)$ is given by summing the deformations due to each actuator,

$$\Phi_{\mathrm{M}}(x, y) = \sum_{j=1}^{N_{\mathrm{A}}} r_j(x, y) c_j \tag{C1.3.1}$$

where $r_j(x, y)$ is the influence function of the jth actuator (with unit control signal) and c_j is the actual value of the control signal applied to that actuator. The influence functions determine how well the mirror can

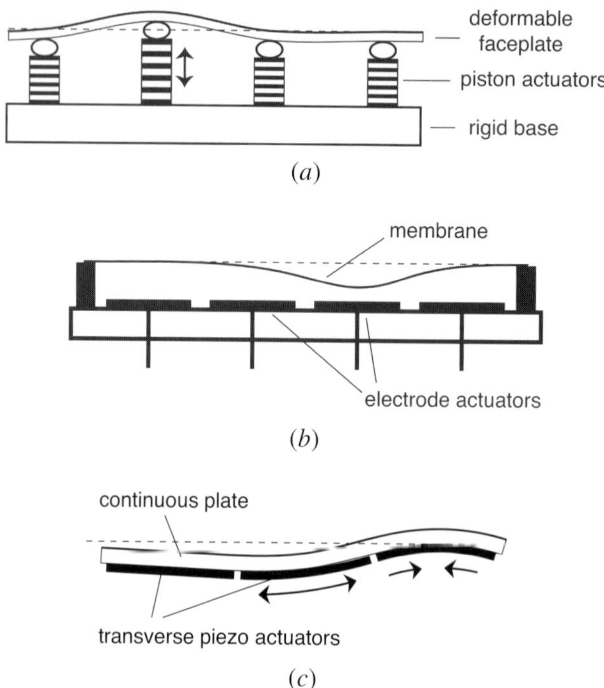

Figure C1.3.1. Deformable mirrors: (*a*) continuous faceplate; (*b*) electrostatic membrane and (*c*) bimorph.

compensate for wavefront errors. Their shape depends on a number of factors including the plate mechanics, the actuator positioning geometry and the flexibility of the actuator-to-mirror mountings. The type of actuator technology also affects the influence functions: piezoelectric stacks control the vertical displacement, whereas voice-coil-type actuators apply a controlled force. The displacement is then determined by boundary conditions and plate mechanics. The influence functions overlap and, in general, will not be orthogonal; however, they are normally linearly independent and an orthogonal set of N_{A} mirror modes can be constructed from them. Expanding in terms of a suitable orthonormal basis, such as Zernike polynomials, $Z_j(x, y)$, the deformation can be written in matrix form as

$$\boldsymbol{\Phi}_{\mathrm{M}} = \boldsymbol{M}\boldsymbol{c}. \tag{C1.3.2}$$

where $\boldsymbol{M}$ is called the influence matrix of the mirror and $\boldsymbol{c}$ is the control vector. Least-squares fitting can be used to calculate the control signals that should be applied to fit the mirror to the required shape,

$$\boldsymbol{c} = \boldsymbol{M}^{\dagger}\boldsymbol{\Phi}_0 = [\boldsymbol{M}\boldsymbol{M}^{\mathrm{T}}]^{-1}\boldsymbol{M}^{\mathrm{T}}\boldsymbol{\Phi}_0 \tag{C1.3.3}$$

where $\boldsymbol{M}^{\dagger}$ is the least-squares inverse of $\boldsymbol{M}$. The difference between the least squares best fit and the required deformation gives the fitting error, which is used to assess the suitability of a given mirror. For the continuous faceplate mirror with piezoelectric stack actuators, the influence functions are smooth and localized near each actuator. The number of actuators needed depends approximately on the space–bandwidth product of the deformations to be produced.

Also important are the dynamic range of the deformations, referred to as the stroke, and the temporal response. For the faceplate mirror, typical values are several micrometres for the stroke and around 1 kHz for the temporal response (limited by mechanical resonance of the device).

Other types of continuous devices include membrane and bimorph mirrors, which are both curvature-type mirrors. Electrostatic membrane mirrors consist of a thin membrane (e.g. silicon nitride) held under tension above an array of electrode pads (figure C1.3.1(*b*)) [3,11]. Applying a voltage between the membrane and one of the electrode pads sets up an electrostatic attractive pressure, deforming the membrane. The deformations obey the membrane equation

$$\nabla^2 z(x, y) = -P(x, y)/T = -\epsilon_0 V^2/Td^2 \tag{C1.3.4}$$

where z is the deformation, T the membrane tension, P the electrostatic pressure due to the voltage V across the gap d between the membrane and the electrode, with the boundary condition of $z(x, y) = 0$ at the fixed edges of the membrane. The influence functions are not localized at the actuators but extend across the whole of the mirror surface. As can be seen from equation (C1.3.4), each actuator controls the local curvature of the mirror above that actuator, for which reason these devices are referred to as curvature mirrors. A consequence of this behaviour is that the stroke is strongly dependent on the spatial frequency of the deformation being produced, the amplitude of the deformation varies as the inverse square of its spatial frequency.

Bimorph mirrors consist of a single plate with piezoelectric actuators bonded onto the back surface such that they can expand or contract laterally causing the plate to bend (figure C1.3.1(*c*)) [10]. More complex sandwiching designs can be used to improve the stroke or make the mirror less susceptible to differential thermal expansion effects. As for the membrane device, each actuator alters the local curvature of the mirror so the influence functions are similar in nature to those of the membrane mirror. Bimorph mirrors with water cooling channels have been used for high-power laser correction.

An alternative to the continuous mirrors are segmented devices. These use an array of separate rigid mirrors, each with its own actuators, which can be used to adjust the height (piston only) or both height and gradient (piston + tilt) of each mirror segment independently. The influence functions of this type of mirror are localized to individual segments and do not overlap. With this type of device, it is possible to introduce discontinuities in the mirror surface at the segment boundaries. This may be undesirable in many applications, requiring constraints to be placed on the mirror operation so that the edges of adjacent segments remain in phase, reducing the number of effective degrees of freedom of the mirror.

A number of different MEMS[1] type mirrors are available, both segmented and monolithic. Although their actuator technologies, which include electrostatic, electrostatic comb and thermal bimorph, may differ from those already described, they are similar in operation, having similar types of influence functions. By their nature MEMS devices are often considerably smaller, which may lead to more stringent limits to their optical power handling. Liquid crystal devices can be used as transmissive wavefront correctors, operating by changing the birefringent properties of a liquid crystal cell electrically. Pixellated devices can be modelled as analogous to segmented piston-only mirrors—the poor fitting error achievable with piston-only correctors being compensated by the large number of pixels available in liquid crystal devices. So called 'modal' devices which have more smoothly tapering influence functions are also available. The drawbacks of liquid crystal devices include the strong polarization dependence of the device and absorption.

C1.3.2 Wavefront sensors, reconstruction and control

To control the deformable mirror in a closed-loop system requires an error signal, which is the purpose of the wavefront sensor. The basic principle will be illustrated with a commonly used wavefront sensor, the Shack–Hartmann (figure C1.3.2). This consists of a lenslet array placed in the beam to be measured. Each lenslet produces its own focal spot, which for an on-axis collimated beam will lie on the optical axis of the lenslet. If, however, the beam is not on-axis and collimated, the position of each focus will be shifted laterally by an amount that depends on the local tilt of the wavefront entering the corresponding lenslet. By measuring

[1] Micro-electromechanical system.

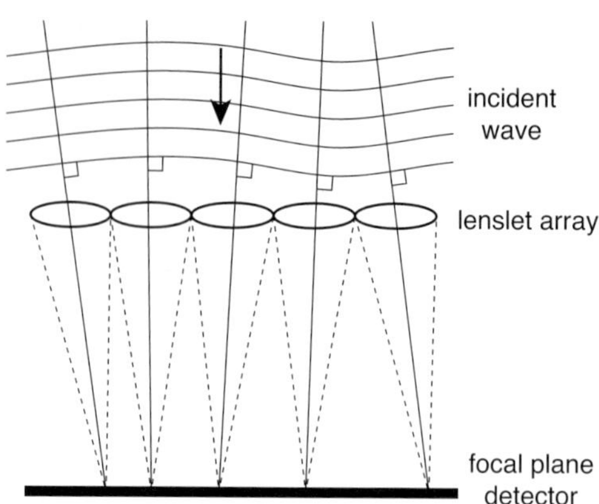

Figure C1.3.2. Shack–Hartmann wavefront sensor.

the transverse displacement of the foci, one can obtain samples of the local gradient of the wavefront at each lenslet. Modelling the wavefront sensor as a linear system, the response of the sensor to an arbitrary input wavefront $\boldsymbol{\phi}_0$ is given by

$$\boldsymbol{s} = \boldsymbol{S}\boldsymbol{\phi}_0. \tag{C1.3.5}$$

$\boldsymbol{S}$ is the wavefront sensor response matrix, the elements of which are calculated from the geometry of the sensor [5]. An estimate of the original wavefront can then be reconstructed from these samples using, for example, a least-squares inverse of $\boldsymbol{S}$ or optimal estimation [12]. In practice, it is possible to dispense with explicit wavefront reconstruction and estimate the required mirror control signals $\boldsymbol{c}$ directly from the sensor output,

$$\boldsymbol{c} = [\boldsymbol{SM}]^{\dagger}\boldsymbol{s} \tag{C1.3.6}$$

For a closed-loop system, this has the advantage that the matrix product $\boldsymbol{SM}$ can be measured directly by varying each mirror actuator control in turn and observing the effect on the wavefront sensor output [7]. Since the wavefront sensor can only sense a finite number of wavefront modes, which for a Shack–Hartmann is twice the number of lenslets, the wavefront cannot be reconstructed completely from the sensor measurements. It is important, therefore, to choose the wavefront sensor geometry with regard to the mirror geometry to ensure that the deformations of the mirror are measurable [9].

There are many alternative wavefront sensor techniques, including wavefront curvature sensing, knife-edge detectors and interferometric techniques: the reader is referred to Hardy or Tyson in further reading for an overview. For some applications, it is possible to dispense with the wavefront sensor. Direct optimization methods rely on monitoring some property of the output (usually the on-axis intensity) while continually adjusting the mirror actuator controls. If a change makes the system worse, the direction of subsequent changes is altered. Optimization methods can work well for systems with small numbers of degrees of freedom; however, they tend to be significantly slower for more spatially complex cases, where they tend to require many iterations to converge. Modulating or dithering the corrective element at a higher frequency than that of the aberrations combined with synchronous detection of the output can be used to lock the system onto a maximum. Multi-dither approaches, where each actuator is modulated at a different frequency, have been used successfully for laser beam correction through turbulence. The number of channels is limited by the finite bandwidth of the mirror and the necessary separation of the dither frequencies.

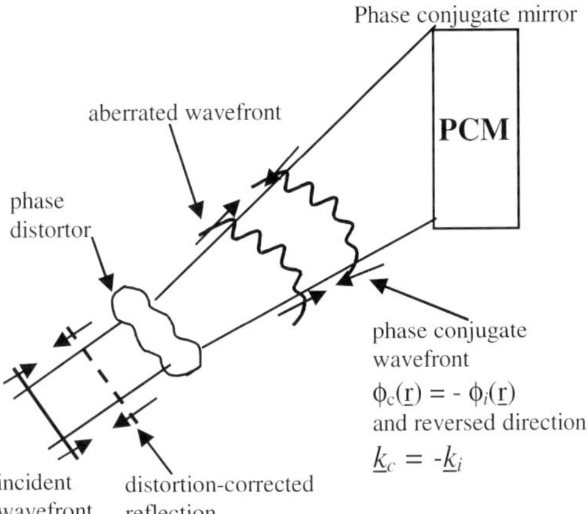

Figure C1.3.3. Phase conjugation as a means for distortion correction.

C1.3.3 Nonlinear optical phase conjugation

Nonlinear optical phase conjugation was first demonstrated in the early 1970s [13]. In its most common form, it is a technique that uses nonlinear optical processes to reverse precisely both the direction of propagation and phase variation of a coherent light wave. Phase conjugation produces true wavefront conjugation in that it preserves both the amplitude distribution while conjugating the phase distribution. It has the advantage over conventional adaptive optics, which relies on discrete active segmented control elements, of being a passive continuous wavefront reversal mechanism and without the requirement for mechanical, magnetic or electro-optical components or software processing. It, therefore, has a potential advantage in terms of resolution, speed, cost, weight and energy efficiency. There are, however, requirements on the input radiation, such as power and temporal coherence for specific nonlinear phase-conjugation techniques that limit its use for certain applications (e.g. astronomy) but it can be used to advantage in laser applications.

The nonlinear device that 'reflects' the input beam to produce the phase conjugate is often referred to as the phase-conjugate mirror. The nonlinear mirror is a real-time processor of the electromagnetic optical field and can respond to the complex amplitude distribution and phase front of the incident field and adapt its reflective properties. Much of the fascination of this technique arises from its ability to restore an aberrated optical beam or image-bearing wavefront to its original, undistorted state (figure C1.3.3). A practical example of this is correction of phase aberrations in a laser amplifier. A high spatial quality (fundamental Gaussian mode) of a laser oscillator is degraded on a single pass through the amplifier by aberrations (e.g. thermally-induced refractive index distortions in a solid state laser rod due to intense pumping by the inversion mechanism). The double pass with a phase-conjugate mirror reflection recovers the high spatial quality of the incident beam, as well as receiving a double-pass amplification in power by the amplifier. A polarization combination (polarizer and quarter-wave retardation plate or Faraday rotator) can act as a simple means of extracting the amplified radiation and isolating the low-power laser oscillator. This was the first application used for phase conjugation [6] and the phase-conjugate mirror was based on the process of stimulated Brillouin scattering. For an introduction to nonlinear optics, see chapter A4.

A surprisingly large number of physical processes in nonlinear optics have the ability to generate a phase conjugate signal [1, 2, 4, 8, 14, 15]. The main mechanisms, illustrated in figure C1.3.4, include:

(i) four-wave mixing (FWM),

(ii) stimulated scattering (especially stimulated Brillouin scattering (SBS)),

(iii) photorefraction and

(iv) self-intersecting loop schemes.

The two most generic phase-conjugating processes are four-wave mixing (FWM) and stimulated Brillouin scattering (SBS). FWM is a real-time holographic process that has the attraction that the beam to be conjugated can have low input power while SBS normally requires a high power input and is especially attractive for high-power pulsed laser applications. FWM is the most versatile of the effects and also plays a major role in the phase-conjugation processes that occur in photorefraction and in self-intersecting loop schemes [1,4]. Other mechanisms also exist but have had more limited applicability. Three-wave mixing in a material with a second-order nonlinear susceptibility has been shown, in principle, to produce phase conjugation, but has a very limited phase-matching bandwidth and this makes it impractical for beams with strong aberrations or divergent wavefronts. Another technique for phase conjugation is surface waves generated at the interface of a material from which the light reflects [2]. In this case it is the counterpropagating surface waves excited by the incident light that produce the phase conjugation.

C1.3.3.1 Four-wave mixing (FWM) [2, 8, 14, 15]

The geometry of the FWM process is shown in figure C1.3.4(*a*) where three waves (E_1, E_2 and E_3) are incident onto a material with a third-order nonlinear susceptibility $\chi^{(3)}$. Different nomenclature is used in the scientific literature for the numbering of the fields. Here E_1 and E_2 are counterpropagating and described as the pump fields and E_3 is the signal field to be phase conjugated. In the nonlinear medium all three waves are coupled in such a manner that a new optical field, the conjugate wave E_4, is generated with a magnitude given by

$$E_4(\boldsymbol{r}) = \left[\mathrm{i}\frac{\mu_0 c\omega}{2n}\chi^{(3)} L E_1 E_2\right] E_3^*(\boldsymbol{r}) = r_C E_3^*(\boldsymbol{r}) \tag{C1.3.7}$$

where $\boldsymbol{r}$ is the transverse spatial coordinate, L is interaction length, $\chi^{(3)}$ is the third-order nonlinear susceptibility of the medium, μ_0 is the permeability of free space, c is the speed of light in vacuum, n is the refractive index and ω is the angular frequency of the optical light. The term r_C is an effective amplitude reflectivity of the phase-conjugate mirror and is dependent on the material parameters, length of the interaction region and strength of pump fields E_1 and E_2.

If the pump fields are constant, e.g. uniform plane-waves, the generated field E_4 is seen to be the complex conjugate of the field E_3 with the phase conjugate wavefront. It is possible to control the conjugate reflectivity $R_C = |r_C|^2$ by varying the intensity of the pump beams and it is even possible to achieve amplified reflectivity $R_C > 1$, since the energy of the conjugate beam is derived from the pump beams. Various types of $\chi^{(3)}$ nonlinearities can be exploited by the use of different materials and physical processes. The highest reflectivities produced to date have been greater than one million (10^6) using Brillouin enhanced FWM with pulsed radiation. The highest reflectivity for continuous-wave radiation has been in saturable gain FWM in a laser amplifier medium, where the radiation is resonant with the laser transition of the medium. High reflectivity can be used as a means of optical image amplification or for boosting weak signals. Many of the FWM processes are non-resonant and provide broad tunability in wavelength often covering the total visible spectrum and even near-UV and IR.

FWM can be thought of as 'real-time' holography—in FWM the writing and reading process can be virtually simultaneous and hence dynamic (e.g. atmospheric turbulence in communications) as well as static aberrations (e.g. in passive optical systems) can be compensated. In the case of transient processes, the response time may be a function of intensity of the writing beams and, in the case of photorefraction, the response time is inversely proportional to intensity.

In its fullest form FWM can be considered as an optical multiplication since the conjugate beam is proportional to the product of the three input fields $E_4 \propto E_1 E_2 E_3^*$. An all-optical processor can be implemented to produce convolution and correlation of images using the Fourier transforming property of lenses and the multiplicative property of FWM.

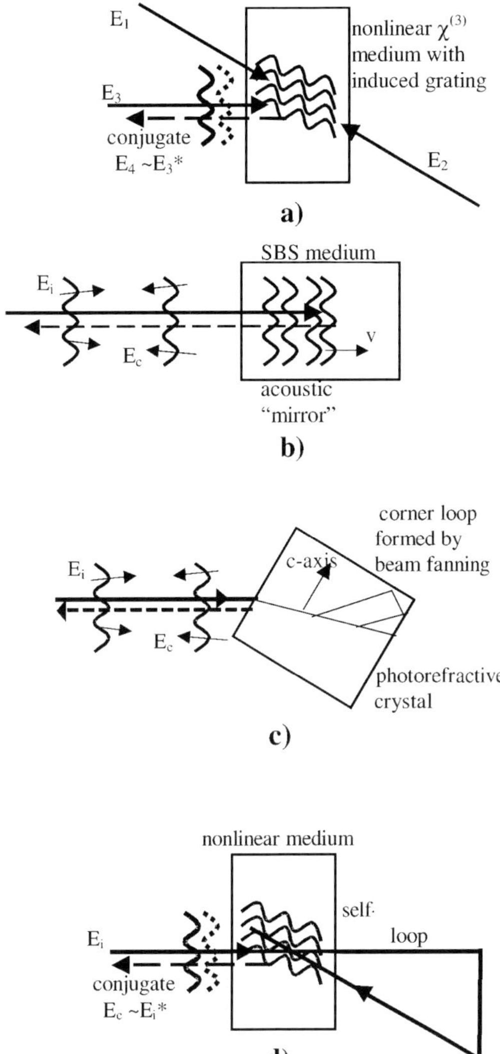

Figure C1.3.4. Nonlinear phase conjugate reflectors include (*a*) four-wave mixing; (*b*) stimulated Brillouin scattering; (*c*) photorefraction in corner-pumped geometry and (*d*) self-intersecting loop conjugators.

Exploitation can even be made of the finite response time of the phase conjugator to produce time-averaged interferometry effects and novelty filters. The novelty filter is effectively a movement detector of speeds faster than the response speed of the phase-conjugating device. In its basic form it is a type of Michelson interferometer with phase-conjugate mirrors in each of the beamsplitter arms. For a static input image, perfect cancellation (destructive interference) results when the two conjugate beams recombine, resulting in zero signal at the output arm. Any movement in the image can be made visible at the output arm if the two conjugators have different speeds. Such a device has been considered for application in surveillance or robotic inspection. Since only the moving object is detected in a large scene, only a small and convenient amount of signal processing is required.

C1.3.3.2 Stimulated Brillouin scattering (SBS) [2, 6, 8, 13–15]

SBS is an alternative method of producing phase conjugation resulting from inelastic nonlinear scattering of high-power radiation from coherent acoustic waves generated by electrostriction in the medium due to the intense optical field (figure C1.3.4(*b*)). A backscattered field is produced which, under many experimental conditions, is found to be the phase conjugate of the incident field. Analysis indicates that when the backscattered field is the phase conjugate of the incident laser field, the growth experienced by the scattered field is greater than any other backscattered field configuration. Physically, one can visualize the acoustic wavefront forming a shape that reproduces the wavefront of the incident light wave. This concept of a flexible mirror has direct analogy to conventional adaptive optics but forming by passive nonlinear optical means. The attraction of SBS compared to FWM is its self-conjugating property, requiring no high-quality additional pump beams.

Typical interaction geometries include focusing into the bulk of a cell containing the Brillouin medium or by launching into a hollow light-pipe filled with a Brillouin-active liquid (e.g. CS_2, n-hexane, acetone) or gas (e.g. CH_4, SF_6). Solids have been less frequently used, since their damage thresholds tend to be comparable to the thresholds for efficient SBS. More recently, there has been renewed interest in solid materials in the form of multi-mode optical fibre whose long interaction length allows reduced power thresholds for SBS as low as $\sim$1 W. In bulk geometries, very high intensities are required for SBS and this means it is most suitable for phase conjugation of high-power pulsed laser radiation. A major motivation of SBS phase conjugation has been directed towards the improvement of beam quality in high-power (solid state, excimer and dye) laser systems. At high-energy and/or high-average-power operation the laser gain medium becomes increasingly distorted by thermally induced refractive index distortions and lensing in solid state and dye lasers and by discharge instabilities in gas excimer lasers. Phase-conjugation is a means to correct for these static and dynamic aberrations.

Other forms of nonlinear inelastic scattering can also lead to phase-conjugation with the phase conjugate wave being selected by a higher growth rate than non-conjugate forms of scattering. The frequency shift of the SBS process determined by the acoustic frequency (acoustic phonon) is typically $\sim$0.1–10 GHz; however, the shift in backward stimulated Raman scattering (SRS) process due to the vibrational transition of the molecular system ('the optical phonon') can be as high as 10% of the optical frequency. The selection of the phase conjugate can be impacted by the difference in diffraction at the shifted frequency leading to an error in correction of aberrations when double passing the distorting system.

C1.3.3.3 Photorefraction [4]

The photorefractive effect is a change in the refractive index due to the optical transfer of charge in the medium leading to change in refractive index. The effect was first observed in lithium niobate ($LiNbO_3$) and was a detrimental effect due to its change in the propagation properties of the light. The basic mechanism is due to the photoexcitation of electrons (or holes) from donor (or acceptor) sites to the conduction (or valence) band, where they are free to migrate by drift or diffusion before recombining into an empty trap site. The dominant rate of charge photoexcitation is in the bright intensity regions with preferential trapping in dark regions. Hence, with an intensity interference pattern the charge is redistributed to dark fringes. The imbalance of charge leads to a space-charge electric field E_{SC} that modulates the refractive index Δn through the electro-optic (Pockels) effect $\Delta n = 1/2n^3 r_{eff} E_{SC}$ where n is the refractive index and r_{eff} is the effective electro-optic coefficient. Hence good choices of photorefractive media exhibit high photoconductivity and have high electro-optic coefficient; examples include barium titanate ($BaTiO_3$) bismuth silicon oxide (BSO) and lithium niobate ($LiNbO_3$). More recently, interest has centred on photorefractive polymer materials where the composite polymer material can be engineered to have particular donor and trapping species but they tend to require high externally-applied electric fields and have high absorption.

An interesting effect known as two-beam coupling occurs when two beams intersect in the photorefractive medium. The non-local displacement of the charge induces a refractive index grating that is $\pi/2$ phase shifted from the light interference pattern (assuming diffusion of charge). This leads to a self-diffraction where one beam is amplified at the expense of the other beam. The direction of energy transfer depends on the sign of the electro-optic coefficient, the orientation of the crystal (that needs to be poled) and the sign of the charge carriers. In highly nonlinear crystals such as barium titanate, the gain of one beam can be by orders of magnitude and, indeed, so high that internal linear scattered light in the crystal can be strongly amplified, a phenomenon known as beam fanning, leading to depletion of the incident light. A new type of phase-conjugate mirror has been demonstrated in this case where the beam fanning is directed to a corner of the crystal acting as a retroreflector (figure C1.3.4(*c*)). A loop of light is observed to evolve and it acts as a counterpropagating pump beam to phase conjugate the input light beam. A phase-conjugate output with reflectivity of 60% or more has been demonstrated. The exact mechanism of the conjugation is rather complex but the internal fanning and beam development is self-organizing and has a preferential tendency to produce phase conjugation, which is the condition for highest spatial overlap and mutual coherence of beams. Like SBS this process is self-conjugating without the requirement of external pump beams (as in FWM). The light required for phase conjugation in photorefractive crystals can be typically at the milliwatt level (or even lower) but the response time of the effect can be very slow. For barium titanate, the response is $\sim$1 s at 1 W cm^{-2} intensity but speed is faster with higher intensity. For other crystals, e.g. BSO, the response is faster ($\sim$1 ms) but the nonlinearity and phase conjugation effects are weaker.

C1.3.3.4 Self-intersecting loop conjugators [1]

Any mechanism that can produce phase conjugation by FWM has the potential to produce a self-intersecting loop conjugator without requiring external pump beams. This geometry was first operated in photorefractive media where the incident beam, after transmission through the crystal, is re-injected to overlap with itself in the crystal (figure C1.3.4(*d*)). In this case, it is the two-beam coupling initiated by beam fanning that causes a build-up of a backward phase-conjugate wave. Reflectivities of many tens of percent can be achieved and the threshold power is generally lower than that required for the corner-pumped phase conjugation. More recently, a number of demonstrations of very high reflectivity loop conjugators have been made with saturable gain as the nonlinear mechanism. Indeed, the self-intersecting input beam can induce a gain grating with sufficient diffraction efficiency to induce a ring laser oscillation in which the lowest-order backward spatial mode is the phase conjugate of the input. Reflectivities approaching 10^5 have been produced by this technique.

References

[1] Damzen M and Crofts G 2002 Wave-mixing and phase conjugation in laser active media *Progress in Photorefractive Nonlinear Optics* ed K Kuroda (London: Taylor and Francis) ch 7

[2] Fisher R A (ed) 1983 *Optical Phase Conjugation* (New York: Academic)

[3] Grosso R P and Yellin M 1977 The membrane mirror as an adaptive optical element *J. Opt. Soc. Am.* **67** 399–406

[4] Gunter P and Huignard J P (ed) 1989 *Photorefractive Materials and their Applications* vols I and II (Berlin: Springer)

[5] Hardy J W 1998 *Adaptive Optics for Astronomical Telescopes* (Oxford: Oxford University Press)

[6] Nosach *et al* D 1972 Cancellation of phase distortions in an amplifying medium with a Brillouin mirror *Sov. Phys.–JETP* **16** 435

[7] Paterson C, Munro I and Dainty J C 2000 A low cost adaptive optics system using a membrane mirror *Opt. Express* **6** 175–85

[8] Pepper D M 1981 Applications of optical phase conjugation *Sci. Am.* **January** 56

[9] Roddier F 1994 The problematic of adaptive optics design *Adaptive Optics for Astronomy* ed D M Alloin and J M Mariotti (Dordrecht: Kluwer Academic) pp 89–111

[10] Steinhaus E and Lipson S G 1979 Bimorph piezoelectric flexible mirror *J. Opt. Soc. Am.* **69** 478–81

[11] Vdovin G and Sarro P M 1995 Flexible mirror micromachined in silicon *Appl. Opt.* **34** 2968–72

[12] Wallner E P 1983 Optimal wave-front correction using slope measurements *J. Opt. Soc. Am.* **73** 1771

[13] Ya Zel'dovich B, Pilipetsky N F and Shkunov V V 1985 *Principles of Phase Conjugation (Springer Optical Sciences 42)* (Berlin: Springer)

[14] Ya Zel'dovich B *et al* 1972 Connection between the wave fronts of the reflected and exciting light in stimulated Mandel'shtam–Brillouin scattering *Sov. Phys.–JETP* **15** 109
[15] Yariv A 1978 Phase conjugate optics and real-time holography *IEEE J. Quantum Electron.* **QE-14** 650

Further reading

Tyson R K 1998 *Principles of Adaptive Optics* 2nd edn (New York: Academic)

An overview of available technologies with an extensive bibliography.

Hardy J W 1998 *Adaptive Optics for Astronomical Telescopes* (Oxford: Oxford University Press)

A comprehensive and detailed account of adaptive optics technologies and methods: although written for astronomy, most of the techniques are directly applicable to lasers.

Love G D (ed) 2000 *Proc. 2nd Int. Workshop on Adaptive Optics for Industry and Medicine* (Singapore: World Scientific)

Contains many examples of adaptive optics applied to lasers.

Fisher R A (ed) 1983 *Optical Phase Conjugation* (New York: Academic)

C1.4
Opto-mechanical parts

Frank Luecke

C1.4.1 Introduction

Art is the triumph over chaos (John Cheever)

The art of opto-mechanical design is to

- put optical surfaces and elements into their proper—not necessarily static—positions;
- keep them there;
- protect them and their surroundings from distortion, damage and stray light; and, in some cases,
- quantify their positions.

Further, it is usually necessary to achieve all this within cost, weight, size, time and numerous other constraints.

To accomplish these objectives requires

- understanding the overall system requirements,
- selecting appropriate materials,
- determining the configuration of the parts,
- anticipating a process for building the product at each phase from feasibility to maximum production and
- testing to assure the objectives are met.

Ideally, the designer pursues a global optimum but reality usually forces compromise among oft-conflicting goals. The design process is seldom simple or linear; most often it is highly iterative as well as interactive with other parts of the bigger picture.

The objective of this chapter is to present some thoughts, information and further resources for creating an opto-mechanical design effectively and efficiently.

C1.4.2 Requirements and specifications

A good design starts with understanding the functional requirements, environmental context and administrative details of the proposed system as completely and accurately as possible. Although it may seem obvious, the importance of fully understanding what a system will be required to do as one starts to design that system cannot be over-emphasized.

Functional requirements:

- What is this device supposed to do?
- How well must it perform: speed, resolution, repeatability, accuracy?
- How long must it continue to operate: operational cycles, number of years?

- How big or small can or should it be?
- How loud or quiet can or should it be?
- What mass and inertia are allowable?
- In what ways are aesthetics important? (Consider all five senses!)
- Which attribute is best to optimize and which attributes are bounds?
- What is the history of the technology? What are the nearest examples presently in existence? In what ways would they be acceptable for the current project and in what ways not? Can you examine, test and dissect some examples?
- Who will use it? Install it? Service it? What skill level will apply? What operator and service interfaces are needed?

Environmental context:

- Where will it be used?
- What conditions of temperature, humidity, pressure/vacuum, vibration/shock, chemical, biological, particulates, ambient light, orientation etc must it withstand during shipment and during operation?
- Are there any EMI, RFI or electrical or magnetic fields considerations?
- Are there any other special materials requirements?
- What energy and supplies are available for input and desired or tolerable for output?
- Are there any safety or health considerations?

Administrative details:

- Are any regulations or certifications applicable to the device, e.g. laser safety, European Compliance (CE), Underwriters' Laboratories Inc (UL)?
- What is the schedule for feasibility, development and production?
- How many units are likely to be needed and when?
- Who is responsible for the rest of the system: optics, electronics, manufacturing?
- What resources are available—funding, skills and facilities—for concept, prototype and production?

Once the functional requirements, environmental context and administrative details of the proposed system are identified and understood as completely and accurately as possible, the creation and refinement of the opto-mechanical design can proceed through the usual steps of layout, analysis, prototyping, testing and preparation for production.

C1.4.3 System considerations

C1.4.3.1 Position description

Many systems consist of an assembly of essentially rigid optical elements that need to be held in a particular spatial relationship with each other. The position of a rigid body with respect to an external frame of reference can be described by six degrees of freedom (DOFs): in a Cartesian coordinate system, these are the X, Y, Z location of a given point on the element and the rotational orientation α_x, α_y, α_z of the element with respect to the X, Y and Z axes. The origin of the coordinate system, i.e. the external frame of reference, can be set at any convenient location; typically the Z-axis is coincident with the optical axis of a linear system. A folded optical path may be described by using an axis-aligned coordinate system for each section with coordinate transforms at each fold. An external ground reference may be used. A polar or spherical coordinate system may be more suitable for some systems.

The term 'kinematic mounting' refers to an approach that provides exactly one constraint for each DOF. This concept was clearly described by Clerk Maxwell in 1876 [1]. Figure C1.4.1 shows under-constrained,

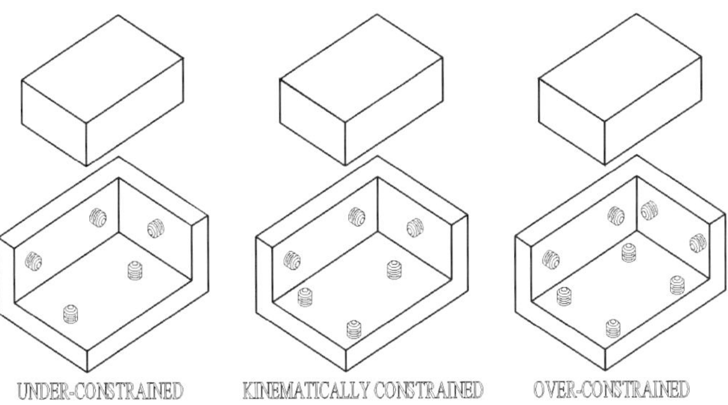

Figure C1.4.1.

Table C1.4.1. Error budget example.

	X (mm)	Y (mm)	Z (mm)	α_X (deg)	α_Y (deg)	α_Z (deg)
Lens	0.2	0.2	0.1	1	1	∞
Shutter	1	1	0.5	5	5	∞
Film	0.3	0.2	0.1	1	1	2

kinematic and over-constrained assemblies. The block, shown separated for clarity, is assumed to be uniformly urged against all the supporting constraints by some method not shown. Many other examples of kinematic mounting appear in the literature [2–4].

C1.4.3.2 Mounting accuracy

Each element in an optical system must be in its correct position within some tolerance for the system to function properly. Table C1.4.1 is an example of an error budget for a fictitious simple camera showing the mounting accuracy required for each optical element. Approximately the same level of performance degradation would result from mis-positioning any element in any one direction by the amount shown. Many programs used to design lenses can supplement judgement and testing in quantifying the positioning tolerances. Tolerances for some DOFs may be unimportant (e.g. the rotation of a circularly symmetric lens about its optic axis), while others will be critical. The inter-relationship between a few elements in a sub-group may be more critical than the alignment of the group to the rest of the system; in such a case it helps to add a separate error budget table for the sub-group. Usually a system need not perform to specification with every element simultaneously at the worst extreme of misalignment; an error probability distribution may apply [4].

C1.4.3.3 Mounting techniques

The methods for placing and holding optical elements fall into three categories:

- deterministic, in which all DOFs of each optical element are adequately constrained by features on the opto-mechanical parts (this category includes innumerable configurations for mounting lenses, prisms, mirrors and other elements as shown in [3, 5, 6].

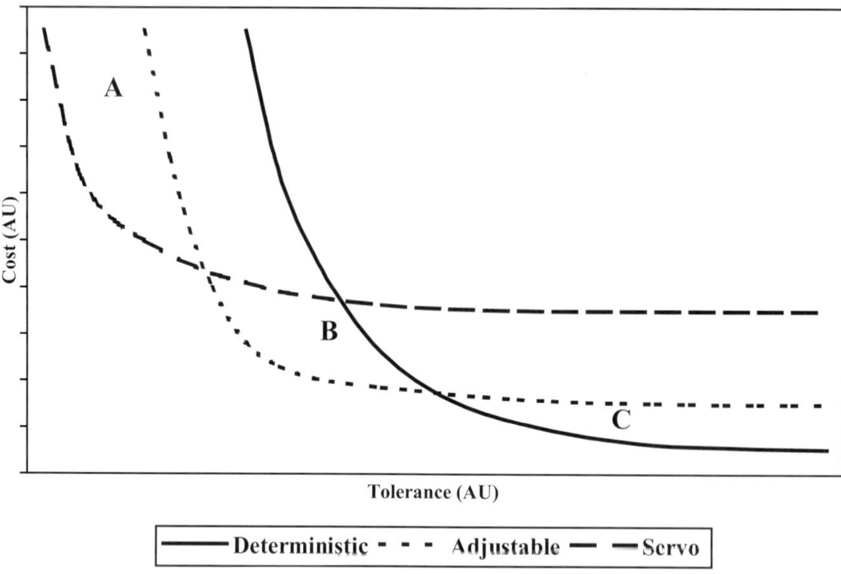

Figure C1.4.2. Cost of holding tolerances.

- adjustable, if deterministic mounting cannot feasibly provide sufficient accuracy; and finally
- servo-actuated, in which elements are continuously and automatically positioned as needed. Such systems are described in more detail in section C1.4.6.3.

The relationship between the cost and achievable accuracy of each of these approaches is shown in figure C1.4.2. The cost of holding a tolerance increases as the tolerance is reduced. In some cases it can be more economical to use a more complex approach, migrating from adjustment to servo (region A) or from deterministic mounting to adjustment (region B). However, including an adjustment or servo system in a product that does not require it (region C) would be a needless complication and a waste of money.

There are hundreds of approaches to holding lenses, mirrors, prisms, windows and other optical elements in place. For a laboratory or prototype system, devices available from suppliers such as those appearing in optical magazines and directories (e.g. *Photonics Directory*) may suffice. For more rigorous requirements and for systems to be produced in quantity, the references show many examples of specific opto-mechanical design elements [1, 3, 4, 6, 7]. Many more can be found by searching through patents and other literature. Dissecting available examples of similar devices can also be instructive.

C1.4.3.4 Optimization

Optimization as a specific discipline is fundamental to the design of modern lens systems as well as many other devices. This discipline can also be used to advantage in certain aspects of opto-mechanical element design [5]. Although literal use of this tool may not be feasible when designing overall opto-mechanical systems due to the difficulty in creating and evaluating a mathematically tractable merit function, it still is helpful to think in optimization terms; e.g. 'Create a gizmo that is as lightweight as possible, while holding performance, cost, size and all other important variables within acceptable limits'.

C1.4.3.5 Design for manufacturability

A design has little value unless it is feasible to actually make and assemble the parts. The processes chosen are influenced by the quantity and quality of each part needed. Basic model-shop equipment including manual mills and lathes may suffice for making one or two copies of relatively simple parts. Rapid prototyping can produce visually representative as well as in some cases completely functional complex parts [8]. Computer-numerically-controlled (CNC) machining can economically produce small to medium quantities of parts, particularly those with complex but machineable shapes or large numbers of features. Production of high volumes must include consideration of other methods such as molding to final or near-finished condition. The references cited in section C1.4.5 describing the characteristics of a variety of materials also describe many parts-forming processes applicable to different materials.

The cost of producing products varies approximately with the negative root of the quantity to be produced: e.g. per-part cost may decline about tenfold when quantities increase 100 fold. This rule of thumb is far from exact or continuous, however, and the discontinuity seems greater in opto-mechanical devices than in many other fields: many optical products resemble prototypes even in volume production.

The considerations for part production apply to assembly and alignment as well. Although it may be appropriate for a PhD to spend hour after hour putting together a first article and coaxing it to life, this probably is not the best way to produce the 10 000th or even the 100th unit. A good design will eliminate as many assembly and alignment steps as possible and will make the remaining steps as easy as possible for a modestly skilled operator, robot or specialized machine to perform correctly.

When adjustments must be performed during assembly, it is better to have excellent manufacturing equipment perform the adjustment on the bench rather than building adjustment devices into each product.

C1.4.3.6 Testing

It may be tempting to minimize development costs by omitting rigorous testing, letting customers discover if the product will work satisfactorily for a reasonable lifetime under the conditions to which it may be subjected. This could be acceptable for very limited production quantities as long as customers knowingly accept the risk. In the long run, though, it is better to verify that the product continues to meet its performance objectives, as outlined in section C1.4.2, when subjected to the various environmental insults also listed in that section. For example, a pair of binoculars might need to maintain alignment, throughput and modulation transfer function (MTF) after being dropped, sprayed with salt water, frozen, irradiated and vibrated. Of course, testing *must* be done in those cases requiring certification by regulatory agencies or industry standards organizations. Numerous suppliers can conduct the needed tests if facilities are not available in-house.

C1.4.4 Materials and finishes

Although black anodized 6061-T6 aluminium may be the most common material and finish now used to build opto-mechanical parts, innumerable other candidates are worthy of consideration [9, 10]. The best choice depends on the combination of functional and esthetic properties required. One very useful approach is to rank the candidate materials using a figure of merit based on the most important functional properties for the system [7]. One example of such a figure of merit is the stiffness-to-density ratio, which can affect the weight and performance of a simple system as well as the bandwidth of a complex, servo-actuated system because of the dependence of the acoustic propagation velocity, as mentioned in section C1.4.6. The most appropriate figure of merit will depend on the specific system requirements and may involve a combination of thermal coefficient of expansion, electrical conductivity, chemical resistance and other factors. The materials palette includes many types of aluminium [11]; other non-ferrous as well as ferrous metals; plastics, elastomers and composites [12]; and structural ceramics [13]. Similarly, many different finishes including paint, powder coating, anodizing, passivation and plating are available to meet the functional as well as aesthetic requirements of the product.

C1.4.5 Parts configuration

C1.4.5.1 Visualization

Since the basic objective of opto-mechanical design is to hold optical elements in their proper positions, a good starting point for creating opto-mechanical parts is to first imagine the complete set of optical elements suspended in space, free of any constraints other than the mutual spatial relationship among the elements (and sometimes between the collection of elements and ground) required for acceptable performance. Ignore for a while the bits and pieces of hardware needed to hold the elements and just picture where the elements need to be. Once that image is clear, add to it the voids and barriers needed to accommodate the light paths. Only then, after contemplating this picture from all angles, use the remaining space as needed for parts to hold and move the optical elements.

A number of companies produce 3D mechanical computer-aided design (CAD) programs that can help in visualization and the more mundane processes of laying out, designing and documenting the mechanical parts. Each of the programs has some merits as well as some shortcomings. Unless the software already standard at your institution restricts the choice of CAD system, there is a daunting array of programs and features to choose from. There may be add-ons and interfaces to ease the process of combining mechanical and optical elements. Use caution in employing the ray-tracing functions of many general-purpose mechanical design programs, because they may be more oriented toward presenting a pleasing image than accurately portraying optical system performance.

C1.4.5.2 Distortion, stress and strain

No thing is perfectly rigid; all real objects are elastic. A force applied to an element will distort both the element providing the force and the one receiving it. Forces will be exerted by gravity, acceleration, vibration, mounting devices and thermal effects, among other agents.

A force applied to any optical element will distort its optical surfaces and will also change the bulk properties of transmissive elements. One example of surface effect is the gravity-induced change in figure of the primary mirror in a large telescope as it moves to view different parts of the sky; this also exemplifies situations in which the structural properties of the optical element itself must be considered. Volume effects include the changes in birefringence of a plastic block subjected to non-uniform external forces, a phenomenon exploited for visualizing stresses in models of structures.

The mechanical elements that are used to hold optics also distort under stress, moving the optical parts attached to them. Such movement must be considered in meeting the overall error budget.

Stress beyond the elastic limit will result in permanent deformation or breakage. It is best to avoid that level of stress in the parts as well as in the personnel on the project. The challenge in creating an acceptable design from a distortion viewpoint, therefore, lies in keeping the stresses and resultant strains within acceptable limits for both the optical and mechanical elements. For simple systems, basic static stress–strain analysis may be more than adequate. For more complex systems or situations, finite-element analysis (FEA) [4,5] can be useful for predicting the effects of applied loads, showing the magnitude of stress at various parts of the element and quantifying the resultant deflections. A skilled practitioner using FEA can help move a design toward optimal or acceptable much faster than would be possible by a 'cut and try' approach alone.

Most opto-mechanical systems will need to function properly over a range of temperatures. Most optical elements change size and shape with changes in temperature. The refractive index of most transmissive optical elements also changes with temperature. Both the sign and magnitude of these changes are dependent on the optical material. Most materials used to support optics expand with increasing temperature. The challenge then is to create a combination of materials and configuration in the support system that closely counteracts the change in optical properties. Many examples exist [5].

Vibration can also perturb an optical system. A system can be partially isolated from externally generated vibrations [14] and modifying the structural stiffness and damping within the system can make it less susceptible to the effects of vibrations from which it cannot readily be isolated. Since a system's natural frequency increases with increasing stiffness and decreasing mass, the figure of merit of stiffness to density ratio is applicable, although not necessarily in a simple way.

C1.4.6 Precision positioning

All optical elements of a system must be positioned to some level of precision. In many cases, deterministic mounting or one-time adjustments provide sufficient accuracy to meet the system specifications as outlined earlier. When they do not or parts of the system need to track moving external objects, a precision positioning mechanism must provide for motion along one or more of the six DOFs.

C1.4.6.1 Stages

Stages nominally constrain motion to follow only the desired path(s). Several companies offer linear and rotary stages for laboratory and prototype applications, and a variety of approaches are built into end products. A well-designed device will allow motion only in the direction(s) of interest while maintaining adequate constraint in the remaining DOFs. A multi-direction-of-motion system ideally will have its coordinate system placed and aligned to minimize the interaction between the functions addressed by motion along each axis; i.e. its axes will be functionally orthogonal.

C1.4.6.2 Actuators

Actuators are the devices that cause motion to occur. They are, in essence, variable links that replace one or more of the six fixed links in a kinematic mounting. Manually operated actuators include the ubiquitous micrometer-type devices. Powered actuator types include electrostatic, electrostrictive, magnetostrictive, piezoelectric, electret, hydraulic, pneumatic and electromagnetic. Electromagnetic actuators can be further subdivided by the magnet, coil and/or iron being the moving element. Each type of actuator has characteristics of moving force, holding force, resolution, range, cost, speed, size and power requirements. The choice of actuator will depend on matching its characteristics with the corresponding set of system requirements [5].

C1.4.6.3 Servo-actuator systems

Servo-actuator systems combine a sensor to detect the difference between an actual and desired position, a processor to interpret these data, an amplifier and an actuator to move parts to reduce the position error. In addition to the limits to performance imposed by the choice of actuator type, several other factors further limit high-speed servo system performance. The acoustic propagation velocity through the structure between the actuator and the payload imposes a fundamental limit to bandwidth. Simply put, the payload will not begin to move until some finite time after the actuator is stimulated. In an ideal system having a nominally solid bar connecting these two elements, this limit is proportional to $(KM^{-1})^{0.5}$, (where K = stiffness and M = mass). In a real system, this limit is usually analytically complex and is established by the combination of material properties and structural configuration. In addition to this fundamental limit, unintentional mechanical resonances can cause the servo loop to oscillate like the squeal in microphone-loudspeaker systems. FEA, as mentioned in section C3.1.4, can help evolve a high-performance system design by predicting structural stiffness and resonant modes. Delays and inaccuracies in position error signal generation, signal processing and amplification may further limit system performance [15].

C1.4.7 Closure

This chapter has touched on the major considerations involved in creating a good opto-mechanical design, emphasizing aspects rarely mentioned in other resources. Remember: it takes a good mechanical system to put optics in their place and keep them there!

References

[1] Evans C 1989 *Precision Engineering: An Evolutionary View* (Bedford: Cranfield)
[2] Blanding D L 1999 *Exact Constraint: Machine Design Using Kinematic Principles* (New York: ASME)
[3] Kamm L J 1993 *Designing Cost-Efficient Mechanisms* (Warrendale, PA: SAE)
[4] Rothbart H A (ed) 1996 *Mechanical Design Handbook* 23rd edn (New York: McGraw-Hill)
[5] Ahmad A (Editor-in-Chief) 1997 *Handbook of Optomechanical Engineering* (Boca Raton, FL: Chemical Rubber Company)
[6] Smith W J 1966 *Modern Optical Engineering: The Design of Optical Systems* (New York: McGraw-Hill)
[7] Smith S T and Chetwynd D G 1992 *Foundations of Ultraprecision Mechanism Design* (Amsterdam: Overseas Publishers Association)
[8] Binstock L (ed) 1994 *Rapid Prototyping Systems: Fast Track to Product Realization* (Dearborn, MI: Society of Manufacturing Engineers)
[9] Brady G S and Clauser H R 1991 *Materials Handbook* (New York: McGraw-Hill)
[10] Lewis G 1990 *Selection of Engineering Materials* (Englewood Cliffs, NJ: Prentice-Hall)
Lewis G 1990 *Photonics Directory* (Pittsfield, MA: Laurin Publishing)
[11] Davis J R (ed) 1993 *Aluminum and Aluminum Alloys* (Materials Park, OH: ASM International)
[12] Harper C A (Editor-in-Chief) 1996 *Handbook of Plastics, Elastomers and Composites* 3rd edn (New York: McGraw-Hill)
[13] Schwartz M M 1992 *Handbook of Structural Ceramics* (New York: McGraw-Hill)
[14] Slocum A H 1992 *Precision Machine Design* (Englewood Cliffs, NJ: Prentice-Hall)
[15] Chang D K 1959 *Analysis of Linear Systems* (Reading, MA: Addison-Wesley)

Further reading

Bouwhuis G, Braat J, Huijser A, Pasman J, van Rosmalen and Schouhamer Immink K 1985 *Principles of Optical Disc Systems* (Bristol: Adam Hilger)

Includes an extensive section on the theory, considerations, approaches and limitations of servo-control mechanisms.

Goldberg N 1992 *Camera Technology: The Dark Side of the Lens* (San Diego, CA: Academic Press)

Presents a good example of the myriad details and design approaches to the multitude of functions incorporated into an everyday opto-mechanical device.

Moore W R 1970 *Foundations of Mechanical Accuracy* (Bridgeport: The Moore Special Tool Company)

Explains the great lengths taken in making and measuring the extremes of flat, straight, and round.

Oberg E, Jones F D, Horton H L and Ryffel H H 1988 *Machinery's Handbook* (New York: Industrial Press)

Includes extensive details of basic machine elements including screw threads.

Reiss R S 1992 *Instrument Design—A Collection of Columns and Features from OE Reports* (SPIE)

Describes the workings of a wide variety of familiar as well as not so familiar devices and designs.

C1.5.1
Power conditioning: supplies for driving semiconductor laser diodes

Ralph Savioli

C1.5.1.1 Introduction

Semiconductor laser diodes are possibly the most fragile semiconductor devices in common usage today. Proper care in their handling and application is essential to prevent damage and failure of the device. Special conductive packaging exists and numerous handling guidelines are well documented to prevent damage due to electrostatic discharge (ESD) [1]. However, not much is written with regard to the circuitry involved in operating or driving a semiconductor laser in a manner that will provide maximum output stability while protecting the laser diode from the various adverse operating conditions that could, and usually do, result in component failure.

The aim of this chapter is to present a conceptual approach to the design of a basic semiconductor laser diode driver. The objective is to describe circuits suitable for continuous wave operation of laser diodes in order to produce a stable output. The designs chosen for this discussion are by no means the only methods for driving a laser diode, but the concepts that will be addressed are fundamental and should be considered regardless of the methodology used. All examples will be specifically geared towards a negatively driven laser diode, typically one that has its anode connection common to the case of the device and in some instances common to the cathode of the integral photodiode of the laser. Anode grounded lasers are common throughout the industry. For a discussion of diode lasers themselves, see section B2.

C1.5.1.2 The basic current source

A laser diode is a current-driven device and the most fundamental method for operating a laser diode is in a constant current mode using automatic current control (ACC). This approach maintains a precise and constant current through the laser diode regardless of impedance changes in the device or other external factors such as variations in the power supplied to the driver itself. One drawback to constant current operation is the fact that the output power of the laser diode changes inversely to its operating temperature [2]. Therefore, ACC works best in conjunction with some sort of temperature control on the laser diode. An advantage of constant current operation is that unlike automated power control (APC) mode, the laser diode is not inside the control loop, making modulation and pulsing of the laser's output power much easier.

Figure C1.5.1.1(*a*) shows a schematic diagram of a very basic current regulator. R_1 is used as a current sense resistor (or shunt resistor), R_2 and R_{set} form a voltage divider that determines the level of the laser drive current, and Q_1 is an NPN transistor used to regulate the laser's current. The magnitude of the laser current is determined from R_1 by the relation $I = V_{R_1}/R_1$, where V_{R_1} is the voltage across R_1. The current is set

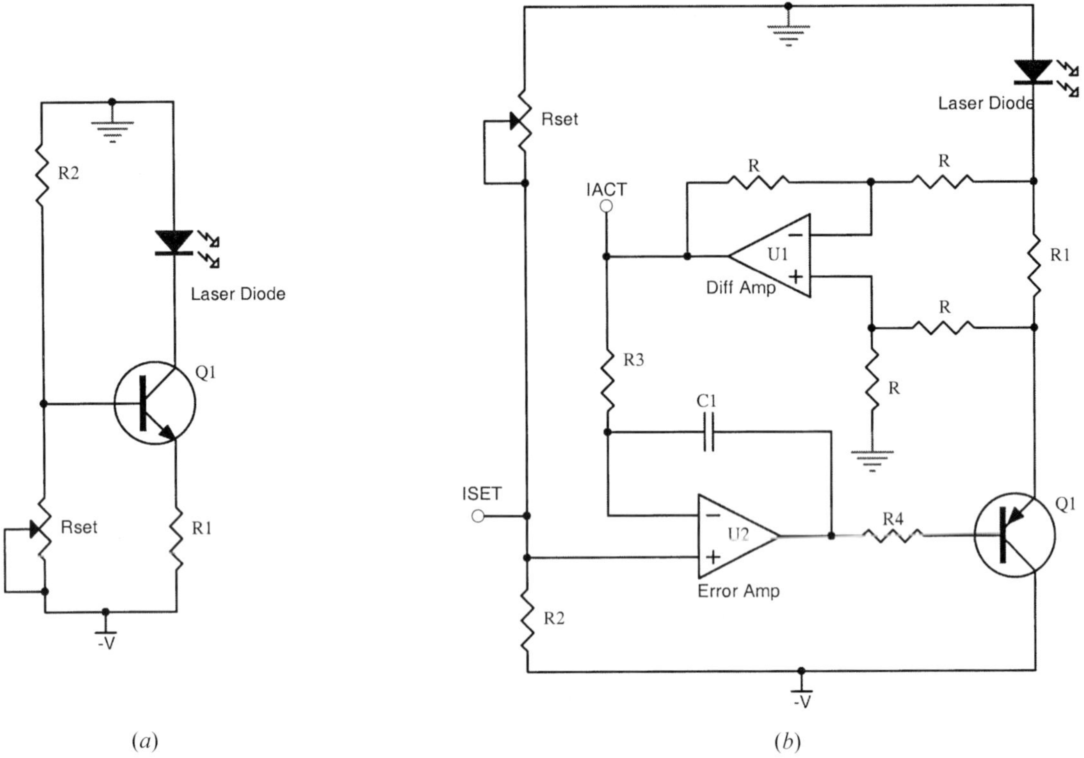

Figure C1.5.1.1. (*a*) A current source; (*b*) a better current source.

using the pot R_{set}. The level of R_{set} will be $V_{R_1} + 0.7$ V for any given current. For example, given a value of 10 Ω for R_1, for a laser current of 100 mA, $V_{R_1} = 1$ V. The voltage across R_{set} must therefore be 1.7 V.

While simple in its design the circuit of C1.5.1.1(*a*) has one major flaw that precludes its use as an aspect of a more complicated driver; the control elements are referenced to the negative power supply ($-V$). From a functional standpoint this practice will work well, but from a practical standpoint it is generally frowned upon. As mentioned earlier the laser will be negatively driven and therefore should use a true negative supply, one in which all functions are referenced to ground.

A better choice for the current regulator circuit is shown in figure C1.5.1.1(*b*). The current set point voltage, I_{set}, is now negative with respect to ground. Two operational amplifiers (op amps) are incorporated. U_1 is configured as a unity gain differential amplifier used to measure the voltage across R_1 and output it with respect to ground (I actual). U_2 is configured as an integrator and used as an 'error amplifier'. Ideally it should have zero potential across its input terminals, therefore I_{act} must equal I_{set}. It achieves this by driving Q_1 accordingly to either increase or decrease the laser diode current. C_1 and R_3 combine to provide stability and define the bandwidth of the closed loop, which is also affected by the individual bandwidth limitations of U_1, U_2 and Q_1.

Lacking a separate current limit circuit it would be possible to adjust R_{set} to a level that could overdrive the laser. An easy solution to this is to chose values for R_2 and R_{set} that only allow adjustment to 10–20% below the maximum operating current of the laser. A current derating is suggested in the event that temperature control is not used and the laser becomes more efficient due to a cooler operating environment, thereby running the risk of exceeding its maximum power rating [2].

Another consideration is the choice in operational amplifiers. While a single supply driver is an attractive feature, it requires the use of "rail to rail" op amps. Since the 'positive' supply voltage in a negative system is ground, the op amps need to be able to sense and source to the positive rail. The other alternatives are to introduce complicated offsets or simply use a bipolar supply [3].

A common specification for any current source is the 'compliance voltage'. The compliance voltage is the maximum voltage a current source can operate into while still maintaining a constant current. That is to say, a current source that is driving a load of resistance R with a current I will develop a voltage V across its output terminals with a magnitude of $V = IR$. As R or I increases, V will also increase, but only to the point of the compliance voltage. Once the compliance voltage is reached the current source can no longer regulate a constant current. The compliance voltage for the circuit of figure C1.5.1.1(b) is calculated using the formula

$$V_{\text{compliance}} = |\text{supply volts}| - |V_{\text{ce}Q_1}| - |V_{R_1}|.$$

The operating voltage of the laser diode, including voltage drops along the output wires, must always be less than the compliance voltage.

Other parameters to consider for a current source are the power dissipation of the components used. For the current sense resistor it is found by the formula $P = I^2 R$ and should be minimized to avoid errors due to heating. If power dissipation becomes an issue the resistance can be lowered and then compensated for with gain added to the current amplifier stage of the control loop. This can also be done to increase the compliance voltage if necessary.

The power dissipation of the regulating transistor is also important to be aware of. Too much power lost in the transistor is inefficient and also increases the requirement for heat sinking. Most semiconductor data books provide tutorials on calculating a safe operating area [4].

C1.5.1.3 Photodiode feedback and APC mode

As previously mentioned, ACC mode has a limitation in the fact that laser diodes are susceptible to changes in operating temperature. This can be overcome by operating the laser diode in constant power mode using APC.

APC mode makes use of a current provided by a photodiode. This current is directly proportional to the output power of the laser diode. The photodiode can either be external to the laser diode or integrated into the rear facet of the device. When operating in APC mode the photodiode current becomes a feedback signal to the current source, thereby placing the laser diode inside of the control loop. Instead of regulating the laser diode current the system regulates the photodiode current and indirectly regulates the output power of the laser diode.

It is necessary to perform a current to voltage conversion of the photodiode current [5]. Figure C1.5.1.2 shows a transimpedance amplifier with a photodiode properly oriented for a negative system.

Figure C1.5.1.3 shows the completed closed loop APC driver. Following the transimpedance amplifier is an integrating error amplifier similar to the one shown in figure C1.5.1.1(b) for the ACC driver. The potentiometer R_{set} is used to set the photodiode regulation level. The value of R_{f} in the transimpedance amplifier is chosen such that at the maximum photodiode current the output of the amplifier is equal to the maximum set point of the R_{set} pot. This is easily determined using the formula $R_{\text{f}} = V_{R_{\text{set}}}/I_{\text{mon}}$, where I_{mon} is the nominal photodiode current found on the laser datasheet or through measurement.

A significant drawback to this APC driver is that there is no means of limiting the current through the laser diode. In cases such as this, simply forgetting to connect the photodiode would cause an open loop situation that would be catastrophic to the laser diode.

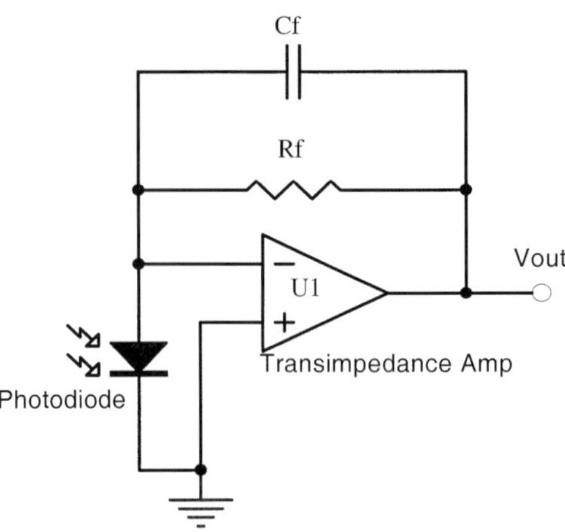

Figure C1.5.1.2. Current to voltage converter.

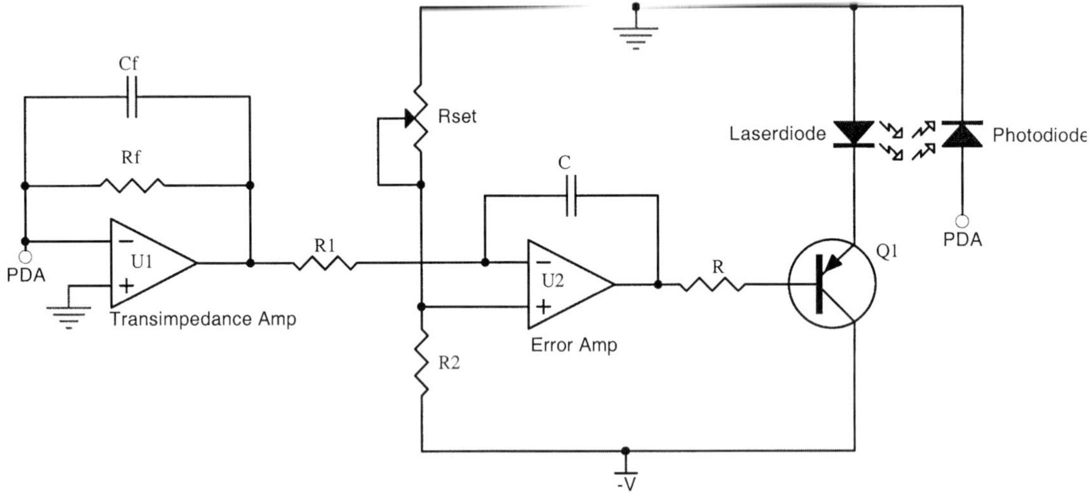

Figure C1.5.1.3. An APC circuit.

C1.5.1.4 A complete APC/ACC cw laser driver

Up to this point we have presented the two forms of laser diode drivers as separate entities, each with their inherent advantages and disadvantages. Figure C1.5.1.4 shows a combination of the two circuits that allows for either constant current operation in ACC mode or constant power operation in APC mode. When properly calibrated this circuit will provide a power limiting function while in ACC mode, as well as a current limiting function while in APC mode.

The final circuit has a number of embellishments that will be briefly explained. A significant improvement in the overall stability of the driver is accomplished by the use of the 2.5 V reference voltage placed across the set point circuits. The switch S1 has been added to allow for turning the laser on and off without powering down the entire circuit. With the switch closed, V_{ref} and all set points are zero voltage. Once the switch is opened the output power will ramp up proportionally to the charging of C_5, providing a soft start function.

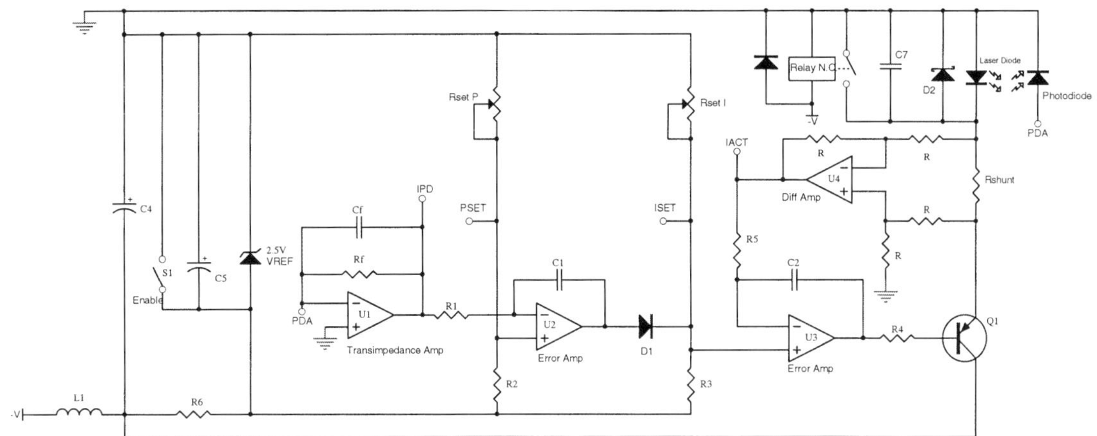

Figure C1.5.1.4. An APC/ACC laser driver.

L_1 and C_4 form an LC filter that will buffer the circuit and laser diode from voltage transients from the power supply, as will C_7, which is across the laser diode itself. C_7 is optional and, in some cases, particularly in APC mode, it might cause loop instabilities. Also connected across the laser diode is D_2. This is typically a Schottky diode used to protect the laser against reverse voltage conditions.

A relay is connected across the laser diode with its coil connected across the negative supply. When the power supply is turned off the normally closed relay will short the laser diode, thereby protecting it from ESD damage.

The APC circuit works by driving the I_{set} node in a positive direction through D_1 to limit and/or regulate the output power. If the power needs to be increased then the APC circuit will reverse bias D_1, thereby allowing the I_{set} node to increase negatively. The addition of D_1 adds the requirement for a bipolar power supply for operating power in order for the APC circuit to drive I_{set} all the way to ground if necessary.

Proper circuit layout will greatly enhance the integrity of the design. In general, it is important to keep high currents and digital signals out of critical analogue signal paths, and, while they are not shown in figure C1.5.1.4, always use decoupling capacitors on the supply pins of most, if not all, integrated circuits.

As previously mentioned, the circuit examples discussed are only a few of the many approaches to implementing a laser diode CW power source. Depending on the complexity and classification of the laser diode system a number of additional circuit requirements may be needed for safety reasons [6]. Using a handful of readily available components though, one should be able to fabricate a simple yet effective laser diode driver.

References

[1] Hodgson D and Olsen B ILX lightwave application note no 3 *Laser Diode Protection Strategies* http://www.ilxlightwave.com/library/index.html

[2] Desmarais L 1998 Diode laser characteristics *Appl. Electro-Opt.* 175–6

[3] Mancini R 2000 Texas Instruments design reference *Op Amps For Everyone* ch 4, Single supply op amp design techniques (Dallas, TX: Texas Instruments)

[4] National Semiconductor Power IC's Databook *Understanding Integrated Circuit Package Power Capabilities* appendix E (New York: McGraw-Hill) pp 8–22

[5] Graeme J 1995 Photodiode amplifiers *Op Amp Solutions* (Santa Clara, CA: National Semiconductor) pp 21–30

[6] International Standard IEC 60825-1 Ed 1.2 2001-08 Safety of Laser Products

Further reading

Coplan M A, Davis C and Moore J H 1991 *Building Scientific Apparatus—A Practical Guide to Design and Construction* 2nd edn (Reading, MA: Perseus)

Extensive sections on electronics principles and construction techniques.

Friauf W S 1998 *Feedback Loop Stability Analysis* (New York: McGraw-Hill)

Includes an informative chapter specifically about operational amplifiers.

Desmarais L 1998 *Diode Laser Characteristics Applied Electro Optics* (Englewood Cliffs, NJ: Prentice-Hall)

Excellent general reference book for many LED and laser diode topics including modulation and driver examples.

Sanyo LCD and LED Website-http://www.sanyo.co.jp/lcd_led/led_drive.html

A number of good ACC and APC driver circuit examples.

C1.5.2
Power conditioning: supplies for driving gas discharges (gas and solid state lasers)

I Smilanski

C1.5.2.1 Introduction: energy, peak power and average power

The gain medium in pulsed gas lasers such as excimer, copper vapour (CVL) etc is gaseous plasma. In these systems, substantial electron energy is required in order to reach and maintain population inversion. Electrons need to be situated in an electrical field E in order to gain energy and the field strength is inversely proportional to the electrodes spacing d:

$$E \sim V/d$$

where V is the voltage between the electrodes. Since the plasma has an impedance Z, supplying the electrons with energy is associated with power P:

$$P = \frac{V^2}{Z}.$$

In a practical CVL (see chapter B3.3), for example, V can be 20 000 V, Z can be 20 Ω so that $P = 20 \times 10^6$ W. If delivered continuously this is the power supplied to a small town. Pulsed gas lasers, however, do not necessarily require continuous (dc) power. Often what is needed are short pulses. In this case, the required power is

$$P = Ef$$

where f is the pulse frequency or the repetition rate and E is the energy per pulse. For the previous example, if the pulse duration is 50 ns and the repetition rate is 5 kHz, then the average power will drop to

$$P = 20 \times 10^6 \times 50 \times 10^{-9} \times 5 \times 10^3 = 5 \text{ kW}.$$

Rather than take a huge power station and slice its output into narrow slices, special compact circuits have been developed to generate these pulses. This technology is called 'pulsed power' and its application to gas lasers is described in this article [1, 2]. The typical power requirements of commercial gas lasers are given in table C1.5.2.1.

C1.5.2.1.1 Capacitive and inductive energy storage

In pulsed power, energy is slowly accumulated in a storage device and then rapidly dumped into a load. The power gain is the ratio between the charging time and the discharging time. The two main energy storage devices are capacitors and inductors.

The energy stored in a capacitor is:

$$E = \tfrac{1}{2}CV^2$$

Table C1.5.2.1. Typical pulse power requirements of commercial gas lasers.

	$\mathrm{J\,l^{-1}}$	τ (ns)	$\mathrm{E\,V\,cm^{-1}}$
Excimer	25	20	10 000
CO_2	50	100	10 000
CVL	1	50	2 000

where E is the energy, C is the capacitance and V is the voltage. The current source charges the capacitor C to a voltage V according to

$$V = \frac{1}{C}\int_0^\tau i_c \, \mathrm{d}t$$

where i_c is the charging current and τ is the charge time.

Once the capacitor is charged to the desired voltage, a switch closes which drives the current into the load. To switch capacitively stored energy, we need a *closing switch.*

Example 1. In the CVL described earlier, 5 kHz at 5 kW means an energy of 1 J pulse. In 20 kV this means a capacitance of

$$\frac{2 \times E}{V^2} = \frac{2}{4 \times 10^8} = 5 \text{ nF}.$$

The energy stored in an inductor is

$$E = \tfrac{1}{2} L I^2$$

where L is the inductance and I is the current through the inductor. Assuming an ideal loss-less inductor, connected to a voltage source V, a current develops according to

$$V = -L\frac{\mathrm{d}I}{\mathrm{d}t}.$$

It follows that, for constant voltage,

$$I = -\int_0^\tau \frac{V}{L}\,\mathrm{d}t = I(0) - \frac{V}{L}\tau.$$

The inductor is linearly charged with current. Consider now a laser connected in parallel to the inductor. Once we disconnect the inductor from the voltage source, current continuity will force the current to continue to flow—with the same magnitude and sign—to the laser, regardless of its impedance or breakdown voltage and, hence, to deliver the energy stored in the inductor to it.

To switch inductively stored energy we need an *opening switch*.

Example 2. For the CVL described earlier, with 1000 A load current and 1 J energy, the storage inductance should be

$$L = \frac{2 \times E}{I^2} = \frac{2}{(10^3)^2} = 2\ \mu\text{H}.$$

The supply voltage in this case would be

$$V = \frac{LI}{\tau} = \frac{2 \times 10^{-6} \times 10^3}{200 \times 10^{-6}} = 10 \text{ V}.$$

Here τ is the inter-pulse duration; and is 200 μs for the 5 kHz CVL. A high-voltage pulse will be automatically sustained on the laser when the switch opens.

C1.5.2.2 Switches

A switch is available for conduction when current carriers (electrons, ions, holes) exist between its electrodes. In some switches (thyratrons, SCRs), these carriers are introduced by diffusion from the control electrode and then multiplied in some positive feedback process. In others (triodes, IGBTs), the charge carriers' capability to move is governed by a control electrode. In this article, we will concentrate on their characteristics as circuit elements, emphasizing the features common to them all.

In closing switches, the switching has four phases: commutation, conduction, turn-off and recovery. They are shown in figure C1.5.2.1 with typical waveforms for a simple RLC circuit.

- *Commutation*: This is the phase in which conducting is developed. It follows the turn-on delay, and is associated with the time required for the charge carriers to propagate and fill the entire region between the electrodes.
- *Conduction*: The phase in which the switch reaches its full conduction capability. In a reasonable circuit, this phase should be much longer than the commutation time.
- *Turn-off*: The time it takes the charge carriers to clear off the active zone.
- *Recovery*: The time required for the switch to regain full holding voltage capability. This is associated with the removal of control charge carriers like plasma from the grid-cathode zone in thyratrons, and its equivalents in semiconductor switches.

An important feature of closing switches is that they cannot be turned off by command. The current through them must first be extinguished—in fact, become smaller than a current I_{h} called the 'holding current'. Figure C1.5.2.2 demonstrates a typical switching situation.

The overall energy loss per pulse is given by $E = \int I \times V \, \mathrm{d}t$, where V is the switch voltage and I is the current through it. To minimize losses, a good switch should have a low conduction voltage but in short pulses the contribution of commutation and turn-off to the losses is dominant.

The high voltage on the switch during turn-off results from the presence of inductance in the circuit. Our intention was to dissipate the energy stored in the capacitor on the resistive load. Due to the unavoidable inductance in the circuit, unless $Q < 1$, the current will not terminate after the first half-cycle. If we had a switch with a zero turn-off time, we could turn the switch off when the current is exactly zero. Due to the finite turn-off time, the switch turns off only some time after the current crosses zero. It actually behaves as an opening switch. The residual energy in the inductor is then split between the load R and the internal resistance of the switch:

$$\tfrac{1}{2}LI^2 = \int (r + R)I^2 \, \mathrm{d}t$$

where R is the load resistance and r is the switch resistance. This energy is larger for longer turn-off times and reaches its maximum at the quarter-cycle time $1/2\pi\sqrt{LC} = \tau$. Since, by definition, during turn-off r increases to infinity, most of the inductor's residual energy will be dissipated on the switch. The little negative hump on the current trace in figure C1.5.2.1(b), following the main positive half-cycle, is created by the finite turn-off time and is responsible for the creation of a huge negative voltage spike shown in figure C1.5.2.1(a), a lot of energy dissipated on the switch, and frequent failures unless preventative measures are taken [3].

A diode–resistor (R,D) combination called a 'snubber', shown in figure C1.5.2.2, can reroute the residual energy and dump it on the resistor instead of the switch.

The value of the resistor is determined by the allowed reversed voltage on the switch:

$$V = I_{\mathrm{turnoff}} R$$

Since this current decays exponentially with a time constant

$$\frac{L_{\mathrm{w}}}{R + R_{\mathrm{L}}} \approx \frac{L_{\mathrm{w}}}{R}$$

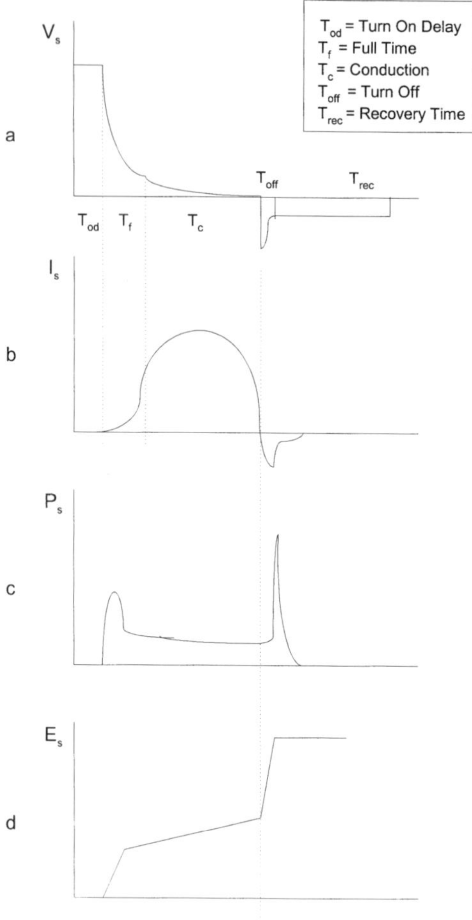

Figure C1.5.2.1. Switching phases: (*a*) switch voltage; (*b*) switch current; (*c*) power dissipated on the switch; and (*d*) energy dissipated on the switch.

where L_W is the wiring inductance (see figure C1.5.2.2), the actual choice of resistor is a compromise between a low value required for a low reverse voltage and a high value required for a short time constant. Snubber resistors should be non-inductive and often need to dissipate high power.

For a half-sine current waveform, the maximum switched energy E_{max} by a given switch will be

$$E_{max} = \frac{\tau V_{max} I_{max}}{2\pi} = \frac{\tau}{2\pi} P_{max}$$

where τ is the pulse duration and V_{hold} and I_{max} are taken from table C1.5.2.2. To switch more energy takes a longer time.

Table C1.5.2.3 shows the minimum pulse duration required for different switch types in order to switch 1, 20 and 100 J pulses—typical excitation energies for commercial gas lasers. These time durations are normalized against the corresponding time in a thyratron. In fact, τ/τ_{th} indicates how many components of that type (in series or in parallel) are required to replace a single thyratron.

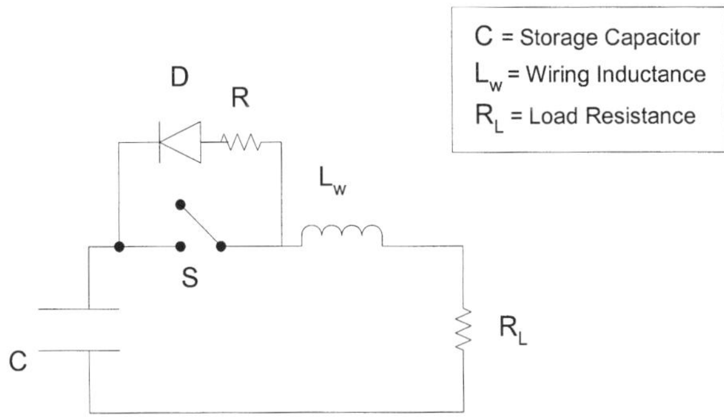

Figure C1.5.2.2. A switch in a capacitor discharge circuit.

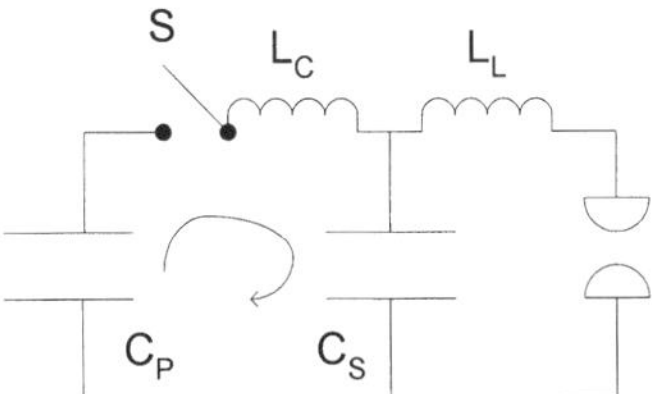

Figure C1.5.2.3. Capacitive transfer circuit: C_P, primary energy storage capacitor; S, switch; L_C, wiring and switch inductance; C_S, secondary capacitance; and L_L, laser inductance.

Table C1.5.2.2. Comparison of switches.

	V_{hold}(kV)	Commutation time	Conduction voltage	Maximum conduction current I_{max}	Turn off time	Recovery time
Thyratron[c] [4]	30	50 ns	250 V	10 kA	20 ns	30 μs
SCR[c] [5, 6]	2	1 μs	2 V	20 kA	500 ns	50 μs
IGBT[c,o] [7]	2	1 μs	2.5 V	1 kA	500 ns	2μs
MOSFET[c,o] [8]	1	50 ns	10 V	100 A	50 ns	500 ns
Triode[c,o] [9]	30	50 ns	2 kV	1 kA	50 ns	50 ns
Spark gap[c] [10]	50	20 ns	500 V	50 kA	10 ms	200 ms
SOS[o] [11]	NA	NA	NA		10 ns	NA

c, closing switch; o, opening switch.

Table C1.5.2.3 maps the boundaries in which a simple RLC circuit can excite fast ($\tau \sim 50$ ns) gas lasers using commonly available switches. In the case of the thyratron with 1 J excitation, the pulse duration equals the commutation time, so the operation will be very inefficient. Some methods to overcome such limitations are described in the next sections of this chapter.

An SOS (semiconductor opening switch) is a relatively new device developed by the Yoffe Institute in St Petersburg, Russia. Its philosophy is, 'If you can't beat them, join them'. It is a diode with a relatively long but abrupt turn-off time. Usually these are highly undesirable features for a switch (shown previously

Table C1.5.2.3. Minimum pulse durations for switches.

	1 J	30 J	100 J	P (MW)	τ/τ_{th}
Thyratron	42 ns	1.25 μs	4.2 μs	24	1
SCR	157 ns	4.7 μs	15.7 μs	6.4	3.75
IGBT	3.14 μs	94 μs	314 μs	0.3	75
MOSFET	63 μs	1.9 ms	6.3 ms	0.015	1500
Spark gap	2.5 ns	75 ns	250 ns	400	0.06

in figure C1.5.2.1) but armed with a very high backward voltage, it is capable of producing high-voltage high-power pulses, in exactly that type of circuit. All you need is to make sure that $1/2\pi\sqrt{LC} = \tau_R$ is the turn-off time of the SOS. Circuits using this device are not very efficient but rather compact and simple [12].

Long turn-off and recovery times usually make the spark gap unsuitable for high-repetition-rate applications (the 'quenched spark gap' is an exception [10].) Common to all spark-gaps is a limited service life, which makes them useful only for special applications and for experimentalists. Thyratrons require bulky services: a couple of hundred watts for the cathode heater and a hydrogen reservoir, grid bias and trigger pulses in the kilovolt range. They require warm-up time and have a limited life expectancy. In fact, when the laser manufacturer 'Lumonics' replaced their thyratron-based CO_2 marking laser with an SCR-based version, they extended their warranty on the switch from 4 months to 5 years. LLNL reported an increase of switch MTBF in their large-scale CVL from 1000 hr for thyratrons to 7000 hr for SCRs [13]. However, thyratrons are very friendly to experimentalists since they are (error) forgiving devices. One needs tens of joules to destroy a thyratron and less than a millijoule to destroy an SCR, the junction being so small.

Almost the same comment holds for triodes compared with MOSFETs. Clearly, the latter require to be arranged in large, carefully designed arrays [14] but for industrial applications they are the only viable solution.

There are many emerging semiconductor switches like the SIT [15], MCT [16] etc but their basic characteristics are still within the aforementioned categories.

C1.5.2.3 Basic circuits

In pulse-forming networks (PFNs), energy is slowly accumulated and switched into the load. PFNs consist of combinations of capacitors and inductors [1]. A unit time constant, one half-cycle, of such a network is $\tau = \pi\sqrt{LC}$ and the typical impedance is $Z = \sqrt{L/C}$ where L is the inductance and C is the capacitance of a single network section. The pulse duration of a network with N sections is N_τ and the shortest pulse duration is for $N = 1$.

Since every capacitance has some residual inductance L_0, the minimum value of τ is $\tau = \pi\sqrt{L_0 C}$ and the minimum impedance will be $Z = \sqrt{L_0/C}$, for any capacitor selected. Therefore, other than in special cases, the PFN of pulsed gas lasers is reduced to a single capacitor.

Between successive pulses, the capacitor is charged to a voltage V and accumulates an amount of energy $E = \frac{1}{2}CV^2$. During the pulse, the capacitor is discharged through the laser. Often this happen in the shortest time possible.

For a first approximation, we can consider the discharge circuit as an *RLC* series resonance circuit, where L is the total residual inductance

$$L = L_{\text{laser}} + L_{\text{switch}} + L_{\text{capacitor}} + L_{\text{wiring}}.$$

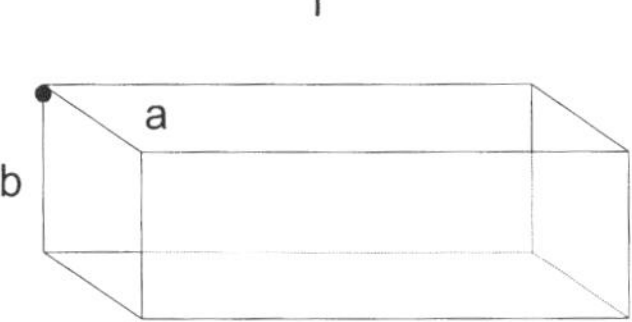

Figure C1.5.2.4. Inductance of a single turn box: $L_{\text{longitudinal}} = \mu_0(lb/a)$; $L_{\text{transversal}} = \mu_0(ba/l)$; $R_{\text{longitudinal}} = \rho(l/ab)$; and $L_{\text{transversal}} = \rho(b/la)$. For a box gain medium of length l and cross section s: $L = \mu_0(1/s)$ and $R = \rho(l/s)$.

The current waveform is:

$$I = I_0 \exp\left(-\mathrm{i}\pi \frac{t}{\tau}\sqrt{1 - \frac{1}{4Q^2}}\right) \exp\left(\frac{\pi}{2}\frac{t}{\tau}\frac{1}{Q}\right)$$

where Q is

$$Q = \frac{L\pi}{R\tau} = \frac{E\pi}{\Delta E} = \frac{1}{R}\sqrt{\frac{L}{C}}$$

or the ratio between the reactive impedance and the resistance as well as the ratio between the stored energy (in the capacitor) and the energy lost per radian. Since we need the energy deposition rate to be as high as possible (= high peak power), we simultaneously need short τ and low Q.

It turns out that the current oscillations will be critically damped only for inductances that satisfy:

$$L \leq \frac{\tau R}{\pi}.$$

Only under these conditions will all of the energy be dissipated in the laser during the time interval τ. As an immediate conclusion, we realize that a smaller inductance in the circuit elements will give a higher speed and energy transfer.

There are two basic situations here:

(1) lasers with self-switching characteristics and relatively high resistance, like excimers and TEA CO_2 lasers; and
(2) lasers that are continuously conducting and have low resistance, like CVLs.

In the first kind of laser, the so-called 'capacitive transfer' circuit is useful (see figure C1.5.2.3).

The branch C_p, S, L_c and C_s is a resonance circuit with a half-period time

$$\tau = \pi\sqrt{L_c \frac{C_p + C_s}{C_p C_s}}.$$

Suppose that we are charging C_p to V_{bd}, which is the break-down voltage of the laser. If we select $C_p = C_s$, then $\pi\sqrt{\frac{1}{2}L_c C_p}$ seconds after closing the switch S all the charge of C_p will be transferred to C_s. At that moment, the laser will break down and C_s will be discharged into it. The only inductance in the laser circuit will now be its self-inductance L_l.

The following example can illuminate the operation of such a circuit. Assume an excimer laser with the following transversal geometry (see figure C1.5.2.4): electrode separation $a = 4$ cm, electrode width $b = 4$ cm, and electrode length $l = 50$ cm.

The breakdown voltage is 25 kV and we want to deposit 20 J into the discharge channel. The storage capacitor will then be:

$$c = \sqrt{\frac{2E}{V}} = 64\ \text{nF} \qquad \text{and} \qquad L_l = \mu_0 \frac{ab}{l} = 10\ \text{nH}$$

so that

$$Z = \sqrt{\frac{L_\mathrm{d}}{C_\mathrm{s}}} = 0.4\ \Omega \qquad \text{and} \qquad \tau = \pi\sqrt{L_\mathrm{d}C_\mathrm{d}} = 80\ \text{ns}$$

A typical value for the external excitation loop (see figure C1.5.2.4) inductance is $L_\mathrm{c} = 0.5\ \mu$H, so that the charge transfer time is $\tau_\mathrm{T} = \pi\sqrt{C_\mathrm{s}L_\mathrm{c}/2} = 430$ ns. Also, the switch peak current is $I_\mathrm{s} = VC_\mathrm{s}\pi/2\tau_\mathrm{T} = 5.8$ kA, and the maximum current in the laser branch is $I_\mathrm{L} = VC_\mathrm{s}\pi/\tau = 62.8$ kA.

We see that the internal switching property enables a considerable compression $(2\tau_\mathrm{T}/\tau)$ of the current in the external circuit. This is why C_s is sometimes called a 'speed-up capacitor'.

Due to the electronegativity of the laser medium, these lasers have a relatively high resistance, so under the previous conditions a good impedance matching is still possible.

Self-switching does not exist in elemental metal vapour lasers. However, in longitudinal excitation, which is favoured in most of these lasers, the capacitive transfer circuit can still be beneficial (shown previously in figure C1.5.2.3). In this case, the self-inductance of the load is sufficiently high so the discharge rate of C_s can be lower than its charge rate. For the sake of simplicity, we will discuss the case in which the two capacitors are equal.

The half-cycle charge time is:

$$\tau_\mathrm{c} = \pi\sqrt{\frac{L_1 C}{2}}.$$

The half-cycle discharge time is:

$$\tau_\mathrm{d} = \pi\sqrt{L_2 C}$$

which, if $L_2 > L_1/2$, is slower. This condition is relatively easy to fulfil. Now, without the speed-up capacitor, the half-cycle discharge time would have been

$$\tau = \pi\sqrt{(L_1 + L_2)C}$$

which is clearly longer.

Trying to use this circuit in a transversal excitation CVL results in severe mismatching and reflections due to the low impedance of the continuously conducting plasma [17].

C1.5.2.4 More circuits

C1.5.2.4.1 Blumlein [18]

A popular circuit is the '*LC* inverter' (also known as the 'Blumlein'), and is shown in figure C1.5.2.5. The idea in both versions is that C_1 and C_2 are charged in parallel and discharged in series. The voltage and current waveforms in a resonance circuit are also shown in figure C1.5.2.5. At $\tau = \pi\sqrt{LC}$, the voltage reversal will be completed. For the best results, the laser tube should then break down. Table C1.5.2.4 compares the two circuits, switching the same energy to the same laser with the same pulse width. We see that halving the voltage results in doubling the current; the circuit designer gets more choice in selecting his or her switch. References [19–21] contain representative applications of this circuit.

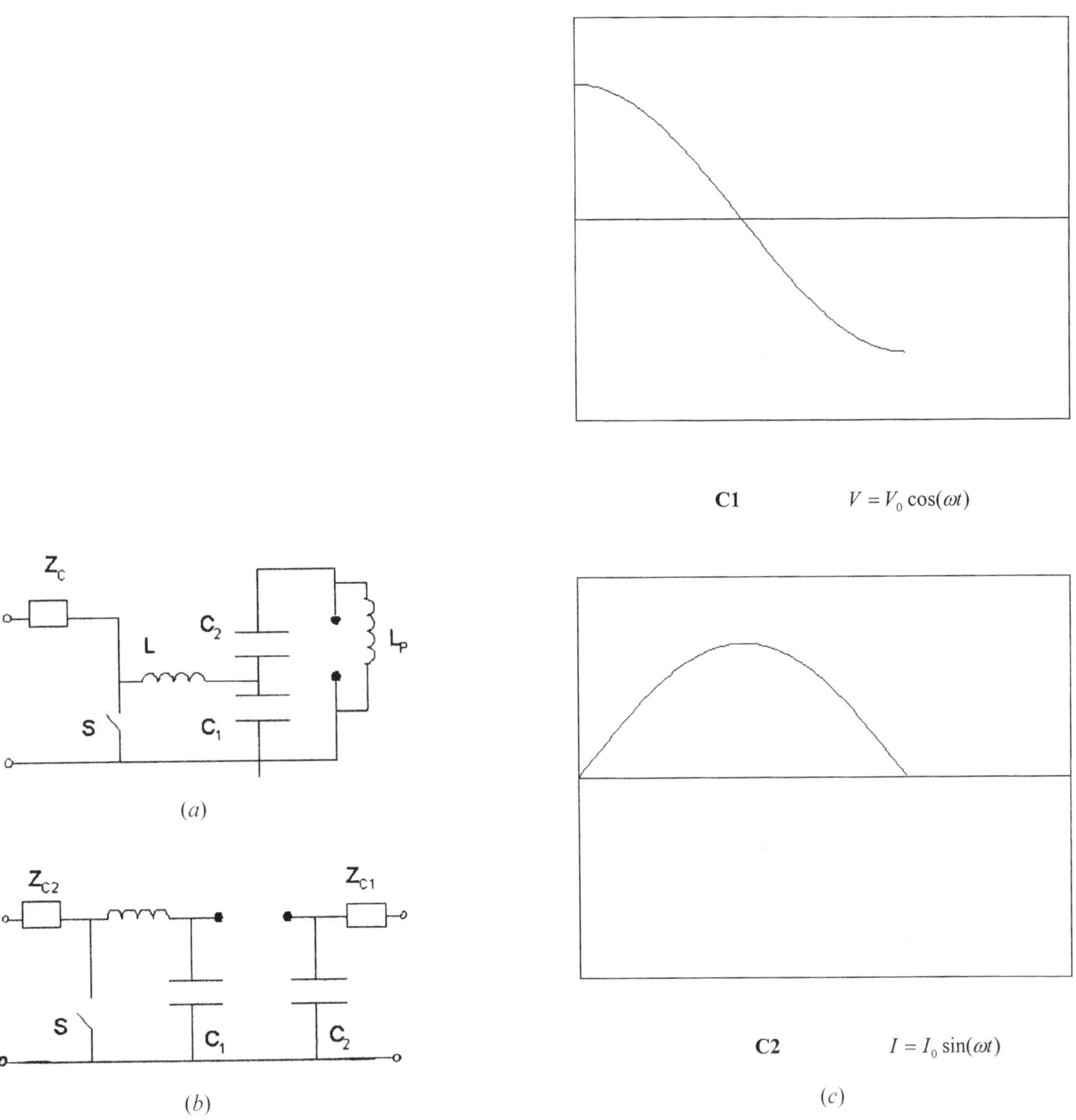

Figure C1.5.2.5. Blumlien's circuits: (*a*) grounded electrode; (*b*) floating electrode and (*c*) voltage (C_1) and current (C_2) on C_1.

C1.5.2.4.2 Spiker sustainer [22]

This is an improvement of the basic self-switching circuit. Often the voltage required to break down a laser gap is higher than the voltage required for sustaining discharge and support lasing. Using the same PFN to do both results in a substantial energy waste, since energy is proportional to the square of the voltage. In spiker sustainer circuits, a separate PFN is devoted to each task—a high-voltage low-energy one for breakdown, and a low-voltage higher-energy one for sustaining the discharge. A switch is required only for the spiker; the sustainer will conduct automatically once plasma is created in the gap. The sustainer PFN, however, should

Table C1.5.2.4. Circuit comparison.

	Switch voltage	Switch current	Storage capacitance	Loop inductance
Capacitive transfer	V	$I_0 = V\sqrt{\frac{C}{2L}}$	C	L
LC inverter	$\frac{1}{2}V$	$2I_0 = V\sqrt{\frac{2C}{L}}$	$2C$	$\frac{1}{2}L$

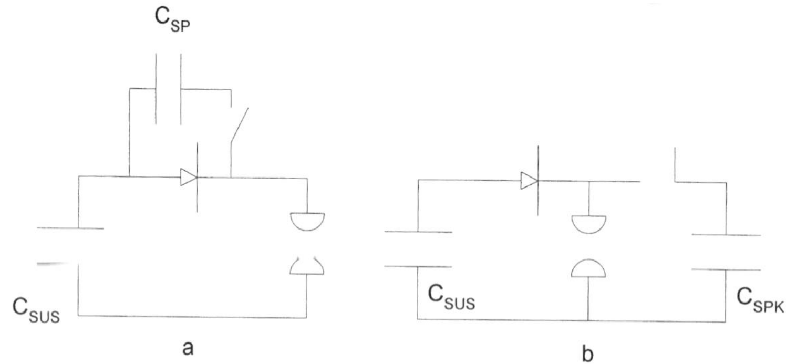

Figure C1.5.2.6. Spiker sustainer circuits.

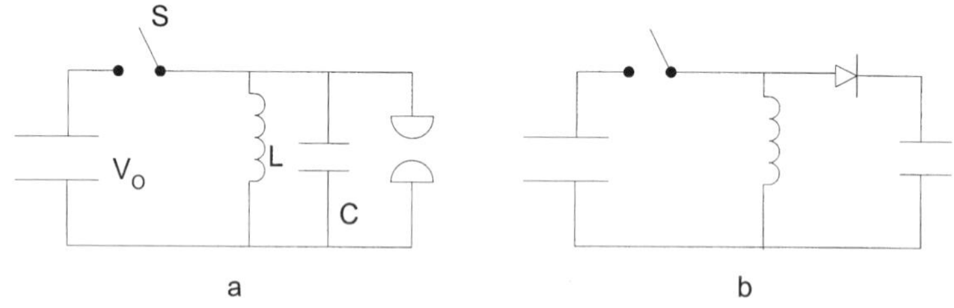

Figure C1.5.2.7. (*a*) Opening switch for creating a high-voltage pulse on C and (*b*) opening switch in a capacitor charging circuit.

be adequately isolated from the spike as suggested schematically in figure C1.5.2.6. The diodes are often replaced with magnetic switches.

C1.5.2.4.3 Interacting

This is a refinement of the capacitive transfer circuit in which two sections peak with each other [23].

C1.5.2.4.4 Opening Switch

Although an opening switch has already been described in section C1.5.2.2, its basic circuit is given in figure C1.5.2.7.

The switch S closes and after a time interval Δt opens again. In $\tau = \pi\sqrt{LC}$, the voltage on the capacitor

C will reach the voltage V:

$$V = \frac{V_0 \Delta T}{\sqrt{LC}} = \pi \frac{\Delta T}{\tau} V_0$$

which means that the voltage gain is proportional to the ratio between the inductor's current charge time Δt and its discharge half-time τ.

C1.5.2.5 Pulse compression

Pulse compression enables us to use a switch in such a way that the commutation and turn-off times will be negligible relative to the conduction time, thus achieving efficient and reliable operation [24–28]. Faraday's law states that for an inductor $V = n\,\mathrm{d}\Phi/\mathrm{d}t$, where V is the voltage induced on it and Φ is the magnetic flux through it. Since $\Phi = AB$ where B is the magnetic flux density and A is the cross section of the inductor we get

$$\int_0^t V\,\mathrm{d}t = nA\int_0^\tau \mathrm{d}\Phi = nA\Delta B$$

where ΔB is the change in flux density. For the circuit in figure C1.5.2.8, the voltage on the magnetic switch inductor (MS) has the form

$$V = \frac{1}{2}V_0\left(1 - \cos\frac{2\pi}{T}t\right)$$

where T is the duration of one cycle. Full charging of C_2 occurs by the end of the first half-cycle ($t = T/2$), therefore

$$\tfrac{1}{2}VT = nA\Delta B.$$

The relationship between the magnetic flux B and the magnetic field H, in a magnetic core, is shown schematically in figure C1.5.2.9.

Far above B_m and below $-B_\mathrm{m}$ the following holds:

$$B = \mu_0 H$$

where μ_0 is the vacuum permeability. Within the region $\pm B_\mathrm{m}$, things are more complicated and depend on the history of the device but for our needs they can be approximated by figure C1.5.2.9(*b*) in which

$$B = \mu\mu_0 H$$

and μ is the relative permeability. There are two distinctive values for μ: The saturated regime $\mu = 1$:

$$B \gg B_\mathrm{m} \qquad B \ll -B_\mathrm{m}.$$

The unsaturated regime $\mu \gg 1$

$$-B_\mathrm{m} \le B \le B_\mathrm{m}.$$

As long as a coil does not change its shape, the flux through it is proportional to its own current. This makes it convenient to define the inductance to be

$$L = \frac{N\Phi}{I}$$

where Φ is the flux through one of the turns of the coil and where N is the number of turns of wire in the coil. Like the capacitance and the resistance, the inductance of a coil depends only on its geometry and on the materials of which it is made. For the case of a long solenoid:

$$L = \mu\mu_0 n^2 Al = \mu\mu_0\frac{N^2}{l}A$$

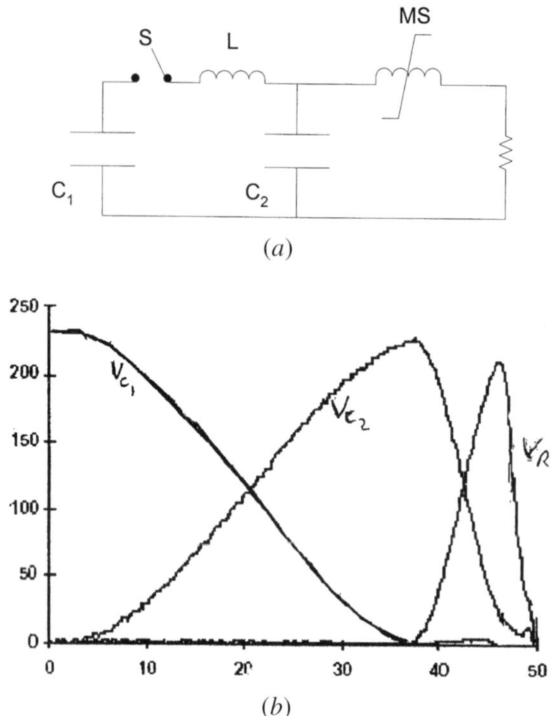

Figure C1.5.2.8. (*a*) Basic magnetic switch circuit and (*b*) C_1, C_2 voltage waveforms.

where n is the number of turns per unit length, N is the number of turns, A is the cross-sectional area of the solenoid and where l is its length. In other inductor geometries, calculating the inductance usually involves complicated integrals.

We see that, depending on the value of B, there are two distinct values of L, orders of magnitude apart, associated with an inductor. This is what enables us to use them as magnetic switches in pulse compressors.

Now consider the circuit shown previously in figure C1.5.2.8. L_{us} is the unsaturated inductance of MS, L_{s} is its saturated inductance and $L_{\mathrm{us}} = \mu L_{\mathrm{s}}$. If we design it so that $L_{\mathrm{us}} \gg L$, then closure of S will cause slow charging of C_2 and a negligible current through MS and R. We remember, however, that $B \sim \int V\,\mathrm{d}t$ and $H \sim I$ so that in the BH plane (shown in figure C1.5.2.9(*b*)), as time passes by, we move towards B_{m}. If we design the inductor correctly, that point will be reached exactly as V_{C_2} reaches maximum. At that moment, μ switches value and drops orders of magnitude down. The ratio ρ between the charging time τ_{c} and the discharging time τ_{d} of C_2 is called the compression ratio:

$$\rho = \frac{\tau_{\mathrm{c}}}{\tau_{\mathrm{d}}} = \frac{\pi\sqrt{\frac{1}{2}L_1 C}}{\pi\sqrt{\frac{1}{2}L_{\mathrm{s}} C}} = \sqrt{\frac{L_1}{L_{\mathrm{s}}}}.$$

In order for ρ to be significant, we need $L \gg L_{\mathrm{s}}$. Overall what we need is:

$$\mu L_{\mathrm{s}} \gg L_1 \gg L_{\mathrm{s}}.$$

For $\rho = 5$, we need $L = 25 L_{\mathrm{s}}$ which means $\mu > 250$.

Table C1.5.2.5. Summary of properties compared to iron.

	ΔB (T)	Resistivity ($\Omega\ m$)	μ_{max}	Hysteresis loss
NiZn	0.7	10^7	1 500	High
MnZn	0.8	1	5 000	Medium
Amorphous alloys	3.5	10^{-4}	400 000	Low
Iron	3.5	10^{-5}	50 000	

In the previous discussion, we assumed that we began our travel in the BH plane at the point A and by doing that guaranteed the highest voltage hold-off ($\int V\,dt$) possible. Let us assume also that after finishing transfering the charge to C_3 the core returns back to point A in the BH plane (the core is 'reset').

The area enclosed by the path ABCD represents the energy lost per pulse in the process of switching μ. Another source of loss arises from eddy currents within the core itself. If the core material is a conductor, it can be viewed as a single-turn secondary in a transformer where the windings are the primary. To reduce the eddy currents, the resistance of the core should be increased by arranging the core in multiple layers of isolated material and not as a single block. The resistance of the conductors by which the turns are made is another source of loss, especially in high frequencies where skin and proximity effects are important.

In conclusion, a suitable magnetic material for pulse compression should have: high μ, high B_m , low hysteresis loss and high resistance in the relevant frequency range. The winding should be made with a low-resistance wire in the relevant frequency.

The volume of the required core is related to the pulse energy and the required compression ratio by

$$\rho^2 = \frac{4A^2\Delta B^2 2l}{\pi^2 E A\mu\mu_0} = \frac{8V_{0l}}{\pi^2\mu\mu_0 E}\Delta B^2$$

where V_{0l} is the volume of the core material. The compression ratio is proportional to ΔB of the material, to the square root of V_{0l} and inversely to the square root of the switched energy.

The magnetic materials most often used for magnetic compressors are amorphous alloys, MgZn and NiZn ferrites.

The main characteristic of amorphous alloys is the absence of periodic crystalline arrangements of atoms: their structure is unordered resembling the distribution of atoms in a melt. As a result, amorphous metals are soft magnets while at the same time mechanically hard and they are not good conductors. Ferrites are ceramic structures which include the oxides of magnetic materials. In that form, their resistance is greater than it would have been in a normal metallic form and therefore the eddy-current loss is reduced. Table C1.5.2.5 summarizes their properties and compares them to iron. Note that the numbers in table C1.5.2.5 indicate the trend, but the actual numbers can vary according to the exact specimen in use.

Both hysteresis and eddy-current loss are frequency dependent and increase as pulses become shorter and shorter. Eddy-current loss can be reduced by slicing the material into thin isolated laminations or by forming it into a thin tape and winding it spirally, keeping the individual turns isolated from each other. For very high frequencies, in spite of lower ΔB and μ_{max} and higher hysteresis loss, the NiZn ferrite is the most suitable.

The properties of magnetic materials are temperature dependent. Above the Curie temperature, ΔB vanishes. In NiZn, ΔB goes down gradually as the temperature rises, so an adequate thermal design to maintain the core temperature in the right range is essential for proper operation.

As a demonstration of pulse compression capabilities, let us return to the 1 J thyratron circuit. The

Table C1.5.2.6. Alternatives for providing L_S.

L_S (nH)	N	m	Nm
220	5	4	20
176	4	5	20

basic capacitive transfer circuit was shown previously in figure C1.5.2.2. The modified circuit was shown in figure C1.5.2.8.

In the basic configuration, if we expect a 50 ns current pulse in the switch then the total wiring inductance L_W should be

$$L_W = \left(\frac{\tau}{\pi}^2\right)\frac{2}{C} = 202 \text{ nH}.$$

The peak current through the thyratron would be

$$I_p = \frac{VC\pi}{2\tau} = 2200 \text{ A}.$$

Now consider a 5:1 pulse compressor. L should be 25 times larger than L_W:

$$L \cong 5\ \mu\text{H}.$$

The current peak will be 404 A, and the 250 ns pulse duration will greatly reduce the effect of commutation and turn-off times.

Suppose we select 9 cm× 3 cm × 1 cm Ni Zn toroids. Then $A = 3 \times 10^{-4}$ m^2 and $\Delta B = 0.7$ T. To hold 28.2 kV for 250 ns, we need

$$Nm = \frac{V\tau}{2\Delta BA} = \frac{28.2 \times 10^3 \times 250 \times 10^{-9}}{2 \times 0.7 \times 3 \times 10^{-4}} \cong 17$$

where N is the turn's number and m is the number of cores. The saturated induction is given by

$$L_S = \frac{\mu_0}{2\pi} N^2 mh \times \ln\left(\frac{D_o}{D_i}\right)$$

where D_o, D_i and h are, respectively, the outer diameter, inner diameter and height of the toroid.

Since L_S and the Nm product are given, we can solve for N and then for m. However, only integers are acceptable solutions, so we are left with two alternatives, as shown in table C1.5..2.6.

The five-toroid four-turn solution is conservative and satisfactory. About 20 mJ per toroid per pulse should be removed from the compressor in order to keep it in normal operating temperature. Returning now to table C1.5.2.3, we realize that a pulse compressor enables us to supply a load with a much higher power than a switch can switch. Using step-up pulse transformers, we can now use relatively slow low-voltage switches like SCRs to switch high-voltage, fast pulses into a gas laser load [5–7].

Finally, saturable inductors (SIs) can significantly reduce the losses in a switch (see figure C1.5.2.9). The SI delays current flow until commutation is over and prevents the current from zero crossing until turn-off is completed, significantly enhancing its capabilities.

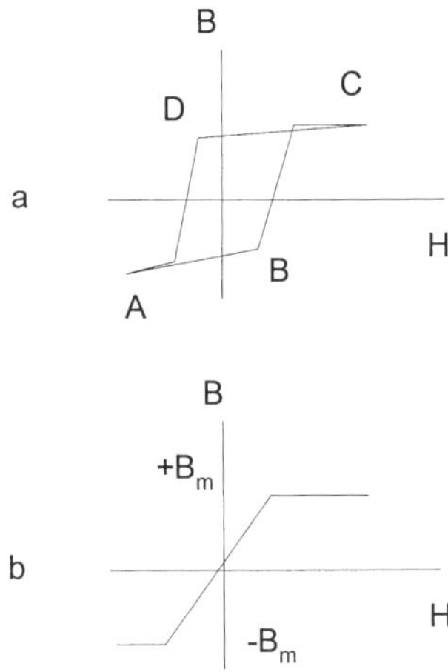

Figure C1.5.2.9. (*a*) Physical hysteresis in the BH plane and (*b*) approximated BH curve.

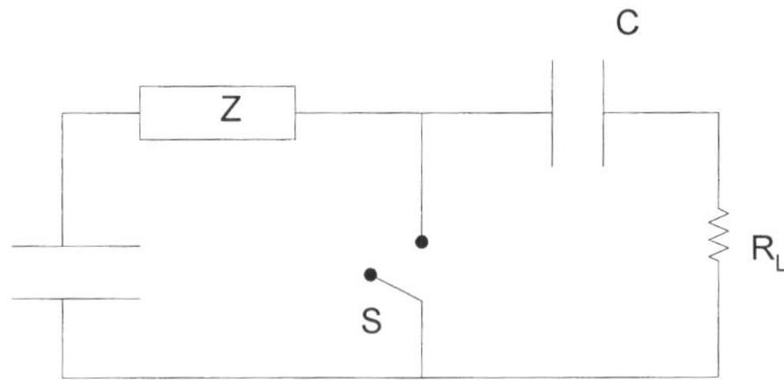

Figure C1.5.2.10. Elementary charging circuit.

C1.5.2.6 Charging

Figure C1.5.2.10 demonstrates a full capacitor discharge circuit—including charging. During recovery, the current through the charging impedance should not exceed the hold current I_h of the switch, otherwise it will latch. For efficient charging, Z needs to be reactive. There are several alternative configurations.

In section C1.5.2.3 we saw that in order to transfer energy efficiently, in just a half-cycle, from a capacitor to a resistive load, we need a low-Q circuit. Just the opposite is true in charging. Here we need a high-Q circuit—the reactive component should be much higher than the resistive one.

The simplest method is just to connect the capacitor to the voltage source by an inductor (resonant charging, figure C1.5.2.11).

Closing the command-charging switch (S_c) will produce (with a reasonable low-loss inductor) the following voltage on the storage capacitor:

$$V_c = (V_{cc} - V_0)\left(1 - \cos\pi\frac{t}{\tau}\right) + V_0$$

where V_0 is the initial voltage on the capacitor and τ is the half-cycle time:

$$\tau = \pi\sqrt{LC}.$$

On $t = \tau$ the voltage on C will be $2V_{cc}$. The voltage can be adjusted or stabilized in the range V_{cc}–$2V_{cc}$ by cutting off the charging current as the voltage on the capacitor reaches the target. Since current through

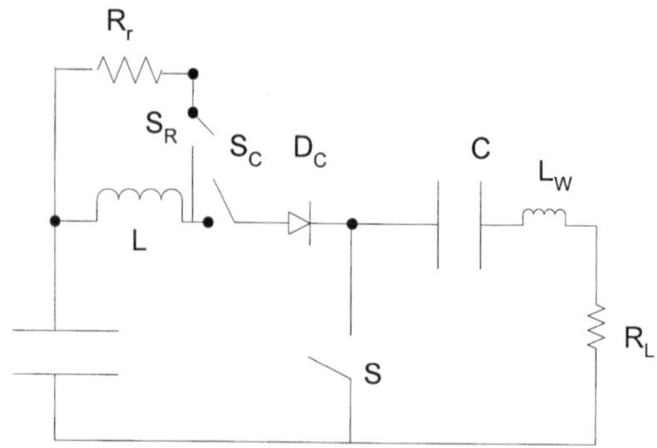

Figure C1.5.2.11. Controlled resonance command charging.

an inductor has to be continuous and, in the moment of interruption, its residual energy should be dumped somewhere, rerouting the current is required. In figure C1.5.2.11, the switch S_R diverts the current into the dump resistor R. If in the moment of switching θ the voltage on R_r will satisfy:

$$V = V_{cc} + I(\theta)R \leq V(\theta).$$

Then the diode D_c will be backward biased, stop conducting and the voltage on C will freeze on target. Many variations of this circuit are described in the literature, for example [1, 29].

The main switch S should be turned on only after the charging current is less then I_h. Otherwise, the continuity of the current through L will make it latch.

If S is uni directional, and C, L_w (the unavoidable circuit inductance) and R_L are not critically damped, then when the current through the switch ends, C will be charged to some negative voltage $-V_f$. The energy dissipated on the load is:

$$E_d = \tfrac{1}{2}C(V^2 - V_f^2).$$

Even if only a small fraction of the energy is reflected, the consequences are severe. when 80% of the energy is dissipated on the load, V_f is:

$$V_f = -\sqrt{0.2}\ \mathrm{V} = -0.45\ \mathrm{V}.$$

Furthermore, on the next charging cycle the target voltage will be 2 V + 0.45 V = 2.45 V. This is an escalating situation that should be avoided by using snubbers or by energy recovery [30]. An advanced approach is resonant inverters. They use the process described earlier deliberately to reach high voltages in a multiple-pulse build-up [31]. They can provide efficient and compact chargers as long as the load does not require a high repetition rate.

Flyback charging is a variation that requires an opening switch (figure C1.5.2.7(b)). The diode prevents the capacitor from discharging back into the inductor. Another degree of freedom is achieved if L is the leakage inductance of a transformer. That enables reversing the voltage, isolating and multiplying it.

C1.5.2.7 Wiring and noise reduction

Pulsed power generators can produce a large amount of EMI, in fact so high that it can impair their own operation. Switching 50 KV in 50 ns means $\mathrm{d}V/\mathrm{d}t$ of 10^{12} V s^{-1}, or driving 1 A of current through each 1 pF coupled to the high voltage side. Large $\mathrm{d}I/\mathrm{d}t$ will also generate considerable voltage on the wiring inductance: 5 kA in 50 ns means 100 V per each nH of induction.

To understand the consequences of this, see figure C1.5.2.12.

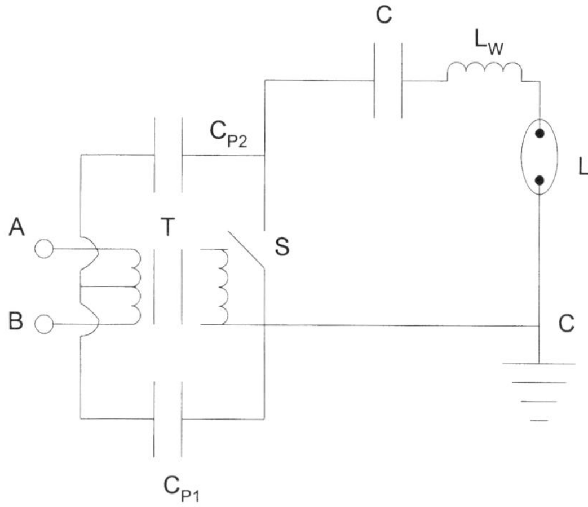

Figure C1.5.2.12. Origin of common-mode voltage on a service transformer.

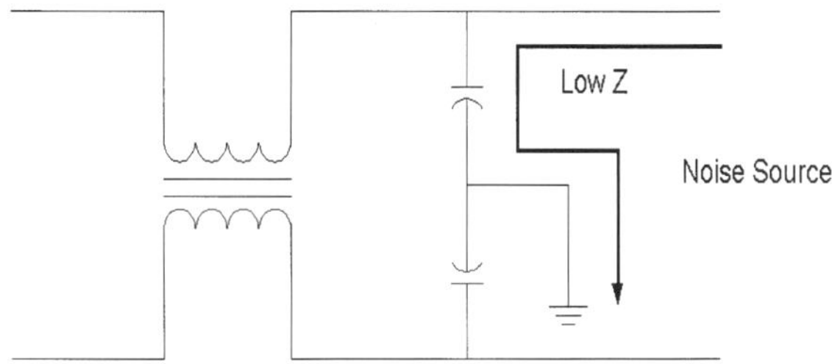

Figure C1.5.2.13. Common-mode noise suppression.

L is a laser tube, grounded at point C, for instance by the dc power supply. T is a service transformer such as a thyratron filament or trigger transformer. C_{p1} and C_{p2} are parasitic capacitances between the primary coil of the transformer and the high and low voltage sides of switch S, respectively. They are drawn as if they are were connected to a centre tap on the primary coil since they equally affect both its terminals, A and B. This shared voltage is called the 'common-mode' voltage, V_{cmd}. If we now ground either A or B we will get a common-mode current through this connection. The common-mode current can reach intolerable values and is the source of many serious interference and reliability problems.

Clearly, avoiding grounding the tube can eliminate the common-mode current. However, this is difficult, not only because it implies that everything connected to the laser should float too—including the power supply, mechanical support and plumbing—but also because any capacitance of the laser to the ground will introduce a common-mode current. The other alternative is to use common-mode filters as shown in figure C1.5.2.13.

A typical filter includes common-mode inductors and capacitors that contribute to the attenuation of the common-mode noise. The inductors become high impedances to the high-frequency noise and either reflect or absorb the noise, while the capacitors become low-impedance paths to ground and redirect the noise away from the main line.

In applications of fast pulse, like trigger pulses, these capacitors should be eliminated. Common-mode inductors are wound with two windings with an equal numbers of turns. The windings are placed on the core so that the line currents in each winding create fluxes that are equal in magnitude but opposite in phase.

These two fluxes 'cancel' each other, thus leaving the core in an unbiased state. The common-mode inductor can use a high permeability material and obtain a very high inductance on a relatively small core.

C1.5.2.7.1 Material selection

Most noise generated by pulse power generators is in the 1–100 MHz range. The impedance of the inductor must be sufficiently high over this frequency range to provide proper attenuation.

Total impedance of the common-mode inductor is composed of two parts: the series inductive reactance (X_s), and the series resistance (R_s). At low frequencies, the reactance is the primary contributor to impedance but as the frequency increases, the real part of the permeability drops and losses within the core rise. These two factors combine to help produce an acceptable impedance (Z_s) over the entire frequency spectrum.

For the most part, ferrites are the material of choice for common-mode inductors. They are divided into two groups: nickel zinc and manganese zinc. Nickel zinc materials are characterized by low initial permeabilities ($\mu < 1000$) but they maintain their permeabilities at very high frequencies ($\mu > 100$ MHz). They are most often used when the majority of unwanted noise is greater than 10 or 20 MHz. Manganese zinc materials can attain permeabilities in excess of 15 000 but may start to 'roll-off' at frequencies as low as 20 kHz. They are very well suited to EMI suppression in the 10 kHz through 50 MHz range.

Common-mode filters are often wound on toroids, since they have the highest effective permeability of any core shape.

The basic parameters needed for common-mode inductor design are input current, differential voltage, impedance and frequency. The input current determines the size of the conductor needed for the windings and the voltage determines the insulation required.

The impedance of the inductor is normally specified as a minimum value at a given frequency. This allows the inductance to be calculated:

$$L_s = X_s/2\pi f$$

Notice that common-mode filters can become saturated if their $nA\Delta B$ product is smaller than the $V\,\mathrm{d}t$ integral of the common-mode pulses or if the leakage magnetic field created by the differential signal exceeds a certain value. The core size and number of turns should be selected accordingly. For reliable operation, the filter should be adequately cooled and wound in such a way that does not introduce stress on the core material.

References

[1] Glasoe G N and Lebacqz J V (ed) 1948 *Pulse Generators (MIT Radiation Laboratory Series 9)* (New York: McGraw Hill)

[2] Sarjeant W J and Dollinger R E 1989 *High Power Electronics* (New York: TAB Books, McGraw Hill)

[3] Goldberg S and Rothstein J 1961 Hydrogen thyratrons *Advances in Electronics and Electron Physics* vol XIV (New York: Academic) pp 207–64

[4] Creedon J 1990 Design Principles and operation characteristics (of thyratrons) *Gas Discharge Closing Switches* ed G Schaefer, M Kristiansen and A Gunther (New York: Plenum) pp 375–407

[5] Fujii T, Nemoto K, Ishikiawa, Hayashi K and Noda E 1991 Development of solid state pulse power supply for copper vapor laser *Proc. SPIE 1412, Gas and Metal Vapor Lasers and Applications* ed J J Kim and F K Tittel, pp 50–8

[6] von Bergmann H M and Swart P H 1992 All-solid-state pulsers for high repetition rate multi-kilowatt lasers *IEE Proc.* B **139** 123–30

[7] Chatroux D, Maury J and Hennevin B 1993 Isolated gate bipolar transistors (IGBT): a solid state switch *Laser Isotope Separation, Proc. SPIE 1859* ed J A Paisner (Bellingham, WA: SPIE), pp 145–53

[8] Vorster A 1991 Replacement of Thyratrons by a MOSMATRIX *8th IEEE Int. Pulsed Power Conf. (San Diego, CA)* (Piscataway, NJ: IEEE) pp 1005–9

[9] Abas-Ogly Ya R, Aboyan S A, Abrosimov G V, Andrianov V A, Vasil'tsov V V, Semykin B I, Shilin A M and Shulakov V N 1981 Copper halide laser pumped with vacuum-tube and thyratron oscillators *Sov. J. Quantum Electron.* **11** 391–3

[10] Morgan A P 1920 *Wireless Telegraphy and Telephony* (New York: Henley)

[11] Kotov Yu A, Mesyats G A, Rukin S N, Filatov A L and Lyubutin S K 1993 A novel nanosecond semiconductor opening switch for megavolt repetitive pulsed power technology: experiment and applications *Proc. IX Int. IEEE Pulsed Power Conf. (Albuquerque, NM)* vol 1 (Piscataway, NJ: IEEE), pp 134–9

[12] Lyubutin S K, Mesyats G A, Rukin S N and Slovikovskii B G 1997 Subnanosecond high-density current interruption in SOS diodes *Proc. XI Int. IEEE Pulsed Power Conf. (Baltimore, MD)* (Piscataway, NJ: IEEE) pp 663–6

Rukin S N, Kotov Yu A, Mesyats G A, Filatov A L, Lyubutin S K, Alichkin E A, Darznek S A, Telnov V A, Slovikovskii B G, Timoshenkov S P, Bushlyakov A I and Turov A M 1994 Pulsed power accelerator technology based on solid-state semiconductor opening switches (SOS) *X Int. Conf. on High Power Particle Beams (San Diego, CA)* (Piscataway, NJ: IEEE) vol 1, pp 33–6

Also see: http://iep.iep.uran.ru/ENGLISH/PPL/Main.htm#Publications

[13] Cook E G, Ball D G, Birx D L, Branum J D, Peluso S E, Langford M D, Speer R D, Sullivan J S and Woods P G 1991 High average power magnetic modulator for copper lasers (invited) *8th IEEE Int. Pulsed Power Conf. (San Diego, CA)* ed R White and K Prestwich (Piscataway, NJ: IEEE) pp 537–42

[14] Bredenkamp G L, Nel J J and Mulder D J 1991 Transient voltage sharing in series coupled high voltage switches *8th IEEE Int. Pulsed Power Conf. (San Diego, CA)* (Piscataway, NJ: IEEE) p 1016

[15] akemoto m, anami s and honma h 2000 high-power switch with si-thyristors for klystron Pulse Modulators *IEEE 24th Int. Power Modulator Symposium (Norfolk, VA)* (Piscataway, NJ: IEEE)

[16] Braun C, Pastore R and Schneider S 1991 Progress towards an MCT-based 100+ kW high-frequency inverter *8th IEEE Int. Pulsed Power Conf. (San Diego, CA)* ed R White and K Prestwich (Piscataway, NJ: IEEE) pp 1009–12

[17] Arte'ev A Yu *et al* 1982 A multi sectional copper vapor laser *Kvant. Electron.* **9** 738–43

[18] Alexander R C 1999 *The Inventor of Stereo: The Life and Works of Alan Dower Blumlein* (Focal Press)

[19] Fitzsimmons W A, Anderson L W, Riedhauser C E and Vrtilek J M 1976 Experimental and theoretical investigation of the nitrogen laser *IEEE J. Quantum Electron.* **QE-12** 624–33

[20] Bokhan P A, Nikolaev V M and Solomonov V I 1975 Sealed copper-vapor laser *Sov. J. Quantum Electron.* **5** 96–8

[21] Pack J L, Liu C S, Feldman D W and Weaver L A 1977 High-average-power pulser design for copper-halide laser systems *Rev. Sci. Instrum.* **48** 1047–9

[22] Taylor R S and Leopold K E 1994 Magnetic-spiker excitation of gas discharge lasers, invited paper *Appl. Phys.* B **59** 479

[23] Vuchkov N K, Astadjov D N and Sabotinov N V 1988 A new circuit for CuBr laser excitation *Opt. Quantum Electron.* **20** 433–8

[24] Melville W S 1951 The use of saturable inductors as discharge device for pulse power generators *Proc. Inst. Electron. Eng.* **98** 185–207

[25] Birx D L, Lauer E J, Reginato L L, Rogers D Jr, Smith M W and Zimmerman T 1981 Experiments in magnetic switching (invited) *3rd IEEE Int. Pulsed Power Conf. (Albuquerque, NM)* ed T H Martin and A H Guenther (Piscataway, NJ: IEEE) p 262

[26] Smilanski I, Byron S R and Burkes T R 1982 Electrical excitation of an XeCl laser using magnetic pulse compression *Appl. Phys. Lett.* **40** 547–8

[27] Smilanski I 1988 Advances in magnetic pulse compression for copper-vapor lasers *18th IEEE Power Modulator Symposium* (Piscataway, NJ: IEEE)

[28] Nel J J, Mulder D J, von Bergmann H M, Swart P H and Tromp H T W 1994 High repetition rate, high voltage pulser for multi-atmosphere CO_2 laser *Proc. 21st Int. Power Modulator Symposium (Costa Mesa, CA)* (Piscataway, NJ: IEEE) pp 63–6

[29] Swart P, Bredenkamp B and von Bergmann H M 1987 A resonant pulsed power supply with pulse-energy regulation *Proc. 6th IEEE Pulsed Power Conf. (Arlington, VA)* (Piscataway, NJ: IEEE) pp 723–6

[30] Birx D E, Das P P, Fomenkov I V, Partlo W N and Watson T A 1998 Pulse power generating circuit with energy recovery, US Patent 5,729,562

[31] McMillan R W 1977 How to pick the best power-supply type for capacitor charging in a pulsed laser *Laser Focus* **13** 62–7

C1.5.3
Power conditioning: supplies for driving flash tubes and arclamps for solid state lasers

Mark Greenwood and D W Miller

C1.5.3.1 Introduction

Solid state lasing material, which is usually in the form of a rod, is optically excited (pumped) using a flashlamp (pulsed applications) or an arclamp (continuous use). Typical laser rods are 10–15 cm long and 4–10 mm in diameter with a matching size for the lamp. The lamp is closely coupled optically to the rod. The output from the laser is primarily controlled by delivering a pre-determined amount of energy per pulse into the flashlamp (pulsed output) or controlling the electrical power into the arclamp (continuous output). Monitoring the laser output and using this as a feedback signal to the power supply can achieve additional control. A power supply is needed to isolate the lamp from input supply variations, ignite the lamp into conduction and control the energy or power into the lamp. For pulse systems, there is an additional requirement to maintain the arc in the lamp between pulses.

C1.5.3.2 Lamp characteristics

The lamp consists of a quartz tube (envelope) with electrodes sealed into the envelope ends. The gas fill is usually either xenon or krypton. Xenon is chosen for pulsed flashlamps because of its higher conversion efficiency. At the lower current densities encountered in cw arclamps, krypton provides a better match to the absorption bands of Nd:YAG (see chapter B1.3). The electrodes are normally made of tungsten with the addition of thorium in the anode to improve machinability. The cathode will also incorporate additional material, such as barium-based compounds, to give a low work function so that it can supply electrons without damage to its surface.

At a current density below about 20 A cm^{-2} the V–I characteristic exhibits a negative resistance region where stable operation is only possible when the supply is high impedance so that the net circuit resistance is positive (see figure C1.5.3.1). At higher current densities the slope resistance becomes positive. The valley point is the transition between these two regions where the lamp voltage is at a minimum. From the valley point up to a current density of about 300 A cm^{-2} the lamp volt drop can be approximated by $V_L = V_0 + I_L \times R_S$ where V_L is the lamp voltage, V_0 a constant, I_L the lamp current and R_S the lamp slope resistance. This is the region where cw arclamps are operated. For a typical 6 kW arclamp with an arc length of 150 mm and an arc diameter of 4 mm $V_0 = 134$ V and $R_s = 3.37\ \Omega$.

At higher current densities the lamp exhibits a nonlinear characteristic given by $V_L = k_0 \times I_L^{0.5}$ where k_0 is the lamp impedance constant. It is this relationship that determines the pulsed flashlamp performance [3].

The lamp impedance constant, k_0, controls the damping factor in pulsed discharge circuits and is given by the equation

$$k_0 = 1.28 \times (L/d) \times (p/x)^{0.2}$$

where L is the arc length in mm, d the bore diameter in mm, p the fill pressure in torr and x a constant, 450 for xenon and 805 for krypton. For example, $k_0 = 31$ for a 1.5 kW xenon-filled flashlamp with arc dimensions of 97 mm by 4 mm diameter and a fill pressure of 450 torr.

In a pulse system, the equation,

$$\text{Pulse-life} = (E_0/E_\text{x})^{-8.5}$$

can be used to estimate the pulse lifetime of a flashlamp where E_0 is the energy per pulse in joules and E_x is the flashlamp explosion energy in joules. The pulse-life is very sensitive to variations in pulse energy due to the exponent value of '−8.5'. This equation should be used with caution for lamp lifetimes in excess of 10^6 pulses because electrode effects and quartz envelope erosion then determine lamp-life. For the 1.5 kW flashlamp described here a lifetime in the range $1–4 \times 10^7$ shots can be expected when the pulsewidth is 100 μs and the input energy per pulse is 50 J.

The explosion energy can be calculated from the equation,

$$E_\text{x} = K_\text{E} \times (T)^{0.5}$$

where T is the one-third pulsewidth in seconds and K_E is the flashlamp single-pulse explosion constant, supplied by the manufacturer [7].

The cooling method depends on the lamp wall loading. Above about 15 w cm^{-2}, water cooling should be used. The water should be de-mineralized so that deposits do not build up on the lamp.

For high-power (>1 kW), high-repetition-rate (>1 kHz) pulsed systems, acoustic resonance in the gas column can cause the arc to become unstable and the laser output to vary. Arc extinction and envelope fracture have been reported [6]. Using a high open-circuit simmer voltage can alleviate arc extinction (see next section). The authors have found envelope fracture in linear flashlamps is not a problem and instability can be avoided by ensuring the pulse operating frequency does not excite the longitudinal resonant modes in the discharge.

Acoustic resonances that can occur in a linear arc discharge are described in [6]. The equations can be applied to flashlamps used in pulse lasers. The manufacturer's technical manual [7] describes the construction and operation of flashlamps in some detail.

C1.5.3.3 Lamp trigger and simmer

To ignite the lamp into conduction the gas fill must first be ionized. This is achieved by generating a high-voltage low-energy trigger pulse. The trigger pulse can be applied either directly to one of the electrodes for series triggering or, for external triggering, to a wire wrapped around the lamp envelope. In both cases the trigger voltage is much higher than the operating voltage. Typically, the voltage pulse will be 10–20 kV, lasting for about 1 μs and with an energy of 0.1 J. External triggering is normally used in low-repetition rate low-energy systems where the lamp is air-cooled. Its advantages are its simplicity and the isolation of the trigger pulse from the rest of the circuit. For industrial systems where the lamp is water-cooled, series triggering is standard because isolating an external trigger wire from the coolant would be difficult. The disadvantage is that the connecting leads from the trigger transformer to the lamp must be rated for the trigger voltage and the circuit is more complex.

A trigger pulse can be generated by discharging a capacitor into the primary winding of a trigger transformer using a thyristor as the switch (see [7] and figure C1.5.3.2 for more details). The trigger capacitor, about 1 μF, is first charged up to about 800 V through the trigger transformer. The inductor in

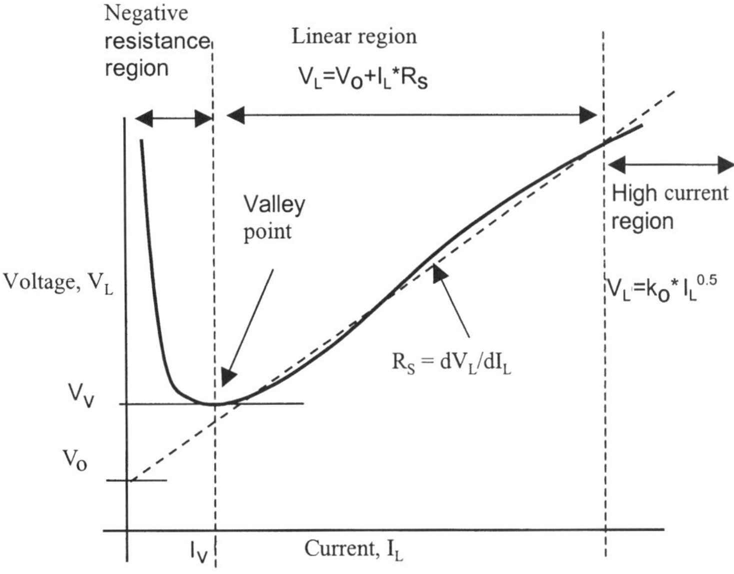

Figure C1.5.3.1. *V–I* lamp characteristic.

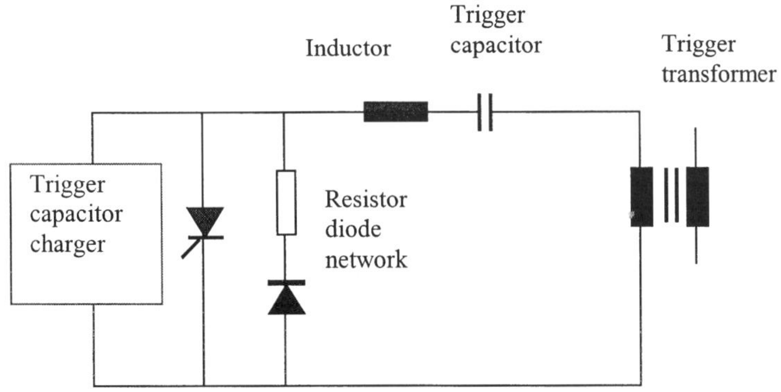

Figure C1.5.3.2. Typical trigger circuit.

series with this capacitor controls the discharge current and the resistor diode network across the thyristor prevents current oscillations. The transformer has a high turn ratio to generate the high voltage necessary to break down the lamp.

The trigger transformer design is important for reliable triggering. It must have a high turn ratio to generate the voltage required and be able to hold off this voltage for long enough to ionize the lamp. The series trigger transformer will be larger than the external trigger transformer because its secondary winding carries the flashlamp current. For pulsed systems, the saturated secondary inductance controls the circuit damping and a large core may be required to obtain the correct inductance, typically 10–40 μH (see section C1.5.3.4).

The simplest way to produce pulses is to charge up a capacitor and then discharge it through the lamp by triggering the lamp into conduction using one of the triggering methods shown in figure C1.5.3.3. They are usually confined to low-repetition-rate and single-shot high-pulse-energy systems.

Instead of breaking down the lamp every time a laser pulse is required, the lamp can be kept in

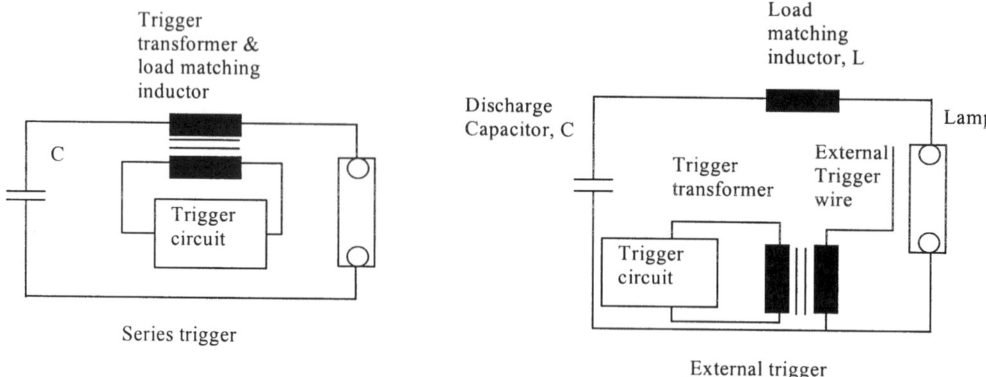

Figure C1.5.3.3. Series and external trigger circuits.

conduction between pulses. This technique is called simmering. A number of worthwhile improvements are then obtained. Lamp-life and pumping efficiency are increased while the laser pulse jitter and electrically generated noise are reduced. The simmer circuit consists of a supply that has an open circuit voltage of about 1400 V dc and a short circuit current between 0.1 to 1.0 A with a lamp simmer voltage of 100–200 V. This can simply be a dc supply with a series resistor or a more efficient current controlled switched-mode supply. The circuit is connected across the lamp. In order to trigger the lamp into conduction, the lamp must be driven down the negative resistance section of its V–I characteristic to a point where the simmer circuit can take over. As the simmer voltage is increased, less trigger energy is required. The high output resistance of the simmer circuit ensures stable operation.

Stable operation below the valley point may not be possible in cw systems where the output impedance of the power supply is low. The lamp must, therefore, be driven above the valley point where the lamp slope resistance is positive and the cw supply can take over. This transition usually requires more energy than the trigger circuit can deliver so another boost stage must be used as shown in figure C1.5.3.8. The initial trigger pulse is used to drive the lamp down to the boost voltage as for the simmer circuit. At this point the capacitor connected across the boost supply discharges into the lamp and drives the lamp into the positive resistance region. A resistor is connected in series with the boost capacitor to control the discharge current time constant. The cw supply must take over the current before the discharge pulse ends otherwise the lamp will revert to its non-conducting state.

C1.5.3.4 Pulsed power circuits

For reasonable efficiency the pumping width pulse should be less than the fluorescent lifetime of the laser rod. For a typical flashlamp this requires a drive voltage of 1–2 kV and peak currents of 1–2 kA when the lasing medium is Nd:YAG, which has a lifetime of 230 μs and an electrical-to-optical conversion efficiency of 1–3% (see chapter B1.3). The circuit elements of a pulsed supply are shown in figure C1.5.3.4 although not all of these may be required.

The basic circuit elements for pulsing the flashlamp are the energy storage capacitor and the matching network as shown in figure C1.5.3.4. The capacitor is first charged up to a pre-determined voltage and then discharged into the flashlamp through the network, which is normally a single inductor. The values of the capacitor and inductor are chosen so that the discharge current into the flashlamp is nearly critically damped without being under damped as this gives the shortest discharge pulse. The energy delivered to the flashlamp is given by $E = 0.5CV^2$ J and the power input is $P = E \times f$ W, where f is the pulse frequency. The required values of capacitor voltage, V, capacitance, C, and inductance, L, are calculated using the following

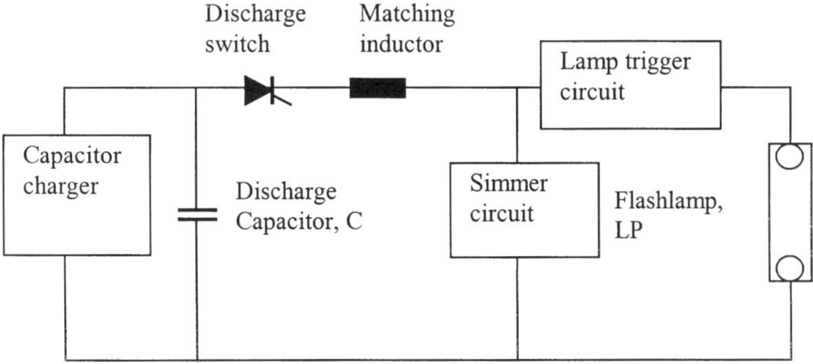

Figure C1.5.3.4. Pulsed supply circuit elements.

equations:

$$\alpha = k_0/(V \times Z)^{0.5}$$

where α is the damping factor and $Z = (L/C)^{0.5}$ Ω. $t = 3(LC)^{0.5}$ where t is the pulsewidth of the current pulse into the flashlamp. The peak discharge current is given by $I_{\mathrm{pk}} = 0.5V/Z$ when $\alpha = 0.8$ (critical damping). (See [3] for the derivation of these equations and [7] for application examples.) For example, with a flashlamp $k_0 = 31$, $C = 25$ μF, $L = 16$ μH and $V = 1400$ V, the energy per pulse, E, is 24.5 J giving 1.23 kW into the flashlamp at 50 Hz. The damping factor α is 0.93 and the pulsewidth is 60 μs, both of which are acceptable. The peak discharge current is 875 A. It should be noted that the value of α is voltage dependent due to the nonlinearity of the flashlamp. When V drops to 600 V, α increases to 1.4.

To take advantage of a simmered system, a semiconductor switch such as a thyristor would be used as shown in figure C1.5.3.4. When the lamp is simmering, the thyristor is turned on and the capacitor is completely discharged into the flashlamp. Once the flashlamp has de-ionized and the thyristor has returned to its off-state the capacitor can be re-charged ready for the next pulse. This system works well for most applications and repetition rates below a few hundred hertz. The upper frequency is limited by the need for the thyristor to recover between pulses before the capacitor can be recharged. Another disadvantage is that the circuit must be well matched to the flashlamp for the best energy transfer [3]. This means that the circuit component values will need to be modified if the flashlamp design, pulse width or pulse energy are changed.

For systems where pulsewidths of less than 10 μs and capacitor values in the region of a few microfarads are required then the charge voltage would be in the region of 5–10 kV. As this voltage approaches the flashlamp self-triggering level a high voltage switch such as a triggered spark gap or a thyratron would have to be put between the capacitor and series connected trigger transformer [7]. Switch life would then be an issue. A simple charging circuit that is suitable for use in the laboratory for low mean-power systems is a constant voltage charger with a resistor to limit the current into the capacitor as shown in figure C1.5.3.5. The maximum pulse repetition frequency is limited to about $1/(5RC)$ and the resistor power loss is equal to the power delivered into the flashlamp so the charger has to supply twice the lamp power.

For high-power high-pulse-rate systems, a more efficient charging method is needed. A series quasi-resonant converter [5] is ideally suited for this application as it is inherently short-circuit and open-circuit proof. The output is constant current so a current-limiting resistor is not required. The complete charger will consist of the converter and its dc voltage source, derived from the mains supply, together with the control electronics.

For simplicity, the ratio of switching frequency to the resonant frequency should be set to $\gamma = 0.5$. The charging current is then determined by Vdc, Lr, Cr and the transformer turn ratio. The control loop is on–off

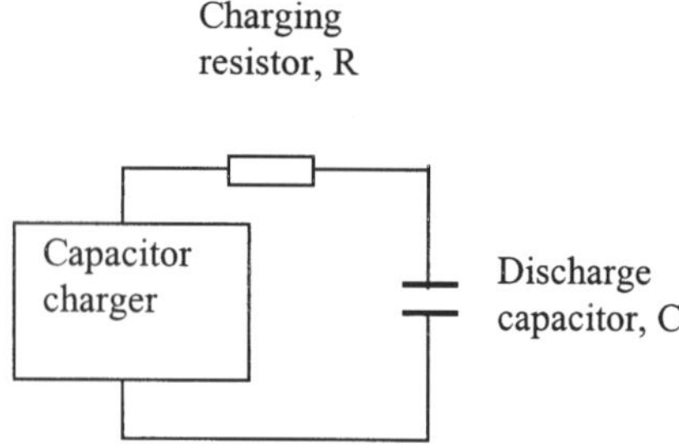

Figure C1.5.3.5. Constant voltage charger.

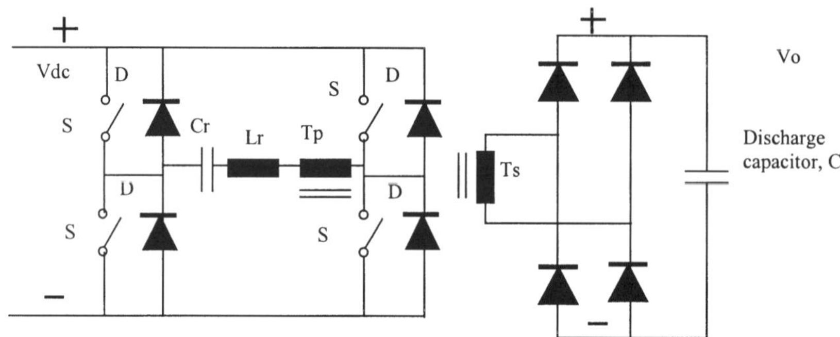

Figure C1.5.3.6. Series quasi-resonant converter.

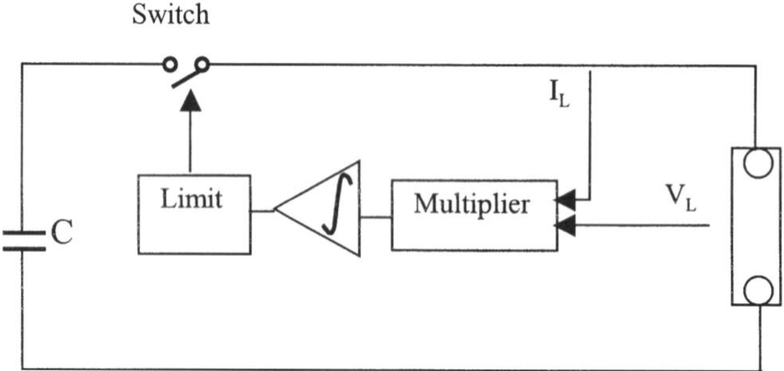

Figure C1.5.3.7. Active energy control.

(as opposed to proportional), that is the charger delivers full current until the output voltage reaches its set value then stops. The charger then turns on for short periods to maintain the capacitor at its set level.

The total discharge system using a thyristor switch and matching inductor is not flexible, as the pulsewidth cannot be varied without component changes. If the thyristor is replaced by a device such as an insulated gate bipolar transistor (IGBT) and the discharge capacitor is increased in value then a more flexible circuit results. The energy delivered into the flashlamp is continuously monitored during each pulse and when the set energy is reached, the control switch is turned off [1]. By this means independent control of pulse length and pulse energy is obtained and operation in the kilohertz region is easily achieved.

The input power factor can be improved and the input current reduced by using either passive or active power factor correction at the power supply input [4]. The active power factor correction is the preferred

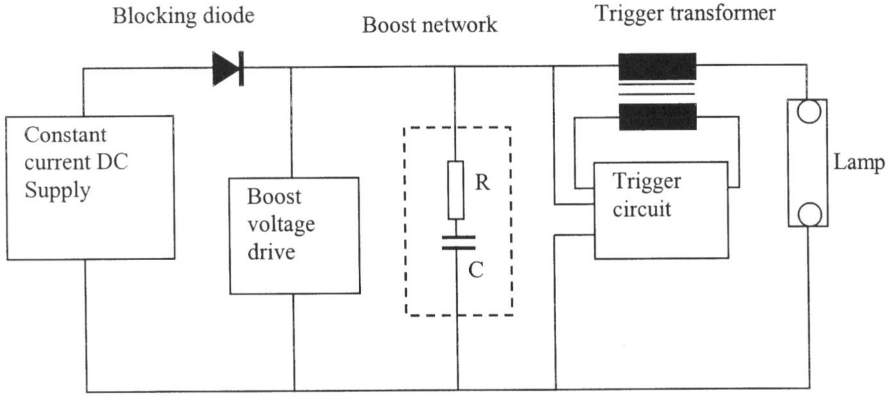

Figure C1.5.3.8. CW supply circuit elements.

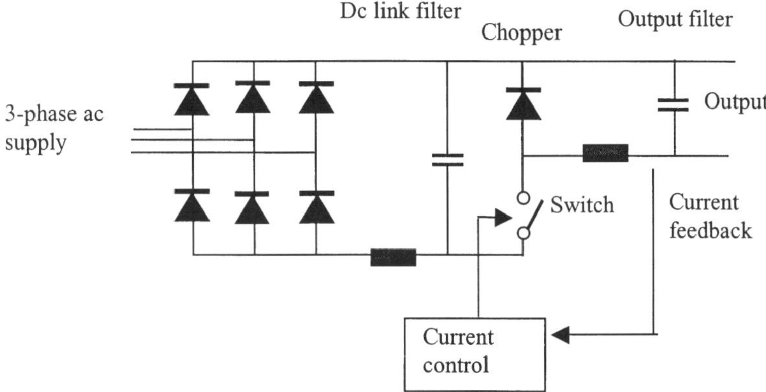

Figure C1.5.3.9. Constant current charger.

choice for single-phase input and is simple to implement using readily available control integrated circuits [8]. For three-phase inputs the more complex 'Vienna' rectifier can be used [2].

C1.5.3.5 CW power circuits

A basic cw supply incorporating the trigger and boost circuits described in section C1.5.3.3 is shown in figure C1.5.3.8. The lamp is ignited into conduction in three stages. The trigger pulse is first applied to drive the lamp voltage down to the boost level of typically 1400 V. The R–C boost network then discharges to take the lamp through the valley point (about 90 V) and into the positive slope region of the V–I curve. The lamp can then be driven from the supply. The boost circuit is a current limited dc supply as for the simmer supply used in pulse systems. However, it can be much smaller as it only has to operate during start-up. The blocking diode prevents the boost voltage from feeding back into the power supply. The lamp power supply is invariably constant current rather than constant voltage, as this gives better control and offers better protection to the lamp. When constant power control is required, the power control loop is the outer one and its error signal is used as the current reference input to the inner current loop. Where the lamp is not required to be electrically isolated from the mains, the basic circuit shown in figure C1.5.3.9 can be used.

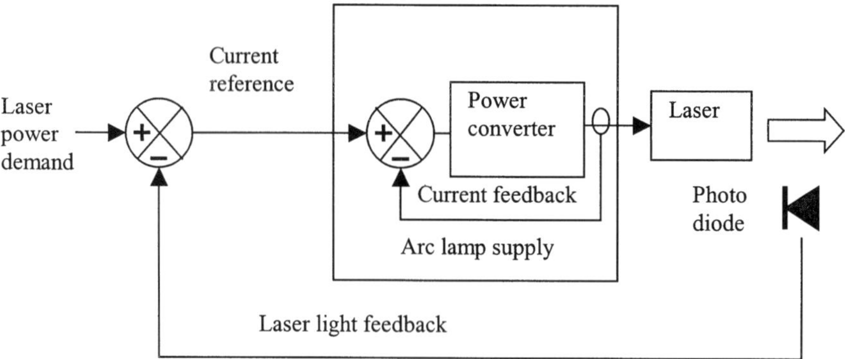

Figure C1.5.3.10. CW laser light stabilization.

The mains supply is rectified and fed into an inductive dc link filter. This L–C filter provides a low impedance source to the pulsewidth modulator (chopper) and improves the mains input power factor. The output L–C filter removes the switching harmonics generated by the modulator to provide a smooth dc output. The current in the output filter inductor is triangular. Monitoring the current in this inductor and comparing it to the required output current achieves control. The modulator switch-on and -off times are controlled to keep the current between two set levels, above and below the required current. The output current is the mean of these two levels. This is known as hysteresis control.

Controlling the current in the output inductor gives a number of benefits compared to using the output current after the filter. The peak current is controlled in the modulator switch so current overshoots are prevented under all operating conditions. The control loop is inherently stable as there are no subharmonics present due to the twice-per-cycle sampling. This makes for a well-behaved supply. When the lamp needs to be isolated from the mains supply, a high-frequency bridge can be used instead of a chopper. The output from the bridge is then transformer-isolated, rectified and smoothed to provide the dc output. In this case, the output current is monitored.

C1.5.3.6 Laser output control

Open-loop control of the lasers output energy is achieved by controlling the energy delivered to the flashlamp (pulse systems) or continuous power input into the arclamp (continuous systems). Where closer control is required, the laser output can be monitored and controlled. For cw systems, a large-area photodiode monitors the laser light and the output of the diode is used as a feedback signal to control the current into the arclamp. A small fraction of light can be picked off from the rear mirror or from the main beam as shown in figure C1.5.3.10. For pulse systems, averaging the energy per pulse using a photodiode detector in closed-loop control can be used.

C1.5.3.7 Specifications and standards

In order to sell mains-driven power supplies within the European Community the equipment has to comply with the EEC Low Voltage Directive, 73/23/EEC, and the Electromagnetic Compatibility (EMC) Directive, 89/336/EEC. It is generally sufficient to demonstrate compliance by ensuring that the equipment meets the appropriate national standards. The main body in the UK for issuing these standards is the British Standards Institution (BSI). Test houses, such as Technische Überwachungsverein (TUV) and Landesgewerbeanstalt Bayern (LGA) in Germany, can advise on the required standards and issue Certificates of Conformity if they carry out the testing. For instance, the standards that could be included for considerations for the electrical safety of laser power supplies are EN60950 covering information technology equipment and EN60601-1 and EN60601-2-22 covering medical electrical equipment. If the equipment is to form part of a machine, then it would make sense to ensure that the appropriate parts of EN60204-1: Safety of Machinery, Electrical Equipment are met. To comply with electromagnetic compatibility (EMC) requirements, the equipment must meet limits on electrical noise and harmonic currents injected into the mains supply and electrical signals radiated into the air. In addition to this, tolerance to injected noise from the mains and to electrostatic discharge are also required. In the first instance, the only way to demonstrate compliance is by testing to the appropriate standards. As the test equipment is expensive and specialist knowledge is required, it is worth considering using a National Measurement Accreditation Service (NAMAS)-accredited test house. The test house will be able to advise on the correct standards to use.

References

[1] Greenwood M and Miller D W 1999 Power electronics for pulsed Nd:YAG laser systems *IEE Colloquium April 1999 Power Electronics for Demanding Applications* (London: IEE) pp 8/1–8/5

[2] Lindemann A and Kolar J E 2001 Single and three phase rectifiers with active power factor correction for enhanced mains power quality *Application Note* (IXYS Corp.) http://www.ixys.com/t062201c.pdf

[3] Markiewicz J P and Emmett J L 1966 Design of flashlamp driving circuits *IEEE J. Quantum Electron.* **QE-2** 707–11

[4] Tarter R E 1993 *Power Factor Correction, Solid State Power Conversion Handbook* 1st edn (New York: Wiley) pp 695–705

[5] Tarter R E 1993 Series resonance *Solid State Power Conversion Handbook* 1st edn (New York: Wiley) pp 448–57

[6] Witting H L 1978 Acoustic resonances in cylindrical high pressure arc discharges *J. Appl. Phys.* **45** 2680–3

[7] High Performance Flash and Arc Lamps. Technical Manual by Perkin Elmer optoelectronics, Saxon Way, Barr Hill, Cambridge, CB3 8SL, UK

[8] UCC3817 BiCMOS 2000 Power factor re-regulator (Texas Instruments) http://www-s.ti.com/sc/ds/ucc3817.pdf

Further reading

Tarter R E 1993 *Solid State Power Conversion Handbook* 1st edn (New York: Wiley)

A good reference book for engineers involved in solid-state power conversion equipment design.

C2
Optical pulse generation

Clive Ireland

The most commonly used categorization of lasers is into continuous wave (cw) or pulsed operation. This comes from the earliest days of laser research investigating the practical possibility of operating three and four-level systems [1, 2]. Subsequently, experiment confirmed that generating sufficient inversion for lasing in three-level systems requires high peak power pulse excitation. However, this is not to say that the majority of pulsed lasers in use today are based on three-level systems. Although there are more than 15 000 laser transitions reported in a broad range of media (the majority only providing weak output under powerful pulsed excitation), even for those that can be operated cw, the great majority finding practical use are employed with modulation or pulsed control of the beam [3].

There can be many applications advantages of lasers producing pulsed output and, indeed, many that just would not work with a cw laser beam. In particular, pulsed operation allows a whole range of applications, including basic research, requiring radiation of high (or very high) peak power that is both spatially and temporally very well defined.

Today, pulsed lasers cover sources producing beams from quasi-cw through to the femtosecond regime. Here, for convenience, they are considered in three time domains; i.e. quasi-cw and modulated beams, beams comprising short pulses and beams of ultrashort pulses. Although this division might initially seem somewhat arbitrary, it has the merit of corresponding reasonably well to that of the different technologies employed to achieve pulsing across the full time regime. Somewhat coincidently, part D of this handbook shows that the division also has parallels for many of the key areas of application of pulsed lasers.

In this section, the time regimes are broadly defined and the technologies used for laser pulse generation in each is described and explained. Where appropriate, examples are given of important types of pulsed laser and the performance that can be achieved. As noted earlier, pulsed performance impacts significantly on applications and this has resulted in a continuing strong R&D effort directed at enlarging the pulse regime envelope, particularly into the attosecond range.

References

[1] Schawlow A L and Townes C H 1958 *Phys. Rev.* **112** 1940

[2] Maiman T H 1960 *Phys. Rev. Lett.* **4** 564

[3] Weber M J (ed) 1999 *Handbook of Laser Wavelengths* (Boca Raton, FL: Chemical Rubber Company)

C2.1
Quasi-cw and modulated beams

K Washio

Laser beams with reproducible and arbitrary controllable pulse shapes are often required for controlling process quality and increasing throughput in material processing. Laser beams with low-noise, single-longitudinal oscillation modes are often required for coherent optical measurements. This section deals with the topics of quasi-cw and modulated beams in the time domain where pumping control or gain dynamics predominantly determine the laser's temporal performance and omits other cases where intracavity active optical components such as Q-switches, modulators, etc, are used.

C2.1.1 Operation of solid state lasers

Solid state lasers are optically pumped by means of either lamps or other lasers, e.g. diode lasers. Therefore, the temporal characteristics of such pumping sources, in addition to the gain dynamics in the laser media, greatly influence the performance of solid state lasers. The relatively long lifetime of the upper laser level (around 0.24 ms for Nd:YAG, for example) limits full-modulation-depth highly repetitive pulsed operation, up to several kHz without utilizing intracavity optical modulators. Adequately long pre-lasing auxiliary excitation can relax the peak power requirement somewhat for pumping sources for fast short-pulse generation and help reduce the laser pulse delay and timing jitter.

C2.1.1.1 Lamp-pumped operation

It has been shown that krypton-filled flashlamps (see chapter C1.5.3) can be efficiently operated at high repetition rates by using simmer current densities of the order of 10 A cm^{-2}. There are several advantages to be gained from using high simmer currents:

(1) it helps prevent lamp misfiring;

(2) it stops the lamp extinguishing;

(3) it removes amplitude and timing jitter from the flashlamp output and;

(4) it improves the energy transfer from the pulse-forming circuit to the laser rod.

A flashlamp-pumped Nd:YAG laser with pulses exceeding 1500 pulse s^{-1} has been obtained [1].

The traditional way of producing high-power high-voltage current pulses was to use a pulse-forming network (PFN), consisting of capacitors and inductors in a network configuration. As the flashlamp impedance is not constant, perfect matching between the PFN and the flashlamp is only possible at one operating voltage. It is very difficult to obtain efficient operation for pulselengths of more than 10 ms. PFNs are very limited in their range of pulse duration adjustment. With the emergence of recently developed high-power solid state switching devices such as gate-turn-off thyristors (GTOs), insulated-gate bipolar transistors (IGBTs) and power MOSFETs, PFNs have almost been replaced by such high-power solid state switching devices. There are a number of advantages in utilizing a transistorized power supply with feedback control of the flashlamp

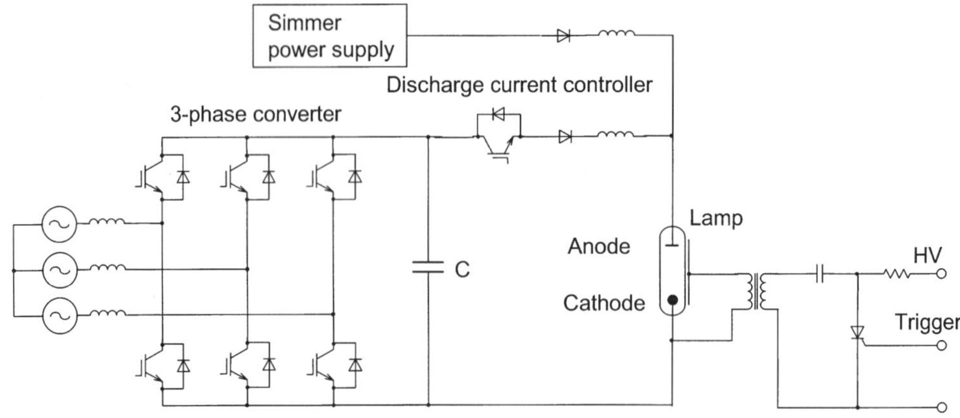

Figure C2.1.1. Schematic diagram of power supply with lamp discharge current controller.

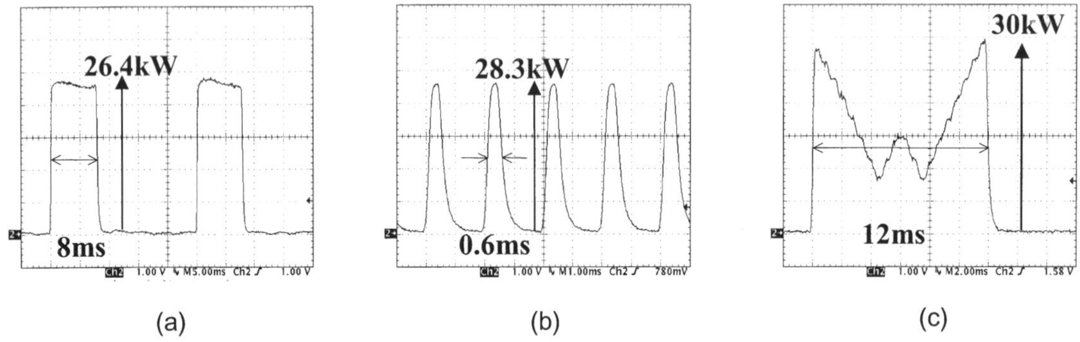

Figure C2.1.2. Waveform examples for a 10 kW average-power Nd:YAG laser system: (*a*) large-pulse energy operation with pulse repetition rate of 40 Hz; (*b*) 500 Hz, high-repetition rate operation; and (*c*) arbitrary pulse shape with 20 segments in a waveform.

current in a Nd:YAG laser for material processing. These include the ability to tailor the pulseshape and modify the pulse parameter (height, width and rate) on a pulse-to-pulse basis. The current is controlled within a feedback control loop, which minimizes the output variation as a result of changes in the supply-line voltage or flashlamp impedance [2]. Various pulses, such as flat-top rectangular, ramp-up or ramp-down ones,with pulse durations in the range 0.1–20 ms are widely used for laser welding (see chapter D1.1). Figure C2.1.1 shows a schematic diagram of a power supply with a lamp discharge current controller.

High-power lamp-pumped Nd:YAG lasers of 6 kW and 10 kW average output power with good beam quality have been developed by using a master oscillator power amplifier (MOPA) arrangement with multi-pump cavities for the oscillator and amplifiers [3]. By pulse modulation control of the power supplies, peak laser powers three times the average power output and fast repetition modulation as high as 500 Hz have been achieved. For a 6 kW average-power Nd:YAG laser, a high peak power exceeding 18 kW has been obtained with a pulsewidth of 0.6 ms and a repletion rate of 500 Hz. For a 10 kW average power laser, a peak power exceeding 26 kW has been obtained with a pulsewidth of 8 ms. Figure C2.1.2 shows examples of waveforms from a 10 kW average-power Nd:YAG laser system [3].

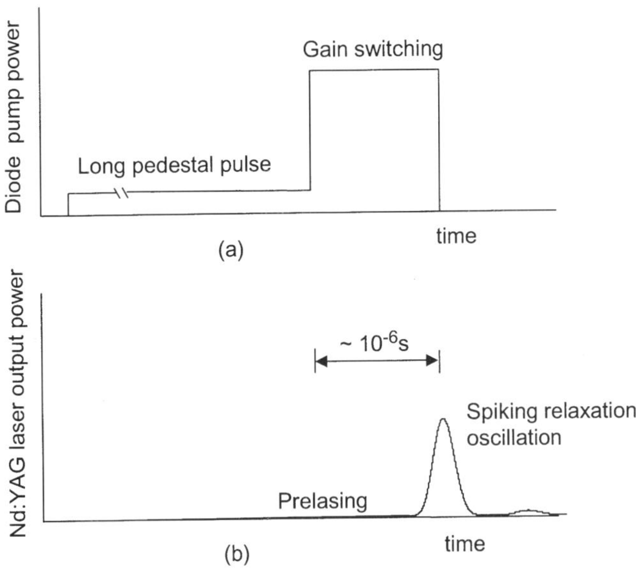

Figure C2.1.3. (*a*) Diode-laser pump power pulse shape; (*b*) gain-switched Nd:YAG laser output pulse shape.

C2.1.1.2 Diode-pumped operation

Recent progress in high-power diode lasers (see chapter B2.4) has so far realized up to multi-kilowatt, high-brightness diode-pumped lasers for precision machining [4]. An average output power of up to 3.6 kW has been realized with good beam quality two to four times diffraction limited, by pumping a zigzag Nd:YAG slab laser by stacked quasi-cw laser diode arrays. Typical diode pulse durations range from 0.1 to 1 ms. In multi-mode operation, 5 kW has been obtained with an optical efficiency of 35%.

The adoption of diode-laser pumping imparts a number of advantages to solid state lasers. Excellent modulation controllability and fast excitation capability realize solid state lasers with higher performances than those attained with lamp pumping.

The low-frequency dynamics of four-level lasers, such as Nd:YAG lasers (see chapter B1.3), have been investigated by many researchers [5, 6, 8]. With gain or loss perturbation, the response of the homogeneously broadened, single-frequency solid state laser can exhibit one sharp peak in the vicinity of the relaxation oscillation frequency together with spiking oscillation or regular pulses of high peak intensity can be obtained. Spiking behaviour becomes chaotic for multi-mode operation.

Gain switching has been shown to be an effective method for obtaining single-frequency short-pulse operation from a miniature monolithic Nd:YAG laser [6]. This method utilizes a low pumping power around 70 mW to select a high-gain single-longitudinal mode of the Nd:YAG laser near its lasing threshold with a long pedestal pulse of about 20 μs duration. This longitudinal mode is then driven by a gain-switching pulse to produce a single-frequency spiking relaxation oscillation. A short (less than 1 μs), high-power (335 mW) gain-switching diode-laser pulse drives the Nd:YAG well above threshold and produces the desired single-frequency emission. Figure C2.1.3 shows schematic diagrams of the pulse shapes for diode-laser pump power and gain-switched spiking Nd:YAG laser output power. With gain switching, the monolithic Nd:YAG laser generates single-frequency pulses with a bandwidth of less than 8 MHz (FWHM) at repetition rates up to 7 kHz. With use of an Nd:glass laser amplifier, single-frequency Nd:YAG pulses can be boosted to powers exceeding 200 W.

In pulsed diode bars used to pump solid state lasers, the temperature rise during the pulse results in a change in the centre wavelength, known as a chirp, and a reduction in efficiency and output power, known as sag [7]. Since many solid state materials such as Nd:YAG have absorption features only a few nanometres in width, efficient system operation requires a correspondingly narrow diode emission including the effects of wavelength chirp. An AlGaAs diode laser shows a shift of about 0.3 nm °C in the emission centre wavelength. In high-peak-power diode-pumped solid state laser systems, diode chirp and sag during the pump can play a major role in determining the pump energy absorbed in the laser medium.

A substantial reduction in intensity noise has been demonstrated for diode-laser-pumped Nd:YAG lasers by application of electronic negative-feedback loops. A fraction of the laser light is detected with a photodiode and ac components are appropriately amplified to generate an error signal. This signal is then fed back to the pump diode. By carefully adjusting the parameters of the feedback loop, suppression of about 25 dB at low frequencies and more than 50 dB at the relaxation oscillation frequency is achieved [8].

C2.1.1.3 Effects of thermal distortion in solid state lasers

One of the big problems in solid state lasers is thermal distortion of the laser medium and beam quality deterioration which are commonly seen in high-power or highly repetitive operation (see chapter C1.2 for correction methods). Detailed interferometric measurements and comparisons have been made of the thermal distortions in xenon-flashlamp- and diode-pumped Nd:YLF laser rods [9]. In both cases, defocus and astigmatism were the dominant thermal distortions. The thermal distortions in the flashlamp-pumped rods were greater than the thermal distortion in the diode-pumped rod when pumped to the same small-signal gain. In the diode-pumped Nd:YLF laser system, a 5 mm-diameter × 54 mm-long Nd:YLF laser rod was used and was pumped with three banks of laser diodes along the length of the rod. The thermal relaxation time constant of their diode-pumped Nd:YLF rod was measured to be approximately 1.5 s.

C2.1.2 Operation of CO_2 lasers

Quasi-cw or modulated CO_2 laser beams are extensively used, particularly in materials processing (see section D1). Among excitation schemes using dc, rf and excitation microwave, rf excitation now seems to be the most popular pumping scheme due to its excellent controllability and ease of incorporation into laser systems. Modulated laser operation has been easily realized by modulating the rf excitation directly. The rf frequencies used for CO_2 lasers (see section B3.1.1) vary from 100 kHz to about 100 MHz, depending on the type of laser systems. The average laser power can be controlled by rf phase modulation or varying rf amplitude in cw mode or by gating rf power in pulsed mode with the appropriate duty cycle and repetition frequency. The average power can be more widely controlled stably in pulsed mode than in cw mode. Figure C2.1.4 shows schematic diagrams of waveforms for gated rf power and CO_2 laser output pulses.

A fast axial flow CO_2 laser system using a 13.56 MHz rf discharge with an output power of up to 1500 W is well known from the early days (around 1986) [10, 11]. Pulse repetition rates between 100 Hz and 12 kHz at duty cycles of 0.1–1 were attained [10]. The rf power was coupled capacitively through an Al_2O_3-plate into the discharge volume of 4 cm × 4 cm × 95 cm. With a maximally available rf-power of 50 kW, a power density of 33 W cm^{-3} has been achieved during pulsed operation [10].

An rf discharge with a frequency below 1 MHz is often called a silent discharge (SD). Transverse flow, as well as fast-axial flow CO_2 lasers with SD excitation have been investigated [12, 13]. When an SD excitation scheme is compared with other rf excitation schemes with higher frequencies (typically 13.56 MHz, 27 MHz, 40.68 MHz, 80.7 MHz etc) or a microwave excitation scheme (typically 2.45 GHz), it has the following major advantages: (1) high-efficiency high-power solid state power sources are readily available and (2) the efficiency of the energy input is high because the power source and the discharge rod are easily coupled effectively. The performance characteristics of the TEM_{00} 2.5 kW SD-excited, transverse-flow CO_2 laser were investigated [12]. Each electrode was a square-shaped metal pipe covered with borosilicate glass 0.8 mm

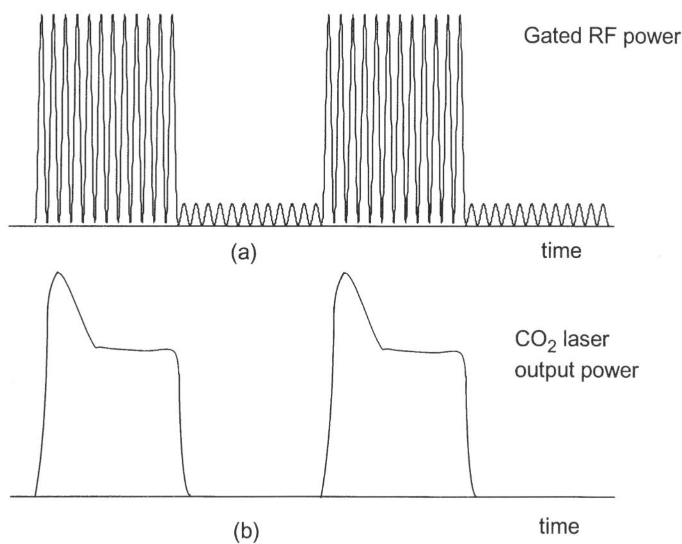

Figure C2.1.4. (*a*) Gated rf power waveform; (*b*) CO_2 laser output pulses.

thick. The discharge space is 45 mm wide in the electric field direction. The power source, which consisted of an inverting circuit switched by a transistor (MOSFET) and a transformer, provided 100 kHz, high-voltage, ac. Pulsed operation of the laser power has been achieved by modulating the applied voltage (around 10 kV) to the electrodes. When the frequency was low, flat-top pulses were obtained. When the frequency was high, the pulse shape lost its flat bottom.

The operation of a pulsed transverse-flow CO_2 laser system with rf excitation has been studied theoretically by computer simulation [14]. The pulse repetition frequency of the CO_2 laser system is restricted by the time delay between the electric field and the laser pulse and by relaxation oscillations. The time delay is roughly inversely proportional to the gas pressure. A pulse repetition frequency over some 30 kHz can be attained only if the excitation pulse period is equal to the delay time but this operation is very sensitive to changes in the duty cycle. Above several 100 mbars, the discharge may become inhomogeneous and can break down in to an arc discharge, which reduces the efficiency.

The response of rf-excited CO_2 lasers to pulsed modulation can be improved by increasing the gas pressure and decreasing the discharge gap distance inversely to keep the gas pressure–discharge gap distance product constant. A fast-axial-flow, rf-excited slab tube CO_2 laser incorporating a high-aspect-ratio rectangular-flow tube and a 5 mm discharge gap has been investigated [15]. The rf frequency was 27 MHz. This laser is expected to be free of thermal filamentation up to 800 mbar. In pulsed operation at pressures of around 400–540 mbar, a repetition frequency of 15 kHz has been obtained with the pulse duration reduced to 15–25 μs, suitable for processing low-thermal-conductivity materials (see section D1.2 and D1.6). Diffusion-cooled, sealed CO_2 slab lasers are advantageous for obtaining efficient high-power operation with smaller inter-electrode distances.

The dependency voltage–current characteristics of a transverse rf-discharge-excited waveguide CO_2 laser have been investigated over a wide range of excitation frequencies (f) (100–160 MHz), for gas pressures of 40–100 Torr and inter-electrode distances (D) of 1–3 mm [16]. It has been found that there is a scaling law—of $fD =$ constant—for such rf-excited CO_2 waveguide lasers.

The trend in rf-excited CO_2 lasers is towards sealed systems and the power of these systems is increasing. There is a commercially available 'quasi-sealed' 3.5 kW laser (basically cw but it can be modulated). Sealed laser discharges have to be narrow to allow gas cooling by diffusion and they should be excited at high frequencies to be efficient, to avoid arcs, etc. The narrower they get, the higher the required gas pressure becomes for stable operation.

A high-power, high-repetition-rate CO_2 laser system can be constructed with the aid of an external modulator. A 1 kW laser capable of full depth modulation in excess of 200 kHz has been developed to satisfy a need within the flexographic printing industry [17]. The laser is based upon a diffusion-cooled CO_2 laser converted into an oscillator–acousto-optic-modulator–amplifier configuration. Switch-on and switch-off both occur within 1 μs, easily achieving full depth modulation at the 200 kHz frequency. Full depth modulation has been achieved up to 500 kHz.

C2.1.3 Examples of quasi-cw or modulated beam applications

A wide variety of pulsed or modulated laser beams are required for high-quality materials processing such as drilling, cutting, welding and hardening (see chapter D1).

In the drilling of microvias for epoxy-glass printed wiring boards (PWBs), for example, the biggest problem to be solved is carbonization of the epoxy resin. The thickness of the heat-affected layer can be reduced from 8 to 2.6 μm by decreasing the pulse duration from 500 to 50 μs and, again, to 1 μm by decreasing the pulse duration to 10–20 μs [18]. On the basis of these data, a new laser drilling system has been developed for PWBs to allow high-quality drilling at the maximum drilling rate of 500 holes s^{-1} with a pulse duration that can be set in the range 16–250 μs [18].

In the cutting process, piercing is performed before cutting is started. To minimize the total piercing time, the laser power is preferably modulated to obtain the desired output waveform during piercing. Ramping is recommended when it is necessary to pierce the work gently to make the first hole. Laser power can be programmed to vary as a function of work-feed rate during cutting. The necessary data for power control can be stored in the laser controller unit. Pulsing is commonly used to reduce burning on corners or detailed work (see chapter D1.2). Furthermore, pulsing is often used to obtain cut surface improvement during cutting. Using the correct pulsing parameters, it is possible to produce cut-edge striations which are very close together and this, in turn, means that the cut surface quality improves because the roughness diminishes.

A mathematical analysis describing keyhole laser welding (see chapter D1.1) by pulsed beams has been investigated [19]. The effect of temporal pulseshape on weld dimensions has been examined over a range of power densities, pulse times and pulse frequencies. Several pulse types (e.g. top-hat, Gaussian, ramp-up and ramp-down) have been considered. Pulse shape had a significant effect on weld dimensions. The predicted width and penetration depth of weld with a ramp-up and ramp-down pulse were greater than those obtained with Gaussian and top-hat pulses.

The cause of penetration instability and counter measures were investigated for butt-welding A3003 aluminium alloys [20]. A slab Nd:YAG laser was used for carrying out the welding experiment in order to exclude instability factors such as a change in the laser beam quality. It was found that the addition of a high peak power, short width pulse superimposed on the early part of the rectangular pulse is very effective in preventing the penetration instability of A3003 alloy.

The weld profile with cw processing commonly shows a wine-cup configuration because a large amount of laser energy is absorbed near the surface. However, when a pulse-modulated beam was used with an increased peak power of up to three times the average power, the weld profile was found to have a larger penetration depth with a wider width near the bottom than those obtained by cw beams [3]. Two and threefold pulse enhancement seems to be very effective for multi-layer deep-penetration overlap welding with low thermal distortion of the workpiece.

References

[1] Corcoran V J, McMillan R W and Barnoske S K 1974 Flashlamp-pumped YAG:Nd^{+3} laser action at kilohertz rates *IEEE J. Quantum Electron.* **QE-10** 618–20

[2] Weedon T M W 1987 Nd–YAG lasers with controlled pulse shape *Proc. LAMP '87 (Osaka, May)* (Osaka: High Temperature Society) pp 75–80

[3] Ishida T, Togawa T, Morita H, Suzuki Y, Okino K, Takenaka H, Kubota K, Washio K and Yamane T 2000 6 kW and 10 kW high power lamp pumped MOPA Nd:YAG laser system *Proc. SPIE* **3888** 568–76

[4] Machan J, Moyer R, Hoffmans D, Zamel J, Burchman D, Tinti R, Holleman G, Marabella L and Injeyan H 1998 Multi-kilowatt, high brightness diode-pumped laser for precision laser machining *Technical Digest of Advanced Solid-State Lasers Topical Meeting (Coeur d'Alene, Idaho, February)* AWA2 (Washington DC: OSA) pp 263–5

[5] Harrison J, Rines G A and Moulton P F 1988 Stable-relaxation-oscillation Nd lasers for long pulse generation *IEEE J. Quantum Electron.* **24** 1181–7

[6] Hadley G R, Owyoung A, Esherick P and Hohimer J P 1998 Numerical simulation and experimental studies of longitudinally excited miniature solid state lasers *Appl. Opt.* **27** 819–27

[7] Honea E C, Skidmore J A, Freitas B L, Utterback E and Emanuel M A 1998 Modelling chirp and sag effects in high-peak power laser-diode-bar pump sources for solid state lasers *Advanced Solid State Lasers (OSA TOPS 19)* ed W R Bosenberg and M M Fejer (Washington, DC: OSA) pp 326–32

[8] Freitag I, Tünnermann A and Welling H 1997 Intensity noise transfer in diode-pumped Nd:YAG lasers *Advanced Solid State Lasers (OSA TOPS 10)* ed C R Pollock and W R Bosenberg (Washington, DC: OSA) pp 380–5

[9] Skeldon M D, Saager R B and Seka W 1999 Quantitative pump-induced wavefront distortions in laser-diode- and flashlamp-pumped Nd:YLF laser rods *IEEE J. Quantum Electron.* **35** 381–6

[10] Hügel H 1986 rf-excitation of high power CO_2 lasers *Proc. SPIE* **650** 2–9

[11] Wollermann-Windgasse R, Akkermann F, Weick J and Brix W 1986 rf-excited high power CO_2-lasers for industrial material processing *Proc. SPIE* **650** 29–35

[12] Yasui K, Kuzumoto M, Ogawa S, Tanaka M and Yagi S 1989 Silent-discharge excited TEM_{00} 2.5 kW CO_2 laser *IEEE J. Quantum Electron.* **25** 836–9

[13] Kuzumoto M, Ogawa S, Tanaka M and Yagi S 1990 Fast axial flow CO_2 laser excited by silent discharge *IEEE J. Quantum Electron.* **26** 1130–4

[14] Offenhäuser F 1988 Theory and experiments on the power modulation of CO_2 lasers *IEEE J. Quantum Electron.* **24** 1289–96

[15] Markillie G A J, Baker H J, Betterton J G and Hall D R 1999 Fast-axial-flow CO_2 slab laser with a narrow-gap rf discharge operating at high pressure *IEEE J. Quantum Electron.* **35** 1134–41

[16] Vitruk P, Baker H J and Hall D R 1994 Similarity and scaling in diffusion-cooled rf-excited carbon dioxide lasers *IEEE J. Quantum Electron.* **30** 1623–34

[17] Wheatley D W 1996 A high power, high modulation bandwidth CO_2 laser *(Proc. XI Int. Symposium on Gas and Chemical Lasers and High-Power Laser Conf. (25–30 August)* Proc. SPIE **3092** 109–13

[18] Takeno S, Moriyasu M, and Kuzumoto M 1997 Laser drilling of epoxy-glass printed circuit boards *Proc. ICALEO 1997* (Orlando, FL: Laser Institute of America) section B-ICALEO, pp 63–72

[19] Mahanty P S, Kar A and Mazumder J 1995 Effects of pulse shaping on keyhole laser welding: A mathematical analysis *Proc. ICALEO 1995* (Orlando, FL: Laser Institute of America) pp 999–1008

[20] Nakamura T, Togawa T, Okino K, Watanabe S and Washio K 1997 Seam welding of A3003 aluminum alloy using high-brightness pulsed slab YAG laser *Proc. ICALEO 1997* (Orlando, FL: Laser Institute of America) section G-ICALEO, pp 130–9

Further reading

Koechner W 1999 *Solid-State Laser Engineering* 5th edn (New York: Springer)

Includes an explanation and illustrations for the characteristics, design, construction and performance of wide variety of solid state lasers for practical use, including pump sources and power supplies.

Iffländer R 2001 *Solid-State Laser for Materials Processing* (Berlin: Springer)

Particularly covers the characteristics, design, construction and performance of high-power solid state lasers for materials processing, including pump sources, power supplies and thermal management.

Steen W M 1991 *Laser Material Processing* (London: Springer)

This book provides an introduction to the industrial application of lasers in cutting, welding and the many new processes of surface treatment.

Metev S M and Veiko V P 1998 *Laser-Assisted Microtechnology* 2nd edn (Berlin: Springer)

This book introduces the principles and techniques of laser-assisted micro-processing of materials, including laser microshaping, laser melting, laser heat treatment, etc.

Powell J 1998 *CO_2 Laser Cutting* 2nd edn (London: Springer).

Wide coverage of CO_2 laser cutting for metals and non-metals with practical advice and processing parameters for pulsed laser cutting of mild steels, piercing of high-reflectivity metals, etc.

C2.2
Short pulses

Andreas Ostendorf

The pulse operation mode of lasers has been of great interest for laser developers and engineers for a long time. The following chapter provides an overview on the different principles of generating short laser pulses. In the case of concrete specifications for the pulse width these are defined by the full width at half maximum (FWHM) method.

C2.2.1 Gain switching

Gain switching can be regarded as the most direct method to generate pulsed laser radiation. During gain switching the pumping process is modulated which results in switching the amplification in the laser medium. In general, after the pumping process has been switched on, the population inversion starts to build up. When the critical inversion is reached, i.e. the gain becomes larger than losses or the loop gain reaches 1.0, the laser starts to oscillate. The oscillation continues until the pumping process is switched off, or until the losses become higher than the amplification. In contrast to the Q-switch mode, which is described in the following section, gain switching is primarily controlled by the pumping process [1]. Gain switching can make use of the transient spiking phenomena in the laser oscillator in order to achieve a high peak power pulse. If a laser medium is pumped at a very fast rate and the population inversion exceeds the threshold significantly before the oscillation starts, the laser reacts by relaxation oscillations, i.e. spiking occurs. If the pump pulse is not only fast but also stops directly after the first peak, a laser pulse is generated which consists of only one spike. The inversion after the first peak is below the threshold and due to the absence of further pump power there is no possibility for further emission.

Gain switching is well established for diode lasers, since for semiconductor lasers the pumping current can easily be modulated. The maximum modulation frequency is mainly limited by the relaxation time constants provided by the active medium and the resonator. Due to the small resonator dimensions within semiconductor lasers these time constants can become very short resulting in modulation frequencies up to the GHz range.

Another example of gain switching are lasers which are excited by diode lasers [2]. Diode lasers can also generate very high optical powers on a short-term basis, and are thus suited for the build-up of high inversions. For diode laser pumped solid-state lasers a spiking phenomenon can be observed which can be used for gain switching. The amplifier gain rises quickly to a high value because of the intense pumping level. This results in a high round trip gain and further relaxation oscillations. These oscillations quickly deplete the population inversion for that particular wavelength, spiking occurs and the lasing process stops. If, at that point, the pump diode current is also turned off, the laser switches itself off momentarily by using up all of its gain.

A representative of gain switched gas lasers is also the transversally excited atmospheric pressure laser (TEA laser). For a TEA-CO_2 laser a discharge voltage is applied to transversally arranged electrodes (see chapter B3.1). If the voltage is pulsed at a duration shorter than 1 μs, instabilities in the discharge will be

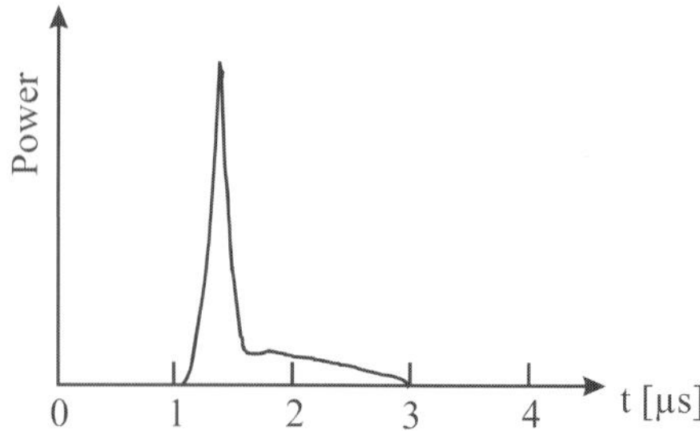

Figure C2.2.1. Typical pulse of a TEA-CO_2 laser.

suppressed and gas pressures of the $CO_2/N_2/He$ mixture can be increased up to 1 bar [3, 4]. This leads to laser pulse energies up to 50 J per litre discharge volume at pulse widths of typically 100 ns. Laser pulses from a TEA laser are characterized by a main pulse of typically 100–500 ns width followed by a pulse tail of approx. 1 μs [5] (see figure C2.2.1). The short main pulse is determined by gain switching. The tail is produced by the inversion recovery which is caused by impacts with N_2 and it is, therefore, dependent on the $N_2:CO_2$ ratio of the active gas.

Because gain switching in pulsed solid-state lasers is related to the spiking in the output, it is difficult to determine the pulse peak power which varies from shot to shot, although pulse energy and overall pulse width remain constant. For these reasons, specifications of pulsed solid-state lasers usually do not include the maximum output power. Instead, pulse energy and pulse width are specified. The peak output power may be approximated by dividing the energy of the output pulse by pulse width as with other pulsed lasers. In gain switching, laser pulse width and peak power are of the same order of the pump pulse. In Q-switching, which is described in the next section, this ratio becomes much more attractive. Moreover, the reproducibility and stability tends to be stochastic, which makes gain switching, in spite of its simplicity, unacceptable for many applications.

C2.2.2 Q-switching

The output of a gain switched, pulsed solid-state laser is generally a train of irregular pulses—irregular in peak power, pulse width and repetition frequency. It is possible to remove these irregularities and at the same time greatly increase the peak power by a technique called Q-switching. Q-switched lasers normally emit only one giant pulse in an operational cycle. The pulse will typically have a time duration less than one microsecond down to several nanoseconds and a peak intensity between 10^6 W and 10^9 W. This technique can also be applied to continuously pumped lasers in order to produce a train of Q-switched pulses with regular width, peak power and repetition rate [6]. Compared to gain switching, a Q-switch converts a relatively long pump pulse with low peak power into a very short pulse with high peak power over several orders of magnitude.

Q-switching is a mode of laser operation in which energy is stored in the laser active material during pumping in the form of excited atoms and suddenly released in a single, short burst. This is accomplished by changing the optical quality of the laser cavity. The quality factor Q is defined as the ratio of the energy stored in the cavity to the energy loss per cycle. During the pumping process the beam path in the Q-switched system is interrupted, resulting in a low Q-factor and preventing the onset of laser emission ($Q\approx1$). After a large amount of energy has been stored in the active medium, the beam path in the resonator is returned to proper alignment, and most of the stored energy emerges in a single, short pulse ($Q\gg1$) [7–11].

In a Q-switched laser a physical change is made in the feedback loop, i.e. in the resonant cavity which drastically lowers the effective reflectivity of the cavity during optical pumping. This prevents the system from oscillating, and allows the population inversion to increase without an output being generated—resulting in an increase in amplifier gain and stored energy in the upper laser level. At the time of maximum population inversion (usually near the end of optical pumping) the quality factor Q is switched back to a high value, causing the system to emit laser radiation and, thereby, releasing its available, stored energy in one, short giant pulse [12]. A Q-switch is essentially a shutter placed between the active medium and the HR mirror. With this shutter closed, the HR mirror is blocked, preventing oscillation. When the amplifier gain reaches a predetermined value, the shutter is opened to increase the cavity quality.

For a further understanding of Q-switched lasers the rate equations have to be solved before and after Q-switching. In the following it is assumed that the Q-switch will take place at $t = t_Q$. The general rate equations for lasers are for the population inversion N:

$$\mathrm{d}N/\mathrm{d}t = 2R(t) - 2B\phi N - 2\tau_{sp}^{-1}N \tag{C2.2.1}$$

and for the photon density ϕ:

$$\mathrm{d}\phi/\mathrm{d}t = -\kappa(t)\phi + BN\phi \tag{C2.2.2}$$

with $R(t)$ describing the pumping rate, B the modified Einstein coefficient, τ_{sp} the characteristic lifetime for spontaneous emission and κ representing the resonator losses. For the description of the Q-switch process, the spontaneous emission can be neglected. The pumping rate $R(t)$ is equal to R for $t < t_P$ and equal to 0 for $t \geq t_P$. According to these definitions, $Q(t < t_Q) \approx 1$, $\kappa(t < t_Q) = \kappa_1 \gg 1$ and $Q(t \geq t_Q) \gg 1$, $\kappa(t \geq t_Q) = \kappa_2 \approx 0$.

Shortly before Q-switching, i.e. $t \approx t_Q$, the population inversion n_Q can be calculated from the steady-state case of equation (C2.2.1) to

$$N(t_Q) = N_Q = \frac{\kappa_1}{B} \tag{C2.2.3}$$

which is much higher than the steady-state population inversion of a laser without Q-switch. This is achieved by keeping the photon density $\phi(t < t_Q)$ at a low level. There is nearly no stimulated emission, which reduces the population inversion within this mode of operation. Also from the steady state case of equation (C2.2.2) the resulting photon density shortly before the switching time can be calculated to

$$\phi(t_Q) = \phi_Q = \frac{R}{\kappa_1} \tag{C2.2.4}$$

which is much lower than the photon density of a laser without Q-switch. Shortly after the Q-switch at $t \geq t_Q$ the following behaviour of the population inversion can be calculated from equation (C2.2.1)

$$\mathrm{d}N/\mathrm{d}t = -2B\phi_Q N_Q = -2R \tag{C2.2.5}$$

or

$$N(t) = N_Q - 2Rt. \tag{C2.2.6}$$

The photon density shortly after the Q-switch can be calculated from equation (C2.2.2) to

$$\mathrm{d}\phi/\phi = BN_Q\mathrm{d}t \approx \kappa_1\mathrm{d}t \tag{C2.2.7}$$

or

$$\phi(t) = \phi_Q \mathrm{e}^{\kappa_1 t} \tag{C2.2.8}$$

which shows that the photon density increases exponentially after the Q-switch. To calculate the pulse peak power the rate equations (C2.2.1) and (C2.2.2) have to be solved after the Q-switch, i.e. with $R(t) = 0$. By taking the quotient of (C2.2.1) and (C2.2.2), time as a parameter can be eliminated.

$$2\frac{\mathrm{d}\phi}{\mathrm{d}N} = \frac{N_{stat}}{N} - 1 \tag{C2.2.9}$$

with N_{stat} the solution of the rate equations in the steady state case. Equation (C2.2.9) can be integrated to give

$$2(\phi - \phi_Q) = (N_Q - N) - N_{stat} \ln(N_Q/N). \tag{C2.2.10}$$

The maximum photon density can be derived from

$$d\phi/dt = 0 = (-\kappa_2 + BN)\phi \tag{C2.2.11}$$

which results in

$$\phi_{max} = \frac{1}{2} N_Q \left(1 - \frac{\kappa_2}{\kappa_1} + \frac{\kappa_2}{\kappa_1} \ln \frac{\kappa_2}{\kappa_1}\right). \tag{C2.2.12}$$

The pulse peak power can be derived directly from the maximum photon density by $P_{max} = \phi_{max} h\nu\kappa_2$.

Figure C2.2.2 provides a time scheme of a laser pulse generated by Q-switched operation of an optically pumped laser.

For Q-switched lasers both the amplifier gain and the deposited energy in the laser active medium reach much higher levels than for the normal gain switching mode. At a time near the end of the optical pumping ($t_4 = t_Q$), the losses are rapidly switched to its minimum value—causing the system to emit a giant laser pulse. During the time of the output pulse, the energy being put back into the laser active medium by external pumping can be neglected [13]. Laser emission starts when the loop gain passes 1.0 at time t_5. The laser pulse builds rapidly up depleting stored energy and reducing the amplifier gain. The peak of the laser output pulse occurs when the loop gain falls below the 1.0 level. After this the output laser power starts to drop.

The pulse width may be controlled in two ways. In the first possibility the Q-switch can be switched back to lower loop gain and terminate laser radiation before the stored energy has been used completely. This can shorten the laser pulse and reduces the pulse energy since some of the energy remains in the active medium. Later on, this energy is depleted through fluorescence. Alternatively, pumping can continue slightly beyond the laser pulse. This extra input energy contributes to fluorescence only. In some cases the Q-switch may remain open with pumping ending during the Q-switch cycle. This usually results in a slightly longer output pulse duration.

The energy of the Q-switched output pulse is dependent of the repetition rate and can be up to 90% of the pulse energy available from normal mode operation. One factor contributing to this is a remainder of energy left in the active medium. Another is that fluorescence begins with pumping, and considerable energy is lost through spontaneous emission before the Q-switch opens. During normal mode operation each atom in the active medium may participate in the emission process several times. In a Q-switched laser each atom contributes only once. This further reduces laser efficiency.

The delay time between the onset of pumping and the opening of the Q-switch influences the efficiency of the laser significantly and should be optimized with respect to the lifetime of the upper level of the active medium. The amplifier gain and the stored energy rise from the beginning of pumping until an optimum has passed. Delaying Q-switch operation beyond this time will not result in more stored energy as energy is lost to fluorescence at the same rate as it is added by pumping. Obviously, the greater the fluorescent lifetime of the laser material, the more energy may be deposited in the active medium. All solid-state laser systems may be effectively Q-switched as their fluorescent lifetimes fall in the range above one microsecond (Nd:YAG 230 μs; Nd:YVO_4 100 μs; Yb:YAG 980 μs; Nd:YLF 480 μs). Molecular gas lasers such as CO_2 are also often Q-switched. Ion lasers cannot be effectively Q-switched because the lifetime of their upper level is too short, allowing insufficient time to build up a large stored energy.

C2.2.2.1 Q-switched cw pumped lasers

Solid-state and molecular gas lasers may be pumped continuously and repetitively Q-switched to produce a regular train of output pulses. The most common examples of this are cw pumped Nd:YAG lasers and cw

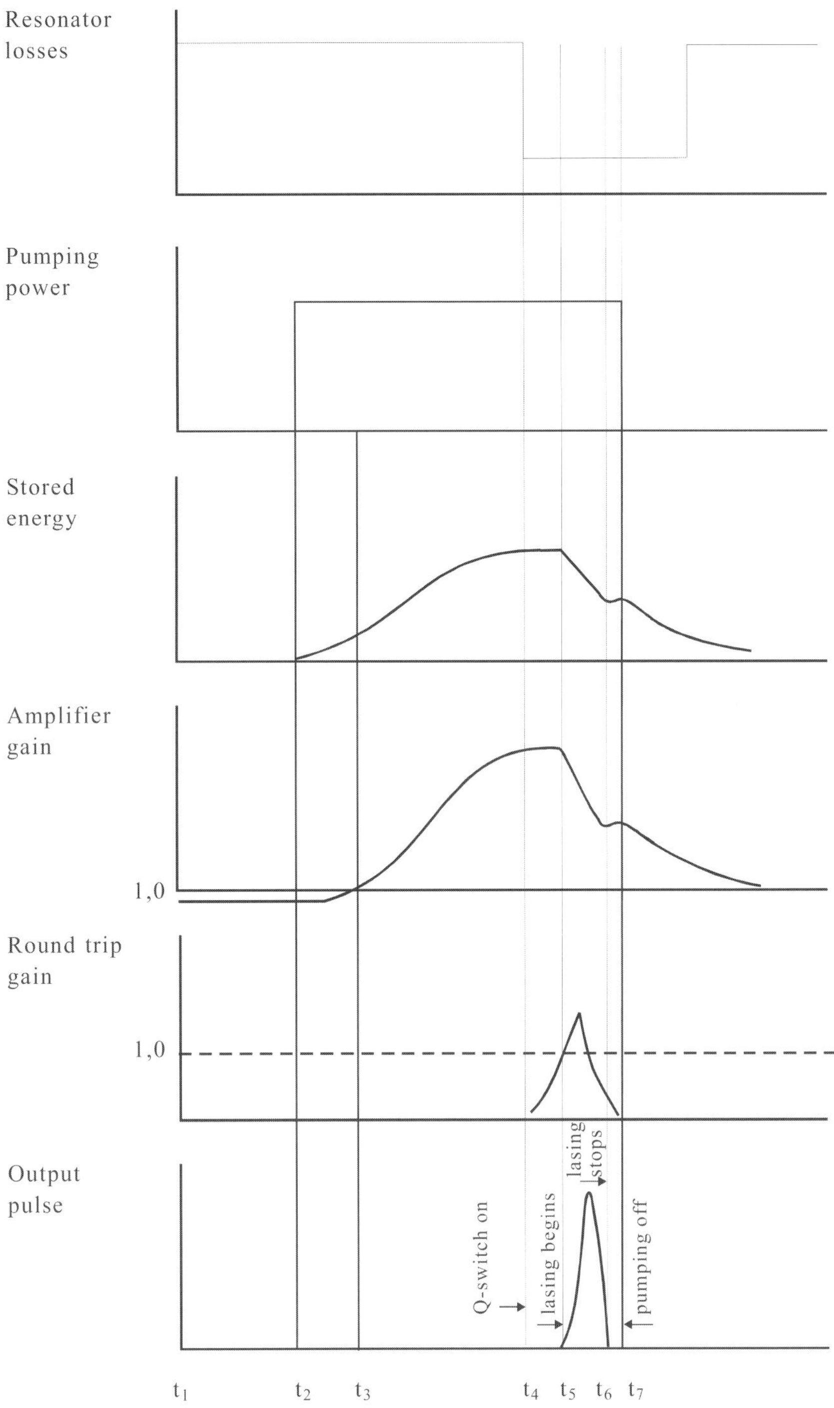

Figure C2.2.2. The time history of laser parameters during pulse pumping and Q-switching.

CO_2 lasers with low output power. Since the pumping rate is much weaker for cw systems, the maximum stored energy is small, and the resulting peak power is lower (typically 10^4 to 10^6 watts).

Nd:YAG is by far the most common Q-switched, cw pumped laser system. Such systems can typically produce several thousand pulses per second without degradation of the pulse energy. Typical pulse widths for such a system range from several tens of nanoseconds to a few microseconds. If the pulse width should be further minimized, it is necessary to use a laser crystal with a larger emission cross section, e.g. Nd:YVO_4. Nd:YVO_4 lasers can deliver pulses with half the duration of Nd:YAG assuming that the same stored energy and resonator configuration is used.

C2.2.2.2 Methods of Q-switching

Several techniques can be used for Q-switching of lasers. Each has its advantages, disadvantages, and specific applications. The four most important Q-switch types are discussed in this section, illustrated in figure C2.2.3, and can be characterized by the following values.

(1) Dynamic loss is the maximum loss introduced in the optical cavity when the Q-switch shutter is closed. Ideally the dynamic loss should be 100% to ensure that lasing does not occur until the Q-switch is opened.

(2) Insertion loss is the minimum loss introduced by the presence of the Q-switch in the open condition. Ideally this is zero, but most Q-switches include optical surfaces that introduce reflection and scattering losses.

(3) Switching time is the time necessary for the Q-switch to open. Faster switching times result in shorter, higher-peak-power pulses because the switch can become fully open before lasing has a significant effect on the population inversion. Slower switches allow significant amounts of the stored energy to be depleted before the Q-switch is fully opened. This lowers the maximum loop gain of the system and stretches out the output laser pulse.

(4) Synchronization is an indication of how well the laser output can be timed with external events. Some Q-switches allow precise control of when the output pulse occurs. Others offer virtually no control at all.

The Q-switch can be carried out by different methods. The most commonly known methods are active switching by mechanical, electro-optical or acousto-optical Q-switches, where the quality control is an explicit function of time, and passive Q-switching by saturable absorbers, where the quality is controlled as a function of the photon density.

C2.2.2.3 Mechanical Q-switches

Two representatives of this category are a light chopper and a spinning mirror. A light chopper is a spinning disc with a hole in it, or a spinning blade (like a fan blade). The chopper is inserted into the optical cavity between the laser rod and the maximum reflectivity mirror. This system provides 100% dynamic losses and 0% insertion losses. A mechanical chopper is relatively slow, however, in that it can Q-switch only a fraction of the beam area at a time as it is swept across the aperture. For this reason, mechanical chopper Q-switches are not practical or effective.

Spinning reflectors are used quite frequently in Q-switched systems where it is not necessary to closely synchronize the output to some other event. Usually the maximum-reflectivity mirror is rotated so that the mirror is tilted out of alignment. The system is Q-switched when the mirror rotates back into alignment (it is in alignment once each revolution). Rotating mirror Q-switches offer 100% dynamic losses and 0% insertion losses. The Q-switch dynamics can be increased to a sufficient level by rotating the mirror at high

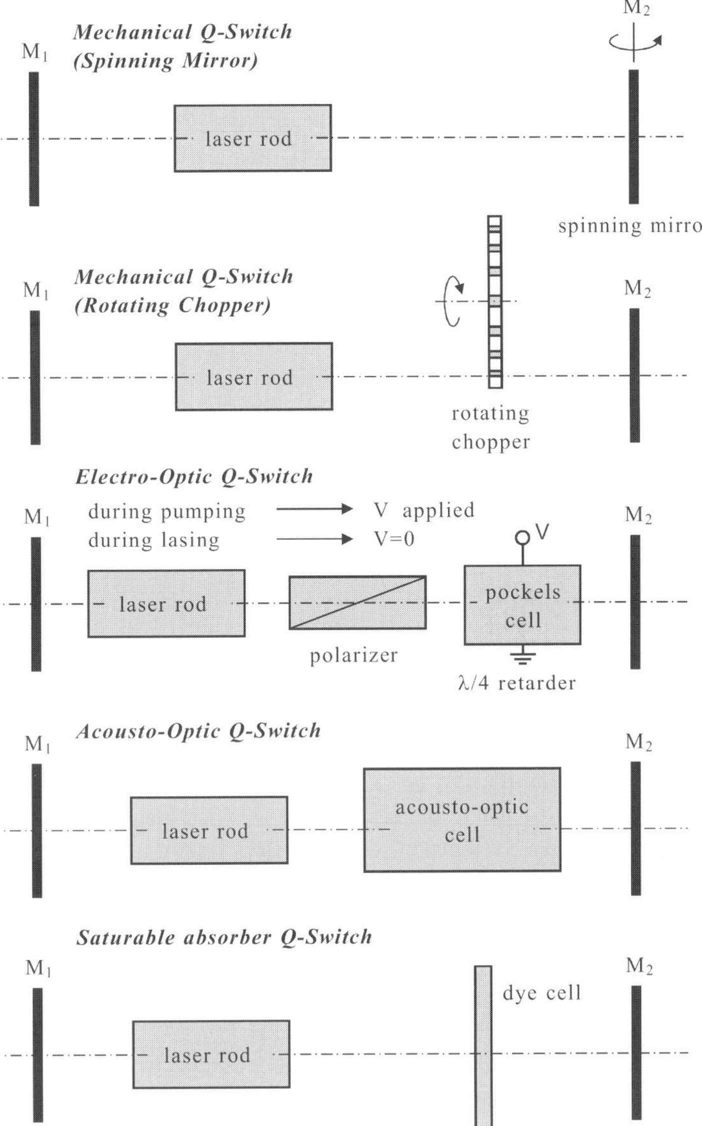

Figure C2.2.3. The principles of Q-switches.

speed (20 000 to 60 000 rpm), or by various optical schemes to multiply the effect of the rotating element. Switching time is typically a few nanoseconds.

For both the chopper and spinning reflector, it is necessary to synchronize the firing of the flashlamps with the position of the spinning element, so that the pumping pulse occurs before the system is Q-switched. Synchronization of the output pulse in mechanical systems is poor. Rotating mirror Q-switches may be used with either pulsed or cw pumped lasers.

C2.2.2.4 Electro-optic Q-switches

These devices usually require the placing of two elements into the reflecting cavity between the laser rod and the maximum reflecting mirror. These elements are a polarization filter (passive) and a polarization rotator (active). Producing a low cavity feedback with these devices involves rotating the polarization vector of the laser beam inside the cavity so that it cannot pass through the polarization filter. When this polarization rotation is terminated, the cavity reflectivity is relatively high and the system will produce a giant pulse. Two of the electro-optic devices used in this application are Kerr cells and Pockels cells. The Pockels effect is a linear electro-optical effect, i.e. the refraction index change in the parallel and orthogonal direction is proportional to the applied voltage ($\Delta n \propto V_{el}$). For Pockels cells usually nonlinear crystals like KD*P, $LiNbO_3$ or $LiTaO_3$ are used for the visible and near-IR region whereas CdTe is used for mid-IR. The Kerr effect is a nonlinear electro-optical effect, i.e. the dependency is a square function of the applied voltage ($\Delta n \propto V_{el}$). Electro-optical Q-switches have high dynamic losses (99%) and acceptable insertion losses (<5%) because of the losses in the optical elements. Switching time is faster compared to competing technologies (mechanical and acousto-optic devices), typically less than a nanosecond, and synchronization is good. Electro-optic devices have some limitations, including their relatively high voltage requirements and limited beam-diameter capabilities.

C2.2.2.5 Acousto-optic Q-switches

This technique involves the use of a transparent element placed in the cavity between the laser rod and the maximum reflectivity mirror. This transparent device, when excited with intense, standing, acoustic waves by piezoelectric crystals, exhibits a diffraction effect on the intracavity laser beam and diffracts part of the beam out of the cavity alignment. This results in a relatively low feedback. When the acoustic wave is removed, the diffraction effect disappears, the cavity is again aligned and the system emits a giant pulse [14]. Acousto-optic devices have low insertion losses (typically less than 1%) and high dynamic losses (90% maximum). Switching time is slow at 100 ns or greater, and the synchronization is good. These devices are ideally suited for use with cw pumped systems or low-gain pulsed lasers.

C2.2.2.6 Saturable-absorber Q-switches

A saturable absorber can be regarded as a passive switch and is the easiest method for a Q-switch [15–20]. The switching time is not externally triggered but defined by the intensity of the beam. Saturable absorbers are available as thin films on glass substrates or as liquids in glass cells. For Q-switching, the absorber is placed in the laser cavity between the amplifier and the maximum reflectivity mirror. It effectively absorbs the laser wavelength at low light intensities, presenting a very high cavity loss to the laser, and preventing lasing until the amplifier has been pumped to a very high gain state. When the irradiance from the active medium becomes intense enough, the energy that is absorbed optically pumps the absorber material, causing it to be transparent at the laser wavelength. The absorber cell is bleached and no longer represents a high cavity loss to the laser, i.e. the quality of the resonator increases. The absorption change is the equivalent of Q-switching in the laser, and it can occur in a period less than a nanosecond. In general, the behaviour of saturable absorbers can be characterized by

$$\alpha = \frac{\alpha_0}{1 + I/I_S} \qquad \text{(C2.2.13)}$$

with $\alpha_0 = \alpha(I \cong 0)$ the initial absorption coefficient and I_S the saturation intensity. For $I = I_S$ the absorption coefficient drops to half of the level without irradiation. For $I \gg I_S$ both energy levels in a two-level system are occupied at the same rate, i.e. the absorption coefficient tends towards zero, the transmission coefficient becomes close to one. In this case the absorber is bleached.

The initial absorption coefficient is determined such that the laser resonator at the time of maximum inversion in its active medium just reaches the threshold. For increasing intensities the transmission coefficient T of the absorber grows, within a very short interval, from $T \ll 1$ to $T \cong 1$. After a relaxation time τ, which is between 1 ps and 1 μs dependent on the absorber material, the molecules fall back to their ground states. The initial absorption coefficient has to be selected so that the remaining intensity in the resonator will not produce a second switching.

Saturable absorber Q-switches provide very high dynamic losses (>99%) and insertion losses (a few per cent at most), and their switching time is fast. Due to their immanent properties, they need virtually no synchronization at all. Saturable absorber Q-switches can be used with pulse pumped systems only because a cw pumped laser never produces sufficient irradiance to bleach the absorber. The main problem with saturable absorbers is to find appropriate materials for the defined absorption wavelength, the saturation intensity, the relaxation time and the susceptibility towards UV-radiation, e.g. from flashlamps in solid-state lasers. For Nd-doped lasers mostly Cr^{4+}:YAG crystals are used as satuarable absorber materials due to their high damage threshold of >500 MW cm^{-2}.

A good example of the use of saturable absorbers is in microchip lasers. Q-switched microlasers are simple, compact and reliable sources that can produce pulses of less than 300 ps duration with peak powers in excess of 25 kW [21,22]. These lasers can produce output pulses as short as large mode-locked lasers with peak powers as high as commercially available Q-switched systems and with a diffraction-limited output beam. Their main fields of application are range finding, micromachining, microsurgery, environmental monitoring, ionization spectroscopy and nonlinear frequency generation [23]. In order to achieve a further decrease of complexity passive Q-switches have also been applied to diode-pumped microchip lasers. The setup is simple and compact consisting of a short piece of a solid-state gain medium Nd:YAG, bonded to a saturable-absorber crystal, for example Cr^{4+}:YAG, each of 0.5 mm thickness [24]. Alternatively, the saturable absorber can be grown epitaxially onto the laser crystal. The facets at the interface between the gain medium and saturable-absorber crystal are coated dielectrically such that the interface is totally transmitting at the oscillating frequency and highly reflecting at the frequency of the pump. The output face of the saturable absorber is coated to be partially reflecting at the oscillating frequency (reflectivity R) and provides the optical output from the device. As the resonator length is only a few hundred micrometres the laser oscillates on a single longitudinal mode which allows Q-switching with pulse widths below 1 ns.

When pumped with one watt from a fibre-coupled diode, microchip lasers can even produce pulses as short as 218 ps, pulse energies as high as 14 mJ, time-averaged powers of up to 120 mW, with pulse repetition rates between 8 and 15 kHz. If semiconductor saturable-absorber mirrors are used as an end mirror with a Nd:YVO_4 laser, crystal pulses down to 37 ps can be realized at different wavelengths by Q-switching the microchip laser [25, 26].

C2.2.3 Cavity dumping

As a result of a high population on the upper long-lifetime laser level, energy is stored in the active medium and subsequently emitted in the form of short laser pulses during Q-switching. In contrast, using cavity dumping, light energy is stored in the resonator. In this case the loss of the cavity is switched from low to high. Cavity dumping is applied for laser media which have an upper laser level with a lifetime that is too short for Q-switching. For cavity dumping the continuously pumped laser medium is placed between two highly reflective mirrors (see figure C2.2.4). The intensity of the laser beam is highly amplified in the resonator. If, e.g. an acousto-optic deflector, which is transparent during intensity build-up, is triggered in the resonator, almost the complete energy in the form of a short laser pulse can leave the resonator. For a cavity-dumping laser the acousto-optic deflector usually operates in an inactive state resulting in extremely high Q-values of the resonant cavity and very intense lasing within the cavity. If an electric pulse is applied to the device, nearly all the laser energy is dumped out of the resonant cavity in the form of a single optical

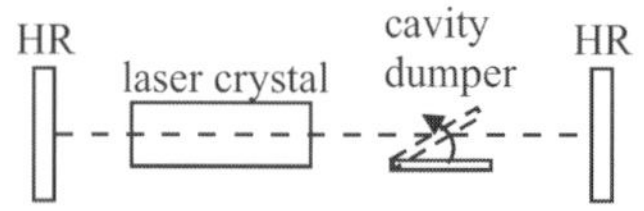

HR laser crystal cavity dumper HR

Figure C2.2.4. The principle of cavity dumping.

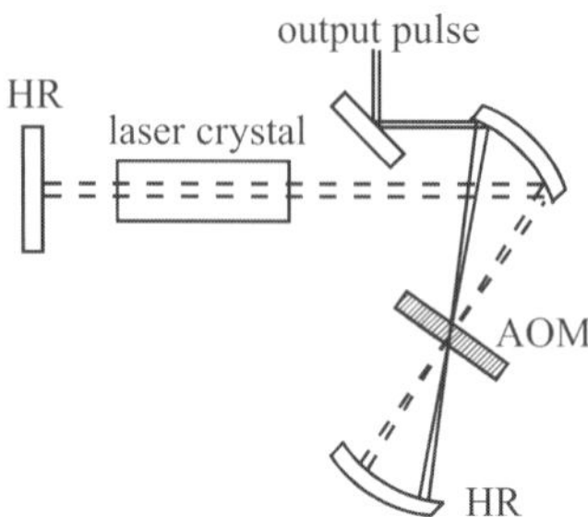

Figure C2.2.5. Cavity dumping by an acousto-optic modulator.

pulse (see figure C2.2.5). Typical pulse widths are of the order of 10 ns, i.e. a round-trip duration to dump out all photons. In order to achieve short pulses the time in which the acoustic wave passes the beam diameter has to be as small as possible. For this reason the beam is focused at the modulator.

Cavity dumping is used, e.g. to generate short laser pulses with a high peak power in continuously pumped argon lasers. The peak power of these laser pulses can be 50 times the output power of the continuous wave mode. Q-switching is not useful in this case, since the lifetime of the upper laser level of these laser media is too short, meaning that a sufficiently high population inversion cannot be generated. Depending on the output coupling mechanism, cavity dumping makes it possible to select pulse repetition rates of up to several tens of MHz at pulse widths of 10–30 ns. The switching elements which are basically used for cavity dumping are electro-optic Pockels cells, acousto-optic devices and polarizing beamsplitters. If used in combination with mode-locking (see next section), cavity dumping can generate sub-nanosecond pulses at repetition rates of several MHz without reducing the average laser power.

C2.2.4 Mode-locking

While Q-switching can be used to generate pulses with high intensities in the ns range, mode-locking is used to generate ultrashort laser pulses with pulse duration in the ps to fs range.

Mode-locking can be used very effectively for lasers with a relatively broad laser transition bandwidth, and thus for lasers with a broad amplification profile, in which numerous longitudinal modes can oscillate simultaneously. The technique is described in the next chapter C2.3.

C2.2.5 Master oscillator with power amplifiers (MOPAs)

If high-energy pulses with high brightness are required, pulse amplifiers have to be used. The generation of high-energy pulses is based on the serial combination of a master oscillator and a multistage power amplifier (MOPA). The principle of the configuration is illustrated in figure C2.2.6.

The oscillator generates the initial low-energy laser pulse with high quality parameters e.g. small beam divergence or narrow spectral width. The power amplifier with its large volume of active material then multiplies the energy up to a factor of 100. With multiple-stage amplifiers even factors of 10^{12} can be

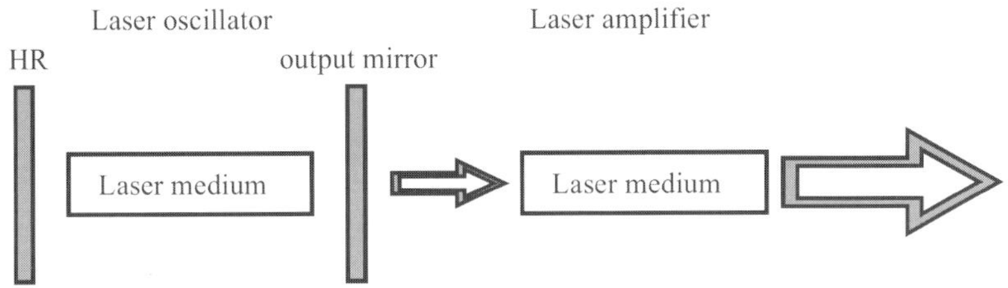

Figure C2.2.6. The schematic setup of a MOPA configuration.

reached. For a MOPA design the pulse width and its influence on the amplification mechanism are of significant importance. If a Q-switched or mode-locked pulse with a pulse width shorter than the pumping rate and the spontaneous emission time of the amplifier enters the amplifier, energy, which was stored in the amplifying medium prior to the arrival of the pulse, will be extracted. As the input pulse passes through the rod, the atoms are stimulated to release the stored energy. The amplifier can be described in this case by the two equations for the population inversion n and the photon flux ϕ:

$$\frac{\mathrm{d}N}{\mathrm{d}t} = -N \cdot c \cdot \sigma \cdot \phi \qquad \text{(C2.2.14)}$$

$$\frac{\mathrm{d}\phi}{\mathrm{d}t} = N \cdot c \cdot \sigma\phi - c \cdot \frac{\mathrm{d}\phi}{\mathrm{d}x} \qquad \text{(C2.2.15)}$$

where σ is the cross section for stimulated emission and c is the speed of light.

The rate at which the photon density changes is equal to the net difference between the generation of photons by simulated emission and the flux of photons leaving that region. The latter term in equation (C2.2.15) is usually absent in basic oscillation processes of laser resonators.

The efficiency of amplifiers, i.e. the ratio of the output energy to the energy stored in the active medium, can be calculated by solving these equations. As a characteristic material-dependent parameter the saturation energy represents the energy which can be extracted from the amplifier divided by the small-signal gain. If the input signal, i.e. the input energy, is very small compared to the saturation energy, no saturation effects occur and the signal gains exponentially with the length of the amplifier medium. For higher energies, e.g. towards the end of the medium, the gain grows linearly. This implies that every excited state contributes its stimulated emission to the beam which obviously represents the most efficient conversion. Usually, lasers are designed to operate within this regime. The efficiency can be further increased by passing the active medium twice. Within a double pass amplifier a mirror at the output returns the beam and a $\lambda/2$ waveplate turns the polarization by 90° [27]. For Nd doped glass laser materials the saturation fluence is between 5–10 J cm^{-2} [28]. For Nd:YAG crystals the saturation fluence is 0.2 J cm^{-2} whereas ruby amplifiers provide a saturation fluence of 11 J cm^{-2}. Due to the inherent advantages of Nd:glass this medium has been chosen to be used for the realization of high power systems for laser fusion experiments (see chapter D10.1). By a superposition of 10 identical beam amplifier lines with pulse energies of 10 kJ each a total pulse energy of 100 kJ has been realized by the NOVA laser system [29].

In very high-peak-power laser systems, the pulse is amplified until it begins to incur one of several nonlinear problems associated with intense light.

At high optical intensities, the index of refraction becomes a function of intensity. The index of refraction determines the phase velocity of light and the optical pathlength experienced and so intense light begins to suffer phase delays, relative to weaker light. The accumulated phase-lag suffered by light in travelling through a medium is given by the B-integral:

$$B = \frac{2\pi}{\lambda} \int_0^L n_2(I) \cdot I(z)\,\mathrm{d}z. \qquad \text{(C2.2.16)}$$

A beam with a non-uniform intensity distribution passing through a transparent medium suffers a delay of phase at the centre of the beam which differs from that at the beam edges. This alteration of the phase of the lightwave is a distortion of the light beam which can seriously degrade the beam quality. If the B-value exceeds values of 3–5 filamentation and self-focusing occur. Besides self-focusing it is possible that a beam which is not perfectly smooth in its intensity profile will break up into beamlets and each of those may self-focus. Between whole-beam self-focusing and filamentation, the difference is only the relative growth rates of different spatial-frequency components of the beam, and the initial amplitude of those components.

For mode-locked sub-ns pulses, however, the incident energy is far below the saturation energy as the high-peak-power would immediately destroy the optical components. For ultrashort laser pulses the energy amplification usually is based on the chirped pulse amplification (CPA) technique (see chapter C2.3).

Regarding the pulse shape, the amplified pulse shapes differ significantly from the incident pulse shapes since the leading edge of the pulse experiences a higher gain. This usually leads to a distortion of the pulse shape which has to be taken into account by laser system designers and users alike.

C2.2.6 Beam characterization and pulse measurement

One of the most important parameters of a laser source is output power and, for pulsed lasers, pulse energy. Temporal measurement of the power of laser pulses down to several nanoseconds can be realized using fast diodes or oscilloscopes. The pulse energy can be calculated by applying the integral function $H = \int P(t)\,\mathrm{d}t$. If the average power is measured by integral devices, the peak pulse power can roughly be estimated, if the repetition rate and pulse width are known.

All sensors can be subdivided into thermal detectors and sensors based on the internal and external photoelectric effect (see chapter A7). For thermal photodetectors, the incident light heats up the sensor, and the temperature rise correlates with the laser pulse energy. The temperature difference can be measured using thermo-elements, temperature-dependent resistors or by using the pyroelectric effect.

In thermo-elements, the incident light of a laser pulse is completely absorbed by a measurement cone. The temperature rise is given by $\Delta T = H/C$, with H being the laser pulse energy and C the heat capacity of the cone. ΔT is measured by thermo-elements and is transformed into $\Delta U = \alpha_{\mathrm{therm}} \Delta T$, with α_{therm} being the thermal force in [V K^{-1}]. Pyroelectric sensors provide a spontaneous electric polarization $P_{\mathrm{e}}(T)$. When the temperature alters, the effective surface charge changes and can be measured at a resistor R by $U(t) = R \cdot \mathrm{d}Q/\mathrm{d}t = A \cdot R \cdot \mathrm{d}P_{\mathrm{e}}/\mathrm{d}T \cdot \mathrm{d}T/\mathrm{d}t$ with A representing the area of the sensor. Consequently, in pyroelectric sensors, voltages can only be measured when the temperature deviates. They are mainly used for determining the laser pulse energy, or as power meters. The energy can then be derived by calculating the time integral. Thermal detectors are nearly independent of the wavelength, which makes them a useful tool for infrared lasers. However, the temporal resolution is not very high. The rising time of pyroelectric sensors is in the range of a nanosecond and the sensitivity is in the range of one μJ.

In addition to sensors based on thermal effects, detectors can exploit the internal photoeffect. Here, electrons are transferred by the incident radiation from the valence to the conduction band, resulting in a free electron and a free hole. The easiest solution is realized by a photo-resistor. Depending on the material, different band gaps and spectral sensitivities can be realized. To avoid free carrier generation by thermal effects, photoresistors are usually cooled. Their time resolution can be up to a few hundred ps. Photodiodes consist of pn junctions applied at a reverse bias. When light hits the depletion zone, free carriers are generated. Compared to photoresistors, the sensitivity of photodiodes is improved, and interference due to thermal effects

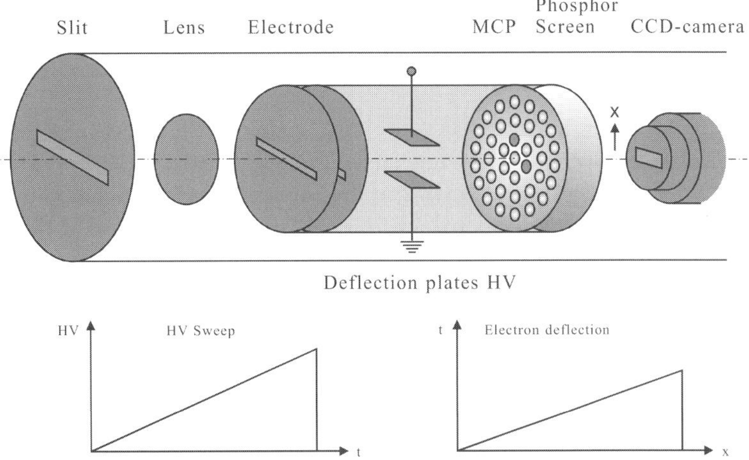

Figure C2.2.7. The principle setup of a streak camera.

can be neglected. Arrays of photodiodes can also measure the spatial and temporal distribution of laser pulses (e.g. CCD cameras).

Sensors which are based on the external photoeffect always operate in a vacuum. The external photoeffect is based on the generation of photoelectrons at a cathode by incident light. The efficiency and spectral sensitivity are strongly dependent on the material and the wavelength. The most simple device of these vacuum photodetectors is the vacuum diode. Here, the free electrons are accelerated and absorbed by an anode located at the opposite side of the diode. By using vacuum diodes, time resolutions down to 100 ps can be achieved. Higher sensitivities, compared to vacuum diodes, can be realized by photomultipliers. The primary photoelectrons from the cathode are accelerated and guided towards a dynode where secondary electrons are generated. By multiple impacts at numerous dynodes, amplification factors of 10^8 can be achieved. The amplification process can be built up within several ns, which makes photomultipliers unique sensors in terms of high amplification, broad bandwidth and low noise.

Characterization of the pulsed beam mainly consists of measuring the spatial, temporal and spectral properties, using appropriate methods. Whereas the spatial and spectral characterization of long and ultrashort laser pulses is identical to cw laser measurement methods (conventional beam profilers and spectrometers), the measurement of the temporal behaviour is more sophisticated for ultrashort pulses in the ps and fs regime. The measurement of ultrashort time intervals has been driven by the need for registration of fast events in chemistry, biology and electronics. Streak cameras and correlation methods are usually used to measure ultrashort time intervals. Both principles are based on mapping short time intervals in a spatial interval that can be easily measured using a camera or a microscope.

Streak cameras convert the time information, *via* the generation and deflection of free electrons, into spatial information. As the sensor can read out the information in two dimensions, the streak camera is able to record the time evolution of one spatial coordinate such as a line focus. Today, streak cameras cover the whole spectral range from the x-ray region to the near-infrared. The time resolution is limited to approximately 0.5 ps. The principle of streak cameras is described later [30]:

The incoming light passes through a one-dimensional slit and is imaged onto the photocathode of the tube. The number of electrons produced by the photoelectric effect is proportional to the number of incident photons and therefore to the intensity. They are accelerated towards the anode by an electric field in the tube. Before the electrons are imaged onto the phosphor screen, they pass a micro channel plate (MCP) where they are amplified with a gain factor of 10^3–10^4. The HV sweep on the deflection plates is synchronized with the incident light pulse. Electrons generated at later times are more strongly deflected compared to electrons generated at earlier times. Therefore, they enter the MCP at different positions and can be detected on the phosphor screen at different x-coordinates.

For laser pulses considerably shorter than 1 ps streak cameras cannot be used [31]. For those short time intervals autocorrelation and cross-correlation methods are usually used [32, 33]. These methods do not resemble the incoming pulse directly, therefore further mathematical transformations have to be applied.

The autocorrelation method is based upon a superposition of the direct incoming pulse E_1 and a delayed replica of the incoming pulse E_2 in a nonlinear crystal. The main requirement for the nonlinear effect is that it has to be instantaneous. The beam E_2 can be shifted in time by varying the beam path. The order of the autocorrelation method is determined by the order of the nonlinear optical conversion in the nonlinear crystal. The beams can either cross the crystal collinearly, which results in an interferometric autocorrelation, or under a certain angle, which results in an intensity autocorrelation. In both cases, however, a signal is recorded depending on the delay between the two incident pulses.

To determine the pulsewidth the delay is gradually increased between successive laser shots and the higher order response signal is measured. This technique is called the multishot method. However, if the pulse to pulse fluctuations are significant the technique provides a relatively high measurement error. Single shot autocorrelators avoid these uncertainties [34]. In single shot autocorrelators the spatial information is also used, e.g. two line foci are superimposed in such a way that the temporal delay varies along the line focus [35]. The resulted signal can be measured by a CCD camera in one shot. An arbitrary point x_0 along the line focus is related to the temporal delay τ by:

$$\tau = \frac{n \cdot x_0 \cdot \sin(\varphi/2)}{c} \tag{C2.2.17}$$

where n is the refractive index of the nonlinear medium, φ is the angle between the two incident beams and c is the speed of light. It can easily be seen that a smaller angle results in an increased time resolution. On the other hand, a certain angle facilitates the measurement as the resulting signal contains no portion of the incident beams and is, therefore, inherently background-free. For the described technique the measured spatial signal is directly proportional to the autocorrelation function. Under the assumption of a certain pulse shape, the pulsewidth can be determined from the measured autocorrelation signal. For rectangular pulse shapes there is obviously no difference between the width of the autocorrelation function and the pulsewidth. For a Gaussian pulse shape, a correction factor of 0.7071 has to be adapted, for the hyperbolic sec shape, the correction factor is 0.6482. If the pulse shape is asymmetrical, an autocorrelation with even order will not reveal any information on the asymmetries as these functions are always symmetrical.

References

[1] Eggleston J M, DeShazer L and Kangas K 1988 Characteristics and kinetics of laser-pumped Ti:Sapphire oscillators *IEEE J. Quantum Electron.* **24** 1009–15

[2] Owyoung A, Hadley G R, Esherick R, Schmitt R L and Rahn L A 1985 Gain switching of a monolithic single-frequency laser-diode-excited Nd:YAG laser *Opt. Lett.* **10** 484

[3] Beaulieu A 1970 Transversely excited atmospheric pressure CO_2 laser *App. Phys. Lett.* **16** 504–5

[4] Dumanchin R and Rocca-Serra J 1970 *High Power Density Pulsed Molecular Laser International Quantum Electronics Conference, Digest of Technical Papers (Kyoto)* p 308

[5] Duley W W 1976 *CO_2 Lasers: Effects and Applications* (New York: Academic)

[6] Degnan J J 1988 Theory of the optimally coupled Q-switched laser *IEEE J. Quantum Electron.* **25** 214–20

[7] Collins R J and Kisliuk P 1962 Control of population inversion in pulsed optical masers by feedback modulation *J. Appl. Phys.* **33** 2009–11

[8] Hellwarth R W 1961 *Advances in Quantum Electronics* (New York: Columbia United Press)

[9] McClung F J and Hellwarth R W 1962 Giant optical pulsations from ruby *J. Appl. Phys.* **33** 828–9

[10] McClung F J and Hellwarth R W 1963 Characteristics of giant optical pulsations from ruby *Proc. IEEE* **51** 46

[11] Stepke E T 1972 *The Key to Q-Switching Electro-Optical Systems Design*
[12] Wang C C 1963 Optical giant pulses from a Q-switched laser *Proc. IEEE* **51** 1767
[13] Wagner W G and Lengyel B A 1963 Evolution of the giant pulse in a laser *J. Appl. Phys.* **34** 2040
[14] Chesler R B, Karr M A and Geusic J E 1970 An experimental and theoretical study of high repetition rate Q-switched Nd:YAG lasers *Proc. IEEE* **58** 1899
[15] Degnan J J 1995 Optimization of passively Q-switched lasers *IEEE J. Quantum Electron.* **31** 1890–901
[16] Erickson L E and Szabo A 1967 Behaviour of saturable-absorber giant-pulse lasers in the limit of large absorber cross section *J. Appl. Phys.* **38** 2540–2
[17] Kafalas P, Masters J I and Murray E M E 1964 Photosensitive liquid used as a nondestructive passive Q-switch in a ruby laser *J. Appl. Phys.* **35** 2349
[18] Soffer B H 1964 Giant pulse laser operation by a passive, reversibly bleachable absorber *J. Appl. Phys.* **35** 2551
[19] Spaeth M L and Sooy W R 1968 Fluorescence and bleaching of organic dyes for a passive Q-switch laser *J. Chem. Phys.* **48** 2315
[20] Szabo A and Stein R A 1965 Theory of laser giant pulsing by a saturable absorber *J. Appl. Phys.* **36** 1562–6
[21] Zayhowski J J 1991 Q-switched operation of a microchip laser *Opt. Lett.* **16** 575–7
[22] Zayhowski J J and Dill C III 1992 Diode-pumped microchip lasers electro-optically Q-switched at high pulse repetition rates *Opt. Lett.* **17** 1201–3
[23] Zayhowski J J 1998 Passively Q-switched microchip lasers and applications *Rev. Laser Eng.* **26** 841–6
[24] Zayhowski J J and Mooradian A 1989 Single-frequency microchip Nd lasers *Opt. Lett.* **14** 24–6
[25] Braun B, Kaertner F X, Moser M, Zhang G and Keller U 1997 56 ps passively Q-switched diode-pumped microchip laser *Opt. Lett.* **22** 381–3
[26] Spuehler G J, Paschotta R, Fluck R, Braun B, Moser M, Zhang G, Gini E and Keller U 1999 Experimentally confirmed design guidelines for passively Q-switched microchip lasers using semiconductor saturable absorbers *J. Opt. Soc. Am.* B **3** 16 376–88
[27] Michon M, Auffret R and Dumanchin R 1970 *J. Appl. Phys.* **41** 2739
[28] Krupke W F 1974 Induced-emission cross-section in neodymium laser glasses *J. Quantum Electron.* **10** 450–7
[29] Brown D C 1981 *High-Peak Power Nd:Glass Laser System* (Berlin: Springer)
[30] Sarger L and Oberlé J 1998 *How to Measure the Characteristics of Laser Pulses Femtosecond Laser Pulses* ed C Rullière (Berlin: Springer) pp 177–201
[31] Sauerbrey R and Feurer T 1996 *Characterization of Short Laser Pulses Methods in Experimental Physics* vol 3 (New York: Academic)
[32] Diels J-C, Fontaine J J, McMichael I C and Simoni F 1985 Control and measurement of ultrashort pulse shape (in amplitude and phase) with femtosecond accuracy *J. Appl. Opt.* **24** 1270–82
[33] Diels J-C, Fontaine J J and Rudolph W 1987 Ultrafast diagnostics revue *Phys. Appl.* **22** 1605–11
[34] Salin F, Georges P, Roger G and Brun A 1987 Single-shot measurement of a 52 fs pulse *Appl. Opt.* **26** 4528
[35] Collier J, Danson C, Johnson C and Mistry, C 1999 Uniaxial single shot autocorrelator *Rev. Sci. Instrum.* **70** 1599–602

Further reading

Diels J C and Rudolph W 1996 *Ultrashort Laser Pulse Phenomena: Fundamentals, Techniques and Applications* (Boston, MA: Academic)
Duling I N III 1995 *Compact Sources of Ultrashort Pulses* (Cambridge: Cambridge University Press)
Herrmann J and Wilhelmi B 1984 *Laser für Ultrakurze Lichtimpulse* (Berlin: Akademie)
Iffländer R 2001 *Solid-State Lasers for Materials Processing* (Berlin: Springer)
Koechner W 1999 *Solid-State Laser Engineering (Springer Series of Optical Sciences 1)* 5th edn (Berlin: Springer)
Ready J F 1978 *Industrial Applications of Lasers* (New York: Academic)
Rullière C 1998 *Femtosecond Laser Pulses* (Berlin: Springer)
Siegman A E 1989 *Lasers* (Mill Valley, CA: University Science Books)
Trebino R 2002 *Frequency-Resolved Optical Gating: The Measurement of Ultrashort Laser Pulses* (Dordrecht: Kluwer)

C2.3
Ultrashort pulses

Derryck T Reid

C2.3.1 Theory of ultrashort pulse generation and modelocking

C2.3.1.1 Active modelocking

In an inhomogeneously broadened laser system (see section A1.8), optical frequencies within the emission bandwidth of the material can experience gain simultaneously at discrete frequencies corresponding to the longitudinal modes of the laser resonator. In the absence of any mechanism to coherently couple energy between these modes, the laser will operate with a continuous-wave (cw) output containing a narrow band of frequencies whose roundtrip cavity gain is highest. An alternative operating regime exists in which each mode is related to its neighbour by a fixed phase relationship and it is this *modelocked* condition that is used to generate a periodic sequence of ultrashort optical pulses.

To understand how modelocking works, consider the frequency separation between adjacent longitudinal resonator modes which can be expressed as

$$\Delta\omega = 2\pi c/l \qquad \text{(C2.3.1)}$$

where l is the optical roundtrip length of the resonator and c is the speed of light. If the nth longitudinal mode has an amplitude E_n then the total optical field can be denoted as

$$E(t) = \sum_n E_n \exp \mathrm{i}[(\omega_o + n\Delta\omega)t + n\Delta\varphi] \qquad \text{(C2.3.2)}$$

where ω_o is the centre frequency of the output and $\Delta\varphi$ is the phase difference between adjacent modes. Figure C2.3.1 illustrates how a simple pulse can be produced by coherently adding together adjacent modes. In this example, three modes with equal amplitudes and a relative phase separation of $\Delta\varphi = 0$ are interfered (figure C2.3.1(*a*)) and, as further modes are added (figure C2.3.1(*b*)), the pulse duration decreases.

Coupling of energy between adjacent cavity modes can be achieved in practice by modulating either the intracavity loss (*AM-modelocking*) or the intracavity phase (*FM-modelocking*). Consider first AM-modelocking where an amplitude modulator with a modulation depth m is driven with a frequency Ω. If the unmodulated field of mode n is $E_n \exp(\mathrm{i}\omega_n t)$ then the modulated field is given by

$$E = E_n \exp(\mathrm{i}\omega_n t) \times (1 + m \cos \Omega t) \qquad \text{(C2.3.3)}$$

which can be written as

$$E = E_n \exp(\mathrm{i}\omega_n t) + \frac{m}{2} E_n \exp(\mathrm{i}(\omega_n + \Omega)t) + \frac{m}{2} E_n \exp(\mathrm{i}(\omega_n - \Omega)t). \qquad \text{(C2.3.4)}$$

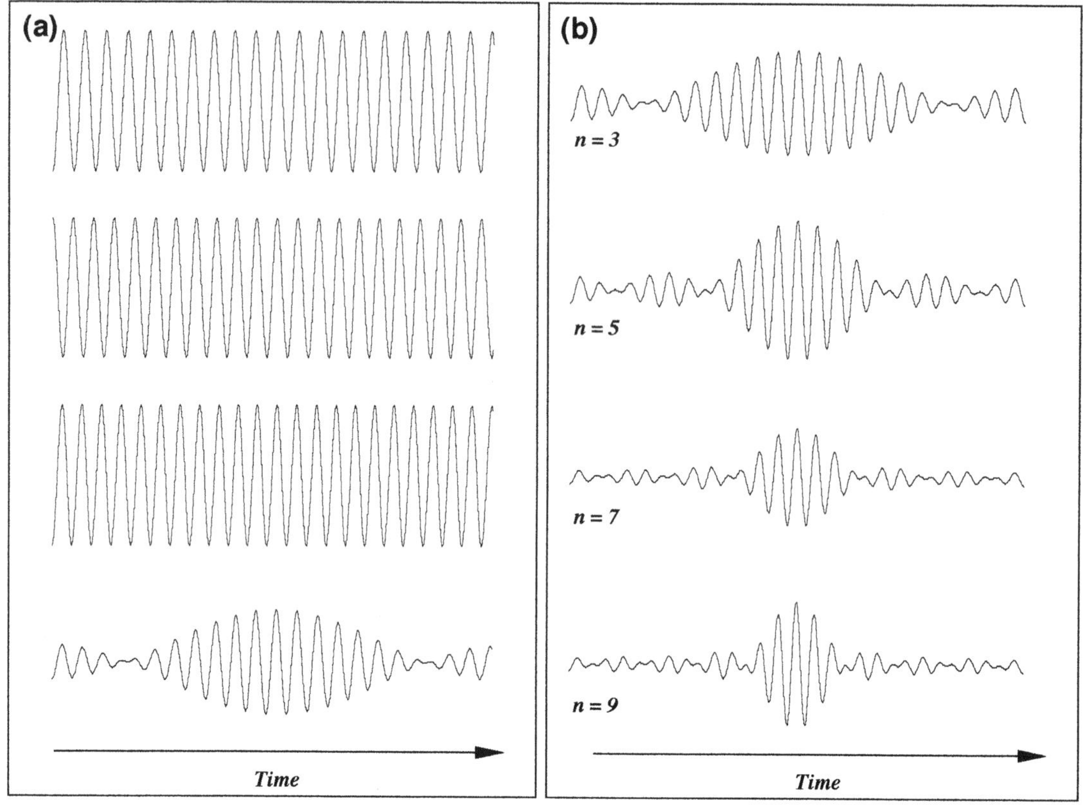

Figure C2.3.1. The principles of modelocking: (*a*) three adjacent modes add coherently to produce a pulse; (*b*) shorter pulses require a larger number of n coupled modes.

The second and third terms in (C2.3.4) represent frequency sidebands added to the mode by amplitude modulation. When the modulation frequency is chosen so that $\Omega = \Delta\omega$, energy is coherently coupled from each cavity mode to its nearest neighbours and modelocking is achieved. Phase modulation can be analysed in a similar way and the phase-modulated field can be written as

$$E = E_n \exp(\mathrm{i}\omega_n t) \times \exp(\mathrm{i}m \sin \Omega t). \tag{C2.3.5}$$

For small modulation depths ($m \ll 1$) the Taylor series approximation $\exp x \approx 1 + x$ can be used to give

$$E = E_n \exp(\mathrm{i}\omega_n t) \times (1 + \mathrm{i}m \sin \Omega t) \tag{C2.3.6}$$

which can be expressed as

$$E = E_n \exp(\mathrm{i}\omega_n t) + \frac{m}{2} E_n \exp(\mathrm{i}(\omega_n + \Omega)t) - \frac{m}{2} E_n \exp(\mathrm{i}(\omega_n - \Omega)t) \tag{C2.3.7}$$

which indicates that phase-modulation can be used to achieve FM-modelocking through the addition of spectral sidebands to each cavity mode.

A common practical amplitude modulator is an acousto-optic modulator in which a standing acoustic wave diffraction grating is formed twice every drive period, implying an electrical drive frequency of $\Delta\omega/2$

for modelocking applications. For operation in the visible and near-infrared, fused silica is a common acousto-optic modulator material. Lithium niobate devices based on the Pockels effect are the most commonly used phase modulators and are attractive for modelocked fibre-laser applications because fibre-pigtailed waveguide devices are readily available.

The modelocking mechanisms discussed so far are examples of *active modelocking* in which the laser resonator is driven by an external reference frequency. Such a system is a classical forced oscillator and, for maximum efficiency, the external frequency must be accurately maintained at the resonant frequency of the laser cavity. Long-term thermal drift in the laser cavity length can make exact synchronization difficult, but a technique known as *regenerative modelocking* can be applied to solve this problem. Using a fast photodiode detector, the modelocked pulse sequence is sampled and the electrical signal bandpass-filtered to produce a clean sinusoidal waveform at the exact cavity frequency. With suitable filtering and amplification the technique is self-starting because random mode-beating detected during cw operation is sufficient to produce the required drive frequency.

Active modelocking has been employed in a wide variety of laser systems including argon ion [1], dye [2], erbium-doped fibre [3], semiconductor [4] and solid-state lasers [5]. Unlike passive modelocking, no optical nonlinearity is required which means that actively-modelocked lasers can be operated with low intracavity powers. The technique is particularly suitable for modelocked telecommunication sources which need to be actively synchronized to an external clock source. Extremely high-frequency operation is possible using the technique of *harmonic modelocking* in which the modulation frequency used is a multiple of the cavity frequency. In this way, sets of widely spaced *supermodes* are coupled together and interfere to generate a modelocked pulse sequence with a repetition frequency equal to a high harmonic of the original cavity frequency. Lasers modelocked using this method exhibit substantial high-frequency noise because the absence of strong coupling between independent sets of supermodes leads to random frequency-beating between them. A form of 'optical' active modelocking exists in which gain modulation resulting from optical pumping using another modelocked laser is the modelocking mechanism. Known as *synchronously-pumped* modelocking, this technique has been used to obtain ultrashort pulses from oscillators incorporating gain media of laser or nonlinear optical materials. The technique is most appropriate to laser gain materials which have a short-lived fluorescence lifetime and whose gain therefore exhibits a significant change when the pump light is modulated at high frequencies. Synchronous-pumping has been used to generate modelocked pulses from colour-centre lasers [6], dye lasers [7] and optical parametric oscillators [8]. In mode-locked lasers one uses the appropriate timing to achieve pulse compression through gain saturation. Mistiming the pump pulses changes the achieved pulse duration by orders of magnitude but has only a small effect on the output power. In optical parametric oscillators (OPOs) the exact matching of the master and slave oscillator pulse repetition frequencies is critical for achieving high conversion efficiency at the desired wavelength but the impact on the pulse duration is generally less noticeable although under certain circumstances substantial compression can result from careful control of the pump pulse timing [9].

C2.3.1.2 Passive modelocking

In the active modelocking technique, pulse shaping is determined directly by the performance (modulation depth, bandwidth, loss etc) of the optical modulator and, consequently, the generation of subpicosecond duration pulses is difficult. To produce the shortest optical pulses, *passive modelocking*, a technique based on exploiting intracavity optical nonlinearities, must be employed. The key difference between active and passive modelocking methods is that, because passive modelocking is a nonlinear technique, the strength of the modelocking action increases as the propagating pulses shorten and their peak intensities rise. One consequence of this nonlinear behaviour is that, unlike active modelocking which generally only locks together the phases of existing laser cavity modes, the passive technique commonly generates and modelocks entirely new cavity frequencies which were not originally present during cw operation.

Passive modelocking has been implemented using a wide range of different nonlinear effects, but in every technique the modelocking process can be described in terms of *dynamic loss saturation* combined, in certain systems, with *dynamic gain saturation*. The principal modelocking element is a *saturable absorber* whose loss decreases as the incident pulse intensity increases. A saturable absorber can take the form either of a physical component such as a solid-state semiconductor device [10] or a dye jet [11], or a virtual component (e.g. a Kerr lens) which replicates a saturable-absorber-like action using self-focusing or self-phase-modulation effects. Saturable absorbers can be categorized as either *slow saturable absorbers* or *fast saturable absorbers*, depending on whether their recovery time is longer or shorter than the pulse duration respectively. The pulse shaping mechanisms are different in each case and we will first describe modelocking using a slow absorber. Most slow saturable absorbers are physical devices, commonly dye molecules in solution [11] or semiconductor-doped glass [12], which rely on resonant excitation to produce an excited electronic state whose transmission is greater than that of the ground state. When an optical pulse is incident on a slow absorber, its leading edge is absorbed and creates an excited state which is relatively transparent to the trailing edge. This effect in isolation is not sufficient to generate ultrashort pulses and must be combined with dynamic gain saturation to achieve modelocking. A laser medium with a gain relaxation time τ_g faster than the cavity roundtrip time, T_{cav}, but slower than the absorber recovery time, τ_a, must be used so that the gain recovers in time to amplify the next roundtrip pulse and the absorber recovers in time to be fully saturated by it. This principle is illustrated in figure C2.3.2 which depicts the individual responses of the absorber and gain medium to an incident ultrashort pulse. A pulse propagating first through the absorber and then through the gain medium experiences a 'gain window' with a finite duration which strongly shapes the pulse. The situation is different when a gain medium with a slow relaxation time is used. Here, dynamic gain saturation cannot be used to shape the resonant pulses and a fast saturable absorber is required to realize modelocking. The time-dependent gain and loss profiles of the absorber and the gain medium are again depicted in figure C2.3.2 and refer to an absorber with both a rapid response and recovery time. In this system, the dynamics of the absorber and gain medium no longer represent the limiting factor influencing the pulse duration, and other effects such as linear and nonlinear dispersion become the dominant pulse shaping mechanisms.

Pulse shaping effects will be discussed in detail in section C2.3.3 but it is useful to explain at this point the key mechanisms which are important in the context of passive modelocking using a near-instantaneous saturable absorber. Pulses propagating in a passively modelocked laser experience strong soliton-like shaping effects associated with anomalous dispersion and self-phase-modulation and can therefore be described in terms of a modified nonlinear Schrödinger equation (NLSE) [13]. Unlike true solitons which propagate in a medium whose nonlinearity and dispersion are distributed uniformly throughout the system, the pulses travelling within a modelocked laser cavity generally experience dispersion and nonlinearity within separate regions of the resonator. The nonlinearity is localized within the gain medium while the (positive) dispersion arises here and within a separate (negative) dispersion-compensating component such as a prism-pair. This departure from the ideal soliton description means that passively modelocked lasers operating under these conditions are better described as *solitary* lasers [14] rather than true soliton lasers [15]. The operating principles of such lasers have been analysed in detail by Krausz *et al* [16], who have shown that, under the weak pulse shaping approximation, the output pulses can be described by a complex amplitude,

$$a_n(t) = \operatorname{sech}(t/\tau)\exp(\mathrm{i}\varphi_n)\left(\frac{W}{2\tau}\right)^{1/2} \tag{C2.3.8}$$

where n is the roundtrip number and φ_n is the linear phase given by,

$$\varphi_n = \varphi_o + nD/\tau^2 \tag{C2.3.9}$$

and D is the roundtrip group-delay dispersion, W is the pulse energy, φ is the roundtrip nonlinear phase shift

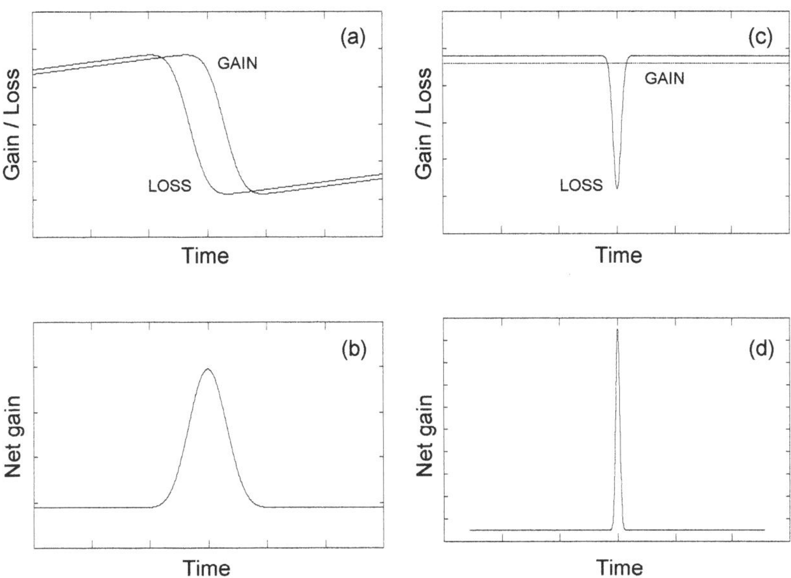

Figure C2.3.2. Saturable absorber action, (*a*) Intracavity gain and loss profiles for a laser comprising a slow saturable absorber and a rapid relaxation time gain medium; (*b*) the net gain (gain – loss) for this system; (*c*) gain and loss profiles for a laser comprising a near-instantaneous saturable absorber and a slow relaxation time gain medium and (*d*) the net gain for this system.

and τ is the pulse duration given by,

$$\tau = 2D/\phi W. \tag{C2.3.10}$$

This result is identical to the solution of the NLSE for a soliton propagating in a uniform medium and justifies the common assertion that passively modelocked lasers produce pulses with a $\text{sech}^2(t)$ intensity profile. In actual modelocked lasers the situation is complicated by the effects of lumped dispersion/nonlinearity, higher order linear dispersion and finite gain bandwidth which mean that spectral and temporal measurements of real lasers rarely conform to the ideal $\text{sech}^2(t)$ solution [17]. Numerous implementations of active and passive modelocking exist and, in this section, we have concentrated on outlining only the general operating principles. Such is the versatility of the passive modelocking technique that the majority of practical ultrafast laser systems are now based exclusively on passive methods, with the exception of hybrid systems which employ multi-stage pulse compression schemes to generate femtosecond pulses from gain-switched or actively-modelocked master oscillators. In the section which follows we will review specific modelocked laser systems and discuss the particular modelocking effect in each case.

C2.3.2 Sources of ultrashort pulses

C2.3.2.1 Dye lasers

Historically, the first practical passively modelocked ultrafast lasers were those based on a gain medium of a thin (~10 μm) organic dye jet (see chapter B5.3) and incorporating a slow saturable absorber jet at a separate cavity focus as described in [18]. The most common combination was R6G in ethylene glycol (gain jet) and DOdcI in ethylene glycol (absorber jet) which produced pulses at a wavelength of 620 nm [10]. One notable refinement of this approach was the introduction of *colliding-pulse modelocking* (CPM) [18] in which two counter-propagating pulses within a ring cavity are synchronized to arrive simultaneously at the absorber jet, so achieving greater saturation of the absorber and allowing the pulses to be transmitted with a loss significantly smaller than in the single-pulse case. CPM enabled the generation of 65 fs pulses directly from the laser [19] and represented the first demonstration of a system producing sub-100 fs pulses [18]. A further important refinement to the CPM dye laser was the compensation of intracavity group-velocity dispersion by incorporating a Brewster-angled 4-prism sequence [20] which resulted in pulse durations of 27 fs [21].

Throughout the 1980s the CPM dye laser was the principal source for ultrafast laser science and, indeed, (after amplification and compression) it was responsible for the record-breaking 6 fs duration pulses [22] which, for an entire decade, remained the shortest pulses ever produced. One drawback of the CPM configuration was the very low efficiency of the system, which required several watts of cw pump power from an Ar^+-ion laser to produce only around 10 mW of average modelocked power. Synchronous pumping using an actively modelocked ion laser allowed R6G dye lasers to produce modelocked pulses without a saturable absorber and enabled the output power to be increased to 100 mW with pulse durations of around 200 fs [23]. Further developments included a hybridly modelocked dye laser based on kiton red and malachite green and operating at 645 nm which, as a commercial system, was available with average output powers of up to 300 mW and pulse durations of 150 fs [24]. All dye lasers are maintenance-intensive systems and this, together with the hazardous nature of many of the chemicals used, led to their gradual replacement by more convenient solid-state alternatives, based principally on Ti:sapphire.

C2.3.2.2 Ti:sapphire lasers

The creation in 1986 [25] of a new broadband laser material—titanium-doped sapphire or Ti:sapphire—began a revolution in ultrafast laser sources which still continues today. Unlike other contemporary solid-state materials for modelocked lasers such as Nd:YAG, Ti:sapphire is a *vibronic* gain medium (see chapter B1.2) which exhibits a strong electron–phonon coupling between the titanium ion transition and the host lattice. This vibronic coupling leads to a broad continuum of possible transition energies and results in a fluorescence bandwidth covering 670—1050 nm, making Ti:sapphire capable of supporting extremely short femtosecond pulses. Furthermore, Ti:sapphire can be grown as high-quality crystals, is mechanically strong with good thermal conductivity and has an absorption band in the blue/green which is compatible with the Ar^+-ion lasers formerly used to pump existing modelocked dye oscillators. The long fluorescence lifetime of Ti:sapphire means that, unlike dye lasers, modelocking requires either active intracavity modulation or a passive method using a fast saturable absorber. Various methods were reported including acousto-optic modelocking [26], intracavity saturable absorbers in the form of a dye jet [27] or semiconductor-doped glass plate [12], additive-pulse modelocking (see section 2.3.2.3) [28–30], semiconductor saturable absorber mirrors [31, 32] and Kerr-lens modelocking [33]. Of the various approaches listed here, only the final two methods have prevailed and these are now discussed in detail.

The first passively modelocked Ti:sapphire lasers were based on additive-pulse modelocking (APM) in which modelocking is achieved using interference between the intracavity pulse and a self-phase-modulated replica propagating in an auxiliary cavity. An important breakthrough was made when in 1990 it was reported that an APM Ti:sapphire laser continued to operate even when the auxiliary cavity was blocked [34]. This discovery was attributed to an entirely new modelocking effect known as *Kerr-lens modelocking* (KLM) or *self-modelocking* in which the presence of a Kerr-lens within the gain medium during modelocked-operation changes the mode focusing within the cavity so that, compared to cw operation, modelocked pulses experience higher gain. Two configurations of KLM exist and are known as *soft aperture modelocking* [35], where the presence of the Kerr-lens increases the gain by improving the overlap between the pump and laser modes, and *hard-aperture modelocking* [36], in which a physical aperture such as a slit or the edge of a prism is adjusted to introduce greater loss for cw operation. Because KLM is based on a non-resonant nonlinearity, Ti:sapphire lasers can be modelocked at any wavelength in their gain bandwidth using this technique. Using specially designed broad-bandwidth mirror sets, femtosecond Ti:sapphire lasers have been demonstrated with continuous tunability across ~300 nm [37]. Independently tunable dual wavelength operation from a single modelocked laser has also been reported [38] and enables two exactly synchronized pulse sequences to be produced which are suitable for spectroscopic pump-probe experiments [39]. Recent advances in controlling the intracavity group-velocity dispersion characteristics have led to the generation of sub-5 fs pulses directly from a Ti:sapphire oscillator [40] which utilize the entire gain bandwidth of the material. The principal drawback of the KLM method is that it is not self-starting because modelocked operation must be seeded by an intense noise spike or other short fluctuation. Different starting methods have been applied successfully including mirror tapping, acousto-optic modulation [41] and mode-dragging [42]; once modelocking has been initiated it is generally stable until some external perturbation disturbs the intracavity beam.

An alternative method for obtaining femtosecond pulses has been pioneered largely in parallel with the KLM efforts and is based on using resonant nonlinearities in semiconductors as fast saturable absorbers. The approach is to fabricate a semiconductor multiple-quantum-well (MQW) device whose bandgap has been engineered to match the laser wavelength and whose absorption can be saturated at high fluences. Initially these devices were utilized in an APM geometry in which the MQW device was situated in an auxiliary cavity [43] but it was later realized that an equivalent monolithic configuration could be obtained by sandwiching the MQW layer between two high-reflectivity mirrors (see figure C2.3.3) to form a Fabry–Pérot structure [44] and the resulting element was known as an anti-resonant Fabry–Pérot saturable absorber (A-FPSA). The first application of a such a *semiconductor saturable absorber mirror (SESAM)* for modelocking Ti:sapphire was reported in 1993 [45] where self-starting operation was demonstrated. A typical modern Ti:sapphire laser design is shown in figure C2.3.4 and illustrates how a saturable absorber mirror can be integrated into the cavity. Further design enhancements such as reducing the finesse of the Fabry–Pérot cavity resulted in SESAMs with a broader wavelength response which enabled sub-10 fs self-starting operation.

C2.3.2.3 Colour-centre lasers

A new source of femtosecond pulses in the near-infrared was reported in 1984. Termed the *soliton laser* [15, 47], the system was a synchronously-pumped KCl:Tl^0 (1) colour-centre laser tunable from 1.4–1.6 μm and modified to allow direct generation of pulses as short as 50 fs, substantially shorter than the 8 ps achieved in synchronously-pumped operation alone. Femtosecond operation was achieved by coupling a length of polarization-preserving anomalously-dispersive fibre to the main laser cavity using a beamsplitter and retroreflector arrangement. This coupled-cavity arrangement later became known as *additive pulse modelocking* (APM) and produced short pulses when the optical length of the fibre was adjusted to be an integral multiple of the main cavity length. In the original demonstration, operation was attributed to bright soliton generation in the fibre arm, but numerical modelling later showed [48] that modelocking was enhanced even when the pulses returned from the fibre were temporally broadened, implying that solitonic effects were not necessary for femtosecond operation. This explanation was later confirmed experimentally when 260 fs pulses were generated from a KCl:Tl colour centre laser modelocked using only normally dispersive fibre [49]. Pulses of 64 fs duration were subsequently produced from the same laser when erbium fibre, which exhibits enhanced nonlinearity, was used [50]. This behaviour was consistent with theories which emphasized the importance of interference between the pulse travelling in the master cavity and the self-phase-modulated pulse returned by the nonlinear fibre [51, 52]. As figure C2.3.5 illustrates, substantial pulse shortening can be achieved by this effect, even after only a single-pass. The long upper-state lifetime of the laser-active centres in the KCl:Tl gain medium allowed particularly stable APM operation, but modelocking of this kind was also successfully employed in an actively stabilized NaCl:OH$^-$ colour-centre laser to produce 110 fs pulses [53]. The need to maintain colour-centre crystals at liquid-nitrogen temperatures (see chapter B1.8) limited the convenience of these lasers but, until the advent of Ti:sapphire, they remained the only solid-state sources capable of femtosecond operation.

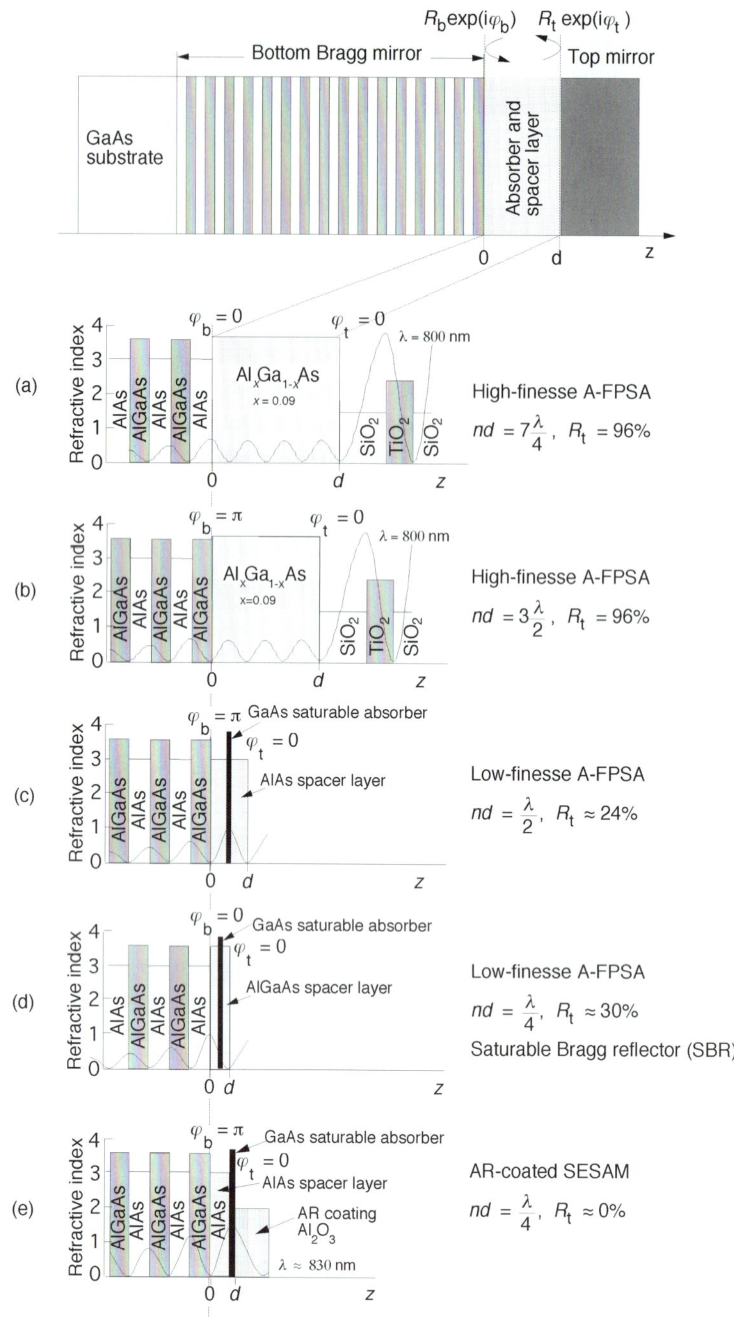

Figure C2.3.3. Different SESAM designs for a centre wavelength of about 800 nm which is suitable for a Ti:sapphire laser. The high-finesse A-FPSA was the first intracavity saturable absorber that stably cw-modelocked solid-state lasers with long (>μs) upper state lifetimes. The saturable Bragg reflector is a special case of a low-finesse A-FPSA for which all layers are quarter-wave layers. Image courtesy of Ursula Keller, Swiss Federal Institute of Technology.

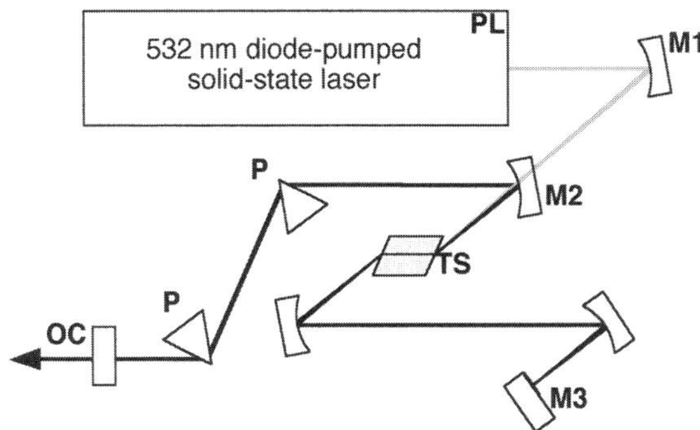

Figure C2.3.4. Typical configuration of a modern femtosecond Ti:sapphire laser. PL, pump laser; M1, pump focusing mirror; M2, high reflectivity cavity mirror (800 nm); TS, Ti:sapphire crystal, typically 2–10 mm long; M3, SBR/SESAM mirror; P, dispersion-compensating prisms; OC, output coupler.

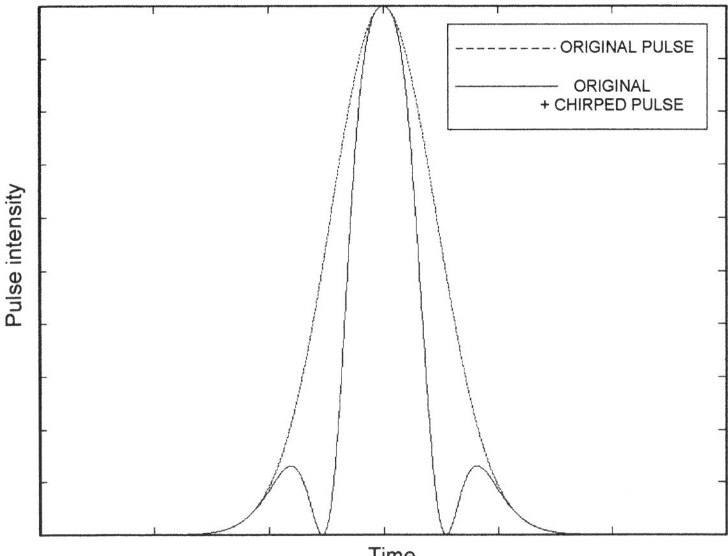

Figure C2.3.5. Additive-pulse modelocking. Figure shows the reduction in pulse duration obtained by interfering a pulse with its self-phase-modulated replica. In this example the replica has been self-phase-modulated with a peak phase shift of 2π.

C2.3.2.4 Fibre lasers

Ultrafast fibre lasers, particularly those based on silica doped with rare-earth Er^{3+}, Nd^{3+} or Yb^{3+} ions (see chapters B4.1 and B4.2), have enjoyed renewed attention in recent years as efforts to scale their average output powers and pulse energies to practical levels have proved successful. Four main modelocking strategies have been applied to these systems: active modelocking [54, 55] which is generally restricted to picosecond generation; intensity-dependent feedback using a nonlinear amplifying loop mirror [56]; polarization APM [57–59]; and semiconductor saturable absorber modelocking [60]. The upper-state lifetime of rare-earth-

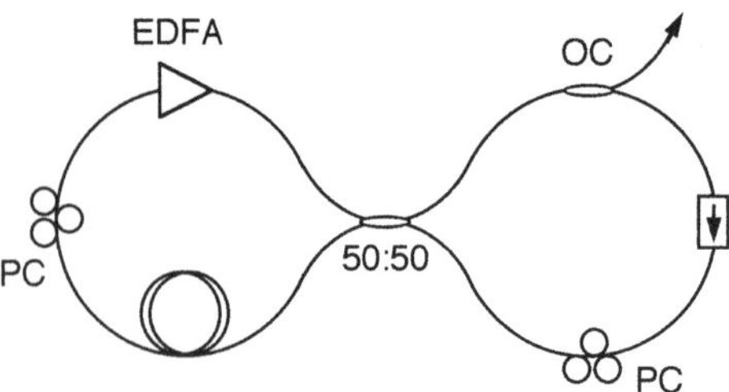

Figure C2.3.6. Schematic diagram of a figure-of-eight femtosecond fibre laser. EDFA, erbium-doped fibre amplifier gain section; OC, output coupler; PC, polarization controller.

doped fibres is long and therefore a fast saturable absorber mechanism is necessary for passive modelocking, limiting the choice of modelocking elements to those based on Kerr or semiconductor nonlinearities.

The development of contemporary actively modelocked fibre sources has concentrated on Er-doped lasers because of the compatibility of their output wavelengths with the 1.5 μm optical communication window in standard silica fibre. One common configuration is the sigma laser which comprises separate gain and phase-modulation sections and has produced 1.5 ps pulses at a repetition frequency of 10 GHz [61]. Passively modelocked designs based on the nonlinear loop mirror work [62] by exploiting the Kerr effect in an intracavity fibre Sagnac interferometer containing a gain section [63]. With suitable control of the intracavity polarization, such a *nonlinear amplifying loop mirror* (NALM) can be set up to transmit light of high intensities but reflect low intensities because of the differential phase shift induced between the two interferometer arms. By adding the loop mirror to a unidirectional ring cavity a *figure-of-eight laser* [64, 65] is formed (figure C2.3.6) which is capable of producing femtosecond soliton pulses with average powers of around 1 mW [66]. Drawbacks of the figure of eight laser include the presence of spectral sidebands at the shortest pulse durations [67] and a tendency for multiple pulse operation.

Modelocking can also be achieved using a variation on the APM technique. An elliptically polarized pulse propagating inside the fibre can be resolved into right- and left-handed components of different intensities which accumulate a differential nonlinear phase shift through the optical Kerr effect. The launch condition into the fibre is adjusted using a waveplate or polarization controller so that interference between the two circular polarization components at a polarizing beamsplitter rejects the wings but transmits the centre of the pulse. The technique can also be understood in terms of nonlinear polarization rotation in which Kerr-induced birefringence in the fibre rotates the centre of the pulse differently from the wings so that only the pulse centre is transmitted by the polarizing beamsplitter. Lasers based on *polarization-APM* can be operated in a soliton mode [68] or in a dispersive mode ('*stretched pulse*') mode [69]. Notably, significant average output powers have been demonstrated from stretched-pulse-APM lasers, with up to 90 mW being reported at pulse energies of 2.25 nJ [70].

Semiconductor saturable absorber modelocking has been demonstrated in fibre lasers using a variety of configurations including bulk InGaAsP on InP [71] and saturable Bragg reflectors based on InGaAs/InP MQWs on a AlAs/GaAs mirror structure [72]. Sub-picosecond pulse durations are readily available from these systems at repetition frequencies in the 10–100 MHz range. Another variation which has been demonstrated is modelocking using a semiconductor optical amplifier and 900 fs pulses at 10 MHz have been reported in a ring laser configuration incorporating a quantum-well diode amplifier. Fibre lasers modelocked using semiconductor saturable absorbers are self-starting and exhibit stable modelocking, normally at the fundamental cavity frequency. The narrow bandwidth of saturable absorber devices at near-infrared wavelengths makes the generation of sub-100 fs pulses difficult.

C2.3.2.5 Semiconductor sources

Diode lasers and external-cavity lasers based on *semiconductor optical amplifiers* (SOAs) are of importance because of their ability to produce ultrashort pulses from a compact system at wavelengths of relevance to optical fibre communications. The two principal methods used to generate ultrashort pulses from such devices are modelocking with an external cavity [73, 74] or *gain-switching* (equivalently known as Q-switching) in a bow-tie waveguide structure [75]. Sub-picosecond pulse durations are now routinely produced using dispersion compensation and/or nonlinear pulse compression techniques which, in fact, pre-empted parallel work in solid-state laser modelocking [76] and, by combining these strategies with Er:fibre amplification, 20 fs pulses have been produced [77]. The modelocking and gain-switching techniques each have their individual advantages. Gain-switching provides a simple and compact approach but significant timing jitter and poor spectral quality can accompany pulses produced in this way. Modelocked systems are physically larger than gain-switched devices because they require an external cavity and additional optical components, but shorter and higher quality pulses are achievable.

Laser action in electrically-pumped semiconductor gain media is unique because the gain can be directly controlled by varying the drive current across the device. As one primary application of semiconductor lasers is as signal sources for optical fibre communications, the ability to modulate their output at high data-bit frequencies is particularly important. The gain-switching method works by biasing a laser diode with a constant dc current slightly below the threshold value and then, using a *bias tee*, superimposing a radio-frequency (rf) current derived from an electrical pulse synthesiser or sine-wave generator. For a short time each period, the rf current brings the diode above threshold and pulsed laser emission occurs, with the pulse duration being sensitive to the relative magnitudes of the dc and rf currents. Increasing the optical gain coefficient and the internal photon density, and decreasing the photon lifetime inside the laser waveguide by minimizing facet reflectivity and separation are ways of increasing the maximum modulation frequency [78]. Maximum modulation frequencies of around 70 GHz have been reported by using short cavity lengths and strong optical confinement, as well as minimizing parasitic effects associated with the rf driver circuitry [78]. Gain-switched Fabry–Pérot diodes are commercially available offering pulse durations of 17 ps at modulation frequencies of 2 GHz [80]. One problematic characteristic of gain-switched laser diodes is their tendency to produce highly-chirped output pulses. Modulating the device current leads to a modulation in the density of electrons and holes which, in turn, perturbs the positions of the Fermi levels lying beside the valence and conduction bands and modifies the oscillation wavelength of the laser. In this way, a nonlinear frequency chirp is imprinted across the laser pulse as the drive current is modulated and this effect places a limit on the minimum pulse duration that can ultimately be produced.

Modelocked pulses have been produced from SOAs using the same range of techniques that are commonly applied to other laser media. Because modelocking requires the mode from the SOA to be efficiently coupled into an external cavity, the SOA must be designed to operate with a fundamental transverse mode and this is achieved using flared-waveguide amplifier structures [81] with facets which are anti-reflection coated to minimize etalon effects. Active modelocking [82] has been demonstrated by using gain modulation near threshold [83] and passive modelocking has been reported using a MQW, saturable Bragg [84] reflector which resulted in 5.6 ps pulses [85]. A theoretical analysis of active modelocking in SOAs has been used to determine the practical limitations of these systems for producing sub-picosecond pulses [86], while similar studies of passively modelocked semiconductor lasers have concentrated on understanding the roles of nonlinearity, dispersion and pulse interaction within the cavity [87]. Active and passive techniques can be combined to implement *hybrid modelocking* [88] and this method has been used to stabilize a passively modelocked system which exhibited fourth-harmonic modelocking due to gain saturation effects in the SOA [85]. A linear-cavity variation of the colliding pulse modelocking technique has been applied in a monolithic form to semiconductor lasers [89], and the technique, which is implemented using laser diodes comprising separate saturable absorber and gain sections [90–92], is now an established method for directly modelocking laser diodes without using an extended cavity.

C2.3.2.6 Other common solid-state laser sources

Femtosecond Ti:sapphire oscillators are widely acknowledged as the modern standard for ultrafast laser systems, but other solid-state sources offer the potential for direct diode pumping, miniaturization and operation at wavelengths or average powers not accessible using Ti:sapphire. Alternative systems based on Nd or Cr ion transitions have also been shown to be capable of ultrashort picosecond or femtosecond operation and a brief review of these sources will now be presented.

Laser materials based on Nd active ions exhibit emission bandwidths enabling sub-picosecond pulse generation, and Nd-based modelocked lasers have existed in a variety of configurations for many years, based either on flashlamp excitation or direct diode-pumping at 808 nm. Actively-modelocked Nd:YAG mainframe systems produce a few picosecond duration pulses at repetition frequencies of around 80 MHz and these have been commonly used for synchronous pumping of other ultrashort lasers [93]. The long upper-state lifetime of Nd:YAG and related materials (see chapter B1.3) means that attempts to passively modelock these lasers using saturable absorbers often result in the production of a modelocked pulse sequence underneath a sub-μs-duration Q-switched envelope [94]. Additive-pulse modelocking has been used successfully to achieve cw-modelocked operation from Nd:YLF [95] and other Nd-based lasers and this technique produces pulse durations of around 1 ps. Passive modelocking of Nd-based lasers using saturable absorber mirrors has been applied to eliminate Q-switching [96] and has enabled the demonstration of picosecond sources with repetition frequencies of 77 GHz [97] and average output powers of >4 W [98]. Femtosecond pulses with durations of 175 fs and an average power of 1 W have been produced using Nd:glass [99] which has one of the broadest fluorescence spectra of the Nd materials.

Cr-ion-doped crystals are promising broadband gain materials for generating femtosecond pulses, and are particularly interesting because many have absorption bands which are suitable for pumping using readily available diode lasers. Crystals can be obtained with active ions of Cr^{2+}, Cr^{3+} and Cr^{4+} which, respectively, have transitions covering wavelengths in the 2.0 μm, 800 nm and 1.4 μm regions (see chapter B1.2). Femtosecond modelocked operation has been demonstrated in Cr^{4+}:YAG at 1.52 μm [100], $Cr^{4+}:Mg_2SiO_4$ (Cr:forsterite) at 1.25 μm [101, 102], $Cr^{3+}:LiSrAlF_4$ (Cr:LiSAF) at 850 nm [103] and $Cr^{3+}:LiSrGaF_6$ (Cr:LiSGAF) [104]. Direct diode pumping of Cr-ion-based lasers is attractive because of the opportunities for producing low-noise, compact and efficient femtosecond sources, but the limited power available from high-quality pump diodes results in low intracavity powers which make KLM operation difficult. As a result, stable operation is better achieved using a physical saturable absorber, and SESAM/SBR modelocking has been applied successfully to many of these lasers [105].

C2.3.2.7 Sources based on nonlinear frequency conversion

Laser sources alone are unable to provide ultrashort pulses continuously tunable from the ultraviolet to the mid-infrared, and large gaps exist at wavelengths where no broadband laser materials are available. Harmonic generation using Ti:sapphire can be used to access some parts of the visible [106] and ultraviolet [107] regions, but infrared wavelengths cannot be generated in this way and coverage is limited by the ability to tune the source itself. The nonlinear production of infrared wavelengths requires parametric downconversion and, as early as 1972, the synchronously-pumped OPO (see chapter B1.6) was proposed as an effective means for generating ultrashort pulses in new wavelength regions [108]. Before the invention of the self-modelocked Ti:sapphire laser, ultrafast pump sources with sufficiently high peak power to enable singly-resonant operation of an OPO were not available. For this reason, the synchronously-pumped parametric oscillator has its origins in frequency-unstable, doubly-resonant configurations such as the one demonstrated by Piskarskas and pumped by the second harmonic of a picosecond cw-modelocked Nd:YAG laser [109]. Although the output of a doubly-resonant OPO is generally characterized by large amplitude and spectral instabilities, Ebrahimzadeh and co-workers exploited the low operating threshold associated with this configuration to demonstrate an all-solid-state oscillator based on KTP and pumped by 15 ps pulses from a frequency-doubled cw-modelocked Nd:YLF laser [110].

The development of synchronously-pumped OPOs had for many years relied on picosecond pump lasers because they represented the only sources capable of providing sufficiently large pulse energies. In 1989, Edelstein *et al* became the first to demonstrate a high-repetition rate continuous-wave femtosecond OPO using an oscillator based on a thin crystal of KTP and pumped at the intracavity focus of a dye laser [111]. The singly-resonant OPO produced 105 fs pulses at a repetition frequency of 100 MHz and tuning was reported from 755–1040 nm (signal) and 1.5–3.2 μm (idler) [112–115]. The low output powers from the CPM-dye laser necessitated the alignment-critical intracavity pumping approach, so an important advance was made when Mak *et al* demonstrated a KTP-based femtosecond OPO pumped extracavity by the output of a hybridly-modelocked dye laser operating at 645 nm [116]. This configuration allowed independent alignment of the laser and the OPO and was an important intermediary to the Ti:sapphire-pumped femtosecond OPO first demonstrated by Pelouch *et al* [117] which produced near-infrared 57 fs pulses at average signal powers of 340 mW. OPO-based visible sources of femtosecond pulses have also been demonstrated, and access to the yellow/red region has been reported using intracavity frequency doubling of KTP [118] and RTA-based [119] oscillators. Direct oscillation in the visible was first demonstrated by Driscoll *et al* in a β-barium-borate (BBO)-based OPO pumped by the second harmonic of a self-modelocked Ti:sapphire laser and resonant at 630 nm [120]. Pulses as short as 14 fs were reported from this oscillator [121] because of its unique phase-matching geometry [122]. New, highly nonlinear materials based on the *quasi-phasematching* and *periodic-poling* techniques emerged in the late-1990s and were soon applied as gain media in femtosecond [123] and picosecond [124] OPOs. Using materials such as PPLN [125] and PPRTA [126], oscillation from Ti:sapphire-pumped OPOs has been achieved at wavelengths extending to the long-wavelength edge of the transmission window at 6.8 μm [127].

C2.3.2.8 Sources of amplified ultrashort pulses

Modern ultrafast oscillators can produce pulses with peak powers of >1 MW [128, 129], average powers of several watts [130] and cavity frequencies from 4 MHz [131] to 2 GHz [132], but many applications (see section C2.3.5) require ultrashort pulses with further enhanced intensities and/or lower pulse repetition frequencies. Additional amplification stages can be used to satisfy these requirements and these are commonly implemented in Nd:glass for picosecond pulses and Ti:sapphire for femtosecond pulses. Other media are also becoming more widely used for amplifier configurations and include Yb:glass [133], Cr:LiSAF [134] and Er/Yb fibre [135]. In order to avoid damage and unwanted nonlinear effects caused by high pulse intensities within the amplifier system, most practical sources now apply the *chirped-pulse amplification* (CPA) approach [136, 137] in which a low-energy femtosecond seed pulse is stretched to sub-ns durations to reduce its peak power is then amplified and is finally compressed to its original duration. Using the CPA technique, pulses with peak powers as high as 1.5 PW have been generated [138]. Two possible amplifier geometries are used which are categorized either as *multi-pass* or *regenerative* amplifiers. In the multi-pass configuration the injected pulse is refocused many times through the same gain crystal using a system of mirrors. The gain crystal is pumped above the saturation fluence in order to extract the maximum stored energy from the medium per pass. The multi-pass geometry is more commonly used for high-average-power operation and critical alignment is required to maximize the overlap between the pump mode profile and the multiple beams intersecting the gain medium. The alternative regenerative amplifier is essentially a stable laser resonator which includes a Pockels cell and a polarizing beamsplitter to enable pulses to be switched into and out of the cavity. The gain per pass is considerably lower than in the multi-pass configuration but regenerative systems produce superior beam quality and their performance can be optimized without affecting the beam pointing of the output. Commercially available Ti:sapphire amplifiers commonly use the regenerative method and are available with pulse energies of up to 1 mJ at repetition frequencies in the 1–300 kHz range.

Several factors affect the quality of the pulse amplification process and the ability of the system to produce output pulses with a similar duration and spectral bandwidth similar to the input pulses. *Gain narrowing* is a fundamental effect caused by the wavelength dependence of the amplifier gain material and, when the seed pulse is centred on the peak of the amplifier gain, leads to greater amplification in the centre of the pulse spectrum than in the wings. The consequent spectral narrowing caused by this effect leads to an increase in the pulse duration. In a similar way, *gain saturation* effects can lead to spectral broadening and introduce chirp on the pulse due to non-uniform amplification across the pulse spectrum. Strategies for minimizing spectral re-shaping include incorporating intra-amplifier frequency filters and using seed pulses whose wavelength is offset from the spectral centre of the amplifier gain. Proper control of second-, third-and fourth-order *group-velocity dispersion* inside an ultrafast amplifier is important and is critically dependent on the design and adjustment of the pulse compressor. Increasingly, exact pulse diagnostics such as FROG and SPIDER (see section C2.3.4.3) are being used to characterize and minimize the high-order spectral phase errors which ultimately limit the maximum peak power which can be achieved.

C2.3.3 Pulse shaping and dispersion in optical systems

C2.3.3.1 Linear material dispersion

Chromatic dispersion, the variation of refractive index with wavelength, leads to a wavelength dependent group velocity which is responsible for pulse broadening and break-up in optical systems. Dispersive effects in materials originate as a result of a frequency-dependent dielectric susceptibility:

$$P(\omega) = \varepsilon_o \chi^{(1)}(\omega) E(\omega) + \varepsilon_o \chi^{(2)}(\omega) E(\omega)^2 + \varepsilon_o \chi^{(3)}(\omega) E(\omega)^3 + \cdots \tag{C2.3.11}$$

which is due to physical resonances (poles) in the electronic response to an applied E-field. Linear dispersion refers to the refractive index associated with the real part of the $\chi^{(1)}$ term, i.e.

$$n_o = \sqrt{1 + \mathrm{Re}\{\chi^{(1)}\}}. \tag{C2.3.12}$$

Dispersive effects are best analysed in terms of optical phase which is related to the refractive index by

$$\varphi(\omega) = \frac{n_o(\omega)\omega L}{c} \tag{C2.3.13}$$

where L is the medium length and c is the vacuum speed of light. The local variation of the spectral phase due to an optical medium can be represented as the Taylor series

$$\varphi(\omega) = \varphi(\omega_o) + (\omega - \omega_o)\left.\frac{\mathrm{d}\varphi}{\mathrm{d}\omega}\right|_{\omega_o} + \frac{1}{2}(\omega - \omega_o)^2 \left.\frac{\mathrm{d}^2\varphi}{\mathrm{d}\omega^2}\right|_{\omega_o} + \frac{1}{6}(\omega - \omega_o)^3 \left.\frac{\mathrm{d}^3\varphi}{\mathrm{d}\omega^3}\right|_{\omega_o} + \cdots \tag{C2.3.14}$$

where ω_o is the centre frequency of the incident light. When an ultrashort pulse propagates through the medium, each term in (C2.3.14) is responsible for a distinct dispersive effect. An arbitrary pulse $E(t)$ with a centre frequency of ω_o can be represented as a carrier field modulated by a complex amplitude $A(t)$ so that

$$E(t) = A(t)\exp(\mathrm{i}\omega_o t). \tag{C2.3.15}$$

The corresponding spectral field is given by the Fourier transform

$$e(\omega) = \frac{1}{\sqrt{2\pi}} \int_{-\infty}^{\infty} E(t) \exp(-\mathrm{i}\omega t)\,\mathrm{d}t. \tag{C2.3.16}$$

After propagating through a dispersive medium, the pulse acquires spectral phase according to

$$e(\omega)' = e(\omega) \exp(\mathrm{i}\varphi(\omega)) \tag{C2.3.17}$$

and the resulting pulse in time is described by the reverse Fourier transform

$$E(t)' = \frac{1}{\sqrt{2\pi}} \int_{-\infty}^{\infty} e(\omega)' \exp(\mathrm{i}\omega t)\,\mathrm{d}\omega. \tag{C2.3.18}$$

The first term in (C2.3.14) is the change in carrier phase after passing through the medium and (within the limits of the *slowly-varying envelope approximation*) has no effect on the pulse. Within the second term, the time spent by the pulse inside the medium is given by the constant

$$\tau = \left.\frac{\mathrm{d}\varphi}{\mathrm{d}\omega}\right|_{\omega_o} \tag{C2.3.19}$$

where τ is known as the *group delay*. In its entirety, the second term in (C2.3.14) describes a linear phase ramp in frequency which corresponds (through the well-known Fourier shift theorem [139]) to a delay of the pulse in time.

The quadratic and higher order terms in (C2.3.14) describe *group-delay dispersion*—the variation of the group delay with frequency—and have a direct influence on the shape of the pulse in time. The $(\omega - \omega_o)^2$ term is responsible for adding *linear chirp*, which refers to a time-varying uniform increase or decrease in the instantaneous optical frequency across the pulse. As an example (neglecting the carrier phase) consider a pulse with a Gaussian amplitude spectrum and quadratic spectral phase represented as

$$e(\omega) = \exp(-(a + \mathrm{i}b)\omega^2). \tag{C2.3.20}$$

The pulse amplitude in time is given by the Fourier transform of (C2.3.20) which, neglecting a constant normalization factor, has the form

$$E(t) \approx \exp(-\alpha t^2/4(a^2 + b^2)) \times \exp(\mathrm{i}bt^2/4(a^2 + b^2)). \tag{C2.3.21}$$

The *full-width half-maximum* (FWHM) intensity duration of the pulse described by (C2.3.21) is

$$\Delta\tau = 2\sqrt{\left(\frac{a^2 + b^2}{a}\right) \ln 4} \tag{C2.3.22}$$

and illustrates that the presence of the spectral quadratic phase of either sign ($\pm b$) will broaden an ultrashort pulse from its unchirped ($b = 0$) duration which is given by

$$\Delta\tau_o = 2\sqrt{a \ln 4}. \tag{C2.3.23}$$

The unchirped duration given in (C2.3.23) is related directly to the FWHM spectral intensity bandwidth of the pulse which is given by

$$\Delta\omega = \sqrt{\frac{\ln 4}{a}}. \tag{C2.3.24}$$

The quantity $\Delta\omega\Delta\tau$ is known as the *duration-bandwidth product* of the pulse and, as comparison of (C2.3.23) and (C2.3.24) confirms, has a constant value for unchirped pulses which is independent of their duration or spectral bandwidth. The duration-bandwidth product is, however, sensitive to the pulse shape: chirp-free Gaussian pulses have $\Delta\omega\Delta\tau = 2.77$ while $\mathrm{sech}^2 t$ pulses have $\Delta\omega\Delta\tau = 1.98$. In experimental measurements of ultrafast lasers the pulse shape is often assumed to be $\mathrm{sech}^2 t$ and the chirp of the pulses is then estimated by comparing their measured duration-bandwidth product with the ideal chirp-free value. Practically, the duration-bandwidth product can therefore be treated as a figure-of-merit for ultrashort pulses.

It was mentioned earlier that linear chirp describes a time-varying uniform increase or decrease in the local carrier frequency across the pulse, and this is evident from the last term in (C2.3.21) which corresponds to a temporal phase of

$$\phi(t) = bt^2/4(a^2 + b^2) \tag{C2.3.25}$$

which leads to a time-dependent variation of the carrier frequency given by

$$\Omega(t) = -\frac{\mathrm{d}\phi(t)}{\mathrm{d}t} = -bt/2(a^2 + b^2). \tag{C2.3.26}$$

All the wavelengths within an unchirped pulse ($b = 0$) travel together and, consequently, $\Omega(t) = 0$ and the local frequency across the entire pulse is the carrier frequency.

The quadratic term in the spectral phase expansion of (C2.3.14) cannot introduce asymmetry into an originally symmetric pulse. In contrast, cubic spectral phase variations associated with the $(\omega - \omega_o)^3$ term are significant because, in the time-domain, they lead to sharpening of one edge of the pulse and broadening of the other, ultimately resulting in pulse break-up. Strategies for generating the shortest pulses from ultrafast lasers rely on controlling the net intracavity dispersion so that the spectral phase associated with one cavity roundtrip is constant across the entire spectral bandwidth of the pulse. Special optical components are used for controlling second-, third- and fourth-order spectral phase and these are described in detail in section C2.3.3.4.

C2.3.3.2 Nonlinear material dispersion

In our treatment of nonlinear dispersion we will only consider effects due to the $\chi^{(3)}$ term in (C2.3.11) because, unlike those associated with the $\chi^{(2)}$ term (e.g. the Pockels effect), $\chi^{(3)}$ effects are present in all materials. Neglecting the $\chi^{(2)}$ term, (C2.3.11) can be re-written as

$$P(\omega) = \varepsilon_o \chi(\omega) E(\omega) \tag{C2.3.27}$$

where

$$\chi(\omega) = \chi^{(1)}(\omega) + \chi^{(3)}(\omega) E^2. \tag{C2.3.28}$$

In a transparent medium, the refractive index due to this net susceptibility is

$$n = \sqrt{1 + \chi^{(1)} + \chi^{(3)} E^2} \tag{C2.3.29}$$

which, by applying the Taylor series expansion $\sqrt{1+x} \approx 1 + x/2$, can be approximated as

$$n \approx n_o + \frac{\chi^{(3)} E^2}{2n_o} \tag{C2.3.30}$$

or simply

$$n \approx n_o + n_2 I. \tag{C2.3.31}$$

The importance of the $\chi^{(3)}$ susceptibility is, therefore, that it causes the refractive index to become intensity-dependent, and significant changes in the refractive index can be induced by the high peak intensities of ultrashort pulses. Two principal effects—*self-phase-modulation* and *self-focusing*—result from the nonlinear refractive index and both are essential in generating the shortest pulses from modern ultrafast lasers.

Self-phase-modulation (SPM) is caused by the time-varying intensity profile of the pulse. The higher-intensity components induce a larger nonlinear refractive index change and therefore experience a greater phase shift than the weaker components. Using (C2.3.26) and (C2.3.31), the resulting frequency shift across the pulse can be shown to be

$$\Omega(t) = -\frac{\mathrm{d}\phi(t)}{\mathrm{d}t} = -\frac{\omega n_2 L}{c}\frac{\mathrm{d}I(t)}{\mathrm{d}t} \tag{C2.3.32}$$

which describes a redshift of the pulse leading edge and a blueshift of the trailing edge. Only the pulse centre frequency remains unchanged and the result of SPM is, therefore, to broaden the pulse spectrum by redistributing energy from the centre to the wings. Figure C2.3.7 illustrates this effect for an initially chirp-free Gaussian pulse. Across the centre of the pulse, SPM leads to an approximately linear positive chirp which, with appropriate optics, can be corrected using an equal but opposite amount of negative second-order dispersion. The combination of spectral broadening using SPM and subsequent chirp removal using linear dispersion is the basis for optical pulse compression techniques which have been applied effectively to create pulses which exist for only a few carrier-wave periods [22, 140]. When the pulse propagates in an environment where SPM and linear dispersion are well balanced, a soliton [13, 141], will evolve which, in the absence of loss, can propagate indefinitely without changing shape.

A spatial analogue to SPM is self-focusing, in which the variation of intensity across the beam profile of an ultrashort pulse leads to a positive *Kerr lens* which retards the centre of the wavefront more than the edges. As already mentioned in section 2.3.2.2, by configuring a laser cavity to properly exploit this effect, the Kerr lens can be used as a versatile means of achieving passive modelocking. A special case exists when beam divergence due to diffraction is exactly balanced by self-focusing. In analogy with temporal solitons, pulses propagating in this way are known as *spatial solitons* [142] and are observed under certain conditions of the beam diameter, pulse intensity and medium nonlinearity. *Catastrophic self-focusing* occurs when the beam power exceeds a critical value [143] given by

$$P_{\mathrm{crit}} = \frac{\pi \varepsilon_0 c^3}{2 n_2 \omega^2}; \tag{C2.3.33}$$

above this value the beam diameter continually decreases until a focal point is formed whose intensity is large enough to damage the medium. Catastrophic self-focusing is of particular concern in ultrafast laser amplifiers and must be avoided by using large beam diameters and long pulses with peak powers below P_{crit}.

C2.3.3.3 Other sources of dispersion

Although dispersion arising from the dielectric response of the propagation medium has the greatest effect on a pulse, other sources also contribute to the overall dispersion profile of an optical medium. The principal remaining source, *modal dispersion*, is relevant only to waveguides and arises because the wave propagating in the guide is split into a travelling forward component and a stationary transverse component [144]. Stronger confinement increases the size of the transverse wavevector component and reduces the size of the travelling component, leading to a lower group-velocity compared to propagation in an equivalent bulk material. In single-mode optical fibres the contribution is significant and shifts the zero-dispersion wavelength from 1.27 μm (bulk silica) to 1.31 μm (typical single-mode silica fibre) and special techniques can be used to achieve even larger shifts [145]. Modal dispersion can be engineered to dominate the dispersion characteristics of a waveguide and this has been demonstrated recently in *photonic-crystal fibres* [146] which, although fabricated from pure silica, can be made to have zero group-velocity dispersion wavelengths as low as 650 nm by varying the core size and void distribution [147].

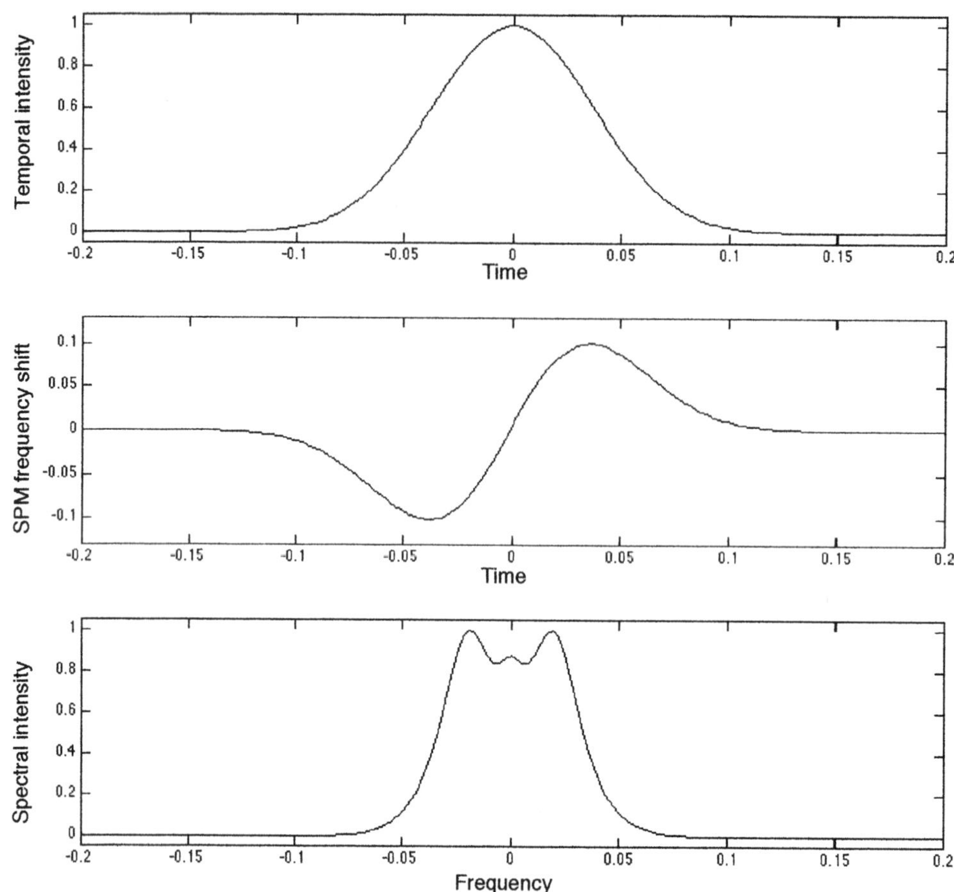

Figure C2.3.7. Self-phase modulation. Intense pulses propagating in a Kerr medium experience a nonlinear phase shift which resembles linear chirp across the centre of the pulse.

Dispersion also arises in multilayer dielectric coatings, particularly high-reflectivity mirrors, and is associated with the wavelength dependence of the transmitted or reflected phase. In Bragg quarter-wave mirror structures the reflected phase is flat in the centre of the stop-band but changes rapidly at the edges, leading to positive dispersion at shorter wavelengths and negative dispersion at longer wavelengths. Interestingly, Bragg mirrors can be understood as one-dimensional *photonic bandgaps* [148], and tunnelling of ultrashort pulses through a Bragg mirror has been observed at superluminal group velocities [149].

C2.3.3.4 Group-velocity dispersion compensation

Obtaining the shortest pulses from a modelocked laser requires the quadratic and higher-order terms in the roundtrip intracavity spectral phase to be zero. In practice this means minimizing the material group-velocity dispersion of the gain medium by using a short but highly-doped laser crystal and then identifying additional optical elements whose group-velocity dispersion is equal in magnitude but opposite in sign to that of the crystal. Optical components for compensating group-velocity dispersion fall into two categories: those which use bulk optics and geometry to achieve a wavelength-dependent path length, and others based on interference effects within a dielectric optical coating.

Pulse compression by dispersion compensation was originally considered by Treacy [150] who discussed in this and a later [151] paper how pairs of diffraction gratings could be configured to compensate for positive quadratic spectral phase. Subsequent work by Fork *et al* [20] showed that prism pairs could be used for the same purpose, thus avoiding the transmission loss associated with diffraction gratings. In both cases the geometry of the systems is such that the group delay is wavelength-dependent and can be made greater for longer wavelengths than for shorter ones. Other authors [22, 152, 153], have refined the analytic expressions describing the quadratic and cubic spectral phase of prism and diffraction grating pairs and their results are summarized in table C2.3.1. Prism pairs and four-prism sequences have been extremely successful in enabling the generation of short pulses from modelocked solid-state lasers, and 11 fs pulses were reported using this approach [154]. Dispersion compensation using prism pairs has fundamental limitations, and the simultaneous compensation of arbitrary second- and third-order spectral phase is difficult and involves identifying a suitable prism material, apex spacing and tip insertion [152].

Dispersion control using mirror coatings is not a new idea and, in 1986, was applied in CPM dye lasers to obtain 50 fs pulses by tuning the laser to a wavelength longer than the mirror centre wavelength. This configuration achieves negative dispersion from a standard Bragg mirror because, towards the long-wavelength edge of the mirror stopband, longer wavelengths penetrate further into the coating than the shorter ones before being reflected. The drawbacks of using a standard Bragg mirror design in this way are that the dispersion response is not readily controllable and the longer wavelengths experience lower reflectivity and so the pulse spectral intensity is modified. A new mirror technology described as *chirped multilayer dielectric coatings* was reported in 1994 [156] and is based on the idea that a Bragg mirror with smaller layer periods at the surface and larger periods deeper in the coating will provide negative group-velocity dispersion because the longer wavelengths experience larger group delays within the coating than the shorter ones. Unlike the earlier approach using simple periodic quarter-wave Bragg layers, 'chirped mirrors' provide constant reflectivity across their stopband and therefore only modify the spectral phase of a pulse, leaving the intensity unchanged. The design of chirped mirrors has been refined in recent years to enhance their operating bandwidth [156] and to reduce undesirable modulations in the group-delay response [157]. The technique has been used to obtain 6.5 fs pulses directly from a Ti:sapphire oscillator [158] and for compression of amplified pulses to 4.9 fs [140].

An alternative approach for obtaining negative group-velocity dispersion in a mirror structure is to exploit the frequency-dependent phase properties of a resonant Fabry–Pérot etalon [159]. A *Gires–Tournois (GTI) mirror* is an etalon fabricated within a dielectric Bragg mirror by including a layer with a thickness of $\lambda/2$ near the surface to form a weakly resonant cavity [160]. The group-delay across the GTI bandwidth increases approximately linearly with wavelength and can therefore be used for dispersion compensation. In comparison to other mirrors used for dispersion compensation, the GTI provides larger amounts of negative dispersion, and a single GTI mirror can therefore provide enough dispersion to compensate the dispersion of all the other components in a laser cavity [161]. GTI mirrors have been applied in both picosecond [162] and femtosecond [163] modelocked lasers.

C2.3.3.5 Fourier-transform pulse shaping

Ultrashort pulses with tailored intensity and phase profiles are of interest for applications in high-energy amplifiers [164], *wavelength-division multiplexed* optical communications [165] and control of chemical reactions [166]. Femtosecond pulse shaping has been studied for a number of years (see the review in [167]), but only recently have diagnostic tools and modulator components become available which enable truly programmable pulse control. Shaping is commonly achieved by placing a spatial light-modulator such as a liquid-crystal display (LCD) array [168] at the Fourier plane of a dispersion-free pulse compressor [169] as illustrated in figure C2.3.8. By modifying the intensity and phase of its spectral components, the pulse is shaped in the time-domain through the Fourier relation given in (C2.3.18). Although LCD-based shaping

Table C2.3.1. Expressions for the second- and third-order dispersion of common optical systems.

Bulk material, thickness t

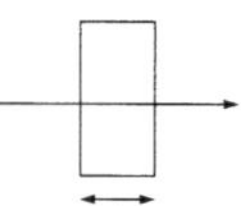

$$\frac{d^2\varphi}{d\omega^2} = t\frac{\lambda^3}{2\pi c^3}\frac{d^2n}{d\lambda^2}$$

$$\frac{d^3\varphi}{d\omega^3} = -t\frac{\lambda^4}{4\pi^2c^3}\left(3\frac{d^2n}{d\lambda^2} + \lambda\frac{d^3n}{d\lambda^3}\right)$$

Double-pass prism-pair, apex separation l, angular deviation within ray of β

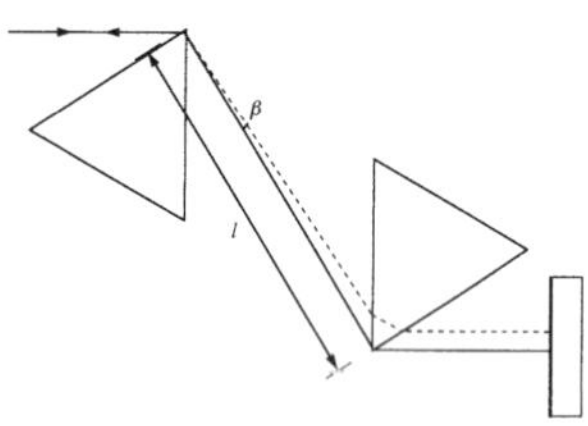

$$\frac{d^2\varphi}{d\omega^2} = \frac{\lambda^3}{2\pi c^2}\frac{d^2P}{d\lambda^2}$$

$$\frac{d^3\varphi}{d\omega^3} = -\frac{\lambda^4}{4\pi^2c^3}\left(3\frac{d^2P}{d\lambda^2} + \lambda\frac{d^3P}{d\lambda^3}\right)$$

$$\frac{d^2P}{d\lambda^2} = 4l\sin\beta\left[\frac{d^2n}{d\lambda^2} + (2n - 1/n^3)\left(\frac{dn}{d\lambda}\right)^2\right] - 8lz\cos\beta\left(\frac{dn}{d\lambda}\right)^2$$

$$\frac{d^3P}{d\lambda^3} = l\cos\beta\left[(24/n^3 - 48n)\left(\frac{dn}{d\lambda}\right)^3 - 24\frac{dn}{d\lambda}\frac{d^2n}{d\lambda^2}\right]$$
$$+l\sin\beta\left[\left(\frac{dn}{d\lambda}\right)^3(12/n^6 + 12/n^4 + 8/n^3 - 16/n^2 + 32n)\right.$$
$$\left.+(24n - 12/n^3)\frac{dn}{d\lambda}\frac{d^2n}{d\lambda^2} + 4\frac{d^3n}{d\lambda^3}\right]$$

Double-pass grating-pair, separation l, angle of incidence γ, line spacing d

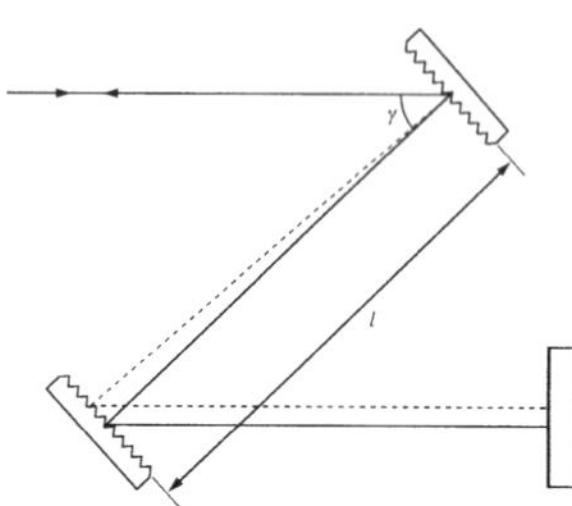

$$\frac{d^2\varphi}{d\omega^2} = l\frac{\lambda^3}{\pi c^2d^2}\left[1 - \left(\frac{\lambda}{d} - \sin\gamma\right)^2\right]^{-3/2}$$

$$\frac{d^3\varphi}{d\omega^3} = -6l\frac{\lambda^4}{c^3d^2}\left(1 + \frac{\lambda}{d}\sin\gamma - \sin^2\gamma\right)\left[1 - \left(\frac{\lambda}{d} - \sin\gamma\right)^2\right]^{-5/2}$$

is the most common approach, other modulation methods have been used successfully and include acousto-optic modulation [170] and deformable mirrors [171]. Pulse shaping is often carried out by using genetic algorithms to condition the modulation signal so that the output pulse maximizes some directly measurable quantity such as second-harmonic generation to optimize pulse peak power [172] or reaction yield in a photochemical process [173]. Feedback can also be applied using exact pulse measurement methods, and a system based on frequency-resolved optical gating and a genetic algorithm has been reported.

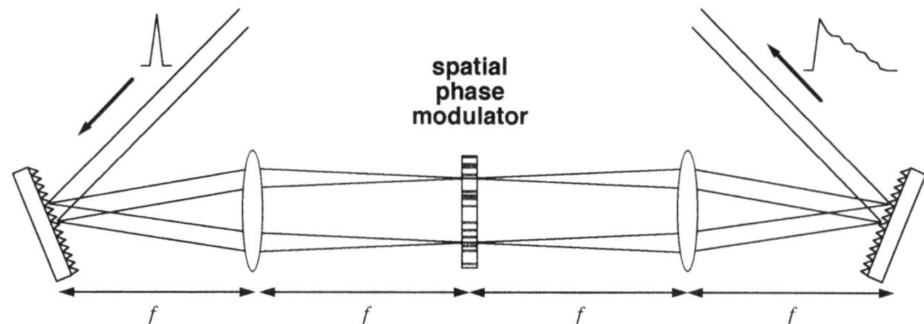

Figure C2.3.8. Femtosecond pulse shaping using a spatial light modulator to modify the spectral phase of a pulse at the Fourier plane of a dispersion-free grating stretcher/compressor.

C2.3.4 Diagnostic techniques

C2.3.4.1 Direct electronic measurements

The principal motivation for using electronic devices such as photodiodes to measure the output from modelocked lasers is to analyse the stability of the modelocked pulse sequence in the radio-frequency (rf) domain. Direct pulse shape measurement using photodiodes is not possible because even the fastest devices and sampling oscilloscopes have finite response times with values in the picosecond range.

RF spectrum analysers are commonly used to determine the presence of multiple-pulsing, amplitude-noise and *phase noise*. Phase noise or *timing jitter* is a form of random noise found in all electronic or optical oscillators and, in modelocked lasers, results in fluctuations in the pulse arrival time. Although in a solid-state modelocked laser the pulse timing error per cavity roundtrip time is generally very small ($\leq$0.01 fs), phase noise prevents the arrival time of any particular future pulse from being predicted exactly. Applications, such as some pump-probe experiments, which rely on synchronization between two ultrafast sources, are particularly vulnerable to the presence of phase noise in the laser oscillator. In solid-state lasers, the major sources of phase noise are acoustic vibrations ($\approx$1–100 Hz) and variations of the refractive index and gain of the laser material due to ripple on the pump laser. Amplitude variations of the intracavity modelocked pulses can also lead to timing fluctuations through nonlinear coupling between intensity and phase.

The rf power spectrum measured using a fast photodiode consists of a series of modes centred on the nth harmonics of the cavity frequency, f_{R}, with the nth mode broadened by a noise sideband S_n [175]:

$$P(\bar{f}) = \bar{P}^2 \sum [\delta(\bar{f} - nf_{\mathrm{R}}) + S_n(\bar{f} - nf_{\mathrm{R}})]. \tag{C2.3.34}$$

The noise sideband surrounding the nth harmonic can be expressed as an energy fluctuation term, S_{E}, whose size is independent of the harmonic number, and a timing fluctuation term, S_J, which dominates at high harmonics because of an n^2 dependence. Ignoring coupling between energy and time fluctuations, the nth sideband is

$$S_n(f) = S_{\mathrm{E}}(f) + n^2(2\pi f_{\mathrm{R}})^2 S_J(f). \tag{C2.3.35}$$

Measurements of the sideband intensity at different harmonics allow the energy and timing fluctuation terms to be separated [176] and the rms fluctuations within a finite rf band from $f_{\min}$ to $f_{\max}$ are obtained using

$$\sigma_{\mathrm{E}(J)} = \sqrt{s \int_{f_{\min}}^{f_{\max}} S_{\mathrm{E}(J)}(f)\,\mathrm{d}f}. \tag{C2.3.36}$$

Quantitative measurements of amplitude and phase noise allow different modelocking methods and laser sources to be compared, and the suitability of any source for an application requiring precise timing synchronization can be determined effectively using this technique.

C2.3.4.2 Approximate methods of pulse shape measurement

Because direct electronic detection is not fast enough to characterize an ultrashort pulse—or even to estimate its duration—it is necessary to use all-optical methods to make the measurement. The ideal measurement device would act like a super-fast mechanical shutter or *gate* which could be opened periodically to sample the energy in a particular time-slice of every pulse. An *autocorrelator* is an optical analogue of this system in which the pulse is overlapped in a nonlinear medium with its time-delayed replica and generates a mixing signal. In the idealized measurement device, an infinitesimally short gate would be able to sample the exact pulse intensity with time as its delay relative to the pulse was scanned. In autocorrelation measurements, the gate pulse is identical to the measurement pulse and results in a broadened signal which, although not a direct map of the pulse envelope, gives a useful estimate of the pulse duration.

Autocorrelation is commonly implemented using second-harmonic generation in a nonlinear crystal to produce a mixing signal which has the form [177]

$$g_2(\tau) = \frac{\int_{-\infty}^{\infty} |\{E(t) + E(t-\tau)\}^2|^2 \, dt}{2\int_{-\infty}^{\infty} |\{E(t)\}^2|^2 \, dt}. \tag{C2.3.37}$$

The interference term in (C2.3.37) leads to $g_2(\tau)$ being known as the *interferometric autocorrelation* profile. When $g_2(\tau)$ is recorded using a measurement system that has insufficient bandwidth to resolve the interference fringes, a modified signal is recorded which is known as the *intensity autocorrelation* profile and is given by

$$G_2(\tau) = 1 + \frac{2\int_{-\infty}^{\infty} I(t) I(t-\tau) \, dt}{\int_{-\infty}^{\infty} I(t)^2 \, dt}. \tag{C2.3.38}$$

Regardless of pulse shape, the autocorrelation functions described always exhibit a characteristic peak-to-background contrast ratio which for the interferometric signal is 8:1 and for the intensity signal is 3:1. A special situation exists when the second-harmonic light generated by each separate pulse (mixing with itself) is not detected and, in this case, a background-free intensity autocorrelation is recorded which is equal to the last term of (C2.3.38).

Separate optical implementations are generally used for amplifier and oscillator sources because of the differences in pulse energy and repetition rate. To characterize pulses from an oscillator, the output is split using a scanning Michelson interferometer (see figure C2.3.9) which then recombines the two beams by focusing into a second-harmonic crystal and detecting the frequency-doubled light using a photomultiplier or other linear detector. The same functionality can be obtained by mixing the pulses in a semiconductor whose bandgap energy, E_g, is related to the pulse photon energy by

$$E_g > E_{\mathrm{photon}} > E_g/2 \tag{C2.3.39}$$

and detecting the photocurrent due to two-photon absorption [178–180]. Scanning autocorrelators are most suitable for high-repetition-frequency ($\sim$ MHz) pulses because each part of the autocorrelation trace results from mixing between distinct pairs of pulses and the measurement is averaged over a time equal to many millions of pulse repetition periods. Mechanical scanning is inconvenient and slow for lower repetition-rate sources such as kHz ultrafast amplifiers and, for these systems, a single-shot autocorrelator [181] is used. Single-shot methods measure the background-free autocorrelation by using non-collinear SHG between two pulses, often by mixing opposite sides of a single beam as shown in figure C2.3.10. When a sufficiently large

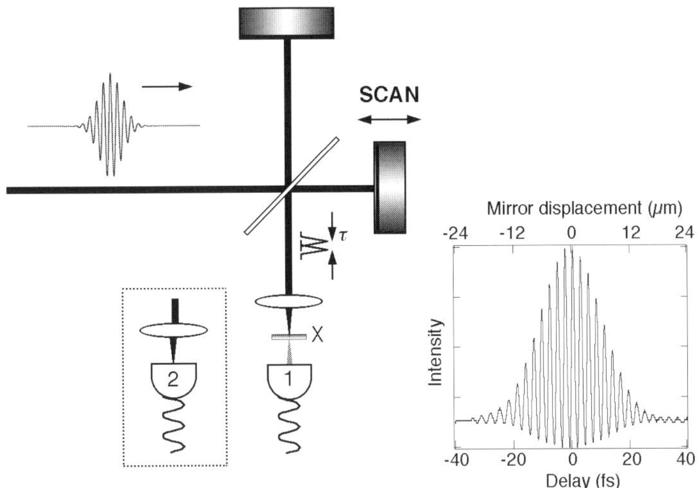

Figure C2.3.9. A scanning Michelson interferometer configured as an autocorrelator. The input pulse is split by the beamsplitter into two replicas which subsequently leave the interferometer with a relative delay τ. A mixing signal is derived either by using a linear (single-photon) detector (1) to record second-harmonic generation in a nonlinear crystal (X) or by focusing the pulses directly onto a nonlinear (two-photon) detector (2).

mixing angle is used the spatial intensity distribution of the frequency-doubled light is proportional to the intensity autocorrelation.

Autocorrelation measurements do not allow the pulse duration to be determined exactly because the width of the autocorrelation function, although directly proportional to the pulse width, is also dependent on the shape of the pulses. Figure C2.3.11 reproduces the intensity and interferometric autocorrelations of five representative pulses with identical FWHM durations and demonstrates the difficulty of trying to infer pulse shape information from autocorrelation measurements. Practically, autocorrelation can be used to estimate the pulse duration from a laser by assuming a pulse shape—normally a Gaussian or $\operatorname{sech}^2 t$ intensity profile—and using a known *deconvolution factor* to infer the pulse duration from the autocorrelation width. For Gaussian pulses $\Delta\tau_{ac}/\Delta\tau = 1.414$ and for pulses $\Delta\tau_{ac}/\Delta\tau = 1.543$.

C2.3.4.3 Exact methods of pulse shape measurement

In many experiments, precise quantitative information about the intensity and phase profile of a pulse is invaluable. Exact pulse measurement methods have been used to tailor pulse chirp to a pre-determined value for quantum-control studies [177]; to test the validity of competing laser theories [182]; and to obtain pulse data for numerical simulations [183]. Throughout the 1990s, various new methods were developed which were able to provide exact shape and chirp information about ultrashort pulses from high-repetition-rate modelocked oscillators or single-shot laser amplifiers. The approaches used can be classified either as two-dimensional methods which measure the pulse spectrogram or sonogram signal in a hybrid time-frequency domain, or one-dimensional methods which measure one or more signals exclusively in either time or frequency. In this section, the principal implementations of each type of technique will be explained and their advantages and limitations discussed.

Although approximate methods for determining the spectral phase of ultrashort laser pulses have existed for some time [177, 184], the first truly exact method was frequency-resolved optical gating (FROG) which was reported in 1993 by Trebino and Kane [185]. Based on measuring a *spectrogram*, FROG records the frequency

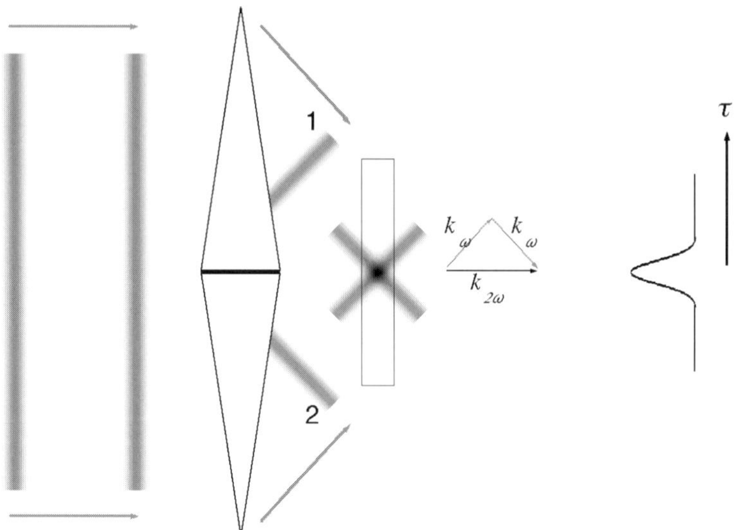

Figure C2.3.10. A single-shot autocorrelator. The input beam is split at a bi-prism and a second-harmonic signal is obtained by non-collinear phasematching. At the top of the crystal, the trailing edge of pulse 1 mixes with the leading edge of pulse 2, while the opposite is true at the bottom of the crystal. In the centre of the crystal the delay between pulses 1 and 2 is zero.

spectrum of different time components of the pulse and uses a two-dimensional (2D) phase-retrieval algorithm (for which a unique solution exists) to obtain from this the pulse characteristics. FROG has been developed extensively since its original demonstration, and a wide variety of implementations has been introduced which allows the measurement of pulses from sources covering a diverse range of wavelengths, pulse durations and energies [186–189]. The mixing of two pulses—the 'probe' and the 'gate'—in a nonlinear medium to obtain a cross- (or auto-) correlation signal is the basis of FROG. The FROG spectrogram is built up by measuring the spectrum of the correlation signal for a continuous range of positive and negative delays between the probe and gate pulses. The specific form of the FROG spectrogram depends on the nonlinearity used to obtain the mixing signal, and common FROG implementations include second-harmonic generation FROG (SHG-FROG) [190], Kerr-effect polarization-gating FROG (PG-FROG) [191], third-harmonic generation FROG (THG-FROG) [192], difference-frequency generation (DFG-FROG) [193] and sum-frequency generation (SFG-FROG) [193]. Table C2.3.2 lists the field of the nonlinear mixing signal, $E_{\mathrm{sig}}(t, \tau)$, associated with the more popular FROG implementations. The FROG spectrogram is related to the mixing signal by

$$S_{\mathrm{FROG}}(\omega, \tau) = \left| \int_{-\infty}^{\infty} E_{\mathrm{sig}}(t, \tau) \exp(-\mathrm{i}\omega t)\,\mathrm{d}t \right|^2 . \tag{C2.3.40}$$

In some FROG geometries, notably SHG-FROG, the roles of the probe and gate pulses are interchangeable and this leads to an ambiguity in the direction of time of the inferred pulse [194]. Figure C2.3.12 shows SHG-FROG traces calculated for some common pulse shapes and illustrates the temporal-symmetry of the measurement. Once the FROG spectrogram has been measured and calibrated, the pulse intensity and phase can be obtained using an iterative *retrieval algorithm* based on an initial guess pulse whose shape is then refined by applying prior knowledge of the nonlinearity and by constraining the FROG trace of the retrieved pulse to match the experimental data. The details of FROG algorithms are beyond the scope of this article but have been described fully by several authors [191, 195, 196]. With the availability of fast computers,

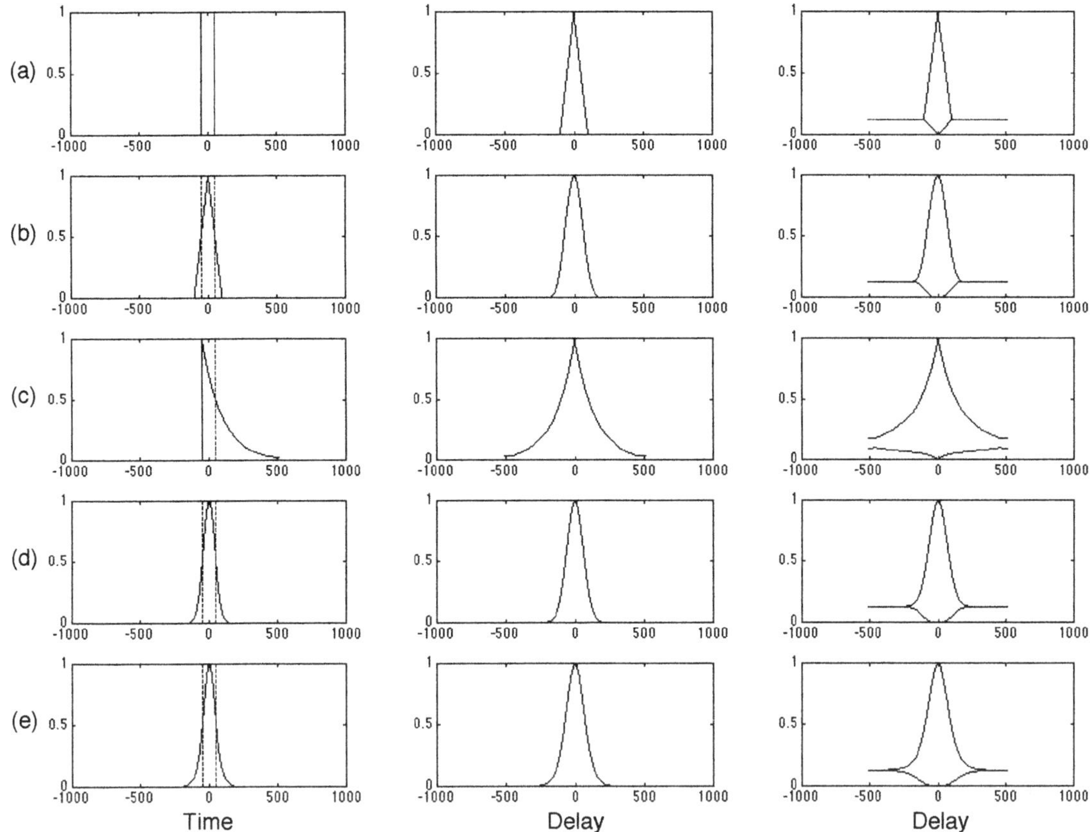

Figure C2.3.11. Illustration of the sensitivity of the second-order autocorrelation profile to pulse shape. Rows: pulses of different shapes but equal FWHM (broken lines) durations—(*a*) square, (*b*) triangular, (*c*) single-side exponential, (*d*) Gaussian, (*e*) $\mathrm{sech}^2(t)$. Columns (left to right): temporal pulse intensity, background-free intensity autocorrelation, envelope of the interferometric autocorrelation.

Table C2.3.2. Expressions for the nonlinear mixing field associated with common FROG geometries.

Second-harmonic	$E_{\mathrm{sig}}(t,\tau) = E(t)E(t-\tau)$
Third-harmonic	$E_{\mathrm{sig}}(t,\tau) = E^2(t)E(t-\tau)$
Polarization gating	$E_{\mathrm{sig}}(t,\tau) = E(t)\vert E(t-\tau)\vert^2$

algorithms retrieving in less than one second have been reported, allowing real-time implementations of FROG to be demonstrated [197].

The pulse *sonogram* can also be used for characterization purposes and is recorded by measuring the cross-correlation signal between different frequency-filtered components of the pulse [198] and a shorter (uncharacterized) reference. Commonly, the reference is the original pulse itself [199, 200] but any available pulse which is synchronous with the pulse to be measured can be used [201]. The sonogram obtained using

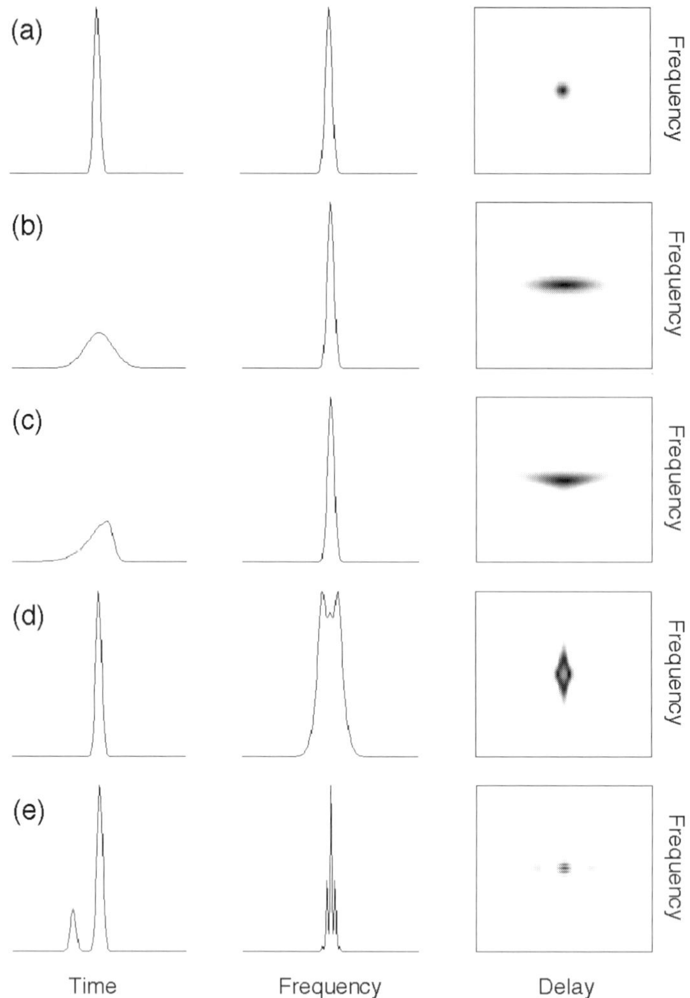

Figure C2.3.12. Examples of SHG-FROG traces for different pulse shapes. Rows: (*a*) transform-limited Gaussian pulse; (*b*) as (*a*) but with quadratic spectral phase added; (*c*) as (*b*) but with cubic spectral phase added; (*d*) as (*a*) but after self-phase-modulation with a peak phase shift of π; (*e*) double pulse. Columns (left to right): temporal pulse intensity, pulse spectrum, SHG-FROG trace.

a zero-dispersion bandpass filter $g(\omega - \Omega)$ with a variable centre frequency Ω is

$$S_{\mathrm{SONO}}(\Omega, t) = \left| \int_{-\infty}^{\infty} E(\omega) g(\omega - \Omega) \exp(\mathrm{i}\omega t)\, \mathrm{d}\omega \right|^2 . \tag{C2.3.41}$$

Sonogram measurements can be made using two-photon-absorption cross-correlation and, using waveguide detectors [202], can be both sensitive and broadband [203]. The requirement for a dispersion-free frequency-filter makes the application of the sonogram technique to sub-50 fs pulses difficult but, unlike SHG-FROG, sonogram measurements have no time ambiguity [200]. Recently, another method known as dispersion-propagation time-resolved optical gating (DP-TROG) has used multiple autocorrelation measurements to record a sonogram using a dispersive filter [204]. In general, 2D techniques are accessible to error-checking methods based on comparing the trace *marginals* (its integral over delay or frequency) with independently measurable quantities. The form and application of the trace marginals has been reported for the FROG [205], sonogram [213] and DP-TROG techniques [206].

Two-dimensional methods are characterized by a high redundancy factor because the measurement of a pulse described by N intensity and N phase samples requires N^2 data points. Such oversampling is desirable because it makes the measurement more resistant to corruption by noise, but the acquisition of 2D datasets often necessitates mechanical scanning and complicated data handling algorithms. It is well known that when a previously-characterized reference pulse exists at the same wavelength as an unknown pulse, the interference signal between the pulses is sufficient for characterization [188, 207]. The use of 2D methods was originally adopted because no well-characterized reference pulse generally exists but, in 1998, a one-dimensional (1D) interferometric method was reported [208] which used a form of spectral shearing interferometry to solve this problem. Known as spectral phase interferometry for direct electric-field reconstruction (SPIDER), the method eliminates the need for a reference pulse by interfering two time-delayed pulses with a relative frequency shift. The SPIDER signal is the interference spectrum between the two pulses,

$$S_{\mathrm{SPIDER}}(\omega) = |e(\omega)|^2 + |e(\omega+\Omega)|^2 + 2|e(\omega)e(\omega+\Omega)|\cos[\varphi(\omega+\Omega) - \varphi(\omega) + \omega\tau] \qquad \text{(C2.3.42)}$$

and, unlike 2D methods, both the time-delay, τ, and the frequency-shift, Ω, are fixed. In practice, the time-delay is achieved using an etalon or a fixed Michelson interferometer and the frequency shift, by sum-frequency mixing the delayed pulses with a quasi-monochromatic wave which is a highly-chirped copy of the original pulse. SPIDER is attractive because it requires no moving parts and its 1D nature means than spatially-resolved measurements of chirp can be recorded [209] but, for optimum performance, the technique requires careful selection of time and frequency shifts [208]. Using SPIDER, pulses as short as 6 fs have been measured [210].

The practicality of measuring pulses using non-phasematched $\chi^{(3)}$ two-photon processes in semiconductors [211] instead of phasematched $\chi^{(2)}$ frequency conversion in bulk nonlinear crystals has motivated efforts to extract intensity and phase information from spectral and auto or cross-correlation measurements alone. This approach is essentially 1D because measurements are made in either the time or frequency domain and the resulting datasets are vectors, not arrays. Various algorithms which use autocorrelation measurements for exact characterization have been reported [212–214] but complex pulse shapes are known to present problems for such techniques. A variation based on the cross-correlation between the measurement pulse and its frequency-chirped replica has been shown to cope successfully with asymmetric and double pulses [215, 216]. The approach uses the spectrum with an initial random phase to estimate the complex spectral field amplitude $e(\omega)$ from which the temporal intensities of the pulse before and after a dispersive element are calculated. By minimizing the RMS error between the measured and calculated cross-correlations of the chirped and unchirped pulses, the algorithm converges to the correct pulse shape. The simplicity of techniques which require only measurements of the spectrum and correlation of the pulse is attractive but the minimization algorithms can be sensitive to the initial guess phase and their convergence can be slow in comparison with iterative spectrogram or sonogram algorithms.

C2.3.5 Applications of ultrashort pulses

C2.3.5.1 Imaging

Beyond the optics research laboratory, many of the applications for low-energy ultrashort pulses concern novel imaging methodologies which, in comparison to conventional techniques, provide greater resolution, three-dimensional (3D) or four-dimensional (4D) sectioning, deeper penetration or other information. Biomedical imaging has been a rapidly developing area and was the first to adopt ultrafast sources for nonlinear microscopy

[217]. Two-photon laser-scanning fluorescence imaging uses a femtosecond laser to achieve confocal-like imaging without an aperture. Because two-photon fluorescence is only generated at the focus of the microscope objective, 3D resolution is obtained and phototoxic and photobleaching effects are minimized. The ability to use longer wavelengths also achieves greater depth penetration with reduced Mie scattering. Similar methods have been applied to image semiconductor devices by mapping the photocurrent across a chip resulting from two-photon absorption [218].

Medical imaging often involves tissues with good optical transmission but high optical scattering cross sections and short mean-free paths. Ultrashort pulse illumination has been combined with time-gated detection to measure only the unscattered *ballistic* light which travels directly through the sample and therefore carries the image information. Experiments by Alfano and others [219, 220] have demonstrated the principle, although for biological samples thicker than a few millimetres the intensity of the ballistic light becomes extremely weak and difficult to detect. Optical coherence tomography (OCT) [221] uses interferometric detection of back-scattered light to provide 3D sectioning of biological samples such as the cornea [222] and has also been implemented at video rates to create movies of living specimens [223]. OCT requires a light source with a short coherence length and, although it can be implemented with a broadband incoherent source, the best results have been obtained by using ultrashort laser pulses. Figure C2.3.13 shows an OCT image through hamster cheek tissue made using a 5 fs Ti:sapphire laser and illustrates the optical sectioning ability of the technique. Another related technique for medical imaging is time-gated fluorescence [224] measurement which, after excitation by a short laser pulse, has the potential to discriminate between cancerous and healthy tissue on the basis of the different fluorescence lifetimes in each region [225]. By implementing this technique in a whole-field configuration, 4D (x, y, z and t) imaging has been demonstrated [226].

Imaging with terahertz radiation is made possible by using femtosecond pulses to trigger a photoconductive antenna [227] or achieve optical rectification [228]. The unique transmission properties of terahertz images have been used to reveal water uptake within plants [229], penetrate optically opaque materials like paper [245] and image dental caries (see chapter D3).

C2.3.5.2 Ultrafast chemistry

Many chemical processes, such as bond breaking and formation; molecular collisions, rotation, vibration and fragmentation; isomerisation and ionization occur with characteristic timescales measurable in femtoseconds. Chemical reactions which are known to proceed from reactants to products *via* an excited transition state can be studied using ultrafast lasers. Commonly an initial pulse is used to excite a molecule into the transition state and a second pulse is used to stimulate *laser-induced fluorescence* (LIF), ionization or *chemiluminescence* whose intensity describes the time-evolution of the product state. One of the first applications of such a scheme was to monitor the vibrational motion of I_2 molecules [230] which can be studied using LIF. Similar approaches have been used to investigate bond breaking in ICN [231], NaI [232] and $C_2F_4I_2$ [233]. Ultrafast laser techniques have also been used extensively to study the effect of the solvent on the dynamics of reactions which occur in solution [234].

A contemporary theme in femtosecond chemistry aims to use femtosecond pulses to drive reactions into particular product channels or prepare molecules in precise quantum states. Work by Kent Wilson and others has applied the FROG pulse measurement technique to prepare negatively-chirped femtosecond pulses with the appropriate duration required to excite I_2 molecules so that, some time after excitation, their vibrational wavepackets had evolved to a state of minimum uncertainty ($\Delta p \Delta x = \hbar/2$) corresponding to a known atomic velocity, separation and momentum [177]. The equivalent positively-chirped pulse was shown, in agreement with theory, to fail to produce localized wavepackets. A related strategy has been used by Gerber and others to optimize the branching ratio of products resulting from the photodissociation of $CpFe(CO)_2Cl$ [235]. The experiment used a genetic algorithm controlling a pulse-shaper to maximize or minimize a feedback signal based on the ratio of $Fe(CO)_5^+/Fe^+$ ions by creating a pulse with an optimized phase profile.

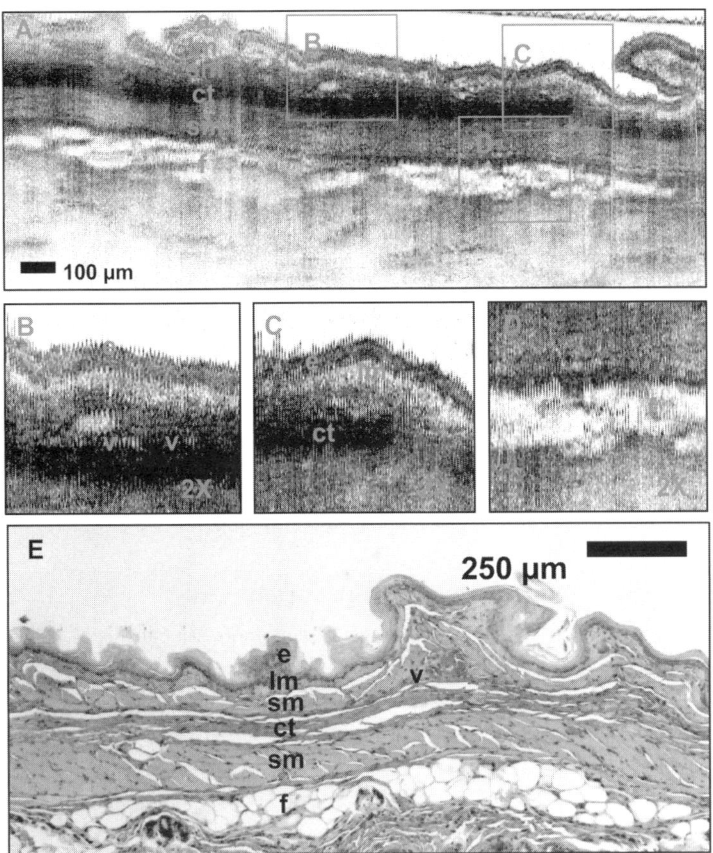

Figure C2.3.13. Optical section through a sample of hamster cheek tissue recorded by OCT using a 5 fs Ti:sapphire laser as the light source. The letters identify distinct microstructures such as blood vessels (v) and fat (f). (Image courtesy of James Fujimoto, MIT, Massachusetts.)

In physical chemistry, x-ray diffraction is a long-established tool for determining atomic configurations and recent emphasis has concentrated on ways of recording how a crystal structure changes after perturbation by an intense laser pulse (see chapter D6). *Femtosecond x-ray diffraction* (figure C2.3.14) is a pump-probe technique in which an optical pulse vibrationally perturbs the atoms in a material and a synchronous x-ray pulse arriving later at the sample is scattered to produce diffraction patterns [236]. Currently the method is being developed to allow the analysis of lattice dynamics and thermalization in bulk semiconductors and to determine the time-evolution of strain within semiconductor heterostructures, but potential exists for the technique to be applied to study complex biological molecules and protein conformations.

C2.3.5.3 Semiconductor spectroscopy

The ability to measure the dynamic behaviour of electrons in semiconductors is important because it enables the speed of modern electronic or opto-electronic devices to be optimized and provides the experimental data necessary to develop theoretical models of these materials. When a semiconductor of bandgap energy E_g absorbs a photon with an energy $\hbar\omega > E_g$, electrons are excited to the conduction band, leaving behind

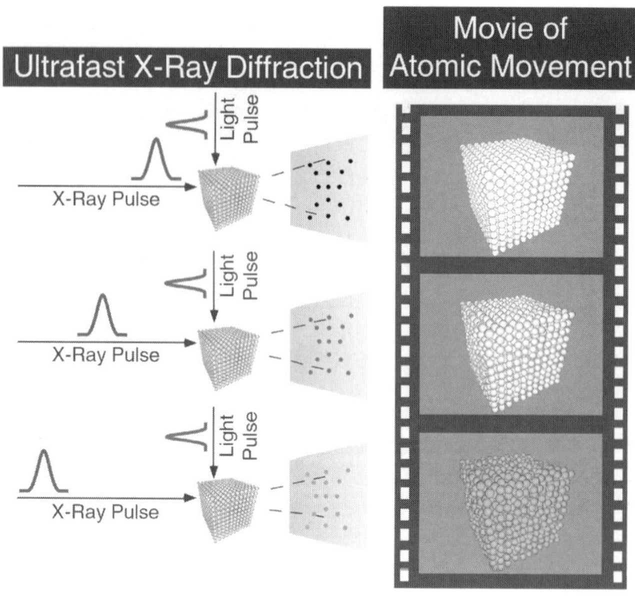

Figure C2.3.14. Intuitive representation of an ultrafast x-ray diffraction experiment. Ideally, after excitation with a short laser pulse, the transient structure of a dynamically evolving system is measured at each time delay. (Image courtesy of Craig W Siders, University of Central Florida.)

holes in the valence band. The scattering and radiative decay mechanisms by which these carriers return to equilibrium determine the electronic properties of the semiconductor. For example, electrons which subsequently become trapped by impurity levels may return to the ground state slowly, with the result that the sample conductivity remains high for a long time after the initial excitation. Intense illumination of the kind available using ultrafast lasers can cause the absorption of a semiconductor to become saturated because the Pauli exclusion principle results in *band-filling*, and full saturation occurs when the populations of the excited and ground states are equal. This effect is enhanced by using (multiple) quantum wells in which strong spatial confinement of the carriers further restricts the allowed transitions to discrete energies. Absorption saturation effects lead to a corresponding change in the refractive index which has been exploited in optical logic devices [237]. The recovery of the absorption can be measured using the *pump-probe* technique in which a strong ultrafast pump pulse induces saturation by exciting carriers, and a time-delayed probe pulse then samples the transmission at a known time after excitation. The rate at which the saturation recovers provides information about the rate at which electrons and holes are scattered out of their initial excited states by non-radiative processes such as phonon emission or electron–electron scattering. Some implementations use a broadband probe pulse created by continuum generation to simultaneously sample absorption across a broad spectral range [238]. Radiative carrier recombination can be measured directly using *time-resolved photoluminescence* techniques [239]. The photoluminescence spectrum is a measure of the carrier energy distribution because the emitted photon energy is the sum of the carrier kinetic energy and the material bandgap energy. By gating the arrival time of the luminescence light using optical sum-frequency mixing, or an electrical instrument such as a streak camera, it is possible to record the time-evolution of the carrier energies and infer the routes by which electrons return to thermal equilibrium.

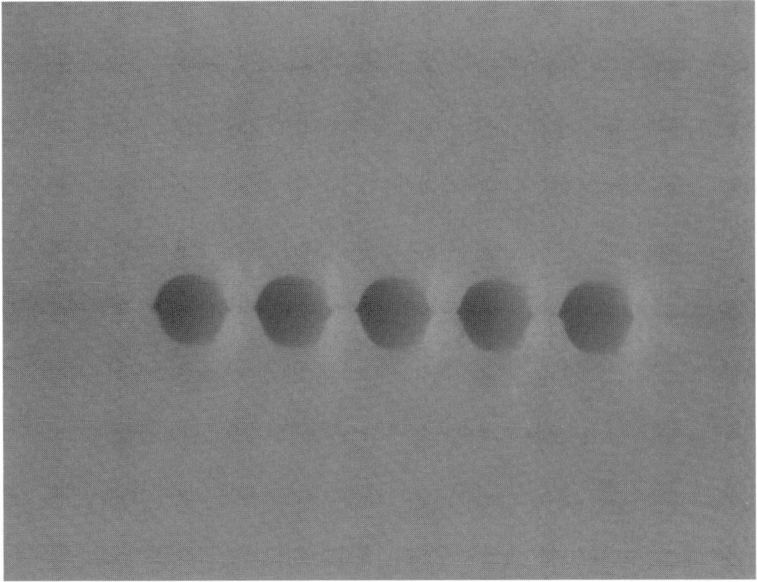

Figure C2.3.15. Cross-section of capillaries in alumina, of 300 μm diameter and 40 mm length. The capillaries are formed from two blocks joined together. The semi-circular grooves in each block are machined by 40 fs laser pulses. (Image courtesy of David R Jones, University of Strathclyde.)

C2.3.5.4 Material processing

Conventional machining of mechanical components with sub-100 μm dimensions is difficult and laser material processing provides a solution which enables the fabrication of features with sizes down to 1 μm. Laser machining is appropriate for metals, plastics, ceramics and glasses sometimes with CW lasers, but often using pulsed radiation. When long pulses are used, thermal processes strongly influence the quality of the resulting feature; in thermal processing, heat is conducted away from the immediate optical absorption region into a wider *heat-affected zone* (see chapter D1.1). In this region, local melting occurs and the ejection of liquid from the interaction region to its edge leads to recast and burr formation. In contrast, femtosecond laser machining is fundamentally non-thermal in nature and the intense electric field of the pulses results in strong multi-photon absorption and a shallow, well-defined absorption region. The material is ablated by ionization and is removed as a plasma, with the result that there is no heat-affected zone, negligible melting and no burr formation. The favourable properties of femtosecond material processing have made it attractive for creating surgical implants [240] and waveguides [242]. Material removal is both precise and predictive and permits accurate 3D profiling of surfaces on the very small scale. A simple example of this capability is shown in figure C2.3.15. Here, 40 fs laser pulses have been used to machine grooves of semi-circular cross section in alumina blocks then joined to form capillaries of 300 μm diameter and length 40 mm. Measurements have implied that femtosecond ablation proceeds by a multi-photon absorption process [241] and this strongly nonlinear behaviour leads to a precisely defined ablation threshold below which no damage occurs. By using tight focusing, this effect has been used to produce structural changes within the bulk of a transparent material without affecting the surface [242]. Further, this process can be controlled to produce only a refractive index modification in plastics and glasses and may thus be utilised to produce a range of 3-D optical devices, including optical circuits, splitters, couplers, sensors, photonic crystals and optical storage.

C2.3.5.5 *High field science*

The intense electromagnetic field available at the beam-focus of powerful ultrafast lasers enables extreme physical conditions to be achieved which are not obtainable in any other way. Using chirped-pulse amplification, 440 fs duration pulses with an energy of 680 J have been demonstrated and focused to an intensity of 6×10^{20} W cm^{-2} [138]. The pulse peak power of 1.5 PW represents the highest laser pulse power ever achieved, and various experiments from laser fusion to experimental astrophysics can exploit the power densities available when the beam is focused on a target (see section D10).

Intense laser pulses have been used to produce relativistic electrons [138] and collimated beams of high-energy protons [243]. The ion beams produced in this way may be used as short-pulse injectors for conventional particle accelerators or as the ignitor for fast-ignition inertial confinement fusion when an indirectly-driven target is used. X-rays and γ-rays can be produced when laser generated ion beams are accelerated by the field in the beam focus onto a solid target, causing bremsstrahlung radiation to be emitted. Laser induced plasmas and ion-beams have become a viable alternative to conventional synchrotron accelerator sources and a review of the physics and applications of the plasmas is given in [244]. Efficient soft x-ray production using moderate energy femtosecond pulses is of interest because of the attractiveness of a 'table-top' source of collimated x-rays for medical and pump-probe experiments (see section C2.3.5.2) and these have been produced by high-harmonic generation from a solid-target using a picosecond Nd:glass amplifier [245] and a gas-cell using a femtosecond Ti:sapphire amplifier [246].

Laser induced nuclear reactions are made possible by the extreme power densities available at the focus of a high-power ultrafast amplifier chain. *Inertial confinement fusion* uses an intense pulse to create a shock wave which compresses a capsule of nuclear fuel to a sufficient density that fusion is initiated in the core of the material and a thermonuclear burn wave travels outward through the fuel releasing energy. The geometry of the fuel and the laser focusing arrangement is critical and both conical [247] and spherical [248] geometries have been used. In the fast igniter concept the fuel is compressed by a pre-pulse and the thermonuclear burn wave is initiated at the outside of the fuel by a second shorter 'ignition' pulse. Nuclear fission is possible by using intense laser pulses to remove neutrons from atomic nuclei to create radioactive isotopes, and fission of uranium has been demonstrated using this approach. The short-lived isotopes which can be produced in this way may have applications in medical therapy.

Conditions at the focus of the most intense amplifier chains are comparable to those found in the centres of stars and planets and, therefore, these lasers are suitable sources with which to carry out experimental simulations of astrophysical events. Using targets of suitable composition, researchers have simulated conditions inside a star as it becomes a supernova [249] and inside a planet where the extreme densities result in novel states of matter with unusual properties not found elsewhere.

C2.3.5.6 *Other applications*

A range of other applications exist for ultrafast laser pulses but the space limitations within this article prevent a full discussion of these here. Amplified femtosecond and picosecond pulses are commonly used for frequency conversion by optical parametric amplifiers (see section 2.3.2.7), particularly to access wavelengths in the visible and mid-infrared not available directly from laser materials [250]. In dense WDM telecommunications, the spectrum of a modelocked femtosecond fibre laser has been sliced to create multiple wavelength channels synchronized to a common clock signal [251]. Attosecond pulse generation in the soft x-ray region will rely on femtosecond lasers as suitable optical pump sources [252] and frequency metrology will benefit from optical self-referencing schemes using phase-locked femtosecond pulses to relate optical frequency standards to conventional microwave reference oscillators [253]. High-power picosecond oscillators are being used to generate the primary colours needed for laser projection cinema [254] which will make possible the digital distribution of cinema films. This list is by no means exhaustive and, as ultrafast lasers become more compact, less expensive and more reliable, other applications beyond the scientific and telecommunication sectors can be expected to emerge.

References

[1] Ruddock I S and Illingworth R 1985 Cavity length optimization in acousto-optically mode-locked argon-ion lasers *J. Phys. E: Sci. Instrum.* **18** 121–3

[2] Marinero E E and Jasny J 1981 An interferometrically tuned and actively modelocked cw dye-laser *Opt. Commun.* **36** 69–74

[3] Bakhshi B and Andrekson P A 1999 Dual-wavelength 10-GHz actively mode-locked erbium fiber laser *IEEE Photon. Tech. Lett.* **11** 1387–9

[4] Dimmick T E, Ho P T and Burdge G L 1984 Coherent pulse generation by active modelocking of a GaAlAs laser in a Selfoc lens extended resonator *Electron. Lett.* **20** 831–3

[5] Albrecht G F 1982 Temporal shape-analysis of Nd-YLif active modelocked/Q-switched oscillator *Opt Commun.* **41** 287–91

[6] Langford N, Smith K and Sibbett W 1987 Subpicosecond-pulse generation in a synchronously mode-locked ring color-center laser *Opt. Lett.* **12** 817-819

[7] May P G, Sibbett W and Taylor J R 1981 Subpicosecond pulse generation in synchronously pumped and hybrid ring dye-lasers *Appl. Phys.* B **26** 179–83

[8] Kafka J D, Watts M L and Pieterse J W 1995 Synchronously pumped optical parametric oscillators with LiB_3O_5 *J. Opt. Soc. Am.* B **12** 2147–57

[9] Lefort L, Puech K, Butterworth S D, Svirko Y P and Hanna D C 1999 Generation of femtosecond pulses from order-of-magnitude pulse compression in a synchronously pumped optical parametric oscillator based on periodically poled lithium niobate *Opt. Lett.* **24** 28–30

[10] Acobovitzveselka G R, Keller U and Asom M T 1992 Broad-band fast semiconductor saturable absorber *Opt. Lett.* **17** 1791–3

[11] Ippen E P, Shank C V and Dienes A 1972 Passive modelocking of the cw dye laser *Appl. Phys. Lett.* **21** 348–50

[12] Bilinsky I P, Prasankuma rR P and Fujimoto J G 1999 Self-starting mode locking and Kerr-lens mode locking of a Ti:Al_2O_3 laser by use of semiconductor-doped glass structures *J. Opt. Soc. Am.* B **16** 546

[13] Agrawal G P 1995 *Nonlinear Fiber Optics* 2nd edn (New York: Academic) p 44

[14] Brabec T, Spielmann C and Krausz F 1991 Mode-locking in solitary lasers *Opt. Lett.* **16** 1961–3

[15] Mollenauer L F and Stolen R H 1984 The soliton laser *Opt. Lett.* **9** 13–15

[16] Krausz F, Ferman M E, Brabec T, Curley P F, Hofer M, Ober M H, Spielmann C, Wintner E and Schmidt A 1992 Femtosecond solid-state lasers *IEEE J. Quantum Electron.* **28** 2097–122

[17] Penman Z E, Schittkowski T, Sleat W, Reid D T and Sibbett W 1998 Experimental comparison of conventional pulse characterization techniques and second-harmonic-generation frequency-resolved optical gating *Opt. Commun.* **155** 297–300

[18] Fork R L, Greene B I and Shank C V 1981 Generation of optical pulses shorter than 0.1 psec by colliding pulse mode locking *Appl. Phys. Lett.* **38** 671

[19] Fork R L, Shank C V, Yen R and Hirlimann C A 1983 Femtosecond optical pulses *IEEE J. Quantum Electron.* **19** 500–5

[20] Fork R L, Martinez O E and Gordon J P 1984 Negative dispersion using pairs of prisms *Opt. Lett.* **9** 150–2

[21] Valdmanis J A, Fork R L and Gordon J P 1985 Generation of optical pulses as short as 27 fs directly from a laser balancing, self-phase modulation, group-velocity dispersion, saturable absorption and saturable gain *Opt. Lett.* **10** 131–3

[22] Fork R L, Brito Cruz C H, Becker P C and Shank C V 1987 Comparison of optical pulses to 6 fs by using cubic phase compensation *Opt. Lett.* **12** 483–5

[23] Johnson A M and Simpson W M 1985 Tunable femtosecond dye laser synchronously pumped by the compressed second harmonic of Nd:YAG *J. Opt. Soc. Am.* B **2** 619

[24] Coherent Satori Hybrid Dye Laser (Kiton Red/Malachite Green)

[25] Moulton P F 1986 Spectroscopic and laser characteristics of Ti:Al_2O_3 *J. Opt. Soc. Am.* B **3** 125

[26] Squier J *et al* 1991 Characteristics of an actively mode-locked 2-psec Ti–sapphire laser operating in the 1 μm wavelength regime *Opt. Lett.* **16** 85–7

[27] Sarukura N, Ishida Y and Nakano H 1991 Generation of 50 fsec pulses from a pulse-compressed, cw passively mode-locked Ti–sapphire laser *Opt. Lett.* **16** 153–5

[28] Goodberlet J *et al* 1989 Femtosecond passively mode-locked Ti-Al_2O_3 laser with a nonlinear external cavity *Opt. Lett.* **14** 1125–7

[29] Goodberlet J *et al* 1990 Starting dynamics of additive-pulse mode-locking in the Ti-Al_2O_3 laser *Opt. Lett.* **15** 1300–1302

[30] French P M W, Williams J A R and Taylor J R 1989 Femtosecond pulse generation from a titanium-doped sapphire laser using nonlinear external cavity feedback *Opt. Lett.* **14** 686–8

[31] Keller U, 't Hooft G W, Knox W H and Cunningham J E 1991 Femtosecond pulses from a continuously self-starting passively-modelocked Ti:sapphire laser *Opt. Lett.* **16** 1022–4

[32] Keller U *et al* 1996 Semiconductor saturable absorber mirrors (SESAMS) for femtosecond to nansecond pulse generation in solid-state lasers *IEEE J. Selected Topics Quantum Electron.* **2** 435

[33] Spence D E, Kean P N and Sibbett W 1991 60 fsec pulse generation from a self-mode-locked Ti–sapphire laser *Opt. Lett.* **16** 42–4

[34] Spence D E *et al* 1990 *Proceedings of the Conference on Lasers and Electro-Optics, Optical Society of America (Anaheim, CA)* Paper CPD10

[35] Juang D-G, Chen Y-C, Hsu S-H, Lin K-H and Hsieh W-F 1997 Differential gain and buildup dynamics of self-starting Kerr lens mode-locked Ti:sapphire laser without an internal aperture *J. Opt. Soc. Am.* B **14** 2116

[36] Brabec T, Curley P F, Spielmann Ch, Wintner E and Schmidt A J 1993 Hard-aperture Kerr-lens mode locking *J. Opt. Soc. Am.* B **10** 1029

[37] Xu L, Spielmann C and Krausz F 1996 Ultrabroadband sub-10 fs Ti:sapphire ring oscillator *Conference on Lasers and Electro-Optics Europe Optical Society of America (Hamburg)* Paper CFF2

[38] Evans J M, Spence D E, Burns D and Sibbett W 1993 Dual-wavelength self-mode-locked Ti:sapphire laser *Opt. Lett.* **18** 1074

[39] Sohn J Y, Ahn Y H, Yee K J and Kim D S 1999 Two-color femtosecond experiments by use of two independently tunable Ti:sapphire lasers with a sample-and-hold switch *Appl. Opt.* **38** 5899

[40] Ell R *et al* 2001 Generation of 5 fs pulses and octave-spanning spectra directly from a Ti:sapphire laser *Opt. Lett.* **26** 373–5

[41] Kafka J D *et al* 1991 *Optical Society of America Annual Meeting Optical Society of America (San Jose, CA)* Paper TUI4

[42] Negus D K *et al* 1991 *Optical Society of America Topical Meeting On Advances in Solid-State Lasers Optical Society of America (Hilton Head, SC)*

[43] Keller U, Knox W H and Roskos H 1990 Coupled-cavity resonant passive mode-locked Ti:sapphire laser *Opt. Lett.* **15** 1377

[44] Keller U, Miller D A B, Boyd G D, Chiu T H, Ferguson J F and Asom M T 1992 Solid-state low-loss intracavity saturable absorber for Nd:YLF lasers: an antiresonant semiconductor Fabry–Pérot saturable absorber *Opt. Lett.* **17** 505

[45] Mellish R, French P M W, Taylor J R, Delfyett P J and Florez L T 1993 Self-starting femtosecond Ti–sapphire laser with intracavity multiquantum-well absorber *Electron. Lett.* **29** 894–6

[46] Jung I D, Kartner F X, Matuschek N, Sutter D H, Morier-Genoud F, Zhang G, Keller U, Scheuer V, Tilsch M and Tschudi T 1997 Self-starting 6.5 fs pulses from a Ti:sapphire laser *Opt. Lett.* **22** 1009

[47] Mitschke F M and Mollenauer L F 1987 *Opt. Lett.* **12** 407

[48] Blow K J and Wood D 1988 *J. Opt. Soc. Am.* B **5** 629

[49] Kean P N, Zhu X, Crust D W, Grant R S, Langford N and Sibbett W 1989 *Opt. Lett.* **14** 39

[50] Zhu X, Kean P N and Sibbett W 1989 *Opt. Lett.* **14** 1192

[51] Mark J, Liu L Y, Hall K L, Haus H A and Ippen E P 1989 *Opt. Lett.* **14** 48

[52] Ouellette F and Piché M 1986 *Opt. Commun.* **60** 99

[53] Kennedy G T, Grant R S, Sleat W E and Sibbett W 1993 *Opt. Lett.* **18** 208

[54] Takara H, Kawanishi S, Saruwatari M and Noguchi K 1992 Generation of highly stable 20 GHz transform-limited optical pulses from actively mode-locked Er^{3+}-doped fibre lasers with an all-polarization maintaining ring cavity *Electron. Lett.* **28** 2095–6

[55] Harvey G T and Mollenauer L F 1993 Harmonically mode-locked fiber ring laser with an internal Fabry–Pérot stabilizer for soliton transmission *Opt. Lett.* **18** 107

[56] Duling I N 1991 All-fiber ring soliton laser mode locked with a nonlinear mirror *Opt. Lett.* **16** 539

[57] Hofer M, Ferman M E, Haberl F, Ober M H and Schmidt A J 1991 *Opt. Lett.* **16** 502

[58] Matsas V J, Newson T P, Richardson D J and Payne D N 1992 *Electron. Lett.* **28** 1391

[59] Tamura K, Haus H A and Ippen E P 1992 *Electron. Lett.* **28** 2226

[60] Zirngibl M, Stulz L W, Stone J, Hugi J, DiGiovanni D and Hansen P B 1991 *Electron. Lett.* **27** 1734

[61] Horowitz M *et al* 2000 Theoretical and experimental study of harmonically modelocked fiber lasers for optical communication systems *J. Lightwave Technol.* **18** 1565–74

[62] Doran N J and Wood D 1988 Nonlinear-optical loop mirror *Opt. Lett.* **13** 56

[63] Fermann M E, Haberl F, Hofer M and Hochreiter H 1990 Nonlinear amplifying loop mirror *Opt. Lett.* **15** 752

[64] Duling I N 1991 Subpicosecond all-fiber erbium laser *Electron. Lett.* **27** 544–5

[65] Richardson D J, Laming R I, Payne D N, Matsas V and Phillips M W 1991 Selfstarting, passively modelocked erbium fiber ring laser based on the amplifying Sagnac switch *Electron. Lett.* **27** 542–4

[66] Nakazawa M, Yoshida E and Kimura Y 1993 Generation of 98 fs optical pulses directly from an erbium-doped fiber ring laser at 1.57 μm *Electron. Lett.* **29** 63

[67] Kelly S M J 1992 *Electron. Lett.* **28** 806

[68] Tamura K, Nelson L E, Haus H A and Ippen E P 1994 Soliton versus nonsoliton operation of fiber ring lasers *Appl. Phys. Lett.* **64** 149–51

[69] Tamura K, Ippen E P, Haus H A and Nelson L E 1993 77 fs pulse generation from a stretched-pulse mode-locked all-fiber ring laser *Opt. Lett.* **18** 1080

[70] Lenz G, Tamura K, Haus H A and Ippen E P 1995 All-solid-state femtosecond source at 1.55 μm *Opt. Lett.* **20** 1289

[71] De Souza E A, Soccolich C E, Pleibel W, Stolen R H, Simpson J R and Digiovanni D J 1993 Saturable absorber modelocked polarization maintaining erbium-doped fiber laser *Electron. Lett.* **29** 447–9

[72] Collings B C *et al* 1997 Short cavity erbium/ytterbium fibre lasers mode-locked with a saturable Bragg reflector *IEEE J. Selected Topics Quantum Electron.* **3** 1065–75

[73] Mar A, Helkey R, Bowers J, Mehuys D and Welch D 1994 Mode-locked operation of a master oscillator power amplifier *IEEE Photon. Technol. Lett.* **6** 1067–9

[74] Goldberg L, Mehuys D and Welch D 1994 High power mode-locked compound laser using a tapered semiconductor amplifier

IEEE Photon. Technol. Lett. **6** 1070–2

[75] Williams K A, Sarma J, White I H, Penty R V, Middlemast I, Ryan T, Laughton F R and Roberts J S 1994 Q-switched bow-tie lasers for high energy picosecond pulse generation *Electron. Lett.* **30** 320–1

[76] Silberberg Y and Smith P W 1986 Sub-picosecond pulses from a modelocked semiconductor laser *IEEE J. Quantum Electron.* **22** 759

[77] Matsui Y, Pelusi M D and Suzuki A 1999 Generation of 20 fs optical pulses from a gain-switched laser diode by a four-stage soliton compression technique *IEEE Photon. Technol. Lett.* **11** 1217–19

[78] Yariv A 1988 *Quantum Electronics* 3rd edn (New York: Wiley) p 259

[79] Vasil'ev P, White I H and Gowar J 2000 Fast phenomena in semiconductor lasers *Rep. Prog. Phys.* **63** 1997–2042

[80] For example, NTT device, Nlk1501

[81] Alphonse G 1997 *Conf. on Lasers and Electro-Optics Optical Society of America (Baltimore, MD)*

[82] Bowers J E, Morton P A, Mar A and Corzine S W 1989 Actively mode-locked semiconductor lasers *IEEE J. Quantum Electron.* **25** 1426–39

[83] Hou A S, Tucker R S and Eisenstein G 1990 Pulse compression of an actively modelocked diode laser using linear dispersion in fibre *IEEE Photon. Technol. Lett.* **2** 322–4

[84] Haus H A and Silberberg Y 1985 Theory of mode locking of a laser diode with a multiple-quantum-well structure *J. Opt. Soc. Am.* B **2** 1237

[85] Gee S, Alphonse G, Connolly J and Delfyett P J 1998 High-power mode-locked external cavity semiconductor laser using inverse bow-tie semiconductor optical amplifiers *IEEE J. Selected Topics Quantum Electron.* **4** 209–15

[86] Schell M, Weber A G, Bottcher E H, Scholl E and Bimberg D 1991 Theory of subpicosecond pulse generation by active modelocking of a semiconductor laser amplifier in an external cavity: limits for the pulsewidth *IEEE J. Quantum Electron.* **27** 402–9

[87] Koumans R G M P and van Roijen R 1996 Theory for passive mode-locking in semiconductor laser structures including the effects of self-phase modulation, dispersion, and pulse collisions *IEEE J. Quantum Electron.* **32** 478–92

[88] Morton P A, Bowers J E, Koszi L A, Soler M, Lopata J and Wilt D P 1990 Monolithic hybrid mode-locked 1.3 μm semiconductor lasers *Appl. Phys. Lett.* **56** 111–13

[89] Young-Kai Chen and Wu M C 1992 Monolithic colliding-pulse mode-locked quantum-well lasers *IEEE J. Quantum Electron.* **28** 2176–85

[90] Derickson D J, Helkey R J, Mar A, Karin J R, Wasserbauer J G and Bowers J E 1992 Short pulse generation using multisegment mode-locked semiconductor lasers *IEEE J. Quantum Electron.* **28** 2186–202

[91] Werner J; Melchior H and Guekos G 1988 Stable optical picosecond pulses from actively mode-locked twin-section diode lasers *Electron. Lett.* **24** 140–1

[92] Martins-Filho J F, Ironside C N and Roberts J S 1993 Quantum well AlGaAs/GaAs monolithic colliding pulse modelocked laser *Electron. Lett.* **29** 1135–6

[93] Sizer T, Mourou G and Rice R R 1981 Picosecond dye laser pulses using a cw frequency doubled Nd:YAG as the pumping source *Opt. Commun.* **37** 207

[94] Jeong T M, Chung C-M, Kim H S, Nam C H and Kim C-J 2000 Generation of passively Q-switched and modelocked pulse from Nd:YVO_4 laser with Cr^{4+}:YAG saturable absorber *Electron. Lett.* **36** 633–4

[95] Malcolm G P A, Curley P F and Ferguson A I 1990 Additive-pulse mode locking of a diode-pumped Nd:YLF laser *Opt. Lett.* **15** 1303

[96] Keller U, Weingarten K J, Kartner F X, Kopf D, Braun B, Jung I D, Fluck R, Honninger C, Matuschek N and Aus Der Au J 1996 Semiconductor saturable absorber mirrors (SESAMs) for femtosecond to nanosecond pulse generation in solid-state lasers *IEEE Selected Topics Quantum Electron.* **2** 435–53

[97] Krainer L, Paschotta R, Moser M and Keller U 2000 77 GHz soliton modelocked Nd:YVO_4 laser *Electron. Lett.* **36** 1846–8

[98] Graf T, Ferguson A I, Bente E, Burns D and Dawson M D 1999, Multi-watt Nd:YVO_4 laser, mode locked by a semiconductor saturable absorber mirror and side-pumped by a diode-laser *Opt. Commun.* **159** 84–7

[99] Aus Der Au J, Loesel F H, Morier-Genoud F, Moser M and Keller U 1998 Femtosecond diode-pumped Nd:glass laser with more than 1 W of average output power *Opt. Lett.* **23** 271

[100] Sennaroglu A, Pollock C R and Nathel H 1994 Continuous-wave self-mode-locked operation of a femtosecond Cr^{4+}:YAG laser *Opt. Lett.* **19** 390

[101] Seas A, Petricevic V and Alfano R R 1992 Generation of sub-100-fs pulses from a cw mode-locked chromium-doped Forsterite laser *Opt. Lett.* **17** 937

[102] Sennaroglu A, Pollock C R and Nathel H 1993 Generation of 48 fs pulses and measurement of crystal dispersion by using a regeneratively initiated self-mode-locked chromium-doped Forsterite laser *Opt. Lett.* **18** 826

[103] Miller A, Likamwa P, Chai B H T and Van Stryland E W Generation Of 150-fs tunable pulses in Cr:$LiSrAlF_6$ *Opt. Lett.* **17** 195

[104] Sorokina I T, Sorokin E, Wintner E, Cassanho A, Jenssen H P and Noginov M A 1996 Efficient continuous-wave TEM_{00} and femtosecond Kerr-lens mode-locked Cr:LiSrGAF laser *Opt. Lett.* **21** 204

[105] Kopf D, Weingarten K J, Zhang G, Moser M, Emanuel M A, Beach R J, Skidmore J A and Keller U 1997 High-average-power diode-pumped femtosecond Cr:LiSAF lasers *Appl. Phys.* B **65** 235–43

[106] Ellingson R J and Tang C L 1992 High-repetition-rate femtosecond pulse generation in the blue *Opt. Lett.* **17** 343–5

[107] Rotermund F and Petrov V 1998 Generation of the fourth harmonic of a femtosecond Ti:sapphire laser *Opt. Lett.* **23** 1040–2

[108] Burneika K, Ignatavicius M, Kabelka V, Piskarskas A and Stabinis A 1972 *IEEE J. Quantum Electron.* **QE-8** 574
[109] Piskarskas A, Smil'gyavichyus V and Umbrasas A 1988 *Sov. J. Quantum Electron.* **18** 155
[110] Ebrahimzadeh M, Malcolm G P A and Ferguson A I 1992 *Opt. Lett.* **17** 183
[111] Edelstein D C, Wachman E S and Tang C L 1989 *Appl. Phys. Lett.* **54** 1728
[112] Wachman E S, Pelouch W S and Tang C L 1991 *J. Appl. Phys.* **70** 1893
[113] Tang C L, Wachman E S and Pelouch W S 1991 *Conf. on Lasers and Electro-Optics Optical Society of America* Paper CFM5
[114] Wachman E S, Edelstein D C and Tang C L 1990 *Opt. Lett.* **15** 136
[115] Moon J A 1993 *IEEE J. Quantum Electron.* **29** 265
[116] Mak G, Fu Q and van Driel H M 1992 *Appl. Phys. Lett.* **60** 542
[117] Pelouch W S, Powers P E and Tang C L 1992 *Opt. Lett.* **17** 1070
[118] Ellingson R J and Tang C L 1993 *Opt. Lett.* **18** 438
[119] Reid D T, Ebrahimzadeh M and Sibbett W 1994 *J. Opt. Soc. Am.* B
[120] Driscoll T J, Gale G M and Hache F 1994 Ti:sapphire second-harmonic-pumped visible range femtosecond optical parametric oscillator *Opt. Commun.* **10** 638
[121] Gale G M, Cavallari M, Driscoll T J and Hache F 1995 Sub20 fs tunable pulses in the visible from an 82 MHz optical parametric oscillator *Opt. Lett.* **20** 1562
[122] Gale G M, Hache F and Cavallari M 1998 Broad-bandwidth parametric amplification in the visible: femtosecond experiments and simulations *IEEE Selected Topics Quantum Electron.* **4** 224–9
[123] Reid D T, Penman Z, Ebrahimzadeh M, Sibbett W, Karlsson H and Laurell F 1997 Broadly tunable infrared femtosecond optical parametric oscillator based on periodically poled $RbTiOAsO_4$ *Opt. Lett.* **22** 1397
[124] Butterworth S D, Pruneri V and Hanna D C 1996 Optical parametric oscillation in periodically poled lithium niobate based on continuous-wave synchronous pumping at 1.047 μm *Opt. Lett.* **21** 1345
[125] Myers L E, Eckardt R C, Fejer M M. Byer R L, Bosenberg W R and Pierce J W 1995 Quasi-phase-matched optical parametric oscillators in bulk periodically poled $LiNbO_3$ *J. Opt. Soc. Am.* B **12** 2102
[126] Karlsson H, Laurell F, Henriksson P and Arvidsson G 1996 Frequency doubling in periodically poled $RbTiOAsO_4$ *Electron. Lett.* **32** 556–7
[127] Loza-Alvarez P, Brown C T A, Reid D T, Sibbett W and Missey M 1999 High-repetition-rate ultrashort-pulse optical parametric oscillator continuously tunable from 2.8 to 6.8 μm *Opt. Lett.* **24** 1523
[128] Beddard T *et al* 1999 High-average-power, 1 MW peak-power self-mode-locked Ti:sapphire oscillator *Opt. Lett.* **24** 163–5
[129] Xu L, Tempea G, Spielmann C, Krausz F, Stingl A, Ferencz K and Takano S 1998 Continuous-wave mode-locked Ti: sapphire laser focusable to 5×10^{13} W Cm^{-2} *Opt. Lett.* **23** 789–91
[130] Liu Z L *et al* , 2000 High-gain, reflection-double pass, Ti:sapphire continuous-wave amplifier delivering 5.77 W average power, 82 MHz repetition rate, femtosecond pulses *Appl. Phys. Lett.* **76** 3182–4
[131] Cho Sh *et al* 2001 Generation of 90 nJ pulses with a 4-MHz repetition-rate Kerr-lens mode-locked Ti:Al_2O_3 laser operating with net positive and negative intracavity dispersion *Opt. Lett.* **26** 560–2
[132] Bartels A, Dekorsy T and Kurz H 1999 Femtosecond Ti:sapphire ring laser with a 2 GHz repetition rate and its application in time-resolved spectroscopy *Opt. Lett.* **24** 996–8
[133] Liu H *et al* 1999 Directly diode-pumped millijoule subpicosecond Yb:glass regenerative amplifier *Opt. Lett.* **24** 917–19
[134] Mellish R *et al* 1997 All-solid-state diode-pumped Cr:LiSAF femtosecond oscillator and regenerative amplifier *Appl. Phys.* B **65** 221–6
[135] Boskovic A *et al* 1995 All-fibre diode-pumped, femtosecond chirped pulse amplification system *Electron. Lett.* **31** 877–9
[136] Fischer R A and Bischel W K 1975, Pulse compression for more efficient operation of solid-state laser amplifier chains II *IEEE J. Quantum Electron.* **11** 46
[137] Strickland D and Mourou G 1985 Compression of amplified chirped optical pulses *Opt. Commun.* **56** 219
[138] Pennington D M *et al* 2000 Petawatt laser system and experiments *IEEE J. Selected Topics Quantum Electron.* **6** 676–88
[139] Riley K F 1987 *Mathematical Methods for the Physical Sciences* (Cambridge: Cambridge University Press) p 213
[140] Baltuska A, Wei Z, Pshenichnikov M S and Wiersma D A 1997 Optical pulse compression to 5 fs at a 1 MHz repetition rate *Opt. Lett.* **22** 102
[141] Dudley J M, Peacock A C and Millot G 2001 The cancellation of nonlinear and dispersive phase components on the fundamental optical fiber soliton: a pedagogical note *Opt. Commun.* **193** 253–9
[142] Akhmediev N N 1998 Spatial solitons in Kerr and Kerr-like media *Opt. Quantum Electron.* **30** 535–69
[143] Yariv A 1988 *Quantum Electronics* 3rd edn (New York: Wiley) p 487
[144] Pain H J 1987 *The Physics of Vibrations And Waves* 3rd edn (New York: Wiley) p 222
[145] Ainslie B J and Day C R 1986 *J. Lightwave Technol.* **4** 967
[146] Knight J C *et al* 1998 Large mode area photonic crystal fibre *Electron. Lett.* **34** 1347–8
[147] Wadsworth W J *et al* 2000 Soliton effects in photonic crystal fibres at 850 nm *Electron. Lett.* **36** 53–5
[148] Yablonovitch E 1994 Photonic crystals *J. Mod. Opt.* **41** 173–94
[149] Spielmann C *et al* 1994 Tunnelling of optical pulses through photonic band-gaps *Phys. Rev. Lett.* **73** 2308–11
[150] Treacy E B 1968 Compression of picosecond light pulses *Phys. Lett.* A **28** 34–5
[151] Treacy E B 1969 Pulse compression with diffraction gratings *IEEE J. Quantum Electron.* **5** 454–8

[152] Lemoff B E and Barty C P J 1993 Cubic-phase-free dispersion compensation in solid-state ultrashort-pulse lasers *Opt. Lett.* **18** 57–9

[153] Sherriff R E 1998 Analytic expressions for group-delay dispersion and cubic dispersion in arbitrary prism sequences *J. Opt. Soc. Am.* B **15** 1224–30

[154] Curley P F, Spielmann C, Brabec T, Krausz F, Wintner E and Schmidt A J 1993 Operation of a femtosecond Ti:sapphire solitary laser in the vicinity of zero group-delay dispersion *Opt. Lett.* **18** 54

[155] Yamashita M, Ishikawa M, Torizuka K and Sato T 1986 Femtosecond-pulse laser chirp compensated by cavity-mirror dispersion *Opt. Lett.* **11** 504

[156] Szipocs R, Ferencz K, Spielmann C and Krausz F 1994 Chirped multilayer coatings for broadband dispersion control in femtosecond lasers *Opt. Lett.* **19** 201

[157] Kartner F X, Matuschek N, Schibli T, Keller U, Haus H A, Heine C, Morf R, Scheuer V, Tilsch M and Tschudi T 1997 Design and fabrication of double-chirped mirrors *Opt. Lett.* **22** 831–3

[158] Jung I D, Kartner F X, Matuschek N, Sutter D H, Morier-Genoud F, Zhang G, Keller U, Scheuer V, Tilsch M and Tschudi T 1997 Self-starting 6.5 fs pulses from a Ti:sapphire laser *Opt. Lett.* **22** 1009

[159] Yariv A 1988 *Optical Electronics* 4th edn (Philadelphia, PA: Saunders College Publishing) p 117

[160] Gires F and Tournois P 1964 *C. R. Acad. Sci. Paris* **258** 6112

[161] For example: Robertson A, Ernst U, Knappe R, Wallenstein R, Scheuer V, Tschudi T, Burns D, Dawson M D and Ferguson A I 1999 Prismless diode-pumped mode-locked femtosecond Cr:LiSAF laser *Opt. Commun.* **163** 38–43

[162] Kuhl J, Serenyi M and Gobel E O 1987 Bandwidth-limited picosecond pulse generation in an actively mode-locked gas laser with intracavity chirp compensation *Opt. Lett.* **12** 334

[163] Heppner J and Kuhl J 1985, Intracavity chirp compensation in a colliding pulse mode-locked laser using thin-film interferometers *Appl. Phys. Lett.* **47** 453–5

[164] Suda A *et al* , 2001 A spatial light modulator based on fused-silica plates for adaptive feedback control of intense femtosecond laser pulses *Opt. Express* **9** 2–6

[165] Patel J S and Silberberg Y 1995 Liquid crystal and grating-based multiple-wavelength cross-connect switch *IEEE Photon. Technol. Lett.* **7** 514–16

[166] Kohler B *et al* 1995 Quantum control of wave-packet evolution with tailored femtosecond pulses *Phys. Rev. Lett.* **74** 3360–3

[167] Weiner A M 1995 Femtosecond optical pulse shaping and processing *Prog. Quantum Electron.* **19** 161–237

[168] Weiner A M, Leaird D E, Patel J S and Wullert J R 1992 Programmable shaping of femtosecond optical pulses by use of 128-element liquid crystal phase modulator *IEEE J. Quantum Electron.* **28** 908–20

[169] Froehly C, Colombeau B and Vampouille M 1983 Shaping and analysis of picosecond light pulses *Prog. Opt.* **20** 65

[170] Hillegas C W, Tull J X, Goswami D, Strickland D and Warren W S 1994 Femtosecond laser pulse shaping by use of microsecond radio-frequency pulses *Opt. Lett.* **19** 737

[171] Chriaux G, Albert O, Wnman V, Chambaret J P, Flix C and Mourou G 2001 Temporal control of amplified femtosecond pulses with a deformable mirror in a stretcher *Opt. Lett.* **26** 169

[172] Brixner T *et al* 2000 Feedback-controlled femtosecond pulse shaping *Appl. Phys. Suppl.* B **70** S119–24

[173] Bergt M, Brixner T, Kiefer B, Strehle M and Gerber G 1999 Controlling the femtochemistry of $Fe(CO)_5$ *J. Phys. Chem.* A. **103** 10 381–7

[174] Zeek E, Maginnis K, Backus S, Russek U, Murnane M, Mourou G, Kapteyn H and Vdovin G 1999 Pulse compression by use of deformable mirrors *Opt. Lett.* **24** 493

[175] Poppe A, Xu L, Krausz F and Spielmann C 1998 Noise characterization of sub-10-fs Ti:sapphire oscillators *IEEE Selected Topics Quantum Electron.* **4** 179–84

[176] Von der Linde D 1986 Characterization of the noise in continuously operating modelocked lasers *Appl. Phys.* B **39** 201

[177] Diels J C M, Fontaine J J, Mcmichael I C and Simoni F 1985, Control and measurement of ultrashort pulse shapes (in amplitude and phase) with femtosecond accuracy *Appl. Opt.* **24** 1270

[178] Laughton F R, Marsh J H and Kean A H 1992 Very sensitive 2-photon absorption GaAs/AlGaAs wave-guide detector for an autocorrelator *Electron. Lett.* **28** 1663–5

[179] Takagi Y *et al* 1992 Multiple-shot and single-shot autocorrelator based on 2-photon conductivity in semiconductors *Opt. Lett.* **17** 658–60

[180] Reid D T *et al* 1997 Light-emitting diodes as measurement devices for femtosecond laser pulses *Opt. Lett.* **22** 233–5

[181] Gyuzalian R N *et al* 1979 Background-free measurement of time behaviour of an individual picosecond pulse *Opt. Commun.* **29** 239

[182] Taft G, Rundquist A, Murnane M M, Kapteyn H C, DeLong K W, Trebino R and Christov I P 1995 Ultrashort optical waveform measurements using frequency resolved optical gating *Opt. Lett.* **20** 743

[183] Loza-Alvarez P, Reid D T, Faller P, Ebrahimzadeh M, Sibbett W, Karlsson H and Laurell F 1999 Simultaneous femtosecond pulse compression and second-harmonic generation in aperiodically poled $KTiOPO_4$ *Opt. Lett.* **24** 1071–3

[184] Treacy E B 1971 Measurement and interpretation of dynamic spectrograms of picosecond light pulses *J. Appl. Phys.* **42** 3848

[185] Kane D J and Trebino R J 1993 Single-shot measurement of the intensity and phase of an arbitrary ultrashort pulse by using frequency-resolved optical gating *Opt. Lett.* **18** 823

[186] Richman B A, Krumbugel M A and Trebino R 1997 Temporal characterization of mid-IR free-electron-laser pulses by frequency-

resolved optical gating *Opt. Lett.* **22** 721–3

[187] Michelmann K, Feurer T, Fernsler R and Sauerbrey R 1996 Frequency resolved optical gating in the UV using the electronic Kerr effect *Appl. Phys. B-Lasers* **63** 485–9

[188] Fittinghoff D N *et al* 1996 Measurement of the intensity and phase of ultraweak, ultrashort laser pulses *Opt. Lett.* **21** 884–6

[189] Baltuska A, Pshenichnikov M S and Wiersma D A 1999 Second-harmonic generation frequency-resolved optical gating in the single-cycle regime *IEEE J. Quantum Electron.* **35** 459–78

[190] Paye J, Ramaswamy M, Fujimoto J G and Ippen E P 1993 Measurement of the amplitude and phase of ultrashort light pulses from spectrally resolved autocorrelation *Opt. Lett.* **18** 1946

[191] Trebino R and Kane D J 1993 Using phase retrieval to measure the intensity and phase of ultrashort pulses: frequency-resolved optical gating *J. Opt. Soc. Am.* A **10** 1101–11

[192] Tsang T *et al* 1996 Frequency-resolved optical-gating measurements of ultrashort pulses using surface third-harmonic generation *Opt. Lett.* **21** 1381–3

[193] Linden S, Kuhl J and Giessen H 1999 Amplitude and phase characterization of weak blue ultrashort pulses by downconversion *Opt. Lett.* **24** 569

[194] Delong K W, Trebino R and Kane D J 1994 Comparison of ultrashort-pulse frequency-resolved-optical-gating traces for three common beam geometries *J. Opt. Soc. Am.* B **11** 1595

[195] Delong K W *et al* 1994 Pulse retrieval in frequency-resolved optical gating based on the method of generalized projections *Opt. Lett.* **19** 2152–4

[196] Kane D J 1998 Real-time measurement of ultrashort laser pulses using principal component generalized projections *IEEE J. Selected Topics Quantum Electron.* **4** 278–84

[197] Kane D J 1999 Recent progress toward real-time measurement of ultrashort laser pulses *IEEE J. Quantum Electron.* **35** 421–31

[198] Chilla J L A and Martinez O E 1991 Direct determination of the amplitude and the phase of femtosecond light pulses *Opt. Lett.* **16** 39–41

[199] Wong V and Walmsley I A 1997 Ultrashort-pulse characterization from dynamic spectrograms by iterative phase retrieval *J. Opt. Soc. Am.* B **14** 944–9

[200] Reid D T 1999 Algorithm for complete and rapid retrieval of ultrashort pulse amplitude and phase from a sonogram *IEEE J. Quantum Electron.* **35** 1584–9

[201] Taira K and Kikuchi K 2001 Optical sampling system at 1.55 μm for the measurement of pulse waveform and phase employing sonogram characterization *IEEE Photon. Technol. Lett.* **13** 505–7

[202] Skovgaard P M W, Mullane R J, Nikogosyan D N and McInerney J G 1998 Two-photon photoconductivity in semiconductor waveguide autocorrelators *Opt. Commun.* **153** 78–82

[203] Reid D T *et al* 2000 Sonogram characterization of picosecond pulses at 1.5 μm using waveguide two photon absorption *Electron. Lett.* **36** 1141–2

[204] Koumans R G M P and Yariv A 2000 Pulse characterization at 1.5 μm using time-resolved optical gating based on dispersive propagation *IEEE Photon. Technol. Lett.* **12** 666–8

[205] Taft G *et al* 1996 Measurement of 10 fs laser pulses *IEEE J. Selected Topics Quantum Electron.* **2** 575–85

[206] Cormack I G, Sibbett W and Reid D T 2001 Measurement of femtosecond optical pulses using time-resolved optical gating, Paper CTuN3 *Conf. on Lasers and Electro-Optics Optical Society of America (Baltimore, MD)*

[207] Rothenberg J E and Grischkowsky D R 1987 Measurement of optical phase with subpicosecond resolution by time-domain interferometry *Opt. Lett.* **12** 99

[208] Iaconis C and Walmsley I A 1998 Spectral phase interferometryfor direct electric-field reconstruction of ultrashort optical pulses *Opt. Lett.* **23** 792

[209] Gallmann L *et al* 2001 Spatially resolved amplitude and phase characterization of femtosecond optical pulses *Opt. Lett.* **26** 96–8

[210] Gallmann L, Sutter D H, Matuschek N, Steinmeyer G, Keller U, Iaconis C and Walmsley I A 1999 Characterization of sub-6 fs optical pulses with spectral phase interferometry for direct electric-field reconstruction *Opt. Lett.* **24** 1314

[211] Reid D T *et al* 1998 Commercial semiconductor devices for two photon absorption autocorrelation of ultrashort light pulses *Appl. Opt.* **37** 8142–4

[212] Peatross J and Rundquist A 1998 Temporal decorrelation of short laser pulses *J. Opt. Soc. Am.* B **15** 216

[213] Baltuska A *et al* 1999 Rapid amplitude-phase reconstruction of femtosecond pulses from intensity autocorrelation and spectrum *Conf. on Lasers and Electro-Optics Optical Society of America (Baltimore, MD)* Paper cwF22

[214] Yau, T-W *et al* 1999,Photodiode-based phase-retrieval ultrafast waveform measurements *Conf. on Lasers and Electro-Optics Optical Society of America (Baltimore, MA)* Paper cwF21

[215] Nicholson J W, Jasapara J, Rudolph W, Omenetto F G and Taylor A J 1999 Full-field characterization of femtosecond pulses by spectrum and cross-correlation measurements *Opt. Lett.* **24** 1774

[216] Nicholson J W, Mero M, Jasapara J and Rudolph W 2000 Unbalanced third-order correlations for full characterization of femtosecond pulses *Opt. Lett.* **25** 1801

[217] Denk W, Strickler J H and Webb W W 1990 2-Photon laser scanning fluorescence microscopy *Science* **248** 73–6

[218] Xu C and Denk W 1997, Two-photon optical beam induced current imaging through the backside of integrated circuits *Appl. Phys. Lett.* **71** 2578–80

[219] Yoo K M, Xing Q and Alfano R R 1991 Imaging objects hidden in highly scattering media using femtosecond second-harmonic-

generation cross-correlation time gating *Opt. Lett.* **16** 1019

[220] Chen H, Chen Y, Dilworth D, Leith E, Lopez J and Valdmanis J 1991 Two-dimensional imaging through diffusing media using 150 fs gated electronic holography techniques *Opt. Lett.* **16** 487

[221] Huang D *et al* 1991 Optical coherence tomography *Science* **254** 1178–81

[222] Drexler W *et al* 2001 Ultrahigh-resolution ophthalmic optical coherence tomography *Nat. Med.* **7** 502–7

[223] Boppart S A *et al* 1997 Non-invasive assessment of the developing *Xenopus* cardiovascular system using optical coherence tomography *Proc. Natl Acad. Sci., USA* **94** 4256–61

[224] Jones R *et al* 1999 Fluorescence lifetime imaging using a diode-pumped all-solid-state laser system *Electron. Lett.* **35** 256–8

[225] Cubeddu R *et al* 1989 Time-gated fluorescence spectroscopy of the tumor localizing fraction of hpd in the presence of cationic surfactant *Photochem. Photobiol.* **50** 157–63

[226] Dowling K *et al* 1998 Whole-field fluorescence lifetime imaging with picosecond resolution using ultrafast 10 KHz solid-state amplifier technology *IEEE J. Selected Topics Quantum Electron.* **4** 370–5

[227] Fattinger C and Grischkowsky D 1989 Terahertz beams *Appl. Phys. Lett.* **54** 490–2

[228] Luo M S C *et al* 1994 Generation of terahertz electromagnetic pulses from quantum-well structures *IEEE J. Quantum Electron.* **30** 1478–88

[229] Mittleman D M, Jacobsen R H and Nuss M C 1996 T-ray imaging *IEEE J. Selected Topics Quantum Electron.* **2** 679–92

[230] Gruebele M, Roberts G, Dantus M, Bowman R M and Zewail A H 1990 Femtosecond temporal spectroscopy and direct inversion to the potential: application to iodine *Chem. Phys. Lett.* **166** 459

[231] Dantus M, Bowman R M and Zewail A H 1990 Femtosecond laser observations of molecular vibration and rotation *Nature* **343** 737

[232] Rose T S, Rosker M J and Zewail A H 1989 Femtosecond real-time probing of reactions. IV. The reactions of alkali halides *J. Chem. Phys.* **91** 7415–36

[233] Khundkar L R and Zewail A H 1990 Picosecond photofragment spectroscopy. IV. Dynamics of consecutive bond breakage in the reaction $C_2F_4I_2C_2F_4$+2I *J. Chem. Phys.* **92** 231–42

[234] Fleming G R and Wolynes P G 1990 Chemical dynamics in solution *Phys. Today* **43** 36–43

[235] Assion A, Baumert T, Bergt M, Brixner T, Kiefer B, Seyfried V, Strehle M and Gerber G 1998 Control of chemical reactions by feedback-optimized phase-shaped femtosecond laser pulses *Science* **282** 919–22

[236] Cavalleri A, Siders cw, Sokolowski-Tinten K, Toth C, Blome C, Squier JA, Von der Linde D, Barty C P J and Wilson K 2001 Femtosecond x-ray diffraction *Opt. Photon. News* **12(5)** 29

[237] Jewell J L, Scherer A, McCall S L, Gossard A C and English J H 1987 *Appl. Phys. Lett.* **51** 94

[238] Shank C V, Fork R L, Leheny R F and Shah J 1979 Dynamics of photoexcited GaAs band-edge absorption with subpicosecond resolution *Phys. Rev. Lett.* **42** 112–15

[239] Shah J, Deveaud B, Damen T C, Tsang W T, Gossard A C and Lugli P 1987 Determination of intervalley scattering rates in GaAs by subpicosecond luminescence spectroscopy *Phys. Rev. Lett.* **59** 2222–5

[240] Nolte S, Momma C, Kamlage G, Chichnov B N, Tunnerman A, von Alvensleben F and Welling H 1998 *Conference on Lasers and Electro-Optics Optical Society of America (San Francisco, CA)* Paper CPD3

[241] Kruger J and Kautek W 1996 Femtosecond-pulse visible laser processing of transparent materials *Appl. Surf. Sci.* **96** 430–8

[242] Schaffer C B *et al* 2001 Micromachining bulk glass by use of femtosecond laser pulses with nanojoule energy *Opt. Lett.* **26** 93–5

[243] Roth M *et al* 2001 Intense ion beams accelerated by petawatt-class lasers *Nucl. Inst. Meth. Phys. Res.* **464** 201

[244] Umstadter D 2001 Review of physics and applications of relativistic plasmas driven by ultra-intense lasers *Phys. Plasmas* **8** 1774–85

[245] Norreys P A, Zepf M, Moustaizis S, Fews A P, Zhang J, Lee P, Bakarezos M, Danson C N, Dyson A, Gibbon P, Loukakos P, Neely D, Walsh F N, Wark J S and Dangor A E 1996 Efficient extreme UV harmonics generated from picosecond laser pulse interactions with solid targets *Phys. Rev. Lett.* **76** 1832–5

[246] Schnurer M *et al* 1998 Guiding and high-harmonic generation of sub-10 fs pulses in hollow-core fibers at 10^{15} W cm^{-2} *Appl. Phys.* B **67** 263–6

[247] Norreys P A *et al* 2000 Experimental studies of the advanced fast ignitor scheme *Phys. Plasmas* **7** 3721–6

[248] Rosen M D 1999 The physics issues that determine inertial confinement fusion target gain and driver requirements: A tutorial *Phys. Plasmas* **6** 1690–9

[249] Kane J *et al* 2000 Supernova experiments on the Nova laser *Astrophys. J. Suppl.* **127** 365–9

[250] Vodopyanov K L and Voevodin V G 1995 Type-I And Type-II $ZnGeP_2$ Travelling-wave optical parametric generator tunable between 3.9 And 10 μm *Opt. Commun.* **117** 277–82

[251] Boivin L *et al* 1999 110 channels $\times$ 2.35 Gb s^{-1} from a single femtosecond laser *IEEE Photon. Tech. Lett.* **11** 466–8

[252] Drescher M, Hentschel M, Kienberger R, Tempea G, Spielmann C, Reider G A, Corkum P B and Krausz F 2001 X-ray pulses approaching the attosecond frontier *Science* **291** 1923–7

[253] Holzwarth R *et al* 2000 Optical frequency synthesizer for precision spectroscopy *Phys. Rev. Lett.* **85** 2264–7

[254] Ruffing B, Nebel A and Wallenstein R 2001 High-power picosecond LiB_3O_5 optical parametric oscillators tunable in the blue spectral range *Appl. Phys.* B **72** 137–49

C3
Frequency conversion and filtering

Terence A King

This section is concerned with wavelength conversion of laser radiation to new wavelengths and the selection and stabilization of the laser frequency. These techniques greatly increase the capabilities of lasers in applications. Methods to extend the range of wavelengths of coherent sources by the use of nonlinear optical mixing techniques based on second harmonic generation and optical parametric oscillation are described. Methods to stabilize laser frequencies for short- and long-term stability are reviewed, with particular consideration to applications in precision measurements.

The operation of lasers directly by stimulated emission from population-inverted media provides discrete wavelengths ranging from the UV to the far-IR from excited gaseous, liquid, solid or plasma media. However, there are broad regions in the UV to far-IR range where laser wavelengths are not directly available. In addition, the wavelength band of almost all laser emissions is very narrow, i.e. highly monochromatic, as a consequence of their emission processes from discrete energy levels and tuning of the output radiation is severely limited. There are a few notable exceptions to this general situation where the emission wavelength band is broad and, hence, can be utilized for tunable laser operation. These include the dye lasers where the laser transition is a broad molecular fluorescence band, solid state lasers such as titanium–sapphire operating on broad dopant vibronic crystal transitions, and colour centre lasers. However, in general, there is substantial incomplete spectral coverage of the UV to far-IR region by direct laser emission.

The range of sources of coherent radiation can be greatly extended by the use of nonlinear optical frequency conversion, in which the nonlinear optical response of an optical medium to an intense electromagnetic field leads to the generation of new wavelengths. Polarization induced in the medium by propagating electromagnetic waves of the same or several frequencies leads to oscillation and emission at the sum and difference frequencies. In the earliest demonstration in 1961 of this frequency conversion process, the newly discovered ruby laser operating at 694.3 nm and propagating in a quartz crystal produced radiation at the second harmonic of 347 nm. Since then, the field of nonlinear optics has flourished in its understanding, the discovery of new processes, the development of new techniques, the availability of nonlinear optical materials and greatly enhanced frequency conversion performance. It has become an important and effective technique to extend the wavelength coverage of coherent sources. In addition, these frequency-converted wavelengths can be generated from those high-power lasers which have superior performance and this increases their applicability. A frequently used example of this is the production of 532 nm radiation by second harmonic generation from the Nd:YAG laser's fundamental wavelength at 1064 nm.

In chapter C3.1, a review is given of the techniques of frequency conversion by second harmonic generation and sum and difference frequency mixing. The methods to achieve efficient frequency conversion by phase-matching and by the use of birefringent anisotropic crystals and periodic poled materials are described. Since the early discovery of nonlinear optical frequency conversion a large number of nonlinear optical materials have been developed and impressive frequency conversion efficiencies are now achievable.

In addition to second harmonic generation and sum and difference frequency mixing the second-order nonlinear optical response of a medium leads to optical parametric oscillation (OPO) and to the linear electro-optic (Pockels) effect. In OPO, using the nonlinear response of the medium, a pump photon (energy $\hbar\omega_p$) produces signal ($\hbar\omega_s$) and idler ($\hbar\omega_i$) photons, such that $\hbar\omega_p = \hbar\omega_s + \hbar\omega_i$. Conservation of photon momentum in the frequency-mixing process produces the phase-matching condition $\hbar k_p = \hbar k_s + \hbar k_i$, and the nonlinear crystal needs to be transparent at the pump, signal and idler wavelengths. Since the first demonstration in 1965 of a pulsed OPO in $LiNbO_3$ and in 1968 a cw OPO, many new materials have been developed giving greatly enhanced performance. The OPO produces signal and idler beams, which are able to be continuously tuned over broad wavelength bands by variation of the crystal angle or temperature.

In many applications, there is a need for selection of the laser frequency by single longitudinal mode operation and accurate stabilization of the frequency. Examples of these applications are in interferometry, precision metrology, gravitational wave detection, high-resolution laser spectroscopy and the measurement of the fundamental physical constants. The stabilization requirements for some of these applications and the techniques to achieve frequency-stabilized lasers are reviewed in chapter C3.3. These methods include those based on stabilization to the laser gain profile, reference to linear atomic or molecular absorption using a gas cell, Doppler-free spectroscopy and frequency locking to a stable optical cavity.

C3.1
Harmonic generation—materials and methods

David J Binks

C3.1.1 Introduction

The fundamental output wavelength of a laser is limited by the linewidth of the atomic transition involved; this linewidth may be very narrow, giving essentially a fixed frequency output as in the case of Nd:YAG, or it may be relatively broad allowing some degree of tunability. Even for Ti:sapphire, however, the most widely tunable visible/near infrared laser gain medium, this tuning range is limited to ~700–1100 nm and for most other laser sources it is significantly less.

The wavelength range of laser radiation can be greatly extended by utilizing the nonlinear optical properties of birefringent crystals to *frequency convert* the output of a laser to other wavelengths. Perhaps, the most common example of frequency conversion is second-harmonic generation, in which part or all of the laser output is converted to a beam of double the frequency. This process can be thought of as being equivalent to adding together two photons from the original beam to produce a single photon of twice the energy. Second-harmonic generation is a special case of the more general process of sum frequency mixing which adds together photons of different energies from separate sources. Difference frequency mixing is the related process by which a wave is produced at a frequency equal to the difference between the frequencies of the beams interacting within the crystal.

In this chapter, the frequency conversion techniques of second-harmonic generation, sum frequency mixing and difference frequency mixing will be described. (The closely related effects of parametric amplification and oscillation will be discussed in the next chapter.) The theoretical analysis of each of these three phenomena will be outlined initially, followed by a discussion of the nonlinear materials commonly used in this type of process. Finally, the design considerations pertinent to the frequency conversion of some common laser systems will be discussed.

C3.1.2 Second-harmonic generation

The optics of anisotropic crystals plays an important part in frequency conversion processes such as second-harmonic generation. The subject of crystal optics is covered in detail in section A5.3.4 of this handbook and also very thoroughly in Born and Wolf (see further reading); a brief outline will be given here, however, to allow the discussion to proceed directly. Light passing through an anisotropic crystal is in general resolved into two beams of orthogonal polarizations, each of which experiences a different refractive index; this is the phenomenon known as birefringence or double refraction. Generally, the value of these two refractive indices will vary in different ways with the direction of light propagation through a crystal and there will be certain directions in which the two values happen to be equal. These special directions are termed the *optic axes* and anisotropic crystals are either uniaxial or biaxial, having one or two of these axes, respectively. In the case of uniaxial crystals, beams with polarization directions perpendicular to the optic axis experience a refractive

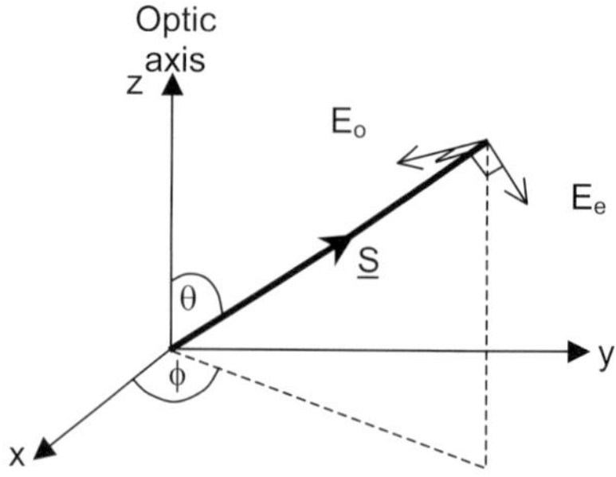

Figure C3.1.1. The orientation of the Poynting vector, S, and the ordinary (E_o) and extraordinary (E_e) polarizations relative to the crystal axis.

index, n_o, that is independent of propagation direction, and are called ordinary or o-rays. In contrast, the beams with a polarization orthogonal to that of the ordinary rays are subject to a direction dependent refractive index, and are hence named extraordinary or e-rays. The refractive index experienced by an e-ray ranges from n_e, when its polarization is parallel to the optic axis, to n_o, when it is perpendicular to the axis. Crystals are termed positive if $n_e > n_o$ and negative if $n_o > n_e$.

This formalism is useful both for the calculation of effective nonlinear coefficients (see the next section) and of *phase-matching* conditions. The phase-evolution of a wave is determined by the refractive index to which it is subject as it propagates through a medium. Moreover, material dispersion usually ensures that the different wavelengths involved in a frequency conversion process experience different refractive indices and therefore become out of phase as they propagate. This generally results in destructive interference between waves generated by the nonlinear process in one part of the crystal and another because the generated wave is produced with the same relationship to the incident wave at each point. However, the birefringent properties of nonlinear crystals can be used to ensure that the refractive index for each wave is equal and hence they remain in phase and interference is constructive. This important technique of phase-matching will be detailed in section C3.1.2.3.

C3.1.2.1 Effective nonlinear coefficient

As was discussed in section A4.1, the amplitude of the nonlinear polarization of the material depends on the product of the interacting fields and on the second-order nonlinear susceptibility, $\chi^{(2)}$, which in general is a tensor quantity. Each tensor element, $\chi^{(2)}_{ijk} = 2d_{ijk}$, contributes differently to the nonlinear material polarization depending on the directions that the fundamental and second-harmonic beams take through the crystal relative to the crystal axes and whether they are ordinary or extraordinary rays. It is usual practice to collect together all the relevant contributions for a particular experimental arrangement and hence calculate an *effective nonlinear coefficient*, d_{eff}, for that case. The value of d_{eff} indicates the strength of the nonlinear interaction that can be expected from a $\chi^{(2)}$ material for a given situation and is useful in selecting appropriate materials for an application and predicting the efficiency of the nonlinear process. This section shows how the effective nonlinear coefficient can be calculated from the experimental geometry and the tensor components. The analysis will be for the particular case of second harmonic generation but can be readily extended to sum and difference mixing by allowing the interacting fields to have different frequencies, i.e. ω_1 and ω_2, and the resulting nonlinear polarization to be at the sum, $\omega_1 + \omega_2$, or difference, $\omega_1 - \omega_2$, frequency.

The effective nonlinear coefficient for second-harmonic generation is defined by the following equation:

$$P(2\omega) = d_{\text{eff}} E(\omega) E(\omega). \tag{C3.1.1}$$

Figure C3.1.1 shows a general geometric arrangement: the Poynting vector, $\boldsymbol{S} = \boldsymbol{E} \times \boldsymbol{H}$, corresponding to the fundamental beam is shown relative to Cartesian axes. The Poynting vector represents the direction of energy transfer (see [102], appendix 5), i.e. the direction in which a beam travels, and in birefringent materials, such as $\chi^{(2)}$ nonlinear crystals, is not in general parallel to the wave propagation vector, k. As will be discussed in section C3.1.5, to ensure that the nonlinear interaction is birefringently phase-matched (see section A4.4) the second-harmonic electric field, $E(2\omega)$, must be of orthogonal polarization to at least one of the fundamental electric fields, $E(\omega)$. (A non-orthogonal fundamental field can be resolved into parallel and orthogonal component fields.) For the purposes of the following example, it will be assumed that the nonlinear medium is a negative uniaxial crystal (i.e. $n_{\mathrm{o}} > n_{\mathrm{e}}$) with the optic axis parallel to the z-axis and that the fundamental wave corresponds to an ordinary ray and the second harmonic to an extraordinary (e-) ray.

Hence, in this case the components of the fundamental electric field along the crystal axes are

$$\begin{aligned} E_x &= E\cos\left(\phi - \frac{\pi}{2}\right) = E\sin(\phi) \\ E_y &= E\sin\left(\phi - \frac{\pi}{2}\right) = -E\cos(\phi) \\ E_z &= 0 \end{aligned} \tag{C3.1.2}$$

As a third rank tensor, $\chi^{(2)}$ has in principle 27 independent components but because no physical significance is attached to the order of the interacting fields i.e. $E_iE_j \equiv E_jE_i$ then this number can be reduced to 18 and the nonlinear polarization, $P(2\omega)$, is found by evaluating:

$$\begin{pmatrix} P_x \\ P_y \\ P_z \end{pmatrix} = \begin{pmatrix} d_{11} & d_{12} & d_{13} & d_{14} & d_{15} & d_{16} \\ d_{21} & d_{22} & d_{23} & d_{24} & d_{25} & d_{26} \\ d_{31} & d_{32} & d_{33} & d_{34} & d_{35} & d_{36} \end{pmatrix} \begin{pmatrix} E_x^2 \\ E_y^2 \\ E_z^2 \\ 2E_xE_y \\ 2E_xE_z \\ 2E_xE_y \end{pmatrix}. \tag{C3.1.3}$$

Furthermore, if for the particular interaction the polarization is due only to electron displacement, rather than including a component from ion displacement, and if there is negligible loss, then the Kleinman conjecture [1] is valid and the following symmetries apply: $d_{21} = d_{16}$, $d_{24} = d_{32}$, $d_{31} = d_{15}$, $d_{13} = d_{35}$, $d_{14} = d_{36} = d_{25}$, $d_{12} = d_{26}$ and $d_{32} = d_{24}$. These conditions apply to most practical cases and so the Kleinman conjecture constitutes a powerful way of simplifying the calculation of the effective nonlinear coefficient. Nonlinear crystals are usually classified according to their point-group and each of these groups has further symmetries, which further reduces the number of independent tensor components (see [29, p 382] for a useful table of crystal symmetries). For a particular nonlinear material, some of the tensor components, whilst non-zero and independent, will be very small in magnitude compared to other components and can be safely neglected. For most materials then, the effective nonlinear coefficient will typically depend on between 1 and 3 of the tensor components only; the relative importance of each of these remaining components is determined by the geometrical arrangement of the second harmonic scheme.

For example, β-barium borate (BBO) is a negative uniaxial crystal and has $d_{22} = -d_{21} = -d_{16}$, $d_{31} = d_{32}$ and $d_{15} = d_{24}$ with all other components zero except d_{35}, therefore

$$\begin{aligned} P_x &= -2d_{16}\sin(\phi)\cos(\phi)E^2 = d_{22}E^2\sin(2\phi) \\ P_y &= [d_{21}\sin^2(\phi) + d_{22}\cos^2(\phi)]E^2 = d_{22}E^2\cos(2\phi) \\ P_z &= [d_{31}\sin^2(\phi) + d_{32}\cos^2(\phi)]E^2 = d_{32}E^2. \end{aligned} \tag{C3.1.4}$$

Since the second-harmonic beam corresponds to an e-ray,

$$\begin{aligned} P(2\omega) &= \sin\left(\theta + \frac{\pi}{2}\right)[P_x \cos(\phi) + P_y \sin(\phi)] + \cos\left(\theta + \frac{\pi}{2}\right) P_z \\ &= [d_{22}\cos(\theta)\sin(3\phi) - d_{31}\sin(\theta)]E^2 \end{aligned} \tag{C3.1.5}$$

Hence, in this case $|d_{\mathrm{eff}}| = d_{22}\cos(\theta)\sin(3\phi) - d_{31}\sin(\theta)$.

C3.1.2.2 Conversion efficiency

In this section, the general solution to the coupled wave equations that describe second harmonic generation for plane waves will be presented. This general solution represents the conversion efficiency achievable for any degree of fundamental wave depletion and for arbitrary phase mismatch between the waves. The limits of non-depletion of the pump beam (i.e. fundamental wave) and of zero phase mismatch will then be examined individually. Note that the field strengths, E, and intensities, I, that are used to describe the interacting waves in the following analysis are *instantaneous* quantities and hence the results that follow apply equally well to both continuous wave and pulsed systems. Total pulse energies can be found by simply integrating over the corresponding instantaneous powers.

General solution. In general, all three-wave mixing phenomena (i.e. second-harmonic generation, sum frequency mixing and difference frequency mixing) can be described by the following coupled wave equations, which relate the Fourier components of the three electric fields propagating collinearly along the z-axis:

$$\begin{aligned} \frac{\partial E_1}{\partial z} &= -\mathrm{i}\frac{\omega_1 d_{\mathrm{eff}}}{n_1 c} E_2^* E_3 \exp(-\mathrm{i}\Delta k z) \\ \frac{\partial E_2}{\partial z} &= -\mathrm{i}\frac{\omega_2 d_{\mathrm{eff}}}{n_2 c} E_1^* E_3 \exp(-\mathrm{i}\Delta k z) \\ \frac{\partial E_3}{\partial z} &= -\mathrm{i}\frac{\omega_3 d_{\mathrm{eff}}}{n_3 c} E_1 E_2 \exp(-\mathrm{i}\Delta k z) \end{aligned} \tag{C3.1.6}$$

where Δk is the phase mismatch defined by

$$\Delta k = k_3 - k_2 - k_1. \tag{C3.1.7}$$

For frequency doubling, the subscript 3 corresponds to the second harmonic and subscripts 1 and 2 both correspond to the fundamental wave, hence, $\omega_1 = \omega_2$ and $\omega_3 = 2\omega_1$.

The following assumptions are now made:

(1) The waves are infinite, unbounded, plane and monochromatic.
(2) The nonlinear material is non-magnetic and lossless.
(3) The slowly varying envelope approximation is valid [2].
(4) The group and phase velocities are equal [3].
(5) Any higher-order nonlinear processes are negligible [4].

Under these assumptions, Milton [5] showed that the second-harmonic conversion efficiency, η, is given by

$$\eta = \frac{I_3}{I_1} = 2B \times sn^2\left(\sqrt{\frac{KL^2 I_1 A}{\hbar\omega_1}}, \gamma\right) \tag{C3.1.8}$$

where L is the interaction length, I_1 is the fundamental intensity, I_3 is the second harmonic intensity and sn is a Jacobian elliptic function [6]. The coupling coefficient, K, is given by

$$K = \frac{2\hbar\omega_1\omega_2\omega_3}{n_1 n_2 n_3 \varepsilon_0 c^3} d_{\text{eff}}^2 \qquad \text{(C3.1.9)}$$

and the parameters A, B and γ are

$$A, B = \tfrac{1}{2}\left[(1+\sigma+\delta) \pm \sqrt{(1+\sigma+\delta)^2 - 4\sigma^2}\right]$$
$$\gamma = \left|\frac{A}{B}\right| \qquad \text{(C3.1.10)}$$

where

$$\sigma = \frac{I_2\omega_2}{I_1\omega_2}$$
$$\delta = \frac{\hbar\omega_1}{K I_1}\left(\frac{\Delta k}{2}\right)^2 \qquad \text{(C3.1.11)}$$

σ is the *photon balance* parameter, i.e. it is unity if the number of photons in the two interacting beams is equal. This is always true for type I phase-matching but only true for types II and III if the angle between the fundamental polarization and the optic axis is exactly 45°. (See section C3.1.2.3 for a definition of phase-matching type.) The parameter δ is the normalized phase error and is equal to zero for a phase-matched interaction.

Non-depleted pump beam and arbitrary phase mismatch. Chapter A4 presented the solution to the coupled wave equations in the case of second-harmonic generation where both the fundamental and second-harmonic waves are plane and phase-matched. To illustrate the importance of phase-matching, the non-phase-matched case will now be considered but, for simplicity, in the limit of non-depletion of the fundamental intensity, i.e. $I_\omega(z) = I_\omega(0)$. Now, the rate of change of the second-harmonic amplitude depends on a constant intensity of the fundamental wave and its z-dependence is only through the phase-matching term $\exp(\mathrm{i}\Delta k z)$:

$$\frac{\mathrm{d}A_3}{\mathrm{d}z} = I_\omega \exp(\mathrm{i}\Delta k z). \qquad \text{(C3.1.12)}$$

The total second-harmonic conversion efficiency over the interaction length, L, (which may be simply limited to the physical length of the nonlinear crystal involved or maybe less than that due to effects such as 'walk-off'—this will be discussed later), may be found by integrating the above expression between the limits of 0 and L to give, in terms of intensity,

$$\eta = \frac{I_3(L)}{I_1} = \frac{A_3^* A_3}{2 Z_3 I_1} = 2\frac{K L^2 I_1}{\hbar\omega_3}\operatorname{sinc}^2\left(\frac{\Delta k L}{2}\right) \qquad \text{(C3.1.13)}$$

where Z_3 is the plane wave impedance for the second-harmonic wavelength. A plot of efficiency as a function of phase mismatch is shown in figure C3.1.2 and shows that a failure to ensure phase matching, i.e. $\Delta k = 0$, results in a much reduced second-harmonic efficiency. Physically, if the fundamental wave gets out of phase with the second-harmonic wave as it progresses through the crystal then the subsequent second-harmonic wavelets it produces interfere destructively with the existing second-harmonic wave, resulting in a reduced amplitude and the oscillatory behaviour shown in figure C3.1.2. The characteristic length over which the

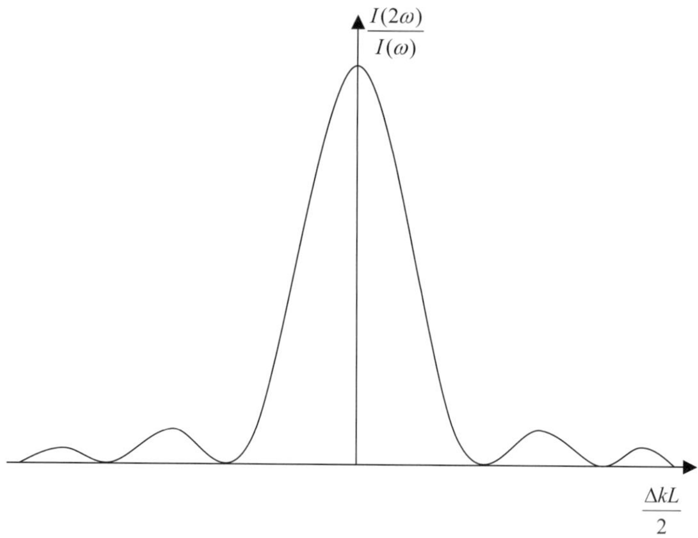

Figure C3.1.2. Second-harmonic conversion efficiency as a function of phase mismatch, Δk.

fundamental and second-harmonic waves stay in phase is known as the coherence length, l_c, and for a particular interaction in a material it is given by

$$l_c = \frac{\pi}{\Delta k}. \tag{C3.1.14}$$

Such is the benefit of ensuring good phase-matching, that applications of the second harmonic that require efficient conversion generally employ a phase-matching technique; these will be described in section C3.1.2.2. Unless otherwise indicated, all analyses from now on will assume that such a technique has been used and therefore that $\Delta k = 0$.

Depleted pump. As was shown in section A4.4, for a phase-matched second-harmonic scheme involving plane waves in which the fractional conversion of the fundamental wave is non-negligible, the coupled wave equations can be solved exactly to give, in terms of intensity, the following expressions for the fundamental and second-harmonic waves [7]:

$$\begin{aligned} \frac{I_1(z)}{I_1(0)} &= \operatorname{sech}^2\left(\frac{z}{l}\right) = 1 - \eta \\ \eta = \frac{I_3(z)}{I_1(0)} &= 2\tanh^2\left(\frac{z}{l}\right) \end{aligned} \tag{C3.1.15}$$

where l is a characteristic length for the interaction and is given here again

$$l = \left(Kz\sqrt{\frac{I_1}{\hbar\omega_1}}\right)^{-1}. \tag{C3.1.16}$$

The evolution of the fundamental and second-harmonic intensities are shown in figure C3.1.3. Hence, given a sufficiently large effective nonlinear coefficient or sufficiently intense fundamental wave or a long

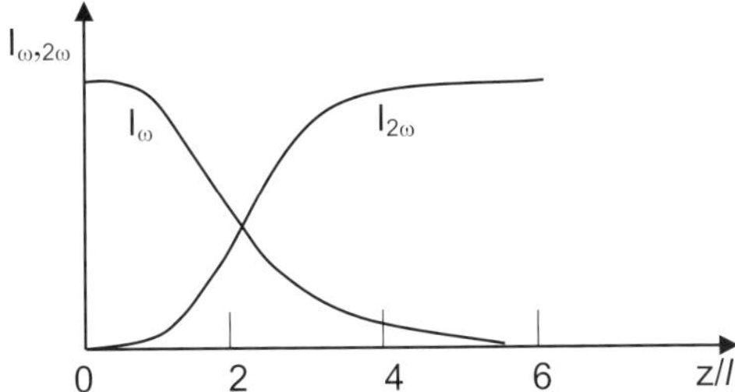

Figure C3.1.3. Evolution of fundamental and second-harmonic intensities as a function of the distance normalized to l.

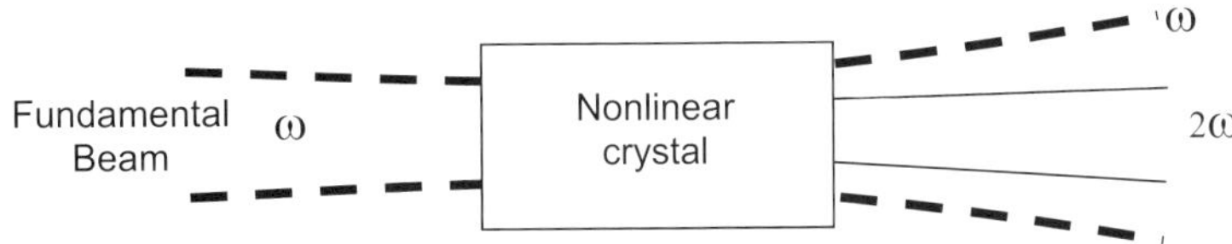

Figure C3.1.4. Second-harmonic generation with Gaussian beams. In general, the fundamental and second harmonic have different beam diameters and divergences.

enough interaction length, complete conversion of the fundamental wave to the second harmonic is possible. Furthermore, note that the nonlinear conversion efficiency function is monotonic with z and so, if the assumptions listed here are valid, then there is no back conversion from the second harmonic to the fundamental. We shall see later that this is a special case for second-harmonic generation when the fundamental is either a pure e-ray or o-ray; it does not necessarily hold for the more general case of sum frequency mixing or for frequency doubling when the fundamental has both e- and o-ray components. In other words, there will be the possibility of back conversion whenever the interacting fields are distinguishable either by their frequency or polarization direction.

Gaussian beams. Due to the weakness of second-order interactions at intensities anything less than those typical of lasers, in general the waves that interact within a nonlinear crystal are usually best described not by plane waves but by Gaussian beams (chapter A2.1). Therefore, for a more accurate description of the second-harmonic process the finite spot size and divergence of the fundamental and second-harmonic beams must be considered, which in general will be different from one another, as in figure C3.1.4.

The plane wave analysis described shows that the second-harmonic efficiency is proportional to the fundamental intensity and so one would expect that greater efficiency could be had from a Gaussian fundamental beam that is tightly focused into the nonlinear crystal. However, two factors mitigate this advantage: the reduced Rayleigh length and increased divergence of the focused beam. If the Rayleigh length, z_0 (see section A2.1.2.2), of the fundamental beam is much shorter than the interaction length (the crystal length in a well-designed scheme) then, even though the peak fundamental intensity is increased by tighter focusing, the average intensity over the interaction length is less and the overall conversion efficiency is reduced. An in-depth analysis by Boyd and Kleinman [8] shows that the optimum conversion efficiency results when the interaction length is $5.68z_0$. Moreover, under these conditions the conversion efficiency increases linearly with interaction length, L, rather than with L^2 as in the plane wave analysis. The greater beam divergence brought about by tight focusing can also have a deleterious effect on efficiency depending on how it compares to the phase-matching acceptance angle for the particular nonlinear material and experimental arrangement, a subject that will be discussed in the next section.

C3.1.2.3 Phase matching

As was illustrated in the previous section, the conversion efficiency of second-harmonic generation and, indeed, all three-wave mixing processes depend crucially on the phase-matching condition being satisfied i.e.

$$\Delta k = k_3 - k_2 - k_1 = 0 \tag{C3.1.17}$$

where k_i represents the magnitude of the propagation vectors of, for $i = 1,2$, the two waves that interact to produce the third, $i = 3$. Since the momentum of a photon is given by $p = \hbar k$ then this statement is equivalent to requiring that momentum is conserved. (Similarly, energy is conserved if $\hbar\omega_3 = \hbar\omega_1 + \hbar\omega_2$.) Dielectric media such as nonlinear crystals are generally dispersive and hence $k = n(\lambda)/\lambda$, where $n(\lambda)$ is the wavelength-dependent refractive index for that material. Hence, the phase-matching condition becomes:

$$\frac{n_3(\lambda_3)}{\lambda_1} = \frac{n_1(\lambda_1)}{\lambda_1} + \frac{n_2(\lambda_2)}{\lambda_2} \tag{C3.1.18}$$

which, for the case of second-harmonic generation, i.e. $\lambda_1 = \lambda_2 = \lambda$ and $2\lambda_3 = \lambda$, means that for exact phase matching, $\Delta k = 0$, the refractive experienced by the second-harmonic wave should be equal to that experienced by the fundamental.

The wavelength dependence of refractive index is usually described by a semi-empirical dispersion relation, called a Sellmeier equation, which would typically be of the form (although other forms are common)

$$n^2 = a + \sum_k \frac{b_k}{\lambda^2 - \lambda_k^2} \tag{C3.1.19}$$

where a and b are constants and the summation is over the k absorption resonances that contribute to dispersion in a particular spectral region; usually dispersion is well described by including only one or two of these resonances. There will be a separate Sellmeier equation for both ordinary and extraordinary refractive indices for uniaxial (i.e. having a single optic axis) crystals, whilst three Sellmeier equations are required to describe refractive index dispersion in biaxial crystals (i.e. having two optic axes), one corresponding to each crystal axis (X, Y, Z) (see section A5.3.4). For example, the ordinary and extraordinary refractive index (n_o and n_e, respectively) of β-barium borate (BBO) is well described over its transparency range ($\sim$0.2–2.6 μm) by

$$\begin{aligned} n_0^2 &= 2.7405 + \frac{0.0184}{\lambda^2 - 0.0179} - 0.00155\lambda^2 \\ n_e^2 &= 2.3730 + \frac{0.0128}{\lambda^2 - 0.0156} - 0.0044\lambda^2 \end{aligned} \tag{C3.1.20}$$

where λ is in units of microns; this is shown in figure C3.1.5.

The refractive index for a second-harmonic and a fundamental wave of the same polarization will in general be different because of this wavelength dispersion. Hence, to achieve phase-matching during frequency doubling, the second-harmonic polarization must be orthogonal to at least one of the fundamental waves. (A single fundamental beam can be regarded as two different waves, one with a polarization parallel and the other with a polarization perpendicular to that of the second harmonic.) Whether the second harmonic ray should be ordinary or extraordinary to achieve phase matching depends on whether the crystal involved is positive uniaxial ($n_e > n_o$) or negative uniaxial ($n_o > n_e$). The various permutations of possible polarization combinations of the three waves involved determine, by convention, the *type* of phase-matching scheme, as outlined in table C3.1.1. A nonlinear interaction between three waves, E_i ($i = 1, 2, 3$), of polarization states $s_i =$ o, e is often described by the notation, $s_1 s_2 \rightarrow s_3$, to indicate which of the waves involved are ordinary (o) or extraordinary (e). For type I phase-matching, waves 1 and 2 are of the same polarization state which

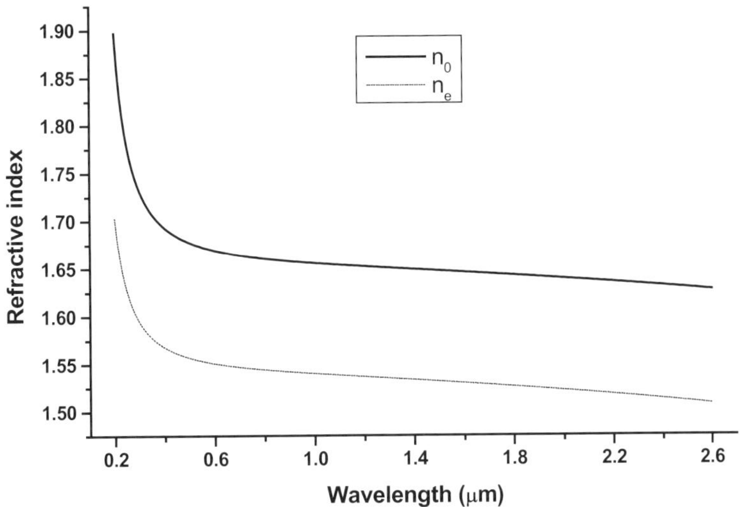

Figure C3.1.5. Ordinary (n_o) and extraordinary (n_e) refractive index change with wavelength for BBO.

Table C3.1.1. Phase matching types. The notation $s_1 s_2 \to s_3$ indicates the polarization states $s_i = o, e$ ($i = 1, 2, 3$) of each of the three waves involved in the interaction.

Phase matching type	Positive crystal	Negative crystal
I	ee → o	oo → e
II	eo → o	oe → e
III	oe → o	eo → e

is orthogonal to that of 3 (i.e. ee → o or oo → e). Similarly, the polarization of waves 1 and 3 is the same and orthogonal to that of 2 for type II phase-matching (i.e. oe → o or eo → e); for type III, 2 and 3 are the same polarization and the state of 1 is orthogonal to the others (i.e. eo → o or oe → e).

Birefringent phase matching. In uniaxial crystals, the extraordinary refractive index depends not only on wavelength but only also on the angle that the e-ray makes with the optic axis (see figure C3.1.6). Calculation of the refractive index experienced by a wave passing through a biaxial crystal at an arbitrary angle is much more complex than for the uniaxial case and there is not an e-ray and an o-ray component for propagation at this arbitrary angle. If, however, the ray is restricted to one of the principal planes (i.e. XZ, XY or YZ) then the calculation of the refractive indices for two orthogonally polarized waves reduces to be the same as for a uniaxial case. For instance, if propagation is restricted to the XY plane then the wave polarized normal to this plane is constant (i.e. becomes the effective o-ray) and has a value of $n_o = n_Z$, where n_X, n_Y and n_Z are the principal refractive indices (see figure C3.1.7). On the other hand, the refractive index experienced by the wave polarized in the XY plane depends on the angle it makes with the X-axis and hence can be regarded as the extraordinary ray for this plane with $n_e = n_X$. In both cases, the loci of n_e values trace out an ellipse as the propagation direction varies between $\theta = 0°$ and 90° (relative to the optic axis for uniaxial crystals and relative to the appropriate principal axis for biaxial crystals). Hence, the refractive index of the extraordinary

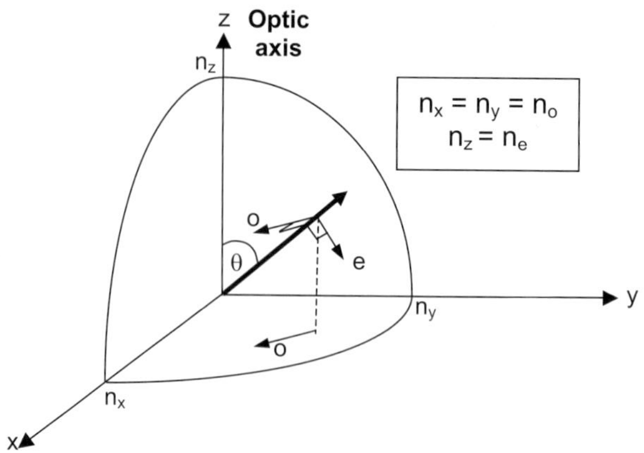

Figure C3.1.6. The index ellipsoid for a uniaxial crystal showing the o- and e-rays.

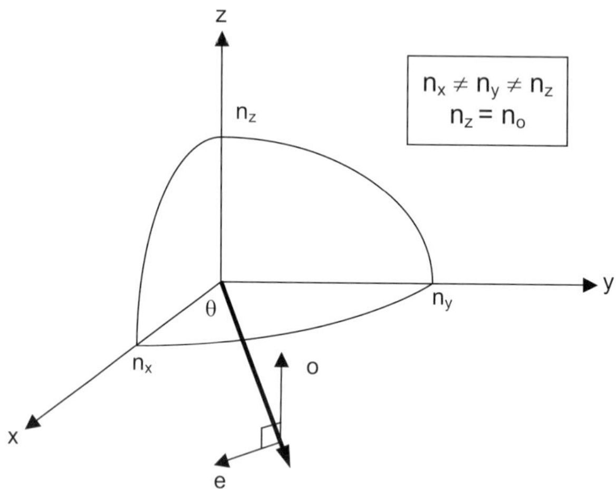

Figure C3.1.7. The index ellipsoid for a biaxial crystal showing the o- and e-rays.

ray, $n(\theta)$, can be calculated from the expression

$$\frac{1}{n^2(\theta)} = \frac{\sin^2(\theta)}{n_e^2} + \frac{\cos^2(\theta)}{n_o^2} \tag{C3.1.21}$$

The combination of equations (C3.1.18), (C3.1.20) and (C3.1.21) allows the angle required for phase matching to be calculated as a function of fundamental wavelength for a particular crystal and phase-matching type. As an example, figure C3.1.8 shows the phase-matching curve for second-harmonic generation using a BBO crystal.

Quasi-phase matching. Quasi-phase matching (QPM) is an alternative method to birefringent phase matching (BPM) of ensuring that the phase mismatch of a particular nonlinear interaction is zero, i.e. $\Delta k = 0$, and thus frequency conversion efficiency is maximized. In contrast to the birefringent technique, which utilizes the natural properties of nonlinear crystals, QPM works by artificially imposing a periodicity

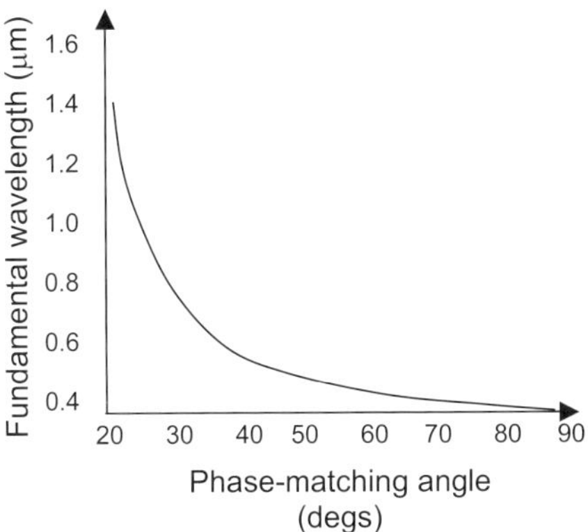

Figure C3.1.8. Phase-matching curve for type I second-harmonic generation in BBO.

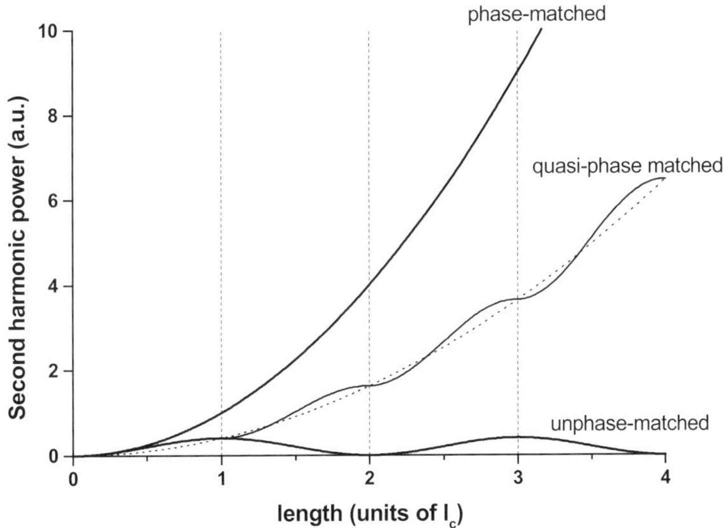

Figure C3.1.9. Second-harmonic power as a function of distance (normalized to coherence length, l_c) for phase-matched, quasi-phase-matched and unphase-matched interactions. The phase-matched interaction experiences monotonic growth whilst, for the unphase-matched situation, energy continuously flows back and forth between the fundamental and the second-harmonic waves. For the quasi-phase-matched interaction, the solid line represents the second-harmonic power variation as the poling direction is alternated every l_c, thereby preventing the onset of power transfer back to the fundamental; the dotted line is the phase-matched growth for an effective nonlinear coefficient of d_Q (see equation (C3.1.23)).

onto the nonlinear coefficient at twice the coherence length, l_c; this resets the accumulated phase difference between the fundamental and second-harmonic wave before destructive interference can cause energy to flow back to the fundamental from the second harmonic (see figure C3.1.9).

This periodic variation of the nonlinear coefficient can be achieved by reversing its sign every l_c to form domains. There are several ways in which such domains can be created, including the stacking of thin plates of the material in alternating orientations [9] and carefully controlled crystal growth [10, 11]. However, over the last decade or so, the technique of imposing a domain structure onto bulk samples of ferroelectric crystals by electric field poling has emerged and proved to be a practical way of producing useful materials for nonlinear optics. The chief benefits of QPM over BPM is that it can be used to ensure that any nonlinear interaction within the transparency range of the material can be phase matched irrespective of angle; hence, the propagation direction is now free to be chosen so that walk-off and angular bandwidth are optimized (see next subsection). Further, given this freedom of phase matching, the beam directions and polarization can be chosen so that the largest component of the nonlinear coefficient tensor can be used for coupling the fields together. This can bring about a considerable improvement in the effective nonlinear coefficient for an interaction; for example, if lithium niobate is used to frequency double the fundamental wavelength of a Nd:YAG laser, $\lambda = 1064$ nm, then the interaction for BPM has $d_{31} = 4.1$ pm V^{-1} whereas the equivalent QPM interaction has $d_{\text{eff}} = 21.6$ pm V^{-1}. This translates to more than an order of magnitude improvement in conversion efficiency in the non-depleted pump limit (see equation (C3.1.13)).

The concept of quasi-phase matching was in fact suggested as long ago as the early 1960s by Armstrong *et al* [12] and independently by Franken and Ward [13] and the analysis of this effect [14] proceeds largely along the same lines as birefringent phase matching, except that now the phase mismatch is given by

$$\Delta k = k_3 - k_1 - k_2 - \frac{2m\pi}{\Lambda} \tag{C3.1.22}$$

where Λ is the repeat length of the nonlinear coefficient variation and m is an integer and represents the order of the process. The nonlinear coefficient of the QPM interaction, d_Q, is related to the effective nonlinear coefficient, d_{eff}, of the equivalent birefringently phase-matched interaction, i.e. the same wavelengths, polarization and propagation direction relative to the crystal axes, *if it could be phase matched* is given by

$$d_Q = \frac{2}{m\pi} d_{\text{eff}} \sin(m\pi D). \tag{C3.1.23}$$

D is the duty cycle, defined as $D = l_{\text{p}}/\Lambda$, where l_{p} is the length of the positive sections of the nonlinear coefficient modulation.

A full analysis of QPM, including tuning, bandwidths and the effects of deviations away from the ideal modulation of the nonlinear coefficient is presented in [14]; the broadening of the QPM second-harmonic bandwidth has been analysed [15].

C3.1.2.4 Phase-matching bandwidth

In the previous section, the propagation direction through a particular nonlinear crystal required to phase-match a *monochromatic plane wave* was calculated. In practice, however, the fundamental wave will have some finite spectral bandwidth and some finite divergence and these deviations from the perfect phase-matching condition will reduce the efficiency of nonlinear conversion because $\Delta k \neq 0$ for all the energy in the fundamental wave. Indeed, effectively no nonlinear conversion will take place for frequencies and angles far from the $\Delta k = 0$ condition and an *angular* and *spectral phase-matching bandwidth* can be defined. Similarly, because refractive index in nonlinear crystals depends on temperature then a *temperature phase-matching bandwidth* can also be defined. The phase mismatch, that is the deviation from $\Delta k = 0$, as a function of angle, θ, temperature, T, and frequency, ν, can be written to the first approximation as

$$\Delta k \approx \frac{\partial(\Delta k)}{\partial\theta}\Delta\theta + \frac{\partial(\Delta k)}{\partial T}\Delta T + \frac{\partial(\Delta k)}{\partial\nu}\Delta\nu \tag{C3.1.24}$$

where $\Delta\nu$, $\Delta\theta$ and ΔT are the spectral, angular and temperature bandwidths, respectively. If the bandwidths are defined as deviation from the $\Delta k = 0$ value that produces a 50% reduction in output power, i.e. when

$\text{sinc}^2(\Delta kL) = 0.5$, then they can be written as

$$\Delta\theta = \frac{1.772\pi}{L\left[\dfrac{\partial(\Delta k)}{\partial\theta}\right]_{\theta=\theta_{pm}}} \qquad \Delta T = \frac{1.772\pi}{L\left[\dfrac{\partial(\Delta k)}{\partial T}\right]_{T=T_{pm}}} \qquad \Delta\nu = \frac{1.772\pi}{L\left[\dfrac{\partial(\Delta k)}{\partial\nu}\right]_{\nu=\nu_{pm}}} \tag{C3.1.25}$$

where the subscript pm indicates the value at $\Delta k = 0$.

The values of these bandwidths will depend upon which phase-matching type is employed (i.e. either I, II or III) and whether the crystals is uniaxial or biaxial because the $n_e = f(\theta)$ but $n_o \neq f(\theta)$ and also because the rate of change with temperature and frequency is different for n_e and n_0. Reference [16] conveniently tabulates the expressions for these bandwidths for the full range of phase-matching schemes.

For a phase-matching angle of $\theta = 90°$, the first derivative $\partial(\Delta k)/\partial\theta = 0$ and the angular bandwidth now depends on the second derivative

$$\Delta\theta_{90} = 2\sqrt{\frac{0.886\pi}{L\left[\dfrac{\partial^2(\Delta k)}{\partial\theta^2}\right]}} \tag{C3.1.26}$$

and is generally much broader than the bandwidth for phase matching at angles significantly different from 90°; in other words, there is enhanced angular tolerance at this angle. For this reason, phase-matching is said to be *non-critical* at $\theta = 90°$ in contrast to *critical phase matching* otherwise. Restricting beam propagation to the $\theta = 90°$ direction of course prevents phase matching being achieved by careful selection of crystal orientation. Hence, second-harmonic generation is usually phase matched at this angle by QPM or by control of crystal temperature; for example, lithium niobate will frequency double 1064 nm radiation from a Nd:YAG with $\theta = 90°$ at a temperature of 120 °C [17].

Walk-off. Another advantage of non-critical phase matching is that at $\theta = 90°$ there is no 'walk-off' between ordinary and extraordinary beams. What is intuitively thought of as the beam direction is the direction of energy transport which is determined by the Poynting vector, $\boldsymbol{S} = \boldsymbol{E} \times \boldsymbol{H}$. In free space and isotropic media the propagation wavevector, $\boldsymbol{k}$, is parallel to $\boldsymbol{S}$ but in anisotropic media this is not generally true because of the tensor nature of the dielectric constant. This means that for collinear phase-matching schemes, i.e. when $\boldsymbol{k}_1$, $\boldsymbol{k}_2$ and $\boldsymbol{k}_3$ are all parallel, the corresponding Poynting vectors, $\boldsymbol{S}_1$, $\boldsymbol{S}_2$ and $\boldsymbol{S}_3$, will not be parallel. For finite beam diameters this results in any extraordinary beams 'walking-off' or separating from the unaffected ordinary beams, as illustrated in figure C3.1.10. For an extraordinary ray propagating through an anisotropic crystal such that its k-vector makes an angle, θ, with the optic axis then the S-vector will be at an angle ρ to the k-vector, where

$$\rho = \pm\tan^{-1}\left[\left(\frac{n_0}{n_e}\right)^2 \tan\theta\right] \mp \theta \tag{C3.1.27}$$

and the upper signs correspond to a negative ($n_o > n_e$) crystal and the lower signs to a positive ($n_e > n_o$) crystal (see [103], section 2.3.) The deleterious effects of walk-off can be largely avoided by the use of large beam diameters, however, if small diameters are necessary then the resulting Poynting vector walk-off length, l_a, can act as the effective interaction length and limit conversion efficiency, where

$$l_a = \frac{w_0\sqrt{\pi}}{\rho} \tag{C3.1.28}$$

where w_0 is the beam waist radius.

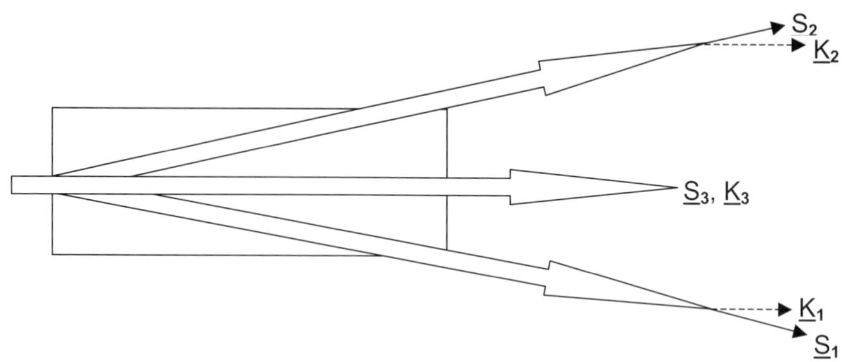

Figure C3.1.10. 'Walk-off' of the Poynting vectors of the phase-matched waves in a nonlinear crystal.

C3.1.2.5 Intracavity and resonant cavity second harmonic generation

As was seen in section C3.1.2.2, the conversion efficiency of second-harmonic generation depends linearly on the intensity of the fundamental beam. For most gain-switched, Q-switched or modelocked lasers the high peak intensities obtained during the pulse (usually kW cm^{-2} or greater) are sufficient to achieve efficient frequency conversion from the nonlinear crystals available. In these cases, simply directing the output beam through an appropriately phase-matched nonlinear crystal is usually adequate. However, the output powers typically available from most continuous wave (cw) lasers are several orders of magnitude less (usually $\sim$1 W or less) and the consequently low intensity becomes a limiting factor to the conversion efficiency. Furthermore, as described in section C3.1.2.2, there is a limit to the efficiency gains to be had from the tight focusing of a cw laser beam because of the associated reduction in Rayleigh length and increase in divergence. One solution is to increase the intensity of the fundamental beam not by focusing but by resonating it inside an optical cavity and the obvious choice for that is the laser cavity itself.

Nonlinear output coupling. A laser cavity consisting of a mirror that is $\sim$100% reflecting at the laser wavelength and an output coupling mirror which transmits a certain fraction, T, at the same wavelength will have a circulating power that that is approximately T^{-1} times larger than the output power available to pump a second-harmonic crystal external to the cavity. This enhancement factor is increased even further if the transmission of the output coupling mirror is made highly reflecting at the laser wavelength as well, i.e. the finesse of the cavity is made as high as possible at this wavelength. In this case, energy escapes the cavity not through transmission at a partially reflecting mirror, as is the case for most laser designs, but by conversion to the second harmonic because the mirror reflectivities have been chosen to be highly transmitting at the doubled frequency as well as highly reflecting at the fundamental wavelength. The frequency doubling crystal now acts as a nonlinear output coupler because the fraction of light that it couples out of the cavity depends on the circulating intensity. An interesting consequence of this nonlinear coupling is that the threshold power for the laser is significantly reduced because the losses due to transmission at the mirrors have been virtually eliminated, and these are usually the chief source of loss in a well-designed laser. There is also minimal output coupling *via* the nonlinear crystal because at threshold the circulating intensities are very low. This arrangement is sometimes called a 'threshold-less' laser even though, of course, there will also be some residual losses and therefore a small but finite threshold. Another benefit of an intracavity scheme is that, whilst for extracavity frequency doubling the coupling between the fundamental and second-harmonic fields should be as large as possible to maximize conversion, the nonlinear coupling at the operating power should be equivalent to the optimum output coupling for the laser. In other words, an extracavity doubling crystal

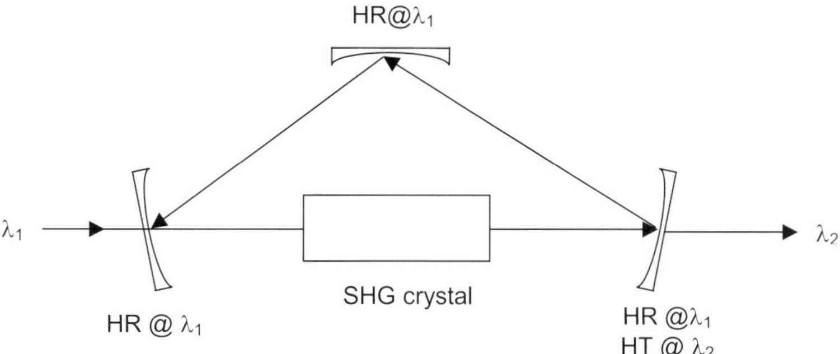

Figure C3.1.11. Optical cavity for resonant second-harmonic generation. (HR=highly reflecting; HT=highly transmitting; SHG=second harmonic generating).

should have a conversion efficiency as near to 100% as possible for a single pass whilst for a laser cavity with an optimum output coupling of, say, 10% then the nonlinear conversion efficiency need only be at this lower level. In this situation, an effective conversion efficiency of 100% has been achieved in the sense that as much second-harmonic power is produced as would have been produced at the fundamental wavelength from a similar laser with a coupling mirror of optimum transmittance.

Disadvantages. Even though there are clear benefits to be had from an intracavity frequency doubling scheme, it is not always the best frequency doubling technique because there are a number of disadvantages with the method. The introduction of any additional optical component into a laser cavity inevitably introduces extra loss, whether it be by absorption in the nonlinear crystal or by reflection from its surfaces, and thereby reduces the efficiency of the system. Also, if the laser normally operates on a number of longitudinal modes, the action of the nonlinear crystal and the gain medium can effectively couple them together resulting in chaotic power fluctuations—often called the 'green problem'. These power fluctuations are most problematic for lasers that operate on a few mode axial modes because if there are a large number of modes then fluctuations tend to average out. Ensuring single mode operation completely obviates the green problem but this can involve introducing lossy optical components into the cavity or finding and securing the resonance of a suitable seed laser. A third disadvantage of intra-cavity second-harmonic generation is that in a standing wave cavity the fundamental circulates into two directions and therefore the second harmonic is also generated into two directions at once. For many laser designs, only the second harmonic generated in one of the two directions is readily extractable and, at the very least, the second harmonic power is split between two beams. A travelling wave cavity solves this problem but usually at the expense of the introduction into the cavity of some component that ensures unidirectional oscillation, such as an optical diode, which inevitably increases the losses.

Resonant cavity second harmonic generation. All of the problems with intracavity frequency doubling can be avoided by putting the nonlinear crystal in a separate optical cavity. Nothing has been introduced into the fundamental laser cavity and so clearly its threshold and slope efficiency have not been compromised. Similarly, separating the laser gain medium and the nonlinear crystal prevents the laser modes from coupling together and producing chaotic power fluctuations. Finally, a ring resonator geometry, such as the one shown in figure C3.1.11, produces a second-harmonic wave only in one direction because the fundamental passes through the nonlinear crystal in that direction only. However, the resonant cavity technique has a drawback in that it requires that the fundamental wave is kept resonant within the cavity containing the frequency doubling

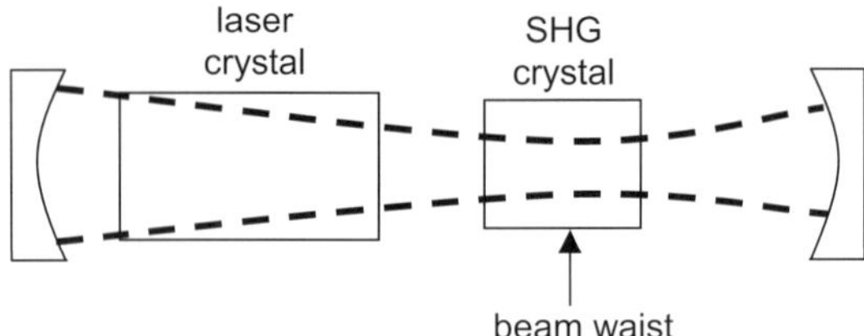

Figure C3.1.12. Intracavity second-harmonic generation configuration, showing the SHG crystal placed at the beam waist and the laser crystal placed where the beam diameter is larger.

crystal. At resonance, the optical impedance of the cavity is well matched to that of the beam path from the laser source and, hence, back reflection is suppressed. Moreover, resonance maximizes the fundamental intensity within the cavity. To achieve resonance, precision control of the cavity length is required, usually by a rapid electronic feedback system and vibration isolation and thermal stabilization. Without these measures, fluctuations in the second-harmonic power will result as the fundamental power inside the resonant cavity varies as the cavity length moves in and out of resonance with the fundamental frequency. These control requirements make for a considerably more complicated and less robust frequency doubling system than the simpler intracavity or external second-harmonic generation schemes.

Theory of intracavity second-harmonic generation

(i) Steady state. The steady-state operation of an intracavity frequency doubled cw laser can be analysed by including a nonlinear loss term into the oscillation condition, i.e. loss = gain, which represents the fraction of light intensity that is lost from the oscillating fundamental wave by conversion to the second harmonic. Hence, the round trip saturated gain is equated to the combined linear and nonlinear losses per round trip (i.e. twice the single-pass second harmonic conversion efficiency, η):

$$\frac{2g_0 L_l I_s}{I_s + I} = F + 2\kappa\eta \tag{C3.1.29}$$

where g_0 is the unsaturated gain coefficient, I is the intensity of the fundamental wave inside the cavity, I_s is the saturation intensity of the laser gain medium, L_l is the length of the gain medium, F is the fractional linear loss per round trip and $\kappa = I_{nonlinear}/I_{gain}$ is the ratio of the fundamental intensity in the nonlinear crystal, $I_{nonlinear}$ and the fundamental intensity in the laser gain medium, I_{gain}. The value of this ratio is typically significantly greater than one because the nonlinear crystal is usually placed at the beam waist of the cavity, where the fundamental intensity is highest, and the gain medium is usually placed elsewhere in the cavity where the cavity mode has a larger diameter, as illustrated in figure C3.1.12. It is advantageous to place the gain medium where the diameter is large because then higher powers can be extracted from the laser before gain saturation.

It can be shown [18, 19] that, in the steady state case, the output coupling through the nonlinear crystal will be optimum and, therefore, the second-harmonic power output will be greatest when

$$\kappa K L^2 = \frac{F}{I_s}. \tag{C3.1.30}$$

Thus, the nonlinear coupling required for maximum second-harmonic power depends only on the cavity losses and the saturation intensity of the gain medium and is independent of the gain. Hence, optimum coupling is achieved for all gain values and consequently all power levels of a given system. This is in contrast to linear output coupling, typically through the partially reflecting mirror, where the optimum value of the coupling is gain and, therefore, power dependent [20]. Moreover, if the frequency doubling crystal can be inserted into the laser cavity with minimal additional loss then the maximum second-harmonic power achieved equals the maximum fundamental power obtainable from the same system, i.e. there is an effective conversion efficiency of 100%.

(ii) Pulsed lasers. The analysis of pulsed intracavity frequency doubled lasers is achieved by the addition of an extra term to the rate equations that represents the energy flow from the fundamental field to the second harmonic field [21]:

$$\begin{aligned}\frac{\mathrm{d}N}{\mathrm{d}t} &= R - \sigma c\phi N - \frac{n}{\tau}\\ \frac{\mathrm{d}\phi}{\mathrm{d}t} &= \frac{-\phi}{\tau_c} + \sigma c N\phi - KL^2\phi^2\end{aligned} \tag{C3.1.31}$$

where N is the population inversion density, ϕ is the photon density, R is the rate at which the laser is pumped; τ is the upper laser level lifetime and τ_c is the lifetime of fundamental photons in the cavity (see chapter C2.2).

These equations cannot be solved analytically but numerical integration is straightforward and can be used to predict second-harmonic pulse energy, peak power and pulse durations; it also reveals that these quantities have an interesting dependence on the magnitude of the nonlinear coupling coefficient. At low nonlinear coupling levels, much less than optimum coupling, the laser can be said to be *undercoupled* in that the second-harmonic pulse energy extracted is less than the maximum possible value and is improved if the coupling coefficient is increased. In this regime, the pulse duration is approximately constant and, hence, the peak power scales with the pulse energy. Once optimum coupling is achieved, however, this behaviour changes. The pulse energy becomes essentially fixed and is unaffected by a further increase in the coupling coefficient. In contrast, additional nonlinear coupling in this regime extends the pulse duration and consequently the peak power decreases; in this situation the laser can be said to be *overcoupled*. Reduced peak powers caused by overcoupling can be problematic if high powers are required, for a subsequent frequency conversion step for instance [22]. Since the nonlinear coupling coefficient can be readily reduced by introducing some degree of phase mismatch, by de-tuning a nonlinear crystal away from the optimum phase-matching angle for example, then this effect can be used to tune the duration and peak power of the second-harmonic pulse whilst maintaining a constant pulse energy. This can be useful for applications such as sum frequency mixing between two pulsed lasers because a good temporal overlap between the two pulses is required for high-conversion efficiency [22].

If the bandwidth of the intracavity fundamental wave is much broader than the phase-matching bandwidth of the nonlinear crystal for intracavity second-harmonic generation, then a form of spectral hole-burning can be observed for pulsed lasers even though they might have a homogeneously broadened gain medium. For instance, this type of behaviour has been reported [23] for a gain-switched Ti:sapphire laser, which is pumped by the frequency doubled output of a Q-switched Nd:YAG laser, and includes a β-barium borate (BBO) nonlinear crystal for intracavity frequency doubling. The bandwidth of the fundamental wave in this free-running laser was ~90 nm whilst the phase-matching bandwidth of the BBO crystal in this particular configuration was ~3 nm. A 3 nm spectral hole is 'burnt' into the fundamental spectrum by the doubling effect; the frequency conversion process is effectively suppressing the fundamental intensity with the phase-matching bandwidth. In this case, if the BBO crystal is rotated in the cavity, changing the phase-matching angle, then this spectral hole can be tuned to different parts of the oscillating fundamental bandwidth. This example also illustrates the importance for conversion efficiency of matching the fundamental bandwidth to the phase-matching bandwidth (PMB): only the radiation within the PMB is converted. This was demonstrated for the present example by introducing a seeding beam from a single mode diode laser into the Ti:sapphire cavity which, when resonant, resulted in a narrowing of the fundamental bandwidth to less than the PMB and a consequent improvement in the second harmonic conversion efficiency from <3% to 17% [23].

C3.1.3 Sum and difference frequency mixing

Second-harmonic generation is a special case of the more general phenomenon of sum frequency mixing (sometimes called 'up-conversion'), in which the two waves that mix together *via* a second-order nonlinear susceptibility to produce a third are at the same frequency. The analysis of the sum frequency mixing (SFM) process essentially proceeds along the same lines as that of second-harmonic generation presented in the previous section but with the two initially present waves allowed to be of different frequencies, i.e. $\omega_1 \neq \omega_2$.

Sum frequency mixing and second-harmonic generation both obey the general phase-matching relations for frequency and propagation vector (corresponding to photon energy and momentum conservation, respectively):

$$\begin{aligned} \omega_3 &= \omega_1 + \omega_2 \\ \boldsymbol{k}_3 &= \boldsymbol{k}_1 + \boldsymbol{k}_2 \end{aligned} \qquad \text{(C3.1.32)}$$

with ω_3 and $\boldsymbol{k}_3$ representing the frequency and propagation vector, respectively.

However, two waves interacting in a nonlinear material also produce a wave corresponding to the difference in their energy and momentum. This process is called difference frequency mixing or 'down-conversion'. That both sum and difference waves are produced concurrently can be seen by considering two plane waves, with entirely real amplitudes A_1 and A_2, interacting *via* an effective nonlinear coefficient, d_{eff}, to produce a plane polarization wave also with an entirely real amplitude, P_3, where

$$\begin{aligned} P_3 \cos(\boldsymbol{k}_3 \cdot \boldsymbol{r} - \omega_3 t) &= d_{\mathrm{eff}} E_1 \cos(\boldsymbol{k}_1 \cdot \boldsymbol{r} - \omega_1 t) E_2 \cos(\boldsymbol{k}_2 \cdot \boldsymbol{r} - \omega_2 t) \\ &= \frac{d_{\mathrm{eff}}}{2} E_1 E_2 \{\cos([\boldsymbol{k}_1 + \boldsymbol{k}_2] \cdot \boldsymbol{r} - [\omega_1 + \omega_2] t) + \cos([\boldsymbol{k}_1 - \boldsymbol{k}_2] \cdot \boldsymbol{r} - [\omega_1 - \omega_2] t)\}. \end{aligned} \qquad \text{(C3.1.33)}$$

Which of the sum or difference frequency light waves actually develops depends on which of them is phase matched; only the phase-matched wave will have significant conversion efficiency. Hence, equating the space and time varying components yields the phase-matching equation for difference frequency mixing:

$$\begin{aligned} \omega_3 &= \omega_1 - \omega_2 \\ \boldsymbol{k}_3 &= \boldsymbol{k}_1 - \boldsymbol{k}_2 \end{aligned} \qquad \text{(C3.1.34)}$$

where $\omega_1 > \omega_2 > \omega_3$ and $k_1 > k_2 > k_3$. Photon energy and momentum are conserved in this case because physically a ω_2 photon stimulates a ω_3 photon to decompose into a ω_3 photon and a further ω_2 photon.

As for sum frequency mixing, the field strengths, E, and intensities, I, used in the analysis of difference frequency mixing are instantaneous quantities and apply equally to both pulsed and steady-state interactions. Again pulse energies can be simply retrieved by integrating the appropriate instantaneous power function. In contrast to second-harmonic generation, however, pulsed sum and difference frequency mixing is further complicated because the two interacting pulses are not necessarily synchronized or of the same duration. Any temporal mismatch between the pulses will reduce the overall conversion efficiency because the magnitude of the generated nonlinear polarization is proportional to the product of the fields corresponding to each pulse. For maximum conversion efficiency, care has to be taken then to maximize the pulse overlap integral:

$$\int_{-\infty}^{\infty} E_1(t) E_2(t - \tau)\, \mathrm{d}t \qquad \text{(C3.1.35)}$$

where E_1 and E_2 are independent fields corresponding to the interacting pulses of generally different functional forms and durations and τ is the delay time between the pulses. It is usually straightforward to ensure that the pulses are synchronized, i.e. $\tau = 0$, by suitable electronic triggering of two separate laser pulses [22]. The pulse durations of the two laser sources are determined by such factors as the gain media used (usually different if two distinct wavelengths are being mixed), the pumping rate and the cavity design. Hence, there is typically less control over the relative durations of the interacting pulses.

C3.1.3.1 Theory

In this section the general theories of the sum and difference frequency mixing of plane waves will be described and some special cases examined including: perfect phase matching (i.e. $\Delta k = 0$), exact photon balance between the interacting waves (i.e. $\sigma = 1$), one interacting field of much larger intensity than the other ($\sigma = 0$ or $\sigma = \infty$), and low total conversion efficiency ($D_{1,2} \to 0$).

C3.1.3.2 Sum frequency mixing

The analysis developed by Milton for the mixing of three plane waves and presented in section C3.1.2.2 for the case of second-harmonic generation can be generalized for sum frequency mixing simply by allowing $\omega_1 \neq \omega_2$ and hence the efficiency of converting wave 1 to wave 3, η_{13}, is

$$\eta_{13} = \frac{I_3}{I_1} = \frac{\omega_3}{\omega_1} B \times sn^2\left(A\sqrt{D_1}, \gamma\right) \tag{C3.1.36}$$

and similarly the efficiency of converting wave 2 to wave 3, η_{23}, is

$$\eta_{23} = \frac{I_3}{I_2} = \frac{\omega_3}{\omega_2} B \times sn^2\left(A\sqrt{D_2}, \gamma\right) \tag{C3.1.37}$$

where D_i is the dimensionless 'nonlinear drive' parameter for wave i, and is defined as

$$D_i = K L^2 \frac{I_i}{\hbar\omega_i} \tag{C3.1.38}$$

D_i is equivalent to the *photon number* conversion efficiency from wave i as $D_i \to 0$. Note that whilst for the special case of second-harmonic generation by type I phase-matching the photon balance parameter, σ, was equal to unity, for sum frequency mixing $\sigma \neq 1$, in general.

Perfect phase matching. For the case of $\Delta k = 0$, the conversion efficiency for the ith wave is considerably simplified and becomes

$$\eta_{i3} = \frac{\omega_3}{\omega_i} \sigma \times sn^2\left(A\sqrt{D_i}, \sqrt{\sigma}\right). \tag{C3.1.39}$$

From now on, it will be assumed that the sum frequency mixing interactions described are all phase matched.

Perfect photon balance. Under the condition that both the interacting fields, $i = 1,2$, are equal in photon number, that is $I_1/\omega_1 = I_2/\omega_2$, then the parameter $\sigma = 1$ and the conversion efficiencies become

$$\begin{aligned} \eta_{13} &= \frac{\omega_3}{\omega_1} \frac{I_2}{\hbar\omega_2} \tanh^2\left(\sqrt{D_1}\right) \\ \eta_{23} &= \frac{\omega_3}{\omega_2} \frac{I_2}{\hbar\omega_2} \tanh^2\left(\sqrt{D_2}\right) \end{aligned} \tag{C3.1.40}$$

which is the same result as for second harmonic if $\omega_1 = \omega_2$. It can be shown that the elliptic function *sn* reduces to a monotonically growing function (i.e *tanh*) only when the condition $\sigma = 1$ is met. In other words, there is no back-conversion from the generated wave ($i = 3$) to the initially interacting waves ($i = 1,2$) only when there is photon number parity between the two initial fields. This is clearly not necessarily true for general sum frequency mixing between waves of different frequencies; it is also not always achieved for second-harmonic generation when a type II or III phase-matching scheme is used. In those cases, the polarization of the fundamental wave is usually orientated at some angle to the optic axis of the nonlinear crystal so that it can be regarded as consisting of two component fields, one parallel to the optic axis and one orthogonal. Only if this angle is exactly 45° will the magnitude of the two component fields be equal and the photon number parity condition be met. Hence, in general, sum frequency mixed interactions can suffer from back conversion of the generated wave.

Large intensity ratio, $I_1 \gg I_2$. If the intensity of one of the interacting fields is much larger than the other, and we can choose $I_1 \gg I_2$ and still retain generality, the conversion efficiency becomes

$$\eta_{23} = \frac{\omega_3}{\omega_2} \sin^2\left(\sqrt{D_1}\right). \tag{C3.1.41}$$

However, caution must be exercised when using this approximate form because its domain of validity is difficult to estimate [5].

Weak nonlinear interaction, $D_1 \to 0$. In the limit of only a weak interaction between the fields the conversion efficiency becomes

$$\begin{aligned} \eta\ 23 &\to \frac{\omega_3}{\omega_2} D_1 \\ \eta\ 23 &= \frac{\omega_3}{\omega_1} D_2. \end{aligned} \tag{C3.1.42}$$

C3.1.3.3 Difference frequency mixing

The general solution to the coupled wave equations for the difference frequency mixing of plane waves yields the following result for nonlinear conversion efficiency

$$\eta_{13} = \frac{I_3}{I_1} = \frac{\omega_3}{\omega_1} B' \left[\frac{\gamma' sn^2(\Lambda, \sqrt{\gamma'})}{1 - \gamma' sn^2(\Lambda, \sqrt{\gamma'})}\right] \tag{C3.1.43}$$

where

$$\begin{aligned} \gamma &= \frac{A'}{A' + B'} \\ \Lambda &= \sqrt{D} \times \sqrt{(\sigma - 1 - \varepsilon)^2 + 4\sigma} \\ A', B' &= \tfrac{1}{2}\left[\pm(\sigma - 1 - \varepsilon) + \sqrt{(\sigma - 1 - \varepsilon)^2 + 4\sigma}\right] \\ \sigma &= \frac{I_2}{I_1}\frac{\omega_1}{\omega_2}. \end{aligned} \tag{C3.1.44}$$

This expression has been derived under the same assumptions as for the sum frequency mixing analysis.

As for sum frequency mixing, approximations can be made to the above expression under different conditions:

Perfect phase-matching, $\Delta k = 0$.

$$\eta_{13} = \frac{\omega_3}{\omega_1}\left[\frac{\gamma' sn^2(\sqrt{D_1(1+\sigma)}, \sqrt{\gamma'})}{1-\gamma' sn^2(\sqrt{D_1(1+\sigma)}, \sqrt{\gamma'})}\right] \tag{C3.1.45}$$

where now

$$\gamma' = \frac{\sigma}{1+\sigma}. \tag{C3.1.46}$$

Large intensity ratio, $I_1 \gg I_2$. If there is large ratio between the intensities of the two interacting waves, then the photon balance parameter $\sigma \to 0$ and the conversion efficiency to the difference frequency becomes:

$$\eta_{13} = \frac{\omega_3}{\omega_2}\frac{I_2}{I_1}\sin^2\left(\sqrt{D_1}\right). \tag{C3.1.47}$$

Weak nonlinear interaction, $D_1 \to 0$. When the mixing between the waves is weak due either to weak beam intensities, small effective nonlinear coefficient, short interaction length or poor phase matching then the conversion efficiency tends to the following expression

$$\eta_{13} = \frac{\omega_3}{\omega_2}\frac{I_2}{I_1}D_1. \tag{C3.1.48}$$

Gaussian beams. The analyses presented before for sum and difference mixing assume that the interacting waves were all plane, which implies that they are infinite in transverse extent (see the first assumption in section C3.1.2.2). As for second harmonic generation, experimentally beams of finite cross section will be used and these will be most likely a TEM_{00} Gaussian mode and hence consideration of the effect of this deviation from the plane wave approximation has to be considered. A comprehensive analysis of three-wave mixing between Gaussian modes has been carried out by Boyd and Kleinman [8]. This analysis reveals that for maximum conversion the *confocal parameter*, b (see section A2.1.2.2), of all three waves must be equal; the confocal parameter for the ith wave is given by

$$b_i = k_i w_0^2 \tag{C3.1.49}$$

where w_0 is the radius of Gaussian beam waist and hence b is twice the Rayleigh length, z_0. Therefore, for optimally efficient sum and difference frequency mixing, the two interacting beams must be focused such that they are collinear and the position of their beam waists overlap and, furthermore, their spot sizes (i.e. beam waist area $= \pi w_{0,i}^2$) must be in the same ratio as their respective wavelengths, λ_i, i.e.

$$\frac{w_{0,1}^2}{w_{0,2}^2} = \frac{\lambda_1}{\lambda_2}. \tag{C3.1.50}$$

Further to this, for maximum conversion efficiency from a nonlinear crystal of length, l, (assuming crystal length limits the interaction length in this case, i.e. $l = L$) then the same optimum focusing condition as was described for second-harmonic generation in section C3.1.2.2 applies: $2.84 \times b = 5.68 \times z_0 = l$.

C3.1.4 Third- and higher-harmonic generation

Third- and higher-order nonlinear effects are generally very much weaker than second-order phenomena. As was argued in section A4.2, this is because the magnitude of the third order nonlinear susceptibility, $\chi^{(3)}$, is

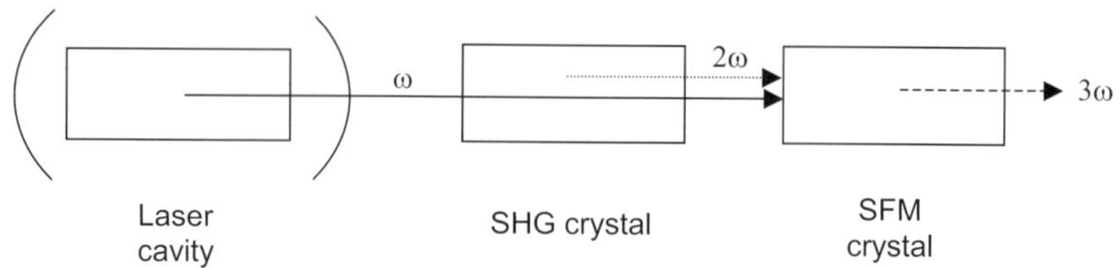

Figure C3.1.13. Third-harmonic generation by cascaded SHG and SFM.

approximately seven orders of magnitude smaller than $\chi^{(2)}$ and the higher order nonlinear susceptibilities, $n > 3$, are much smaller than that. Hence, the efficient generation of third- and higher-order harmonics is not done by utilizing the corresponding order of nonlinear response of the material but rather by using a sequence of second-order processes. For instance, laser radiation can be converted to its third harmonic by initially frequency doubling part of the output and then sum frequency mixing this second-harmonic wave with the residual fundamental wave, as illustrated in figure C3.1.13.

Each individual second-order frequency conversion step of an overall high-order harmonic generation scheme can be treated separately and hence the analyses presented in the previous section can be used to describe the whole process. One step interacts with the next simply by determining the photon balance parameter, σ, of the radiation incident on the subsequent step. As noted in the previous section, the conversion efficiency is only a monotonic function of distance through the crystal for a perfect photon balance, i.e. $\sigma = 1$, and hence only in this case is the mixing process free from the possibility of back conversion. For perfect photon balance, complete conversion of the fundamental to the desired harmonic can be achieved, in principle. In all other cases, there will be an optimum distance into the crystal up to which the generated wave grows steadily in amplitude and after which the amplitude decreases or oscillates. Efficient high-harmonic generation schemes, therefore, ensure that there is a perfect photon balance before each constituent frequency conversion step.

For third-harmonic generation, Craxton [24, 25], demonstrated that the requirement of perfect photon balance can be achieved automatically by an appropriate choice of polarization angle in the frequency doubling crystal (the first step). In fact, it was shown that there are several possible ways of achieving this, depending on the type of phase matching used in both steps and the polarization angle. In each case, however, exactly 2 out of 3 of the fundamental photons are converted into second-harmonic photons after the first step, leaving an equal number of second-harmonic and residual fundamental photons for the final sum frequency mixing step, thus ensuring a theoretical 100% conversion to the third harmonic. In practice, the use of each of these schemes has resulted in up to 80% conversion efficiency to the third harmonic [26] but, for simplicity, those schemes that obviate the need for rotation of the fundamental polarization by waveplates in between stages are preferred. The need for a waveplate is removed by careful choice of the polarization angle of the fundamental beam relative to the optic axis of the first crystal. For a negative uniaxial crystal for example, a polarization angle of 55° results in two thirds of the photons contributing to the o-ray and one third to the e-ray. After the frequency doubling stage, maximum conversion efficiency produces equal numbers of second-harmonic and fundamental photons, with the polarization angle for the residual fundamental being either 0° or 90° depending on the phase-matching type. This perfect photon balance and the alignment of the residual fundamental polarization so that it is entirely an o- or e-ray allows complete conversion to the third harmonic. In contrast, at other initial polarization angles the fundamental after the first stage is neither purely an o- nor an e-ray. Hence, even though the photon numbers may be balanced, complete conversion to the third harmonic is not possible.

A notable example of multiple harmonic generation is Zayhowski's work using microchip lasers [27]. In that case a passively Q-switched, microchip laser (a 1 mm^3 dielectrically coated Nd:YAG crystal) was used to produce 8.0, 3.5, 0.3, 0.7 and 0.01 μJ of pulse energy at the first to fifth harmonic, respectively; KTP and BBO were used as the nonlinear crystals (see the next section for a description of these materials). The excellent beam quality and short (~200 ps) pulse duration obtained from microchip lasers make them well suited to nonlinear frequency conversion at such modest pulse energies.

Very high harmonic generation that produces light in the extreme ultraviolet (XUV) or x-ray regions of the spectrum can be achieved at very high intensities (typically in the terawatt range or higher) in, usually, a noble gas medium (see section B5.2.4). In this process, atoms are ionized by the laser field and the resulting free electrons accelerated by the same field before recombination. The high-energy photons produced in this way can interfere constructively to produce a coherent XUV or x-ray beam. Further details can be founds in a recent review [28].

C3.1.5 Nonlinear materials for frequency conversion

Since the first demonstration of harmonic generation in the early sixties just after the invention of the laser, a great number of nonlinear crystals have been studied. This section will not attempt a comprehensive survey of these materials; such reviews already exist. For instance, the book by Dmitriev *et al* [103] provides an extensive review of over fifty nonlinear crystals and lists all the parameters necessary to calculate the phase-matching angles and conversion efficiencies for each material. Yariv [29] provides a useful table on nonlinear coefficients for over 20 materials. In contrast, Koechner [30] provides a more selective survey of nonlinear materials but with more commentary on each (see further reading for details of these books). Similarly, this section will concentrate on those few materials whose superior characteristics has meant that they have come to prominence. These materials can be divided into two classes: *birefringent*, for which phase matching is ensured by using the natural properties of the crystal, and *quasi-phase matched*, which rely on an imposed periodic poling of the crystal domains to achieve phase matching. Self-doubling and self-summing crystals will also be described, which are an interesting, though not yet commonplace, class of materials.

There are many questions regarding the properties of nonlinear crystals that can have a bearing on their selection for a particular application; some important ones are listed below:

(1) Is it possible to phase match the interaction in question?

(2) What will the effective nonlinear coefficient be?

(3) Do all the wavelengths concerned lie within the transparency range of the crystal?

(4) Will the peak intensity damage the crystal?

(5) What are the absorption coefficients at the relevant wavelengths?

(6) Is the divergence of the pump laser larger than the angular bandwidth?

(7) Is the spectral width of the pump laser larger than the frequency bandwidth?

(8) If the interaction has a small temperature bandwidth, will temperature stabilization of the crystal be required?

(9) What crystal sizes are available?

(10) What is the optical quality of the crystal?

(11) Is the crystal chemically stable?

(12) Is the crystal mechanically stable?

The relative importance of these questions will depend on the details of the application. For instance, for cw interactions a large effective nonlinear coefficient and a low absorption coefficient are usually desirable whilst for the frequency conversion of high-energy pulses a high damage threshold and good optical quality are likely to be more important. Moreover, if the beam path within the material limits the interaction length then large crystal size can compensate for a lower effective nonlinear coefficent.

C3.1.5.1 *Birefringent materials*

The phase matching of three-wave mixing interactions in this class of materials relies on the natural angle, polarization, temperature or field dependence of the refractive indices experienced by the interacting waves as they pass through the medium, as was described in section C3.1.1.2. The materials used for three-wave mixing in the UV, visible and near and mid-infrared are described in this section; those materials that are used to mix far-infrared radiation, however, will be discussed in the section on carbon dioxide lasers, C3.1.6.3. Direct comparisons of some of the crystals described here have been reported elsewhere [31, 32]. In this work, the properties of the materials discussed are compared in table C3.1.2. Since many of the important material performance parameters, such as effective nonlinear coefficient and absorption coefficient, are determined by the particular nonlinear interaction and the wavelengths of the beams involved; the relative merit of each type of crystal depends on the application. Nevertheless, a comparison based on a particular application does indicate relative performance; the parameters listed in table C3.1.2 are based, where possible, on perhaps the most common application of nonlinear crystals: the second-harmonic generation of the fundamental wavelength of a Nd:YAG laser (1064 nm); the other two materials listed are compared on the basis of their performance in frequency doubling the fundamental of a CO_2 laser.

*KDP, KD*P, ADP and CDA.* Good optical quality crystals of potassium dihydrogen phosphate, KH_2PO_4, and its deuterated analogue KD*P, can be grown from aqueous solution at room temperature up to diameters of 27 cm, the largest size for any nonlinear crystal [33]. However, as well as being water soluble, these crystals must be used below 100 °C and heated at a rate no greater than 5 °C min^{-1}. The nonlinear coefficients of KDP are rather low compared to other leading nonlinear materials ($d_{36} = 0.44$ pm V^{-1}; $d_{36} = 0.40$ pm V^{-1} for KD*P), however, this is largely compensated for by the large size and good optical quality of the crystals, and high conversion efficiencies have been achieved. Since KDP and another of its analogues, ammonium dihydrogen phosphate (ADP), were the crystals used in the first demonstration of phase-matched second harmonic generation [34], the nonlinear coefficients of other materials are often given relative to that of KDP. A comprehensive calculation of the properties of KDP relevant to the generation of harmonics of 1 μm radiation has been reported [35]. Two further analogues of KDP, cesium dihydrogen phosphate (CDA) and cesium dideuterium phosphate (CD*A) have also been studied and are interesting because they can non-critically phase match the fundamental wavelength of both Nd:YAG and Nd:glass [36, 37].

KTP and KTA. Potassium titanyl phosphate, $KTiOPO_4$, (KTP) has a very favourable set of characteristics when compared to other birefringent materials; other materials might score more highly in particular categories but KTP has superior overall properties. It has large nonlinear coefficients ($d_{31} = 6.5$ pm V^{-1} and $d_{32} \sim 5.0$ pm V^{-1}), phase matches over a wide wavelength range (for type II), and has broad angular and temperature bandwidths; its thermal properties are also good and it has a high damage threshold. These properties often make it the nonlinear crystal of choice for the important application of the frequency doubling of Nd:lasers. However, growing KTP is a difficult process, which limits the crystal size and availability makes it an expensive material. Even with the relatively small size of crystals available the high nonlinear coefficients of KTP mean that the high-frequency doubling efficiencies can be readily achieved. KTP is also susceptible to photochemical degradation or 'grey tracking', which is a cumulative effect dependent on total exposure; it leads to increased absorption and eventual crystal cracking. Fortunately, operation at high temperature (>65 °C) prevents and reverses the degradation process. Potassium titanyl arsenate (KTA) is an isomorph of KTP that has largely similar properties. It differs from KTP in that it cannot phase-match SHG for a Nd:YAG laser, although it will phase-match 1053 nm radiation and so can be used to frequency double Nd:YLF lasers and for this application exhibits a 1.6× higher effective nonlinear coefficient [38]. It also absorbs less than KTP in the 3–5 μm part of the infrared spectrum and has found use as a material for optical parametric oscillators (see the following chapter C3.2).

Table C3.1.2. Properties of some nonlinear crystals pertinent to SHG.

Material	Phase-matching type[a]	Phase-matching angle[a]	d_{eff} (pm V^{-1})[a]	Transparency range (μm)	Damage threshold (GW cm^{-2})[b]	Absorption coeffiecient (cm^{-1})[a]	Angular bandwidth (mrad)[c]	Temperature bandwidth (°C)[c]	Spectral bandwidth (nm)[c]
KD*P	II	53.5°	0.38	0.2–2.0	0.5	0.005	5.0	13	6.5
KTP	II	25°	5.27	0.35–4.5	1.0	0.05	15–68	25	0.56
$LiNbO_3$	I	78°	5.89	0.33–5.5	0.06	0.08	3.1	1.1	0.3
BBO	I	21°	2.01	0.2–2.6	23	0.005[d]	1.8[e]	50[e]	0.66[d]
LBO	I	90°	1.17	0.16–2.6	19	0.000 35[f]	70[f]	5.8[f]	4.37[g]
Te[†]	I	14.5°	629	3.8–32	0.045	0.96	0.7	—	0.001
$ZnGeP_2^{\dagger}$	I	76°	35.4	0.74–12	0.06	0.9	24.5	44[h]	0.05

Notes: All the data in the table has been taken from Dmitriev *et al* (see further reading), unless otherwise indicated:
[a] for 1064 nm radiation, except for † which are for 10 μm radiation;
[b] for 1064 nm, ~10 ns pulses, except for † which are for 10 μm and ~200 ns;
[c] for a 1 cm interaction length;
[d] [30, table 10.1];
[e] [41];
[f] [43];
[g] Velsko S P, Webb M, Davis L and Huang C 1991 *IEEE J. Quantum Electron.* **27** 2182;
[h] Kato K 1997 *Appl. Opt.* **36** 2506.

Lithium niobate. Lithium niobate, $LiNbO_3$, has favourable chemical and mechanical properties, being non-hygroscopic and hard, thus making the processing and polishing of these crystals straightforward. It also has relatively large nonlinear coefficients ($d_{31} = -5.4$ pm V^{-1} and $d_{22} = 2.8$ pm V^{-1};) and can be temperature-tuned when non-critically phase-matched. However, a low damage threshold and a particular susceptibility to degradation by photorefractive damage offset these advantageous properties. As with KTP, this degradation is reversed by increasing the temperature but, for the lithium niobate, the temperatures required are much higher, at $\sim 170\,^{\circ}$C. The susceptibility to damage is reduced by doping lithium niobate with magnesium oxide, MgO, to a typical level of 5%. Due to these damage problems and emergence of newer materials without these limitations, lithium niobate had been receiving less attention until the invention of the quasi-phase matching technique. It was the first material to be successfully periodically poled and, hence, once again became the focus of much interest. Quasi-phase matched materials will be covered in more detail in the next section.

BBO. Barium borate, BaB_2O_4, (BBO) was discovered in 1984 by Chen *et al* [39] and has two crystalline phases, α and β, with the β-phase being non-centrosymmetric. BBO has nonlinear coefficients ($d_{22} =$ 2.2 pm V^{-1} and $d_{31} = 0.16$ pm V^{-1}) that are somewhat smaller than some of the other materials discussed but it is its other properties that make it interesting. It has a wide temperature phase-matching bandwidth, low absorption ($\sim$0.01 cm^{-1} at $\lambda = 532$ nm [40]) and a very high damage threshold. Of particular note, is that the transparency range of BBO extends down to wavelengths of just 189 nm thus allowing second-harmonic generation and sum-frequency mixing into the ultraviolet. The only major weakness of BBO as a nonlinear material is that it has a small angular phase-matching bandwidth. A comprehensive review of the properties and applications of BBO has been published [41].

LBO. Lithium triborate, LiB_3O_5, (LBO) was discovered in 1988 by the same group who had earlier developed BBO [42]. At 155 nm, the short wavelength limit of its transparency range extends even further into the important UV part of the spectrum than does that of BBO. It is a hard, chemically stable, non-hygroscopic crystal making it easy to handle and able to take a good polish. It can also be non-critically phase matched and temperature-tuned. However, it has nonlinear coefficients that are even smaller than those of BBO ($d_{31} = 1.1$ pm V^{-1} and $d_{32} = 1.2$ pm V^{-1}) but does not suffer from the same narrow angular bandwidth. Its properties and applications have been comprehensively reviewed [43].

C3.1.5.2 Quasi-phase-matched materials

The physical properties of quasi-phase-matched materials are identical to their birefringently phase-matched counterparts except with regard to their phase-matching behaviour and the effective nonlinear coefficients. Nowadays, the domain structure is imposed onto bulk ferroelectric materials by lithographically writing a pattern of electrodes on to the crystals and then applying large electric fields (typically 25 kV mm^{-1} for lithium niobate); hence, these materials are often termed 'periodically poled'. A more detailed description of poling process is given in [44]. Another advantage of the poling technique is that domain structures other than the simple periodic grating can be implemented and these can determine, for example, the spectral dependence of the conversion [45], the tuning range [46] and the beam profile [47].

Periodically poled lithium niobate. Periodically poled lithium niobate (PPLN) was the subject of much of the initial development of QPM materials fabricated by this technique because it is ferroelectric and because of its favourable chemical and mechanical properties, making it available in large, easily processable samples. The PPLN crystal is usually poled so that phase-matching occurs for all three waves polarized parallel to the crystal z-axis. This enables the largest nonlinear coefficient component, $d_{33} = 34$ pm V^{-1}, to be utilized; hence, theoretically $d_Q = 2d_{33}/\pi = 21.6\,\mathrm{pm\,V^{-1}}$ but imperfections in the domain grating lead to a reduction in the effective value of d_Q to typically [48] 15–17 pm $\mathrm{V^{-1}}$. Experimentally, QPM typically leads to a factor ~10–20 improvement in the conversion efficiency over the birefrigent phase-matched case (which can only use the smaller d_{31} component) [49]. This high nonlinearity enables cw and low energy pulses to be efficiently converted to the second harmonic, even for an extracavity, single pass arrangement. For example, approximately 40% frequency doubling conversion efficiency has been obtained in such an arrangement both for a 6.5 W cw 1064 nm Nd:YAG laser [50] and a 7 nJ pulse^{-1}, 946 nm Nd:YAG laser [51]. A poling period could be chosen such that NCPM occurred at room temperature but generally PPLN is operated at an elevated temperature, typically ~150 °C, to mitigate the photorefractive effect.

Periodically poled KTP. The periodic poling of KTP was first demonstrated in 1994 and has developed rapidly since. Periodically poled KTP (PPKTP) has a number of advantages over PPLN: the field required for poling is approximately an order of magnitude less, the higher crystal anisotropy prevents lateral domain growth and so allows smaller domain sizes to be fabricated more consistently, its transparency extends further into the UV and, finally, KTP is less susceptible to photorefractive damage allowing room-temperature operation [52]. On the other hand, the value of the nonlinear coefficient tensor component used in PPKTP interactions, also d_{33}, is at 16.9 pm $\mathrm{V^{-1}}$ less then half that of PPLN. Nevertheless, conversion efficiencies of approximately 65% have been reported for the extracavity frequency doubling of sub-millijoule pulses in both the picosecond and nanosecond [53] regimes for Nd:YAG (twice that for BPM KTP). For a cw Ti:sapphire laser, a doubling efficiency of 10% has been reported recently [54] for a 1 cm long crystal. The greater UV transparency of KTP compared to lithium niobate has been utilized to generate 390 nm radiation from a mode-locked femtosecond Ti:sapphire laser [54] and 460 nm light from a mode-locked diode laser [55].

Periodically poled lithium tantalate. Periodically poled lithium tantalate (PPLT) has a smaller nonlinear coefficient than either PPLN or PPKTP: $d_{33} = 13.8$ pm $\mathrm{V^{-1}}$ for a wavelength of 1064 nm, corresponding to a theoretical, and experimentally observed, effective nonlinear coefficient for first-order QPM of $d_Q = 9$ pm $\mathrm{V^{-1}}$ [56]. However, it does have a photorefractive damage threshold nearly twice that of PPLN [57] and is transparent to wavelengths as low as 280 nm [58]. For pumping with low-power cw diodes, conversion efficiencies of 1.2% $\mathrm{W^{-1}\,cm^{-1}}$ have been demonstrated [59]. For higher peak powers, conversion efficiencies of 13% and 35% have been reported for pulsed diode [60] and fibre lasers [61]. For doubling into the UV region, however, the effective d_Q achievable drops to ~2 pm $\mathrm{V^{-1}}$ because of the difficulty in fabricating grating periods of ~2 μm and the necessity of using second-order ($m = 2$) QPM in some cases [60, 61].

Periodically poled rubidium titanyl arsenate. Periodically poled rubidium titanyl arsenate, $RbTiOAsO_4$, or PPRTA is an isomorph of KTP and shares many of its advantageous properties such as large nonlinear coefficients (e.g. $d_{33} = 15.6$ pm $\mathrm{V^{-1}}$ [62]) and a high-photorefractive damage threshold permitting room-temperature operation; like KTP, RTA requires poling fields approximately one order of magnitude less than lithium niobate. Furthermore, the absorption coefficient for RTA is smaller than that for KTP in the mid-infrared region [63, 64], which is advantageous for DFM; and RTA is flux grown, a technique that often allows crystals to be produced more cheaply and in larger sizes than the hydrothermal technique used for KTP. A normalized conversion efficiency of 6.5% $\mathrm{W^{-1}\,cm^{-1}}$ for the frequency doubling of cw laser diode at a wavelength of 875 nm has been reported [62].

C3.1.5.3 Self-doubling and summing materials

Certain nonlinear materials can be doped with laser active ions to produce crystals that act as both a laser gain medium and an intracavity frequency doubler or summer. The requirement that the crystal must both accept the dopant ion and also be phase matchable for the laser emission wavelength in practice restricts this phenomenon to only a few possible combinations of ions and crystals. A thorough review of self-doubling and self-summing lasers, including a survey of materials and performance results, has recently been published [65]; it also contains a theoretical analysis of the self-doubling and self-processes.

The first demonstration of a self-doubled laser [66] utilized Tm doped in lithium niobate (doped also with MgO to reduce photorefractive damage) although more recently Nd and Yb have been the more common dopant ions. Apart from lithium niobate (LNB), other nonlinear crystals that been successfully doped with laser active ions include: $LaBeGeO_5$ (LBG), $Ba_2NaNb_5O_{15}$ (BNN), $Gd_2(MoO_4)_3$, $YAl_3(BO_3)_4$ (YAB), $CaY_4(BO_3)_3O$ (YCOB) and $CaGd_4(BO_3)_3O$ (GdCOB). Of the Nd-doped crystals, YAB has produced the best results so far [67,68] and the best reported efficiency for a Yb-doped laser has been for a LNB:MgO crystal [69]. Self-frequency doubling has also been reported in some quasi-phase-matched materials: periodically poled Nd-doped lithium tantalate [70] and periodically poled Ti-doped LN [71].

C3.1.6 Frequency conversion of particular lasers

This section surveys the materials and techniques used in the frequency conversion of a few common laser types. Each laser will be examined in turn and the nonlinear techniques utilized for different relevant regimes of operation, such as cw, Q-switched or mode-locked and high- or low-energy/power. Further information of this type can be found in the books listed in further reading; in particular, Koechner [30] describes a variety of laser cavity designs for the intracavity frequency doubling of Nd:YAG lasers and Dmitriev *et al* [103] have produced a very comprehensive survey of the techniques used for harmonic generation and sum and difference mixing of a wide range of important lasers, including all the major solid state and gas types.

C3.1.6.1 Nd lasers

KTP is probably the most important nonlinear material for the SHG of Nd lasers (chapter B1.3) in both the cw regime (see, [72] for example) and the pulsed regime up to high powers [73] and for pulse durations greater than a few hundred picoseconds [74]. Its large nonlinearity and large angular and temperature phase-matching bandwidths for the SHG of 1.06 μm radiation (~0.5° and 20 °C, respectively, for a 1 cm long crystal [75]) are a compelling combination for most applications; it also has a high damage threshold and is non-hygroscopic. In the pulsed regime, extracavity conversion efficiencies of 55–60% have been obtained [73, 74].

However, for very high-energy picosecond Nd lasers, such as the ones used in thermonuclear fusion research, KDP may be a more appropriate material. KDP has a number of advantageous properties in this regime: it is resistant to optical damage by sub-nanosecond pulses of intensities up to 10 GW cm^{-2}, it exhibits low one- and two-photon absorption at the harmonics of Nd lasers and both picosecond continuum generation and stimulated Raman scattering do not occur in this material for intensities less than 100 GW cm^{-2}. Moreover, high-quality KDP crystals can be grown to large sizes, which enables the large beam cross sections necessary to prevent damage to optical components to be accommodated. The lower value of the effective nonlinear coefficient for KDP compared to KTP is not an issue in this high peak power regime and second-harmonic conversion efficiencies of 90% have been reported [76].

Higher-harmonic generation for Nd lasers has been achieved for a number of different nonlinear materials including KDP, ADP and BBO. For instance, KDP has been used to convert 0.15 ns pulses at the fundamental wavelength of Nd:YAG to the third harmonic with 32% efficiency [77] and a 85% second- to fourth-harmonic efficiency has been achieved for 30 ps pulses in ADP [78]. For very high intensity lasers, the overall conversion efficiency (fundamental to harmonic) rises to 80% for the third [76] and 70% for the fourth harmonic [79]. A combination of BBO and KTP has also been used to produce tunable UV radiation (230–330 nm) from a Q-switched Nd:YAG by a combination of harmonic generation and sum-frequency mixing [80] with an overall efficiency of 5%.

C3.1.6.2 Ti:sapphire lasers

Ti:sapphire (i.e. the Ti^{3+} ion doped into Al_2O_3) has the widest tunability of any gain medium in the visible and near-infrared optical region, lasing having been demonstrated from 645 nm to 1178 nm [81] (see chapter B1.2). This impressive tunability can be extended even further, however, by the use of nonlinear frequency conversion techniques, to cover the range from ~200 nm to 3000 nm. Given the wide range of possible fundamental wavelengths from a Ti:sapphire laser, it is not surprising that there are a number of nonlinear crystals that will phase match over at least part of its tuning range. In fact, three of the leading nonlinear materials described in the previous section will frequency double the output of a Ti:sapphire laser over its full tuning range, and have low absorption at the second-harmonic wavelength. These three are LBO, BBO and KDP and a comparison of the relative performance of these materials for frequency conversion applications (second-, third- and fourth-harmonic generation, and both sum and difference frequency mixing) is reported in [70]. The use of NCPM, temperature tuned LBO for the frequency doubling of Ti:sapphire output has also been studied elsewhere [82] and includes details of temperature tuning curves and phase-matching bandwidths. For SHG, LBO proved to have the lowest threshold of the three materials for both type I and type II phase-matching, with the latter being significantly the lowest. However, LBO will only phase match for the entire Ti:sapphire tuning range in a type I arrangement, for type II there is a short wavelength cut-off at 792 nm. Of the other two materials, BBO demonstrated the lowest threshold and highest conversion efficiency, and, importantly, because of its sensitivity to change in phase-matching angle, a single crystal could be used to frequency double the entire tuning range of Ti:sapphire. (Other materials need to have several crystals, each cut at different angles to the crystal axes, to cover the entire range.) The disadvantage of this angular sensitivity, however, is small phase-matching tolerance angles, necessitating low divergence pump beams for efficient frequency conversion. On the other hand, the phase-matching temperature for BBO was the greatest of the three at ~20 °C.

C3.1.6.3 Carbon dioxide lasers

The long fundamental wavelength, $\lambda = 10\ \mu m$, of the CO_2 laser (chapter B3.1) lies outside the transparency range of all of the major nonlinear crystals described in the previous section. Frequency conversion of CO_2 laser output is achieved by using one of a number of semiconductor materials that are non-centrosymmetric and are transparent deep into the infrared; these include: proustite (Ag_3AsS_3), pyrargyrite (Ag_3SbS_3), silver thiogallate ($AgGaS_2$), $AgGaSe_2$, $ZnGeP_2$, $CdGeAs_2$, gallium selenide (GaSe), mercury sulphide (HgS), selenium and tellurium. The important properties of these for frequency conversion can be found in Dmitriev *et al* [103], and the properties of ZnGeP2 and tellurium are compared in table C3.1.2. The nonlinear coefficients for these materials can be very large compared to those of the crystals used for frequency conversion in the UV, visible and near-infrared, e.g. at 10 μm tellerium has $d_{11} = 650$ pm V^{-1}. However, the advantage of the consequent factor of $\sim 10^4$ increase in $(d_{eff})^2$ is offset by the factor of ~10 decrease in the frequency of all three of the interacting waves and the rise in the values of refractive index experienced by each wave (e.g. at 10 μm $n_0 \approx 5$ and $n_e \approx 6.5$ for tellurium). Hence, despite the very large nonlinear coefficients of this type of material, the coupling coefficient, K, (see equation (C3.1.9)) achievable in the 10 μm region is comparable to that obtained at shorter wavelengths.

For SHG, conversion efficiencies as high as 80% inside a 3 mm long crystal of zinc germanium phosphide ($ZnGeP_2$; $d_{36} = 75.4$ pm V^{-1}) have been reported [83] for 10 mJ, 2 ns pulses. The high refractive index of these materials means that Fresnel reflection losses are high and so the corresponding conversion efficiency outside the crystal was ~50%.

Sum frequency mixing between CO_2 and near-infrared and visible lasers has also been reported. For instance, in the nanosecond pulse regime conversion efficiencies of between of 30–40% have been obtained by mixing 10 μm radiation with the output from a Nd:YAG laser in proustite [84] and silver thiogallate [85] and a dye laser in silver thiogallate [86] to obtain 976 nm and 564 nm radiation, respectively.

C3.1.7 Developing and growth areas

Current research into the frequency conversion of laser sources is continuing across a broad range of topics covering basic physics, new materials and applications. In this section, a few of the areas that are presently receiving attention will be briefly discussed. The success of the quasi-phase-matching technique over the past decade has encouraged further investigations into structured nonlinear materials. This structuring can take the form of variations from the simple periodically poled grating used for QPM such as fanning and chirping of the grating [87, 88], to produce, for instance, near-transform limited SHG pulses. The formation of QPM materials into waveguide structures is also being actively studied [89–91] and recent highlights include the observation of 99% single-pass conversion efficiency to the second harmonic of a quasi-cw beam in a PPLN waveguide [92]. The use of the periodic structure of photonic bandgap materials for harmonic generation has also recently been investigated [93, 94].

Some of the physical phenomena associated with frequency conversion that have lately attracted interest are the vortices and singularities that can form in the optical field [95, 96]. Other processes investigated recently include quantum noise and squeezing of light in SHG [97, 98], and optical pattern formation [99]. Finally, the increasingly powerful diode lasers that have become available over the last few years, coupled with the emergence of QPM materials with high effective nonlinearities in the same period, has made the combination of these two technologies a practical prospect. Diode lasers have the benefits of being compact and efficient but are only generally available (at the moment) economically and at high powers for emission in the red and near infrared part of the spectrum. Combination with an efficient nonlinear material extends this wavelength range to the blue and near-UV part of the spectrum, which is an important region for various sensing applications. For instance, several groups have recently used periodically poled KTP to double the frequency of picosecond pulses from a diode laser [100, 101].

References

[1] Kleinman D A 1962 Nonlinear dielectric polarization in optical media *Phys. Rev.* **126** 1977
[2] Byer R L and Herbst R L 1977 Parametric oscillation and mixing *Nonlinear Infrared Generation* ed Y R Shen (Berlin: Springer)
[3] Armstrong J A, Jha S S and Shiren N S 1970 *IEEE J. Quantum Electron.* **6** 123
[4] Harter D J and Brown D C 1982 *IEEE J. Quantum Electron.* **18** 1146
[5] Milton M J 1992 *IEEE J. Quantum Electron.* **28** 739
[6] Byrd P F and Friedman M D 1954 *Handbook of Elliptic Integrals for Engineers and Scientists* (Berlin: Springer)
[7] Boyd G D, Ashkin A, Dziedzic J M and Kleinman D A 1965 *Phys. Rev.* **137** 1305
[8] Boyd G D and Kleinman D A 1968 Parametric interaction of focused Gaussian light beams *J. Appl. Phys.* **39**
[9] Okada M, Takizawa K and Ieiri S 1976 *Opt. Commun.* **18** 331
[10] Feisst A and Koidl P 1985 *Appl. Phys. Lett.* **47** 1125
[11] Wang W S, Zhou Q, Geng Z H and Feng D 1986 *J. Cryst. Growth* **79** 706
[12] Armstrong J A, Bloembergen N, Ducuing J and Pershan P S 1962 *Phys. Rev.* **127** 1918
[13] Franken P A and Ward J F 1963 *Rev. Mod. Phys.* **35** 23
[14] Fejer M M, Magel G A, Jundt D H and Byer R L 1992 *IEEE J. Quantum Electron.* **28** 2631
[15] Mizuuchi K, Yamamoto K, Kato M and Sato H 1994 *IEEE J. Quantum Electron.* **30** 1596
[16] Dmitriev V G, Gurzadyan G G and Nikogosyan D N 1991 *Handbook of Nonlinear Optical Crystals* (Berlin: Springer) section 2.10, p 33
[17] Byer R L, Park Y K, Feigelson R S and Kway W L 1981 *Appl. Phys. Lett.* **39** 17
[18] Guesic J E, Levinson H J, Singh S, Smith R G and van Uitert L G 1968 *Appl. Phys. Lett.* **12** 306
[19] Smith R G 1970 *IEEE J. Quantum Electron.* **6** 215
[20] Degnan J J 1989 *IEEE J. Quantum Electron.* **25** 214
[21] Murray J E and Harris S E 1970 *J. Appl. Phys.* **41** 609
[22] Binks D J, Golding P S and King T A 2000 *J. Mod. Opt.* **47** 1899
[23] Mohammed A K, Pruvost J-A, Ribet I, Lefebvre M, Rosencher E and Binks D J 2001 *IEEE J. Quantum Electron.* **37** 290
[24] Craxton R S 1981 *IEEE J. Quantum Electron.* **17** 1771
[25] Craxton R S 1980 *Opt. Comm.* **34** 474

[26] Seka W, Jacobs S D, Rizzo J E, Boni R and Craxton R S 1980 *Opt. Commun.* **34** 469
[27] Zayhowski J J 1996 *Opt. Lett.* **21** 588
Zayhowski J J 1996 *Opt. Lett.* **21** 1618
[28] Balcou P *et al* 2002 *Appl. Phys.* B **74** 509
[29] Yariv A *Quantum Electronics* 3rd edn (New York: Wiley) p 387
[30] Koechner W 1996 *Solid State Laser Engineering* (Berlin: Springer) section 10.1.3
[31] Eckardt R C, Masuda H, Fan Y X and Byer R L 1990 *IEEE J. Quantum Electron.* **26** 921
[32] Borsutzky A, Brünger R, Huang Ch and Wallenstein R 1991 *Appl. Phys.* B **52** 52
[33] Koechner W 1996 *Solid State Laser Engineering* (Berlin: Springer) p 580
[34] Ashkin A, Boyd G D and Dziedzic J M 1963 *Phys. Rev. Lett.* **11** 14
[35] Craxton R S, Jacobs S D, Rizzo J E and Boni R 1981 *IEEE J. Quantum Electron.* **17** 1782
[36] Kato K 1973 *IEEE J. Quantum Electron.* **10** 616
[37] Kato K 1973 *Opt. Commun.* **9** 249
[38] Bierlein J O, Vanherzeele H and Ballman A A 1989 *Appl. Phys. Lett.* **54** 783
[39] Chen C, Wu B, Jiang A and You G 1985 *Sci. Sin.* B **28** 235
[40] Fan Y X, Eckardt R C, Byer R L, Chen C and Jiang A D 1989 *IEEE J. Quantum Electron.* **25** 1196
[41] Nikogosyan D N 1991 *Appl. Phys.* A **52** 359
[42] Chen C, Wu Y, Jiang A, Wu B, You G, Li R and Lin S 1989 *J. Opt. Soc. Am.* B **6** 616
[43] Nikogosyan D N 1994 *Appl. Phys.* A **58** 181
[44] Myers L E, Miller G D, Eckhardt R C, Fejer M M, Byer R L and Bosenberg W R 1995 *Opt. Lett.* **20** 52
[45] Bortz M L, Fujimara M and Mejer M M 1997 *Opt. Lett.* **22** 1341
[46] Ishigame Y, Suhara T and Ishihara H 1991 *Opt. Lett.* **3** 3597
[47] Imeshev G, Proctor M and Fejer M M 1998 *Opt. Lett.* **23** 673
[48] Myers L E, Eckardt R C, Fejer M M, Byer R L. Bosenberg W R and Pierce J W 1995 *J. Opt. Soc. Am.* B **12** 2102
[49] Myers L E, Miller G D, Eckardt R C, Fejer M M, Byer R L and Bosenberg W R 1995 *Opt. Lett.* **20** 52
[50] Miller G D, Batchko R G, Tulloch W M, Weise D R, Fejer M M and Byer R L 1997 *Opt. Lett.* **22** 1834
[51] Ross G W, Pollnau M, Smith P G R, Clarkson W A, Britton P E and Hanna D C 1998 *Opt. Lett.* **23** 171
[52] Pasiskevicius V, Wang S, Tellefsen J A, Laurell F and Karlsson H 1998 *Appl. Opt.* **37** 7116
[53] Englander A, Lavi R, Katz M, Oron M, Eger D, Lebiush E, Rosenman G and Sklair A 1997 *Opt. Lett.* **22** 1598
[54] Wang S, Pasiskevicius V, Laurell F and Karlsson H 1998 *Opt. Lett.* **23** 1883
[55] Woll D, Schumacher J, Robertson A, Tremont M A, Wallenstein R, Katz M, Eger D and Englander A 2002 *Opt. Lett.* **27** 1055
[56] Meyn J-P and Fejer M M 1997 *Opt. Lett.* **22** 1214
[57] Skvortsov L A and Stepantsov E S 1993 *Kvantovaya Electron.* **20** 1127
[58] Kase S and Ohi K 1974 *Ferroelectrics* **8** 1201
[59] Mizuuchi K, Yamamoto K and Kato M 1997 *Appl. Phys. Lett.* **70** 1201
[60] Yamamoto K, Mizuuchi K, Litaoka Y and Kato M 1995 *Opt. Lett.* **20** 273
[61] Champert P A, Popov S V, Taylor J R and Meyn J P 2000 *Opt. Lett.* **25** 1252
[62] Risk W P and Loiacono G M 1996 *Appl. Phys. Lett.* **69** 311
[63] Scheidt M, Beier B, Boller K-J and Wallenstein R 1997 *Opt. Lett.* **22** 1287
[64] Karlsson H, Olson M, Arvidsson G, Laurell F, Bader U, Botsuzky A, Wallenstein R, Wickstrom S and Gustafsson M 1999 *Opt. Lett.* **24** 330
[65] Brenier A 2000 *J. Lumin.* **91** 121
[66] Johnson L F and Ballman A A 1969 *J. Appl. Phys.* **40** 297
[67] Amano S and Mochizuki T 1991 *Nonlinear Opt.* **1** 297
[68] Hemmati 1992 *IEEE J. Quantum Electron.* **28** 1169
[69] Montoya E, Capmany J, Bausa L E, Kellner T, Diener A and Huber G 1999 *Appl. Phys. Lett.* **74** 3113
[70] Adedin K S, Tsuritani T, Sato M and Ito H 1997 *Appl. Phys. Lett.* **70** 10
[71] Amin J, Pruneri V, Webjorn J, Russell P S, Hanna D C and Wilkinson J S 1997 *Opt. Commun.* **135** 41
[72] Jiangrui G, Hai W, Maoquan H, Changde X, Kunchi P, Zhenggang Y, Changqin M and Xuning W 1995 *Appl. Opt.* **34** 1519
[73] Driscoll T A, Hoffman H J, Stone R E and Perkins P E 1986 *J. Opt. Soc. Am.* B **3** 683
[74] Moody S E, Eggleston J M and Seamans J F 1987 *IEEE J. Quantum Electron.* **23** 335
[75] Lavorovskaya O I, Pavlova N I and Tarasov A V 1986 *Sov. Phys. Crystallogr.* **31** 678
[76] Gulamov A A, Ibragimov E A, Redkorechev V I and Usmanov T 1983 *Sov. J. Quantum Electron.* **13** 1054
[77] Attwood D, Pierce E L and Coleman L W 1975 *Opt. Commun.* **15** 10
[78] Reintjes J J and Eckhardt R C 1977 *Appl. Phys. Lett.* **30** 91
[79] Linford G J, Johnson B C, Hildum J S, Martin W E, Snyder K, Boyd R D, Smith W L, Vercimak C L, Eimerl D and Hunt J T 1982 *Appl. Opt.* **21** 3633
[80] Fix A and Ehret G 1998 *Appl. Phys.* B **67** 331
[81] Rines G A, Zenzie H H, Schwarz R A, Isyanova Y and Moulton P F 1995 *IEEE J. Selected Topics Quantum Electron.* **1** 50
[82] Chen D W and Lin J T 1992 *IEEE J. Quantum Electron.* **29** 307

[83] Andreev Y M, Baranov V Yu, Voevodin V G, Geiko P P, Gribenyukov A A, Izyumov S V, Kozochkin S M, Pismenny V D, Sato Yu A and Streltsov A P 1987 *Sov. J. Quantum Electron.* **17** 1435

[84] Jaanimagi P A, Richardson M C and Isenor N R 1979 *Opt. Lett.* **4** 45

[85] Voronin E S, Solomatin V S, Cherepov N I, Shuvalov Y V, Badikov V V and Pivovarov O N 1975 *Sov. J. Quantum Electron.* **2** 597

[86] Jantz W and Koidl P 1977 *Appl. Phys. Lett.* **31** 99

[87] Zimmermann J, Struckmeier J, Hofmann M R and Meyn J P 2002 *Opt. Lett.* **27** 604

[88] Schober A M, Imeshev G and Fejer M M 2002 *Opt. Lett.* **27** 1129

[89] Rafailov E U, Birkin D J L, Sibbett W, Battle P, Fry T and Mohatt D 2002 *Opt. Lett.* **26** 1961

[90] Parameswaran K R, Route R K, Kurz J R, Roussev R V, Fejer M M and Fujimura M 2002 *Opt. Lett.* **27** 179

[91] Rafailov E U, Loza-Alvarez P, Brown C T A, Sibbett W, De La Rue R M, Millar P, Yanson D A, Roberts J S and Houston P A 2001 *Opt. Lett.* **26** 1984

[92] Parameswaran K R, Kurz J R, Roussev R V and Fejer M M 2002 *Opt. Lett.* **27** 43

[93] Shin K C, Hoshi H, Chung D H, Ishikawa K and Takezoe H 2002 *Opt. Lett.* **27** 128

[94] Norton A H and de Sterke C M 2003 *Opt. Lett.* **28** 188

[95] Frend I 2002 *Opt. Lett.* **27** 1640

[96] Loina-Terriza G, Petrov D V, Recolons J and Torner L 2002 *Opt. Lett.* **27** 625

[97] Lodahl P and Saffman M 2002 *Opt. Lett.* **27** 110

[98] Lodahl P and Saffman M 2002 *Opt. Lett.* **27** 551

[99] Mamaev A V, Lodahl P and Saffman M 2003 *Opt. Lett.* **28** 31

[100] Rafailov E U, Birkin D J L, Sibbett W, Battle P, Fry T and Mohatt D 2001 *Opt. Lett.* **26** 1961

[101] Woll D, Schumacher J, Robertson A, Tremont M A, Wallenstein R, Katz M, Eger D and Englander A 2002 *Opt. Lett.* **27** 1055

[102] Davis C 1996 *Lasers and Electro-optics* (Cambridge: Cambridge University Press)

[103] Dmitriev V G, Gurzadyan G G and Nikogosyan D N 1991 *Handbook of Nonlinear Optical Crystals* (Berlin: Springer)

Further reading

Dmitriev V G, Gurzadyan G G and Nikogosyan D N 1991 *Handbook of Nonlinear Optical Crystals* (Berlin: Springer)

A very thorough survey of over 50 nonlinear crystals, it is also includes a detailed description of the theory of three-wave mixing interactions and describes the results of harmonic generation experiments performed on a wide range of different lasers.

Yariv A 1989 *Quantum Electronics* 3rd edn (New York: Wiley)

First published in 1967, this is an introduction to general laser physics that includes chapters on second harmonic generation and other frequency conversion techniques. This work approaches the subject from a rigorous theoretical standpoint, often returning to the quantum mechanical basis of many phenomena.

Koechner W 1996 *Solid State Laser Engineering* (New York: Springer)

A detailed and practical-minded overview of modern techniques in laser engineering, which includes descriptions of many actual laser systems and comprehensively references the literature.

Davis C 1996 *Lasers and Electro-optics* (Cambridge: Cambridge University Press)

This is a wide-ranging text that covers many areas of laser physics, nonlinear optics and photonics at a level suitable for senior undergraduate or graduate students.

Saleh B E A and Teich M C 1991 *Fundamentals of Photonics* (New York: Wiley)

This book covers a broad of topics that includes lasers; the principles of nonlinear optics are described.

Born M and Wolf E 1999 *Principles of Optics* (Cambridge: Cambridge University Press) 7th edition

This book describes classical optics in a very thorough and authoritative way. Although it does not cover the physics of lasers or their applications, it does describe the optics of crystals in great detail.

C3.2
Optical parametric devices

M Ebrahimzadeh

C3.2.1 Introduction

Since its invention in 1960 [1], the *laser* has become an indispensable tool that has transformed optical science and technology over the past four decades. The important characteristics of laser light with regard to intensity, spectral purity, spatial coherence and directionality have led to the establishment of a new branch of optical science and technology now referred to as *photonics* and many new applications in previously unexplored areas of pure and applied science have become possible with the advent of the laser.

The laser, however, is not without its limitations. By virtue of its unique coherence properties, light emission from the laser can occur over only a confined band of frequencies determined by the energy gap in the gain material, and emission over extended spectral bands is generally difficult. It is possible to develop lasers with more extended spectral emission by using gain materials with broadened energy levels (or energy *bands*). In this case, transitions from the upper to the lower energy bands can occur over a correspondingly wide band of frequencies, resulting in laser emission over a broad spectral range. Examples of this type of *tunable* laser are the liquid dye or solid-state vibronic lasers such as the Ti:sapphire laser (see chapter B1.2). However, even in such cases, the maximum spectral coverage available to most prominent tunable lasers (e.g. Ti:sapphire) is limited to at best 300–400 nm. Moreover, the restricted availability of suitable laser gain materials has confined the wavelength coverage of existing tunable lasers mainly to the visible and near-infrared spectrum, as shown in figure C3.2.1. These limitations have left substantial portions of the optical spectrum, particularly in the infrared, inaccessible to lasers and alternative methods for the generation of coherent light in these important regions have had to be devised. One of the most effective techniques to address the spectral limitations of conventional lasers is to exploit *nonlinear optics*. In particular, *optical parametric generation, amplification* and *oscillation* in a nonlinear material have long been recognized as versatile techniques for the generation of widely tunable radiation in spectral regions inaccessible to traditional lasers. The aim of this chapter is to provide an overview of optical parametric devices, including their basic operating principles, material and pump source requirements, and the latest developments in the field. While many of the fundamental principles are universal to parametric devices of all types, the main focus of the discussion will be on *optical parametric oscillators* (OPOs), which represent the majority of the systems developed to date.

C3.2.2 Nonlinear frequency conversion

The potential of nonlinear optical techniques as a means of providing coherent light in new spectral regions was recognized soon after the invention of laser and many of the early ideas were formulated more than forty years ago [2, 3]. Nonlinear optical phenomena are based on a fundamentally different principle from conventional lasers for the generation of coherent radiation. While the process of light generation in a laser is

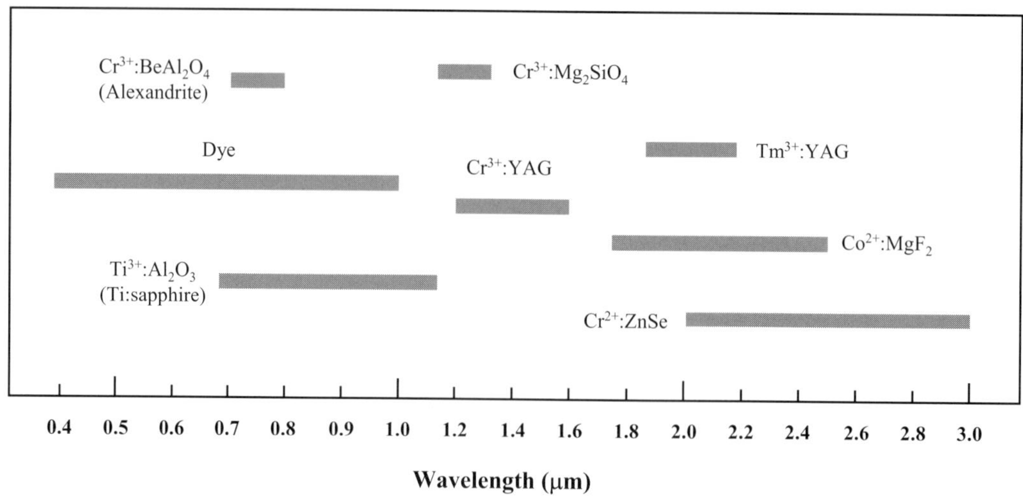

Figure C3.2.1. The spectral coverage of the most prominent tunable lasers, including the traditional dye laser and a number of solid-state vibronic lasers. The tuning range corresponding to the dye laser requires up to 20 different dyes. The vertical scale has no significance.

a direct consequence of the transitions (absorption, spontaneous and stimulated emission) between the energy levels in the gain material, nonlinear processes rely on an alternative mechanism for light generation, namely, electric dipole oscillations in a dielectric medium.

Nonlinear optical processes can readily be understood in terms of a simple classical picture [4]. When a dielectric material is subjected to an external optical field, the constituent dipoles within the material can be set into oscillations through interaction with the oscillating electric field of the incoming light wave. The dipole oscillations can result in the emission of new light waves with an intensity and frequency determined by the dipole moment and the dipole oscillation frequency. In the case when the input optical intensity is small, the dipole displacement follows a linear dependence on the input field strength. The result is dipole oscillation at the same frequency as that of the input optical wave. This situation corresponds to the regime of *linear optics*. On the other hand, when the input optical intensity is large, the dipole displacement from rest will become nonlinear towards the higher field strengths, as shown in figure C3.2.2. This is the regime of *nonlinear optics*. The exact form of the nonlinear response will depend on the structural symmetry of the dielectric material[1]. However, regardless of the crystallographic structure the dipoles will oscillate not only at the input frequency, but also over an infinite range of frequencies, above and below the input frequency[2]. The resulting light emission from the oscillating dipoles will therefore also be over an infinite band of frequencies. In the photon picture, this process is equivalent to the generation of large numbers of new photons with energies extending over an infinite range, below and above the input photon energy, but subject to the conservation of energy.

The potential of nonlinear optical processes is thus immediately clear—they provide a mechanism for the generation of new frequencies (wavelengths) from an already existing input frequency. In other words, they provide a convenient technique for *frequency conversion* of light from an old to a new spectral range.

[1] In a centrosymmetric material, only the odd harmonics of the input wave are generated through dipole oscillations, whereas in a non-centrosymmetric material all frequency harmonics, odd and even, are emitted.

[2] Mathematically, the range of frequencies emitted by the nonlinear oscillating dipoles can be viewed as the Fourier components of the input frequency. In a centrosymmetric material, only odd Fourier components of the input wave are generated, whereas in a non-centrosymmetric material all frequency components are available.

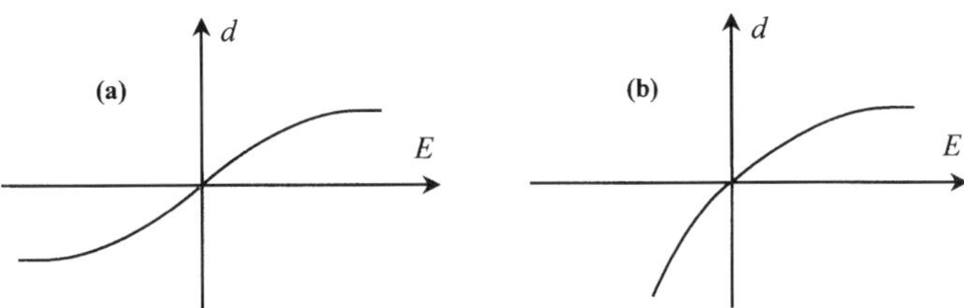

Figure C3.2.2. Dipole displacement from rest, d, as a function of the input electric field amplitude, E, for (*a*) centrosymmetric and (*b*) non-centrosymmetric dielectric materials. The regime of linear optics corresponds to the small and linear displacements of the dipole at low values of field. The regime of nonlinear optics corresponds to the high values of field, where the dipole response is distorted.

This is one of the most fundamental principles of nonlinear optics, first suggested more than four decades ago [2, 3].

Nonlinear optical processes can take a variety of forms. The most important in the context of frequency conversion are second harmonic generation (SHG), sum- and difference-frequency mixing (SFM and DFM) and, central to this discussion, optical parametric generation and amplification (chapter A4). The underlying physical principles responsible for all such processes are essentially the same and relate to the second-order susceptibility $\chi^{(2)}$ in a dielectric crystalline medium. In such a medium, the induced polarization, $\vec{P}$, due to an input optical field, $\vec{E}$, is in its most general form given by

$$\vec{P} = \varepsilon_0[\chi^{(1)} \cdot \vec{E} + \chi^{(2)} \cdot \vec{E} \cdot \vec{E} + \chi^{(3)} \cdot \vec{E} \cdot \vec{E} \cdot \vec{E} + \ldots] \tag{C3.2.1}$$

where $\chi^{(1)}$ is the linear susceptibility and $\chi^{(2)}, \chi^{(3)}, \ldots$ are the nonlinear susceptibilities of the medium. The linear susceptibility is related to the refractive index through $\chi^{(1)} = n^2 - 1$ and is responsible for the linear optical properties of the material such as refraction, dispersion, absorption and birefringence. On the other hand, the second-order susceptibility $\chi^{(2)}$ is the component that gives rise to the processes of SHG, SFM and DFM, linear electro-optic (Pockels) effect, and most importantly in the context of this discussion, optical parametric generation. The third-order nonlinear susceptibility $\chi^{(3)}$ is responsible for the phenomena of third harmonic generation, optical bistability, phase conjugation and the optical Kerr effect including self-focusing and self-phase-modulation. The susceptibilities $\chi^{(1)}, \chi^{(2)}, \chi^{(3)}, \ldots$ are tensors of the second, third, fourth and higher rank, respectively. In writing (C3.2.1), we have used a tensor notation, because most crystalline media that exhibit second-order nonlinearity are optically anisotropic, so that $\vec{P}$ and $\vec{E}$ are generally not parallel in such media.

For the onset of nonlinear optical processes an essential prerequisite is clearly a large input optical intensity. In general, nonlinear optical effects can be observed only when the input electric field approaches the electric field strength binding the material dipoles themselves (of the order of $\sim 10^8$–10^9 V cm^{-1}). The attainment of such high electric field strengths requires optical intensities as large as $\sim 10^{13}$ W cm^{-2}, which can be provided only by a laser. Moreover, the magnitude of the susceptibility tensors decreases rapidly with increasing rank of nonlinearity, so that higher-order nonlinear effects become increasingly difficult to induce and require higher input intensities. Thus, it is not surprising that the observation of nonlinear optical effects only became possible after the invention of the laser.

In this discussion, we are concerned with the second-order nonlinear susceptibility $\chi^{(2)}$ and focus on the implications of this property only in relation to the parametric process. More extensive treatments of other $\chi^{(2)}$ processes and higher-order nonlinear effects can be found in numerous texts and review articles in the literature [5–8]. In its most general form, the $\chi^{(2)}$ tensor has 27 elements, but in practice many of the components vanish under certain symmetry conditions, so the total number of independent components is

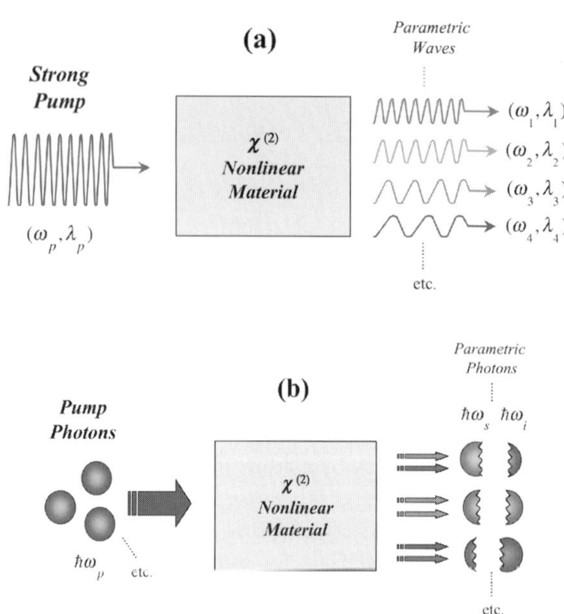

Figure C3.2.3. (*a*) The simple classical wave picture and (*b*) the photon picture of the optical parametric generation. In (*a*) strong dipole oscillations induced by an intense optical pump field results in the generation of new optical waves at lower frequencies. In (*b*) the pump photons are spontaneously broken up into a pair of constituent photons (signal and idler), whose energies add up to the energy of the pump photon, due to energy conservation.

generally far fewer. The tensor is non-zero only in non-centrosymmetric media that lack inversion symmetry in their crystalline structure. In centrosymmetric materials, $\chi^{(2)}$ and all other even-order susceptibilities reduce to zero, and so second-order nonlinear frequency conversion processes, including parametric generation, are not attainable in such crystals. In the definition of the induced polarization through (C3.2.1), the units of $\chi^{(2)}$ are m V^{-1} in MKS units, which is the system of units used throughout this chapter.

C3.2.2.1 Optical parametric generation

As highlighted above, nonlinear frequency conversion processes can take a variety forms. For light emission over extended spectral regions, optical parametric generation [9–11] is the process of primary interest. It corresponds to the most fundamental nonlinear effect where, in the classical picture, a single input *pump* frequency is converted to a broad range of lower optical frequencies in a new spectral range, as illustrated in figure C3.2.3(*a*). In other words, it provides the mechanism for tunable light generation in a new and longer wavelength range using an existing laser pump source at a fixed wavelength. The process is thus highly effective for the generation of tunable coherent radiation in spectral regions where conventional lasers are not available. In the photon picture, figure C3.2.3(*b*), the process is equivalent to the spontaneous break-up of each high-energy pump photon into two constituent parts of lower energy (termed *signal* and *idler*), subject to the conservation of energy. Since energy must be conserved, the sum of energies of the constituent photons has to equal the pump photon energy. Given the large number of pump photons, there are statistically an infinite number of ways in which the break-up can occur, so that an infinite range of signal and idler energies (frequencies) will be emitted during the process. However, out of the infinite number of generated signal and idler photons, only a single pair determined by the *phase-matching* condition and satisfying the energy conservation condition will be emitted in practice. The concept of phase-matching is discussed in more detail in section C3.2.3 (see also chapter C3.1).

Analytically, the parametric generation process can therefore be treated by considering the interaction of three optical fields at frequencies ω_3, ω_2 and ω_1, such that $\omega_3 = \omega_2 + \omega_1$. The field at ω_3 corresponds to the intense input optical pump field, giving rise to a pair of signal and idler fields at ω_2 and ω_1, respectively. The generated field at the higher frequency, ω_2, say, is usually designated the signal, while the field at the lower frequency, ω_1, is designated the idler, but variations in this nomenclature are also frequently used in the literature. Although more rigorous treatments of the parametric generation process can be found elsewhere [3], here we restrict discussion to the essential features of the treatment and focus on the main results of the analysis.

At a fundamental level, the parametric process can be described by considering Maxwell's wave equation for the propagation of three optical fields, the pump, signal, and idler, at respective frequencies ω_3, ω_2 and ω_1, in a non-centrosymmetric dielectric exhibiting second-order susceptibility, $\boldsymbol{\chi}^{(2)}$. The propagation of the three optical fields in such a medium involves the solution of Maxwell's nonlinear wave equation with the second-order nonlinear polarization as the source term, namely

$$\frac{\partial^2 \boldsymbol{E}}{\partial z^2} = \mu\varepsilon \frac{\partial^2 \boldsymbol{E}}{\partial t^2} + \mu \frac{\partial^2 \boldsymbol{P}^{(2)}}{\partial t^2} \tag{C3.2.2}$$

where $\boldsymbol{P}^{(2)} = \varepsilon_0 \boldsymbol{\chi}^{(2)} \boldsymbol{E}^2$ is the second-order polarization and $\boldsymbol{E}$ is the electric field of the propagating wave, given by

$$\boldsymbol{E} = \tfrac{1}{2}[E(z)\mathrm{e}^{\mathrm{i}(kz-\omega t)} + \mathrm{cc}] \tag{C3.2.3}$$

with $E(z)$ representing the complex field amplitude. In writing (C3.2.2), we have ignored nonlinear polarizations terms higher than second-order. We have also reverted to a scalar notation for convenience and taken the propagation to be along the z-axis. In addition, we have assumed that the medium is lossless, non-conducting and non-magnetic, as is generally the case in practice. Since the parametric process involves the interaction of three optical fields, the total field $\boldsymbol{E}$ will comprise three harmonic waves representing the pump, signal and idler, so that

$$\boldsymbol{E} = \boldsymbol{E}_1(\omega_1) + \boldsymbol{E}_2(\omega_2) + \boldsymbol{E}_3(\omega_3) \tag{C3.2.4}$$

To simplify the analysis further, we neglect the effects of focusing and assume that the optical fields are infinite uniform plane waves. We also assume that the fields are monochromatic and neglect any effects due to double-refraction.

The parametric interaction process can then be understood by seeking solution to the nonlinear wave equation (C3.2.2). This is done by separating (C3.2.2) into three components at the three different frequencies $\omega_1, \omega_2, \omega_3$, each of which must separately satisfy the wave equation. Then, by considering the three separate wave equations at each frequency and assuming that the field amplitudes vary only slowly over distances compared to a wavelength, after some manipulation, we obtain the variations of three field amplitudes with propagation as

$$\frac{\partial E_1(z)}{\partial z} = \mathrm{i}\kappa_1 E_3(z) E_2^*(z) \mathrm{e}^{\mathrm{i}\Delta kz} \tag{C3.2.5a}$$

$$\frac{\partial E_2(z)}{\partial z} = \mathrm{i}\kappa_2 E_3(z) E_1^*(z) \mathrm{e}^{\mathrm{i}\Delta kz} \tag{C3.2.5b}$$

$$\frac{\partial E_3(z)}{\partial z} = \mathrm{i}\kappa_3 E_1(z) E_2(z) \mathrm{e}^{-\mathrm{i}\Delta kz} \tag{C3.2.5c}$$

where $\kappa_j = (\omega_j d_{\mathrm{eff}}/n_j c)$, with $j = 1, 2, 3$, n is the refractive index, $\Delta k = k_3 - k_2 - k_1$ is the *phase-mismatch* parameter, and $d_{\mathrm{eff}} = \chi^{(2)}_{\mathrm{eff}}/2$ is the *effective nonlinear coefficient* representing the appropriate combination of the nonlinear tensor elements taking part in the parametric process. These are the *coupled-wave equations* governing the parametric interaction of the pump, signal, and idler ($\omega_3 \rightarrow \omega_2 + \omega_1$) in a dielectric medium exhibiting second-order nonlinear susceptibility.

The coupled-wave equations are the starting point in the analysis of a wide range of nonlinear optical effects and apply universally to any three-wave mixing process involving the second-order susceptibility, where the frequencies of the fields satisfy $\omega_3 = \omega_2 + \omega_1$. Similar equations to (C3.2.5a)–(C3.2.5c) apply to other mixing processes including SFM ($\omega_1 + \omega_2 \rightarrow \omega_3$), DFM ($\omega_3 - \omega_2 \rightarrow \omega_1$) and SHG ($\omega + \omega \rightarrow 2\omega$) [8]. We notice from (C3.2.5a)–(C3.2.5c) that the amplitudes of the three optical fields are coupled to one another through d_{eff}. Physically, this coupling provides the mechanism for the exchange of energy among the interacting fields as they propagate through the nonlinear material. The direction of energy flow in a given three-wave mixing process depends on the relative phase and the intensity of the input fields.

C3.2.2.2 Optical parametric gain

In practice, the parametric process is initiated by a single intense pump field at frequency ω_3 at the input to a nonlinear crystal. This field, which is provided by a laser, in turn mixes, through the nonlinear susceptibility, with a signal field at ω_2 to give rise to an idler field at $\omega_1 = \omega_3 - \omega_2$. The idler field so generated in turn mixes back with the pump to produce additional signal and the re-generated signal re-mixes with the pump to produce more idler. Under suitable phase-matching conditions (see section C3.2.3), the process can continue in this way until power is gradually transferred from the strong pump to the initially weak signal and idler fields. The generated signal and idler fields can therefore grow to macroscopic levels by draining power from the input pump field as they propagate through the nonlinear crystal.

In the absence of a coherent source of signal and idler beams at the input, the initial supply of photons at ω_1 and ω_2 for mixing with the input pump is provided by the spontaneous break-up of pump photons through *spontaneous parametric fluorescence*. This process, also referred to as *parametric noise* or *parametric luminescence*, may be viewed to arise from the mixing of the zero-point flux of the electromagnetic field at the signal and idler frequency, quantized within the volume of the crystal, with the incoming pump photons, through the nonlinear polarization [12, 13]. The effective zero-point flux at the signal and idler is obtained by allowing one half-photon of energy at both ω_2 and ω_1 or one photon of energy at either frequency to be present in each black-body mode of the quantizing volume.

We can obtain the gain and amplification factor for the growth of the signal and idler fields in the parametric process from the solution of the coupled wave equations (C3.2.5a)–(C3.2.5c). The general solution is beyond the scope of the present discussion and can be found elsewhere [3]. However, if it is assumed that the input pump field does not undergo strong depletion with propagation through the medium, then $\partial E_3/\partial z = 0$ in (C3.2.5c). The coupled-wave equations are reduced to two, with E_3 independent of z in both (C3.2.5a) and (C3.2.5b). Subject to the initial condition of no input idler field, $E_1(z = 0) = 0$, and finite input signal, $E_2(z = 0) \neq 0$, the fractional gain in signal intensity with propagation through the nonlinear crystal is obtained as

$$G_2(\ell) = \frac{I_2(z=\ell)}{I_2(z=0)} = 1 + \Gamma^2\ell^2 \frac{\sinh^2[\Gamma^2\ell^2 - (\Delta k\ell/2)^2]^{1/2}}{[\Gamma^2\ell^2 - (\Delta k\ell/2)^2]} \qquad \text{(C3.2.6)}$$

where ℓ is the interaction length, $I = nc\varepsilon_0 EE^*/2$ is the intensity or flux (in W m^{-2}) and Γ is the *gain factor* defined as

$$\Gamma^2 = \frac{8\pi^2 d_{\mathrm{eff}}^2}{c\varepsilon_0 n_1 n_2 n_3 \lambda_1 \lambda_2} I_3(z=0). \qquad \text{(C3.2.7)}$$

Here, n and λ are the refractive index and wavelength of the respective waves, $I_3(z = 0)$ is the input pump intensity and d_{eff} has units of m V^{-1}. From the same analysis, an expression similar to (C3.2.6) may be derived for the growth of the idler field from its initial zero value at the input to the nonlinear crystal [7]. The case of non-zero input idler as well as signal can also be treated similarly [14] and results analogous to (C3.2.6) can be derived for the amplification of the generated fields. It is sometimes useful to express the gain factor in the form

$$\Gamma^2 = \frac{8\pi^2 d_{\mathrm{eff}}^2}{c\varepsilon_0 n_0^2 n_3 \lambda_0^2}(1-\delta^2) I_3(z=0) \tag{C3.2.8}$$

where δ is the *degeneracy factor* defined as

$$1+\delta = \frac{\lambda_0}{\lambda_2} \qquad 1-\delta = \frac{\lambda_0}{\lambda_1} \qquad 0 \le \delta \le 1 \tag{C3.2.9}$$

where $\lambda_0 (= 2\lambda_3)$ is the degenerate wavelength and n_0 is the refractive index at degeneracy, with $n_0 \sim n_1 \sim n_2$. The factor δ is a measure of how close the signal and wavelengths are to degeneracy. It is clear from (C3.2.8) that parametric gain has a maximum value at degeneracy, where $\delta \sim 0$, and decreases for operation away from degeneracy as $\delta \to 1$.

C3.2.2.3 Optical parametric amplification

It can be seen from (C3.2.6) and (C3.2.7) that the magnitude of nonlinear gain in the parametric process depends on material parameters such as the refractive index, interaction length and nonlinear coefficient, the signal and idler wavelengths, as well as the input pump intensity. It is also seen that amplification is strongly dependent on the phase-mismatch parameter, Δk. For $\Delta k = 0$, the generated fields experience maximum gain, whereas the growth of the parametric waves is severely hampered by an increase in the magnitude of Δk. In practice, therefore, it is imperative to ensure maximum gain and amplification for the parametric waves by setting $\Delta k = 0$. This is achieved through the technique of phase-matching, as discussed in more detail in section C3.2.3. Under this condition, $\Delta k \sim 0$ and (C3.2.6) is reduced to

$$G_2(\ell) = \cosh^2(\Gamma\ell) \tag{C3.2.10}$$

Parametric devices often operate under different gain conditions, depending on the magnitude of the gain factor, Γ. In the low-gain regime, corresponding to $\Gamma\ell \lesssim 1$, (C3.2.6) can be simplified to

$$G_2(\ell) \approx 1 + \Gamma^2\ell^2 \tag{C3.2.11}$$

On the other hand, in the high-gain regime, corresponding to $\Gamma\ell \gg 1$, (C3.2.6) can be approximated by

$$G_2(\ell) \approx \tfrac{1}{4}\mathrm{e}^{2\Gamma\ell} \tag{C3.2.12}$$

It can be seen from (C3.2.11) and (C3.2.12) that, under the phase-match condition, the single-pass intensity gain has a quadratic dependence on Γl in the low-gain limit, whereas it increases exponentially with $2\Gamma l$ in the high-gain limit. Experimentally, the low-gain limit corresponds to parametric generation when using continuous-wave (cw) or low- to moderate-peak-power pulsed pump sources. This regime is pertinent to most parametric devices. On the other hand, the high-gain limit corresponds to pumping with high-intensity, pulsed and amplified laser pump sources.

C3.2.3 Phase-matching

From the preceding discussion, it is clear that in the presence of a strong input pump the signal (and idler) field in the parametric process can experience growth with propagation through the crystal, provided the phase-matching condition, $\Delta k = 0$, is satisfied (chapter C3.1). However, because of normal dispersion,

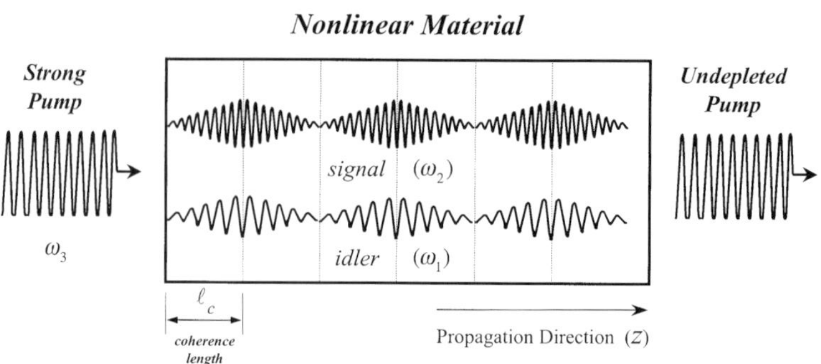

Figure C3.2.4. Wave velocity mismatch and coherence length. The signal and idler wave amplitudes undergo oscillations in propagating through the material, without experiencing any growth. The pump amplitude is unaffected by the generation and propagation of the parametric waves. The amplitudes of the generated waves are grossly exaggerated relative to that of the pump.

optical fields at different frequencies cannot generally travel with the same phase velocity and maintain synchronism in propagating through the material due to different refractive indices. The degree of phase velocity synchronism among the pump, signal and idler is determined by the phase-mismatch parameter $\Delta k = k_3 - k_2 - k_1$. Under normal conditions, $\Delta k \neq 0$, so that the optical waves at different frequencies slip out of phase after travelling a short distance through the medium. This distance is known as the *coherence length*, given by $\ell_c = \pi/|\Delta k|$, and is illustrated in figure C3.2.4. With propagation beyond a coherence length the intensities of the generated waves begin to fall back to zero and with further propagation the waves step in and out of phase periodically, with a period determined by the coherence length. As a result, the optical waves cannot maintain any phase unison beyond a coherence length. The pump and the parametric fields exchange energy back and forth, and no net transfer of energy from the pump to the generated waves can occur in travelling through the medium. The coherence length is thus a measure of the maximum interaction length over which amplification of the parametric fields can be sustained in the presence of dispersion. It is the distance after which the relative phase of the pump, signal and idler waves slips by π and the generated intensities begin to fall back to zero. Typical values of ℓ_c are a few to tens of microns in most nonlinear materials. Given the direct dependence of $G_2(\ell)$ on the interaction length, ℓ, using (C3.2.6), it is clear that over negligible distances corresponding to a coherence length only trivial gains are available.

The general functional dependence of $G_2(\ell)$ on Δk may be derived from (C3.2.6) and can be found in a number of reference texts [7, 8, 14, 15]. In most situations of practical interest, however, we are concerned with small gains, for which $\Gamma^2\ell^2 \ll (\Delta k\ell/2)^2$. Under this condition, the fractional single-pass signal gain, given by (C3.2.6), is modified to

$$G_2(\ell) \approx 1 + \Gamma^2\ell^2 \left[\frac{\sin(\Delta k\ell/2)}{(\Delta k\ell/2)}\right]^2 \qquad \text{(C3.2.13)}$$

which has a sinc^2 dependence on the $\Delta k\ell/2$, as shown in figure C3.2.5. As expected, the gain has a maximum value at $\Delta k = 0$ and is reduced with increasing Δk, dropping to zero when $\Delta k = 2\pi/\ell$.

Clearly, for any meaningful growth of the parametric waves, phase velocity synchronism must be maintained over interaction lengths significantly longer than the coherence length. In fact, in order to exploit the full length of the nonlinear crystal, the necessary length for velocity synchronism must be comparable or longer than the crystal length. This requirement can be met by exploiting the technique of phase-matching, which permits the condition $\Delta k = 0$ to be satisfied, hence allowing the coherence length to become infinite,

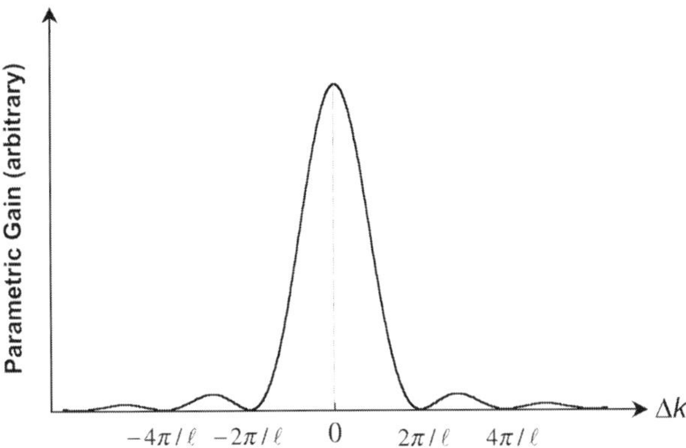

Figure C3.2.5. The dependence of parametric gain on the phase-mismatch parameter $\Delta k\ell/2$ in the small gain limit, $\Gamma^2\ell^2 \ll (\Delta k\ell/2)^2$.

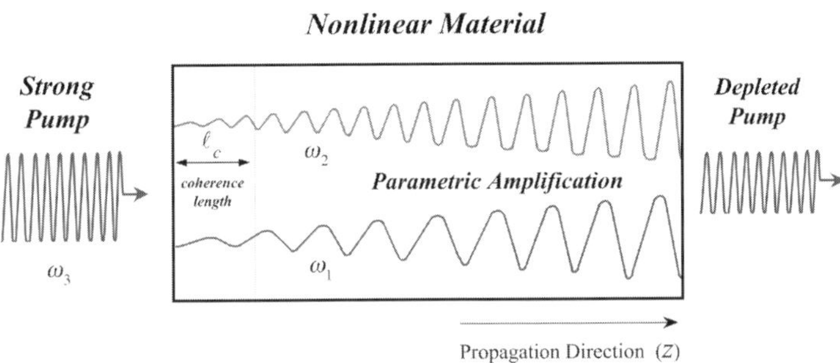

Figure C3.2.6. Optical parametric amplification. Phase-matching provides the velocity synchronism for a particular pair of signal and idler waves, out of the infinite number of waves generated, resulting in amplification. Power transfer occurs from the pump to the parametric waves, resulting in a monotonic increase in the signal and idler intensity and a drop in the transmitted pump intensity with propagation through the material. The amplitudes of the generated waves are grossly exaggerated relative to that of the pump.

and phase velocity synchronism between the interacting fields to be maintained indefinitely. Under this condition, the signal and idler fields undergo amplification over the full interaction length of crystal by continually draining power from the pump, as shown in figure C3.2.6, and coherent macroscopic output can be extracted at ω_2 and ω_1.

The phase-match condition, $\Delta k = 0$, may also be usefully expressed as $n_3\omega_3 - n_2\omega_2 - n_2\omega_2 = 0$, which highlights one of the most attractive features of the parametric process—its tunability. If for a given pump frequency, ω_3 the refractive indices n_3, n_2 and n_1 are made to vary through, for example, the change in crystal angle or temperature, the phase-match condition will be satisfied for a new pair of signal and idler frequencies, ω_2 and ω_1. As a consequence, the generated frequencies will be 'forced' to shift to new values for which phase-matched amplification is available. In addition to *angle-* and *temperature-tuning*, there are also other tuning methods including *pump-tuning* and, in the case of *quasi-phase-matched* interactions, *grating-tuning* (see section C3.2.3.2).

C3.2.3.1 Birefringent phase-matching

The traditional method for achieving phase-matching is to use the birefringence of optically anisotropic media to compensate for dispersion [16, 17]. Most non-centrosymmetric crystals in which second-order nonlinear processes are attainable are optically anisotropic and exhibit the phenomenon of birefringence. In such media, the index of refraction (and hence the phase velocity) for a wave at a given frequency depends on its state of polarization as well as the direction of propagation. It can be shown that for an arbitrary propagation direction in a birefringent crystal, two orthogonal linear polarization states are permitted [18]. This means that an optical wave at given frequency can exhibit two different phase velocities depending on its state of polarization. Alternatively, two optical waves at different frequencies may propagate with the same phase velocity in a birefringent crystal by suitable choice of orthogonal polarization vectors and propagation direction. It is this important property that can be used to obtain phase velocity synchronism between optical waves at different frequencies in nonlinear processes, including parametric generation. The technique is known as *birefringent phase-matching* (BPM).

Detailed discussion of crystal optics and the general procedures for determining the allowed polarization states and phase-matching directions in BPM are not within the scope of this treatment and can be found in other reference texts [6, 8, 19] (see also chapters A5 and C3.1). Here, we focus on the essential elements of the approach and provide a brief summary of procedures involved for the attainment of BPM.

The orientation and phase velocities of the two allowed linear polarization states in a birefringent crystal can be obtained from the so-called *normal index surface*, which uniquely describes the optical properties of the particular crystal. This surface is defined in such a way that the index of refraction (or the phase velocity) of a wave propagating in a given direction is equal to the distance between the intersection of the wave normal with the surface in that direction and the origin. The principal axes of the surface are defined by the principal refractive indices n_x, n_y, n_z, which lie along the principal dielectric axes of the crystal, ε_x, ε_y and ε_z. Because of dispersion, the length of the principal axes of the normal index surface, and thus its shape, vary with the frequency of the optical radiation.

With regard to their optical properties, birefringent crystals may be classified into two distinct categories: optically *uniaxial* crystals in which $n_x = n_y \neq n_z$; and optically *biaxial* crystals in which $n_x \neq n_y \neq n_z$. In uniaxial crystals, one may find a unique direction along which the two allowed orthogonal polarizations have the same refractive index (i.e. travel with the same phase velocity). In biaxial crystals, two such directions may be identified. These directions define the *optic axes* of the crystal. In uniaxial materials, the optic axis coincides with the z-axis of the index surface, while in biaxial materials the two optic axes generally lie symmetrically about the z-axis, in the xz-plane. In both crystal classes it is in principle possible, by an appropriate choice of polarizations and propagation direction, for the optical waves at different frequencies to achieve phase-matching with $\Delta k = 0$.

In uniaxial crystals, of the two allowed polarization waves, one experiences the same refractive index regardless of the direction of propagation in the medium. This wave is known as the *ordinary wave* (or o-wave). If θ is the angle between the wave normal direction and the z-axis of the normal index surface (figure C3.2.7), then the refractive index of the o-wave is a constant for all θ, given by $n_o(\theta) = n_o$. The wave with orthogonal polarization to the o-wave, however, experiences a refractive index that varies with the propagation direction in the crystal. This wave is referred to as the *extraordinary wave* (or e-wave). The refractive index of the e-wave, $n_e(\theta)$, varies from n_o to n_e as θ varies from 0° to 90°. The refractive indices n_o and n_e, therefore, represent the principal axes of the normal index surfaces for the uniaxial crystal. The normal index surface for the o-wave is a sphere, whereas that for the e-wave is an ellipsoid of revolution about the z-axis, as shown in figure C3.2.7.

We may distinguish between two classes of uniaxial crystals: *positive* uniaxial for which $n_e > n_o$; and *negative* uniaxial for which $n_e < n_o$. The normal index surfaces for a positive crystal are those shown in figure C3.2.7, and the surfaces for a negative crystal are shown in figure C3.2.8. In either case, it is often

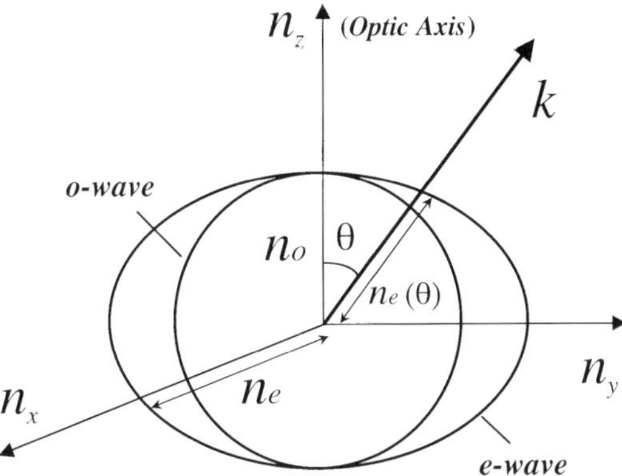

Figure C3.2.7. The normal index surfaces, representing the refractive indices for the o-wave and the e-wave in a uniaxial crystal. The surfaces correspond to a positive uniaxial crystal.

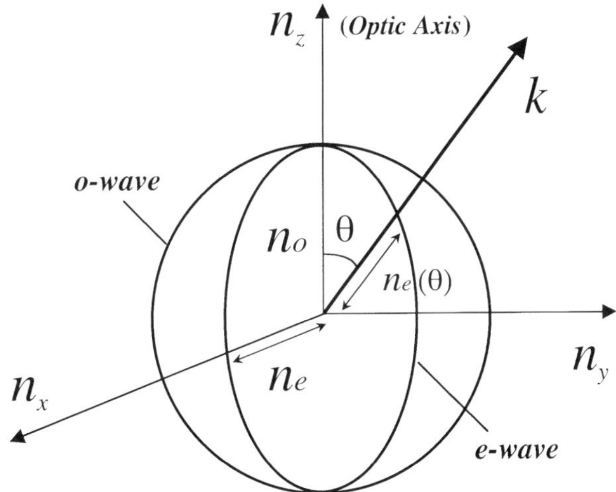

Figure C3.2.8. The normal index surfaces for a negative uniaxial crystal.

possible to find a direction along which suitably polarized pump and parametric waves have refractive indices (and phase velocities) that satisfy the phase-match condition $\Delta k = k_3 - k_2 - k_1 = 0$. In a positive crystal, two phase-matching schemes are possible, both involving the propagation of the pump as an o-wave, with either the signal (or idler) or both as e-waves. In a negative crystal, the pump is polarized as an e-wave, with either the signal (or idler) or both polarized as o-waves. The case where the signal and idler have the same polarization is referred to as *type I* (or *parallel*) phase-matching. When the signal and idler have orthogonal polarizations, it is known as *type II* (or *orthogonal*) phase-matching. Table C3.2.1 summarizes the possible phase-matching schemes for parametric generation in uniaxial crystals.

We may also identify two classes of biaxial crystal depending on whether $n_x < n_y < n_z$ or $n_x > n_y > n_z$. In both cases, the normal index surface for the allowed polarizations does not follow a simple spherical or spheroidal symmetry as in a uniaxial crystal, but has a more complex bilayer structure with four points of interlayer contact through which the two optic axes pass [6]. The determination of the phase-match condition

Table C3.2.1. Possible birefringent phase-matching schemes and corresponding field polarizations for parametric generation ($\omega_3 \rightarrow \omega_2 + \omega_1$) in uniaxial crystals. $\boldsymbol{E}$ represents the electric field vector of the optical waves. The subscripts o and e refer to *ordinary* and *extraordinary* polarization, respectively.

	$\omega_3 \rightarrow \omega_2 + \omega_1$			
Crystal class	Positive uniaxial		Negative uniaxial	
Phase-matching scheme	Type I	Type II	Type I	Type II
Field polarizations	$\boldsymbol{E}_o \rightarrow \boldsymbol{E}_e + \boldsymbol{E}_e$	$\boldsymbol{E}_o \rightarrow \boldsymbol{E}_o + \boldsymbol{E}_e$ $\boldsymbol{E}_o \rightarrow \boldsymbol{E}_e + \boldsymbol{E}_o$	$\boldsymbol{E}_e \rightarrow \boldsymbol{E}_o + \boldsymbol{E}_o$	$\boldsymbol{E}_e \rightarrow \boldsymbol{E}_o + \boldsymbol{E}_e$ $\boldsymbol{E}_e \rightarrow \boldsymbol{E}_e + \boldsymbol{E}_o$

and allowed polarizations for an arbitrary propagation direction (θ, ϕ) is thus considerably more complex, since the refractive indices of not only one, but both allowed polarizations in general vary with propagation direction. Therefore, one cannot strictly consider the allowed polarizations as an o-wave or an e-wave, as in uniaxial crystals. The situation may, however, be greatly simplified if propagation is confined to one of the principal planes xy, xz or yz. In this case, the refractive index of one of the allowed waves becomes independent of the direction of propagation, so that one may think of this wave as an o-wave in analogy with the uniaxial case. The wave with orthogonal polarization to the o-wave has a refractive index that varies with the propagation direction in the phase-match plane and may thus be treated as an e-wave. The situation thus simplifies to that in a uniaxial crystal.

As an example, if we consider propagation in the xy-plane ($\theta = 90°$), for a wave polarized along the z-axis the refractive index does not vary with propagation angle ϕ. This wave can, therefore, be considered as an o-wave with a refractive index $n_1(\phi) = n_z \equiv n_o$. However, the refractive index of a wave with an orthogonal polarization vector in the xy-plane varies with the phase-match angle ϕ. This wave can then be treated as an e-wave with a refractive index $n_2(\phi) = n_{xy} \equiv n_e(\phi)$, say, where n_{xy} varies from n_y to n_x as ϕ varies from 0° to 90°. It thus becomes possible to obtain phase matching in the xy-plane in the same way as in a uniaxial crystal using similar type I and type II schemes. Phase-matching in the other principal planes xz and yz can also be treated in a similar way. Whether the phase-matching scheme in a particular plane corresponds to a positive or negative uniaxial crystal depends on relative indices of the biaxial crystal. The possible phase-matching schemes for parametric generation in the principal planes of the two classes of biaxial crystals with $n_x < n_y < n_z$ and $n_x > n_y > n_z$ are summarized in tables C3.2.2(*a*) and C3.2.2(*b*), respectively. More general treatments of BPM and calculation of phase-matching directions in biaxial crystals can be found elsewhere [6, 20].

It is therefore clear that by judicial choice of wave polarizations and propagation direction in a particular crystal, it is possible to obtain phase-velocity synchronism in the parametric generation process using the BPM and to achieve amplification of the generated waves.

C3.2.3.2 Quasi-phase-matching

An alternative technique for the attainment of phase velocity synchronism in nonlinear processes is *quasi-phase-matching* (QPM) [21, 22]. The technique was first proposed as early as four decades ago when the original ideas on nonlinear frequency conversion were first formulated [3]. However, its practical implementation did not become possible until the recent development of reliable fabrication methods in ferroelectric materials [22].

In the QPM process, the orientation of the electric dipoles in the nonlinear material is periodically reversed by 180° along the pump propagation direction (figure C3.2.9). A practical method to achieve

Table C3.2.2. (*a*) Possible birefringent phase-matching schemes and corresponding field polarizations for parametric generation ($\omega_3 \rightarrow \omega_2 + \omega_1$) in the three principal planes of biaxial crystals with $n_x < n_y < n_z$. The angles θ and ϕ are the polar and azimuthal angles, measured from the principal optical axes z and x, respectively. The angle θ_0 is the angle between the optic axis and the z-axis, measured in the xz-plane [6]. $\boldsymbol{E}$ represents the electric field vector of the optical waves. The subscripts represent the polarization of the electric vector, with x, y, z representing polarization along the respective axes and xy, xz and yz designating polarization in the respective planes.

Crystal class	$\omega_3 \rightarrow \omega_2 + \omega_1$ Biaxial ($n_x < n_y < n_z$)			
	xy-plane ($\theta = 90$; $0 < \phi < 90$) (Negative uniaxial)		yz-plane ($\phi = 90$; $0 < \theta < 90$) (Positive uniaxial)	
Phase-matching scheme	Type I	Type II	Type I	Type II
Field polarizations	$\boldsymbol{E}_{xy} \rightarrow \boldsymbol{E}_z + \boldsymbol{E}_z$	$\boldsymbol{E}_{xy} \rightarrow \boldsymbol{E}_z + \boldsymbol{E}_{xy}$ $\boldsymbol{E}_{xy} \rightarrow \boldsymbol{E}_{xy} + \boldsymbol{E}_z$	$\boldsymbol{E}_x \rightarrow \boldsymbol{E}_{yz} + \boldsymbol{E}_{yz}$	$\boldsymbol{E}_x \rightarrow \boldsymbol{E}_x + \boldsymbol{E}_{yz}$ $\boldsymbol{E}_x \rightarrow \boldsymbol{E}_{yz} + \boldsymbol{E}_x$
	xz-plane			
	($\phi = 0$; $0 < \theta < \theta_0$) (Negative uniaxial)		($\phi = 0$; $\theta_0 < \theta < 90$) (Positive uniaxial)	
Phase-matching scheme	Type I	Type II	Type I	Type II
Field polarizations	$\boldsymbol{E}_{xz} \rightarrow \boldsymbol{E}_y + \boldsymbol{E}_y$	$\boldsymbol{E}_{xz} \rightarrow \boldsymbol{E}_y + \boldsymbol{E}_{xz}$ $\boldsymbol{E}_{xz} \rightarrow \boldsymbol{E}_{xz} + \boldsymbol{E}_y$	$\boldsymbol{E}_y \rightarrow \boldsymbol{E}_{xz} + \boldsymbol{E}_{xz}$	$\boldsymbol{E}_y \rightarrow \boldsymbol{E}_y + \boldsymbol{E}_{xz}$ $\boldsymbol{E}_y \rightarrow \boldsymbol{E}_{xz} + \boldsymbol{E}_y$

such domain reversal is through *periodic poling* of ferroelectric materials during the fabrication process by applying a high electrical field (several kV) across the crystal using patterned electrodes. The result of periodic poling is that the radiated optical waves from the oscillating dipoles in consecutive domains become out of phase by π in both time and space. The domain reversal is equivalent to a periodic change in the sign of the effective nonlinear coefficient between $+d_{\mathrm{eff}}$ and $-d_{\mathrm{eff}}$. If the poling period, Λ, is made to correspond to two coherence lengths for the nonlinear interaction ($\Lambda = 2\ell_{\mathrm{c}}$), then the phase of the generated waves is periodically reversed by π every two coherence lengths. This periodic re-adjustment of phase preserves a constructive relative phase between the pump and parametric waves (albeit in a quasi-continuous manner) as they propagate through the material. This prevents the waves from slipping out of phase after only one coherence length, as would be the case under normal dispersion.

The end result of periodic poling is that the generated waves can undergo quasi-continuous growth as they propagate with the pump through the material, as illustrated in figure C3.2.10. A qualitative comparison of the generated intensity between QPM and BPM, as well as the non-phase-matched process with propagation through the crystal is shown in figure C3.2.11. It can be seen that, in the QPM process, the growth of optical waves is not monotonic, as it is in BPM. However, because QPM does not rely on a prescribed combination of wave polarizations for phase-matching, it can provide access to the highest nonlinear tensor coefficients in the material, which are not generally available under BPM. This flexibility more than compensates for the quasi-monotonic amplification of the generated waves, so that the QPM

Table C3.2.2. (*b*) Possible birefringent phase-matching schemes and corresponding field polarizations for parametric generation ($\omega_3 \rightarrow \omega_2 + \omega_1$) in the three principal planes of biaxial crystals with $n_x > n_y > n_z$. The angles θ and ϕ are the polar and azimuthal angles, measured from the principal optical axes z and x, respectively. The angle θ_0 is the angle between the optic axis and the z-axis, measured in the xz-plane [6]. $\boldsymbol{E}$ represents the electric field vector of the optical waves. The subscripts represent the polarization of the electric vector, with x, y, z representing polarization along the respective axes and xy, xz and yz designating polarization in the respective planes.

Crystal class	$\omega_3 \rightarrow \omega_2 + \omega_1$ Biaxial ($n_x > n_y > n_z$)			
	xy-plane ($\theta = 90$; $0 < \phi < 90$) (Positive uniaxial)		yz-plane ($\phi = 90$; $0 < \theta < 90$) (Negative uniaxial)	
Phase-matching scheme	Type I	Type II	Type I	Type II
Field polarizations	$\boldsymbol{E}_z \rightarrow \boldsymbol{E}_{xy} + \boldsymbol{E}_{xy}$	$\boldsymbol{E}_z \rightarrow \boldsymbol{E}_z + \boldsymbol{E}_{xy}$ $\boldsymbol{E}_z \rightarrow \boldsymbol{E}_{xy} + \boldsymbol{E}_z$	$\boldsymbol{E}_{yz} \rightarrow \boldsymbol{E}_x + \boldsymbol{E}_x$	$\boldsymbol{E}_{yz} \rightarrow \boldsymbol{E}_x + \boldsymbol{E}_{yz}$ $\boldsymbol{E}_{yz} \rightarrow \boldsymbol{E}_{yz} + \boldsymbol{E}_x$
	xz-plane			
	($\phi = 0$; $0 < \theta < \theta_0$) (Positive uniaxial)		($\phi = 0$; $\theta_0 < \theta < 90$) (Negative uniaxial)	
Phase-matching scheme	Type I	Type II	Type I	Type II
Field polarizations	$\boldsymbol{E}_y \rightarrow \boldsymbol{E}_{xz} + \boldsymbol{E}_{xz}$	$\boldsymbol{E}_y \rightarrow \boldsymbol{E}_y + \boldsymbol{E}_{xz}$ $\boldsymbol{E}_y \rightarrow \boldsymbol{E}_{xz} + \boldsymbol{E}_y$	$\boldsymbol{E}_{xz} \rightarrow \boldsymbol{E}_y + \boldsymbol{E}_y$	$\boldsymbol{E}_{xz} \rightarrow \boldsymbol{E}_y + \boldsymbol{E}_{xz}$ $\boldsymbol{E}_{xz} \rightarrow \boldsymbol{E}_{xz} + \boldsymbol{E}_y$

process can ultimately provide substantially higher gains for the same crystal length and operating conditions than BPM. In addition, the direction of propagation in QPM can also be freely chosen to provide optimized phase-matching geometries such as *noncritical phase-matching* (NCPM), where the effects of *spatial walkoff* are minimized (see section C3.2.6.1). This is again not possible in BPM, where the choice of propagation direction in the crystal is dictated by the birefringence properties of the material and the type of phase-matching permitted. The NCPM capability is particularly important in the presence of low parametric gains, since it allows for the use of long crystal interaction lengths, to maximize device efficiency. Another important merit of QPM is that it can be implemented in optically isotropic as well as anisotropic materials, as it does not rely on birefringence for phase-matching. With regard to tuning capability, QPM also has an overriding advantage over BPM in that it can be freely engineered to provide maximum spectral coverage for the generated waves. By fabricating the correct poling (or *grating*) period, it is in principle possible to generate any desired wavelength within the entire transparency range of the material.

The phase-matching condition under QPM can be expressed in a similar way to BPM, but is modified by the periodic phase re-adjustments introduced by the poling process. As a result, the phase-mismatch parameter under QPM is given by $\Delta k_Q = k_3 - k_2 - k_1 - k_Q$, where $k_Q = 2\pi/\Lambda$ is the so-called *grating vector* corresponding to the periodic grating. The effective nonlinear coefficient in QPM is also similarly modified to take account of the periodic phase reversal of the optical waves. The form of the coefficient depends on the order of the QPM process and the poling cycle [21, 22] and is given by $d_Q = G_m d_{\text{eff}}$, where

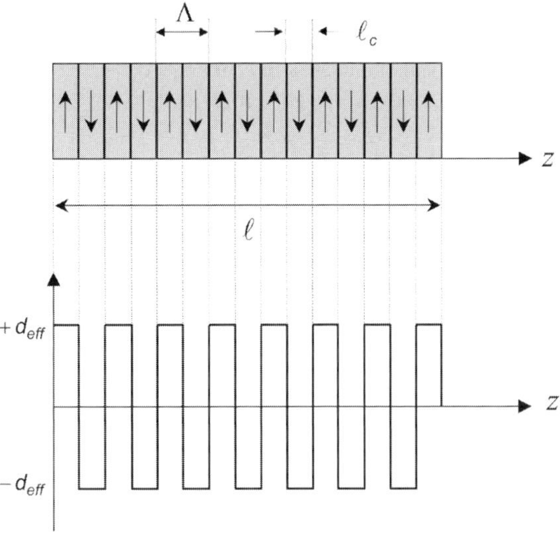

Figure C3.2.9. Quasi-phase-matching through periodic poling. The orientation of the electric dipoles is periodically flipped through the material, resulting in a corresponding modulation in the effective nonlinear coefficient.

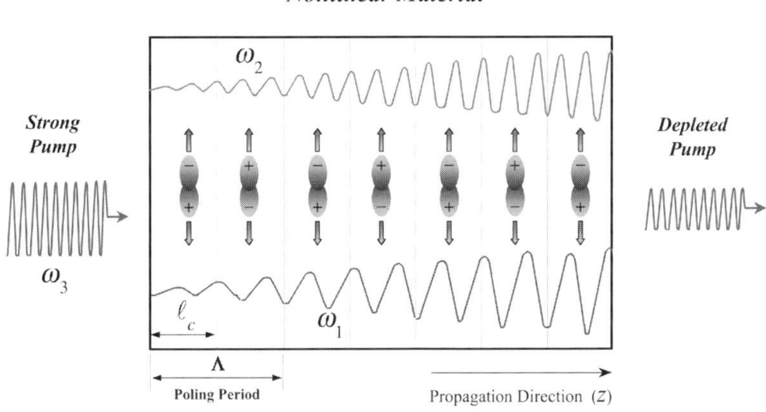

Figure C3.2.10. Optical parametric amplification under quasi-phase-matching. The generated waves undergo phase re-adjustment every coherence length, preserving a constructive relative phase, and thus experiencing quasi-continuous growth with propagation through the material. Power transfer occurs from the pump to the parametric waves, resulting in the depletion of the transmitted pump. The amplitudes of the generated waves are grossly exaggerated relative to that of the pump.

d_{eff} would be the effective coefficient for the same wave polarizations and propagation direction in the absence of the periodic grating. The factor G_m is given by

$$G_m = \frac{2}{m\pi} \sin(m\pi D) \tag{C3.2.14}$$

where m is the order of QPM and $D = l/\Lambda$ is the poling duty factor, with l being the length of the reversed

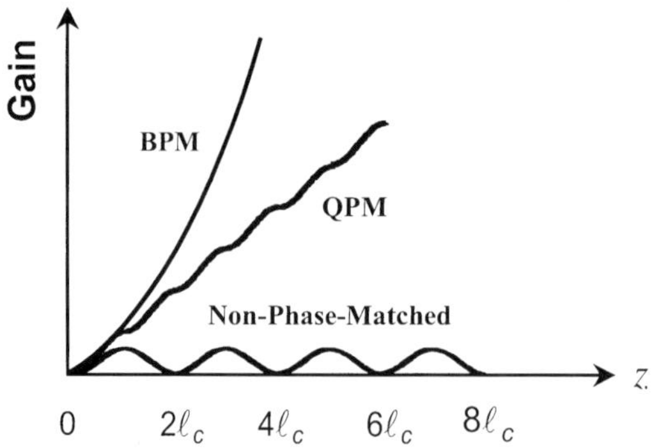

Figure C3.2.11. Qualitative comparison of gain under birefringent phase-matching (BPM), quasi-phase-matching (QPM), and non-phase-matched interaction with propagation through the nonlinear crystal. The plot assumes the same effective nonlinear coefficient for QPM as for BPM. The vertical scale, hence the relative gain, is arbitrary. The intensity of the non-phase-matched interaction is grossly exaggerated relative to that of BPM and QPM.

domain. The QPM order, m, represents the periodicity of the poling in terms of coherence length. For an mth-order process, the poling period is given by $\Lambda = 2m\ell_c$, so that for a first-order grating ($m = 1$) the domains are reversed every coherence length, for a second-order grating ($m = 2$) every two coherence lengths, and so on. The existing analyses of parametric amplification and gain used in section C3.2.2 equally apply to QPM processes. For a given QPM order, grating period and poling cycle, the appropriate expressions for Δk_Q and d_Q may be substituted into equations (C3.2.7) and (C3.2.6) to obtain the gain factor and growth in signal intensity.

For a first-order QPM process ($\Lambda = 2\ell_c$) and a 50% poling duty cycle, the effective nonlinear coefficient is maximized to $d_Q = (2/\pi)d_{\text{eff}}$. This is therefore the configuration of most practical interest, as it minimizes device threshold and maximizes output power and efficiency. Higher-order processes and poling duty factors other than 50% are of interest for novel applications including spectral and spatial tailoring, temporal pulse shaping, expansion and compression, tailoring of phase-matching line-shape and bandwidth modification in frequency conversion processes [21–25]. The unique flexibility and important advantages of QPM have now established this technique as a highly effective approach to the development of a new class of frequency conversion sources, particularly parametric devices [26–28], with exceptional versatility and unprecedented performance capabilities with regard to wavelength coverage, output power and efficiency and spectral and temporal characteristics (see sections C3.2.7–C3.2.9).

C3.2.4 Optical parametric devices

From the preceding discussion, it may be concluded that on the attainment of the phase-match condition ($\Delta k = 0$) in the parametric process, the generated waves will experience optical gain and amplification, and coherent macroscopic output can be obtained at the signal and idler wavelengths. To obtain practical estimates of the magnitude of parametric gain under different pumping conditions, we can consider the main operating regimes of parametric devices. These are summarized in table C3.2.3, where the gain factor, Γ, and the fractional single-pass intensity gain, $G_2(\ell)$, have been calculated under the phase-matched condition ($\Delta k = 0$) for different pumping regimes, using equations (C3.2.8) and (C3.2.10). We have taken typical values for pump pulse energy, duration, power, focused beam waist radius (w_3) and crystal length (ℓ) appropriate to each mode of operation in practice. We have also assumed a typical nonlinear coefficient of

Table C3.2.3. The parametric gain factor and the fractional single-pass intensity gain for different pumping regimes, calculated from (C3.2.8) and (C3.2.10). The calculations are based on typical experimental values for pump laser and nonlinear material parameters for each operating regime and assume phase-matched interaction ($\Delta k = 0$) and near-degenerate operation at $\sim 2\ \mu$m.

		Pulsed		
	CW	Q-switched	Mode-locked	Mode locked amplified
Pump pulse energy	—	10 mJ	15 nJ	10 μJ
Pump pulse duration	—	10 ns	100 fs	200 fs
Peak pump power	5 W	1 MW	150 kW	50 MW
Focused waist radius (w_3)	20 μm	1 mm	15 μm	15 μm
Peak intensity (I_3)	400 kW cm^{-2}	30 MW cm^{-2}	20 GW cm^{-2}	7 TW cm^{-2}
Crystal length (ℓ)	10 mm	10 mm	1 mm	1 mm
$\Gamma\ell$	0.09	0.77	1.99	37
$G_2(\ell)$	1.008	1.72	13.88	3.4×10^{31}

$d_{\text{eff}} \sim 3$ pm V^{-1} and have considered degenerate operation at 2 μm ($\lambda_2 \sim \lambda_1 \sim 2\ \mu$m) with $n_3 \sim n_0 \sim 1.5$, for convenience.

We can see from table C3.2.3 that the net percentage single-pass parametric gain varies from a mere 0.8% under cw pumping with a relatively powerful 5 W laser to as much as $\sim 3.4 \times 10^{31}$ when using high-energy ultrashort pump pulses. The latter would correspond to the use of high-intensity, mode-locked and amplified laser systems based on, for example, regenerative amplification and cavity-dumping schemes. In the intermediate regime of pumping with Q-switched nanosecond pulses, net optical gains of $\sim$70% are available, whereas with ultrashort pulses of relatively low energy, corresponding to typical cw mode-locked lasers (e.g. Ti:sapphire), single-pass intensity gains of the order of $\sim$1300% are expected. Given that the initial signal intensity at the crystal input is provided by spontaneous parametric noise, it is clear that any meaningful amplification to macroscopic levels over a single pass of the crystal is not practicable, except in the high-energy ultrashort pump pulse regime where large exponential gains are available. This is the configuration corresponding to *optical parametric generator* (OPG) and *amplifier* (OPA) devices. In all other operating regimes, discernible output can only be made available by enclosing the nonlinear material within an optical resonator to provide feedback at the generated parametric wave(s). This is the configuration corresponding to OPOs, which is the most common architecture for parametric devices. In both architectures, there is an operation threshold associated with the device, which in practice would correspond to the detection and extraction of coherent output at the signal and idler wavelengths. In this treatment, we focus mainly on a description of OPO devices and their operating characteristics. While many of the design criteria and operating principles are equally applicable to OPG and OPAs, more extensive discussion of these single-pass devices can be found elsewhere in the literature [29].

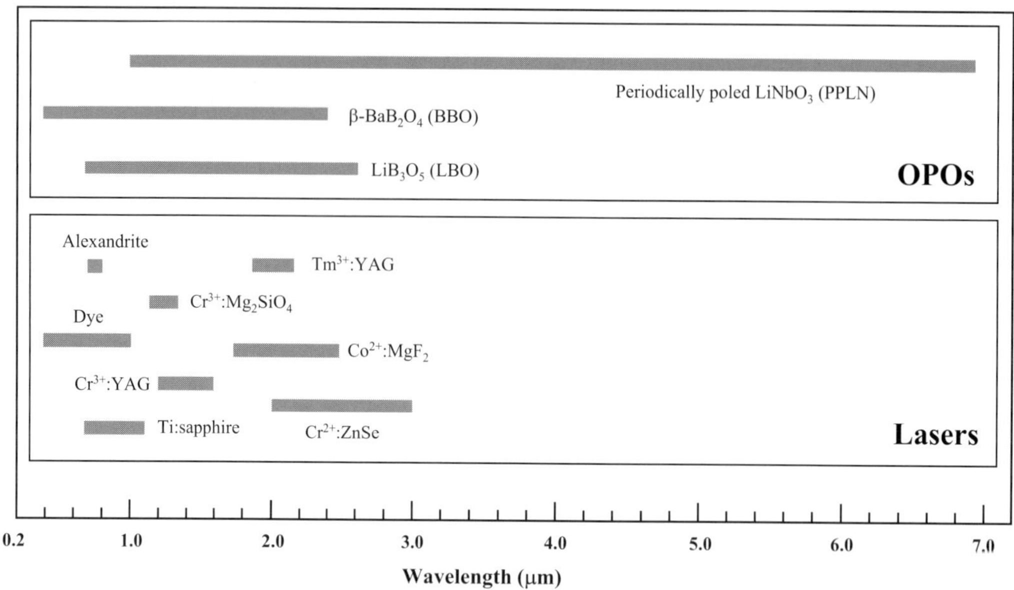

Figure C3.2.12. Comparison of the spectral coverage of prominent conventional tunable lasers (in figure C3.2.1) with that of a number of OPO devices demonstrated to date. The vertical scale has no significance.

C3.2.5 Optical parametric oscillators

Of the different types of parametric devices, the large majority conform to the low- to moderate-gain operating limit. This is the regime most frequently encountered in practice when commonly available laser pump sources and nonlinear materials are deployed. From the previous discussion, it is seen that the small-signal gains available in such devices are of the order of 0.1–1000% for typical nonlinear materials and pumping intensities. Such gains are clearly insufficient to achieve macroscopic amplification of parametric waves from noise in a single pass through the nonlinear crystal. Therefore, parametric devices of this type are operated in OPO configuration, as in a conventional laser, by enclosing the nonlinear gain medium within an optical cavity to provide feedback at the generated waves. In this way, the amplification of parametric waves is achieved to macroscopic levels by successive transits through the nonlinear crystal and coherent output can be extracted from the oscillator.

The potential of OPOs derives from their exceptional wavelength flexibility, which allows access to spectral regions unavailable to conventional lasers. The OPO can readily provide widely tunable radiation across substantial portions of the optical spectrum by suitable choice of nonlinear material and laser pump source. Figure C3.2.12 shows the wavelength tuning range of a number of OPO devices developed to date. The spectral range available to several conventional tunable lasers (as in figure C3.2.1) is also shown for comparison. It is seen that many wavelength regions unavailable to lasers are readily accessible to OPOs. Moreover, spectral regions far more extensive than any tunable laser can be accessed by a single device using one nonlinear crystal. In addition to its spectral versatility, the OPO can be configured as a highly compact device, has a simple tuning mechanism and offers high efficiencies in converting the input pump energy into useful output. It also has a practical solid-state design unlike, for example, tunable dye lasers employing liquids as the gain medium. Another important characteristic of OPOs is their temporal flexibility, which allows these devices to operate across all temporal regimes. This property is a consequence of the instantaneous nature of electronic polarization, the origin of nonlinear gain. In contrast to conventional lasers,

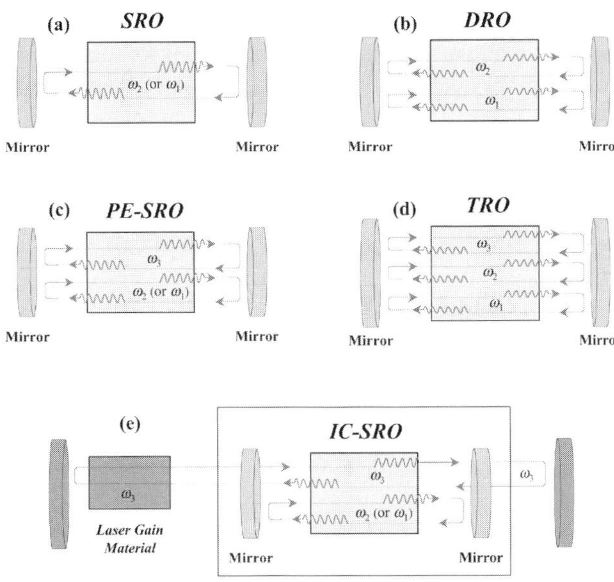

Figure C3.2.13. The principal cavity configurations for OPOs.

where the generation of the shortest optical pulses is limited by the upper state lifetime of the laser transition, OPOs can provide output in all temporal regimes, from the cw to ultrafast femtosecond time scales, by the suitable choice of laser pump source.

As in a conventional laser oscillator, the OPO is characterized by a *threshold* condition, defined by the pumping intensity at which the growth of the parametric waves in one complete round-trip of the optical cavity just balances the total loss in that round-trip. However, in contrast to the laser, amplification of the generated waves in an OPO occurs only along the pump beam, so that optical gain is generally available only in one direction, unless the pump beam is returned to the crystal by back-reflection (see section C3.2.5.1). The gain is thus single-pass, whereas the losses occur in both passes of the nonlinear crystal. Once threshold has been surpassed, coherent light at macroscopic levels can be extracted from the oscillator.

In order to provide feedback in an OPO, a variety of resonance configurations may be deployed by the suitable choice of mirrors forming the optical cavity. The principal cavity geometries are summarized in figure C3.2.13. The mirrors may be highly reflecting at only one of the parametric waves (signal or idler, but not both), as in figure C3.2.13(a), in which case the device is known as a *singly resonant oscillator* (SRO). This configuration is characterized by the highest operation threshold. In order to reduce threshold, alternative resonator schemes may be employed where additional optical waves are resonated in the optical cavity. These include the *doubly resonant oscillator* (DRO), figure C3.2.13(b), in which both the signal and idler waves are resonant in the optical cavity, and the *pump-resonant* or *pump-enhanced* (PE) SRO, figure C3.2.13(c), in which the pump as well as one of the generated waves (signal or idler) is resonant in the optical cavity. In an alternative scheme, the pump may be resonated together with both parametric waves, as in figure C3.2.13(d), in which case the device is known as a PE-DRO or a *triply resonant oscillator* (TRO). Such schemes effect substantial reductions in threshold from the SRO configuration, with the TRO offering the lowest operation threshold. On the other hand, the threshold reduction is achieved at the expense of increased spectral and power instability in the OPO output arising from the difficulty in maintaining resonance for more than one optical wave in a single optical cavity. For this reason, the SRO offers the most direct route to the attainment of high output stability and spectral control without stringent demands on the frequency stability of the laser pump source. On the other hand, multiple resonant oscillators (DRO, PE-SRO, and TRO) require active stabilization techniques to achieve output power and spectral stability, with the TRO representing the most difficult configuration in practice. In addition, the implementation of practical OPOs in multiple resonant cavities can only be achieved by using stable, single-frequency pump lasers and such devices also require more complex protocols for frequency tuning and control than the SRO [30].

An alternative technique to obtain major reductions in threshold pump power requirement in cw SROs is the use of intracavity pumping [31]. The *intracavity* SRO (IC-SRO) takes advantage of the large circulating intensities within the pump laser oscillator itself to reach operation threshold. In the scheme, shown in figure C3.2.13(*e*), the OPO is placed inside the cavity of the pump laser to allow access to high circulating pump intensities, which would otherwise not be available external to the laser. The technique has proved highly effective in bringing the operation of cw SROs within the reach of commonly available pump lasers and birefringent nonlinear materials and enabling the development of practical OPO devices with minimal requirements on pump power and frequency stability.

More detailed descriptions of the different resonance and pumping schemes for OPOs and analytical treatment of tuning mechanisms, spectral behaviour, frequency control and stabilization can be found elsewhere [30]. Here, we focus on the main results of the analysis and provide a summary of the main operating parameters for OPOs in the different resonance configurations and temporal regimes from the cw to ultrafast femtosecond time scales.

Because of the high operation threshold, successful operation of SROs in conventional external pumping configurations using commonly available nonlinear materials generally necessitates large input pumping intensities, which are often beyond the reach of many existing cw laser sources. Such high intensities can, however, be made readily available with pulsed lasers operating in nanosecond, picosecond or femtosecond regimes. As a result, OPOs operating in the cw regime generally require multiple resonant cavities based on the DRO and PE-SRO to reach threshold, whereas pulsed OPOs can be readily implemented in the SRO configuration. On the other hand, practical implementation of cw OPOs in the SRO configuration is readily attainable using the intracavity pumping approach.

C3.2.5.1 Continuous-wave OPOs

This class of OPO devices conforms to a steady-state operating regime as in a cw laser. The successful operation of the OPO is achieved by increasing the cw pumping intensity until the optical gain at the parametric waves overcomes the parasitic losses in the cavity and the threshold is surpassed. Once above threshold, steady-state oscillation is established and parametric down-conversion from the pump to the signal and idler waves occurs, resulting in macroscopic output from the device. Starting from the coupled-wave equations (C3.2.5a)–(C3.2.4c), analytical expressions for threshold pumping intensity and conversion efficiency in cw OPOs can be obtained under steady-state conditions by considering the balance between parametric gain and parasitic loss within the OPO cavity [30].

C3.2.5.1.1 Steady-state threshold

The analysis is based on the same assumptions as used in the derivation of the coupled-wave equations and parametric gain in section C3.2.2. All three optical waves, the pump, signal and idler, are assumed to be infinite plane waves with uniform intensity across the beams and the effects of focusing and double refraction are ignored. The interaction is also assumed to be phase-matched ($\Delta k = 0$), with the pump making a single pass through the nonlinear crystal, so that the gain is only in the forward direction. In addition, the analysis assumes that the OPO cavity losses at the resonant wave(s) are small, which is generally true in practice. Pump depletion is also considered to be small, a condition which is true close to threshold. Both these assumptions allow E_3 to be constant throughout the length of the nonlinear crystal, making $\partial E_3/\partial z$ = constant in the third coupled-wave equation (C3.2.5c), hence simplifying the analysis.

Under these assumptions, steady-state oscillation thresholds can be obtained for both SRO and DRO cavities. For the SRO, the threshold pumping intensity is derived as

$$I_{\text{th}} = \alpha_2 \left(\frac{c\varepsilon_0 n_1 n_2 n_3 \lambda_1 \lambda_2}{8\pi^2 \ell^2 d_{\text{eff}}^2} \right) \tag{C3.2.15}$$

where $\alpha_2 \ll 1$ is the total fractional round-trip power loss for the resonant wave (in this case, the signal). The loss at the idler wave is taken as 100% ($\alpha_1 = 1$), corresponding to an ideal SRO. For the DRO, the threshold condition is similarly obtained as

$$I_{\text{th}} = \frac{\alpha_1 \alpha_2}{4} \left(\frac{c\varepsilon_0 n_1 n_2 n_3 \lambda_1 \lambda_2}{8\pi^2 \ell^2 d_{\text{eff}}^2} \right) \tag{C3.2.16}$$

where α_2 and α_1 are the total fractional round-trip power losses for resonant signal and idler waves, respectively. Both signal and idler losses are assumed to be small ($\ll$ 100%), so that α_2 and α_1 are both $\ll 1$. It can be seen from equations (C3.2.15) and (C3.2.16) that a DRO with signal and idler losses, α_1 and α_2, will have a threshold which is a factor $\alpha_1/4$ lower than a SRO with the same loss at the resonant signal and 100% loss at the idler. Therefore, a DRO with an idler loss of 2% will have a pump power threshold that is 200 times lower than the equivalent SRO. The threshold advantage of the DRO is therefore immediately clear.

In both SRO and DRO configurations, the pump threshold can be substantially reduced by returning the pump beam back to the crystal after its first pass in the forward direction [32]. The double-pass pumping ensures that optical gain is available in both directions, as in a laser oscillator, hence increasing the round-trip optical gain relative to loss. For a SRO, the threshold is reduced by a factor of two using this technique. Further reductions in threshold by as much as four times can also be obtained in a SRO if, in addition to the pump, the non-resonant wave is also double-passed through the crystal. However, to achieve the maximum four-fold reduction, it is necessary to maintain the appropriate relative phase between the pump and the parametric waves on reflection. In a DRO, the threshold reduction can be as much as four times, but this again requires that the correct relative phase among the three waves be preserved on reflection.

For a PE-SRO, the threshold condition can be derived in a similar manner to SRO, but by considering gain in both directions through the nonlinear crystal, as the pump is now also resonant within the optical cavity. On resonance, the internal pump power threshold for the PE-SRO will, therefore, be the same as the incident threshold of the SRO with double-pass pump. The external threshold pumping intensity can, however, be shown to be [30]

$$I_{\text{th}} = \alpha \left(\frac{c\varepsilon_0 n_1 n_2 n_3 \lambda_1 \lambda_2}{16\pi^2 \ell^2 d_{\text{eff}}^2} \right) \left[\frac{(1 - \sqrt{r_1 r_2 t t'})^2}{t_1} \right] \tag{C3.2.17}$$

where $\alpha \ll 1$ is the round-trip power loss of the parametric wave (signal or idler) that is resonant together with the pump in the cavity. Hence, α is equal to α_2 or α_1, depending on whether the resonant wave is the signal or the idler. r_1 and t_1 are the pump power reflectivity and transmission of the input mirror through which the pump is coupled into the cavity and r_2 is the pump power reflectivity of the output mirror. t and t' are the effective round-trip pump transmission of the crystal to linear and nonlinear loss (due to parametric conversion), respectively. Clearly, for $r_1 = 0$ and $t_1 = 1$, (C3.2.17) reduces to the SRO threshold condition (C3.2.15). It can also been seen from (C3.2.17) that resonating the pump has the effect of reducing the SRO threshold by a factor $(1 - \sqrt{r_1 r_2 t t'})^2/t_1$. To achieve maximum threshold reduction, this factor must thus be minimized through the interplay and optimization of the controllable parameters r_1, r_2, t_1 and t. The nonlinear transmission of the crystal, t', must also be maximized through the use of long crystal lengths, and optimization of phase-matching and focusing parameters in order to achieve the lowest threshold.

In an IC-SRO, where the OPO is placed within the pump laser cavity, the steady-state threshold corresponds to a value of circulating pump intensity within the laser cavity, which is equal to the external pump intensity for an externally pumped SRO with double-pass pumping. The required external input power to the laser will in turn be determined by the internal slope efficiency of the pump laser in the presence of the SRO within the cavity [30, 31].

C3.2.5.1.2 Conversion efficiency

The steady-state analysis of cw OPOs can be similarly extended to operation above threshold to determine the maximum conversion efficiency from the pump to the parametric waves in the different resonance configurations [30]. The plane-wave analysis is again based on the same assumptions as used in the threshold calculations. These include small cavity losses, optimum relative phase for parametric generation, and phase-matched interaction. For a SRO with a single pass of the pump, the maximum conversion efficiency can be derived from the relation

$$I_{\text{out}} = I_{\text{in}} \cos^2 \left[\left(\frac{I_{\text{in}} - I_{\text{out}}}{I_{\text{th}}} \right)^2 \right] \qquad \text{(C3.2.18)}$$

where I_{in} and I_{out} are the steady-state pump intensities at the input and output of the nonlinear crystal, respectively, and I_{th} is the threshold pumping intensity defined by (C3.2.15). Form (C3.2.18), we can find the condition for maximum conversion efficiency in the SRO. This occurs when the pump is completely depleted in passing through the crystal, so that $I_{\text{out}} = 0$. This results in the relation

$$I_{\text{in}} = \left(\frac{\pi}{2} \right)^2 I_{\text{th}} = 2.47 I_{\text{th}} \qquad \text{(C3.2.19)}$$

which implies that in the SRO a maximum efficiency of 100% is attainable when the input pump intensity is 2.47 times that at threshold. The analysis, as presented, holds up only to the point where the pump intensity just reaches zero at the output face of the nonlinear crystal. If the input pump intensity is further increased, then the pump wave will be totally depleted at some point within the crystal. In the remaining length of the crystal, the signal and idler waves are then able to regenerate the pump through sum-frequency mixing. As a result, pump intensity once again now exits the crystal, and the SRO conversion efficiency is reduced. There is hence an optimum value of the pump intensity, at 2.47 times threshold, for which 100% efficiency is obtained. Above and below this value, lower conversion efficiencies are available.

Interestingly, it can also be shown that if the pump is double-passed through the SRO, a conversion efficiency of 100% is still attainable, but will require the same input pump intensity, namely 2.47 times the single-pass threshold intensity [32]. This is despite the two-fold reduction in SRO threshold under double-pass pumping. The analysis of conversion efficiency in the SRO leads to the same results, (C3.2.18) and (C3.2.19), regardless of whether the SRO is configured in a standing wave or ring cavity. This is because the non-resonant wave is not present on the return pass through the crystal and so regeneration of the pump wave from mixing between the signal and idler waves does not arise here. Since the non-resonant wave is generated from zero at the input of the crystal, the appropriate relative phase for maximum conversion efficiency is acquired among the optical waves from the start and, under phase matching, is maintained throughout the process.

In a DRO, the analysis of conversion efficiency must include the effects of pump wave regeneration, which can limit the maximum attainable efficiency [32, 33]. In a standing-wave cavity with a single-pass pump only in the forward direction, the pump wave is regenerated through the mixing of signal and idler waves on their backward propagation through the nonlinear crystal. If it is assumed that all waves maintain the optimum relative phase for maximum conversion efficiency, then the pump intensity at the output of the crystal is derived as

$$I_{\mathrm{out}} = I_{\mathrm{in}} - \left[2\sqrt{I_{\mathrm{th}}}\left(\sqrt{I_{\mathrm{in}}} - \sqrt{I_{\mathrm{th}}}\right)\right] \tag{C3.2.20}$$

where I_{in} is the steady-state pump intensity at the input of the crystal, and I_{th} is the threshold pump intensity defined by equation (C3.2.16). From the expression, we see that the maximum conversion efficiency in a DRO (corresponding to $I_{\mathrm{out}} = 0$) occurs at four times threshold, instead of the 2.47 times in the SRO. The maximum efficiency is also now limited to only 50% due to pump wave back-conversion. It can be readily shown that the back-converted pump intensity is given by

$$I_{\mathrm{BC}} = \left(\sqrt{I_{\mathrm{in}}} - \sqrt{I_{\mathrm{th}}}\right)^2 \tag{C3.2.21}$$

from which we can calculate the transmitted intensity of the pump under steady-state operation. If we define the down-converted pump intensity as $I_{\mathrm{DC}} = I_{\mathrm{in}} - I_{\mathrm{out}}$, then the transmitted intensity can be obtained from the relation $I_{\mathrm{out}} = I_{\mathrm{in}} - I_{\mathrm{BC}} - I_{\mathrm{DC}}$. Substituting for I_{BC} and I_{DC} using equations (C3.2.20) and (C3.2.21), we find $I_{\mathrm{out}} = I_{\mathrm{th}}$. This is an interesting result as it implies that, under steady-state operation above threshold, the transmitted pump intensity through the DRO is clamped at its threshold value. In other words, the DRO acts as an optical intensity or power limiter for the pump and regardless of the magnitude of the input pump intensity above threshold, the transmitted intensity remains fixed at I_{th}. Any increase in the input intensity above the threshold value has the effect of increasing the back-generation of the pump, without a rise in either the down-converted or the transmitted pump intensity.

The effects of back-generation in a DRO can, however, be eliminated by allowing the pump to be present with the parametric waves on the return pass through the crystal using double-pass pumping [32]. Under this condition, 100% conversion efficiency becomes attainable in the DRO. As noted earlier, double passing of the pump also has the effect of reducing the DRO threshold by as much as four times. However, unlike the SRO, the required pump intensity for 100% conversion efficiency in the DRO is four times the new and lowered threshold. At the same time, the attainment of the maximum four-fold reduction in DRO threshold requires that the optimum relative phase among the pump, signal, and idler is preserved on the return pass through the crystal. This may, in practice, be best achieved by interferometric control of the reflected pump wave using a separate mirror.

On the other hand, it is also possible to achieve 100% conversion efficiency in the DRO without double passing of the pump, but by using a travelling-wave rather than a standing-wave cavity for the OPO. In this geometry, the pump wave always accompanies the signal and idler waves in every pass of the nonlinear crystal, so that no back-generation of the pump can occur. The steady-state analysis of conversion efficiency in a travelling-wave DRO is similar to a standing-wave DRO, but by ignoring back-generation of the pump. The result of the analysis is the following expression for the steady-state output pump intensity at the exit of the crystal [30]

$$I_{\mathrm{out}} = \left(\sqrt{4I_{\mathrm{th}}} - \sqrt{I_{\mathrm{in}}}\right)^2 \tag{C3.2.22}$$

where I_{in} is the pump intensity at the input of the crystal and I_{th} is the threshold intensity. From the above, it is thus clear that the maximum DRO conversion efficiency (corresponding to $I_{\mathrm{out}} = 0$) still occurs at four times threshold. However, if we consider the down-converted pump intensity intensity, $I_{\mathrm{DC}} = I_{\mathrm{in}} - I_{\mathrm{out}}$, we find from equation (C3.2.22) that

$$I_{\mathrm{DC}} = 4\sqrt{I_{\mathrm{th}}}\left(\sqrt{I_{\mathrm{in}}} - \sqrt{I_{\mathrm{th}}}\right) \tag{C3.2.23}$$

which implies that a DRO conversion efficiency of 100% is now attainable at four times threshold because of the absence of pump back-conversion. It is, however, important to note that back-conversion of the pump can still occur if the pump is fully depleted within the crystal. In this case, the signal and idler fields can mix in the remaining length of the crystal to regenerate the pump, hence reducing the conversion efficiency. This would occur when the input pump intensity increases above the optimum value of four times threshold. The back-conversion of pump within the crystal can also arise in the SRO, where 100% depletion of the pump is possible.

In a PE-SRO, the conversion efficiency may be analysed by considering the down-converted pump intensity, defined as [30]

$$I_{\mathrm{DC}} = (1 - t')I_{\mathrm{cc}} \tag{C3.2.24}$$

where I_{cc} is the circulating pump intensity within the PE-SRO cavity at threshold. It is equal to the incident threshold of the SRO with double-pass pump. The down-converted intensity may then be shown to be

$$I_{\mathrm{DC}} = I_{\mathrm{cc}} \left[1 - \left(\frac{1}{r_1 r_2 t} \right) \left(1 - \sqrt{\frac{t_1 I_{\mathrm{ext}}}{I_{\mathrm{cc}}}} \right)^2 \right] \tag{C3.2.25}$$

where I_{ext} is the external pump power incident on the input mirror and all other parameters as defined previously (see section C3.2.5.1.1). For a perfect input mirror with $r_1 + t_1 = 1$, the optimum transmission of the input mirror can be found from equation (C3.2.25) to be $t_1^{\mathrm{opt}} = I_{\mathrm{ext}}/I_{\mathrm{cc}}$. The optimum down-converted pump intensity can similarly be obtained as

$$I_{\mathrm{DC}}^{\mathrm{opt}} = I_{\mathrm{cc}} \left[1 - \left(\frac{1}{r_2 t} \right) \left(1 - \frac{I_{\mathrm{ext}}}{I_{\mathrm{cc}}} \right) \right] \tag{C3.2.26}$$

We can see from this that, for zero parasitic loss of pump through the output mirror and no linear transmission through the crystal ($rt = 1$), the optimum down-converted intensity in the PE-SRO under optimized input coupling will equal the external pump intensity at the input. In other words, the PE-SRO is capable of 100% conversion efficiency from the external pump to the generated parametric waves in the absence of any linear parasitic loss of pump from the cavity.

For an IC-SRO, the analysis of conversion efficiency leads to the following relation for down-converted pump intensity [30, 31]

$$I_{\mathrm{DC}} = \sigma_{\mathrm{max}} (I_{\mathrm{in}} - I_{\mathrm{th}}) \left(1 - \frac{I_{\mathrm{th}}^{\mathrm{L}}}{I_{\mathrm{th}}} \right) \tag{C3.2.27}$$

where σ_{max} is the output slope efficiency of the pump laser with optimum output coupling for a given parasitic loss. I_{in} is the external input intensity to the pump laser, $I_{\mathrm{th}}^{\mathrm{L}}$ and I_{th} are the external input intensities corresponding to the pump laser and IC-SRO threshold, respectively. From equation (C3.2.27) the condition for maximum down-converted intensity can be obtained as

$$I_{\mathrm{th}} = \sqrt{I_{\mathrm{th}}^{\mathrm{L}} I_{\mathrm{in}}} \tag{C3.2.28}$$

Under this condition, the total down-converted power can be shown to be

$$(I_{\mathrm{DC}})_{\mathrm{max}} = \sigma_{\mathrm{max}} \left(\sqrt{I_{\mathrm{in}}} - \sqrt{I_{\mathrm{th}}^{\mathrm{L}}} \right)^2 \equiv (I_{\mathrm{out}}^{\mathrm{L}})_{\mathrm{max}} \tag{C3.2.29}$$

which also identically describes the condition for the maximum attainable output intensity from the pump laser, $(I_{\mathrm{out}}^{\mathrm{L}})_{\mathrm{max}}$, at a given input intensity, when subject to optimum output coupling. This implies that when the pump laser and SRO input intensity thresholds satisfy equation (C3.2.28), then 100% of the maximum output intensity potentially extractable from the pump laser itself can be down-converted into the parametric waves. In other words, equation (C3.2.28) describes the condition for optimization of the nonlinear coupling loss presented to the pump laser by the IC-SRO to achieve 100% down-conversion at a given input power. Under this condition, the IC-SRO can be viewed as an optimum output coupler to the pump laser. Increasing the external input intensity to the pump laser leads to the down-converted intensity from the IC-SRO approaching the maximum available output from the optimally out-coupled pump laser, reaching 100% efficiency when the input intensity reaches a value $I_{in} = I_{th}^2/I_{th}^L$. Above this value, the efficiency gradually tails off as the IC-SRO begins to over-couple the laser field. Back-conversion of the pump is generally not relevant in influencing the efficiency of IC-SRO, since one of the generated waves is non-resonant in the OPO cavity and the pump also always accompanies the parametric waves in every pass of the crystal. In addition, full pump depletion within the crystal is not possible because the single-pass conversion is at best only a few per cent (equivalent to the loss presented to the pump laser by an optimized output coupler). Equation (C3.2.28) provides a practical design criterion for the attainment of maximum power conversion in the IC-SRO for a given available input power. In practice, however, it is often difficult to achieve the maximum 100% efficiency, but down-conversion efficiencies in excess of 90% can be obtained experimentally through suitable design of the pump laser and ICSRO threshold parameters, as described in section C3.2.7.

C3.2.5.2 Pulsed OPOs

As highlighted earlier, the SRO cavity configuration in which only one of the generated waves is resonant represents the simplest and most practical device architecture for an OPO, because of its minimal demands on pump beam coherence, high passive stability and reduced complexity in spectral and temporal behaviour. On the other hand, the SRO yields the highest threshold of all cavity configurations, so that the attainment of oscillation threshold using conventional external pumping schemes is generally beyond the reach of commonly available cw pump lasers, particularly when conventional birefringent materials are employed. Given this limitation, practical operation of SROs has historically necessitated the use of pulsed lasers, where the high available peak powers readily permit device operation without resort to intracavity pumping or any form of pump enhancement. Such devices can be readily implemented in the nanosecond regime by deploying Q-switched pump lasers or in the ultrafast picosecond and femtosecond regime by using mode-locked laser sources. In the following discussion, we provide an overview of the main operating characteristics of pulsed OPOs, with an emphasis on the SRO configuration, as the majority of practical devices developed to date conform to this mode of operation.

C3.2.5.2.1 Nanosecond OPOs

Because of their high peak powers, nanosecond pulsed lasers represent the most viable choice of pump for SROs. Combined with the widespread availability of such laser sources, OPOs operating in the nanosecond pulsed regime have traditionally been the most extensively developed of all parametric devices. The basic operation principles in nanosecond OPOs are the same as in cw oscillators. However, the steady-state threshold analysis used earlier in the treatment of cw OPOs is not strictly applicable here, because the instantaneous nature of parametric gain does not allow steady-state conditions to prevail within the finite duration of the pump pulse. This situation is markedly different from that in a conventional laser, where under pulsed excitation the gain storage capability of the gain medium generally permits amplification after the pump pulse excitation for a time determined by the upper state lifetime of the transition. In the parametric process, the electronic susceptibility is effectively instantaneous, so that the nonlinear medium has no gain storage capacity. This means that coherent amplification of the parametric wave(s) has to occur in the presence of the pump and that no gain is available outside the temporal window of the pump pulse. Given the finite duration of the pump pulse, only a limited number of round-trips for the parametric wave(s) can be made available through the nonlinear crystal in OPO cavities of practical length, thus preventing the establishment of steady-state conditions. Therefore, nanosecond pulsed OPOs generally operate in a transient regime and a modified model taking account of the dynamic behaviour of OPO is necessary for the analysis of such devices.

By using a time-dependent gain analysis, it is possible to derive the threshold condition in pulsed SROs in terms of pump energy fluence (energy/area) [34]. The model assumes a Gaussian temporal profile for the incident pump pulse and a Gaussian spatial distribution for the pump and the resonant signal, with an idler wave that is unconstrained by the optical cavity. The model also assumes a single pass pump and includes the effects of mode overlap and spatial walkoff. The threshold condition is then derived from the solution of the coupled-wave equations (C3.2.5a)–(C3.2.5c) in the limit of low pump depletion and zero input idler field, by allowing the resonant signal wave to be amplified from the initial parametric noise power to a detectable level by successive transits of the SRO cavity. By defining the SRO threshold as a signal energy of $\sim$100 μJ, corresponding to a signal power to parametric noise power of $\ln(P_{\rm s}^{\rm th}/P_{\rm n}) = 33$, the threshold energy fluence is derived as

$$J_{\rm th} \cong \frac{2.25}{\gamma g_{\rm s}\zeta}\tau\left[\frac{L}{2\tau c}\ln\frac{P_{\rm s}^{\rm th}}{P_{\rm n}} + 2\alpha_2\ell + \ln\frac{1}{\sqrt{R}} + \ln 2\right]^2 \qquad \text{(C3.2.30)}$$

where τ is the $1/e^2$ intensity half-width of the pump pulse, L is the cavity length, α_2 is the signal field amplitude loss coefficient within the crystal, R is the mirror reflectivity coefficient, ℓ is the crystal length. The parameter γ is a modified gain coefficient, which is related to the gain factor Γ defined in equation (C3.2.7) by

$$\gamma = \frac{\Gamma^2}{I_3(z=0)} = \frac{8\pi^2 d_{\rm eff}^2}{c\varepsilon_0 n_1 n_2 n_3 \lambda_1 \lambda_2}. \qquad \text{(C3.2.31)}$$

The factor $g_{\rm s}$ is the spatial coupling coefficient describing the mode overlap between the resonant signal and the pump field. It is defined as

$$g_{\rm s} = \frac{w_3^2}{w_3^2 + w_2^2} \qquad \text{(C3.2.32)}$$

where w_3 and w_2 are the Gaussian mode electric field radii of the pump and signal, respectively. The parameter ζ is the effective parametric gain length given by

$$\zeta = \ell_{\rm w}\,{\rm erf}\left(\frac{\sqrt{\pi}}{2}\frac{\ell}{\ell_{\rm w}}\right) \qquad \text{(C3.2.33)}$$

where $\ell_{\rm w}$ is the *walkoff length* defined as

$$\ell_{\rm w} = \frac{\sqrt{\pi}}{2}\frac{w_3}{\rho}\sqrt{\frac{w_3^2 + w_2^2}{w_3^2 + w_2^2/2}} \qquad \text{(C3.2.34)}$$

where ρ is the double-refraction angle (see section C3.2.6.1). The walkoff length, $\ell_{\rm w}$, is closely related to the aperture length, $\ell_{\rm a}$ (see section C3.2.6.1), and accounts for spatial walkoff between the Gaussian pump and signal fields. In the absence of spatial walkoff, $\ell_{\rm w} \to \infty$, and the crystal length ℓ becomes the effective parametric gain length.

The main conclusions of the time-dependent model relate to the strong dependence of pulsed OPO threshold on the characteristic *rise time* of the oscillator. This is a measure of the time required for the parametric gain to build up from noise to oscillation threshold [35]. In order for the OPO to switch on, it is essential that the oscillator rise time is shorter than the pump pulse duration. For efficient operation, however, the rise time must be minimized so that the parametric waves are rapidly amplified above the threshold level in a time significantly shorter than the pump pulse interval. Therefore, the rise time represents an effective loss for pulsed OPOs, with a direct impact on oscillation threshold. To exploit maximum gain, it is essential that the parametric waves are amplified in as short a time interval as possible in the presence of the pump pulse. In practice, this can be achieved by minimizing the OPO cavity length to allow the maximum number of round-trips over the pump pulse interval. This is also evident from the direct dependence of J_{th} on the cavity length L in equation (C3.2.30). In nanosecond OPOs, cavity lengths of a few centimetres or shorter are often practicable, yielding as many as one hundred round-trips for pump pulses of 1–50 ns duration. As can be verified from equation (C3.2.30), the pulsed OPO threshold can be further reduced by using longer pump pulse durations, τ (and maintaining the same peak intensity), and by lowering the intracavity parasitic losses through minimizing α and maximizing R. Above threshold, the operation of pulsed OPOs is characterized by the extraction of energy from the pump pulse and saturation of nonlinear gain, in a similar manner to cw OPOs discussed earlier. This is often observed in the form of depletion towards the trailing edge of pump pulse. The signal (and idler) pulse is then amplified with a characteristic delay from the pump pulse by a time interval determined by the OPO rise time [36–38].

We can obtain a measure of the threshold fluence in pulsed SROs by considering a typical example based on a 10 mm long KTP crystal, pumped by 10 ns pulses from a Q-switched Nd:YAG laser 1.064 μm. We assume type II NCPM (see section C3.2.6.1), so that the full crystal length is utilized for the nonlinar interaction. We take a practical cavity length $L = 20$ mm, mirror reflectivities $R \sim 99.5\%$, an effective nonlinear coefficient $d_{eff} \sim 3$ pm V^{-1}, refractive indices $n_3 \sim n_2 \sim n_1 \sim 1.5$, and mode radii $w_3 \sim w_2$. For near-degenerate operation with $\lambda_1 \sim \lambda_2 \sim 2$ μm and for $\alpha \sim 2\%$ and $\rho = 0$, substitution of these parameters into equations (C3.2.31)–(C3.2.34) and then into (C3.2.30) yields a threshold pump energy fluence $J_{th} \sim 1.2$ J cm^{-2}. This translates into a threshold pump pulse energy of $E_{th} \sim 37$ mJ for a focused pump mode radius $w_3 \sim 1$ mm. The threshold energy can, of course, be substantially reduced by using tighter focusing. For example, the use of a pump beam waist of $w_3 \sim 0.5$ mm will lead to a nearly four-fold reduction in threshold energy to ~9 mJ, whereas reducing the focused radius to $w_3 \sim 0.1$ mm will result in nearly a ten-fold threshold reduction to ~0.4 mJ. It should, however, be noted that considerations of material damage in the presence of nanosecond pulses present a practical upper limit to the minimum focused pump spot radii that can be used in such devices (see section C3.2.6.1). Therefore, the optimum choice of experimental parameters for the attainment of minimum operation threshold in pulsed OPOs is often compromised by material damage threshold. Material damage remained a major limitation to the operation of early devices, even at moderate power levels, and for many years hampered the development of practical nanosecond OPOs.

C3.2.5.2.2 Picosecond and femtosecond OPOs

The high peak pulse intensities available with mode-locked picosecond and femtosecond pulses can facilitate the attainment of sufficient nonlinear gain in an OPO to overcome threshold, even in SRO configurations exhibiting the highest operation threshold. Moreover, the low-energy fluence associated with ultrashort pulses results in increased damage tolerance for the material, thereby enabling reliable operation of the OPO at high average powers using high-power input pump sources. On the other hand, unlike cw and pulsed nanosecond devices, operation of ultrafast OPOs is based on the principle of *synchronous pumping*. As noted in section C3.2.5.2.1, due the instantaneous nature of nonlinear polarization, optical gain in an OPO is available only in the presence of the pump pulse, during which macroscopic amplification of the parametric waves from noise to coherent output must take place. However, in contrast to nanosecond pulses, the temporal window in picosecond and femtosecond pulses is too narrow to allow a finite number of cavity round-trips for the build-up of the parametric waves over the pump pulse duration, even for practical OPO cavity lengths as short as a few millimetres. To overcome this difficulty, the OPO resonator length is matched to the length of the pump laser, so that the round-trip transit time in the OPO cavity is equal to the repetition period of the pump pulse train. In this way, the resonated parametric pulses experience amplification with consecutive coincidences with the input pump pulses as they make successive transits through the nonlinear crystal. In practice, the technique is practical only at relatively high pulse repetition rates (>50 MHz). At lower repetition frequencies, the required OPO cavity lengths for synchronous pumping become too long and cumbersome to be useful in practice.

In general, synchronously-pumped OPOs may be classified into either cw or pulsed devices. In cw oscillators, the input pump radiation comprises a continuous train of ultrashort pulses, whereas in pulsed devices the pump consists of trains of ultrashort pulses contained within a nanosecond or microsecond envelope. With regard to their operating characteristics, cw synchronously-pumped OPOs may be treated as steady-state devices in the same way as cw OPOs, but with the peak pump pulse intensity determining the nonlinear gain. As such, the steady-state analysis of cw OPOs is similarly applicable to cw synchronously-pumped OPOs by using the peak pulse intensity as the incident pump intensity. On the other hand, the operating dynamics of pulsed synchronously-pumped OPOs is similar to nanosecond OPOs, where a transient analysis taking account of rise-time effects is necessary to adequately describe the OPO behaviour [84]. In pulsed synchronously-pumped OPOs, the nonlinear gain is similarly determined by the peak pulse intensity, but rise-time effects arising from the finite duration of the pulse envelope lead to an additional loss mechanism. In either case, however, further effects pertinent to the operation of synchronously-pumped OPOs including *temporal walkoff* and group-velocity dispersion (see section C3.2.9.2) also have to be included in the model. In picosecond OPOs, where pulse durations of >1 ps are involved, such temporal effects can generally be ignored for practical crystal lengths of up to 20 mm. However, they become increasingly important in femtosecond OPOs, where pulses of 100 fs or shorter are involved (see section C3.2.9.2).

The analysis of synchronously-pumped OPOs under steady-state and transient operation has been performed for plane waves [39] and for cw synchronously-pumped SROs with Gaussian beams [40]. One of the main conclusions of the analyses is that in the absence of group velocity dispersion and temporal walkoff, the output pulses from a synchronously-pumped OPO are always shorter than the input pump pulses, with the parametric pulses broadening towards the pump pulse length with increasing pump depletion. They also show that it is possible to obtain conversion efficiencies as high as $\sim$70% in such devices with a suitable choice of design parameters. The analysis of cw synchronously-pumped SROs also shows that, for high efficiency and short output pulses, the oscillator should be designed to operate with a peak pump intensity that is approximately twice the steady-state threshold value. Thus, for a given pump intensity and nonlinear material parameters, the output coupling mirror can be adjusted to allow this condition to be met. Near threshold, the effects of temporal walkoff can be almost perfectly compensated for by fine tuning the cavity length.

C3.2.6 OPO design issues

The principal design criterion for optimum operation of parametric devices of all types, including OPOs, is the maximization of parametric gain through the suitable choice of nonlinear material and laser pump sources, optimization of phase-matching, cavity design and focusing parameters.

C3.2.6.1 Nonlinear material

The selection of nonlinear material for OPOs is governed by a number of universal factors that are equally applicable to all types of parametric devices. A fundamental material parameter is a broad transparency and the ability to be phase-matched in the wavelength range of interest. As evident from equation (C3.2.7), it is also important for the material to have a large effective nonlinear coefficient, d_{eff}, for the attainment of maximum gain and highest efficiency. Other material requirements include a high optical damage threshold, good optical quality and low absorption loss at the pump, signal and idler wavelengths, mechanical, chemical, and thermal stability, and availability in bulk form and sufficiently large size.

In addition to the fundamental material parameters, there are also a number of general material requirements that are desirable for optimum operation of OPOs. These include favourable phase-matching geometries, small double-refraction angles and spatial walkoff and, in the case of femtosecond OPOs, low temporal walkoff and group velocity dispersion. It is also desirable for the material to display large tolerances to possible deviations in the spectral and spatial quality of the pump beam, which result in an increase in the magnitude of the phase mismatch, Δk, from zero. Such deviations arise in practice from the finite spectral bandwidth and divergence of the pump beam. The tolerance of the nonlinear crystal to such effects is measured in terms of the so-called *spectral* and *angular acceptance bandwidths*, which can be calculated from the rate of change of phase mismatch with the wavelength spread ($\Delta\lambda$) and angular divergence ($\Delta\theta$) at the pump, using a series expansion of Δk [6]. The acceptance bandwidths can then be obtained by solving for the quantities $\Delta\lambda \cdot \ell$ and $\Delta\theta \cdot \ell$, using the boundary condition $|\Delta k \ell| \sim 2\pi$. For a given crystal length, ℓ, the acceptance bandwidths set an upper limit to the maximum allowable pump linewidth and divergence before parametric gain is severely degraded. Equivalently, for a given pump linewidth and angular divergence, the acceptance bandwidths determine the maximum length of the nonlinear crystal that can effectively contribute to the nonlinear gain. Since the crystal length follows an inverse relationship with the maximum allowable pump bandwidth and divergence, for shorter crystal lengths larger deviations in pump beam quality can be tolerated and *vice versa*. For materials that exhibit temperature-dependent refractive indices and, hence, a temperature-tuning capability, one can similarly define a *temperature acceptance bandwidth*, $\Delta T \cdot \ell$, which is a measure of the sensitivity of phase-matching and, thus, parametric gain, to changes in the crystal temperature. As a general rule, for the attainment of maximum nonlinear gain and minimum OPO threshold, it is advantageous to use materials which exhibit large spectral, angular and temperature acceptance bandwidths.

Another important material parameter is double-refraction, which leads to spatial walkoff among the interacting fields. Under this condition, the strength of nonlinear coupling can rapidly diminish and the interaction becomes ineffective after a finite length through the medium, known as the *aperture length*, and given by $\ell_a = w_3\sqrt{\pi}/\rho$, where w_3 is the pump beam waist radius and ρ is the double-refraction angle. It is clear that for longer aperture lengths, it is desirable to use nonlinear crystals with small double-refraction and larger pump beam waists. In pulsed nanosecond OPOs, the high peak intensities available from Q-switched pump lasers allow for the use of large beam waists. However, in cw and synchronously-pumped OPOs, it is often necessary to use tightly focused beams because of the substantially lower available pumping intensities.

In the operating regime of cw and synchronously-pumped OPOs, therefore, it is imperative to use materials that possess a NCPM capability, which allows phase-matched interaction along a principal optical axis with no spatial walkoff. Given the generally low intensities available from cw and mode-locked pump lasers, the NCPM geometry allows tight focusing of the pump beam to achieve the necessary intensities without the deleterious effects of beam walkoff. Therefore, NCPM is generally a highly important material property in the context of cw and synchronously-pumped OPOs, particularly picosecond OPOs, where long interaction lengths are imperative for the attainment of threshold. This requirement is somewhat less stringent when deploying high-power (multi-watt) pump sources for picosecond OPOs, or in femtosecond OPOs, where the high peak powers and short interaction lengths of the material allow for the use of critical phase-matching (CPM) in collinear or non-collinear schemes (see section C3.2.9.2). However, even in these cases NCPM still offers the advantage of zero spatial walkoff and hence higher nonlinear gain. Another important advantage of NCPM is that under this geometry the angular acceptance bandwidth of the material is maximized, because of the small sensitivity of refractive indices to beam propagation angle along a principal optical axis. Therefore, the NCPM geometry is also highly beneficial when using pump beams of poor spatial quality.

In addition to these requirements, one of the most important properties that can ultimately limit operation of parametric devices in practice is material damage threshold. Because of the high pumping intensities in the parametric process, the materials must withstand power densities typically >10 MW cm^{-2} and often many times this value. Optical damage may be caused by several different mechanisms and can be of various types. The most predominant type is the irreversible damage, which can take the form of surface or bulk damage. As a general trend, material damage tolerance is found to decrease for increasing pump energy fluence and with shorter pumping wavelengths. As a result, material damage becomes more important in nanosecond OPOs than in synchronously-pumped and cw devices and also in the presence of ultraviolet and visible radiation. Another type of optical damage that is prevalent in PPLN is a reversible form, which results

Table C3.2.4. The relative significance of the key material parameters for different modes of OPO operation, from cw and pulsed nanosecond to cw synchronously-pumped picosecond and femtosecond regimes. The universal material requirements of transparency, phase-matchability, high optical quality and low absorption loss, and mechanical and chemical stability are essential for all modes of operation.

	OPOs			
	CW	Nanosecond	Picosecond	Femtosecond
High d_{eff} (>3 pm V^{-1})	Essential	Desirable	Essential	Desirable
Low spatial walkoff	Essential	Desirable	Essential	Desirable
Noncritical phase-matching	Essential	Desirable	Essential	Desirable
Long interaction length (>10 mm)	Essential	Desirable	Essential	Undesirable
High damage threshold	Desirable	Essential	Desirable	Desirable
Low group velocity dispersion	Unnecessary	Unnecessary	Desirable	Highly desirable
Low temporal walkoff	Unnecessary	Unnecessary	Desirable	Highly desirable
Temperature phase-matching	Highly desirable	Unnecessary	Highly desirable	Unnecessary
Angle phase-matching	Generally not permitted	Desirable	Generally not permitted	Unnecessary
Large phase-matching acceptance bandwidths	Desirable	Desirable	Desirable	Desirable

from illumination by visible radiation. This type of damage, known as photorefractive damage, manifests itself as inhomogeneities in the refractive index, which can lead to beam distortions, non-uniformities, and instabilities in the output beam. The induced damage can be reversed by ultraviolet illumination of the crystal or by elevating its temperature.

The relative importance of the various material parameters discussed is critically dependent on the particular mode of operation. In table C3.2.4, we summarize the key material requirements for each OPO mode of operation, from the cw to the synchronously-pumped femtosecond regime. The effective nonlinearity, transparency and potential tuning range of a number of prominent birefringent and periodically-poled materials used in the development of OPOs are also summarized in figure C3.2.14. A more comprehensive survey of a wide range of birefringent nonlinear materials can be found elsewhere [6].

C3.2.6.2 Pump laser

As with the nonlinear material, the choice of laser pump source is also governed by a number of universal factors relating to the maximization of nonlinear gain. A principal requirement is set by phase-matching to access the wavelength regions of interest, and the pump wavelength should obviously be within the transparency range of the material. The pump source must also be of sufficient intensity, as evident from (C3.2.7), to provide appreciable nonlinear gain for the OPO to reach operation threshold. This, in turn, necessitates a sufficiently low beam divergence to allow high focused intensities and depth of focus within the nonlinear material. Considerations of phase-mismatch, Δk, place further demands on the spectral and spatial coherence of the pump beam. As can be seen from (C3.2.6), parametric gain is strongly dependent on the

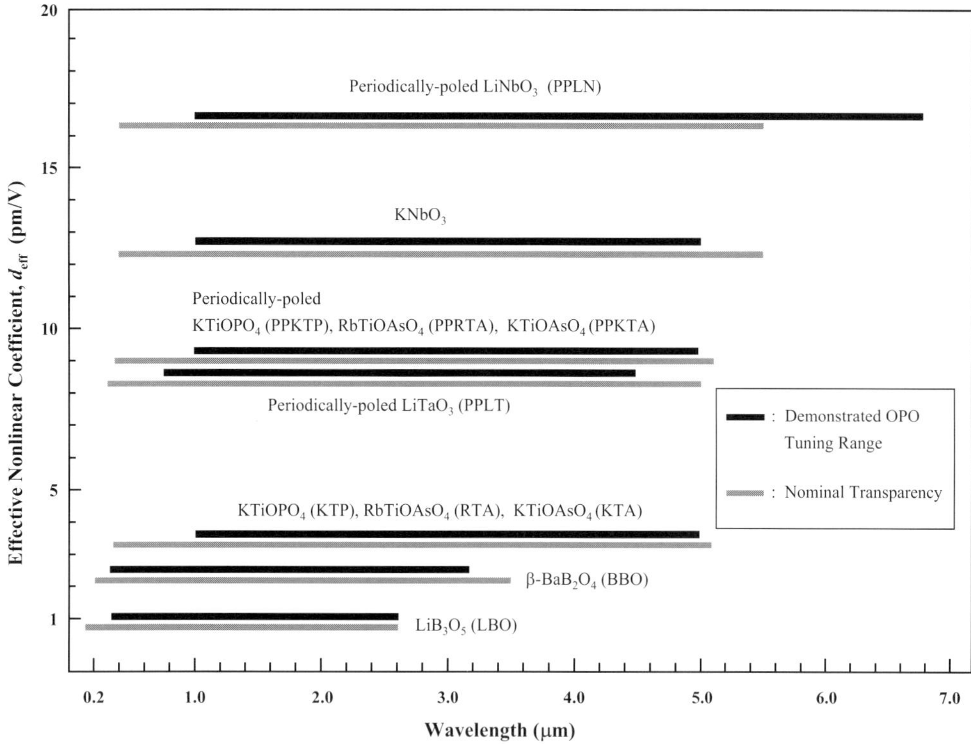

Figure C3.2.14. The characteristics of a number of important birefringent and periodically-poled nonlinear materials used in the development of OPOs. The indicated spectral coverage may not in all cases correspond to a single device, but is potentially available to the material in different device configurations.

phase-mismatch parameter, Δk. Maximum gain occurs for $\Delta k = 0$, while increases in the magnitude of Δk result in severe reductions in nonlinear gain. Despite the use of phase-matching techniques, the attainment of perfect phase-matching ($\Delta k = 0$) is generally not possible in practice, because of finite spectral bandwidth and spatial divergence of the pump beam. This leads to an increase in the magnitude of Δk, with the net result that the parametric gain is severely reduced from its peak value at $\Delta k = 0$. The maximum allowable pump bandwidth and divergence can be calculated from considerations of phase-mismatch [41]. As discussed in section C3.2.6.1, such limitations to pump beam quality can be equivalently considered in terms of the spatial and spectral acceptance bandwidths for the nonlinear process. To maintain high parametric gains, it is important to employ laser pump sources of narrow linewidth and low beam divergence.

C3.2.7 Continuous-wave OPO devices

Of the different types of OPO devices, the development of practical OPOs in the cw regime has been traditionally more difficult than pulsed and synchronously-pumped devices because of the substantially lower nonlinear gains available under cw pumping. The recent development of novel birefringent and periodically-poled materials, new innovations in resonance and pumping techniques, and the advent of stable, high-power solid-state laser pump sources, have led to important breakthroughs in cw OPOs in the past few years, particularly for the infrared spectral range at wavelengths >1 μm. Because of low spatial walkoff, NCPM capability, moderate nonlinear optical coefficients, an extended transparency window to 5 μm, the birefringent material KTP and its arsenate isomorphs, KTA and RTA, have shown promise as candidates for cw OPOs. Configured in IC-SRO cavities and pumped by Ti:sapphire lasers, devices based on KTP and KTA have been shown to be practical sources of cw radiation, providing total output powers of up to 1.46 W at up to 90% conversion efficiency, with as much as 840 mW of idler power available in the 2.4–2.9 μm spectral range [42, 43]. In alternative TRO and PE-SRO resonance configurations, low-threshold operation of cw OPOs has also been achieved using semiconductor diode lasers as the pump source [44, 45]. By using KTP and RTA under NCPM and GaAlAs diode lasers at ~800 nm, pump power thresholds <100 mW and tuning in the 2.1–2.7 μm range have been demonstrated through the tunability of the diode laser pump source. Because of a lack of significant temperature tuning in KTP and its arsenate analogues, wavelength tuning in cw OPOs based on these materials is generally achieved through static pump tuning while maintaining the NCPM condition. With the available laser sources based on cw Ti:sapphire, semiconductor and Nd:YAG lasers, the potential infrared spectral coverage of cw OPOs based on the KTP family of crystals under NCPM is limited to ~3.5 μm.

In contrast, the most important recent advances in cw OPOs have been brought about by periodically-poled materials. In particular, the advent of PPLN with high effective nonlinearity and long interaction lengths has brought the threshold of cw SROs within the reach of cw solid-state lasers and enabled the operation of these devices at unprecedented power levels and conversion efficiencies in simple external pumping schemes. Using a 50 mm long PPLN crystal pumped with a high-power diode-pumped Nd:YAG laser, infrared output powers of as much as 3.6 W have been generated from a cw SRO over a spectral range from 3.25 to 3.95 μm, for 13.5 W of input pump power [46]. Wavelength tuning was achieved through temperature tuning or by using different grating periods incorporated onto the single crystal. Alternative tuning schemes include the use of PPLN with fanned gratings, where the grating period is progressively varied across the crystal [47]. With the advent of PPLN, the operation of cw SROs in external pumping configuration has also extended to high-power semiconductor diode pump lasers [48]. Pumped by an InGaAs laser at 925 nm in a master oscillator-power amplifier configuration and a 38 mm PPLN crystal, SRO operation thresholds of 1.7 W and idler output powers of 480 mW were generated over a tuning range of 2.03–2.29 μm, for 2.5 W of pump power. Appropriate synchronization of pump tuning and cavity length scanning results in a wide, continuous, single-frequency tuning range. More recently, high-power fibre lasers (see chapter B4.2) have been successfully used to pump cw PPLN-based SROs [49]. Using an Yb-doped cw fibre laser tunable over 1.031–1.100 μm and delivering 8.3 W of pump power into a 40 mm PPLN crystal, mid-IR idler powers of up to 1.9 W over a spectral range of 2.98–3.7 μm have been generated, with a SRO power threshold at 3.5 W.

At the same time, the use of intracavity pumping schemes in combination with PPLN have, for the first time, permitted cw SRO operation with minimal operation threshold based on commonly available, low-power, diode-pumped solid-state lasers [50]. Using a Nd:YVO_4 laser pumped by a 1 W diode laser, practical mid-IR output powers of 70 mW over a spectral range from 3.16–4.02 μm have been generated from an IC-SRO in a simple, compact, all-solid-state design, with a diode pump power threshold of only 310 mW. Operation of mid-IR cw OPOs based on PPLN has also been achieved in PE-SRO configurations. Pumped by a miniature diode-pumped single-frequency Nd:YAG laser, 140 mW of idle power over the range 2.29–2.96 μm has been obtained from a PE-SRO, for 800 mW of pump power. The device had an external pump power threshold of 250 mW and wavelength tuning was available by a combination of temperature and grating period tuning. More recently, operation of a mid-IR PPLN PE-SRO pumped by a cw single-frequency Ti:sapphire laser and tunable from 4.07–5.26 μm was also reported [51]. By using a twin-cavity arrangement, mode-hop-free tuning of the single-frequency idler over 10.8 GHz was demonstrated by fine-tuning the pump laser over 12.3 GHz.

The high optical nonlinearity of PPLN and long available interaction lengths has also permitted the development of PE-SRO devices pumped directly by single-stripe semiconductor diode lasers. Operation of a device based on a 50 mm PPLN crystal and pumped at ~810 nm with a single-mode extended-cavity AlGaAs laser has been achieved with a power threshold of only around 25–30 mW [52]. This device could provide up to 4 mW of unidirectional idler power, with tuning coverage of from 2.58–3.44 μm. Using alternative DRO configurations, similar cw OPOs based on PPLN have also been demonstrated by the use of solitary, single-stripe diode lasers [53, 54]. These devices can operate with pump powers of less than 20 mW and can provide practical mid-IR output powers of up to 5 mW over a spectral range 2.2–3.7 μm. It is important to note that the potential tuning range available to all PPLN-based cw OPOs under Ti:sapphire, Nd:YAG, or direct diode laser pumping extends across a continuous range of ~1.5–5 μm, limited by the increased absorption in the material beyond ~5 μm. By suitable choice of OPO mirror and crystal coatings, it is possible to access this entire tuning range with one device incorporating a single PPLN crystal.

At present, substantial progress continues in the development of cw OPOs based on PPLN. These devices have now been developed to the point where they can provide widely-tunable, stable, single-frequency coherent radiation at practical power levels in the mid-IR. However, effective frequency control and power stabilization of PPLN devices, particularly under high-intensity pumping regimes in SRO configurations is impeded by the photorefractive effect, thermal phase-mismatching and green-induced infrared absorption; practical operation of these devices requires careful control of these parameters. In this context, alternative periodically-poled materials, particularly phosphates and arsenates of KTP, represent attractive material candidates for mid-IR cw OPOs, because of the absence of such detrimental effects, an equally extended transparency range to >5 μm, and lower coercive fields for poling. However, the shorter available interaction lengths (typically <20 mm) and lower effective nonlinear coefficients lead to higher operation thresholds than in equivalent PPLN devices. This precludes SRO operation in external pumping configurations even with high-power multiwatt laser pump sources, and so other approaches based on the DRO, PE-SRO and IC-SRO are the most viable route to the development of mid-IR cw OPOs based on these materials [55,56]. Currently, the KTP family of periodically-poled materials also offer more limited tuning flexibility than PPLN, because of the lack of significant temperature tuning and the difficulty in incorporating several different grating periods onto a single crystal due to limited apertures. However, the continuing advances in poling and fabrication technology are expected to pave the way for the practical development of stable and widely tunable cw OPOs for the mid-IR based on these materials. Figure C3.2.15 provides a summary of the spectral coverage and power performance of several cw OPOs developed to date.

C3.2.8 Nanosecond OPO devices

Advances in OPO devices have traditionally been most deliberate in the nanosecond regime, where the high peak pump intensities can be readily made available, even for SRO operation, with the use of Q-switched laser sources delivering pulses between 1 to 100 ns. Indeed, the first operation of an OPO was achieved with a pulsed nanosecond laser [57] and the majority of the early devices conform to this mode of operation. The use of this device configuration was also driven by the easier accessibility to such pulsed lasers in the early days of nonlinear optics. This operating regime has continued to represent the most widespread and extensively developed of all OPO device architectures.

In contrast, practical operation of nanosecond OPOs is more demanding on material damage tolerance than other device configurations, because of the relatively high optical energy associated with nanosecond pulses. Material damage is generally exacerbated at higher pump energy fluence, and this was perhaps the single most important factor hampering the practical development of early nanosecond OPOs that followed the demonstration of the very first device in 1965 [57]. With the emergence of a new generation of nonlinear materials in the 1980s, exhibiting unprecedented damage thresholds, the development of truly practical nanosecond OPOs became a reality for the first time. Of the new materials, BBO, LBO and KTP have been the prime candidates for visible and near-IR parametric generation. Other nonlinear crystals such as KNB and the arsenate isomorphs of KTP, namely KTA, CTA, RTA, have been of particular interest for wavelength generation in the difficult 3–5 μm spectral range in the mid-IR, because of their longer infrared transmission beyond 5 μm.

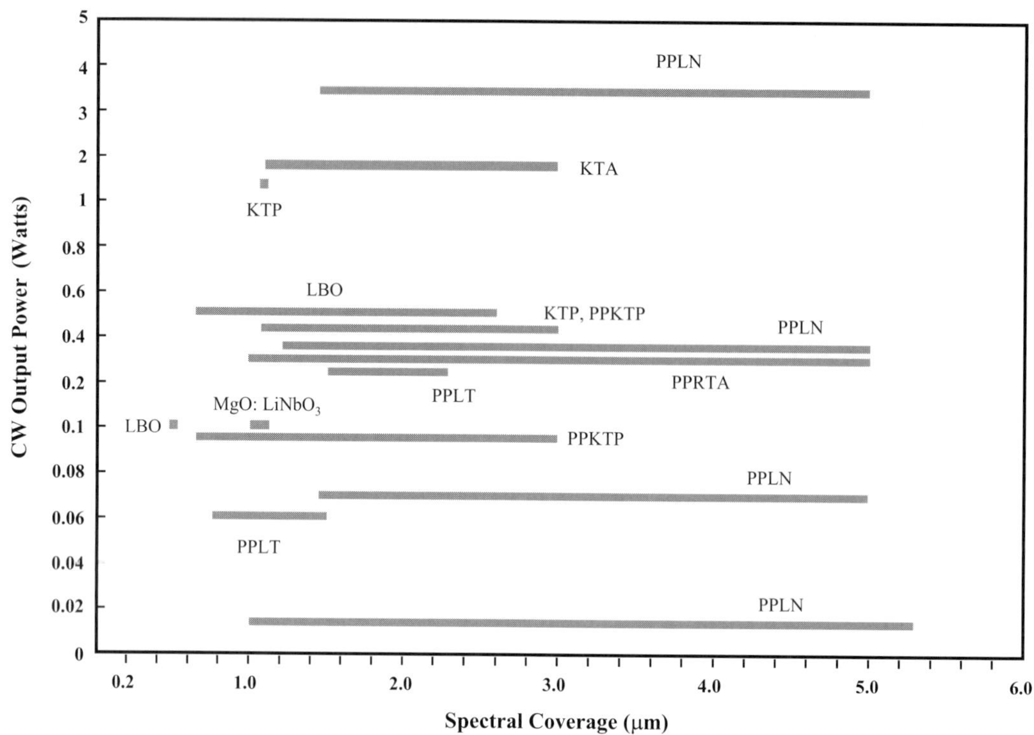

Figure C3.2.15. Survey of tuning range and output power of a number of cw OPOs demonstrated to date. The indicated tuning limits correspond to the potential coverage available to the particular crystal in the given phase-matching and pumping configuration, although the experimental tuning range may not be continuous or may have been limited by the available mirrors and crystal coatings. The output powers correspond to the maximum combined signal and idler power generated within the demonstrated tuning range.

The majority of nanosecond OPOs have been pumped by the well-established Q-switched Nd:YAG laser [58–68]; some in diode-pumped all-solid-state configurations [69–71]. Due to their high average power capability, ultraviolet excimer lasers (see chapter B3.2) have also been used as optical pumps [72–74]. The upper limit to the conversion efficiency and output pulse energy in nanosecond OPOs is generally set by optical damage to the mirror and crystal coatings, which can be caused either by the pump pulses or by the circulating signal intensities. This can be minimized by using novel pumping and crystal configurations [60], longer pump pulses and shorter cavity lengths. For enhancement of spectral and spatial coherence and output efficiency, novel design concepts and new resonator architectures have also been effectively deployed [75–80]. These have enabled the generation of optical pulses with high spectral purity, excellent spatial beam quality, and broad spectral tunability using nanosecond OPOs. These devices have now been established as viable sources of coherent light pulses from ~400 nm in the near-UV to ~ 5 μm the mid-IR. Optical energies from typically a few mJ to as much as 200 mJ can now be made available from such OPOs over a range of pulse durations from 1 to 50 ns. For wavelengths beyond 5 μm, classical materials such as $ZnGeP_2$ have proved highly effective. By using a *cascaded* pumping approach, spectral coverage to ~12 μm in spectrally narrow output pulses has been achieved [81, 82]. The important advances in this area have also led to the commercial realization of several OPO devices based on the nanosecond pumping approach. The recent development of wide-aperture periodically-poled materials PPKTP and PPRTA also promises further progress in high-energy OPOs with more flexible operating parameters [83–85]. The large nonlinear gain of

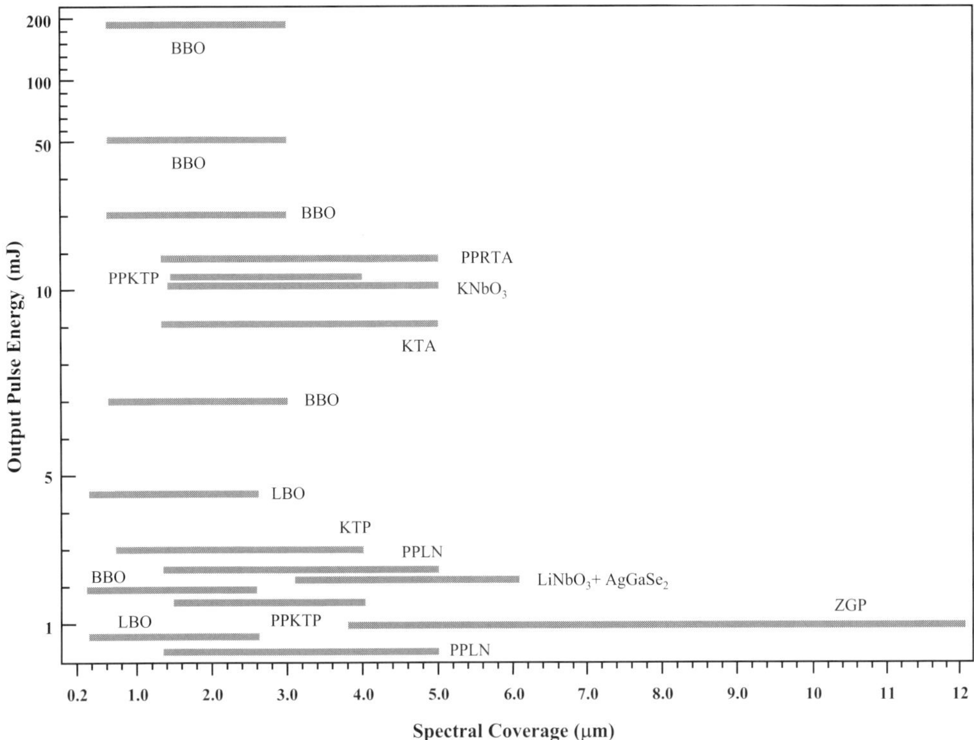

Figure C3.2.16. Survey of the tuning range and output pulse energy of several pulsed nanosecond OPOs developed to date. The indicated tuning limits correspond to the potential tuning range available to the particular crystal in the given phase-matching and pumping geometry, although not necessarily demonstrated. The indicated energies correspond to the maximum pulse energy obtained over the demonstrated tuning range.

PPLN has also enabled fibre laser pumping of nanosecond OPOs [86] and the development of new types of single-pass parametric generators using nanosecond pulse pumping [87]. Figure C3.2.16 provides a summary of the main performance characteristics of several nanosecond OPOs developed to date.

At the same time, the availability of PPLN, PPKTP and its arsenate isomorphs such as PPRTA and PPKTA, with large effective nonlinearity ($d_{\mathrm{eff}} \sim 8$–16 pm V^{-1}, see figure C3.2.14) and long interaction lengths (up to 60 mm in PPLN) has enabled the realization of pulsed OPOs at considerably lower pump pulse energies than practicable with birefringent materials. Combined with the advances in pump laser technology, this has facilitated the development of nanosecond OPOs using high-repetition-rate or long-pulse pump sources with relatively low pump pulse energies (few μJ to few mJ) [88, 89]. While avoiding material damage due to the lower energy fluence, this approach has permitted the successful development of OPOs that can deliver a continuous train of nanosecond pulses at high repetition rates up to 30 kHz, with average powers as high as 10 W. Almost all devices in this operating regime have been pumped by the Nd:YAG laser or its variations, many in diode-pumped all-solid-state configurations. The main operating characteristics of a number of high-repetition-rate nanosecond OPOs are summarized in figure C3.2.17.

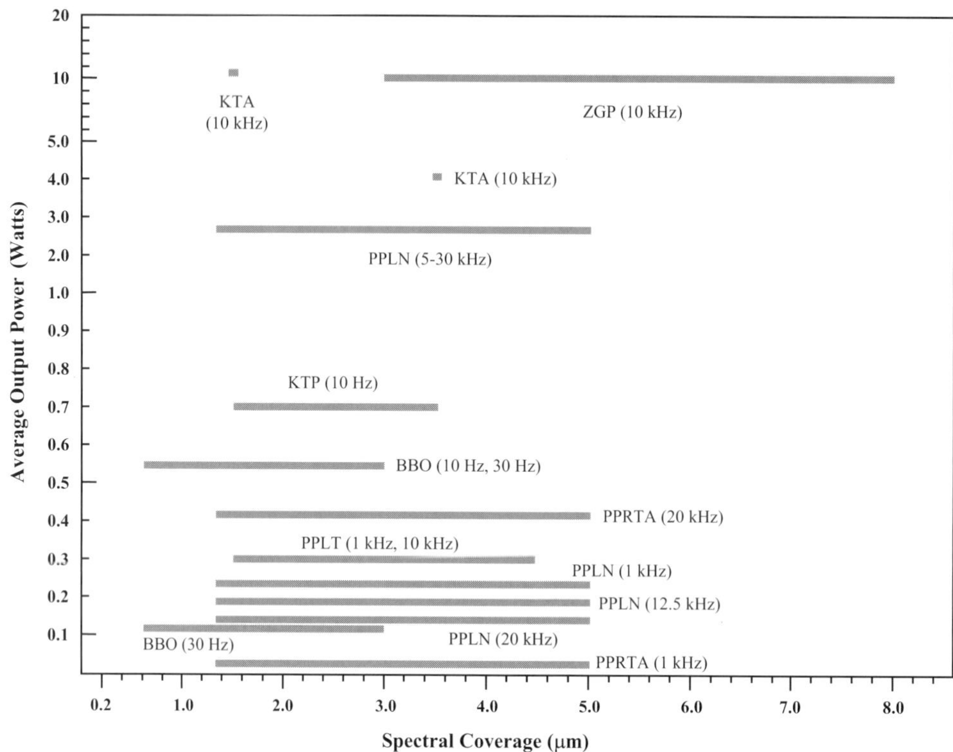

Figure C3.2.17. Survey of performance characteristics of high-repetition-rate nanosecond OPOs. The indicated tuning limits correspond to the potential tuning range available to the particular crystal in the given phase-matching and pumping geometry, although not necessarily demonstrated. The indicated average powers correspond to the maximum power generated over the demonstrated tuning range.

C3.2.9 Synchronously-pumped OPO devices

As noted in section C3.2.5.2.2, synchronously-pumped OPOs may be operated either as steady-state cw or pulsed (quasi-cw) devices. For most practical applications, cw operation is the desired configuration because the output consists of a truly continuous train of identical pulses. In pulsed oscillators, the amplitude, intensity and duration of the output pulses can vary across the pulse envelope, and the output does not constitute a truly repetitive pulse train. However, because of the significantly higher peak powers available from pulsed mode-locked than the equivalent cw mode-locked pump lasers, operation of synchronously-pumped OPOs is more readily attainable under pulsed conditions, particularly in SRO configurations of practical interest. For this reason, most of the early synchronously-pumped OPOs based on birefringent materials have been pulsed oscillators. With the availability of high-power mode-locked laser sources and novel nonlinear materials in recent years, operation of cw synchronously-pumped OPOs has become a practical reality, circumventing the need for pulsed pumping or the use of DRO schemes. We, therefore, focus on a description of synchronously-pumped OPOs in the cw SRO configuration, as this represents the most practical and stable mode of operation for picosecond and femtosecond OPOs.

C3.2.9.1 Picosecond OPO devices

As with nanosecond OPOs, the majority of picosecond OPOs demonstrated to date have been based on the cw mode-locked Nd laser or its variations as the pump source. These include all-solid-state cw picosecond SROs based on KTP [90] and LBO [91, 92], pumped by coupled-cavity mode-locked Nd:YLF lasers, providing output pulses with durations of 1–5 ps at repetition rates of up to 130 MHz. The LBO-based picosecond SRO can generate continuously tunable output from <650 nm to >2.5 μm with a single crystal, while the KTP device is tunable over 900 nm to 1.2 μm. Average output powers of tens of milliwatts at efficiencies of up to 30% are routinely available from these devices. High-power operation of cw picosecond SROs has also been achieved with the use of flashlamp-pumped cw mode-locked neodymium lasers (see chapter C2.3). By using the fundamental output of a Nd:YLF laser at 1.053 μm, total average output powers of up to 2.8 W have been obtained from a KTP picosecond SRO for 14 W of pump, with 800 mW in the idler beam near 3.2 μm [93]. This device generated signal pulses of 12 ps duration for 40 ps input pump pulses at 76 MHz repetition rate. A similar mid-IR oscillator was demonstrated by using temporally compressed pulses from a Nd:YAG laser at 1.064 μm to pump a KTP SRO [94]. Average output powers of up to 350 mW in pulses of 2–3 ps at 75 MHz were generated for 4 W of pump power. More recently, operation of a cw picosecond SRO at unprecedented power levels was achieved by using a diode-pumped, cw mode-locked Nd:YVO_4 oscillator-amplifier laser system at 1.064 μm to pump the crystal of KTA in a NCPM geometry [95]. With the pump radiation consisting of 7 ps pulses at 83 MHz repetition rate and an average power of 29 W, the OPO delivered as much as 6.4 W of idler output at 3.47 μm in the mid-IR. The combined signal and idler output power from this OPO was as high as 21 W, corresponding to an external efficiency in excess of 70%. High-repetition-rate operation of the KTA OPO at frequencies as high as 1.33 GHz was also demonstrated. Picosecond pulses in the visible have also been generated in LBO using the third harmonic of a cw mode-locked Nd:YLF laser [96]. Average powers of 275 mW in the blue spectral range 453–472 nm were obtained in 15 ps pulses at 75 MHz repetition rate.

The operation of cw picosecond SROs has also been extended to the Kerr-lens mode-locked (KLM) Ti:sapphire laser as the pump source (see chapter C2.3). This approach is attractive because the tunability of the Ti:sapphire pump source allows wavelength tuning in the OPO without resort to angle phase-matching. This minimizes intracavity reflection losses caused by crystal rotation, thus maintaining maximum efficiency across the available tuning range. By using type II NCPM in KTP, picosecond pulses in the 1–1.2 μm and 2.3–2.9 μm spectral range have been generated by tuning the Ti:sapphire laser over 720–853 nm [97]. Average output powers of up to 700 mW in 1.2 ps pulses at 82 MHz repetition rate were produced for 1.6 W of pump. The combination of tunable Ti:sapphire laser with temperature-tuned LBO under type I NCPM has also been shown to provide picosecond pulses with continuous tunability over 1–2.4 μm [98]. This SRO could generate average powers of up to 325 mW for 1.2 W of pump in pulses of ~1–2 ps durations at 81 MHz repetition rate. For wavelength coverage further into the infrared, the arsenate isomorphs of KTP, namely KTA and RTA, represent excellent material candidates because of their extended transparency beyond that of KTP and LBO as well as their NCPM capability under Ti:sapphire pump tuning. The use of type II NCPM in KTA has enabled the generation of picosecond pulses out to 3.6 μm, with average powers of up to 400 mW in 1–3 ps pulses [99]. The extension of tuning range of Ti:sapphire-pumped picosecond OPOs to the visible has also been demonstrated by intracavity frequency doubling of signal pulses. By using LBO as the nonlinear material, visible pulses of ~1 ps duration in the 584–771 nm spectral range have been obtained at 320 mW average power with this technique [100].

The availability of periodically-poled materials has also brought about new opportunities for the development of cw picosecond SROs with improved performance characteristics. The large nonlinearity of these materials combined with NCPM capability and broad infrared transparency has enabled the implementation of picosecond SROs with minimal pump power requirement, high output efficiency, practical output powers, and extended tunability into the difficult mid-IR spectral range. In particular, compact all-solid-state SROs based on PPLN or PPRTA and pumped by mode-locked Nd:YLF and Ti:sapphire lasers have been shown to be versatile sources of picosecond pulses with tunability across the entire 1.5–6 μm range [101–103]. These devices exhibit average pump power thresholds as low as 10 mW, and can provide signal output powers of up to ~400 mW in pulses of ~1–5 ps duration, with >10 mW of idler available beyond 5 μm.

Power scaling of cw picosecond SROs to unprecedented levels has become possible by using flashlamp-pumped Nd lasers in combination with PPLN. By using a picosecond Nd:YLF pump laser, an average mid-IR output power of 2.4 W in pulses of ~45 ps at ~76 MHz repetition rate was generated in PPLN, with a mid-IR tuning range from ~2.15 to ~2.6 μm [104]. The SRO generated a total output power of ~5 W for 7.4 W of input pump power. Major enhancements in the mid-infrared output power and photon conversion efficiency have been shown to be possible by using intracavity DFM techniques [105]. In a separate experiment, the use of 80 ps pulses at 76 MHz from a high-power Nd:YAG laser has permitted the generation of more than 4 W of mid-IR idler power from a PPLN picosecond OPO, with a tunable range over 2.2–2.8 μm [106]. With an average input pump power of 18 W, the OPO provided a combined signal and idler output power in excess of 12 W at ~65% extraction efficiency. Operation of cw picosecond OPOs has recently also been extended to novel mode-locked semiconductor pump lasers operating at GHz repetition rates. By using an InGaAs oscillator-amplifier system delivering 7.8 ps pulses at 2.5 GHz, up to 78 mW of mid-IR idler power over a tunable range of 1.99–2.35 μm was generated from a PPLN OPO for 900 mW of input pump power [107]. More recently, operation of a cw picosecond SROs has been achieved at repetition rates as high as 10 GHz by using novel all-solid-state Nd lasers in combination with monolithic PPLN [108]. Figure C3.2.18 provides a summary of the performance characteristics of cw picosecond OPOs developed to date.

C3.2.9.2 Femtosecond OPO devices

Optical parametric oscillators operating in the femtosecond time domain represent the latest class of parametric devices. The high peak intensities available from ultrashort femtosecond pulses are particularly suited to the exploitation of the small nonlinear gain in parametric devices, enabling the practical development of femtosecond OPOs in cw SRO configurations. However, because of the short temporal duration (~100 fs) and large spectral content (~10 nm) of femtosecond pulses, additional effects such as group velocity dispersion, temporal walk-off and spectral acceptance bandwidths play an important role in the operation of these devices. Group velocity dispersion often leads to pulse broadening in femtosecond OPOs, while temporal walk-off can degrade nonlinear gain or even modify the temporal characteristics of the output pulses. The spectral acceptance bandwidth of the crystal can also reduce gain as well as set a lower limit to the minimum attainable pulse duration from the OPO. The influence of temporal walk-off and spectral bandwidth can be minimized by using short crystal lengths of typically 1–2 mm, while the effects of group velocity dispersion can be overcome by the inclusion of dispersion compensation in the OPO cavity, as in conventional femtosecond lasers. The choice of nonlinear crystal, therefore, requires a trade-off with the crystal length. The need for high optical nonlinearity for low operation threshold must be traded against the desire for large phase-matching bandwidth and low temporal walkoff for optimum design. While crystal lengths of 1–2 mm are too short to provide sufficient gain in picosecond and nanosecond oscillators, the substantially higher peak intensities available with femtosecond pulses can adequately compensate for this shortfall in gain. At the same time, the high in-crystal peak intensities (a few GW cm^{-2}) can introduce higher-order nonlinear effects such as self-phase-modulation in femtosecond OPOs. Such effects, which are not generally present in picosecond and nanosecond OPOs, often lead to spectral broadening and chirping of the output pulses and become more pronounced with longer crystal lengths. The spectral broadening due to self-phase-modulation can, however, be exploited for subsequent compression of output pulses. Therefore, in addition to the general phase-matching and material requirements for parametric devices, the design of femtosecond OPOs involves consideration of the pertinent temporal and spectral effects.

The first successful operation of a cw femtosecond OPO was achieved in a 1.4 mm crystal of KTP,

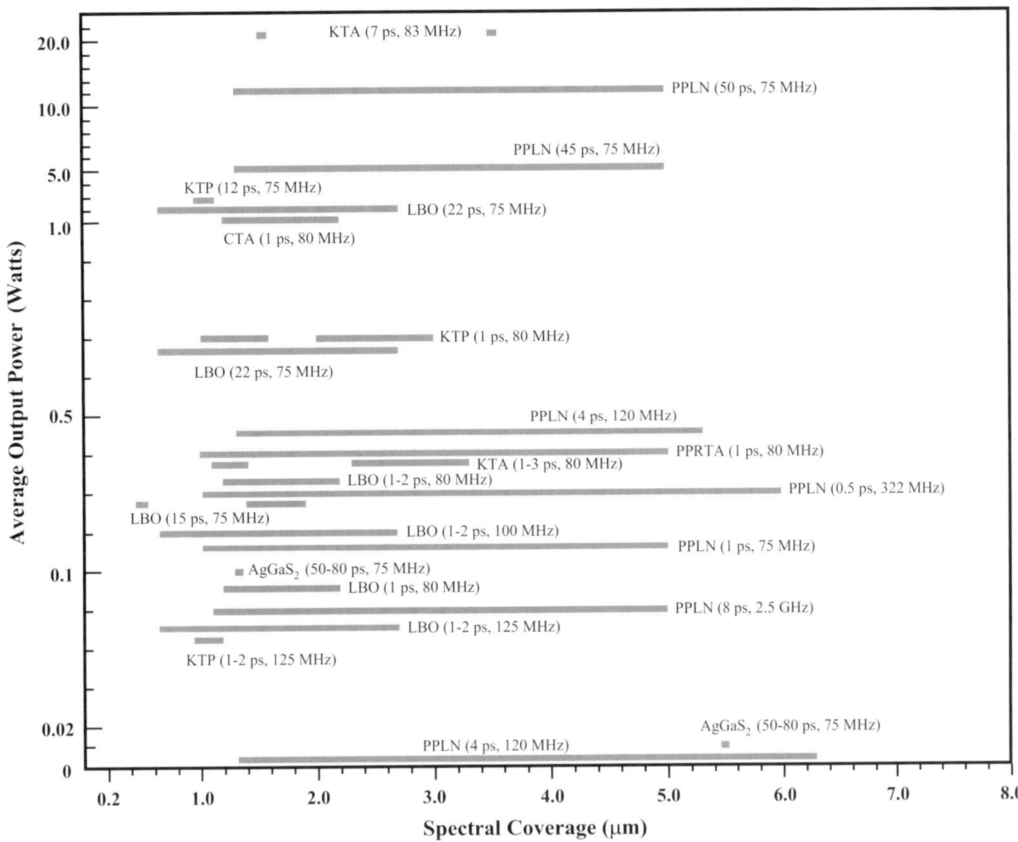

Figure C3.2.18. Survey of the performance characteristics of several cw picosecond OPOs demonstrated to date. The average output powers represent the maximum combined power in the signal and idler beams. The indicated tuning limits correspond to the potential tuning range available to the particular crystal in the given phase-matching geometry and pumping configuration, although the experimental tuning range may have been limited by the available mirrors and crystal coatings. The pulse durations correspond to the measured signal or idler pulse width within the tuning range.

pumped by 170 fs pulses from a colliding-pulse mode-locked dye laser at 620 nm [109]. To access the high peak intensities necessary for oscillation, the KTP crystal was pumped at the intracavity focus of the dye laser in a non-collinear phase-matching geometry. The SRO produced 220 fs signal pulses at milliwatt average power levels in the near-IR. The approach was subsequently extended to an externally-pumped SRO where increased signal powers of up to 30 mW were generated [110].

Soon after, the availability of the KLM Ti:sapphire laser provided a laser source capable of providing cw femtosecond pulses at far higher intensities and many times above SRO threshold. This laser has since become the primary pump source for femtosecond OPOs, enabling the development of a KTP-based device capable of producing total average output powers of up to 750 mW at wavelength $<2\ \mu$m in the near-IR using non-collinear CPM or collinear NCPM schemes [111–113]. Pulse durations of <100 fs at repetition rates of 75–85 MHz are routinely available from these devices. The non-collinear CPM technique was soon extended to other KTP-based systems [114], as well as to KTA [115, 116], CTA [117, 118], RTA [119] and KNB [120], providing femtosecond pulses into the 2–5 μm spectral range using pump and angle tuning schemes. Total average powers of hundreds of milliwatts and mid-IR output powers of tens of milliwatts in

100–200 fs pulses have been obtained from these devices. By using alternative collinear NCPM schemes in combination with Ti:sapphire pump tuning, operation of a femtosecond near-IR OPO based on KTP was achieved, generating 40 fs output pulses [121], while a similar device based on RTA was reported as having a pump power threshold as low as 50 mW [122]. This device provided mid-IR pulses in the 2–3 μm spectral range at 50–100 mW of average power. A femtosecond OPO using type I temperature-tuned NCPM in LBO was also demonstrated, providing $\sim$500 mW in $\sim$100 fs pulses from 1.1–2.4 μm range [123].

The spectral range of Ti:sapphire-pumped femtosecond OPOs has also been extended to the visible by using two different approaches. The first relies on SHG of near-IR signal pulses internal to the OPO cavity. By using BBO as the doubling crystal, output powers of up to 240 mW in 70–100 fs pulses have been generated over a spectral range of 580–660 nm in KTP- and RTA-based OPOs [124, 125]. Intracavity SFM between the pump and signal pulses in BBO has also been used in a KTP OPO to provide sub-100 fs pulses in the blue spectral range from 426–483 nm at $\sim$200 mW average power [126]. Internal SHG and parametric generation has also been simultaneously achieved in an OPO based on a single KTP crystal, providing visible femtosecond pulses in the 530–580 nm range with up to 200 mW average power [127]. The second approach to visible pulse generation has been the use of frequency-doubled output of the Ti:sapphire laser at $\sim$400 nm to directly pump a visible OPO based on BBO as the nonlinear crystal [128]. This approach provided 30 fs signal pulses with average powers of up to 100 mW from 566–676 nm. Subsequently, femtosecond pulses with durations as short as 13 fs were generated from such a device [129].

More recently, the introduction of periodically-poled nonlinear materials has led to further advances in femtosecond OPOs. Devices based on PPLN, PPRTA and PPKTP have been shown to be highly versatile sources of femtosecond pulses, offering minimum pump thresholds, high output powers and vast spectral coverage from a single device. Operation of femtosecond OPOs based on periodically-poled materials has been demonstrated over a wide spectral range from $\sim$1 to $\sim$ 6.8 μm using non-collinear CPM or collinear NCPM schemes combined with temperature, grating, or angle-tuning [130–134]. Because of the large nonlinear gains available, these oscillators exhibit pump power thresholds typically well below 100 mW, enabling the use of all-solid-state Ti:sapphire lasers as the pump source [130, 132]. These devices can readily provide practical powers in excess of 100 mW in the 1–4 μm range, with milliwatt level output attainable in the difficult spectral regions beyond 5 μm. Output pulse durations of 100–200 fs at $\sim$80 MHz repetition rate are typically available. Because of the large nonlinear gain bandwidth of periodically poled crystals, combined with the short interaction lengths used in femtosecond OPOs, devices based on such materials have been shown to be particularly flexible in providing extensive wavelength coverage only through cavity length tuning of the OPO [131–133]. This provides a highly convenient method for wavelength tuning and without the need for pump, angle, grating or temperature tuning. This tuning mechanism occurs because the cavity length detuning introduces a loss at the signal wavelength by reducing the synchronism between the pump and signal pulses. To maintain synchronism and optimize gain, the signal shifts to a more favourable wavelength with a group velocity that satisfies a constant round-trip time. This cavity length tuning, which was observed in the first demonstration of a femtosecond OPO [109], is a useful mechanism for tuning the output wavelength, often by as much as 50 nm in birefringent crystals. In periodically-poled materials, the large phase-matching bandwidths can provide cavity length tuning over hundreds of nanometres, limited by the bandwidth of OPO mirrors.

Femtosecond pulses in the 5–10 μm spectral range have also been generated by using a cascaded, two-stage pumping arrangement, where the output of a Ti:sapphire-pumped CTA OPO has been used as the pump for a second oscillator based on the classical nonlinear material $AgGaSe_2$ [135]. The wavelength flexibility of the primary OPO provides a suitable pump wavelength above the absorption edge of the material, which avoids material absorption and at the same time facilitates the required phase-matching condition for mid-IR generation. Using this technique, femtosecond pulses in the 4–8 μm spectral range with average powers of up to 35 mW and pulse durations of of 300–640 fs have been obtained at 82 MHz repetition rate. Successful implementation of this technique, however, necessitates the use of high-power input laser pump sources.

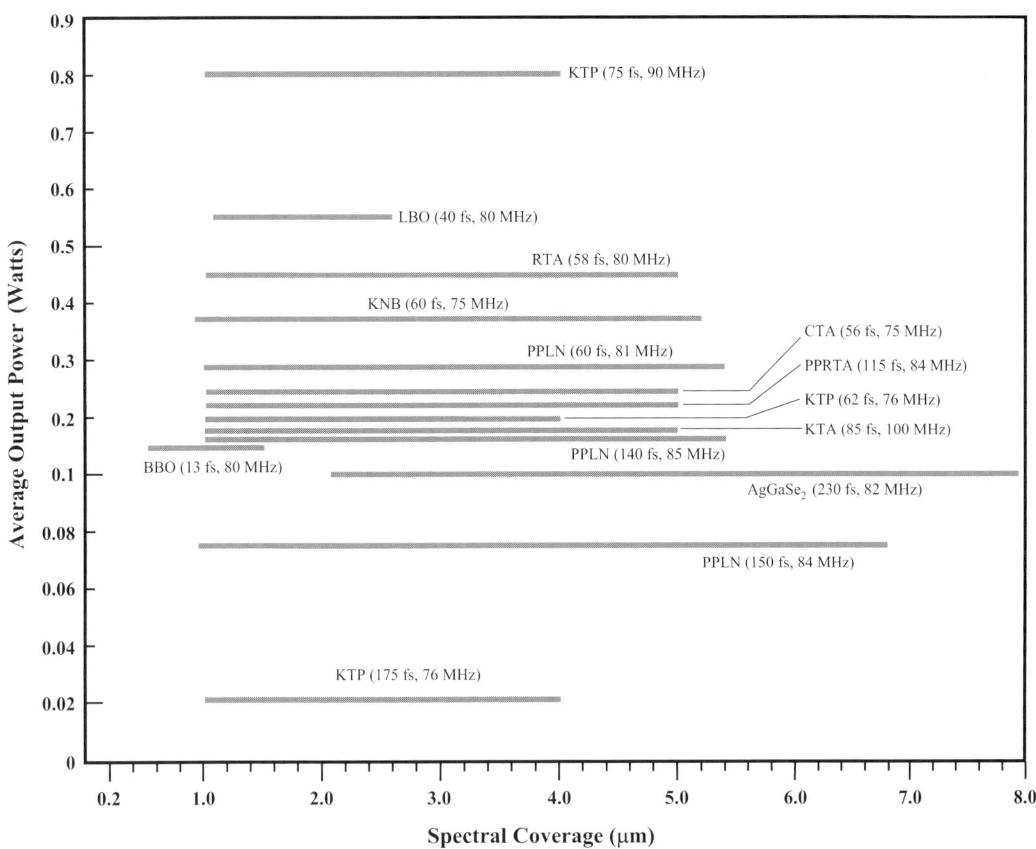

Figure C3.2.19. Survey of the performance characteristics of Ti:sapphire-pumped cw femtosecond OPOs demonstrated to date. The indicated tuning limits correspond to the potential tuning range available to the particular crystal in the given phase-matching geometry and pumping configuration, although the experimental tuning range may have been limited by the available mirrors and crystal coatings. The output powers correspond to the maximum combined signal and idler power generated within the demonstrated tuning range. The pulse durations correspond to the measured signal or idler pulse width within the tuning range.

More recently, operation of Ti:sapphire-pumped femtosecond OPOs at repetition rates as high as 1 GHz has been achieved [136, 137] and the use of mode-locked fibre lasers as pump sources for femtosecond OPOs has been demonstrated [138]. Figure C3.2.19 provides a summary of the performance characteristics of several Ti:sapphire-pumped cw femtosecond OPOs demonstrated to date.

C3.2.10 Summary

This chapter has provided a description of optical parametric devices, with particular emphasis on OPOs operating from the cw to the ultrafast femtosecond time scales. The discussion has included the basic elements of nonlinear frequency conversion, the main operating principles of parametric devices, and a survey of the most recent developments in OPOs. A key element in the rapid advancement of OPO technology in recent years has been the emergence of new nonlinear materials, replacing the more traditional nonlinear crystals, as well as the development of novel resonator and pumping techniques. The advent of PPLN, PPRTA and other periodically-poled materials, as well as birefringent materials such as BBO, LBO, KTP and its arsenate isomorphs have led to the realization of practical OPOs covering spectral regions from <400 nm in the near-UV to >12 μm in mid-IR across all temporal time scales and with unprecedented performance characteristics.

In cw operation, OPOs have been shown to be capable of providing output powers of up to $\sim$4 W and wavelength coverage to >5 μm, while in pulsed nanosecond operation pulse energies as high as 200 mJ and spectral access to >12 μm have been achieved. In picosecond operation, average output powers of more than 20 W have been generated and tuning to >6 μm demonstrated. In the femtosecond regime, average powers in excess of 1.5 W, spectral coverage to $\sim$8 μm and pulse durations as short as $\sim$ 13 fs have been obtained from OPO devices. With the continuing progress in nonlinear materials, laser pump sources, novel phase-matching, pumping and resonator techniques, and the need for new coherent light sources in new spectral regions, particularly the mid-IR, the potential for further advancement of OPO device technology remains strong.

References

[1] Maiman T H 1960 *Nature* **187** 493

[2] Franken P A, Hill A E, Peters C W and Weinreich G 1961 Generation of optical harmonics *Phys. Rev. Lett.* **7** 118–19

[3] Armstrong J A, Bloembergen N, Ducuing J and Pershan P S 1962 Interaction between light waves in a nonlinear dielectric *Phys. Rev.* **127** 1918–39

[4] Ebrahimzadeh M 2003 Parametric light generation *Phil. Trans. R. Soc.* A in press

[5] Shen Y R 1984 *Principles of Nonlinear Optics* (New York: Wiley)

[6] Dimitriev V G, Gurzadyan G G and Nikogosyan D N 1991 *Handbook of Nonlinear Optical Crystals* (Berlin: Springer)

[7] Boyd R W 1992 *Nonlinear Optics* (New York: Academic)

[8] Sutherland R L 1996 *Handbook of Nonlinear Optics* (New York: Marcel Dekker)

[9] Kingston R H 1962 Parametric amplification and oscillation at optical frequencies *Proc. IRE* **50** 472–3

[10] Kroll N M 1962 Parametric amplification in spatially extended media and application to the design of tuneable oscillators at optical frequencies *Phys. Rev.* **127** 1207–11

[11] Akhmanov S A and Khokhlov R V 1963 Concerning one possibility of amplification of light waves *Sov. Phys. J. Exp. Theor. Phys.* **16** 252–4

[12] Louisell W H, Yariv A and Siegman A E 1961 Quantum fluctuations and noise in parametric processes *Phys. Rev.* **124** 1646

[13] Yariv A 1989 *Quantum Electronics* (New York: Wiley)

[14] Smith R G 1976 *Optical Parametric Oscillators (Lasers)* ed A K Levine and A J DeMaria (New York: Marcel Dekker) pp 189–307

[15] Harris S E 1969 Tunable optical parametric oscillators *Proc. IEEE* **57** 2096–113

[16] Giordmaine J A 1962 Mixing of light beams in crystals *Phys. Rev. Lett.* **8** 19–20

[17] Maker P D, Terhune R W, Nissenoff M and Savage C M 1962 Effects of dispersion and focusing on the production of optical harmonics *Phys. Rev. Lett.* **8** 21–2

[18] Born M and Wolf E 1984 *Principles of Optics* (Oxford: Pergamon)

[19] Ebrahimzadeh M and Ferguson A I 1993 *Novel Nonlinear Crystals in Principles and Applications of Nonlinear Optical Materials* (London: Chapman and Hall) pp 99–142

[20] Hobden M V 1967 *J. Appl. Phys.* **38** 4365

[21] Fejer M M, Magel G A, Jundt D H and Byer R L 1992 Quasi-phase-matched second harmonic generation: tuning and tolerances *IEEE J. Quantum Electron.* **28** 2631–54

[22] Myers L E, Eckardt R C, Fejer M M, Byer R L, Bosenberg W R and Pierce J W 1995 Quasi-phase-matched optical parametric oscillators in bulk periodically poled $LiNbO_3$ *J. Opt. Soc. Am.* B **12** 2102–16

[23] Arbore M A, Marco O and Fejer M M 1997 Pulse compression during second-harmonic generation in aperiodic quasi-matching gratings *Opt. Lett.* **22** 865–7

[24] Imeshev G, Galvanauskas A, Harter D, Arbore M A, Procter M and Fejer M M 1998 Engineerable femtosecond pulse shaping by second-harmonic generation with synthetic quasi-phase-matching gratings *Opt. Lett.* **23** 864–6

[25] Beddard T, Ebrahimzadeh M, Reid D T and Sibbett W 2000 Five-optical-cycle pulse generation in the mid-infrared from an optical parametric oscillator based on aperiodically-poled lithium niobate *Opt. Lett.* **25** 1052–54

[26] Byer R L and Piskarskas A S (ed) 1993 Optical parametric oscillation and amplification *J. Opt. Soc. Am.* B **10** 1656–791, 2148–243

[27] Bosenberg W R and Eckardt R C (ed) 1995 Optical parametric devices *J. Opt. Soc. Am.* B **12** 2084–322

[28] Ebrahimzadeh M, Eckardt R C and Dunn M H (ed) 1999 Optical parametric devices and processes *J. Opt. Soc. Am.* B **16** 1477–597

[29] Danielius R, Piskarskas A, Stabinis A, Banfi G P, Di Trapani P and Righini R 1995 Travelling-wave parametric generation of widely tunable, highly coherent femtosecond light pulses *J. Opt. Soc. Am.* B **10** 2222–31

[30] Ebrahimzadeh M and Dunn M H 2000 Optical parametric oscillators *Handbook of Optics* vol IV (New York: McGraw-Hill) pp 2201–72

[31] Ebrahimzadeh M, Turnbull G A, Edwards T J, Stothard D J M, Lindsay I D and Dunn M H 1999 Intracavity continuous-wave singly resonant optical parametric oscillators *J. Opt. Soc. Am.* B **16** 1499–511

[32] Bjorkholm J E, Ashkin A and Smith R G 1970 *IEEE J. Quantum Electron.* **QE-6** 797

[33] Siegman A E 1962 Nonlinear optical effects: an optical power limiter *Appl. Opt.* **1** 739–44

[34] Brosnan S J and Byer R L 1979 Optical parametric oscillator threshold and linewidth studies *IEEE J. Quantum Electron.* **QE-15** 415–31

[35] Pearson J E, Ganiel U and Yariv A 1972 Rise time of pulsed parametric oscillators *IEEE J. Quantum Electron.* **QE-8** 433–40

[36] Bjorkholm J E 1968 Efficient optical parametric oscillation using doubly and singly resonant cavities *Appl. Phys. Lett.* **13** 53–5

[37] Kreuzer L B 1968 High-efficiency optical parametric oscillation and power limiting in $LiNbO_3$ *Appl. Phys. Lett.* **13** 57–9

[38] Bjorkholm J E 1971 Some effects of spatially nonuniform pumping in pulsed optical parametric oscillators *IEEE J. Quantum Electron.* **QE-7** 109–18

[39] Becker M F, Kuizanga D J, Phillion D W and Siegman A E 1974 Analytic expressions for ultrashort pulse generation in mode-locked optical parametric oscillators *J. Appl. Phys.* **45** 3996–4005

[40] Cheung E C and Liu J M 1991 Theory of a synchronously pumped optical parametric oscillator in steady-state operation *J. Opt. Soc. Am.* B **7** 1385–401

Cheung E C and Liu J M 1991 Efficient generation of ultrashort, wavelength-tunable infrared pulses *J. Opt. Soc. Am.* B **8** 1491–506

[41] Ebrahimzadeh M, Henderson A J and Dunn M H 1990 *IEEE J. Quantum Electron.* **26** 1241–52

[42] Colville F G, Dunn M H and Ebrahimzadeh M 1997 Continuous-wave, singly resonant intracavity parametric oscillator *Opt. Lett.* **22** 75–9

[43] Edwards T J, Turnbull G A, Dunn M H and Ebrahimzadeh M and Colville F G 1998 High-power, continuous-wave, singly resonant intracavity optical parametric oscillator *Appl. Phys. Lett.* **72** 1527–9

[44] Scheidt M, Beier B, Knappe R, Boller K-J and Wallenstein R 1995 Diode-laser-pumped continuous-wave KTP optical parametric oscillator *J. Opt. Soc. Am.* B **12** 2087–94

[45] Scheidt M, Beier B, Boller K-J and Wallenstein R 1997 Frequency-stable operation of a diode-pumped continuous-wave $RbTiOAsO_4$ optical parametric oscillator *Opt. Lett.* **22** 1287–9

[46] Bosenberg W R, Drobshoff A, Alexander J I, Myers L E and Byer R L 1996 93% pump depletion, 3.5 W continuous-wave, singly resonant optical parametric oscillator *Opt. Lett.* **21** 1336–8

[47] Powers P E, Kulp T J and Bisson S E 1998 Continuous tuning of a continuous-wave periodically poled lithium niobate optical parametric oscillator by use of a fan-out grating design *Opt. Lett.* **23** 159–61

[48] Klein M E, Lee D H, Meyn J-P, Boller K-J and Wallenstein R 1999 Singly resonant continuous-wave optical parametric oscillator pumped by a diode laser *Opt. Lett.* **24** 1142–4

[49] Gross P, Klein M E, Walde T, Boller K-J, Auerbach M, Wessels P and Fallnich C 2002 Fiber-laser-pumped continuous-wave singly-resonant optical parametric oscillator *Opt. Lett.* **27** 418–20

[50] Stothard D J M, Ebrahimzadeh M and Dunn M H 1998 Low-pump-threshold, continuous-wave, singly resonant optical parametric oscillator *Opt. Lett.* **23** 1895–7

[51] Turnbull G A, McGloin D, Lindsay I D, Ebrahimzadeh M and Dunn M H 2000 Extended mode-hop-free tuning using a dual-cavity, pump-enhanced optical parametric oscillator *Opt. Lett.* **25** 341–3

[52] Lindsay I D, Petridis C, Dunn M H and Ebrahimzadeh M 2001 Continuous-wave pump-enhanced singly-resonant optical parametric oscillator pumped by an external-cavity diode laser *Appl. Phys. Lett.* **78** 871–3

[53] Lindsay I D, Turnbull G A, Dunn M H and Ebrahimzadeh M 1998 Doubly-resonant continuous-wave optical parametric oscillator pumped by a single-mode laser diode *Opt. Lett.* **23** 1889–91

[54] Henderson A J, Roper P M, Borschowa L A and Mead R D 2000 Stable, continuously tunable operation of a diode-pumped doubly resonant optical parametric oscillator *Opt. Lett.* **25** 1264–6

[55] Edwards T J, Turnbull G A, Dunn M H and Ebrahimzadeh M 1998 Continuous-wave, singly-resonant optical parametric oscillator based on periodically-poled $RbTiOAsO_4$ *Opt. Lett.* **23** 837–9

[56] Edwards T J, Turnbull G A, Dunn M H and Ebrahimzadeh M 2000 Continuous-wave, singly-resonant optical parametric oscillator based on periodically-poled $KTiOPO_4$ *Opt. Exp.* **6** 58–63

[57] Giordmaine J A and Miller R C 1965 Tunable coherent parametric oscillation in $LiNbo_3$ at optical frequencies *Phys. Rev. Lett.* **14** 973–6

[58] Cheng L K, Bosenberg W R and Tang C L 1988 Broadly tunable optical parametric oscillation in β-BaB_2O_4 *Appl. Phys. Lett.* **53** 175–7

[59] Fan Y X, Eckardt R C and Byer R L 1988 *Appl. Phys. Lett.* **50** 2014

[60] Bosenberg W R, Cheng L K and Tang C L 1989 Ultraviolet optical parametric oscillation in β-BaB_2O_4 *Appl. Phys. Lett.* **54** 13–15

[61] Fan Y X, Eckardt R C, Byer R L, Chen C and Jiang A D 1989 *IEEE J. Quantum Electron.* **25** 1196

[62] Bosenberg W R, Pelouch W S and Tang C L 1989 High-efficiency and narrow-linewidth operation of a two-crystal β-BaB_2O_4 optical parametric oscillator *Appl. Phys. Lett.* **55** 1952–4
[63] Bosenberg W R and Tang C L 1990 Type II phase matching in β-barium borate optical parametric oscillator *Appl. Phys. Lett.* **56** 1819–21
[64] Kato K 1991 Parametric oscillation at 3.2 μm in KTP pumped at 1.064 μm *IEEE J. Quantum Electron.* **27** 1137–40
[65] Hanson F and Dick D 1991 Blue parametric generation from temperature-tuned LiB_3O_5 *Opt. Lett.* **16** 205–7
[66] Bosenberg W R, Cheng L K and Bierlein J D 1994 Optical parametric frequency-conversion properties of $KTiOAsO_4$ *Appl. Phys. Lett.* **65** 2765–7
[67] Gloster L A W, Jiang Z X and King T A 1994 *IEEE J. Quantum Electron.* **30** 2961
[68] Urschel R, Fix A, Wallenstein R, Rytz D and Zysset B 1995 *J. Opt. Soc. Am.* B **12** 726
[69] Cui Y, Dunn M H, Norrie C J, Sibbett W, Sinclair B D, Tang Y and Terry J A C 1992 *Opt. Lett.* **17** 646
[70] Cui Y, Withers D E, Rae C F, Norrie C J, Tang Y, Sinclair B D, Sibbett W and Dunn M H 1993 *Opt. Lett.* **18** 122–4
[71] Marshall L R, Kasinski J and Burnham R L 1991 Efficient optical parametric oscillation at 1.6 μm *Opt. Lett.* **16** 1680
[72] Komine H 1988 Optical parametric oscillation in beta-barium borate crystal pumped by an XeCl excimer laser *Opt. Lett.* **13** 643–5
[73] Ebrahimzadeh M, Henderson A J and Dunn M H 1990 An excimer-pumped β-BaB_2O_4 optical parametric oscillator tunable from 354 nm to 2.370 μm *IEEE J. Quantum Electron.* **26** 1241–52
[74] Ebrahimzadeh M, Robertson G and Dunn M H 1991 *Opt. Lett.* **16** 767
[75] Bosenberg W R and Guyer D R 1993 *J. Opt. Soc. Am.* B **10** 1716
[76] Fix A, Schroder T, Wallenstein R, Haub J G, Johnson M J and Orr B J 1993 *J. Opt. Soc. Am.* B **10** 1744
[77] Boon-Engering J M, van der Veer W E and Gerritsen J W 1995 Bandwidth studies of an injection-seeded β-barium borate optical parametric oscillator *Opt. Lett.* **20** 380–2
[78] Haub J G, Johnson M J, Powell A J and Orr B J 1995 Bandwidth characteristics of a pulsed optical parametric oscillator: application to degenerate four-wave mixing spectroscopy 1995 *Opt. Lett.* **20** 1637–9
[79] Urschel R, Bader U, Borzutsky A and Wallenstein R *J. Opt. Soc. Am.* B
[80] Ribet I, Drag C, Lefebvre M and Rosencher E 2002 *Opt. Lett.* **27** 255
[81] Vodopyanov K L, Ganikhanov F, Maffetone J P, Zwieback I and Ruderman W 2000 *Opt. Lett.* **25** 841
[82] Ganikhanov F, Caughey T and Vodopyanov K L 2001 *Opt. Lett.* **25** 818
[83] Karlsson H, Olson M, Arvidsson G, Laurell F, Bader U, Borsutzky A, Wallenstein R, Wickstrom S and Gustafsson M 1999 *Opt. Lett.* **24** 330–2
[84] Hellstrom J, Pasiskevicius V, Karlsson H and Laurell F 2000 *Opt. Lett.* **25** 174
[85] Feve J-P, Pacaud O, Boulanger B, Menaert B, Hellstrom J. Pasiskevicius V and Laurell F 2001 *Opt. Lett.* **26** 1882
[86] Britton P E, Offerhaus H L, Richardson D J, Smith P G R, Ross G W and Hanna D C 1999 *Opt. Lett.* **24** 975–7
[87] Missey M J, Dominic V, Powers P E and Schepler K L 1999 *Opt. Lett.* **24** 1227–9
[88] Myers L E and Bosenberg W R 1997 *IEEE J. Quantum Electron.* **33** 1663–72
[89] Bader U, Bartschke J, Klimov I, Orsutzky A B and Wallenstein R 1998 *Opt. Commun.* **147** 95–8
[90] McCarthy M J and Hanna D C 1992 *Opt. Lett.* **17** 402–4
[91] Hall G J, Ebrahimzadeh M, Robertson A, Malcolm G P A and Ferguson A I 1993 *J. Opt. Soc. Am.* B **10** 2168–79
[92] Butterworth S D, McCarthy M J and Hanna D C 1993 *Opt. Lett.* **18** 1429–31
[93] Grasser Ch, Wang D, Beigang R and Wallenstein R 1993 *J. Opt. Soc. Am.* B **10** 2218–21
[94] Chung J and Siegman A E 1993 Singly resonant continuous-wave mode-locked $KTiOPO_4$ optical parametric oscillator pumped by a Nd:YAG Laser *J. Opt. Soc. Am.* B **10** 2201–10
[95] Ruffing B, Nebel A and Wallenstein R 1998 All-solid-state cw mode-locked picosecond $KTiOAsO_4$ (KTA) optical parametric oscillator *Appl. Phys.* B **67** 537
[96] Wang D, Grasser C, Beigang R and Wallenstein R 1997 The generation of tunable blue ps-light-pulses from a cw mode-locked LBO optical parametric oscillator *Opt. Commun.* **138** 87–90
[97] Nebel A, Fallnich C, Beigang R and Wallenstein R 1993 Noncritically phase-matched continuous-wave mode-locked singly resonant optical parametric oscillator synchronously pumped by a Ti:sapphire laser *J. Opt. Soc. Am.* B **10** 2195–200
[98] Ebrahimzadeh M, French S and Miller A 1995 Design and performance of a singly resonant picosecond LiB_3O_5 optical parametric oscillator synchronously pumped by a self-mode-locked Ti:sapphire laser *J. Opt. Soc. Am.* B **12** 2180–91
[99] French S, Ebrahimzadeh M and Miller A 1996 High-power, high-repetiton-rate picosecond optical parametric oscillator for the near- to mid-infrared *Opt. Lett.* **21** 131–3
[100] French S, Ebrahimzadeh M and Miller A 1996 High-power, high-repetition-rate picosecond optical parametric oscillator tunable in the visible *Opt. Lett.* **21** 976–8
[101] Kennedy G T, Reid D T, Miller A, Ebrahimzadeh M, Karlsson H, Arvidsson G and Laurell F 1998 Broadly tunable mid-infrared picosecond optical parametric oscillator based on periodically-poled $RbTiOAsO_4$ *Opt. Lett.* **23** 503–5
[102] Lefort L, Peuch K, Ross G W, Svirko Y P and Hanna D C 1998 Optical parametric oscillation out to 6.3 μm in periodically poled lithium niobate under strong idler absorption *Appl. Phys. Lett.* **73** 1610–12
[103] Phillips P J, Das S and Ebrahimzadeh M 2000 *Appl. Phys. Lett.* **77** 469
[104] Finsterbusch K, Urschel R and Zakarias H 2000 Fourier-transform-limited, high-power picosecond optical parametric oscillator

based on periodically poled lithium niobate *Appl. Phys.* B **70** 741–6

[105] Dearborn M E, Koch K, Moore G T and Diels J C 1998 Greater than 100% photon-conversion efficiency from an optical parametric oscillator with intracavity difference-frequency mixing *Opt. Lett.* **23** 759–61

[106] Hoyt C W, Sheik-Bahae M and Ebrahimzadeh M 2002 High-power picosecond optical parametric oscillator based on periodically poled lithium niobate *Opt. Lett.* **27** 1543

[107] Robertson A, Klein M E, Tremont M A, Boller K-J and Wallesnstein R 2000 2.5-GHz repetition-rate singly resonant optical parametric oscillator synchronously pumped by a mode-locked diode oscillator amplifier system *Opt. Lett.* **25** 657

[108] Lecomte S, Krainer L, Paschotta R, Dymott M J P, Weingarten K J and Keller U 2002 *Opt. Lett.* **27** 1714

[109] Edelstein D C, Wachman E S and Tang C L 1989 Broadly tunable high repetition rate femtosecond optical parametric oscillator *Appl. Phys. Lett.* **54** 1728–30

[110] Mak G, Fu Q and van Driel H M 1992 Externally pumped high repetition rate femtosecond infrared optical parametric oscillator *Appl. Phys. Lett.* **60** 542–4

[111] Fu Q, Mak G and van Driel H M 1992 High-power, 62-fs infrared optical parametric oscillator synchronously pumped by a 76-MHz Ti:sapphire laser *Opt. Lett.* **17** 1006–8

[112] Pelouch W S, Powers P E and Tang C L 1992 Ti:sapphire-pumped, high-repetition-rate femtosecond optical parametric oscillator *Opt. Lett.* **17** 1070–2

[113] Dudley J M, Reid D T, Ebrahimzadeh M and Sibbett W 1994 Characteristics of a noncritically phase-matched Ti:sapphire-pumped femtosecond optical parametric oscillator *Opt. Commun.* **104** 419–30

[114] McCahon S W, Anson S A, Jang D-J and Boggess T F 1995 Generation of 3–4 mm femtosecond pulses from a synchronously pumped, critically phase-matched $KTiOPO_4$ optical parametric oscillator *Opt. Lett.* **22** 2309–11

[115] Powers P E, Ramakrishna S, Tang C L and Cheng L K 1993 Optical parametric oscillation with $KTiOAsO_4$ *Opt. Lett.* **18** 1171–3

[116] Reid D T, McGowan C, Ebrahimzadeh M and Sibbett W 1997 Characterization and modelling of a noncollinearly phase-matched femtosecond optical parametric oscillator based on KTA and operating beyond 4 μm *IEEE J. Quantum Electron.* **33** 1–9

[117] Powers P E, Tang C L and Cheng L K 1994 High-repetition-rate femtosecond optical parametric oscillator based on $CsTiOAsO_4$ *Opt. Lett.* **19** 37–9

[118] Holtom G R, Crowell R A and Cheng L K 1995 Femtosecond mid-infrared optical parametric oscillator based on $CsTiOAsO_4$ *Opt. Lett.* **20** 1880–2

[119] Powers P E, Tang C L and Cheng L K 1994 High-repetition-rate femtosecond optical parametric oscillator based on $RbTiOAsO_4$ *Opt. Lett.* **19** 1439–41

[120] Spence D E, Wielandy S and Tang C L 1996 High average power, high repetition rate femtosecond pulse generation in the 1–5 μm region using an optical parametric oscillator *Appl. Phys. Lett.* **68** 452–4

[121] Dudley J M, Reid D T, Ebrahimzadeh M and Sibbett W 1994 Characteristics of a noncritically phase-matched Ti:sapphire-pumped femtosecond optical parametric oscillator *Opt. Commun.* **104** 419–30

[122] Reid D T, Ebrahimzadeh M and Sibbett W 1995 Noncritically phase-matched Ti:sapphire-pumped femtosecond optical parametric oscillator based on $RbTiOAsO_4$ *Opt. Lett.* **20** 55–7

[123] Kafka J D, Watts M L and Pieterse J W 1995 Synchronously pumped optical parametric oscillators with LiB_3O_5 *J. Opt. Soc. Am.* B **12** 2147–57

[124] Ellingson R J and Tang C L 1993 High-power, high-repetition-rate femtosecond pulses tunable in the visible *Opt. Lett.* **18** 438–40

[125] Reid D T, Ebrahimzadeh M and Sibbett W 1995 Efficient femtosecond pulse generation in the visible in a frequency-doubled optical parametric oscillator based on $RbTiOAsO_4$ *J. Opt. Soc. Am.* B **12** 1157–63

[126] Shirakawa A, Mao H W and Kobayashi T 1996 Highly efficient generation of blue-orange femtosecond pulses from intracavity-frequency-mixed optical parametric oscillator *Opt. Commun.* **123** 121–8

[127] Kartaloglu T, Koprulu K G and Aytur O 1997 Phase-matched self-doubling optical parametric oscillator *Opt. Lett.* **22** 280–2

[128] Driscoll T J, Gale G M and Hache F 1994 Ti:sapphire second-harmonic-pumped visible range femtosecond optical parametric oscillator *Opt. Commun.* **110** 638–44

[129] Gale G M, Hache F and Cavallari M 1998 Broad-bandwidth parametric amplification in the visible: femtosecond experiments and simulations *IEEE J. Sel. Top. Quantum Electron.* **4** 224–9

[130] Burr K C, Tang C L, Arbore M A and Fejer M M 1997 Broadly tunable mid-infrared femtosecond optical parametric oscillator using all-solid-state-pumped periodically poled lithium niobate *Opt. Lett.* **22** 1458–60

[131] Reid D T, Penman Z, Ebrahimzadeh M, Sibbett W, Karlsson H and Laurell F 1997 Broadly tunable infrared femtosecond optical parametric oscillator based on periodically poled $RbTiOAsO_4$ *Opt. Lett.* **22** 1397–9

[132] Reid D T, Kennedy G T, Miller A, Sibbett W and Ebrahimzadeh M 1998 Widely tunable near- to mid-infrared femtosecond and picosecond optical parametric oscillators using periodically poled $LiNbO_3$ and $RbTiOAsO_4$ *IEEE J. Selected Topics Quantum Electron.* **4** 238–48

[133] McGowan C, Reid D T, Penman Z E, Ebrahimzadeh M, Sibbett W and Jundt D T 1998 Femtosecond optical parametric oscillator based on periodically poled lithium niobate *J. Opt. Soc. Am.* B **15** 694–701

[134] Kartaloglu T, Koprulu K G, Aytur O, Sundheimer M and Risk W P 1998 Femtosecond optical parametric oscillator based on periodically poled $KTiOPO_4$ *Opt. Lett.* **23** 61–3

[135] Marzenell S, Beigang R and Wallenstein R 1999 Synchronously pumped femtosecond optical parametric oscillator based on $AgGaSe_2$ tunable from 2 μm to 8 μm *Appl. Phys.* B **69** 423–8

[136] Zhang X P, Hebling J, Bartels A, Nau D, Kuhl J, Ruhle W W and Giessen H 2002 *Appl. Phys. Lett.* **80** 1873–5
[137] Jiang J and Hasama T 2003 *Opt. Commun.* **220** 193–202
[138] O'Connor M V, Watson M A, Shepherd D P, Hanna D C, Price J H V, Malinowski A, Nilsson J, Broderick N G R, Richardson D J and Lefort L 2002 *Opt. Lett.* **27** 1052

C3.3 Laser stabilization for precision measurements

G P Barwood and P Gill

C3.3.1 Basic spatial and spectral characteristics of lasers

Lasers are uniquely suited to applications in precision measurement due to their output beam spatial and spectral characteristics. They emit light in a near diffraction-limited beam and may also emit in a single longitudinal mode, with a narrow linewidth. By way of introduction to this chapter, a few comments covering spatial characteristics are given, in so far as they can affect the stabilization and application of narrow linewidth lasers. Subsequently, we concentrate solely on spectral characteristics.

Ideally, the emitted light from a laser has a Gaussian spatial profile (the TEM_{00} mode), although more complicated spatial mode patterns are possible. The intensity profile is then of the form $I = I_0 \exp(-2r^2/w_0^2)$, where r is the distance from the beam centre and w_0 is the $1/e$ amplitude radius of the beam [1]. HeNe and YAG lasers are examples of systems that emit near-Gaussian beams. However, optical elements in the cavity, for example Brewster windows or even the gain medium itself, can make the beam slightly astigmatic and, therefore, non-Gaussian. Diode lasers are a common example of a system with a poor spatial beam profile [2]. In this case, the beam profile is determined by the shape of the semiconductor junction. Since the junction's cross section tends to be rectangular, even for low-power lasers, the far-field beam pattern is elliptical. Special beam-shaping optics are often used in order to achieve a more circular Gaussian beam profile. Near-Gaussian beam quality is important for frequency stabilization to transitions in gas or vapour cells, where higher-order spatial modes can give rise to small frequency shifts. It is also important when frequency stabilizing lasers to optical cavities, otherwise the input coupling efficiency will be low. Poor spatial mode quality can also limit the use of lasers in interferometry, causing beam distortion when propagating over long distances.

Spectrally, lasers may emit light in either a single longitudinal mode or via several modes throughout their spectral gain profile. Some applications in precision measurement, for example displacement measuring interferometry, can use either single or multi-mode laser sources. The latter can only be used where the optical path difference is less than a few tens of cms. Many other applications in spectroscopy and optical frequency metrology inevitably require the use of a single-mode source. There is a variety of methods available for forcing multi-mode lasers to emit in only a single mode. A common technique is to use a low-finesse etalon inside the cavity, creating a small loss at frequencies other than at the desired mode. Competition between adjacent modes in the laser then ensures that only this mode lases.

With a single longitudinal mode laser, there are a number of factors which can determine the spectral linewidth. In many common cw single longitudinal mode lasers, such as HeNe devices, the linewidth is often determined by the level of air or ground borne vibrations in the laboratory, perturbing the cavity length and, hence, the emitted frequency. If these vibration levels are reduced, then in HeNe lasers, there remains the problem of noise due to the discharge. In optically-pumped solid state lasers, where there is no discharge, the laser linewidth is far smaller. In particular, the diode-pumped Nd:YAG laser at 1064 nm can have a linewidth in the region of 10 kHz. The frequency-doubled Nd:YAG laser at 532 nm is thus well suited to frequency stabilization, since the narrow linewidth should ensure good short-term frequency stability. Laser diodes, which have a much larger tuning range (several THz for a given device), however, have a frequency noise problem associated with their short cavity length. In this case, the dominant problem is with the very short, low-finesse cavity of a diode. This dominant noise source arises from noise due to spontaneous emission in the cavity. The linewidth of a diode laser is further increased because, following each spontaneous emission event, there is a small change $\Delta n''$ in the imaginary part of the refractive index of the semiconductor. This is caused by a change in the carrier density, which also changes the real part of the refractive index by $\Delta n'$. Defining $\alpha = \Delta n'/\Delta n''$, the formula for the full-width-at-half-maximum (FWHM) linewidth, γ, is

$$\gamma = \frac{2\pi h f \eta \Delta f_{1/2}^2 (1+\alpha^2)}{P}. \tag{C3.3.1}$$

This is the modified Schawlow–Townes formula: the original formula has $\alpha = 0$ but, for semiconductor lasers, $\alpha \simeq 5$ is more appropriate [3, 4]. Equation (C3.3.1) shows that the FWHM depends on the cavity half-width half maximum (HWHM) ($\Delta f_{1/2}$) and the emitted power (P). In equation (C3.3.1), h is the Planck constant, f the optical frequency and η is the inversion factor ($\eta = N_2/(N_2 - N_1)$), with N denoting the occupancy of the two levels. For a diode laser, with a cavity length of 1 mm and a mirror reflectivity of 30%, the limit is in the region of a few tens of MHz and, therefore, this dominates over other effects. In contrast, for a typical HeNe laser, this Schawlow–Townes limit is in the region of a few mHz and so other effects, principally acoustic vibration or HeNe discharge noise, always dominate.

C3.3.2 Advantages of frequency stabilization

For many applications in precision measurements, it is important to have good laser frequency stability and to know the absolute optical frequency (chapter D2.1). Knowledge of the absolute frequency at a particular level of accuracy implies that the laser also needs to be reproducible at that level from day to day. In order to develop a frequency-stabilized laser, it is generally necessary to use a single-mode system. Common choices of single-mode laser include the HeNe, Nd:YAG, diode or titanium–sapphire (Ti:S) laser. The Ti:S laser has now almost completely superseded the dye laser [5], where a tunable high-power source is required. These lasers have been listed in order of increasing tuning range. HeNe lasers, for example, emit only within a Doppler-broadened gain profile of typically 1 GHz, at very specific wavelengths determined by the neon transitions. In the visible part of the spectrum, these wavelengths are 543, 594, 612, 633 and 640 nm. The optically pumped neodymium-doped YAG tunes over a few tens of GHz. The output wavelength is determined by transitions in Nd, the most common being at 1064 nm. This output is often doubled to produce 532 nm and higher harmonics are possible. Laser diodes can be produced to cover most wavelengths between 630 and 1000 nm, with other wavelengths available at 400, 1300 and 1500–1600 nm. Individual units can tune several THz. However, in order to develop a stabilized laser at a particular frequency, the most difficult problem is generally to source a diode which can tune to exactly the desired frequency. Manufacturers may be able to provide this service, and some companies hold considerable stock to allow customer selection. Finally, a Ti:S laser can be used to cover the entire region between 700 and 1000 nm. The high output power also allows frequency doubling from 350 to 500 nm. All these lasers can be designed to emit in a single longitudinal mode but, as can readily be appreciated, this single mode can emit anywhere within a considerable range. For many applications it is advantageous to restrict laser operation within a much tighter bandwidth by frequency stabilization.

C3.3.3 Applications of frequency-stabilized lasers

Stabilized lasers can be used in a number of applications, such as interferometry (including dimensional metrology and gravitational wave detection), measurements of fundamental constants, including investigations regarding their possible variation with time, frequency standards and optical clocks. In the following sections (C3.3.3.1–C3.3.3.3), some of these applications are discussed in outline.

C3.3.3.1 Frequency-stabilized lasers as sources for dimensional interferometry

Interferometry is a widely used technique for the precision measurement of length. One of the most common interferometer systems for length (or, more strictly, displacement) measurement is the Michelson interferometer, shown in figure C3.3.1 (e.g. [6]). In this system, light from the laser is split, with one beam going to a fixed (or reference) reflector (usually a corner cube) and the other beam travelling along the path to be measured. On return, the beams are re-combined at the beamsplitter and produce optical fringes. The resulting intensity is given by

$$I = |E_{\mathrm{meas}} \exp(\mathrm{i}(\omega t + 2kL)) + E_{\mathrm{ref}} \exp(\mathrm{i}\omega t)|^2 \tag{C3.3.2}$$

where L is the distance between the beamsplitter and the measurement corner cube. The reference path and measurement path electric field amplitudes at the detector are denoted by E_{ref} and E_{meas} respectively. It is easily shown that this equation becomes:

$$I = E_{\mathrm{meas}}^2 + E_{\mathrm{ref}}^2 + 2E_{\mathrm{ref}}E_{\mathrm{meas}} \cos 2kL. \tag{C3.3.3}$$

The intensity function, therefore, has a periodicity given by $kL = n\pi$, where n is an integer and the wavenumber k is $2\pi/\lambda$. The detected periodicity or 'fringe' can be electronically counted as a function of displacement of the measurement retro-reflector, using the initial position of the retro-reflector as the zero reference position. The displacement is then given by $L = \frac{1}{2}n\lambda$, where n is the number of fringes counted and the wavelength is λ. For a 1 m displacement, for example, this corresponds to around three million fringes. Provided we know the wavelength very well, then this gives a simple but accurate means to determine displacement. However, we should note that if this measurement is performed in air, then the air refractive index must also be taken into account [7–9], so that a knowledge of the laser vacuum wavelength (and, hence, the frequency) is only part of the measurement problem. In most systems, the air refractive index is calculated from the air temperature, pressure and relative humidity. System measurement capabilities of 1 part in 10^6 over the range of 1 m to tens of metres are typical for systems where the refractive index correction is automatically compensated for. This capability can be improved to approximately 1 part in 10^7, provided more accurate environmental measurements are made, and the laser is stabilized to a few parts in 10^8 or better. Alternatively, interferometric techniques may also be used either to measure the refractive index absolutely [10] or track changes in the refractive index. Although we require a known and stable laser frequency in order to obtain the most accurate measurements, a multi-mode HeNe laser may be used if we only require an accuracy of no better than a few in 10^5. This is because the gain profile of the laser is only around 1 GHz (two parts in 10^6 of the optical frequency).

From equation (C3.3.3), the intensity output of the Michelson interferometer varies between $I_{\mathrm{max}} = (E_{\mathrm{meas}} + E_{\mathrm{ref}})^2$ and $I_{\mathrm{min}} = (E_{\mathrm{meas}} - E_{\mathrm{ref}})^2$ and we may define a fringe visibility function as follows

$$V = \frac{(I_{\mathrm{max}} - I_{\mathrm{min}})}{(I_{\mathrm{max}} + I_{\mathrm{min}})}. \tag{C3.3.4}$$

In this case, considering only the effect of unequal beam intensities in the two arms of the interferometer, we have

$$V = \frac{2E_{\mathrm{ref}}E_{\mathrm{meas}}}{(E_{\mathrm{ref}}^2 + E_{\mathrm{meas}}^2)}. \tag{C3.3.5}$$

The fringe visibility is, therefore, unity in the ideal case where the measurement and reference amplitudes are equal. However, there are other factors which can reduce this visibility from unity and the most important

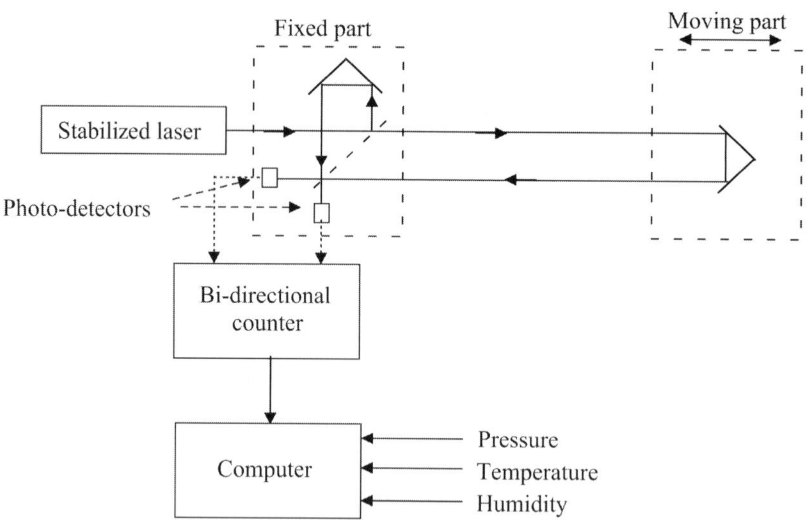

Figure C3.3.1. Michelson interferometer system for measuring displacement.

of these is probably the laser linewidth. It can be shown that the fringe visibility is proportional to the Fourier transform of the laser spectral profile. For example, for a laser diode with a Lorentzian profile and FWHM of $c/\pi L_c$, the fringe visibility varies as $V \propto \exp(-L/L_c)$, where L_c is the laser coherence length. This demonstrates another reason for frequency stabilizing the laser, namely to prevent the laser from tuning to a region where the spectrum is multi-mode with a corresponding reduction in fringe visibility. This is particularly the case with HeNe lasers. With a narrow-linewidth source (for example, <1 MHz), the distance that can be measured is generally determined by air turbulence, which causes rapid changes in the fringe visibility as the return-beam position varies spatially. For long distances, the optical beam must also be fairly large (typically 10 mm in diameter) in order to prevent too much beam divergence over the return distance. Commercial systems are available that can measure displacements of up to 80 m. Further discussion on the use of inteferometry in dimensional metrology is given in sections D2.1.1.3 and D2.1.1.4.

It is important to emphasize that the Michelson system described in the previous paragraph is measuring the displacement of the corner cube and not an absolute distance. For distance measurement, other techniques are available—for example swept frequency interferometry [11, 12 also references therein], where the interferometer path-lengths are kept fixed but the laser frequency is scanned over a known frequency range in order to obtain scanning fringes. This will require the frequency at both ends of the scan to be accurately known.

In order to calibrate material length standard artefacts, such as gauge blocks, the method of excess fringe fractions is used (for a diagram showing a gauge block interferometer system, see figure D2.1.18). With this method, one end of the gauge block is wrung to a polished base-plate and wedged fringes are formed between the base-plate and a reference flat, and between the top end of the gauge block and the reference flat[1]. For a single fixed wavelength, a fringe discontinuity (i.e. fringe fraction) between the two sets of wedged fringes is observed. However, this does not provide information on the difference in order number corresponding to the length of the gauge. In order to determine this, fringe fractions are measured at a number of wavelengths and, together with a nominal value for gauge length determined by initial lower-accuracy

[1] For details of a commercially available gauge block interferometer system, see http://www.npl.co.uk/length/dmet/services/equipment/gbi.html.

methods, the order number can be unambiguously assigned and a higher accuracy measurement of length achieved. Length measurement accuracies of approximately one part in 10^7 are possible using this technique, limited by environmental and surface features, rather than the laser frequency uncertainty. More details on gauge block calibration are given in section D2.1.4.1.

C3.3.3.2 Interferometry for gravitational wave detection

Gravitational waves, originally predicted by Einstein, are estimated to produce distortions in space, although at levels well below that which are achieved by conventional interferometry [13]. For example, violent astrophysical events, such as supernova explosions, require strain sensitivity measurements of around a part in 10^{21}. Although this represents a major difficulty, it is only necessary to achieve this in a certain bandwidth. For ground-based interferometers, this is typically a few tens of Hz in the hundreds of Hz region. A number of novel 'recycling' interferometer systems has been developed for this purpose and shown to have sensitivities close to this required value. For example, in the power recycling mode [14], light returning towards the laser is reflected back into the interferometer in order to assist the power build-up. It can be shown [13] that for powers in the region of 7×10^7 W optical sensitivity close to the standard quantum limit is achieved. For more details of specific systems, see the websites cited in the footnotes, which have extensive lists of recent publications. There are a number of groups worldwide with experiments targeted on gravity wave detection and developing sensitive ground-based interferometers for this purpose. Interferometer arm dimensions are typically 600 m to 3 km, with evacuated tubes and interferometer optics and mirrors suspended from seismically damped supports. The UK–German ground-based interferometer GEO600, based around a 600 m interferometer system, is currently being tested near Hanover. This is a recycling interferometer system, based on a 10 W single-mode YAG laser at 1064 nm and has a sensitivity of 10^{-20}–10^{-21}. Power recycling increases the power in the interferometer to 10 kW. It is designed to operate in a 60 Hz bandwidth between 50 Hz and 1.5 kHz. Space-based interferometers offer improved resolution and can detect gravity waves at much lower frequencies. Other gravitational wave detector facilities include, example, VIRGO (France–Italy), TAMA (Japan) and LIGO (USA)[2]. There is a proposal for a future space-based interferometer (LISA), with pathlengths of around 5×10^9 m. This is around $1/30$ of the distance between the earth and the sun.

C3.3.3.3 The determination of fundamental constants

Frequency-stabilized lasers also have a key role to play in the determination of some fundamental constants, through their use as references for tunable sources in high-resolution spectroscopy. Rydberg originally observed that the wavelengths of known lines in hydrogen fitted a simple pattern. As is now well known, these lines are, at vacuum wavelengths λ_0, given by

$$\frac{1}{\lambda} = R\left(\frac{1}{p^2} - \frac{1}{q^2}\right) \tag{C3.3.6}$$

where p and q are integers and R is the Rydberg constant. The Bohr theory of the hydrogen atom allows the determination of R:

$$R_\infty = \frac{m_e e^4}{8\varepsilon_0^2 h^3 c} \tag{C3.3.7}$$

[2] More information on gravitational waves is available at http:www.physics.gla.ac.uk/gwg/index.html (Institute for Gravitational Research, Glasgow); http://www.geo600.uni-hannover.de/ for the GEO600 project in Hanover,http://www.ligo.caltech.edu/ for the US Ligo project; http://www.virgo.infn.it/ for the joint French-Italian VIRGO project; http//tamago.mtk.nao.ac.jp/ for 'Tama300' in Japan; http://lisa.jpl.nasa.goc/ for 'LISA' and http://www.anu.edu.au/Physics/ACIGA/ for a project based in Australia.

where m_e and e are the electron mass and charge, ε_0 is the permittivity of free space, h the Planck constant and c the speed of light. The subscript used for R indicates that the Bohr model assumes that the proton mass is infinite. The finite proton mass is just one of a large number of corrections that need applying to the simple formula to explain various frequency shifts and splittings that are observed in hydrogen. In recent years, there has been considerable activity in the precision measurements of frequencies of transitions both in hydrogen and hydrogen-like species (i.e. deuterium, positronium and muonium). Local frequency standards were developed based on tellurium, in order to provide local high-accuracy references for hydrogenic work [15]. Within a few years, measurements of hydrogenic frequency standards had progressed to the point where the narrow 1S–2S two-photon transition at 243 nm had been observed and measured to a precision of two parts in 10^{14} [16]. High-accuracy comparison of measured and calculated hydrogen-like frequencies now provide stringent tests of theories such as quantum electrodynamics (QED). More details of this subject can be found in [17, 18].

The long-term reproducibility of some high-accuracy lasers and frequency standards can also be used to investigate the constancy of the spectroscopic fine structure constant, $\alpha = (\mu_0 ce^2/2h)$. Since this constant is dimensionless, any change is independent of the units in which it is expressed. Both laboratory-based and astrophysical observations have been published in recent years: recent astrophysical data appear to indicate a real change as a function of the redshift [19], although earlier measurements failed to detect any significant change (see, for example, [20]). Laboratory experiments (see, for example, [21]) have so far been unable to establish any evidence for a change with time. These experiments rely on different standards, whether optical or microwave, having different α dependences, so that measurements of ratios between them will vary. In the optical case, the recent availability of femtosecond laser comb technology to measure optical frequency ratios to high accuracy between very stable lasers and optical frequency standards should help make significant advances in this area. More details on the use of femtosecond laser combs to measure optical frequencies is to be found in chapter D6.2. For more details on fundamental constants, a recent review has been published [18] and a brief introduction to recent measurements on variations in α is discussed by [22].

C3.3.4 Gas-cell-absorption-based stabilization techniques

This section describes some of the more common techniques available for frequency-stabilizing lasers, using transitions in gases, vapours or discharges. Typically, these transitions have linewidths of a few MHz but the observed feature width is broadened due to the movement of the atoms and the resulting Doppler shift (see section A1.8). Since the velocity distribution of the atoms is Gaussian, the transition profile of the Doppler-broadened feature will also be Gaussian and typically has a linewidth of a few hundred MHz. Where the spectral widths are limited in this way, the resolution is said to be Doppler-limited. Fortunately, techniques are available to observe the transitions without the Doppler broadening, giving Doppler-free resolution, and these methods will be discussed in sections 3.3.4.3 and 3.3.4.4.

Initially, we discuss Doppler-limited methods of laser frequency stabilization, as applied in the case of HeNe and diode lasers. In the case of the HeNe laser, the Doppler-broadened gain profile is around 1 GHz wide (see section B3.6.7) and techniques are available to frequency stabilize the laser to the centre of the gain profile. These provide an easy-to-use system with reproducibilities in the range of one part in 10^8 to one part in 10^7. These methods are so useful and straightforward that systems are commercially available.

C3.3.4.1 Frequency stabilization of a HeNe laser to the gain curve

One of the simplest and most widely used methods of frequency stabilization is to use a HeNe laser with a gain tube of around 30 cm in length [23, 24]. After turn-on, the laser plasma tube gradually heats up and the cavity expands. The frequency of the nth mode of the cavity is given by $f_n = nc/2L$, for a cavity of length L. As the cavity expands, the frequencies of individual modes therefore decrease, drifting through

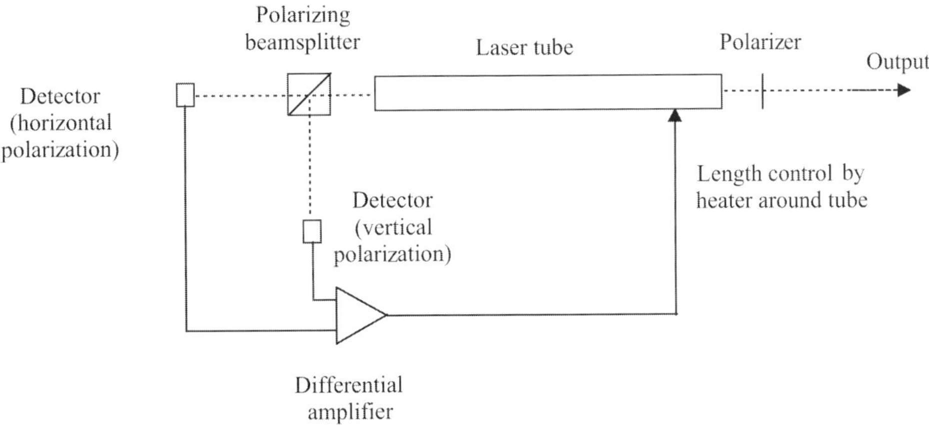

Figure C3.3.2. Schematic diagram for the frequency stabilization of a two-mode HeNe laser by balancing the two-mode intensities.

the gain profile. Typically, for a 30 cm cavity length, the gain profile will support two or three modes. The three-mode case occurs with one dominant mode close to the gain centre, whereas the two-mode case occurs with both modes nominally equidistant from gain centre. Due to mode competition, each mode is orthogonally polarized to the adjacent one. As a result, for the two-mode case, the modes can be split with a polarizing beamsplitter, as shown in figure C3.3.2. Two detectors are then used to measure the power in these two modes and the difference in power can be used to generate an error signal. This difference signal can then be integrated and fed back onto heaters which control the cavity length. Integration of the error signal ensures that, at low frequencies, the servo system gain corresponds to the maximum open-loop gain of the amplifier. This minimizes the error signal, which will inevitably be present, if the laser cavity length is slowly drifting. An output polarizer is used to select one of the two modes as the main output beam. Relatively straightforward circuitry allows the design of a stabilized laser system which turns on from cold and automatically frequency stabilizes in this way, generally in around 20 min. This is the basis of a number of 633-nm frequency-stabilized HeNe lasers, currently on the market. A slightly more elaborate algorithm is employed for lasers at the lower gain lines at 543 and 612 nm, which generally emit in three longitudinal modes at the centre of the gain profile [25].

This two-mode balanced stabilization technique generates an error signal based on the Doppler-broadened gain profile of the HeNe laser. Two further schemes exist which make use of the Doppler-free neon feature at the centre of the gain profile. The first of these is based on the Zeeman effect, and the method requires a laser tube sufficiently short that it lases on only one longitudinal mode at the centre of the gain profile. The tube should also have no polarization-selective optics, such as a Brewster-angled window. The HeNe discharge is placed in an axial magnetic field [26, 27], which splits the gain curve into two oppositely-circularly-polarized components, σ_+ and σ_-. Both these Zeeman components provide gain but their cavity mode frequencies are non-degenerate on account of their different refractive index dependences on the magnetic field. In low fields, the frequency difference is so small that the frequencies of the two modes usually lock together. However, in a magnetic field of 10 mT, the refractivity difference becomes sufficiently different that the two modes separate in frequency and a Zeeman beat of a few hundred kHz between the modes is observed. The difference in this refractivity changes as the laser is tuned through the central portion of the gain curve, resulting in a Zeeman beat maximum at line centre. This detected Zeeman beat frequency signal is used to stabilize the laser.

An early frequency stabilization technique, using the Lamb dip, made use of the gain competition experienced by the two travelling-wave components of the standing-wave cavity mode. In general, atoms with velocity $+v$ and $-v$ can contribute to the gain experienced by a cavity mode at the frequency corresponding to either of these velocity classes. However, at line centre, only one set of neon atoms (those with a zero longitudinal velocity component) can contribute to the gain, giving rise to a slight dip in the output power of the mode at this point. In order to stabilize to this Lamb dip, the laser cavity length and, hence, the frequency is modulated by means of a piezoelectrically driven cavity mirror, at a frequency of a few hundred Hz. The technique of frequency modulating the laser and subsequent de-modulation of the power change is now so widely used that we should consider it here in some detail. In general, we have a situation where the power varies with frequency, i.e. $P = g(f)$, and we are particularly interested in the behaviour of this function around some maximum or minimum value at $f = f_0$. We then modulate the laser frequency at an angular frequency Ω with amplitude f_m, so that the power output varies with time and frequency de-tuning as

$$P(t) = g(f + f_m \sin \Omega t). \tag{C3.3.8}$$

Expanding this function to first order in f_m gives:

$$P(t) = g(f) + f_m \sin(\Omega t) g'(f) + O(f_m^2) \tag{C3.3.9}$$

where the prime denotes a derivative and the terminology O( ) indicates the order of higher terms in the expansion. A phase-sensitive detector (or double balanced mixer) will pick out the signal at Ω, which is $f_m g'(f)$, so giving the first derivative of the spectral profile. This is an ideal feature shape to produce an error signal, since it is zero at the centre of the Lamb dip and is of different sign either side of the centre. The Doppler-broadened gain curve will also give a slowly varying derivative signal, and this is also zero at the centre of the Lamb dip. As we shall see later, for Doppler-free absorption spectroscopy, this is not generally the case, and so we will need to use additional techniques to remove this Doppler-limited background.

C3.3.4.2 Frequency stabilization based on Doppler-limited absorption

An alternative method of providing frequency stabilization, giving modest levels of stability, is by referencing to the linear absorption of a molecule or atom in a cell. This technique is advantageous in that the frequency reference providing the stabilization discriminant is decoupled from the laser gain medium. As a result, it is much more widely applicable to different types of laser, whereas the gain-related techniques already described are applicable, in the main, to the HeNe laser. A secondary advantage is that the stabilization signal is derived from a quiescent absorption rather than an excited discharge. In absorption spectroscopy, the laser intensity (I) transmitted through a cell of length L is given by the Beer law, $I = I_o \exp(-\alpha L)$, where I_o is the incident intensity and the absorbance is αL [28]. The absorption coefficient α is given by the product of the molecular number density, N, and the absorption cross section, σ,

$$\alpha = N\sigma \tag{C3.3.10}$$

As the laser frequency f is scanned through a single Doppler-broadened linear absorption, the cross section, σ, varies as

$$\sigma = \frac{S}{\gamma\sqrt{\pi}} \exp(-(f - f_0)^2/\gamma^2) \tag{C3.3.11}$$

where $2\gamma\sqrt{\ln 2}$ is the FWHM of the absorption and f_o is the laser frequency at line centre. The line-strength S of the transition is the total cross section integrated over all frequencies. It follows from equation (C3.3.11), that the absorption cross section σ_o, at line centre is given by

$$\sigma_0 = \frac{S}{\gamma\sqrt{\pi}} \tag{C3.3.12}$$

Although linear absorption Doppler-limited spectroscopy is less widely applied than the Doppler-free techniques described in section C3.3.4.3, a recent example of this method is the stabilization of laser diodes at 1500 nm to acetylene and other molecules [29, 30]. A frequency-doubled Nd:YAG laser has also been stabilized to linear absorptions in iodine at 532 nm [31] and a stability of 2.3×10^{-8} was reported. Linear absorption spectroscopy was also recently used to frequency stabilize a frequency-doubled diode laser system at 422 nm, using Doppler-broadened transitions in rubidium [32].

C3.3.4.3 Frequency-stabilization-based Doppler-free spectroscopy

In order to achieve better frequency stability and reproducibility from absorptions in molecules or atoms in a cell, we need to use Doppler-free spectroscopy. This produces far narrower features for frequency locking—typically a few MHz or less compared to the 500 MHz observed for Doppler-limited features. When an absorption cell is placed within a laser cavity, the laser gain profile is modified by any Doppler-limited linear absorptions coincident with the lasing wavelength. In addition, a series of weak hyperfine saturated spectra (inverted Lamb dips) can be observed as the laser is tuned. These features arise since the standing-wave cavity mode sees a narrow region of decreased absorption loss when it is within a natural width of the hyperfine absorption line centre. In effect, the cavity mode standing wave comprises of two oppositely directed travelling-wave components, each of which interacts with a particular positive or negative velocity component (dependent on the direction) of the atom's or molecule's thermal velocity distribution which gives rise to the Doppler width. As the laser mode is tuned within a natural linewidth of the hyperfine component centre frequency, it interacts only with the zero-velocity part of the distribution and there are reduced numbers of absorbing molecules within this velocity group (see section A1.15.1). As a result, the absorption is reduced within this narrow region and the laser output increases accordingly, resulting in the saturated spectra superimposed on the tuning curve. This is a standard technique for the removal of Doppler broadening due to absorber thermal motion, and together with the use of a low-pressure absorption cell, provide the means for a long-term narrow spectral reference which is relatively perturbation-free and very reproducible with time.

In order to observe these weak saturated features with good signal-to-noise ratios and then frequency stabilize, standard techniques of signal recovery using phase-sensitive detection (PSD) are used. The laser cavity length (frequency) is modulated by a piezoelectric tube (PZT) to a depth of a few MHz at a rate of typically around 500 Hz. The third derivative of the laser-power tuning curve (obtained by PZT tuning of the cavity length) is recovered by phase-sensitive detection at the third harmonic of the modulation frequency. This is readily appreciated from the earlier discussions on observation of the first derivative of the Lamb dip. If the expansion of equation (C3.3.9) is taken further, we obtain:

$$P(t) = g(f) + f_{\mathrm{m}} g'(f) \sin \Omega t + \tfrac{1}{2} f_{\mathrm{m}}^2 g''(f) \sin^2 \Omega t + \tfrac{1}{6} f_{\mathrm{m}}^3 g'''(f) \sin^3 \Omega t + O(f_{\mathrm{m}}^4). \qquad \text{(C3.3.13)}$$

If we now modify our PSD electronics, so that it looks for signals at the third harmonic of the modulation frequency, we see that it will predominantly be observing the third derivative of the profile, since $\sin^3 \Omega t$ itself contains a term in $\sin 3\Omega t$. Although the third derivative does contain terms relating to the Doppler-limited changes in the iodine absorption and neon gain profile, these are very much smaller than observed in the case of first harmonic detection and the offsets introduced are usually negligible. So, this technique provides a zero-crossing discriminant on a flat background free from large offsets and suitable for stabilization purposes.

However, in modulating the laser within the frequency response range of a PZT (typically a few hundreds of Hz to a few kHz), we are modulating the laser at a frequency where much of the frequency and intensity noise is. This noise may be acoustic or arise from plasma noise in the case of a HeNe laser. With a diode laser, the frequency noise is $1/f$, and so arises predominantly at low frequencies. In many cases, therefore, it is desirable to modulate the laser at frequencies of several MHz or above. This has the added benefit of allowing us to increase the laser servo bandwidth, which has to be less than the modulation frequency. This

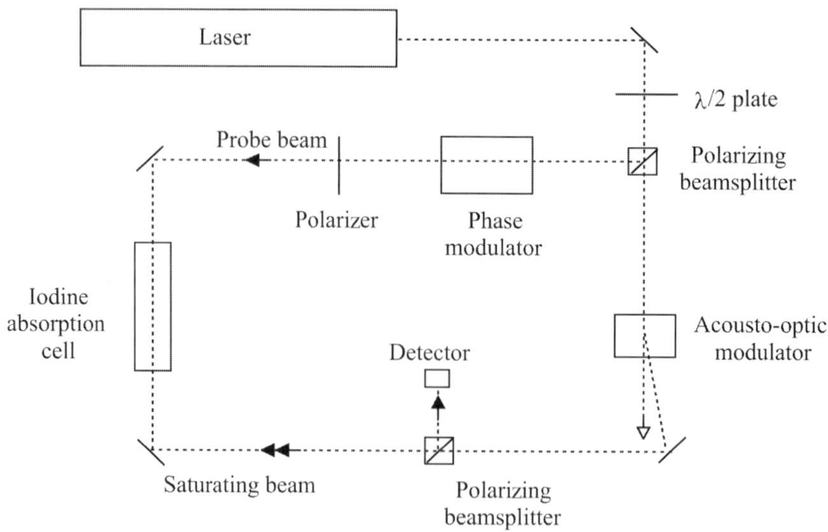

Figure C3.3.3. Experimental arrangement for FM Doppler-free spectroscopy.

enables significant reduction of the laser frequency noise. Frequency modulation (FM) spectroscopy was first demonstrated in the 1980s [33–35] and a typical system or the Doppler-free case in iodine is shown in figure C3.3.3.

The probe beam through the iodine cell is phase modulated at an angular frequency Ω, so that the output spectrum is:

$$E = E_0 \exp(\mathrm{i}(\omega t + \beta \sin \Omega t)) \tag{C3.3.14}$$

where β is termed the modulation index. This expression is mathematically identical to

$$E = E_0 \exp(\mathrm{i}\omega t) \sum_n J_n(\beta) \cos n\Omega t \tag{C3.3.15}$$

so that, provided β is less than 1, this may be approximated as a carrier with a pair of sidebands at frequencies $\pm\Omega/2\pi$. In the absence of any interaction with the absorption cell, equation (C3.3.15) represents pure phase modulation—no amplitude modulation is involved. Each sideband forms a beat with the carrier but the pairs of beats are in anti-phase and so cancel out. In figure C3.3.3, the probe and saturating beam overlap in the cell and so the probe beam transmission will be slightly greater as the laser is tuned through the natural Lorentzian width of an iodine transition. The mechanism producing this is nominally the same as that for an inverted Lamb dip using an absorption cell within the laser cavity. As the phase-modulated probe beam is scanned through the transition, the three frequencies (carrier and two sidebands) are absorbed by differing amounts. Therefore, some amplitude modulation is introduced and this is detected and demodulated by a double balanced mixer. The profile that is produced is of a characteristic shape, which may be calculated [36] and has a zero cross-over at line centre and so is suitable for frequency locking. The shape is also dependent upon the relative phases of the reference signal entering the double balanced mixer and the phase modulator. Although this method produces FM features suitable for frequency control, it produces them on a slowly varying background which arises from the Doppler-limited absorption. In figure C3.3.3, this is removed by amplitude modulation of the saturating beam, which amplitude modulates the detected saturated signal but not the Doppler-limited background. A phase-sensitive detector, operating at the amplitude chopping frequency, demodulates the double balanced mixer output and produces the FM features on a flat background.

C3.3.4.4 Frequency-stabilized lasers referenced to iodine, rubidium and acetylene

The question of which particular atomic or molecular absorber to use depends on the wavelength of the laser. For fixed-frequency lasers, such as HeNe lasers, operating in the 500–650 nm region, iodine is usually chosen [37]. The electronic transition of the iodine diatomic molecule has associated with it a vast number of molecular rotational and vibrational absorption lines, distributed throughout this visible region, with separations between neighbouring absorptions of a few GHz or less. Each of these lines has an underlying hyperfine structure, which arises from the 5/2 spin of the ^{127}I nucleus. For an even value of J (the rotational quantum number) the total nuclear spin takes the values of 1, 3 or 5 and the transition is, therefore, split into 21 hyperfine levels [38]. For an odd value of J, the total nuclear spin takes the values 0, 2 or 4 and the transition is split into 15 hyperfine levels. There is, therefore, a good probability that any laser that can be tuned across a few hundred MHz can access at least some of these hyperfine states. Iodine vapour can also conveniently be confined in a short silica cell, and its pressure controlled through the use of a cold finger, whilst the vapour temperature can be controlled by heaters around the cell body. At wavelengths longer than 650 nm, the absorptions become weaker as the transitions have to involve ground states with a high vibrational quantum number and low occupancy. The lower state, however, can be better populated by heating the cell, perhaps to 100 °C, and using longer cells to enhance the absorption.

With extensive possibilities for frequency stabilizing lasers to various absorbers, there is international agreement regarding frequency values and uncertainties adopted for such lasers in respect of their application to the practical realization of the metre. This list is the *mise en pratique* of the Comité International des Poids et Mesures or CIPM [37] and includes a number of iodine-stabilized lasers, a diode laser based on the Rb two-photon absorption at 778 nm, an acetylene-stabilized diode at 1500 nm, together with some frequency references based on cooled atoms and ions. These latter examples, which provide the most reproducible standards, are discussed in more detail in chapter D6.2. In this chapter, we will limit discussions to vapour-or gas-cell-based standards based on iodine, the 778 nm rubidium absorption and 1.5 μm acetylene absorptions.

The earliest iodine-stabilized HeNe laser was based on the 11–5 R(127) absorption line in $^{127}I_2$, coincident with the HeNe laser line [39]. This has been extensively characterized, so that various frequency shifts connected with iodine cell pressure, wall temperature, depth of frequency modulation and cavity geometry are well documented. The *mise en pratique* contains various parameter ranges within which the laser should be operated in order to obtain the specified reproducibility. In general, for a well-designed laser, around one part in 10^{11} reproducibility can be expected, although this is critically dependent upon the iodine cell quality. Systems can be made relatively 'user-friendly' through automatic in-lock monitoring[3], because the $2f$ signal, which is at a maximum at line centre, can be used to indicate whether the laser is in lock. The $1f$ signal, which varies with the Doppler-limited background, can be used to indicate which of the main iodine components (d, e, f or g) the laser is locked to. An overall schematic diagram of a full system is shown in figure C3.3.4. Third harmonic 633 nm iodine features are shown in figure D2.1.9. There have been frequency chain measurements of this laser [40–42], as well as more recent measurements based on femtosecond laser combs [43, 44]. Also, there have been direct international intercomparisons between a number of countries on various different occasions (e.g. [45, 46]). HeNe lasers at 543, 612 and 640 nm can also be stabilized to Doppler-free iodine spectra, generally using iodine cells external to the cavity.

Solid state lasers and diode lasers have also been stabilized to iodine. By far the most impressive results in terms of frequency stability have been obtained with the iodine stabilized frequency-doubled Nd:YAG laser at 532 nm [47]. It can tune to a number of strong iodine transitions with narrow ($\approx$1 MHz) linewidths and so this system can provide excellent frequency stability. The laser-to-laser reproducibility is usually determined by the iodine cell quality; background gases in the cell can cause small pressure shifts. An experimental arrangement similar to figure C3.3.3 can be used, with an iodine cell length of a few tens of cms. There have

[3] For a example of an automatically monitored iodine stabilised laser system at 633 nm see, http://www.npl.co.uk/length/wss/services/equipment/i2henelaser.html.

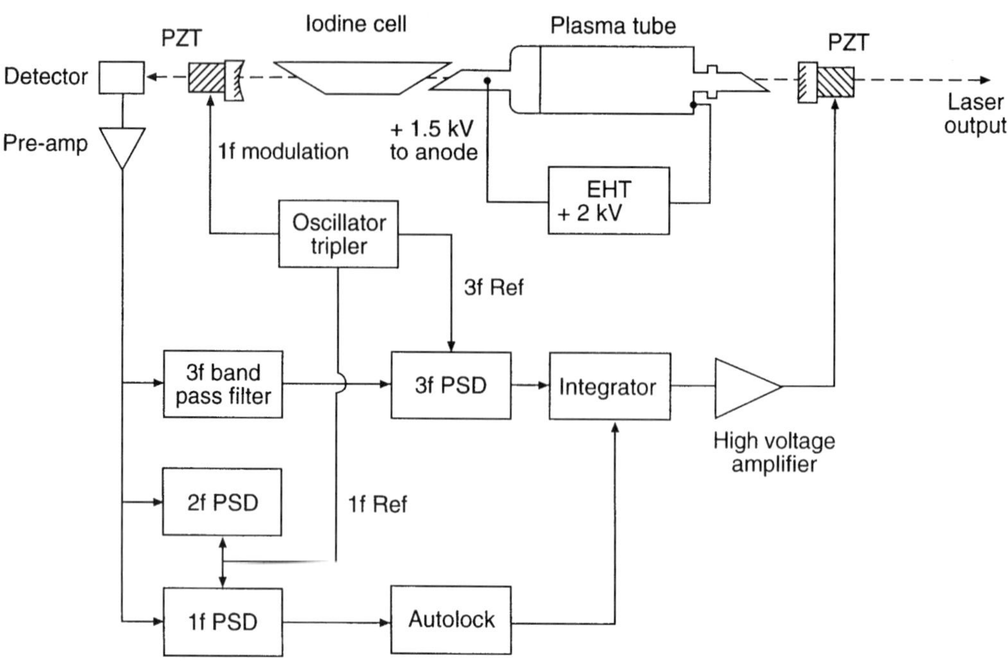

Figure C3.3.4. Overall schematic diagram for stabilizing a HeNe laser to iodine, including the $1f$ signal required for component monitoring and the $2f$ signal used to advise the operator that the laser is in lock. PSD, phase-sensitive detector; PZT, piezoelectric.

been international intercomparisons of this laser system, together with a number of frequency measurements based on femtosecond laser combs [48–51]. These femtosecond comb techniques are discussed in more detail in section D6.2.

Laser diodes, operating in the range 630–690 nm, can also be frequency stabilized to iodine but with increasing difficulty for longer wavelengths. In principle, the tunability of laser diodes means that any one of hundreds of absorption lines could be used for frequency control. However, most interest has centred on those close in frequency to the 11–5 R(127) line used to stabilized the 633 nm HeNe laser. One line often used is the 6–3 P(33) line [52], which has a stronger linear absorption than 11–5 R(127). Work has also been published on the R(60)8–4, R(125)9–4 and P(54)8–4 transitions. A number of countries took part in a recent international intercomparison of eight lasers at the Bureau International des Poids et Mesures (BIPM) [53]. System reproducibility of a few tens of kHz was recorded, with best stabilities at 1 s of better than one part in 10^{11}.

The optical frequency standard at 778 nm is based on a two-photon transition in rubidium, whose partial energy level structure is shown in figure C3.3.5. This standard has importance since it is a convenient reference at twice the frequency of the telecommunications region at 1556 nm. Two-photon spectroscopy is a Doppler-free technique, which can be used for some absorptions which are forbidden as conventional one-photon transitions. In two-photon spectroscopy, the two photons are absorbed, one from each of two counter-propagating beams. This method was first demonstrated in sodium [54]. In rubidium, the two-photon transition rate is enhanced because there is an intermediate level (the $^2P_{3/2}$ state) which is nearly halfway between the two levels involved in the two-photon transition. This is close to resonance with the 'virtual state' (shown dotted in figure C3.3.5) reached after the absorption of just one photon. Detection of the transition

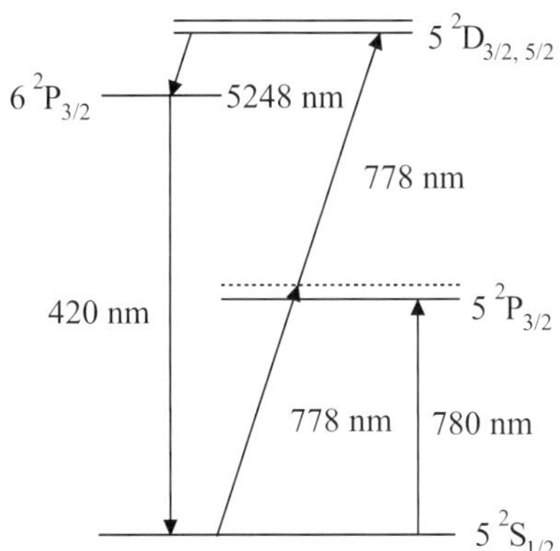

Figure C3.3.5. Partial-term scheme of rubidium, showing the two-photon transition at 778 nm, the decay route back to the ground state and the D-line transition at 780 nm.

is achieved by observing 420 nm fluorescence using a photomultiplier, which arises from the decay from the $P_{3/2}$ to the ground state. One difficulty with two-photon absorptions is that the signal is weak, although it increases as the square of the laser power. In order to maximize the signal, the cell is therefore placed inside a resonant power build-up cavity (figure C3.3.6). Two servo systems are required: one to lock the cavity to the laser and the second to lock the laser to the transition peak. The cell is also heated to around 100 °C in order to increase the rubidium vapour density. The frequency was first measured using a frequency chain [55, 56], although femtosecond laser measurements are also planned. A summary table showing the various frequency stabilized laser wavelengths discussed is given in the table D2.1.1 in section D2.1.2.

The final system discussed here is a stabilized diode laser system based on saturated absorptions in acetylene [57]. A schematic diagram of a typical system is shown in figure C3.3.7 and is also being developed at a number of standards laboratories (e.g. PTB, NPL, and NRC). Acetylene has a number of molecular rotational and vibrational transitions between 1510 and 1540 nm for $^{12}C_2H_2$ and 1520–1550 nm for $^{13}C_2H_2$ and has importance through the provision of a number of standards covering the telecommunications window at 1500 nm. With the ever-increasing demand for data transfer using fibre-based telecommunications, it is anticipated that fibres will eventually utilize the whole of the 1300–1600 nm region. Other molecular absorbers, such as CO, CH_4 or HCN, are able to provide frequency references in various other parts of this spectral region. Frequency standards based on some of these absorbers are being developed, with reproducibilities at a few parts in 10^{10} to one in 10^9 level for Doppler-free stabilization techniques. The acetylene system is a further variation of saturated absorption spectroscopy. In this implementation, the cell is inside a separate cavity to the laser. Again, the cavity transmission demonstrates an inverted Lamb dip at the centre of the Doppler absorption profile. In figure C3.3.7, the laser is locked to the cavity and the cavity is then locked to the Doppler-free acetylene signal, using a third harmonic lock to remove the Doppler-limited background.

In acetylene, there is no hyperfine structure and so we only observe one peak within each Doppler-broadened transition. The $\nu_1 + \nu_3$ combination band of acetylene has more than 50 strong absorption lines in the 1510–1540 nm region.

All the frequency-stabilized laser systems discussed here have reproducibility limitations at the one part in 10^{11} level. There are a number of factors contributing to this limit and these are briefly discussed. The first is connected with the Doppler shift. All the techniques discussed reduce or eliminate the first-order Doppler shift but are unable to reduce the second-order term, which arises from special relativity and varies as $v^2/2c^2$. In one dimension (for example for motion along the laser beam), this relative frequency shift

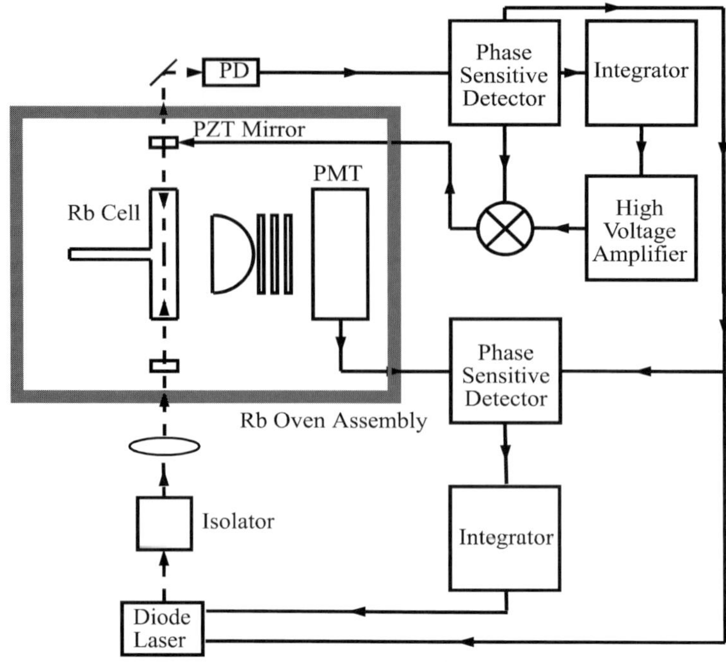

Figure C3.3.6. Schematic diagram of a system used to observe the two-photon transition at 778 nm. The photomultiplier (PMT) is used to record the 420 nm fluorescence [58].

becomes $kT/2mc^2$, where k is the Boltzman constant, T the absolute temperature, m the mass of the atom and c is the speed of light. Typically, this amounts to a few parts in 10^{12} of the optical frequency at room temperature, although this depends upon the mass of the absorbing atom or molecule. There will also be residual first-order Doppler shifts arising, for example, from the probe and saturating beams not being exactly co-linear. A further limitation will be frequency shifts caused by collisions, either with other molecules of the same type or with foreign (background) gases, produced in the cell preparation and sealing-off processes. These effects are difficult to quantify to better than the one part in 10^{11} level. There are also frequency broadening and shifts which can arise since the molecule only spends a finite time in the laser beam. These 'transit time' effects arise from the uncertainty principle: if the molecule only spends a time Δt in the beam, then the observed linewidth must be limited at the $1/\Delta t$ level in frequency. Finally, there are frequency stability limitations arising from the signal-to-noise ratio of the saturated features. For a typical iodine linewidth of 5 MHz, a signal-to-noise ratio of around a thousand is required to achieve a stability of one part in 10^{11} (5 kHz). These limitations point to a requirement to have the frequency standard based on a transition in an atom which is cold (to remove Doppler effects and to keep the atom in the laser beam indefinitely), to be in a vacuum on its own (to remove collisional effects) and to have a transition with a low linewidth (to reduce the problems associated with the signal-to-noise ratio affecting the line centre). All these requirements point to the ideal of an ultimate frequency standard based on a cooled trapped atom or ion. This is discussed further in chapter D6.2.

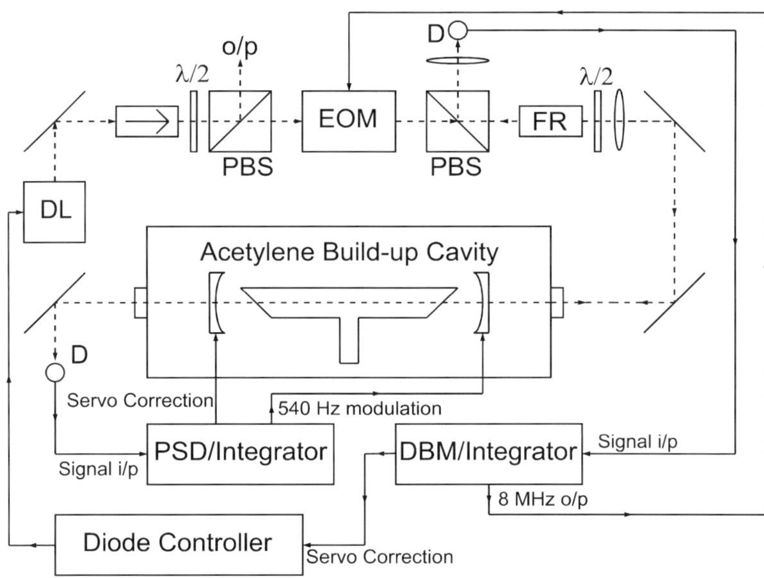

Figure C3.3.7. Overall schematic diagram of the experimental arrangement used to observe Doppler-free transitions in acetylene Key: DL, diode laser, D, detector; EOM, electro-optic modulator; FR, Faraday rotator; PBS, polarizing beamsplitter, i/p, input; o/p, output and $\lambda/2$ denotes a half-wave plate [59].

C3.3.5 Evaluation of frequency stability and reproducibility

The question arises of how to evaluate the performance of a stabilized laser system. The normal approach is to build two similar systems and then evaluate the difference (beat) frequency. This is typically arranged to be in the few tens of MHz to few GHz region, within the response range of commercially available detectors and amplifiers. The results are usually collected by a computer-controlled frequency counter; the computer is then also used for subsequent data analysis. A typical beat frequency monitoring system is shown in figure C3.3.8.

The output beams from the two lasers should have the same (linear) polarization and approximately the same beam diameters and divergence in order to provide a strong rf signal. The two beams are overlapped at the beam combiner and focused onto the detector. The time-dependent part of the electric field of the laser beam may be expressed as $E\exp(\mathrm{i}\omega t)$, where the laser frequency is $\omega/2\pi$ and E is the electric field amplitude. The amplitude of the combined beams may, therefore, be represented by the expression $(E_1\exp(\mathrm{i}\omega_1 t) + E_2\exp(\mathrm{i}\omega_2 t))$. The detector measures the intensity of the combined beam:

$$I = |E_1\exp(\mathrm{i}\omega_1 t) + E_2\exp(\mathrm{i}\omega_2 t)|^2 = E_1^2 + E_2^2 + 2E_1E_2\cos(\omega_1 - \omega_2)t \qquad \text{(C3.3.16)}$$

and so an rf beat frequency is produced at the difference frequency. The beat signal size depends not only on the laser intensities (as indicated earlier) but also upon the mode matching, alignment and relative polarization of the two lasers. The beat is produced as a photocurrent in the detector which needs subsequent amplification.

Although one would expect the laser frequency noise to obey normal (Gaussian) statistics, with the standard deviation reducing as $1/\sqrt{(\text{measurement time})}$, this is not always the case. Often, the mean can vary over time as the laser drifts and a more appropriate statistic is a variance based on successive differences. In frequency metrology, this parameter is called the Allan variance, $\sigma^2(2, T, \tau)$ [60,61]. Here, τ is the averaging time (counter gate time) and T is the time between the start of successive readings. Ideally it is expected

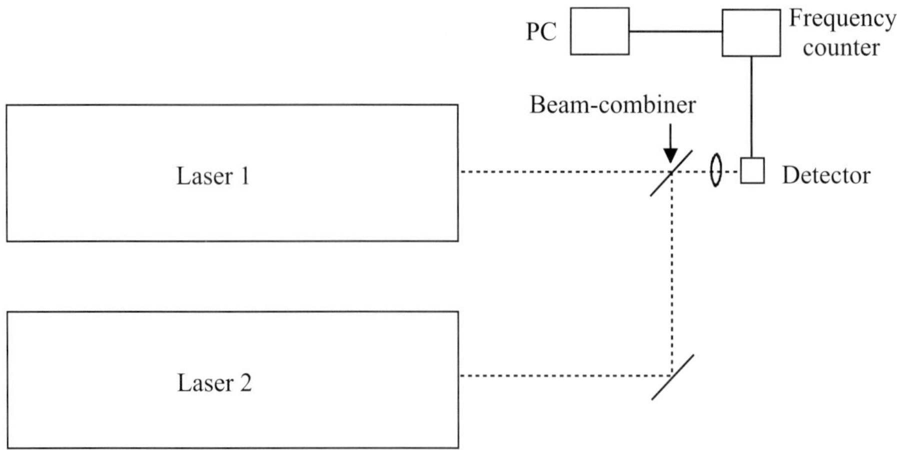

Figure C3.3.8. Measurement of the frequency difference between two lasers using rf beats.

Table C3.3.1. The main types of laser frequency noise and their characteristic Allan deviation slopes.

Slope	Source
−1	White phase noise or discrete frequency modulation
−1/2	White frequency noise
0	Flicker (1/f) noise
+1/2	Random frequency walk
+1	Steady frequency drift

that $T = \tau$, but all frequency counters exhibit a 'dead time' between readings, especially if data need to be transferred to a computer. The statistic σ^2 is the variance of successive samples of the frequency averaged over time τ. If the output of the laser is represented by

$$E = E_o \exp\{\mathrm{j}(\Omega t + \phi(t))\} \tag{C3.3.17}$$

then the instantaneous departure from the nominal frequency is $\dot{\phi}(t)/2\pi$. Hence, at time t, we have

$$\bar{f_k} = \frac{1}{2\pi\tau}\int_t^{t+\tau} \dot{\phi}(t)\,\mathrm{d}t. \tag{C3.3.18}$$

If $\bar{f}_{k+1}$ and $\bar{f_k}$ are successive values of the frequency, averaged over time τ, then

$$\sigma^2(2, T, \tau) = \langle \tfrac{1}{2}(\bar{f}_{k+1} - \bar{f_k})^2 \rangle \tag{C3.3.19}$$

where < > denotes a time average. It may be shown that a log–log plot of σ *versus* τ (Allan deviation *versus* time), has various slopes, depending upon the type of frequency noise present. Investigation of the type of laser frequency noise present is one of the most important uses of this statistic. The most important σ *versus* τ slopes are listed in table C3.3.1, together with the characteristic noise source yielding this slope.

Finally, the effect of counter 'dead time' should be noted. This can be very significant when $\tau \leq 30$ ms. However, during these periods, we are usually in a region where the slope is −1 (pure frequency modulation)

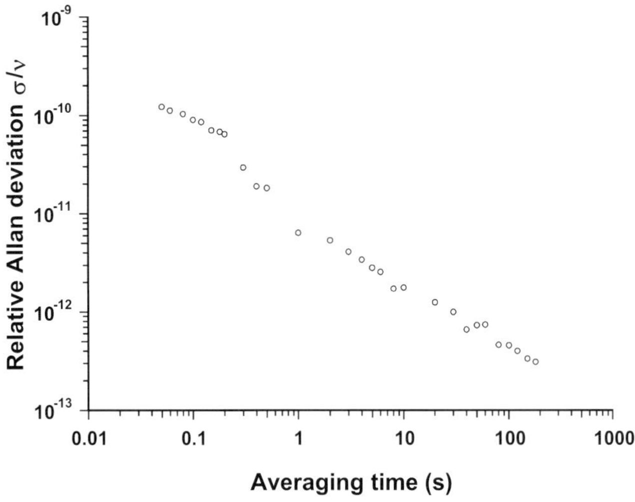

Figure C3.3.9. Plot of the Allan deviation of the beat frequency between two iodine-stabilized HeNe lasers at 633 nm. The slope is -1, resulting from the frequency modulation of the lasers.

or $-1/2$ (white frequency noise). In these cases, σ is unaffected by a non-zero counter dead time. A plot of a typical Allan deviation, in this case between two iodine-stabilized HeNe lasers at 633 nm, is shown in figure C3.3.9. The slope is approximately -1, typical for lasers which need to be frequency modulated in order to produce the discriminants suitable for frequency stabilization.

In addition to the measurement of frequency stability, we need to measure the mean offset between two laser systems to determine the long-term frequency reproducibility. If both lasers are emitting at nominally the same frequency, then the most convenient method is probably to shift one frequency by a known amount using an acousto-optic modulator and then compare the beat with the modulator drive frequency. However, if the lasers can be locked to different components in, for example, iodine, that are a few tens of MHz apart, then there is an alternative method. Let us consider two components 'a' and 'b' and assume that $f_{\mathrm{b}} > f_{\mathrm{a}}$. We also assume that the components accessed by laser 1 are offset from the true value by Δf_1 and those accessed by laser 2 are offset by Δf_2. We lock laser 1 to 'a' and laser 2 to 'b', and the beat frequency (B) is then

$$B = (f_b + \Delta f_2) - (f_a + \Delta f_1). \tag{C3.3.20}$$

If the components are exchanged, so that laser 2 is locked to 'a' and laser 1 to 'b', then the observed beat will become

$$B' = (f_b + \Delta f_1) - (f_a - \Delta f_2). \tag{C3.3.21}$$

The unshifted interval is, therefore, $(B + B')/2$ and the difference between the laser offsets is given by $(B' - B)/2$. This difference between the offsets gives a measure of the long-term frequency reproducibility. This is usually a higher figure than the short-term frequency stability shown in figure C3.3.9 and may result from small design differences between the two systems.

C3.3.6 Cavity stabilization techniques

Although atomic and molecular transitions provide the best long-term frequency stability and reproducibility for laser stabilization, an optical cavity can provide better short-term stability. This is because of the high signal-to-noise ratios obtainable with frequency discriminants based on optical cavity resonances observed either in reflection or transmission. Atomic transitions that are narrow are difficult to interrogate and lock onto if the laser linewidth is broad and the laser drifts through the transition too quickly. Typically, forbidden transitions in trapped ions, which are being developed as optical frequency standards (see chapter D6.2), have linewidths of the order of 1 Hz and the observed linewidth is usually limited by the laser. A very high-finesse cavity that is acoustically well isolated and has a narrow fringe width and a low drift rate has a pivotal role in developing these standards. Lower finesse cavities provide an important means of stabilizing many different types of laser without the need for the lasing wavelength to be coincident with a quantum absorber. This is particularly relevant for tunable laser systems, such as dye lasers or Ti:sapphire solid state lasers or many different diode lasers.

As with Doppler-free spectroscopy, there are a number of techniques available for frequency locking to optical cavities. The simplest method is either to frequency modulate the laser or to length modulate the cavity at a few kHz and use a PSD to provide a first derivative feature in the same way as described earlier for locking to transitions in iodine. This error signal is then used, together with an integrator, to feed back onto the laser frequency. However, this method requires either the laser to be modulated or the cavity length to be dithered. Neither of these is desirable, since one often requires an unmodulated laser output and, similarly, modulating the cavity length can cause instabilities in the length caused by heating in the piezoelectrics used to effect tuning.

A method of frequency stabilization which does not require modulation is side-of-fringe locking [62]. With this technique, two signals are differenced in order to produce the error signal. The first is the transmitted fringe and the second is the power monitor. The result is a Fabry–Pérot fringe signal that is offset so that zero volts corresponds to a frequency halfway up the side of the fringe. This lock point is immune to small power changes in the laser. However, this method can only give a small 'capture range'—the frequency range that the laser has to be within in order to acquire lock. This range will be extremely narrow for the case of ultrahigh-finesse cavities. Ideally, we require a lock feature that has a sharp change in signal with frequency at the cavity resonance line centre (that depends upon the cavity finesse) but still has a large capture range in order to acquire lock easily.

Another method which does not involve frequency modulation of the laser was first demonstrated by Hänsch and Coulliard [63]. In this method, linearly polarized light is reflected by a confocal reference cavity with an internal Brewster plate (or another optical element to induce a loss for one polarization). The incoming light comprises light polarized both parallel and perpendicular to the axis of the Brewster plate. One polarization, therefore, sees a cavity of low loss and experiences a frequency-dependent phase shift on reflection. The other component, to first order, is merely reflected by the input mirror. At resonance, both reflected beams are in phase and the reflected beam remains linearly polarized. Away from resonance, however, the parallel component acquires a phase shift and the reflected beam becomes elliptically polarized. This ellipticity can be detected with a simple polarization analyser to give an error signal which has a dispersion shape, but with wings that extend the 'capture range' of the lock.

By far the best cavity lock method uses phase modulation of the laser to produce sidebands equation (C3.3.15). Since the modulation is applied externally to the laser, we can pick off part of the beam before the modulator to provide a narrow-linewidth unmodulated source for high-resolution spectroscopy or frequency standards work. This method is normally referred to as the 'Pound–Drever–Hall' method [64] and produces a feature for locking similar to that obtained in FM spectroscopy. Close to cavity resonance, an imbalance is produced in the sidebands in both the reflected and transmitted powers and this produces amplitude modulation of the light at the detector. This signal is demodulated in a double balanced mixer (DBM) to produce electronic signals suitable for locking. In order to achieve the narrowest linewidths, cavities with very high finesses are required. Typically, these cavities have a free spectral range of the order of 1 GHz and a finesse of 2×10^5. The mirrors must have very good surface quality, in order to reduce losses caused by scattering. If the mirrors cause significant scatter, then the cavity may have a high finesse but the transmission on resonance will be unacceptably low. The technology for producing these 'super-polished'

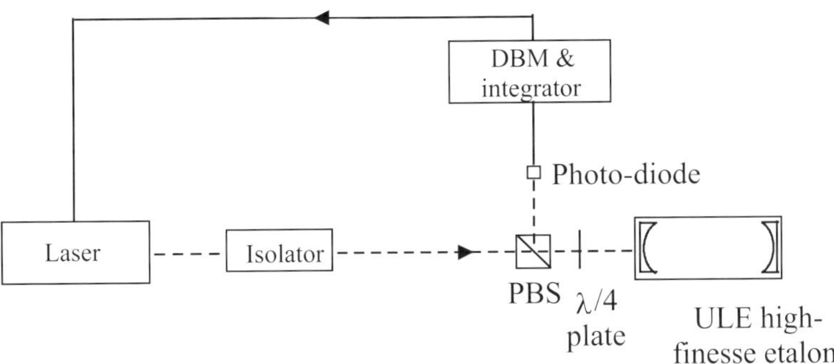

Figure C3.3.10. Typical arrangement for locking a laser to a cavity using the Pound–Drever–Hall method: PBS, polarizing beamsplitter; ULE, ultralow-expansion.

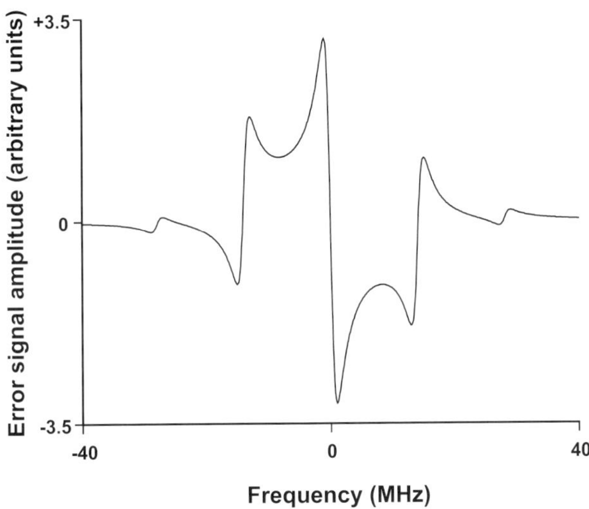

Figure C3.3.11. Calculated frequency discriminant, using a Pound–Drever–Hall lock using a 14 MHz modulation frequency and a cavity linewidth of 1 MHz and modulation index of 0.8. Note that this allows the laser to acquire lock over a frequency range determined by the modulation frequency (i.e. ± 14 MHz); the central slope is determined by the cavity linewidth.

mirrors was developed initially for aerospace applications and the production of laser gyros. Typical cavities have linewidths of a few kHz. This poses a problem in that the cavity has a characteristic response time, which is of the order of the reciprocal of the linewidth. If the laser frequency changes too quickly, then the transmitted power through the cavity will take time to change. The normal method of stabilization to these high-finesse cavities is, therefore, to use them in reflection. On time scales faster than the cavity response time, the reflected power is a combination of the light directly reflected from the cavity and the stored power inside the cavity. On these and faster time scales, the Pound–Drever–Hall lock, therefore, produces a phase error signal, rather than the frequency error signal relevant to longer time scales. A typical system is shown in figure C3.3.10 and a typical lineshape observed from the DBM is shown in figure C3.3.11.

There are two critical elements in producing a narrow-linewidth laser source. The first is to ensure that

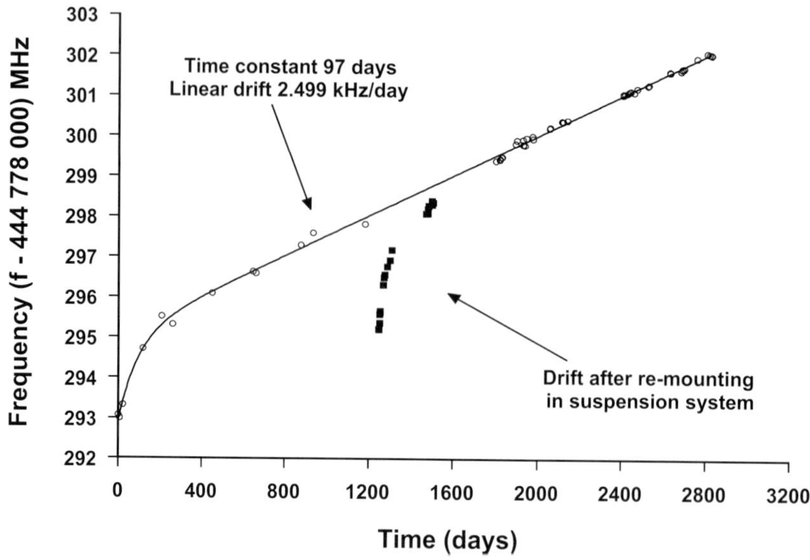

Figure C3.3.12. Long-term drift of a ULE cavity measured over several years.

the electronic servocontrol scheme is capable of producing a narrow-linewidth laser. The usual method to determine the linewidth is to lock two lasers to adjacent modes of the same cavity and monitor the beat, which is, of course, at a frequency equivalent to one free spectral range. This usually produces very narrow linewidths, for example 50 mHz has been reported [65], but there is a degree of common-mode noise rejection, so that the beat width is artificially narrowed. The more demanding task is to demonstrate narrow linewidths by locking two lasers to two independent cavities and measure the linewidth of the beat between them. It is necessary to acoustically isolate the cavity sufficiently in order to achieve this. A laser linewidth of only 0.2 Hz for averaging times up to 32 s, with a frequency stability of 3×10^{-16} at 1 s for a laser at 563 nm, was developed at NIST in order to interrogate their mercury trapped-ion standard [66, 67].

Although acoustic isolation is important to obtain short-term frequency stability, thermal stability and isothermal mechanical stability determine the longer-term drift. Typically, the ULE material used in their cavities has a thermal expansion coefficient of around a $1 \times 10^{-8}\,{}^{\circ}\mathrm{C}^{-1}$. This coefficient is temperature dependent and crosses through zero, at some temperature close to room temperature dependent upon the batch of material. If the cavity can be controlled close to this zero thermal expansion temperature, then this is clearly advantageous. In addition to thermally induced drift, longer-term isothermal drifts of cavities have been reported [68]. Over several years, a cavity isothermal drift rate of only 2.5 kHz/day ($\approx$0.03 Hz s^{-1}) has been reported, as shown in figure C3.3.12. These cavities thus provide excellent 'optical flywheels', providing good short-term frequency stability, whilst a quantum standard is used to provide the long-term frequency control. The use of ULE cavities in this way is discussed in more detail in chapter D6.2 (section D6.2.4).

C3.3.7 Summary

This chapter has described a number of methods for frequency stabilizing lasers. Relatively simple methods are available for 633-nm HeNe lasers. Some systems are commercially available, which provide a reproducibility of one to a few parts in 10^8 of the optical frequency. For other lasers, modest frequency control is also possible using Doppler-limited absorption but the best frequency control for gas-cell-based reference lasers uses Doppler-free spectroscopy. If good short-term frequency stability is required but long-term reproducibility is less critical, then an ultrastable cavity provides a convenient solution. Applications for these systems include dimensional metrology, gravitational wave detection, spectroscopy and the determination of fundamental constants.

References

[1] Kogelnik H and Li T 1966 *Proc. IEEE* **54** 1312–29
[2] Camparo J C 1985 *Contemp. Phys.* **26** 443–77
[3] Henry C H 1982 *IEEE J. Quantum Electron.* **QE-18** 259–64
[4] Welford D and Mooradian A 1982 *Appl. Phys. Lett.* **40** 865–7
[5] Johnston T F Jr, Brady R H and Profitt W 1982 *Appl. Opt.* **21** 2307–16
[6] Gill P 1993 Laser interferometry for precision engineering metrology *Optical Methods in Engineering Metrology* ed D C Williams (London: Chapman and Hall) ch 5, p 153
[7] Birch K P and Downs M J 1994 *Metrologia* **31** 315–16
[8] Bönsch G and Potulski E 1998 *Metrologia* **35** 133–9
[9] Ciddor P E 1996 *Appl. Opt.* **35** 1566–73
[10] Birch K P and Downs M J 1993 *Metrologia* **30** 155–62
[11] Barwood G P, Gill P and Rowley W R C 1993 *Meas. Sci. Technol.* **4** 988–94
[12] Barwood G P, Gill P and Rowley W R C 1998 *Meas. Sci. Technol.* **9** 1036–41 and references therein
[13] Ju L, Blair D G and Zhao C 2000 *Rep. Prog. Phys.* **63** 1317–427
[14] Drever R W P, Ford G M, Hough J, Kerr I, Munley A J, Pugh J R, Robertson N A and Ward H 1983 *Proc. 9th Int. Conf. on General Relativity and Gravitation* ed E Shmutzer (Cambridge: Cambridge University Press)
[15] Barwood G P, Rowley W R C, Gill P, Flowers J L and Petley B W 1991 *Phys. Rev.* A **43** 4783–90
[16] Niering M *et al* 2000 *Phys. Rev. Lett.* **84** 5496–9
[17] Flowers J L, Klein H A, Knight D J E and Margolis H S 2001 Hydrogenic systems for calculable frequency standards: status and options *NPL Report* CBTLM 11. This can be found at http://www.npl.co.uk/~jlf/refs/H_report_final_gs_processed.pdf
[18] Flowers J L and Petley B W 2001 *Rep. Prog. Phys.* **64** 1191–246
[19] Webb J K *et al* 2001 *Phys. Rev. Lett.* **89** 091301
[20] Carilli C L *et al* 2000 *Phys. Rev. Lett.* **85** 5511–14
[21] Prestage J D, Tjoelker R L and Maleki L 1995 *Phys. Rev. Lett.* **74** 3511
[22] Carilli C L 2001 *Physics World* **14** 26–7
[23] Balhorn, Kunzmann H and Lebowsky F 1972 *Appl. Opt.* **11** 742
[24] Bennett S J, Ward R E and Wilson D C 1973 *Appl. Opt.* **12** 1406
[25] Rowley W R C and Gill P 1990 *Appl. Phys.* B **51** 421–6
[26] Baer T, Kowalski F V and Hall J L 1980 *Appl. Opt.* **19** 3173–7
[27] Rowley W R C 1990 *Meas. Sci. Technol.* **1** 348–351
[28] Brassington D J 1995 Tunable diode laser absorption spectroscopy for the measurement of atmospheric species *Spectroscopy in Environment Science (Advances in Spectroscopy 24)* ed R J H Clark and R E Hester (Chichester: Wiley) ch 3
[29] Gilbert S L, Swann W C and Dennis T 2001 *Proc. SPIE* **4269** 184–91
[30] Swann W C and Gilbert S L 2000 *J. Opt. Soc. Am.* B **17** 1263–70
[31] Schulthess M and Giggenbach D 1998 *Electron. Lett.* **34** 1854
[32] Madej A A, Marmet L and Bernard J E 1998 *Appl. Phys.* B **67** 229–34
[33] Bjorklund G C 1980 *Opt. Lett.* **5** 15–17
[34] Drever R W P, Hall J L, Kowalski, Hough J, Ford G M, Manley A J and Ward H 1983 *Appl. Phys.* B **31** 97
[35] Hall J L, Hollberg J L, Baer T and Robinson H G 1981 *Appl. Phys. Lett.* **39** 680
[36] Hall J L, Robinson H G, Baer T and Hollberg L 1983 The lineshapes of sub-Doppler resonances observable with FM sideband (optical heterodyne) laser techniques *Advances in Laser Spectroscopy (NATO Advanced Science Institutes Series)* ed F T Arrechi, F Strumia and H Walther (New York: Plenum)
[37] Quinn T J 2003 *Metrologia* **40** 103–33
[38] Broyer M, Lehmann J-C and Vigue J 1975 *J. Physique* **36** 235–41
[39] Hanes G R and Dahlstrom C E 1969 *Appl. Phys. Lett.* **14** 362
[40] Acef O *et al* 1993 *Opt. Commun.* **97** 29–34
[41] Bernard J E, Madej A A, Siemsen K J and Marmet L 2001 *Opt. Commun.* **187** 211–18
[42] Jennings D A *et al* 1983 *Opt. Lett.* **8** 136–8
[43] Lea S N *et al* 2003 *Metrologia* **40** 84–8
[44] Ye J *et al* 2000 *Phys. Rev. Lett.* **85** 3797–800
[45] Chartier J-M and Chartier A 2001 *Proc. SPIE* **4269** 123–33
[46] Darnedde H *et al* 1999 *Metrologia* **36** 199–206

[47] Ye J, Ma L-S and Hall J L 2001 *Phys. Rev. Lett.* **87** 270801
[48] Robertsson L, Picard S, Hong F L, Millerioux Y, Juncar P and Ma L-S 2001 *Metrologia* **38** 567–72
[49] Zhang Y, Ishikawa J and Hong F-L 2001 *Opt. Commun.* **200** 209–15
[50] Holzwarth R *et al* 2001 *Appl. Phys.* B **73** 269–71
[51] Rovera G D, Ducos F, Zondy J-J, Acef O, Wallerand J-J, Knight J C and Russell P St J 2002 *Meas. Sci. Technol.* **13** 918–22
[52] Edwards C S, Barwood G P, Gill P and Rowley W R C 1999 *Metrologia* **36** 41–5
[53] Zarka A *et al* 2000 *Metrologia* **37** 329–39
[54] Pritchard D, Apt J and Ducas T W 1974 *Phys. Rev. Lett.* **32** 641–2
[55] Bernard J E *et al* 2000 *Opt. Commun.* **173** 357–64
[56] Poulin M, Latrasse C, Touahri D and Tetu M 2002 *Opt. Commun.* **207** 233–42
[57] Onae A *et al* 1999 *IEEE Trans. Instrum. Meas.* **48** 563–6
[58] Edwards C S, Barwood G P, Gill P and Rowley W R C 2002 *Proc. 6th Symposium on Frequency Standards and Metrology* ed P Gill (River Edge, NJ: World Scientific) pp 643–5
[59] Edwards C S, Barwood G P, Gill P and Rowley W R C 2002 *Technical Digest of the LEOS 2002 Conference and Annual Meeting (Glasgow)* vol 1, p 281–2 (IEEE cat. no. 02CH37369)
[60] Allan D W 1966 *Proc. IEEE* **54** 221–30
[61] Allan D W 1987 *IEEE Trans. Ultrason. Ferroelectr. Freq. Control* **UFFC-34** 647–54
[62] Barger R L, Sorem M S and Hall J L 1973 *Appl. Phys. Lett.* **22** 573
[63] Hänsch T W and Coulliard B 1980 *Opt. Commun.* **35** 441
[64] Hall J L, Baer T, Hollberg L and Robinson H G 1981 *Laser Spectroscopy (Springer Series in Optical Sciences 30)* vol V, ed A R W McKellar, T Oka and B P Stoicheff (Berlin: Springer) p 15
[65] Hils D and Hall J L 1988 *Proc. 4th Symposium on Frequency Standards and Metrology (Ancona, Italy)* (New Jersey: World Scientific) pp 162–73
[66] Young B C, Cruz F C, Itano W M and Bergquist J C 1999 *Phys. Rev. Lett.* **82** 3799–802
[67] Young B C, Rafac R J, Beall J A, Cruz F C, Itano W M, Wineland D J, and Bergquist J C 1999 *Proc. XIV Int. Conf. on Laser Spectroscopy* ed R Blatt *et al* (Singapore: World Scientific) pp 61–70
[68] Barwood G P, Gao K, Gill P, Huang G and Klein H A 2001 *Proc. SPIE* **4269** 134–42

Further reading

Camparo J C 1985 *Contemp. Phys.* **26** 443–77

A useful early review article, describing the general properties of laser diodes, for spectroscopic applications

Brassington D J 1995 Tunable diode laser absorption spectroscopy for the measurement of atmospheric species *Spectroscopy in Environment Science (Advances in Spectroscopy 24)* ed R J H Clark and R E Hester (Chichester: Wiley) ch 3

A review of linear absorption spectroscopy, from an environmental monitoring viewpoint.

Gill P 1993 Laser interferometry for precision engineering metrology *Optical Methods in Engineering Metrology* ed D C Williams (London: Chapman and Hall) ch 5, p 153

A review of the use of laser interferometry in dimensional metrology.

Flowers J L and Petley B W 2001 *Rep. Prog. Phys.* **64** 1191–246

A comprehensive and recent review of fundamental constants (not only those such as the Rydberg and fine structure constants, which can be measured spectroscopically).

C4
Beam delivery

Julian Jones

The process of conveying a laser beam from the laser to the point at which the beam is to be applied is *beam delivery*, the subject of section C4. Delivery may be either by conventional optics in free space, C4.2, or in waveguides, C4.3. In either case, the point of delivery must be controllable and accurately defined in space, either statically or dynamically—hence the related topics of positioning and scanning, the subjects of chapter C4.4.

The basic principles of beam delivery—the focusing, shaping and manipulation of the beam relative to the workpiece—are expounded in chapter C4.1 by Duncan Hand. Focusing and shaping are dependent on the properties of the laser beam, the required beam profile at the workpiece and whether the beam is to be imaged onto the workpiece and scanned or projected via a mask. The laser wavelength and power densities set the choice of materials to be used (and whether fibre-optic beam delivery would be an option).

Chapter C4.2 (Leo Beckmann) is concerned with the design of delivery systems using free-space optics and builds on the foundations of chapter C1.1 in which the optical components to be used were described. Much depends on the wavelength of the laser. For CO_2 optics at the infrared wavelength of 10.6 μm the choice of transmissive optical materials is limited, with ZnSe a popular choice and artificial diamond a modern alternative in some situations: reflective optics are appropriate too. For the Nd:YAG and diode lasers of the visible/near infrared range, more familiar materials are appropriate and fibre delivery (cf the next section) is a promising option. Diode lasers create special problems because their output beam is of relatively low quality, elliptical and can be astigmatic: more complex optics are required when an array of diode lasers is to be coupled or even a stack of arrays. Excimer lasers have UV wavelengths, requiring careful material choice and have low beam quality, favouring techniques of mask projection combined with beam homogenization.

For Nd:YAG lasers, fibre-optic beam delivery has become the technique of choice in many situations and such is the topic of chapter C4.3 (Duncan Hand). For the 1.06 μm wavelength of the Nd:YAG laser the familiar solid-core fused silica fibre, of step or graded core refractive index profile, guiding by total internal reflection, is most commonly used. However, materials such as chalcogenide, fluoride and germanate glasses are all used to extend operation to longer wavelengths—as are the crystalline materials of the silver halide class (polycrystalline) and sapphire (single crystal). The most successful mid-infrared wavelength fibres currently in use are those based on hollow fibres, where the waveguiding mechanism is relatively lossy but the radiation need not pass through significant depths of the waveguide materials. Causing great excitement are a new class of optical fibres, often called microstructured, photonic crystal or ‘holey’ fibres, which rely on their structure rather than material properties in order to guide radiation. One type of photonic crystal fibre, the *photonic bandgap*, is capable of providing guidance in a hollow-core structure with orders of magnitude less attenuation than a conventional hollow-core fibre. Fibre systems, whilst flexible, have distinct limitations in beam quality, power and energy-handling capacity, all of which are discussed.

Jürgen Koch’s chapter C4.4, on scanning and positioning, is essentially *dynamic* beam delivery, covering the basic actuation mechanisms: mechanical, electromagnetic, electro-optic, magneto-optic, acousto-optic and piezoelectric—building on chapter C1.2. The chapter covers the basic design issues in positioning—

motion, accuracy, repeatability and drift and dynamic effects (essentially, the speed of operation)—as well as discussing practical systems with several axes of motion. Positioning is mainly concerned with stepping between fixed positions: scanning is more dynamic, progressing from stepping through vector scanning (stepping at rates beyond system bandwidth) to raster scanning. Scanning is often based on moving mirrors of which the galvanometer and polygon are special cases. More rapid, although with limited amplitude, are piezoelectric scanners and faster still are those based on acousto-optics.

C4.1
Basic principles

D P Hand

The beam delivery of laser light is an important consideration for any high-power laser application, whether for manufacturing (section D1) or medical applications (section D3). Such lasers are normally large and heavy and so cannot be easily moved. In addition, in many applications such as welding (chapter D1.1) in automotive manufacture or ship-building, the workpiece is also of considerable size and mass. This means that the ideal solution is to manipulate the laser *beam* relative to the workpiece.

In addition to beam manipulation issues, most processes require a focused beam and so the beam delivery system must provide a correctly focused beam at, or close to, the surface of the workpiece. This can be achieved using either transmissive or reflective optics. In some cases, simple focusing of the laser beam is not sufficient; instead different and more complex beam shapes at the workpiece may be required. For example there is evidence that in processes such as 'laser direct casting' and 'concrete scabbling', alternative beam shapes are more efficient [1]. Also, with excimer lasers a 'mask-imaging' system is normally used, due to the poor beam quality and low coherence.

The high capital cost of high-power lasers means that there is often a requirement for 'multiplexing' arrangements, where a single laser beam can either be switched 'time-shared' or split 'power-shared' into different paths to different machines. This is particularly useful where workpiece fixturing and positioning is more time-consuming than the laser processing itself or where it is important to process more than one point at a time on a component to ensure symmetrical heat input and, therefore, prevent distortion problems.

Beam delivery, therefore, encompasses optics and associated mechanics for beam manipulation, multiplexing, focusing and beam shaping. The main lasers used for industrial applications are Nd:YAG (1.06 μm) and CO_2 (10.6 μm), and so I will concentrate on beam delivery for these lasers here. However, where appropriate I will also address specific issues with other lasers currently used, including excimer (248 nm, 193 nm), diode ($\sim$800 nm, 980 nm), copper vapour (510.6 nm, 578.2 nm), and femtosecond Ti:Sapphire ($\sim$800 nm), and frequency-multiplied versions of Nd:YAG and copper vapour.

C4.1.1 Beam manipulation

There are three main ways of manipulating the beam: (i) flying optics, (ii) articulated arms and (iii) fibre optic delivery. Examples of these are shown in figures C4.1.1, C4.1.2 and C4.1.3, respectively. The choice of technique depends on the laser wavelength, peak power and the application. For example, CO_2 lasers for cutting flat plates typically use a flying optics configuration, since suitable fibre optics are not available for this wavelength and movement is only required in two dimensions. Alternatively, for the delivery of high peak power, short pulse Nd:YAG light for dentistry applications, an articulated arm is normally used, due to the damage problems experienced with optical fibres at such high peak powers and the requirement for true 3D manipulation. Fibre optics are used in many applications, however, including the delivery of high cw Nd:YAG power for laser welding. Indeed, Nd:YAG welding lasers are often *only* supplied with fibre optic beam delivery.

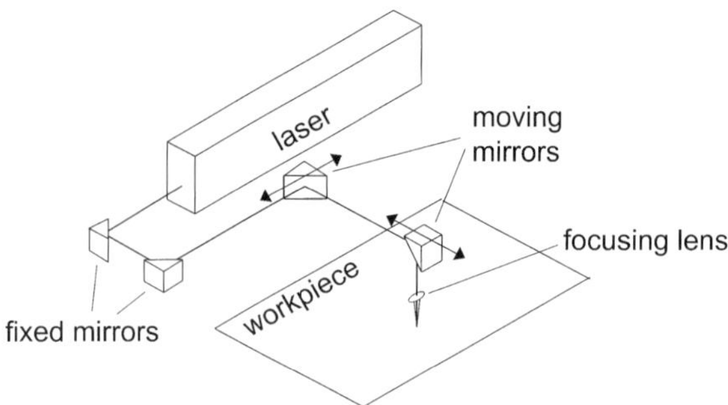

Figure C4.1.1. Flying optics configuration, suitable for moving laser beam around surface of large workpiece.

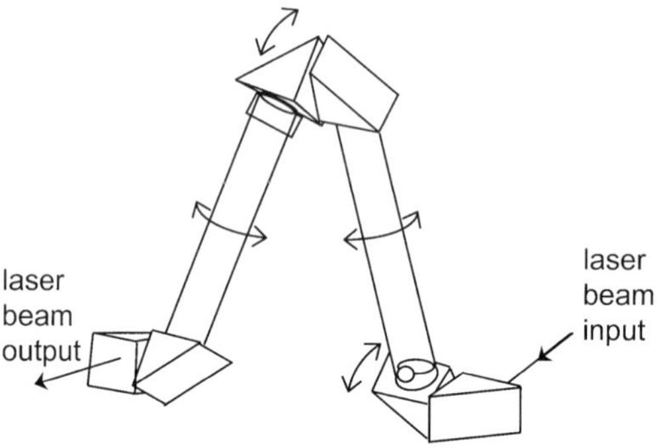

Figure C4.1.2. Articulated arm for flexible delivery of high power laser beams.

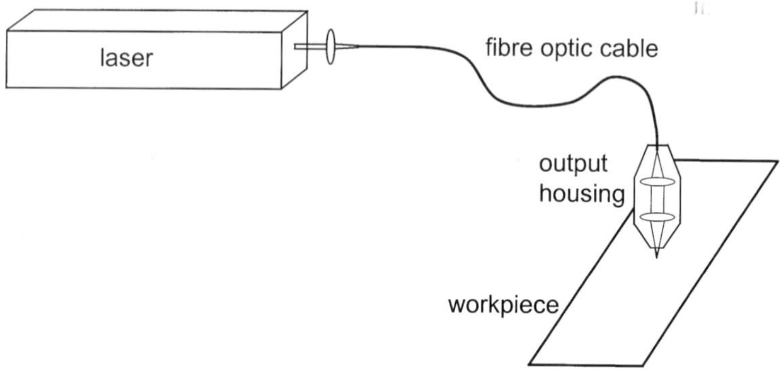

Figure C4.1.3. Fibre optic delivery system.

C4.1.2 Materials for transmissive and reflective optics

The choice of material for beam delivery optics is strongly dependent on wavelength. With visible and NIR lasers (including Nd:YAG), a number of different optical glasses may be used, with fused silica particularly useful due to its very low absorption and suitability for the UV (excimer or frequency-tripled Nd:YAG) as well as the visible and NIR parts of the spectrum. Materials for UV, visible and NIR laser optics are discussed in more detail in section C4.2.7.1.

With CO_2 lasers (chapter B3.1) however, standard optical glasses cannot be used due to the long wavelength (10.6 μm). A number of other materials can be used instead, with the most popular being ZnSe and GaAs, see sections C4.2.5.2.1 and C4.2.5.2.4. ZnSe, in particular, has low absorption and, hence, good thermal properties; also its good transmission in the visible means that it can be used with on-axis monitoring alignment systems, which view through the laser focusing optics (see section C4.3.3 and figure C4.3.8). GaAs has particular application as a final focusing lens, with its high thermal conductivity reducing the likelihood of thermal runaway damage if spatter from the workpiece deposits onto the surface. An alternative possibility is artificial diamond, as discussed in section C4.2.5.2.5.

For high-power CO_2 applications (>5 kW), reflective rather than transmissive optics are normally used due to their higher damage resistance, see section C4.2.5.3 for more details. Copper and silicon mirrors are popular, since they offer advantages in terms of reflectivity (>99.3% for copper) and thermal conductivity. Copper is also easy to machine, with diamond machining used to produce mirror surfaces directly. This greatly simplifies the production of aspheric surfaces for focusing the beam–the most common design being the off-axis parabola described in section C.4.2.5.3.2. In welding and heat treating applications, there is no high-pressure coaxial gas flow to protect the optics from spatter (unlike cutting and drilling) and so molybdenum is sometimes used for the final focusing optic, due to its extremely high resistance to laser damage [2].

C4.1.3 Beam quality

In general, high-power lasers do not produce a single transverse mode, due to thermal effects in the laser gain medium. Heating gives rise to a non-uniform temperature profile and, hence, thermal lensing. Conventional lenses are often placed within the laser resonator to compensate for this lensing effect but residual effects remain, which lead to the generation of higher order modes. This is particularly an issue for lamp-pumped solid state lasers, where only a small percentage of the broad-band pump light has the correct wavelength to pump the laser, with the remainder being converted into heat. This means that a typical Nd:YAG welding laser will have a highly multi-mode beam with a beam profile which is roughly 'top-hat'. Spatial filtering can be used to reduce the number of modes; however, this also reduces the average power. Narrow-band pumping can significantly increase efficiency and, hence, reduce heating effects, and so high-power laser diode arrays are being increasingly used as pump sources for Nd:YAG lasers. The capital cost of these arrays is significantly higher than flashlamps; however, they do have advantages in terms of maintenance cost, with typical lifetimes of 10 000 h compared with 1000 h for flashlamps. With gas lasers, thermal lensing effects are significantly reduced since the dependence of refractive index on temperature is several orders of magnitude lower than with solid state lasers; also novel resonator designs such as the hybrid unstable resonator [3] have been used to remove heat more efficiently.

The modal characteristics of a laser beam are important because they directly affect its 'focusability', or beam quality. This depends on both the wavelength and the transverse modes produced by a laser. For a given laser beam, the product of the focused spot-size (w_0) and the focusing angle (α_0) are a constant (see figure C4.1.4):

$$w_0\alpha_0 = M^2\frac{\lambda}{\pi}. \tag{C4.1.1}$$

The M^2 parameter [4] is often quoted as a measure of the beam quality of a particular laser. A 'perfect' laser beam has a single mode with a fundamental Gaussian profile, which has an M^2 value of 1. In reality, even a single-mode laser will have an M^2 value of $\geq$1.1, and a high-power Nd:YAG welding laser may have an M^2 value of 100 or more. However, the M^2 parameter is not the only term used to describe beam quality; in some countries (particularly in Germany), it is more common to use the beam propagation factor K to

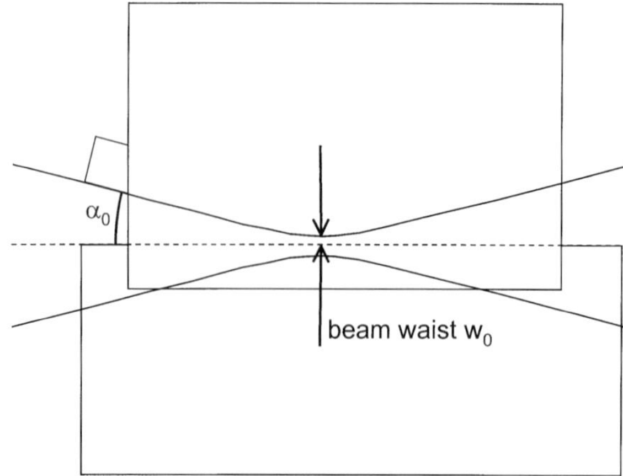

Figure C4.1.4. Focused laser beam, showing beam waist w_0 and focusing angle α_0

describe beam quality, where

$$K = \frac{1}{M^2}. \tag{C4.1.2}$$

In addition, many laser manufacturers describe beam quality using the 'beam parameter product', normally $w_0\alpha_0$, expressed in mm mrad. This has the advantage of incorporating the effect of wavelength but the disadvantage that sometimes focus *diameter* $2w_0$ and *full-angle* $2\alpha_0$ are used for this product instead of focus *radius* w_0 and *half-angle* α_0, giving a factor of four difference in the numerical value quoted.

There is generally a trade-off between average power and beam quality, particularly in solid state lasers, due to thermal lensing effects as discussed earlier; or at least a trade-off between beam quality and cost. It is, therefore, clearly undesirable that any beam delivery system should reduce the beam quality. There are three main ways in which beam quality can be lost: (i) lens aberrations, (ii) thermal distortions in mirrors and lenses and (iii) mode coupling in optical fibres. Lens aberrations can be avoided by using appropriate lens designs, as discussed in section C4.21.3. Thermal distortions are a particular problem with CO_2 optics but can be minimized by using effective cooling, e.g. water cooling of mirrors. However, there is always some loss of beam quality (or laser power) when a multi-mode laser beam is coupled into an optical fibre. In order to achieve coupling of light into the fibre, a focused spot radius of typically 90% of the fibre core radius is used. This will 'grow' to 100% after a short distance within the fibre. Furthermore, mode coupling occurs in the fibre, due to imperfections and bending, which means that the output divergence angle will always be greater than the focusing angle at the input [5]. For a typical fibre beam delivery system using 600 μm diameter optical fibre, a reduction in beam quality of about 16% would be expected for an input M^2 value of $\sim$100.

C4.1.4 Beam requirements at the workpiece

There are two basic types of processing: (i) a focused spot is moved relative to the workpiece (e.g. figure C4.1.1 and C4.1.3); and (ii) a mask-imaging system (figure C4.1.5). CO_2, Nd:YAG, copper vapour and diode lasers all use the first technique, whereas excimer lasers use the second. With mask imaging, the main issues are beam uniformity and optical aberrations, with such aberrations being particularly critical in high-resolution photolithographic applications in the electronics industry—see section C4.2.7.

Any focused spot process has particular requirements in terms of focused spot size, depth of field, beam shape and, in some cases, polarization. As shown in equation (C4.1.1), the spot size varies with both the M^2

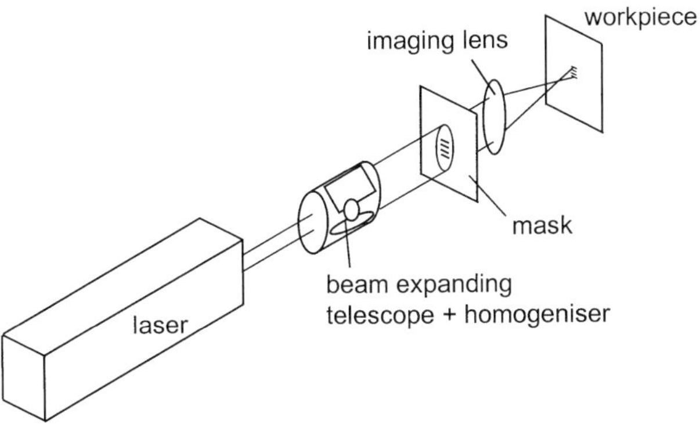

Figure C4.1.5. Mask projection arrangement.

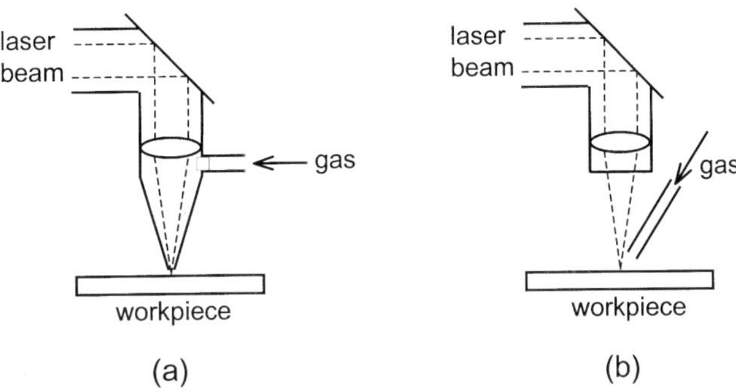

Figure C4.1.6. Laser processing head with (*a*) co-axial gas assist and (*b*) off-axis gas shielding.

parameter and wavelength for a given focusing angle. The equations which describe the focusing of a laser beam are developed in C4.2.2. It is important to realize that a small spot size carries the penalty of a short depth of field, which has a quadratic dependence on spot size, for a given wavelength and M^2 parameter. Although there are significant differences between the required focused beam for different processes, there are some general principles which can be defined.

First, there is a requirement for a large stand-off (several tens of mm) between the focusing optics and the workpiece. This is to prevent damage to these optics during processing. Such damage can occur due to direct thermal radiation from the process, reflected laser light or debris from the process (e.g. spatter from laser drilling) contaminating the lens and, hence, causing runaway thermal damage due to subsequent heating from the transmitted laser beam. Gas jets are normally used to help prevent such contamination, which can be either co-axial (as used in cutting) or off-axis (commonly used for welding)—see figure C4.1.6.

Second, many processes (e.g. precision cutting) require a small focused spot (tens to hundreds of micrometers in diameter, dependent on the process) at the surface of the workpiece. This combination of small focused spot and large stand-off means that relatively large diameter lenses must be used, particularly if the beam quality is poor, and so 50 mm diameter optics are commonly employed, although some systems are based around 25 mm diameter optics. The high numerical aperture used means that aberrations must be considered, see section C.4.2.3.3. Spherical aberration is the main problem, which is commonly addressed in the visible-NIR by using achromatic doublets; aspheric optics are used at CO_2 wavelengths. Although achromatic doublets are optimized to deal with chromatic aberration, they also provide good correction for spherical aberration, are relatively low cost, and have the additional advantage of providing a high-quality imaging route for the on-axis camera systems which are often used for workpiece setup and monitoring.

The final issue is depth of field. This is normally defined as being twice the Rayleigh range (see section A2.1.2.2), and is the distance over which the laser beam area (and hence intensity) is within a factor of two of its value at the beam waist. However, such a variation in intensity would have a significant effect on most high-power laser processes, so the effective depth of field is somewhat less than this. For example, a 10% change in intensity occurs at a distance after the waist of roughly one-third of the Rayleigh range. Some processes, such as deep hole drilling, require a long depth of field [6], whilst with others it is simply important to minimize the tolerancing requirements of the beam-positioning system. For example, a typical Nd:YAG beam used for drilling cooling holes in turbine blades has an M^2 parameter of 20 and a focused spot diameter of 0.5 mm, giving a Rayleigh range of 9 mm. This Rayleigh range is necessary since holes of several mm depth are required, due to the combination of material thickness and the shallow angle of the holes relative to the surface of workpiece. A typical Nd:YAG beam used for welding, meanwhile, may have an M^2 value of 125, with a focused spot diameter of 0.6 mm and Rayleigh range of 2.1 mm.

In processes which do not require a standard focused beam but instead demand a special beam shape at the workpiece, other types of optics may be necessary. For example, diffractive optical elements (DOEs) may be used, which can be designed to produce an almost infinite variety of beam shapes [1]. Such DOEs have, for example, been applied to useful effect in the laser scabbling of concrete. Other techniques, including kaleidoscope-type reflective cavities, have been developed for beam homogenization, as described in C4.2.5.4. Another optical element which has been applied to great effect is the 'dual-focus' lens [7]. This produces a beam which is particularly suited to deep section cutting, with the outside of the beam focusing at the workpiece surface and the central part a few mm below the surface.

C4.1.5 Attenuation

An ideal beam delivery system would be perfectly transparent. In reality that can never be the case, primarily due to reflections from the surfaces of the optical elements but also due to absorption, particularly in optical fibres, and incomplete reflection from mirrors. The reflectivity of a transmissive optic is dependent on its refractive index. For NIR, visible and UV wavelengths (Nd:YAG, copper vapour, diode, excimer), fused silica or BK7 optics are normally used, which have a refractive index of 1.46 and so each surface reflects about 3.5%. For CO_2 laser light, however, ZnSe is commonly used. It has a refractive index of 2.4 and, hence, 17% of the beam is reflected at each surface. Anti-reflection coatings are used to considerably reduce these surface reflectivities; however, they reduce the damage threshold of these optics and so cannot always be used with very high-power beams. Mirror reflectivities are normally very high, up to 99.9%, and so do not significantly affect the overall attenuation of the beam delivery system. Metal mirrors have slightly lower reflectivities but are still typically $>99\%$.

Attenuation in large-core fused silica optical fibres is typically 2.5 dB km^{-1} at 1064 nm (see figure C4.3.5), which means that it only becomes significant when lengths of 200 m or more are used and so is not normally an issue with industrial applications. However, the absorption increases at shorter wavelengths, giving a loss of 2.4 dB in 10 m at 266 nm (frequency-tripled Nd:YAG). The dominant effect with fibres is end-face reflection; here, it is not practical to use anti-reflection coatings due to the high intensity incident on the fibre end-face which would quickly result in damage. In addition, some light is normally lost when coupled into the fibre, as the focused spot from the laser is not a perfect 'top-hat' intensity distribution and so a small percentage (typically 1–2%) is not coupled into the core.

For a typical Nd:YAG fibre optic delivery system, consisting of two mirrors and a single doublet lens at the input end of the 10 m fibre, with two doublets at the output to image the light onto the workpiece the overall loss is about 11% if the lenses have anti-reflection coatings, rising to about 26% if uncoated.

C4.1.6 Optical damage

Beam delivery systems for industrial applications must be able to deliver light with sufficient power to melt and vapourize a wide range of materials. This has the potential to cause catastrophic damage to the various optical elements in the delivery system. The threshold for damage at the surface of a material is usually lower than in the bulk, due to small imperfections at or near the surface, dependent on the quality of the polish. Care must be taken to keep all surfaces clean, as any dust can be burnt onto the surface resulting in localized heating and, hence, damage. The risk of damage is reduced if the beam size is as large as possible, although, in some cases, the damage has been shown to scale with diameter rather than area [8]. Thin film optical coatings, such as those used to make mirrors or anti-reflection coatings, significantly reduce the damage threshold.

C4.1.7 Safety

Safety must always be a key consideration, since the high power (average and/or peak) which must be delivered could easily cause damage if even a small fraction was directed somewhere other than intended—hazards include eye damage, skin burns and fire, as discussed in section C6. Delivery systems are designed to prevent the beam from being accidentally intercepted, by enclosing the beam either within tubing or in an optical fibre. In addition, optical fibre cables for high-power laser light normally incorporate some kind of continuity sensor, to detect if fibre breakage occurs. One such example is a fine wire coiled around the fibre along its length, whose continuity is continually checked. If the fibre breaks, the light which emerges will very quickly burn through the wire, breaking the continuity. Sensors have also been developed to detect any misalignment of the laser beam relative to the input end of the fibre by monitoring the power coupled into the fibre cladding [9].

C4.1.8 Summary

Laser beam delivery is an intrinsic part of any high-power laser processing system, in order to transport, focus and manipulate the beam as required. The components of a beam delivery system include lenses, mirrors and optical fibres, together with appropriate safety enclosures or continuity monitors and process gas delivery. The particular components used (and the material from which they are made) primarily depend on the laser wavelength, peak power and beam quality required for a particular application.

References

[1] Taghizadeh M R, Blair P, Balluder K, Waddie A J, Rudman P and Ross N 2000 Design and fabrication of diffractive elements for laser material processing applications *Opt. Lasers Eng.* **34** 289–307

[2] Danielewicz E, von Der Ahe T and King S 2000 Reflective optics add focus *Industrial Laser Solutions* April

[3] Jackson P E, Baker H J and Hall D R 1992 CO_2 large-area discharge laser using an unstable-waveguide hybrid resonator *Appl. Phys. Lett.* **54** 1950

[4] Sasnett M W 1989 Propagation of Multimode Laser Beams—The M^2 Factor *The Physics and Technology of Laser Resonators* ed D R Hall and P E Jackson (Bristol: Adam Hilger)

[5] Kuhn A, Blewett I J, Hand D P and Jones J D C 2000 Beam quality after propagation of Nd:YAG laser light through large-core optical fibers *Appl. Opt.* **39** 6754–60

[6] Kudesia S S, Rodden W S O, Hand D P and Jones J D C 2001 Effect of beam quality on single pulse laser drilling *Proc. ICALEO 2001 Jacksonville, USA* (Orlando FL: Laser Institute of America)

[7] Powell J, Tan W K, Maclennan P, Rudd D, Wykes C and Engstrom H 2000 Laser cutting stainless steel with dual focus lenses *J. Laser Applications* **12** 224–31

[8] Wood R M 1997 Laser induced damage thresholds and laser safety levels. Do the units of measurement matter? *Opt. Laser Technol.* **29** 517–22

[9] Boechat A A P, Su D and Jones J D C 1992 Bidirectional cladding power monitor for fiberoptic beam delivery systems *Meas. Sci. Technol.* **3** 897–901

C4.2
Free-space optics

Leo H J F Beckmann

C4.2.1 Introduction

The subject of this chapter is free-space optics, that is the optical means and devices which transfer and transform the radiation output from a laser to a workpiece in such a way that it activates the desired processes. Thus it entails both radiation beam transport (relaying) and beam manipulation. The latter is mostly in the form of focusing to a spot of some desired size and shape but an application may also require a radiation beam transformation, which produces a homogeneous irradiation over a certain area of the workpiece. The optical methods and devices for achieving these goals vary considerably with the type of laser used and, more specifically its wavelength, mainly due to restrictions set by wavelength-dependent material properties.

As we saw in chapter A1, the unique properties of laser radiation, in particular its spatial coherence in conjunction with monochromaticity and scalability to high power levels, have given lasers a firm, often exclusive, place in applications ranging from information processing, transport and storage to thermal machining and material processing. The usefulness of lasers of medium and high power stems largely from the exceptionally high radiance of the laser radiation source. A (rotationally symmetrical) laser which delivers a radiation beam of radius w_0, and has a radiating surface $A = \pi w_0^2$ at the point of smallest beam radius, or beam 'waist', will radiate its power P of wavelength λ into a cone with a divergence half-angle $\vartheta = M^2\lambda/(\pi w_0)$, where M^2 is the beam quality, which is equal to or greater than 1. The corresponding (small) solid angle is $\Omega = \pi\vartheta^2$, so that the radiance $N = P/(A\Omega)$ of the laser is found to be $N = P(M^2\lambda)^{-2}$, a noteworthy result in that $M^2$and λ are the only variables besides the power P and radiance depends on the squares of both the beam quality and the wavelength. The radiance found by this simplified calculation is an average quantity—the actual radiance is even higher in a central peak or for pulsed lasers.

Since radiance is a field quantity, invariant to (loss and aberration-free) optical manipulation, optical systems allow the transformation of laser radiation to (very high) power density (or irradiation) $P/A' = N\Omega'$, by suitably choosing the solid angle Ω' at which the laser radiation impinges on a surface A'. This solid angle can—and most often will—be much greater than that into which the laser radiates. A parameter, more commonly used in optics, is the numerical aperture (NA), $NA = n \sin U'$, where U' is the half-angle of the radiation cone, which emerges from the optical system. It is related to the solid angle (in air, and for a Lambertian absorber) by $\Omega' = \pi(NA)^2$, so that

$$P/A' = [\pi P/(M^2\lambda)](NA)^2. \tag{C4.2.1}$$

It should be noted that, with the exception of numerical aperature, all parameters in this equation are fixed by the choice of laser source, so that the numerical aperture is the only quantity available to the system designer for controlling the power density on the workpiece and, thus, the process.

For a numerical example, consider a CO_2 laser of 1 KW with a beam quality of $M^2 = 2$. Its radiance is approximately 2.2×10^8 W cm^{-2} sr^{-1} and it is noteworthy that this level corresponds to blackbody radiation at a temperature of more than 100 000 K. It allows power densities well in excess of 1 MW cm^{-2} to be achieved as is necessary for such processes as the cutting and drilling of metals, even with optics of moderate numerical aperture.

In many respects, laser radiation behaves similarly to non-coherent radiation from classical radiation sources and the design of devices for its manipulation can, therefore, often benefit from classical optical expertise which has accumulated over more than a century but there are important differences and hitherto unknown challenges and traps. Aside from coherence and its optical consequences, treated in the next section, it is the power of a laser beam, the high power density—whether wanted or unwanted—it creates in foci and the associated thermal effects, which require careful systems consideration. Lasers which deliver high power outputs span a wide range of wavelengths, from ultraviolet (less than 200 nm) to infrared (more than 10 μm), and their optical manipulation faces serious, unavoidable and partly unsolvable problems rooted in basic materials properties, specific to the appropriate wavelength region. The wide range has also consequences with respect to the manufacturing tolerances of the optics, those for the ultraviolet region being roughly 50 times more severe—and costly to achieve—than for the infrared. The following sections will address these problems in conjunction with the laser types which have found broad application in materials processing or otherwise, notably the CO_2 laser, the Nd:YAG laser, the diode laser and the family of excimer lasers.

C4.2.2 Laser beam propagation and its optical consequences

The theory behind the propagation of Gaussian laser beams is given in chapter A2.1. However, we summarize the principal results in this section. Spatial coherence is probably the most characteristic and most consequential property of a laser beam. It causes a beam, which simultaneously has a Gaussian beam profile, to propagate such that the beam radius $w(z)$ along the axis of propagation expands in a nonlinear fashion [1]:

$$w(z) = w_0\{1 + [\lambda z/(\pi w_0^2)]^2\}^{1/2} \tag{C4.2.2}$$

so that the beam boundaries do not form straight lines as in classical optics. Instead, the beam radius attains a minimum of w_0 at $z = 0$, known as the beam waist and fans out to a radiation cone with a limiting half-angle (also called the half-beam divergence):

$$\vartheta(z \to \infty) = \lambda/(\pi w_0). \tag{C4.2.3}$$

The axial distance over which the beam has grown to $\sqrt{2}$ times its size at the waist is known as the Rayleigh range and

$$z_R = \pi w_0^2/\lambda. \tag{C4.2.4}$$

At the site of the waist, the beam wavefront is flat and its radius of curvature, $R(z)$, changes along the axis of propagation:

$$R(z) = z[1 + (z_R/z)^2]. \tag{C4.2.5}$$

These relations differ from those for classical ray optics, yet the concept of optical rays retains some usefulness when treating laser beams, provided the rays are strictly seen as the normal to the wavefront of the beam. The well-established methods for optical design and analysis, which are essentially based on geometrical ray tracing, can, with some modifications, be applied to optics for laser-beam handling. The modifications concern the nonlinear change in the wavefront curvature, as given by equation (C4.2.5), when the beam propagates from one optical surface to the next but the difference and, thus, the necessary correction is small except for near-telescopic situations. The laws of refraction and reflection apply unchanged.

The axial region between the beam waist and the Rayleigh range is referred to as the near field. It is in this region that the difference between laser and classical optics, notably the curvature of the wavefront, is greatest. At distances much greater than the Rayleigh range, called the far field, the laser-beam wavefront curvature approaches that described by classical theory. As the differences are all tied to the Rayleigh range, classical optics may well be considered to be a special—or limiting—case of general (coherent) optics for the Rayleigh range becoming zero.

The analogy between coherent and incoherent optics may be extended to the generation of an image, which is, by definition, the formation of a diffraction pattern from a converging wavefront at some point in space and which may be simulated by calculating the (Huygens) diffraction integral. Quantitative differences result from differences in the amplitude and phase distribution over the wavefront, however, and these affect the shape and size of the diffraction pattern as well as the axial point of minimum size (or maximum energy concentration). Most marked is the absence of outer rings in the case of an unapertured coherent beam with a Gaussian profile, as contrasted to the Airy pattern which results from non-coherent illumination. Indeed, the Gaussian profile of the laser beam is preserved throughout all optical-beam manipulations by aberration-free optics [1].

C4.2.2.1 Beam size

The previous relations apply to purely Gaussian or fundamental mode (TEM_{00}) laser beams and have to be modified for real beams, which are characterized by higher and mixed modes [2].

Before elaborating on real beams, the question of beam-size definition has to be addressed. While non-coherent beams are usually limited by hard aperture stops to well-defined beam widths, the definition of that quantity for a laser beam is much less obvious. For pure Gaussian (TEM_{00}) beams, the diameter at which the power has dropped to $1/e^2$ of its central peak has been quite generally used to describe the beam width but for higher and mixed mode beams, this is no longer a meaningful quantity and must, therefore, be replaced by a more rigorous and fundamental definition [3, 4]. A suitable and generally applicable description of the beam diameter is based on the second moments of its power distribution function and this has been adopted for both the relevant ISO and CEN standards [5]. Using this beam size definition, the previous formulas for Gaussian beams can also be applied to real beams, if the wavelength λ is replaced by the quantity

$$M^2\lambda = \lambda/K \tag{C4.2.6}$$

which could be looked at as an 'effective' wavelength. M^2 (see section A2.1.4.3) is called the beam quality or, more specifically, the 'times diffraction-limited factor' and was introduced by Siegman [6]. The ability to express the most important performance parameters of any real laser by a single additional quantity—where (in accordance with the previously mentioned standards) $1/K$ and M^2 may be used interchangeably—is the main reason and advantage of its introduction but it must be borne in mind that it does not describe the shape of the power distribution within the beam and, consequentially, the focal spot.

The beam quality is directly related to the product of the beam-waist radius (W_0) and the half-beam divergence (ϑ):

$$\vartheta w_0 = M^2\lambda/\pi. \tag{C4.2.7}$$

This quantity is also known as the beam parameter product [7]; it is invariant throughout the beam propagation, as long as aberration-free optics are involved and it attains a minimum for a 'diffraction-limited', i.e. purely TEM_{00} beam, for which $M^2 \rightarrow 1$. In fact, the relation (C4.2.7) is typically used to determine M^2 from measured values of the beam-waist radius and beam divergence [8]. This definition of the beam parameter product is not adhered to consistently throughout the literature and occasionally the product of beam-waist diameter and the full beam divergence is used instead.

C4.2.3 Computation of laser optical systems

C4.2.3.1 Location of the laser beam image

Chapter A2.1 developed a matrix formalism to describe the geometrical propagation of paraxial rays and the propagation with diffraction of Gaussian beams. In the following sections, we describe how to calculate the position and dimensions of laser beam waists when transformed by lenses, as introduced in chapter A3. In geometrical optics, an object (loosely something that can be imaged) is considered to be the point of origin of a spherical wavefront, the radius of which equals the distance from the object. Given a lens (or, more generally, an optical imaging system) of focal length f, object (s) and image (s') distances from the (thin) lens are given by the well-known 'lens formula' [9]:

$$1/s + 1/s' = 1/f. \tag{C4.2.8}$$

The waist of a laser beam is not an object in the same sense, as the wavefront varies with distance in accordance with equation (C4.2.5), but the lens formula can be applied in expanded form:

$$\frac{1}{s+q} + \frac{1}{s'+q'} = \frac{1}{f} \tag{C4.2.9}$$

where the beam parameters q, q', at the object and image waists, respectively, are purely imaginary [1]:

$$q = j\pi w_0^2/\lambda \qquad q' = j\pi w_0'^2/\lambda \tag{C4.2.10}$$

Equating the imaginary parts, one obtains

$$\frac{s-f}{s'-f} = \frac{w_0^2}{w_0'^2}. \tag{C4.2.11}$$

Equating the real parts results in

$$(s-f)(s'-f) = f^2 - f_0^2 \tag{C4.2.12}$$

with

$$f_0 = \pi w_0 w_0'/\lambda. \tag{C4.2.13}$$

From this we can derive an expression for the position of the image waist (s') as a function of object-waist position (s) and focal length [9]:

$$\frac{1}{s + b^2/(s-f)} + \frac{1}{s'} = \frac{1}{f} \tag{C4.2.14}$$

with

$$b = \pi w_0^2/\lambda. \tag{C4.2.15}$$

It will be noted that b is the Rayleigh range of the source beam. The nonlinear dependence of s' on s, together with the variation of beam diameter with distance s from the beam waist (equation (C4.2.2)), affects both the focal position and the numerical aperature of the focused beam in systems where the laser is stationary and flying optics are used to concentrate its radiation on the workpiece (such as in gantry-type machining stations). To minimize these dependencies, it is advisable to laterally expand the source beam, which effectively increases the source-beam Rayleigh range and, thus, reduces the contribution of s. If this is not done, the resulting variation of focal point distance from the lens may need to be compensated for at the optical focusing head.

If the expression (C4.2.14) is rewritten in normalized form [9]:

$$(s'/f) - 1 = \frac{[(s/f) - 1]}{[(s/f) - 1]^2 + (b/f)^2} \tag{C4.2.16}$$

and compared to the equally rewritten geometrical lens formula

$$(s'/f) - 1 = \frac{1}{[(s/f) - 1]} \tag{C4.2.17}$$

it is clear, that the appearance of the term $(b/f)^2$ removes the pole at $(s/f) = 1$, the case for which the object is at the (forward) focal point and which causes the image of a classical object to move to infinity. Instead, the image position (s') of a laser beam waist is always finite: its maximum distance from the lens is

$$s'(\max) = f[1 + (f/b)/2] \tag{C4.2.18}$$

for a source-beam-waist distance of $s = f + b$. Furthermore, unlike the case of classical geometrical objects where there is a minimum object-to-image distance of $4f$, there is no minimum separation between input (source) and output (image) beam waist for Gaussian beams.

The magnification m, i.e. the ratio of the image- and source-waist radii, is found to be

$$m = w_0'/w_0 = \frac{1}{\{[1 - (s/f)]^2 + (b/f)^2\}^{1/2}}. \tag{C4.2.19}$$

As is easily seen, equations (C4.2.14), (C4.2.16) and (C4.2.19) reduce to the corresponding geometrical optics formulae when $b \to 0$, which again emphasizes that classical geometrical optics is the special case at the $b \to 0$ limit.

It is interesting to discuss another special case: in many applications, the focusing lens will be situated at or near (an image of) the source waist and $s \to 0$. Then

$$s' = \frac{f}{1 + (f/b)^2} \tag{C4.2.20}$$

that is the image waist is always closer to the lens than the geometrical image. Since the focal length is, however, in most cases much smaller than the Rayleigh range, the deviation is usually very small. In other words, if a laser beam is focused by an optical system located well within the near field, the position of the focus will be near the geometrical focal point.

C4.2.3.1.1 Multiple optical elements and the use of ray tracing

The previous discussion has tacitly assumed a single (thin) optical element and the distances from it. In practice, however, we will often encounter optical systems with a succession of optical components as needed to perform a given task. For this situation, the previous treatment with its use of the imaginary beam parameters q and q', although mathematically quite elegant, is less suitable. In practice, we prefer an approach which is closer to the well-established methods of ray tracing but where a 'ray' is strictly seen as the normal to a wavefront and ray propagation is computed in accordance with equation (C4.2.2). Such a calculation will involve the following steps:

(1) Calculation of the wavefront size w and radius of curvature $R(z)$ at the point of intersection with the first optical surface, using the laser beam waist radius and waist distance. These data result in a 'ray' of height w and slope $w/R(z)$ on the first optical surface.

(2) Refraction of the ray, which results in the slope of the exiting ray (there is no change of the ray height). Conversion to the corresponding wavefront curvature of the exiting beam.
(3) Calculation of the waist size (w_0') and position (z') of this exiting beam from that wavefront curvature and size.
(4) Calculation of the wavefront curvature and size for that beam at the intersection with the next optical surface and conversion to a 'ray' of appropriate slope and height.
(5) Repetition of steps (2) through (4) for as many optical surfaces as there are in the system, ending with a step (3), which yields the size and position of the image waist.

The formulas for calculating the beam waist size and position from a point along the axis, where the wavefront has radius w and radius of curvature R, as needed for step (3), are [10]:

$$w_0 = w/\{[1 + (\pi w^2/\lambda R)^2]\}^{1/2} \tag{C4.2.21}$$
$$z = R/[1 + (\lambda R/\pi w^2)^2]. \tag{C4.2.22}$$

For use in a computer program, it is preferable to replace these formulas by the equivalent expressions:

$$w_0 = \{R^2(\lambda/\pi)^2/[w^2 + R^2(\lambda/\pi)^2/w^2]\}^{1/2} \tag{C4.2.23}$$
$$z = R[1 - w_0^2/w^2] \tag{C4.2.24}$$

This procedure is easily extended to include real beams of a given beam quality M^2 by multiplying λ with M^2. It should further be noted that the calculations can be (and often are) carried out with 'paraxial formulas' for the refraction part, which means approximating the sine of an angle by the angle itself and the cosine by 1. The final result is the ideal image waist size and position, i.e. neglecting aberrations and beam aperturing (to be discussed later).

C4.2.3.2 Focal spot size

As in geometrical optics, the size of and energy distribution within the focus of a laser beam is governed by diffraction. With aberration-free optics, the size of a real laser focal spot, actually the radius w_0' of an image beam waist, for a laser of beam quality M^2 is

$$w_0' = M^2\lambda/[\pi(NA)] \tag{C4.2.25}$$

and since λ and M^2 are determined by the choice of the laser used, the only parameter left to the user for control of the spot size is the NA of the focusing optics—quite in analogy to classical optics. It is also apparent that a laser of superior beam quality (lower M^2) allows one to achieve a desired spot size at lower NA. This is important for three reasons: (1) a higher NA generally requires more sophisticated optics, which (2) has typically a shorter back focus (i.e. the last optical surface will be closer to the workpiece). It (3) also has a shorter depth of focus (i.e. the radiation cone has a wider angle at the image). The latter is quantified by the Rayleigh range (R_i) of the image,

$$R_i = \pi(w_0')^2/(\lambda M^2) \tag{C4.2.26}$$

or

$$R_i = w_0'/(NA) \tag{C4.2.27}$$

which shows that, for any given spot size w_0', the Rayleigh range will be inversely proportional to both M^2 and to the NA which is required to achieve that spot size.

C4.2.3.3 *Effect of aberrations on image size and shape*

The aberrations of a non-ideal optical system will, in general, cause a deformation of the wavefront with respect to its—originally spherical—shape. The equally occurring (small) changes in the amplitude distribution over the wavefront are normally—and in by far the most cases justifiably—neglected. The effect of wavefront deformations in laser beams is quite analogous to that in non-coherent optics: a reduction in the power density in the central peak of the diffraction image, associated with a shift of optical power to surrounding rings. There is also some broadening of the central peak.

A quantitative comparison between the diffraction images obtained from non-coherent and coherent wavefronts relies on a suitable definition of the image boundary. For a non-coherent diffraction image, it is the diameter of the first dark ring of the well known Airy pattern [11, p 182 ff] which provides a suitable measure of the image diameter but the diffraction pattern of a perfect Gaussian wavefront is Gaussian itself, has no such ring features and (theoretically) extends to infinity. The generally accepted [5] diameter definition of laser-beam images—actually image waists—again uses the second moments of the power distribution function, completely analogous to the definition of beam diameters. This value corresponds to the $1/e^2$ diameter for the limiting case of a TEM_{00} beam; that diameter is $2\lambda/(\pi NA)$ or $0.64\lambda/(NA)$. We note that the diameter of the first dark ring of the Airy pattern (for homogeneous illumination by non-coherent radiation) is $1.22\lambda/(NA)$, seemingly much bigger for equal numerical apertures but it must be borne in mind that focusing a Gaussian beam will require that the free diameter of the optics be larger than that of the beam to avoid beam aperturing and associated image degradation (to be discussed shortly). Thus the difference in NA essentially offsets the numerical factors of these expressions. The diffraction image of a laser beam is, however, more compact than the Airy pattern with its extended ring structure and more than 99% of the energy falls within two diffraction diameters as long as the aberrations are small.

What constitutes a ‘small’ aberration is somewhat arbitrary [12]. In non-coherent optics, it is customary to consider an optical system to be ‘diffraction limited’ if the wavefront error due to, for instance, spherical aberration is less than one-quarter wavelength, which causes a reduction in the Strehl ratio (the peak intensity in relation to the theoretical maximum) to 80%. Much smaller (1/14–1/20 wavelength) random peak-to-peak wavefront errors, such as those resulting from manufacturing tolerances, cause a deterioration of the same order of magnitude [13] and make the 80% limit, which was originally proposed by Lord Rayleigh [14], widely accepted. There is no such commonly agreed acceptance limit in laser applications and, since peak power density is quite often the quantity which determines the effectiveness of a laser process (for example cutting and drilling), low aberration optical systems, with a Strehl ratio greater than 0.9 are preferable. The effect of spherical aberration on the diffraction image of a Gaussian beam—essentially a shift of power into an outer ring structure—is shown in figure C4.2.1.

C4.2.3.4 *Beam aperturing requirements and effects*

The already mentioned requirement that the free diameter of a focusing optics well exceeds the beam diameter arises for two reasons: One, of particular importance for high-power beams [15], is to ensure that no appreciable portion of the incident radiation falls on the lens-holder and, is converted to heat hence causing unforeseen thermal problems. The second, which is valid for all power levels, is a change of shape of the central maximum of the diffraction image with a corresponding reduction in the peak power density. The increase in focal diameter from aperturing a TEM_{00} beam at 1.5 times the $1/e^2$ diameter amounts to about 10% for an aberration-free lens. As a practical rule, the free optical diameter for laser-beam handling should be at least 1.5—and preferably 2.0—times the beam diameter. Lower factors are acceptable for very low quality beams. Although an even larger factor would be theoretically beneficial, it is hardly useful in practice, as some aperturing has always already occurred within the laser source. Formulas describing the effect of beam aperturing on the far-field beam divergence have been derived by Drége [16].

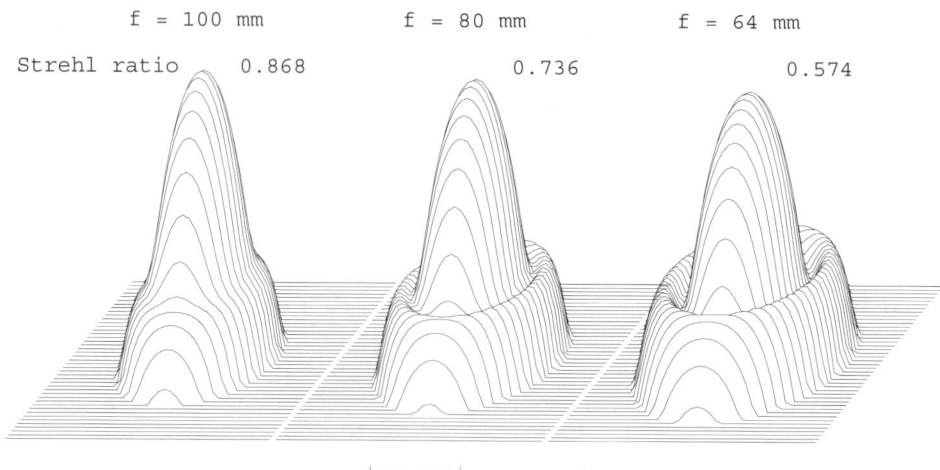

Figure C4.2.1. Diffraction patterns of foci of TEM_{00}-laser radiation obtained with lenses of increasingly shorter focal length and associated spherical aberration. Logarithmic vertical scale, 0.002 threshold. Spherical aberration reduces peak intensity and broadens the image as power is shifted to outer rings.

C4.2.3.5 Low beam quality sources

For low-quality laser beams—$M^2 > 10$—as is typical for most high-power solid state lasers, the focal spot size approaches that of the classical image of an extended object. This is easily demonstrated for the case of the radiation output from, for instance, an Nd:YAG laser, which is delivered via an optical fibre of diameter D. The fibre output forms a source of radiant surface

$$A = \pi D^2/4 \tag{C4.2.28}$$

and if the beam quality is M^2, it will radiate all its power into a cone of half-divergence

$$\sin V = \frac{\lambda M^2}{\pi D/2}. \tag{C4.2.29}$$

Now let this (divergent) radiation beam enter an optical system of magnification m; then the exiting radiation cone will have a half-beam divergence of $\sin V' = (1/m) \sin V$ and the geometrical image diameter will be $d = mD$. The diffraction image of a Gaussian source, however, would have a diameter of

$$\delta = \frac{2\lambda}{\pi \sin V'} = \frac{2\lambda m}{\pi \sin V} = mD/M^2 \tag{C4.2.30}$$

and is seen to be smaller by a factor $1/M^2$. In other words, the diffraction contribution is only a fraction $1/M^2$ of the image size.

C4.2.3.6 Non-rotationally-symmetrical laser beams

Throughout most of the previous discussion, it has been tacitly assumed that the laser beams are essentially rotationally symmetrical, since this is the preferred shape for most applications and generally striven for by laser manufacturers. A number of real lasers, notably diode lasers, however, emit beams of different width and/or beam quality in two transverse, mutually perpendicular, directions. Focusing such beams generally results in more or less elliptical spots and, in accordance with the previous formulas, adaptation of the numerical aperture in one direction will be needed to form a circular spot (see chapter A3 and C4.2.6.4.3).

C4.2.4 General practical guidelines for optics for laser applications

As in any other application of optics, the optical layout is dictated by system requirements but there are additional and important considerations specific to laser applications, mostly related to the typically high (peak) powers. For lasers operating outside the visible and near-infrared spectrum, applications rely on few, often costly, transmissive materials, suitable for use as lenses or windows. Mirror optics can be used rather universally at any wavelength and throughout the laser optical train but obviously cannot reproduce the function of a (pressure) window.

Lenses are often preferred over mirrors because they allow straight optical paths and, thus, make the mechanical layout, assembly and alignment easier. Mirror systems, however, are always 'off-axis', causing folded paths, and the mechanics must occasionally accommodate odd angles as determined by the optical design.

At very high powers, mirrors are the only choice: metal-based mirrors can withstand very high powers and power densities (up to, say, 10^5 W cm^{-2}), if channels for circulating cooling water are integrated into the mirror body, a widely applied technique. Metal-based mirrors are typically manufactured by computer-controlled single-point diamond machining and the generation of various non-spherical shapes is well within the capability of that technique, though often at higher cost.

When designing an opto-mechanical system for laser powers in excess of, say, 500 W, due attention has to be given to thermal problems, which derive from stray radiation, residual absorption in transmissive materials and thermal expansion and resultant stress. Forced cooling ensures thermal stability and avoids thermal runaway, an effect which otherwise can occur if absorption increases with temperature. Since laser power is often applied abruptly, the risk of thermal shock and rupture is to be considered and this favours the use, if possible, of materials with low thermal expansion and high thermal conductivity. Lens surfaces exposed to possible contamination from process fumes or even spatters are particularly vulnerable and are often protected by an (expendable) window made from lower-cost material. Purging with a well-directed stream of a clean shielding gas is helpful and is implemented more easily for optics with a comparatively long back focus (the distance between the last lens surface and the focal spot), and corresponding lens designs of the 'retrofocus' type are, therefore, preferable [17]. Thermal aspects also put restrictions on the optical design of lens systems in general: the refractive surfaces, even if coated for maximum transmission, will always have some residual reflection. In complex lens systems consisting of several lenses at appreciable distances, such residual reflections may cause a focus in or, even worse, on the surface of one of the earlier lens elements. This situation, which will inevitably cause lens damage, must be avoided at all cost, in particular for applications with pulsed lasers, due to their inherently very high peak powers.

It is good practice in optical design to strive to achieve a specified performance with a minimum number of optical elements. For laser applications, this aspect is emphasized by the often high cost of the materials, the need to minimize the loss of (expensive) laser power and the complexity of cooling. What helps is that many laser focusing optics need to perform only over a small angular field, typically less than one degree and this often permits relatively unsophisticated designs with low element counts. An exception is the optics for excimer laser applications, which use mask projection techniques rather than single-point imaging. These must cover angular fields of several degrees and require excellent field flatness. The problems are compounded by the short wavelength, the scarcity and low refractive index of the available materials and the requirement for a relatively long back focus.

Finally, the user of laser optics has to face the fact that laser optics are easily damaged, be it due to low hardness of the material, burn-in of accumulated surface contaminations or irreversible changes within the material, a known effect in excimer laser optics used at high power levels. Cleanliness and special care in handling laser optics, in particular lenses, are absolutely essential.

C4.2.5 Optics for CO_2 laser systems

The relatively high power efficiency of the CO_2 laser, combined with scalability to high powers (greater than 10 kW), with only a gradual decrease in beam quality from, typically, $M^2 < 2$ at powers below 3 kW to $M^2 > 5$ well above 10 kW, has given this type of laser a strong position as a 'work horse' for materials processing, such as cutting, welding and surface treatment. Its long wavelength, 10.6 μm, is a drawback in that there are very few environmentally stable optical materials which have high transmission—i.e. low absorption losses—at that wavelength, none of them even near-perfect, a fact which has frustrated hopes for the feasibility of beam transport through an optical fibre (short hollow waveguides are a limited alternative [18]). At the higher power levels, beam handling, including focusing, is by means of mirror systems rather than lenses. The limit is somewhat elastic, depending on the effort spent on lens cooling and ranges between 3 and 5 kW.

C4.2.5.1 Beam transport

High-power CO_2 lasers (see chapter B3.1) are bulky and lend themselves to stationary installation, while applications such as cutting and welding of large workpieces often require bridging long distances to the point of processing. When relaying a CO_2 laser beam from the source to the focusing optics near the point of application, two aspects have to be considered. One is the position of the source beam waist: This is typically located near the out-coupling window of the laser and, thus, often far away from the location of the focusing optics, at which the beam may have expanded to an undesirably large diameter. The second is the beam divergence in the far field (equation (C4.2.2)), which causes a variation the beam diameter with varying distance from the source, as typically occurs with gantry-type installations. The answer to these problems is the use of a telescope, operating as a beam expander and, if adjustable from slightly converging to slightly diverging, as a means for relocating the beam waist to an axial position near the focusing optics, together with an adaptation of the beam size. CO_2 laser beam telescopes are mostly based on the Galilean type, using a positive–negative combination of optical elements, usually mirrors, because it avoids an intermediate focus and is shorter (figure C4.2.2). If the mirrors are paraboloidal, the ingoing and outgoing beams are parallel (figure C4.2.2(*a*)). If spherical mirrors are applied, the angles of deflection must be different (figure C4.2.2(*b*)) to compensate for astigmatism [19]. Waist-position control is achieved by varying the distance between the two mirrors. This is, of course, associated with a lateral beam shift, which must be compensated by a third mirror, often arranged so that the unit has a total deflection of exactly 90 degrees (figure C4.2.2(*c*)). The use of adaptive mirrors for relocating a source waist has also been proposed [20, 21].

C4.2.5.1.1 Articulated arms

An alternative approach to beam positioning in multiple dimensions is the use of an 'articulated arm', a hinged tubular structure with mirrored rotational joints and sufficient degrees of freedom of motion, to allow it to passively follow a three-dimensional path plus angular orientation if gripped at the end by, for instance, an industrial robot. The necessary very precise alignment, and freedom from mechanical play, are difficult to achieve (and maintain) in practice and the required large number of mirrors and their cooling demands are further drawbacks. Applications of CO_2 laser systems requiring articulated arms have now been largely superseded by installations based on fibre-optically transported Nd:YAG laser radiation but, since the latter

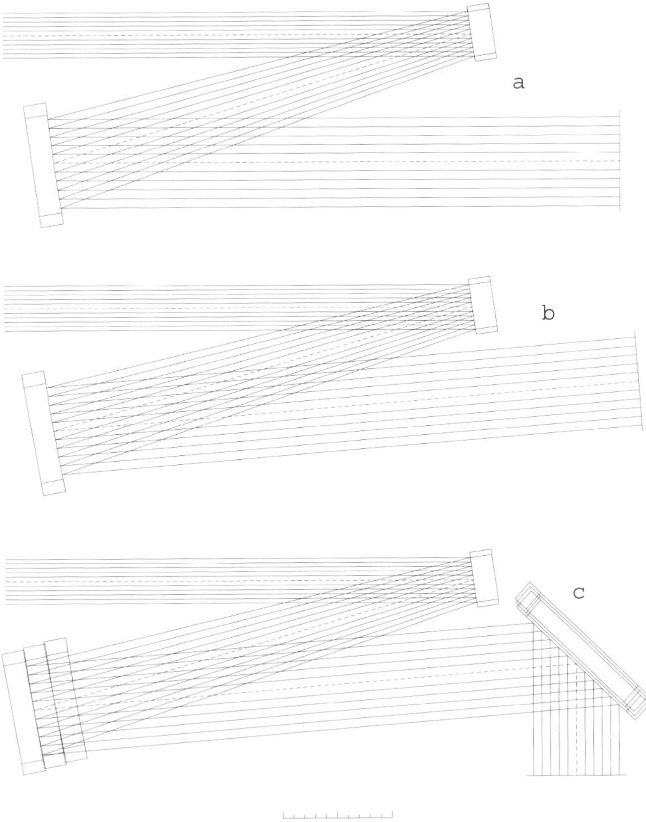

Figure C4.2.2. Galilean telescopes for beam expansion and/or waist relocation of CO_2 laser beams, using off-axis paraboloidal (*a*) or spherical mirrors (*b*). Note different tilt angles in the latter case to compensate for astigmatism. Mirror separation may be adjusted to control the location of the image waist, and the associated lateral beam shift is compensated by synchronously moving a third, flat, mirror, often arranged for a total deflection of 90° (*c*).

approach is typically more costly at power levels in the multi-kilowatt range, the use of articulated arms for positioning CO_2 laser beams and foci by industrial robots is still a viable option.

C4.2.5.2 Lenses for focusing CO_2 laser radiation

C4.2.5.2.1 The single (ZnSe) lens

CO_2 laser radiation can be conveniently focused by means of lenses at power levels up to, say, 3 kW and, with proper measures for forced cooling, at even higher power levels. In applications such as cutting and welding, which are by far the most common applications of this type of laser, the lens simultaneously serves as a pressure window for the cutting or shielding gas used with these processes. These lenses are almost exclusively made from ZnSe, as this is the only environmentally stable material, which, in pure form—referred to as 'laser grade'—has the required low absorption. GaAs has also been considered in the past (for a comparison of the properties see Reedy [22]). A new and upcoming possibility is the use of artificial diamond, to be briefly discussed in section C4.2.5.5. High-quality ZnSe is available from several vendors

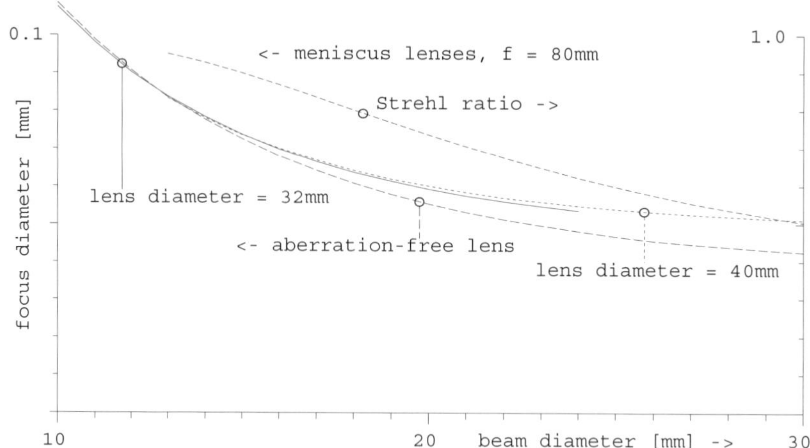

Figure C4.2.3. Performance of ZnSe meniscus lenses for focusing TEM_{00} CO_2 laser radiation in comparison with an aberration-free lens, showing broadening of the $(1/e^2)$ focal spot and reduction of the peak intensity (Strehl ratio). Deterioration becomes noticeable at a beam diameter of 16 mm ($f/5$) but is acceptable up to 18 mm ($f/4.4$). The aberration-free lens used as a reference is the one shown in figure C4.2.4.

with a bulk absorption of less than 0.0005 cm^{-1}. Due to the relatively high refractive index of ZnSe (2.4028), it is mandatory to apply anti-reflection coatings and, in most situations, it is the quality of the coating which determines the overall absorption of laser radiation by a ZnSe lens. For a good new lens, this may run as low as 0.15% of the incident radiation.

The relatively high refractive index of ZnSe helps the lens designer: it keeps spherical aberration low and, thus, allows diffraction-limited performance to be achieved with a single lens up to useful numerical apertures. The lens shape for which spherical aberration is minimum is a meniscus with the stronger curved side directed towards the incident beam [23, p 25 ff]. Figure C4.2.3 shows the performance of such a lens with a focal length of 80 mm over a range of $(1/e^2)$ beam diameters in comparison with an aberration-free lens. The diameter of the focus will be somewhat larger than that obtained with an aberration-free lens but the deviation is below 10% for an 18 mm beam ($f/4.4$, Strehl ratio 0.8) and gradually grows to about 20% at, say $f/2.8$, where the Strehl ratio has dropped to 0.6, which means that the peak intensity at the focus is only 60% of what it would be with perfect optics. If a (cheaper) plano-convex lens is used instead of the meniscus lens, the performance drops by another 15%. In practice, where peak intensity is often the controlling factor for process speed, a plano-convex ZnSe lens is adequate up to $NA = 0.1(f/5)$.

While the above numbers were evaluated for a TEM_{00} laser beam, they apply equally well to low-order mixed mode beams such as a combination TEM_{01*} (ring shaped) plus TEM_{00} beam, a beam type often generated by medium-power fast axial flow CO_2 lasers.

C4.2.5.2.2 Extending the limits of the single lens

When a higher NA at the focal spot is desired, for instance to cut very thin material at maximum speed, there are several options to shorten the focal length: One is the use of diffractive optics, an approach obviously favoured by the relatively long wavelength, and diffractive optical elements with the required high diffraction efficiency are available commercially. A second option is to use an aspherical rather than a spherical lens surface shape. Both these techniques have the disadvantage of a relatively small back focus: the distance between the last lens surface and the workpiece is typically only 90% of the focal length of a single lens.

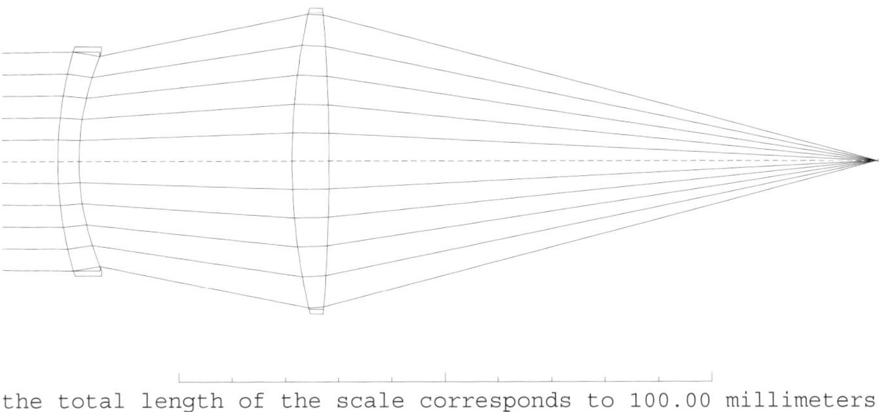

Figure C4.2.4. 2-element ZnSe lens of the retrofocus type for aberration-free focusing of CO_2-laser radiation at high numerical apertures (up to 0.25 or $f/1.96$ at the lens rim). Note that the back focus (100 mm) is substantially larger than the focal length (80 mm).

It is, therefore, much more desirable to use dual lens systems of the retrofocus type [17], exemplified by figure C4.2.4, which can be considered to combine a beam widening and focusing function and creates a long back focus albeit at the expense of a larger lens diameter. This lens type allows excellent aberration correction, with diffraction-limited performance extended to NA greater than 0.25 ($f/1.9$) at the lens rim.

C4.2.5.2.3 Use of zoom optics

The NA is the quantity of choice for adapting the focal spot size, on one hand, and the depth of focus (the Rayleigh range), on the other, to the needs of a process. Thus, a zoom lens system which allows a continuous adjustment of the NA can optimize processes like sheet metal cutting by selecting an appropriate NA for a given material thickness. Such zoom systems are comparable to zoom lenses which are widely used in photography, except that here they are operated at constant entrance pupil diameter rather than exit pupil diameter, and the NA varies accordingly—which is here the reason for zooming. They are also much simpler since they only need to function near the axis: A range of 3:1 in focal lengths (and NA) can be covered with optics using only two lenses (figure C4.2.5) with one aspheric surface or three lenses—all spherical [24]. Performance is diffraction limited. Note that the coverage of the depth of focus is nearly one order of magnitude as the Rayleigh length goes by the square of the focal length. The zoom lens approach has hardly been exploited commercially but deserves more attention as its feasibility and usefulness have been well demonstrated.

C4.2.5.2.4 Thermal effects in ZnSe lenses

The unavoidable residual absorption of CO_2 laser power in ZnSe lenses, of which the largest part normally occurs at the surfaces due to imperfect coatings or contamination, causes lens heating and a radial temperature gradient, since the lens is heated in the centre and cooled by heat conduction to the lens rim and mounting [25–27]. In a new and perfectly clean lens, the effect is small enough to be neglected in practice but it increases strongly if lens surfaces are contaminated and/or damaged. A radial temperature gradient is associated with a radial gradient of the refractive index and the primary effect of this is that radiation paths inside the lens are bent inwardly, thereby shortening the focal length. If the actual temperature increase in the centre of the lens is known, the effect can be computed. A simple model which describes the situation in thermal

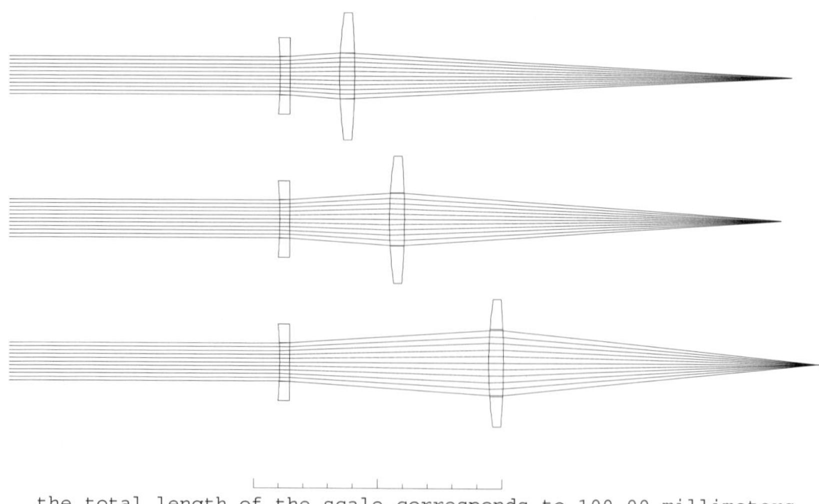

Figure C4.2.5. Zoom lens for focusing CO_2 laser radiation at variable focal lengths for adaptation to material thickness in laser cutting. The lens is of the retrofocus type and has only two (ZnSe) lens elements, surface 3 is (weakly) aspheric. Note the relatively long back focus at the position of shortest focal length (below). The lens covers a factor of >3 in focal length (and spot diameter) which corresponds to nearly one order of magnitude in Rayleigh range and, thus, depth of focus.

equilibrium—reached typically after half a minute—has been shown to match experimental data within about 10% [28, 29]. A more sophisticated model based on finite-element calculations leads to similar results and also covers the time dependency of the effect [30]. In an example situation of a (contaminated) lens, which absorbs 20 W (i.e. 1% of a 2 kW beam), the centre of the lens will, in thermal equilibrium, run about 25 K warmer than the lens rim and the focal shift is of the order of one Rayleigh range. For an application like cutting, this is normally beyond acceptable limits and, as a rule of thumb, the limit of the lifetime of a ZnSe lens is reached when it absorbs more than 10 W.

Interestingly, the warm-up of the lens centre hardly affects its aberrations; in fact, the spherical aberration is even slightly reduced [28]. Lens heating and its associated focal shift have had surprisingly little attention in the literature on laser materials processing and are probably largely underestimated as a source of poor process reproducibilty.

C4.2.5.2.5 Use of artificial diamond

Diamond has long been recognized as an excellent optical material for use at wavelengths in excess of 10 μm but natural diamond is scarce in large enough sizes and prohibitively expensive. Artificial diamond, synthesised by a chemical vapour deposition (CVD) process, has recently become available with a quality and size which allows its use in industrial CO_2 laser systems. Due to the manufacturing process, CVD-diamond comes in plates of up to 2 mm thick and up to, say, 120 mm in diameter. Highly transparent samples are sold under the trade name DIAFILM and are polycrystalline with grain sizes below 0.1 mm, and a predominant crystallographic orientation of $\langle 110 \rangle$ perpendicular to the growth plane. The material closely approaches the properties of natural diamond type IIb, which is the purest form, and likewise exhibits intrinsic multi-phonon absorption bands in the medium infrared region near 5 μm. The wings of these bands are responsible for a residual absorption at the CO_2 laser wavelength, where the absorption coefficient is typically 0.05 cm^{-1}, which is close to the theoretical limit. Although this coefficient is substantially higher than that of ZnSe, there are two main factors that make (CVD-) diamond superior for high-power CO_2 laser applications:

(1) the thermal conductivity is more than two orders of magnitude higher than that of ZnSe (2000 *versus* 17 W m^{-1} K^{-1}); and

(2) the temperature dependence of the refractive index is only about one-sixth (10 *versus* 57 $\times$ 10^{-6} K^{-1}).

In addition, the high strength of diamond allows its use at much lower thickness for a given mechanical strength requirement. The total absorption—bulk plus surface coatings—is, thus, somewhat higher than that of the corresponding ZnSe component but the optical effects of heat absorbed in a lens (or window) previously discussed in C4.2.5.2.4 are much reduced with diamond, as the higher thermal conductivity keeps the thermal gradients low and the lower temperature coefficient of the refractive index reduces the optical effect of the remaining gradient. Theoretical modelling has shown a gain by a factor of between 40 and 240 depending on lens (or window) thickness [31]. Diamond has about the same refractive index as ZnSe (2.38 at 10.6 μm), which likewise makes it attractive for lens applications but since the material is best available in plates typically 1 mm thick or less, its actual use in CO_2 laser systems has so far been predominantly for out-coupling windows in high-power CO_2 lasers above, say, 3 kW.

C4.2.5.3 Mirror systems for use with CO_2 lasers

The unavoidable residual absorption of CO_2 laser radiation by lenses makes mirror systems an attractive and at very high power levels, the only feasible alternative for radiation focusing. Since there is no pressure window function as with lenses, the use of all-mirror systems also requires alternative nozzle designs for applying cutting or shielding gas. Owing to the excellent thermal conductivity of the relevant mirror substrate materials and efficient cooling geometries, which are essentially independent of the mirror size, reflective optics can withstand high overall power levels and local power densities up to 10^5 W cm^{-2}. At large beam diameters, typically associated with high-power beams, mirrors also offer substantial cost advantages over refractive optics.

The two most common substrate materials for CO_2-laser mirrors are silicon and copper. Copper has a higher thermal conductivity by a factor of 2.5, which favours its use at very high power levels but its thermal expansion is more than six times higher so that, under equal thermal loading, silicon will deform less [32], and is preferred for reflectors inside a CO_2 laser. Both materials are normally used with reflection-enhancing coatings, which can raise reflectivity to up to $>$99.5%. Suitability for precision machining and integration of cooling channels into the mirror body have made (oxygen-free) copper the most popular substrate material for mirrors in CO_2 laser applications. Its main drawback is the low hardness and susceptibility to scratching during cleaning. In applications where a mirror is likely to be easily contaminated by spatters or fumes and may require frequent cleaning, a material of higher durability and hardness is preferable. These properties are offered by (uncoated) molybdenum, albeit at the expense of reduced reflectivity (typically 98%) [33]. In view of their application at high power levels, it is common practice to have cooling water channels integrated into the mirror body with a geometry that minimizes surface distortion by unavoidable thermal gradients.

C4.2.5.3.1 Angular field considerations

Since laser beam focusing normally involves a very small angular field centred about the optical axis, it is tempting to consider only the axial aberrations of such systems. It must be borne in mind, however, that mirror optics for laser-beam handling are of necessity applied in some off-axis arrangement (figure C4.2.6) in order to avoid central obscuration and that such systems are notoriously difficult to align such that they are truly operating with their optical axis parallel to that of the ingoing laser beam. Consequently, the optical system should tolerate a sensible angular deviation from the theoretical axis of at least, say, 2 mrad. Only if

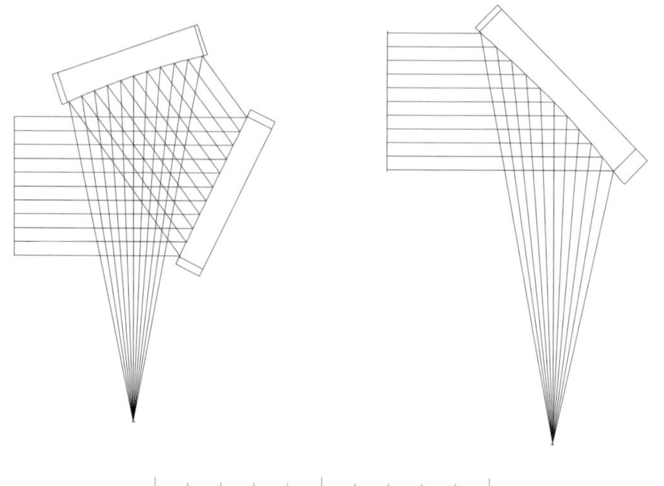

Figure C4.2.6. Off-axis paraboloids for 90° beam deviation.

the optical aberrations at such a field angle, notably coma, are negligible, can an optical system be considered fit for general applications.

C4.2.5.3.2 The off-axis paraboloid

The only axially corrected one-mirror focusing system, the paraboloid, is particularly troublesome in this respect. Due to the impossibility of fulfilling the optical sine condition, the paraboloid suffers from severe coma, when used at an angle to the (theoretical) axis [11, p 95 ff]; the effect rapidly increases when an off-axis section of the paraboloid is used, which is the only way to avoid obscuration (figure C4.2.6). Figure C4.2.7 shows the deformations of the wavefront at angular deviations from the optical axis of 2 and 3 mrad for paraboloids used under 40° and 90° off-axis, respectively. It will be noted that the wavefront deformation is of a saddle shape, quite similar to that caused by astigmatism, but it is actually a segment of the unsymmetrical comatic wavefront and its magnitude is proportional to the field angle as is characteristic for (third-order) coma terms [10, p 118 ff]. From calculations of the diffraction image, it can be concluded that a 40° off-axis paraboloid requires alignment to better than 2 mrad, and a 90° off-axis paraboloid to better than half a milliradian. Thus, in spite of its apparent simplicity, the one-mirror 90° off-axis paraboloid is the least recommendable solution, while the two-mirror folded variety with the paraboloid under about 40° degrees, is just acceptable for practice.

C4.2.5.3.3 Coma-corrected optics

There is no cure for coma of single mirrors other than the use of designs involving two or more curved surfaces. This has, of course, been known by optical scientists for a long time and has led to well-known designs such as, for instance, the Ritchey–Chretien system which dominates modern astronomical telescopes. Unfortunately, this and related systems do not lend themselves well to off-axis derivatives in that (1) the numerical aperture of an off-axis cut-out will be much too small for most laser applications and (2) the back focus will be much shorter than the focal length, as is true in all systems with a concave–convex sequence of mirrors, regardless

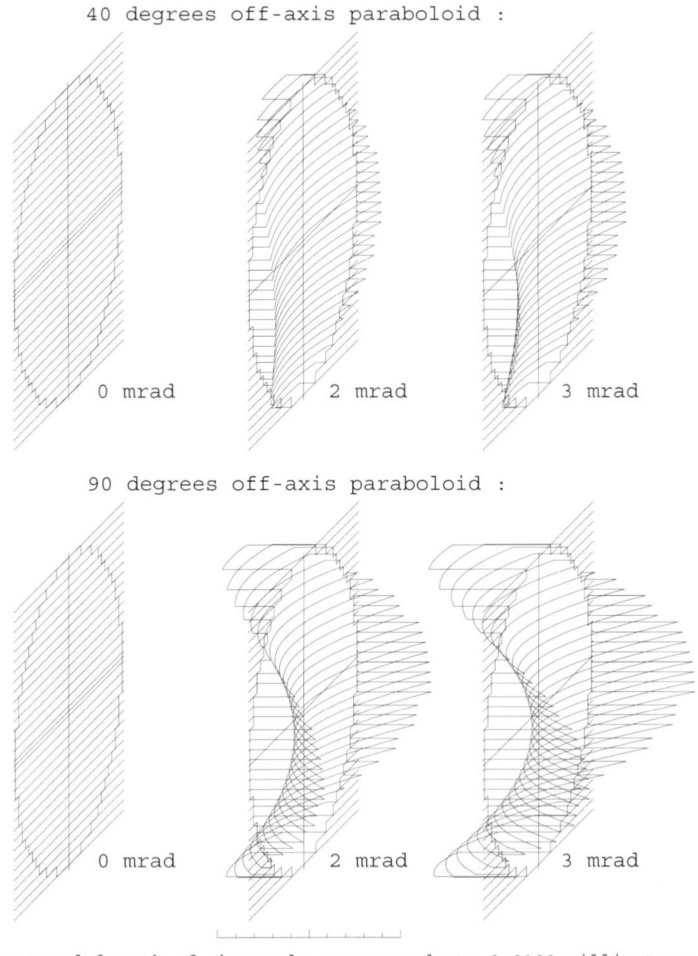

Figure C4.2.7. Phase-front distortion of off-axis paraboloid mirrors used under 40 and 90° off-axis (figure C4.2.6) for beams entering under angles of 0, 2 and 3 mrad with respect to the optical axis. The phase distortions are saddle-shaped and represent segments of the off-axis comatic phasefront which increase linearly with the angle but they result in astigmatic images.

of their shape. Thus, for use with lasers, the opposite configuration, a convex–concave sequence of mirrors, i.e. the well-known Schwarzschild design [34] is a better starting point for creating an off-axis laser focusing system. One family of designs derived on this basis uses four mirrors—three spherical, one flat—in a variety of arrangements with the focus either in line with the incoming laser beam or under 90°. Such systems have negligible aberrations over a field of more than one degree which makes them truly alignment tolerant [35].

C4.2.5.3.4 General remarks on aspherical mirrors

Some general remarks about the use of aspheric mirrors may be in order. While in the past, aspheric surfaces were known to be costly to produce by classical grinding and polishing techniques and, thus, likely to be avoided by designers, mirrors using copper substrates are currently manufactured by single-point diamond turning which allows machining of aspherical shapes of high precision, albeit at sensibly higher cost for off-axis segments. There are two aspects that make aspheric mirrors less attractive than spherics:

(1) An off-axis section of an aspheric has no rotational symmetry, which makes shape measurement and verification, and positioning in a mechanical mount, quite demanding and time consuming, unless sophisticated measuring tools are available.

(2) All aspheric mirrors contribute strong coma, which drastically limits the angular field of use. This was discussed earlier for the off-axis paraboloid but it applies equally well to the ellipsoid and the hyperboloid, which are candidates in conjunction with divergent or convergent beams. Contrary to common belief, off-axis aspheric surfaces are no universal problem-solvers and should be used selectively and with caution.

C4.2.5.4 CO_2 laser beam integration for homogeneous illumination

CO_2 laser applications such as cutting and welding use beam focusing to produce small circular spots. A quite different radiation pattern is desirable for applications in various forms of surface treatments such as transformation hardening, cladding and alloying. In these applications, the surface is scanned with a spot of—preferably—rectangular shape and a flat power profile, in order to create a stroke of approximately homogeneously elevated temperature. The required power densities are in the 10^4–10^5 W cm^{-2} range and typical stroke widths range from a few millimetres for lasers in the low kilowatt class to several centimetres for laser powers above 10 kW. Several techniques have evolved for creating rectangular flat-topped beam profiles, referred to as 'beam homogenization' or 'beam integration'. These methods split the original beam into smaller rectangular parts and overlap them to form the desired rectangular pattern in a given plane. This can, for instance, be achieved by a mirror having rectangular flat areas [36], the size of, say, 1/10 to 1/50 of the beam area, or by just folding two halves of the beam [37]. If these areas are properly tilted with respect to each other, the beamlets reflected from such a mirror overlap and add up in one plane. Note that the recombination of beam parts stemming from the same source invariably leads to interference which shows up as a rather disturbing modulation of the power density within the integrated spot. A way to bring this modulation down to acceptable limits is to have the beamlets recombine at angles as large as are compatible with the system configuration, which at least raises the spatial frequency of the interference pattern [38].

Another approach to beam integration uses a kaleidoscope-type rectangular reflective cavity, where different parts of the original wavefront undergo different reflections so that the outcome is a more or less randomized sum. More reflections result in better randomization but also higher losses due to the high angles of incidence. Incoupling of the radiation by means of an axicon is helpful to minimize the number of internal reflections needed to achieve a given homogeneity [39]. The effective randomization of the optical pathlengths reduces the modulation due to interference to an unproblematic level.

A predetermined irradiation pattern over a given stroke width can also be achieved by rapid scanning or oscillation of a more or less focused beam in a direction perpendicular to the (slow) motion of the beam over the workpiece, creating an irradiation profile by time averaging. To avoid excessive temperature fluctuations, the scan frequency must be very high, of the order of 1000 Hz for transformation hardening, somewhat lower for processes, where a melt pool is formed. The method is quite flexible and with little extra effort allows variation of the stroke in width and temperature profile by scanning with nonlinear speed but the latter is difficult to combine with high repetition rates.

C4.2.5.5 Power distribution shaping by phase modulation

A radically different concept for beam shaping, be it beam homogenization or other, makes use of phase modulation, exploiting in effect the spatial coherence of the laser beam and manipulating the phase distribution of the original beam so that the power distribution in the diffraction image attains a predetermined shape and profile. Although known for some time as a viable principle [40–43], the method has been slow to be implemented in practice, partly because of technical hurdles that need to be overcome and partly because the theoretical understanding and modelling have so far been limited to TEM_{00} beams. Computer simulations have shown that phase modulation has the potential to shape a three-dimensional power distribution in the focal region in such a way that an axially elongated, thread-like region is formed, within which the power density exceeds a certain value (Beckmann, unpublished). Although that value will be a little less than that at the peak of a diffraction-limited image, it stays high over an extended axial range due to the formation of multiple waist-like concentrations along the axis. This particular behaviour is obtained with a very simple phase element, having a single phase step of $\pi (= 1/2$ wavelength) somewhere near the $1/e^2$ diameter of the (TEM_{00}) beam. The exact diameter of that phase step—the amount of outer beam radiation which interferes with the π phase difference with the inner part—determines the exact nature of the three-dimensional power pattern. It is mandatory that beam aperturing is virtually absent. The method still awaits experimental verification.

C4.2.6 Optics for lasers operating in the visible or near-IR

A common advantage of all lasers which operate in the visible or near-infrared spectral region, that is, between, say, 0.40 and 1.5 μm, is that they can make use of the wide range of optical glasses of high quality, precision of definition in terms of refractive index and dispersion, stability and workability, which have been available for many decades. At least some of the more popular glasses have low residual absorption in this spectral region. It must be noted though that the data are, by tradition, given in, for instance, the Schott glass catalogues, as bulk transmissions for samples respectively 5 and 25 mm thick and with a precision of only three figures. More precise data would be highly desirable for (cw) laser applications in view of the ever increasing powers at which, for instance, Nd:YAG-lasers have become available over the last decade. Quite a different matter is the handling of radiation from Q-switched solid state lasers, where power densities well above 10^{10} W cm^{-2} are routine and the possibility of laser damage to optical glasses has to be considered. There are two mechanisms that cause internal damage to optical glasses at high peak powers: one is due to absorption by foreign particles in the glass, such as platinum, which entered during the manufacturing process; the other is self-focusing of the laser radiation due to an increase in the effective refractive index with the local power density as characterized by the nonlinear refractive index n_2 of the glass. A further concern is surface damage, which is largely independent of the glass type, i.e. its chemical composition, but related to the polishing agent and, of course, surface cleanliness. Schott [44] published a list of optical glasses for which damage-related quantities have been measured. Low-index glasses typically perform better. Data for fused silica were reviewed recently [45].

C4.2.6.1 Optical handling of fibre-delivered Nd:YAG laser radiation

The transport of radiation by an optical fibre has important advantages from a systems standpoint as it allows laser machining to be combined with positioning by (standard) industrial robots (fibre-optic beam delivery is considered in detail in chapter C4.3). It is also beneficial in that the fibre output constitutes a homogeneous radiation source with sharply defined boundaries. It is customary to combine the fibre output with collimating optics into one hermetically sealed unit, protecting the fibre end from contamination and ensuing catastrophic failure. The beam quality at the fibre output of a (lamp pumped) state-of-the-art laser may typically run around $M^2 = 100$ but the advantages of the (collimated) fibre output readily outweigh that low beam quality in typical applications such as welding. Diode-laser-pumped high-power (>2 kW) Nd:YAG lasers have recently become available with beam qualities $M^2 < 50$ at the fibre output. Focusing the radiation onto the workpiece should be by (near) diffraction-limited optics, if the sharpness of the fibre boundary and its

Table C4.2.1. Air-spaced doublet lenses for focusing collimated Nd:YAG radiation, made from different materials. All lenses are of the type exemplified by figure C4.2.8, with the lens elements in edge contact. Optical diameter is 50 mm. The lens systems are optimized (and diffraction limited) for 1060 nm and obviously not colour corrected. They nevertheless also show diffraction-limited performance at the HeNe laser wavelength (633 nm) for the indicated focal offset, a feature that is helpful for testing and adjustment.

Glass	Refractive index	Focal length [mm]	Back focus [mm]	Numerical aperture	F-number	focal offset [mm]
SiO_2	1.44967	177.0	156.3	0.1412	3.505	−0.295
BK7	1.50669	163.0	142.2	0.1534	3.222	−0.274
LF5	1.56594	152.0	131.0	0.1645	3.000	−0.336
SF2	1.62766	143.0	121.9	0.1748	2.816	−0.361
SF15	1.67516	137.0	115.8	0.1825	2.694	−0.378
SF14	1.73313	131.0	109.7	0.1908	2.572	−0.397

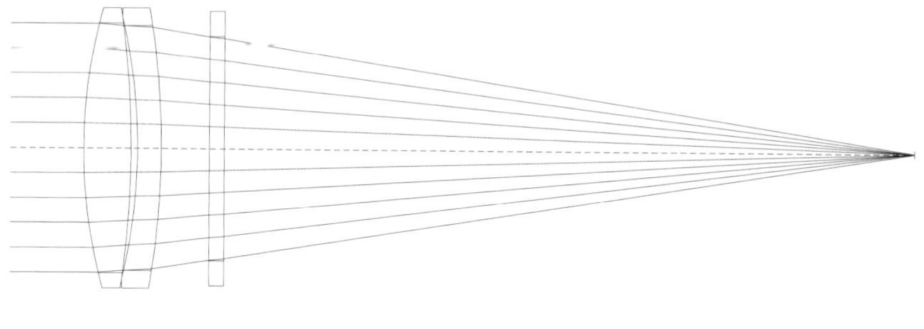

the total length of the scale corresponds to 100.00 millimeters

Figure C4.2.8. Air-spaced doublet for aberration-free focusing of collimated high-power Nd:YAG radiation at numerical apertures up to 0.16 ($f/3.1$). The two lenses are in edge contact to ease manufacturing. This type of lens can be made from different materials (see table C4.2.1) and covers a wide spectral range, albeit at varying focal position. A protective window, typically made from low-cost glass, is optional.

homogeneous (= top-hat) power profile are to be preserved in the image, which is usually the preferred mode of operation [46].

C4.2.6.2 Lens design

The design of focusing optics for visible- and near-infrared lasers follows essentially the general design rules and practice known from classical geometrical optics [23, 47, 48], with noteworthy exceptions: cemented lenses are better avoided where appreciable powers are to be handled and the glass catalogues should be checked for (bulk) optical transmission at the wavelength of use. To be on the safe side when designing optics for high-power Nd:YAG lasers above, say, 2 kW, the use of quartz glass of the kind specifically made for that spectral region by fusion of natural quartz [49, 50] has merits, as it greatly reduces the risk of thermal stress and rupture. A typical design is shown in figure C4.2.8. It is diffraction limited at a focal length of 177 mm for an optical diameter of 50 mm, corresponding to a numerical aperture of 0.14 ($f/3.5$) at the lens rim. Table C4.2.1 shows what can be achieved with higher refractive index materials. All these lenses

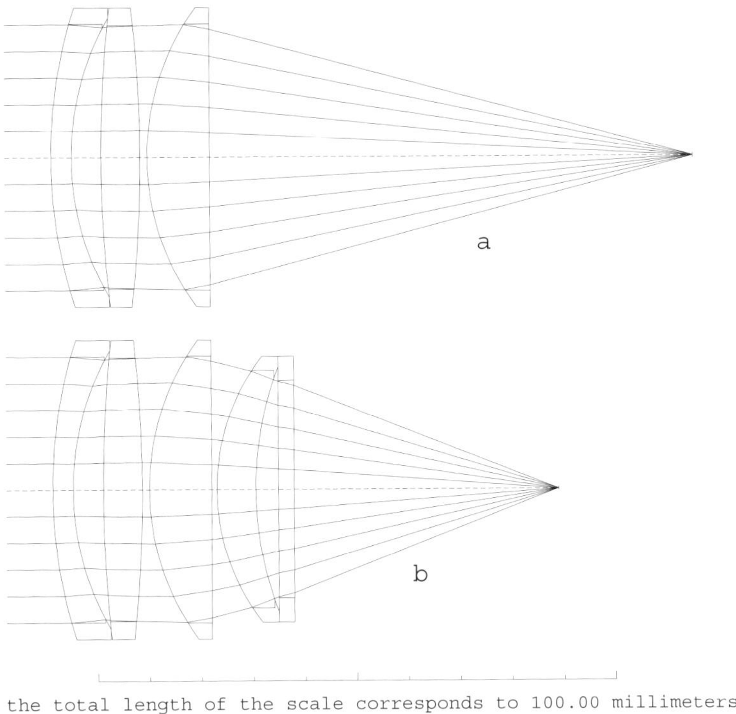

Figure C4.2.9. High-aperture focusing optics for high-power Nd:YAG radiation using SiO_2. The basic lens (*a*) has 3 elements and 100 mm focal length for a numerical aperture of up to 0.25 ($f/2.0$). Adding an aplanatic lens (b) shortens the focal length to 70 mm for a numerical aperture of up to 0.36 ($f/1.3$), but at the expense of a much shorter back focus, so that concurrent use of an (expendable) protective window is recommended. Note that the aberrations caused by this window at high aperture have to be taken care of in the design of the added lens element. To ease manufacturing, lenses 1 and 2 are in edge contact, lens 2 is symmetrically biconvex and lens 3 has a flat rear surface.

are air-spaced doublets with the positive element in front and the separation precisely set by edge contact (figure C4.2.8). More lens elements will be needed for higher numerical apertures; an example is shown in figure C4.2.9. The basic three-element lens (figure C4.2.9(*a*)) has a 100 mm focal length, an optical diameter of 50 mm and is diffraction limited up to a numerical aperture of 0.25 ($f/2.0$). Note that this system has a negative front lens followed by two positive lenses, a configuration which maximizes the back focus and, at the same time, minimizes spherical aberration. Addition of a (near aplanatic) lens element (figure C4.2.9(*b*)) shortens the focal length to 70 mm, increases the numerical aperture to 0.36 ($f/1.3$), while maintaining diffraction limited performance (Strehl ratio >0.95). The back focus is much shorter, however, and use of a protective window made from a less expensive material is highly recommended. Note that, at this high numerical aperture, such a window has to be accounted for in the design of the last lens.

C4.2.6.3 Colour correction

System requirements may demand that a lens for handling Nd:YAG laser radiation be colour corrected to have the same focal position for the wavelength of a pilot laser emitting visible light, often a HeNe laser. In this case, the classical and well-established methods for correcting colour for visible light optics by a suitable

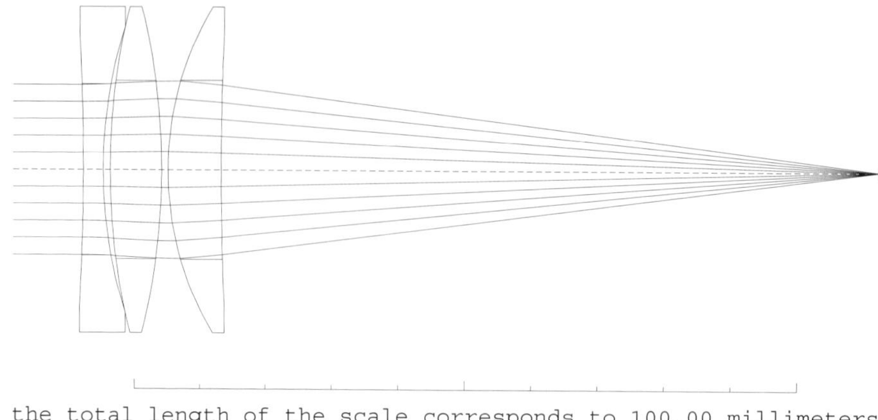

Figure C4.2.10. Three element lens simultaneously corrected for Nd:YAG and HeNe laser radiation. The lens uses BK7 glass in the two positive lenses and SF10 for the negative front element to match the change of dispersion when shifting from visible light to the near infrared. For use in the visible spectral region, the front lens would be made from SF2 with near identical surface curvatures.

choice of glasses come into play but have to be modified for the change in wavelength region. The dispersion of the high index—and, in the visible, high dispersion—glasses, becomes progressively smaller when moving towards the near infrared [51, p 142 ff]. Thus, expertise and experience about glass combinations which allow colour correction in the visible spectrum and simultaneously result in low monochromatic aberrations cannot be transferred to the longer wavelengths [52]. Figure C4.2.10 shows an example. For use in the visible spectrum (C,d,F light), the positive lens elements are made from BK7, the negative lens from SF2, a popular choice. To correct for use with Nd:YAG and HeNe laser light, however, the negative lens is better made from SF10, a glass of much higher index and higher dispersion for visible light but a good match for the longer wavelengths. The system is diffraction limited (up to NA $= 0.155$ or $f/3.2$) and has exactly the same focal position for both laser wavelengths but there is strong (and uncorrectable) secondary colour, i.e. the focal position deviates sensibly at wavelengths in between. Full colour correction is not always desired: a well-chosen amount of chromatic aberration can be used to image different radiation sources at different axial positions back into the fibre for process control in laser welding [53].

C4.2.6.4 Optics for diode lasers

Diode lasers differ from all other laser types in several important aspects. They (1) are inherently simple devices in that they convert electrical energy directly to (laser) light in a properly configured semiconductor crystal, which makes them very suitable for (low-cost) mass production, and (2) allows high energy conversion efficiencies but they (3) are not scalable to high output powers from a single diode. The process of generating laser light in a diode does (4) create a beam with different properties in two mutually perpendicular directions and, due to the very small size of the laser cavity, with (5) small waist sizes and accordingly large far-field divergence. The last three aspects govern all optics for diode laser applications. Diode lasers themselves are considered in chapter B2.

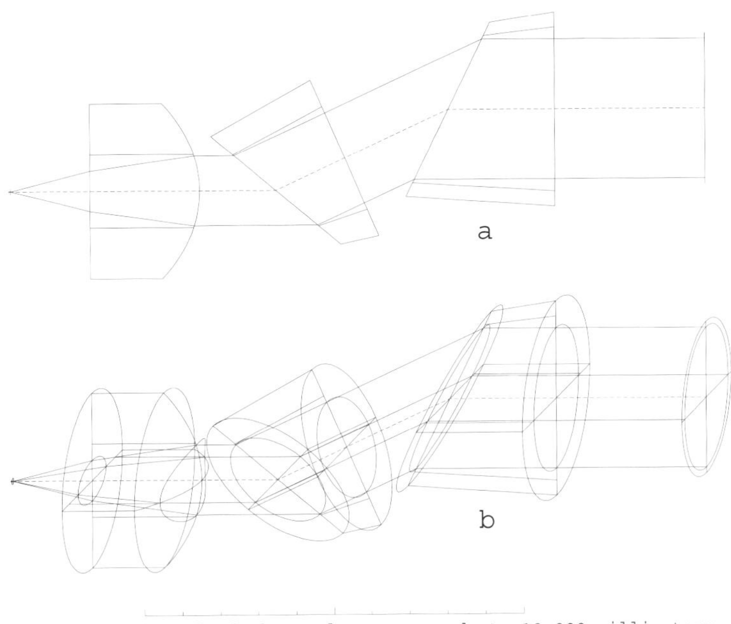

Figure C4.2.11. Collimation and one-dimensional beam expansion for diode laser beam equalization using a prism pair in the collimated beam: (*a*) meridional cross section; (*b*) 3D a view showing elliptical input and circular output cross section.

C4.2.6.4.1 Classes of diode lasers and applications

There are three major areas of diode laser applications and corresponding classes of laser diodes and associated optics:

(1) telecommunication via optical fibres, which is mainly done at wavelengths in the 1.3–1.5 μm range and requires diodes of medium power and highest reliability;

(2) optical information readout and storage, which applies low power (<100 mW) diodes at near infrared (CD) and red (DVD) wavelengths with a projected switchover in the near future to shorter wavelengths; and

(3) a cluster of applications which demand the highest available powers from individual diodes or from (stacks of) linear arrays of diodes, for such widely differing applications as pumping of solid state lasers, material processing and a number of medical uses.

Due to their compactness, low power consumption and low operating voltages, diode lasers have also largely replaced gas lasers such as the HeNe laser in various types of sensors (e.g. bar code readers) and are gradually displacing ion lasers in applications such as printing, but these applications fall into one or the other of the previous classes.

C4.2.6.4.2 Diode laser optical output properties

Optically speaking, the diode laser is an astigmatic light source with, in general, different beam properties in two orthogonal directions [54]. In the direction perpendicular to the diode junction, the beam quality is high, with M^2 typically running less than 1.4, and close to 1 at low powers, and the source waist width

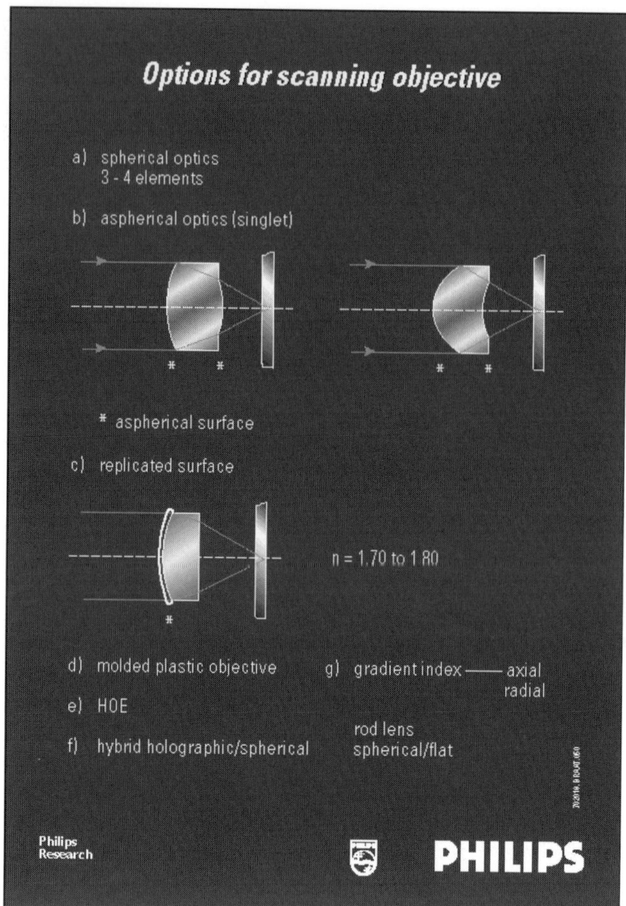

Figure C4.2.12. Options for the collimating optics for use in optical storage devices. (Courtesy J Braat, Philips Research Laboratories, Eindhoven, The Netherlands.)

w_s is typically less than 1 μm, so that in accordance with equation (C4.2.3), the corresponding half-beam divergence can exceed 0.5 rad. This direction is, therefore, usually referred to as the 'fast' direction. In the other direction the beam quality and beam width vary widely, depending on diode size and power. For low-power diodes, with outputs below, say, 100 mW, a beam quality close to 1 is achieved in this direction as well but high power diodes may show a $M^2 > 50$ and a waist width w_p up to 0.25 mm, with a typical half-beam divergence of 0.08 rad. This direction is therefore called the 'slow' direction. Radiances of high-power diode lasers well exceed 10^7 W cm^{-2}sr^{-1}.

C4.2.6.4.3 Optical handling of single diode laser beams

The strong divergence of the diode laser beam in the 'fast' direction (if not in both) suggests the need to intercept that radiation by some optics as close to the diode emitter as possible. In the most simple case—coupling of the diode laser output straightaway into a single-mode fibre—a single, small ball lens is often used, which offers by far the most economic, albeit optically imperfect solution. Methods for calculating and optimizing the coupling efficiency have been described by Ratowsky [55] and Wilson [56]. In all other

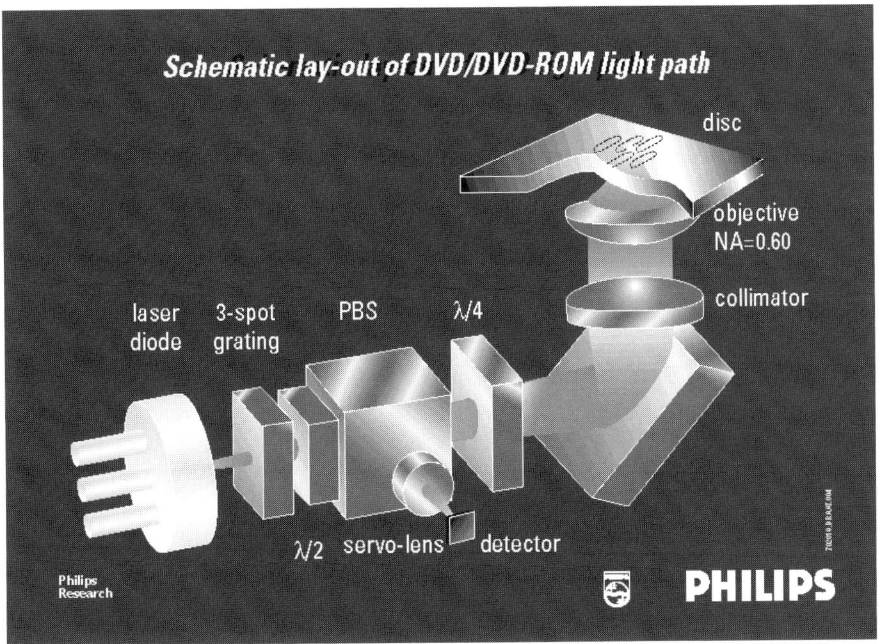

Figure C4.2.13. Schematic layout of the readout optics for the DVD optical storage system. (Courtesy J Braat, Philips Research Laboratories, Eindhoven The Netherlands.)

cases an aspheric lens with a short focal length is typically used for collimating the beam. Alignment is critical because of the very limited angular field of such lenses. Since a diode laser beam nearly always has strongly differing divergence in the directions parallel and perpendicular to the diode junction, the collimated beam is, in general, elliptical and will, for most applications, require equalization of the beam width by beam expansion in one direction so that subsequent focusing yields a circular spot. Figure C4.2.11 shows how this is achieved with a pair of prisms. Alternatively, the required beam shaping can be performed in a single step by a properly designed thick lens [57] with non-spherical surfaces that can be produced by plastic moulding. This is currently the preferred approach for use in high-volume consumer products (CD readers), which use diode lasers of 785 nm wavelength with beam divergences of 7° and 20° (FWHM), respectively, and require beam transformation to a wavefront of about 3 mm diameter with less than 1/50 wavelength distortion.

C4.2.6.4.4 Single diode laser focusing optics

As with all other laser types, the focusing requirements, especially the NA, depend on the application. In optical storage, which is the largest diode laser application by numbers, diffraction-limited performance is required at NA = 0.45 for standard CD readout and NA = 0.50 for writing. Since the focusing lens forms part of the servo-controlled optical head, its mass must be as small as possible, so a single aspheric lens is used exclusively. Several options are shown in figure C4.2.12. The design has to take care of the thickness (= 0.12 mm) and thickness tolerance, of the (refractive) CD substrate ($n = 1.58$) in the convergent beam [58]. To ensure a stable product, this lens (option d of figure C4.2.12) is typically manufactured from a basic plano-convex spherical glass lens of high refractive index, with a thin plastic aspheric shell moulded onto the convex surface. Higher density optical data storage (on the DVD, figure C4.2.13) uses a somewhat shorter wavelength (635–670 nm) and a substantially higher NA = 0.60, in order to generate a smaller diffraction spot, imposing correspondingly higher demands on the quality of the optical components. For the next generation of optical storage, a wavelength of 400–405 nm and a yet higher NA $= 0.85$ are envisaged.

C4.2.6.4.5 Optics for applications of diode laser arrays

A single diode laser is only scalable to powers around a few watts and, for higher output powers, a series of diodes is arranged in a linear array, also referred to as a diode bar, with a length of typically 10 mm [59]. The individual diodes have widths varying between 0.5 mm and some tens of micrometres, they are multi-mode in the 'slow' direction; the divergence of each emitted beam will typically be 10–25° (FWHM). As the thickness of the diode junction is as small as in any single diode, the beam quality in the 'fast' direction is high (typically $M^2 < 1.4$) as is the divergence. For just about all applications, it is common practice to place a collimating optic for the fast direction very close (<0.2 mm) to the emitting surface. This is mostly in the form of a cylindrical lens of short focal length, with a non-circular cross section and a planar surface at the side of the diode emitter. Because of the relatively high thermal load, such lenses are made from (high-index) glass. Lenses of this type, with focal lengths in the 1–2 mm range, are available commercially. Alignment is critical and unless both the diode bar and the cylinder lens are strictly straight within, say, a few micrometres, the centre lines of the individual collimated beams will not lie in one plane, a deviation referred to as 'smile'.

Behind the fast-axis collimator, the beam divergence will be reduced to a few milliradians but the beams fan out in the slow direction. They also start to overlap in that direction at some distance. Depending on the fast axis collimator focal length, the beams can, in principle, be collimated in the slow direction before they overlap but it should be noted that this requires an array of cylindrical lenses with the lens distances exactly matched to the positions of the diodes in the bar. Alternatively, and depending on the application, the beams can be jointly manipulated with common optics. Without any further imaging, they are, for instance, suitable for side-pumping of a solid state laser [60]. When it is desired to concentrate the total optical power from the bar into one (circular) spot or an optical fibre, one has to bear in mind that the beam quality of the individual diodes is low in the slow and high in the fast direction. Thus, to combine them into one common beam with reasonably equal quality in both directions, they should be added (concatenated) in the fast direction. This is, however, not the direction in which they are lined up behind the fast axis collimator and for concatenation in the proper (=fast) direction, the individual beams have to be rearranged, i.e. effectively rotated by 90°. A number of optical schemes for this purpose has been proposed in the patent literature [61].

C4.2.6.4.6 Stacks of diode laser arrays

In practice, diode laser bars as described earlier are mounted (soldered) onto a heat sink which has provisions for water cooling and forms a package of, typically, 2 mm in height. Such packages, complete with collimating cylinder lenses, can be stacked to form larger units with output powers of hundreds of watts. It should be noted that collimation in the slow direction is virtually impossible for a stack since the bars are not in line within the required small tolerances. Stacks of diode laser bars are suitable for pumping (high-power) solid state lasers and for direct materials processing, the latter so far limited to applications which require power densities below, say, 10^5 W cm^{-2}, such as soldering or surface heat treatment. With increasing power output from the individual bars and through the use of more sophisticated schemes for combining diode laser outputs, such as polarization and wavelength multiplexing, the achievable power densities are expected to rise to levels which allow keyhole-based welding of metals, within just a few years.

C4.2.7 Optics for excimer and other UV lasers

Excimer lasers and their applications differ strongly from all other lasers in several respects. The radiation spans the whole ultraviolet region, the exact wavelength being determined by the gas combination used (see chapter B3.2). The output is in the form of short pulses (typically 10–20 ns), pulse energies run up to, say, 1 J and create peak powers of tens of megawatts with average power levels of tens of watts. In typical commercial excimer lasers, beam quality is poor, with M^2 well in excess of 100, so an excimer laser beam behaves largely like (collimated) incoherent radiation. Direct focusing does not make much sense but the capability for high-precision machining, which results from the short (UV) wavelength, can nevertheless be exploited by using the laser for illuminating a mask, which is imaged on the workpiece so as to replicate the mask pattern. Although the mask can be in direct contact with the workpiece, it is more common to project a demagnified image onto the workpiece. This then requires essentially diffraction-limited optics with a very flat image plane of, typically, several square millimetres for simple applications (simultaneous drilling of small holes in a small area) and up to some square centimetres for higher end systems. Due to the high photon energy of excimer laser radiation, the interaction with matter, in particular with organic matter (plastics), is largely non-thermal at the shorter wavelengths and referred to as ablation; it allows the machining of sharply defined features in the micrometre and sub-micrometre range without causing thermal damage. There is a threshold of laser energy density for the ablation process (typically less than 0.5 J cm^{-2} for plastics, higher for metals and ceramics). This limits the image area over which the radiation may be spread out and, indirectly, sets a lower limit for the numerical aperture at the mask image. Material removal is minute (several 10 μg per pulse) but care has to be taken that the resulting debris does not deposit on the imaging lens. A long back focus of, say, 70 mm and preferably more is, therefore, mandatory. Finally, the typical excimer laser beam is rectangular and inhomogeneous in at least one direction, so it requires homogenization before it is fit for mask illumination. The technique of homogeneously illuminating a mask and imaging the mask pattern makes, in general, only inefficient use of the laser light. When the mask pattern is simple and fixed, the laser energy can better be concentrated on the open parts of the mask [62].

These comments also largely apply to handling of radiation from other lasers operating in the ultraviolet, such as frequency-multiplied solid state lasers.

C4.2.7.1 Material aspects

The ultraviolet region below, say, 300 nm is known for the very small number of available transparent optical materials. SiO_2 in the form of fused synthetic silica is most widely used, as are some of the crystalline fluorides, notably CaF_2 and LiF and, to a lesser extent, MgF_2, as it suffers from birefringence. At the shorter UV-wavelengths, the bulk absorption coefficient of fused silica depends on the fabrication process. Manufacturers offer certified material with an absorption coefficient of <0.0009 cm^{-1} at 248 nm (for KrF laser applications) and <0.003 cm^{-1} at 193 nm (for ArF laser applications) [63]. The high peak powers and high photon energy of excimer laser pulses cause irreversible changes [64], the mechanisms of which are not yet fully understood [65]. Thus, optics for excimer lasers have a limited lifetime. Under prolonged irradiation by, for instance, KrF excimer laser pulses, fused silica develops an absorption band around 210 nm [66, 67], which affects the transmission at both the wavelengths of the ArF and the KrF excimer lasers. Although this absorption vanishes for low intensities within minutes after the irradiation, it re-occurs at the previous level if high-energy pulses are again applied [66]. Excimer laser irradiation also increases the density and, thus, the refractive index of fused silica, depending upon the fluence and number of pulses but the index change is only of the order of 10^{-6}–10^{-5} and, thus, not harmful in most applications.

All UV-transmitting materials have low refractive indices, so single lenses are rarely adequate. The spherical aberration of a single lens is strongly dependent on the refractive index [23, p 25 ff] and optical systems virtually always have multiple lens elements. SiO_2 and CaF_2 have a sufficient difference in dispersion to allow for colour correction at two wavelengths. Durable anti-reflection coatings are available for both materials [68, 69].

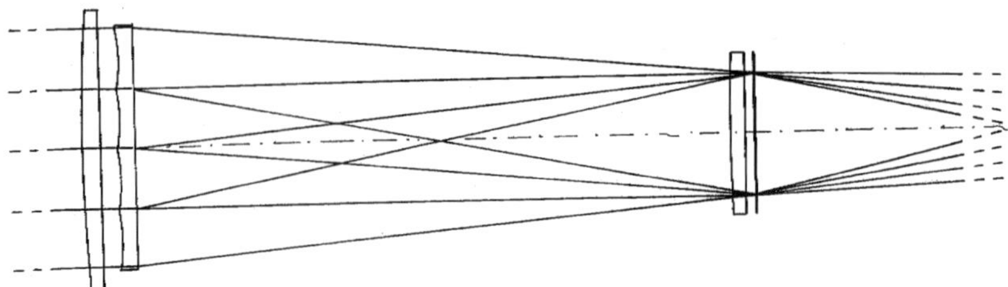

Figure C4.2.14. Beam homogenization system for excimer laser radiation (schematic diagram). The beam is split by means of an array of cylindrical lenses into four lightly diverging beamlets which are overlapped by a common lens in the plane of the mask, which is then imaged onto the workpiece (not shown). Two mutually perpendicular cylindrical lens arrays provide homogenization in both directions of the beam.

C4.2.7.2 Beam homogenization

A typical beam homogenizer for excimer lasers has two mutually perpendicular arrays of cylindrical lenses, which slice the original beam into four to six strips and diverge and—by means of a common lens—redirect each strip of radiation so that they all overlap with the desired size in one common plane, which then becomes the plane of the mask (figure C4.2.14). Diffractive microlenses have also been proposed for this purpose [70]. The homogenized beam tends to diverge behind the mask and a (low-power) field lens is needed to direct the radiation into the opening of the lens system that images the mask onto the workpiece. In fact, the field lens has a dual function: it ensures, on one hand, that all laser radiation is utilized and, on the other, that the entrance pupil of the imaging lens is properly filled so that the lens can perform to its full diffraction-limited capability corresponding to the aperture for which it has been designed.

C4.2.7.3 Optics for imaging a mask

Most of the requirements which have to be met by optics for imaging a mask for excimer laser machining have already been mentioned. For processing of plastics, they translate typically to a demagnifying lens (or mirror system) with a NA at the image of about 0.1 ($f/5$). For machining of edges well perpendicular to the surface over a sizeable field and to reduce sensitivity to axial alignment, (near) telecentricity in image space is advantageous and often mandatory. To achieve a flat image plane over a sizeable field calls for optics with a low Petzval sum, achievable only with lens system types with some length, and this, in turn, causes problems with foci from reflections at later surfaces (see section C4.2.4). This seriously limits the design options.

Optics for UV radiation are costly to manufacture because all tolerances tend to go down proportionally to the wavelength. This, and the already mentioned limited lifetime, justifies every effort to keep the number of optical elements as small as possible.

C4.2.7.3.1 Design examples

Figure C4.2.15 shows an example of an 'economic' lens with only three elements for use in micromachining of small areas (less than 2 mm circular): demagnification is 10, object-to-image distance 1048 mm, the back focus a comfortable 88 mm. The system is corrected for 248 nm (KrF) excimer laser radiation and, for ease of focusing and alignment, simultaneously for HeNe laser radiation. The two outer lenses are made from CaF_2, the inner (negative) lens from fused silica. The lens has sensible field curvature, and this limits the Strehl ratio in the best image plane for the total field to not more than 0.33 for a numerical aperture of just over 0.11 ($f/4.7$). The image of an edge has a steepness, defined here as the distance over which the energy rises from 15 to 85% of the total, of better than 2 μm over the total field and the overall performance, in practice,

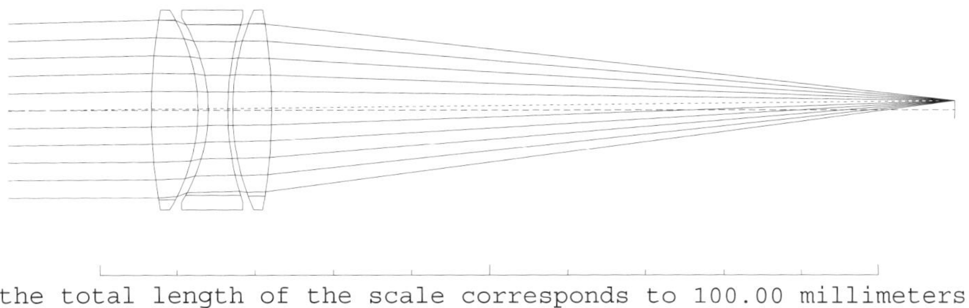

Figure C4.2.15. 3-element lens simultaneously corrected for 248 nm (KrF) excimer and HeNe laser radiation for 10:1 mask imaging on a workpiece for micromachining with feature sizes down to about 2 μm over an area of a few millimetres diameter. Note the long back focus, intended to protect the lens against debris from the process.

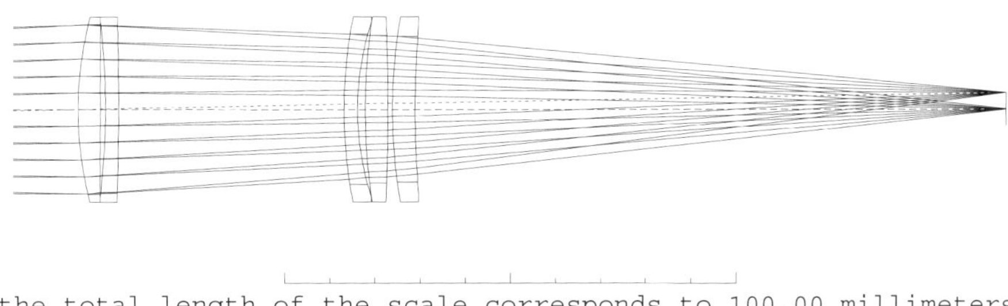

Figure C4.2.16. Five-element lens of the Petzval type for 4:1 mask imaging for micromachining with 248 nm KrF excimer laser radiation. The lens covers an image field of 7 mm diameter with better than 1.25 μm resolution and is near-telecentric in image space. The positive–negative lens pairs are in edge contact.

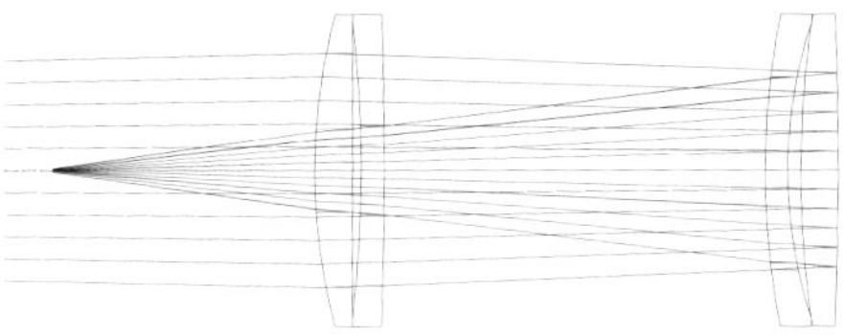

Figure C4.2.17. Focus caused by (residual) reflected radiation at surface 8 of the lens shown in figure C4.2.16. Such foci must lie outside the lens elements if the lenses are to be used at the high (peak) power levels needed for micromachining with excimer lasers.

has been quite satisfactory. It should be mentioned, though, that the design is difficult to manufacture as the strongly curved surfaces imply tight tolerances for lens separation and centration.

Figure C4.2.16 shows a more sophisticated system with five lenses, made exclusively from SiO_2 and thus not colour corrected, and optimized for 248 nm (KrF laser). Demagnification is only 4 and the long object-to-image distance, a system requirement in this case, actually 891.7 mm, produces a long back focus of 131.8 mm. An image field of up to 7 mm diameter is covered with an NA of 0.11 ($f/4.7$). The system is very well corrected with a Strehl ratio of >0.8 over most of the field and near telecentric. The steepness, as previously defined, is better than 1.25 μm. The system is basically of the Petzval lens type (with the last lens functioning as an aplanatic amplifier) which is known for its excellent definition over a limited angular field.

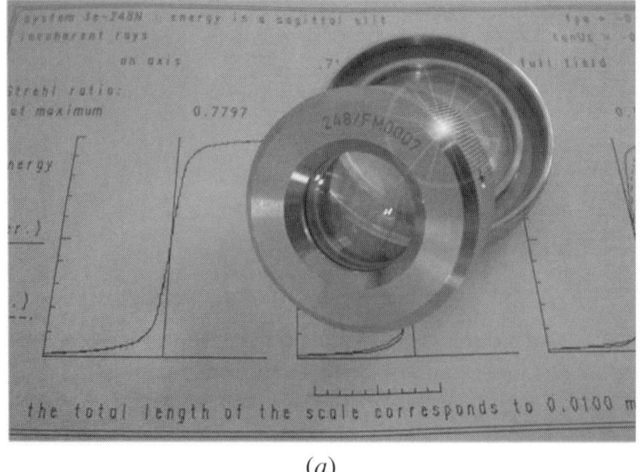

(*a*)

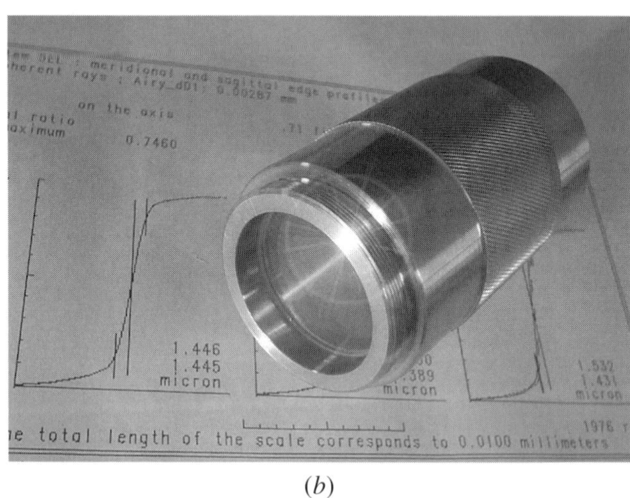

(*b*)

Figure C4.2.18. Photograph of the mask imaging lens systems of (*a*) figure C4.2.15 and (*b*) figure C4.2.16 (Courtesy OPTEC s.a.,Hornu, Belgium).

To ease manufacturing, all lens curvatures are weak and the components of the two positive–negative lens groups are in edge contact. As figure C4.2.17 demonstrates, the only critical reflected image (from surface 8) is well outside the front lens group. Although designed and optimized for operation at 248 nm, the lens is surprisingly perfect at the wavelength of the HeNe laser, yet with a much longer back focus of 154.7 mm, and this can be exploited to (1) mark the focal position with a properly convergent HeNe laser beam or (2) check for manufacturing errors with visible (HeNe) light. Figure C4.2.18 shows photographs of these two lens systems.

A final example, shown in figure C4.2.19 is again a lens for the 248 nm KrF laser with, again, five elements (all SiO_2) and $\times 4$ demagnification but with an even larger field of 14 mm diameter (10 mm square). The object-to-image distance, 376.2 mm, is much shorter than in all previous examples, yet combined with a long back focus of 109.5 mm. The lens is (perfectly) telecentric in image space, and fully diffraction limited at NA $= 0.106$ ($f/4.7$), with a Strehl ratio >0.9 and the previously defined steepness a nearly constant

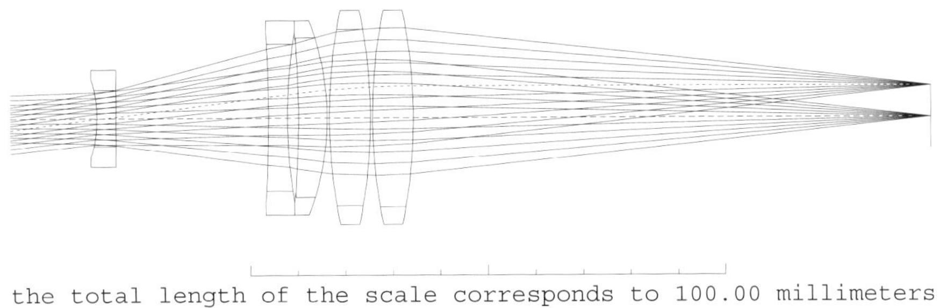

Figure C4.2.19. Five-element retrofocus-type lens for 4:1 mask imaging for micromachining with 248 nm KrF excimer laser radiation. The lens covers a flat image field of 13 mm diameter with better than 1.2 μm resolution, and is perfectly telecentric in image space. The back focus is 109.5 mm for a mask-to-image distance of only 376.2 mm. To ease manufacturing, lenses 2 and 3 are in edge contact, lenses 4 and 5 are identical and symmetrical, and lens 1 has a flat rear surface.

1.2 μm over the whole field. Efforts to reduce manufacturing costs led to (1) a plane rear surface of the first lens, (2) edge contact between the second and the third lens and (3) identical and symmetrical biconvex lenses 4 and 5. The lens is again diffraction limited for HeNe laser light as well, (on-axis only). Reflections of concern are from surfaces 6, 8 and 10, the last just outside the troublesome region (figure C4.2.20).

An interesting mirror system for 1:1 imaging is shown in figure C4.2.21. It dates back to the early days of photolithography [71] and has a remarkable set of properties, achieved with only two spherical mirrors in a concentric set-up. The Petzval sum is zero and, thus, the image field essentially flat, at least in the sagittal direction, with some field curvature in the meridional direction and resulting astigmatism. Thus, imaging is best over a strip-shaped (or better, an annular segment) image field of, say, 6×30 mm in the example layout, at NA $= 0.1$ ($f/5$), the strip extending in the sagittal direction. The convex secondary mirror acts as aperture stop and the system is telecentric in both object and image space. The large distances between the (concave) mirror and the object and image planes, respectively, allow to freely rearrange their positions with flat folding mirrors, if so desired. Derivatives of this system are presently under study for use at 157 nm (F_2 excimer laser). Applications of excimer lasers in micromachining are discussed in chapter D1.6.

C4.2.7.3.2 Photolithography

Photolithography, a key technology for the fabrication of semiconductor integrated circuits ('chips') also uses UV (excimer laser) radiation to image a mask on, in this case, a semiconductor wafer but, in contrast to laser machining, the laser power density is small and geared to a chemical change in the photoresist which covers the surface. As there is no material removal, the back focus may be short (beneficial for the designer) but there are ever increasing demands on resolution, numerical aperture and field size and flatness, which puts photolithographic optics into a class of its own and outside the scope of this treatment. Photolithography forms the subject of chapter D1.5.

C4.2.8 Optics for other laser sources

The previously described lasers and their optics are representative for the three main spectral regions and their respective materials limitations. There is virtually only one material for handling CO_2 laser radiation; when shifting to shorter infrared wavelengths of, say, about 5 μm, several materials with excellent transmission and durability become usable, most of them crystalline such as Al_2O_3 (sapphire), MgO, ZnS and spinel; and these materials also have favourable (= high) refractive indices and thermal conductivities. Yet, the CO laser

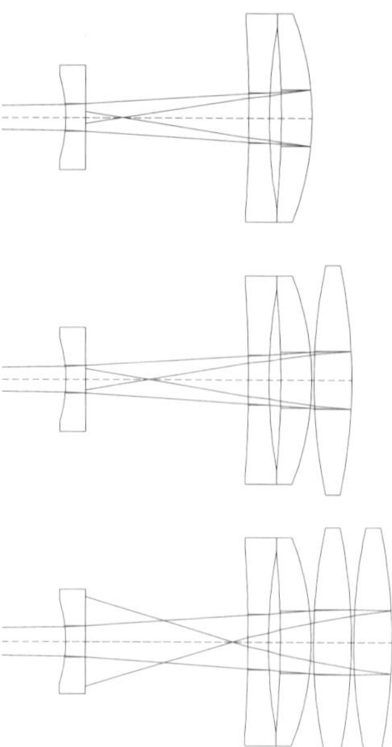

Figure C4.2.20. Foci caused by (residual) reflected radiation at surfaces 6, 8 and 10 of the lens shown in figure C4.2.19. Note that the reflection from surface 10 barely meets the requirement.

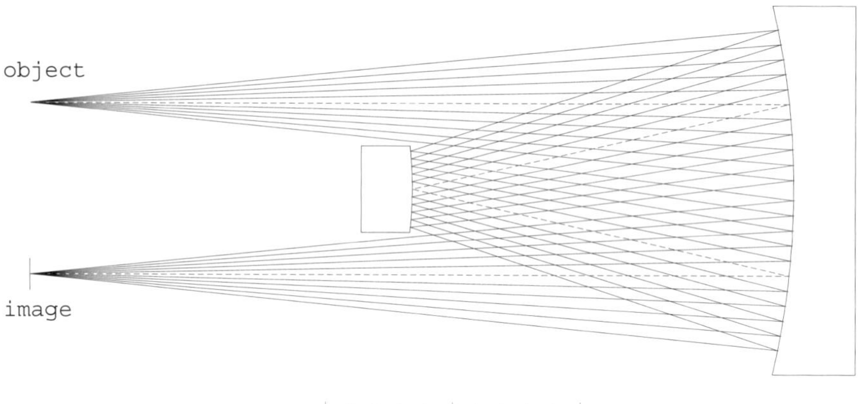

Figure C4.2.21. Two-mirror 'Offner'-type 1:1 imaging system. The system is telecentric in both object and image space and remarkably free of aberrations over a sizeable field. The mirrors are concentric and there is ample space for folding the system to match to mechanical constraints.

which operates in this wavelength region and could, in principle, benefit from these facts has not attained any sensible market shares, clearly for other than optical (or materials) reasons.

Ion lasers are visible light sources with excellent beam quality and this fact has accounted for their popularity in applications which need that property, such as scanners. Needless to say, optical systems must be of a corresponding (i.e. diffraction-limited) quality level. Design aspects correspond to those discussed before for the more widespread visible and near-infrared lasers (C4.2.6). There is a tendency to replace ion lasers with diode lasers wherever possible, for reasons of efficiency and life expectancy.

The copper vapour laser which has become available at powers of 200 W also has high beam quality which makes it a candidate for precision machining [72], allowing focal diameters down to 10 μm at a wavelength favourable for both the radiation–material interaction and optical handling. Optics have to reckon with high peak powers resulting from the short pulse duration.

Most of the lasers considered in C4.2.5 through C4.2.7 are typical high-power lasers, so possible thermal problems with the optics require adequate attention. Other lasers rarely exceed 10 W average output and the thermal aspects are much more relaxed. With the great effort spent on ultrashort laser pulses down to the femtosecond region, however, extremely high peak powers become increasingly common. Little is known, and less published, about the performance of optics at these extremes. Theory predicts a different diffraction pattern [73].

C4.2.9 Conclusions

Free-space optics play a key role in achieving optimum—even adequate—results in a laser application. To select the right optics for each task within a laser system is far from trivial and requires a full understanding of the applicable optical design principles, materials properties and power-related thermal aspects [74]. Optical expertise which has accumulated over centuries has not always found its way into the laser community and this suggests a need for a stronger interaction between optical engineers, laser system designers and laser users [75].

References

[1] Kogelnik H and Li T 1966 Laser beams and resonators *Appl. Opt.* **5** 1550–66

[2] Jabczynski J 1992 Quasi-geometrical model of partially coherent beam propagation in real axially symmetric optical systems *Proc. SPIE* **1780** 584–9

[3] Siegman A E, Sasnett M W and Johnston jr T F 1991 Choice of clip levels for beam width measurements using knife-edge techniques *IEEE J. Quantum Electron.* **27** 1098–104

[4] Siegman A E 1991 Defining the effective radius of curvature for a nonideal optical beam *IEEE J. Quantum Electron.* **27** 1146–8

[5] CEN/TC 123 N79, 1996 Optics and optical instruments, Lasers and laser related equipment, vocabulary and symbols, Secretariat DIN Aussenstelle Pforzheim, Westliche 56, D-75172 Pforzheim

[6] Siegman A E 1990 New developments in laser resonators *Proc. SPIE* **1224** 2–14

[7] Borik S, Wittig K and Zoske U 1994 Zur Bedeutung des Strahlparameterprodukts für das Propagationsverhalten von Hochleistungslaserstrahlen *Laser Optoelektron.* **26** 50–7

[8] Johnston T F Jr 1998 Beam propagation (M^2) measurement made as aesy as it gets: the four-cuts method *Appl. Opt.* **37** 4840–50

[9] Self S A 1983 Focusing of spherical Gaussian beams *Appl. Opt.* **22** 658–61

[10] Welford W T 1986 *Aberrations of Optical Systems* (Bristol: Adam Hilger)

[11] Schroeder D J 1987 *Astronomical Optics* (San Diego, CA: Academic)

[12] Borik S 1994 Einfluss optischer Komponenten auf die Fokussierbarkeit *Laser Optoelektron.* **26** 58–63

[13] Smith W J 1989 Optical design *The Infrared Handbook* ed W L Wolfe and G I Zissis 3rd printing (Washington, DC: Office of Naval Research) ch 8

[14] Rayleigh Lord 1879 *Phil. Mag.* **8** 403

[15] Yura H T and Rose T S 1995 Gaussian beam transfer through hard-aperture optics *Appl. Opt.* **34** 6826–8

[16] Drége E M, Skinner N G and Byrne D M 2000 Analytical far-field divergence angle of a truncated Gaussian beam *Appl. Opt.* **39** 4918–25

[17] Beckmann L H J F and De Meijere J L F 1988 Retrofocus optics for the focusing of CO_2-laser radiation; a comparison between designs using spherical optics and aspherics *Proc. SPIE* **1020** 192–5

[18] Matuura Y, Miura D and Miyagi M 1999 Fabrication of copper oxide-coated hollow waveguides for CO_2 laser radiation *Appl. Opt.* **38** 1700–3

[19] Hello P and Man C N 1996 Design of a low-loss off-axis beam expander *Appl. Opt.* **35** 2534–6

[20] Bea M, Giesen A and Huegel H 1994 Gezielte Steuerung der Fokusgeometrie durch gekoppelte adaptive systeme *Laser Optoelektron.* **26** 43–9

[21] Jarosch U K 1996 Adaptive metal mirror for high power CO_2 lasers *Proc. SPIE* **2774** 457–67

[22] Reedy H E and Herrit G L 1988 Comparison of GaAs and ZnSe for high power CO_2 laser optics *Proc. SPIE* **1020** 180–91

[23] Smith W J 1992 *Modern Lens Design* (New York: McGraw-Hill)

[24] Beckmann L H J F and Märten O 1992, Zoom lens designs for use in sheet metal cutting by high power CO_2-lasers *Proc. SPIE* **1780** 765–76

[25] Miyamoto I, Nanba H and Maruo H 1990 Analysis of thermally induced optical distortion in lens during focusing high power CO_2 laser beam *Proc. SPIE* **1276** 112–21

[26] Miyamoto I, Horiguchi Y and Maruo H 1990 Novel shaping optics of CO_2 laser beam: LSV optics—principle and applications *Proc. SPIE* **1276** 202–16

[27] Tangelder R J, Beckmann L H J F and Meijer J 1992 Influence of temperature gradients on the performance of ZnSe-lenses *Proc. SPIE* **1780** 294–302

[28] Beckmann L H J F 1994 Modelling of, and design for, thermal radial gradients in lenses for use with high power laser radiation *Proc. Int. Optical Design Conf. (Rochester)* p 11–15

[29] Beckmann L H J F and Meijer J 1998 Modelling of and design for thermal radial gradients in lenses for use with high power laser radiation *Welding in the World* **41** 97–104

[30] Weck M, Hermanns C, Ostendarp H and Wermeyer K 1994 Betriebsverhalten transmissiver Optiken bei der Laser-Materialbearbeitung *Laser Optoelektron.* **26** 67–72

[31] Godfried H P, Coe S E, Hall C E, Pickles C S J, Sussmann R S, Tang X and van der Voorden W K L 2000 Use of CVD diamond in high-power CO_2 lasers and laser diode arrays *Proc. SPIE* **3889** 553–63

[32] Herrit G L and Reedy H E 1989 Advanced figure of merit evaluation for CO_2 laser optics using finite element analysis *Proc. SPIE* **1047** 33-42

[33] Plansee 1987 Wolfram- und Molybdänlaserspiegel (datasheet) Metallwerk Plansee GmbH, A-6600 Reutte

[34] Erdös P 1959 Mirror anastigmat with two concentric spherical surfaces *J. Opt. Soc. Am.* **49** 877–82

[35] Beckmann L H J F and Ehrlichmann D 1994 Three mirror off-axis systems for laser applications *Proc. Int. Optical Design Conf. (Rochester)* pp 340–8

[36] Henning T, Scholl M, Unnebrink L, Habich U, Lenert R and Herziger G 1997 Beam shaping for laser materials processing with non-rotationally symmetric optical elements *Proc. SPIE* **3092** 126–9

[37] Armengol J, Lupon N and Laguarta F 1997 Two-faceted mirror for active integration of coherent high-power laser beams *Appl. Opt.* **36** 658–61

[38] Schneider M 1998 Laser cladding *PhD Thesis* Univerity of Twente (Enschede, The Netherlands)

[39] Beckmann L H J F 1990 A small-computer program for optical design and analysis written in 'C' *Proc. SPIE* **1354** 254–61

[40] Bett T H, Danson C N, Jinks P, Pepler DA, Ross I N and Stevenson R M 1995 Binary Phase zone-plate arrays for laser-beam spatial-intensity distribution conversion *Appl. Opt.* **34** 4025–36

[41] Dickey F M and Holswade S C 1996 Gaussian laser beam profile shaping *Opt. Eng.* **35** 3285–95

[42] Kuittinen M, Vahimaa P, Honkanen M and Turunen J 1997 Beam shaping in the nonparaxial domain of diffractive optics *Appl. Opt.* **36** 2034–41

[43] Zhang G Q, Gu B Y and Yang G Z 1995 Design of diffractive phase elements that produce focal annuli: a new method *Appl. Opt.* **34** 8110–15

[44] Schott 1988 Resistance of optical glasses to short laser pulses, Technical Information—Optical Glass, No 21, 4/1988 Schott Glaswerke, Postfach 2480, 55014 Mainz, Germany

[45] Milam D 1998 Review and assessment of measured values of the nonlinear refractive- index coefficient of fused silica *Appl. Opt.* **37** 546–50

[46] Verboven P E 1995 Beam delivery for Nd:YAG lasers *Opt. Eng.* **34** 2683–6

[47] Kingslake R 1978 *Lens Design Fundamentals* (New York: Academic)

[48] Smith W J 1990 *Modern Optical Engineering* 2nd edn (New York: McGraw-Hill)

[49] Heraeus 1994, Infrasil 301, 302 and 303 POL Information, POL-O/439M-E, 06/94 and: HOQ 310, POL Information, POL-O/424M-E, 06/94, Heraeus Quarzglas GmbH & Co KG, Division POL, PO Box 1554, 63405 Hanau, Germany

[50] Heraeus 2000, Homosil 101 and Herasil 102, POL Information, POL-O/439M-E,Heraeus Quarzglas GmbH & Co KG, Division POL, PO Box 1554, 63405 Hanau, Germany

[51] Shannon R R 1997 *The Art and Science of Optical Design* (New York: Cambridge University Press)

[52] Walker B H 1995 Lens design for the near IR... correction of primary chromatic aberration *Appl. Opt.* **34** 8072–3

[53] Haran F M, Hand D P, Peters C and Jones J D C 1997 Focus control system for laser welding *Appl. Opt.* **36** 5246–51

[54] Sun H 1997 Measurement of laser diode astigmatism *Opt. Eng.* **36** 1082–7

[55] Ratowsky R P, Long Y, Deri R J, Chang K W, Kallman J S and Trott G 1997 Laser diode to single-mode fiber ball lens coupling efficiency: full-wave calculation and measurements *Appl. Opt.* **36** 3435–8

[56] Wilson R G 1998 Ball-lens coupling efficiency for laser-diode to single-mode fiber: comparison of independent studies by distinct methods *Appl. Opt.* **37** 3201–5

[57] Braat J 1995 Design of beam-shaping optics *Appl. Opt.* **34** 2665–70

[58] Braat J 1997 Influence of substrate thickness on optical disk readout *Appl. Opt.* **36** 8056–62

[59] Muller N G, Weber R and Weber H P 1995 Output beam characteristics of high-power continuous-wave diode laser bars *Opt. Eng.* **34** 2384–9

[60] Du K, Zhang J, Quade M, Liao Y, Falter S, Baumann M, Loosen P and Poprawe R 1998 Neodymium:YAG 30 W cw laser side pumped by three diode laser bars *Appl. Opt.* **37** 2361–4
[61] see, for example, US 5,784,203; UK 2319630; DE 19752416.8
[62] Holmer A-K and Hård S 1995 Laser-machining experiment with an excimer laser and a kinoform *Appl. Opt.* **34** 7718–23
[63] Heraeus 1999, Quartz Glass for Excimer Laser Applications, Application Note, June 1999, Heraeus Quarzglas GmbH & Co KG, Division POL, PO Box 1554, 63405 Hanau, Germany
[64] Schenker R and Oldham W 1998 Damage-limited lifetime of 193 nm lithography tools as a function of system variables *Appl. Opt.* **37** 733–8
[65] Rothschild M 1993 Optical materials for excimer laser applications *Opt. Photon. News* **May** 8–15
[66] Leclerc N, Pfleiderer C, Hitzler H, Wolfrum J, Greulich K O, Thomas St Fabian H, Takke R and Englisch W Transient 210-nm absorption in fused silica induced by high-power UV laser irradiation *Opt. Lett.* **16** 940–2
[67] Thomas Th and Kuehn B 1996 KrF Laser induced absorption in synthtic fused silica *Proc. SPIE* **2966** 56–64
[68] Krajnovich D J, Kulkarni M, Leung W, Tam A C, Spool A and York B 1992 Testing of the durability of single-crystal calcium fluoride with and without antireflection coatings for use with high-power KrF excimer lasers *Appl. Opt.* **31** 6062–75
[69] Laux S, Mann K, Granitza B, Kaiser U and Richter W 1996 Antireflection coatings for UV radiation obtained by molecular-beam deposition *Appl. Opt.* **35** 6216–18
[70] Nocolajeff F, Hård S and Curtis B 1997 Diffractive microlenses replicated in fused silica for excimer laser-beam homogenizing *Appl. Opt.* **36** 8481–9
[71] Offner A 1975 New concepts in projection mask aligners *Opt. Eng.* **14** 131–7
[72] Hartmann M, Koch J, Lang A, Schutte K and Bergmann H W 1997 Industrial aspects of precision machining with copper vapour lasers *Proc. SPIE* **3097** 260–6
[73] Aleshkevich V, Kartashov Y and Vysloukh V 1999 Diffraction and focusing of extremely short optical pulses: generalization of the Sommerfeld integral *Appl. Opt.* **38** 1677–81
[74] Klein C A 1997 Materials for high-power laser optics: figures of merit for thermally induced beam distortions *Opt. Eng.* **36** 1586–95
[75] Beckmann L H J F and Ehrlichmann D 1995 Optical systems for high-power laser applications: principles and design aspects *Opt. Quantum Electron.* **27** 1407–25

Further reading

There are no specific books on the subject of this chapter, but useful information can be found in most texts on Laser materials processing, such as:

Caird Luxon J T and Parker E 1992 *Industrial Lasers and Their Application* (Englewood Cliffs, NJ: Prentice-Hall)
Crafer R C and Oakley P J 1993 *Laser Processing in Manufacturing* (London: Chapman and Hall)
Herziger G and Loosen P 1993 *Werkstoffbearbeitung mit Laserstrahlung* (München: Hanser)

The design of optical systems in general is treated in a number of textbooks, such as those cited for this chapter, but without adressing laser-specific subjects, with the exception of Smith (1990, 1992) and Welford (1986) (in a 2-page appendix). The subject of power distribution shaping of laser beams (see section C4.2.5.5) has been treated in detail in a recent book:

Dickey F M and Holswade S C (ed) 2000 *Laser Beam Shaping* (New York: Marcel Dekker)

C4.3
Fibre optic beam delivery

D P Hand

Fibre optics are a particularly attractive means of high-power laser beam delivery, offering the highest degree of flexibility and, hence, full directional control of light, coupled with transmission over long distances if required (hundreds of metres). If integrated into suitable robots and positioning systems, they provide multi-axis processing (important for complex workpieces) and remote working capabilities and allow one laser to be easily time-shared between many different workstations.

The most widely used type of optical fibre is made from fused silica. The telecommunications industry has driven the development of these fibres, which have since been adapted for use in laser materials processing. These fibres have excellent transmission throughout the visible and near infrared (NIR) part of the spectrum, and are ideally suited to the 1.06 μm wavelength of Nd:YAG lasers, with losses as low as 1 to 2 dB km^{-1}. Unfortunately, they are not suitable for the 10.6 μm wavelength produced by CO_2 lasers where they are opaque. Alternative fibres do exist for such longer wavelengths but these have more severe limitations in terms of power handling, attenuation, beam profile preservation and available lengths. However, they are used in a limited range of power delivery applications, in particular for laser surgery.

One example process where fibre optic delivery is often used is laser welding in the automotive industry e.g. for welding car body parts (see chapter D1.1). The 3D shapes being welded require a highly manoeuvrable laser beam, and so a fibre is ideal. Typically, in such an application, 2–4 kW of cw laser power is required, focused down to a spot of 0.3–0.5 mm diameter. A suitable beam from a Nd:YAG laser can be readily delivered through a 600 μm core diameter fused silica optical fibre. Indeed, many high-power Nd:YAG welding lasers are *only* available with fibre optic delivery. Nd:YAG laser drilling (see chapter D1.4), meanwhile, requires high peak power pulses *and* a high beam quality (necessitating a smaller diameter optical fibre, 200 μm) which means that optical fibre beam delivery cannot normally be used due to optical damage problems.

C4.3.1 Fibre operation

An optical fibre is an example of an optical waveguide. Most optical waveguides work on the principle of total internal reflection (TIR), where the light is guided within a high-index core region surrounded by a lower index cladding (figure C4.3.1). Light which is incident on the core/cladding interface at an angle θ greater than the critical angle θ_c is 100% reflected. The angle θ_c is dependent on the refractive index difference between the core and cladding, according to the equation $\sin\theta_c = n_1/n_2$. It is often more useful to express this angle in terms of the fibre's numerical aperture (NA). Light which is incident on the front face of the fibre at an angle less than that defined by the fibre's NA will satisfy $\theta > \theta_c$, and so be guided by the fibre. The theory of such waveguides is discussed in detail in chapter A6. Optical fibres based on this principle are normally made from fused silica, with a suitable dopant used to alter the refractive index of either the core or cladding region. Fused silica is suitable over a wide wavelength spectrum, from UV at 200 nm up to NIR at 2 μm but other glasses have been used for transmission of mid-IR light, for which fused silica is opaque.

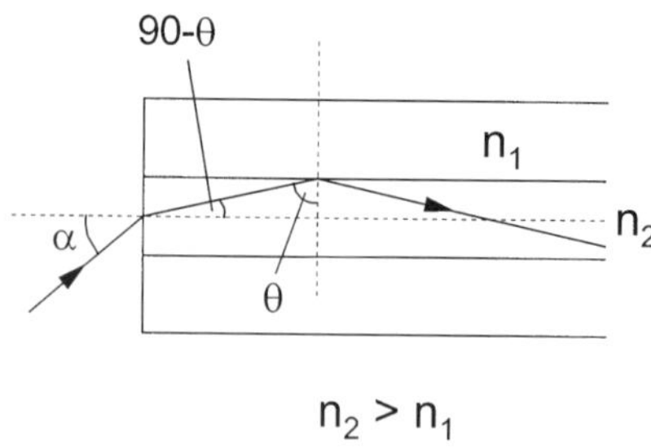

Figure C4.3.1. Step-index fibre optic.

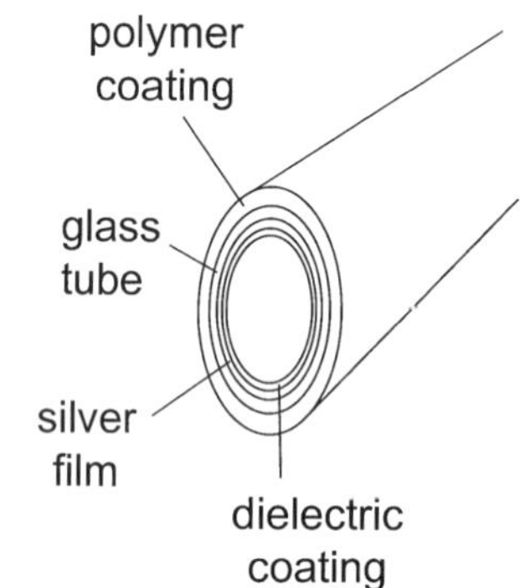

Figure C4.3.2. Hollow glass fibre optic.

Plastic optical fibres are also available, as are plastic-clad silica fibres but these cannot be used with large optical powers due to their low damage thresholds.

It is also possible to provide guidance using Fresnel reflection (rather than total internal reflection), which is the technique normally used with hollow fibres for CO_2 laser light. An example of such a fibre is shown in figure C.4.3.2, consisting of a very fine glass capillary tube with an internal dielectric coating to provide a high reflectivity at the appropriate wavelength. These hollow fibres are described in more detail later, but for an in-depth review see Harrington [1].

Both TIR and hollow waveguide fibres guide a number of discrete waveguide modes (see section A6.3.1). The number of modes depends on the guide's dimensions and NA. If the core region is made sufficiently small ($\sim 7\ \mu$m at 1.064 μm for a 0.11 NA fibre), only one mode can propagate and the fibre is said to be single mode (see section A6.4.1). Single mode fibres are normally used in telecommunications applications, where intermodal dispersion would otherwise cause problems (see section A6.5.2). For power delivery, however, multi-mode fibres with much larger core diameters are normally used. This clearly reduces the intensity in the core and, hence, the likelihood of optical damage. In addition, high-power laser beams, particularly from solid state lasers such as Nd:YAG, are normally highly multi-moded due to thermal lensing effects in the laser. It is impossible to couple such a beam efficiently into a single-mode fibre; instead only the lowest order mode will be coupled efficiently. A typical silica fibre used for high power Nd:YAG beam delivery has a core diameter of between 200 and 1000 μm, with an NA of 0.22, and can support many tens to hundreds of thousands of modes.

Finally, there is a new type of optical fibre which has been developed in recent years. This is the microstructured fibre, often called a 'photonic crystal fibre' or PCF. PCF is an optical fibre with an ordered array of air holes running along its length—typical fibre cross sections are shown in figure C4.3.3. There

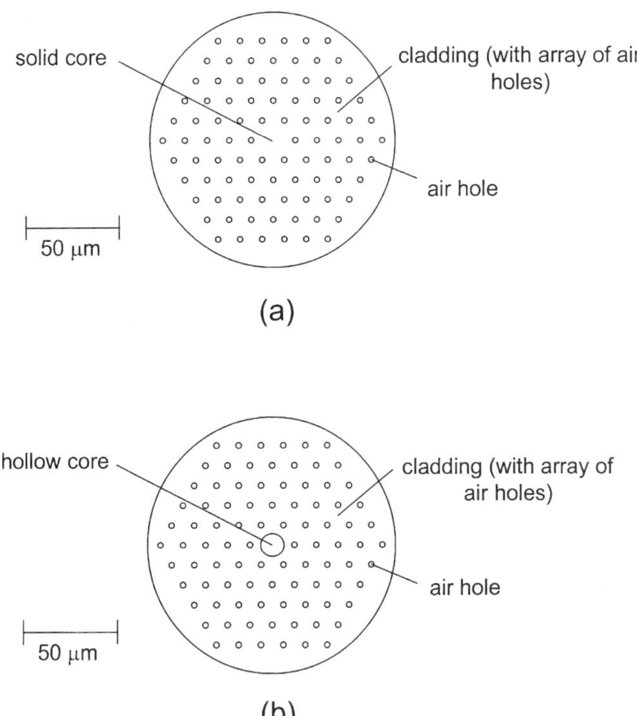

Figure C4.3.3. Photonic crystal fibre (*a*) index-guiding and (*b*) air-guiding 'photonic bandgap fibre'.

are two main categories of PCF: 'index-guiding' PCF in which a solid core is surrounded by a cladding of the same material but including an array of air holes (figure C4.3.3(*a*)); and air-guiding 'photonic bandgap fibres' where a hollow core is surrounded by a cladding with an array of holes (figure C4.3.3(*b*)). With the index-guiding PCF structure, the core index is greater than the average index of the cladding, which allows guiding in much the same way as conventional optical fibres (by a modified form of TIR). Photonic bandgap fibres, meanwhile, use the special optical properties of the *periodic pattern* of the air holes to provide guidance by Bragg reflection.

The performance of these various types of optical fibre is now described and summarized in table C4.3.1.

C4.3.1.1 Total internal reflectance fibres

Total internal reflectance (TIR) fibres can be either step-index or graded-index, with refractive index profiles as shown in figure C.4.3.4. Graded-index fibres were developed for telecommunications, since they provide multi-mode operation with greatly reduced intermodal dispersion. They have only been used to a very limited extent for high-power beam delivery. Step-index fibres, meanwhile, have been specifically developed for power delivery applications. These normally have a pure silica core to maximize the damage threshold and a fluorine-doped cladding to provide the requisite refractive index difference. Telecommunication fibres, by contrast, normally have a germanium-doped core and pure silica cladding. The attenuation spectrum is dependent on the OH content of the fibre and, typically, manufacturers offer both 'low OH' and 'high OH' fibre. Attenuation spectra are shown in figure C4.3.5, courtesy of Polymicro Technologies, Inc. High OH fibres are most suitable for the UV/visible part of the spectrum, whereas low OH fibres are used for the longer

Table C4.3.1. Details and performance of the various types of optical fibre.

Fibre type	Operating wavelengths (μm)	Power handling	Typical core diameter (μm)	Attenuation (dependent on wavelength)	Usable length	Typical beam quality (M^2)
TOTAL INTERNAL REFLECTANCE FIBRES						
Fused silica (large core multimode)	0.2–2	> 4 kW cw in 600 μm 55 J/pulse in 400 μm (1 ms pulse)	100–1000	<0.1 dB/m from 0.5 to 1.9 μm	Hundreds of metres	20–300
Chalcogenide	1–6		200–300	0.2–1 dB	Hundreds of metres	15–200
Fluoride	1–4	970 mJ in 200 μs	70–600	< 0.1 dB/m from 2 to 3 μm	Hundreds of metres	20–200
Germanate	1–3	20 J/pulse		0.1–1 dB/m	Hundreds of metres	
Silver halide	4–16	100 W cw (1000 μm ∅)	500–1000	0.5–2 dB	15 m	20–200
Sapphire (solid)	0.5–3.5	1 J/pulse in 400 μm	150–400	0.25–2 dB/m	3 m	15–250
Sapphire (hollow)	10–16	1.9 kW (water cooled)	250–1000	1–10 dE/m (dependent on ∅)		
HOLLOW WAVEGUIDE FIBRES	Internal coating optimized for wavelength					
Metal tube	1–25	2.7 kW cw	700–1000	0.1–10 dB/m (dependent on ∅ and bend radii)	Few metres	
Plastic tube	1–25	65 W cw (1800 μm ∅	800–1800	0.1–2 dB/m		
Glass tube	1–25	1 kW cw (water cooled) 75 mJ in 7 ns	250–1000	0.1–0.3 dB	13 m	
MICROSTRUCTURED OPTICAL FIBRES						
Fused silica	0.2–2		5–150	Down to 0.05 dB/m	Hundreds of metres	

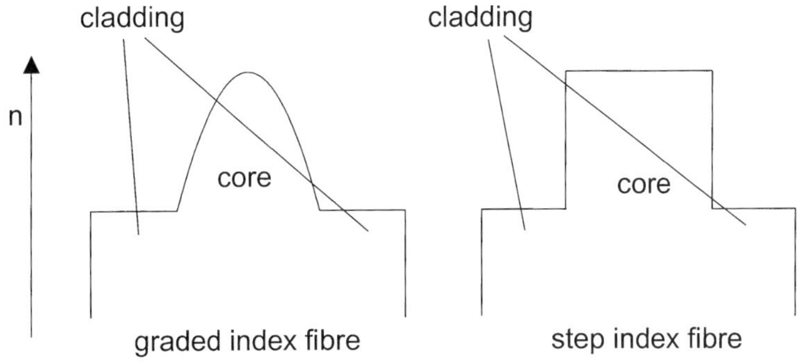

Figure C4.3.4. Refractive index profiles of graded and step index fibre optics.

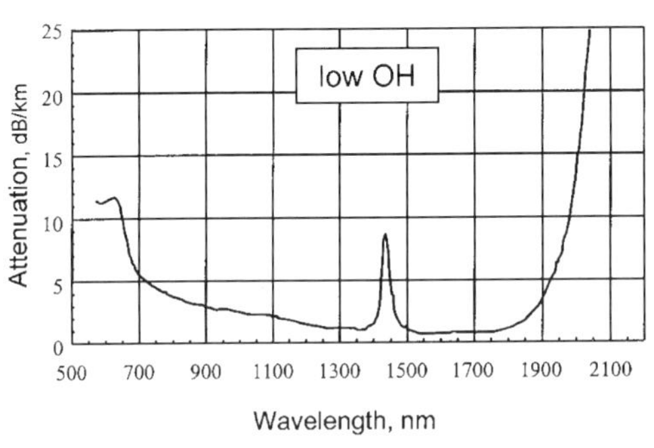

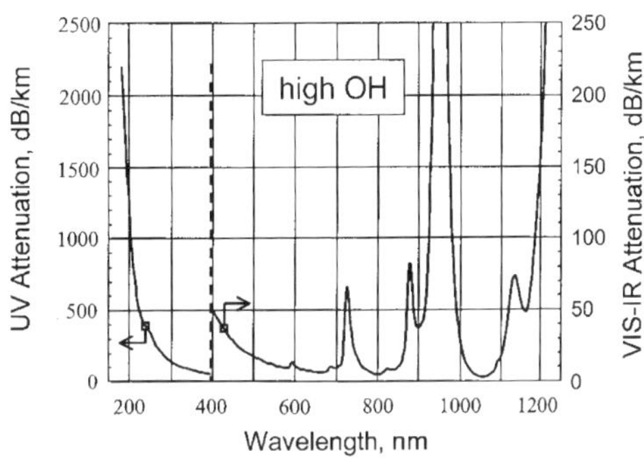

Figure C4.3.5. Attenuation spectra of low OH and high OH fibre optics. Figures courtesy of Polymicro Technologies, Inc.

wavelength end of the visible and the NIR. In most power delivery applications, it is unusual to use more than about 10 m of fibre. This means that an attenuation of ~0.05 dB m^{-1} is acceptable (10% loss in 10 m).

Other materials have been developed for TIR optical fibres, specifically aimed at the transmission of longer wavelength light. These can be either glasses or crystalline materials. The glasses are suitable for the mid-IR, whereas only crystalline fibres can be used at the important CO_2 wavelength of 10.6 μm.

The main glasses used are fluorides (e.g. fluorozirconate, fluoroaluminate), chalcogenides [2] (sulphide, selenide or telluride) and germanates. Fluoride fibres transmit light from 1–4 μm, with attenuation <1 dB m^{-1}, and so are suitable for delivery of Er:YAG laser light at 2.94 μm, used for medical applications. Germanate fibres, however, offer a much higher damage threshold (20 J pulse from Er:YAG), even though their attenuation is slightly higher. Chalcogenide fibres, meanwhile, may be used to transmit light from 1 to 6 μm, with an attenuation of 0.2–1 dB m^{-1}. Typical sizes available are 200–300 μm core diameter, with an NA of 0.3 to 0.5. These fibres are useful with CO lasers, which emit a number of lines between 5 and 6 μm. The maximum power which can be transmitted, however, is limited to about 10 W at present.

Two crystalline materials are commonly used for optical fibres: silver halide (polycrystalline) and sapphire (single crystal). Silver halide fibres can be used to transmit light in the mid to far IR, with an attenuation at 10.6 μm as low as 0.5 dB m^{-1} [3]. These fibres do not normally have a cladding region; instead, the polycrystalline silver halide core is simply surrounded by air and, hence, they have a high NA of 0.7. The fibre must be held within an opaque tube, in order to prevent UV light from reaching the photosensitive core which causes premature failure. Diameters of 200–1000 μm are typical and the damage threshold is about 100 W cw with the 1000 μm fibre. The fibres are mechanically quite weak. Single-crystal sapphire fibres, meanwhile, can be used for wavelengths of up to 3.5 μm and are mechanically very strong but are limited to lengths of 3 m. Sapphire can also be used to guide 10.6 μm light, although in this case it must be fabricated into a hollow tube. Anomalous dispersion means that sapphire has a refractive index of less than one between 10 and 16.7 μm [4] and so light is guided in the air core by TIR. These fibres have significant loss (about 10 dB m^{-1} for a 200 μm core fibre) but this varies with $1/R$, with the attenuation for 1000 μm fibre $\sim$1 dB. Hollow sapphire fibres have been used with CO_2 lasers for surgical applications and, with a water-cooled jacket, have delivered 1.9 kW of laser power [5].

C4.3.1.2 Hollow waveguide fibres

Hollow waveguide fibres are made by depositing a high-reflectivity layer inside a glass [6], metal [7] or plastic [8] tube. For example, hollow glass fibre lengths of up 13 m have been manufactured with bore sizes varying from 250 to 1300 μm. One interesting feature of hollow fibres is the differential modal attenuation, where higher order modes are strongly attenuated. This means that the output will consist of only a few low-order modes and, hence, has a high beam quality *but* a large speckle pattern due to intermodal interference, which changes as the fibre is bent [9]. This can be a significant problem for many laser materials processing applications, where a more uniform spot is required. Matsuura *et al* [10] have developed a silver halide beam homogenizer at the output end of the fibre in order to address this, but this, of course, will reduce the beam quality. The attenuation of higher order modes in hollow waveguide fibres means that bend loss becomes significant, since bends cause coupling to higher order modes [9]. Indeed, bend loss has been shown to vary with the inverse of the bend radius [1]. The loss also depends on the bore diameter, with a loss of 2 dB m^{-1} for a 250 μm diameter bore dropping to 0.25 dB m^{-1} for a 530 μm bore for 10.6 μm light in a hollow glass waveguide [11]. Designs based on a glass tube generally have the best performance, due to the superior interior surface quality in comparison with metal or plastic tubes.

The significant attenuation of hollow waveguide fibres means that water cooling is essential for very high power transmission. The largest CO_2 laser power reported to have been delivered through a water-cooled hollow glass waveguide with a bore of 700 μm is 1040 W [5].

C4.3.2 Fabrication

C4.3.2.1 Fused silica optical fibres

Fused silica optical fibres are made by first manufacturing a very high purity glass rod or tube called the *preform* (typically 10–25 mm in diameter and 60–120 cm long) which is then pulled into a very thin filament in a drawing furnace. There are two main types of preform manufacture process used: one where the glass is deposited in layers and in the other the glass for the core (rod) and cladding (tube) are made separately and subsequently combined. The layered technique uses vapour phase deposition and allows complex radial refractive index profiles to be constructed and is the process normally used for small core fibres, such as the single-mode fibre used in telecommunications. The rod-in-tube technique, meanwhile, is quicker (radial heating is applied to the tube in order to collapse it onto the rod) and is particularly suitable for the large-core step-index fibres used for power delivery (200–1000 μm core diameter). However, it is more likely to create inherent problems such as core/cladding interface irregularities.

Vapour phase deposition involves an oxidation process where highly pure vapours of $SiCl_4$ react with oxygen to form silica (SiO_2) particles. Other vapours, such as $GeCl_4$, are included in the gas mixture to adjust the refractive index of the deposited layers through the Ge content. The deposition process can be performed in two basic ways, namely external or internal. In the external lateral deposition process [12] as illustrated in figure C4.3.6(*a*), the glass particles (or soot) are formed on the outside of a rotating graphite or ceramic mandrel which is also translated in the lateral direction. The desired index profile is built up layer

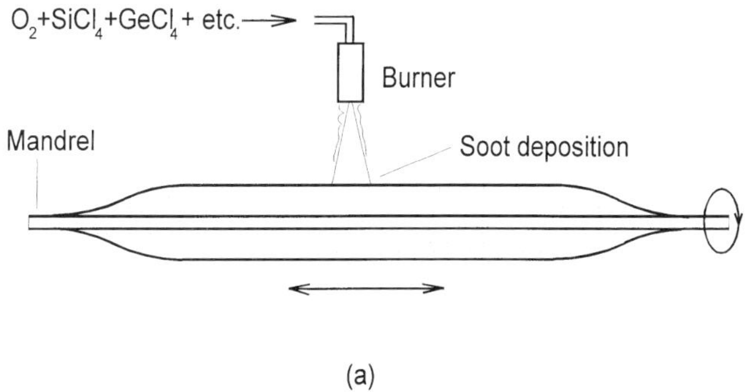

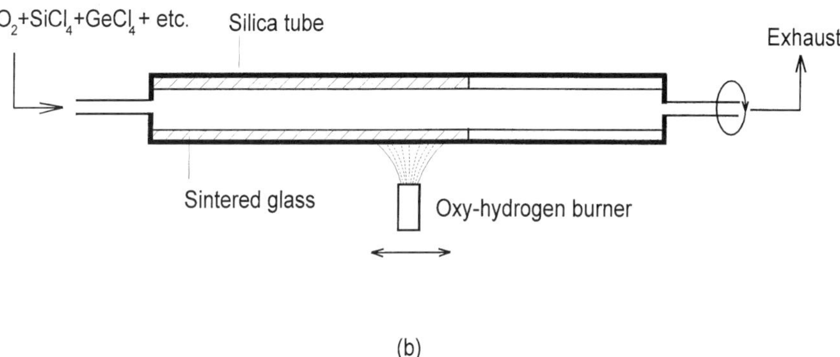

Figure C4.3.6. Manufacture of a fibre preform by vapour phase deposition (*a*) Outside vapour phase oxidation (OVPO) (*b*) Modified chemical vapour deposition (MCVD).

by layer. After deposition, the mandrel is removed and the preform is vitrified at high temperature to form a clear glass preform. (The small central hole from the mandrel disappears during the fibre drawing process). An alternative external technique is axial deposition [13], or VAD (vapour phase axial deposition), where both core and cladding layers are deposited simultaneously on the end of a fused silica seed rod using two oxy-hydrogen burners. Because of axial deposition, the preform can be made in continuous lengths. An internal deposition method, the modified chemical vapour deposition (MCVD) process [14], is illustrated in figure C4.3.6(*b*). Here, the constituent vapours, along with oxygen, flow through the inside of a rotating silica tube and, as the particles are deposited, they are sintered to glass by the oxy-hydrogen flame traversing back and forth along the tube. In plasma-activated CVD [15], the oxy-hydrogen flame is replaced with a moving microwave resonator which generates a plasma within the tube (held at 1100 °C) and activates the chemical reactions. The process produces a clear glass preform directly without the need for sintering.

C4.3.2.2 Chalcogenide, fluoride and germanate glasses

These fibres are normally fabricated using rod-in-tube techniques similar to that previously described for large-core fused silica fibres, although in some cases extrusion techniques are used [16].

C4.3.2.3 *Crystalline optical fibres*

There are both polycrystalline and single-crystal fibres which are made in different ways. Silver halide fibres are polycrystalline and are made into fibres using a hot extrusion process, where a single-crystal billet or preform is placed in a heated chamber and the fibre extruded through a diamond or tungsten die at a temperature of about half the melting point, resulting in a polycrystalline structure with a grain size of about 10 μm. Sapphire fibres, meanwhile, are single crystal and are made by melting some or all of the starting sapphire material and slowly drawing a fibre, either from a melt through a capillary tube or directly from a sapphire, using a CO_2 laser for local melting.

C4.3.2.4 *Hollow waveguide optical fibres*

A number of different techniques have been used in the fabrication of hollow waveguide fibres. The smoothness of the internal surface is particularly important, with a very smooth surface essential to prevent coupling to higher order modes and, hence, high losses.

(a) *Metal tube waveguides.* In 1983, Miyagi *et al* [17] developed a technique where a dielectric layer and then a metallic film are deposited on the outside of an aluminium tube. The metal film is subsequently electroplated with nickel, before etching away the aluminium tube. A different technique was developed by Bhardwaj *et al* [7] in 1993 which used a silver tube as a starting point. This tube is first internally etched to make it smooth, before depositing an AgBr film on the inside using wet chemistry.

(b) *Plastic tube waveguides.* Croitoru *et al* [18] have developed a technique where a silver film is deposited on the inside of Teflon and polyethylene tubing, which is then overcoated with AgI using wet chemistry. Haan and Harrington [19] have used similar techniques to deposit Ag/AgI films inside polycarbonate tubing.

(c) *Hollow glass waveguides.* Hollow glass waveguide fibres are normally based on fused silica. A fused silica tube is heated and pulled down to a small diameter in the same way as a TIR optical fibre and a polyimide coating is deposited on the outside for mechanical protection. A conventional electroless plating technique is then used to deposit a silver film internally. A dielectric layer is then deposited on top of the silver. One technique is to flow iodine through the tube, which reacts with the silver surface and, hence, forms a thin uniform dielectric layer of AgI [20]. An alternative is to deposit a suitable polymer layer inside the tube, by flowing a polymer solution through and subsequently heating [21].

C4.3.2.5 *Microstructured optical fibres*

Microstructured optical fibres are made by stacking fused silica capillary tubes in order to give the required cross section. In the ‘index-guiding’ case, the core is defined by substituting one of the tubes with a solid rod, whereas in the ‘photonic bandgap’(hollow core) case, a tube with a larger bore diameter is used. These stacks are then heated and fused together to form a preform, from which the optical fibre is pulled.

C4.3.3 Implementation

Focusing optics are necessary to couple the light from the laser into the fibre. These are chosen to give a focal spot diameter only slightly smaller than that of the fibre core. However, with some lasers, the beam quality can change with operating parameters, which can lead to an increased focal spot size, so this must be taken into account. One way to avoid this is to use an imaging system where the cone angle rather than spot size changes with beam quality. This can be achieved, for example, by imaging an aperture within the laser resonator.

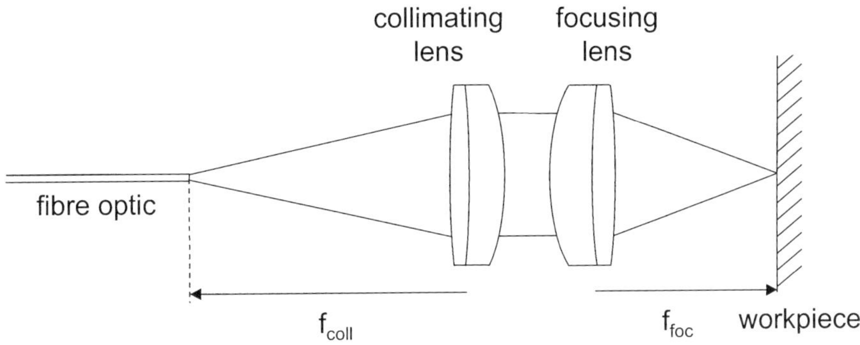

Figure C4.3.7. Imaging optics used at the output end of a fibre optic delivery system.

With solid TIR fibres, the end-faces must be very flat and clean to avoid damage on coupling the light in and so are usually prepared by careful polishing to give a scratch-free surface. With small diameter fibres (outer diameter $<220\ \mu m$), however, high quality end-faces can be produced simply by cleaving with a special diamond tool—this technique is much quicker and easier and is the standard one used with telecommunications.

At the output end of the fibre, the light is imaged onto the workpiece with typical magnification ratios between 1:1 and 1:2. A standard arrangement is to use a pair of lenses, both operating at infinity—see figure C4.3.7. The fibre end-face is placed at the focal point of the first (collimating) lens, giving an image at the focal point of the second (focusing) lens with a magnification that is simply dependent on the ratio of the two focal lengths. The lenses are normally held in a rigid assembly (output housing) and the fibre is held in a connector which plugs into the back of this. For many machining operations, high-pressure gases are used to assist with material removal and these are introduced through a nozzle coaxial with the laser beam, which is also integrated into the output housing.

For alignment and inspection purposes it is often beneficial to image the workpiece co-axially with the laser beam. This can be achieved by placing a laser turning mirror or beamsplitter between the two lenses (figure C4.3.8). For this reason, although the laser light has a very narrow bandwidth, achromats are often used in the output housing. Thus allowing white light imaging to be used. In addition, they are well corrected for spherical aberration and are available at reasonable cost.

The main problem which can arise at the fibre output end is one of back-reflection of laser light from the workpiece. This is refocused by the optics onto the fibre end and can cause the fibre and mounting to heat up and even damage when using high average powers. To avoid this, the fibre output housing that contains these optics is often placed at an angle of perhaps 10° to the workpiece normal.

C4.3.4 Beam division and combination

Fibre optic beamsplitters are often used with single-mode optical fibres, in order to route light down more than one path and for (re-)combination of light from separate paths. These are normally fused tapered couplers, where a taper section is used to expand the optical field out of the core, in order that coupling can occur between the two fibres [22]. It is possible to make multi-mode versions but they have high loss and so are easily damaged with high powers. An alternative design is to use two polished 'half-couplers' in close contact but again losses are likely to be too high for high-power applications. If a high-power laser is required to be coupled into a number of fibres, it is, therefore, normal to split it into separate beams using bulk optics before coupling into the optical fibres. Three techniques are commonly used: (i) high reflectivity mirrors partially

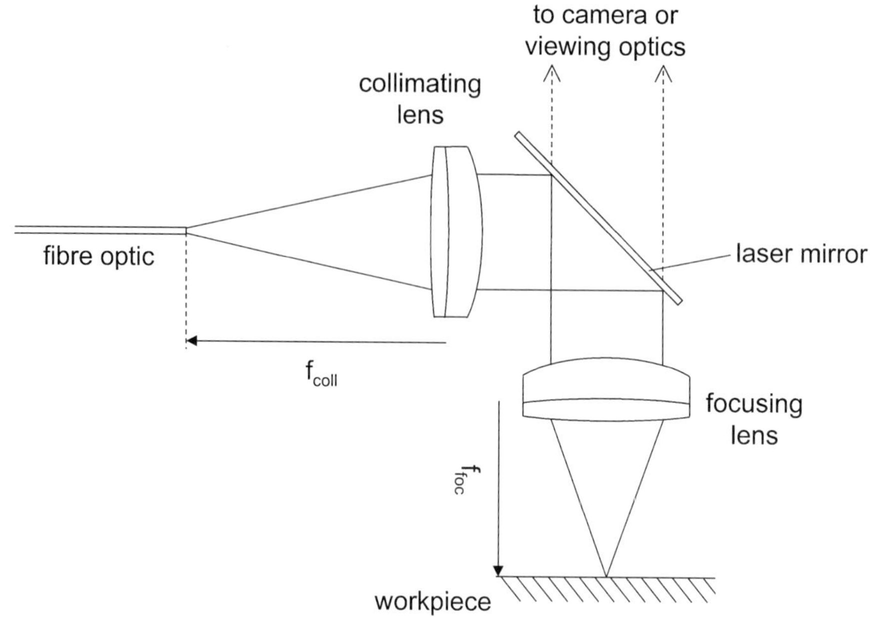

Figure C4.3.8. Combined high power laser imaging and viewing optics at the output end of a fibre optic delivery system.

inserted into the beam; (ii) partial reflectors; or (iii) diffractive optical elements. In addition to splitting light into several fibres, there are some very high power applications where two or three lasers have been combined into a single fibre. Such a device has been developed [23] to combine light from three 4 kW Nd:YAG lasers.

C4.3.5 Limitations

Despite the advantages of fibre optic beam delivery described here (flexibility, multiplexing), the implementation of fibre optics for delivery is often limited by optical damage and degradation of beam quality and/or beam profile. Changes in beam quality or profile are strongly dependent on the type of fibre (TIR or hollow waveguide), its dimensions and the coupling optics used at the input end. Optical damage, meanwhile, is dependent both on the peak power and the temporal profile of the laser; for long pulses (milliseconds up to cw), the main damage problems are thermal, associated with the high average laser power, whereas with short pulses (nanoseconds), damage occurs at much lower average powers and is limited by non linear processes in the fibre such as self-focusing.

C4.3.5.1 Beam quality and profile

It is important to consider how the properties of the beam are preserved through the fibre. As indicated in the introduction, there is really no problem with silica fibres for low beam quality applications such as welding, where the loss of beam quality on transmission through the fibre is very small. Unfortunately, for applications requiring higher beam quality, such as cutting, drilling and marking, degradation of both beam quality and profile can be a severe problem. Beam quality degradation arises partly because such beams do not match well to the mode structure of a step-index fibre (whereas the 'top-hat' profile typical of low beam quality is quite suitable) and also because mode coupling occurs as the light propagates through the fibre.

A rough measure of beam quality is the M^2 parameter, which as described in section C4.1.3 is proportional to the product of beam waist and divergence angle. The beam waist at the fibre output will always be equal to the fibre radius, irrespective of the spot size at the input of the fibre. Similarly, even if a low NA beam is launched into the fibre, the process of mode coupling (e.g. when the fibre is bent) will cause the NA of the guided beam to increase, giving an increased divergence at the fibre exit [24]. Thus, the quality of the output beam is always less than that of the input. Consequently, high quality beams must be guided by small core fibres.

Light propagates through a fibre in a number of discrete modes and laser light output from these modes will interfere at the output of the fibre, creating a 'speckle pattern'. With a low beam quality, large core fibre (600 μm diameter), there are so many modes excited (>75 000) that these speckles are too small to be resolved from the perspective of the process. However, the smaller fibres required by higher beam quality lasers have significantly fewer modes and so the speckle interference can become quite severe (and random!) to the extent of affecting the process. Such problems have been observed both with silica fibres for Nd:YAG precision machining [25] and hollow waveguide fibres for CO_2 marking [9]. One way of improving this is to excite equally all the modes within a particular numerical aperture, which can be achieved by increasing the fibre length or introducing controlled bends. A better solution is to restrict the fibre core size sufficiently so that only one mode will propagate; such fibres are described as 'single mode' and are commonly used in telecommunications applications. Unfortunately, however, to obtain efficient coupling from the laser into the fibre, the laser must have an $M^2 \sim 1$. As a demonstration, light delivered through a 200 m length of single-mode fibre has been used to cut 0.2 mm thick stainless steel sheet at 4 mm s^{-1} [26].

C4.3.5.2 Thermal damage

With hollow waveguide fibres, thermal damage can occur simply due to the attenuation of the fibre, since a small fraction of the light is lost and converted to heat on each Fresnel reflection. Since higher order modes are attenuated more quickly, this heating effect is normally greatest close to the input end. In addition, care must be taken to ensure that the focused laser beam does not strike the tube end-face, so an aperture is normally placed just in front of the fibre to ensure that light can only go *inside* the tube. An alternative solution proposed by Matsuura *et al* [27] is a launching coupler consisting of a tapered section of hollow waveguide. In order to allow high power transmission, hollow waveguide fibres are often water cooled [5], and, in this way, average powers of up to 1040 W have been transmitted through a hollow glass fibre.

In TIR fibres, by contrast, the guiding mechanism is loss-less, and fused silica fibres have very low intrinsic absorption and so no heating problems occur due to guided light. Problems can arise, however, on coupling the light into the fibre. The laser must be accurately focused onto the fibre core: it is important not to overfill either the core diameter or the NA of the fibre. If this is allowed to happen, a proportion of the light is coupled into the cladding, and damage can then occur by absorption in a material in optical contact with the cladding, for example the glue used to mount the fibre. Provided the focusing condition can be satisfied, however, there is the potential to significantly increase the average power transmitted, compared to that used in current applications. Experiments with an 8 μm core diameter fibre [26] have shown that the maximum long-pulse (0.1 ms) power which can be transmitted is 300 W. If this is scaled up to a more typical 600 μm fibre, and assuming that the power handling capacity scales with area, it implies that a power level of 1.7 MW is transmissible. It may, however, be necessary to take into account the optical damage $1/d$ scaling law described by Wood [28] (in some regimes optical damage threshold scales with $1/d$ rather than $1/d^2$ as might be expected) but even so a transmitted power of at least 23 kW should be possible.

With both hollow and TIR fibres, very sharp bends can also lead to damage. Sharp bends in a TIR fibre will couple light into the cladding, whereas in a hollow fibre they will couple light to higher order modes, which are then strongly attenuated. Appropriate cabling designs are, therefore, used to prevent this.

C4.3.5.3 *Pulsed laser damage*

Optical damage in silica fibres is much more of a limiting factor with short pulses (tens of ns). In this case the damage threshold can be orders of magnitude lower than expected from measurements with silica blocks—nominally the same material! Rather than the damage occurring at the fibre end-face, it typically occurs a few mm inside the fibre and is believed to be associated with self-focusing. The damage threshold can be significantly increased if the laser light is coupled to a greater range of fibre modes, reducing the likelihood of self-focusing [29]. This may be achieved by using a diffractive optical element (DOE), which is designed to reduce controllably the laser beam quality and match it to that of the fibre. By doing this, the damage threshold may be significantly increased and coupled with careful end-face polishing, the damage threshold becomes similar to that predicted by measurements on fused silica blocks—7.3 GW cm^{-2}, corresponding to 51 J cm^{-2} in a 7 ns pulse [30]. However, in a 200 μm fibre, this is still only equivalent to 16 mJ.

With the longer pulses (ms) used for laser percussion drilling, a different kind of damage is observed, which originates at the input end-face. The threshold for this damage has been shown to be related to the cleaving and polishing processes used to prepare the fibre end, which can leave surface scratches and sub-surface damage. Kuhn *et al* [31] developed a technique using a CO_2 laser to 'anneal' this damage, and demonstrated an increase in damage threshold intensity of almost two orders of magnitude. This means that it is now possible to deliver suitable pulse energy and beam quality through 400 μm diameter fibres for drilling holes of a few hundred micrometres diameter in a thickness of a few millimetres.

C4.3.5.4 *Nonlinear effects*

In addition to the self-focusing effect described earlier, two other non linear effects, stimulated Raman scattering (SRS) and stimulated Brillouin scattering (SBS), are also observed in fused silica optical fibres with a high peak power. For a full explanation see '*Nonlinear Fiber Optics*' by Agrawal [32]. Damage is not a problem with either of these processes, instead the light is strongly coupled to other wavelengths. Brillouin scattering is an interaction of the laser beam with an acoustic wave, which couples the power into a frequency-shifted beam travelling in the opposite direction to the laser, thus greatly attenuating the light emerging from the output end of the fibre. Raman scattering, meanwhile, is the scattering of a photon from a molecule, with a change in vibrational energy level of the molecule and a consequent increase in the wavelength of the light. With SRS, the scattered beam normally travels in the same direction as the original beam. If a sufficient length of fibre is used, further scattering of the scattered light to longer wavelengths is observed, resulting in a spectrum with a number of broad peaks [33]. Since the scattered light propagates in the same direction as the original laser beam, SRS is not such a problem as SBS in wavelength-sensitive applications, as the power will still emerge from the far end of the fibre. However, the focusing optics at the output end must be free from chromatic aberration over the range of wavelengths produced.

The intensity thresholds for both SRS and SBS processes are dependent on the fibre composition, with a linear dependence on the fibre length. SBS is also dependent on the linewidth and pulsewidth of the laser. SBS has a lower threshold with cw laser light, whereas SRS dominates for short pulses—with Nd:YAG pulses, SRS has been shown to dominate even with a pulse length of 0.1 ms in a single-mode fibre [26]. In current fibre optic delivery systems, however, the fibre length is normally quite short, e.g. 10 m, so the threshold intensities are very high, $\sim 8.6 \times 10^8$ W cm^{-2}. With a 600 μm core diameter fibre this corresponds to a peak power of 2.4 MW, much lower than the 4 kW cw laser power transmitted for welding applications. Based on these results, even for a drilling application where 30 kW peak power pulses are required through a 200 μm diameter fibre, SRS is only likely to become an issue if the fibre length exceeds 90 m.

C4.3.5.5 Mechanical damage

A fused silica optical fibre is very strong provided that there are no scratches on its surface. To prevent such mechanical damage, a coating is placed on the fibre as it is pulled from the preform, before it comes into contact with any other object. Such a coated fibre can be bent to a very tight radius without damage (50 mm with a 1000 μm diameter fibre). Hollow glass waveguides are also based on fused silica and so have similar properties. To make the fibres suitably robust for an engineering environment, the fibres are encased inside a suitable armoured cable structure.

C4.3.6 Summary and future directions

Fibre optics have found wide acceptance for high power laser beam delivery and, in applications such as Nd:YAG laser welding, have become standard components. Hollow waveguide fibres are also becoming a practical proposition for delivery of high-power CO_2 light, with low-loss glass hollow waveguides now available. However, there is a range of applications where current fibres are still unsuitable, including welding using high-power (multi-kW) CO_2 lasers and Nd:YAG micromachining where high power short pulses and high beam quality are required. These applications are still limited by damage problems.

One exciting advance in recent years which could have implications for the delivery of short pulse Nd:YAG light is the development of microstructured or PCF optical fibres, as described in C4.3.1. In particular, the air-guiding version holds promise as a means of delivering high peak power Nd:YAG laser pulses with good beam quality. New versions of hollow waveguide fibres designed for operation at 1.064 μm also have potential for this application, with low loss delivery of 158 mJ, 10 ns pulses already demonstrated [21, 34].

There have been significant developments in hollow waveguide fibres in recent years in the mid-IR, and research in this area remains very active. Current fibres have a loss which is acceptable for many applications but their sensitivity to bending remains a problem—I would suggest that the major challenge that remains to be addressed with these fibres, therefore, is that of reducing differential modal attenuation, which is the primary reason for this sensitivity.

References

[1] Harrington J A 2000 A review of IR transmitting, hollow waveguides *Fiber and Integrated Optics* **19** 211–17

[2] Nishii J, Morimoto S, Inagawa I, Iizuka R, Yamashita T, Yamagashi T 1992 Recent advances and trends in chalcogenide glass fiber technology: a review *J. Non-cryst. solids* **140** 199–208

[3] Artjushenko V G, Butvina L N, Vojtsekhovsky V V, Dianov E M, Kolesnikov J G 1986 Mechanisms of optical losses in polycrystalline KRS-5 fibers *J. Lightwave Technol.* **LT-4** 461–5

[4] Gregory C C, Harrington J A 1993 Attenuation, modal, polarization properties of $n < 1$, hollow dielectric waveguides *Appl. Opt.* **32** 5302–9

[5] Nubling R K, Harrington J A 1996 Hollow-waveguide delivery systems for high-power, industrial CO_2 lasers *Appl. Opt.* **34** 372–80

[6] Abel T, Hirsch J, Harrington J A 1994 Hollow glass waveguides for broadband infrared transmission *Opt. Lett.* **19** 1034–6

[7] Bhardwaj P, Gregory O J, Morrow C, Gu G, Burbank K 1993 Performance of a dielectric-coated monolithic hollow metallic waveguide *Mater. Lett.* **16** 150–6

[8] R Dahan, J Dror, N Croitoru 1992 Characterization of chemically formed silver iodide layers from hollow infrared guides *Mater. Res. Bull.* **27** 761–6

[9] Su D, Somkuarnpanit S, Hall D R, Jones J D C 1995 Hollow core waveguide for high quality CO_2 laser beam delivery: exploitation of bend-induced mode coupling *Opt. Commun.* **114** 255–61

[10] Matsuura Y, Miyagi M, German A, Nagli L, Katzir A 1997 Silver-halide fiber tip as a beam homogenizer for infrared hollow waveguides *Opt. Lett.* **22** 1308–10

[11] Matsuura Y, Abel T, Harrington J A 1995 Optical properties of small-bore hollow glass waveguides *Appl. Opt.* **34** 6842–7

[12] van Dewoestine R V and Morrow A J 1986 Developments in optical waveguide fabrication by the outside vapor deposition *Proc. IEEE J. Lightwave Technol.* **4** 1020–5

[13] Murata H 1986 Recent developments in vapor phase axial deposition *IEEE J. Lightwave Technol.* **4** 1026–33

[14] Nagel S R, MacChesney J B and Walker K L 1985 *Modified Chemical Vapor Deposition Optical Fiber Communications: Fiber Fabrication* vol 1, ed T Li (New York: Academic)

[15] Lydtin H 1986 PCVD: A technique suitable for large scale fabrication of optical fibers *IEEE J. Lightwave Technol.* **4** 1034–8

[16] Itoh K, Miura K, Masuda M, Ikwakura M and Yamagashi T 1991 Low-loss fluorozirco-aluminate glass fiber *Proc. 7th Int. Symp. Halide Glass* pp 2.7–2.12 (North Holland: Elsevier)

[17] Miyagi M, Hongo A, Aizawa Y and Kawakami S 1983 Fabrication of germanium-coated nickel hollow waveguides for infrared transmission *Appl. Phys. Lett.* **43** 430–2

[18] Croitoru N, Dror J and Gannot I 1990 Characterization of hollow fibers for the transmission of infrared radiation *Appl. Opt.* **29** 1805–9

[19] Haan D J and Harrington J A 1999 Hollow waveguides for gas sensing and near-IR applications specialty fiber optics for medical applications *Proc. SPIE* **3596** 43–9

[20] Matsuura Y, Abel T and Harrington J A 1995 Optical properties of small-bore hollow glass waveguides *Appl. Opt.* **34** 6842–7

[21] Abe Y, Matsuura Y, Shi Y, Wang Y, Uyama H and Miyagi M 1998 Polymer-coated hollow fiber for CO_2 laser delivery *Opt. Lett.* **23** 89–90

[22] Payne F P, Hussey C D and Yataki M S 1985 Modelling fused single-mode-fibre couplers *Electron. Lett.* **21** 461–2

[23] Olivier C A, Hilton P A and Russell J D 1999 Materials processing with a 10kW Nd:YAG laser facility *Proc. ICALEO '99* pp D233–41 (Orlando FL: Laser Inst. of America)

[24] Kuhn A, Blewett I J, Hand D P and Jones J D C 2000 Beam quality after propagation of Nd:YAG laser light through large-core optical fibers *Appl. Opt.* **39** 6754–60

[25] Hand D P, Su D, Naeem M and Jones J D C 1996 Fibre optic high quality Nd:YAG beam delivery for materials processing *Opt. Eng.* **35** 502–6

[26] Hand D P and Jones J D C 1998 Single-mode fibre delivery of Nd:YAG light for precision machining applications *Appl. Opt.* **37** 1602–6

[27] Matsuura Y, Hiraga H, Wang Y, Kato Y, Miyagi M, Abe S and Onodera S 1997 Lensed taper launching coupler for small-bore, infrared hollow fibers *Appl. Opt.* **36** 7818–21

[28] Wood RM 1997 Laser induced damage thresholds and laser safety levels. Do the units of measurement matter? *Opt. Laser Technol.* **29** 517–22

[29] Sweatt W C and Farn M W 1993 Kinoform/lens system for injecting a high power laser beam into an optical fiber *Proc. SPIE* **2114** 82–6

[30] Maier R R J, Hand D P, Kuhn A, Blair P, Taghizadeh M R and Jones J D C 1999 Fibre optic beam delivery of nano-second Nd:YAG laser pulses for micro-machining *Proc ICALEO '99, Laser Microfabrication Conf. (San Diego, CA)* pp 204–18 (Orlando FL: Laser Inst. of America)

[31] Kuhn A, French P, Hand D P, Blewett I J, Richmond M and Jones J D C 2000 Preparation of fibre optics for the delivery of high-energy, high-beam-quality Nd:YAG laser pulses *Appl. Opt.* **39** 6136–43

[32] Agrawal G P 1989 *Nonlinear Fiber Optics* (London: Academic) chs 8 and 9

[33] Cohen L G and Lin C 1978 A universal fiber-optic (UFO) measurement system based on a near-IR fiber Raman laser *IEEE J. Quantum Electron.* **QE-14** 855–9

[34] Sato S, Ashida H and Arai T 2000 Vacuum-cored hollow waveguide for transmission of high-energy, nanosecond Nd:YAG laser pulses and its application to biological tissue ablation *Opt. Lett.* **25** 49–51

C4.4
Positioning and scanning systems

Jürgen Koch

C4.4.1 Introduction

In the context of industrial material processing, the 'positioning system' incorporates the optical and mechanical techniques for focusing, directing and controlling a high-powered laser beam, focused to a spot, to the coordinates of a specific location [position] on a workpiece. Either the beam or the workpiece is moved to achieve this objective; both can also be moved relative to each other to provide a more compact or a more flexible system. The 'scanning system' comprises the optical, acousto-optical, electro-optical and opto-mechanical techniques for controlling and scanning a laser beam, focused to a spot, across a surface that is usually one- or two-dimensional, such as a bar code or a display screen. Such systems are referred to as beam delivery systems. Besides the positioning and scanning systems, the free-space and fibre optics mentioned in previous chapters are also part of beam delivery systems but predominantly their purpose is to locate the laser beam at a workstation and not at a workpiece.

C4.4.2 General requirements

The general requirements for positioning and scanning systems derive from the previous definitions. Pointing out the character of each system reveals its main property: positioning a laser beam on the coordinates of a specific location demands precision, while laser beam scanning necessitates speed. According to the specific application, these two properties are characterized by different parameters—see table C4.4.1. Furthermore, both systems involve motion, so the dynamical behaviour caused by certain forces also affects these parameters. Further demands are associated with the type of drive used in the positioning system or are related to the method of steering a laser beam in scanning systems. In addition, specific requirements are associated with particular applications. Table C4.4.1 lists several parameters classified according to the main application. This table is not exhaustive. Each developer of a beam delivery system for a specific application has to bear in mind the important parameters, but engineers may use it as a checklist.

It should be noted that different parameters cannot be chosen arbitrarily, because they are mostly mutually dependent. Furthermore, positioning and scanning systems are usually equipped with specific controllers, which principally influence the system performance by their settings. For a specific application, optimization of these settings is necessary to achieve the best performance. Thereby, the whole assembly has to be taken into account. Given the number of possible different set-ups, specific types of controllers cannot be mentioned in this chapter but general remarks related to this subject are presented where necessary.

An example taken from laboratory practice will illuminate some aspects regarding the general requirements for beam delivery systems for laser material processing—some of this has been reported in [1]:

> A feasibility study on the manufacture of multiple-hole metal sheets, which form part of an active system for the reduction of aerodynamic resistance in aircrafts, has been carried out. The

Table C4.4.1. Considerations for beam delivery systems.

Purpose of application	Main parameters
Motion	Traverse path, angle of deviation, velocity.
Accuracy	Stepwidth, resolution, accuracy of position, drift hysteresis.
Dynamic	Accuracy of path, acceleration, deceleration, rise time, resonance frequency, transient response, heating.
Repetition	Repeatability, drift.
Forces	Load and stress, stiffness.

application's requirements were to drill thousands of holes perpendicular to the surface of flat aluminium or titanium 1 mm thick in an area of up to 200 mm × 200 mm with a distance of 0.5 mm between adjacent holes. Each hole had to have a diameter of 50 μm and processing time should not exceed 20 ms, including positioning.

A high-frequency (pulse repetition rate of approximately 10 kHz), short-pulsed (pulsewidth 30 ns) copper vapour laser (CVL) system was applied to perform the task. Hence, the drilling of a single hole had to be done by percussion drilling, i.e. the use of several consecutive pulses for each drilling and, hence, the positioning of each hole had to be done step by step.

This example—we will come back to it later—demonstrates only a few of the possible requirements placed upon a beam delivery system by a specific application.

First, *the laser system*: drilling holes with a diameter of 50 μm necessitates the use of a laser source with high beam quality and the pulses should also be of high energy, according to the amount of material that has to be ablated. For the processing time, a highly repetitive system has to be chosen. Nowadays, pulsed Nd:YAG systems would fit very well but at the time the investigations were carried out the copper vapour laser (CVL) was the best choice. More details about laser systems are given in part B of this handbook.

Second, *workpiece positioning*: if we neglect the necessity of obtaining the focus, i.e. placing the focused spot of the laser beam on the surface of the workpiece, the whole problem is two-dimensional. For this, taking into account the maximum size of the metal sheets, a coordinate table would do but, considering the distance between the adjacent holes and the processing time allowed for each drilling, a device with high speed—not velocity, but high values of acceleration and deceleration—is required. In addition, wear can play a role because of the many thousand repetitive positionings. Finally, vibrations caused by acceleration and deceleration have to be avoided.

As we can see, developing beam delivery systems starts with the laser source. The laser selected for an application limits the choice of usable positioning or scanning devices and *vice versa*. The wavelength, power, energy, mode of operation—continuous wave (cw) or pulsed—pulse duration and beam width of the designated laser have to be considered. Using mirror-based deflection systems, a wide range of different laser sources can be selected for many applications. Only the optical materials available for mirrors restricts our choice. The same applies to transmissive optics such as acousto-optical ones. The limitations on workpiece handling are even fewer because the positioning of the workpieces is independent of the laser. Hence, in the following, different types of beam delivery systems for specific laser sources are not treated in detail, rather certain devices are introduced and their advantages and limitations discussed. The basic principles of laser optics, techniques for focusing and projection as well as the choice of material for specific wavelengths are explained in parts A, and chapters C1.1 and C4.1, respectively. In addition, specific examples of beam delivery systems used in laser applications are given in part D of this handbook.

C4.4.3 Positioning systems

Positioning in the context of laser material processing means controlling and directing the focused spot of a laser beam to the coordinates of a specific location on a workpiece. Either the beam or the workpiece is moved to achieve this objective—both can also be moved relative to each other to provide a more complex or a more flexible system. Hence, the main purpose of a positioning system is motion, whether it be translation or rotation.

C4.4.3.1 Motion devices

The technique most commonly used in laser material processing, e.g. cutting, welding, structuring and micromachining, to perform translation is mechanical in type. Usually, a rotation is translated into linear motion by a ball screw. For this purpose, the rotation is achieved by mounting an electromagnetic drive to the shaft. The accuracy of such systems is determined by the precision of the screw shaft and the ball bearings used to fix and guide the shaft. Mounting two such axes together, one perpendicular to the other, facilitates two-dimensional motion. These systems are called coordinate tables due to the possibility of addressing a specific position by its coordinates (X and Y). The same set-up can be realized with linear drives, where magnetic forces move the axes. The use of magnetic drives solves an additional problem for two-dimensional motion: a parallel set-up of the two axes overcomes the problem of a different load on each axis if they are mounted one on the other. The system is driven by moving coils and is also equipped with air bearings. A comparison of the two different driving techniques reveals the specific advantages and disadvantages.

Back to the example. Different techniques were used to move the metal sheets linearly:

> At first, the stepping for positioning each hole was achieved by a screw-shaft-driven coordinate table. After optimization of the motion parameters, the positioning of a single hole already exceeded the required processing time by a factor of five. In addition, repetitive positioning of many thousand holes resulted in wear of the screw shaft, in turn causing unreasonable deviations in position.
>
> In the next step, a magnetically driven motion system with a parallel set-up for the axes was used. Hence, the stepping time was reduced by half but the drillings appeared somewhat elliptical. A close investigation of the whole system revealed that vibrations in one axis of the coordinate table's base frame, caused by high values of the acceleration and deceleration in combination with a weakness in the substructure in one direction, were responsible for the anomaly in shape. By just changing the main direction of motion, circular holes were once again processed but time was still a problem.

The differences in positioning time results from different driving forces. A mechanical system is very stiff because of the use of a screw shaft and ball bearings. Hence, high velocity can be achieved but as a serial assembly—the Y-axis is mounted on the X-axis— a large mass has to be moved and friction becomes noticeable. Hence, acceleration and deceleration, the main parameters for stepping small distances, are limited. In addition, repetitive positioning often results in wear of the screw shaft or bearings.

Using a parallel set-up of magnetically driven axes with air bearings reduces friction. Thus, wear is not observable; moreover, higher values of acceleration and an increase in deceleration are achieved. However, due to the mechanically contact-free assembly, the system shows no ideal stiffness—see also [5]. Hence, the load on the axes has to be taken into account because the resulting transient response, i.e. the behaviour of the drive until it reaches the desired position, depends on the mass mounted on the system. Furthermore, the noticeable vibrations at least demonstrate that at elevated rates of acceleration and deceleration, the environment also has to be taken into account.

Table C4.4.2. Comparison of the accuracy of different driving techniques.

System	Drive mechanism/ bearing	Field size (mm^2)	Position accuracy full stroke (μm) at	Velocity ($mm\ s^{-1}$)	Angular deviation (μrad)	Resolution (μm)
[6]	Ball screw/hydrostatic	200 × 200	0.05	—[b]	2.4	0.005
LPKF HS 8 GP[a]	Moving coil/air bearing	200 × 200	<2	up to 300	—	0.25
LPKF XY 60 G[a]	Ball screw/air bearing	600 × 600	±3	155	24	0.5

[a] Taken from supplier's catalogue.

[b] According to its use as a high-precision device, Kami did not report on velocity investigations. Besides, an increase in accuracy can also be achieved by sensor technology adapted to the specific device—see further reading [14, 15].

Table C4.4.2 presents a comparison of the achievable accuracy of different driving techniques—hydrostatic ball screw, moving coil with air bearings and a ball screw with air bearings—for use in micromachining, e.g. with excimer lasers, Q-switched Nd:YAG lasers and ultrashort-pulsed laser systems.

Coordinate tables are not limited to linear motion: curved lines can also be realized, dividing the path into small consecutive parts of linear motion, thus limiting the accuracy of curved lines to the stepwidth of the system, i.e. the smallest range of motion possible. Hence, applications necessitating rotation only employ a rotation axis. The power transmission of these systems is performed directly, as in ball-screw-driven devices, or indirectly with cogwheels, chains or belts.

Finally, the combination of several axes facilitates processing in three dimensions. However, the motion sequences necessary to locate a distinctive point in space have first to be determined.

C4.4.3.2 Kinematics

In general, the positioning of an optic, hence a laser beam, for 3D processing tasks requires a minimum of five degrees of freedom regarding motion. The definition of a particular point in a workspace needs at least three coordinates of translation (usually called the X-, Y- and Z-axes) and another two of rotation (B- and C-axes) with respect to an orientation at that point—compare figure C4.4.1. Differences in the realization of 3D motions are correlated to the type of movement chosen (translation, rotation) and to the sequence in which the axes move.

The gantry systems with translation axes for the main directions of motion as shown in figure C4.4.1 are an example for a 3D assembly with a Cartesian solution. In a Cartesian coordinate system, it is possible to compute all the axial positions necessary to locate a point in space. The coordinate axes are linearly independent.

In contrast, industrial robots which use a combination of rotation axes, as shown in figure C4.4.2, demand a minimum of six rotation axes to achieve the same performance as that of gantry systems. Furthermore, the coordinate axes of the kinematic chain are not linearly independent due to the superposition of non-Cartesian coordinate systems. The same point in space can be accessed with the elbow up or elbow down. Thus, increased effort has to put into the controlling tools—see [16]—and position measuring becomes inevitable.

C4.4.3.3 Measuring devices

The measurement of positions in linear devices, such as gantry systems and coordinate tables, is carried out by linear measuring scales. The advantage of these measuring systems is that the actual value of the

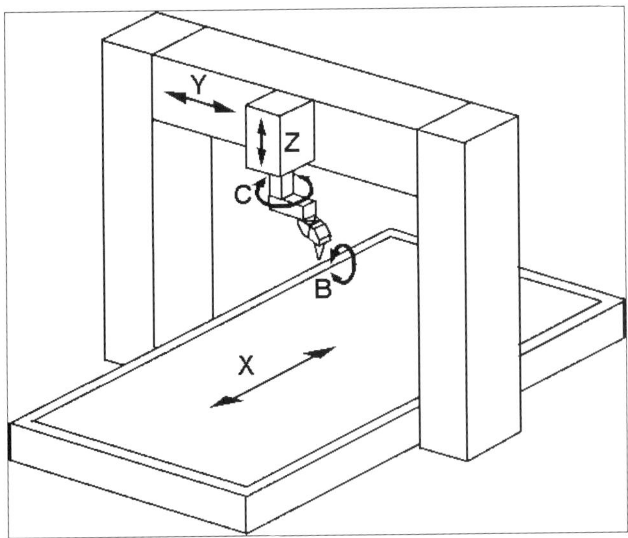

Figure C4.4.1. Gantry system (reproduced by permission of Trumpf, Germany).

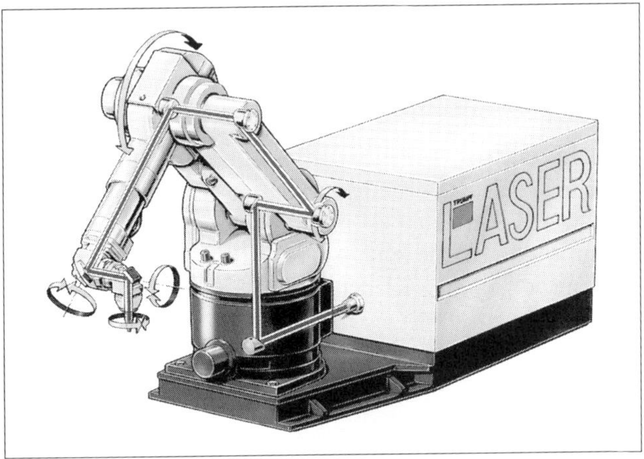

Figure C4.4.2. Robot (reproduced by permission of Trumpf, Germany).

position is measured directly. For very precise requirements, e.g. in micromachining with excimer lasers, the measuring scale is made of glass due to its lower thermal expansion coefficient. When linear motion is performed by a ball screw, encoders are usually applied. Their purpose is to transform the number of rotations into linear scaling and *vice versa*. Therefore, the measurement of the position of the axes is indirect. Comparable systems are used in rotation axes. Errors due to the low stiffness of the axis, fetch in the power train or overload of the axis are not registered. Moreover, the errors add up when chain-linking several axes, for example to achieve advanced systems for three-dimensional processing.

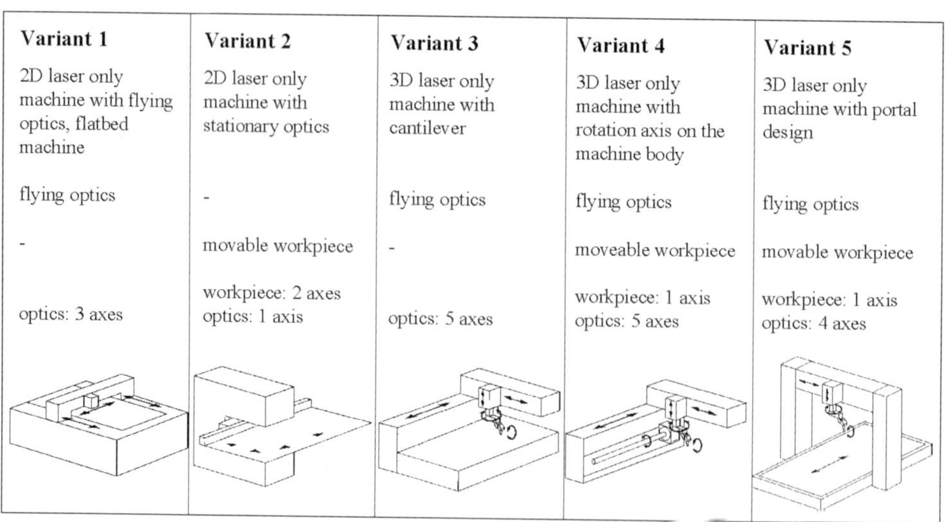

Variant 1	Variant 2	Variant 3	Variant 4	Variant 5
2D laser only machine with flying optics, flatbed machine	2D laser only machine with stationary optics	3D laser only machine with cantilever	3D laser only machine with rotation axis on the machine body	3D laser only machine with portal design
flying optics	-	flying optics	flying optics	flying optics
-	movable workpiece	-	moveable workpiece	movable workpiece
optics: 3 axes	workpiece: 2 axes optics: 1 axis	optics: 5 axes	workpiece: 1 axis optics: 5 axes	workpiece: 1 axis optics: 4 axes

Figure C4.4.3. Overview on gantry systems (reproduced by permission of Trumpf, Germany).

C4.4.3.4 Advanced positioning systems and optical set-ups

The introduction of relative motion between workpiece and tool—here a laser beam—can be accomplished by moving the workpiece or the tool; both can also be moved relative to each other.

Gantry systems feature large mass as well as vibration-restricted constructions. Therefore, high processing and positioning velocities are achieved, even when considering the motion of large mass. A set-up with a fixed workpiece—examples 1 and 3 in figure C4.4.3—denotes a definite position for the workpiece, but a complex beam delivery system with movable optical parts—also called flying optics, see chapter C4.2—has to be used to locate the radiation at the optics needed for processing. Moving the workpiece during processing results in a reduction in the specifications for the beam delivery system and saves space but these are mostly limited to small and lightweight components. For precision processing, a hybrid solution where the workpiece moves along one direction—example 5 in figure C4.4.3—is more convenient. Figure C4.4.3 also demonstrates some assembly examples for 2D and 3D machining, equipped with different numbers of axes.

Fixed workpieces can also be accessed by industrial robots. Such robots are available with internal beam delivery systems, as shown in figure C4.4.2, or external beam supplies, comparable to articulated arms—see chapter C4.1—thus increasing flexibility. The internal beam supply demands careful adjustment of the optical components, because the beam has to be directed exactly along the rotation axes. Therefore, fibre optics—see chapter C4.3—are in use where applicable.

A current method to widen workspaces is to combine translation and rotation motion units. Hence, additional axes, for example, an assembly of industrial robots on linear motion devices and workpiece holders which are rotate-and-pivot units, allow processing at areas with difficult access but beam delivery and system control are more expensive.

Furthermore, the growing use of process control elements, e.g. [3], controlling accuracy during processing [2], optical focus [3] and feedback control systems, increases the precision of positioning systems and is steadily enhancing the complexity of beam delivery systems. Already available means—like adaptive optics, see chapter C1.3—designed for this purpose can also be used.

Table C4.4.3. Comparison of 3D-positioning systems.

Assembly	Accuracy of position (mm)	Accuracy of path (mm)	Investment costs (€)
Robot	±0.05–±0.2	±0.05–±1.4	50 000–70 000
Gantry system	±0.03–±0.05	±0.1–±0.2	>200 000

C4.4.3.5 Influences on beam propagation

The dimensions of the workspace and, therefore, the length of free-space beam propagation (the sum of the X-, Y- and Z-directions) influence the focability of the beam. In particular, the movable parts in the flying optics (see chapter C4.2) need careful adjustment, because they can greatly modify the length of beam propagation, whereby the quality of the processing varies across the workspace. Affected by the divergence, the beam width changes and, hence, the irradiance distribution which depends on the position of the working optics along the beam axis. Deviations in the position of the focus relative to the working optics can be corrected with telescopes, adaptive optics or mechanical systems for compensating the beam propagation length.

C4.4.3.6 Comparison of 3D-positioning systems

Hoffmann recently compared the accuracy and investment cost of 3D positioning systems, e.g. those used for hard soldering in the automobile industry [4]—see table C4.4.3.

Differences in investment costs derive from the complexity and accuracy of the various systems. For example, robots, such as those shown in figure C4.4.2, compared to gantry systems are a less expensive option but they do not give such good guidance. In addition, robots are more sensitive to misalignment and the loss in laser power is higher because of the necessary number of mirrors. The main advantage of robots is higher flexibility in positioning. With these systems, even working areas which are difficult to access can be reached, especially if the less expensive internal beam delivery is exchanged for an external beam supply.

C4.4.4 Scanning systems

In general, two different groups of scanning systems can be distinguished. In reflective systems the laser beam scans by means of mirrors, which are usually mounted to torsion or rotation drives. Galvanometer scanners, polygonal scanners and piezoelectrically driven trepanning heads belong to this group. In contrast, transmissive systems, like acousto-optical (AO) and electro-optical (EO) devices, deflect the laser beam while it is being transmitted through a crystal. The optical properties of the crystal are changed by means of acoustic or electric waves applied to the crystal. Some essential aspects of AOs are mentioned in the corresponding section of this chapter. EOs are neglected here as they are rarely used in beam delivery systems for laser material processing. However, they are considered in other literature.

In principle, positioning systems can be applied to scanning operations. However, the nature of scanning processes is a high velocity with an adapted accuracy. So, some positioning devices may also be used in scanning systems and *vice versa*. For this to happen, only the accent has to be changed. Piezoelectric actuators are the prime example here, as will be shown later. In general, scanning devices are designed to meet a particular task. Reflective scanning systems are usually used in industrial laser material processing. The basic requirements concerning reflective optics were explained in chapter C1.

The benefit of scanning devices for use in industrial applications, such as materials processing or microstructuring, can be shown with the example of perforating metal sheets, mentioned previously.

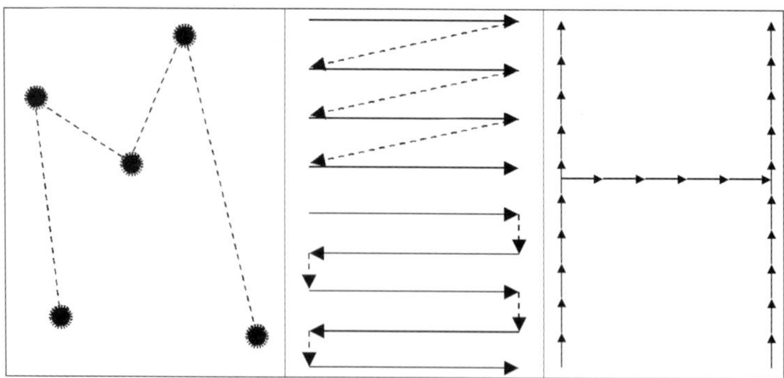

Figure C4.4.4. Step (a), raster (b) and vector (c) scanning.

After several optimization steps, first with a ball-screw-driven coordinate table, and then applying a moving-coil driving mechanism, the time necessary to position each hole was still a problem. It was finally solved by the implementation of a fast laser beam deflection system. To be more precise a galvanometer scanner due to its large aperture for beam deflection equipped with a telecentric f–θ lens was chosen to position the holes accurately in an area 25 mm × 25 mm and to achieve appropriate focusing of the radiation. After processing such a patch, the metal sheet was translated to be worked on the next patch. That way, the pre-determined requirement (a processing time of less than 20 ms, including positioning) was fulfilled.

This example shows that scanning systems can be successfully applied to laser material processing. Hence, as their use is not restricted to small step distances, several scanning methods are available.

C4.4.4.1 Scanning methods

Three different scanning methods are used in laser material processing according to the various applications (figure C4.4.4(a)–(c)).

The perforation of sheet metal in our example was finally carried out by *step scanning*: the laser beam is active at predefined positions and during positioning operations laser processing is suppressed—see figure C4.4.4(a).

In *raster scanning*, the scanning system sweeps the laser beam at a constant velocity over a fixed angular range in a repetitive manner, thus scanning a full image with a series of parallel lines. At the end of each line, the laser beam is turned off and the scanning device returns to the starting position. Where applicable, raster scanning is done bi-directionally, i.e. processing from the left to the right and *vice versa*—see the lower part of figure C4.4.4(b).

In *vector scanning*, the motion is separated into many small steps, indicated by the arrows drawing an 'H' in figure C4.4.4(c). A continuous track on the workpiece surface is made when the command waveform's frequency exceeds the scanning system's bandwidth. This technique enables writing (with a laser beam), e.g. letters or symbols, on the object surface when the scanning device moves the beam during laser operation. A new starting position is reached while the laser beam is deactivated (figure C4.4.4(c)).

The various scanning methods can be performed by transmissive devices as well as by reflective ones. However, in laser material processing, the laser beam is usually focused to a spot to perform the processing tasks. Hence, a lens is part of the scanning system's set-up. Due to the position of the focusing lens relative to the scanning device, two optical configurations can be specified.

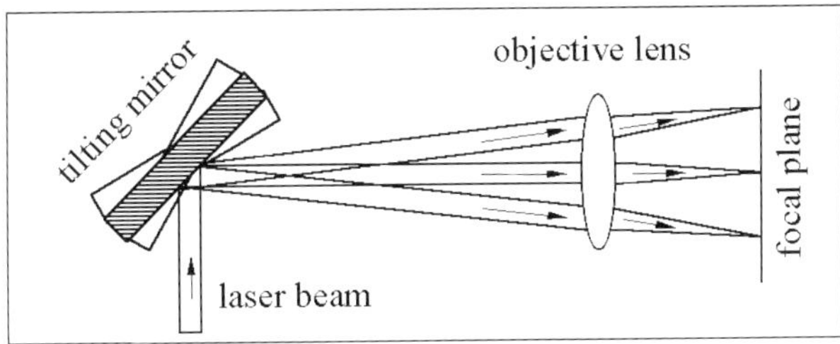

Figure C4.4.5. Pre-objective configuration.

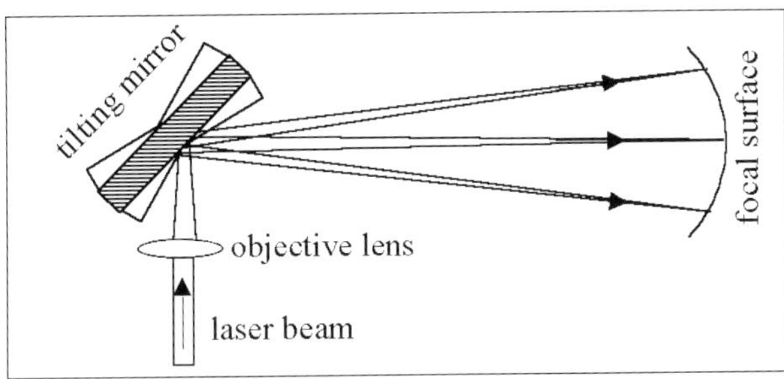

Figure C4.4.6. Post-objective configuration.

C4.4.4.2 Optical configurations of scanning systems

The focal spot size necessary for laser material processing, can be achieved by two types of optical configurations: a pre-objective one (figure C4.4.5), which is also applicable to AO devices and not restricted to reflective systems as depicted here; and a post-objective one (figure C4.4.6), which depends on the relative position of the scanning device with respect to the focusing objective lens.

In the pre-objective configuration, the laser beam scans before it is focused by the objective lens, while in post-objective systems the beam is focused first. When necessary, an additional lens for field correction is applied, as is the case with wire marking. Here, a mask is projected by means of a telescope, settled before a rotating mirror, and a specific lens corrects the focal plane afterwards, in order to perform sharp images of the bar codes, for example, when the wire is moved during processing.

The pre-objective configuration allows only limited angles due to the aperture of the objective; the post-objective configuration, in principle, can reach a full circle.

Two-dimensional scanning is achieved by the use of a second scanning device and the optical configuration can be pre-objective, as shown by the sketch in figure C4.4.7, or post-objective; a pre–post-objective configuration is also possible, when the objective lens is positioned between the two devices. The most common arrangement is one in which both scan units are in a pre-objective configuration. In general, f–θ lenses are applied to achieve a maximum flat field size h [7].

$$h = 2F\tan(\Theta) \qquad \text{(C4.4.1)}$$

with F the focal distance and Θ the optical scan half-angle.

As previously indicated, scanning operations are performed by transmissive or reflective devices. Hence, the means for deflecting a laser beam depend on the type of device chosen. Moreover, if using mirrors, the different types of driving systems have to be considered.

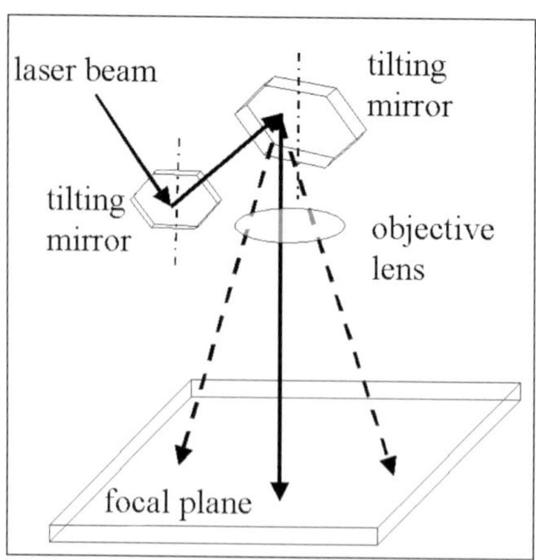

Figure C4.4.7. Pre-objective 2D scanning system.

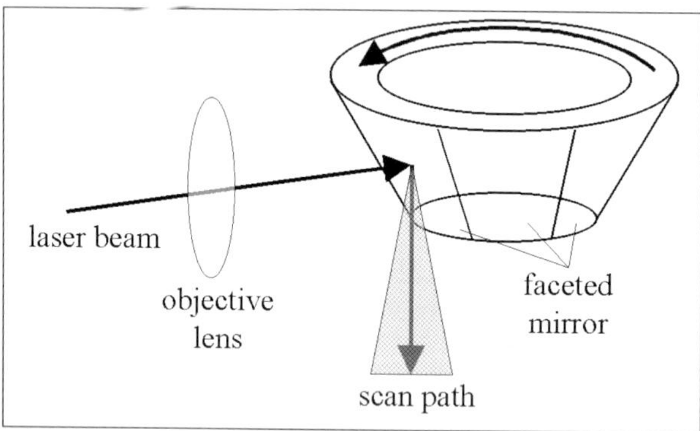

Figure C4.4.8. Diagram of a pyramid polygonal scanning system (post-objective configuration).

In industrial applications where large apertures due to the beam width of the applied laser systems, e.g. CO_2 lasers, are important, as in the heat treatment of metals, regular prismatic polygonal scanners or pyramid polygonal systems [8] move the laser beam across the workpiece's surface. Galvanometer-driven systems are used in laser welding applications, e.g. in the automotive industry [10], and are also employed for microstructure surfaces [9]. Furthermore, piezoelectric actuators are applied to drilling operations, e.g. in trepanning heads [11].

C4.4.4.3 Polygonal scanners

Polygonal scanners deflect the laser beam by means of a faceted mirror. Usually regular prismatic polygonal scanners, with the mirror facets placed on the periphery of a cylindrical wheel, are used but pyramid polygonal scanners, i.e. ones in which the mirror facets are placed on the surface of a truncated cone (figure C4.4.8), are applied as well, for example for heat treatment applications, although they are difficult to manufacture [8]. The advantage of the pyramid polygonal scanner compared to the regular prismatic scanner is that the scan width can be varied due to the height chosen for the incident beam on the cone.

Two-dimensional scanning is achieved by inserting an additional beam-deflecting device, e.g. a gal-

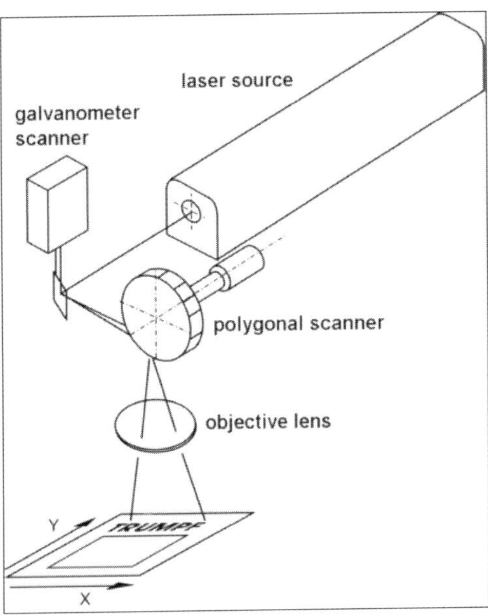

Figure C4.4.9. Regular prismatic polygonal scanners in a 2D assembly (pre-objective configuration) (reproduced by permission of Trumpf, Germany).

vanometer scanner. Hence, it is obvious that polygonal scanners are limited to raster scanning, as used in laser projection systems and laser-marking applications (figure C4.4.9).

The performance of polygonal scanners, i.e. the scan frequency (representing the number of repetitive scans), depends on the rotation drives of these scanners. Usually, they provide high scan frequencies f according to the rate N of the rotation axis and the number m of facets on the polygonal mirror:

$$f = Nm. \tag{C4.4.2}$$

Power transmission is performed directly, when the mirror wheel is fixed to the shaft of the axis, or indirectly, e.g. by belt.

Another way to perform raster scanning is by the use of scanning galvanometers but these are not limited to raster scanning.

C4.4.4.4 Galvanometer scanners

The scanning galvanometer is an electromagnetic actuator that moves a flat mirror (see figure C4.4.7) in order to deflect a laser beam in a controlled way and with high speed. Different galvanometer drives are used for laser beam scanning for industrial processes.

- *Resonant galvanometers* are driven at their resonant frequency, which limits their use to high-speed raster scanning. Drawing arbitrary paths or step scanning is not possible with these systems.
- *Moving magnet galvanometers* feature very low rotor momentum of inertia, in connection with the torque generated by them. Hence, they offer high accelerations and, in combination with efficient position sensors, they are applicable to all scan methods mentioned earlier.
- *Moving coil galvanometers* are mainly used due to their high and stable torque when the beam diameter exceeds 30 mm and, therefore, the mirrors have to be very large.

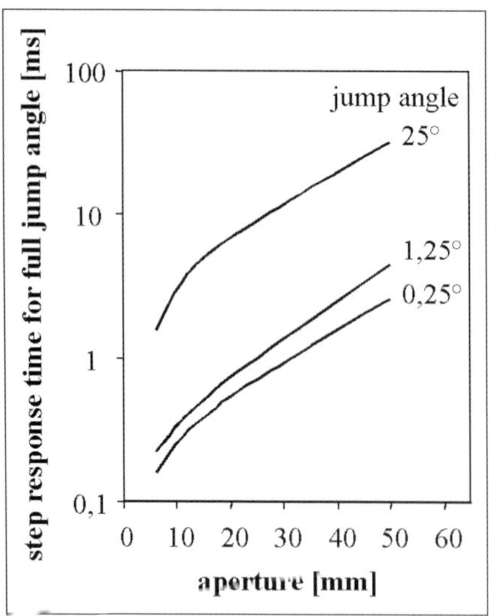

Figure C4.4.10. Step performance *versus* aperture size (reproduced by permission of Arges, Germany).

In any case, because of the high values of the tangential acceleration performed by a galvanometer, its shaft is mounted on high-precision, pre-loaded ball bearings. The shaft rotation has to be very smooth with very low friction to minimize positioning and tracking errors, at all ranges of working temperature. Indeed, some types of inertia effects cannot be neglected, as figure C4.4.10 shows. In step scanning, the time of response depends on the desired angle of deviation (jump angle) as expected and also on the diameter of the aperture. Larger apertures are used with larger beam widths; therefore, the mirrors applied to the scanner have to be larger—being heavier, the momentum of inertia is, thus, increased. Obviously, at smaller apertures (up to approximately 11 mm) when the mirrors are small and lightweight, the friction from the drive and the inertia of the rotation axis prevail over the inertia effects of the mirrors.

The limitation of the measurement of the response time to a maximum jump angle of 25° in figure C4.4.10 indicates the maximum mechanical deflection angle preferred in industrial material processing tasks.

A galvanometer drive for scanning purposes is usually equipped with a position sensor intimately linked to the shaft. The motor moves by means of a feedback amplifier controlled by the position sensor signal. In this way, the galvanometer shaft (which means the laser beam deflection) can be accurately positioned by injecting an electrical signal into the amplifier input or, if the electrical signal is time-dependent, the laser beam moves precisely following the applied signal. However, system performance and the accuracy of the galvanometer scanners depend on their driving force. The parameters reflect the oscillatory character of the electromagnetic drive.

C4.4.4.5 Performance and accuracy of scanning galvanometers

Usually, the performance and accuracy of scanning galvanometers are dependent on the scanning method used for a particular application. In addition, the focal distance of the objective lens also influences accuracy. Therefore, in the following, only a general view of these parameters with respect to the applied scanning method is given. It should be noted that the definitions of the parameters are valid for any scanning system but some are only specific to the use of galvanometers.

The performance parameter for step scanning is the *step time*, measured from the initial motion of the scanning device to a pre-defined position, within a settling tolerance, known as the transient response of the

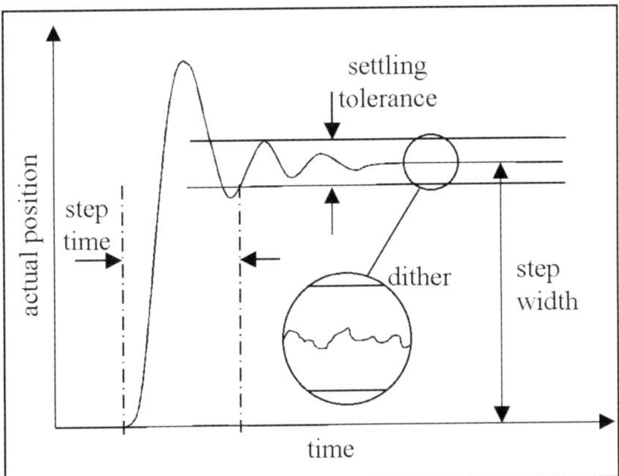

Figure C4.4.11. Step scan performance.

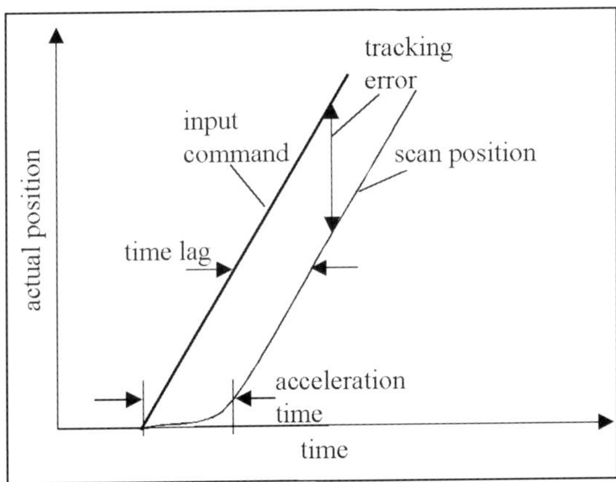

Figure C4.4.12. Vector scan performance.

device—see figure C4.4.11.

Besides heat, *dither*, i.e. a slight vibration of the shaft, caused by the noise of the power supply affects the accuracy of the position. Therefore, the power supplies used are specially designed for low noise output.

In raster scanning, the *scan frequency* is the important parameter which represents the number of repetitive scans. The retrace time, which is the time necessary to reposition the system from the end of one scan to the beginning of the next, limits the scan frequency.

Figure C4.4.12 describes the performance with vector scanning. Generally, a *tracking error* occurs due to acceleration of the scanning device. To minimize the tracking error, state-of-the-art systems offer a programmable 'head time' to overcome the acceleration effect in a particular application.

According to the desired accuracy in the position or path, three effects—overshoot, cross-scan error (also known as across-scan error or wobble) and in-scan error (also known as along-scan error or jitter)—should also be mentioned. Figure C4.4.13 displays the latter two errors in brief. Overshoot describes a movement that is farther than necessary. Hence, some applications overcome deceleration effects, such as an increase in the overlap of consecutive pulses at the end of a scan, to achieve higher performance or accuracy but only when the overshoot is programmable by an additional 'tail time' and appropriate laser control. Cross-scan

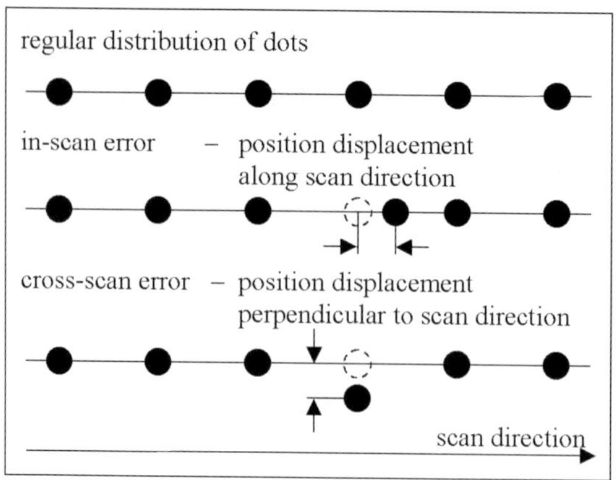

Figure C4.4.13. Position errors in scanning applications.

error and in-scan error are two positioning errors, perpendicular and parallel to the direction of motion, respectively. These arise from heating effects, calibration failures and power supply noise.

It should be mentioned that the performance of galvanometer scanners is generally affected by heating in dynamic applications. Therefore, most systems are available in a temperature-stabilized design where the system is heated to a value above the ambient temperature. However, working within a fixed coordinate system requires calibration of the scanning devices before a new process is started.

The *resolution* of galvanometer scanners as well as that of piezoelectric drives depends on the lenses used with the system; accuracy and linearity are also linked to this. In addition, the position sensor applied to the system is a very important item that has a considerable influence on the response time and accuracy.

The accuracy and electromechanical characteristics of a galvanometer–amplifier assembly enable high-speed positioning of the deflecting mirror with a resolution of the order of a micro-radian and a reaction time of a microsecond (figure C4.4.10), if we take into account the previously mentioned limitations. However, the lateral resolution depends on the focal distance of the lenses used for focusing the laser beam. Employing equation (C4.4.1), the lateral resolution r can be approximated by exchanging the optical scan half-angle α with half the minimum step width $\Delta\alpha$:

$$r = 2F\tan(\Delta\alpha). \qquad \text{(C4.4.3)}$$

To drive the motors, specialized software which runs on a PC is used. The input to this software is a datafile of a drawing. The motors are driven by means of an electronic card (path generator) that also switches the laser on or off.

Another aspect which has to be considered with respect to galvanometer scanners and their oscillatory nature and high values of acceleration is its mirrors.

C4.4.4.6 Mirrors in oscillatory scanning systems

Both the static and dynamic characteristics of scanning mirrors are important, especially in galvanometer scanners, where a flat mirror is fixed to a shaft. Inadequate mirror flatness increases the laser beam divergence. In addition, the mirror flatness can be degraded by rapid rotation of the mirror. In high-speed scanning systems, the mirror rotates with very high tangential accelerations (180×10^3 rad s^{-2}) and frequencies up to 500 Hz. Such accelerations strongly stress the mirror and elastically distort it or even make it vibrate. Therefore, the stiffness of the mirror is important. It has to be high enough to minimize distortion of the mirror. Furthermore, appropriate mirror stiffness prevents vibrations of the mirror as it rotates. Otherwise, the feedback electronic system is affected adversely and, especially when the vibrations match one of the mirror's resonant frequencies, the mirror will be destroyed. However, the mirror dimensions should not limit the laser beam diameter over the full angular field needed for the application.

In a pre-objective configuration, the width of the mirror along the rotation axis is determined by the width of the laser beam. The minimum height H of the mirror—perpendicular to the rotation axis—can be estimated with the assumptions that a laser beam of a width W completely fills the mirror at the desired maximum of the angle of rotation α:

$$H = W/\cos\alpha. \tag{C4.4.4}$$

However, to overcome the diffraction effects caused by the mirror, both of its dimensions—parallel and perpendicular to the rotation axis—usually exceed the minimum requirements by 10%.

C4.4.4.7 Piezoelectric devices

Among many other applications, piezoelectric devices are also employed in high-resolution positioning and dynamic applications, e.g. in trepanning heads they provide the laser beam with circular motion by moving a single mirror [11] but their use is limited to the generation of Lissajou patterns, comparable to resonantly excited 2D scanners [12].

According to Piezojena's handbook, [13], there are many parameters which have to be taken into account with piezoelectric devices. In the following, some additional aspects to those already mentioned, concerning scanning and positioning applications, are described. For further information see [13].

The driving force of piezoelements is the piezoelectric or inverse piezoelectric effect.

A voltage applied to piezoelectric material (PZT or lead–zirconium–titanate) changes the dimensions of the material, thereby generating motion—the inverse piezoelectric effect. Hence, the driving force of a piezoelectric actuator is *expansion*. The relative expansion $S = \Delta l/L_0$ (without any external forces) of a piezoelement is proportional to the applied electrical field strength. Typical values for ceramic materials are $S = 0.1$–0.13% (field strength $E = 2$ kV mm^{-1}):

$$\Delta l/L_0 = S = d \times E. \tag{C4.4.5}$$

Here S is the relative stretch (dimensionless), $d = d_{ij}$ is the piezomodule or parameter of the material, $E(= U/d_s)$ is the electrical field strength, where U is the applied voltage and d_s is the thickness of a single disk.

The maximum expansion that can be achieved by using normal stacks, i.e. an assembly of several discs of piezoelectric material, is up to 300 μm. The length of such a stack will be 300 mm! Typical piezostacks have a motion of 20–60 μm.

According to equation (C4.4.5) the maximum expansion will rise with increasing voltage but the relation is not linear as predicted by the equation. Because of their ferroelectric nature, piezoelements show typical hysteresis behaviour. A butterfly-like curve reflects the inherent hysteresis. To reduce the hysteresis effect, usually unipolar voltage is applied, thus leading to the characteristic expansion curve of piezoelements, (shown in figure C4.4.14).

The typical width of hysteresis is 10–15% of the full motion. Working in a small voltage range, the hysteresis is also smaller.

Apart from hysteresis, the piezoelectrical effect as a solid-state effect has a very high resolution. Investigations carried out with typical piezoelements, e.g. PX 38 from Piezosysteme Jena, show a resolution of 1/100 nm. Therefore, the resolution is limited by the noise characteristics of the power supply.

Another characteristic of piezoelectric actuators is a short dimensional stabilization known as *creep*. A step change in the applied voltage will produce an initial response in a fraction of a millisecond, followed by

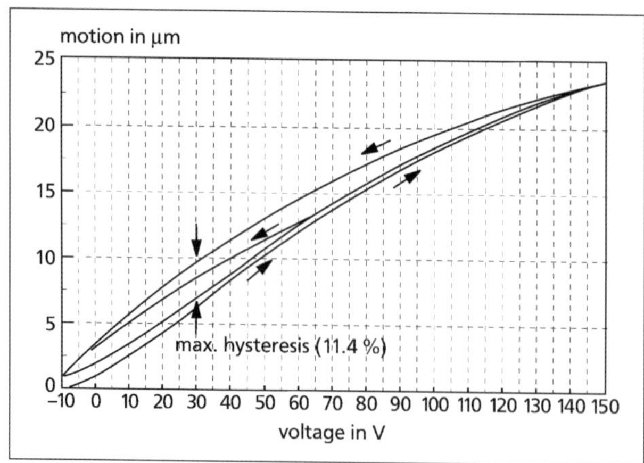

Figure C4.4.14. Hysteresis curve at unipolar voltage (reproduced by permission of Piezosysteme Jena, Germany).

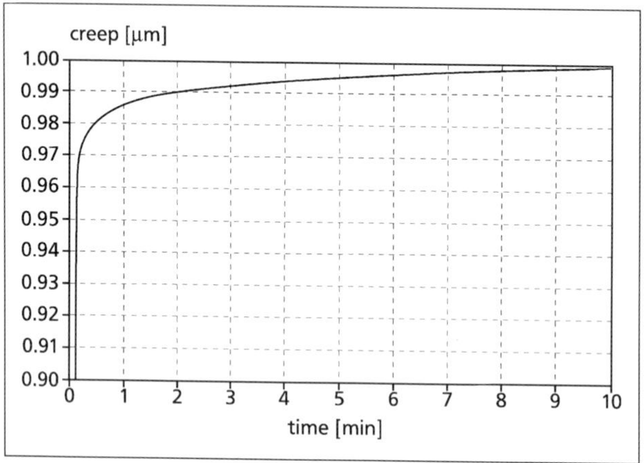

Figure C4.4.15. Typical creep of a PZT (reproduced by permission of Piezosysteme Jena, Germany).

a smaller change in a much longer time scale as shown in figure C4.4.15. Creep will be within 0.5–0.8% in the next decade.

The creep depends on the expansion, Δl, of the ceramic material, on the external loads, and on time. The creep can also be shown to have a logarithmic dependence on time:

$$\Delta l(t) = \Delta l_{0.1} \times (1 + \gamma \lg(t/0.1 \text{ s})). \tag{C4.4.6}$$

Here, $\Delta l_{0.1}$ is the stroke 0.1 s after the end of the voltage's risetime.

In this case, a value of 0.015 is reached for γ. The value of γ depends on the material, the construction and the environmental conditions (e.g. forces). When the motion (voltage) stops, the creep takes a few seconds longer to stop.

When working with periodic signals, as is the case in step scanning, the *repeatability* of a position will not be affected by creep. Because of the strong time dependence of the motion, creep occurs in all oscillations of the same order.

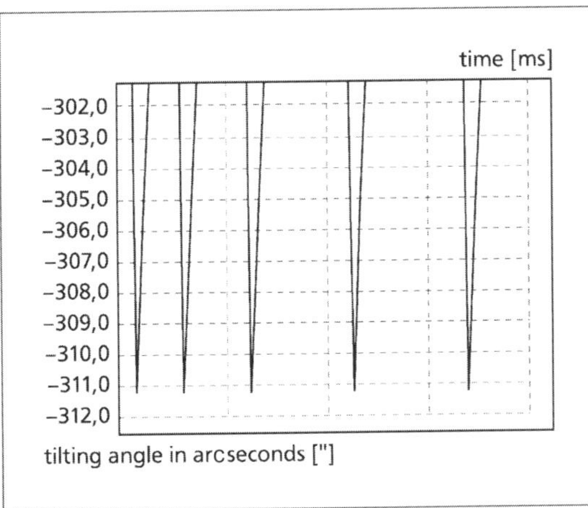

Figure C4.4.16. Repeatability of a mirror mount with periodic motion (reproduced by permission of Piezosysteme Jena, Germany).

The periodic oscillation of a mirror mount is shown in figure C4.4.16. The full tilting angle is approximately 380 arcsec. In the figure, only a 10 arcsec section (302–312 arcsec) is displayed—the repeatability is better than 0.1 arcsec.

A result of this experiment is that high repeatability within the system is reached without the use of closed-loop control. For such experiments, the repeatability is only determined by the quality of the power supply.

Taking scanning operations into account, the dynamic properties also have to be considered. Piezoactuators are oscillating mechanical systems characterized by their *resonant frequency* f_{res}. The resonant frequency is determined by the stiffness and the mass distribution (effective mass moved) within the actuator. Therefore, the resonant frequency of a complete system can deviate considerably from the resonant frequency of a single actuator. This is an important fact, for example, when using a mirror for fast tilting.

Reputable suppliers provide data sheets not only with the resonant frequency but also with the effective mass. Knowing the effective mass, it is possible to estimate the resonant frequency f^1_{res} with an additional mass (using formula (C4.4.7)):

$$f^1_{res} = 1/2\pi \times \sqrt{(c_T/(m_{eff} + M))} = f^0_{res} \times \sqrt{(m_{eff}/(m_{eff} + M))}. \tag{C4.4.7}$$

Here, f^0_{res} is the resonant frequency of the single actuator, m_{eff} the mass off the actuator system and M the additional mass, e.g. a mirror.

The *risetime*, previously mentioned in general, can be treated in a special way for piezoactuators. The shortest risetime, t_{min}, an actuator needs for expansion is determined by its resonant frequency f_{res}:

$$t_{min} = (3f_{res})^{-1}. \tag{C4.4.8}$$

If a short electrical pulse is applied to an actuator, the actuator expands within its risetime t_{min}. The piezoelement's resonant frequency will be simultaneously excited. Therefore, it begins to oscillate with a damped oscillation relative to its position. A shorter electrical pulse can result in a higher super-elevation but the risetime will not be shortened.

A typical response to a short electrical excitation of a piezoactuator (PAHL 40/20 from Piezosysteme Jena) is shown in figure C4.4.17. Although the excitation pulse has a duration of approximately 8 μs, the risetime of the actuator is only 20 μs. This value agrees with a resonant frequency of 16 kHz.

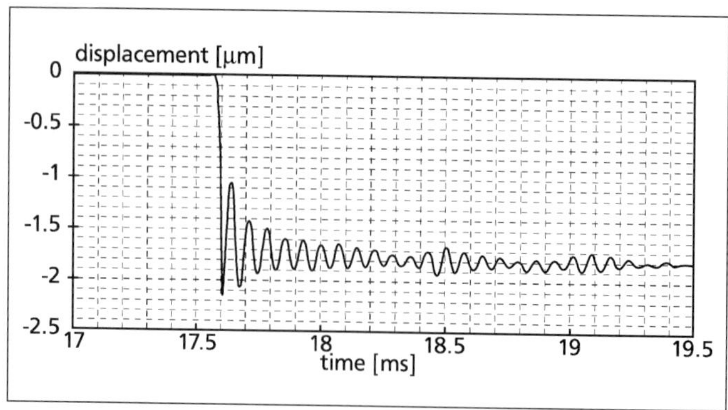

Figure C4.4.17. Response of a piezoelement (reproduced by permission of Piezosysteme Jena, Germany).

Table C4.4.4. Typical values of compressive stress and tensile forces on piezoelements during dynamic operation.

Compressive stress forces (kg cm^{-2})	Tensile forces (kg cm^{-2})
~9000	~50

While working in the dynamic regime, compressive stress and tensile forces act on the piezoactuators. The compressive strength of piezoactuators is very high but they are very sensitive to tensile forces. However, both forces occur to the same order while moving dynamically. Typical tensile and compressive stress forces for piezoactuators without pre-loads are shown in table C4.4.4.

According to this, *dynamic forces* have to be considered while in dynamic motion. They also appear without external loads! Actuators without pre-loads can only be used for small motions in special cases. Hence, it is necessary to use pre-loaded actuators for dynamic applications.

C4.4.4.8 Acousto-optic deflectors

Acousto-optic (AO) deflectors are completely different from mirror-based scanning devices with respect to the physical mechanism for beam deflection. The mirror systems mentioned before reflect the radiation, AO deflectors deviate the beam while it is being transmitted through a crystal. According to the optical materials used in the AO deflectors, they can be applied to every application that necessitates beam manipulation but the parameters differ appropriately with the material.

The principle in brief is as follows. An acoustic wave, generated by an rf signal, is applied to a piezoelectric transducer, bonded to a suitable crystal, which promotes a phase grating in the crystal, while it travels through the crystal at the acoustic velocity of the material with the acoustic wavelength being dependent on the frequency of the rf signal. Any incident laser beam will be diffracted by this grating, generally giving a number of diffracted beams. The basic principles of AO devices are given in chapter C1.2.

Using the *Bragg regime* (figure C4.4.18), beam deflection can be achieved. In this case, at one particular angle of incidence, only one diffraction order is produced—all the others are annihilated by destructive interference. Most AO devices operate in the Bragg regime, e.g. modulators and deflectors—the common exception being AO mode-lockers and Q-switches.

For specific use as deflectors, the resolution, efficiency and bandwidth are of interest in most applications.

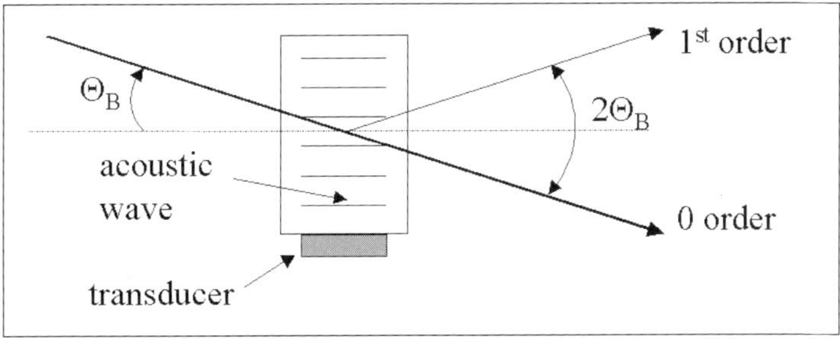

Figure C4.4.18. Bragg regime.

A high *resolution* is provided by the use of large-sized crystals (up to 30 mm or more) in order to work with large beam diameters, decrease optical divergence and increase resolution. The *static resolution*, r, is defined as the number of distinct directions that the diffracted beam can have. The centre of two consecutive points will be separated by the laser beam diameter (at $1/e^2$) in the case of a TEM_{00} beam:

$$r = (\pi/4) \times \Delta f \times (d/v) \tag{C4.4.9}$$

where Δf is the AO frequency range (f_{min}–f_{max}); this is also called the *bandwidth*. d describes the beam diameter at an intensity $1/e^2$ and v is the acoustic velocity.

To avoid first- and second-order overlapping, the bandwidth is limited to an octave and, therefore, the maximum scan angle is also limited.

More information on TEM_{00} is given in section A2.1.2 and more detailed information on static resolution can be found in textbooks on optics.

Due to the use of AO devices in scanning applications as well, the *access time* t_a—a synonym for transient response but more common when referring to AOs—of the deflector is of interest, too. This corresponds to the time necessary for the acoustic wave to travel through the laser beam and, thus, to the time needed by the deflector to move from one position to another:

$$t_a = d/v \tag{C4.4.10}$$

where d is the beam width and v the acoustic velocity of the material.

Hence, beam width and acoustic velocity work arbitrarily. Therefore, a high static resolution is achieved by the use of a large-sized crystal with a high access time. It should be noted that the acoustic velocity depends on the temperature of the material. Therefore, even the AO characteristics are influenced by heating.

A deflector is often characterized with the access time and bandwidth product ($t_a \times \Delta f$). Typical values are summarized in table C4.4.6.

When the field of frequencies no longer consists of discrete values, as in step scanning, but continuous sweeping as in raster scanning, it is necessary to define the *dynamic resolution* r_d, which takes account of the 'gradient' of the frequencies.

In the case of linear frequency sweeping, at $z = 0$ (i.e. on entry into the crystal), the frequency f is equal to

$$f = f_0 + (\Delta f/T) \times t \tag{C4.4.11}$$

where f_0 is the frequency at the beginning of the sweeping and T the time to sweep from f_{min} to f_{max}. At z,

$$f = f_0 + (\Delta f/T) \times t - (\Delta f/T) \times (z/v). \tag{C4.4.12}$$

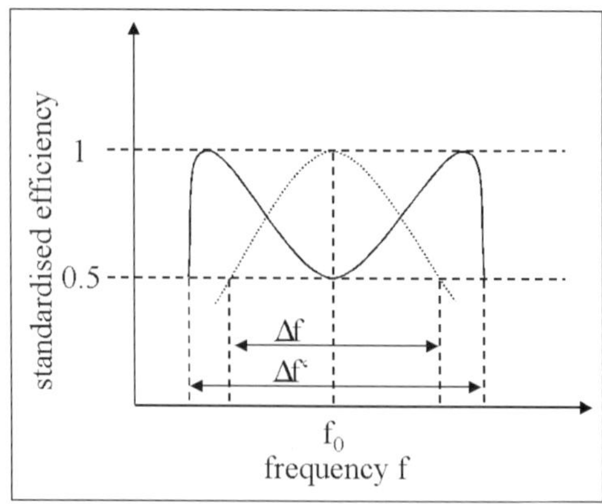

Figure C4.4.19. Efficiency of AO deflectors.

Table C4.4.5. Selected properties of acousto-optic materials.

Material	Refractive index	At λ (nm)	Max cw laser power density (W cm^{-2})	Acoustic velocity (m s^{-1})	M^2 (AO figure of merit) ($\times 10^{-15}$ S^3 kg^{-1})	At λ (nm)
Ge	4	10 600	5	5500	180	10 600
Doped glass	2.09	633	1	3400	24	633
$Ge_{33}As_{12}Se_{55}$	2.59	1064	1	2520	248	1064
As_2S_3	2.46	1150	1	2600	433	633
$PbMoO_4$	2.26/2.38	633	0.5	3630	36	633
TeO_2	2.26	633	5	4200	34	633
TeO_2	2.26	633	5	620	1200	633
SiO_2 (fused silica)	1.46	633	>100	5960	1.5	633
SiO_2 (fused silica)	1.46	633	>100	3760	0.5	633

The angle of deviation, $\Theta = \lambda f/v$, is now a function of the distance z and of time t. For z and $z + \mathrm{d}z$, the angle of deviation is not the same. Focusing occurs in only one plane of the diffracted beam. It is important to note this cylindrical lens effect, which intervenes during sequential sweeping, e.g. in televisions with a raster scan, printing and so on.

An equivalent cylindrical focal length is given by

$$F_{\mathrm{cyl}} = \alpha^2 (v/\lambda \times (\mathrm{d}f/\mathrm{d}t)^{-1}) \tag{C4.4.13}$$

with $\mathrm{d}f/\mathrm{d}t$ the frequency modulation slope and α a parameter depending on the beam profile (1 for a rectangular shape, ~1.34 for TEM_{00}).

The *efficiency*, i.e. the amount of irradiance at a distinct angle of deviation with respect to the irradiance of the incident laser beam, usually shows a dependence on the frequency as can be seen in figure C4.4.19 (dotted line). However, some applications require a quasi-constant efficiency, on all bandwidths. This can be obtained by decreasing the width of the ultrasonic beam but with a resulting loss in maximum efficiency, except for an anisotropic interaction. In this case, it is possible to match the Bragg condition at specific interaction angles with two synchronous frequencies. Therefore, the deflection angle range can be broadened, providing a good efficiency—see figure C4.4.19 (full line).

Table C4.4.6. Property dependency of TeO_2 regarding maximum aperture[a].

Aperture ∅ (mm)	Efficiency (%)	Access time T_a (μs)	Bandwidth Δf (MHz)	$T \times \Delta f$	Deflection angle Θ (mrad)
1.7[b]	70	2	50	100	48
4.2[b]	70	5	50	250	48
6.7[b]	70	8	50	400	48
1.7[c]	40	2	50 × 50	100 × 100	48 × 48
4.2[c]	40	5	50 × 50	250 × 250	48 × 48
6.7[c]	40	8	50 × 50	400 × 400	48 × 48

[a] spectral range (nm) depends on the specific anti-reflection coating: 514–532, 633–670, 780–820 or 360–1100.
[b] Single axis.
[c] Dual axes.

Table C4.4.5 gives some examples of AO materials including specific parameters and table C4.4.6 shows some specific properties for TeO_2 with respect to different beam diameters.

The values displayed in tables C4.4.5 and C4.4.6 are taken from consumer information supplied by AA Acousto-optics, France.

Due to their optical properties, AO devices affect the incident radiation. Hence, optical influences, e.g. on the polarization, should be mentioned but are no matter for this chapter. Additional information can be found in other literature, for example [17].

C4.4.5 Conclusion

Given the broad distribution of laser applications, whether it be household, medicine, measuring techniques or material processing, it is not possible to mention all aspects linked to positioning and scanning systems. Limiting ourselves to industrial material processing has successfully reduced the work. However, the responsibility of finding appropriate solutions for beam delivery systems lies in the hands of those involved in developing them, aided by their imagination and knowledge of the possibilities. Hopefully, the latter has been amplified here.

Acknowledgments

The author would like to thank Dr Braun (Fraunhofer IPM), Fr Schmid (Trumpf GmbH), Dr Jovanovic (Pegasus Optik), Mr Hartmann (Arges), Dr Blank (LPKF Motion Control), Mr Elbinger (Piezosysteme Jena), Mr Höppe (Reis-robotics) and Crisel Instruments for support, assistance and discussion.

References

[1] Koch J, Wolf M and Sepold G 1999 Mikrobohrungen in Dünnblech-hergestellt mit dem Laser *wt werkstattstechnik* **89** 492–4

[2] Liu J S, Li L J and Jin X Z 1999 Accuracy control of three-dimensional Nd:YAG laser shaping by ablation *Opt. Laser Technol.* **31** 419–23

[3] Hand D P, Fox M D T, Haran F M, Peters C, Morgan S A, McLean M A, Steen W M and Jones J D C 2000 Optical focus control system for laser welding and direct casting *Opt. Lasers Eng.* **34** 415–27

[4] Hoffmann P, Schwab J and Kugler P 2002 Laserstrahlhartlöten-Systemtechnik, Anwendungen und Potenziale *Laserstrahlfügen: Prozesse, Systeme, Anwendungen, Trends Strahltechnik* Band 19, ed G Sepold and T Seefeld (Bremen: BIAS) pp 315–24

[5] Weck M, Kruger P and Brecher C 2001 Limits for controller settings with electric linear direct drives *Int. J. Machine Tools Manufacture* **41** 65–88

[6] Kami Y, Yabuya M and Shimizu T 1995 Research and development of an ultraprecision positioning system *Nanotechnology* **5** 127–34

[7] Grey D J 1997 Design of high performance CO_2 laser beam scanning objectives *Lasers in Materials Processing (Proc. SPIE 3097)* ed L H J F Beckmann (Bellingham, WA: SPIE)

[8] Kim J-D, Jung J-K, Jeon B-C and Cho C-D 2001 Wide band laser heat treatment using pyramid polygon mirror *Opt. Lasers Eng.* **35** 285–97

[9] Rizvi N and Apte P 2002 Developments in laser micro-machining technique *J. Mater. Process. Technol.* **127** 206–10

[10] Hornig 2002 Laserstrahlbearbeitung bei BMW: Anwendungen und Trends *Laserstrahlfügen: Prozesse, Systeme, Anwendungen, Trends Strahltechnik* Band 19, ed G Sepold and T Seefeld (Bremen: BIAS) pp 351–65

[11] Koch J and Lang A 1997 Feinstbohren mit dem Kupferdampflaser *Präzisionslaserstrahl-fertigungstechnik für den Maschinenbau* ed M Geiger *et al* (Erlangen: Bayerisches Laserzentrum) pp 127–44

[12] Schenk H, Dürr P, Kunze D, Lakner H and Kück H 2001 A resonantly excited 2D-micro-scanning-mirror with large deflection *Sensors Actuators* A **89** 104–11

[13] Piezosysteme 2001 *Piezoline* part of catalogue (Jena: Piezosysteme Jena GmbH) which is also available on the internet: www.piezojena.com.

[14] Chang S B, Wu S H and Hu Y C 1997 Submicrometre overshoot control of rapid and precise positioning *Precision Eng.* **20** 161–70

[15] Petiot J-F, Chedmail P and Hascoet J-Y 1998 Contribution to the scheduling of trajectories in robotics *Robot. Computer-Integrated Manufacturing* **14** 237–51

[16] Chen J S and Dwang I C 2000 A ballscrew drive mechanism with piezo-electric nut for preload and motion control International *J. Machine Tool Manufacture* **40** 513–26

[17] Blomme E, Gondek G, Katkowski T, Kwiek P, Sliwinski A and Leroy O 2000 On the polarization of light diffracted by ultrasound *Ultrasonics* **38** 575–80

Further reading

n.n. 1996 *The Fascinating World of Sheet Metal* ed TRUMPF GmbH & Co. (Dietzingen: Trumpf GmbH)

Overview on sheet metal with general information related to laser processing of sheet metal. Includes CO_2- and Nd:YAG high power lasers as well as workpiece positioning and beam delivery systems. Also available on the Internet: www.trumpf.com.

Petiot J-F, Chedmail P and Hascoet J-Y 1998 Contribution to the scheduling of trajectories in robotics *Robot. Computer-Integrated Manufacturing* **14** 237–51

Written from a scientific point of view. Considers a general problem that comes along with robotics, the travelling salesman, with briefer mention of commonly used solutions. Introduces a new approach to optimization. With detailed reference listing.

Chang S B, Wu S H and Hu Y C 1997 Submicrometre overshoot control of rapid and precise positioning *Precision Eng.* **20** 161–70

Written from a scientific point of view. An optimized controller for high precision positioning of ball guide tables is introduced. Considers several possible errors, like stick-slip motion and overshoot.

Chen J S and Dwang I C 2000 A ballscrew drive mechanism with piezo-electric nut for preload and motion control *International J. Machine Tool Manufacture* **40** 513–26

Another scientific presentation. Introduces a long stroke, high precision positioning system performed by combination of a ball screw drive unit additionally equipped with a piezo-electric nut. Considers stick-slip motion at low speed and mentions friction as well.

Montague J and DeWeerd H 1996 Optomechanical Scanning Application, Techniques, and Devices *The Infrared and Electro-Optical Systems Handbook* 2nd edn, vol 3, ed J S Accetta and D L Shumaker (Bellingham, WA: SPIE)

A good review on scanning techniques and applications.

Beiser L 2003 *Unified Optical Scanning Technology* (New York: Wiley) ISBN 0-471-31654-7

A cohesive view of the expanding field of optical scanning, in a single compact volume.

Waynant R and Edinger M 2000 *Electro-optics Handbook* (New York: McGraw)

Master the latest electro-optical sources, materials, detectors, and applications. Mentions the wide array of gas, solid state, semiconductor, dye, chemical and free electron lasers.

Agullo-Lopez F and Cabrera J M 1994 *Electrooptics: Phenomena, Materials and Applications* (San Diego, CA: Academic)

Introduces the fundamental concepts and phenomena of electro optics at an undergraduate and graduate level. Gives a broad overview of inorganic materials, including ceramics and novel single crystals.

Goutzoulis A and Pape D 1994 *Design and Fabrication of Acousto-optic Devices* (New York: Marcel Dekker)

Offers in-depth discussions on all aspects of acousto-optic deflectors, modulators, and tenable filters-emphasizing hands-on procedures for design, fabrication, and testing.

C5
Laser beam measurement

Julian Jones

To characterize a laser beam requires complete knowledge of its mutual coherence function—the power (or energy) distribution and the relative phase distribution over a single transverse plane—achieved by complex measurement and subsequent analysis. In chapter 5.1, Brooke Ward describes a useful approximation to the ideal, based on measuring the beam diameter and divergence at a set of transverse planes along the axis of the laser beam, considering how the diameter is to be defined (using a second-moment technique) for different types of beams; how the measurements are to be made, using sensor arrays or scanning slits; and considering the propagation characteristics of the beam, encountered earlier in chapter C1.1.

Intrinsic to the characterization of a beam is the ability to be able to measure optical power or energy. Optical detectors were introduced in chapter A7. In chapter C5.2, Kenny Weir advances the discussion to consider the various practical devices that are required to cover the range of wavelengths of the lasers described elsewhere in this Handbook, extending from ultraviolet to infrared. In the ultraviolet, there remains a place for the vacuum-tube photomultiplier, although photodiodes (notably of the Schottky type), using semiconductor materials such as Si, SiC, GaP and GaN, are useful. More recently, semiconductor artificial diamond has come forward as a promising material on account of its very wide bandgap. For many purposes, the silicon photodiode is ubiquitous for measurements in the visible wavelength range. Infrared detection poses more of a challenge. In the nearest wavelength range, say up to the 1.55 μm typical of optical communications, InGaAsP approaches the ideal. At longer wavelengths, where the thermal quantum approaches the optical quantum energy, more complex structures (such as quantum wells) and different materials are needed, such as the lead salts in the mid-infrared; perhaps HgCdTe is the most flexible material at these wavelengths. Thermal detectors, or *bolometers*, have their place for the detection of the longest wavelengths and development continues, e.g. with the silicon microbolometer.

International standards exist for the measurement of the radiant power and energy of lasers. Techniques for the transfer of these standards are discussed by Robert Tyson in chapter C5.3, spanning the wide wavelengths (IR to UV) and pulselengths (fs to cw) required for practical lasers, noting the standard calibration categories, and describing some standard test configurations. Returning to the complete characterization of a laser beam in terms of the irradiance and phase distributions in a single transverse plane and which are often written as $I(x, y)$ and $\phi(x, y)$, Bernd Schäfer in chapter 5.4 reviews the relevant international standards ISO13694 and ISO15367. The irradiance (power density) is measured by the techniques of C5.1 with array or scanning detectors. The phase distribution is measured with wavefront sensors, such as those used in adaptive optics (chapter C1.3), e.g. the Shack–Hartmann sensor, or by using an interferometer (chapter A5). Alternatively, the complete mutual coherence function can be measured using an optical arrangement similar to that used for producing Young's fringes.

C5.1
Beam propagation

B A Ward

C5.1.1 Introduction

The propagation behaviour of a highly coherent laser beam can be predicted from a detailed knowledge of both the power/energy density distribution and the relative phase distribution in a single transverse plane of the beam. Complete characterization of a more general beam requires knowledge of the mutual coherence function combined with complex measurement and analysis procedures. A much simpler and equally useful procedure that can be used to predict much of the propagation behaviour of laser beams, as well as some incoherent broadband beams of light, involves measurement of the beam diameter and divergence. The measurements are made using second-order moment analysis of the transverse power/energy density distribution at a number of positions along a beam. This section of the handbook is devoted to describing this latter technique. In what follows, the word 'irradiance' is used to mean the power/energy density in a transverse plane of a beam.

C5.1.2 Beam types

The analytic methods adopted in this section can usually be applied to beams of radiation whose full divergence angle is less than 30°. The simplest type is the circular cross-section beam whose transverse irradiance distribution is radially symmetric. This is a *stigmatic* beam. If the radial symmetry exists at all planes along a beam, it is classified as *intrinsically stigmatic*. If, on the other hand, the beam is circular at only one or two locations along the beam then the beam is classified as astigmatic.

If the axes of radial asymmetry of a propagating beam remain parallel to each other then the beam is classified as *simple astigmatic* (see figure C5.1.1). If the axes of asymmetry rotate in the transverse plane of a propagating beam then that beam is classified as *general astigmatic*.

C5.1.3 Beam diameter

There are a number of alternative definitions [2] of the diameter of a beam of electromagnetic radiation. Many are based on selecting the diameter where the irradiance is a given fraction of that at the peak. Others are based on the diameter of the contour enclosing a fraction of the total beam power/energy. Perhaps the most common definition is known as the $1/e^2$ diameter (figure C5.1.2). This is based on the diameter of the fundamental TEM_{00} single Gaussian mode emitted by a stable resonator where the radial irradiance distribution takes the form

$$I = I_0 \cdot \exp\left(-2\left(\frac{r}{w}\right)^2\right)$$

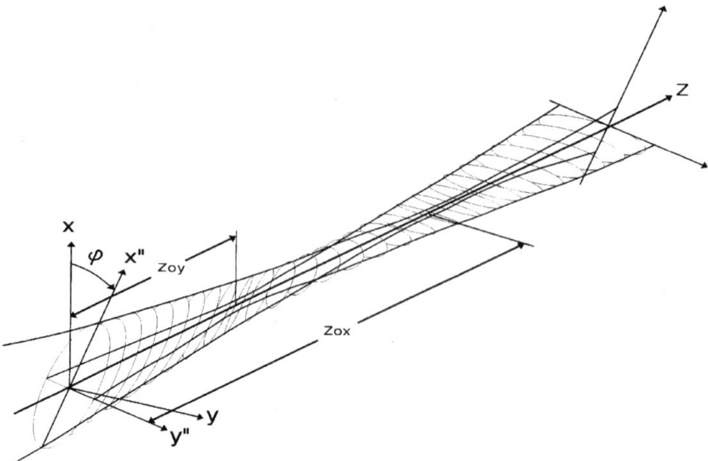

Figure C5.1.1. Beam axis coordinate system showing an astigmatic irradiance distribution and axially separated orthogonal waists at Z_{0x} and Z_{0y}

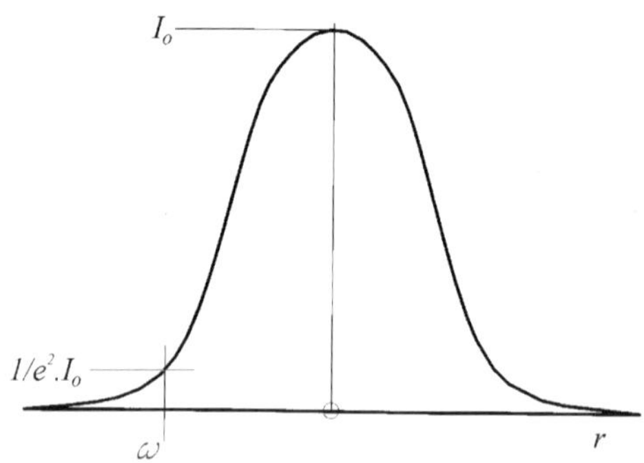

Figure C5.1.2. The $1/e^2$ radius of a lowest order Gaussian irradiance distribution

and w is the radius where the irradiance is $1/e^2$ of the peak irradiance I_0. Gaussian beams were introduced in chapters A2 and A3 and were discussed in the context of optical components in section C1.1.

The fundamental Gaussian mode can be compared with the statistical normal probability or Gaussian error distribution

$$P = P_0 \exp\left(-\frac{1}{2}\left(\frac{r}{\sigma}\right)^2\right)$$

where σ is the standard deviation of the distribution. The relationship results in the convenient definition of the $1/e^2$ diameter to be that of the contour that contains 87% of the total beam power. It is referred to as the second moment diameter, $d_{87} \equiv d_\sigma = 2w = 4\sigma$.

This statistical definition of beam diameter is convenient since it does not require identification of a peak irradiance and is not restricted to Gaussian beam forms.

Evaluation of the beam diameter at location z along the beam axis in the direction of the transverse x and y axes is performed by taking the first moments of the irradiance distribution $I(x, y, z)$ to determine the centroid $\bar{x}$, $\bar{y}$ of the beam.

$$\bar{x} = \frac{\sum_x \sum_y x \cdot I(x, y, z)}{\sum_x \sum_y I(x, y, z)} \qquad \bar{y} = \frac{\sum_x \sum_y y \cdot I(x, y, z)}{\sum_x \sum_y I(x, y, z)}. \tag{C5.1.1}$$

The diameters of a beam are identified as 'second moment' diameters by using σ to suggest the similarity with statistically determined 'standard deviation' of a variable. The orthogonal diameters are given by:

$$d_{\sigma x}(z) = 4\sigma_x(z) \qquad \text{and} \qquad d_{\sigma y}(z) = 4\sigma_y(z) \tag{C5.1.2}$$

where the second moments of the power density distribution of the beam at location z are given by:

$$\begin{aligned} \sigma_x^2(z) &= \frac{\sum_x \sum_y (x - \bar{x})^2 \cdot I(x, y, z)}{\sum_x \sum_y I(x, y, z)} \equiv \langle x^2 \rangle \\ \sigma_y^2(z) &= \frac{\sum_x \sum_y (y - \bar{y})^2 \cdot I(x, y, z)}{\sum_x \sum_y I(x, y, z)} \equiv \langle y^2 \rangle \end{aligned} \tag{C5.1.3}$$

C5.1.4 Practical measurements

The previous expressions have been written with summation symbols rather than integrals to underline the fact that the irradiance data will most likely be acquired by array detectors and the analysis involved is simple processing of digitized pixel values. However, a number of precautions and approximations have to be made before second moment measurements can be used for propagation predictions.

The raw data collected from an array detector will require processing to reduce the uncertainty in beam diameter measurement that can be introduced by optical noise and sharp-edge diffraction effects. Sources of optical noise include stray light from pump lamps or discharge fluorescence as well as background illumination effects. A high background level can lead to a 'baseline offset' that can cause large errors in the evaluation of beam characteristics.

There are a number of methods for reducing the influence of optical noise. The simplest technique is to record, calculate and subtract an average background illumination pattern from the power density distribution that includes the beam pattern of interest as well as the interfering background. The background pattern can be obtained by carefully blocking and absorbing just the beam under study as it leaves the laser and recording the rest. The averaging procedure is necessary to reduce any statistical fluctuations that might be present in the background pattern. However, it is possible that some of the background light is coherent and capable of producing speckle interference patterns when combined with the main laser beam. If this is the case, more sophisticated compensation techniques may be required [3].

One result of proper background subtraction will be the presence of negative noise values in the corrected distribution. These negative values should be included in the further evaluation procedures to compensate for the presence of residual positive noise signals.

A further uncertainty is introduced by the fact that equations C5.1.1 and C5.1.2 imply that the summations have to be carried out over an infinitely wide aperture. This means that the measurements are sensitive to widely scattered light. Beams passing through sharp-edged apertures will suffer diffraction and a certain amount of large-angle scattering and may not be amenable to second moment measurements.

Infinite measurement apertures are obviously impractical but a solution does exist that preserves the usefulness of second moment width measurements [3]. The total number of pixels used should cover an area whose width is more than four times that of the beam diameter. There should be sufficient numbers of pixels across the beam to permit an accurate estimate to be made of the second moment width. It is suggested that there should be more than 400 kpixels in a suitable array detector. The technique involves making an initial estimate of the beam widths using all available pixels. The summations for both the centroid and width estimates are then repeated but the pixels included are limited to those that are within a rectangle whose sides are three times the estimated beam widths along each axis and centred on the beam centroid. The procedure results in a new estimate of the beam width which is then used to repeat the process. The iteration continues until the width estimate converges to a steady value. In practice, five to seven cycles are generally sufficient to produce a value that is precise, reproducible and useful for propagation prediction. The procedure is summarized by a modification of equation C5.1.3 so that:

$$\sigma_x^2(z) = \frac{\sum_{y_1}^{y_2}\sum_{x_1}^{x_2}(x-\bar{x})^2 \cdot I(x,y,z)}{\sum_{y_1}^{y_2}\sum_{x_1}^{x_2} I(x,y,z)} \equiv \langle x^2 \rangle$$

$$\sigma_y^2(z) = \frac{\sum_{y_1}^{y_2}\sum_{x_1}^{x_2}(y-\bar{y})^2 \cdot I(x,y,z)}{\sum_{y_1}^{y_2}\sum_{x_1}^{x_2} I(x,y,z)} \equiv \langle y^2 \rangle. \tag{C5.1.4}$$

The summations are carried out over a rectangle parallel to the x- and y-axes where:

$$x_1 = \bar{x} - \tfrac{3}{2}d_{\sigma x} \qquad x_2 = \bar{x} + \tfrac{3}{2}d_{\sigma x}$$

$$y_1 = \bar{y} - \tfrac{3}{2}d_{\sigma y} \qquad y_2 = \bar{y} + \tfrac{3}{2}d_{\sigma y}. \tag{C5.1.5}$$

This procedure for the analysis of beam diameter can also be applied to some incoherent and broadband beams as well as monochromatic laser beams. It has been used for the examination of a number of beam types. For many sources, it is found to provide a reproducible estimate of beam width and demonstrates useful precision. However, the absolute accuracy of the beam propagation parameters derived using this method may not be high enough for some of the more complex beams. Nevertheless, the procedure appears to provide a robust and transportable protocol for optical beam analysis that might be used in different laboratories to examine the same source and to give similar results.

C5.1.5 Propagation characteristics

Investigations of the propagation behaviour of an arbitrary optical beam have intensified over the last ten years [4, 5, 7]. They have revealed that two of the most useful parameters for predicting the diameters of a propagating beam are the dimensions of the beam waist $d_{\sigma 0}$ and the far-field divergence of the beam Θ_σ. If the diameter of the beam is measured using the second moment of its irradiance distribution, it will vary in a hyperbolic manner and be symmetrical about the beam waist [6].

The usefulness of the waist width and far-field divergence measurements stems from the fact that their product is invariant when passing through an aberration-free optical system of sufficient aperture to avoid detectable truncation of the beam. The far-field divergence can be determined by inserting an aberration-free lens (focal length f) on the axis of the beam under investigation and measuring the second moment diameter $d_{\sigma f}$ of the irradiance distribution in the focal plane of the transforming lens. The far-field divergence is $\Theta_\sigma = d_{\sigma f}/f$. The far-field is that part of a beam that is a significant distance (>5 Rayleigh lengths) from the waist of the beam (figure C5.1.3)

Locations close to the beam waist are said to be in the near-field. A useful parameter for describing distances from a beam waist is the Rayleigh length $z_R = d_{0\sigma}/\Theta_\sigma$ (synonymous with the Rayleigh range (see section A2.1.2.2)). It is the distance from the waist over which the cross-sectional area of the beam has doubled.

Figure C5.1.3 indicates the 87% enclosed power/energy contour. This contour will have a symmetrical shape with an hyperbolic form $d_{\sigma x}^2(z) = a + bz + cz^2$. This fact indicates a convenient method for measuring the main propagation parameters of a beam. The second moment diameter is measured at ten or more

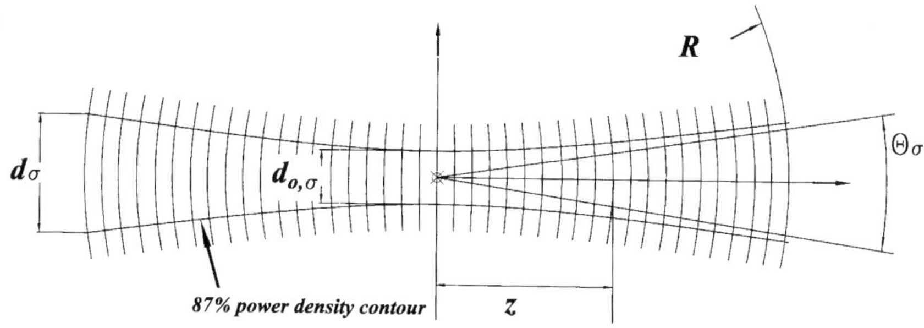

Figure C5.1.3. Characteristics of a beam of electromagnetic radiation

positions around the waist and statistical procedures are used to determine the most probable hyperbola coefficients [8]. The location and diameter of the beam waist is then given by:

$$z_0 = \frac{-b}{2c} \qquad \text{and} \qquad d_{\sigma 0} = \sqrt{a - \frac{b^2}{4c}}. \tag{C5.1.6}$$

The other propagation constants, i.e. divergence, propagation ratio and Rayleigh length, are given by:

$$\Theta_\sigma = \sqrt{c} \qquad M^2 = \frac{\pi}{4\lambda}\sqrt{a \cdot c - \frac{b^2}{4}} \qquad \text{and} \qquad z_R = \frac{1}{c}\sqrt{a \cdot c - \frac{b^2}{4}}. \tag{C5.1.7}$$

The propagation ratio M^2 is a quantity popularly known as the 'beam quality factor'. It is now called the 'beam propagation ratio'. Its value is based on the fact that the product of the waist diameter and far-field divergence of a beam is an invariant of propagation. In other words, this value is not changed by passing the beam through an aberration-free optical system. The smallest value of the diameter/divergence product is that achieved by the fundamental TEM_{00} Gaussian beam and has the value $\frac{4}{\pi}\lambda^2$, where λ is the wavelength of the monochromatic radiation. M^2 is the ratio of the diameter/divergence product of a beam being characterized to this theoretical minimum product.

C5.1.6 Beam transformation by a lens

When a beam enters an aberration-free lens, the emerging beam can also be described by a hyperbola. As we have seen in sections C1.1 and C4.2, the new hyperbola will have its own Rayleigh length, waist diameter and asymptotic divergence (figure C5.1.4). Prediction of the width and position of the waist of the laser beam after it has passed through a lens is relatively simple.

The expressions for the diameter and location of the transformed beam are best considered in terms of the displacement of the input and output beam waists from the front and rear focal planes of the lens, respectively (see figure C5.1.4).

The relationship between the propagation characteristics of the input and output beams is summarized as follows:

$$\text{if} \qquad d_{02} = V_1 \cdot d_{01} \qquad \text{where } V_1^2 = \frac{f^2}{x^2 + z_{R1}^2} \tag{C5.1.8}$$

$$\text{then} \qquad y = V_1^2 \cdot x \qquad \Theta_2 = \frac{1}{V_1} \cdot \Theta_1 \qquad \text{and} \qquad z_{R2} = V_1^2 \cdot z_{R1}. \tag{C5.1.9}$$

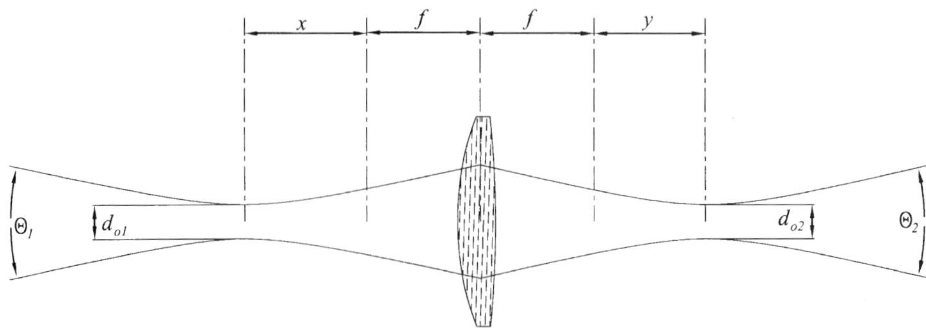

Figure C5.1.4. Beam transformation by a lens. Waist distances are measured from the focal planes of the lens

It should be noted that as the Rayleigh length of the input beam approaches zero (i.e. the waist becomes a point source) then the relationship between the locations of the input and output waists of the beams approaches the Newtonian expression for geometrical image formation, $x \cdot y = f^2$. This is an interesting result since it enables examination of the relationship between the image-forming properties in a beam and the contrasting behaviour of propagation envelopes.

C5.1.7 Alternative measurement methods

Sensor arrays provide the greatest detail and analysis capability when measuring beam properties. However, simpler detectors and technologies do exist. An effective method for measuring the power content diameter of a stigmatic beam is the use of a sequence of variable circular apertures to locate that diameter that contains (say) 87% of the total beam power. Alternative methods that can accommodate astigmatic beams are based on the scanning slit or knife edge sensor. These instruments, their limitations and correction methods are discussed in detail in the international standard [3]. They are very useful devices for characterizing good quality beams but mixed beam modes and sharp-edge diffraction effects can lead to significant measurement uncertainties.

It is possible to test the measurement capability of an instrument when examining a beam of unknown quality. The technique is based on measuring the propagation characteristics of the raw subject beam and using them to predict the location and size of the new waist produced by insertion of a low-aberration (long focal length) transforming lens. The location and size of the transformed waist can then be confirmed by measurements taken with the same instrument on the actual transformed beam.

C5.1.8 Propagation of astigmatic beams

Most of the previous discussion has been aimed at the propagation properties of stigmatic or simple astigmatic beams. In fact, if the ratio $\sigma_x/\sigma_y \leq 1.15$ then the beam may be considered circular with a diameter $d_\sigma = 2\sqrt{\sigma_x^2 + \sigma_y^2}$ at the measurement location. If the azimuth angle φ (figure C5.1.1) varies by no more than 10° over a distance of two Rayleigh lengths, the beam may be simple astigmatic. In this case, an effective beam propagation ratio can be defined to be $M_{\text{eff}}^2 = \sqrt{M_x^2 \cdot M_y^2}$. However, if neither of these two conditions apply, the beam is probably general astigmatic.

Analysis and propagation prediction of a general astigmatic beam is obviously more complex than the study of simple astigmatic beams. Nevertheless, the measurement and evaluation principles are virtually the same for both beam types. The concept of second moment measurements is now extended to include the 'mixed moments' of the spatial and divergence properties of the beam. For example, the spatial mixed moment is:

$$\sigma_{xy}^2(z) = \frac{\sum_{y_1}^{y_2}\sum_{x_1}^{x_2}(x-\bar{x})(y-\bar{y})\cdot I(x,y,z)}{\sum_{y_1}^{y_2}\sum_{x_1}^{x_2} I(x,y,z)} \equiv \langle xy\rangle. \tag{C5.1.10}$$

This measurement can be combined with the second moment diameter measurements to reveal the mean azimuth angle φ of the irradiance distribution in the laboratory coordinate system. The alternative nomenclature used in equations (C5.1.4) and (C5.1.10) for second moment measurements is now used to simplify the equations.

$$\varphi = \frac{1}{2}\arctan\left[\frac{2\langle xy\rangle}{\langle x^2\rangle - \langle y^2\rangle}\right] \tag{C5.1.11}$$

There are ten second moment values that can be used to specify the propagation properties of a general astigmatic beam. Those moments can be used to form an ABCD matrix (see chapter A2) that can be used to compute the changes in the beam characteristics as it passes through an idealized optical system [1]. The form of the matrix and its application is summarized in considerable detail in the forthcoming revision of the international standard on beam propagation [3]. The overall beam matrix has the form:

$$P = \begin{pmatrix} \langle x^2\rangle & \langle xy\rangle & \langle x\Theta_x\rangle & \langle x\Theta_y\rangle \\ \langle xy\rangle & \langle y^2\rangle & \langle y\Theta_x\rangle & \langle y\Theta_y\rangle \\ \langle x\Theta_x\rangle & \langle y\Theta_x\rangle & \langle\Theta_x^2\rangle & \langle\Theta_x\Theta_y\rangle \\ \langle x\Theta_y\rangle & \langle y\Theta_y\rangle & \langle\Theta_x\Theta_y\rangle & \langle\Theta_y^2\rangle \end{pmatrix} \tag{C5.1.12}$$

In spite of the complexity of the definition of a general astigmatic beam it is still possible to define a beam propagation ratio as an invariant of propagation

$$M_{\text{eff}}^2 = \frac{4\pi}{\lambda}(\det(P))^{1/4} \tag{C5.1.13}$$

Other invariants of propagation through a homogeneous system of spherical optical elements are the intrinsic astigmatism and the twist parameter. This first parameter results from a complex combination of the ten second moment properties of the beam. The twist parameter $t = \langle x\Theta_y\rangle - \langle y\Theta_x\rangle$ is a parameter that can be changed by propagation through cylindrical lenses.

C5.1.9 Summary

Many of the measurement techniques and analysis methods discussed in this section are specified and described in the latest revised international standard on beam parameter measurements [3]. This standard has been produced and developed following eight workshops on laser beam characterization and two Eurolaser Eureka programmes conducted over the last ten years. The standard has been subject to international voting and has been passed as a final draft. Full publication can be anticipated during 2003.

References

[1] Belanger P A 1991 Beam propagation and the ABCD ray matrices *Opt. Lett.* **16** 196

[2] ISO 11145:2001 Optics and optical instruments—Lasers and laser related equipment—Vocabulary and symbols. Available as BS EN ISO 11145:2001 from the British Standards Institute

[3] Revised ISO 11146 ~2003 Lasers and laser related equipment—Test methods for laser beam widths, divergence angle and beam propagation factor—Part 1: Stigmatic and simple astigmatic beams. Part 2: General astigmatic beams. Part 3: Alternative test methods and geometrical laser beam classification and propagation. To be published

[4] Nemes G and Serna J 1997 The ten physical parameters associated with a full general astigmatic beam *4th Int, Workshop on Laser Beam and Optics Characterization* ed A Giesen and M Morin (Munich: IFSW/VDI Messe) pp 92–105

[5] Sasnett M W and Johnston T F 1991 *Conf. on Laser Beam Diagnostics (Proc. SPIE)* **1414** 1–32

[6] Siegman A E 1990 New developments in laser resonators *Optical Resonators (Proc SPIE)* **1224** 2–14
[7] Siegman A E 1993 *Laser Beam Characterization* ed P M Mejias, H Weber, R Martinez-Herrero and A Gonzalez-Urena (Madrid: SEDO) pp 1–22
[8] Ward B A 1993 *Laser Beam Characterization* ed P M Mejias, H Weber, R Martinez-Herrero and A Gonzalez-Urena (Madrid: SEDO) pp 62–4

C5.2
Detectors

Kenny Weir

C5.2.1 Introduction

The principles of the reliable detection of light, in any region of the spectrum, depends on careful consideration of a large number of factors. The importance of signal level, noise and signal to noise ratio cannot be underestimated, and these general considerations are discussed in detail elsewhere (see chapter A7), together with details of the operating principles of many generic detectors. However, detection in the extremes of the ultraviolet (UV) and infrared (IR) introduces a new range of considerations in these areas, and in the material characteristics that are required to construct reliable detectors.

For the purposes of the current discussion, the optical spectrum is considered in three distinct regions. The visible is considered to extend from a wavelength of 400 nm to 1.5 μm. Although visible to the eye generally only extends a little above 650 nm, this range is grouped together as the same operating principles can be applied to detectors out to 1.5 μm. The details of visible detectors appear elsewhere (see chapter A7) and are not considered further here. The ultraviolet below 400 nm is considered down to 193 nm, to include the output of the ArF laser. The infrared above 1.5 μm is considered up to 14 μm. Although there are only distinct laser outputs in these region, they are considered as a whole as the detection principles apply across a wide range and, indeed, there is widespread interest in detectors in the ultraviolet and infrared, for example in spectroscopy and sensors.

C5.2.2 Operating principles

Detection of optical radiation relies on either detecting the light due to the absorption of its energy by a body and observing the change in its temperature (photothermal), or observing the effect on the detector when individual photons are detected (photoelectric). Further detailed discussion on these approaches can be found in chapter A7.

The issues relating to semiconductor solid state detector construction and operation in the two extremes of UV and IR (and indeed across the visible) are in many cases quite similar. In considering solid-state devices the relationship between the photon energy and the bandgap energy of the intrinsic semiconductor material is important. The energy of the photon, $h\nu$, where h is Planck's constant and ν is the frequency of the radiation, must be greater than the bandgap energy of the intrinsic semiconductor, E_{g}. This may be written as

$$h\nu = \frac{hc}{\lambda} > E_{\mathrm{g}} \qquad \text{(C5.2.1)}$$

where c is the speed of light and λ is the wavelength of the light.

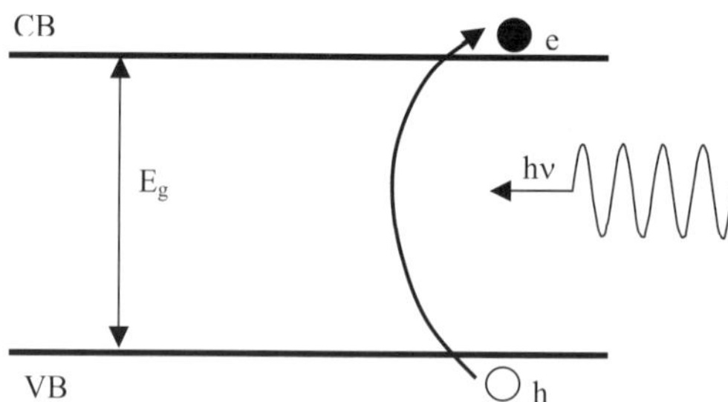

Figure C5.2.1. Absorption of a photon in an intrinsic semiconductor promotes an electron (e) from the valence band (VB) to the conduction band (CB) leaving a hole (h) in the valence band. The energy of the photon must be greater than the bandgap energy (E_g).

If this is satisfied an absorbed photon may promote an electron from the valence band to the conduction band, leaving a hole in the valence band, as shown schematically in figure C5.2.1. It also means there is a cut-off wavelength λ_c given by the relation

$$\lambda_c[\mu\mathrm{m}] = \frac{1.24}{E_g[\mathrm{eV}]} \tag{C5.2.2}$$

which holds for wavelengths given in μm, and bandgap energy given in eV. Above this wavelength the photon does not have sufficient energy to promote an electron to the conduction band.

The actual way in which the detector is constructed is also critical to the way in which it is used. Such intrinsic semiconductor detectors may be used in one of two modes. In the photoconductive mode the electron-hole pair contribute to an increase in the conductivity of the semiconductor material. In the photovoltaic mode the semiconductor structure has an internal potential barrier (arising due to a p–n junction or other heterostructures) which causes a separation of the electron-hole pair to create a potential difference across the device. There are additional advantages in using photovoltaic devices in that the electrons and holes can be more carefully controlled through the band structure and the associated dark current is generally lower. However, they do require a more complicated structure of differently doped, or physically different, semiconductors to introduce the potential barriers in the band structure. A more extended discussion on the theory of operation of photoconductive and photovoltaic detectors may be found in chapter A7.

C5.2.3 Responsivity, noise and figures of merit

The responsivity of a detector relates the output from the detector as a function of the incident optical power. If we consider the output to be an electric current, i, this is summarized as:

$$i = \Re P \tag{C5.2.3}$$

where $\Re$ is the responsivity and P is the incident optical power.

For photoelectric detectors (such as photovoltaic and photoconductive detectors) the responsivity relates the number of incident photons to the number of charge carriers promoted to the conduction band and may be written:

$$\Re = G\frac{\eta e}{h\nu} \tag{C5.2.4}$$

where $h\nu$ is the photon energy, e is the charge on an electron, η is the quantum efficiency of the detector and G is the gain in a photoconductor detector ($G = 1$ for a photovoltaic detector).

The performance of a detector is not just defined by the responsivity and how big the signal is. The quality of the signal is important and therefore origins of noise on the signal must be considered. We summarize here the expressions for the most important sources of noise (the derivations can be found in any of the books on detectors in further reading).

Thermal noise (also known as Johnson noise) arises from the fluctuations in the thermal motion of charge carriers in a resistive element. The thermal current noise, i_j, this is written as

$$i_j = \sqrt{\frac{4kT\Delta f}{R}} \tag{C5.2.5}$$

where k is Boltzmann's constant, Δf is the bandwidth of the detector system, T is the temperature and R is the resistance of the resistive element.

Shot noise is associated with the discrete nature of the charge carriers. The current shot noise, i_{sn}, may be written as

$$i_{sn} = \sqrt{2ei_T\Delta f} \tag{C5.2.6}$$

where e is the electronic charge, i_T is the total current flowing in the circuit and Δf is the bandwidth of the detector system. Note that the total current may consist of several components including the signal current. For example, there may be contributions due to the dark current of the device (the current that flows even when the detector is not illuminated) and the current due to radiation following on from the target that is not from the object of interest—background radiation.

In a photoconductor there is an additional noise source which arises due to the random generation and recombination of carriers. If the detector is operated in the white noise region this results in an expression similar to that for the shot noise but with an increase in noise by a factor of $\sqrt{2}$ (assuming $G = 1$).

There are other contributing sources of noise, such as $1/f$ noise and electronic amplifier noise, but the user has the option to reduce their effect (by modulating the signal away from dc, or using different electronic configurations), but the expressions given here represent the fundamental limits in what may be achieved. If there is more than one noise source, they combine as the sum of the mean square currents.

Note that the shot noise, or the generation recombination noise in photoconductors, depends on the optical radiation incident on the detector. Thus, even if the detector is operating in a regime where the other sources of noise are minimized, there is still a noise component dependent on the incident radiation. This places a fundamental limit on the noise performance that can be achieved and is often termed the shot noise limit, or signal limit.

These expressions then allow the overall performance of a detector system to be quantified in a single number or figure of merit. The most obvious is by looking at the signal to noise ratio (formally this is the ratio of the signal and noise powers) which is very specific to the particular manner in which a detector is used. This can be extended to define the noise equivalent power (NEP) which is the optical radiation incident on the detector that results in a signal to noise ratio of unity. The NEP is often quoted by manufacturers, but it includes parameters which depend on the actual detector, in particular the bandwidth of the system and the active area of the detector. NEP can be normalized to system bandwidth, but the specific detectivity, D^* (called 'Dee star') is a quantity which allows more direct comparison between detectors and is defined as

$$D^* = \frac{\sqrt{A}\sqrt{\Delta f}}{NEP}. \tag{C5.2.7}$$

where A is the active area of the detector in cm^2. Note the unusual unit of D^* which is cm $\mathrm{Hz}^{1/2}\ \mathrm{W}^{-1}$. When comparing detectors using D^* the measurement conditions must be considered, as it does depend on these conditions which include the wavelength used, the detector temperature, the modulation of the incident radiation and biasing of the detector. Manufacturers will specify in their data sheets the conditions under which D^* is determined.

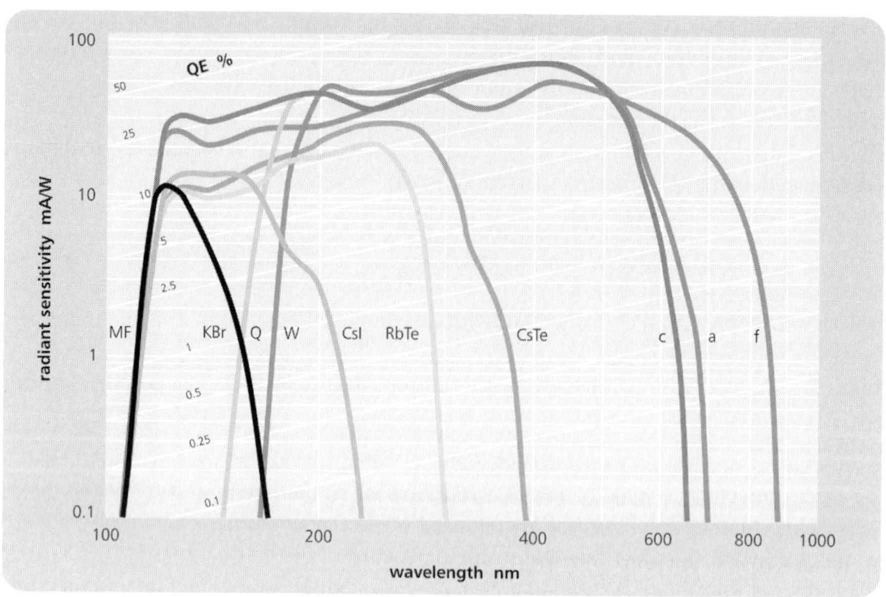

Figure C5.2.2. Photocathode responses in photomultipliers. Photocathodes represented by curves Q-a, W-a and MF-c represent bialkalis. Curve MF-f is the S20 multialkali. (Reproduced by permission of Electron Tubes Ltd, Bury St, Ruislip UK.)

C5.2.4 Detectors for the UV

Detection in the ultraviolet tends to rely on the high energy of the individual photons. The photoelectric effect and the photomultiplier are still as applicable to this particular problem as they have ever been [1]. Of course the photocathode needs to be appropriate to the region of the spectrum of interest, such that the energy of the photon is greater that the work function of the material forming the cathode. Typical spectral responses of photocathode materials are shown in figure C5.2.2. There are a wide range of materials in this spectral region because of the high energy of the individual photons (most of the work on photocathode materials has been to extend the red response). Care must also be taken to consider the material used to construct the window of the photomultiplier due to the high absorption coefficient of many materials in the UV. The quantum efficiency of the photocathode is usually higher in the UV than the visible region of the spectrum. The photomultiplier structure then provides gain through the multiplication process as electrons are accelerated across a potential difference and incident on a dynode where secondary electrons are generated. This process continues along the dynode chain, each successive dynode being held at a higher potential before being collected at the anode and forming the current in an external circuit.

Solid state semiconductor detectors for the UV [2] rely on the same internal process as visible and near infrared detectors, and generally offer a higher quantum efficiency than photoemissive devices (though do not necessarily offer the overall sensitivity of a photomultiplier). The incident photon must have an energy greater than the intrinsic bandgap energy. UV photons have high energies, a wavelength of 200 nm corresponds to a photon energy of 7.25 eV. This energy implies that any of the solid state detectors used in the visible or near-infrared can be used for the UV. This is indeed the case, though there is one main difficulty that has to be overcome—absorption. The high absorption coefficient for these materials in the UV means the active

Table C5.2.1. Properties of semiconductors for UV detectors.

Material	Bandgap energy at 300 K (eV)	Cut-off wavelength (nm)
Si	1.1	1100
GaAs	1.4	880
GaP	2.3	540
GaN	3.39	360
SiC	2.2	560
Diamond	5.5	230

area of the device has to be very close to the surface. In addition, the thin oxide layer that is produced on the surface often has high absorption in the UV and there are often radiation induced effects with extended exposure to UV which reduce performance.

This means that a silicon photodiode may be used as a UV detector but needs special consideration for this application. The depletion layer must be very near the surface for the absorption of the photon to be in this layer and for it to be detected. There are approaches to help arrange this such as the Schottky junction (see, for example [2]). Alternatively, the substrate of the photodiode material may be etched to bring the depletion layer much closer to the surface. This may be used to enhance the UV sensitivity of CCD arrays [3].

Other solid state detectors for the UV were developed to help overcome these problems and offer further advantages. Using material with a higher bandgap energy means the detector would be more wavelength selective. This has a number of advantages, in that the range of wavelengths to which it is sensitive is reduced (e.g. it may be 'solar blind') which in turn may reduce the noise. (Detectors are often limited by background noise from other parts of the spectrum.) A number of materials have been identified and developed, their properties relevant to detection in the UV are given in table C5.2.1. Obviously bandgap is important in considering the detector performance, but so are the physical properties of the material (e.g. how easily it can be used to make detectors and its compatibility with other materials to form heterostructures) and the electrical characteristics for controlling electron and hole transport, biasing considerations and bandwidth potential.

The material characteristics of silicon carbide (SiC) are very attractive for detectors. Not only has it a suitable bandgap, but the processing of the material and its stability are good. Its peak responsivity is around 280 nm (it exhibits some temperature dependence) and is effective down to around 220 nm and up to about 380 nm. Both photoconductive and photovoltaic structures have been developed.

Gallium phosphide is a standard III–V semiconductor and is often used for the blue/green sensitivity it offers. Its sensitivity also extends into the UV. It has usable response from 200 nm to 520 nm with a peak sensitivity around 440 nm. The nitride III–V semiconductor materials also have large bandgaps suggesting suitability for application in the UV. Gallium nitride is the most developed and is used in the 240–380 nm range. Other III–V nitrides have been studies in detail [4]. For example, AlN has a bandgap corresponding to the UV ($E_g = 6.2$ eV at 300 K). Compounds of AlGaN may be used to tailor the peak in responsivity.

Semiconductor diamond as a UV detector is attractive as it has the largest bandgap and has good UV optical properties. In addition it has high thermal conductivity and a small dielectric constant, bringing together properties that are ideal for UV optical detectors. The higher bandgap energy means that it is particularly useful for shorter wavelengths. However, getting high quality devices has meant a poor yield. Recent progress in thin-film diamond technology means there is still work being done in this area [2], although at present there are no commercial devices.

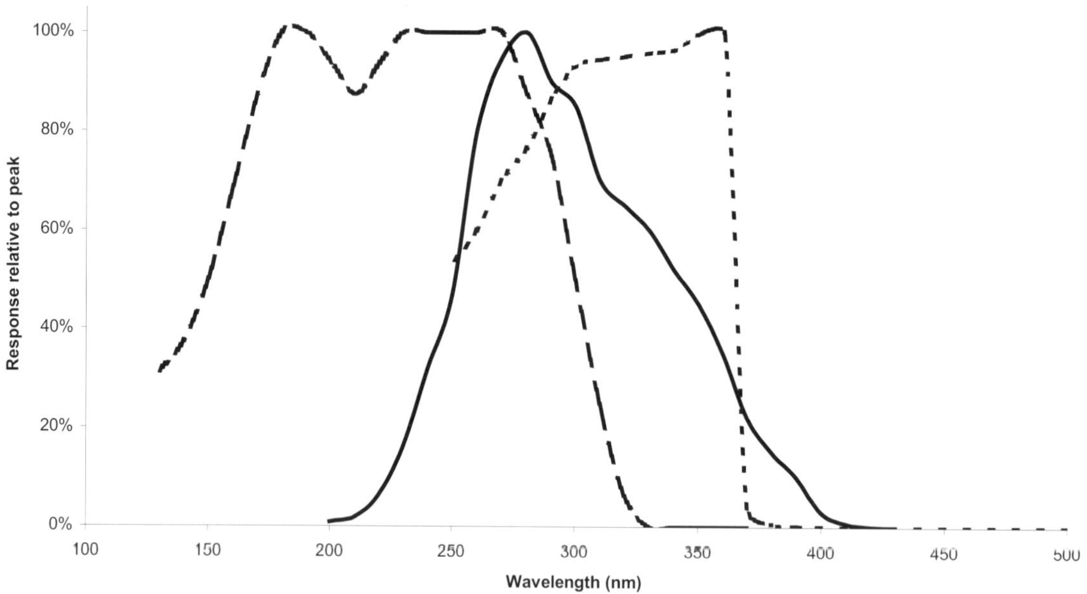

Figure C5.2.3. Manufacturer's data showing the spectral sensitivity of semiconductor detectors in the UV showing SiC (full curve), GaN (dotted curve) and AlGaN (broken curve). (Reproduced by permission of Boston Electronics Corporation, Brookline, MA USA.)

A Schottky barrier can be used to create photovoltaic devices for each of the materials discussed. The Schottky barrier creates a depletion layer between the metal contacting the semiconductor, and the semiconductor material. Thus, the sensitive region of the device is brought very close to the surface. The transmission of the UV through a thin metal layer can be very high.

The response of a selection of commercially available semiconductor detectors for the UV is summarized in the data shown in figure C5.2.3.

C5.2.5 Detectors for the IR

For the purposes of detectors, IR radiation is considered to extend above a wavelength of 1.5 μm up to 14 μm. The measurement of radiation up to the lower end of this region can be satisfied by standard Si, Ge and InGaAs semiconductor technology (see chapter A7). Thermal detectors are used to detect in this region of the spectrum, with the associated trade-off in sensitivity and response time. There are however, a range of semiconductor materials appropriate for radiation up to 14 μm where the photon energy is 0.089 eV. Special consideration must be taken to account for background radiation–more than for other regions of the spectrum, as black body considerations tell us that a body of temperature 300 K radiates with a peak wavelength around 10 μm and the detector may well have a significant response in this region. This is specifically important for infrared sensors where objects in the field of view, by virtue of their equilibrium temperature, will radiate in the spectral region of the detector sensitivity.

This introduces other considerations in considering the limiting performance of the detector. In discussing noise we noted that the shot and generation recombination noise were dependent on the total current in the device. The situation here is that a major contribution to the total current arises from the background radiation and this is the dominant noise source. This leads to the condition where the detector is described as a background limited infrared photodetector or BLIP. If the detector is operating in this regime it establishes

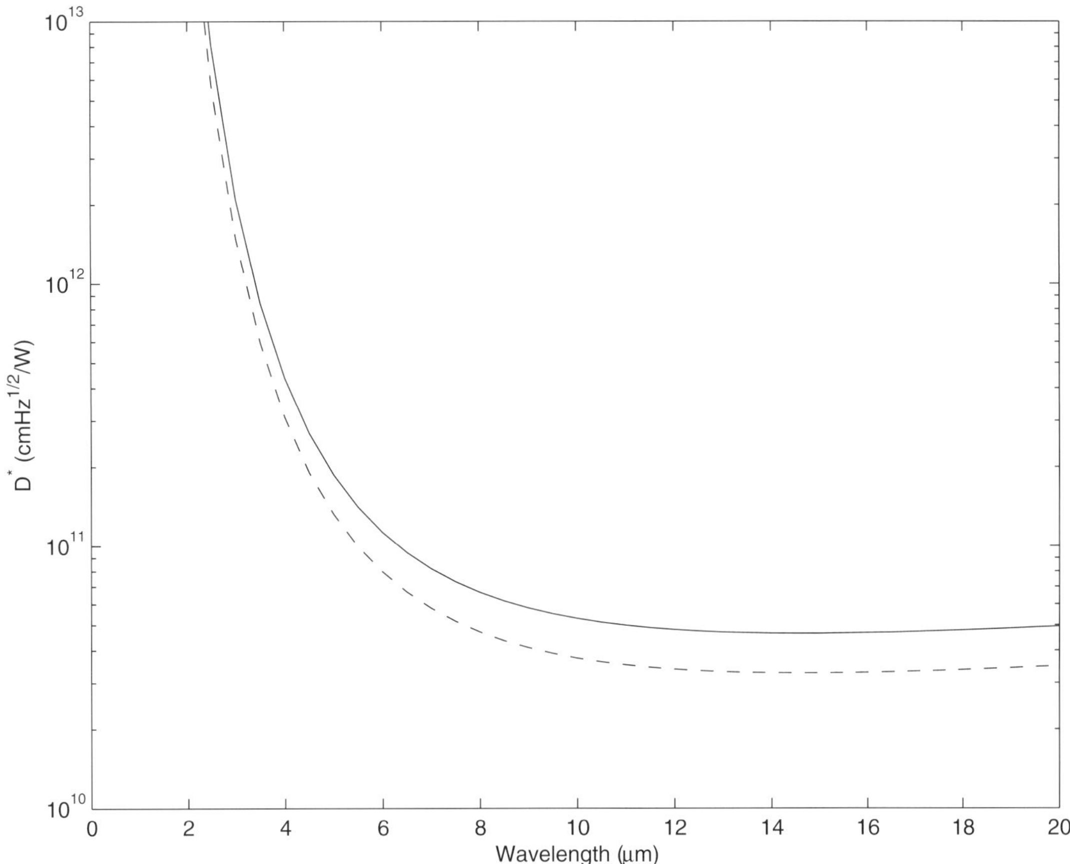

Figure C5.2.4. Calculated BLIP performance limits for photovoltaic (full curve) and photoconductive (broken curve) detectors with a field of view of 180°, and a background temperature of 300 K.

a limit, independent of the actual detector, which cannot be improved upon. If we consider the specific detectivity for a BLIP detector we find, for a photovoltaic detector

$$D^*_{\mathrm{BLIP}} = \frac{\lambda}{hc}\sqrt{\frac{h\nu\eta}{2(\text{Total background power})}} \tag{C5.2.8}$$

where 'total background power' is the incident optical power from the field of view which falls within the spectral response of the detector. For a photoconductive detector, the BLIP specific detectivity is reduced by a factor of $\sqrt{2}$.

As black body radiation tells us the spectral characteristics of objects as a function of temperature, the specific detectivity of an ideal detector as a function of cut-off wavelength of the detector may be calculated for specific fields of view and background temperatures. Calculated BLIP performance curves for a field of view of 180°, and a temperature of 300 K are shown in figure C5.2.4

The detector must be used with these points in mind. It is usual for detectors to be held in a mount that defines the angular field of view of the detector. The mount can then be cooled to minimize the contribution it makes to the background radiation. An example of this type of arrangement is illustrated in figure C5.2.5.

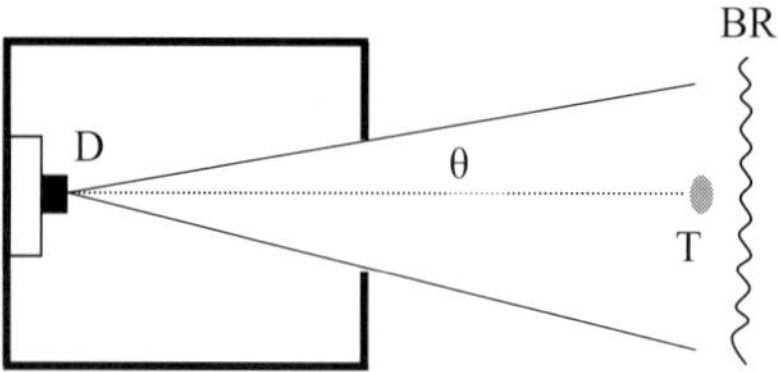

Figure C5.2.5. Illustration of a detector mount to reduce the field of view. D represents the detector, T is the target, BR is the background radiation and θ is the angular field of view limited by the aperture. The mount and aperture are usually cooled to reduces the their thermal background, this also reduces the temperature of the detector, reducing the number of thermally generated carriers.

Table C5.2.2. Extrinsic semiconductors—host material, dopant (d, donor or a, acceptor) and associated ionization energies for IR detectors.

Dopant	Host	Ionization energy at 300 K (eV)	Cut-off wavelength (μm)
B (acceptor)	Ge	0.0104	119
	Si	0.045	27
Al (a)	Ge	0.0102	121
	Si	0.059	22
Ga (a)	Ge	0.0108	115
	Si	0.065	19
P (donor)	Ge	0.0120	103
	Si	0.039	31
As (d)	Ge	0.0127	98
	Si	0.049	25
Sb (d)	Ge	0.0097	127
	Si	0.039	31

This means that the detector will also be cooled (as it is in thermal contact with the mount) which also reduces the thermally generated carriers in the device. The thermal energy (kT) at a room temperature of 300 K is only 0.025 eV. It is not uncommon for devices to be cooled down to liquid nitrogen temperatures to optimize performance.

Photoconductive or photovoltaic techniques based on the intrinsic semiconductor detectors already discussed are not possible in the 2–14 μm region of the spectrum as the individual photons have insufficient energy to promote an electron from the valence to the conduction band. To use semiconductor detectors in the infrared [5] requires new materials that have a small intrinsic bandgap energy. One alternative to this is to use extrinsic semiconductor materials. Extrinsic semiconductors are semiconductors doped with impurities which introduce energy levels in the forbidden region of the bandgap. The generation of a free carrier is then by promoting an electron from a donor energy level to the conduction band, or promoting an electron from the valence band to the acceptor energy level leaving behind a hole. This only requires the absorbed photon to have the energy associated with the energy difference between the edge of the conduction band and the donor energy level, or between the edge of the valence band and the acceptor energy level. This is illustrated in figure C5.2.6.

Some typical dopants together with their ionization energy and corresponding cut-off wavelength are

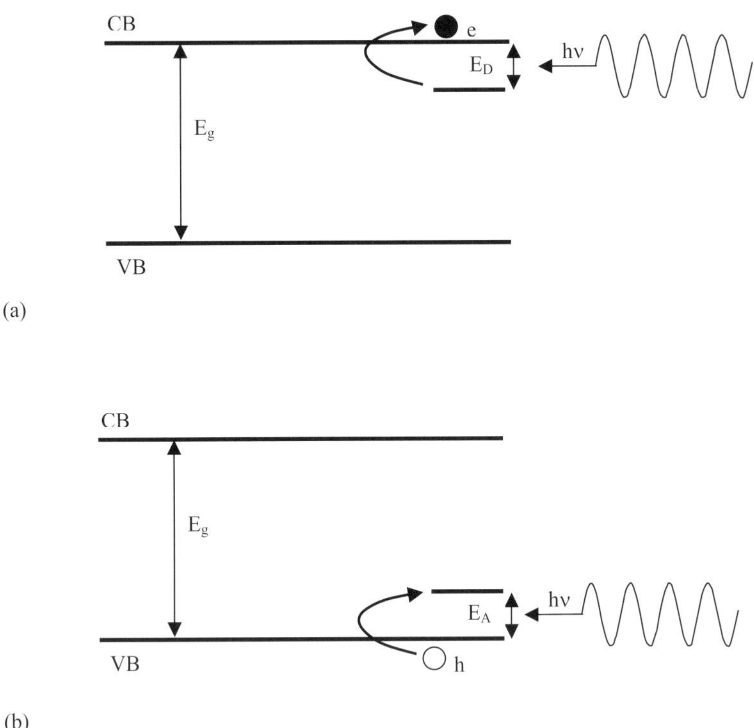

Figure C5.2.6. Absorption of a photon in an extrinsic semiconductor. (*a*) The semiconductor is doped with donor impurities, which introduces an energy level E_{D} below the conduction band (CB) edge. Absorption of a photon of energy greater than this energy ionizes the donor and promotes an electron to the conduction band. (*b*) The semiconductor is doped with acceptor impurities, which introduces an energy level E_{A} above the valence band (VB) edge. Absorption of a photon of energy greater than this energy promotes an electron from the valence band, leaving a hole in the valence band.

shown in table C5.2.2. A large number of materials could potentially be available, but a difficulty is the thermal generation of carriers. To approach the BLIP limit the generation recombination noise (or shot noise) associated with thermal transitions must be negligible compared with the generation recombination noise associated with the background radiation. For intrinsic materials the thermal carrier generation rate is usually lower at a given temperature and bandgap, than for an extrinsic material at the same temperature and with an ionization energy the same as the bandgap of the intrinsic material. Extrinsic semiconductors have impurities in the forbidden region of the bandgap which require lower energies to excite to or from these levels. This is why intrinsic detectors do not need to be cooled as much as extrinsic detectors. This is also one reason why other intrinsic materials are the subject of continued research.

The so-called lead salts, PbS, PbSe and PbTe, and the antimonide detector InSb, all have intrinsic bandgaps appropriate to detection of the medium infrared (typically 1–8 μm). These detectors have their own respective advantages and disadvantages (see, for example, Dereniak in further reading). Indium antimonide is easily prepared as a single crystal, and controlling the ratio of In to Sb dictates whether it is intrinsic, n-type or p-type semiconductor allowing it to be used as a photovoltaic detector (though the absorption is still across the intrinsic bandgap). In contrast, it is difficult to achieve high purity in the lead salts, and these impurities give rise to energy levels in the bandgap. This in turn increases the noise due to thermally generated carriers. Thus, for a given operating condition the InSb detector has a better D^* value. Generally it also offers faster response.

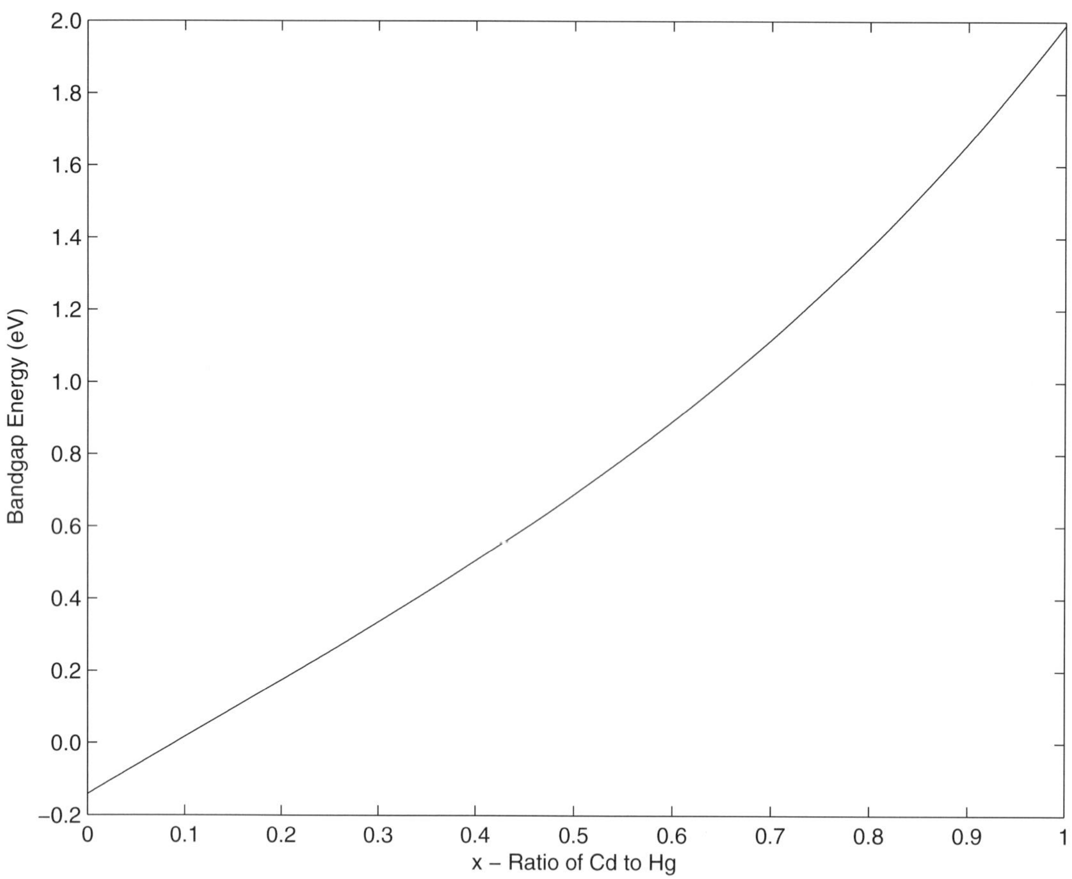

Figure C5.2.7. Variation in bandgap energy of $Hg_{1-x}Cd_xTe$ with x.

A more adaptable material for detection in the infrared is mercury cadmium telluride, $Hg_{1-x}Cd_xTe$, or MCT [6]. MCT is a semiconductor that operates as a infrared detector across its intrinsic bandgap and is a widely used device because of its good detectivity [7]. The response can also be tailored, as the bandgap energy varies with the ratio of mercury to cadmium. The variation in intrinsic bandgap with ratio of mercury to cadmium is shown in figure C5.2.7. With a value of $x = 0.2$ the bandgap is 0.1 eV and the sensitive range extends out to 14 μm. Thus, MCT offers the user flexibility in controlling the responsivity by altering the composition. Since the lattice constant changes very little as the composition is changed, it is also possible to make heterostructures. However, it is hard to grow with uniform x, and is difficult to handle. It is widely accepted that it is more difficult to process low bandgap material than high bandgap (such as GaAs).

A MCT detector can approach BLIP performance at higher temperature (compared with extrinsic semiconductor materials) as it operates as an intrinsic device. Much of the current effort on these devices is directed at improving frequency response, constructing arrays, reducing noise and reducing cooling requirements [6]. These last two requirements are related as the dark current increases as temperature rises. One area which is being examined is to use a quaternary semiconductor such as HgCdZnTe [8]. Arrays can also be produced [9] which exploit the use of heterostructures to reduce the cooling requirements.

A further area which has attracted significant development is to use quantum wells [10, 11] to create a

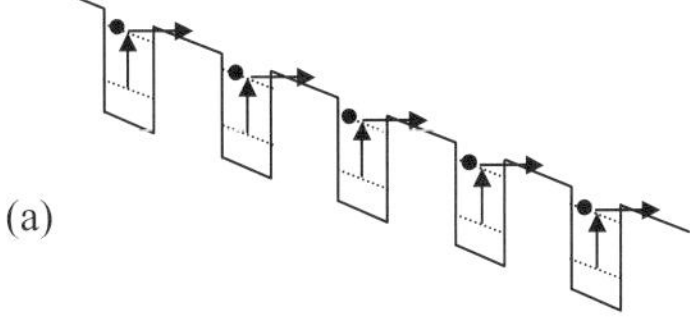

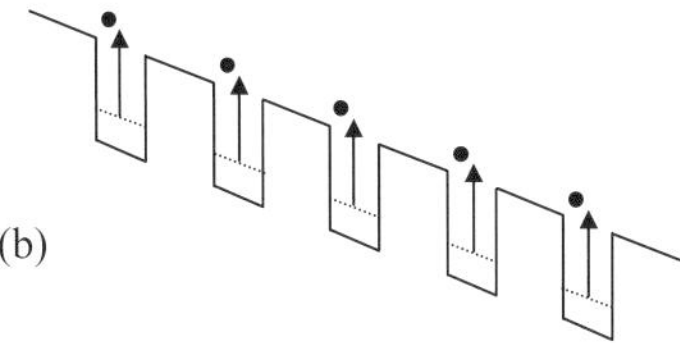

Figure C5.2.8. Representation of the conduction band structure for quantum well detectors. (*a*) Absorption of a photon promotes the electron to the upper bound state of the quantum well from which it tunnels into the continuum. (*b*) Absorption of a photon promotes the electron directly to the continuum of the conduction band.

quantum well infrared photodetector (QWIP). The quantum wells are designed such that the energy required to promote a carrier from a lower energy level of the well to a second bound state near the top of the well, is close to the energy of the incident photons. Biasing of the device then allows the excited electrons to tunnel out to the continuum of the conduction band. Alternatively, the well can be designed so that absorption of the photon promotes a carrier directly to the continuum of the conduction band. Both of these situations are shown in figure C5.2.8. The second construction does tend to have a slightly higher absorption [5]. The two situations also have different spectral responses. For a transition from a bound state of the well to a quasibound state the ratio of the spectral width of the responsivity, $\Delta\lambda$, to the peak wavelength, λ, is typically 10%. For a transition from a bound state to the continuum the same ratio is $\sim$24%. This is due to the fact that the excited state is above the barrier and so the energy associated with this is broad [10].

The use of quantum well devices is an attractive alternative to extrinsic devices as the probability of absorption (the dipole of the transition) is greater for the quantum well than the donor or acceptor site. They also are built on GaAs and AlGaAs semiconductor structures, which have a higher bandgap and so are easier to process. A large number of wells may be stacked to increase the sensitive area—50 wells is not uncommon. This may seem a demanding requirement, but it should be remembered that this is a very mature technology and the geometry and dimensions of the wells can be accurately controlled. This then allows the energy levels of the quantum wells and the spectral response to be tailored. As an example, in some of the first examples of these devices [12], a series of 50 GaAs wells of 6.5 nm separated by barriers of $Al_{0.25}Ga_{0.75}As$ of 9.5 nm should give rise to wells with two bound states separated by an energy close to 10 μm. Experimental examination of the first such device found a peak wavelength of 10.9 μm. Extending the design to combine structures which exhibit two wavelength response, multiple wavelength or broadband response is relatively straightforward. A full review and comparison of various types of QWIPs has been carried out by Gunapala and Bandara [13]. QWIP devices compete directly with MCT detectors, and are currently only just behind in terms of detectivity. Recent work is now extending these original ideas to create

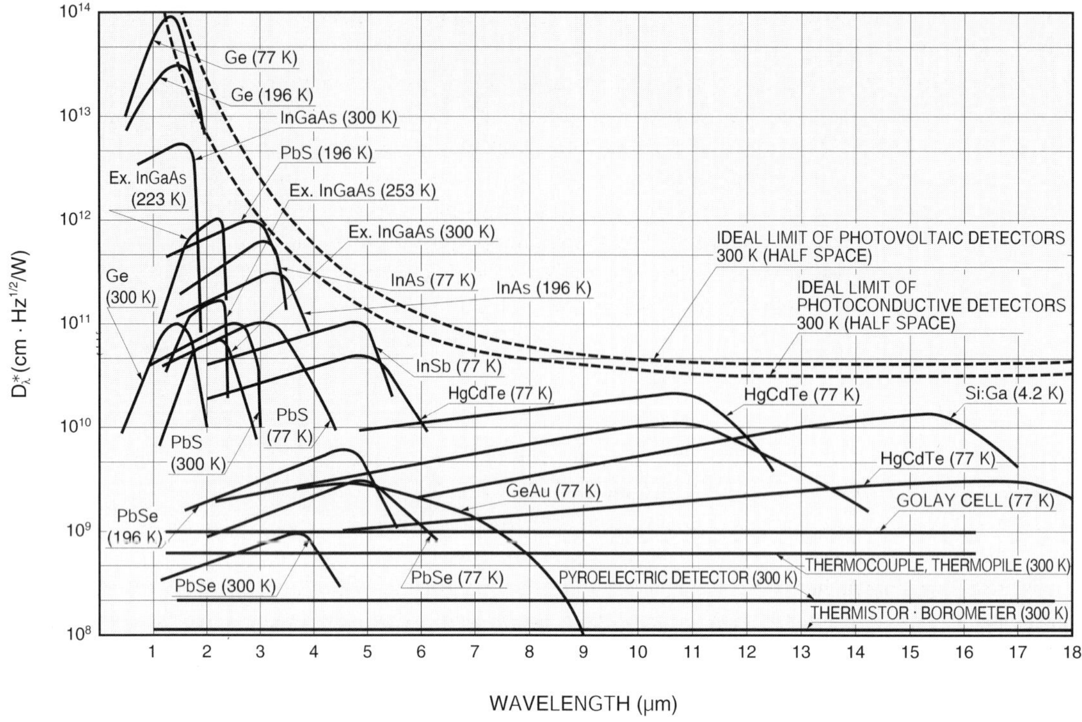

Figure C5.2.9. Specific detectivity for a range of detectors across the visible and IR, with the background limit for a photoconductive and photovoltaic detector indicated for a background temperature of 300 K. (Reproduced by permission of Hamamatsu Photonics KK, Japan.)

devices based on superlattices (multiple quantum wells with a thin barrier layer), for example, work has been reported on a device in which the responsivity is tunable by adjusting the external bias [14].

Moving from a single element detector to a detector array allows images to be recorded in the infrared. The mature technology of semiconductor material processing applied to GaAs and AlGaAs means extending the growth process to create QWIP arrays reveals a host of new applications, and at relatively low cost. There are many potential applications of such devices. For example the absorption lines of many gas molecules lie in the 2–14 μm range which opens up many applications in atmospheric and space-based spectroscopy. Recent reports have described a 640 $\times$ 486 QWIP focal plane array with a pixel pitch of 25 μm, and a pixel size of 18 $\times$ 18 μm. The peak responsivity of the device was 8.5 μm. The noise equivalent temperature of the device (the temperature difference across the array which would produce a signal ratio of unity) was 36 mK [13, 15].

Thermal detectors (bolometers) are still of relevance as they span the entire spectrum without concern for the photon energy. The development of new devices and materials discussed previously, and the long time response of bolometers mean they may no longer be the first choice. Work has recently been reported on a silicon microbolometer array. The silicon is etched to leave a thermally isolated (50 by 50) μm absorber. The small size of the device means the time constant is small and using mature silicon processing technology arrays can be produced. A 'heat balancing' process, where electrical energy is delivered to each bolometer to balance the absorbed radiation means extensive calibration is not necessary [16, 17]. These devices are available commercially and are widely used in thermal imaging cameras (see, for example, Raytheon Commercial Infrared, Dallas USA).

Another thermal detector is the pyroelectric detector. These are the detectors that are widely used in the passive infrared detectors of intruder alarms, though they are not common in laser systems. The operation of this detector relies on the change in the electric polarization of the crystal material as the temperature increases due to the absorbed optical radiation. This in turn gives an output current. As the device responds to temperature changes it does not show any DC response, only giving output as the temperature rises or falls. If such a detector is to be used in a DC system, chopping of the signal is required. These devices are thermal, so the spectral response is dictated by the window material enclosing the detector. As a thermal detector there is a trade-off between sensitivity and response time.

C5.2.6 Summary

The detection of optical radiation across the spectrum relies not only on having a detector with the appropriate spectral response, but also on a careful consideration of noise. Noise can arise due to processes intrinsic to the detection system and also arises due to spectral sensitivity to other sources of radiation, and the manner in which the detector is used is critical to the final performance.

Figure C5.2.9 shows the specific detectivity of detectors across the spectrum in relation to the BLIP limit for both photoconductive and photovoltaic devices. This illustrates that commercial sensors are approaching the background limits. Work continues to improve detector performance and to push performance even closer to the limits.

The spectral regions of the UV and the IR present special challenges to detectors. Detector technology has risen to these challenges and detector performance continues to improve. Material developments open new regions of sensitivity, and new structures reduce noise and cooling requirements which mean the user benefits from increased performance and a wide range of devices to choose from across the spectrum.

References

[1] Lerner E J 2001 Photomultiplier tubes stay the course *Laser Focus* **37** 151–3

[2] Razeghi M and Rogalski A 1996 Semiconductor ultraviolet detectors *J. Appl. Phys.* **79** 7433–73

[3] Schaefer A R and Janesick J A 1994 *Charge Coupled Devices Sensor Technology and Devices* ed L Ristic (Boston, MA: Artech House) pp 341–76

[4] Morkoc H, Strite S, Gao G B, Lin M E, Sverdlov B and Burns M 1994 *J. Appl. Phys.* **76** 1363–98

[5] Elliot C T and Gordon N T 1993 *Infrared Detectors Handbook on Semiconductors* vol 4, ed C Hilsum (Amsterdam: North-Holland) pp 841–936

[6] Piotrowski J and Rogalski A 1998 New generation of infrared photodetectors *Sensors and Actuators* A **67** 146–52

[7] Piotrowski J and Perry F 1997 Designers still choose mercury cadmium telluride *Laser Focus* **33** 135–41

[8] Piotrowski J 1994 New generation of near-room temperature photodetectors *Opt. Eng.* **33** 1413–21

[9] Elliott C T, Gordon N T, White M A 1999 Towards background limited, room-temperature, infrared photon detectors in the 3–13 μm wavelength range *Appl. Phys. Lett.* **74** 2881–3

[10] Bandara S V, Gunapala S D, Liu J K Luong E M, Mumolo J M, Hong W Sengupta D K and McKelvey M J 1998 *Appl. Phys. Lett.* **72** 2427–9

[11] Levine B P 1993 Quantum-well infrared photodetectors *J. Appl. Phys.* **74** R1–81

[12] Levine B P, Choi K K, Bethea C G, Walker J and Malik R J 1987 New 10 μm infrared detector using intersubband absorption in resonant tunnelling GaAlAs superlattices *Appl. Phys. Lett.* **50** 1092

[13] Gunapala S D and Bandara S V 2000 *Semicond. Semimetals* **62** 197–282

[14] Chen C C, Chen H C, Hsu M C, Hsieh W H, Kuan C H, Wany S Y and Lee C P 2002 *J. Appl. Phys.* **91** 943–8

[15] Gunapala S D, Bandara S V, Singh A, Liu J K, Luong E M, Mumolo J M and LeVan P D 2000 Recent developments and applications of quantum well infrared photodetector focal plane arrays *Physica* E **7** 108–11

[16] Liu C C and Mastrangelo C H 2000 A CMOS uncooled heat-balancing infrared imager *IEEE J. Solid-State Circuits* **35** 527–35

[17] Lerner E J 2001 Uncooled IR detectors move into the mainstream *Laser Focus* **37** 201–4

Further reading

Boyd R W 1983 *Radiometry and the Detection of Optical Radiation* (New York: Wiley)

A good introduction to the theory of detectors and consideration of the parameters affecting detector performance.

Dereniak E L and Boreman G D 1996 *Infrared Detectors and Systems* (New York: Wiley)

Comprehensive coverage of infrared detectors.

Keyes R J 1980 *Optical and Infrared Detectors* 2nd edn (Berlin: Springer)

Good discussion on the operating principles and applications.

Kingston R H 1978 *Detection of Optical and Infrared Radiation* (Berlin: Springer)

Quite theoretical but covers techniques in the visible and infrared, though not as up-to-date as the others.

Rieke G H 1994 *Detection of Light from the Ultraviolet to the Submillimeter* (Cambridge: Cambridge University Press)

Covers a range of detector principles across the spectrum.

Sze S M 1981 *The Physics of Semiconductor Devices* (New York: Wiley)

A good introduction to the semiconductor of devices with some discussion on detectors.

Wood D 1994 *Optoelectronic Semiconductor Devices* (London: Prentice-Hall)

A more general book on devices with a good background on solid state physics and three chapters on detectors for the visible and infrared.

C5.3
Laser energy and power measurement

Robert K Tyson

To fully characterize a laser one must measure the energy and the power or, more specifically, the *radiant energy* in joules (J) and *radiant power* or *energy flux* in J s^{-1}. Due to the nature and diversity of lasers, which operate from the deep ultraviolet to the infrared, from continuous output to picosecond pulses and from microwatts to megawatts, energy-measuring devices encompass a wide range of possibilities [1–3]. Detectors are sensitive to the spectrum of the beam and the speed of energy deposition in their conversion of energy to a useful readable or recordable quantity that can be interpreted as energy or power.

A detector is any device in which optical radiation produces a measurable physical effect. Laser energy measurement devices are divided into general groups: calorimetric, photoelectric, photochemical and mechanical. Detector methods are covered in detail in chapter A7. A specific method is chosen depending upon the duration of the energy pulse, the spectral response of the detector and the dynamic range of the incident energy. Absolute power and energy measurement and laser dose (energy density) measurement is dependent upon the accurate calibration of the detectors. Calibration, data collection and data interpretation are three necessary steps of the *measurement* process.

C5.3.1 Measurement technique selection

The choice of measurement technique is governed by the application and generally when the measurement technique closely resembles the application the measurement is most useful. Lasers must meet standards dictated by their application in manufacturing, electronics, medicine, communications and the military. The first characteristic dictating the measurement is whether the laser is continuous wave (cw) or pulsed. In terms of modern energy measurement, applications for cw lasers include optical (lightwave) communications, laser-based medical instrumentation, materials processing, photolithography, data storage and laser safety equipment. Accurate measurement is required for pulsed lasers such as ultraviolet excimer lasers (optical lithography for semiconductors, corneal sculpting for photorefractive keratectomy (PRK) or laser *in situ* keratomileusis (LASIK), and micro-machining of small structures) and deep-UV lasers used for lithography.

These applications have various power levels and wavelengths and can be grouped into standard calibration categories [3].

- cw laser power below 2 W,
- cw laser power at 1064 nm above 2 W,
- cw laser power at 10.6 μm,
- pulsed laser energy (Q-switched YAG at 1064 nm) and
- pulsed laser energy (KrF excimer at 248 nm and ArF at 193 nm).

For any of the methods, despite their variation in employed physical principles, the measuring device must respond in a predictable fashion to the actual incident energy [4]. This responsivity $\Re$ is expressed as the ratio

$$\Re = C\frac{\text{output}}{\text{power incident}} \tag{C5.3.1}$$

where the output is the measurable voltage, current, resistance, pressure etc of the detector and C is a calibration factor. Because C can be, and almost always is, dependent upon wavelength, pressure, temperature, quantum efficiency, frequency, detector area, load resistance or noise, there are always important steps in calibrating the detectors and determining their useful ranges all the while relating them to radiometric standards.

Whenever the desired range of operation for a device is exceeded, it is necessary to employ electronic or optical techniques to keep the responsivity known [2]. Sometimes it is necessary to match the angular response of detectors with integrating spheres or diffusers. When the source energy or power exceeds the linear region of a detector, calibrated neutral density filters can be used to bring the incident energy onto the detector into a usable regime. Filters can also be used to match the spectral response of the detector to the spectrum of the incident radiation. The addition of choppers for attenuation or ac amplification is often employed. In all cases of photoelectric measurement of power and energy it is necessary to maintain calibration and international standardization [5].

C5.3.2 Test configuration

For laser power and energy measurement, it is necessary to maintain repeatable and calibrated environments. For example, isoperibol (constant temperature environment) calorimeters are used in many cases. Solid state and thermal detectors are used with well-characterized transfer standards (calibrated against primary standards). Ultra-high accuracy can be achieved using a cryogenically cooled electrical substitution laser power meter [3].

Laboratories often use beamsplitter-based calibration systems that allow various power and energy detectors. The critical parameters such as absorptivity and transmittance are well characterized at all the operating wavelengths. Small angles of incidence are used to minimize polarization effects.

Optical fibre power meters are calibrated using a system in which the test meter and the laboratory standard are both exposed to a stable diode laser source. For many commercial and standards laboratory measurements, the reported results include both the test data and the calibration data.

References

[1] Budde W 1983 *Physical Detectors of Optical Radiation* (New York: Academic)

[2] Heard H G 1968 *Laser Parameter Measurements Handbook* (New York: Wiley)

[3] National Institute of Standards and Technology (NIST), 2001, Electronics and Electrical Engineering Laboratory, Optoelectronics Division Programs, Activities, and Accomplishments, NISTIR 6602, January

[4] Grum F and Becherer R J 1979 *Optical Radiation Measurements* vol 1 (New York: Academic)

[5] Commission Internationale de l'Eclairage (CIE) 1982 *Methods for Characterizing the Performance of Radiometers (Photometers) (Publ. No. 53)* (Paris: CIE)

Further reading

Acetta J S and Shumaker D L (ed) 1993 *The Infrared and Electro-Optical Systems Handbook* (Ann Arbor, MI: ERIM) and (Bellingham, WA: SPIE)

Bode D E 1976 Infrared detector technology at the Santa Barbara Research Center *Proc. Electro-opt. Syst. Des. Conf.* p 63

Dowell M L 2001 Choosing the right detector is key to accurate beam power measurements *SPIE OE Magazine* **56**

Lehman J and Li X 1999 A transfer standard for optical fiber power metrology *Opt. Photon. News* **10** 44f–h

Leong K H, Holdridge D J and Sabo K R Characteristics of power meters for high-power CO_2 lasers 1994 *J. Laser Appl.* **6** 231–6

Putley E H 1980 Thermal detectors *Optical and Infrared Detectors* ed R J Keyes (Berlin: Springer) ch 3

Vayshenker I, Li X and Scott T R 1994 Optical power meter calibration using tunable laser diodes *Proc. Natl Conf. Stud. Lab. (Jul 31–Aug 4, Chicago, IL)* pp 337–52

West E D and Schmidt L B 1987 A system for calibrating laser power meters for the range 5–1000 W *Natl Bur. Stand.* (US) Technical Note 685

C5.4
Irradiance and phase distribution measurement

B Schäfer

This section concerns the measurement of the irradiance distribution $I(x, y)$ [1] and the phase distribution $\phi(x, y)$ [2] of laser beams in xy-planes perpendicular to the direction of beam propagation. The topic intrinsically includes aspects of spatial coherence measurement which are, therefore, considered in the last section.

C5.4.1 Basic concepts and definitions

In this context irradiance terms, i.e. either the relative power density or the relative energy density depending on whether cw or pulsed laser beams are considered. A convenient phase definition, which applies to fully and partially coherent beams, is based on the time averaged Poynting vector distribution $\boldsymbol{S}(x, y)$ according to [3]

$$\frac{\boldsymbol{S}_\perp}{|\boldsymbol{S}|} = \frac{\lambda}{2\pi}\nabla_\perp\phi(x, y) \qquad \text{(C5.4.1)}$$

where λ denotes the mean wavelength of light and $\nabla_\perp$ the two-dimensional gradient[1]. Defined in this way, $\phi(x, y)$ represents the deterministic part of the fluctuating phase associated with partially coherent fields and for coherent beams it coincides with the familiar definition as a surface of constant phase. However, equation (C5.4.1) is consistent only if $\nabla \times \boldsymbol{S} = 0$.

Whenever this condition is violated as, e.g., for some higher-order Gauss–Laguerre modes, an adequate description of the beam requires the full Poynting vector distribution.

C5.4.2 Principal measurement set-up

The principal experimental arrangement for electronic laser beam analysis (figure C5.4.1) consists of a beam sampler for producing a high-quality low-energy replica of the original beam, a filter set for further attenuation, an optional re-imaging optics in order to adjust the beam diameter and sensor aperture, an irradiance or phase distribution sensor and a readout device, normally a personal computer system equipped with a frame grabber and beam analysis software.

C5.4.3 Irradiance distribution measurement

C5.4.3.1 Scanning devices

Mechanical scanning devices are commonly based on a rotating drum or a moveable opaque screen, with slits, knife edges or pinholes mounted on it and a single-element detector positioned behind (figure C5.4.2).

[1] The quantity $w(x, y) = \lambda\phi(x, y)/2\pi$ is often called the wavefront of the beam. If there is no risk of confusion, both terms—phase and wavefront—will be used simultaneously.

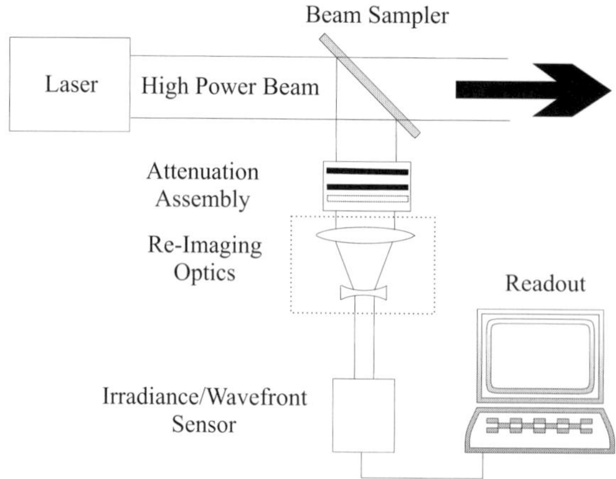

Figure C5.4.1. Setup for electronic irradiance and phase distribution measurement.

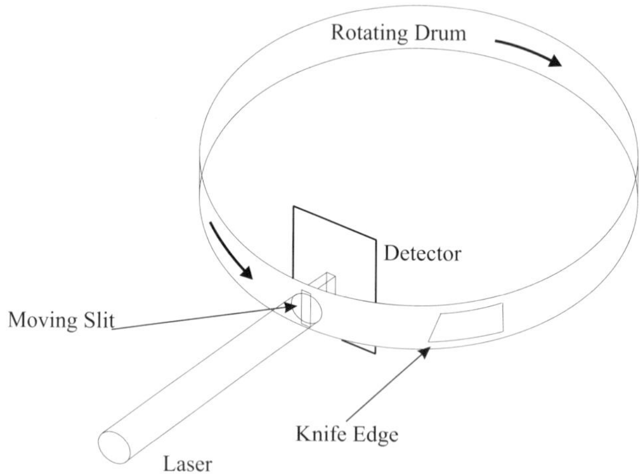

Figure C5.4.2. Principal arrangement of a scanning laser beam profiler.

These methods work only with cw beams but the requirements on beam attenuation are rather modest, as only a small part of the total beam power hits the detector. The spatial resolution can be as good as 0.5 μm but only a few ($\sim$10) one-dimensional profile scans are delivered at a time, thus yielding quite limited off-axis information.

C5.4.3.2 Camera based systems

Camera-based systems deliver the complete two-dimensional irradiance distribution, covering a wavelength range from the far-IR to the extreme UV [4]. They are suitable for cw as well as for pulsed beams and, by means of fast electronic shutters, they can pick up single pulses from a 10 kHz pulse train. In general, very strong beam attenuation is needed to account for their high sensitivity. Without re-imaging, the spatial resolution is limited by the pixel size, which ranges between 5 and 15 μm for CCD detectors and approximately 30 μm for pyroelectric cameras. For optimum re-imaging conditions, the detector diameter d_{Det} should accommodate

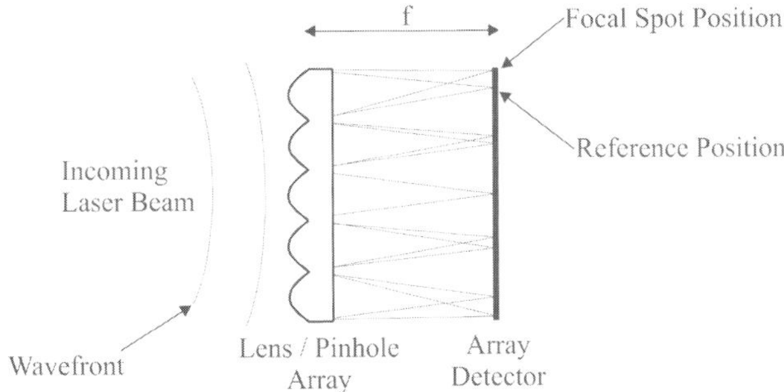

Figure C5.4.3. Principle of the Hartman and Hartmann–Shack wavefront sensors (see text).

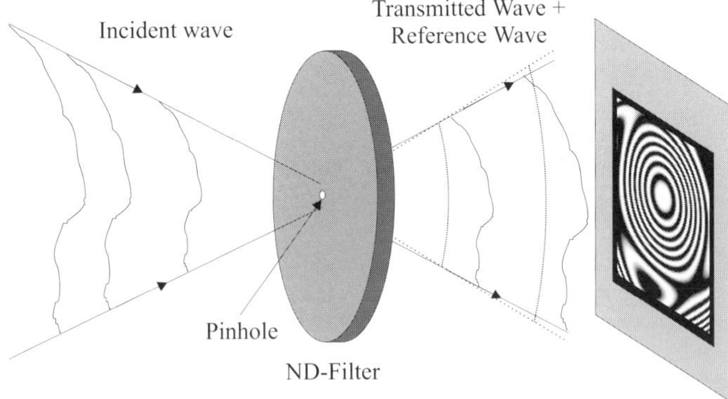

Figure C5.4.4. Basic setup of the point diffraction interferometer (see text).

the beam profile in a way that less than 1% of the total beam power is lost by clipping [5]. The corresponding magnification, M, of the beam yields the effective resolution, δ:

$$\delta = \frac{d_{\mathrm{Det}}}{\sqrt{N_{\mathrm{tot}}}M}.$$

The total number of pixels , N_{tot}, varies between 128 × 128 say for pyroelectric cameras up to 4k × 4k for digital CCD cameras. Qualitative profile inspection as well as low-accuracy beam parameter estimation can be performed by standard 8-bit cameras but higher demands require 12- or even 16-bit dynamic resolution. A special problem occurring for vidicon-type cameras and for some interline transfer CCD chips is a spatially varying baseline offset which must be compensated by subtraction of a background map if reliable quantitative results are desired.

C5.4.4 Phase distribution measurement

C5.4.4.1 Hartmann–Shack wavefront sensor

The Hartmann principle [6–10] is based on an orthogonal or hexagonal array of lenses (Hartmann–Shack) or pinholes (Hartmann), which divides the incoming beam into a large number of sub-rays (figure C5.4.3). The total irradiance and the position of the individual spots are monitored with a position-sensitive detector placed at a distance f behind the array. The displacement of the spot centroid $\boldsymbol{x}^{\mathrm{c}}$ with respect to a plane-wave

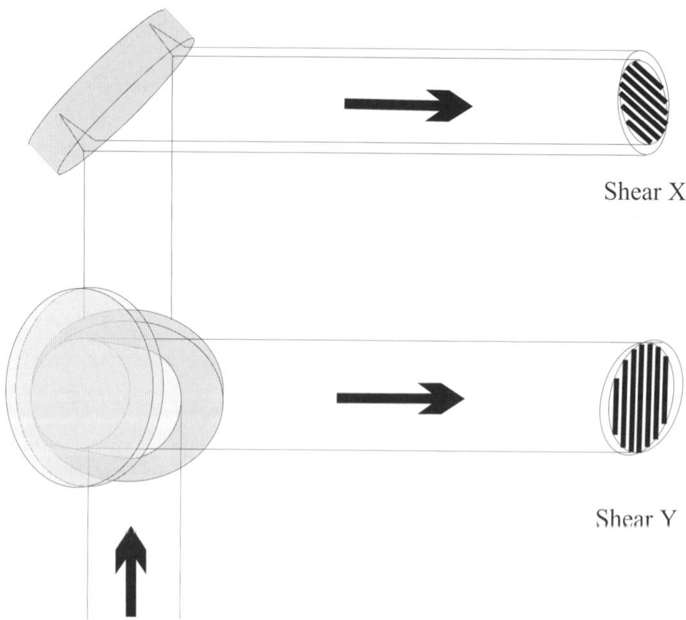

Figure C5.4.5. Operation principle of an etalon-based lateral shearing interferometer (see text).

reference position $\boldsymbol{x}^{\mathrm{r}}$ measures, for some sub-aperture (i, j), the local phase gradient distribution according to[2]

$$\begin{pmatrix} \partial\phi/\partial x \\ \partial\phi/\partial y \end{pmatrix}_{ij} = \frac{2\pi}{f\cdot\lambda}\begin{pmatrix} x^{\mathrm{c}} - x^{\mathrm{r}} \\ y^{\mathrm{c}} - y^{\mathrm{r}} \end{pmatrix}_{ij} \tag{C5.4.2}$$

Obviously, the accuracy in phase estimation depends on the quality of the reference wavefront and on how accurate the centroid estimation is performed. The latter is not only influenced by detector properties as pixel size and dynamic range but suffers from diffraction-induced cross-talk between neighbouring sub-apertures. However, crosstalk can be reduced efficiently in several ways [11, 12], so relative rms wavefront errors as low as $\lambda/20$ at 633 nm are achievable even with quite moderate effort [11]. If compared to interferometrical techniques, the spatial resolution ($\sim 50\,\mu$m) is lower by an order of magnitude but, on the credit side, Hartmann–Shack systems are capable of sensing white light, they are suited for pulsed laser beams and there is no phase ambiguity.

C5.4.4.2 Interferometers

In principle, there are many interferometer designs which are capable of laser beam phase determination. In this section two particular attractive approaches utilizing the method of self-referencing are briefly considered.

The point-diffraction interferometer (PDI) is a common path interferometer exhibiting a simple design [13] (figure C5.4.4). A laser beam is directed onto a neutral density (ND) filter containing a small pinhole. Diffraction at this pinhole creates a virtually undisturbed spherical reference wave which interferes with the beam. Furthermore, a configuration working in reflection rather than transmission is also possible. Except for a quadratic lag, the interference pattern in the detector plane gives directly the phase distribution with

[2] The wavefront reconstruction from the gradient data is performed by standard techniques of linear theory and the reader is referred to [14] and [15] as well as references therein for a detailed analysis.

reference to the pinhole position. Obviously, the spherical reference wave has to interfere with the whole beam, so the latter must be globally coherent.

The principle of a lateral shearing interferometer [16] consists of amplitude division of the beam and introducing a small relative amount of transverse shear, s, which is produced, for example by dual frequency gratings [17] or precision etalons (figure C5.4.5). Interference of the two beams produces fringes which represent, in the limiting case of small shear, curves of constant phase gradient. Complete gradient information requires shear in two orthogonal directions. The extraction of the phase differences $\Delta\phi$ from the detector signal N_{d}^{ij} at pixel (ij):

$$\begin{aligned} N_{\mathrm{d}}^{ij} = \int_{\text{pixel}} & \mathrm{d}x\,\mathrm{d}y\, I(x, y) + I(x + s_{\mathrm{s}}, y + s_y) \\ & + 2\sqrt{I(x, y) I(x + s_{\mathrm{s}}, y + s_y)}\, g(x, y, s_x, s_y, \tau) \cos(\Delta\phi(x, y, s_x, s_y, \tau)) \end{aligned} \qquad \text{(C5.4.3)}$$

with g denoting the modulus of the complex degree of coherence and $c\tau$ the optical path difference (OPD) between both sheared beams, is performed either by ac modulating the phase with angular frequency ω (ac-shearing interferometer) or by introducing three or four fixed delays (three- or four-bin shearing interferometer) in one of the beams before detection. Finally, the phase gradient is obtained as

$$\frac{\partial\phi^{ij}}{\partial x} \approx \frac{\Delta\phi(x_i, y_j, s_x, 0)}{s_x} \qquad \frac{\partial\phi^{ij}}{\partial y} \approx \frac{\Delta\phi(x_i, y_i, 0, s_y)}{s_y}. \qquad \text{(C5.4.4)}$$

From equation (C5.4.3) it is evident that a reliable phase gradient measurement requires a certain amount of temporal and spatial beam coherence. However, these restriction are much weaker than for the PDI and even white light can be sensed by some modifications of the principal design [18]. Regarding the measurement accuracy, commercial devices yielding rms wavefront errors of $\lambda/50$ at 633 nm are already available.

C5.4.4.3 Phase retrieval from intensity transport equations

The law of energy conservation applied to equation (C5.4.1) yields with $|\boldsymbol{S}(x, y)| \approx I(x, y)$:

$$\nabla \cdot (I \cdot \nabla\phi) = 0 \qquad \text{(C5.4.5)}$$

and, finally, the paraxial approximation gives the intensity transport equation:

$$k \cdot \frac{\partial I(\boldsymbol{x}, z)}{\partial z} = -\nabla_{\perp} \cdot (I(\boldsymbol{x}, z) \cdot \nabla_{\perp}\phi(\boldsymbol{x}, z)) \qquad \text{(C5.4.6)}$$

It is, therefore, possible to determine the phase distribution from the derivative of the irradiance distribution in the direction of beam propagation by solving the intensity transport equation [19, 20]. The advantage of this approach is that only irradiance distribution measurements in two or three neighbouring z-planes are needed and that there are no requirements on beam coherence. However, a partial differential equation has to be solved and real-time operation requires the synchronization and simultaneous evaluation of several detectors.

C5.4.5 Coherence measurement

A possible way of determining the coherence properties of laser beams, which is still frequently employed in practice, is Young's experiment employing pairs of slits or pinholes in an opaque screen. Indeed, each term in the expression for the mutual coherence function Γ,

$$\Gamma(x_1, y_1, x_2, y_2; \tau) = \sqrt{I(x_1, y_1)}\sqrt{I(x_2, y_2)}\, \underbrace{g(x_1, y_1, x_2, y_2; \tau) \exp[\mathrm{i}\phi(x_1, y_1, x_2, y_2; \tau)]}_{\gamma(x_1, y_1, x_2, y_2; \tau)} \qquad \text{(C5.4.7)}$$

has its experimental Young counterpart [21]. In particular, the coherence function g determines the contrast in the centre of the interference pattern but, unfortunately, as g depends on four variables, even highly automated measurements become very tedious. A much more efficient way to determine the beam coherence uses a lateral shearing interferometer [22, 23]. It is striking from equation (C5.4.3), that the variation of the local fringe contrast with the applied shear enables the evaluation of g across the whole beam profile and, moreover, the temporal coherence can be estimated from the variation of the fringe visibility with the OPD between both beams.

References

[1] ISO/DIS 13694, Optics and optical instruments—Lasers and laser-related equipment—Test methods for laser beam power (energy) distribution

[2] ISO/CD 15367-1, Lasers and laser related equipment—Test methods for determination of the shape of a laser beam wavefront—Part I: Terminology and fundamental aspects

[3] Paganin D and Nugent K A 1998 Noninterferometric phase imaging with partially coherent light *Phys. Rev. Lett.* **80** 2586–9

[4] Hoover R B and Tat M W (ed) 1994 X-ray and UV detectors *SPIE* **2278**

[5] ISO/FDIS 11146, Lasers and laser related equipment—Test methods for laser beam parameters—Beam widths divergence angle and beam propagation factor

[6] Flath L, An J, Brase J, Carrano C, Brent Dane C, Fochs S, Hurd R, Kartz M and Sawvel R 2000 Development of adaptive resonator techniques for high power lasers *Proc. 2nd Int. Workshop on Adaptive Optics for Industry and Medicine* ed G D Love (Durham) (Singapore: World Scientific) pp 163–8

[7] Hartmann J 1900 Bemerkungen über den Bau und die Justierung von Spektrographen, *Z. Instrumentenkunde* **8** 2

[8] Neal D R, Alford W J, Gruetzner J K and Warren M E 1996 Amplitude and phase beam characterization using a two-dimensional wavefront sensor *SPIE* **2870** 72–82

[9] Schäfer B and Mann K 2000 Investigation of the propagation characteristics of excimer lasers using a Hartmann–Shack sensor *Rev. Sci. Instrum.* **71** 2663–8

[10] Shack R B 1971 Lenticular Hartmann screen *Opt. Sci. Center Newsletters* **5** 15

[11] Laude V, Olivier S, Dirson C and Huignard J P 1999 Hartmann wave-front scanner *Opt. Lett.* **24** 1796–8

[12] Mansel J D, Byer R L and Neal D R 2000 Apodized micro-lenses for Hartmann wavefront sensing *Proc. 2nd Int. Workshop on Adaptive Optics for Industry and Medicine* ed G D Love (Durham) (Singapore: World Scientific) pp 203–8

[13] Smartt R N and Steel W H 1975 Theory and application of point-diffraction interferometers *Japan. J. Appl. Phys.* **14** 351–6

[14] Tyson R K 1998 *Principles of Adaptive Optics* 2nd edn (New York: Academic)

[15] Press W H, Teukolsky S A, Vetterling W T and Flannery P B 1992 *Numerical Recipes in C* 2nd edn (Cambridge: Cambridge University Press)

[16] Bates W J 1947 A wavefront shearing interferometer *Proc. Phys. Soc.* **59** 940–50

[17] Wyant J C 1973 Double frequency grating lateral shearing interferometer *Appl. Opt.* **12** 2057–60

[18] Wyant J C 1974 White light extended source shearing interferometer *Appl. Opt.* **13** 200–2

[19] Streibel N 1984 Phase imaging by the transport equation of intensity *Opt. Commun.* **49** 6–10

[20] Teague M R 1983 Deterministic phase retrieval: a Greens function solution *J. Opt. Soc. Am.* **73** 1434–41

[21] Born M and Wolf E 1982 *Principles of Optics* 6th edn (New York: Academic)

[22] Kawata S, Hikima I, Ichihara Y and Watanabe S 1992 Spatial coherence of excimer lasers *Appl. Opt.* **31** 385–96

[23] Omatsu T, Kuroda K and Takase T 1992 Time resolved measurement of spatial coherence of a copper vapor laser using a reversal shear interferometer *Opt. Commun.* **87** 278–86

Further reading

Geary J M 1995 *Introduction to Wavefront Sensors (Tutorial Texts in Optical Engineering TT18)* (Bellingham, WA: SPIE Optical Engineering Press)

A tutorial text with main emphasis on the optical principles and the hardware implementation of wavefront sensors as well as their application in different fields of optical engineering.

Mandel L and Wolf E 1994 *Optical Coherence and Quantum Optics* 2nd edn (Cambridge: Cambridge University Press)

A standard textbook on optical coherence theory including a comprehensible mathematical introduction to stochastic processes and a couple of applications.

Roundy C B 2000 Current technology of beam profile measurement *Laser Beam Shaping* ed F M Dickey and S C Holswade (New York: Marcel Dekker)

Written from a practitioners point of view. Includes a lot of valuable advices for experimenters engaged in the field of laser beam profile and laser beam parameter measurement.

Tyson R K 1998 *Principles of Adaptive Optics* 2nd edn (New York: Academic)

Thorough treatment of wavefront sensing, wavefront sensors and wavefront reconstruction, considering both theoretical and practical aspects of the field. Many references.

C6
Laser safety

Colin Webb

Although many low-power lasers in everyday use as barcode readers at supermarket checkout tills or as display screen pointers present little or no safety hazard, there are many lasers whose output power capability and beam collimation make them potentially very dangerous. It is important that laser users are aware of the dangers associated with these devices and take adequate steps to ensure that neither they, nor anyone else, can be exposed to harmful levels of radiation.

There are, of course, other types of hazards associated with laser use. Such general hazards include those connected with the materials and chemicals used in lasers (see also chapters B5.3 and B5.4), exhaust fumes and gases and, not least, high-voltage power supplies. The subject of laser safety, then, is a vast one, and the reader is advised to consult appropriate texts or qualified safety consultants on the subject of these more general hazards. We have chosen in this section to highlight the hazards which are specific to lasers. The types of injury which can arise, the precautions necessary to prevent them and the classification of lasers according to the degree of hazard they present are described by two of the leading experts in this field.

C6.1
Laser safety

J Michael Green and Karl Schulmeister

C6.1.1 Introduction

This chapter describes the hazard posed by laser radiation, how it is assessed and the principles of how to minimize the risk. The nature of the laser radiation hazard is well known, yet safety standards and guidance documents must continually evolve in response to the challenges of new laser applications and laser source developments; in this chapter the major revisions to the international laser safety standard introduced in 2001 have been included.

The general trend in laser safety standards is a relaxation of what was historically a highly cautious approach to the setting of threshold limits and classifying of laser products. These subjects form the more complex and technical aspects of laser safety, yet from a user point of view the subject remains an essentially practical one, dealing primarily with the design and selection of appropriate engineering controls, administrative controls and protective eyewear.

Wavelength is the principal determinant of the nature of the laser injury. This is particularly true for the eyes and, after a review in section C6.1.2 of the wavelength dependence of absorption in the various components of the eye and the various injury mechanisms, section C6.1.3 addresses the setting of threshold limits, with particular emphasis on recent developments in the international laser safety standard. The hazard classification scheme, the bedrock of manufacturer requirements and user guidance, is described in section C6.1.4. Finally, section C6.1.5 reviews the principal control measures for safe laser use.

This chapter is intended only to provide a basic overview of the subject[1]. It highlights the principal hazards and how to achieve safe use. However, it should not be regarded as a safety manual. In particular, the assessment of laser hazards, classification of laser products and the implementation of laser safety controls requires detailed considerations beyond the scope of this chapter.

C6.1.2 Laser injuries to the eyes and skin

C6.1.2.1 Injury mechanims

The full range of radiation wavelengths available from laser sources extends from the ultraviolet through to the far-infrared. Over this range the skin is a strong absorber, thereby providing primary protection for all the bodily organs, except for the eyes. When absorption of radiation occurs, the temperature of the absorbing tissue rises, and if the temperature exceeds a critical temperature of approximately 60 °C, proteins in cells denature and the cell dies. If the temperatures exceeds 100 °C, water in the tissue begins to boil and further temperature increases lead to a carbonization of the tissue. This is the *thermal* damage process. It is characterized by a sharp threshold for injury and is non-cumulative in that, if thermal injury does not occur

[1] For a discussion of laser safety in biomedical and clinical environments see chapter D3.5.

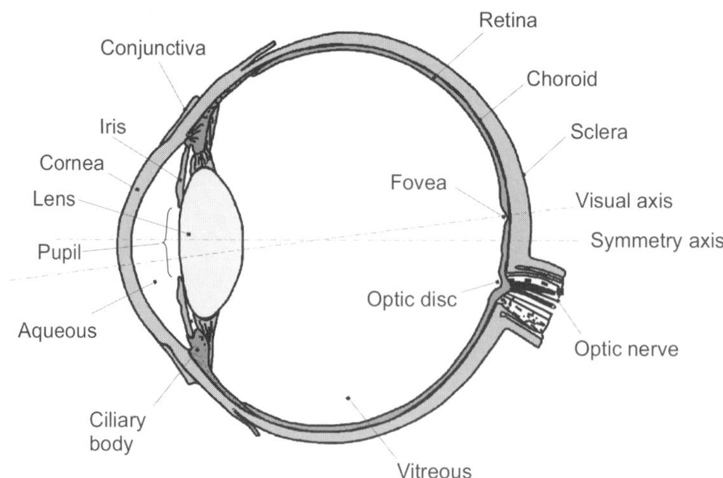

Figure C6.1.1. A schematic horizontal cross section of a human eye. Central sharp vision is only possible in the central part of the retina, the macula (yellow spot), and particularly in the even smaller centre of the macula, the fovea. The main refractive power of the eye is provided by the cornea, while the lens is needed for accommodation (imaging) of distant and close objects.

during the first approximately 10 s during which the temperature of the exposed tissue will be rising to some steady-state value, then it will not occur for prolonged or repeated exposure at that level.

For radiation in the ultraviolet and the blue/green end of the visible range of wavelengths, the thermal competes with the photochemical damage process. The capability of radiation to initiate photochemical effects is strongly wavelength-dependent and generally increases with decreasing wavelength. Chemical changes in the exposed tissue induced by this shorter wavelength radiation is cumulative, certainly over a working day, so the degree of damage depends on the radiant exposure (J m^{-2} on the exposed surface) i.e. the accumulated energy. Unlike the threshold for thermal damage, which varies with wavelength only in response to the changes in the absorption depth in the tissue, the photochemical 'action curve' has a strong wavelength dependence and photochemical injury can be the dominant mechanism, but only for prolonged or repeated exposures at levels that are too small to cause a sufficient temperature rise for thermal damage. Conversely, thermal damage will always be the dominant injury mechanism for short time (less than 1 s) single exposures.

C6.1.2.2 Principal components and operation of the eye

There are many reference sources describing the structure and operation of the eye in sufficient detail for laser safety purposes. Figure C6.1.1 shows the structural elements of the human eye, which is equipped to transmit electromagnetic radiation in the restricted wavelength range of 400 nm (blue light) through to 1400 nm (near infrared):

- through the cornea, which provides the primary means of focusing of the radiation;
- through the pupil at the centre of the iris, whose partial closure in response to bright light provides a degree of regulation;
- through the somewhat pliable crystalline lens, which under muscular control provides the remaining amount of focusing of the radiation;

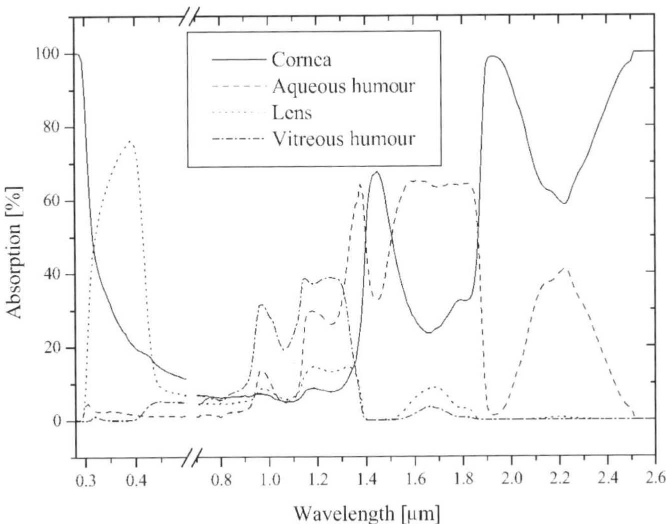

Figure C6.1.2. The spectral absorption of the ocular media for the eye. The break in the curves corresponds to the visible 0.4–0.7 μm region where there is very little absorption through to the retina. Of particular note is the strong UV and near-IR absorption of the lens and the 0.86–1.4 μm region (only) where the vitreous absorbs a significant fraction of the incident radiation. Corneal absorption of 100% at both wavelength extremes of the figure extends to longer and shorter wavelengths.

- creating an image of the radiation source on the foveal region of the retina, where photoreceptors provide a signal to the brain from which the structure and colour of the image are interpreted.

C6.1.2.3 Laser injuries to the eye

The absorption of radiation incident on the eye has a strong and complex wavelength dependence. This is shown in figure C6.1.2, which plots the percentage absorption of radiation incident on the cornea within the different components of the ocular media. Various wavelength bands can be discerned over which the cornea, the anterior region (aqueous and lens) and the posterior (vitreous) become the major absorbing components. Between 0.4 and 1.4 μm the accumulated absorption of these components is less than 100%, and the retina is therefore at risk of injury.

Table C6.1.1 delineates the principal types and locations of injury according to spectral band and exposure duration. Note that damage to the cornea or lens can result in serious, though in principle correctable, vision loss. By contrast, injuries to the retina do not heal and are not repairable.

C6.1.2.4 Retinal injuries

Within the ‘retinal hazard range’ (0.4–1.4 μm) a laser beam entering the eye and passing through the pupil may be focused to a spot of the order of 20 μm diameter on the retina, resulting in a concentration of the exposure level at the cornea of up to approximately 100 000 times, i.e. the ratio of areas of a 7 mm diameter pupil to a 20 μm image. In practice, factors during laser exposure including pupil size and accommodation of the eye (i.e. the distance at which the eye is focused), combined with involuntary eye movements and, in the wavelength range of light (0.4–0.7 μm), aversion responses (including the blink reflex), lessen the retinal

Table C6.1.1. Summary of principal ocular injuries for radiation from ultraviolet to infrared wavelengths. The cornea has a highly effective repair mechanism for minor injuries, but for the lens and retina the injury is permanent. Note that the injury mechanisms apply to both laser and non-laser sources, except for short pulse (typically Q-switched lasers with pulse durations less than 1 μs) injury mechanisms, which are unique to lasers.

Wavelength range	Tissue affected	Single-pulse injury	Long exposure (several seconds and more) cw and repetitively pulsed lasers
Ultraviolet (180–400 nm)	Cornea lens	Thermal damage dominates, leading to denaturation (clouding) of the cornea and (0.28–0.4 μm) lens.	Photochemical damage dominates, leading to photokeratitis ('arc-eye' or 'snow blindness') of the cornea and photochemical cataract (0.28–0.4 μm).
		Short high power pulses can photoablate corneal tissue.	
Visible and near-infrared (400–1400 nm)	Retina	Thermal damage dominates, leading to a retinal burn (protein denaturation) with potential severe vision loss	400–550 nm (blue to green) may cause photochemical damage for prolonged exposure.
		Short pulses can cause photomechanical damage (i.e. rupture of tissue) leading to extensive retinal damage and bleeding into the inner eye.	Exposure to visible radiation is normally limited to less than 0.25 s by the natural aversion response to bright light. More generally, the saccadic eye movement reduces the retinal hazard for longer durations.
Far-infrared (1400 nm to 1 mm)	Cornea lens	Thermal damage dominates, leading to denaturation (clouding) of the cornea and (1.4–1.9 μm) lens.	Exposure for thermal damage is normally limited to a few seconds by reaction to pain due to heating of the cornea.
		Short high-power pulses can photoablate corneal tissue.	Long-term exposure can lead to infrared cataract (1.4–3 μm).

exposure. Nevertheless, quoted safe exposure limits at the cornea undergo a step function change of several orders of magnitude at 400 nm.

Figure C6.1.3 shows the overall transmission curve for the eye together with the retinal visual response curve. It shows the retina as sensitive only to wavelengths in the approximate range 0.4–0.7 μm, though a limited retinal sensitivity persists into the near-infrared (0.7–1.4 μm). For laser safety considerations this sensitivity is regarded as insufficient to evoke an aversion response to hazardous levels of near-infrared radiation and there are many cases of such near-infrared exposures causing damage without the victim being aware.

As confirmed by accident statistics [1], the near-infrared is the most hazardous wavelength range for lasers in general and near-infrared pulsed lasers are the most hazardous sources. A fundus photograph in figure C6.1.4 shows several types of thermal and thermo-mechanical retinal injuries in a rhesus monkey eye produced by a laser in this category. The white lesions result from thermally-induced denaturation of proteins and are characteristic of exposures longer than 1 μs. The area of the injury is confined to the irradiated area,

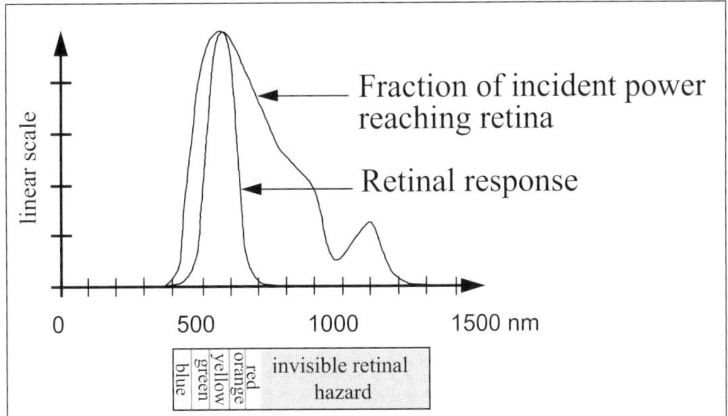

Figure C6.1.3. Semi-qualitative comparison of the photopic response curve and the transmission curve for the eye. The sensitivity of the eye in respect to visual stimulus is highest in the green, around 550 nm, and falls off towards red on one side and blue on the other. There is no sharp dividing line between visible and non-visible. Whereas the limits of the visible spectrum are defined as 380 nm and 780 nm [CIE], in laser safety, in the context of inducing an aversion response to bright light, they are taken as 400 nm and 700 nm.

extended by thermal conduction. However, even minimal retinal lesions may result in permanent serious vision loss if the damage is located in the foveal region of the retina or the optic disc. For laser pulses of duration less than 1 μs, rapid heating of the absorbing tissue can cause a micro-explosion at the back of the eye, possibly leading to haemorrhaging and damage extending well beyond the irradiated area, the flow of blood into the vitreous humour greatly increasing the impairment of vision and the psychological impact of the injury on the victim. Examples of these injuries are illustrated in figure C6.1.4.

C6.1.3 Exposure limits

C6.1.3.1 Establishing a threshold level

The level of exposure or irradiance which can be thought of as the border between safe and potentially harmful is called maximum permissible exposure (MPE). MPE values are mostly derived from animal experiments and a very limited number of human exposures. In such threshold experiments, a number of exposures are delivered to the eye or the skin for some fixed set of conditions (laser wavelength, exposure duration, spot size). After each exposure at a given energy level, the exposed site is examined for a detectable lesion. For obvious reasons, the exposed site can only be exposed once and therefore a range of sites per animal and also a number of animals have to be exposed. As a result, and in combination with experimental uncertainties, there is not a sharply defined threshold exposure value below which no lesions are found and above which all exposures lead to damage [2].

Figure C6.1.5 is a typical dose-response curve, often called a 'probit plot'. The exposure dose at which 50% of the exposures lead to a lesion, ED-50, is generally referred to as the threshold, even though there is a finite probability for damage at energies somewhat below the ED-50. The MPE is set well below the ED-50 point to ensure that the probability of a detectable minimal lesion is practically zero.

The safety factor between ED-50 and MPE is often set at 10. However, where there is less variability and small experimental uncertainty, such as for corneal photochemical injury in the UV range, a safety factor less than 10 has been agreed by the international committee. To cover bands of wavelength and pulse duration,

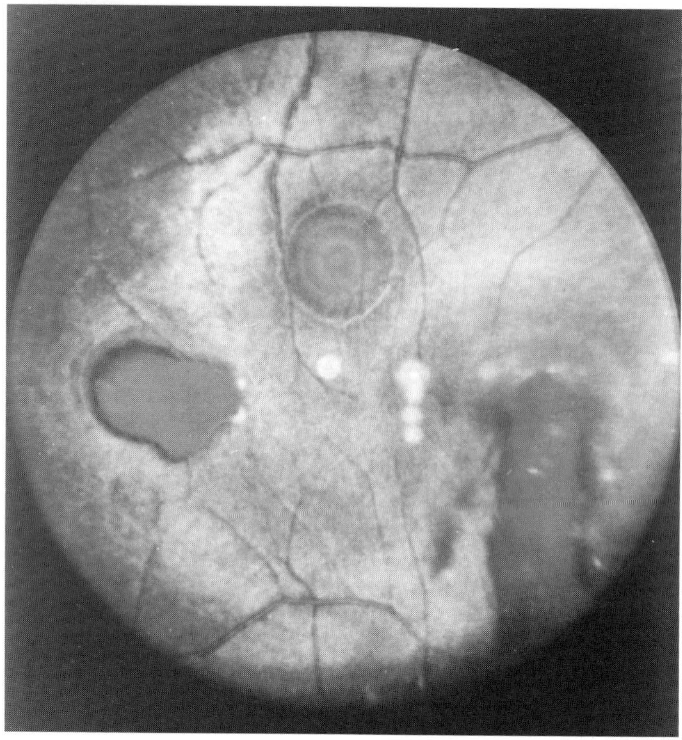

Figure C6.1.4. A range of injuries induced with a Nd:YAG laser on a monkey retina. The white spots in the centre are thermal burns, i.e. coagulation of retinal layers. With larger pulse energies, holes are produced in the retina which result in bleeding either into the vitreous or contained within the retina, both of which result in functional loss in the affected area. (Photograph courtesy of J Zuclich, TASC Litton, TX, USA.)

MPE values are expressed as single values or by a simple analytical expression and, as a result, safety margins greater than 10 are often encountered.

C6.1.3.2 Calculating MPE from tables

Internationally-approved MPE values are to be found in the IEC 60825-1 standard [3]. Values expressed as an irradiance (W m^{-2}) or as a radiant exposure (J m^{-2}), averaged over an area of specified diameter or 'limiting aperture'. The importance of the limiting aperture is primarily in hazard assessment. For example, the ocular MPE for a 0.25 s intra-beam exposure to radiation in the 0.4–0.7 μm wavelength range is quoted as 26 W m^{-2} and the corresponding limiting aperture size is 7 mm. In this case, multiplying the MPE by the area of the limiting aperture gives a result of 1 mW. If a 0.25 s laser exposure is not to exceed the MPE, then the power passing through a 7 mm diameter must be less than 1 mW, even if the actual beam diameter is less than this. Separate MPE and limiting aperture values are assigned for the eye and the skin, though they broadly coincide outside the retinal hazard range.

MPE values vary principally with wavelength, exposure duration and, in the retinal hazard range, retinal image size and are quoted as irradiance and radiant exposure levels at the cornea, even in the retinal hazard range. Annex B 'Biophysical considerations' of [3] provides a more comprehensive explanation of the biophysical considerations behind the derivation of MPE values.

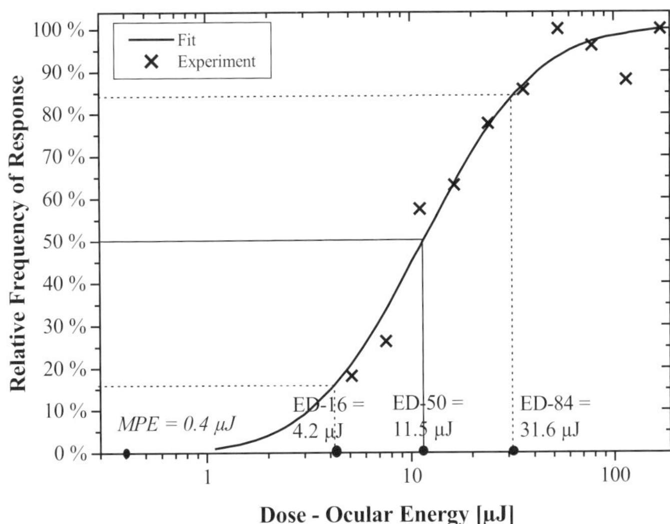

Figure C6.1.5. A typical dose–response curve as obtained in threshold experiments (here for a 850 nm laser, 180 ns pulse duration, minimal retinal spot size). For a given laser wavelength, pulse duration and spot size, the energy is varied. Due to variability of the threshold within the sampled population, a distribution of percentages of exposures which lead to a lesion results (in this example the sample population consisted of a total of 191 exposures on four animals). Also shown on the plot is the exposure corresponding to the MPE, in this case, 3.5% of the ED-50 value to ensure that there is practically zero risk when exposed at this level [2].

C6.1.3.2.1 Assessing the exposure duration

Current MPE values are defined for wavelengths between 180 nm and 1 mm and for exposure durations from 10^{-13} s to 30 000 s. For single pulsed lasers the appropriate exposure duration is the duration of the laser pulse but for exposure to cw and repetitively-pulsed lasers the exposure duration has to be assessed. Table C6.1.2 lists 'typical' values according to potential exposure scenarios. Among the situations considered in this table is unexpected exposure to bright light. Such an exposure evokes natural aversion responses, including the blink reflex and turning away of the eyes and head, for which laser safety committees have established a figure of 0.25 s. At the other extreme, 30 000 s corresponds to almost 8 hr, a full working day.

C6.1.3.2.2 Dealing with multiple pulse exposures

Assessing the MPE for exposure to a repetitively pulsed laser beam involves selecting the most restrictive of at least two approaches: (i) exposure to each pulse within the train taken separately and (ii) exposure to a cw beam of the same average power. This approach is satisfactory for wavelengths less than 400 nm, where the injury mechanism is photochemical. However, for wavelengths exceeding 400 nm a third approach to the assessment of MPE is required to take into account the cumulative heating effect of multiple pulses, the chosen MPE being the most restrictive of the three. For a train of pulses of the same magnitude and duration this third MPE is the single pulse MPE reduced by $N^{-0.25}$, where N is the number of pulses in the assessed duration of exposure. A more flexible approach, however, is to evaluate the MPE for a single total-on-time-pulse (TOTP).

The TOTP is a hypothetical pulse with an energy equal to the sum of the energy of all pulses and a duration given by summing their individual durations. The time frame recommended in the standard [3] for

Table C6.1.2. Suggested exposure times for calculating MPE values for cw and repetitively-pulsed lasers.

Wavelength	Situation assessed	Duration*	Comment
180–400 nm	Eye and skin exposure	30 000 s	Accumulated photochemical injury over a full working day.
400–700 nm	Accidental eye exposure	0.25 s	Aversion response to bright light
700–1400 nm	Accidental eye exposure	10 s**	Involuntary eye movement
400–1400 nm	Accidental skin exposure	10 s	Aversion response to local heating.
400–1400 nm	Continuous intentional exposure of eyes and skin	30 000 s	Full working day
>1400 nm	Accidental exposure of skin and eyes	10 s	Aversion response to local heating.

* If the laser emission duration is less than the suggested duration then, of course, the former value should be used.

** For small sources. A larger duration is recommended where the image size corresponds to an angular subtense greater than 1.5 mrad (see discussion on apparent source size)

this summation is generally limited to 10 s (eye movement limit) but is longer for extended source viewing. However, if within this time frame there are laser pulses of duration less than a specified wavelength-dependent value T_i, typically 1 ms or less, such pulses are for the sake of calculation given a duration T_i. This accounts for the time-independence of ocular MPE for exposure duration values below T_i.

C6.1.3.2.3 Small and extended sources

The minimum image size that can be produced by the eye is estimated to be of order 20 μm diameter, a value strongly influenced by diffraction, aberrations and scattering. The cornea/lens optical system of a normal eye has a focal length of 17 mm, so the minimum spot corresponds to a linear angle of 1.2 mrad. In laser safety standards a source is referred to as a 'point' or 'small' source if the angular size of a source as perceived by the eye during exposure (a value known as the 'angular subtense') is less than 1.5 mrad. This 'minimum angular subtense' value applies to intra-beam viewing of all but the poorest quality laser beams and can apply to the viewing of small near-monochromatic incoherent sources (e.g. LEDs).

Ocular MPE values in the 400–1400 nm wavelength range relate directly to exposure to 'point' or 'small' sources. Exposure conditions where the image size corresponds to an angular subtense greater than 1.5 mrad is referred to as 'extended source viewing' and applies to most conditions of viewing incoherent sources or diffuse reflections of laser beams, when viewed close enough for shape to be geometrically resolved by the eye (as illustrated in figure C6.1.6). Although the area of the retinal image has an α^2 dependence, where α is the angular subtense, thermal diffusion from the centre of the retinal image will be slower and, as a result, thermal retinal MPE values are quoted with a multiplication factor $\alpha/1.5$ (α in milliradians) for $\alpha \geq 1.5$ mrad. A different procedure is required to separately assess the photochemical retinal injury for cw exposure in the 400—600 nm wavelength range.

C6.1.3.3 Nominal ocular hazard distance

The minimum distance from the laser source at which the beam is so large that the MPE for ocular exposure is no longer exceeded is termed the nominal ocular hazard distance (NOHD). Class 1 and class 2 lasers (see section C6.1.4) do not have a NOHD, but a well collimated class 3B or class 4 laser could have a NOHD measured in kilometres. For indoor applications involving class 3B and class 4 lasers, the NOHD will often

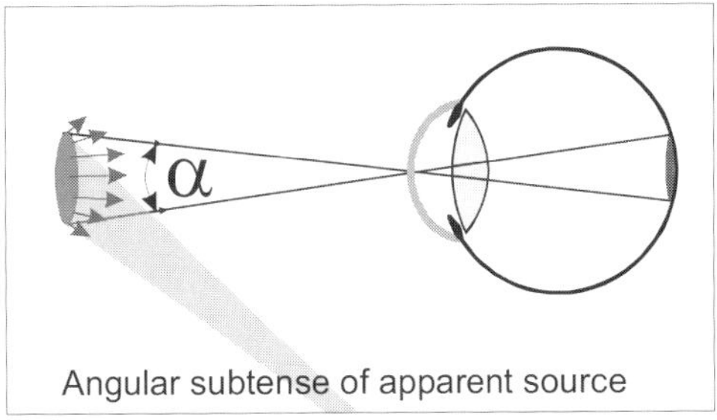

Figure C6.1.6. Definition of the angular subtense: the linear angle subtended by the retinal image. In this illustration a laser beam is incident on a matt surface and the eye 'sees' the illuminated area as it would a conventional light source. More generally, the angular subtense of a source is determined by placing a lens at the desired point of measurement and dividing the image size in the focal plane by the focal length of the lens.

exceed the dimensions of the room or enclosure that it is in, and the calculation is of little value except if the laser radiation is highly divergent e.g. is expanding from focus or from the end of an optical fibre. NOHD calculations for diffuse reflections generally assume a Lambertian cosine law for angle of reflection and an inverse square law for irradiance, though for extended source viewing the calculation is complicated by the image size dependence of MPE as discussed in section C6.1.3.2.

NOHD calculations are particularly useful in assessing the safety of outdoor applications of lasers and in most cases an approximate answer based on a simple conical expansion of the beam (i.e. expansion at the far field divergence angle from the position of the beam waist) and a Gaussian beam profile is sufficient. Outdoor assessments may need to take account of the increased hazard created by the use of binoculars or telescopes, use of which may be hazardous at distances greater than the NOHD. Such considerations form part of the test procedures for laser classification and are addressed quantitatively in section C6.1.4.2. The extended nominal ocular hazard distance (ENOHD) is defined on this basis.

C6.1.4 Safety in product design

C6.1.4.1 The classification scheme

The laser hazard classification scheme groups laser products according to increasing hazard potential and to recommended safety control measures such as interlocked access to laser areas (engineering controls) and restricted use of optical instruments (administrative controls). The IEC laser product safety standard [3], approved in Europe as EN 60825-1 and reproduced by national standards bodies in the EU member states (e.g. in the UK as BS EN 60825-1, in Germany as DIN EN 60825-1), specifies requirements for manufacturers on how to classify a laser product and the safety features and warning labels to be provided. At the present time there is a convergence but not an equivalence between the IEC standard and the standard adopted in the USA [4].

In essence, the classification scheme divides laser products into four major classes.

Class 1: No risk to eyes or skin. Lasers that are safe in normal operations under reasonably foreseeable conditions, including direct intra-beam viewing.

Class 2: Low risk to eyes. No risk to skin. Lasers emitting visible radiation in the wavelength range 400–700 nm for which the natural aversion response to bright light (including the blink reflex) prevents retinal injury for direct intra-beam viewing. These lasers do, however, present a dazzle hazard.

Table C6.1.3. Qualitative description of laser safety classes.

Class	Sub-division	Meaning	Warning label and safety features	Examples	Relationship of AEL to MPE	Typical AEL for cw lasers	Comparison with 1993 scheme
1	Intrinsic	Safe by virtue of the intrinsic low power of the laser, even with the use of optical instruments	No warning label. No additional safety features	Laser scanning ophthalmoscopes, low-power rangefinders, most LEDs	(i) Output does not exceed MPE (time base 30 000 s for UV sources and products intended to be intentionally viewed, otherwise 100 s). (ii) Output collected by 7 mm aperture 14 mm from apparent source does not exceed AEL. (iii) Output collected by 50 mm aperture 2 m from source does not exceed AEL.	10 mW in wavelength range 1.4–4 μm 40 μW for blue light	AELs are higher than previous class 1 by virtue of higher MPE values and some changes in limiting aperture sizes
	Engineering	Embedded laser products, safe by virtue of engineering controls e.g. total enclosure guarding, scan failure mechanism.	No warning label No additional safety features (but see requirements for protective housings in C6.4.3)	CD players, laser printers, some fully enclosed cutting, drilling, welding and marking machines			
1M	Collimated	Well-collimated beam, output in range 302.5–4000 nm, with large diameter that is safe for unaided viewing but potentially hazardous when a telescope or binoculars are used.	LASER RADIATION DO NOT VIEW DIRECTLY WITH BINOCULARS OR TELESCOPES* No additional safety features	Some LIDAR and large aperture infrared optical test instruments	Relationship (i) and (ii) above are satisfied but not (iii).	As for class 1 but, additionally, for a collimated beam, output collected by 50 mm aperture 2 m from source less than class 3B AEL.	New, previously was non-visible part of class 3A
	High divergence	Output in range 302.5–4000 nm. Safe for unaided viewing but potentially hazardous when an eye loupe or magnifier is used.	LASER RADIATION DO NOT VIEW DIRECTLY WITH MAGNIFIERS* No additional safety features	Some low to medium power lasers with optical fibre outputs, raw beam laser diodes, line lasers, higher power infrared LEDs	Relationship (i) and (iii) above are satisfied but not (ii).	As for class 1 but additionally, for a high divergence beam output collected by 7 mm aperture 14 mm from source less than class 3B AEL	
2	—	Output in range 400–700 nm. Safe for unintended exposure, even with the use of optical instruments, by virtue of natural aversion response to bright light.	DO NOT STARE INTO THE BEAM No additional safety features	Visible alignment laser, laser pointer	(i) Output does not exceed MPE (time base 0.25 s). (ii) Output collected by 7 mm aperture 14 mm from source does not exceed AEL. (iii) Output collected by 50 mm aperture 2 m from source does not exceed AEL.	1 mW for cw beam.	Unchanged

Table C6.1.3. (Continued.)

Class	Sub-division	Meaning	Warning label and safety features	Examples	Relationship of AEL to MPE	Typical AEL for cw lasers	Comparison with 1993 scheme
2M	Collimated	Well collimated beam in range 400–700 nm, with large diameter that is safe for unaided viewing by virtue of natural aversion response to bright light but potentially hazardous when a telescope or binoculars are used.	LASER RADIATION DO NOT STARE INTO THE BEAM OR VIEW DIRECTLY WITH BINOCULARS OR TELESCOPES* No additional safety features	Some visible LIDAR and large aperture optical test instruments	Relationship (i) and (ii) above are satisfied but not (iii).	As for class 2 but, additionally, for a collimated beam, output collected by 50 mm aperture 2 m from source less than class 3B AEL.	New, previously was visible part of class 3A, 5 mW power cap is lifted
	High divergence	High divergence source with output in range 400–700 nm. Safe for unaided viewing by virtue of natural aversion response to bright light but potentially hazardous when an eye loupe or magnifier is used.	LASER RADIATION DO NOT STARE INTO THE BEAM OR VIEW DIRECTLY WITH MAGNIFIERS* No additional safety features	Some visible lasers with optical fibre output, some visible LEDs and raw beam laser diodes.	Relationship (i) and (iii) above are satisfied but not (ii).	As for class 2 but, for a high divergence beam, output collected by 7 mm aperture 14 mm from source less than class 3B AEL.	
3R	Visible	Output in range 400–700 nm. Direct intra-beam viewing is potentially hazardous, but by virtue of natural aversion response to bright light the risk is lower than for class 3B.	AVOID DIRECT EYE EXPOSURE No additional safety features	Alignment lasers such as HeNe lasers and laser diodes, which are used widely in industry, science and medicine.	(i) Output does not exceed MPE (time base 0.25 s) by more than 5 times. (ii) Output collected by 7 mm aperture 14 mm from source does not exceed class 2 AEL by more than 5 times. (iii) Output collected by 50 mm aperture 2 m from source does not exceed class 2 AEL by more than 5 times.	5 times the limit for class 2 i.e. 5 mW.	Equivalent to the existing class 3B with less than 5 mW (often referred to as 3B star).
	Non-visible	Output in UV (302.5–400 nm) or IR (700 nm–1 mm). Direct intrabeam viewing is potentially hazardous, but the risk is lower than for Class 3B.	AVOID DIRECT EYE EXPOSURE (700–1400 nm) AVOID EXPOSURE TO THE BEAM (outside the range 400–1400 nm Emission warning device	Some low power IR range finders.	(i) Output does not exceed MPE (time base 30 000 s for UV sources and products intended to be intentionally viewed, otherwise 100 s) by more than 5 times. (ii) Output collected by 7 mm aperture 14 mm from source does not exceed class 1 AEL by more than 5 times. (iii) Output collected by 50 mm aperture 2 m from source does not exceed class 1 AEL by more than 5 times.	5 times the limit for class 1.	New, no equivalence in previous system.

Table C6.1.3. (Continued.)

Class	Sub-division	Meaning	Warning label and safety features	Examples	Relationship of AEL to MPE	Typical AEL for cw lasers	Comparison with 1993 scheme
3B	—	Medium power laser. Direct ocular exposure is hazardous, even taking into account aversion responses, but diffuse reflections are usually safe.	AVOID EXPOSURE TO THE BEAM Key switch, emission warning device, external interlock connection and beam stop	Large area illuminators (e.g. industrial holography) and laser displays.	Ocular MPE (10 s) not exceeded at 130 mm from normal incidence viewing of beam-off diffuse (Lambertian) reflector.	500 mW for cw laser with a wavelength >315 nm.	Meaning unchanged, but 1993 class 3B included some lasers now moved to lower classes.
4	—	High-power laser. Direct exposure is hazardous to eye and diffuse reflection may also be hazardous. Skin and potential fire hazard.	AVOID EYE OR SKIN EXPOSURE TO DIRECT OR SCATTERED RADIATION Safety features as for 3B	Lasers for materials processing. High power diode arrays. Professional laser displays.	No AEL.	No limit.	Unchanged.

* The general phrase 'optical instruments' can be used in place of 'binoculars or telescopes' or 'magnifiers'. However, generally only one or other group of optical instruments leads to an increase in the hazard for a given laser product. Therefore, at the discretion of the manufacturer, a specific wording can be added to the warning label. This distinction and information should help to place the hazard into perspective: for instance, the use of class 1M range finders (highly collimated large diameter beam) might have to be restricted in outdoor applications where binoculars are in frequent use (military base, bird sanctuary), while class 1M visual LED displays are safe to be viewed with binoculars.

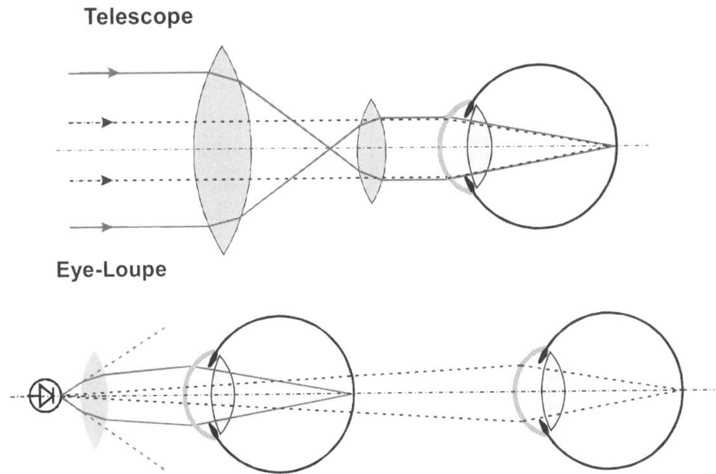

Figure C6.1.7. Use of viewing aids increasing the ocular hazard. Upper: A telescope or binocular increases the hazard of a well-collimated beam if it has a diameter exceeding the pupil size. Lower: An eye loupe or magnifier increases the hazard of highly divergent sources by allowing the source to be imaged at a closer distance (full lines) than would be the case for the unaided eye, where the closest distance of accommodation is about 100–200 mm (broken lines).

Class 3: Medium risk to eyes. Low risk to skin. Lasers for which intra-beam viewing is hazardous, but for which the viewing of diffuse reflections is normally safe. Natural aversion response to localized heating prevents serious skin injury.

Class 4: High risk to eyes and skin. Lasers for which intra-beam viewing and skin exposure is hazardous and for which the viewing of diffuse reflections may be hazardous. These lasers are also a fire hazard.

Each class has associated with it an accessible emission limit (AEL), generally expressed in watts and joules, except class 4 for which there is no upper limit. The AEL is a maximum level of 'accessible' laser radiation for the class, according to prescribed measurement conditions, measurement detector positions and collection aperture sizes. The measurement procedures are intended to take account of many of the worst-case assumptions of exposure conditions for which the product could be used i.e. exposure duration, closeness of viewing and the use of optical instruments.

The current scheme introduces relaxed forms of class 1, 2 and 3 in a systematic manner and, for the first time, addresses measurement conditions for classification more appropriate to high divergence sources.

Table C6.1.3 summarizes the current classification scheme. A sub-division of classes has been included to reveal how, on specifics of warning signs, manufacturer-installed safety measures and (in section C6.1.5) user guidance, a total of 7 classes and 11 'sub-classes' can be identified from the original four.

C6.1.4.2 Optical viewing aids

Measurements for classification for class 1, 2, 3R and 3B in the wavelength range 302.5–4000 nm (the transmission window of silica) take into account the use of optical viewing aids, as illustrated in figure C6.1.7. For example, for outputs in the 400–1400 nm wavelength range (only), three measurement conditions apply:

(i) the maximum output collected through a 50 mm aperture at 2 m from the product must not exceed the appropriate AEL. This measurement simulates the output that could enter the eye if a ×7 telescope or binocular of 50 mm acceptance aperture is used to view the laser output at this distance;

(ii) the appropriate AEL must not be exceeded through a 7 mm aperture at 14 mm from the apparent source (this term is used to include virtual as well as real source positions). This measurement simulates the output that could enter the eye if a magnifier or eye loupe is used close to the laser source, but different measurement distances and aperture sizes are required to assess the photochemical and thermal hazard for extended sources;

(iii) For class 1 and 2 only, the AEL must not be exceeded through a 7 mm aperture placed 100 mm away from the apparent source and simulates the output that could enter the eye in 'worst case' naked eye viewing i.e. the shortest distance of accommodation. However, if (i) and (ii) are satisfied, then this condition is redundant.

Outside the retinal hazard range the limiting aperture for MPE values reduces from 7 mm to 3.5 mm and the collection apertures referred to in the measurement conditions are reduced accordingly. The reader should refer to the standard for full measurement details [3].

For class 1M and 2M laser products, defined as safe only for unaided viewing, (iii) only applies, with the caveat that the output collected under measurement conditions (i) and (ii) must not exceed the class 3B AEL. It follows that lasers with high divergence or large area outputs can be class 1M or 2M yet have total output powers greatly exceeding the class 1 and 2 AEL limits. Although these measurements for classification are designed to be performed by the manufacturers of lasers, they can usefully be exploited by users wishing to assess situations where optical viewing aids may be used.

C6.1.4.3 Engineering safety features on laser products

The product safety standard provides details of general requirements for warning labels and engineering safety features. These are summarized in table C6.1.3. There are additional requirements for medical laser equipment [5] and optical fibre communication equipment [6] and for machines that use lasers to process materials [7, 8].

C6.1.4.3.1 Protective housing

All classes of laser product are required to have a protective housing to limit human access to levels of laser radiation below the AEL assigned to the product. The design of this primary engineered safety feature on a laser product becomes particularly important for class 1 (embedded) laser products, particularly those enclosing class 4 laser sources.

Laser safety standards include requirements for labelling on access and service panels that form part of the protective housing and provide access to hazardous levels of laser radiation and, in addition, interlocking for such panels that are intended to be removed during normal use or maintenance.

Thin walled enclosures will 'keep fingers out' and block any weakly scattered laser radiation but insofar as they are incapable of withstanding a high power laser beam they rely heavily on maintaining the direction of the beam path. Regular alignment checks and the maintenance of optics and the environment (e.g. vibration, humidity, temperature, dust) can therefore be important safety considerations in the context of high power laser beams [9]. Indeed, the subject of laser guards, a 'machine' term but essentially simply a component of a protective housing, is the subject of a separate standard [8].

C6.1.5 Safety in practice

C6.1.5.1 Class-based user guidance

The laser hazard classification scheme is designed primarily as the basis for specifying manufacturer requirements but as the basis for user guidance it is widely recognized as insufficient, in particular since severity of injury plays no part in the classification scheme. For example, most user guidance relates to the safe use of class 3B and 4 lasers but class 3B lasers which threaten permanent visual impairment are in the same class as class 3B lasers for which the maximum injury is mild corneal irritation lasting a few days, and class 4 lasers capable of burning through firebricks in seconds are placed together with lasers capable of causing no more serious a skin burn than could be inflicted by momentary contact with a soldering iron.

Section C6.1.5.2 addresses considerations by which appropriate control measures can be selected and imposed to reflect the severity and probability of injury based on wavelength, power and use of the laser. Nevertheless, there is a relatively small number of safety control measures to choose from to deal with the laser hazard and these can generally be divided according to laser class, as summarized in table C6.1.4. The particular layout of the table has been chosen to emphasize the progressive adding of appropriate control measures as the class of laser increases. Also, the table separates engineering controls (generally the most effective and therefore to be preferred, but also the most expensive and restrictive) from administrative controls (less effective, but lower cost and more flexible), and personal protective equipment. As described in section C6.1.5.3, PPE is generally recognized as the measure of last resort and certainly not an equivalent alternative to the others.

Table C6.1.4. Summary of recommended control measures by laser class.

		Appropriate controls during normal operation*		
Class	Sub-division	Engineering controls	Administrative controls [NOHD, ENOHD]	Eye protection
1	Intrinsic	None.	None. [0,0]	None.
	Engineering	None.	Operation procedures. [0*,0*]	None.*
1M	Collimated	Employ basic beam path design principles**.	Safety training of laser user. Control use within NHZ of binoculars and telescope and specularly reflecting surfaces that may focus the beam. [0, >2 m]	None required. Filters should be fitted to binoculars or telescope within NOHA.
	High div.	None.	Control use of magnifiers close to source and specularly reflecting surfaces that may focus the beam. [0,0]	None required. Filters should be fitted to magnifiers if used.
2	—	Employ basic beam path design principles**.	Restrict use around activities where dazzle could cause an accident. Control use of specularly reflecting surfaces in vicinity of unenclosed beam path. [0,0] for accidental exposure.	None required.
2M	Collimated	Same controls as appropriate for class 1M (collimated) products plus dealing with dazzle hazard		
	High div.	Same controls as appropriate for class 1M (high divergence) products		

3R	Visible	Same controls as appropriate for class 2 products, but some training of laser user is appropriate (to enforce the importance of avoiding intentional exposure.) NOHD >0 but risk of injury during short time exposure is very small.		
	Invisible	Except for dazzle hazard, same controls as appropriate for class 3R (visible) products plus:		
		Employ restricted beam path design principles ***.	Appoint a Laser Safety Officer Provide basic safety training to users and other personnel in the NHZ appropriate to the small but finite risk of injury.	Recommended.
3B	—	Same controls as appropriate for Class 3R (visible) products plus:		
		Employ remote interlock connector.	Post warning signs at approaches to the hazard area. Provide training to users and other personnel in the NHZ appropriate to the severity of potential injury. Control access to key switch for laser use. Interpose the beam stop or attenuator when the beam is not required.	
4	—	Same controls as appropriate for Class 3B products, plus those below:		
		Design exposed surfaces for power handling capability. Shroud diffuse reflections.	Potentially large NOHD	Skin protection recommended.

* For embedded laser products, controls appropriate to the embedded laser may be required during maintenance and service operations and the NOHD is potentially large.

** Basic beam path design: locate horizontal beams above or below eye level, avoid upwardly directed beam paths and terminate beam path at end of its useful range.

*** Restricted beam path design: beam enclosed and secondary beams blocked (for class 3B and 4) as far as reasonably practicable; robust, secure mounting of laser and beamline optics; open beam paths as short as reasonably practicable with a minimum number of directional changes, clear of surfaces producing hazardous reflections, not crossing walkways.

C6.1.5.2 Application of control measures

The international safety standard presents control measures in the form of guidance (this includes parts of the IEC 60825 series that provide application-specific laser safety guidance [9, 11]), leaving it to the user to decide if, for example, the 'should' in 'the remote interlock connector of class 3B and 4 lasers *should* be connected to . . . a room door' should be interpreted as a 'shall'. Where, for example within a particular organization or profession, the conditions of laser use are well defined and a common approach to laser safety is desirable, then working codes based on the hazard classification system can be devised to take away some or all of the decision from the user. The most comprehensive example of working codes is the USAs ANSI Z136 series [12] with additional parts dealing with optical fiber communication systems (part 2), health care facilities (part 3), measurement (part 4), education (part 5) and outdoor lasers (part 6). This prescriptive approach can only work, however, where the scope of the code is restricted to specific laser applications and/or types of users and environments.

A more general and more flexible approach to laser safety is provided by risk assessment and is the basis by which safety in general is dealt with in the various EU directives for both product and workplace safety. The approach involves first identifying potentially hazardous events and then, for each such event, assessing the likelihood of occurrence, the severity of harm and thereby, the risk. For each unacceptable risk so identified, and with reference to applicable standards and best practice guidelines, appropriate measures are identified and implemented, beginning with reasonably practicable engineering controls, followed by safe working practices and other administrative controls, until the risk is as low as reasonably practicable. Any residual risk is dealt with by the use of personal protective equipment.

The imposition of reasonable practicability and reference to applicable guidelines leads to different solutions depending on application and environment. For example, engineering controls are used in industrial laser processing (full enclosure of the laser radiation) whereas laser displays invariably rely heavily on administrative controls (crowd control) and many medical procedures require close proximity to open beams and therefore rely heavily on the use of laser safety eyewear. The flexibility often required for laser use in research may make full beam enclosure impractical (though partial enclosure is often feasible and is strongly recommended), so a mix of controls including interlocked access to laser laboratories and training of laser operators in safe working procedures is generally required, including the use of laser safety eyewear.

C6.1.5.3 Personal protective equipment

C6.1.5.3.1 Eye protection

The primary requirement to laser safety eyewear is a filter attenuation at the laser wavelength(s) sufficient to ensure that the transmitted radiation will not exceed the MPE at the maximum foreseeable laser exposure, usually but not necessarily chosen to be the exposure level at the laser aperture. The product standards for safety eyewear [13] specifies a 30 000 s time basis for MPE values at UV wavelengths, to address cumulative photochemical damage from scattered light over a working day, otherwise a value of 10 s, based on head movement and other behavioural considerations. For providing protection at visible laser wavelengths whilst allowing sufficient visible transmission for beam alignment purposes a related standard [14] specifies a time basis of 0.25 s.

Where laser safety eyewear is required, every reasonable means should be taken to ensure that it is properly used. This includes the choice of an appropriate model, i.e. high visible light transmission, a comfortable and appropriate design, convenient protective storage, clear labelling and regular maintenance. In Europe, this advice is embodied in the Personal Protective Equipment at Work Directive. This Directive also requires that all PPE be type tested and CE marked. The eyewear product standards [13, 14] include a test requirement for 'stability to laser radiation' (i.e. resistance to laser damage for 10 s exposure) which non-CE marked eyewear would generally not meet.

Beam alignment is the most hazardous laser operation, but it also provides the greatest temptation for the user to remove protective eyewear, especially if the beam to be aligned is of a visible wavelength. Consequently, at the same time as selecting safety eyewear it is important to consider the implications of its use. Beam visualization techniques involving fluorescent or scintillation cards and cameras and co-linear class 2 laser beams should always be investigated. Equally, depending on the visible transmission of the eyewear, it may be important to provide strong room lighting and to pay particular attention to trip hazards.

C6.1.5.3.2 Skin protection

Except for exposure to focused beams, skin injuries inflicted by lasers at the high end of class 3B and the low end of class 4 are generally regarded as minor (equivalent to a soldering iron burn) and skin protection is not worn. This practice can generally be justified on the grounds that the wearing of protective gloves, face shields and/or suits induces discomfort and can result in a greater accident as a result of the reduced ability of the wearer to carry out delicate operations. However, the wearing of gloves and other protective clothing should certainly be considered for open beam use of the higher power industrial lasers and for work with

Table C6.1.5. Summary of the main associated (i.e. non-beam) hazards and their control.

Origin	Hazard	Typical hazardous situation	Typical control measures
Laser	High voltage	Laser head and power supply exposed during servicing.	Proper screening of exposed HV. Restricted access to qualified persons. Use of earthing stick.
	Explosion	During changes of high pressure flashlamps in laser.	Gloves and face shield. Training.
	Collateral radiation	RF and UV accessible during servicing.	Proper screening combined with access restricted to service engineers.
	Mechanical	Unloading and positioning of laser power supplies.	Attachment points provided for use of lifting equipment by qualified persons. Training and use of gloves.
	Chemical	Dyes for dye lasers, halogen gas in excimer lasers, zinc selenide optics.	Procedures for storing, handling and disposing of hazardous materials. Training and use of gloves.
Process	Fume	Material removed during laser materials processing.	Fume extraction and filtration. Face mask and gloves worn during cleaning operations.
	Fire	Improperly terminated class 4 laser radiation.	Control of flammable materials and beam path. Provision of fire extinguisher.
	Secondary radiation	X-ray, UV and blue light during laser-workpiece/target interactions.	Enclosure of target area and monitoring of hazard. PPE for exposure to UV and blue light.
	Mechanical	Manipulators for laser and/or workpiece/target.	Guarding of traps. Restricted access to moving parts.

open beam UV lasers in general, though, to the author's knowledge, proprietary laser-specified skin-wear is not generally available.

C6.1.5.4 Accident reports

Reports of laser injuries are notoriously unreliable and incomplete and the available statistics are poor, with relatively few accidents reported and recorded per year [1]. The value of these reports lies primarily in identifying potentially hazardous activities and here they clearly imply that:

- research workers and service engineers are in the highest risk category for laser injuries;
- the most hazardous laser activity is laser beam alignment;
- pulsed class 3B and 4 lasers inflict the worst injuries;
- the near-infrared (700–1400 nm) is the most hazardous part of the laser spectrum;
- the most common cause of accidents is an over-reliance on laser safety eyewear combined with poor enforcement in its use;
- the lowest power eye injuries inflicted by visible cw lasers have occurred at 3 mW and involve use by untrained people and deliberate viewing of several seconds.

C6.1.6 Associated hazards

In addition to the laser radiation hazard, there are other general hazards associated with laser use. Hazards can be divided into two categories; those generally associated with the laser source (including high voltage, high pressure gases, collateral radiation) and those generally associated with the laser application (including fume, fire and mechanical hazards) though the list is not intended to be exhaustive or exclusive. Table C6.1.5 summarizes the main laser-related hazards and control measures, but details go beyond the scope of this chapter.

Manufacturers of machines for laser materials processing should be aware that the international standard for these products [7] requires measures to deal with associated hazards, the fume hazard in particular.

C6.1.7 Summary

Whilst the potential for injury exists, lasers have an excellent track record for safety. The vulnerability of the eye to laser injury has always been well appreciated, but as the nature of the injury has become more precisely defined so the setting of safe exposure limits has become more complex, especially for retinal injuries. The classification scheme provides the basis for manufacturers' requirements and goes some way to assisting users on precautions, but the appropriateness of control measures in a given situation can really only be decided on the basis of a well-informed risk assessment.

References

[1] Rockwell Laser Industries web site http://www.rli.com provides up-to-date accident statistics.
[2] Sliney D H, Mellerio J, Gabel V-P and Schulmeister K 2002 What is the meaning of thresholds in laser injury experiments? Implications for human exposure limits *Health Physics* **82** 335–47
[3] IEC 60825-1: 2001 Safety of laser products. Part 1. Equipment classification, requirements and user's guide
[4] Federal Laser Products Performance Standard 21 CFR 1040.10 and 1040.11
[5] IEC 601-2-22: 1991 Medical laser equipment. Part 2: Particular requirements for the safety of diagnostic and therapeutic laser equipment
[6] IEC 60825-2: 2000 Safety of laser products. Part 2. Safety of optical fibre communication systems
[7] ISO 11553: 1996 Safety of machinery—Laser processing machines—Safety requirements (equivalent to EN 12626)
[8] IEC 60825-4: 1997 Safety of laser products. Part 4. Laser guards
[9] Green J M 1989 High power laser safety *Opt. Laser Technol.* **21** 244–8
[10] IEC 60825-3: 1997 Safety of laser products. Part 5. Guidance for laser displays and shows
[11] IEC 60825-8: 1997 Safety of laser products. Part 8. Guidelines for the use of medical laser equipment
[12] ANSI Z136.1-2000 American National Standard for Safe Use of Lasers
[13] EN 207: 1998 Personal eye-protection—Filters and eye-protectors against laser radiation (laser eye-protectors)
[14] EN 208: 1998 Personal eye-protection—Eye-protectors for adjustment work on lasers and laser systems (laser adjustment eye-protectors)

Further reading

Sliney D H and Wolbarsht M L 1980 *Safety with Lasers and other Optical Sources* (New York: Plenum)
Sliney D H 1995 (ed) Selected papers on laser safety *SPIE Milestone Series* **MS-117** 1995
Henderson A R and Schulmeister K 2003 *Laser Safety* (Bristol: Institute of Physics)